Introductory & Intermediate Algebra

Fourth Edition

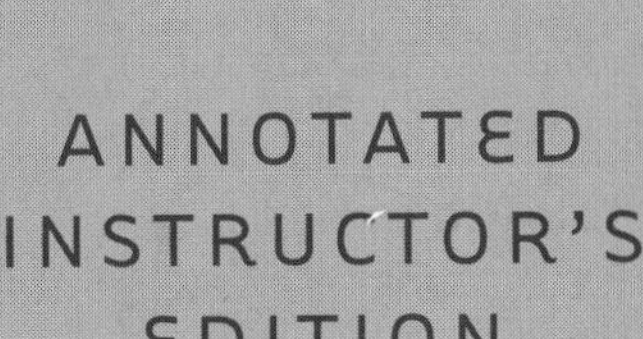

ANNOTATED INSTRUCTOR'S EDITION

Introductory & Intermediate Algebra

Fourth Edition

ANNOTATED INSTRUCTOR'S EDITION

Margaret L. Lial
American River College

John Hornsby
University of New Orleans

Terry McGinnis

Addison-Wesley
Boston • New York • San Francisco
London • Toronto • Sydney • Tokyo • Singapore • Madrid
Mexico City • Munich • Paris • Cape Town • Hong Kong • Montreal

Editorial Director	Christine Hoag
Editor in Chief	Maureen O'Connor
Executive Project Manager	Kari Heen
Project Editor	Courtney Slade
Editorial Assistant	Mary Gallagher
Senior Managing Editor	Karen Wernholm
Senior Production Supervisor	Kathleen A. Manley
Senior Designer	Barbara T. Atkinson
Photo Researcher	Beth Anderson
Supplements Production	Marianne Groth and Kayla Smith-Tarbox
Media Producers	Ceci Fleming, Lin Mahoney, and Jean Choe
Software Development	Eric Gregg, MathXL; Mary Durnwald, TestGen
Senior Marketing Manager	Michelle Renda
Marketing Assistant	Alicia Frankel
Senior Author Support/Technology Specialist	Joe Vetere
Senior Prepress Supervisor	Caroline Fell
Senior Media Buyer	Ginny Michaud
Rights and Permissions Advisor	Shannon Barbe
Senior Manufacturing Buyer	Carol Melville
Composition/Production Coordination	Nesbitt Graphics, Inc.
Cover Image	Golden Autumn Copyright © Lorraine Cota Manley

Library of Congress Cataloging-in-Publication Data

Lial, Margaret L.

Introductory and intermediate algebra.—4th ed. / Margaret L. Lial, John Hornsby, Terry McGinnis.

p. cm.

ISBN 13: 978-0-321-57569-2 (student edition) ISBN 13: 978-0-321-57605-7 (annotated instructor's edition)

ISBN 10: 0-321-57569-5 (student edition) ISBN 10: 0-321-57605-5 (annotated instructor's edition)

1. Algebra—Textbooks. I. Hornsby, E. John. II. McGinnis, Terry. III. Title.

QA152.3.L56 2010

512.9—dc22 2008024478

For permission to use copyrighted material, grateful acknowledgment is made to the copyright holders on page P-1, which is hereby made part of this copyright page.

1 2 3 4 5 6 7 8 9 10—CRK—12 11 10 09

Addison-Wesley
is an imprint of

www.pearsonhighered.com

ISBN 10: 0-321-57569-5
ISBN 13: 978-0-321-57569-2

To my two "scholarly" buddies Wayne and BK, and in memory of Bill

Johnny

To my sons, Andrew and Tyler, for blessing each of my days

Mom

CONTENTS

PREFACE

It is with pleasure that we offer the fourth edition of *Introductory and Intermediate Algebra*. With each new edition, the text has been shaped and adapted to meet the changing needs of both students and educators, and this edition faithfully continues that process. As always, we have taken special care to respond to the specific suggestions of users and reviewers through enhanced discussions, new and updated examples and exercises, helpful features, updated figures and graphs, and an extensive package of supplements and study aids. We believe the result is an easy-to-use, comprehensive text that is the best edition yet.

Students who have never studied algebra—as well as those who require further review of basic algebraic concepts before taking additional courses in mathematics, business, science, nursing, or other fields—will benefit from the text's student-oriented approach. Of particular interest to students and instructors will be the **NEW** pointers in examples, Study Skills activities, Math in the Media feature, and Solutions section.

This text is part of a series that also includes the following books:

- *Essential Mathematics*, Third Edition, by Lial and Salzman
- *Basic College Mathematics*, Eighth Edition, by Lial, Salzman, and Hestwood
- *Prealgebra*, Fourth Edition, by Lial and Hestwood
- *Introductory Algebra*, Ninth Edition, by Lial, Hornsby, and McGinnis
- *Intermediate Algebra*, Ninth Edition, by Lial, Hornsby, and McGinnis
- *Prealgebra and Introductory Algebra*, Third Edition, by Lial, Hestwood, Hornsby, and McGinnis
- *Developmental Mathematics: Basic Mathematics and Algebra,* Second Edition, by Lial, Hornsby, McGinnis, Salzman, and Hestwood

Hallmark Features

We are pleased to offer the following features, each of which is designed to increase ease-of-use by students and actively engage them in learning mathematics.

▶ ***Chapter Openers*** New and updated chapter openers feature real-world applications of mathematics that are relevant to students and tied to specific material within the chapters. Examples of topics include Americans' personal savings rate, the Olympics, and student credit card debt. (See pp. 29, 123, and 221—Chapters 1, 2, and 3.)

▶ ***Real-Life Applications*** We are always on the lookout for interesting data to use in real-life applications. As a result, we have included new or updated examples and exercises from fields such as business, pop culture, sports, the life sciences, and technology that show the relevance of algebra to daily life. (See pp. 223, 243, and 244.)

▶ ***Figures, Photos, and* NEW *Hand-Drawn Graphs*** Today's students are more visually oriented than ever. Thus, we have made a concerted effort to include mathematical figures, diagrams, tables, and graphs, including the new "hand-drawn" style of graphs, whenever possible. (See pp. 237, 238, and 286.) Many of the graphs use a style similar to that seen by students in today's print and electronic media. Even more photos have been incorporated to enhance applications in examples and exercises. (See pp. 154 and 162.)

▶ ***Emphasis on Problem Solving*** Introduced in Chapter 2, our six-step problem-solving method is integrated throughout the text. The six steps, *Read, Assign a Variable, Write an Equation, Solve, State the Answer*, and *Check*, are emphasized in boldface type and repeated in examples and exercises to reinforce the problem-solving process for students. (See pp. 353 and 354.) **Problem-Solving Hint** boxes provide students with helpful problem-solving tips and strategies. (See pp. 156 and 502.)

Appendix B: Strategies for Problem Solving provides examples of additional problem-solving techniques, such as working backward, using trial and error, and looking for patterns. A wide variety of applications are included. (See pp. 995–1004.)

▶ ***Learning Objectives*** Each section begins with clearly stated, numbered objectives, and the included material is directly keyed to these objectives so that students know exactly what is covered in each section. (See pp. 222 and 237.)

▶ ***Examples*** The new edition of the text features a multitude of step-by-step, worked-out examples that include pedagogical color, helpful side comments, and **NEW** pointers. We give increased attention to checking example solutions—more checks, designated using a special **Check** tag, are included than in past editions. (See pp. 125 and 194.)

▶ ***Margin Problems*** Margin problems, with answers immediately available at the bottom of the page, are found in every section of the text. (See pp. 227 and 228.) This popular feature allows students to immediately practice the material covered in the examples in preparation for the exercise sets. We have added more margin problems in this edition.

▶ ***Cautions and Notes*** One of the most popular features of previous editions, **CAUTION** and **Note** boxes warn students about common errors and emphasize important ideas throughout the exposition. (See pp. 228 and 230.) The text design makes them easy to spot: Cautions are highlighted in bright yellow and Notes are highlighted in purple.

▶ ***Calculator Tips*** Optional Calculator Tips, marked with calculator icons, offer basic information and instruction for students using calculators in the course. (See pp. 32 and 438.)

▶ ***Ample and Varied Exercise Sets*** One of the most commonly mentioned strengths of this text is its exercise sets. The text contains a wealth of exercises to provide students with opportunities to practice, apply, connect, and extend the algebraic concepts and skills they are learning. Numerous illustrations, tables, graphs, and photos have been added to the exercise sets to help students visualize the problems they are solving. Problem types include writing, estimation, and calculator exercises as well as applications and multiple-choice, matching, true/false, and fill-in-the-blank problems. In the *Annotated Instructor's Edition* of the text, writing exercises are marked with icons so that teachers may assign these problems at their discretion. Exercises suitable for calculator work are marked in both the student and teacher editions with icons. (See pp. 341, 342, and 350.) Students can watch an instructor work through the solutions for exercises marked with the DVD icon on the Videos on DVD or in MyMathLab.

▶ ***Relating Concepts Exercises*** These sets of exercises help students tie together topics and develop problem-solving skills as they compare and contrast ideas, identify and describe patterns, and extend concepts to new situations. (See pp. 342 and 350.) These exercises make great collaborative activities for pairs or small groups of students.

▶ ***Summary Exercises*** These special exercise sets provide students with the all-important *mixed* review problems they need to master topics. Summaries of solution methods or additional examples are often included. (See pp. 351 and 352.)

▶ ***Test Your Word Power*** To help students understand and master mathematical vocabulary, this feature can be found in each chapter summary. Key terms from the chapter are presented along with four possible definitions in a multiple-choice format. Answers and examples illustrating each term are provided. (See p. 371.)

▶ ***Ample Opportunity for Review*** Each chapter concludes with a Chapter Summary that features Key Terms with definitions and helpful graphics, New Symbols, Test Your Word Power, and a Quick Review of each section's content with additional examples. A comprehensive set of Chapter Review Exercises, keyed to individual sections, is included, as are Mixed Review Exercises and a Chapter Test. Beginning with Chapter 2, each chapter concludes with a set of Cumulative Review Exercises that cover material going back to Chapters R and 1. (See pp. 371–382.) Students can watch an instructor work through the full solutions for all Chapter Test exercises on the Chapter Test Prep Video CD that accompanies each new copy of the text.

What's New in This Edition?

Throughout this edition of the text, we are pleased to offer the following new student-oriented features:

NEW ***Math in the Media*** These new one-page activities provide a relevant application of mathematics as it is found in various media forms, such as newspapers, movies, and TV. Designed to help teachers answer the often-asked question, "Why do I need to learn this?", these activities are well-suited for individual or collaborative work, as well as class discussions. We hope both students and instructors will enjoy them. They include the following:

- The Magic Number in Sports (p. 104)
- Interpreting (and Misinterpreting) Line Graphs (p. 276)
- At What Temperature Does Paper Burn? (p. 368)
- Connecting Graphs in the Media with Systems of Equations (p. 374)
- More Power to You, Captain Kirk (p. 420)
- Floods, Hurricanes, and Earthquakes, Oh My! (p. 444)
- To Play Baseball, You Must *Work* At It! (p. 576)
- Learn Math, Lose Weight (p. 614)
- Prime Numbers in Prime Time (p. 696)
- So, Did the Scarecrow Really Get a Brain? (p. 718)

NEW ***Study Skills*** Poor study skills are a major reason why students do not succeed in mathematics. These new two-page activities provide helpful information, tips, and strategies on a variety of essential study skills, including *Using Your Textbook, Taking Lecture Notes, Tips for Taking Math Tests*, and *Managing Your Time*. While most of the activities are concentrated in the early chapters of the text, each has been designed independently and can be used at most any point in your course with individuals or small groups of students, or as a source of material for in-class discussions. (See pp. 321 and 322.)

NEW ***Solutions to Selected Exercises*** Exercise numbers enclosed in a blue square, such as **11.**, indicate that a step-by-step, worked-out solution for the problem is included at the back of the text. These solutions are given for selected exercises that extend the skills and concepts presented in the section examples—actually providing students with a pool of examples for exercises that include some kind of twist or are a bit more difficult. (See pp. S-1 through S-21.)

NEW ***Pointers*** Pointers from the authors have been added to examples and provide students with important on-the-spot reminders and warnings about common pitfalls. (See pp. 327 and 337.)

NEW ***Chapter Test Prep Video CD*** The Chapter Test Prep Video CD provides students with the opportunity to watch instructors work through step-by-step solutions to all the Chapter Test exercises from the textbook. The Chapter Test Prep Video CD is included with each new student text.

A primary focus of this revision of the text was to polish and enhance individual presentations of topics and exercise sets, based on user and reviewer feedback, and we have worked hard to do this throughout the book. Some of the specific content changes you may notice include the following:

- The exercise sets received special attention. There are approximately 400 new and updated exercises, including problems that check conceptual understanding, focus on skill development, and provide review.
- Real-world data in over 200 applications in the examples and exercises have been updated.
- There is an increased emphasis on the difference between expressions and equations, including a new Caution at the beginning of Section 2.1. Throughout the text, we have reformatted many example solutions to use a "drop down" layout in order to further emphasize for students the difference between simplifying expressions and solving equations.
- There is an increased emphasis on checking solutions and answers, as indicated by the new **Check** tag in the exposition and examples.
- The presentation on solving linear equations in Sections 2.1–2.3 includes five new examples, with new margin problems and corresponding exercises.
- When a new type of graph is introduced (Sections 3.2, 3.5, 7.4, 9.1, 10.6, 10.7, 11.2, 11.3, 12.1–12.3, and 12.5), a new "hand-drawn" graph style is used to simulate what a student might actually sketch on graph paper.
- The *Summary Exercises on Factoring* (from the third edition) in Chapter 6 have been expanded into new Section 6.6 that now features examples and a comprehensive review of factoring strategies, plus additional randomly organized exercises.
- Presentations of the following topics have also been enhanced and expanded:

  Review of fractions and decimals (Chapter R)
  Applications from geometry (Sections 2.4 and 2.5)
  Slopes of lines and average rate of change (Section 3.3)
  Slope-intercept form of the equation of a line (Section 3.4)
  Solving systems of equations with decimal coefficients (Section 4.2)
  Scientific notation (Section 5.8)
  Solving quadratic equations by factoring (Section 6.7)
  Solving work applications (Section 7.5)
  Solving systems of linear equations in three variables (Section 8.5)
  Solving equations quadratic in form (Section 10.4)
  Graphing logarithmic functions (Section 11.3)
  Applying the laws of logarithms (Section 11.4)

What Supplements Are Available?

For a comprehensive list of the supplements and study aids that accompany *Introductory and Intermediate Algebra*, Fourth Edition, see pages xvii and xviii.

Acknowledgments

The comments, criticisms, and suggestions of users, nonusers, instructors, and students have positively shaped this textbook over the years, and we are most grateful for the many responses we have received. We especially wish to thank the following reviewers whose valuable contributions have helped to refine this edition of this text.

Mary Kay Abbey, *Montgomery College*
Randall Allbritton, *Daytona Beach Community College*
Theresa Allen, *University of Idaho*

Sonya Armstrong, *West Virginia State College*
Jannette Avery, *Monroe Community College*
Linda Beattie, *Western New Mexico University*
Linda Beller, *Brevard Community College*
Carla J. Bissell, *University of Nebraska at Omaha*
Jean Bolyard, *Fairmont State University*
Vernon Bridges, *Durham Technical Community College*
Tim C. Caldwell, *Meridian Community College*
Russell Campbell, *Fairmont State University*
Dawn Cox, *Cochise College*
Julie Dewan, *Mohawk Valley Community College*
Bill Dunn, *Las Positas College*
Lucy Edwards, *Las Positas College*
Rob Farinelli, *Community College of Allegheny–Boyce Campus*
Scott Fallstrom, *Shoreline Community College*
J. Lloyd Harris, *Gulf Coast Community College*
Terry Haynes, *Eastern Oklahoma State College*
Edith Hays, *Texas Woman's University*
Anthony Hearn, *Community College of Philadelphia*
Karen Heavin, *Morehead State University*
Elizabeth Heston, *Monroe Community College*
Sharon Jackson, *Brookhaven College*
Harriet Kiser, *Floyd College*
Jeffrey Kroll, *Brazosport College*
Barbara Krueger, *Cochise College*
Valerie Lazzara, *Palm Beach Community College*
Christine Heinecke Lehmann, *Purdue University—North Central*
Sandy Lofstock, *California Lutheran University*
Valerie H. Maley, *Cape Fear Community College*
Susan McClory, *San Jose State University*
Pam Miller, *Phoenix College*
Jeffrey Mills, *Ohio State University*
Linda J. Murphy, *Northern Essex Community College*
Celia Nippert, *Western Oklahoma State College*
Elizabeth Olgilvie, *Horry-Georgetown Technical College*
Ted Panitz, *Cape Cod Community College*
Claire Peacock, *Chattanooga State Technical Community College*
Faith Peters, *Miami Dade Community College*
Larry Pontaski, *Pueblo Community College*
Serban Raianu, *California State University—Dominguez Hills*
Janice Rech, *University of Nebraska at Omaha*
Diann Robinson, *Ivy Tech State College—Lafayette*
Rachael Schettenhelm, *Southern Connecticut State University*
Dwight Smith, *Prestonburg Community College*
Lee Ann Spahr, *Durham Technical Community College*
Theresa Stalder, *University of Illinois–Chicago*
Carol Stewart, *Fairmont State University*
Mark Tom, *College of the Sequoias*
Cora S. West, *Florida Community College at Jacksonville*
Leigh Ann Wheeler, *Greenville Technical College*
Johanna Windmueller, *Seminole Community College*
Gabriel Yimesghen, *Community College of Philadelphia*

Over the years, we have come to rely on an extensive team of experienced professionals. Our sincere thanks go to these dedicated individuals at Addison-Wesley, who worked long and hard to make this revision a success: Greg Tobin, Maureen O'Connor, Michelle Renda, Kari Heen, Courtney Slade, Kathy Manley, Barbara Atkinson, Beth Anderson, Lin Mahoney, Ceci Fleming, Alicia Frankel, and Mary Gallagher.

Abby Tanenbaum did an outstanding job helping us with manuscript preparation. We are truly grateful for her contributions to so many of our books over the years. Janette Krauss, Bonnie Boehme, and Nesbitt Graphics, Inc. provided excellent production work on the challenging format of these books. Special thanks are due Diana Hestwood and Linda Russell for their hard work on the Study Skills; Jeff Cole, who continues to provide accurate, helpful solutions manuals; Barb Brown, who helped us update the real-data applications; and Lucie Haskins for another useful index. Janis Cimperman, Shannon d'Hemecourt, Perian Herring, Paul Lorczak, and Sarah Sponholz did a wonderful and timely job accuracy checking.

As an author team, we are committed to providing the best possible text and supplements package to help students succeed and instructors teach. As we continue to work toward this goal, we would welcome any comments or suggestions you might have via e-mail to *math@pearson.com.*

Margaret L. Lial
John Hornsby
Terry McGinnis

Student Supplements

Student's Solutions Manual

- By Jeffery A. Cole, *Anoka-Ramsey Community College*
- Provides detailed solutions to the odd-numbered, section-level exercises and to all margin, Relating Concepts, Summary, Chapter Review, Chapter Test, and Cumulative Review Exercises

ISBNs: 0-321-57612-8, 978-0-321-57612-5

Videos on DVD

- Feature an engaging team of lecturers
- Include a complete set of lectures for each section of the text on DVD for student use at home or on campus
- Ideal for distance learning or supplemental instruction
- Watch an instructor work through the complete solution for all exercises marked with a DVD icon
- Include optional English and Spanish subtitles for lectures

ISBNs: 0-321-57611-X, 978-0-321-57611-8

Worksheets for Classroom or Lab Practice

- Provide extra practice exercises for every section of the text with ample space for students to show their work
- List the learning objectives and key vocabulary terms for every text section, along with vocabulary practice problems

ISBNs: 0-321-57618-7, 978-0-321-57618-7

InterAct Math Tutorial Website *www.interactmath.com*

- Offers online practice and tutorial help
- Allows students to retry an exercise with new values each time for unlimited practice and mastery
- Every exercise accompanied by an interactive guided solution that gives helpful feedback when an incorrect answer is entered
- Allows students to view steps of a worked-out sample problem similar to those in the text

Chapter Test Prep Video CD

- Watch instructors work through step-by-step solutions to all the Chapter Test exercises from the textbook
- Included with each new student text
- Available with optional English subtitles

Instructor Supplements

Annotated Instructor's Edition

- Provides answers to all text exercises in color next to the corresponding problem
- Includes icons to identify writing and calculator exercises

ISBNs: 0-321-57605-5, 978-0-321-57605-7

Instructor's Solutions Manual

- By Jeffery A. Cole, *Anoka-Ramsey Community College*
- Provides complete answers to all the exercises in the text

ISBNs: 0-321-57606-3, 978-0-321-57606-4

Additional Teaching Resources

Includes resources to help both new and adjunct faculty with course preparation and classroom management by offering helpful teaching tips correlated to the sections of the text

Available for download at *www.pearsonhighered.com*

Instructor's Resource Manual with Tests

- By James Ball, *Indiana State University*
- Contains a test bank with two diagnostic pretests, six free-response and two multiple-choice test forms per chapter, and two final exams
- Also contains a mini-lecture for each section of the text with objectives, key examples, and teaching tips
- Includes a correlation guide from the third to the fourth edition and phonetic spellings for all key terms in the text

ISBNs: 0-321-57607-1, 978-0-321-57607-1

PowerPoint® Lecture Slides

- Present key concepts and definitions from the text
- Available for download at *www.pearsonhighered.com*

TestGen® *www.pearsonhighered.com/testgen*

- Enables instructors to build, edit, print, and administer tests using a computerized bank of questions developed to cover all text objectives
- Allows instructors to create multiple but equivalent versions of the same question or test with the click of a button
- Allows instructors to modify test bank questions or add new questions
- Tests can be printed or administered online

Pearson Math Adjunct Support Center

http://www.pearsontutorservices.com/math-adjunct.html

Staffed by qualified instructors with more than 50 years of combined experience at both the community college and university levels. Assistance provided for faculty in the following areas:

- Suggested syllabus consultation
- Tips on using materials packed with your book
- Book-specific content assistance
- Teaching suggestions, including advice on classroom strategies

Available for Students and Instructors

MyMathLab® MyMathLab is a series of text-specific, easily customizable online courses for Pearson Education's textbooks in mathematics and statistics. Powered by CourseCompass™ (our online teaching and learning environment) and MathXL® (our online homework, tutorial, and assessment system), MyMathLab provides the tools needed to deliver all or a portion of a course online, whether students are in a lab setting or working from home. MyMathLab provides a rich and flexible set of course materials, featuring free-response exercises that are algorithmically generated for unlimited practice and mastery. Students can also use online tools, such as video lectures, animations, and a multimedia textbook, to independently improve their understanding and performance. Instructors can use MyMathLab's homework and test managers to select and assign online exercises correlated directly to the textbook, and they can also create and assign their own online exercises and import TestGen tests for added flexibility. MyMathLab's online gradebook—designed specifically for mathematics and statistics—automatically tracks students' homework and test results and gives the instructor control over how to calculate final grades. Instructors can also add offline (paper-and-pencil) grades to the gradebook. MyMathLab also includes access to the **Pearson Tutor Center** (*www.pearsontutorservices.com*). The Tutor Center is staffed by qualified mathematics instructors who provide textbook-specific tutoring for students via toll-free phone, fax, email, and interactive Web sessions. MyMathLab is available to qualified adopters. For more information, visit our Web Site at *www.mymathlab.com,* or contact your sales representative.

MathXL® MathXL is a powerful online homework, tutorial, and assessment system that accompanies Pearson Education's textbooks in mathematics or statistics. With MathXL, instructors can create, edit, and assign online homework and tests using algorithmically generated exercises correlated at the objective level to the textbook. They can also create and assign their own online exercises and import TestGen tests for added flexibility. All student work is tracked in MathXL's online gradebook. Students can take chapter tests in MathXL and receive personalized study plans based on their test results. The study plan diagnoses weaknesses and links students directly to tutorial exercises for the objectives they need to study and retest. Students can also access supplemental animations and video clips directly from selected exercises. MathXL is available to qualified adopters. For more information, visit our Web Site at *www.mathxl.com,* or contact your sales representative.

MathXL® Tutorials on CD This interactive tutorial CD-ROM provides algorithmically generated practice exercises that are correlated at the objective level to the exercises in the textbook. Every practice exercise is accompanied by an example and a guided solution designed to involve students in the solution process. Selected exercises may also include a video clip to help students visualize concepts. The software provides helpful feedback for incorrect answers and can generate printed summaries of students' progress.
ISBNs:
0-321-57610-1
978-0-321-57610-1

Study Skills

YOUR BRAIN *CAN* LEARN MATHEMATICS

OBJECTIVES

1. **Describe how practice fosters dendrite growth.**
2. **Explain the effect of anxiety on the brain.**

Your brain knows how to learn, just as your lungs know how to breathe; however, there are important things you can do to maximize your brain's ability to do its work. This short introduction will help you choose effective strategies for learning mathematics. This is a simplified explanation of a complex process.

Your brain's outer layer, called the **neocortex,** is where higher level thinking, language, reasoning, and purposeful behavior occur. The neocortex has about 100 billion (100,000,000,000) brain cells called **neurons.**

Learning Something New

- As you learn something new, threadlike branches grow out of each neuron. These branches are called **dendrites.**
- When the dendrite from one neuron grows close enough to the dendrite from another neuron, a connection is made. There is a small gap at the connection point called a **synapse.** One dendrite sends an electrical signal across the gap to another dendrite.
- ***Learning = growth and connecting of dendrites.***

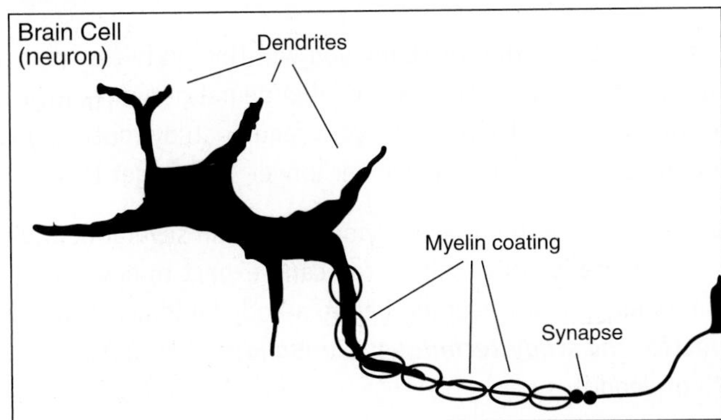

A neuron with several dendrites: one dendrite has developed a myelin coating through repeated practice

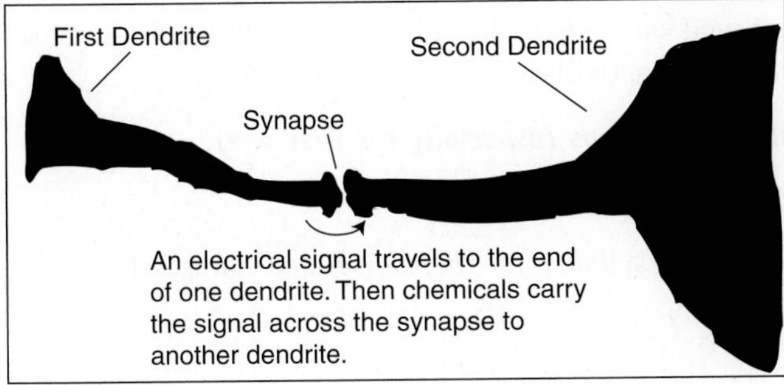

A close up view of the connection (synapse) between two dendrites

Remembering New Skills

- When you practice a skill just once or twice, the connections between neurons are very weak. If you do not practice the skill again, the dendrites at the connection points wither and die back. You have forgotten the new skill.
- If you practice a new skill many times, the dendrites for that skill become coated with a fatty protein called **myelin.** Each time one dendrite sends a signal to another dendrite, the myelin coating becomes thicker and smoother, allowing the signals to move faster and with less interference. Thinking can now occur more quickly and easily, and you will remember the skill for a long time because the dendrite connections are strong.

Become An Effective Student

- You grow dendrites specifically for the topic you are studying. ***So, be sure you are actively learning and practicing.***
- If you practice something the wrong way, you will develop connections for doing it the wrong way. ***So, as you study, check frequently that you are getting correct answers.***
- As you study a new topic that is related to things you already know, your brain sends signals throughout the network of dendrites for the related topics. In this way, you build a complex **neural network** that allows you to apply concepts, see differences and similarities between ideas, and understand relationships between concepts.

In the first few chapters of this textbook you will find activities that are designed to help you grow and develop your own reliable neural networks for mathematics. When you incorporate the activities into your regular study routine, you will discover that you understand better, remember longer, and forget less.

It takes time to learn something new. Trying to cram in several new concepts and skills at the last minute is not possible. You can't expect to develop huge muscles by lifting weights for just one evening before a body building competition. In the same way, ***practice the study techniques throughout the course*** to facilitate strong growth of dendrites.

When Anxiety Strikes

If you are under stress or feeling anxious, such as during a test, your body secretes **adrenaline** into your system. Adrenaline in the brain blocks connections between neurons, so you can't think clearly. If you've ever experienced "blanking out" on a test, you know what adrenaline does. You'll learn several solutions to that problem in later activities.

Start Your Course Right!

- ***Attend all class sessions (especially the first one).***
- ***Gather the necessary supplies.***
- ***Carefully read the syllabus for the course, and ask questions if you don't understand.***

Prealgebra Review

R.1 Fractions

OBJECTIVES

1. Identify prime numbers.
2. Write numbers in prime factored form.
3. Write fractions in lowest terms.
4. Convert between improper fractions and mixed numbers.
5. Multiply and divide fractions.
6. Add and subtract fractions.
7. Solve applied problems that involve fractions.

The numbers used most often in everyday life are the **whole numbers,**

$$0, 1, 2, 3, 4, 5, \ldots$$

and **fractions,** such as

$$\frac{1}{3}, \quad \frac{5}{4}, \quad \text{and} \quad \frac{11}{12}.$$

The parts of a fraction are named as follows.

$$\text{Fraction bar} \rightarrow \frac{4}{7} \begin{array}{l} \leftarrow \text{Numerator} \\ \leftarrow \text{Denominator} \end{array}$$

If the numerator of a fraction is less than the denominator, we call it a **proper fraction.** A proper fraction has a value less than 1. If the numerator is greater than or equal to the denominator, the fraction is an **improper fraction.** Some examples follow.

$$\frac{1}{5}, \quad \frac{2}{7}, \quad \frac{9}{10}, \quad \frac{23}{25} \qquad \text{Proper fractions}$$

$$\frac{3}{2}, \quad \frac{5}{5}, \quad \frac{11}{7}, \quad \frac{28}{4} \qquad \text{Improper fractions}$$

OBJECTIVE 1 Identify prime numbers. In work with fractions, we will need to write the numerators and denominators as products. A **product** is the answer to a multiplication problem. When 12 is written as the product 2×6, for example, 2 and 6 are called **factors** of 12. Other factors of 12 are 1, 3, 4, and 12. A whole number is **prime** if it has exactly two different factors (itself and 1). The first dozen primes are listed here.

2, 3, 5, 7, 11, 13, 17, 19, 23, 29, 31, 37 Prime numbers

A whole number greater than 1 that is not prime is called a **composite number.** Some examples follow.

4, 6, 8, 9, 10, 12 Composite numbers

The number 1 is neither prime nor composite.

EXAMPLE 1 Distinguishing between Prime and Composite Numbers

Decide whether each number is *prime* or *composite*.

(a) 33
33 has factors of 3 and 11, as well as 1 and 33, so it is composite.

(b) 43
Since there are no numbers other than 1 and 43 itself that divide *evenly* into 43, the number 43 is prime.

(c) 9832
Since 9832 can be divided by 2, giving 2×4916, it is composite.

◀ Work Problem 1 at the Side.

1 Tell whether each number is *prime* or *composite*.

(a) 12

(b) 13

(c) 27

(d) 59

(e) 1806

OBJECTIVE 2 Write numbers in prime factored form. To factor a number means to write it as the product of two or more numbers. Factoring is the reverse of multiplying two numbers to get the product.

Multiplication	Factoring
$6 \cdot 3 = 18$	$18 = 6 \cdot 3$
↑ ↑ ↑	↑ ↑ ↑
Factors Product	Product Factors

In algebra, a dot $\cdot$ is used instead of the $\times$ symbol to indicate multiplication because $\times$ may be confused with the letter x. A composite number written using factors that are all prime numbers is in **prime factored form.**

EXAMPLE 2 Writing Numbers in Prime Factored Form

Write each number in prime factored form.

(a) 35
Factor 35 as the product of the prime factors 5 and 7, or as $35 = 5 \cdot 7$.

(b) 24
We use a factor tree, as shown below. The prime factors are circled.

Divide by the smallest prime, 2, to get $24 = 2 \cdot 12$.

Now divide 12 by 2 to find factors of 12. $24 = 2 \cdot 2 \cdot 6$

Since 6 can be factored as $2 \cdot 3$, $24 = 2 \cdot 2 \cdot 2 \cdot 3$, where all factors are prime. $24 = 2 \cdot 2 \cdot 2 \cdot 3$

24
(2) · 12
(2) · 6
(2) · (3)

◀ Work Problem 2 at the Side.

2 Write each number in prime factored form.

(a) 70

(b) 72

(c) 693

(d) 97

OBJECTIVE 3 Write fractions in lowest terms. A fraction is in **lowest terms** when the numerator and denominator have no factors in common (other than 1). The following properties are useful.

Properties of 1

Any nonzero number divided by itself is equal to 1; for example, $\frac{3}{3} = 1$.

Any number multiplied by 1 remains the same; for example, $7 \cdot 1 = 7$.

ANSWERS

1. (a) composite (b) prime (c) composite (d) prime (e) composite
2. (a) $2 \cdot 5 \cdot 7$ (b) $2 \cdot 2 \cdot 2 \cdot 3 \cdot 3$ (c) $3 \cdot 3 \cdot 7 \cdot 11$ (d) 97 is prime.

Writing a Fraction in Lowest Terms

Step 1 Write the numerator and denominator in prime factored form.

Step 2 Replace each pair of factors common to the numerator and denominator with 1.

Step 3 Multiply the remaining factors in the numerator and in the denominator.

(This procedure is sometimes called "simplifying the fraction.")

EXAMPLE 3 **Writing Fractions in Lowest Terms**

Write each fraction in lowest terms.

(a) $\frac{10}{15} = \frac{2 \cdot 5}{3 \cdot 5} = \frac{2}{3} \cdot \frac{5}{5} = \frac{2}{3} \cdot 1 = \frac{2}{3}$

Since 5 is a common factor of 10 and 15, we use the first property of 1 to replace $\frac{5}{5}$ with 1.

(b) $\frac{15}{45} = \frac{3 \cdot 5}{3 \cdot 3 \cdot 5} = \frac{1 \cdot 3 \cdot 5}{3 \cdot 3 \cdot 5} = \frac{1}{3} \cdot \frac{3}{3} \cdot \frac{5}{5} = \frac{1}{3} \cdot 1 \cdot 1 = \frac{1}{3}$

Multiplying by 1 in the numerator does not change the value of the numerator and makes it possible to rewrite the expression as the product of three fractions in the next step.

(c) $\frac{150}{200}$

It is not always necessary to factor into *prime* factors in Step 1. Here, if you see that 50 is a common factor of the numerator and the denominator, factor as follows:

$$\frac{150}{200} = \frac{3 \cdot 50}{4 \cdot 50} = \frac{3}{4} \cdot 1 = \frac{3}{4}.$$

Note

When writing a fraction in lowest terms, look for the largest common factor in the numerator and the denominator. If none is obvious, factor the numerator and the denominator into prime factors. *Any* common factor can be used and the fraction can be simplified in stages.

For example, $\frac{150}{200} = \frac{15 \cdot 10}{20 \cdot 10} = \frac{3 \cdot 5 \cdot 10}{4 \cdot 5 \cdot 10} = \frac{3}{4}.$

Work Problem 3 at the Side. ▶

OBJECTIVE 4 Convert between improper fractions and mixed numbers. A **mixed number** is a single number that represents the sum of a whole number and a fraction. For example, $5\frac{3}{4} = 5 + \frac{3}{4}$. Mixed numbers are commonly used in everyday life. For example, a child's age might be given as $2\frac{1}{2}$ years old, or a cookie recipe might call for $3\frac{1}{4}$ cups of flour.

3 Write each fraction in lowest terms.

(a) $\frac{8}{14}$

(b) $\frac{35}{42}$

(c) $\frac{72}{120}$

Answers

3. (a) $\frac{4}{7}$ **(b)** $\frac{5}{6}$ **(c)** $\frac{3}{5}$

4 Write $\frac{92}{5}$ as a mixed number.

Any improper fraction whose value is not a whole number can be rewritten as a mixed number, and any mixed number can be rewritten as an improper fraction. It is convenient to be able to convert between these two forms of the same number.

EXAMPLE 4 Converting an Improper Fraction to a Mixed Number

Write $\frac{59}{8}$ as a mixed number.

To convert an improper fraction to a mixed number, divide the numerator by the denominator. Here, divide 59 by 8. Use the quotient and remainder to form the mixed number.

Denominator of fraction (divisor) → 8)59 ← Numerator of fraction (dividend)

$$\begin{array}{r} 7 \\ 8\overline{)59} \\ \underline{56} \\ 3 \end{array}$$

7 ← Quotient; 3 ← Remainder

Thus, $\frac{59}{8} = 7\frac{3}{8}$.

◀ Work Problem 4 at the Side.

EXAMPLE 5 Converting a Mixed Number to an Improper Fraction

5 Write $11\frac{2}{3}$ as an improper fraction.

Write $6\frac{4}{7}$ as an improper fraction.

To convert a mixed number to an improper fraction, multiply the denominator of the fraction by the whole number and add the numerator of the fraction to get the numerator of the improper fraction. To write $6\frac{4}{7}$ as an improper fraction, the numerator is

$$7 \cdot 6 + 4 = 42 + 4 = 46.$$

The denominator of the improper fraction is the same as the denominator in the mixed number. The denominator is 7. Thus, $6\frac{4}{7} = \frac{46}{7}$.

◀ Work Problem 5 at the Side.

OBJECTIVE 5 Multiply and divide fractions.

Multiplying Fractions

To multiply two fractions, multiply the numerators to get the numerator of the product, and multiply the denominators to get the denominator of the product. ***The product should be written in lowest terms.***

EXAMPLE 6 Multiplying Fractions

Find each product, and write it in lowest terms.

(a) $\frac{3}{8} \cdot \frac{4}{9} = \frac{3 \cdot 4}{8 \cdot 9}$ Multiply numerators. Multiply denominators.

$= \frac{3 \cdot 4}{2 \cdot 4 \cdot 3 \cdot 3}$ Factor.

$= \frac{1}{2 \cdot 3} = \frac{1}{6}$ Write in lowest terms.

Continued on Next Page

ANSWERS

4. $18\frac{2}{5}$

5. $\frac{35}{3}$

(b) $2\frac{1}{3} \cdot 5\frac{1}{2} = \frac{7}{3} \cdot \frac{11}{2}$ Write as improper fractions.

$= \frac{77}{6}$, or $12\frac{5}{6}$ Multiply numerators and denominators; write as a mixed number.

Work Problem 6 at the Side. ▶

Two fractions are **reciprocals** of each other if their product is 1. For example, $\frac{3}{4}$ and $\frac{4}{3}$ are reciprocals because

$$\frac{3}{4} \cdot \frac{4}{3} = 1.$$

The numbers $\frac{7}{11}$ and $\frac{11}{7}$ are reciprocals also. Other examples are $\frac{1}{5}$ and 5 (which can be written as $\frac{5}{1}$), $\frac{4}{9}$ and $\frac{9}{4}$, and 16 and $\frac{1}{16}$.

Because division is the opposite, or inverse, of multiplication, we use reciprocals to divide fractions.

Dividing Fractions

To divide two fractions, multiply the first fraction by the reciprocal of the second. The result, called the **quotient,** should be written in lowest terms.

The reason this method works will be explained in **Section 1.6.** However, as an example, we know that $20 \div 10 = 2$, and $20 \cdot \frac{1}{10} = 2$.

EXAMPLE 7 Dividing Fractions

Find each quotient, and write it in lowest terms.

(a) $\frac{3}{4} \div \frac{8}{5} = \frac{3}{4} \cdot \frac{5}{8} = \frac{3 \cdot 5}{4 \cdot 8} = \frac{15}{32}$

Multiply by the reciprocal of the second fraction.

(b) $\frac{3}{4} \div \frac{5}{8} = \frac{3}{4} \cdot \frac{8}{5} = \frac{3 \cdot 8}{4 \cdot 5} = \frac{3 \cdot 4 \cdot 2}{4 \cdot 5} = \frac{6}{5}$, or $1\frac{1}{5}$

The answer can be written as an improper fraction or as a mixed number.

(c) $\frac{5}{8} \div 10 = \frac{5}{8} \div \frac{10}{1} = \frac{5}{8} \cdot \frac{1}{10} = \frac{5 \cdot 1}{8 \cdot 10} = \frac{5 \cdot 1}{8 \cdot 2 \cdot 5} = \frac{1}{16}$

Write 10 as $\frac{10}{1}$.

Remember to write 1 in the numerator.

(d) $1\frac{2}{3} \div 4\frac{1}{2} = \frac{5}{3} \div \frac{9}{2}$ Write as improper fractions.

$= \frac{5}{3} \cdot \frac{2}{9}$ Multiply by the reciprocal of the second fraction.

$= \frac{10}{27}$ Multiply numerators and denominators.

Always check to make sure your final answer is in lowest terms.

6 Find each product, and write it in lowest terms.

(a) $\frac{5}{8} \cdot \frac{2}{10}$

(b) $\frac{1}{10} \cdot \frac{12}{5}$

(c) $\frac{7}{9} \cdot \frac{12}{14}$

(d) $3\frac{1}{3} \cdot 1\frac{3}{4}$

ANSWERS

6. (a) $\frac{1}{8}$ **(b)** $\frac{6}{25}$ **(c)** $\frac{2}{3}$ **(d)** $\frac{35}{6}$, or $5\frac{5}{6}$

7 Find each quotient, and write it in lowest terms.

(a) $\frac{3}{10} \div \frac{2}{7}$

(b) $\frac{3}{4} \div \frac{7}{16}$

(c) $\frac{4}{3} \div 6$

(d) $3\frac{1}{4} \div 1\frac{2}{5}$

8 Add. Write sums in lowest terms.

(a) $\frac{3}{5} + \frac{4}{5}$

(b) $\frac{5}{14} + \frac{3}{14}$

Answers

7. (a) $\frac{21}{20}$, or $1\frac{1}{20}$ (b) $\frac{12}{7}$, or $1\frac{5}{7}$ (c) $\frac{2}{9}$ (d) $\frac{65}{28}$, or $2\frac{9}{28}$

8. (a) $\frac{7}{5}$, or $1\frac{2}{5}$ (b) $\frac{4}{7}$

CAUTION

Notice in Example 7 that *only* the second fraction (the divisor) is replaced by its reciprocal in the multiplication.

◀ Work Problem 7 at the Side.

OBJECTIVE 6 Add and subtract fractions. The result of adding two numbers is called the **sum** of the numbers. For example, since $2 + 3 = 5$, the sum of 2 and 3 is 5.

Adding Fractions

To find the sum of two fractions with the *same* denominator, add their numerators and ***keep the same denominator.***

EXAMPLE 8 Adding Fractions with the Same Denominator

Add. Write sums in lowest terms.

(a) $\frac{3}{7} + \frac{2}{7} = \frac{3+2}{7} = \frac{5}{7}$ Add numerators; keep the same denominator.

(b) $\frac{2}{10} + \frac{3}{10} = \frac{2+3}{10} = \frac{5}{10} = \frac{1}{2}$ Write in lowest terms.

◀ Work Problem 8 at the Side.

If the fractions to be added do not have the same denominator, the procedure above can still be used, but only *after* the fractions are rewritten with a common denominator. For example, to rewrite $\frac{3}{4}$ as a fraction with a denominator of 32,

$$\frac{3}{4} = \frac{?}{32},$$

we must find the number that can be multiplied by 4 to give 32. Since $4 \cdot 8 = 32$, we use the number 8. By the second property of 1, we can multiply the numerator and the denominator by 8.

$$\frac{3}{4} = \frac{3}{4} \cdot 1 = \frac{3}{4} \cdot \frac{8}{8} = \frac{3 \cdot 8}{4 \cdot 8} = \frac{24}{32}$$

This process is the reverse of writing a fraction in lowest terms.

Finding the Least Common Denominator (LCD)

Step 1 Factor all denominators to prime factored form.

Step 2 The LCD is the product of every (different) factor that appears in any of the factored denominators. If a factor is repeated, use the greatest number of repeats as factors of the LCD.

Step 3 Write each fraction with the LCD as the denominator, using the second property of 1.

EXAMPLE 9 Adding Fractions with Different Denominators

Add. Write sums in lowest terms.

(a) $\frac{4}{15} + \frac{5}{9}$

Step 1 To find the LCD, factor the denominators to prime factored form.

$$15 = 5 \cdot 3 \quad \text{and} \quad 9 = 3 \cdot 3$$

3 is a factor of both denominators.

Step 2

$$\text{LCD} = \underbrace{5 \cdot 3}_{15} \cdot 3 = 45 \quad (9 = 3 \cdot 3)$$

In this example, the LCD needs one factor of 5 and two factors of 3 because the second denominator has two factors of 3.

Step 3 Now we can use the second property of 1 to write each fraction with 45 as the denominator.

$$\frac{4}{15} = \frac{4}{15} \cdot \frac{3}{3} = \frac{12}{45} \quad \text{and} \quad \frac{5}{9} = \frac{5}{9} \cdot \frac{5}{5} = \frac{25}{45}$$

At this stage, the fractions are *not* in lowest terms.

Now add the two equivalent fractions to get the required sum.

$$\frac{4}{15} + \frac{5}{9} = \frac{12}{45} + \frac{25}{45} = \frac{37}{45}$$

(b) $3\frac{1}{2} + 2\frac{3}{4} = \frac{7}{2} + \frac{11}{4}$ Change to improper fractions.

$= \frac{14}{4} + \frac{11}{4}$ Get a common denominator.

$= \frac{25}{4}$, or $6\frac{1}{4}$ Add; write as a mixed number.

(c) $45\frac{2}{3} + 73\frac{1}{2}$

We use a vertical method here.

$$\begin{array}{r} 45\frac{2}{3} = 45\frac{4}{6} \\ +\ 73\frac{1}{2} = 73\frac{3}{6} \\ \hline \end{array}$$

Add the whole numbers and the fractions separately.

$$118\frac{7}{6} = 118 + \left(1 + \frac{1}{6}\right) = 119\frac{1}{6}$$

Work Problem **9** *at the Side.* ▶

The **difference** between two numbers is found by subtracting the numbers. For example, $9 - 5 = 4$, so the difference between 9 and 5 is 4. We find the difference between two fractions as given in the box on the next page.

9 Add. Write sums in lowest terms.

(a) $\frac{7}{30} + \frac{2}{45}$

(b) $\frac{17}{10} + \frac{8}{27}$

(c) $2\frac{1}{8} + 1\frac{2}{3}$

(d) $132\frac{4}{5} + 28\frac{3}{4}$

Answers

9. (a) $\frac{5}{18}$ **(b)** $\frac{539}{270}$, or $1\frac{269}{270}$
(c) $\frac{91}{24}$, or $3\frac{19}{24}$ **(d)** $161\frac{11}{20}$

10 Subtract.

(a) $\frac{9}{11} - \frac{3}{11}$

(b) $\frac{13}{15} - \frac{5}{6}$

(c) $2\frac{3}{8} - 1\frac{1}{2}$

(d) $50\frac{1}{4} - 32\frac{2}{3}$

Subtracting Fractions

To find the difference between two fractions with the *same* denominator, subtract their numerators and ***keep the same denominator.***

If the fractions have *different* denominators, write them with a common denominator first.

EXAMPLE 10 Subtracting Fractions

Subtract. Write differences in lowest terms.

(a) $\frac{15}{8} - \frac{3}{8} = \frac{15 - 3}{8}$ Subtract numerators; keep the same denominator.

$= \frac{12}{8}$

$= \frac{3}{2}$, or $1\frac{1}{2}$ Write in lowest terms or as a mixed number.

(b) $\frac{15}{16} - \frac{4}{9}$

Since $16 = 2 \cdot 2 \cdot 2 \cdot 2$ and $9 = 3 \cdot 3$ have no common factors, the LCD is $16 \cdot 9 = 144$.

$\frac{15}{16} - \frac{4}{9} = \frac{15 \cdot 9}{16 \cdot 9} - \frac{4 \cdot 16}{9 \cdot 16}$ Get a common denominator.

$= \frac{135}{144} - \frac{64}{144}$

$= \frac{71}{144}$ Subtract numerators; keep the same denominator.

(c) $2\frac{1}{2} - 1\frac{3}{4} = \frac{5}{2} - \frac{7}{4}$ Write as improper fractions.

$= \frac{10}{4} - \frac{7}{4}$ Get a common denominator.

$= \frac{3}{4}$ Subtract.

Alternatively, we could use a vertical method.

$$\begin{array}{rcrcr} 2\frac{1}{2} & = & 2\frac{2}{4} & = & 1\frac{6}{4} \\ -\,1\frac{3}{4} & = & 1\frac{3}{4} & = & 1\frac{3}{4} \\ \hline & & & & \frac{3}{4} \end{array}$$

◀ Work Problem 10 at the Side.

OBJECTIVE 7 Solve applied problems that involve fractions.

We often see mixed numbers used in applications of mathematics, as shown in Examples 11 and 12 on the next page.

ANSWERS

10. (a) $\frac{6}{11}$ (b) $\frac{1}{30}$ (c) $\frac{7}{8}$ (d) $17\frac{7}{12}$

EXAMPLE 11 **Solving an Applied Problem with Fractions**

The diagram in Figure 1 appears in the book *Woodworker's 39 Sure-Fire Projects.* It is a view of a bookcase/desk. Add the fractions in the diagram to find the height of the bookcase/desk to the top of the writing surface.

We must add the following measures (″ means inches):

$$\frac{3}{4},\quad 4\frac{1}{2},\quad 9\frac{1}{2},\quad \frac{3}{4},\quad 9\frac{1}{2},\quad \frac{3}{4},\quad 4\frac{1}{2}.$$

We begin by changing $4\frac{1}{2}$ to $4\frac{2}{4}$ and $9\frac{1}{2}$ to $9\frac{2}{4}$, since the common denominator is 4. Then we use the method of Example 9(c).

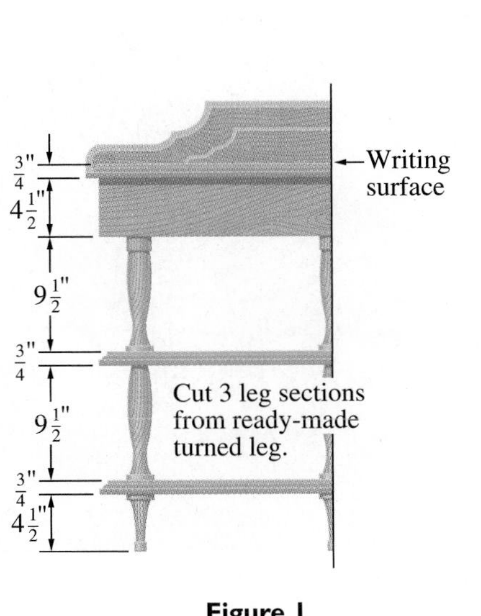

Figure 1

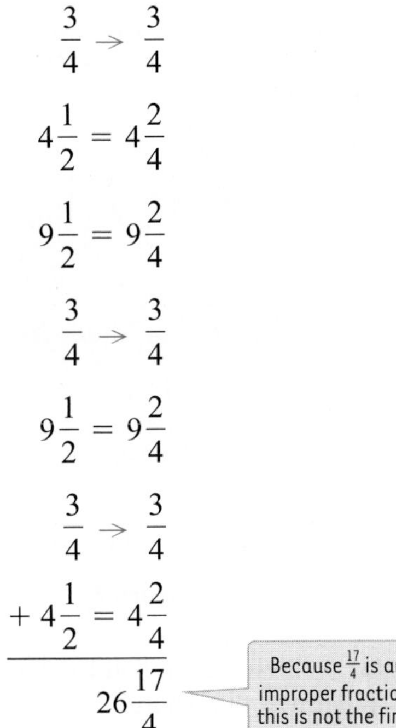

$$\begin{array}{rcl} \frac{3}{4} & \rightarrow & \frac{3}{4} \\ 4\frac{1}{2} & = & 4\frac{2}{4} \\ 9\frac{1}{2} & = & 9\frac{2}{4} \\ \frac{3}{4} & \rightarrow & \frac{3}{4} \\ 9\frac{1}{2} & = & 9\frac{2}{4} \\ \frac{3}{4} & \rightarrow & \frac{3}{4} \\ +\,4\frac{1}{2} & = & 4\frac{2}{4} \\ \hline & & 26\frac{17}{4} \end{array}$$

Because $\frac{17}{4}$ is an improper fraction, this is not the final answer.

Since $\frac{17}{4} = 4\frac{1}{4}$, $26\frac{17}{4} = 26 + 4\frac{1}{4} = 30\frac{1}{4}$. The height is $30\frac{1}{4}$ in.

Work Problem **11** *at the Side.* ▶

EXAMPLE 12 **Solving an Applied Problem with Fractions**

An upholsterer needs $2\frac{1}{4}$ yd from a bolt of fabric to cover a chair. How many chairs can be covered with $23\frac{2}{3}$ yd of fabric?

To better understand the problem, we replace the fractions with whole numbers. Suppose each chair requires 2 yd, and we have 24 yd of fabric. Dividing 24 by 2 gives the number of chairs (12) that can be covered. To solve the original problem, we must divide $23\frac{2}{3}$ by $2\frac{1}{4}$.

$$23\frac{2}{3} \div 2\frac{1}{4} = \frac{71}{3} \div \frac{9}{4}$$
$$= \frac{71}{3} \cdot \frac{4}{9}$$
$$= \frac{284}{27}, \quad \text{or} \quad 10\frac{14}{27}$$

Thus, 10 chairs can be covered, with some fabric left over.

Work Problem **12** *at the Side.* ▶

11 Solve the problem.

To make a three-piece outfit from the same fabric, Wei Jen needs $1\frac{1}{4}$ yd for the blouse, $1\frac{2}{3}$ yd for the skirt, and $2\frac{1}{2}$ yd for the jacket. How much fabric does she need?

12 Solve the problem.

A gallon of paint covers 500 ft^2. (ft^2 means square feet.) To paint his house, Tram needs enough paint to cover 4200 ft^2. How many gallons of paint should he buy?

Answers

11. $5\frac{5}{12}$ yd

12. $8\frac{2}{5}$ gal are needed, so he should buy 9 gal.

A **circle graph,** or **pie chart,** is often used to give a pictorial representation of data. A circle is used to indicate the total of all the categories represented. The circle is divided into sectors, or wedges (like pieces of pie), whose sizes show the relative magnitudes of the categories. The sum of all the fractional parts must be 1 (for 1 whole circle).

13 Refer to the circle graph in Figure 2.

(a) Which region had the second-largest number of Internet users in November 2005?

(b) Estimate the number of Internet users in Europe.

(c) How many actual Internet users were there in Europe?

EXAMPLE 13 Using a Circle Graph to Interpret Information

In November 2005, there were about 970 million Internet users worldwide. The circle graph in Figure 2 shows the approximate fractions of these users living in various regions of the world.

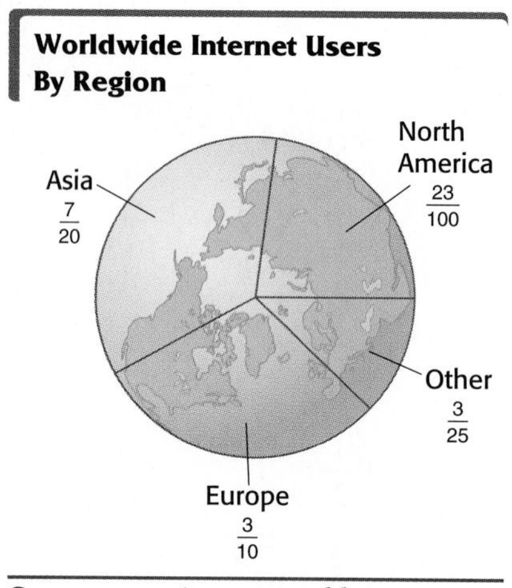

Figure 2

(a) Which region had the largest share of Internet users in November 2005? What was that share?

In the circle graph, the sector for Asia is the largest, so Asia had the largest share of Internet users, $\frac{7}{20}$.

(b) Estimate the number of Internet users in North America in November 2005.

A share of $\frac{23}{100}$ can be rounded to $\frac{25}{100}$, or $\frac{1}{4}$, and the total number of Internet users, 970 million, can be rounded to 1000 million (1 billion). We multiply $\frac{1}{4}$ by 1000. The number of Internet users in North America would be about

$$\frac{1}{4} \cdot 1000 = 250 \text{ million.}$$

(c) How many actual Internet users were there in North America in November 2005?

To find the answer, we multiply the actual fraction from the graph for North America, $\frac{23}{100}$, by the number of users, 970 million:

$$\frac{23}{100} \cdot 970 = \frac{23}{100} \cdot \frac{970}{1} = \frac{22{,}310}{100} = 223\frac{1}{10}.$$

Thus, $223\frac{1}{10}$ million, or 223,100,000 (since $\frac{1}{10}$ million $= \frac{1}{10} \cdot 1{,}000{,}000 = 100{,}000$) people in North America used the Internet. This number is reasonable, given our estimate in part (b).

◀ Work Problem 13 at the Side.

ANSWERS

13. **(a)** Europe **(b)** 300 million **(c)** 291 million

R.1 Exercises

FOR EXTRA HELP

PRACTICE
WATCH

DOWNLOAD

READ

REVIEW

Decide whether each statement is true *or* false. *If it is* false, *say why.*

1. In the fraction $\frac{3}{7}$, 3 is the numerator and 7 is the denominator.

true

2. The mixed number equivalent of $\frac{41}{5}$ is $8\frac{1}{5}$.

true

3. The fraction $\frac{17}{51}$ is in lowest terms.

false; The fraction $\frac{17}{51}$ can be simplified to $\frac{1}{3}$.

4. The reciprocal of $\frac{8}{2}$ is $\frac{4}{1}$.

false; The reciprocal of $\frac{8}{2} = 4$ is $\frac{2}{8} = \frac{1}{4}$.

5. The product of 8 and 2 is 10.

false; *Product* indicates multiplication, so the product of 8 and 2 is 16.

6. The difference between 12 and 2 is 6.

false; *Difference* indicates subtraction, so the difference between 12 and 2 is 10.

Identify each number as prime, composite, *or* neither. *See Example 1.*

7. 19 — **prime**

8. 99 — **composite**

9. 52 — **composite**

10. 61 — **prime**

11. 2468 — **composite**

12. 3125 — **composite**

13. 1 — **neither**

14. 83 — **prime**

Write each number in prime factored form. See Example 2.

15. 30 — $2 \cdot 3 \cdot 5$

16. 40 — $2 \cdot 2 \cdot 2 \cdot 5$

17. 252 — $2 \cdot 2 \cdot 3 \cdot 3 \cdot 7$

18. 168 — $2 \cdot 2 \cdot 2 \cdot 3 \cdot 7$

19. 124 — $2 \cdot 2 \cdot 31$

20. 165 — $3 \cdot 5 \cdot 11$

21. 29 — 29

22. 31 — 31

Write each fraction in lowest terms. See Example 3.

23. $\frac{8}{16}$ — $\frac{1}{2}$

24. $\frac{4}{12}$ — $\frac{1}{3}$

25. $\frac{15}{18}$ — $\frac{5}{6}$

26. $\frac{16}{20}$ — $\frac{4}{5}$

27. $\frac{15}{75}$ — $\frac{1}{5}$

28. $\frac{24}{64}$ — $\frac{3}{8}$

29. $\frac{144}{120}$ — $\frac{6}{5}$

30. $\frac{132}{77}$ — $\frac{12}{7}$

Write each improper fraction as a mixed number. See Example 4.

31. $\frac{12}{7}$ $1\frac{5}{7}$

32. $\frac{28}{5}$ $5\frac{3}{5}$

33. $\frac{77}{12}$ $6\frac{5}{12}$

34. $\frac{101}{15}$ $6\frac{11}{15}$

35. $\frac{83}{11}$ $7\frac{6}{11}$

36. $\frac{67}{13}$ $5\frac{2}{13}$

Write each mixed number as an improper fraction. See Example 5.

37. $2\frac{3}{5}$ $\frac{13}{5}$

38. $5\frac{6}{7}$ $\frac{41}{7}$

39. $10\frac{3}{8}$ $\frac{83}{8}$

40. $12\frac{2}{3}$ $\frac{38}{3}$

41. $10\frac{4}{5}$ $\frac{54}{5}$

42. $18\frac{1}{6}$ $\frac{109}{6}$

43. For the fractions $\frac{p}{q}$ and $\frac{r}{s}$, which can serve as a common denominator? **A**

A. $q \cdot s$

B. $q + s$

C. $p \cdot r$

D. $p + r$

44. Which fraction is *not* equal to $\frac{5}{9}$? **C**

A. $\frac{15}{27}$ **B.** $\frac{30}{54}$

C. $\frac{40}{74}$ **D.** $\frac{55}{99}$

Find each product or quotient, and write it in lowest terms. See Examples 6 and 7.

45. $\frac{4}{5} \cdot \frac{6}{7}$ $\frac{24}{35}$

46. $\frac{5}{9} \cdot \frac{10}{7}$ $\frac{50}{63}$

47. $\frac{1}{10} \cdot \frac{12}{5}$ $\frac{6}{25}$

48. $\frac{6}{11} \cdot \frac{2}{3}$ $\frac{4}{11}$

49. $\frac{15}{4} \cdot \frac{8}{25}$ $\frac{6}{5}$, or $1\frac{1}{5}$

50. $\frac{4}{7} \cdot \frac{21}{8}$ $\frac{3}{2}$, or $1\frac{1}{2}$

51. $2\frac{2}{3} \cdot 5\frac{4}{5}$ $\frac{232}{15}$, or $15\frac{7}{15}$

52. $3\frac{3}{5} \cdot 7\frac{1}{6}$ $\frac{129}{5}$, or $25\frac{4}{5}$

53. $\frac{5}{4} \div \frac{3}{8}$ $\frac{10}{3}$, or $3\frac{1}{3}$

54. $\frac{7}{6} \div \frac{9}{10}$ $\frac{35}{27}$, or $1\frac{8}{27}$

55. $\frac{32}{5} \div \frac{8}{15}$ 12

56. $\frac{24}{7} \div \frac{6}{21}$ 12

57. $\frac{3}{4} \div 12$ $\frac{1}{16}$

58. $\frac{2}{5} \div 30$ $\frac{1}{75}$

59. $2\frac{5}{8} \div 1\frac{15}{32}$ $\frac{84}{47}$, or $1\frac{37}{47}$

60. $2\frac{3}{10} \div 7\frac{4}{5}$ $\frac{23}{78}$

Find each sum or difference, and write it in lowest terms. See Examples 8–10.

61. $\frac{7}{12}+\frac{1}{12}$

$\frac{2}{3}$

62. $\frac{3}{16}+\frac{5}{16}$

$\frac{1}{2}$

63. $\frac{5}{9}+\frac{1}{3}$

$\frac{8}{9}$

64. $\frac{4}{15}+\frac{1}{5}$

$\frac{7}{15}$

65. $3\frac{1}{8}+\frac{1}{4}$

$\frac{27}{8}$, or $3\frac{3}{8}$

66. $5\frac{3}{4}+\frac{2}{3}$

$\frac{77}{12}$, or $6\frac{5}{12}$

67. $\frac{7}{12}-\frac{1}{9}$

$\frac{17}{36}$

68. $\frac{11}{16}-\frac{1}{12}$

$\frac{29}{48}$

69. $6\frac{1}{4}-5\frac{1}{3}$

$\frac{11}{12}$

70. $8\frac{4}{5}-7\frac{4}{9}$

$\frac{61}{45}$, or $1\frac{16}{45}$

71. $\frac{5}{3}+\frac{1}{6}-\frac{1}{2}$

$\frac{4}{3}$, or $1\frac{1}{3}$

72. $\frac{7}{15}+\frac{1}{6}-\frac{1}{10}$

$\frac{8}{15}$

Solve each applied problem. See Examples 11 and 12.

Use the chart, which appears on a package of Quaker Quick Grits, to answer the questions in Exercises 73 and 74.

73. How many cups of water would be needed for eight microwave servings?

6 cups

74. How many teaspoons of salt would be needed for five stove top servings? (*Hint:* 5 is halfway between 4 and 6.)

$\frac{3}{8}$ tsp

	Microwave	Stove Top		
Servings	**1**	**1**	**4**	**6**
Water	$\frac{3}{4}$ cup	1 cup	3 cups	4 cups
Grits	3 Tbsp	3 Tbsp	$\frac{3}{4}$ cup	1 cup
Salt (optional)	Dash	Dash	$\frac{1}{4}$ tsp	$\frac{1}{2}$ tsp

75. A motel owner has decided to expand his business by buying a piece of property next to the motel. The property has an irregular shape, with five sides as shown in the figure. Find the total distance around the piece of property. This is called the **perimeter** of the figure.

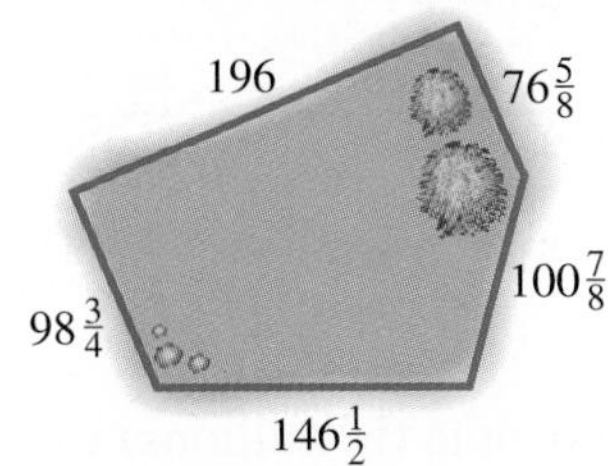

Measurements in feet

$618\frac{3}{4}$ ft

76. A triangle has sides of lengths $5\frac{1}{4}$ ft, $7\frac{1}{2}$ ft, and $10\frac{1}{8}$ ft. Find the perimeter of the triangle. See Exercise 75.

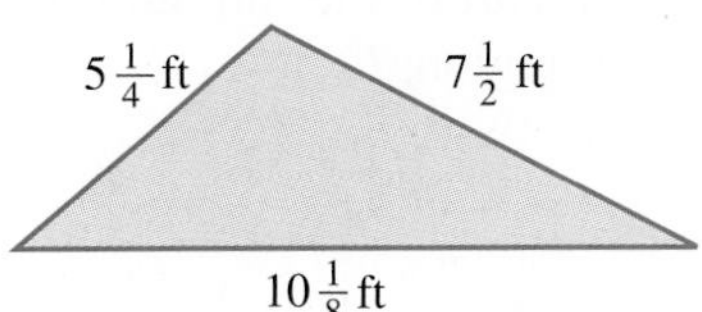

$22\frac{7}{8}$ ft

77. A hardware store sells a 40-piece socket wrench set. The measure of the largest socket is $\frac{3}{4}$ in., while the measure of the smallest socket is $\frac{3}{16}$ in. What is the difference between these measures?

$\frac{9}{16}$ in.

78. Two sockets in a socket wrench set have measures of $\frac{9}{16}$ in. and $\frac{3}{8}$ in. What is the difference between these two measures?

$\frac{3}{16}$ in.

79. Under existing standards, most of the holes in Swiss cheese must have diameters between $\frac{11}{16}$ and $\frac{13}{16}$ in. To accommodate new high-speed slicing machines, the USDA wants to reduce the minimum size to $\frac{3}{8}$ in. How much smaller is $\frac{3}{8}$ in. than $\frac{11}{16}$ in.? (*Source:* U.S. Department of Agriculture.)

$\frac{5}{16}$ in.

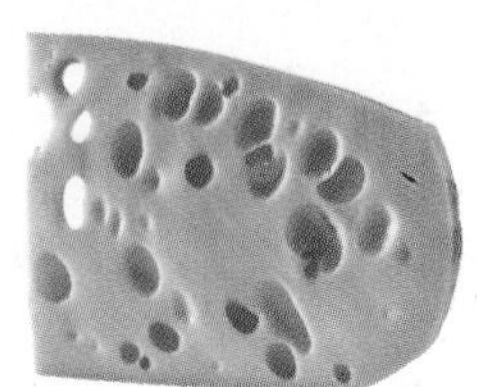

80. Loretta's favorite recipe for barbecue sauce calls for $2\frac{1}{3}$ cups of tomato sauce. The recipe makes enough barbecue sauce to serve 7 people. How much tomato sauce is needed for 1 serving?

$\frac{1}{3}$ cup

81. It takes $2\frac{3}{8}$ yd from a bolt of fabric to make a costume for a school play. How much fabric would be needed for 7 costumes?

$16\frac{5}{8}$ yd

82. A cake recipe calls for $1\frac{3}{4}$ cups of sugar. A caterer has $15\frac{1}{2}$ cups of sugar on hand. How many cakes can he make?

8 cakes (There will be some sugar left over.)

Approximately 34 *million people living in the United States in 2004 were born in other countries. The circle graph gives the fractional number from each region of birth for these people. Use the graph to answer each question. See Example 13.*

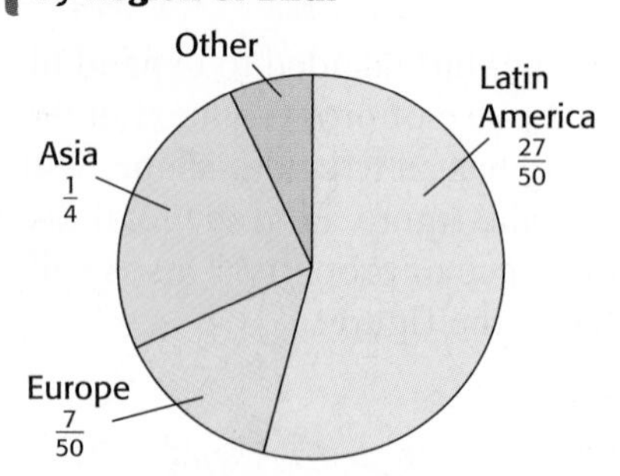

Source: U.S. Census Bureau.

83. What fractional part of the foreign-born population was from other regions?

$\frac{7}{100}$

84. What fractional part of the foreign-born population was from Latin America or Asia?

$\frac{79}{100}$

85. How many people (in millions) were born in Europe?

about $4\frac{19}{25}$ million, or 4,760,000

86. How many more people (in millions) were born in Latin America than in Asia?

about $9\frac{43}{50}$ million, or 9,860,000

R.2 Decimals and Percents

OBJECTIVES

1. Write decimals as fractions.
2. Add and subtract decimals.
3. Multiply and divide decimals.
4. Write fractions as decimals.
5. Convert percents to decimals and decimals to percents.
6. Use fraction, decimal, and percent equivalents.

Fractions are one way to represent parts of a whole. Another way is with a **decimal fraction** or **decimal,** a number written with a decimal point, such as 9.4. Each digit in a decimal number has a place value, as shown below.

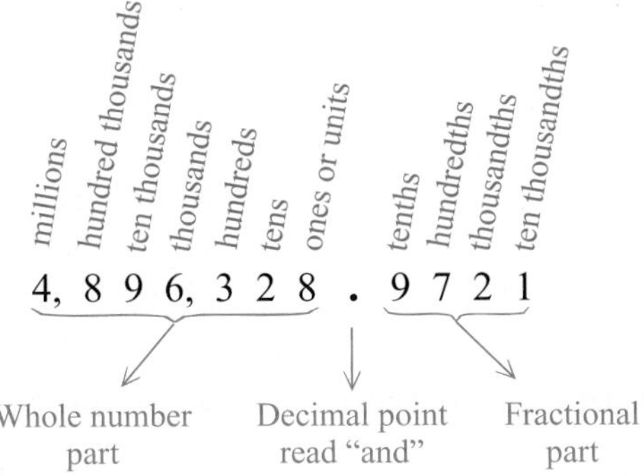

Each successive place value is ten times greater than the place value to its right and is one-tenth as great as the place value to its left.

Prices are often written as decimals. The price \$14.75 means 14 dollars and 75 cents, or 14 dollars and $\frac{75}{100}$ of a dollar.

OBJECTIVE 1 Write decimals as fractions. Place value is used to write a decimal number as a fraction. For example, since the last digit (that is, the digit farthest to the right) of 0.67 is in the *hundredths* place,

$$0.67 = \frac{67}{100}.$$

Similarly, $0.9 = \frac{9}{10}$ and $0.25 = \frac{25}{100}$. Digits to the left of the decimal point indicate whole numbers, so 12.342 is the sum of 12 and 0.342, and

$$12.342 = 12 + 0.342 = 12 + \frac{342}{1000} = \frac{12{,}000}{1000} + \frac{342}{1000} = \frac{12{,}342}{1000}.$$

These examples suggest the following rule.

Converting a Decimal to a Fraction

Read the name using the correct place value. Write it in fraction form just as you read it. The denominator will be a **power of 10,** a number like 10, 100, 1000, and so on.

For example, we read 0.16 as "sixteen hundredths" and write it in fraction form as $\frac{16}{100}$. The same thing is accomplished by counting the number of digits to the right of the decimal point, then writing the given number without a decimal point over a denominator of 1 followed by that number of zeros.

EXAMPLE 1 Writing Decimals as Fractions

Write each decimal as a fraction. Do not write in lowest terms.

(a) 0.95

We read 0.95 as 95 hundredths, so the fraction form is $\frac{95}{100}$. Using the shortcut method, since there are two places to the right of the decimal point, there will be two zeros in the denominator.

Continued on Next Page

$$0.95 = \frac{95}{100}$$

2 places 2 zeros

(b) $0.056 = \frac{56}{1000}$

3 places 3 zeros

(c) $4.2095 = 4 + 0.2095 = 4 + \frac{2095}{10{,}000} = \frac{42{,}095}{10{,}000}$

4 places 4 zeros

◀ Work Problem 1 at the Side.

1 Write each decimal as a fraction. Do not write in lowest terms.

(a) 0.8

(b) 0.431

(c) 20.58

OBJECTIVE 2 Add and subtract decimals.

EXAMPLE 2 Adding and Subtracting Decimals

Add or subtract as indicated.

(a) $6.92 + 14.8 + 3.217$

Place the digits of the numbers in columns, so that tenths are in one column, hundredths in another column, and so on.

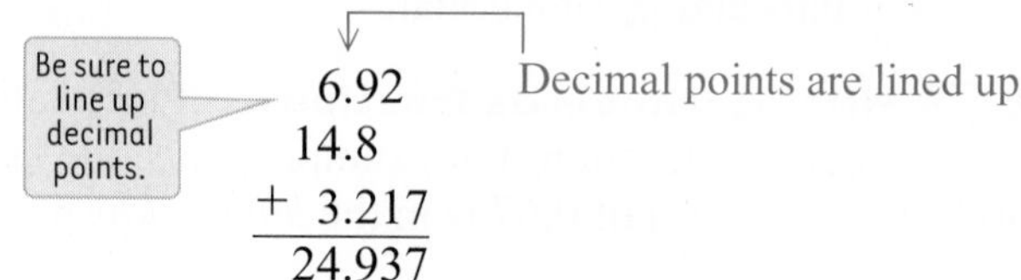

A good way to avoid errors is to attach zeros to make all the numbers the same length.

```
  6.92               6.920   Attach zeros.
 14.8     becomes   14.800
+ 3.217            + 3.217
                    24.937
```

(b) $47.6 - 32.509$

Write the numbers in columns, attaching zeros to 47.6.

```
  47.6                47.600
- 32.509   becomes  - 32.509
                      15.091
```

(c) $3 - 0.253$

A whole number is assumed to have the decimal point at the right of the number. Write 3 as 3.000; then subtract.

```
  3.000
- 0.253
  2.747
```

◀ Work Problem 2 at the Side.

2 Add or subtract as indicated.

(a)
```
  68.9
  42.72
+ 8.973
```

(b)
```
  32.5
- 21.72
```

(c) $42.83 + 71 + 3.074$

(d) $351.8 - 2.706$

OBJECTIVE 3 Multiply and divide decimals. We multiply decimals by slightly modifying multiplication of whole numbers. We will sometimes use the times symbol, $\times$, instead of a dot, to avoid confusion with the decimal point.

ANSWERS

1. (a) $\frac{8}{10}$ (b) $\frac{431}{1000}$ (c) $\frac{2058}{100}$
2. (a) 120.593 (b) 10.78 (c) 116.904 (d) 349.094

Multiplying Decimals

Ignore the decimal points and multiply as if the numbers were whole numbers. Then add together the number of **decimal places** (digits to the *right* of the decimal point) in each number being multiplied. Place the decimal point in the answer that many digits from the right.

EXAMPLE 3 Multiplying Decimals

Multiply.

(a) 29.3 × 4.52

Multiply as if the numbers were whole numbers.

$$\begin{array}{r} 29.3 \\ \times \quad 4.52 \\ \hline 586 \\ 1465 \\ 1172 \\ \hline 132.436 \end{array}$$

1 decimal place in first number
2 decimal places in second number
1 + 2 = 3
3 decimal places in answer

(b) 7.003 × 55.8

$$\begin{array}{r} 7.003 \\ \times \quad 55.8 \\ \hline 56024 \\ 35015 \\ 35015 \\ \hline 390.7674 \end{array}$$

3 decimal places
1 decimal place
3 + 1 = 4
4 decimal places

(c) 31.42 × 65

$$\begin{array}{r} 31.42 \\ \times \quad 65 \\ \hline 15710 \\ 18852 \\ \hline 2042.30 \end{array}$$

2 decimal places
0 decimal places
2 + 0 = 2
2 decimal places

The final 0 here can be dropped and the result can be expressed as 2042.3.

Work Problem **3** *at the Side.* ▶

To divide decimals, convert the divisor to a whole number.

Dividing Decimals

Change the **divisor** (the number you are dividing *by*) into a whole number by moving the decimal point as many places as necessary to the right. Move the decimal point in the **dividend** (the number you are dividing *into*) to the right by the same number of places. Move the decimal point straight up and then divide as with whole numbers.

$$\text{Divisor} \longrightarrow 25\overline{)125}$$

5 ← Quotient; 125 ↑ Dividend

3 Multiply.

(a) 2.13 × 0.05

(b) 9.32 × 1.4

(c) 300.2 × 0.052

(d) 42,001 × 0.012

ANSWERS

3. **(a)** 0.1065 **(b)** 13.048 **(c)** 15.6104 **(d)** 504.012

4 Divide.

(a) $14.9\overline{)451.47}$

(b) $0.37\overline{)5.476}$

(c) $375.1 \div 3.001$

ANSWERS

4. (a) 30.3 (b) 14.8
(c) 124.99 (rounded)

EXAMPLE 4 Dividing Decimals

Divide.

(a) $233.45 \div 11.5$

Write the problem as follows.

$$11.5\overline{)233.45}$$

To change 11.5 into a whole number, move the decimal point one place to the right. Move the decimal point in 233.45 the *same* number of places to the right, to get 2334.5.

$$11.5\overline{)233.4\,5}$$ Move each decimal point one place to the right.

To see why this works, write the division in fraction form and multiply by $\frac{10}{10}$, or 1.

$$\frac{233.45}{11.5} \cdot \frac{10}{10} = \frac{2334.5}{115}$$

The result is the same as when we moved the decimal point one place to the right in the divisor and the dividend.

Move the decimal point straight up and divide as with whole numbers.

```
       20.3
115)2334.5     Move the decimal point straight up.
    230
     345
     345
       0
```

In the second step of the division, 115 does not divide into 34, so we used zero as a placeholder in the quotient.

(b) $73.85\overline{)1852.882}$ (Round the answer to two decimal places.)

Move the decimal point two places to the right in 73.85, to get 7385. Do the same thing with 1852.882, to get 185288.2.

$$73.85\overline{)1852.88\,2}$$

Move the decimal point straight up and divide as with whole numbers.

```
          25.089
7385)185288.200
     14770
      37588
      36925
        66320
        59080
         72400
         66465
          5935
```

We carried out the division to three decimal places so that we could round to two decimal places, obtaining the quotient 25.09.

◀ Work Problem 4 at the Side.

A shortcut can be used when multiplying or dividing by powers of 10.

Multiplying or Dividing by Powers of 10

To ***multiply*** by a power of 10, ***move the decimal point to the right*** as many places as the number of zeros.

To ***divide*** by a power of 10, ***move the decimal point to the left*** as many places as the number of zeros.

In both cases, insert 0s as placeholders if necessary.

EXAMPLE 5 **Multiplying and Dividing by Powers of 10**

Multiply or divide as indicated.

(a) $48.731 \times 100 = 48.73\,1 = 4873.1$

We moved the decimal point two places to the right because 100 has two zeros.

(b) $48.7 \div 1000 = 048.7 = 0.0487$

We moved the decimal point three places to the left because 1000 has three zeros. We needed to insert a zero in front of the 4 to do this.

Work Problem **5** *at the Side.* ▶

5 Multiply or divide as indicated.

(a) 294.72×10

(b) 19.5×1000

(c) $4.793 \div 100$

(d) $960.1 \div 10$

To avoid misplacing the decimal point, check your work by estimating the answer. ***To get a quick estimate, round the numbers so that only the first digit is not zero,*** using the rule for rounding. For more accurate estimates, the numbers could be rounded to the first two or even three nonzero digits.

Rule for Rounding

If the digit to become 0 or be dropped is 5 or more, round up by adding 1 to the final digit to be kept.

If the digit to become 0 or be dropped is 4 or less, do not round up.

For example, to estimate the answer to Example 2(a), round

6.92 to 7, 14.8 to 10, and 3.217 to 3.

(5 or more; 4 or less; 4 or less)

Since $7 + 10 + 3 = 20$, the answer of 24.937 is reasonable. In Example 4(a), round 233.45 to 200 and 11.5 to 10. Since $200 \div 10 = 20$, the answer of 20.3 is reasonable.

OBJECTIVE 4 **Write fractions as decimals.**

Writing a Fraction as a Decimal

Because a fraction bar indicates division, write a fraction as a decimal by dividing the denominator into the numerator.

Answers

5. (a) 2947.2 **(b)** 19,500 **(c)** 0.04793 **(d)** 96.01

6 Convert to decimals. For repeating decimals, write the answer two ways: using the bar notation and rounding to the nearest thousandth.

(a) $\frac{2}{9}$

(b) $\frac{17}{20}$

(c) $\frac{1}{11}$

EXAMPLE 6 Writing Fractions as Decimals

Write each fraction as a decimal.

(a) $\frac{19}{8}$

$$\begin{array}{r} 2.375 \\ 8\overline{)19.000} \\ \underline{16} \\ 3\,0 \\ \underline{2\,4} \\ 60 \\ \underline{56} \\ 40 \\ \underline{40} \\ 0 \end{array}$$

Divide 19 by 8. Add a decimal point and as many 0s as necessary.

$$\frac{19}{8} = 2.375$$

(b) $\frac{2}{3}$

$$\begin{array}{r} 0.6666\ldots \\ 3\overline{)2.0000\ldots} \\ \underline{1\,8} \\ 20 \\ \underline{18} \\ 20 \\ \underline{18} \\ 20 \end{array}$$

The remainder in the division in part (b) is never 0. Because 2 is always left after the subtraction, this quotient is a **repeating decimal.** A convenient notation for a repeating decimal is a bar over the digit (or digits) that repeats. For instance, the decimal for $\frac{2}{3}$ can be written as either of the following.

$$0.6666\ldots \quad \text{or} \quad 0.\overline{6}$$

We often round repeating decimals to as many places as needed. Rounding to the *nearest thousandth,*

$$\frac{2}{3} \approx 0.667. \qquad \text{An approximation } (\approx \text{ means "approximately equal to"})$$

CAUTION

When rounding, be careful to distinguish between *thousandths* and *thousands* or between *hundredths* and *hundreds,* and so on.

◀ *Work Problem 6 at the Side.*

OBJECTIVE 5 Convert percents to decimals and decimals to percents. An important application of decimals is in work with percents. The word **percent** means "per one hundred." Percent is written with the sign **%**. ***One percent means "one per one hundred," or "one one-hundredth."***

$$\mathbf{1\% = 0.01}, \quad \text{or} \quad \mathbf{1\% = \frac{1}{100}}$$

EXAMPLE 7 Converting Percents and Decimals

(a) Write 73% as a decimal.

Since $1\% = 0.01$, we convert as follows.

$$73\% = 73 \cdot \mathbf{1\%} = 73 \times \mathbf{0.01} = 0.73$$

Also, 73% can be written as a decimal using the fraction form $1\% = \frac{1}{100}$.

$$73\% = 73 \cdot \mathbf{1\%} = 73 \cdot \mathbf{\frac{1}{100}} = \frac{73}{100} = 0.73$$

Continued on Next Page

ANSWERS

6. (a) $0.\overline{2}$, 0.222 **(b)** 0.85 **(c)** $0.\overline{09}$, 0.091

(b) Write 125% as a decimal.

$$125\% = 125 \cdot 1\% = 125 \times 0.01 = 1.25$$

(c) Write $3\frac{1}{2}\%$ as a decimal.
First write the fractional part as a decimal.

$$3\frac{1}{2}\% = (3 + 0.5)\% = 3.5\%$$

Now change the percent to decimal form.

$$3.5\% = 3.5 \times 0.01 = 0.035 \qquad 1\% = 0.01$$

(d) Write 0.32 as a percent.
Since 0.32 means 32 hundredths, write 0.32 as 32×0.01. Finally, replace 0.01 with 1%.

$$0.32 = 32 \times 0.01 = 32 \times 1\% = 32\%$$

(e) Write 2.63 as a percent.

$$2.63 = 263 \times 0.01 = 263 \times 1\% = 263\%$$

A number greater than 1 is more than 100%.

Note

A quick way to change from a percent to a decimal is to move the decimal point two places to the left and drop the % symbol. To change from a decimal to a percent, move the decimal point two places to the right, and attach a percent symbol.

Divide by 100; Move 2 places left. (← from Percent to Decimal)

Decimal **Percent**

Multiply by 100; Move 2 places right. (→ from Decimal to Percent)

EXAMPLE 8 Converting Percents and Decimals by Moving the Decimal Point.

Convert each percent to a decimal and each decimal to a percent.

(a) $45\% = 0.45$

(b) $250\% = 2.50$

(c) $0.57 = 57\%$

(d) $1.5 = 1.50 = 150\%$

(e) $0.327 = 32.7\%$

Work Problem 7 at the Side. ▶

In this book, we use 0 in the ones place for decimals between 0 and 1, such as 0.5 and 0.72. Some calculators, including graphing calculators, do *not* show 0 in the ones place. Either way is correct.

OBJECTIVE 6 Use fraction, decimal, and percent equivalents. The fraction $\frac{1}{2}$ can be written as the decimal 0.5, or as the percent 50%. These are three different ways of writing the same number. It is convenient to know the equivalent forms of commonly used fractions, decimals, and percents.

7 Convert as indicated.

(a) 23% to a decimal

(b) 310% to a decimal

(c) 0.71 to a percent

(d) 1.32 to a percent

(e) 0.685 to a percent

Answers

7. **(a)** 0.23 **(b)** 3.10 **(c)** 71% **(d)** 132% **(e)** 68.5%

8 Solve each problem.

(a) Chris found a pair of jeans, regularly priced at \$50, on sale at 20% off. If he buys the jeans on sale, how much will he save? What is the sale price?

(b) Miguel was earning \$7.50 per hour on his job and received a $33\frac{1}{3}\%$ raise. How much was his hourly raise? What was his new hourly pay?

FRACTION, DECIMAL, AND PERCENT EQUIVALENTS

Fraction (or whole number)	Decimal	Percent
$\frac{1}{100}$	0.01	1%
$\frac{1}{50}$	0.02	2%
$\frac{1}{20}$	0.05	5%
$\frac{1}{10}$	0.1	10%
$\frac{1}{8}$	0.125	$12\frac{1}{2}\%$, or 12.5%
$\frac{1}{6}$	$0.1\overline{6}$	$16\frac{2}{3}\%$, or $16.\overline{6}\%$
$\frac{1}{5}$	0.2	20%
$\frac{1}{4}$	0.25	25%
$\frac{1}{3}$	$0.\overline{3}$	$33\frac{1}{3}\%$, or $33.\overline{3}\%$
$\frac{1}{2}$	0.5	50%
$\frac{2}{3}$	$0.\overline{6}$	$66\frac{2}{3}\%$, or $66.\overline{6}\%$
$\frac{3}{4}$	0.75	75%
1	1.0	100%

These equivalents can be used with mental mathematics to solve some everyday problems.

EXAMPLE 9 Using Percent and Fraction Equivalents

Solve each problem.

(a) Marissa found a jacket with a regular price of \$80, on sale at 25% off. If she buys the jacket on sale, how much money will she save? What is the sale price?

From the table, $25\% = \frac{1}{4}$, and $\frac{1}{4}$ of \$80 is \$20, so she will save \$20. The sale price is

$$\$80 - \$20 = \$60.$$

(b) Caleb is paying \$600 per month in rent for his apartment. If his landlord raises the rent by 15%, how much more will he pay each month? What will his new rent be?

Although 15% is not listed in the table, it is still useful. We know that

$$15\% = 10\% + 5\%,$$

and $10\% = \frac{1}{10}$. Thus, 10%, or $\frac{1}{10}$, of \$600 is \$60. Also, 5% is half of 10%, so if 10% of \$600 is \$60, then 5% of \$600 is half of \$60, or \$30. Caleb's new monthly rent will be

$$\$600 + \underbrace{\underset{10\%}{\underset{\uparrow}{\$60}} + \underset{5\%}{\underset{\uparrow}{\$30}}}_{15\%} = \$690.$$

He will pay \$90 more each month.

◀ *Work Problem* **8** *at the Side.*

We will solve other types of problems involving percent in **Section 2.6.**

ANSWERS

8. **(a)** \$10; \$40 **(b)** \$2.50; \$10.00

R.2 Exercises

FOR EXTRA HELP

Math XL PRACTICE

WATCH

DOWNLOAD

READ

REVIEW

1. In the decimal 367.9412, name the digit that has each place value.

(a) tens **6** **(b)** tenths **9** **(c)** thousandths **1**

(d) ones or units **7** **(e)** hundredths **4**

2. Write a number that has 5 in the thousands place, 0 in the tenths place, and 4 in the ten thousandths place.

Answers will vary. One example is 5243.0164.

3. For the decimal number 46.249, round to the place value indicated.

(a) hundredths **46.25** **(b)** tenths **46.2**

(c) ones or units **46** **(d)** tens **50**

4. Round each decimal to the nearest thousandth.

(a) $0.\overline{8}$ **0.889** **(b)** $0.\overline{5}$ **0.556**

(c) 0.9762 **0.976** **(d)** 0.8642 **0.864**

5. For the sum 35.89 + 24.1, which is the best estimate? **C**

A. 40 **B.** 50 **C.** 60 **D.** 70

6. For the difference 119.83 − 52.4, which is the best estimate? **D**

A. 40 **B.** 50 **C.** 60 **D.** 70

7. For the product 84.9 × 98.3, which is the best estimate? **B**

A. 7000 **B.** 8000 **C.** 80,000 **D.** 70,000

8. For the quotient 9845.3 ÷ 97.2, which is the best estimate? **C**

A. 10 **B.** 1000 **C.** 100 **D.** 10,000

Write each decimal as a fraction. Do not write in lowest terms. See Example 1.

9. 0.4 $\frac{4}{10}$ **10.** 0.6 $\frac{6}{10}$ **11.** 0.64 $\frac{64}{100}$ **12.** 0.82 $\frac{82}{100}$

13. 0.138 $\frac{138}{1000}$ **14.** 0.104 $\frac{104}{1000}$ **15.** 3.805 $\frac{3805}{1000}$ **16.** 5.166 $\frac{5166}{1000}$

Add or subtract as indicated. Make sure that your answer is reasonable by estimating first. Give your estimate, and then give the exact answer. See Example 2 and the rules for rounding on page 21.

17. 25.32 + 109.2 + 8.574 **139; 143.094**

18. 90.527 + 32.43 + 589.83 + 399.327 **1120; 1112.114**

19. 28.73 − 3.12 **27; 25.61**

20. 46.88 − 13.45 **40; 33.43**

21. 43.5 − 28.17 **10; 15.33**

22. 345.1 − 56.31 **240; 288.79**

23. 32.56 + 47.356 + 1.8 **82; 81.716**

24. 75.2 + 123.96 + 3.897 **184; 203.057**

25. 18 − 2.789 **17; 15.211**

26. 29 − 8.582 **21; 20.418**

Multiply or divide as indicated. Make sure that your answer is reasonable by estimating first. Give your estimate, and then give the exact answer. See Examples 3–5 and the rules for rounding on page 21.

27. 0.2 × 0.03 **0.006; 0.006**

28. 0.07 × 0.004 **0.00028; 0.00028**

29. 12.8 × 9.1 **90; 116.48**

30. 34.04 × 0.56 **18; 19.0624**

31. 57.2 ÷ 8 **6; 7.15**

32. 73.36 ÷ 14 **7; 5.24**

33. 19.967 ÷ 9.74 **2; 2.05**

34. 44.4788 ÷ 5.27 **8; 8.44**

35. 57.116×100 **6000; 5711.6**

36. 0.094×1000 **100; 94**

37. $1.62 \div 10$ **0.2; 0.162**

38. $24.03 \div 100$ **0.2; 0.2403**

Write each fraction as a decimal. For repeating decimals, write the answer two ways: using the bar notation and rounding to the nearest thousandth. See Example 6.

39. $\frac{1}{8}$ **0.125**

40. $\frac{7}{8}$ **0.875**

41. $\frac{5}{4}$ **1.25**

42. $\frac{9}{5}$ **1.8**

43. $\frac{5}{9}$ **$0.\overline{5}$; 0.556**

44. $\frac{8}{9}$ **$0.\overline{8}$; 0.889**

45. $\frac{1}{6}$ **$0.1\overline{6}$; 0.167**

46. $\frac{5}{6}$ **$0.8\overline{3}$; 0.833**

47. In your own words, explain how to convert a decimal to a percent.

To convert a decimal to a percent, move the decimal point two places to the right and attach a percent symbol (%).

48. In your own words, explain how to convert a percent to a decimal.

To convert a percent to a decimal, move the decimal point two places to the left and drop the percent symbol (%).

Convert each percent to a decimal. See Examples 7(a)–(c), 8(a), and 8(b).

49. 54% **0.54**

50. 39% **0.39**

51. 117% **1.17**

52. 189% **1.89**

53. 2.4% **0.024**

54. 3.1% **0.031**

55. $6\frac{1}{4}\%$ **0.0625**

56. $5\frac{1}{2}\%$ **0.055**

57. 0.8% **0.008**

58. 0.9% **0.009**

Convert each decimal to a percent. See Examples 7(d), 7(e), and 8(c)–(e).

59. 0.73 **73%**

60. 0.83 **83%**

61. 0.004 **0.4%**

62. 0.005 **0.5%**

63. 1.28 **128%**

64. 2.35 **235%**

65. 0.3 **30%**

66. 0.6 **60%**

One method of converting a fraction to a percent is to first convert the fraction to a decimal, as shown in Example 6, and then convert the decimal to a percent, as shown in Examples 7 and 8. Convert each fraction to a percent in this way.

67. $\frac{4}{5}$ **80%**

68. $\frac{3}{25}$ **12%**

69. $\frac{2}{11}$ **$18.\overline{18}\%$**

70. $\frac{4}{9}$ **$44.\overline{4}\%$**

71. $\frac{7}{4}$ **175%**

72. $\frac{11}{8}$ **137.5%**

Solve each problem mentally. See Example 9.

73. Liam, Shannon, Chris, and Kayla went out for dinner together. The total cost of their meals was $75. If they decide to leave a 20% tip, how much should the tip be? **$15**

74. Where Lan lives, a sales tax of 5% is added to all purchases. If he buys a DVD player that costs $80, how much will he pay for the DVD player, including the tax? **$84**

75. Nicole took an algebra test that was worth 50 points. If she earned 35 points on the test, what percent did she score? **70%**

76. In a survey of 2400 eligible voters, 400 said that they planned to vote in an upcoming election. What percent of those surveyed planned to vote?

$16\frac{2}{3}\%$, or $16.\overline{6}\%$

Study Skills

USING YOUR TEXTBOOK

Your textbook is a valuable resource. To find out what it has to offer, we look at some general features that will help in all chapters.

OBJECTIVES

1 **Become familiar with the features of this text.**

2 **Locate the Index, Answers, and Solutions sections.**

Look at the very front of the book for the Table of Contents. Use it to locate topics in your text. Each chapter is divided into sections, and each section has a number, such as 3.5.

Chapter 3 → **3.5** ← Section 5 within Chapter 3

Your instructor will use these numbers to assign readings and homework.

Chapter Features

Pay special attention to these features as you work through the chapters in your book.

- **Objectives** Each section lists the objectives in the upper corner of the page and again within the section as the corresponding material is presented. An objective tells you what you will be able to do after you complete the section. To check your learning, go back to the list of objectives when you are finished with a section and ask yourself if you can do them all.
- **Margin Exercises** The exercises in the margins in your textbook give you immediate practice and actively involve you in learning the material.
- **Cautions** The yellow boxes titled CAUTION provide warnings about common errors that students often make or trouble spots you will want to avoid.
- **Pointers** These comments in small shaded boxes provide on-the-spot warnings and reminders, point out key steps, and give other helpful tips.
- **Notes** These light purple boxes provide additional explanations or emphasize important ideas in the discussion.
- **Problem-Solving Hints** These green boxes provide helpful problem-solving tips and strategies to use when you are working applications.
- **Calculator Tips** A small red calculator indicates a Calculator Tip, which provides helpful information about using your calculator. A calculator beside an exercise is a recommendation to use your calculator to work that exercise.

List a page number from Chapter 1 or 2 for each of these features.

A *Caution* appears on page __________.

A *Pointer* appears on page __________.

A *Note* appears on page __________.

A *Problem-Solving Hint* appears on page __________.

A *Calculator Tip* appears on page __________.

End of Chapter Features

At the end of most chapters, there are four important sections.

- **Chapter Summary** Turn to the Summary for Chapter 2. It lists the chapter's **Key Terms** (arranged by section), as well as any **New Symbols. Test Your Word Power** checks your understanding of the math vocabulary. The **Quick Review** lists the main concepts in each section of the chapter with corresponding worked-out examples. **Answers to Test Your Word Power** are given right after the Quick Review.
- **Review Exercises** Use these exercises as a way to check your understanding of all the concepts in the chapter. If you have trouble, the red numbers in brackets tell you which section of the chapter to go back to for more explanations. Make sure you do the **Mixed Review Exercises** to practice for tests.
- **Chapter Test** Take the test as a practice exam to be sure you know how to work all types of problems without looking back at the chapter.
- **Cumulative Review** (starting with Chapter 2) These exercises help you maintain the skills you've learned in all previous chapters. Working on previous skills throughout the course will be a big help on the final exam.

Answers

To find out if you've worked the exercises correctly, your textbook provides many of the answers. **Answers to the margin problems** are at the bottom of each page in the margin area. For homework, you can find the **answers to all of the odd-numbered section exercises** in the **Answers to Selected Exercises** section near the end of your textbook. *All* of the answers are given for the Chapter Review Exercises, Chapter Tests, and Cumulative Reviews.

Flag the Answers section with a sticky note or other device, so that you can turn to it quickly.

Solutions

The **Solutions to Selected Exercises** section is located near the end of the text and provides step-by-step, worked-out solutions to some exercises that have a twist or are a bit more difficult. Exercises that have a blue square highlighting the exercise number, such as **11.** , have a complete solution in the Solutions section.

Index

Now find the Index. All of the topics, vocabulary, and concepts are listed in alphabetical order in the Index. There may be several subheadings listed under the main word or several page numbers listed. Usually, the *first* place that a word appears in the textbook is where it is introduced and defined.

Look up the following in the index. Write down the page that introduces or defines each one.

Double negative rule ______________________________

Pythagorean formula ______________________________

Reciprocals ______________________________

1

The Real Number System

The personal savings rate of Americans has declined steadily since 1984, when it stood at 10.8% of after-tax income. In 2006, Americans spent everything they earned and more, causing the personal savings rate to drop to -1%, the worst personal savings rate since 1933. That year, Americans depleted their savings to cope with the job layoffs and business failures of the Great Depression. (*Source:* Commerce Department.)

In this chapter, we examine *positive* and *negative numbers* and apply them to situations such as the personal savings rate of Americans in Exercises 77 and 78 of Section 1.5.

1.1 Exponents, Order of Operations, and Inequality

OBJECTIVES

1. Use exponents.
2. Use the rules for order of operations.
3. Use more than one grouping symbol.
4. Know the meanings of $\neq$, $<$, $>$, $\leq$, and $\geq$.
5. Translate word statements to symbols.
6. Write statements that change the direction of inequality symbols.

OBJECTIVE 1 Use exponents. In algebra, we use a raised dot for multiplication as shown in **Chapter R,** where we factored a number as the product of its prime factors. For example, 81 is written in prime factored form as

$$81 = 3 \cdot 3 \cdot 3 \cdot 3,$$

where the factor 3 appears four times. Repeated factors are written in an abbreviated form by using an *exponent.* The prime factored form of 81 is written with an exponent as

$$\underbrace{3 \cdot 3 \cdot 3 \cdot 3}_{\text{4 factors of 3}} = 3^4.$$

Exponent (points to 4); Base (points to 3)

The number 4 is the **exponent,** or **power,** and 3 is the **base** in the **exponential expression** 3^4. The exponent always follows the base and tells how many times the base is used as a factor. We read 3^4 as "3 to the fourth power," or simply "3 to the fourth." ***A number raised to the first power is simply that number.*** For example, $6^1 = 6$ and $(2.5)^1 = 2.5$.

EXAMPLE 1 Evaluating Exponential Expressions

Find the value of each exponential expression.

(a) $5^2 = \underbrace{5 \cdot 5} = 25$

5 is used as a factor 2 times.

Read 5^2 as "5 to the second power" or, more commonly, "5 squared."

(b) $6^3 = \underbrace{6 \cdot 6 \cdot 6} = 216$

6 is used as a factor 3 times.

Read 6^3 as "6 to the third power" or, more commonly, "6 cubed."

(c) $2^5 = 2 \cdot 2 \cdot 2 \cdot 2 \cdot 2 = 32$ 2 is used as a factor 5 times.

Read 2^5 as "2 to the fifth power."

(d) $7^4 = 7 \cdot 7 \cdot 7 \cdot 7 = 2401$ 7 is used as a factor 4 times.

Read 7^4 as "7 to the fourth power."

(e) $\left(\frac{2}{3}\right)^3 = \frac{2}{3} \cdot \frac{2}{3} \cdot \frac{2}{3} = \frac{8}{27}$ $\frac{2}{3}$ is used as a factor 3 times.

◀ Work Problem 1 at the Side.

CAUTION

Squaring, or raising a number to the second power, is not the same as doubling the number. For example,

$$3^2 \text{ means } 3 \cdot 3, \text{ not } 2 \cdot 3.$$

Thus $3^2 = 9$, not 6. Similarly, cubing, or raising a number to the third power, does *not* mean tripling the number.

1 Find the value of each exponential expression.

(a) 6^2

(b) 3^5

(c) $\left(\frac{3}{4}\right)^2$

(d) $\left(\frac{1}{2}\right)^4$

(e) $(0.4)^3$

ANSWERS

1. (a) 36 (b) 243 (c) $\frac{9}{16}$ (d) $\frac{1}{16}$ (e) 0.064

OBJECTIVE 2 Use the rules for order of operations. Many problems involve more than one operation. To indicate the order in which the operations should be performed, we often use **grouping symbols.** If no grouping symbols are used, we apply the rules for order of operations.

Consider the expression $5 + 2 \cdot 3$. To show that the multiplication should be performed before the addition, parentheses can be used to write

$$5 + (2 \cdot 3), \quad \text{which equals} \quad 5 + 6, \quad \text{or} \quad 11.$$

If addition is to be performed first, the parentheses should group $5 + 2$ as follows:

$$(5 + 2) \cdot 3, \quad \text{which equals} \quad 7 \cdot 3, \quad \text{or} \quad 21.$$

Other grouping symbols used in more complicated expressions are brackets [ ], braces { }, and fraction bars. (For example, in $\frac{8-2}{3}$, the expression $8 - 2$ is considered to be grouped in the numerator.)

To work problems with more than one operation, use the following **order of operations.** This order is used by most calculators and computers.

Order of Operations

If grouping symbols are present, simplify within them, innermost first (and above and below fraction bars separately), in the following order.

Step 1 Apply all **exponents.**

Step 2 Do any **multiplications** or **divisions** in the order in which they occur, working from left to right.

Step 3 Do any **additions** or **subtractions** in the order in which they occur, working from left to right.

If no grouping symbols are present, start with Step 1.

A dot has been used to show multiplication; another way to show multiplication is with parentheses. For example, 3(7) means $3 \cdot 7$, or 21. Also, $3(4 + 5)$ means "3 times the sum of 4 and 5." By the order of operations, the sum in parentheses must be found first, then the product.

EXAMPLE 2 Using the Rules for Order of Operations

Find the value of each expression.

(a) $24 - 12 \div 3$

Use the order of operations given in the box.

$$24 - 12 \div 3$$

Be careful! Divide first.

$= 24 - 4$ Divide.

$= 20$ Subtract.

(b) $9(6 + 11)$

$= 9(17)$ Add inside parentheses.

$= 153$ Multiply.

Continued on Next Page

(c) $6 \cdot 8 + 5 \cdot 2$

$= 48 + 10$ Multiply, working from left to right.

$= 58$ Add.

(d) $2(5 + 6) + 7 \cdot 3$

Start here.

$= 2(11) + 7 \cdot 3$ Add inside parentheses.

$= 22 + 21$ Multiply.

$= 43$ Add.

(e) $9 + 2^3 - 5$

$= 9 + 8 - 5$ Apply the exponent.

$= 12$ Add, and then subtract.

◀ *Work Problem* **2** *at the Side.*

2 Find the value of each expression.

(a) $7 + 3 \cdot 8$

(b) $2 \cdot 9 + 7 \cdot 3$

(c) $7 \cdot 6 - 3(8 + 1)$

(d) $2 + 3^2 - 5$

OBJECTIVE 3 Use more than one grouping symbol. An expression with double parentheses, such as the expression $2(8 + 3(6 + 5))$, can be confusing. For clarity, we often use brackets, [ ], in place of the outer pair of parentheses.

EXAMPLE 3 Using Brackets and Fraction Bars as Grouping Symbols

Find the value of each expression.

Start here.

(a) $2[8 + 3(6 + 5)]$

$= 2[8 + 3(11)]$ Add inside parentheses.

$= 2[8 + 33]$ Multiply inside brackets.

$= 2[41]$ Add inside brackets.

$= 82$ Multiply.

(b) $\dfrac{4(5 + 3) + 3}{2(3) - 1}$ Simplify the numerator and denominator separately.

$= \dfrac{4(8) + 3}{2(3) - 1}$ Add inside parentheses.

$= \dfrac{32 + 3}{6 - 1}$ Multiply.

$= \dfrac{35}{5}$ Add and subtract.

$= 7$ Divide.

◀ *Work Problem* **3** *at the Side.*

3 Find the value of each expression.

(a) $9[(4 + 8) - 3]$

(b) $\dfrac{2(7 + 8) + 2}{3 \cdot 5 + 1}$

Calculator Tip Calculators follow the order of operations given in this section. Try some of the examples to see that your calculator gives the same answers. Be sure to use the parentheses keys to insert parentheses where they are needed. To work Example 3(b) with a calculator, you must put parentheses around the numerator and the denominator.

ANSWERS

2. **(a)** 31 **(b)** 39 **(c)** 15 **(d)** 6
3. **(a)** 81 **(b)** 2

OBJECTIVE 4 Know the meanings of $\neq$, $<$, $>$, $\leq$, and $\geq$. So far, we have used only the symbols of arithmetic, such as $+$, $-$, $\cdot$, and $\div$ and the equality symbol $=$. The equality symbol with a slash through it means "is *not* equal to." For example,

$$7 \neq 8 \qquad \text{7 is not equal to 8.}$$

indicates that 7 is not equal to 8.

If two numbers are not equal, then one of the numbers must be less than the other. The symbol $<$ represents "is less than," so "7 is less than 8" is written as

$$7 < 8. \qquad \text{7 is less than 8.}$$

The symbol $>$ means "is greater than." For example,

$$8 > 2. \qquad \text{8 is greater than 2.}$$

The statement "17 is greater than 11" becomes $17 > 11$.

To keep the meanings of the symbols $<$ and $>$ clear, remember that the symbol always points to the lesser number.

Lesser number $\longrightarrow$ **8** < 15

$15 >$ **8** $\longleftarrow$ Lesser number

Work Problem 4 at the Side. ▶

Two other symbols, $\leq$ and $\geq$, also represent the idea of inequality. The symbol $\leq$ means "is less than or equal to," so

$$5 \leq 9 \qquad \text{5 is less than or equal to 9.}$$

means "5 is less than or equal to 9." ***If either the $<$ part or the $=$ part is true, then the inequality $\leq$ is true.*** The statement $5 \leq 9$ is true because $5 < 9$ is true. Also, $8 \leq 8$ is true because $8 = 8$ is true. But $13 \leq 9$ is not true because neither $13 < 9$ nor $13 = 9$ is true.

The symbol $\geq$ means "is greater than or equal to," so

$$9 \geq 5 \qquad \text{9 is greater than or equal to 5}$$

is true because $9 > 5$ is true.

EXAMPLE 4 Using the Symbols $\leq$ and $\geq$

Determine whether each statement is *true* or *false*.

(a) $15 \leq 20$ The statement $15 \leq 20$ is true because $15 < 20$.

(b) $12 \geq 12$ Since $12 = 12$, this statement is true.

(c) $\frac{6}{15} \geq \frac{2}{3}$

To compare fractions, write them with a common denominator. Here, 15 is a common denominator and $\frac{2}{3} = \frac{10}{15}$. Now decide whether $\frac{6}{15} \geq \frac{10}{15}$ is true or false. Both statements $\frac{6}{15} > \frac{10}{15}$ and $\frac{6}{15} = \frac{10}{15}$ are false; therefore, $\frac{6}{15} \geq \frac{2}{3}$ is false.

Work Problem 5 at the Side. ▶

OBJECTIVE 5 Translate word statements to symbols. An important part of algebra deals with translating words into algebraic notation.

4 Write each statement in words. Then decide whether it is *true* or *false*.

(a) $7 < 5$

(b) $12 > 6$

(c) $4 \neq 10$

(d) $28 \neq 4 \cdot 7$

5 Tell whether each statement is *true* or *false*.

(a) $30 \leq 40$

(b) $25 \geq 10$

(c) $40 \leq 10$

(d) $21 \leq 21$

(e) $3 \geq 3$

Answers

4. (a) Seven is less than five. False
(b) Twelve is greater than six. True
(c) Four is not equal to ten. True
(d) Twenty-eight is not equal to four times seven. False

5. (a) true **(b)** true **(c)** false
(d) true **(e)** true

6 Write in symbols.

(a) Nine is equal to eleven minus two.

(b) Seventeen is less than thirty.

(c) Eight is not equal to ten.

(d) Fourteen is greater than twelve.

(e) Thirty is less than or equal to fifty.

(f) Two is greater than or equal to two.

7 Write each statement with the inequality symbol reversed.

(a) $8 < 10$

(b) $3 > 1$

(c) $9 \leq 15$

(d) $6 \geq 2$

ANSWERS

6. **(a)** $9 = 11 - 2$ **(b)** $17 < 30$ **(c)** $8 \neq 10$ **(d)** $14 > 12$ **(e)** $30 \leq 50$ **(f)** $2 \geq 2$

7. **(a)** $10 > 8$ **(b)** $1 < 3$ **(c)** $15 \geq 9$ **(d)** $2 \leq 6$

EXAMPLE 5 **Converting Words to Symbols**

Write each word statement in symbols.

(a) Twelve is equal to ten plus two. $12 = 10 + 2$

(b) Nine is less than ten. $9 < 10$
Compare this with "9 less than 10," which is written $10 - 9$.

(c) Fifteen is not equal to eighteen. $15 \neq 18$

(d) Seven is greater than four. $7 > 4$

(e) Thirteen is less than or equal to forty. $13 \leq 40$

(f) Six is greater than or equal to six. $6 \geq 6$

◀ Work Problem 6 at the Side.

OBJECTIVE 6 **Write statements that change the direction of inequality symbols.** Any statement with $<$ can be converted to one with $>$, and any statement with $>$ can be converted to one with $<$. ***We do this by reversing both the order of the numbers and the direction of the symbol.*** For example, the statement $6 < 10$ can be written as $10 > 6$.

Interchange numbers.

$6 < 10$ becomes $10 > 6$

Reverse symbol.

EXAMPLE 6 **Converting between < and >**

Parts (a)–(d) show the same statements written in two equally correct ways. In each inequality, the symbol points toward the lesser number.

(a) $9 < 16, \quad 16 > 9$ **(b)** $5 > 2, \quad 2 < 5$

(c) $3 \leq 8, \quad 8 \geq 3$ **(d)** $12 \geq 5, \quad 5 \leq 12$

◀ Work Problem 7 at the Side.

Here is a summary of the symbols of equality and inequality.

Symbol	Meaning	Example
$=$	Is equal to	$0.5 = \frac{1}{2}$ means 0.5 is equal to $\frac{1}{2}$.
$\neq$	Is not equal to	$3 \neq 7$ means 3 is not equal to 7.
$<$	Is less than	$6 < 10$ means 6 is less than 10.
$>$	Is greater than	$15 > 14$ means 15 is greater than 14.
$\leq$	Is less than or equal to	$4 \leq 8$ means 4 is less than or equal to 8.
$\geq$	Is greater than or equal to	$1 \geq 0$ means 1 is greater than or equal to 0.

CAUTION

The symbols of equality and inequality are used to write mathematical ***sentences.*** They differ from the symbols for operations ($+$, $-$, $\cdot$, and $\div$), discussed earlier, which are used to write mathematical ***expressions*** that represent a number. For example, compare the sentence $4 < 10$, which gives the relationship between 4 and 10, with the expression $4 + 10$, which tells how to operate on 4 and 10 to get the number 14.

1.1 Exercises

FOR EXTRA HELP

MyMathLab

Math XL PRACTICE

WATCH

DOWNLOAD

READ

REVIEW

Decide whether each statement is true *or* false. *If it is* false, *explain why.*

1. An exponent tells how many times its base is used as a factor.

true

2. Some grouping symbols are $+$, $-$, $\cdot$, and $\div$.

false; $+$, $-$, $\cdot$, and $\div$ are operation symbols.

3. When evaluated, $4 + 3(8 - 2)$ is equal to 42.

false; Using the guidelines for order of operations gives $4 + 3(8 - 2) = 4 + 3(6) = 4 + 18 = 22$.

4. $3^3 = 9$

false; $3^3 = 3 \cdot 3 \cdot 3 = 27$.

5. The statement "4 is 12 less than 16" is interpreted $4 = 12 - 16$.

false; The correct translation is $4 = 16 - 12$.

6. The statement "6 is 4 less than 10" is interpreted $6 < 10 - 4$.

false; The correct translation is $6 = 10 - 4$.

Find the value of each exponential expression. See Example 1.

7. 7^2 **49**

8. 4^2 **16**

9. 12^2 **144**

10. 14^2 **196**

11. 4^3 **64**

12. 5^3 **125**

13. 10^3 **1000**

14. 11^3 **1331**

15. 3^4 **81**

16. 6^4 **1296**

17. 4^5 **1024**

18. 3^5 **243**

19. $\left(\frac{2}{3}\right)^4$ $\frac{16}{81}$

20. $\left(\frac{3}{4}\right)^3$ $\frac{27}{64}$

21. $(0.04)^3$ **0.000064**

22. $(0.05)^4$ **0.00000625**

23. When evaluating $(4^2 + 3^3)^4$, what is the *last* exponent that would be applied? Explain your answer.

The 4 would be applied last because we work first inside the parentheses.

24. Which are not grouping symbols—parentheses, brackets, fraction bars, exponents?

exponents

Find the value of each expression. See Examples 2 and 3.

25. $13 + 9 \cdot 5$ **58**

26. $11 + 7 \cdot 6$ **53**

27. $20 - 4 \cdot 3 + 5$ **13**

28. $18 - 7 \cdot 2 + 6$ **10**

29. $9 \cdot 5 - 13$ **32**

30. $7 \cdot 6 - 11$ **31**

31. $18 - 2 + 3$ **19**

32. $22 - 8 + 9$ **23**

33. $\frac{1}{4} \cdot \frac{2}{3} + \frac{2}{5} \cdot \frac{11}{3}$ $\frac{49}{30}$

34. $\frac{9}{4} \cdot \frac{2}{3} + \frac{4}{5} \cdot \frac{5}{3}$ $\frac{17}{6}$

35. $9 \cdot 4 - 8 \cdot 3$ **12**

36. $11 \cdot 4 + 10 \cdot 3$ **74**

37. $2.5(1.9) + 4.3(7.3)$ **36.14**

38. $4.3(1.2) + 2.1(8.5)$ **23.01**

39. $10 + 40 \div 5 \cdot 2$ **26**

40. $12 + 8^2 \div 8 - 4$ **16**

41. $18 - 2(3 + 4)$ **4**

42. $30 - 3(4 + 2)$ **12**

43. $5[3 + 4(2^2)]$ **95**

44. $6\left[\frac{3}{4} + 8\left(\frac{1}{2}\right)^3\right]$ $\frac{21}{2}$

45. $\left(\frac{3}{2}\right)^2\left[\left(11 + \frac{1}{3}\right) - 6\right]$ **12**

46. $4^2[(13 + 4) - 8]$ **144**

47. $\frac{8 + 6(3^2 - 1)}{3 \cdot 2 - 2}$ **14**

48. $\frac{8 + 2(8^2 - 4)}{4 \cdot 3 - 10}$ **64**

49. $\dfrac{4(7+2)+8(8-3)}{6(4-2)-2^2}$ **$\dfrac{19}{2}$**

50. $\dfrac{6(5+1)-9(1+1)}{5(8-4)-2^3}$ **$\dfrac{3}{2}$**

Tell whether each statement is true *or* false. *In Exercises 53–62, first simplify each expression involving an operation. See Example 4.*

51. $8 \geq 17$ **false**

52. $10 \geq 41$ **false**

53. $17 \leq 18 - 1$ **true**

54. $12 \geq 10 + 2$ **true**

55. $6 \cdot 8 + 6 \cdot 6 \geq 0$ **true**

56. $4 \cdot 20 - 16 \cdot 5 \geq 0$ **true**

57. $6[5 + 3(4 + 2)] \leq 70$ **false**

58. $6[2 + 3(2 + 5)] \leq 135$ **false**

59. $\dfrac{9(7-1)-8 \cdot 2}{4(6-1)} > 3$ **false**

60. $\dfrac{2(5+3)+2 \cdot 2}{2(4-1)} > 1$ **true**

61. $8 \leq 4^2 - 2^2$ **true**

62. $10^2 - 8^2 > 6^2$ **false**

Write each word statement in symbols. See Example 5.

63. Fifteen is equal to five plus ten. **$15 = 5 + 10$**

64. Twelve is equal to twenty minus eight. **$12 = 20 - 8$**

65. Nine is greater than five minus four. **$9 > 5 - 4$**

66. Ten is greater than six plus one. **$10 > 6 + 1$**

67. Sixteen is not equal to nineteen. **$16 \neq 19$**

68. Three is not equal to four. **$3 \neq 4$**

69. Two is less than or equal to three. **$2 \leq 3$**

70. Five is less than or equal to nine. **$5 \leq 9$**

Write each statement in words and decide whether it is true *or* false. *(Hint: To compare fractions, write them with the same denominator.)*

71. $7 < 19$
Seven is less than nineteen; true

72. $9 < 10$
Nine is less than ten; true

73. $\dfrac{1}{3} \neq \dfrac{3}{10}$
One-third is not equal to three-tenths; true

74. $\dfrac{10}{7} \neq \dfrac{3}{2}$
Ten-sevenths is not equal to three-halves; true

75. $8 \geq 11$
Eight is greater than or equal to eleven; false

76. $4 \leq 2$
Four is less than or equal to two; false

Write each statement with the inequality symbol reversed. See Example 6.

77. $5 < 30$
$30 > 5$

78. $8 > 4$
$4 < 8$

79. $12 \geq 3$
$3 \leq 12$

80. $25 \leq 41$
$41 \geq 25$

The table shows the number of pupils per teacher in U.S. public schools in selected states in a recent year. Use this table to answer the questions in Exercises 81–84.

81. Which states had a number greater than 13.9?
Alaska, Texas, California, Idaho

82. Which states had a number that was at most 14.7?
Texas, Wyoming, Maine, Missouri

83. Which states had a number not less than 13.9?
Alaska, Texas, California, Idaho, Missouri

84. Which states had a number greater than 20.5?
none

State	Pupils per Teacher
Alaska	16.7
Texas	14.7
California	20.5
Wyoming	12.5
Maine	12.3
Idaho	17.8
Missouri	13.9

Source: National Center for Education Statistics.

Study Skills

HOMEWORK: HOW, WHY, AND WHEN

It is best for success in mathematics if you keep up with the reading and homework in your math class. The more times you work with the information, the more you learn. So, give yourself every opportunity to read, work problems, and review your mathematics.

Here are two options for reading your math textbook. Read the short descriptions below and decide which will be best for you.

OBJECTIVES

1 Select an appropriate strategy for homework.

2 Use textbook features effectively.

Preview before Class; Read Carefully *after* Class

Abby learns best by listening to her teacher explain things. She "gets it" when she sees the instructor work problems on the board. She likes to ask questions in class and put the information in her notes. She has learned that it helps if she has *previewed* the section before the lecture, so she knows generally what to expect in class. *But after the class instruction*, when Abby gets home, she finds that she can better understand the math textbook. She remembers what her teacher said, and she can double-check her notes if she gets confused. So, **Abby carefully reads the section in her text *after* she hears the classroom lecture on the topic.**

Read Carefully *before* Class

De'Lore, on the other hand, feels he learns well by reading on his own. He prefers to read the section and try working the example problems before coming to class. That way, he already knows what the teacher is going to talk about. Then, he can follow the teacher's examples more easily. It is also easier for him to take notes in class. De'Lore likes to have his questions answered right away, which he can do if he has already read the chapter section. So, **De'Lore carefully reads the section in his text *before* he hears the classroom lecture on the topic.**

Notice that there is no one right way to work with your textbook. You must always figure out what works best for you. Note also that both Abby and De'Lore work with one section at a time. ***The key is that you read the textbook regularly.*** The rest of this activity will give you some ideas of how to make the most of your reading.

Try the following steps as you read your math textbook.

- **Read slowly.** Read only one section—or even part of a section—at a time.
- **Do the sample problems in the margins as you go.** Check them right away. The answers are at the bottom of the page.
- If your mind wanders, **work problems on separate paper and write explanations in your own words.**
- **Make study cards as you read each section.** Pay special attention to the colored boxes in the book. Make cards for new vocabulary, rules, procedures, formulas, and sample problems.
- **NOW,** you are ready to do your homework assignment.

Why Are These Techniques Brain Friendly?

The steps here encourage you to be actively working with the material in your text. You learn best when you are actively doing something.

These methods require you to try several different techniques, not just the same thing over and over.

Also, the techniques allow you to take small breaks in your learning. Those rest periods are crucial for good learning.

Which two or three steps will be most helpful for you?

1. ______________ 2. ______________ 3. ______________

Homework

Teachers assign homework so you can learn the material and then remember the material through practice. In learning, you get good at what you practice. So, completing homework every day will build your confidence, strengthen your skills, and prepare you for exams.

If you have read each section in your textbook according to the steps above, you will probably encounter few difficulties with the exercises in the homework. Here are some additional suggestions that will help you succeed with the homework.

- **If you have trouble with a problem,** find a similar worked example problem in the section. Pay attention to *every line* of the worked example to see how to get from step to step. Work it yourself too, on separate paper; don't just look at it.
- **If it is hard to remember the steps** to follow for certain procedures, write the steps on a separate card. Then write a short explanation of each step. Keep the card nearby while you do the exercises, but try *not* to look at it.
- **If you aren't sure you are working the assigned exercises correctly,** choose two or three odd-numbered problems that are a similar type and work them. Then check the answers in the answer section of your book and see if you are doing them correctly. If you aren't, go back to the section in the text and review the examples and find out how to correct your errors. Finally, when you are sure you understand, try the assigned problems again.
- **If the problem or a similar problem has a blue screen around the problem number,** such as **11.** , there is a worked-out solution in the selected solutions section at the back of the book. Study this solution.
- **Make sure you do some homework every day,** even if your math class does not meet each day.

What are your biggest homework concerns?

List your two main concerns below. Then write a solution for each one.

1. Concern: ______________ **Solution:** ______________

2. Concern: ______________ **Solution:** ______________

Why Are These Suggestions Brain Friendly?

You will learn mathematics as you study the worked examples in the text and try doing them yourself on separate paper. Then, when you see similar problems in the homework, you will already have experience to work from.

Giving yourself a practice test by trying to remember the steps (without looking at your study cards or notes) is an excellent way to reinforce what you are learning.

Correcting errors right away is how you learn and reinforce the correct procedures. It is hard to unlearn a mistake, so always check frequently to see that you are on the right track.

1.2 Variables, Expressions, and Equations

A **variable** is a symbol, usually a letter such as x, y, or z, used to represent an unknown number. Different numbers can replace the variables to form specific statements. For example, in **Section 1.7** we will see that

$$a + b = b + a.$$

This statement is true for any replacements of the variables a and b, such as 2 for a and 5 for b, which gives the true statement

$$2 + 5 = 5 + 2.$$

An **algebraic expression** is a collection of numbers, variables, operation symbols, and grouping symbols, such as parentheses, square brackets, or fraction bars.

$x + 5$, $2m - 9$, and $8p^2 + 6(p - 2)$ Algebraic expressions

In $2m - 9$, the $2m$ means $2 \cdot m$, the product of 2 and m; $8p^2$ represents the product of 8 and p^2. Also, $6(p - 2)$ means the product of 6 and $p - 2$.

OBJECTIVES

1. **Evaluate algebraic expressions, given values for the variables.**
2. **Translate phrases from words to algebraic expressions.**
3. **Identify solutions of equations.**
4. **Translate sentences to equations.**
5. **Distinguish between *expressions* and *equations*.**

OBJECTIVE 1 Evaluate algebraic expressions, given values for the variables. An algebraic expression can have different numerical values for different values of the variables.

EXAMPLE 1 Evaluating Expressions

Find the value of each algebraic expression if $m = 5$ and then if $m = 9$.

(a) $8m$

$8m$		$8m$	
$= 8 \cdot 5$	Let $m = 5$.	$= 8 \cdot 9$	Let $m = 9$.
$= 40$	Multiply.	$= 72$	Multiply.

(b) $3m^2$

$3m^2$		$3m^2$	
$= 3 \cdot 5^2$	Let $m = 5$.	$= 3 \cdot 9^2$	Let $m = 9$.
$= 3 \cdot 25$	Square 5.	$= 3 \cdot 81$	Square 9.
$= 75$	Multiply.	$= 243$	Multiply.

CAUTION
In Example 1(b), $3m^2$ means $3 \cdot m^2$, ***not*** $3m \cdot 3m$. ***Unless parentheses are used, the exponent refers only to the variable or number just before it.*** Use parentheses to write $3m \cdot 3m$ with exponents as $(3m)^2$.

Work Problem 1 at the Side. ▶

1 Find the value of each expression if $p = 3$.

(a) $6p$

(b) $p + 12$

(c) $5p^2$

EXAMPLE 2 Evaluating Expressions

Find the value of each expression if $x = 5$ and $y = 3$.

(a)

$2x + 5y$

$= 2 \cdot 5 + 5 \cdot 3$ Replace x with 5 and y with 3.

$= 10 + 15$ Multiply.

$= 25$ Add.

Follow the rules for order of operations.

Continued on Next Page

ANSWERS

1. (a) 18 (b) 15 (c) 45

2 Find the value of each expression if $x = 6$ and $y = 9$.

(a) $4x + 7y$

(b) $\dfrac{4x - 2y}{x + 1}$

(c) $2x^2 + y^2$

(b) $\dfrac{9x - 8y}{2x - y}$

$= \dfrac{9 \cdot 5 - 8 \cdot 3}{2 \cdot 5 - 3}$ Replace x with 5 and y with 3.

$= \dfrac{45 - 24}{10 - 3}$ Multiply.

$= \dfrac{21}{7}$ Subtract.

$= 3$ Divide.

(c) $x^2 - 2y^2$

$3^2 = 3 \cdot 3$

$5^2 = 5 \cdot 5$

$= 5^2 - 2 \cdot 3^2$ Replace x with 5 and y with 3.

$= 25 - 2 \cdot 9$ Apply the exponents.

$= 25 - 18$ Multiply.

$= 7$ Subtract.

Work Problem 2 at the Side.

OBJECTIVE 2 Translate phrases from words to algebraic expressions.

Problem-Solving Hint

Sometimes variables must be used to change word phrases into algebraic expressions. This process will be important later for solving applied problems.

EXAMPLE 3 Using Variables to Write Word Phrases as Algebraic Expressions

Write each word phrase as an algebraic expression, using x as the variable.

(a) The **sum** of a number and 9

"Sum" is the answer to an addition problem. This phrase translates as

$$x + 9, \quad \text{or} \quad 9 + x.$$

(b) 7 **minus** a number

"Minus" indicates subtraction, so the translation is

$$7 - x.$$

Note that $x - 7$ would* not *be correct because we cannot subtract in either order and get the same results.

(c) A number **subtracted from 12**

Since a number is subtracted *from* 12, write this as

$$12 - x.$$

Compare this result with "12 subtracted from a number," which is $x - 12$.

(d) The **product** of 11 and a number

$$11 \cdot x, \quad \text{or} \quad 11x$$

Continued on Next Page

Answers

2. (a) 87 (b) $\frac{6}{7}$ (c) 153

(e) 5 **divided by** a number

$$5 \div x, \quad \text{or} \quad \frac{5}{x}$$

$\frac{x}{5}$ is not correct here.

(f) The **product of** 2 and the **difference** between a number and 8

$$2(x - 8)$$

We are multiplying 2 times another number. This number is the difference between some number and 8, written $x - 8$. Using parentheses around this difference, the final expression is $2(x - 8)$.

CAUTION

Notice that in translating the words "the difference between a number and 8" in Example 3(f), the order is kept the same: $x - 8$. "The difference between 8 and a number" would be written $8 - x$.

Work Problem 3 at the Side. ▶

OBJECTIVE 3 Identify solutions of equations. An **equation** is a statement that two expressions are equal. ***Therefore, an equation always includes the equality symbol, =.***

$$x + 4 = 11, \quad 2y = 16, \quad \text{and} \quad 4p + 1 = 25 - p \qquad \text{Equations}$$

To **solve** an equation, we must find all values of the variable that make the equation true. Such values of the variable are called the **solutions** of the equation.

EXAMPLE 4 Deciding Whether a Number Is a Solution of an Equation

Decide whether the given number is a solution of the equation.

(a) Is 7 a solution of $5p + 1 = 36$?

$$5p + 1 = 36$$

$$5 \cdot 7 + 1 \stackrel{?}{=} 36 \qquad \text{Replace } p \text{ with 7.}$$

$$35 + 1 \stackrel{?}{=} 36 \qquad \text{Multiply.}$$

$$36 = 36 \qquad \text{True}$$

The number 7 is a solution of the equation.

(b) Is $\frac{14}{3}$ a solution of $9m - 6 = 32$?

$$9m - 6 = 32$$

$$9 \cdot \frac{14}{3} - 6 \stackrel{?}{=} 32 \qquad \text{Replace } m \text{ with } \tfrac{14}{3}.$$

$$42 - 6 \stackrel{?}{=} 32 \qquad \text{Multiply.}$$

$$36 = 32 \qquad \text{False}$$

The number $\frac{14}{3}$ is not a solution of the equation.

Work Problem 4 at the Side. ▶

3 Write each word phrase as an algebraic expression. Use x as the variable.

(a) The sum of 5 and a number

(b) A number minus 4

(c) A number subtracted from 48

(d) The product of 6 and a number

(e) 9 multiplied by the sum of a number and 5

4 Decide whether the given number is a solution of the equation.

(a) $p - 1 = 3$; 2

(b) $2k + 3 = 15$; 7

(c) $8p - 11 = 5$; 2

Answers

3. **(a)** $5 + x$ **(b)** $x - 4$ **(c)** $48 - x$ **(d)** $6x$ **(e)** $9(x + 5)$

4. **(a)** no **(b)** no **(c)** yes

OBJECTIVE 4 Translate sentences to equations. We have seen how to translate phrases from words to expressions. Sentences given in words are translated as equations.

EXAMPLE 5 Translating Sentences to Equations

Write each word sentence as an equation. Use x as the variable.

(a) Twice the sum of a number and four is six.

"Twice" means two times. The word *is* suggests equals. With x representing the number, translate as follows.

Twice	the sum of a number and four	is	six.
↓	↓	↓	↓
$2 \cdot$	$(x + 4)$	$=$	6

$$2(x + 4) = 6$$

(b) Nine more than five times a number is 49.

Use x to represent the unknown number. Start with $5x$ and then add 9 to it. The word *is* translates as $=$.

$$5x + 9 = 49$$

(c) Seven less than three times a number is eleven.

Here, 7 is *subtracted* from three times a number to get 11.

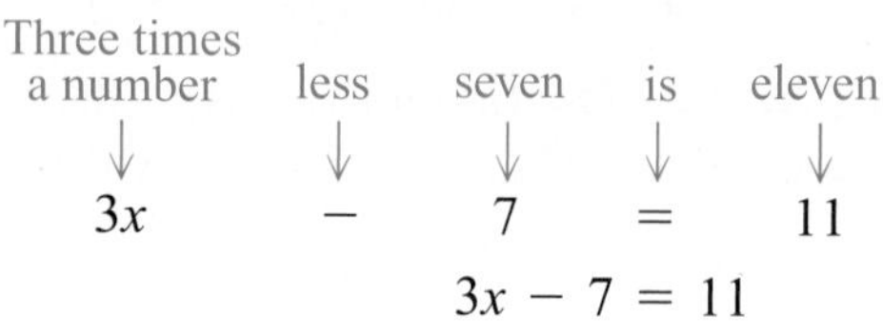

$$3x - 7 = 11$$

◀ Work Problem **5** at the Side.

OBJECTIVE 5 Distinguish between *expressions* and *equations*. Students often have trouble distinguishing between equations and expressions. ***Remember that an equation is a sentence (with an = symbol); an expression is a phrase that represents a number.***

$$4x + 5 = 9 \qquad\qquad 4x + 5$$

Left Side ↑ Right side — Equation (to solve)

↑ Expression (to simplify or evaluate)

EXAMPLE 6 Distinguishing between Equations and Expressions

Decide whether each is an *equation* or an *expression*.

(a) $2x - 5y$

There is no equals symbol, so this is an expression.

(b) $2x = 5y$

Because there is an equals symbol with something on either side of it, this is an equation.

◀ Work Problem **6** at the Side.

5 Write each word sentence as an equation. Use x as the variable.

(a) Three times the sum of a number and 13 is 19.

(b) Five times a number is subtracted from 21, giving 15.

6 Decide whether each is an *equation* or an *expression*.

(a) $2x + 5y - 7$

(b) $\dfrac{3x - 1}{5}$

(c) $2x + 5 = 7$

(d) $\dfrac{x}{y - 3} = 4x$

ANSWERS

5. **(a)** $3(x + 13) = 19$ **(b)** $21 - 5x = 15$
6. **(a)** expression **(b)** expression **(c)** equation **(d)** equation

1.2 Exercises

FOR EXTRA HELP MyMathLab Math XL PRACTICE  WATCH DOWNLOAD READ REVIEW

Fill in each blank with the correct response.

1. If $x = 3$, then the value of $x + 8$ is **11**.

2. If $x = 1$ and $y = 2$, then the value of $5xy$ is **10**.

3. "The sum of 13 and x" is represented by the expression **13 + x**. If $x = 3$, the value of that expression is **16**.

4. Will the equation $x = x + 5$ ever have a solution? **no**.

5. $2x + 6$ is an **expression** (equation/expression), while $2x + 6 = 8$ is an **equation** (equation/expression).

Exercises 6–10 cover some of the concepts introduced in this section. Give a short explanation for each.

6. Why is $2x^3$ not the same as $2x \cdot 2x \cdot 2x$? Explain, using an exponent to write $2x \cdot 2x \cdot 2x$.

$2x^3 = 2 \cdot x \cdot x \cdot x$, while $2x \cdot 2x \cdot 2x = (2x)^3$.

7. If the words *more than* in Example 5(b) were changed to *less than,* how would the equation be changed?

The equation would be $5x - 9 = 49$.

8. Explain in your own words why, when evaluating the expression $4x^2$ for $x = 3$, 3 must be squared *before* multiplying by 4.

The exponent 2 applies only to the base x, and exponentials must be evaluated before products.

9. There are many pairs of values of x and y for which $2x + y$ will equal 6. Name two such pairs and describe how you determined them.

Answers will vary. Two such pairs are $x = 0, y = 6$ and $x = 1, y = 4$. To find a pair, choose one number, substitute it for a variable, then calculate the value for the other variable.

10. Suppose that for the equation $3x - y = 9$, the value of x is given as 4. What would be the corresponding value of y? How do you know this?

The value for y is 3. If $x = 4$, then we have $12 - y = 9$, and 3 subtracted from 12 equals 9.

Find the numerical value of each expression if ***(a)*** $x = 4$ *and* ***(b)*** $x = 6$. *See Example 1.*

11. $4x^2$
(a) 64 (b) 144

12. $5x^2$
(a) 80 (b) 180

13. $\frac{3x - 5}{2x}$
(a) $\frac{7}{8}$ (b) $\frac{13}{12}$

14. $\frac{4x - 1}{3x}$
(a) $\frac{5}{4}$ (b) $\frac{23}{18}$

15. $\frac{6.459x}{2.7}$ (to the nearest thousandth)
(a) 9.569 (b) 14.353

16. $\frac{0.74x^2}{0.85}$ (to the nearest thousandth)
(a) 13.929 (b) 31.341

17. $3x^2 + x$
(a) 52 (b) 114

18. $2x + x^2$
(a) 24 (b) 48

Find the value of each expression if ***(a)*** $x = 2$ *and* $y = 1$ *and* ***(b)*** $x = 1$ *and* $y = 5$. *See Example 2.*

19. $3(x + 2y)$
(a) 12 (b) 33

20. $2(2x + y)$
(a) 10 (b) 14

21. $x + \frac{4}{y}$
(a) 6 (b) $\frac{9}{5}$

22. $y + \frac{8}{x}$
(a) 5 (b) 13

23. $\frac{x}{2} + \frac{y}{3}$
(a) $\frac{4}{3}$ (b) $\frac{13}{6}$

24. $\frac{x}{5} + \frac{y}{4}$
(a) $\frac{13}{20}$ (b) $\frac{29}{20}$

25. $\frac{2x + 4y - 6}{5y + 2}$
(a) $\frac{2}{7}$ (b) $\frac{16}{27}$

26. $\frac{4x + 3y - 1}{2x + y}$
(a) 2 (b) $\frac{18}{7}$

27. $2y^2 + 5x$
(a) 12 (b) 55

28. $6x^2 + 4y$
(a) 28 (b) 26

29. $\frac{3x + y^2}{2x + 3y}$
(a) 1 (b) $\frac{28}{17}$

30. $\frac{x^2 + 1}{4x + 5y}$
(a) $\frac{5}{13}$ (b) $\frac{2}{29}$

31. $0.841x^2 + 0.32y^2$

(a) 3.684 **(b)** 8.841

32. $0.941x^2 + 0.2y^2$

(a) 3.964 **(b)** 5.941

Write each word phrase as an algebraic expression, using x as the variable. See Example 3.

33. Twelve times a number

$12x$

34. Thirteen added to a number

$x + 13$

35. Two subtracted from a number

$x - 2$

36. Eight subtracted from a number

$x - 8$

37. One-third of a number, subtracted from seven

$7 - \frac{1}{3}x$

38. One-fifth of a number, subtracted from fourteen

$14 - \frac{1}{5}x$

39. The difference between twice a number and 6

$2x - 6$

40. The difference between 6 and half a number

$6 - \frac{1}{2}x$

41. 12 divided by the sum of a number and 3

$\frac{12}{x + 3}$

42. The difference between a number and 5, divided by 12

$\frac{x - 5}{12}$

43. The product of 6 and four less than a number

$6(x - 4)$

44. The product of 9 and five more than a number

$9(x + 5)$

45. In the phrase "four more than the product of a number and 6," does the word *and* signify the operation of addition? Explain.

The word *and* does not signify addition here. In the phrase "the product of a number and 6," *and* connects two quantities to be multiplied.

46. Suppose that the directions on a test read "Solve the following expressions." How would you politely correct the person who wrote these directions?

An expression cannot be solved; it merely indicates a series of operations to be performed. An equation can be solved.

Decide whether the given number is a solution of the equation. See Example 4.

47. Is 7 a solution of $x - 5 = 12$?

no

48. Is 10 a solution of $x + 6 = 15$?

no

49. Is 1 a solution of $5x + 2 = 7$?

yes

50. Is 1 a solution of $3x + 5 = 8$?

yes

51. Is $\frac{1}{5}$ a solution of $6x + 4x + 9 = 11$?

yes

52. Is $\frac{12}{5}$ a solution of $2x + 3x + 8 = 20$?

yes

53. Is 3 a solution of $2y + 3(y - 2) = 14$?

no

54. Is 2 a solution of $6a + 2(a + 3) = 14$?

no

55. Is $\frac{1}{3}$ a solution of $\frac{z + 4}{2 - z} = \frac{13}{5}$?

yes

56. Is $\frac{13}{4}$ a solution of $\frac{x + 6}{x - 2} = \frac{37}{5}$?

yes

57. Is 4.3 a solution of $3r^2 - 2 = 53.47$?

yes

58. Is 3.7 a solution of $2x^2 + 1 = 28.38$?

yes

Write each word sentence as an equation. Use x as the variable. See Example 5.

59. The sum of a number and 8 is 18.

$x + 8 = 18$

60. A number minus three equals 1.

$x - 3 = 1$

61. Five more than twice a number is 5.

$2x + 5 = 5$

62. The product of 2 and the sum of a number and 5 is 14.

$2(x + 5) = 14$

63. Sixteen minus three-fourths of a number is 13.

$16 - \frac{3}{4}x = 13$

64. The sum of six-fifths of a number and 2 is 14.

$\frac{6}{5}x + 2 = 14$

65. Three times a number is equal to 8 more than twice the number.

$3x = 2x + 8$

66. Twelve divided by a number equals $\frac{1}{3}$ times that number.

$\frac{12}{x} = \frac{1}{3}x$

Identify each as an expression or an equation. See Example 6.

67. $3x + 2(x - 4)$

expression

68. $5y - (3y + 6)$

expression

69. $7t + 2(t + 1) = 4$

equation

70. $9r + 3(r - 4) = 2$

equation

Relating Concepts (Exercises 71–74) For Individual or Group Work

*A **mathematical model** is an equation that describes the relationship between two quantities. For example, the life expectancy of Americans at birth can be approximated by the equation*

$$y = 0.212x - 347,$$

where x is a year between 1943 and 2005 and y is age in years. (Source: Centers for Disease Control and Prevention.)

Use this model to approximate life expectancy (to the nearest tenth of a year) in each of the following years.

71. 1943

64.9 yr

72. 1960

68.5 yr

73. 1980

72.8 yr

74. 2005

78.1 yr

Study Skills

TAKING LECTURE NOTES

Study the set of sample math notes below, and read the comments about them. Then try to incorporate the techniques into your own math note taking in class.

OBJECTIVE

1 Identify and apply note taking strategies.

- **Always include the date and title** of the day's lecture topic at the top of every page. **Always begin a new day with a new page.**
- **Note the definitions** of base and exponent are written in parentheses—don't trust your memory!
- **Skipping lines** makes the notes easier to read.
- **See how the direction word** (*simplify*) **is emphasized** and explained.
- **A star marks an important concept.** This is a warning to avoid future mistakes. **Note the underlining,** too, which highlights the importance.
- Notice the two columns, which allow for the example and its explanation to be close together. **Whenever you know you'll be given a series of steps to follow, try the two-column method.**
- **Note the brackets and arrows,** which clearly show how the problem is set up to be simplified.

January 2

Exponents

Exponents used to show repeated multiplication.

$3 \cdot 3 \cdot 3 \cdot 3$ can be written 3^4 — *exponent (how many times it's multiplied)*; *base (the number being multiplied)*

Read 3^2 as 3 to the 2nd power or 3 squared

3^3 as 3 to the 3rd power or 3 cubed

3^4 as 3 to the 4th power

etc.

Simplifying an expression with exponents → *actually do the repeated multiplication*

2^3 means $2 \cdot 2 \cdot 2$ and $2 \cdot 2 \cdot 2 = 8$

☆ *Careful!* [*5^2 means $5 \cdot 5$ NOT $5 \cdot 2$*
so $5^2 = 5 \cdot 5 = 25$ BUT $5^2 \neq 10$]

Example	*Explanation*
Simplify $2^4 \cdot 3^2$	*Exponents mean multiplication.*
$2 \cdot 2 \cdot 2 \cdot 2 \cdot 3 \cdot 3$	*Use 2 as a factor 4 times.* *Use 3 as a factor 2 times.*
$16 \cdot 9$	*$2 \cdot 2 \cdot 2 \cdot 2$ is 16* *$3 \cdot 3$ is 9* → *$16 \cdot 9$ is 144*
144	*Simplified result is 144 (no exponents left)*

Now Try This ▶▶▶

Find one or two people in your math class to work with. Compare each other's lecture notes over a period of a week or so. Ask yourself the following questions as you examine the notes.

1. **What are you doing to show the main points** or larger concepts in your notes (such as underlining, boxing, using stars or capital letters, etc.)?

   ____

   ____

2. **In what ways do you set off the explanations** for worked problems, examples, and subpoints or smaller concepts (such as indenting, using arrows, circling or boxing, etc.)?

   ____

   ____

3. **What does your instructor do** to show that he or she is moving from one idea to the next (such as saying "Next," "Any questions," "Now," or erasing the board, etc.)?

   ____

   ____

4. **How do you mark a change in ideas** or topics in your notes (such as skipping lines, using dashes or numbers, etc.)?

   ____

   ____

5. **What explanations (in words) do you include** in your notes, so that you can read over them and still remember new concepts when doing your homework?

   ____

   ____

6. **What new ideas did you learn** by examining your classmates' notes?

   ____

   ____

7. **What new techniques will you try in your own note taking?** List several that you will use next time you take notes in math class.

   ____

   ____

Why Are These Notes Brain Friendly?

The notes are easy to look at. Other techniques that are visually memorable are the use of spacing (the two columns), stars, underlining, and circling. All of these methods allow you to take note of important concepts and steps.

The notes are also systematic, which means that they use certain techniques regularly. This way, you can easily recognize the topic of the day, the signals that show an important point, and the steps to follow for procedures. When you develop a system that you always use in your notes, they are easy to understand later when you are reviewing for a test.

1.3 Real Numbers and the Number Line

OBJECTIVES

1. Classify numbers and graph them on number lines.
2. Tell which of two real numbers is less than the other.
3. Find the opposite of a real number.
4. Find the absolute value of a real number.

A **set** is a collection of objects. In mathematics, these objects are usually numbers. The objects that belong to the set, called **elements** of the set, are written between braces. For example, the set of numbers 1, 2, 3, 4, 5 is written

$$\{1, 2, 3, 4, 5\}.$$

OBJECTIVE 1 Classify numbers and graph them on number lines. The set of numbers used for counting is called the **natural numbers.**

Natural Numbers

$$\{1, 2, 3, 4, 5, \ldots\}$$

In **Chapter R,** we introduced the set of **whole numbers.**

Whole Numbers

$$\{0, 1, 2, 3, 4, 5, \ldots\}$$

These numbers, along with many others, can be represented on a **number line** like the one in Figure 1. We draw a number line by choosing any point on the line and labeling it 0. Choose any point to the right of 0 and label it 1. The distance between 0 and 1 gives a unit of measure used to locate other points, as shown in Figure 1. The points labeled in Figure 1 correspond to the first few whole numbers.

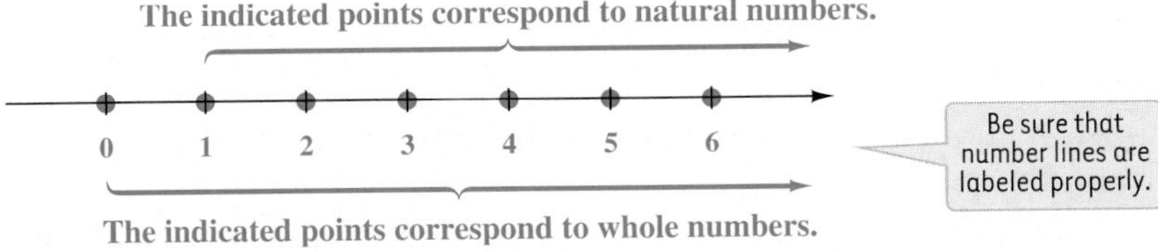

Figure 1

The natural numbers are located to the right of 0 on the number line. But numbers may also be placed to the left of 0. For each natural number we can place a corresponding number to the left of 0. These numbers, written $-1, -2, -3, -4$, and so on, are shown in Figure 2 on the next page. Each is the **opposite,** or **negative,** of a natural number. The natural numbers, their opposites, and 0 form a new set of numbers called the **integers.**

Integers

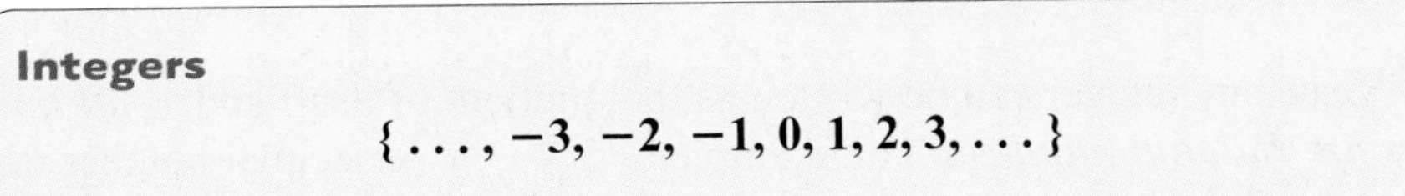

$$\{\ldots, -3, -2, -1, 0, 1, 2, 3, \ldots\}$$

1 Use an integer to express the boldface italic number(s) in each application.

(a) Erin discovers that she has spent $***53*** more than she has in her checking account.

(b) The record high Fahrenheit temperature in the United States was ***134***° in Death Valley, California, on July 10, 1913. (*Source: World Almanac and Book of Facts.*)

(c) A football team gained ***5*** yd, then lost ***10*** yd on the next play.

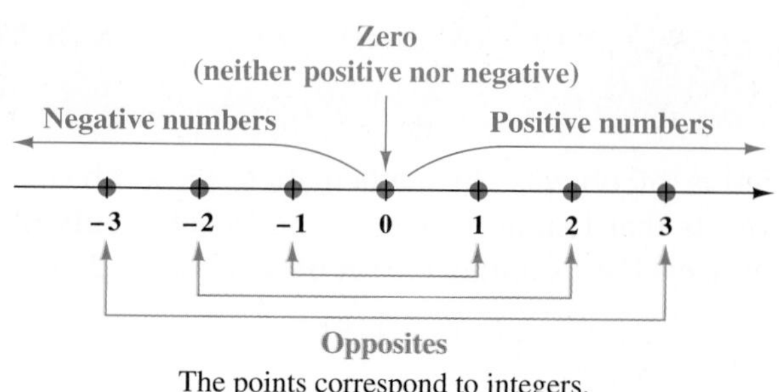

Figure 2

There are many practical applications of negative numbers. For example, a Fahrenheit temperature on a cold January day might be $-10°$, and a business that spends more than it takes in has a negative "profit" (a loss).

EXAMPLE 1 Using Negative Numbers in Applications

Use an integer to express the boldface italic number in each application.

(a) The lowest Fahrenheit temperature ever recorded in meteorological records was ***129***° below zero at Vostok, Antarctica, on July 21, 1983. (*Source: World Almanac and Book of Facts.*)
Use -129 because "below zero" indicates a negative number.

(b) The shore surrounding the Dead Sea is ***1348*** ft below sea level. (*Source: World Almanac and Book of Facts.*)
Again, "below sea level" indicates a negative number, -1348.

◀ Work Problem 1 at the Side.

Not all numbers are integers. For example, $\frac{1}{2}$ is not; it is a number halfway between the integers 0 and 1. Also, $3\frac{1}{4}$ is not an integer. These numbers and others that are quotients of integers are *rational numbers.* (The name comes from the word *ratio,* which indicates a quotient.)

Rational Numbers

$\{x \mid x$ is a quotient of two integers, with denominator not $0\}$ is the set of **rational numbers.**

(Read the part in the braces as "the set of all numbers x such that x is a quotient of two integers, with denominator not 0.")

Note

The set symbolism used in the definition of rational numbers,

$$\{x \mid x \textbf{ has a certain property}\},$$

is called **set-builder notation.** This notation is convenient to use when it is not possible to list all the elements of a set.

Since any integer can be written as the quotient of itself and 1, ***all integers are rational numbers.*** For example, $-5 = \frac{-5}{1}$. A decimal number that comes to an end (terminates), such as 0.23, is a rational number. For example, $0.23 = \frac{23}{100}$. Decimal numbers that repeat in a fixed block of digits, such as $0.3333\ldots = 0.\overline{3}$ and $0.454545\ldots = 0.\overline{45}$, are also rational numbers. For example, $0.\overline{3} = \frac{1}{3}$.

ANSWERS

1. **(a)** -53 **(b)** 134 **(c)** $5, -10$

As shown in Figures 1 and 2 on the preceding pages, to **graph** a number, we place a dot on the number line at the point that corresponds to the number. The number is called the **coordinate** of the point. Think of the graph of a set of numbers as a picture of the set.

EXAMPLE 2 Graphing Rational Numbers

Graph each number on the number line.

$$-\frac{3}{2},\quad -\frac{2}{3},\quad \frac{1}{2},\quad 1\frac{1}{3},\quad \frac{23}{8},\quad 3\frac{1}{4}$$

To locate the improper fractions on the number line, write them as mixed numbers or decimals. The graph is shown in Figure 3.

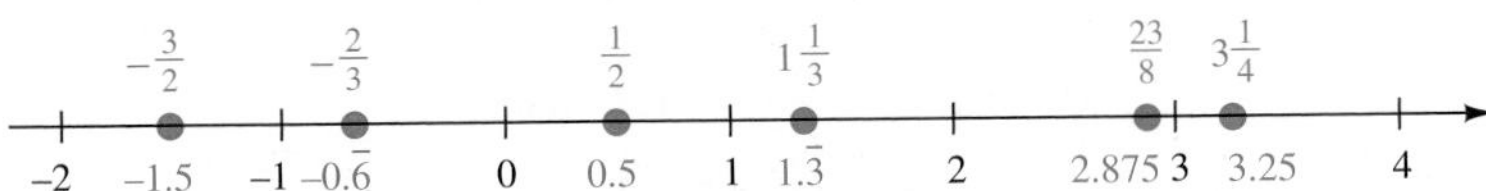

Figure 3

Work Problem 2 at the Side. ▶

2 Graph each number on the number line.

$$-3,\ \frac{17}{8},\ -2.75,\ 1\frac{1}{2},\ -\frac{3}{4}$$

−3 −2 −1 0 1 2 3

The number system contains numbers that are not rational. For example, a square that measures one unit on a side has a diagonal whose length is the square root of 2, written $\sqrt{2}$. See Figure 4. It can be shown that $\sqrt{2}$ cannot be written as a quotient of integers. Because of this, $\sqrt{2}$ is not rational; it is *irrational*. Other examples of irrational numbers are $\sqrt{3}, \sqrt{7}, -\sqrt{10}$, and π (the ratio of the *circumference* of a circle to its *diameter*).

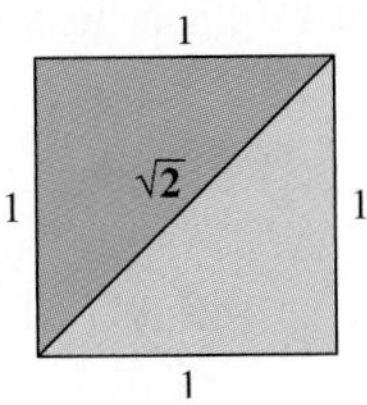

Figure 4

Irrational Numbers

$\{x \mid x$ is a nonrational number represented by a point on the number line$\}$ is the set of **irrational numbers.**

The decimal form of an irrational number neither terminates nor repeats. Irrational numbers are discussed in **Chapter 9.**

Both rational and irrational numbers can be represented by points on the number line and are called **real numbers.**

Real Numbers

$\{x \mid x$ is a rational or an irrational number$\}$ is the set of **real numbers.**

All the numbers mentioned so far are real numbers.* The relationships between the various types of numbers are shown in Figure 5 on the next page. Every real number is either a rational number or an irrational number.

ANSWER

2. -2.75 $-\frac{3}{4}$ $1\frac{1}{2}$ $\frac{17}{8}$
−3 −2 −1 0 1 2 3

* An example of a number that is not real is the square root of a negative number, such as $\sqrt{-4}$.

3 Tell whether each statement is *true* or *false*.

(a) $-2 < 4$

(b) $6 > -3$

(c) $-9 < -12$

(d) $-4 \geq -1$

(e) $-6 \leq 0$

ANSWERS

3. **(a)** true **(b)** true **(c)** false **(d)** false **(e)** true

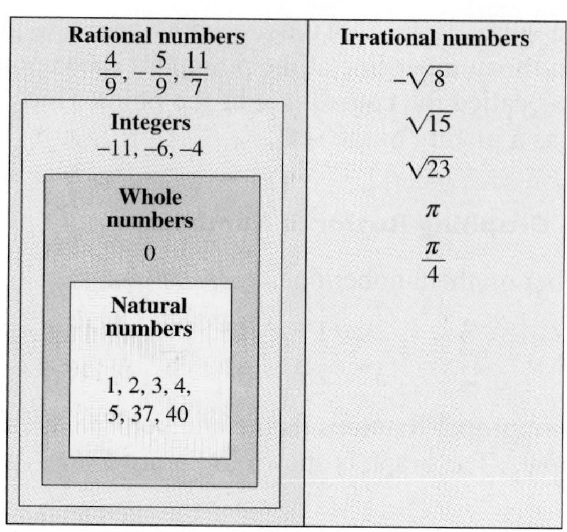

All numbers shown are real numbers.

Figure 5

OBJECTIVE 2 Tell which of two real numbers is less than the other. Given any two whole numbers, we can tell which number is less than the other. But what about two negative numbers, as in the set of integers? Moving from 0 to the right along a number line, the positive numbers corresponding to the points on the number line *increase.* For example, $8 < 12$, and 8 is to the left of 12 on a number line. We extend this ordering to all real numbers.

Ordering of the Real Numbers

For any two real numbers a and b, ***a* is less than *b*** if a is to the left of b on a number line.

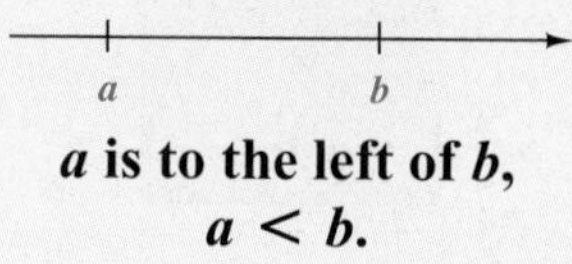

***a* is to the left of *b*,**
$a < b$.

Thus, any negative number is less than 0, and any negative number is less than any positive number. Also, 0 is less than any positive number.

EXAMPLE 3 Determining the Order of Real Numbers

Is it true that $-3 < -1$?

To find out, locate -3 and -1 on a number line, as shown in Figure 6. Because -3 is to the left of -1 on the number line, -3 is less than -1. The statement $-3 < -1$ is true.

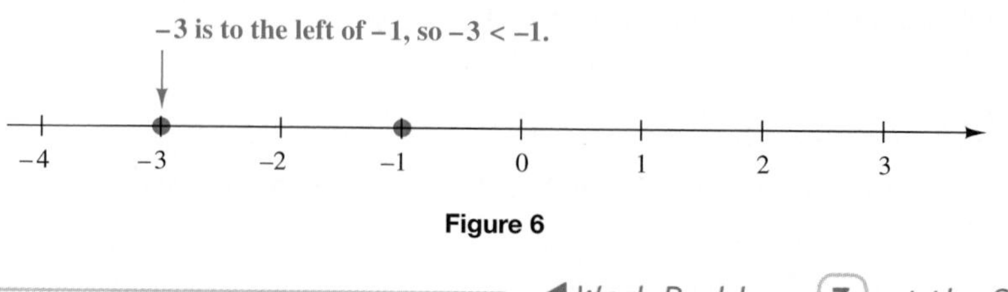

Figure 6

◀ *Work Problem* **3** *at the Side.*

We can also say that, for any real numbers a and b, ***a* is greater than *b*** if a is to the right of b on the number line.

OBJECTIVE 3 Find the opposite of a real number. Earlier, we saw that every positive integer has a negative integer that is its opposite, or negative. This is true for every real number except 0, which is its own opposite.* A characteristic of pairs of opposites is that they are the same distance from 0 on the number line but in opposite directions. See Figure 7.

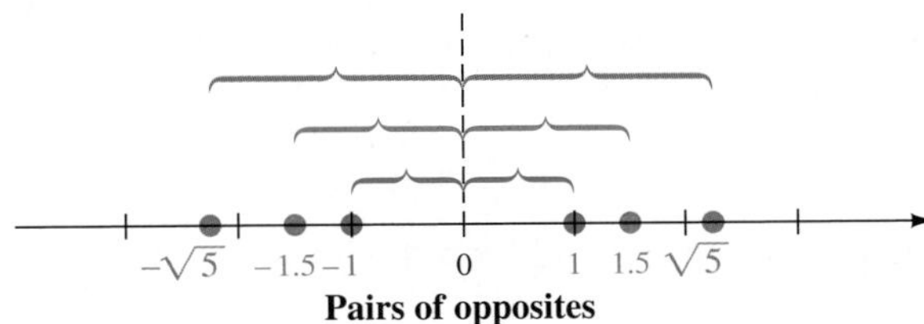

Pairs of opposites

Figure 7

We indicate the opposite of a number by writing the symbol $-$ in front of the number. For example, the opposite of 7 is -7 (read "negative 7"). We could write the opposite of -4 as $-(-4)$, but we know that 4 is the opposite of -4. Since a number can have only one opposite, $-(-4)$ and 4 must represent the same number, so

$$-(-4) = 4.$$

This idea can be generalized.

Double Negative Rule

For any real number a,

$$-(-a) = a.$$

The following chart shows several numbers and their opposites.

Number	Opposite
-4	$-(-4)$, or 4
0	0
5	-5
$-\frac{2}{3}$	$\frac{2}{3}$
0.52	-0.52

The chart suggests the following rule.

Except for 0, the opposite of a number is found by changing the sign of the number.

Work Problem 4 at the Side. ▶

4 Find the opposite of each number.

(a) 6

(b) 15

(c) -9

(d) -12

(e) 0

*The opposite (or negative) of a number is also called the *additive inverse* of the number, as we shall see in **Section 1.7**.

ANSWERS

4. (a) -6 **(b)** -15 **(c)** 9 **(d)** 12 **(e)** 0

OBJECTIVE 4 Find the absolute value of a real number. As previously mentioned, opposites are numbers the same distance from 0 on the number line but on opposite sides of 0. Another way to say this is to say that opposites have the same *absolute value*. The **absolute value** of a number is the undirected distance between 0 and the number on the number line. The symbol for the absolute value of the number a is $|a|$, which is read **"the absolute value of *a*."** For example, the distance between 2 and 0 on the number line is 2 units, so

$$|2| = 2.$$

Also, the distance between -2 and 0 on the number line is 2, so

$$|-2| = 2.$$

Because distance is a physical measurement, which is never negative, we can make the following statement.

The absolute value of a number can never be negative.

For example,

$$|12| = 12 \quad \text{and} \quad |-12| = 12$$

because both 12 and -12 lie at a distance of 12 units from 0 on the number line. Since the distance of 0 from 0 is 0 units, we have

$$|0| = 0.$$

EXAMPLE 4 Evaluating Absolute Value

Simplify by finding the absolute value.

(a) $|5| = 5$

(b) $|-5| = 5$

(c) $-|5| = -(5) = -5$

(d) $-|-5| = -(5) = -5$ Replace $|-5|$ with 5.

(e) $|8 - 5|$

Simplify within the absolute value bars first.

$$|8 - 5| = |3| = 3$$

(f) $-|8 - 5| = -|3| = -3$

(g) $-|12 - 3| = -|9| = -9$

Parts (e)–(g) in Example 4 show that absolute value bars also act as grouping symbols. We must perform any operations within absolute value bars before finding the absolute value.

Work Problem 5 at the Side.

5 Simplify.

(a) $|-6|$

(b) $|9|$

(c) $-|15|$

(d) $-|-9|$

(e) $|9 - 4|$

(f) $-|32 - 2|$

Answers

5. (a) 6 **(b)** 9 **(c)** -15 **(d)** -9 **(e)** 5 **(f)** -30

In Exercises 1–6, give an example of a number that satisfies each given condition.

1. An integer between 3.6 and 4.6

4

2. A rational number between 2.8 and 2.9

One example is 2.85. There are others.

3. A whole number that is not positive and is less than 1

0

4. A whole number greater than 3.5

One example is 4. There are others.

5. An irrational number that is between $\sqrt{12}$ and $\sqrt{14}$

One example is $\sqrt{13}$. There are others.

6. A real number that is neither negative nor positive

0

*List all numbers from each set that are **(a)** natural numbers, **(b)** whole numbers, **(c)** integers, **(d)** rational numbers, **(e)** irrational numbers, **(f)** real numbers.*

7. $\left\{-9, -\sqrt{7}, -1\frac{1}{4}, -\frac{3}{5}, 0, \sqrt{5}, 3, 5.9, 7\right\}$

(a) 3, 7 (b) 0, 3, 7 (c) −9, 0, 3, 7

(d) $-9, -1\frac{1}{4}, -\frac{3}{5}, 0, 3, 5.9, 7$

(e) $-\sqrt{7}, \sqrt{5}$ (f) All are real numbers.

8. $\left\{-5.3, -5, -\sqrt{3}, -1, -\frac{1}{9}, 0, 1.2, 4, \sqrt{12}\right\}$

(a) 4 (b) 0, 4 (c) −5, −1, 0, 4

(d) $-5.3, -5, -1, -\frac{1}{9}, 0, 1.2, 4$

(e) $-\sqrt{3}, \sqrt{12}$ (f) All are real numbers.

Use an integer to express each boldface italic number representing a change in the following applications. See Example 1.

9. Between July 1, 2004, and July 1, 2005, the population of the United States increased by approximately ***2,845,000.*** (*Source:* U.S. Census Bureau.)

2,845,000

10. From 2000 to 2005, the mean SAT verbal score for Massachusetts residents increased by ***9,*** while the mathematics score increased by ***14.*** (*Source:* The College Board.)

9; 14

11. From 1995 to 2005, the number of cable TV systems in the United States went from 11,218 to 8409, representing a decrease of ***2809.*** (*Source: Television and Cable Factbook.*)

−2809

12. In 1935, there were 15,295 banks in the United States. By 2004, the number was 8975, representing a decrease of ***6320*** banks. (*Source:* Federal Deposit Insurance Corporation.)

−6320

Graph each group of numbers on a number line. See Example 2.

13. $0, 3, -5, -6$

[Number line with points at −6, −5, 0, 3; labels: −6 −5 0 3]

14. $2, 6, -2, -1$

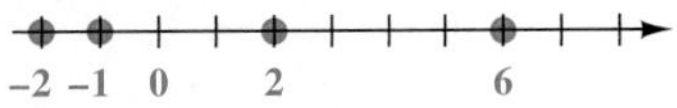

[Number line with points at −2, −1, 2, 6; labels: −2 −1 0 2 6]

15. $-2, -6, -4, 3, 4$

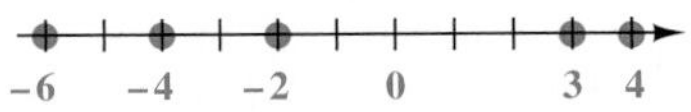

[Number line with points at −6, −4, −2, 3, 4; labels: −6 −4 −2 0 3 4]

16. $-5, -3, -2, 0, 4$

[Number line with points at −5, −3, −2, 0, 4; labels: −5 −3 −2 0 4]

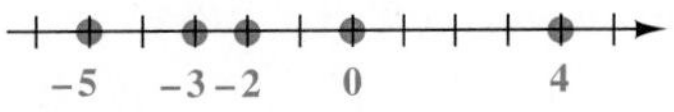

17. $\frac{1}{4}, 2\frac{1}{2}, -3\frac{4}{5}, -4, -\frac{13}{8}$

[Number line with points labeled $-3\frac{4}{5}$, $-\frac{13}{8}$, $\frac{1}{4}$, $2\frac{1}{2}$; axis labels: −4 0 3]

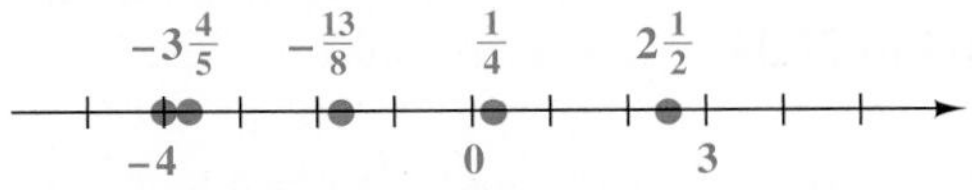

18. $5\frac{1}{4}, \frac{41}{9}, -2\frac{1}{3}, 0, -3\frac{2}{5}$

[Number line with points labeled $-3\frac{2}{5}$, $-2\frac{1}{3}$, $\frac{41}{9}$, $5\frac{1}{4}$; axis labels: −3 0 3]

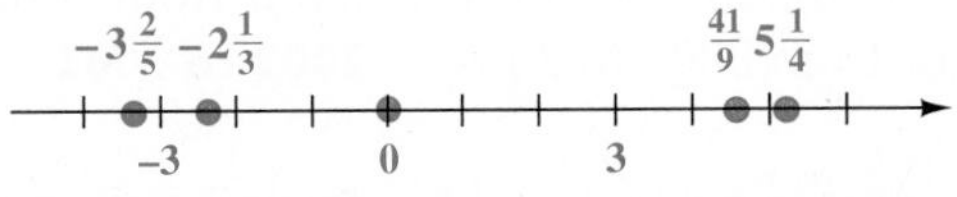

Select the lesser number in each pair. See Example 3.

19. $-11, -4$ **−11**

20. $-9, -16$ **−16**

21. $-21, 1$ **−21**

22. $-57, 3$ **−57**

23. $0, -100$ **−100**

24. $-215, 0$ **−215**

25. $-\frac{2}{3}, -\frac{1}{4}$ $-\frac{2}{3}$

26. $-\frac{3}{8}, -\frac{9}{16}$ $-\frac{9}{16}$

Decide whether each statement is true *or* false. *See Example 3.*

27. $8 < -16$ **false**

28. $12 < -24$ **false**

29. $-3 < -2$ **true**

30. $-10 < -9$ **true**

For each number, **(a)** *find its opposite and* **(b)** *find its absolute value.*

31. -2 **(a)** 2 **(b)** 2

32. -8 **(a)** 8 **(b)** 8

33. 6 **(a)** −6 **(b)** 6

34. 11 **(a)** −11 **(b)** 11

35. $-\frac{3}{4}$ **(a)** $\frac{3}{4}$ **(b)** $\frac{3}{4}$

36. $-\frac{1}{3}$ **(a)** $\frac{1}{3}$ **(b)** $\frac{1}{3}$

Simplify. See Example 4.

37. $|-7|$ **7**

38. $|-3|$ **3**

39. $-|12|$ **−12**

40. $-|23|$ **−23**

41. $-\left|-\frac{2}{3}\right|$ $-\frac{2}{3}$

42. $-\left|-\frac{4}{5}\right|$ $-\frac{4}{5}$

43. $|13 - 4|$ **9**

44. $|8 - 7|$ **1**

Decide whether each statement is true *or* false.

45. $|-8| < 7$ **false**

46. $|-6| \geq -|6|$ **true**

47. $4 \leq |4|$ **true**

48. $-|-3| > 2$ **false**

49. Students often say "The absolute value of a number is always positive." Is this true? If not, explain.
No; the statement is false for one number, 0.

50. If the absolute value of a number is equal to the number itself, what must be true about the number?
It must be greater than or equal to 0.

To answer the questions in Exercises 51–54, refer to the table, which gives the changes in producer price indexes for two recent periods.

Commodity	Change from 2002 to 2003	Change from 2003 to 2004
Farm products	12.6	10.6
Gasoline	23.3	24.7
Machinery and equipment	−0.8	0.2
Iron and steel	6.5	33.7
Electronic components and accessories	−1.7	−2.2

Source: U.S Bureau of Labor Statistics.

51. What commodity for which period represents the greatest decrease?
electronic components and accessories, 2003 to 2004

52. What commodity for which period represents the least change?
machinery and equipment, 2003 to 2004

53. Which has lesser absolute value, the change for electronic components and accessories from 2002 to 2003 or from 2003 to 2004? **2002 to 2003**

54. Which has greater absolute value, the change for machinery and equipment from 2002 to 2003 or from 2003 to 2004? **2002 to 2003**

Study Skills

USING STUDY CARDS

OBJECTIVES

1. **Create study cards for new terms.**
2. **Create study cards for new procedures.**

You may have used "flash cards" in other classes. In math, "study cards" can be helpful. The main things to remember in math besides terms and definitions are *sets of steps to follow* to solve problems (and how to know which set of steps to follow) and *concepts about how math works* (principles). So, the cards may look different but will be just as useful.

In this two-part activity, you will find four types of study cards to use in math. Look carefully at what kinds of information to put on them and where to put it. Then use them the way you would any flash card:

- to quickly review when you have a few minutes,
- for daily reviews,
- to review before a quiz or test.

Remember, the most helpful thing about study cards is making them. After each card description you will find an assignment to try, marked **Now Try This.**

New Vocabulary Cards

To make a new vocabulary card, put the word (spelled correctly) and the page number where it is found on the front of the card. On the back, write:

- the definition (in your own words if possible),
- an example, including any exceptions or other special information,
- any related words, and
- a sample problem (if appropriate).

Front of Card

Integers *p. 49*

Back of Card

Def: The natural numbers {1, 2, 3, 4, ...}
their opposites {-1, -2, -3, -4, ...}
and 0.

Integers {..., -3, -2, -1, 0, 1, 2, 3, ...}

→ No fractions, decimals, roots
→ Related word: rational numbers

Now Try This

List 4 new vocabulary words/concepts you need to learn right now. Make a card for each one.

__________ __________ __________ __________

Procedure ("Steps") Cards

To make a procedure (steps) card, write the name of the procedure at the top on the front of the card. Then write each step *in words*. If you need to learn abbreviations for some words, include them along with the whole words written out. On the back of the card, put an example of the procedure, showing each step you need to take. You can review by looking at the front and practicing a new worked example, or by looking at the back and remembering the procedure and its steps.

Front of Card

Evaluating Absolute Value (Simplifying)

1. *Work inside absolute value bars first (like working inside parentheses).*
2. *Find the absolute value (<u>never</u> negative).*
3. *A negative sign <u>in front</u> of the absolute value bar is NOT affected, so keep it!*

Back of Card

Examples:

Simplify $|10 - 6|$ → $|4| = 4$ — *Work inside: 10 − 6 = 4. Absolute value of 4 is 4*

Simplify $-|-12|$ → -12 — *Absolute value of −12 is 12. Keep negative sign that was in front.*

Now Try This ▶▶▶

What procedure are you learning right now? Write below the steps that you will put on your study card.

Procedure: ______________________

Step 1 ______________________

Step 2 ______________________

Step 3 ______________________

Step 4 ______________________

Step 5 ______________________

1.4 Adding Real Numbers

OBJECTIVES

1. Add two numbers with the same sign.
2. Add numbers with different signs.
3. Add mentally.
4. Use the rules for order of operations with real numbers.
5. Translate words and phrases that indicate addition.

OBJECTIVE 1 Add two numbers with the same sign. We can use the number line to explain addition of real numbers. Later, we will give the rules for addition. Recall that the answer to an addition problem is called the **sum.**

EXAMPLE 1 Adding Two Positive Numbers on a Number Line

Use a number line to find the sum $2 + 3$.

Add the positive numbers 2 and 3 by starting at 0 and drawing an arrow two units to the *right,* as shown in Figure 8. This arrow represents the number 2 in the sum $2 + 3$. Next, from the right end of this arrow draw another arrow three units to the right. The number below the end of this second arrow is 5, so $2 + 3 = 5$.

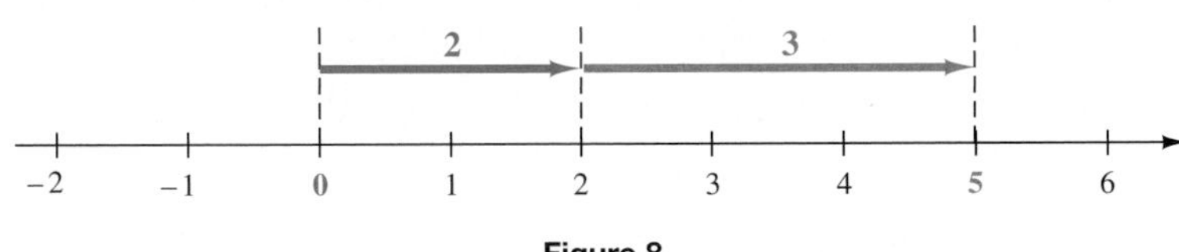

Figure 8

EXAMPLE 2 Adding Two Negative Numbers on a Number Line

Use a number line to find the sum $-2 + (-4)$. (Parentheses are placed around the -4 to avoid the confusing use of + and − next to each other.)

To add the negative numbers -2 and -4 on the number line, we start at 0 and draw an arrow two units to the *left,* as shown in Figure 9. From the left end of this first arrow, we draw a second arrow four units to the left. We draw the arrow to the left to represent the addition of the *negative* number, -4. The number below the end of this second arrow is -6, so $-2 + (-4) = -6$.

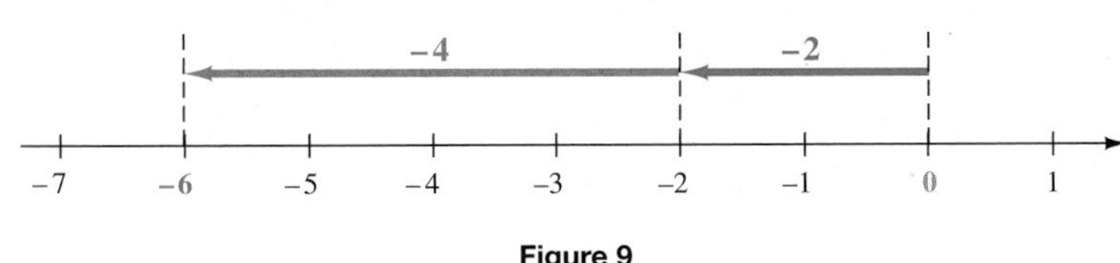

Figure 9

Work Problem 1 at the Side. ▶

1 Use a number line to find each sum.

(a) $1 + 4$

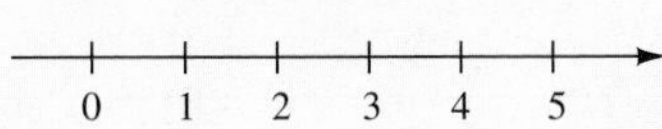

(b) $-2 + (-5)$

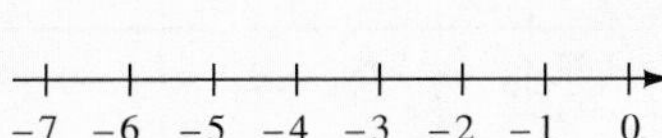

In Example 2, we found that the sum of the two negative numbers -2 and -4 is a negative number whose distance from 0 is the sum of the distance of -2 from 0 and the distance of -4 from 0. ***That is, the sum of two negative numbers is the opposite of the sum of their absolute values.***

$$-2 + (-4) = -(|-2| + |-4|) = -(2 + 4) = -6$$

Adding Real Numbers with the Same Sign

To add two numbers with the same sign, add the absolute values of the numbers. Give the result the same sign as the numbers being added.

Example: $-4 + (-3) = -7$

ANSWERS

1. (a) $1 + 4 = 5$

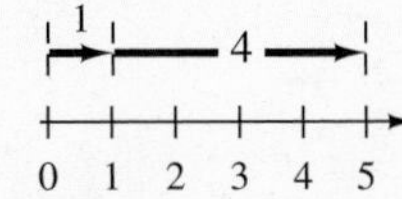

(b) $-2 + (-5) = -7$

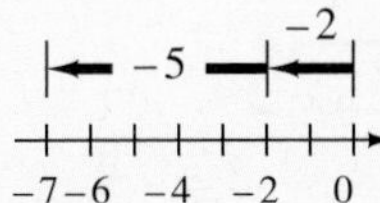

2 Find each sum.

(a) $-7 + (-3)$

(b) $-12 + (-18)$

(c) $-15 + (-4)$

3 Use a number line to find each sum.

(a) $6 + (-3)$

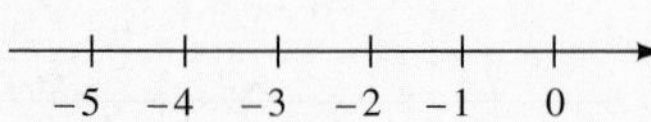

(b) $-5 + 1$

-5 -4 -3 -2 -1 0

Answers

2. **(a)** -10 **(b)** -30 **(c)** -19

3. **(a)** $6 + (-3) = 3$

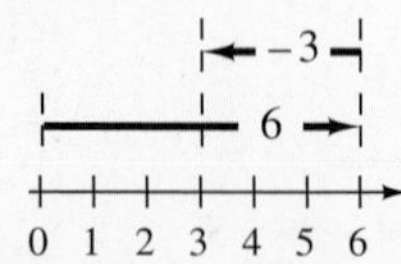

(b) $-5 + 1 = -4$

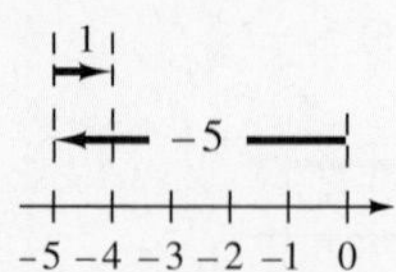

EXAMPLE 3 **Adding Two Negative Numbers**

Find each sum.

(a) $-2 + (-9) = -11$ The sum of two negative numbers is negative.

(b) $-8 + (-12) = -20$

(c) $-15 + (-3) = -18$

Work Problem **2** *at the Side.*

OBJECTIVE 2 Add numbers with different signs. We use the number line again to illustrate the sum of a positive number and a negative number.

EXAMPLE 4 **Adding Numbers with Different Signs**

Use the number line to find the sum $-2 + 5$.

We find the sum $-2 + 5$ on the number line by starting at 0 and drawing an arrow two units to the left. From the left end of this arrow, we draw a second arrow five units to the right, as shown in Figure 10. The number below the end of this second arrow is 3, so $-2 + 5 = 3$.

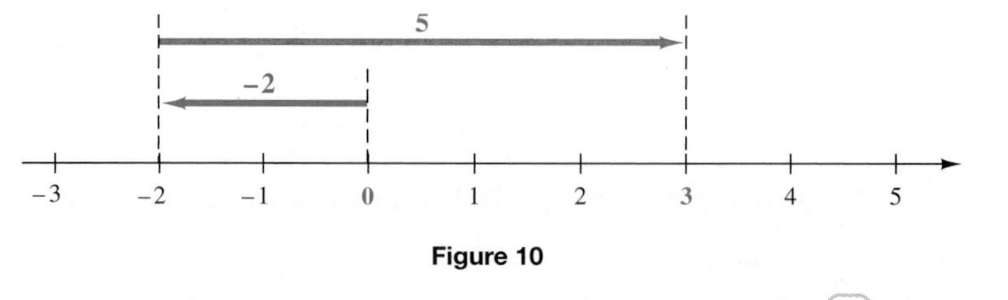

Figure 10

Work Problem **3** *at the Side.*

Addition of numbers with different signs also can be defined using absolute value.

> **Adding Real Numbers with Different Signs**
> To add numbers with different signs, find the absolute values of the numbers, and subtract the lesser absolute value from the greater. Give the answer the same sign as the number with the greater absolute value.
> *Example:* $-12 + 6 = -6$

For example, to add -12 and 5, we find their absolute values:

$$|-12| = 12 \quad \text{and} \quad |5| = 5.$$

Then we find the difference between these absolute values: $12 - 5 = 7$. Since $|-12| > |5|$, the sum will be negative, so

$$-12 + 5 = -7.$$

> **Calculator Tip** The (−) or (+/−) key is used to input a negative number in some scientific calculators. Try using your calculator to add negative numbers.

OBJECTIVE 3 Add mentally. While a number line is useful in showing the rules for addition, it is important to be able to find sums mentally.

EXAMPLE 5 Adding a Positive Number and a Negative Number

Check each answer, trying to work the addition mentally. If you have trouble, use a number line.

(a) $7 + (-4) = 3$

(b) $-8 + 12 = 4$

(c) $-\frac{1}{2} + \frac{1}{8} = -\frac{4}{8} + \frac{1}{8} = -\frac{3}{8}$

Remember to find a common denominator.

(d) $\frac{5}{6} + \left(-1\frac{1}{3}\right) = \frac{5}{6} + \left(-\frac{4}{3}\right) = \frac{5}{6} + \left(-\frac{8}{6}\right) = -\frac{3}{6} = -\frac{1}{2}$

(e) $-4.6 + 8.1 = 3.5$

(f) $-16 + 16 = 0$

Work Problem 4 at the Side. ▶

The rules for adding signed numbers are summarized below.

Adding Signed Numbers

Same sign Add the absolute values of the numbers. Give the sum the same sign as the numbers being added.

Different signs Find the absolute values of the numbers, and subtract the lesser absolute value from the greater. Give the answer the sign of the number having the greater absolute value.

OBJECTIVE 4 Use the rules for order of operations with real numbers. Sometimes a problem involves square brackets, [ ]. As we mentioned earlier, brackets are treated just like parentheses. We do the calculations inside the brackets until a single number is obtained. Remember to use the rules for order of operations given in **Section 1.1** for adding more than two numbers.

EXAMPLE 6 Adding with Brackets

Find each sum.

Start here.

(a) $-3 + [4 + (-8)]$
$= -3 + (-4)$
$= -7$

(b) $8 + [(-2 + 6) + (-3)]$
$= 8 + [4 + (-3)]$
$= 8 + 1$
$= 9$

Work Problem 5 at the Side. ▶

OBJECTIVE 5 Translate words and phrases that indicate addition. We now interpret words and phrases that involve addition. Problem solving often requires translating such words and phrases into symbols. We began this process in **Section 1.1.**

4 Check each answer, trying to work the addition mentally. If you have trouble, use a number line.

(a) $-8 + 2 = -6$

(b) $-15 + 4 = -11$

(c) $17 + (-10) = 7$

(d) $\frac{3}{4} + \left(-1\frac{3}{8}\right) = -\frac{5}{8}$

(e) $-9.5 + 3.8 = -5.7$

(f) $42 + (-42) = 0$

5 Find each sum.

(a) $2 + [7 + (-3)]$

(b) $6 + [(-2 + 5) + 7]$

(c) $-9 + [-4 + (-8 + 6)]$

ANSWERS

4. All are correct.
5. (a) 6 (b) 16 (c) −15

The word *sum* indicates addition. The table lists other key words and phrases that also indicate addition.

Word or Phrase	Example	Numerical Expression and Simplification
Sum of	The *sum of* -3 and 4	$-3 + 4 = 1$
Added to	5 *added to* -8	$-8 + 5 = -3$
More than	12 *more than* -5	$(-5) + 12 = 7$
Increased by	-6 *increased by* 13	$-6 + 13 = 7$
Plus	3 *plus* 14	$3 + 14 = 17$

EXAMPLE 7 Translating Words and Phrases (Addition)

Write a numerical expression for each phrase, and simplify the expression.

(a) The **sum of** -8 and 4 and 6

$$-8 + 4 + 6$$

To simplify, add in order from left to right, to obtain

$$-4 + 6, \quad \text{or} \quad 2.$$

(b) 3 **more than** -5, **increased by** 12

$$-5 + 3 + 12 \quad \text{simplifies to} \quad -2 + 12, \quad \text{or} \quad 10.$$

Work Problem 6 at the Side.

Gains (or increases) and losses (or decreases) sometimes appear in applied problems. When they do, the gains may be interpreted as positive numbers and the losses as negative numbers.

EXAMPLE 8 Interpreting Gains and Losses

The Carolina Panthers football team gained 3 yd on first down, lost 12 yd on second down, and then gained 13 yd on third down. How many yards did the team gain or lose altogether on these plays?

The gains are represented by positive numbers and the loss by a negative number.

$$3 + (-12) + 13$$
$$= [3 + (-12)] + 13 \quad \text{Add from left to right.}$$
$$= (-9) + 13$$
$$= 4$$

The team gained 4 yd altogether on these plays.

Work Problem 7 at the Side.

6 Write a numerical expression for each phrase, and simplify the expression.

(a) 4 more than -12

(b) The sum of 6 and -7

(c) -12 added to -31

(d) 7 increased by the sum of 8 and -3

7 Solve the problem.

A football team lost 8 yd on first down, lost 5 yd on second down, and then gained 7 yd on third down. How many yards did the team gain or lose altogether on these plays?

Answers

6. **(a)** $-12 + 4$; -8 **(b)** $6 + (-7)$; -1 **(c)** $-31 + (-12)$; -43 **(d)** $7 + [8 + (-3)]$; 12

7. The team lost 6 yd.

1.4 Exercises

FOR EXTRA HELP

MyMathLab

Math XL PRACTICE

WATCH

DOWNLOAD

READ

REVIEW

By the order of operations, what is the first step you would use to simplify each expression?

1. $4[3(-2 + 5) - 1]$ **Add −2 and 5.**

2. $[-4 + 7(-6 + 2)]$ **Add −6 and 2.**

3. $9 + ([-1 + (-3)] + 5)$ **Add −1 and −3.**

4. $[(-8 + 4) + (-6)] + 5$ **Add −8 and 4.**

Find each sum. See Examples 1–6.

5. $6 + (-4)$ **2**

6. $8 + (-5)$ **3**

7. $12 + (-15)$ **−3**

8. $4 + (-8)$ **−4**

9. $-7 + (-3)$ **−10**

10. $-11 + (-4)$ **−15**

11. $-10 + (-3)$ **−13**

12. $-16 + (-7)$ **−23**

13. $-12.4 + (-3.5)$ **−15.9**

14. $-21.3 + (-2.5)$ **−23.8**

15. $10 + [-3 + (-2)]$ **5**

16. $13 + [-4 + (-5)]$ **4**

17. $5 + [14 + (-6)]$ **13**

18. $7 + [3 + (-14)]$ **−4**

19. $-3 + [5 + (-2)]$ **0**

20. $-7 + [10 + (-3)]$ **0**

21. $-8 + [3 + (-1) + (-2)]$ **−8**

22. $-7 + [5 + (-8) + 3]$ **−7**

23. $\frac{9}{10} + \left(-\frac{3}{5}\right)$ $\mathbf{\frac{3}{10}}$

24. $\frac{5}{8} + \left(-\frac{17}{12}\right)$ $\mathbf{-\frac{19}{24}}$

25. $-\frac{1}{6} + \frac{2}{3}$ $\mathbf{\frac{1}{2}}$

26. $-\frac{6}{25} + \frac{19}{20}$ $\mathbf{\frac{71}{100}}$

27. $2\frac{1}{2} + \left(-3\frac{1}{4}\right)$ $\mathbf{-\frac{3}{4}}$

28. $-4\frac{3}{8} + 6\frac{1}{2}$ $\mathbf{\frac{17}{8}}$**, or** $\mathbf{2\frac{1}{8}}$

29. $7.8 + (-9.4)$ **−1.6**

30. $14.7 + (-10.1)$ **4.6**

31. $-7.1 + [3.3 + (-4.9)]$ **−8.7**

32. $-9.5 + [-6.8 + (-1.3)]$ **−17.6**

33. $[-8 + (-3)] + [-7 + (-7)]$ **−25**

34. $[-5 + (-4)] + [9 + (-2)]$ **−2**

35. $\left(-\frac{1}{2} + 0.25\right) - \left(-\frac{3}{4} + 0.75\right)$ $\mathbf{-\frac{1}{4}}$**, or −0.25**

36. $\left(-\frac{3}{2} - 0.75\right) - \left(2.25 - \frac{1}{2}\right)$ **−4**

Perform each operation, and then determine whether the statement is true *or* false. *Try to do all work mentally. See Examples 5 and 6.*

37. $-11 + 13 = 13 + (-11)$ **true**

38. $16 + (-9) = -9 + 16$ **true**

39. $-10 + 6 + 7 = -3$ **false**

40. $-12 + 8 + 5 = -1$ **false**

41. $\frac{7}{3} + \left(-\frac{1}{3}\right) + \left(-\frac{6}{3}\right) = 0$ **true**

42. $-\frac{3}{2} + 1 + \frac{1}{2} = 0$ **true**

43. $|-8 + 10| = -8 + (-10)$ **false**

44. $|-4 + 6| = -4 + (-6)$ **false**

45. $2\frac{1}{5} + \left(-\frac{6}{11}\right) = -\frac{6}{11} + 2\frac{1}{5}$ **true**

46. $-1\frac{1}{2} + \frac{5}{8} = \frac{5}{8} + \left(-1\frac{1}{2}\right)$ **true**

47. $-7 + [-5 + (-3)] = [(-7) + (-5)] + 3$ **false**

48. $6 + [-2 + (-5)] = [(-4) + (-2)] + 5$ **true**

Relating Concepts (Exercises 49–52) For Individual or Group Work

Recall the rules for adding signed numbers introduced in this section, and ***work Exercises 49–52 in order.***

49. Suppose that the sum of two numbers is negative, and you know that one of the numbers is positive. What can you conclude about the other number?

It must be negative and have the greater absolute value.

50. If you are solving the equation $x + 5 = -7$ from a set of numbers, why could you immediately eliminate any positive numbers as possible solutions? (Remember how you answered Exercise 49.)

The sum of a positive number and 5 cannot be −7.

51. Suppose that the sum of two numbers is positive, and you know that one of the numbers is negative. What can you conclude about the other number?

It must be positive and have the greater absolute value.

52. If you are solving the equation $x + (-8) = 2$ from a set of numbers, why could you immediately eliminate any negative numbers as possible solutions? (Remember how you answered Exercise 51.)

The sum of a negative number and −8 cannot be 2.

53. In your own words, explain how to add two negative numbers.

Add the absolute values of the numbers. The sum will be negative.

54. In your own words, explain how to add a positive number and a negative number. Give two cases.

Subtract the lesser absolute value from the greater absolute value. The number with greater absolute value determines the sign of the sum. If that number is positive, the sum will be positive. If that number is negative, the sum will be negative.

Write a numerical expression for each phrase, and simplify the expression. See Example 7.

55. The sum of -5 and 12 and 6

$-5 + 12 + 6; 13$

56. The sum of -3 and 5 and -12

$-3 + 5 + (-12); -10$

57. 14 added to the sum of -19 and -4

$[-19 + (-4)] + 14; -9$

58. -2 added to the sum of -18 and 11

$(-18 + 11) + (-2); -9$

59. The sum of -4 and -10, increased by 12

$[-4 + (-10)] + 12; -2$

60. The sum of -7 and -13, increased by 14

$[-7 + (-13)] + 14; -6$

61. $\frac{2}{7}$ more than the sum of $\frac{5}{7}$ and $-\frac{9}{7}$

$\left[\frac{5}{7} + \left(-\frac{9}{7}\right)\right] + \frac{2}{7}; -\frac{2}{7}$

62. 0.85 more than the sum of -1.25 and -4.75

$[-1.25 + (-4.75)] + 0.85; -5.15$

Solve each problem. See Example 8.

63. Nathaniel owed his older sister Jenna \$24 for his share of the bill when they took their mother out to dinner for her birthday. He later borrowed \$38 from his younger sister Ilana to buy two DVDs. What positive or negative number represents Nathaniel's financial situation with his siblings?

−\$62

64. Bonika's checking account balance is \$54.00. She then takes a gamble by writing a check for \$89.00. What is her new balance? (Write the balance as a signed number.)

−\$35.00

65. The surface, or rim, of a canyon is at altitude 0. On a hike down into the canyon, a party of hikers stops for a rest at 130 m below the surface. They then descend another 54 m. What is their new altitude? (Write the altitude as a signed number.) **−184 m**

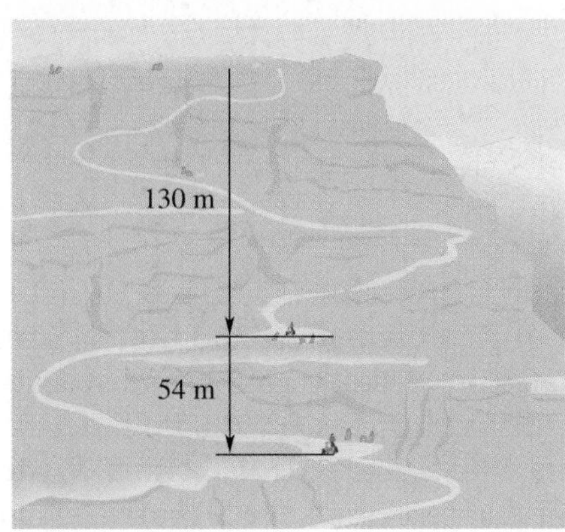

66. A pilot announces to the passengers that the current altitude of their plane is 34,000 ft. Because of some unexpected turbulence, the pilot is forced to descend 2100 ft. What is the new altitude of the plane? (Write the altitude as a signed number.) **31,900 ft**

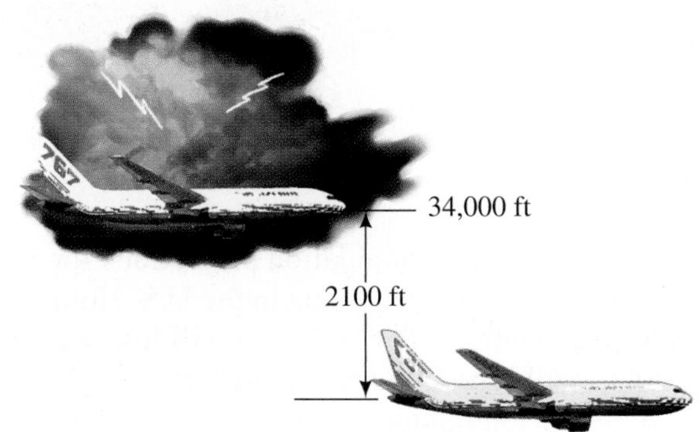

67. On three consecutive passes, Drew Brees of the New Orleans Saints passed for a gain of 6 yd, was sacked for a loss of 12 yd, and passed for a gain of 43 yd. What positive or negative number represents the total net yardage for the plays?

37 yd

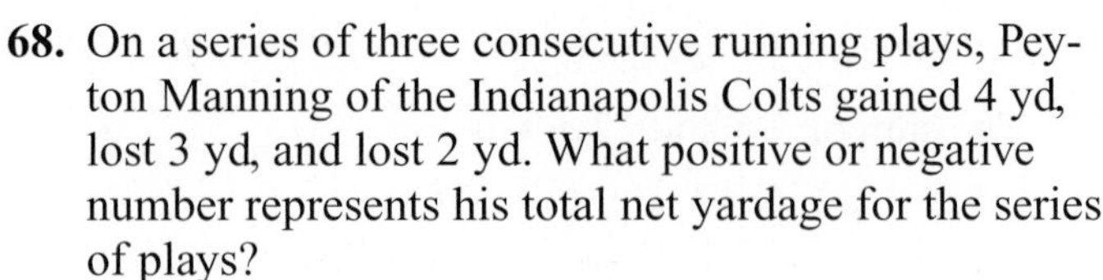

68. On a series of three consecutive running plays, Peyton Manning of the Indianapolis Colts gained 4 yd, lost 3 yd, and lost 2 yd. What positive or negative number represents his total net yardage for the series of plays?

−1 yd

69. The lowest temperature ever recorded in Arkansas was −29°F. The highest temperature ever recorded there was 149°F more than the lowest. What was this highest temperature? (*Source: World Almanac and Book of Facts.*) **120°F**

70. On January 23, 1943, the temperature rose 49°F in two minutes in Spearfish, South Dakota. If the starting temperature was −4°F, what was the temperature two minutes later? **45°F**

71. Dana Weightman owes $153 to a credit card company. She makes a $14 purchase with the card, and then pays $60 on the account. What is her current balance as a signed number? **−$107**

72. A female polar bear weighed 660 lb when she entered her winter den. She lost 45 lb during each of the first two months of hibernation, and another 205 lb before leaving the den with her two cubs in March. How much did she weigh when she left the den? **365 lb**

73. Based on census population projections for 2020, New York will lose 5 seats in the U.S. House of Representatives, Pennsylvania will lose 4 seats, and Ohio will lose 3. Write a signed number that represents the total number of seats these three states are projected to lose. (*Source:* Population Reference Bureau.)

−12

74. Michigan is projected to lose 3 seats in the U.S. House of Representatives and Illinois 2 in 2020. The states projected to gain the most seats are California with 9, Texas with 5, Florida with 3, Georgia with 2, and Arizona with 2. Write a signed number that represents the algebraic sum of these changes. (*Source:* Population Reference Bureau.)

+16

1.5 Subtracting Real Numbers

OBJECTIVE 1 Find a difference. In the operation $a - b$, a is called the **minuend** and b is called the **subtrahend.** As we mentioned earlier, the answer to a subtraction problem is called a **difference.**

Differences between signed numbers can be found by using a number line. Addition and subtraction are opposite operations. Thus, because *addition* of a positive number on the number line is shown by drawing an arrow to the *right, subtraction* of a positive number is shown by drawing an arrow to the *left.*

EXAMPLE 1 Subtracting with the Number Line

Use a number line to find the difference $7 - 4$.

To find the difference $7 - 4$ on the number line, begin at 0 and draw an arrow 7 units to the *right*. From the right end of this arrow, draw an arrow 4 units to the *left,* as shown in Figure 11. The number at the end of the second arrow shows that $7 - 4 = 3$.

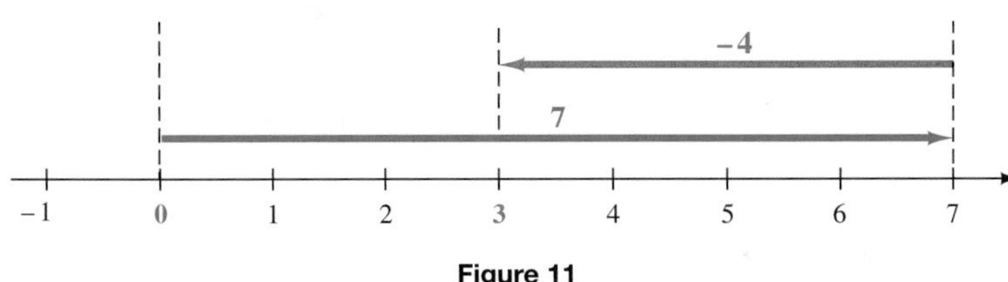

Figure 11

Work Problem 1 at the Side. ▶

OBJECTIVE 2 Use the definition of subtraction. The procedure used in Example 1 to find $7 - 4$ is exactly the same procedure that would be used to find $7 + (-4)$, so

$$7 - 4 = 7 + (-4).$$

This shows that *subtracting* a positive number from a larger positive number is the same as *adding* the opposite of the smaller number to the larger. We use this idea to define subtraction for all real numbers.

> **Subtraction**
>
> For any real numbers a and b,
>
> $$a - b = a + (-b).$$
>
> *Example:* $4 - 9 = 4 + (-9) = -5$

To subtract b from a, add the opposite (or negative) of b to a. In other words, change the subtrahend to its opposite and add.

> **Subtracting Signed Numbers**
>
> ***Step 1*** Change the subtraction symbol to addition and change the sign of the subtrahend.
>
> ***Step 2*** Add, as in the previous section.

OBJECTIVES

1. Find a difference.
2. Use the definition of subtraction.
3. Work subtraction problems that involve brackets.
4. Translate words and phrases that indicate subtraction.

1 Use the number line to find each difference.

(a) $5 - 1$

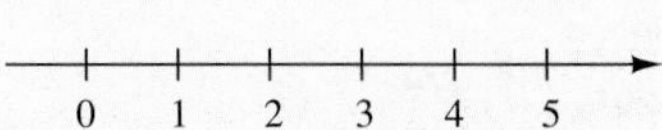

(b) $6 - 2$

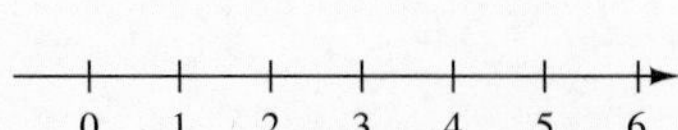

ANSWERS

1. (a) $5 - 1 = 4$

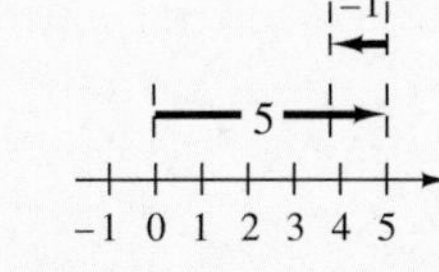

(b) $6 - 2 = 4$

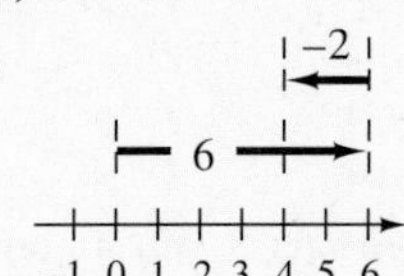

2 Subtract.

(a) $6 - 10$

(b) $-2 - 4$

(c) $3 - (-5)$

(d) $-8 - (-12)$

(e) $\frac{5}{4} - \left(-\frac{3}{7}\right)$

EXAMPLE 2 Using the Definition of Subtraction

Subtract.

No change; Change − to +.; Opposite of 3

(a) $12 - 3 = 12 + (-3) = 9$ **(b)** $5 - 7 = 5 + (-7) = -2$

−7 has the greater absolute value, so the sum is negative.

(c) $-8 - 15 = -8 + (-15) = -23$

No change; Change − to +.; Opposite of −5

(d) $-3 - (-5) = -3 + (5) = 2$

(e) $\frac{3}{8} - \left(-\frac{4}{5}\right) = \frac{15}{40} - \left(-\frac{32}{40}\right) = \frac{15}{40} + \frac{32}{40} = \frac{47}{40}$

◀ *Work Problem* **2** *at the Side.*

Subtraction can be used to reverse the result of an addition problem. For example, if 4 is added to a number and then subtracted from the sum, the original number is the result.

$$12 + 4 = 16 \quad \text{and} \quad 16 - 4 = 12$$

Uses of the Symbol −

The symbol − has now been used for three purposes:

1. ***to represent subtraction,*** as in $9 - 5 = 4$;
2. ***to represent negative numbers,*** such as -10, -2, and -3;
3. ***to represent the opposite (or negative) of a number,*** as in "the opposite (or negative) of 8 is -8."

We may see more than one use in the same problem, such as $-6 - (-9)$, where -9 is subtracted from -6. The meaning of the symbol depends on its position in the algebraic expression.

OBJECTIVE 3 Work subtraction problems that involve brackets. As before, with problems that have both parentheses and brackets, first do any operations inside the parentheses and brackets. Work from the inside out.

EXAMPLE 3 Subtracting with Grouping Symbols

Perform each operation.

Start here.

(a) $-6 - [2 - (8 + 3)]$

$= -6 - [2 - 11]$ Add.

$= -6 - [2 + (-11)]$ Change − to +.

$= -6 - (-9)$ Add.

$= -6 + (9)$ Change − to +.

$= 3$

Continued on Next Page

ANSWERS

2. (a) -4 (b) -6 (c) 8 (d) 4 (e) $\frac{47}{28}$

(b) $5 - \left[\left(-\frac{1}{3} - \frac{1}{2}\right) - (4 - 1)\right]$

$= 5 - \left[\left(-\frac{1}{3} + \left(-\frac{1}{2}\right)\right) - 3\right]$ Work within the parentheses inside the brackets.

$= 5 - \left[\left(-\frac{5}{6}\right) - 3\right]$ Use 6 as the common denominator; $-\frac{1}{3} + (-\frac{1}{2}) = -\frac{2}{6} + (-\frac{3}{6}) = -\frac{5}{6}$

$= 5 - \left[\left(-\frac{5}{6}\right) + (-3)\right]$ Definition of subtraction

$= 5 - \left[\left(-\frac{5}{6}\right) + \left(-\frac{18}{6}\right)\right]$ $-\frac{3}{1} \cdot \frac{6}{6} = -\frac{18}{6}$

Find a common denominator.

$= 5 - \left(-\frac{23}{6}\right)$

$= 5 + \frac{23}{6}$ Definition of subtraction

$= \frac{53}{6}$ $5 = \frac{30}{6}$

Work Problem 3 at the Side. ▶

3 Perform each operation.

(a) $2 - [(-3) - (4 + 6)]$

(b) $[(5 - 7) + 3] - 8$

(c) $6 - [(-1 - 4) - 2]$

OBJECTIVE 4 Translate words and phrases that indicate subtraction. *Difference* is one word that indicates subtraction of real numbers. Some others are given in the table.

Word or Phrase	Example	Numerical Expression and Simplification
Difference between	The *difference between* −3 and −8	$-3 - (-8) = -3 + 8 = 5$
Subtracted from	12 *subtracted from* 18	$18 - 12 = 6$
Less than	6 *less than* 5	$5 - 6 = 5 + (-6) = -1$
Decreased by	9 *decreased by* −4	$9 - (-4) = 9 + 4 = 13$
Minus	−8 *minus* 5	$-8 - 5 = -8 + (-5) = -13$

CAUTION

When subtracting two numbers, it is important to write them in the correct order, because, in general, $a - b \neq b - a$. For example, $5 - 3 \neq 3 - 5$. ***Think carefully before interpreting an expression involving subtraction!*** Subtracting a larger number from a smaller number *always* produces a negative number.

EXAMPLE 4 Translating Words and Phrases (Subtraction)

Write a numerical expression for each phrase, and simplify the expression.

(a) The **difference between** −8 and 5

When "difference between" is used, write the numbers in the order they are given. The expression is

$-8 - 5$, which simplifies to $-8 + (-5)$, or -13.

Continued on Next Page

ANSWERS

3. (a) 15 **(b)** −7 **(c)** 13

4 Write a numerical expression for each phrase, and simplify the expression.

(a) The difference between -5 and -12

(b) -2 subtracted from the sum of 4 and -4

(c) 7 less than -2

(d) 9, decreased by 10 less than 7

5 Solve the problem.

The highest elevation in Argentina is Mt. Aconcagua, which is 6960 m above sea level. The lowest point in Argentina is the Valdes Peninsula, 40 m below sea level. Find the difference between the highest and lowest elevations.

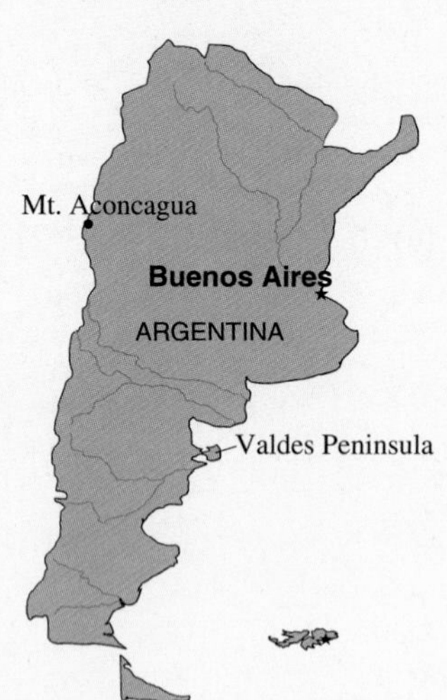

Answers

4. (a) $-5-(-12)$; 7
(b) $[4+(-4)]-(-2)$; 2
(c) $-2-7$; -9
(d) $9-(7-10)$; 12

5. 7000 m

(b) 4 **subtracted from** the sum of 8 and -3

Here the operation of addition is also used, as indicated by the word *sum*. First, add 8 and -3. Next, subtract 4 from this sum. The expression is

$$[8+(-3)]-4, \quad \text{which simplifies to} \quad 5-4, \quad \text{or} \quad 1.$$

(c) 4 **less than** -6

Here 4 must be taken *from* -6, so write -6 first.

Be careful with order.

$$-6-4 \quad \text{simplifies to} \quad -6+(-4), \quad \text{or} \quad -10.$$

Notice that "4 less than -6" differs from "4 *is less than* -6." The statement "4 is less than -6" is symbolized as $4<-6$ (which is a false statement).

(d) 8, **decreased by** 5 **less than** 12

First, write "5 less than 12" as $12-5$. Next, subtract $12-5$ from 8.

$$8-(12-5) \quad \text{simplifies to} \quad 8-7, \quad \text{or} \quad 1.$$

◀ *Work Problem* **4** *at the Side.*

We have seen a few applications of signed numbers in earlier sections. The next example involves subtraction of signed numbers.

EXAMPLE 5 Solving a Problem Involving Subtraction

The record high temperature of 134°F in the United States was recorded at Death Valley, California, in 1913. The record low was -80°F, at Prospect Creek, Alaska, in 1971. See Figure 12. What is the difference between these highest and lowest temperatures? (*Source: World Almanac and Book of Facts.*)

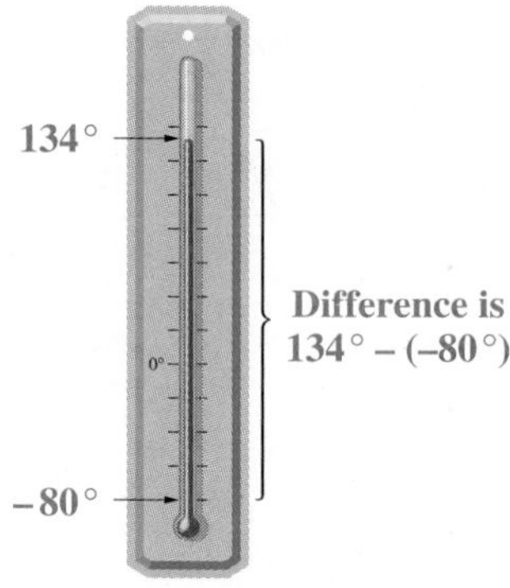

Figure 12

We must subtract the lowest temperature from the highest temperature.

Order of numbers matters in subtraction.

$$\begin{aligned} &134-(-80) \\ &=134+80 \quad \text{Use the definition of subtraction.} \\ &=214 \quad \text{Add.} \end{aligned}$$

The difference between the two temperatures is 214°F.

◀ *Work Problem* **5** *at the Side.*

1.5 Exercises

FOR EXTRA HELP

MyMathLab

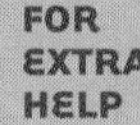
Math XL PRACTICE

WATCH

DOWNLOAD

READ

REVIEW

Fill in each blank with the correct response.

1. By the definition of subtraction, in order to perform the subtraction $-6 - (-8)$, we must add the opposite of $\underline{-8}$ to $\underline{-6}$.

2. By the rules for order of operations, to simplify $8 - [3 - (-4 - 5)]$, the first step is to subtract $\underline{5}$ from $\underline{-4}$.

3. "The difference between 7 and 12" translates as $\underline{7 - 12}$, while "the difference between 12 and 7" translates as $\underline{12 - 7}$.

4. $-9 - (-3) = -9 + \underline{3}$

5. $-8 - 4 = -8 + \underline{-4}$

6. $-19 - 22 = -19 + \underline{-22}$

Find each difference. See Examples 1–3.

7. $-7 - 3$ **−10**

8. $-12 - 5$ **−17**

9. $-10 - 6$ **−16**

10. $-13 - 16$ **−29**

11. $7 - (-4)$ **11**

12. $9 - (-6)$ **15**

13. $6 - (-13)$ **19**

14. $13 - (-3)$ **16**

15. $-7 - (-3)$ **−4**

16. $-8 - (-6)$ **−2**

17. $3 - (4 - 6)$ **5**

18. $6 - (7 - 14)$ **13**

19. $-3 - (6 - 9)$ **0**

20. $-4 - (5 - 12)$ **3**

21. $\frac{1}{2} - \left(-\frac{1}{4}\right)$ $\frac{3}{4}$

22. $\frac{1}{3} - \left(-\frac{4}{3}\right)$ $\frac{5}{3}$

23. $-\frac{3}{4} - \frac{5}{8}$ $-\frac{11}{8}$

24. $-\frac{5}{6} - \frac{1}{2}$ $-\frac{4}{3}$

25. $\frac{5}{8} - \left(-\frac{1}{2} - \frac{3}{4}\right)$ $\frac{15}{8}$

26. $\frac{9}{10} - \left(\frac{1}{8} - \frac{3}{10}\right)$ $\frac{43}{40}$

27. $4.4 - (-9.2)$ **13.6**

28. $6.7 - (-12.6)$ **19.3**

29. $-7.4 - 4.5$ **−11.9**

30. $-5.4 - 9.6$ **−15.0**

31. $-5.2 - (8.4 - 10.8)$ **−2.8**

32. $-9.6 - (3.5 - 12.6)$ **−0.5**

33. $[(-3.1) - 4.5] - (0.8 - 2.1)$ **−6.3**

34. $[(-7.8) - 9.3] - (0.6 - 3.5)$ **−14.2**

35. $-12 - [(9 - 2) - (-6 - 3)]$ **−28**

36. $-4 + [(-6 - 9) - (-7 + 4)]$ **−16**

37. $-8 + [(-3 - 10) - (-4 + 1)]$ **−18**

38. $\left(-\frac{3}{4} - \frac{5}{2}\right) - \left(-\frac{1}{8} - 1\right)$ $-\frac{17}{8}$

39. $\left(-\frac{3}{8} - \frac{2}{3}\right) - \left(-\frac{9}{8} - 3\right)$ $\frac{37}{12}$

40. $[-34.99 + (6.59 - 12.25)] - 8.33$ **−48.98**

41. $[-12.25 - (8.34 + 3.57)] - 17.88$ **−42.04**

42. Explain in your own words how to subtract signed numbers.

To subtract signed numbers, add the opposite of the subtrahend to the minuend.

43. We know that, in general, $a - b \neq b - a$. Find two pairs of values for a and b so that $a - b = b - a$.

For example, let $a = 1$, $b = 1$ or let $a = 2$, $b = 2$. In general, choose $a = b$.

Simplify each expression. Use the rules for order of operations.

44. $-3 - (-4) - 5$ **−4**

45. $8 - (-3) - 9 + 6$ **8**

46. $-5 - 2 + 4 - 8 - (-6)$ **−5**

47. Make up a subtraction problem so that the difference between two negative numbers is a negative number.

For example, $-8 - (-2) = -6$.

48. Make up a subtraction problem so that the difference between two negative numbers is a positive number.

For example, $-2 - (-8) = 6$.

Write a numerical expression for each phrase and simplify. See Example 4.

49. The difference between 4 and -8

$4 - (-8)$; 12

50. The difference between 7 and -14

$7 - (-14)$; 21

51. 8 less than -2

$-2 - 8$; -10

52. 9 less than -13

$-13 - 9$; -22

53. The sum of 9 and -4, decreased by 7

$[9 + (-4)] - 7$; -2

54. The sum of 12 and -7, decreased by 14

$[12 + (-7)] - 14$; -9

55. 12 less than the difference between 8 and -5

$[8 - (-5)] - 12$; 1

56. 19 less than the difference between 9 and -2

$[9 - (-2)] - 19$; -8

Solve each problem. See Example 5.

57. The coldest temperature recorded in Chicago, Illinois, was -35°F in 1996. The record low in South Dakota was set in 1936 and was 23°F lower than -35°F. What was the record low in South Dakota? (*Source: World Almanac and Book of Facts.*)

-58°F

58. No one knows just why humpback whales love to heave their 45-ton bodies out of the water, but leap they do. This is called *breaching*. Chantelle, a researcher based on the island of Maui, noticed that one of her favorite whales, "Pineapple," breached 15 ft above the surface of the ocean while her mate cruised 12 ft below the surface. What is the difference between these two heights? **27 ft**

59. The top of Mount Whitney, visible from Death Valley, has an altitude of 14,494 ft above sea level. The bottom of Death Valley is 282 ft below sea level. Using 0 as sea level, find the difference between these two elevations. (*Source: World Almanac and Book of Facts.*)

14,776 ft

60. A chemist is running an experiment under precise conditions. At first, she runs it at −174.6°F. She then lowers the temperature by 2.3°F. What is the new temperature for the experiment?

−176.9°F

61. Samir owed his brother $10. He later borrowed $70. What positive or negative number represents his present financial status?

−$80

62. Francesca has $15 in her purse, and Emilio has a debt of $12. Find the difference between these amounts.

$27

63. For the year 2007, one health club showed a profit of $86,000, while another showed a loss of $19,000. Find the difference between these amounts.

$105,000

64. At 2:00 A.M., a plant worker found that a dial reading was 7.904. At 3:00 A.M., she found the reading to be −3.291. Find the difference between these two readings.

11.195

65. J. D. Patin enjoys playing Triominoes every Wednesday night. Last Wednesday, on four successive turns, his scores were −19, 28, −5, and 13. What was his final score for the four turns?

17

66. Gay Aguillard also enjoys playing Triominoes. On five successive turns, her scores were −13, 15, −12, 24, and 14. What was her total score for the five turns?

28

67. In August, Kari Heen began with a checking account balance of $904.89. Her checks and deposits for August are given below:

Checks	Deposits
$35.84	$85.00
$26.14	$120.76
$3.12	

Assuming no other transactions, what was her account balance at the end of August?

$1045.55

68. In September, Derek Bowen began with a checking account balance of $904.89. His checks and deposits for September are given below:

Checks	Deposits
$41.29	$80.59
$13.66	$276.13
$84.40	

Assuming no other transactions, what was his account balance at the end of September?

$1122.26

69. A certain Greek mathematician was born in 426 B.C. His father was born 43 years earlier. In what year was his father born?

469 B.C.

70. A certain Roman philosopher was born in 325 B.C. Her mother was born 35 years earlier. In what year was her mother born?

360 B.C.

71. Kim Falgout owes \$870.00 on her MasterCard account. She returns two items costing \$35.90 and \$150.00 and receives credits for these on the account. Next, she makes a purchase of \$82.50, and then two more purchases of \$10.00 each. She makes a payment of \$500.00. She then incurs a finance charge of \$37.23. How much does she still owe?

\$323.83

72. Charles Vosburg owes \$679.00 on his Visa account. He returns three items costing \$36.89, \$29.40, and \$113.55 and receives credits for these on the account. Next, he makes purchases of \$135.78 and \$412.88, and two purchases of \$20.00 each. He makes a payment of \$400. He then incurs a finance charge of \$24.57. How much does he still owe?

\$712.39

73. José Martinez enjoys scuba diving. He dives to 34 ft below the surface of a lake. His partner, Sean O'Malley, dives to 40 ft below the surface, but then ascends 20 ft. What is the vertical distance between José and Sean?

14 ft

74. Rhonda Alessi also enjoys diving. She dives to 12 ft below the surface of False River. Her sister, Sandy, dives to 20 ft below the surface, but then ascends 10 ft. What is the vertical distance between Rhonda and Sandy?

2 ft

75. The height of Mt. Foraker is 17,400 ft, while the depth of the Java Trench is 23,376 ft. What is the vertical distance between the top of Mt. Foraker and the bottom of the Java Trench? (*Source: World Almanac and Book of Facts.*)

40,776 ft

76. The height of Mt. Wilson is 14,246 ft, while the depth of the Cayman Trench is 24,721 ft. What is the vertical distance between the top of Mt. Wilson and the bottom of the Cayman Trench? (*Source: World Almanac and Book of Facts.*)

38,967 ft

77. In 1984, Americans saved 10.8% of their after-tax incomes. In 2005, they saved −0.5%, the first negative personal savings rate since 1933. Find the difference between those two amounts.

11.3%

78. Refer to Exercise 77. How is it possible that Americans had a negative personal savings rate in 2005?

Americans spent more money than they earned, which means that they had to dip into savings or borrow money.

79. In 2000, the federal budget had a surplus of $236 billion. In 2004, the federal budget had a deficit of $413 billion. Find the difference between these amounts. (*Source:* Treasury Department.)

$649 billion

80. In 1998, undergraduate college students had an average credit card balance of $1879. The average balance increased $869 by 2000 and then dropped $579 by 2004. What was the average credit card balance of undergraduate college students in 2004? (*Source:* Nellie Mae.)

$2169

Median sales prices for existing single-family homes in the United States for the years 2003 through 2007 are shown in the table. Complete the table, determining the change from one year to the next by subtraction.

	Year	Median Sales Price	Change from Previous Year
	2003	$180,200	—
81.	2004	$195,200	**$15,000**
82.	2005	$219,000	**$23,800**
83.	2006	$221,900	**$2900**
84.	2007*	$217,800	**−$4100**

*Projected.
Source: National Association of Realtors.

In Exercises 85–88, suppose that x represents a positive number and y represents a negative number. Determine whether the given expression must represent a positive number or a negative number.

85. $x - y$

positive

86. $y - x$

negative

87. $x + |y|$

positive

88. $y - |x|$

negative

1.6 Multiplying and Dividing Real Numbers

The result of multiplication is called the **product.** We already know how to multiply positive numbers and that the product of two positive numbers is positive. We also know that the product of 0 and any positive number is 0, so we extend that property to all real numbers.

Multiplication Property of 0

For any real number a,

$$a \cdot 0 = 0 \cdot a = 0.$$

OBJECTIVE 1 Find the product of a positive number and a negative number. To define the product of numbers with different signs so that the result is consistent with multiplication of positive numbers, look at the following pattern.

$$\begin{aligned} 3 \cdot 5 &= 15 \\ 3 \cdot 4 &= 12 \\ 3 \cdot 3 &= 9 \\ 3 \cdot 2 &= 6 \\ 3 \cdot 1 &= 3 \\ 3 \cdot 0 &= 0 \\ 3 \cdot (-1) &= ? \end{aligned}$$

The products decrease by 3.

What should $3(-1)$ equal? Since multiplication can also be considered repeated addition, the product $3(-1)$ represents the sum

$$-1 + (-1) + (-1) = -3,$$

so the product should be -3, which fits the pattern. Also,

$$3(-2) = -2 + (-2) + (-2) = -6.$$

Work Problem 1 at the Side. ▶

These results suggest the following rule.

Multiplying Numbers with Different Signs

The product of a positive number and a negative number is negative.

Examples: $6(-3) = -18$ and $-6(3) = -18$

EXAMPLE 1 Multiplying a Positive Number and a Negative Number

Find each product using the multiplication rule.

(a) $8(-5) = -(8 \cdot 5) = -40$

(b) $-7(2) = -(7 \cdot 2) = -14$

(c) $-9\left(\frac{1}{3}\right) = -3$

(d) $-6.2(4.1) = -25.42$

Work Problem 2 at the Side. ▶

OBJECTIVES

1. Find the product of a positive number and a negative number.
2. Find the product of two negative numbers.
3. Use the reciprocal of a number to apply the definition of division.
4. Use the rules for order of operations when multiplying and dividing signed numbers.
5. Evaluate expressions involving variables.
6. Translate words and phrases involving multiplication and division.
7. Translate simple sentences into equations.

1 Find each product by finding the sum of three numbers.

(a) $3(-3)$

(b) $3(-4)$

(c) $3(-5)$

2 Find each product.

(a) $2(-6)$

(b) $7(-8)$

(c) $-9(2)$

(d) $-16\left(\frac{5}{32}\right)$

(e) $4.56(-10)$

ANSWERS

1. (a) -9 **(b)** -12 **(c)** -15

2. (a) -12 **(b)** -56 **(c)** -18 **(d)** $-\frac{5}{2}$ **(e)** -45.6

3 Find each product.

(a) $-5(-6)$

(b) $-7(-3)$

(c) $-8(-5)$

(d) $-11(-2)$

(e) $-17(3)(-7)$

(f) $-41(2)(-13)$

OBJECTIVE 2 Find the product of two negative numbers. The product of two positive numbers is positive, and the product of a positive number and a negative number is negative. What about the product of two negative numbers? Look at another pattern.

$$\begin{aligned} -5(4) &= -20 \\ -5(3) &= -15 \\ -5(2) &= -10 \\ -5(1) &= -5 \\ -5(0) &= 0 \\ -5(-1) &= ? \end{aligned}$$

The products increase by 5.

The numbers on the left of the equals signs (in color) decrease by 1 for each step down the list. The products on the right increase by 5 for each step down the list. To maintain this pattern, $-5(-1)$ should be 5 more than $-5(0)$, or 5 more than 0, so

$$-5(-1) = 5.$$

The pattern continues with

$$\begin{aligned} -5(-2) &= 10 \\ -5(-3) &= 15 \\ -5(-4) &= 20 \\ -5(-5) &= 25, \end{aligned}$$

and so on. This pattern suggests the next rule.

Multiplying Two Negative Numbers

The product of two negative numbers is positive.

Example: $-5(-4) = 20$

EXAMPLE 2 Multiplying Two Negative Numbers

Find each product using the multiplication rule.

(a) $-9(-2) = 18$

(b) $-6(-12) = 72$

(c) $-2(4)(-1) = -8(-1) = 8$

(d) $3(-5)(-2) = -15(-2) = 30$

◀ Work Problem 3 at the Side.

Multiplying Signed Numbers

The product of two numbers having the *same* sign is *positive.*

The product of two numbers having *different* signs is *negative.*

OBJECTIVE 3 Use the reciprocal of a number to apply the definition of division. Recall that the result of division is called the **quotient.** In **Section 1.5** we saw that the difference between two numbers is found by adding the opposite of the subtrahend to the minuend. Similarly, the *quotient* of two numbers involves multiplying by the *reciprocal* of the second number, which is the *divisor.*

ANSWERS

3. (a) 30 (b) 21 (c) 40 (d) 22 (e) 357 (f) 1066

Reciprocals

Pairs of numbers whose product is 1 are called **reciprocals** of each other.

Since $$8 \cdot \frac{1}{8} = \frac{8}{8} = 1 \quad \text{and} \quad \frac{5}{4} \cdot \frac{4}{5} = \frac{20}{20} = 1,$$

the reciprocal of 8 is $\frac{1}{8}$, and the reciprocal of $\frac{5}{4}$ is $\frac{4}{5}$. The following table shows several numbers and their reciprocals.

Number	Reciprocal
4	$\frac{1}{4}$
-5	$\frac{1}{-5}$, or $-\frac{1}{5}$
0.3, or $\frac{3}{10}$	$\frac{10}{3}$
$-\frac{5}{8}$	$-\frac{8}{5}$
0	None

By definition, the product of a number and its reciprocal is 1. But the multiplication property of 0 says that the product of 0 and any number is 0. Thus, **0 has no reciprocal.**

Work Problem 4 *at the Side.* ▶

By definition, the quotient of a and b is the product of a and the reciprocal of b.

Division

The quotient $\frac{a}{b}$ of real numbers a and b, with $b \neq 0$, is

$$\frac{a}{b} = a \cdot \frac{1}{b}.$$

Example: $\frac{8}{-4} = 8\left(-\frac{1}{4}\right) = -2$

This definition indicates that b, the divisor, cannot be 0. Since 0 has no reciprocal,

$\frac{a}{0}$ is not a number and *division by 0 is undefined.* If a division problem requires division by 0, write "undefined."

To illustrate, $\frac{6}{2} = 3$ since $2 \cdot 3 = 6$, but there is no number to represent $\frac{6}{0}$, since there is no number that when multiplied by 0 gives 6 as a product.

Note

Although *division by zero is undefined, we may divide zero by any nonzero real number,* to obtain the quotient 0.

$$\text{If} \quad b \neq 0, \quad \text{then} \quad \frac{0}{b} = 0.$$

4 Complete the table.

Number	Reciprocal
(a) 6	
(b) -2	
(c) $\frac{2}{3}$	
(d) $-\frac{1}{4}$	
(e) 0.75	
(f) 0	

Answers

4. (a) $\frac{1}{6}$ (b) $\frac{1}{-2}$, or $-\frac{1}{2}$ (c) $\frac{3}{2}$ (d) -4 (e) $\frac{4}{3}$ (f) none

5 Find each quotient.

(a) $\frac{42}{7}$

(b) $\frac{-36}{(-2)(-3)}$

(c) $\frac{-12.56}{-0.4}$

(d) $\frac{10}{7} \div \left(-\frac{24}{5}\right)$

(e) $\frac{-3}{0}$

(f) $\frac{0}{-53}$

6 Find each quotient.

(a) $\frac{-8}{-2}$

(b) $\frac{-16.4}{2.05}$

(c) $\frac{1}{4} \div \left(-\frac{2}{3}\right)$

(d) $\frac{12}{-4}$

Answers

5. (a) 6 (b) -6 (c) 31.4 (d) $-\frac{25}{84}$ (e) undefined (f) 0

6. (a) 4 (b) -8 (c) $-\frac{3}{8}$ (d) -3

EXAMPLE 3 Using the Definition of Division

Find each quotient.

(a) $\frac{12}{3} = 12 \cdot \frac{1}{3} = 4$

(b) $\frac{5(-2)}{2} = -10 \cdot \frac{1}{2} = -5$

(c) $\frac{-1.47}{-7} = -1.47 \cdot \left(-\frac{1}{7}\right) = 0.21$

(d) $-\frac{2}{3} \div \left(-\frac{5}{4}\right) = -\frac{2}{3} \cdot \left(-\frac{4}{5}\right) = \frac{8}{15}$

(e) $\frac{-10}{0}$ Undefined

(f) $\frac{0}{13} = 0$ $\frac{0}{b} = 0 \quad (b \neq 0)$

◀ Work Problem 5 at the Side.

When dividing fractions, multiplying by the reciprocal of the divisor works well. However, using the definition of division directly with integers is awkward. It is easier to divide in the usual way, and then determine the sign of the answer.

Dividing Signed Numbers

The quotient of two numbers having the *same* sign is *positive.*

The quotient of two numbers having *different* signs is *negative.*

Examples: $\frac{-15}{-5} = 3$, $\frac{15}{-5} = -3$, and $\frac{-15}{5} = -3$

EXAMPLE 4 Dividing Signed Numbers

Find each quotient.

(a) $\frac{8}{-2} = -4$

(b) $\frac{-10}{2} = -5$

(c) $\frac{-4.5}{-0.09} = 50$

(d) $-\frac{1}{8} \div \left(-\frac{3}{4}\right) = -\frac{1}{8} \cdot \left(-\frac{4}{3}\right) = \frac{1}{6}$ (Remember to write in lowest terms.)

◀ Work Problem 6 at the Side.

From the definitions of multiplication and division of real numbers,

$$\frac{-40}{8} = -40 \cdot \frac{1}{8} = -5 \quad \text{and} \quad \frac{40}{-8} = 40\left(\frac{1}{-8}\right) = -5, \text{ so}$$

$$\frac{-40}{8} = \frac{40}{-8}.$$

Based on this example, the quotient of a positive number and a negative number can be written in any of the following three forms.

Equivalent Forms

For any positive real numbers a and b,

$$\frac{-a}{b} = \frac{a}{-b} = -\frac{a}{b}.$$

Similarly, the quotient of two negative numbers can be expressed as the quotient of two positive numbers.

Equivalent Forms

For any positive real numbers a and b,

$$\frac{-a}{-b} = \frac{a}{b}.$$

OBJECTIVE 4 Use the rules for order of operations when multiplying and dividing signed numbers.

EXAMPLE 5 Using the Rules for Order of Operations

Simplify.

(a) $-9(2) - (-3)(2)$

$= -18 - (-6)$ Multiply.

$= -18 + 6$ Definition of subtraction

$= -12$ Add.

(b) $-6(-2) - 3(-4)$

$= 12 - (-12)$

$= 12 + 12$

$= 24$

(c) $\dfrac{5(-2) - 3(4)}{2(1 - 6)}$

$= \dfrac{-10 - 12}{2(-5)}$ Simplify the numerator and denominator separately.

$= \dfrac{-22}{-10}$

$= \dfrac{11}{5}$ Remember to write in lowest terms.

Work Problem 7 at the Side. ▶

The rules for operations with signed numbers are summarized here.

Operations with Signed Numbers

Addition

Same sign Add the absolute values of the numbers. The sum has the same sign as the numbers.

$$-4 + (-6) = -10$$

Different signs Find the absolute values of the numbers, and subtract the lesser absolute value from the greater. Give the sum the sign of the number having the greater absolute value.

$$4 + (-6) = -(6 - 4) = -2$$

(continued)

7 Perform the indicated operations.

(a) $-3(4) - 2(6)$

(b) $-8[-1 - (-4)(-5)]$

(c) $\dfrac{6(-4) - 2(5)}{3(2 - 7)}$

(d) $\dfrac{-6(-8) + 3(9)}{-2[4 - (-3)]}$

ANSWERS

7. (a) -24 **(b)** 168 **(c)** $\frac{34}{15}$ **(d)** $-\frac{75}{14}$

Operations with Signed Numbers (continued)

Subtraction

Add the opposite of the subtrahend to the minuend.

$$8 - (-3) = 8 + 3 = 11$$

Multiplication and Division

Same sign The product or quotient of two numbers with the same sign is positive.

$$-5(-6) = 30 \quad \text{and} \quad \frac{-36}{-12} = 3$$

Different signs The product or quotient of two numbers with different signs is negative.

$$-5(6) = -30 \quad \text{and} \quad \frac{18}{-6} = -3$$

Division by 0 is undefined.

OBJECTIVE 5 Evaluate expressions involving variables. To *evaluate* an expression means to find its *value*.

EXAMPLE 6 Evaluating Expressions for Numerical Values

Evaluate each expression, given that $x = -1$, $y = -2$, and $m = -3$.

(a) $(3x + 4y)(-2m)$ Use parentheses around substituted negative values to avoid errors.

$= [3(-1) + 4(-2)][-2(-3)]$ Substitute the given values for the variables.

$= [-3 + (-8)]\,[6]$ Multiply.

$= [-11]6$ Add inside the brackets.

$= -66$ Multiply.

(b) $2x^2 - 3y^2$

$= 2(-1)^2 - 3(-2)^2$ Substitute.

$= 2(1) - 3(4)$ Apply the exponents.

$= 2 - 12$ Multiply.

$= -10$ Subtract.

(c) $\dfrac{4y^2 + x}{m}$

$= \dfrac{4(-2)^2 + (-1)}{-3}$ Substitute.

$= \dfrac{4(4) + (-1)}{-3}$ Apply the exponent.

$= \dfrac{16 + (-1)}{-3}$ Multiply.

$= \dfrac{15}{-3}$, or -5 Add, and then divide.

◀ Work Problem 8 at the Side.

8 Evaluate each expression.

(a) $2x - 7(y + 1)$, if $x = -4$ and $y = 3$

(b) $2x^2 - 4y^2$, if $x = -2$ and $y = -3$

(c) $\dfrac{4x - 2y}{-3x}$, if $x = 2$ and $y = -1$

ANSWERS

8. (a) -36 (b) -28 (c) $-\frac{5}{3}$

OBJECTIVE 6 Translate words and phrases involving multiplication and division. Just as there are words and phrases that indicate addition or subtraction, certain words and phrases indicate multiplication or division. The chart gives some phrases indicating multiplication.

Word or Phrase	Example	Numerical Expression and Simplification
Product of	The *product of* -5 and -2	$-5(-2) = 10$
Times	13 *times* -4	$13(-4) = -52$
Twice (meaning "2 times")	*Twice* 6	$2(6) = 12$
Of (used with fractions)	$\frac{1}{2}$ *of* 10	$\frac{1}{2}(10) = 5$
Percent of	12% *of* -16	$0.12(-16) = -1.92$

EXAMPLE 7 Translating Words and Phrases (Multiplication)

Write a numerical expression for each phrase, and simplify the expression.

(a) The **product of** 12 and the sum of 3 and -6
Here, 12 is multiplied by "the sum of 3 and -6." The expression is

$$12[3 + (-6)], \quad \text{which simplifies to} \quad 12[-3], \quad \text{or} \quad -36.$$

(b) **Twice** the difference between 8 and -4

$$2[8 - (-4)], \quad \text{simplifies to} \quad 2[12], \quad \text{or} \quad 24.$$

(c) Two-thirds **of** the sum of -5 and -3

$$\frac{2}{3}[-5 + (-3)] \quad \text{simplifies to} \quad \frac{2}{3}[-8], \quad \text{or} \quad -\frac{16}{3}.$$

(d) 15% **of** the difference between 14 and -2

$$0.15[14 - (-2)] \quad \text{simplifies to} \quad 0.15[16], \quad \text{or} \quad 2.4.$$

Remember that 15% = 0.15.

Work Problem 9 at the Side. ▶

9 Write a numerical expression for each phrase, and simplify the expression.

(a) The product of 6 and the sum of -5 and -4

(b) Three times the difference between 4 and -6

(c) Three-fifths of the sum of 2 and -7

(d) 20% of the sum of 9 and -4

The word *quotient* refers to the answer in a division problem. In algebra, a quotient is usually represented with a fraction bar; the symbol ÷ is seldom used. The chart gives some phrases associated with division.

Word or Phrase	Example	Numerical Expression and Simplification
Quotient of	The *quotient of* -24 and 3	$\frac{-24}{3} = -8$
Divided by	-16 *divided by* -4	$\frac{-16}{-4} = 4$
Ratio of	The *ratio of* 2 to 3	$\frac{2}{3}$

When translating a phrase involving division, we write the first number named as the numerator and the second as the denominator.

Answers

9. **(a)** $6[(-5) + (-4)]$; -54
(b) $3[4 - (-6)]$; 30
(c) $\frac{3}{5}[2 + (-7)]$; -3
(d) $0.20[9 + (-4)]$; 1

EXAMPLE 8 Translating Words and Phrases (Division)

Write a numerical expression for each phrase, and simplify the expression.

(a) The **quotient of** 14 and the sum of -9 and 2

"Quotient" indicates division. The number 14 is the numerator and "the sum of -9 and 2" is the denominator. The expression is

$$\frac{14}{-9+2}, \quad \text{which simplifies to} \quad \frac{14}{-7}, \quad \text{or} \quad -2.$$

(b) The product of 5 and -6, **divided by** the difference between -7 and 8

The numerator of the fraction representing the division is found by multiplying 5 and -6. The denominator is found by subtracting -7 and 8. The expression is

$$\frac{5(-6)}{-7-8}, \quad \text{which simplifies to} \quad \frac{-30}{-15}, \quad \text{or} \quad 2.$$

◀ Work Problem 10 at the Side.

10 Write a numerical expression for each phrase, and simplify the expression.

(a) The quotient of 20 and the sum of 8 and -3

(b) The product of -9 and 2, divided by the difference between 5 and -1

11 Write each sentence in symbols, using x to represent the number.

(a) Twice a number is -6.

(b) The difference between -8 and a number is -11.

(c) The sum of 5 and a number is 8.

(d) The quotient of a number and -2 is 6.

OBJECTIVE 7 Translate simple sentences into equations. In this section and the previous two sections, important words and phrases involving the four operations of arithmetic have been introduced. We can use these words and phrases to translate sentences into equations.

EXAMPLE 9 Translating Sentences into Equations

Write each sentence in symbols, using x to represent the number.

(a) Three **times** a number **is** -18.

The word *times* indicates multiplication, and the word *is* translates as the equals sign (=).

$$3x = -18 \qquad 3 \cdot x = 3x$$

(b) The **sum** of a number and 9 **is** 12.

$$x + 9 = 12$$

(c) The **difference between** a number and 5 **is** 0.

$$x - 5 = 0$$

(d) The **quotient of** 24 and a number **is** -2.

$$\frac{24}{x} = -2$$

◀ Work Problem 11 at the Side.

CAUTION

It is important to recognize the distinction between the types of problems found in Examples 7 and 8 and those in Example 9. In Examples 7 and 8, the phrases translate as *expressions,* while in Example 9, the sentences translate as *equations.* ***Remember that an expression is a phrase, while an equation is a sentence.***

$$\frac{5(-6)}{-7-8} \qquad\qquad 3x = -18$$

↑ Expression ↑ Equation

ANSWERS

10. (a) $\frac{20}{8+(-3)}$; 4

(b) $\frac{-9(2)}{5-(-1)}$; -3

11. (a) $2x = -6$

(b) $-8 - x = -11$

(c) $5 + x = 8$

(d) $\frac{x}{-2} = 6$

1.6 Exercises

FOR EXTRA HELP

Fill in each blank with one of the following: greater than 0, less than 0, equal to 0.

1. The product or the quotient of two numbers with the same sign is **greater than 0**.

2. The product or the quotient of two numbers with different signs is **less than 0**.

3. If three negative numbers are multiplied together, the product is **less than 0**.

4. If two negative numbers are multiplied and then their product is divided by a negative number, the result is **less than 0**.

5. If a negative number is squared and the result is added to a positive number, the final answer is **greater than 0**.

6. The reciprocal of a negative number is **less than 0**.

Find each product. See Examples 1 and 2.

7. $-7(4)$ **−28**

8. $-8(5)$ **−40**

9. $-5(-6)$ **30**

10. $-4(-20)$ **80**

11. $-8(0)$ **0**

12. $0(-12)$ **0**

13. $-\frac{3}{8}\left(-\frac{20}{9}\right)$ $\frac{5}{6}$

14. $-\frac{5}{4}\left(-\frac{6}{25}\right)$ $\frac{3}{10}$

15. $-6.8(0.35)$ **−2.38**

16. $-4.6(0.24)$ **−1.104**

17. $-6\left(-\frac{1}{4}\right)$ $\frac{3}{2}$

18. $-8\left(-\frac{1}{2}\right)$ **4**

Find each quotient. See Examples 3 and 4.

19. $\frac{-15}{5}$ **−3**

20. $\frac{-18}{6}$ **−3**

21. $\frac{20}{-10}$ **−2**

22. $\frac{28}{-4}$ **−7**

23. $\frac{-160}{-10}$ **16**

24. $\frac{-260}{-20}$ **13**

25. $\frac{0}{-3}$ **0**

26. $\frac{0}{-5}$ **0**

27. $\frac{-10.252}{0}$ **undefined**

28. $\frac{-29.584}{0}$ **undefined**

29. $\left(-\frac{3}{4}\right) \div \left(-\frac{1}{2}\right)$ $\frac{3}{2}$

30. $\left(-\frac{3}{16}\right) \div \left(-\frac{5}{8}\right)$ $\frac{3}{10}$

31. Which expression is undefined?

A. $\frac{5-5}{5+5}$ **B.** $\frac{5+5}{5+5}$ **C.** $\frac{5-5}{5-5}$ **D.** $\frac{5-5}{5}$ **C**

32. What is the reciprocal of 0.4? **2.5, or $\frac{5}{2}$**

Perform each indicated operation. See Example 5.

33. $\frac{-5(-6)}{9-(-1)}$ **3**

34. $\frac{-12(-5)}{7-(-5)}$ **5**

35. $\frac{-21(3)}{-3-6}$ **7**

36. $\frac{-40(3)}{-2-3}$ **24**

37. $\frac{-10(2)+6(2)}{-3-(-1)}$ **4**

38. $\frac{8(-1)+6(-2)}{-6-(-1)}$ **4**

39. $\frac{-27(-2)-(-12)(-2)}{-2(3)-2(2)}$ **−3**

40. $\frac{-13(-4)-(-8)(-2)}{(-10)(2)-4(-2)}$ **−3**

41. $\frac{3^2-4^2}{7(-8+9)}$ **−1**

42. $\frac{5^2-7^2}{2(3+3)}$ **−2**

43. $\frac{4(2^3-5)-5(-3^3+21)}{3[6-(-2)]}$ **$\frac{7}{4}$**

44. $\frac{-3(-2^4+10)+4(2^5-12)}{-2[8-(-7)]}$ **$-\frac{49}{15}$**

Evaluate each expression if $x = 6$, $y = -4$, and $a = 3$. See Example 6.

45. $6x - 5y + 4a$ **68**

46. $5x - 2y + 3a$ **47**

47. $(5x - 2y)(-2a)$ **−228**

48. $(2x + y)(3a)$ **72**

49. $\left(\frac{5}{6}x + \frac{3}{2}y\right)\left(-\frac{1}{3}a\right)$ **1**

50. $\left(\frac{1}{3}x - \frac{4}{5}y\right)\left(-\frac{1}{5}a\right)$ **$-\frac{78}{25}$**

51. $(6 - x)(5 + y)(3 + a)$ **0**

52. $(-5 + x)(-3 + y)(3 - a)$ **0**

53. $5x - 4a^2$ **−6**

54. $-2y^2 + 3a$ **−23**

55. $\frac{xy + 9a}{x + y - 2}$ **undefined**

56. $\frac{2y^2 - x}{a - 3}$ **undefined**

Write a numerical expression for each phrase and simplify. See Examples 7 and 8.

57. The product of 4 and -7, added to -12

$-12 + 4(-7); -40$

58. The product of -9 and 2, added to 9

$9 + (-9)(2); -9$

59. Twice the product of -8 and 2, subtracted from -1

$-1 - 2(-8)(2); 31$

60. Twice the product of -1 and 6, subtracted from -4

$-4 - 2(-1)(6); 8$

61. The product of -3 and the difference between 3 and -7

$-3[3 - (-7)]; -30$

62. The product of 12 and the difference between 9 and -8

$12[9 - (-8)]; 204$

63. Three-tenths of the sum of -2 and -28

$\frac{3}{10}[-2 + (-28)]; -9$

64. Four-fifths of the sum of -8 and -2

$\frac{4}{5}[-8 + (-2)]; -8$

65. The quotient of -20 and the sum of -8 and -2

$\frac{-20}{-8 + (-2)}; 2$

66. The quotient of -12 and the sum of -5 and -1

$\frac{-12}{-5 + (-1)}; 2$

67. The sum of -18 and -6, divided by the product of 2 and -4

$\frac{-18 + (-6)}{2(-4)}; 3$

68. The sum of 15 and -3, divided by the product of 4 and -3

$\frac{15 + (-3)}{4(-3)}; -1$

69. The product of $-\frac{2}{3}$ and $-\frac{1}{5}$, divided by $\frac{1}{7}$

$\frac{-\frac{2}{3}\left(-\frac{1}{5}\right)}{\frac{1}{7}}$; $\frac{14}{15}$

70. The product of $-\frac{1}{2}$ and $\frac{3}{4}$, divided by $-\frac{2}{3}$

$\frac{-\frac{1}{2}\left(\frac{3}{4}\right)}{-\frac{2}{3}}$; $\frac{9}{16}$

Write each sentence with symbols, using x to represent the number. See Example 9.

71. Nine times a number is -36.

$9x = -36$

72. Seven times a number is -42.

$7x = -42$

73. The quotient of a number and 4 is -1.

$\frac{x}{4} = -1$

74. The quotient of a number and 3 is -3.

$\frac{x}{3} = -3$

75. $\frac{9}{11}$ less than a number is 5.

$x - \frac{9}{11} = 5$

76. $\frac{1}{2}$ less than a number is 2.

$x - \frac{1}{2} = 2$

77. When 6 is divided by a number, the result is -3.

$\frac{6}{x} = -3$

78. When 15 is divided by a number, the result is -5.

$\frac{15}{x} = -5$

Relating Concepts (Exercises 79–84) For Individual or Group Work

*To find the **average** of a group of numbers, we add the numbers and then divide the sum by the number of terms added. **Work Exercises 79–82 in order,** to find the average of* 23, 18, 13, −4, *and* −8. *Then find the averages in Exercises 83 and 84.*

79. Find the sum of the given group of numbers.

42

80. How many numbers are in the group?

5

81. Divide your answer for Exercise 79 by your answer for Exercise 80. Give the quotient as a mixed number.

$8\frac{2}{5}$

82. What is the average of the given group of numbers?

$8\frac{2}{5}$

83. What is the average of all integers between -10 and 14, including both -10 and 14?

2

84. What is the average of the integers between -15 and -10, including -15 and -10?

$-12\frac{1}{2}$

Summary Exercises on Operations with Real Numbers

Operations with Signed Numbers

Addition

Same sign Add the absolute values of the numbers. The sum has the same sign as the numbers.

Different signs Find the absolute values of the numbers, and subtract the lesser absolute value from the greater. Give the sum the sign of the number having the greater absolute value.

Subtraction

Add the opposite of the subtrahend to the minuend.

Multiplication and Division

Same sign The product or quotient of two numbers with the same sign is positive.

Different signs The product or quotient of two numbers with different signs is negative.

Division by 0 is undefined.

Perform each indicated operation.

1. $14 - 3 \cdot 10$

-16

2. $-3(8) - 4(-7)$

4

3. $(3 - 8)(-2) - 10$

0

4. $-6(7 - 3)$

-24

5. $7 - (-3)(2 - 10)$

-17

6. $-4[(-2)(6) - 7]$

76

7. $(-4)(7) - (-5)(2)$

-18

8. $-5[-4 - (-2)(-7)]$

90

9. $40 - (-2)[8 - 9]$

38

10. $\dfrac{5(-4)}{-7 - (-2)}$

4

11. $\dfrac{-3 - (-9 + 1)}{-7 - (-6)}$

-5

12. $\dfrac{5(-8 + 3)}{13(-2) + (-7)(-3)}$

5

13. $\dfrac{6^2 - 8}{-2(2) + 4(-1)}$

$-\frac{7}{2}$, or $-3\frac{1}{2}$

14. $\dfrac{16(-8 + 5)}{15(-3) + (-7 - 4)(-3)}$

4

15. $\dfrac{9(-6) - 3(8)}{4(-7) + (-2)(-11)}$

13

16. $\frac{2^2 + 4^2}{5^2 - 3^2}$
$\frac{5}{4}$, or $1\frac{1}{4}$

17. $\frac{(2 + 4)^2}{(5 - 3)^2}$
9

18. $\frac{4^3 - 3^3}{-5(-4 + 2)}$
$\frac{37}{10}$, or $3\frac{7}{10}$

19. $\frac{-9(-6) + (-2)(27)}{3(8 - 9)}$
0

20. $|-4(9)| - |-11|$
25

21. $\frac{6(-10 + 3)}{15(-2) - 3(-9)}$
14

22. $\frac{(-9)^2 - 9^2}{3^2 - 5^2}$
0

23. $\frac{(-10)^2 + 10^2}{-10(5)}$
-4

24. $-\frac{3}{4} \div \left(-\frac{5}{8}\right)$
$\frac{6}{5}$, or $1\frac{1}{5}$

25. $\frac{1}{2} \div \left(-\frac{1}{2}\right)$
-1

26. $\frac{8^2 - 12}{(-5)^2 + 2(6)}$
$\frac{52}{37}$, or $1\frac{15}{37}$

27. $\left[\frac{5}{8} - \left(-\frac{1}{16}\right)\right] + \frac{3}{8}$
$\frac{17}{16}$, or $1\frac{1}{16}$

28. $\left(\frac{1}{2} - \frac{1}{3}\right) - \frac{5}{6}$
$-\frac{2}{3}$

29. $-0.9(-3.7)$
3.33

30. $-5.1(-0.2)$
1.02

31. $-3^2 - 2^2$
-13

32. $|-2(3) + 4| - |-2|$
0

33. $40 - (-2)[-5 - 3]$
24

Evaluate each expression if $x = -2$, $y = 3$, and $a = 4$.

34. $-x + y - 3a$
-7

35. $(x + 6)^3 - y^3$
37

36. $(x - y) - (a - 2y)$
-3

37. $\left(\frac{1}{2}x + \frac{2}{3}y\right)\left(-\frac{1}{4}a\right)$
-1

38. $\frac{2x + 3y}{a - xy}$
$\frac{1}{2}$

39. $\frac{x^2 - y^2}{x^2 + y^2}$
$-\frac{5}{13}$

40. $-x^2 + 3y$
5

41. $\frac{-x + 2y}{2x + a}$
undefined

42. $\frac{2x + a}{-x + 2y}$
0

1.7 Properties of Real Numbers

If you are asked to find the sum

$$3 + 89 + 97,$$

you might mentally add $3 + 97$ to get 100, and then add $100 + 89$ to get 189. While the rules for order of operations say to add (or multiply) from left to right, the fact is we may change the order of the terms (or factors) and group them in any way we choose without affecting the sum (or product).

This is an example of a shortcut we use in everyday mathematics that is justified by the properties of real numbers introduced in this section. In the following statements, a, b, and c represent real numbers.

OBJECTIVES

1. **Use the commutative properties.**
2. **Use the associative properties.**
3. **Use the identity properties.**
4. **Use the inverse properties.**
5. **Use the distributive property.**

OBJECTIVE 1 Use the commutative properties. The word *commute* means to go back and forth. Many people commute to work or to school. If you travel from home to work and follow the same route from work to home, you travel the same distance each time. The **commutative properties** say that if two numbers are added or multiplied in any order, they give the same result.

$$a + b = b + a \quad \text{Addition}$$
$$ab = ba \quad \text{Multiplication}$$

EXAMPLE 1 Using the Commutative Properties

Use a commutative property to complete each statement.

(a) $-8 + 5 = 5 +$ ______

By the commutative property for addition, the missing number is -8 because $-8 + 5 = 5 + (-8)$.

(b) $-2(7) =$ ______ (-2)

By the commutative property for multiplication, the missing number is 7, since $-2(7) = 7(-2)$.

Work Problem 1 at the Side. ▶

1 Complete each statement. Use a commutative property.

(a) $x + 9 = 9 +$ ______

(b) $-12(4) =$ ______ (-12)

(c) $5x = x \cdot$ ______

OBJECTIVE 2 Use the associative properties. When we *associate* one object with another, we tend to think of those objects as being grouped together. The **associative properties** say that when we add or multiply three numbers, we can group them in any manner and get the same answer.

$$(a + b) + c = a + (b + c) \quad \text{Addition}$$
$$(ab)c = a(bc) \quad \text{Multiplication}$$

EXAMPLE 2 Using the Associative Properties

Use an associative property to complete each statement.

(a) $8 + (-1 + 4) = (8 +$ ______ $) + 4$

The missing number is -1.

(b) $[2 \cdot (-7)] \cdot 6 = 2 \cdot$ ______

The missing expression on the right should be $[(-7) \cdot 6]$.

Work Problem 2 at the Side. ▶

2 Complete each statement. Use an associative property.

(a) $(9 + 10) + (-3)$
$= 9 + [$ ______ $+ (-3)]$

(b) $-5 + (2 + 8)$
$= ($ ______ $) + 8$

(c) $10 \cdot [-8 \cdot (-3)] =$ ______

ANSWERS

1. (a) x (b) 4 (c) 5
2. (a) 10 (b) $-5 + 2$ (c) $[10 \cdot (-8)] \cdot (-3)$

By the associative property of addition, the sum of three numbers will be the same no matter how the numbers are "associated" in groups. For this reason, parentheses can be left out in many addition problems. For example, both

$$(-1 + 2) + 3 \quad \text{and} \quad -1 + (2 + 3)$$

can be written as

$$-1 + 2 + 3.$$

In the same way, parentheses also can be left out of many multiplication problems.

EXAMPLE 3 Distinguishing between Associative and Commutative Properties

(a) Is $(2 + 4) + 5 = 2 + (4 + 5)$ an example of an associative or a commutative property?

The order of the three numbers is the same on both sides of the equals sign. The only change is in the *grouping,* or association, of the numbers. Therefore, this is an example of an associative property.

(b) Is $6(3 \cdot 10) = 6(10 \cdot 3)$ an example of an associative or a commutative property?

The same numbers, 3 and 10, are grouped on each side. On the left, however, 3 appears first in $(3 \cdot 10)$. On the right, 10 appears first. Since the only change involves the *order* of the numbers, this statement is an example of a commutative property.

(c) Is $(8 + 1) + 7 = 8 + (7 + 1)$ an example of an associative or a commutative property, or both?

In the statement, both the order and the grouping are changed. On the left, the order of the three numbers is 8, 1, and 7. On the right, it is 8, 7, and 1. On the left, 8 and 1 are grouped, and on the right, 7 and 1 are grouped. Therefore, both associative and commutative properties are used.

◀ Work Problem 3 at the Side.

EXAMPLE 4 Using Commutative and Associative Properties

Find each sum or product.

(a) $23 + 41 + 2 + 9 + 25$

$= (41 + 9) + (23 + 2) + 25$ Commutative and associative properties

$= 50 + 25 + 25$

$= 100$

(b) $25(69)(4)$

$= 25(4)(69)$

$= 100(69)$

$= 6900$

◀ Work Problem 4 at the Side.

3 Decide whether each statement is an example of a commutative property, an associative property, or both.

(a) $2(4 \cdot 6) = (2 \cdot 4)6$

(b) $(2 \cdot 4)6 = (4 \cdot 2)6$

(c) $(2 + 4) + 6 = 4 + (2 + 6)$

4 Find each sum or product.

(a) $5 + 18 + 29 + 31 + 12$

(b) $5(37)(20)$

ANSWERS

3. (a) associative **(b)** commutative **(c)** both

4. (a) 95 **(b)** 3700

OBJECTIVE 3 Use the identity properties. If a child wears a costume on Halloween, the child's appearance is changed, but his or her *identity* is unchanged. In the same way, the identity, or value, of a real number is left unchanged when identity properties are applied. The **identity properties** say that the sum of 0 and any number equals that number, and the product of 1 and any number equals that number.

$$\boldsymbol{a + 0 = a} \quad \text{and} \quad \boldsymbol{0 + a = a} \qquad \text{Addition}$$

$$\boldsymbol{a \cdot 1 = a} \quad \text{and} \quad \boldsymbol{1 \cdot a = a} \qquad \text{Multiplication}$$

The number 0 leaves the *identity,* or value, of any real number unchanged by addition. For this reason, 0 is called the **identity element for addition,** or the **additive identity.** Since multiplication by 1 leaves any real number unchanged, 1 is the **identity element for multiplication,** or the **multiplicative identity.**

EXAMPLE 5 Using Identity Properties

These statements are examples of identity properties.

(a) $-3 + 0 = -3$ Addition

(b) $1 \cdot 25 = 25$ Multiplication

Work Problem 5 at the Side. ▶

We use the identity property for multiplication to write fractions in lowest terms and to find common denominators.

EXAMPLE 6 Using the Identity Element for Multiplication to Simplify Expressions

Simplify each expression.

(a) $\dfrac{49}{35}$

$= \dfrac{7 \cdot 7}{5 \cdot 7}$ Factor.

$= \dfrac{7}{5} \cdot \dfrac{7}{7}$ Write as a product.

$= \dfrac{7}{5} \cdot 1$ Property of 1

$= \dfrac{7}{5}$ Identity property

(b) $\dfrac{3}{4} + \dfrac{5}{24} = \dfrac{3}{4} \cdot 1 + \dfrac{5}{24}$ Identity property

$= \dfrac{3}{4} \cdot \dfrac{6}{6} + \dfrac{5}{24}$ Use $1 = \frac{6}{6}$ to get a common denominator.

$= \dfrac{18}{24} + \dfrac{5}{24}$ Multiply.

$= \dfrac{23}{24}$ Add.

Work Problem 6 at the Side. ▶

5 Use an identity property to complete each statement.

(a) $9 + 0 =$ ______

(b) ______ $+ (-7) = -7$

(c) ______ $\cdot 1 = 5$

6 Use an identity property to simplify each expression.

(a) $\dfrac{85}{105}$

(b) $\dfrac{9}{10} - \dfrac{53}{50}$

ANSWERS

5. (a) 9 **(b)** 0 **(c)** 5

6. (a) $\dfrac{17}{21}$ **(b)** $-\dfrac{4}{25}$

OBJECTIVE 4 **Use the inverse properties.** Each day before you go to work or school, you probably put on your shoes before you leave. When you get home or before you go to sleep at night, you probably take them off, and this leads to the same situation that existed before you put them on. These operations from everyday life are examples of inverse operations.

The **inverse properties** of addition and multiplication lead to the additive and multiplicative identities, respectively. The *opposite* of a, $-a$, is the **additive inverse** of a and the *reciprocal* of a, $\frac{1}{a}$, is the **multiplicative inverse** of the nonzero number a. The sum of the numbers a and $-a$ is 0, and the product of the nonzero numbers a and $\frac{1}{a}$ is 1.

$$\mathbf{a + (-a) = 0 \quad \text{and} \quad -a + a = 0} \qquad \text{Addition}$$

$$\mathbf{a \cdot \frac{1}{a} = 1 \quad \text{and} \quad \frac{1}{a} \cdot a = 1 \quad (a \neq 0)} \qquad \text{Multiplication}$$

EXAMPLE 7 **Using the Inverse Properties**

The following statements are examples of the inverse properties.

(a) $(-5)\left(-\frac{1}{5}\right) = 1$ Multiplication **(b)** $4 + (-4) = 0$ Addition

(c) $\frac{1}{2} + (-0.5) = 0$ Addition **(d)** $\frac{2}{3} \cdot \frac{3}{2} = 1$ Multiplication

(e) $-\frac{1}{2} + \frac{1}{2} = 0$ Addition **(f)** $4(0.25) = 1$ Multiplication

◀ *Work Problem 7 at the Side.*

OBJECTIVE 5 **Use the distributive property.** The everyday meaning of the word *distribute* is "to give out from one to several." An important property of real number operations involves this idea.

Look at the value of the following expressions:

$$2(5 + 8), \quad \text{which equals} \quad 2(13), \quad \text{or} \quad 26;$$
$$2(5) + 2(8), \quad \text{which equals} \quad 10 + 16, \quad \text{or} \quad 26.$$

Since both expressions equal 26,

$$2(5 + 8) = 2(5) + 2(8).$$

This result is an example of the *distributive property of multiplication with respect to addition,* the only property involving *both* addition and multiplication. With this property, a product can be changed to a sum or difference. This idea is illustrated by the divided rectangle in Figure 13.

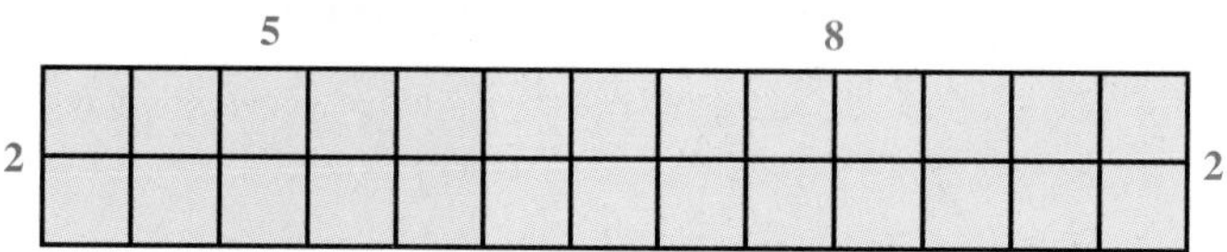

The area of the left part is 2(5) = 10.
The area of the right part is 2(8) = 16.
The total area is 2(5 + 8) = 26 or the total area is
2(5) + 2(8) = 10 + 16 = 26.
Thus, 2(5 + 8) = 2(5) + 2(8).

Figure 13

7 Complete each statement so that it is an example of either an identity property or an inverse property. Tell which property is used.

(a) $-6 + \_\_\_\_ = 0$

(b) $\frac{4}{3} \cdot \_\_\_\_ = 1$

(c) $-\frac{1}{9} \cdot \_\_\_\_ = 1$

(d) $275 + \_\_\_\_ = 275$

(e) $-0.75 + \frac{3}{4} = \_\_\_\_$

(f) $0.2(5) = \_\_\_\_$

ANSWERS

7. **(a)** 6; inverse **(b)** $\frac{3}{4}$; inverse **(c)** -9; inverse **(d)** 0; identity **(e)** 0; inverse **(f)** 1; inverse

The **distributive property** says that multiplying a number a by a sum of numbers $b + c$ gives the same result as multiplying a by b and a by c and then adding the two products.

$$\boldsymbol{a(b + c) = ab + ac} \quad \text{and} \quad \boldsymbol{(b + c)a = ba + ca}$$

As the arrows show, the a outside the parentheses is "distributed" over the b and c inside. The distributive property is also valid for subtraction.

$$\boldsymbol{a(b - c) = ab - ac} \quad \text{and} \quad \boldsymbol{(b - c)a = ba - ca}$$

The distributive property also can be extended to the sum (or difference) of more than two numbers.

$$\boldsymbol{a(b + c + d) = ab + ac + ad}$$

The distributive property can also be written "in reverse."

$$\boldsymbol{ab + ac = a(b + c)}$$

EXAMPLE 8 **Using the Distributive Property**

Use the distributive property to rewrite each expression.

(a) $5(9 + 6)$

$= 5 \cdot 9 + 5 \cdot 6$ Distributive property

$= 45 + 30$ Multiply.

$= 75$ Add.

Multiply first.

(b) $4(x + 5 + y)$

$= 4x + 4 \cdot 5 + 4y$ Distributive property

$= 4x + 20 + 4y$ Multiply.

(c) $-2(x + 3)$

$= -2x + (-2)(3)$ Distributive property

$= -2x + (-6)$ Multiply.

$= -2x - 6$

(d) $3(k - 9)$

$= 3[k + (-9)]$ Definition of subtraction

$= 3k + 3(-9)$ Distributive property

$= 3k - 27$ Multiply.

(e) $8(3r + 11t + 5z)$

$= 8(3r) + 8(11t) + 8(5z)$ Distributive property

$= (8 \cdot 3)r + (8 \cdot 11)t + (8 \cdot 5)z$ Associative property

$= 24r + 88t + 40z$ Multiply.

(f) $6 \cdot 8 + 6 \cdot 2$

$= 6(8 + 2)$ Distributive property in reverse

$= 6(10)$ Add.

$= 60$ Multiply.

(g) $4x - 4m$

$= 4(x - m)$ Distributive property in reverse

Work Problem **8** at the Side. ▶

8 Use the distributive property to rewrite each expression.

(a) $2(p + 5)$

(b) $-4(y + 7)$

(c) $5(m - 4)$

(d) $7(2y + 7k - 9m)$

(e) $9 \cdot k + 9 \cdot 5$

(f) $3a - 3b$

ANSWERS

8. **(a)** $2p + 10$ **(b)** $-4y - 28$ **(c)** $5m - 20$ **(d)** $14y + 49k - 63m$ **(e)** $9(k + 5)$ **(f)** $3(a - b)$

The symbol $-a$ may be interpreted as $-1 \cdot a$. Similarly, when a negative sign precedes an expression within parentheses, it may also be interpreted as a factor of -1. Thus, we can use the distributive property to remove (or clear) the parentheses from expressions such as $-(2y + 3)$.

$$\begin{aligned} &-(2y+3) \\ &= -1 \cdot (2y+3) && -a = -1 \cdot a \\ &= -1 \cdot (2y) + (-1) \cdot (3) && \text{Distributive property} \\ &= -2y - 3 && \text{Multiply.} \end{aligned}$$

EXAMPLE 9 Using the Distributive Property to Remove (Clear) Parentheses

Write each expression without parentheses.

(a) $-(7r - 8)$

$$\begin{aligned} &= -1(7r - 8) && -a = -1 \cdot a \\ &= -1(7r) + (-1)(-8) && \text{Distributive property} \\ &= -7r + 8 && \text{Multiply.} \end{aligned}$$

(b) $-(-9w + 2)$

$= -1(-9w + 2)$

$= 9w - 2$

We can interpret the $-$ sign in front of the parentheses to mean -1, yielding the opposite of each term inside the parentheses after it is distributed.

◀ Work Problem 9 at the Side.

Here is a summary of the basic properties of real numbers.

Properties of Addition and Multiplication

For any real numbers a, b, and c, the following properties hold.

Commutative properties $\quad a + b = b + a \qquad ab = ba$

Associative properties $\quad (a + b) + c = a + (b + c)$

$(ab)c = a(bc)$

Identity properties There is a real number 0 such that

$a + 0 = a$ and $0 + a = a.$

There is a real number 1 such that

$a \cdot 1 = a$ and $1 \cdot a = a.$

Inverse properties For each real number a, there is a single real number $-a$ such that

$a + (-a) = 0$ and $(-a) + a = 0.$

For each nonzero real number a, there is a single real number $\frac{1}{a}$ such that

$$a \cdot \frac{1}{a} = 1 \quad \text{and} \quad \frac{1}{a} \cdot a = 1.$$

Distributive property $\quad a(b + c) = ab + ac$

$(b + c)a = ba + ca$

9 Write each expression without parentheses.

(a) $-(3k - 5)$

(b) $-(2 - r)$

(c) $-(-5y + 8)$

(d) $-(-z + 4)$

ANSWERS

9. **(a)** $-3k + 5$ **(b)** $-2 + r$ **(c)** $5y - 8$ **(d)** $z - 4$

1.7 Exercises

FOR EXTRA HELP

Math XL PRACTICE

WATCH

DOWNLOAD

READ

REVIEW

Match each item in Column I with the correct choice from Column II. Choices may be used once, more than once, or not at all.

I

1. Identity element for addition B
2. Identity element for multiplication F
3. Additive inverse of a C
4. Multiplicative inverse, or reciprocal, of the nonzero number a I
5. The only number that has no multiplicative inverse B
6. An example of an associative property A
7. An example of a commutative property G
8. An example of the distributive property H

II

A. $(5 \cdot 4) \cdot 3 = 5 \cdot (4 \cdot 3)$
B. 0
C. $-a$
D. -1
E. $5 \cdot 4 \cdot 3 = 60$
F. 1
G. $(5 \cdot 4) \cdot 3 = 3 \cdot (5 \cdot 4)$
H. $5(4 + 3) = 5 \cdot 4 + 5 \cdot 3$
I. $\frac{1}{a}$

Decide whether each statement is an example of a commutative, associative, identity, or inverse property, or of the distributive property. See Examples 1, 2, 3, and 5–8.

9. $\frac{2}{3}(-4) = -4\left(\frac{2}{3}\right)$
commutative property

10. $6\left(-\frac{5}{6}\right) = \left(-\frac{5}{6}\right)6$
commutative property

11. $-6 + (12 + 7) = (-6 + 12) + 7$
associative property

12. $(-8 + 13) + 2 = -8 + (13 + 2)$
associative property

13. $-6 + 6 = 0$
inverse property

14. $12 + (-12) = 0$
inverse property

15. $\left(\frac{2}{3}\right)\left(\frac{3}{2}\right) = 1$
inverse property

16. $\left(\frac{5}{8}\right)\left(\frac{8}{5}\right) = 1$
inverse property

17. $2.34 \cdot 1 = 2.34$
identity property

18. $-8.456 \cdot 1 = -8.456$
identity property

19. $(4 + 17) + 3 = 3 + (4 + 17)$
commutative property

20. $(-8 + 4) + (-12) = -12 + (-8 + 4)$
commutative property

21. $6(x + y) = 6x + 6y$
distributive property

22. $14(t + s) = 14t + 14s$
distributive property

23. $-\frac{5}{9} = -\frac{5}{9} \cdot \frac{3}{3} = -\frac{15}{27}$

identity property

24. $\frac{13}{12} = \frac{13}{12} \cdot \frac{7}{7} = \frac{91}{84}$

identity property

25. $5(2x) + 5(3y) = 5(2x + 3y)$

distributive property

26. $3(5t) - 3(7r) = 3(5t - 7r)$

distributive property

27. What number(s) satisfy each condition? **(a)** a number that is its own additive inverse **(b)** two numbers that are their own multiplicative inverses **(a) 0 (b) 1, −1**

28. The distributive property holds for multiplication with respect to addition. Is there a distributive property for addition with respect to multiplication? That is, does $a + b \cdot c = (a + b)(a + c)$? If not, give an example to show why.

No. For example, $2 + (3 \cdot 4) \neq (2 + 3) \cdot (2 + 4)$.

29. Evaluate $25 - (6 - 2)$ and $(25 - 6) - 2$. Use the results to explain why subtraction is or is not associative.

$25 - (6 - 2) = 25 - 4 = 21$ and $(25 - 6) - 2 = 19 - 2 = 17$. Since these results are different, subtraction is not associative.

30. Suppose that a classmate shows you the following work.

$$\begin{aligned} &-2(5 - 6) \\ &= -2(5) - 2(6) \\ &= -10 - 12 \\ &= -22 \end{aligned}$$

The classmate made a very common error. ***WHAT WENT WRONG?***

In the first step, using the distributive property, the second product should be −2 times −6 (not 6). The correct steps are $-2(5 - 6) = (-2)(5) + (-2)(-6) = -10 + 12 = 2$.

Write a new expression that is equal to the given expression, using the given property. Then simplify the new expression if possible. See Examples 1, 2, 5, 7, and 8.

31. $r + 7$; commutative

$7 + r$

32. $t + 9$; commutative

$9 + t$

33. $s + 0$; identity

s

34. $w + 0$; identity

w

35. $-6(x + 7)$; distributive

$-6x + (-6)7$; $-6x - 42$

36. $-5(y + 2)$; distributive

$-5y + (-5)2$; $-5y - 10$

37. $(w + 5) + (-3)$; associative

$w + [5 + (-3)]$; $w + 2$

38. $(b + 8) + (-10)$; associative

$b + [8 + (-10)]$; $b - 2$

39. Explain how the procedure of changing $\frac{3}{4}$ to $\frac{9}{12}$ requires the use of the multiplicative identity element, 1.

We must multiply $\frac{3}{4}$ by 1 in the form $\frac{3}{3}$: $\frac{3}{4} \cdot \frac{3}{3} = \frac{9}{12}$.

Use the properties of this section to simplify each expression. See Example 4.

40. $26 + 8 - 26 + 12$

20

41. $-\frac{3}{8} + \frac{2}{5} + \frac{8}{5} + \frac{3}{8}$

2

42. $\frac{9}{7}(-0.38)\left(\frac{7}{9}\right)$

−0.38

Use the distributive property to rewrite each expression. Simplify if possible. See Example 8.

43. $4(t + 3)$

$4t + 12$

44. $5(w + 4)$

$5w + 20$

45. $-8(r + 3)$

$-8r - 24$

46. $-11(x + 4)$

$-11x - 44$

47. $-5(y - 4)$

$-5y + 20$

48. $-9(g - 4)$

$-9g + 36$

49. $-\frac{4}{3}(12y + 15z)$

$-16y - 20z$

50. $-\frac{2}{5}(10b + 20a)$

$-4b - 8a$

51. $8 \cdot z + 8 \cdot w$

$8(z + w)$

52. $4 \cdot s + 4 \cdot r$

$4(s + r)$

53. $5 \cdot 3 + 5 \cdot 17$

$5(3 + 17)$; 100

54. $15 \cdot 6 + 5 \cdot 6$

$(15 + 5)6$; 120

55. $7(2v) + 7(5r)$

$7(2v + 5r)$

56. $13(5w) + 13(4p)$

$13(5w + 4p)$

57. $8(3r + 4s - 5y)$

$24r + 32s - 40y$

58. $2(5u - 3v + 7w)$

$10u - 6v + 14w$

59. $-3(8x + 3y + 4z)$

$-24x - 9y - 12z$

60. $-5(2x - 5y + 6z)$

$-10x + 25y - 30z$

Use the distributive property to write each expression without parentheses. See Example 9.

61. $-(4t + 5m)$

$-4t - 5m$

62. $-(9x + 12y)$

$-9x - 12y$

63. $-(-5c - 4d)$

$5c + 4d$

64. $-(-13x - 15y)$

$13x + 15y$

65. $-(-3q + 5r - 8s)$

$3q - 5r + 8s$

66. $-(-4z + 5w - 9y)$

$4z - 5w + 9y$

67. "Starting a car" and "driving away in a car" are not commutative. Give an example of another pair of everyday activities that are not commutative.

Answers will vary. For example, "putting on your socks" and "putting on your shoes"

68. Are "undressing" and "taking a shower" commutative?

no

69. *True* or *false:* "preparing a meal" and "eating a meal" are commutative.

false

70. The phrase "dog biting man" has two different meanings, depending on how the words are associated.

(dog biting) man or dog (biting man)

Give another example of a three-word phrase that has different meanings depending on how the words are associated.

Answers will vary. For example, "defective merchandise counter"

71. Use parentheses to show how the associative property can be used to give two different meanings to "foreign sales clerk."

(foreign sales) clerk; foreign (sales clerk)

72. Use parentheses to show two different meanings for "hot pink pants."

(hot pink) pants; hot (pink pants)

Relating Concepts (Exercises 73–76) For Individual or Group Work

In **Section 1.6** *we used a pattern to see that the product of two negative numbers is a positive number. In the exercises that follow, we show another justification for determining the sign of the product of two negative numbers.* ***Work Exercises 73–76 in order.***

73. Evaluate the expression $-3[5 + (-5)]$ by using the order of operations.

0

74. Write the expression in Exercise 73 using the distributive property. Do not simplify the products.

$\mathbf{-3(5) + (-3)(-5)}$

75. The product $-3(5)$ should be one of the terms you wrote when answering Exercise 74. Based on the results in **Section 1.6,** what is this product?

−15

76. In Exercise 73, you should have obtained 0 as the answer. Now, consider the following, using the results of Exercises 73 and 75.

$$-3[5 + (-5)] = -3(5) + (-3)(-5)$$
$$0 = -15 + ?$$

The question mark represents the product $-3(-5)$. When added to -15, it must give a sum of 0. Therefore, $-3(-5)$ must equal what?

The product −3(−5) must equal 15, since it is the additive inverse of −15.

1.8 Simplifying Expressions

OBJECTIVE 1 Simplify expressions. We now simplify expressions using the properties of addition and multiplication introduced in **Section 1.7.**

EXAMPLE 1 Simplifying Expressions

Simplify each expression.

(a) $4x + 8 + 9$ simplifies to $4x + 17$.

(b) $4(3m - 2n)$

$= 4(3m) - 4(2n)$ Distributive property

$= 12m - 8n$ Associative property

(c) $6 + 3(4k + 5)$

Don't start by adding.

$= 6 + 3(4k) + 3(5)$ Distributive property

$= 6 + 12k + 15$ Multiply.

$= 21 + 12k$ Add.

(d) $5 - (2y - 8)$

Be careful with signs.

$= 5 - 1(2y - 8)$ $-a = -1 \cdot a$

$= 5 - 2y + 8$ Distributive property

$= 13 - 2y$ Add.

Note

In Examples 1(c) and 1(d), we mentally used the commutative and associative properties to add in the last step. In practice, these steps are usually left out, but we should realize that they are used whenever the ordering and grouping in a sum are rearranged.

Work Problem 1 at the Side. ▶

OBJECTIVE 2 Identify terms and numerical coefficients. A **term** is a number, a variable, or a product or quotient of a number and one or more variables raised to powers. Examples of terms include

$$-9x^2, \quad 15y, \quad -3, \quad 8m^2n, \quad \frac{2}{p}, \quad \text{and} \quad k. \qquad \text{Terms}$$

The **numerical coefficient,** or simply **coefficient,** of the term $9m$ is 9; the numerical coefficient of $-15x^3y^2$ is -15; the numerical coefficient of x is 1; and the numerical coefficient of 8 is 8. In the expression $\frac{x}{3}$, the numerical coefficient of x is $\frac{1}{3}$ since $\frac{x}{3} = \frac{1x}{3} = \frac{1}{3}x$.

CAUTION

It is important to be able to distinguish between ***terms*** and ***factors.*** For example, in the expression $8x^3 + 12x^2$, there are two *terms*, $8x^3$ and $12x^2$. Terms are separated by a + or − sign. On the other hand, in the one-term expression $(8x^3)(12x^2)$, $8x^3$ and $12x^2$ are *factors*. Factors are multiplied.

OBJECTIVES

1. Simplify expressions.
2. Identify terms and numerical coefficients.
3. Identify like terms.
4. Combine like terms.
5. Simplify expressions from word phrases.

1 Simplify each expression.

(a) $9k + 12 - 5$

(b) $7(3p + 2q)$

(c) $2 + 5(3z - 1)$

(d) $-3 - (2 + 5y)$

ANSWERS

1. (a) $9k + 7$ (b) $21p + 14q$ (c) $15z - 3$ (d) $-5 - 5y$

2 Give the numerical coefficient of each term.

(a) $15q$

(b) $-2m^3$

(c) $-18m^7q^4$

(d) $-r$

(e) $\frac{5x}{4}$

3 Identify each pair of terms as *like* or *unlike*.

(a) $9x, 4x$

(b) $-8y^3, 12y^2$

(c) $5x^2y^4, 5x^4y^2$

(d) $7x^2y^4, -7x^2y^4$

(e) $13kt, 4tk$

4 Combine like terms.

(a) $4k + 7k$

(b) $4r - r$

(c) $5z + 9z - 4z$

(d) $8p + 8p^2$

(e) $5x - 3y + 2x - 5y - 3$

ANSWERS

1. **(a)** 15 **(b)** -2 **(c)** -18 **(d)** -1 **(e)** $\frac{5}{4}$

2. **(a)** like **(b)** unlike **(c)** unlike **(d)** like **(e)** like

3. **(a)** $11k$ **(b)** $3r$ **(c)** $10z$ **(d)** cannot be combined **(e)** $7x - 8y - 3$

Here are some examples of terms and their numerical coefficients.

Term	Numerical Coefficient
$-7y$	-7
$34r^3$	34
$-26x^5yz^4$	-26
$-k$, or $-1 \cdot k$	-1
r, or $1r$	1
$\frac{3x}{8} = \frac{3}{8}x$	$\frac{3}{8}$

◀ *Work Problem* **2** *at the Side.*

OBJECTIVE 3 Identify like terms. Terms with exactly the same variables (including the same exponents) are called **like terms.** For example, $9m$ and $4m$ have the same variables and are like terms. Also, $6x^3$ and $-5x^3$ are like terms. The terms $-4y^3$ and $4y^2$ have different exponents and are unlike terms. Here are some additional examples:

$5x$ and $-12x$, $3x^2y$ and $5x^2y$ Like terms

$4xy^2$ and $5xy$, $8x^2y^3$ and $7x^3y^2$. Unlike terms

◀ *Work Problem* **3** *at the Side.*

OBJECTIVE 4 Combine like terms. Recall the distributive property:

$$x(y + z) = xy + xz.$$

As seen in **Section 1.7,** this statement can also be written as

$$xy + xz = x(y + z) \quad \text{or} \quad yx + zx = (y + z)x.$$

Thus, the distributive property may be used to find the sum or difference of like terms. For example,

$$3x + 5x = (3 + 5)x = 8x.$$

This process is called **combining like terms.**

EXAMPLE 2 Combining Like Terms

Combine like terms in each expression.

(a) $9m + 5m$
$= (9 + 5)m$
$= 14m$

(b) $6r + 3r + 2r$
$= (6 + 3 + 2)r$
$= 11r$

(c) $\frac{3}{4}x + x$
$= \frac{3}{4}x + 1x$
$= \left(\frac{3}{4} + 1\right)x$
$= \frac{7}{4}x$ $\quad 1 = \frac{4}{4}$

(d) $16y^2 - 9y^2$
$= (16 - 9)y^2$
$= 7y^2$

(e) $32y + 10y^2$ cannot be combined because $32y$ and $10y^2$ are unlike terms. The distributive property cannot be used here to combine coefficients.

◀ *Work Problem* **4** *at the Side.*

CAUTION

Remember that only like terms may be combined.

EXAMPLE 3 **Simplifying Expressions Involving Like Terms**

Simplify each expression.

(a) $14y + 2(6 + 3y)$

$= 14y + 2(6) + 2(3y)$ Distributive property

$= 14y + 12 + 6y$ Multiply.

$= 20y + 12$ Combine like terms.

(b) $9k - 6 - 3(2 - 5k)$ Be careful with signs.

$= 9k - 6 - 3(2) - 3(-5k)$ Distributive property

$= 9k - 6 - 6 + 15k$ Multiply.

$= 24k - 12$ Combine like terms.

(c) $-(2 - r) + 10r$

$= -1(2 - r) + 10r$ $\quad -(2 - r) = -1(2 - r)$

$= -1(2) - 1(-r) + 10r$ Distributive property

Be careful with signs.

$= -2 + r + 10r$ Multiply.

$= -2 + 11r$ Combine like terms.

(d) $5(2a^2 - 6a) - 3(4a^2 - 9)$

$= 10a^2 - 30a - 12a^2 + 27$ Distributive property

$= -2a^2 - 30a + 27$ Combine like terms.

Work Problem **5** *at the Side.* ▶

OBJECTIVE 5 **Simplify expressions from word phrases.** We now can simplify translated expressions by combining like terms.

EXAMPLE 4 **Translating Words into a Mathematical Expression**

Write the phrase as a mathematical expression and simplify.

Four times a number, subtracted from the sum of twice the number and 4

Let x represent the number. The expression is

The sum of twice the number and 4 ↓ ; Four times the number ↓

$(2x + 4) - 4x,$ Write with symbols.

which simplifies to

$-2x + 4.$ Combine like terms.

Work Problem **6** *at the Side.* ▶

CAUTION

In Example 4, we are dealing with an expression to be simplified, *not* an equation to be solved.

5 Simplify.

(a) $10p + 3(5 + 2p)$

(b) $7z - 2 - (1 + z)$

(c) $-(3k^2 + 5k) + 7(k^2 - 4k)$

6 Write each phrase as a mathematical expression, and simplify by combining like terms.

(a) Three times a number, subtracted from the sum of the number and 8

(b) Twice a number added to the sum of 6 and the number

ANSWERS

5. **(a)** $16p + 15$ **(b)** $6z - 3$ **(c)** $4k^2 - 33k$

6. **(a)** $(x + 8) - 3x; -2x + 8$

(b) $2x + (6 + x); 3x + 6$

Math in the Media

THE MAGIC NUMBER IN SPORTS

The climax of any sports season is the playoffs. Baseball fans eagerly debate predictions of which team will win the pennant for their division. The *magic number* for each first-place team is often reported in media outlets. The **magic number** (sometimes called the **elimination number**) is the combined number of wins by the first-place team and losses by the second-place team that would clinch the title for the first-place team.

American League

East	W	L	PCT	GB
Boston	90	63	.588	—
New York	88	64	.579	1.5
Toronto	77	75	.507	12.5
Baltimore	65	87	.428	24.5
Tampa Bay	63	90	.412	27.0

Central	W	L	PCT	GB
Cleveland	90	62	.592	—
Detroit	83	70	.542	7.5
Minnesota	75	77	.493	15.0
Kansas City	66	86	.434	24.0
Chicago	66	87	.431	24.5

West	W	L	PCT	GB
Los Angeles	91	62	.595	—
Seattle	81	71	.533	9.5
Oakland	74	80	.481	17.5
Texas	70	83	.458	21.0

Source: mlb.com

To calculate the magic number, consider the following conditions.

The number of wins for the first-place team (W_1) plus the magic number (M) is one more than the sum of the number of wins to date (W_2) and the number of games remaining in the season (N_2) for the second-place team.

1. First, use the variable definitions to write an equation involving the magic number. Second, solve the equation for the magic number. Write the formula for the magic number.

$W_1 + M = W_2 + N_2 + 1$;

$M = W_2 + N_2 + 1 - W_1$

2. The American League standings on September 10, 2007, are shown above. There were 162 regulation games in the 2007 season. Find the magic number for each team. The number of games remaining in the season for the second-place team is calculated as

$$N_2 = 162 - (W_2 + L_2),$$

where L_2 represents the number of losses for the second-place team.

(a) AL East: Boston vs New York

Magic Number 9

(b) AL Central: Cleveland vs Detroit

Magic Number 3

(c) AL West: Los Angeles vs Seattle

Magic Number 1

3. Try to calculate the magic number for Toronto vs Boston. (Treat Toronto as if it were the second-place team.) How can you interpret the result?

$M = -2$; A negative magic number indicates that Toronto has been eliminated from winning the division.

1.8 Exercises

FOR EXTRA HELP

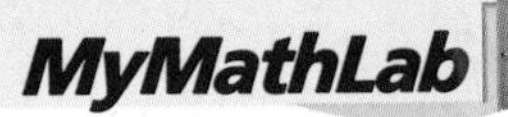

Math XL PRACTICE

WATCH

DOWNLOAD

READ

REVIEW

In Exercises 1–4, choose the letter of the correct response.

1. Which is true for all real numbers x? C

A. $6 + 2x = 8x$ **B.** $6 - 2x = 4x$

C. $6x - 2x = 4x$ **D.** $3 + 8(4x - 6) = 11(4x - 6)$

2. Which is an example of a pair of like terms? C

A. $6t, 6w$ **B.** $-8x^2y, 9xy^2$

C. $5ry, 6yr$ **D.** $-5x^2, 2x^3$

3. Which is an example of a term with numerical coefficient 5? A

A. $5x^3y^7$ **B.** x^5 **C.** $\frac{x}{5}$ **D.** 5^2xy^3

4. Which is a correct translation for "six times a number, subtracted from the product of eleven and the number" (if x represents the number)? B

A. $6x - 11x$ **B.** $11x - 6x$

C. $(11 + x) - 6x$ **D.** $6x - (11 + x)$

Simplify each expression. See Example 1.

5. $3x + 12x$ $15x$

6. $4y + 9y$ $13y$

7. $8t - 5t + 2t$ $5t$

8. $6s - 9s + 4s$ s

9. $4r + 19 - 8$ $4r + 11$

10. $7t + 18 - 4$ $7t + 14$

11. $5 + 2(x - 3y)$ $5 + 2x - 6y$

12. $8 + 3(s - 6t)$ $8 + 3s - 18t$

13. $-2 - (5 - 3p)$ $-7 + 3p$

14. $-10 - (7 - 14r)$ $-17 + 14r$

Give the numerical coefficient of each term.

15. $-12k$ -12

16. $-23y$ -23

17. $5m^2$ 5

18. $-3n^6$ -3

19. xw 1

20. pq 1

21. $-x$ -1

22. $-t$ -1

23. 74 74

24. 98 98

25. Give an example of a pair of like terms with the variable x, such that one of them has a negative numerical coefficient, one has a positive numerical coefficient, and their sum has a positive numerical coefficient.

Answers will vary. For example, $-3x$ and $4x$

26. Give an example of a pair of unlike terms such that each term has x as the only variable.

Answers will vary. For example, x^2 and x^3

Identify each group of terms as like *or* unlike.

27. $8r, -13r$ like

28. $-7a, 12a$ like

29. $5z^4, 9z^3$ unlike

30. $8x^5, -10x^3$ unlike

31. $4, 9, -24$ like

32. $7, 17, -83$ like

33. x, y unlike

34. t, s unlike

35. There is an old saying "You can't add apples and oranges." Explain how this saying can be applied to Objective 3 in this section.

We cannot "add" two unlike terms to obtain a single term, so we must be able to identify like terms in order to combine them.

36. Explain how the distributive property is used in combining $6t + 5t$ to get $11t$.

The variable t appears in both terms, allowing the use of the distributive property: $6t + 5t = (6 + 5)t = 11t$.

Simplify each expression. See Examples 2 and 3.

37. $-5 - 2(x - 3)$ **$1 - 2x$**

38. $-8 - 3(2x + 4)$ **$-20 - 6x$**

39. $-\frac{4}{3} + 2t + \frac{1}{3}t - 8 - \frac{8}{3}t$ **$-\frac{1}{3}t - \frac{28}{3}$**

40. $-\frac{5}{6} + 8x + \frac{1}{6}x - 7 - \frac{7}{6}$ **$\frac{49}{6}x - 9$**

41. $-5.3r + 4.9 - (2r + 0.7) + 3.2r$ **$-4.1r + 4.2$**

42. $2.7b + 5.8 - (3b + 0.5) - 4.4b$ **$-4.7b + 5.3$**

43. $2y^2 - 7y^3 - 4y^2 + 10y^3$ **$-2y^2 + 3y^3$**

44. $9x^4 - 7x^6 + 12x^4 + 14x^6$ **$21x^4 + 7x^6$**

45. $13p + 4(4 - 8p)$ **$-19p + 16$**

46. $5x + 3(7 - 2x)$ **$-x + 21$**

47. $-\frac{4}{3}(y - 12) - \frac{1}{6}y$ **$-\frac{3}{2}y + 16$**

48. $-\frac{7}{5}(t - 15) - \frac{3}{2}$ **$-\frac{7}{5}t + \frac{39}{2}$**

49. $-5(5y - 9) + 3(3y + 6)$ **$-16y + 63$**

50. $-3(2t + 4) + 8(2t - 4)$ **$10t - 44$**

Write each phrase as a mathematical expression. Use x to represent the number. Combine like terms when possible. See Example 4.

51. Five times a number, added to the sum of the number and three

$(x + 3) + 5x; 6x + 3$

52. Six times a number, added to the sum of the number and six

$(x + 6) + 6x; 7x + 6$

53. A number multiplied by -7, subtracted from the sum of 13 and six times the number

$(13 + 6x) - (-7x); 13 + 13x$

54. A number multiplied by 5, subtracted from the sum of 14 and eight times the number

$(14 + 8x) - 5x; 14 + 3x$

55. Six times a number added to -4, subtracted from twice the sum of three times the number and 4

$2(3x + 4) - (-4 + 6x); 12$

56. Nine times a number added to 6, subtracted from triple the sum of 12 and 8 times the number

$3(12 + 8x) - (6 + 9x); 30 + 15x$

57. Write the expression $9x - (x + 2)$ using words, as in Exercises 51–56.

Wording may vary. One example is "the difference between 9 times a number and the sum of the number and 2."

58. Write the expression $2(3x + 5) - 2(x + 4)$ using words, as in Exercises 51–56.

Wording may vary. One example is "the difference between twice the sum of three times a number and 5, and twice the sum of the number and 4."

Relating Concepts (Exercises 59–62) For Individual or Group Work

A manufacturer has fixed costs of \$1000 *to produce widgets. Each widget costs* \$5 *to make. The fixed cost to produce gadgets is* \$750, *and each gadget costs* \$3 *to make.* ***Work Exercises 59–62 in order.***

59. Write an expression for the cost to make x widgets. (*Hint:* The cost will be the sum of the fixed cost and the cost per item times the number of items.)

$1000 + 5x$ (dollars)

60. Write an expression for the cost to make y gadgets.

$750 + 3y$ (dollars)

61. Write an expression for the total cost to make x widgets and y gadgets.

$1000 + 5x + 750 + 3y$ (dollars)

62. Simplify the expression you wrote in Exercise 61.

$1750 + 5x + 3y$ (dollars)

Study Skills

REVIEWING A CHAPTER

This activity is really about preparing for tests. Some of the suggestions are ones that you will learn to use a little later in the term, but it is a good idea to start trying them out now.

OBJECTIVES

1. **Use the Chapter Summary to practice every type of problem.**
2. **Create study cards for key terms.**
3. **Practice by doing review and mixed review exercises.**
4. **Take the Chapter Test as a practice test.**

Use These Chapter Reviewing Techniques

- **Make a study card for each key term and concept.** Include a definition, an example, a sketch, and a page reference. Include a symbol or formula if there is one. See the *Using Study Cards* activity for a quick look at some sample study cards.

- **Go back to the section** to find more explanations or information about any new vocabulary, formulas, or symbols.

- **Use the Chapter Summary** to practice each type of problem. Do not expect the Summary to substitute for reading and working through the whole chapter. First, take the "Test Your Word Power" quiz to check your understanding of new vocabulary. The answers are at the end of the Quick Review. Then read the Quick Review. **Pay special attention to the headings.** Study the explanations and examples given for each concept. **Try to think about the whole chapter.**

- **Reread your lecture notes** to see what your instructor has emphasized in class. Then review that material in your text.

- **Do the Review Exercises.**
  - ✓ Check your answers after you're done with each section of exercises.
  - ✓ If you get stuck on a problem, first check the Chapter Summary. If that doesn't clear up your confusion, then check the section and your lecture notes.
  - ✓ Pay attention to direction words for the problems, such as *simplify, round, solve*, and *estimate*.
  - ✓ Make study cards for especially difficult problems.

- **Do the Mixed Review Exercises.** This is a good check to see if you can still do the problems when they are in mixed-up order. **Check your answers carefully** in the answer section in the back of your book. Are your answers exact and complete? Make sure you are labeling answers correctly, using the right units. For example, does your answer need to include \$, cm^2, ft, and so on?

Why Are These Techniques Brain Friendly?

You have already become familiar with the features of your textbook. This activity requires you to make good use of them. Your brain needs repetition in order to learn. By following the steps outlined here, you will be reinforcing the concepts, procedures, and skills you need to use for tests (and for the next chapters).

This combination of techniques also provides repetition in different ways. A thorough review of each chapter will help you be sure that you understand the concepts *completely and accurately*. Also, taking the Chapter Test will *simulate the testing situation*, which gives you practice in test taking conditions.

Now Try This ▶▶▶

- **Take the Chapter Test as if it were a real test.** If your instructor has skipped sections in the chapter, figure out which problems to skip on the test before you start.
  - ✓ **Time yourself** just as you would for a real test.
  - ✓ **Use a calculator or notes** just as you would be permitted to (or not) on a real test.
  - ✓ **Take the test in one sitting,** just like a real test is given in one sitting.
  - ✓ **Show all your work.** Practice showing your work just the way your instructor has asked you to show it.
  - ✓ **Practice neatness.** Can someone else follow your steps?
  - ✓ **Check your answers** in the back of the book.

Notice that reviewing a chapter will take some time. You cannot fully learn the material by rushing through a review in one night. But if you use the suggestions over a few days or evenings, you will notice that you understand the material more thoroughly and remember it longer.

Follow the reviewing techniques listed above for your next test. For each technique, write a comment about how it worked for you in the spaces below.

1. **Make a study card for each key term and concept.**

2. **Go back to the section** to find more explanations or information.

3. **Take the Test Your Word Power Quiz and use the Quick Review** to review each concept in the chapter.

4. **Study your lecture notes** to see what your instructor has emphasized in class.

5. **Do the Review Exercises,** following the specific suggestions on the previous page.

6. **Do the Mixed Review Exercises.**

7. **Take the Chapter Test** as if it were a real test.

Chapter 1 Summary

Key Terms

1.1	**exponent**	An exponent, or **power**, is a number that indicates how many times a factor is repeated. (3^4 ← Exponent; 3 ← Base; 3^4 Exponential expression)
	base	The base is the number that is a repeated factor when written with an exponent.
	exponential expression	A number written with an exponent is an exponential expression.
1.2	**variable**	A variable is a symbol, usually a letter, used to represent an unknown number.
	algebraic expression	An algebraic expression is a collection of numbers, variables, operation symbols, and grouping symbols.
	equation	An equation is a statement that says two expressions are equal.
	solution	A solution of an equation is any value of the variable that makes the equation true.
1.3	**natural numbers**	The set of natural numbers is $\{1, 2, 3, 4, \ldots\}$.
	whole numbers	The set of whole numbers is $\{0, 1, 2, 3, 4, \ldots\}$.
	number line	The number line shows the ordering of the real numbers on an infinite line.
	opposite	The opposite of a number a is the number that is the same distance from 0 on the number line as a, but on the opposite side of 0. This number is also called the **negative** of a or the **additive inverse** of a.
	integers	The set of integers is $\{\ldots, -3, -2, -1, 0, 1, 2, 3, \ldots\}$.
	negative number	A negative number is located to the *left* of 0 on the number line.
	positive number	A positive number is located to the *right* of 0 on the number line.
	rational numbers	A rational number is a number that can be written as the quotient of two integers, with denominator not 0.
	set-builder notation	Set builder notation uses a variable and a description to describe a set. It is often used to describe sets whose elements cannot easily be listed.
	coordinate	The number that corresponds to a point on the number line is the coordinate of that point.
	irrational numbers	An irrational number is a real number that is not a rational number.
	real numbers	Real numbers are numbers that can be represented by points on the number line (that is, all rational and irrational numbers).
	absolute value	The absolute value of a number is the distance between 0 and the number on the number line.
1.4	**sum**	The answer to an addition problem is called the sum.
1.5	**minuend**	In the operation $a - b$, a is called the minuend.
	subtrahend	In the operation $a - b$, b is called the subtrahend.
	difference	The answer to a subtraction problem is called the difference.
1.6	**product**	The answer to a multiplication problem is called the product.
	quotient	The answer to a division problem is called the quotient.
	reciprocal	Pairs of numbers whose product is 1 are called reciprocals, or **multiplicative inverses,** of each other.
1.7	**identity element for addition**	When the identity element for addition, which is 0, is added to a number, the number is unchanged.
	identity element for multiplication	When a number is multiplied by the identity element for multiplication, which is 1, the number is unchanged.

Number line
Negative numbers | Positive numbers
−3 −2 −1 0 1 2 3
Opposites

(continued)

1.8	**term**	A term is a number, a variable, or a product or quotient of a number and one or more variables raised to powers.
	numerical coefficient	The numerical factor in a term is its numerical coefficient, or **coefficient.**
	like terms	Terms with exactly the same variables (including the same exponents) are called like terms.

▶ New Symbols

a^n	n factors of a	$a(b)$, $(a)b$, $(a)(b)$, $a \cdot b$, or ab	a times b
$=$	is equal to	$\frac{a}{b}$, a/b, or $a \div b$	a divided by b
$\neq$	is not equal to	$\{\ \}$	set braces
$<$	is less than	$\{x \mid x$ **has a certain property**$\}$	set-builder notation
$\leq$	is less than or equal to	$[\ ]$	square brackets
$>$	is greater than	$\lvert x \rvert$	absolute value of x
$\geq$	is greater than or equal to	$-x$	additive inverse, or opposite, of x
		$\frac{1}{x}$	multiplicative inverse, or reciprocal, of x $(x \neq 0)$

▶ Test Your Word Power

See how well you have learned the vocabulary in this chapter. Answers, with examples, follow the Quick Review.

1. The **product** is
- **A.** the answer in an addition problem
- **B.** the answer in a multiplication problem
- **C.** one of two or more numbers that are added to get another number
- **D.** one of two or more numbers that are multiplied to get another number.

2. A number is **prime** if
- **A.** it cannot be factored
- **B.** it has just one factor
- **C.** it has exactly two different factors (itself and 1)
- **D.** it has at least two different factors.

3. An **exponent** is
- **A.** a symbol that tells how many numbers are being multiplied
- **B.** a number raised to a power
- **C.** a number that tells how many times a factor is repeated
- **D.** one of two or more numbers that are multiplied.

4. A **variable** is
- **A.** a symbol used to represent an unknown number
- **B.** a value that makes an equation true
- **C.** a solution of an equation
- **D.** the answer in a division problem.

5. An **integer** is
- **A.** a positive or negative number
- **B.** a natural number, its opposite, or zero
- **C.** any number that can be graphed on a number line
- **D.** the quotient of two numbers.

6. A **coordinate** is
- **A.** the number that corresponds to a point on a number line
- **B.** the graph of a number
- **C.** any point on a number line
- **D.** the distance from 0 on a number line.

7. The **absolute value** of a number is
- **A.** the graph of the number
- **B.** the reciprocal of the number
- **C.** the opposite of the number
- **D.** the distance between 0 and the number on a number line.

8. A **term** is
- **A.** a numerical factor
- **B.** a number, a variable, or a product or quotient of numbers and variables raised to powers
- **C.** one of several variables with the same exponents
- **D.** a sum of numbers and variables raised to powers.

9. A **numerical coefficient** is
- **A.** the numerical factor of the variables in a term
- **B.** the number of terms in an expression
- **C.** a variable raised to a power
- **D.** the variable factor in a term.

10. The **subtrahend** in $a - b = c$ is
- **A.** a
- **B.** b
- **C.** c
- **D.** $a - b$.

▸ Quick Review

Concepts	Examples

1.1 Exponents, Order of Operations, and Inequality

Order of Operations

Simplify within any parentheses or brackets and above and below fraction bars, using the following steps.

Step 1 Apply all exponents.

Step 2 Multiply or divide from left to right.

Step 3 Add or subtract from left to right.

$$\frac{9(2+6)}{2} - 2(2^3+3)$$
$$= 36 - 2(8+3)$$
$$= 36 - 2(11)$$
$$= 36 - 22$$
$$= 14$$

1.2 Variables, Expressions, and Equations

Evaluate an expression with a variable by substituting a given number for the variable.

Evaluate $2x + y^2$ if $x = 3$ and $y = -4$.

$$2x + y^2$$
$$= 2(3) + (-4)^2$$
$$= 6 + 16$$
$$= 22$$

Values of a variable that make an equation true are solutions of the equation.

Is 2 a solution of $5x + 3 = 18$?

$$5(2) + 3 \stackrel{?}{=} 18$$
$$13 = 18 \quad \text{False}$$

2 is not a solution.

1.3 Real Numbers and the Number Line

Ordering Real Numbers

a is less than b if a is to the left of b on the number line.

The opposite, or additive inverse, of a is $-a$.

The absolute value of a, written $|a|$, is the distance between a and 0 on the number line.

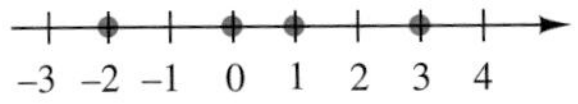

$-2 < 3$ $\quad$ $3 > 0$ $\quad$ $1 < 3$

$-(5) = -5$ $\quad$ $-(-7) = 7$ $\quad$ $-0 = 0$

$|13| = 13$ $\quad$ $|0| = 0$ $\quad$ $|-5| = 5$

1.4 Adding Real Numbers

To add two numbers with the *same sign*, add their absolute values. The sum has that same sign.

To add two numbers with *different signs*, subtract their absolute values. The sum has the sign of the number with greater absolute value.

$$9 + 4 = 13$$
$$-8 + (-5) = -13$$
$$7 + (-12) = -5$$
$$-5 + 13 = 8$$

1.5 Subtracting Real Numbers

For any real numbers a and b,

$$\boldsymbol{a - b = a + (-b).}$$

$$5 - (-2) = 5 + 2 = 7$$
$$-3 - 4 = -3 + (-4) = -7$$
$$-2 - (-6) = -2 + 6 = 4$$

Concepts	Examples
1.6 Multiplying and Dividing Real Numbers	
The product (or quotient) of two numbers having the *same sign* is *positive*.	$6 \cdot 5 = 30$ $\quad -7(-8) = 56 \quad \frac{-24}{-6} = 4$
The product (or quotient) of two numbers having *different signs* is *negative*.	$-6(5) = -30 \quad 6(-5) = -30$ $-18 \div 9 = \frac{-18}{9} = -2 \quad 49 \div (-7) = \frac{49}{-7} = -7$
To divide a by b, multiply a by the reciprocal of b.	$\frac{10}{\frac{2}{3}} = 10 \div \frac{2}{3} = 10 \cdot \frac{3}{2} = 15$
0 divided by a nonzero number is 0. Division by 0 is undefined.	$\frac{0}{5} = 0 \quad \frac{5}{0}$ is undefined.
1.7 Properties of Real Numbers	
Commutative Properties $a + b = b + a$ $ab = ba$	$7 + (-1) = -1 + 7$ $5(-3) = (-3)5$
Associative Properties $(a + b) + c = a + (b + c)$ $(ab)c = a(bc)$	$(3 + 4) + 8 = 3 + (4 + 8)$ $[-2(6)]4 = -2[6(4)]$
Identity Properties $a + 0 = a \quad 0 + a = a$ $a \cdot 1 = a \quad 1 \cdot a = a$	$-7 + 0 = -7 \quad 0 + (-7) = -7$ $9 \cdot 1 = 9 \quad 1 \cdot 9 = 9$
Inverse Properties $a + (-a) = 0 \quad -a + a = 0$ $a \cdot \frac{1}{a} = 1 \quad \frac{1}{a} \cdot a = 1 \quad (a \neq 0)$	$7 + (-7) = 0 \quad -7 + 7 = 0$ $-2\left(-\frac{1}{2}\right) = 1 \quad -\frac{1}{2}(-2) = 1$
Distributive Properties $a(b + c) = ab + ac$ $(b + c)a = ba + ca$ $a(b - c) = ab - ac$	$5(4 + 2) = 5(4) + 5(2)$ $(4 + 2)5 = 4(5) + 2(5)$ $9(5 - 4) = 9(5) - 9(4)$
1.8 Simplifying Expressions	
Only like terms may be combined. We use the distributive property.	$4(3 + 2x) - 6(5 - x)$ $= 12 + 8x - 30 + 6x$ $= 14x - 18$

ANSWERS TO TEST YOUR WORD POWER

1. B; *Example:* The product of 2 and 5, or 2 times 5, is 10. **2.** C; *Examples:* 2, 3, 11, 41, 53
3. C; *Example:* In 2^3, the number 3 is the exponent (or power), so 2 is a factor three times; $2^3 = 2 \cdot 2 \cdot 2 = 8$.
4. A; *Examples:* a, b, c **5.** B; *Examples:* $-9, 0, 6$
6. A; *Example:* The point graphed three units to the right of 0 on a number line has coordinate 3.
7. D; *Examples:* $|2| = 2$ and $|-2| = 2$ **8.** B; *Examples:* $6, \frac{x}{2}, -4ab^2$
9. A; *Examples:* The term 3 has numerical coefficient 3, $8z$ has numerical coefficient 8, and $-10x^4y$ has numerical coefficient -10.
10. B; *Example:* In $5 - 3 = 2$, 5 is the minuend, 3 is the subtrahend, and 2 is the difference.

Chapter 1 Review Exercises

If you need help with any of these Review Exercises, look in the section indicated in brackets.

[1.1] *Find the value of each exponential expression.*

1. 5^4 **625**

2. $(0.03)^4$ **0.00000081**

3. 0.21^3 **0.009261**

4. $\left(\frac{5}{2}\right)^3$ $\frac{125}{8}$

Find the value of each expression.

5. $8 \cdot 5 - 13$ **27**

6. $5[4^2 + 3(2^3)]$ **200**

7. $\dfrac{7(3^2 - 5)}{16 - 2 \cdot 6}$ **7**

8. $\dfrac{3(9 - 4) + 5(8 - 3)}{2^3 - (5 - 3)}$ $\frac{20}{3}$

Write each word sentence in symbols.

9. Thirteen is less than seventeen. **$13 < 17$**

10. Five plus two is not equal to ten. **$5 + 2 \neq 10$**

11. Write $6 < 15$ in words. **Six is less than fifteen.**

12. Construct a false statement that involves addition on the left side, the symbol $\geq$, and division on the right side. **Answers will vary. One example is $2 + 5 \geq \frac{16}{2}$.**

[1.2] *Evaluate each expression if $x = 6$ and $y = 3$.*

13. $2x + 6y$ **30**

14. $4(3x - y)$ **60**

15. $\dfrac{x}{3} + 4y$ **14**

16. $\dfrac{x^2 + 3}{3y - x}$ **13**

Change each word phrase to an algebraic expression. Use x to represent the number.

17. Six added to a number **$x + 6$**

18. A number subtracted from eight **$8 - x$**

19. Nine subtracted from six times a number **$6x - 9$**

20. Three-fifths of a number added to 12 **$12 + \frac{3}{5}x$**

Decide whether the given number is a solution of the equation.

21. $5x + 3(x + 2) = 22; 2$ **yes**

22. $\dfrac{x + 5}{3x} = 1; 6$ **no**

Change each word sentence to an equation. Use x to represent the number.

23. Six less than twice a number is 10. **$2x - 6 = 10$**

24. The product of a number and 4 is 8. **$4x = 8$**

Identify each of the following as either an equation *or an* expression.

25. $5r - 8(r + 7) = 2$ **equation**

26. $2y + (5y - 9) + 2$ **expression**

[1.3] *Graph each group of numbers on a number line.*

27. $-4, -\frac{1}{2}, 0, 2.5, 5$

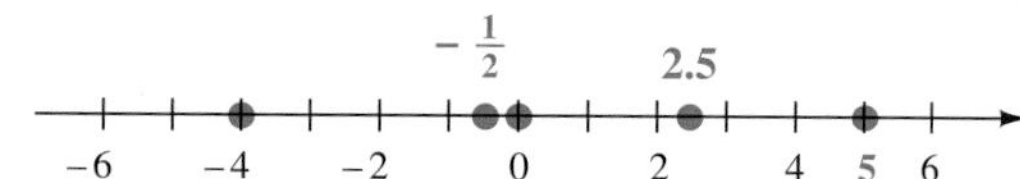

28. $-2, -3, |-3|, |-1|$

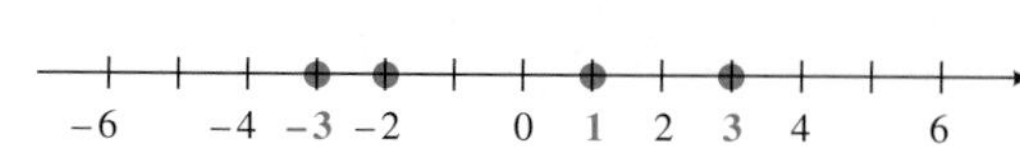

29. $-3\frac{1}{4}, \frac{14}{5}, -1\frac{1}{8}, \frac{5}{6}$

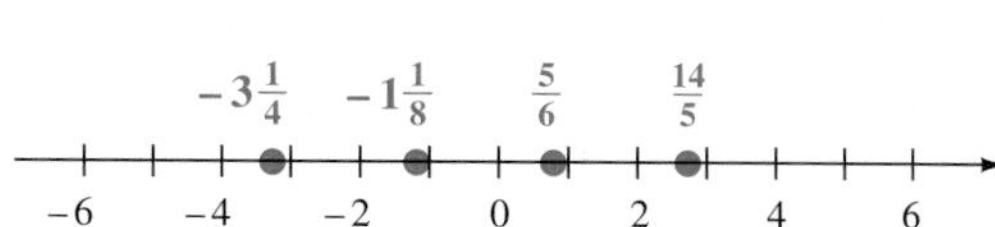

30. $|-4|, -|-3|, -|-5|, -6$

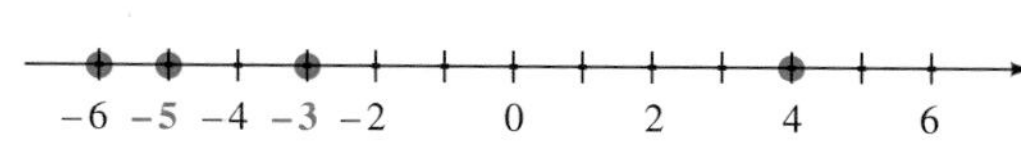

Select the lesser number in each pair.

31. $-10, 5$ **-10**

32. $-8, -9$ **-9**

33. $-\frac{2}{3}, -\frac{3}{4}$ **$-\frac{3}{4}$**

34. $0, -|23|$ **$-|23|$**

Decide whether each statement is true *or* false.

35. $12 > -13$ **true**

36. $0 > -5$ **true**

37. $-9 < -7$ **true**

38. $-13 > -13$ **false**

Simplify by finding the absolute value.

39. $-|3|$ **-3**

40. $-|-19|$ **-19**

41. $-|9 - 2|$ **-7**

42. $|15 - 6|$ **9**

[1.4] *Find each sum.*

43. $-10 + 4$ **-6**

44. $14 + (-18)$ **-4**

45. $-8 + (-9)$ **-17**

46. $\frac{4}{9} + \left(-\frac{5}{4}\right)$ **$-\frac{29}{36}$**

47. $[-6 + (-8) + 8] + [9 + (-13)]$ **-10**

48. $(-4 + 7) + (-11 + 3) + (-15 + 1)$ **-19**

Write a numerical expression for each phrase, and simplify the expression.

49. 19 added to the sum of -31 and 12

$(-31 + 12) + 19; 0$

50. 13 more than the sum of -4 and -8

$[-4 + (-8)] + 13; 1$

Solve each problem.

51. Like many people, Otis Taylor neglects to keep up his checkbook balance. When he finally balanced his account, he found that the balance was $-\$23.75$, so he deposited $50.00. What is his new balance?

$26.25

52. The low temperature in Yellowknife, in the Canadian Northwest Territories, one January day was -26°F. It rose 16° that day. What was the high temperature?

-10°F

[1.5] *Find each difference.*

53. $-7 - 4$ -11

54. $-12 - (-11)$ -1

55. $5 - (-2)$ 7

56. $-\frac{3}{7} - \frac{4}{5}$ $-\frac{43}{35}$

57. $2.56 - (-7.75)$ 10.31

58. $(-10 - 4) - (-2)$ -12

59. $(-3 + 4) - (-1)$ 2

60. $|5 - 9| - |-3 + 6|$ 1

Write a numerical expression for each phrase, and simplify the expression.

61. The difference between -4 and -6

$-4 - (-6); 2$

62. Five less than the sum of 4 and -8

$[4 + (-8)] - 5; -9$

63. The difference between 18 and -23, decreased by 15

$[18 - (-23)] - 15; 26$

64. Nineteen, decreased by 12 less than -7

$19 - (-7 - 12); 38$

Solve each problem.

65. Peyton Manning of the Indianapolis Colts passed for a gain of 8 yd, was sacked for a loss of 12 yd, and then threw a 42 yd touchdown pass. What positive or negative number represents the total net yardage for the plays?

38

66. On Friday, February 22, 2008, the Dow Jones Industrial Average closed at 12,381.02, up 96.72 from the previous day. What was the closing price on Thursday, February 21, 2008? (*Source: The Washington Post.*)

12,284.30

The table shows the number of people naturalized in the United States (that is, made citizens of the United States) for the years 1999 through 2006. In Exercises 67–70, use a signed number to represent the change in the number of people naturalized for each time period.

Year	Number of People (in thousands)
1999	843
2000	899
2001	606
2002	574
2003	463
2004	337
2005	604
2006	702

Source: U.S. Department of Homeland Security.

67. 1999–2000 56 thousand

68. 2000–2001 -293 thousand

69. 2003–2004 -126 thousand

70. 2005–2006 98 thousand

[1.6] *Perform the indicated operations.*

71. $(-12)(-3)$

36

72. $15(-7)$

-105

73. $-\frac{4}{3}\left(-\frac{3}{8}\right)$

$\frac{1}{2}$

74. $-4.8(-2.1)$

10.08

75. $5(8 - 12)$

-20

76. $(5 - 7)(8 - 3)$

-10

77. $2(-6) - (-4)(-3)$

-24

78. $3(-10) - 5$

-35

79. $\frac{-36}{-9}$

4

80. $\frac{220}{-11}$

-20

81. $-\frac{1}{2} \div \frac{2}{3}$

$-\frac{3}{4}$

82. $-33.9 \div (-3)$

11.3

83. $\frac{-5(3) - 1}{8 - 4(-2)}$

-1

84. $\frac{5(-2) - 3(4)}{-2[3 - (-2)] + 10}$

undefined

85. $\frac{10^2 - 5^2}{8^2 + 3^2 - (-2)}$

1

86. $\frac{4^2 - 8 \cdot 2}{(-1.2)^2 - (-0.56)}$

0

Evaluate each expression if $x = -5$, $y = 4$, *and* $z = -3$.

87. $6x - 4z$

-18

88. $5x + y - z$

-18

89. $5x^2$

125

90. $z^2(3x - 8y)$

-423

Write a numerical expression for each phrase, and simplify the expression.

91. Nine less than the product of -4 and 5

$-4(5) - 9; -29$

92. Five-sixths of the sum of 12 and -6

$\frac{5}{6}[12 + (-6)]; 5$

93. The quotient of 12 and the sum of 8 and -4

$\frac{12}{8 + (-4)}; 3$

94. The product of -20 and 12, divided by the difference between 15 and -15

$\frac{-20(12)}{15 - (-15)}; -8$

Translate each sentence to an equation, using x to represent the number.

95. The quotient of a number and the sum of the number and 5 is -2.

$\frac{x}{x + 5} = -2$

96. 3 less than 8 times a number is -7.

$8x - 3 = -7$

[1.7] *Decide whether each statement is an example of a commutative, associative, identity, or inverse property, or of the distributive property.*

97. $6 + 0 = 6$

identity property

98. $5 \cdot 1 = 5$

identity property

99. $-\frac{2}{3}\left(-\frac{3}{2}\right) = 1$

inverse property

100. $17 + (-17) = 0$

inverse property

101. $5 + (-9 + 2) = [5 + (-9)] + 2$

associative property

102. $w(xy) = (wx)y$

associative property

103. $3x + 3y = 3(x + y)$

distributive property

104. $(1 + 2) + 3 = 3 + (1 + 2)$

commutative property

Use the distributive property to rewrite each expression. Simplify if possible.

105. $7y + y$

$(7 + 1)y$; $8y$

106. $-12(4 - t)$

$-12 \cdot 4 - 12(-t)$; $-48 + 12t$

107. $3(2s) + 3(4y)$

$3(2s + 4y)$; $6s + 12y$

108. $-(-4r + 5s)$

$-1(-4r) + (-1)(5s)$; $4r - 5s$

[1.8] *Use the distributive property as necessary and combine like terms.*

109. $16p^2 - 8p^2 + 9p^2$

$17p^2$

110. $4r^2 - 3r + 10r + 12r^2$

$16r^2 + 7r$

111. $-8(5k - 6) + 3(7k + 2)$

$-19k + 54$

112. $2s - (-3s + 6)$

$5s - 6$

113. $-7(2t - 4) - 4(3t + 8) - 19(t + 1)$

$-45t - 23$

114. $3.6t^2 + 9t - 8.1(6t^2 + 4t)$

$-45t^2 - 23.4t$

Translate each phrase into a mathematical expression. Use x to represent the number, and combine like terms when possible.

115. Seven times a number, subtracted from the product of −2 and three times the number

$-2(3x) - 7x$; $-13x$

116. The quotient of 9 more than a number and 6 less than the number

$\frac{x + 9}{x - 6}$

117. In Exercise 115, does the word *and* signify addition? Explain.

No. The use of *and* there indicates the two quantities that are to be multiplied.

118. Write the expression $3(4x - 6)$ using words, as in Exercises 115 and 116.

Answers may vary. For example, "3 times the difference between 4 times a number and 6"

Mixed Review Exercises*

Perform the indicated operations.

119. $[(-2) + 7 - (-5)] + [-4 - (-10)]$ **16**

120. $\left(-\frac{5}{6}\right)^2$ $\frac{25}{36}$

121. $-|(-7)(-4)| - (-2)$ **−26**

122. $\frac{6(-4) + 2(-12)}{5(-3) + (-3)}$ $\frac{8}{3}$

123. $\frac{3}{8} - \frac{5}{12}$ $-\frac{1}{24}$

124. $\frac{12^2 + 2^2 - 8}{10^2 - (-4)(-15)}$ $\frac{7}{2}$

125. $\frac{8^2 + 6^2}{7^2 + 1^2}$ **2**

126. $-16(-3.5) - 7.2(-3)$ **77.6**

127. $2\frac{5}{6} - 4\frac{1}{3}$ $-1\frac{1}{2}$

128. $-8 + [(-4 + 17) - (-3 - 3)]$ **11**

129. $-\frac{12}{5} \div \frac{9}{7}$ $-\frac{28}{15}$

130. $(-8 - 3) - 5(2 - 9)$ **24**

131. $[-7 + (-2) - (-3)] + [8 + (-13)]$ **−11**

132. $\frac{15}{2} \cdot \left(-\frac{4}{5}\right)$ **−6**

Solve each problem.

133. The highest temperature ever recorded in Iowa was 118°F at Keokuk on July 20, 1934. The lowest temperature ever recorded in the state was at Elkader on February 3, 1996, and was 165° lower than the highest temperature. What is the record low temperature for Iowa? (*Source:* National Climatic Data Center.)

−47°F

134. For a certain system of rating relief pitchers, 3 points are awarded for a save, 3 points are awarded for a win, 2 points are subtracted for a loss, and 2 points are subtracted for a blown save. If Mariano Rivera of the New York Yankees has 4 saves, 3 wins, 2 losses, and 1 blown save, how many points does he have?

15 points

*The order of exercises in this final group does not correspond to the order in which topics occur in the chapter. This random ordering should help you prepare for the chapter test in yet another way.

Chapter 1 Test

Use the Chapter Test Prep Video CD to see fully worked-out solutions to any of the exercises you want to review.

Decide whether each statement is true *or* false.

1. $4[-20 + 7(-2)] \leq -135$

1. true

2. $\left(\frac{1}{2}\right)^2 + \left(\frac{2}{3}\right)^2 = \left(\frac{1}{2} + \frac{2}{3}\right)^2$

2. false

3. Graph the numbers -1, -3, $|-4|$, and $|-1|$ on the number line.

3. 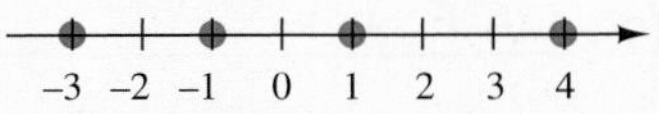

Select the lesser number from each pair.

4. $6, -|-8|$

4. $-|-8|$ (or -8)

5. $-0.742, -1.277$

5. -1.277

6. Write in symbols: The quotient of -6 and the sum of 2 and -8. Simplify the expression.

6. $\frac{-6}{2 + (-8)}$; 1

7. If a and b are both negative, is $\frac{a + b}{a \cdot b}$ positive or negative?

7. negative

Perform the indicated operations whenever possible.

8. $-2 - (5 - 17) + (-6)$

8. 4

9. $-5\frac{1}{2} + 2\frac{2}{3}$

9. $-2\frac{5}{6}$

10. $-6.2 - [-7.1 + (2.0 - 3.1)]$

10. 2

11. $4^2 + (-8) - (2^3 - 6)$

11. 6

12. $(-5)(-12) + 4(-4) + (-8)^2$

12. 108

13. $\frac{-7 - |-6 + 2|}{-5 - (-4)}$

13. 11

14. $\frac{30}{7}$

14. $\dfrac{30(-1 - 2)}{-9[3 - (-2)] - 12(-2)}$

In Exercises 15 and 16, evaluate each expression if $x = -2$ and $y = 4$.

15. -70

15. $3x - 4y^2$

16. 3

16. $\dfrac{5x + 7y}{3(x + y)}$

17. 178°F

17. The highest Fahrenheit temperature ever recorded in Idaho was 118°F, while the lowest was −60°F. What is the difference between these highest and lowest temperatures? (*Source: World Almanac and Book of Facts.*)

Match each example in Column I with a property in Column II.

18. D
19. A
20. E
21. B
22. C

	I	II
18.	$3x + 0 = 3x$	**A.** Commutative
19.	$(5 + 2) + 8 = 8 + (5 + 2)$	**B.** Associative
20.	$-3(x + y) = -3x + (-3y)$	**C.** Inverse
21.	$-5 + (3 + 2) = (-5 + 3) + 2$	**D.** Identity
22.	$-\frac{5}{3}\left(-\frac{3}{5}\right) = 1$	**E.** Distributive

23. $-9x^2 - 6x - 8$

23. Simplify $-2(3x^2 + 4) - 3(x^2 + 2x)$ by using the distributive property and combining like terms.

24. identity and distributive properties

24. Which properties are used to show that $-(3x + 1) = -3x - 1$?

25. (a) -18
(b) -18
(c) The distributive property tells us that the two methods produce equal results.

25. Consider the expression $-6[5 + (-2)]$.
(a) Evaluate it by first working within the brackets.
(b) Evaluate it by using the distributive property.
(c) Why must the answers in parts (a) and (b) be the same?

Study Skills

PREPARING FOR TESTS

Many things besides studying can improve your test scores. You may not realize that eating the right foods, and getting enough exercise and sleep can also improve your scores. Your ability to think is affected by the condition of your whole body. So, part of your preparation for tests includes keeping yourself in good physical shape as well as spending time on the actual course material. Try these suggestions and see the difference.

OBJECTIVES

1. **Restate the importance of sleep and good nutrition as it affects learning.**
2. **Explain the effect of anxiety and stress on learning.**

Performance Health Tips to Improve Your Test Score	Explanation
Get seven to eight hours of sleep the night before the exam. (It's helpful to get that much sleep every night.)	***Fatigue and exhaustion reduce efficiency.*** They also cause poor memory and recall. If you didn't sleep much the night before a test, 20 minutes of relaxation or meditation can help. (Also see the comments below about eating carbohydrates to help you sleep.)
Eat a small, high-energy meal about two hours before the test. Start the meal with a small amount of protein such as fish, chicken, or nonfat yogurt. Include carbohydrates if you like, but no high-fat foods.	Just 3 to 4 ounces of protein increases the amount of a chemical in the brain called tyrosine, which ***improves your alertness, accuracy, and motivation.*** High-fat foods dull your mind and slow down your ability to think.
Drink plenty of water. Don't wait until you feel thirsty; your body is already dehydrated by the time you feel it.	Research suggests that staying well hydrated improves the electrochemical communications in your brain.
Give your brain the time it needs to learn the material.	***Cramming doesn't work;*** your brain cannot learn that quickly. ***Studying every day*** using these study skills techniques is the way to give your brain the time it needs.

Techniques to Prevent Anxiety	Explanation
Practice slow, deep breathing for five minutes each day. Then do a minute or two of deep breathing right before the test. Also, if you feel your anxiety building during the test, stop for a minute, close your eyes, and do some deep breathing.	When ***test anxiety*** hits, you breathe more quickly and shallowly, which causes hyperventilation. Symptoms may be confusion, inability to concentrate, shaking, dizziness, and more. ***Slow, deep breathing will calm you and prevent panic.***

(continued)

Techniques to Prevent Anxiety	Explanation
Do 15 to 20 minutes of moderate exercise (like walking) shortly before the test. Daily exercise is even better.	***Exercise reduces stress*** and will help prevent "blanking out" on a test. Exercise also increases your alertness, clear thinking, and energy.
To help you sleep the night before the test, or any time you need to calm down, ***eat high carbohydrate foods*** such as popcorn, bread, rice, crackers, muffins, bagels, pasta, corn, baked potatoes (not fries or chips), and cereals.	Carbohydrates increase the level of a chemical in the brain called serotonin, which has a ***calming effect on the mind.*** It reduces feelings of tension and stress and improves your ability to concentrate. You only need to eat a small amount, like half a bagel, to get this effect.
Before the test, ***go easy on caffeinated beverages*** such as coffee, tea, and soft drinks. Do not eat candy bars or other sugary snacks.	Extra caffeine can ***make you jittery,*** "hyper," and shaky for the test. It can increase the tendency to panic. Too much sugar causes negative emotional reactions in some people.

Now Try This ▶▶▶

What will you do to improve your next test score? List the three or four tips you think will help you the most.

1. ______________________________
2. ______________________________
3. ______________________________
4. ______________________________

What changes will you have to make in order to try the tips you chose?

____

____

____

See *Tips for Taking Math Tests* and *Preparing for Your Math Final Exam* for more ideas about managing anxiety. (Check the Table of Contents to find their locations.)

2

Equations, Inequalities, and Applications

In 1896, 241 competitors from 14 countries gathered in Athens, Greece, for the first modern Olympic Games. What began as a small, mainly European, sports competition has become the world's largest global sporting event. The Games of the XXIX Olympiad, hosted in 2008 by Beijing, China, attracted 10,500 athletes from over 200 countries. First introduced at the 1920 Games in Antwerp, Belgium, the five interlocking rings on the Olympic flag symbolize unity among the nations of Africa, the Americas, Asia, Australia, and Europe. (*Source:* www.olympic.org, *Microsoft Encarta Encyclopedia.*)

Throughout this chapter we use *linear equations* to solve applications about the Olympics.

2.1 The Addition Property of Equality

OBJECTIVES

1. **Identify linear equations.**
2. **Use the addition property of equality.**
3. **Simplify, and then use the addition property of equality.**

An *equation* is a statement that two algebraic expressions are equal.

CAUTION
Remember that an equation includes an equals sign.

Equation (to solve): $\underbrace{x - 5}_{\text{Left side}} = \underbrace{2}_{\text{Right side}}$

Expression (to simplify or evaluate): $x - 5$

OBJECTIVE 1 Identify linear equations.

Linear Equation in One Variable

A **linear equation in one variable** can be written in the form

$$Ax + B = C,$$

where A, B, and C are real numbers, with $A \neq 0$.

Some examples of linear and *non*linear equations follow.

$$4x + 9 = 0, \quad 2x - 3 = 5, \quad \text{and} \quad x = 7 \qquad \text{Linear equations}$$

$$x^2 + 2x = 5, \quad \frac{1}{x} = 6, \quad \text{and} \quad |2x + 6| = 0 \qquad \text{Nonlinear equations}$$

As we saw in **Section 1.2,** a *solution* of an equation is a number that makes the equation true when it replaces the variable. An equation is solved by finding its **solution set,** the set of all solutions. Equations that have exactly the same solution sets are **equivalent equations.** A linear equation in x is solved by using a series of steps to produce a simpler equivalent equation of the form

$$\mathbf{x = a\ number} \quad \text{or} \quad \mathbf{a\ number = x}.$$

OBJECTIVE 2 Use the addition property of equality. In the equation $x - 5 = 2$, both $x - 5$ and 2 represent the same number because that is the meaning of the equals sign. To solve the equation, we change the left side from $x - 5$ to just x. We do this by adding 5 to $x - 5$. We use 5 because 5 is the opposite (additive inverse) of -5, and $-5 + 5 = 0$. To keep the two sides equal, we must also add 5 to the right side.

$$\begin{aligned} x - 5 &= 2 & \\ x - 5 + 5 &= 2 + 5 & \text{Add 5 to each side.} \\ x + 0 &= 7 & \text{Additive inverse property} \\ x &= 7 & \text{Additive identity property} \end{aligned}$$

The solution is 7. We check by replacing x with 7 in the original equation.

Check

$$\begin{aligned} x - 5 &= 2 & \text{Original equation} \\ 7 - 5 &\stackrel{?}{=} 2 & \text{Let } x = 7. \\ 2 &= 2 & \text{True} \end{aligned}$$

Since the final equation is true, 7 checks as the solution and $\{7\}$ is the solution set. We write a solution set using set braces $\{\ \}$.

To solve the equation $x - 5 = 2$, we added the same number, 5, to each side. The **addition property of equality** justifies this step.

Addition Property of Equality

If A, B, and C represent real numbers, then the equations

$$A = B \quad \text{and} \quad A + C = B + C$$

are equivalent equations.

In words, we can add the same number to each side of an equation without changing the solution.

In this property, C represents a real number. Any quantity that represents a real number can be added to each side of an equation to obtain an equivalent equation.

Note

Equations can be thought of in terms of a balance. Thus, adding the same quantity to each side does not affect the balance. See Figure 1.

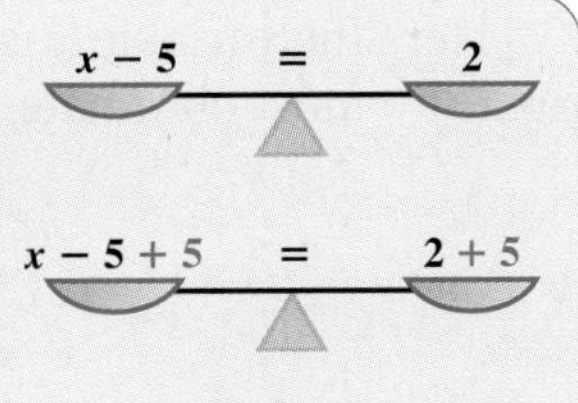

Figure 1

EXAMPLE 1 **Using the Addition Property of Equality**

Solve $x - 16 = 7$.

Our goal is to get an equivalent equation of the form **x = a number.**

$$x - 16 = 7$$

$$x - 16 + 16 = 7 + 16 \qquad \text{Add 16 to each side.}$$

$$x = 23 \qquad \text{Combine like terms.}$$

Check Substitute 23 for x in the *original* equation.

$$x - 16 = 7 \qquad \text{Original equation}$$

$$23 - 16 \stackrel{?}{=} 7 \qquad \text{Let } x = 23.$$

$$7 = 7 \qquad \text{True}$$

7 is *not* the solution.

Since a true statement results, 23 is the solution and $\{23\}$ is the solution set.

Work Problem 1 at the Side. ▶

EXAMPLE 2 **Using the Addition Property of Equality**

Solve $x - 2.9 = -6.4$.

Our goal is to isolate x.

$$x - 2.9 = -6.4$$

$$x - 2.9 + 2.9 = -6.4 + 2.9 \qquad \text{Add 2.9 to each side.}$$

$$x = -3.5$$

Check

$$x - 2.9 = -6.4 \qquad \text{Original equation}$$

$$-3.5 - 2.9 \stackrel{?}{=} -6.4 \qquad \text{Let } x = -3.5.$$

$$-6.4 = -6.4 \qquad \text{True}$$

Since a true statement results, the solution set is $\{-3.5\}$.

Work Problem 2 at the Side. ▶

1 Solve.

(a) $x - 12 = 9$

(b) $x - 25 = -18$

2 Solve.

(a) $x - 3.7 = -8.1$

(b) $a - 4.1 = 6.3$

Answers

1. **(a)** $\{21\}$ **(b)** $\{7\}$
2. **(a)** $\{-4.4\}$ **(b)** $\{10.4\}$

The addition property of equality says that the same number may be *added* to each side of an equation. In **Section 1.5,** subtraction was defined as addition of the opposite. Thus, we can also use the following rule when solving an equation.

The same number may be *subtracted* from each side of an equation without changing the solution.

For example, to solve the equation $x + 4 = 10$, we *subtract* 4 from each side, which is the same as adding -4. The result is $x = 6$.

3 Solve.

(a) $-3 = a + 2$

(b) $22 = -16 + r$

EXAMPLE 3 Using the Addition Property of Equality

Solve $-7 = x + 22$.

Here the variable x is on the right side of the equation. To isolate x on the right, we must eliminate the 22 by subtracting 22 from each side.

$$-7 = x + 22$$ The variable can be isolated on *either* side.

$$-7 - 22 = x + 22 - 22$$ Subtract 22 from each side.

$$-29 = x, \quad \text{or} \quad x = -29$$ Rewrite; a number $= x$, or $x =$ a number.

Check

$$-7 = x + 22$$ Original equation

$$-7 \stackrel{?}{=} -29 + 22$$ Let $x = -29$.

$$-7 = -7$$ True

The check confirms that the solution set is $\{-29\}$.

Work Problem 3 at the Side.

CAUTION

The final line of the check does not give the solution to the problem, only a confirmation that the solution found is correct.

EXAMPLE 4 Subtracting a Variable Term

Solve $\frac{3}{5}k + 15 = \frac{8}{5}k$.

To get all terms with variables on the same side of the equation, subtract $\frac{3}{5}k$ from each side.

$$\frac{3}{5}k + 15 = \frac{8}{5}k$$

$$\frac{3}{5}k + 15 - \frac{3}{5}k = \frac{8}{5}k - \frac{3}{5}k$$ Subtract $\frac{3}{5}k$ from each side.

$$15 = 1k$$ $\frac{3}{5}k - \frac{3}{5}k = 0$; $\frac{8}{5}k - \frac{3}{5}k = \frac{5}{5}k = 1k$

$$15 = k$$ Multiplicative identity property

From now on we will skip the step that changes $1k$ to k. Check the solution by replacing k with 15 in the original equation. The solution set is $\{15\}$.

ANSWERS

3. (a) $\{-5\}$ **(b)** $\{38\}$

What happens if we solve the equation in Example 4 by first subtracting $\frac{8}{5}k$ from each side?

$$\frac{3}{5}k + 15 = \frac{8}{5}k \qquad \text{Equation from Example 4}$$

$$\frac{3}{5}k + 15 - \frac{8}{5}k = \frac{8}{5}k - \frac{8}{5}k \qquad \text{Subtract } \tfrac{8}{5}k \text{ from each side.}$$

$$15 - k = 0 \qquad \tfrac{3}{5}k - \tfrac{8}{5}k = -\tfrac{5}{5}k = -1k = -k;\ \tfrac{8}{5}k - \tfrac{8}{5}k = 0$$

$$15 - k - 15 = 0 - 15 \qquad \text{Subtract 15 from each side.}$$

$$-k = -15 \qquad \text{Combine like terms; additive inverse}$$

This result gives the value of $-k$, but not of k itself. However, it does say that the additive inverse of k is -15, which means that k must be 15.

$$-k = -15$$

$$k = 15 \qquad \text{Same result as in Example 4}$$

(This result can also be justified using the multiplication property of equality, covered in **Section 2.2.**) We can make the following generalization.

If a is a number and $-x = a$, then $x = -a$.

Work Problem 4 *at the Side.* ▶

EXAMPLE 5 Using the Addition Property of Equality Twice

Solve $8 - 6p = -7p + 5$.

We must get all terms with variables on the same side of the equation and all terms without variables on the other side of the equation.

$$8 - 6p = -7p + 5$$

$$8 - 6p + 7p = -7p + 5 + 7p \qquad \text{Add } 7p \text{ to each side.}$$

$$8 + p = 5 \qquad \text{Combine like terms.}$$

$$8 + p - 8 = 5 - 8 \qquad \text{Subtract 8 from each side.}$$

$$p = -3 \qquad \text{Combine like terms.}$$

Check Substitute -3 for p in the original equation.

$$8 - 6p = -7p + 5 \qquad \text{Original equation}$$

$$8 - 6(-3) \stackrel{?}{=} -7(-3) + 5 \qquad \text{Let } p = -3.$$

$$8 + 18 \stackrel{?}{=} 21 + 5 \qquad \text{Multiply.}$$

$$26 = 26 \qquad \text{True}$$

Use parentheses when substituting to avoid errors.

The check results in a true statement, so the solution set is $\{-3\}$.

Work Problem 5 *at the Side.* ▶

Note

There are often several equally correct ways to solve an equation. In Example 5, we could begin by adding $6p$, instead of $7p$, to each side. Combining like terms and subtracting 5 from each side gives $3 = -p$. (Try this.) If $3 = -p$, then $-3 = p$, and the variable has been isolated on the right side of equation. The same solution results.

4 **(a)** Solve $5m + 4 = 6m$.

(b) Solve $\frac{7}{2}m + 1 = \frac{9}{2}m$.

(c) What is the solution set of $-x = 6$?

(d) What is the solution set of $-x = -12$?

5 Solve.

(a) $10 - a = -2a + 9$

(b) $6x - 8 = 12 + 5x$

Answers

4. (a) $\{4\}$ **(b)** $\{1\}$ **(c)** $\{-6\}$ **(d)** $\{12\}$
5. (a) $\{-1\}$ **(b)** $\{20\}$

OBJECTIVE 3 Simplify, and then use the addition property of equality. Sometimes the terms of an equation must be simplified as a first step in its solution.

EXAMPLE 6 Combining Like Terms before Solving

Solve $3t - 12 + t + 2 = 5 + 3t + 2$.

Begin by combining like terms on each side of the equation.

$$3t - 12 + t + 2 = 5 + 3t + 2$$

$$4t - 10 = 7 + 3t \quad \text{Combine like terms.}$$

$$4t - 10 - 3t = 7 + 3t - 3t \quad \text{Subtract } 3t \text{ from each side.}$$

$$t - 10 = 7 \quad \text{Combine like terms.}$$

$$t - 10 + 10 = 7 + 10 \quad \text{Add 10 to each side.}$$

$$t = 17 \quad \text{Combine like terms.}$$

Check Substitute 17 for t in the original equation.

$$3t - 12 + t + 2 = 5 + 3t + 2 \quad \text{Original equation}$$

$$3(17) - 12 + 17 + 2 \stackrel{?}{=} 5 + 3(17) + 2 \quad \text{Let } t = 17.$$

$$51 - 12 + 17 + 2 \stackrel{?}{=} 5 + 51 + 2 \quad \text{Multiply.}$$

$$58 = 58 \quad \text{True}$$

The check results in a true statement, so the solution set is $\{17\}$.

◀ *Work Problem 6 at the Side.*

6 Solve.

(a) $4x + 6 + 2x - 3 = 9 + 5x - 4$

(b) $9r + 4r + 6 - 2 = 9r + 4 + 3r$

EXAMPLE 7 Using the Distributive Property before Solving

Solve $3(2 + 5x) - (1 + 14x) = 6$.

$$3(2 + 5x) - (1 + 14x) = 6$$

Be careful here!

$$3(2 + 5x) - 1(1 + 14x) = 6 \quad -(1 + 14x) = -1(1 + 14x)$$

$$3(2) + 3(5x) - 1(1) - 1(14x) = 6 \quad \text{Distributive property}$$

$$6 + 15x - 1 - 14x = 6 \quad \text{Multiply.}$$

$$x + 5 = 6 \quad \text{Combine like terms.}$$

$$x + 5 - 5 = 6 - 5 \quad \text{Subtract 5 from each side.}$$

$$x = 1 \quad \text{Combine like terms.}$$

Check by substituting 1 for x in the original equation. The solution set is $\{1\}$.

CAUTION

Be careful to apply the distributive property correctly in a problem like that in Example 7, or a sign error may result.

◀ *Work Problem 7 at the Side.*

7 Solve.

(a) $4(r + 1) - (3r + 5) = 1$

(b) $-3(m - 4) + 2(5 + 2m) = 29$

ANSWERS

6. (a) $\{2\}$ (b) $\{0\}$
7. (a) $\{2\}$ (b) $\{7\}$

2.1 Exercises

FOR EXTRA HELP

Math XL PRACTICE

WATCH

DOWNLOAD

READ

REVIEW

1. Decide whether each is an *expression* or an *equation*. If it is an expression, simplify it. If it is an equation, solve it.

(a) $5x + 8 - 4x + 7$
expression; $x + 15$

(b) $-6m + 12 + 7m - 5$
expression; $m + 7$

(c) $5x + 8 - 4x = 7$
equation; $\{-1\}$

(d) $-6m + 12 + 7m = -5$
equation; $\{-17\}$

2. Which of the pairs of equations are equivalent equations?

A. $x + 2 = 6$ and $x = 4$
B. $10 - x = 5$ and $x = -5$
C. $x + 3 = 9$ and $x = 6$
D. $4 + x = 8$ and $x = -4$

A and C

3. Which of the following are not linear equations in one variable?

A. $x^2 - 5x + 6 = 0$
B. $x^3 = x$
C. $3x - 4 = 0$
D. $7x - 6x = 3 + 9x$

A and B

4. Explain how to check a solution of an equation.

Replace the variable(s) in the *original* equation with the proposed solution. A true statement will result if the proposed solution is correct.

Solve each equation, and check your solution. See Examples 1–5.

5. $x - 4 = 8$
$\{12\}$

6. $x - 8 = 9$
$\{17\}$

7. $x - 5 = -8$
$\{-3\}$

8. $x - 7 = -9$
$\{-2\}$

9. $r + 9 = 13$
$\{4\}$

10. $t + 6 = 10$
$\{4\}$

11. $x + 26 = 17$
$\{-9\}$

12. $x + 45 = 24$
$\{-21\}$

13. $x + \frac{1}{4} = -\frac{1}{2}$ $\left\{-\frac{3}{4}\right\}$

14. $x + \frac{2}{3} = -\frac{1}{6}$ $\left\{-\frac{5}{6}\right\}$

15. $x - 8.4 = -2.1$
$\{6.3\}$

16. $z - 15.5 = -5.1$
$\{10.4\}$

17. $t + 12.3 = -4.6$
$\{-16.9\}$

18. $x + 21.5 = -13.4$
$\{-34.9\}$

19. $7 + r = -3$
$\{-10\}$

20. $8 + k = -4$
$\{-12\}$

21. $2 = p + 15$
$\{-13\}$

22. $3 = z + 17$
$\{-14\}$

23. $-\frac{1}{3} = x - \frac{3}{5}$ $\left\{\frac{4}{15}\right\}$

24. $-\frac{1}{4} = x - \frac{2}{3}$ $\left\{\frac{5}{12}\right\}$

25. $3x = 2x + 7$
$\{7\}$

26. $5x = 4x + 9$
$\{9\}$

27. $10x + 4 = 9x$
$\{-4\}$

28. $8t + 5 = 7t$
$\{-5\}$

29. $\frac{9}{7}r - 3 = \frac{2}{7}r$
$\{3\}$

30. $\frac{8}{5}w - 6 = \frac{3}{5}w$
$\{6\}$

31. $5.6x + 2 = 4.6x$
$\{-2\}$

32. $9.1x - 5 = 8.1x$
$\{5\}$

33. $3p + 6 = 10 + 2p$
$\{4\}$

34. $8x - 4 = -6 + 7x$
$\{-2\}$

35. $5 - x = -2x - 11$
$\{-16\}$

36. $3 - 8x = -9x - 1$
$\{-4\}$

37. $-4z + 7 = -5z + 9$
$\{2\}$

38. $-6q + 3 = -7q + 10$
$\{7\}$

Solve each equation, and check your solution. See Examples 6 and 7.

39. $3x + 6 - 10 = 2x - 2$
$\{2\}$

40. $8k - 4 + 6 = 7k + 1$
$\{-1\}$

41. $6x + 5 + 7x + 3 = 12x + 4$
$\{-4\}$

42. $4x - 3 - 8x + 1 = -5x + 9$
$\{11\}$

43. $10x + 5x + 7 - 8 = 12x + 3 + 2x$
$\{4\}$

44. $7p + 4p + 13 - 7 = 7p + 9 + 3p$
$\{3\}$

45. $5.2q - 4.6 - 7.1q = -0.9q - 4.6$
$\{0\}$

46. $-4.0x + 2.7 - 1.6x = -4.6x + 2.7$
$\{0\}$

47. $\frac{5}{7}x + \frac{1}{3} = \frac{2}{5} - \frac{2}{7}x + \frac{2}{5}$ $\left\{\frac{7}{15}\right\}$

48. $\frac{6}{7}s - \frac{3}{4} = \frac{4}{5} - \frac{1}{7}s + \frac{1}{6}$ $\left\{\frac{103}{60}\right\}$

49. $(5x + 6) - (3 + 4x) = 10$
$\{7\}$

50. $(8r - 3) - (7r + 1) = -6$
$\{-2\}$

51. $2(p + 5) - (9 + p) = -3$
$\{-4\}$

52. $4(k - 6) - (3k + 2) = -5$
$\{21\}$

53. $-6(2x + 1) + (13x - 7) = 0$
$\{13\}$

54. $-5(3w - 3) + (1 + 16w) = 0$
$\{-16\}$

55. $10(-2x + 1) = -19(x + 1)$
$\{29\}$

56. $2(2 - 3r) = -5(r - 3)$
$\{-11\}$

57. $-2(8p + 2) - 3(2 - 7p) = 2(4 + 2p)$
$\{18\}$

58. $-5(1 - 2z) + 4(3 - z) = 7(3 + z)$
$\{-14\}$

59. Write an equation that requires the use of the addition property of equality, in which 6 must be added to each side to solve the equation and the solution is a negative number.
Answers will vary. One example is $x - 6 = -8$.

60. Write an equation that requires the use of the addition property of equality, in which $\frac{1}{2}$ must be subtracted from each side and the solution is a positive number.
Answers will vary. One example is $x + \frac{1}{2} = 1$.

2.2 The Multiplication Property of Equality

OBJECTIVES

1 **Use the multiplication property of equality.**

2 **Simplify, and then use the multiplication property of equality.**

OBJECTIVE 1 Use the multiplication property of equality. The addition property of equality alone is not enough to solve some equations, such as $3x + 2 = 17$.

$$3x + 2 = 17$$

$$3x + 2 - 2 = 17 - 2 \quad \text{Subtract 2 from each side.}$$

$$3x = 15 \quad \text{Combine like terms.}$$

Notice that the coefficient of x on the left side is 3, not 1 as desired. Another property, the **multiplication property of equality,** is needed to change $3x = 15$ to an equation of the form

$$\mathbf{x = a\ number.}$$

Since $3x = 15$, both $3x$ and 15 must represent the same number. Multiplying both $3x$ and 15 by the same number will also result in an equality.

Multiplication Property of Equality

If A, B, and C $(C \neq 0)$ represent real numbers, then the equations

$$A = B \quad \text{and} \quad AC = BC$$

are equivalent equations.

In words, we can multiply each side of an equation by the same nonzero number without changing the solution.

In $3x = 15$, we must change $3x$ to $1x$, or x. To isolate x, we multiply each side by $\frac{1}{3}$, the reciprocal of 3, because $\frac{1}{3} \cdot 3 = \frac{3}{3} = 1$.

$$3x = 15$$

$$\frac{1}{3}(3x) = \frac{1}{3}(15) \quad \text{Multiply each side by } \tfrac{1}{3}.$$

$$\left(\frac{1}{3} \cdot 3\right)x = \frac{1}{3}(15) \quad \text{Associative property}$$

$$1x = 5 \quad \text{Multiplicative inverse property}$$

$$x = 5 \quad \text{Multiplicative identity property}$$

The product of a number and its reciprocal is 1.

The solution is 5. We can check this result in the original equation.

Work Problem 1 at the Side. ▶

Just as the addition property of equality permits *subtracting* the same number from each side of an equation, the multiplication property of equality permits *dividing* each side of an equation by the same nonzero number.

$$3x = 15$$

$$\frac{3x}{3} = \frac{15}{3} \quad \text{Divide each side by 3.}$$

$$x = 5 \quad \text{Same result as above}$$

We can divide each side of an equation by the same nonzero number without changing the solution. ***Do not, however, divide each side by a variable, since the variable might be equal to 0.***

1 Check that 5 is the solution of $3x = 15$, and write the solution set.

ANSWER

1. Since $3(5) = 15$, the solution of $3x = 15$ is 5, and the solution set is $\{5\}$.

Note

In practice, it is usually easier to multiply on each side if the coefficient of the variable is a fraction, and divide on each side if the coefficient is an integer or a decimal. For example, to solve

$$-\frac{3}{4}x = 12,$$

it is easier to multiply by $-\frac{4}{3}$, the reciprocal of $-\frac{3}{4}$, than to divide by $-\frac{3}{4}$. On the other hand, to solve

$$-5x = -20,$$

it is easier to divide by -5 than to multiply by $-\frac{1}{5}$.

EXAMPLE 1 Dividing Each Side of an Equation by a Nonzero Number

Solve $5x = 60$.

Isolate x on the left by using the multiplication property of equality. Divide each side of the equation by 5, the coefficient of x.

$$5x = 60$$

$$\frac{5x}{5} = \frac{60}{5} \qquad \text{Divide each side by 5.}$$

$$x = 12 \qquad \frac{5x}{5} = \frac{5}{5}x = 1x = x$$

Check Substitute 12 for x in the original equation.

$$5x = 60 \qquad \text{Original equation}$$

$$5(12) \stackrel{?}{=} 60 \qquad \text{Let } x = 12.$$

$$60 = 60 \qquad \text{True}$$

Since a true statement results, the solution set is $\{12\}$.

◀ *Work Problem* 2 *at the Side.*

EXAMPLE 2 Using the Multiplication Property of Equality

Solve $-25p = 30$.

$$-25p = 30$$

$$\frac{-25p}{-25} = \frac{30}{-25} \qquad \text{Divide by } -25.$$

$$p = \frac{30}{-25} = -\frac{6}{5} \qquad \frac{a}{-b} = -\frac{a}{b}; \text{ Write in lowest terms.}$$

Check

$$-25p = 30 \qquad \text{Original equation}$$

$$\frac{-25}{1}\left(-\frac{6}{5}\right) \stackrel{?}{=} 30 \qquad \text{Let } p = -\frac{6}{5}.$$

$$30 = 30 \qquad \text{True}$$

The check confirms that the solution set is $\{-\frac{6}{5}\}$.

◀ *Work Problem* 3 *at the Side.*

2 Solve.

(a) $6p = 60$

(b) $3r = -12$

(c) $15x = 75$

3 Solve.

(a) $2m = 15$

(b) $-6x = 14$

(c) $10z = -45$

ANSWERS

2. (a) $\{10\}$ **(b)** $\{-4\}$ **(c)** $\{5\}$

3. (a) $\left\{\frac{15}{2}\right\}$ **(b)** $\left\{-\frac{7}{3}\right\}$ **(c)** $\left\{-\frac{9}{2}\right\}$

EXAMPLE 3 Solving an Equation with Decimals

Solve $6.09 = 2.1x$.

$$6.09 = 2.1x \quad \text{Isolate } x \text{ on the right.}$$

$$\frac{6.09}{2.1} = \frac{2.1x}{2.1} \quad \text{Divide each side by 2.1.}$$

$$2.9 = x, \quad \text{or} \quad x = 2.9 \quad \text{Divide; you may use a calculator.}$$

Check that the solution set is $\{2.9\}$.

Work Problem 4 at the Side. ▶

EXAMPLE 4 Using the Multiplication Property of Equality

Solve $\frac{x}{4} = 3$.

Replace $\frac{x}{4}$ by $\frac{1}{4}x$, since dividing by 4 is the same as multiplying by $\frac{1}{4}$. To isolate x, multiply each side by 4, the *reciprocal* of the coefficient of x.

$$\frac{x}{4} = 3$$

$$\frac{1}{4}x = 3 \quad \text{Change } \tfrac{x}{4} \text{ to } \tfrac{1}{4}x.$$

$$4 \cdot \frac{1}{4}x = 4 \cdot 3 \quad \text{Multiply each side by 4.}$$

$$x = 12 \quad \text{Multiplicative inverse property; multiplicative identity property}$$

($4 \cdot \frac{1}{4}x = 1x = x$)

Check

$$\frac{x}{4} = 3 \quad \text{Original equation}$$

$$\frac{12}{4} \stackrel{?}{=} 3 \quad \text{Let } x = 12.$$

$$3 = 3 \quad \text{True}$$

Since a true statement results, the solution set is $\{12\}$.

Work Problem 5 at the Side. ▶

EXAMPLE 5 Using the Multiplication Property of Equality

Solve $\frac{3}{4}h = 6$.

To isolate h on the left, multiply each side of the equation by $\frac{4}{3}$, the reciprocal of $\frac{3}{4}$, since $\frac{4}{3} \cdot \frac{3}{4}h = 1 \cdot h = h$.

$$\frac{3}{4}h = 6$$

$$\frac{4}{3} \cdot \frac{3}{4}h = \frac{4}{3} \cdot 6 \quad \text{Multiply each side by } \tfrac{4}{3}.$$

$$1 \cdot h = \frac{4}{3} \cdot \frac{6}{1} \quad \text{Multiplicative inverse property}$$

$$h = 8 \quad \text{Multiplicative identity property; multiply fractions.}$$

Check that the solution set is $\{8\}$.

Work Problem 6 at the Side. ▶

4 Solve.

(a) $-0.7m = -5.04$

(b) $-63.75 = 12.5k$

5 Solve.

(a) $\frac{x}{5} = 5$

(b) $\frac{p}{4} = -6$

6 Solve.

(a) $-\frac{5}{6}t = -15$

(b) $\frac{3}{5}k = -21$

Answers

4. **(a)** $\{7.2\}$ **(b)** $\{-5.1\}$
5. **(a)** $\{25\}$ **(b)** $\{-24\}$
6. **(a)** $\{18\}$ **(b)** $\{-35\}$

7 Solve.

(a) $-m = 2$

(b) $-p = -7$

In **Section 2.1,** we obtained the equation $-k = -15$ in our alternate solution to Example 4. We reasoned that since this equation says that the additive inverse (or opposite) of k is -15, then k must equal 15. We can also use the multiplication property of equality to obtain the same result, as shown in the next example.

EXAMPLE 6 Using the Multiplication Property of Equality When the Coefficient of the Variable Is −1

Solve $-k = -15$.

Work on the left side, first writing $-k$ as $-1 \cdot k$.

$-k = -15$	
$-1 \cdot k = -15$	$-k = -1 \cdot k$
$-1(-1 \cdot k) = -1(-15)$	Multiply by -1, since $-1(-1) = 1$.
$[-1(-1)] \cdot k = 15$	Associative property; multiply.
$1 \cdot k = 15$	Multiplicative inverse property
$k = 15$	Multiplicative identity property

Check

$-k = -15$	Original equation
$-(15) \stackrel{?}{=} -15$	Let $k = 15$.
$-15 = -15$	True

The solution, 15, checks, so the solution set is $\{15\}$.

◀ *Work Problem* 7 *at the Side.*

8 Solve.

(a) $7m - 5m = -12$

(b) $4r - 9r = 20$

OBJECTIVE 2 Simplify, and then use the multiplication property of equality. In the next example, it is necessary to combine like terms before using the multiplication property of equality.

EXAMPLE 7 Combining Like Terms before Solving

Solve $5m + 6m = 33$.

$5m + 6m = 33$	
$11m = 33$	Combine like terms.
$\frac{11m}{11} = \frac{33}{11}$	Divide each side by 11.
$m = 3$	Multiplicative identity property

Check

$5m + 6m = 33$	Original equation
$5(3) + 6(3) \stackrel{?}{=} 33$	Let $m = 3$.
$15 + 18 \stackrel{?}{=} 33$	Multiply.
$33 = 33$	True

The solution, 3, checks, so the solution set is $\{3\}$.

◀ *Work Problem* 8 *at the Side.*

ANSWERS

7. (a) $\{-2\}$ (b) $\{7\}$
8. (a) $\{-6\}$ (b) $\{-4\}$

2.2 Exercises

FOR EXTRA HELP

By what number is it necessary to multiply each side of each equation in order to isolate x on the left side? Do not actually solve.

1. $\frac{2}{3}x = 8$ **$\frac{3}{2}$**

2. $\frac{4}{5}x = 6$ **$\frac{5}{4}$**

3. $\frac{x}{10} = 3$ **10**

4. $\frac{x}{100} = 8$ **100**

5. $-\frac{9}{2}x = -4$ **$-\frac{2}{9}$**

6. $-\frac{8}{3}x = -11$ **$-\frac{3}{8}$**

7. $-x = 0.36$ **-1**

8. $-x = 0.29$ **-1**

By what number is it necessary to divide each side of each equation in order to isolate x on the left side? Do not actually solve.

9. $6x = 5$ **6**

10. $7x = 10$ **7**

11. $-4x = 13$ **-4**

12. $-13x = 6$ **-13**

13. $0.12x = 48$ **0.12**

14. $0.21x = 63$ **0.21**

15. $-x = 23$ **-1**

16. $-x = 49$ **-1**

17. Tell whether you would use the addition or multiplication property of equality to solve each equation. *Do not actually solve.*

(a) $3x = 12$ **(b)** $3 + x = 12$ **(c)** $-x = 4$ **(d)** $-12 = 6 + x$

(a) and (c): multiplication property of equality

(b) and (d): addition property of equality

18. Which equation does *not* require the use of the multiplication property of equality?

A. $3x - 5x = 6$ **B.** $-\frac{1}{4}x = 12$ **C.** $5x - 4x = 7$ **D.** $\frac{x}{3} = -2$ **C**

19. In the solution of a linear equation, the next-to-the-last step reads "$-x = -\frac{3}{4}$." Which of the following would be the solution to this equation?

A. $-\frac{3}{4}$ **B.** $\frac{3}{4}$ **C.** -1 **D.** $\frac{4}{3}$ **B**

20. A student tried to solve the equation $4x = 8$ by dividing each side by 8. What is a better approach?

To isolate x on the left side, divide each side by 4, the coefficient of x.

Solve each equation, and check your solution. See Examples 1–7.

21. $5x = 30$ **{6}**

22. $7x = 56$ **{8}**

23. $2m = 15$ **$\left\{\frac{15}{2}\right\}$**

24. $3m = 10$ **$\left\{\frac{10}{3}\right\}$**

25. $3a = -15$ **{−5}**

26. $5k = -70$ **{−14}**

27. $10t = -36$ **$\left\{-\frac{18}{5}\right\}$**

28. $4s = -34$ **$\left\{-\frac{17}{2}\right\}$**

29. $-6x = -72$ **{12}**

30. $-8x = -64$ **{8}**

31. $2r = 0$ **{0}**

32. $5x = 0$ **{0}**

33. $-x = 12$
{−12}

34. $-t = 14$
{−14}

35. $0.2t = 8$
{40}

36. $0.9x = 18$
{20}

37. $-2.1m = 25.62$
{−12.2}

38. $-3.9a = 31.2$
{−8}

39. $\frac{1}{4}x = -12$
{−48}

40. $\frac{1}{5}p = -3$
{−15}

41. $\frac{z}{6} = 12$
{72}

42. $\frac{x}{5} = 15$
{75}

43. $\frac{x}{7} = -5$
{−35}

44. $\frac{k}{8} = -3$
{−24}

45. $\frac{2}{7}p = 4$
{14}

46. $\frac{3}{8}x = 9$
{24}

47. $-\frac{7}{9}c = \frac{3}{5}$
$\left\{-\frac{27}{35}\right\}$

48. $-\frac{5}{6}d = \frac{4}{9}$
$\left\{-\frac{8}{15}\right\}$

49. $4x + 3x = 21$
{3}

50. $9x + 2x = 121$
{11}

51. $3r - 5r = 10$
{−5}

52. $9p - 13p = 24$
{−6}

53. $\frac{2}{5}x - \frac{3}{10}x = 2$
{20}

54. $\frac{2}{3}x - \frac{5}{9}x = 4$
{36}

55. $x + x - 3x = 12$
{−12}

56. $z - 3z + z = -16$
{16}

57. $5m + 6m - 2m = 63$
{7}

58. $11r - 5r + 6r = 168$
{14}

59. $-6x + 4x - 7x = 0$
{0}

60. $-5x + 4x - 8x = 0$
{0}

61. $0.9w - 0.5w + 0.1w = -3$
{−6}

62. $0.5x - 0.6x + 0.3x = -1$
{−5}

63. Write an equation that requires the use of the multiplication property of equality, where each side must be multiplied by $\frac{2}{3}$ and the solution is a negative number.
Answers will vary. One example is $\frac{3}{2}x = -6$.

64. Write an equation that requires the use of the multiplication property of equality, where each side must be divided by 100 and the solution is not an integer.
Answers will vary. One example is $100x = 17$.

Write an equation using the information given in the problem. Use x as the variable. Then solve the equation.

65. When a number is multiplied by −4, the result is 10. Find the number.
$-4x = 10; -\frac{5}{2}$

66. When a number is divided by −5, the result is 2. Find the number.
$\frac{x}{-5} = 2; -10$

2.3 More on Solving Linear Equations

In this section, we solve linear equations using *both* properties of equality introduced in **Sections 2.1 and 2.2.**

Work Problem 1 at the Side. ▶

OBJECTIVE 1 Learn and use the four steps for solving a linear equation. ***Remember that when we solve an equation, our primary goal is to isolate the variable on one side of the equation.*** We use the following four-step method.

Solving a Linear Equation

Step 1 **Simplify each side separately.** Clear (eliminate) parentheses, fractions, and decimals, using the distributive property as needed, and combine like terms.

Step 2 **Isolate the variable term on one side.** Use the addition property so that the variable term is on one side of the equation and a number is on the other.

Step 3 **Isolate the variable.** Use the multiplication property to get the equation in the form $x =$ a number, or a number $= x$. (Other letters may be used for the variable.)

Step 4 **Check.** Substitute the proposed solution into the *original* equation to see if a true statement results.

EXAMPLE 1 Using Both Properties of Equality to Solve an Equation

Solve $-6x + 5 = 17$.

Step 1 There are no parentheses, fractions, or decimals in this equation, so this step is not necessary.

$$-6x + 5 = 17$$

Step 2 $-6x + 5 - 5 = 17 - 5$ Subtract 5 from each side.

$-6x = 12$ Combine like terms.

Step 3 $\frac{-6x}{-6} = \frac{12}{-6}$ Divide each side by -6.

$$x = -2$$

Step 4 Check by substituting -2 for x in the original equation.

Check $-6x + 5 = 17$ Original equation

$-6(-2) + 5 \stackrel{?}{=} 17$ Let $x = -2$.

$12 + 5 \stackrel{?}{=} 17$ Multiply.

$17 = 17$ True

The solution, -2, checks, so the solution set is $\{-2\}$.

Work Problem 2 at the Side. ▶

OBJECTIVES

1. **Learn and use the four steps for solving a linear equation.**
2. **Solve equations with fractions or decimals as coefficients.**
3. **Solve equations that have no solution or infinitely many solutions.**
4. **Write expressions for two related unknown quantities.**

1 As a review, tell whether you would use the addition or multiplication property of equality to solve each equation. *Do not actually solve.*

(a) $7 + x = -9$

(b) $-13x = 26$

(c) $-x = \frac{3}{4}$

(d) $-12 = x - 4$

2 Solve.

(a) $-5p + 4 = 19$

(b) $7 + 2m = -3$

ANSWERS

1. **(a)** and **(d)**: addition property of equality
   (b) and **(c)**: multiplication property of equality
2. **(a)** $\{-3\}$ **(b)** $\{-5\}$

3 Solve.

(a) $2q + 3 = 4q - 9$

(b) $5 - 8k = 2k - 5$

Answers

3. (a) $\{6\}$ (b) $\{1\}$

EXAMPLE 2 **Using Both Properties of Equality to Solve an Equation**

Solve $3x + 2 = 5x - 8$.

Step 1 Again, there are no parentheses, fractions, or decimals in the equation, so we begin with Step 2.

$$3x + 2 = 5x - 8$$ Our goal is to isolate x.

Step 2

$$3x + 2 - 5x = 5x - 8 - 5x$$ Subtract $5x$ from each side.

$$-2x + 2 = -8$$ Combine like terms.

$$-2x + 2 - 2 = -8 - 2$$ Subtract 2 from each side.

$$-2x = -10$$ Combine like terms.

Step 3

$$\frac{-2x}{-2} = \frac{-10}{-2}$$ Divide each side by -2.

$$x = 5$$

Step 4 Check by substituting 5 for x in the original equation.

Check

$$3x + 2 = 5x - 8$$ Original equation

$$3(5) + 2 \stackrel{?}{=} 5(5) - 8$$ Let $x = 5$.

$$15 + 2 \stackrel{?}{=} 25 - 8$$ Multiply.

$$17 = 17$$ True

The solution, 5, checks, so the solution set is $\{5\}$.

Note

Remember that the variable can be isolated on either side of the equation. In Example 2, x will be isolated on the right if we begin by subtracting $3x$, instead of $5x$, from each side of the equation.

$$3x + 2 = 5x - 8$$ Equation from Example 2

$$3x + 2 - 3x = 5x - 8 - 3x$$ Subtract $3x$ from each side.

$$2 = 2x - 8$$ Combine like terms.

$$2 + 8 = 2x - 8 + 8$$ Add 8 to each side.

$$10 = 2x$$ Combine like terms.

$$\frac{10}{2} = \frac{2x}{2}$$ Divide each side by 2.

$$5 = x$$ The same solution results.

There are often several equally correct ways to solve an equation.

Work Problem 3 at the Side.

EXAMPLE 3 **Using the Four Steps to Solve an Equation**

Solve $4(k - 3) - k = k - 6$.

Step 1 Clear the parentheses using the distributive property.

$$4(k - 3) - k = k - 6$$

$$4(k) + 4(-3) - k = k - 6 \quad \text{Distributive property}$$

$$4k - 12 - k = k - 6 \quad \text{Multiply.}$$

$$3k - 12 = k - 6 \quad \text{Combine like terms.}$$

Step 2

$$3k - 12 - k = k - 6 - k \quad \text{Subtract } k.$$

$$2k - 12 = -6 \quad \text{Combine like terms.}$$

$$2k - 12 + 12 = -6 + 12 \quad \text{Add 12.}$$

$$2k = 6 \quad \text{Combine like terms.}$$

Step 3

$$\frac{2k}{2} = \frac{6}{2} \quad \text{Divide by 2.}$$

$$k = 3$$

Step 4 Check by substituting 3 for k in the original equation.

Check

$$4(k - 3) - k = k - 6 \quad \text{Original equation}$$

$$4(3 - 3) - 3 \stackrel{?}{=} 3 - 6 \quad \text{Let } k = 3.$$

$$4(0) - 3 \stackrel{?}{=} 3 - 6 \quad \text{Work inside the parentheses.}$$

$$-3 = -3 \quad \text{True}$$

The solution is 3, so $\{3\}$ is the solution set.

Work Problem **4** *at the Side.* ▶

EXAMPLE 4 **Using the Four Steps to Solve an Equation**

Solve $8a - (3 + 2a) = 3a + 1$.

Step 1

$$8a - (3 + 2a) = 3a + 1$$

$$8a - 1(3 + 2a) = 3a + 1 \quad \text{Multiplicative identity property}$$

$$8a - 3 - 2a = 3a + 1 \quad \text{Distributive property}$$

Be careful with signs.

$$6a - 3 = 3a + 1 \quad \text{Combine like terms.}$$

Step 2

$$6a - 3 - 3a = 3a + 1 - 3a \quad \text{Subtract } 3a.$$

$$3a - 3 = 1 \quad \text{Combine like terms.}$$

$$3a - 3 + 3 = 1 + 3 \quad \text{Add 3.}$$

$$3a = 4 \quad \text{Combine like terms.}$$

Step 3

$$\frac{3a}{3} = \frac{4}{3} \quad \text{Divide by 3.}$$

$$a = \frac{4}{3}$$

Step 4 Check that the solution set is $\{\frac{4}{3}\}$.

---

4 Solve.

(a) $7(p - 2) + p = 2p + 4$

(b) $11 + 3(x + 1) = 5x + 16$

ANSWERS

4. **(a)** $\{3\}$ **(b)** $\{-1\}$

CAUTION

Be very careful with signs when solving an equation like the one in Example 4. When clearing parentheses in the expression

$$8a - (3 + 2a),$$

remember that the $-$ sign acts like a factor of -1 and affects the sign of *every* term inside the parentheses. Thus,

$$\begin{aligned} &8 - (3 + 2a) \\ &= 8 - 1(3 + 2a) \\ &= 8a + (-1)(3 + 2a) \\ &= 8a - 3 - 2a. \end{aligned}$$

Change to $-$ in *both* terms.

◀ *Work Problem* 5 *at the Side.*

5 Solve.

(a) $7m - (2m - 9) = 39$

(b) $4x - (x + 7) = 9$

EXAMPLE 5 Using the Four Steps to Solve an Equation

Solve $4(8 - 3t) = 32 - 8(t + 2)$.

Step 1	$4(8 - 3t) = 32 - 8(t + 2)$	Be careful with signs.
	$32 - 12t = 32 - 8t - 16$	Distributive property
	$32 - 12t = 16 - 8t$	Combine like terms.
Step 2	$32 - 12t + 8t = 16 - 8t + 8t$	Add $8t$.
	$32 - 4t = 16$	Combine like terms.
	$32 - 4t - 32 = 16 - 32$	Subtract 32.
	$-4t = -16$	Combine like terms.
Step 3	$\frac{-4t}{-4} = \frac{-16}{-4}$	Divide by -4.
	$t = 4$	

Step 4 Check this solution in the original equation.

Check	$4(8 - 3t) = 32 - 8(t + 2)$	Original equation
	$4[8 - 3(4)] \stackrel{?}{=} 32 - 8(4 + 2)$	Let $t = 4$.
	$4[8 - 12] \stackrel{?}{=} 32 - 8(6)$	Simplify.
	$4[-4] \stackrel{?}{=} 32 - 48$	
	$-16 = -16$	True

The solution, 4, checks. The solution set is $\{4\}$.

◀ *Work Problem* 6 *at the Side.*

6 Solve.

(a) $2(4 + 3r) = 3(r + 1) + 11$

(b) $2 - 3(2 + 6z) = 4(z + 1) + 18$

OBJECTIVE 2 Solve equations with fractions or decimals as coefficients. We clear an equation of fractions by multiplying each side by the least common denominator (LCD) of all the fractions in the equation. It is a good idea to do this in Step 1 to avoid messy computations, as shown in Examples 6–8.

ANSWERS

5. (a) $\{6\}$ (b) $\left\{\frac{16}{3}\right\}$

6. (a) $\{2\}$ (b) $\left\{-\frac{13}{11}\right\}$

EXAMPLE 6 **Solving an Equation with Fractions as Coefficients**

Solve $\frac{2}{3}x - \frac{1}{2}x = -\frac{1}{6}x - 2$.

Step 1 The LCD of all the fractions in the equation is 6, so we multiply each side by 6 to clear the fractions.

$$\frac{2}{3}x - \frac{1}{2}x = -\frac{1}{6}x - 2$$

Pay particular attention here.

$$6\left(\frac{2}{3}x - \frac{1}{2}x\right) = 6\left(-\frac{1}{6}x - 2\right) \quad \text{Multiply by 6.}$$

$$6\left(\frac{2}{3}x\right) + 6\left(-\frac{1}{2}x\right) = 6\left(-\frac{1}{6}x\right) + 6(-2) \quad \text{Distributive property}$$

$$4x - 3x = -x - 12 \quad \text{Multiply.}$$

$$x = -x - 12 \quad \text{Combine like terms.}$$

Step 2

$$x + x = -x - 12 + x \quad \text{Add } x.$$

$$2x = -12 \quad \text{Combine like terms.}$$

Step 3

$$\frac{2x}{2} = \frac{-12}{2} \quad \text{Divide by 2.}$$

$$x = -6$$

Step 4 Check by substituting -6 for x in the original equation.

Check

$$\frac{2}{3}x - \frac{1}{2}x = -\frac{1}{6}x - 2 \quad \text{Original equation}$$

$$\frac{2}{3}(-6) - \frac{1}{2}(-6) \stackrel{?}{=} -\frac{1}{6}(-6) - 2 \quad \text{Let } x = -6.$$

$$-4 + 3 \stackrel{?}{=} 1 - 2$$

$$-1 = -1 \quad \text{True}$$

The solution, -6, checks. The solution set is $\{-6\}$.

Work Problem **7** *at the Side.* ▶

7 Solve

$$\frac{1}{4}x - 4 = \frac{3}{2}x + \frac{3}{4}x.$$

EXAMPLE 7 **Solving an Equation with Fractions as Coefficients**

Solve $\frac{1}{3}(x + 5) - \frac{3}{5}(x + 2) = 1$.

Step 1 We first clear the fractions by multiplying by the LCD, 15.

$$\frac{1}{3}(x + 5) - \frac{3}{5}(x + 2) = 1$$

$$15\left[\frac{1}{3}(x + 5) - \frac{3}{5}(x + 2)\right] = 15(1) \quad \text{Multiply by 15.}$$

$$15\left[\frac{1}{3}(x + 5)\right] + 15\left[-\frac{3}{5}(x + 2)\right] = 15 \quad \text{Distributive property}$$

$$5(x + 5) - 9(x + 2) = 15 \quad \text{Multiply.}$$

$$15\left[\tfrac{1}{3}(x + 5)\right] = 15 \cdot \tfrac{1}{3} \cdot (x + 5) = 5(x + 5)$$

$$5x + 25 - 9x - 18 = 15 \quad \text{Distributive property}$$

$$-4x + 7 = 15 \quad \text{Combine like terms.}$$

Continued on Next Page

ANSWER

7. $\{-2\}$

8 Solve

$$\frac{1}{4}(x+3) - \frac{2}{3}(x+1) = -2.$$

Step 2	$-4x + 7 - 7 = 15 - 7$	Subtract 7.
	$-4x = 8$	Combine like terms.
Step 3	$\frac{-4x}{-4} = \frac{8}{-4}$	Divide by -4.
	$x = -2$	

Step 4 Check to confirm that $\{-2\}$ is the solution set.

◀ *Work Problem* **8** *at the Side.*

CAUTION
Be sure you understand how to multiply by the LCD to clear an equation of fractions. Study Step 1 in Examples 6 and 7 carefully.

EXAMPLE 8 **Solving an Equation with Decimals as Coefficients**

Solve $0.1t + 0.05(20 - t) = 0.09(20)$.

Step 1 The decimals here are expressed as tenths (0.1) and hundredths (0.05 and 0.09). We choose the least exponent on 10 needed to eliminate the decimals; in this case, we use $10^2 = 100$.

	$\mathbf{0.1}t + 0.05(20 - t) = 0.09(20)$	
	$\mathbf{0.10}t + 0.05(20 - t) = 0.09(20)$	$0.1 = 0.10$
	$\mathbf{100}[0.10t + 0.05(20 - t)] = \mathbf{100}[0.09(20)]$	Multiply by 100.
	$\mathbf{100}(0.10t) + \mathbf{100}[0.05(20 - t)] = \mathbf{100}[0.09(20)]$	Distributive property
	$10t + 5(20 - t) = 9(20)$	Multiply.
	$10t + 5(20) + 5(-t) = 180$	Distributive property
	$10t + 100 - 5t = 180$	Multiply.
	$5t + 100 = 180$	Combine like terms.
Step 2	$5t + 100 - \mathbf{100} = 180 - \mathbf{100}$	Subtract 100.
	$5t = 80$	Combine like terms.
Step 3	$\frac{5t}{5} = \frac{80}{5}$	Divide by 5.
	$t = 16$	

Step 4 Check to confirm that $\{16\}$ is the solution set.

◀ *Work Problem* **9** *at the Side.*

9 Solve $0.06(100 - x) + 0.04x = 0.05(92)$.

Note
In Example 8, multiplying by 100 is the same as moving the decimal point two places to the right.

$$0.10t + 0.05(20 - t) = 0.09(20)$$

$$10t + 5(20 - t) = 9(20) \qquad \text{Multiply by 100.}$$

ANSWERS

8. $\{5\}$
9. $\{70\}$

OBJECTIVE 3 Solve equations that have no solution or infinitely many solutions. Every equation we have solved so far has had exactly one solution. Such an equation, which is true for some values of the variable and false for others, is a **conditional equation**—it is true only under certain conditions. Some equations have no solution or infinitely many solutions.

EXAMPLE 9 Solving an Equation That Has Infinitely Many Solutions

Solve $5x - 15 = 5(x - 3)$.

$$5x - 15 = 5(x - 3)$$

$$5x - 15 = 5x - 15 \qquad \text{Distributive property}$$

$$5x - 15 - \mathbf{5x} = 5x - 15 - \mathbf{5x} \qquad \text{Subtract } 5x.$$

$$-15 = -15$$

$$-15 + \mathbf{15} = -15 + \mathbf{15} \qquad \text{Add 15.}$$

$$\mathbf{0 = 0} \qquad \text{True}$$

Solution set: {all real numbers}

The variable has "disappeared." Since the last statement $(0 = 0)$ is true, *any* real number is a solution. (We could have predicted this from the line in the solution that says $5x - 15 = 5x - 15$, which is true for *any* value of x.) Try several values for x in the original equation to see that they all satisfy it.

An equation with both sides exactly the same, like $0 = 0$, is called an **identity.** An identity is true for all replacements of the variables. As shown above, we write the solution set as **{all real numbers}.**

CAUTION
In Example 9, do not write $\{0\}$ as the solution set of the equation. While 0 is a solution, there are infinitely many other solutions.

EXAMPLE 10 Solving an Equation That Has No Solution

Solve $2x + 3(x + 1) = 5x + 4$.

$$2x + 3(x + 1) = 5x + 4$$

$$2x + 3x + 3 = 5x + 4 \qquad \text{Distributive property}$$

$$5x + 3 = 5x + 4 \qquad \text{Combine like terms.}$$

$$5x + 3 - \mathbf{5x} = 5x + 4 - \mathbf{5x} \qquad \text{Subtract } 5x.$$

$$3 = 4 \qquad \text{False}$$

There is no solution. Solution set: $\emptyset$

Again, the variable has disappeared, but this time a false statement $(3 = 4)$ results. This is a signal that the equation, called a **contradiction,** has no solution. Its solution set is the **empty set,** or **null set,** symbolized $\boldsymbol{\emptyset}$.

CAUTION
Do not write $\{\emptyset\}$ to represent the empty set.

Work Problem 10 at the Side. ▶

10 Solve each equation.

(a) $2(x - 6) = 2x - 12$

(b) $3x + 6(x + 1) = 9x - 4$

ANSWERS

10. (a) {all real numbers} **(b)** $\emptyset$

OBJECTIVE 4 Write expressions for two related unknown quantities.

Problem-Solving Hint

Often we are given a problem in which the sum of two quantities is a particular number, and we are asked to find the values of the two quantities. Example 11 shows how to express the unknown quantities in terms of a single variable.

EXAMPLE 11 Translating a Phrase into an Algebraic Expression

Two numbers have a sum of 23. If one of the numbers is represented by k, write an expression for the other number.

First, suppose that the sum of two numbers is 23, and one of the numbers is 10. How would you find the other number? You would subtract 10 from 23 to get 13.

$$23 - 10 = 13$$

So instead of using 10 as one of the numbers, use k as stated in the problem. The other number would be obtained in the same way. You must subtract k from 23. Therefore, an expression for the other number is

$$23 - k.$$

CAUTION

Since the sum of the two numbers in Example 11 is 23, the expression for the other number must be $23 - k$, *not* $k - 23$. (Subtraction is *not* commutative.) To check, find the sum of the two numbers:

$$k + (23 - k) = 23, \quad \text{as required.}$$

◀ Work Problem 11 at the Side.

11 Perform each translation.

(a) Two numbers have a sum of 36. One of the numbers is r. Write an expression for the other number.

(b) The product of two numbers is -6. One of the numbers is q. Write an expression for the other number.

ANSWERS

11. (a) $36 - r$ **(b)** $\frac{-6}{q}$

2.3 Exercises

FOR EXTRA HELP

MyMathLab

Math XL PRACTICE

WATCH

DOWNLOAD

READ

REVIEW

Using the methods of this section, what should we do first when solving each equation? Do not actually solve.

1. $7x + 8 = 1$

Use the addition property of equality to subtract 8 from each side.

2. $7x - 5x + 15 = 8 + x$

Combine like terms on the left.

3. $3(2t - 4) = 20 - 2t$

Clear parentheses by using the distributive property.

4. $\frac{3}{4}z = -15$

Use the multiplication property of equality to multiply each side by $\frac{4}{3}$.

5. $\frac{2}{3}x - \frac{1}{6} = \frac{3}{2}x + 1$

Clear fractions by multiplying by the LCD, 6.

6. $0.9x + 0.3(x + 12) = 6$

Clear decimals by multiplying each side by 10.

Solve each equation, and check your solution. See Examples 1–5, 9, and 10.

7. $3x + 2 = 14$

$\{4\}$

8. $4x + 3 = 27$

$\{6\}$

9. $-5z - 4 = 21$

$\{-5\}$

10. $-7w - 4 = 10$

$\{-2\}$

11. $4p - 5 = 2p$

$\left\{\frac{5}{2}\right\}$

12. $6q - 2 = 3q$

$\left\{\frac{2}{3}\right\}$

13. $5m + 8 = 7 + 3m$

$\left\{-\frac{1}{2}\right\}$

14. $4r + 2 = r - 6$

$\left\{-\frac{8}{3}\right\}$

15. $10p + 6 = 12p - 4$

$\{5\}$

16. $-5x + 8 = -3x + 10$

$\{-1\}$

17. $7r - 5r + 2 = 5r - r$

$\{1\}$

18. $9p - 4p + 6 = 7p - 3p$

$\{-6\}$

19. $x + 3 = -(2x + 2)$

$\left\{-\frac{5}{3}\right\}$

20. $2x + 1 = -(x + 3)$

$\left\{-\frac{4}{3}\right\}$

21. $4(2x - 1) = -6(x + 3)$

$\{-1\}$

22. $6(3w + 5) = 2(10w + 10)$

$\{5\}$

23. $6(4x - 1) = 12(2x + 3)$

$\emptyset$

24. $6(2x + 8) = 4(3x - 6)$

$\emptyset$

25. $3(2x - 4) = 6(x - 2)$

{all real numbers}

26. $3(6 - 4x) = 2(-6x + 9)$

{all real numbers}

27. $3(4x - 2) + 5x = 30 - x$

{2}

28. $5(2m + 3) - 4m = 8m + 27$

{−6}

29. $-2p + 7 = 3 - (5p + 1)$

$\left\{-\frac{5}{3}\right\}$

30. $4x + 9 = 3 - (x - 2)$

$\left\{-\frac{4}{5}\right\}$

31. Which linear equation does *not* have all real numbers as solutions?

A. $5x = 4x + x$ **B.** $2(x + 6) = 2x + 12$ **C.** $\frac{1}{2}x = 0.5x$ **D.** $3x = 2x$ D

32. The expression $100[0.03(x - 10)]$ is equivalent to which of the following?

A. $0.03x - 0.3$ **B.** $3x - 3$ **C.** $3x - 10$ **D.** $3x - 30$ D

Solve each equation, and check your solution. See Examples 6–8.

33. $-\frac{2}{7}r + 2r = \frac{1}{2}r + \frac{17}{2}$

{7}

34. $\frac{3}{5}t - \frac{1}{10}t = t - \frac{5}{2}$

{5}

35. $\frac{3}{4}x - \frac{1}{3}x + 5 = \frac{5}{6}x$

{12}

36. $\frac{1}{5}x - \frac{2}{3}x - 2 = -\frac{2}{5}x$

{−30}

37. $\frac{1}{7}(3x + 2) - \frac{1}{5}(x + 4) = 2$

{11}

38. $\frac{1}{4}(3x - 1) + \frac{1}{6}(x + 3) = 3$

{3}

39. $\frac{1}{9}(x + 18) + \frac{1}{3}(2x + 3) = x + 3$

$\{0\}$

40. $-\frac{1}{4}(x - 12) + \frac{1}{2}(x + 2) = x + 4$

$\{0\}$

41. $-\frac{5}{6}q - \left(q - \frac{1}{2}\right) = \frac{1}{4}(q + 1)$

$\left\{\frac{3}{25}\right\}$

42. $\frac{2}{3}k - \left(k + \frac{1}{4}\right) = \frac{1}{12}(k + 4)$

$\left\{-\frac{7}{5}\right\}$

43. $0.3(30) + 0.15x = 0.2(30 + x)$

$\{60\}$

44. $0.2(60) + 0.05x = 0.1(60 + x)$

$\{120\}$

45. $0.92x + 0.98(12 - x) = 0.96(12)$

$\{4\}$

46. $1.00x + 0.05(12 - x) = 0.10(63)$

$\{6\}$

47. $0.02(5000) + 0.03x = 0.025(5000 + x)$

$\{5000\}$

48. $0.06(10{,}000) + 0.08x = 0.072(10{,}000 + x)$

$\{15{,}000\}$

Solve each equation, and check your solution. See Examples 1–10.

49. $-3(5z + 24) + 2 = 2(3 - 2z) - 4$

$\left\{-\frac{72}{11}\right\}$

50. $-2(2s - 4) - 8 = -3(4s + 4) - 1$

$\left\{-\frac{13}{8}\right\}$

51. $-(6k - 5) - (-5k + 8) = -3$

$\{0\}$

52. $-(4x + 2) - (-3x - 5) = 3$

$\{0\}$

53. $8(t-3)+4t=6(2t+1)-10$

$\emptyset$

54. $9(v+1)-3v=2(3v+1)-8$

$\emptyset$

55. $4(x+3)=2(2x+8)-4$

{all real numbers}

56. $4(x+8)=2(2x+6)+20$

{all real numbers}

57. $\frac{1}{3}(x+3)+\frac{1}{6}(x-6)=x+3$

$\{-6\}$

58. $\frac{1}{2}(x+2)+\frac{3}{4}(x+4)=x+5$

$\{4\}$

59. $0.3(x+15)+0.4(x+25)=25$

$\{15\}$

60. $0.1(x+80)+0.2x=14$

$\{20\}$

Write the answer to each problem in terms of the variable. See Example 11.

61. Two numbers have a sum of 12. One number is q. What expression represents the other number?

$12-q$

62. Two numbers have a sum of 26. One of the numbers is r. What expression represents the other number?

$26-r$

63. The product of two numbers is 9. One of the numbers is z. What expression represents the other number?

$\frac{9}{z}$

64. The product of two numbers is 13. One number is k. What expression represents the other number?

$\frac{13}{k}$

65. Monica is a years old. What expression represents her age 12 yr from now? 2 yr ago?

$a+12$; $a-2$

66. Chandler is b years old. What expression represents his age 3 yr ago? 5 yr from now?

$b-3$; $b+5$

67. Tom has r quarters. Express the value of the quarters in cents.

$25r$

68. Jean has y dimes. Express the value of the dimes in cents.

$10y$

Study Skills

USING STUDY CARDS REVISITED

This is the second part of the Study Cards activity. As you get further into a chapter, you can choose particular problems that will serve as a good test review. Here are two more types of study cards that will help you.

OBJECTIVES

1. **Create study cards for difficult problems.**
2. **Create study cards of quiz problems.**

Tough Problems Card

When you are doing your homework and find yourself saying, "This is really hard," or "I'm having trouble with this," make a tough problem study card. On the front, write out the procedure to work the type of problem *in words*. If there are special notes (like what *not* to do), include them. On the back, work at least one example. Make sure you label what you are doing.

Front of Card

When solving a linear equation, be careful when clearing parentheses if there is a minus sign in front.

$$6x - (x + 3)$$

The minus sign acts like -1, so change the sign of every term inside the parentheses.

Back of Card

Solve.

$$6x - (x + 3) = 7$$
$$6x - 1(x + 3) = 7$$

Change both terms.

$6x - x - 3 = 7$	*Distributive property*
$5x - 3 = 7$	*Combine terms.*
$5x - 3 + 3 = 7 + 3$	*Add 3.*
$5x = 10$	*Combine terms.*
$\frac{5x}{5} = \frac{10}{5}$	*Divide by 5.*
$x = 2$	

Solution set: $\{2\}$

Now Try This

Choose three types of difficult problems, and work them out on study cards. Be sure to put the words for solving the problem on one side and the worked problem on the other side.

Practice Quiz Cards

Quiz study cards cover each type of problem you learn. They are useful when you prepare for a test. To make a quiz card, put the problem with the direction words (like *solve, simplify*) on the front of the card, and work the problem on the back. If you like, include the page number from the text. When you review, work the problem on a separate paper and check it by looking at the back of your quiz card.

Front of Card

Solve $4(3a - 4) = 2(6a - 9) + 2$.

Back of Card

$4(3a - 4)$	$= 2(6a - 9) + 2$	
$12a - 16$	$= 12a - 18 + 2$	Distributive property
$12a - 16$	$= 12a - 16$	Combine terms.
$12a - 16 + 16$	$= 12a - 16 + 16$	Add 16.
$12a$	$= 12a$	Combine terms.
$12a - 12a$	$= 12a - 12a$	Subtract 12a.
0	$= 0$	True

When both sides of an equation are the same, it is called an identity.

Any real number will work, so the solution set is {all real numbers} (not just {0}).

Now Try This ▶▶▶

Choose three problems from different sections of this chapter, and work them on study cards. Be sure you don't just choose the easiest problems.

Why Are Study Cards Brain Friendly?

First, making the study cards is an active technique. You have to make decisions about what is most important and how to put it on a card. This kind of thinking is more involved than just memorizing, and as a result, you will understand the concepts better and remember them longer.

Second, the cards are visually appealing (if you write neatly and try some color). You will remember a visual image longer and may even be able to "picture in your mind" how your cards look. This will help you during tests.

Third, because study cards are small and portable, you can review them easily whenever you have a few minutes. Even while you're waiting for a bus or have a few minutes between classes, you can take out your cards and read over them. After a while, the information will become automatic and easier to remember.

Summary Exercises on Solving Linear Equations

This section of miscellaneous linear equations provides practice in solving all the types introduced in **Sections 2.1–2.3.** Refer to the examples in these sections to review the various solution methods.

Solve each equation, and check your solution.

1. $a + 2 = -3$
$\{-5\}$

2. $2m + 8 = 16$
$\{4\}$

3. $16.5k = -84.15$
$\{-5.1\}$

4. $-x = -25$
$\{25\}$

5. $\frac{4}{5}x = -20$
$\{-25\}$

6. $9x - 7x = -12$
$\{-6\}$

7. $5x - 9 = 4(x - 3)$
$\{-3\}$

8. $\frac{a}{-2} = 8$
$\{-16\}$

9. $-3(t - 5) + 2(7 + 2t) = 36$
$\{7\}$

10. $\frac{2}{3}x + 8 = \frac{1}{4}x$
$\left\{-\frac{96}{5}\right\}$

11. $0.08x + 0.06(x + 9) = 1.24$
$\{5\}$

12. $x - 16.2 = 7.5$
$\{23.7\}$

13. $4x + 2(3 - 2x) = 6$
{all real numbers}

14. $-0.3x + 2.1(x - 4) = -6.6$
$\{1\}$

15. $-x = 16$
$\{-16\}$

16. $3(m + 5) - 1 + 2m = 5(m + 2)$
$\emptyset$

17. $10m - (5m - 9) = 39$

{6}

18. $7(p - 2) + p = 2(p + 2)$

{3}

19. $-2t + 5t - 9 = 3(t - 4) - 5$

∅

20. $-9z = -21$

$\left\{\frac{7}{3}\right\}$

21. $0.02(50) + 0.08r = 0.04(50 + r)$

{25}

22. $2.3x + 13.7 = 1.3x + 2.9$

{−10.8}

23. $2(3 + 7x) - (1 + 15x) = 2$

{3}

24. $6q - 9 = 12 + 3q$

{7}

25. $2(5 + 3x) = 3(x + 1) + 13$

{2}

26. $r + 9 + 7r = 4(3 + 2r) - 3$

{all real numbers}

27. $\frac{5}{6}x + \frac{1}{3} = 2x + \frac{2}{3}$

$\left\{-\frac{2}{7}\right\}$

28. $0.06x + 0.09(15 - x) = 0.07(15)$

{10}

29. $\frac{3}{4}(a - 2) - \frac{1}{3}(5 - 2a) = -2$

$\left\{\frac{14}{17}\right\}$

30. $2 - (m + 4) = 3m + 8$

$\left\{-\frac{5}{2}\right\}$

31. $5.2x - 4.6 - 7.1x = -2.1 - 1.9x - 2.5$

{all real numbers}

32. $9(2m - 3) - 4(5 + 3m) = 5(4 + m) - 3$

{64}

2.4 An Introduction to Applications of Linear Equations

OBJECTIVES

1. **Learn the six steps for solving applied problems.**
2. **Solve problems involving unknown numbers.**
3. **Solve problems involving sums of quantities.**
4. **Solve problems involving supplementary and complementary angles.**
5. **Solve problems involving consecutive integers.**

OBJECTIVE 1 Learn the six steps for solving applied problems. We now look at how algebra is used to solve applied problems. Some of the problems may seem contrived, but the skills you develop in solving them will help you solve more realistic problems in chemistry, biology, business, and other fields.

While there is not one specific method that enables you to solve all kinds of applied problems, we suggest the following six-step method.*

Solving an Applied Problem

Step 1 **Read** the problem, several times if necessary, until you *understand* what is given and what is to be found.

Step 2 **Assign a variable** to represent the unknown value, using diagrams or tables as needed. Write down what the variable represents. Express any other unknown values in terms of the variable.

Step 3 **Write an equation** using the variable expression(s).

Step 4 **Solve** the equation.

Step 5 **State the answer.** Does it seem reasonable?

Step 6 **Check** the answer in the words of the *original* problem.

OBJECTIVE 2 Solve problems involving unknown numbers.

Problem-Solving Hint

The third step in solving an applied problem is often the hardest. To translate the problem into an equation, write the given phrases as mathematical expressions. Replace any words that mean *equal* or *same* with an = sign. Other forms of the verb "to be," such as *is, are, was,* and *were,* also translate as an = sign. The = sign leads to an equation to be solved.

EXAMPLE 1 Finding the Value of an Unknown Number

The product of 4, and a number decreased by 7, is 100. What is the number?

Step 1 **Read** the problem carefully. We are asked to find a number.

Step 2 **Assign a variable** to represent the unknown quantity. In this problem, we are asked to find a number, so we write

$$\text{Let } x = \text{the number.}$$

There are no other unknown quantities to find.

Continued on Next Page

*__Appendix B__ *Strategies for Problem Solving* introduces additional methods and tips for solving applied problems.

1 Use the six steps to solve the problem. Give the equation, using x as the variable, and give the answer.

If 5 is added to the product of 9 and a number, the result is 19 less than the number. Find the number.

Step 3 **Write an equation.**

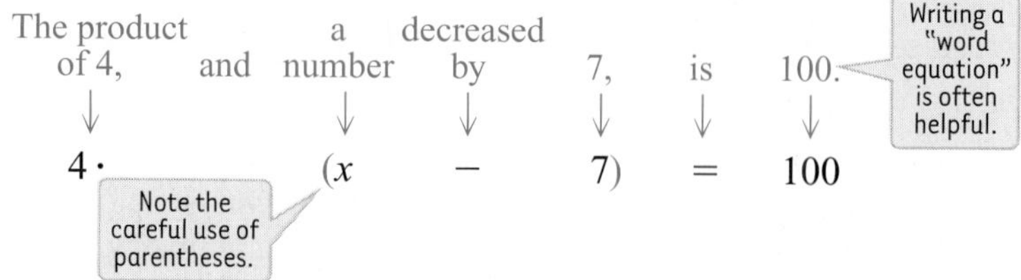

Because of the commas in the given problem, writing the equation as $4x - 7 = 100$ is *incorrect.* The equation $4x - 7 = 100$ corresponds to the statement "The product of 4 and a number, decreased by 7, is 100."

Step 4 **Solve** the equation.

$$4(x - 7) = 100$$

$$4x - 28 = 100 \quad \text{Distributive property}$$

$$4x - 28 + 28 = 100 + 28 \quad \text{Add 28.}$$

$$4x = 128 \quad \text{Combine like terms.}$$

$$\frac{4x}{4} = \frac{128}{4} \quad \text{Divide by 4.}$$

$$x = 32$$

Step 5 **State the answer.** The number is 32.

Step 6 **Check** When 32 is decreased by 7, we get $32 - 7 = 25$. If 4 is multiplied by 25, we get 100, as required. The answer, 32, is correct.

Work Problem **1** *at the Side.*

OBJECTIVE 3 Solve problems involving sums of quantities.

Problem-Solving Hint

In general, to solve problems involving sums of quantities, choose a variable to represent one of the unknowns and then ***represent the other quantity in terms of the same variable.*** (See Example 11 in **Section 2.3.**)

EXAMPLE 2 Finding Numbers of Olympic Medals

In the 2006 Winter Olympics in Torino, Italy, the United States won 11 more medals than Sweden. The two countries won a total of 39 medals. How many medals did each country win? (*Source:* U.S. Olympic Committee.)

Step 1 **Read** the problem. We are given information about the total number of medals and asked to find the number each country won.

Step 2 **Assign a variable.**

Let $x =$ the number of medals Sweden won.

Then $x + 11 =$ the number of medals the U.S. won.

Continued on Next Page

Answer

1. $9x + 5 = x - 19$; -3

Step 3 **Write an equation.**

The total	is	the number of medals Sweden won	plus	the number of medals the U.S. won.
↓	↓	↓	↓	↓
39	=	x	+	$(x + 11)$

Step 4 **Solve** the equation.

$$39 = 2x + 11 \qquad \text{Combine like terms.}$$

$$39 - \mathbf{11} = 2x + 11 - \mathbf{11} \qquad \text{Subtract 11.}$$

$$28 = 2x \qquad \text{Combine like terms.}$$

$$\frac{28}{2} = \frac{2x}{2} \qquad \text{Divide by 2.}$$

$$14 = x, \quad \text{or} \quad x = 14$$

Step 5 **State the answer.** The variable x represents the number of medals Sweden won, so Sweden won 14 medals. Then the number of medals the United States won is $x + 11 = 14 + 11 = 25$.

Step 6 **Check.** Since the United States won 25 medals and Sweden won 14, the total number of medals was $25 + 14 = 39$. Because $25 - 14 = 11$, the United States won 11 more medals than Sweden. This information agrees with what is given in the problem, so the answer checks.

Problem-Solving Hint

The problem in Example 2 could also be solved by letting x represent the number of medals the United States won. Then $x - 11$ would represent the number of medals Sweden won. The equation would be

$$39 = x + (x - 11).$$

The solution of this equation is 25, which is the number of U. S. medals. The number of Swedish medals would be $25 - 11 = 14$. ***The answers are the same,*** whichever approach is used, even though the equation and its solution are different.

Work Problem 2 *at the Side.* ▶

2 Solve the problem.
In the 2006 Winter Olympics in Torino, Italy, Canada won 5 more medals than Norway. The two countries won a total of 43 medals. How many medals did each country win? (*Source:* U.S. Olympic Committee.)

EXAMPLE 3 **Analyzing a Gasoline/Oil Mixture**

A lawn trimmer uses a mixture of gasoline and oil. The mixture contains 16 oz of gasoline for each ounce of oil. If the tank holds 68 oz of the mixture, how many ounces of oil and how many ounces of gasoline does it require when it is full?

Step 1 **Read** the problem. We must find how many ounces of oil and gasoline are needed to fill the tank.

Step 2 **Assign a variable.**

Let $x =$ the number of ounces of oil required.

Then $16x =$ the number of ounces of gasoline required.

Continued on Next Page

ANSWER

2. Norway: 19; Canada: 24

A diagram like the following is sometimes helpful.

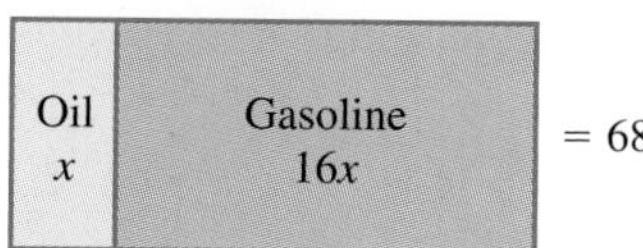

Step 3 **Write an equation.**

Amount of gasoline	plus	amount of oil	is	total amount in tank.
↓	↓	↓	↓	↓
$16x$	$+$	x	$=$	68

Step 4 **Solve.**

$$17x = 68 \quad \text{Combine like terms.}$$

$$\frac{17x}{17} = \frac{68}{17} \quad \text{Divide by 17.}$$

$$x = 4$$

Step 5 **State the answer.** The lawn trimmer requires 4 oz of oil and $16(4) = 64$ oz of gasoline when full.

Step 6 **Check.** Since $4 + 64 = 68$, and 64 is 16 times 4, the answer checks.

◀ *Work Problem* 3 *at the Side.*

Problem-Solving Hint

Sometimes it is necessary to find three unknown quantities in an applied problem. Frequently the three unknowns are compared in *pairs.* When this happens, it is usually best to ***let the variable represent the unknown found in both pairs.*** The next example illustrates this.

EXAMPLE 4 **Dividing a Board into Pieces**

The instructions for a woodworking project call for three pieces of wood. The longest piece must be twice the length of the middle-sized piece, and the shortest piece must be 10 in. shorter than the middle-sized piece. Maria Gonzales has a board 70 in. long that she wishes to use. How long must each piece be?

Step 1 **Read** the problem. Three lengths must be found.

Step 2 **Assign a variable.** Since the middle-sized piece appears in both pairs of comparisons, let x represent the length, in inches, of the middle-sized piece. We have

x = the length of the middle-sized piece,

$2x$ = the length of the longest piece, and

$x - 10$ = the length of the shortest piece.

Continued on Next Page

3 Solve the problem.

At a meeting of the local coin club, each member brought two nonmembers. If a total of 27 people attended, how many were members and how many were nonmembers?

Meeting

Members x	Nonmembers $2x$	= 27

ANSWER

3. members: 9; nonmembers: 18

A sketch is helpful here. See Figure 2.

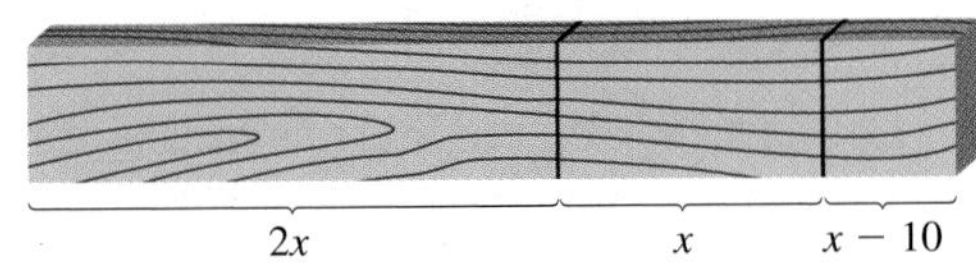

Figure 2

Step 3 **Write an equation.**

Longest	plus	middle-sized	plus	shortest	is	total length.
↓	↓	↓	↓	↓	↓	↓
$2x$	$+$	x	$+$	$(x - 10)$	$=$	70

Step 4 **Solve.**

$$4x - 10 = 70 \quad \text{Combine like terms.}$$

$$4x - 10 + \mathbf{10} = 70 + \mathbf{10} \quad \text{Add 10.}$$

$$4x = 80 \quad \text{Combine like terms.}$$

$$\frac{4x}{4} = \frac{80}{4} \quad \text{Divide by 4.}$$

$$x = 20$$

Step 5 **State the answer.** The middle-sized piece is 20 in. long, the longest piece is $2(20) = 40$ in. long, and the shortest piece is $20 - 10 = 10$ in. long.

Step 6 **Check.** The sum of the lengths is 70 in. All conditions of the problem are satisfied.

Work Problem **4** *at the Side.* ▶

OBJECTIVE 4 **Solve problems involving supplementary and complementary angles.** An angle can be measured by a unit called the degree (°), which is $\frac{1}{360}$ of a complete rotation. Two angles whose sum is 90° are said to be **complementary**, or *complements* of each other. An angle that measures 90° is a **right angle.** Two angles whose sum is 180° are said to be **supplementary**, or supplements of each other. One angle *supplements* the other to form a **straight angle** of 180°. See Figure 3.

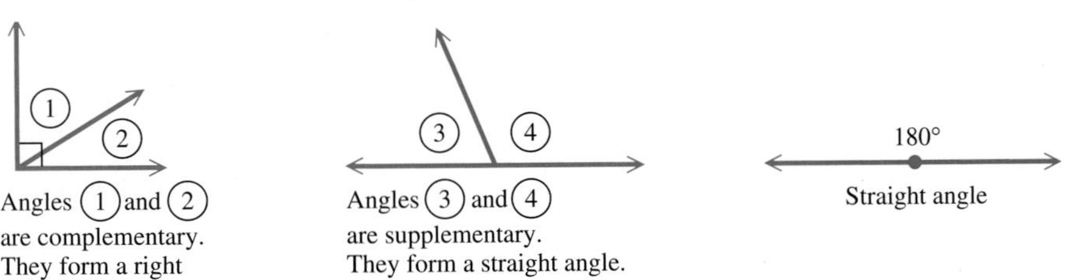

Figure 3

Problem-Solving Hint

If x represents the degree measure of an angle, then

90 − x represents the degree measure of its complement, and

180 − x represents the degree measure of its supplement.

Work Problem **5** *at the Side.* ▶

4 Solve the problem.

A piece of pipe is 50 in. long. It is cut into three pieces. The longest piece is 10 in. longer than the middle-sized piece, and the shortest piece measures 5 in. less than the middle-sized piece. Find the lengths of the three pieces.

5 Find each angle measure.

(a) Fill in the blank below the figure. Then find the complement of an angle that measures 26°.

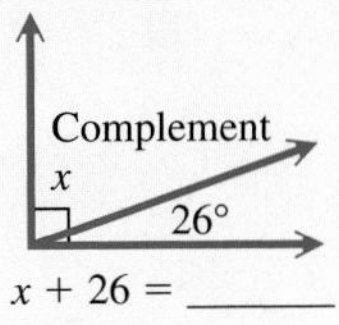

(b) Fill in the blank below the figure. Then find the supplement of an angle that measures 92°.

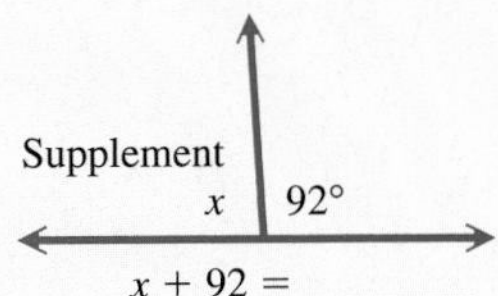

ANSWERS

4. longest: 25 in.; middle: 15 in.; shortest: 10 in.
5. **(a)** 90; 64° **(b)** 180; 88°

6 Solve the problem.
Find the measure of an angle whose complement is eight times its measure.

EXAMPLE 5 Finding the Measure of an Angle

Find the measure of an angle whose complement is five times its measure.

Step 1 **Read** the problem. We must find the measure of an angle, given information about the measure of its complement.

Step 2 **Assign a variable.**

Let $x =$ the degree measure of the angle.

Then $90 - x =$ the degree measure of its complement.

Step 3 **Write an equation.**

Measure of the complement	is	5 times the measure of the angle.
↓	↓	↓
$90 - x$	$=$	$5x$

Step 4 **Solve.**

$$90 - x + x = 5x + x \qquad \text{Add } x.$$

$$90 = 6x \qquad \text{Combine like terms.}$$

$$\frac{90}{6} = \frac{6x}{6} \qquad \text{Divide by 6.}$$

$$15 = x, \quad \text{or} \quad x = 15$$

Step 5 **State the answer.** The measure of the angle is 15°.

Step 6 **Check.** If the angle measures 15°, then its complement measures $90° - 15° = 75°$, which is equal to five times 15°, as required.

◀ *Work Problem* **6** *at the Side.*

EXAMPLE 6 Finding the Measure of an Angle

Find the measure of an angle whose supplement is 10° more than twice its complement.

Step 1 **Read** the problem. We are to find the measure of an angle, given information about its complement and its supplement.

Step 2 **Assign a variable.**

Let $x =$ the degree measure of the angle.

Then $90 - x =$ the degree measure of its complement;

$180 - x =$ the degree measure of its supplement.

We can visualize this information using a sketch. See Figure 4.

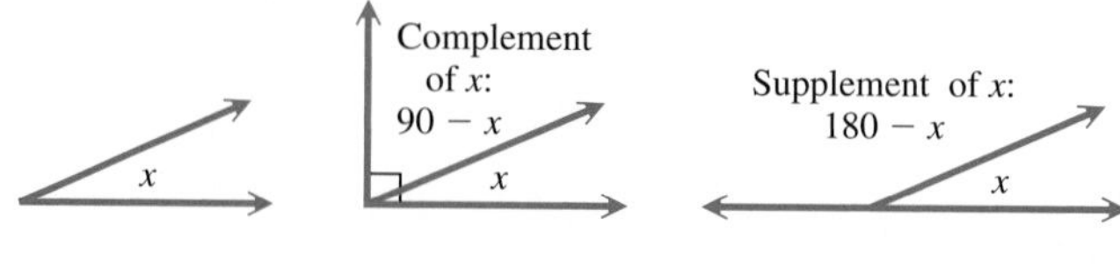

Figure 4

Continued on Next Page

Answer

6. 10°

Step 3 **Write an equation.**

Supplement	is	10	more than	twice		its complement.
↓	↓	↓	↓	↓		↓
$180 - x$	$=$	10	$+$	2	$\cdot$	$(90 - x)$

Be sure to use parentheses here.

Step 4 **Solve.**

$$180 - x = 10 + 2(90 - x)$$

$$180 - x = 10 + 180 - 2x \quad \text{Distributive property}$$

$$180 - x = 190 - 2x \quad \text{Combine like terms.}$$

$$180 - x + \mathbf{2x} = 190 - 2x + \mathbf{2x} \quad \text{Add } 2x.$$

$$180 + x = 190 \quad \text{Combine like terms.}$$

$$180 + x - \mathbf{180} = 190 - \mathbf{180} \quad \text{Subtract 180.}$$

$$x = 10$$

Step 5 **State the answer.** The measure of the angle is 10°.

Step 6 **Check.** The complement of 10° is 80° and the supplement of 10° is 170°. Also, 170° is equal to 10° more than twice 80° (that is, $170 = 10 + 2(80)$ is true). Therefore, the answer is correct.

Work Problem 7 at the Side. ▶

7 Solve the problem.

Find the measure of an angle such that twice its complement is 30° less than its supplement.

OBJECTIVE 5 Solve problems involving consecutive integers. Two integers that differ by 1 are called **consecutive integers.** For example, 3 and 4, 6 and 7, and −2 and −1 are pairs of consecutive integers. ***In general, if x represents an integer, x + 1 represents the next greater consecutive integer.***

8 Solve the problem.

Two back-to-back page numbers in this book have a sum of 569. What are the page numbers?

EXAMPLE 7 Finding Consecutive Integers

Two pages that face each other in this book have 317 as the sum of their page numbers. What are the page numbers?

Step 1 **Read** the problem. Because the two pages face each other, they must have page numbers that are consecutive integers.

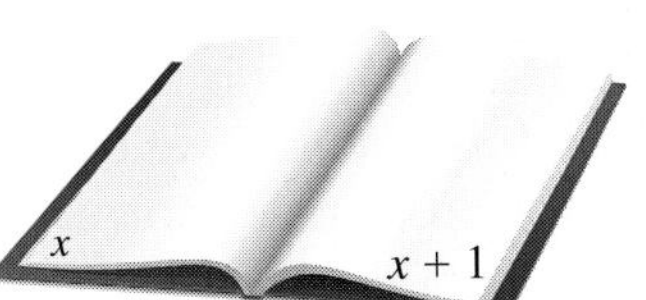

Step 2 **Assign a variable.**

Let $x =$ the lesser page number.

Then $x + 1 =$ the greater page number.

Step 3 **Write an equation.** The sum of the page numbers is 317, so

$$x + (x + 1) = 317.$$

Step 4 **Solve.**

$$2x + 1 = 317 \quad \text{Combine like terms.}$$

$$2x = 316 \quad \text{Subtract 1.}$$

$$x = 158 \quad \text{Divide by 2.}$$

Step 5 **State the answer.** The lesser page number is 158, and the greater page number is $158 + 1 = 159$. (Your book is opened to these two pages!)

Step 6 **Check.** The sum of 158 and 159 is 317. The answer is correct.

Work Problem 8 at the Side. ▶

ANSWERS

7. 30°
8. 284, 285

9 Solve the problem.

Find two consecutive even integers such that six times the lesser added to the greater gives a sum of 86.

Consecutive *even* integers, such as 8 and 10, differ by 2. Similarly, **consecutive *odd* integers,** such as 9 and 11, also differ by two. In general, if x represents an even integer, $x + 2$ represents the next greater consecutive even integer. The same holds true for odd integers; that is, if x is an odd integer, $x + 2$ is the next greater odd integer. In this book, we list consecutive integers in increasing order when solving applications.

Problem-Solving Hint

When solving consecutive integer problems, if $x =$ the lesser integer, then for any

two consecutive integers, use	$\mathbf{x},$	$\mathbf{x + 1;}$
two consecutive *even* integers, use	$\mathbf{x},$	$\mathbf{x + 2;}$
two consecutive *odd* integers, use	$\mathbf{x},$	$\mathbf{x + 2.}$

In Example 8, see if you can identify the six steps.

EXAMPLE 8 Finding Consecutive Odd Integers

If the lesser of two consecutive odd integers is doubled, the result is 7 more than the greater of the two integers. Find the two integers.

Let x be the lesser integer. Since the two numbers are consecutive *odd* integers, then $x + 2$ is the greater. Now write an equation.

If the lesser is doubled,	the result is	7	more than	the greater.
↓	↓	↓	↓	↓
$2x$	$=$	7	$+$	$x + 2$

$$2x = 9 + x \quad \text{Combine like terms.}$$

$$x = 9 \quad \text{Subtract } x.$$

The lesser integer is 9 and the greater is $9 + 2 = 11$. To check, we see that when 9 is doubled, we get 18, which is 7 more than the greater odd integer, 11. The answers are correct.

◀ Work Problem **9** at the Side.

Answer

9. 12, 14

2.4 Exercises

FOR EXTRA HELP MyMathLab Math XL PRACTICE WATCH DOWNLOAD READ REVIEW

1. In your own words, write the general procedure for solving applications as outlined in this section.

   The procedure should include the following steps: read the problem carefully; assign a variable to represent the unknown to be found; write down variable expressions for any other unknown quantities; translate into an equation; solve the equation; state the answer; check your solution.

2. List some of the words that translate as "=" when writing an equation to solve an applied problem.

   Some examples are *is, are, was,* and *were*.

3. Suppose that a problem requires you to find the number of cars on a dealer's lot. Which one of the following would not be a reasonable answer? Justify your answer.

   A. 0 **B.** 45 **C.** 1 **D.** $6\frac{1}{2}$ **D; There cannot be a fractional number of cars.**

4. Suppose that a problem requires you to find the number of hours a light bulb is on during a day. Which one of the following would not be a reasonable answer? Justify your answer.

   A. 0 **B.** 4.5 **C.** 13 **D.** 25 **D; A day cannot have more than 24 hr.**

5. Suppose that a problem requires you to find the distance traveled in miles. Which one of the following would not be a reasonable answer? Justify your answer.

   A. -10 **B.** 1.8 **C.** $10\frac{1}{2}$ **D.** 50 **A; Distance cannot be negative.**

6. Suppose that a problem requires you to find the time in minutes. Which one of the following would not be a reasonable answer? Justify your answer.

   A. 0 **B.** 10.5 **C.** -5 **D.** 90 **C; Time cannot be negative.**

Solve each problem. See Example 1.

7. The product of 8, and a number increased by 6, is 104. What is the number? **7**

8. The product of 5, and 3 more than twice a number, is 85. What is the number? **7**

9. Two less than three times a number is equal to 14 more than five times the number. What is the number? **-8**

10. Nine more than five times a number is equal to 3 less than seven times the number. What is the number? **6**

11. If 2 is subtracted from a number and this difference is tripled, the result is 6 more than the number. Find the number. **6**

12. If 3 is added to a number and this sum is doubled, the result is 2 more than the number. Find the number. **-4**

13. The sum of three times a number and 7 more than the number is the same as the difference between -11 and twice the number. What is the number? **-3**

14. If 4 is added to twice a number and this sum is multiplied by 2, the result is the same as if the number is multiplied by 3 and 4 is added to the product. What is the number? **-4**

Solve each problem. See Example 2.

15. The number of drive-in movie screens has declined steadily in the United States. Pennsylvania and Ohio had the most remaining drive-in movie screens in 2007. Pennsylvania had 2 more screens than Ohio, and there were 68 screens total in the two states. How many drive-in movie screens remained in each state? (*Source:* Drive-Ins.com)

Pennsylvania: 35 screens; Ohio: 33 screens

16. Two of the most watched episodes in television were the final episode of *M*A*S*H,* broadcast on February 23, 1983, and the "Who Shot J. R.?" episode of *Dallas*, broadcast on November 21, 1980. The total number of viewers for these two episodes was about 91 million, with 9 million more people watching the *M*A*S*H* episode than the *Dallas* one. How many people watched each show? (*Source:* Nielsen Media Research.) ***M*A*S*H*:* 50 million; *Dallas:* 41 million**

17. During the 109th session (2005–2006), the U.S. Senate had a total of 99 Democrats and Republicans. There were 11 more Republicans than Democrats. How many Democrats and Republicans were there in the Senate? (*Source: World Almanac and Book of Facts.*)

Democrats: 44; Republicans: 55

18. The total number of Democrats and Republicans in the U.S. House of Representatives during the 109th session was 434. There were 30 more Republicans than Democrats. How many members of each party were there? (*Source: World Almanac and Book of Facts.*)

Democrats: 202; Republicans: 232

19. The Police and Kenny Chesney had the two top-grossing North American concert tours in 2007, together generating $204.3 million in ticket sales. If Kenny Chesney took in $62.1 million less than The Police, how much did each tour generate? (*Source:* Pollstar.)

The Police: $133.2 million; Kenny Chesney: $71.1 million

20. The Toyota Camry was the top-selling passenger car in the United States in 2005, followed by the Honda Accord. Honda Accord sales were 65 thousand less than Toyota Camry sales, and 803 thousand of these two cars were sold. How many of each make of car were sold? (*Source:* www.wikipedia.org)

Camry: 434 thousand; Accord: 369 thousand

21. In the 2006–2007 NBA regular season, the Phoenix Suns won 19 more than twice as many games as they lost. The Suns played 82 games. How many wins and losses did the team have? (*Source:* nba.com)

wins: 61; losses: 21

22. In the 2007 regular baseball season, the Boston Red Sox won 36 less than twice as many games as they lost. They played 162 regular season games. How many wins and losses did the team have? (*Source:* www.mlb.com)

wins: 96; losses: 66

23. A one-cup serving of orange juice contains 3 mg less than four times the amount of vitamin C as a one-cup serving of pineapple juice. Servings of the two juices contain a total of 122 mg of vitamin C. How many milligrams of vitamin C are in a serving of each type of juice? (*Source:* U.S. Agriculture Department.)

orange: 97 mg; pineapple: 25 mg

24. A one-cup serving of pineapple juice has 9 more than three times as many calories as a one-cup serving of tomato juice. Servings of the two juices contain a total of 173 calories. How many calories are in a serving of each type of juice? (*Source:* U.S. Agriculture Department.)

tomato: 41 calories; pineapple: 132 calories

Solve each problem. See Example 3.

25. The value of a "Mint State-63" (uncirculated) 1950 Jefferson nickel minted at Denver is $\frac{4}{3}$ the value of a similar condition 1944 nickel minted at Philadelphia. Together, the value of the two coins is \$28.00. What is the value of each coin? (*Source:* Yeoman, R., *A Guide Book of United States Coins,* edited by K. Bressett, 61st edition, 2008.)

1950 Denver nickel: \$16.00; 1944 Philadelphia nickel: \$12.00

26. In one day, a store sold $\frac{8}{5}$ as many DVDs as CDs. The total number of DVDs and CDs sold that day was 273. How many DVDs were sold?

168 DVDs

27. The world's largest taco was made in the city of Mexicali, Mexico. The taco contained approximately 1 kg of onion for every 6.6 kg of grilled steak. The total weight of these two ingredients was 617.6 kg. To the nearest tenth of a kilogram, how many kilograms of onions and how many kilograms of grilled steak were used to make the taco? (*Source: Guinness World Records.*)

onions: 81.3 kg; grilled steak: 536.3 kg

28. The world's most populous countries are China and India. As of mid-2005, the combined population of these two countries was estimated at 2.4 billion. If there were about $\frac{4}{5}$ as many people living in India as China, what was the population of each country, to the nearest tenth of a billion? (*Source:* U.S. Census Bureau.)

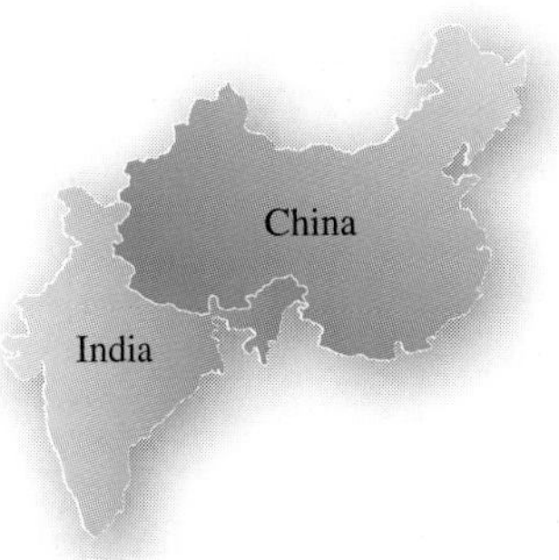

China: 1.3 billion; India: 1.1 billion

29. U.S. five-cent coins are made from a combination of two metals: nickel and copper. For every pound of nickel, 3 lb of copper are used. How many pounds of copper would be needed to make 560 lb of five-cent coins? (*Source:* The United States Mint.)

420 lb

30. A bakery makes a special whole-grain bread using two kinds of flour: whole wheat and rye. The recipe for this bread calls for 1 oz of rye flour for every 4 oz of whole-wheat flour. How many ounces of each kind of flour should be used to make a loaf of bread weighing 32 oz?

whole wheat: 25.6 oz; rye: 6.4 oz

Solve each problem. See Example 4.

31. Al Moser, an office manager, books airline tickets for business trips that employees need to make. In one week, he booked 55 tickets. He booked 7 more tickets on American Airlines than United Airlines. On Southwest Airlines, he booked 4 more than twice as many tickets as on United. How many tickets did he book on each airline?

American: 18; United: 11; Southwest: 26

32. Lauren Morse, a mathematics textbook editor, works 7.5 hr a day. She spent a recent day making telephone calls, writing e-mails, and attending meetings. She spent twice as much time attending meetings as making telephone calls and 0.5 hr longer writing e-mails than making telephone calls. How many hours did she spend on each task?

telephone calls: 1.75 hr; e-mails: 2.25 hr; meetings: 3.5 hr

33. The United States earned 103 medals at the 2004 Summer Olympics in Athens. The number of silver medals earned was 4 more than the number of gold medals. The number of bronze medals earned was 6 less than the number of gold medals. How many of each kind of medal did the United States earn? (*Source: The Gazette,* August 30, 2004.)

gold: 35; silver: 39; bronze: 29

34. Nagaraj Nanjappa has a party-length submarine sandwich 59 in. long. He wants to cut it into three pieces so that the middle piece is 5 in. longer than the shortest piece and the shortest piece is 9 in. shorter than the longest piece. How long should the three pieces be?

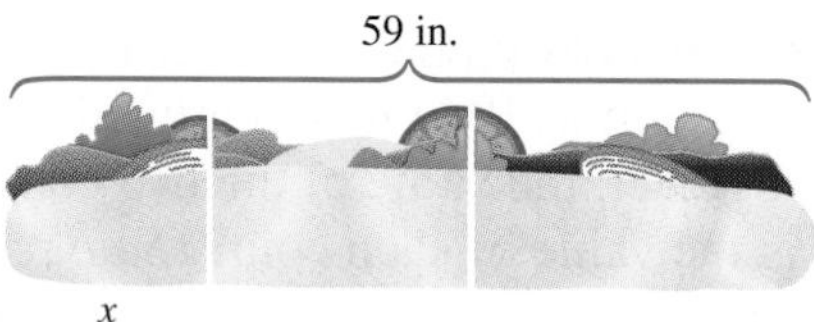

shortest piece: 15 in.; middle piece: 20 in.; longest piece: 24 in.

35. Venus is 31.2 million mi farther from the sun than Mercury, while Earth is 57 million mi farther from the sun than Mercury. If the total of the distances from these three planets to the sun is 196.2 million mi, how far away from the sun is Mercury? (All distances given here are mean (*average*) distances.) (*Source: The New York Times Almanac.*)

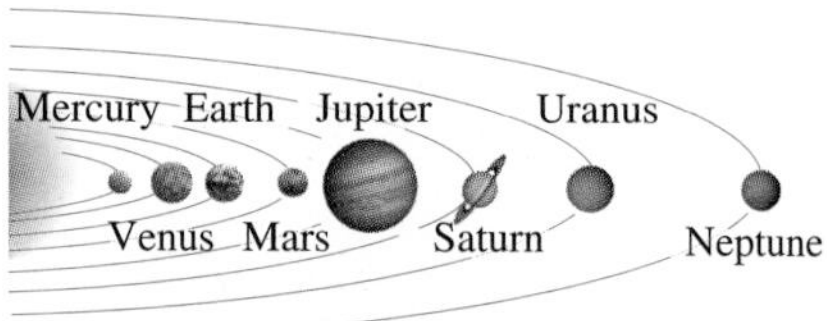

36 million mi

36. Saturn, Jupiter, and Uranus together have a total of 137 known satellites (moons). Jupiter has 16 more satellites than Saturn, and Uranus has 20 fewer satellites than Saturn. How many known satellites does Uranus have? (*Source: The New York Times Almanac.*) **27**

37. The sum of the measures of the angles of any triangle is 180°. In triangle *ABC,* angles *A* and *B* have the same measure, while the measure of angle *C* is 60° greater than each of *A* and *B.* What are the measures of the three angles?

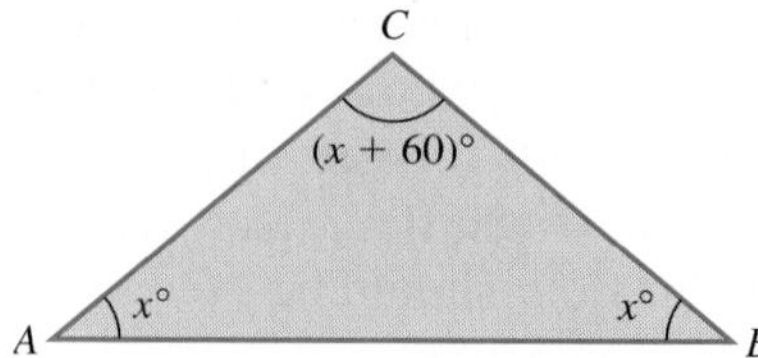

***A* and *B*: 40°; *C*: 100°**

38. In triangle *ABC,* the measure of angle *A* is 141° more than the measure of angle *B.* The measure of angle *B* is the same as the measure of angle *C.* Find the measure of each angle. (*Hint:* See Exercise 37.)

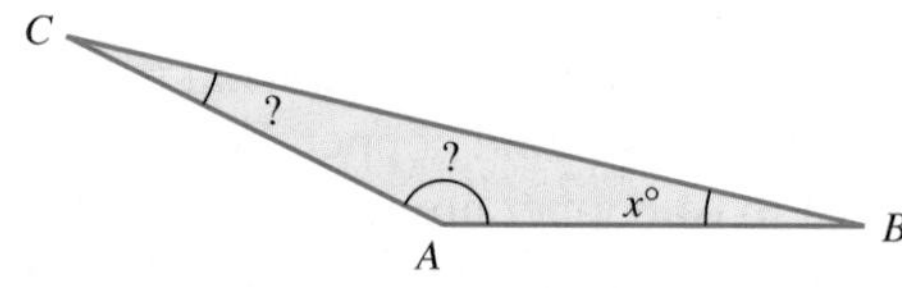

***A*: 154°; *B* and *C*: 13°**

Use the concepts of this section to answer each question.

39. Is there an angle that is equal to its supplement? Is there an angle that is equal to its complement? If the answer is yes to either question, give the measure of the angle.

yes, 90°; yes, 45°

40. If x represents an integer, how can you express the next smaller consecutive integer in terms of x? The next smaller even integer?

$x - 1$; $x - 2$

Solve each problem. See Examples 5 and 6.

41. Find the measure of an angle whose complement is four times its measure. **18°**

42. Find the measure of an angle whose complement is five times its measure. **15°**

43. Find the measure of an angle whose supplement is eight times its measure. **20°**

44. Find the measure of an angle whose supplement is three times its measure. **45°**

45. Find the measure of an angle whose supplement measures 39° more than twice its complement.

39°

46. Find the measure of an angle whose supplement measures 38° less than three times its complement.

26°

47. Find the measure of an angle such that the difference between the measures of its supplement and three times its complement is 10°. **50°**

48. Find the measure of an angle such that the sum of the measures of its complement and its supplement is 160°. **55°**

Solve each problem. See Examples 7 and 8.

49. The numbers on two consecutively numbered gym lockers have a sum of 137. What are the locker numbers? **68, 69**

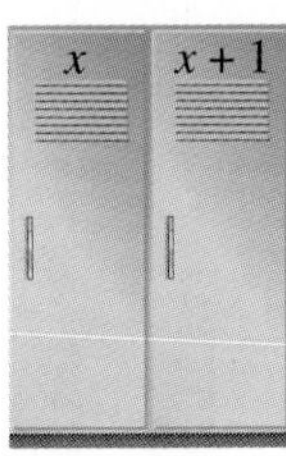

50. The sum of two consecutive check numbers is 357. Find the numbers. **178, 179**

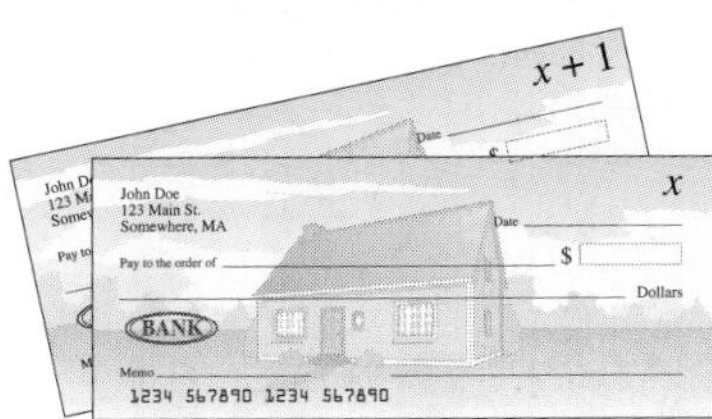

51. Two pages that are back-to-back in this book have 293 as the sum of their page numbers. What are the page numbers? **146, 147**

52. Two houses on the same side of the street have house numbers that are consecutive even integers. The sum of the integers is 58. What are the two house numbers? **28, 30**

53. Find two consecutive even integers such that the lesser added to three times the greater gives a sum of 46. **10, 12**

54. Find two consecutive odd integers such that twice the greater is 17 more than the lesser. **13, 15**

55. When the lesser of two consecutive integers is added to three times the greater, the result is 43. Find the integers. **10, 11**

56. If five times the lesser of two consecutive integers is added to three times the greater, the result is 59. Find the integers. **7, 8**

57. If the sum of three consecutive even integers is 60, what is the first of the three even integers? (*Hint:* If x and $x + 2$ represent the first two consecutive even integers, how would you represent the third consecutive even integer?) **18**

58. If the sum of three consecutive odd integers is 69, what is the third of the three odd integers? **25**

Apply the ideas of this section to solve Exercises 59 and 60, which are based on the graphs.

59. In 2003, federal funding for Head Start programs increased by \$0.13 billion from the previous year. The increase from 2003 to 2004 was \$0.10 billion. Over the three-year period 2002–2004, the total funding was \$19.98 billion. What was federal Head Start funding for each of these years? (*Source:* U.S. Department of Health and Human Services.)

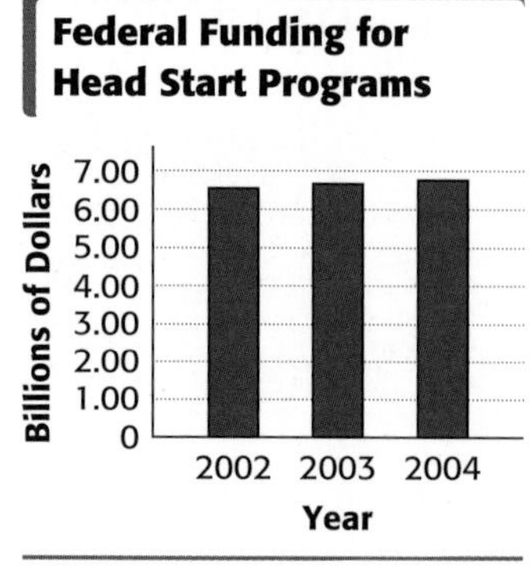

2002: \$6.54 billion; 2003: \$6.67 billion; 2004: \$6.77 billion

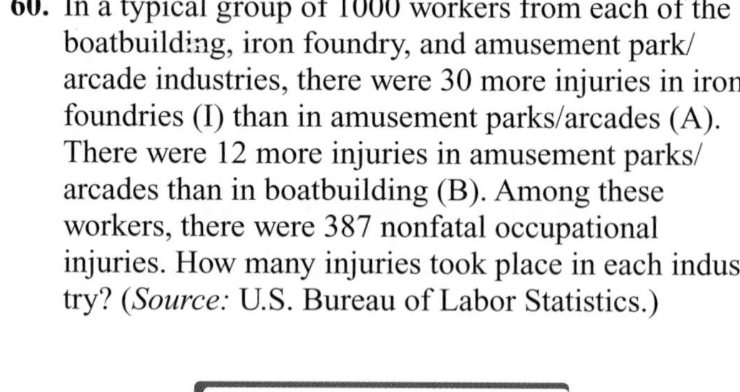

60. In a typical group of 1000 workers from each of the boatbuilding, iron foundry, and amusement park/arcade industries, there were 30 more injuries in iron foundries (I) than in amusement parks/arcades (A). There were 12 more injuries in amusement parks/arcades than in boatbuilding (B). Among these workers, there were 387 nonfatal occupational injuries. How many injuries took place in each industry? (*Source:* U.S. Bureau of Labor Statistics.)

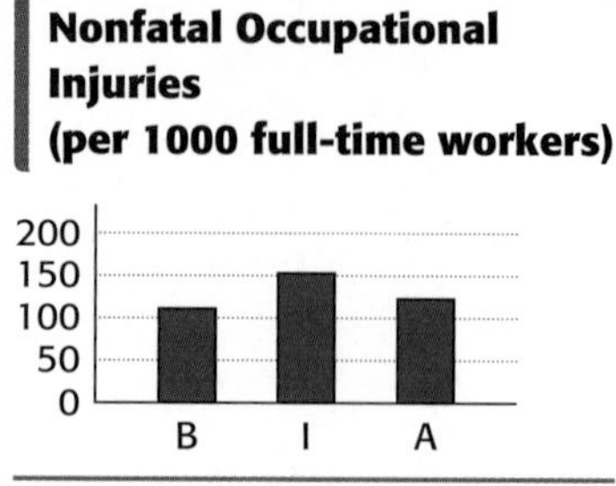

boatbuilding: 111; iron foundries: 153; amusement parks/arcades: 123

2.5 Formulas and Additional Applications from Geometry

A **formula** is an equation in which variables are used to describe a relationship. For example, formulas exist for finding perimeters and areas of geometric figures such as squares and circles, for calculating money earned on bank savings, and for converting among measurements.

$$P = 4s, \quad A = \pi r^2, \quad I = prt, \quad F = \frac{9}{5}C + 32 \qquad \text{Formulas}$$

The formulas used in this book are given on the inside covers.

OBJECTIVES

1. **Solve a formula for one variable, given the values of the other variables.**
2. **Use a formula to solve an applied problem.**
3. **Solve problems involving vertical angles and straight angles.**
4. **Solve a formula for a specified variable.**

OBJECTIVE 1 Solve a formula for one variable, given the values of the other variables. In Example 1, we use the idea of *area*. The **area** of a plane (two-dimensional) geometric figure is a measure of the surface covered by the figure.

EXAMPLE 1 Using Formulas to Evaluate Variables

Find the value of the remaining variable in each formula.

(a) $A = LW$; $A = 64$, $L = 10$

As shown in Figure 5, this formula gives the area of a rectangle with length L and width W. Substitute the given values into the formula.

$$A = LW \qquad \text{Solve for } W.$$

$$64 = 10W \qquad \text{Let } A = 64 \text{ and } L = 10.$$

$$\frac{64}{10} = \frac{10W}{10} \qquad \text{Divide by 10.}$$

$$6.4 = W$$

The width is 6.4. Since $10(6.4) = 64$, the given area, the answer checks.

(b) $A = \frac{1}{2}h(b + B)$; $A = 210$, $B = 27$, $h = 10$

This formula gives the area of a trapezoid with parallel sides of lengths b and B and distance h between the parallel sides. See Figure 6.

$$A = \frac{1}{2}h(b + B) \qquad \text{Solve for } b.$$

$$210 = \frac{1}{2}(10)(b + 27) \qquad \text{Let } A = 210, h = 10, B = 27.$$

$$210 = 5(b + 27) \qquad \text{Multiply.}$$

$$210 = 5b + 135 \qquad \text{Distributive property}$$

$$210 - 135 = 5b + 135 - 135 \qquad \text{Subtract 135.}$$

$$75 = 5b \qquad \text{Combine like terms.}$$

$$\frac{75}{5} = \frac{5b}{5} \qquad \text{Divide by 5.}$$

$$15 = b$$

The length of the shorter parallel side, b, is 15. Since $\frac{1}{2}(10)(15 + 27) = 210$, the given area, the answer checks.

Work Problem 1 at the Side. ▶

1 Find the value of the remaining variable in each formula.

(a) $I = prt$; $I = \$246$, $r = 0.06$, $t = 2$

(b) $P = 2L + 2W$; $P = 126$, $W = 25$

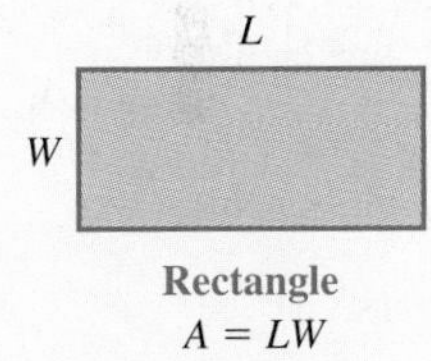

Rectangle
$A = LW$

Figure 5

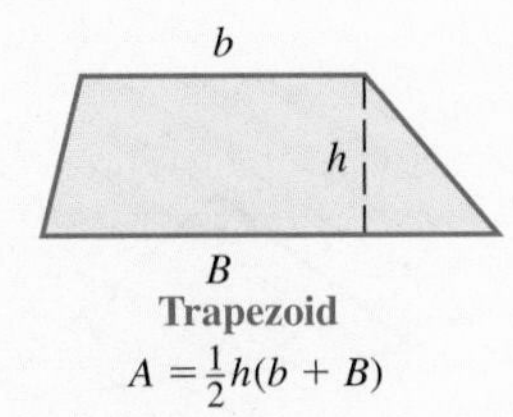

Trapezoid
$A = \frac{1}{2}h(b + B)$

Figure 6

ANSWERS

1. **(a)** $p = \$2050$ **(b)** $L = 38$

2 Solve the problem.

A farmer has 800 m of fencing material to enclose a rectangular field. The width of the field is 175 m. Find the length of the field.

OBJECTIVE 2 Use a formula to solve an applied problem. Formulas are often used to solve applied problems. ***It is a good idea to draw a sketch when a geometric figure is involved.***

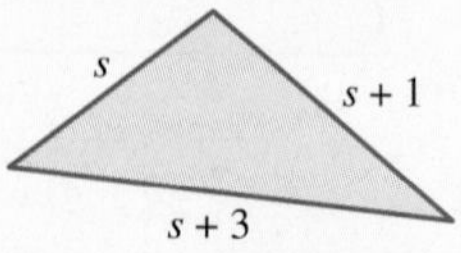

Examples 2 and 3 use the idea of *perimeter*. The **perimeter** of a plane (two-dimensional) geometric figure is the distance around the figure. For a polygon (such as a rectangle, square, or triangle), the perimeter is the sum of the lengths of the sides.

EXAMPLE 2 Finding the Dimensions of a Rectangular Yard

Kari Heen's backyard is in the shape of a rectangle. The length is 5 m less than twice the width, and the perimeter is 80 m. Find the dimensions of the yard.

Step 1 **Read** the problem. We must find the dimensions of the yard.

Step 2 **Assign a variable.** Let $W =$ the width of the lot, in meters. Since the length is 5 m less than twice the width, the length is given by $L = 2W - 5$. See Figure 7.

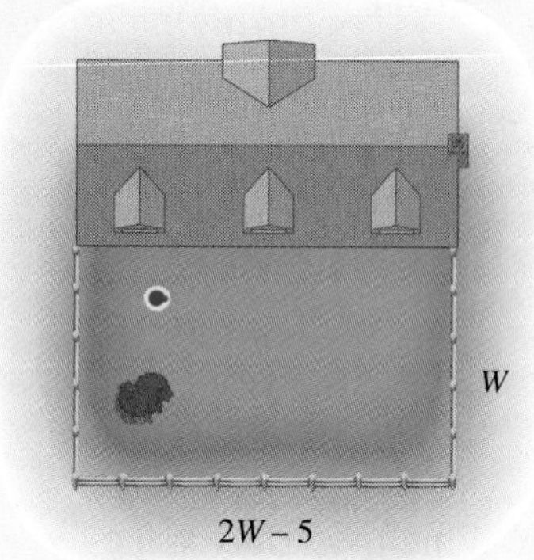

Figure 7

Step 3 **Write an equation.** The formula for the perimeter of a rectangle is

$$P = 2L + 2W.$$

$$\underset{\displaystyle 80}{\overset{\text{Perimeter}}{\downarrow}} = \underset{\displaystyle 2(2W-5)}{\overset{2\cdot\text{Length}}{\downarrow}} + \underset{\displaystyle 2W}{\overset{2\cdot\text{Width}}{\downarrow}} \qquad \text{Substitute } 2W - 5 \text{ for length } L.$$

Step 4 **Solve.**

$$80 = 4W - 10 + 2W \qquad \text{Distributive property}$$

$$80 = 6W - 10 \qquad \text{Combine like terms.}$$

$$80 + \mathbf{10} = 6W - 10 + \mathbf{10} \qquad \text{Add 10.}$$

$$90 = 6W \qquad \text{Combine like terms.}$$

$$\frac{90}{\mathbf{6}} = \frac{6W}{\mathbf{6}} \qquad \text{Divide by 6.}$$

$$15 = W$$

Step 5 **State the answer.** The width is 15 m and the length is $2(\mathbf{15}) - 5 = 25$ m.

Step 6 **Check.** If the width of the yard is 15 m and the length is 25 m, the perimeter is $2(25) + 2(15) = 50 + 30 = 80$ m, as required.

◀ Work Problem **2** at the Side.

EXAMPLE 3 Finding the Dimensions of a Triangle

The longest side of a triangle is 3 ft longer than the shortest side. The medium side is 1 ft longer than the shortest side. If the perimeter of the triangle is 16 ft, what are the lengths of the three sides?

Step 1 **Read** the problem. We are given the perimeter of a triangle and want to find the lengths of the three sides.

Step 2 **Assign a variable.**

Let $s =$ the length of the shortest side, in feet,

$s + 1 =$ the length of the medium side, in feet, and,

$s + 3 =$ the length of the longest side, in feet.

See Figure 8.

Figure 8

Continued on Next Page

ANSWER

2. 225 m

Step 3 **Write an equation.** Use the formula for the perimeter of a triangle.

$$P = a + b + c$$

$$16 = s + (s + 1) + (s + 3) \qquad \text{Substitute.}$$

Step 4 **Solve.**

$$16 = 3s + 4 \qquad \text{Combine like terms.}$$

$$12 = 3s \qquad \text{Subtract 4.}$$

$$4 = s \qquad \text{Divide by 3.}$$

Step 5 **State the answer.** Since s represents the length of the shortest side, its measure is 4 ft. The medium side measures

$$s + 1 = 4 + 1 = 5 \text{ ft},$$

and the longest side measures

$$s + 3 = 4 + 3 = 7 \text{ ft}.$$

Step 6 **Check.** The medium side, 5 ft, is 1 ft longer than the shortest side, and the longest side, 7 ft, is 3 ft longer than the shortest side. Furthermore, the perimeter is $4 + 5 + 7 = 16$ ft, as required.

  Work Problem 3 at the Side. ▶

3 Solve the problem.

The longest side of a triangle is 1 in. longer than the medium side. The medium side is 5 in. longer than the shortest side. If the perimeter is 32 in., what are the lengths of the three sides?

EXAMPLE 4 Finding the Height of a Triangular Sail

The area of a triangular sail of a sailboat is 126 ft^2. (Recall that ft^2 means "square feet.") The base of the sail is 12 ft. Find the height of the sail.

Step 1 **Read.** We must find the height of the triangular sail.

Step 2 **Assign a variable.** Let $h =$ the height of the sail, in feet. See Figure 9.

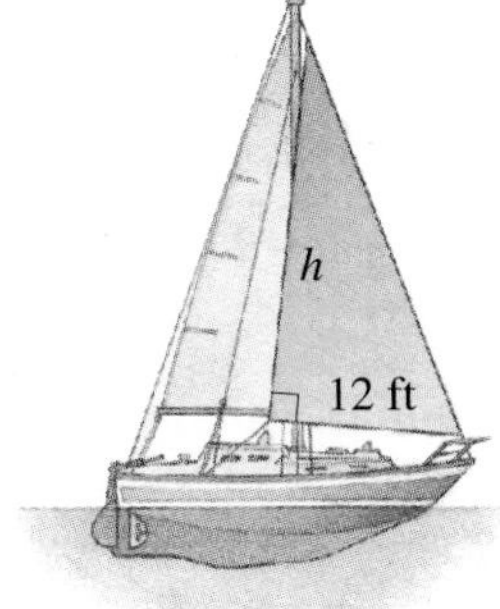

Figure 9

Step 3 **Write an equation.** The formula for the area of a triangle is $A = \frac{1}{2}bh$, where A is the area, b is the base, and h is the height.

$$A = \frac{1}{2}bh$$

$$126 = \frac{1}{2}(12)h \qquad \text{Substitute } A = 126,\ b = 12.$$

Step 4 **Solve.**

$$126 = 6h \qquad \text{Multiply.}$$

$$21 = h \qquad \text{Divide by 6.}$$

Step 5 **State the answer.** The height of the sail is 21 ft.

Step 6 **Check** to see that the values $A = 126$, $b = 12$, and $h = 21$ satisfy the formula for the area of a triangle.

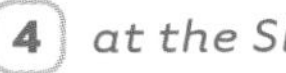 Work Problem 4 at the Side. ▶

4 Solve the problem.

The area of a triangle is 120 m^2. The height is 24 m. Find the length of the base of the triangle.

Answers

3. 7 in.; 12 in.; 13 in.
4. 10 m

OBJECTIVE 3 Solve problems involving vertical angles and straight angles. Figure 10 shows two intersecting lines forming angles that are numbered ①, ②, ③, and ④. Angles ① and ③ lie "opposite" each other. They are called **vertical angles.** Another pair of vertical angles is ② and ④. ***Vertical angles have equal measures.***

Now look at angles ① and ②. When their measures are added, we get the measure of a **straight angle,** which is **180°**. There are three other such pairs of angles: ② and ③, ③ and ④, and ① and ④.

The next example uses these ideas.

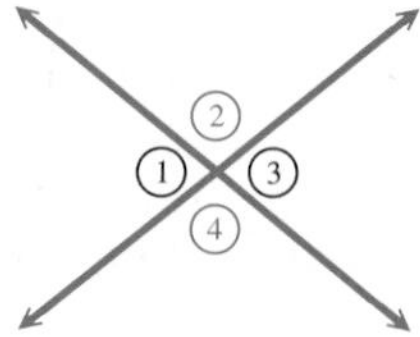

Figure 10

EXAMPLE 5 Finding Angle Measures

Refer to the appropriate figure in each part.

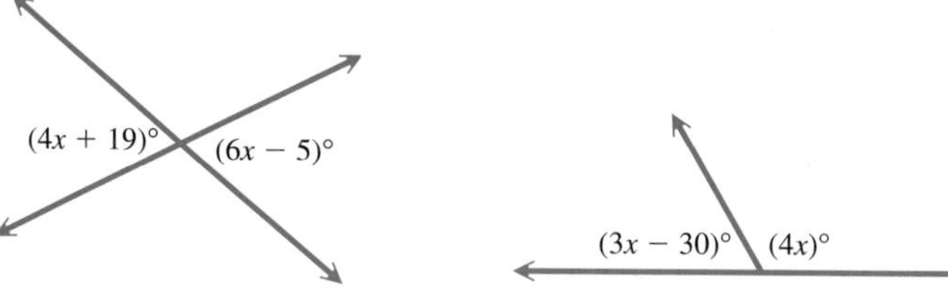

Figure 11 Figure 12

(a) Find the measure of each marked angle in Figure 11.

Since the marked angles are vertical angles, they have equal measures.

$$4x + 19 = 6x - 5 \quad \text{Set } 4x + 19 \text{ equal to } 6x - 5.$$
$$19 = 2x - 5 \quad \text{Subtract } 4x.$$
$$24 = 2x \quad \text{Add 5.}$$
$$12 = x \quad \text{Divide by 2.}$$

Don't stop here!

Since $x = 12$, one angle has measure $4(12) + 19 = 67$ degrees. The other has the same measure, since $6(12) - 5 = 67$ as well. Each angle measures 67°.

(b) Find the measure of each marked angle in Figure 12.

The measures of the marked angles must add to 180° because together they form a straight angle. (They are also supplements of each other.)

$$(3x - 30) + 4x = 180$$
$$7x - 30 = 180 \quad \text{Combine like terms.}$$
$$7x = 210 \quad \text{Add 30.}$$
$$x = 30 \quad \text{Divide by 7.}$$

To find the measures of the angles, replace x with 30 in the two expressions.

$$3x - 30 = 3(30) - 30 = 90 - 30 = 60$$
$$4x = 4(30) = 120$$

The two angle measures are 60° and 120°.

CAUTION

In Example 5, the answer is *not* the value of x. ***Remember to substitute the value of the variable into the expression given for each angle.***

◀ *Work Problem 5 at the Side.*

5 Find the measure of each marked angle.

(a)

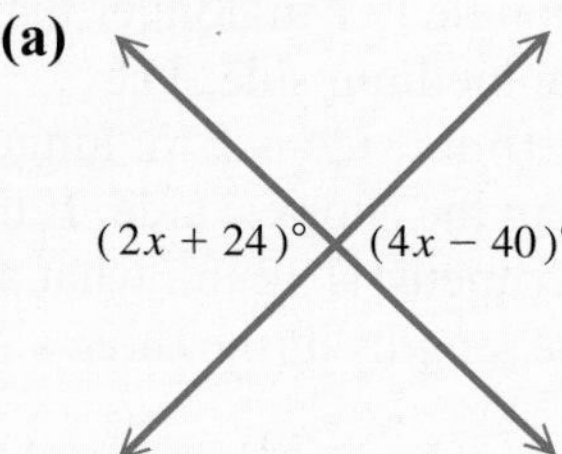

(b)

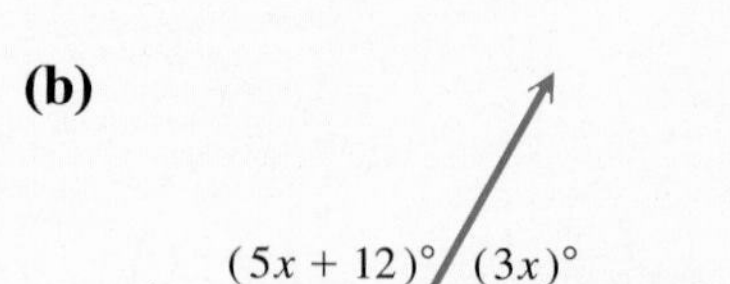

(c)

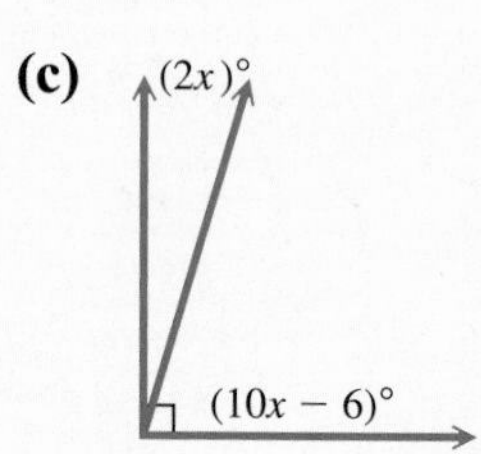

ANSWERS

5. **(a)** Both measure 88°. **(b)** 117° and 63° **(c)** 16° and 74°

OBJECTIVE 4 Solve a formula for a specified variable. Sometimes it is necessary to solve a number of problems that use the same formula. For example, a surveying class might need to solve several problems that involve the formula for the area of a rectangle, $A = LW$. Suppose that in each problem the area (A) and the length (L) of a rectangle are given, and the width (W) must be found. Rather than solving for W each time the formula is used, it would be simpler to *rewrite the formula* so that it is solved for W. This process is called **solving for a specified variable** or **solving a literal equation.**

We use the *same* steps to solve a formula for a specified variable that we used to solve an equation with just one variable. Consider the parallel reasoning to solve the following for x.

$3x + 4 = 13$		$ax + b = c$	
$3x + 4 - 4 = 13 - 4$	Subtract 4.	$ax + b - b = c - b$	Subtract b.
$3x = 9$		$ax = c - b$	
$\frac{3x}{3} = \frac{9}{3}$	Divide by 3.	$\frac{ax}{a} = \frac{c - b}{a}$	Divide by a.
$x = 3$		$x = \frac{c - b}{a}$	

In solving a formula for a specified variable, we treat the specified variable as if it were the ONLY variable in the equation, and treat the other variables as if they were numbers.

EXAMPLE 6 Solving for a Specified Variable

Solve $A = LW$ for W.

Think of undoing what has been done to W. Since W is multiplied by L, undo the multiplication by dividing each side of $A = LW$ by L.

$$A = LW$$ Our goal is to isolate W.

$$\frac{A}{L} = \frac{LW}{L}$$ Divide by L.

$$\frac{A}{L} = W, \quad \text{or} \quad W = \frac{A}{L}$$ $\frac{LW}{L} = \frac{L}{L} \cdot W = 1 \cdot W = W$

Work Problem 6 at the Side. ▶

EXAMPLE 7 Solving for a Specified Variable

Solve $P = 2L + 2W$ for L.

$$P = 2L + 2W$$ Our goal is to isolate L.

$$P - 2W = 2L + 2W - 2W$$ Subtract $2W$.

$$P - 2W = 2L$$ Combine like terms.

$$\frac{P - 2W}{2} = \frac{2L}{2}$$ Divide by 2.

$$\frac{P - 2W}{2} = L, \quad \text{or} \quad L = \frac{P - 2W}{2}$$

Work Problem 7 at the Side. ▶

6 Solve $I = prt$ for t.

7 Solve $P = a + b + c$ for a.

ANSWERS

6. $t = \frac{I}{pr}$

7. $a = P - b - c$

8 (a) Solve $A = p + prt$ for t.

(b) Solve $Ax + B = C$ for x.

ANSWERS

8. (a) $t = \frac{A - p}{pr}$ (b) $x = \frac{C - B}{A}$

EXAMPLE 8 Solving for a Specified Variable

Solve $F = \frac{9}{5}C + 32$ for C. (This is the formula for converting temperatures from Celsius to Fahrenheit.)

We need to isolate C on one side of the equation. First undo the addition of 32 to $\frac{9}{5}C$ by subtracting 32 from each side.

$$F = \frac{9}{5}C + 32$$

$$F - 32 = \frac{9}{5}C + 32 - 32 \qquad \text{Subtract 32.}$$

$$F - 32 = \frac{9}{5}C$$

$$\frac{5}{9}(F - 32) = \frac{5}{9} \cdot \frac{9}{5}C \qquad \text{Multiply by } \tfrac{5}{9}.$$

Be sure to use parentheses.

$$\frac{5}{9}(F - 32) = C, \quad \text{or} \quad C = \frac{5}{9}(F - 32)$$

This last result is the formula for converting temperatures from Fahrenheit to Celsius.

Work Problem 8 at the Side.

2.5 Exercises

FOR EXTRA HELP MyMathLab Math XL PRACTICE WATCH DOWNLOAD READ REVIEW

1. In your own words, explain what is meant by each term.
   (a) Perimeter of a plane geometric figure
   The perimeter of a plane geometric figure is the distance around the figure.
   (b) Area of a plane geometric figure
   The area of a plane geometric figure is the measure of the surface covered or enclosed by the figure.

2. In parts (a)–(c), choose one of the following words to make the statement true: *linear, square,* or *cubic.*
   (a) If the dimensions of a plane geometric figure are given in feet, then the **area** is given in **square** feet.
   (b) If the dimensions of a rectangle are given in yards, then the **perimeter** is given in **linear** yards.
   (c) If the dimensions of a pyramid are given in meters, then the **volume** (which is defined on the next page) is given in **cubic** meters.

3. If a formula has exactly five variables, how many values would you need to be given in order to find the value of any one variable? **four**

4. The formula for changing Celsius temperature to Fahrenheit is given in Example 8 as $F = \frac{9}{5}C + 32$. Sometimes it is seen as $F = \frac{9C}{5} + 32$. These are both correct. Why is it true that $\frac{9}{5}C$ is equal to $\frac{9C}{5}$?
   $\frac{9}{5}C$ **equals** $\frac{9C}{5}$ **because** $\frac{9}{5}C = \frac{9}{5} \cdot \frac{C}{1} = \frac{9C}{5}$.

Decide whether perimeter or area would be used to solve a problem concerning the measure of the quantity.

5. Sod for a lawn **area**
6. Carpeting for a bedroom **area**
7. Baseboards for a living room **perimeter**
8. Fencing for a yard **perimeter**
9. Fertilizer for a garden **area**
10. Tile for a bathroom **area**
11. Determining the cost of planting rye grass in a lawn for the winter **area**
12. Determining the cost of replacing a linoleum floor with a wood floor **area**

In the following exercises a formula is given, along with the values of all but one of the variables in the formula. Find the value of the variable that is not given. (When necessary, use 3.14 as an approximation for π (pi).) See Example 1.

13. $P = 2L + 2W$ (perimeter of a rectangle); $L = 8$, $W = 5$
    $P = 26$

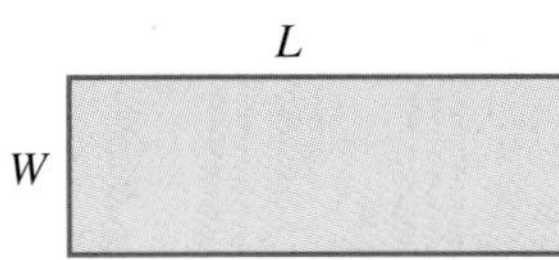

14. $P = 2L + 2W$; $L = 6$, $W = 4$
    $P = 20$

15. $A = \frac{1}{2}bh$ (area of a triangle); $b = 8$, $h = 16$

$A = 64$

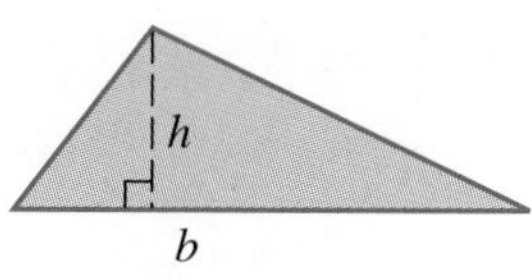

16. $A = \frac{1}{2}bh$; $b = 10$, $h = 14$

$A = 70$

17. $P = a + b + c$ (perimeter of a triangle); $P = 12$, $a = 3$, $c = 5$

$b = 4$

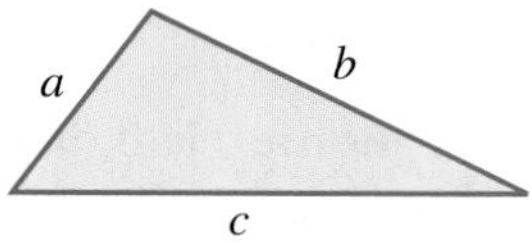

18. $P = a + b + c$; $P = 15$, $a = 3$, $b = 7$

$c = 5$

19. $d = rt$ (distance formula); $d = 252$, $r = 45$

$t = 5.6$

20. $d = rt$; $d = 100$, $t = 2.5$

$r = 40$

21. $I = prt$ (simple interest); $p = 7500$, $r = 0.035$, $t = 6$

$I = 1575$

22. $I = prt$; $p = 5000$, $r = 0.025$, $t = 7$

$I = 875$

23. $C = 2\pi r$ (circumference of a circle); $C = 16.328$

$r = 2.6$

24. $C = 2\pi r$; $C = 8.164$

$r = 1.3$

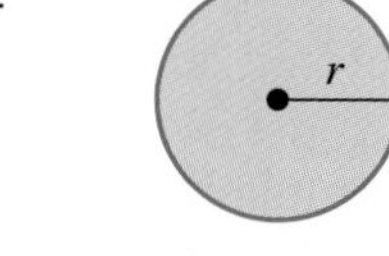

25. $A = \pi r^2$ (area of a circle); $r = 4$

$A = 50.24$

26. $A = \pi r^2$; $r = 12$

$A = 452.16$

The ***volume*** *of a three-dimensional object is a measure of the space occupied by the object. For example, we would need to know the volume of a gasoline tank in order to know how many gallons of gasoline it would take to completely fill the tank. In the following exercises, a formula for the volume (V) of a three-dimensional object is given, along with values for the other variables. Evaluate V. (Use 3.14 as an approximation for π.) See Example 1.*

27. $V = LWH$ (volume of a rectangular box); $L = 10$, $W = 5$, $H = 3$

$V = 150$

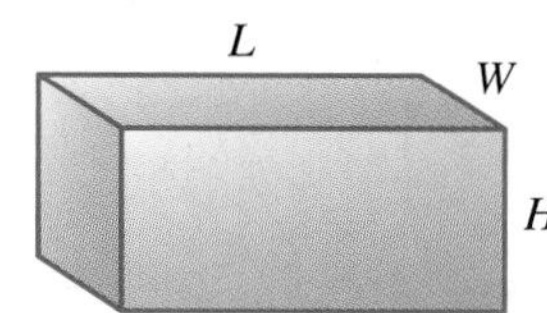

28. $V = LWH$; $L = 12$, $W = 8$, $H = 4$

$V = 384$

29. $V = \frac{1}{3}Bh$ (volume of a pyramid); $B = 12$, $h = 13$

$V = 52$

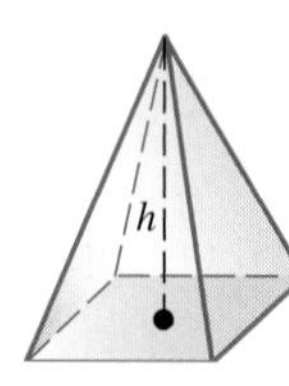

30. $V = \frac{1}{3}Bh$; $B = 36$, $h = 4$

$V = 48$

31. $V = \frac{4}{3}\pi r^3$ (volume of a sphere); $r = 12$

$V = 7234.56$

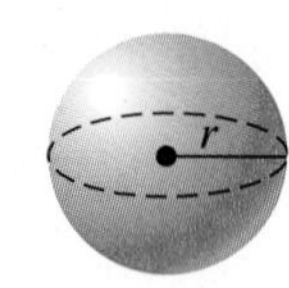

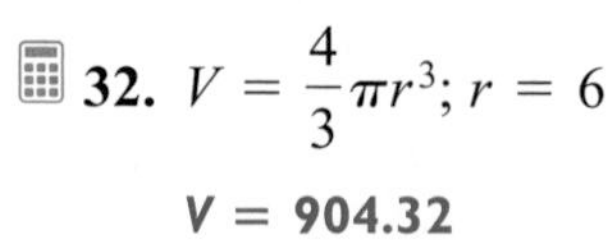

32. $V = \frac{4}{3}\pi r^3$; $r = 6$

$V = 904.32$

Solve each perimeter problem. See Examples 2 and 3.

33. The length of a rectangle is 9 in. more than the width. The perimeter is 54 in. Find the length and the width of the rectangle.

length: 18 in.; width: 9 in.

34. The width of a rectangle is 3 ft less than the length. The perimeter is 62 ft. Find the length and the width of the rectangle.

length: 17 ft; width: 14 ft

35. The perimeter of a rectangle is 36 m. The length is 2 m more than three times the width. Find the length and the width of the rectangle.

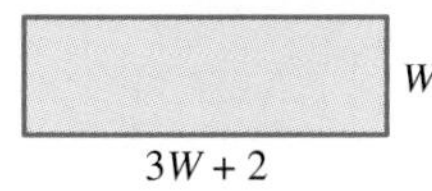

length: 14 m; width: 4 m

36. The perimeter of a rectangle is 36 yd. The width is 18 yd less than twice the length. Find the length and the width of the rectangle.

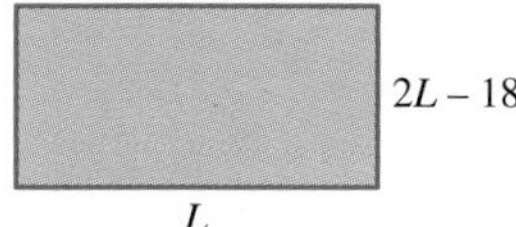

length: 12 yd; width: 6 yd

37. The longest side of a triangle is 3 in. longer than the shortest side. The medium side is 2 in. longer than the shortest side. If the perimeter of the triangle is 20 in., what are the lengths of the three sides?

shortest: 5 in.; medium: 7 in.; longest: 8 in.

38. The perimeter of a triangle is 28 ft. The medium side is 4 ft longer than the shortest side, while the longest side is twice as long as the shortest side. What are the lengths of the three sides?

shortest: 6 ft; medium: 10 ft; longest: 12 ft

39. Two sides of a triangle have the same length. The third side measures 4 m less than twice that length. The perimeter of the triangle is 24 m. Find the lengths of the three sides.

two equal sides: 7 m; third side: 10 m

40. A triangle is such that its medium side is twice as long as its shortest side and its longest side is 7 yd less than three times its shortest side. The perimeter of the triangle is 47 yd. What are the lengths of the three sides?

shortest 9 yd; medium: 18 yd; longest: 20 yd

Use a formula to write an equation for each application, and then use the problem-solving method of ***Section 2.4*** *to solve. (Use* 3.14 *as an approximation for* π*.)* ***Formulas are found on the inside covers of this book.*** *See Examples 2–4.*

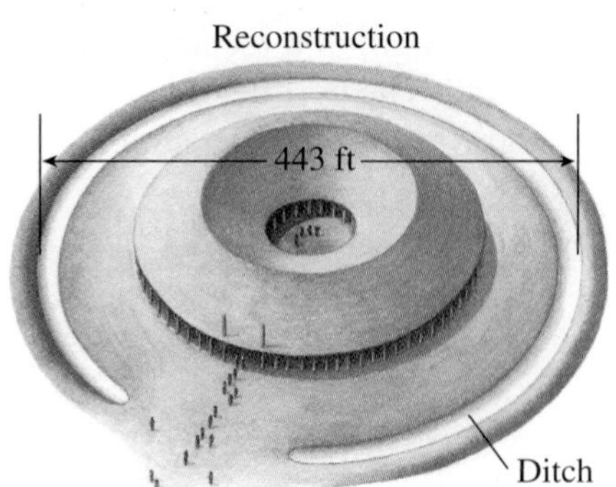

41. A prehistoric ceremonial site dating to about 3000 B.C. was discovered at Stanton Drew in southwestern England. The site, which is larger than Stonehenge, is a nearly perfect circle, consisting of nine concentric rings that probably held upright wooden posts. Around this timber temple is a wide, encircling ditch enclosing an area with a diameter of 443 ft. Find this enclosed area to the nearest thousand square feet. (*Source: Archaeology,* vol. 51, no. 1, Jan./Feb. 1998.) **about 154,000 ft^2**

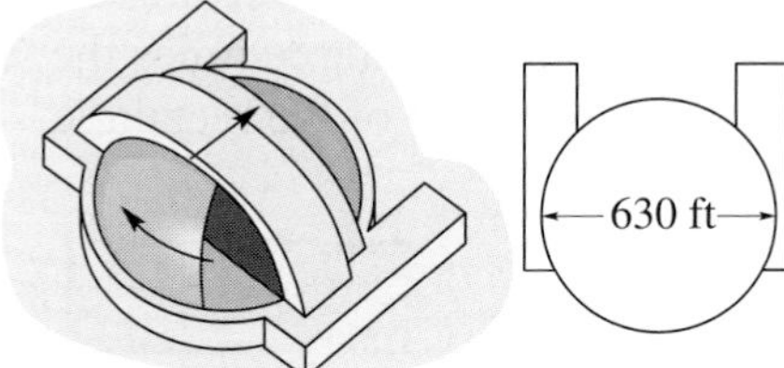

42. The Rogers Centre in Toronto, Canada, is the first stadium with a hard-shell, retractable roof. The steel dome is 630 ft in diameter. To the nearest foot, what is the circumference of this dome? (*Source:* www.ballparks.com)

about 1978 ft

43. The largest fashion catalogue in the world was published in Hamburg, Germany. Each of the 212 pages in the catalogue measured 1.2 m by 1.5 m. What was the perimeter of a page? What was the area? (*Source: Guinness World Records.*)

perimeter: 5.4 m; area: 1.8 m^2

44. The world's largest sand painting was created by Buddhist monks in the Singapore Expo Hall. The painting measured 12.24 m by 12.24 m. What was the perimeter of the sand painting? To the nearest hundredth of a square meter, what was the area? (*Source: Guinness World Records.*)

perimeter: 48.96 m; area: 149.82 m^2

45. The largest drum ever constructed was made from Japanese cedar and cowhide, with diameter 15.74 ft. What was the area of the circular face of the drum? Round your answer to the nearest hundredth of a square foot. (*Hint:* Use $A = \pi r^2$.) (*Source: Guinness World Records.*)

194.48 ft^2

46. What was the circumference of the drum described in Exercise 45? Round your answer to the nearest hundredth of a foot. (*Hint:* Use $C = 2\pi r$.)

49.42 ft

47. The area of a triangular road sign is 70 ft^2. If the base of the sign measures 14 ft, what is the height of the sign?

10 ft

48. The area of a triangular advertising banner is 96 ft^2. If the height of the banner measures 12 ft, find the measure of the base.

16 ft

49. The survey plat depicted here shows two lots that form a trapezoid. The measures of the parallel sides are 115.80 ft and 171.00 ft. The height of the trapezoid is 165.97 ft. Find the combined area of the two lots. Round your answer to the nearest hundredth of a square foot.

23,800.10 ft^2

50. Lot A in the figure is in the shape of a trapezoid. The parallel sides measure 26.84 ft and 82.05 ft. The height of the trapezoid is 165.97 ft. Find the area of Lot A. Round your answer to the nearest hundredth of a square foot.

9036.24 ft^2

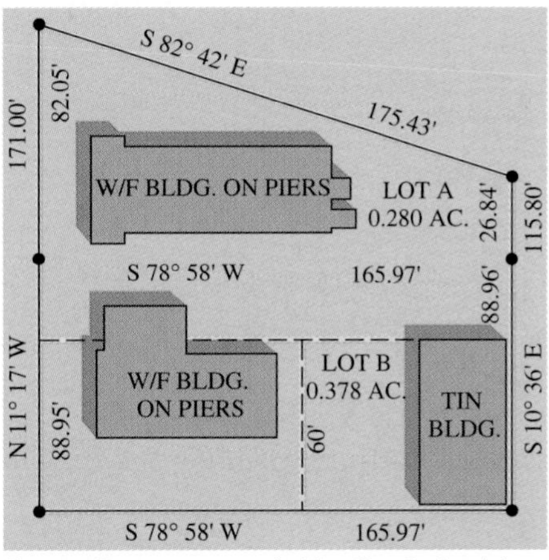

Source: Property survey in New Roads, Louisiana.

51. The U.S. Postal Service requires that any box sent by Priority Mail® have length plus girth (distance around) totaling no more than 108 in. The maximum volume that meets this condition is contained by a box with a square end 18 in. on each side. What is the length of the box? What is the maximum volume? (*Source:* United States Postal Service.)

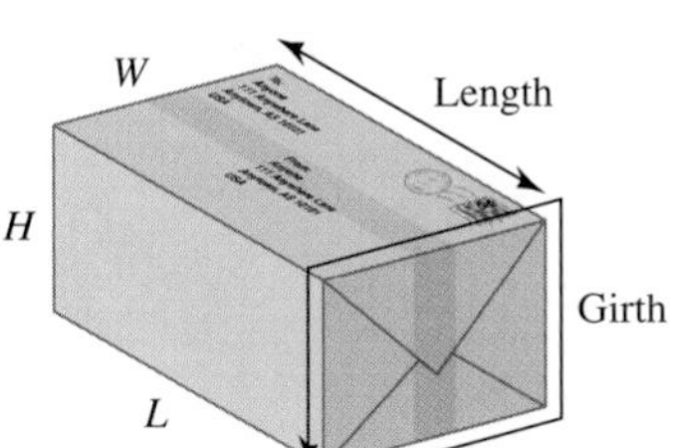

length: 36 in.; maximum volume: 11,664 in.^{3}

52. On March 17, 2005, a new record was set for the world's largest sandwich. The fillings of this sandwich were corned beef, cheese, lettuce, and mustard. The sandwich, made by Wild Woody's Chill and Grill in Roseville, Michigan, was 12 ft long, 12 ft wide, and $17\frac{1}{2}$ in. ($1\frac{11}{24}$ ft) thick. What was the volume of the sandwich? (*Source: Guinness World Records.*) **210 ft^3**

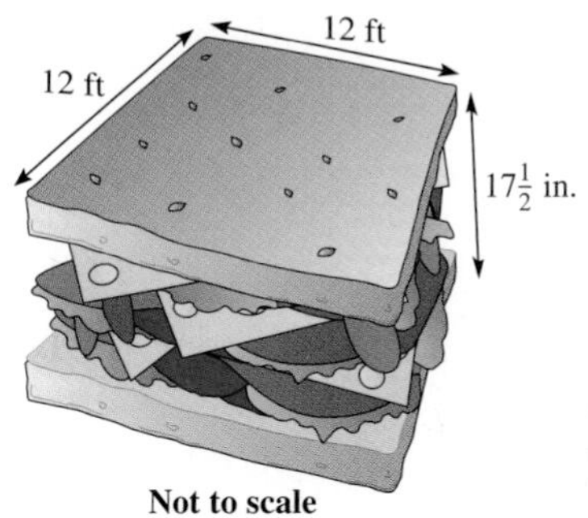

Find the measure of each marked angle. See Example 5.

53.

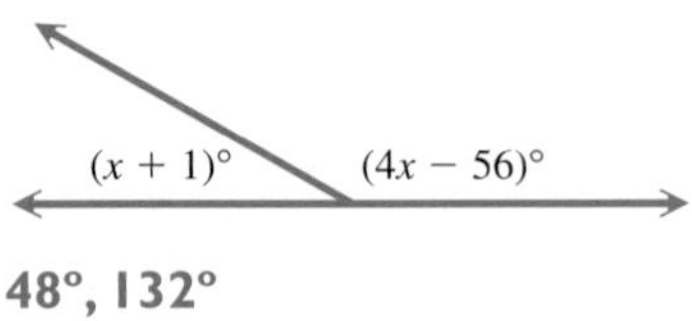

48°, 132°

54.

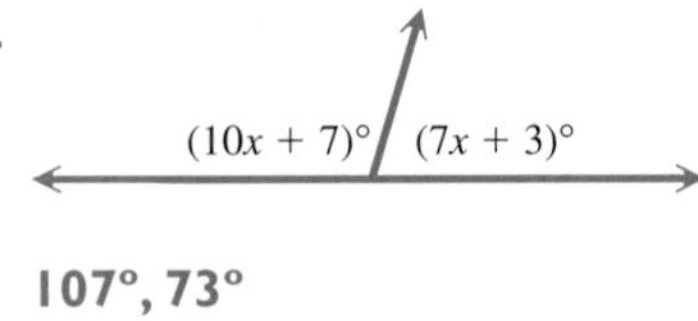

107°, 73°

55.

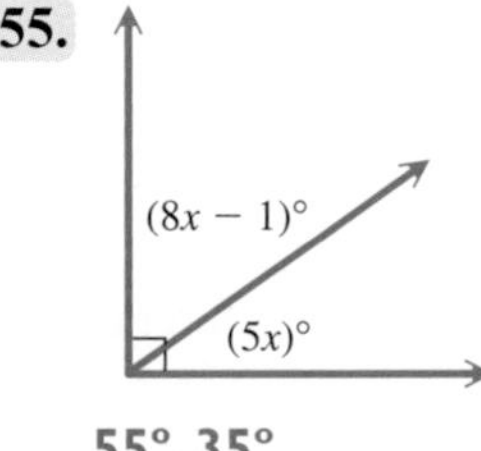

55°, 35°

56.

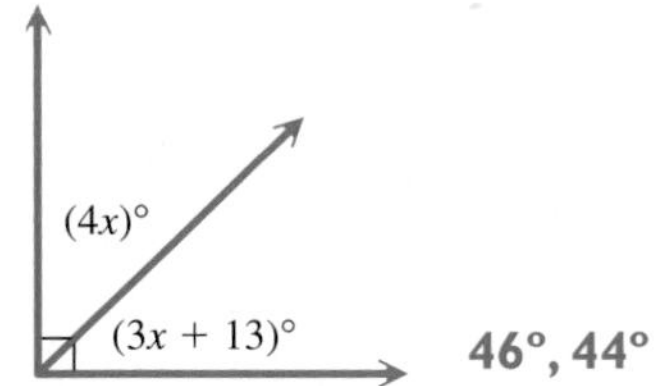

46°, 44°

57.

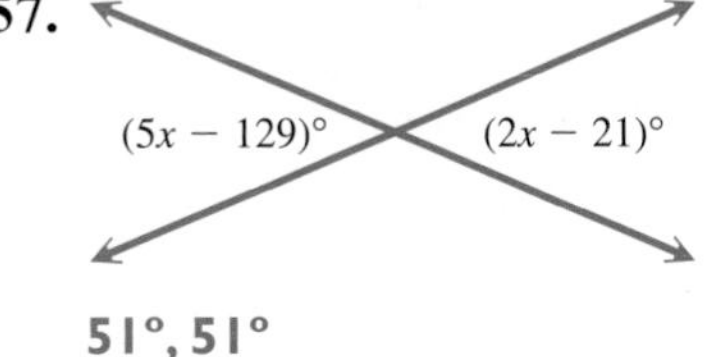

51°, 51°

58.

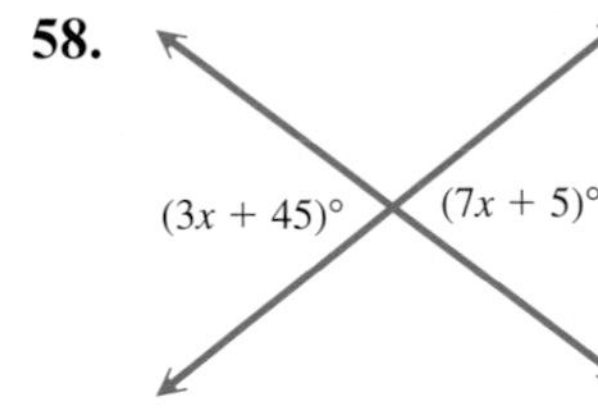

75°, 75°

59.

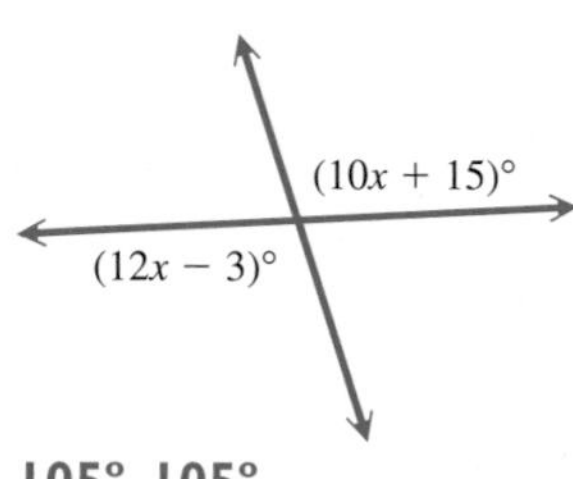

105°, 105°

60.

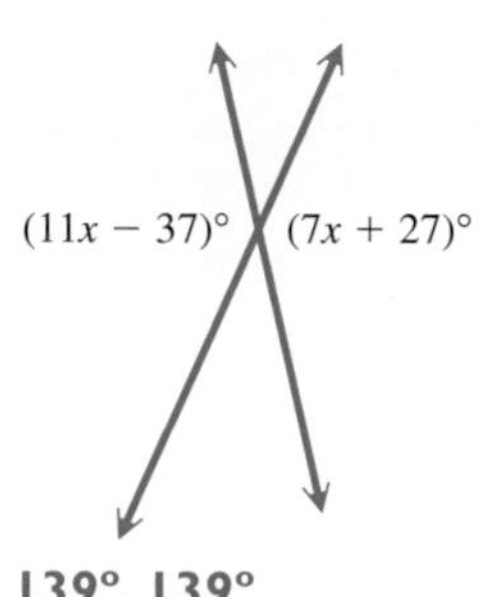

139°, 139°

Solve each formula for the specified variable. See Examples 6–8.

61. $d = rt$ for t

$t = \dfrac{d}{r}$

62. $d = rt$ for r

$r = \dfrac{d}{t}$

63. $V = LWH$ for H

$H = \dfrac{V}{LW}$

64. $V = LWH$ for L

$L = \dfrac{V}{WH}$

65. $P = a + b + c$ for b

$b = P - a - c$

66. $P = a + b + c$ for c

$c = P - a - b$

67. $C = 2\pi r$ for r

$r = \dfrac{C}{2\pi}$

68. $C = \pi d$ for d

$d = \dfrac{C}{\pi}$

69. $I = prt$ for r

$r = \dfrac{I}{pt}$

70. $I = prt$ for p

$p = \dfrac{I}{rt}$

71. $A = \dfrac{1}{2}bh$ for h

$h = \dfrac{2A}{b}$

72. $A = \dfrac{1}{2}bh$ for b

$b = \dfrac{2A}{h}$

73. $V = \dfrac{1}{3}\pi r^2 h$ for h

$h = \dfrac{3V}{\pi r^2}$

74. $V = \pi r^2 h$ for h

$h = \dfrac{V}{\pi r^2}$

75. $P = 2L + 2W$ for W

$W = \dfrac{P - 2L}{2}$

76. $A = p + prt$ for r

$r = \dfrac{A - p}{pt}$

77. $y = mx + b$ for m

$m = \dfrac{y - b}{x}$

78. $y = mx + b$ for x

$x = \dfrac{y - b}{m}$

79. $Ax + By = C$ for y

$y = \dfrac{C - Ax}{B}$

80. $Ax + By = C$ for x

$x = \dfrac{C - By}{A}$

81. $M = C(1 + r)$ for r

$r = \dfrac{M - C}{C}$

82. $C = \dfrac{5}{9}(F - 32)$ for F

$F = \dfrac{9}{5}C + 32$

83. $P = 2(a + b)$ for a

$a = \dfrac{P - 2b}{2}$

84. $P = 2(a + b)$ for b

$b = \dfrac{P - 2a}{2}$

2.6 Ratio, Proportion, and Percent

OBJECTIVES

1. **Write ratios.**
2. **Solve proportions.**
3. **Solve applied problems using proportions.**
4. **Find percentages and percents.**

OBJECTIVE 1 Write ratios. A **ratio** is a comparison of two quantities using a quotient.

Ratio

The ratio of the number a to the number b is written

$$a \text{ to } b, \quad a:b, \quad \text{or} \quad \frac{a}{b}.$$

This last way of writing a ratio is most common in algebra.

EXAMPLE 1 Writing a Word Phrase as a Ratio

Write a ratio for each word phrase.

(a) The ratio of 5 hr to 3 hr is

$$\frac{5 \text{ hr}}{3 \text{ hr}} = \frac{5}{3}.$$

(b) To find the ratio of 6 hr to 3 days, first convert 3 days to hours.

$$3 \text{ days} = 3 \cdot 24$$
$$3 \text{ days} = 72 \text{ hr}$$

The ratio of 6 hr to 3 days is thus

$$\frac{\mathbf{6\ hr}}{\mathbf{3\ days}} = \frac{\mathbf{6\ hr}}{\mathbf{72\ hr}} = \frac{6}{72} = \frac{1}{12}.$$

Work Problem 1 at the Side. ▶

1 Write each ratio.

(a) 9 women to 5 women

(b) 4 in. to 1 ft

An application of ratios is in unit pricing, to see which size of an item offered in different sizes produces the best price per unit. To do this, set up the ratio of the price of the item to the number of units on the label. Then divide to obtain the price per unit.

EXAMPLE 2 Finding the Price per Unit

A Cub Foods supermarket charges the following prices for a jar of extra crunchy peanut butter.

PEANUT BUTTER

Size	Price
18 oz	\$1.78
28 oz	\$2.97
40 oz	\$3.98

Which size is the best buy? That is, which size has the lowest unit price?

Continued on Next Page

ANSWERS

1. (a) $\frac{9}{5}$ (b) $\frac{4}{12} = \frac{1}{3}$

2 Solve the problem.

The local supermarket charges the following prices for a popular brand of pork and beans.

PORK AND BEANS

Size	Price
8 oz	$0.76
28 oz	$1.00
53 oz	$1.99

Which size is the best buy? What is the unit cost for that size?

To find the best buy, write ratios comparing the price for each size jar to the number of units (ounces) per jar. The results in the following table are rounded to the nearest thousandth.

Size	Unit Cost (dollars per ounce)	
18 oz	$\frac{\$1.78}{18} = \mathbf{\$0.099}$ ←	The best buy
28 oz	$\frac{\$2.97}{28} = \$0.106$	
40 oz	$\frac{\$3.98}{40} = \$0.100$	

Because the 18-oz size produces the lowest unit cost, it is the best buy. This example shows that buying the largest size does not always provide the best buy, although it often does.

◀ Work Problem 2 at the Side.

OBJECTIVE 2 Solve proportions. A ratio is used to compare two numbers or amounts. A **proportion** says that two ratios are equal, so it is a special type of equation. For example,

$$\frac{3}{4} = \frac{15}{20} \quad \text{Proportion}$$

is a proportion that says that the ratios $\frac{3}{4}$ and $\frac{15}{20}$ are equal. In the proportion

$$\frac{a}{b} = \frac{c}{d} \quad (b, d \neq 0),$$

a, b, c, and d are the **terms** of the proportion. The a and d terms are called the **extremes,** and the b and c terms are called the **means.** We read the proportion $\frac{a}{b} = \frac{c}{d}$ as **"*a* is to *b* as *c* is to *d*."** Multiplying each side of this proportion by the common denominator, bd, gives

$$bd \cdot \frac{a}{b} = bd \cdot \frac{c}{d}$$

$$\frac{b}{b}(d \cdot a) = \frac{d}{d}(b \cdot c) \quad \text{Associative and commutative properties}$$

$$ad = bc. \quad \text{Commutative and identity properties}$$

We can also find the products ad and bc by multiplying diagonally.

$$\frac{a}{b} = \frac{c}{d}$$

$$ad = bc$$

For this reason, ad and bc are called **cross products.**

Cross Products

If $\frac{a}{b} = \frac{c}{d}$, then the cross products ad and bc are equal—***the product of the extremes equals the product of the means.***

Also, if $ad = bc$, then $\frac{a}{b} = \frac{c}{d}$ (where $b, d \neq 0$).

ANSWER

2. 28 oz; $0.036 per oz

Note

If $\frac{a}{c} = \frac{b}{d}$, then $ad = cb$, or $ad = bc$. This means that the two proportions are equivalent, and

$$\text{the proportion } \frac{a}{b} = \frac{c}{d} \text{ can also be written as } \frac{a}{c} = \frac{b}{d}.$$

Sometimes one form is more convenient to work with than the other.

Four numbers are used in a proportion. If any three of these numbers are known, the fourth can be found.

EXAMPLE 3 Finding an Unknown in a Proportion

Solve the proportion $\frac{5}{9} = \frac{x}{63}$.

$$\frac{5}{9} = \frac{x}{63}$$

$5 \cdot 63 = 9 \cdot x$ — Cross products must be equal.

$315 = 9x$ — Multiply.

$35 = x$ — Divide by 9.

Check by substituting 35 for x in the proportion. The solution set is $\{35\}$.

Work Problem 3 at the Side. ▶

CAUTION

The cross product method cannot be used directly if there is more than one term on either side.

EXAMPLE 4 Solving an Equation by Using Cross Products

Solve the equation $\frac{m-2}{5} = \frac{m+1}{3}$.

$$\frac{m-2}{5} = \frac{m+1}{3}$$

$3(m-2) = 5(m+1)$ — Find the cross products. (Be sure to use parentheses.)

$3m - 6 = 5m + 5$ — Distributive property

$3m = 5m + 11$ — Add 6.

$-2m = 11$ — Subtract $5m$.

$m = -\frac{11}{2}$ — Divide by -2.

A check confirms that the solution is $-\frac{11}{2}$, so the solution set is $\{-\frac{11}{2}\}$.

Work Problem 4 at the Side. ▶

Note

When you set cross products equal to each other, you are really multiplying each ratio in the proportion by a common denominator.

3 Solve each proportion.

(a) $\frac{y}{6} = \frac{35}{42}$

(b) $\frac{a}{24} = \frac{15}{16}$

4 Solve each equation.

(a) $\frac{z}{2} = \frac{z+1}{3}$

(b) $\frac{p+3}{3} = \frac{p-5}{4}$

Answers

3. (a) $\{5\}$ **(b)** $\left\{\frac{45}{2}\right\}$

4. (a) $\{2\}$ **(b)** $\{-27\}$

5 Solve the problem.

Twelve gal of gasoline costs \$57.48. To the nearest cent, how much would 16.5 gal of the same fuel cost?

OBJECTIVE 3 Solve applied problems using proportions. Proportions are useful in many practical applications. We continue to use the six-step method, although the steps are not numbered here.

EXAMPLE 5 Applying Proportions

After Lee Ann Spahr pumped 5.0 gal of gasoline, the display showing the price read \$23.20. When she finished pumping the gasoline, the price display read \$67.28. How many gallons did she pump?

To solve this problem, set up a proportion, with prices in the numerators and gallons in the denominators.

Let $x =$ the number of gallons she pumped.

$$\begin{array}{r} \text{Price} \rightarrow \\ \text{Gallons} \rightarrow \end{array} \frac{\$23.20}{5.0} = \frac{\$67.28}{x} \begin{array}{l} \leftarrow \text{Price} \\ \leftarrow \text{Gallons} \end{array}$$

Notice how the labels correspond.

$$23.20x = 5.0(67.28) \quad \text{Cross products}$$

$$23.20x = 336.40 \quad \text{Multiply.}$$

$$x = 14.5 \quad \text{Divide by 23.20.}$$

She pumped a total of 14.5 gal. Check this answer. Notice that the way the proportion was set up uses the fact that the unit price is the same, no matter how many gallons are purchased.

Calculator Tip Using a calculator to perform the arithmetic in Example 5 reduces the possibility of errors.

◀ *Work Problem* **5** *at the Side.*

OBJECTIVE 4 Find percentages and percents. ***A percent is a ratio where the second number is always 100.*** For example, 50% represents the ratio of 50 to 100, 27% represents the ratio of 27 to 100, and so on. We can use the techniques for solving proportions to solve percent problems.

Recall from **Section R.2** that the decimal point is moved two places to the left to change a percent to a decimal number. For example, 75% can be written as the decimal 0.75.

Calculator Tip Many calculators have a percent key that does this automatically.

We can solve a percent problem by writing it as the proportion

$$\frac{\textit{amount}}{\textit{base}} = \frac{\textit{percent}}{100}, \quad \text{or} \quad \frac{a}{b} = \frac{p}{100}.$$

The amount, or **percentage,** is compared to the **base** (the whole amount). Since ***percent* means *per 100,*** we compare the numerical value of the percent to 100. Thus, we write 50% as

$$\frac{p}{100} = \frac{50}{100}. \quad p = 50$$

ANSWER

5. \$79.04

EXAMPLE 6 **Finding Amounts (Percentages)**

Solve each problem.

(a) Find 15% of 600.

Here, the base b is 600, the percent p is 15, and we must find the amount a (the percentage).

$$\frac{\text{amount } a}{\text{base } b} = \frac{\text{percent } p}{100}$$

Solve for a.

$$\frac{a}{600} = \frac{15}{100} \quad \text{Substitute known values.}$$

$$100a = 600(15) \quad \text{Cross products}$$

$$a = \frac{600(15)}{100} \quad \text{Divide by 100.}$$

$$a = 90 \quad \text{Simplify.}$$

Thus, 15% of 600 is 90.

(b) A DVD with a regular price of \$18 is on sale this week at 22% off. Find the amount of the discount and the sale price of the disc.

The discount is 22% of \$18. We want to find a, given b is 18 and p is 22.

$$\frac{a}{b} = \frac{p}{100}$$

$$\frac{a}{18} = \frac{22}{100} \quad \text{Substitute known values.}$$

$$100a = 18(22) \quad \text{Cross products}$$

$$a = \frac{18(22)}{100} \quad \text{Divide by 100.}$$

$$a = 3.96 \quad \text{Simplify.}$$

The amount of the discount on the DVD is \$3.96, and the sale price is \$18.00 − \$3.96 = \$14.04.

Work Problem 6 *at the Side.* ▶

EXAMPLE 7 **Finding a Base (Whole)**

32% of what number is 64?

The amount is 64, the percent is 32, and we must find the base (whole).

$$\frac{\text{amount } a}{\text{base } b} = \frac{\text{percent } p}{100}$$

Solve for b.

$$\frac{64}{b} = \frac{32}{100} \quad \text{Substitute known values.}$$

$$64(100) = 32b \quad \text{Cross products}$$

$$\frac{64(100)}{32} = b \quad \text{Divide by 32.}$$

$$200 = b \quad \text{Simplify.}$$

Thus, 32% of 200 is 64.

Work Problem 7 *at the Side.* ▶

6 Solve each problem.

(a) Find 20% of 70.

(b) Find the discount on a television set with a regular price of \$950 if the set is on sale at 25% off. Find the sale price of the set.

7 Solve each problem.

(a) 40% of what number is 130?

(b) Mark scored 34 points on a test, which was 85% of the possible points. How many possible points were on the test?

ANSWERS

6. (a) 14 **(b)** \$237.50; \$712.50

7. (a) 325 **(b)** 40 points

8 Solve each problem.

(a) 90 is what percent of 360?

(b) The interest in 1 yr on a deposit of $11,000 was $682. What percent interest was paid?

EXAMPLE 8 **Solving an Applied Percent Problem**

A newspaper ad offered a set of tires at a sale price of $258. The regular price was $300. What percent of the regular price was the savings?

The savings amounted to $\$300 - \$258 = \$42$. We can now restate the problem: What percent of 300 is 42? Substitute into the percent proportion. We have $a = 42$, $b = 300$, and p is to be found.

$$\frac{a}{b} = \frac{p}{100}$$ Solve for p.

$$\frac{42}{300} = \frac{p}{100}$$ Substitute known values.

$$4200 = 300p$$ Cross products

$$14 = p$$ Divide by 300.

The sale price represented a 14% savings.

◀ Work Problem **8** at the Side.

Note

Another way to solve percent problems uses the following percent equation, which can be obtained from the proportion we used in Examples 6–8.

percent (as a decimal) · base (whole) = amount

In Example 7, we were asked to answer the question "32% of what number is 64?" To use the percent equation to solve this problem, we let x represent the number. The word "of" translates as multiplication. Then

32%	of	a number	is	64.	
↓	↓	↓	↓	↓	
0.32	·	x	=	64	Write the percent as a decimal.

$$x = \frac{64}{0.32}$$ Divide by 0.32.

$$x = 200.$$ Simplify.

The result is the same.

Answers

8. (a) 25% **(b)** 6.2%

2.6 Exercises

FOR EXTRA HELP

Math XL PRACTICE

WATCH

DOWNLOAD

READ

REVIEW

1. Match each ratio in Column I with the ratio equivalent to it in Column II.

I	II
(a) 75 to 100	**A.** 80 to 100
(b) 5 to 4	**B.** 50 to 100
(c) $\frac{1}{2}$	**C.** 3 to 4
(d) 4 to 5	**D.** 15 to 12

(a) C (b) D (c) B (d) A

2. Which one of the following represents a ratio of 3 days to 2 weeks?

A. $\frac{3}{2}$ **B.** $\frac{3}{7}$

C. $\frac{1.5}{1}$ **D.** $\frac{3}{14}$

D

Write a ratio for each word phrase. In Exercises 7–10, first write the amounts with the same units. Write fractions in lowest terms. See Example 1.

3. 60 ft to 70 ft
$\frac{6}{7}$

4. 40 mi to 30 mi
$\frac{4}{3}$

5. 72 dollars to 220 dollars
$\frac{18}{55}$

6. 120 people to 90 people
$\frac{4}{3}$

7. 30 in. to 8 ft
$\frac{5}{16}$

8. 20 yd to 8 ft
$\frac{15}{2}$

9. 16 min to 1 hr
$\frac{4}{15}$

10. 24 min to 2 hr
$\frac{1}{5}$

11. 2 yd to 60 in.
$\frac{6}{5}$

12. 5 days to 40 hr
$\frac{3}{1}$, or 3

Find the best buy (based on price per unit) for each item. Give the unit price to the nearest thousandth for that size. See Example 2. (Source: Cub Foods.*)*

13. GRANULATED SUGAR

Size	Price
4 lb	\$1.79
10 lb	\$4.29

10 lb; \$0.429

14. GROUND COFFEE

Size	Price
15 oz	\$3.43
34.5 oz	\$6.98

34.5 oz; \$0.202

15. BLACK PEPPER

Size	Price
2 oz	\$2.23
4 oz	\$2.49
8 oz	\$6.59

4 oz; \$0.623

16. SALAD DRESSING

Size	Price
16 oz	\$2.44
32 oz	\$2.98
48 oz	\$4.95

32 oz; \$0.093

17. VEGETABLE OIL

Size	Price
16 oz	\$1.66
32 oz	\$2.59
64 oz	\$4.29
128 oz	\$6.49

128 oz; \$0.051

18. MOUTHWASH

Size	Price
8.5 oz	\$0.99
16.9 oz	\$1.87
33.8 oz	\$2.49
50.7 oz	\$2.99

50.7 oz; \$0.059

19. TOMATO KETCHUP

Size	Price
14 oz	\$1.39
24 oz	\$1.55
36 oz	\$1.78
64 oz	\$3.99

36 oz; \$0.049

20. GRAPE JELLY

Size	Price
12 oz	\$1.05
18 oz	\$1.73
32 oz	\$1.84
48 oz	\$2.88

32 oz; \$0.058

Solve each equation. See Examples 3 and 4.

21. $\frac{k}{4} = \frac{175}{20}$

{35}

22. $\frac{x}{6} = \frac{18}{4}$

{27}

23. $\frac{49}{56} = \frac{z}{8}$

{7}

24. $\frac{20}{100} = \frac{z}{80}$

{16}

25. $\frac{z}{4} = \frac{z+1}{6}$

{2}

26. $\frac{m}{5} = \frac{m-2}{2}$

$\left\{\frac{10}{3}\right\}$

27. $\frac{3y-2}{5} = \frac{6y-5}{11}$

{−1}

28. $\frac{2r+8}{4} = \frac{3r-9}{3}$

{10}

29. $\frac{5k+1}{6} = \frac{3k-2}{3}$

{5}

30. $\frac{x+4}{6} = \frac{x+10}{8}$

{14}

31. $\frac{2p+7}{3} = \frac{p-1}{4}$

$\left\{-\frac{31}{5}\right\}$

32. $\frac{3m-2}{5} = \frac{4-m}{3}$

$\left\{\frac{13}{7}\right\}$

Solve each problem. See Example 5.

33. If 16 candy bars cost $20.00, how much do 24 candy bars cost?

$30.00

34. If 12 ring tones cost $30.00, how much do 8 ring tones cost?

$20.00

35. If 6 gal of premium gasoline cost $22.74, how much would it cost to completely fill a 15-gal tank?

$56.85

36. If sales tax on a $16.00 DVD is $1.32, how much would the sales tax be on a $120.00 DVD player?

$9.90

37. In a recent year, there were 19 vehicles for every 10 U.S. households. How many vehicles were there for the 107 million U.S. households that year? (*Source:* U.S. Transportation Department.)

203.3 million vehicles

38. If there are 18 licensed drivers for every 10 U.S. households, how many licensed drivers are there in the 107 million U.S. households? (*Source:* U.S. Transportation Department.)

192.6 million drivers

39. Biologists tagged 500 fish in Grand Bay on October 12. At a later date, they found 7 tagged fish in a sample of 700. Estimate the total number of fish in Grand Bay to the nearest hundred.

50,000 fish

40. On May 25, researchers at an oxbow lake called Old River tagged 840 fish. When they returned a few weeks later, their sample of 1000 fish contained 18 that were tagged. Give an approximation of the fish population in Old River to the nearest hundred.

46,700 fish

41. The distance between Kansas City, Missouri, and Denver is 600 mi. On a certain wall map, this is represented by a length of 2.4 ft. On the map, how many feet would there be between Memphis and Philadelphia, two cities that are actually 1000 mi apart?

4 ft

42. The distance between Singapore and Tokyo is 3300 mi. On a certain wall map, this distance is represented by 11 in. The actual distance between Mexico City and Cairo is 7700 mi. How far apart are they on the same map?

$25\frac{2}{3}$ in.

43. According to the directions on the label of a bottle of Armstrong® Concentrated Floor Cleaner, for routine cleaning, $\frac{1}{4}$ cup of cleaner should be mixed with 1 gal of warm water. How much cleaner should be mixed with $10\frac{1}{2}$ gal of water?

$2\frac{5}{8}$ cups

44. The directions on the bottle mentioned in Exercise 43 also specify that, for extra-strength cleaning, $\frac{1}{2}$ cup of cleaner should be used for each gallon of water. For extra-strength cleaning, how much cleaner should be mixed with $15\frac{1}{2}$ gal of water?

$7\frac{3}{4}$ cups

45. The euro is the common currency used by most European countries, including Italy. On January 29, 2006, the exchange rate between euros and U.S. dollars was 1 euro to $1.2128. Ashley went to Rome and exchanged her U.S. currency for euros, receiving 300 euros. How much in U.S. dollars did she exchange? (*Source*: www.xe.com/ucc)

$363.84

46. If 8 U.S. dollars can be exchanged for 84.3 Mexican pesos, how many pesos can be obtained for $65? (Round to the nearest tenth.)

684.9 pesos

Answer each question about percent. See Examples 6–8.

47. What is 48.6% of 19? 9.234

48. What is 26% of 480? 124.8

49. What percent of 48 is 96? 200%

50. What percent of 30 is 36? 120%

51. 12% of what number is 3600? 30,000

52. 25% of what number is 150? 600

53. 78.84 is what percent of 292? 27%

54. 0.392 is what percent of 28? 1.4%

Use mental techniques to answer the questions in Exercises 55 and 56.

55. A recent census showed that the population of Alabama was 4,447,000, with 26.0% represented by African-Americans. What is the best estimate of the African-American population in Alabama? (*Source:* U.S. Census Bureau.) C

A. 500,000 **B.** 750,000

C. 1,000,000 **D.** 1,500,000

56. The same census showed that the population of New Mexico was 1,819,000, with 42.1% being Hispanic. What is the best estimate of the Hispanic population of New Mexico? (*Source:* U.S. Census Bureau.) A

A. 720,000 **B.** 72,000

C. 650,000 **D.** 36,000

Work each problem. Round all money amounts to the nearest dollar and percents to the nearest tenth, as needed. See Examples 6–8.

57. A family of four with a monthly income of $3800 plans to spend 8% of this amount on entertainment. How much will be spent on entertainment? $304

58. Quinhon Dac Ho earns $3200 per month. He saves 12% of this amount. How much does he save?

$384

59. In 2005, the U.S. civilian labor force consisted of 149,320,000 persons. Of this total, 7,615,320 were unemployed. What was the percent of unemployment? (*Source:* U.S. Bureau of Labor Statistics.)

5.1%

60. In 2006, the U.S. labor force (excluding agricultural employees, self-employed persons, and the unemployed) consisted of 128,300,000 persons. Of this total, 15,400,000 were union members. What percent of this labor force belonged to unions? (*Source:* U.S. Bureau of Labor Statistics.)

12.0%

61. As of February 2007, U.S. households owned 358,800,000 pets. Of these, 73,900,000 were dogs. What percent of the pets were *not* dogs? (*Source*: American Pet Product Manufacturers Association.) **79.4%**

62. During the 2005–2006 school year, the total public and private school enrollment in the United States was 54,993,000. Of this total, 11.7% of the students were enrolled in private schools. How many students were enrolled in public schools? (*Source:* National Center for Education Statistics.) **48,558,819**

63. The 1916 dime minted in Denver is quite rare. The 1979 edition of *A Guide Book of United States Coins* listed its value in Extremely Fine condition as \$625. The 2008 value had increased to \$6200. What was the percent increase in the value of this coin? **892%**

64. Here is a common business problem: If the sales tax rate is 6.5% and I have collected \$3400 in sales tax, how much were my sales? **\$52,308**

The Consumer Price Index provides a means of determining the purchasing power of the U.S. dollar from one year to the next. Using the period from 1982 to 1984 as a measure of 100.0, *the Consumer Price Index for selected years from 1995 to 2007 is shown here. To use the Consumer Price Index to predict a price in a particular year, we can set up a proportion and compare it with a known price in another year, as follows:*

$$\frac{\text{price in year } A}{\text{index in year } A} = \frac{\text{price in year } B}{\text{index in year } B}.$$

Use the Consumer Price Index figures in the table to find the amount that would be charged for the use of the same amount of electricity that cost \$225 in 1995. Give your answer to the nearest dollar.

Year	Consumer Price Index
1995	152.4
1997	160.5
1999	166.6
2001	177.1
2003	184.0
2005	195.3
2007	207.3

Source: U.S. Bureau of Labor Statistics.

65. in 1997 **\$237** **66.** in 1999 **\$246** **67.** in 2003 **\$272** **68.** in 2007 **\$306**

Relating Concepts (Exercises 69–72) For Individual or Group Work

Work Exercises 69–72 in order. *The steps justify the method of solving a proportion by cross products.*

69. What is the LCD of the fractions in the equation $\frac{x}{6} = \frac{2}{5}$? **30**

70. Solve the equation in Exercise 69 as follows.

(a) Multiply each side by the LCD. What equation do you get? $5x = 12$

(b) Solve the equation from part (a) by dividing each side by the coefficient of x. $\left\{\frac{12}{5}\right\}$

71. Solve the equation in Exercise 69 using cross products. $\left\{\frac{12}{5}\right\}$

72. Compare your answers from Exercises 70(b) and 71. What do you notice?
Both methods give the same solution set.

Summary Exercises on Solving Applied Problems

The following problems are of the various types discussed in this chapter. Solve each problem.

1. On an algebra test, the highest grade was 42 points more than the lowest grade. The sum of the two grades was 138. Find the lowest grade.
   48

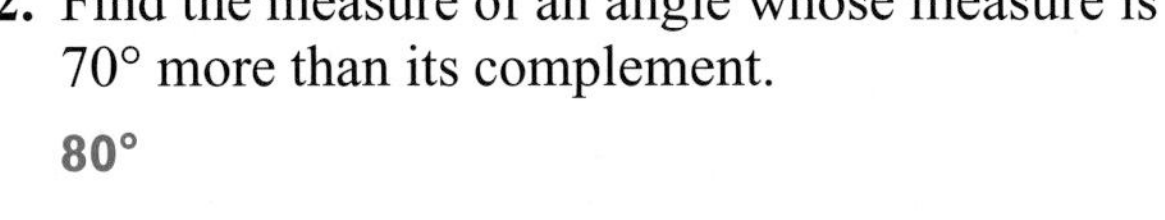

2. Find the measure of an angle whose measure is 70° more than its complement.
   80°

3. The perimeter of a certain square is seven times the length of a side, decreased by 12. Find the length of a side.

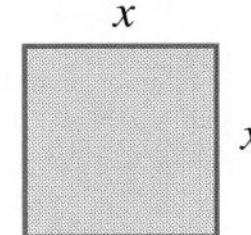

   4

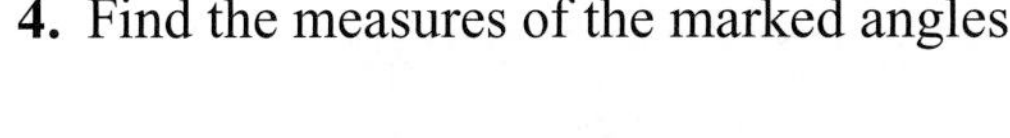

4. Find the measures of the marked angles.

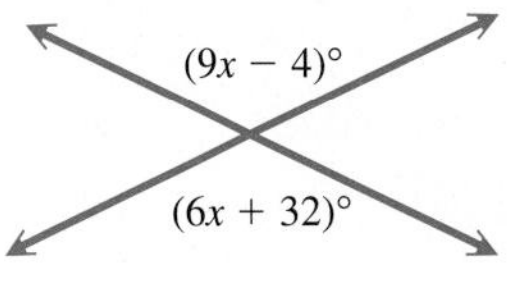

   104°, 104°

5. If 2 is added to five times a number, the result is equal to 5 more than four times the number. Find the number.
   3

6. Find two consecutive even integers such that four times the greater added to the lesser is 98.
   18, 20

7. Find the measures of the marked angles.

$(10x + 50)°$ $(4x + 4)°$

   140°, 40°

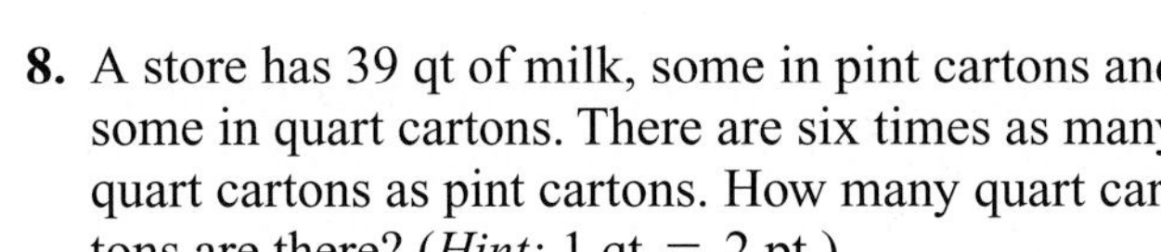

8. A store has 39 qt of milk, some in pint cartons and some in quart cartons. There are six times as many quart cartons as pint cartons. How many quart cartons are there? (*Hint:* 1 qt = 2 pt.)
   36 quart cartons

9. Grant Wood painted his most famous work, *American Gothic,* in 1930 on composition board with perimeter 108.44 in. If the width of the painting is 29.88 in., find the length. Then find the area. Round your answers to the nearest hundredth. (*Source: The Gazette.*) **24.34 in.; 727.28 in.^{2}**

10. A fully inflated professional basketball has a circumference of 78 cm. What is the radius of a circular cross section through the center of the ball? (Use 3.14 as the approximation for π.) Round your answer to the nearest hundredth. **12.42 cm**

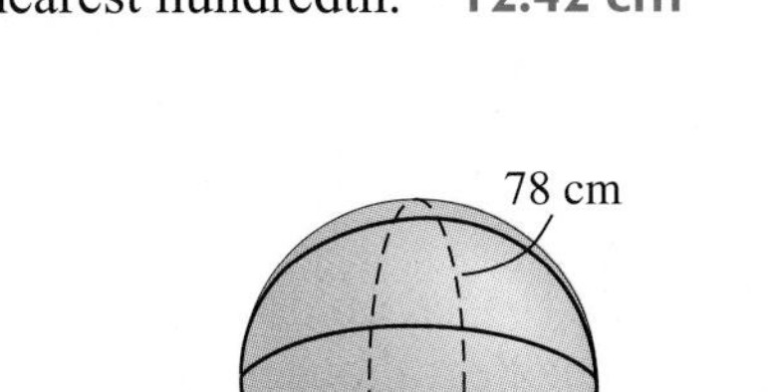

11. An MP3 player that normally sells for \$90 is on sale for \$75. What is the percent discount on the player?

$16\frac{2}{3}\%$

12. Two slices of bacon contain 85 calories. How many calories are there in twelve slices of bacon?

510 calories

13. Athletes in vigorous training programs can eat 50 calories per day for every 2.2 lb of body weight. To the nearest hundred, how many calories can a 175 lb athlete consume per day? (*Source: The Gazette.*)

4000 calories

14. In the 2007 Masters Golf Tournament in Augusta, Georgia, Phil Mickelson finished with a score of 10 more than the winner, Zach Johnson. The sum of their scores was 588. Find their scores. (*Source:* www.masters.org)

Zach Johnson: 289; Phil Mickelson: 299

15. The average cost of a traditional Thanksgiving dinner for 10 people, featuring turkey, stuffing, cranberries, pumpkin pie, and trimmings, was \$36.78 in 2005. This price increased 3.6% in 2006. What was the price of this traditional Thanksgiving dinner in 2006? Round your answer to the nearest cent. (*Source:* American Farm Bureau.)

\$38.10

16. Refer to Exercise 15. The cost of a traditional Thanksgiving dinner in 1987, the year this information was first collected, was \$26.24. What percent increase is the 2006 cost over the 1987 cost? Round your answer to the nearest tenth. (*Source:* American Farm Bureau.)

45.2%

17. Find the best buy (based on price per unit). Give the unit price to the nearest thousandth for that size. (*Source:* Cub Foods.)

SPAGHETTI SAUCE

Size	Price
$15\frac{1}{2}$ oz	\$1.19
32 oz	\$1.69
48 oz	\$2.69

32 oz; \$0.053

18. In the 2006 Winter Olympics in Torino, Italy, athletes from the Russian Federation earned 22 medals. Four of every 11 medals were gold. How many gold medals did the Russian Federation earn? (*Source*: www.torino2006.org)

8 gold medals

2.7 Solving Linear Inequalities

OBJECTIVES

1. Graph intervals on a number line.
2. Use the addition property of inequality.
3. Use the multiplication property of inequality.
4. Solve inequalities using both properties of inequality.
5. Solve linear inequalities with three parts.
6. Use inequalities to solve applied problems.

Inequalities are algebraic expressions related by

$<$	"is less than,"	$\leq$	"is less than or equal to,"
$>$	"is greater than,"	$\geq$	"is greater than or equal to."

We solve an inequality by finding all real number solutions for it. For example, the solutions of $x \leq 2$ include all *real numbers* that are less than or equal to 2, not just the *integers* less than or equal to 2.

OBJECTIVE 1 Graph intervals on a number line. Graphing is a good way to show the solution set of an inequality. To graph all real numbers satisfying $x \leq 2$, we place a square bracket at 2 on a number line and draw an arrow extending from the bracket to the left (since all numbers less than 2 are also part of the graph). The graph is shown in Figure 13.

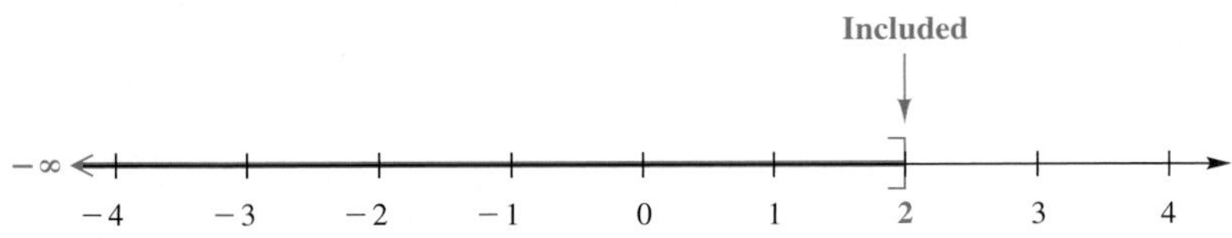

Figure 13 Graph of the interval $(-\infty, 2]$

The set of numbers less than or equal to 2 is an example of an **interval** on the number line. To write intervals, we use **interval notation**. For example, the interval of all numbers less than or equal to 2 is written $(-\infty, 2]$. The **negative infinity** symbol $-\infty$ does not indicate a number but shows that the interval includes all real numbers less than 2. As on the number line, the square bracket indicates that 2 is part of the solution.

1 Write each inequality in interval notation and graph the interval.

(a) $x \leq 3$

(b) $x > -4$

(c) $x \leq -\dfrac{3}{4}$

EXAMPLE 1 Graphing an Interval on a Number Line

Graph $x > -5$.

The statement $x > -5$ says that x can represent any value greater than -5 but cannot equal -5. The interval is written $(-5, \infty)$. We show this interval on a graph by placing a parenthesis at -5 and drawing an arrow to the right, as in Figure 14. The parenthesis at -5 indicates that -5 is *not* part of the graph.

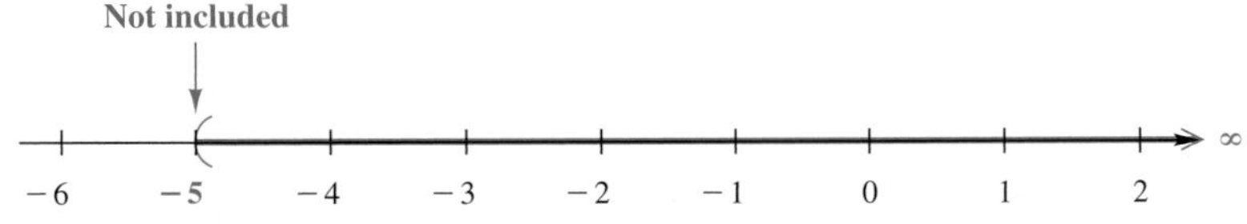

Figure 14 Graph of the interval $(-5, \infty)$

Keep the following important concepts regarding interval notation in mind:

1. A parenthesis indicates that an endpoint is *not included* in a solution set.
2. A bracket indicates that an endpoint is *included* in a solution set.
3. A parenthesis is *always* used next to an infinity symbol, $-\infty$ or ∞.
4. The set of all real numbers is written in interval notation as $(-\infty, \infty)$.

Work Problem 1 at the Side. ▶

ANSWERS

1. (a) $(-\infty, 3]$

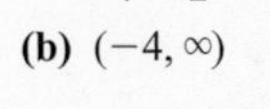

(b) $(-4, \infty)$

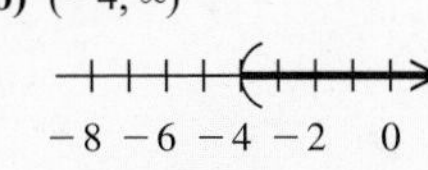

(c) $\left(-\infty, -\frac{3}{4}\right]$

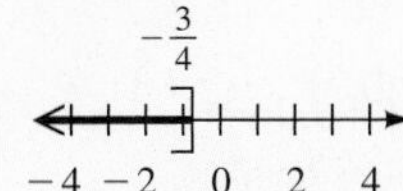

2 Write each inequality in interval notation and graph the interval.

(a) $-4 \geq x$

(b) $0 < x$

3 Write each inequality in interval notation and graph the interval.

(a) $-7 < x < -2$

(b) $-6 < x \leq -4$

ANSWERS

2. (a) $(-\infty, -4]$

(b) $(0, \infty)$

3. (a) $(-7, -2)$

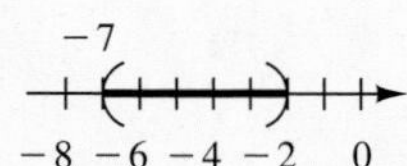

(b) $(-6, -4]$

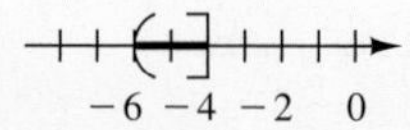

EXAMPLE 2 Graphing an Interval on a Number Line

Graph $3 > x$.

The statement $3 > x$ means the same as $x < 3$. The graph of $x < 3$, written in interval notation as $(-\infty, 3)$, is shown in Figure 15.

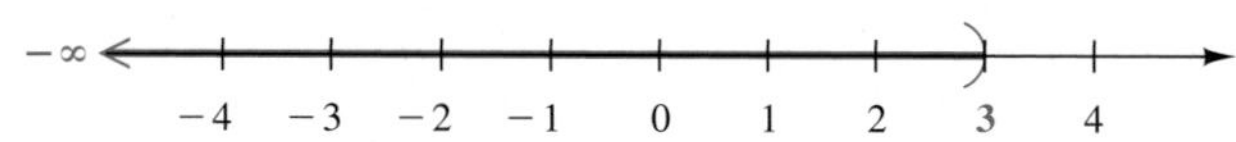

Figure 15

◀ Work Problem **2** at the Side.

EXAMPLE 3 Graphing an Interval on a Number Line

Graph $-3 \leq x < 2$.

The statement $-3 \leq x < 2$ is read "-3 is less than or equal to x *and* x is less than 2." We want the set of numbers that are *between* -3 and 2, with -3 included and 2 excluded. In interval notation, we write $[-3, 2)$, using a square bracket at -3 because -3 is part of the graph and a parenthesis at 2 because 2 is not part of the graph. The graph is shown in Figure 16.

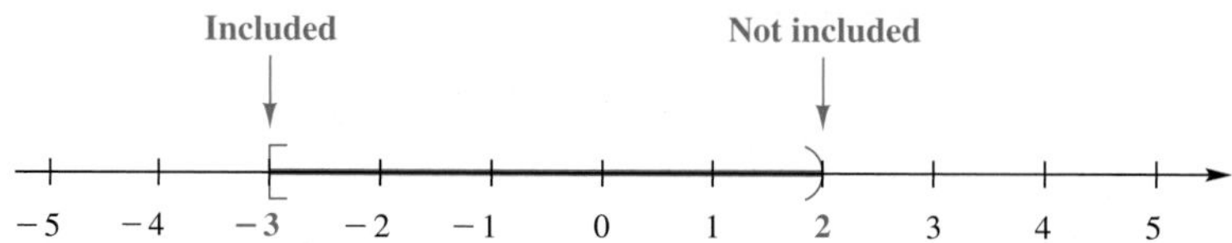

Figure 16 Graph of the interval $[-3, 2)$

◀ Work Problem **3** at the Side.

We now summarize the various types of intervals.

Type of Interval	Set-Builder Notation	Interval Notation	Graph
Open interval	$\{x \mid a < x < b\}$	(a, b)	a b
Closed interval	$\{x \mid a \leq x \leq b\}$	$[a, b]$	a b
Half-open (or half-closed) interval	$\{x \mid a \leq x < b\}$	$[a, b)$	a b
	$\{x \mid a < x \leq b\}$	$(a, b]$	a b
Disjoint interval*	$\{x \mid x < a \text{ or } x > b\}$	$(-\infty, a) \cup (b, \infty)$	a b
Infinite interval	$\{x \mid x > a\}$	(a, ∞)	a
	$\{x \mid x \geq a\}$	$[a, \infty)$	a
	$\{x \mid x < a\}$	$(-\infty, a)$	a
	$\{x \mid x \leq a\}$	$(-\infty, a]$	a
	$\{x \mid x \text{ is a real number}\}$	$(-\infty, \infty)$	0

*We will work with disjoint intervals in **Section 8.2** when we study *set operations* and *compound inequalities*.

Recall from **Section 1.3** that the notation used in the second column of the table, such as $\{x \mid a < x < b\}$, is called **set-builder notation.**

OBJECTIVE 2 Use the addition property of inequality. Solving an inequality is similar to solving an equation.

Linear Inequality in One Variable

A **linear inequality in one variable** can be written in the form

$$Ax + B < C,$$

where A, B, and C are real numbers, with $A \neq 0$.

(All definitions and rules are also valid for $>$, $\leq$, and $\geq$.) Examples of linear inequalities in one variable include

$x + 5 < 2$, $t - 3 \geq 5$, and $2k + 5 \leq 10$. Linear inequalities

Consider the inequality $2 < 5$. If 4 is added to each side, the result is

$$2 + 4 < 5 + 4 \quad \text{Add 4.}$$
$$6 < 9, \quad \text{True}$$

a true sentence. This example suggests the **addition property of inequality.**

Addition Property of Inequality

For any real numbers A, B, and C, the inequalities

$$A < B \quad \text{and} \quad A + C < B + C$$

have exactly the same solutions.

In words, the same number may be added to each side of an inequality without changing the solutions.

As with the addition property of equality, the same number may be subtracted from each side of an inequality.

EXAMPLE 4 Using the Addition Property of Inequality

Solve $7 + 3k \geq 2k - 5$, and graph the solution set.

$$7 + 3k \geq 2k - 5$$
$$7 + 3k - 2k \geq 2k - 5 - 2k \quad \text{Subtract } 2k.$$
$$7 + k \geq -5 \quad \text{Combine like terms.}$$
$$7 + k - 7 \geq -5 - 7 \quad \text{Subtract 7.}$$
$$k \geq -12 \quad \text{Combine like terms.}$$

The solution set is $[-12, \infty)$. Its graph is shown in Figure 17.

−13 −12 −11 −10 −9 −8 −7 −6 −5 −4 −3 −2 −1 0

Figure 17

Work Problem 4 at the Side. ▶

4 Solve each inequality, and graph the solution set.

(a) $-1 + 8r < 7r + 2$

(b) $5m - \frac{4}{3} \leq 4m$

ANSWERS

4. (a) $(-\infty, 3)$

−4 −2 0 2 3 4

(b) $\left(-\infty, \frac{4}{3}\right]$

$\frac{4}{3}$

−2 −1 0 1 2

5 **(a)** Multiply each side of

$$-2 < 8$$

by 6 and then by -5. Reverse the direction of the inequality symbol if necessary to make a true statement.

(b) Multiply each side of

$$-4 > -9$$

by 2 and then by -8. Reverse the direction of the inequality symbol if necessary to make a true statement.

Note

Because an inequality has many solutions, we cannot check all of them by substitution as we did with the single solution of an equation. To check the solutions of

$$7 + 3k \geq 2k - 5$$

from Example 4, which we found to be in the interval $[-12, \infty)$, we use a multistep process. First, we substitute -12 for k in the related *equation*.

Check

$7 + 3k = 2k - 5$	Related equation
$7 + 3(-12) \stackrel{?}{=} 2(-12) - 5$	Let $k = -12$.
$7 - 36 \stackrel{?}{=} -24 - 5$	Multiply.
$-29 = -29$	True

A true statement results, so -12 is indeed the "boundary" point. Now we test a number other than -12 from the interval $[-12, \infty)$. We choose 0 since it is easy to substitute.

Check

$7 + 3k \geq 2k - 5$	Original inequality
$7 + 3(0) \stackrel{?}{\geq} 2(0) - 5$	Let $k = 0$.
$7 \geq -5$	True

Again, a true statement results, so the checks confirm that solutions to the inequality are in the interval $[-12, \infty)$. Any number "outside" the interval $[-12, \infty)$, that is, any number in $(-\infty, -12)$, will give a false statement when tested. (Try this.)

OBJECTIVE 3 Use the multiplication property of inequality. The addition property of inequality cannot be used to solve an inequality such as $4x \geq 28$. The *multiplication property of inequality* is required. To see how this property works, consider the true statement $3 < 7$.

Multiply each side of the inequality $3 < 7$ by the positive number 5.

$3 < 7$	
$5(3) < 5(7)$	Multiply by 5.
$15 < 35$	True

Now multiply each side of $3 < 7$ by the negative number -5.

$3 < 7$	
$-5(3) < -5(7)$	Multiply by -5.
$-15 < -35$	False

To get a true statement when multiplying each side by −5, we must reverse the direction of the inequality symbol.

$3 < 7$	
$-5(3) > -5(7)$	Multiply by -5; reverse the direction of the symbol.
$-15 > -35$	True

◀ *Work Problem* 5 *at the Side.*

Answers

5. **(a)** $-12 < 48$; $10 > -40$
(b) $-8 > -18$; $32 < 72$

In summary, the **multiplication property of inequality** has two parts.

> **Multiplication Property of Inequality**
>
> For any real numbers A, B, and C $(C \neq 0)$,
>
> **1.** if C is ***positive***, then the inequalities
>
> $$A < B \quad \text{and} \quad AC < BC$$
>
> have exactly the same solutions;
>
> **2.** if C is ***negative***, then the inequalities
>
> $$A < B \quad \text{and} \quad AC > BC$$
>
> have exactly the same solutions.
>
> In words, each side of an inequality may be multiplied by the same positive number without changing the solutions. ***If the multiplier is negative, we must reverse the direction of the inequality symbol.***

As with the multiplication property of equality, the same nonzero number may be divided into each side.

It is important to remember the differences in the multiplication property for positive and negative numbers.

1. When each side of an inequality is multiplied or divided by a ***positive number,*** the direction of the inequality symbol *does not change.* (Adding or subtracting terms on each side also does not change the symbol.)
2. When each side of an inequality is multiplied or divided by a ***negative number,*** the direction of the symbol *does change.* ***Reverse the direction of the inequality symbol only when multiplying or dividing each side by a negative number.***

EXAMPLE 5 Using the Multiplication Property of Inequality

Solve $3r < -18$, and graph the solution set.

Using the multiplication property of inequality, we divide each side by 3. Since 3 is a positive number, the direction of the inequality symbol *does not* change. ***It does not matter that the number on the right side of the inequality is negative.***

$$3r < -18$$

$$\frac{3r}{3} < \frac{-18}{3} \quad \text{Divide by 3.}$$

$$r < -6$$

3 is *positive.* Do NOT reverse the direction of the symbol.

The graph of the solution set, $(-\infty, -6)$, is shown in Figure 18.

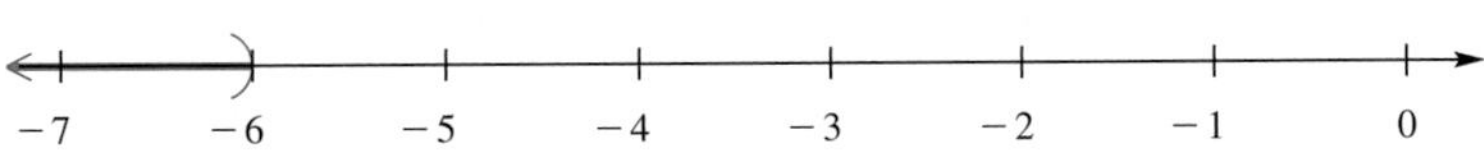

Figure 18

6 Solve each inequality. Graph the solution set.

(a) $9x < -18$

(b) $-2r > -12$

(c) $-5p \leq 0$

EXAMPLE 6 Using the Multiplication Property of Inequality

Solve $-4t \geq 8$, and graph the solution set.

Here each side of the inequality must be divided by -4, a negative number, which *does* require changing the direction of the inequality symbol.

$$-4t \geq 8$$

$$\frac{-4t}{-4} \leq \frac{8}{-4} \qquad \text{Divide by } -4.$$

(-4 is *negative*. Change $\geq$ to $\leq$.)

$$t \leq -2$$

The solution set, $(-\infty, -2]$, is graphed in Figure 19.

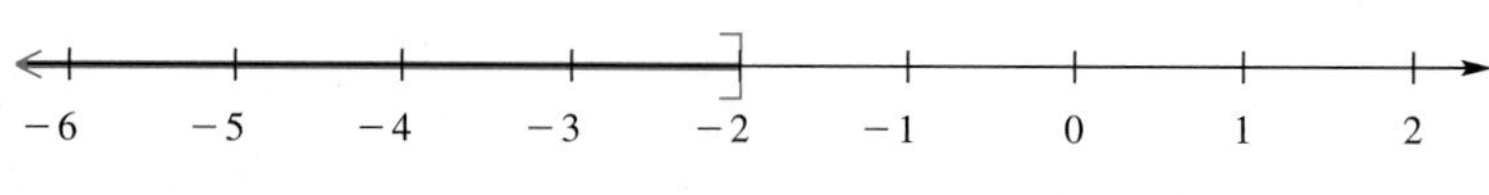

Figure 19

◀ Work Problem **6** at the Side.

OBJECTIVE 4 Solve inequalities using both properties of inequality. The steps to solve a linear inequality are summarized below.

> **Solving a Linear Inequality**
>
> ***Step 1*** **Simplify each side separately.** Use the distributive property to clear parentheses and combine like terms on each side as needed.
>
> ***Step 2*** **Isolate the variable term on one side.** Use the addition property to get all terms with variables on one side of the inequality and all numbers on the other side.
>
> ***Step 3*** **Isolate the variable.** Use the multiplication property to write the inequality in the form $x < c$ or $x > c$.
>
> ***Remember: Reverse the direction of the inequality symbol only when multiplying or dividing each side of an inequality by a negative number.***

EXAMPLE 7 Solving a Linear Inequality

Solve $3x + 2 - 5 > -x + 7 + 2x$. Graph the solution set.

Step 1 Combine like terms and simplify.

$$3x + 2 - 5 > -x + 7 + 2x$$

$$3x - 3 > x + 7$$

Step 2 Use the addition property of inequality.

$$3x - 3 - x > x + 7 - x \qquad \text{Subtract } x.$$

$$2x - 3 > 7 \qquad \text{Combine like terms.}$$

$$2x - 3 + 3 > 7 + 3 \qquad \text{Add 3.}$$

$$2x > 10 \qquad \text{Combine like terms.}$$

Step 3 Use the multiplication property of inequality.

$$\frac{2x}{2} > \frac{10}{2} \qquad \text{Divide by 2.}$$

(Because 2 is positive, keep the symbol $>$.)

$$x > 5$$

Continued on Next Page

ANSWERS

6. (a) $(-\infty, -2)$

(b) $(-\infty, 6)$

(c) $[0, \infty)$

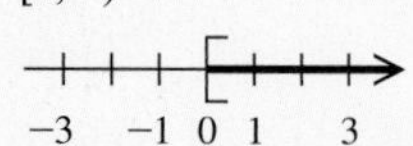

The solution set, $(5, \infty)$, is graphed in Figure 20.

Figure 20

Work Problem 7 at the Side. ▶

EXAMPLE 8 Solving a Linear Inequality

Solve $5(k-3)-7k \geq 4(k-3)+9$. Graph the solution set.

Step 1 Clear parentheses; then combine like terms.

$$5(k-3)-7k \geq 4(k-3)+9$$

$$5k-15-7k \geq 4k-12+9 \quad \text{Distributive property}$$

$$-2k-15 \geq 4k-3 \quad \text{Combine like terms.}$$

Step 2 Use the addition property.

$$-2k-15-\mathbf{4k} \geq 4k-3-\mathbf{4k} \quad \text{Subtract } 4k.$$

$$-6k-15 \geq -3 \quad \text{Combine like terms.}$$

$$-6k-15+\mathbf{15} \geq -3+\mathbf{15} \quad \text{Add 15.}$$

$$-6k \geq 12 \quad \text{Combine like terms.}$$

Step 3 Divide each side by -6, a negative number. Change the direction of the inequality symbol.

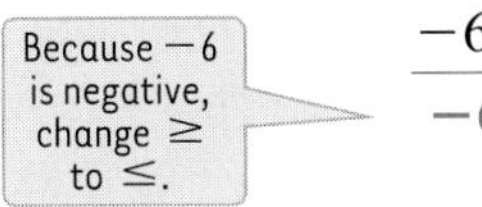

$$\frac{-6k}{\mathbf{-6}} \leq \frac{12}{\mathbf{-6}} \quad \text{Divide by } -6.$$

$$k \leq -2$$

A graph of the solution set, $(-\infty, -2]$, is shown in Figure 21.

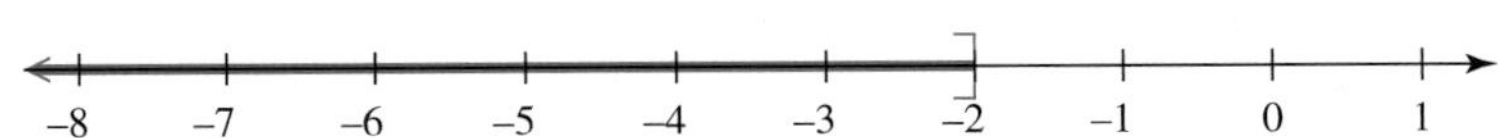

Figure 21

Work Problem 8 at the Side. ▶

EXAMPLE 9 Solving a Linear Inequality with Fractions

Solve $-\frac{2}{3}(r-3)-\frac{1}{2}<\frac{1}{2}(5-r)$, and graph the solution set.

Step 1 To clear fractions, multiply by the least common denominator, 6.

$$-\frac{2}{3}(r-3)-\frac{1}{2}<\frac{1}{2}(5-r)$$

$$6\left[-\frac{2}{3}(r-3)-\frac{1}{2}\right]<6\left[\frac{1}{2}(5-r)\right] \quad \text{Multiply by 6, the LCD.}$$

Be careful here.

$$6\left[-\frac{2}{3}(r-3)\right]-6\left(\frac{1}{2}\right)<6\left[\frac{1}{2}(5-r)\right] \quad \text{Distributive property}$$

$$-4(r-3)-3<3(5-r) \quad \text{Multiply.}$$

$$-4r+12-3<15-3r \quad \text{Distributive property}$$

$$-4r+9<15-3r$$

Continued on Next Page

7 Solve

$7x-6+1 \geq 5x-x+2.$

Graph the solution set.

8 Solve

$-15-(2x+1) \geq 4(x-1)-3x.$

Graph the solution set.

ANSWERS

7. $\left[\frac{7}{3}, \infty\right)$

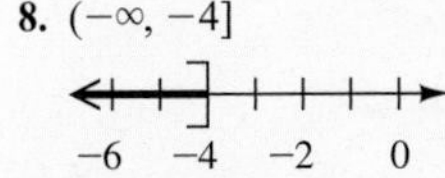

8. $(-\infty, -4]$

−6 −4 −2 0

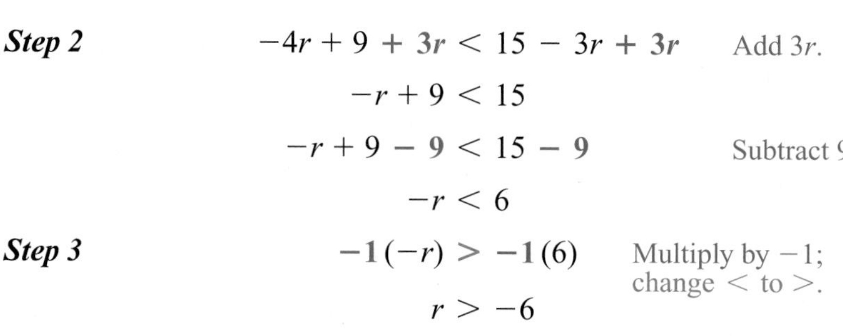

Step 2 $-4r + 9 + 3r < 15 - 3r + 3r$ Add $3r$.

$$-r + 9 < 15$$

$$-r + 9 - 9 < 15 - 9 \quad \text{Subtract 9.}$$

$$-r < 6$$

Step 3 $-1(-r) > -1(6)$ Multiply by -1; change $<$ to $>$.

$$r > -6$$

Check that the solution set is $(-6, \infty)$. See Figure 22.

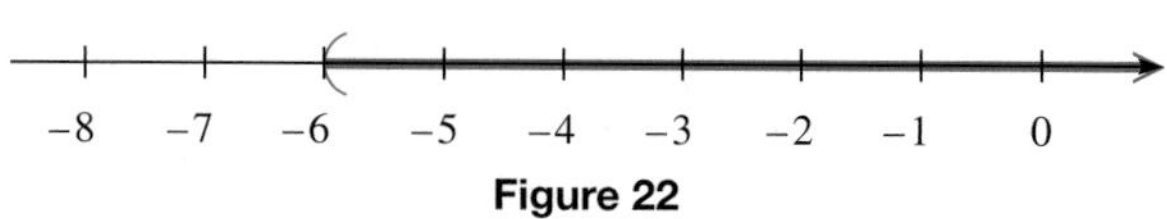

Figure 22

◀ Work Problem **9** at the Side.

9 Solve, check, and graph the solution set of each inequality.

(a) $5 - 3(m - 1) \leq 2(m + 3) + 1$

(b) $\frac{1}{4}(m + 3) + 2 \leq \frac{3}{4}(m + 8)$

OBJECTIVE 5 Solve linear inequalities with three parts. An inequality that says that one number is *between* two other numbers is a **three-part inequality.** For example, $-3 < 5 < 7$ says that 5 is between -3 and 7. Similarly, the three-part inequality

$$3 < x + 2 < 8$$

says that $x + 2$ is between 3 and 8. We solve this inequality as follows.

$$3 - 2 < x + 2 - 2 < 8 - 2 \quad \text{Subtract 2 from each part.}$$

$$1 < x < 6$$

Thus, x must be between 1 and 6 so that $x + 2$ will be between 3 and 8. The idea is to get the inequality in the form

a number $< x <$ another number.

10 Solve

$$2 \leq 3x - 1 \leq 8.$$

Graph the solution set.

CAUTION

Three-part inequalities are written so that the symbols point in the same direction and toward the lesser number. It would be *wrong* to write $8 < x + 2 < 3$, since this would imply that $8 < 3$, a false statement.

EXAMPLE 10 Solving a Three-Part Inequality

Solve $4 \leq 3x - 5 < 10$. Graph the solution set.

$$4 \leq 3x - 5 < 10$$

$$4 + 5 \leq 3x - 5 + 5 < 10 + 5 \quad \text{Add 5 to each part.}$$

$$9 \leq 3x < 15$$

$$\frac{9}{3} \leq \frac{3x}{3} < \frac{15}{3} \quad \text{Divide each part by 3.}$$

Remember to divide all *three* parts by 3.

$$3 \leq x < 5$$

The solution set, $[3, 5)$, is graphed in Figure 23.

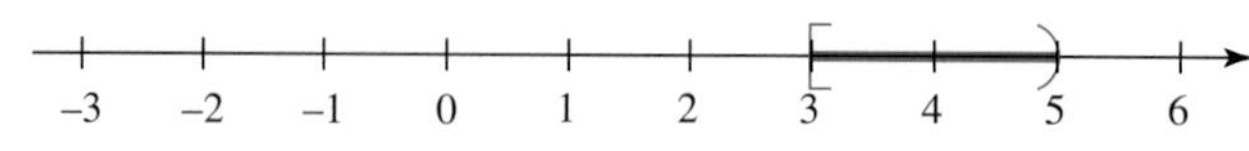

Figure 23

◀ Work Problem **10** at the Side.

ANSWERS

9. (a) $\left[\frac{1}{5}, \infty\right)$

$\frac{1}{5}$

0 1 2 3 4 5

(b) $\left[-\frac{13}{2}, \infty\right)$

$-\frac{13}{2}$

−8 −6 −4 −2 0 2

10. $[1, 3]$

0 1 3

Examples of the types of solution sets to be expected from solving linear equations and linear inequalities are shown below.

SOLUTIONS OF LINEAR EQUATIONS AND INEQUALITIES

Equation or Inequality	Typical Solution Set	Graph of Solution Set
Linear equation $5x + 4 = 14$	$\{2\}$	(point at 2)
Linear inequality $5x + 4 < 14$	$(-\infty, 2)$	(ray left of 2)
$5x + 4 > 14$	$(2, \infty)$	(ray right of 2)
Three-part inequality $-1 \leq 5x + 4 \leq 14$	$[-1, 2]$	(segment from −1 to 2)

OBJECTIVE 6 Use inequalities to solve applied problems. The table below gives some of the more common phrases that suggest inequality.

Phrase/Word	Example	Inequality
Is greater than	A number *is greater than* 4	$x > 4$
Is less than	A number *is less than* -12	$x < -12$
Exceeds	A number *exceeds* 3.5	$x > 3.5$
Is at least	A number *is at least* 6	$x \geq 6$
Is at most	A number *is at most* 8	$x \leq 8$

Work Problem 11 at the Side. ▶

11 Translate each statement into an inequality, using x as the variable.

(a) The total cost is less than \$10.

(b) Chicago received at most 5 in. of snow.

(c) The car's speed exceeded 60 mph.

(d) You must be at least 18 yr old to vote.

The next example uses the idea of finding the average of a number of scores. ***In general, to find the average of n numbers, add the numbers and divide by n.*** We use the six problem-solving steps from **Section 2.4,** changing Step 3 to "Write an inequality."

EXAMPLE 11 Finding an Average Test Score

Brent has grades of 86, 88, and 78 on his first three tests in geometry. If he wants an average of at least 80 after his fourth test, what are the possible scores he can make on that test?

Step 1 **Read** the problem again.

Step 2 **Assign a variable.** Let x = Brent's score on his fourth test.

Step 3 **Write an inequality.** To find his average after four tests, add the test scores and divide by 4.

Average ↓ is at least ↓ 80. ↓

$$\frac{86 + 88 + 78 + x}{4} \geq 80$$

Continued on Next Page

ANSWERS

11. (a) $x < 10$ (b) $x \leq 5$
(c) $x > 60$ (d) $x \geq 18$

12 Solve the problem.

Maggie has scores of 98, 86, and 88 on her first three tests in algebra. If she wants an average of at least 90 after her fourth test, what score must she make on her fourth test?

Step 4 **Solve.** $\frac{252 + x}{4} \geq 80$ Add the known scores.

$4\left(\frac{252 + x}{4}\right) \geq 4(80)$ Multiply by 4.

$252 + x \geq 320$

$252 + x - 252 \geq 320 - 252$ Subtract 252.

$x \geq 68$ Combine like terms.

Step 5 **State the answer.** He must score 68 or more on the fourth test to have an average of *at least* 80.

Step 6 **Check.** $\frac{86 + 88 + 78 + 68}{4} = \frac{320}{4} = 80$

(Also show that any number in (68, 100] makes the average greater than 80.)

◀ Work Problem **12** at the Side.

EXAMPLE 12 **Using a Linear Inequality to Solve a Rental Problem**

A rental company charges \$20 to rent a chain saw, plus \$9 per hr. Tom Ruhberg can spend no more than \$65 to clear some logs from his yard. What is the *maximum* amount of time he can use the rented saw?

Step 1 **Read** the problem again.

Step 2 **Assign a variable.** Let h = the number of hours he can rent the saw.

Step 3 **Write an inequality.** He must pay \$20, plus \$9h, to rent the saw for h hours, and this amount must be *no more than* \$65.

Cost of renting	is no more than	65 dollars.
$20 + 9h$	$\leq$	65

Step 4 **Solve.** $9h \leq 45$ Subtract 20.

$h \leq 5$ Divide by 9.

Step 5 **State the answer.** He can use the saw for a maximum of 5 hr. (He may use it for less time, as indicated by the inequality $h \leq 5$.)

Step 6 **Check.** If Tom uses the saw for 5 hr, he will spend

$20 + 9(5) = 65$ dollars, the maximum amount.

◀ Work Problem **13** at the Side.

13 Solve the problem.

A rental company charges \$10 to rent a leaf blower, plus \$7.50 per hr. Marge Ruhberg can spend no more than \$40 to blow leaves from her driveway and pool deck. What is the *maximum* amount of time she can use the rented leaf blower?

ANSWERS

12. 88 or more
13. 4 hr

2.7 Exercises

FOR EXTRA HELP

MyMathLab

Math XL PRACTICE

WATCH

DOWNLOAD

READ

REVIEW

Write an inequality using the variable x that corresponds to each graph of solutions on a number line.

1.

$x > -4$

2.

$x \geq -4$

3.

$x \leq 4$

4. 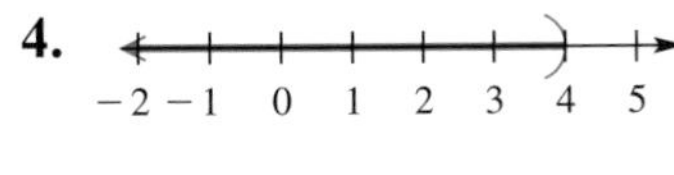

$x < 4$

5. 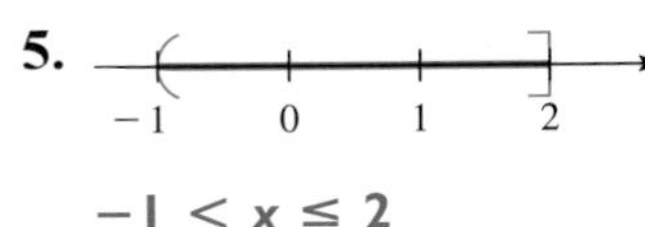

$-1 < x \leq 2$

6. 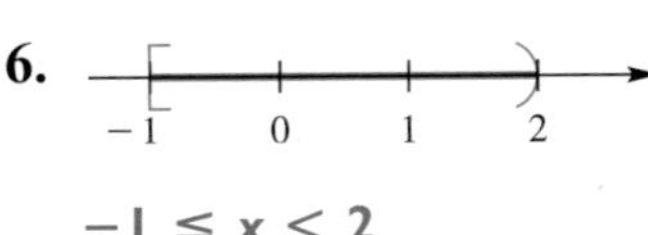

$-1 \leq x < 2$

Write each inequality in interval notation, and graph the interval. See Examples 1–3.

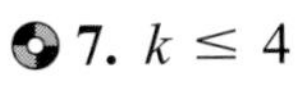
7. $k \leq 4$

$(-\infty, 4]$

8. $r \leq -10$ 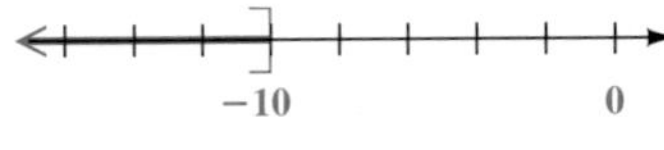

$(-\infty, -10]$

9. $x > -3$ 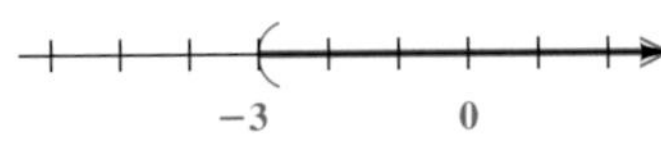

$(-3, \infty)$

10. $x > 3$ 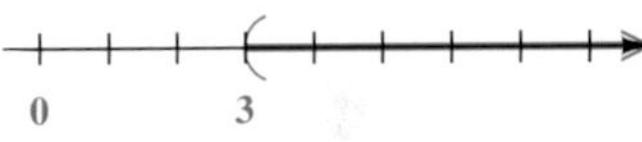

$(3, \infty)$

11. $8 \leq x \leq 10$ 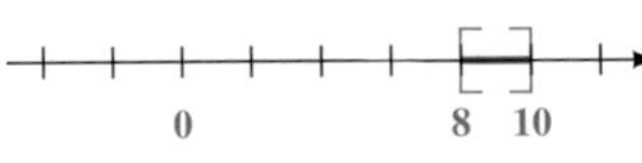

$[8, 10]$

12. $3 \leq x \leq 5$ 

$[3, 5]$

13. $0 < x \leq 10$ 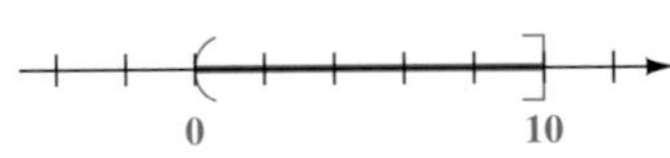

$(0, 10]$

14. $-3 \leq x < 5$ 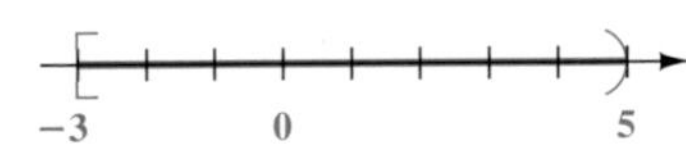

$[-3, 5)$

Solve each inequality. Write the solution set in interval notation, and graph it. See Example 4.

15. $z - 8 \geq -7$ 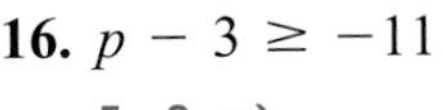

$[1, \infty)$

16. $p - 3 \geq -11$ 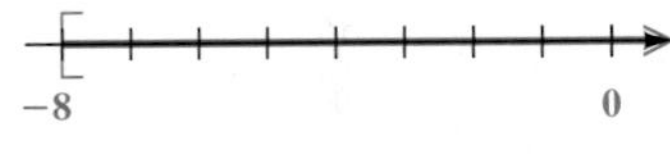

$[-8, \infty)$

17. $2k + 3 \geq k + 8$

$[5, \infty)$

18. $3x + 7 \geq 2x + 11$ 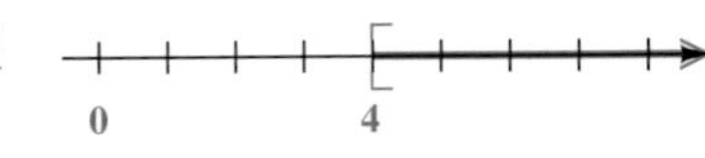

$[4, \infty)$

19. $3n + 5 < 2n - 1$ 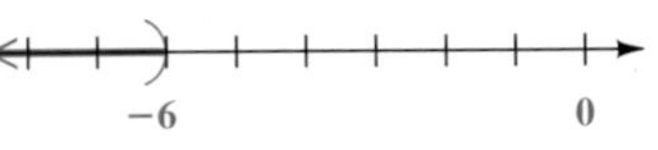

$(-\infty, -6)$

20. $5x - 2 < 4x - 5$

$(-\infty, -3)$

21. Under what conditions must the inequality symbol be reversed when using the multiplication property of inequality?

It must be reversed when multiplying or dividing by a negative number.

22. Explain the steps you would use to solve the inequality $-5x > 20$.

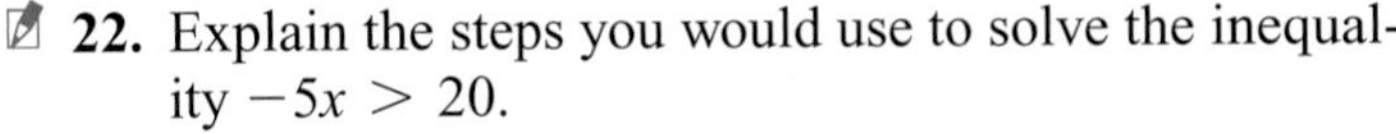

Divide by -5 and reverse the inequality symbol to get $x < -4$.

Solve each inequality. Write the solution set in interval notation, and graph it. See Examples 5 and 6.

23. $3x < 18$

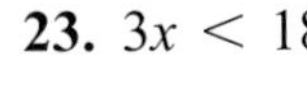
$(-\infty, 6)$

24. $5x < 35$

$(-\infty, 7)$

25. $2x \geq -20$ 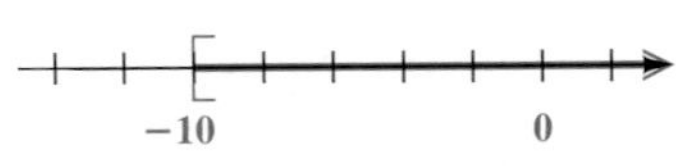

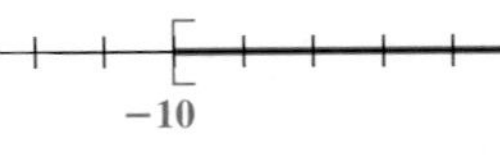

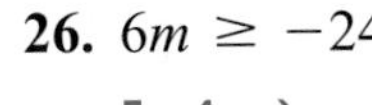

$[-10, \infty)$

26. $6m \geq -24$ 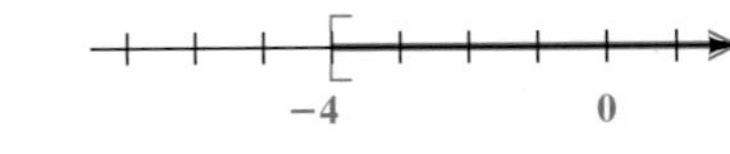

$[-4, \infty)$

27. $-8t > 24$

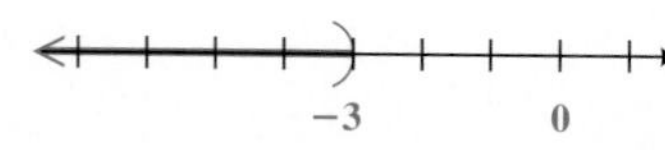

$(-\infty, -3)$

28. $-7x > 49$

$(-\infty, -7)$

29. $-x \geq 0$

$(-\infty, 0]$

30. $-k < 0$

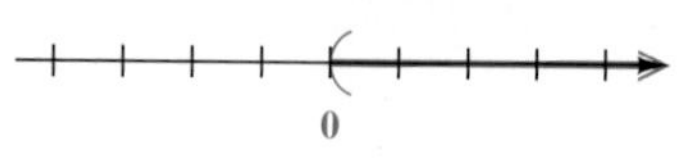

$(0, \infty)$

31. $-\frac{3}{4}r < -15$

$(20, \infty)$

32. $-\frac{7}{8}t < -14$

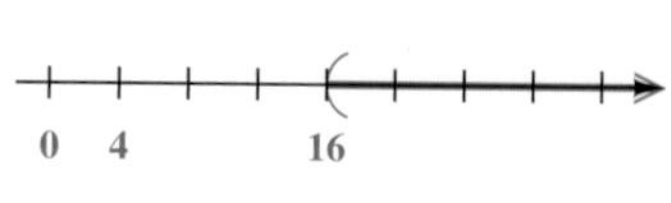

$(16, \infty)$

33. $-0.02x \leq 0.06$

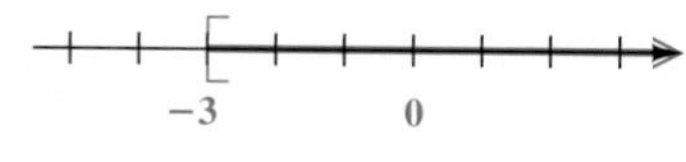

$[-3, \infty)$

34. $-0.03v \geq -0.12$

$(-\infty, 4]$

Solve each inequality. Write the inequality in interval notation, and graph it. See Examples 4–9.

35. $8x + 9 \leq -15$

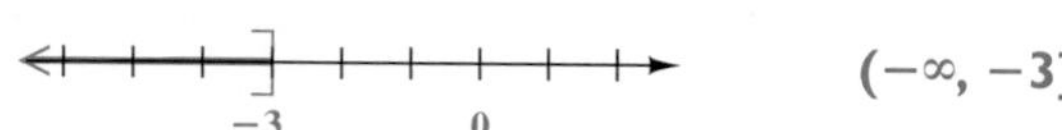

$(-\infty, -3]$

36. $6x + 7 \leq -17$

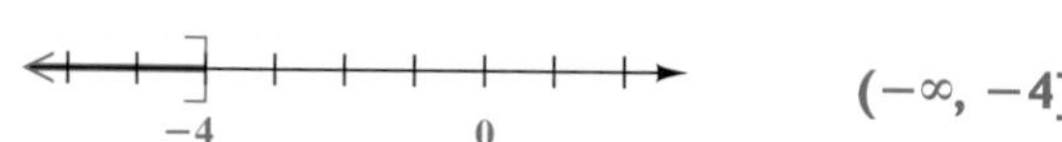

$(-\infty, -4]$

37. $-4x - 3 < 1$

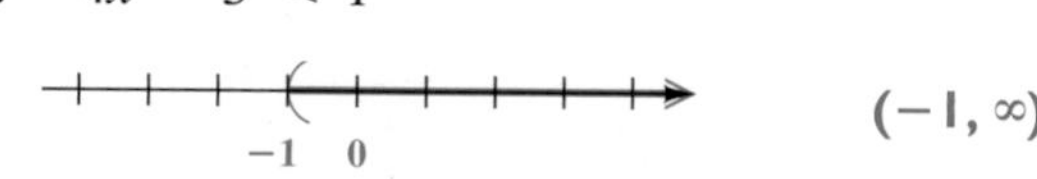

$(-1, \infty)$

38. $-5x - 4 < 6$

$(-2, \infty)$

39. $5r + 1 \geq 3r - 9$

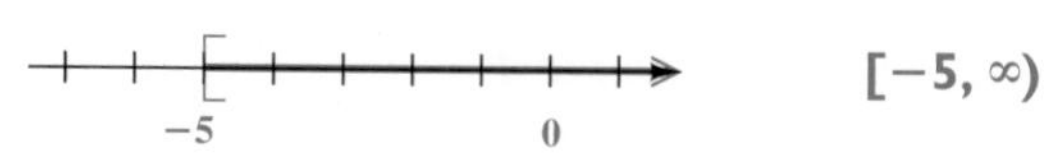

$[-5, \infty)$

40. $6t + 3 < 3t + 12$

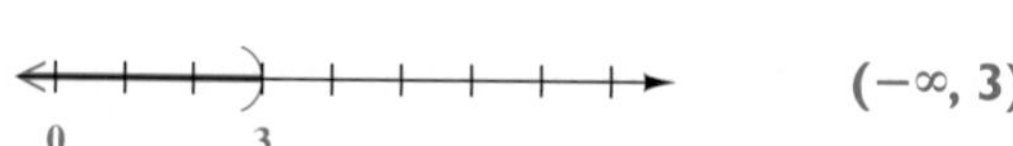

$(-\infty, 3)$

41. $6x + 3 + x < 2 + 4x + 4$

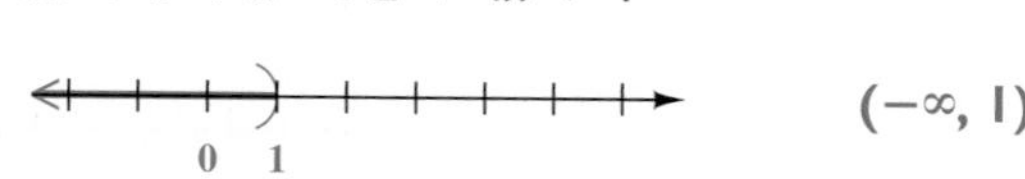

$(-\infty, 1)$

42. $-4w + 12 + 9w \geq w + 9 + w$

$[-1, \infty)$

43. $-x + 4 + 7x \leq -2 + 3x + 6$

$(-\infty, 0]$

44. $14x - 6 + 7x > 4 + 10x - 10$

$(0, \infty)$

45. $5(t - 1) > 3(t - 2)$

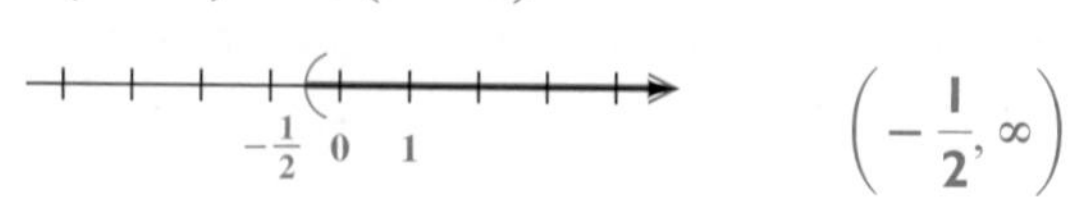

$\left(-\frac{1}{2}, \infty\right)$

46. $7(m - 2) < 4(m - 4)$

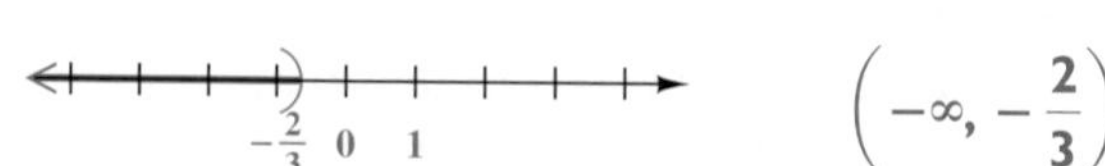

$\left(-\infty, -\frac{2}{3}\right)$

47. $5(x + 3) - 6x \leq 3(2x + 1) - 4x$

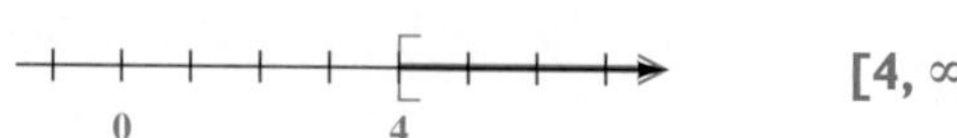

$[4, \infty)$

48. $2(x - 5) + 3x < 4(x - 6) + 1$

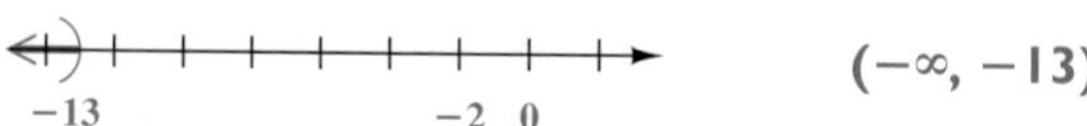

$(-\infty, -13)$

49. $\frac{2}{3}(p+3) > \frac{5}{6}(p-4)$

$(-\infty, 32)$

50. $\frac{7}{9}(n-4) \le \frac{4}{3}(n+5)$

$\left[-\frac{88}{5}, \infty\right)$

51. $-\frac{1}{4}(p+6) + \frac{3}{2}(2p-5) < 10$

$\left(-\infty, \frac{76}{11}\right)$

52. $\frac{3}{5}(k-2) - \frac{1}{4}(2k-7) \le 3$

$\left(-\infty, \frac{49}{2}\right]$

53. $4x - (6x+1) \le 8x + 2(x-3)$

$\left[\frac{5}{12}, \infty\right)$

54. $2x - (4x+3) < 6x + 3(x+4)$

$\left(-\frac{15}{11}, \infty\right)$

Solve each inequality. Write the solution set in interval notation, and graph it. See Example 10.

55. $-5 \le 2x - 3 \le 9$

$[-1, 6]$

56. $-7 \le 3x - 4 \le 8$

$[-1, 4]$

57. $10 < 7p + 3 < 24$

$(1, 3)$

58. $-8 \le 3r - 1 \le -1$

$\left[-\frac{7}{3}, 0\right]$

59. $-12 < -1 + 6m \le -5$

$\left(-\frac{11}{6}, -\frac{2}{3}\right]$

60. $-14 \le 1 + 5q < 3$

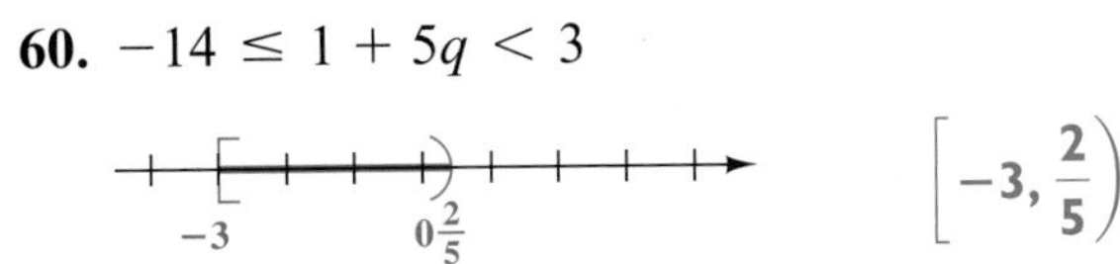

$\left[-3, \frac{2}{5}\right)$

61. $4 < -9x + 5 \le 8$

$\left[-\frac{1}{3}, \frac{1}{9}\right)$

62. $4 < -2x + 3 \le 8$

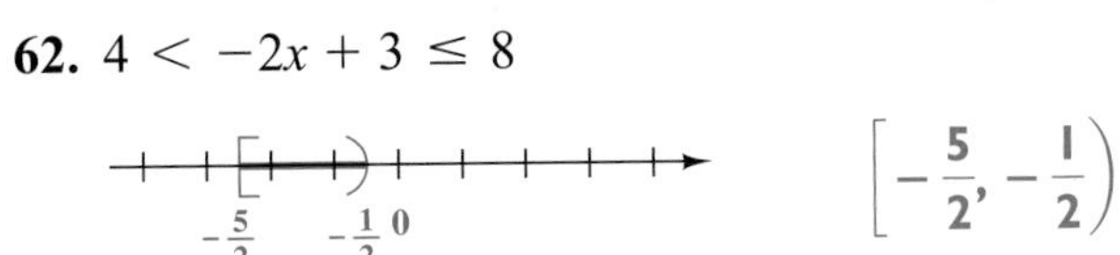

$\left[-\frac{5}{2}, -\frac{1}{2}\right)$

Solve each problem. See Examples 11 and 12.

63. John Douglas has grades of 84 and 98 on his first two history tests. What must he score on his third test so that his average is at least 90?

88 or more

64. Elizabeth Gainey has scores of 74 and 82 on her first two algebra tests. What must she score on her third test so that her average is at least 80?

84 or more

65. When 2 is added to the difference between six times a number and 5, the result is greater than 13 added to 5 times the number. Find all such numbers.

all numbers greater than 16

66. When 8 is subtracted from the sum of three times a number and 6, the result is less than 4 more than the number. Find all such numbers.

all numbers less than 3

67. The formula for converting Celsius temperature to Fahrenheit is

$$F = \frac{9}{5}C + 32.$$

The Fahrenheit temperature of Providence, Rhode Island, has never exceeded 104°. How would you describe this using Celsius temperature?

It has never exceeded 40°C.

68. The formula for converting Fahrenheit temperature to Celsius is

$$C = \frac{5}{9}(F - 32).$$

If the Celsius temperature on a certain day in San Diego, California, is never more than 25°, how would you describe the corresponding Fahrenheit temperature?

It is never more than 77°F.

69. For what values of x would the rectangle have perimeter of at least 400? **32 or greater**

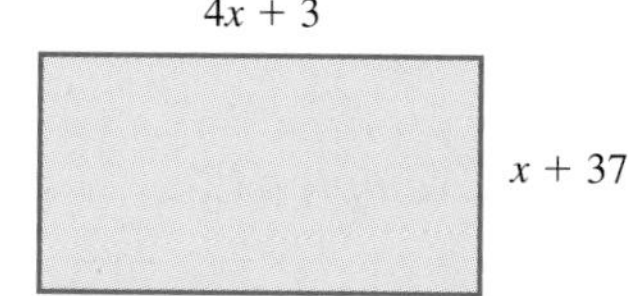

70. For what values of x would the triangle have perimeter of at least 72? **14 or greater**

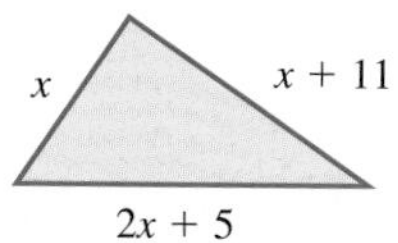

71. An international phone call costs \$2.00 for the first three minutes plus \$0.30 per minute for each minute or fractional part of a minute after the first three minutes. If x represents the number of minutes of the length of the call after the first three minutes, then $2 + 0.30x$ represents the cost of the call. If Jorge has \$5.60 to spend on a call, what is the maximum total time he can use the phone?

15 min

72. At the Speedy Gas 'n Go, a car wash costs \$4.50, and gasoline is selling for \$4.20 per gal. Terri Hoelker has \$48.60 to spend, and her car is so dirty that she must have it washed. What is the maximum number of gallons of gasoline that she can purchase?

10.5 gal

Relating Concepts (Exercises 73–77) For Individual or Group Work

Work Exercises 73–77 in order, *to see the connection between the solution of an equation and the solutions of the corresponding inequalities. Graph the solutions in Exercises 73–75.*

73. $3x + 2 = 14$

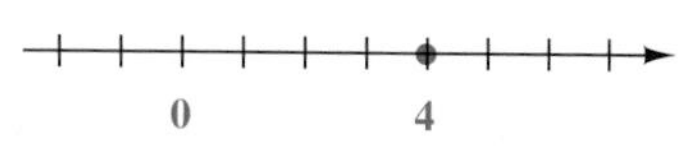

74. $3x + 2 < 14$

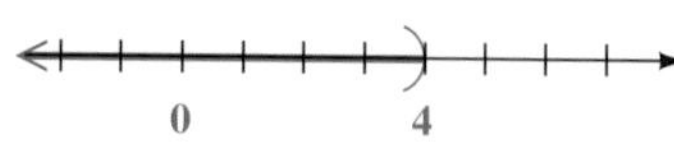

75. $3x + 2 > 14$

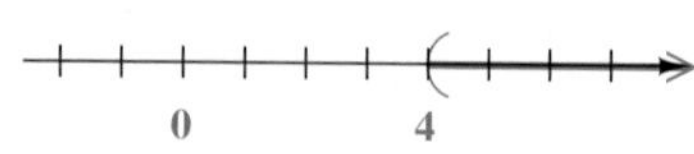

76. Now graph all the solutions together on the following number line.

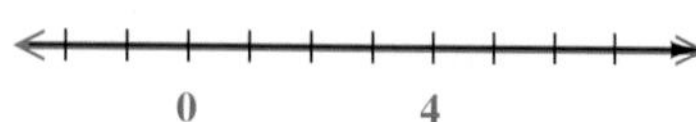

How would you describe the graph?

It is the set of all real numbers.

77. Based on your results from Exercises 73–76, if you were to graph the solutions of

$$-4x + 3 = -1, \quad -4x + 3 > -1,$$
$$\text{and} \quad -4x + 3 < -1$$

on the same number line, what do you think the graph would be?

The graph would be the set of all real numbers.

Study Skills

MANAGING YOUR TIME

Many college students find themselves juggling a difficult schedule and multiple responsibilities. Perhaps you are going to school, working part time, and managing family demands. Here are some tips to help you develop good time management skills and habits.

- **Read the syllabus for each class.** Check on class policies, such as attendance, late homework, and make-up tests. Find out how you are graded. Keep the syllabus in your notebook.
- **Make a semester or quarter calendar.** Put test dates and major due dates for *all* your classes on the same calendar. That way you will see which weeks are the really busy ones. Try using a different color pen for each class. A sample semester calendar is given on the next page.
- **Make a weekly schedule.** After you fill in your classes and other regular responsibilities (such as work, picking up kids from school, etc.), block off some study periods during the day that you can guarantee you will use for studying. Aim for 2 hours of study for each 1 hour you are in class.
- **Make "to-do" lists.** Then use them by crossing off the tasks as you complete them. You might even number them in the order they need to be done (most important ones first).
- **Break big assignments into smaller chunks.** They won't seem so big that way. Make deadlines for each small part so that you stay on schedule.
- **Give yourself small breaks in your studying.** Do not try to study for hours at a time. Your body needs rest between periods of learning. Try to give yourself a 10 minute break each hour or so. You will learn more and remember it longer.
- **If you get off schedule, just try to get back on schedule tomorrow.** We all slip from time to time. Make a new "to-do" list and start doing the most important things first.
- **Get help when you need it.** Talk with your instructor during office hours. Also, most colleges have some kind of learning center, tutoring center, or counseling office. If you feel lost or overwhelmed, ask for help. Someone can help you prioritize and decide what to spend your time on right away.

What two or three of the suggestions above will you try this week? How do you think they will help you?

1. ____________________

2. ____________________

3. ____________________

OBJECTIVES

1 Create a semester schedule.

2 Create a "to do" list.

Why Are These Techniques Brain Friendly?

We are creatures of habit. We enjoy a little routine. For example, by choosing the same study time and place each day, you will find that you quickly settle in to your reading or homework.

You function better when you are calm. Too much rushing around at the last minute to get your homework and studying done makes it more difficult for you to learn and remember. So, a little planning can really pay off.

Building rest into your schedule is also good for you. Many people benefit from short "power" naps.

We've suggested using color on your calendars. Messy and hard to read calendars will not be helpful, and you probably won't look at them often.

SEMESTER CALENDAR

WEEK	MON	TUES	WED	THUR	FRI	SAT	SUN
1							
2							
3							
4							
5							
6							
7							
8							
9							
10							
11							
12							
13							
14							
15							
16							

Chapter 2 Summary

▶ Key Terms

2.1

linear equation A linear equation in one variable is an equation that can be written in the form $Ax + B = C$, where A, B, and C are real numbers, with $A \neq 0$.

solution set The set of all solutions of an equation is its solution set.

equivalent equations Equations that have exactly the same solution sets are equivalent equations.

2.3

conditional equation A conditional equation is an equation that is true for some values of the variable and false for others.

identity An identity is an equation that is true for all values of the variable.

contradiction A contradiction is an equation with no solution.

2.4

complementary angles Two angles whose measures have a sum of 90° are complementary angles.

right angle A right angle measures 90°.

supplementary angles Two angles whose measures have a sum of 180° are supplementary angles.

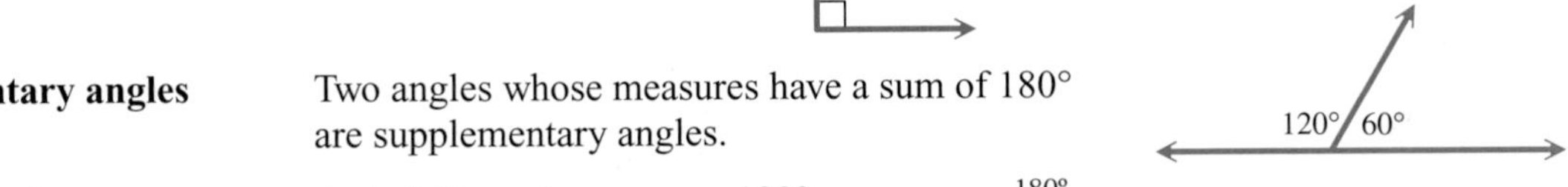

straight angle A straight angle measures 180°.

consecutive integers Two integers that differ by 1 are consecutive integers.

2.5

formula A formula is an equation in which variables are used to describe a relationship.

area The area of a plane geometric figure is a measure of the surface covered by the figure.

perimeter The perimeter of a plane geometric figure is the distance around the figure.

vertical angles Vertical angles are angles formed by intersecting lines. They have the same measure.

2.6

ratio A ratio is a comparison of two quantities using a quotient.

proportion A proportion is a statement that two ratios are equal.

cross products The method of cross products provides a way of determining whether a proportion is true.

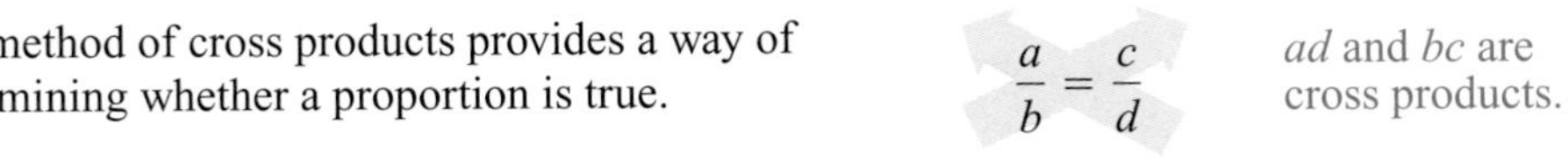

terms In the proportion $\frac{a}{b} = \frac{c}{d}$, a, b, c, and d are the terms. The a and d terms are called the **extremes,** and the b and c terms are called the **means.**

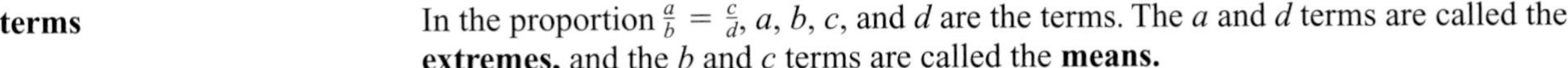

2.7

inequality Inequalities are algebraic expressions related by $<$, $\leq$, $>$, or $\geq$.

interval An interval is a portion of a number line.

The interval $(-1, \infty)$

interval notation Interval notation is a special notation that uses parentheses ( ) and/or brackets [ ] to describe an interval on a number line.

linear inequality A linear inequality in one variable can be written in the form $Ax + B < C$, $Ax + B \leq C$, $Ax + B > C$, or $Ax + B \geq C$, where A, B, and C are real numbers, with $A \neq 0$.

three-part inequality An inequality that says that one number is between two other numbers is a three-part inequality.

New Symbols

$\emptyset$	empty (null) set	(a, b)	interval notation for $a < x < b$	∞	infinity
$1°$	one degree	$[a, b]$	interval notation for $a \leq x \leq b$	$-\infty$	negative infinity
a to b, $a:b$, or $\frac{a}{b}$	the ratio of a to b			$(-\infty, \infty)$	set of all real numbers

Test Your Word Power

See how well you have learned the vocabulary in this chapter. Answers, with examples, follow the Quick Review.

1. A **solution** of an equation is a number that
   A. makes an expression undefined
   B. makes the equation false
   C. makes the equation true
   D. makes an expression equal to 0.

2. **Complementary angles** are angles
   A. formed by two parallel lines
   B. whose sum is 90°
   C. whose sum is 180°
   D. formed by perpendicular lines.

3. **Supplementary angles** are angles
   A. formed by two parallel lines
   B. whose sum is 90°
   C. whose sum is 180°
   D. formed by perpendicular lines.

4. A **ratio**
   A. compares two quantities using a quotient
   B. says that two quotients are equal
   C. is a product of two quantities
   D. is a difference between two quantities.

5. A **proportion**
   A. compares two quantities using a quotient
   B. says that two ratios are equal
   C. is a product of two quantities
   D. is a difference between two quantities.

6. An **inequality** is
   A. a statement that two algebraic expressions are equal
   B. a point on a number line
   C. an equation with no solutions
   D. a statement with algebraic expressions related by $<, \leq, >,$ or $\geq$.

Quick Review

Concepts	Examples
2.1 The Addition Property of Equality	
The same number may be added to (or subtracted from) each side of an equation without changing the solution.	Solve. $x - 6 = 12$ $x - 6 + 6 = 12 + 6$ Add 6. $x = 18$ Combine like terms. Solution set: $\{18\}$
2.2 The Multiplication Property of Equality	
Each side of an equation may be multiplied (or divided) by the same nonzero number without changing the solution.	Solve. $\frac{3}{4}x = -9$ $\frac{4}{3} \cdot \frac{3}{4}x = \frac{4}{3}(-9)$ Multiply by $\frac{4}{3}$. $x = -12$ Solution set: $\{-12\}$

Concepts	Examples
2.3 More on Solving Linear Equations	
Step 1 Simplify each side separately.	Solve. $2x + 2(x + 1) = 14 + x$ $2x + 2x + 2 = 14 + x$ Distributive property $4x + 2 = 14 + x$ Combine like terms.
Step 2 Isolate the variable term on one side.	$4x + 2 - x - 2 = 14 + x - x - 2$ Subtract x; subtract 2. $3x = 12$ Combine like terms.
Step 3 Isolate the variable.	$\frac{3x}{3} = \frac{12}{3}$ Divide by 3. $x = 4$
Step 4 Check.	To check, substitute 4 for x in the original equation. $2(4) + 2(4 + 1) \stackrel{?}{=} 14 + 4$ Let $x = 4$. $18 = 18$ True Solution set: $\{4\}$
2.4 An Introduction to Applications of Linear Equations	
Step 1 Read.	One number is 5 more than another. Their sum is 21. What are the numbers? We are looking for two numbers.
Step 2 Assign a variable.	Let x represent the lesser number. Then $x + 5$ represents the greater number.
Step 3 Write an equation.	$x + (x + 5) = 21$
Step 4 Solve the equation.	$2x + 5 = 21$ Combine like terms. $2x = 16$ Subtract 5. $x = 8$ Divide by 2.
Step 5 State the answer.	The numbers are 8 and 13.
Step 6 Check.	13 is 5 more than 8, and $8 + 13 = 21$. It checks.
2.5 Formulas and Additional Applications from Geometry	
To find the value of one of the variables in a formula, given values for the others, substitute the known values into the formula.	Find L if $A = LW$, given that $A = 24$ and $W = 3$. $24 = L \cdot 3$ $A = 24, W = 3$ $\frac{24}{3} = \frac{L \cdot 3}{3}$ Divide by 3. $8 = L$
To solve a formula for one of the variables, isolate that variable by treating the other variables as numbers and using the steps for solving equations.	Solve $P = 2a + 2b$ for b. $P - 2a = 2a + 2b - 2a$ Subtract $2a$. $P - 2a = 2b$ Combine like terms. $\frac{P - 2a}{2} = \frac{2b}{2}$ Divide by 2. $\frac{P - 2a}{2} = b$, or $b = \frac{P - 2a}{2}$

Concepts	Examples

2.6 Ratio, Proportion, and Percent

To write a ratio, express quantities in the same units.

4 ft to 8 in. = 48 in. to 8 in. $= \frac{48}{8} = \frac{6}{1}$

To solve a proportion, use the method of cross products.

Solve $\frac{x}{12} = \frac{35}{60}$.

$60x = 12 \cdot 35$ Cross products

$60x = 420$ Multiply.

$\frac{60x}{60} = \frac{420}{60}$ Divide by 60.

$x = 7$

Solution set: $\{7\}$

To solve a percent problem, use the proportion

$$\frac{\textbf{amount}}{\textbf{base}} = \frac{\textbf{percent}}{\textbf{100}}.$$

2.7 Solving Linear Inequalities

Step 1 Simplify each side separately.

Step 2 Isolate the variable term on one side.

Step 3 Isolate the variable.

Be sure to reverse the direction of the inequality symbol when multiplying or dividing by a negative number.

Solve and graph the solution set.

$3(1 - x) + 5 - 2x > 9 - 6$

$3 - 3x + 5 - 2x > 9 - 6$ Distributive property

$8 - 5x > 3$ Combine like terms.

$8 - 5x - 8 > 3 - 8$ Subtract 8.

$-5x > -5$ Combine like terms.

$\frac{-5x}{-5} < \frac{-5}{-5}$ Divide by -5; change $>$ to $<$.

$x < 1$

Solution set: $(-\infty, 1)$

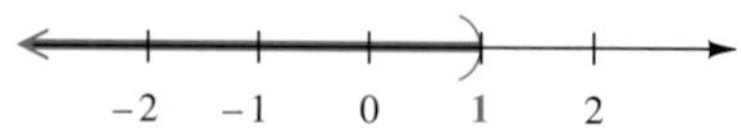

To solve a three-part inequality such as

$$4 < 2x + 6 < 8,$$

work with all three parts at the same time.

Solve.

$4 < 2x + 6 < 8$

$4 - 6 < 2x + 6 - 6 < 8 - 6$ Subtract 6.

$-2 < 2x < 2$ Combine like terms.

$\frac{-2}{2} < \frac{2x}{2} < \frac{2}{2}$ Divide by 2.

$-1 < x < 1$

Solution set: $(-1, 1)$

▸ Answers to Test Your Word Power

1. C; *Example:* 8 is the solution of $2x + 5 = 21$.
2. B; *Example:* Angles with measures 35° and 55° are complementary angles.
3. C; *Example:* Angles with measures 112° and 68° are supplementary angles.
4. A; *Example:* $\frac{7 \text{ in.}}{12 \text{ in.}} = \frac{7}{12}$
5. B; *Example:* $\frac{2}{3} = \frac{8}{12}$
6. D; *Examples:* $x < 5$, $7 + 2y \geq 11$

Chapter 2 Review Exercises

[2.1–2.3] *Solve each equation. Check the solution.*

1. $x - 7 = 2$

   {9}

2. $4r - 6 = 10$

   {4}

3. $5x + 8 = 4x + 2$

   {−6}

4. $8t = 7t + \frac{3}{2}$

   $\left\{\frac{3}{2}\right\}$

5. $(4r - 8) - (3r + 12) = 0$

   {20}

6. $7(2x + 1) = 6(2x - 9)$

   $\left\{-\frac{61}{2}\right\}$

7. $-\frac{6}{5}y = -18$

   {15}

8. $\frac{1}{2}r - \frac{1}{6}r + 3 = 2 + \frac{1}{6}r + 1$

   {0}

9. $3x - (-2x + 6) = 4(x - 4) + x$

   $\emptyset$

10. $0.10(x + 80) + 0.20x = 8 + 0.30x$

    {all real numbers}

[2.4] *Solve each problem.*

11. If 7 is added to five times a number, the result is equal to three times the number. Find the number.

    $-\frac{7}{2}$

12. If 4 is subtracted from twice a number, the result is 36. Find the number.

    20

13. The land area of Hawaii is 5213 mi^2 greater than that of Rhode Island. Together, the areas total 7637 mi^2. What is the area of each state?

    Hawaii: 6425 mi^2; Rhode Island: 1212 mi^2

14. The height of Seven Falls in Colorado is $\frac{5}{2}$ the height (in feet) of Twin Falls in Idaho. The sum of the heights is 420 ft. Find the height of each.

    Seven Falls: 300 ft; Twin Falls: 120 ft

15. The supplement of an angle measures 10 times the measure of its complement. What is the measure of the angle (in degrees)?

    80°

16. Find two consecutive odd integers such that when the lesser is added to twice the greater, the result is 24 more than the greater integer.

    11, 13

[2.5] *A formula is given in each exercise, along with the values for all but one of the variables. Find the value of the variable that is not given. (For Exercises 19 and 20, use 3.14 as an approximation for π.)*

17. $A = \frac{1}{2}bh$; $A = 44, b = 8$

    $h = 11$

18. $A = \frac{1}{2}h(b + B)$; $b = 3, B = 4, h = 8$

    $A = 28$

19. $C = 2\pi r$; $C = 29.83$

    $r = 4.75$

20. $V = \frac{4}{3}\pi r^3$; $r = 9$

    $V = 3052.08$

Solve each formula for the specified variable.

21. $A = bh$ for h $\quad h = \frac{A}{b}$

22. $A = \frac{1}{2}h(b + B)$ for h $\quad h = \frac{2A}{b + B}$

Find the measure of each marked angle.

23.

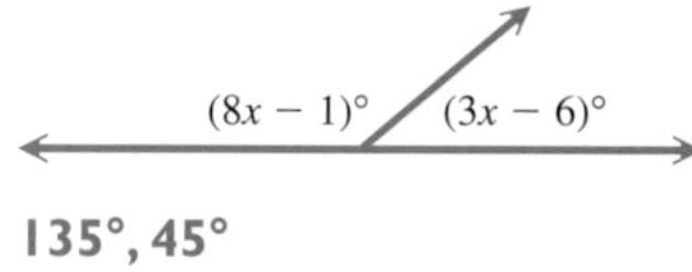

135°, 45°

24.

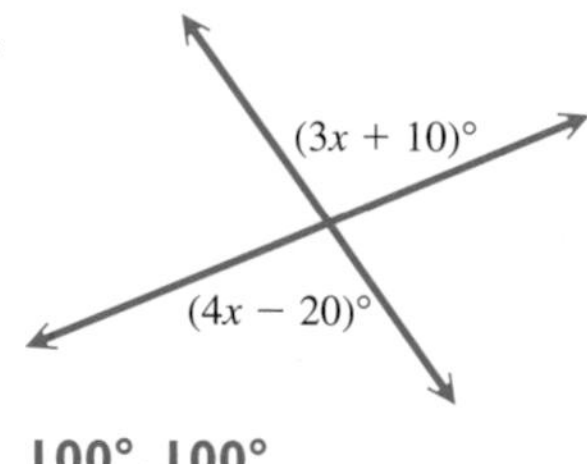

100°, 100°

Solve each application of geometry.

25. A cinema screen in Indonesia has length 92.75 ft and width 70.5 ft. What is the perimeter? What is the area? (*Source: Guinness World Records.*)

perimeter: 326.5 ft; area: 6538.875 ft^2

26. There is a Montezuma cypress in Mexico that is 137 ft tall and has a circumference of about 146.9 ft. What is the diameter of the tree? What is the radius of the tree? Use 3.14 as an approximation for π. Round your answers to the nearest hundredth. (*Source: Guinness World Records.*)

diameter: 46.78 ft; radius: 23.39 ft

[2.6] *Write a ratio for each word phrase. Write fractions in lowest terms.*

27. 60 cm to 40 cm $\quad \frac{3}{2}$

28. 5 days to 2 weeks $\quad \frac{5}{14}$

29. 90 in. to 10 ft $\quad \frac{3}{4}$

30. 3 months to 3 yr $\quad \frac{1}{12}$

Solve each proportion.

31. $\frac{p}{21} = \frac{5}{30}$ $\quad \left\{\frac{7}{2}\right\}$

32. $\frac{5 + x}{3} = \frac{2 - x}{6}$ $\quad \left\{-\frac{8}{3}\right\}$

33. $\frac{y}{5} = \frac{6y - 5}{11}$ $\quad \left\{\frac{25}{19}\right\}$

34. Explain how 40% can be expressed as a ratio of two whole numbers.

40% means $\frac{40}{100}$, or $\frac{2}{5}$. It is the same as the ratio of 2 to 5.

Solve each problem.

35. If 2 lb of fertilizer will cover 150 ft^2 of lawn, how many pounds would be needed to cover 500 ft^2?

$6\frac{2}{3}$ lb

36. If 8 oz of medicine must be mixed with 20 oz of water, how many ounces of medicine must be mixed with 90 oz of water?

36 oz

37. The distance between two cities on a road map is 32 cm. The two cities are actually 150 km apart. The distance on the map between two other cities is 80 cm. How far apart are these cities?

375 km

38. Find the best buy. Give the unit price to the nearest thousandth for that size. (*Source:* Cub Foods.)

CEREAL

Size	Price
15 oz	\$2.69
20 oz	\$3.29
25.5 oz	\$3.49

25.5 oz; \$0.137

39. What is 23% of 76? $\quad$ 17.48

40. What percent of 12 is 21? $\quad$ 175%

41. 6 is what percent of 18? $\quad 33\frac{1}{3}\%$

42. 36% of what number is 900? $\quad$ 2500

43. Terry paid \$28,191, including tax, for her 2007 Honda CR-V. The sales tax rate was 6%. What was the actual price of the car to the nearest dollar? (*Source:* Author's sales receipt.)

\$26,595

44. Courtney, from the mathematics editorial division of Addison-Wesley, took the mathematics faculty from a community college out to dinner. The bill was \$304.75. Courtney added a 15% tip, and paid for the meal with her corporate credit card. What was the total price she paid to the nearest cent?

\$350.46

[2.7] *Write each inequality in interval notation, and graph it.*

45. $p \geq -4$ **$[-4, \infty)$**

46. $x < 7$ **$(-\infty, 7)$**

47. $-5 \leq k < 6$ **$[-5, 6)$**

48. $r \geq \frac{1}{2}$ **$\left[\frac{1}{2}, \infty\right)$**

Solve each inequality. Write the solution set in interval notation, and graph it.

49. $x + 6 \geq 3$ **$[-3, \infty)$**

50. $5t < 4t + 2$ **$(-\infty, 2)$**

51. $-6x \leq -18$ **$[3, \infty)$**

52. $8(k - 5) - (2 + 7k) \geq 4$ **$[46, \infty)$**

53. $4x - 3x > 10 - 4x + 7x$ **$(-\infty, -5)$**

54. $3(2w + 5) + 4(8 + 3w) < 5(3w + 2) + 2w$ **$(-\infty, -37)$**

55. $-3 \leq 2x + 1 < 4$ **$\left[-2, \frac{3}{2}\right)$**

56. $8 < 3x + 5 \leq 20$ **$(1, 5]$**

57. Justin Sudak has grades of 94 and 88 on his first two calculus tests. What possible scores on a third test will give him an average of at least 90?

88 or more

58. If nine times a number is added to 6, the result is at most 3. Find all such numbers.

all numbers less than or equal to $-\frac{1}{3}$

Mixed Review Exercises

Solve.

59. $\frac{x}{7} = \frac{x - 5}{2}$ **{7}**

60. $d = 2r$ for r **$r = \frac{d}{2}$**

61. $-2x > -4$ **$(-\infty, 2)$**

62. $2k - 5 = 4k + 13$ **{−9}**

63. $0.05x + 0.02x = 4.9$ **{70}**

64. $2 - 3(t - 5) = 4 + t$ **$\left\{\frac{13}{4}\right\}$**

65. $9x - (7x + 2) = 3x + (2 - x)$ **∅**

66. $\frac{1}{3}s + \frac{1}{2}s + 7 = \frac{5}{6}s + 5 + 2$ **{all real numbers}**

67. Pizza Hut and Domino's, the top-selling pizza restaurants in 2006, together had sales of $14.45 billion. Domino's sales were $4.25 billion less than Pizza Hut's. What were sales in billions for each restaurant? (*Source:* Pizza Today.)

Pizza Hut: $9.35 billion; Domino's: $5.1 billion

68. On a world globe, the distance between Capetown and Bangkok, two cities that are actually 10,080 km apart, is 12.4 in. The actual distance between Moscow and Berlin is 1610 km. How far apart are Moscow and Berlin on this globe, to the nearest inch?

2.0 in.

69. Of the 29 medals earned by Germany during the 2006 Winter Olympic games, there were two times as many silver as bronze medals and 5 more gold than bronze medals. How many of each medal did Germany earn? (*Source:* www.torin2006.org)

gold: 11; silver: 12; bronze: 6

70. In triangle *DEF*, the measure of angle *E* is twice the measure of angle *D*. Angle *F* has measure 18 less than six times the measure of angle *D*. Find the measure of each angle.

D: 22°; E: 44°; F: 114°

71. The perimeter of a triangle is 96 m. One side is twice as long as another, and the third side is 30 m long. What is the length of the longest side?

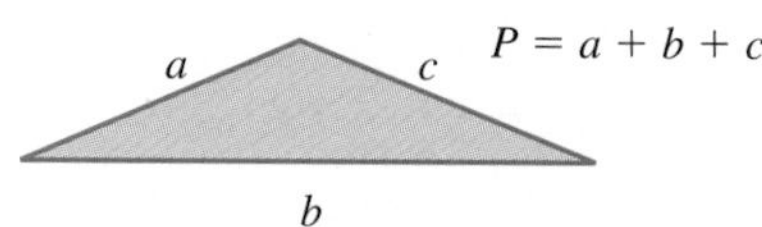

44 m

72. The perimeter of a rectangle is 288 ft. The length is 4 ft longer than the width. Find the width.

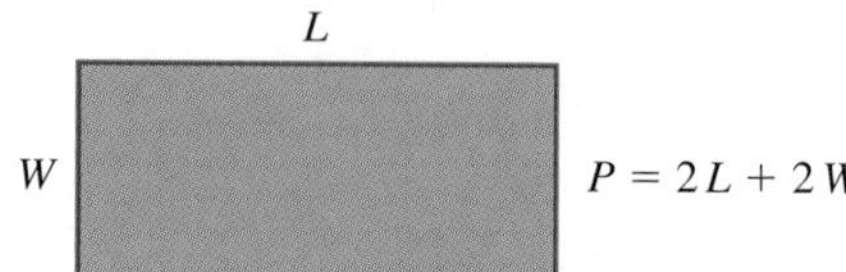

70 ft

73. Find the best buy. Give the unit price to the nearest thousandth for that size. (*Source:* Cub Foods.)

LAUNDRY DETERGENT

Size	Price
50 oz	$4.69
100 oz	$5.98
200 oz	$13.68

100 oz; $0.060

74. Find the measure of each marked angle.

24°, 66°

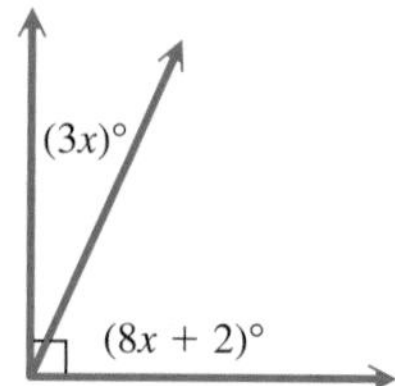

75. Latarsha has grades of 82 and 96 on her first two English tests. What must she make on her third test so that her average will be at least 90?

92 or more

76. If nine pairs of jeans cost $121.50, find the cost of five pairs. (Assume all are equally priced.)

$67.50

Chapter 2 Test

Use the Chapter Test Prep Video CD to see fully worked-out solutions to any of the exercises you want to review.

Solve each equation, and check the solution.

1. $3x - 7 = 11$ — **1.** $\{6\}$

2. $5x + 9 = 7x + 21$ — **2.** $\{-6\}$

3. $2 - 3(x - 5) = 3 + (x + 1)$ — **3.** $\left\{\frac{13}{4}\right\}$

4. $2.3x + 13.7 = 1.3x + 2.9$ — **4.** $\{-10.8\}$

5. $7 - (m - 4) = -3m + 2(m + 1)$ — **5.** $\emptyset$

6. $-\frac{4}{7}x = -12$ — **6.** $\{21\}$

7. $0.06(x + 20) + 0.08(x - 10) = 4.6$ — **7.** $\{30\}$

8. $-8(2x + 4) = -4(4x + 8)$ — **8.** {all real numbers}

Solve each problem.

9. The Dallas Mavericks finished with the best record for the 2006–2007 NBA regular season. They won 7 more than four times as many games as they lost. They played 82 games. How many games did they win and lose? (*Source:* www.sports.yahoo.com)

9. 67 wins, 15 losses

10. Three islands in the Hawaiian island chain are Hawaii (the Big Island), Maui, and Kauai. Together, their areas total 5300 mi^2. The island of Hawaii is 3293 mi^2 larger than the island of Maui, and Maui is 177 mi^2 larger than Kauai. What is the area of each island?

Kauai
Oahu
Molokai
Lanai
Maui
The Big Island
HAWAII

10. Hawaii: 4021 mi^2; Maui: 728 mi^2; Kauai: 551 mi^2

11. Find the measure of an angle if its supplement measures 10° more than three times its complement.

11. 50°

12. The formula for the perimeter of a rectangle is $P = 2L + 2W$.

(a) Solve for W.

(b) If $P = 116$ and $L = 40$, find the value of W.

12. (a) $W = \frac{P - 2L}{2}$

(b) 18

Find the measure of each marked angle.

13. 100°, 80°

14. 75°, 75°

13.

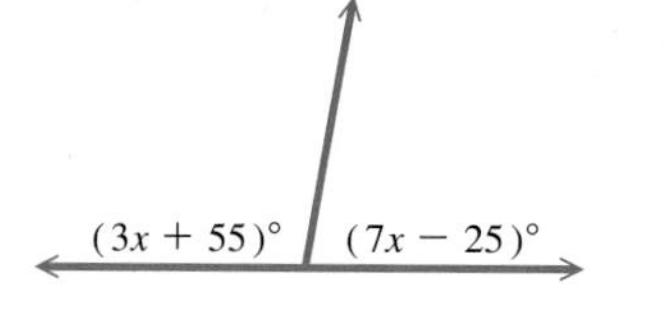

14.

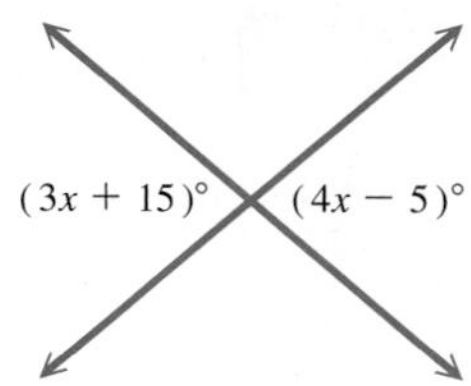

Solve each proportion.

15. {6}

16. {−29}

15. $\frac{z}{8} = \frac{12}{16}$

16. $\frac{x + 5}{3} = \frac{x - 3}{4}$

17. 32 oz; $0.250

17. Find the best buy. Give the unit price to the nearest thousandth for that size.

PROCESSED CHEESE SLICES

Size	Price
8 oz	$2.79
16 oz	$4.99
32 oz	$7.99

18. 2300 mi

18. The distance between Milwaukee and Boston is 1050 mi. On a certain map, this distance is represented by 42 in. On the same map, Seattle and Cincinnati are 92 in. apart. What is the actual distance between Seattle and Cincinnati?

19. 4.2%

19. Average tuition at two-year colleges was $2361 in 2007, up from $2265 in 2006. What percent increase in cost was this? Round your answer to the nearest tenth of a percent. (*Source:* College Board.)

20. (a) $x < 0$
(b) $-2 < x \leq 3$

20. Write an inequality involving x that describes the numbers graphed.

(a) −2 −1 0 1 2 3

(b)

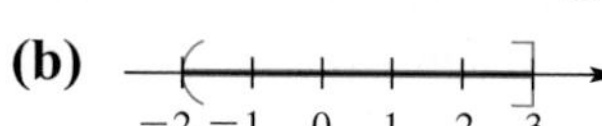

Solve each inequality. Write the solution set in interval notation, and graph it.

21. $(-\infty, 11)$ — 0 11

21. $-3x > -33$

22. $[-3, \infty)$ — −3 0

22. $-0.04x \leq 0.12$

23. $(-\infty, 4]$ — 0 4

23. $-4x + 2(x - 3) \geq 4x - (3 + 5x) - 7$

24. $(-2, 6]$ — −2 0 6

24. $-10 < 3x - 4 \leq 14$

25. 83 or more

25. Shania Johnson has scores of 76 and 81 on her first two algebra tests. If she wants an average of at least 80 after her third test, what score must she make on her third test?

Study Skills

TIPS FOR TAKING MATH TESTS

Techniques To Improve Your Test Score	Comments
Come prepared with a pencil, eraser, and calculator, if allowed. If you are easily distracted, sit in the corner farthest from the door.	***Working in pencil lets you erase,*** keeping your work neat and readable.
Scan the entire test, note the point value of different problems, and plan your time accordingly. Allow at least five minutes to check your work at the end of the testing time.	If you have 50 minutes to do 20 problems, $50 \div 20 = 2.5$ minutes per problem. ***Spend less time on easy ones,*** and more time on problems with higher point values.
Read directions carefully, and circle any significant words. When you finish a problem, read the directions again to make sure you did what was asked.	***Pay attention to announcements*** written on the board or made by your instructor. Ask if you don't understand. You don't want to get problems wrong because you misread the directions.
Show your work. Most math teachers give partial credit if some of the steps in your work are correct, even if the final answer is wrong. ***Write neatly.***	***If your teacher can't read your writing, you won't get credit for it.*** If you need more space to work, ask if you can use extra pieces of paper that you hand in with your test paper.
Check that the answer to an application problem is reasonable and makes sense. Read the problem again to make sure you've answered the question.	***Use common sense.*** Can the father really be seven years old? Would a month's rent be \$32,140? Label your answer: \$, years, inches, etc.
To check for careless errors, rework the problem without looking at your previous work. Cover your work with a piece of scratch paper, and pretend you are doing the problem for the first time. Then compare the two answers.	If you just "look over" your work, your mind can make the same mistake again without noticing it. ***Reworking the problem from the beginning forces you to rethink it.*** If possible, use a different method to solve the problem the second time.

Technique To Reduce Anxiety	Comments
Do not try to review up until the last minute before the test. Instead, go for a walk, do some deep breathing, and arrive just in time for the test. Ignore other students.	Listening to anxious classmates before the test ***may cause you to panic.*** Moderate exercise and deep breathing will calm your mind.

(continued)

OBJECTIVES

1 **Apply suggestions to tests and quizzes.**

2 **Develop a set of "best practices" to apply while testing.**

Why Are These Suggestions Brain Friendly?

Several of the suggestions address anxiety. Reducing anxiety allows you to think clearly.

Remember that your subconscious continues to work on a difficult problem even if you skip it and go on to the next one. Your mind will often come through for you if you are open to the idea.

Some of the suggestions ask you to use your common sense. Follow the directions, show your work, write neatly, and pay attention to whether your answers really make sense.

Techniques To Reduce Anxiety	Comments
Do a "knowledge dump" as soon as you get the test. Write important notes to yourself in a corner of the test paper, such as formulas or common errors you want to watch out for.	***Writing down tips and things that you've memorized lets you relax.*** You won't have to worry about forgetting those things and can refer to them as needed.
Do the easy problems first in order to build confidence. If you feel your anxiety starting to build, *immediately* stop for a minute, close your eyes, and take several slow, deep breaths.	***Greater confidence helps you get the easier problems correct.*** Anxiety causes shallow breathing, which leads to confusion and reduced concentration. Deep breathing calms you.
As you work on more difficult problems, you may have negative thoughts such as, "I can't do it," or "Who cares about this test anyway?" In your mind, think, "STOP" and take several deep, slow breaths. Or, replace the negative thoughts with positive ones.	***Try writing one of these positive statements on the top of your test paper.*** • I know I can do it. • I can do this one step at a time. • I've studied hard, and I'll do the best I can. • This test is a positive challenge for me to show what I've learned.
Read the harder problems twice. Write down *anything* that might help solve the problem: a formula, a diagram, etc. If you still can't get it, circle the problem and ***come back to it later***. Do *not* erase any of the things you wrote down.	If you know even a little bit about the problem, write it down. ***The answer may come to you as you work on it, or you may get partial credit.*** Don't spend too long on any one problem. Your subconscious mind will work on the tough problem while you go on with the test.
If you still can't solve a difficult problem when you come back to it the second time, ***make a guess and do not change it.*** In this situation, your first guess is your best bet. On any problem, do not change the answer just because you're a little unsure. ***Change it only if you find an obvious mistake.***	If you are thinking about changing an answer, be sure you have a good reason for changing it. If you cannot find a specific error, leave your first answer alone. ***When the test is returned, check to see if changing answers helped or hurt you.***
Ignore students who finish early. Use the entire test time. You do not get extra credit for finishing early. Use the extra time to rework problems and correct careless errors.	Students who leave early are often the ones who didn't study or who are too anxious to continue working. If they bother you, ***sit as far from the door as possible.***

Cumulative Review Exercises ▶▶▶ Chapters R–2

Write each fraction in lowest terms.

1. $\frac{15}{40}$ $\frac{3}{8}$

2. $\frac{108}{144}$ $\frac{3}{4}$

Perform the indicated operations.

3. $\frac{5}{6}+\frac{1}{4}+\frac{7}{15}$
$\frac{31}{20}$

4. $16\frac{7}{8}-3\frac{1}{10}$
$\frac{551}{40}$, or $13\frac{31}{40}$

5. $\frac{9}{8}\cdot\frac{16}{3}$
6

6. $\frac{3}{4}\div\frac{5}{8}$
$\frac{6}{5}$

7. $4.8+12.5+16.73$
34.03

8. $56.3-28.99$
27.31

9. $67.8(0.45)$
30.51

10. $236.46\div4.2$
56.3

11. In making dresses, Earth Works uses $\frac{5}{8}$ yd of trim per dress. How many yards of trim would be used to make 56 dresses?
35 yd

12. A cook wants to increase a recipe for Quaker Quick Grits that serves 4 to make enough for 10 people. The recipe calls for 3 cups of water. How much water will be needed to serve 10?
$7\frac{1}{2}$ cups

13. First published in 1953, the digest-sized *TV Guide* has recently been changed to a full-sized magazine. See the figure. The new magazine is 3 in. wider than the old guide. What is the difference in their heights? (*Source: TV Guide.*)
$3\frac{3}{8}$ in.

14. A small business owner bought 3 computer workstations of various sizes for \$329.99, \$379.99, and \$439.99 and 3 ergonomic office chairs for \$539.97 each. What was the final bill (without tax)? (*Source:* Staples "Furniture Values" catalog, 2007.)
\$2769.88

Tell whether each inequality is true *or* false.

15. $\frac{8(7)-5(6+2)}{3\cdot5+1}\geq1$
true

16. $\frac{4(9+3)-8(4)}{2+3-3}\geq2$
true

Perform the indicated operations.

17. $-11+20+(-2)$ **7**

18. $13+(-19)+7$ **1**

19. $9-(-4)$ **13**

20. $-2(-5)(-4)$ **−40**

21. $\frac{4\cdot9}{-3}$ **−12**

22. $\frac{8}{7-7}$ **undefined**

23. $(-5+8)+(-2-7)$ **−6**

24. $(-7-1)(-4)+(-4)$ **28**

25. $\frac{-3-(-5)}{1-(-1)}$ **1**

26. $\frac{6(-4)-(-2)(12)}{3^2+7^2}$ **0**

27. $\frac{(-3)^2-(-4)(2^4)}{5\cdot2-(-2)^3}$ **$\frac{73}{18}$**

28. $\frac{-2(5^3)-6}{4^2+2(-5)+(-2)}$ **−64**

Find the value of each expression when $x = -2$, $y = -4$, and $z = 3$.

29. $xz^3 - 5y^2$ **−134**

30. $\dfrac{xz - y^3}{-4z}$ $\mathbf{-\dfrac{29}{6}}$

Name the property illustrated by each equation.

31. $7(k + m) = 7k + 7m$
distributive property

32. $3 + (5 + 2) = 3 + (2 + 5)$
commutative property

33. $7 + (-7) = 0$
inverse property

34. $3.5(1) = 3.5$
identity property

Simplify each expression.

35. $4p - 6 + 3p - 8$ **7p − 14**

36. $-4(k + 2) + 3(2k - 1)$ **2k − 11**

Solve each equation, and check the solution.

37. $2r - 6 = 8$
{7}

38. $2(p - 1) = 3p + 2$
{−4}

39. $4 - 5(a + 2) = 3(a + 1) - 1$
{−1}

40. $2 - 6(z + 1) = 4(z - 2) + 10$
$\left\{-\dfrac{3}{5}\right\}$

41. $-(m - 1) = 3 - 2m$
{2}

42. $\dfrac{x - 2}{3} = \dfrac{2x + 1}{5}$
{−13}

43. $\dfrac{2x + 3}{5} = \dfrac{x - 4}{2}$
{26}

44. $\dfrac{2}{3}x + \dfrac{3}{4}x = -17$
{−12}

Solve each formula for the indicated variable.

45. $P = a + b + c + B$ for c $\quad c = P - a - b - B$

46. $P = 4s$ for s $\quad s = \dfrac{P}{4}$

Solve each inequality. Write the solution set in interval notation, and graph it.

47. $-5z \geq 4z - 18$

(−∞, 2]

48. $6(r - 1) + 2(3r - 5) < -4$

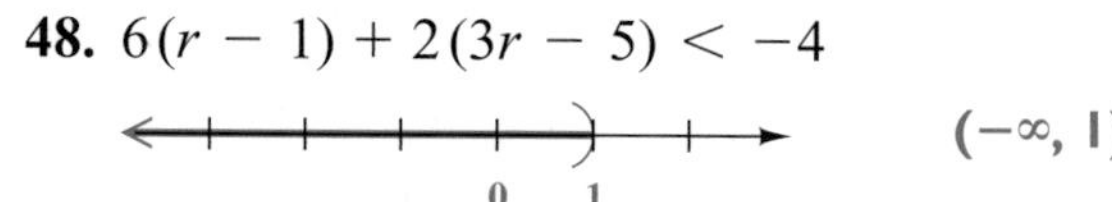

(−∞, 1)

Solve each problem.

49. The small business owner in Exercise 14 paid a sales tax of $6\frac{1}{4}\%$ on his purchase. What was the final bill, including tax, to the nearest cent? **$2943.00**

50. A car has a price of $5000. For trading in her old car, Shannon d'Hemecourt will get 25% off. Find the price of the car with the trade-in. **$3750**

51. Abby Tanenbaum bought textbooks at the college bookstore for $244.33, including 6% sales tax. What did the books cost before tax? **$230.50**

52. Stuart Sudak received a bill from his credit card company for $104.93. The bill included interest at $1\frac{1}{2}\%$ per month for one month and a $5.00 late charge. How much did his purchases amount to? **$98.45**

53. The perimeter of a rectangle is 98 cm. The width is 19 cm. Find the length. **30 cm**

?
19 cm

54. The area of a triangle is 104 in.^{2}. The base is 13 in. Find the height.

16 in.

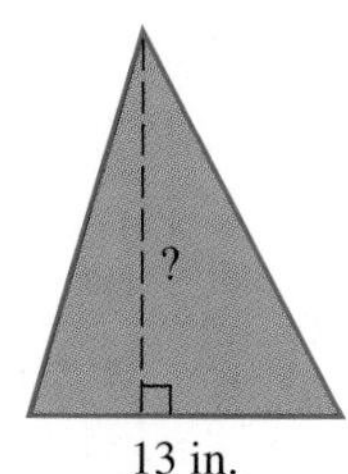

3

Graphs of Linear Equations and Inequalities; Functions

While U.S. consumers in general continue to pile up credit card debt, fewer undergraduate college students are carrying credit cards, and those with cards are using them less. In 2004, 76% of undergraduates carried at least one credit card, down from a peak of 83% in 2001. The average outstanding balance also dropped to \$2169, from a high of \$2748 in 2000. These declines are attributed to increased financial education aimed specifically at high school and college students. (*Source:* Nellie Mae.)

In Example 7 of Section 3.2, we examine a *linear equation in two variables* that models credit card debt in the United States.

3.1 Reading Graphs; Linear Equations in Two Variables

OBJECTIVES

1. Interpret graphs.
2. Write a solution as an ordered pair.
3. Decide whether a given ordered pair is a solution of a given equation.
4. Complete ordered pairs for a given equation.
5. Complete a table of values.
6. Plot ordered pairs.

As we saw in **Section R.1,** circle graphs (pie charts) provide a convenient way to organize and communicate information. Along with *bar graphs* and *line graphs,* they can be used to analyze data, make predictions, or simply to entertain us.

OBJECTIVE 1 Interpret graphs. A **bar graph** is used to show comparisons. It consists of a series of bars (or simulations of bars) arranged either vertically or horizontally. In a bar graph, values from two categories are paired with each other.

EXAMPLE 1 Interpreting a Bar Graph

The bar graph in Figure 1 shows U.S. sales of motor scooters, which have gained popularity due to their fuel efficiency. The graph compares sales in thousands.

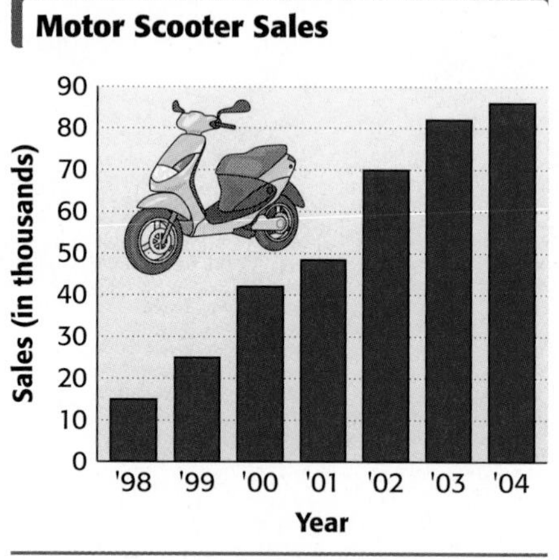

Source: Motorcycle Industry Council.

Figure 1

(a) In what years were sales greater than 50 thousand?

Locate 50 on the vertical scale and follow the line across to the right. Three years—2002, 2003, and 2004—have bars that extend above the line for 50, so sales were greater than 50 thousand in those years.

(b) Estimate sales in 2000 and 2004.

Locate the top of the bar for 2000, and move horizontally across to the vertical scale to see that it is about 40. Sales in 2000 were about 40 thousand. Follow the top of the bar for 2004 across to the vertical scale to see that it lies about halfway between 80 and 90 thousand, so sales in 2004 were about 85,000.

(c) Describe the change in sales as the years progressed.

As the years progressed, sales increased steadily, from about 15 thousand in 1998 to about 85 thousand in 2004.

Work Problem 1 at the Side.

1 Refer to the bar graph in Figure 1.

(a) Which years had sales less than 50 thousand?

(b) Estimate sales of motor scooters in 1999 and 2001.

(c) Describe the change in sales of motor scooters from 1999 to 2001.

ANSWERS

1. **(a)** 1998, 1999, 2000, 2001
   (b) 1999: about 25 thousand; 2001: about 50 thousand
   (c) Sales approximately doubled from 1999 to 2001.

A **line graph** is used to show changes or trends in data over time. To form a line graph, we connect a series of points representing data with line segments.

EXAMPLE 2 Interpreting a Line Graph

Current projections indicate that funding for Medicare will not cover its costs unless the program changes. The line graph in Figure 2 shows Medicare funds in billions of dollars for the years 2004 through 2013.

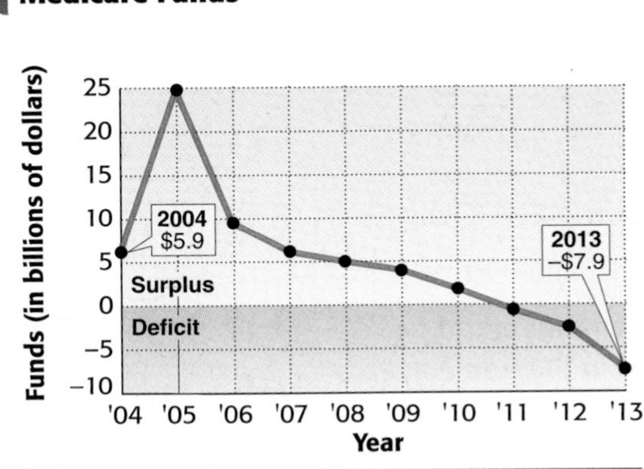

Source: Centers for Medicare and Medicaid Services.
*Projected

Figure 2

(a) Which is the only period in which Medicare funds increased?
Because the graph *rises* from 2004 to 2005 and falls in every other case, funds increased between these two years.

(b) What is the projected trend from 2005 to 2013?
Funds will decrease, since the graph *falls* during this period.

(c) In which year is it projected that funds will first show a deficit?
From 2004 to 2010, the graph is always above 0, but in 2011, it falls slightly below 0 for the first time, indicating a deficit.

(d) Based on the figures shown in the graph, what is the difference in Medicare funds from 2004 to 2013?

$$\underbrace{-\$7.9 \text{ billion}}_{\text{2013 amount}} - \underbrace{\$5.9 \text{ billion}}_{\text{2004 amount}} = \underbrace{-\$13.8 \text{ billion}}_{\text{Difference}}$$

The fund amount will have *decreased* $13.8 billion (as indicated by the negative sign in $-\$13.8$).

Work Problem 2 at the Side. ▶

The line graph in Figure 2 relates years to Medicare funds. We can also represent these two related quantities using a table of data, as shown at the side. Notice that in table form, we can see specific data rather than estimating it. Trends in the data are easier to see from the graph, however, which gives us a "picture" of the data.

We can extend these ideas to the subject of this chapter, *linear equations in two variables*. A linear equation in two variables, one for each of the quantities being related, can be used to represent the data in the table or graph. ***The graph of a linear equation in two variables is a line.***

2 Refer to the line graph in Figure 2.

(a) Which year has the greatest amount of Medicare funds?

(b) Estimate projected Medicare funds for 2010. Is there a surplus or a deficit in 2010?

(c) About how much is it projected that funds will decrease from 2006 to 2011?

Year	Medicare Funds (in billions of dollars)*
2004	5.9
2005	25.0
2006	9.5
2007	6.0
2008	5.0
2009	4.0
2010	2.0
2011	−0.5
2012	−2.5
2013	−7.9

*Projected

ANSWERS

2. (a) 2005
(b) $2 billion; surplus
(c) $10 billion

3 Write each solution as an ordered pair.

(a) $x = 5$ and $y = 7$

(b) $y = 6$ and $x = -1$

(c) $y = 4$ and $x = -3$

(d) $x = \frac{2}{3}$ and $y = -12$

(e) $y = 1.5$ and $x = -2.4$

(f) $x = 0$ and $y = 0$

Answers

3. (a) $(5, 7)$ **(b)** $(-1, 6)$ **(c)** $(-3, 4)$ **(d)** $\left(\frac{2}{3}, -12\right)$ **(e)** $(-2.4, 1.5)$ **(f)** $(0, 0)$

Linear Equation in Two Variables

A **linear equation in two variables** is an equation that can be written in the form

$$Ax + By = C,$$

where A, B, and C are real numbers and A and B are not both 0.

Some examples of linear equations in two variables in this form, called *standard form,* are

$$3x + 4y = 9, \quad x - y = 0, \quad \text{and} \quad x + 2y = -8. \quad \text{Linear equations in two variables}$$

Note

Other linear equations in two variables, such as

$$y = 4x + 5 \quad \text{and} \quad 3x = 7 - 2y,$$

are not written in standard form but could be. We discuss the forms of linear equations in more detail in **Section 3.4.**

OBJECTIVE 2 Write a solution as an ordered pair. Recall from **Section 1.2** that a *solution* of an equation is a number that makes the equation true when it replaces the variable. For example, the linear equation in one variable $x - 2 = 5$ has solution 7, since replacing x with 7 gives a true statement.

A solution of a linear equation in two variables requires two numbers, one for each variable. For example, a true statement results when we replace x with 2 and y with 13 in the equation $y = 4x + 5$ since

$$13 = 4(2) + 5. \quad \text{Let } x = 2, y = 13.$$

The pair of numbers $x = 2$ and $y = 13$ gives one solution of the equation $y = 4x + 5$. The phrase "$x = 2$ and $y = 13$" is abbreviated

x-value → (2, 13) ← y-value

Ordered pair

with the x-value, 2, and the y-value, 13, given as a pair of numbers written inside parentheses. ***The x-value is always given first.*** A pair of numbers such as (2, 13) is called an **ordered pair.** As the name indicates, the order in which the numbers are written is important. The ordered pairs (2, 13) and (13, 2) are *not* the same. The second pair indicates that $x = 13$ and $y = 2$. ***For two ordered pairs to be equal, their x-values must be equal and their y-values must be equal.***

◀ *Work Problem* **3** *at the Side.*

OBJECTIVE 3 Decide whether a given ordered pair is a solution of a given equation. We substitute the x- and y-values of an ordered pair into a linear equation in two variables to see whether the ordered pair is a solution.

EXAMPLE 3 Deciding Whether Ordered Pairs Are Solutions of an Equation

Decide whether each ordered pair is a solution of the equation $2x + 3y = 12$.

(a) (3, 2)

To see whether (3, 2) is a solution of the given equation $2x + 3y = 12$, substitute 3 for x and 2 for y in the equation.

$$2x + 3y = 12$$

$$2(3) + 3(2) \stackrel{?}{=} 12 \quad \text{Let } x = 3; \text{ let } y = 2.$$

$$6 + 6 \stackrel{?}{=} 12 \quad \text{Multiply.}$$

$$12 = 12 \quad \text{True}$$

This result is true, so (3, 2) is a solution of $2x + 3y = 12$.

(b) $(-2, -7)$

$$2x + 3y = 12$$

$$2(-2) + 3(-7) \stackrel{?}{=} 12 \quad \text{Let } x = -2; \text{ let } y = -7.$$

Use parentheses to avoid errors.

$$-4 + (-21) \stackrel{?}{=} 12 \quad \text{Multiply.}$$

$$-25 = 12 \quad \text{False}$$

This result is false, so $(-2, -7)$ is *not* a solution of $2x + 3y = 12$.

Work Problem 4 *at the Side.* ▶

OBJECTIVE 4 Complete ordered pairs for a given equation. Choosing a number for one variable in a linear equation makes it possible to find the value of the other variable.

EXAMPLE 4 Completing Ordered Pairs

Complete each ordered pair for the equation $y = 4x + 5$.

(a) (7, __)

The x-value always comes first.

In this ordered pair, $x = 7$. To find the corresponding value of y, replace x with 7 in the equation.

$$y = 4x + 5$$

$$y = 4(7) + 5 \quad \text{Let } x = 7.$$

$$y = 28 + 5 \quad \text{Multiply.}$$

$$y = 33 \quad \text{Add.}$$

The ordered pair is (7, 33).

(b) $(\_\_, -3)$

In this ordered pair, $y = -3$. Find the value of x by replacing y with -3 in the equation; then solve for x.

$$y = 4x + 5$$

$$-3 = 4x + 5 \quad \text{Let } y = -3.$$

$$-8 = 4x \quad \text{Subtract 5 from each side.}$$

$$-2 = x \quad \text{Divide each side by 4.}$$

The ordered pair is $(-2, -3)$.

Work Problem 5 *at the Side.* ▶

4 Decide whether each ordered pair is a solution of the equation $5x + 2y = 20$.

(a) (0, 10)

$$5x + 2y = 20$$

$$5(\_\_) + 2(\_\_) \stackrel{?}{=} 20$$

$$\_\_\_\_ + 20 \stackrel{?}{=} 20$$

$$\_\_\_\_ = 20$$

Is (0, 10) a solution?

(b) $(2, -5)$

(c) (3, 2)

(d) $(-4, 20)$

5 Complete each ordered pair for the equation $y = 2x - 9$.

(a) (5, __)

$$y = 2(\_\_) - 9$$

$$y = \_\_\_\_ - 9$$

$$y = \_\_\_\_$$

The ordered pair is _____.

(b) (2, __)

(c) (__, 7)

(d) $(\_\_, -13)$

ANSWERS

4. **(a)** 0; 10; 0; 20; yes **(b)** no **(c)** no **(d)** yes

5. **(a)** 5; 10; 1; (5, 1) **(b)** $(2, -5)$ **(c)** (8, 7) **(d)** $(-2, -13)$

OBJECTIVE 5 Complete a table of values. Ordered pairs are often displayed in a **table of values.** The table may be written either vertically or horizontally.

EXAMPLE 5 Completing Tables of Values

Complete the table of values for each equation. Then write the results as ordered pairs.

(a) $x - 2y = 8$

x	y
2	
10	
	0
	-2

To complete the first two ordered pairs of the table, let $x = 2$ and $x = 10$, respectively.

If	$x = 2,$	If	$x = 10,$
then	$x - 2y = 8$	then	$x - 2y = 8$
becomes	$2 - 2y = 8$	becomes	$10 - 2y = 8$
	$-2y = 6$		$-2y = -2$
	$y = -3.$		$y = 1.$

Now complete the last two ordered pairs by letting $y = 0$ and $y = -2$, respectively.

If	$y = 0,$	If	$y = -2,$
then	$x - 2y = 8$	then	$x - 2y = 8$
becomes	$x - 2(0) = 8$	becomes	$x - 2(-2) = 8$
	$x - 0 = 8$		$x + 4 = 8$
	$x = 8.$		$x = 4.$

The completed table of values follows.

x	y
2	-3
10	1
8	0
4	-2

Write y-values here.

Write x-values here.

The corresponding ordered pairs are $(2, -3)$, $(10, 1)$, $(8, 0)$, and $(4, -2)$. Each ordered pair is a solution of the given equation $x - 2y = 8$.

(b) $x = 5$

x	y
	-2
	6
	3

The given equation is $x = 5$. No matter which value of y is chosen, the value of x is *always* 5.

x	y
5	-2
5	6
5	3

The corresponding ordered pairs are $(5, -2)$, $(5, 6)$, and $(5, 3)$.

◀ *Work Problem* 6 *at the Side.*

6 Complete the table of values for each equation.

(a) $2x - 3y = 12$

x	y
0	
	0
3	
	-3

(b) $x = -1$

x	y
	-4
	0
	2

(c) $y = 4$

x	y
-3	
2	
5	

ANSWERS

6. (a)

x	y
0	-4
6	0
3	-2
$\frac{3}{2}$	-3

(b)

x	y
-1	-4
-1	0
-1	2

(c)

x	y
-3	4
2	4
5	4

Note

We can think of $x = 5$ in Example 5(b) as an equation in two variables by rewriting $x = 5$ as $x + 0y = 5$. This form of the equation shows that for any value of y, the value of x is 5. Similarly, $y = 4$ in Problem 6(c) in the margin on the preceding page is the same as $0x + y = 4$.

7 Name the quadrant in which each point in the figure is located.

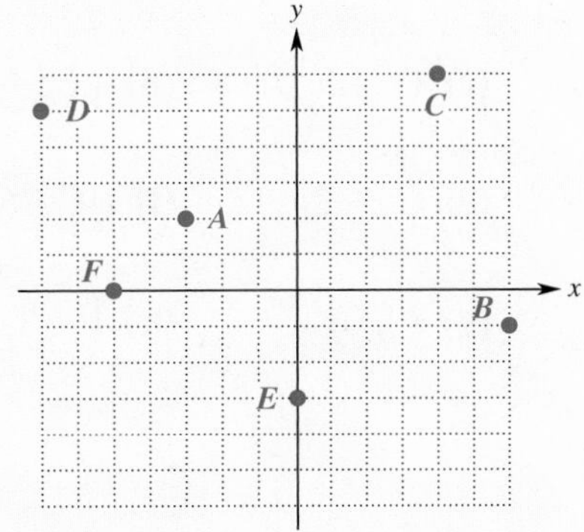

OBJECTIVE 6 Plot ordered pairs. In **Section 2.3,** we saw that linear equations in *one* variable had either one, zero, or an infinite number of real number solutions. These solutions could be graphed on *one* number line. For example, the linear equation in one variable $x - 2 = 5$ has solution 7, which is graphed on the number line in Figure 3.

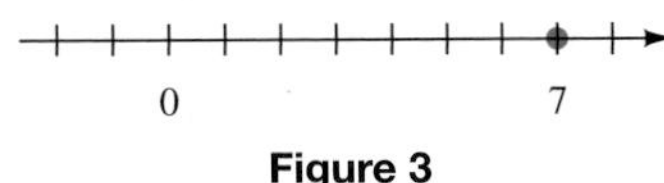

Figure 3

Every linear equation in *two* variables has an infinite number of ordered pairs as solutions. Each choice of a number for one variable leads to a particular real number for the other variable.

To graph these solutions, represented as the ordered pairs (x, y), we need *two* number lines, one for each variable, as drawn in Figure 4. The horizontal number line is called the **x-axis,** and the vertical line is called the **y-axis.** Together, the x-axis and y-axis form a **rectangular coordinate system,** also called the **Cartesian coordinate system,** in honor of René Descartes, the French mathematician who is credited with its invention.

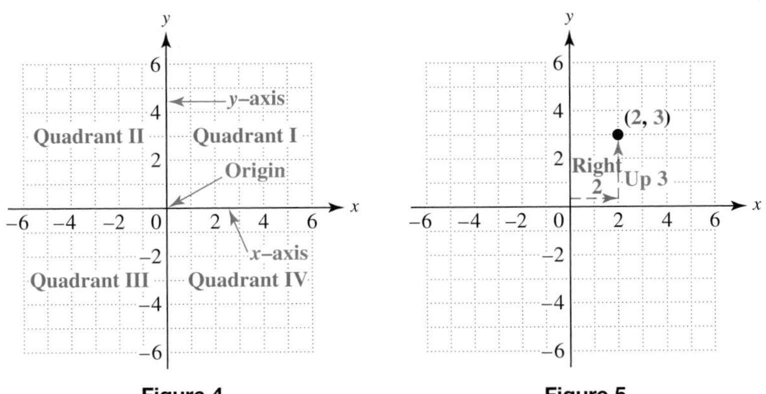

Figure 4 **Figure 5**

René Descartes (1596–1650)

The coordinate system is divided into four regions, called **quadrants.** These quadrants are numbered counterclockwise, as shown in Figure 4. ***Points on the axes themselves are not in any quadrant.*** The point at which the x-axis and y-axis meet is called the **origin.** The origin, which is labeled 0 in Figure 4, is the point corresponding to $(0, 0)$.

Work Problem **7** *at the Side.* ▶

The x-axis and y-axis determine a **plane**—a flat surface illustrated by a sheet of paper. By referring to the two axes, every point in the plane can be associated with an ordered pair. The numbers in the ordered pair are called the **coordinates** of the point.

For example, we locate the point associated with the ordered pair (2, 3) by starting at the origin. Since the x-coordinate is 2, we go 2 units to the right along the x-axis. Then, since the y-coordinate is 3, we turn and go up 3 units on a line parallel to the y-axis. The point (2, 3) is **plotted** in Figure 5. From now on, we will refer to the point with x-coordinate 2 and y-coordinate 3 as the point (2, 3).

ANSWER

7. *A*: II; *B*: IV; *C*: I; *D*: II; *E*: no quadrant; *F*: no quadrant

8 Plot each ordered pair on a coordinate system.

(a) $(3, 5)$ **(b)** $(-2, 6)$

(c) $(-4.5, 0)$ **(d)** $(-5, -2)$

(e) $(6, -2)$ **(f)** $(0, -6)$

(g) $(0, 0)$ **(h)** $\left(-3, \frac{5}{2}\right)$

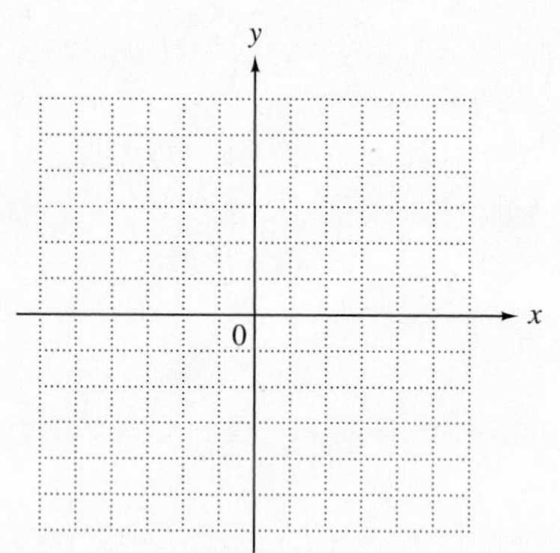

Note

When we graph on a number line (as in Figure 3), one number corresponds to each point. On a plane, however, *both* numbers in an ordered pair are needed to locate a point (as in Figure 5). The ordered pair is a name for the point.

We mentioned that René Descartes is credited with inventing the Cartesian coordinate system. Legend has it that Descartes, who was lying in bed ill, was watching a fly crawl about on the ceiling near a corner of the room. It occurred to him that the location of the fly could be described by determining its distances from the two adjacent walls. See the figure.

Locating a fly on a ceiling

EXAMPLE 6 Plotting Ordered Pairs

Plot each ordered pair on a coordinate system.

(a) $(1, 5)$ **(b)** $(-2, 3)$ **(c)** $(-1, -4)$ **(d)** $(3, -2)$

(e) $\left(\frac{3}{2}, 2\right)$ **(f)** $(5, 0)$ **(g)** $(0, -3)$ **(h)** $(4, -3.75)$

See Figure 6. In each case, begin at the origin. Move right or left the number of units that corresponds to the x-coordinate in the ordered pair—***right if the x-coordinate is positive or left if it is negative.*** Then turn and move up or down the number of units that corresponds to the y-coordinate—***up if the y-coordinate is positive or down if it is negative.*** So in part (c), locate the point $(-1, -4)$ by first going 1 unit to the *left* along the x-axis. Then turn and go 4 units *down,* parallel to the y-axis.

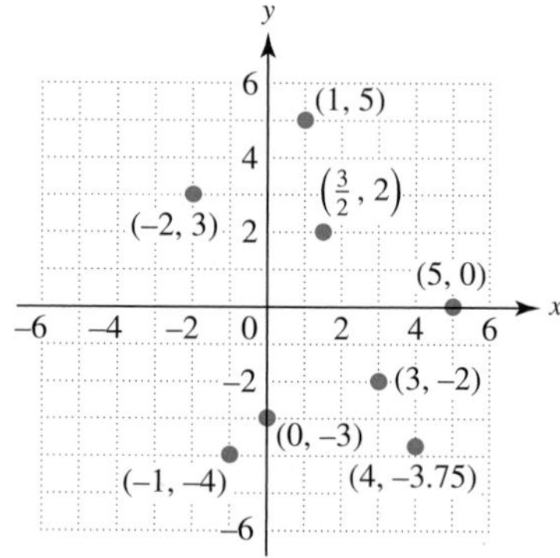

Figure 6

Notice the difference in the locations of the points $(-2, 3)$ and $(3, -2)$ in parts (b) and (d). The point $(-2, 3)$ is in quadrant II, whereas the point $(3, -2)$ is in quadrant IV. ***The order of the coordinates is important. Remember that the x-coordinate is always given first in an ordered pair.***

To plot the point $(\frac{3}{2}, 2)$ in part (e), think of the improper fraction $\frac{3}{2}$ as the mixed number $1\frac{1}{2}$ and move $\frac{3}{2}$ (or $1\frac{1}{2}$) units to the right along the x-axis. Then turn and go 2 units up, parallel to the y-axis. The point $(4, -3.75)$ in part (h) is plotted similarly, by approximating the location of the decimal y-coordinate.

In part (f), the point $(5, 0)$ lies on the x-axis since the y-coordinate is 0. In part (g), the point $(0, -3)$ lies on the y-axis since the x-coordinate is 0.

◀ Work Problem **8** at the Side.

ANSWERS

8.

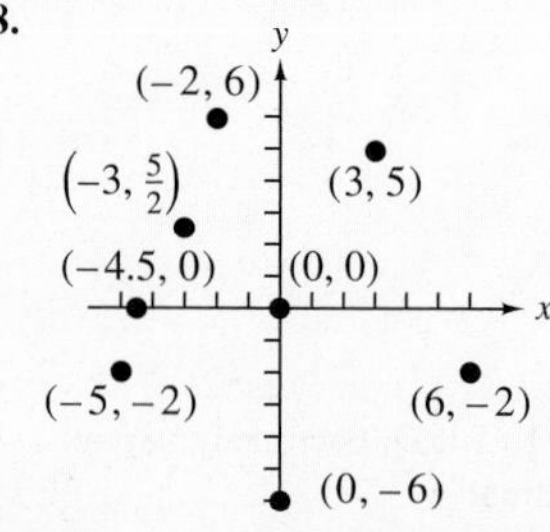

Sometimes we can use a linear equation in two variables to mathematically describe, or *model*, a real-life situation, as shown in the next example.

EXAMPLE 7 Completing Ordered Pairs to Estimate the Number of Twin Births

The number of twin births in the United States has increased steadily in recent years. The annual number of twin births from 2000 through 2005 can be closely approximated by the linear equation

Number of twin births → y; Year → x

$$y = 3.074x - 6029.7,$$

which relates x, the year, and y, the number of twin births in thousands. (*Source: National Vital Statistics Reports,* Vol. 56, No. 6, December 5, 2007.)

(a) Complete the table of values for the given linear equation.

x (Year)	y (Number of Twin Births, in thousands)
2000	
2002	
2005	

To find y when $x = 2000$, substitute into the equation.

$$y = 3.074x - 6029.7$$

$$y = 3.074(\mathbf{2000}) - 6029.7 \quad \text{Let } x = 2000.$$

$$y \approx 118 \quad \text{Use a calculator.}$$

$\approx$ means "is approximately equal to."

This means that in 2000, there were about 118 thousand (or 118,000) twin births in the United States.

Work Problem 9 *at the Side.* ▶

Including the results from Problem 9 at the side gives the completed table that follows.

x (Year)	y (Number of Twin Births, in thousands)
2000	118
2002	124
2005	134

We can write the results from the table of values as ordered pairs (x, y). Each year x is paired with its number of twin births y (in thousands):

$$(2000, 118), \quad (2002, 124), \quad \text{and} \quad (2005, 134).$$

Continued on Next Page

9 Refer to the linear equation in Example 7.

(a) Find the y-value for $x = 2002$. Round to the nearest whole number.

(b) Find the y-value for $x = 2005$. Interpret your result.

ANSWERS

9. (a) 124
(b) 134; In 2005, there were about 134 thousand (or 134,000) twin births in the United States.

(b) Graph the ordered pairs found in part (a).

The ordered pairs (2000, 118), (2002, 124), and (2005, 134) are graphed in Figure 7. This graph of ordered pairs of data is called a **scatter diagram.** Notice how the axes are labeled: x represents the year, and y represents the number of twin births in thousands. Different scales are used on the two axes. Here, each square represents one unit in the horizontal direction and 5 units in the vertical direction. Because the numbers in the first ordered pair are large, we show a break in the axes near the origin.

x (Year)	y (Number of Twin Births, in thousands)
2000	118
2002	124
2005	134

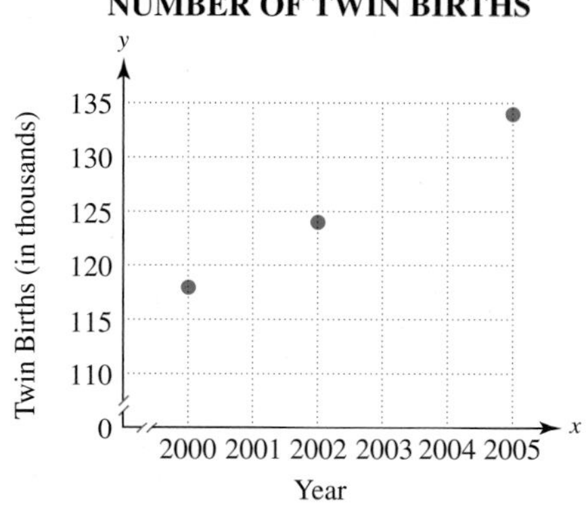

Figure 7

A scatter diagram enables us to tell whether two quantities are related to each other. In Figure 7, the plotted points could be connected to closely approximate a straight ***line,*** so the variables x (year) and y (number of twin births) have a ***line***ar relationship. The increase in the number of twin births is also reflected.

CAUTION

The equation in Example 7 is valid only for the years 2000 through 2005 because it was based on data for those years. ***Do not assume that this equation would provide reliable data for other years since the data for those years may not follow the same pattern.***

3.1 Exercises

The bar graph shows total U.S. milk production in billions of pounds for the years 2001 through 2007. Use the bar graph to work Exercises 1–4. See Example 1.

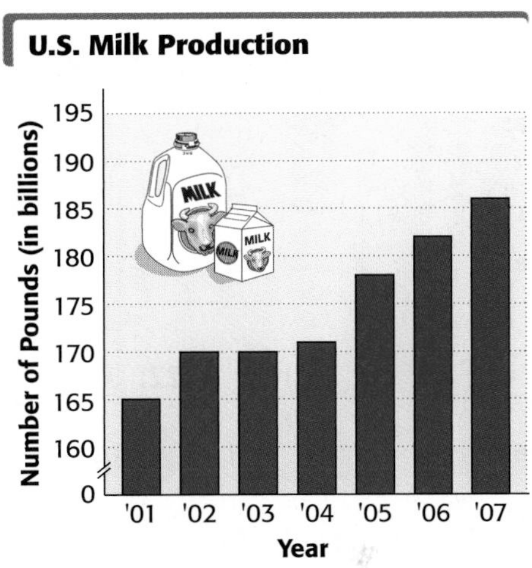

Source: U.S. Department of Agriculture.

1. In what years was U.S. milk production greater than 175 billion pounds?

2005, 2006, 2007

2. In what years was U.S. milk production about the same?

2002 and 2003

3. Estimate U.S. milk production in 2001 and 2007.

2001: about 165 billion lb; 2007: about 185 billion lb

4. Describe the change in U.S. milk production from 2001 to 2007.

U.S. milk production increased about 20 billion lb during these years, from about 165 billion lb to about 185 billion lb.

The line graph shows the average price, adjusted for inflation, that Americans have paid for a gallon of gasoline for selected years since 1970. Use the line graph to work Exercises 5–8. See Example 2.

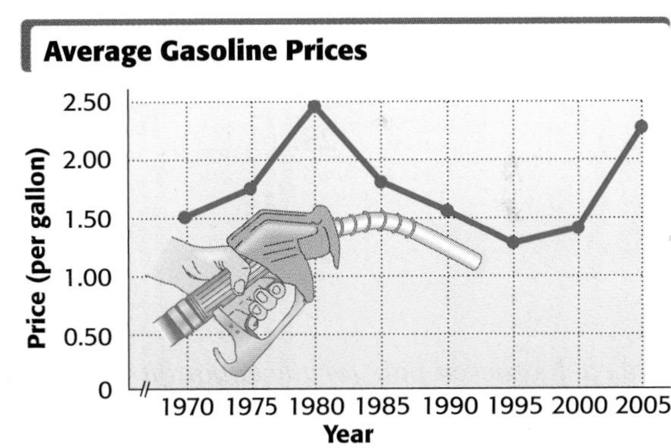

Source: Energy Information Administration.

5. Over which period of years did the greatest increase in the price of a gallon of gas occur? About how much was this increase?

from 2000 to 2005; about $0.85

6. Estimate the price of a gallon of gas during 1985, 1990, 1995, and 2000.

1985: $1.80; 1990: $1.55; 1995: $1.30; 2000: $1.40

7. Describe the trend in gas prices from 1980 to 1995.

The price of a gallon of gas was decreasing.

8. During which year(s) did a gallon of gas cost approximately $1.50?

1970, 1991, 2001

Use the concepts of this section to fill in each blank with the correct response.

9. The symbol (x, y) **does** _____ (does/does not) represent an ordered pair, while the symbols $[x, y]$ and $\{x, y\}$ **do not** _____ (do/do not) represent ordered pairs.

10. The point whose graph has coordinates $(-4, 2)$ is in quadrant **II**.

11. The point whose graph has coordinates $(0, 5)$ lies on the **y**-axis.

12. The ordered pair $(4, $ **3** $)$ is a solution of the equation $y = 3$.

13. The ordered pair $($ **6** $, -2)$ is a solution of the equation $x = 6$.

14. The ordered pair $(3, 2)$ is a solution of the equation $2x - 5y =$ **−4**.

Decide whether each ordered pair is a solution of the given equation. See Example 3.

15. $x + y = 9$; $(0, 9)$
yes

16. $x + y = 8$; $(0, 8)$
yes

17. $2x - y = 6$; $(4, 2)$
yes

18. $2x + y = 5$; $(3, -1)$
yes

19. $4x - 3y = 6$; $(2, 1)$
no

20. $5x - 3y = 15$; $(5, 2)$
no

21. $y = \frac{2}{3}x$; $(-6, -4)$
yes

22. $y = -\frac{1}{4}x$; $(-8, 2)$
yes

23. $x = -6$; $(5, -6)$
no

24. $y = 2$; $(2, 4)$
no

25. Do $(4, -1)$ and $(-1, 4)$ represent the same ordered pair? Explain.

No. For two ordered pairs (x, y) to be equal, the x-values must be equal and the y-values must be equal. Here we have $4 \neq -1$ and $-1 \neq 4$.

26. Explain why it would be easier to find the corresponding y-value for $x = \frac{1}{3}$ in the equation $y = 6x + 2$ than it would be for $x = \frac{1}{7}$.

Substituting $\frac{1}{3}$ for x in $y = 6x + 2$ gives $y = 6(\frac{1}{3}) + 2 = 2 + 2 = 4$. Because $6(\frac{1}{7}) = \frac{6}{7}$, calculating y requires working with fractions.

Complete each ordered pair for the equation $y = 2x + 7$. See Example 4.

27. $(2, \_\_)$
11

28. $(0, \_\_)$
7

29. $(\_\_, 0)$
$-\frac{7}{2}$

30. $(\_\_, -3)$
−5

Complete each ordered pair for the equation $y = -4x - 4$. See Example 4.

31. $(0, \_\_)$
−4

32. $(\_\_, 0)$
−1

33. $(\_\_, 16)$
−5

34. $(\_\_, 24)$
−7

Complete each table of values. In Exercises 35–38, write the results as ordered pairs. See Example 5.

35. $2x + 3y = 12$

x	y
0	4
6	0
−6	8

(0, 4); (6, 0); (−6, 8)

36. $4x + 3y = 24$

x	y
0	8
6	0
3	4

(0, 8); (6, 0); (3, 4)

37. $3x - 5y = -15$

x	y
0	3
−5	0
−15	−6

(0, 3); (−5, 0); (−15, −6)

38. $4x - 9y = -36$

x	y
−9	0
0	**4**
9	8

(−9, 0); (0, 4); (9, 8)

39. $x = -9$

x	y
−9	6
−9	2
−9	−3

40. $x = 12$

x	y
12	3
12	8
12	0

41. $y = -6$

x	y
8	**−6**
4	**−6**
−2	**−6**

42. $y = -10$

x	y
4	**−10**
0	**−10**
−4	**−10**

43. $x - 8 = 0$

x	y
8	8
8	3
8	0

44. $y + 2 = 0$

x	y
9	**−2**
2	**−2**
0	**−2**

Give the ordered pairs for the points labeled A–F in the figure. Tell the quadrant in which each point is located.

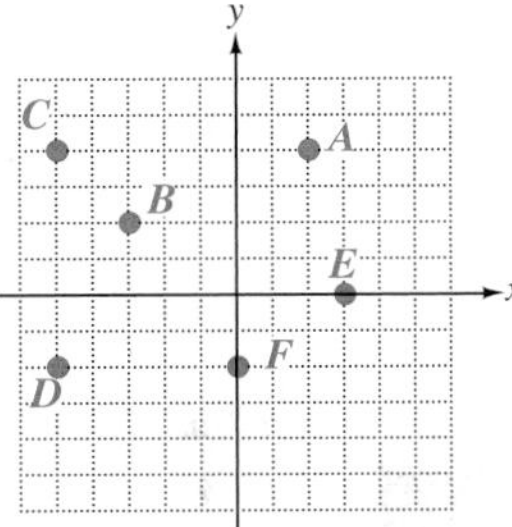

45. A
(2, 4); I

46. B
(−3, 2); II

47. C
(−5, 4); II

48. D
(−5, −2); III

49. E
(3, 0); no quadrant

50. F
(0, −2); no quadrant

Fill in each blank with the word positive *or the word* negative.

The point with coordinates (x, y) is in

51. quadrant III if x is **negative** and y is **negative**.

52. quadrant II if x is **negative** and y is **positive**.

53. quadrant IV if x is **positive** and y is **negative**.

54. quadrant I if x is **positive** and y is **positive**.

55. A point (x, y) has the property that $xy < 0$. In which quadrant(s) must the point lie? Explain.
If xy < 0, then either x < 0 and y > 0 or x > 0 and y < 0. If x < 0 and y > 0, then the point lies in quadrant II. If x > 0 and y < 0, then the point lies in quadrant IV.

56. A point (x, y) has the property that $xy > 0$. In which quadrant(s) must the point lie? Explain.
If xy > 0, then either x > 0 and y > 0 or x < 0 and y < 0. If x > 0 and y > 0, then the point lies in quadrant I. If x < 0 and y < 0, then the point lies in quadrant III.

Plot each ordered pair on the rectangular coordinate system provided. See Example 6.

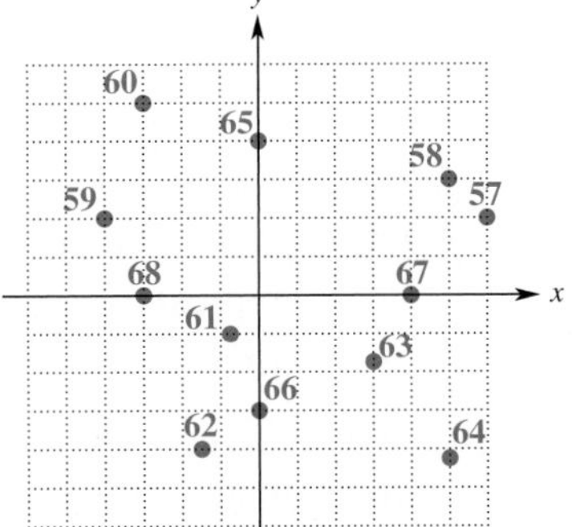

57. (6, 2)

58. (5, 3)

59. (−4, 2)

60. (−3, 5)

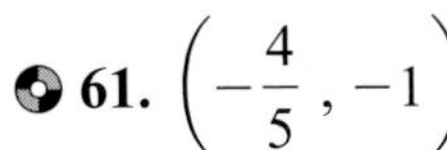

61. $\left(-\frac{4}{5}, -1\right)$

62. $\left(-\frac{3}{2}, -4\right)$

63. (3, −1.75)

64. (5, −4.25)

65. (0, 4)

66. (0, −3)

67. (4, 0)

68. (−3, 0)

Complete each table of values, and then plot the ordered pairs. See Examples 5 and 6.

69. $x - 2y = 6$

x	y
0	**−3**
6	0
2	**−2**
4	**−1**

70. $2x - y = 4$

x	y
0	**−4**
2	0
1	**−2**
−1	−6

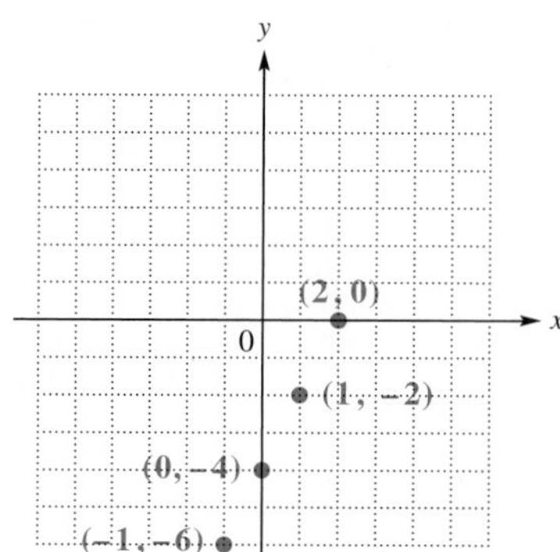

71. $3x - 4y = 12$

x	y
0	**−3**
4	0
−4	**−6**
$-\frac{4}{3}$	−4

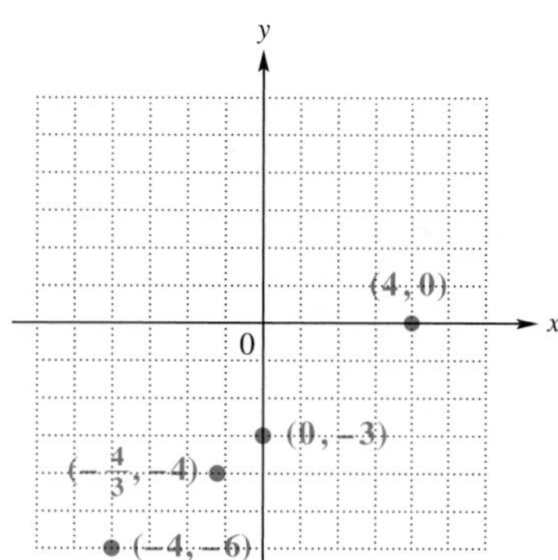

72. $2x - 5y = 10$

x	y
0	**−2**
5	0
−5	**−4**
$-\frac{5}{2}$	−3

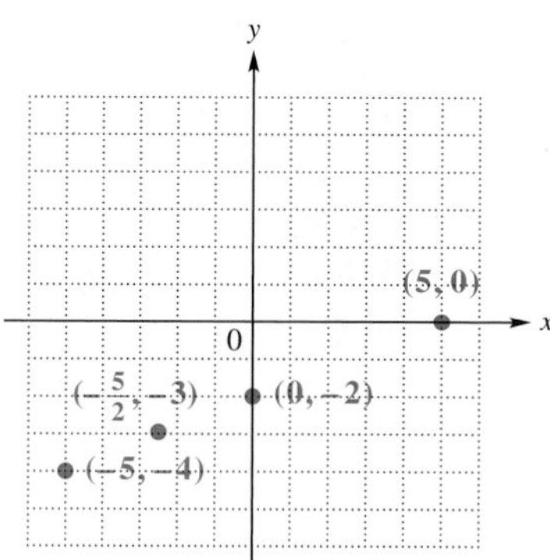

73. $y + 4 = 0$

x	y
0	**−4**
5	**−4**
−2	**−4**
−3	**−4**

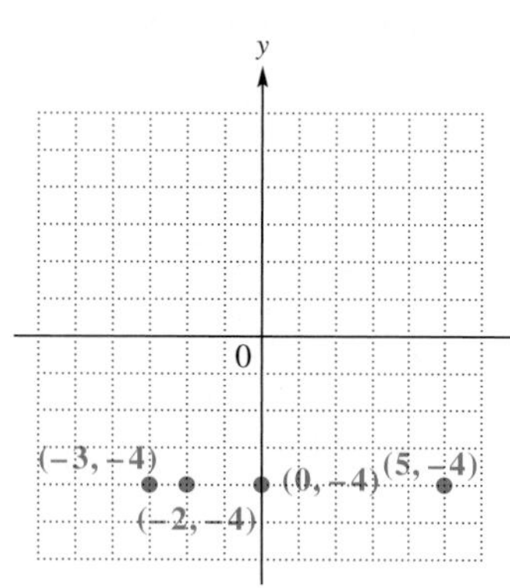

74. $x - 5 = 0$

x	y
5	1
5	0
5	6
5	−4

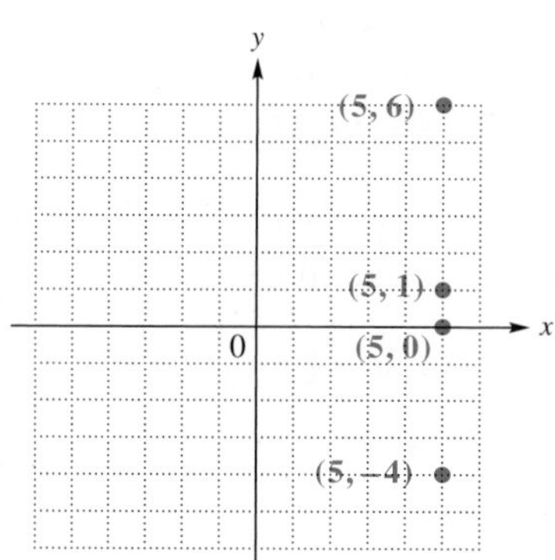

75. Look at the graphs of the ordered pairs in Exercises 69–74. Describe the pattern indicated by the plotted points.

The points in each graph appear to lie on a straight line.

Work each problem. See Example 7.

76. Suppose that it costs \$5000 to start up a business selling snow cones. Furthermore, it costs \$0.50 per cone in labor, ice, syrup, and overhead. Then the cost to make x snow cones is given by y dollars, where

$$y = 0.50x + 5000.$$

Express each of the following as an ordered pair.

(a) When 100 snow cones are made, the cost is \$5050. (*Hint:* What does x represent? What does y represent?)

(100, 5050)

(b) When the cost is \$6000, the number of snow cones made is 2000.

(2000, 6000)

77. It costs a flat fee of \$20 plus \$5 per day to rent a pressure washer. Therefore, the cost to rent the pressure washer for x days is given by

$$y = 5x + 20,$$

where y is in dollars. Express each of the following as an ordered pair.

(a) When the washer is rented for 5 days, the cost is \$45. (*Hint:* What does x represent? What does y represent?)

(5, 45)

(b) I paid \$50 when I returned the washer, so I must have rented it for 6 days.

(6, 50)

78. The table shows the number of U.S. students studying abroad (in thousands) for several academic years.

Academic Year	Number of Students (in thousands)
2000	154
2001	161
2002	175
2003	191
2004	206
2005	224

Source: Institute of International Education.

(a) Write the data from the table as ordered pairs (x, y), where x represents the year and y represents the number of U.S. students studying abroad.

(2000, 154), (2001, 161), (2002, 175), (2003, 191), (2004, 206), (2005, 224)

(b) What does the ordered pair (2004, 206) mean in the context of this problem?

(2004, 206) means that in 2004, about 206 thousand (or 206,000) U.S. students studied abroad.

(c) Make a scatter diagram of the data using the ordered pairs from part (a). **See the graph.**

U.S. STUDENTS STUDYING ABROAD

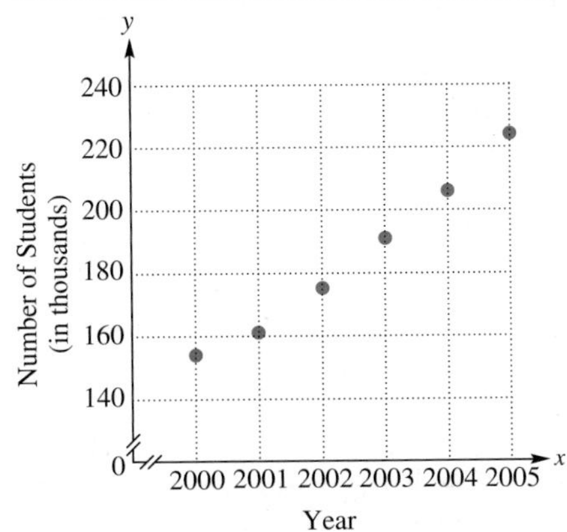

(d) Describe the pattern indicated by the points on the scatter diagram. What is the trend in the number of U.S. students studying abroad?

The points lie approximately on a straight line. The number of U.S. students studying abroad is increasing year by year.

79. The table shows the rate (in percent) at which 2-year college students (public) complete a degree within 3 years.

Year	Percent
2000	32.4
2001	31.6
2002	31.6
2003	30.1
2004	29.0
2005	27.5

Source: ACT.

(a) Write the data from the table as ordered pairs (x, y), where x represents the year and y represents the percent.

(2000, 32.4), (2001, 31.6), (2002, 31.6), (2003, 30.1), (2004, 29.0), (2005, 27.5)

(b) What would the ordered pair (2007, 27.1) mean in the context of this problem?

(2007, 27.1) means that 27.1 percent of 2-year college students in 2007 received a degree within 3 years.

(c) Make a scatter diagram of the data using the ordered pairs from part (a). **See the graph.**

2-YEAR COLLEGE STUDENTS COMPLETING A DEGREE WITHIN 3 YEARS

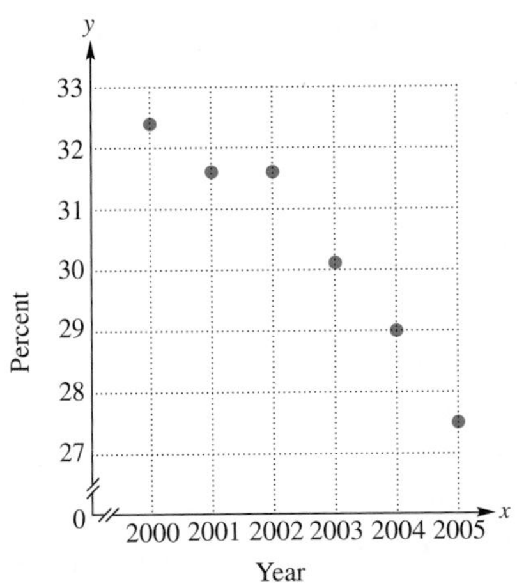

(d) Describe the pattern indicated by the points on the scatter diagram. What is happening to the rates at which 2-year college students complete a degree within 3 years?

With the exception of the point for 2002, the points lie approximately on a straight line. Rates at which 2-year college students complete a degree within 3 years are generally decreasing.

80. The maximum benefit for the heart from exercising occurs if the heart rate is in the target heart rate zone. The lower limit of this target zone can be approximated by the linear equation

$$y = -0.5x + 108,$$

where x represents age and y represents heartbeats per minute. (*Source:* www.fitresource.com)

(a) Complete the table of values for this linear equation.

Age	Heartbeats (per minute)
20	**98**
40	**88**
60	**78**
80	**68**

See the table.

(b) Write the data from the table of values as ordered pairs.

(20, 98), (40, 88), (60, 78), (80, 68)

(c) Make a scatter diagram of the data. Do the points lie in a linear pattern?

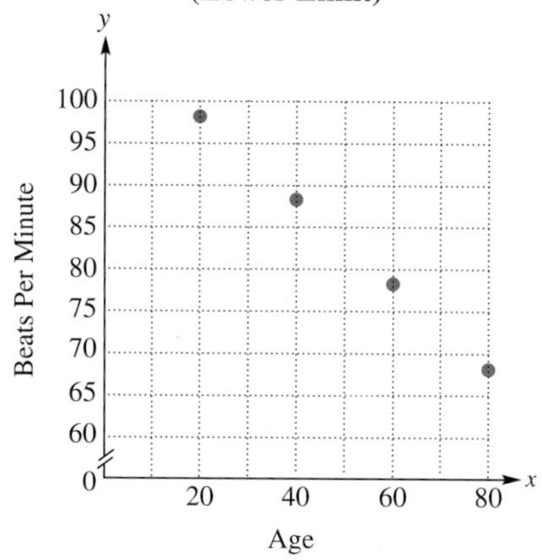

See the graph; yes

81. (See Exercise 80.) The upper limit of the target heart rate zone can be approximated by the linear equation

$$y = -0.8x + 173,$$

where x represents age and y represents heartbeats per minute. (*Source:* www.fitresource.com)

(a) Complete the table of values for this linear equation.

Age	Heartbeats (per minute)
20	**157**
40	**141**
60	**125**
80	**109**

See the table.

(b) Write the data from the table of values as ordered pairs.

(20, 157), (40, 141), (60, 125), (80, 109)

(c) Make a scatter diagram of the data. Describe the pattern indicated by the data.

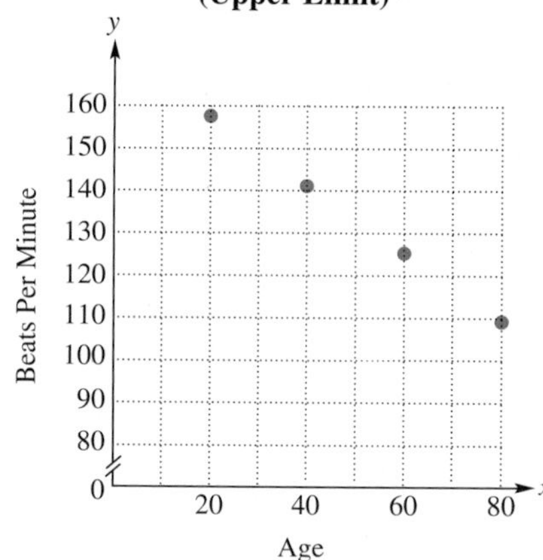

See the graph; The points lie in a linear pattern.

82. Refer to Exercises 80 and 81. What is the target heart rate zone for age 20? age 40?

between 98 and 157 beats per minute; between 88 and 141 beats per minute

3.2 Graphing Linear Equations in Two Variables

OBJECTIVES

1. **Graph linear equations by plotting ordered pairs.**
2. **Find intercepts.**
3. **Graph linear equations of the form $Ax + By = 0$.**
4. **Graph linear equations of the form $y = k$ or $x = k$.**
5. **Use a linear equation to model data.**

OBJECTIVE 1 Graph linear equations by plotting ordered pairs. There are infinitely many ordered pairs that satisfy an equation in two variables. We find these ordered-pair solutions by choosing as many values of x (or y) as we wish and then completing each ordered pair.

For example, consider the equation $x + 2y = 7$. If we choose $x = 1$, then $y = 3$, so the ordered pair $(1, 3)$ is a solution of the equation $x + 2y = 7$.

$$1 + 2(3) = 7$$

Work Problem 1 at the Side. ▶

1 Complete each ordered pair for the equation $x + 2y = 7$.

(a) $(-3, \_\_)$

(b) $(-1, \_\_)$

(c) $(3, \_\_)$

(d) $(5, \_\_)$

(e) $(7, \_\_)$

Figure 8 shows a graph of all the ordered-pair solutions found above and in Problem 1 at the side for $x + 2y = 7$.

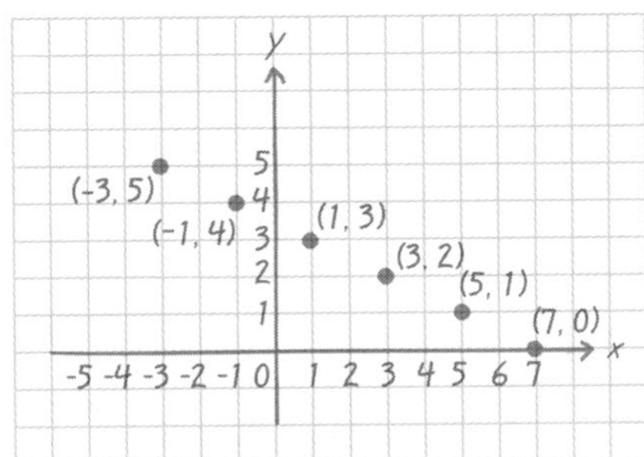

Figure 8

Notice that the points plotted in Figure 8 all appear to lie on a straight line, as shown in Figure 9. In fact, the following is true.

Every point on the line represents a solution of the equation $x + 2y = 7$, and every solution of the equation corresponds to a point on the line.

The line gives a "picture" of all the solutions of the equation $x + 2y = 7$. Only a portion of the line is shown here, but it extends indefinitely in both directions, as suggested by the arrowhead on each end.

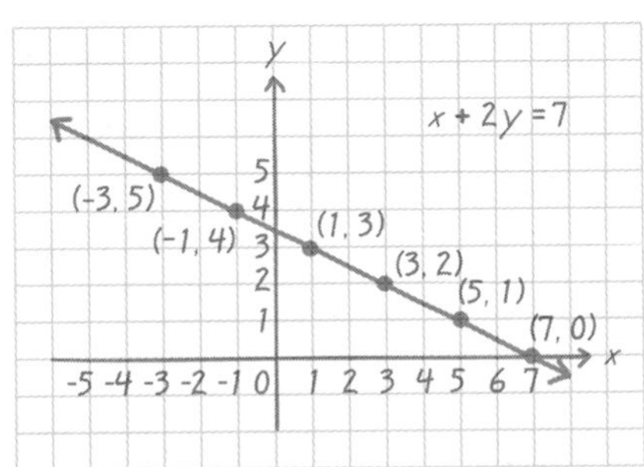

Figure 9

The line in Figure 9 is called the **graph** of the equation $x + 2y = 7$, and the process of plotting the ordered pairs and drawing the line through the corresponding points is called **graphing.**

ANSWERS

1. (a) $(-3, 5)$ (b) $(-1, 4)$ (c) $(3, 2)$ (d) $(5, 1)$ (e) $(7, 0)$

The preceding discussion can be generalized.

Graph of a Linear Equation

The graph of any linear equation in two variables is a straight line.

(Notice that the word ***line*** appears in the term "***line***ar equation.")

Because two distinct points determine a line, a straight line can be graphed by finding any two different points on the line. However, it is a good idea to plot a third point as a check.

2 Complete the table of values, and graph the linear equation.

$x + y = 6$

x	y
0	
	0
2	

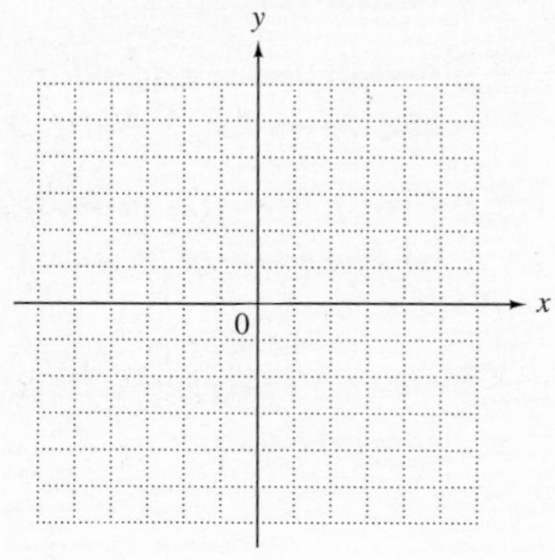

EXAMPLE 1 Graphing a Linear Equation

Graph the linear equation $4x - 5y = 20$.

At least two different points are needed to draw the graph. First let $x = 0$ and then let $y = 0$ to complete two ordered pairs.

$$
\begin{aligned}
4x - 5y &= 20 \\
4(0) - 5y &= 20 \quad \text{Let } x = 0. \\
0 - 5y &= 20 \\
-5y &= 20 \\
y &= -4
\end{aligned}
\qquad
\begin{aligned}
4x - 5y &= 20 \\
4x - 5(0) &= 20 \quad \text{Let } y = 0. \\
4x - 0 &= 20 \\
4x &= 20 \\
x &= 5
\end{aligned}
$$

Write each x-value first.

The ordered pairs are $(0, -4)$ and $(5, 0)$. Find a third ordered pair (as a check) by choosing a number other than 0 for x or y. We choose $y = 2$.

$$
\begin{aligned}
4x - 5y &= 20 \\
4x - 5(2) &= 20 \qquad \text{Let } y = 2. \\
4x - 10 &= 20 \\
4x &= 30 \qquad \text{Add 10.} \\
x &= \frac{30}{4}, \text{ or } \frac{15}{2} \qquad \text{Divide by 4; lowest terms}
\end{aligned}
$$

This gives the ordered pair $(\frac{15}{2}, 2)$, or $(7\frac{1}{2}, 2)$. Plot the three ordered pairs $(0, -4)$, $(5, 0)$, and $(7\frac{1}{2}, 2)$, and draw a line through them. This line, shown in Figure 10, is the graph of $4x - 5y = 20$.

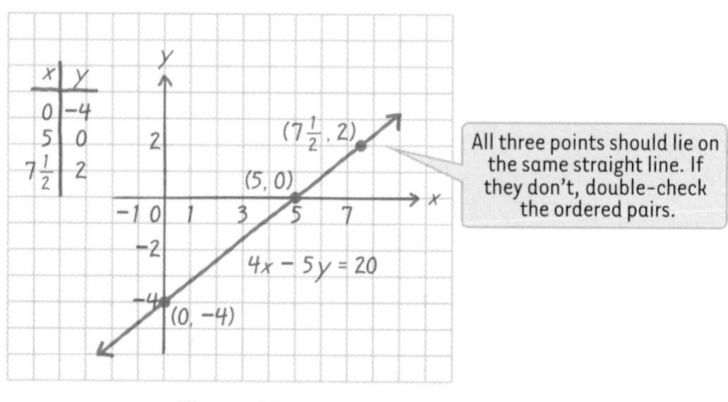

Figure 10

◀ Work Problem **2** at the Side.

ANSWER

2.

x	y
0	6
6	0
2	4

(0, 6), (2, 4), (6, 0); $x + y = 6$

EXAMPLE 2 Graphing a Linear Equation

Graph the linear equation $y = -\frac{3}{2}x + 3$.

Although this equation is not in the form $Ax + By = C$, it *could* be written in that form, so it is a linear equation. Two different points on the graph can be found by first letting $x = 0$ and then letting $y = 0$.

If $x = 0$, then

$$y = -\frac{3}{2}x + 3$$

$$y = -\frac{3}{2}(0) + 3 \quad \text{Let } x = 0.$$

$$y = 0 + 3 \quad \text{Multiply.}$$

$$y = 3. \quad \text{Add.}$$

If $y = 0$, then

$$y = -\frac{3}{2}x + 3$$

$$0 = -\frac{3}{2}x + 3 \quad \text{Let } y = 0.$$

$$\frac{3}{2}x = 3 \quad \text{Add } \tfrac{3}{2}x.$$

$$x = 2. \quad \text{Multiply by } \tfrac{2}{3}.$$

This gives the ordered pairs (0, 3) and (2, 0). We find a third point (as a check) by letting x or y equal some other number. For example, let $x = -2$.

$$y = -\frac{3}{2}x + 3$$

$$y = -\frac{3}{2}(-2) + 3 \quad \text{Let } x = -2.$$

$$y = 3 + 3 \quad \text{Multiply.}$$

$$y = 6 \quad \text{Add.}$$

Choosing a multiple of 2 makes multiplying by $-\frac{3}{2}$ easier.

This gives the ordered pair (−2, 6). These three ordered pairs are shown in the table with Figure 11. Plot the corresponding points, and then draw a line through them. This line, shown in Figure 11, is the graph of $y = -\frac{3}{2}x + 3$.

x	y
0	3
2	0
−2	6

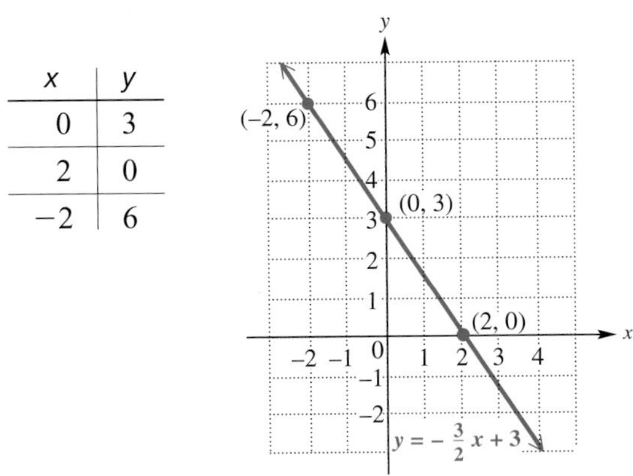

Figure 11

Work Problem 3 at the Side. ▶

3 Make a table of values, and graph the linear equation.

$$y = \frac{2}{3}x - 2$$

x	y

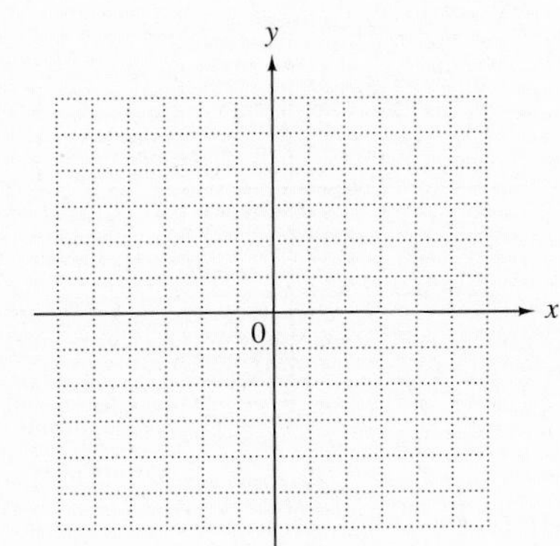

ANSWER

3.

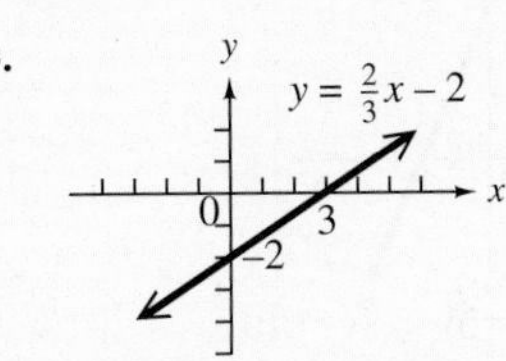

4 Find the intercepts for the graph of $5x + 2y = 10$. Then draw the graph. (Be sure to get a third point as a check.)

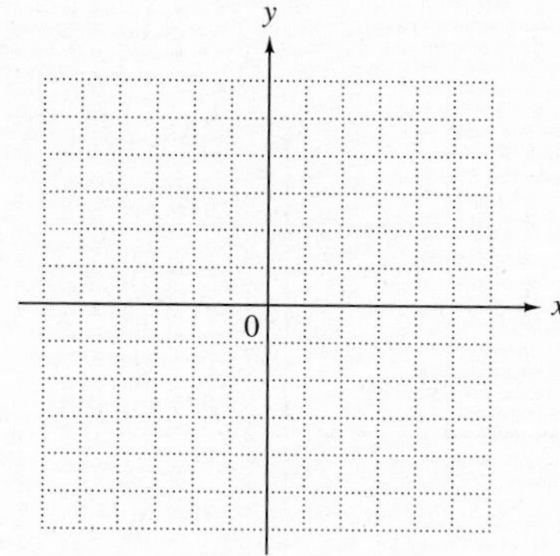

OBJECTIVE 2 Find intercepts. In Figure 11, the graph crosses, or intersects, the y-axis at (0, 3) and the x-axis at (2, 0). For this reason, (0, 3) is called the ***y*-intercept,** and (2, 0) is called the ***x*-intercept** of the graph.

The intercepts are particularly useful for graphing linear equations. The intercepts are found by replacing, in turn, each variable with 0 in the equation and solving for the value of the other variable.

Finding Intercepts

To find the x-intercept, let $y = 0$ in the given equation and solve for x. Then $(x, 0)$ is the x-intercept.

To find the y-intercept, let $x = 0$ in the given equation and solve for y. Then $(0, y)$ is the y-intercept.

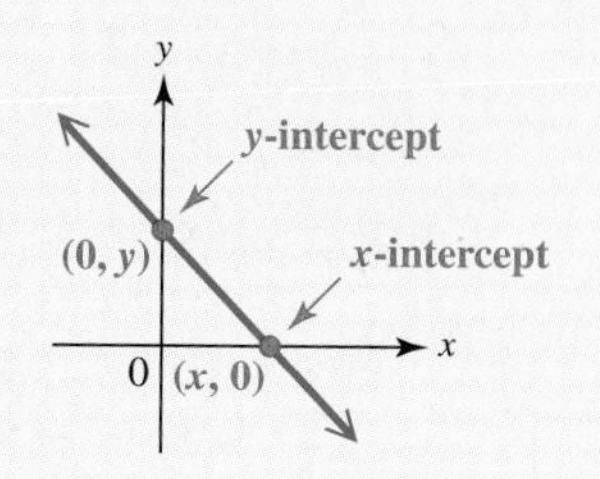

EXAMPLE 3 Finding Intercepts

Find the intercepts for the graph of $2x + y = 4$. Then draw the graph.

To find the y-intercept, let $x = 0$; to find the x-intercept, let $y = 0$.

$$2x + y = 4$$
$$2(\mathbf{0}) + y = 4 \quad \text{Let } x = 0.$$
$$0 + y = 4$$
$$y = 4$$

$$2x + y = 4$$
$$2x + \mathbf{0} = 4 \quad \text{Let } y = 0.$$
$$2x = 4$$
$$x = 2$$

The y-intercept is (0, 4). The x-intercept is (2, 0). Find a third point as a check. For example, choosing $x = 1$ gives $y = 2$. Plot (0, 4), (2, 0), and (1, 2) and draw the line through them. This line, shown in Figure 12, is the graph.

x	y
0	4
2	0
1	2

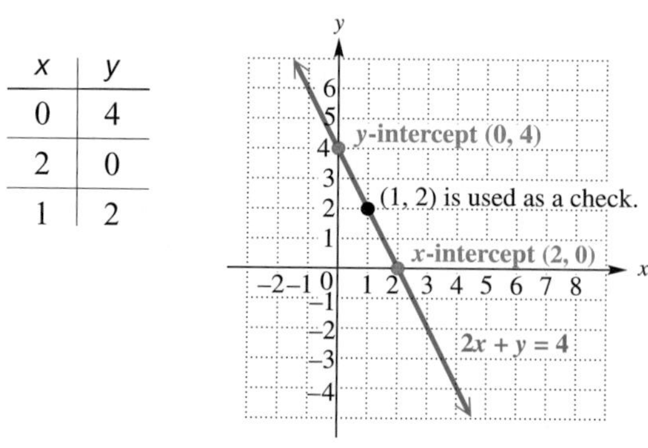

Figure 12

◀ Work Problem **4** at the Side.

ANSWER

4. x-intercept (2, 0); y-intercept (0, 5)

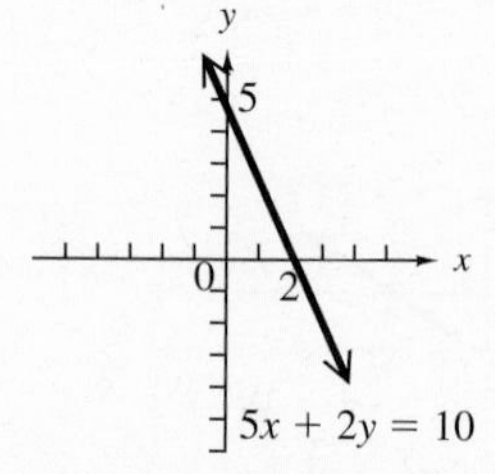

OBJECTIVE 3 Graph linear equations of the form $Ax + By = 0$. In the preceding examples, the x- and y-intercepts were used to help draw the graphs. This is not always possible. Example 4 shows what to do when the x- and y-intercepts are the same point (that is, coincide).

EXAMPLE 4 **Graphing an Equation of the Form $Ax + By = 0$**

Graph the linear equation $x - 3y = 0$.

If we let $x = 0$, then $y = 0$, giving the ordered pair $(0, 0)$. Letting $y = 0$ also gives $(0, 0)$. This is the same ordered pair, so we choose two *other* values for x or y. Choosing 2 for y gives $x - 3 \cdot 2 = 0$, leading to $x = 6$, so another ordered pair is $(6, 2)$. Choosing -2 for y gives $x - 3(-2) = 0$, leading to $x = -6$, so a third ordered pair is $(-6, -2)$. We use the ordered pairs $(-6, -2)$, $(0, 0)$, and $(6, 2)$ to sketch the graph in Figure 13.

x	y
0	0
6	2
−6	−2

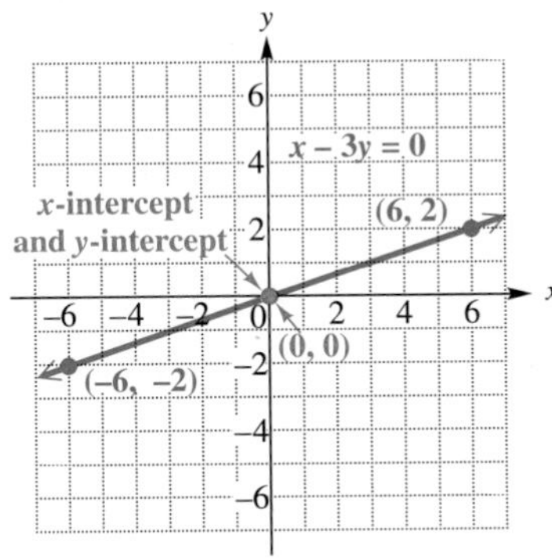

Figure 13

Work Problem 5 at the Side. ▶

Line through the Origin

If A and B are nonzero real numbers, the graph of a linear equation of the form

$$Ax + By = 0$$

passes through the origin $(0, 0)$.

OBJECTIVE 4 **Graph linear equations of the form $y = k$ or $x = k$.** The equation $y = -4$ is a linear equation in which the coefficient of x is 0. (To see this, write $y = -4$ as $0x + y = -4$.) Also, $x = 3$ is a linear equation in which the coefficient of y is 0. These equations lead to horizontal or vertical straight lines, as the next examples show.

EXAMPLE 5 **Graphing an Equation of the Form $y = k$**

Graph $y = -4$.

As the equation states, for any value of x, y is always equal to -4. Three ordered pairs that satisfy the equation are shown. The graph is the horizontal line in Figure 14. The y-intercept is $(0, -4)$; there is no x-intercept.

x	y
−2	−4
0	−4
3	−4

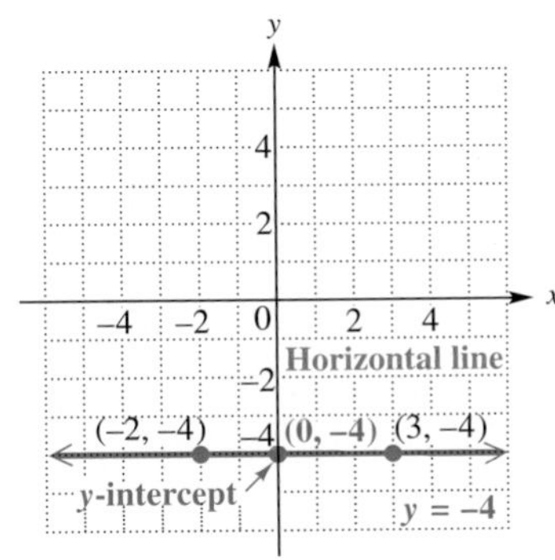

Figure 14

Work Problem 6 at the Side. ▶

5 Graph $2x - y = 0$.

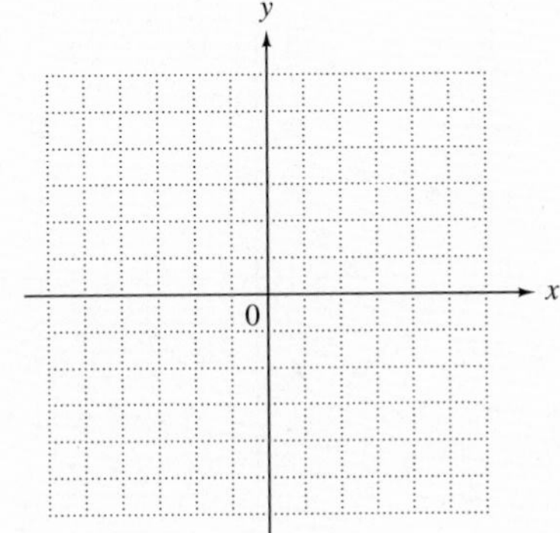

6 Graph $y = -5$.

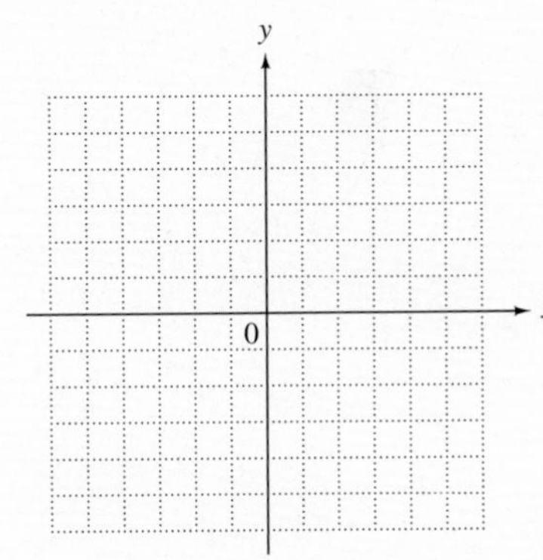

ANSWERS

5.

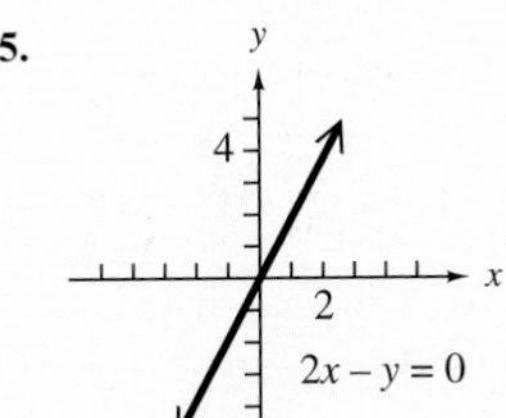

6.

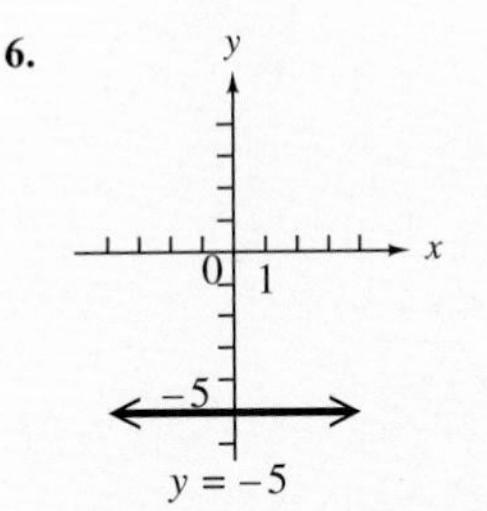

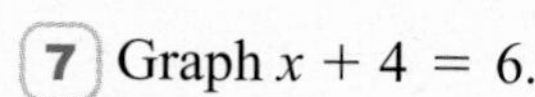

7 Graph $x + 4 = 6$.

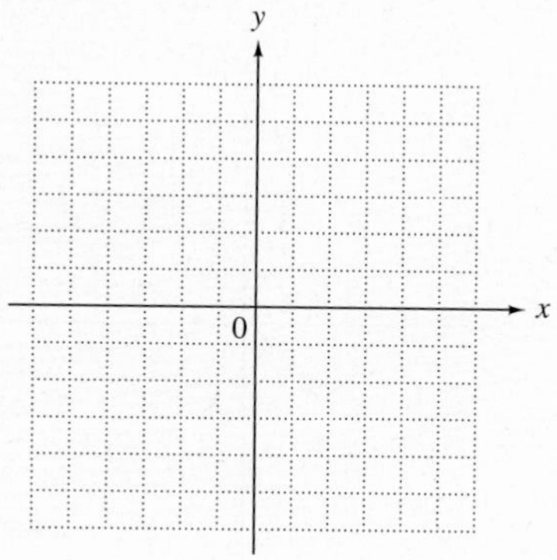

EXAMPLE 6 Graphing an Equation of the Form $x = k$

Graph $x - 3 = 0$.

First add 3 to each side of $x - 3 = 0$ to get $x = 3$. All the ordered pairs that satisfy this equation have x-coordinate 3. Any number can be used for y. See Figure 15 for the graph of this vertical line, along with a table of values. The x-intercept is (3, 0); there is no y-intercept.

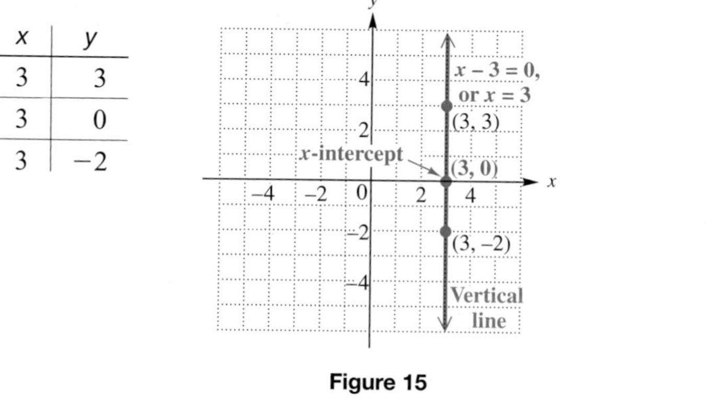

x	y
3	3
3	0
3	−2

Figure 15

◀ Work Problem 7 at the Side.

From the results in Examples 5 and 6, we make the following observations.

Horizontal and Vertical Lines

The graph of the linear equation $y = k$, where k is a real number, is the horizontal line with y-intercept $(0, k)$ and no x-intercept.

The graph of the linear equation $x = k$, where k is a real number, is the vertical line with x-intercept $(k, 0)$ and no y-intercept.

In particular, notice that the horizontal line $y = 0$ is the x-axis and the vertical line $x = 0$ is the y-axis. The different forms of linear equations from this section and the methods of graphing them are summarized below.

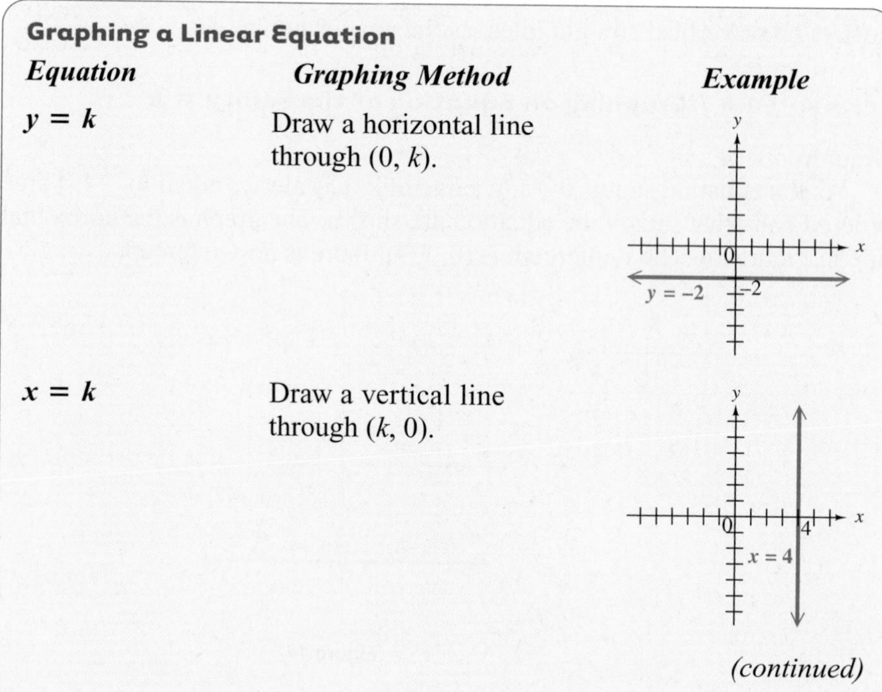

Graphing a Linear Equation

Equation	*Graphing Method*	*Example*
$y = k$	Draw a horizontal line through $(0, k)$.	
$x = k$	Draw a vertical line through $(k, 0)$.	

(continued)

ANSWER

7. $x + 4 = 6$, or $x = 2$

Equation	*Graphing Method*	*Example*
$Ax + By = 0$	Graph passes through (0, 0). To get additional points that lie on the graph, choose any values for x or y, except 0.	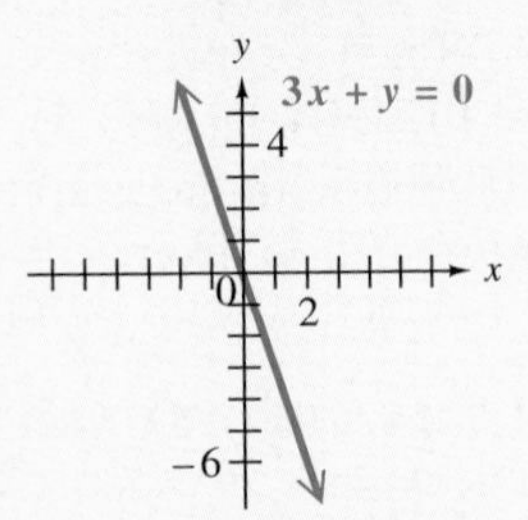
$Ax + By = C$ **(but not of the types above)**	Find any two points on the line. A good choice is to find the intercepts. Let $x = 0$, and find the corresponding value of y; then let $y = 0$, and find x. As a check, get a third point by choosing a value of x or y that has not yet been used.	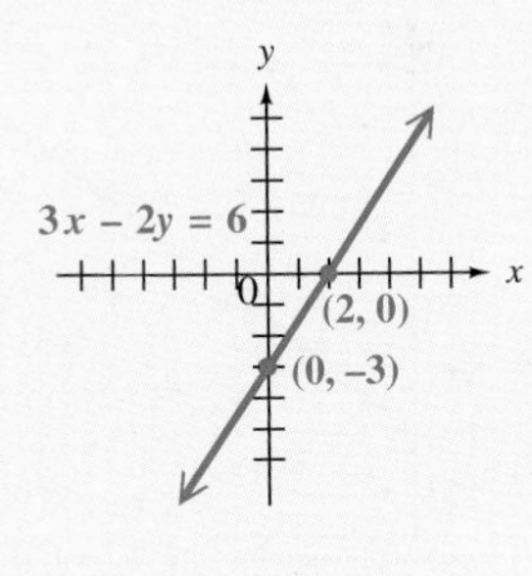

Work Problem **8** at the Side. ▶

8 Match the information about the graphs in parts (a)–(d) with the linear equations in A–D.

A. $x = 5$
B. $2x - 5y = 8$
C. $y - 2 = 3$
D. $x + 4y = 0$

(a) The graph of the equation is a horizontal line.

(b) The graph of the equation passes through the origin.

(c) The graph of the equation is a vertical line.

(d) The graph of the equation passes through (9, 2).

Note

Another method of graphing linear equations, using the concepts of slope and y-intercept, will be covered in Objective 2 of **Section 3.4.**

OBJECTIVE 5 Use a linear equation to model data.

EXAMPLE 7 Using a Linear Equation to Model Credit Card Debt

Credit card debt in the United States has increased steadily during recent years. The amount of debt y in billions of dollars can be modeled by the linear equation

$$y = 38.7x + 450,$$

where $x = 0$ represents the year 1995, $x = 1$ represents 1996, and so on. (*Source:* Board of Governors of the Federal Reserve System.)

(a) Use the equation to approximate credit card debt in the years 1995, 2000, and 2003.

Substitute the appropriate value for each year x to find credit card debt in that year.

For 1995: $y = 38.7(0) + 450$ Replace x with 0.
$y = 450$ billion dollars

For 2000: $y = 38.7(5) + 450$ $2000 - 1995 = 5$; Replace x with 5.
$y = 643.5$ billion dollars

For 2003: $y = 38.7(8) + 450$ $2003 - 1995 = 8$; Replace x with 8.
$y = 759.6$ billion dollars

Continued on Next Page

Answers

8. **(a)** C **(b)** D **(c)** A **(d)** B

9 Use the graph and then the equation in Example 7 to approximate credit card debt in 1997.

(b) Write the information from part (a) as three ordered pairs, and use them to graph the given linear equation.

Since x represents the year and y represents the debt, the ordered pairs are (0, 450), (5, 643.5), and (8, 759.6). See Figure 16. (Arrowheads are not included with the graphed line, since the data are for the years 1995 to 2003 only—that is, from $x = 0$ to $x = 8$.)

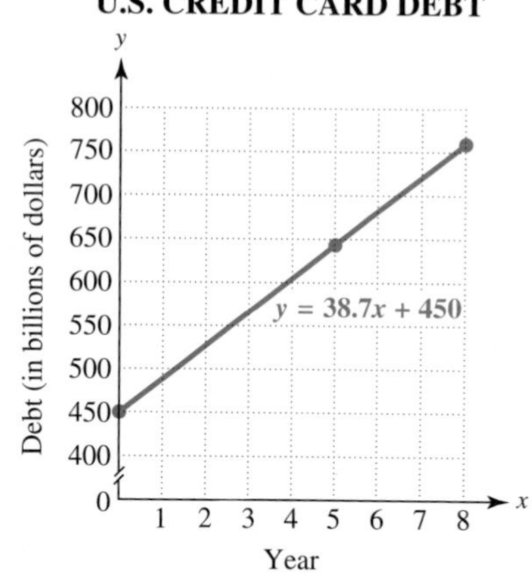

Figure 16

(c) Use the graph and then the equation to approximate credit card debt in 2002.

For 2002, $x = 7$. On the graph, find 7 on the horizontal axis, move up to the graphed line and then across to the vertical axis. It appears that credit card debt in 2002 was about 725 billion dollars. To use the equation, substitute 7 for x.

$$y = 38.7x + 450$$

$$y = 38.7(7) + 450 \qquad \text{Let } x = 7.$$

$$y = 720.9 \text{ billion dollars}$$

This result for 2002 is close to our estimate of 725 billion dollars from the graph.

◀ *Work Problem* **9** *at the Side.*

Answer

9. about 525 billion dollars; 527.4 billion dollars

3.2 Exercises

FOR EXTRA HELP

MyMathLab

Math XL PRACTICE — WATCH — DOWNLOAD — READ — REVIEW

Complete the given ordered pairs for each equation. Then graph each equation by plotting the points and drawing the line through them. See Examples 1 and 2.

1. $x + y = 5$ **5; 5; 3**

$(0, \_\_), (\_\_, 0), (2, \_\_)$

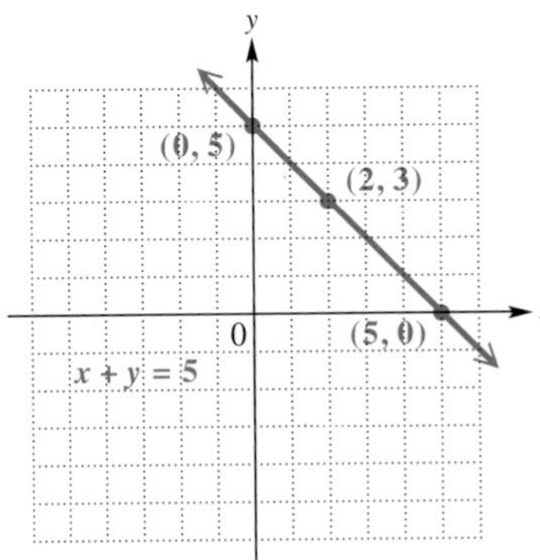

2. $x - y = 2$ **−2; 2; 3**

$(0, \_\_), (\_\_, 0), (5, \_\_)$

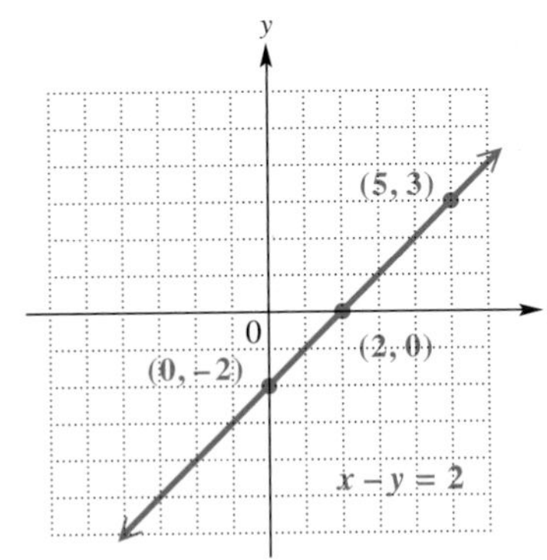

3. $y = \frac{2}{3}x + 1$ **1; 3; −1**

$(0, \_\_), (3, \_\_), (-3, \_\_)$

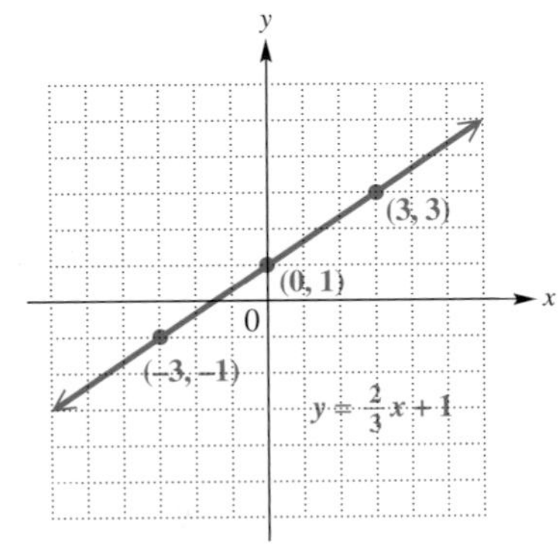

4. $y = -\frac{3}{4}x + 2$ **2; −1; 5**

$(0, \_\_), (4, \_\_), (-4, \_\_)$

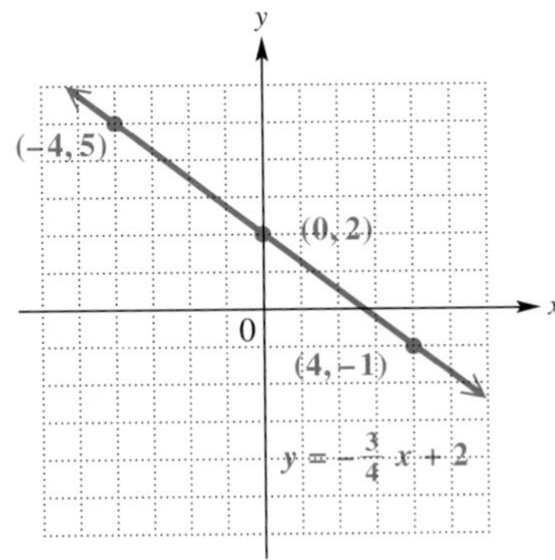

5. $3x = -y - 6$ **−6; −2; −5**

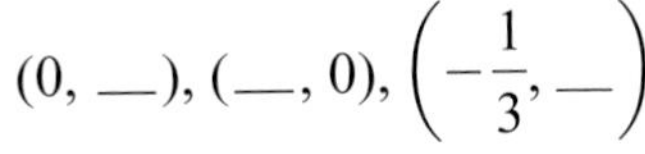

$(0, \_\_), (\_\_, 0), \left(-\frac{1}{3}, \_\_\right)$

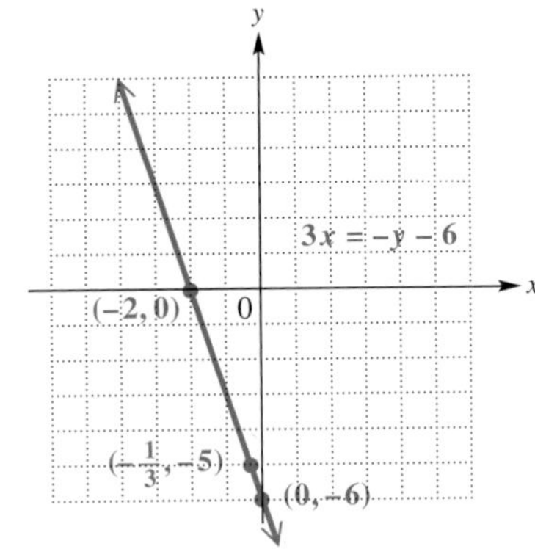

6. $x = 2y + 3$ **3; $-\frac{3}{2}$; 4**

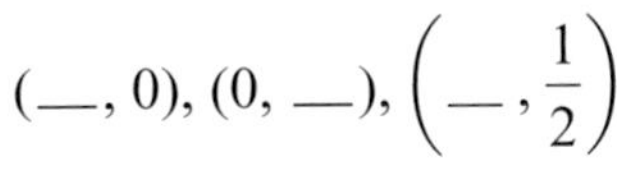

$(\_\_, 0), (0, \_\_), \left(\_\_, \frac{1}{2}\right)$

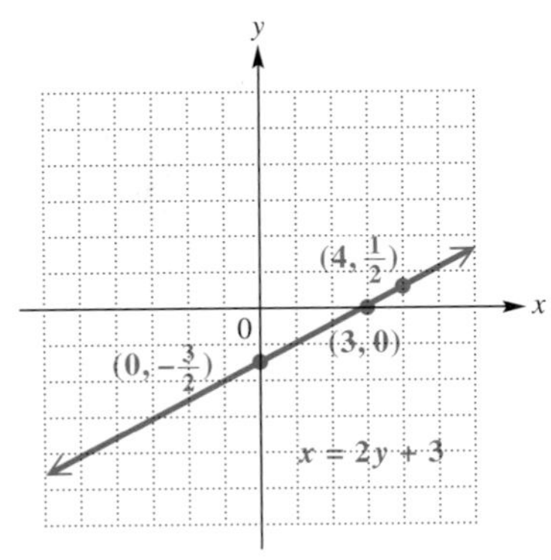

7. Match the information about each graph in Column I with the correct linear equation in Column II.

I	II
(a) The graph of the equation has y-intercept $(0, -4)$. **A**	**A.** $3x + y = -4$
(b) The graph of the equation has $(0, 0)$ as x-intercept and y-intercept. **C**	**B.** $x - 4 = 0$
(c) The graph of the equation does not have an x-intercept. **D**	**C.** $y = 4x$
(d) The graph of the equation has x-intercept $(4, 0)$. **B**	**D.** $y = 4$

8. Write a few sentences summarizing how to graph a linear equation in two variables.

Find two ordered pairs that satisfy the equation. (The intercepts are good choices.) As a check, find a third ordered pair. Plot the corresponding points on a coordinate system. Draw a straight line through the points.

Find the intercepts for the graph of each equation. See Example 3.

9. $2x - 3y = 24$
x-intercept: **(12, 0)**
y-intercept: **(0, −8)**

10. $-3x + 8y = 48$
x-intercept: **(−16, 0)**
y-intercept: **(0, 6)**

11. $x + 6y = 0$
x-intercept: **(0, 0)**
y-intercept: **(0, 0)**

12. $3x - y = 0$
x-intercept: **(0, 0)**
y-intercept: **(0, 0)**

Graph each linear equation. See Examples 1–6.

13. $y = x - 2$

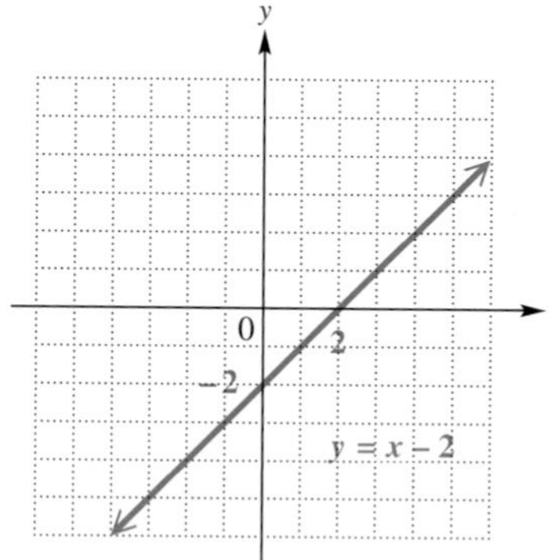

14. $y = -x + 6$

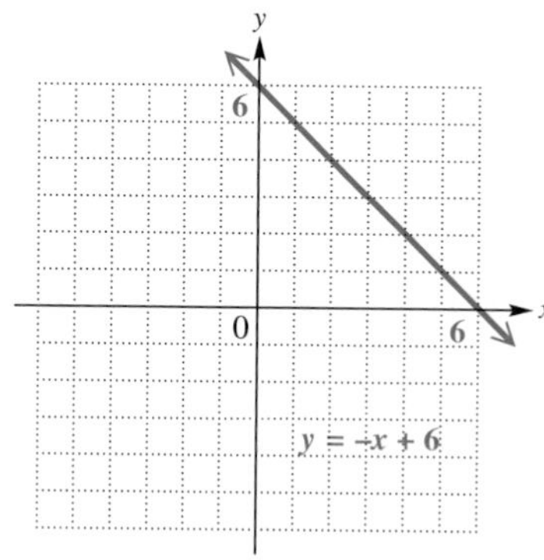

15. $x - y = 4$

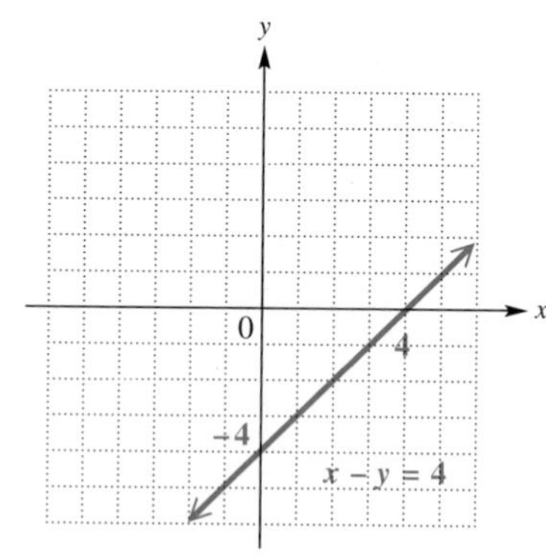

16. $x - y = 5$

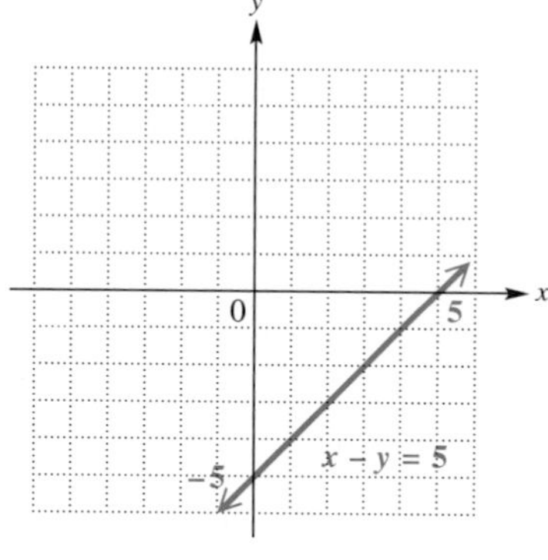

17. $2x + y = 6$

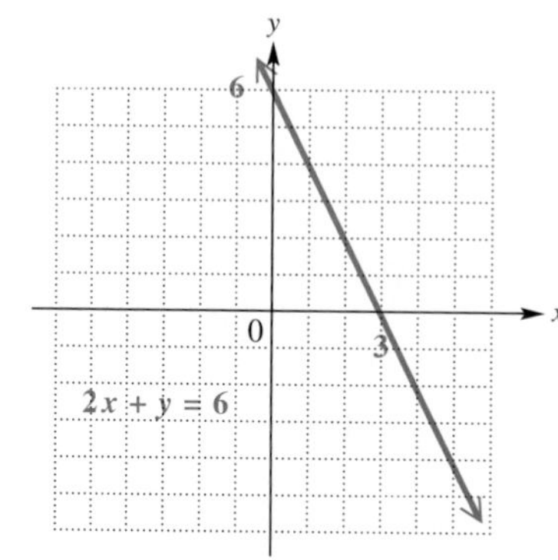

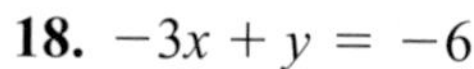

18. $-3x + y = -6$

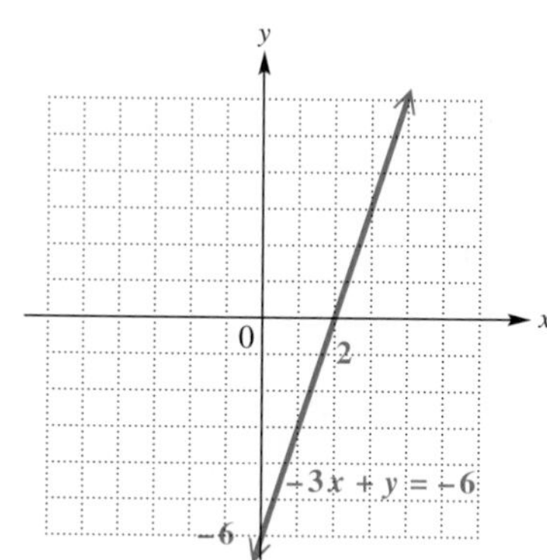

19. $3x + 7y = 14$

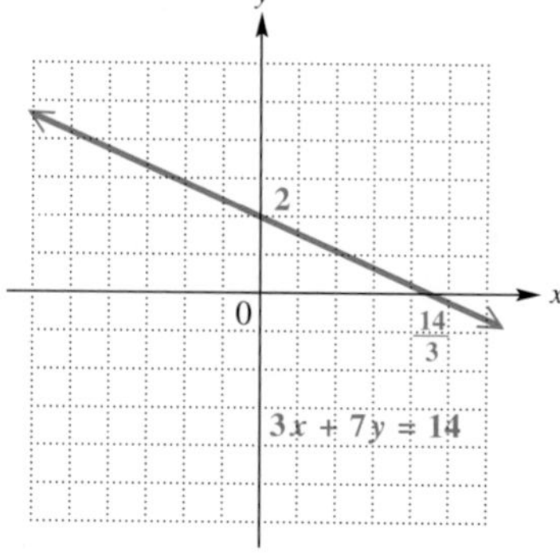

20. $6x - 5y = 18$

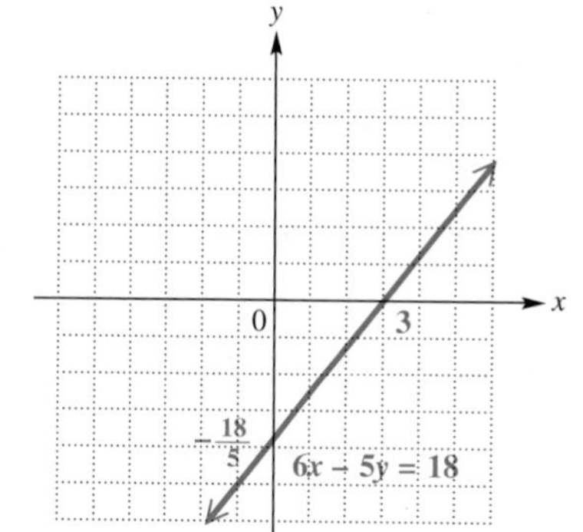

21. $y - 2x = 0$

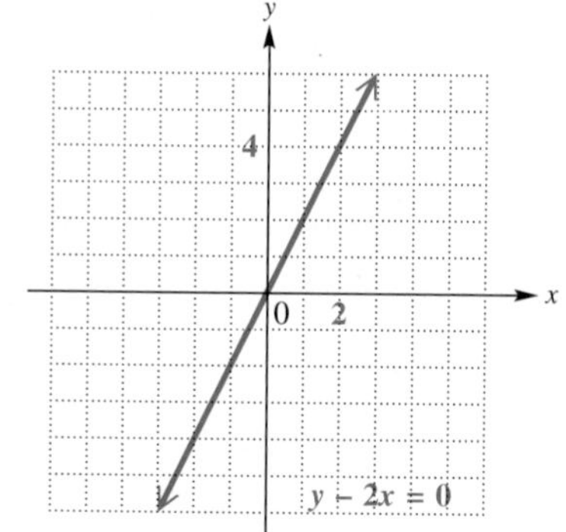

22. $y + 3x = 0$

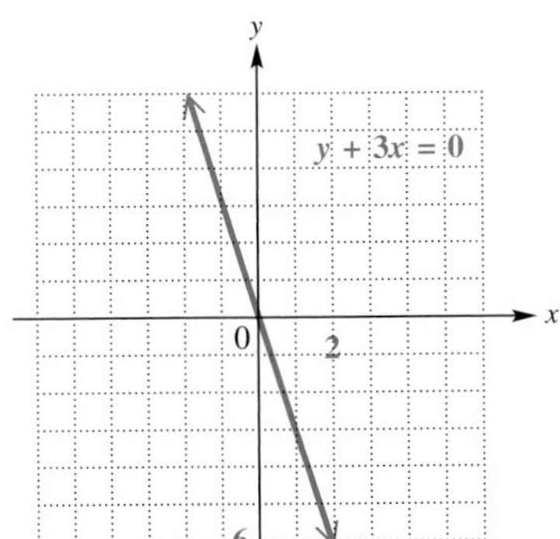

23. $y = -6x$

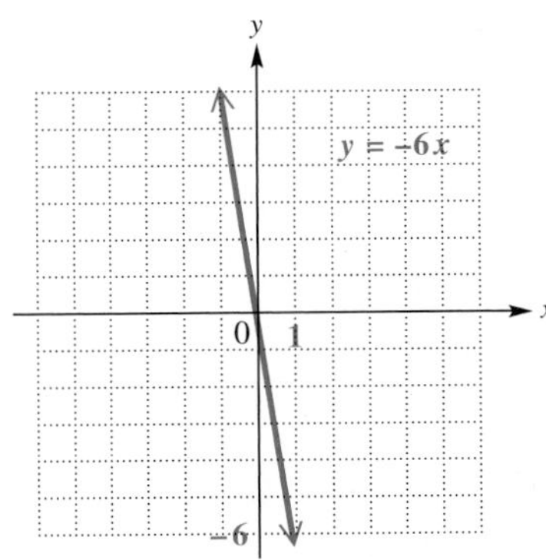

24. $y = 4x$

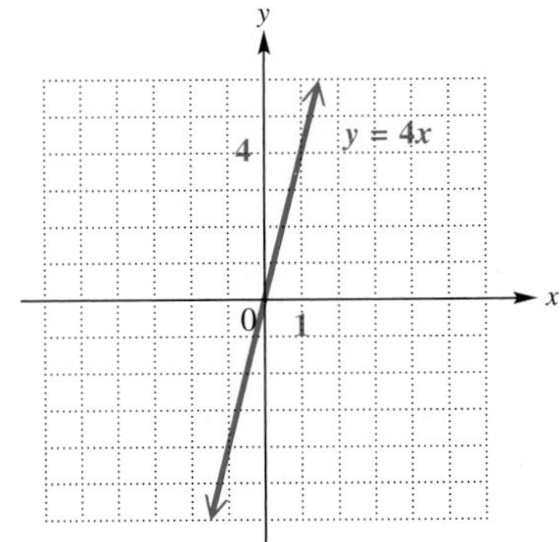

25. $x = -2$

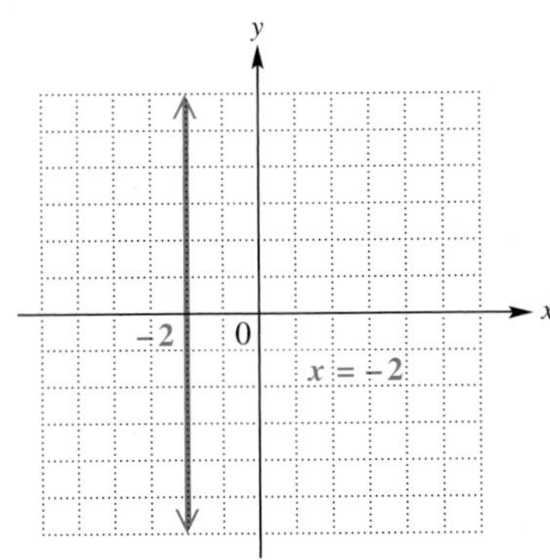

26. $x = 4$

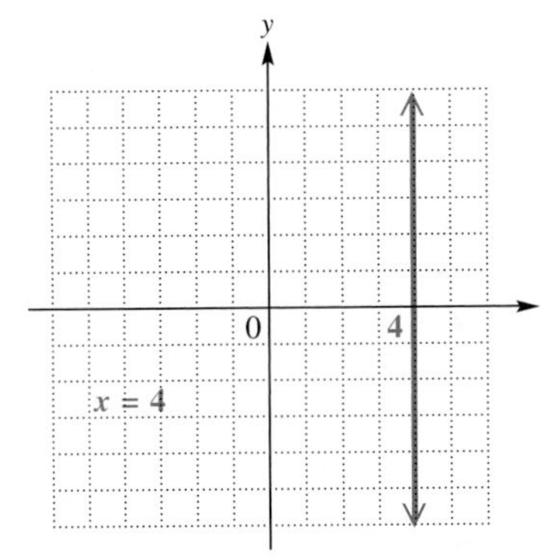

27. $y - 3 = 0$

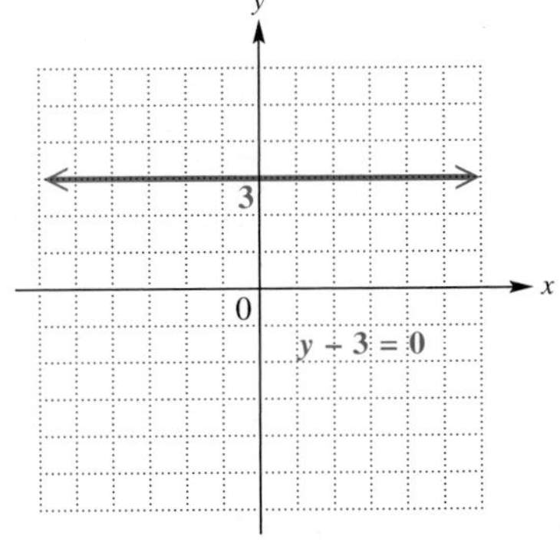

28. $y + 1 = 0$

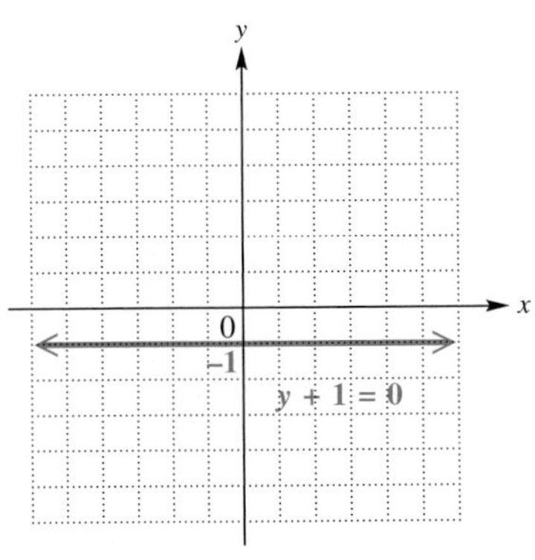

29. $-3y = 15$

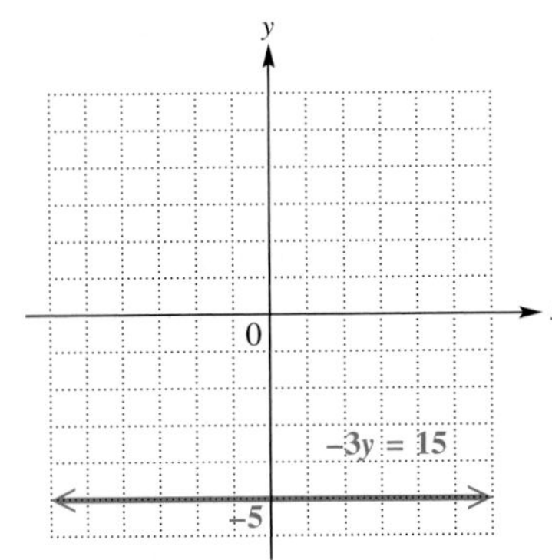

30. $-2y = 12$

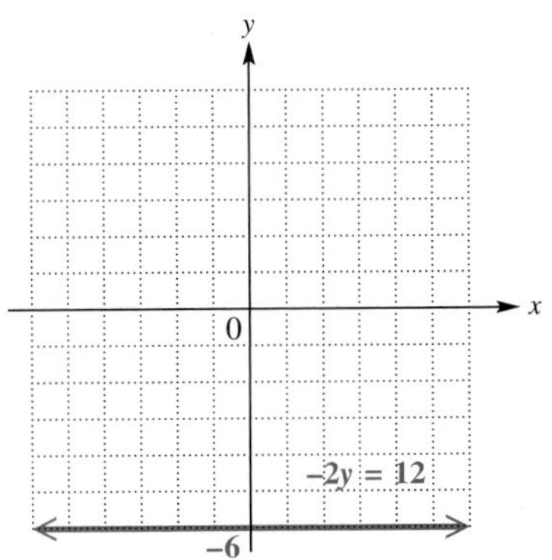

In Exercises 31–34, describe what the graph of each linear equation will look like on the coordinate plane. (Hint: Rewrite the equation if necessary so that it is in a more recognizable form.) **In Exercises 31–34, descriptions may vary.**

31. $3x = y - 9$

The graph is a line with x-intercept $(-3, 0)$ and y-intercept $(0, 9)$.

32. $x - 10 = 1$

The graph is a vertical line with x-intercept $(11, 0)$.

33. $3y = -6$

The graph is a horizontal line with y-intercept $(0, -2)$.

34. $2x = 4y$

The graph passes through the origin $(0, 0)$. It also passes through the points $(2, 1)$ and $(4, 2)$.

35. A student attempted to graph $4x + 5y = 0$ by finding intercepts. She first let $x = 0$ and found y; then she let $y = 0$ and found x. In both cases, the resulting point was (0, 0). She knew that she needed at least two points to graph the line, but was unsure what to do next because finding intercepts gave her only one point. How would you explain to her what to do next?

Choose a value *other than* 0 for either x or y. For example, if $x = -5$, $y = 4$.

36. What is the equation of the x-axis? What is the equation of the y-axis?

$y = 0$; $x = 0$

Solve each problem. See Example 7.

37. The height y (in centimeters) of a woman is related to the length of her radius bone x (from the wrist to the elbow) and is approximated by the linear equation

$$y = 3.9x + 73.5.$$

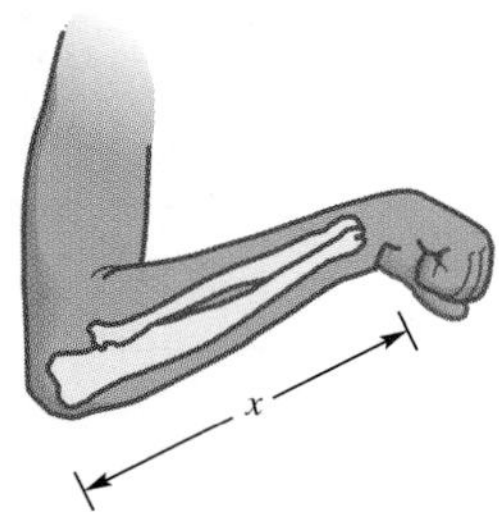

(a) Use the equation to find the approximate heights of women with radius bones of lengths 20 cm, 22 cm, and 26 cm.

151.5 cm, 159.3 cm, 174.9 cm

(b) Write the information from part (a) as three ordered pairs.

(20, 151.5), (22, 159.3), (26, 174.9)

(c) Graph the equation using the data from part (b).

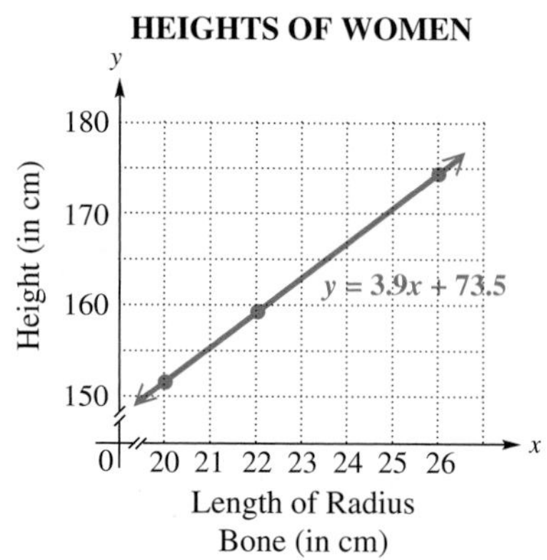

See the graph.

(d) Use the graph to estimate the length of the radius bone in a woman who is 167 cm tall. Then use the equation to find the length of this radius bone to the nearest centimeter. (*Hint:* Substitute for y in the equation.)

24 cm; 24 cm

38. The weight y (in pounds) of a man taller than 60 in. can be roughly approximated by the linear equation

$$y = 5.5x - 220,$$

where x is the height of the man in inches.

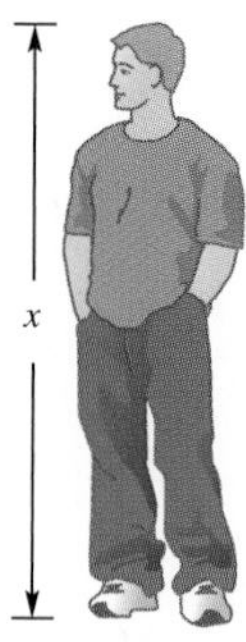

(a) Use the equation to approximate the weights of men whose heights are 62 in., 66 in., and 72 in.

121 lb, 143 lb, 176 lb

(b) Write the information from part (a) as three ordered pairs.

(62, 121), (66, 143), (72, 176)

(c) Graph the equation using the data from part (b).

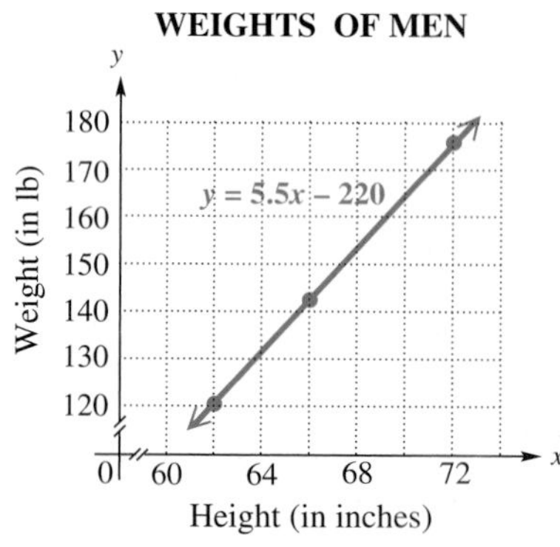

See the graph.

(d) Use the graph to estimate the height of a man who weighs 155 lb. Then use the equation to find the height of this man to the nearest inch. (*Hint:* Substitute for y in the equation.)

68 in.; 68 in.

39. As a fundraiser, a school club is selling posters. The printer charges a \$25 set-up fee, plus \$0.75 for each poster. Then the cost y in dollars to print x posters is given by the linear equation

$$y = 0.75x + 25.$$

(a) What is the cost y in dollars to print 50 posters? to print 100 posters?

\$62.50; \$100

(b) Find the number of posters x if the printer billed the club for costs of \$175.

200

(c) Write the information from parts (a) and (b) as three ordered pairs.

(50, 62.50), (100, 100), (200, 175)

(d) Use the data from part (c) to graph the equation.

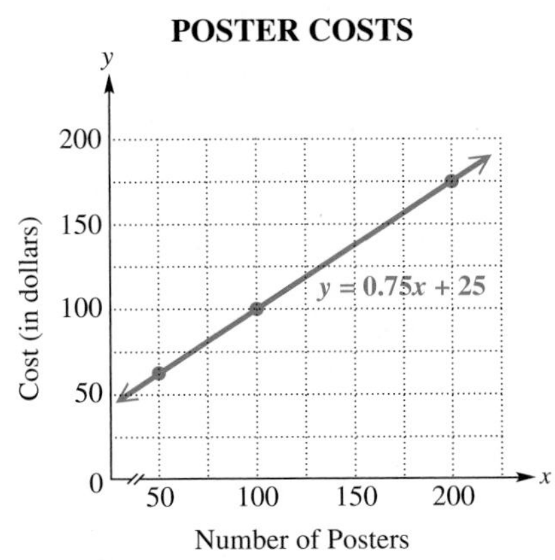

See the graph.

40. A gas station is selling gasoline for \$4.50 per gallon and charges \$7 for a car wash. Then the cost y in dollars for x gallons of gasoline and a car wash is given by the linear equation

$$y = 4.50x + 7.$$

(a) What is the cost y in dollars for 9 gallons of gasoline and a car wash? for 4 gallons of gasoline and a car wash?

\$47.50; \$25.00

(b) Find the number of gallons of gasoline x if the cost for the gasoline and a car wash is \$43.00.

8 gal

(c) Write the information from parts (a) and (b) as three ordered pairs.

(9, 47.50), (4, 25.00), (8, 43.00)

(d) Use the data from part (c) to graph the equation.

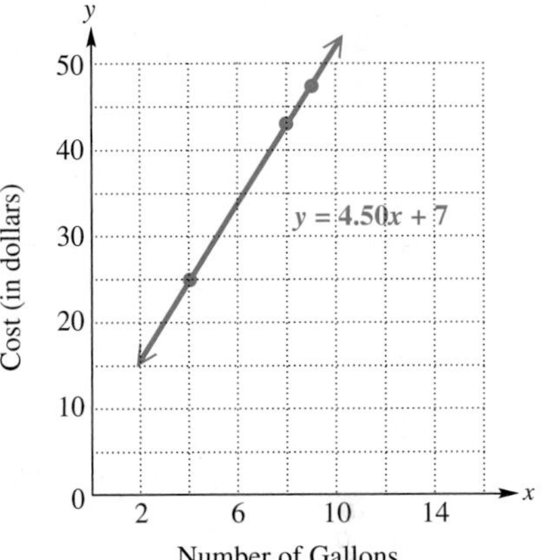

See the graph.

41. The graph shows the value of a certain sport-utility vehicle over the first 5 yr of ownership. Use the graph to do the following.

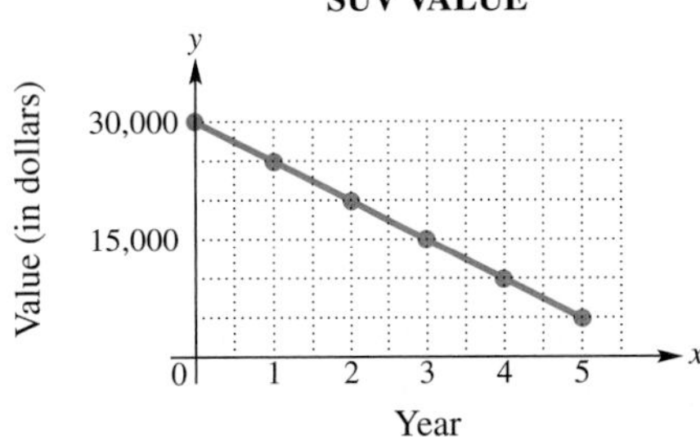

(a) Determine the initial value of the SUV.

\$30,000

(b) Find the **depreciation** (loss in value) from the original value after the first 3 yr.

\$15,000

(c) What is the annual or yearly depreciation in each of the first 5 yr?

\$5000

(d) What does the ordered pair (5, 5000) mean in the context of this problem?

After 5 yr, the SUV has a value of \$5000.

42. Demand for an item is often closely related to its price. As price increases, demand decreases, and as price decreases, demand increases. Suppose demand for a video game is 2000 units when the price is \$40, and demand is 2500 units when the price is \$30.

(a) Let x be the price and y be the demand for the game. Graph the two given pairs of prices and demands.

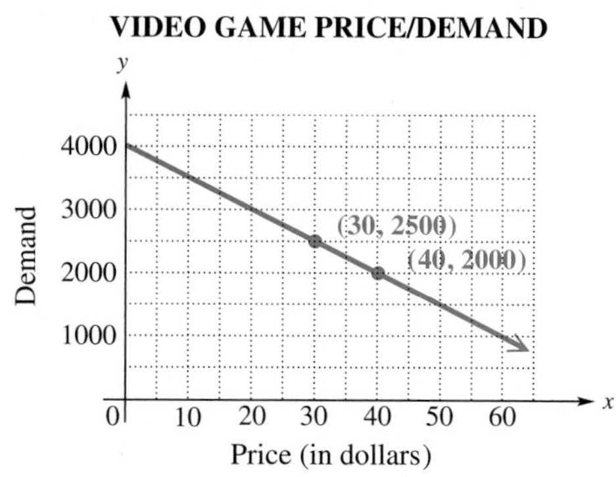

See the indicated points on the graph.

(b) Assume the relationship is linear. Draw a line through the two points from part (a). From your graph, estimate the demand if the price drops to \$20.

See the graph; 3000 units

(c) Use the graph to estimate the price if the demand is 3500 units.

\$10

(d) Write the prices and demands from parts (b) and (c) as ordered pairs.

(20, 3000), (10, 3500)

43. U.S. per capita consumption of cheese increased for the years 1980 through 2005 as shown in the graph. If $x = 0$ represents 1980, $x = 5$ represents 1985, and so on, per capita consumption y in pounds can be modeled by the linear equation

$$y = 0.5383x + 18.74.$$

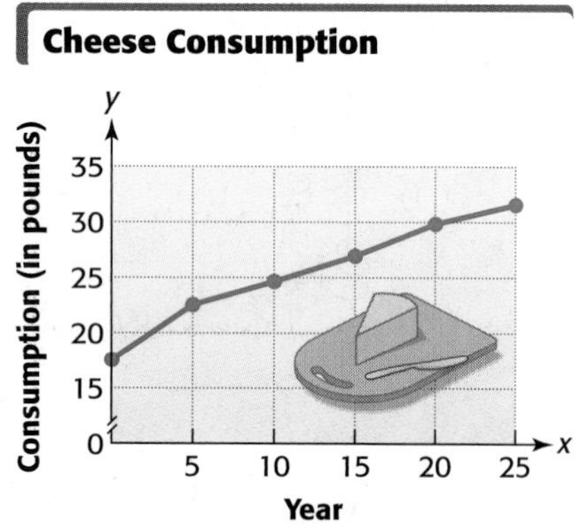

Source: U.S. Department of Agriculture.

(a) Use the equation to approximate consumption in 1990, 2000, and 2005 to the nearest tenth.

1990: 24.1 lb; 2000: 29.5 lb; 2005: 32.2 lb

(b) Use the graph to estimate consumption for the same years.

1990: 25 lb; 2000: 30 lb; 2005: 32 lb

(c) How do the approximations using the equation compare to the estimates from the graph?

The values are quite close.

44. In the United States, sporting goods sales y (in billions of dollars) from 2000 through 2006 are shown in the graph and modeled by the linear equation

$$y = 3.018x + 72.52,$$

where $x = 0$ corresponds to 2000, $x = 1$ corresponds to 2001, and so on.

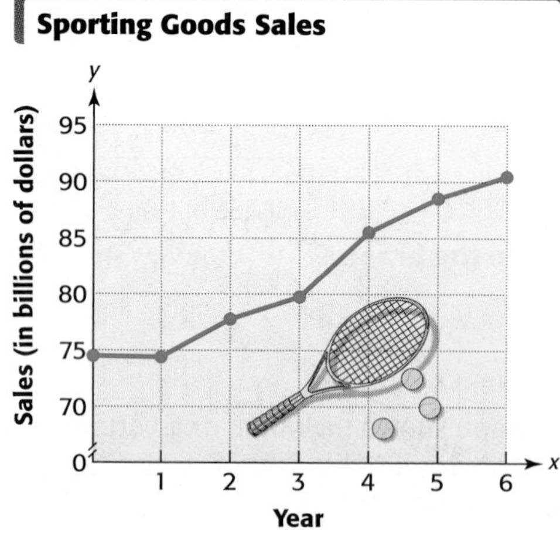

Source: National Sporting Goods Association.

(a) Use the equation to approximate sporting goods sales in 2000, 2004, and 2006. Round your answers to the nearest billion dollars.

2000: \$73 billion; 2004: \$85 billion;
2006: \$91 billion

(b) Use the graph to estimate sales for the same years.

2000: \$74 billion; 2004: \$86 billion;
2006: \$91 billion

(c) How do the approximations using the equation compare to the estimates using the graph?

The values are quite close.

3.3 Slope of a Line

OBJECTIVES

1. **Find the slope of a line given two points.**
2. **Find the slope from the equation of a line.**
3. **Use slope to determine whether two lines are parallel, perpendicular, or neither.**
4. **Solve problems involving average rate of change.**

An important characteristic of the lines we graphed in the previous section is their slant or "steepness", as viewed from *left to right*. See Figure 17.

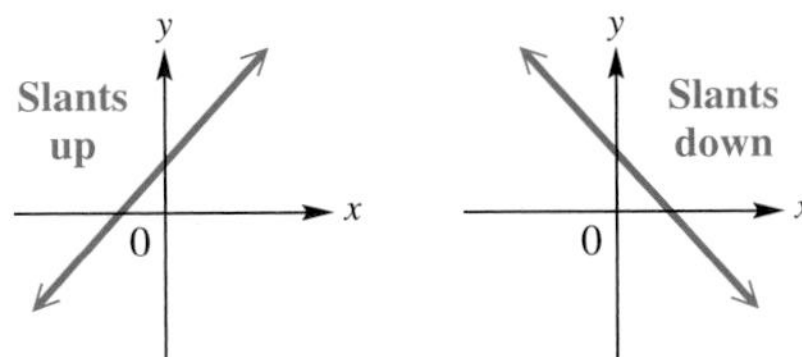

Figure 17

One way to measure the steepness of a line is to compare the vertical change in the line to the horizontal change while moving along the line from one fixed point to another. This measure of steepness is called the *slope* of the line.

OBJECTIVE 1 Find the slope of a line given two points. To find the steepness, or slope, of the line in Figure 18, we begin at point Q and move to point P. The vertical change, or **rise,** is the change in the y-values, which is the difference $6 - 1 = 5$ units. The horizontal change, or **run,** from Q to P is the change in the x-values, which is the difference $5 - 2 = 3$ units.

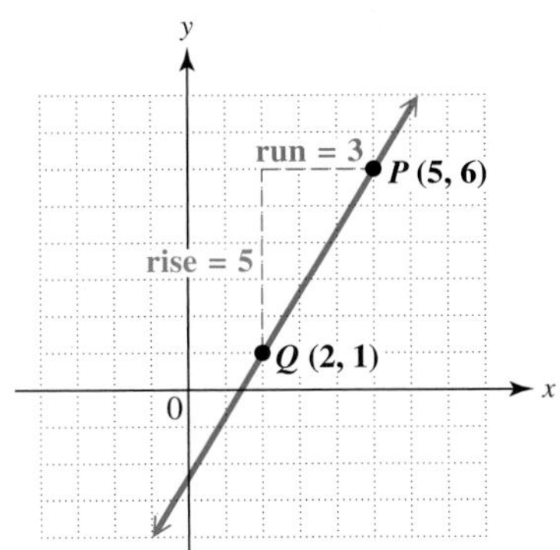

Figure 18

Remember from **Section 2.6** that one way to compare two numbers is by using a ratio. The **slope** is the ratio of the vertical change in y to the horizontal change in x. The line in Figure 18 has

$$\text{slope} = \frac{\text{vertical change in } y \text{ (rise)}}{\text{horizontal change in } x \text{ (run)}} = \frac{5}{3}.$$

To confirm this ratio, we can count grid squares. We start at point Q in Figure 18 and count *up* 5 grid squares to find the vertical change (rise). To find the horizontal change (run) and arrive at point P, we count to the *right* 3 grid squares. The slope is $\frac{5}{3}$, as found above. ***Slope is a single number that allows us to determine the direction in which a line is slanting from left to right, as well as how much slant there is to the line.***

 Find the slope ratio of each line.

(a)

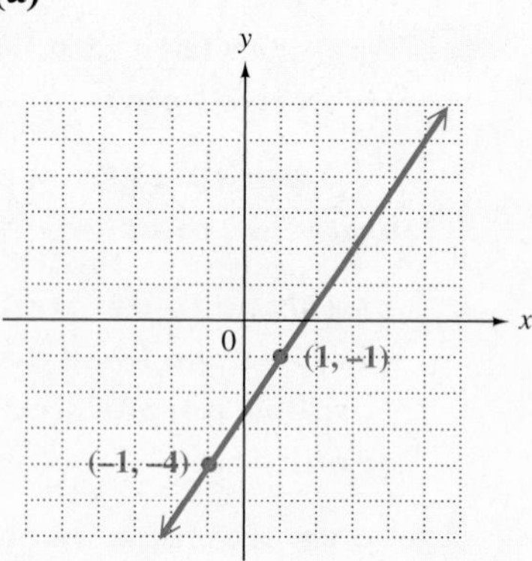

(b)

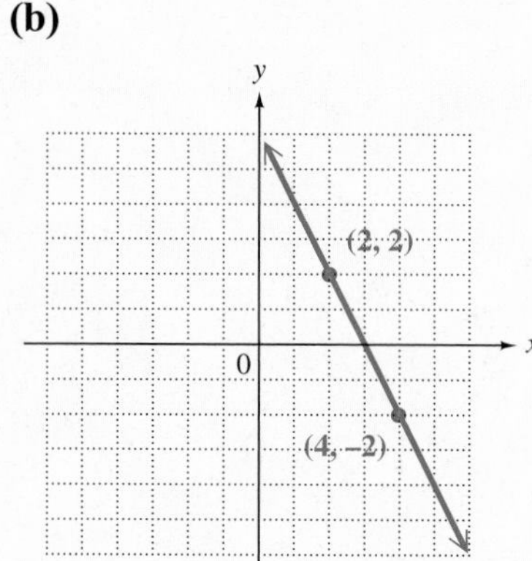

EXAMPLE 1 Finding the Slope of a Line

Find the slope of the line in Figure 19.

We use the two points shown on the line. The vertical change is the difference in the y-values, or $-1 - 3 = -4$, and the horizontal change is the difference in the x-values, or $6 - 2 = 4$. Thus, the line has

$$\text{slope} = \frac{\text{change in } y \text{ (rise)}}{\text{change in } x \text{ (run)}} = \frac{-4}{4}, \quad \text{or} \quad -1.$$

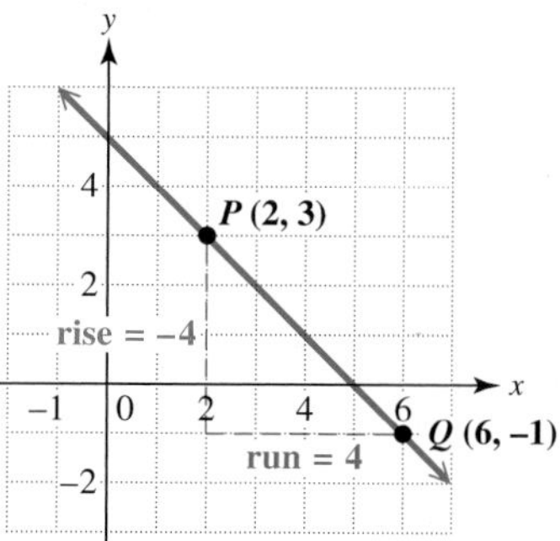

Figure 19

Counting grid squares, we begin at point P and count *down* 4 grid squares. Because we counted down, we write the vertical change as a negative number, -4 here. Then we count to the *right* 4 grid squares to reach point Q. The slope is $\frac{-4}{4}$, or -1.

Work Problem 1 at the Side.

Note

The slope of a line is the same for any two points on the line. To see this, refer to Figure 19. Find the points (3, 2) and (5, 0), which also lie on the line. If we start at (3, 2) and count *down* 2 units and then to the *right* 2 units, we arrive at (5, 0). The slope is $\frac{-2}{2}$, or -1, the same slope we found in Example 1.

The concept of slope is used in many everyday situations. See Figure 20. For example, a highway with a 10%, or $\frac{1}{10}$, grade (or slope) rises 1 m for every 10 m horizontally. Architects specify the pitch of a roof by using slope; a $\frac{5}{12}$ roof means that the roof rises 5 ft for every 12 ft that it runs in the horizontal direction. The slope of a stairwell also indicates the ratio of the vertical rise to the horizontal run. In the figure, the slope of the stairwell is $\frac{8}{12}$, or $\frac{2}{3}$.

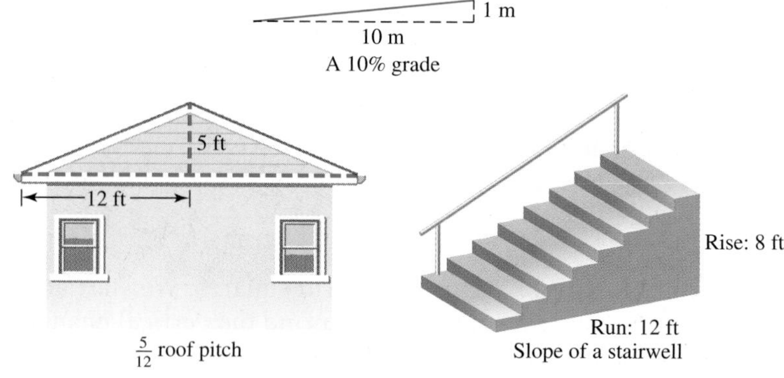

Figure 20

We can generalize the preceding discussion and find the slope of a line through two nonspecific points (x_1, y_1) and (x_2, y_2). (This notation is called **subscript notation.** Read x_1 as **"x-sub-one"** and x_2 as **"x-sub-two."**) See Figure 21.

ANSWERS

1. (a) $\frac{3}{2}$ (b) -2

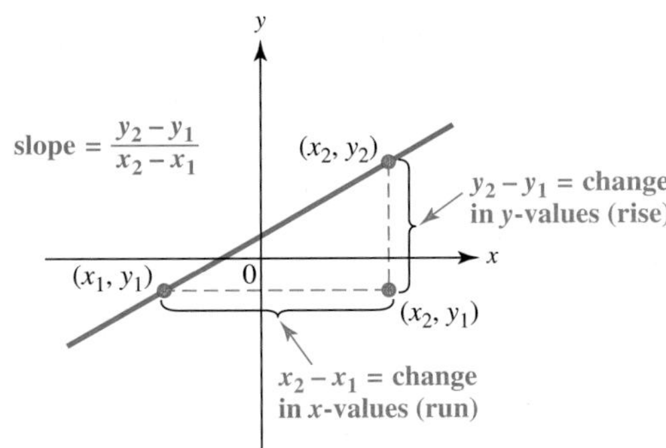

Figure 21

Moving along the line from the point (x_1, y_1) to the point (x_2, y_2), we see that y changes by $y_2 - y_1$ units. This is the vertical change (rise). Similarly, x changes by $x_2 - x_1$ units, which is the horizontal change (run). The slope of the line is the ratio of $y_2 - y_1$ to $x_2 - x_1$.

> **Note**
> Subscript notation is used to identify a point. It does *not* indicate any operation. Note the difference between x_2, a nonspecific value, and x^2, which means $x \cdot x$. Read x_2 as "x-sub-two," *not* "x squared."

Traditionally, the letter m represents slope. The slope m of a line is defined as follows.

> **Slope Formula**
> The **slope** of the line through the points (x_1, y_1) and (x_2, y_2) is
>
> $$m = \frac{\textbf{change in } y}{\textbf{change in } x} = \frac{y_2 - y_1}{x_2 - x_1} \quad (x_1 \neq x_2).$$

The slope gives the change in y for each unit of change in x.

Work Problem **2** *at the Side.* ▶

EXAMPLE 2 Finding Slopes of Lines

Find the slope of each line.

(a) The line through $(-4, 7)$ and $(1, -2)$

Use the slope formula. Let $(-4, 7) = (x_1, y_1)$ and $(1, -2) = (x_2, y_2)$.

$$\text{slope } m = \frac{\text{change in } y}{\text{change in } x} = \frac{y_2 - y_1}{x_2 - x_1} = \frac{-2 - 7}{1 - (-4)} = \frac{-9}{5} = -\frac{9}{5}$$

Substitute carefully here.

Continued on Next Page

2 Find $\frac{y_2 - y_1}{x_2 - x_1}$ for the following values.

(a) $y_2 = 4, y_1 = -1,$
$x_2 = 3, x_1 = 4$

(b) $x_1 = 3, x_2 = -5,$
$y_1 = 7, y_2 = -9$

(c) $x_1 = 2, x_2 = 7,$
$y_1 = 4, y_2 = 9$

Answers

2. (a) -5 **(b)** 2 **(c)** 1

3 Find the slope of each line.

(a) The line through $(6, -2)$ and $(5, 4)$

(b) The line through $(-3, 5)$ and $(-4, -7)$

(c) The line through $(6, -8)$ and $(-2, 4)$
(Find this slope in two different ways as in Example 2(b).)

Count grid squares in Figure 22 to confirm that the slope is $-\frac{9}{5}$.

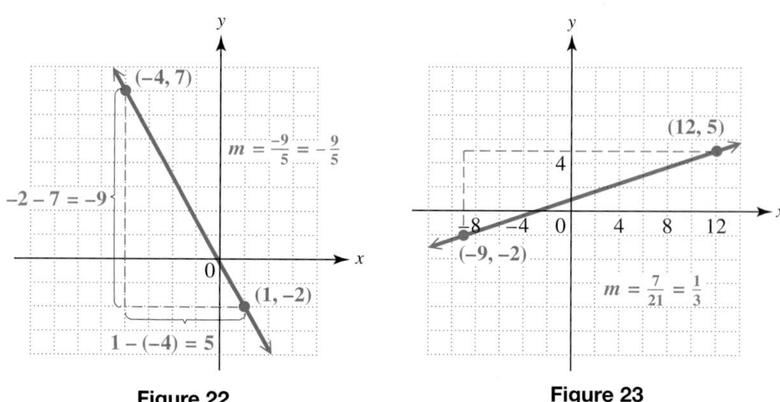

Figure 22 Figure 23

(b) The line through $(-9, -2)$ and $(12, 5)$

$$\text{slope } m = \frac{5-(-2)}{12-(-9)} = \frac{7}{21} = \frac{1}{3}$$

(y-value on top; x-value from the *same* ordered pair on the bottom)

See Figure 23. Note that the same slope is obtained by subtracting in reverse order.

$$\text{slope } m = \frac{-2-5}{-9-12} = \frac{-7}{-21} = \frac{1}{3}$$

(y-value on top; x-value from the *same* ordered pair on the bottom)

CAUTION
It makes no difference which point is (x_1, y_1) or (x_2, y_2); however, be consistent. Start with the x- and y-values of one point (either one), and subtract the corresponding values of the other point.

◀ *Work Problem 3 at the Side.*

The slopes we found for the lines in Figures 22 and 23 suggest the following generalization.

Positive and Negative Slopes
A line with positive slope rises (slants up) from left to right.
A line with negative slope falls (slants down) from left to right.

ANSWERS
3. (a) -6 (b) 12 (c) $-\frac{3}{2}$; $-\frac{3}{2}$

EXAMPLE 3 Showing that the Slope of a Horizontal Line Is Zero

Find the slope of the line through $(-8, 4)$ and $(2, 4)$.

$$m = \frac{y_2 - y_1}{x_2 - x_1} = \frac{4 - 4}{-8 - 2} = \frac{0}{-10} = 0 \quad \text{Zero slope}$$

As shown in Figure 24, the line through the given points is horizontal, with equation $y = 4$. ***All horizontal lines have slope* 0** since the difference in their y-values is always 0.

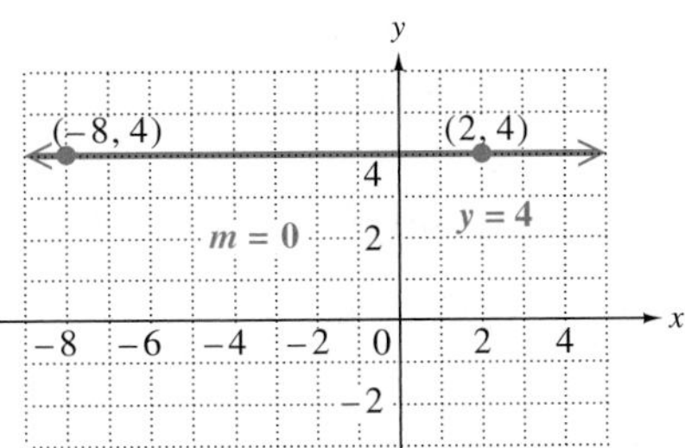

Figure 24

EXAMPLE 4 Showing that a Vertical Line Has Undefined Slope

Find the slope of the line through $(6, 2)$ and $(6, -4)$.

$$m = \frac{y_2 - y_1}{x_2 - x_1} = \frac{2 - (-4)}{6 - 6} = \frac{6}{0} \quad \text{Undefined slope}$$

Because division by 0 is undefined, this line has undefined slope. (This is why the slope formula at the beginning of this section had the restriction $x_1 \neq x_2$.) The graph in Figure 25 shows that this line is vertical, with equation $x = 6$. All points on a vertical line have the same x-value, so ***all vertical lines have undefined slope.***

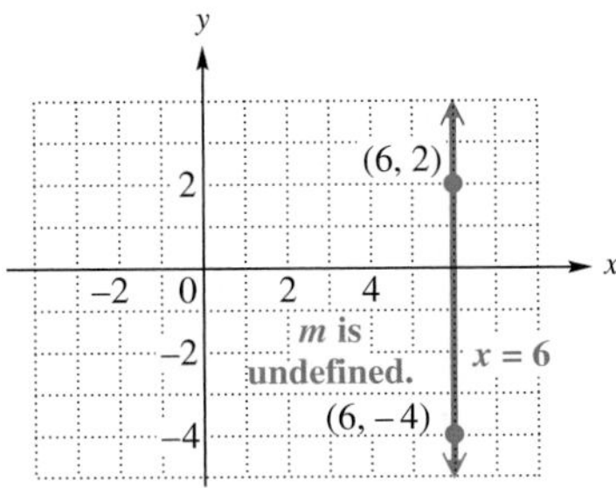

Figure 25

> **Slopes of Horizontal and Vertical Lines**
>
> **Horizontal lines,** which have equations of the form $y = k$, have **slope 0.**
>
> **Vertical lines,** which have equations of the form $x = k$, have **undefined slope.**

Work Problem 4 at the Side. ▶

4 Find the slope of each line.

(a) The line through $(2, 5)$ and $(-1, 5)$

(b) The line through $(3, 1)$ and $(3, -4)$

(c) The line with equation $y = -1$

(d) The line with equation $x - 4 = 0$

Answers

4. (a) 0 **(b)** undefined **(c)** 0 **(d)** undefined

Figure 26 summarizes the four cases for slopes of lines.

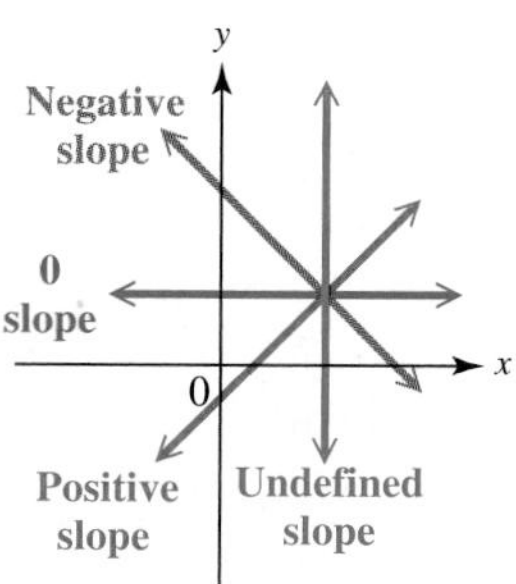

Slopes of lines

Figure 26

OBJECTIVE 2 Find the slope from the equation of a line. Consider the equation

$$y = -3x + 5.$$

We can find its slope using any two points on the line. We get these two points by first choosing two different values of x and then finding the corresponding values of y. For example, choose $x = -2$ and $x = 4$.

$y = -3x + 5$		$y = -3x + 5$	
$y = -3(-2) + 5$	Let $x = -2$.	$y = -3(4) + 5$	Let $x = 4$.
$y = 6 + 5$		$y = -12 + 5$	
$y = 11$		$y = -7$	

The ordered pairs are $(-2, 11)$ and $(4, -7)$. Now use the slope formula.

$$m = \frac{11 - (-7)}{-2 - 4} = \frac{18}{-6} = -3$$

The slope, $m = -3$, is the same number as the coefficient of x in the equation $y = -3x + 5$. It can be shown that this always happens, *as long as the equation is solved for y.* This fact is used to find the slope of a line from its equation.

Finding the Slope of a Line from Its Equation

Step 1 Solve the equation for y.

Step 2 The slope is given by the coefficient of x.

Note

We will see in **Section 3.4** that the equation $y = -3x + 5$ is written using a special form of the equation of a line, called *slope-intercept form,*

$$y = mx + b.$$

EXAMPLE 5 Finding Slopes from Equations

Find the slope of each line.

(a) $2x - 5y = 4$

Step 1 Solve the equation for y.

$$2x - 5y = 4 \quad \text{Isolate } y \text{ on one side.}$$
$$-5y = -2x + 4 \quad \text{Subtract } 2x.$$
$$y = \frac{2}{5}x - \frac{4}{5} \quad \text{Divide by } -5.$$

Step 2 The slope is given by the coefficient of x, so the slope is $\frac{2}{5}$.

(b)
$$8x + 4y = 1$$
$$4y = -8x + 1 \quad \text{Subtract } 8x. \quad \text{Solve for } y.$$
$$y = -2x + \frac{1}{4} \quad \text{Divide by } 4.$$

The slope of this line is given by the coefficient of x, which is -2.

Work Problem 5 at the Side. ▶

5 Find the slope of each line.

(a) $y = -\frac{7}{2}x + 1$

(b) $3x + 2y = 9$

(c) $y + 4 = 0$

(d) $x + 3 = 7$

OBJECTIVE 3 Use slope to determine whether two lines are parallel, perpendicular, or neither. Two lines in a plane that never intersect are **parallel.** We use slopes to tell whether two lines are parallel. For example, Figure 27 shows the graphs of $x + 2y = 4$ and $x + 2y = -6$. These lines appear to be parallel. Solving $x + 2y = 4$ for y gives $y = -\frac{1}{2}x + 2$. Solving $x + 2y = -6$ for y gives $y = -\frac{1}{2}x - 3$. Both lines have slope $-\frac{1}{2}$. ***Nonvertical parallel lines always have equal slopes.***

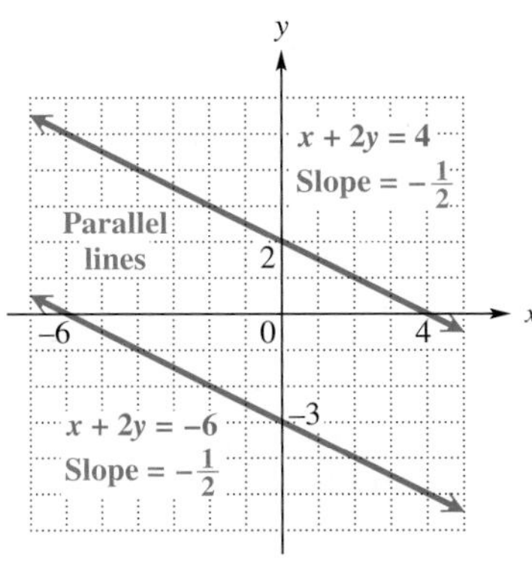

Figure 27

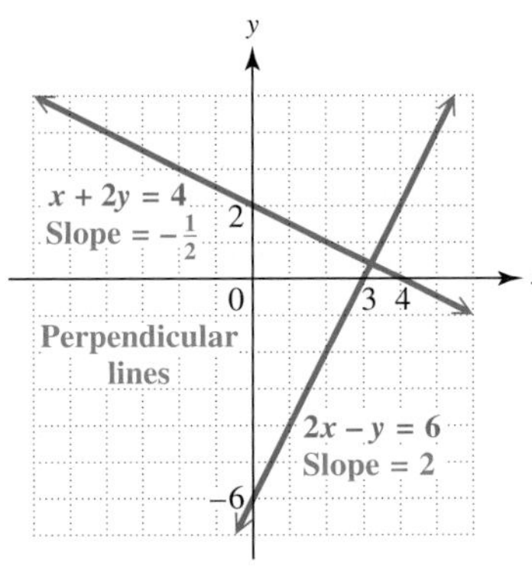

Figure 28

Figure 28 shows the graphs of $x + 2y = 4$ and $2x - y = 6$. These lines appear to be **perpendicular** (that is, they intersect at a 90° angle). Solving $x + 2y = 4$ for y gives $y = -\frac{1}{2}x + 2$, with slope $-\frac{1}{2}$. Solving $2x - y = 6$ for y gives $y = 2x - 6$, with slope 2. The product of $-\frac{1}{2}$ and 2 is

$$-\frac{1}{2}(2) = -1.$$

This condition is true in general. ***The product of the slopes of two perpendicular lines, neither of which is vertical, is always*** -1. This means that the slopes of perpendicular lines are negative (or opposite) reciprocals—if one slope is the nonzero number a, then the other is $-\frac{1}{a}$. The table in the margin shows several examples.

Number	Negative Reciprocal
$\frac{3}{4}$	$-\frac{4}{3}$
$\frac{1}{2}$	$-\frac{2}{1}$, or -2
-6, or $-\frac{6}{1}$	$\frac{1}{6}$
-0.4, or $-\frac{4}{10}$	$\frac{10}{4}$, or 2.5

The product of each number and its negative reciprocal is -1.

Answers

5. **(a)** $-\frac{7}{2}$ **(b)** $-\frac{3}{2}$ **(c)** 0 **(d)** undefined

Slopes of Parallel and Perpendicular Lines

Two lines with the same slope are parallel.

Two lines whose slopes have a product of -1 are perpendicular.

EXAMPLE 6 Deciding Whether Lines Are Parallel, Perpendicular, or Neither

Decide whether each pair of lines is *parallel, perpendicular,* or *neither.*

(a) $x + 2y = 7$

$-2x + y = 3$

Find the slope of each line by first solving each equation for y.

$x + 2y = 7$		$-2x + y = 3$	
$2y = -x + 7$	Subtract x.	$y = 2x + 3$	Add $2x$.
$y = -\frac{1}{2}x + \frac{7}{2}$	Divide by 2.		
Slope is $-\frac{1}{2}$.		Slope is 2.	

Because the slopes are not equal, the lines are not parallel. Check the product of the slopes: $-\frac{1}{2}(2) = -1$. The two lines are perpendicular because the product of their slopes is -1. See Figure 29.

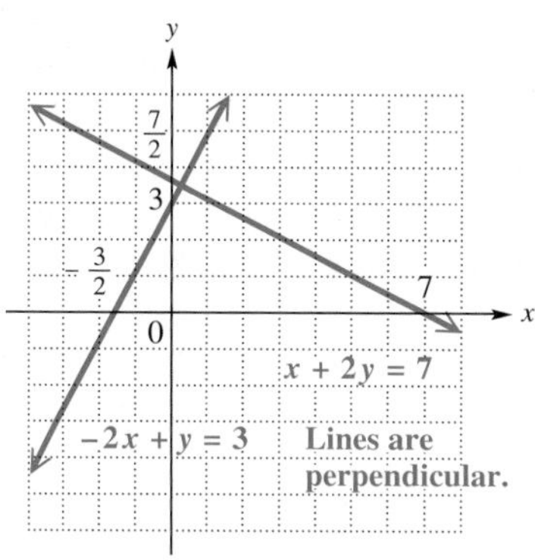

Figure 29

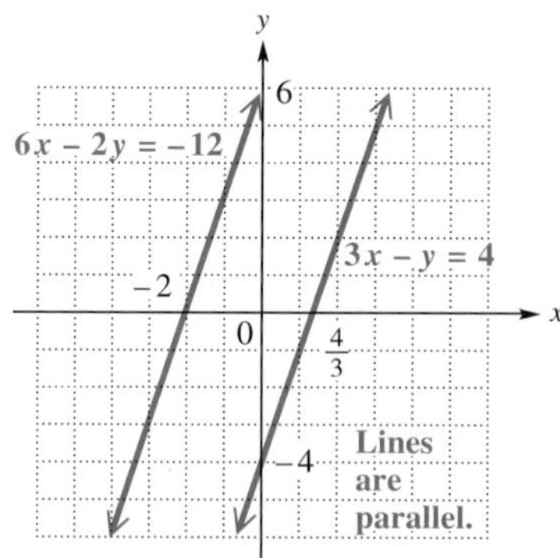

Figure 30

(b) $3x - y = 4$ $\quad$ Solve for y. $\quad$ $y = 3x - 4$

$6x - 2y = -12$ $\quad\quad$ $y = 3x + 6$

Both lines have slope 3, so the lines are parallel. See Figure 30.

(c) $4x + 3y = 6$ $\quad$ Solve for y. $\quad$ $y = -\frac{4}{3}x + 2$

$2x - y = 5$ $\quad\quad$ $y = 2x - 5$

Here the slopes are $-\frac{4}{3}$ and 2. Because $-\frac{4}{3} \neq 2$ and $-\frac{4}{3}(2) \neq -1$, these lines are neither parallel nor perpendicular.

(d) $5x - y = 1$ $\quad$ Solve for y. $\quad$ $y = 5x - 1$

$x - 5y = -10$ $\quad\quad$ $y = \frac{1}{5}x + 2$

The slopes are 5 and $\frac{1}{5}$. The lines are not parallel, nor are they perpendicular. ***(Be careful!*** $\mathbf{5(\frac{1}{5}) = 1}$, ***not*** $\mathbf{-1}$.**)**

◀ *Work Problem* 6 *at the Side.*

6 Decide whether each pair of lines is *parallel, perpendicular,* or *neither*.

(a) $x + y = 6$

$x + y = 1$

(b) $3x - y = 4$

$x + 3y = 9$

(c) $2x - y = 5$

$2x + y = 3$

(d) $3x - 7y = 35$

$7x - 3y = -6$

ANSWERS

6. **(a)** parallel **(b)** perpendicular **(c)** neither **(d)** neither

OBJECTIVE 4 Solve problems involving average rate of change. We know that the slope of a line is the ratio of the vertical change in y to the corresponding horizontal change in x. Thus, the slope formula applied to any two points on the line gives the **average rate of change** in y per unit change in x, where the value of y depends on the value of x.

For example, if the height of a boy increased from 60 to 68 in. between the ages of 12 and 16, as shown in the graph in Figure 31, then the boy's average growth rate (or average change in height) from ages 12 to 16 was

$$\frac{\text{Change in height } y \longrightarrow 68 - 60}{\text{Change in age } x \longrightarrow 16 - 12} = \frac{8}{4} = 2 \text{ in. per year.}$$

The boy may actually have grown more than 2 in. during some years and less than 2 in. during other years. If we plotted ordered pairs (age, height) for those years, the points would be a little above or below the line shown in the graph. If we drew a line connecting any two of those points and calculated the average rate of change, it would likely be slightly different than that found above. However using the height data for ages 12 and 16, the boy's *average* change in height was 2 in. per year over these years.

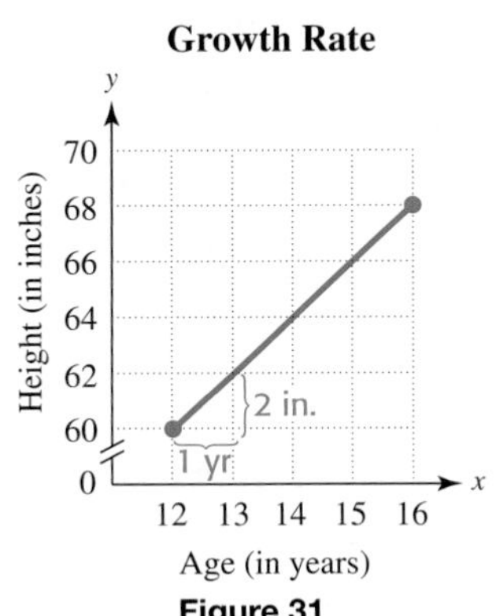

Figure 31

7 Americans spent an average of 886 hr in 2003 watching cable and satellite TV. (*Source:* Veronis Suhler Stevenson.)

(a) Using this data for 2003 and the data for 2000 from the graph in Figure 32, find the average rate of change to the nearest tenth of an hour from 2000 to 2003.

(b) How does the average rate of change from part (a) compare to the average rate of change from 2000 to 2005 found in Example 7?

EXAMPLE 7 Interpreting Slope as Average Rate of Change

The graph in Figure 32 approximates the average number of hours per year spent watching cable and satellite TV for each person in the United States during the years 2000 through 2005. Find the average rate of change in number of hours per year.

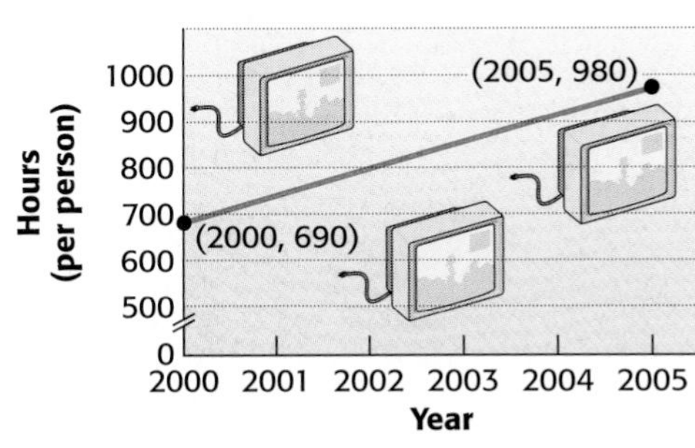

Source: Veronis Suhler Stevenson.
Figure 32

To determine the average rate of change, we need two pairs of data. From the graph, if $x = 2000$, then $y = 690$ and if $x = 2005$, then $y = 980$. Thus, we have the ordered pairs (2000, 690) and (2005, 980). By the slope formula,

$$\text{average rate of change} = \frac{980 - 690}{2005 - 2000} = \frac{290}{5} = 58$$

A positive slope indicates an increase.

This means that the average time per person spent watching cable and satellite TV *increased* by 58 hr per year from 2000 through 2005.

Work Problem 7 at the Side. ▶

Answers

7. **(a)** 65.3 hr per yr
(b) It is greater than the average rate of change for 2000–2005.

8 In 2000, 942.5 million compact discs were sold in the United States. In 2006, 614.9 million CDs were sold. Find the average rate of change in CDs sold per year. (*Source:* Recording Industry Association of America.)

EXAMPLE 8 Interpreting Slope as Average Rate of Change

During the year 2000, the average person in the United States spent 866 hr watching broadcast TV. In 2004, the average number of hours per person spent watching broadcast TV was 678. Find the average rate of change in number of hours per year. (*Source:* Veronis Suhler Stevenson.)

To use the slope formula, we need two ordered pairs. Here, we let one ordered pair be (2000, 866) and the other be (2004, 678).

$$\text{average rate of change} = \frac{678 - 866}{2004 - 2000} = \frac{-188}{4} = -47$$

A negative slope indicates a decrease.

The graph in Figure 33 confirms that the line through the ordered pairs falls from left to right and, therefore, has negative slope. Thus, the average time per person spent watching broadcast TV *decreased* by 47 hr per year from 2000 through 2004.

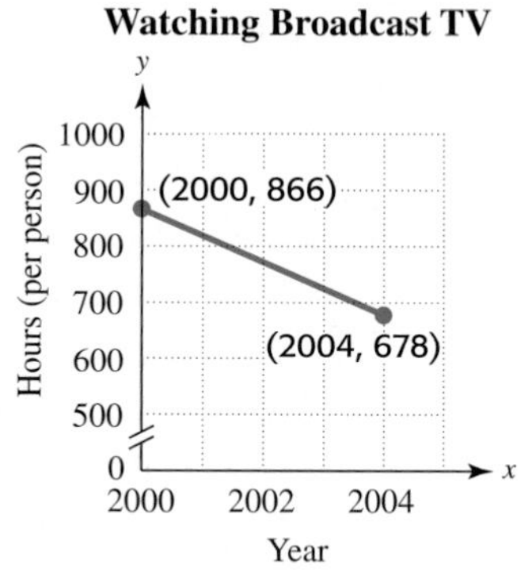

Figure 33

Work Problem 8 at the Side.

ANSWER

8. −54.6 million CDs per yr

3.3 Exercises

FOR EXTRA HELP

READ

Use the coordinates of the indicated points to find the slope of each line. See Example 1.

1. $\frac{3}{2}$

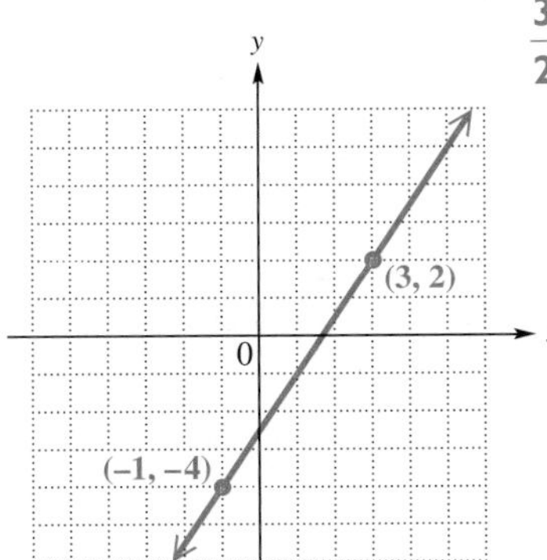

2. 4

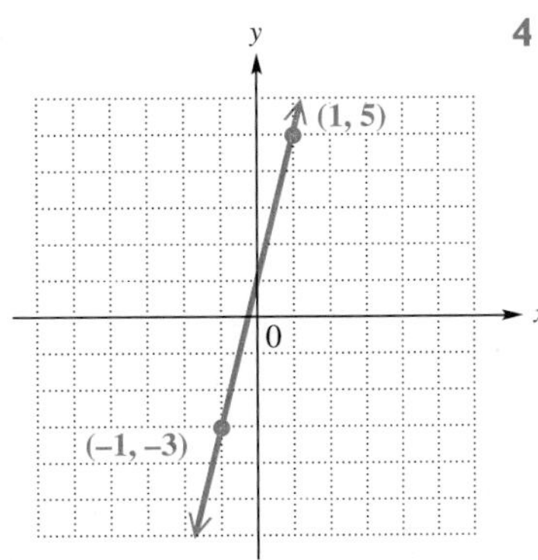

3. $-\frac{7}{4}$

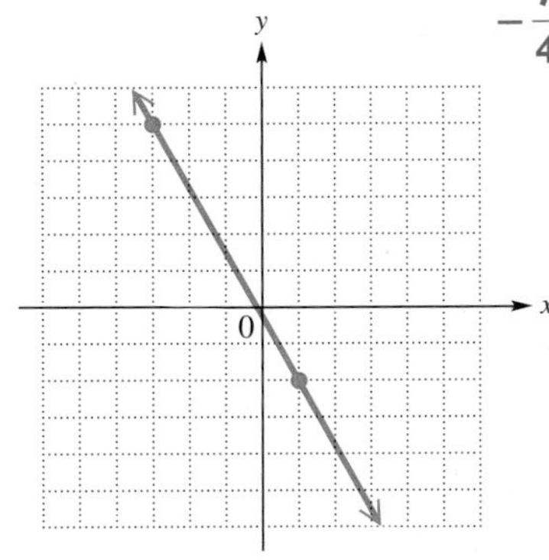

4. $-\frac{1}{2}$

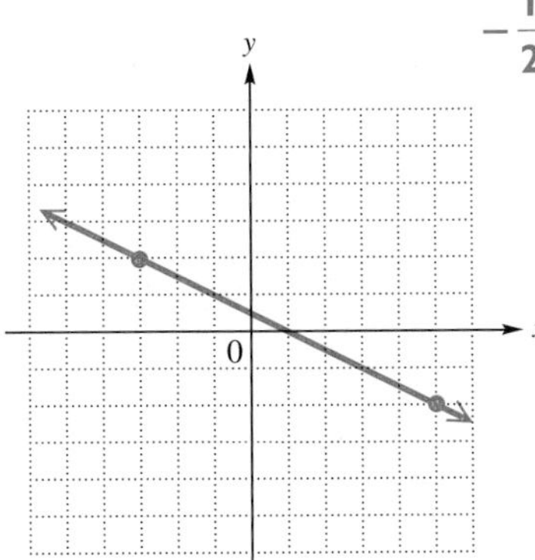

5. 0

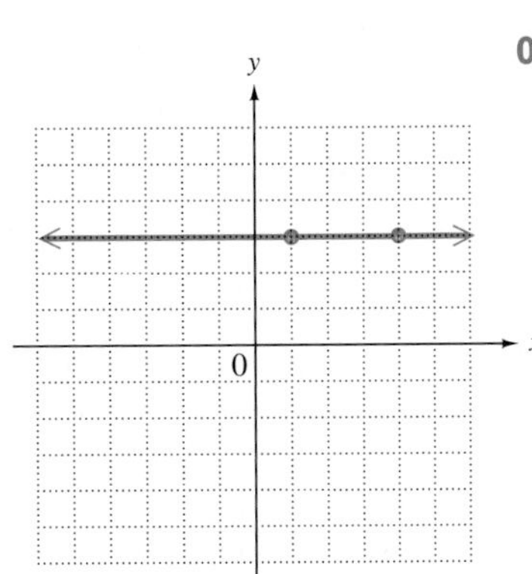

6. 0

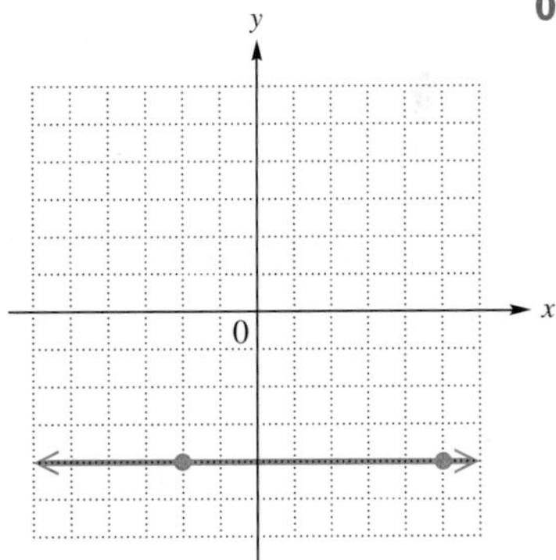

7. In the context of the graph of a straight line, what is meant by "rise"? What is meant by "run"?

Rise is the vertical change between two different points on a line. Run is the horizontal change between two different points on a line.

8. Look at the graph in Exercise 1, and answer the following.

(a) Start at the point $(-1, -4)$ and count vertically up to the horizontal line that goes through the other plotted point. What is this vertical change? (Remember: "up" means positive, "down" means negative.) <u>6</u>

(b) From this new position, count horizontally to the other plotted point. What is this horizontal change? (Remember: "right" means positive, "left" means negative.) <u>4</u>

(c) What is the quotient of the numbers found in parts (a) and (b)? $\frac{6}{4}$, or $\frac{3}{2}$
What do we call this number? <u>slope of the line</u>

(d) If we were to *start* at the point (3, 2) and *end* at the point $(-1, -4)$, would the answer to part (c) be the same? Explain why or why not.

Yes, the answer would be the same. It doesn't matter which point you start with. The slope would be expressed as the quotient of -6 and -4, which simplifies to $\frac{3}{2}$.

On the given coordinate system, sketch the graph of a straight line with the indicated slope.

9. Negative

10. Positive

11. Undefined

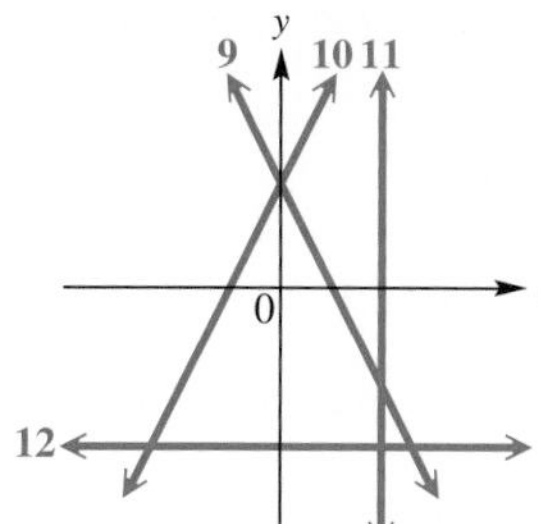

12. Zero

Answers will vary.

13. Decide whether the line with the given slope rises from left to right, falls from left to right, is horizontal, or is vertical.

(a) $m = -4$ **(b)** $m = 0$ **(c)** m is undefined. **(d)** $m = \frac{3}{7}$

falls from left to right **horizontal** **vertical** **rises from left to right**

14. Explain in your own words what is meant by the *slope* of a line.

The slope of a line is the ratio (or quotient) of the rise, the change in y, and the run, the change in x.

15. A student found the slope of the line through the points (2, 5) and (−1, 3) and got $-\frac{2}{3}$ as his answer. He showed his work as

$$\frac{3-5}{2-(-1)} = \frac{-2}{3} = -\frac{2}{3}.$$

WHAT WENT WRONG? Give the correct slope.

Because he found the difference $3 - 5 = -2$ in the numerator, he should have subtracted in the same order in the denominator to get $-1 - 2 = -3$. The correct slope is $\frac{-2}{-3} = \frac{2}{3}$.

Find the slope of the line through each pair of points. See Examples 2–4.

16. (4, −1) and (−2, −8)
$\frac{7}{6}$

17. (1, −2) and (−3, −7)
$\frac{5}{4}$

18. (−8, 0) and (0, −5)
$-\frac{5}{8}$

19. (0, 3) and (−2, 0)
$\frac{3}{2}$

20. (−4, −5) and (−5, −8)
3

21. (−2, 4) and (−3, 7)
−3

22. (6, −5) and (−12, −5)
0

23. (4, 3) and (−6, 3)
0

24. (−8, 6) and (−8, −1)
undefined

25. (−12, 3) and (−12, −7)
undefined

26. (3.1, 2.6) and (1.6, 2.1)
$\frac{1}{3}$

27. $\left(-\frac{7}{5}, \frac{3}{10}\right)$ and $\left(\frac{1}{5}, -\frac{1}{2}\right)$
$-\frac{1}{2}$

Find the slope of each line. See Example 5.

28. $y = 2x - 3$
2

29. $y = 5x + 12$
5

30. $2y = -x + 4$
$-\frac{1}{2}$

31. $4y = x + 1$
$\frac{1}{4}$

32. $-6x + 4y = 4$ $\frac{3}{2}$

33. $3x - 2y = 3$ $\frac{3}{2}$

34. $y = 4$ 0

35. $y = 6$ 0

36. $x = 5$ undefined

37. $x = -2$ undefined

38. $x + y = 0$ -1

39. $x - y = 0$ 1

The figure at the right shows a line that has a positive slope (because it rises from left to right) and a positive y-value for the y-intercept (because it intersects the y-axis above the origin).

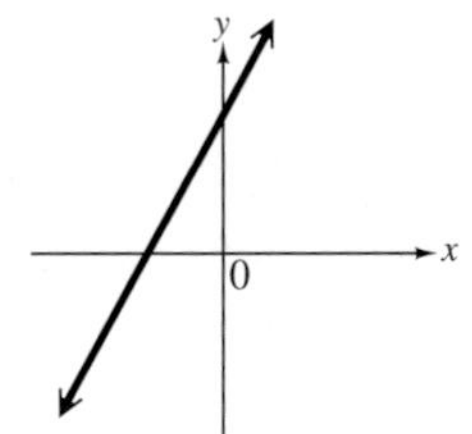

For each figure in Exercises 40–45, decide whether ***(a)*** *the slope is* positive, negative, *or* 0 *and whether* ***(b)*** *the y-value of the y-intercept is* positive, negative, *or* 0.

40. (a) negative
(b) negative

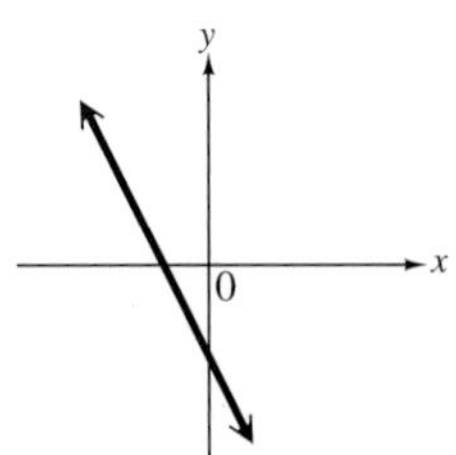

41. (a) negative
(b) 0

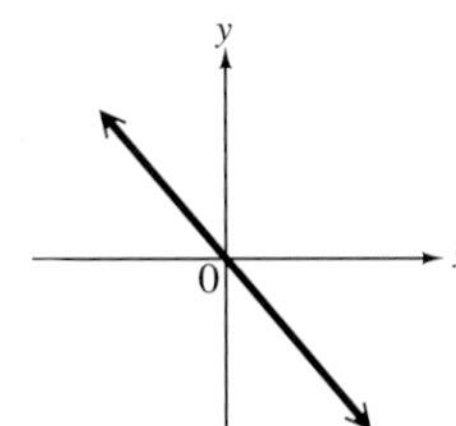

42. (a) positive
(b) 0

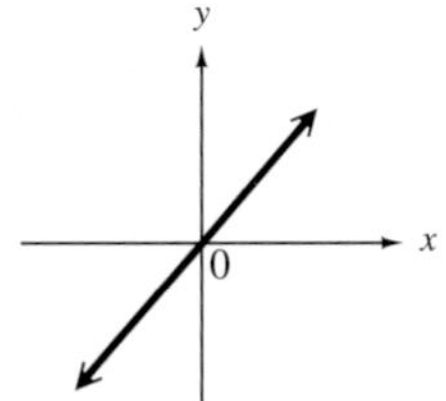

43. (a) positive
(b) negative

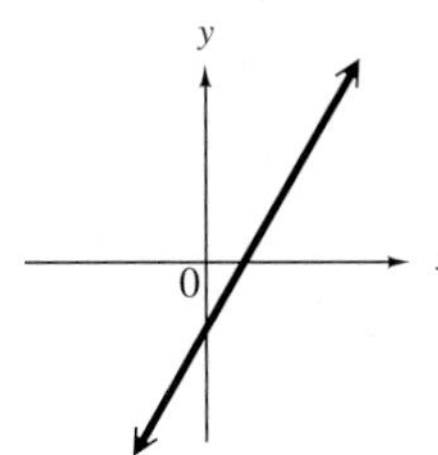

44. (a) 0
(b) positive

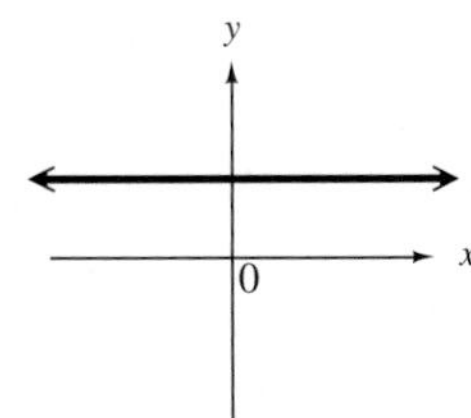

45. (a) 0
(b) negative

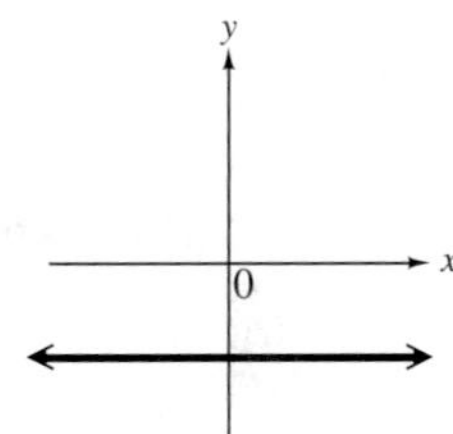

In each pair of equations, give the slope of each line, and then determine whether the two lines are parallel, perpendicular, *or* neither *parallel nor perpendicular. See Example 6.*

46. $2x + 5y = 4$
$4x + 10y = 1$
$-\frac{2}{5}; -\frac{2}{5}$; parallel

47. $-4x + 3y = 4$
$-8x + 6y = 0$
$\frac{4}{3}; \frac{4}{3}$; parallel

48. $8x - 9y = 6$
$8x + 6y = -5$
$\frac{8}{9}; -\frac{4}{3}$; neither

49. $5x - 3y = -2$
$3x - 5y = -8$
$\frac{5}{3}; \frac{3}{5}$; neither

50. $3x - 2y = 6$
$2x + 3y = 3$
$\frac{3}{2}; -\frac{2}{3}$; perpendicular

51. $3x - 5y = -1$
$5x + 3y = 2$
$\frac{3}{5}; -\frac{5}{3}$; perpendicular

52. What is the slope (or pitch) of this roof? $\frac{3}{10}$

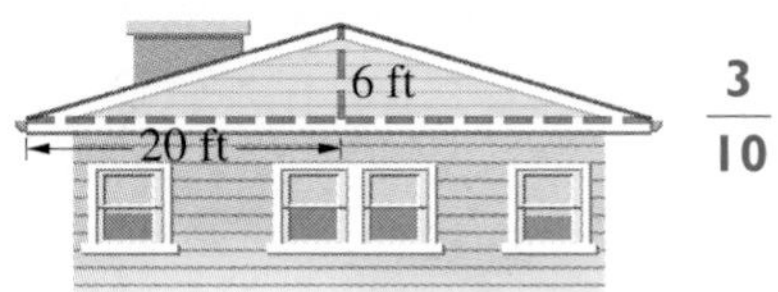

53. What is the slope (or grade) of this hill? $\frac{8}{27}$

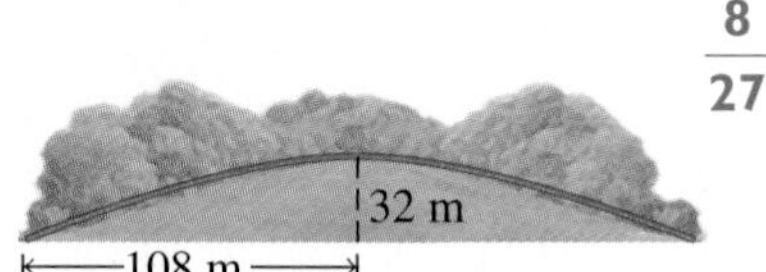

54. What is the slope (or grade) of this ski slope? $-\frac{1}{4}$

Relating Concepts (Exercises 55–60) For Individual or Group Work

Figure A gives public school enrollment (in thousands) in grades 9–12 in the United States. Figure B gives the (average) number of public school students per computer.

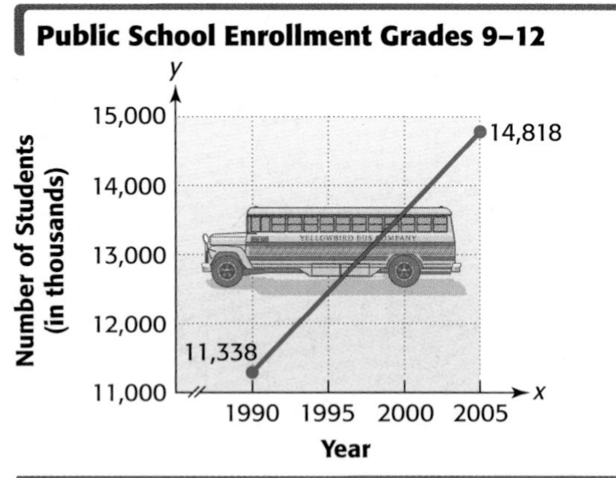

Source: U.S. Department of Education.

Figure A

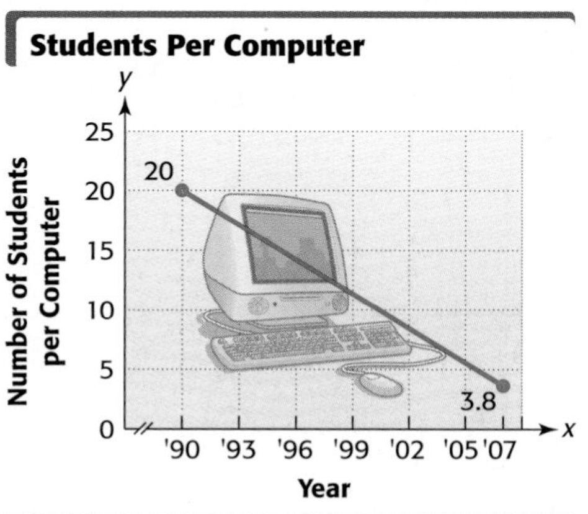

Source: Quality Education Data, Inc.

Figure B

Work Exercises 55–60 in order.

55. Use the ordered pairs (1990, 11,338) and (2005, 14,818) to find the slope of the line in Figure A. **232 thousand, or 232,000**

56. The slope of the line in Figure A is **positive** (positive/negative). This means that during the period represented, enrollment **increased** (increased/decreased).

57. The slope of a line represents its *rate of change*. Based on Figure A, what was the increase in students *per year* during the period shown? **232,000 students**

58. Use the given information to find the slope, to the nearest hundredth, of the line in Figure B. **−0.95**

59. The slope of the line in Figure B is **negative** (positive/negative). This means that during the period represented, the number of students per computer **decreased** (increased/decreased).

60. Based on Figure B, what was the decrease in students per computer *per year* during the period shown? **0.95 student per computer**

Use the concept of slope to solve each problem.

61. When designing the TD Bank North Garden arena in Boston, architects designed the ramps leading up to the entrances so that circus elephants would be able to march up the ramps. The maximum grade (or slope) that an elephant will walk on is 13%. Suppose that such a ramp were constructed with a horizontal run of 150 ft. What would be the maximum vertical rise the architects could use?

19.5 ft

62. The upper deck at U.S. Cellular Field (formerly Comiskey Park) in Chicago has produced, among other complaints, displeasure with its steepness. It is 160 ft from home plate to the front of the upper deck and 250 ft from home plate to the back. The top of the upper deck is 63 ft above the bottom. What is its slope? (Consider the slope as a positive number.)

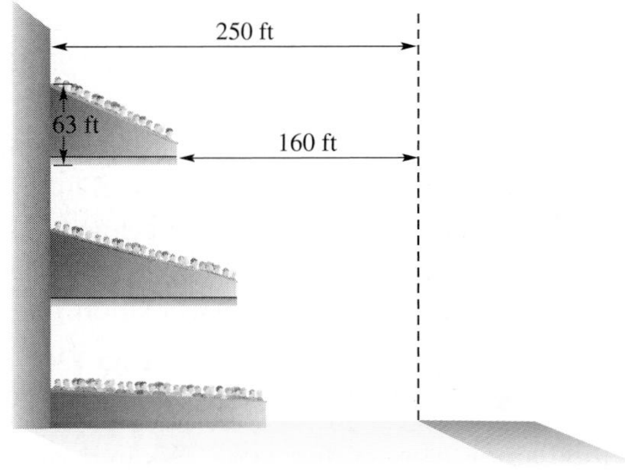

$\mathbf{\frac{7}{10}}$

Find and interpret the average rate of change illustrated in each graph.

63.

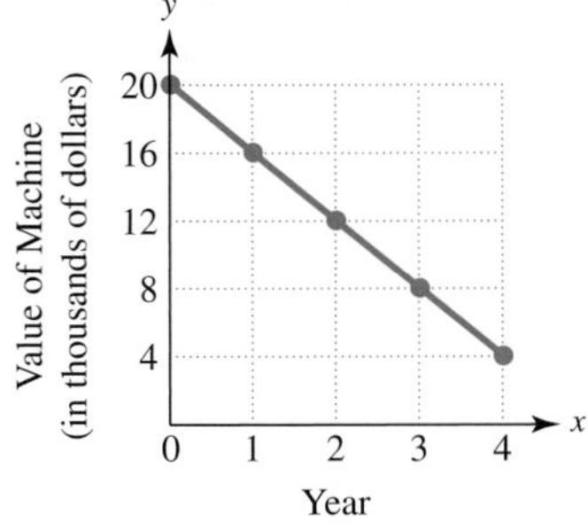

−$4000 per yr; The value of the machine is decreasing $4000 each year during these years.

64.

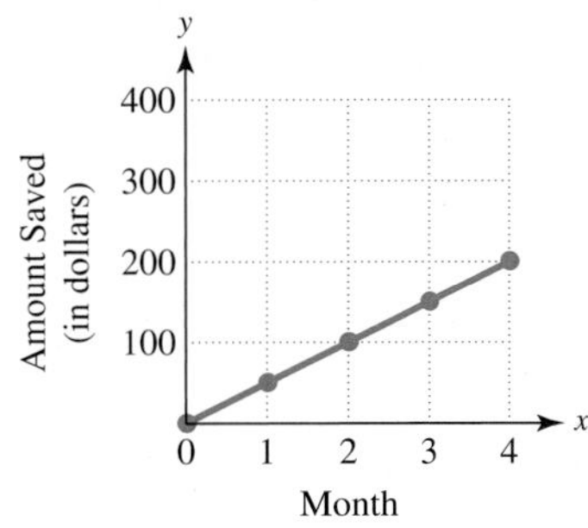

$50 per month; The amount saved is increasing $50 each month during these months.

65.

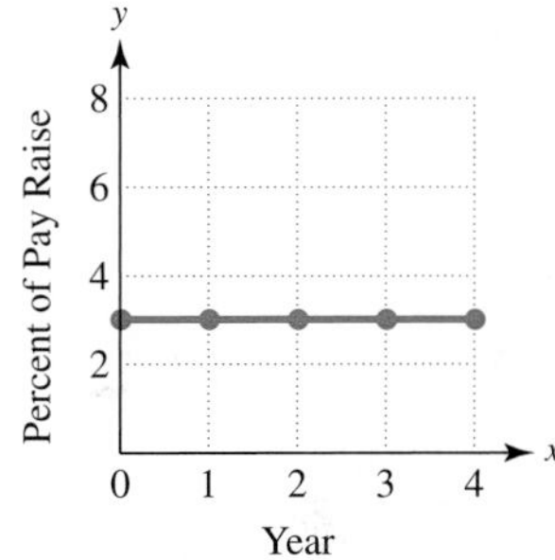

0% per yr (or no change); The percent of pay raise is not changing—it is 3% each year during these years.

Use the idea of average rate of change to solve each problem. Round answers to the nearest thousandth. See Examples 7 and 8.

66. The graph provides a good approximation of the number of mobile homes (in thousands) placed in use in the United States during 2000–2006.

(a) Use the given ordered pairs to find the average rate of change in the number of mobile homes per year during this period.

−28.27 thousand mobile homes per yr

(b) Interpret what a negative slope means in this situation.

The negative slope means that the number of mobile homes *decreased* by an average of 28.27 thousand each year from 2000 to 2006.

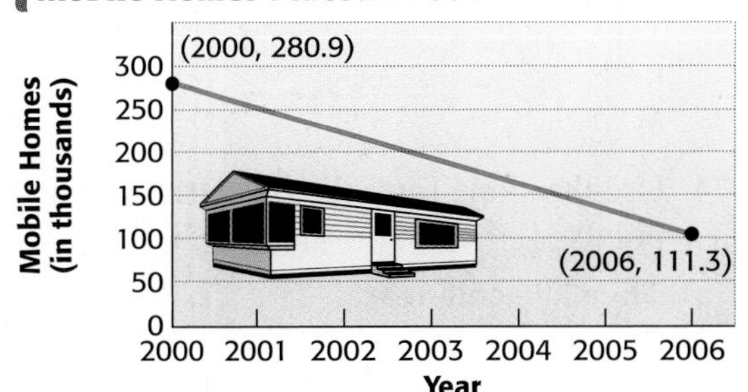

Source: U.S. Census Bureau.

67. Personal spending on recreation in the United States (in billions of dollars) in recent years is closely approximated by the graph.

(a) Use the given ordered pairs to determine the average rate of change in these expenditures per year.

$18.78 billion per yr

(b) Explain how a positive slope is interpreted in this situation.

The positive slope means that personal spending on recreation in the United States *increased* by an average of $18.78 billion each year from 2000 to 2006.

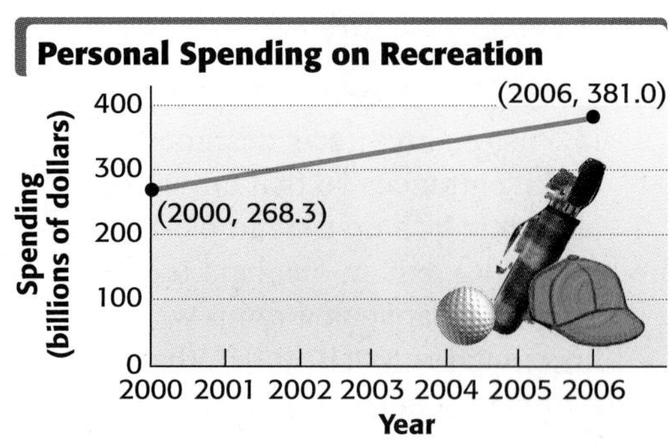

Source: U.S. Department of Commerce.

68. The total amount spent on plasma TVs in the United States changed from $1590 million in 2003 to $5705 million in 2006. Find and interpret the average rate of change in sales, in millions of dollars per year. Round your answer to the nearest hundredth. (*Source:* Consumer Electronics Association.)

$1371.67 million per yr; Sales of plasma TVs *increased* by an average of $1371.67 million each year from 2003 to 2006.

69. The total amount spent on analog TVs in the United States changed from $5836 million in 2003 to $1424 million in 2006. Find and interpret the average rate of change in sales, in millions of dollars per year. Round your answer to the nearest hundredth. (*Source:* Consumer Electronics Association.)

−$1470.67 million per yr; Sales of analog TVs *decreased* by an average of $1470.67 million each year from 2003 to 2006.

Relating Concepts (Exercises 70–75) For Individual or Group Work

In these exercises we investigate a method of determining whether three points lie on the same straight line. (Such points are said to be **collinear.***) The points we consider are* $A(3, 1)$, $B(6, 2)$, *and* $C(9, 3)$. ***Work Exercises 70–75 in order.***

70. Find the slope of segment AB.

$\frac{1}{3}$

71. Find the slope of segment BC.

$\frac{1}{3}$

72. Find the slope of segment AC.

$\frac{1}{3}$

73. If slope of AB = slope of BC = slope of AC, then A, B, and C are collinear. Use the results of Exercises 70–72 to show that this statement is satisfied.

$\frac{1}{3} = \frac{1}{3} = \frac{1}{3}$ **is true.**

74. Use the slope formula to determine whether the points $(1, -2)$, $(3, -1)$, and $(5, 0)$ are collinear.

They are collinear.

75. Repeat Exercise 74 for the points $(0, 6)$, $(4, -5)$, and $(-2, 12)$.

They are not collinear.

3.4 Equations of Lines

OBJECTIVE 1 Write an equation of a line given its slope and y-intercept. In **Section 3.3** we found the slope of a line from the equation of the line by solving the equation for y. For example, we found that the slope of the line with equation

$$y = -3x + 5$$

is -3, the coefficient of x. What does the number 5 represent?

To find out, suppose a line has slope m and y-intercept $(0, b)$. We can find an equation of this line by choosing another point (x, y) on the line, as shown in Figure 34. Using the slope formula,

$$m = \frac{y - b}{x - 0}$$

$$m = \frac{y - b}{x}$$

$$mx = y - b \qquad \text{Multiply by } x.$$

$$mx + b = y \qquad \text{Add } b.$$

$$y = mx + b. \qquad \text{Rewrite.}$$

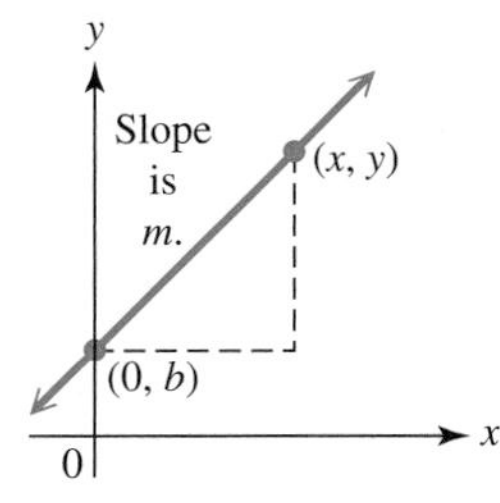

Figure 34

This last equation is called the **slope-intercept form** of the equation of a line, because we can identify the slope m and y-intercept $(0, b)$ at a glance. Thus, in $y = -3x + 5$, the number 5 indicates that the y-intercept is $(0, 5)$.

Slope-Intercept Form

The slope-intercept form of the equation of a line with slope m and y-intercept $(0, b)$ is

$$y = mx + b.$$

↑ Slope ↑ y-intercept is $(0, b)$.

REMEMBER: The intercept in slope-intercept form is the y-intercept.

EXAMPLE 1 Writing an Equation of a Line

Write an equation of the line with slope $\frac{2}{3}$ and y-intercept $(0, -1)$.

Here $m = \frac{2}{3}$ and $b = -1$, so an equation is

Slope → m; y-intercept $(0, b)$ → b

$$y = mx + b$$

$$y = \frac{2}{3}x + (-1), \quad \text{or} \quad y = \frac{2}{3}x - 1.$$

Work Problem 1 at the Side. ▶

Note

The slope-intercept form is the most useful form for a linear equation because of the information we can determine from it. It is also the form that describes a *linear function*, which we introduce in **Section 3.6.**

OBJECTIVES

1. Write an equation of a line given its slope and y-intercept.
2. Graph a line given its slope and a point on the line.
3. Write an equation of a line given its slope and any point on the line.
4. Write an equation of a line given two points on the line.
5. Write an equation of a line parallel or perpendicular to a given line.
6. Write an equation of a line that models real data.

1 Write an equation of the line with the given slope and y-intercept.

(a) slope $\frac{1}{2}$; y-intercept $(0, -4)$

(b) slope -1; y-intercept $(0, 8)$

(c) slope 3; y-intercept $(0, 0)$

(d) slope 0; y-intercept $(0, 2)$

(e) slope 1; y-intercept $(0, 0.75)$

ANSWERS

1. **(a)** $y = \frac{1}{2}x - 4$ **(b)** $y = -x + 8$ **(c)** $y = 3x$ **(d)** $y = 2$ **(e)** $y = x + 0.75$

2 Graph $3x - 4y = 8$ by using the slope and y-intercept.

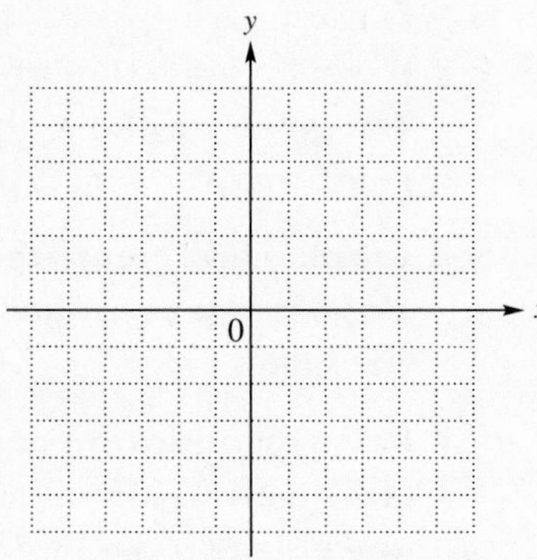

OBJECTIVE 2 Graph a line given its slope and a point on the line. We can use the slope and y-intercept to graph a line.

> **Graphing a Line by Using the Slope and y-Intercept**
>
> ***Step 1*** Write the equation in slope-intercept form, if necessary, by solving for y.
>
> ***Step 2*** Identify the y-intercept. Graph the point $(0, b)$.
>
> ***Step 3*** Identify slope m of the line. Use the geometric interpretation of slope ("rise over run") to find another point on the graph by counting from the y-intercept.
>
> ***Step 4*** Join the two points with a line to obtain the graph.

EXAMPLE 2 Graphing a Line by Using the Slope and y-Intercept

Graph $2x - 3y = 3$ by using the slope and y-intercept.

Step 1 Solve for y to write the equation in slope-intercept form.

Isolate y on one side.

$$2x - 3y = 3 \quad \text{Given equation}$$

$$-3y = -2x + 3 \quad \text{Subtract } 2x.$$

$$y = \frac{2}{3}x - 1 \quad \text{Divide by } -3.$$

Slope m ($\frac{2}{3}$); y-intercept $(0, b)$ (-1)

Step 2 The y-intercept is $(0, -1)$. Graph this point. See Figure 35.

Step 3 The slope is $\frac{2}{3}$. By the definition of slope,

$$m = \frac{\textbf{change in } y}{\textbf{change in } x} = \frac{2}{3}.$$

Counting from the y-intercept 2 units up and 3 units to the right, we obtain another point on the graph, $(3, 1)$.

Step 4 Draw the line through the points $(0, -1)$ and $(3, 1)$ to obtain the graph of the given equation $2x - 3y = 3$. See Figure 35.

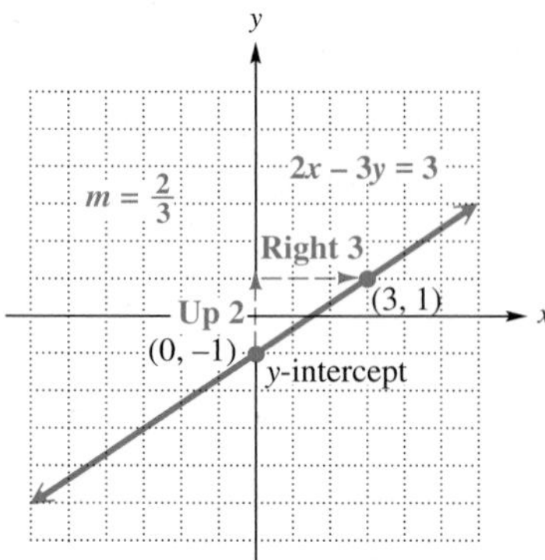

Figure 35

◀ *Work Problem* **2** *at the Side.*

ANSWER

2.

y
x
3x – 4y = 8
(4, 1)
(0, –2)

The method of Example 2 can be extended to graph a line given its slope and *any* point on the line.

EXAMPLE 3 Graphing a Line by Using the Slope and a Point

Graph the line passing through the point $(-2, 3)$, with slope -4.

First, locate the point $(-2, 3)$. See Figure 36. Then write the slope -4 as

$$\text{slope } m = \frac{\text{change in } y}{\text{change in } x} = -4 = \frac{-4}{1}.$$

Locate another point on the line by counting 4 units *down* (because of the negative sign) from $(-2, 3)$ and then 1 unit to the right. Finally, draw the line through this new point P and the given point $(-2, 3)$. See Figure 36.

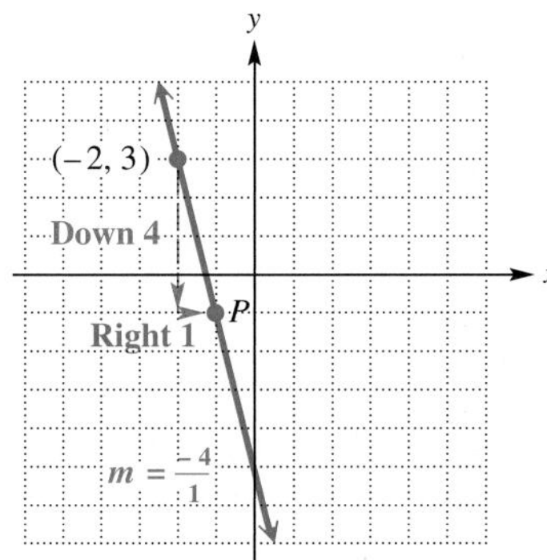

Figure 36

Note

In Example 3, we could have written the slope as $\frac{4}{-1}$ instead. In this case, we would move 4 units up from $(-2, 3)$ and then 1 unit to the *left* (because of the negative sign). Verify that this produces the same line.

Work Problem 3 at the Side. ▶

3 Graph the line passing through the point $(2, -3)$, with slope $-\frac{1}{3}$.

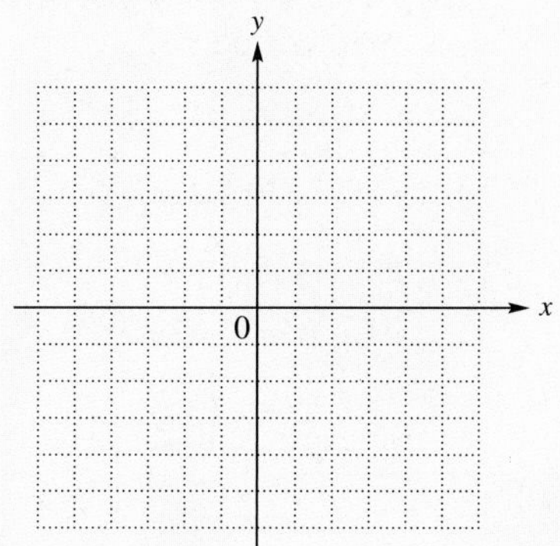

OBJECTIVE 3 Write an equation of a line given its slope and any point on the line. We can use the slope-intercept form to write the equation of a line if we know the slope and any point on the line.

4 Write an equation, in slope-intercept form, of the line having slope -2 and passing through the point $(-1, 4)$.

EXAMPLE 4 Using the Slope-Intercept Form to Write an Equation of a Line

Write an equation, in slope-intercept form, of the line having slope 4 passing through the point $(2, 5)$.

Since the line passes through the point $(2, 5)$, we can substitute $x = 2, y = 5$, and the given slope $m = 4$ into $y = mx + b$ and solve for b.

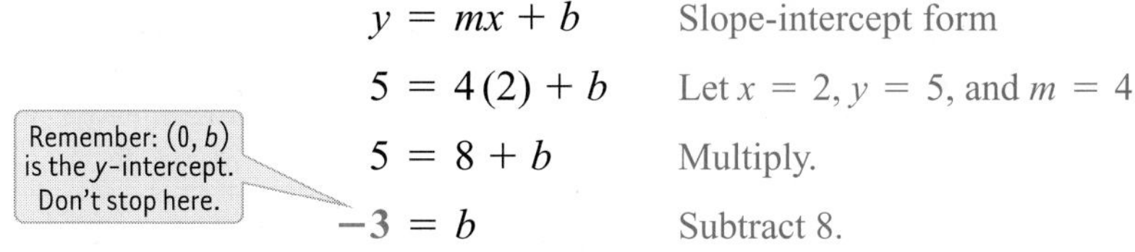

$$y = mx + b \quad \text{Slope-intercept form}$$

$$5 = 4(2) + b \quad \text{Let } x = 2, y = 5, \text{ and } m = 4.$$

$$5 = 8 + b \quad \text{Multiply.}$$

$$-3 = b \quad \text{Subtract 8.}$$

Remember: $(0, b)$ is the y-intercept. Don't stop here.

The y-intercept is $(0, -3)$. Using the given slope, 4, an equation of the line is

$$y = 4x - 3. \quad \text{Slope-intercept form}$$

Work Problem 4 at the Side. ▶

ANSWERS

3.

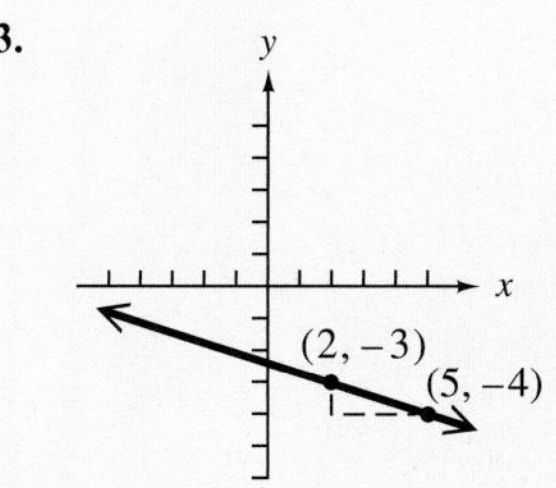

4. $y = -2x + 2$

There is another form that can be used to write the equation of a line. To develop this form, let m represent the slope of a line and let (x_1, y_1) represent a given point on the line. Let (x, y) represent any other point on the line. See Figure 37. Then,

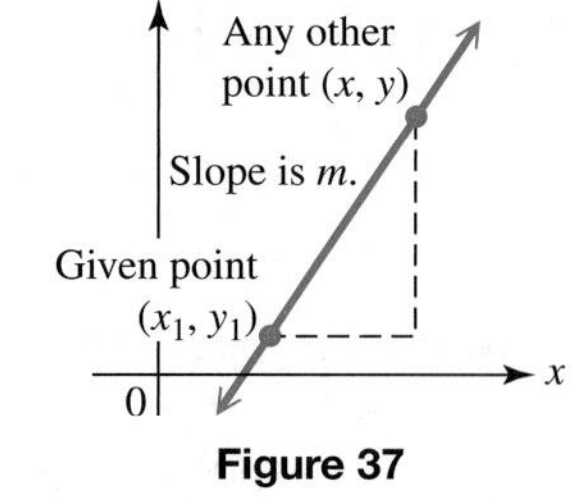

Figure 37

$$m = \frac{y - y_1}{x - x_1} \quad \text{Definition of slope}$$

$$m(x - x_1) = y - y_1 \quad \text{Multiply each side by } x - x_1.$$

$$y - y_1 = m(x - x_1). \quad \text{Rewrite.}$$

This result is the **point-slope form** of the equation of a line.

Point-Slope Form

The point-slope form of the equation of a line with slope m passing through the point (x_1, y_1) is

$$y - y_1 = m(x - x_1).$$

(Slope: m; Given point: (x_1, y_1))

EXAMPLE 5 Using the Point-Slope Form to Write Equations

Write an equation of each line. Give the final answer in slope-intercept form.

(a) The line through $(-2, 4)$, with slope -3

The given point is $(-2, 4)$ so $x_1 = -2$ and $y_1 = 4$. Also, $m = -3$.

$$y - y_1 = m(x - x_1) \quad \text{Point-slope form}$$

$$y - 4 = -3[x - (-2)] \quad \text{Let } x_1 = -2, y_1 = 4, m = -3.$$

$$y - 4 = -3(x + 2) \quad \text{Be careful substituting.}$$

$$y - 4 = -3x - 6 \quad \text{Distributive property}$$

$$y = -3x - 2 \quad \text{Add 4.}$$

(b) The line through $(4, 2)$, with slope $\frac{3}{5}$

$$y - y_1 = m(x - x_1) \quad \text{Point-slope form}$$

$$y - 2 = \frac{3}{5}(x - 4) \quad \text{Let } x_1 = 4, y_1 = 2, m = \tfrac{3}{5}.$$

$$y - 2 = \frac{3}{5}x - \frac{12}{5} \quad \text{Distributive property}$$

$$y = \frac{3}{5}x - \frac{12}{5} + \frac{10}{5} \quad \text{Add } 2 = \tfrac{10}{5}.$$

$$y = \frac{3}{5}x - \frac{2}{5} \quad \text{Combine like terms.}$$

We did not clear fractions after the substitution step because we want the equation in slope-intercept form—that is, solved for y.

◀ *Work Problem* **5** *at the Side.*

5 Write an equation of each line. Give the final answer in slope-intercept form.

(a) The line through $(-1, 3)$, with slope -2

$$y - y_1 = m(x - x_1)$$

$$y - \_\_\_\_ = \_\_\_\_ [x - (\_\_\_\_)]$$

$$y - 3 = -2(x + \_\_\_\_)$$

$$y - 3 = -2x - \_\_\_\_$$

$$y = \_\_\_\_\_\_$$

(b) The line through $(5, 2)$, with slope $-\frac{1}{3}$

Answers

5. (a) $3; -2; -1; 1; 2; -2x + 1$

(b) $y = -\frac{1}{3}x + \frac{11}{3}$

OBJECTIVE 4 Write an equation of a line given two points on the line. We can also use the point-slope form to find an equation of a line when two points on the line are known.

EXAMPLE 6 Writing an Equation of a Line Given Two Points

Write an equation of the line through the points $(-2, 5)$ and $(3, 4)$. Give the final answer in slope-intercept form.

First, find the slope of the line, using the slope formula.

$$\text{slope } m = \frac{y_2 - y_1}{x_2 - x_1} = \frac{5 - 4}{-2 - 3} = \frac{1}{-5} = -\frac{1}{5}$$

Now use either $(-2, 5)$ or $(3, 4)$ and the point-slope form. Using $(3, 4)$ gives

$$y - y_1 = m(x - x_1) \quad \text{Point-slope form}$$

$$y - 4 = -\frac{1}{5}(x - 3) \quad \text{Let } x_1 = 3, y_1 = 4, m = -\tfrac{1}{5}.$$

$$y - 4 = -\frac{1}{5}x + \frac{3}{5} \quad \text{Distributive property}$$

$$y = -\frac{1}{5}x + \frac{3}{5} + \frac{20}{5} \quad \text{Add } 4 = \tfrac{20}{5}.$$

$$y = -\frac{1}{5}x + \frac{23}{5}. \quad \text{Combine like terms.}$$

The same result would be found using $(-2, 5)$ for (x_1, y_1).

Work Problem 6 at the Side. ▶

6 Write an equation in slope-intercept form of the line through each pair of points.

(a) $(-3, 1)$ and $(2, 4)$

(b) $(2, 5)$ and $(-1, 6)$

Note

In Example 6, the same result would also be found by substituting the slope and either given point in slope-intercept form $y = mx + b$ and then solving for b, as in Example 4. Try this.

Many of the linear equations in **Sections 3.1–3.3** were given in the form

$$Ax + By = C,$$

called **standard form,** where A, B, and C are real numbers and A and B are not both 0. In most cases, A, B, and C are rational numbers. For consistency in this book, we give answers so that A, B, and C are integers with greatest common factor 1 and $A \geq 0$.

Note

The definition of standard form is not the same in all texts. A linear equation can be written in many different, equally correct, ways. For example,

$$3x + 4y = 12, \quad 6x + 8y = 24, \quad \text{and} \quad -9x - 12y = -36$$

all represent the same set of ordered pairs. When giving answers, let us agree that $3x + 4y = 12$ is preferable to the other forms because the greatest common factor of 3, 4, and 12 is 1 and $A \geq 0$.

ANSWERS

6. (a) $y = \frac{3}{5}x + \frac{14}{5}$

(b) $y = -\frac{1}{3}x + \frac{17}{3}$

OBJECTIVE 5 Write an equation of a line parallel or perpendicular to a given line. As mentioned in the previous section, parallel lines have the same slope and perpendicular lines have slopes with product -1.

EXAMPLE 7 Finding Equations of Parallel or Perpendicular Lines

Find the equation in slope-intercept form of the line passing through the point $(-4, 5)$ and **(a)** parallel to the line $2x + 3y = 6$; **(b)** perpendicular to the line $2x + 3y = 6$.

(a) We must first find the slope of the line $2x + 3y = 6$ by solving for y.

$$2x + 3y = 6$$

$$3y = -2x + 6 \quad \text{Subtract } 2x.$$

$$y = -\frac{2}{3}x + 2 \quad \text{Divide by 3.}$$

↑ Slope

The slope is given by the coefficient of x, so $m = -\frac{2}{3}$. See Figure 38.

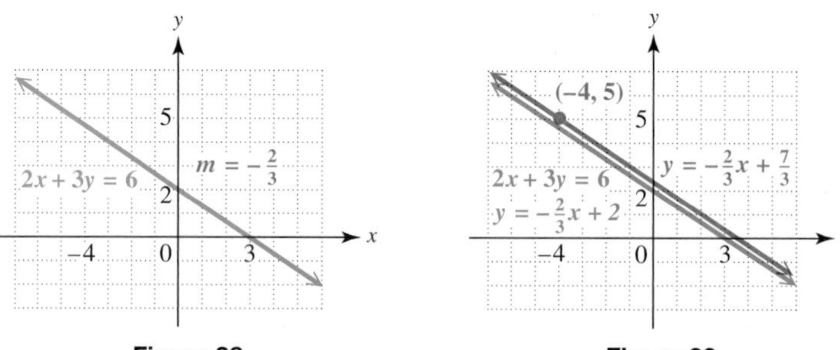

Figure 38

Figure 39

Since parallel lines have the same slope, the required equation of the line through $(-4, 5)$ and parallel to $2x + 3y = 6$ must also have slope $-\frac{2}{3}$. To find this equation, we use the point-slope form, with $(x_1, y_1) = (-4, 5)$ and $m = -\frac{2}{3}$.

$$y - y_1 = m(x - x_1) \quad \text{Point-slope form}$$

$$y - 5 = -\frac{2}{3}[x - (-4)] \quad y_1 = 5, m = -\tfrac{2}{3}, x_1 = -4$$

$$y - 5 = -\frac{2}{3}(x + 4) \quad \text{Definition of subtraction}$$

$$y - 5 = -\frac{2}{3}x - \frac{8}{3} \quad \text{Distributive property}$$

$$y = -\frac{2}{3}x - \frac{8}{3} + \frac{15}{3} \quad \text{Add } 5 = \tfrac{15}{3}.$$

$$y = -\frac{2}{3}x + \frac{7}{3} \quad \text{Combine like terms.}$$

We did not clear fractions after the substitution step here because we want the equation in slope-intercept form—that is, solved for y. Both lines are shown in Figure 39.

Continued on Next Page

(b) The given line $2x + 3y = 6$ can be written in slope-intercept form as

$$y = -\frac{2}{3}x + 2, \quad \text{From part (a)}$$

so the line has slope $-\frac{2}{3}$. To be perpendicular to the line $2x + 3y = 6$, a line must have a slope that is the negative reciprocal of $-\frac{2}{3}$, which is $\frac{3}{2}$. We use the point $(-4, 5)$ and slope $\frac{3}{2}$ in the point-slope form to get the equation of the perpendicular line shown in Figure 40.

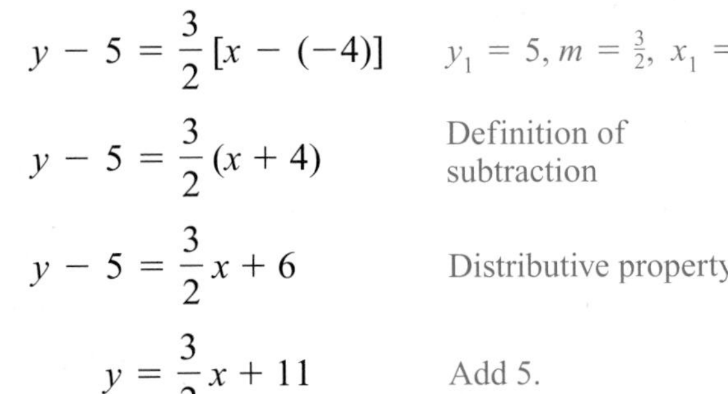

$$y - 5 = \frac{3}{2}[x - (-4)] \quad y_1 = 5, m = \frac{3}{2}, x_1 = -4$$

$$y - 5 = \frac{3}{2}(x + 4) \quad \text{Definition of subtraction}$$

$$y - 5 = \frac{3}{2}x + 6 \quad \text{Distributive property}$$

$$y = \frac{3}{2}x + 11 \quad \text{Add 5.}$$

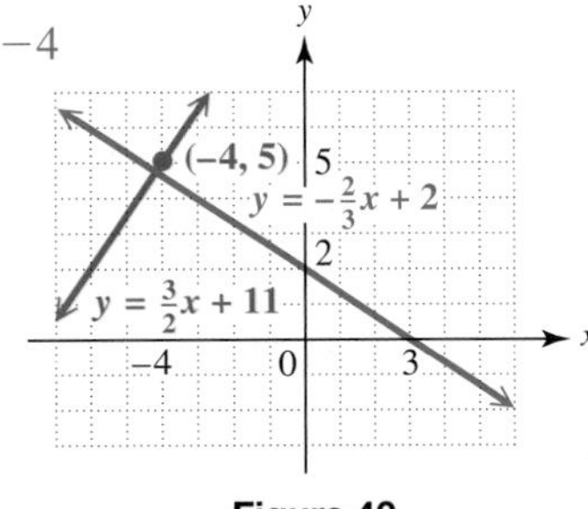

Figure 40

Work Problem 7 at the Side. ▶

7 Write an equation in slope-intercept form of the line passing through the point $(-8, 3)$ and

(a) parallel to the line $2x - 3y = 10$.

(b) perpendicular to the line $2x - 3y = 10$.

A summary of the various forms of linear equations follows.

FORMS OF LINEAR EQUATIONS

Equation	Description	When to Use
$y = mx + b$	**Slope-Intercept Form** Slope is m. y-intercept is $(0, b)$.	The slope and y-intercept can be easily identified and used to quickly graph the equation.
$y - y_1 = m(x - x_1)$	**Point-Slope Form** Slope is m. Line passes through (x_1, y_1).	This form is ideal for finding the equation of a line if the slope and a point on the line or two points on the line are known.
$Ax + By = C$	**Standard Form** Slope is $-\frac{A}{B}$ $(B \neq 0)$. x-intercept is $(\frac{C}{A}, 0)$ $(A \neq 0)$. y-intercept is $(0, \frac{C}{B})$ $(B \neq 0)$.	The x- and y-intercepts can be found quickly and used to graph the equation. The slope must be calculated.
$y = k$	**Horizontal Line** Slope is 0. y-intercept is $(0, k)$.	If the graph intersects only the y-axis, then y is the only variable in the equation.
$x = k$	**Vertical Line** Slope is undefined. x-intercept is $(k, 0)$.	If the graph intersects only the x-axis, then x is the only variable in the equation.

OBJECTIVE 6 Write an equation of a line that models real data. We can use the information presented in this section to write equations of lines that mathematically describe, or *model,* real data if the given set of data changes at a fairly constant rate. In this case, the data fit a linear pattern, and the rate of change is the slope of the line.

ANSWERS

7. **(a)** $y = \frac{2}{3}x + \frac{25}{3}$ **(b)** $y = -\frac{3}{2}x - 9$

EXAMPLE 8 **Determining a Linear Equation to Describe Real Data**

Suppose it is time to fill your car with gasoline. At your local station, 89-octane gas is selling for \$4.50 per gal.

(a) Write an equation that describes the cost y to buy x gallons of gas.

Experience has taught you that the total price you pay is determined by the number of gallons you buy multiplied by the price per gallon (in this case, \$4.50). As you pump the gas, two sets of numbers flash by: the number of gallons pumped and the price for that number of gallons.

The table uses ordered pairs to illustrate this situation.

Number of Gallons Pumped	Price of This Number of Gallons
0	$0(\$4.50) = \$\ 0.00$
1	$1(\$4.50) = \$\ 4.50$
2	$2(\$4.50) = \$\ 9.00$
3	$3(\$4.50) = \$13.50$
4	$4(\$4.50) = \$18.00$

If we let x denote the number of gallons pumped, then the total price y in dollars can be found by using the linear equation

Total price → $y = 4.50x.$ ← Number of gallons

Theoretically, there are infinitely many ordered pairs (x, y) that satisfy this equation, but here we are limited to nonnegative values for x, since we cannot have a negative number of gallons. There is also a practical maximum value for x in this situation, which varies from one car to another. What determines this maximum value?

(b) You can also get a car wash at the gas station if you pay an additional \$3.00. Write an equation that defines the price for gas and a car wash.

Since an additional \$3.00 will be charged, you pay $4.50x + 3.00$ dollars for x gallons of gas and a car wash, described by

$$y = 4.5x + 3. \quad \text{Delete unnecessary 0s.}$$

(c) Interpret the ordered pairs (5, 25.5) and (10, 48) in relation to the equation from part (b).

The ordered pair (5, 25.5) indicates that the price of 5 gal of gas and a car wash is \$25.50. Similarly, (10, 48) indicates that the price of 10 gal of gas and a car wash is \$48.00.

◀ *Work Problem* **8** *at the Side.*

8 **(a)** Suppose it costs \$0.10 per minute to make a long-distance call. Write an equation to describe the cost y to make an x-minute call.

(b) Suppose there is a flat rate of \$0.20 plus a charge of \$0.10 per minute to make a call. Write an equation that gives the cost y for a call of x minutes.

(c) Interpret the ordered pair (15, 1.7) in relation to the equation from part (b).

> **Note**
>
> In Example 8(a), the ordered pair (0, 0) satisfied the equation, so the linear equation has the form $y = mx$, where $b = 0$. If a situation involves an initial charge b plus a charge per unit m as in Example 8(b), the equation has the form $y = mx + b$, where $b \neq 0$.

ANSWERS

8. **(a)** $y = 0.1x$ (Note: $0.10x = 0.1x$)
(b) $y = 0.1x + 0.2$
(c) The ordered pair (15, 1.7) indicates that the price of a 15-minute call is \$1.70.

EXAMPLE 9 Finding an Equation of a Line That Models Data

Average annual tuition and fees for in-state students at public four-year colleges are shown in the table for selected years and graphed as ordered pairs of points in the scatter diagram in Figure 41, where $x = 0$ represents 1990, $x = 4$ represents 1994, and so on, and y represents the cost in dollars.

Year	Cost (in dollars)
1990	2035
1994	2820
1996	3151
1998	3486
2000	3774
2002	4273
2004	4920

Source: U.S. National Center for Education Statistics.

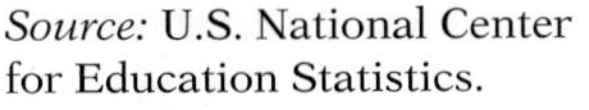

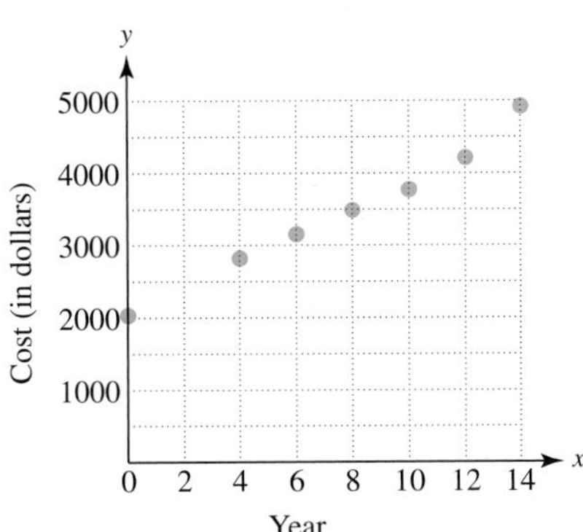

Figure 41

(a) Find an equation that models the data.

Since the points in Figure 41 lie approximately on a straight line, we can write a linear equation that models the relationship between year x and cost y. We choose two data points, (0, 2035) and (12, 4273), to find the slope of the line.

$$m = \frac{4273 - 2035}{12 - 0} = \frac{2238}{12} = 186.5$$

Start with the x- and y-values of the same point.

The slope 186.5 indicates that the cost of tuition and fees increased by about \$186.50 per year from 1990 to 2002. We use this slope, the y-intercept (0, **2035**), and the slope-intercept form to write an equation of the line. Thus,

$$y = 186.5x + 2035.$$

(b) Use the equation from part (a) to approximate the cost of tuition and fees at public four-year colleges in 2006.

The value $x = 16$ corresponds to the year 2006.

$$y = 186.5x + 2035$$

$$y = 186.5(16) + 2035 \quad \text{Substitute 16 for } x.$$

$$y = 5019$$

According to the model, average tuition and fees for in-state students at public four-year colleges in 2006 were about \$5019.

Note

In Example 9, if we had chosen different data points, we would have found a slightly different equation. However, all such equations should yield similar results, since the data points are approximately linear.

Work Problem 9 at the Side. ▶

9 The percentage of the U.S. population 25 yr and older with at least a high school diploma is shown in the table for selected years.

Year	Percent
1950	34.3
1960	41.1
1970	52.3
1980	66.5
1990	77.6
2000	84.1
2005	85.2

Source: U.S. Census Bureau.

(a) Let $x = 0$ represent 1950, $x = 10$ represent 1960, and so on. Use the data for 1950 and 2000 to find an equation that models the data.

(b) Use the equation from part (a) to approximate the percentage, to the nearest tenth, of the U.S. population 25 yr and older who were at least high school graduates in 1995.

ANSWERS

9. (a) $y = 0.996x + 34.3$ **(b)** 79.1%

Math in the Media

INTERPRETING (AND MISINTERPRETING) LINE GRAPHS

The graph shown here is typical of many graphs that appear in magazines and newspapers. This one shows that between 1996 and 2006, the number of McDonald's restaurants worldwide rose from 20,000 to 31,000. This is depicted by the line segment joining the two points labeled A and B.

Use the graph to answer each of the following.

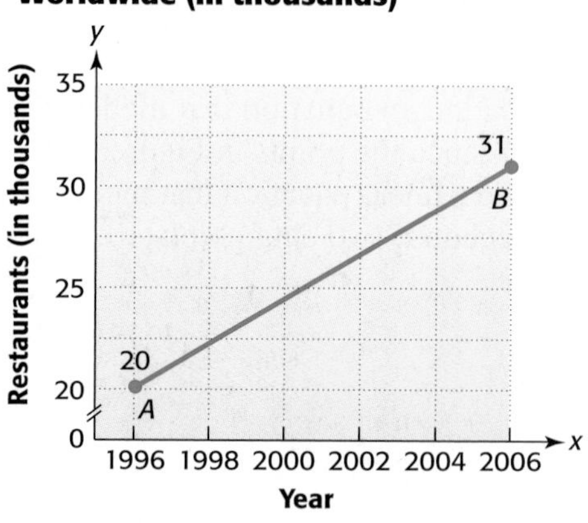

Source: McDonald's Corp.; Hoovers.

1. To represent point A, write an ordered pair in the form

(year, number of restaurants in thousands).

Do this for point B also.
(1996, 20); (2006, 31)

2. The points A and B as well as the points on line AB between them make up line segment AB. We can find the coordinates of a point M on line segment AB that is exactly halfway between A and B. (This point is called the **midpoint** of the segment.)

(a) To find the x-coordinate of M, we find the average of the x-coordinates of A and B by adding them and dividing by 2. What is the x-coordinate of M? **2001**

(b) To find the y-coordinate of M, we find the average of the y-coordinates of A and B. What is the y-coordinate of M? **25.5**

3. Fill in the blanks with the appropriate responses: The ordered pair that represents M, the midpoint of segment AB, is (_____, _____). This suggests that in the year _____, there were _____ thousand McDonald's restaurants worldwide.
2001; 25.5; 2001; 25.5

4. Use the points (1996, 20) and (2006, 31) to find the $y = mx + b$ form of the equation of the line containing A and B. $y = 1.1x - 2175.6$

5. Use the result of Exercise 4 to find the value of y when $x = 2001$. Does this correspond to the result you found in Exercise 2(b)?
25.5; yes

6. The actual number of McDonald's restaurants worldwide was 30 thousand in 2001. How does this compare to your answers in Exercises 3 and 5? Explain how a line graph such as this one can be misleading. Use the concept of *slope* in your explanation.
The actual number is greater. A line graph can be misleading if it contains only the endpoints, as some intermediate points may not lie on the line. The change in y compared to the change in x (the slope) may not be constant for all intermediate points.

3.4 Exercises

FOR EXTRA HELP 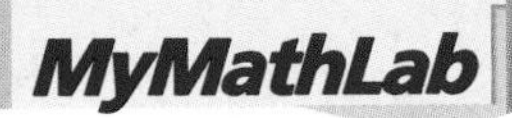  PRACTICE  WATCH  DOWNLOAD READ  REVIEW

1. Match the correct equation in Column II with the description given in Column I.

I		II
(a) Slope -2, the line through the point $(4, 1)$	**D**	A. $y = 4x$
(b) Slope -2, y-intercept $(0, 1)$	**C**	B. $y = \frac{1}{4}x$
(c) The line through the points $(0, 0)$ and $(4, 1)$	**B**	C. $y = -2x + 1$
(d) The line through the points $(0, 0)$ and $(1, 4)$	**A**	D. $y - 1 = -2(x - 4)$

2. The equations $y = \frac{3}{2}x - 6$ and $y + 3 = \frac{3}{2}(x - 2)$ are examples of equations in slope-intercept form and point-slope form, respectively. Write each of these equations in standard form. What do you notice?

In standard form, both equations are written $3x - 2y = 12$. They are different, but equivalent, forms of the same equation.

Use the geometric interpretation of slope (rise divided by run, from **Section 3.3***) to find the slope of each line. Then, by identifying the y-intercept from the graph, write the slope-intercept form of the equation of the line.*

3.
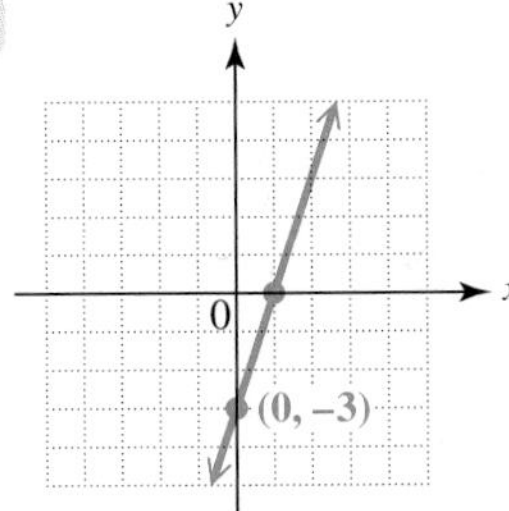

$y = 3x - 3$

4.
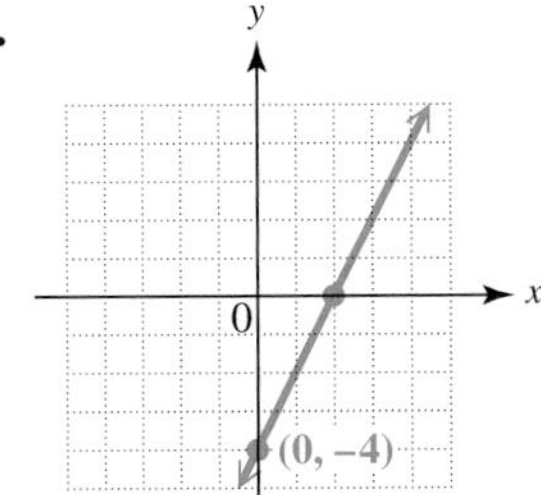

$y = 2x - 4$

5.
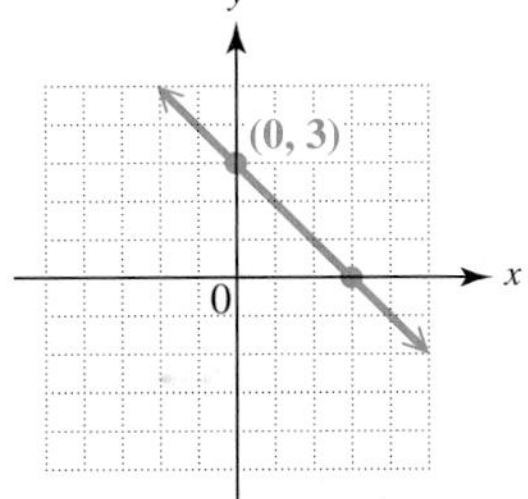

$y = -x + 3$

6.
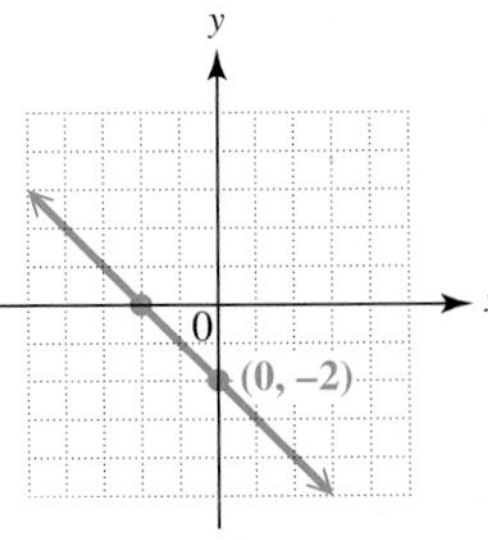

$y = -x - 2$

7.
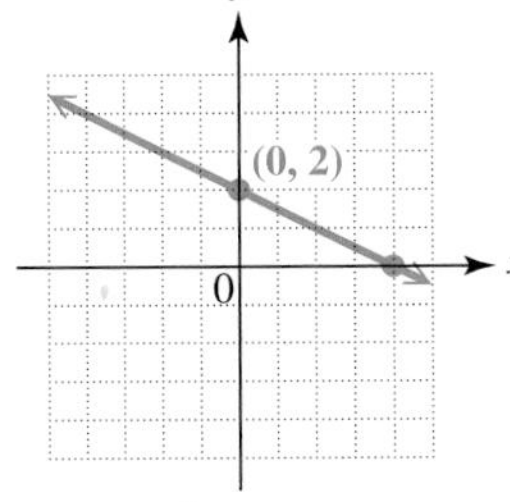

$y = -\frac{1}{2}x + 2$

8.
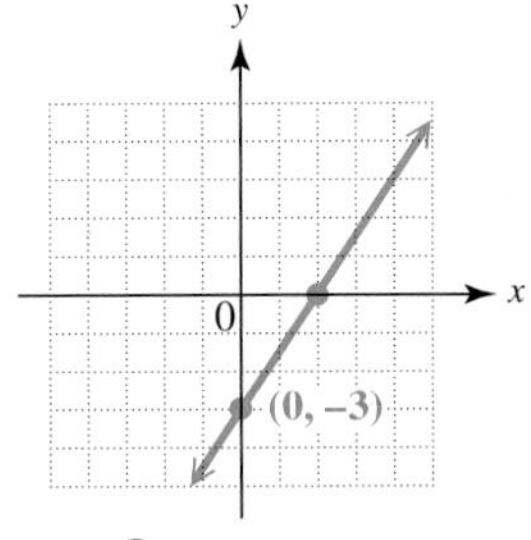

$y = \frac{3}{2}x - 3$

Write the equation of the line with the given slope and y-intercept. See Example 1.

9. slope 4; y-intercept $(0, -3)$

$y = 4x - 3$

10. slope -5; y-intercept $(0, 6)$

$y = -5x + 6$

11. slope 0; y-intercept $(0, 3)$

$y = 3$

12. slope 3; y-intercept $(0, 0)$

$y = 3x$

13. Match each equation with the graph that would most closely resemble its graph.

(a) $y = x + 3$ C

(b) $y = -x + 3$ B

(c) $y = x - 3$ A

(d) $y = -x - 3$ D

A.
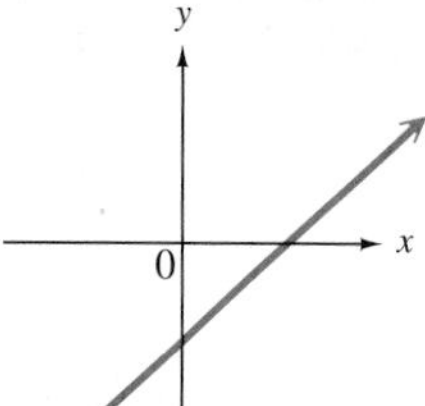

B.
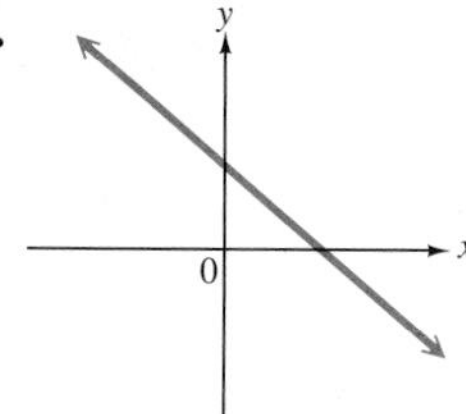

C.
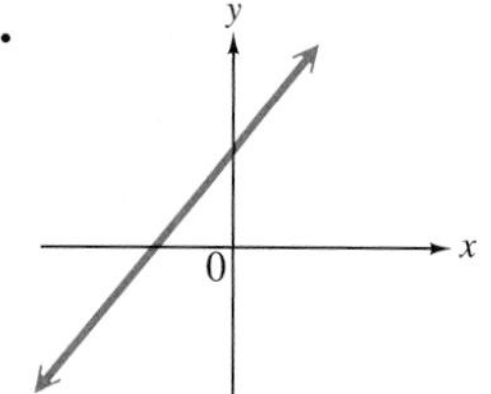

D.
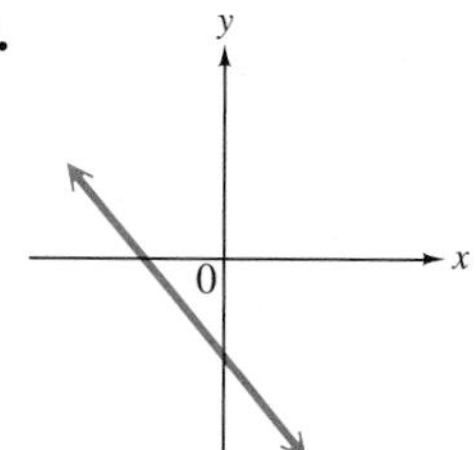

14. Explain why the equation of a vertical line cannot be written in the form $y = mx + b$.

A vertical line has undefined slope, so there is no value for *m*. Also, there is no y-intercept, so there can be no value for *b*.

Graph each equation by finding the slope and y-intercept, and using their definitions to find two points on the line. See Example 2.

15. $y = 3x + 2$

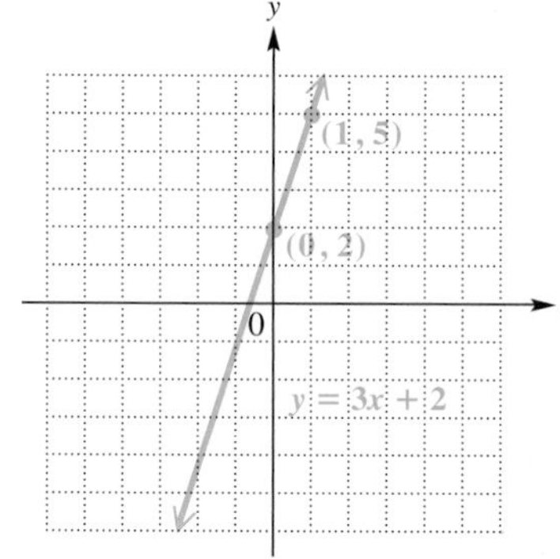

16. $y = 4x - 4$

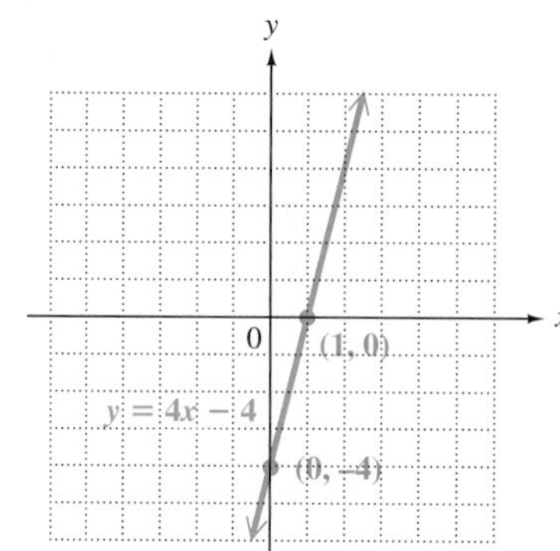

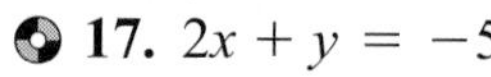

17. $2x + y = -5$

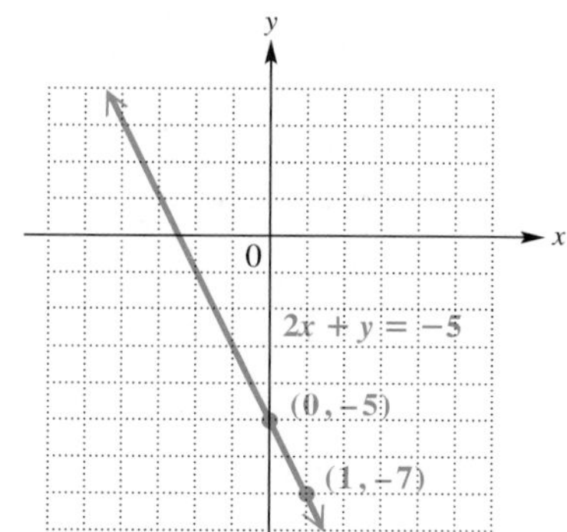

18. $3x + y = -2$

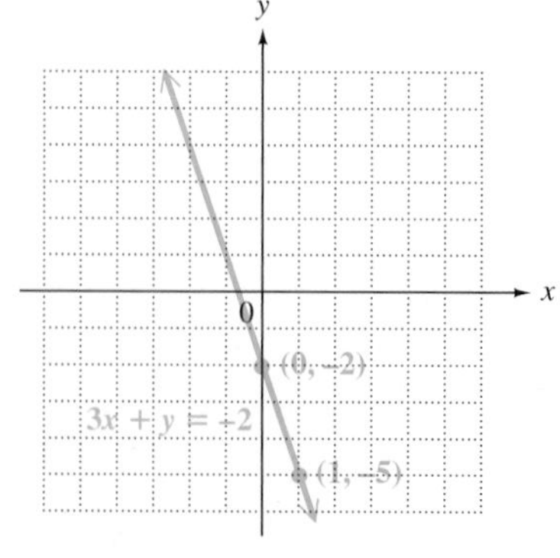

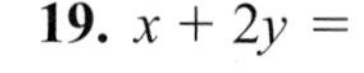

19. $x + 2y = 4$

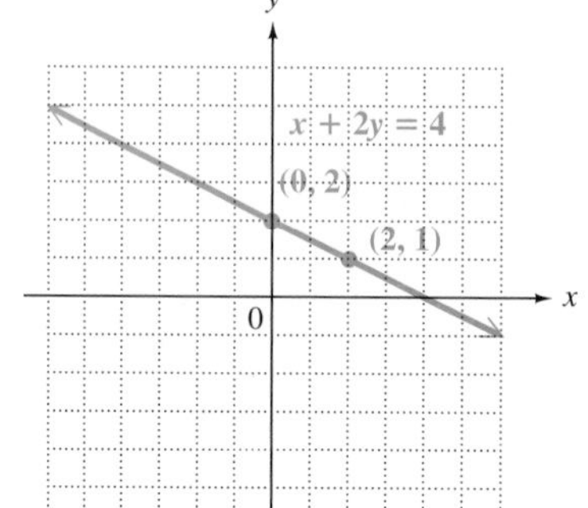

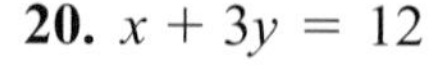

20. $x + 3y = 12$

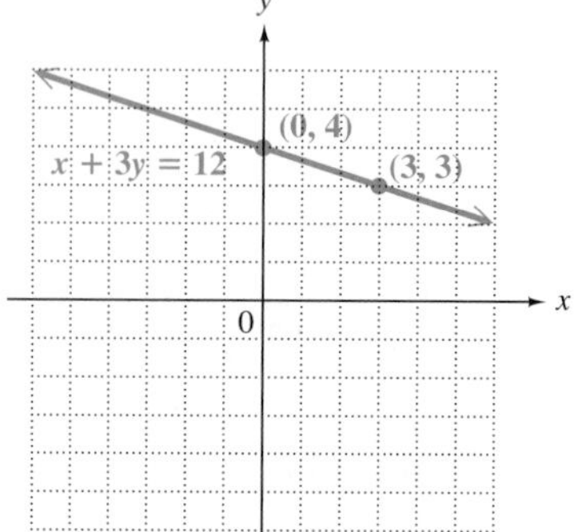

Graph each line passing through the given point and having the given slope. (In Exercises 25–28, recall the types of lines having slope 0 and undefined slope.) Give the slope-intercept form of the equation of the line if possible. See Example 3.

21. $(-2, 3)$, $m = \frac{1}{2}$

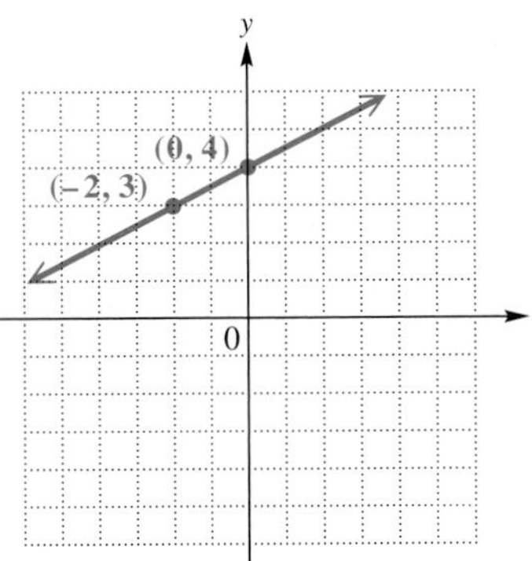

$y = \frac{1}{2}x + 4$

22. $(-4, -1)$, $m = \frac{3}{4}$

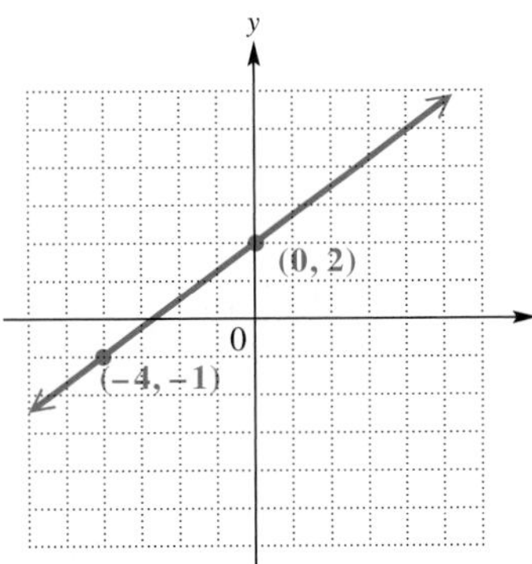

$y = \frac{3}{4}x + 2$

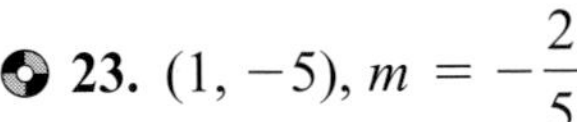

23. $(1, -5)$, $m = -\frac{2}{5}$

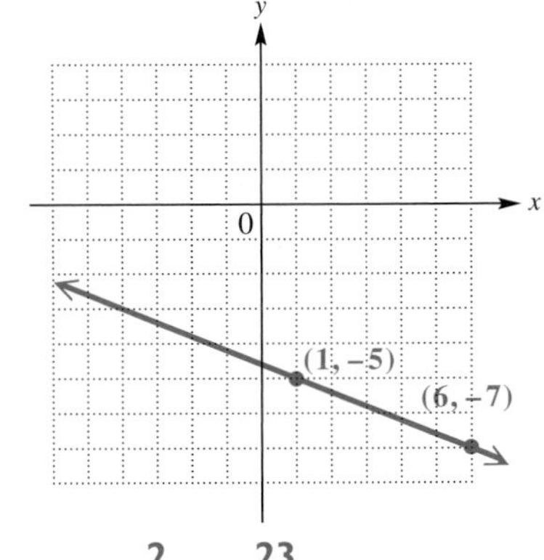

$y = -\frac{2}{5}x - \frac{23}{5}$

24. $(2, -1)$, $m = -\frac{1}{3}$

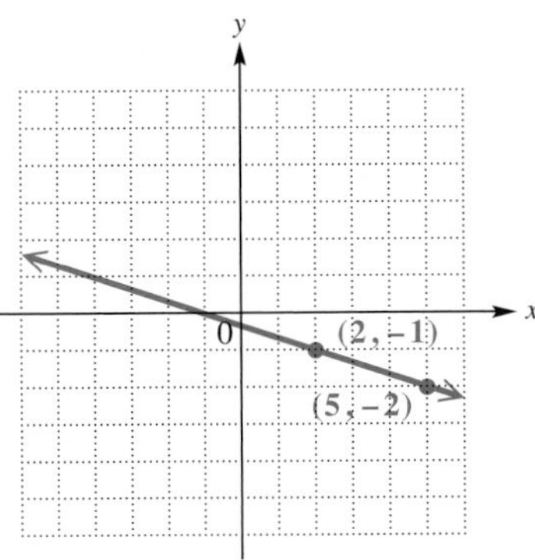

$y = -\frac{1}{3}x - \frac{1}{3}$

25. $(3, 2)$, $m = 0$

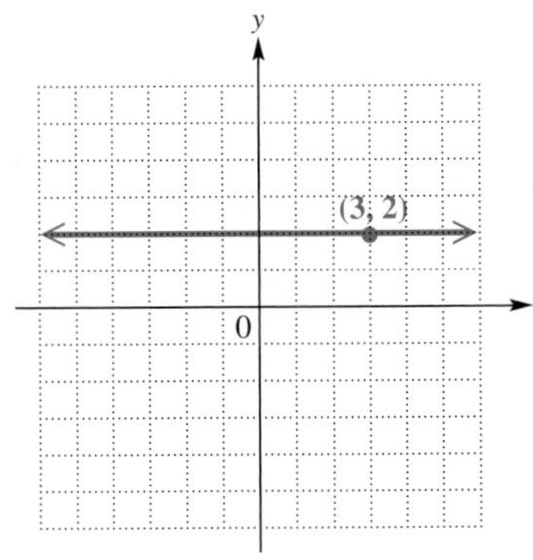

$y = 2$

26. $(-2, 3)$, $m = 0$

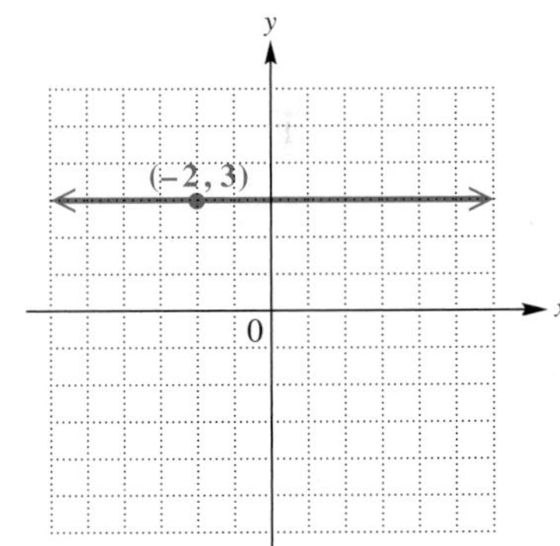

$y = 3$

27. $(3, -2)$, undefined slope

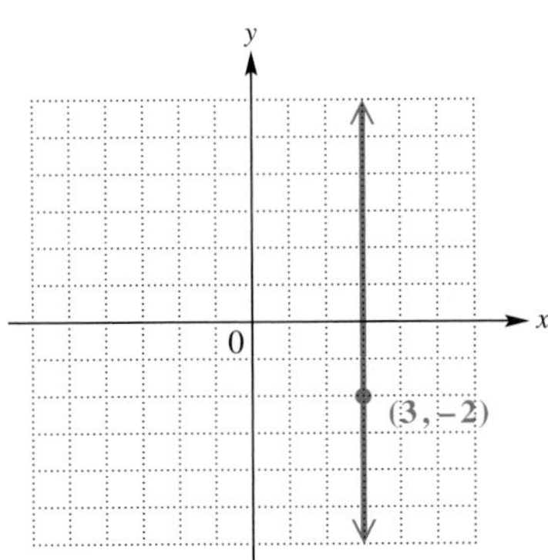

$x = 3$ (no slope-intercept form)

28. $(2, 4)$, undefined slope

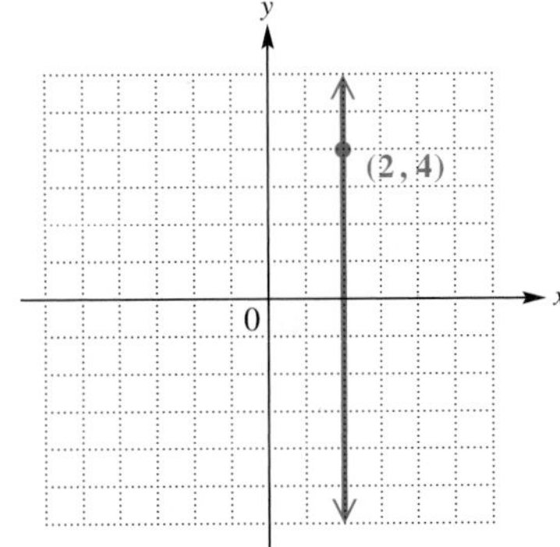

$x = 2$ (no slope-intercept form)

29. $(0, 0)$, $m = \frac{2}{3}$

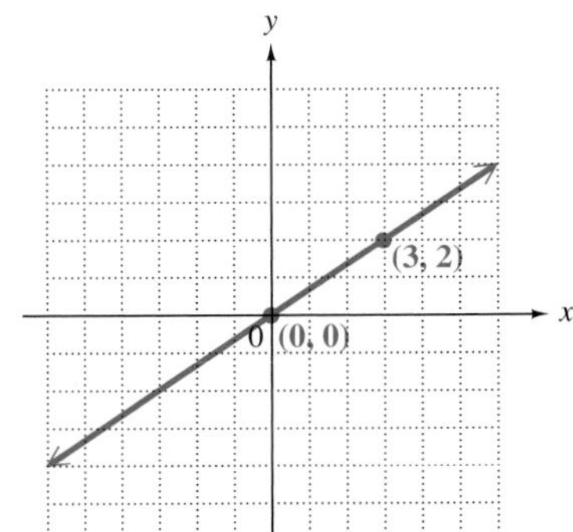

$y = \frac{2}{3}x$

30. (a) What is the common name given to the vertical line whose x-intercept is the origin? **the y-axis**

(b) What is the common name given to the line with slope 0 whose y-intercept is the origin? **the x-axis**

Write an equation of the line passing through the given point and having the given slope. Give the final answer in slope-intercept form. See Examples 4 and 5.

31. $(4, 1)$, $m = 2$
$y = 2x - 7$

32. $(2, 7)$, $m = 3$
$y = 3x + 1$

33. $(3, -10)$, $m = -2$
$y = -2x - 4$

34. $(2, -5)$, $m = -4$
$y = -4x + 3$

35. $(-2, 5)$, $m = \frac{2}{3}$
$y = \frac{2}{3}x + \frac{19}{3}$

36. $(-4, 1)$, $m = \frac{3}{4}$
$y = \frac{3}{4}x + 4$

Write an equation of the line passing through each pair of points. Give the final answer in slope-intercept form, if possible. See Example 6.

37. $(8, 5)$ and $(9, 6)$
$y = x - 3$

38. $(4, 10)$ and $(6, 12)$
$y = x + 6$

39. $(-1, -7)$ and $(-8, -2)$
$y = -\frac{5}{7}x - \frac{54}{7}$

40. $(-2, -1)$ and $(3, -4)$
$y = -\frac{3}{5}x - \frac{11}{5}$

41. $(0, -2)$ and $(-3, 0)$
$y = -\frac{2}{3}x - 2$

42. $(-4, 0)$ and $(0, 2)$
$y = \frac{1}{2}x + 2$

43. $(3, 5)$ and $(3, -2)$
$x = 3$ **(no slope-intercept form)**

44. $(3, -5)$ and $(-1, -5)$
$y = -5$

45. $\left(\frac{1}{2}, \frac{3}{2}\right)$ and $\left(-\frac{1}{4}, \frac{5}{4}\right)$
$y = \frac{1}{3}x + \frac{4}{3}$

46. $\left(-\frac{2}{3}, \frac{8}{3}\right)$ and $\left(\frac{1}{3}, \frac{7}{3}\right)$
$y = -\frac{1}{3}x + \frac{22}{9}$

Write an equation of the line satisfying the given conditions. Give the final answer in slope-intercept form. See Example 7.

47. The line through $(2, -3)$, parallel to $3x = 4y + 5$
$y = \frac{3}{4}x - \frac{9}{2}$

48. The line through $(-1, 4)$, perpendicular to $2x + 3y = 8$
$y = \frac{3}{2}x + \frac{11}{2}$

49. The line through $(8, 5)$, perpendicular to $2x - y = 7$
$y = -\frac{1}{2}x + 9$

50. The line through $(2, -7)$, perpendicular to $5x + 2y = 18$
$y = \frac{2}{5}x - \frac{39}{5}$

51. The line perpendicular to $x - 2y = 7$, y-intercept $(0, -3)$
$y = -2x - 3$

52. The line parallel to $5x = 2y + 10$, y-intercept $(0, 4)$
$y = \frac{5}{2}x + 4$

Write an equation in the form $y = mx$ for each situation. Then give the three ordered pairs associated with the equation for x-values of 0, 5, *and* 10. *See Example 8(a).*

53. x represents the number of hours traveling at 45 mph, and y represents the distance traveled (in miles).

$y = 45x$; $(0, 0), (5, 225), (10, 450)$

54. x represents the number of compact discs sold at $16 each, and y represents the total cost of the discs (in dollars).

$y = 16x$; $(0, 0), (5, 80), (10, 160)$

55. x represents the number of gallons of gas sold at $5.00 per gal, and y represents the total cost of the gasoline (in dollars).

$y = 5.00x$; $(0, 0), (5, 25.00), (10, 50.00)$

56. x represents the number of days a DVD movie is rented at $2.50 per day, and y represents the total charge for the rental (in dollars).

$y = 2.50x$; $(0, 0), (5, 12.50), (10, 25.00)$

For each situation, **(a)** *write an equation in the form $y = mx + b$,* **(b)** *find and interpret the ordered pair associated with the equation for $x = 5$, and* **(c)** *answer the question. See Examples 8(b) and 8(c).*

57. A membership to the Midwest Athletic Club costs $99 plus $41 per month. (*Source:* Midwest Athletic Club.) Let x represent the number of months and y represent the cost. How much does the first year's membership cost?

(a) $y = 41x + 99$

(b) $(5, 304)$; The cost of a 5-month membership is $304.

(c) $591

58. A cell phone plan includes 900 anytime minutes for $60 per month, plus a one-time activation fee of $36. (*Source:* AT&T.) Let x represent the number of months of service and y represent the cost. If you sign a 2-yr contract, how much will this cell phone plan cost? (Assume that you never use more than the allotted number of minutes.)

(a) $y = 60x + 36$

(b) $(5, 336)$; The cost of the plan for 5 months is $336.

(c) $1476

59. A rental car costs $50 plus $0.20 per mile. Let x represent the number of miles driven and y represent the total charge to the renter. How many miles was the car driven if the renter paid $84.60?

(a) $y = 0.20x + 50$

(b) $(5, 51)$; The charge for driving 5 mi is $51.

(c) 173 mi

60. There is a $30 fee to rent a chain saw, plus $6 per day. Let x represent the number of days the saw is rented and y represent the charge to the user in dollars. If the total charge is $138, for how many days is the saw rented?

(a) $y = 6x + 30$

(b) $(5, 60)$; It costs $60 to rent the saw for 5 days.

(c) 18 days

The cost to produce x items is, in some cases, expressed as $y = mx + b$. The number b gives the ***fixed cost*** *(the cost that is the same no matter how many items are produced), and the number m is the* ***variable cost*** *(the cost to produce an additional item). Use this information to work Exercises 61 and 62.*

61. It costs $400 to start up a business selling campaign buttons. Each button costs $0.25 to produce.

(a) What is the fixed cost? $400

(b) What is the variable cost? $0.25

(c) Write the cost equation. $y = 0.25x + 400$

(d) What will be the cost to produce 100 campaign buttons, based on the cost equation? $425

(e) How many campaign buttons will be produced if total cost is $775? 1500

62. It costs $2000 to purchase a copier, and each copy costs $0.02 to make.

(a) What is the fixed cost? $2000

(b) What is the variable cost? $0.02

(c) Write the cost equation. $y = 0.02x + 2000$

(d) What will be the cost to produce 10,000 copies, based on the cost equation? $2200

(e) How many copies will be produced if total cost is $2600? 30,000

Solve each problem. In part (a), give equations in slope-intercept form. (Round the slope to the nearest tenth.) See Example 9. (Source for Exercises 63 and 64: Consumer Electronics Association.)

63. Total sales of digital cameras in the United States (in millions of dollars) are shown in the graph, where the year 2003 corresponds to $x = 0$.

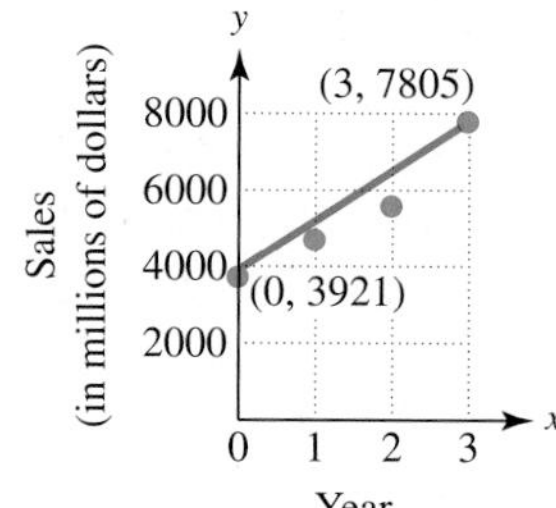

(a) Use the ordered pairs from the graph to write an equation that models the data. What does the slope tell us in the context of this problem?

$y = 1294.7x + 3921$; Sales of digital cameras in the United States increased by $1294.7 million per yr from 2003 to 2006.

(b) Use the equation from part (a) to approximate the sales of digital cameras in the United States in 2007.

$9099.8 million

64. Total sales of fax machines in the United States (in millions of dollars) are shown in the graph, where the year 2003 corresponds to $x = 0$.

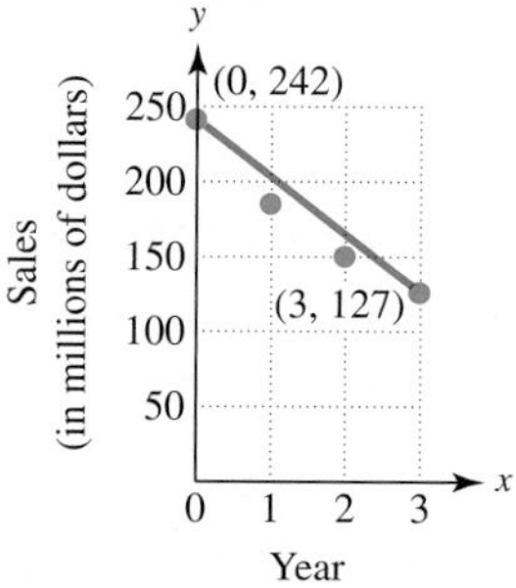

(a) Use the ordered pairs from the graph to write an equation that models the data. What does the slope tell us in the context of this problem?

$y = -38.3x + 242$; Sales of fax machines in the United States decreased by $38.3 million per yr from 2003 to 2006.

(b) Use the equation from part (a) to approximate the sales of fax machines in the United States in 2007.

$88.8 million

Solve the problem. See Example 9.

65. The table gives heavy-metal nuclear waste (in thousands of metric tons) from spent reactor fuel now stored temporarily at reactor sites, awaiting permanent storage. (*Source:* "Burial of Radioactive Nuclear Waste Under the Seabed," *Scientific American*, January 1998.)

Year x	Waste y
1995	32
2000	42
2010*	61
2020*	76

*Estimates by the U.S. Department of Energy.

HEAVY-METAL NUCLEAR WASTE AWAITING STORAGE

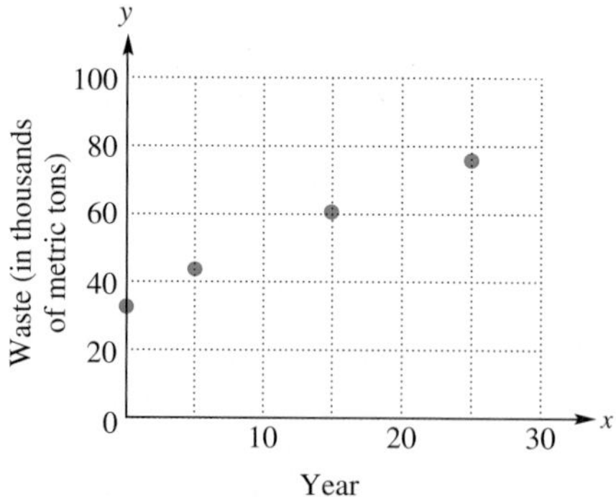

Let $x = 0$ represent 1995, $x = 5$ represent 2000 (since $2000 - 1995 = 5$), and so on.

(a) For 1995, the ordered pair is (0, 32). Write ordered pairs for the data for the other years given in the table.

(5, 42), (15, 61), (25, 76)

(b) Plot the ordered pairs (x, y) on the grid provided at the right. Do the points lie approximately in a straight line?

See the graph; yes

(c) Use the ordered pairs (0, 32) and (25, 76) to find the equation of a line that approximates the data. Write the equation in slope-intercept form.

$y = 1.76x + 32$

(d) Use the equation from part (c) to estimate the amount of nuclear waste in 2015. (*Hint:* What is the value of x for 2015?)

67.2 thousand (or 67,200) metric tons

Summary Exercises on Linear Equations and Graphs

Identify the slope and the y-intercept of the graph of each equation.

1. $3x + y = -6$ **−3; (0, −6)**

2. $2x + y = -4$ **−2; (0, −4)**

3. $-4x - y = 3$ **−4; (0, −3)**

4. $-5x - y = 8$ **−5; (0, −8)**

5. $-3x + 2y = 12$ $\mathbf{\frac{3}{2}}$**; (0, 6)**

6. $-5x + 3y = 15$ $\mathbf{\frac{5}{3}}$**; (0, 5)**

Graph each line, using the given information or equation.

7. $m = 1, b = -2$

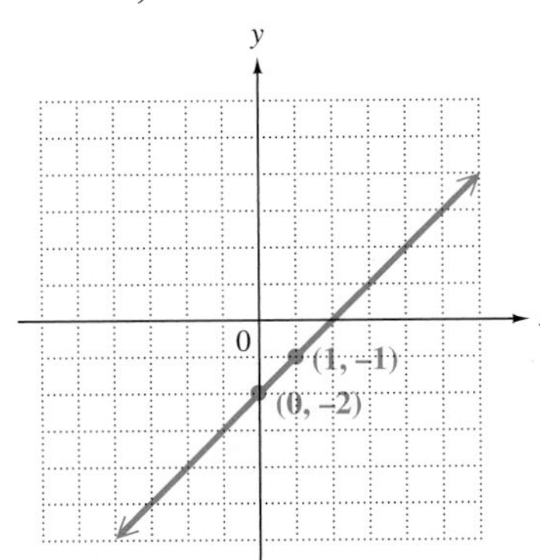

8. $m = 1$, y-intercept $(0, -4)$

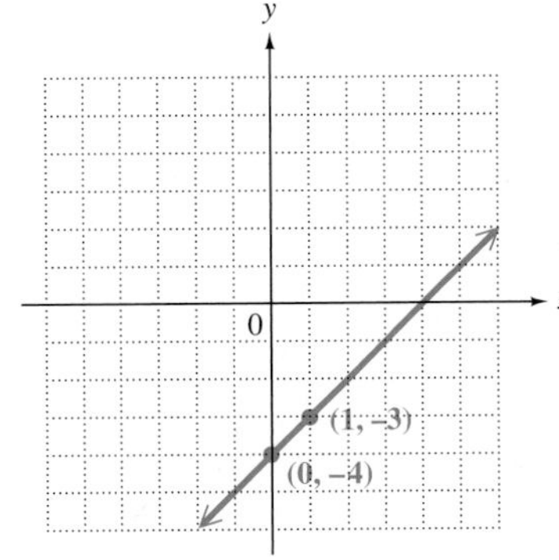

9. $y = -2x + 6$

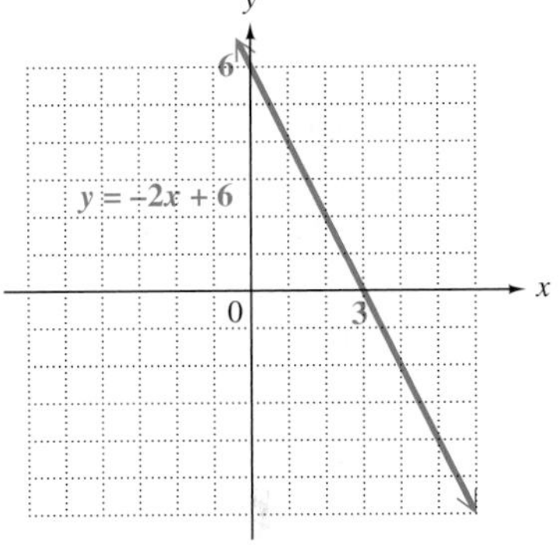

10. $x + 4 = 0$

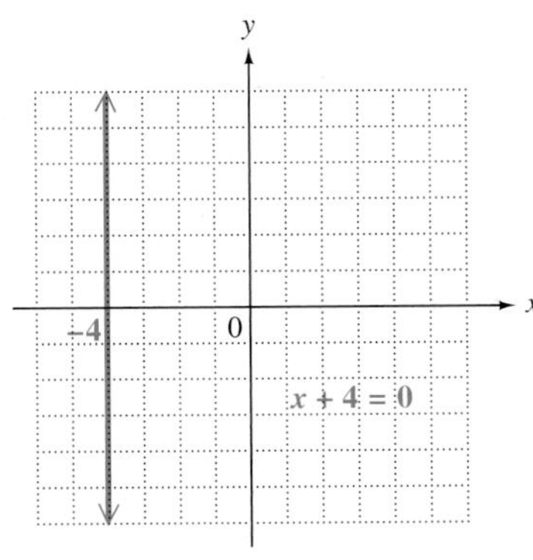

11. $m = -\frac{2}{3}$, passes through $(3, -4)$

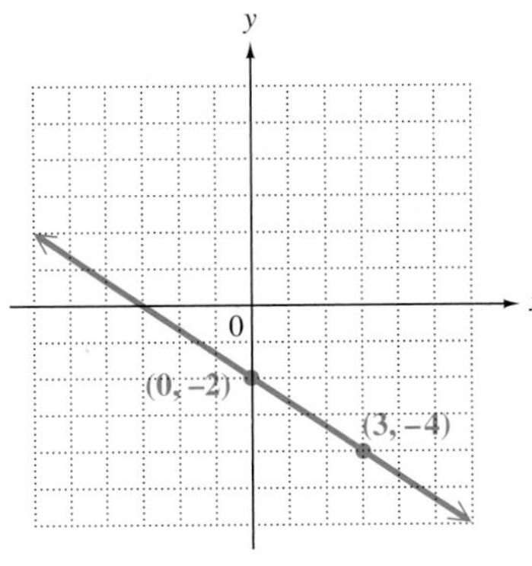

12. $y = -\frac{1}{2}x + 2$

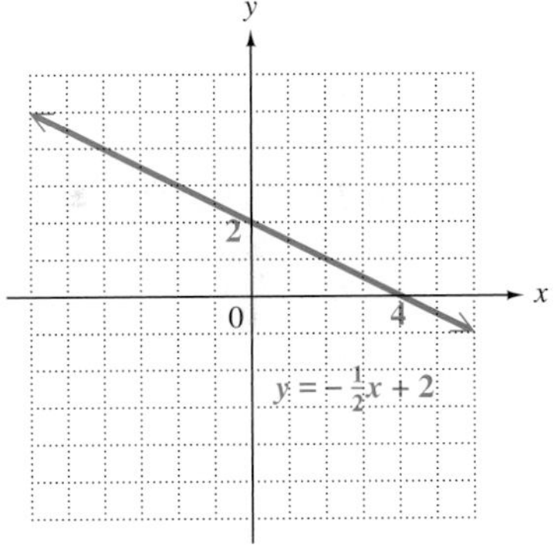

13. $y - 4 = -9$

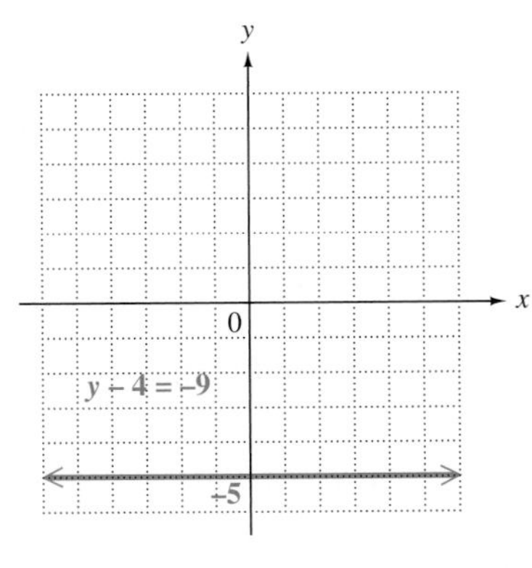

14. $m = -\frac{3}{4}$, passes through $(4, -4)$

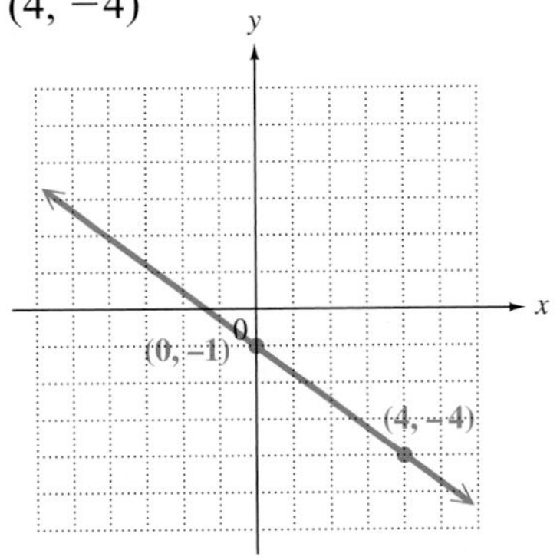

15. Undefined slope, passes through $(3.5, 0)$

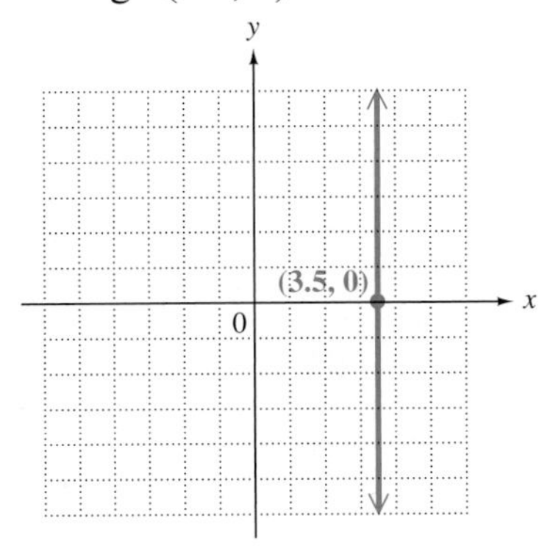

16. Slope $-\frac{1}{5}$, passes through $(0, 0)$

17. $4x - 5y = 20$

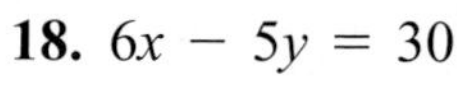

18. $6x - 5y = 30$

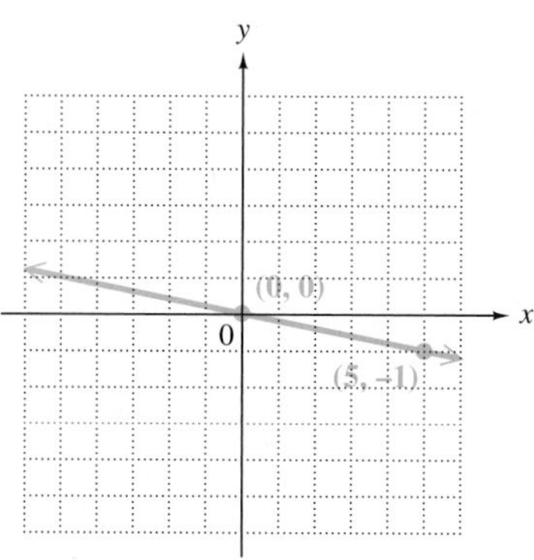

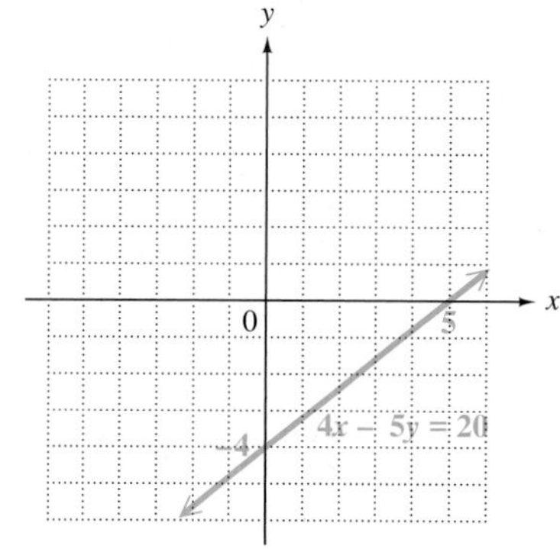

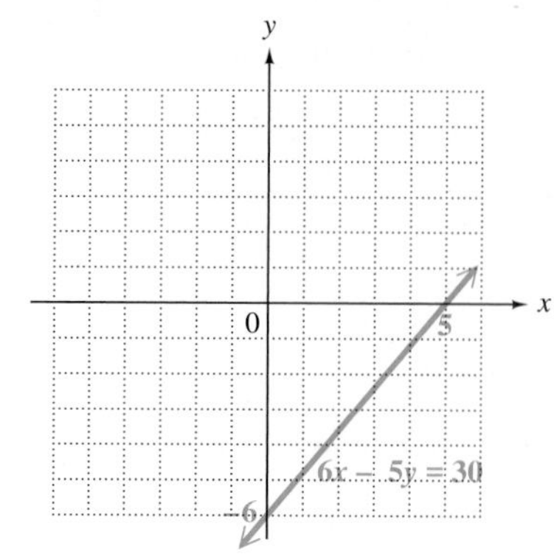

19. $x - 4y = 0$

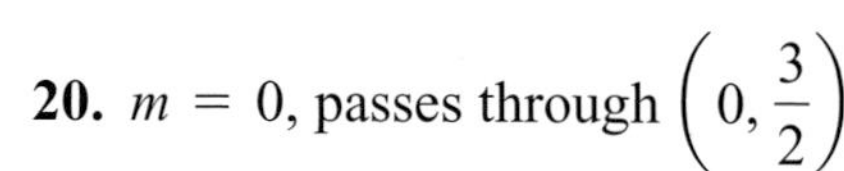

20. $m = 0$, passes through $\left(0, \frac{3}{2}\right)$

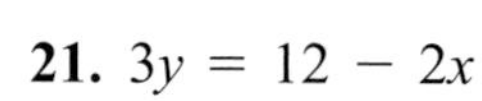

21. $3y = 12 - 2x$

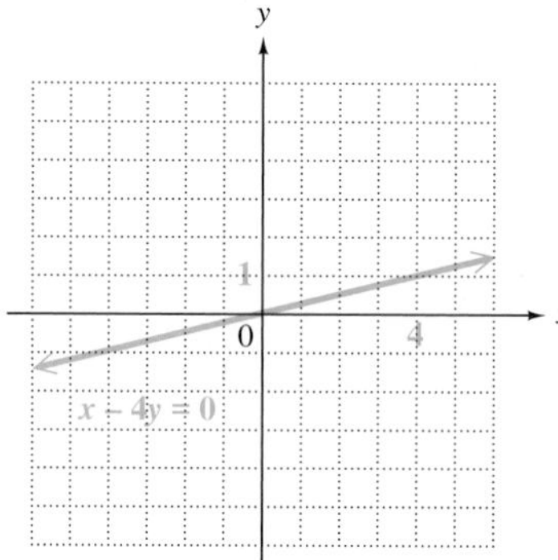

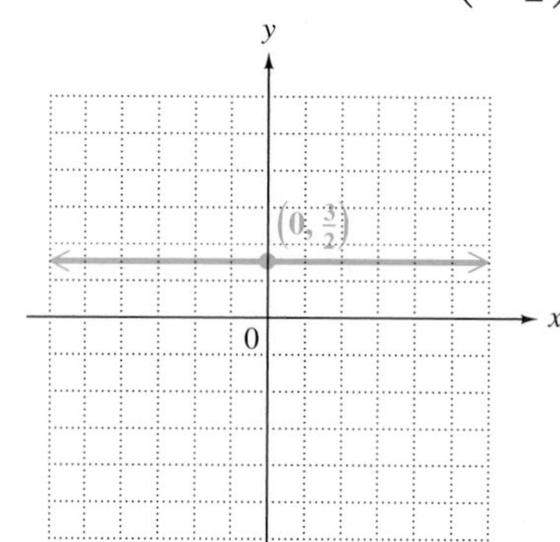

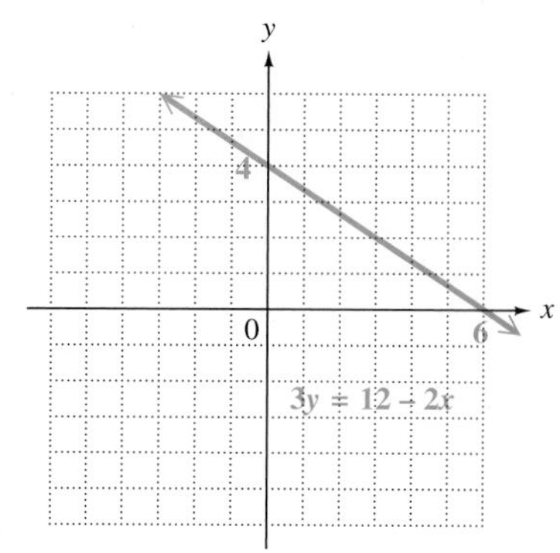

Write an equation of each line. Give the final answer in slope-intercept form if possible.

22. $m = -3, b = -6$

$y = -3x - 6$

23. The line through $(1, -7)$ and $(-2, 5)$

$y = -4x - 3$

24. The line through $(0, 0)$ and $(5, 3)$

$y = \frac{3}{5}x$

25. The line through $(0, 0)$, undefined slope

$x = 0$

26. The line through $(0, 0)$, $m = 0$

$y = 0$

27. $m = -2$, y-intercept $(0, -4)$

$y = -2x - 4$

28. $m = \frac{5}{3}$, through $(-3, 0)$

$y = \frac{5}{3}x + 5$

3.5 Graphing Linear Inequalities in Two Variables

OBJECTIVES

1 **Graph linear inequalities in two variables.**

2 **Graph a linear inequality with boundary line through the origin.**

OBJECTIVE 1 Graph linear inequalities in two variables. In **Section 2.7** we graphed linear inequalities in one variable on the number line. We now graph linear inequalities in two variables on a rectangular coordinate system.

Linear Inequality in Two Variables

An inequality that can be written as

$$Ax + By < C \quad \text{or} \quad Ax + By > C,$$

where A, B, and C are real numbers and A and B are not both 0, is a **linear inequality in two variables.**

The symbols $\leq$ and $\geq$ may replace $<$ and $>$ in the definition.

Consider the graph in Figure 42. The graph of the line $x + y = 5$ divides the points in the rectangular coordinate system into three sets:

1. Those points that lie on the line itself and satisfy the equation $x + y = 5$ [like (0, 5), (2, 3), and (5, 0)];
2. Those that lie in the half-plane above the line and satisfy the inequality $x + y > 5$ [like (5, 3) and (2, 4)];
3. Those that lie in the half-plane below the line and satisfy the inequality $x + y < 5$ [like (0, 0) and $(-3, -1)$].

The graph of the line $x + y = 5$ is called the **boundary line** for the inequalities $x + y > 5$ and $x + y < 5$. Graphs of linear inequalities in two variables are *regions* in the real number plane that may or may not include boundary lines.

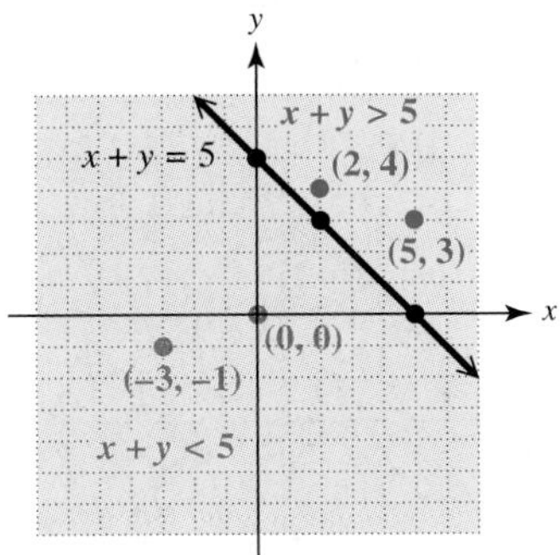

Figure 42

Graphing a Linear Inequality

Step 1 **Draw the graph of the straight line that is the boundary.** Make the line solid if the inequality involves $\leq$ or $\geq$; make the line dashed if the inequality involves $<$ or $>$.

Step 2 **Choose a test point.** Choose any point not on the line, and substitute the coordinates of this point in the inequality. Simplify to determine whether the inequality is true or false.

Step 3 **Shade the appropriate region.** Shade the region that includes the test point if it satisfies the original inequality; otherwise, shade the region on the other side of the boundary line.

1 Graph each inequality.

(a) $x + y \leq 4$

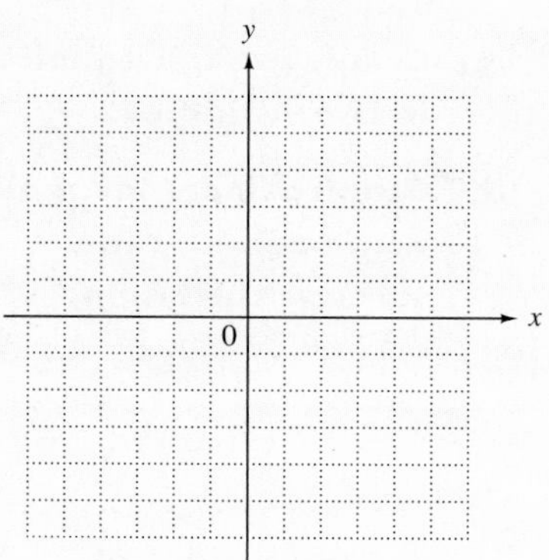

(b) $3x + y \geq 6$

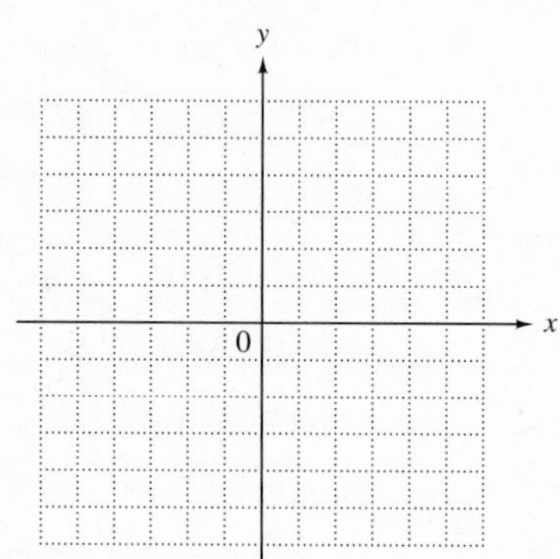

ANSWERS

1. (a)

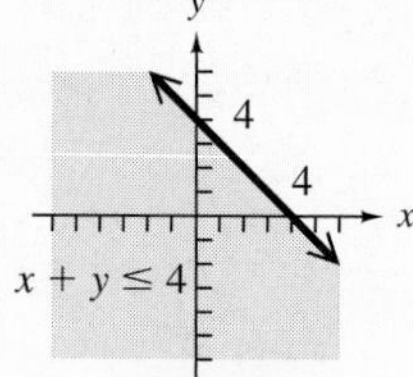

(b)

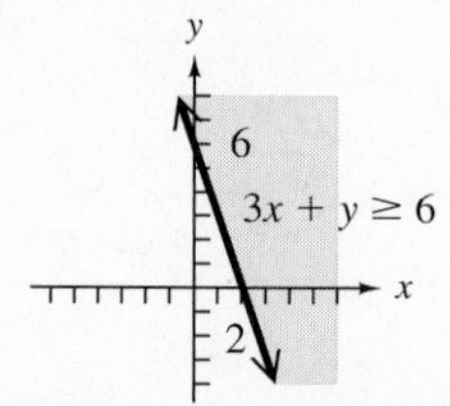

EXAMPLE 1 Graphing a Linear Inequality

Graph $3x + 2y \geq 6$.

Step 1 First graph the line $3x + 2y = 6$, which has intercepts (2, 0) and (0, 3). The graph of this line, the boundary of the graph of the inequality, is shown in Figure 43.

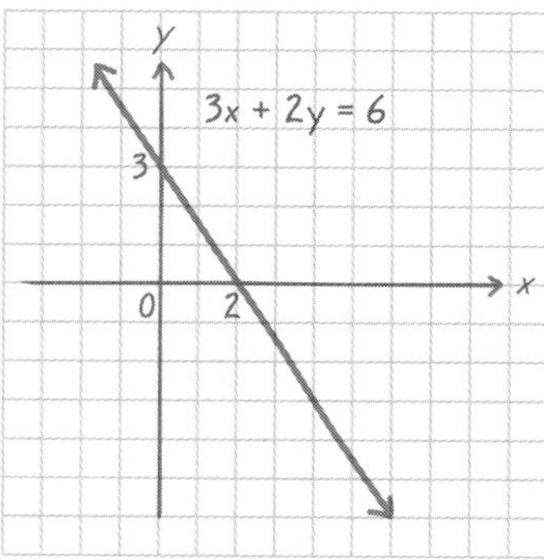

Figure 43

Step 2 The graph of the inequality $3x + 2y \geq 6$ includes the points of the boundary line $3x + 2y = 6$ and either the points *above* the line $3x + 2y = 6$ or the points *below* that line. To decide which, select any point not on the line $3x + 2y = 6$ as a test point. The origin, (0, 0), is often a good choice. Substitute the values from the test point (0, 0) for x and y in the inequality.

$$3x + 2y > 6$$

$$3(\mathbf{0}) + 2(\mathbf{0}) \overset{?}{>} 6 \quad \text{Let } x = 0 \text{ and } y = 0.$$

$$0 > 6 \quad \text{False}$$

Step 3 Because the result is false, (0, 0) does *not* satisfy the inequality, and so the solution set includes all points on the other side of the line. This region is shaded in Figure 44.

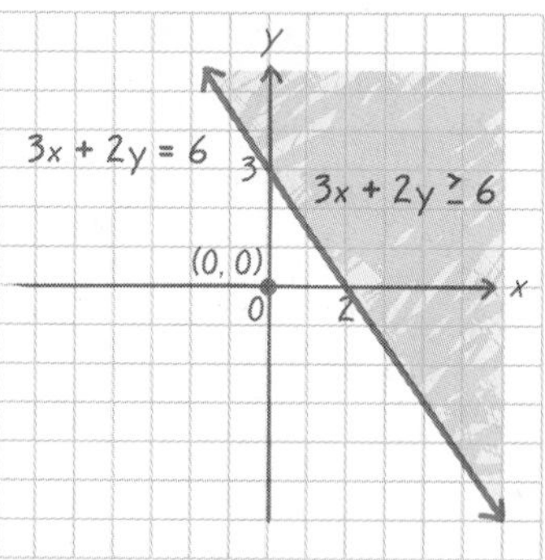

Figure 44

◀ Work Problem 1 at the Side.

If the inequality is written in the form $y > mx + b$ or $y < mx + b$, the inequality symbol indicates which half-plane to shade.

If $y > mx + b$, shade **above** the boundary line.

If $y < mx + b$, shade **below** the boundary line.

This method works only if the inequality is solved for y.

EXAMPLE 2 Graphing a Linear Inequality

Graph $x - 3y < 4$.

First graph the boundary line $x - 3y = 4$ through $(4, 0)$ and $(0, -\frac{4}{3})$, as shown in Figure 45. The points of the boundary line do not belong to the inequality $x - 3y < 4$ (because the inequality symbol is $<$, not $\leq$). For this reason, the line is dashed. Now solve the inequality for y.

Solve for y.

$$x - 3y < 4$$
$$-3y < -x + 4 \quad \text{Subtract } x.$$
$$y > \frac{x}{3} - \frac{4}{3} \quad \text{Multiply by } -\tfrac{1}{3}\text{; change } < \text{ to } >.$$

Because of the *is greater than* symbol, shade *above* the line. As a check, choose a test point not on the line, say $(1, 2)$, and substitute for x and y in the original inequality.

$$x - 3y < 4 \quad \text{Original inequality}$$
$$1 - 3(2) \overset{?}{<} 4 \quad \text{Let } x = 1 \text{ and } y = 2.$$
$$-5 < 4 \quad \text{True}$$

This result agrees with the decision to shade above the line. The solution set, graphed in Figure 45, includes only those points in the shaded half-plane (not those on the line).

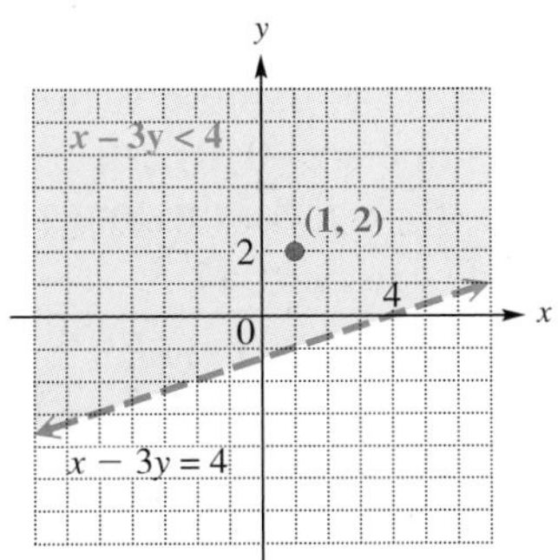

Figure 45

Work Problem 2 at the Side. ▶

2 Graph each inequality.

(a) $x - y > 2$

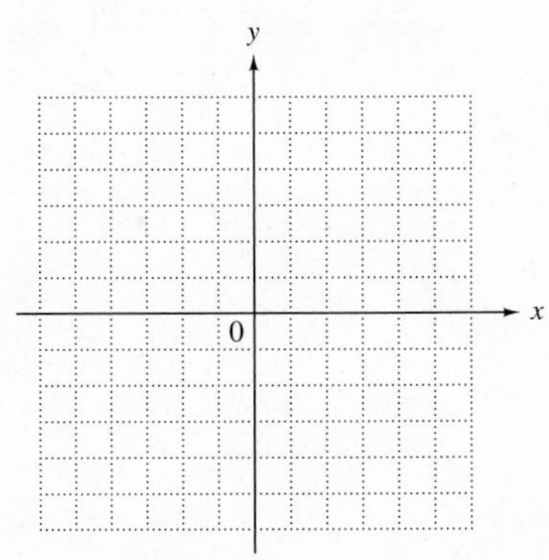

(b) $3x + 4y < 12$

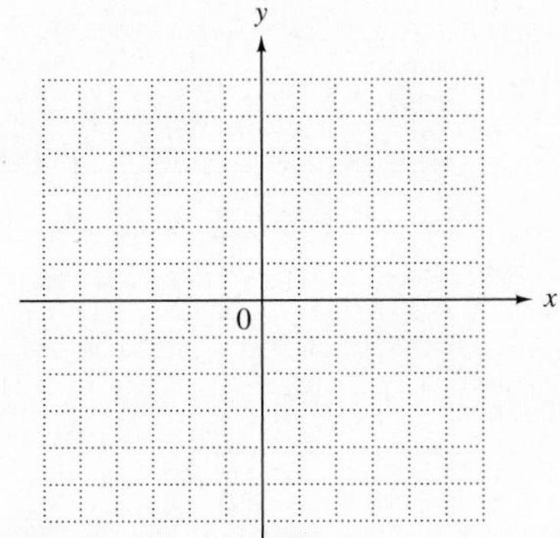

EXAMPLE 3 Graphing a Linear Inequality with a Vertical Boundary Line

Graph the inequality $x < 3$.

First graph $x = 3$, a vertical line through the point $(3, 0)$. Use a dashed line (why?) and choose $(0, 0)$ as a test point.

$$x < 3 \quad \text{Original inequality}$$
$$0 \overset{?}{<} 3 \quad \text{Let } x = 0.$$
$$0 < 3 \quad \text{True}$$

Because $0 < 3$ is true, we shade the region containing $(0, 0)$, as in Figure 46 on the next page.

Continued on Next Page

ANSWERS

2. (a)

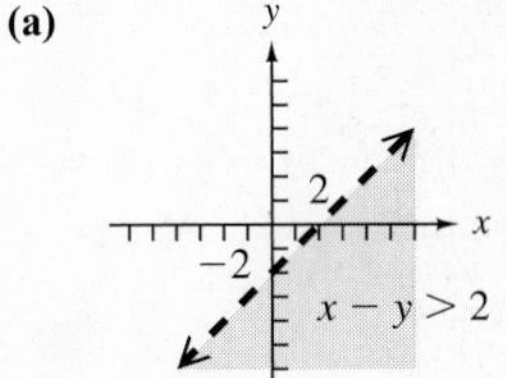

(b)

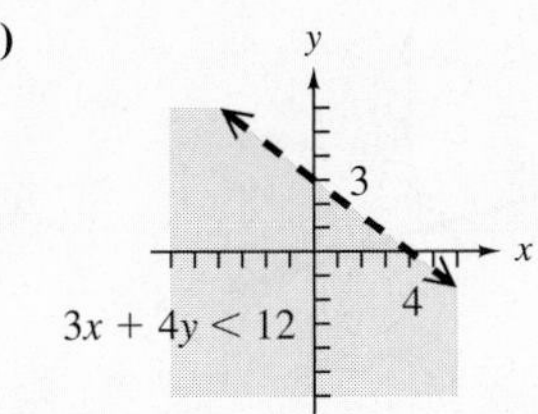

3 Graph $y < 4$.

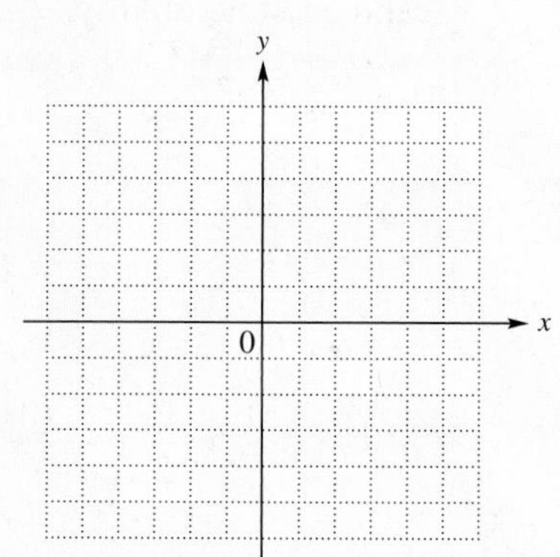

Be careful to draw a dashed line.

Figure 46

◀ *Work Problem* **3** *at the Side.*

OBJECTIVE 2 Graph a linear inequality with boundary line through the origin. ***If the graph of an inequality has a boundary line through the origin, (0, 0) cannot be used as a test point.***

4 Graph $x \geq -3y$.

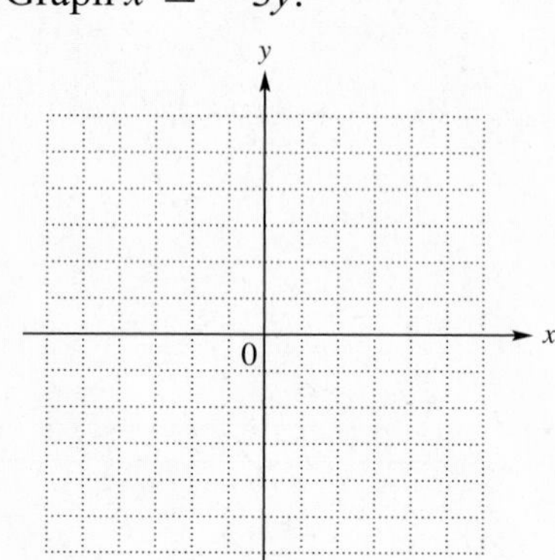

EXAMPLE 4 Graphing a Linear Inequality with a Boundary Line through the Origin

Graph the inequality $x \leq 2y$.

We begin by graphing $x = 2y$, using a solid line. Some ordered pairs that can be used to graph this line are (0, 0), (6, 3), and (4, 2). We cannot use (0, 0) as a test point because (0, 0) is *on* the line $x = 2y$. Instead, we choose a test point *off* the line, say (1, 3).

$x \leq 2y$	Original inequality
$1 \overset{?}{\leq} 2(3)$	Let $x = 1$ and $y = 3$.
$1 \leq 6$	True

Because $1 \leq 6$ is true, we shade the side of the graph containing the test point (1, 3). See Figure 47.

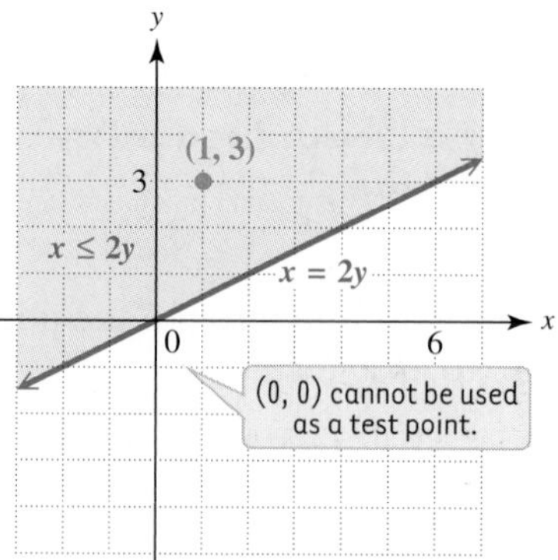

Figure 47

◀ *Work Problem* **4** *at the Side.*

Answers

3.

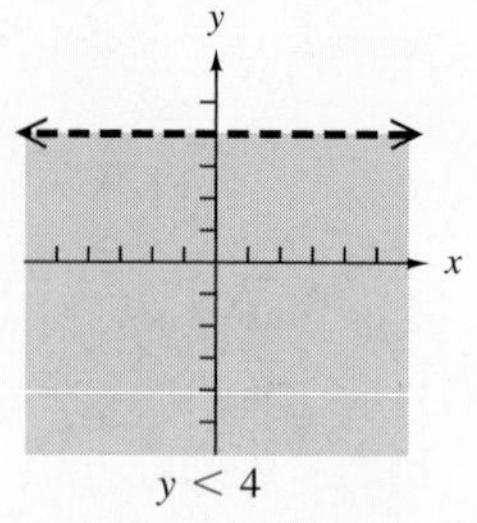

4.

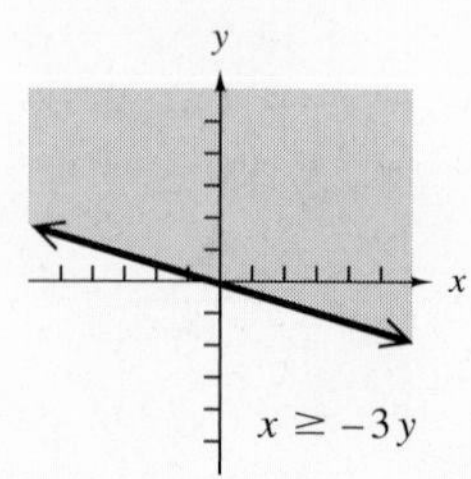

3.5 Exercises

FOR EXTRA HELP 

Math XL PRACTICE  WATCH  DOWNLOAD  READ  REVIEW

1. Explain how to determine whether to use a dashed line or a solid line when graphing a linear inequality in two variables.

   Use a dashed line if the symbol is $<$ or $>$. Use a solid line if the symbol is $\leq$ or $\geq$.

2. Explain why the point (0, 0) is not an appropriate choice for a test point when graphing an inequality whose boundary goes through the origin.

   A test point cannot lie on the boundary line. It must lie on one side of the boundary.

Graph each linear inequality. See Examples 1–4.

3. $x + y \leq 5$

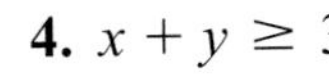
4. $x + y \geq 3$

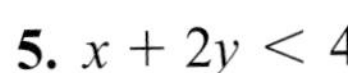
5. $x + 2y < 4$

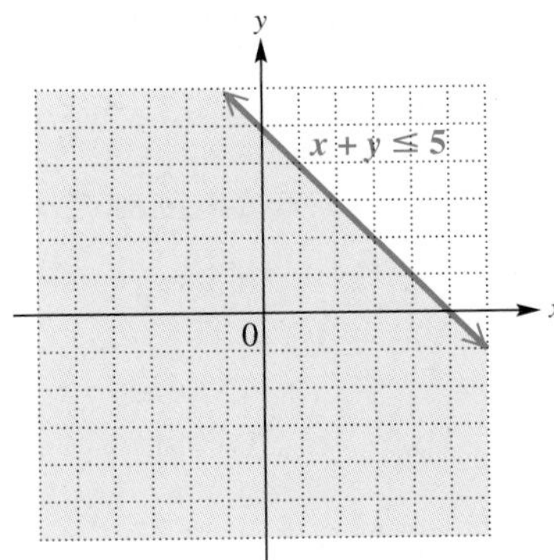

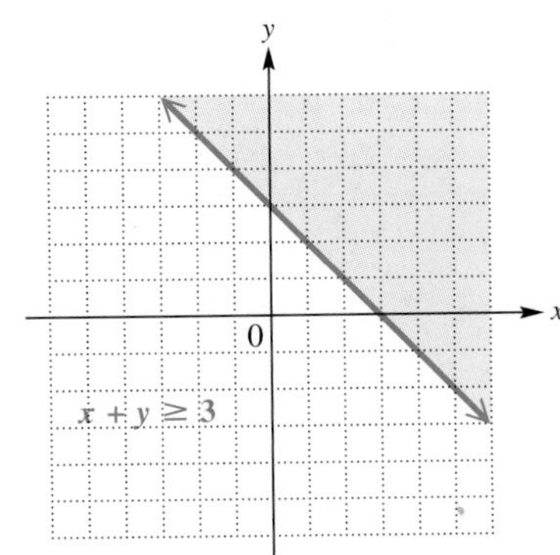

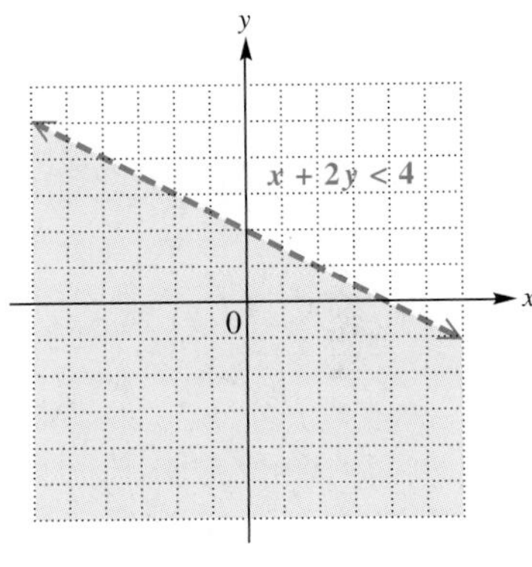

6. $x + 3y > 6$

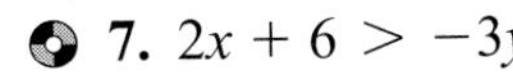
7. $2x + 6 > -3y$

8. $-4y > 3x - 12$

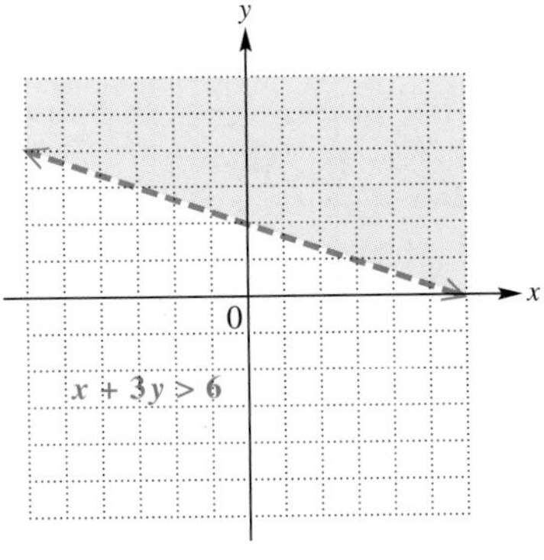

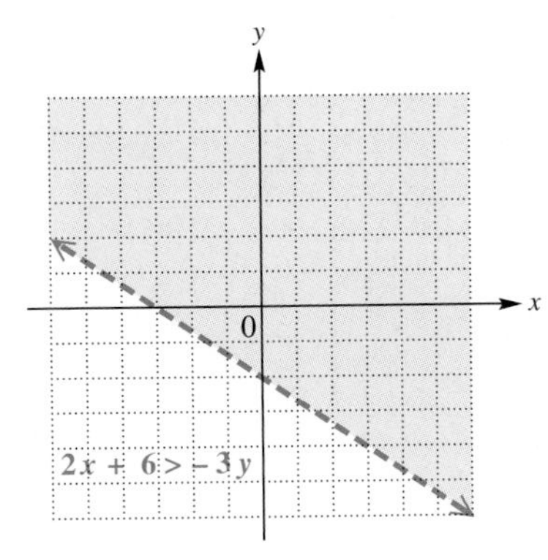

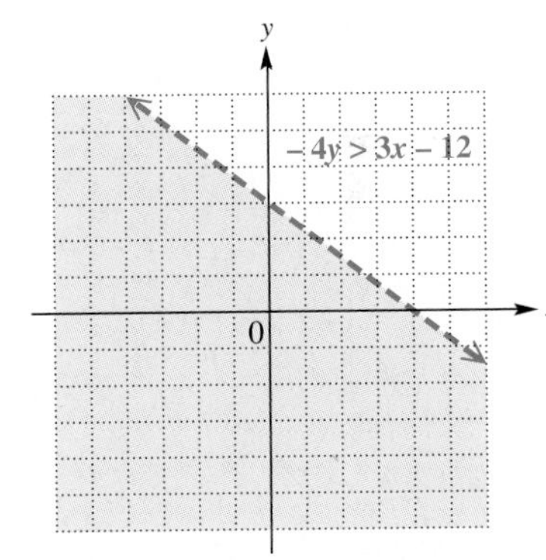

9. $y \geq 2x + 1$

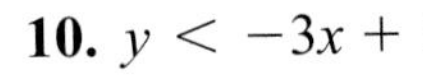
10. $y < -3x + 1$

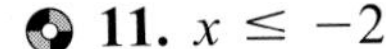
11. $x \leq -2$

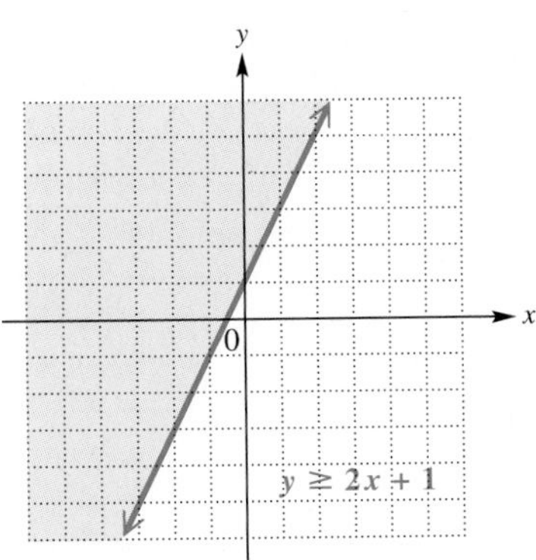

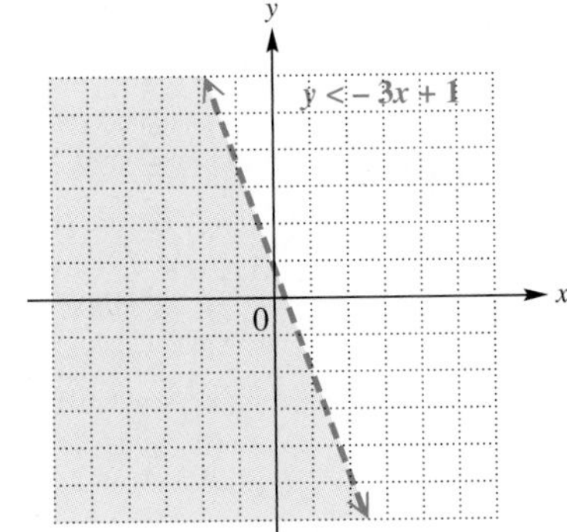

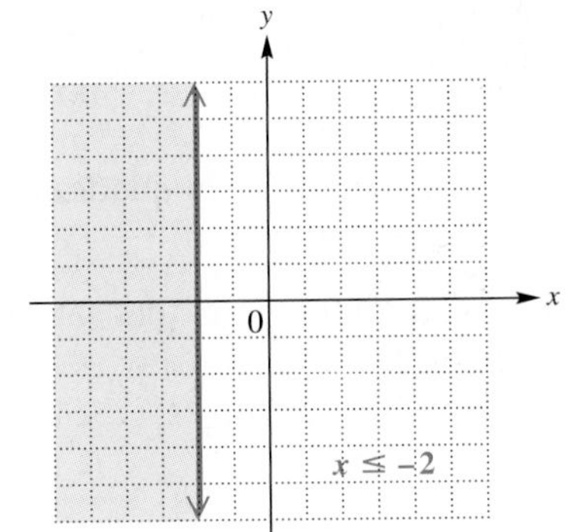

12. $x \geq 1$

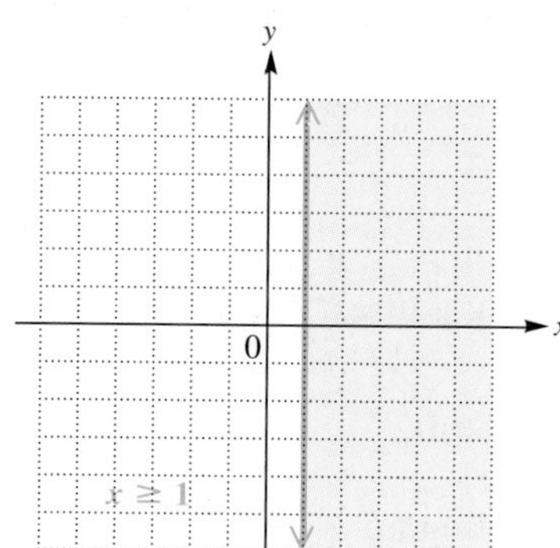

13. $y < 5$

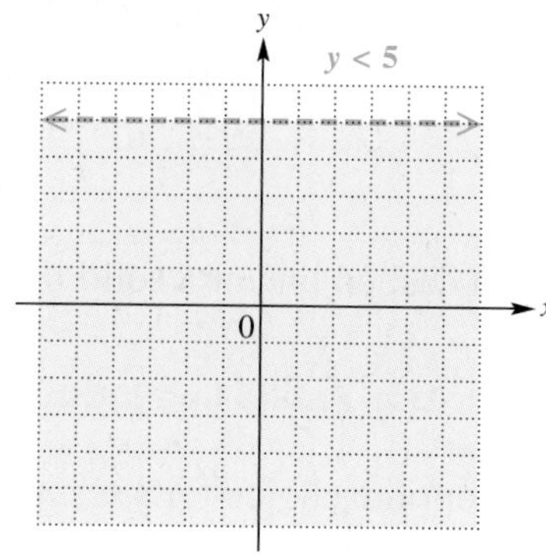

14. $y < -3$

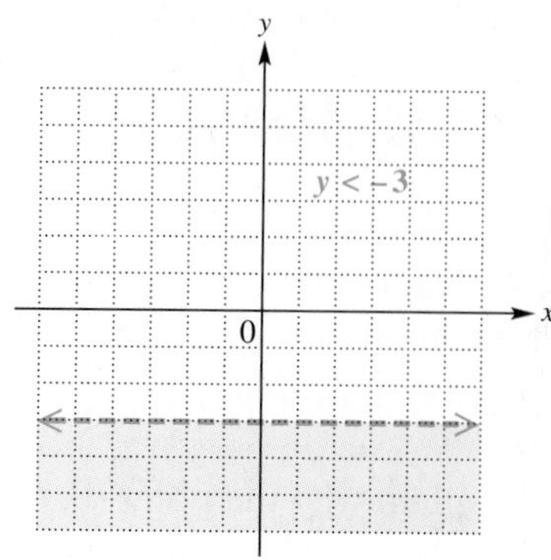

15. $y \geq 4x$

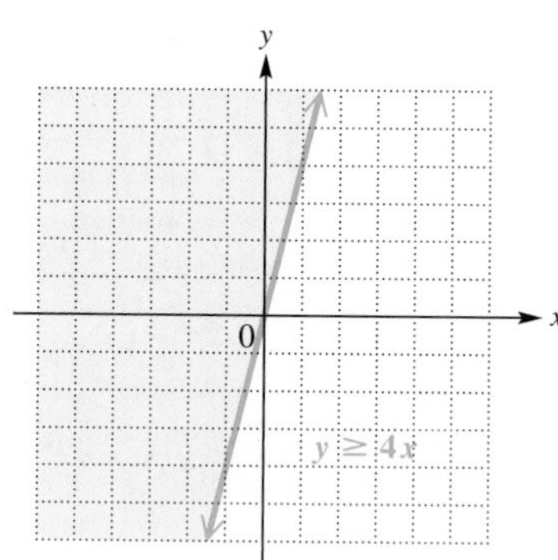

16. $y \leq 2x$

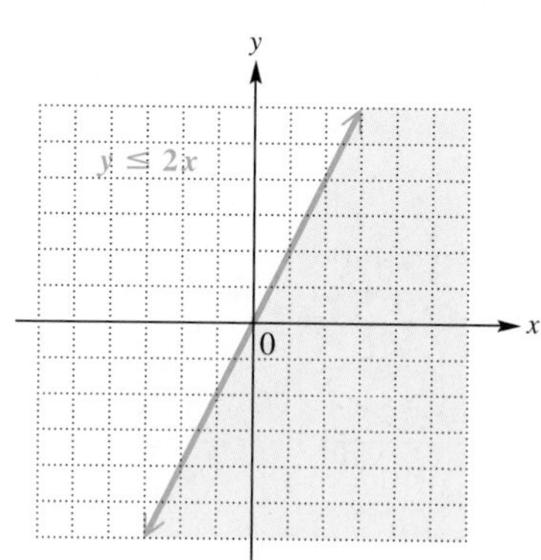

17. $x + y > 0$

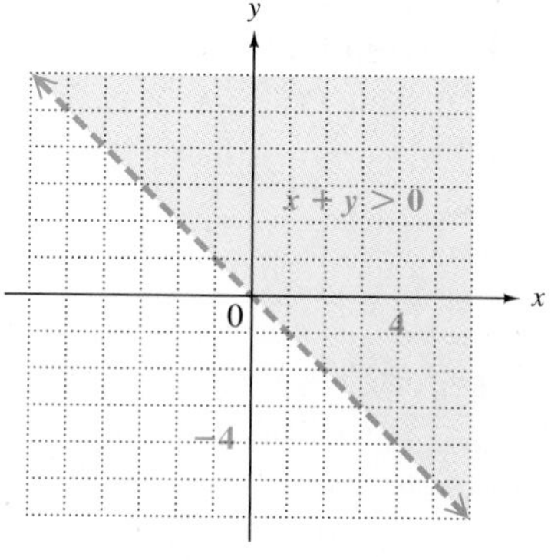

18. $x + 2y > 0$

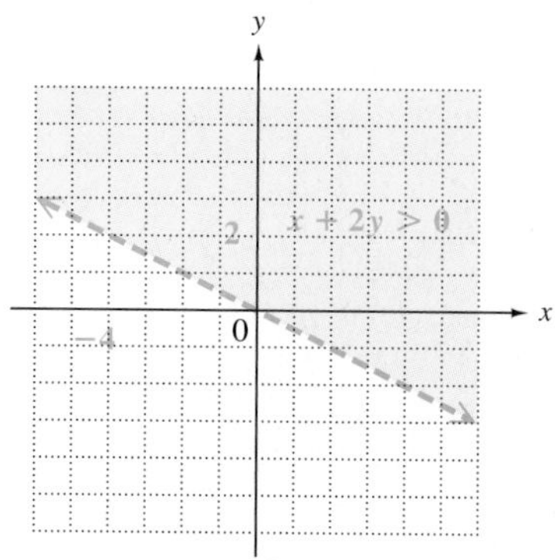

19. $x - 3y \leq 0$

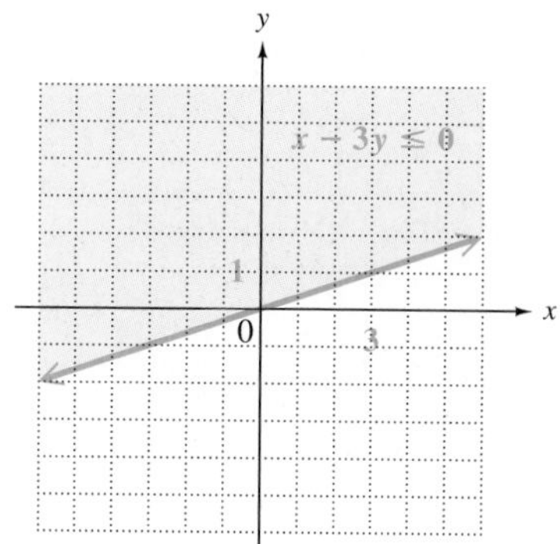

20. $x - 5y \leq 0$

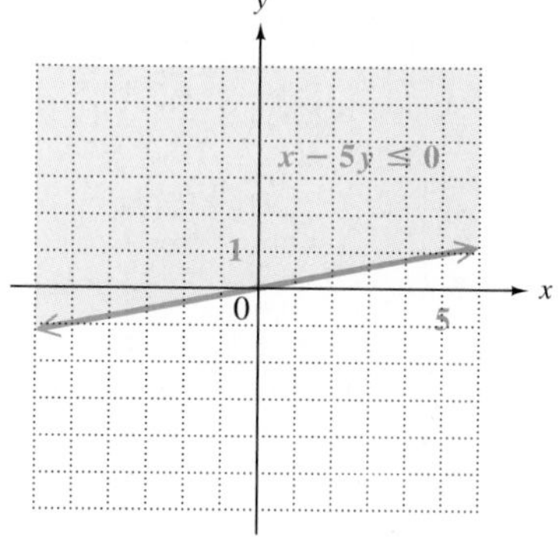

A company will ship x units of merchandise to outlet I and y units of merchandise to outlet II. The company must ship a total of at least 500 *units to these two outlets. This can be expressed by writing*

$$x + y \geq 500.$$

21. Graph the inequality in quadrant I on the given grid since $x \geq 0$ and $y \geq 0$.

22. Give two ordered pairs that satisfy the inequality in Exercise 21.

(500, 0) and (200, 400); Other answers are possible.

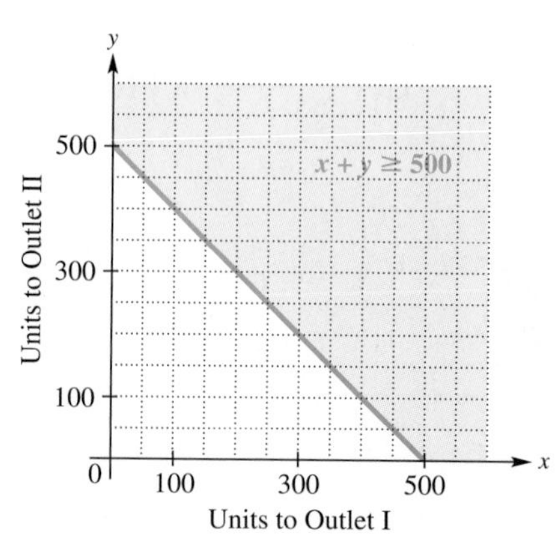

3.6 Introduction to Functions

We often describe one quantity in terms of another. Consider the following.

- The amount of your paycheck if you are paid hourly depends on the number of hours you worked.
- The cost at the gas station depends on the number of gallons of gas you pumped into your car.
- The distance traveled by a car moving at a constant speed depends on the time traveled.

We can use ordered pairs to represent these corresponding quantities. For example, we indicate the relationship between the amount of your paycheck and hours worked by writing ordered pairs in which the first number represents hours worked and the second number represents paycheck amount in dollars. Then the ordered pair (5, 40) indicates that when you work 5 hr, your paycheck is \$40. Similarly, the ordered pairs (10, 80) and (20, 160) show that working 10 hr results in an \$80 paycheck and working 20 hr results in a \$160 paycheck.

Work Problem 1 *at the Side.* ▶

Since the amount of your paycheck *depends* on the number of hours worked, your paycheck amount is called the *dependent variable,* and the number of hours worked is called the *independent variable*. Generalizing, if the value of the variable y depends on the value of the variable x, then y is the **dependent variable** and x is the **independent variable.**

Independent variable ↓ ↓ Dependent variable

$$(x, y)$$

OBJECTIVE 1 Define and identify relations and functions.

Since we can write related quantities using ordered pairs, a set of ordered pairs such as

$$\{(5, 40), (10, 80), (20, 160), (40, 320)\}$$

is called a *relation*.

> **Relation**
>
> A **relation** is any set of ordered pairs.

A special kind of relation, called a *function,* is very important in mathematics and its applications.

> **Function**
>
> A **function** is a relation in which, for each value of the first component of the ordered pairs, there is *exactly one value* of the second component.

OBJECTIVES

1. Define and identify relations and functions.
2. Find domain and range.
3. Identify functions defined by graphs and equations.
4. Use function notation.
5. Graph linear and constant functions.

1 What would the ordered pair (40, 320) in the correspondence between number of hours worked and paycheck amount (in dollars) indicate?

ANSWER

1. It indicates that when you work 40 hr, your paycheck is \$320.

2 Determine whether each relation defines a function.

(a) $\{(0, 3), (-1, 2), (-1, 3)\}$

(b) $\{(2, -2), (4, -4), (6, -6)\}$

(c) $\{(-1, 5), (0, 5)\}$

EXAMPLE 1 Determining Whether Relations Are Functions

Tell whether each relation defines a function.

$$F = \{(1, 2), (-2, 4), (3, -1)\}$$
$$G = \{(-2, -1), (-1, 0), (0, 1), (1, 2), (2, 2)\}$$
$$H = \{(-4, 1), (-2, 1), (-2, 0)\}$$

Relations F and G are functions, because for each different x-value there is exactly one y-value. Notice that in G, the last two ordered pairs have the same y-value (1 is paired with 2, and 2 is paired with 2). This does not violate the definition of function, since the first components (x-values) are different and each is paired with only one second component (y-value).

In relation H, however, the last two ordered pairs have the *same* x-value paired with *two different* y-values (-2 is paired with both 1 and 0), so H is a relation but not a function. ***In a function, no two ordered pairs can have the same first component and different second components.***

Different y-values

$$H = \{(-4, 1), (-2, 1), (-2, 0)\}$$ Not a function

Same x-value

◀ Work Problem **2** at the Side.

In a function, there is* exactly one *value of the dependent variable, the second component, for each value of the independent variable, the first component. This is what makes functions so important in applications.

Relations and functions can also be expressed as a correspondence or *mapping* from one set to another, as shown in Figure 48 for function F and relation H from Example 1. The arrow from 1 to 2 indicates that the ordered pair (1, 2) belongs to F—each first component is paired with exactly one second component. In the mapping for set H, which is not a function, the first component -2 is paired with two different second components, 1 and 0.

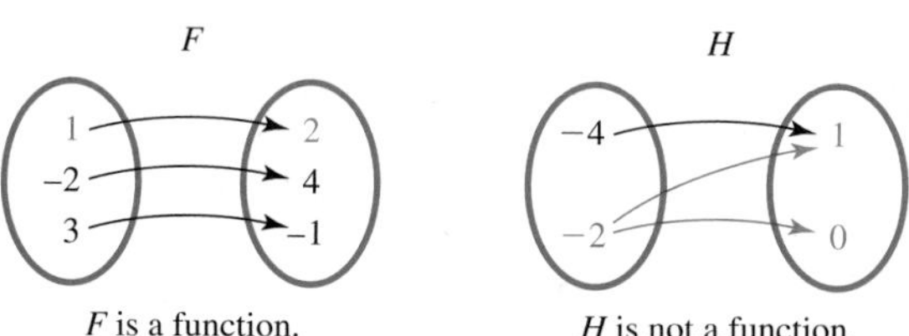

F is a function. H is not a function.

Figure 48

Since relations and functions are sets of ordered pairs, we can represent them using tables and graphs. A table and graph for function F is shown in Figure 49.

x	y
1	2
-2	4
3	-1

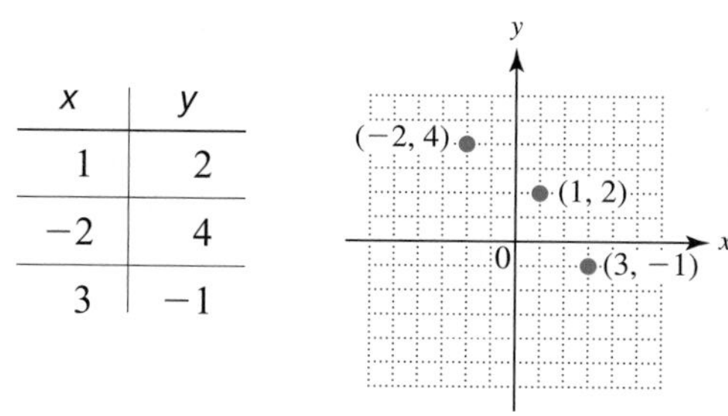

Graph of F

Figure 49

ANSWERS

2. (a) not a function **(b)** function **(c)** function

Finally, we can describe a relation or function using a rule that tells how to determine the dependent variable for a specific value of the independent variable. The rule may be given in words, such as "the dependent variable is twice the independent variable." Usually, however, the rule is given as an equation:

$$y = 2x.$$

Dependent variable (y) — Independent variable (x)

An equation is the most efficient way to define a relation or function.

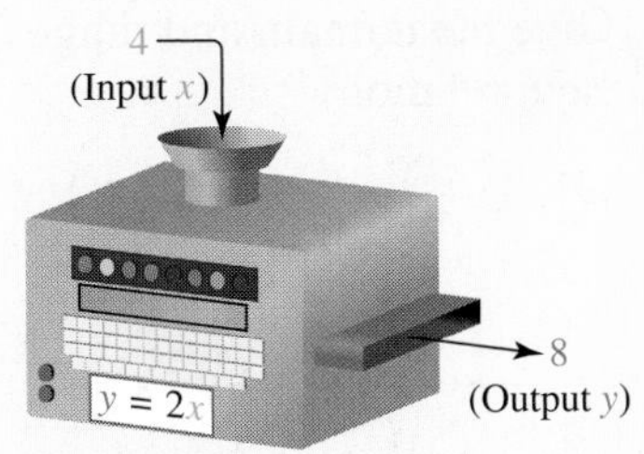

Function machine

> **Note**
>
> Another way to think of a function relationship is to think of the independent variable as an input and the dependent variable as an output. This is illustrated by the input-output (function) machine in the margin for the function defined by $y = 2x$.

OBJECTIVE 2 Find domain and range. For every relation, there are two important sets of elements called the *domain* and *range*.

> **Domain and Range**
>
> In a relation, the set of all values of the independent variable (x) is the **domain.** The set of all values of the dependent variable (y) is the **range.**

EXAMPLE 2 Finding Domains and Ranges of Relations

Give the domain and range of each relation. Tell whether the relation defines a function.

(a) $\{(3, -1), (4, 2), (4, 5), (6, 8)\}$

The domain, the set of x-values, is $\{3, 4, 6\}$; the range, the set of y-values, is $\{-1, 2, 5, 8\}$. This relation is not a function because the same x-value 4 is paired with two different y-values, 2 and 5.

(b)

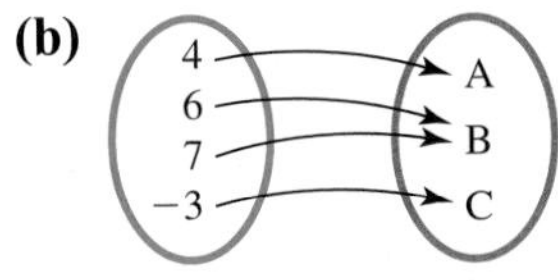

The domain of this relation is

$$\{4, 6, 7, -3\}.$$

The range is

$$\{A, B, C\}.$$

This mapping defines a function—each x-value corresponds to exactly one y-value.

(c)

x	y
−5	2
0	2
5	2

This is a table of ordered pairs, so the domain is the set of x-values, $\{-5, 0, 5\}$, and the range is the set of y-values, $\{2\}$. The table defines a function because each different x-value corresponds to exactly one y-value (even though it is the same y-value).

Work Problem 3 at the Side. ▶

3 Give the domain and range of each relation. Does the relation define a function?

(a) $\{(4, 0), (4, 1), (4, 2)\}$

(b)

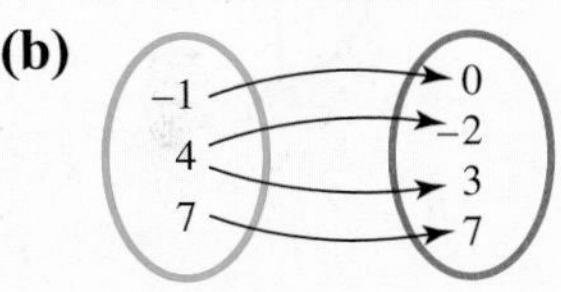

(c)

Year	Cell Phone Subscribers (in thousands)
2002	140,766
2003	158,722
2004	182,140
2005	207,896
2006	233,041

Source: CTIA-The Wireless Association.

ANSWERS

3. (a) domain: $\{4\}$; range: $\{0, 1, 2\}$; No, the relation does not define a function.
(b) domain: $\{-1, 4, 7\}$; range: $\{0, -2, 3, 7\}$; No, the relation does not define a function.
(c) domain: $\{2002, 2003, 2004, 2005, 2006\}$; range: $\{140{,}766, 158{,}722, 182{,}140, 207{,}896, 233{,}041\}$; Yes, the relation defines a function.

4 Give the domain and range of each relation.

(a)

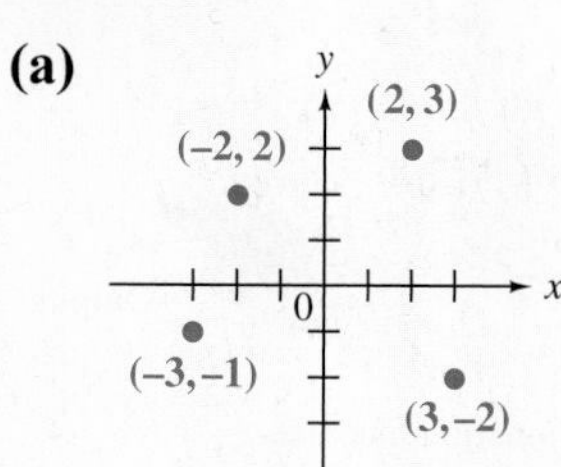

(b)

(c)

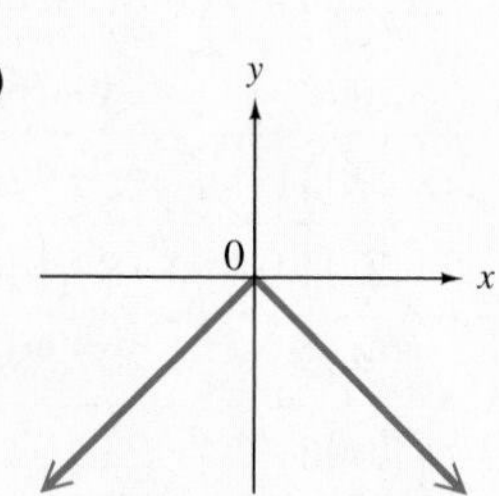

The graph of a relation gives a picture of the relation, which can be used to determine its domain and range.

EXAMPLE 3 Finding Domains and Ranges from Graphs

Give the domain and range of each relation.

(a)

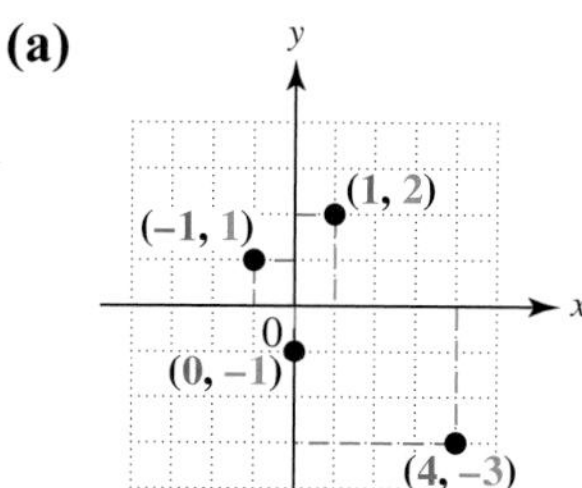

The domain is the set of x-values,

$$\{-1, 0, 1, 4\}.$$

The range is the set of y-values,

$$\{-3, -1, 1, 2\}.$$

(b)

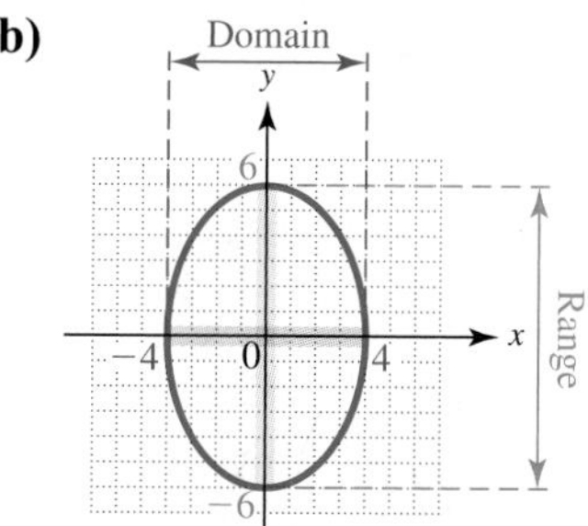

The x-values of the points on the graph include all numbers between -4 and 4, inclusive. The y-values include all numbers between -6 and 6, inclusive. Using interval notation,

the domain is $[-4, 4]$;

the range is $[-6, 6]$.

(c)

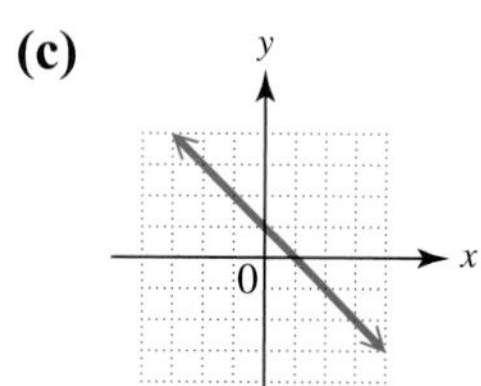

The arrowheads indicate that the line extends indefinitely left and right, as well as up and down. Therefore, both the domain and the range include all real numbers, written $(-\infty, \infty)$.

(d)

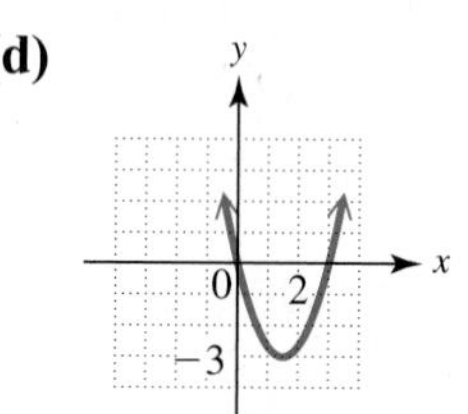

The arrowheads indicate that the graph extends indefinitely left and right, as well as upward. The domain is $(-\infty, \infty)$, and the range includes all numbers greater than or equal to -3, written $[-3, \infty)$.

◀ Work Problem 4 at the Side.

Since relations are often defined by equations, such as $y = 2x + 3$ and $y^2 = x$, we must sometimes determine the domain of a relation from its equation. We assume the following agreement on the domain of a relation.

Agreement on Domain

The domain of a relation is assumed to be all real numbers that produce real numbers when substituted for the independent variable.

To illustrate this agreement, since any real number can be used as a replacement for x in $y = 2x + 3$, the domain of this function is the set of all real numbers. The function defined by $y = \frac{1}{x}$ has all real numbers except 0 as domain, since y is undefined if $x = 0$. ***In general, the domain of a function defined by an algebraic expression is all real numbers, except those numbers that lead to division by 0 or, as we will see in Chapter 9, an even root of a negative number.***

ANSWERS

4. (a) domain: $\{-3, -2, 2, 3\}$; range: $\{-2, -1, 2, 3\}$
(b) domain: $[-2, \infty)$; range: $(-\infty, \infty)$
(c) domain: $(-\infty, \infty)$; range: $(-\infty, 0]$

OBJECTIVE 3 Identify functions defined by graphs and equations. Since each value of x in a function corresponds to only one value of y, any vertical line drawn through the graph of a function must intersect the graph in at most one point. This is the *vertical line test* for a function.

Vertical Line Test

If every vertical line intersects the graph of a relation in no more than one point, then the relation represents a function.

For example, the graph shown in Figure 50(a) is not the graph of a function since a vertical line intersects the graph in more than one point. The graph in Figure 50(b) does represent a function.

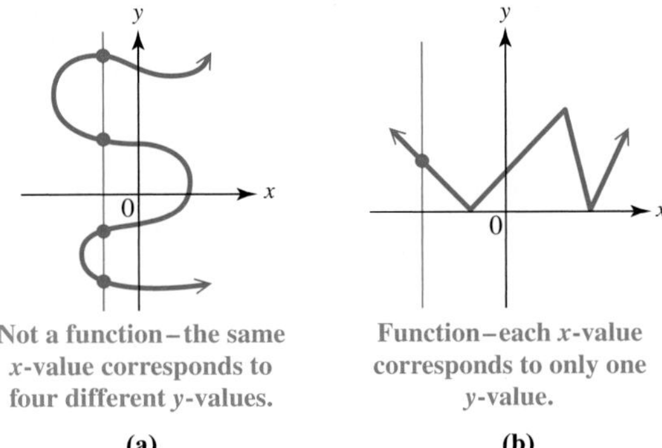

Figure 50

EXAMPLE 4 Using the Vertical Line Test

Use the vertical line test to determine whether each relation graphed in Example 3 is a function.

(a)

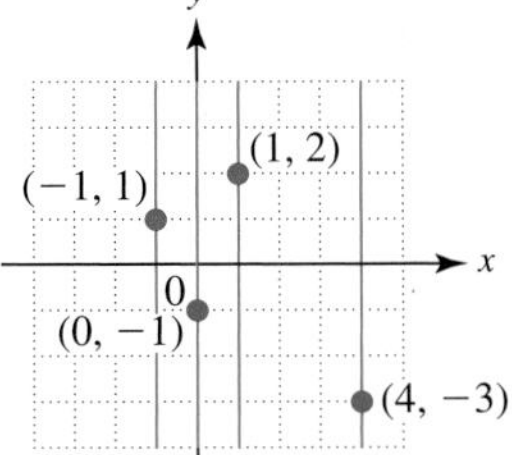

Function

(b)

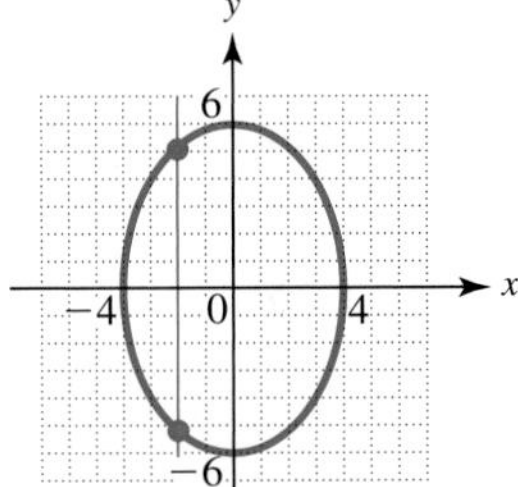

Not a function

(c)

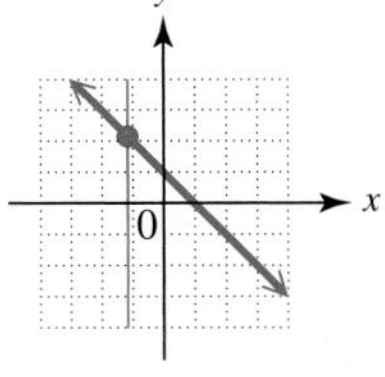

Function

(d)

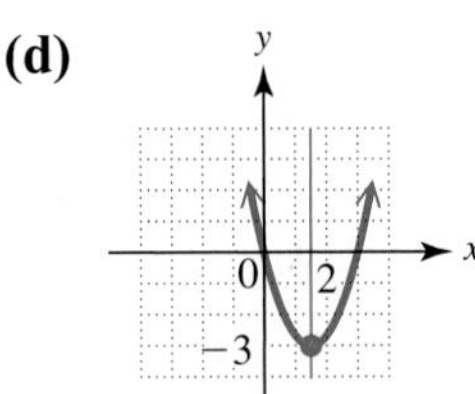

Function

The graphs in (a), (c), and (d) represent functions. The graph of the relation in (b) fails the vertical line test, since the same x-value corresponds to two different y-values; therefore, it is not the graph of a function.

Work Problem  *at the Side.* ▶

5 Use the vertical line test to decide which graphs represent functions.

A.

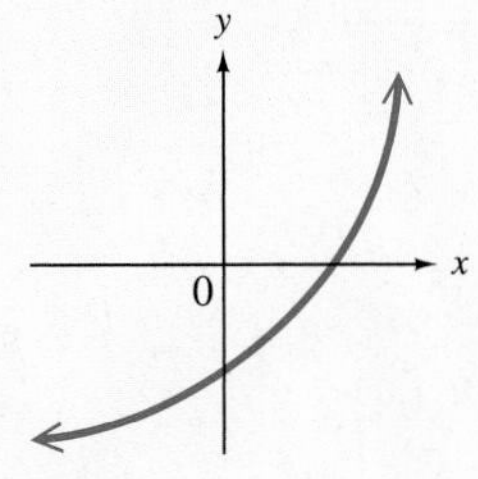

B.

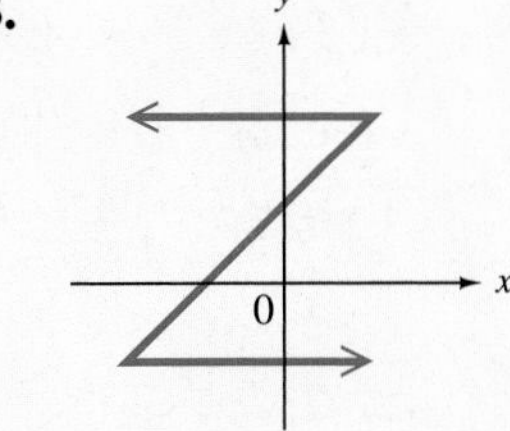

C.

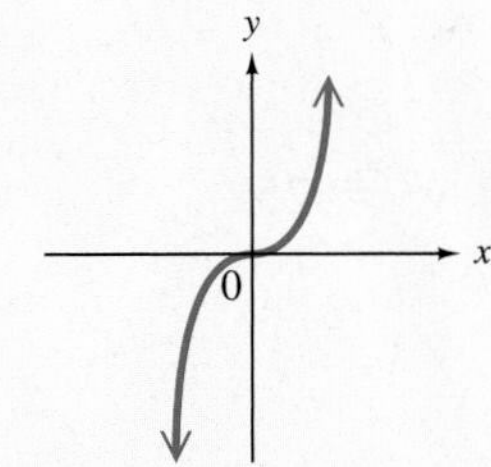

ANSWER

5. A and C are graphs of functions.

6 Decide whether each relation defines a function, and give the domain.

(a) $y = 6x + 12$

(b) $y \le 4x$

(c) $y = \dfrac{1}{3x - 2}$

(d) $y^2 = 25x$

Answers

6. (a) yes; $(-\infty, \infty)$ **(b)** no; $(-\infty, \infty)$

(c) yes; $\left(-\infty, \frac{2}{3}\right) \cup \left(\frac{2}{3}, \infty\right)$

(d) no; $[0, \infty)$

Note

Graphs that do not represent functions are still relations. ***All equations and graphs represent relations and all relations have a domain and range.***

It can be more difficult to decide whether a relation defined by an equation or an inequality is a function. The next example gives some hints.

EXAMPLE 5 Identifying Functions from Their Equations

Decide whether each relation defines a function and give the domain.

(a) $y = x + 4$

In the defining equation, $y = x + 4$, y is always found by adding 4 to x. Thus, each value of x corresponds to just one value of y and the relation defines a function; x can be any real number, so the domain is $(-\infty, \infty)$.

(b) $y^2 = x$

The ordered pairs (16, 4) and (16, −4) both satisfy this equation. Since one value of x, 16, corresponds to two values of y, 4 and −4, this equation does not define a function. Because x is equal to the square of y, the values of x must always be nonnegative. The domain of the relation is $[0, \infty)$.

(c) $y \le x - 1$

By definition, y is a function of x if every value of x leads to exactly one value of y. Here a particular value of x, say 1, corresponds to many values of y. The ordered pairs (1, 0), (1, −1), (1, −2), (1, −3), and so on, all satisfy the inequality. Thus, this relation does not define a function. Any number can be used for x, so the domain is the set of real numbers, $(-\infty, \infty)$.

(d) $y = \dfrac{5}{x - 1}$

Given any value of x in the domain, we find y by subtracting 1, then dividing the result into 5. This process produces exactly one value of y for each value in the domain, so this equation defines a function. The domain includes all real numbers except those that make the denominator 0. We find these numbers by setting the denominator equal to 0 and solving for x.

$$x - 1 = 0$$
$$x = 1 \quad \text{Add 1.}$$

The domain includes all real numbers *except* 1, written $(-\infty, 1) \cup (1, \infty)$.*

◀ Work Problem 6 at the Side.

In summary, three variations of the definition of function are given here.

Variations of the Definition of Function

1. A **function** is a relation in which, for each value of the first component of the ordered pairs, there is exactly one value of the second component.
2. A **function** is a set of ordered pairs in which no first component is repeated.
3. A **function** is a rule or correspondence that assigns exactly one range value to each distinct domain value.

*The **union** of two sets A and B, written $A \cup B$, is the set of all elements of A together with the set of all elements of B.

OBJECTIVE 4 Use function notation. When a function f is defined with a rule or an equation using x and y for the independent and dependent variables, we say "y is a function of x" to emphasize that *y depends on x*. We use the notation

$$y = f(x),$$

called **function notation,** to express this and read $f(x)$ as "f of x." (In this special notation the parentheses do not indicate multiplication.) The letter f stands for *function*. For example, if $y = 9x - 5$, we can name this function f and write

$$f(x) = 9x - 5.$$

Note that ***$f(x)$ is just another name for the dependent variable y.*** For example, if $y = f(x) = 9x - 5$ and $x = 2$, then we find y, or $f(2)$, by replacing x with 2.

$$\begin{aligned} y &= f(2) & & \\ &= 9 \cdot 2 - 5 & & \text{Let } x = 2. \\ &= 18 - 5 & & \text{Multiply.} \\ &= 13. & & \text{Subtract.} \end{aligned}$$

For function f, the statement "if $x = 2$, then $y = 13$" is represented by the ordered pair $(2, 13)$ and is abbreviated with function notation as

$$f(2) = 13.$$

Read $f(2)$ as "f of 2" or "f at 2." Also,

$$f(0) = 9 \cdot 0 - 5 = -5 \quad \text{and} \quad f(-3) = 9(-3) - 5 = -32.$$

These ideas can be illustrated as follows.

Name of the function

Defining expression

$$y = f(x) = 9x - 5$$

Value of the function

Name of the independent variable

CAUTION

The symbol $f(x)$ *does not* indicate "f times x," but represents the y-value for the indicated x-value. As just shown, $f(2)$ is the y-value that corresponds to the x-value 2.

EXAMPLE 6 Using Function Notation

Let $f(x) = -x^2 + 5x - 3$. Find the following.

(a) $f(2)$

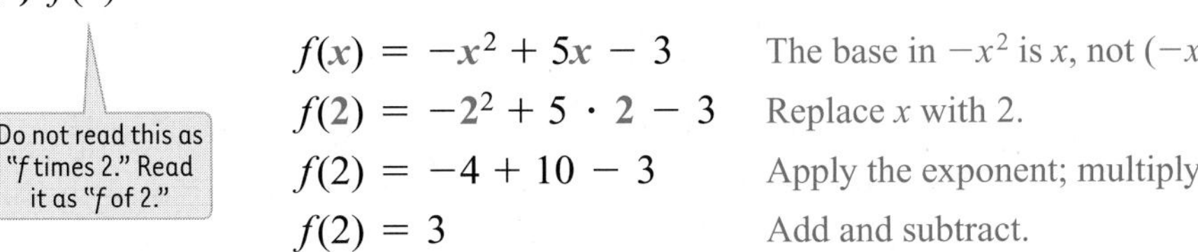

$$\begin{aligned} f(x) &= -x^2 + 5x - 3 & & \text{The base in } -x^2 \text{ is } x, \text{ not } (-x). \\ f(2) &= -2^2 + 5 \cdot 2 - 3 & & \text{Replace } x \text{ with 2.} \\ f(2) &= -4 + 10 - 3 & & \text{Apply the exponent; multiply.} \\ f(2) &= 3 & & \text{Add and subtract.} \end{aligned}$$

Do not read this as "f times 2." Read it as "f of 2."

Since $f(2) = 3$, the ordered pair $(2, 3)$ belongs to f.

Continued on Next Page

7 Find $f(-3)$, $f(p)$, and $f(m+1)$.

(a) $f(x) = 6x - 2$

(b) $f(x) = \dfrac{-3x + 5}{2}$

(c) $f(x) = \dfrac{1}{6}x - 1$

8 For each function, find $f(-2)$.

(a) $f(x) = -4x - 8$

(b) $f = \{(0, 5), (-1, 3), (-2, 1)\}$

(c)

x	$f(x)$
-4	16
-2	4
0	0
2	4
4	16

Answers

7. (a) -20; $6p - 2$; $6m + 4$

(b) 7; $\dfrac{-3p + 5}{2}$; $\dfrac{-3m + 2}{2}$

(c) $-\dfrac{3}{2}$; $\dfrac{1}{6}p - 1$; $\dfrac{1}{6}(m + 1) - 1$, or $\dfrac{1}{6}m - \dfrac{5}{6}$

8. (a) 0 **(b)** 1 **(c)** 4

(b) $f(q)$

$$f(x) = -x^2 + 5x - 3 \quad \text{Given function}$$
$$f(q) = -q^2 + 5q - 3 \quad \text{Replace } x \text{ with } q.$$

The replacement of one variable with another is important in later courses.

Sometimes letters other than f, such as g, h, or capital letters F, G, and H, are used to name functions.

EXAMPLE 7 Using Function Notation

Let $g(x) = 2x + 3$. Find and simplify $g(a + 1)$.

$$g(x) = 2x + 3$$
$$g(a + 1) = 2(a + 1) + 3 \quad \text{Replace } x \text{ with } a + 1.$$
$$g(a + 1) = 2a + 2 + 3 \quad \text{Distributive property}$$
$$g(a + 1) = 2a + 5 \quad \text{Add.}$$

◀ Work Problem **7** at the Side.

Functions can be evaluated in a variety of ways, as shown in Example 8.

EXAMPLE 8 Using Function Notation

For each function, find $f(3)$.

(a) $f(x) = 3x - 7$

$f(3) = 3(3) - 7$

$f(3) = 9 - 7$

$f(3) = 2$

(b) $f = \{(-3, 5), (0, 3), (3, 1), (6, -1)\}$

We want $f(3)$, the y-value of the ordered pair where $x = 3$. As indicated by the ordered pair $(3, 1)$, when $x = 3$, $y = 1$, so $f(3) = 1$.

(c)

Domain	f	Range
-2	→	6
3	→	5
10	→	12

The domain element 3 is paired with 5 in the range, so $f(3) = 5$.

(d)

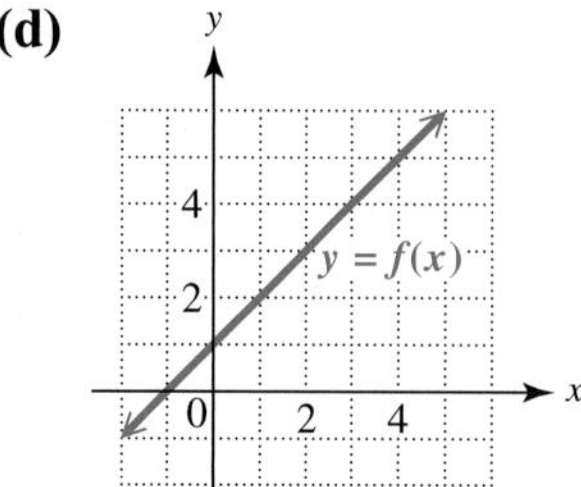

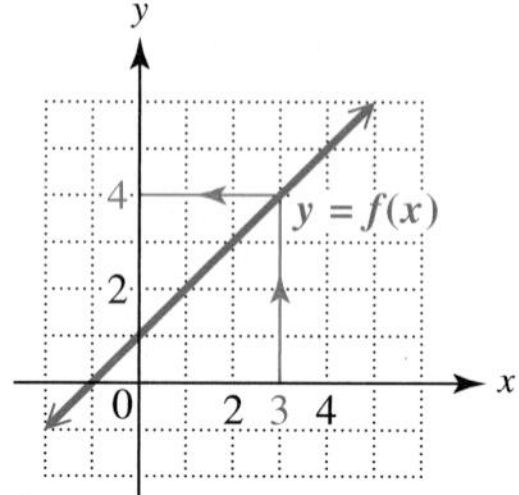

Figure 51

To evaluate $f(3)$, find 3 on the x-axis. See Figure 51. Then move up until the graph of f is reached. Moving horizontally to the y-axis gives 4 for the corresponding y-value. Thus, $f(3) = 4$.

◀ Work Problem **8** at the Side.

If a function f is defined by an equation with x and y instead of with function notation, use the following steps to find $f(x)$.

Writing an Equation Using Function Notation

Step 1 Solve the equation for y.

Step 2 Replace y with $f(x)$.

EXAMPLE 9 Writing Equations Using Function Notation

Rewrite each equation using function notation $f(x)$. Then find $f(-2)$ and $f(a)$.

(a) $y = x^2 + 1$

This equation is already solved for y. Since $y = f(x)$,

$$f(x) = x^2 + 1.$$

To find $f(-2)$, let $x = -2$.

$$f(-2) = (-2)^2 + 1 \quad \text{Let } x = -2.$$
$$f(-2) = 4 + 1 \quad \text{Apply the exponent.}$$
$$f(-2) = 5 \quad \text{Add.}$$

Find $f(a)$ by letting $x = a$: $f(a) = a^2 + 1$.

(b) $x - 4y = 5$

First solve $x - 4y = 5$ for y. Then replace y with $f(x)$.

$$x - 4y = 5$$
$$x - 5 = 4y \quad \text{Add } 4y\text{; subtract 5.}$$
$$y = \frac{x - 5}{4}, \quad \text{so} \quad f(x) = \frac{1}{4}x - \frac{5}{4}$$

Now find $f(-2)$ and $f(a)$.

$$f(-2) = \frac{1}{4}(-2) - \frac{5}{4} = -\frac{7}{4} \quad \text{Let } x = -2.$$
$$f(a) = \frac{1}{4}a - \frac{5}{4} \quad \text{Let } x = a.$$

Work Problem 9 at the Side. ▶

OBJECTIVE 5 Graph linear and constant functions. Our first two-dimensional graphing was of straight lines. Linear equations (except for vertical lines with equations $x = a$) define *linear functions*.

Linear Function

A function that can be defined by

$$\boldsymbol{f(x) = ax + b,}$$

for real numbers a and b, is a **linear function.** The value of a is the slope m of the graph of the function.

A linear function defined by $\boldsymbol{f(x) = b}$ (whose graph is a horizontal line) is sometimes called a **constant function.** The domain of any linear function is $(-\infty, \infty)$. The range of a nonconstant linear function is $(-\infty, \infty)$, while the range of the constant function defined by $f(x) = b$ is $\{b\}$.

9 Rewrite each equation using function notation $f(x)$. Then find $f(-1)$.

(a) $y = \sqrt{x + 2}$

(b) $x^2 - 4y = 3$

ANSWERS

9. (a) $f(x) = \sqrt{x + 2}$; 1

(b) $f(x) = \frac{x^2 - 3}{4}$, or $f(x) = \frac{1}{4}x^2 - \frac{3}{4}$; $-\frac{1}{2}$

10 Graph each linear function. Give the domain and range.

(a) $f(x) = \frac{3}{4}x - 2$

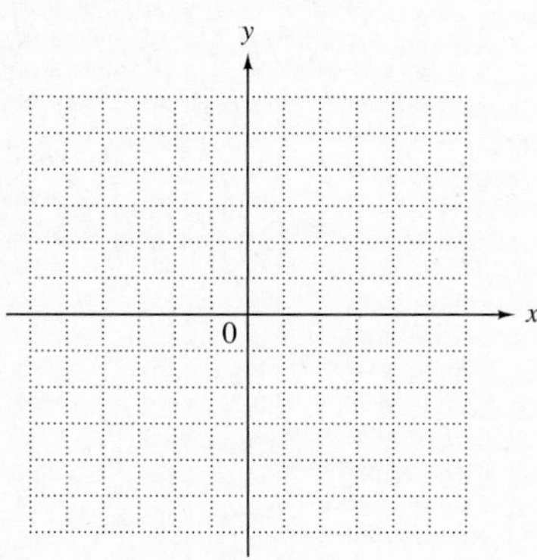

(b) $g(x) = 3$

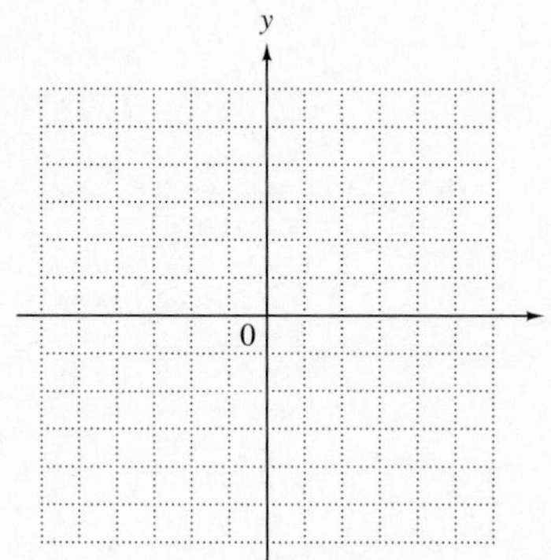

EXAMPLE 10 Graphing Linear and Constant Functions

Graph each function. Give the domain and range.

(a) $f(x) = \frac{1}{4}x - \frac{5}{4}$

Recall from **Section 3.4** that m is the slope of the line and $(0, b)$ is the y-intercept. In Example 9(b), we wrote the equation $x - 4y = 5$ as the linear function defined by

$$f(x) = \frac{1}{4}x - \frac{5}{4}.$$

Slope ↑ (the $\frac{1}{4}$); y-intercept is $(0, -\frac{5}{4})$.

To graph this function, plot the y-intercept $(0, -\frac{5}{4})$ and use the definition of slope as $\frac{\text{rise}}{\text{run}}$ to find a second point on the line. Since the slope is $\frac{1}{4}$, move 1 unit up from $(0, -\frac{5}{4})$ and 4 units to the right to the point $(4, -\frac{1}{4})$. Draw the straight line through these points to obtain the graph shown in Figure 52. The domain and range are both $(-\infty, \infty)$.

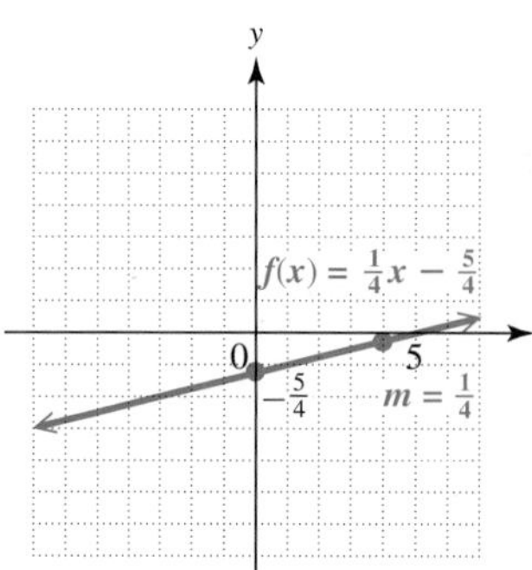

Figure 52

Figure 53

(b) $f(x) = 4$

This is a constant function. Its graph is the horizontal line containing all points with y-coordinate equal to 4. See Figure 53. The domain is $(-\infty, \infty)$ and the range is $\{4\}$.

◀ *Work Problem* **10** *at the Side.*

Answers

10. (a)

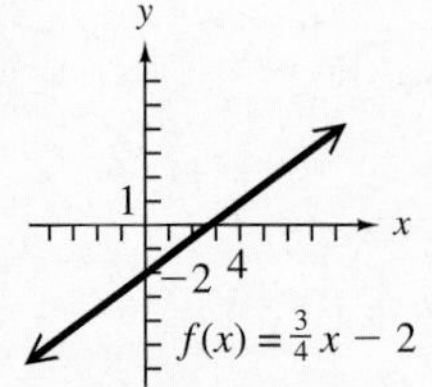

domain: $(-\infty, \infty)$; range: $(-\infty, \infty)$

(b)

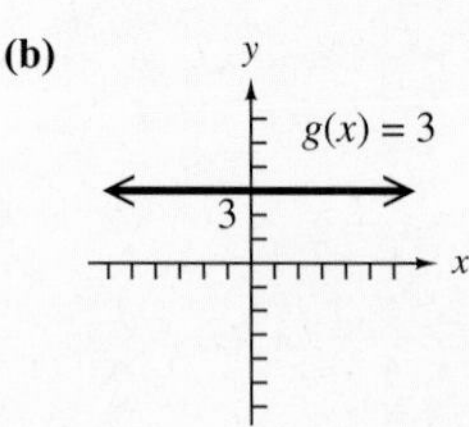

domain: $(-\infty, \infty)$; range: $\{3\}$

3.6 Exercises

FOR EXTRA HELP MyMathLab Math XL PRACTICE WATCH DOWNLOAD READ REVIEW

1. In an ordered pair of a relation, is the first element the independent or the dependent variable?

   independent variable

2. Give an example of a relation that is not a function, having domain $\{-3, 2, 6\}$ and range $\{4, 6\}$. (There are many possible correct answers.)

   One example is $\{(-3, 4), (2, 4), (2, 6), (6, 4)\}$.

3. Explain what is meant by each term.
   (a) Relation **(b)** Domain of a relation
   (c) Range of a relation **(d)** Function

   (a) A relation is a set of ordered pairs.
   (b) The domain is the set of all first components (x-values).
   (c) The range is the set of all second components (y-values).
   (d) A function is a relation in which each domain element is paired with one and only one range element.

4. Describe the use of the vertical line test.

   The vertical line test is used to determine whether a graph is that of a function. Any vertical line will intersect the graph of a function in at most one point.

Decide whether each relation is a function, and give the domain and the range. Use the vertical line test in Exercises 17–22. See Examples 1–4.

5. $\{(5, 1), (3, 2), (4, 9), (7, 3)\}$

   function; domain: $\{5, 3, 4, 7\}$; range: $\{1, 2, 9, 3\}$

6. $\{(8, 0), (5, 4), (9, 3), (3, 9)\}$

   function; domain: $\{8, 5, 9, 3\}$; range: $\{0, 4, 3, 9\}$

7. $\{(2, 4), (0, 2), (2, 6)\}$

   not a function; domain: $\{2, 0\}$; range: $\{4, 2, 6\}$

8. $\{(9, -2), (-3, 5), (9, 1)\}$

   not a function; domain: $\{9, -3\}$; range: $\{-2, 5, 1\}$

9. $\{(-3, 1), (4, 1), (-2, 7)\}$

   function; domain: $\{-3, 4, -2\}$; range: $\{1, 7\}$

10. $\{(-12, 5), (-10, 3), (8, 3)\}$

    function; domain: $\{-12, -10, 8\}$; range: $\{5, 3\}$

11. $\{(1, 1), (1, -1), (0, 0), (2, 4), (2, -4)\}$

    not a function; domain: $\{1, 0, 2\}$; range: $\{1, -1, 0, 4, -4\}$

12. $\{(2, 5), (3, 7), (4, 9), (5, 11)\}$

    function; domain: $\{2, 3, 4, 5\}$; range: $\{5, 7, 9, 11\}$

13. 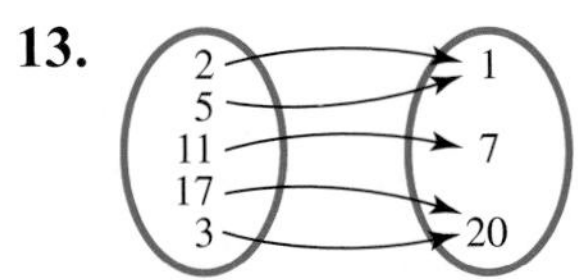

    function; domain: $\{2, 5, 11, 17, 3\}$; range: $\{1, 7, 20\}$

14. 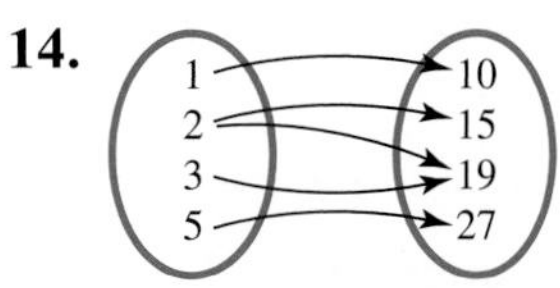

    not a function; domain: $\{1, 2, 3, 5\}$; range: $\{10, 15, 19, 27\}$

15.

x	y
1	5
1	2
1	−1
1	−4

not a function;
domain: {1};
range: {5, 2, −1, −4}

16.

x	y
4	−3
2	−3
0	−3
−2	−3

function;
domain: {4, 2, 0, −2};
range: {−3}

17.

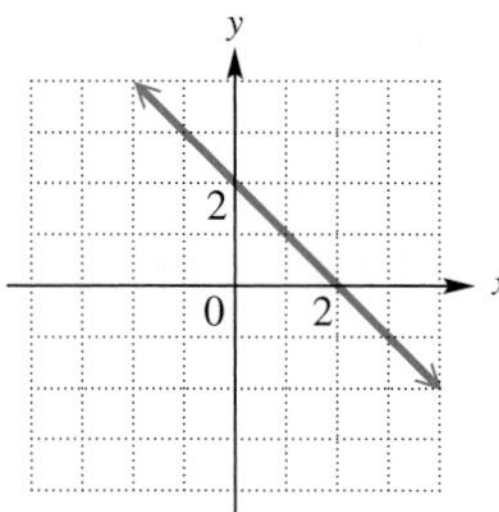

function;
domain: $(-\infty, \infty)$;
range: $(-\infty, \infty)$

18.

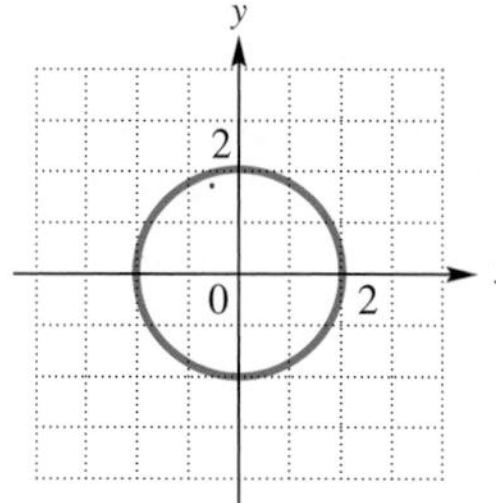

not a function;
domain: $[-2, 2]$;
range: $[-2, 2]$

19.

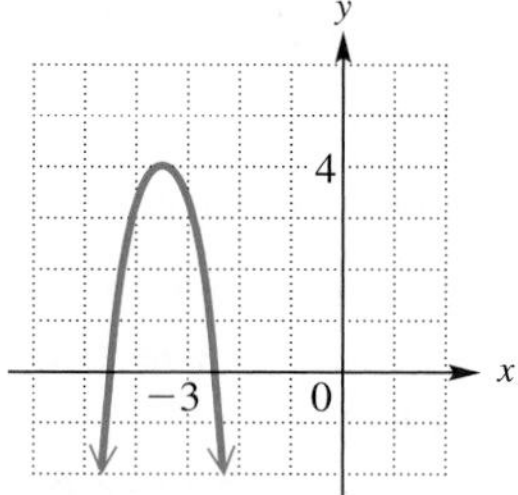

function;
domain: $(-\infty, \infty)$;
range: $(-\infty, 4]$

20.

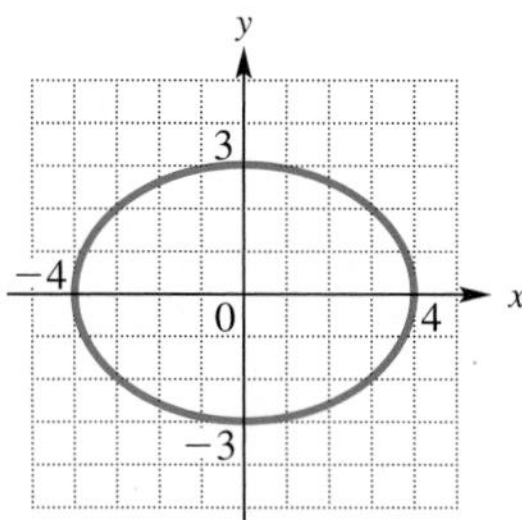

not a function;
domain: $[-4, 4]$;
range: $[-3, 3]$

21.

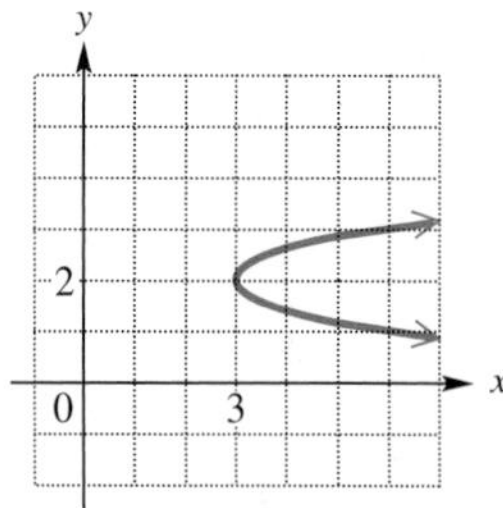

not a function;
domain: $[3, \infty)$;
range: $(-\infty, \infty)$

22.

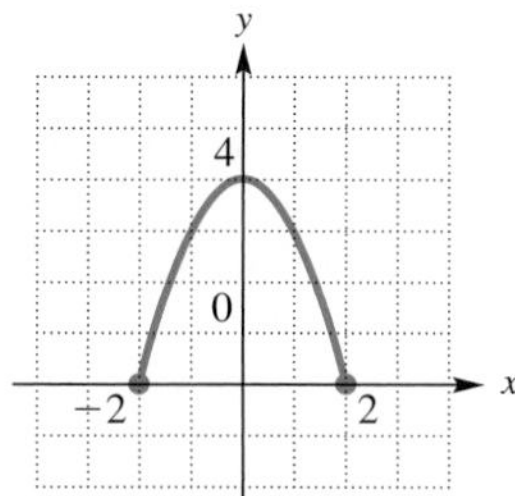

function;
domain: $[-2, 2]$;
range: $[0, 4]$

Decide whether each relation defines y as a function of x. Give the domain. See Example 5.

23. $y = x^2$
function;
domain: $(-\infty, \infty)$

24. $y = x^3$
function;
domain: $(-\infty, \infty)$

25. $x = y^6$
not a function;
domain: $[0, \infty)$

26. $x = y^4$
not a function;
domain: $[0, \infty)$

27. $y = 2x - 6$
function;
domain: $(-\infty, \infty)$

28. $y = -6x + 8$
function;
domain: $(-\infty, \infty)$

29. $x + y < 4$
not a function;
domain: $(-\infty, \infty)$

30. $x - y < 3$
not a function;
domain: $(-\infty, \infty)$

31. $y = x$

function; domain: $(-\infty, \infty)$

32. $y = -x$

function; domain: $(-\infty, \infty)$

33. $xy = 1$

function; domain: $(-\infty, 0) \cup (0, \infty)$

34. $xy = -3$

function; domain: $(-\infty, 0) \cup (0, \infty)$

35. $y = \dfrac{2}{x - 9}$

function; domain: $(-\infty, 9) \cup (9, \infty)$

36. $y = \dfrac{-7}{x - 16}$

function; domain: $(-\infty, 16) \cup (16, \infty)$

37. $y = \dfrac{1}{4x + 2}$

function; domain: $\left(-\infty, -\frac{1}{2}\right) \cup \left(-\frac{1}{2}, \infty\right)$

38. $y = \dfrac{1}{9 - 2x}$

function; domain: $\left(-\infty, \frac{9}{2}\right) \cup \left(\frac{9}{2}, \infty\right)$

39. Refer to the graph to answer the questions.

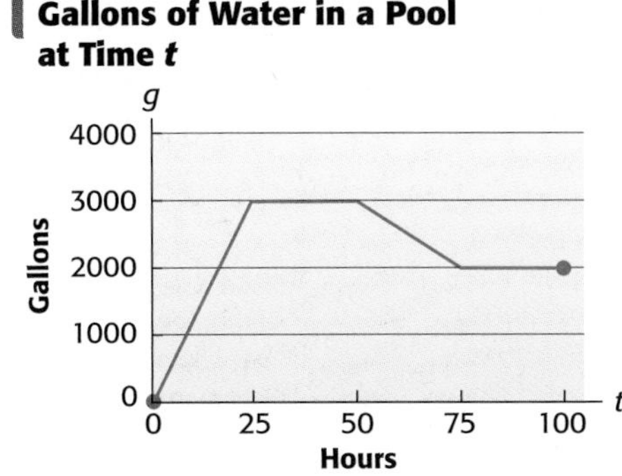

(a) What numbers are possible values of the dependent variable? **[0, 3000]**

(b) For how long is the water level increasing? decreasing? **25 hr; 25 hr**

(c) How many gallons are in the pool after 90 hr? **2000 gal**

(d) Call this function g. What is $g(0)$? What does it mean in this example? **$g(0) = 0$; The pool is empty at time 0.**

40. The graph shows the daily megawatts of electricity used on a record-breaking summer day in Sacramento, California.

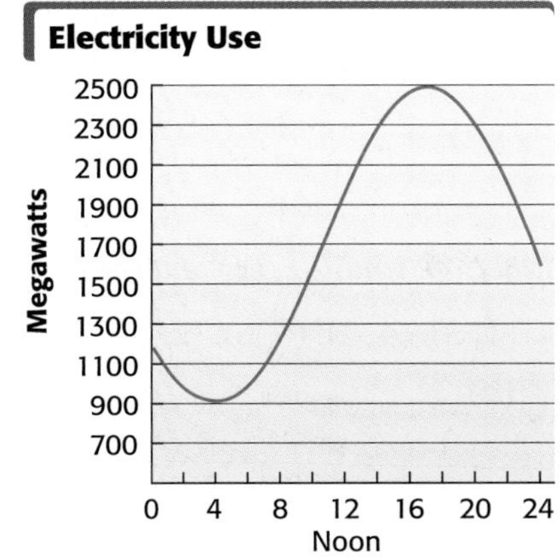

Source: Sacramento Municipal Utility District.

(a) Is this the graph of a function? **yes**

(b) What is the domain? **[0, 24]**

(c) Estimate the number of megawatts used at 8 A.M. **1200 megawatts**

(d) At what time was the most electricity used? the least electricity? **at 17 hr or 5 P.M.; at 4 hr or 4 A.M.**

41. Give an example of a function from everyday life. (*Hint:* Fill in the blanks: _______ depends on _______, so _______ is a function of _______.)

Here is one example. The cost of gasoline; number of gallons purchased; cost; number of gallons

42. Choose the correct response: The notation $f(3)$ means

A. the variable f times 3 or $3f$

B. the value of the dependent variable when the independent variable is 3

C. the value of the independent variable when the dependent variable is 3

D. f equals 3.

B

Let $f(x) = -3x + 4$ *and* $g(x) = -x^2 + 4x + 1$. *Find the following. See Examples 6 and 7.*

43. $f(0)$
4

44. $f(-3)$
13

45. $g(-2)$
−11

46. $g(10)$
−59

47. $f(p)$
$-3p + 4$

48. $g(k)$
$-k^2 + 4k + 1$

49. $f(-x)$
$3x + 4$

50. $g(-x)$
$-x^2 - 4x + 1$

51. $f(x + 2)$
$-3x - 2$

52. $g\left(-\frac{1}{x}\right)$
$-\frac{1}{x^2} - \frac{4}{x} + 1$

53. $g\left(\frac{p}{3}\right)$
$-\frac{p^2}{9} + \frac{4p}{3} + 1$

54. $f(3t - 2)$
$-9t + 10$

For each function, find ***(a)*** $f(2)$ *and* ***(b)*** $f(-1)$. *See Example 8.*

55. $f = \{(-1, 3), (4, 7), (0, 6), (2, 2)\}$
(a) 2 **(b)** 3

56. $f = \{(2, 5), (3, 9), (-1, 11), (5, 3)\}$
(a) 5 **(b)** 11

57.

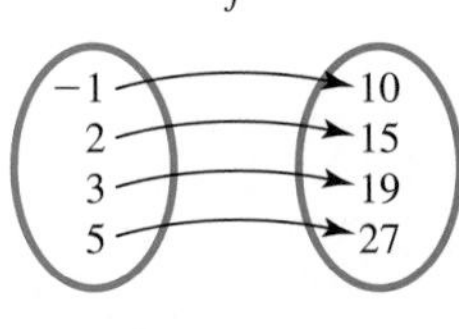

(a) 15 **(b)** 10

58.

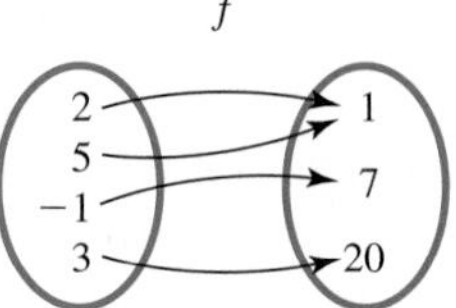

(a) 1 **(b)** 7

59.

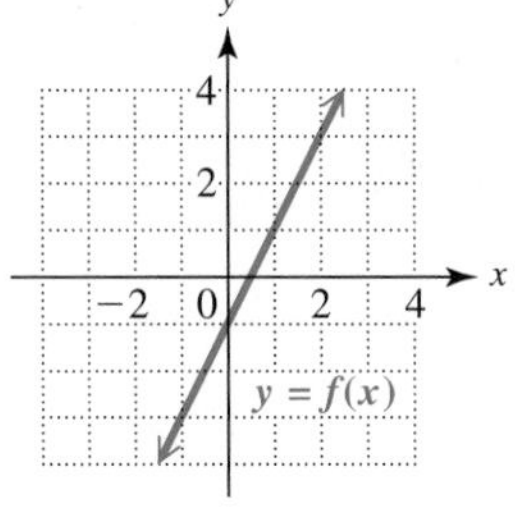

(a) 3 **(b)** −3

60.

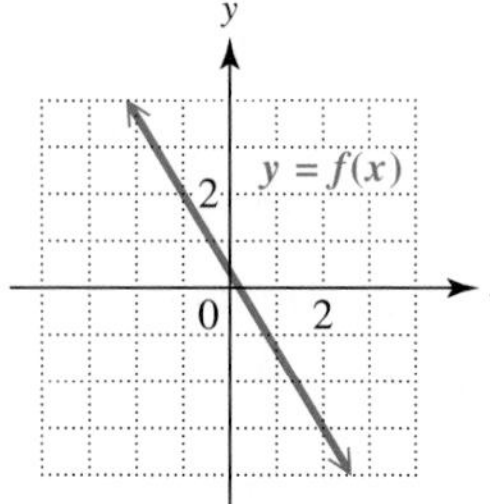

(a) −3 **(b)** 2

61. Fill in each blank with the correct response.

The equation $2x + y = 4$ has a straight **line** as its graph. One point that lies on the graph is (3, **−2**). If we solve the equation for y and use function notation, we have a **linear** function defined by $f(x) =$ **−2x + 4**. For this function, $f(3) =$ **−2**, meaning that the point (**3**, **−2**) lies on the graph of the function.

62. Which of the following defines a linear function?

A. $y = \frac{2}{5}x - 3$ **B.** $y = \frac{1}{x}$

C. $y = x^2$ **D.** $x = 1$

A

An equation that defines y as a function f of x is given. ***(a)*** *Solve for y in terms of x, and replace y with the function notation f(x).* ***(b)*** *Find f(3). See Example 9.*

63. $x + 3y = 12$

(a) $f(x) = -\frac{1}{3}x + 4$ **(b)** 3

64. $x - 4y = 8$

(a) $f(x) = \frac{1}{4}x - 2$ **(b)** $-\frac{5}{4}$

65. $y + 2x^2 = 3$

(a) $f(x) = 3 - 2x^2$ **(b)** −15

66. $y - 3x^2 = 2$

(a) $f(x) = 2 + 3x^2$ **(b)** 29

67. $4x - 3y = 8$

(a) $f(x) = \frac{4}{3}x - \frac{8}{3}$ **(b)** $\frac{4}{3}$

68. $-2x + 5y = 9$

(a) $f(x) = \frac{2}{5}x + \frac{9}{5}$ **(b)** 3

Graph each linear or constant function. Give the domain and range. See Example 10.

69. $f(x) = -2x + 5$

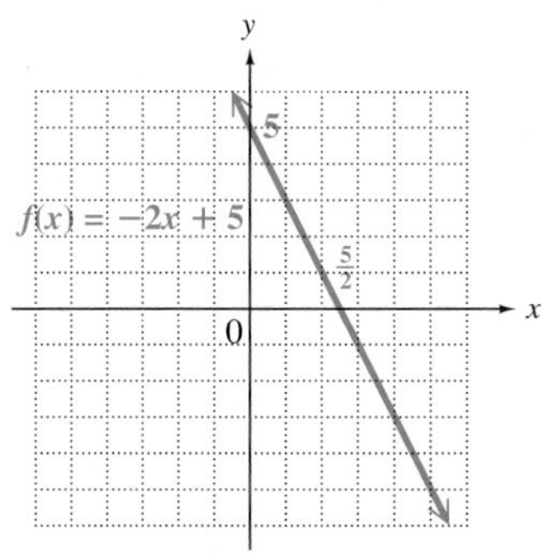

domain: $(-\infty, \infty)$; **range:** $(-\infty, \infty)$

70. $g(x) = 4x - 1$

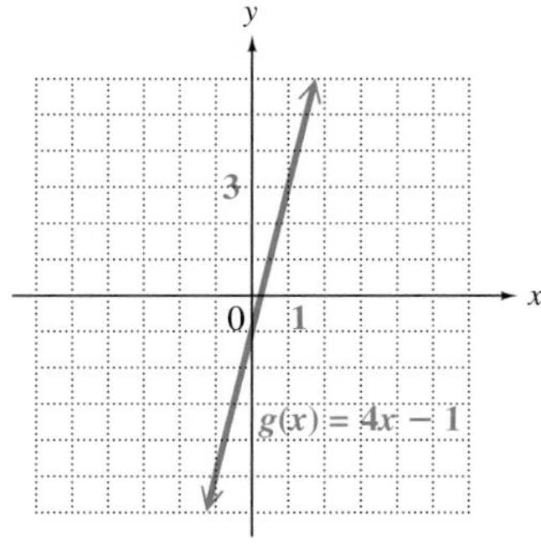

domain: $(-\infty, \infty)$; **range:** $(-\infty, \infty)$

71. $h(x) = \frac{1}{2}x + 2$

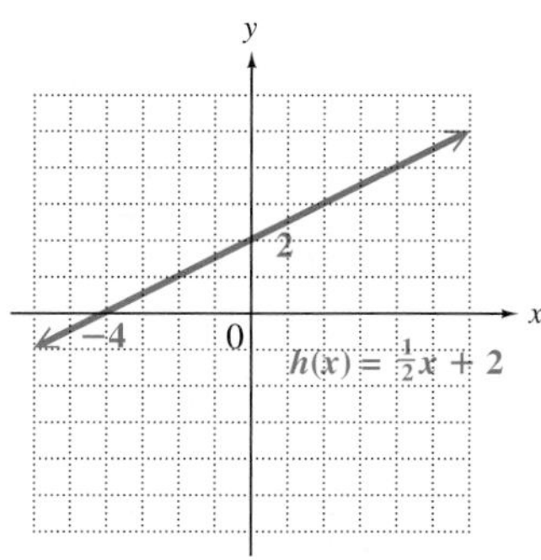

domain: $(-\infty, \infty)$; **range:** $(-\infty, \infty)$

72. $F(x) = -\frac{1}{4}x + 1$

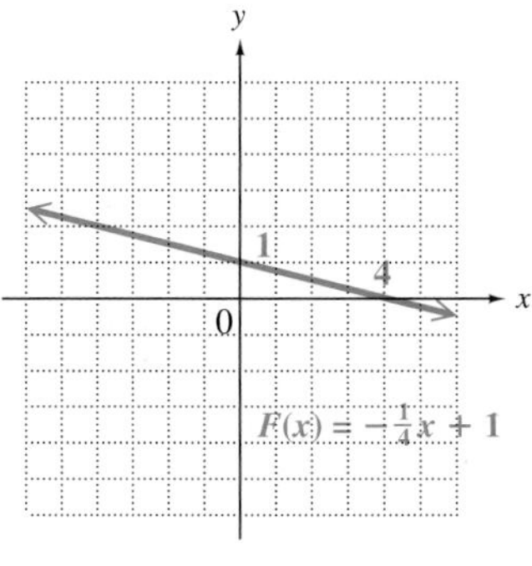

domain: $(-\infty, \infty)$; **range:** $(-\infty, \infty)$

73. $g(x) = -4$

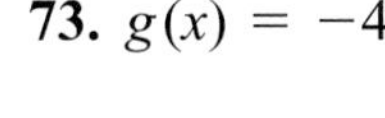
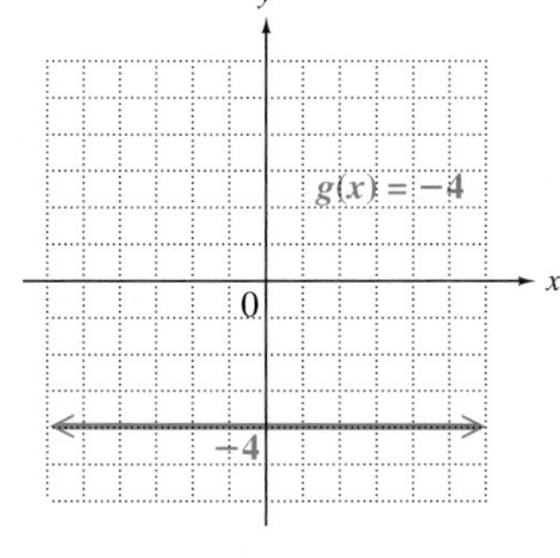

domain: $(-\infty, \infty)$; **range:** $\{-4\}$

74. $f(x) = 5$

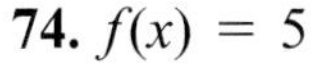
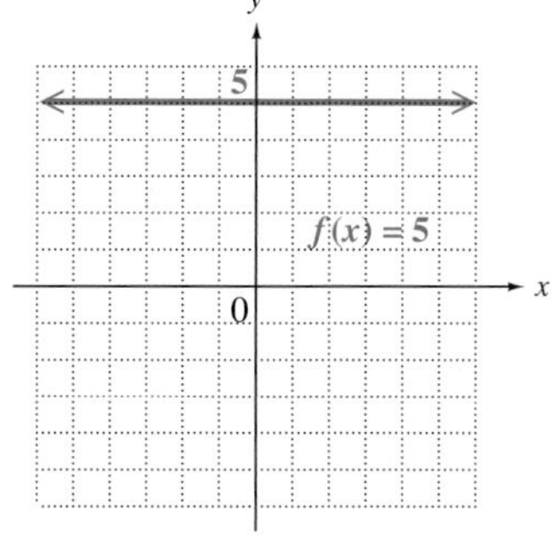

domain: $(-\infty, \infty)$; **range:** $\{5\}$

Solve each problem.

75. Suppose that a taxicab driver charges \$2.50 per mi.

(a) Fill in the table with the correct response for the price $f(x)$ he charges for a trip of x miles.

x	$f(x)$
0	**\$0**
1	**\$2.50**
2	**\$5.00**
3	**\$7.50**

(b) The linear function that gives a rule for the amount charged is $f(x) =$ **2.50x**.

(c) Graph this function for the domain $\{0, 1, 2, 3\}$.

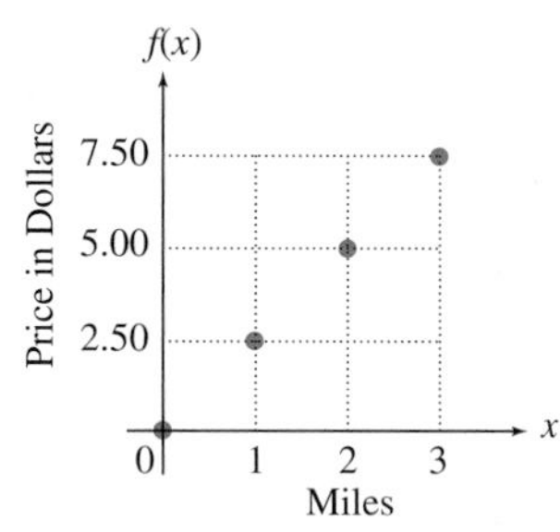

76. Suppose that a package weighing x pounds costs $f(x)$ dollars to ship to a given location, where $f(x) = 3.75x$.

(a) What is the value of $f(3)$?

11.25 (dollars)

(b) In your own words, describe what 3 and the value $f(3)$ mean in part (a), using the terms *independent variable* and *dependent variable.*

3 is the value of the independent variable, which represents a package weight of 3 lb; $f(3) = \$11.25$ is the value of the dependent variable representing the cost to mail a 3-lb package.

(c) How much would it cost to mail a 5-lb package? Write the answer using function notation.

\$18.75; $f(5) = 18.75$

Forensic scientists use the lengths of certain bones to calculate the height of a person. Two bones often used are the tibia (t), the bone from the ankle to the knee, and the femur (r), the bone from the knee to the hip socket. A person's height (h) is determined from the lengths of these bones using functions defined by the following formulas. All measurements are in centimeters.

Functions for men: $h(r) = 69.09 + 2.24r$ or $h(t) = 81.69 + 2.39t$

Functions for women: $h(r) = 61.41 + 2.32r$ or $h(t) = 72.57 + 2.53t$

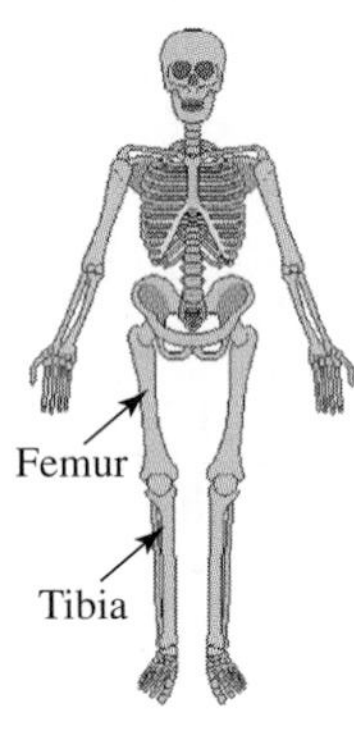

77. Find the height of a man with a femur measuring 56 cm.

194.53 cm

78. Find the height of a man with a tibia measuring 40 cm.

177.29 cm

79. Find the height of a woman with a femur measuring 50 cm.

177.41 cm

80. Find the height of a woman with a tibia measuring 36 cm.

163.65 cm

Chapter 3 Summary

▶ Key Terms

3.1

bar graph A bar graph is a series of bars used to show comparisons between two categories of data.

line graph A line graph consists of a series of points that are connected with line segments and is used to show changes or trends in data.

linear equation in two variables An equation that can be written in the form $Ax + By = C$ is a linear equation in two variables. (A and B are real numbers that cannot both be 0.)

ordered pair A pair of numbers written between parentheses in which order is important is called an ordered pair.

table of values A table showing selected ordered pairs of numbers that satisfy an equation is called a table of values.

x	y
0	4
2	0
1	2

Table of values for $2x + y = 4$

x-axis The horizontal axis in a coordinate system is called the x-axis.

y-axis The vertical axis in a coordinate system is called the y-axis.

rectangular (Cartesian) coordinate system An x-axis and y-axis at right angles form a coordinate system.

Quadrant II | Quadrant I | Quadrant III | Quadrant IV | y-axis | x-axis | Origin

quadrants A coordinate system divides the plane into four regions called quadrants.

origin The point at which the x-axis and y-axis intersect is called the origin.

plane A flat surface determined by two intersecting lines is a plane.

coordinates The numbers in an ordered pair are called the coordinates of the corresponding point.

plot To plot an ordered pair is to find the corresponding point on a coordinate system.

scatter diagram A graph of ordered pairs of data is a scatter diagram.

3.2

graph The graph of an equation is the set of all points that correspond to the ordered pairs that satisfy the equation.

graphing The process of plotting the ordered pairs that satisfy a linear equation and drawing a line through them is called graphing.

y-intercept If a graph intersects the y-axis at k, then the y-intercept is $(0, k)$.

x-intercept If a graph intersects the x-axis at k, then the x-intercept is $(k, 0)$.

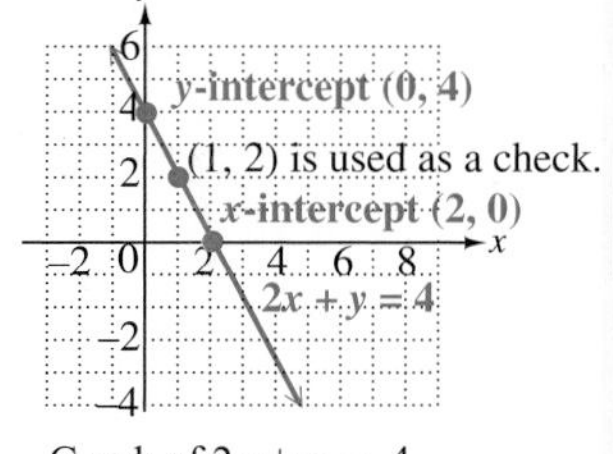

Graph of $2x + y = 4$

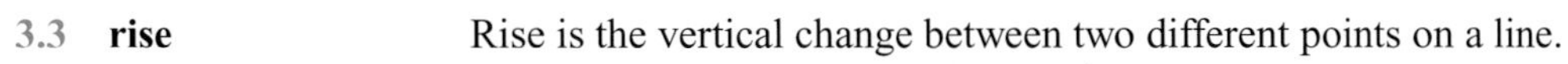

3.3

rise Rise is the vertical change between two different points on a line.

run Run is the horizontal change between two different points on a line.

slope The slope of a line is the ratio of the change in y compared to the change in x when moving along the line from one point to another.

parallel lines Two lines in a plane that never intersect are parallel.

perpendicular lines Perpendicular lines intersect at a 90° angle.

3.5

linear inequality in two variables An inequality that can be written in the form $Ax + By < C$, $Ax + By > C$, $Ax + By \leq C$, or $Ax + By \geq C$ is a linear inequality in two variables.

boundary line In the graph of a linear inequality, the boundary line separates the region that satisfies the inequality from the region that does not satisfy the inequality.

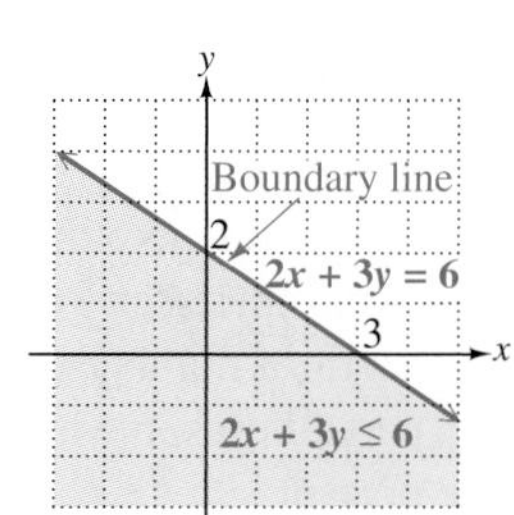

3.6

dependent variable If the quantity y depends on x, then y is called the dependent variable in a relation between x and y.

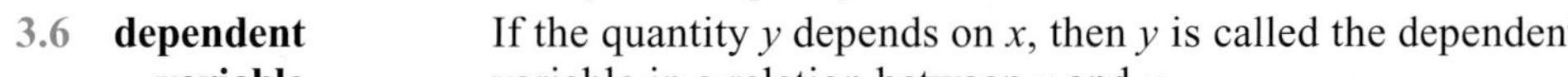

(*continued*)

▸ Key Terms (continued)

independent variable If y depends on x, then x is the independent variable in a relation between x and y.

relation A relation is a set of ordered pairs of real numbers.

function A function is a set of ordered pairs in which each value of the first component, x, corresponds to exactly one value of the second component, y.

domain The domain of a relation is the set of first components (x-values) of the ordered pairs of the relation.

range The range of a relation is the set of second components (y-values) of the ordered pairs of the relation.

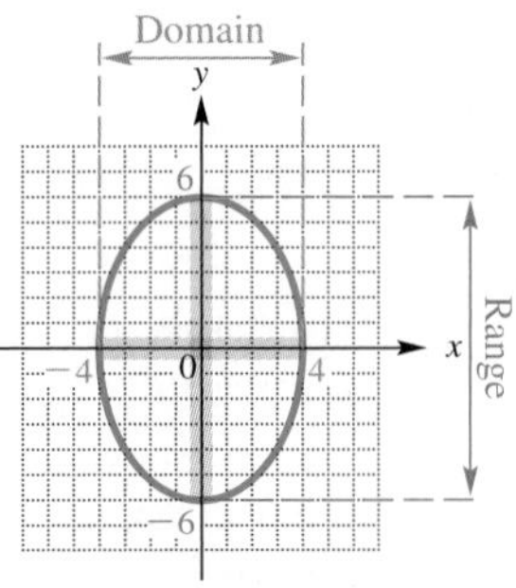

Graph of a relation

function notation The function notation $f(x)$ is another way to represent the dependent variable y for the function f.

linear function A function that is defined by $f(x) = mx + b$ is a linear function.

constant function A constant function is a linear function of the form $f(x) = b$, for a real number b.

▸ New Symbols

(x, y)	ordered pair	(x_1, y_1)	subscript notation; x-sub-one, y-sub-one	m	slope
				$f(x)$	function f of x (read "f of x")

▸ Test Your Word Power

See how well you have learned the vocabulary in this chapter. Answers, with examples, follow the Quick Review.

1. An **ordered pair** is a pair of numbers written
   - **A.** in numerical order between brackets
   - **B.** between parentheses or brackets
   - **C.** between parentheses in which order is important
   - **D.** between parentheses in which order does not matter.

2. A **linear equation in two variables** is an equation that can be written in the form
   - **A.** $Ax + By < C$
   - **B.** $ax = b$
   - **C.** $y = x^2$
   - **D.** $Ax + By = C$.

3. An **intercept** is
   - **A.** the point where the x-axis and y-axis intersect
   - **B.** a pair of numbers written in parentheses in which order is important
   - **C.** one of the four regions determined by a rectangular coordinate system
   - **D.** the point where a graph intersects the x-axis or the y-axis.

4. The **slope** of a line is
   - **A.** the measure of the run over the rise of the line
   - **B.** the distance between two points on the line
   - **C.** the ratio of the change in y to the change in x along the line
   - **D.** the horizontal change compared to the vertical change of two points on the line.

5. A **relation** is
   - **A.** a set of ordered pairs
   - **B.** the ratio of the change in y to the change in x along a line
   - **C.** the set of all possible values of the independent variable
   - **D.** all the second components of a set of ordered pairs.

6. A **function** is
   - **A.** the numbers in an ordered pair
   - **B.** a set of ordered pairs in which each x-value corresponds to exactly one y-value
   - **C.** a pair of numbers written between parentheses in which order matters
   - **D.** the set of all ordered pairs that satisfy an equation.

7. The **domain** of a function is
   - **A.** the set of all possible values of the dependent variable y
   - **B.** a set of ordered pairs
   - **C.** the difference between the x-values
   - **D.** the set of all possible values of the independent variable x.

8. The **range** of a function is
   - **A.** the set of all possible values of the dependent variable y
   - **B.** a set of ordered pairs
   - **C.** the difference between the y-values
   - **D.** the set of all possible values of the independent variable x.

▶ Quick Review

Concepts	Examples

3.1 Reading Graphs; Linear Equations in Two Variables

Bar graphs and line graphs are ways to "picture," or represent, the relationship between two variables.

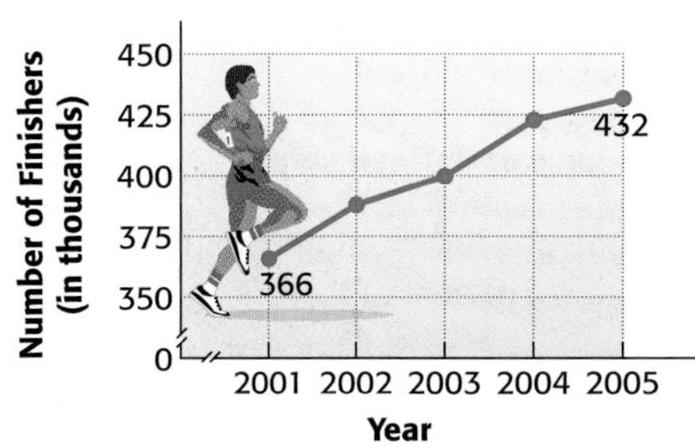

Source: Running USA.

The line graph illustrates the number of U.S. runners in thousands who finished marathons in the years 2001–2005.

An ordered pair is a solution of an equation if it makes the equation a true statement.

Is $(2, -5)$ or $(0, -6)$ a solution of $4x - 3y = 18$?

$4(2) - 3(-5) = 23 \neq 18$ | $4(0) - 3(-6) = 18$

$(2, -5)$ is not a solution. | $(0, -6)$ is a solution.

If a value of either variable in an equation is given, the value of the other variable can be found by substitution.

Complete the ordered pair $(0, \_\_)$ for $3x = y + 4$.

$3(0) = y + 4$ Let $x = 0$.

$0 = y + 4$ Multiply.

$-4 = y$ Subtract 4.

The ordered pair is $(0, -4)$.

To plot the ordered pair $(-3, 4)$, start at the origin, go 3 units to the left, and from there go 4 units up.

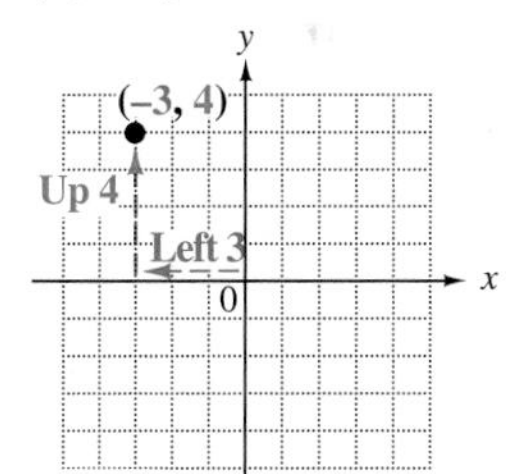

3.2 Graphing Linear Equations in Two Variables

To graph a linear equation:

Step 1 Find at least two ordered pairs that are solutions of the equation.

Step 2 Plot the corresponding points.

Step 3 Draw a straight line through the points.

The graph of $y = k$ is a horizontal line through $(0, k)$.

The graph of $x = k$ is a vertical line through $(k, 0)$.

Graph $x - 2y = 4$.

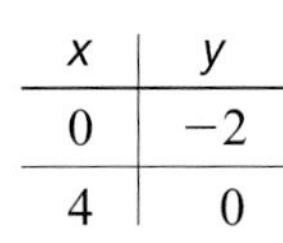

x	y
0	−2
4	0

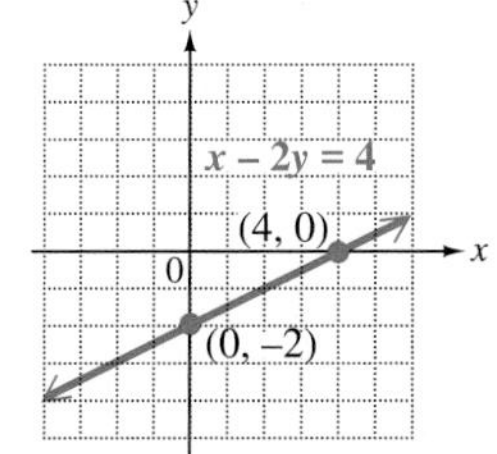

Concepts	Examples

3.3 Slope of a Line

The slope of the line through (x_1, y_1) and (x_2, y_2) is

$$m = \frac{\textbf{change in } y}{\textbf{change in } x} = \frac{y_2 - y_1}{x_2 - x_1} \quad (x_1 \neq x_2).$$

The line through $(-2, 3)$ and $(4, -5)$ has slope

$$m = \frac{-5 - 3}{4 - (-2)} = \frac{-8}{6} = -\frac{4}{3}.$$

Horizontal lines have slope 0.

The line $y = -2$ has slope 0.

Vertical lines have undefined slope.

The line $x = 4$ has undefined slope.

To find the slope of a line from its equation, solve for y. The slope is the coefficient of x.

Find the slope of the graph of $3x - 4y = 12$.

$$-4y = -3x + 12 \qquad \text{Add } -3x.$$

$$y = \frac{3}{4}x - 3 \qquad \text{Divide by } -4.$$

↑ Slope (the coefficient $\frac{3}{4}$)

Parallel lines have equal slopes.

The lines $y = 2x + 3$ and $4x - 2y = 6$ are **parallel**; both have $m = 2$.

$y = 2x + 3$	$4x - 2y = 6$
$m = 2$	$-2y = -4x + 6$
	$y = 2x - 3$
	$m = 2$

The slopes of perpendicular lines are negative reciprocals (with a product of -1).

The lines $y = 3x - 1$ and $x + 3y = 4$ are **perpendicular**; their slopes are negative reciprocals.

$y = 3x - 1$	$x + 3y = 4$
$m = 3$	$3y = -x + 4$
	$y = -\frac{1}{3}x + \frac{4}{3}$
	$m = -\frac{1}{3}$

3.4 Equations of Lines

Slope-Intercept Form
$y = mx + b$
m is the slope.
$(0, b)$ is the y-intercept.

Write an equation of the line with slope 2 and y-intercept $(0, -5)$.

$$y = 2x - 5$$

Point-Slope Form
$y - y_1 = m(x - x_1)$
m is the slope.
(x_1, y_1) is a point on the line.

Write an equation of the line with slope $-\frac{1}{2}$ through $(-4, 5)$.

$$y - 5 = -\frac{1}{2}[x - (-4)] \qquad \text{Substitute.}$$

$$y - 5 = -\frac{1}{2}(x + 4)$$

$$y - 5 = -\frac{1}{2}x - 2 \qquad \text{Distributive property}$$

$$y = -\frac{1}{2}x + 3 \qquad \text{Add 5.}$$

Concepts	Examples

3.4 Equations of Lines *(continued)*

Standard Form
$Ax + By = C$

The equation $y = -\frac{1}{2}x + 3$ is written in standard form as

$$x + 2y = 6,$$

with $A = 1$, $B = 2$, and $C = 6$.

Horizontal Line
$y = k$

$y = 4$ Horizontal line

Vertical Line
$x = k$

$x = -1$ Vertical line

3.5 Graphing Linear Inequalities in Two Variables

Graphing a Linear Inequality

Step 1 Draw the graph of the line that is the boundary. Make the line solid if the inequality involves $\leq$ or $\geq$; make the line dashed if the inequality involves $<$ or $>$.

Graph $2x - 3y \leq 6$.
Draw the graph of $2x - 3y = 6$. Use a solid line because of the inclusion of equality in the symbol $\leq$.

Step 2 Choose any point not on the line as a test point. Substitute the coordinates in the inequality. Simplify to determine whether the inequality is true or false.

Choose (0, 0), for example.

$2(0) - 3(0) = 0$, and $0 \leq 6$ True

Step 3 Shade the region that includes the test point if the test point satisfies the original inequality; otherwise, shade the region on the other side of the boundary line.

Shade the side of the line that includes (0, 0).

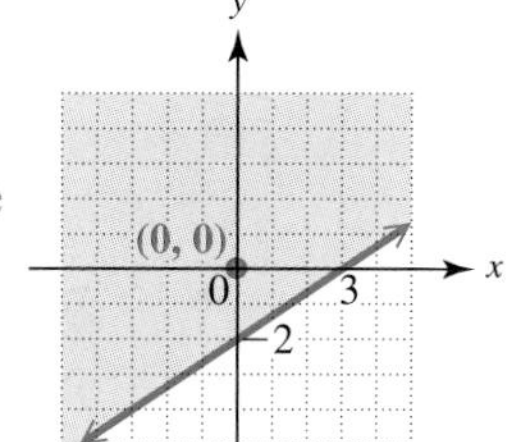

3.6 Introduction to Functions

A **function** is a set of ordered pairs such that, for each first component, there is one and only one second component. The set of first components is called the **domain,** and the set of second components is called the **range.**

$y = f(x) = x^2$ defines a function f with domain $(-\infty, \infty)$ and range $[0, \infty)$.

To evaluate a function using function notation (that is, $f(x)$ notation) for a given value of x, substitute the value wherever x appears.

If $f(x) = x^2 - 7x + 12$, then

$$f(1) = 1^2 - 7(1) + 12$$
$$= 6.$$

To write the equation that defines a function f in function notation, follow these steps.

Step 1 Solve the equation for y.

Step 2 Replace y with $f(x)$.

Write $2x + 3y = 12$ in function notation for function f.

$3y = -2x + 12$ Subtract $2x$.

$y = -\frac{2}{3}x + 4$ Divide by 3.

$f(x) = -\frac{2}{3}x + 4$

ANSWERS TO TEST YOUR WORD POWER

1. C; *Examples:* (0, 3), (3, 8), (4, 0)
2. D; *Examples:* $3x + 2y = 6$, $x = y - 7$, $4x = y$
3. D; *Example:* The graph of the equation $4x - 3y = 12$ has x-intercept (3, 0) and y-intercept (0, −4).
4. C; *Example:* The line through (3, 6) and (5, 4) has slope $\frac{4-6}{5-3} = \frac{-2}{2} = -1$.
5. A; *Example:* The set {(2, 0), (4, 3), (6, 6), (8, 9)} defines a relation.
6. B; *Example:* The relation given in Answer 5 is a function since each x-value corresponds to exactly one y-value.
7. D; *Example:* In the function in Answer 5, the domain is the set of x-values, {2, 4, 6, 8}.
8. A; *Example:* In the function in Answer 5, the range is the set of y-values, {0, 3, 6, 9}.

Chapter 3 Review Exercises

[3.1] *The percent of first-year college students at two-year public institutions who returned for a second year for the years 2001 through 2007 are shown in the graph.*

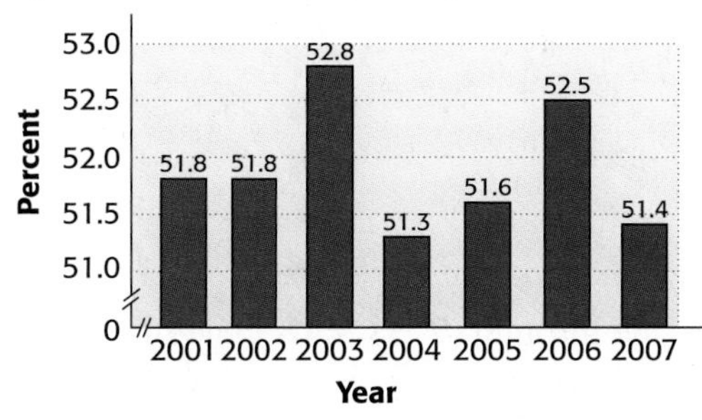

Source: ACT.

1. Write ordered pairs of the form (year, percent) for the data shown in the graph.

   (2001, 51.8), (2002, 51.8), (2003, 52.8), (2004, 51.3), (2005, 51.6), (2006, 52.5), (2007, 51.4)

2. What does the ordered pair (2006, 52.5) mean in the context of these problems?

   In the year 2006, 52.5% of first-year college students at two-year public institutions returned for a second year.

3. In what year did the percent show the greatest decrease from the previous year? What was this decrease?

   2004; 1.5%

4. In what year did the percent show the greatest increase from the previous year? What was this increase?

   2003; 1.0%

Complete the given ordered pairs for each equation.

5. $y = 3x + 2$ $(-1, \_\_), (0, \_\_), (\_\_, 5)$

   -1; 2; 1

6. $4x + 3y = 6$ $(0, \_\_), (\_\_, 0), (-2, \_\_)$

   2; $\frac{3}{2}$; $\frac{14}{3}$

Decide whether each ordered pair is a solution of the given equation.

7. $2x + y = 5; (-1, 3)$

   no

8. $3x - y = 4; \left(\frac{1}{3}, -3\right)$

   yes

9. If $x > 0$ and $y < 0$, in what quadrant(s) must (x, y) lie? Explain.

   x is positive in quadrants I and IV; y is negative in quadrants III and IV. Thus, if x is positive and y is negative, (x, y) must lie in quadrant IV.

10. On what axis does the point $(k, 0)$ lie for any real value of k? the point $(0, k)$? Explain.

    In the ordered pair (k, 0), the y-value is 0, so the point lies on the x-axis. In the ordered pair (0, k), the x-value is 0, so the point lies on the y-axis.

[3.2] *Find the intercepts for the graph of each equation.*

11. $y = 2x + 5$

    x-intercept: $\left(-\frac{5}{2}, 0\right)$

    y-intercept: $(0, 5)$

12. $2x + y = -7$

    x-intercept: $\left(-\frac{7}{2}, 0\right)$

    y-intercept: $(0, -7)$

13. $3x + 2y = 8$

    x-intercept: $\left(\frac{8}{3}, 0\right)$

    y-intercept: $(0, 4)$

Graph each linear equation.

14. $2x - y = 3$

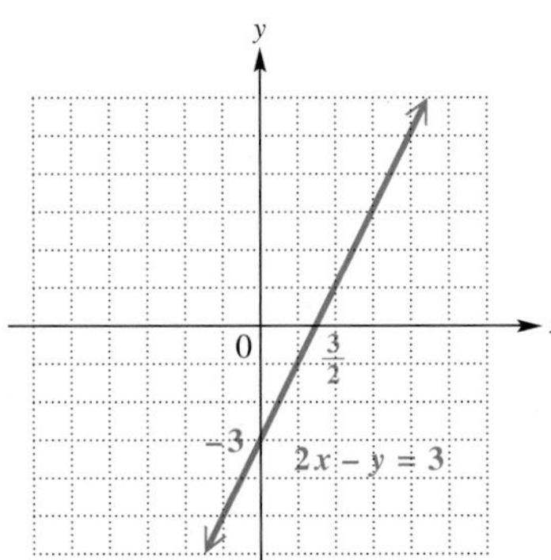

15. $x + 2y = -4$

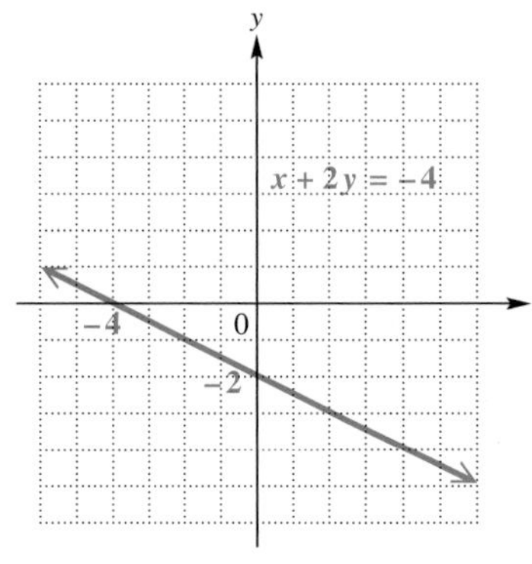

16. $x + y = 0$

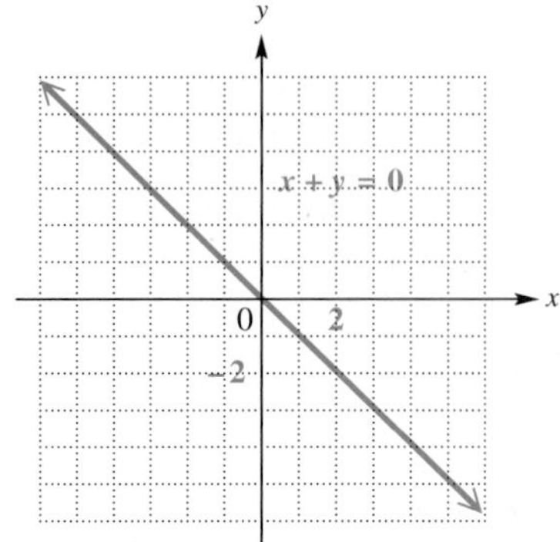

[3.3] *Find the slope of each line.*

17. The line through (2, 3) and (−4, 6)

$-\frac{1}{2}$

18. The line through (2, 5) and (2, 8)

undefined

19. $y = 3x - 4$

3

20. $y = 4$

0

21.

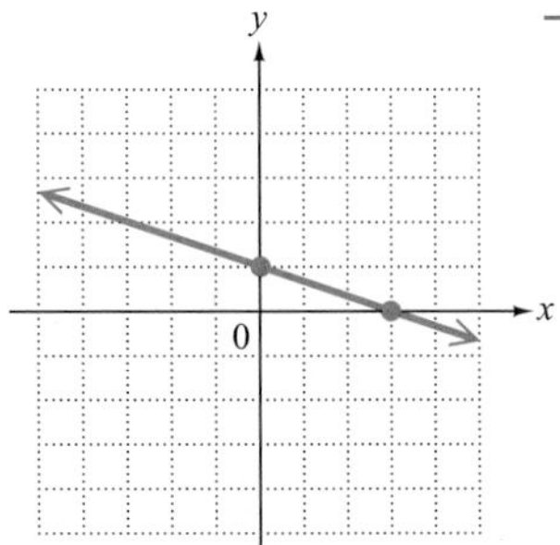

$-\frac{1}{3}$

22. The line having these points $\frac{3}{2}$

x	y
0	1
2	4
6	10

23. **(a)** A line parallel to the graph of $y = 2x + 3$ **2**

(b) A line perpendicular to the graph of $y = -3x + 3$

$\frac{1}{3}$

Decide whether each pair of lines is parallel, perpendicular, *or* neither.

24. $3x + 2y = 6$
$6x + 4y = 8$

parallel

25. $x - 3y = 1$
$3x + y = 4$

perpendicular

26. $x - 2y = 8$
$x + 2y = 8$

neither

27. If the pitch of a roof is $\frac{1}{4}$, how many feet in the horizontal direction correspond to a rise of 3 ft?

12 ft

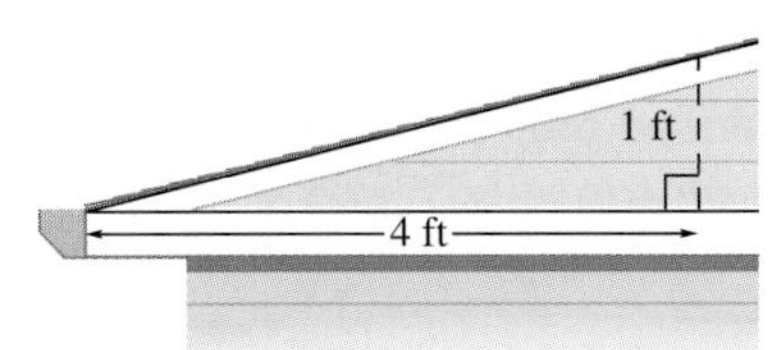

28. In 1980, the median family income in the United States was about \$21,000 per yr. In 2005, it was about \$56,200 per yr. Find the average rate of change of median family income over that period. (*Source:* U.S. Census Bureau.)

\$1408 per yr

[3.4] *Write an equation of each line. Give the final answer in slope-intercept form (if possible).*

29. $m = -1, b = \frac{2}{3}$

$y = -x + \frac{2}{3}$

30. The line in Exercise 21

$y = -\frac{1}{3}x + 1$

31. The line through (−1, 4), $m = \frac{2}{3}$

$y = \frac{2}{3}x + \frac{14}{3}$

32. The line through (2, 1) and (−2, 2)

$y = -\frac{1}{4}x + \frac{3}{2}$

33. The line through (−4, 1), slope 0

$y = 1$

34. The line through $\left(\frac{1}{3}, -\frac{3}{4}\right)$, undefined slope

$x = \frac{1}{3}$

*Write an equation for each line **(a)** in slope-intercept form and **(b)** in standard form.*

35. Parallel to $4x - y = 3$ and through $(6, -2)$

(a) $y = 4x - 26$ **(b)** $4x - y = 26$

36. Perpendicular to $2x - 5y = 7$ and through $(0, 1)$

(a) $y = -\frac{5}{2}x + 1$ **(b)** $5x + 2y = 2$

37. The Midwest Athletic Club offers two special membership plans. (*Source:* Midwest Athletic Club.) For each plan, write a linear equation in slope-intercept form and give the cost y in dollars of a 1-yr membership. Let x represent the number of months.

(a) Executive VIP/Gold membership: $159 fee plus $57 per month

$y = 57x + 159$; $843

(b) Executive Regular/Silver membership: $159 fee plus $47 per month

$y = 47x + 159$; $723

[3.5] *Graph each linear inequality.*

38. $2x - 3y > -6$

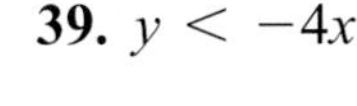

39. $y < -4x$

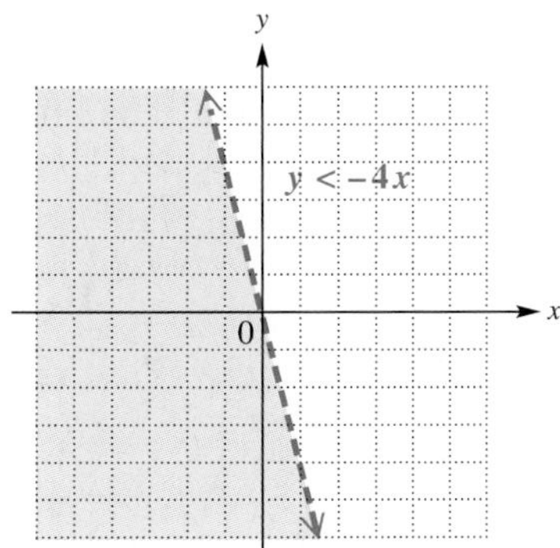

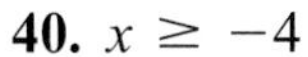

40. $x \geq -4$

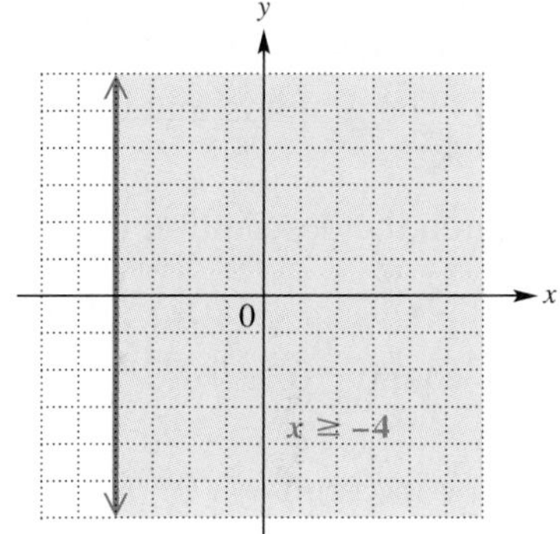

[3.6] *Give the domain and range of each relation. Identify any functions.*

41. $\{(-4, 2), (-4, -2), (1, 5), (1, -5)\}$

domain: $\{-4, 1\}$;
range: $\{2, -2, 5, -5\}$;
not a function

42.

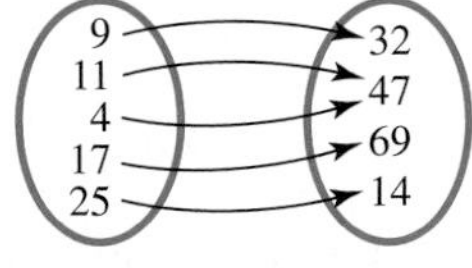

domain: $\{9, 11, 4, 17, 25\}$;
range: $\{32, 47, 69, 14\}$;
function

43.

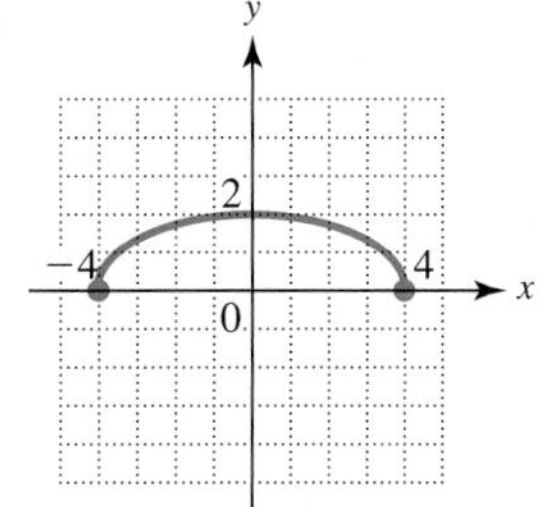

domain: $[-4, 4]$; **range:** $[0, 2]$;
function

Given $f(x) = -2x^2 + 3x - 6$, *find each of the following.*

44. $f(0)$

-6

45. $f(3)$

-15

46. $f(-k)$

$-2k^2 - 3k - 6$

Determine whether each relation defines y as a function of x. Identify any linear functions. Give the domain in each case.

47. $y = 3x - 3$

function; linear function; domain: $(-\infty, \infty)$

48. $x = y^2$

not a function; domain: $[0, \infty)$

49. $y = \dfrac{7}{x - 36}$

function; domain: $(-\infty, 36) \cup (36, \infty)$

50. The equation $2x^2 - y = 0$ defines y as a function of x. Rewrite it using $f(x)$ notation, and find $f(3)$.

$f(x) = 2x^2$; 18

51. Describe the graph of a constant function.

It is a horizontal line.

Mixed Review Exercises

52. Match each statement in (a)–(f) to the appropriate graph or graphs in A–D. Graphs may be used more than once.

A.

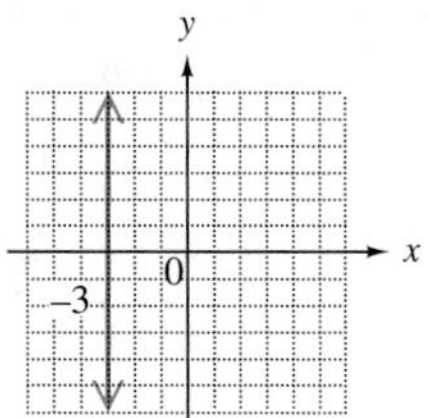

B.

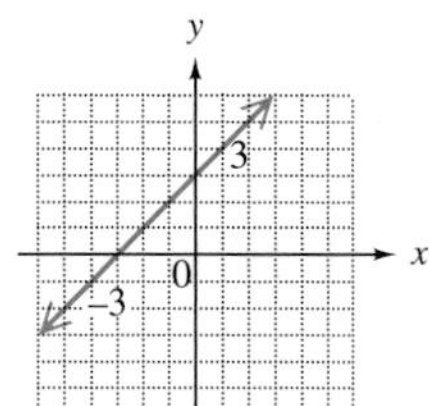

C.

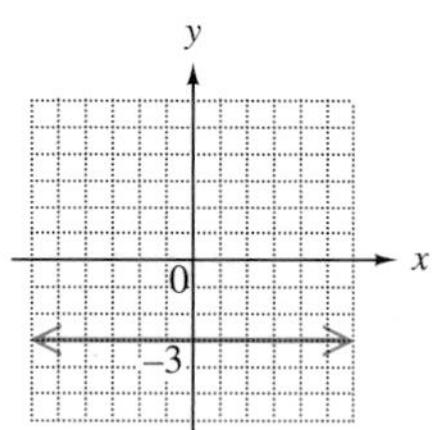

D.

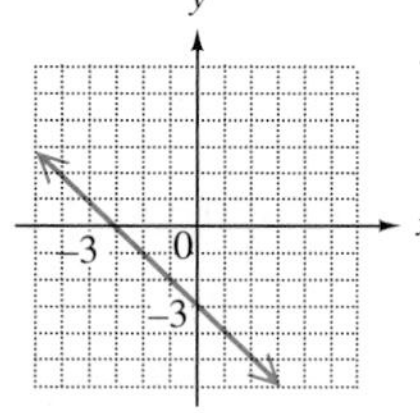

(a) The line shown in the graph has undefined slope.
A

(b) The graph of the equation has y-intercept $(0, -3)$.
C, D

(c) The graph of the equation has x-intercept $(-3, 0)$.
A, B, D

(d) The line shown in the graph has negative slope.
D

(e) The graph is that of the equation $y = -3$.
C

(f) The line shown in the graph has slope 1.
B

53. Match the description in Column I with the correct equation in Column II.

I	II
(a) Slope -0.5, $b = -2$ **B**	**A.** $y = -\frac{1}{2}x$
(b) x-intercept $(4, 0)$, y-intercept $(0, 2)$ **D**	**B.** $y = -\frac{1}{2}x - 2$
(c) The line through $(4, -2)$ and $(0, 0)$ **A**	**C.** $x - 2y = 2$
(d) $m = \frac{1}{2}$, passes through $(-2, -2)$ **C**	**D.** $x + 2y = 4$
	E. $x = 2y$

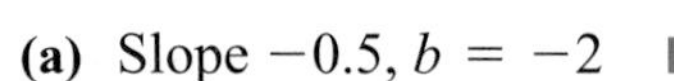

54. The table shows life expectancy at birth in the United States for selected years.

Year	Life Expectancy at Birth (in years)
1943	63.3
1953	68.8
1963	69.9
1973	71.4
1983	74.6
1993	75.5
2003	77.6

Source: Centers for Disease Control and Prevention.

(a) Does the data in the table define a function? What are the domain and range?

yes; domain: {1943, 1953, 1963, 1973, 1983, 1993, 2003}; range: {63.3, 68.8, 69.9, 71.4, 74.6, 75.5, 77.6}

(b) Call this function f. Give two ordered pairs that belong to f.

Answers will vary. Two possible ordered pairs are (1953, 68.8) and (1973, 71.4).

(c) Find $f(2003)$. What does it mean?

77.6; In 2003, life expectancy at birth was 77.6 yr.

(d) If $f(x) = 75.5$, what does x equal? **1993**

Chapter 3 Test

Use the Chapter Test Prep Video CD to see fully worked-out solutions to any of the exercises you want to review.

The line graph shows the overall unemployment rate in the U.S. civilian labor force for the years 1998 through 2005. Use the graph to work Exercises 1–3.

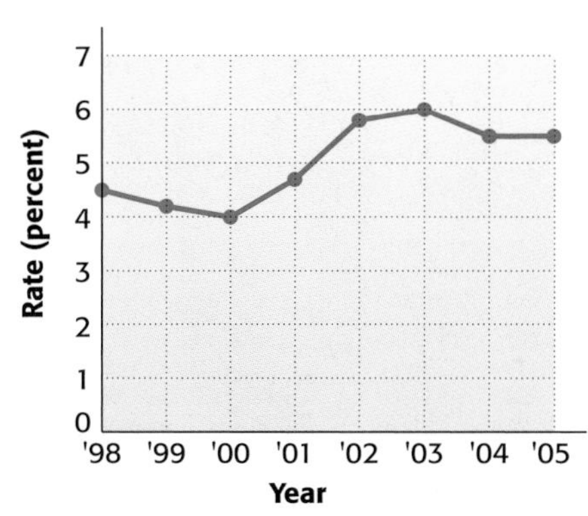

Source: U.S. Department of Labor.

1. Between which pairs of consecutive years did the unemployment rate decrease?

2. What was the general trend in the unemployment rate between 2000 and 2003?

3. Estimate the overall unemployment rate in 2003 and 2004. About how much did the unemployment rate decline between 2003 and 2004?

Graph each linear equation. Give the x- and y-intercepts.

4. $3x + y = 6$

5. $y - 2x = 0$

1. **between 1998 and 1999, 1999 and 2000, and 2003 and 2004**

2. **The unemployment rate was increasing.**

3. **2003: 6.0%; 2004: 5.5%; decline: 0.5%**

4. x-intercept: **(2, 0)**
   y-intercept: **(0, 6)**

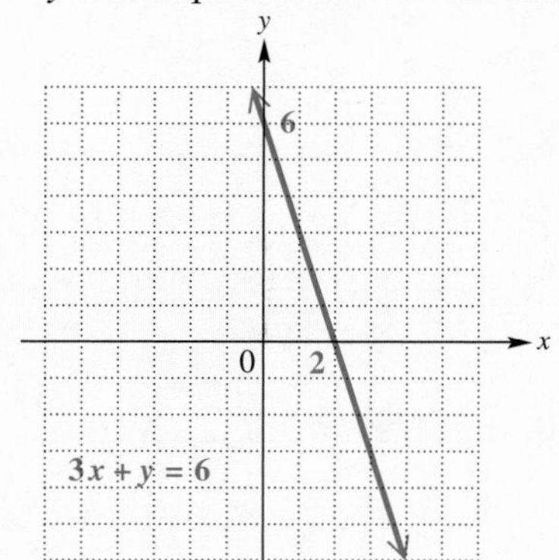

5. x-intercept: **(0, 0)**
   y-intercept: **(0, 0)**

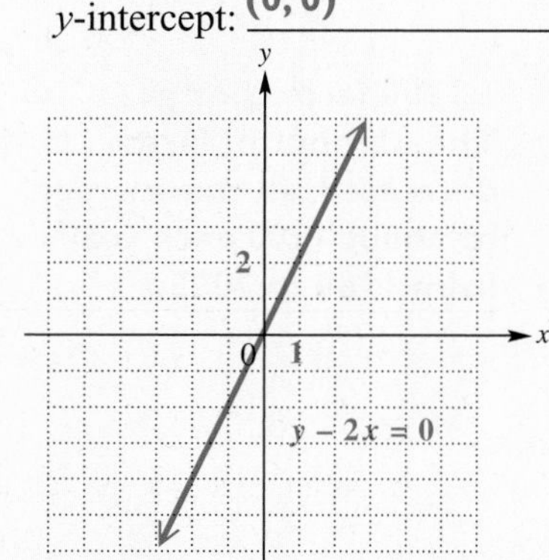

6. x-intercept: **(−3, 0)**

y-intercept: **none**

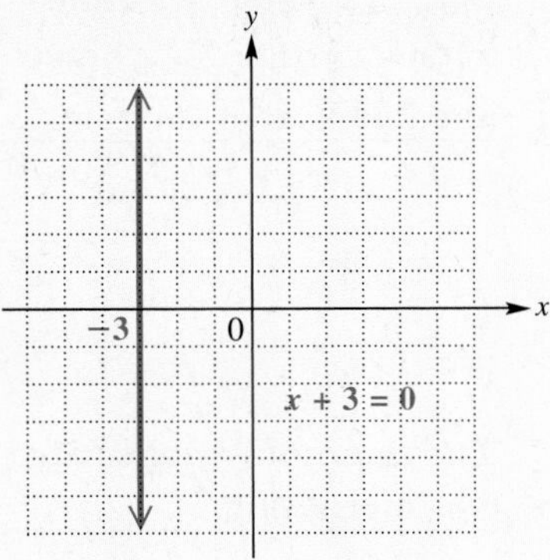

6. $x + 3 = 0$

7. x-intercept: **(4, 0)**

y-intercept: **(0, −4)**

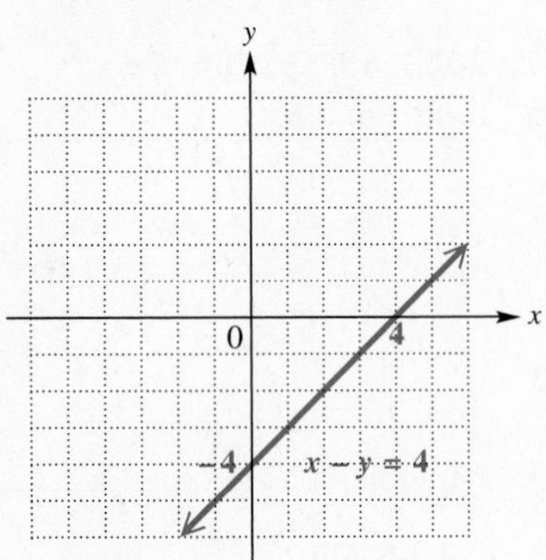

7. $x - y = 4$

Find the slope of each line.

8. $-\frac{8}{3}$

8. The line through $(-4, 6)$ and $(-1, -2)$

9. **−2**

9. $2x + y = 10$

10. **undefined**

10. $x + 12 = 0$

11. $\frac{5}{2}$

11.

y

1

0 2

−4

x

12. **−1200 farms per yr; The number of farms decreased, on the average, by about 1200 each year from 1980 to 2005.**

12. In 1980, there were 119,000 farms in Iowa. As of 2005, there were 89,000. Find and interpret the average rate of change in the number of farms per year. (*Source:* U.S. Department of Agriculture.)

Write an equation for each line. Give the final answer in slope-intercept form.

13. The line through $(-1, 4)$, $m = 2$

13. $y = 2x + 6$

14. The line in Exercise 11

14. $y = \frac{5}{2}x - 4$

15. Write the equation of the line through $(-3, 14)$ and $(-6, 9)$ in standard form.

15. $5x - 3y = -57$

16. Write the equation in slope-intercept form for the line through $(-7, 2)$ and

(a) parallel to $3x + 5y = 6$. **(b)** perpendicular to $y = 2x$.

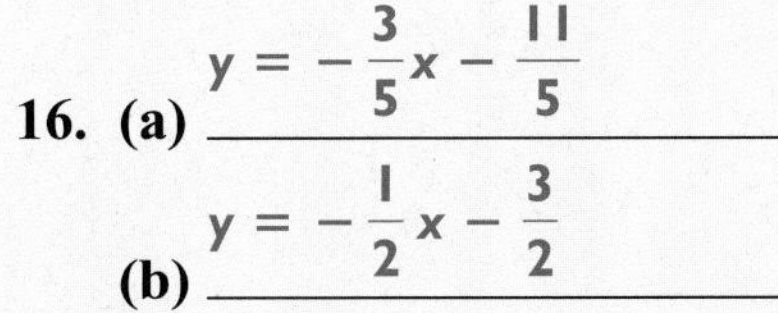

16. (a) $y = -\frac{3}{5}x - \frac{11}{5}$

(b) $y = -\frac{1}{2}x - \frac{3}{2}$

Graph each linear inequality.

17. $x + y \leq 3$

17.

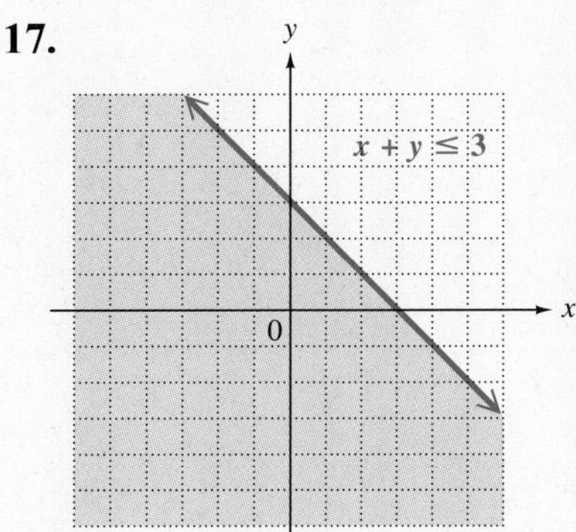

18. $3x - y > 0$

18.

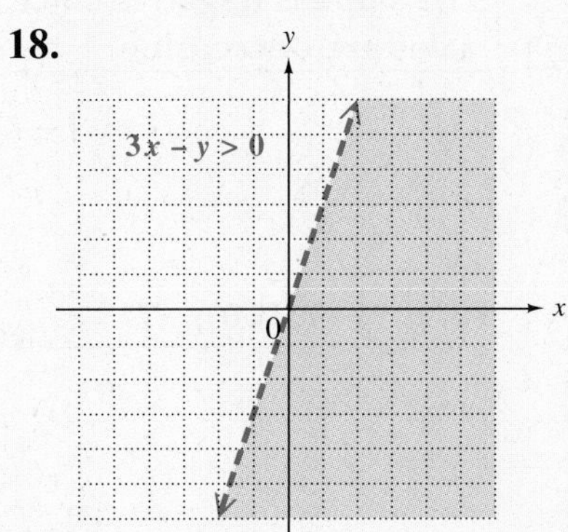

19. Which of the following is the graph of a function? Give its domain and range.

A.

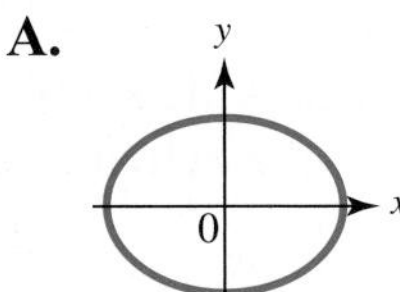

B.

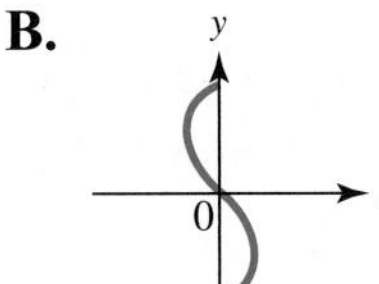

C.

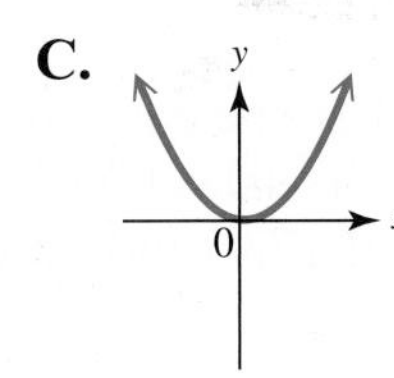

19. C; domain: $(-\infty, \infty)$; range: $[0, \infty)$

20. C; domain: $\{0, 3, 6\}$; range: $\{1, 2, 3\}$

20. Which of the following does not define a function? Give its domain and range.

A. $\{(0, 1), (-2, 3), (4, 8)\}$ **B.** $y = 2x - 6$ **C.**

x	y
0	1
3	2
0	2
6	3

21. 0; $-a^2 + 2a - 1$

21. If $f(x) = -x^2 + 2x - 1$, find $f(1)$ and $f(a)$.

The graph shows worldwide snowmobile sales from 2000 through 2007, where 2000 corresponds to $x = 0$. Use the graph to work Exercises 22–25.

Source: www.snowmobile.org

22. The slope is negative since sales are decreasing.

22. Is the slope of the line in the graph positive or negative? Explain.

23. (0, 209), (7, 160); −7

23. Write two ordered pairs for the data points shown in the graph. Use them to find the slope of the line.

24. $y = -7x + 209$

24. Use the ordered pairs and slope from Exercise 23 to find an equation of a line that models the data. Write the equation in slope-intercept form.

25. 174 thousand; The equation gives a good approximation of the actual sales.

25. Use the equation from Exercise 24 to approximate worldwide snowmobile sales for 2005. How does your answer compare to the actual sales of 173.7 thousand?

Study Skills

ANALYZING YOUR TEST RESULTS

OBJECTIVES

1. Determine the reason for errors.
2. Develop a plan to avoid test taking errors.
3. Review material to correct misunderstandings.

After taking a test, many students heave a big sigh of relief and try to forget it ever happened. Don't fall into this trap. ***An exam is a learning opportunity.*** It gives you clues about *what your instructor thinks is important, what concepts and skills are valued* in mathematics, and *if you are on the right track.*

After the test is returned, do the following:

- **Jot down problems that caused you trouble.** Find out how to solve them by checking your textbook or notes, or asking your instructor. You might see those same problems again on a final exam.
- **Find out what you got wrong and why you had points deducted.** Write down the problem so you can learn how to do it correctly. Sometimes you only have a short time in class to review your test. If you need more time, ask your instructor if you can look at the test in his or her office.

Find Out Why You Made the Errors You Made

Here is a list of typical reasons you might make errors on math tests.

1. You read the directions wrong.
2. You read the question wrong or skipped over something.
3. You made a computation error (maybe even a minor one).
4. Your answer is not accurate.
5. Your answer is not complete.
6. You labeled your answer wrong. For example, you labeled it "ft" and it should have been "ft^2."
7. You didn't show your work.

8.* You didn't understand the concept.

9.* You were unable to go from words (in an application) to setting up the problem.

10.* You were unable to apply a procedure to a new situation.

11. You were anxious and made errors even when you knew the material.

The first seven errors are test taking errors. They are easy to correct if you carefully read test questions and directions, proofread or rework the problems, show all your work, and double-check units and labels every time.

The three starred errors (*) are test preparation errors. Remember that you need to practice the kinds of problems that you will see on tests. So, for example, if application problems are difficult for you, you must do more application problems. If you have practiced the various techniques outlined in the Study Skills, you will be less likely to make these kinds of errors on tests.

The last item isn't really an error. **Anxiety can play a big part in your test results.** Go back to the *Preparing for Tests* Study Skill, and read the suggestions for managing anxiety. Five minutes of brisk walking before your test can help you relax. Also, practicing a relaxation technique while you do your homework will make it more likely that you will benefit from using it during a test. Deep breathing is one such technique. When you are anxious, you tend to breathe more shallowly, which can make you feel confused and easily distracted.

Make a Plan for the Next Test

Make this plan based on your results from a test. You might review the Chapter Summary and work the problems in the Chapter Review Exercises or the Chapter Test. Perhaps ask your instructor for more help.

Now Try This ▶▶▶

Below is a record sheet to use to track your progress in test taking. Use it to find out if you tend to make particular kinds of errors on tests. Then you can work specifically on correcting them. For each category of errors, check the appropriate box when you made one of the errors.

Test Taking Errors

Test	Read directions wrong	Read question wrong	Computation error	Not exact or accurate	Not complete	Labeled wrong	Didn't show work
1							
2							
3							
4							
5							

What will you do to avoid these kinds of errors on your next test?

____

____

Test Preparation Errors

Test	Didn't understand concept	Didn't set up problem correctly	Couldn't apply concept to new situation
1			
2			
3			
4			
5			

What will you do to avoid these kinds of errors on your next test?

____

____

Anxiety

Test	Felt anxious *before* the exam	Felt anxious *during* the exam	Blanked out on questions	Got questions wrong that I knew how to do
1				
2				
3				
4				
5				

What will you do to reduce your anxiety before the next test? ________

____

____

Cumulative Review Exercises Chapters R–3

Perform the indicated operations.

1. $10\frac{5}{8} - 3\frac{1}{10}$ **$\frac{301}{40}$, or $7\frac{21}{40}$**

2. $\frac{3}{4} \div \frac{1}{8}$ **6**

3. $5 - (-4) + (-2)$ **7**

4. $\frac{(-3)^2 - (-4)(2^4)}{5(2) - (-2)^3}$ **$\frac{73}{18}$, or $4\frac{1}{18}$**

5. *True* or *false*? $\frac{4(3 - 9)}{2 - 6} \geq 6$ **true**

6. Find the value of $xz^3 - 5y^2$ when $x = -2$, $y = -3$, and $z = -1$. **−43**

7. What property does $3(-2 + x) = -6 + 3x$ illustrate? **distributive property**

8. Simplify $-4p - 6 + 3p + 8$ by combining like terms. **$-p + 2$**

Solve.

9. $V = \frac{1}{3}\pi r^2 h$ for h **$h = \frac{3V}{\pi r^2}$**

10. $6 - 3(1 + a) = 2(a + 5) - 2$ **{−1}**

11. $-(m - 3) = 5 - 2m$ **{2}**

12. $\frac{y - 2}{3} = \frac{2y + 1}{5}$ **{−13}**

Solve each inequality. Write the solution set in interval notation and graph it.

13. $-2.5x < 6.5$

−2.6 0

(−2.6, ∞)

14. $4(x + 3) - 5x < 12$

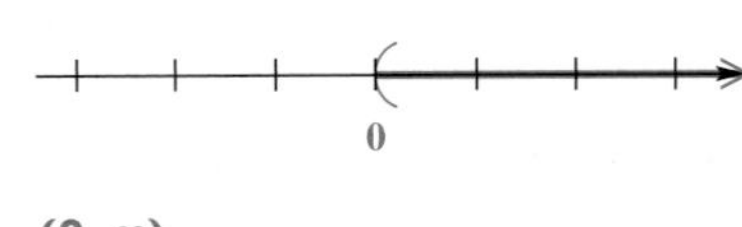

(0, ∞)

15. $\frac{2}{3}t - \frac{1}{6}t \leq -2$

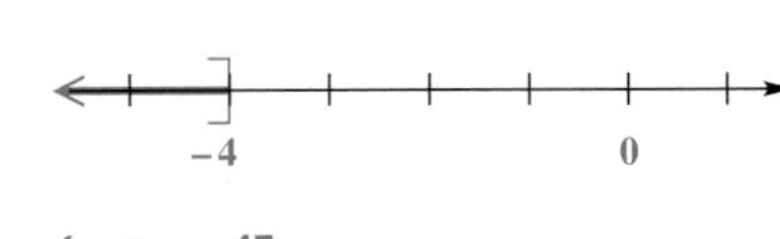

(−∞, −4]

Solve each problem.

16. The gap in average annual earnings by level of education continues to increase. Based on the most recent statistics available, a person with a bachelor's degree can expect to earn $31,120 more each year than someone with a high school diploma. Together the individuals would earn $107,892. How much can a person at each level of education expect to earn? (*Source*: U.S. Census Bureau.)

high school diploma: $38,386; bachelor's degree: $69,506

17. Mount Mayon in the Philippines is the most perfectly shaped conical volcano in the world. Its base is a circle with circumference 80 mi, and it has a height of about 8200 ft. (One mile is 5280 ft.) Find the radius of the circular base to the nearest mile. (*Hint*: This problem has some unneeded information.) (*Source: Microsoft Encarta Encyclopedia.*) **13 mi**

Circumference = 80 mi

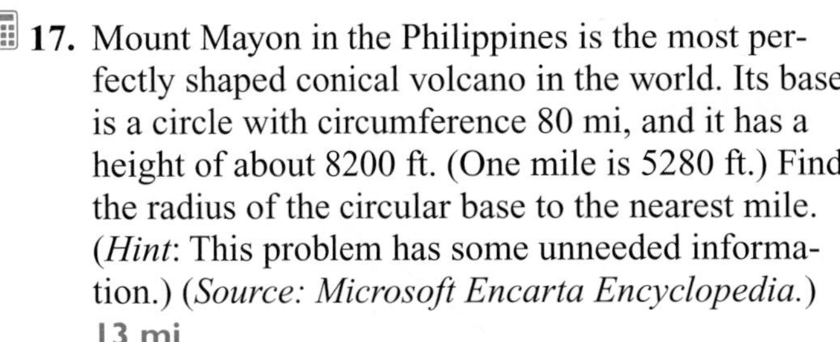

18. The winning times in seconds for the women's 1000 m speed skating event in the Winter Olympics for the years 1960 through 2006 can be closely approximated by the linear equation

$$y = -0.4787x + 95.10$$

where x is the number of years since 1960. That is, $x = 4$ represents 1964, $x = 8$ represents 1968, and so on. (*Source: World Almanac and Book of Facts.*)

(a) Use this equation to complete the table of values. Round times to the nearest hundredth of a second.

x	y
20	**85.53**
38	**76.91**
42	**74.99**

(b) Based on this equation, what does the ordered pair (32, 79.78) mean?

In 1992, the winning time was approximately 79.78 sec.

19. Baby boomers are expected to inherit \$10.4 trillion from their parents over the next 45 yr, an average of \$50,000 each. The circle graph shows how they plan to spend their inheritances.

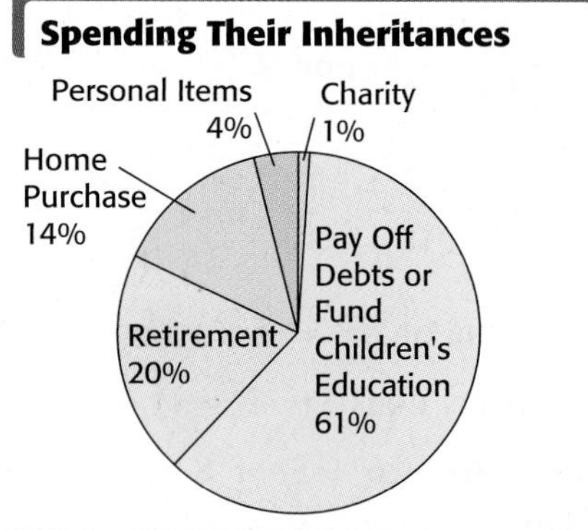

Source: First Interstate Bank Trust and Private Banking Group.

(a) How much of the \$50,000 is expected to go toward home purchase?

\$7000

(b) How much is expected to go toward retirement?

\$10,000

(c) Use the answer from part (b) to estimate the amount expected to go toward paying off debts or funding children's education.

about \$30,000

Consider the linear equation $-3x + 4y = 12$. *Find the following.*

20. The x- and y-intercepts

(−4, 0); (0, 3)

21. The slope

$\frac{3}{4}$

22. The graph

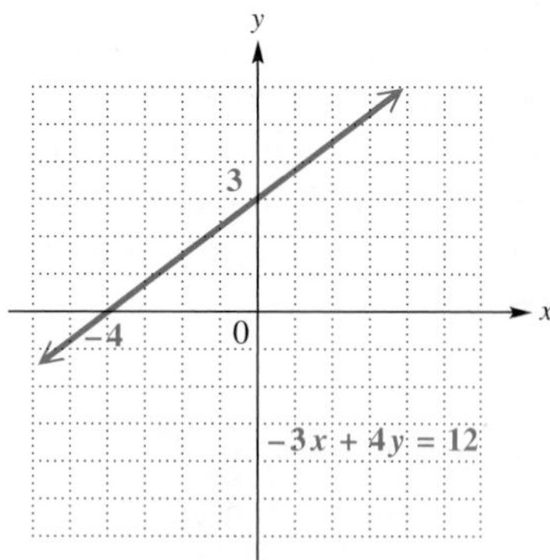

23. The y-value of the point having x-value 4

6

24. Are the lines with equations $x + 5y = -6$ and $y = 5x - 8$ *parallel, perpendicular,* or *neither?* **perpendicular**

Write an equation of each line. Give the final answer in slope-intercept form if possible.

25. The line through $(2, -5)$ with slope 3 **$y = 3x - 11$**

26. The line through $(0, 4)$ and $(2, 4)$ **$y = 4$**

4

Systems of Linear Equations and Inequalities

Over the years, Americans have continued their fascination with Hollywood and the movies. Although many people now prefer to watch movies at home on DVD, the number of tickets sold at domestic movie theaters increased in 2007 to about 1.4 billion, with a total gross of over $9.6 billion. The top box office draws of that year—*Spider-Man 3* and *Shrek the Third*—attracted millions of adults and children wishing to get away from it all for a few hours. (*Source:* www.boxofficemojo.com)

In Exercises 13 and 14 of Section 4.4, we use *systems of equations* to find out how much money these top films earned in total and on their opening weekends.

4.1 Solving Systems of Linear Equations by Graphing

OBJECTIVES

1. Decide whether a given ordered pair is a solution of a system.
2. Solve linear systems by graphing.
3. Solve special systems by graphing.
4. Identify special systems without graphing.

A **system of linear equations,** often called a **linear system,** consists of two or more linear equations with the same variables.

$$\begin{aligned} 2x + 3y &= 4 \\ 3x - y &= -5 \end{aligned} \qquad \begin{aligned} x + 3y &= 1 \\ -y &= 4 - 2x \end{aligned} \qquad \begin{aligned} x - y &= 1 \\ y &= 3 \end{aligned} \qquad \text{Linear systems}$$

In the system on the right, think of $y = 3$ as an equation in two variables by writing it as $0x + y = 3$.

OBJECTIVE 1 Decide whether a given ordered pair is a solution of a system. A **solution of a system** of linear equations is an ordered pair that makes both equations true at the same time. A solution of an equation is said to *satisfy* the equation.

EXAMPLE 1 Determining Whether an Ordered Pair Is a Solution

Is $(4, -3)$ a solution of each system?

(a) $x + 4y = -8$
$3x + 2y = 6$

To decide whether $(4, -3)$ is a solution of the system, substitute 4 for x and -3 for y in each equation.

$x + 4y = -8$		$3x + 2y = 6$	
$4 + 4(-3) \stackrel{?}{=} -8$		$3(4) + 2(-3) \stackrel{?}{=} 6$	
$4 + (-12) \stackrel{?}{=} -8$	Multiply.	$12 + (-6) \stackrel{?}{=} 6$	Multiply.
$-8 = -8$	True	$6 = 6$	True

Because $(4, -3)$ satisfies both equations, it is a solution of the system.

(b) $2x + 5y = -7$
$3x + 4y = 2$

Again, substitute 4 for x and -3 for y in both equations.

$2x + 5y = -7$		$3x + 4y = 2$	
$2(4) + 5(-3) \stackrel{?}{=} -7$		$3(4) + 4(-3) \stackrel{?}{=} 2$	
$8 + (-15) \stackrel{?}{=} -7$	Multiply.	$12 + (-12) \stackrel{?}{=} 2$	Multiply.
$-7 = -7$	True	$0 = 2$	False

The ordered pair $(4, -3)$ is not a solution of this system because it does not satisfy the second equation.

◀ Work Problem 1 at the Side.

1 Fill in the blanks, and decide whether the given ordered pair is a solution of the system.

(a) $(2, 5)$

$3x - 2y = -4$
$5x + y = 15$

$3x - 2y = -4$
$3(\_\_\_) - 2(\_\_\_) \stackrel{?}{=} -4$

$5x + y = 15$
$5(2) + \_\_\_ \stackrel{?}{=} \_\_\_$

$(2, 5)$ ______ a solution. (is/is not)

(b) $(1, -2)$

$x - 3y = 7$
$4x + y = 5$

$(1, -2)$ ______ a solution. (is/is not)

OBJECTIVE 2 Solve linear systems by graphing. The set of all ordered pairs that are solutions of a system is its **solution set.** One way to find the solution set of a system of two linear equations is to graph both equations on the same axes. The graph of each line shows points whose coordinates satisfy the equation of that line. Any intersection point would be on both lines and would therefore be a solution of *both* equations. ***Thus, the coordinates of any point where the lines intersect give a solution of the system.***

ANSWERS

1. (a) 2; 5; 5; 15; is **(b)** is not

The graph in Figure 1 shows that the solution of the system in Example 1(a) is the intersection point $(4, -3)$. Because *two different* straight lines can intersect at no more than one point, there can never be more than one solution for such a system.

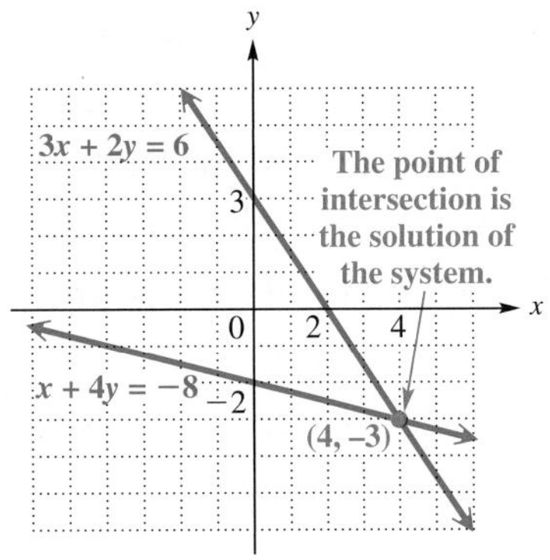

Figure 1

2 Solve each system of equations by graphing both equations on the same axes. Check your solutions.

(a) $5x - 3y = 9$

$x + 2y = 7$

(One of the lines is already graphed.)

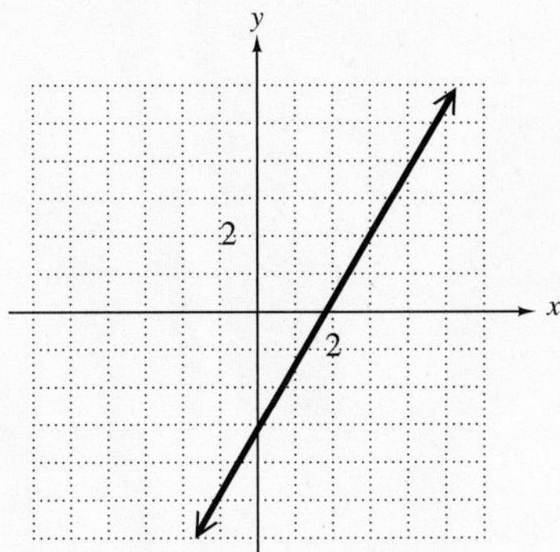

(b) $x + y = 4$

$2x - y = -1$

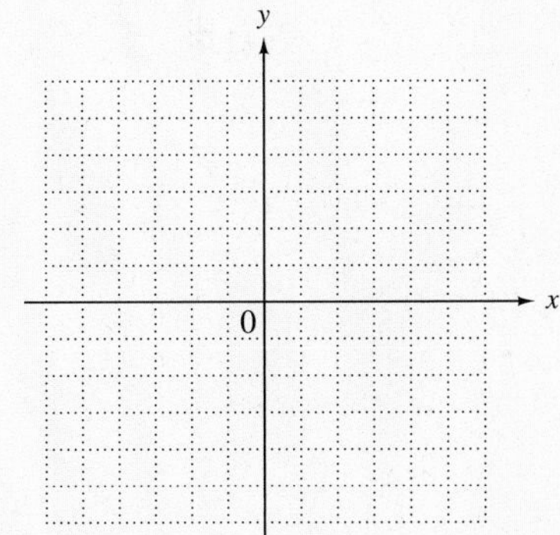

EXAMPLE 2 Solving a System by Graphing

Solve the system of equations by graphing both equations on the same axes.

$$2x + 3y = 4$$
$$3x - y = -5$$

We graph these two equations by plotting several points for each line. Recall from **Section 3.2** that the intercepts are often convenient choices.

$2x + 3y = 4$

x	y
0	$\frac{4}{3}$
2	0
-2	$\frac{8}{3}$

Find a third ordered pair as a check.

$3x - y = -5$

x	y
0	5
$-\frac{5}{3}$	0
-2	-1

The lines in Figure 2 suggest that the graphs intersect at the point $(-1, 2)$. We check this by substituting -1 for x and 2 for y in both equations.

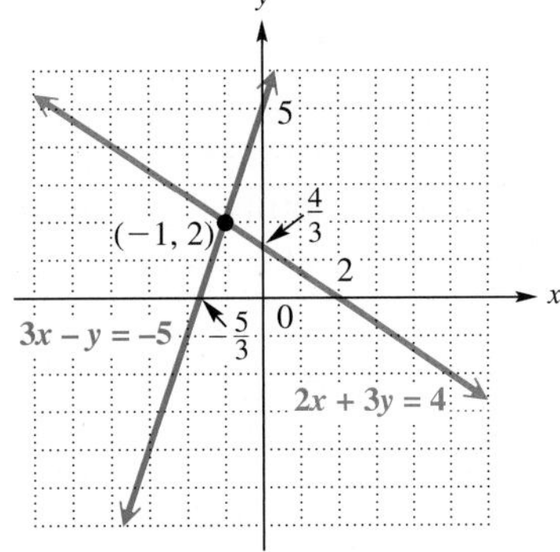

Figure 2

Check

$$2x + 3y = 4$$
$$2(-1) + 3(2) \stackrel{?}{=} 4$$
$$4 = 4 \quad \text{True}$$
$$3x - y = -5$$
$$3(-1) - 2 \stackrel{?}{=} -5$$
$$-5 = -5 \quad \text{True}$$

Because $(-1, 2)$ satisfies both equations, the solution set of this system is $\{(-1, 2)\}$.

Work Problem **2** *at the Side.* ▶

Note

We can also graph a linear system by writing each equation in the system in slope-intercept form and using the slope and y-intercept to graph each line. For Example 2,

$2x + 3y = 4$ becomes $y = -\frac{2}{3}x + \frac{4}{3}$ y-intercept $(0, \frac{4}{3})$; slope $-\frac{2}{3}$

$3x - y = -5$ becomes $y = 3x + 5$. y-intercept $(0, 5)$; slope 3, or $\frac{3}{1}$

Confirm that graphing these equations results in the same lines and the same solution shown in Figure 2.

Answers

2. (a) $\{(3, 2)\}$ **(b)** $\{(1, 3)\}$

To solve a linear system by graphing, follow these steps.

Solving a Linear System by Graphing

Step 1 **Graph each equation** of the system on the same coordinate axes.

Step 2 **Find the coordinates of the point of intersection** of the graphs if possible. This is the solution of the system.

Step 3 **Check** the solution in *both* of the original equations. Then write the solution set.

CAUTION

A difficulty with the graphing method of solution is that it may not be possible to determine from the graph the exact coordinates of the point that represents the solution, particularly if these coordinates are not integers. For this reason, algebraic methods of solution are explained later in this chapter. The graphing method does, however, show geometrically how solutions are found and is useful when approximate answers will do.

OBJECTIVE 3 Solve special systems by graphing. Sometimes the graphs of the two equations in a system either do not intersect at all or are the same line, as in the systems in Example 3.

EXAMPLE 3 Solving Special Systems

Solve each system by graphing.

(a) $2x + y = 2$

$2x + y = 8$

The graphs of these lines are shown in Figure 3. The two lines are parallel and have no points in common. For such a system, there is no solution. We write the solution set as $\emptyset$.

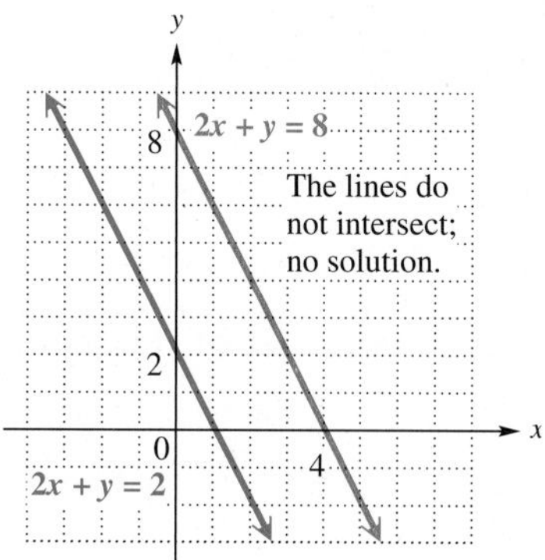

Figure 3

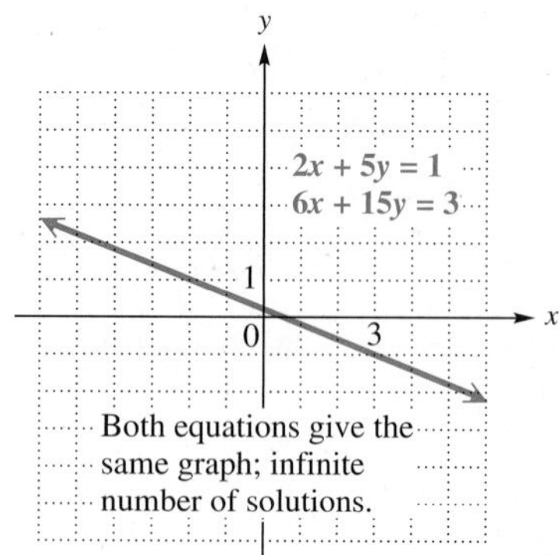

Figure 4

(b) $2x + 5y = 1$

$6x + 15y = 3$

The graphs of these two equations are the same line. See Figure 4. The second equation can be obtained by multiplying each side of the first equation by 3. In this case, every point on the line is a solution of the system, and the solution set contains an infinite number of ordered pairs that satisfy the equations.

Continued on Next Page

We write the solution set as

$$\{(x, y) \mid 2x + 5y = 1\},$$

This is the first equation in the system. See the Note below.

read "the set of ordered pairs (x, y) such that $2x + 5y = 1$." Recall from **Section 1.3** that this notation is called **set-builder notation.**

Note

When a system has an infinite number of solutions, as in Example 3(b), either equation of the system could be used to write the solution set. ***We prefer to use the equation in standard form with coefficients that are integers having greatest common factor 1.*** If neither of the given equations of the system is in this form, use an *equivalent* equation that is in standard form with coefficients that are integers having greatest common factor 1 to write the solution set with set-builder notation.

Work Problem 3 *at the Side.* ▶

The system in Example 2 has exactly one solution. A system with at least one solution is called a **consistent system.** A system of equations with no solution, such as the one in Example 3(a), is called an **inconsistent system.** The equations in Example 2 are **independent equations** with different graphs. The equations of the system in Example 3(b) have the same graph and are equivalent. Because they are different forms of the same equation, these equations are called **dependent equations.**

Examples 2 and 3 show the three cases that may occur when solving a system of two equations with two variables.

Three Cases for Solutions of Systems

1. The graphs intersect at exactly one point, which gives the (single) ordered-pair solution of the system. The **system is consistent** and the **equations are independent.** See Figure 5(a).
2. The graphs are parallel lines, so there is no solution and the solution set is $\emptyset$. The **system is inconsistent** and the **equations are independent.** See Figure 5(b).
3. The graphs are the same line. There is an infinite number of solutions, and the solution set is written in set-builder notation as $\{(x, y) \mid \_\_\_\_\_\_\}$, where one of the equations is written after the | symbol. The **system is consistent** and the **equations are dependent.** See Figure 5(c).

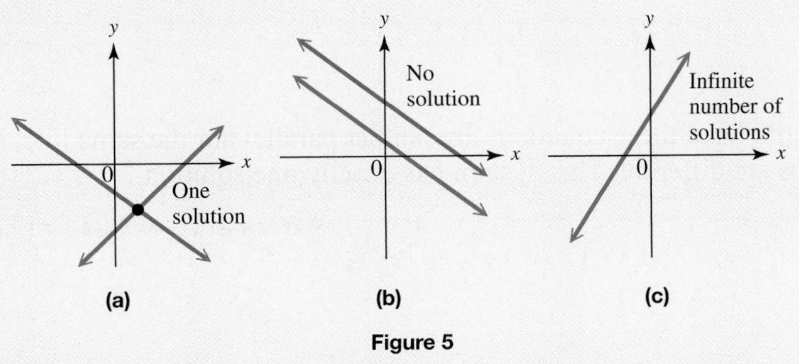

Figure 5

3 Solve each system of equations by graphing both equations on the same axes.

(a) $3x - y = 4$
$6x - 2y = 12$

(One of the lines is already graphed.)

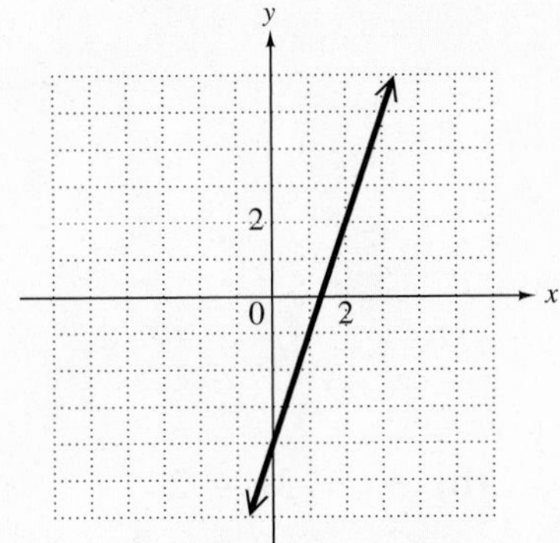

(b) $x - 3y = -2$
$2x - 6y = -4$

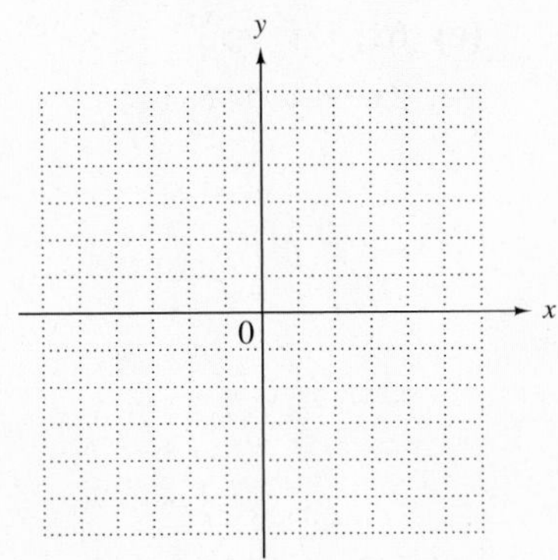

ANSWERS

3. (a) $\emptyset$ **(b)** $\{(x, y) \mid x - 3y = -2\}$

OBJECTIVE 4 Identify special systems without graphing. Example 3 showed that the graphs of an inconsistent system are parallel lines and the graphs of a system of dependent equations are the same line. We can recognize these special kinds of systems without graphing by using slopes.

EXAMPLE 4 Identifying the Three Cases by Using Slopes

Describe each system without graphing. State the number of solutions.

(a) $3x + 2y = 6$

$-2y = 3x - 5$

Write each equation in slope-intercept form, $y = mx + b$, by solving for y.

$3x + 2y = 6$			$-2y = 3x - 5$	
$2y = -3x + 6$	Subtract $3x$.		$y = -\frac{3}{2}x + \frac{5}{2}$	Divide by -2.
$y = -\frac{3}{2}x + 3$	Divide by 2.			

Both equations have slope $-\frac{3}{2}$ but they have different y-intercepts, 3 and $\frac{5}{2}$. In **Section 3.3,** we found that lines with the same slope are parallel, so these equations have graphs that are parallel lines. Thus, the system has no solution.

(b) $2x - y = 4$

$x = \frac{y}{2} + 2$

Again, write the equations in slope-intercept form.

$$2x - y = 4$$
$$-y = -2x + 4$$
$$y = 2x - 4$$

$$x = \frac{y}{2} + 2$$
$$\frac{y}{2} + 2 = x$$
$$\frac{y}{2} = x - 2$$
$$y = 2x - 4$$

The equations are exactly the same—their graphs are the same line. Thus, the system has an infinite number of solutions.

(c) $x - 3y = 5$

$2x + y = 8$

In slope-intercept form, the equations are as follows.

$$x - 3y = 5$$
$$-3y = -x + 5$$
$$y = \frac{1}{3}x - \frac{5}{3}$$

$$2x + y = 8$$
$$y = -2x + 8$$

The graphs of these equations are neither parallel nor the same line, since the slopes are different. This system has exactly one solution.

◀ Work Problem 4 at the Side.

4 Describe each system without graphing. State the number of solutions.

(a) $2x - 3y = 5$
$3y = 2x - 7$

(b) $-x + 3y = 2$
$2x - 6y = -4$

(c) $6x + y = 3$
$2x - y = -11$

ANSWERS

4. (a) The equations represent parallel lines. The system has no solution.
(b) The equations represent the same line. The system has an infinite number of solutions.
(c) The equations represent lines that are neither parallel nor the same line. The system has exactly one solution.

4.1 Exercises

FOR EXTRA HELP MyMathLab Math XL PRACTICE  WATCH DOWNLOAD  READ  REVIEW

1. Which ordered pair could be a solution of the system graphed? Why is it the only valid choice?

   A. (2, 2)
   B. (−2, 2)
   C. (−2, −2)
   D. (2, −2)

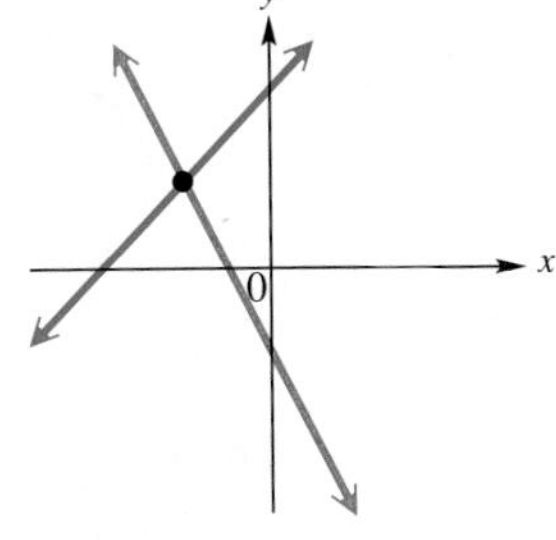

B, because the ordered pair must be in quadrant II.

2. Which ordered pair could be a solution of the system graphed? Why is it the only valid choice?

   A. (2, 0)
   B. (0, 2)
   C. (−2, 0)
   D. (0, −2)

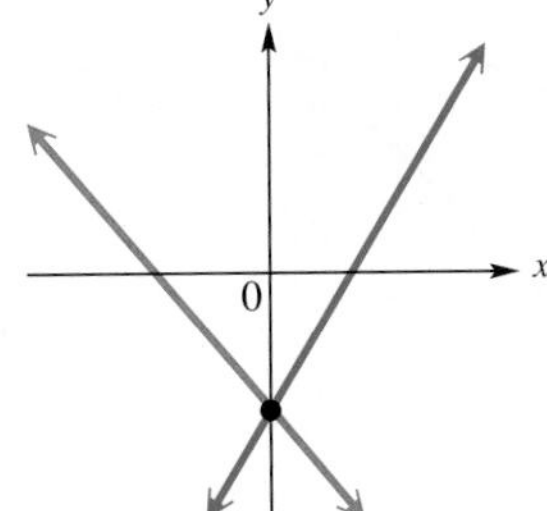

D, because the ordered pair must be on the y-axis, with y < 0.

3. How can you tell without graphing that this system has no solution?

$$x + y = 2$$
$$x + y = 4$$

There is no way that the sum of two numbers can be both 2 and 4 at the same time.

4. Explain why a system of two linear equations cannot have exactly two solutions.

Two lines will intersect in at most one point (if they are different) or infinitely many points (if they are the same) or no points (if they are parallel). They cannot intersect in exactly two points.

Decide whether the given ordered pair is a solution of the given system. See Example 1.

5. (2, −3)
$$x + y = -1$$
$$2x + 5y = 19$$
no

6. (4, 3)
$$x + 2y = 10$$
$$3x + 5y = 3$$
no

7. (−1, −3)
$$3x + 5y = -18$$
$$4x + 2y = -10$$
yes

8. (−9, −2)
$$2x - 5y = -8$$
$$3x + 6y = -39$$
yes

9. (7, −2)
$$4x = 26 - y$$
$$3x = 29 + 4y$$
yes

10. (9, 1)
$$2x = 23 - 5y$$
$$3x = 24 + 3y$$
yes

11. (6, −8)
$$-2y = x + 10$$
$$3y = 2x + 30$$
no

12. (−5, 2)
$$5y = 3x + 20$$
$$3y = -2x - 4$$
no

Solve each system of equations by graphing. If the system is inconsistent or the equations are dependent, say so. See Examples 2 and 3. **We show the graphs here only for Exercises 13–18.**

13. $x - y = 2$
$x + y = 6$

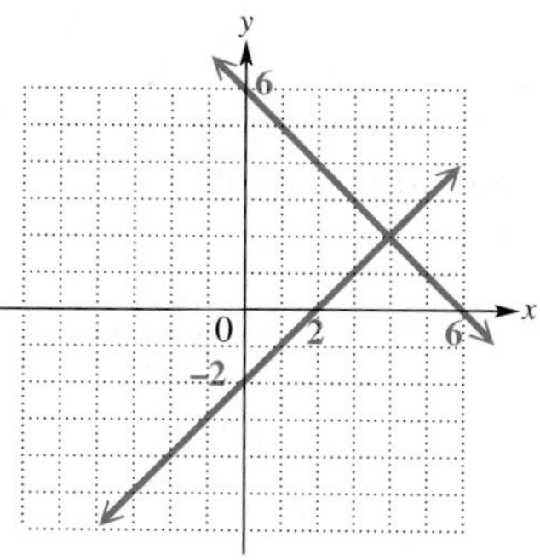

{(4, 2)}

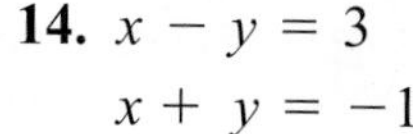

14. $x - y = 3$
$x + y = -1$

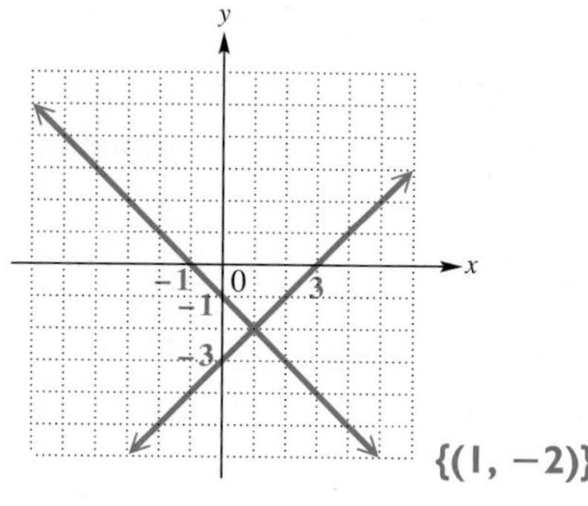

{(1, −2)}

15. $x + y = 4$
$y - x = 4$

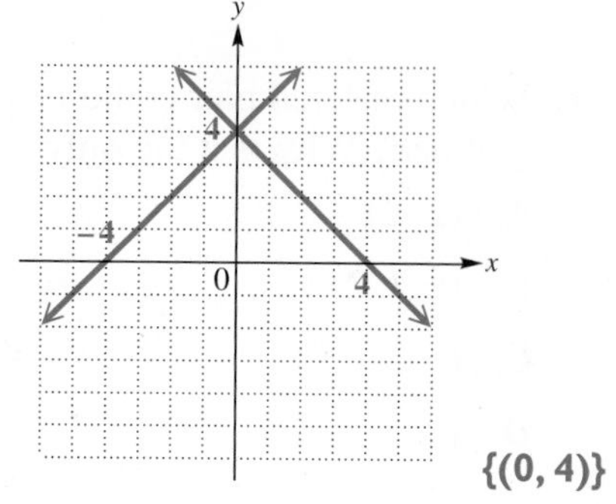

{(0, 4)}

16. $x + y = -5$
$x - y = 5$

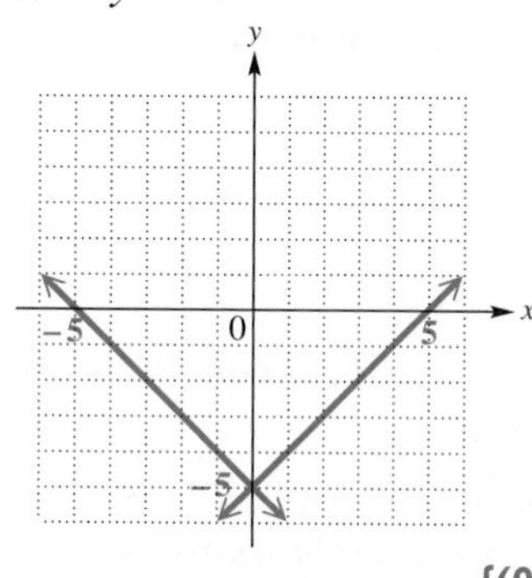

{(0, −5)}

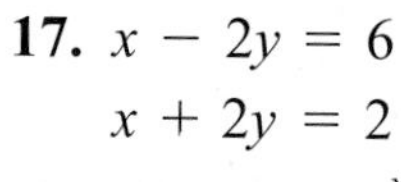

17. $x - 2y = 6$
$x + 2y = 2$

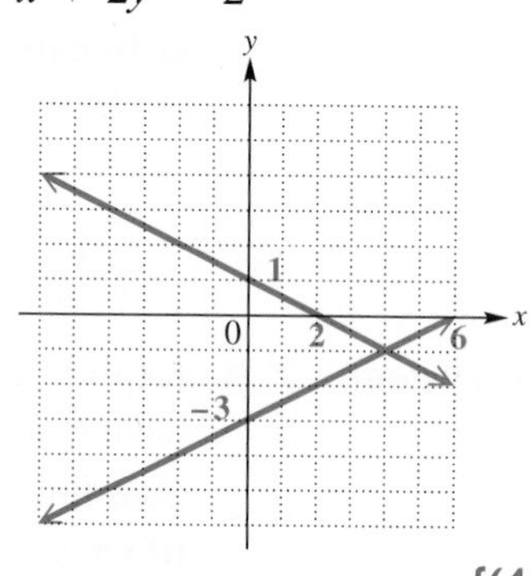

{(4, −1)}

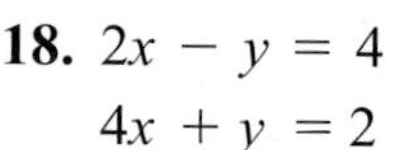

18. $2x - y = 4$
$4x + y = 2$

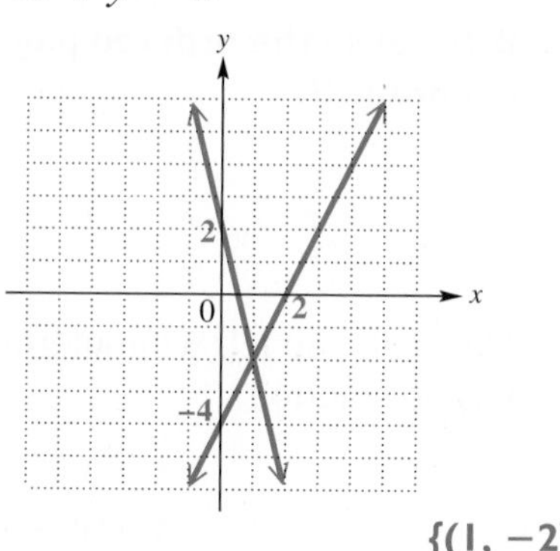

{(1, −2)}

19. $3x - 2y = -3$
$-3x - y = -6$

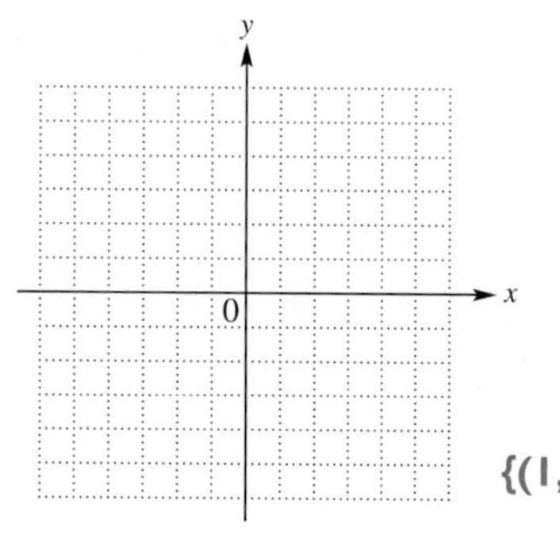

{(1, 3)}

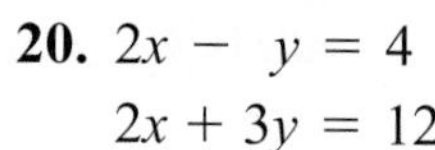

20. $2x - y = 4$
$2x + 3y = 12$

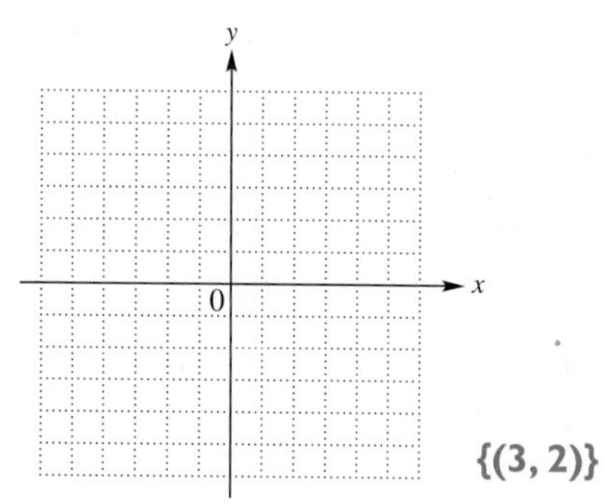

{(3, 2)}

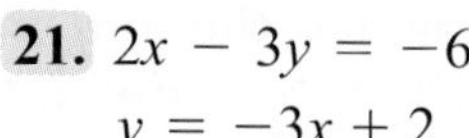

21. $2x - 3y = -6$
$y = -3x + 2$

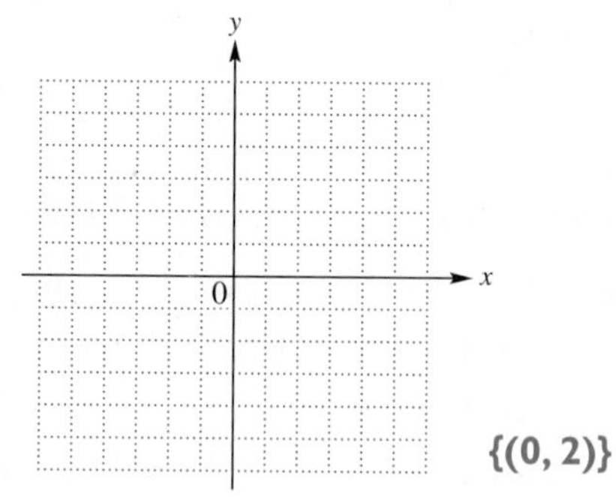

{(0, 2)}

22. $-3x + y = -3$
$y = x - 3$

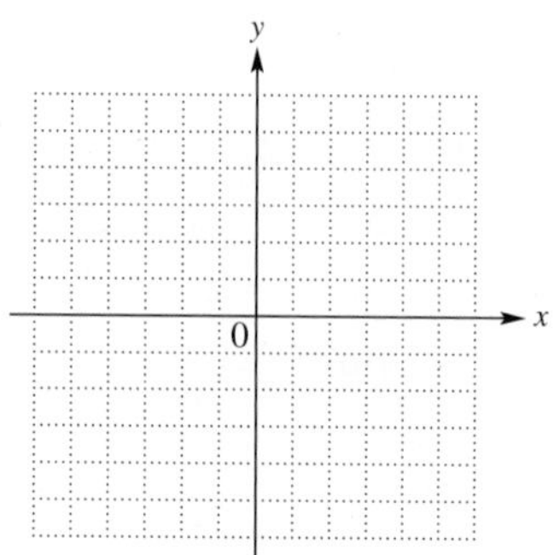

{(0, −3)}

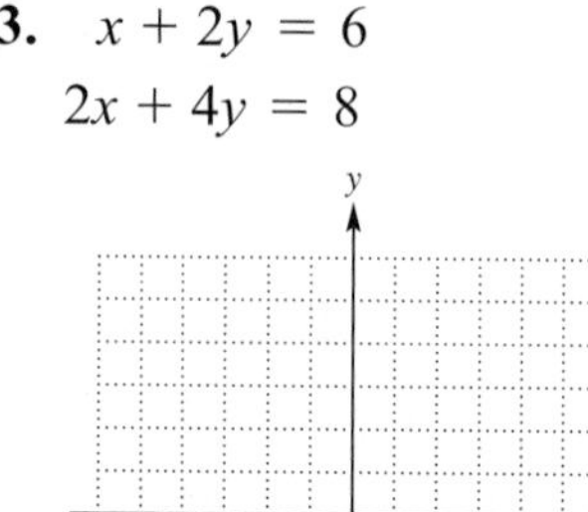

23. $x + 2y = 6$
$2x + 4y = 8$

24. $2x - y = 6$
$6x - 3y = 12$

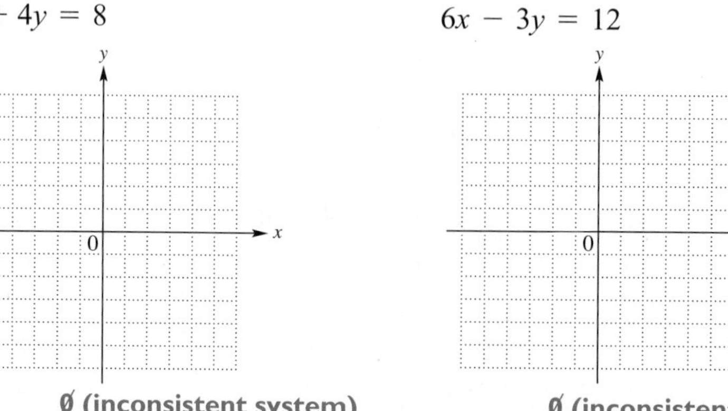

∅ (inconsistent system)

∅ (inconsistent system)

25. $4x - 2y = 8$
$2x = y + 4$

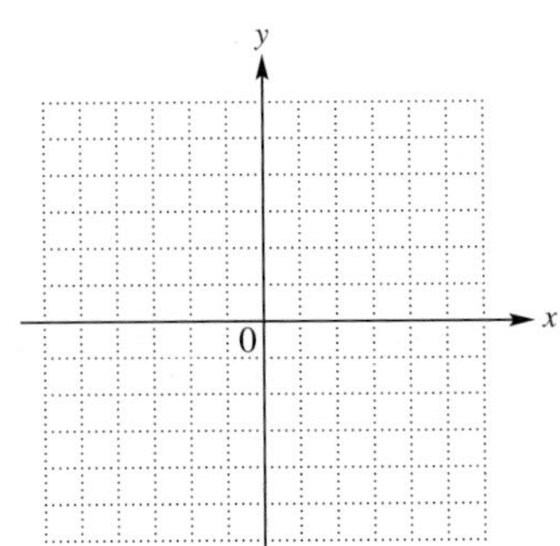

$\{(x, y) \mid 2x - y = 4\}$
(dependent equations)

26. $3x = 5 - y$
$6x + 2y = 10$

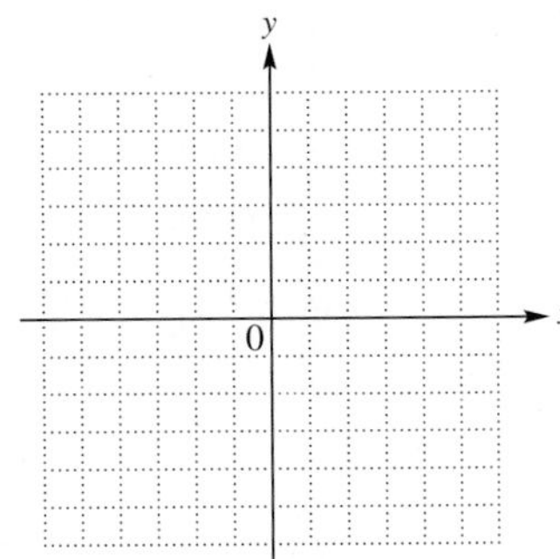

$\{(x, y) \mid 3x + y = 5\}$
(dependent equations)

27. $3x - 4y = 24$
$y = -\frac{3}{2}x + 3$

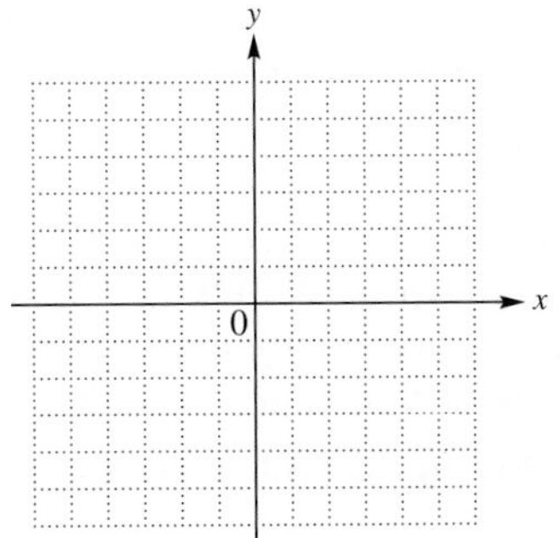

$\{(4, -3)\}$

28. $3x - 2y = 12$
$y = -4x + 5$

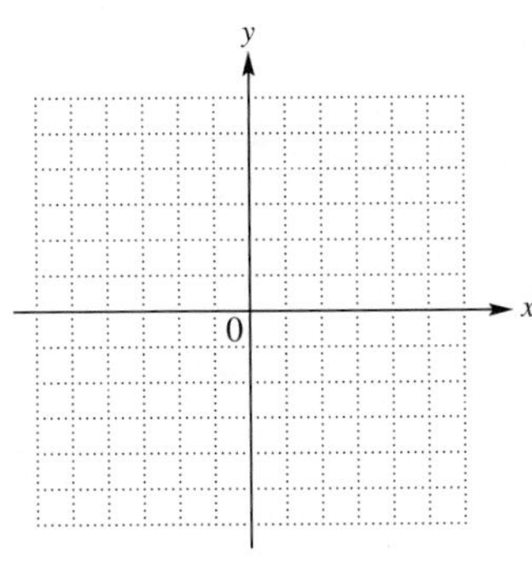

$\{(2, -3)\}$

29. $3x = y + 5$
$6x - 5 = 2y$

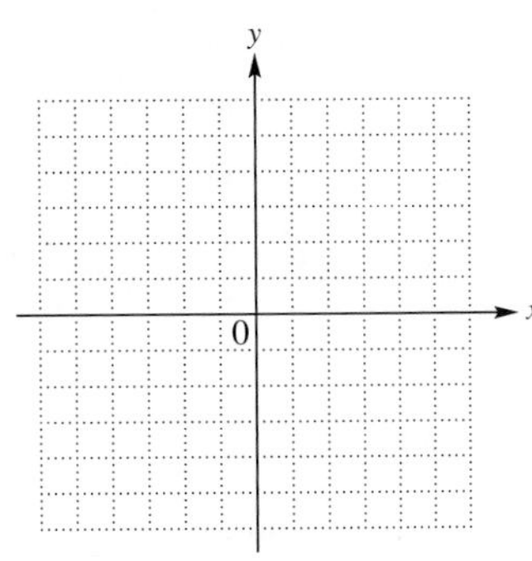

∅ (inconsistent system)

30. $2x = y - 4$
$4x - 2y = -4$

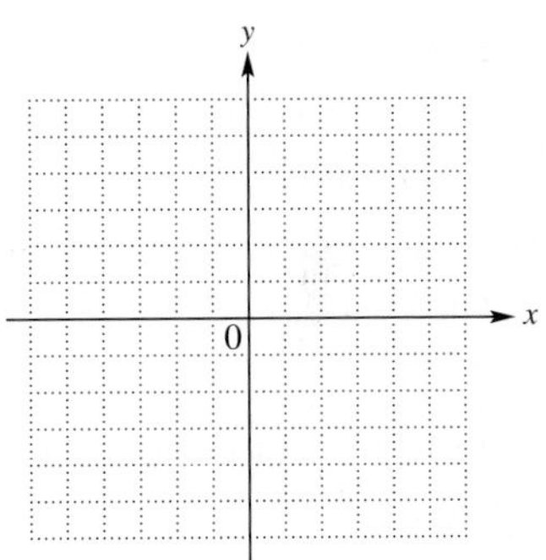

∅ (inconsistent system)

Without graphing, answer the following questions for each linear system. See Example 4.

(a) *Is the system inconsistent, are the equations dependent, or neither?*
(b) *Is the graph a pair of intersecting lines, a pair of parallel lines, or one line?*
(c) *Does the system have one solution, no solution, or an infinite number of solutions?*

31. $y - x = -5$
$x + y = 1$

(a) neither
(b) intersecting lines
(c) one solution

32. $2x + y = 6$
$x - 3y = -4$

(a) neither
(b) intersecting lines
(c) one solution

33. $x + 2y = 0$
$4y = -2x$

(a) dependent
(b) one line
(c) infinite number of solutions

34. $y = 3x$
$y + 3 = 3x$

(a) inconsistent
(b) parallel lines
(c) no solution

35. $5x + 4y = 7$
$10x + 8y = 4$

(a) inconsistent
(b) parallel lines
(c) no solution

36. $4x - 6y = 10$
$-6x + 9y = -15$

(a) dependent
(b) one line
(c) infinite number of solutions

The numbers of daily morning and evening newspapers in the United States in selected years are shown in the graph. Use the graph to work Exercises 37 and 38.

Number of Daily Newspapers

Evening dailies
Morning dailies
Number of Newspapers
Year

Source: Editor & Publisher International Year Book.

37. For which years were there more evening dailies than morning dailies?

1980–2000

38. Estimate the year in which the number of evening and morning dailies was closest to the same. About how many newspapers of each type were there in that year?

2001; about 800 newspapers

Work Exercises 39 and 40 using the graphs provided.

39. The graph shows how college students managed their money during the period 1997–2004.

(a) During what period did ATM use dominate both credit card *and* debit card use?

1997–2002

(b) In what year did debit card use overtake credit card use?

2001

(c) In what year did debit card use overtake ATM use?

2002

(d) Write an ordered pair for the debit card use data in the year 2004.

(2004, 70) (The y-value is approximate.)

(e) Describe the trend in debit card use over this period.

During the period 1997–2004, debit card use went from least popular to most popular of the three methods depicted.

How College Students Manage Their Money

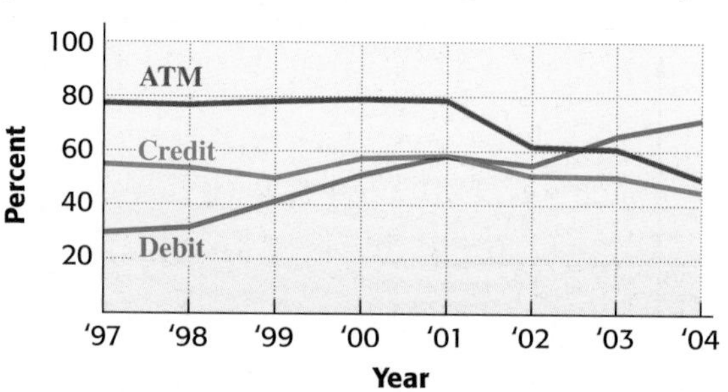

Source: Georgetown University Credit Research Center.

40. The graph shows how the average viewing hours for broadcast TV and cable/satellite TV in the United States has changed during the period 1998–2004.

(a) In approximately what year did Americans spend almost the same number of hours watching broadcast and cable/satellite TV? How many hours per year was this?

2001; about 840 hr

(b) Express the point of intersection of the two graphs as an ordered pair of the form (year, hours).

(2001, 840)

(c) During what period was the time spent watching broadcast TV almost constant?

1999–2000

Watching the Tube

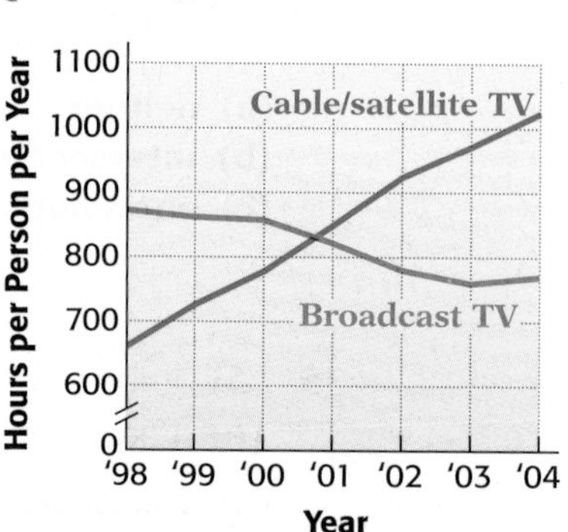

Source: Veronis Suhler Stevenson.

4.2 Solving Systems of Linear Equations by Substitution

OBJECTIVES

1. Solve linear systems by substitution.
2. Solve special systems by substitution.
3. Solve linear systems with fractions and decimals by substitution.

OBJECTIVE 1 Solve linear systems by substitution.

Work Problem 1 at the Side. ▶

As we saw in Problem 1 at the side, graphing to solve a system of equations has a serious drawback: It is difficult to accurately find a solution such as $\left(\frac{11}{3}, -\frac{4}{9}\right)$ from a graph. One algebraic method for solving a system of equations is the **substitution method.**

EXAMPLE 1 Using the Substitution Method

Solve the system by the substitution method.

$$3x + 5y = 26$$
$$y = 2x$$

The second equation is already solved for y. This equation says that $y = 2x$. Substituting $2x$ for y in the first equation gives

$$\begin{aligned} 3x + 5y &= 26 \\ 3x + 5(2x) &= 26 && \text{Let } y = 2x. \\ 3x + 10x &= 26 && \text{Multiply.} \\ 13x &= 26 && \text{Combine like terms.} \\ x &= 2. && \text{Divide by 13.} \end{aligned}$$

Because $x = 2$, we find y from the equation $y = 2x$ by substituting 2 for x.

$$y = 2(2) = 4 \qquad \text{Let } x = 2.$$

Check that the solution set of the given system is $\{(2, 4)\}$ by substituting 2 for x and 4 for y in *both* equations.

Work Problem 2 at the Side. ▶

EXAMPLE 2 Using the Substitution Method

Solve the system by the substitution method.

$$2x + 5y = 7$$
$$x = -1 - y$$

The second equation gives x in terms of y. Substitute $-1 - y$ for x in the first equation.

$$\begin{aligned} 2x + 5y &= 7 \\ 2(-1 - y) + 5y &= 7 && \text{Let } x = -1 - y. \\ -2 - 2y + 5y &= 7 && \text{Distributive property} \\ -2 + 3y &= 7 && \text{Combine like terms.} \\ 3y &= 9 && \text{Add 2.} \\ y &= 3 && \text{Divide by 3.} \end{aligned}$$

Distribute 2 to *both* -1 and $-y$.

To find x, substitute 3 for y in the equation $x = -1 - y$ to get

$$x = -1 - 3 = -4.$$

Write the x-coordinate first.

Check that the solution set of the given system is $\{(-4, 3)\}$.

Work Problem 3 at the Side. ▶

1 Solve the system by graphing.

$$2x + 3y = 6$$
$$x - 3y = 5$$

Can you determine the answer? Why or why not?

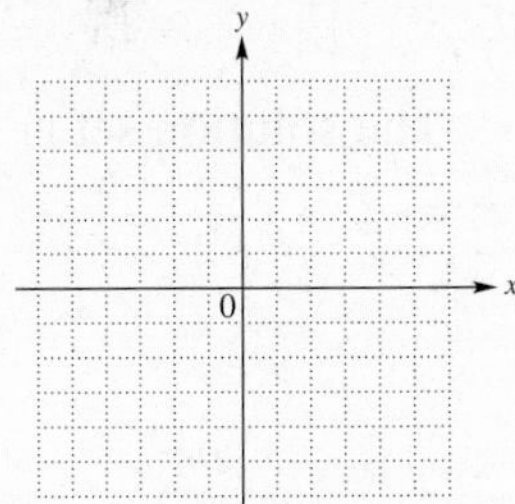

2 Fill in the blanks to solve by the substitution method. Check your solution.

$$3x + 5y = 69$$
$$y = 4x$$

$3x + 5(\_\_\_\_) = 69$

$\_\_\_\_ = 69$

$x = \_\_\_\_$

$y = 4(\_\_\_\_) = \_\_\_\_$

The solution set is ____.

3 Solve by the substitution method. Check your solution.

$$2x + 7y = -12$$
$$x = 3 - 2y$$

ANSWERS

1. The answer cannot be determined from the graph because it is too difficult to read the exact coordinates.
2. $4x$; $23x$; 3; 3; 12; $\{(3, 12)\}$
3. $\{(15, -6)\}$

CAUTION

Even though we found y first in Example 2, ***the x-coordinate is always written first in the ordered-pair solution of a system.***

To solve a system by substitution, follow these steps.

Solving a Linear System by Substitution

Step 1 **Solve one equation for either variable.** If one of the variables has coefficient 1 or -1, choose it, since it usually makes the substitution easier.

Step 2 **Substitute** for that variable in the other equation. The result should be an equation with just one variable.

Step 3 **Solve** the equation from Step 2.

Step 4 **Substitute** the result from Step 3 into the equation from Step 1 to find the value of the other variable.

Step 5 **Check** the solution in *both* of the original equations. Then write the solution set.

EXAMPLE 3 Using the Substitution Method

Use substitution to solve the system.

$$2x = 4 - y \quad (1)$$
$$5x + 3y = 10 \quad (2)$$

Step 1 For the substitution method, we must solve one of the equations for either x or y. Because the coefficient of y in equation (1) is -1, we choose equation (1) and solve for y.

$$2x = 4 - y \quad (1)$$
$$y + 2x = 4 \quad \text{Add } y.$$
$$y = -2x + 4 \quad \text{Subtract } 2x.$$

Step 2 Now substitute $-2x + 4$ for y in equation (2).

$$5x + 3y = 10 \quad (2)$$
$$5x + 3(-2x + 4) = 10 \quad \text{Let } y = -2x + 4.$$

Step 3 Now solve the equation from Step 2.

$$5x - 6x + 12 = 10 \quad \text{Distributive property}$$
$$-x + 12 = 10 \quad \text{Combine like terms.}$$
$$-x = -2 \quad \text{Subtract 12.}$$
$$x = 2 \quad \text{Multiply by } -1.$$

Distribute 3 to both $-2x$ and 4.

Step 4 Since $y = -2x + 4$ and $x = 2$, $y = -2(2) + 4 = 0$.

Step 5 Check that $(2, 0)$ is the solution.

Check

$2x = 4 - y$ (1)	$5x + 3y = 10$ (2)
$2(2) \stackrel{?}{=} 4 - 0$	$5(2) + 3(0) \stackrel{?}{=} 10$
$4 = 4$ True	$10 = 10$ True

Since both results are true, the solution set of the system is $\{(2, 0)\}$.

◀ Work Problem 4 at the Side.

4 Solve each system by substitution. Check each solution.

(a) Fill in the blanks to solve

$$x + 4y = -1$$
$$2x - 5y = 11.$$

Solve the first equation for x.

$$x = -1 - \_\_\_\_$$

Substitute into the second equation to find y.

$$2(\_\_\_\_) - 5y = 11$$
$$-2 - 8y - 5y = 11$$
$$-2 - \_\_\_\_ y = 11$$
$$\_\_\_\_ y = 13$$
$$y = \_\_\_\_$$

Find x.

$$x = -1 - \_\_\_\_$$
$$x = \_\_\_\_$$

The solution set is ____.

(b) $2x + 5y = 4$
$x + y = -1$

ANSWERS

4. (a) $4y$; $-1 - 4y$; 13; -13; -1; -4; 3; $\{(3, -1)\}$ **(b)** $\{(-3, 2)\}$

EXAMPLE 4 **Using the Substitution Method**

Use substitution to solve the system.

$$2x + 3y = 10 \quad (1)$$
$$-3x - 2y = 0 \quad (2)$$

Step 1 To use the substitution method, we must solve one of the equations for one of the variables. We choose equation (1) and solve for x.

$$2x + 3y = 10 \quad (1)$$
$$2x = 10 - 3y \quad \text{Subtract } 3y.$$
$$x = 5 - \frac{3}{2}y \quad \text{Divide by 2.}$$

Step 2 Substitute this expression for x in equation (2).

$$-3x - 2y = 0 \quad (2)$$
$$-3\left(5 - \frac{3}{2}y\right) - 2y = 0 \quad \text{Let } x = 5 - \tfrac{3}{2}y.$$

Step 3

$$-15 + \frac{9}{2}y - 2y = 0 \quad \text{Distributive property}$$

$-3(-\frac{3}{2}) = (-\frac{3}{1})(-\frac{3}{2}) = \frac{9}{2}$

$$-15 + \frac{5}{2}y = 0 \quad \text{Combine like terms.}$$
$$\frac{5}{2}y = 15 \quad \text{Add 15.}$$

$\frac{2}{5}(\frac{5}{2}y) = (\frac{2}{5} \cdot \frac{5}{2})y = 1 \cdot y = y$

$$y = \frac{30}{5} = 6 \quad \text{Multiply by } \tfrac{2}{5}.$$

Step 4 Find x by substituting 6 for y in $x = 5 - \frac{3}{2}y$.

$$x = 5 - \frac{3}{2}(6) = -4$$

Step 5 Check that $(-4, 6)$ is the solution.

Check

$2x + 3y = 10$ (1)		$-3x - 2y = 0$ (2)
$2(-4) + 3(6) \stackrel{?}{=} 10$		$-3(-4) - 2(6) \stackrel{?}{=} 0$
$-8 + 18 \stackrel{?}{=} 10$		$12 - 12 \stackrel{?}{=} 0$
$10 = 10$ True		$0 = 0$ True

Both results are true, so the solution set of the system is $\{(-4, 6)\}$.

Note

In Example 4, we could have started the solution by solving the second equation for either x or y and then substituting the result into the first equation. The solution would be the same.

Work Problem 5 at the Side. ▶

5 Solve the system by substitution. Check your solution.

$$3x + 2y = 1$$
$$3x - 4y = -11$$

Answer

5. $\{(-1, 2)\}$

OBJECTIVE 2 Solve special systems by substitution. We can solve inconsistent systems with graphs that are parallel lines and systems of dependent equations with graphs that are the same line using the substitution method.

EXAMPLE 5 Solving an Inconsistent System by Substitution

Use substitution to solve the system.

$$x = 5 - 2y \quad (1)$$
$$2x + 4y = 6 \quad (2)$$

Substitute $5 - 2y$ for x in equation (2).

$$2x + 4y = 6 \quad (2)$$
$$2(5 - 2y) + 4y = 6 \quad \text{Let } x = 5 - 2y.$$
$$10 - 4y + 4y = 6 \quad \text{Distributive property}$$
$$10 = 6 \quad \text{False}$$

This false result means that the equations in the system have graphs that are parallel lines. The system is inconsistent, and the solution set is $\emptyset$. See Figure 6.

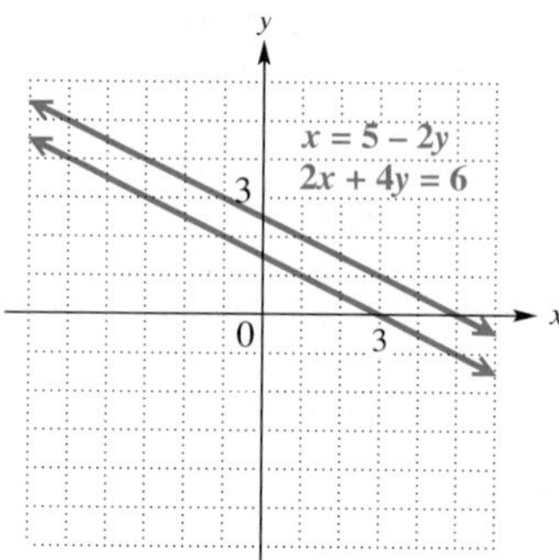

Figure 6

CAUTION
It is a common error to give "false" as the answer to an inconsistent system. The correct response is $\emptyset$.

EXAMPLE 6 Solving a System with Dependent Equations by Substitution

Solve the system by the substitution method.

$$3x - y = 4 \quad (1)$$
$$-9x + 3y = -12 \quad (2)$$

Begin by solving equation (1) for y to get $y = 3x - 4$. Substitute $3x - 4$ for y in equation (2) and solve the resulting equation.

$$-9x + 3y = -12 \quad (2)$$
$$-9x + 3(3x - 4) = -12 \quad \text{Let } y = 3x - 4.$$
$$-9x + 9x - 12 = -12 \quad \text{Distributive property}$$
$$0 = 0 \quad \text{Add 12; combine like terms.}$$

Continued on Next Page

This true result means that every solution of one equation is also a solution of the other, so the system has an infinite number of solutions—all the ordered pairs corresponding to points that lie on the common graph. The solution set is $\{(x, y) | 3x - y = 4\}$. A graph of the equations of this system is shown in Figure 7.

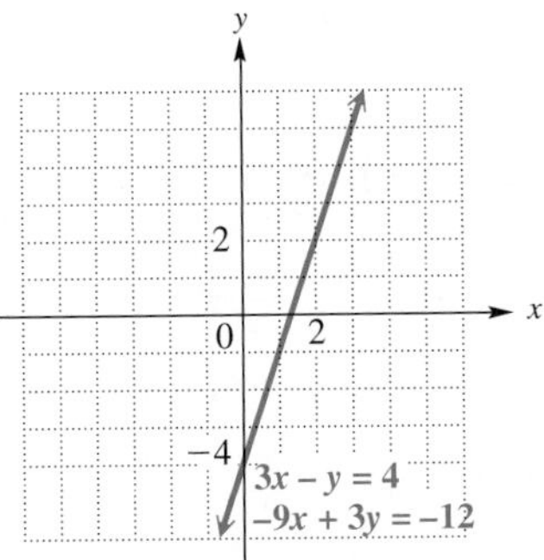

Figure 7

CAUTION

It is a common error to give "true" as the solution of a system of dependent equations. Remember that we give the solution set in set-builder notation using an equation that is in standard form, with integer coefficients having greatest common factor 1.

Work Problem 6 at the Side. ▶

6 Solve each system by substitution.

(a) $8x - y = 4$
$y = 8x + 4$

(b) $7x - 6y = 10$
$-14x + 20 = -12y$

OBJECTIVE 3 Solve linear systems with fractions and decimals by substitution. When a system includes an equation with fractions as coefficients, eliminate the fractions by multiplying each side of the equation by a common denominator. Then solve the resulting system.

EXAMPLE 7 Using the Substitution Method with Fractions as Coefficients

Solve the system by the substitution method.

$$3x + \frac{1}{4}y = 2 \qquad (1)$$

$$\frac{1}{2}x + \frac{3}{4}y = -\frac{5}{2} \qquad (2)$$

Clear equation (1) of fractions by multiplying each side by 4.

$$4\left(3x + \frac{1}{4}y\right) = 4(2) \qquad \text{Multiply by 4.}$$

$$4(3x) + 4\left(\frac{1}{4}y\right) = 4(2) \qquad \text{Distributive property}$$

$$12x + y = 8 \qquad (3)$$

Continued on Next Page

ANSWERS

6. (a) $\emptyset$ **(b)** $\{(x, y) | 7x - 6y = 10\}$

7 Solve the system by substitution. First clear all fractions.

$$\frac{2}{3}x + \frac{1}{2}y = 6$$
$$\frac{1}{2}x - \frac{3}{4}y = 0$$

Now clear equation (2) of fractions by multiplying each side by the common denominator 4.

$$\frac{1}{2}x + \frac{3}{4}y = -\frac{5}{2} \quad (2)$$
$$4\left(\frac{1}{2}x + \frac{3}{4}y\right) = 4\left(-\frac{5}{2}\right) \quad \text{Multiply by 4.}$$
$$4\left(\frac{1}{2}x\right) + 4\left(\frac{3}{4}y\right) = 4\left(-\frac{5}{2}\right) \quad \text{Distributive property}$$
$$2x + 3y = -10 \quad (4)$$

The given system of equations has been simplified to the equivalent system

$$12x + y = 8 \quad (3)$$
$$2x + 3y = -10. \quad (4)$$

To solve this system by substitution, equation (3) can be solved for y.

$$12x + y = 8 \quad (3)$$
$$y = -12x + 8 \quad \text{Subtract } 12x.$$

Now substitute this result for y in equation (4).

$$2x + 3y = -10 \quad (4)$$
$$2x + 3(-12x + 8) = -10 \quad \text{Let } y = -12x + 8.$$
$$2x - 36x + 24 = -10 \quad \text{Distributive property}$$
$$-34x = -34 \quad \text{Combine like terms; subtract 24.}$$
$$x = 1 \quad \text{Divide by } -34.$$

Distribute 3 to both $-12x$ and 8.

Substitute 1 for x in $y = -12x + 8$ to get

$$y = -12(1) + 8 = -4.$$

Check by substituting 1 for x and -4 for y in both of the original equations. The solution set is $\{(1, -4)\}$.

◀ *Work Problem* 7 *at the Side.*

8 Complete the process of solving the system given in Example 8 by substitution. (*Hint:* Solve equation (4) for x. Then substitute this result for x in equation (3) to find y.)

If any of the coefficients in the equations of a system are decimals, we can eliminate the decimals by multiplying by a power of 10, as we did when solving linear equations with decimal coefficients in **Section 2.3.**

EXAMPLE 8 Using the Substitution Method with Decimals as Coefficients

Solve the system by the substitution method.

$$0.5x + 2.4y = 4.2 \quad (1)$$
$$-0.1x + 1.5y = 5.1 \quad (2)$$

To eliminate (or "clear") decimals, multiply each equation by 10.

$$5x + 24y = 42 \quad (3)$$
$$-x + 15y = 51 \quad (4) \quad (-0.1x) \cdot 10 = -1x = -x$$

Now we can solve this equivalent system by substitution.

◀ *Work problem* 8 *at the Side.*

ANSWERS

7. $\{(6, 4)\}$
8. $\{(-6, 3)\}$

4.2 Exercises

FOR EXTRA HELP

MyMathLab

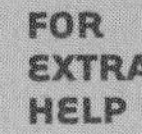
Math XL PRACTICE

WATCH

DOWNLOAD

READ

REVIEW

1. A student solves the system

$$5x - y = 15$$
$$7x + y = 21$$

and finds that $x = 3$, which is the correct value for x. The student gives the solution set as $\{3\}$. ***WHAT WENT WRONG?***

The y-value must also be determined. The solution set is {(3,0)}.

2. When you use the substitution method, how can you tell that a system has

(a) no solution?

A false statement, such as 0 = 3, occurs.

(b) an infinite number of solutions?

A true statement, such as 0 = 0, occurs.

Solve each system by the substitution method. Check each solution. See Examples 1–8.

3. $x + y = 12$
$y = 3x$
$\{(3, 9)\}$

4. $x + 3y = -28$
$y = -5x$
$\{(2, -10)\}$

5. $3x + 2y = 27$
$x = y + 4$
$\{(7, 3)\}$

6. $4x + 3y = -5$
$x = y - 3$
$\{(-2, 1)\}$

7. $3x + 5y = 14$
$x - 2y = -10$
$\{(-2, 4)\}$

8. $5x + 2y = -1$
$2x - y = -13$
$\{(-3, 7)\}$

9. $3x + 4 = -y$
$2x + y = 0$
$\{(-4, 8)\}$

10. $2x - 5 = -y$
$x + 3y = 0$
$\{(3, -1)\}$

11. $7x + 4y = 13$
$x + y = 1$
$\{(3, -2)\}$

12. $3x - 2y = 19$
$x + y = 8$
$\{(7, 1)\}$

13. $3x - y = 5$
$y = 3x - 5$
$\{(x, y) \mid 3x - y = 5\}$

14. $4x - y = -3$
$y = 4x + 3$
$\{(x, y) \mid 4x - y = -3\}$

15. $6x - 8y = 6$
$2y = -2 + 3x$
$\left\{\left(\frac{1}{3}, -\frac{1}{2}\right)\right\}$

16. $3x + 2y = 6$
$6x = 8 + 4y$
$\left\{\left(\frac{5}{3}, \frac{1}{2}\right)\right\}$

17. $2x + 8y = 3$
$x = 8 - 4y$
$\emptyset$

18. $2x + 10y = 3$
$x = 1 - 5y$
$\emptyset$

19. $12x - 16y = 8$
$3x = 4y + 2$
$\{(x, y) | 3x - 4y = 2\}$

20. $6x + 9y = 6$
$2x = 2 - 3y$
$\{(x, y) | 2x + 3y = 2\}$

21. $\frac{1}{5}x + \frac{2}{3}y = -\frac{8}{5}$
$3x - y = 9$
$\{(2, -3)\}$

22. $\frac{1}{3}x - \frac{1}{2}y = \frac{1}{6}$
$3x - 2y = 9$
$\{(5, 3)\}$

23. $\frac{x}{2} - \frac{y}{3} = 9$
$\frac{x}{5} - \frac{y}{4} = 5$
$\{(10, -12)\}$

24. $\frac{1}{6}x + \frac{1}{6}y = 2$
$-\frac{1}{2}x - \frac{1}{3}y = -8$
$\{(24, -12)\}$

25. $\frac{x}{5} + 2y = \frac{16}{5}$
$\frac{3x}{5} + \frac{y}{2} = -\frac{7}{5}$
$\{(-4, 2)\}$

26. $\frac{x}{3} - \frac{3y}{4} = -\frac{1}{2}$
$\frac{x}{6} + \frac{y}{8} = \frac{3}{4}$
$\{(3, 2)\}$

27. $0.1x + 0.9y = -2$
$0.5x - 0.2y = 4.1$
$\{(7, -3)\}$

28. $0.2x - 1.3y = -3.2$
$-0.1x + 2.7y = 9.8$
$\{(10, 4)\}$

29. $0.08x - 0.01y = 1.3$
$0.22x + 0.15y = 8.9$
$\{(20, 30)\}$

Relating Concepts (Exercises 30–33) For Individual or Group Work

A system of linear equations can be used to model the cost and the revenue of a business. ***Work Exercises 30–33 in order.***

30. Suppose that you start a business manufacturing and selling bicycles, and it costs you \$5000 to get started. You determine that each bicycle will cost \$400 to manufacture. Explain why the linear equation $y_1 = 400x + 5000$ gives your *total* cost to manufacture x bicycles (y_1 in dollars).

To find the total cost, multiply the number of bicycles (x) by the cost per bicycle (400 dollars) and add the fixed cost (5000 dollars). Thus, $y_1 = 400x + 5000$ gives this total cost (in dollars).

31. You decide to sell each bike for \$600. What expression in x represents the revenue you will take in if you sell x bikes? Write an equation using y_2 to express your revenue when you sell x bikes (y_2 in dollars).

$y_2 = 600x$

32. Form a system from the two equations in Exercises 30 and 31, and then solve the system, assuming $y_1 = y_2$, that is, cost = revenue.

$y_1 = 400x + 5000$
$y_2 = 600x$; solution set: $\{(25, 15{,}000)\}$

33. The value of x from Exercise 32 is the number of bikes it takes to *break even.* Fill in the blanks: When ______ bikes are sold, the break-even point is reached. At that point, you have spent ______ dollars and taken in ______ dollars.

25; 15,000; 15,000

4.3 Solving Systems of Linear Equations by Elimination

OBJECTIVE 1 Solve linear systems by elimination. An algebraic method that depends on the addition property of equality can be used to solve systems. As mentioned earlier, adding the same quantity to each side of an equation results in equal sums.

$$\text{If} \quad A = B, \quad \text{then} \quad A + C = B + C.$$

This addition can be taken a step further. Adding *equal* quantities, rather than the *same* quantity, to both sides of an equation also results in equal sums.

$$\text{If} \quad A = B \quad \text{and} \quad C = D, \quad \text{then} \quad A + C = B + D.$$

Using the addition property to solve systems is called the **elimination method.** When using this method, the idea is to *eliminate* one of the variables. ***To do this, one of the variables in the two equations must have coefficients that are opposites.***

OBJECTIVES

1. **Solve linear systems by elimination.**
2. **Multiply when using the elimination method.**
3. **Use an alternative method to find the second value in a solution.**
4. **Use the elimination method to solve special systems.**

EXAMPLE 1 Using the Elimination Method

Use the elimination method to solve the system.

$$x + y = 5$$
$$x - y = 3$$

Each equation in this system is a statement of equality, so the sum of the left sides equals the sum of the right sides. Adding in this way gives

$$(x + y) + (x - y) = 5 + 3.$$

Combine like terms and simplify to get

$$2x = 8$$
$$x = 4. \quad \text{Divide by 2.}$$

Notice that y has been eliminated. The result, $x = 4$, gives the x-value of the solution of the given system. To find the y-value of the solution, substitute 4 for x in either of the two equations of the system.

Work Problem 1 at the Side. ▶

Check the solution set found at the side, $\{(4, 1)\}$, by substituting 4 for x and 1 for y in both equations of the given system.

Check

$x + y = 5$	$x - y = 3$
$4 + 1 \stackrel{?}{=} 5$	$4 - 1 \stackrel{?}{=} 3$
$5 = 5$ True	$3 = 3$ True

Since both results are true, the solution set of the system is $\{(4, 1)\}$.

1 (a) Substitute 4 for x in the equation $x + y = 5$ to find the value of y.

(b) Give the solution set of the system.

CAUTION
A system is not completely solved until values for both x and y are found. Do not stop after finding the value of only one variable. Remember to write the solution set as a set containing an ordered pair.

ANSWERS

1. (a) $y = 1$ **(b)** $\{(4, 1)\}$

◀ Work Problem 2 at the Side.

In general, to solve a system by elimination, follow these steps.

Solving a Linear System by Elimination

Step 1 **Write both equations in standard form** $Ax + By = C$.

Step 2 **Transform so that the coefficients of one pair of variable terms are opposites.** Multiply one or both equations by appropriate numbers so that the sum of the coefficients of either the x- or y-terms is 0.

Step 3 **Add** the new equations to eliminate a variable. The sum should be an equation with just one variable.

Step 4 **Solve** the equation from Step 3 for the remaining variable.

Step 5 **Substitute** the result from Step 4 into *either* of the original equations and solve for the other variable.

Step 6 **Check** the solution in *both* of the original equations. Then write the solution set.

It does not matter which variable is eliminated first. Usually we choose the one that is more convenient to work with.

EXAMPLE 2 **Using the Elimination Method**

Solve the system.

$$y + 11 = 2x$$
$$5x = y + 26$$

Step 1 Rewrite both equations in the form $Ax + By = C$ to get the system

$$-2x + y = -11 \quad \text{Subtract } 2x \text{ and } 11.$$
$$5x - y = 26. \quad \text{Subtract } y.$$

Step 2 Because the coefficients of y are 1 and -1, adding will eliminate y. It is not necessary to multiply either equation by a number.

Step 3 Add the two equations. This time we use vertical addition.

$$\begin{array}{r} -2x + y = -11 \\ 5x - y = 26 \\ \hline 3x \phantom{{}-y} = 15 \end{array} \quad \text{Add in columns.}$$

Step 4 Solve the equation.

$$3x = 15$$
$$x = 5 \quad \text{Divide by 3.}$$

Don't stop here.

Step 5 Find the value of y by substituting 5 for x in either of the original equations. Choosing the first equation gives

$$y + 11 = 2x$$
$$y + 11 = 2(5) \quad \text{Let } x = 5.$$
$$y + 11 = 10$$
$$\mathbf{y = -1}. \quad \text{Subtract 11.}$$

Continued on Next Page

2 Solve each system by the elimination method. Check each solution.

(a) Fill in the blanks to solve the following system.

$$x + y = 8$$
$$x - y = 2$$

Add.

$$(x + y) + (x - y) = 8 + \_\_\_$$
$$2\_\_\_ = \_\_\_$$
$$x = \_\_\_$$

Find y.

$$x - y = 2$$
$$\_\_\_ - y = 2$$
$$-y = \_\_\_$$
$$y = \_\_\_$$

The solution set is ___.

(b) $3x - y = 7$
$2x + y = 3$

ANSWERS

2. (a) 2; x; 10; 5; 5; -3; 3; $\{(5, 3)\}$
(b) $\{(2, -1)\}$

Step 6 Check the solution by substituting $x = 5$ and $y = -1$ into both of the original equations.

Check

$y + 11 = 2x$	$5x = y + 26$
$-1 + 11 \stackrel{?}{=} 2(5)$	$5(5) \stackrel{?}{=} -1 + 26$
$10 = 10$ True	$25 = 25$ True

Since $(5, -1)$ is a solution of *both* equations, the solution set is $\{(5, -1)\}$.

Work Problem **3** *at the Side.* ▶

OBJECTIVE 2 Multiply when using the elimination method. Sometimes we need to multiply each side of one or both equations in a system by some number before adding the equations will eliminate a variable.

EXAMPLE 3 Multiplying Both Equations When Using the Elimination Method

Solve the system.

$$2x + 3y = -15 \quad (1)$$
$$5x + 2y = 1 \quad (2)$$

Adding the two equations gives $7x + 5y = -14$, which does not eliminate either variable. However, we can multiply each equation by a suitable number so that the coefficients of one of the two variables are opposites. For example, to eliminate x, multiply each side of equation (1) by 5, and each side of equation (2) by -2.

$$10x + 15y = -75 \quad \text{Multiply equation (1) by 5.}$$
$$-10x - 4y = -2 \quad \text{Multiply equation (2) by } -2.$$
$$11y = -77 \quad \text{Add.}$$
$$y = -7 \quad \text{Divide by 11.}$$

Substituting -7 for y in either equation (1) or (2) gives $x = 3$. Check that the solution set of the system is $\{(3, -7)\}$.

Work Problem **4** *at the Side.* ▶

OBJECTIVE 3 Use an alternative method to find the second value in a solution. Sometimes it is easier to find the value of the second variable in a solution by using the elimination method twice.

EXAMPLE 4 Finding the Second Value Using an Alternative Method

Solve the system.

$$4x = 9 - 3y \quad (1)$$
$$5x - 2y = 8 \quad (2)$$

Rearrange the terms in equation (1) so that like terms are aligned in columns. To do this, add $3y$ to each side to get the following system.

$$4x + 3y = 9 \quad (3)$$
$$5x - 2y = 8 \quad (2)$$

One way to proceed is to eliminate y by multiplying each side of equation (3) by 2 and each side of equation (2) by 3, and then adding.

Continued on Next Page

3 Solve each system by the elimination method. Check each solution.

(a) $2x - y = 2$
$4x + y = 10$

(b) $8x - 5y = 32$
$4x + 5y = 4$

4 **(a)** Solve the system in Example 3 by first eliminating the variable y. Check your solution.

(b) Solve

$$6x + 7y = 4$$
$$5x + 8y = -1,$$

and check your solution.

ANSWERS

3. (a) $\{(2, 2)\}$ **(b)** $\left\{\left(3, -\frac{8}{5}\right)\right\}$

4. (a) $\{(3, -7)\}$ **(b)** $\{(3, -2)\}$

$$\begin{aligned} 8x + 6y &= 18 \quad &&\text{Multiply equation (3) by 2.} \\ 15x - 6y &= 24 &&\text{Multiply equation (2) by 3.} \\ \hline 23x &= 42 &&\text{Add.} \\ x &= \frac{42}{23} &&\text{Divide by 23.} \end{aligned}$$

Substituting $\frac{42}{23}$ for x in one of the given equations would give y, but the arithmetic involved would be messy. Instead, solve for y by starting again with the original equations and eliminating x. Multiply each side of equation (3) by 5 and each side of equation (2) by -4, and then add.

$$\begin{aligned} 20x + 15y &= 45 \quad &&\text{Multiply equation (3) by 5.} \\ -20x + 8y &= -32 &&\text{Multiply equation (2) by } -4. \\ \hline 23y &= 13 &&\text{Add.} \\ y &= \frac{13}{23} &&\text{Divide by 23.} \end{aligned}$$

Check that the solution set is $\{(\frac{42}{23}, \frac{13}{23})\}$.

◀ Work Problem **5** at the Side.

5 Solve each system of equations.

(a) $5x = 7 + 2y$
$5y = 5 - 3x$

(b) $3y = 8 + 4x$
$6x = 9 - 2y$

When the value of the first variable is a fraction, the method used in Example 4 helps avoid arithmetic errors. Of course, this method could be used to solve any system of equations.

OBJECTIVE 4 **Use the elimination method to solve special systems.**

EXAMPLE 5 **Using the Elimination Method for an Inconsistent System or Dependent Equations**

Solve each system by the elimination method.

(a) $2x + 4y = 5$
$4x + 8y = -9$

Multiply each side of $2x + 4y = 5$ by -2; then add to $4x + 8y = -9$.

$$\begin{aligned} -4x - 8y &= -10 \\ 4x + 8y &= -9 \\ \hline 0 &= -19 \quad \text{False} \end{aligned}$$

The false statement $0 = -19$ indicates that the solution set is $\emptyset$.

(b) $3x - y = 4$
$-9x + 3y = -12$

Multiply each side of the first equation by 3; then add the two equations.

$$\begin{aligned} 9x - 3y &= 12 \\ -9x + 3y &= -12 \\ \hline 0 &= 0 \quad \text{True} \end{aligned}$$

A true statement occurs when the equations are equivalent. As before, this indicates that every solution of one equation is also a solution of the other. The solution set is $\{(x, y) \mid 3x - y = 4\}$. (See **Section 4.2,** Example 6, where the same system was solved using substitution.)

◀ Work Problem **6** at the Side.

6 Solve each system by the elimination method.

(a) $4x + 3y = 10$
$2x + \frac{3}{2}y = 12$

(b) $4x - 6y = 10$
$-10x + 15y = -25$

Answers

5. (a) $\left\{\left(\frac{45}{31}, \frac{4}{31}\right)\right\}$ **(b)** $\left\{\left(\frac{11}{26}, \frac{42}{13}\right)\right\}$

6. (a) $\emptyset$
(b) $\{(x, y) \mid 2x - 3y = 5\}$ (*Note:* To write the solution set, we divided the equation $4x - 6y = 10$ by 2 so that the coefficients would have greatest common factor 1.)

4.3 Exercises

FOR EXTRA HELP MyMathLab Math XL PRACTICE  WATCH  DOWNLOAD  READ  REVIEW

In Exercises 1–4, answer true *or* false *for each statement. If* false, *tell why.*

1. The ordered pair (0, 0) *must* be a solution of a system of the form

$$Ax + By = 0$$
$$Cx + Dy = 0.$$

true

2. To eliminate the y-terms in the system

$$2x + 12y = 7$$
$$3x + 4y = 1,$$

we should multiply the bottom equation by 3 and then add.

false; Multiply by −3.

3. The system

$$x + y = 1$$
$$x + y = 2$$

has $\emptyset$ as its solution set.

true

4. The ordered pair (4, −5) cannot be a solution of a system that contains the equation $5x - 4y = 0$.

true

Solve each system by the elimination method. Check each solution. See Examples 1 and 2.

5. $x + y = 2$
$2x - y = -5$
{(−1, 3)}

6. $3x - y = -12$
$x + y = 4$
{(−2, 6)}

7. $2x + y = -5$
$x - y = 2$
{(−1, −3)}

8. $2x + y = -15$
$-x - y = 10$
{(−5, −5)}

9. $3x + 2y = 0$
$-3x - y = 3$
{(−2, 3)}

10. $5x - y = 5$
$-5x + 2y = 0$
{(2, 5)}

11. $6x - y = -1$
$5y = 17 + 6x$
$\left\{\left(\frac{1}{2}, 4\right)\right\}$

12. $y = 9 - 6x$
$-6x + 3y = 15$
$\left\{\left(\frac{1}{2}, 6\right)\right\}$

Solve each system by the elimination method. Check each solution. See Examples 3–5.

13. $2x - y = 12$
$3x + 2y = -3$
$\{(3, -6)\}$

14. $x + y = 3$
$-3x + 2y = -19$
$\{(5, -2)\}$

15. $x + 3y = 19$
$2x - y = 10$
$\{(7, 4)\}$

16. $4x - 3y = -19$
$2x + y = 13$
$\{(2, 9)\}$

17. $x + 4y = 16$
$3x + 5y = 20$
$\{(0, 4)\}$

18. $2x + y = 8$
$5x - 2y = -16$
$\{(0, 8)\}$

19. $5x - 3y = -20$
$-3x + 6y = 12$
$\{(-4, 0)\}$

20. $4x + 3y = -28$
$5x - 6y = -35$
$\{(-7, 0)\}$

21. $2x - 8y = 0$
$4x + 5y = 0$
$\{(0, 0)\}$

22. $3x - 15y = 0$
$6x + 10y = 0$
$\{(0, 0)\}$

23. $x + y = 7$
$x + y = -3$
$\emptyset$

24. $x - y = 4$
$x - y = -3$
$\emptyset$

25. $-x + 3y = 4$
$-2x + 6y = 8$
$\{(x, y) \mid x - 3y = -4\}$

26. $6x - 2y = 24$
$-3x + y = -12$
$\{(x, y) \mid 3x - y = 12\}$

27. $4x - 3y = -19$
$3x + 2y = 24$
$\{(2, 9)\}$

28. $5x + 4y = 12$
$3x + 5y = 15$
$\{(0, 3)\}$

29. $3x - 7 = -5y$
$5x + 4y = -10$
$\{(-6, 5)\}$

30. $2x + 3y = 13$
$6 + 2y = -5x$
$\{(-4, 7)\}$

31. $2x + 3y = 0$
$4x + 12 = 9y$
$\left\{\left(-\frac{6}{5}, \frac{4}{5}\right)\right\}$

32. $-4x + 3y = 2$
$5x + 3 = -2y$
$\left\{\left(-\frac{13}{23}, -\frac{2}{23}\right)\right\}$

33. $24x + 12y = -7$
$16x - 17 = 18y$
$\left\{\left(\frac{1}{8}, -\frac{5}{6}\right)\right\}$

34. $9x + 4y = -3$
$6x + 7 = -6y$
$\left\{\left(\frac{1}{3}, -\frac{3}{2}\right)\right\}$

35. $3x = 3 + 2y$
$-\frac{4}{3}x + y = \frac{1}{3}$
$\{(11, 15)\}$

36. $3x = 27 + 2y$
$x - \frac{7}{2}y = -25$
$\{(17, 12)\}$

37. $5x - 2y = 3$
$10x - 4y = 5$
$\varnothing$

38. $3x - 5y = 1$
$6x - 10y = 4$
$\varnothing$

39. $6x + 3y = 0$
$-18x - 9y = 0$
$\{(x, y) \mid 2x + y = 0\}$

40. $3x - 5y = 0$
$9x - 15y = 0$
$\{(x, y) \mid 3x - 5y = 0\}$

Relating Concepts (Exercises 41–46) For Individual or Group Work

Attending the movies is one of America's favorite forms of entertainment. The graph shows U.S. movie attendance from 1996 through 2004. In 1996, attendance was 1339 *million, as represented by the point P*(1996, 1339). *In 2004, attendance was* 1536 *million, as represented by the point Q*(2004, 1536). *We can find an equation of line segment PQ by using a system of equations. Then we use the equation we found to approximate the attendance in any of the years between 1996 and 2004.* **Work Exercises 41–46 in order.**

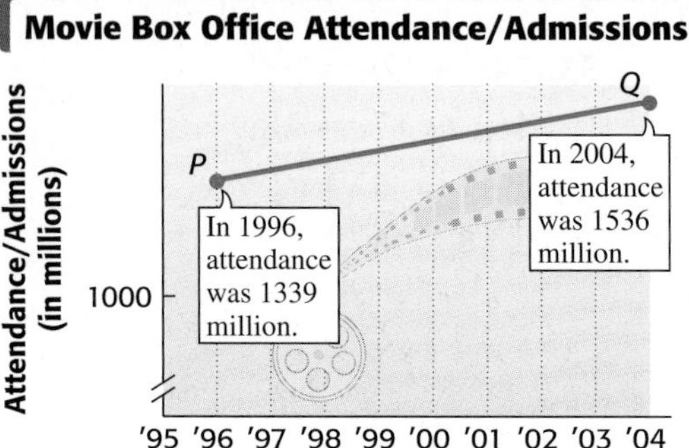

Source: Motion Picture Association of America.

41. The line segment has an equation that can be written in the form $y = ax + b$. Using the coordinates of point P with $x = 1996$ and $y = 1339$, write an equation in the variables a and b. **$1339 = 1996a + b$**

42. Using the coordinates of point Q with $x = 2004$ and $y = 1536$, write a second equation in the variables a and b. **$1536 = 2004a + b$**

43. Write the system of equations formed from the two equations in Exercises 41 and 42, and solve the system using the elimination method.

$1996a + b = 1339$

$2004a + b = 1536$;

solution set: $\{(24.625, -47{,}812.5)\}$

44. What is the equation of the segment PQ?

$y = 24.625x - 47{,}812.5$

45. Let $x = 2002$ in the equation of Exercise 44, and solve for y to the nearest tenth. How does the result compare with the actual figure of 1639 million?

1486.8 (million); This is quite a bit less than the actual figure.

46. The actual data points for the years 1996 through 2004 do not lie in a perfectly straight line. Explain the pitfalls of relying too heavily on using the equation in Exercise 44 to predict attendance.

Since the data do not lie in a perfectly straight line, the quantity obtained from an equation determined in this way will probably be "off" a bit. We cannot put too much faith in models such as this one, because not all sets of data points are linear in nature.

Summary Exercises on Solving Systems of Linear Equations

The exercises in this summary include a variety of problems on solving systems of linear equations. Since we do not usually specify the method of solution, use the following guidelines to help you decide whether to use substitution or elimination.

Guidelines for Choosing a Method to Solve a System of Linear Equations

1. If one of the equations of the system is already solved for one of the variables, as in the systems

$$\begin{aligned} 3x + 4y &= 9 \\ y &= 2x - 6 \end{aligned} \qquad \text{or} \qquad \begin{aligned} -5x + 3y &= 9 \\ x &= 3y - 7, \end{aligned}$$

the substitution method is the better choice.

2. If both equations are in standard $Ax + By = C$ form, as in

$$\begin{aligned} 4x - 11y &= 3 \\ -2x + 3y &= 4, \end{aligned}$$

and none of the variables has coefficient -1 or 1, the elimination method is the better choice.

3. If one or both of the equations are in standard form and the coefficient of one of the variables is -1 or 1, as in the systems

$$\begin{aligned} 3x + y &= -2 \\ -5x + 2y &= 4 \end{aligned} \qquad \text{or} \qquad \begin{aligned} -x + 3y &= -4 \\ 3x - 2y &= 8, \end{aligned}$$

either method is appropriate.

Use the preceding guidelines to solve each problem.

1. Assuming you want to minimize the amount of work required, tell whether you would use the substitution or elimination method to solve each system. Explain your answers. *Do not actually solve.*

(a) $3x + 2y = 18$

$y = 3x$

Use substitution since the second equation is solved for y.

(b) $3x + y = -7$

$x - y = -5$

Use elimination since the coefficients of the y-terms are opposites.

(c) $3x - 2y = 0$

$9x + 8y = 7$

Use elimination since the equations are in standard form with no coefficients of 1 or −1. Solving by substitution would involve fractions.

2. Which one of the following systems would be easier to solve using the substitution method? Why?

$$\begin{aligned} 5x - 3y &= 7 \\ 2x + 8y &= 3 \end{aligned} \qquad \begin{aligned} 7x + 2y &= 4 \\ y &= -3x + 1 \end{aligned}$$

The system on the right is easier to solve by substitution because the second equation is already solved for y.

*In Exercises 3 and 4, **(a)** solve the system by the elimination method, **(b)** solve the system by the substitution method, and **(c)** tell which method you prefer for that particular system and why.*

3. $4x - 3y = -8$
$x + 3y = 13$

(a) $\{(1, 4)\}$
(b) $\{(1, 4)\}$
(c) **Answers will vary.**

4. $2x + 5y = 0$
$x = -3y + 1$

(a) $\{(-5, 2)\}$
(b) $\{(-5, 2)\}$
(c) **Answers will vary.**

Solve each system by the method of your choice. (For Exercises 5–7, see your answers for Exercise 1.)

5. $3x + 2y = 18$
$y = 3x$
$\{(2, 6)\}$

6. $3x + y = -7$
$x - y = -5$
$\{(-3, 2)\}$

7. $3x - 2y = 0$
$9x + 8y = 7$
$\left\{\left(\frac{1}{3}, \frac{1}{2}\right)\right\}$

8. $x + y = 7$
$x = -3 - y$
$\emptyset$

9. $5x - 4y = 15$
$-3x + 6y = -9$
$\{(3, 0)\}$

10. $4x + 2y = 3$
$y = -x$
$\left\{\left(\frac{3}{2}, -\frac{3}{2}\right)\right\}$

11. $3x = 7 - y$
$2y = 14 - 6x$
$\{(x, y) \mid 3x + y = 7\}$

12. $3x - 5y = 7$
$2x + 3y = 30$
$\{(9, 4)\}$

13. $3y = 4x + 2$
$5x - 2y = -3$
$\left\{\left(-\frac{5}{7}, -\frac{2}{7}\right)\right\}$

14. $4x + 3y = 1$
$3x + 2y = 2$
$\{(4, -5)\}$

15. $2x - 3y = 7$
$-4x + 6y = 14$
$\emptyset$

16. $0.2x + 0.3y = 1.0$
$-0.3x + 0.1y = 1.8$
$\{(-4, 6)\}$

17. $6x + 5y = 13$
$3x + 3y = 4$
$\left\{\left(\frac{19}{3}, -5\right)\right\}$

18. $x - 3y = 7$
$4x + y = 5$
$\left\{\left(\frac{22}{13}, -\frac{23}{13}\right)\right\}$

19. $\frac{1}{4}x - \frac{1}{5}y = 9$
$y = 5x$
$\{(-12, -60)\}$

20. $\frac{1}{2}x + \frac{1}{3}y = -\frac{1}{3}$
$\frac{1}{2}x + 2y = -7$
$\{(2, -4)\}$

21. $-\frac{1}{2}x - \frac{1}{3}y = -5$
$-\frac{1}{2}x - \frac{1}{3}y = -5$
$\{(18, -12)\}$

22. $\frac{x}{5} + 2y = \frac{8}{5}$
$\frac{3x}{5} + \frac{y}{2} = -\frac{7}{10}$
$\{(-2, 1)\}$

23. $\frac{x}{5} + y = \frac{6}{5}$
$\frac{x}{10} + \frac{y}{3} = \frac{5}{6}$
$\left\{\left(13, -\frac{7}{5}\right)\right\}$

24. $\frac{2}{5}x + \frac{4}{3}y = -8$
$\frac{7}{10}x - \frac{2}{9}y = 9$
$\{(10, -9)\}$

25. $0.5x + 0.2y = 0.2$
$x - 0.6y = -0.5$
$\{(0.04, 0.9)\}$

4.4 Applications of Linear Systems

Recall from **Section 2.4** the six-step method for solving applied problems. We modify those steps slightly to allow for two variables and two equations.

OBJECTIVES

1. Solve problems about unknown numbers.
2. Solve problems about quantities and their costs.
3. Solve problems about mixtures.
4. Solve problems about distance, rate (or speed), and time.

Solving an Applied Problem with Two Variables

Step 1 **Read** the problem, several times if necessary, until you understand what is given and what is to be found.

Step 2 **Assign variables** to represent the unknown values, using diagrams or tables as needed. Write down what each variable represents.

Step 3 **Write two equations** using both variables.

Step 4 **Solve** the system of two equations.

Step 5 **State the answer** to the problem. Is the answer reasonable?

Step 6 **Check** the answer in the words of the original problem.

OBJECTIVE 1 Solve problems about unknown numbers.

EXAMPLE 1 Solving a Problem about Two Unknown Numbers

In 2004, sales of athletic/sports footwear were $3551 million more than sales of sports clothing. Together, total sales for these items were $25,953 million. (*Source:* National Sporting Goods Association.) What were the sales for each?

Step 1 **Read** the problem carefully. We must find the 2004 sales (in millions of dollars) for athletic/sports footwear and clothing. We know how much more footwear sales were than clothing sales. Also, we know the total sales.

Step 2 **Assign variables.**

Let x = sales of footwear in millions of dollars,

and y = sales of clothing in millions of dollars.

Step 3 **Write two equations.**

$x = 3551 + y$ Sales of footwear were $3551 million more than sales of clothing.

$x + y = 25{,}953$ Total sales were $25,953 million.

Step 4 **Solve** the system for x and y from Step 3. The substitution method works well here since the first equation is already solved for x.

Work Problem 1 at the Side. ▶

Step 5 **State the answer.** Footwear sales were $14,752 million, and clothing sales were $11,201 million.

Step 6 **Check** the answer in the original problem. Since

$$14{,}752 - 11{,}201 = 3551 \quad \text{and} \quad 14{,}752 + 11{,}201 = 25{,}953,$$

the answer satisfies the information in the problem.

1 Solve the system.

$$x = 3551 + y$$
$$x + y = 25{,}953$$

ANSWER

1. $x = 14{,}752$, $y = 11{,}201$

2 Set up a system of equations for the following problem. Do not solve the system.

Two of the most popular movies of 2007 were *Ratatouille* and *The Simpsons Movie*. Together, their domestic gross was \$389.5 million. *The Simpsons Movie* grossed \$23.3 million less than *Ratatouille*. How much did each movie gross? (*Source:* www.boxofficemojo.com)

Let x = the amount (in millions) that *Ratatouille* grossed, and y = the amount (in millions) that __________ grossed.

CAUTION

If an applied problem asks for *two* values as in Example 1, be sure to give both of them in your answer.

◀ *Work Problem* **2** *at the Side.*

OBJECTIVE 2 Solve problems about quantities and their costs. We can also use a linear system to solve an applied problem involving two quantities and their costs.

EXAMPLE 2 Solving a Problem about Quantities and Costs

Musicals have long been the most popular shows on Broadway. The musical *Wicked,* based on a "re-imagining" of *The Wizard of Oz*, has played to sold-out houses in cities around the world. (*Source:* www.broadway.com)

For the production playing at the Ford Center in Chicago, orchestra (main floor) seats cost \$148, while the best balcony tickets cost \$65. (*Source*: www.ticketmaster.com) Suppose that the members of a club spent a total of \$2614 for 30 tickets to *Wicked*. How many tickets of each kind did they buy?

Step 1 **Read** the problem several times.

Step 2 **Assign variables.**

Let x = the number of orchestra seats,

and y = the number of balcony seats.

Summarize the information given in the problem in a table. The entries in the first two rows of the Total Value column were found by multiplying the number of tickets sold by the price per ticket.

	Number of Tickets	Price per Ticket (in dollars)	Total Value
Orchestra	x	148	$\mathbf{148x}$
Balcony	y	65	$\mathbf{65y}$
Total	30	XXXXXX	**2614**

Step 3 **Write two equations.** The total number of tickets was 30, so

$$\mathbf{x + y = 30.} \quad \text{Total number of tickets}$$

Since the total value was \$2614, the final column leads to

$$\mathbf{148x + 65y = 2614.} \quad \text{Total value of tickets}$$

These two equations form the system

$$\begin{aligned} x + \quad y &= 30 \qquad (1) \\ 148x + 65y &= 2614. \qquad (2) \end{aligned}$$

Step 4 **Solve** the system using the elimination method. To eliminate the y-terms, multiply each side of equation (1) by -65 to get

$$-65x - 65y = -1950.$$

Continued on Next Page

ANSWER

2. *The Simpsons Movie*; $x + y = 389.5$ $y = x - 23.3$

Then add this result to equation (2).

$$\begin{aligned} -65x - 65y &= -1950 \\ 148x + 65y &= 2614 \quad (2) \\ \hline 83x &= 664 \quad \text{Add.} \\ x &= 8 \quad \text{Divide by 83.} \end{aligned}$$

Substitute 8 for x in equation (1) to get

$$\begin{aligned} x + y &= 30 \quad (1) \\ 8 + y &= 30 \quad \text{Let } x = 8. \\ y &= 22 \end{aligned}$$

Step 5 **State the answer.** The club members bought 8 orchestra tickets and 22 balcony tickets.

Step 6 **Check.** The sum of 8 and 22 is 30, so the total number of tickets is correct. Since 8 tickets were purchased at \$148 each and 22 at \$65 each, the total of all the ticket prices is

$$\$148(8) + \$65(22) = \$2614,$$

which agrees with the total amount stated in the problem.

Work Problem 3 at the Side. ▶

OBJECTIVE 3 Solve problems about mixtures. In **Section 2.6** we solved percent problems using one variable. Many problems about mixtures that involve percent can be solved using a system of two equations in two variables.

EXAMPLE 3 Solving a Mixture Problem Involving Percent

A pharmacist needs 100 L of 50% alcohol solution. She has on hand 30% alcohol solution and 80% alcohol solution, which she can mix. How many liters of each will be required to make the 100 L of 50% alcohol solution?

Step 1 **Read** the problem. Note the percent of each solution and of the mixture.

Step 2 **Assign variables.**

Let x = the number of liters of 30% alcohol needed,

and y = the number of liters of 80% alcohol needed.

Summarize the information given in the problem in a table. Percents are written as decimals.

Liters of Mixture	Percent	Liters of Pure Alcohol
x	0.30	$0.30x$
y	0.80	$0.80y$
100	0.50	$0.50(100)$

Continued on Next Page

3 For the production of *Wicked* playing at the Pantages Theatre in Los Angeles, orchestra seats cost \$96 and mid-priced mezzanine tickets cost \$58. (*Source:* www.ticketmaster.com) If a group of 18 people attended the show and spent a total of \$1234 for their tickets, how many of each kind of ticket did they buy?

(a) Complete the table.

	Number of Tickets Sold	Price (in dollars)	Total Value
Orchestra	x		
Mezzanine	y		
Total		XXXXXX	

(b) Write a system of equations.

(c) Solve the system and check your answer in the words of the original problem.

Answers

3. (a)

	Number of Tickets Sold	Price (in dollars)	Total Value
Orchestra	x	96	$96x$
Mezzanine	y	58	$58y$
Total	18	XXXXXX	1234

(b) $x + y = 18$
$96x + 58y = 1234$

(c) orchestra: 5; mezzanine: 13

4 How many liters of 25% alcohol solution must be mixed with 12% solution to get 13 L of 15% solution?

(a) Complete the table.

Liters	Percent	Liters of Pure Alcohol
x	0.25	$0.25x$
y	0.12	
13	0.15	

(b) Write a system of equations, and solve it.

5 Solve the problem.

Joe needs 60 milliliters (mL) of 20% acid solution for a chemistry experiment. The lab has on hand only 10% and 25% solutions. How much of each should he mix to get the desired amount of 20% solution?

6 Solve using the formula $d = rt$.

A small plane traveled from Stockholm, Sweden, to Oslo, Norway, averaging 244 km per hr. The trip took 1.7 hr. To the nearest kilometer, what is the distance between the two cities?

Figure 8 gives an idea of what is actually happening in this problem.

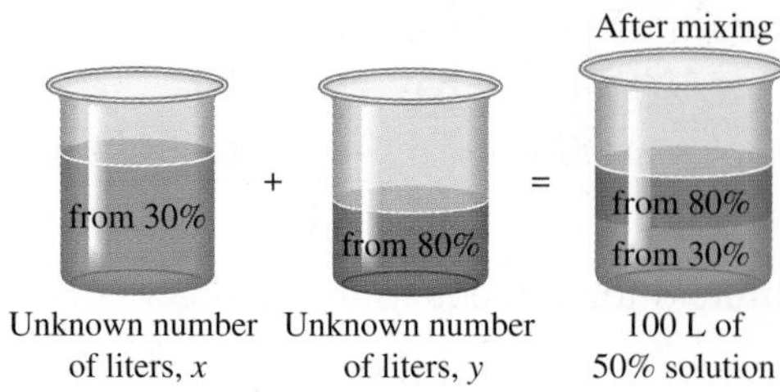

Figure 8

Step 3 **Write two equations.** Since the total number of liters in the final mixture will be 100, the first equation is

$$x + y = 100.$$

To find the amount of pure alcohol in each mixture, multiply the number of liters by the concentration. The amount of pure alcohol in the 30% solution added to the amount of pure alcohol in the 80% solution will equal the amount of pure alcohol in the final 50% solution. This gives the second equation,

$$0.30x + 0.80y = 0.50(100).$$

These two equations form the system

Be sure to write two equations.

$$\begin{aligned} x + \quad y &= 100 \\ 0.30x + 0.80y &= 50. \qquad 0.50(100) = 50 \end{aligned}$$

Step 4 **Solve** this system by the substitution method. Solving the first equation of the system for x gives $x = 100 - y$. Substitute $100 - y$ for x in the second equation.

$$\begin{aligned} 0.30x + 0.80y &= 50 \\ 0.30(\mathbf{100 - y}) + 0.80y &= 50 \qquad \text{Let } x = 100 - y. \\ 30 - 0.30y + 0.80y &= 50 \qquad \text{Distributive property} \\ 30 + 0.50y &= 50 \qquad \text{Combine like terms.} \\ 0.50y &= 20 \qquad \text{Subtract 30.} \\ y &= 40 \qquad \text{Divide by 0.50.} \end{aligned}$$

Then $x = 100 - y = 100 - \mathbf{40} = 60$.

Step 5 **State the answer.** The pharmacist should use 60 L of the 30% solution and 40 L of the 80% solution.

Step 6 Since $60 + 40 = 100$ and $0.30(60) + 0.80(40) = 50$, this mixture will give the 100 L of 50% solution, as required in the original problem.

◀ *Work Problems* 4 *and* 5 *at the Side.*

OBJECTIVE 4 Solve problems about distance, rate (or speed), and time. If an automobile travels at an average rate of 50 mph for 2 hr, then it travels $50 \times 2 = 100$ mi. This is an example of the basic relationship between distance, rate, and time:

distance = rate × time, given by the formula $d = rt$.

◀ *Work Problem* 6 *at the Side.*

ANSWERS

4. (a)

Liters	Percent	Liters of Pure Alcohol
x	0.25	$0.25x$
y	0.12	$0.12y$
13	0.15	$0.15(13)$

(b) $x + y = 13$
$0.25x + 0.12y = 0.15(13)$
3 L of 25%, 10 L of 12%

5. 20 mL of 10%, 40 mL of 25%

6. 415 km

EXAMPLE 4 Solving a Problem about Distance, Rate, and Time

Two executives in cities 400 mi apart drive to a business meeting at a location on the line between their cities. They meet after 4 hr. Find the speed of each car if one car travels 20 mph faster than the other.

Step 1 **Read** the problem carefully.

Step 2 **Assign variables.** Let $x =$ the speed of the faster car, and $y =$ the speed of the slower car.

We use the formula $d = rt$. Since each car travels for 4 hr, the time, t, for each car is 4. The distance is found by using the formula $d = rt$ and the expressions already entered in the table.

	r	t	d
Faster Car	x	4	$4x$
Slower Car	y	4	$4y$

Find d from $d = rt$.

Figure 9 shows what is happening in the problem.

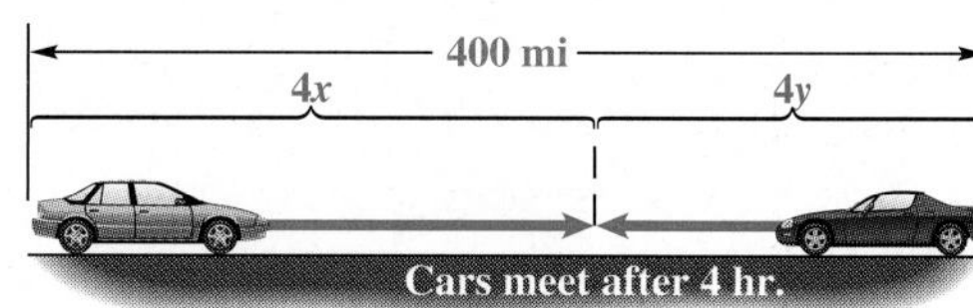

Figure 9

Step 3 **Write two equations.** As shown in the figure, since the total distance traveled by both cars is 400 mi, one equation is

$$4x + 4y = 400.$$

Because the faster car goes 20 mph faster than the slower car, the second equation is

$$x = 20 + y.$$

Step 4 **Solve** the system of equations,

$$4x + 4y = 400 \quad (1)$$
$$x = 20 + y, \quad (2)$$

by substitution. Replace x with $20 + y$ in equation (1) and then solve for y.

$$4(20 + y) + 4y = 400 \quad \text{Let } x = 20 + y.$$
$$80 + 4y + 4y = 400 \quad \text{Distributive property}$$
$$80 + 8y = 400 \quad \text{Combine like terms.}$$
$$8y = 320 \quad \text{Subtract 80.}$$
$$y = 40 \quad \text{Divide by 8.}$$

Since $x = 20 + y$ and $y = 40$,

$$x = 20 + 40 = 60.$$

Step 5 **State the answer.** The speeds of the cars are 40 mph and 60 mph.

Step 6 **Check** the answer. Since each car travels for 4 hr, total distance is

$$4(60) + 4(40) = 240 + 160 = 400 \text{ mi}, \quad \text{as required.}$$

Work Problem 7 at the Side. ▶

7 Two cars that were 450 mi apart traveled toward each other. They met after 5 hr. If one car traveled twice as fast as the other, what were their speeds?

(a) Complete this table.

	r	t	d
Faster Car	x	5	
Slower Car	y	5	

(b) Write a system, and solve it.

Answers

7. (a)

	r	t	d
Faster Car	x	5	$5x$
Slower Car	y	5	$5y$

(b) $5x + 5y = 450$
$x = 2y$
faster car: $x = 60$ mph;
slower car: $y = 30$ mph

8 Solve the system.

$$x + y = 320$$
$$x - y = 280$$

9 Solve the problem.

In 1 hr, Gigi can row 2 mi against the current or 10 mi with the current. Find the speed of the current and Gigi's speed in still water. (*Hint:* Let $x =$ the speed of the current and $y =$ Gigi's speed in still water. Then her rate against the current is $(y - x)$ mph, and her rate with the current is $(y + x)$ mph.)

CAUTION

Be careful! ***When you use two variables to solve a problem, you must write two equations.***

EXAMPLE 5 Solving a Problem about Distance, Rate, and Time

A plane flies 560 mi in 1.75 hr traveling with the wind. The return trip against the same wind takes the plane 2 hr. Find the speed of the plane and the speed of the wind.

Step 1 **Read** the problem several times.

Step 2 **Assign variables.**

Let $x =$ the speed of the plane,

and $y =$ the speed of the wind.

The speed (rate) of the plane *with* the wind is $(x + y)$ mph, and the speed (rate) of the plane *against* the wind is $(x - y)$ mph. See Figure 10.

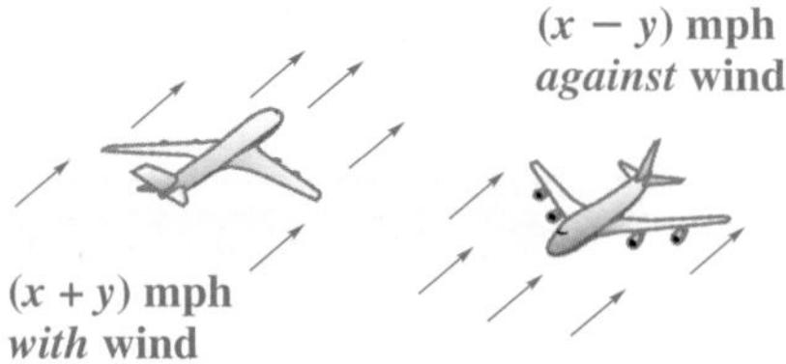

Figure 10

We use this information and the formula $d = rt$ (or $rt = d$) to complete a table.

	r	t	d
With Wind	$x + y$	1.75	560
Against Wind	$x - y$	2	560

Step 3 **Write two equations.** From the table,

$$1.75(x + y) = 560 \xrightarrow{\text{Divide by 1.75.}} x + y = 320 \quad (1)$$
$$2(x - y) = 560 \xrightarrow{\text{Divide by 2.}} x - y = 280. \quad (2)$$

Step 4 **Solve** the system of equations (1) and (2).

◀ Work Problem **8** at the Side.

Step 5 **State the answer.** From Problem 8 at the side, the speed of the plane is 300 mph and the speed of the wind is 20 mph.

Step 6 **Check.** The answer seems reasonable, and true statements result when the values are substituted into the equations of the system.

◀ Work Problem **9** at the Side.

ANSWERS

8. (300, 20)
9. current: 4 mph;
Gigi's speed in still water: 6 mph

Choose the correct response in Exercises 1–7.

1. Which expression represents the monetary value of x 20-dollar bills?

A. $\frac{x}{20}$ dollars **B.** $\frac{20}{x}$ dollars **C.** $(20 + x)$ dollars **D.** $20x$ dollars D

2. Which expression represents the cost of t pounds of candy that sells for \$1.95 per lb?

A. $\$1.95t$ **B.** $\frac{\$1.95}{t}$ **C.** $\frac{t}{\$1.95}$ **D.** $\$1.95 + t$ A

3. Which expression represents the amount of interest earned on d dollars at an interest rate of 2%?

A. $2d$ dollars **B.** $0.02d$ dollars **C.** $0.2d$ dollars **D.** $200d$ dollars B

4. Suppose that x liters of a 40% acid solution are mixed with y liters of a 35% solution to obtain 100 L of a 38% solution. One equation in a system for solving this problem is $x + y = 100$. Which one of the following is the other equation?

A. $0.35x + 0.40y = 0.38(100)$ **B.** $0.40x + 0.35y = 0.38(100)$

C. $35x + 40y = 38$ **D.** $40x + 35y = 0.38(100)$ B

5. According to *Natural History* magazine, the speed of a cheetah is 70 mph. If a cheetah runs for x hours, how many miles does the cheetah cover?

A. $(70 + x)$ miles **B.** $(70 - x)$ miles **C.** $\frac{70}{x}$ miles **D.** $70x$ miles D

6. What is the speed of a plane that travels at a rate of 560 mph *against* a wind of r mph?

A. $(560 + r)$ mph **B.** $\frac{560}{r}$ mph **C.** $(560 - r)$ mph **D.** $(r - 560)$ mph C

7. What is the speed of a plane that travels at a rate of 560 mph *with* a wind of r mph?

A. $\frac{r}{560}$ mph **B.** $(560 - r)$ mph **C.** $(560 + r)$ mph **D.** $(r - 560)$ mph C

8. Using the list of steps for solving an applied problem with two variables, describe the general procedure you will use to solve the problems that follow in this exercise set.

Choose a variable that represents one of the unknown quantities, and choose another variable that represents the other unknown quantity. Write a system of two equations in these two variables, and then solve the system. Check to see that the two values satisfy all conditions of the problem.

Exercises 9 and 10 are good warm-up problems. In each case, refer to the six-step problem-solving method, fill in the blanks for Steps 2 and 3, and then complete the solution by applying Steps 4–6.

9. The sum of two numbers is 98 and the difference between them is 48. Find the two numbers.

Step 1 **Read** the problem carefully.

Step 2 **Assign variables.**

Let $x =$ the first number and let

$y =$ **the second number**.

Step 3 **Write two equations.**

First equation: $x + y = 98$

Second equation: **$x - y = 48$**

The two numbers are 73 and 25.

10. The sum of two numbers is 201 and the difference between them is 11. Find the two numbers.

Step 1 **Read** the problem carefully.

Step 2 **Assign variables.**

Let $x =$ the first number and let

$y =$ **the second number**.

Step 3 **Write two equations.**

First equation: $x + y = 201$

Second equation: **$x - y = 11$**

The two numbers are 106 and 95.

Write a system of equations for each problem, and then solve the problem. See Example 1.

11. As of 2008, the two longest-running shows in Broadway history were *The Phantom of the Opera* and *Cats*. As of September 26, 2007, there had been a total of 15,682 Broadway performances of the two shows, with 712 more performances of *The Phantom of the Opera* than *Cats*. How many performances were there of each show? (*Source:* The Broadway League.)

***The Phantom of the Opera:* 8197; *Cats:* 7485**

12. Two other musicals that had very long Broadway runs were *A Chorus Line* and *Beauty and the Beast*. During their runs, there were 676 fewer performances of *Beauty and the Beast* than of *A Chorus Line,* and a total of 11,598 performances of the two shows. How many performances were there of each show? (*Source:* The Broadway League.)

***A Chorus Line:* 6137; *Beauty and the Beast:* 5461**

13. The two domestic top-grossing movies of 2007 were *Spider-Man 3* and *Shrek the Third. Shrek the Third* grossed \$13.8 million less than *Spider-Man 3*, and together the two films took in \$659.2 million. How much did each of these movies earn? (*Source:* www.boxofficemojo.com)

***Spider-Man 3:* \$336.5 million; *Shrek the Third:* \$322.7 million**

14. During their opening weekends, *Spider-Man 3* and *Shrek the Third* grossed a total of \$272.7 million, with *Spider-Man 3* grossing \$29.5 million more than *Shrek the Third*. How much did each of these movies earn during their opening weekends? (*Source:* www.boxofficemojo.com)

***Spider-Man 3:* \$151.1 million; *Shrek the Third:* \$121.6 million**

15. The Terminal Tower in Cleveland, Ohio, is 242 ft shorter than the Key Tower, also in Cleveland. The total of the heights of the two buildings is 1658 ft. Find the heights of the buildings. (*Source: World Almanac and Book of Facts.*)

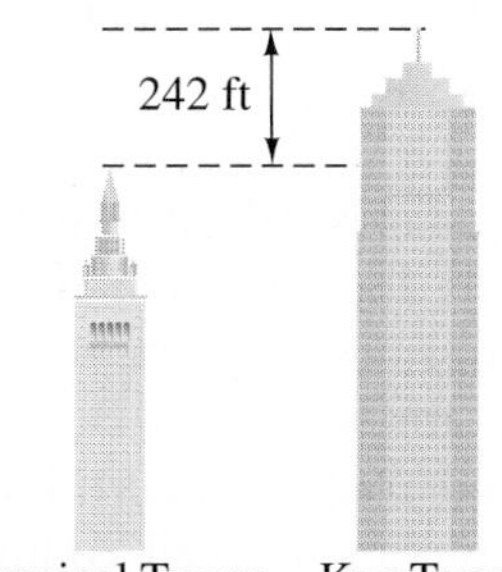

Terminal Tower: 708 ft; Key Tower: 950 ft

16. In 2006, a total of 646.3 thousand people lived in the Twin Cities of Minneapolis and St. Paul, Minnesota. Minneapolis had 99.3 thousand more residents than St. Paul. What was the population of each city? (*Source:* U.S. Census Bureau.)

Minneapolis: 372.8 thousand; St. Paul: 273.5 thousand

If x units of a product cost C dollars to manufacture and earn revenue of R dollars, the value of x where the expressions for C and R are equal is called the ***break-even quantity,*** *the number of units that produce* 0 *profit. In Exercises 17 and 18,* ***(a)*** *find the break-even quantity, and* ***(b)*** *decide whether the product should be produced based on whether it will earn a profit. (Profit equals revenue minus cost.)*

17. $C = 85x + 900$; $R = 105x$; no more than 38 units can be sold.

(a) 45 units (b) Do not produce; the product will lead to a loss.

18. $C = 105x + 6000$; $R = 255x$; no more than 400 units can be sold.

(a) 40 units (b) Produce; the product will earn a profit.

Write a system of equations for each problem, and then solve the system. See Example 2.

19. A motel clerk counts his \$1 and \$10 bills at the end of a day. He finds that he has a total of 74 bills having a combined monetary value of \$326. Find the number of bills of each denomination that he has.

Number of Bills	Denomination of Bill	Total Value
x	\$1	
y	\$10	
74	XXXXXX	\$326

46 ones; 28 tens

20. Carly is a bank teller. At the end of a day, she has a total of 69 \$5 and \$10 bills. The total value of the money is \$590. How many of each denomination does she have?

Number of Bills	Denomination of Bill	Total Value
x	\$5	\$5x
y	\$10	
	XXXXXX	

20 fives; 49 tens

21. A newspaper advertised DVDs and CDs. Tracy Sudak went shopping and bought each of her seven nephews a gift, either a DVD of the movie *Night at the Museum* or the latest Linkin Park CD. The DVD cost \$14.95 and the CD cost \$16.88, and she spent a total of \$114.30. How many DVDs and how many CDs did she buy?

2 DVDs of *Night at the Museum*; 5 Linkin Park CDs

22. Terry Wong saw the ad (see Exercise 21) and he, too, went shopping. He bought each of his five nieces a gift, either a DVD of *Hairspray* or the CD soundtrack to *High School Musical 2*. The DVD cost \$14.99 and the soundtrack cost \$13.88, and he spent a total of \$70.51. How many DVDs and CDs did he buy?

1 DVD of *Hairspray*; 4 *High School Musical 2* CD soundtracks

23. Maria Lopez has twice as much money invested at 5% simple annual interest as she does at 4%. If her yearly income from these two investments is \$350, how much does she have invested at each rate?

\$2500 at 4%; \$5000 at 5%

24. Charles Miller invested his textbook royalty income in two accounts, one paying 3% annual simple interest and the other paying 2% interest. He earned a total of \$11 interest. If he invested three times as much in the 3% account as he did in the 2% account, how much did he invest at each rate?

\$300 at 3%; \$100 at 2%

25. The two top-grossing North American concert tours in 2007 were The Police and Van Halen. Based on the average ticket prices for these tours, it cost a total of \$1217 to buy six tickets for The Police and five tickets to a Van Halen concert. Three tickets for The Police and four tickets for Van Halen cost a total of \$781. How much did an average ticket cost for each tour? (*Source:* Pollstar.)

The Police: \$107; Van Halen: \$115

26. Two other popular North American concert tours in 2007 were Billy Joel and Neil Young. Based on the average ticket prices for these tours, it cost a total of \$986 to buy eight tickets for Billy Joel and three tickets to a Neil Young concert. Four tickets for Billy Joel and five tickets for Neil Young cost a total of \$878. How much did an average ticket cost for each tour? (*Source:* Pollstar.)

Billy Joel: \$82; Neil Young: \$110

Write a system of equations for each problem, and then solve the system. See Example 3.

27. A 40% dye solution is to be mixed with a 70% dye solution to get 120 L of a 50% solution. How many liters of the 40% and 70% solutions will be needed?

Liters of Solution	Percent (as a Decimal)	Liters of Pure Dye
x	0.40	
y	0.70	
120	0.50	

80 L of 40% solution; 40 L of 70% solution

28. A 90% antifreeze solution is to be mixed with a 75% solution to make 120 L of a 78% solution. How many liters of the 90% and 75% solutions will be used?

Liters of Solution	Percent (as a Decimal)	Liters of Pure Antifreeze
x	0.90	
y	0.75	
120	0.78	

24 L of 90% solution; 96 L of 75% solution

29. Ahmad Hashemi wishes to mix coffee worth \$6 per lb with coffee worth \$3 per lb to get 90 lb of a mixture worth \$4 per lb. How many pounds of the \$6 and the \$3 coffees will be needed?

Pounds	Dollars per Pound	Cost
x	6	
y		
90		

30 lb at \$6 per lb; 60 lb at \$3 per lb

30. Mariana Coanda wishes to blend candy selling for \$1.20 per lb with candy selling for \$1.80 per lb to get a mixture that will be sold for \$1.40 per lb. How many pounds of the \$1.20 and the \$1.80 candies should be used to get 45 lb of the mixture?

Pounds	Dollars per Pound	Cost
x		
y	1.80	
45		

30 lb at \$1.20 per lb; 15 lb at \$1.80 per lb

31. How many pounds of nuts selling for \$6 per lb and raisins selling for \$3 per lb should Kelli Hammer combine to obtain 60 lb of a trail mix selling for \$5 per lb?

nuts: 40 lb; raisins: 20 lb

32. Avis Proctor works at a gourmet delicatessen. She is preparing cheese trays for a large reception. She is using some cheeses that sell for \$8 per lb and others that sell for \$12 per lb. How many pounds of cheese at each price should she use in order for the mixed cheeses on the trays to weigh a total of 56 lb and sell for \$10.50 per lb?

21 lb at \$8 per lb; 35 lb at \$12 per lb

Write a system of equations for each problem, and then solve the system. See Examples 4 and 5.

33. **RAGBRAI**®, the Des Moines **R**egister's **A**nnual **G**reat **B**icycle **R**ide **A**cross **I**owa, is the longest and oldest touring bicycle ride in the world. Suppose a cyclist began the 471 mi ride on July 20, 2008, in western Iowa at the same time that a car traveling toward it left eastern Iowa. If the bicycle and the car met after 7.5 hr and the car traveled 35.8 mph faster than the bicycle, find the average speed of each. (*Source:* www.ragbrai.org)

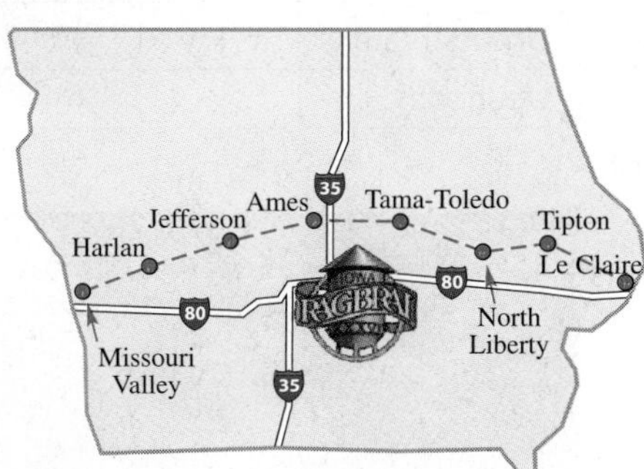

bicycle: 13.5 mph; car: 49.3 mph

34. In 2006, Atlanta's Hartsfield Airport was the nation's busiest. Suppose two planes leave the airport at the same time, one traveling east and the other traveling west. If the planes are 2100 mi apart after 2 hr and one plane travels 50 mph faster than the other, find the speed of each plane. (*Source:* Airports Council International.)

500 mph and 550 mph

35. Toledo and Cincinnati are 200 mi apart. A car leaves Toledo traveling toward Cincinnati, and another car leaves Cincinnati at the same time, traveling toward Toledo. The car leaving Toledo averages 15 mph faster than the other, and they meet after 1 hr and 36 min. What are the rates of the cars?

car leaving Cincinnati: 55 mph;
car leaving Toledo: 70 mph

36. Kansas City and Denver are 600 mi apart. Two cars start from these cities, traveling toward each other. They meet after 6 hr. Find the rate of each car if one travels 30 mph slower than the other.

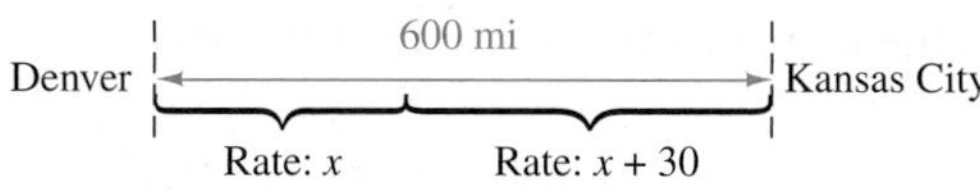

35 mph and 65 mph

37. At the beginning of a bicycle ride for charity, Roberto and Juana are 30 mi apart. If they leave at the same time and ride in the same direction, Roberto overtakes Juana in 6 hr. If they ride toward each other, they meet in 1 hr. What are their speeds?

Roberto: 17.5 mph; Juana: 12.5 mph

38. Mr. Abbot left Farmersville in a plane at noon to travel to Exeter. Mr. Baker left Exeter in his automobile at 2 P.M. to travel to Farmersville. It is 400 mi from Exeter to Farmersville. If the sum of their speeds was 120 mph, and if they crossed paths at 4 P.M., find the speed of each.

Abbot: 80 mph; Baker: 40 mph

39. A boat takes 3 hr to go 24 mi upstream. It can go 36 mi downstream in the same time. Find the speed of the current and the speed of the boat in still water if $x =$ the speed of the boat in still water and $y =$ the speed of the current.

	r	t	d
Downstream	$x + y$		36
Upstream	$x - y$		24

boat: 10 mph; current: 2 mph

40. It takes a boat $1\frac{1}{2}$ hr to go 12 mi downstream, and 6 hr to return. Find the speed of the boat in still water and the speed of the current. Let $x =$ the speed of the boat in still water and $y =$ the speed of the current.

	r	t	d
Downstream	$x + y$	$\frac{3}{2}$	12
Upstream		6	

boat: 5 mph; current: 3 mph

41. If a plane can travel 440 mph against the wind and 500 mph with the wind, find the speed of the wind and the speed of the plane in still air.

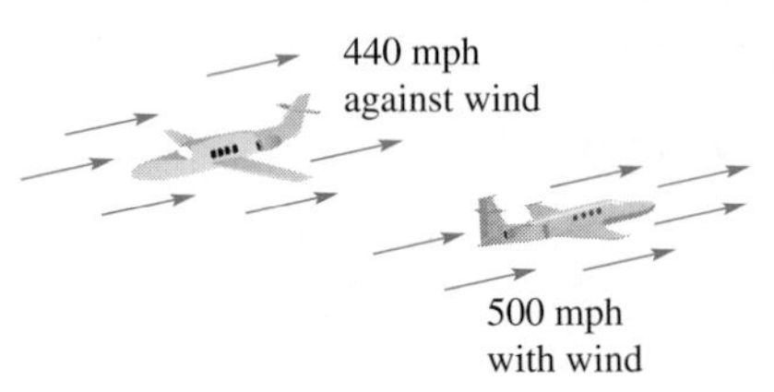

plane: 470 mph; wind: 30 mph

42. A small plane travels 200 mph with the wind and 120 mph against it. Find the speed of the wind and the speed of the plane in still air.

plane: 160 mph; wind: 40 mph

4.5 Solving Systems of Linear Inequalities

OBJECTIVE

1 Solve systems of linear inequalities by graphing.

We graphed the solutions of a linear inequality in **Section 3.5.** Recall that to graph the solutions of $x + 3y > 12$, for example, we first graph $x + 3y = 12$ by finding and plotting a few ordered pairs that satisfy the equation. Because the points on the line do *not* satisfy the inequality, we use a dashed line. To decide which side of the line includes the points that are solutions, we choose a test point not on the line, such as (0, 0). Substituting these values for x and y in the inequality gives

$$x + 3y > 12$$
$$0 + 3(0) \overset{?}{>} 12$$
$$0 > 12. \quad \text{False}$$

This false result indicates that the solutions are those points on the side of the line that does not include (0, 0), as shown in Figure 11.

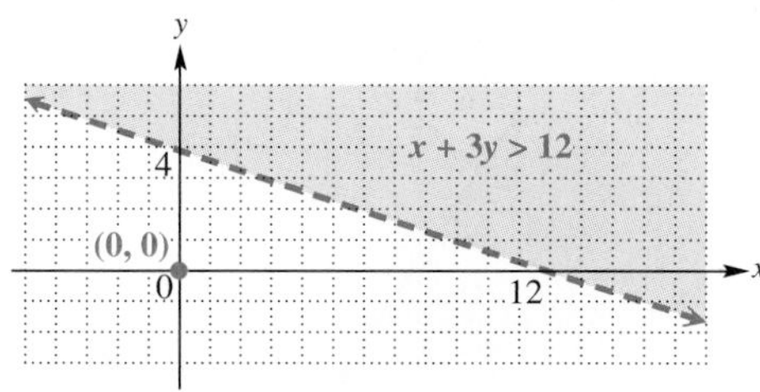

Figure 11

Now we use the same techniques to solve systems of linear inequalities.

OBJECTIVE 1 Solve systems of linear inequalities by graphing. A **system of linear inequalities** consists of two or more linear inequalities. The **solution set of a system of linear inequalities** includes all points that make all inequalities of the system true at the same time. To solve a system of linear inequalities, use the following steps.

Solving a System of Linear Inequalities

Step 1 **Graph the inequalities.** Graph each inequality using the method of **Section 3.5.**

Step 2 **Choose the intersection.** Indicate the solution set of the system by shading the intersection of the graphs (the region where the graphs overlap).

EXAMPLE 1 Solving a System of Two Linear Inequalities

Graph the solution set of the system.

$$3x + 2y \le 6$$
$$2x - 5y \ge 10$$

To graph $3x + 2y \le 6$, graph the solid boundary line $3x + 2y = 6$ and shade the region containing (0, 0), as shown in Figure 12(a) on the next page. Then graph $2x - 5y \ge 10$ with the solid boundary line $2x - 5y = 10$. The test point (0, 0) makes this inequality false, so shade the region on the other side of the boundary line. See Figure 12(b).

Continued on Next Page

1 Graph the solution set of the system.

$$x - 2y \leq 8$$
$$3x + y \geq 6$$

To get you started, the graphs of $x - 2y = 8$ and $3x + y = 6$ are shown.

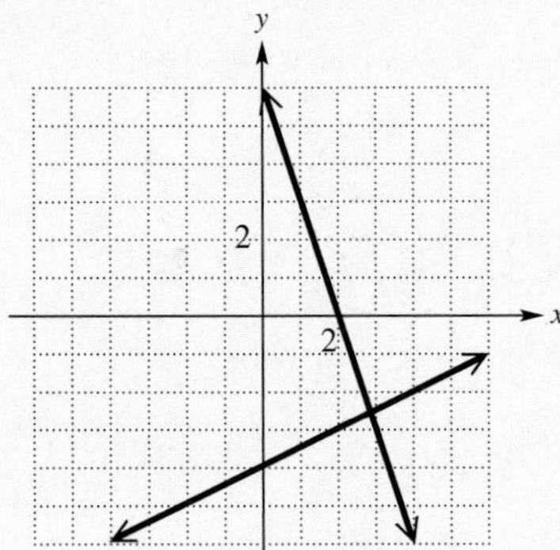

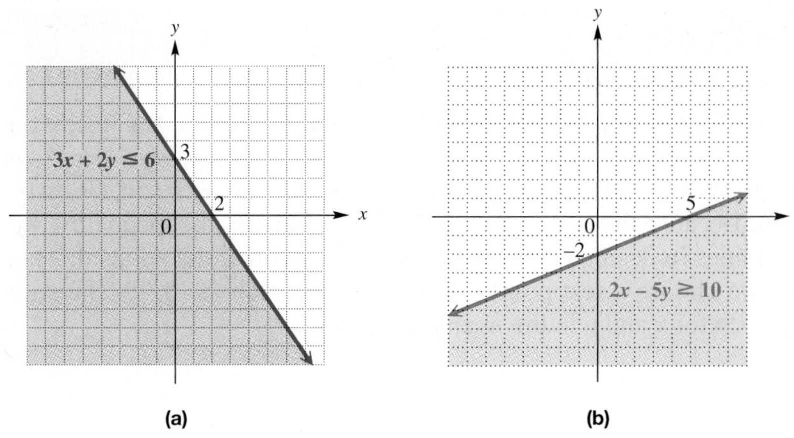

Figure 12

The solution set of this system includes all points in the intersection (overlap) of the graphs of the two inequalities. It includes the shaded region and portions of the two boundary lines shown in Figure 13.

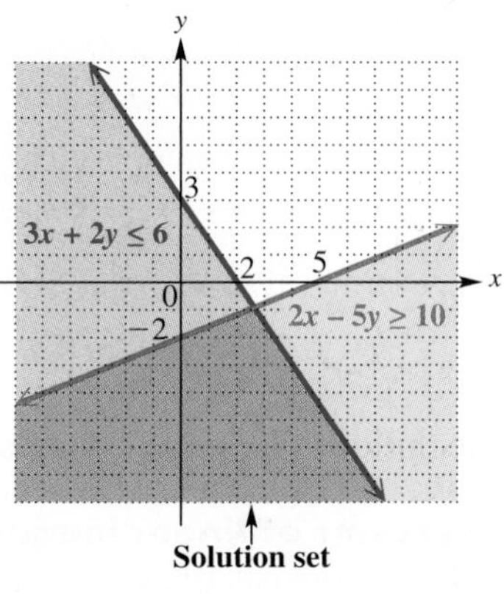

Figure 13

◀ *Work Problem*  *at the Side.*

> **Note**
>
> We usually do all the work on one set of axes. In the following examples, only one graph is shown. Be sure that the region of the final solution set is clearly indicated.

EXAMPLE 2 Solving a System of Two Linear Inequalities

Graph the solution set of the system.

$$x - y > 5$$
$$2x + y < 2$$

Figure 14 shows the graphs of both $x - y > 5$ and $2x + y < 2$. Dashed lines show that the graphs of the inequalities do not include their boundary lines. The solution set of the system is the region with the darkest shading. The solution set does not include either boundary line.

Continued on Next Page

ANSWER

1.

y
x
$x - 2y \leq 8$
$3x + y \geq 6$

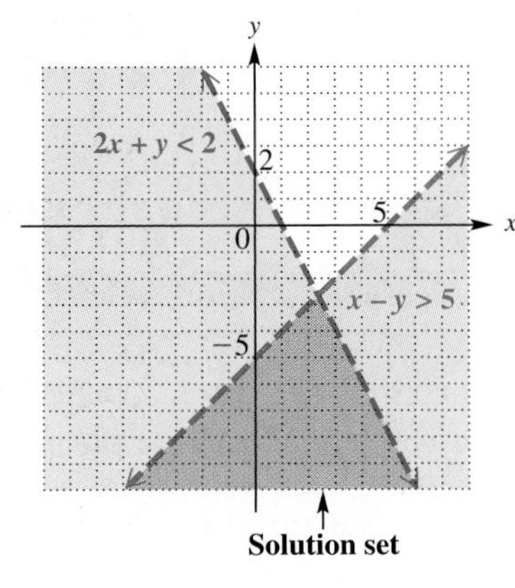

Figure 14

EXAMPLE 3 **Solving a System of Three Linear Inequalities**

Graph the solution set of the system.

$$4x - 3y \leq 8$$
$$x \geq 2$$
$$y \leq 4$$

Recall that $x = 2$ is a vertical line through the point (2, 0), and $y = 4$ is a horizontal line through (0, 4). The graph of the solution set is the shaded region in Figure 15, including all boundary lines.

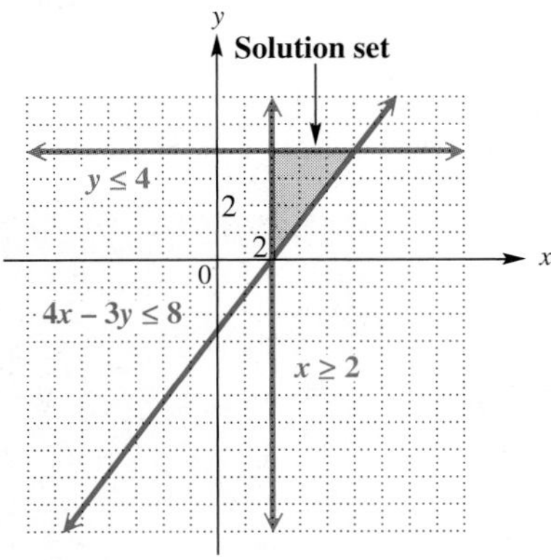

Figure 15

Work Problem **2** *at the Side.* ▶

2 Graph the solution set of each system.

(a) $x + 2y < 0$
$3x - 4y < 12$

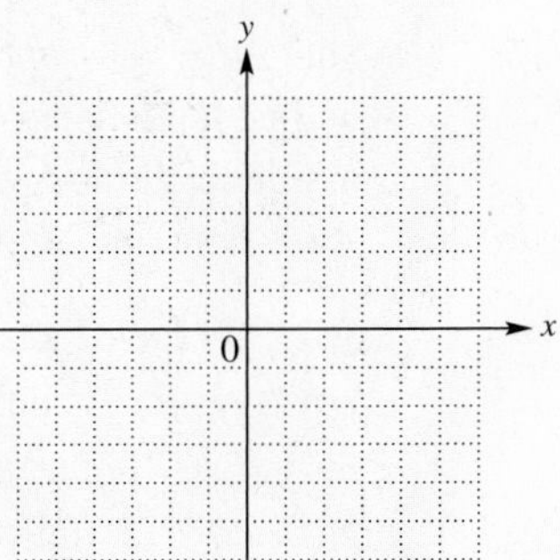

(b) $3x + 2y \leq 12$
$x \leq 2$
$y \leq 4$

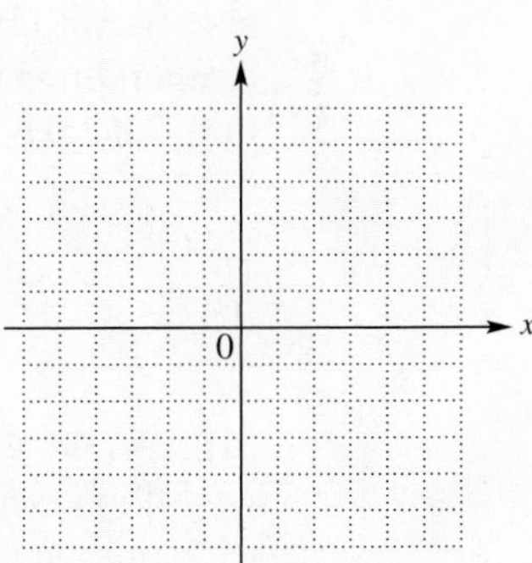

ANSWERS

2. (a)

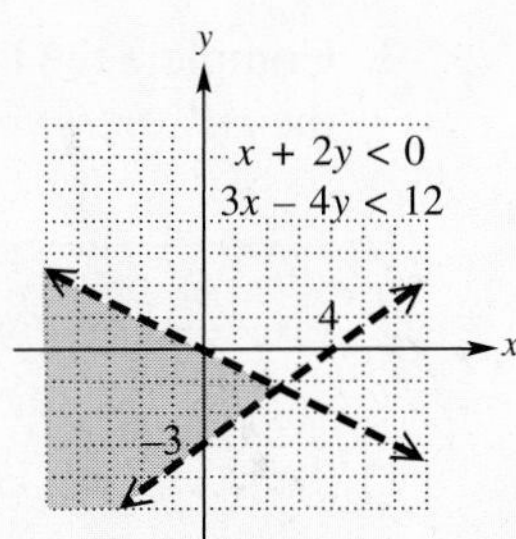

(b)

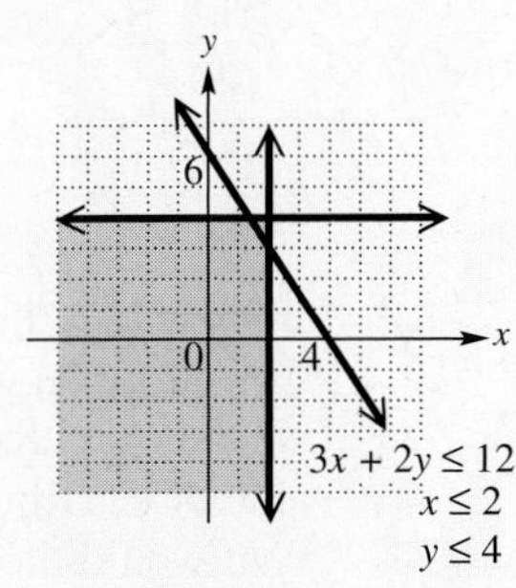

Math in the Media

AT WHAT TEMPERATURE DOES PAPER BURN?

The 1966 film *Fahrenheit 451*, directed by François Truffaut and based on the Ray Bradbury book of the same name, is a science fiction classic. In a future dominated by oppression, a fireman is assigned the task of burning books to discourage independent thinking. He eventually begins to read them and questions the motive of the government.

The Fahrenheit scale for temperature is used in the United States, but in many other countries temperature is routinely reported in Celsius. The formula for converting Celsius to Fahrenheit is $F = \frac{9}{5}C + 32$, but a quick "rule of thumb" given by travel books is "*Double the Celsius temperature and add 30,*" which is mathematically stated by the approximation formula

$$F = 2C + 30.$$

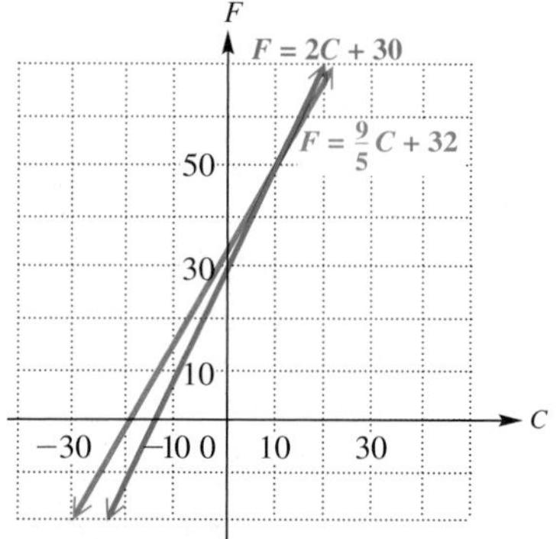

1. Suppose you are interested in knowing for what temperature the rule of thumb and the actual formulas give the same result. You also want to know if the rule of thumb formula is predicting temperatures that are lower or higher than the actual temperature. The two formulas can be written as the system of equations

$$F = \frac{9}{5}C + 32$$

$$F = 2C + 30.$$

(a) Use the graph of the system of equations to find the point of intersection. (*Hint:* To check your answer, use substitution to see if it satisfies both formulas.) **(10, 50)**

(b) For what temperature in degrees Celsius do the two formulas agree? **10°C**

(c) For what temperature in degrees Fahrenheit do the two formulas agree? **50°F**

2. Complete the table of values to compare the *actual* and the *rule of thumb* formulas for temperature conversion.

°C	°F (*Actual*)	°F (*Rule of Thumb*)
0	**32**	**30**
5	**41**	**40**
10	**50**	**50**
15	**59**	**60**
20	**68**	**70**
30	**86**	**90**

3. Suppose that the movie and book title *Fahrenheit 451* was given in Celsius rather than Fahrenheit. Use the exact formula $F = \frac{9}{5}C + 32$ to solve for this value of C. (Round to the nearest whole number.) What would the title then be? ***Celsius 233***

4.5 Exercises

FOR EXTRA HELP 

Math XL PRACTICE  WATCH  DOWNLOAD  READ  REVIEW

Match each system of inequalities with the correct graph from choices A–D.

1. $x \geq 5$ $y \leq -3$ C

2. $x \leq 5$ $y \geq -3$ A

3. $x > 5$ $y < -3$ B

4. $x < 5$ $y > -3$ D

A.

B.

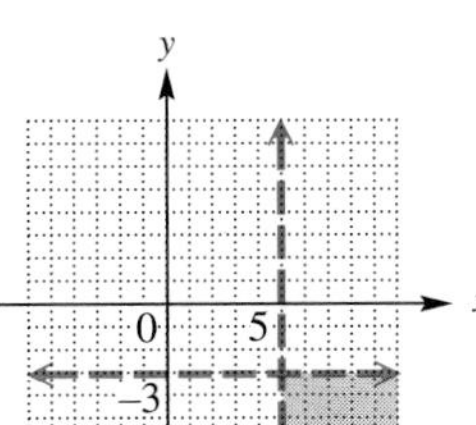

C.

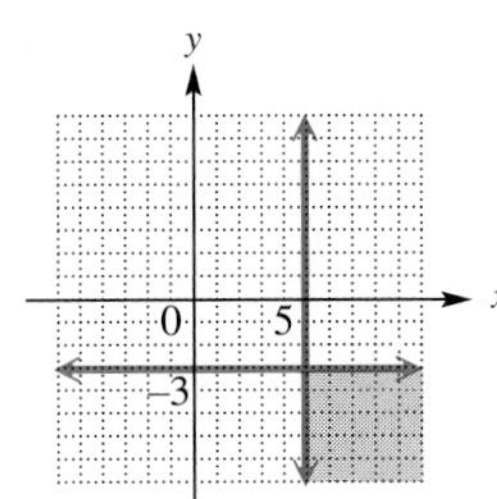

D.

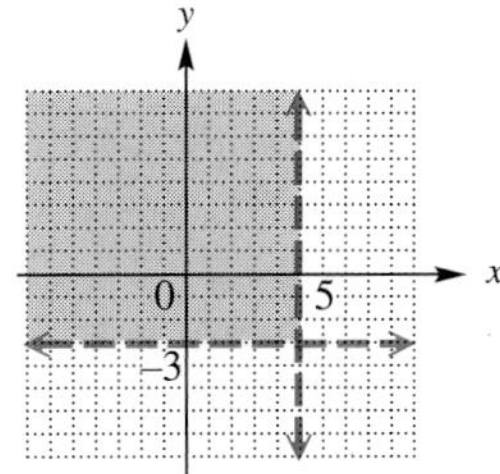

Graph the solution set of each system of linear inequalities. See Examples 1–3.

5. $x + y \leq 6$
$x - y \geq 1$

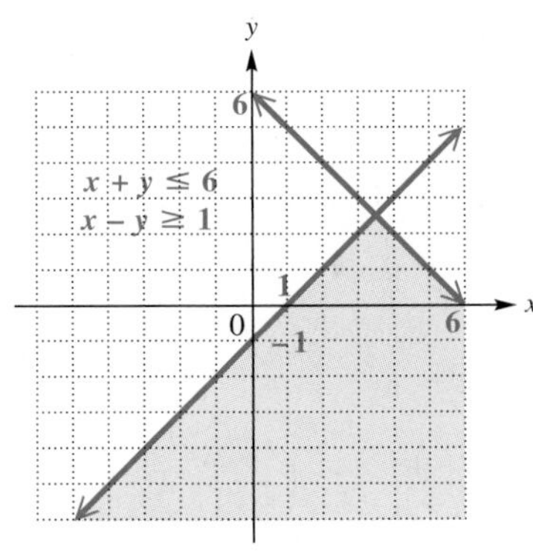

6. $x + y \leq 2$
$x - y \geq 3$

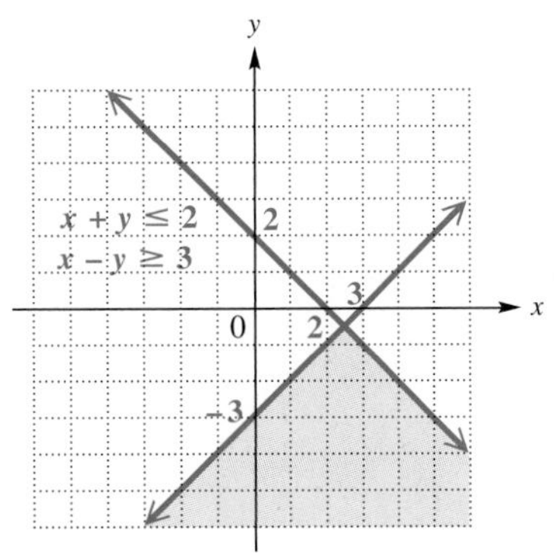

7. $4x + 5y \geq 20$
$x - 2y \leq 5$

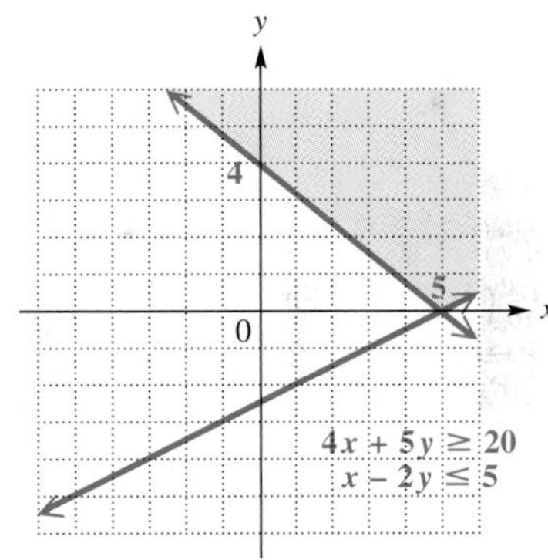

8. $x + 4y \leq 8$
$2x - y \geq 4$

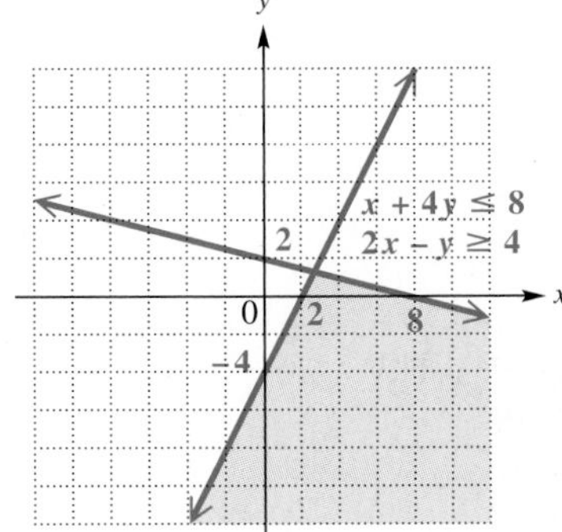

9. $2x + 3y < 6$
$x - y < 5$

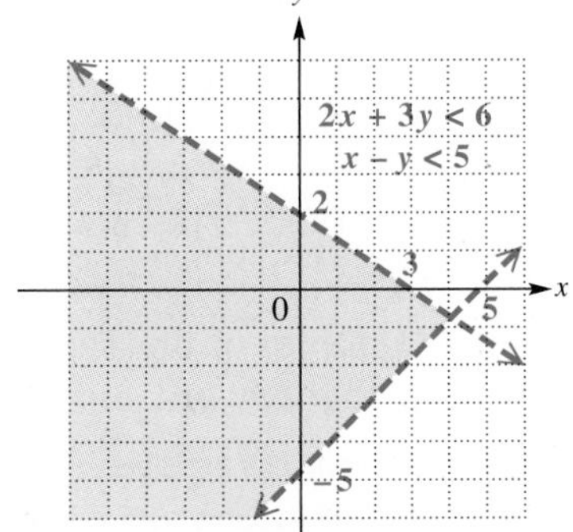

10. $x + 2y < 4$
$x - y < -1$

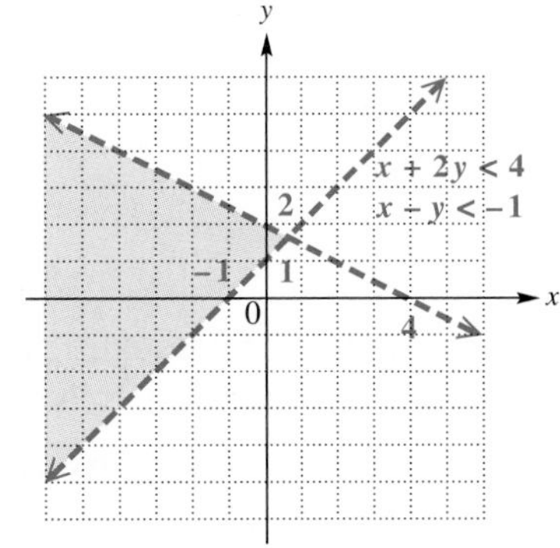

11. $y \le 2x - 5$
$x < 3y + 2$

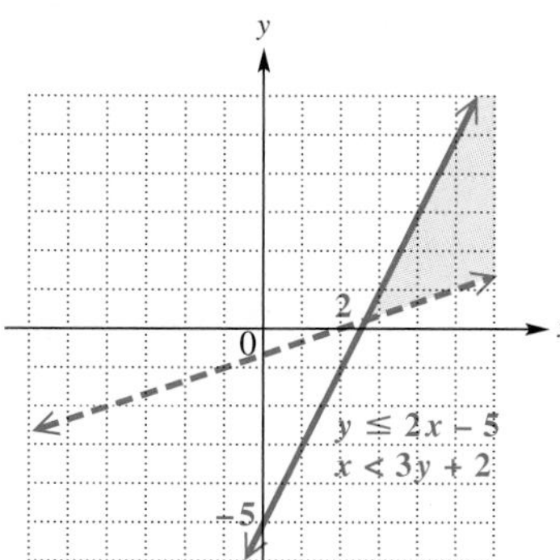

12. $x \ge 2y + 6$
$y > -2x + 4$

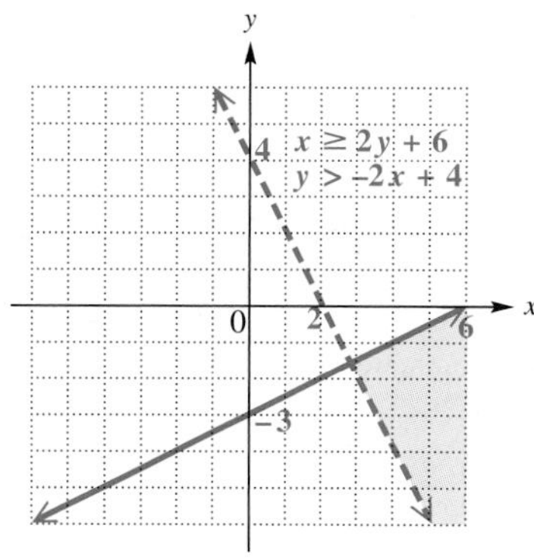

13. $4x + 3y < 6$
$x - 2y > 4$

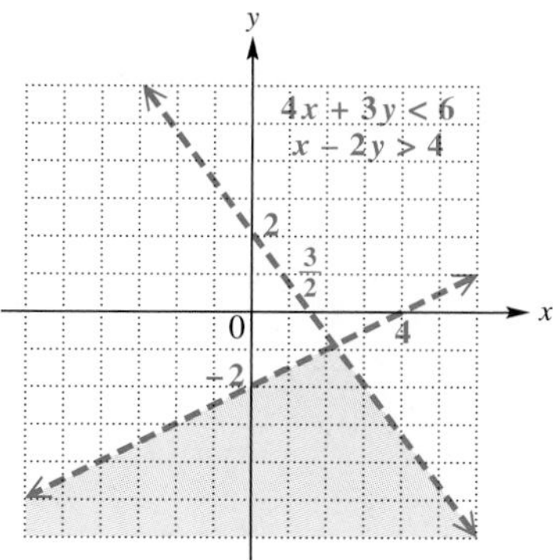

14. $3x + \ y > 4$
$x + 2y < 2$

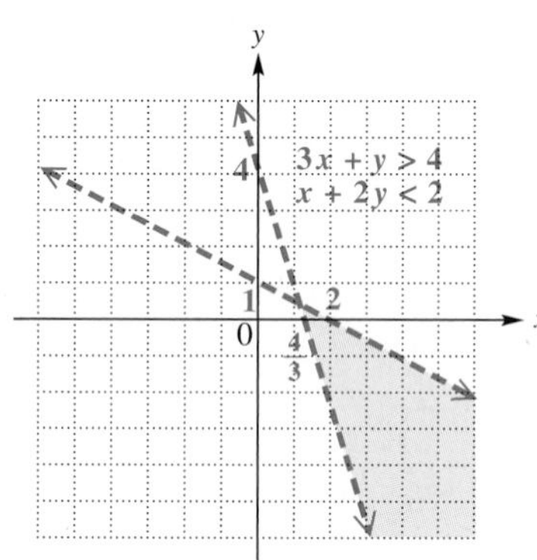

15. $x \le 2y + 3$
$x + y < 0$

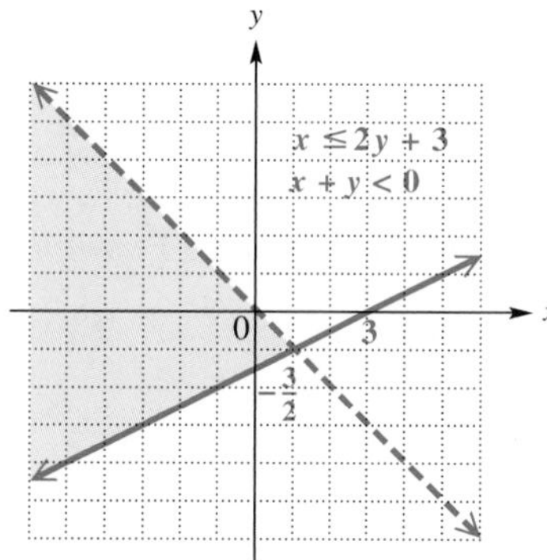

16. $x \le 4y + 3$
$x + y > 0$

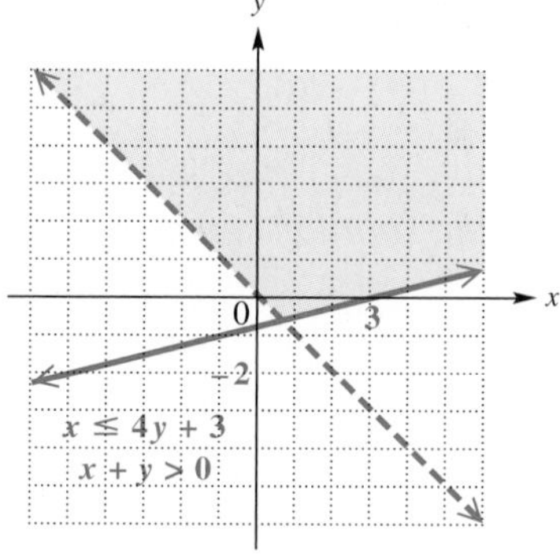

17. $4x + 5y < 8$
$y > -2$
$x > -4$

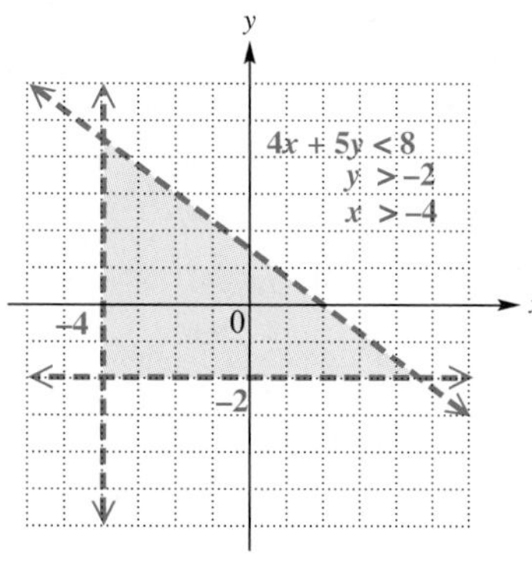

18. $x + y \ge -3$
$x - y \le 3$
$y \le 3$

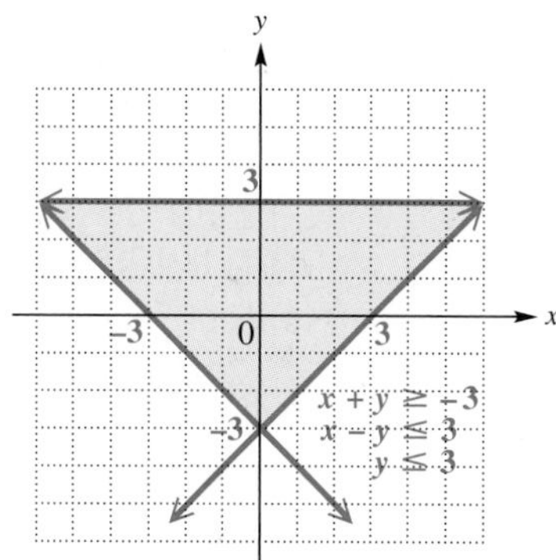

19. $3x - 2y \ge 6$
$x + y \le 4$
$x \ge 0$
$y \ge -4$

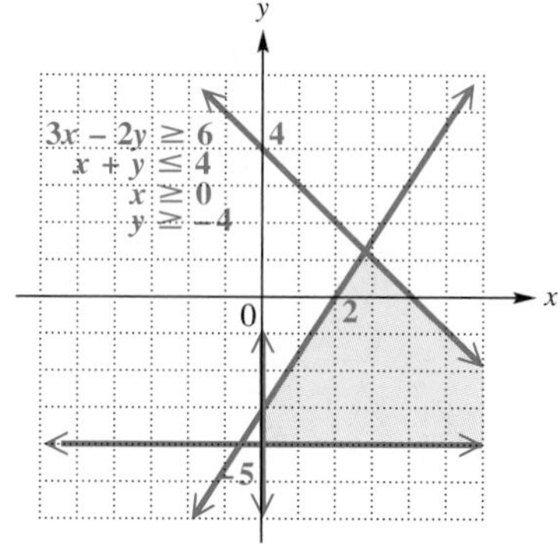

20. Every system of inequalities illustrated in the examples of this section has infinitely many solutions. Explain why this is so. Does this mean that *any* ordered pair is a solution?

There are infinitely many points in any of the shaded regions. There are, however, points that are not in the shaded regions, and thus there are ordered pairs that are not solutions.

Chapter 4 Summary

▶ Key Terms

4.1

system of linear equations A system of linear equations (or **linear system**) consists of two or more linear equations with the same variables.

solution of a system The solution of a system of linear equations includes all the ordered pairs that make all the equations of the system true at the same time.

solution set of a system The set of all ordered pairs that are solutions of a system is its solution set.

consistent system A system of equations with at least one solution is a consistent system.

inconsistent system An inconsistent system of equations is a system with no solution.

independent equations Equations of a system that have different graphs are called independent equations.

dependent equations Equations of a system that have the same graph (because they are different forms of the same equation) are called dependent equations.

4.5

system of linear inequalities A system of linear inequalities contains two or more linear inequalities (and no other kinds of inequalities).

solution set of a system of linear inequalities The solution set of a system of linear inequalities includes all points that make all inequalities of the system true at the same time.

▶ Test Your Word Power

See how well you have learned the vocabulary in this chapter. Answers, with examples, follow the Quick Review.

1. A **system of linear equations** consists of
   - **A.** at least two linear equations with different variables
   - **B.** two or more linear equations that have an infinite number of solutions
   - **C.** two or more linear equations with the same variables
   - **D.** two or more linear inequalities.

2. A **solution of a system** of linear equations is
   - **A.** an ordered pair that makes one equation of the system true
   - **B.** an ordered pair that makes all the equations of the system true at the same time
   - **C.** any ordered pair that makes one or the other or both equations of the system true
   - **D.** the set of values that make all the equations of the system false.

3. A **consistent system** is a system of equations
   - **A.** with at least one solution
   - **B.** with no solution
   - **C.** with an infinite number of solutions
   - **D.** that have the same graph.

4. An **inconsistent system** is a system of equations
   - **A.** with one solution
   - **B.** with no solution
   - **C.** with an infinite number of solutions
   - **D.** that have the same graph.

5. **Dependent equations**
   - **A.** have different graphs
   - **B.** have no solution
   - **C.** have one solution
   - **D.** are different forms of the same equation.

▶ Quick Review

Concepts	Examples

4.1 Solving Systems of Linear Equations by Graphing

An ordered pair is a solution of a system if it makes all equations of the system true at the same time.

Is $(4, -1)$ a solution of the system $\begin{aligned} x + y &= 3 \\ 2x - y &= 9 \end{aligned}$?

Because $4 + (-1) = 3$ and $2(4) - (-1) = 9$ are both true, $(4, -1)$ is a solution.

If the graphs of the equations of a system are both sketched on the same axes, then the points of intersection, if any, are solutions of the system.

Solve by graphing.

$$x + y = 5$$
$$2x - y = 4$$

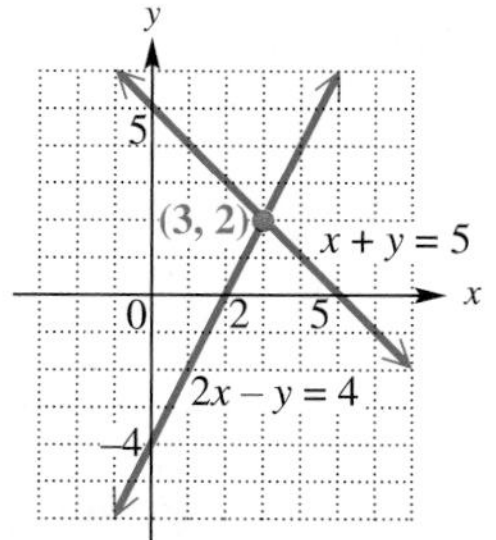

If the graphs of the equations do not intersect (that is, the lines are parallel), then the system has no solution and the solution set is $\emptyset$.

If the graphs of the equations are the same line, then the system has an infinite number of solutions. Use set-builder notation to write the solution set: $\{(x, y) \mid \_\_\_\_\_\}$.

The ordered pair $(3, 2)$ satisfies both equations, so $\{(3, 2)\}$ is the solution set.

4.2 Solving Systems of Linear Equations by Substitution

Step 1 Solve one equation for either variable.

Solve by substitution.

$$x + 2y = -5 \quad (1)$$
$$y = -2x - 1 \quad (2)$$

Equation (2) is already solved for y.

Step 2 Substitute for that variable in the other equation to get an equation in one variable.

Substitute $-2x - 1$ for y in equation (1).

Step 3 Solve the equation from Step 2.

$$\begin{aligned} x + 2(-2x - 1) &= -5 \\ x - 4x - 2 &= -5 \\ -3x - 2 &= -5 \\ -3x &= -3 \\ x &= 1 \end{aligned}$$

Step 4 Substitute the result into the equation from Step 1 to get the value of the other variable.

To find y, let $x = 1$ in equation (2):

$$y = -2(1) - 1 = -3.$$

Step 5 Check. Write the solution set.

The solution $(1, -3)$ checks, so $\{(1, -3)\}$ is the solution set.

4.3 Solving Systems of Linear Equations by Elimination

Step 1 Write both equations in standard form $Ax + By = C$.

Solve by elimination.

$$x + 3y = 7 \quad (1)$$
$$3x - y = 1 \quad (2)$$

Step 2 If necessary, multiply one or both equations by appropriate numbers so that the sum of the coefficients of either the x- or y-terms is 0.

Multiply equation (1) by -3 to eliminate the x-terms.

Concepts	Examples
4.3 Solving Systems of Linear Equations by Elimination *(continued)*	
Step 3 Add the equations to get an equation with only one variable (or no variable).	$\begin{aligned} -3x - 9y &= -21 \\ 3x - y &= 1 \\ \hline -10y &= -20 \quad \text{Add.} \end{aligned}$
Step 4 Solve the equation from Step 3.	$y = 2$ Divide by -10.
Step 5 Substitute the solution from Step 4 into either of the original equations to find the value of the remaining variable.	Substitute to get the value of x. $x + 3(2) = 7$ (1) $x + 6 = 7$ $x = 1$
Step 6 Check. Write the solution set.	Since $1 + 3(2) = 7$ and $3(1) - 2 = 1$, the solution $(1, 2)$ checks, so the solution set is $\{(1, 2)\}$.
4.4 Applications of Linear Systems	
Use the modified six-step method. *Step 1* **Read** the problem carefully. *Step 2* **Assign variables** for each unknown value. Use diagrams or tables as needed.	The sum of two numbers is 30. Their difference is 6. Find the numbers. Let x represent one number. Let y represent the other number.
Step 3 **Write two equations** using both variables.	$x + y = 30$ $x - y = 6$
Step 4 **Solve** the system.	$2x = 36$ Add. $x = 18$ Divide by 2.
Step 5 **State the answer.**	Let $x = 18$ in the first equation: $18 + y = 30$. Solve to get $y = 12$. The numbers are 18 and 12.
Step 6 **Check** the answer in the words of the original problem.	The sum of 18 and 12 is 30, and the difference between 18 and 12 is 6, so the answer checks.
4.5 Solving Systems of Linear Inequalities	
To solve a system of two or more linear inequalities, graph the inequalities on the same axes. (This was explained in **Section 3.5.**) The solution of the system is the intersection (overlap) of the regions of the graphs. The portions of the boundary lines that bound the region of solutions are included for a $\leq$ or $\geq$ inequality and excluded for a $<$ or $>$ inequality.	The shaded region is the solution of the system $2x + 4y \geq 5$ $x \geq 1$. 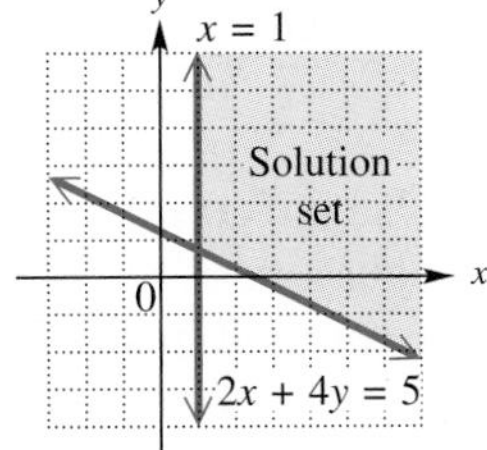

ANSWERS TO TEST YOUR WORD POWER

1. C; *Example:* $2x + y = 7$, $3x - y = 3$
2. B; *Example:* The ordered pair (2, 3) satisfies both equations of the system in the Answer 1 example, so it is a solution of the system.
3. A; *Example:* The system in the Answer 1 example is consistent. The graphs of the equations intersect at exactly one point, in this case the solution (2, 3).
4. B; *Example:* The equations of two parallel lines make up an inconsistent system; their graphs never intersect, so there is no solution to the system.
5. D; *Example:* The equations $4x - y = 8$ and $8x - 2y = 16$ are dependent because their graphs are the same line.

Math in the Media

CONNECTING GRAPHS IN THE MEDIA WITH SYSTEMS OF EQUATIONS

The March 16, 2008, headline and accompanying graph shown here indicate the Port of New Orleans cargo volume declined since 1998, after a period of growth between 1991 and 1998. The graph shows year-to-year fluctuations, as it consists of line segments describing the data from each year to the next.

N.O. port fights to grow as cargo lags

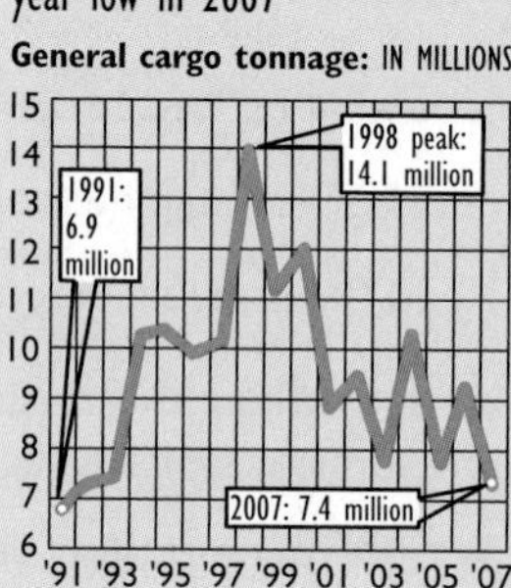

1. Suppose that we wish to depict the basic idea but provide a less detailed graph of the data consisting of only three data points: the point for the year 1991, represented by $A(1991, 6.9)$; the point for the year 1998, represented by $B(1998, 14.1)$; and the point for the year 2007, represented by $C(2007, 7.4)$. We could simply graph the segment AB and then graph the segment BC. Do this on the axes provided.

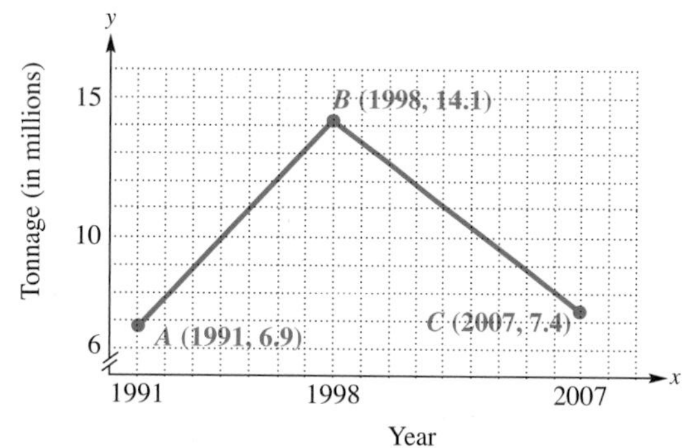

2. Use the coordinates for points A and B to find the equation of the line on which segment AB lies. Express it in slope-intercept form.

   $$y = \frac{36}{35}x - \frac{142{,}869}{70}$$

3. Use the coordinates for points B and C to find the equation of the line on which segment BC lies. Express it in slope-intercept form.

   $$y = -\frac{67}{90}x + \frac{3003}{2}$$

4. Consider the two equations you found in Exercises 3 and 4 as a system of linear equations. Solve the system, and confirm that the solution of the system is (1998, 14.1).

   The point (1998, 14.1) does satisfy both equations.

Chapter 4 Review Exercises

[4.1] *Decide whether the given ordered pair is a solution of the given system.*

1. $(3, 4)$

$4x - 2y = 4$

$5x + y = 19$ **yes**

2. $(-5, 2)$

$x - 4y = -13$

$2x + 3y = 4$ **no**

Solve each system by graphing.

3. $x + y = 4$

$2x - y = 5$ **{(3, 1)}**

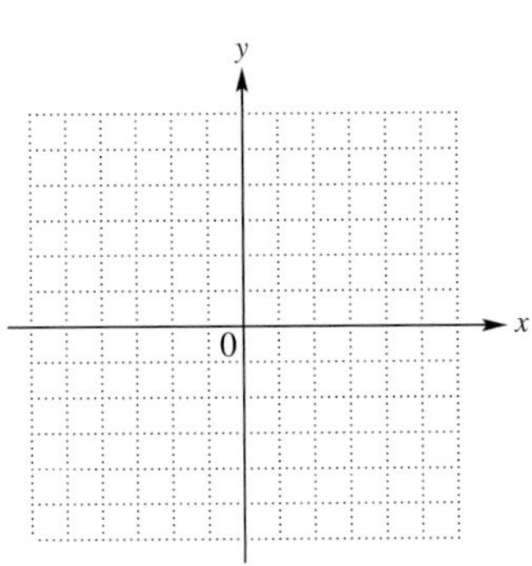

4. $x - 2y = 4$

$2x + y = -2$ **{(0, −2)}**

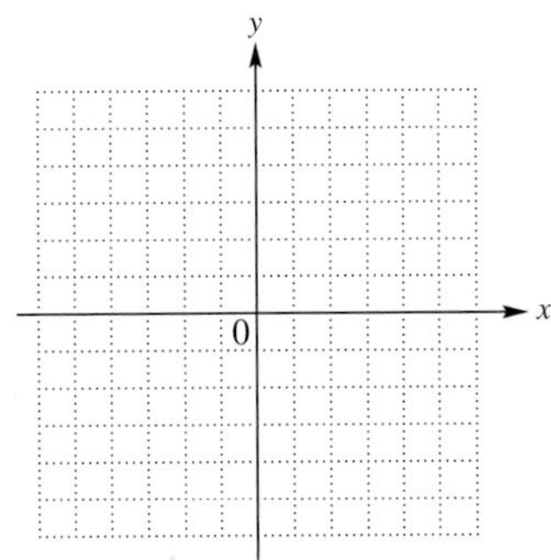

5. $x - 2 = 2y$

$2x - 4y = 4$ **{(x, y) | x − 2y = 2}**

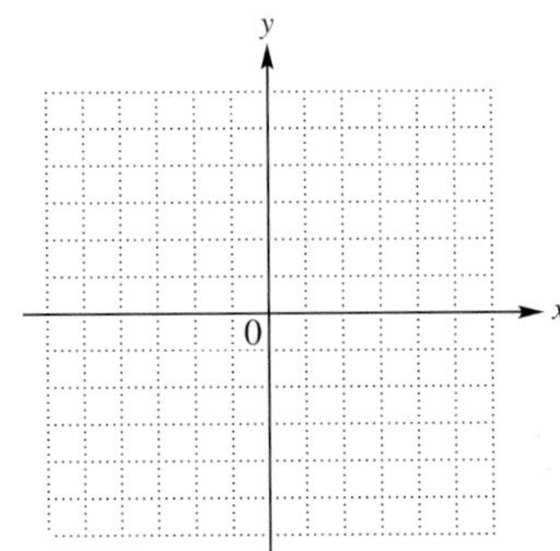

6. $2x + 4 = 2y$

$y - x = -3$ **∅**

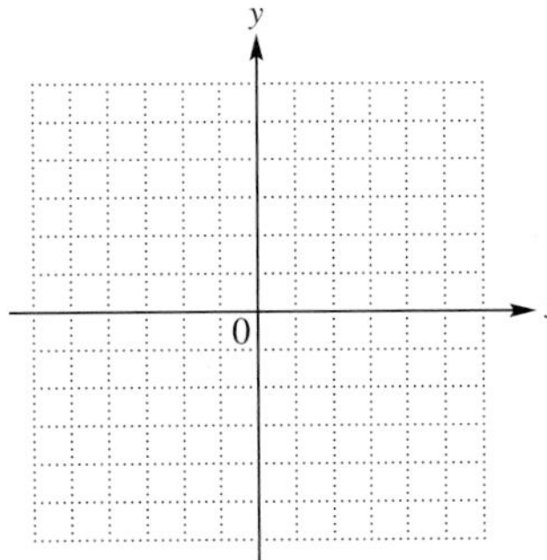

7. When a student was asked to determine whether the ordered pair $(1, -2)$ is a solution of the system

$$x + y = -1$$
$$2x + y = 4,$$

he answered "yes." His reasoning was that the ordered pair satisfies the equation $x + y = -1$; that is, $1 + (-2) = -1$ is true. Why is his answer wrong?

It is not a solution of the system because it is not also a solution of the second equation, 2x + y = 4.

[4.2] *Solve each system by the substitution method.*

8. $3x + y = 7$

$x = 2y$

{(2, 1)}

9. $2x - 5y = -19$

$y = x + 2$

{(3, 5)}

10. $4x + 5y = 44$
$x + 2 = 2y$
$\{(6, 4)\}$

11. $5x + 15y = 3$
$x + 3y = 2$
$\emptyset$

[4.3] *Solve each system by the elimination method.*

12. $2x - y = 13$
$x + y = 8$
$\{(7, 1)\}$

13. $3x - y = -13$
$x - 2y = -1$
$\{(-5, -2)\}$

14. $-4x + 3y = 25$
$6x - 5y = -39$
$\{(-4, 3)\}$

15. $3x - 4y = 9$
$6x - 8y = 18$
$\{(x, y) \mid 3x - 4y = 9\}$

16. For the system

$$2x + 12y = 7$$
$$3x + 4y = 1,$$

if we were to multiply the first (top) equation by -3, by what number would we have to multiply the second (bottom) equation in order to

(a) eliminate the x-terms when solving by the elimination method? 2

(b) eliminate the y-terms when solving by the elimination method? 9

Solve each system by any method.

17. $x - 2y = 5$
$y = x - 7$
$\{(9, 2)\}$

18. $5x - 3y = 11$
$2y = x - 4$
$\left\{\left(\frac{10}{7}, -\frac{9}{7}\right)\right\}$

19. $\frac{x}{2} + \frac{y}{3} = 7$
$\frac{x}{4} + \frac{2y}{3} = 8$
$\{(8, 9)\}$

20. $\frac{3x}{4} - \frac{y}{3} = \frac{7}{6}$
$\frac{x}{2} + \frac{2y}{3} = \frac{5}{3}$
$\{(2, 1)\}$

21. $2.4x + 1.7y = 7.6$
$1.2x - 0.5y = 9.2$
$\{(6, -4)\}$

22. $0.5x + 3.4y = 13$
$1.5x - 2.6y = -25$
$\{(-8, 5)\}$

[4.4] *Solve each problem by using a system of equations.*

23. At the end of 2006, Subway topped McDonald's as the largest restaurant chain in the United States. Subway operated 6981 more restaurants than McDonald's, and together the two chains had 34,529 restaurants. How many restaurants did each company operate? (*Source:* Technomic.)

Subway: 20,755 restaurants; McDonald's: 13,774 restaurants

24. In 2006, the two magazines with the largest circulations in the United States were *AARP The Magazine* and *Reader's Digest*. Together, the average total circulation of these two magazines was 33.5 million copies. The circulation of *Reader's Digest* was 13.3 million less than that of *AARP The Magazine.* What were the circulation figures for each magazine? (*Source:* www.myjobsource.com)

AARP *The Magazine*: 23.4 million; *Reader's Digest*: 10.1 million

25. The perimeter of a rectangle is 90 m. Its length is $1\frac{1}{2}$ times its width. Find the length and width of the rectangle.

length: 27 m; width: 18 m

26. A cashier has 20 bills, all of which are \$10 or \$20 bills. The total value of the money is \$330. How many of each type does the cashier have?

Number of Bills	Denomination of Bills	Total Value
x	\$10	$10x$
	\$20	
	XXXXXX	\$330

13 twenties; 7 tens

27. Candy that sells for \$1.30 per lb is to be mixed with candy selling for \$0.90 per lb to get 100 lb of a mix that will sell for \$1 per lb. How much of each type should be used?

25 lb of \$1.30 candy; 75 lb of \$0.90 candy

28. A certain plane flying with the wind travels 540 mi in 2 hr. Later, flying against the same wind, the plane travels 690 mi in 3 hr. Find the speed of the plane in still air and the speed of the wind.

plane: 250 mph; wind: 20 mph

29. After taxes, Ms. Cesar's game show winnings were \$18,000. She invested part of it at 3% annual simple interest and the rest at 4%. Her interest income for the first year was \$650. How much did she invest at each rate?

Amount of Principal	Percent (as a Decimal)	Interest
x	0.03	
y	0.04	
\$18,000	XXXXXX	

\$7000 at 3%; \$11,000 at 4%

30. A 40% antifreeze solution is to be mixed with a 70% solution to get 90 L of a 50% solution. How many liters of the 40% and 70% solutions will be needed?

Number of Liters	Percent (as a Decimal)	Amount of Pure Antifreeze
x	0.40	
y	0.70	
90	0.50	XXXXXX

60 L of 40% solution;
30 L of 70% solution

[4.5] *Graph the solution set for each system of linear inequalities.*

31. $x + y \geq 2$
$x - y \leq 4$

32. $y \geq 2x$
$2x + 3y \leq 6$

33. $x + y < 3$
$2x > y$

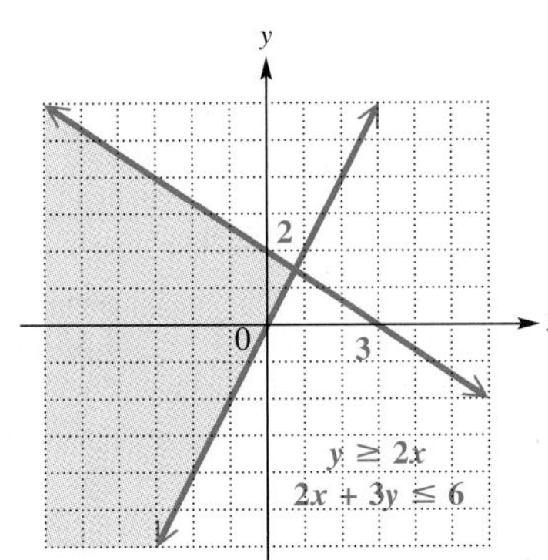

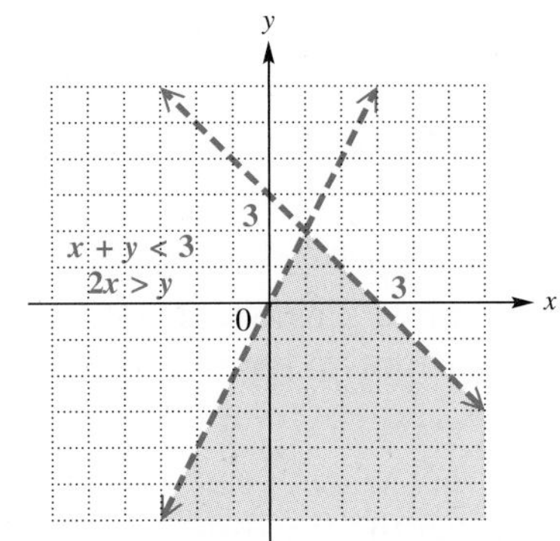

34. Which system of linear inequalities is graphed in the figure?

A. $x \leq 3$, $y \leq 1$ **B.** $x \leq 3$, $y \geq 1$ **C.** $x \geq 3$, $y \leq 1$ **D.** $x \geq 3$, $y \geq 1$

B

35. Without actually graphing, determine which system of inequalities has no solution.

A. $x \geq 4$, $y \leq 3$ **B.** $x + y > 4$, $x + y < 3$ **C.** $x > 2$, $y < 1$ **D.** $x + y < 4$, $x - y < 3$

B

Mixed Review Exercises

Solve each system.

36. $3x + 4y = 6$
$4x - 5y = 8$ **{(2, 0)}**

37. $\frac{3x}{2} + \frac{y}{5} = -3$
$4x + \frac{y}{3} = -11$ **{(−4, 15)}**

38. $x + 6y = 3$
$2x + 12y = 2$ **∅**

39. $x + y < 5$
$x - y \geq 2$

40. $y \leq 2x$
$x + 2y > 4$

41. $y < -4x$
$y < -2$

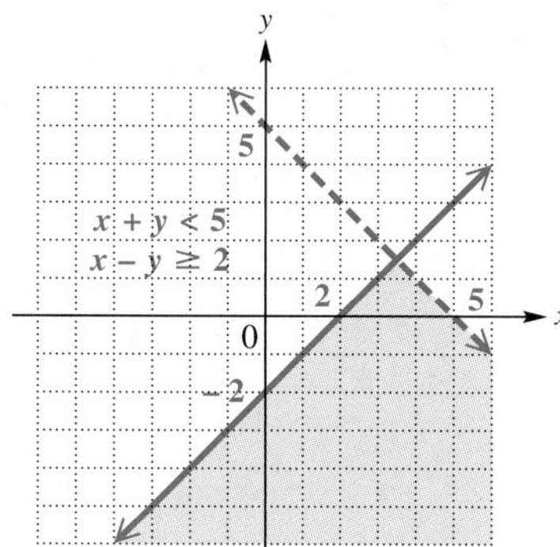

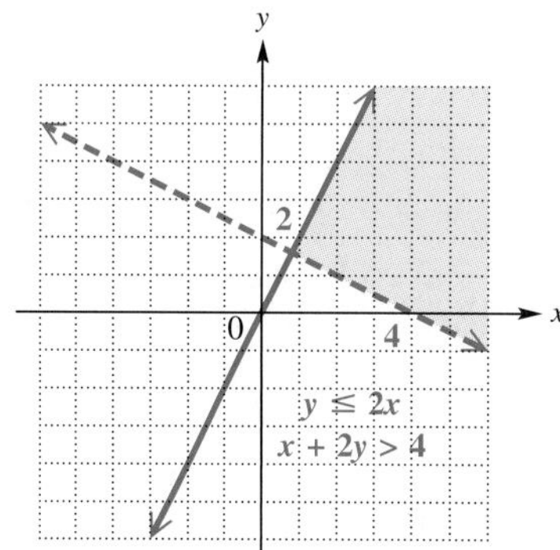

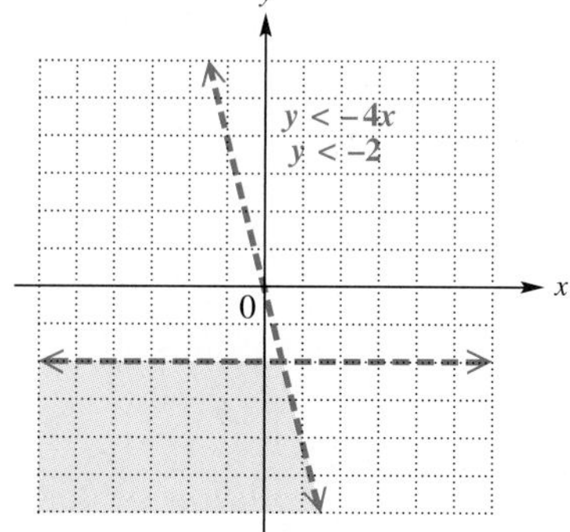

42. The perimeter of an isosceles triangle is 29 in. One side of the triangle is 5 in. longer than each of the two equal sides. Find the lengths of the sides of the triangle.

8 in., 8 in., and 13 in.

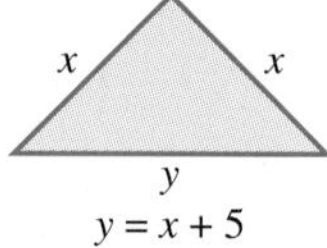

43. Super Bowl XLII was played in Glendale, Arizona, on February 3, 2008. The New York Giants beat the New England Patriots by 3 points, and the winning score was 11 points less than twice the losing score. What was the final score of the game? (*Source:* NFL.)

Giants: 17; Patriots: 14

44. Eboni Perkins compared the monthly payments she would incur for two types of mortgages: fixed-rate and variable-rate. Her observations led to the following graph.

(a) For which years would the monthly payment be more for the fixed-rate mortgage than for the variable-rate mortgage? **years 0–6**

(b) In what year would the payments be the same, and what would those payments be?

year 6; about $650

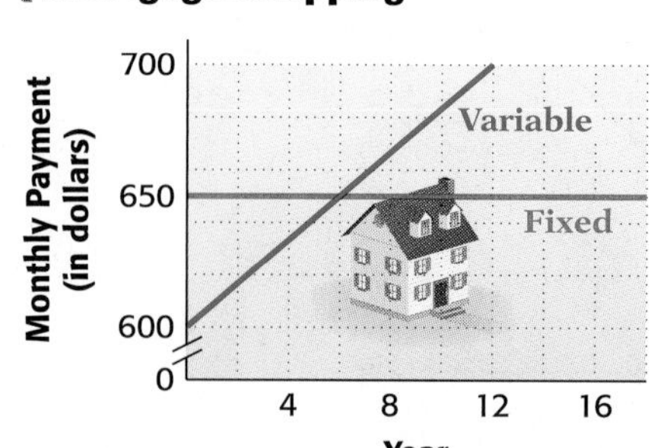

Chapter 4 Test

Use the Chapter Test Prep Video CD to see fully worked-out solutions to any of the exercises you want to review.

1. Solve the system by graphing.

$$2x + y = 1$$
$$3x - y = 9$$

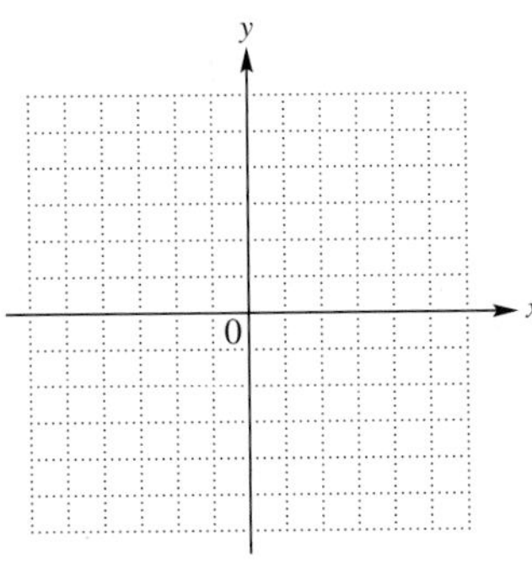

2. Suppose that the graph of a system of two linear equations consists of lines that have the same slope but different y-intercepts. How many solutions does the system have?

Solve each system by the substitution method.

3. $2x + y = -4$
   $x = y + 7$

4. $4x + 3y = -35$
   $x + y = 0$

Solve each system by the elimination method.

5. $2x - y = 4$
   $3x + y = 21$

6. $4x + 2y = 2$
   $5x + 4y = 7$

7. $6x - 5y = 0$
   $-2x + 3y = 0$

8. $4x + 5y = 2$
   $-8x - 10y = 6$

Solve each system by any method.

9. $3x = 6 + y$
   $6x - 2y = 12$

10. $\frac{x}{2} - \frac{y}{4} = 7$
    $\frac{2x}{3} + \frac{5y}{4} = 3$

1. $\{(2, -3)\}$

2. It has no solution.

3. $\{(1, -6)\}$

4. $\{(-35, 35)\}$

5. $\{(5, 6)\}$

6. $\{(-1, 3)\}$

7. $\{(0, 0)\}$

8. $\emptyset$

9. $\{(x, y) \mid 3x - y = 6\}$

10. $\{(12, -4)\}$

Solve each problem.

11. Memphis and Atlanta: 394 mi; Minneapolis and Houston: 1176 mi

11. The distance between Memphis and Atlanta is 782 mi less than the distance between Minneapolis and Houston. Together, the two distances total 1570 mi. How far is it between Memphis and Atlanta? How far is it between Minneapolis and Houston? (*Source: Rand McNally Road Atlas.*)

12. Statue of Liberty: 3.6 million; National World War II Memorial: 5.4 million

12. In 2004, a total of 9.0 million people visited the Statue of Liberty and the National World War II Memorial, two popular tourist attractions. The Statue of Liberty had 1.8 million fewer visitors than the National World War II Memorial. How many visitors did each of these attractions have? (*Source:* National Park Service, Department of the Interior.)

13. 20 L of 15% solution; 30 L of 40% solution

13. A 15% solution of alcohol is to be mixed with a 40% solution to get 50 L of a final mixture that is 30% alcohol. How much of each of the original solutions should be used?

14. slower car: 45 mph; faster car: 60 mph

14. Two cars leave from Perham, Minnesota, and travel in the same direction. One car travels $1\frac{1}{3}$ times as fast as the other. After 3 hr they are 45 mi apart. What are the speeds of the cars?

15.

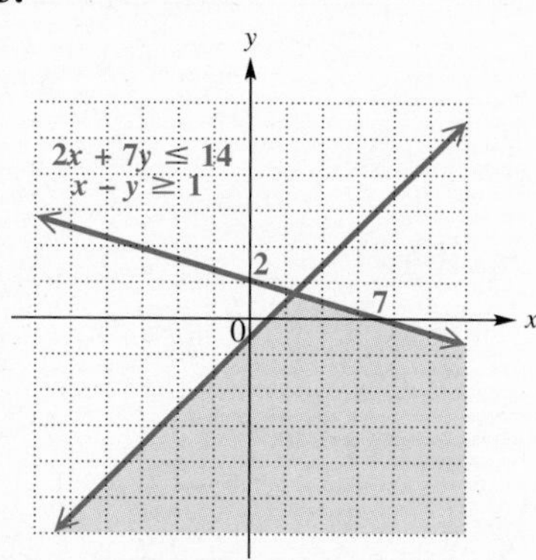

16.

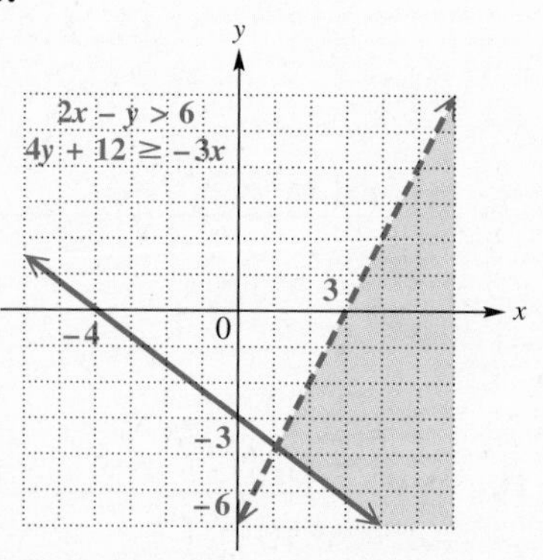

Graph the solution set of each system of inequalities.

15. $2x + 7y \leq 14$
$x - y \geq 1$

16. $2x - y > 6$
$4y + 12 \geq -3x$

Cumulative Review Exercises Chapters R–4

1. List all integer factors of 40.

−1, 1, −2, 2, −4, 4, −5, 5, −8, 8, −10, 10, −20, 20, −40, 40

2. Find the value of the expression if $x = 1$ and $y = 5$.

1

$$\frac{3x^2 + 2y^2}{10y + 3}$$

Name the property that justifies each statement.

3. $5 + (-4) = (-4) + 5$

commutative property

4. $r(s - k) = rs - rk$

distributive property

5. $-\frac{2}{3} + \frac{2}{3} = 0$

inverse property

6. Evaluate $-2 + 6[3 - (4 - 9)]$.

46

Solve each linear equation.

7. $2 - 3(6x + 2) = 4(x + 1) + 18$

$\left\{-\frac{13}{11}\right\}$

8. $\frac{3}{2}\left(\frac{1}{3}x + 4\right) = 6\left(\frac{1}{4} + x\right)$

$\left\{\frac{9}{11}\right\}$

Solve each linear inequality. Write the solution set in interval notation.

9. $-\frac{5}{6}x < 15$

$(-18, \infty)$

10. $-8 < 2x + 3$

$\left(-\frac{11}{2}, \infty\right)$

11. No baseball fan should be without a copy of *The Sports Encyclopedia: Baseball 2006* by David S. Neft and Richard M. Cohen. It provides the history of every player, team, and season from 1902–2005, and includes exhaustive statistics. The book has a perimeter of 37.8 in., and its width measures 2.58 in. less than its length. What are its dimensions? (*Source:* www.amazon.com)

width: 8.16 in.; length: 10.74 in.

Graph each linear equation.

12. $x - y = 4$

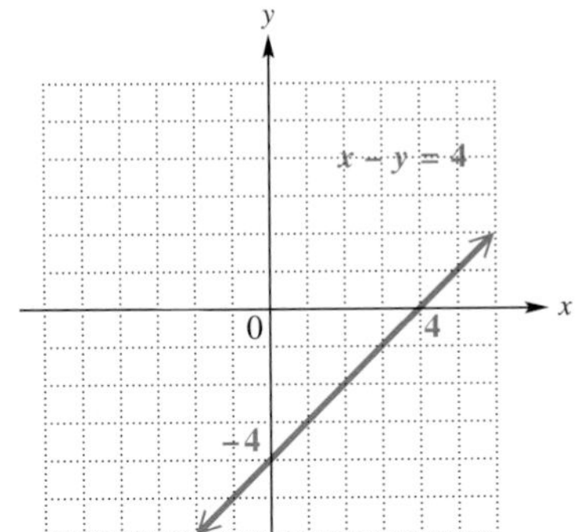

13. $3x + y = 6$

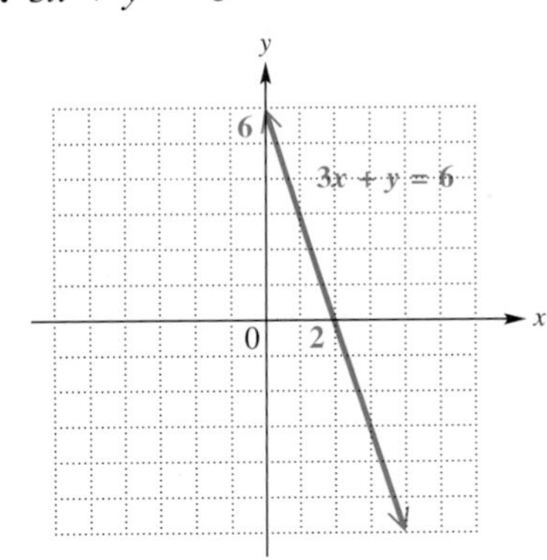

Find the slope of each line.

14. Through $(-5, 6)$ and $(1, -2)$

$-\frac{4}{3}$

15. Perpendicular to the line $y = 4x - 3$

$-\frac{1}{4}$

Write an equation for each line in slope-intercept form.

16. Through $(-4, 1)$ with slope $\frac{1}{2}$

$y = \frac{1}{2}x + 3$

17. Through the points $(1, 3)$ and $(-2, -3)$

$y = 2x + 1$

18. (a) Write an equation of the vertical line through $(9, -2)$. $x = 9$

(b) Write an equation of the horizontal line through $(4, -1)$. $y = -1$

Solve each system by any method.

19. $2x - y = -8$
$x + 2y = 11$
$\{(-1, 6)\}$

20. $4x + 5y = -8$
$3x + 4y = -7$
$\{(3, -4)\}$

21. $3x + 4y = 2$
$6x + 8y = 1$
$\emptyset$

Use a system of equations to solve each problem.

22. Admission prices at a football game were \$6 for adults and \$2 for children. The total value of the tickets sold was \$2528, and 454 tickets were sold. How many adults and how many children attended the game?

Kind of Ticket	Number Sold	Cost of Each (in dollars)	Total Value (in dollars)
Adult	x	6	$6x$
Child	y		
Total	454	XXXXXX	

405 adults and 49 children

23. The perimeter of a triangle is 53 in. If two sides are of equal length, and the third side measures 4 in. less than each of the equal sides, what are the lengths of the three sides?

19 in., 19 in., 15 in.

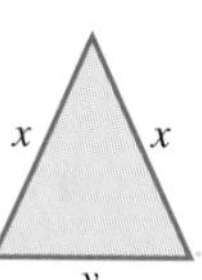

24. Graph the solution set of the system.

$$x + 2y \le 12$$
$$2x - y \le 8$$

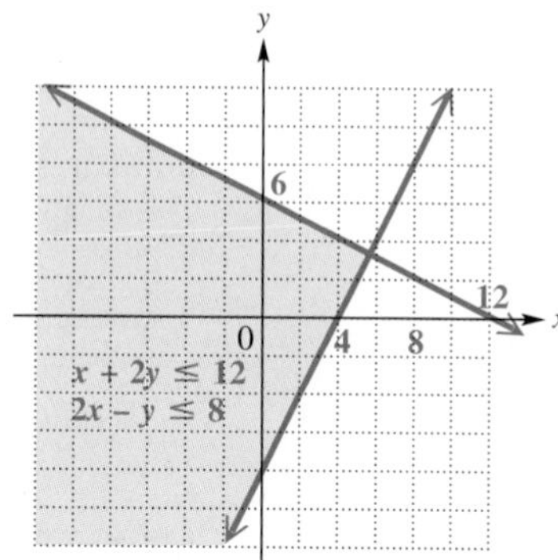

5

Exponents and Polynomials

Just how much is a *trillion*? A trillion, written 1,000,000,000,000, is a million million, or a thousand billion. A trillion seconds would last more than 31,000 years—that is, 310 centuries. By 2017, the U.S. government projects that consumers and taxpayers will spend more than \$4 trillion on health care, accounting for \$1 of every \$5 spent. (*Source:* Centers for Medicare and Medicaid Services.)

In Section 5.8, we use *exponents* and *scientific notation* to write and calculate with large numbers, such as the national debt, tax revenue, and the distances of a double-helix nebula and the star Pollux from Earth.

5.1 Adding and Subtracting Polynomials

OBJECTIVES

1. Review combining like terms.
2. Know the vocabulary for polynomials.
3. Evaluate polynomials.
4. Add polynomials.
5. Subtract polynomials.
6. Add and subtract polynomials with more than one variable.

Recall from **Section 1.8** that in an expression such as

$$4x^3 + 6x^2 + 5x + 8,$$

the quantities that are added, $4x^3$, $6x^2$, $5x$, and 8, are called **terms.** In the term $4x^3$, the number 4 is called the **numerical coefficient,** or simply the **coefficient,** of x^3. In the same way, 6 is the coefficient of x^2 in the term $6x^2$, 5 is the coefficient of x in the term $5x$, and 8 is the **constant** term. Other examples are given in the table at the side.

Term	Numerical Coefficient
$-7y$	-7
$34r^3$	34
$-26x^5yz^4$	-26
$-k = -1k$	-1
$r = 1r$	1
$\frac{3x}{8} = \frac{3}{8}x$	$\frac{3}{8}$
$\frac{x}{3} = \frac{1x}{3} = \frac{1}{3}x$	$\frac{1}{3}$

OBJECTIVE 1 **Review combining like terms.** Recall from **Section 1.8** that **like terms** have exactly the same combination of variables, with the same exponents on the variables. ***Only the coefficients may differ.***

$19m^5$ and $14m^5$
$-37y^9$ and y^9
$3pq$ and $-2pq$
$2xy^2$ and $-xy^2$
} Examples of like terms

$7x$ and $7y$
z^4 and z
$2pq$ and $2p$
$-4xy^2$ and $5x^2y$
} Examples of unlike terms

Using the distributive property, we combine, or add, like terms by adding their coefficients.

EXAMPLE 1 **Adding Like Terms**

Simplify each expression by adding like terms.

(a) $-4x^3 + 6x^3$
$= (-4 + 6)x^3$ Distributive property
$= 2x^3$

(b) $9x^6 - 14x^6 + x^6$
$= (9 - 14 + 1)x^6$ $x^6 = 1x^6$
$= -4x^6$

(c) $12m^2 + 5m + 4m^2$
$= (12 + 4)m^2 + 5m$
$= 16m^2 + 5m$

(d) $3x^2y + 4x^2y - x^2y$
$= (3 + 4 - 1)x^2y$
$= 6x^2y$

In Example 1(c), we cannot combine $16m^2$ and $5m$. These two terms are unlike because the exponents on the variables are different. ***Unlike terms have different variables or different exponents on the same variables.***

◀ Work Problem 1 at the Side.

1 Add like terms.

(a) $5x^4 + 7x^4$

(b) $9pq + 3pq - 2pq$

(c) $r^2 + 3r + 5r^2$

(d) $x + \frac{1}{2}x$

(e) $8t + 6w$

(f) $3x^4 - 3x^2$

OBJECTIVE 2 **Know the vocabulary for polynomials.** A **polynomial in** x is a term or the sum of a finite number of terms of the form ax^n, for any real number a and any whole number n. For example,

$$16x^8 - 7x^6 + 5x^4 - 3x^2 + 4 \quad \text{Polynomial}$$

is a polynomial in x. This polynomial is written in **descending powers,** because the exponents on x decrease from left to right.

ANSWERS

1. (a) $12x^4$ (b) $10pq$ (c) $6r^2 + 3r$ (d) $\frac{3}{2}x$
(e) These are unlike terms. They cannot be added.
(f) These are unlike terms. They cannot be added.

On the other hand, $2x^3 - x^2 + \frac{4}{x}$ is not a polynomial, since a variable appears in a denominator. We can define a *polynomial* using any variable, not just x, as in Example 1(c). Polynomials may have terms with more than one variable, as in Example 1(d).

Work Problem **2** *at the Side.* ▶

The **degree of a term** is the sum of the exponents on the variables. A constant term has degree 0. For example, $3x^4$ has degree 4, while $6x^{17}$ has degree 17. The term $5x$ (or $5x^1$) has degree 1, -7 has degree 0, and $2x^2y$ has degree $2 + 1 = 3$ (y has an exponent of 1).

The **degree of a polynomial** is the greatest degree of any nonzero term of the polynomial. For example, $3x^4 - 5x^2 + 6$ is of degree 4, the polynomial $5x + 7$ is of degree 1, 3 is of degree 0, and $x^2y + xy - 5xy^2$ is of degree 3.

Three types of polynomials are very common and are given special names. A polynomial with only one term is called a **monomial.** (*Mono-* means "one," as in *mono*rail.) Examples are

$$9m, \quad -6y^5, \quad a^2, \quad \text{and} \quad 6. \qquad \text{Monomials}$$

A polynomial with exactly two terms is called a **binomial.** (*Bi-* means "two," as in *bi*cycle.) Examples are

$$-9x^4 + 9x^3, \quad 8m^2 + 6m, \quad \text{and} \quad 3m^5 - 9m^2. \qquad \text{Binomials}$$

A polynomial with exactly three terms is called a **trinomial.** (*Tri-* means "three," as in *tri*angle.) Examples are

$$9m^3 - 4m^2 + 6, \quad \frac{19}{3}y^2 + \frac{8}{3}y + 5, \quad \text{and} \quad -3m^5 - 9m^2 + 2. \qquad \text{Trinomials}$$

EXAMPLE 2 Classifying Polynomials

Simplify each polynomial if possible. Then give the degree and tell whether the polynomial is a *monomial,* a *binomial,* a *trinomial,* or *none of these.*

(a) $2x^3 + 5$ We cannot simplify further. This is a binomial of degree 3.

(b) $4x - 5x + 2x$

Add like terms to simplify: $4x - 5x + 2x = x$. The degree is 1 (since $x = x^1$). The simplified polynomial is a monomial.

Work Problem **3** *at the Side.* ▶

OBJECTIVE 3 Evaluate polynomials. A polynomial usually represents different numbers for different values of the variable.

EXAMPLE 3 Evaluating a Polynomial

Find the value of $3x^4 + 5x^3 - 4x - 4$ when $x = -2$ and when $x = 3$.

First, substitute -2 for x.

$$3x^4 + 5x^3 - 4x - 4$$

$$= 3(-2)^4 + 5(-2)^3 - 4(-2) - 4 \qquad \text{Let } x = -2.$$

Use parentheses to avoid errors.

$$= 3(16) + 5(-8) - 4(-2) - 4 \qquad \text{Apply the exponents.}$$

$$= 48 - 40 + 8 - 4 \qquad \text{Multiply.}$$

$$= 12 \qquad \text{Add and subtract.}$$

Continued on Next Page

2 Choose all descriptions that apply for each of the expressions in parts (a)–(d).

A. Polynomial
B. Polynomial written in descending powers
C. Not a polynomial

(a) $3m^5 + 5m^2 - 2m + 1$

(b) $2p^4 + p^6$

(c) $\frac{1}{x} + 2x^2 + 3$

(d) $x - 3$

3 Simplify each polynomial if possible. Then give the degree and tell whether the polynomial is a *monomial, binomial, trinomial,* or *none of these.*

(a) $3x^2 + 2x - 4$

(b) $x^3 + 4x^3$

(c) $x^8 - x^7 + 2x^8$

ANSWERS

2. (a) A and B **(b)** A **(c)** C **(d)** A and B
3. (a) degree 2; trinomial
(b) degree 3; monomial (simplify to $5x^3$)
(c) degree 8; binomial (simplify to $3x^8 - x^7$)

Next, replace x with 3.

$$\begin{aligned} &3x^4 + 5x^3 - 4x - 4 \\ &= 3(3)^4 + 5(3)^3 - 4(3) - 4 && \text{Let } x = 3. \\ &= 3(81) + 5(27) - 4(3) - 4 && \text{Apply the exponents.} \\ &= 243 + 135 - 12 - 4 && \text{Multiply.} \\ &= 362 && \text{Add and subtract.} \end{aligned}$$

CAUTION
Use parentheses around the numbers that are substituted for the variable in Example 3, particularly when substituting a negative number for a variable that is raised to a power. Otherwise, a sign error may result.

◀ *Work Problem* **4** *at the Side.*

4 Find the value of $2x^3 + 8x - 6$ in each case.

(a) When $x = -1$

(b) When $x = 4$

OBJECTIVE 4 Add polynomials. Polynomials may be added, subtracted, multiplied, and divided.

Adding Polynomials
To add two polynomials, add like terms.

5 Add each pair of polynomials.

(a) $4x^3 - 3x^2 + 2x$ and $6x^3 + 2x^2 - 3x$

(b) $x^2 - 2x + 5$ and $4x^2 - 2$

EXAMPLE 4 Adding Polynomials Vertically

(a) Add $6x^3 - 4x^2 + 3$ and $-2x^3 + 7x^2 - 5$.
Write like terms in columns.

$$\begin{array}{r} 6x^3 - 4x^2 + 3 \\ \underline{-2x^3 + 7x^2 - 5} \end{array}$$

Now add, column by column.

$$\begin{array}{rrr} 6x^3 & -4x^2 & 3 \\ \underline{-2x^3} & \underline{7x^2} & \underline{-5} \\ 4x^3 & 3x^2 & -2 \end{array}$$

Add the three sums together.

$$4x^3 + 3x^2 + (-2) = 4x^3 + 3x^2 - 2$$

(b) Add $2x^2 - 4x + 3$ and $x^3 + 5x$.
Write like terms in columns and add column by column.

$$\begin{array}{r} 2x^2 - 4x + 3 \\ \underline{x^3 \qquad\quad + 5x \quad\;} \\ x^3 + 2x^2 + \;\; x + 3 \end{array}$$

Leave spaces for missing terms.

◀ *Work Problem* **5** *at the Side.*

The polynomials in Example 4 also could be added horizontally.

ANSWERS

4. **(a)** -16 **(b)** 154
5. **(a)** $10x^3 - x^2 - x$ **(b)** $5x^2 - 2x + 3$

EXAMPLE 5 Adding Polynomials Horizontally

(a) Add $6x^3 - 4x^2 + 3$ and $-2x^3 + 7x^2 - 5$.
Combine like terms.

$$(6x^3 - 4x^2 + 3) + (-2x^3 + 7x^2 - 5) = 4x^3 + 3x^2 - 2$$ Same answer found in Example 4(a)

(b) Add $2x^2 - 4x + 3$ and $x^3 + 5x$.

$$(2x^2 - 4x + 3) + (x^3 + 5x)$$
$$= x^3 + 2x^2 - 4x + 5x + 3 \quad \text{Commutative property}$$
$$= x^3 + 2x^2 + x + 3 \quad \text{Combine like terms.}$$

Work Problem 6 at the Side. ▶

OBJECTIVE 5 Subtract polynomials. In **Section 1.5,** the difference $\boldsymbol{x - y}$ was defined as $\boldsymbol{x + (-y)}$. (We find the difference $x - y$ by adding x and the opposite of y.) For example,

$$7 - 2 = 7 + (-2) = 5 \quad \text{and} \quad -8 - (-2) = -8 + 2 = -6.$$

A similar method is used to subtract polynomials.

Subtracting Polynomials

To subtract two polynomials, change all the signs of the second polynomial and add the result to the first polynomial.

EXAMPLE 6 Subtracting Polynomials

(a) Perform the subtraction $(5x - 2) - (3x - 8)$.
Change the signs in the second polynomial and add.

$$(5x - 2) - (3x - 8)$$
$$= (5x - 2) + (-3x + 8)$$
$$= 2x + 6$$

(b) Subtract $6x^3 - 4x^2 + 2$ from $11x^3 + 2x^2 - 8$.

$$(11x^3 + 2x^2 - 8) - (6x^3 - 4x^2 + 2)$$ Write the problem in the correct order.
$$= (11x^3 + 2x^2 - 8) + (-6x^3 + 4x^2 - 2)$$
$$= 5x^3 + 6x^2 - 10$$

To check a subtraction problem, use the following fact:

$$\text{If} \quad a - b = c, \quad \text{then} \quad a = b + c.$$

For example, $6 - 2 = 4$, so we check by writing $6 = 2 + 4$, which is correct. We check the polynomial subtraction above as follows:

$$(6x^3 - 4x^2 + 2) + (5x^3 + 6x^2 - 10)$$
$$= 11x^3 + 2x^2 - 8.$$

Since the sum is $11x^3 + 2x^2 - 8$, the subtraction was performed correctly.

Work Problem 7 at the Side. ▶

6 Find each sum.

(a) $(2x^4 - 6x^2 + 7)$
$+ (-3x^4 + 5x^2 + 2)$

(b) $(3x^2 + 4x + 2)$
$+ (6x^3 - 5x - 7)$

7 Subtract, and check your answers by addition.

(a) $(14y^3 - 6y^2 + 2y - 5)$
$- (2y^3 - 7y^2 - 4y + 6)$

(b) Subtract
$\left(-\frac{3}{2}y^2 + \frac{4}{3}y + 6\right)$
from $\left(\frac{7}{2}y^2 - \frac{11}{3}y + 8\right)$.

ANSWERS

6. (a) $-x^4 - x^2 + 9$
(b) $6x^3 + 3x^2 - x - 5$

7. (a) $12y^3 + y^2 + 6y - 11$
(b) $5y^2 - 5y + 2$

Subtraction also can be done in columns. We use vertical subtraction in **Section 5.7** when we study polynomial division.

EXAMPLE 7 Subtracting Polynomials Vertically

Subtract by columns: $(14y^3 - 6y^2 + 2y - 5) - (2y^3 - 7y^2 - 4y + 6)$.

$$\begin{array}{r} 14y^3 - 6y^2 + 2y - 5 \\ \underline{2y^3 - 7y^2 - 4y + 6} \end{array}$$ Arrange like terms in columns.

Change all signs in the second row, and then add.

$$\begin{array}{r} 14y^3 - 6y^2 + 2y - 5 \\ \underline{-2y^3 + 7y^2 + 4y - 6} \\ 12y^3 + y^2 + 6y - 11 \end{array}$$ Change signs. Add.

◀ Work Problem 8 at the Side.

EXAMPLE 8 Adding and Subtracting More Than Two Polynomials

Perform the indicated operations to simplify the expression

$$(4 - x + 3x^2) - (2 - 3x + 5x^2) + (8 + 2x - 4x^2).$$

Rewrite, changing the subtraction to adding the opposite.

$(4 - x + 3x^2) - (2 - 3x + 5x^2) + (8 + 2x - 4x^2)$

$= (4 - x + 3x^2) + (-2 + 3x - 5x^2) + (8 + 2x - 4x^2)$

$= (2 + 2x - 2x^2) + (8 + 2x - 4x^2)$ Combine like terms.

$= 10 + 4x - 6x^2$ Combine like terms.

◀ Work Problem 9 at the Side.

OBJECTIVE 6 Add and subtract polynomials with more than one variable. Polynomials in more than one variable are added and subtracted by combining like terms, just as with single-variable polynomials.

EXAMPLE 9 Adding and Subtracting Multivariable Polynomials

Add or subtract as indicated.

(a) $(4a + 2ab - b) + (3a - ab + b)$

$= 4a + 2ab - b + 3a - ab + b$

$= 7a + ab$ Combine like terms.

(b) $(2x^2y + 3xy + y^2) - (3x^2y - xy - 2y^2)$

$= 2x^2y + 3xy + y^2 - 3x^2y + xy + 2y^2$

$= -x^2y + 4xy + 3y^2$ Be careful with signs.

◀ Work Problem 10 at the Side.

8 Subtract by columns.

$(4y^3 - 16y^2 + 2y)$
$- (12y^3 - 9y^2 + 16)$

9 Perform the indicated operations.

$(6p^4 - 8p^3 + 2p - 1)$
$- (-7p^4 + 6p^2 - 12)$
$+ (p^4 - 3p + 8)$

10 Add or subtract.

(a) $(3mn + 2m - 4n)$
$+ (-mn + 4m + n)$

(b) $(5p^2q^2 - 4p^2 + 2q)$
$- (2p^2q^2 - p^2 - 3q)$

ANSWERS

8. $-8y^3 - 7y^2 + 2y - 16$
9. $14p^4 - 8p^3 - 6p^2 - p + 19$
10. (a) $2mn + 6m - 3n$
(b) $3p^2q^2 - 3p^2 + 5q$

5.1 Exercises

FOR EXTRA HELP

Math XL PRACTICE

WATCH

DOWNLOAD

READ

REVIEW

Fill in each blank with the correct response.

1. In the term $7x^5$, the coefficient is **7** and the exponent is **5**.
2. The expression $5x^3 - 4x^2$ has **two** (how many?) term(s).
3. The degree of the term $-4x^8$ is **8**.
4. The polynomial $4x^2 - y^2$ **is not** (is/is not) an example of a trinomial.
5. When $x^2 + 10$ is evaluated for $x = 4$, the result is **26**.
6. **$5x^9$** is an example of a monomial with coefficient 5, in the variable x, having degree 9.

For each polynomial, determine the number of terms, and name the coefficient of each term.

7. $6x^4$ **1; 6**
8. $-9y^5$ **1; −9**
9. t^4 **1; 1**
10. s^7 **1; 1**
11. $\frac{x}{5}$ **1; $\frac{1}{5}$**
12. $\frac{z}{8}$ **1; $\frac{1}{8}$**
13. $-19r^2 - r$ **2; −19, −1**
14. $2y^3 - y$ **2; 2, −1**
15. $x - 8x^2 + \frac{2}{3}x^3$ **3; 1, −8, $\frac{2}{3}$**
16. $v - 2v^3 + \frac{3}{4}v^2$ **3; 1, −2, $\frac{3}{4}$**

In each polynomial, combine like terms whenever possible. Write the result with descending powers. See Example 1.

17. $-3m^5 + 5m^5$ **$2m^5$**
18. $-4y^3 + 3y^3$ **$-y^3$**
19. $2r^5 + (-3r^5)$ **$-r^5$**
20. $-19y^2 + 9y^2$ **$-10y^2$**
21. $\frac{1}{2}x^4 + \frac{1}{6}x^4$ **$\frac{2}{3}x^4$**
22. $\frac{3}{10}x^6 + \frac{1}{5}x^6$ **$\frac{1}{2}x^6$**
23. $0.2m^5 - 0.5m^2$ **cannot be simplified; $0.2m^5 - 0.5m^2$**
24. $-0.9y + 0.9y^2$ **cannot be simplified; $0.9y^2 - 0.9y$**
25. $-3x^5 + 2x^5 - 4x^5$ **$-5x^5$**
26. $6x^3 - 8x^3 + 9x^3$ **$7x^3$**
27. $-4p^7 + 8p^7 + 5p^9$ **$5p^9 + 4p^7$**
28. $-3a^8 + 4a^8 - 3a^2$ **$a^8 - 3a^2$**
29. $-4y^2 + 3y^2 - 2y^2 + y^2$ **$-2y^2$**
30. $3r^5 - 8r^5 + r^5 + 2r^5$ **$-2r^5$**

For each polynomial, first simplify, if possible, and write it with descending powers. Then give the degree of the resulting polynomial, and tell whether it is a monomial, *a* binomial, *a* trinomial, *or* none of these. *See Example 2.*

31. $6x^4 - 9x$ **already simplified; 4; binomial**
32. $7t^3 - 3t$ **already simplified; 3; binomial**
33. $5m^4 - 3m^2 + 6m^5 - 7m^3$ **already simplified; $6m^5 + 5m^4 - 7m^3 - 3m^2$; 5; none of these**

34. $6p^5 + 4p^3 - 8p^4 + 10p^2$
already simplified;
$\mathbf{6p^5 - 8p^4 + 4p^3 + 10p^2}$**; 5;**
none of these

35. $\frac{5}{3}x^4 - \frac{2}{3}x^4 + \frac{1}{3}x^2 - 4$
$x^4 + \frac{1}{3}x^2 - 4$**; 4; trinomial**

36. $\frac{4}{5}r^6 + \frac{1}{5}r^6 - r^4 + \frac{2}{5}r$
$r^6 - r^4 + \frac{2}{5}r$**; 6; trinomial**

37. $0.8x^4 - 0.3x^4 - 0.5x^4 + 7$
7; 0; monomial

38. $1.2t^3 - 0.9t^3 - 0.3t^3 + 9$
9; 0; monomial

39. $2.5x^2 + 0.5x + x^2 - x - 2x^2$
$\mathbf{1.5x^2 - 0.5x}$**; 2; binomial**

Find the value of each polynomial ***(a)*** *when $x = 2$ and* ***(b)*** *when $x = -1$. See Example 3.*

40. $5x - 4$
(a) 6 (b) −9

41. $-2x + 3$
(a) −1 (b) 5

42. $-3x^2 + 14x - 2$
(a) 14 (b) −19

43. $2x^2 + 5x + 1$
(a) 19 (b) −2

44. $x^4 - 6x^3 + x^2 + 1$
(a) −27 (b) 9

45. $2x^5 - 4x^4 + 5x^3 - x^2$
(a) 36 (b) −12

46. $2x^6 - 4x$
(a) 120 (b) 6

47. $-4x^5 + x^2$
(a) −124 (b) 5

Relating Concepts (Exercises 48–52) For Individual or Group Work

A polynomial can model the distance in feet that a car going approximately 68 mph will skid in t seconds. If we let D represent this distance, then using function notation from **Section 3.6,** *we have*

$$D(t) = 100t - 13t^2.$$

Each time we evaluate this function (called a ***polynomial function****) for a value of t, we get one and only one output value D(t). Exercises 48–52 illustrate this idea with this polynomial function and three others.* ***Work them in order.***

48. Evaluate the given function when $t = 5$. Use the result to fill in the blanks: In ______ seconds, the car will skid ______ feet.
5;175

49. Use the given polynomial function to find the distance the car will skid in 1 sec. Write an ordered pair of the form $(t, D(t))$.
87 ft; (1, 87)

50. If gasoline costs \$4.00 per gal, then the function defined by $C(x) = 4.00x$ gives the cost, in dollars, of x gallons. How much would 4 gal cost?
\$16.00

51. If it costs \$15 plus \$2 per day to rent a chain saw, then $C(x) = 2x + 15$ gives the cost in dollars to rent the chain saw for x days. How much would it cost to rent the saw for 6 days?
\$27

52. If an object is projected upward under certain conditions, its height in feet is given by $h(t) = -16t^2 + 60t + 80$, where t is in seconds. Evaluate this function for $t = 2.5$, and then use the result to fill in the blanks: If ______ seconds have elapsed, the height of the object is ______ feet.
2.5; 130

Add or subtract as indicated. See Examples 4 and 7.

53. Add.
$$\begin{array}{r} 3m^2 + 5m \\ 2m^2 - 2m \\ \hline \end{array}$$
$5m^2 + 3m$

54. Add.
$$\begin{array}{r} 4a^3 - 4a^2 \\ 6a^3 + 5a^2 \\ \hline \end{array}$$
$10a^3 + a^2$

55. Subtract.
$$\begin{array}{r} 12x^4 - x^2 \\ 8x^4 + 3x^2 \\ \hline \end{array}$$
$4x^4 - 4x^2$

56. Subtract.
$$\begin{array}{r} 13y^5 - y^3 \\ 7y^5 + 5y^3 \\ \hline \end{array}$$
$6y^5 - 6y^3$

57. Add.
$$\begin{array}{r} \frac{2}{3}x^2 + \frac{1}{5}x + \frac{1}{6} \\ \frac{1}{2}x^2 - \frac{1}{3}x + \frac{2}{3} \\ \hline \end{array}$$
$\frac{7}{6}x^2 - \frac{2}{15}x + \frac{5}{6}$

58. Add.
$$\begin{array}{r} \frac{4}{7}y^2 - \frac{1}{5}y + \frac{7}{9} \\ \frac{1}{3}y^2 - \frac{1}{3}y + \frac{2}{5} \\ \hline \end{array}$$
$\frac{19}{21}y^2 - \frac{8}{15}y + \frac{53}{45}$

59. Subtract.
$$\begin{array}{r} 12m^3 - 8m^2 + 6m + 7 \\ 5m^2 \qquad\quad - 4 \\ \hline \end{array}$$
$12m^3 - 13m^2 + 6m + 11$

60. Subtract.
$$\begin{array}{r} 5a^4 - 3a^3 + 2a^2 - a + 6 \\ -6a^4 \qquad\quad - a^2 + a - 1 \\ \hline \end{array}$$
$11a^4 - 3a^3 + 3a^2 - 2a + 7$

61. Subtract.
$$\begin{array}{r} 4.3x^3 - 6.1x^2 - 3.0x - 5 \\ 1.4x^3 - 2.6x^2 - 1.5x + 4 \\ \hline \end{array}$$
$2.9x^3 - 3.5x^2 - 1.5x - 9$

Perform the indicated operations. See Examples 5, 6, and 8.

62. $(3r^2 + 5r - 6) + (2r - 5r^2)$

$-2r^2 + 7r - 6$

63. $(2r^2 + 3r - 12) + (6r^2 + 2r)$

$8r^2 + 5r - 12$

64. $(x^2 + x) - (3x^2 + 2x - 1)$

$-2x^2 - x + 1$

65. $(8m^2 - 7m) - (3m^2 + 7m - 6)$

$5m^2 - 14m + 6$

66. $(-2b^6 + 3b^4 - b^2) + (b^6 + 2b^4 + 2b^2)$

$-b^6 + 5b^4 + b^2$

67. $(16x^3 - x^2 + 3x) + (-12x^3 + 3x^2 + 2x)$

$4x^3 + 2x^2 + 5x$

68. $(8t^5 + 3t^3 + 5t) - (19t^4 - 6t^2 + t)$

$8t^5 - 19t^4 + 3t^3 + 6t^2 + 4t$

69. $(7y^4 + 3y^2 + 2y) - (18y^5 - 5y^3 + y)$

$-18y^5 + 7y^4 + 5y^3 + 3y^2 + y$

70. $[(9b^3 - 4b^2 + 3b + 2) - (-2b^3 + b)] - (8b^3 + 6b + 4)$

$3b^3 - 4b^2 - 4b - 2$

71. $[(8m^2 + 4m - 7) - (2m^3 - 5m + 2)] - (m^2 + m)$

$-2m^3 + 7m^2 + 8m - 9$

72. Subtract $-5w^3 + 5w^2 - 7$ from $6w^3 + 8w + 5$.

$11w^3 - 5w^2 + 8w + 12$

73. Subtract $9x^2 - 3x + 7$ from $-2x^2 - 6x + 4$.

$-11x^2 - 3x - 3$

74. Find the difference when $9x^4 + 3x^2 + 5$ is subtracted from $8x^4 - 2x^3 + x - 1$.

$-x^4 - 2x^3 - 3x^2 + x - 6$

Find a polynomial that represents the perimeter of each square, rectangle, or triangle.

75.

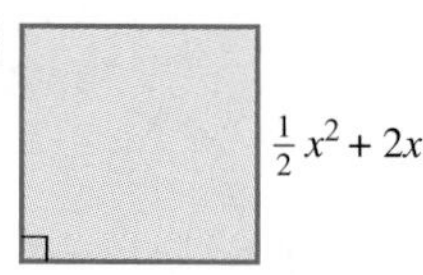

$2x^2 + 8x$

76.

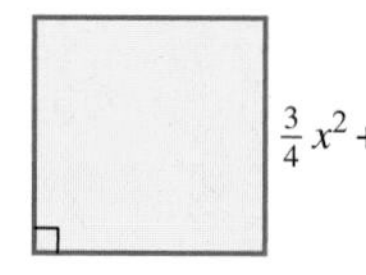

$3x^2 + 4x$

77.

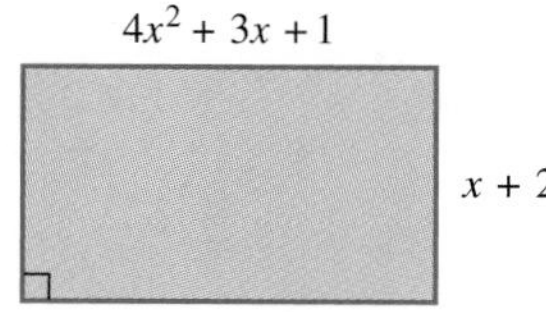

$8x^2 + 8x + 6$

78.

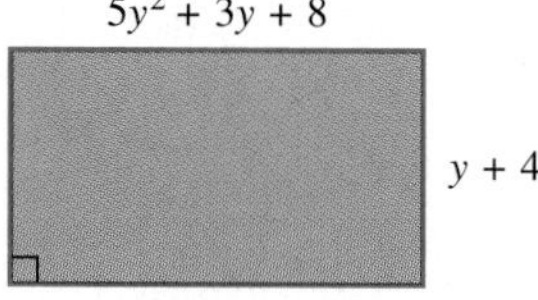

$10y^2 + 8y + 24$

79.

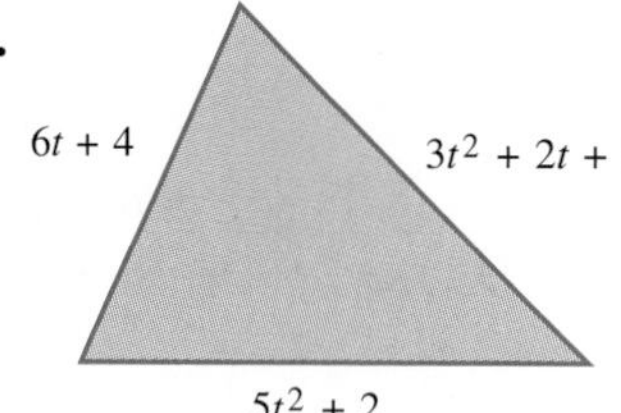

$8t^2 + 8t + 13$

80.

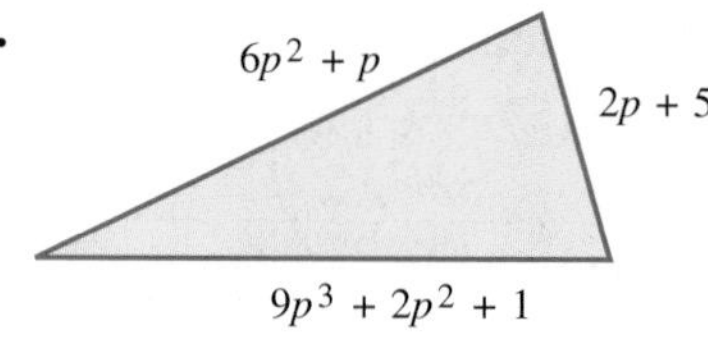

$9p^3 + 8p^2 + 3p + 6$

Add or subtract as indicated. See Example 9.

81. $(9a^2b - 3a^2 + 2b) + (4a^2b - 4a^2 - 3b)$

$13a^2b - 7a^2 - b$

82. $(4xy^3 - 3x + y) + (5xy^3 + 13x - 4y)$

$9xy^3 + 10x - 3y$

83. $(2c^4d + 3c^2d^2 - 4d^2) - (c^4d + 8c^2d^2 - 5d^2)$

$c^4d - 5c^2d^2 + d^2$

84. $(3k^2h^3 + 5kh + 6k^3h^2) - (2k^2h^3 - 9kh + k^3h^2)$

$k^2h^3 + 14kh + 5k^3h^2$

85. Subtract.

$$\begin{array}{r} 9m^3n - 5m^2n^2 + 4mn^2 \\ \underline{-3m^3n + 6m^2n^2 + 8mn^2} \end{array}$$

$12m^3n - 11m^2n^2 - 4mn^2$

86. Subtract.

$$\begin{array}{r} 12r^5t + 11r^4t^2 - 7r^3t^3 \\ \underline{-8r^5t + 10r^4t^2 + 3r^3t^3} \end{array}$$

$20r^5t + r^4t^2 - 10r^3t^3$

*Find **(a)** a polynomial that represents the perimeter of each triangle and **(b)** the measures of the angles of the triangle. (Hint: In part (b), the sum of the measures of the angles of any triangle is 180°.)*

87.

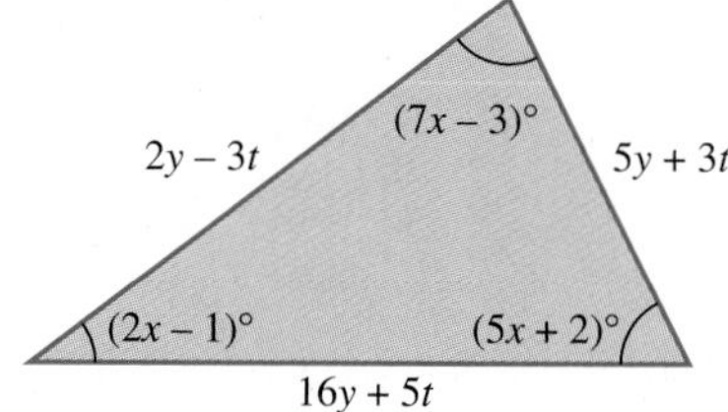

(a) $23y + 5t$ (b) 25°, 67°, 88°

88.

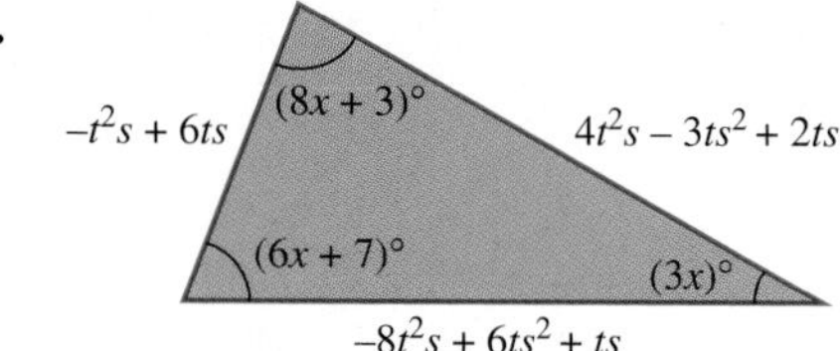

(a) $-5t^2s + 3ts^2 + 9ts$ (b) 30°, 67°, 83°

5.2 The Product Rule and Power Rules for Exponents

OBJECTIVES

1. Use exponents.
2. Use the product rule for exponents.
3. Use the rule $(a^m)^n = a^{mn}$.
4. Use the rule $(ab)^m = a^m b^m$.
5. Use the rule $\left(\frac{a}{b}\right)^m = \frac{a^m}{b^m}$.
6. Use combinations of the rules for exponents.
7. Use the rules for exponents in a geometry application.

OBJECTIVE 1 Use exponents. In **Section 1.1,** we used exponents to write repeated products. Recall that in the expression 5^2, the number 5 is called the **base** and 2 is called the **exponent,** or **power.** The expression 5^2 is called an **exponential expression.** Although we do not usually write a quantity with an exponent of 1, in general, for any quantity a, $\boldsymbol{a = a^1}$.

EXAMPLE 1 Using Exponents

Write $3 \cdot 3 \cdot 3 \cdot 3 \cdot 3$ in exponential form and evaluate.

Since 3 occurs as a factor five times, the base is 3 and the exponent is 5. The exponential expression is 3^5, read "3 to the fifth power," or simply "3 to the fifth."

$$\underbrace{3 \cdot 3 \cdot 3 \cdot 3 \cdot 3}_{\text{5 factors of 3}} \quad \text{means} \quad 3^5, \quad \text{or} \quad 243.$$

Work Problem 1 at the Side. ▶

EXAMPLE 2 Evaluating Exponential Expressions

Evaluate. Name the base and the exponent.

	Base	Exponent
(a) $5^4 = 5 \cdot 5 \cdot 5 \cdot 5 = 625$	5	4
(b) $-5^4 = -1 \cdot 5^4 = -1 \cdot (5 \cdot 5 \cdot 5 \cdot 5) = -625$	5	4
(c) $(-5)^4 = (-5)(-5)(-5)(-5) = 625$	-5	4

CAUTION

Look at Examples 2(b) and (c). In -5^4, the absence of parentheses shows that the exponent 4 applies only to the base 5, and not -5. In $(-5)^4$, the parentheses show that the exponent 4 applies to the base -5. In summary, $-a^n$ and $(-a)^n$ are not necessarily the same.

Expression	Base	Exponent	Example
$-a^n$	a	n	$-3^2 = -(3 \cdot 3) = -9$
$(-a)^n$	$-a$	n	$(-3)^2 = (-3)(-3) = 9$

Work Problem 2 at the Side. ▶

OBJECTIVE 2 Use the product rule for exponents. To develop the product rule, we use the definition of an exponent.

$$2^4 \cdot 2^3 = \overbrace{(2 \cdot 2 \cdot 2 \cdot 2)}^{\text{4 factors}}\overbrace{(2 \cdot 2 \cdot 2)}^{\text{3 factors}}$$
$$= \underbrace{2 \cdot 2 \cdot 2 \cdot 2 \cdot 2 \cdot 2 \cdot 2}_{4 + 3 = 7 \text{ factors}}$$
$$= 2^7$$

1 Write $2 \cdot 2 \cdot 2 \cdot 2$ in exponential form and evaluate.

2 Evaluate. Name the base and the exponent.

(a) $(-2)^5$ **(b)** -2^5

(c) -4^2 **(d)** $(-4)^2$

ANSWERS

1. 2^4, or 16
2. **(a)** -32; -2; 5 **(b)** -32; 2; 5 **(c)** -16; 4; 2 **(d)** 16; -4; 2

Also, $6^2 \cdot 6^3 = (6 \cdot 6)(6 \cdot 6 \cdot 6)$

$$= 6 \cdot 6 \cdot 6 \cdot 6 \cdot 6$$

$$= 6^5.$$

Generalizing from these examples,

$$2^4 \cdot 2^3 = 2^{4+3} = 2^7 \quad \text{and} \quad 6^2 \cdot 6^3 = 6^{2+3} = 6^5.$$

In each case, adding the exponents gives the exponent of the product, suggesting the **product rule for exponents.**

Product Rule for Exponents

For any positive integers m and n, $a^m \cdot a^n = a^{m+n}$.
(Keep the same base and add the exponents.)

Example: $6^2 \cdot 6^5 = 6^{2+5} = 6^7$

CAUTION

Do not multiply the bases when using the product rule. ***Keep the same base and add the exponents.*** For example,

$$6^2 \cdot 6^5 = 6^7, \quad \textit{not} \quad 36^7.$$

EXAMPLE 3 Using the Product Rule

Use the product rule for exponents to simplify, if possible.

(a) $6^3 \cdot 6^5 = 6^{3+5} = 6^8$ (Keep the same base.)

(b) $(-4)^7(-4)^2 = (-4)^{7+2} = (-4)^9$

(c) $x^2 \cdot x = x^2 \cdot x^1 = x^{2+1} = x^3$

(d) $m^4m^3m^5 = m^{4+3+5} = m^{12}$

(e) $2^3 \cdot 3^2$

The product rule does not apply to the product $2^3 \cdot 3^2$ because the bases are different.

$$2^3 \cdot 3^2 = 8 \cdot 9 = 72 \quad \text{Evaluate } 2^3 \text{ and } 3^2\text{; then multiply.}$$

Think: 2^3 means $2 \cdot 2 \cdot 2$. Think: 3^2 means $3 \cdot 3$.

(f) $2^3 + 2^4$

The product rule does not apply to $2^3 + 2^4$ because it is a *sum,* not a *product*.

$$2^3 + 2^4 = 8 + 16 = 24 \quad \text{Evaluate } 2^3 \text{ and } 2^4\text{; then add.}$$

CAUTION

The bases of the factors must be the same before we can apply the product rule for exponents.

◀ Work Problem 3 at the Side.

3 Simplify by using the product rule, if possible.

(a) $8^2 \cdot 8^5$

(b) $(-7)^5(-7)^3$

(c) $y^3 \cdot y$

(d) $z^2z^5z^6$

(e) $4^2 \cdot 3^5$

(f) $6^4 + 6^2$

ANSWERS

3. **(a)** 8^7 **(b)** $(-7)^8$ **(c)** y^4 **(d)** z^{13} **(e)** The product rule does not apply. (product: 3888) **(f)** The product rule does not apply. (sum: 1332)

EXAMPLE 4 **Using the Product Rule**

Multiply $2x^3$ and $3x^7$.

$$2x^3 \cdot 3x^7 \qquad 2x^3 = 2 \cdot x^3;\ 3x^7 = 3 \cdot x^7$$

$$= (2 \cdot 3) \cdot (x^3 \cdot x^7) \quad \text{Commutative and associative properties}$$

$$= 6x^{3+7} \quad \text{Multiply; product rule}$$

$$= 6x^{10} \quad \text{Add the exponents.}$$

CAUTION
Be sure you understand the difference between *adding* and *multiplying* exponential expressions. For example,

$$8x^3 + 5x^3 \text{ means } (8+5)x^3, \text{ or } 13x^3,$$

but $$(8x^3)(5x^3) \text{ means } (8 \cdot 5)x^{3+3}, \text{ or } 40x^6.$$

Work Problem **4** at the Side. ▶

OBJECTIVE 3 **Use the rule $(a^m)^n = a^{mn}$.** We can simplify an expression such as $(8^3)^2$ with the product rule for exponents, as follows.

$$(8^3)^2 = (8^3)(8^3) = 8^{3+3} = 8^6$$

The product of the exponents in $(8^3)^2$, $3 \cdot 2$, gives the exponent in 8^6. Also,

$$(5^2)^4 = 5^2 \cdot 5^2 \cdot 5^2 \cdot 5^2 \quad \text{Definition of exponent}$$

$$= 5^{2+2+2+2} \quad \text{Product rule}$$

$$= 5^8, \quad \text{Add the exponents.}$$

and $2 \cdot 4 = 8$. These examples suggest **power rule (a) for exponents.**

Power Rule (a) for Exponents
For any positive integers m and n, $(a^m)^n = a^{mn}$.
(Raise a power to a power by multiplying exponents.)
Example: $(3^2)^4 = 3^{2 \cdot 4} = 3^8$

EXAMPLE 5 **Using Power Rule (a)**

Use power rule (a) for exponents to simplify.

(a) $(2^5)^3 = 2^{5 \cdot 3} = 2^{15}$ **(b)** $(5^7)^2 = 5^{7 \cdot 2} = 5^{14}$ **(c)** $(x^2)^5 = x^{2 \cdot 5} = x^{10}$

Work Problem **5** at the Side. ▶

OBJECTIVE 4 **Use the rule $(ab)^m = a^m b^m$.** We can rewrite the expression $(4x)^3$ as shown below.

$$(4x)^3 = (4x)(4x)(4x) \quad \text{Definition of exponent}$$

$$= 4 \cdot 4 \cdot 4 \cdot x \cdot x \cdot x \quad \text{Commutative and associative properties}$$

$$= 4^3x^3 \quad \text{Definition of exponent}$$

This example suggests **power rule (b) for exponents.**

4 Multiply.

(a) $5m^2 \cdot 2m^6$

(b) $3p^5 \cdot 9p^4$

(c) $-7p^5 \cdot (3p^8)$

5 Simplify.

(a) $(5^3)^4$

(b) $(6^2)^5$

(c) $(3^2)^4$

(d) $(a^6)^5$

ANSWERS

4. (a) $10m^8$ **(b)** $27p^9$ **(c)** $-21p^{13}$
5. (a) 5^{12} **(b)** 6^{10} **(c)** 3^8 **(d)** a^{30}

Power Rule (b) for Exponents

For any positive integer m, $(ab)^m = a^m b^m$.
(Raise a product to a power by raising each factor to the power.)

Example: $(2p)^5 = 2^5 p^5$

EXAMPLE 6 Using Power Rule (b)

Use power rule (b) for exponents to simplify.

(a) $(3xy)^2$

$= 3^2x^2y^2$ Power rule (b)

$= 9x^2y^2$ $3^2 = 3 \cdot 3 = 9$

(b) $9(pq)^2$

$= 9(p^2q^2)$ Power rule (b)

$= 9p^2q^2$ Multiply.

(c) $5(2m^2p^3)^4$

$= 5[2^4(m^2)^4(p^3)^4]$ Power rule (b)

$= 5(2^4m^8p^{12})$ Power rule (a)

$= 5 \cdot 2^4m^8p^{12}$

$= 80m^8p^{12}$ $5 \cdot 2^4 = 5 \cdot 16 = 80$

(d) $(-5^6)^3$

$= (-1 \cdot 5^6)^3$ $-a = -1 \cdot a$

$= (-1)^3(5^6)^3$ Power rule (b)

$= -1 \cdot 5^{18}$ Power rule (a)

$= -5^{18}$

Raise −1 to the designated power.

CAUTION

Power rule (b) does not apply to a sum:

$$(4x)^2 = 4^2x^2, \quad \text{but} \quad (4 + x)^2 \neq 4^2 + x^2.$$

◀ *Work Problem* 6 *at the Side.*

OBJECTIVE 5 Use the rule $\left(\frac{a}{b}\right)^m = \frac{a^m}{b^m}$. Since the quotient $\frac{a}{b}$ can be written as $a \cdot \frac{1}{b}$, we can use power rule (b), together with some of the properties of real numbers, to get **power rule (c) for exponents.**

Power Rule (c) for Exponents

For any positive integer m, $\left(\frac{a}{b}\right)^m = \frac{a^m}{b^m}$ $(b \neq 0)$.

(Raise a quotient to a power by raising both the numerator and the denominator to the power.)

Example: $\left(\frac{5}{3}\right)^2 = \frac{5^2}{3^2}$

---

6 Simplify.

(a) $(2ab)^4$

(b) $5(mn)^3$

(c) $(3a^2b^4)^5$

(d) $(-5m^2)^3$

ANSWERS

6. **(a)** $16a^4b^4$ **(b)** $5m^3n^3$ **(c)** $243a^{10}b^{20}$ **(d)** $-125m^6$

EXAMPLE 7 Using Power Rule (c)

Use power rule (c) for exponents to simplify.

(a) $\left(\frac{2}{3}\right)^5 = \frac{2^5}{3^5} = \frac{32}{243}$

(b) $\left(\frac{m}{n}\right)^4 = \frac{m^4}{n^4}, \quad n \neq 0$

(c) $\left(\frac{1}{5}\right)^4 = \frac{1^4}{5^4} = \frac{1}{5^4} = \frac{1}{625}$ $\quad 1^4 = 1 \cdot 1 \cdot 1 \cdot 1 = 1$

Note

In Example 7(c), we used the fact that $1^4 = 1$.

In general, $\boldsymbol{1^n = 1,}$ ***for any integer n.***

Work Problem 7 at the Side. ▶

The rules for exponents discussed in this section are basic to the study of algebra and should be *memorized*.

Rules for Exponents

For positive integers m and n: *Examples*

Product rule	$a^m \cdot a^n = a^{m+n}$	$6^2 \cdot 6^5 = 6^{2+5} = 6^7$
Power rules (a)	$(a^m)^n = a^{mn}$	$(3^2)^4 = 3^{2 \cdot 4} = 3^8$
(b)	$(ab)^m = a^m b^m$	$(2p)^5 = 2^5 p^5$
(c)	$\left(\frac{a}{b}\right)^m = \frac{a^m}{b^m}$ $(b \neq 0)$.	$\left(\frac{5}{3}\right)^2 = \frac{5^2}{3^2}$

OBJECTIVE 6 Use combinations of the rules for exponents. More than one rule may be needed to simplify an exponential expression.

EXAMPLE 8 Using Combinations of Rules

Simplify each expression.

(a) $\left(\frac{2}{3}\right)^2 \cdot 2^3$

$= \frac{2^2}{3^2} \cdot \frac{2^3}{1}$ Power rule (c)

$= \frac{2^2 \cdot 2^3}{3^2 \cdot 1}$ Multiply fractions.

$= \frac{2^{2+3}}{3^2}$ Product rule

$= \frac{2^5}{3^2}$

$= \frac{32}{9}$

(b) $(5x)^3(5x)^4$

$= (5x)^7$ Product rule

$= 5^7x^7$ Power rule (b)

Continued on Next Page

7 Simplify. Assume that all variables represent nonzero real numbers.

(a) $\left(\frac{5}{2}\right)^4$

(b) $\left(\frac{p}{q}\right)^2$

(c) $\left(\frac{r}{t}\right)^3$

(d) $\left(\frac{1}{3}\right)^5$

(e) $\left(\frac{1}{x}\right)^{10}$

Answers

7. (a) $\frac{625}{16}$ (b) $\frac{p^2}{q^2}$ (c) $\frac{r^3}{t^3}$ (d) $\frac{1}{243}$ (e) $\frac{1}{x^{10}}$

(c) $(2x^2y^3)^4\,(3xy^2)^3$

$= 2^4(x^2)^4(y^3)^4 \cdot 3^3x^3(y^2)^3$ Power rule (b)

$= 2^4x^8y^{12} \cdot 3^3x^3y^6$ Power rule (a)

$= 2^4 \cdot 3^3x^8x^3y^{12}y^6$ Commutative and associative properties

$= 16 \cdot 27x^{11}y^{18}$ Product rule

$= 432x^{11}y^{18}$ Multiply.

Notice that $(2x^2y^3)^4$ means $2^4x^{2\cdot4}y^{3\cdot4}$, ***not*** $(2\cdot4)x^{2\cdot4}y^{3\cdot4}$.

Do *not* multiply the coefficient 2 and the exponent 4.

(d) $(-x^3y)^2(-x^5y^4)^3$

$= (-1x^3y)^2(-1x^5y^4)^3$ $-a = -1 \cdot a$

Think of the negative sign in each factor as -1.

$= (-1)^2(x^3)^2(y^2) \cdot (-1)^3(x^5)^3(y^4)^3$ Power rule (b)

$= (-1)^2(x^6)(y^2) \cdot (-1)^3(x^{15})(y^{12})$ Power rule (a)

$= (-1)^5(x^{6+15})(y^{2+12})$ Product rule

$= -1x^{21}y^{14}$

$= -x^{21}y^{14}$

◀ *Work Problem* 8 *at the Side.*

OBJECTIVE 7 Use the rules for exponents in a geometry application.

EXAMPLE 9 Using Area Formulas

Find a polynomial that represents the area of each geometric figure.

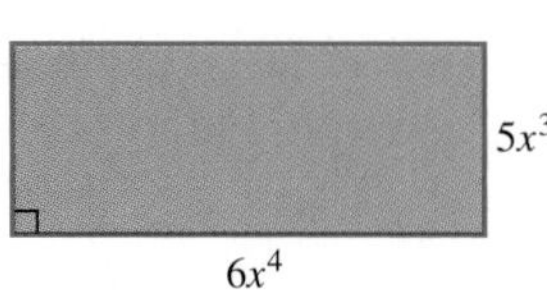

Figure 1

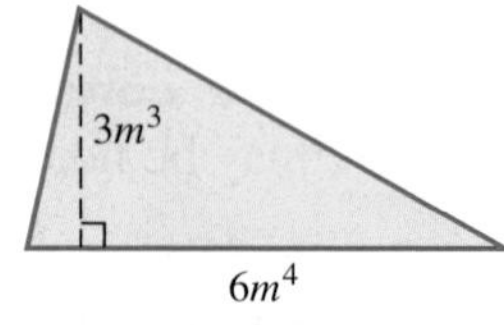

Figure 2

For Figure 1, use the formula for the area of a rectangle, $A = LW$.

$A = (6x^4)(5x^3)$ Area formula

$A = 6 \cdot 5 \cdot x^{4+3}$ Product rule

$A = 30x^7$

Figure 2 is a triangle with base $6m^4$ and height $3m^3$. Substitute into the formula for the area of a triangle and simplify.

$A = \frac{1}{2}bh$ Area formula

$A = \frac{1}{2}(6m^4)(3m^3)$ Substitute.

$A = \frac{1}{2}(18m^7)$, or $9m^7$ Product rule; multiply.

◀ *Work Problem* 9 *at the Side.*

8 Simplify.

(a) $(2m)^3(2m)^4$

(b) $\left(\frac{5k^3}{3}\right)^2$

(c) $\left(\frac{1}{5}\right)^4(2x)^2$

(d) $(-3xy^2)^3(x^2y)^4$

9 Find a polynomial that represents the area of the figure.

$4x^2$

$8x^4$

ANSWERS

8. **(a)** 2^7m^7, or $128m^7$ **(b)** $\frac{5^2k^6}{3^2}$, or $\frac{25k^6}{9}$

(c) $\frac{2^2x^2}{5^4}$, or $\frac{4x^2}{625}$

(d) $-3^3x^{11}y^{10}$, or $-27x^{11}y^{10}$

9. $32x^6$

5.2 Exercises

FOR EXTRA HELP

Math XL PRACTICE WATCH DOWNLOAD READ REVIEW

1. What exponent is understood on the base x in the expression xy^2? **1**

2. How are the expressions 3^2, 5^3, and 7^4 read? **three squared; five cubed; seven to the fourth power**

Decide whether each statement is true *or* false.

3. $3^3 = 9$ **false**

4. $(-2)^4 = 2^4$ **true**

5. $(a^2)^3 = a^5$ **false**

6. $\left(\frac{1}{4}\right)^2 = \frac{1}{4^2}$ **true**

Write each expression using exponents. See Example 1.

7. $t \cdot t \cdot t \cdot t \cdot t \cdot t \cdot t$ $\mathbf{t^7}$

8. $w \cdot w \cdot w \cdot w \cdot w \cdot w$ $\mathbf{w^6}$

9. $\left(\frac{1}{2}\right)\left(\frac{1}{2}\right)\left(\frac{1}{2}\right)\left(\frac{1}{2}\right)\left(\frac{1}{2}\right)$ $\left(\frac{1}{2}\right)^5$

10. $\left(-\frac{1}{4}\right)\left(-\frac{1}{4}\right)\left(-\frac{1}{4}\right)\left(-\frac{1}{4}\right)$ $\left(-\frac{1}{4}\right)^4$

11. $(-8p)(-8p)$ $\mathbf{(-8p)^2}$

12. $(-7x)(-7x)(-7x)$ $\mathbf{(-7x)^3}$

13. Explain how the expressions $(-3)^4$ and -3^4 are different.

The expression $(-3)^4$ means $(-3)(-3)(-3)(-3) = 81$, while -3^4 means $-(3 \cdot 3 \cdot 3 \cdot 3) = -81$.

14. Explain how the expressions $(5x)^3$ and $5x^3$ are different.

The expression $(5x)^3$ means $(5x)(5x)(5x) = 125x^3$, while $5x^3$ means $5x \cdot x \cdot x = 5x^3$.

Identify the base and the exponent for each exponential expression. In Exercises 15–18, also evaluate the expression. See Example 2.

15. 3^5
base: 3; exponent: 5; 243

16. 2^7
base: 2; exponent: 7; 128

17. $(-3)^5$
base: −3; exponent: 5; −243

18. $(-2)^7$
base: −2; exponent: 7; −128

19. $(-6x)^4$
base: $-6x$; exponent: 4

20. $(-8x)^4$
base: $-8x$; exponent: 4

21. $-6x^4$
base: x; exponent: 4

22. $-8x^4$
base: x; exponent: 4

23. Explain why the product rule does not apply to the expression $5^2 + 5^3$. Then evaluate the expression.

The product rule does not apply to $5^2 + 5^3$ because it is a *sum*, not a product.

$5^2 + 5^3 = 25 + 125 = 150$

24. Explain why the product rule does not apply to the expression $3^2 \cdot 4^3$. Then evaluate the expression.

The product rule applies only to exponential expressions with the same base.

$3^2 \cdot 4^3 = 9 \cdot 64 = 576$

Use the product rule for exponents to simplify each expression, if possible. Write each answer in exponential form. See Examples 3 and 4.

25. $5^2 \cdot 5^6$ $\mathbf{5^8}$

26. $3^6 \cdot 3^7$ $\mathbf{3^{13}}$

27. $4^2 \cdot 4^7 \cdot 4^3$ $\mathbf{4^{12}}$

28. $5^3 \cdot 5^8 \cdot 5^2$ $\mathbf{5^{13}}$

29. $(-7)^3(-7)^6$ $\mathbf{(-7)^9}$

30. $(-9)^8(-9)^5$ $\mathbf{(-9)^{13}}$

31. $t^3t^8t^{13}$ $\mathbf{t^{24}}$

32. $n^5n^6n^9$ $\mathbf{n^{20}}$

33. $(-8r^4)(7r^3)$
$\mathbf{-56r^7}$

34. $(10a^7)(-4a^3)$
$\mathbf{-40a^{10}}$

35. $(-6p^5)(-7p^5)$
$\mathbf{42p^{10}}$

36. $(-5w^8)(-9w^8)$
$\mathbf{45w^{16}}$

37. $3^8 + 3^9$

38. $4^{12} + 4^5$

39. $5^8 \cdot 3^8$

40. $6^3 \cdot 8^3$

In Exercises 37–40, the product rule does not apply.

Use the power rules for exponents to simplify each expression. See Examples 5–7.

41. $(4^3)^2$ 4^6

42. $(8^3)^6$ 8^{18}

43. $(t^4)^5$ t^{20}

44. $(y^6)^5$ y^{30}

45. $(7r)^3$ $343r^3$

46. $(11x)^4$ 11^4x^4

47. $(-5^2)^6$ 5^{12}

48. $(-9^4)^8$ 9^{32}

49. $(-8^3)^5$ -8^{15}

50. $(-7^5)^7$ -7^{35}

51. $(5xy)^5$ $5^5x^5y^5$

52. $(9pq)^6$ $9^6p^6q^6$

53. $8(qr)^3$ $8q^3r^3$

54. $4(vw)^5$ $4v^5w^5$

55. $\left(\frac{1}{2}\right)^3$ $\frac{1}{8}$

56. $\left(\frac{1}{3}\right)^5$ $\frac{1}{243}$

57. $\left(\frac{a}{b}\right)^3$, $b \neq 0$ $\frac{a^3}{b^3}$

58. $\left(\frac{r}{t}\right)^4$, $t \neq 0$ $\frac{r^4}{t^4}$

59. $\left(\frac{9}{5}\right)^8$ $\frac{9^8}{5^8}$

60. $\left(\frac{12}{7}\right)^6$ $\frac{12^6}{7^6}$

61. $(-2x^2y)^3$ $-8x^6y^3$

62. $(-5m^4p^2)^3$ $-125m^{12}p^6$

63. $(3a^3b^2)^2$ $9a^6b^4$

64. $(4x^3y^5)^4$ $256x^{12}y^{20}$

Simplify each expression. See Example 8.

65. $\left(\frac{5}{2}\right)^3 \cdot \left(\frac{5}{2}\right)^2$ $\frac{5^5}{2^5}$

66. $\left(\frac{3}{4}\right)^5 \cdot \left(\frac{3}{4}\right)^6$ $\frac{3^{11}}{4^{11}}$

67. $\left(\frac{9}{8}\right)^3 \cdot 9^2$ $\frac{9^5}{8^3}$

68. $\left(\frac{8}{5}\right)^4 \cdot 8^3$ $\frac{8^7}{5^4}$

69. $(2x)^9(2x)^3$ $2^{12}x^{12}$

70. $(6y)^5(6y)^8$ $6^{13}y^{13}$

71. $(-6p)^4(-6p)$ -6^5p^5

72. $(-13q)^3(-13q)$ 13^4q^4

73. $(6x^2y^3)^5$ $6^5x^{10}y^{15}$

74. $(5r^5t^6)^7$ $5^7r^{35}t^{42}$

75. $(x^2)^3(x^3)^5$ x^{21}

76. $(y^4)^5(y^3)^5$ y^{35}

77. $(2w^2x^3y)^2(x^4y)^5$ $4w^4x^{26}y^7$

78. $(3x^4y^2z)^3(yz^4)^5$ $27x^{12}y^{11}z^{23}$

79. $(-r^4s)^2(-r^2s^3)^5$ $-r^{18}s^{17}$

80. $(-ts^6)^4(-t^3s^5)^3$ $-t^{13}s^{39}$

81. $\left(\frac{5a^2b^5}{c^6}\right)^3$, $c \neq 0$ $\frac{125a^6b^{15}}{c^{18}}$

82. $\left(\frac{6x^3y^9}{z^5}\right)^4$, $z \neq 0$ $\frac{1296x^{12}y^{36}}{z^{20}}$

83. $(-5m^3p^4q)^2(p^2q)^3$ $25m^6p^{14}q^5$

84. $(-a^4b^5)(-6a^3b^3)^2$ $-36a^{10}b^{11}$

85. $(2x^2y^3z)^4(xy^2z^3)^2$ $16x^{10}y^{16}z^{10}$

Find a polynomial that represents the area of each figure. See Example 9.

86.

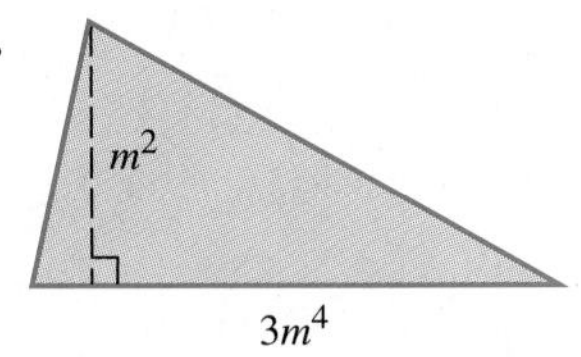

$\frac{3}{2}m^6$

87.

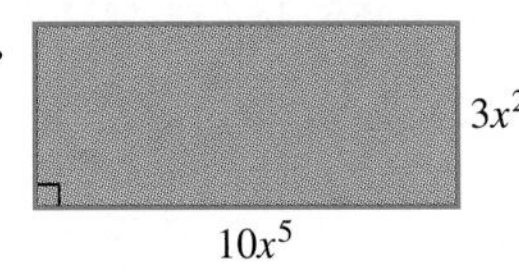

$30x^7$

88.

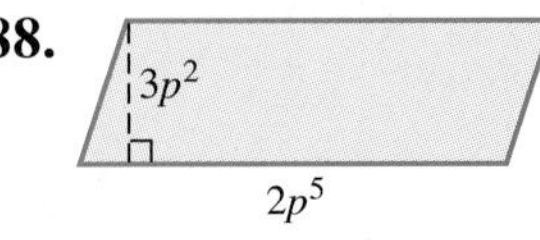

$6p^7$

5.3 Multiplying Polynomials

OBJECTIVES

1. Multiply a monomial and a polynomial.
2. Multiply two polynomials.
3. Multiply binomials by the FOIL method.

OBJECTIVE 1 Multiply a monomial and a polynomial. As shown in **Section 5.2,** we find the product of two monomials by using the rules for exponents and the commutative and associative properties. For example,

$$\begin{aligned}&(-8m^6)(-9n^6)\\&=(-8)(-9)(m^6)(n^6)\\&=72m^6n^6.\end{aligned}$$

CAUTION

Do not confuse addition of terms with multiplication of terms.

$$7q^5+2q^5=9q^5, \quad \text{but} \quad (7q^5)(2q^5)=7\cdot 2q^{5+5}=14q^{10}.$$

To find the product of a monomial and a polynomial with more than one term, we use the distributive property and multiplication of monomials.

EXAMPLE 1 Multiplying Monomials and Polynomials

Find each product.

(a) $4x^2(3x+5)$

$$\begin{aligned}4x^2(3x+5)&=4x^2(3x)+4x^2(5) &&\text{Distributive property}\\&=12x^3+20x^2 &&\text{Multiply monomials.}\end{aligned}$$

(b) $-8m^3(4m^3+3m^2+2m-1)$

$$\begin{aligned}&=-8m^3(4m^3)+(-8m^3)(3m^2)\\&\quad+(-8m^3)(2m)+(-8m^3)(-1) &&\text{Distributive property}\\&=-32m^6-24m^5-16m^4+8m^3 &&\text{Multiply monomials.}\end{aligned}$$

Work Problem 1 at the Side. ▶

1 Find each product.

(a) $5m^3(2m+7)$

(b) $2x^4(3x^2+2x-5)$

(c) $-4y^2(3y^3+2y^2-4y+8)$

OBJECTIVE 2 Multiply two polynomials. We can use the distributive property repeatedly to find the product of any two polynomials. For example, to find the product of the polynomials x^2+3x+5 and $x-4$, think of $x-4$ as a single quantity and use the distributive property as follows.

$$\begin{aligned}&(x^2+3x+5)(x-4)\\&=x^2(x-4)+3x(x-4)+5(x-4) &&\text{Distributive property}\\&=x^2(x)+x^2(-4)+3x(x)+3x(-4)+5(x)+5(-4)\\& &&\text{Distributive property again}\\&=x^3-4x^2+3x^2-12x+5x-20 &&\text{Multiply monomials.}\\&=x^3-x^2-7x-20 &&\text{Combine like terms.}\end{aligned}$$

This example suggests the following rule.

Multiplying Polynomials

To multiply two polynomials, multiply each term of the second polynomial by each term of the first polynomial and add the products.

ANSWERS

1. (a) $10m^4+35m^3$
   (b) $6x^6+4x^5-10x^4$
   (c) $-12y^5-8y^4+16y^3-32y^2$

2 Multiply.

(a) $(m + 3)(m^2 - 2m + 1)$

(b) $(6p^2 + 2p - 4)(3p^2 - 5)$

EXAMPLE 2 Multiplying Two Polynomials

Multiply $(m^2 + 5)(4m^3 - 2m^2 + 4m)$.

Multiply each term of the second polynomial by each term of the first.

$$
\begin{aligned}
&(m^2 + 5)(4m^3 - 2m^2 + 4m)\\
&= m^2(4m^3) + m^2(-2m^2) + m^2(4m) + 5(4m^3) + 5(-2m^2) + 5(4m)\\
&= 4m^5 - 2m^4 + 4m^3 + 20m^3 - 10m^2 + 20m\\
&= 4m^5 - 2m^4 + 24m^3 - 10m^2 + 20m \quad \text{Combine like terms.}
\end{aligned}
$$

◀ Work Problem **2** at the Side.

3 Find the product.

$$
\begin{array}{r}
3x^2 + 4x - 5\\
x + 4\\
\hline
\end{array}
$$

EXAMPLE 3 Multiplying Polynomials Vertically

Multiply $(x^3 + 2x^2 + 4x + 1)(3x + 5)$ vertically.

Write the polynomials as follows.

$$
\begin{array}{r}
x^3 + 2x^2 + 4x + 1\\
3x + 5\\
\hline
\end{array}
$$

Begin by multiplying each of the terms in the top row by 5.

$$
\begin{array}{rl}
x^3 + 2x^2 + 4x + 1 & \\
3x + 5 & \\
\hline
5x^3 + 10x^2 + 20x + 5 & \quad 5(x^3 + 2x^2 + 4x + 1)
\end{array}
$$

Notice how this process is similar to multiplication of whole numbers. Now multiply each term in the top row by $3x$. Then add like terms.

Place *like* terms in columns so they can be added.

$$
\begin{array}{rl}
x^3 + 2x^2 + 4x + 1 & \\
3x + 5 & \\
\hline
5x^3 + 10x^2 + 20x + 5 & \\
3x^4 + 6x^3 + 12x^2 + 3x \phantom{{}+5} & \quad 3x(x^3 + 2x^2 + 4x + 1)\\
\hline
3x^4 + 11x^3 + 22x^2 + 23x + 5 & \quad \text{Add.}
\end{array}
$$

The product is $3x^4 + 11x^3 + 22x^2 + 23x + 5$.

◀ Work Problem **3** at the Side.

4 Use the rectangle method to find each product.

(a) $(4x + 3)(x + 2)$

	x	2
$4x$	$4x^2$	
3		

(b) $(x + 5)(x^2 + 3x + 1)$

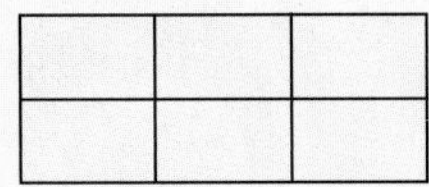

We can use a rectangle to model polynomial multiplication. For example, to find the product

$$(2x + 1)(3x + 2),$$

label a rectangle with each term as shown below on the left. Then put the product of each pair of monomials in the appropriate box as shown on the right.

	$3x$	2
$2x$		
1		

	$3x$	2
$2x$	$6x^2$	$4x$
1	$3x$	2

The product of the binomials is the sum of these four monomial products.

$$
\begin{aligned}
&(2x + 1)(3x + 2)\\
&= 6x^2 + 4x + 3x + 2\\
&= 6x^2 + 7x + 2
\end{aligned}
$$

◀ Work Problem **4** at the Side.

Answers

2. (a) $m^3 + m^2 - 5m + 3$
   (b) $18p^4 + 6p^3 - 42p^2 - 10p + 20$
3. $3x^3 + 16x^2 + 11x - 20$
4. (a) $4x^2 + 11x + 6$
   (b) $x^3 + 8x^2 + 16x + 5$

OBJECTIVE 3 Multiply binomials by the FOIL method. In algebra, many of the polynomials to be multiplied are both binomials (with just two terms). For these products, the **FOIL method** reduces the rectangle method to a systematic approach without the rectangle. To develop the FOIL method, we use the distributive property to find $(x + 3)(x + 5)$.

$$
\begin{aligned}
&(x+3)(x+5) \\
&= (x+3)x + (x+3)5 && \text{Distributive property} \\
&= x(x) + 3(x) + x(5) + 3(5) && \text{Distributive property again} \\
&= x^2 + 3x + 5x + 15 && \text{Multiply.} \\
&= x^2 + 8x + 15 && \text{Combine like terms.}
\end{aligned}
$$

Here is where the letters of the word FOIL originate.

$(x + 3)(x + 5)$ Multiply the **First terms**: $x(x)$. **F**

$(x + 3)(x + 5)$ Multiply the **Outer terms**: $x(5)$. **O**
This is the **outer product.**

$(x + 3)(x + 5)$ Multiply the **Inner terms**: $3(x)$. **I**
This is the **inner product.**

$(x + 3)(x + 5)$ Multiply the **Last terms**: $3(5)$. **L**

The outer product, $5x$, and the inner product, $3x$, should be added mentally so that the three terms of the answer can be written without extra steps.

$$
\begin{aligned}
&(x+3)(x+5) \\
&= x^2 + 8x + 15
\end{aligned}
$$

A summary of the steps in the FOIL method follows.

Multiplying Binomials by the FOIL Method

Step 1 Multiply the two **F**irst terms of the binomials to get the first term of the answer.

Step 2 Find the **O**uter product and the **I**nner product and add them (when possible) to get the middle term of the answer.

Step 3 Multiply the two **L**ast terms of the binomials to get the last term of the answer.

F $= x^2$ **L** $= 15$

$(x + 3)(x + 5)$

I $= 3x$
O $= 5x$
$8x$ Add.

Work Problem **5** *at the Side.* ▶

5 For the product

$$(2p - 5)(3p + 7),$$

find the following.

(a) Product of first terms

(b) Outer product

(c) Inner product

(d) Product of last terms

(e) Complete product in simplified form

ANSWERS

5. (a) $2p(3p) = 6p^2$ **(b)** $2p(7) = 14p$ **(c)** $-5(3p) = -15p$ **(d)** $-5(7) = -35$ **(e)** $6p^2 - p - 35$

6 Use the FOIL method to find each product.

(a) $(m + 4)(m - 3)$

(b) $(y + 7)(y + 2)$

(c) $(r - 8)(r - 5)$

EXAMPLE 4 Using the FOIL Method

Use the FOIL method to find the product $(x + 8)(x - 6)$.

Step 1 F Multiply the **first** terms: $x(x) = x^2$.

Step 2 O Find the **outer** product: $x(-6) = -6x$.

I Find the **inner** product: $8(x) = 8x$.

Add the outer and inner products mentally: $-6x + 8x = 2x$.

Step 3 L Multiply the **last** terms: $8(-6) = -48$.

The product $(x + 8)(x - 6)$ is $x^2 + 2x - 48$, the sum of the terms found in Steps 1–3. As a shortcut, this product can be found as follows.

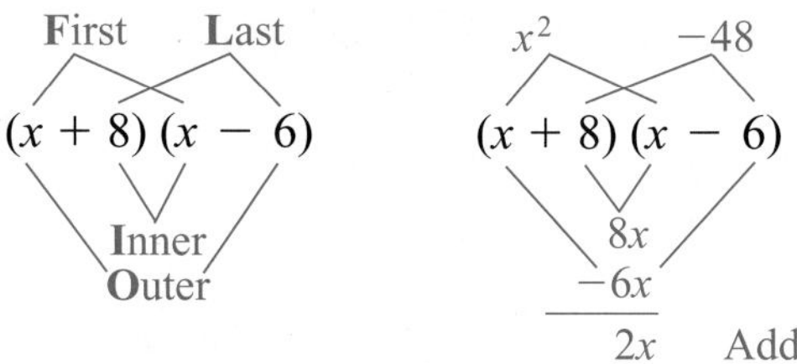

◀ Work Problem 6 at the Side.

7 Find the product.

$(4x - 3)(2y + 5)$

EXAMPLE 5 Using the FOIL Method

Multiply $(9x - 2)(3y + 1)$.

First	$(9x - 2)(3y + 1)$	$27xy$
Outer	$(9x - 2)(3y + 1)$	$9x$
Inner	$(9x - 2)(3y + 1)$	$-6y$
Last	$(9x - 2)(3y + 1)$	-2

These unlike terms *cannot* be added.

The product $(9x - 2)(3y + 1)$ is $\overset{F}{27xy} + \overset{O}{9x} - \overset{I}{6y} - \overset{L}{2}$.

◀ Work Problem 7 at the Side.

8 Find each product.

(a) $(6m + 5)(m - 4)$

(b) $(3r + 2t)(3r + 4t)$

(c) $y^2(8y + 3)(2y + 1)$

EXAMPLE 6 Using the FOIL Method

Find each product.

(a) $(2k + 5y)(k + 3y)$

$$= \overset{F}{2k(k)} + \overset{O}{2k(3y)} + \overset{I}{5y(k)} + \overset{L}{5y(3y)}$$
$$= 2k^2 + 6ky + 5ky + 15y^2$$
$$= 2k^2 + 11ky + 15y^2$$

(b) $(7p + 2q)(3p - q)$

$$= 21p^2 - 7pq + 6pq - 2q^2$$
$$= 21p^2 - pq - 2q^2$$

(c) $2x^2(x - 3)(3x + 4)$

$$= 2x^2(3x^2 - 5x - 12)$$
$$= 6x^4 - 10x^3 - 24x^2$$

◀ Work Problem 8 at the Side.

Note

In Example 6(c), we could have multiplied $2x^2$ and $x - 3$ first.

$$2x^2(x - 3)(3x + 4)$$
$$= (2x^3 - 6x^2)(3x + 4)$$
$$= 6x^4 - 10x^3 - 24x^2 \quad \text{Same answer}$$

Answers

6. **(a)** $m^2 + m - 12$ **(b)** $y^2 + 9y + 14$ **(c)** $r^2 - 13r + 40$
7. $8xy + 20x - 6y - 15$
8. **(a)** $6m^2 - 19m - 20$ **(b)** $9r^2 + 18rt + 8t^2$ **(c)** $16y^4 + 14y^3 + 3y^2$

FOR EXTRA HELP

Math XL PRACTICE

WATCH

DOWNLOAD

READ

REVIEW

Find each product using the rectangle method shown in the text.

1. $(x + 3)(x + 4)$
$x^2 + 7x + 12$

2. $(x + 5)(x + 2)$
$x^2 + 7x + 10$

3. $(2x + 1)(x^2 + 3x + 2)$
$2x^3 + 7x^2 + 7x + 2$

4. $(x + 4)(3x^2 + 2x + 1)$
$3x^3 + 14x^2 + 9x + 4$

5. In multiplying a monomial by a polynomial, such as in $4x(3x^2 + 7x^3) = 4x(3x^2) + 4x(7x^3)$, the first property that is used is the distributive property.

6. Match each product in parts (a)–(d) with the correct polynomial in choices A–D.

(a) $(x - 5)(x + 3)$ C

(b) $(x + 5)(x + 3)$ A

(c) $(x - 5)(x - 3)$ B

(d) $(x + 5)(x - 3)$ D

A. $x^2 + 8x + 15$ **B.** $x^2 - 8x + 15$ **C.** $x^2 - 2x - 15$ **D.** $x^2 + 2x - 15$

Find each product. See Example 1.

7. $-2m(3m + 2)$
$-6m^2 - 4m$

8. $-5p(6 + 3p)$
$-30p - 15p^2$

9. $\frac{3}{4}p(8 - 6p + 12p^3)$
$6p - \frac{9}{2}p^2 + 9p^4$

10. $\frac{4}{3}x(3 + 2x + 5x^3)$
$4x + \frac{8}{3}x^2 + \frac{20}{3}x^4$

11. $2y^5(3 + 2y + 5y^4)$
$6y^5 + 4y^6 + 10y^9$

12. $2m^4(3m^2 + 5m + 6)$
$6m^6 + 10m^5 + 12m^4$

Find each product. See Examples 2 and 3.

13. $(6x + 1)(2x^2 + 4x + 1)$
$12x^3 + 26x^2 + 10x + 1$

14. $(9y - 2)(8y^2 - 6y + 1)$
$72y^3 - 70y^2 + 21y - 2$

15. $(2r - 1)(3r^2 + 4r - 4)$
$6r^3 + 5r^2 - 12r + 4$

16. $(9a + 2)(9a^2 + a + 1)$
$81a^3 + 27a^2 + 11a + 2$

17. $(4m + 3)(5m^3 - 4m^2 + m - 5)$
$20m^4 - m^3 - 8m^2 - 17m - 15$

18. $(y + 4)(3y^3 - 2y^2 + y + 3)$
$3y^4 + 10y^3 - 7y^2 + 7y + 12$

19. $(5x^2 + 2x + 1)(x^2 - 3x + 5)$
$5x^4 - 13x^3 + 20x^2 + 7x + 5$

20. $(2m^2 + m - 3)(m^2 - 4m + 5)$
$2m^4 - 7m^3 + 3m^2 + 17m - 15$

Find each product. See Examples 4–6.

21. $(m + 7)(m + 5)$
$m^2 + 12m + 35$

22. $(x + 4)(x + 7)$
$x^2 + 11x + 28$

23. $(n - 2)(n + 3)$
$n^2 + n - 6$

24. $(r - 6)(r + 8)$
$r^2 + 2r - 48$

25. $(4r + 1)(2r - 3)$
$8r^2 - 10r - 3$

26. $(5x + 2)(2x - 7)$
$10x^2 - 31x - 14$

27. $(3x + 2)(3x - 2)$
$9x^2 - 4$

28. $(7x + 3)(7x - 3)$
$49x^2 - 9$

29. $(3q + 1)(3q + 1)$
$9q^2 + 6q + 1$

30. $(4w + 7)(4w + 7)$
$16w^2 + 56w + 49$

31. $(5x + 7)(3y - 8)$
$15xy - 40x + 21y - 56$

32. $(4x + 3)(2y - 1)$
$8xy - 4x + 6y - 3$

33. $(3t + 4s)(2t + 5s)$
$6t^2 + 23st + 20s^2$

34. $(8v + 5w)(2v + 3w)$
$16v^2 + 34vw + 15w^2$

35. $(-0.3t + 0.4)(t + 0.6)$
$-0.3t^2 + 0.22t + 0.24$

36. $(-0.5x + 0.9)(x - 0.2)$
$-0.5x^2 + x - 0.18$

37. $\left(x - \frac{2}{3}\right)\left(x + \frac{1}{4}\right)$
$x^2 - \frac{5}{12}x - \frac{1}{6}$

38. $\left(-\frac{8}{3} + 3k\right)\left(-\frac{2}{3} - k\right)$
$\frac{16}{9} + \frac{2}{3}k - 3k^2$

39. $\left(-\frac{5}{4} + 2r\right)\left(-\frac{3}{4} - r\right)$
$\frac{15}{16} - \frac{1}{4}r - 2r^2$

40. $2m^3(4m - 1)(2m + 3)$
$16m^5 + 20m^4 - 6m^3$

41. $x(2x - 5)(x + 3)$
$2x^3 + x^2 - 15x$

42. $5t^4(t + 3)(3t - 1)$
$15t^6 + 40t^5 - 15t^4$

43. $3y^3(2y + 3)(y - 5)$
$6y^5 - 21y^4 - 45y^3$

Relating Concepts (Exercises 44–48) For Individual or Group Work

Work Exercises 44–48 in order. *(All units are in yards.)*

44. Find a polynomial that represents the area of the rectangle.

$3x + 6$ (length), 10 (width)

$(30x + 60)$ yd^2

45. Suppose you know that the area of the rectangle is 600 yd^2. Use this information and the polynomial from Exercise 44 to write an equation in x, and solve it.
$30x + 60 = 600$; $\{18\}$

46. (a) What are the dimensions of the rectangle?
10 yd by 60 yd

(b) Use the result of part (a) to find the perimeter of the lawn.
140 yd

47. Suppose the rectangle represents a lawn and it costs \$3.50 per square yard to lay sod on the lawn. How much will it cost to sod the entire lawn? \$2100

48. Again, suppose the rectangle represents a lawn and it costs \$9.00 per yard to fence the lawn. How much will it cost to fence the lawn? \$1260

---

49. Perform the following multiplications: $(x + 4)(x - 4)$; $(y + 2)(y - 2)$; $(r + 7)(r - 7)$. Observe your answers, and explain the pattern that can be found in the answers.
The answers are $x^2 - 16$, $y^2 - 4$, and $r^2 - 49$. Each product is the difference of the square of the first term and the square of the last term of the binomials.

50. Repeat Exercise 49 for the following: $(x + 4)(x + 4)$; $(y - 2)(y - 2)$; $(r + 7)(r + 7)$.
The answers are $x^2 + 8x + 16$, $y^2 - 4y + 4$, and $r^2 + 14r + 49$. Each product is the sum of the square of the first term, twice the product of the two terms, and the square of the last term of the binomials.

5.4 Special Products

In this section, we develop shortcuts to find certain binomial products.

OBJECTIVE 1 Square binomials. The square of a binomial can be found quickly by using the method shown in Example 1.

EXAMPLE 1 Squaring a Binomial

Find $(m + 3)^2$.

$(m + 3)^2$ means $(m + 3)(m + 3)$.

$$\begin{aligned}&(m + 3)(m + 3)\\ &= m^2 + 3m + 3m + 9 && \text{FOIL}\\ &= m^2 + 6m + 9 && \text{Combine like terms.}\end{aligned}$$

This result has the squares of the first and the last terms of the binomial:

$$m^2 = m^2 \quad \text{and} \quad 3^2 = 9.$$

The middle term, 6m, is twice the product of the two terms of the binomial, since the outer and inner products are $m(3)$ and $3(m)$, and

$$m(3) + 3(m) = 2(m)(3) = 6m.$$

Work Problem 1 at the Side. ▶

Example 1 suggests the following rules.

Square of a Binomial

The square of a binomial is a trinomial consisting of the square of the first term, plus twice the product of the two terms, plus the square of the last term of the binomial. For a and b,

$$(a + b)^2 = a^2 + 2ab + b^2.$$

Also, $$(a - b)^2 = a^2 - 2ab + b^2.$$

EXAMPLE 2 Squaring Binomials

Square each binomial.

$$(a - b)^2 = a^2 - 2 \cdot a \cdot b + b^2$$

(a) $(5z - 1)^2 = (5z)^2 - 2(5z)(1) + (1)^2$

$= 25z^2 - 10z + 1$ $\qquad (5z)^2 = 5^2z^2 = 25z^2$

(b) $(3b + 5r)^2$

$= (3b)^2 + 2(3b)(5r) + (5r)^2$

$= 9b^2 + 30br + 25r^2$

(c) $(2a - 9x)^2$

$= (2a)^2 - 2(2a)(9x) + (9x)^2$

$= 4a^2 - 36ax + 81x^2$

Continued on Next Page

OBJECTIVES

1. Square binomials.
2. Find the product of the sum and difference of two terms.
3. Find greater powers of binomials.

1 Consider the binomial $x + 4$.

(a) What is the first term of the binomial? Square it.

(b) What is the last term of the binomial? Square it.

(c) Find twice the product of the two terms of the binomial.

(d) Find $(x + 4)^2$.

ANSWERS

1. **(a)** x; x^2 **(b)** 4; 16 **(c)** $8x$ **(d)** $x^2 + 8x + 16$

2 Square each binomial.

(a) $(t - 6)^2$

(b) $(2m - p)^2$

(c) $(4p + 3q)^2$

(d) $(5r - 6s)^2$

(e) $\left(3k - \frac{1}{2}\right)^2$

(f) $x(2x + 7)^2$

Answers

2. **(a)** $t^2 - 12t + 36$
(b) $4m^2 - 4mp + p^2$
(c) $16p^2 + 24pq + 9q^2$
(d) $25r^2 - 60rs + 36s^2$
(e) $9k^2 - 3k + \frac{1}{4}$
(f) $4x^3 + 28x^2 + 49x$

(d) $\left(4m + \frac{1}{2}\right)^2$

$= (4m)^2 + 2(4m)\left(\frac{1}{2}\right) + \left(\frac{1}{2}\right)^2 \qquad (a + b)^2 = a^2 + 2ab + b^2$

$= 16m^2 + 4m + \frac{1}{4}$

(e) $x(4x - 3)^2$ — Remember the middle term.

$= x(16x^2 - 24x + 9)$ Square the binomial.

$= 16x^3 - 24x^2 + 9x$ Distributive property

Notice that in the square of a sum, all of the terms are positive, as in Examples 2(b) and (d). ***In the square of a difference, the middle term is negative,*** as in Examples 2(a) and (c).

CAUTION

A common error when squaring a binomial is to forget the middle term of the product. In general,

$$(a + b)^2 = a^2 + 2ab + b^2, \quad \textbf{not} \quad a^2 + b^2,$$

and

$$(a - b)^2 = a^2 - 2ab + b^2, \quad \textbf{not} \quad a^2 - b^2.$$

◀ *Work Problem* **2** *at the Side.*

OBJECTIVE 2 Find the product of the sum and difference of two terms. In binomial products of the form $(a + b)(a - b)$, one binomial is the sum of two terms, and the other is the difference of the *same* two terms. For example, the product of $x + 2$ and $x - 2$ is

$(x + 2)(x - 2)$

$= x^2 - 2x + 2x - 4$ FOIL

$= x^2 - 4.$ Combine like terms.

As the above example suggests, the product of $a + b$ and $a - b$ is the difference of two squares.

Product of the Sum and Difference of Two Terms

$$(a + b)(a - b) = a^2 - b^2$$

Note

The expressions $a + b$ and $a - b$, the sum and difference of the *same* two terms, are called **conjugates.** In the example above, $x + 2$ and $x - 2$ are conjugates.

EXAMPLE 3 Finding the Product of the Sum and Difference of Two Terms

Find each product.

(a) $(x + 4)(x - 4)$

Use the rule for the product of the sum and difference of two terms.

$$(x + 4)(x - 4)$$
$$= x^2 - 4^2$$
$$= x^2 - 16$$

(b) $\left(\frac{2}{3} - w\right)\left(\frac{2}{3} + w\right)$

$= \left(\frac{2}{3} + w\right)\left(\frac{2}{3} - w\right)$ Commutative property

$= \left(\frac{2}{3}\right)^2 - w^2$ Multiply.

$= \frac{4}{9} - w^2$ Square $\frac{2}{3}$.

(c) $x(x + 2)(x - 2)$

$= x(x^2 - 4)$ Find the product of the sum and difference of two terms.

$= x^3 - 4x$ Distributive property

EXAMPLE 4 Finding the Product of the Sum and Difference of Two Terms

Find each product.

$(a \quad + b)\ (a \quad - b)$

$\downarrow \quad \downarrow \quad \downarrow \quad \downarrow$

(a) $(5m + 3)(5m - 3)$

Use the rule for the product of the sum and difference of two terms.

$$(5m + 3)(5m - 3)$$

$= (5m)^2 - 3^2$ $(a + b)(a - b) = a^2 - b^2$

$= 25m^2 - 9$ Apply the exponents.

(b) $(4x + y)(4x - y)$

$= (4x)^2 - y^2$

$= 16x^2 - y^2$

(c) $\left(z - \frac{1}{4}\right)\left(z + \frac{1}{4}\right)$

$= z^2 - \left(\frac{1}{4}\right)^2$

$= z^2 - \frac{1}{16}$

(d) $2p(p^2 + 3)(p^2 - 3)$

$= 2p(p^4 - 9)$ Multiply the conjugates.

$= 2p^5 - 18p$ Distributive property

Work Problem 3 at the Side. ▶

3 Find each product.

(a) $(y + 3)(y - 3)$

(b) $(10m + 7)(10m - 7)$

(c) $(7p + 2q)(7p - 2q)$

(d) $\left(3r - \frac{1}{2}\right)\left(3r + \frac{1}{2}\right)$

(e) $3x(x^3 - 4)(x^3 + 4)$

ANSWERS

3. (a) $y^2 - 9$ **(b)** $100m^2 - 49$ **(c)** $49p^2 - 4q^2$ **(d)** $9r^2 - \frac{1}{4}$ **(e)** $3x^7 - 48x$

4 Find each product.

(a) $(m + 1)^3$

(b) $(3k - 2)^4$

(c) $-3x(x - 4)^3$

The product rules of this section will be important in **Chapters 6** and **7** and should be *memorized.*

OBJECTIVE 3 **Find greater powers of binomials.** The methods used in the previous section and this section can be combined to find greater powers of binomials.

EXAMPLE 5 **Finding Greater Powers of Binomials**

Find each product.

(a) $(x + 5)^3$

$= (x + 5)^2(x + 5)$ — $a^3 = a^2 \cdot a$

$= (x^2 + 10x + 25)(x + 5)$ — Square the binomial.

$= x^3 + 10x^2 + 25x + 5x^2 + 50x + 125$ — Multiply polynomials.

$= x^3 + 15x^2 + 75x + 125$ — Combine like terms.

(b) $(2y - 3)^4$

$= (2y - 3)^2(2y - 3)^2$ — $a^4 = a^2 \cdot a^2$

$= (4y^2 - 12y + 9)(4y^2 - 12y + 9)$ — Square each binomial.

$= 16y^4 - 48y^3 + 36y^2 - 48y^3 + 144y^2$ — Multiply polynomials.
$- 108y + 36y^2 - 108y + 81$

$= 16y^4 - 96y^3 + 216y^2 - 216y + 81$ — Combine like terms.

(c) $-2r(r + 2)^3$

$= -2r(r + 2)(r + 2)^2$

$= -2r(r + 2)(r^2 + 4r + 4)$

$= -2r(r^3 + 4r^2 + 4r + 2r^2 + 8r + 8)$

$= -2r(r^3 + 6r^2 + 12r + 8)$

$= -2r^4 - 12r^3 - 24r^2 - 16r$

◀ *Work Problem* **4** *at the Side.*

ANSWERS

4. (a) $m^3 + 3m^2 + 3m + 1$
(b) $81k^4 - 216k^3 + 216k^2 - 96k + 16$
(c) $-3x^4 + 36x^3 - 144x^2 + 192x$

5.4 Exercises

FOR EXTRA HELP

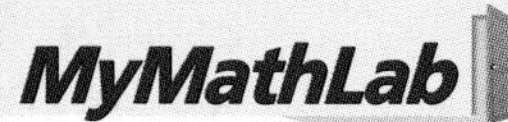
Math XL PRACTICE

WATCH

DOWNLOAD

READ

REVIEW

1. Consider the square $(2x + 3)^2$.

(a) What is the square of the first term, $(2x)^2$? $4x^2$

(b) What is twice the product of the two terms, $2(2x)(3)$? $12x$

(c) What is the square of the last term, 3^2? 9

(d) Write the final product, which is a trinomial, using your results from parts (a)–(c). $4x^2 + 12x + 9$

2. Repeat Exercise 1 for the square $(3x - 2)^2$.

(a) $9x^2$ (b) $-12x$ (c) 4 (d) $9x^2 - 12x + 4$

Find each square. See Examples 1 and 2.

3. $(p + 2)^2$ — $p^2 + 4p + 4$

4. $(r + 5)^2$ — $r^2 + 10r + 25$

5. $(z - 5)^2$ — $z^2 - 10z + 25$

6. $(x - 3)^2$ — $x^2 - 6x + 9$

7. $(4x - 3)^2$ — $16x^2 - 24x + 9$

8. $(5y + 2)^2$ — $25y^2 + 20y + 4$

9. $(2p + 5q)^2$ — $4p^2 + 20pq + 25q^2$

10. $(8a - 3b)^2$ — $64a^2 - 48ab + 9b^2$

11. $(0.8t + 0.7s)^2$ — $0.64t^2 + 1.12ts + 0.49s^2$

12. $(0.7z - 0.3w)^2$ — $0.49z^2 - 0.42zw + 0.09w^2$

13. $\left(5x + \frac{2}{5}y\right)^2$ — $25x^2 + 4xy + \frac{4}{25}y^2$

14. $\left(6m - \frac{4}{5}n\right)^2$ — $36m^2 - \frac{48}{5}mn + \frac{16}{25}n^2$

15. $t(3t - 1)^2$ — $9t^3 - 6t^2 + t$

16. $x(2x + 5)^2$ — $4x^3 + 20x^2 + 25x$

17. $-(4r - 2)^2$ — $-16r^2 + 16r - 4$

18. $-(3y - 8)^2$ — $-9y^2 + 48y - 64$

19. Consider the product $(7x + 3y)(7x - 3y)$.

(a) What is the product of the first terms, $7x(7x)$? $49x^2$

(b) Multiply the outer terms, $7x(-3y)$. Then multiply the inner terms, $3y(7x)$. Add the results. What is this sum? 0

(c) What is the product of the last terms, $3y(-3y)$? $-9y^2$

(d) Write the complete product using your answers in parts (a) and (c). $49x^2 - 9y^2$ Why is the sum found in part (b) omitted here?

Because 0 is the identity element for addition, it is not necessary to write "+0."

20. Repeat Exercise 19 for the product $(5x + 7y)(5x - 7y)$.

(a) $25x^2$ (b) 0 (c) $-49y^2$ (d) $25x^2 - 49y^2$; Because 0 is the identity element for addition, it is not necessary to write "+0."

Find each product. See Examples 3 and 4.

21. $(q + 2)(q - 2)$ — $q^2 - 4$

22. $(x + 8)(x - 8)$ — $x^2 - 64$

23. $(2w + 5)(2w - 5)$ — $4w^2 - 25$

24. $(3z + 8)(3z - 8)$ — $9z^2 - 64$

25. $(10x + 3y)(10x - 3y)$ — $100x^2 - 9y^2$

26. $(13r + 2z)(13r - 2z)$ — $169r^2 - 4z^2$

27. $(2x^2 - 5)(2x^2 + 5)$ — $4x^4 - 25$

28. $(9y^2 - 2)(9y^2 + 2)$ — $81y^4 - 4$

29. $\left(7x + \frac{3}{7}\right)\left(7x - \frac{3}{7}\right)$ — $49x^2 - \frac{9}{49}$

30. $\left(9y + \frac{2}{3}\right)\left(9y - \frac{2}{3}\right)$ — $81y^2 - \frac{4}{9}$

31. $p(3p + 7)(3p - 7)$ — $9p^3 - 49p$

32. $q(5q - 1)(5q + 1)$ — $25q^3 - q$

Relating Concepts (Exercises 33–42) For Individual or Group Work

Special products can be illustrated by using areas of rectangles. Use the figure and **work Exercises 33–38 in order,** *to justify the special product* $(a + b)^2 = a^2 + 2ab + b^2$.

33. Express the area of the large square as the square of a binomial. $(a + b)^2$

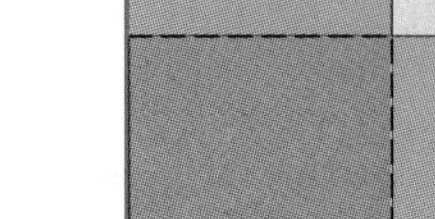

34. Give the monomial that represents the area of the red square. a^2

35. Give the monomial that represents the sum of the areas of the blue rectangles.

$2ab$

36. Give the monomial that represents the area of the yellow square. b^2

37. What is the sum of the monomials you obtained in Exercises 34–36?

$a^2 + 2ab + b^2$

38. Explain why the binomial square you found in Exercise 33 must equal the polynomial you found in Exercise 37.

They both represent the area of the entire large square.

To understand how the special product $(a + b)^2 = a^2 + 2ab + b^2$ *can be applied to a purely numerical problem,* **work Exercises 39–42 in order.**

39. Evaluate 35^2 using either traditional paper-and-pencil methods or a calculator. 1225

40. The number 35 can be written as $30 + 5$. Therefore, $35^2 = (30 + 5)^2$. Use the special product for squaring a binomial with $a = 30$ and $b = 5$ to write an expression for $(30 + 5)^2$. Do not simplify at this time. $30^2 + 2(30)(5) + 5^2$

41. Use the order of operations to simplify the expression you found in Exercise 40. 1225

42. How do the answers in Exercises 39 and 41 compare? They are equal.

Find each product. See Example 5.

43. $(m - 5)^3$

$m^3 - 15m^2 + 75m - 125$

44. $(p + 3)^3$

$p^3 + 9p^2 + 27p + 27$

45. $(y + 2)^3$

$y^3 + 6y^2 + 12y + 8$

46. $(x - 7)^3$

$x^3 - 21x^2 + 147x - 343$

47. $(2a + 1)^3$

$8a^3 + 12a^2 + 6a + 1$

48. $(3m - 1)^3$

$27m^3 - 27m^2 + 9m - 1$

49. $(3r - 2t)^4$

$81r^4 - 216r^3t + 216r^2t^2 - 96rt^3 + 16t^4$

50. $(2z + 5y)^4$

$16z^4 + 160z^3y + 600z^2y^2 + 1000zy^3 + 625y^4$

51. $3x^2(x - 3)^3$

$3x^5 - 27x^4 + 81x^3 - 81x^2$

52. $4p^3(p + 4)^3$

$4p^6 + 48p^5 + 192p^4 + 256p^3$

53. $-8x^2y(x + y)^4$

$-8x^6y - 32x^5y^2 - 48x^4y^3 - 32x^3y^4 - 8x^2y^5$

In Exercises 54 and 55, refer to the figure shown here.

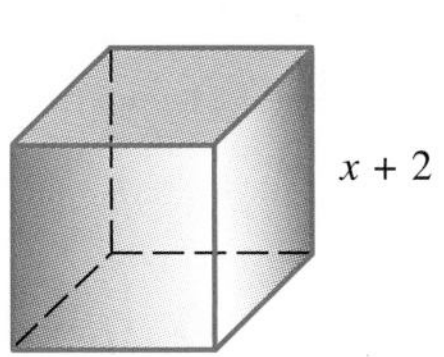

54. Find a polynomial that represents the volume of the cube.

$x^3 + 6x^2 + 12x + 8$

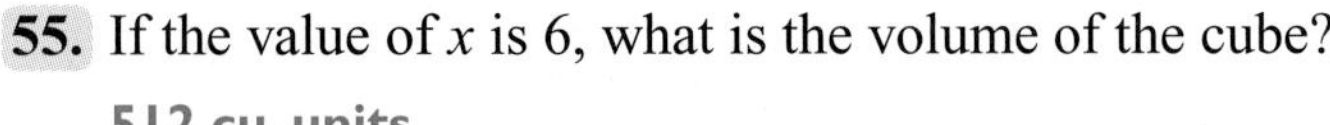

55. If the value of x is 6, what is the volume of the cube?

512 cu. units

5.5 Integer Exponents and the Quotient Rule

In all our earlier work, exponents were positive integers. Now we want to develop meaning for exponents that are *not* positive integers.

Consider the following list.

$$2^4 = 16$$
$$2^3 = 8$$
$$2^2 = 4$$

Do you see the pattern in the values? Each time we reduce the exponent by 1, the value is divided by 2 (the base). Using this pattern, we can continue the list to smaller and smaller integer exponents.

$$2^1 = 2$$
$$2^0 = 1$$
$$2^{-1} = \frac{1}{2}$$

Work Problem 1 at the Side. ▶

From the preceding list and the answers to Problem 1 at the side, it appears that we should define 2^0 as 1 and negative exponents as reciprocals.

OBJECTIVE 1 Use 0 as an exponent. We want the definitions of 0 and negative exponents to satisfy the rules for exponents from **Section 5.2.** For example, if $6^0 = 1$,

$$6^0 \cdot 6^2 = 1 \cdot 6^2 = 6^2 \quad \text{and} \quad 6^0 \cdot 6^2 = 6^{0+2} = 6^2,$$

so the product rule is satisfied. Check that the power rules are also valid for a 0 exponent. Thus, we define a 0 exponent as follows.

Zero Exponent

For any nonzero real number a, $\quad \mathbf{a^0 = 1.}$

Example: $17^0 = 1$

EXAMPLE 1 Using Zero Exponents

Evaluate.

(a) $60^0 = 1$

(b) $(-60)^0 = 1$

(c) $-60^0 = -(1) = -1$

(d) $y^0 = 1, \quad y \neq 0$

(e) $6y^0 = 6(1) = 6, \quad y \neq 0$

(f) $(6y)^0 = 1, \quad y \neq 0$

CAUTION

Look again at Examples 1(b) and (c). In $(-60)^0$, the base is -60 and the exponent is 0. Any nonzero base raised to the exponent 0 is 1. In -60^0, the base is 60. Then $60^0 = 1$, and $-60^0 = -1$.

Work Problem 2 at the Side. ▶

OBJECTIVES

1. Use 0 as an exponent.
2. Use negative numbers as exponents.
3. Use the quotient rule for exponents.
4. Use combinations of rules.

1 Continue the list of exponentials using -2, -3, and -4 as exponents.

$2^{-2} =$ ______

$2^{-3} =$ ______

$2^{-4} =$ ______

2 Evaluate.

(a) 28^0

(b) $(-16)^0$

(c) -7^0

(d) $m^0, \quad m \neq 0$

(e) $-p^0, \quad p \neq 0$

ANSWERS

1. $2^{-2} = \frac{1}{4}$; $2^{-3} = \frac{1}{8}$; $2^{-4} = \frac{1}{16}$

2. (a) 1 (b) 1 (c) -1 (d) 1 (e) -1

OBJECTIVE 2 Use negative numbers as exponents. From the lists at the beginning of this section and margin Problem 1, since $2^{-2} = \frac{1}{4}$ and $2^{-3} = \frac{1}{8}$, we can deduce that 2^{-n} should equal $\frac{1}{2^n}$. Is the product rule valid in such cases? For example, if we multiply 6^{-2} by 6^2, we get

$$6^{-2} \cdot 6^2 = 6^{-2+2} = 6^0 = 1.$$

The expression 6^{-2} behaves as if it were the reciprocal of 6^2, because their product is 1. The reciprocal of 6^2 may be written $\frac{1}{6^2}$, leading us to define 6^{-2} as $\frac{1}{6^2}$. This is a particular case of the definition of negative exponents.

Negative Exponents

For any nonzero real number a and any integer n, $\boldsymbol{a^{-n} = \frac{1}{a^n}}$.

Example: $3^{-2} = \frac{1}{3^2}$

By definition, a^{-n} and a^n are reciprocals, since

$$a^n \cdot a^{-n} = a^n \cdot \frac{1}{a^n} = 1.$$

Since $1^n = 1$, the definition of a^{-n} can also be written

$$a^{-n} = \frac{1}{a^n} = \frac{1^n}{a^n} = \left(\frac{1}{a}\right)^n.$$

For example, $6^{-3} = \left(\frac{1}{6}\right)^3$ and $\left(\frac{1}{3}\right)^{-2} = 3^2$.

EXAMPLE 2 Using Negative Exponents

Simplify by writing with positive exponents. Assume that all variables represent nonzero real numbers.

(a) $3^{-2} = \frac{1}{3^2} = \frac{1}{9}$ $\quad a^{-n} = \frac{1}{a^n}$

(b) $5^{-3} = \frac{1}{5^3} = \frac{1}{125}$

(c) $\left(\frac{1}{2}\right)^{-3} = 2^3 = 8$ $\quad \frac{1}{2}$ and 2 are reciprocals.

Notice that we can change the base to its reciprocal if we also change the sign of the exponent.

(d) $\left(\frac{2}{5}\right)^{-4} = \left(\frac{5}{2}\right)^4 = \frac{5^4}{2^4} = \frac{625}{16}$

$\frac{2}{5}$ and $\frac{5}{2}$ are reciprocals.

(e) $\left(\frac{4}{3}\right)^{-5} = \left(\frac{3}{4}\right)^5 = \frac{3^5}{4^5} = \frac{243}{1024}$

(f) $4^{-1} - 2^{-1}$

$= \frac{1}{4} - \frac{1}{2} = \frac{1}{4} - \frac{2}{4} = -\frac{1}{4}$ $\quad$ Apply the exponents first; then subtract.

Continued on Next Page

(g) $p^{-2} = \frac{1}{p^2}$

(h) $\frac{1}{x^{-4}} = \frac{1^{-4}}{x^{-4}}$ $\quad 1^n = 1$, for any integer n

$= \left(\frac{1}{x}\right)^{-4}$ $\quad$ Power rule (c)

$= x^4$ $\quad \frac{1}{x}$ and x are reciprocals.

Notice that, in general, $\frac{\mathbf{1}}{\boldsymbol{a^{-n}}} = \boldsymbol{a^n}$.

(i) $x^3y^{-4} = \frac{x^3}{1} \cdot \frac{1}{y^4} = \frac{x^3}{y^4}$

CAUTION

A negative exponent does not indicate a negative number. Negative exponents lead to reciprocals.

Expression	**Example**	
a^{-n}	$3^{-2} = \frac{1}{3^2} = \frac{1}{9}$	Not negative
$-a^{-n}$	$-3^{-2} = -\frac{1}{3^2} = -\frac{1}{9}$	Negative

Work Problem **3** at the Side. ▶

Consider the following:

$$\frac{2^{-3}}{3^{-4}} = \frac{\frac{1}{2^3}}{\frac{1}{3^4}} = \frac{1}{2^3} \div \frac{1}{3^4} = \frac{1}{2^3} \cdot \frac{3^4}{1} = \frac{3^4}{2^3}.$$

To divide by a fraction, multiply by its reciprocal.

Therefore,

$$\frac{2^{-3}}{3^{-4}} = \frac{3^4}{2^3}.$$

Changing from Negative to Positive Exponents

For any nonzero numbers a and b, and any integers m and n,

$$\frac{a^{-m}}{b^{-n}} = \frac{b^n}{a^m} \quad \text{and} \quad \left(\frac{a}{b}\right)^{-m} = \left(\frac{b}{a}\right)^m.$$

Examples: $\frac{3^{-5}}{2^{-4}} = \frac{2^4}{3^5}$ and $\left(\frac{4}{5}\right)^{-3} = \left(\frac{5}{4}\right)^3$

3 Simplify by writing with positive exponents. Assume that all variables represent nonzero real numbers.

(a) 4^{-3}

(b) 6^{-2}

(c) $\left(\frac{1}{4}\right)^{-2}$

(d) $\left(\frac{2}{3}\right)^{-2}$

(e) $2^{-1} + 5^{-1}$

(f) m^{-5}

(g) $\frac{1}{z^{-4}}$

(h) p^2q^{-5}

Answers

3. (a) $\frac{1}{4^3} = \frac{1}{64}$ **(b)** $\frac{1}{6^2} = \frac{1}{36}$ **(c)** $4^2 = 16$

(d) $\left(\frac{3}{2}\right)^2 = \frac{9}{4}$ **(e)** $\frac{1}{2} + \frac{1}{5} = \frac{7}{10}$

(f) $\frac{1}{m^5}$ **(g)** z^4 **(h)** $\frac{p^2}{q^5}$

4 Simplify. Assume that all variables represent nonzero real numbers.

(a) $\frac{7^{-1}}{5^{-4}}$

(b) $\frac{x^{-3}}{y^{-2}}$

(c) $\frac{4h^{-5}}{m^{-2}k}$

(d) $\left(\frac{3m}{p}\right)^{-2}$

EXAMPLE 3 Changing from Negative to Positive Exponents

Simplify. Assume that all variables represent nonzero real numbers.

(a) $\frac{4^{-2}}{5^{-3}} = \frac{5^3}{4^2} = \frac{125}{16}$

(b) $\frac{m^{-5}}{p^{-1}} = \frac{p^1}{m^5} = \frac{p}{m^5}$

(c) $\frac{a^{-2}b}{3d^{-3}} = \frac{bd^3}{3a^2}$ Notice that b in the numerator and the coefficient 3 in the denominator are not affected.

(d) $\left(\frac{x}{2y}\right)^{-4}$

$= \left(\frac{2y}{x}\right)^4$ Negative-to-positive rule

$= \frac{2^4y^4}{x^4}$ Power rule (c)

$= \frac{16y^4}{x^4}$

◀ Work Problem **4** at the Side.

CAUTION

Be careful. We cannot use the rule $\frac{a^{-m}}{b^{-n}} = \frac{b^n}{a^m}$ to change negative exponents to positive exponents if the exponents occur in a *sum* or *difference* of terms. For example,

$\frac{5^{-2} + 3^{-1}}{7 - 2^{-3}}$ would be written with positive exponents as $\frac{\frac{1}{5^2} + \frac{1}{3}}{7 - \frac{1}{2^3}}$.

OBJECTIVE 3 Use the quotient rule for exponents. Consider a quotient of two exponential expressions with the same base.

$$\frac{6^5}{6^3} = \frac{6 \cdot 6 \cdot 6 \cdot 6 \cdot 6}{6 \cdot 6 \cdot 6} = 6^2$$

Notice that the difference between the exponents, $5 - 3 = 2$, is the exponent in the quotient. Also,

$$\frac{6^2}{6^4} = \frac{6 \cdot 6}{6 \cdot 6 \cdot 6 \cdot 6} = \frac{1}{6^2} = 6^{-2}.$$

Here, $2 - 4 = -2$. These examples suggest the **quotient rule for exponents.**

Quotient Rule for Exponents

For any nonzero real number a and any integers m and n,

$$\frac{a^m}{a^n} = a^{m-n}.$$

(Keep the same base and subtract the exponents.)

Example: $\frac{5^8}{5^4} = 5^{8-4} = 5^4$

ANSWERS

4. (a) $\frac{625}{7}$ (b) $\frac{y^2}{x^3}$ (c) $\frac{4m^2}{h^5k}$ (d) $\frac{p^2}{9m^2}$

CAUTION

A common **error** is to write $\frac{5^8}{5^4} = 1^{8-4} = 1^4$. ***This is incorrect.*** By the quotient rule, the quotient must have the *same base,* 5, so

$$\frac{5^8}{5^4} = 5^{8-4} = 5^4.$$

We can confirm this by using the definition of exponents to write out the factors:

$$\frac{5^8}{5^4} = \frac{5 \cdot 5 \cdot 5 \cdot 5 \cdot 5 \cdot 5 \cdot 5 \cdot 5}{5 \cdot 5 \cdot 5 \cdot 5} = 5^4.$$

EXAMPLE 4 Using the Quotient Rule

Simplify. Assume that all variables represent nonzero real numbers.

(a) $\frac{5^8}{5^6} = 5^{8-6} = 5^2 = 25$ (Keep the same base.)

(b) $\frac{4^2}{4^9} = 4^{2-9} = 4^{-7} = \frac{1}{4^7}$

(c) $\frac{5^{-3}}{5^{-7}} = 5^{-3-(-7)} = 5^4 = 625$ (Be careful with signs.)

(d) $\frac{q^5}{q^{-3}} = q^{5-(-3)} = q^8$

(e) $\frac{3^2x^5}{3^4x^3}$

$$= \frac{3^2}{3^4} \cdot \frac{x^5}{x^3}$$

$$= 3^{2-4} \cdot x^{5-3}$$

$$= 3^{-2}x^2$$

$$= \frac{x^2}{3^2}$$

$$= \frac{x^2}{9}$$

(f) $\frac{(m+n)^{-2}}{(m+n)^{-4}}$

$$= (m+n)^{-2-(-4)}$$

$$= (m+n)^{-2+4}$$

$$= (m+n)^2, \quad m \neq -n$$

The restriction $m \neq -n$ is necessary to prevent a denominator of 0 in the original expression. Division by 0 is undefined.

(g) $\frac{7x^{-3}y^2}{2^{-1}x^2y^{-5}}$

$$= \frac{7 \cdot 2^1y^2y^5}{x^2x^3}$$ Definition of negative exponent

$$= \frac{14y^7}{x^5}$$ Multiply; product rule

Work Problem **5** *at the Side.* ▶

The definitions and rules for exponents given in this section and **Section 5.2** are summarized on the next page.

5 Simplify. Assume that all variables represent nonzero real numbers.

(a) $\frac{5^{11}}{5^8}$

(b) $\frac{4^7}{4^{10}}$

(c) $\frac{6^{-5}}{6^{-2}}$

(d) $\frac{8^4m^9}{8^5m^{10}}$

(e) $\frac{3^{-1}(x+y)^{-3}}{2^{-2}(x+y)^{-4}}, \quad x \neq -y$

ANSWERS

5. (a) 125 **(b)** $\frac{1}{64}$ **(c)** $\frac{1}{216}$ **(d)** $\frac{1}{8m}$ **(e)** $\frac{4}{3}(x+y)$

Definitions and Rules for Exponents

For any integers m and n:

		Examples
Product rule	$a^m \cdot a^n = a^{m+n}$	$7^4 \cdot 7^5 = 7^{4+5} = 7^9$
Zero exponent	$a^0 = 1 \quad (a \neq 0)$	$(-3)^0 = 1$
Negative exponent	$a^{-n} = \frac{1}{a^n} \quad (a \neq 0)$	$5^{-3} = \frac{1}{5^3}$
Quotient rule	$\frac{a^m}{a^n} = a^{m-n} \quad (a \neq 0)$	$\frac{2^2}{2^5} = 2^{2-5} = 2^{-3} = \frac{1}{2^3}$
Power rules (a)	$(a^m)^n = a^{mn}$	$(4^2)^3 = 4^{2 \cdot 3} = 4^6$
(b)	$(ab)^m = a^m b^m$	$(3k)^4 = 3^4 k^4$
(c)	$\left(\frac{a}{b}\right)^m = \frac{a^m}{b^m} \quad (b \neq 0)$	$\left(\frac{2}{3}\right)^2 = \frac{2^2}{3^2}$
Negative-to-positive rules	$\frac{a^{-m}}{b^{-n}} = \frac{b^n}{a^m} \quad (a, b \neq 0)$	$\frac{2^{-4}}{5^{-3}} = \frac{5^3}{2^4}$
	$\left(\frac{a}{b}\right)^{-m} = \left(\frac{b}{a}\right)^m.$	$\left(\frac{4}{7}\right)^{-2} = \left(\frac{7}{4}\right)^2$

OBJECTIVE 4 Use combinations of rules. We sometimes need to use more than one rule to simplify an expression.

EXAMPLE 5 Using a Combination of Rules

Simplify each expression. Assume that all variables represent nonzero real numbers.

(a) $\frac{(4^2)^3}{4^5}$

$= \frac{4^6}{4^5}$ Power rule (a)

$= 4^{6-5}$ Quotient rule

$= 4^1$

$= 4$

(b) $(2x)^3(2x)^2$

$= (2x)^5$ Product rule

$= 2^5x^5$ Power rule (b)

$= 32x^5$

(c) $\left(\frac{2x^3}{5}\right)^{-4}$

$= \left(\frac{5}{2x^3}\right)^4$ Negative-to-positive rule

$= \frac{5^4}{2^4x^{12}}$ Power rules (a)–(c)

$= \frac{625}{16x^{12}}$

(d) $\left(\frac{3x^{-2}}{4^{-1}y^3}\right)^{-3}$

$= \frac{3^{-3}x^6}{4^3y^{-9}}$ Power rules (a)–(c)

$= \frac{x^6y^9}{4^3 \cdot 3^3}$ Negative-to-positive rule

$= \frac{x^6y^9}{1728}$ $4^3 \cdot 3^3 = 64 \cdot 27 = 1728$

Continued on Next Page

(e) $\dfrac{(4m)^{-3}}{(3m)^{-4}}$

$= \dfrac{4^{-3}m^{-3}}{3^{-4}m^{-4}}$ Power rule (b)

$= \dfrac{3^4m^4}{4^3m^3}$ Negative-to-positive rule

$= \dfrac{3^4m^{4-3}}{4^3}$ Quotient rule

$= \dfrac{3^4m}{4^3}$

$= \dfrac{81m}{64}$

Note

Since the steps can be done in several different orders, there are many equally correct ways to simplify expressions like those in Examples 5(c) through 5(e).

Work Problem **6** *at the Side.* ▶

6 Simplify each expression. Assume that all variables represent nonzero real numbers.

(a) $\dfrac{(3^4)^2}{3^3}$

(b) $(4x)^2(4x)^4$

(c) $\dfrac{(6x)^{-1}}{(3x^2)^{-2}}$

(d) $\dfrac{3^9 \cdot (x^2y)^{-2}}{3^3 \cdot x^{-4}y}$

Answers

6. (a) 243 **(b)** 4^6x^6, or $4096x^6$ **(c)** $\dfrac{3x^3}{2}$

(d) $\dfrac{729}{y^3}$

Math in the Media

MORE POWER TO YOU, CAPTAIN KIRK

The original *Star Trek* series first aired during the 1966 to 1967 television season and started the phenomenon that continues today. There have been five different television series and 10 feature movies with the *Star Trek* theme.

Captain James T. Kirk, portrayed by William Shatner, led the Starship Enterprise during its first three seasons. During the first season, the February 2, 1967, episode "Court Martial" told the story of Kirk being put on trial. He was accused of negligence in the death of a crewmember, because the computer records of the ship contradicted Kirk's logs. As the trial begins, Kirk explains how the sounds on the ship can be recorded and magnified:

Kirk: *Gentlemen, this computer has an auditory sensor. It can, in effect, hear sounds. By installing a booster we can increase that capability on an order of one to the fourth power. The computer should be able to bring us every sound occurring on the ship.*

1. Read Captain Kirk's statement carefully. What error did he make?

   "One to the fourth power" is $1^4 = 1$. So there is no boost, because the sound level will remain the same.

2. What is the result if we raise the number 1 to any whole number power?

   1

3. It is possible that Kirk meant "10 to the fourth power." Express 10^4 in expanded form.

   10,000

4. The word **googol** was invented to express a very large power of 10. The search engine Google was named in honor of it. Look up the meaning of googol, and write it in exponential form.

   A googol is 10 to the 100th power: 10^{100}.

5. Investigate the meaning of the word **googolplex.**

   A googolplex is 10 to the googol power: $10^{\text{googol}} = 10^{10^{100}}$.

5.5 Exercises

FOR EXTRA HELP

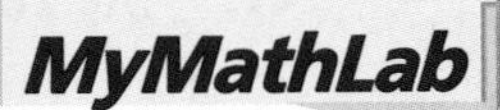

Decide whether each expression is positive, negative, or 0.

1. $(-2)^{-3}$ negative

2. $(-3)^{-2}$ positive

3. -2^4 negative

4. -3^6 negative

5. $\left(\frac{1}{4}\right)^{-2}$ positive

6. $\left(\frac{1}{5}\right)^{-2}$ positive

7. $1 - 5^0$ 0

8. $1 - 7^0$ 0

Decide whether each expression is equal to either 0, 1, *or* −1. *See Example 1.*

9. 9^0 1

10. 5^0 1

11. $(-4)^0$ 1

12. $(-10)^0$ 1

13. -9^0 −1

14. -5^0 −1

15. $(-2)^0 - 2^0$ 0

16. $(-8)^0 - 8^0$ 0

17. $\frac{0^{10}}{10^0}$ 0

18. $\frac{0^5}{5^0}$ 0

Evaluate each expression. See Examples 1 and 2.

19. $7^0 + 9^0$ 2

20. $8^0 + 6^0$ 2

21. 4^{-3} $\frac{1}{64}$

22. 5^{-4} $\frac{1}{625}$

23. $\left(\frac{1}{2}\right)^{-4}$ 16

24. $\left(\frac{1}{3}\right)^{-3}$ 27

25. $\left(\frac{6}{7}\right)^{-2}$ $\frac{49}{36}$

26. $\left(\frac{2}{3}\right)^{-3}$ $\frac{27}{8}$

27. $(-3)^{-4}$ $\frac{1}{81}$

28. $(-4)^{-3}$ $-\frac{1}{64}$

29. $5^{-1} + 3^{-1}$ $\frac{8}{15}$

30. $6^{-1} + 2^{-1}$ $\frac{2}{3}$

31. $-2^{-1} + 3^{-2}$ $-\frac{7}{18}$

32. $(-3)^{-2} + (-4)^{-1}$ $-\frac{5}{36}$

Relating Concepts (Exercises 33–36) For Individual or Group Work

In Objective 1, we used the product rule to motivate the definition of a 0 *exponent. We can also use the quotient rule. To see this,* ***work Exercises 33–36 in order.***

33. Consider the expression $\frac{25}{25}$. What is its simplest form? 1

34. Write the quotient in Exercise 33 using the fact that $25 = 5^2$. $\frac{5^2}{5^2}$

35. Apply the quotient rule for exponents to your answer for Exercise 34. Give the answer as a power of 5.

$5^{2-2} = 5^0$

36. Because your answers for Exercises 33 and 35 both represent $\frac{25}{25}$, they must be equal. Write this equality. What definition does it support?

$5^0 = 1$; This supports the definition of a 0 exponent.

Simplify by writing each expression with positive exponents. Assume that all variables represent nonzero real numbers. See Examples 2–4.

37. $\frac{9^4}{9^5}$ $\frac{1}{9}$

38. $\frac{7^3}{7^4}$ $\frac{1}{7}$

39. $\frac{6^{-3}}{6^2}$ $\frac{1}{6^5}$, or $\frac{1}{7776}$

40. $\frac{4^{-2}}{4^3}$ $\frac{1}{4^5}$, or $\frac{1}{1024}$

41. $\frac{1}{6^{-3}}$ 216

42. $\frac{1}{5^{-2}}$ 25

43. $\frac{2}{r^{-4}}$ $2r^4$

44. $\frac{3}{s^{-8}}$ $3s^8$

45. $\frac{4^{-3}}{5^{-2}}$ $\frac{25}{64}$

46. $\frac{6^{-2}}{5^{-4}}$ $\frac{625}{36}$

47. p^5q^{-8} $\frac{p^5}{q^8}$

48. $x^{-8}y^4$ $\frac{y^4}{x^8}$

49. $\frac{r^5}{r^{-4}}$ r^9

50. $\frac{a^6}{a^{-4}}$ a^{10}

51. $\frac{6^4x^8}{6^5x^3}$ $\frac{x^5}{6}$

52. $\frac{3^8y^5}{3^{10}y^2}$ $\frac{y^3}{9}$

53. $\frac{6y^3}{2y}$ $3y^2$

54. $\frac{5m^2}{m}$ $5m$

55. $\frac{3x^5}{3x^2}$ x^3

56. $\frac{10p^8}{2p^4}$ $5p^4$

57. $\frac{x^{-3}y}{4z^{-2}}$ $\frac{yz^2}{4x^3}$

58. $\frac{p^{-5}q^{-4}}{9r^{-3}}$ $\frac{r^3}{9p^5q^4}$

59. $\frac{(a+b)^{-3}}{(a+b)^{-4}}$ $a+b$

60. $\frac{(x+y)^{-8}}{(x+y)^{-9}}$ $x+y$

Simplify by writing each expression with positive exponents. Assume that all variables represent nonzero real numbers. See Example 5.

61. $\frac{(7^4)^3}{7^9}$ 343

62. $\frac{(5^3)^2}{5^2}$ 625

63. $x^{-3} \cdot x^5 \cdot x^{-4}$ $\frac{1}{x^2}$

64. $y^{-8} \cdot y^5 \cdot y^{-2}$ $\frac{1}{y^5}$

65. $\frac{(3x)^{-2}}{(4x)^{-3}}$ $\frac{64x}{9}$

66. $\frac{(2y)^{-3}}{(5y)^{-4}}$ $\frac{625y}{8}$

67. $\left(\frac{x^{-1}y}{z^2}\right)^{-2}$ $\frac{x^2z^4}{y^2}$

68. $\left(\frac{p^{-4}q}{r^{-3}}\right)^{-3}$ $\frac{p^{12}}{r^9q^3}$

69. $(6x)^4(6x)^{-3}$ $6x$

70. $(10y)^9(10y)^{-8}$ $10y$

71. $\frac{(m^7n)^{-2}}{m^{-4}n^3}$ $\frac{1}{m^{10}n^5}$

72. $\frac{(m^8n^{-4})^2}{m^{-2}n^5}$ $\frac{m^{18}}{n^{13}}$

73. $\frac{5x^{-3}}{(4x)^2}$ $\frac{5}{16x^5}$

74. $\frac{-3k^5}{(2k)^2}$ $\frac{-3k^3}{4}$

75. $\left(\frac{2p^{-1}q}{3^{-1}m^2}\right)^2$ $\frac{36q^2}{m^4p^2}$

76. $\left(\frac{4xy^2}{x^{-1}y}\right)^{-2}$ $\frac{1}{16x^4y^2}$

Summary Exercises on the Rules for Exponents

Simplify each expression. Assume that all variables represent nonzero real numbers.

1. $\left(\frac{6x^2}{5}\right)^{12}$

$\frac{6^{12}x^{24}}{5^{12}}$

2. $\left(\frac{rs^2t^3}{3t^4}\right)^6$

$\frac{r^6s^{12}}{729t^6}$

3. $(10x^2y^4)^2(10xy^2)^3$

$100{,}000x^7y^{14}$

4. $(-2ab^3c)^4(-2a^2b)^3$

$-128a^{10}b^{15}c^4$

5. $\left(\frac{9wx^3}{y^4}\right)^3$

$\frac{729w^3x^9}{y^{12}}$

6. $(4x^{-2}y^{-3})^{-2}$

$\frac{x^4y^6}{16}$

7. $\frac{c^{11}(c^2)^4}{(c^3)^3(c^2)^{-6}}$

c^{22}

8. $\left(\frac{k^4t^2}{k^2t^{-4}}\right)^{-2}$

$\frac{1}{k^4t^{12}}$

9. $5^{-1}+6^{-1}$

$\frac{11}{30}$

10. $\frac{(3y^{-1}z^3)^{-1}(3y^2)}{(y^3z^2)^{-3}}$

$y^{12}z^3$

11. $\frac{(2xy^{-1})^3}{2^3x^{-3}y^2}$

$\frac{x^6}{y^5}$

12. $-8^0+(-8)^0$

0

13. $(z^4)^{-3}(z^{-2})^{-5}$

$\frac{1}{z^2}$

14. $\left(\frac{r^2st^5}{3r}\right)^{-2}$

$\frac{9}{r^2s^2t^{10}}$

15. $\frac{(3^{-1}x^{-3}y)^{-1}(2x^2y^{-3})^2}{(5x^{-2}y^2)^{-2}}$

$\frac{300x^3}{y^3}$

16. $\left(\frac{5x^2}{3x^{-4}}\right)^{-1}$

$\frac{3}{5x^6}$

17. $\left(\frac{-2x^{-2}}{2x^2}\right)^{-2}$

x^8

18. $\frac{(x^{-4}y^2)^3(x^2y)^{-1}}{(xy^2)^{-3}}$

$\frac{y^{11}}{x^{11}}$

19. $\frac{(a^{-2}b^3)^{-4}}{(a^{-3}b^2)^{-2}(ab)^{-4}}$

$\frac{a^6}{b^4}$

20. $(2a^{-30}b^{-29})(3a^{31}b^{30})$

$6ab$

21. $5^{-2}+6^{-2}$

$\frac{61}{900}$

22. $\left(\dfrac{(x^{47}y^{23})^2}{x^{-26}y^{-42}}\right)^0$

1

23. $\left(\dfrac{7a^2b^3}{2}\right)^3$

$\mathbf{\dfrac{343a^6b^9}{8}}$

24. $-(-12^0)$

1

25. $-(-12)^0$

$\mathbf{-1}$

26. $\dfrac{0^{12}}{12^0}$

0

27. $\dfrac{(2xy^{-3})^{-2}}{(3x^{-2}y^4)^{-3}}$

$\mathbf{\dfrac{27y^{18}}{4x^8}}$

28. $\left(\dfrac{a^2b^3c^4}{a^{-2}b^{-3}c^{-4}}\right)^{-2}$

$\mathbf{\dfrac{1}{a^8b^{12}c^{16}}}$

29. $(6x^{-5}z^3)^{-3}$

$\mathbf{\dfrac{x^{15}}{216z^9}}$

30. $(2p^{-2}qr^{-3})(2p)^{-4}$

$\mathbf{\dfrac{q}{8p^6r^3}}$

31. $\dfrac{(xy)^{-3}(xy)^5}{(xy)^{-4}}$

$\mathbf{x^6y^6}$

32. $42^0 - (-12)^0$

0

33. $\dfrac{(7^{-1}x^{-3})^{-2}(x^4)^{-6}}{7^{-1}x^{-3}}$

$\mathbf{\dfrac{343}{x^{15}}}$

34. $\left(\dfrac{3^{-4}x^{-3}}{3^{-3}x^{-6}}\right)^{-2}$

$\mathbf{\dfrac{9}{x^6}}$

35. $(5p^{-2}q)^{-3}(5pq^3)^4$

$\mathbf{5p^{10}q^9}$

36. $8^{-1} + 6^{-1}$

$\mathbf{\dfrac{7}{24}}$

37. $\left(\dfrac{4r^{-6}s^{-2}t}{2r^8s^{-4}t^2}\right)^{-1}$

$\mathbf{\dfrac{r^{14}t}{2s^2}}$

38. $(13x^{-6}y)(13x^{-6}y)^{-1}$

1

39. $\dfrac{(8pq^{-2})^4}{(8p^{-2}q^{-3})^3}$

$\mathbf{8p^{10}q}$

40. $\left(\dfrac{mn^{-2}p}{m^2np^4}\right)^{-2}\left(\dfrac{mn^{-2}p}{m^2np^4}\right)^3$

$\mathbf{\dfrac{1}{mn^3p^3}}$

41. $-(-3^0)^0$

$\mathbf{-1}$

42. $5^{-1} - 8^{-1}$

$\mathbf{\dfrac{3}{40}}$

43. A student simplified $(10^2)^3$ as 1000^6. ***WHAT WENT WRONG?*** Give the correct answer.

Using the product rule, simplify as follows:

$$(10^2)^3 = 10^{2\cdot 3} = 10^6 = 1{,}000{,}000.$$

44. A student simplified -5^4 as shown:

$$-5^4 = (-5)^4 = 625.$$

WHAT WENT WRONG? Give the correct answer.

The negative sign is not part of the base:

$$-5^4 = -1(5^4) = -1 \cdot 625 = -625.$$

5.6 Dividing a Polynomial by a Monomial

OBJECTIVE 1 Divide a polynomial by a monomial. We add two fractions with a common denominator as follows.

$$\frac{a}{c}+\frac{b}{c}=\frac{a+b}{c}$$

In reverse, this statement gives a rule for dividing a polynomial by a monomial.

Dividing a Polynomial by a Monomial

To divide a polynomial by a monomial, divide each term of the polynomial by the monomial:

$$\frac{a+b}{c}=\frac{a}{c}+\frac{b}{c} \quad (c \neq 0).$$

Examples: $\frac{2+5}{3}=\frac{2}{3}+\frac{5}{3}$ and $\frac{x+3z}{2y}=\frac{x}{2y}+\frac{3z}{2y} \quad (y \neq 0)$

The parts of a division problem are named here.

$$\text{Dividend} \rightarrow \frac{12x^2+6x}{6x} = 2x+1 \leftarrow \text{Quotient}$$
$$\text{Divisor} \rightarrow 6x$$

EXAMPLE 1 Dividing a Polynomial by a Monomial

Divide $5m^5 - 10m^3$ by $5m^2$.

$$\frac{5m^5-10m^3}{5m^2}$$

$$=\frac{5m^5}{5m^2}-\frac{10m^3}{5m^2} \quad \text{Use the preceding rule, with + replaced by −.}$$

$$=m^3-2m \quad \text{Quotient rule}$$

Check Multiply: $5m^2 \cdot (m^3 - 2m) = 5m^5 - 10m^3$. ← Original polynomial (Dividend)

($5m^2$ ↑ Divisor; $(m^3 - 2m)$ ↑ Quotient)

Because division by 0 is undefined, the quotient $\frac{5m^5-10m^3}{5m^2}$ is undefined if $m = 0$. From now on, we assume that no denominators are 0.

Work Problem 1 at the Side. ▶

EXAMPLE 2 Dividing a Polynomial by a Monomial

Divide $\frac{16a^5-12a^4+8a^2}{4a^3}$.

$$\frac{16a^5-12a^4+8a^2}{4a^3}$$

$$=\frac{16a^5}{4a^3}-\frac{12a^4}{4a^3}+\frac{8a^2}{4a^3} \quad \text{Divide each term by } 4a^3.$$

$$=4a^2-3a+\frac{2}{a} \quad \text{Quotient rule}$$

Continued on Next Page

OBJECTIVE

1 **Divide a polynomial by a monomial.**

1 Divide.

(a) $\frac{6p^4+18p^7}{3p^2}$

(b) $\frac{12m^6+18m^5+30m^4}{6m^2}$

(c) $(18r^7 - 9r^2) \div (3r)$

ANSWERS

1. (a) $2p^2 + 6p^5$ (b) $2m^4 + 3m^3 + 5m^2$ (c) $6r^6 - 3r$

The quotient $4a^2 - 3a + \frac{2}{a}$ is not a polynomial because of the expression $\frac{2}{a}$, which has a variable in the denominator. While the sum, difference, and product of two polynomials are always polynomials, the quotient of two polynomials may not be.

Check $4a^3\left(4a^2 - 3a + \frac{2}{a}\right)$ Divisor × Quotient should equal Dividend.

$= 4a^3(4a^2) + 4a^3(-3a) + 4a^3\left(\frac{2}{a}\right)$ Distributive property

$= 16a^5 - 12a^4 + 8a^2$ Dividend

◀ *Work Problem* 2 *at the Side.*

EXAMPLE 3 Dividing a Polynomial by a Monomial with a Negative Coefficient

Divide $-7x^3 + 12x^4 - 4x$ by $-4x$.

Write the polynomial in descending powers as $12x^4 - 7x^3 - 4x$ before dividing.

$$\frac{12x^4 - 7x^3 - 4x}{-4x}$$ Write in descending powers.

$= \frac{12x^4}{-4x} - \frac{7x^3}{-4x} - \frac{4x}{-4x}$ Divide each term by $-4x$.

$= -3x^3 - \frac{7x^2}{-4} - (-1)$ Quotient rule

$= -3x^3 + \frac{7}{4}x^2 + 1$ Be sure to include the 1 in the answer.

Check by multiplying.

◀ *Work Problem* 3 *at the Side.*

EXAMPLE 4 Dividing a Polynomial by a Monomial

Divide $180x^4y^{10} - 150x^3y^8 + 120x^2y^6 - 90xy^4 + 100y$ by $-30xy^2$.

$$\frac{180x^4y^{10} - 150x^3y^8 + 120x^2y^6 - 90xy^4 + 100y}{-30xy^2}$$

$$= \frac{180x^4y^{10}}{-30xy^2} - \frac{150x^3y^8}{-30xy^2} + \frac{120x^2y^6}{-30xy^2} - \frac{90xy^4}{-30xy^2} + \frac{100y}{-30xy^2}$$

$$= -6x^3y^8 + 5x^2y^6 - 4xy^4 + 3y^2 - \frac{10}{3xy}$$

◀ *Work Problem* 4 *at the Side.*

2 Divide.

(a) $\frac{20x^4 - 25x^3 + 5x}{5x^2}$

(b) $\frac{50m^4 - 30m^3 + 20m}{10m^3}$

3 Divide.

(a) $\frac{-9y^6 + 8y^7 - 11y - 4}{y^2}$

(b) $\frac{-8p^4 - 6p^3 - 12p^5}{-3p^3}$

4 Divide.

$\frac{45x^4y^3 + 30x^3y^2 - 60x^2y}{-15x^2y}$

Answers

2. (a) $4x^2 - 5x + \frac{1}{x}$ **(b)** $5m - 3 + \frac{2}{m^2}$

3. (a) $8y^5 - 9y^4 - \frac{11}{y} - \frac{4}{y^2}$

(b) $4p^2 + \frac{8p}{3} + 2$

4. $-3x^2y^2 - 2xy + 4$

5.6 Exercises

FOR EXTRA HELP

MyMathLab

Math XL PRACTICE

WATCH

DOWNLOAD

READ

REVIEW

Fill in each blank with the correct response.

1. In the statement $\frac{6x^2 + 8}{2} = 3x^2 + 4$, **$6x^2 + 8$** is the dividend, **2** is the divisor, and **$3x^2 + 4$** is the quotient.

2. The expression $\frac{3x + 12}{x}$ is undefined if $x =$ **0**.

3. To check the division shown in Exercise 1, multiply **$3x^2 + 4$** by **2** and show that the product is **$6x^2 + 8$**.

4. The expression $5x^2 - 3x + 6 + \frac{2}{x}$ **is not** (is/is not) a polynomial.

5. Explain why the division problem $\frac{16m^3 - 12m^2}{4m}$ can be performed using the method of this section, while the division problem $\frac{4m}{16m^3 - 12m^2}$ cannot.

To use the method of this section, the divisor must be just one term. This is true of the first problem, but not the second.

6. Evaluate $\frac{5y + 6}{2}$ when $y = 2$. Evaluate $5y + 3$ when $y = 2$. Does $\frac{5y + 6}{2}$ equal $5y + 3$? **8; 13; no**

Perform each division. See Examples 1–4.

7. $\frac{60x^4 - 20x^2 + 10x}{2x}$

$30x^3 - 10x + 5$

8. $\frac{120x^6 - 60x^3 + 80x^2}{2x}$

$60x^5 - 30x^2 + 40x$

9. $\frac{20m^5 - 10m^4 + 5m^2}{-5m^2}$

$-4m^3 + 2m^2 - 1$

10. $\frac{12t^5 - 6t^3 + 6t^2}{-6t^2}$

$-2t^3 + t - 1$

11. $\frac{8t^5 - 4t^3 + 4t^2}{2t}$

$4t^4 - 2t^2 + 2t$

12. $\frac{8r^4 - 4r^3 + 6r^2}{2r}$

$4r^3 - 2r^2 + 3r$

13. $\frac{4a^5 - 4a^2 + 8}{4a}$

$a^4 - a + \frac{2}{a}$

14. $\frac{5t^8 + 5t^7 + 15}{5t}$

$t^7 + t^6 + \frac{3}{t}$

15. $\frac{12x^5 - 4x^4 + 6x^3}{-6x^2}$

$-2x^3 + \frac{2x^2}{3} - x$

16. $\frac{24x^6 - 12x^5 + 30x^4}{-6x^2}$

$-4x^4 + 2x^3 - 5x^2$

17. $\frac{4x^2 + 20x^3 - 36x^4}{4x^2}$

$-9x^2 + 5x + 1$

18. $\frac{5x^2 - 30x^4 + 30x^5}{5x^2}$

$6x^3 - 6x^2 + 1$

19. $\frac{-3x^3 - 4x^4 + 2x}{-3x^2}$

$\frac{4x^2}{3} + x - \frac{2}{3x}$

20. $\frac{-8x + 6x^3 - 5x^4}{-3x^2}$

$\frac{5x^2}{3} - 2x + \frac{8}{3x}$

21. $\frac{27r^4 - 36r^3 - 6r^2 + 3r - 2}{3r}$

$9r^3 - 12r^2 - 2r + 1 - \frac{2}{3r}$

22. $\frac{8k^4 - 12k^3 - 2k^2 - 2k - 3}{2k}$

$4k^3 - 6k^2 - k - 1 - \frac{3}{2k}$

23. $\dfrac{2m^5 - 6m^4 + 8m^2}{-2m^3}$ $-m^2 + 3m - \dfrac{4}{m}$

24. $\dfrac{6r^5 - 8r^4 + 10r^2}{-2r^4}$ $-3r + 4 - \dfrac{5}{r^2}$

25. $(120x^{11} - 60x^{10} + 140x^9 - 100x^8) \div (10x^{12})$

$\dfrac{12}{x} - \dfrac{6}{x^2} + \dfrac{14}{x^3} - \dfrac{10}{x^4}$

26. $(120x^{12} - 84x^9 + 60x^8 - 36x^7) \div (12x^9)$

$10x^3 - 7 + \dfrac{5}{x} - \dfrac{3}{x^2}$

27. $(20a^4b^3 - 15a^5b^2 + 25a^3b) \div (-5a^4b)$

$-4b^2 + 3ab - \dfrac{5}{a}$

28. $(16y^5z - 8y^2z^2 + 12yz^3) \div (-4y^2z^2)$

$-\dfrac{4y^3}{z} + 2 - \dfrac{3z}{y}$

29. What polynomial represents the length of the rectangle?

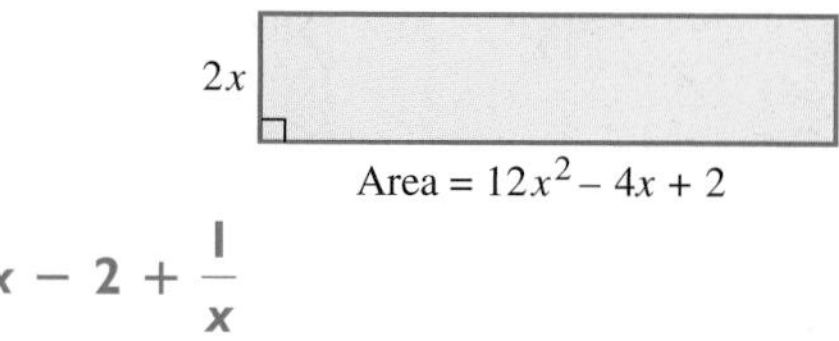

$6x - 2 + \dfrac{1}{x}$

30. What polynomial represents the length of the base of the triangle?

m

Area = $24m^3 + 48m^2 + 12m$

$48m^2 + 96m + 24$

31. What polynomial, when divided by $5x^3$, yields $3x^2 - 7x + 7$ as a quotient?

$15x^5 - 35x^4 + 35x^3$

32. The quotient of a certain polynomial and $-12y^3$ is $6y^3 - 5y^2 + 2y - 3 + \frac{7}{y}$. Find the polynomial.

$-72y^6 + 60y^5 - 24y^4 + 36y^3 - 84y^2$

Relating Concepts (Exercises 33–36) For Individual or Group Work

Our system of numeration is called a decimal system. It is based on powers of ten. In a whole number such as 2846, each digit is understood to represent the number of powers of ten for its place value. The 2 represents two thousands (2×10^3), *the 8 represents eight hundreds* (8×10^2), *the 4 represents four tens* (4×10^1), *and the 6 represents six ones (or units)* (6×10^0). *In expanded form we write*

$$2846 = (2 \times 10^3) + (8 \times 10^2) + (4 \times 10^1) + (6 \times 10^0).$$

Keeping this information in mind, ***work Exercises 33–36 in order.***

33. Divide 2846 by 2, using paper-and-pencil methods: $2\overline{)2846}$. 1423

34. Write your answer in Exercise 33 in expanded form.

$(1 \times 10^3) + (4 \times 10^2) + (2 \times 10^1) + (3 \times 10^0)$

35. Use the methods of this section to divide the polynomial $2x^3 + 8x^2 + 4x + 6$ by 2.

$x^3 + 4x^2 + 2x + 3$

36. Compare your answers in Exercises 34 and 35. How are they similar? How are they different? For what value of x does the answer in Exercise 35 equal the answer in Exercise 34?

They are similar in that the coefficients of the powers of ten are equal to the coefficients of the powers of x. They are different in that one is a number while the other is a polynomial. They are equal if $x = 10$.

5.7 Dividing a Polynomial by a Polynomial

OBJECTIVES

1. Divide a polynomial by a polynomial.
2. Apply division to a geometry problem.

OBJECTIVE 1 Divide a polynomial by a polynomial. We use a method of "long division" to divide a polynomial by a polynomial (other than a monomial). ***Both polynomials must be written in descending powers.***

Dividing Whole Numbers	Dividing Polynomials
Step 1	
Divide 6696 by 27. $27\overline{)6696}$	Divide $8x^3 - 4x^2 - 14x + 15$ by $2x + 3$. $2x + 3\overline{)8x^3 - 4x^2 - 14x + 15}$
Step 2	
66 divided by 27 $= 2$; $2 \cdot 27 = 54$. $\begin{array}{r} 2 \\ 27\overline{)6696} \\ \underline{54} \end{array}$	$8x^3$ divided by $2x = 4x^2$; $4x^2(2x + 3) = 8x^3 + 12x^2$. $\begin{array}{rl} & 4x^2 \\ 2x + 3 & \overline{)8x^3 - 4x^2 - 14x + 15} \\ & \underline{8x^3 + 12x^2} \end{array}$
Step 3	
Subtract; then bring down the next digit. $\begin{array}{r} 2 \\ 27\overline{)6696} \\ \underline{54}\downarrow \\ 129 \end{array}$	Subtract; then bring down the next term. $\begin{array}{rl} & 4x^2 \\ 2x + 3 & \overline{)8x^3 - 4x^2 - 14x + 15} \\ & \underline{8x^3 + 12x^2} \quad \downarrow \\ & \quad -16x^2 - 14x \end{array}$ ***(To subtract two polynomials, change the signs of the second and then add.)***
Step 4	
129 divided by 27 $= 4$; $4 \cdot 27 = 108$. $\begin{array}{r} 24 \\ 27\overline{)6696} \\ \underline{54} \\ 129 \\ \underline{108} \end{array}$	$-16x^2$ divided by $2x = -8x$; $-8x(2x + 3) = -16x^2 - 24x$. $\begin{array}{rl} & 4x^2 - 8x \\ 2x + 3 & \overline{)8x^3 - 4x^2 - 14x + 15} \\ & \underline{8x^3 + 12x^2} \\ & \quad -16x^2 - 14x \\ & \quad \underline{-16x^2 - 24x} \end{array}$
Step 5	
Subtract; then bring down the next digit. $\begin{array}{r} 24 \\ 27\overline{)6696} \\ \underline{54} \\ 129 \\ \underline{108}\downarrow \\ 216 \end{array}$	Subtract; then bring down the next term. $\begin{array}{rl} & 4x^2 - 8x \\ 2x + 3 & \overline{)8x^3 - 4x^2 - 14x + 15} \\ & \underline{8x^3 + 12x^2} \\ & \quad -16x^2 - 14x \\ & \quad \underline{-16x^2 - 24x} \quad \downarrow \\ & \qquad\qquad 10x + 15 \end{array}$

(continued)

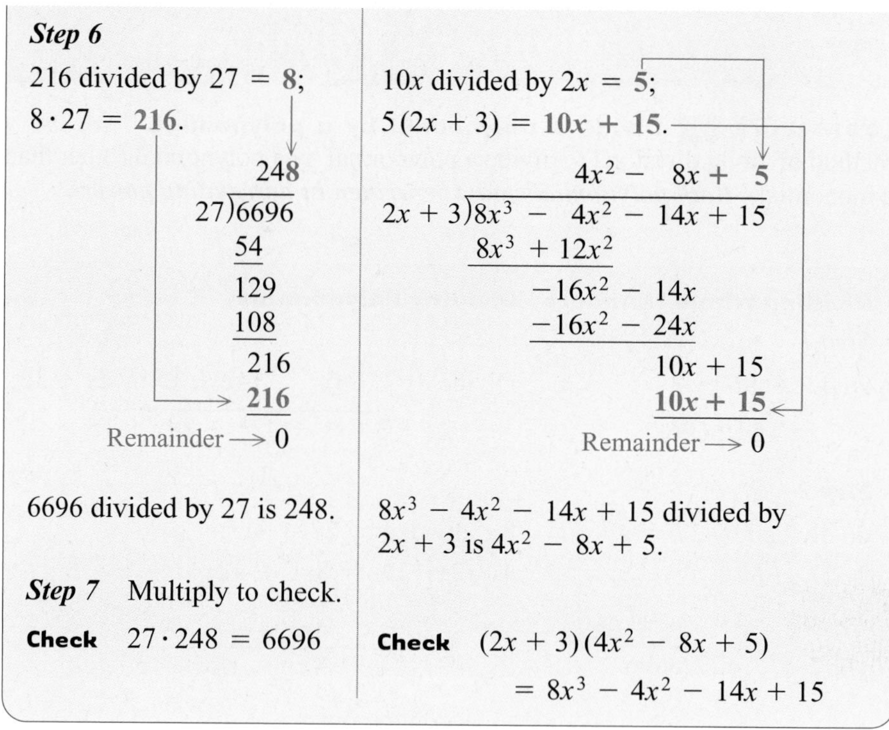

EXAMPLE 1 Dividing a Polynomial by a Polynomial

Divide $5x + 4x^3 - 8 - 4x^2$ by $2x - 1$.

The first polynomial must be written with the exponents in descending powers as $4x^3 - 4x^2 + 5x - 8$. Then divide by $2x - 1$.

$$\begin{array}{r} 2x^2 - x + 2 \\ 2x - 1\overline{)4x^3 - 4x^2 + 5x - 8} \\ \underline{4x^3 - 2x^2} \qquad\qquad \\ -2x^2 + 5x \qquad \\ \underline{-2x^2 + x} \qquad \\ 4x - 8 \\ \underline{4x - 2} \\ -6 \end{array}$$

−6 ← Remainder

To subtract, add the opposite.

Write in descending powers.

Step 1 $4x^3$ divided by $2x = 2x^2$; $2x^2(2x - 1) = 4x^3 - 2x^2$.

Step 2 Subtract; bring down the next term.

Step 3 $-2x^2$ divided by $2x = -x$; $-x(2x - 1) = -2x^2 + x$.

Step 4 Subtract; bring down the next term.

Step 5 $4x$ divided by $2x = 2$; $2(2x - 1) = 4x - 2$.

Step 6 Subtract. The remainder is -6. Write the remainder as the numerator of a fraction that has $2x - 1$ as its denominator. The answer is not a polynomial because of the nonzero remainder.

$$\frac{4x^3 - 4x^2 + 5x - 8}{2x - 1} = \underbrace{2x^2 - x + 2}_{\text{Quotient polynomial}} + \frac{-6}{2x - 1}$$

Dividend → (numerator); Divisor → (denominator)

Continued on Next Page

Step 7 Multiply to check.

Check $(2x-1)\left(2x^2-x+2+\dfrac{-6}{2x-1}\right)$

$= (2x-1)(2x^2)+(2x-1)(-x)+(2x-1)(2)$

$+ (2x-1)\left(\dfrac{-6}{2x-1}\right)$

$= 4x^3-2x^2-2x^2+x+4x-2-6$

$= 4x^3-4x^2+5x-8$

Work Problem **1** at the Side. ▶

EXAMPLE 2 Dividing into a Polynomial with Missing Terms

Divide x^3-1 by $x-1$.

Here the polynomial x^3-1 is missing the x^2-term and the x-term. When terms are missing, use **0** as the coefficient for each missing term. (Zero acts as a placeholder here, just as it does in our numeration system.) Thus, $x^3-1=x^3+\mathbf{0}x^2+\mathbf{0}x-1$. Now divide.

$$
\begin{array}{r|l}
 & x^2+x+1 \\
x-1 & x^3+0x^2+0x-1 \\
 & \underline{x^3-x^2} \\
 & x^2+0x \\
 & \underline{x^2-x} \\
 & x-1 \\
 & \underline{x-1} \\
 & 0
\end{array}
$$

Insert placeholders for the missing terms.

The remainder is 0. The quotient is x^2+x+1.

Check $(x-1)(x^2+x+1)$

$= x^3+x^2+x-x^2-x-1$

$= x^3-1$

Work Problem **2** at the Side. ▶

EXAMPLE 3 Dividing by a Polynomial with Missing Terms

Divide $x^4+2x^3+2x^2-x-1$ by x^2+1.

Since x^2+1 has a missing x-term, write it as $x^2+\mathbf{0}x+1$.

$$
\begin{array}{r|l}
 & x^2+2x+1 \\
x^2+0x+1 & x^4+2x^3+2x^2-x-1 \\
 & \underline{x^4+0x^3+x^2} \\
 & 2x^3+x^2-x \\
 & \underline{2x^3+0x^2+2x} \\
 & x^2-3x-1 \\
 & \underline{x^2+0x+1} \\
 & -3x-2 \leftarrow \text{Remainder}
\end{array}
$$

Insert a placeholder for the missing term.

Continued on Next Page

1 Divide.

(a) $(x^3+x^2+4x-6) \div (x-1)$

(b) $\dfrac{p^3-2p^2-5p+9}{p+2}$

2 Divide.

(a) $\dfrac{r^2-5}{r+4}$

(b) $(x^3-8) \div (x-2)$

ANSWERS

1. (a) x^2+2x+6
   (b) $p^2-4p+3+\dfrac{3}{p+2}$
2. (a) $r-4+\dfrac{11}{r+4}$
   (b) x^2+2x+4

When the result of subtracting ($-3x - 2$, in this case) is a constant or a polynomial of degree less than the divisor ($x^2 + 0x + 1$), that constant or polynomial is the remainder. We write the answer as

$$x^2 + 2x + 1 + \frac{-3x - 2}{x^2 + 1}.$$

Remember to include "$+ \frac{\text{remainder}}{\text{divisor}}$."

Multiply to check that this is the correct quotient.

◀ Work Problem **3** at the Side.

3 Divide.

(a) $(2x^4 + 3x^3 - x^2 + 6x + 5) \div (x^2 - 1)$

(b) $\dfrac{2m^5 + m^4 + 6m^3 - 3m^2 - 18}{m^2 + 3}$

EXAMPLE 4 Dividing a Polynomial when the Quotient Has Fractional Coefficients

Divide $4x^3 + 2x^2 + 3x + 2$ by $4x - 4$.

$$\begin{array}{r} x^2 + \frac{3}{2}x + \frac{9}{4} \\ 4x - 4 \overline{)4x^3 + 2x^2 + 3x + 2} \\ \underline{4x^3 - 4x^2} \qquad\qquad \\ 6x^2 + 3x \qquad \\ \underline{6x^2 - 6x} \qquad \\ 9x + 2 \\ \underline{9x - 9} \\ 11 \end{array}$$

$\frac{6x^2}{4x} = \frac{3}{2}x$

$\frac{9x}{4x} = \frac{9}{4}$

The answer is $x^2 + \frac{3}{2}x + \frac{9}{4} + \frac{11}{4x - 4}$.

◀ Work Problem **4** at the Side.

4 Divide $3x^3 + 7x^2 + 7x + 10$ by $3x + 6$.

OBJECTIVE 2 Apply division to a geometry problem.

EXAMPLE 5 Using an Area Formula

The area of the rectangle in Figure 3 is given by $x^3 + 4x^2 + 8x + 8$ sq. units and the width by $x + 2$ units. What is its length?

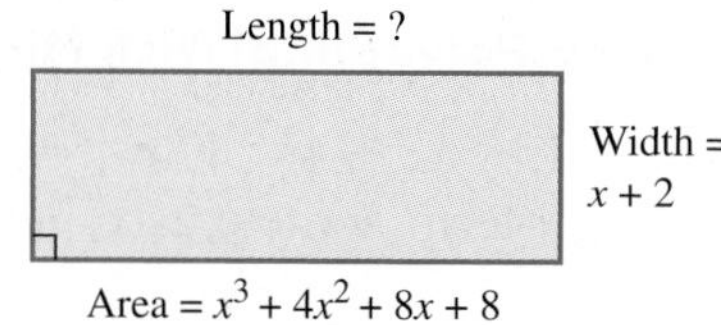

Figure 3

Since $A = LW$, solving for L gives $L = \frac{A}{W}$. Divide $x^3 + 4x^2 + 8x + 8$ by the width, $x + 2$.

◀ Work Problem **5** at the Side.

5 Divide $x^3 + 4x^2 + 8x + 8$ by $x + 2$.

The quotient from Problem 5 at the side, $x^2 + 2x + 4$, represents the length of the rectangle in units.

Answers

3. (a) $2x^2 + 3x + 1 + \frac{9x + 6}{x^2 - 1}$

(b) $2m^3 + m^2 - 6$

4. $x^2 + \frac{1}{3}x + \frac{5}{3}$

5. $x^2 + 2x + 4$

5.7 Exercises

FOR EXTRA HELP

MyMathLab

Math XL PRACTICE

WATCH

DOWNLOAD

READ

REVIEW

1. In the division problem $(4x^4 + 2x^3 - 14x^2 + 19x + 10) \div (2x + 5) = 2x^3 - 4x^2 + 3x + 2$, which polynomial is the divisor? Which is the quotient?

   The divisor is $2x + 5$; the quotient is $2x^3 - 4x^2 + 3x + 2$.

2. When dividing one polynomial by another, how do you know when to stop dividing?

   Stop when the degree of the remainder is less than the degree of the divisor, or when the remainder is 0.

3. In dividing $12m^2 - 20m + 3$ by $2m - 3$, what is the first step?

   Divide $12m^2$ by $2m$ to get $6m$.

4. In the division in Exercise 3, what is the second step?

   Multiply $6m$ by $2m - 3$ to get $12m^2 - 18m$.

Perform each division. See Example 1.

5. $\dfrac{x^2 - x - 6}{x - 3}$ **$x + 2$**

6. $\dfrac{m^2 - 2m - 24}{m - 6}$ **$m + 4$**

7. $\dfrac{2y^2 + 9y - 35}{y + 7}$ **$2y - 5$**

8. $\dfrac{2y^2 + 9y + 7}{y + 1}$ **$2y + 7$**

9. $\dfrac{p^2 + 2p + 20}{p + 6}$ **$p - 4 + \dfrac{44}{p + 6}$**

10. $\dfrac{x^2 + 11x + 16}{x + 8}$ **$x + 3 + \dfrac{-8}{x + 8}$**

11. $(r^2 - 8r + 15) \div (r - 3)$ **$r - 5$**

12. $(t^2 + 2t - 35) \div (t - 5)$ **$t + 7$**

13. $\dfrac{4a^2 - 22a + 32}{2a + 3}$ **$2a - 14 + \dfrac{74}{2a + 3}$**

14. $\dfrac{9w^2 + 6w + 10}{3w - 2}$ **$3w + 4 + \dfrac{18}{3w - 2}$**

15. $\dfrac{8x^3 - 10x^2 - x + 3}{2x + 1}$ **$4x^2 - 7x + 3$**

16. $\dfrac{12t^3 - 11t^2 + 9t + 18}{4t + 3}$ **$3t^2 - 5t + 6$**

Perform each division. See Examples 2–4.

17. $\dfrac{3y^3 + y^2 + 2}{y + 1}$ **$3y^2 - 2y + 2$**

18. $\dfrac{2r^3 - 6r - 36}{r - 3}$ **$2r^2 + 6r + 12$**

19. $\dfrac{2x^3 + x + 2}{x + 1}$ **$2x^2 - 2x + 3 + \dfrac{-1}{x + 1}$**

20. $\dfrac{3x^3 + x + 5}{x + 1}$ **$3x^2 - 3x + 4 + \dfrac{1}{x + 1}$**

21. $\dfrac{3k^3 - 4k^2 - 6k + 10}{k^2 - 2}$

$3k - 4 + \dfrac{2}{k^2 - 2}$

22. $\dfrac{5z^3 - z^2 + 10z + 2}{z^2 + 2}$

$5z - 1 + \dfrac{4}{z^2 + 2}$

23. $(x^4 - x^2 - 2) \div (x^2 - 2)$

$x^2 + 1$

24. $(r^4 + 2r^2 - 3) \div (r^2 - 1)$

$r^2 + 3$

25. $\dfrac{x^4 - 1}{x^2 - 1}$

$x^2 + 1$

26. $\dfrac{y^3 + 1}{y + 1}$

$y^2 - y + 1$

27. $\dfrac{6p^4 - 15p^3 + 14p^2 - 5p + 10}{3p^2 + 1}$

$2p^2 - 5p + 4 + \dfrac{6}{3p^2 + 1}$

28. $\dfrac{6r^4 - 10r^3 - r^2 + 15r - 8}{2r^2 - 3}$

$3r^2 - 5r + 4 + \dfrac{4}{2r^2 - 3}$

29. $\dfrac{2x^5 + x^4 + 11x^3 - 8x^2 - 13x + 7}{2x^2 + x - 1}$

$x^3 + 6x - 7$

30. $\dfrac{4t^5 - 11t^4 - 6t^3 + 5t^2 - t + 3}{4t^2 + t - 3}$

$t^3 - 3t^2 - 1$

31. $(10x^3 + 13x^2 + 4x + 1) \div (5x + 5)$

$2x^2 + \dfrac{3}{5}x + \dfrac{1}{5}$

32. $(6x^3 - 19x^2 - 19x - 4) \div (2x - 8)$

$3x^2 + \dfrac{5}{2}x + \dfrac{1}{2}$

Work each problem. See Example 5.

33. Give the length of the rectangle.

The area is $5x^3 + 7x^2 - 13x - 6$ sq. units.

$(x^2 + x - 3)$ units

34. Find the measure of the base of the parallelogram.

The area is $2x^3 + 2x^2 - 3x - 1$ sq. units.

$(2x^2 + 4x + 1)$ units

Relating Concepts (Exercises 35–38) For Individual or Group Work

We can find the value of a polynomial in x for a given value of x by substituting that number for x. Surprisingly, we can accomplish the same thing by division. For example, to find the value of $2x^2 - 4x + 3$ *for* $x = -3$, *we would divide* $2x^2 - 4x + 3$ *by* $x - (-3)$. *The remainder will give the value of the polynomial for* $x = -3$. ***Work Exercises 35–38 in order.***

35. Find the value of $2x^2 - 4x + 3$ for $x = -3$ by substitution. 33

36. Divide $2x^2 - 4x + 3$ by $x + 3$. Give the remainder. 33

37. Compare your answers to Exercises 35 and 36. What do you notice?

They are the same.

38. Choose another polynomial and evaluate it both ways for some value of the variable. Do the answers agree?

The answers should agree.

5.8 An Application of Exponents: Scientific Notation

OBJECTIVES

1. Express numbers in scientific notation.
2. Convert numbers in scientific notation to numbers without exponents.
3. Use scientific notation in calculations.

OBJECTIVE 1 Express numbers in scientific notation. Numbers occurring in science are often extremely large (such as the distance from Earth to the sun, 93,000,000 mi) or extremely small (the wavelength of yellow-green light, approximately 0.0000006 m). Because of the difficulty of working with many zeros, scientists often express such numbers with exponents, using a form called *scientific notation.*

Scientific Notation

A number is written in **scientific notation** when it is expressed in the form

$$a \times 10^n,$$

where $1 \leq |a| < 10$ and n is an integer.

In **scientific notation,** there is always one nonzero digit before the decimal point. This is shown in the following examples.

$3.19 \times 10^1 = 3.19 \times 10 = 31.9$ Decimal point moves 1 place to the right.

$3.19 \times 10^2 = 3.19 \times 100 = 319.$ Decimal point moves 2 places to the right.

$3.19 \times 10^3 = 3.19 \times 1000 = 3190.$ Decimal point moves 3 places to the right.

$3.19 \times 10^{-1} = 3.19 \times 0.1 = 0.319$ Decimal point moves 1 place to the left.

$3.19 \times 10^{-2} = 3.19 \times 0.01 = 0.0319$ Decimal point moves 2 places to the left.

$3.19 \times 10^{-3} = 3.19 \times 0.001 = 0.00319$ Decimal point moves 3 places to the left.

Note

In scientific notation, the times symbol, $\times$, is commonly used.

A number in scientific notation is always written with the decimal point after the first nonzero digit and then multiplied by the appropriate power of 10. For example, 56,200 is written 5.62×10^4, since

$$56{,}200 = 5.62 \times \mathbf{10{,}000} = 5.62 \times \mathbf{10^4}.$$

Other examples include

	42,000,000	written	4.2×10^7,
	0.000586	written	5.86×10^{-4},
and	2,000,000,000	written	2×10^9.

It is not necessary to write 2.0.

To write a number in scientific notation, follow the steps given on the next page. (For a negative number, follow these steps using the *absolute value* of the number; then make the result negative.)

Writing a Number in Scientific Notation

Step 1 Move the decimal point to the right of the first nonzero digit.

Step 2 Count the number of places you moved the decimal point.

Step 3 The number of places in Step 2 is the absolute value of the exponent on 10.

Step 4 The exponent on 10 is positive if the original number is greater than the number in Step 1; the exponent is negative if the original number is less than the number in Step 1. If the decimal point is not moved, the exponent is 0.

1 Write each number in scientific notation.

(a) 63,000

(b) 5,870,000

(c) 7.0065

(d) 0.0571

(e) -0.00062

EXAMPLE 1 Using Scientific Notation

Write each number in scientific notation.

(a) 93,000,000

Move the decimal point to follow the first nonzero digit (the 9). Count the number of places the decimal point was moved.

$$\underbrace{93{,}000{,}000.}_{\text{7 places}} \leftarrow \text{Decimal point}$$

The number will be written in scientific notation as 9.3×10^n. To find the value of n, first compare the original number, 93,000,000, with 9.3. Since 93,000,000 is *greater* than 9.3, we must multiply by a *positive* power of 10 so that the product 9.3×10^n will equal the larger number.

Since the decimal point was moved 7 places, and since n is positive,

$$93{,}000{,}000 = 9.3 \times 10^7.$$

(b) $63{,}200{,}000{,}000 = 6.\underbrace{3200000000}_{\text{10 places}} = 6.32 \times 10^{10}$

(c) $3.021 = 3.021 \times 10^0$

(d) 0.00462

Move the decimal point to the right of the first nonzero digit and count the number of places the decimal point was moved.

$$0.00462 \quad \text{3 places}$$

Since 0.00462 is *less* than 4.62, the exponent must be *negative*.

$$0.00462 = 4.62 \times 10^{-3}$$

(e) $-0.\underbrace{0000762}_{\text{5 places}} = -7.62 \times 10^{-5}$ Remember the negative sign.

◀ Work Problem **1** at the Side.

Note

To choose the exponent when you write a number in scientific notation, think: If the original number is "large," like 93,000,000, use a *positive* exponent on 10, since positive is greater than negative. However, if the original number is "small," like 0.00462, use a *negative* exponent on 10, since negative is less than positive.

ANSWERS

1. **(a)** 6.3×10^4 **(b)** 5.87×10^6 **(c)** 7.0065×10^0 **(d)** 5.71×10^{-2} **(e)** -6.2×10^{-4}

OBJECTIVE 2 Convert numbers in scientific notation to numbers without exponents. To convert a number written in scientific notation to a number without exponents, work in reverse. ***Multiplying a number by a positive power of 10 will make the number greater; multiplying by a negative power of 10 will make the number less.***

EXAMPLE 2 Writing Numbers without Exponents

Write each number without exponents.

(a) 6.2×10^3

Since the exponent is positive, make 6.2 greater by moving the decimal point 3 places to the right. It is necessary to attach two 0s.

$$6.2 \times 10^3 = 6.200 = 6200$$

(b) $4.283 \times 10^5 = 4.28300 = 428,300$ Move 5 places to the right; attach 0s as necessary.

(c) $-9.73 \times 10^{-2} = -09.73 = -0.0973$ Move 2 places to the left.

The exponent tells the number of places and the direction that the decimal point is moved.

Work Problem 2 at the Side. ▶

OBJECTIVE 3 Use scientific notation in calculations. The next example uses scientific notation with products and quotients.

EXAMPLE 3 Multiplying and Dividing with Scientific Notation

Perform each calculation. Write answers in scientific notation and also without exponents.

(a)

$$(7 \times 10^3)(5 \times 10^4)$$

$= (7 \times 5)(10^3 \times 10^4)$ Commutative and associative properties

$= 35 \times 10^7$ Multiply; product rule

Don't stop! This number is *not* in scientific notation, since 35 is not between 1 and 10.

$= (3.5 \times 10^1) \times 10^7$ Write 35 in scientific notation.

$= 3.5 \times (10^1 \times 10^7)$ Associative property

$= 3.5 \times 10^8$ Product rule

$= 350,000,000$ Write without exponents.

(b) $\dfrac{4 \times 10^{-5}}{2 \times 10^3} = \dfrac{4}{2} \times \dfrac{10^{-5}}{10^3} = 2 \times 10^{-8} = 0.00000002$

Work Problem 3 at the Side. ▶

Note

Multiplying or dividing numbers written in scientific notation may produce an answer in the form $a \times 10^0$. Since $10^0 = 1$, $a \times 10^0 = a$. For example,

$$(8 \times 10^{-4})(5 \times 10^4) = 40 \times 10^0 = 40. \qquad 10^0 = 1$$

Also, if $a = 1$, then $a \times 10^n = 10^n$. For example, we could write 1,000,000 as 10^6 instead of 1×10^6.

2 Write without exponents.

(a) 4.2×10^3

(b) 8.7×10^5

(c) 6.42×10^{-3}

3 Perform each calculation. Write answers in scientific notation and also without exponents.

(a) $(2.6 \times 10^4)(2 \times 10^{-6})$

(b) $(3 \times 10^5)(5 \times 10^{-2})$

(c) $\dfrac{4.8 \times 10^2}{2.4 \times 10^{-3}}$

Answers

2. (a) 4200 **(b)** 870,000 **(c)** 0.00642

3. (a) 5.2×10^{-2}; 0.052

(b) 1.5×10^4; 15,000

(c) 2×10^5; 200,000

4 The speed of light is approximately 3.0×10^5 km per sec. How far does light travel in 6.0×10^1 sec? (*Source: World Almanac and Book of Facts.*)

Calculator Tip Calculators usually have a key labeled EE or EXP for scientific notation. See your owner's manual for more information.

EXAMPLE 4 Using Scientific Notation to Solve an Application

A *nanometer* is a very small unit of measure that is equivalent to about 0.00000003937 in. About how much would 700,000 nanometers measure in inches? (*Source: World Almanac and Book of Facts.*)

Write each number in scientific notation, and then multiply.

$$700{,}000(0.00000003937)$$

$$= (7 \times 10^5)(3.937 \times 10^{-8}) \qquad \text{Write in scientific notation.}$$

$$= (7 \times 3.937)(10^5 \times 10^{-8}) \qquad \text{Properties of real numbers}$$

$$= \mathbf{27.559} \times 10^{-3} \qquad \text{Multiply; product rule}$$

Don't stop here.

$$= (\mathbf{2.7559 \times 10^1}) \times 10^{-3} \qquad \text{Write 27.559 in scientific notation.}$$

$$= 2.7559 \times 10^{-2} \qquad \text{Product rule}$$

$$= 0.027559 \qquad \text{Write without exponents.}$$

Thus, 700,000 nanometers would measure

$$2.7559 \times 10^{-2} \text{ in., or } 0.027559 \text{ in.}$$

◀ *Work Problem* **4** *at the Side.*

5 If the speed of light is approximately 3.0×10^5 km per sec, how many seconds does it take light to travel approximately 1.5×10^8 km from the sun to Earth? (*Source: World Almanac and Book of Facts.*)

EXAMPLE 5 Using Scientific Notation to Solve an Application

In 2003, the national debt was $\$3.9136 \times 10^{12}$ (which is more than \$3 trillion). The population of the United States was approximately 290 million that year. About how much would each person have had to contribute in order to pay off the national debt? (*Source:* U.S. Office of Management and Budget; U.S. Census Bureau.)

Write the population in scientific notation. Then divide to obtain the per person contribution.

$$\frac{3.9136 \times 10^{12}}{\mathbf{290{,}000{,}000}}$$

$$= \frac{3.9136 \times 10^{12}}{\mathbf{2.9 \times 10^8}} \qquad \text{Write 290 million in scientific notation.}$$

$$= \frac{3.9136}{2.9} \times 10^4 \qquad \text{Quotient rule}$$

$$\approx 1.3495 \times 10^4 \qquad \text{Divide; round to 4 decimal places.}$$

$$\approx 13{,}495 \qquad \text{Write without exponents.}$$

Each person would have to pay about \$13,495.

◀ *Work Problem* **5** *at the Side.*

ANSWERS

4. 1.8×10^7 km, or 18,000,000 km
5. 5×10^2 sec, or 500 sec

5.8 Exercises

FOR EXTRA HELP MyMathLab Math XL PRACTICE WATCH DOWNLOAD READ REVIEW

Write the numbers (other than dates) mentioned in the following statements in scientific notation.

1. NASA has budgeted \$6,130,900,000 for 2003 and \$5,868,900,000 for 2004 for the international space station. (*Source:* U.S. National Aeronautics and Space Administration.)

6.1309×10^9; 5.8689×10^9

2. The mass of Pluto is 0.0021 times that of Earth; the mass of Jupiter is 317.83 times that of Earth. (*Source: World Almanac and Book of Facts.*)

2.1×10^{-3}; 3.1783×10^2

Determine whether or not the given number is written in scientific notation as defined in Objective 1. If it is not, write it as such.

3. 4.56×10^3
in scientific notation

4. 7.34×10^5
in scientific notation

5. 5,600,000
not in scientific notation; 5.6×10^6

6. 34,000
not in scientific notation; 3.4×10^4

7. 0.004
not in scientific notation; 4×10^{-3}

8. 0.0007
not in scientific notation; 7×10^{-4}

9. 0.8×10^2
not in scientific notation; 8×10^1

10. 0.9×10^3
not in scientific notation; 9×10^2

11. Explain in your own words what it means for a number to be written in scientific notation.

A number is written in scientific notation if it is the product of a number whose absolute value is between 1 and 10 (inclusive of 1) and a power of 10.

12. Explain how to multiply a number by a positive power of ten. Then explain how to multiply a number by a negative power of ten.

To multiply by a positive power of 10, move the decimal point to the right as many places as the exponent on 10. With a negative power of 10, move the decimal point to the left as many places as the absolute value of the exponent on 10.

Write each number in scientific notation. See Example 1.

13. 5,876,000,000
5.876×10^9

14. 9,994,000,000
9.994×10^9

15. 82,350
8.235×10^4

16. 78,330
7.833×10^4

17. 0.000007
7×10^{-6}

18. 0.0000004
4×10^{-7}

19. -0.00203
-2.03×10^{-3}

20. -0.0000578
-5.78×10^{-5}

Write each number without exponents. See Example 2.

21. 7.5×10^5
750,000

22. 8.8×10^6
8,800,000

23. 5.677×10^{12}
5,677,000,000,000

24. 8.766×10^9
8,766,000,000

25. 1×10^{12}
1,000,000,000,000

26. 1×10^7
10,000,000

27. -6.21×10^0
-6.21

28. -8.56×10^0
-8.56

29. 7.8×10^{-4}
0.00078

30. 8.9×10^{-5}
0.000089

31. 5.134×10^{-9}
0.000000005134

32. 7.123×10^{-10}
0.0000000007123

Perform the indicated operations. Write the answers in scientific notation and then without exponents. See Example 3.

33. $(2 \times 10^8)(3 \times 10^3)$
6×10^{11}; 600,000,000,000

34. $(3 \times 10^7)(3 \times 10^3)$
9×10^{10}; 90,000,000,000

35. $(5 \times 10^4)(3 \times 10^2)$
1.5×10^7; 15,000,000

36. $(8 \times 10^5)(2 \times 10^3)$
1.6×10^9; 1,600,000,000

37. $(4 \times 10^{-6})(2 \times 10^3)$
8×10^{-3}; 0.008

38. $(3 \times 10^{-7})(2 \times 10^2)$
6×10^{-5}; 0.00006

39. $(6 \times 10^3)(4 \times 10^{-2})$
2.4×10^2; 240

40. $(7 \times 10^5)(3 \times 10^{-4})$
2.1×10^2; 210

41. $(9 \times 10^4)(7 \times 10^{-7})$
6.3×10^{-2}; 0.063

42. $(6 \times 10^4)(8 \times 10^{-8})$
4.8×10^{-3}; 0.0048

43. $(3.15 \times 10^{-4})(2.04 \times 10^8)$
6.426×10^4; 64,260

44. $(4.92 \times 10^{-3})(2.25 \times 10^7)$
1.107×10^5; 110,700

45. $\dfrac{9 \times 10^{-5}}{3 \times 10^{-1}}$
3×10^{-4}; 0.0003

46. $\dfrac{12 \times 10^{-4}}{4 \times 10^{-3}}$
3×10^{-1}; 0.3

47. $\dfrac{8 \times 10^3}{2 \times 10^2}$
4×10^1; 40

48. $\dfrac{15 \times 10^4}{3 \times 10^3}$
5×10^1; 50

49. $\dfrac{2.6 \times 10^{-3}}{2 \times 10^2}$
1.3×10^{-5}; 0.000013

50. $\dfrac{9.5 \times 10^{-1}}{5 \times 10^3}$
1.9×10^{-4}; 0.00019

51. $\dfrac{4 \times 10^5}{8 \times 10^2}$
5×10^2; 500

52. $\dfrac{3 \times 10^9}{6 \times 10^5}$
5×10^3; 5000

53. $\dfrac{2.6 \times 10^{-3} \times 7.0 \times 10^{-1}}{2 \times 10^2 \times 3.5 \times 10^{-3}}$
2.6×10^{-3}; 0.0026

54. $\dfrac{9.5 \times 10^{-1} \times 2.4 \times 10^4}{5 \times 10^3 \times 1.2 \times 10^{-2}}$
3.8×10^2; 380

55. $\dfrac{(1.65 \times 10^8)(5.24 \times 10^{-2})}{(6 \times 10^4)(2 \times 10^7)}$
7.205×10^{-6}; 0.000007205

Work each problem. In Exercises 58–60, give answers without exponents. See Examples 4 and 5.

56. Pollux, one of the brightest stars in the night sky, is 33.7 light-years from Earth. If one light-year is about 6,000,000,000,000 mi (that is, 6 trillion mi), about how many miles is Pollux from Earth? (*Source: World Almanac and Book of Facts.*)
2.022×10^{14} mi

57. In March 2006, astronomers using the Spitzer Space Telescope discovered a twisted double-helix nebula, a conglomeration of dust and gas stretching across the center of the Milky Way galaxy. This nebula is 25,000 light-years from Earth. If one light-year is about 6,000,000,000,000 mi, about how many miles is the twisted double-helix nebula from Earth? (*Source*: http://articles.news.aol.com)
1.5×10^{17} mi

58. In 2003, the U.S. government collected about $6730 per person in taxes. If the population at that time was 290,000,000, how much did the government collect in taxes for 2003? (*Source*: U.S. Internal Revenue Service.)
$1,951,700,000,000

59. In 2000, the population of the United States was about 281.4 million. To the nearest dollar, calculate how much each person in the United States would have had to contribute in order to make one lucky person a trillionaire (that is, to give that person $1,000,000,000,000). (*Source*: U.S. Census Bureau.)
$3554

60. In 2006, Congress raised the government's debt limit to $\$9 \times 10^{12}$. When this national debt limit is reached, about how much is it for every man, woman, and child in the country? Use 300 million as the population of the United States. (*Source: The Gazette*, Cedar Rapids, Iowa, March 17, 2006.)
$30,000

Chapter 5  Summary

▸ Key Terms

Section	Term	Definition
5.1	**term**	A term is a number, a variable, or a product or quotient of a number and one or more variables raised to powers.
	like terms	Terms with exactly the same variables (including the same exponents) are called like terms.
	polynomial	A polynomial is a term or the sum of a finite number of terms with whole number exponents.
	descending powers	A polynomial in x is written in descending powers if the exponents on x in its terms are in decreasing order.
	degree of a term	The degree of a term is the sum of the exponents on the variables.
	degree of a polynomial	The degree of a polynomial is the greatest degree of any term of the polynomial.
	monomial	A monomial is a polynomial with exactly one term.
	binomial	A binomial is a polynomial with exactly two terms.
	trinomial	A trinomial is a polynomial with exactly three terms.
5.2	**exponential expression**	A number written with an exponent is an exponential expression. 3^4 ← Exponent, ↑ Base } Exponential expression
5.3	**FOIL**	FOIL is a shortcut method for finding the product of two binomials. The letters of the word **FOIL** originate as follows: Multiply the **F**irst terms, multiply the **O**uter terms (to get the outer product), multiply the **I**nner terms (to get the inner product), and multiply the **L**ast terms.
	outer product	The outer product of $(2x + 3)(x - 5)$ is $2x(-5)$.
	inner product	The inner product of $(2x + 3)(x - 5)$ is $3x$.
5.4	**conjugate**	The conjugate of $a + b$ is $a - b$.
5.8	**scientific notation**	A number written as $a \times 10^n$, where $1 \leq \lvert a \rvert < 10$ and n is an integer, is in scientific notation.

▸ New Symbols

x^{-n} x to the negative n power

▸ Test Your Word Power

See how well you have learned the vocabulary in this chapter. Answers, with examples, follow the Quick Review.

1. A **polynomial** is an algebraic expression made up of
   A. a term or a finite product of terms with positive coefficients and exponents
   B. a term or a finite sum of terms with real coefficients and whole number exponents
   C. the product of two or more terms with positive exponents
   D. the sum of two or more terms with whole number coefficients and exponents.

2. The **degree of a term** is
   A. the number of variables in the term
   B. the product of the exponents on the variables
   C. the least exponent on the variables
   D. the sum of the exponents on the variables.

3. A **trinomial** is a polynomial with
   A. only one term
   B. exactly two terms
   C. exactly three terms
   D. more than three terms.

4. A **binomial** is a polynomial with
   A. only one term
   B. exactly two terms
   C. exactly three terms
   D. more than three terms.

5. A **monomial** is a polynomial with
   A. only one term
   B. exactly two terms
   C. exactly three terms
   D. more than three terms.

6. **FOIL** is a method for
   A. adding two binomials
   B. adding two trinomials
   C. multiplying two binomials
   D. multiplying two trinomials.

▶ Quick Review

Concepts	Examples

5.1 Adding and Subtracting Polynomials

Addition
Add like terms.

Add.
$$\begin{array}{r} 2x^2 + 5x - 3 \\ 5x^2 - 2x + 7 \\ \hline 7x^2 + 3x + 4 \end{array}$$

Subtraction
Change the signs of the terms in the second polynomial and add to the first polynomial.

Subtract. $(2x^2 + 5x - 3) - (5x^2 - 2x + 7)$
$$= (2x^2 + 5x - 3) + (-5x^2 + 2x - 7)$$
$$= -3x^2 + 7x - 10$$

5.2 The Product Rule and Power Rules for Exponents

For any integers m and n:

Product rule $a^m \cdot a^n = a^{m+n}$

Power rules (a) $(a^m)^n = a^{mn}$

(b) $(ab)^m = a^m b^m$

(c) $\left(\frac{a}{b}\right)^m = \frac{a^m}{b^m}$ $(b \neq 0)$.

Simplify.
$$2^4 \cdot 2^5 = 2^{4+5} = 2^9$$
$$(3^4)^2 = 3^{4 \cdot 2} = 3^8$$
$$(6a)^5 = 6^5 a^5$$
$$\left(\frac{2}{3}\right)^4 = \frac{2^4}{3^4}$$

5.3 Multiplying Polynomials

Multiply each term of the first polynomial by each term of the second polynomial. Then add like terms.

Multiply.
$$\begin{array}{rrrrr} & 3x^3 & -\ 4x^2 & +\ 2x & -\ 7 \\ & & & 4x & +\ 3 \\ \hline & 9x^3 & -\ 12x^2 & +\ 6x & -\ 21 \\ 12x^4 & -\ 16x^3 & +\ 8x^2 & -\ 28x & \\ \hline 12x^4 & -\ 7x^3 & -\ 4x^2 & -\ 22x & -\ 21 \end{array}$$

FOIL Method

Step 1 Multiply the two **F**irst terms to get the first term of the answer.

Step 2 Find the **O**uter product and the **I**nner product and mentally add them, when possible, to get the middle term of the answer.

Step 3 Multiply the two **L**ast terms to get the last term of the answer.

Add the terms found in Steps 1–3.

Multiply $(2x + 3)(5x - 4)$.
$$2x(5x) = 10x^2$$
$$2x(-4) + 3(5x) = 7x$$
$$3(-4) = -12$$
The product is $10x^2 + 7x - 12$.

5.4 Special Products

Square of a Binomial
$$(a + b)^2 = a^2 + 2ab + b^2$$
$$(a - b)^2 = a^2 - 2ab + b^2$$

Product of the Sum and Difference of Two Terms
$$(a + b)(a - b) = a^2 - b^2$$

Multiply.

$(3x + 1)^2$
$$= (3x)^2 + 2(3x)(1) + 1^2$$
$$= 9x^2 + 6x + 1$$

$(2m - 5n)^2$
$$= (2m)^2 - 2(2m)(5n) + (5n)^2$$
$$= 4m^2 - 20mn + 25n^2$$

$(4a + 3)(4a - 3)$
$$= (4a)^2 - 3^2$$
$$= 16a^2 - 9$$

Concepts	Examples

5.5 Integer Exponents and the Quotient Rule

If $a, b \neq 0$, for integers m and n:

Zero exponent $a^0 = 1$

Negative exponent $a^{-n} = \frac{1}{a^n}$

Quotient rule $\frac{a^m}{a^n} = a^{m-n}$

Negative-to-positive rules $\frac{a^{-m}}{b^{-n}} = \frac{b^n}{a^m}$ $\left(\frac{a}{b}\right)^{-m} = \left(\frac{b}{a}\right)^m$.

Simplify.

$15^0 = 1$

$5^{-2} = \frac{1}{5^2} = \frac{1}{25}$

$\frac{4^8}{4^3} = 4^{8-3} = 4^5$

$\frac{6^{-2}}{7^{-3}} = \frac{7^3}{6^2}$ $\left(\frac{5}{3}\right)^{-4} = \left(\frac{3}{5}\right)^4$

5.6 Dividing a Polynomial by a Monomial

Divide each term of the polynomial by the monomial:

$$\frac{a+b}{c} = \frac{a}{c} + \frac{b}{c}.$$

Divide. $\frac{4x^3 - 2x^2 + 6x - 8}{2x}$

$$= \frac{4x^3}{2x} - \frac{2x^2}{2x} + \frac{6x}{2x} - \frac{8}{2x}$$

$$= 2x^2 - x + 3 - \frac{4}{x}$$

5.7 Dividing a Polynomial by a Polynomial

Use "long division."

$$\begin{array}{r} 2x - 5 + \frac{-1}{3x+4} \\ 3x+4\,\overline{)\,6x^2 - 7x - 21} \\ \underline{6x^2 + 8x} \\ -15x - 21 \\ \underline{-15x - 20} \\ -1 \end{array}$$

Divide. (The final -1 is the ← Remainder.)

5.8 An Application of Exponents: Scientific Notation

To write a number in scientific notation (as $a \times 10^n$, where $1 \leq |a| < 10$), move the decimal point to the right of the first nonzero digit. If the decimal point is moved n places, and this makes the number smaller, n is positive; if it makes the number larger, n is negative. If the decimal point is not moved, n is 0.

Write in scientific notation.

$247 = 2.47 \times 10^2$

$0.0051 = 5.1 \times 10^{-3}$

Write without exponents.

$3.25 \times 10^5 = 325{,}000$

$8.44 \times 10^{-6} = 0.00000844$

ANSWERS TO TEST YOUR WORD POWER

1. B; *Example:* $5x^3 + 2x^2 - 7$
2. D; *Examples:* The term 6 has degree 0, $3x$ has degree 1, $-2x^8$ has degree 8, and $5x^2y^4$ has degree 6.
3. C; *Example:* $2a^2 - 3ab + b^2$
4. B; *Example:* $3t^3 + 5t$
5. A; *Examples:* -5 and $4xy^5$
6. C; *Example:* $(m+4)(m-3)$

$$\overset{\textbf{F}}{=m(m)} \overset{\textbf{O}}{-3m} \overset{\textbf{I}}{+4m} \overset{\textbf{L}}{+4(-3)}$$

$$= m^2 + m - 12$$

Math in the Media

FLOODS, HURRICANES, AND EARTHQUAKES, OH MY!

In recent years, the number of natural disasters seems to be on the increase. Charles F. Richter devised a scale in 1935 to compare the intensities, or relative power, of earthquakes. The **intensity** of an earthquake (often mentioned in media reports) is measured relative to the intensity of a standard **zero-level** earthquake of intensity I_0. The relationship is equivalent to $I = I_0 \times 10^R$, where R is the **Richter scale** measure. For example, if an earthquake has magnitude 5.0 on the Richter scale, then its intensity is calculated as $I = I_0 \times 10^{5.0} = I_0 \times 100{,}000$, which is 100,000 times as intense as a zero-level earthquake.

To compare an earthquake that measures 8.0 on the Richter scale to one that measures 5.0, find the ratio of the intensities:

$$\frac{\text{intensity } 8.0}{\text{intensity } 5.0} = \frac{I_0 \times 10^{8.0}}{I_0 \times 10^{5.0}} = \frac{10^8}{10^5} = 10^{8-5} = 10^3 = 1000.$$

Therefore, an earthquake that measures 8.0 is 1000 times as intense as one that measures 5.0.

The Gazette TUESDAY April 19, 2008

Quake rattles Iowans

Chances of 'big one' happening here remote UNI professor says

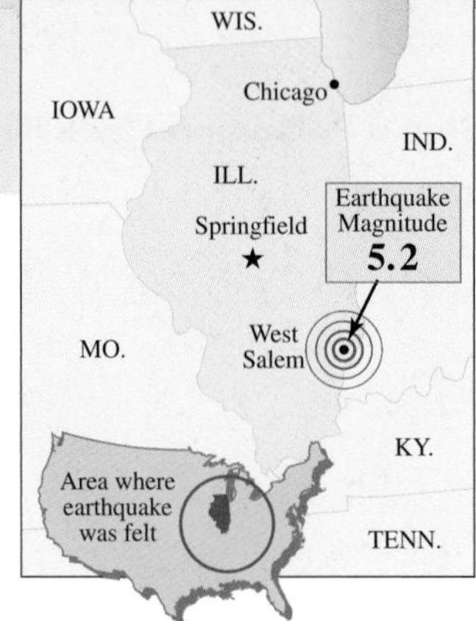

Source: ESRI; USGS.

The table gives Richter scale measurements for several earthquakes.

Year	Earthquake	Richter Scale Measurement
2008	West Salem, IL	5.2
2005	Northern Sumatra, Indonesia	8.6
2004	West coast of Northern Sumatra	9.1
2003	Southeastern Iran	6.6
1998	Balleny Islands region	8.1
1906	San Francisco, CA	7.7

Source: U.S. Geological Survey.

1. Compare the intensity of the 2004 west coast of northern Sumatra earthquake to that of the 1998 Balleny Islands region earthquake.

   The 2004 quake was 10 times as intense as the 1998 quake.

2. Compare the intensity of the 2005 northern Sumatra, Indonesia earthquake to that of the 2003 southeastern Iran earthquake.

   The 2005 quake was 100 times as intense as the 2003 quake.

3. Compare the intensity of the 1906 San Francisco earthquake, the most powerful to strike the United States, to that of the 2008 West Salem earthquake. (*Hint*: Use the exponential key of a scientific calculator to compute the required power of 10.)

   The 1906 quake was 316 times as intense as the 2008 quake.

4. Suppose an earthquake measures a value of x on the Richter scale. How would the intensity of a second earthquake compare if its Richter scale measure is $x + 4.0$? How would it compare if its Richter scale measure is $x - 1.0$?

   $10^4 = 10{,}000$ times as intense; $10^{-1} = \frac{1}{10}$ times as intense.

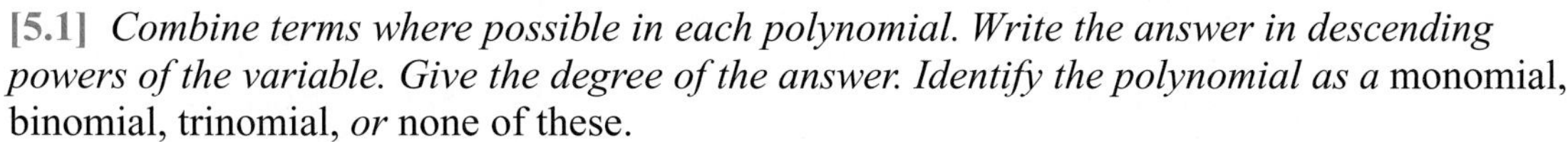

Chapter 5 Review Exercises

[5.1] *Combine terms where possible in each polynomial. Write the answer in descending powers of the variable. Give the degree of the answer. Identify the polynomial as a* monomial, binomial, trinomial, *or* none of these.

1. $9m^2 + 11m^2 + 2m^2$

$22m^2$; degree 2; monomial

2. $-4p + p^3 - p^2 + 8p + 2$

$p^3 - p^2 + 4p + 2$; degree 3; none of these

3. $12a^5 - 9a^4 + 8a^3 + 2a^2 - a + 3$

already in descending powers; degree 5; none of these

4. $-7y^5 - 8y^4 - y^5 + y^4 + 9y$

$-8y^5 - 7y^4 + 9y$; degree 5; trinomial

Add or subtract as indicated.

5. Add.

$$\begin{array}{r} -2a^3 + 5a^2 \\ \underline{-3a^3 - a^2} \\ -5a^3 + 4a^2 \end{array}$$

6. Add.

$$\begin{array}{r} 4r^3 - 8r^2 + 6r \\ \underline{-2r^3 + 5r^2 + 3r} \\ 2r^3 - 3r^2 + 9r \end{array}$$

7. Subtract.

$$\begin{array}{r} 6y^2 - 8y + 2 \\ \underline{-5y^2 + 2y - 7} \\ 11y^2 - 10y + 9 \end{array}$$

8. Subtract.

$$\begin{array}{r} -12k^4 - 8k^2 + 7k - 5 \\ \underline{k^4 + 7k^2 + 11k + 1} \\ -13k^4 - 15k^2 - 4k - 6 \end{array}$$

9. $(2m^3 - 8m^2 + 4) + (8m^3 + 2m^2 - 7)$

$10m^3 - 6m^2 - 3$

10. $(-5y^2 + 3y + 11) + (4y^2 - 7y + 15)$

$-y^2 - 4y + 26$

11. $(6p^2 - p - 8) - (-4p^2 + 2p + 3)$

$10p^2 - 3p - 11$

12. $(12r^4 - 7r^3 + 2r^2) - (5r^4 - 3r^3 + 2r^2 + 1)$

$7r^4 - 4r^3 - 1$

[5.2] *Simplify each expression.*

13. $4^3 \cdot 4^8$

4^{11}

14. $(-5)^6(-5)^5$

-5^{11}

15. $(-8x^4)(9x^3)$

$-72x^7$

16. $(2x^2)(5x^3)(x^9)$

$10x^{14}$

17. $(19x)^5$

19^5x^5

18. $(-4y)^7$

-4^7y^7

19. $5(pt)^4$

$5p^4t^4$

20. $\left(\frac{7}{5}\right)^6$

$\frac{7^6}{5^6}$

21. $(3x^2y^3)^3$

$27x^6y^9$

22. $(t^4)^8(t^2)^5$

t^{42}

23. $(6x^2z^4)^2(x^3yz^2)^4$

$36x^{16}y^4z^{16}$

24. $\left(\frac{2m^3n}{p^2}\right)^3$

$\frac{8m^9n^3}{p^6}$

25. Find a polynomial that represents the volume of the figure. (If necessary, refer to the formulas on the inside covers.)

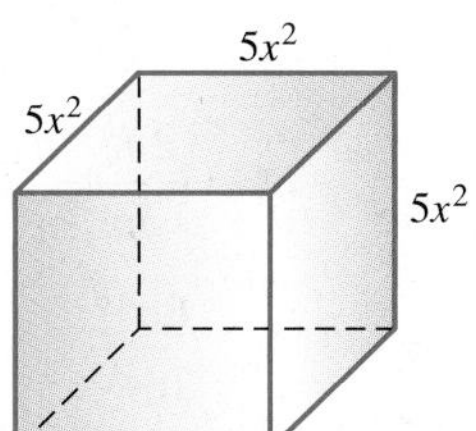

$125x^6$

26. Explain why the product rule for exponents does not apply to the expression $7^2 + 7^4$.

The product rule for exponents does not apply here because we want the sum of 7^2 and 7^4, not their product.

[5.3] *Find each product.*

27. $5x(2x + 14)$
$10x^2 + 70x$

28. $-3p^3(2p^2 - 5p)$
$-6p^5 + 15p^4$

29. $(3r - 2)(2r^2 + 4r - 3)$
$6r^3 + 8r^2 - 17r + 6$

30. $(2y + 3)(4y^2 - 6y + 9)$
$8y^3 + 27$

31. $(5p^2 + 3p)(p^3 - p^2 + 5)$
$5p^5 - 2p^4 - 3p^3 + 25p^2 + 15p$

32. $(x + 6)(x - 3)$
$x^2 + 3x - 18$

33. $(3k - 6)(2k + 1)$
$6k^2 - 9k - 6$

34. $(6p - 3q)(2p - 7q)$
$12p^2 - 48pq + 21q^2$

35. $(m^2 + m - 9)(2m^2 + 3m - 1)$
$2m^4 + 5m^3 - 16m^2 - 28m + 9$

[5.4] *Find each product.*

36. $(a + 4)^2$
$a^2 + 8a + 16$

37. $(3p - 2)^2$
$9p^2 - 12p + 4$

38. $(2r + 5s)^2$
$4r^2 + 20rs + 25s^2$

39. $(r + 2)^3$
$r^3 + 6r^2 + 12r + 8$

40. $(2x - 1)^3$
$8x^3 - 12x^2 + 6x - 1$

41. $(2z + 7)(2z - 7)$
$4z^2 - 49$

42. $(6m - 5)(6m + 5)$
$36m^2 - 25$

43. $(5a + 6b)(5a - 6b)$
$25a^2 - 36b^2$

44. $(2x^2 + 5)(2x^2 - 5)$
$4x^4 - 25$

45. The square of a binomial leads to a polynomial with how many terms? The product of the sum and difference of two terms leads to a polynomial with how many terms?
three; two

46. Explain why $(a + b)^2$ is not equal to $a^2 + b^2$.
$(a + b)^2 = (a + b)(a + b) = a^2 + 2ab + b^2$. The term $2ab$ is not in $a^2 + b^2$.

[5.5] *Evaluate each expression.*

47. $5^0 + 8^0$ 2

48. 2^{-5} $\frac{1}{32}$

49. $\left(\frac{6}{5}\right)^{-2}$ $\frac{25}{36}$

50. $4^{-2} - 4^{-1}$ $-\frac{3}{16}$

Simplify each expression. Assume that all variables represent nonzero numbers.

51. $\frac{6^{-3}}{6^{-5}}$ 36

52. $\frac{x^{-7}}{x^{-9}}$ x^2

53. $\frac{p^{-8}}{p^4}$ $\frac{1}{p^{12}}$

54. $\frac{r^{-2}}{r^{-6}}$ r^4

55. $(2^4)^2$ 2^8

56. $(9^3)^{-2}$ $\frac{1}{9^6}$

57. $(5^{-2})^{-4}$ 5^8

58. $(8^{-3})^4$ $\frac{1}{8^{12}}$

59. $\frac{(m^2)^3}{(m^4)^2}$ $\frac{1}{m^2}$

60. $\frac{y^4 \cdot y^{-2}}{y^{-5}}$ y^7

61. $\frac{r^9 \cdot r^{-5}}{r^{-2} \cdot r^{-7}}$ r^{13}

62. $(-5m^3)^2$ $25m^6$

63. $(2y^{-4})^{-3}$
$\frac{y^{12}}{8}$

64. $\frac{ab^{-3}}{a^4b^2}$
$\frac{1}{a^3b^5}$

65. $\frac{(6r^{-1})^2 \cdot (2r^{-4})}{r^{-5}(r^2)^{-3}}$
$72r^5$

66. $\frac{(2m^{-5}n^2)^3(3m^2)^{-1}}{m^{-2}n^{-4}(m^{-1})^2}$
$\frac{8n^{10}}{3m^{13}}$

[5.6] *Perform each division.*

67. $\dfrac{-15y^4}{-9y^2}$

$\dfrac{5y^2}{3}$

68. $\dfrac{-12x^3y^2}{6xy}$

$-2x^2y$

69. $\dfrac{6y^4 - 12y^2 + 18y}{-6y}$

$-y^3 + 2y - 3$

70. $\dfrac{2p^3 - 6p^2 + 5p}{2p^2}$

$p - 3 + \dfrac{5}{2p}$

71. $(5x^{13} - 10x^{12} + 20x^7 - 35x^5) \div (-5x^4)$

$-x^9 + 2x^8 - 4x^3 + 7x$

72. $(-10m^4n^2 + 5m^3n^3 + 6m^2n^4) \div (5m^2n)$

$-2m^2n + mn^2 + \dfrac{6n^3}{5}$

[5.7] *Perform each division.*

73. $(2r^2 + 3r - 14) \div (r - 2)$

$2r + 7$

74. $\dfrac{12m^2 - 11m - 10}{3m - 5}$

$4m + 3 + \dfrac{5}{3m - 5}$

75. $\dfrac{10a^3 + 5a^2 - 14a + 9}{5a^2 - 3}$

$2a + 1 + \dfrac{-8a + 12}{5a^2 - 3}$

76. $\dfrac{2k^4 + 4k^3 + 9k^2 - 8}{2k^2 + 1}$

$k^2 + 2k + 4 + \dfrac{-2k - 12}{2k^2 + 1}$

[5.8] *Write each number in scientific notation.*

77. 48,000,000

4.8×10^7

78. 28,988,000,000

2.8988×10^{10}

79. 0.000065

6.5×10^{-5}

80. 0.0000000824

8.24×10^{-8}

Write each number without exponents.

81. 2.4×10^4

24,000

82. 7.83×10^7

78,300,000

83. 8.97×10^{-7}

0.000000897

84. 9.95×10^{-12}

0.00000000000995

Perform the indicated operations. Write the answers in scientific notation and then without exponents.

85. $(2 \times 10^{-3})(4 \times 10^5)$

8×10^2; 800

86. $\dfrac{8 \times 10^4}{2 \times 10^{-2}}$

4×10^6; 4,000,000

87. $\dfrac{12 \times 10^{-5} \times 5 \times 10^4}{4 \times 10^3 \times 6 \times 10^{-2}}$

2.5×10^{-2}; 0.025

88. $\dfrac{2.5 \times 10^5 \times 4.8 \times 10^{-4}}{7.5 \times 10^8 \times 1.6 \times 10^{-5}}$

1×10^{-2}; 0.01

89. A computer can perform 466,000,000 calculations per second. How many calculations can it perform per minute? Per hour?

2.796×10^{10} calculations;

1.6776×10^{12} calculations

90. There are 1×10^9 Social Security numbers. The population of the United States is about 3×10^8. How many Social Security numbers are available for each person? (*Source*: U.S. Census Bureau.)

about 3.3

Mixed Review Exercises

Perform each indicated operation. Assume that all variables represent nonzero real numbers.

91. $19^0 - 3^0$ **0**

92. $(3p)^4(3p^{-7})$ $\frac{243}{p^3}$

93. 7^{-2} $\frac{1}{49}$

94. $(-7 + 2k)^2$

$49 - 28k + 4k^2$

95. $\dfrac{2y^3 + 17y^2 + 37y + 7}{2y + 7}$

$y^2 + 5y + 1$

96. $\left(\dfrac{6r^2s}{5}\right)^4$ $\frac{1296r^8s^4}{625}$

97. $-m^5(8m^2 + 10m + 6)$

$-8m^7 - 10m^6 - 6m^5$

98. $\left(\dfrac{1}{2}\right)^{-5}$

32

99. $(25x^2y^3 - 8xy^2 + 15x^3y) \div (5x)$

$5xy^3 - \frac{8y^2}{5} + 3x^2y$

100. $(6r^{-2})^{-1}$

$\frac{r^2}{6}$

101. $(2x + y)^3$

$8x^3 + 12x^2y + 6xy^2 + y^3$

102. $2^{-1} + 4^{-1}$

$\frac{3}{4}$

103. $(a + 2)(a^2 - 4a + 1)$

$a^3 - 2a^2 - 7a + 2$

104. $(5y^3 - 8y^2 + 7) - (-3y^3 + y^2 + 2)$

$8y^3 - 9y^2 + 5$

105. $(2r + 5)(5r - 2)$

$10r^2 + 21r - 10$

106. $(12a + 1)(12a - 1)$

$144a^2 - 1$

107. What polynomial represents the area of this rectangle?

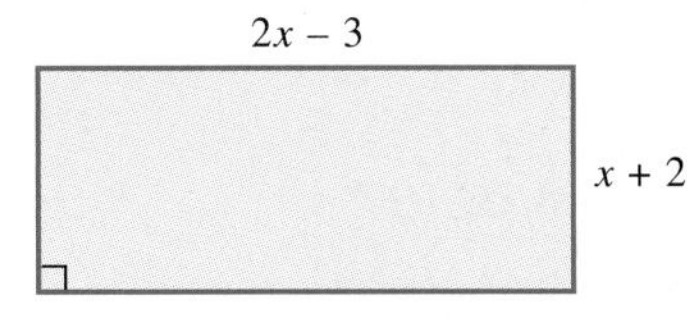

$2x^2 + x - 6$

108. What polynomial represents the perimeter of this square? The area?

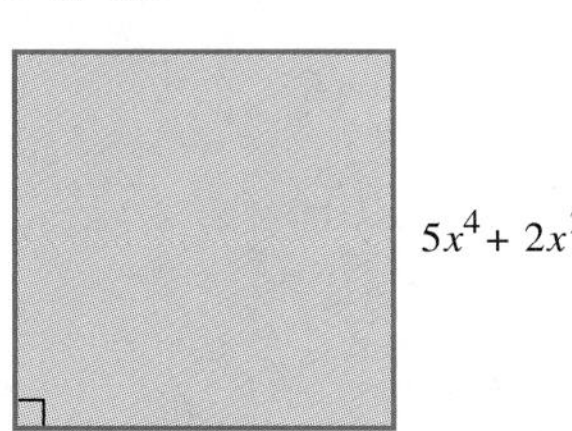

$20x^4 + 8x^2$; $25x^8 + 20x^6 + 4x^4$

Chapter 5 Test

Use the Chapter Test Prep Video CD to see fully worked-out solutions to any of the exercises you want to review.

Perform the indicated operations.

1. $(5t^4 - 3t^2 + 7t + 3) - (t^4 - t^3 + 3t^2 + 8t + 3)$

2. $(2y^2 - 8y + 8) + (-3y^2 + 2y + 3) - (y^2 + 3y - 6)$

3. Subtract.

$$\begin{array}{r} 9t^3 - 4t^2 + 2t + 2 \\ \underline{9t^3 + 8t^2 - 3t - 6} \end{array}$$

Simplify.

4. $(-2)^3(-2)^2$

5. $\left(\frac{6}{m^2}\right)^3, \quad m \neq 0$

6. $3x^2(-9x^3 + 6x^2 - 2x + 1)$

7. $(2r - 3)(r^2 + 2r - 5)$

8. $(t - 8)(t + 3)$

9. $(4x + 3y)(2x - y)$

10. $(5x - 2y)^2$

11. $(10v + 3w)(10v - 3w)$

12. $(x + 1)^3$

13. What polynomial represents the perimeter of this square? The area?

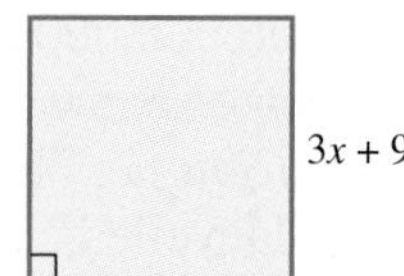

Evaluate each expression.

14. 5^{-4}

15. $(-3)^0 + 4^0$

16. $4^{-1} + 3^{-1}$

1. $4t^4 + t^3 - 6t^2 - t$

2. $-2y^2 - 9y + 17$

3. $-12t^2 + 5t + 8$

4. -32

5. $\frac{216}{m^6}$

6. $-27x^5 + 18x^4 - 6x^3 + 3x^2$

7. $2r^3 + r^2 - 16r + 15$

8. $t^2 - 5t - 24$

9. $8x^2 + 2xy - 3y^2$

10. $25x^2 - 20xy + 4y^2$

11. $100v^2 - 9w^2$

12. $x^3 + 3x^2 + 3x + 1$

13. $12x + 36;\ 9x^2 + 54x + 81$

14. $\frac{1}{625}$

15. 2

16. $\frac{7}{12}$

Perform the indicated operations. In Exercises 17 and 18, write each answer using only positive exponents. Assume that variables represent nonzero numbers.

17. $\frac{8^{-1} \cdot 8^4}{8^{-2}}$

17. 8^5

18. $\frac{(x^{-3})^{-2}(x^{-1}y)^2}{(xy^{-2})^2}$

18. x^2y^6

19. $\frac{8y^3 - 6y^2 + 4y + 10}{2y}$

19. $4y^2 - 3y + 2 + \frac{5}{y}$

20. $(-9x^2y^3 + 6x^4y^3 + 12xy^3) \div (3xy)$

20. $-3xy^2 + 2x^3y^2 + 4y^2$

21. $\frac{2x^2 + x - 36}{x - 4}$

21. $2x + 9$

22. $(3x^3 - x + 4) \div (x - 2)$

22. $3x^2 + 6x + 11 + \frac{26}{x - 2}$

Write each number in scientific notation.

23. (a) 344,000,000,000
(b) 0.00000557

23. (a) 3.44×10^{11}
(b) 5.57×10^{-6}

Write each number without exponents.

24. (a) 2.96×10^7
(b) 6.07×10^{-8}

24. (a) 29,600,000
(b) 0.0000000607

25. A satellite galaxy of our own Milky Way, known as the Large Magellanic Cloud, is 1000 light-years across. If a light-year is equal to 5,890,000,000,000 mi, how many miles across is the Large Magellanic Cloud? (*Source:* "Images of Brightest Nebula Unveiled," *USA Today*, June 12, 2002.)

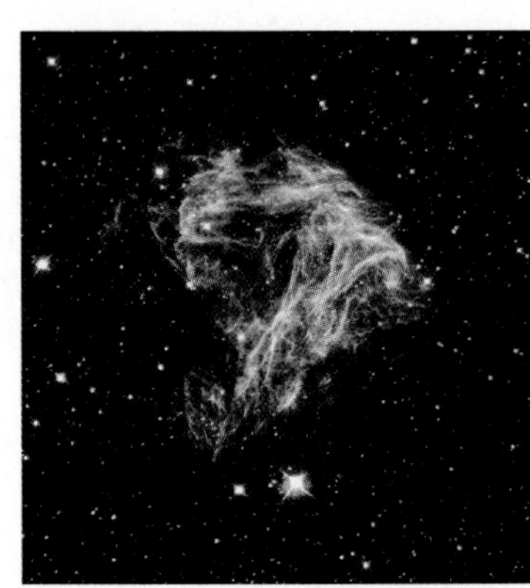

25. 5.89×10^{15} mi

Cumulative Review Exercises Chapters R–5

Work each problem.

1. $\frac{2}{3} + \frac{1}{8}$

   $\frac{19}{24}$

2. $\frac{7}{4} - \frac{9}{5}$

   $-\frac{1}{20}$

3. $8.32 - 4.6$

   3.72

4. 7.21×8.6

   62.006

5. A retailer has \$34,000 invested in her business. She finds that last year she earned 5.4% on this investment. How much did she earn?

   \$1836

Find the value of each expression if $x = -2$ and $y = 4$.

6. $\frac{4x - 2y}{x + y}$

   -8

7. $x^3 - 4xy$

   24

Perform the indicated operations.

8. $\frac{(-13 + 15) - (3 + 2)}{6 - 12}$ $\frac{1}{2}$

9. $-7 - 3[2 + (5 - 8)]$ -4

Decide what property justifies each statement.

10. $(9 + 2) + 3 = 9 + (2 + 3)$

    associative property

11. $-7 + 7 = 0$

    inverse property

12. $6(4 + 2) = 6(4) + 6(2)$

    distributive property

Solve each equation.

13. $2x - 7x + 8x = 30$

    $\{10\}$

14. $2 - 3(t - 5) = 4 + t$

    $\left\{\frac{13}{4}\right\}$

15. $2(5h + 1) = 10h + 4$

    $\emptyset$

16. $d = rt$ for r

    $r = \frac{d}{t}$

17. $\frac{x}{5} = \frac{x - 2}{7}$

    $\{-5\}$

18. $\frac{1}{3}p - \frac{1}{6}p = -2$

    $\{-12\}$

19. $0.05x + 0.15(50 - x) = 5.50$

    $\{20\}$

20. $4 - (3x + 12) = (2x - 9) - (5x - 1)$

    {all real numbers}

Solve each problem.

21. A 1-oz mouse takes about 16 times as many breaths as does a 3-ton elephant. (*Source: Dinosaurs, Spitfires, and Sea Dragons*, McGowan, C., Harvard University Press, 1991.) If the two animals take a combined total of 170 breaths per minute, how many breaths does each take during that time period?

    mouse: 160; elephant: 10

22. If a number is subtracted from 8 and this difference is tripled, the result is three times the number. Find this number, and you will learn how many times a dolphin rests during a 24-hr period.

    4

Solve each inequality. Write the solution set in interval notation.

23. $-8x \le -80$

$[10, \infty)$

24. $-2(x + 4) > 3x + 6$

$\left(-\infty, -\frac{14}{5}\right)$

25. $-3 \le 2x + 5 < 9$

$[-4, 2)$

Given $2x - 3y = -6$, *find the following.*

26. The intercepts of the graph

$(0, 2)$ and $(-3, 0)$

27. The graph

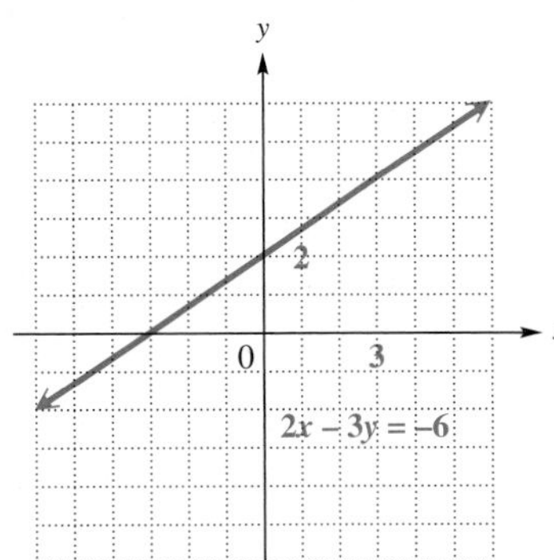

28. The slope of the line

$\frac{2}{3}$

Solve the system using the method indicated.

29. $y = 2x + 5$
$x + y = -4$ (Substitution)

$\{(-3, -1)\}$

30. $3x + 2y = 2$
$2x + 3y = -7$ (Elimination)

$\{(4, -5)\}$

Evaluate each expression.

31. $4^{-1} + 3^0$ $\frac{5}{4}$, or $1\frac{1}{4}$

32. $2^{-4} \cdot 2^5$ 2

33. $\frac{8^{-5} \cdot 8^7}{8^2}$ 1

34. Write with positive exponents only: $\frac{(a^{-3}b^2)^2}{(2a^{-4}b^{-3})^{-1}}$.

$\frac{2b}{a^{10}}$

35. Write in scientific notation: 34,500.

3.45×10^4

Perform the indicated operations.

36. $(7x^3 - 12x^2 - 3x + 8) + (6x^2 + 4) - (-4x^3 + 8x^2 - 2x - 2)$ $11x^3 - 14x^2 - x + 14$

37. $6x^5(3x^2 - 9x + 10)$

$18x^7 - 54x^6 + 60x^5$

38. $(7x + 4)(9x + 3)$

$63x^2 + 57x + 12$

39. $(5x + 8)^2$

$25x^2 + 80x + 64$

40. $\frac{y^3 - 3y^2 + 8y - 6}{y - 1}$

$y^2 - 2y + 6$

6

Factoring and Applications

Wireless communication uses radio waves to carry signals and messages across distances. Cellular phones, one of the most popular forms of wireless communication, have become an invaluable tool for people to stay connected to family, friends, and work while on the go. In 2007, there were about 243 million cell phone subscribers in the United States, with 81% of the population having cell phone service. Total revenue from this service was about \$133 billion. (*Source:* CITA–The Wireless Association.)

In Exercise 31 of Section 6.8, we use a *quadratic equation* to model the number of cell phone subscribers in the United States in recent years.

6.1 Factors; The Greatest Common Factor

OBJECTIVES

1. Find the greatest common factor of a list of numbers.
2. Find the greatest common factor of a list of variable terms.
3. Factor out the greatest common factor.
4. Factor by grouping.

Recall from **Section R.1** that to **factor** a number means to write it as the product of two or more numbers. The product is called the **factored form** of the number. For example,

$$12 = \underbrace{6 \cdot 2}_{\text{Factored form}}.$$

(Factors: 6 and 2)

Factoring is a process that "undoes" multiplying. We multiply $6 \cdot 2$ to get 12, but we factor 12 by writing it as $6 \cdot 2$.

OBJECTIVE 1 Find the greatest common factor of a list of numbers. An integer that is a factor of two or more integers is a **common factor** of those integers. For example, 6 is a common factor of 18 and 24 because 6 is a factor of both 18 and 24. Other common factors of 18 and 24 are 1, 2, and 3. The **greatest common factor (GCF)** of a list of integers is the largest common factor of those integers. This means 6 is the greatest common factor of 18 and 24, since it is the largest of their common factors.

Note

Factors of a number are also divisors of the number. The greatest common factor is the same as the greatest common divisor.

EXAMPLE 1 Finding the Greatest Common Factor for Numbers

Find the greatest common factor for each list of numbers.

(a) 30, 45

First write each number in prime factored form.

$$30 = 2 \cdot 3 \cdot 5$$
$$45 = 3 \cdot 3 \cdot 5$$

Use each prime the *least* number of times it appears in *all* the factored forms. There is no 2 in the prime factored form of 45, so there will be no 2 in the greatest common factor. The least number of times 3 appears in all the factored forms is 1. The least number of times 5 appears is also 1. From this, the

$$\text{GCF} = 3^1 \cdot 5^1 = 3 \cdot 5 = 15.$$

(b) 72, 120, 432

Find the prime factored form of each number.

$$72 = 2 \cdot 2 \cdot 2 \cdot 3 \cdot 3$$
$$120 = 2 \cdot 2 \cdot 2 \cdot 3 \cdot 5$$
$$432 = 2 \cdot 2 \cdot 2 \cdot 2 \cdot 3 \cdot 3 \cdot 3$$

The least number of times 2 appears in all the factored forms is 3, and the least number of times 3 appears is 1. There is no 5 in the prime factored form of either 72 or 432, so the

$$\text{GCF} = 2^3 \cdot 3^1 = 24.$$

Continued on Next Page

(c) 10, 11, 14

Write the prime factored form of each number.

$$10 = 2 \cdot 5$$
$$11 = 11$$
$$14 = 2 \cdot 7$$

There are no primes common to all three numbers, so the GCF is 1.

Work Problem **1** *at the Side.* ▶

OBJECTIVE 2 Find the greatest common factor of a list of variable terms. The terms x^4, x^5, x^6, and x^7 have x^4 as the greatest common factor because the least exponent on the variable x is 4.

$$x^4 = 1 \cdot x^4, \quad x^5 = x \cdot x^4, \quad x^6 = x^2 \cdot x^4, \quad x^7 = x^3 \cdot x^4$$

Note

The exponent on a variable in the GCF is the *least* exponent that appears on that variable in *all* the terms.

EXAMPLE 2 Finding the Greatest Common Factor for Variable Terms

Find the greatest common factor for each list of terms.

(a) $21m^7, -18m^6, 45m^8$

$$21m^7 = 3 \cdot 7 \cdot m^7$$
$$-18m^6 = -1 \cdot 2 \cdot 3 \cdot 3 \cdot m^6$$
$$45m^8 = 3 \cdot 3 \cdot 5 \cdot m^8$$

First, 3 is the greatest common factor of the coefficients 21, −18, and 45. The least exponent on m is 6, so the

$$\text{GCF} = 3m^6.$$

(b) $x^4y^2, x^7y^5, x^3y^7, y^{15}$

$$x^4y^2, \quad x^7y^5, \quad x^3y^7, \quad y^{15}$$

There is no x in the last term, y^{15}, so x will not appear in the greatest common factor. There is a y in each term, however, and 2 is the least exponent on y. The GCF is y^2.

(c) $-a^2b, -ab^2$

$$-a^2b = -1a^2b = -1 \cdot 1 \cdot a^2b$$
$$-ab^2 = -1ab^2 = -1 \cdot 1 \cdot ab^2$$

The factors of −1 are −1 and 1. Since $1 > -1$, the GCF is $1ab$, or ab.

Note

In a list of negative terms, sometimes a negative common factor is preferable (even though it is not the greatest common factor). In Example 2(c), for instance, we might prefer $-ab$ as the common factor. In factoring exercises like this, either answer will be acceptable.

1 Find the greatest common factor for each list of numbers.

(a) 30, 20, 15

$$30 = 2 \cdot 3 \cdot 5$$
$$20 = 2 \cdot \_\_\_\_ \cdot \_\_\_\_$$
$$15 = 3 \cdot \_\_\_\_$$
$$\text{GCF} = \_\_\_\_$$

(b) 42, 28, 35

(c) 12, 18, 26, 32

(d) 10, 15, 21

Answers

1. (a) 2; 5; 5; 5 **(b)** 7 **(c)** 2 **(d)** 1

2 Find the greatest common factor for each list of terms.

(a) $6m^4, 9m^2, 12m^5$

$6m^4 = 2 \cdot \_\_\_\_ \cdot m^4$

$9m^2 = 3 \cdot \_\_\_\_ \cdot \_\_\_\_$

$12m^5 = 2 \cdot 2 \cdot \_\_\_\_ \cdot \_\_\_\_$

GCF = ____

(b) $-12p^5, -18q^4$

(c) y^4z^2, y^6z^8, z^9

(d) $12p^{11}, 17q^5$

Finding the Greatest Common Factor (GCF)

Step 1 **Factor.** Write each number in prime factored form.

Step 2 **List common factors.** List each prime number or each variable that is a factor of every term in the list. (If a prime does not appear in one of the prime factored forms, it cannot appear in the greatest common factor.)

Step 3 **Choose least exponents.** Use as exponents on the common prime factors the *least* exponents from the prime factored forms.

Step 4 **Multiply.** Multiply the primes from Step 3. If there are no primes left after Step 3, the greatest common factor is 1.

*◀ Work Problem **2** at the Side.*

OBJECTIVE 3 Factor out the greatest common factor. The polynomial

$$3m + 12$$

has two terms, $3m$ and 12. The greatest common factor of these two terms is 3. We can write $3m + 12$ so that each term is a product with 3 as one factor.

$$\begin{aligned} &3m + 12 \\ &= 3 \cdot m + 3 \cdot 4 \\ &= 3(m + 4) \qquad \text{Distributive property} \end{aligned}$$

The factored form of $3m + 12$ is $3(m + 4)$. This process is called **factoring out the greatest common factor.**

CAUTION

The polynomial $3m + 12$ is *not* in factored form when written as the *sum*

$$3 \cdot m + 3 \cdot 4. \qquad \text{Not in factored form}$$

The terms are factored, but the polynomial is not. The factored form of $3m + 12$ is the *product*

$$3(m + 4). \qquad \text{In factored form}$$

Writing a polynomial as a product, that is, in factored form, is called **factoring** the polynomial.

EXAMPLE 3 Factoring Out the Greatest Common Factor

Factor out the greatest common factor.

(a) $5y^2 + 10y$

$$\begin{aligned} &= 5y(y) + 5y(2) \qquad \text{GCF} = 5y \\ &= 5y(y + 2) \qquad \text{Distributive property} \end{aligned}$$

Check Multiply the factored form.

$$\begin{aligned} &5y(y + 2) \\ &= 5y(y) + 5y(2) \qquad \text{Distributive property} \\ &= 5y^2 + 10y \qquad \text{Original polynomial} \end{aligned}$$

Continued on Next Page

ANSWERS

2. **(a)** 3; 3; m^2; 3; m^5; $3m^2$
(b) 6 **(c)** z^2 **(d)** 1

(b) $20m^5 + 10m^4 - 15m^3$

$= 5m^3(4m^2) + 5m^3(2m) - 5m^3(3)$ GCF $= 5m^3$

$= 5m^3(4m^2 + 2m - 3)$ Factor out $5m^3$.

Check $5m^3(4m^2 + 2m - 3)$

$= 20m^5 + 10m^4 - 15m^3$ Original polynomial

(c) $x^5 + x^3$

$= x^3(x^2) + x^3(1)$

$= x^3(x^2 + 1)$ Don't forget the 1.

(d) $20m^7p^2 - 36m^3p^4$

$= 4m^3p^2(5m^4) - 4m^3p^2(9p^2)$ GCF $= 4m^3p^2$

$= 4m^3p^2(5m^4 - 9p^2)$ Factor out $4m^3p^2$.

(e) $\frac{1}{6}n^2 + \frac{5}{6}n$

$= \frac{1}{6}n(n) + \frac{1}{6}n(5)$ GCF $= \frac{1}{6}n$

$= \frac{1}{6}n(n + 5)$

CAUTION
Be sure to include the 1 in a problem like Example 3(c). ***Check that the factored form can be multiplied out to give the original polynomial.***

Work Problem 3 at the Side. ▶

EXAMPLE 4 Factoring Out a Negative Common Factor

Factor $-8x^4 + 16x^3 - 4x^2$.

We can factor out either $4x^2$ or $-4x^2$ here. We factor out $-4x^2$ so that the coefficient of the first term in the trinomial factor will be positive.

$-8x^4 + 16x^3 - 4x^2$ Be careful with signs.

$= -4x^2(2x^2) - 4x^2(-4x) - 4x^2(1)$ $-4x^2$ is a common factor.

$= -4x^2(2x^2 - 4x + 1)$ Factor out $-4x^2$.

Check $-4x^2(2x^2 - 4x + 1)$

$= -4x^2(2x^2) - 4x^2(-4x) - 4x^2(1)$ Distributive property

$= -8x^4 + 16x^3 - 4x^2$ Original polynomial

Work Problem 4 at the Side. ▶

Note
Whenever we factor a polynomial in which the coefficient of the first term of a polynomial is negative, we will factor out the negative common factor, even if it is just -1. However, it would also be correct to factor out $4x^2$ in Example 4 to obtain $4x^2(-2x^2 + 4x - 1)$.

3 Factor out the greatest common factor.

(a) $4x^2 + 6x$

(b) $10y^5 - 8y^4 + 6y^2$

(c) $m^7 + m^9$

(d) $8p^5q^2 + 16p^6q^3 - 12p^4q^7$

(e) $\frac{1}{3}b^2 - \frac{2}{3}b$

4 Factor

$-14a^3b^2 - 21a^2b^3 + 7ab$

by factoring out a negative common factor.

Answers

3. **(a)** $2x(2x + 3)$
(b) $2y^2(5y^3 - 4y^2 + 3)$
(c) $m^7(1 + m^2)$
(d) $4p^4q^2(2p + 4p^2q - 3q^5)$
(e) $\frac{1}{3}b(b - 2)$

4. $-7ab(2a^2b + 3ab^2 - 1)$

EXAMPLE 5 Factoring Out a Common Binomial Factor

Factor out the greatest common factor.

(a) $a(a + 3) + 4(a + 3)$

Sometimes the GCF has a factor with more than one term. The binomial $a + 3$ is the greatest common factor here.

Same

$$a(a + 3) + 4(a + 3)$$
$$= (a + 3)(a + 4)$$

(b) $x^2(x + 1) - 5(x + 1)$

$= (x + 1)(x^2 - 5)$ Factor out $x + 1$.

◀ *Work Problem* 5 *at the Side.*

5 Factor out the greatest common factor.

(a) $r(t - 4) + 5(t - 4)$

(b) $y^2(y + 2) - 3(y + 2)$

(c) $x(x - 1) - 5(x - 1)$

OBJECTIVE 4 Factor by grouping. ***When a polynomial has four terms, common factors can sometimes be used to factor by grouping.***

EXAMPLE 6 Factoring by Grouping

Factor by grouping.

(a) $2x + 6 + ax + 3a$

Group the first two terms and the last two terms, since the first two terms have a common factor of 2 and the last two terms have a common factor of a.

$$2x + 6 + ax + 3a$$
$$= (2x + 6) + (ax + 3a)$$
$$= 2(x + 3) + a(x + 3)$$

The expression is still not in factored form because it is the ***sum*** of two terms. Now, however, $x + 3$ is a common factor and can be factored out.

$2x + 6 + ax + 3a$

$= (2x + 6) + (ax + 3a)$ Group the terms.

$= 2(x + 3) + a(x + 3)$ Factor each group.

$= (x + 3)(2 + a)$ Factor out $x + 3$.

The final result is in factored form because it is a ***product.*** Note that the goal in factoring by grouping is to get a common factor, $x + 3$ here, so that the last step is possible. Check by multiplying the binomials using the FOIL method from **Section 5.3.**

Check $(x + 3)(2 + a)$

$= 2x + ax + 6 + 3a$ FOIL

$= 2x + 6 + ax + 3a,$ Rearrange terms.

which is the original polynomial.

Continued on Next Page

ANSWERS

5. (a) $(t - 4)(r + 5)$
(b) $(y + 2)(y^2 - 3)$
(c) $(x - 1)(x - 5)$

(b) $6ax + 24x + a + 4$

$= (6ax + 24x) + (a + 4)$ Group the terms.

$= 6x(a + 4) + 1(a + 4)$ Factor each group.

Remember the 1.

$= (a + 4)(6x + 1)$ Factor out $a + 4$.

Check $(a + 4)(6x + 1)$

$= 6ax + a + 24x + 4$ FOIL

$= 6ax + 24x + a + 4,$ Rearrange terms.

which is the original polynomial.

(c) $2x^2 - 10x + 3xy - 15y$

$= (2x^2 - 10x) + (3xy - 15y)$ Group the terms.

$= 2x(x - 5) + 3y(x - 5)$ Factor each group.

$= (x - 5)(2x + 3y)$ Factor out the common factor, $x - 5$.

Check $(x - 5)(2x + 3y)$

$= 2x^2 + 3xy - 10x - 15y$ FOIL

$= 2x^2 - 10x + 3xy - 15y$ Original polynomial

(d) $t^3 + 2t^2 - 3t - 6$

Be sure to write a + sign between the groups.

$= (t^3 + 2t^2) + (-3t - 6)$ Group the terms.

$= t^2(t + 2) - 3(t + 2)$ Factor out -3 so there is a common factor, $t + 2$; $-3(t + 2) = -3t - 6$.

Be careful with signs.

$= (t + 2)(t^2 - 3)$ Factor out $t + 2$.

Check by multiplying.

CAUTION

Be careful with signs when grouping in a problem like Example 6(d). It is wise to check the factoring in the second step, as shown in the example side comment, before continuing.

Work Problem 6 at the Side. ▶

Factoring a Polynomial with Four Terms by Grouping

Step 1 **Group terms.** Collect the terms into two groups so that each group has a common factor.

Step 2 **Factor within groups.** Factor out the greatest common factor from each group.

Step 3 **Factor the entire polynomial.** Factor a common binomial factor from the results of Step 2.

Step 4 **If necessary, rearrange terms.** If Step 2 does not result in a common binomial factor, try a different grouping.

6 Factor by grouping.

(a) $pq + 5q + 2p + 10$

(b) $2xy + 3y + 2x + 3$

(c) $2a^2 - 4a + 3ab - 6b$

(d) $x^3 + 3x^2 - 5x - 15$

ANSWERS

6. **(a)** $(p + 5)(q + 2)$
(b) $(2x + 3)(y + 1)$
(c) $(a - 2)(2a + 3b)$
(d) $(x + 3)(x^2 - 5)$

7 Factor by grouping.

(a) $6y^2 - 20w + 15y - 8yw$

(b) $9mn - 4 + 12m - 3n$

EXAMPLE 7 Rearranging Terms Before Factoring by Grouping

Factor by grouping.

(a) $10x^2 - 12y + 15x - 8xy$

Factoring out the common factor of 2 from the first two terms and the common factor of x from the last two terms gives

$$10x^2 - 12y + 15x - 8xy$$
$$= 2(5x^2 - 6y) + x(15 - 8y).$$

This does not lead to a common factor, so we try rearranging the terms. There is usually more than one way to do this. We try the following.

$10x^2 - 12y + 15x - 8xy$

$= 10x^2 - 8xy - 12y + 15x$ Commutative property

$= (10x^2 - 8xy) + (-12y + 15x)$ Group the terms.

$= 2x(5x - 4y) + 3(-4y + 5x)$ Factor each group.

$= 2x(5x - 4y) + 3(5x - 4y)$ Rewrite $-4y + 5x$.

$= (5x - 4y)(2x + 3)$ Factor out $5x - 4y$.

Check $(5x - 4y)(2x + 3)$

$= 10x^2 + 15x - 8xy - 12y$ FOIL

$= 10x^2 - 12y + 15x - 8xy$ Original polynomial

(b) $2xy + 12 - 3y - 8x$

We need to rearrange these terms to get two groups that each have a common factor. Trial and error suggests the following grouping.

$2xy + 12 - 3y - 8x$ (Always write a + sign between the two groups.)

$= (2xy - 3y) + (-8x + 12)$ Group the terms.

$= y(2x - 3) - 4(2x - 3)$ Factor each group. (Be careful with signs.)

$= (2x - 3)(y - 4)$ Factor out $2x - 3$.

Since the quantities in parentheses in the second step must be the same, we factored out -4 rather than 4.

Check $(2x - 3)(y - 4)$

$= 2xy - 8x - 3y + 12$ FOIL

$= 2xy + 12 - 3y - 8x$ Original polynomial

CAUTION

Use negative signs carefully when grouping, as in Example 7(b), or a sign error will occur. ***Always check by multiplying.***

◀ *Work Problem 7 at the Side.*

ANSWERS

7. (a) $(2y + 5)(3y - 4w)$
(b) $(3m - 1)(3n + 4)$

6.1 Exercises

FOR EXTRA HELP   PRACTICE  WATCH  DOWNLOAD  READ  REVIEW

Find the greatest common factor for each list of numbers. See Example 1.

1. 12, 16 **4**
2. 18, 24 **6**
3. 40, 20, 4 **4**
4. 50, 30, 5 **5**
5. 18, 24, 36, 48 **6**
6. 15, 30, 45, 75 **15**
7. 4, 9, 12 **1**
8. 9, 16, 24 **1**

Find the greatest common factor for each list of terms. See Example 2.

9. $16y, 24$ **8**
10. $18w, 27$ **9**
11. $30x^3, 40x^6, 50x^7$ $\mathbf{10x^3}$
12. $60z^4, 70z^8, 90z^9$ $\mathbf{10z^4}$
13. $-x^4y^3, -xy^2$ $\mathbf{xy^2}$
14. $-a^4b^5, -a^3b$ $\mathbf{a^3b}$
15. $42ab^3, -36a, 90b, -48ab$ **6**
16. $45c^3d, 75c, 90d, -105cd$ **15**

Complete each factoring.

17. $9m^4 = 3m^2(\quad)$ $\mathbf{3m^2}$
18. $12p^5 = 6p^3(\quad)$ $\mathbf{2p^2}$
19. $-8z^9 = -4z^5(\quad)$ $\mathbf{2z^4}$
20. $-15k^{11} = -5k^8(\quad)$ $\mathbf{3k^3}$
21. $6m^4n^5 = 3m^3n(\quad)$ $\mathbf{2mn^4}$
22. $27a^3b^2 = 9a^2b(\quad)$ $\mathbf{3ab}$
23. $12y + 24 = 12(\quad)$ $\mathbf{y + 2}$
24. $18p + 36 = 18(\quad)$ $\mathbf{p + 2}$
25. $10a^2 - 20a = 10a(\quad)$ $\mathbf{a - 2}$
26. $15x^2 - 30x = 15x(\quad)$ $\mathbf{x - 2}$
27. $8x^2y + 12x^3y^2 = 4x^2y(\quad)$ $\mathbf{2 + 3xy}$
28. $18s^3t^2 + 10st = 2st(\quad)$ $\mathbf{9s^2t + 5}$

Factor out the greatest common factor, or a negative common factor if the coefficient of the term of greatest degree is negative. See Examples 3–5.

29. $x^2 - 4x$
$\mathbf{x(x - 4)}$

30. $m^2 - 7m$
$\mathbf{m(m - 7)}$

31. $6t^2 + 15t$
$\mathbf{3t(2t + 5)}$

32. $8x^2 + 6x$
$\mathbf{2x(4x + 3)}$

33. $\frac{1}{4}d^2 - \frac{3}{4}d$
$\mathbf{\frac{1}{4}d(d - 3)}$

34. $\frac{1}{5}z^2 + \frac{3}{5}z$
$\mathbf{\frac{1}{5}z(z + 3)}$

35. $-12x^3 - 6x^2$
$\mathbf{-6x^2(2x + 1)}$

36. $-21b^3 + 7b^2$
$\mathbf{-7b^2(3b - 1)}$

37. $65y^{10} + 35y^6$
$\mathbf{5y^6(13y^4 + 7)}$

38. $100a^5 + 16a^3$
$\mathbf{4a^3(25a^2 + 4)}$

39. $11w^3 - 100$
no common factor (except 1)

40. $13z^5 - 80$
no common factor (except 1)

41. $8m^2n^3 + 24m^2n^2$
$\mathbf{8m^2n^2(n + 3)}$

42. $19p^2y - 38p^2y^3$
$\mathbf{19p^2y(1 - 2y^2)}$

43. $-4x^3 + 10x^2 - 6x$
$\mathbf{-2x(2x^2 - 5x + 3)}$

44. $-9z^3 + 6z^2 - 12z$
$\mathbf{-3z(3z^2 - 2z + 4)}$

45. $13y^8 + 26y^4 - 39y^2$
$\mathbf{13y^2(y^6 + 2y^2 - 3)}$

46. $5x^5 + 25x^4 - 20x^3$
$5x^3(x^2 + 5x - 4)$

47. $45q^4p^5 + 36qp^6 + 81q^2p^3$
$9qp^3(5q^3p^2 + 4p^3 + 9q)$

48. $125a^3z^5 + 60a^4z^4 - 85a^5z^2$
$5a^3z^2(25z^3 + 12az^2 - 17a^2)$

49. $c(x + 2) + d(x + 2)$
$(x + 2)(c + d)$

50. $r(5 - x) + t(5 - x)$
$(5 - x)(r + t)$

51. $a^2(2a + b) - b(2a + b)$
$(2a + b)(a^2 - b)$

52. $3x(x^2 + 5) - y(x^2 + 5)$
$(x^2 + 5)(3x - y)$

53. $q(p + 4) - 1(p + 4)$
$(p + 4)(q - 1)$

54. $y^2(x - 4) + 1(x - 4)$
$(x - 4)(y^2 + 1)$

Factor by grouping. See Examples 6 and 7.

55. $5m + mn + 20 + 4n$
$(5 + n)(m + 4)$

56. $ts + 5t + 2s + 10$
$(s + 5)(t + 2)$

57. $6xy - 21x + 8y - 28$
$(2y - 7)(3x + 4)$

58. $2mn - 8n + 3m - 12$
$(m - 4)(2n + 3)$

59. $3xy + 9x + y + 3$
$(y + 3)(3x + 1)$

60. $6n + 4mn + 3 + 2m$
$(3 + 2m)(2n + 1)$

61. $7z^2 + 14z - az - 2a$
$(z + 2)(7z - a)$

62. $2b^2 + 3b - 8ab - 12a$
$(2b + 3)(b - 4a)$

63. $18r^2 + 12ry - 3xr - 2xy$
$(3r + 2y)(6r - x)$

64. $5m^2 + 15mp - 2mr - 6pr$
$(m + 3p)(5m - 2r)$

65. $w^3 + w^2 + 9w + 9$
$(w + 1)(w^2 + 9)$

66. $y^3 + y^2 + 6y + 6$
$(y + 1)(y^2 + 6)$

67. $3a^3 + 6a^2 - 2a - 4$
$(a + 2)(3a^2 - 2)$

68. $10x^3 + 15x^2 - 8x - 12$
$(2x + 3)(5x^2 - 4)$

69. $16m^3 - 4m^2p^2 - 4mp + p^3$
$(4m - p^2)(4m^2 - p)$

70. $10t^3 - 2t^2s^2 - 5ts + s^3$
$(5t - s^2)(2t^2 - s)$

71. $y^2 + 3x + 3y + xy$
$(y + 3)(y + x)$

72. $m^2 + 14p + 7m + 2mp$
$(m + 7)(m + 2p)$

73. $2z^2 + 6w - 4z - 3wz$
$(z - 2)(2z - 3w)$

74. $2a^2 + 20b - 8a - 5ab$
$(a - 4)(2a - 5b)$

Relating Concepts (Exercises 75–78) For Individual or Group Work

In many cases, the choice of which pairs of terms to group when factoring by grouping can be made in different ways. To see this for Example 7(b), ***work Exercises 75–78 in order.***

75. Start with the polynomial from Example 7(b), $2xy + 12 - 3y - 8x$, and rearrange the terms as follows: $2xy - 8x - 3y + 12$. What property from **Section 1.7** allows this?
commutative property

76. Group the first two terms and the last two terms of the rearranged polynomial in Exercise 75. Then factor each group.
$2x(y - 4) - 3(y - 4)$

77. Is your result from Exercise 76 in factored form? Explain your answer.
No, because it is not a product. It is the difference between $2x(y - 4)$ and $3(y - 4)$.

78. If your answer to Exercise 77 is *no*, factor the polynomial. Is the result the same as the one shown for Example 7(b)?
$(2x - 3)(y - 4)$; yes

6.2 Factoring Trinomials

Using FOIL, the product of the binomials $k - 3$ and $k + 1$ is

$$(k - 3)(k + 1) = k^2 - 2k - 3. \quad \text{Multiplying}$$

Suppose instead that we are given the polynomial $k^2 - 2k - 3$ and want to rewrite it as the product $(k - 3)(k + 1)$. That is,

$$k^2 - 2k - 3 = (k - 3)(k + 1). \quad \text{Factoring}$$

Recall from **Section 6.1** that this process is called *factoring* the polynomial. Factoring reverses or, "undoes," multiplying.

OBJECTIVES

1. **Factor trinomials with a coefficient of 1 for the squared term.**
2. **Factor trinomials after factoring out the greatest common factor.**

OBJECTIVE 1 Factor trinomials with a coefficient of 1 for the squared term. When factoring polynomials with integer coefficients, we use only integers in the factors. For example, we can factor $x^2 + 5x + 6$ by finding integers m and n such that

$$x^2 + 5x + 6 \quad \text{is written as} \quad (x + m)(x + n).$$

To find these integers m and n, we first use FOIL to multiply the two binomials on the right above:

$$\begin{aligned} &(x + m)(x + n) \\ &= x^2 + nx + mx + mn \\ &= x^2 + (n + m)x + mn. \quad \text{Distributive property} \end{aligned}$$

Comparing this result with $x^2 + 5x + 6$ shows that we must find integers m and n having a sum of 5 and a product of 6.

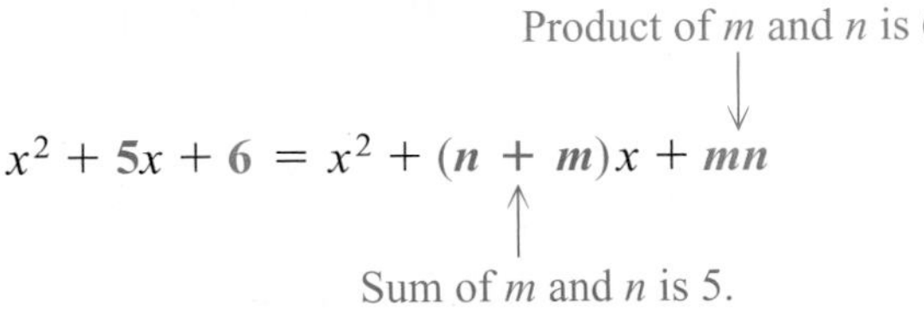

Because many pairs of integers have a sum of 5, it is best to begin by listing those pairs of integers whose product is 6. Both 5 and 6 are positive, so we consider only pairs in which both integers are positive.

Work Problem 1 at the Side. ▶

1 (a) List all pairs of positive integers whose product is 6.

(b) Find the pair from part (a) whose sum is 5.

From Problem 1 at the side, we see that the numbers 1 and 6 and the numbers 2 and 3 both have a product of 6, but only the pair 2 and 3 has a sum of 5. So 2 and 3 are the required integers, and

$$x^2 + 5x + 6 \quad \text{is factored as} \quad (x + 2)(x + 3).$$

Check by multiplying the binomials using FOIL. ***Make sure that the sum of the outer and inner products produces the correct middle term.***

Check $(x + 2)(x + 3) = x^2 + 5x + 6$ Correct

$2x$
$3x$
$5x$ Add.

This method of factoring can be used only for trinomials that have 1 as the coefficient of the squared term.

ANSWERS

1. (a) 1, 6; 2, 3 **(b)** 2, 3

2 Factor each trinomial.

(a) $y^2 + 12y + 20$

First complete the given list of numbers.

Factors of 20	Sums of Factors
20, 1	$20 + 1 = 21$
10, ___	$10 +$ ___ = ___
5, ___	$5 +$ ___ = ___

(b) $x^2 + 9x + 18$

3 Factor each trinomial.

(a) $t^2 - 12t + 32$

First complete the given list of numbers.

Factors of 32	Sums of Factors
$-32, -1$	$-32 + (-1) = -33$
-16, ___	$-16 +$ (___) = ___
-8, ___	$-8 +$ (___) = ___

(b) $y^2 - 10y + 24$

ANSWERS

2. **(a)** 2; 2; 12; 4; 4; 9; $(y + 10)(y + 2)$
   (b) $(x + 3)(x + 6)$
3. **(a)** -2; -2; -18; -4; -4; -12; $(t - 8)(t - 4)$
   (b) $(y - 6)(y - 4)$

EXAMPLE 1 Factoring a Trinomial with All Positive Terms

Factor $m^2 + 9m + 14$.

Look for two integers whose product is 14 and whose sum is 9. List the pairs of integers whose products are 14. Then examine the sums. Only positive integers are needed since all signs in $m^2 + 9m + 14$ are positive.

Factors of 14	Sums of Factors	
14, 1	$14 + 1 = 15$	
7, 2	$7 + 2 = 9$	Sum is 9.

From the list, 7 and 2 are the required integers, since $7 \cdot 2 = 14$ and $7 + 2 = 9$. Thus,

$$m^2 + 9m + 14 \quad \text{factors as} \quad (m + 2)(m + 7).$$

Check $(m + 2)(m + 7)$

$= m^2 + 7m + 2m + 14$ FOIL

$= m^2 + 9m + 14$ Original polynomial

Note

In Example 1, the answer $(m + 2)(m + 7)$ also could have been written

$$(m + 7)(m + 2).$$

Because of the commutative property of multiplication, the order of the factors does not matter. ***Always check by multiplying.***

◀ Work Problem **2** at the Side.

EXAMPLE 2 Factoring a Trinomial with a Negative Middle Term

Factor $x^2 - 9x + 20$.

Find two integers whose product is 20 and whose sum is -9. Since the numbers we are looking for have a *positive product* and a *negative sum*, we consider only pairs of negative integers.

Factors of 20	Sums of Factors	
$-20, -1$	$-20 + (-1) = -21$	
$-10, -2$	$-10 + (-2) = -12$	
$-5, -4$	$-5 + (-4) = -9$	Sum is -9.

The required integers are -5 and -4, so

$$x^2 - 9x + 20 \quad \text{factors as} \quad (x - 5)(x - 4).$$

Check $(x - 5)(x - 4)$

$= x^2 - 4x - 5x + 20$ FOIL

$= x^2 - 9x + 20$ Original polynomial

◀ Work Problem **3** at the Side.

EXAMPLE 3 **Factoring a Trinomial with Two Negative Terms**

Factor $p^2 - 2p - 15$.

Find two integers whose product is -15 and whose sum is -2. If these numbers do not come to mind right away, find them (if they exist) by listing all the pairs of integers whose product is -15. Because the last term, -15, is negative, we need pairs of integers with different signs.

Factors of -15	Sums of Factors	
$15, -1$	$15 + (-1) = 14$	
$-15, 1$	$-15 + 1 = -14$	
$5, -3$	$5 + (-3) = 2$	
$-5, 3$	$-5 + 3 = -2$	Sum is -2.

The required integers are -5 and 3, so

$$p^2 - 2p - 15 \quad \text{factors as} \quad (p - 5)(p + 3).$$

Check Multiply $(p - 5)(p + 3)$ to obtain $p^2 - 2p - 15$.

Note

In Examples 1–3, notice that we listed factors in descending order (disregarding sign) when we were looking for the required pair of integers. This helps avoid skipping the correct combination.

Work Problem **4** *at the Side.* ▶

As shown in the next example, some trinomials cannot be factored using only integers. We call such trinomials **prime polynomials.**

EXAMPLE 4 **Deciding Whether Polynomials Are Prime**

Factor each trinomial.

(a) $x^2 - 5x + 12$

As in Example 2, both factors must be negative to give a positive product and a negative sum. First, list all the pairs of negative integers whose product is 12. Then examine the sums.

Factors of 12	Sums of Factors
$-12, -1$	$-12 + (-1) = -13$
$-6, -2$	$-6 + (-2) = -8$
$-4, -3$	$-4 + (-3) = -7$

None of the pairs of integers has a sum of -5. Therefore, the trinomial $x^2 - 5x + 12$ *cannot be factored using only integers;* it is a *prime polynomial.*

(b) $k^2 - 8k + 11$

There is no pair of integers whose product is 11 and whose sum is -8, so $k^2 - 8k + 11$ is a prime polynomial.

Work Problem **5** *at the Side.* ▶

4 Factor each trinomial.

(a) $a^2 - 9a - 22$

(b) $r^2 - 6r - 16$

5 Factor each trinomial, if possible.

(a) $r^2 - 3r - 4$

(b) $m^2 - 2m + 5$

Answers

4. **(a)** $(a - 11)(a + 2)$ **(b)** $(r - 8)(r + 2)$
5. **(a)** $(r - 4)(r + 1)$ **(b)** prime

Guidelines for factoring a trinomial of the form $x^2 + bx + c$ are summarized here.

Factoring $x^2 + bx + c$

Find two integers whose product is c and whose sum is b.

1. Both integers must be positive if b and c are positive.
2. Both integers must be negative if c is positive and b is negative.
3. One integer must be positive and one must be negative if c is negative.

EXAMPLE 5 Factoring a Trinomial with Two Variables

Factor $z^2 - 2bz - 3b^2$.

Here, the coefficient of z in the middle term is $-2b$, so we need to find two expressions whose product is $-3b^2$ and whose sum is $-2b$. The expressions are $-3b$ and b, so

$$z^2 - 2bz - 3b^2 \quad \text{factors as} \quad (z - 3b)(z + b).$$

Check $(z - 3b)(z + b)$

$= z^2 + zb - 3bz - 3b^2$ FOIL

$= z^2 + 1bz - 3bz - 3b^2$ Identity and commutative properties

$= z^2 - 2bz - 3b^2$ Combine like terms.

◀ Work Problem 6 at the Side.

OBJECTIVE 2 Factor trinomials after factoring out the greatest common factor. The trinomial in the next example does not have a coefficient of 1 for the squared term. (In fact, there is no squared term.) However, there may be a common factor.

EXAMPLE 6 Factoring a Trinomial with a Common Factor

Factor $4x^5 - 28x^4 + 40x^3$.

First, factor out the greatest common factor, $4x^3$.

$$4x^5 - 28x^4 + 40x^3$$
$$= 4x^3(x^2 - 7x + 10)$$

Now factor $x^2 - 7x + 10$. The integers -5 and -2 have a product of 10 and a sum of -7. The complete factored form is

Include $4x^3$. $\quad 4x^3(x - 5)(x - 2).$

Check $4x^3(x - 5)(x - 2)$

$= 4x^3(x^2 - 7x + 10)$ FOIL

$= 4x^5 - 28x^4 + 40x^3$ Distributive property

◀ Work Problem 7 at the Side.

CAUTION

When factoring, ***always look for a common factor first.*** Remember to include the common factor as part of the answer. As a check, multiplying out the complete factored form should give the original polynomial.

---

6 Factor each trinomial.

(a) $b^2 - 3ab - 4a^2$

(b) $r^2 - 6rs + 8s^2$

7 Factor each trinomial completely.

(a) $2p^3 + 6p^2 - 8p$

(b) $-3x^4 + 15x^3 - 18x^2$

Answers

6. (a) $(b - 4a)(b + a)$
   (b) $(r - 4s)(r - 2s)$
7. (a) $2p(p + 4)(p - 1)$
   (b) $-3x^2(x - 3)(x - 2)$

6.2 Exercises

FOR EXTRA HELP

MyMathLab

Math XL PRACTICE

WATCH

DOWNLOAD

READ

REVIEW

1. When factoring a trinomial in x as $(x + a)(x + b)$, what must be true of a and b, if the last term of the trinomial is negative?

   a and b must have different signs.

2. In Exercise 1, what must be true of a and b if the last term is positive?

   a and b must have the same sign.

3. What is meant by a *prime polynomial?*

   A prime polynomial is one that cannot be factored using only integers in the factors.

4. How can you check your work when factoring a trinomial? Does the check ensure that the trinomial is *completely* factored?

   To check a factored form, multiply the factors. You should get the trinomial you started with. If you forget to factor out any common factors, the check will not indicate it.

In Exercises 5–8, list all pairs of integers with the given product. Then find the pair whose sum is given. See the tables in Examples 1–4.

5. Product: 12; Sum: 7

   1 and 12, −1 and −12, 2 and 6, −2 and −6, 3 and 4, −3 and −4; The pair with a sum of 7 is 3 and 4.

6. Product: 18; Sum: 9

   1 and 18, −1 and −18, 2 and 9, −2 and −9, 3 and 6, −3 and −6; The pair with a sum of 9 is 3 and 6.

7. Product: −24; Sum: −5

   1 and −24, −1 and 24, 2 and −12, −2 and 12, 3 and −8, −3 and 8, 4 and −6, −4 and 6; The pair with a sum of −5 is 3 and −8.

8. Product: −36; Sum: −16

   1 and −36, −1 and 36, 2 and −18, −2 and 18, 3 and −12, −3 and 12, 4 and −9, −4 and 9, 6 and −6; The pair with a sum of −16 is 2 and −18.

9. Which one of the following is the correct factored form of $x^2 - 12x + 32$?

   A. $(x - 8)(x + 4)$ **B.** $(x + 8)(x - 4)$

   C. $(x - 8)(x - 4)$ **D.** $(x + 8)(x + 4)$

   C

10. What would be the first step in factoring

$$2x^3 + 8x^2 - 10x?$$

   Factor out the greatest common factor, $2x$.

Complete each factoring.

11. $x^2 + 15x + 44 = (x + 4)(\quad)$

    $x + 11$

12. $r^2 + 15r + 56 = (r + 7)(\quad)$

    $r + 8$

13. $x^2 - 9x + 8 = (x - 1)(\quad)$

    $x - 8$

14. $t^2 - 14t + 24 = (t - 2)(\quad)$

    $t - 12$

15. $y^2 - 2y - 15 = (y + 3)(\quad)$

    $y - 5$

16. $t^2 - t - 42 = (t + 6)(\quad)$

    $t - 7$

17. $x^2 + 9x - 22 = (x - 2)(\quad)$

    $x + 11$

18. $x^2 + 6x - 27 = (x - 3)(\quad)$

    $x + 9$

19. $y^2 - 7y - 18 = (y + 2)(\quad)$

    $y - 9$

20. $y^2 - 2y - 24 = (y + 4)(\quad)$

    $y - 6$

Factor completely. If a polynomial cannot be factored, write prime. See Examples 1–4.

21. $y^2 + 9y + 8$
$(y + 8)(y + 1)$

22. $a^2 + 9a + 20$
$(a + 4)(a + 5)$

23. $b^2 + 8b + 15$
$(b + 3)(b + 5)$

24. $x^2 + 6x + 8$
$(x + 4)(x + 2)$

25. $m^2 + m - 20$
$(m + 5)(m - 4)$

26. $p^2 + 4p - 5$
$(p + 5)(p - 1)$

27. $x^2 + 3x - 40$
$(x + 8)(x - 5)$

28. $d^2 + 4d - 45$
$(d + 9)(d - 5)$

29. $y^2 - 8y + 15$
$(y - 5)(y - 3)$

30. $y^2 - 6y + 8$
$(y - 4)(y - 2)$

31. $z^2 - 15z + 56$
$(z - 8)(z - 7)$

32. $x^2 - 13x + 36$
$(x - 9)(x - 4)$

33. $r^2 - r - 30$
$(r - 6)(r + 5)$

34. $q^2 - q - 42$
$(q - 7)(q + 6)$

35. $a^2 - 8a - 48$
$(a - 12)(a + 4)$

36. $m^2 - 10m - 24$
$(m - 12)(m + 2)$

37. $x^2 + 4x + 5$
prime

38. $t^2 + 11t + 12$
prime

Factor completely. See Examples 5 and 6.

39. $r^2 + 3ra + 2a^2$
$(r + 2a)(r + a)$

40. $x^2 + 5xa + 4a^2$
$(x + 4a)(x + a)$

41. $x^2 + 4xy + 3y^2$
$(x + y)(x + 3y)$

42. $p^2 + 9pq + 8q^2$
$(p + q)(p + 8q)$

43. $t^2 - tz - 6z^2$
$(t + 2z)(t - 3z)$

44. $a^2 - ab - 12b^2$
$(a + 3b)(a - 4b)$

45. $v^2 - 11vw + 30w^2$
$(v - 5w)(v - 6w)$

46. $v^2 - 11vx + 24x^2$
$(v - 8x)(v - 3x)$

47. $4x^2 + 12x - 40$
$4(x + 5)(x - 2)$

48. $5y^2 - 5y - 30$
$5(y + 2)(y - 3)$

49. $2t^3 + 8t^2 + 6t$
$2t(t + 1)(t + 3)$

50. $3t^3 + 27t^2 + 24t$
$3t(t + 8)(t + 1)$

51. $-2x^6 - 8x^5 + 42x^4$
$-2x^4(x - 3)(x + 7)$

52. $-4y^5 - 12y^4 + 40y^3$
$-4y^3(y - 2)(y + 5)$

53. $a^5 + 3a^4b - 4a^3b^2$
$a^3(a + 4b)(a - b)$

54. $z^{10} - 4z^9y - 21z^8y^2$
$z^8(z - 7y)(z + 3y)$

55. $m^3n - 10m^2n^2 + 24mn^3$
$mn(m - 6n)(m - 4n)$

56. $y^3z + 3y^2z^2 - 54yz^3$
$yz(y - 6z)(y + 9z)$

57. Use the FOIL method from **Section 5.3** to show that $(2x + 4)(x - 3) = 2x^2 - 2x - 12$. Why, then, is it incorrect to completely factor $2x^2 - 2x - 12$ as $(2x + 4)(x - 3)$?
The factored form $(2x + 4)(x - 3)$ is incorrect because $2x + 4$ has a common factor of 2, which must be factored out for the trinomial to be completely factored.

58. Why is it incorrect to completely factor $3x^2 + 9x - 12$ as the product $(x - 1)(3x + 12)$?
The factored form $(x - 1)(3x + 12)$ is incorrect because $3x + 12$ has a common factor of 3, which must be factored out for the trinomial to be completely factored.

6.3 Factoring Trinomials by Grouping

Trinomials like $2x^2 + 7x + 6$, in which the coefficient of the squared term is *not* 1, are factored with extensions of the methods from the previous sections. One such method uses factoring by grouping from **Section 6.1.**

OBJECTIVE 1 Factor trinomials by grouping when the coefficient of the squared term is not 1. Recall that a trinomial such as $m^2 + 3m + 2$ is factored by finding two integers whose product is 2 and whose sum is 3. To factor $2x^2 + 7x + 6$, we look for two integers whose product is $2 \cdot 6 = 12$ and whose sum is 7.

$$2x^2 + 7x + 6$$

Sum is 7. (coefficient 7)

Product is $2 \cdot 6 = 12$. (2 and 6)

By considering pairs of positive integers whose product is 12, the necessary integers are found to be 3 and 4. We use these integers to write the middle term, $7x$, as $7x = 3x + 4x$. The trinomial $2x^2 + 7x + 6$ becomes

$$2x^2 + 7x + 6$$

$$= 2x^2 + \underbrace{3x + 4x}_{7x} + 6$$

$$= (2x^2 + 3x) + (4x + 6) \quad \text{Group the terms.}$$

$$= x(2x + 3) + 2(2x + 3) \quad \text{Factor each group.}$$

Must be the same

$$= (2x + 3)(x + 2). \quad \text{Factor out } 2x + 3.$$

Check Multiply $(2x + 3)(x + 2)$ to obtain $2x^2 + 7x + 6$.

In the preceding example, we could have written $7x$ as $4x + 3x$. Factoring by grouping this way would give the same answer.

Work Problem 1 at the Side. ▶

EXAMPLE 1 Factoring Trinomials by Grouping

Factor each trinomial.

(a) $6r^2 + r - 1$

We must find two integers with a product of $6(-1) = -6$ and a sum of 1. The integers are -2 and 3. We write the middle term, r, as $-2r + 3r$.

$$6r^2 + r - 1$$

$$= 6r^2 - 2r + 3r - 1 \quad r = -2r + 3r$$

$$= (6r^2 - 2r) + (3r - 1) \quad \text{Group the terms.}$$

$$= 2r(3r - 1) + 1(3r - 1) \quad \text{The binomials must be the same.}$$

Remember the 1.

$$= (3r - 1)(2r + 1) \quad \text{Factor out } 3r - 1.$$

Check Multiply $(3r - 1)(2r + 1)$ to obtain $6r^2 + r - 1$.

Continued on Next Page

OBJECTIVE

1 **Factor trinomials by grouping when the coefficient of the squared term is not 1.**

1 **(a)** Factor $2x^2 + 7x + 6$ by writing $7x$ as $4x + 3x$. Complete the following.

$$2x^2 + 7x + 6$$
$$= 2x^2 + 4x + 3x + 6$$
$$= (2x^2 + \_\_\_) + (3x + \_\_\_)$$
$$= 2x(x + \_\_\_) + 3(x + \_\_\_)$$
$$= (\_\_\_\_)(2x + 3)$$

(b) Is the answer in part (a) the same as in the example? (Remember that the order of the factors does not matter.)

ANSWERS

1. (a) $4x$; 6; 2; 2; $x + 2$ **(b)** yes

(b) $12z^2 - 5z - 2$

Look for two integers whose product is $12(-2) = -24$ and whose sum is -5. The required integers are 3 and -8, so

$$\begin{aligned} &12z^2 - 5z - 2 \\ &= 12z^2 + 3z - 8z - 2 && -5z = 3z - 8z \\ &= (12z^2 + 3z) + (-8z - 2) && \text{Group the terms.} \\ &= 3z(4z + 1) - 2(4z + 1) && \text{Factor each group.} \\ &= (4z + 1)(3z - 2). && \text{Factor out } 4z + 1. \end{aligned}$$

Be careful with signs.

Check Multiply $(4z + 1)(3z - 2)$ to obtain $12z^2 - 5z - 2$.

(c) $10m^2 + mn - 3n^2$

Two integers whose product is $10(-3) = -30$ and whose sum is 1 are -5 and 6. Rewrite the trinomial with four terms.

$$\begin{aligned} &10m^2 + mn - 3n^2 \\ &= 10m^2 - 5mn + 6mn - 3n^2 && mn = -5mn + 6mn \\ &= 5m(2m - n) + 3n(2m - n) && \text{Group the terms; factor each group.} \\ &= (2m - n)(5m + 3n) && \text{Factor out } 2m - n. \end{aligned}$$

Check by multiplying.

◀ Work Problem **2** at the Side.

2 Factor each trinomial by grouping.

(a) $2m^2 + 7m + 3$

(b) $5p^2 - 2p - 3$

(c) $15k^2 - km - 2m^2$

EXAMPLE 2 **Factoring a Trinomial with a Common Factor by Grouping**

Factor $28x^5 - 58x^4 - 30x^3$.

First factor out the greatest common factor, $2x^3$.

$$\begin{aligned} &28x^5 - 58x^4 - 30x^3 \\ &= 2x^3(14x^2 - 29x - 15) \end{aligned}$$

To factor $14x^2 - 29x - 15$, find two integers whose product is $14(-15) = -210$ and whose sum is -29. Factoring 210 into prime factors gives

$$210 = 2 \cdot 3 \cdot 5 \cdot 7.$$

Combine these prime factors in pairs in different ways, using one positive factor and one negative factor to get -210. The factors 6 and -35 have the correct sum, -29. Now rewrite the given trinomial and factor it.

$$\begin{aligned} &28x^5 - 58x^4 - 30x^3 \\ &= 2x^3(14x^2 - 29x - 15) \\ &= 2x^3(14x^2 + 6x - 35x - 15) \\ &= 2x^3[(14x^2 + 6x) + (-35x - 15)] \\ &= 2x^3[2x(7x + 3) - 5(7x + 3)] \\ &= 2x^3[(7x + 3)(2x - 5)] \\ &= 2x^3(7x + 3)(2x - 5) \end{aligned}$$

Remember the common factor.

◀ Work Problem **3** at the Side.

3 Factor each trinomial completely.

(a) $-4x^2 + 2x + 30$

(b) $18p^4 + 63p^3 + 27p^2$

(c) $6a^2 + 3ab - 18b^2$

Answers

2. **(a)** $(2m + 1)(m + 3)$
   (b) $(5p + 3)(p - 1)$
   (c) $(5k - 2m)(3k + m)$
3. **(a)** $-2(2x + 5)(x - 3)$
   (b) $9p^2(2p + 1)(p + 3)$
   (c) $3(2a - 3b)(a + 2b)$

6.3 Exercises

FOR EXTRA HELP MyMathLab Math XL PRACTICE  WATCH  DOWNLOAD  READ  REVIEW

The middle term of each trinomial has been rewritten. Now factor by grouping. See Example 1.

1. $m^2 + 8m + 12$
$= m^2 + 6m + 2m + 12$
$(m + 6)(m + 2)$

2. $x^2 + 9x + 14$
$= x^2 + 7x + 2x + 14$
$(x + 7)(x + 2)$

3. $a^2 + 3a - 10$
$= a^2 + 5a - 2a - 10$
$(a + 5)(a - 2)$

4. $y^2 - 2y - 24$
$= y^2 + 4y - 6y - 24$
$(y + 4)(y - 6)$

5. $10t^2 + 9t + 2$
$= 10t^2 + 5t + 4t + 2$
$(2t + 1)(5t + 2)$

6. $6x^2 + 13x + 6$
$= 6x^2 + 9x + 4x + 6$
$(2x + 3)(3x + 2)$

7. $15z^2 - 19z + 6$
$= 15z^2 - 10z - 9z + 6$
$(3z - 2)(5z - 3)$

8. $12p^2 - 17p + 6$
$= 12p^2 - 9p - 8p + 6$
$(4p - 3)(3p - 2)$

9. $8s^2 + 2st - 3t^2$
$= 8s^2 - 4st + 6st - 3t^2$
$(2s - t)(4s + 3t)$

10. $3x^2 - xy - 14y^2$
$= 3x^2 - 7xy + 6xy - 14y^2$
$(3x - 7y)(x + 2y)$

11. $15a^2 + 22ab + 8b^2$
$= 15a^2 + 10ab + 12ab + 8b^2$
$(3a + 2b)(5a + 4b)$

12. $25m^2 + 25mn + 6n^2$
$= 25m^2 + 15mn + 10mn + 6n^2$
$(5m + 3n)(5m + 2n)$

13. Which pair of integers would be used to rewrite the middle term when factoring $12y^2 + 5y - 2$ by grouping?

A. $-8, 3$ **B.** $8, -3$ **C.** $-6, 4$ **D.** $6, -4$

B

14. Which pair of integers would be used to rewrite the middle term when factoring $20b^2 - 13b + 2$ by grouping?

A. $10, 3$ **B.** $-10, -3$ **C.** $8, 5$ **D.** $-8, -5$

D

Complete the steps to factor each trinomial by grouping.

15. $2m^2 + 11m + 12$

(a) Find two integers whose product is 2 · 12 = 24 and whose sum is 11.

(b) The required integers are 3 and 8. (Order is irrelevant.)

(c) Write the middle term $11m$ as $3m$ + $8m$.

(d) Rewrite the given trinomial using four terms.
$2m^2 + 3m + 8m + 12$

(e) Factor the polynomial in part (d) by grouping.
$(2m + 3)(m + 4)$

(f) Check by multiplying.
$(2m + 3)(m + 4)$
$= 2m^2 + 11m + 12$

16. $6y^2 - 19y + 10$

(a) Find two integers whose product is 6 · 10 = 60 and whose sum is -19.

(b) The required integers are -4 and -15. (Order is irrelevant.)

(c) Write the middle term $-19y$ as $-4y$ + $(-15y)$.

(d) Rewrite the given trinomial using four terms.
$6y^2 - 4y - 15y + 10$

(e) Factor the polynomial in part (d) by grouping.
$(3y - 2)(2y - 5)$

(f) Check by multiplying.
$(3y - 2)(2y - 5)$
$= 6y^2 - 19y + 10$

Factor each trinomial by grouping. See Examples 1 and 2. **The order of the factors is irrelevant in Exercises 17–40.**

17. $2x^2 + 7x + 3$
$(2x + 1)(x + 3)$

18. $3y^2 + 13y + 4$
$(3y + 1)(y + 4)$

19. $4r^2 + r - 3$
$(4r - 3)(r + 1)$

20. $4r^2 + 3r - 10$
$(4r - 5)(r + 2)$

21. $8m^2 - 10m - 3$
$(4m + 1)(2m - 3)$

22. $20x^2 - 28x - 3$
$(10x + 1)(2x - 3)$

23. $21m^2 + 13m + 2$
$(3m + 1)(7m + 2)$

24. $38x^2 + 23x + 2$
$(2x + 1)(19x + 2)$

25. $6b^2 + 7b + 2$
$(2b + 1)(3b + 2)$

26. $6w^2 + 19w + 10$
$(2w + 5)(3w + 2)$

27. $12y^2 - 13y + 3$
$(4y - 3)(3y - 1)$

28. $15a^2 - 16a + 4$
$(3a - 2)(5a - 2)$

29. $24x^2 - 42x + 9$
$3(4x - 1)(2x - 3)$

30. $48b^2 - 74b - 10$
$2(8b + 1)(3b - 5)$

31. $2m^3 + 2m^2 - 40m$
$2m(m - 4)(m + 5)$

32. $3x^3 + 12x^2 - 36x$
$3x(x - 2)(x + 6)$

33. $-32z^5 + 20z^4 + 12z^3$
$-4z^3(z - 1)(8z + 3)$

34. $-18x^5 - 15x^4 + 75x^3$
$-3x^3(2x + 5)(3x - 5)$

35. $12p^2 + 7pq - 12q^2$
$(3p + 4q)(4p - 3q)$

36. $6m^2 - 5mn - 6n^2$
$(3m + 2n)(2m - 3n)$

37. $6a^2 - 7ab - 5b^2$
$(3a - 5b)(2a + b)$

38. $25g^2 - 5gh - 2h^2$
$(5g + h)(5g - 2h)$

39. $5 - 6x + x^2$
$(5 - x)(1 - x)$

40. $7 + 8x + x^2$
$(7 + x)(1 + x)$

41. On a quiz, a student factored $16x^2 - 24x + 5$ by grouping as follows.

$$16x^2 - 24x + 5$$
$$= 16x^2 - 4x - 20x + 5$$
$$= 4x(4x - 1) - 5(4x - 1) \quad \text{His answer}$$

He thought his answer was correct since it checked by multiplying. Why was his answer marked wrong? What is the correct factored form?

The student stopped too soon. He needs to factor out the common factor $4x - 1$ to get $(4x - 1)(4x - 5)$ as the correct answer.

42. On the same quiz, another student factored $3k^3 - 12k^2 - 15k$ by first factoring out the common factor $3k$ to get $3k(k^2 - 4k - 5)$. Then she wrote

$$k^2 - 4k - 5$$
$$= k^2 - 5k + k - 5$$
$$= k(k - 5) + 1(k - 5)$$
$$= (k - 5)(k + 1). \quad \text{Her answer}$$

Why was her answer marked wrong? What is the correct factored form?

The student forgot to include the common factor $3k$ in her answer. The correct answer is $3k(k - 5)(k + 1)$.

6.4 Factoring Trinomials Using FOIL

OBJECTIVE 1 Factor trinomials using FOIL. This section shows an alternative method of factoring trinomials in which the coefficient of the squared term is not 1. This method uses trial and error.

To factor $2x^2 + 7x + 6$ (the same trinomial factored at the beginning of **Section 6.3**) by trial and error, we use FOIL backwards. We want to write $2x^2 + 7x + 6$ as the product of two binomials.

$$(\qquad)(\qquad)$$

The product of the two first terms of the binomials is $2x^2$. The possible factors of $2x^2$ are $2x$ and x or $-2x$ and $-x$. Since all terms of the trinomial are positive, we consider only positive factors. Thus, we have

$$(2x \qquad)(x \qquad).$$

The product of the two last terms, 6, can be factored as $1 \cdot 6$, $6 \cdot 1$, $2 \cdot 3$, or $3 \cdot 2$. Try each pair to find the pair that gives the correct middle term, $7x$.

Work Problem 1 *at the Side.* ▶

In part (b) at the side, since $2x + 6 = 2(x + 3)$, the binomial $2x + 6$ has a common factor of 2, while $2x^2 + 7x + 6$ has no common factor other than 1. The product $(2x + 6)(x + 1)$ cannot be correct. (Part (c) also has one binomial factor with a common factor.)

> **Note**
> If the original polynomial has no common factor, then none of its binomial factors will either.

Now try the remaining numbers 3 and 2 as factors of 6.

$$(2x + 3)(x + 2) = 2x^2 + 7x + 6 \qquad \text{Correct}$$

$3x$
$4x$
$7x$ Add.

Finally, we see that $2x^2 + 7x + 6$ factors as $(2x + 3)(x + 2)$.

Check Multiply $(2x + 3)(x + 2)$ to obtain $2x^2 + 7x + 6$.

EXAMPLE 1 Factoring a Trinomial with All Positive Terms Using FOIL

Factor $8p^2 + 14p + 5$.

The number 8 has several possible pairs of factors, but 5 has only 1 and 5 or -1 and -5. For this reason, it is easier to begin by considering the factors of 5. Ignore the negative factors, since all coefficients in the trinomial are positive. If $8p^2 + 14p + 5$ can be factored, the factors will have the form

$$(\quad + 5)(\quad + 1).$$

Continued on Next Page

OBJECTIVE

1 Factor trinomials using FOIL.

1 Multiply to decide whether each factored form is correct or incorrect for

$$2x^2 + 7x + 6.$$

(a) $(2x + 1)(x + 6)$

(b) $(2x + 6)(x + 1)$

(c) $(2x + 2)(x + 3)$

Answers

1. (a) incorrect **(b)** incorrect **(c)** incorrect

When factoring $8p^2 + 14p + 5$, the possible pairs of factors of $8p^2$ are $8p$ and p, or $4p$ and $2p$. Try various combinations, checking to see if the middle term is $14p$ in each case.

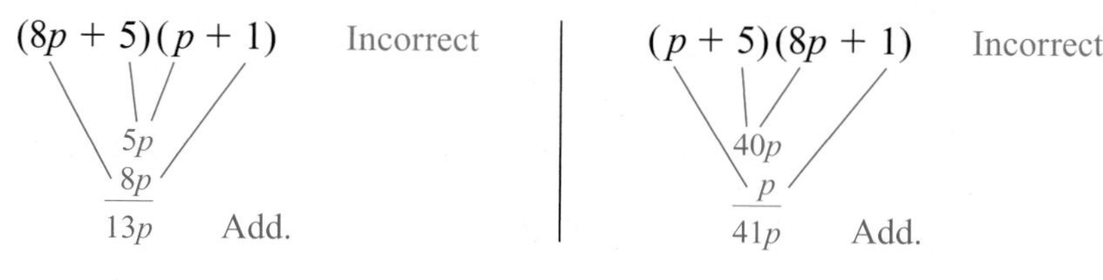

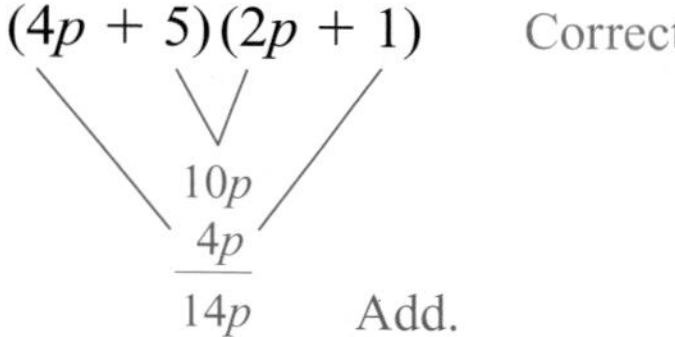

Since $14p$ is the correct middle term,

$$8p^2 + 14p + 5 \quad \text{factors as} \quad (4p + 5)(2p + 1).$$

Check Multiply $(4p + 5)(2p + 1)$ to obtain $8p^2 + 14p + 5$.

◀ *Work Problem* 2 *at the Side.*

2 Factor each trinomial.

(a) $2p^2 + 9p + 9$

(b) $6p^2 + 19p + 10$

(c) $8x^2 + 14x + 3$

EXAMPLE 2 Factoring a Trinomial with a Negative Middle Term Using FOIL

Factor $6x^2 - 11x + 3$.

Since 3 has only 1 and 3 or -1 and -3 as factors, it is better here to begin by factoring 3. The last term of the trinomial $6x^2 - 11x + 3$ is positive and the middle term has a negative coefficient, so we consider only negative factors. We need two negative factors because the *product* of two negative factors is positive and their *sum* is negative, as required.

Try -3 and -1 as factors of 3:

$$(\quad - 3)(\quad - 1).$$

The factors of $6x^2$ may be either $6x$ and x, or $2x$ and $3x$.

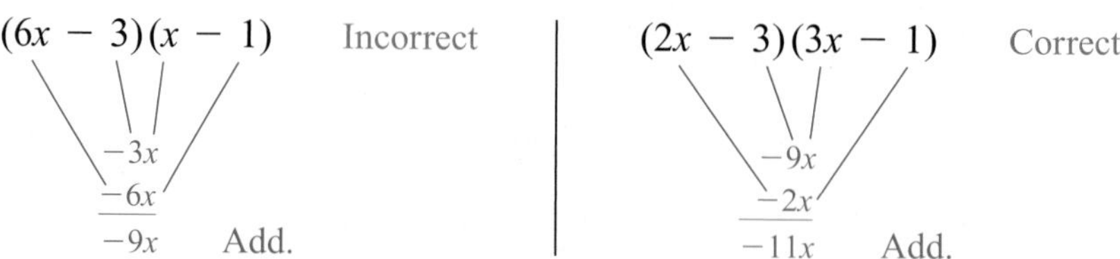

The factors $2x$ and $3x$ produce $-11x$, the correct middle term, so

$$6x^2 - 11x + 3 \quad \text{factors as} \quad (2x - 3)(3x - 1).$$

Check Multiply $(2x - 3)(3x - 1)$ to obtain $6x^2 - 11x + 3$.

3 Factor each trinomial.

(a) $4y^2 - 11y + 6$

(b) $9x^2 - 21x + 10$

Note

In Example 2, we might also realize that our initial attempt to factor $6x^2 - 11x + 3$ as $(6x - 3)(x - 1)$ *cannot* be correct since $6x - 3$ has a common factor of 3 and the original polynomial does not.

◀ *Work Problem* 3 *at the Side.*

Answers

2. (a) $(2p + 3)(p + 3)$
(b) $(3p + 2)(2p + 5)$
(c) $(4x + 1)(2x + 3)$

3. (a) $(4y - 3)(y - 2)$
(b) $(3x - 5)(3x - 2)$

EXAMPLE 3 **Factoring a Trinomial with a Negative Last Term Using FOIL**

Factor $8x^2 + 6x - 9$.

The integer 8 has several possible pairs of factors, as does -9. Since the last term is negative, one positive factor and one negative factor of -9 are needed. Since the coefficient of the middle term is small, it is wise to avoid large factors such as 8 or 9. We try 4 and 2 as factors of 8, and 3 and -3 as factors of -9, and check the middle term.

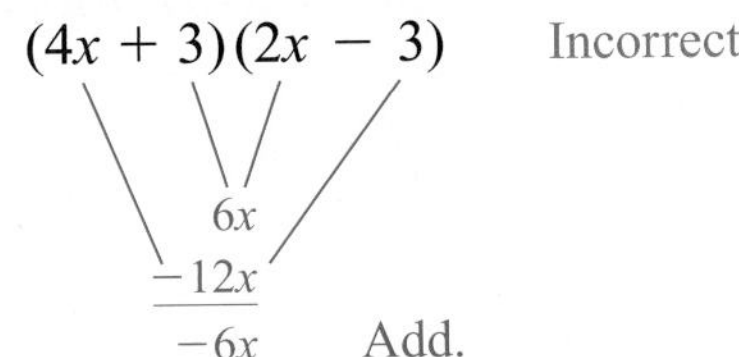

Now we try interchanging 3 and -3, since only the sign of the middle term is incorrect.

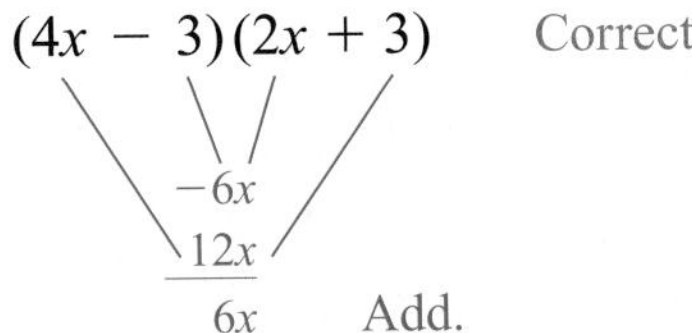

This combination produces $6x$, the correct middle term, so

$$8x^2 + 6x - 9 \quad \text{factors as} \quad (4x - 3)(2x + 3).$$

Work Problem **4** *at the Side.* ▶

EXAMPLE 4 **Factoring a Trinomial with Two Variables**

Factor $12a^2 - ab - 20b^2$.

There are several pairs of factors of $12a^2$, including

$$12a \text{ and } a, \quad 6a \text{ and } 2a, \quad \text{and} \quad 4a \text{ and } 3a,$$

just as there are many possible pairs of factors of $-20b^2$, including

$$20b \text{ and } -b, \quad -20b \text{ and } b, \quad 10b \text{ and } -2b,$$
$$-10b \text{ and } 2b, \quad 4b \text{ and } -5b, \quad \text{and} \quad -4b \text{ and } 5b.$$

Once again, since the coefficient of the middle term is small, avoid the larger factors. Try the factors $6a$ and $2a$ and $4b$ and $-5b$.

$$(6a + 4b)(2a - 5b)$$

This cannot be correct, as mentioned before, since $6a + 4b$ has 2 as a common factor, while the given trinomial does not. Try $3a$ and $4a$ with $4b$ and $-5b$.

$$(3a + 4b)(4a - 5b)$$
$$= 12a^2 + ab - 20b^2 \quad \text{Incorrect}$$

Here the middle term is ab, rather than $-ab$. Interchange the signs of the last two terms in the factors.

$$(3a - 4b)(4a + 5b)$$
$$= 12a^2 - ab - 20b^2 \quad \text{Correct}$$

Work Problem **5** *at the Side.* ▶

4 Factor each trinomial, if possible.

(a) $6x^2 + 5x - 4$

(b) $6m^2 - 11m - 10$

(c) $4x^2 - 3x - 7$

(d) $3y^2 + 8y - 6$

5 Factor each trinomial.

(a) $2x^2 - 5xy - 3y^2$

(b) $8a^2 + 2ab - 3b^2$

ANSWERS

4. **(a)** $(3x + 4)(2x - 1)$
(b) $(2m - 5)(3m + 2)$
(c) $(4x - 7)(x + 1)$
(d) prime

5. **(a)** $(2x + y)(x - 3y)$
(b) $(4a + 3b)(2a - b)$

6 Factor each trinomial.

(a) $36z^3 - 6z^2 - 72z$

(b) $-24x^3 + 32x^2y + 6xy^2$

EXAMPLE 5 Factoring Trinomials with Common Factors

Factor each trinomial.

(a) $15y^3 + 55y^2 + 30y$

First factor out the greatest common factor, $5y$.

$$15y^3 + 55y^2 + 30y$$
$$= 5y(3y^2 + 11y + 6)$$

Now factor $3y^2 + 11y + 6$. Try $3y$ and y as factors of $3y^2$ and 2 and 3 as factors of 6.

$$(3y + 2)(y + 3)$$
$$= 3y^2 + 11y + 6 \quad \text{Correct}$$

The complete factored form of $15y^3 + 55y^2 + 30y$ is

$$5y(3y + 2)(y + 3).$$

Remember the common factor.

Check $5y(3y + 2)(y + 3)$

$$= 5y(3y^2 + 11y + 6) \quad \text{FOIL}$$
$$= 15y^3 + 55y^2 + 30y \quad \text{Distributive property}$$

(b) $-24a^3 - 42a^2 + 45a$

The common factor could be $3a$ or $-3a$. If we factor out $-3a$, the first term of the trinomial will be positive, which makes it easier to factor.

$$-24a^3 - 42a^2 + 45a$$
$$= -3a(8a^2 + 14a - 15) \quad \text{Factor out } -3a.$$
$$= -3a(4a - 3)(2a + 5) \quad \text{Use trial and error.}$$

Check $-3a(4a - 3)(2a + 5)$

$$= -3a(8a^2 + 14a - 15)$$
$$= -24a^3 - 42a^2 + 45a$$

CAUTION

This caution bears repeating: ***Remember to include the common factor in the final factored form.***

◀ Work Problem **6** at the Side.

ANSWERS

6. (a) $6z(2z - 3)(3z + 4)$
(b) $-2x(6x + y)(2x - 3y)$

6.4 Exercises

FOR EXTRA HELP

Math XL PRACTICE

WATCH

DOWNLOAD

READ

REVIEW

Decide which is the correct factored form of the given polynomial.

1. $2x^2 - x - 1$
A. $(2x - 1)(x + 1)$ **B.** $(2x + 1)(x - 1)$ **B**

2. $3a^2 - 5a - 2$
A. $(3a + 1)(a - 2)$ **B.** $(3a - 1)(a + 2)$ **A**

3. $4y^2 + 17y - 15$
A. $(y + 5)(4y - 3)$ **B.** $(2y - 5)(2y + 3)$ **A**

4. $12c^2 - 7c - 12$
A. $(6c - 2)(2c + 6)$ **B.** $(4c + 3)(3c - 4)$ **B**

5. $4k^2 + 13mk + 3m^2$
A. $(4k + m)(k + 3m)$ **B.** $(4k + 3m)(k + m)$ **A**

6. $2x^2 + 11x + 12$
A. $(2x + 3)(x + 4)$ **B.** $(2x + 4)(x + 3)$ **A**

Complete each factoring.

7. $6a^2 + 7ab - 20b^2$
$= (3a - 4b)(\quad)$ **$2a + 5b$**

8. $9m^2 - 3mn - 2n^2$
$= (3m + n)(\quad)$ **$3m - 2n$**

9. $2x^2 + 6x - 8$
$= 2(\quad)$
$= 2(\quad)(\quad)$
$x^2 + 3x - 4$; $x + 4$, $x - 1$, or $x - 1$, $x + 4$

10. $3x^2 - 9x - 30$
$= 3(\quad)$
$= 3(\quad)(\quad)$
$x^2 - 3x - 10$; $x - 5$, $x + 2$, or $x + 2$, $x - 5$

11. $4z^3 - 10z^2 - 6z$
$= 2z(\quad)$
$= 2z(\quad)(\quad)$
$2z^2 - 5z - 3$; $2z + 1$, $z - 3$, or $z - 3$, $2z + 1$

12. $15r^3 - 39r^2 - 18r$
$= 3r(\quad)$
$= 3r(\quad)(\quad)$
$5r^2 - 13r - 6$; $5r + 2$, $r - 3$, or $r - 3$, $5r + 2$

13. For the polynomial $12x^2 + 7x - 12$, 2 is not a common factor. Explain why the binomial $2x - 6$, then, cannot be a factor of the polynomial.

The binomial $2x - 6$ cannot be a factor because it has a common factor of 2, but the polynomial does not.

14. How are the signs of the last terms of the two binomial factors of a trinomial determined?

If the first term of the trinomial is positive and the last term is negative, the signs of the last two terms of the factors must be different. The signs are chosen so the middle term has the correct sign. If the first term of the trinomial is positive and the last term is positive, then the signs of the two last terms of the factors must be the same, the sign of the middle term of the trinomial.

Factor each trinomial completely. See Examples 1–5. **The order of the factors is irrelevant in Exercises 15–50.**

15. $3a^2 + 10a + 7$
$(3a + 7)(a + 1)$

16. $7r^2 + 8r + 1$
$(7r + 1)(r + 1)$

17. $2y^2 + 7y + 6$
$(2y + 3)(y + 2)$

18. $5z^2 + 12z + 4$
$(5z + 2)(z + 2)$

19. $15m^2 + m - 2$
$(3m - 1)(5m + 2)$

20. $6x^2 + x - 1$
$(2x + 1)(3x - 1)$

21. $12s^2 + 11s - 5$
$(3s - 1)(4s + 5)$

22. $20x^2 + 11x - 3$
$(4x + 3)(5x - 1)$

23. $10m^2 - 23m + 12$
$(5m - 4)(2m - 3)$

24. $6x^2 - 17x + 12$
$(3x - 4)(2x - 3)$

25. $8w^2 - 14w + 3$
$(4w - 1)(2w - 3)$

26. $9p^2 - 18p + 8$
$(3p - 2)(3p - 4)$

27. $20y^2 - 39y - 11$
$(4y + 1)(5y - 11)$

28. $10x^2 - 11x - 6$
$(2x - 3)(5x + 2)$

29. $3x^2 - 15x + 16$
prime

30. $2t^2 + 13t - 18$
prime

31. $20x^2 + 22x + 6$
$2(5x + 3)(2x + 1)$

32. $36y^2 + 81y + 45$
$9(4y + 5)(y + 1)$

33. $-40m^2q - mq + 6q$
$-q(5m + 2)(8m - 3)$

34. $-15a^2b - 22ab - 8b$
$-b(3a + 2)(5a + 4)$

35. $15n^4 - 39n^3 + 18n^2$
$3n^2(5n - 3)(n - 2)$

36. $24a^4 + 10a^3 - 4a^2$
$2a^2(3a + 2)(4a - 1)$

37. $-15x^2y^2 + 7xy^2 + 4y^2$
$-y^2(5x - 4)(3x + 1)$

38. $-14a^2b^3 - 15ab^3 + 9b^3$
$-b^3(7a - 3)(2a + 3)$

39. $5a^2 - 7ab - 6b^2$
$(5a + 3b)(a - 2b)$

40. $6x^2 - 5xy - y^2$
$(6x + y)(x - y)$

41. $12s^2 + 11st - 5t^2$
$(4s + 5t)(3s - t)$

42. $25a^2 + 25ab + 6b^2$
$(5a + 2b)(5a + 3b)$

43. $6m^6n + 7m^5n^2 + 2m^4n^3$
$m^4n(3m + 2n)(2m + n)$

44. $12k^3q^4 - 4k^2q^5 - kq^6$
$kq^4(6k + q)(2k - q)$

If a trinomial has a negative coefficient for the squared term, such as $-2x^2 + 11x - 12$, it may be easier to factor by first factoring out the common factor -1:

$$\begin{aligned} &-2x^2 + 11x - 12 \\ &= -1(2x^2 - 11x + 12) \\ &= -1(2x - 3)(x - 4). \end{aligned}$$

Use this method to factor the trinomials in Exercises 45–50.

45. $-x^2 - 4x + 21$
$-1(x + 7)(x - 3)$

46. $-x^2 + x + 72$
$-1(x + 8)(x - 9)$

47. $-3x^2 - x + 4$
$-1(3x + 4)(x - 1)$

48. $-5x^2 + 2x + 16$
$-1(5x + 8)(x - 2)$

49. $-2a^2 - 5ab - 2b^2$
$-1(a + 2b)(2a + b)$

50. $-3p^2 + 13pq - 4q^2$
$-1(p - 4q)(3p - q)$

Relating Concepts (Exercises 51–56) For Individual or Group Work

One of the most common problems that beginning algebra students face is this: If an answer obtained doesn't look exactly like the one given in the back of the book, is it necessarily incorrect? Often there are several different equivalent forms of an answer that are all correct. **Work Exercises 51–56 in order,** *to see how and why this is possible for factoring problems.*

51. Factor the integer 35 as the product of two prime numbers. **$5 \cdot 7$**

52. Factor the integer 35 as the product of the negatives of two prime numbers. **$(-5)(-7)$**

53. Verify that $6x^2 - 11x + 4$ factors as $(3x - 4)(2x - 1)$.
The product of $3x - 4$ and $2x - 1$ is $6x^2 - 11x + 4$.

54. Verify that $6x^2 - 11x + 4$ factors as $(4 - 3x)(1 - 2x)$.
The product of $4 - 3x$ and $1 - 2x$ is $6x^2 - 11x + 4$.

55. Compare the two valid factored forms in Exercises 53 and 54. How do the factors in each case compare?
The factors in Exercise 53 are the opposites of the factors in Exercise 54.

56. Suppose you know that the correct factored form of a particular trinomial is $(7t - 3)(2t - 5)$. Based on your observations in Exercises 51–55, what is another valid factored form?
$(3 - 7t)(5 - 2t)$

6.5 Special Factoring Techniques

OBJECTIVES

1. Factor a difference of squares.
2. Factor a perfect square trinomial.
3. Factor a difference of cubes.
4. Factor a sum of cubes.

By reversing the rules for multiplication of binomials from **Section 5.4,** we obtain rules for factoring polynomials in certain forms.

OBJECTIVE 1 Factor a difference of squares. The formula for the product of the sum and difference of the same two terms is

$$(a + b)(a - b) = a^2 - b^2.$$

Reversing this rule leads to the following special factoring rule.

Factoring a Difference of Squares

$$a^2 - b^2 = (a + b)(a - b)$$

For example,
$$m^2 - 16$$
$$= m^2 - 4^2$$
$$= (m + 4)(m - 4).$$

As the next examples show, the following conditions must be true for a binomial to be a difference of squares.

1. Both terms of the binomial must be squares, such as
$$x^2, \quad 9y^2, \quad 25, \quad 1, \quad m^4.$$
2. The terms of the binomial must have different signs (one positive and one negative).

EXAMPLE 1 Factoring Differences of Squares

Factor each binomial, if possible. (In part (c), use fractions.)

$$a^2 - b^2 = (a + b)(a - b)$$

(a) $x^2 - 49 = x^2 - 7^2 = (x + 7)(x - 7)$

(b) $y^2 - m^2$
$= (y + m)(y - m)$

(c) $z^2 - \frac{9}{16}$
$= \left(z + \frac{3}{4}\right)\left(z - \frac{3}{4}\right)$ $\quad \frac{9}{16} = \left(\frac{3}{4}\right)^2$

(d) $x^2 - 8$
Because 8 is not the square of an integer, this binomial is not a difference of squares. It is a prime polynomial.

(e) $p^2 + 16$
Since $p^2 + 16$ is a *sum* of squares, it is not equal to $(p + 4)(p - 4)$. Also, using FOIL,

$$(p - 4)(p - 4)$$
$$= p^2 - 8p + 16, \quad \text{not} \quad p^2 + 16,$$

and
$$(p + 4)(p + 4)$$
$$= p^2 + 8p + 16, \quad \text{not} \quad p^2 + 16,$$

so $p^2 + 16$ is a prime polynomial.

CAUTION

As Example 1(e) suggests, ***after any common factor is removed, a sum of squares cannot be factored.***

EXAMPLE 2 Factoring Differences of Squares

Factor each difference of squares.

$$a^2 - b^2 = (a+b)(a-b)$$

(a) $25m^2 - 16 = (5m)^2 - 4^2 = (5m+4)(5m-4)$

(b) $49z^2 - 64$

$= (7z)^2 - 8^2$

$= (7z+8)(7z-8)$

◀ Work Problem **1** at the Side.

Note

Always check a factored form by multiplying.

EXAMPLE 3 Factoring More Complex Differences of Squares

Factor completely.

(a) $81y^2 - 36$

$= 9(9y^2 - 4)$ Factor out the GCF, 9.

$= 9[(3y)^2 - 2^2]$

$= 9(3y+2)(3y-2)$ Difference of squares

(b) $9x^2 - 4z^2$

$= (3x)^2 - (2z)^2$

$= (3x+2z)(3x-2z)$

(c) $p^4 - 36$

$= (p^2)^2 - 6^2$

$= (p^2+6)(p^2-6)$

(d) $m^4 - 16$

$= (m^2)^2 - 4^2$

$= (m^2+4)(m^2-4)$ Difference of squares

Don't stop here.

$= (m^2+4)(m+2)(m-2)$ Difference of squares again

CAUTION

Factor again when any of the factors is a difference of squares, as in Example 3(d). Check by multiplying.

◀ Work Problem **2** at the Side.

1 Factor, if possible. (In part (b), use fractions.)

(a) $p^2 - 100$

(b) $x^2 - \frac{25}{36}$

(c) $x^2 + y^2$

(d) $9m^2 - 49$

(e) $64a^2 - 25$

2 Factor completely.

(a) $50r^2 - 32$

(b) $27y^2 - 75$

(c) $25a^2 - 64b^2$

(d) $k^4 - 49$

(e) $81r^4 - 16$

ANSWERS

1. **(a)** $(p+10)(p-10)$
(b) $\left(x+\frac{5}{6}\right)\left(x-\frac{5}{6}\right)$
(c) prime
(d) $(3m+7)(3m-7)$
(e) $(8a+5)(8a-5)$

2. **(a)** $2(5r+4)(5r-4)$
(b) $3(3y+5)(3y-5)$
(c) $(5a+8b)(5a-8b)$
(d) $(k^2+7)(k^2-7)$
(e) $(9r^2+4)(3r+2)(3r-2)$

OBJECTIVE 2 Factor a perfect square trinomial. The expressions 144, $4x^2$, and $81m^6$ are called *perfect squares* because

$$144 = 12^2, \quad 4x^2 = (2x)^2, \quad \text{and} \quad 81m^6 = (9m^3)^2.$$

A **perfect square trinomial** is a trinomial that is the square of a binomial. For example, $x^2 + 8x + 16$ is a perfect square trinomial because it is the square of the binomial $x + 4$:

$$\begin{aligned} &x^2 + 8x + 16 \\ &= (x + 4)(x + 4) \\ &= (x + 4)^2. \end{aligned}$$

On the one hand, a necessary condition for a trinomial to be a perfect square is that *two of its terms must be perfect squares*. For this reason, $16x^2 + 4x + 15$ is not a perfect square trinomial because only the term $16x^2$ is a perfect square.

On the other hand, even if two of the terms are perfect squares, the trinomial may not be a perfect square trinomial. For example, $x^2 + 6x + 36$ has two perfect square terms, but it is not a perfect square trinomial. (Try to find a binomial that can be squared to give $x^2 + 6x + 36$.)

We can multiply to see that the square of a binomial gives one of the following perfect square trinomials.

Factoring Perfect Square Trinomials

$$a^2 + 2ab + b^2 = (a + b)^2$$

$$a^2 - 2ab + b^2 = (a - b)^2$$

The middle term of a perfect square trinomial is always twice the product of the two terms in the squared binomial. (See **Section 5.4.**) Use this to check any attempt to factor a trinomial that appears to be a perfect square.

EXAMPLE 4 Factoring a Perfect Square Trinomial

Factor $x^2 + 10x + 25$.

The term x^2 is a perfect square, and so is 25. Try to factor the trinomial

$$x^2 + 10x + 25 \quad \text{as} \quad (x + 5)^2.$$

To check, take twice the product of the two terms in the squared binomial.

$$2 \cdot x \cdot 5 = 10x$$

Twice First term of binomial Last term of binomial

Since $10x$ is the middle term of the trinomial, the trinomial is a perfect square and can be factored as $(x + 5)^2$. Thus,

$$x^2 + 10x + 25 \quad \text{factors as} \quad (x + 5)^2.$$

Work Problem 3 at the Side. ▶

3 Factor each trinomial.

(a) $p^2 + 14p + 49$

(b) $m^2 + 8m + 16$

(c) $x^2 + 2x + 1$

ANSWERS

3. (a) $(p + 7)^2$ (b) $(m + 4)^2$ (c) $(x + 1)^2$

4 Factor each trinomial.

(a) $p^2 - 18p + 81$

(b) $16a^2 + 56a + 49$

(c) $121p^2 + 110p + 100$

(d) $64x^2 - 48x + 9$

(e) $27y^3 + 72y^2 + 48y$

EXAMPLE 5 Factoring Perfect Square Trinomials

Factor each trinomial.

(a) $x^2 - 22x + 121$

The first and last terms are perfect squares ($121 = 11^2$ or $(-11)^2$). Check to see whether the middle term of $x^2 - 22x + 121$ is twice the product of the first and last terms of the binomial $x - 11$.

$$2 \cdot x \cdot (-11) = -22x$$

Twice — First term — Last term

Since twice the product of the first and last terms of the binomial is the middle term, $x^2 - 22x + 121$ is a perfect square trinomial and

$$x^2 - 22x + 121 \quad \text{factors as} \quad (x - 11)^2.$$

Same sign

Notice that the sign of the second term in the squared binomial is the same as the sign of the middle term in the trinomial.

(b) $9m^2 - 24m + 16 = (3m)^2 + 2(3m)(-4) + (-4)^2 = (3m - 4)^2$

Twice — First term — Last Term

(c) $25y^2 + 20y + 16$

The first and last terms are perfect squares.

$$25y^2 = (5y)^2 \quad \text{and} \quad 16 = 4^2$$

However, twice the product of the first and last terms of the binomial $5y + 4$ is $2 \cdot 5y \cdot 4 = 40y$, which is not the middle term of $25y^2 + 20y + 16$. This trinomial is not a perfect square. In fact, the trinomial cannot be factored, even with the methods of the previous sections. It is a prime polynomial.

(d) $12z^3 + 60z^2 + 75z$

$= 3z(4z^2 + 20z + 25)$ Factor out $3z$.

$= 3z[(2z)^2 + 2(2z)(5) + 5^2]$ $4z^2 + 20z + 25$ is a perfect square trinomial.

$= 3z(2z + 5)^2$ Factor.

Note

1. The sign of the second term in the squared binomial is always the same as the sign of the middle term in the trinomial.
2. The first and last terms of a perfect square trinomial must be *positive,* because they are squares. For example, the polynomial $x^2 - 2x - 1$ cannot be a perfect square because the last term is negative.
3. Perfect square trinomials can also be factored using grouping or FOIL, although using the method of this section is often easier.

◀ Work Problem **4** at the Side.

ANSWERS

4. **(a)** $(p - 9)^2$ **(b)** $(4a + 7)^2$ **(c)** prime **(d)** $(8x - 3)^2$ **(e)** $3y(3y + 4)^2$

OBJECTIVE 3 Factor a difference of cubes. We factored a difference of squares at the beginning of this section. We can also factor a **difference of cubes.** Use the following pattern.

Difference of Cubes

$$a^3 - b^3 = (a - b)(a^2 + ab + b^2)$$

This pattern should be memorized. Multiply on the right to see that the pattern gives the correct factors, as shown in the margin.

Notice the pattern of the terms in the factored form of $a^3 - b^3$.

- $a^3 - b^3 =$ (a binomial factor) (a trinomial factor)
- The binomial factor has the difference of the cube roots of the given terms.
- The terms in the trinomial factor are all positive.
- What you write in the binomial factor determines the trinomial factor.

$$a^3 - b^3 = (a - b)(\underbrace{a^2}_{\text{First term squared}} + \underbrace{ab}_{\text{positive product of the terms}} + \underbrace{b^2}_{\text{second term squared}})$$

$$\begin{array}{r} a^2 + ab + b^2 \\ a - b \\ \hline -a^2b - ab^2 - b^3 \\ a^3 + a^2b + ab^2 \quad\quad \\ \hline a^3 \quad\quad\quad\quad - b^3 \end{array}$$

EXAMPLE 6 Factoring Differences of Cubes

Factor each difference of cubes.

(a) $m^3 - 125$

Use the pattern for a difference of cubes.

$$\begin{aligned} a^3 - b^3 &= (a - b)(a^2 + ab + b^2) \\ m^3 - 125 = m^3 - 5^3 &= (m - 5)(m^2 + 5m + 5^2) \\ &= (m - 5)(m^2 + 5m + 25) \end{aligned}$$

(b) $8p^3 - 27$

$= (2p)^3 - 3^3$ $\quad 8p^3 = (2p)^3;\ 27 = 3^3$

$= (2p - 3)[(2p)^2 + (2p)3 + 3^2]$ $\quad a^3 - b^3 = (a - b)(a^2 + ab + b^2)$

$= (2p - 3)(4p^2 + 6p + 9)$ $\quad (ab)^2 = a^2b^2$

(c) $4m^3 - 32n^3$

$= 4(m^3 - 8n^3)$ Factor out the common factor.

$= 4[m^3 - (2n)^3]$ $\quad 8n^3 = (2n)^3$

$= 4(m - 2n)[m^2 + m(2n) + (2n)^2]$

$= 4(m - 2n)(m^2 + 2mn + 4n^2)$

Work Problem 5 at the Side. ▶

5 Factor each difference of cubes.

(a) $t^3 - 64$

(b) $2x^3 - 54$

(c) $8k^3 - y^3$

CAUTION

A common error in factoring $a^3 - b^3 = (a - b)(a^2 + ab + b^2)$ is to try to factor $a^2 + ab + b^2$. This particular trinomial cannot be factored.

ANSWERS

5. (a) $(t - 4)(t^2 + 4t + 16)$
(b) $2(x - 3)(x^2 + 3x + 9)$
(c) $(2k - y)(4k^2 + 2ky + y^2)$

OBJECTIVE 4 Factor a sum of cubes. A sum of squares, such as $m^2 + 25$, cannot be factored using real numbers, but a **sum of cubes** can.

Sum of Cubes

$$a^3 + b^3 = (a + b)(a^2 - ab + b^2)$$

Observe the positive and negative signs in these factoring patterns.

$$a^3 - b^3 = (a - b)(a^2 + ab + b^2) \quad \text{Difference of cubes}$$

(Same sign: the $-$ in $a^3 - b^3$ and in $(a - b)$; Opposite sign: the $+$ before ab; Positive: the $+$ before ab and before b^2.)

$$a^3 + b^3 = (a + b)(a^2 - ab + b^2) \quad \text{Sum of cubes}$$

(Same sign: the $+$ in $a^3 + b^3$ and in $(a + b)$; Opposite sign: the $-$ before ab; Positive: the $+$ before b^2.)

EXAMPLE 7 Factoring Sums of Cubes

Factor each sum of cubes.

(a) $k^3 + 27$

$= k^3 + 3^3$

$= (k + 3)(k^2 - 3k + 3^2)$

$= (k + 3)(k^2 - 3k + 9)$

(b) $8m^3 + 125p^3$

$= (2m)^3 + (5p)^3$ $\quad 8m^3 = (2m)^3;\ 125p^3 = (5p)^3$

$= (2m + 5p)[(2m)^2 - (2m)(5p) + (5p)^2]$

$= (2m + 5p)(4m^2 - 10mp + 25p^2)$

◀ Work Problem 6 at the Side.

6 Factor each sum of cubes.

(a) $x^3 + 8$

(b) $64y^3 + 1$

(c) $27m^3 + 343n^3$

The methods of factoring discussed in this section are summarized here.

Special Factoring Rules

Difference of squares	$a^2 - b^2 = (a + b)(a - b)$
Perfect square trinomials	$a^2 + 2ab + b^2 = (a + b)^2$
	$a^2 - 2ab + b^2 = (a - b)^2$
Difference of cubes	$a^3 - b^3 = (a - b)(a^2 + ab + b^2)$
Sum of cubes	$a^3 + b^3 = (a + b)(a^2 - ab + b^2)$

Remember that a sum of squares cannot be factored using real numbers unless the terms have a common factor.

ANSWERS

6. (a) $(x + 2)(x^2 - 2x + 4)$
(b) $(4y + 1)(16y^2 - 4y + 1)$
(c) $(3m + 7n)(9m^2 - 21mn + 49n^2)$

6.5 Exercises

FOR EXTRA HELP MyMathLab  Math XL PRACTICE  WATCH  DOWNLOAD  READ  REVIEW

1. To help you factor a difference of squares, complete the following list of squares.

$1^2 =$ **1** $2^2 =$ **4** $3^2 =$ **9** $4^2 =$ **16** $5^2 =$ **25**

$6^2 =$ **36** $7^2 =$ **49** $8^2 =$ **64** $9^2 =$ **81** $10^2 =$ **100**

$11^2 =$ **121** $12^2 =$ **144** $13^2 =$ **169** $14^2 =$ **196** $15^2 =$ **225**

$16^2 =$ **256** $17^2 =$ **289** $18^2 =$ **324** $19^2 =$ **361** $20^2 =$ **400**

2. To use the factoring techniques described in this section, you will sometimes need to recognize fourth powers of integers. Complete the following list of fourth powers.

$1^4 =$ **1** $2^4 =$ **16** $3^4 =$ **81** $4^4 =$ **256** $5^4 =$ **625**

3. The following powers of x are all perfect squares: $x^2, x^4, x^6, x^8, x^{10}$. Based on this observation, we may make a conjecture (an educated guess) that if the power of a variable is divisible by **2** (with 0 remainder), then it is a perfect square.

4. Which of the following are differences of squares? **A, D**

A. $x^2 - 4$ **B.** $y^2 + 9$ **C.** $2a^2 - 25$ **D.** $9m^2 - 1$

Factor each binomial completely. In Exercises 7, 8, 13, and 14, use fractions. See Examples 1–3.

5. $y^2 - 25$ — $(y + 5)(y - 5)$

6. $t^2 - 16$ — $(t + 4)(t - 4)$

7. $p^2 - \frac{1}{9}$ — $\left(p + \frac{1}{3}\right)\left(p - \frac{1}{3}\right)$

8. $q^2 - \frac{1}{4}$ — $\left(q + \frac{1}{2}\right)\left(q - \frac{1}{2}\right)$

9. $m^2 - 12$ — prime

10. $k^2 - 18$ — prime

11. $9r^2 - 4$ — $(3r + 2)(3r - 2)$

12. $4x^2 - 9$ — $(2x + 3)(2x - 3)$

13. $4m^2 - \frac{9}{25}$ — $\left(2m + \frac{3}{5}\right)\left(2m - \frac{3}{5}\right)$

14. $100b^2 - \frac{49}{81}$ — $\left(10b + \frac{7}{9}\right)\left(10b - \frac{7}{9}\right)$

15. $36x^2 - 16$ — $4(3x + 2)(3x - 2)$

16. $32a^2 - 8$ — $8(2a + 1)(2a - 1)$

17. $196p^2 - 225$ — $(14p + 15)(14p - 15)$

18. $361q^2 - 400$ — $(19q + 20)(19q - 20)$

19. $16r^2 - 25a^2$ — $(4r + 5a)(4r - 5a)$

20. $49m^2 - 100p^2$ — $(7m + 10p)(7m - 10p)$

21. $100x^2 + 49$ — prime

22. $81w^2 + 16$ — prime

23. $p^4 - 49$ — $(p^2 + 7)(p^2 - 7)$

24. $r^4 - 25$ — $(r^2 + 5)(r^2 - 5)$

25. $x^4 - 1$ — $(x^2 + 1)(x + 1)(x - 1)$

26. $y^4 - 10{,}000$ — $(y^2 + 100)(y + 10)(y - 10)$

27. $p^4 - 256$ — $(p^2 + 16)(p + 4)(p - 4)$

28. $16k^4 - 1$ — $(4k^2 + 1)(2k + 1)(2k - 1)$

29. When a student was directed to factor $x^4 - 81$ completely, his teacher did not give him full credit for the answer $(x^2 + 9)(x^2 - 9)$. The student argued that because his answer does indeed give $x^4 - 81$ when multiplied out, he should be given full credit. ***WHAT WENT WRONG?*** Give the correct factored form.

The teacher was justified, because it was not factored *completely*; $x^2 - 9$ can be factored as $(x + 3)(x - 3)$. The complete factored form is $(x^2 + 9)(x + 3)(x - 3)$.

30. The binomial $4x^2 + 16$ is a sum of squares that *can* be factored. How is this binomial factored? When can a sum of squares be factored?

$4x^2 + 16$ is factored as $4(x^2 + 4)$. A sum of squares can be factored if the terms have a common factor.

31. In the polynomial $9y^2 + 14y + 25$, the first and last terms are perfect squares. Can the polynomial be factored? If it can, factor it. If it cannot, explain why it is not a perfect square trinomial.

No, it is not a perfect square since the middle term would have to be $30y$.

32. Which of the following are perfect square trinomials?

A. $y^2 - 13y + 36$ **B.** $x^2 + 6x + 9$ **C.** $4z^2 - 4z + 1$ **D.** $16m^2 + 10m + 1$

B, C

Factor each trinomial completely. It may be necessary to factor out the greatest common factor first. In Exercises 37–40, use fractions or decimals, as appropriate. See Examples 4 and 5.

33. $w^2 + 2w + 1$ **$(w + 1)^2$**

34. $p^2 + 4p + 4$ **$(p + 2)^2$**

35. $x^2 - 8x + 16$ **$(x - 4)^2$**

36. $x^2 - 10x + 25$ **$(x - 5)^2$**

37. $t^2 + t + \frac{1}{4}$ **$\left(t + \frac{1}{2}\right)^2$**

38. $m^2 + \frac{2}{3}m + \frac{1}{9}$ **$\left(m + \frac{1}{3}\right)^2$**

39. $x^2 - 1.0x + 0.25$ **$(x - 0.5)^2$**

40. $y^2 - 1.4y + 0.49$ **$(y - 0.7)^2$**

41. $2x^2 + 24x + 72$ **$2(x + 6)^2$**

42. $3y^2 - 48y + 192$ **$3(y - 8)^2$**

43. $16x^2 - 40x + 25$ **$(4x - 5)^2$**

44. $36y^2 - 60y + 25$ **$(6y - 5)^2$**

45. $49x^2 - 28xy + 4y^2$ **$(7x - 2y)^2$**

46. $4z^2 - 12zw + 9w^2$ **$(2z - 3w)^2$**

47. $64x^2 + 48xy + 9y^2$ **$(8x + 3y)^2$**

48. $9t^2 + 24tr + 16r^2$ **$(3t + 4r)^2$**

49. $-50h^3 + 40h^2y - 8hy^2$ **$-2h(5h - 2y)^2$**

50. $-18x^3 - 48x^2y - 32xy^2$ **$-2x(3x + 4y)^2$**

51. To help you factor a sum or difference of cubes, complete the following list of cubes.

$1^3 =$ 1 $\quad 2^3 =$ 8 $\quad 3^3 =$ 27 $\quad 4^3 =$ 64 $\quad 5^3 =$ 125

$6^3 =$ 216 $\quad 7^3 =$ 343 $\quad 8^3 =$ 512 $\quad 9^3 =$ 729 $\quad 10^3 =$ 1000

52. The following powers of x are all perfect cubes: $x^3, x^6, x^9, x^{12}, x^{15}$. Based on this observation, we may make a conjecture that if the power of a variable is divisible by 3 (with 0 remainder), then we have a perfect cube.

53. Which of the following are differences of cubes?

A. $9x^3 - 125$ **B.** $x^3 - 16$ **C.** $x^3 - 1$ **D.** $8x^3 - 27y^3$ **C, D**

54. Which of the following are sums of cubes?

A. $x^3 + 1$ **B.** $x^3 + 36$ **C.** $12x^3 + 27$ **D.** $64x^3 + 216y^3$ **A, D**

Factor. Use your answers in Exercises 51 and 52 as necessary. See Examples 6 and 7.

55. $a^3 + 1$

$(a + 1)(a^2 - a + 1)$

56. $m^3 + 8$

$(m + 2)(m^2 - 2m + 4)$

57. $a^3 - 1$

$(a - 1)(a^2 + a + 1)$

58. $m^3 - 8$

$(m - 2)(m^2 + 2m + 4)$

59. $p^3 + q^3$

$(p + q)(p^2 - pq + q^2)$

60. $w^3 + z^3$

$(w + z)(w^2 - wz + z^2)$

61. $y^3 - 216$

$(y - 6)(y^2 + 6y + 36)$

62. $x^3 - 343$

$(x - 7)(x^2 + 7x + 49)$

63. $k^3 + 1000$

$(k + 10)(k^2 - 10k + 100)$

64. $p^3 + 512$

$(p + 8)(p^2 - 8p + 64)$

65. $27x^3 - 1$

$(3x - 1)(9x^2 + 3x + 1)$

66. $64y^3 - 27$

$(4y - 3)(16y^2 + 12y + 9)$

67. $125a^3 + 8$

$(5a + 2)(25a^2 - 10a + 4)$

68. $216b^3 + 125$

$(6b + 5)(36b^2 - 30b + 25)$

69. $y^3 - 8x^3$

$(y - 2x)(y^2 + 2xy + 4x^2)$

70. $w^3 - 216z^3$

$(w - 6z)(w^2 + 6wz + 36z^2)$

71. $27a^3 - 64b^3$

$(3a - 4b)(9a^2 + 12ab + 16b^2)$

72. $125m^3 - 8n^3$

$(5m - 2n)(25m^2 + 10mn + 4n^2)$

73. $8p^3 + 729q^3$

$(2p + 9q)(4p^2 - 18pq + 81q^2)$

74. $27x^3 + 1000y^3$

$(3x + 10y)(9x^2 - 30xy + 100y^2)$

75. $16t^3 - 2$

$2(2t - 1)(4t^2 + 2t + 1)$

76. $3p^3 - 81$

$3(p - 3)(p^2 + 3p + 9)$

77. $40w^3 + 135$

$5(2w + 3)(4w^2 - 6w + 9)$

78. $32z^3 + 500$

$4(2z + 5)(4z^2 - 10z + 25)$

79. $x^3 + y^6$

$(x + y^2)(x^2 - xy^2 + y^4)$

80. $p^9 + q^3$

$(p^3 + q)(p^6 - p^3q + q^2)$

81. $125k^3 - 8m^9$

$(5k - 2m^3)(25k^2 + 10km^3 + 4m^6)$

82. $125c^6 - 216d^3$

$(5c^2 - 6d)(25c^4 + 30c^2d + 36d^2)$

Relating Concepts (Exercises 83–90) For Individual or Group Work

A binomial may be both a difference of squares and a difference of cubes. One example of such a binomial is $x^6 - 1$. Using the techniques of this section, one factoring method will give the complete factored form, while the other will not. ***Work Exercises 83–90 in order,*** *to determine the method to use.*

83. Factor $x^6 - 1$ as the difference of two squares.

$(x^3 - 1)(x^3 + 1)$

84. The factored form obtained in Exercise 83 consists of a difference of cubes multiplied by a sum of cubes. Factor each binomial further.

$(x - 1)(x^2 + x + 1)(x + 1)(x^2 - x + 1)$

85. Now start over and factor $x^6 - 1$ as a difference of cubes.

$(x^2 - 1)(x^4 + x^2 + 1)$

86. The factored form obtained in Exercise 85 consists of a binomial that is a difference of squares and a trinomial. Factor the binomial further.

$(x - 1)(x + 1)(x^4 + x^2 + 1)$

87. Compare your results in Exercises 84 and 86. Which one of these is the completely factored form?

The result in Exercise 84 is completely factored.

88. Verify that the trinomial in the factored form in Exercise 86 is the product of the two trinomials in the factored form in Exercise 84.

Show that $x^4 + x^2 + 1 = (x^2 + x + 1)(x^2 - x + 1)$.

89. Use the results of Exercises 83–88 to complete the following statement:

In general, if I must choose between factoring first using the method for a difference of squares or the method for a difference of cubes, I should choose the **difference of squares** method to eventually obtain the complete factored form.

90. Find the *complete* factored form of $x^6 - 729$ using the knowledge you have gained in Exercises 83–89.

$(x - 3)(x^2 + 3x + 9)(x + 3)(x^2 - 3x + 9)$

6.6 A General Approach to Factoring

A polynomial is completely factored when **(1)** it is written as a *product* of prime polynomials with integer coefficients, and **(2)** none of the polynomial factors can be factored further.

OBJECTIVES

1. Factor out any common factor.
2. Factor binomials.
3. Factor trinomials.
4. Factor polynomials with more than three terms.

Factoring a Polynomial

Step 1 **Factor out any common factor.**

Step 2 **If the polynomial is a binomial,** check to see if it is a difference of squares, a difference of cubes, or a sum of cubes.

If the polynomial is a trinomial, check to see if it is a perfect square trinomial. If it is not, factor as in **Sections 6.2–6.4.**

If the polynomial has more than three terms, try to factor by grouping as in **Section 6.1.**

Step 3 ***Check the factored form by multiplying.***

OBJECTIVE 1 Factor out any common factor. ***This step is always the same, regardless of the number of terms in the polynomial.***

EXAMPLE 1 Factoring Out a Common Factor

Factor each polynomial.

(a) $9p + 45$

$= 9(p + 5)$ GCF $= 9$

(b) $8m^2p^2 + 4mp$

$= 4mp(2mp + 1)$

Check the factored form by multiplying.

(c) $5x(a + b) - y(a + b)$

$= (a + b)(5x - y)$ Factor out $a + b$.

Work Problem 1 at the Side. ▶

1 Factor each polynomial.

(a) $8x - 80$

(b) $2x^3 + 10x^2 - 2x$

(c) $12m(p - q) - 7n(p - q)$

OBJECTIVE 2 Factor binomials. Check for the following patterns.

Factoring a Binomial (Two Terms)

Difference of squares	$x^2 - y^2 = (x + y)(x - y)$
Difference of cubes	$x^3 - y^3 = (x - y)(x^2 + xy + y^2)$
Sum of cubes	$x^3 + y^3 = (x + y)(x^2 - xy + y^2)$

EXAMPLE 2 Factoring Binomials

Factor each binomial if possible.

(a) $64m^2 - 9n^2$

$= (8m)^2 - (3n)^2$ Difference of squares

$= (8m + 3n)(8m - 3n)$

(b) $27x^3 - 8$ $27x^3 = (3x)^3$; $8 = 2^3$

$= (3x - 2)[(3x)^2 + (3x)(2) + 2^2]$

$= (3x - 2)(9x^2 + 6x + 4)$

(c) $1000m^3 + 1$

$= (10m)^3 + 1^3$ Sum of cubes

$= (10m + 1)[(10m)^2 - (10m)(1) + 1^2]$

$= (10m + 1)(100m^2 - 10m + 1)$ Check by multiplying.

(d) $25m^2 + 121$ is prime. It is a *sum* of squares.

Work Problem 2 at the Side. ▶

2 Factor each binomial if possible.

(a) $36x^2 - y^2$

(b) $4t^2 + 1$

(c) $125x^3 - 27y^3$

(d) $x^3 + 343y^3$

Answers

1. **(a)** $8(x - 10)$
   (b) $2x(x^2 + 5x - 1)$
   (c) $(p - q)(12m - 7n)$
2. **(a)** $(6x + y)(6x - y)$
   (b) prime
   (c) $(5x - 3y)(25x^2 + 15xy + 9y^2)$
   (d) $(x + 7y)(x^2 - 7xy + 49y^2)$

Note

The binomial $25m^2 + 625$ is a sum of squares. It *can* be factored, however, as $25(m^2 + 25)$ because it has a common factor, 25.

OBJECTIVE 3 Factor trinomials. Consider the following.

Factoring a Trinomial (Three Terms)

Decide whether it is a perfect square trinomial of the form

$$x^2 + 2xy + y^2 = (x + y)^2 \quad \text{or} \quad x^2 - 2xy + y^2 = (x - y)^2.$$

If not, use the general factoring methods of **Sections 6.2–6.4.**

EXAMPLE 3 Factoring Trinomials

Factor each trinomial.

(a) $p^2 + 10p + 25$
$= (p + 5)^2$ $\quad 2(p)(5) = 10p$

(b) $49z^2 - 42z + 9$
$= (7z - 3)^2$ $\quad 2(7z)(3) = 42z$

(c) $y^2 - 5y - 6$
$= (y - 6)(y + 1)$ $\quad$ The integers -6 and 1 have a product of -6 and a sum of -5. **(Section 6.2)**

(d) $2k^2 - k - 6$
$= (2k + 3)(k - 2)$
(Section 6.4)

(e) $28z^2 + 6z - 10$
$= 2(14z^2 + 3z - 5)$
$= 2(7z + 5)(2z - 1)$

◀ Work Problem 3 at the Side.

OBJECTIVE 4 Factor polynomials with more than three terms. Consider factoring by grouping from **Section 6.1.**

EXAMPLE 4 Factoring Polynomials with More than Three Terms

Factor each polynomial.

(a) $4 - 2q - 6p + 3pq$

$= (4 - 2q) + (-6p + 3pq)$	Group the terms.
$= 2(2 - q) - 3p(2 - q)$	Factor each group.
$= (2 - q)(2 - 3p)$	Factor out $2 - q$.

(b) $20k^3 + 4k^2 - 45k - 9$

$= (20k^3 + 4k^2) + (-45k - 9)$	Group the terms.
$= 4k^2(5k + 1) - 9(5k + 1)$	Factor each group.
$= (5k + 1)(4k^2 - 9)$	$5k + 1$ is a common factor.
$= (5k + 1)(2k + 3)(2k - 3)$	Difference of squares

(c) $4a^2 + 4a + 1 - b^2$

$= (4a^2 + 4a + 1) - b^2$	Group the first three terms.
$= (2a + 1)^2 - b^2$	Perfect square trinomial
$= (2a + 1 + b)(2a + 1 - b)$	Difference of squares

◀ Work Problem 4 at the Side.

---

3 Factor each trinomial.

(a) $16m^2 + 56m + 49$

(b) $r^2 + 18r + 72$

(c) $8t^2 - 13t + 5$

(d) $6x^2 - 3x - 63$

4 Factor each polynomial.

(a) $20 - 5m - 12n + 3mn$

(b) $p^3 - 2pq^2 + p^2q - 2q^3$

(c) $9x^2 + 24x + 16 - y^2$

Answers

3. **(a)** $(4m + 7)^2$
(b) $(r + 6)(r + 12)$
(c) $(8t - 5)(t - 1)$
(d) $3(2x - 7)(x + 3)$

4. **(a)** $(4 - m)(5 - 3n)$
(b) $(p + q)(p^2 - 2q^2)$
(c) $(3x + 4 + y)(3x + 4 - y)$

6.6 Exercises

FOR EXTRA HELP

Math XL PRACTICE

WATCH

DOWNLOAD

READ

REVIEW

Match each polynomial in Column I with the method you would use to factor it in Column II. The choices in Column II may be used once, more than once, or not at all.

I	II
1. $12x^2 + 20x + 8$ **F**	**A.** Factor out the GCF; no further factoring is possible.
2. $x^2 - 17x + 72$ **G**	**B.** Factor a difference of squares.
3. $-16m^2n + 24mn - 40mn^2$ **A**	**C.** Factor a difference of cubes.
4. $64a^2 - 121b^2$ **B**	**D.** Factor a perfect square trinomial.
5. $36p^2 - 60pq + 25q^2$ **D**	**E.** Factor by grouping.
6. $z^2 - 4z + 6$ **H**	**F.** Factor out the GCF; then factor a trinomial by grouping or trial and error.
7. $8p^3 - 1$ **C**	**G.** Factor into two binomials by finding two integers whose product is the constant in the trinomial and whose sum is the coefficient of the middle term.
8. $x^6 + 4x^4 - 3x^2 - 12$ **E**	
9. $4w^2 + 49$ **H**	
10. $144 - 24z + z^2$ **D**	**H.** The polynomial is prime.

Factor each polynomial completely. We have randomly included all the different types of factoring exercises here to give you practice in applying factoring strategies. See Examples 1–4.

11. $100a^2 - 9b^2$
$(10a + 3b)(10a - 3b)$

12. $10r^2 + 13r - 3$
$(5r - 1)(2r + 3)$

13. $18p^5 - 24p^3 + 12p^6$
$6p^3(3p^2 - 4 + 2p^3)$

14. $15x^2 - 20x$
$5x(3x - 4)$

15. $x^2 + 2x - 35$
$(x + 7)(x - 5)$

16. $108m^2 - 36m + 3$
$3(6m - 1)^2$

17. $225p^2 + 256$
prime

18. $x^3 + 1000$
$(x + 10)(x^2 - 10x + 100)$

19. $6b^2 - 17b - 3$
$(6b + 1)(b - 3)$

20. $k^2 - 6k + 16$
prime

21. $18m^3n + 3m^2n^2 - 6mn^3$
$3mn(3m + 2n)(2m - n)$

22. $6t^2 + 19tu - 77u^2$
$(3t - 7u)(2t + 11u)$

23. $m^3 + 4m^2 - 6m - 24$
$(m + 4)(m^2 - 6)$

24. $9m^2 - 45m + 18m^3$
$9m(m - 5 + 2m^2)$

25. $4k^2 + 28kr + 49r^2$
$(2k + 7r)^2$

26. $54m^3 - 2000$
$2(3m - 10)(9m^2 + 30m + 100)$

27. $mn - 2n + 5m - 10$
$(m - 2)(n + 5)$

28. $9m^2 - 30mn + 25n^2 - p^2$
$(3m - 5n + p)(3m - 5n - p)$

29. $x^3 + 3x^2 - 9x - 27$
$(x + 3)^2(x - 3)$

30. $56k^3 - 875$
$7(2k - 5)(4k^2 + 10k + 25)$

31. $9r^2 + 100$
prime

32. $6k^2 - k - 1$
$(3k + 1)(2k - 1)$

33. $8p^3 - 125$
$(2p - 5)(4p^2 + 10p + 25)$

34. $27m^2 + 144mn + 192n^2$
$3(3m + 8n)^2$

35. $x^4 - 625$
$(x^2 + 25)(x + 5)(x - 5)$

36. $p^3 + 64$
$(p + 4)(p^2 - 4p + 16)$

37. $ab + 6b + ac + 6c$
$(a + 6)(b + c)$

38. $12z^3 - 6z^2 + 18z$
$6z(2z^2 - z + 3)$

39. $14z^2 - 3zk - 2k^2$
$(7z + 2k)(2z - k)$

40. $z^2 - zp + 20p^2$
prime

41. $256b^2 - 400c^2$
$16(4b + 5c)(4b - 5c)$

42. $64m^2 - 25n^2$
$(8m + 5n)(8m - 5n)$

43. $1000z^3 + 512$
$8(5z + 4)(25z^2 - 20z + 16)$

44. $y^4 - 81$
$(y^2 + 9)(y + 3)(y - 3)$

45. $10r^2 + 23rs - 5s^2$
$(5r - s)(2r + 5s)$

46. $2a^3 + a^2 - 14a - 7$
$(2a + 1)(a^2 - 7)$

47. $32x^2 + 16x^3 - 24x^5$
$8x^2(4 + 2x - 3x^3)$

48. $y^2 + 3y - 10$
$(y + 5)(y - 2)$

49. $14x^2 - 25xq - 25q^2$
$(2x - 5q)(7x + 5q)$

50. $b^2 - 7ba - 18a^2$
$(b - 9a)(b + 2a)$

51. $4(p + 2) + m(p + 2)$
$(p + 2)(4 + m)$

52. $kq - 9q + kr - 9r$
$(k - 9)(q + r)$

53. $50p^2 - 162$
$2(5p + 9)(5p - 9)$

54. $25x^2 - 20xy + 4y^2$
$(5x - 2y)^2$

55. $16a^2 + 8ab + b^2$
$(4a + b)^2$

6.7 Solving Quadratic Equations by Factoring

Galileo Galilei developed theories to explain physical phenomena and set up experiments to test his ideas. According to legend, Galileo dropped objects of different weights from the Leaning Tower of Pisa to disprove the belief that heavier objects fall faster than lighter objects. He developed a formula for freely falling objects described by $\mathbf{d = 16t^2}$, where d is the distance in feet that an object falls (disregarding air resistance) in t seconds, regardless of weight.

The equation $d = 16t^2$ is a *quadratic equation.* A quadratic equation contains a squared term and no terms of higher degree.

OBJECTIVES

1. Solve quadratic equations by factoring.
2. Solve other equations by factoring.

Quadratic Equation

A **quadratic equation** is an equation that can be written in the form

$$ax^2 + bx + c = 0,$$

where a, b, and c are real numbers, with $a \neq 0$. The given form is called **standard form.**

Galileo Galilei (1564–1642)

$x^2 + 5x + 6 = 0, \quad 2t^2 - 5t = 3, \quad y^2 = 4$ Quadratic equations

In these examples, only $x^2 + 5x + 6 = 0$ is in standard form.

Work Problems 1 and 2 at the Side. ▶

Up to now, we have factored *expressions,* including many quadratic expressions of the form $ax^2 + bx + c$. In this section, we use factored quadratic expressions to solve quadratic *equations.*

OBJECTIVE 1 Solve quadratic equations by factoring. We use the **zero-factor property** to solve a quadratic equation by factoring.

Zero-Factor Property

If a and b are real numbers and $ab = 0$, then $a = 0$ or $b = 0$.

In words, if the product of two numbers is 0, then at least one of the numbers must be 0. One number *must* be 0, but both *may* be 0.

EXAMPLE 1 Using the Zero-Factor Property

Solve each equation.

(a) $(x + 3)(2x - 1) = 0$

The product $(x + 3)(2x - 1)$ is equal to 0. By the zero-factor property, the only way that the product of these two factors can be 0 is if at least one of the factors equals 0. Therefore, either $x + 3 = 0$ or $2x - 1 = 0$.

$x + 3 = 0 \quad$ or $\quad 2x - 1 = 0$ Zero-factor property

$x = -3 \quad$ or $\quad 2x = 1$ Solve each equation.

$$x = \frac{1}{2}$$

Continued on Next Page

1 Which of the following equations are quadratic equations?

A. $y^2 - 4y - 5 = 0$

B. $x^3 - x^2 + 16 = 0$

C. $2z^2 + 7z = -3$

D. $x + 2y = -4$

2 Write each quadratic equation in standard form.

(a) $x^2 - 3x = 4$

(b) $y^2 = 9y - 8$

ANSWERS

1. A, C
2. (a) $x^2 - 3x - 4 = 0$
   (b) $y^2 - 9y + 8 = 0$

The given equation, $(x + 3)(2x - 1) = 0$, has two solutions, -3 and $\frac{1}{2}$. *Check* these solutions by substituting -3 for x in the original equation, $(x + 3)(2x - 1) = 0$. Then start over and substitute $\frac{1}{2}$ for x.

Check If $x = -3$, then

$$(x + 3)(2x - 1) = 0$$
$$(-3 + 3)[2(-3) - 1] \stackrel{?}{=} 0$$
$$0(-7) = 0. \quad \text{True}$$

If $x = \frac{1}{2}$, then

$$(x + 3)(2x - 1) = 0$$
$$\left(\frac{1}{2} + 3\right)\left(2 \cdot \frac{1}{2} - 1\right) \stackrel{?}{=} 0$$
$$\frac{7}{2}(1 - 1) \stackrel{?}{=} 0$$
$$\frac{7}{2} \cdot 0 = 0. \quad \text{True}$$

Both -3 and $\frac{1}{2}$ result in true equations, so the solution set is $\{-3, \frac{1}{2}\}$.

(b)
$$y(3y - 4) = 0$$
$$y = 0 \quad \text{or} \quad 3y - 4 = 0 \qquad \text{Zero-factor property}$$
$$3y = 4$$
$$y = \frac{4}{3}$$

Don't forget that 0 is a solution.

Check these solutions by substituting each one in the original equation. The solution set is $\{0, \frac{4}{3}\}$.

◀ Work Problem **3** at the Side.

3 Solve each equation. Check your solutions.

(a) $(x - 5)(x + 2) = 0$

(b) $(3x - 2)(x + 6) = 0$

(c) $z(2z + 5) = 0$

Note

The word *or* as used in Example 1 means "one or the other or both."

In Example 1, each equation to be solved was given with the polynomial in factored form. If the polynomial in an equation is not already factored, first make sure that the equation is in standard form. Then factor and solve.

EXAMPLE 2 Solving Quadratic Equations

Solve each equation.

(a) $x^2 - 5x = -6$

First, write the equation in standard form by adding 6 to each side.

$$x^2 - 5x = -6$$
$$x^2 - 5x + 6 = 0 \qquad \text{Add 6.}$$

Don't factor x out at this step.

Now factor $x^2 - 5x + 6$. Find two numbers whose product is 6 and whose sum is -5. These two numbers are -2 and -3, so the equation becomes

$$(x - 2)(x - 3) = 0. \qquad \text{Factor the trinomial.}$$
$$x - 2 = 0 \quad \text{or} \quad x - 3 = 0 \qquad \text{Zero-factor property}$$
$$x = 2 \quad \text{or} \quad x = 3 \qquad \text{Solve each equation.}$$

Continued on Next Page

ANSWERS

3. **(a)** $\{-2, 5\}$ **(b)** $\left\{-6, \frac{2}{3}\right\}$ **(c)** $\left\{-\frac{5}{2}, 0\right\}$

Check If $x = 2$, then

$$2^2 - 5(2) \stackrel{?}{=} -6$$
$$4 - 10 \stackrel{?}{=} -6$$
$$-6 = -6. \quad \text{True}$$

If $x = 3$, then

$$3^2 - 5(3) \stackrel{?}{=} -6$$
$$9 - 15 \stackrel{?}{=} -6$$
$$-6 = -6. \quad \text{True}$$

Both solutions check, so the solution set is $\{2, 3\}$.

(b) $y^2 = y + 20$

$y^2 - y - 20 = 0$ Write in standard form.

$(y - 5)(y + 4) = 0$ Factor the trinomial.

$y - 5 = 0$ or $y + 4 = 0$ Zero-factor property

$y = 5$ or $y = -4$ Solve each equation.

Check by substituting in the original equation. The solution set is $\{-4, 5\}$.

Work Problem 4 at the Side. ▶

4 Solve each equation. Check your solutions.

(a) $m^2 - 3m - 10 = 0$

(b) $r^2 + 2r = 8$

Solving a Quadratic Equation by Factoring

Step 1 **Write the equation in standard form,** that is, with all terms on one side of the equals sign in descending powers of the variable and 0 on the other side.

Step 2 **Factor** completely.

Step 3 **Use the zero-factor property** to set each factor with a variable equal to 0.

Step 4 **Solve** the resulting equations.

Step 5 **Check** each solution in the original equation.

5 Solve each equation. Check your solutions.

(a) $10a^2 - 5a - 15 = 0$

(b) $4x^2 - 2x = 42$

EXAMPLE 3 Solving a Quadratic Equation (Common Factor)

Solve $4p^2 + 40 = 26p$.

$4p^2 - 26p + 40 = 0$ Standard form

$2(2p^2 - 13p + 20) = 0$ Factor out 2.

$2p^2 - 13p + 20 = 0$ Divide each side by 2.

$(2p - 5)(p - 4) = 0$ Factor.

$2p - 5 = 0$ or $p - 4 = 0$ Zero-factor property

$2p = 5$ or $p = 4$ Solve each equation.

$p = \frac{5}{2}$

Check that the solution set is $\{\frac{5}{2}, 4\}$ by substituting in the original equation.

CAUTION

A common error is to include the common factor 2 as a solution in Example 3. ***Only factors containing variables lead to solutions.***

Work Problem 5 at the Side. ▶

ANSWERS

4. (a) $\{-2, 5\}$ **(b)** $\{-4, 2\}$

5. (a) $\left\{-1, \frac{3}{2}\right\}$ **(b)** $\left\{-3, \frac{7}{2}\right\}$

6 Solve each equation. Check your solutions.

(a) $49m^2 - 9 = 0$

(b) $p(4p + 7) = 2$

(c) $m^2 = 3m$

EXAMPLE 4 Solving Quadratic Equations

Solve each equation.

(a) $16m^2 - 25 = 0$

We can factor the left side of the equation as the difference of squares (**Section 6.5**).

$$16m^2 - 25 = 0$$

$$(4m + 5)(4m - 5) = 0 \quad \text{Factor the difference of squares.}$$

$$4m + 5 = 0 \quad \text{or} \quad 4m - 5 = 0 \quad \text{Zero-factor property}$$

$$4m = -5 \quad \text{or} \quad 4m = 5 \quad \text{Solve each equation.}$$

$$m = -\frac{5}{4} \quad \text{or} \quad m = \frac{5}{4}$$

Check the solutions $-\frac{5}{4}$ and $\frac{5}{4}$ in the original equation. The solution set is $\{-\frac{5}{4}, \frac{5}{4}\}$.

(b) $k(2k + 5) = 3$

We need to write this equation in standard form.

$$k(2k + 5) = 3$$

To be in standard form, 0 must be on one side.

$$2k^2 + 5k = 3 \quad \text{Multiply.}$$

$$2k^2 + 5k - 3 = 0 \quad \text{Subtract 3.}$$

$$(2k - 1)(k + 3) = 0 \quad \text{Factor.}$$

$$2k - 1 = 0 \quad \text{or} \quad k + 3 = 0 \quad \text{Zero-factor property}$$

$$2k = 1 \quad \text{or} \quad k = -3$$

$$k = \frac{1}{2}$$

The solution set is $\{-3, \frac{1}{2}\}$.

(c) $y^2 = 2y$

$$y^2 - 2y = 0 \quad \text{Standard form}$$

Don't forget to set the variable factor y equal to 0.

$$y(y - 2) = 0 \quad \text{Factor.}$$

$$y = 0 \quad \text{or} \quad y - 2 = 0 \quad \text{Zero-factor property}$$

$$y = 2$$

The solution set is $\{0, 2\}$.

CAUTION

In Example 4(b), the zero-factor property could not be used to solve the equation $k(2k + 5) = 3$ in its given form because of the 3 on the right. ***The zero-factor property applies only to a product that equals 0.***

In Example 4(c), it is tempting to begin by dividing each side of the equation $y^2 = 2y$ by y to get $y = 2$. Note that we do not get the other solution, 0, if we divide by a variable. (We *may* divide each side of an equation by a *nonzero* real number, however. For instance, in Example 3 we divided each side by 2.)

◀ Work Problem 6 at the Side.

ANSWERS

6. (a) $\left\{-\frac{3}{7}, \frac{3}{7}\right\}$ **(b)** $\left\{-2, \frac{1}{4}\right\}$ **(c)** $\{0, 3\}$

EXAMPLE 5 Solving a Quadratic Equation with a Double Solution

Solve each equation.

(a)

$$z^2 + 121 = 22z$$

$$z^2 - 22z + 121 = 0 \quad \text{Standard form}$$

Because $121 = 11^2$ and $22z = 2 \cdot z \cdot 11$, the trinomial on the left is a perfect square.

$$(z - 11)^2 = 0 \quad \text{Factor.}$$

To apply the zero-product property, write $(z - 11)^2$ as two separate factors.

$$(z - 11)(z - 11) = 0 \quad a^2 = a \cdot a$$

$$z - 11 = 0 \quad \text{or} \quad z - 11 = 0 \quad \text{Zero-factor property}$$

Because the two factors are identical, they both lead to the same solution. (This is called a **double solution.**) Thus,

$$z - 11 = 0$$

$$z = 11. \quad \text{Add 11.}$$

Check

$$z^2 + 121 = 22z$$

$$\mathbf{11}^2 + 121 \stackrel{?}{=} 22(\mathbf{11}) \quad \text{Let } z = 11.$$

$$121 + 121 \stackrel{?}{=} 242$$

$$242 = 242 \quad \text{True}$$

The solution set is $\{11\}$.

(b)

$$9t^2 - 30t = -25$$

$$9t^2 - 30t + 25 = 0 \quad \text{Standard form}$$

$$(3t - 5)^2 = 0 \quad \text{Factor the perfect square trinomial.}$$

$$3t - 5 = 0 \quad \text{or} \quad 3t - 5 = 0 \quad \text{Zero-factor property}$$

$$3t = 5 \quad \text{Solve the equation.}$$

$$t = \frac{5}{3}$$

Check the double solution by substituting $\frac{5}{3}$ in the original equation. The solution set is $\{\frac{5}{3}\}$.

Work Problem 7 at the Side. ▶

7 Solve each equation. Check your solutions.

(a) $x^2 + 16x = -64$

(b) $4x^2 - 4x + 1 = 0$

CAUTION

When a trinomial has two identical factors (a perfect square trinomial), as in Examples 5(a) and (b), it is common for students to write the solution of the corresponding quadratic equation twice in the solution set. Each of these equations has only *one* distinct solution. ***There is no need to write the same number more than once in a solution set.***

Note

Not all quadratic equations can be solved by factoring. A more general method for solving such equations is given in **Chapter 10.**

ANSWERS

7. (a) $\{-8\}$ **(b)** $\left\{\frac{1}{2}\right\}$

OBJECTIVE 2 Solve other equations by factoring. We can extend the zero-factor property to solve equations that involve more than two factors with variables, as shown in Examples 6 and 7. (These equations are *not* quadratic equations. Why not?)

8 Solve each equation. Check your solutions.

(a) $r^3 - 16r = 0$

(b) $x^3 - 3x^2 - 18x = 0$

EXAMPLE 6 Solving an Equation with More Than Two Variable Factors

Solve $6z^3 - 6z = 0$.

$$6z^3 - 6z = 0$$

$$6z(z^2 - 1) = 0 \quad \text{Factor out } 6z.$$

$$6z(z + 1)(z - 1) = 0 \quad \text{Factor } z^2 - 1.$$

By an extension of the zero-factor property, this product can equal 0 only if at least one of the factors equals 0. Write and solve three equations, one for each factor with a variable.

$$6z = 0 \quad \text{or} \quad z + 1 = 0 \quad \text{or} \quad z - 1 = 0$$

$$z = 0 \quad \text{or} \quad z = -1 \quad \text{or} \quad z = 1$$

Check by substituting, in turn, 0, -1, and 1 in the original equation. The solution set is $\{-1, 0, 1\}$.

◀ *Work Problem* **8** *at the Side.*

9 Solve each equation. Check your solutions.

(a) $(m + 3)(m^2 - 11m + 10) = 0$

(b) $(2x + 5)(4x^2 - 9) = 0$

EXAMPLE 7 Solving an Equation with a Quadratic Factor

Solve $(2x - 1)(x^2 - 9x + 20) = 0$.

$$(2x - 1)(x^2 - 9x + 20) = 0$$

$$(2x - 1)(x - 5)(x - 4) = 0 \quad \text{Factor } x^2 - 9x + 20.$$

$$2x - 1 = 0 \quad \text{or} \quad x - 5 = 0 \quad \text{or} \quad x - 4 = 0 \quad \text{Zero-factor property}$$

$$x = \frac{1}{2} \quad \text{or} \quad x = 5 \quad \text{or} \quad x = 4$$

Check to verify that the solution set is $\{\frac{1}{2}, 4, 5\}$.

◀ *Work Problem* **9** *at the Side.*

CAUTION
In Example 7, it would be unproductive to begin by multiplying the two factors together. Keep in mind that the zero-factor property and its extension requires the product of two or more factors to equal 0. ***Always consider first whether an equation is given in the appropriate form to apply the zero-factor property.***

ANSWERS

8. (a) $\{-4, 0, 4\}$ **(b)** $\{-3, 0, 6\}$

9. (a) $\{-3, 1, 10\}$ **(b)** $\left\{-\frac{5}{2}, -\frac{3}{2}, \frac{3}{2}\right\}$

6.7 Exercises

FOR EXTRA HELP

Math XL PRACTICE

WATCH

DOWNLOAD

READ

REVIEW

Solve each equation, and check your solutions. See Example 1.

1. $(x + 5)(x - 2) = 0$
$\{-5, 2\}$

2. $(x - 1)(x + 8) = 0$
$\{-8, 1\}$

3. $(2m - 7)(m - 3) = 0$
$\left\{3, \frac{7}{2}\right\}$

4. $(6k + 5)(k + 4) = 0$
$\left\{-4, -\frac{5}{6}\right\}$

5. $t(6t + 5) = 0$
$\left\{-\frac{5}{6}, 0\right\}$

6. $w(4w + 1) = 0$
$\left\{-\frac{1}{4}, 0\right\}$

7. $2x(3x - 4) = 0$
$\left\{0, \frac{4}{3}\right\}$

8. $6x(4x + 9) = 0$
$\left\{-\frac{9}{4}, 0\right\}$

9. $\left(x + \frac{1}{2}\right)\left(2x - \frac{1}{3}\right) = 0$
$\left\{-\frac{1}{2}, \frac{1}{6}\right\}$

10. $\left(a + \frac{2}{3}\right)\left(5a - \frac{1}{2}\right) = 0$
$\left\{-\frac{2}{3}, \frac{1}{10}\right\}$

11. $(x - 9)(x - 9) = 0$
$\{9\}$

12. $(2x + 1)(2x + 1) = 0$
$\left\{-\frac{1}{2}\right\}$

13. Look at this "solution." ***WHAT WENT WRONG?***

$$2x(3x - 4) = 0$$

$$x = 2 \quad \text{or} \quad x = 0 \quad \text{or} \quad 3x - 4 = 0$$

$$x = \frac{4}{3}$$

The solution set is $\{2, 0, \frac{4}{3}\}$.

Set each *variable* factor equal to 0, to get $2x = 0$ or $3x - 4 = 0$. The solution set is $\{0, \frac{4}{3}\}$.

14. Look at this "solution." ***WHAT WENT WRONG?***

$$x(7x - 1) = 0$$

$$7x - 1 = 0 \quad \text{Zero-factor property}$$

$$x = \frac{1}{7}$$

The solution set is $\{\frac{1}{7}\}$.

The variable x is another factor to set equal to 0, so the solution set is $\{0, \frac{1}{7}\}$.

Solve each equation, and check your solutions. See Examples 2–7.

15. $y^2 + 3y + 2 = 0$
$\{-2, -1\}$

16. $p^2 + 8p + 7 = 0$
$\{-7, -1\}$

17. $y^2 - 3y + 2 = 0$
$\{1, 2\}$

18. $r^2 - 4r + 3 = 0$
$\{1, 3\}$

19. $x^2 = 24 - 5x$
$\{-8, 3\}$

20. $t^2 = 2t + 15$
$\{-3, 5\}$

21. $x^2 = 3 + 2x$
$\{-1, 3\}$

22. $m^2 = 4 + 3m$
$\{-1, 4\}$

23. $z^2 + 3z = -2$
$\{-2, -1\}$

24. $p^2 - 2p = 3$
$\{-1, 3\}$

25. $m^2 + 8m + 16 = 0$
$\{-4\}$

26. $b^2 - 6b + 9 = 0$
$\{3\}$

27. $3x^2 + 5x - 2 = 0$
$\left\{-2, \frac{1}{3}\right\}$

28. $6r^2 - r - 2 = 0$
$\left\{-\frac{1}{2}, \frac{2}{3}\right\}$

29. $6p^2 = 4 - 5p$
$\left\{-\frac{4}{3}, \frac{1}{2}\right\}$

30. $6x^2 = 4 + 5x$

$\left\{-\frac{1}{2}, \frac{4}{3}\right\}$

31. $9s^2 + 12s = -4$

$\left\{-\frac{2}{3}\right\}$

32. $36x^2 + 60x = -25$

$\left\{-\frac{5}{6}\right\}$

33. $y^2 - 9 = 0$

$\{-3, 3\}$

34. $m^2 - 100 = 0$

$\{-10, 10\}$

35. $16k^2 - 49 = 0$

$\left\{-\frac{7}{4}, \frac{7}{4}\right\}$

36. $4w^2 - 9 = 0$

$\left\{-\frac{3}{2}, \frac{3}{2}\right\}$

37. $n^2 = 121$

$\{-11, 11\}$

38. $x^2 = 400$

$\{-20, 20\}$

39. $x^2 = 7x$

$\{0, 7\}$

40. $t^2 = 9t$

$\{0, 9\}$

41. $6r^2 = 3r$

$\left\{0, \frac{1}{2}\right\}$

42. $10y^2 = -5y$

$\left\{-\frac{1}{2}, 0\right\}$

43. $g(g - 7) = -10$

$\{2, 5\}$

44. $r(r - 5) = -6$

$\{2, 3\}$

45. $z(2z + 7) = 4$

$\left\{-4, \frac{1}{2}\right\}$

46. $b(2b + 3) = 9$

$\left\{-3, \frac{3}{2}\right\}$

47. $2(y^2 - 66) = -13y$

$\left\{-12, \frac{11}{2}\right\}$

48. $3(t^2 + 4) = 20t$

$\left\{\frac{2}{3}, 6\right\}$

49. $5x^3 - 20x = 0$

$\{-2, 0, 2\}$

50. $3x^3 - 48x = 0$

$\{-4, 0, 4\}$

51. $9y^3 - 49y = 0$

$\left\{-\frac{7}{3}, 0, \frac{7}{3}\right\}$

52. $16r^3 - 9r = 0$

$\left\{-\frac{3}{4}, 0, \frac{3}{4}\right\}$

53. $(2r + 5)(3r^2 - 16r + 5) = 0$

$\left\{-\frac{5}{2}, \frac{1}{3}, 5\right\}$

54. $(3m + 4)(6m^2 + m - 2) = 0$

$\left\{-\frac{4}{3}, -\frac{2}{3}, \frac{1}{2}\right\}$

55. $(2x + 7)(x^2 + 2x - 3) = 0$

$\left\{-\frac{7}{2}, -3, 1\right\}$

56. $(x + 1)(6x^2 + x - 12) = 0$

$\left\{-\frac{3}{2}, -1, \frac{4}{3}\right\}$

57. Galileo's formula for freely falling objects, $\mathbf{d = 16t^2}$, was given at the beginning of this section. The distance d in feet an object falls depends on the time elapsed t in seconds. (This is an example of a **function,** introduced in **Section 3.6.**)

(a) Use Galileo's formula and complete the following table. (*Hint:* Substitute each given value into the formula and solve for the unknown value.)

t in seconds	0	1	2	3	4	6
d in feet	0	16	64	144	256	576

(b) When $t = 0$, $d = 0$. Explain this in the context of the problem.

No time has elapsed, so the object hasn't fallen (been released) yet.

58. In Exercise 57, when you substituted 256 for d and solved for t, you should have found two solutions: 4 and -4. Why doesn't -4 make sense as an answer?

Time cannot be negative.

6.8 Applications of Quadratic Equations

OBJECTIVES

1. Solve problems about geometric figures.
2. Solve problems about consecutive integers.
3. Solve problems using the Pythagorean formula.
4. Solve problems using given quadratic models.

We can use factoring to solve quadratic equations that arise in applications. We follow the same six problem-solving steps given in **Section 2.4.**

Solving an Applied Problem

Step 1 **Read** the problem, several times if necessary, until you *understand* what is given and what is to be found.

Step 2 **Assign a variable** to represent the unknown value, using diagrams or tables as needed. Write a statement that tells what the variable represents. Express any other unknown values in terms of the variable.

Step 3 **Write an equation** using the variable expression(s).

Step 4 **Solve** the equation.

Step 5 **State the answer.** Does it seem reasonable?

Step 6 **Check** the answer in the words of the original problem.

OBJECTIVE 1 Solve problems about geometric figures. Refer to the formulas given on the inside covers of the text, if necessary.

EXAMPLE 1 Solving an Area Problem

The Monroes want to plant a rectangular garden in their yard. The width of the garden will be 4 ft less than its length, and they want it to have an area of 96 ft^2. (ft^2 means square feet.) Find the length and width of the garden.

Step 1 **Read** the problem carefully. We need to find the dimensions of a garden with area 96 ft^2.

Step 2 **Assign a variable.**

Let x = the length of the garden.

Then $x - 4$ = the width. (The width is 4 ft less than the length.)

See Figure 1.

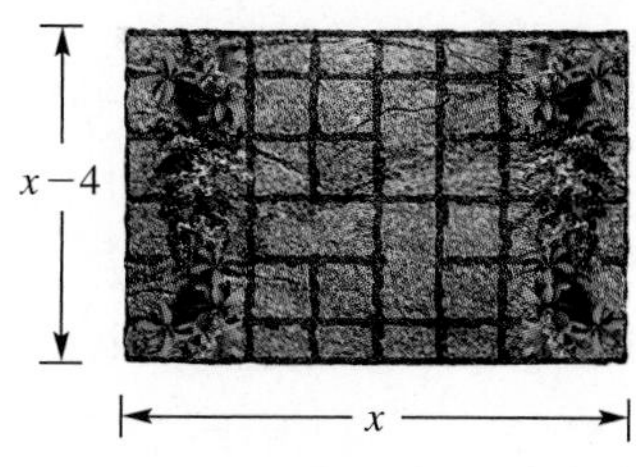

Figure 1

Step 3 **Write an equation.** The area of a rectangle is given by

$$\text{Area} = LW = \text{Length} \times \text{Width}. \quad \text{Area formula}$$

Substitute 96 for area, x for length, and $x - 4$ for width.

$$A = LW$$

$$96 = x(x - 4) \quad \text{Let } A = 96,\ L = x,\ W = x - 4.$$

Continued on Next Page

1 Solve each problem.

(a) The length of a rectangular room is 2 m more than the width. The area of the floor is 48 m^2. Find the length and width of the room.

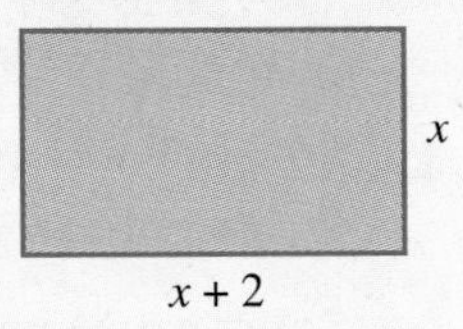

(b) The length of each side of a square is increased by 4 in. The sum of the areas of the original square and the larger square is 106 in^2. What is the length of a side of the original square?

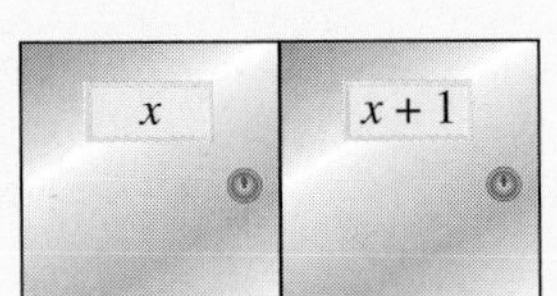

Figure 2

Answers

1. **(a)** length: 8 m; width: 6 m **(b)** 5 in.

Step 4 **Solve.**

$$96 = x(x - 4)$$

$$96 = x^2 - 4x \qquad \text{Distributive property}$$

$$x^2 - 4x - 96 = 0 \qquad \text{Standard form}$$

$$(x - 12)(x + 8) = 0 \qquad \text{Factor.}$$

$$x - 12 = 0 \quad \text{or} \quad x + 8 = 0 \qquad \text{Zero-factor property}$$

$$x = 12 \quad \text{or} \quad x = -8$$

Step 5 **State the answer.** The solutions are 12 and -8. A rectangle cannot have a side of negative length, so discard -8. The length of the garden will be 12 ft. The width will be $12 - 4 = 8$ ft.

Step 6 **Check.** The width of the garden is 4 ft less than the length; the area is $12 \cdot 8 = 96$ ft^2.

Problem-Solving Hint

When solving applied problems, ***always check solutions against physical facts*** and discard any answers that are not appropriate.

◀ Work Problem **1** *at the Side.*

OBJECTIVE 2 Solve problems about consecutive integers. Recall from **Section 2.4** that **consecutive integers** are integers that are next to each other on a number line, such as 5 and 6, or -11 and -10. **Consecutive odd integers** are *odd* integers that are next to each other, such as 5 and 7, or -13 and -11. **Consecutive even integers** are defined similarly; 4 and 6 are consecutive even integers, as are -10 and -8. (In this book, we will list consecutive integers in increasing order from left to right.)

Problem-Solving Hint

In consecutive integer problems, if x represents the first integer, then for

two consecutive integers, use	$x, \; x + 1;$
three consecutive integers, use	$x, \; x + 1, \; x + 2;$
two consecutive even or odd integers, use	$x, \; x + 2;$
three consecutive even or odd integers, use	$x, \; x + 2, \; x + 4.$

EXAMPLE 2 Solving a Consecutive Integer Problem

The product of the numbers on two consecutive post-office boxes is 210. Find the box numbers.

Step 1 **Read** the problem. Note that the boxes are numbered consecutively.

Step 2 **Assign a variable.**

Let x = the first box number.

Then $x + 1$ = the next consecutive box number.

See Figure 2.

Step 3 **Write an equation.** The product of the box numbers is 210, so

$$x(x + 1) = 210.$$

Continued on Next Page

Step 4 **Solve.**

$$x(x+1) = 210$$
$$x^2 + x = 210 \quad \text{Distributive property}$$
$$x^2 + x - 210 = 0 \quad \text{Standard form}$$
$$(x+15)(x-14) = 0 \quad \text{Factor.}$$
$$x + 15 = 0 \quad \text{or} \quad x - 14 = 0 \quad \text{Zero-factor property}$$
$$x = -15 \quad \text{or} \quad x = 14$$

Step 5 **State the answer**. The solutions are -15 and 14. Discard the solution -15 since a box number cannot be negative. When $x = 14$, then $x + 1 = 15$, so the post-office boxes have the numbers 14 and 15.

Step 6 **Check.** The numbers 14 and 15 are consecutive and their product is $14 \cdot 15 = 210$, as required.

Work Problem **2** *at the Side.* ▶

EXAMPLE 3 **Solving a Consecutive Integer Problem**

The product of two consecutive odd integers is 1 less than five times their sum. Find the integers.

Step 1 **Read** carefully. This problem is a little more complicated.

Step 2 **Assign a variable.** We must find two consecutive *odd* integers.

Let x = the lesser integer.

Then $x + 2$ = the next greater odd integer.

Step 3 **Write an equation.** According to the problem, the product is 1 less than five times the sum.

The product	is	five times the sum	less 1.
↓	↓	↓	↓
$x(x+2)$	$=$	$5(x+x+2)$	$-\ 1$

Step 4 **Solve.**

$$x^2 + 2x = 5x + 5x + 10 - 1 \quad \text{Distributive property}$$
$$x^2 + 2x = 10x + 9 \quad \text{Combine like terms.}$$
$$x^2 - 8x - 9 = 0 \quad \text{Standard form}$$
$$(x-9)(x+1) = 0 \quad \text{Factor.}$$
$$x - 9 = 0 \quad \text{or} \quad x + 1 = 0 \quad \text{Zero-factor property}$$
$$x = 9 \quad \text{or} \quad x = -1$$

Step 5 **State the answer.** We need to find two consecutive odd integers.

If $x = \mathbf{9}$ is the lesser, then $x + 2 = 9 + 2 = \mathbf{11}$ is the greater.

If $x = \mathbf{-1}$ is the lesser, then $x + 2 = -1 + 2 = \mathbf{1}$ is the greater.

There are two sets of answers here since integers can be positive or negative.

Step 6 **Check.** The product of the first pair of integers is $9 \cdot 11 = 99$. One less than five times their sum is $5(9 + 11) - 1 = 99$. Thus 9 and 11 satisfy the problem. Repeat the check with -1 and 1.

Work Problem **3** *at the Side.* ▶

2 Solve the problem.

The product of the numbers on two consecutive lockers at a health club is 132. Find the locker numbers.

3 Solve each problem.

(a) The product of two consecutive even integers is 4 more than two times their sum. Find the integers.

(b) Find three consecutive odd integers such that the product of the least and greatest is 16 more than the middle integer.

Answers

2. 11 and 12
3. (a) 4 and 6 or -2 and 0 **(b)** 3, 5, 7

CAUTION

Do *not* use $x, x + 1, x + 3$, and so on to represent consecutive odd integers. To see why, let $x = 3$. Then $x + 1 = 3 + 1 = 4$ and $x + 3 = 3 + 3 = 6$, and 3, 4, and 6 are not consecutive odd integers.

OBJECTIVE 3 Solve problems using the Pythagorean formula.

The next example requires the Pythagorean formula from geometry.

Pythagorean Formula

If a right triangle (a triangle with a 90° angle) has longest side of length c and two other sides of lengths a and b, then

$$a^2 + b^2 = c^2.$$

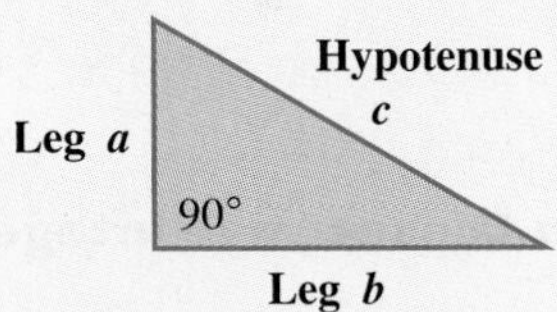

The longest side, the **hypotenuse,** is opposite the right angle. The two shorter sides are the **legs** of the triangle.

EXAMPLE 4 Using the Pythagorean Formula

Amy and Kevin leave their office, with Amy traveling north and Kevin traveling east. When Kevin is 1 mi farther than Amy from the office, the distance between them is 2 mi more than Amy's distance from the office. Find their distances from the office and the distance between them.

Step 1 **Read** the problem again. We must find three distances.

Step 2 **Assign a variable.** Let x represent Amy's distance from the office, $x + 1$ represent Kevin's distance from the office, and $x + 2$ represent the distance between them. Place these on a right triangle, as in Figure 3.

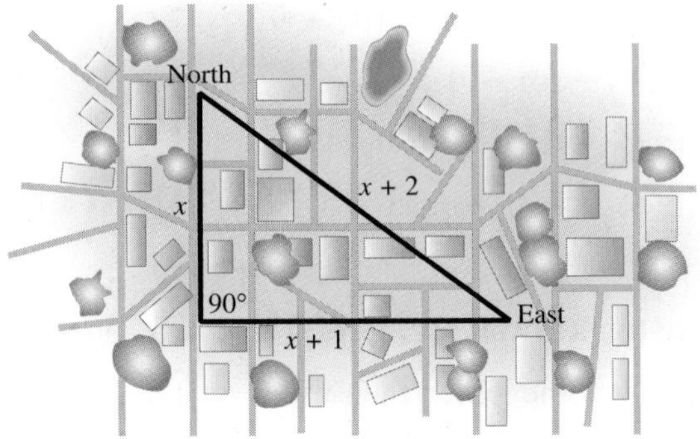

Figure 3

Step 3 **Write an equation.** Substitute into the Pythagorean formula.

$$a^2 + b^2 = c^2$$

$$x^2 + (x + 1)^2 = (x + 2)^2$$

Be careful to substitute properly.

Continued on Next Page

Step 4 **Solve.**

$$x^2 + x^2 + 2x + 1 = x^2 + 4x + 4$$
$$x^2 - 2x - 3 = 0 \quad \text{Standard form}$$
$$(x - 3)(x + 1) = 0 \quad \text{Factor.}$$
$$x - 3 = 0 \quad \text{or} \quad x + 1 = 0 \quad \text{Zero-factor property}$$
$$x = 3 \quad \text{or} \quad x = -1$$

Step 5 **State the answer.** Since -1 cannot represent a distance, 3 is the only possible answer. Amy's distance is 3 mi, Kevin's distance is $3 + 1 = 4$ mi, and the distance between them is $3 + 2 = 5$ mi.

Step 6 **Check.** Since $3^2 + 4^2 = 5^2$, the answer is correct.

CAUTION

When solving a problem involving the Pythagorean formula, be sure that the expressions for the sides are properly placed.

$$\textbf{leg}^2 + \textbf{leg}^2 = \textbf{hypotenuse}^2$$

Work Problem **4** *at the Side.* ▶

OBJECTIVE 4 Solve problems using given quadratic models. In Examples 1–4, we wrote quadratic equations to model, or mathematically describe, various situations and then solved the equations. Now we are given the quadratic models and must use them to determine data.

EXAMPLE 5 Finding the Height of a Ball

A tennis player can hit a ball 180 ft per sec. If she hits a ball directly upward, the height h of the ball in feet at time t in seconds is modeled by the quadratic equation

$$h = -16t^2 + 180t + 6.$$

When will the ball be 206 ft above the ground?

A height of 206 ft means $h = 206$, so we substitute 206 for h in the equation and then solve for t.

$$\mathbf{206} = -16t^2 + 180t + 6 \quad \text{Let } h = 206.$$
$$-16t^2 + 180t + 6 = 206 \quad \text{Interchange sides.}$$
$$-16t^2 + 180t - 200 = 0 \quad \text{Standard form}$$
$$4t^2 - 45t + 50 = 0 \quad \text{Divide by } -4.$$
$$(4t - 5)(t - 10) = 0 \quad \text{Factor.}$$
$$4t - 5 = 0 \quad \text{or} \quad t - 10 = 0 \quad \text{Zero-factor property}$$
$$t = \frac{5}{4} \quad \text{or} \quad t = 10$$

Since we found two acceptable answers, the ball will be 206 ft above the ground twice (once on its way up and once on its way down)—at $\frac{5}{4}$ sec and at 10 sec after it is hit. See Figure 4.

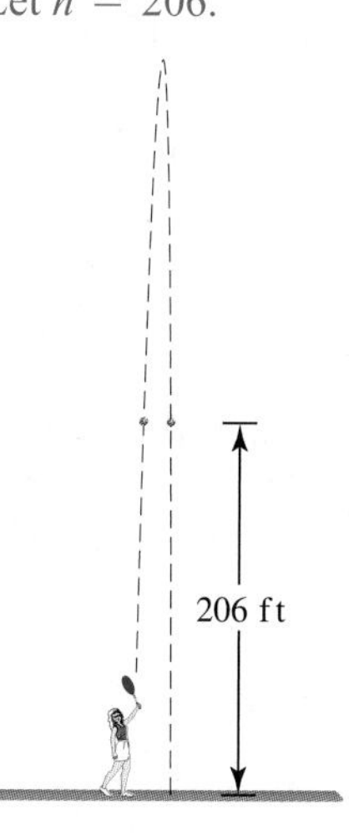

Figure 4

Work Problem **5** *at the Side.* ▶

4 Solve the problem.

The hypotenuse of a right triangle is 3 in. longer than the longer leg. The shorter leg is 3 in. shorter than the longer leg. Find the lengths of the sides of the triangle.

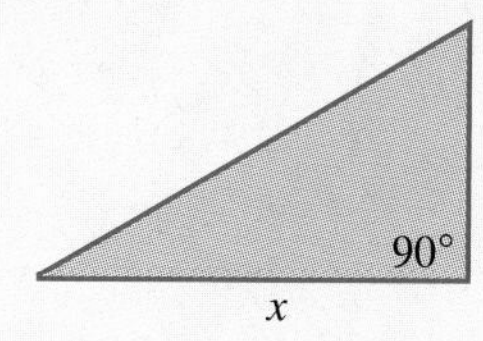

5 Solve the problem.

The number of impulses fired after a nerve has been stimulated is modeled by

$$I = -x^2 + 2x + 60,$$

where x is in milliseconds (ms) after the stimulation. When will 45 impulses occur? Do you get two solutions? Why is only one answer given?

ANSWERS

4. 9 in., 12 in., 15 in.
5. After 5 ms; There are two solutions, -3 and 5; Only one answer makes sense here because a negative answer is not appropriate.

6 Use the model in Example 6 to find the foreign-born population of the United States in 1990. Give your answer to the nearest tenth of a million. How does it compare to the actual value from the table?

EXAMPLE 6 Modeling the Foreign-Born Population of the United States

After decreasing in the middle of the 20th century, the foreign-born population of the United States started to increase in the later part of the century and is now increasing rapidly. The foreign-born population over the years 1930–2004 can be modeled by the quadratic equation

$$y = 0.01036x^2 - 0.5316x + 15.36,$$

where $x = 0$ represents 1930, $x = 10$ represents 1940, and so on, and y is the number of people in millions. (*Source:* U.S. Census Bureau.)

(a) Use the model to find the foreign-born population in 1980 to the nearest tenth of a million.

Since $x = 0$ represents 1930, $x = 50$ represents 1980. Substitute 50 for x in the equation.

$$y = 0.01036(50)^2 - 0.5316(50) + 15.36 \qquad \text{Let } x = 50.$$

$$y = 14.7 \qquad \text{Round to the nearest tenth.}$$

In 1980, the foreign-born population of the United States was about 14.7 million.

(b) Repeat part (a) for 2004.

$$y = 0.01036(74)^2 - 0.5316(74) + 15.36 \qquad \text{For 2004, let } x = 74.$$

$$y = 32.8 \qquad \text{Round to the nearest tenth.}$$

In 2004, the foreign-born population of the United States was about 32.8 million.

(c) The model used in parts (a) and (b) was developed using the data in the table below. How do the results in parts (a) and (b) compare to the actual data from the table?

Year	Foreign-Born Population (millions)
1930	14.2
1940	11.6
1950	10.3
1960	9.7
1970	9.6
1980	14.1
1990	19.8
2000	28.4
2004	34.2

From the table, the actual value for 1980 is 14.1 million. Our answer in part (a), 14.7 million, is slightly high. For 2004, the actual value is 34.2 million, so our answer of 32.8 million in part (b) is somewhat low.

 *Work Problem 6 at the Side.*

ANSWER

6. 20.8 million; The actual value is 19.8 million, so our answer using the model is somewhat high.

6.8 Exercises

FOR EXTRA HELP MyMathLab Math XL PRACTICE WATCH DOWNLOAD READ REVIEW

1. To review the six problem-solving steps first introduced in **Section 2.4,** complete each statement.

Step 1: **Read** the problem, several times if necessary, until you understand what is given and what must be found.

Step 2: Assign a **variable** to represent the unknown value.

Step 3: Write a(n) **equation** using the variable expression(s).

Step 4: **Solve** the equation.

Step 5: State the **answer**.

Step 6: **Check** the answer in the words of the **original** problem.

2. A student solves an applied problem and gets 6 or -3 for the length of the side of a square. Which of these answers is reasonable? Explain.

Only 6 is reasonable since a square cannot have a side of negative length.

In Exercises 3–6, a figure and a corresponding geometric formula are given. Using x as the variable, complete Steps 3–6 for each problem. (Refer to the steps in Exercise 1 as needed.)

3.

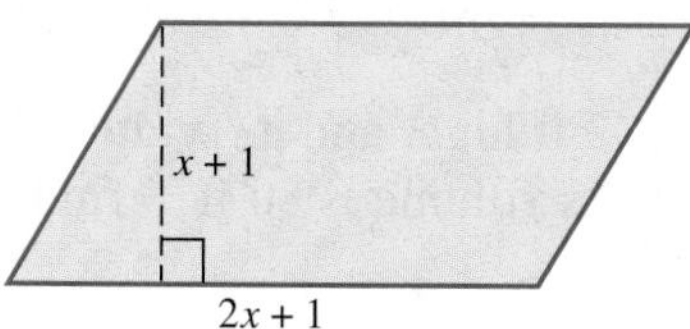

Area of a parallelogram: $A = bh$

The area of this parallelogram is 45 sq. units. Find its base and height.

Step 3: $45 = (2x + 1)(x + 1)$

Step 4: $x = 4$ **or** $x = -\frac{11}{2}$

Step 5: **base: 9 units; height: 5 units**

Step 6: $9 \cdot 5 = 45$

4.

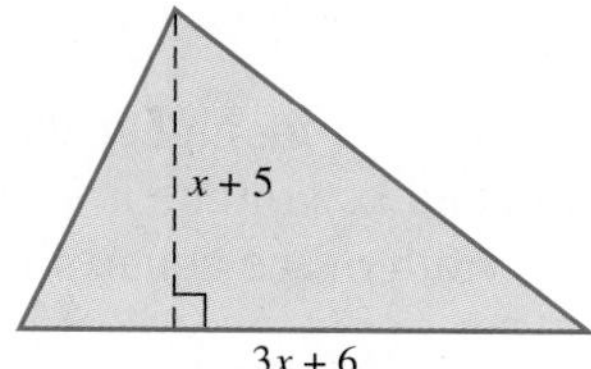

Area of a triangle: $A = \frac{1}{2}bh$

The area of this triangle is 60 sq. units. Find its base and height.

Step 3: $60 = \frac{1}{2}(3x + 6)(x + 5)$

Step 4: $x = 3$ **or** $x = -10$

Step 5: **base: 15 units; height: 8 units**

Step 6: $\frac{1}{2}(15)(8) = 60$

5.

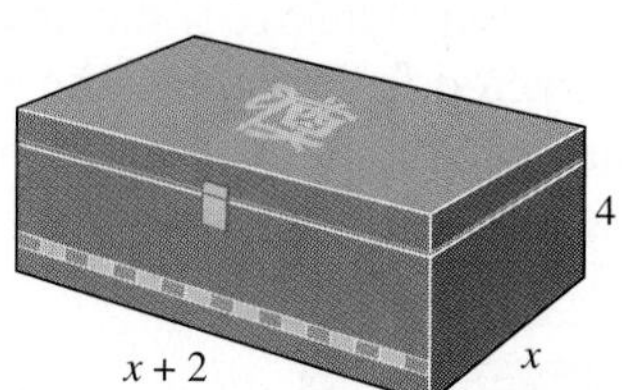

Volume of a rectangular Chinese box: $V = LWH$

The volume of this box is 192 cu. units. Find its length and width.

Step 3: $192 = 4x(x + 2)$

Step 4: $x = 6$ **or** $x = -8$

Step 5: **length: 8 units; width: 6 units**

Step 6: $8 \cdot 6 \cdot 4 = 192$

6.

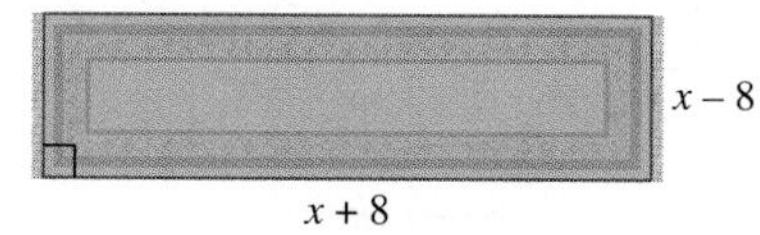

Area of a rectangular rug: $A = LW$

The area of this rug is 80 sq. units. Find its length and width.

Step 3: $80 = (x + 8)(x - 8)$

Step 4: $x = 12$ **or** $x = -12$

Step 5: **length: 20 units; width: 4 units**

Step 6: $20 \cdot 4 = 80$

Solve each problem. Check your answers to be sure they are reasonable. Refer to the formulas on the inside covers. See Example 1.

7. The length of a standard jewel case is 2 cm more than its width. The area of the rectangular top of the case is 168 cm^2. Find the length and width of the jewel case.

   length: 14 cm; width: 12 cm

8. A standard DVD case is 6 cm longer than it is wide. The area of the rectangular top of the case is 247 cm^2. Find the length and width of the case.

   length: 19 cm; width: 13 cm

9. The dimensions of an HPf1905 flat-panel monitor are such that its length is 3 in. more than its width. If the length were doubled and if the width were decreased by 1 in., the area would be increased by 150 in.^{2}. What are the length and width of the flat panel?

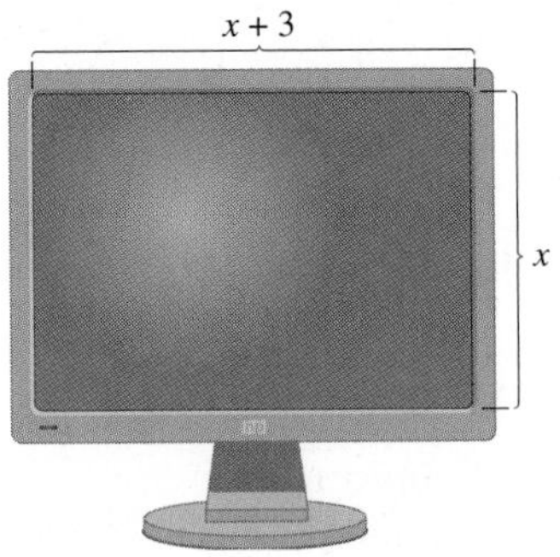

   length: 15 in.; width: 12 in.

10. The keyboard of the computer in Exercise 9 is 11 in. longer than it is wide. If both its length and width are increased by 2 in., the area of the top of the keyboard is increased by 54 in.^{2}. Find the length and width of the keyboard. (*Source:* Author's computer.)

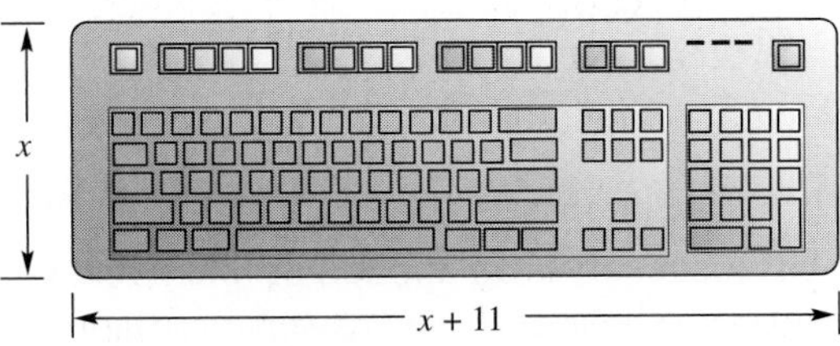

   length: 18 in.; width: 7 in.

11. A 10-gal aquarium is 3 in. higher than it is wide. Its length is 21 in., and its volume is 2730 in.^{3}. What are the height and width of the aquarium?

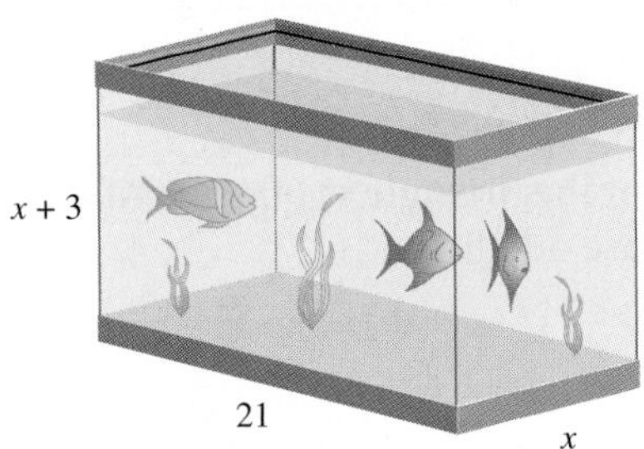

   height: 13 in.; width: 10 in.

12. A toolbox is 2 ft high, and its width is 3 ft less than its length. If its volume is 80 ft^3, find the length and width of the box.

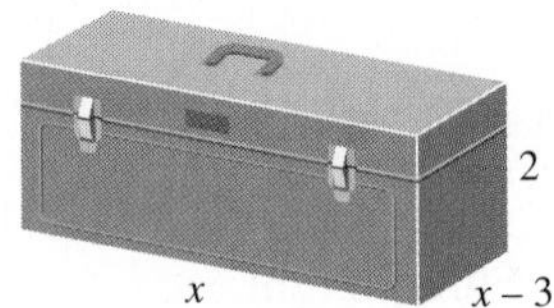

   length: 8 ft; width: 5 ft

13. A square mirror has sides measuring 2 ft less than the sides of a square painting. If the difference between their areas is 32 ft^2, find the lengths of the sides of the mirror and the painting.

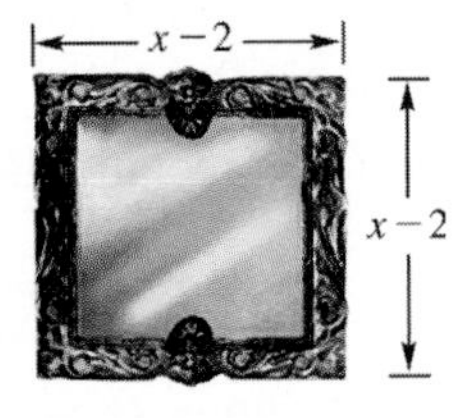

   mirror: 7 ft; painting: 9 ft

14. The sides of one square have length 3 m more than the sides of a second square. If the area of the larger square is subtracted from 4 times the area of the smaller square, the result is 36 m^2. What are the lengths of the sides of each square?

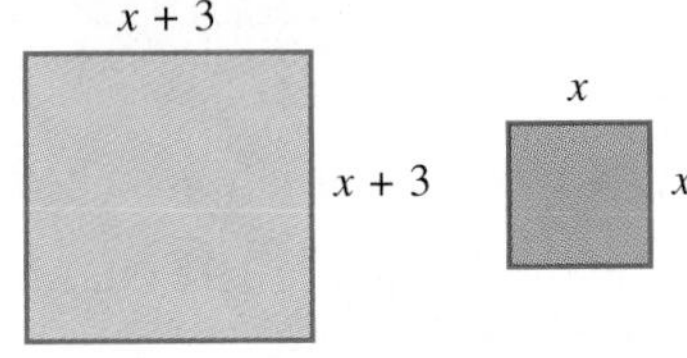

   larger square: 8 m; smaller square: 5 m

Solve each problem about consecutive integers. See Examples 2 and 3.

15. The product of the numbers on two consecutive volumes of research data is 420. Find the volume numbers.

20, 21

16. The product of the page numbers on two facing pages of a book is 600. Find the page numbers.

24, 25

17. The product of two consecutive integers is 11 more than their sum. Find the integers.

−3, −2 or 4, 5

18. The product of two consecutive integers is 4 less than four times their sum. Find the integers.

0, 1 or 7, 8

19. Find two consecutive odd integers such that their product is 15 more than three times their sum.

−3, −1 or 7, 9

20. Find two consecutive odd integers such that five times their sum is 23 less than their product.

−3, −1 or 11, 13

21. Find three consecutive even integers such that the sum of the squares of the lesser two is equal to the square of the greatest.

−2, 0, 2 or 6, 8, 10

22. Find three consecutive even integers such that the square of the sum of the lesser two is equal to twice the greatest.

−2, 0, 2

Use the Pythagorean formula to solve each problem. See Example 4.

23. The hypotenuse of a right triangle is 1 cm longer than the longer leg. The shorter leg is 7 cm shorter than the longer leg. Find the length of the longer leg of the triangle.

12 cm

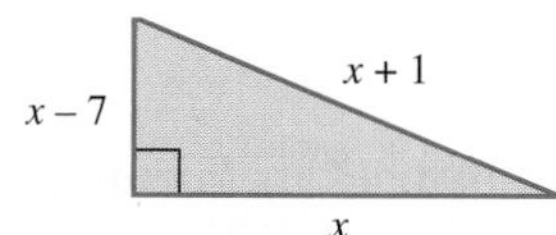

24. The longer leg of a right triangle is 1 m longer than the shorter leg. The hypotenuse is 1 m shorter than twice the shorter leg. Find the length of the shorter leg of the triangle.

3 m

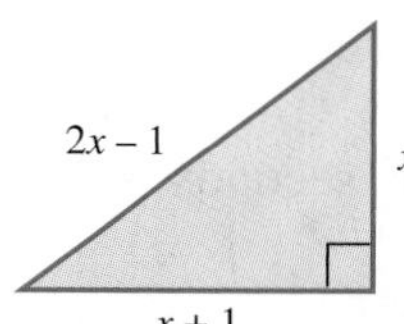

25. Terri works due north of home. Her husband Denny works due east. They leave for work at the same time. By the time Terri is 5 mi from home, the distance between them is 1 mi more than Denny's distance from home. How far from home is Denny?

12 mi

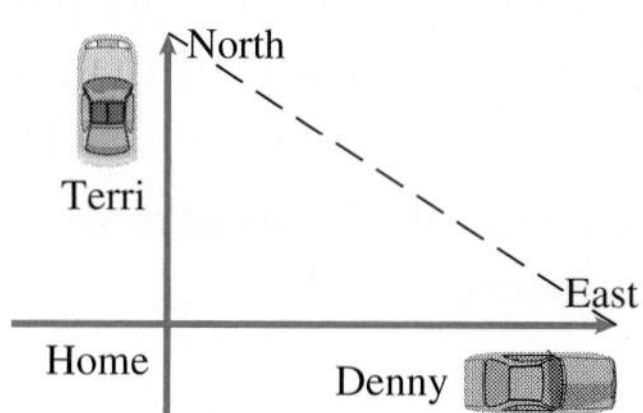

26. Two cars left an intersection at the same time. One traveled north. The other traveled 14 mi farther, but to the east. How far apart were they then, if the distance between them was 4 mi more than the distance traveled east?

34 mi

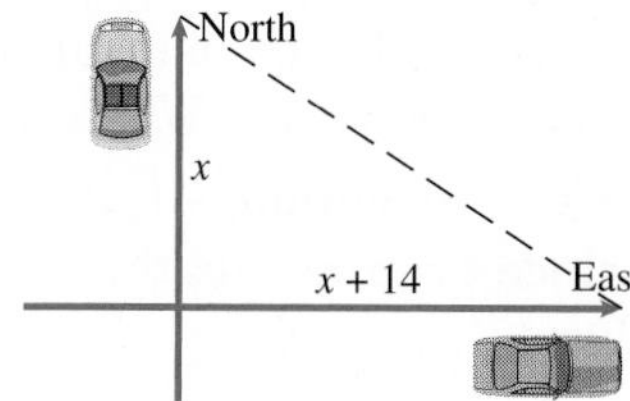

27. A ladder is leaning against a building. The distance from the bottom of the ladder to the building is 4 ft less than the length of the ladder. How high up the side of the building is the top of the ladder if that distance is 2 ft less than the length of the ladder?

8 ft

28. A lot has the shape of a right triangle with one leg 2 m longer than the other. The hypotenuse is 2 m less than twice the length of the shorter leg. Find the length of the shorter leg.

6 m

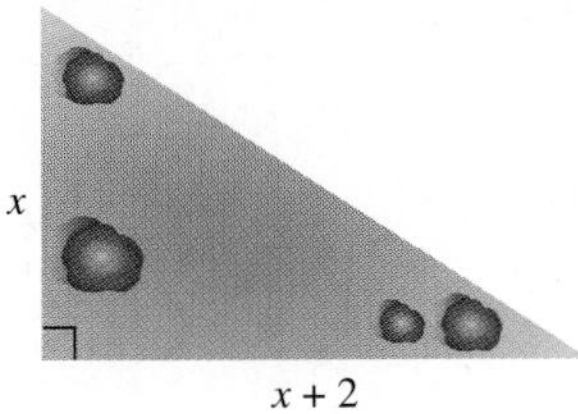

Solve each problem. See Examples 5 and 6.

29. An object projected from a height of 48 ft with an initial velocity of 32 ft per sec after t seconds has height

$$h = -16t^2 + 32t + 48.$$

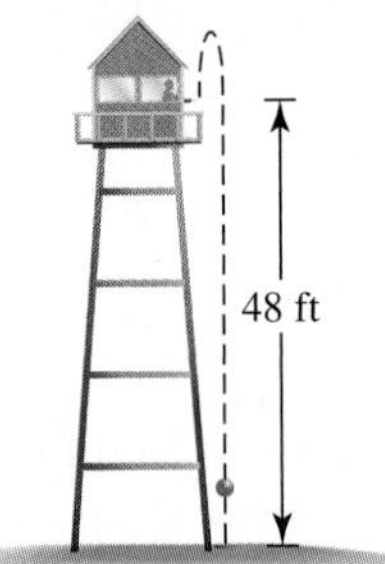

(a) After how many seconds is the height 64 ft? (*Hint:* Let $h = 64$ and solve.)

1 sec

(b) After how many seconds is the height 60 ft?

$\frac{1}{2}$ sec and $1\frac{1}{2}$ sec

(c) After how many seconds does the object hit the ground? (*Hint:* When the object hits the ground, $h = 0$.)

3 sec

(d) The quadratic equation from part (c) has two solutions, yet only one of them is appropriate for answering the question. Why is this so?

The negative solution, −1, does not make sense since t represents time, which cannot be negative.

30. If an object is projected upward from ground level with an initial velocity of 64 ft per sec, its height h in feet t seconds later is

$$h = -16t^2 + 64t.$$

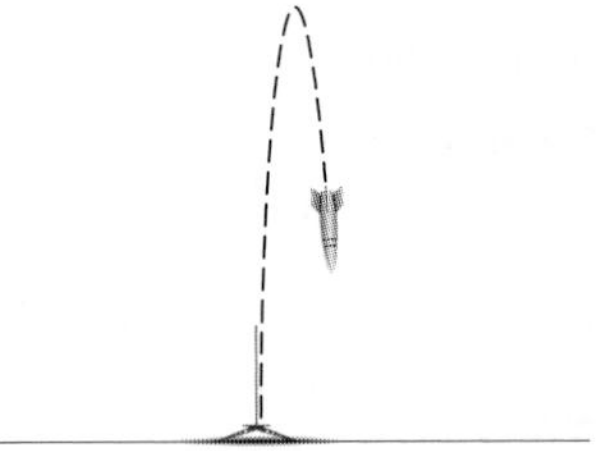

(a) After how many seconds is the height 48 ft?

1 sec and 3 sec

(b) The object reaches its maximum height 2 sec after it is projected. What is this maximum height?

64 ft

(c) After how many seconds does the object hit the ground?

4 sec

(d) The quadratic equation from part (c) has two solutions, yet only one of them is appropriate for answering the question. Why is this so?

The solution 0 represents the time at which it was projected.

31. The table shows the number of cellular phone subscribers (in millions) in the United States.

Year	Subscribers (in millions)
1990	5
1992	11
1994	24
1996	44
1998	69
2000	109
2002	141
2004	182
2006	233

Source: CTIA-The Wireless Association.

We used the data to develop the quadratic equation

$$y = 0.734x^2 + 2.62x + 3.37,$$

which models the number of cellular phone subscribers y (in millions) in the year x, where $x = 0$ represents 1990, $x = 2$ represents 1992, and so on.

(a) Use the model to find the number of cellular phones in 1996 to the nearest million. How does the result compare to the actual data in the table?

46 million; The result using the model is a little more than 44 million, the actual number for 1996.

(b) What value of x corresponds to 2004?

14

(c) Use the model to find the number of cellular phones in 2004 to the nearest million. How does the result compare to the actual data in the table?

184 million; The result is a little more than 182 million, the actual number for 2004.

(d) Assuming that the trend in the data continues, use the quadratic equation to estimate the number of cellular phones in 2009 to the nearest million.

318 million

Relating Concepts (Exercises 32–40) For Individual or Group Work

The U.S. trade deficit represents the amount by which exports are less than imports. It provides not only a sign of economic prosperity but also a warning of potential decline. The data in the table shows the U.S. trade deficit in goods and services for 2001 through 2005.

Year	Deficit (in billions of dollars)
2001	365.1
2002	423.7
2003	496.9
2004	612.1
2005	714.4

Source: U.S. Census Bureau.

*Use the data to **work Exercises 32–40 in order.***

32. How much did the trade deficit in goods and services increase from 2001 to 2002? What percent increase is this (to the nearest percent)?

\$58.6 billion; 16%

33. The U.S. trade deficit might be approximated by the linear equation

$$y = 88.7x + 256,$$

where y is the deficit in billions of dollars. Here $x = 1$ represents 2001, $x = 2$ represents 2002, and so on. Use this equation to approximate the trade deficits in 2003, 2004, and 2005.

2003; $522.1 billion; 2004: $610.8 billion; 2005: $699.5 billion

34. How do your answers from Exercise 33 compare to the actual data in the table?

The answer using the linear equation is close to the actual data for 2004, but not for the other years.

35. The trade deficit y (in billions of dollars) might also be approximated by the quadratic equation

$$y = 9.24x^2 + 33.24x + 321,$$

where $x = 1$ again represents 2001, $x = 2$ represents 2002, and so on. Use this equation to approximate the trade deficits in 2003, 2004, and 2005.

2003: $503.9 billion; 2004: $601.8 billion; 2005: $718.2 billion

36. Compare your answers from Exercise 35 to the actual data in the table. Which equation, the linear one in Exercise 33 or the quadratic one in Exercise 35, models the data better?

The answers in Exercise 35 are fairly close to the actual data. The quadratic equation models the data better.

37. Write the data from the table as a set of ordered pairs (x, y), where x represents the years starting with 2001, such that $x = 1$ for 2001, $x = 2$ for 2002, and so on, and y represents the trade deficit in billions of dollars.

Year	Deficit (in billions of dollars)
2001	365.1
2002	423.7
2003	496.9
2004	612.1
2005	714.4

Source: U.S. Census Bureau.

(1, 365.1), (2, 423.7), (3, 496.9), (4, 612.1), (5, 714.4)

38. Plot the ordered pairs from Exercise 37 on the graph.

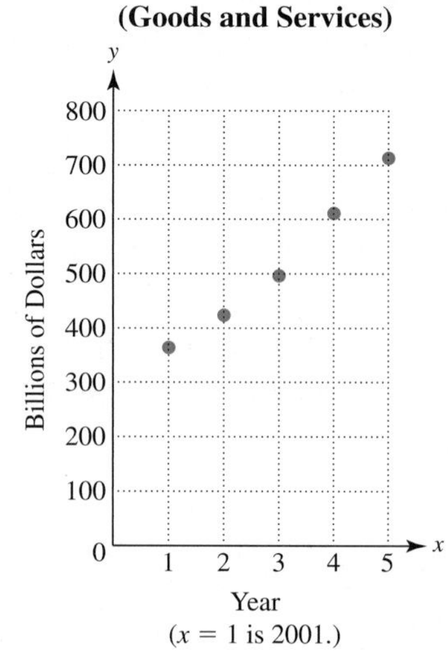

39. Assuming that the trend in the data continues, use the quadratic equation from Exercise 35 to estimate the trade deficit for the year 2006.

$853.1 billion

40. The actual trade deficit for 2006 was 758.2 billion dollars.

(a) How does the actual deficit for 2006 compare to your estimate from Exercise 39?

The actual deficit is quite a bit less than the estimate.

(b) Should the quadratic equation be used to estimate the U.S. trade deficit for years after 2005? Explain.

No, data for later years might not follow the same pattern.

Chapter 6 Summary

Key Terms

6.1	**factor**	An expression A is a factor of an expression B if B can be divided by A with 0 remainder.
	factored form	An expression is in factored form when it is written as a product.
	greatest common factor (GCF)	The greatest common factor is the largest quantity that is a factor of each of a group of quantities.
	factoring	The process of writing a polynomial as a product is called factoring.
6.2	**prime polynomial**	A prime polynomial is a polynomial that cannot be factored using only integers.
6.5	**perfect square trinomial**	A perfect square trinomial is a trinomial that can be factored as the square of a binomial.
6.7	**quadratic equation**	A quadratic equation is an equation that can be written in the form $ax^2 + bx + c = 0$, with $a \neq 0$.
	standard form	The form $ax^2 + bx + c = 0$ is the standard form of a quadratic equation.
6.8	**hypotenuse**	The longest side of a right triangle, opposite the right angle, is the hypotenuse.
	legs	The two shorter sides of a right triangle are the legs.

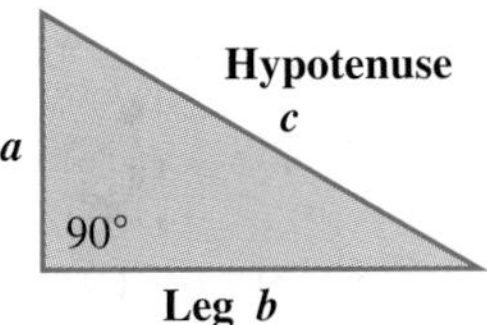

Test Your Word Power

See how well you have learned the vocabulary in this chapter. Answers, with examples, follow the Quick Review.

1. **Factoring** is
   A. a method of multiplying polynomials
   B. the process of writing a polynomial as a product
   C. the answer in a multiplication problem
   D. a way to add the terms of a polynomial.

2. A polynomial is in **factored form** when
   A. it is prime
   B. it is written as a sum
   C. the squared term has a coefficient of 1
   D. it is written as a product.

3. The **greatest common factor** of a polynomial is
   A. the least integer that divides evenly into all the terms of the polynomial
   B. the least term that is a factor of all the terms in the polynomial
   C. the greatest term that is a factor of all the terms in the polynomial
   D. the variable that is common to all the terms in the polynomial.

4. A **perfect square trinomial** is a trinomial
   A. that can be factored as the square of a binomial
   B. that cannot be factored
   C. that is multiplied by a binomial
   D. where all terms are perfect squares.

5. A **quadratic equation** is an equation that can be written in the form
   A. $y = mx + b$
   B. $ax^2 + bx + c = 0 \, (a \neq 0)$
   C. $Ax + By = C$
   D. $x = k$.

6. A **hypotenuse** is
   A. either of the two shorter sides of a triangle
   B. the shortest side of a right triangle
   C. the side opposite the right angle in a right triangle
   D. the longest side in any triangle.

▶ Quick Review

Concepts	Examples

6.1 Factors; The Greatest Common Factor

Finding the Greatest Common Factor (GCF)

Step 1 Write each number in prime factored form.

Step 2 List each prime number or each variable that is a factor of every term in the list.

Step 3 Use as exponents on the common prime factors the least exponents from the prime factored forms.

Step 4 Multiply the primes from Step 3.

Find the greatest common factor of $4x^2y$, $-6x^2y^3$, and $2xy^2$.

$$4x^2y = 2 \cdot 2 \cdot x^2 \cdot y$$
$$-6x^2y^3 = -1 \cdot 2 \cdot 3 \cdot x^2 \cdot y^3$$
$$2xy^2 = 2 \cdot x \cdot y^2$$

The greatest common factor is $2xy$.

Factoring by Grouping

Step 1 Group the terms.

Step 2 Factor out the greatest common factor from each group.

Step 3 Factor a common binomial factor from the results of Step 2.

Step 4 If necessary, rearrange terms and try a different grouping.

Factor by grouping.

$$\begin{aligned} &2a^2 + 2ab + a + b \\ &= (2a^2 + 2ab) + (a + b) \\ &= 2a(a + b) + 1(a + b) \\ &= (a + b)(2a + 1) \end{aligned}$$

6.2 Factoring Trinomials

To factor $x^2 + bx + c$, find m and n such that $mn = c$ and $m + n = b$.

$$\overset{mn = c}{\downarrow}$$
$$x^2 + bx + c$$
$$\underset{m + n = b}{\uparrow}$$

Then $x^2 + bx + c$ factors as $(x + m)(x + n)$.

Check by multiplying.

Factor $x^2 + 6x + 8$.

$$\overset{mn = 8}{\downarrow}$$
$$x^2 + 6x + 8$$
$$\underset{m + n = 6}{\uparrow}$$

$m = 2$ and $n = 4$

$x^2 + 6x + 8$ factors as $(x + 2)(x + 4)$.

Check $(x + 2)(x + 4)$

$$= x^2 + 4x + 2x + 8$$
$$= x^2 + 6x + 8$$

6.3 Factoring Trinomials by Grouping

To factor $ax^2 + bx + c$, find m and n such that $mn = ac$.

$$\overset{m + n = b}{\downarrow}$$
$$ax^2 + bx + c$$
$$\underset{mn = ac}{\uparrow \qquad \uparrow}$$

Then factor $ax^2 + mx + nx + b$ by grouping.

Factor $3x^2 + 14x - 5$.

$$\underset{-15}{\uparrow \qquad \uparrow}$$

Find two integers with a product of $3(-5) = -15$ and a sum of 14. The integers are -1 and 15.

$$\begin{aligned} &3x^2 + 14x - 5 \\ &= 3x^2 - x + 15x - 5 \\ &= (3x^2 - x) + (15x - 5) \\ &= x(3x - 1) + 5(3x - 1) \\ &= (3x - 1)(x + 5) \end{aligned}$$

Concepts	Examples

6.4 Factoring Trinomials Using FOIL

To factor $ax^2 + bx + c$ by trial and error, use FOIL backwards.

By trial and error,

$$3x^2 + 14x - 5 \quad \text{factors as} \quad (3x - 1)(x + 5).$$

6.5 Special Factoring Techniques

Difference of Squares

$$a^2 - b^2 = (a + b)(a - b)$$

Perfect Square Trinomials

$$a^2 + 2ab + b^2 = (a + b)^2$$

$$a^2 - 2ab + b^2 = (a - b)^2$$

Difference of Cubes

$$x^3 - y^3 = (x - y)(x^2 + xy + y^2)$$

Sum of Cubes

$$x^3 + y^3 = (x + y)(x^2 - xy + y^2)$$

Factor.

$$4x^2 - 9$$
$$= (2x + 3)(2x - 3)$$

$$9x^2 + 6x + 1 \qquad 4x^2 - 20x + 25$$
$$= (3x + 1)^2 \qquad = (2x - 5)^2$$

$$8 - 27a^3$$
$$= (2 - 3a)(4 + 6a + 9a^2)$$

$$64z^3 + 1$$
$$= (4z + 1)(16z^2 - 4z + 1)$$

6.6 A General Approach to Factoring

See pages 489–490 for guidelines and examples.

6.7 Solving Quadratic Equations by Factoring

Zero-Factor Property

If a and b are real numbers and $ab = 0$, then $a = 0$ or $b = 0$.

Solving a Quadratic Equation by Factoring

Step 1 Write the equation in standard form.

Step 2 Factor.

Step 3 Use the zero-factor property.

Step 4 Solve the resulting equations.

Step 5 Check.

If $(x - 2)(x + 3) = 0$, then $x - 2 = 0$ or $x + 3 = 0$.

Solve $2x^2 = 7x + 15$.

$$2x^2 - 7x - 15 = 0$$

$$(2x + 3)(x - 5) = 0$$

$$2x + 3 = 0 \quad \text{or} \quad x - 5 = 0$$

$$2x = -3 \qquad x = 5$$

$$x = -\frac{3}{2}$$

The solutions $-\frac{3}{2}$ and 5 satisfy the original equation.

The solution set is $\{-\frac{3}{2}, 5\}$.

Concepts	Examples

6.8 Applications of Quadratic Equations

Pythagorean Formula

In a right triangle, the square of the hypotenuse equals the sum of the squares of the legs.

$$a^2 + b^2 = c^2$$

Hypotenuse c

Leg a

$90°$

Leg b

In a right triangle, one leg measures 2 ft longer than the other. The hypotenuse measures 4 ft longer than the shorter leg. Find the lengths of the three sides of the triangle.

Let $x =$ the length of the shorter leg. Then

$$x^2 + (x + 2)^2 = (x + 4)^2.$$

Solve this equation to get $x = 6$ or $x = -2$. Discard -2 as a solution. Check that the sides measure 6 ft, $6 + 2 = 8$ ft, and $6 + 4 = 10$ ft.

ANSWERS TO TEST YOUR WORD POWER

1. B; *Example:* $x^2 - 5x - 14$ factors as $(x - 7)(x + 2)$.
2. D; *Example:* The factored form of $x^2 - 5x - 14$ is $(x - 7)(x + 2)$.
3. C; *Example:* The greatest common factor of $8x^2$, $22xy$, and $16x^3y^2$ is $2x$.
4. A; *Example:* $a^2 + 2a + 1$ is a perfect square trinomial; its factored form is $(a + 1)^2$.
5. B; *Examples:* $y^2 - 3y + 2 = 0$, $x^2 - 9 = 0$, $2m^2 = 6m + 8$
6. C; *Example:* See the triangle included in the Quick Review above for **Section 6.8.**

Chapter 6 Review Exercises

[6.1] *Factor out the greatest common factor or factor by grouping.*

1. $15t + 45$
   $15(t + 3)$
2. $60z^3 + 30z$
   $30z(2z^2 + 1)$
3. $44x^3 + 55x^2$
   $11x^2(4x + 5)$
4. $100m^2n^3 - 50m^3n^4 + 150m^2n^2$
   $50m^2n^2(2n - mn^2 + 3)$
5. $2xy - 8y + 3x - 12$
   $(x - 4)(2y + 3)$
6. $6y^2 + 9y + 4xy + 6x$
   $(2y + 3)(3y + 2x)$

[6.2] *Factor completely.*

7. $x^2 + 10x + 21$
   $(x + 3)(x + 7)$
8. $y^2 - 13y + 40$
   $(y - 5)(y - 8)$
9. $x^2 + x + 1$
   prime
10. $r^2 - r - 56$
    $(r - 8)(r + 7)$
11. $r^2 - 4rs - 96s^2$
    $(r + 8s)(r - 12s)$
12. $p^2 + 2pq - 120q^2$
    $(p + 12q)(p - 10q)$
13. $-8p^3 + 24p^2 + 80p$
    $-8p(p + 2)(p - 5)$
14. $3x^4 + 30x^3 + 48x^2$
    $3x^2(x + 2)(x + 8)$
15. $p^7 - p^6q - 2p^5q^2$
    $p^5(p - 2q)(p + q)$

[6.3–6.4] *Factor completely.*

16. $3r^2 + 11r - 4$
    $(3r - 1)(r + 4)$
17. $6r^2 - 5r - 6$
    $(3r + 2)(2r - 3)$
18. $5t^2 - 11t + 12$
    prime
19. $24x^5 - 20x^4 + 4x^3$
    $4x^3(3x - 1)(2x - 1)$
20. $-3m^3n + 19m^2n + 40mn$
    $-mn(3m + 5)(m - 8)$
21. $14a^2 - 27ab - 20b^2$
    $(2a - 5b)(7a + 4b)$

[6.5]

22. Which one of the following is a difference of squares?
    A. $32x^2 - 1$ B. $4x^2y^2 - 25z^2$
    C. $x^2 + 36$ D. $25y^3 - 1$
    B
23. Which one of the following is a perfect square trinomial?
    A. $x^2 + x + 1$ B. $y^2 - 4y + 9$
    C. $4x^2 + 10x + 25$ D. $x^2 - 20x + 100$
    D

Factor completely. In Exercise 28, use fractions.

24. $n^2 - 64$
    $(n + 8)(n - 8)$
25. $25b^2 - 121$
    $(5b + 11)(5b - 11)$
26. $144p^2 - 36q^2$
    $36(2p + q)(2p - q)$

27. $x^2 + 100$

prime

28. $x^2 - \frac{49}{100}$

$\left(x + \frac{7}{10}\right)\left(x - \frac{7}{10}\right)$

29. $r^2 - 12r + 36$

$(r - 6)^2$

30. $16m^2 + 40mn + 25n^2$

$(4m + 5n)^2$

31. $125x^3 - 1$

$(5x - 1)(25x^2 + 5x + 1)$

32. $1000p^3 + 27$

$(10p + 3)(100p^2 - 30p + 9)$

[6.7] *Solve each equation, and check the solutions.*

33. $(4t + 3)(t - 1) = 0$

$\left\{-\frac{3}{4}, 1\right\}$

34. $x(2x - 5) = 0$

$\left\{0, \frac{5}{2}\right\}$

35. $z^2 + 4z + 3 = 0$

$\{-3, -1\}$

36. $x^2 = -15 + 8x$

$\{3, 5\}$

37. $3z^2 - 11z - 20 = 0$

$\left\{-\frac{4}{3}, 5\right\}$

38. $81t^2 - 64 = 0$

$\left\{-\frac{8}{9}, \frac{8}{9}\right\}$

39. $y^2 = 8y$

$\{0, 8\}$

40. $n(n - 5) = 6$

$\{-1, 6\}$

41. $t^2 = 12(t - 3)$

$\{6\}$

42. $(5z + 2)(z^2 + 3z + 2) = 0$

$\left\{-\frac{2}{5}, -2, -1\right\}$

43. $49x^3 - 9x = 0$

$\left\{-\frac{3}{7}, 0, \frac{3}{7}\right\}$

44. $25w^2 - 90w + 81 = 0$

$\left\{\frac{9}{5}\right\}$

[6.8] *Solve each problem.*

45. The length of a rug is 6 ft more than the width. The area is 40 ft^2. Find the length and width of the rug. **length: 10 ft; width: 4 ft**

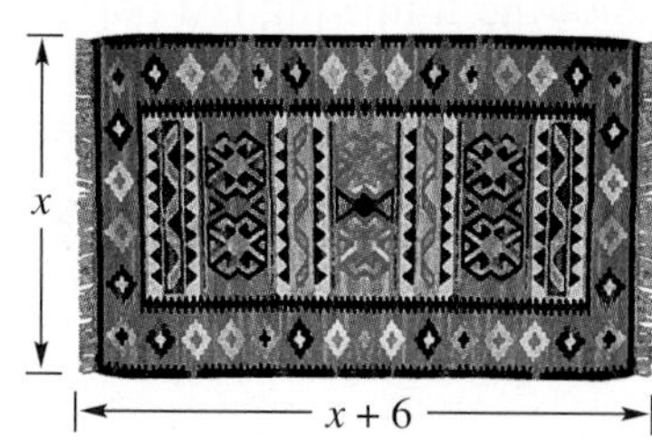

46. The surface area S of a box is given by

$$S = 2WH + 2WL + 2LH.$$

A treasure chest from a sunken galleon has dimensions as shown in the figure. Its surface area is 650 ft^2. Find its width. **5 ft**

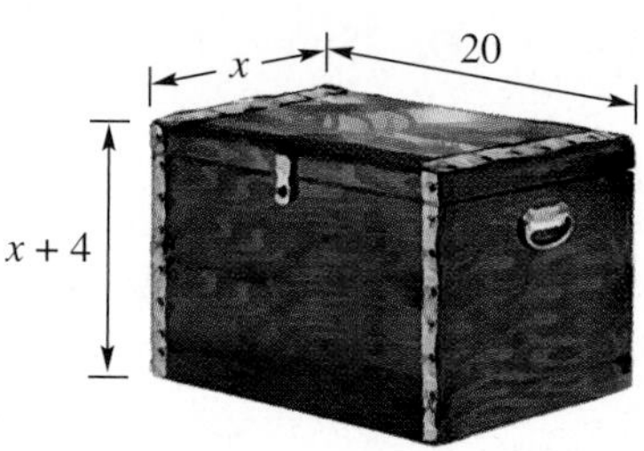

47. The product of two consecutive integers is 29 more than their sum. What are the integers?

6, 7 or −5, −4

48. Two cars left an intersection at the same time. One traveled west, and the other traveled 14 mi less, but to the south. How far apart were they then, if the distance between them was 16 mi more than the distance traveled south?

26 mi

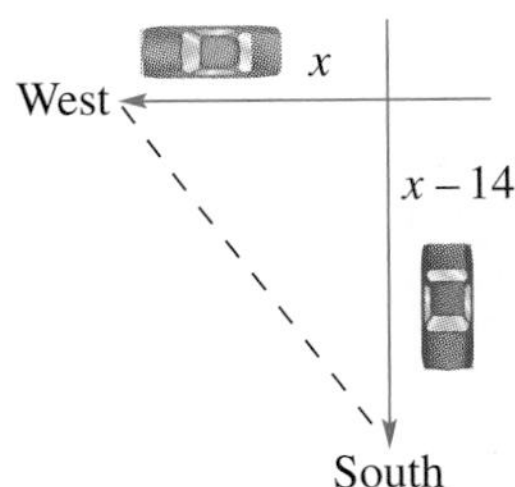

If an object is projected upward with an initial velocity of 128 *ft per sec, its height h in feet after t seconds is*

$$h = 128t - 16t^2.$$

Find the height of the object after each period of time.

49. 1 sec

112 ft

50. 2 sec

192 ft

51. 4 sec

256 ft

52. For the object described above, when does it return to the ground?

after 8 sec

53. Annual revenue in billions of dollars for eBay is shown in the table.

Year	Annual Revenue (in billions of dollars)
2002	1.21
2003	2.17
2004	3.27
2005	4.55
2006	5.77

Source: eBay.

Using the data, we developed the quadratic equation

$$y = 0.05x^2 + 0.95x + 1.19$$

to model eBay revenues y in year x, where $x = 0$ represents 2002, $x = 1$ represents 2003, and so on.

(a) Use the model to find eBay revenue (to the nearest hundredth) in 2005. How does your answer compare to the actual data from the table?

$4.49 billion; The answer using the model is very close.

(b) Use the model to estimate annual revenue (to the nearest hundredth) for eBay in 2007.

$7.19 billion

Mixed Review Exercises

54. Which of the following is *not* factored completely?

A. $3(7t)$ **B.** $3x(7t + 4)$ **C.** $(3 + x)(7t + 4)$ **D.** $3(7t + 4) + x(7t + 4)$

D

55. Although $(2x + 8)(3x - 4) = 6x^2 + 16x - 32$ is a true statement, the polynomial is not factored completely. Explain why and give the complete factored form.

The factor $2x + 8$ has a common factor of 2. The complete factored form is $2(x + 4)(3x - 4)$.

Factor completely.

56. $z^2 - 11zx + 10x^2$

$(z - x)(z - 10x)$

57. $3k^2 + 11k + 10$

$(3k + 5)(k + 2)$

58. $15m^2 + 20mp - 12m - 16p$

$(3m + 4p)(5m - 4)$

59. $y^4 - 625$

$(y^2 + 25)(y + 5)(y - 5)$

60. $6m^3 - 21m^2 - 45m$

$3m(2m + 3)(m - 5)$

61. $24ab^3c^2 - 56a^2bc^3 + 72a^2b^2c$

$8abc(3b^2c - 7ac^2 + 9ab)$

62. $25a^2 + 15ab + 9b^2$

prime

63. $12x^2yz^3 + 12xy^2z - 30x^3y^2z^4$

$6xyz(2xz^2 + 2y - 5x^2yz^3)$

64. $2a^5 - 8a^4 - 24a^3$

$2a^3(a + 2)(a - 6)$

65. $-12r^2 - 8rq + 15q^2$

$-(2r + 3q)(6r - 5q)$

66. $8z^3 + 64y^3$

$8(z + 2y)(z^2 - 2zy + 4y^2)$

67. $49t^2 + 56t + 16$

$(7t + 4)^2$

Solve.

68. $t(t - 7) = 0$

$\{0, 7\}$

69. $x^2 + 3x = 10$

$\{-5, 2\}$

70. $25x^2 + 20x + 4 = 0$

$\left\{-\frac{2}{5}\right\}$

Solve each problem.

71. The floor plan for a house is a rectangle with length 7 m more than its width. The area is 170 m². Find the width and length of the house.

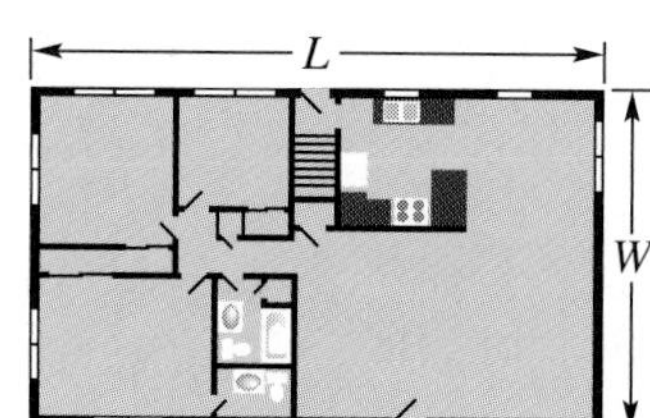

width: 10 m; length: 17 m

72. The triangular sail of a schooner has an area of 30 m². The height of the sail is 4 m more than the base. Find the length of the base of the sail. **6 m**

Chapter 6 Test

CHAPTER Test Prep VIDEO CD — Use the Chapter Test Prep Video CD to see fully worked-out solutions to any of the exercises you want to review.

1. Which one of the following is the correct, completely factored form of $2x^2 - 2x - 24$?

   A. $(2x + 6)(x - 4)$ **B.** $(x + 3)(2x - 8)$

   C. $2(x + 4)(x - 3)$ **D.** $2(x + 3)(x - 4)$

Factor each polynomial completely.

2. $12x^2 - 30x$
3. $2m^3n^2 + 3m^3n - 5m^2n^2$
4. $2ax - 2bx + ay - by$
5. $x^2 - 9x + 14$
6. $2x^2 + x - 3$
7. $6x^2 - 19x - 7$
8. $3x^2 - 12x - 15$
9. $10z^2 - 17z + 3$
10. $t^2 + 6t + 10$
11. $x^2 + \frac{1}{36}$
12. $y^2 - 49$
13. $81a^2 - 121b^2$
14. $x^2 + 16x + 64$
15. $4x^2 - 28xy + 49y^2$
16. $-2x^2 - 4x - 2$
17. $x^3 - 512$
18. $8k^3 + 64$

Answers:

1. D
2. $6x(2x - 5)$
3. $m^2n(2mn + 3m - 5n)$
4. $(2x + y)(a - b)$
5. $(x - 7)(x - 2)$
6. $(2x + 3)(x - 1)$
7. $(3x + 1)(2x - 7)$
8. $3(x + 1)(x - 5)$
9. $(5z - 1)(2z - 3)$
10. prime
11. prime
12. $(y + 7)(y - 7)$
13. $(9a + 11b)(9a - 11b)$
14. $(x + 8)^2$
15. $(2x - 7y)^2$
16. $-2(x + 1)^2$
17. $(x - 8)(x^2 + 8x + 64)$
18. $8(k + 2)(k^2 - 2k + 4)$

19. $4t(t + 4)^2$

20. $(x^2 + 9)(x + 3)(x - 3)$

21. $\{-3, 9\}$

22. $\left\{\frac{1}{2}, 6\right\}$

23. $\left\{-\frac{2}{5}, \frac{2}{5}\right\}$

24. $\{10\}$

25. $\{0, 3\}$

26. $\left\{-8, -\frac{5}{2}, \frac{1}{3}\right\}$

27. 6 ft by 9 ft

28. $-2, -1$

29. 17 ft

30. 49 million

19. $4t^3 + 32t^2 + 64t$

20. $x^4 - 81$

Solve each equation.

21. $(x + 3)(x - 9) = 0$

22. $2r^2 - 13r + 6 = 0$

23. $25x^2 - 4 = 0$

24. $x(x - 20) = -100$

25. $t^2 = 3t$

26. $(s + 8)(6s^2 + 13s - 5) = 0$

Solve each problem.

27. The length of a rectangular flower bed is 3 ft less than twice its width. The area of the bed is 54 ft^2. Find the dimensions of the flower bed.

28. Find two consecutive integers such that the square of the sum of the two integers is 11 more than the lesser integer.

29. A carpenter needs to cut a brace to support a wall stud, as shown in the figure. The brace should be 7 ft less than three times the length of the stud. If the brace will be anchored on the floor 15 ft away from the stud, how long should the brace be?

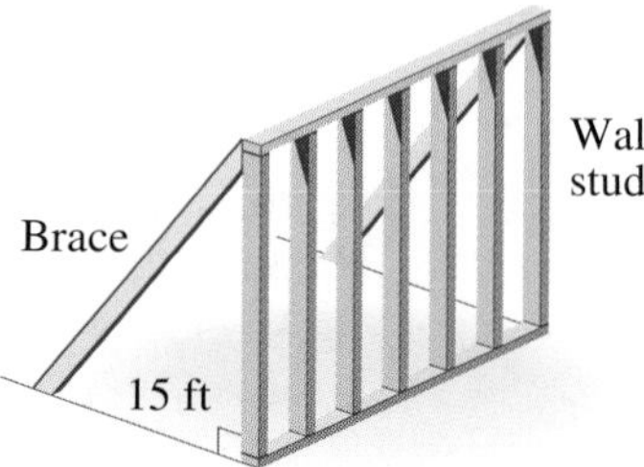

30. The number of Americans using broadband Internet access at home from 2001 through 2006 can be approximated by the quadratic equation

$$y = 2.36x^2 + 4.81x + 13.81,$$

where $x = 0$ represents 2001, $x = 1$ represents 2002, and so on, and y is number in millions. (*Source:* Pew Internet Project, Nielsen/Net Ratings.) Use the model to estimate the number of Americans using broadband Internet at home in 2004. Round your answer to the nearest million.

Cumulative Review Exercises Chapters R–6

Solve each equation.

1. $3x + 2(x - 4) = 4(x - 2)$
{0}

2. $0.3x + 0.9x = 0.06$
{0.05}

3. $\frac{2}{3}n - \frac{1}{2}(n - 4) = 3$
{6}

4. Solve for t: $A = P + Prt$
$t = \frac{A - P}{Pr}$

5. From a list of "everyday items" often taken for granted, adults were recently surveyed as to those items they wouldn't want to live without. Complete the results shown in the table if 500 adults were surveyed.

Item	Percent That Wouldn't Want to Live Without	Number That Wouldn't Want to Live Without
Toilet paper	69%	**345**
Zipper	42%	**210**
Frozen foods	**38%**	190
Self-stick note pads	**15%**	75

(Other items included tape, hairspray, paper clips, and Velcro.)
Source: Market Facts for Kleenex Cottonelle.

Solve each problem.

6. At the 2006 Winter Olympics in Torino, Italy, the top medal winner was Germany, which won a total of 29 medals. Germany won 1 more silver medal than gold and 5 more gold medals than bronze. Find the number of each type of medal won. (*Source:* www.infoplease.com)
gold: 11; silver: 12; bronze: 6

7. In 2004, American women working full time earned, on average, 80.8 cents for every dollar earned by men working full time. The median weekly earnings for full-time male workers were \$743. To the nearest dollar, what were the median weekly earnings for full-time female workers? (*Source:* U.S. Bureau of Labor Statistics.)
\$600

8. Find the measures of the marked angles.
110° and 70°

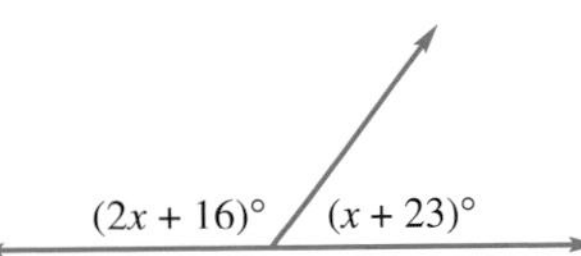

9. Fill in each blank with *positive* or *negative*. The point with coordinates (a, b) is in

(a) quadrant II if a is **negative** and b is **positive**.

(b) quadrant III if a is **negative** and b is **negative**.

Consider the equation $y = 12x + 3$. *Find the following.*

10. The x- and y-intercepts
$\left(-\frac{1}{4}, 0\right)$, $(0, 3)$

11. The slope
12

12. The graph

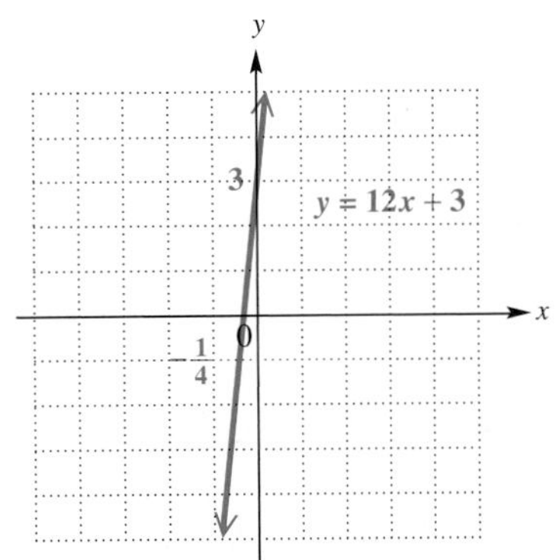

13. The points on the graph show the total retail sales of prescription drugs in the United States in the years 2000–2006, along with a graph of a linear equation that models the data.

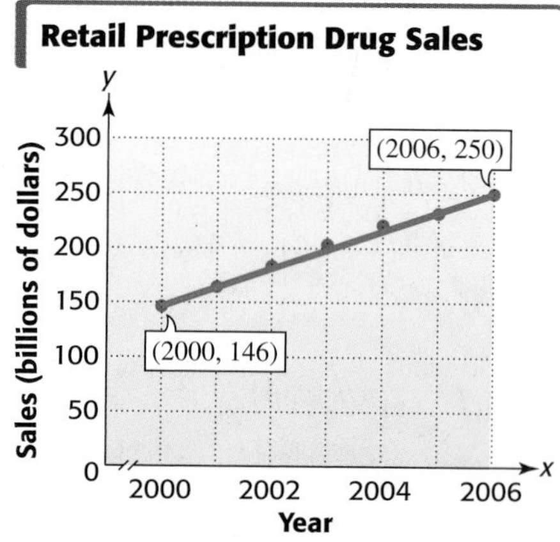

Source: National Association of Chain Drug Stores.

(a) Use the ordered pairs shown on the graph to find the slope of the line to the nearest whole number. Interpret the slope.

17; A slope of (approximately) 17 means that the retail sales of prescription drugs increased by about \$17 billion per year.

(b) Use the graph to estimate sales in the year 2005. Write your answer as an ordered pair of the form (year, sales in billions of dollars).

(2005, 230)

Solve each system of equations.

14. $4x - y = -6$ **{(−1, 2)}**
$2x + 3y = 4$

15. $5x + 3y = 10$ **∅**
$2x + \frac{6}{5}y = 5$

Evaluate each expression.

16. $2^{-3} \cdot 2^5$ **4**

17. $\left(\frac{3}{4}\right)^{-2}$ $\mathbf{\frac{16}{9}}$

18. $\frac{6^5 \cdot 6^{-2}}{6^3}$ **1**

19. $\left(\frac{4^{-3} \cdot 4^4}{4^5}\right)^{-1}$ **256**

Simplify each expression and write the answer using only positive exponents. Assume no denominators are 0.

20. $\frac{(p^2)^3 p^{-4}}{(p^{-3})^{-1} p}$ $\mathbf{\frac{1}{p^2}}$

21. $\frac{(m^{-2})^3 m}{m^5 m^{-4}}$ $\mathbf{\frac{1}{m^6}}$

Perform the indicated operations.

22. $(2k^2 + 4k) - (5k^2 - 2) - (k^2 + 8k - 6)$
$\mathbf{-4k^2 - 4k + 8}$

23. $(9x + 6)(5x - 3)$
$\mathbf{45x^2 + 3x - 18}$

24. $(3p + 2)^2$
$\mathbf{9p^2 + 12p + 4}$

25. $\frac{8x^4 + 12x^3 - 6x^2 + 20x}{2x}$
$\mathbf{4x^3 + 6x^2 - 3x + 10}$

Factor completely.

26. $2a^2 + 7a - 4$
$\mathbf{(2a - 1)(a + 4)}$

27. $10m^2 + 19m + 6$
$\mathbf{(2m + 3)(5m + 2)}$

28. $8t^2 + 10tv + 3v^2$
$\mathbf{(4t + 3v)(2t + v)}$

29. $4p^2 - 12p + 9$
$\mathbf{(2p - 3)^2}$

30. $25r^2 - 81t^2$
$\mathbf{(5r + 9t)(5r - 9t)}$

31. $2pq + 6p^3q + 8p^2q$
$\mathbf{2pq(3p + 1)(p + 1)}$

Solve each equation.

32. $6m^2 + m - 2 = 0$
$\mathbf{\left\{-\frac{2}{3}, \frac{1}{2}\right\}}$

33. $8x^2 = 64x$
{0, 8}

34. $49x^2 - 56x + 16 = 0$
$\mathbf{\left\{\frac{4}{7}\right\}}$

35. The length of the hypotenuse of a right triangle is 3 m more than twice the length of the shorter leg. The longer leg is 7 m longer than the shorter leg. Find the lengths of the sides.
5 m, 12 m, 13 m

Study Skills

PREPARING FOR YOUR MATH FINAL EXAM

OBJECTIVES

1. **Create a final exam week plan.**
2. **Break studying into chunks and study over several days.**
3. **Practice all types of problems.**

Your math final exam is likely to be a **comprehensive exam.** This means that it will cover material from the entire term. The end of the term will be less stressful if you make a plan for how you will prepare for each of your exams.

First, figure out the grade you need to earn on the final exam to get the course grade you are aiming for. Check your course syllabus for grading policies, or ask your instructor if you are not sure of them. This allows you to set a goal for yourself.

> How many points do you need to earn on your mathematics final exam to get the grade you want? ____________________

Second, create a final exam week plan for your work and personal life. If you need to make an adjustment in your work schedule, do it in advance. If you have family members to care for, perhaps enlist some help from others so you can spend extra time studying. Try to plan in advance so you don't create additional stress for yourself. You will have to set some priorities, and studying has to be at the top of the list. Although life doesn't stop for finals, some things can be ignored for a short time. **Get enough sleep and healthy food so you can perform your best.**

> What adjustments in your personal life do you need to make for final exam week? ____________________

Third, use the following suggestions to guide your studying and reviewing.

- **Know exactly which chapters and sections will be on the final exam.**
- **Divide up the chapters,** and decide how much you will review each day.
- **Begin your reviewing several days before the exam.**
- **Use returned quizzes and tests to review earlier material** (if you have them).
- **Practice all types of problems,** but emphasize the types that are most difficult for you. Use the Cumulative Reviews that are at the end of each chapter in your textbook.
- **Rewrite your notes or make mind maps** to create summaries of important information.
- **Make study cards for all types of problems.** Carry the cards with you, and review them whenever you have a few spare minutes.

Managing Stress

Of course, a week of final exams produces stress. **Students who develop skills for reducing and managing stress do better on their final exams.** You will feel better if you make a conscious effort to reduce your stress level. Even if it takes you away from studying for a little while each day, the time will be well spent.

Reducing Physical Stress

Examples of ways to reduce physical stress are listed below.

- *Laugh*. Watch your favorite funny movie, exchange a joke with a friend, or view a comedy bit on the Internet.
- *Exercise for 20 to 30 minutes.* If you normally exercise regularly, do NOT stop during final exam week.
- *Practice deep breathing.* Several minutes of deep breathing will calm you.
- *Visualize a relaxing scene.* Choose something that you find peaceful and picture it. Imagine what it feels like and sounds like. Try to put yourself in the picture.
- If you feel stress in your muscles, such as your shoulders or back, *slowly squeeze the muscles as much as you can, and then release them.* Sometimes we don't realize we are clenching our teeth or holding tension in our shoulders until we consciously work with them. Try to notice what it feels like when they are relaxed and loose.

Reducing Mental Stress

Mental stress reduction is also a powerful tool both before and during an exam.

- *Talk positively to yourself.* Tell yourself you will get through it.
- *Reward yourself.* Give yourself small breaks, a treat—something that makes you feel cared for—every day of final exam week.
- *Make a list of things to do* and feel the sense of accomplishment when you cross each item off.
- *Relax your mind* by using it for something *completely* different from the kind of thinking you do when you study. Play your favorite music, walk your dog, read a good book.
- *Visualize.* Picture yourself completing exams and projects successfully. Picture yourself taking tests calmly and confidently.

Finally, DON'T stay up all night the night before an exam—get a good night's sleep.

Now Try This ▶▶▶

Choose three techniques for reducing stress from the lists above to try during final exam week. Include other ideas as well.

1. ______________________________

2. ______________________________

3. ______________________________

Other ideas ______________________________

7

Rational Expressions and Functions

Americans have been car crazy ever since the first automobiles hit the road early in the twentieth century. Nowhere is this more apparent than in Michigan, long the home of automobile manufacturing in the United States. To coincide with the 100th anniversary of the Model T, the 100th birthday of General Motors Corporation, and the first Autopalooza, a summer auto tourism festival expected to draw over one million visitors, Michigan officially designated 2008 as the Year of the Car. (*Source: The Gazette*, May 25, 2008.)

In Exercises 67 and 68 of Section 7.2, we use a *rational expression* to determine the cost of restoring a vintage automobile.

7.1 Rational Expressions and Functions; Multiplying and Dividing

OBJECTIVES

1. Define rational expressions.
2. Define rational functions and describe their domains.
3. Write rational expressions in lowest terms.
4. Multiply rational expressions.
5. Find reciprocals for rational expressions.
6. Divide rational expressions.

OBJECTIVE 1 Define rational expressions. In arithmetic, a rational number is the quotient of two integers, with the denominator not 0. In algebra, a **rational expression** or *algebraic fraction* is the quotient of two polynomials, again with the denominator not 0. For example,

$$\frac{x}{y}, \quad \frac{-a}{4}, \quad \frac{m+4}{m-2}, \quad \frac{8x^2-2x+5}{4x^2+5x}, \quad \text{and} \quad x^5\left(\text{or } \frac{x^5}{1}\right) \qquad \text{Rational expressions}$$

are rational expressions. Rational expressions are the elements of the set

$$\left\{\frac{P}{Q} \,\middle|\, P \text{ and } Q \text{ are polynomials, with } Q \neq 0\right\}.$$

OBJECTIVE 2 Define rational functions and describe their domains. A function that is defined by a rational expression is called a **rational function** and has the form

$$f(x) = \frac{P(x)}{Q(x)}, \quad \text{where } Q(x) \neq 0.$$

The domain of a rational function includes all real numbers except those that make $Q(x)$, that is, the denominator, equal to 0. For example, the domain of

$$f(x) = \frac{2}{\underbrace{x-5}_{\text{Cannot equal 0}}}$$

includes all real numbers except 5, because 5 would make the denominator equal to 0. Figure 1 shows a graph of the function defined by

$$f(x) = \frac{2}{x-5}.$$

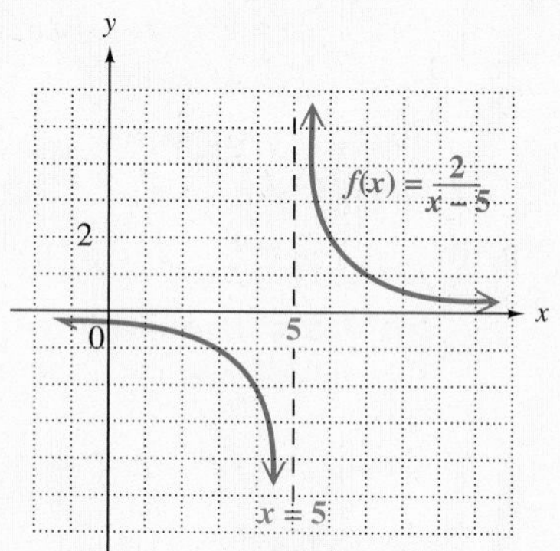

Figure 1

Notice that the graph does not exist when $x = 5$. It does not intersect the dashed vertical line whose equation is $x = 5$. This line is an **asymptote.** We will discuss graphs of rational functions in more detail in **Section 7.4.**

EXAMPLE 1 Finding Numbers That Are Not in the Domains of Rational Functions

Find all numbers that are not in the domain of each rational function. Then give the domain using set notation.

(a) $f(x) = \frac{3}{7x-14}$

The only values that cannot be used are those that make the denominator 0. To find these values, set the denominator equal to 0 and solve the resulting equation.

$$7x - 14 = 0$$

$$7x = 14 \qquad \text{Add 14.}$$

$$x = 2 \qquad \text{Divide by 7.}$$

The number 2 cannot be used as a replacement for x. The domain of f includes all real numbers except 2, written using set notation as $\{x \mid x \neq 2\}$.

Continued on Next Page

(b) $g(x) = \dfrac{3 + x}{x^2 - 4x + 3}$ Values that make the denominator 0 must be excluded.

$$x^2 - 4x + 3 = 0 \quad \text{Set the denominator equal to 0.}$$

$$(x - 1)(x - 3) = 0 \quad \text{Factor.}$$

$$x - 1 = 0 \quad \text{or} \quad x - 3 = 0 \quad \text{Zero-factor property}$$

$$x = 1 \quad \text{or} \quad x = 3 \quad \text{Solve each equation.}$$

The domain of g includes all real numbers except 1 and 3, written $\{x \mid x \neq 1, 3\}$.

(c) $h(x) = \dfrac{8x + 2}{3}$

The denominator, 3, can never be 0, so the domain of h includes all real numbers, written $(-\infty, \infty)$.

(d) $f(x) = \dfrac{2}{x^2 + 4}$

Setting $x^2 + 4$ equal to 0 leads to $x^2 = -4$. There is no real number whose square is -4. Therefore, any real number can be used as a replacement for x. As in part (c), the domain of f consists of all real numbers $(-\infty, \infty)$.

Work Problem 1 at the Side. ▶

OBJECTIVE 3 Write rational expressions in lowest terms. In arithmetic, we write the fraction $\frac{15}{20}$ in lowest terms by dividing the numerator and denominator by 5 to get $\frac{3}{4}$. We write rational expressions in lowest terms in a similar way, using the **fundamental property of rational numbers.**

Fundamental Property of Rational Numbers

If $\frac{a}{b}$ is a rational number and if c is any nonzero real number, then

$$\frac{a}{b} = \frac{ac}{bc}.$$

In words, the numerator and denominator of a rational number may either be multiplied or divided by the same nonzero number without changing the value of the rational number.

Because $\frac{c}{c}$ is equivalent to 1, the fundamental property is based on the identity property of multiplication.

A rational expression is a quotient of two polynomials. Since the value of a polynomial is a real number for every value of the variable for which it is defined, any statement that applies to rational numbers will also apply to rational expressions.

We use the following steps to write rational expressions in lowest terms.

Writing a Rational Expression in Lowest Terms

Step 1 **Factor** both numerator and denominator to find their greatest common factor (GCF).

Step 2 **Apply the fundamental property.**

1 Find all numbers that are not in the domain of each rational function. Then give the domain using set notation.

(a) $f(x) = \dfrac{x + 4}{x - 6}$

(b) $f(x) = \dfrac{x + 6}{x^2 - x - 6}$

(c) $f(x) = \dfrac{3 + 2x}{5}$

(d) $f(x) = \dfrac{2}{x^2 + 1}$

Answers

1. (a) 6; $\{x \mid x \neq 6\}$ **(b)** $-2, 3$; $\{x \mid x \neq -2, 3\}$
(c) none; The domain consists of all real numbers $(-\infty, \infty)$.
(d) none; The domain consists of all real numbers $(-\infty, \infty)$.

EXAMPLE 2 **Writing Rational Expressions in Lowest Terms**

Write each rational expression in lowest terms.

(a) $\dfrac{8k}{16} = \dfrac{k \cdot 8}{2 \cdot 8} = \dfrac{k}{2} \cdot 1 = \dfrac{k}{2}$ Factor; apply the fundamental property.

(b) $\dfrac{8 + k}{16}$ Be careful. The numerator cannot be factored.

This expression cannot be simplified further and is in lowest terms.

(c) $\dfrac{a^2 - a - 6}{a^2 + 5a + 6}$

$= \dfrac{(a - 3)(a + 2)}{(a + 3)(a + 2)}$ Factor the numerator and the denominator.

$= \dfrac{a - 3}{a + 3} \cdot 1$ Fundamental property

$= \dfrac{a - 3}{a + 3}$ Lowest terms

(d) $\dfrac{y^2 - 4}{2y + 4}$

$= \dfrac{(y + 2)(y - 2)}{2(y + 2)}$ Factor the difference of squares in the numerator; factor the denominator.

$= \dfrac{y - 2}{2}$ Lowest terms

(e) $\dfrac{x^3 - 27}{x - 3}$

$= \dfrac{(x - 3)(x^2 + 3x + 9)}{x - 3}$ Factor the difference of cubes.

$= x^2 + 3x + 9$ Lowest terms

(f) $\dfrac{pr + qr + ps + qs}{pr + qr - ps - qs}$

$= \dfrac{(pr + qr) + (ps + qs)}{(pr + qr) - (ps + qs)}$ Group the terms.

$= \dfrac{r(p + q) + s(p + q)}{r(p + q) - s(p + q)}$ Factor within the groups.

$= \dfrac{(p + q)(r + s)}{(p + q)(r - s)}$ Factor by grouping.

$= \dfrac{r + s}{r - s}$ Lowest terms

CAUTION

Be careful! ***When using the fundamental property of rational numbers, only common factors may be divided.*** For example,

$$\frac{y-2}{2} \neq y \quad \text{and} \quad \frac{y-2}{2} \neq y-1$$

because the 2 in $y - 2$ is not a *factor* of the numerator. ***Remember to factor before writing a fraction in lowest terms.***

Work Problem 2 at the Side. ▶

In the rational expression from Example 2(c),

$$\frac{a^2-a-6}{a^2+5a+6}, \quad \text{or} \quad \frac{(a-3)(a+2)}{(a+3)(a+2)},$$

a can take any value except -3 or -2 since these values make the denominator 0. In the simplified rational expression

$$\frac{a-3}{a+3},$$

a cannot equal -3. Because of this,

$$\frac{a^2-a-6}{a^2+5a+6} = \frac{a-3}{a+3}$$

for all values of a except -3 or -2. From now on such statements of equality will be made with the understanding that they apply only for those real numbers that make neither denominator equal 0. We will no longer state such restrictions.

EXAMPLE 3 **Writing Rational Expressions in Lowest Terms**

Write each rational expression in lowest terms.

(a) $\frac{m-3}{3-m}$

Here, the numerator and denominator are opposites. This expression can be written in lowest terms by writing the denominator as $-1(m-3)$.

$$\frac{m-3}{3-m} = \frac{m-3}{-1(m-3)} = \frac{1}{-1} = -1$$

The numerator could have been rewritten instead to get the same result.

(b) $\frac{r^2-16}{4-r}$

$= \frac{(r+4)(r-4)}{4-r}$ Factor the difference of squares in the numerator.

$= \frac{(r+4)(r-4)}{-1(r-4)}$ Write $4 - r$ as $-1(r-4)$.

$= \frac{r+4}{-1}$ Fundamental property

$= -(r+4)$ or $-r-4$ Lowest terms

2 Write each rational expression in lowest terms.

(a) $\frac{y^2+2y-3}{y^2-3y+2}$

(b) $\frac{3y+9}{y^2-9}$

(c) $\frac{y+2}{y^2+4}$

(d) $\frac{1+p^3}{1+p}$

(e) $\frac{3x+3y+rx+ry}{5x+5y-rx-ry}$

ANSWERS

2. (a) $\frac{y+3}{y-2}$ **(b)** $\frac{3}{y-3}$
(c) already in lowest terms
(d) $1-p+p^2$ **(e)** $\frac{3+r}{5-r}$

3 Write each rational expression in lowest terms.

(a) $\frac{y-2}{2-y}$

(b) $\frac{8-b}{8+b}$

(c) $\frac{p-2}{4-p^2}$

ANSWERS

3. (a) -1 (b) already in lowest terms
(c) $\frac{-1}{2+p}$

As shown in Example 3, the quotient $\frac{a}{-a}$ $(a \neq 0)$ can be simplified as

$$\frac{a}{-a} = \frac{a}{-1(a)} = \frac{1}{-1} = -1.$$

Quotient of Opposites

In general, if the numerator and the denominator of a rational expression are opposites, the expression equals -1.

Based on this result, the following are true:

$$\frac{q-7}{7-q} = -1 \quad \text{and} \quad \frac{-5a+2b}{5a-2b} = -1.$$

Numerator and denominator in each expression are opposites.

However, the following expression cannot be simplifed further.

$$\frac{r-2}{r+2}$$ Numerator and denominator are *not* opposites.

◀ *Work Problem* **3** *at the Side.*

OBJECTIVE 4 Multiply rational expressions. To multiply rational expressions, follow these steps. (In practice, we usually simplify before multiplying.)

Multiplying Rational Expressions

Step 1 **Factor** all numerators and denominators as completely as possible.

Step 2 **Apply the fundamental property.**

Step 3 **Multiply** remaining factors in the numerator and remaining factors in the denominator. Leave the denominator in factored form.

Step 4 **Check** to be sure the product is in lowest terms.

EXAMPLE 4 Multiplying Rational Expressions

Multiply.

(a) $\frac{5p-5}{p} \cdot \frac{3p^2}{10p-10}$

$= \frac{5(p-1)}{p} \cdot \frac{3p \cdot p}{2 \cdot 5(p-1)}$ Factor.

$= \frac{5(p-1)}{5(p-1)} \cdot \frac{p}{p} \cdot \frac{3p}{2}$ Commutative property

$= \frac{1}{1} \cdot \frac{1}{1} \cdot \frac{1}{1} \cdot \frac{3p}{2}$ Fundamental property

$= \frac{3p}{2}$ Lowest terms

Continued on Next Page

(b) $\dfrac{k^2 + 2k - 15}{k^2 - 4k + 3} \cdot \dfrac{k^2 - k}{k^2 + k - 20}$

$= \dfrac{(k + 5)(k - 3)}{(k - 3)(k - 1)} \cdot \dfrac{k(k - 1)}{(k + 5)(k - 4)}$ Factor.

$= \dfrac{k}{k - 4}$ Lowest terms

(c) $(p - 4) \cdot \dfrac{3}{5p - 20}$

$= \dfrac{p - 4}{1} \cdot \dfrac{3}{5p - 20}$ Write $p - 4$ as $\frac{p - 4}{1}$.

$= \dfrac{p - 4}{1} \cdot \dfrac{3}{5(p - 4)}$ Factor.

$= \dfrac{3}{5}$ Multiply; lowest terms

(d) $\dfrac{x^2 + 2x}{x + 1} \cdot \dfrac{x^2 - 1}{x^3 + x^2}$

$= \dfrac{x(x + 2)}{x + 1} \cdot \dfrac{(x + 1)(x - 1)}{x^2(x + 1)}$ Factor.

$= \dfrac{(x + 2)(x - 1)}{x(x + 1)}$ Multiply; lowest terms.

(e) $\dfrac{x - 6}{x^2 - 12x + 36} \cdot \dfrac{x^2 - 3x - 18}{x^2 + 7x + 12}$

$= \dfrac{x - 6}{(x - 6)^2} \cdot \dfrac{(x + 3)(x - 6)}{(x + 3)(x + 4)}$ Factor.

$= \dfrac{1}{x + 4}$ Lowest terms

Remember to include 1 in the numerator when all other factors are eliminated.

Work Problem **4** *at the Side.* ▶

4 Multiply.

(a) $\dfrac{2r + 4}{5r} \cdot \dfrac{3r}{5r + 10}$

(b) $\dfrac{c^2 + 2c}{c^2 - 4} \cdot \dfrac{c^2 - 4c + 4}{c^2 - c}$

(c) $\dfrac{m^2 - 16}{m + 2} \cdot \dfrac{1}{m + 4}$

(d) $\dfrac{x - 3}{x^2 + 2x - 15} \cdot \dfrac{x^2 - 25}{x^2 + 3x - 40}$

OBJECTIVE 5 Find reciprocals for rational expressions. The rational numbers $\frac{a}{b}$ and $\frac{c}{d}$ are reciprocals of each other if they have a product of 1. The **reciprocal** of a rational expression is defined in the same way: ***Two rational expressions are reciprocals of each other if they have a product of 1. Recall that 0 has no reciprocal.*** The table shows several rational expressions and their reciprocals.

Rational Expression	Reciprocal
$\dfrac{5}{k}$	$\dfrac{k}{5}$
$\dfrac{m^2 - 9m}{2}$	$\dfrac{2}{m^2 - 9m}$
$\dfrac{0}{4}$	undefined

ANSWERS

4. (a) $\dfrac{6}{25}$ (b) $\dfrac{c - 2}{c - 1}$ (c) $\dfrac{m - 4}{m + 2}$ (d) $\dfrac{1}{x + 8}$

The examples in the table on the previous page suggest the following.

Finding the Reciprocal

To find the reciprocal of a nonzero rational expression, interchange the numerator and denominator of the expression.

◀ Work Problem 5 at the Side.

OBJECTIVE 6 Divide rational expressions. Dividing rational expressions is like dividing rational numbers.

Dividing Rational Expressions

To divide two rational expressions, multiply the first (the dividend) by the reciprocal of the second (the divisor).

EXAMPLE 5 Dividing Rational Expressions

Divide.

(a) $\frac{2z}{9} \div \frac{5z^2}{18}$

$= \frac{2z}{9} \cdot \frac{18}{5z^2}$ Multiply by the reciprocal of the divisor.

$= \frac{2z}{9} \cdot \frac{2 \cdot 9}{5z^2}$ Factor.

$= \frac{4}{5z}$ Multiply; lowest terms

(b) $\frac{8k - 16}{3k} \div \frac{3k - 6}{4k^2}$

$= \frac{8k - 16}{3k} \cdot \frac{4k^2}{3k - 6}$ Multiply by the reciprocal.

$= \frac{8(k - 2)}{3k} \cdot \frac{4k^2}{3(k - 2)}$ Factor.

$= \frac{32k}{9}$ Multiply; lowest terms

(c) $\frac{5m^2 + 17m - 12}{3m^2 + 7m - 20} \div \frac{5m^2 + 2m - 3}{15m^2 - 34m + 15}$

$= \frac{5m^2 + 17m - 12}{3m^2 + 7m - 20} \cdot \frac{15m^2 - 34m + 15}{5m^2 + 2m - 3}$ Definition of division

$= \frac{(5m - 3)(m + 4)}{(m + 4)(3m - 5)} \cdot \frac{(3m - 5)(5m - 3)}{(5m - 3)(m + 1)}$ Factor.

$= \frac{5m - 3}{m + 1}$ Lowest terms

◀ Work Problem 6 at the Side.

5 Find each reciprocal.

(a) $\frac{-3}{r}$

(b) $\frac{7}{y + 8}$

(c) $\frac{a^2 + 7a}{2a - 1}$

(d) $\frac{0}{-5}$

6 Divide.

(a) $\frac{16k^2}{5} \div \frac{3k}{10}$

(b) $\frac{5p + 2}{6} \div \frac{15p + 6}{5}$

(c) $\frac{y^2 - 2y - 3}{y^2 + 4y + 4} \div \frac{y^2 - 1}{y^2 + y - 2}$

Answers

5. (a) $\frac{r}{-3}$ (b) $\frac{y + 8}{7}$ (c) $\frac{2a - 1}{a^2 + 7a}$
(d) There is no reciprocal.

6. (a) $\frac{32k}{3}$ (b) $\frac{5}{18}$ (c) $\frac{y - 3}{y + 2}$

7.1 Exercises

FOR EXTRA HELP

Math XL PRACTICE

WATCH

DOWNLOAD

READ

REVIEW

Rational expressions can often be written in lowest terms in seemingly *different ways. For example,*

$$\frac{y-3}{-5} \quad \textit{and} \quad \frac{-y+3}{5}$$

look different, but we get the second expression by multiplying the first by -1 *in both the numerator and denominator. To practice recognizing equivalent rational expressions, match the expressions in Exercises 1–6 with their equivalents in Choices A–F.*

1. $\dfrac{x-3}{x+4}$ C

2. $\dfrac{x+3}{x-4}$ A

3. $\dfrac{x-3}{x-4}$ D

4. $\dfrac{x+3}{x+4}$ B

5. $\dfrac{3-x}{x+4}$ E

6. $\dfrac{x+3}{4-x}$ F

A. $\dfrac{-x-3}{4-x}$

B. $\dfrac{-x-3}{-x-4}$

C. $\dfrac{3-x}{-x-4}$

D. $\dfrac{-x+3}{-x+4}$

E. $\dfrac{x-3}{-x-4}$

F. $\dfrac{-x-3}{x-4}$

7. In Example 1(a), we showed that the domain of the rational function defined by $f(x) = \dfrac{3}{7x-14}$ does not include 2. Explain in your own words why this is so. In general, how do we find the value or values excluded from the domain of a rational function?

Replacing x with 2 makes the denominator 0 and the value of the expression undefined. To find the values excluded from the domain, set the denominator equal to 0 and solve the equation. All solutions of the equation are excluded from the domain.

8. The domain of the rational function defined by $g(x) = \dfrac{x+1}{x^2+3}$ includes all real numbers. Explain.

There is no number that makes the denominator 0. In fact, the denominator is greater than or equal to 3 for any replacement of x. Thus, the domain includes all real numbers.

Find all numbers that are not in the domain of each function. Then give the domain using set notation. See Example 1.

9. $f(x) = \dfrac{x}{x-7}$

$7; \{x \mid x \neq 7\}$

10. $f(x) = \dfrac{x}{x+3}$

$-3; \{x \mid x \neq -3\}$

11. $f(x) = \dfrac{6x-5}{7x+1}$

$-\frac{1}{7}; \left\{x \mid x \neq -\frac{1}{7}\right\}$

12. $f(x) = \dfrac{8x-3}{2x+7}$

$-\frac{7}{2}; \left\{x \mid x \neq -\frac{7}{2}\right\}$

13. $f(x) = \dfrac{12x+3}{x}$

$0; \{x \mid x \neq 0\}$

14. $f(x) = \dfrac{9x+8}{x}$

$0; \{x \mid x \neq 0\}$

15. $f(x) = \dfrac{3x+1}{2x^2+x-6}$

$-2, \frac{3}{2}; \left\{x \mid x \neq -2, \frac{3}{2}\right\}$

16. $f(x) = \dfrac{2x+4}{3x^2+11x-42}$

$-6, \frac{7}{3}; \left\{x \mid x \neq -6, \frac{7}{3}\right\}$

17. $f(x) = \dfrac{x+2}{14}$

none; $(-\infty, \infty)$

18. $f(x) = \dfrac{x-9}{26}$

none; $(-\infty, \infty)$

19. $f(x) = \dfrac{2x^2-3x+4}{3x^2+8}$

none; $(-\infty, \infty)$

20. $f(x) = \dfrac{9x^2-8x+3}{4x^2+1}$

none; $(-\infty, \infty)$

21. **(a)** Identify the two *terms* in the numerator and the two *terms* in the denominator of the rational expression $\frac{x^2 + 4x}{x + 4}$.

numerator: x^2, $4x$; denominator: x, 4

(b) Describe the steps you would use to write this rational expression in lowest terms. (*Hint:* It simplifies to x.)

First factor the numerator, getting $x(x + 4)$, then divide the numerator and denominator by the common factor of $x + 4$ to get $\frac{x}{1}$, or x.

22. Only one of the following rational expressions can be simplified. Which one is it?

A. $\frac{x^2 + 2}{x^2}$ **B.** $\frac{x^2 + 2}{2}$

C. $\frac{x^2 + y^2}{y^2}$ **D.** $\frac{x^2 - 5x}{x}$

D

23. Only one of the following rational expressions is *not* equivalent to $\frac{x - 3}{4 - x}$. Which one is it?

A. $\frac{3 - x}{x - 4}$ **B.** $\frac{x + 3}{4 + x}$

C. $-\frac{3 - x}{4 - x}$ **D.** $-\frac{x - 3}{x - 4}$

B

24. Which two of the following rational expressions equal -1?

A. $\frac{2x + 3}{2x - 3}$ **B.** $\frac{2x - 3}{3 - 2x}$

C. $\frac{2x + 3}{3 + 2x}$ **D.** $\frac{2x + 3}{-2x - 3}$

B, D

Write each rational expression in lowest terms. See Example 2.

25. $\frac{x^2(x + 1)}{x(x + 1)}$

x

26. $\frac{y^3(y - 4)}{y^2(y - 4)}$

y

27. $\frac{(x + 4)(x - 3)}{(x + 5)(x + 4)}$

$\frac{x - 3}{x + 5}$

28. $\frac{(2x + 7)(x - 1)}{(2x + 3)(2x + 7)}$

$\frac{x - 1}{2x + 3}$

29. $\frac{4x(x + 3)}{8x^2(x - 3)}$

$\frac{x + 3}{2x(x - 3)}$

30. $\frac{5y^2(y + 8)}{15y(y - 8)}$

$\frac{y(y + 8)}{3(y - 8)}$

31. $\frac{3x + 7}{3}$

already in lowest terms

32. $\frac{4x - 9}{4}$

already in lowest terms

33. $\frac{6m + 18}{7m + 21}$

$\frac{6}{7}$

34. $\frac{5r - 20}{3r - 12}$

$\frac{5}{3}$

35. $\frac{3z^2 + z}{18z + 6}$

$\frac{z}{6}$

36. $\frac{2x^2 - 5x}{16x - 40}$

$\frac{x}{8}$

37. $\frac{2t + 6}{t^2 - 9}$

$\frac{2}{t - 3}$

38. $\frac{5s - 25}{s^2 - 25}$

$\frac{5}{s + 5}$

39. $\frac{x^2 + 2x - 15}{x^2 + 6x + 5}$

$\frac{x - 3}{x + 1}$

40. $\dfrac{y^2 - 5y - 14}{y^2 + y - 2}$

$\dfrac{y - 7}{y - 1}$

41. $\dfrac{8x^2 - 10x - 3}{8x^2 - 6x - 9}$

$\dfrac{4x + 1}{4x + 3}$

42. $\dfrac{12x^2 - 4x - 5}{8x^2 - 6x - 5}$

$\dfrac{6x - 5}{4x - 5}$

43. $\dfrac{a^3 + b^3}{a + b}$

$a^2 - ab + b^2$

44. $\dfrac{r^3 - s^3}{r - s}$

$r^2 + rs + s^2$

45. $\dfrac{2c^2 + 2cd - 60d^2}{2c^2 - 12cd + 10d^2}$

$\dfrac{c + 6d}{c - d}$

46. $\dfrac{3s^2 - 9st - 54t^2}{3s^2 - 6st - 72t^2}$

$\dfrac{s + 3t}{s + 4t}$

47. $\dfrac{ac - ad + bc - bd}{ac - ad - bc + bd}$

$\dfrac{a + b}{a - b}$

48. $\dfrac{2xy + 2xw + y + w}{2xy + y - 2xw - w}$

$\dfrac{y + w}{y - w}$

Write each rational expression in lowest terms. See Example 3.

49. $\dfrac{7 - b}{b - 7}$

-1

50. $\dfrac{r - 13}{13 - r}$

-1

51. $\dfrac{x^2 - y^2}{y - x}$

In Exercises 51–56, there are other acceptable ways to express each answer.

$-(x + y)$

52. $\dfrac{m^2 - n^2}{n - m}$

$-(m + n)$

53. $\dfrac{(a - 3)(x + y)}{(3 - a)(x - y)}$

$-\dfrac{x + y}{x - y}$

54. $\dfrac{(8 - p)(x + 2)}{(p - 8)(x - 2)}$

$-\dfrac{x + 2}{x - 2}$

55. $\dfrac{5k - 10}{20 - 10k}$

$-\dfrac{1}{2}$

56. $\dfrac{7x - 21}{63 - 21x}$

$-\dfrac{1}{3}$

57. $\dfrac{a^2 - b^2}{a^2 + b^2}$

already in lowest terms

58. $\dfrac{p^2 + q^2}{p^2 - q^2}$

already in lowest terms

Multiply or divide as indicated. See Examples 4 and 5.

59. $\dfrac{(x + 2)(x + 1)}{(x + 3)(x - 2)} \cdot \dfrac{(x + 3)(x + 4)}{(x + 2)(x + 1)}$

$\dfrac{x + 4}{x - 2}$

60. $\dfrac{(x + 3)(x - 4)}{(x - 4)(x + 2)} \cdot \dfrac{(x + 5)(x - 6)}{(x + 3)(x - 6)}$

$\dfrac{x + 5}{x + 2}$

61. $\dfrac{(2x + 3)(x - 4)}{(x + 8)(x - 4)} \div \dfrac{(x - 4)(x + 2)}{(x - 4)(x + 8)}$

$\dfrac{2x + 3}{x + 2}$

62. $\dfrac{(6x + 5)(x - 3)}{(x + 9)(x - 1)} \div \dfrac{(x - 3)(2x + 7)}{(x - 1)(x + 9)}$

$\dfrac{6x + 5}{2x + 7}$

63. $\frac{7t + 7}{-6} \div \frac{4t + 4}{15}$

$-\frac{35}{8}$

64. $\frac{8z - 16}{-20} \div \frac{3z - 6}{40}$

$-\frac{16}{3}$

65. $\frac{4x}{8x + 4} \cdot \frac{14x + 7}{6}$

$\frac{7x}{6}$

66. $\frac{12x - 20}{5x} \cdot \frac{6}{9x - 15}$

$\frac{8}{5x}$

67. $\frac{p^2 - 25}{4p} \cdot \frac{2}{5 - p}$

$-\frac{p + 5}{2p}$ (There are other ways.)

68. $\frac{a^2 - 1}{4a} \cdot \frac{2}{1 - a}$

$-\frac{a + 1}{2a}$ (There are other ways.)

69. $\frac{m^2 - 49}{m + 1} \div \frac{7 - m}{m}$

$\frac{-m(m + 7)}{m + 1}$ (There are other ways.)

70. $\frac{k^2 - 4}{3k^2} \div \frac{2 - k}{11k}$

$\frac{-11(k + 2)}{3k}$ (There are other ways.)

71. $\frac{12x - 10y}{3x + 2y} \cdot \frac{6x + 4y}{10y - 12x}$

-2

72. $\frac{9s - 12t}{2s + 2t} \cdot \frac{3s + 3t}{4t - 3s}$

$-\frac{9}{2}$

73. $\frac{x^2 - 25}{x^2 + x - 20} \cdot \frac{x^2 + 7x + 12}{x^2 - 2x - 15}$

$\frac{x + 4}{x - 4}$

74. $\frac{t^2 - 49}{t^2 + 4t - 21} \cdot \frac{t^2 + 8t + 15}{t^2 - 2t - 35}$

$\frac{t + 3}{t - 3}$

75. $\frac{6x^2 + 5xy - 6y^2}{12x^2 - 11xy + 2y^2} \div \frac{4x^2 - 12xy + 9y^2}{8x^2 - 14xy + 3y^2}$

$\frac{2x + 3y}{2x - 3y}$

76. $\frac{8a^2 - 6ab - 9b^2}{6a^2 - 5ab - 6b^2} \div \frac{4a^2 + 11ab + 6b^2}{9a^2 + 12ab + 4b^2}$

$\frac{3a + 2b}{a + 2b}$

77. $\frac{3k^2 + 17kp + 10p^2}{6k^2 + 13kp - 5p^2} \div \frac{6k^2 + kp - 2p^2}{6k^2 - 5kp + p^2}$

$\frac{k + 5p}{2k + 5p}$

78. $\frac{16c^2 + 24cd + 9d^2}{16c^2 - 16cd + 3d^2} \div \frac{16c^2 - 9d^2}{16c^2 - 24cd + 9d^2}$

$\frac{4c + 3d}{4c - d}$

79. $\left(\frac{6k^2 - 13k - 5}{k^2 + 7k} \div \frac{2k - 5}{k^3 + 6k^2 - 7k}\right) \cdot \frac{k^2 - 5k + 6}{3k^2 - 8k - 3}$

$(k - 1)(k - 2)$

80. $\left(\frac{2x^3 + 3x^2 - 2x}{3x - 15} \div \frac{2x^3 - x^2}{x^2 - 3x - 10}\right) \cdot \frac{5x^2 - 10x}{3x^2 + 12x + 12}$

$\frac{5(x - 2)}{9}$

Study Skills

MAKING A MIND MAP

Mind mapping is a visual way to show information that you have learned. It is an excellent way to review. Mapping is flexible and can be personalized, which is helpful for your memory. You remember information better if it is pleasing to look at, colorful, and shows connections between ideas. Take advantage of that by creating maps that are

- easy to read,
- use color in a systematic way, and
- clearly show you how different concepts are related (using arrows or dotted lines, for example).

OBJECTIVES

1. **Create mind maps for appropriate concepts.**
2. **Visually show how concepts relate to each other using arrows or lines.**

Directions for Making a Mind Map

Here are some general directions for making a mind map. After you read them, work on completing the map that has been started for you on the next page. It is from **Section 7.1.**

- To begin a mind map, write the concept in the center of a piece of paper and either circle it or draw a box around it.
- Make a line out from the center concept, and draw a box large enough to write the definition of the concept.
- Think of the other aspects (subpoints) of the concept that you have learned, such as procedures to follow or formulas. Make a separate line and box connecting each subpoint to the center.
- From each of the new boxes, add the information you've learned. You can continue making new lines and boxes or circles, or you can list items below the new information.
- Use color to highlight the major points. For example, everything related to one subpoint might be the same color. That way you can easily see related ideas.
- Use arrows, underlining, or small drawings to help you remember.

Why Is Mapping Brain Friendly?

Remember that you learn better when you are actively thinking about and working with information. Making a map requires you to think hard about *how to place the information, how to show connections* between parts of the map, and *how color will be useful*. It also takes a lot of thinking to fill in all related details and *show how those details connect* to the larger concept. It is time well spent.

Try This Mind Map Using Sections 7.1 and 7.2

On a separate sheet of paper, make a map that summarizes *Multiplying and Dividing Rational Expressions* (**Section 7.1**). Follow the directions and use the starter map below.

- Notice that *dividing rational expressions requires two steps first*, then you follow the same steps as you do when multiplying. Notice the placement of those two steps and the arrow leading you to the next steps.
- Notice under "multiplying" that the written steps are on the left, and the example is on the right.
- **Your job is to complete the map by writing an example for dividing rational expressions. Then make another map for adding and subtracting rational expressions when you get to Section 7.2.**

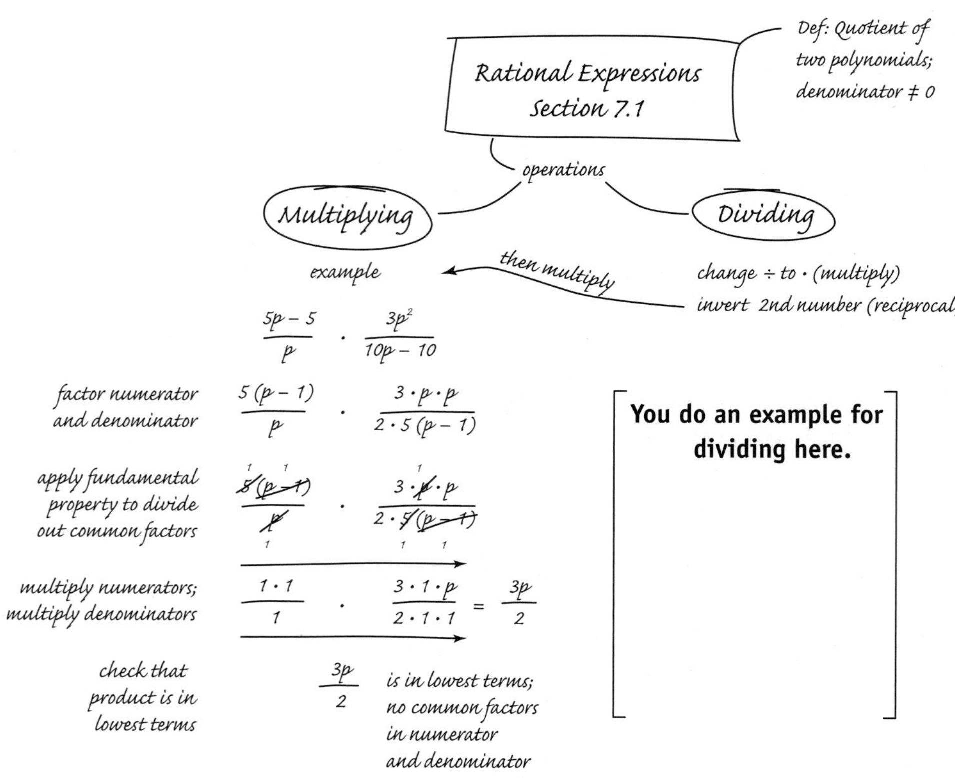

7.2 Adding and Subtracting Rational Expressions

OBJECTIVES

1. Add and subtract rational expressions with the same denominator.
2. Find a least common denominator.
3. Add and subtract rational expressions with different denominators.

OBJECTIVE 1 Add and subtract rational expressions with the same denominator. We do so as we would with rational numbers.

Adding or Subtracting Rational Expressions

Step 1 **If the denominators are the same,** add or subtract the numerators. Place the result over the common denominator.

If the denominators are different, first find the least common denominator. Write all rational expressions with this LCD, and then add or subtract the numerators. Place the result over the common denominator.

Step 2 **Simplify.** Write all answers in lowest terms.

EXAMPLE 1 Adding and Subtracting Rational Expressions with the Same Denominator

Add or subtract as indicated.

(a) $\frac{3y}{5} + \frac{x}{5} = \frac{3y + x}{5}$ ← Add the numerators. ← Keep the common denominator.

(b) $\frac{7}{2r^2} - \frac{11}{2r^2}$

$= \frac{7 - 11}{2r^2}$ Subtract the numerators; keep the common denominator.

$= \frac{-4}{2r^2}$

$= -\frac{2}{r^2}$ Lowest terms

(c) $\frac{m}{m^2 - p^2} + \frac{p}{m^2 - p^2}$

$= \frac{m + p}{m^2 - p^2}$ Add the numerators; keep the common denominator.

$= \frac{m + p}{(m + p)(m - p)}$ Factor.

$= \frac{1}{m - p}$ Lowest terms (Remember to write 1 in the numerator.)

(d) $\frac{4}{x^2 + 2x - 8} + \frac{x}{x^2 + 2x - 8}$

$= \frac{4 + x}{x^2 + 2x - 8}$ Add.

$= \frac{4 + x}{(x - 2)(x + 4)}$ Factor.

$= \frac{1}{x - 2}$ Lowest terms

Work Problem 1 at the Side. ▶

1 Add or subtract.

(a) $\frac{3m}{8} + \frac{5n}{8}$

(b) $\frac{7}{3a} + \frac{10}{3a}$

(c) $\frac{2}{y^2} - \frac{5}{y^2}$

(d) $\frac{a}{a + b} + \frac{b}{a + b}$

(e) $\frac{2y - 1}{y^2 + y - 2} - \frac{y}{y^2 + y - 2}$

ANSWERS

1. (a) $\frac{3m + 5n}{8}$ (b) $\frac{17}{3a}$ (c) $-\frac{3}{y^2}$ (d) 1 (e) $\frac{1}{y + 2}$

OBJECTIVE 2 Find a least common denominator. We add or subtract rational expressions with different denominators by first writing them with a common denominator, usually the **least common denominator (LCD).**

Finding the Least Common Denominator

Step 1 **Factor** each denominator.

Step 2 **Find the least common denominator.** The LCD is the product of all different factors from each denominator, with each factor raised to the *greatest* power that occurs in any denominator.

EXAMPLE 2 Finding Least Common Denominators

Assume that the given expressions are denominators of fractions. Find the LCD for each group.

(a) $5xy^2$, $2x^3y$

Each denominator is already factored.

$$5xy^2 = 5 \cdot x \cdot y^2$$
$$2x^3y = 2 \cdot x^3 \cdot y$$
$$\text{LCD} = 5 \cdot 2 \cdot x^3 \cdot y^2 = 10x^3y^2$$

Greatest exponent on x is 3.
Greatest exponent on y is 2.

(b) $k - 3$, k

Each denominator is already factored. The LCD, an expression divisible by *both* $k - 3$ and k, is

$$k(k - 3).$$

Don't forget the factor k.

It is usually best to leave a least common denominator in factored form.

(c) $y^2 - 2y - 8$, $y^2 + 3y + 2$

Factor the denominators.

$$y^2 - 2y - 8 = (y - 4)(y + 2)$$
$$y^2 + 3y + 2 = (y + 2)(y + 1)$$
Factor.

The LCD, divisible by both polynomials, is $(y - 4)(y + 2)(y + 1)$.

(d) $8z - 24$, $5z^2 - 15z$

$$8z - 24 = 8(z - 3)$$
$$5z^2 - 15z = 5z(z - 3)$$
Factor.

The LCD is $8 \cdot 5z \cdot (z - 3) = 40z(z - 3)$.

(e) $m^2 + 5m + 6$, $m^2 + 4m + 4$, $2m + 6$

$$m^2 + 5m + 6 = (m + 3)(m + 2)$$
$$m^2 + 4m + 4 = (m + 2)^2$$
$$2m + 6 = 2(m + 3)$$
Factor.

The LCD is $2(m + 3)(m + 2)^2$.

◀ Work Problem 2 at the Side.

2 Find the LCD for each group of denominators.

(a) $5k^3s$, $10ks^4$

(b) $3 - x$, $9 - x^2$

(c) z, $z + 6$

(d) $2y^2 - 3y - 2$, $2y^2 + 3y + 1$

(e) $x^2 - 2x + 1$, $x^2 - 4x + 3$, $4x - 4$

ANSWERS

2. (a) $10k^3s^4$ (b) $(3 + x)(3 - x)$ (c) $z(z + 6)$ (d) $(y - 2)(2y + 1)(y + 1)$ (e) $4(x - 3)(x - 1)^2$

OBJECTIVE 3 Add and subtract rational expressions with different denominators. Before adding or subtracting two rational expressions, we write each expression with the least common denominator by multiplying its numerator and denominator by the factors needed to get the LCD. This procedure is valid because we are multiplying each rational expression by a form of 1, the identity element for multiplication.

Consider the sum $\frac{7}{15} + \frac{5}{12}$. The LCD for 15 and 12 is 60. Multiply $\frac{7}{15}$ by $\frac{4}{4}$ (a form of 1) and multiply $\frac{5}{12}$ by $\frac{5}{5}$ so that each fraction has denominator 60. Then add the numerators.

$$\frac{7}{15} + \frac{5}{12}$$

$$= \frac{7 \cdot 4}{15 \cdot 4} + \frac{5 \cdot 5}{12 \cdot 5} \quad \text{Fundamental property}$$

$$= \frac{28}{\mathbf{60}} + \frac{25}{\mathbf{60}}$$

$$= \frac{28 + 25}{\mathbf{60}} \quad \text{Add the numerators; keep the common denominator.}$$

$$= \frac{53}{60}$$

EXAMPLE 3 Adding and Subtracting Rational Expressions with Different Denominators

Add or subtract as indicated.

(a) $\frac{5}{2p} + \frac{3}{8p}$ The LCD for $2p$ and $8p$ is $8p$.

$$= \frac{5 \cdot 4}{2p \cdot 4} + \frac{3}{8p} \quad \text{Fundamental property}$$

$$= \frac{20}{\mathbf{8p}} + \frac{3}{\mathbf{8p}}$$

$$= \frac{20 + 3}{\mathbf{8p}} \quad \text{Add the numerators; keep the common denominator.}$$

$$= \frac{23}{8p}$$

(b) $\frac{6}{r} - \frac{5}{r - 3}$ The LCD is $r(r - 3)$.

$$= \frac{6(r - 3)}{r(r - 3)} - \frac{r \cdot 5}{r(r - 3)} \quad \text{Fundamental property}$$

$$= \frac{6r - 18}{r(r - 3)} - \frac{5r}{r(r - 3)} \quad \text{Distributive and commutative properties}$$

$$= \frac{6r - 18 - 5r}{r(r - 3)} \quad \text{Subtract the numerators.}$$

$$= \frac{r - 18}{r(r - 3)} \quad \text{Combine terms in the numerator.}$$

Work Problem 3 at the Side. ▶

3 Add or subtract.

(a) $\frac{6}{7} + \frac{1}{5}$

(b) $\frac{8}{3k} - \frac{2}{9k}$

(c) $\frac{2}{y} - \frac{1}{y + 4}$

ANSWERS

3. (a) $\frac{37}{35}$ (b) $\frac{22}{9k}$ (c) $\frac{y + 8}{y(y + 4)}$

4 Subtract.

(a) $\frac{5x+7}{2x+7} - \frac{-x-14}{2x+7}$

(b) $\frac{2}{r-2} - \frac{r}{r-1}$

CAUTION

Sign errors can easily occur when a rational expression with two or more terms in the numerator is being subtracted. In this case, ***the subtraction sign must be distributed to every term in the numerator of the fraction that follows it.*** Study Example 4 carefully to see how this is done.

EXAMPLE 4 Using the Distributive Property When Subtracting Rational Expressions

Subtract.

(a) $\frac{7x}{3x+1} - \frac{x-2}{3x+1}$

The denominators are the same for both rational expressions. The subtraction sign must be applied to *both* terms in the numerator of the second rational expression. Notice the careful use of the distributive property here.

$$\frac{7x}{3x+1} - \frac{x-2}{3x+1}$$

Use parentheses to avoid errors.

$$= \frac{7x-(x-2)}{3x+1}$$ Subtract the numerators; keep the common denominator.

Be careful with signs.

$$= \frac{7x-x+2}{3x+1}$$ Distributive property

$$= \frac{6x+2}{3x+1}$$ Combine terms in the numerator.

$$= \frac{2(3x+1)}{3x+1}$$ Factor the numerator.

$$= 2$$ Lowest terms

(b) $\frac{1}{q-1} - \frac{1}{q+1}$ The LCD is $(q-1)(q+1)$.

$$= \frac{1(q+1)}{(q-1)(q+1)} - \frac{1(q-1)}{(q+1)(q-1)}$$ Fundamental property

$$= \frac{(q+1)-(q-1)}{(q-1)(q+1)}$$ Subtract.

Be careful with signs.

$$= \frac{q+1-q+1}{(q-1)(q+1)}$$ Distributive property

$$= \frac{2}{(q-1)(q+1)}$$ Combine terms in the numerator.

◀ *Work Problem* **4** *at the Side.*

In some problems, rational expressions to be added or subtracted have denominators that are opposites of each other, such as

$$\frac{y}{y-2} + \frac{8}{2-y}.$$ Denominators are opposites.

The next example illustrates how to proceed in such a problem.

ANSWERS

4. (a) 3 **(b)** $\frac{-r^2+4r-2}{(r-2)(r-1)}$

EXAMPLE 5 Adding Rational Expressions with Denominators That Are Opposites

Add.

$$\frac{y}{y-2}+\frac{8}{2-y}$$

$$=\frac{y}{y-2}+\frac{8(-1)}{(2-y)(-1)}$$ Multiply the second expression by $\frac{-1}{-1}$.

$$=\frac{y}{y-2}+\frac{-8}{y-2}$$ The LCD is $y-2$.

$$=\frac{y-8}{y-2}$$ Add the numerators.

We could use $2-y$ as the common denominator and rewrite the first expression.

$$\frac{y}{y-2}+\frac{8}{2-y}$$

$$=\frac{y(-1)}{(y-2)(-1)}+\frac{8}{2-y}$$ Multiply the first expression by $\frac{-1}{-1}$.

$$=\frac{-y+8}{2-y}, \text{ or equivalently } \frac{8-y}{2-y}$$ The LCD is $2-y$.

Work Problem **5** *at the Side.* ▶

EXAMPLE 6 Adding and Subtracting Three Rational Expressions

Add and subtract as indicated.

$$\frac{3}{x-2}+\frac{5}{x}-\frac{6}{x^2-2x}$$

$$=\frac{3}{x-2}+\frac{5}{x}-\frac{6}{x(x-2)}$$ Factor the third denominator.

$$=\frac{3x}{x(x-2)}+\frac{5(x-2)}{x(x-2)}-\frac{6}{x(x-2)}$$ The LCD is $x(x-2)$; fundamental property

$$=\frac{3x+5(x-2)-6}{x(x-2)}$$ Add and subtract the numerators.

$$=\frac{3x+5x-10-6}{x(x-2)}$$ Distributive property

$$=\frac{8x-16}{x(x-2)}$$ Combine terms in the numerator.

$$=\frac{8(x-2)}{x(x-2)}$$ Factor the numerator.

$$=\frac{8}{x}$$ Lowest terms

Work Problem **6** *at the Side.* ▶

5 Add or subtract as indicated.

(a) $\frac{8}{x-4}+\frac{2}{4-x}$

(b) $\frac{9}{2x-9}-\frac{4}{9-2x}$

6 Add and subtract as indicated.

$$\frac{4}{x-5}+\frac{-2}{x}-\frac{10}{x^2-5x}$$

Answers

5. (a) $\frac{6}{x-4}$, or $\frac{-6}{4-x}$ **(b)** $\frac{13}{2x-9}$, or $\frac{-13}{9-2x}$

6. $\frac{2}{x-5}$

7 Subtract.

$$\frac{-a}{a^2 + 3a - 4} - \frac{4a}{a^2 + 7a + 12}$$

8 Add.

$$\frac{4}{p^2 - 6p + 9} + \frac{1}{p^2 + 2p - 15}$$

EXAMPLE 7 Subtracting Rational Expressions

Subtract.

$$\frac{m + 4}{m^2 - 2m - 3} - \frac{2m - 3}{m^2 - 5m + 6}$$

$$= \frac{m + 4}{(m - 3)(m + 1)} - \frac{2m - 3}{(m - 3)(m - 2)}$$ Factor each denominator.

$$= \frac{(m + 4)(m - 2)}{(m - 3)(m + 1)(m - 2)} - \frac{(2m - 3)(m + 1)}{(m - 3)(m - 2)(m + 1)}$$ Fundamental property

The LCD is $(m - 3)(m + 1)(m - 2)$.

$$= \frac{(m + 4)(m - 2) - (2m - 3)(m + 1)}{(m - 3)(m + 1)(m - 2)}$$ Subtract the numerators.

Note the careful use of parentheses.

$$= \frac{m^2 + 2m - 8 - (2m^2 - m - 3)}{(m - 3)(m + 1)(m - 2)}$$ Multiply in the numerator.

Be careful with signs.

$$= \frac{m^2 + 2m - 8 - 2m^2 + m + 3}{(m - 3)(m + 1)(m - 2)}$$ Distributive property

$$= \frac{-m^2 + 3m - 5}{(m - 3)(m + 1)(m - 2)}$$ Combine terms in the numerator.

If we try to factor the numerator, we find that this rational expression is in lowest terms.

◀ Work Problem **7** at the Side.

EXAMPLE 8 Adding Rational Expressions

Add.

$$\frac{5}{x^2 + 10x + 25} + \frac{2}{x^2 + 7x + 10}$$

$$= \frac{5}{(x + 5)^2} + \frac{2}{(x + 5)(x + 2)}$$ Factor each denominator.

$$= \frac{5(x + 2)}{(x + 5)^2(x + 2)} + \frac{2(x + 5)}{(x + 5)^2(x + 2)}$$ The LCD is $(x + 5)^2(x + 2)$; fundamental property

$$= \frac{5(x + 2) + 2(x + 5)}{(x + 5)^2(x + 2)}$$ Add the numerators.

$$= \frac{5x + 10 + 2x + 10}{(x + 5)^2(x + 2)}$$ Distributive property

$$= \frac{7x + 20}{(x + 5)^2(x + 2)}$$ Combine terms in the numerator.

◀ Work Problem **8** at the Side.

ANSWERS

7. $\frac{-5a^2 + a}{(a + 4)(a - 1)(a + 3)}$

8. $\frac{5p + 17}{(p - 3)^2(p + 5)}$

7.2 Exercises

FOR EXTRA HELP MyMathLab

Math XL PRACTICE  WATCH  DOWNLOAD  READ  REVIEW

1. Write an explanation for adding or subtracting rational expressions that have a common denominator.

To add or subtract rational expressions that have a common denominator, first add or subtract the numerators. Then place the result over the common denominator. Write the answer in lowest terms.

2. Write an explanation for adding or subtracting rational expressions that have different denominators.

To add or subtract rational expressions that have different denominators, first write each expression as an equivalent expression with the least common denominator. Then follow the procedure described in the answer to Exercise 1.

Add or subtract as indicated. Write all answers in lowest terms. See Example 1.

3. $\frac{7}{t}+\frac{2}{t}$ **$\frac{9}{t}$**

4. $\frac{5}{r}+\frac{9}{r}$ **$\frac{14}{r}$**

5. $\frac{11}{5x}-\frac{1}{5x}$ **$\frac{2}{x}$**

6. $\frac{7}{4y}-\frac{3}{4y}$ **$\frac{1}{y}$**

7. $\frac{5x+4}{6x+5}+\frac{x+1}{6x+5}$ **1**

8. $\frac{6y+12}{4y+3}+\frac{2y-6}{4y+3}$ **2**

9. $\frac{x^2}{x+5}-\frac{25}{x+5}$ **$x-5$**

10. $\frac{y^2}{y+6}-\frac{36}{y+6}$ **$y-6$**

11. $\frac{4}{p^2+7p+12}+\frac{p}{p^2+7p+12}$ **$\frac{1}{p+3}$**

12. $\frac{5}{x^2+x-20}+\frac{x}{x^2+x-20}$ **$\frac{1}{x-4}$**

13. $\frac{a^3}{a^2+ab+b^2}-\frac{b^3}{a^2+ab+b^2}$ **$a-b$**

14. $\frac{p^3}{p^2-pq+q^2}+\frac{q^3}{p^2-pq+q^2}$ **$p+q$**

Assume that the expressions given are denominators of fractions. Find the least common denominator (LCD) for each group. See Example 2.

15. $18x^2y^3$, $24x^4y^5$ **$72x^4y^5$**

16. $24a^3b^4$, $18a^5b^2$ **$72a^5b^4$**

17. $z-2$, z **$z(z-2)$**

18. $k+3$, k **$k(k+3)$**

19. $2y+8$, $y+4$ **$2(y+4)$**

20. $3r-21$, $r-7$ **$3(r-7)$**

21. x^2-81, $x^2+18x+81$ **$(x+9)^2(x-9)$**

22. y^2-16, $y^2-8y+16$ **$(y-4)^2(y+4)$**

23. $m+n$, $m-n$, m^2-n^2 **$(m+n)(m-n)$**

24. $r+s$, $r-s$, r^2-s^2 **$(r+s)(r-s)$**

25. x^2-3x-4, $x+x^2$ **$x(x-4)(x+1)$**

26. $y^2-8y+12$, y^2-6y **$y(y-2)(y-6)$**

27. $2t^2+7t-15$, $t^2+3t-10$ **$(t+5)(t-2)(2t-3)$**

28. s^2-3s-4, $3s^2+s-2$ **$(s+1)(s-4)(3s-2)$**

29. $2y+6$, y^2-9, y **$2y(y+3)(y-3)$**

30. $9x+18$, x^2-4, x **$9x(x+2)(x-2)$**

31. One student added two rational expressions and obtained the answer $\frac{3}{5-y}$. Another student obtained the answer $\frac{-3}{y-5}$ for the same problem. Is it possible that both answers are correct? Explain.

Yes, they could both be correct because the expressions are equivalent. Multiplying $\frac{3}{5-y}$ by 1 in the form $\frac{-1}{-1}$ gives $\frac{-3}{y-5}$.

32. Consider the following incorrect work. ***WHAT WENT WRONG?***

$$\frac{x}{x+2} - \frac{4x-1}{x+2} = \frac{x-4x-1}{x+2} = \frac{-3x-1}{x+2}$$

The expression $\frac{x-4x-1}{x+2}$ is incorrect. The third term in the numerator should be $+1$, since the $-$ sign should be distributed to both $4x$ and -1. The answer should be $\frac{-3x+1}{x+2}$.

Add or subtract as indicated. Write all answers in lowest terms. See Examples 3–8.

33. $\frac{8}{t} + \frac{7}{3t}$ **$\frac{31}{3t}$**

34. $\frac{5}{x} + \frac{9}{4x}$ **$\frac{29}{4x}$**

35. $\frac{5}{12x^2y} - \frac{11}{6xy}$ **$\frac{5-22x}{12x^2y}$**

36. $\frac{7}{18a^3b^2} - \frac{2}{9ab}$ **$\frac{7-4a^2b}{18a^3b^2}$**

37. $\frac{1}{x-1} - \frac{1}{x}$ **$\frac{1}{x(x-1)}$**

38. $\frac{3}{x-3} - \frac{1}{x}$ **$\frac{2x+3}{x(x-3)}$**

39. $\frac{3a}{a+1} + \frac{2a}{a-3}$ **$\frac{5a^2-7a}{(a+1)(a-3)}$**

40. $\frac{2x}{x+4} + \frac{3x}{x-7}$ **$\frac{5x^2-2x}{(x+4)(x-7)}$**

41. $\frac{17y+3}{9y+7} - \frac{-10y-18}{9y+7}$ **3**

42. $\frac{7x+8}{3x+2} - \frac{x+4}{3x+2}$ **2**

43. $\frac{2}{4-x} + \frac{5}{x-4}$ **$\frac{3}{x-4}$, or $\frac{-3}{4-x}$**

44. $\frac{3}{2-t} + \frac{1}{t-2}$ **$\frac{2}{2-t}$, or $\frac{-2}{t-2}$**

45. $\frac{w}{w-z} - \frac{z}{z-w}$ **$\frac{w+z}{w-z}$, or $\frac{-w-z}{z-w}$**

46. $\frac{a}{a-b} - \frac{b}{b-a}$ **$\frac{a+b}{a-b}$, or $\frac{-a-b}{b-a}$**

47. $\frac{5}{12+4x} - \frac{7}{9+3x}$ **$\frac{-13}{12(3+x)}$**

48. $\frac{3}{10x+15} - \frac{8}{12x+18}$ **$\frac{-11}{15(2x+3)}$**

49. $\frac{4x}{x-1} - \frac{2}{x+1} - \frac{4}{x^2-1}$ **$\frac{2(2x-1)}{x-1}$**

50. $\frac{4}{x+3} - \frac{x}{x-3} - \frac{18}{x^2-9}$ **$\frac{-x^2+x-30}{(x+3)(x-3)}$**

51. $\dfrac{15}{y^2 + 3y} + \dfrac{2}{y} + \dfrac{5}{y + 3}$

$\dfrac{7}{y}$

52. $\dfrac{7}{t - 2} - \dfrac{6}{t^2 - 2t} - \dfrac{3}{t}$

$\dfrac{4}{t - 2}$

53. $\dfrac{5}{x - 2} + \dfrac{1}{x} + \dfrac{2}{x^2 - 2x}$

$\dfrac{6}{x - 2}$

54. $\dfrac{5x}{x - 3} + \dfrac{2}{x} + \dfrac{6}{x^2 - 3x}$

$\dfrac{5x + 2}{x - 3}$

55. $\dfrac{3x}{x + 1} + \dfrac{4}{x - 1} - \dfrac{6}{x^2 - 1}$

$\dfrac{3x - 2}{x - 1}$

56. $\dfrac{5x}{x + 3} + \dfrac{x + 2}{x} - \dfrac{6}{x^2 + 3x}$

$\dfrac{6x + 5}{x + 3}$

57. $\dfrac{4}{x + 1} + \dfrac{1}{x^2 - x + 1} - \dfrac{12}{x^3 + 1}$

$\dfrac{4x - 7}{x^2 - x + 1}$

58. $\dfrac{5}{x + 2} + \dfrac{2}{x^2 - 2x + 4} - \dfrac{60}{x^3 + 8}$

$\dfrac{5x - 18}{x^2 - 2x + 4}$

59. $\dfrac{2x + 4}{x + 3} + \dfrac{3}{x} - \dfrac{6}{x^2 + 3x}$

$\dfrac{2x + 1}{x}$

60. $\dfrac{4x + 1}{x + 5} - \dfrac{2}{x} + \dfrac{10}{x^2 + 5x}$

$\dfrac{4x - 1}{x + 5}$

61. $\dfrac{3}{x^2 - 5x + 6} - \dfrac{2}{x^2 - 4x + 4}$

$\dfrac{x}{(x - 2)^2(x - 3)}$

62. $\dfrac{2}{m^2 - 4m + 4} + \dfrac{3}{m^2 + m - 6}$

$\dfrac{5m}{(m - 2)^2(m + 3)}$

63. $\dfrac{3}{x^2 + 4x + 4} + \dfrac{7}{x^2 + 5x + 6}$

$\dfrac{10x + 23}{(x + 2)^2(x + 3)}$

64. $\dfrac{5}{x^2 + 6x + 9} - \dfrac{2}{x^2 + 4x + 3}$

$\dfrac{3x - 1}{(x + 3)^2(x + 1)}$

65. $\dfrac{5x}{x^2 + xy - 2y^2} - \dfrac{3x}{x^2 + 5xy - 6y^2}$

$\dfrac{2x(x + 12y)}{(x + 2y)(x - y)(x + 6y)}$

66. $\dfrac{6x}{6x^2 + 5xy - 4y^2} - \dfrac{2y}{9x^2 - 16y^2}$

$\dfrac{2(9x^2 - 14xy + y^2)}{(3x + 4y)(2x - y)(3x - 4y)}$

*A **concours d'elegance** is a competition in which a maximum of 100 points is awarded to a car based on its general attractiveness. The function defined by the rational expression*

$$c(x) = \frac{1010}{49(101 - x)} - \frac{10}{49}$$

approximates the cost, in thousands of dollars, of restoring a car so that it will win x points.

Use this information to work Exercises 67 and 68.

67. Simplify the expression for $c(x)$ by performing the indicated subtraction.

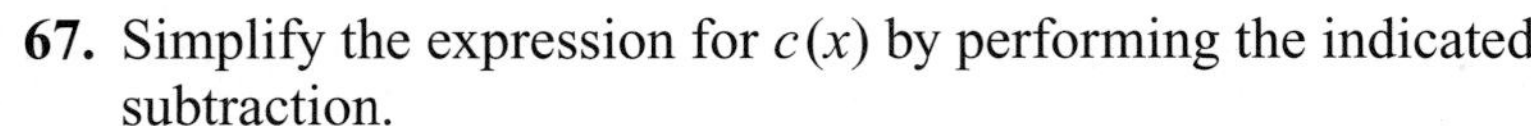

$c(x) = \dfrac{10x}{49(101 - x)}$

68. Use the simplified expression to determine how much it would cost to win 95 points.

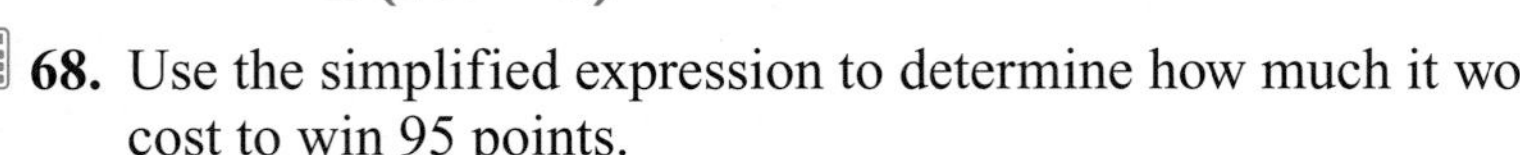

approximately 3.23 thousand dollars

Relating Concepts (Exercises 69–74) For Individual or Group Work

In Example 6 we showed that

$$\frac{3}{x-2} + \frac{5}{x} - \frac{6}{x^2 - 2x} \quad \textit{simplifies to} \quad \frac{8}{x}.$$

*Algebra is, in a sense, a generalized form of arithmetic. **Work Exercises 69–74 in order,** to see how the algebra in this example is related to the arithmetic of common fractions.*

69. Perform the following operations, and express your answer in lowest terms.

$$\frac{3}{7} + \frac{5}{9} - \frac{6}{63}$$

$\frac{8}{9}$

70. Substitute 9 for x in the given problem from Example 6. Compare this problem to the one given in Exercise 69. What do you notice?

$\frac{3}{7} + \frac{5}{9} - \frac{6}{63}$; **They are the same.**

71. Now substitute 9 for x in the answer given in Example 6. Do your results agree with the result you obtained in Exercise 69?

$\frac{8}{9}$; **yes**

72. Replace x in the problem from Example 6 with the number of letters in your last name, assuming that this number is not 2. If your last name has two letters, let $x = 3$. Now predict the answer to your problem. Verify that your prediction is correct.

Answers will vary. Suppose the name is Sosa, so that $x = 4$. The problem is $\frac{3}{2} + \frac{5}{4} - \frac{6}{8}$. The predicted answer is $\frac{8}{4} = 2$, which is correct.

73. Why will $x = 2$ not work for the problem from Example 6?

It causes $\dfrac{3}{x-2}$ and $\dfrac{6}{x^2 - 2x}$ to be undefined, since 0 appears in the denominators.

74. What other value of x is not allowed in the problem given from Example 6?

0

7.3 Complex Fractions

OBJECTIVES

1. Simplify complex fractions by simplifying the numerator and denominator (Method 1).
2. Simplify complex fractions by multiplying by a common denominator (Method 2).
3. Compare the two methods of simplifying complex fractions.
4. Simplify rational expressions with negative exponents.

A **complex fraction** is an expression having a fraction in the numerator, denominator, or both. Examples of complex fractions include

$$\frac{1+\frac{1}{x}}{2}, \quad \frac{\frac{4}{y}}{6-\frac{3}{y}}, \quad \text{and} \quad \frac{\frac{m^2-9}{m+1}}{\frac{m+3}{m^2-1}}. \quad \text{Complex fractions}$$

OBJECTIVE 1 Simplify complex fractions by simplifying the numerator and denominator (Method 1). There are two different methods for simplifying complex fractions.

Simplifying a Complex Fraction: Method 1

Step 1 Simplify the numerator and denominator separately.

Step 2 Divide by multiplying the numerator by the reciprocal of the denominator.

Step 3 Simplify the resulting fraction, if possible.

In Step 2, we are treating the complex fraction as a quotient of two rational expressions and dividing. Before performing this step, be sure that both the numerator and denominator are single fractions.

EXAMPLE 1 Simplifying Complex Fractions by Method 1

Use Method 1 to simplify each complex fraction.

(a) $\dfrac{\frac{x+1}{x}}{\frac{x-1}{2x}}$ Both the numerator and the denominator are already simplified. (Step 1)

$= \dfrac{x+1}{x} \div \dfrac{x-1}{2x}$ Write as a division problem.

$= \dfrac{x+1}{x} \cdot \dfrac{2x}{x-1}$ Multiply by the reciprocal of $\frac{x-1}{2x}$. (Step 2)

$= \dfrac{2x(x+1)}{x(x-1)}$ Multiply.

$= \dfrac{2(x+1)}{x-1}$ Simplify. (Step 3)

Continued on Next Page

1 Use Method 1 to simplify each complex fraction.

(a) $\dfrac{\dfrac{a+2}{5a}}{\dfrac{a-3}{7a}}$

(b) $\dfrac{2+\dfrac{1}{k}}{2-\dfrac{1}{k}}$

(c) $\dfrac{\dfrac{r^2-4}{4}}{1+\dfrac{2}{r}}$

ANSWERS

1. (a) $\dfrac{7(a+2)}{5(a-3)}$ (b) $\dfrac{2k+1}{2k-1}$ (c) $\dfrac{r(r-2)}{4}$

(b) $\dfrac{2+\dfrac{1}{y}}{3-\dfrac{2}{y}}$

$= \dfrac{\dfrac{2y}{y}+\dfrac{1}{y}}{\dfrac{3y}{y}-\dfrac{2}{y}}$ Simplify the numerator and denominator. (Step 1)

$= \dfrac{\dfrac{2y+1}{y}}{\dfrac{3y-2}{y}}$ $\dfrac{\frac{2y+1}{y}}{\frac{3y-2}{y}}$ means $\dfrac{2y+1}{y} \div \dfrac{3y-2}{y}$.

$= \dfrac{2y+1}{y} \cdot \dfrac{y}{3y-2}$ Multiply by the reciprocal of $\frac{3y-2}{y}$. (Step 2)

$= \dfrac{2y+1}{3y-2}$ Multiply and simplify. (Step 3)

◀ *Work Problem* **1** *at the Side.*

OBJECTIVE 2 Simplify complex fractions by multiplying by a common denominator (Method 2). The second method for simplifying complex fractions uses the identity property of multiplication.

Simplifying a Complex Fraction: Method 2

Step 1 Multiply the numerator and denominator of the complex fraction by the least common denominator of the fractions in the numerator and the fractions in the denominator of the complex fraction.

Step 2 Simplify the resulting fraction, if possible.

EXAMPLE 2 Simplifying Complex Fractions by Method 2

Use Method 2 to simplify each complex fraction.

(a) $\dfrac{2+\dfrac{1}{y}}{3-\dfrac{2}{y}}$ This is the same fraction as in Example 1(b) above. Compare the solution methods.

$= \dfrac{\left(2+\dfrac{1}{y}\right)\cdot y}{\left(3-\dfrac{2}{y}\right)\cdot y}$ Multiply the numerator and denominator by the LCD, y. (Step 1)

$= \dfrac{2\cdot y+\dfrac{1}{y}\cdot y}{3\cdot y-\dfrac{2}{y}\cdot y}$ Distributive property

$= \dfrac{2y+1}{3y-2}$ Simplify. (Step 2)

Continued on Next Page

(b) $\dfrac{2p + \dfrac{5}{p-1}}{3p - \dfrac{2}{p}}$

$= \dfrac{\left(2p + \dfrac{5}{p-1}\right) \cdot p(p-1)}{\left(3p - \dfrac{2}{p}\right) \cdot p(p-1)}$ Multiply the numerator and denominator by the LCD, $p(p-1)$.

$= \dfrac{2p[p(p-1)] + \dfrac{5}{p-1} \cdot p(p-1)}{3p[p(p-1)] - \dfrac{2}{p} \cdot p(p-1)}$ Distributive property

$= \dfrac{2p[p(p-1)] + 5p}{3p[p(p-1)] - 2(p-1)}$ Multiply.

$= \dfrac{2p^3 - 2p^2 + 5p}{3p^3 - 3p^2 - 2p + 2}$ Multiply; lowest terms

Work Problem 2 at the Side. ▶

OBJECTIVE 3 Compare the two methods of simplifying complex fractions. Choosing whether to use Method 1 or Method 2 to simplify a complex fraction is usually a matter of preference. Some students prefer one method over the other, while other students feel comfortable with both methods and rely on practice with many examples to determine which method they will use on a particular problem. In the next example, we illustrate how to simplify a complex fraction using both methods so that you can observe the processes and decide for yourself the pros and cons of each method.

EXAMPLE 3 Simplifying Complex Fractions Using Both Methods

Use both Method 1 and Method 2 to simplify each complex fraction.

Method 1

(a) $\dfrac{\dfrac{2}{x-3}}{\dfrac{5}{x^2-9}}$

$= \dfrac{\dfrac{2}{x-3}}{\dfrac{5}{(x-3)(x+3)}}$

$= \dfrac{2}{x-3} \div \dfrac{5}{(x-3)(x+3)}$

$= \dfrac{2}{x-3} \cdot \dfrac{(x-3)(x+3)}{5}$

$= \dfrac{2(x+3)}{5}$

Method 2

(a) $\dfrac{\dfrac{2}{x-3}}{\dfrac{5}{x^2-9}}$

$= \dfrac{\dfrac{2}{x-3}}{\dfrac{5}{(x-3)(x+3)}}$

$= \dfrac{\dfrac{2}{x-3} \cdot (x-3)(x+3)}{\dfrac{5}{(x-3)(x+3)} \cdot (x-3)(x+3)}$

$= \dfrac{2(x+3)}{5}$

Continued on Next Page

2 Use Method 2 to simplify each complex fraction.

(a) $\dfrac{\dfrac{5}{y} + 6}{\dfrac{8}{3y} - 1}$

(b) $\dfrac{\dfrac{1}{y} + \dfrac{1}{y-1}}{\dfrac{1}{y} - \dfrac{2}{y-1}}$

ANSWERS

2. (a) $\dfrac{15 + 18y}{8 - 3y}$ **(b)** $\dfrac{2y-1}{-y-1}$, or $\dfrac{1-2y}{y+1}$

3 Use both methods to simplify each complex fraction.

(a) $\dfrac{\dfrac{5}{y+2}}{\dfrac{-3}{y^2-4}}$

(b) $\dfrac{\dfrac{1}{a}-\dfrac{1}{b}}{\dfrac{1}{a^2}-\dfrac{1}{b^2}}$

4 Simplify each expression, using only positive exponents in the answer.

(a) $\dfrac{r^{-2}-s^{-1}}{4r^{-1}+s^{-2}}$

(b) $\dfrac{b^{-4}}{b^{-5}+2}$

Method 1

(b) $\dfrac{\dfrac{1}{x}+\dfrac{1}{y}}{\dfrac{1}{x^2}-\dfrac{1}{y^2}}$

$$= \frac{\dfrac{y}{xy}+\dfrac{x}{xy}}{\dfrac{y^2}{x^2y^2}-\dfrac{x^2}{x^2y^2}}$$

$$= \frac{\dfrac{y+x}{xy}}{\dfrac{y^2-x^2}{x^2y^2}}$$

$$= \frac{y+x}{xy} \div \frac{y^2-x^2}{x^2y^2}$$

$$= \frac{y+x}{xy} \cdot \frac{x^2y^2}{(y-x)(y+x)}$$

$$= \frac{xy}{y-x}$$

Method 2

(b) $\dfrac{\dfrac{1}{x}+\dfrac{1}{y}}{\dfrac{1}{x^2}-\dfrac{1}{y^2}}$

$$= \frac{\left(\dfrac{1}{x}+\dfrac{1}{y}\right)\cdot x^2y^2}{\left(\dfrac{1}{x^2}-\dfrac{1}{y^2}\right)\cdot x^2y^2}$$

$$= \frac{\left(\dfrac{1}{x}\right)x^2y^2+\left(\dfrac{1}{y}\right)x^2y^2}{\left(\dfrac{1}{x^2}\right)x^2y^2-\left(\dfrac{1}{y^2}\right)x^2y^2}$$

$$= \frac{xy^2+x^2y}{y^2-x^2}$$

$$= \frac{xy(y+x)}{(y+x)(y-x)}$$

$$= \frac{xy}{y-x}$$

◀ Work Problem **3** at the Side.

OBJECTIVE 4 Simplify rational expressions with negative exponents. To simplify such expressions, we begin by rewriting the expressions with only positive exponents.

EXAMPLE 4 Simplifying a Rational Expression with Negative Exponents

Simplify, using only positive exponents in the answer.

$$\frac{m^{-1}+p^{-2}}{2m^{-2}-p^{-1}}$$

$$= \frac{\dfrac{1}{m}+\dfrac{1}{p^2}}{\dfrac{2}{m^2}-\dfrac{1}{p}}$$ Write with positive exponents.

The base of $2m^{-2}$ is m, not $2m$: $2m^{-2}=\frac{2}{m^2}$.

$$= \frac{m^2p^2\left(\dfrac{1}{m}+\dfrac{1}{p^2}\right)}{m^2p^2\left(\dfrac{2}{m^2}-\dfrac{1}{p}\right)}$$ Simplify by Method 2, multiplying the numerator and denominator by the LCD, m^2p^2.

$$= \frac{m^2p^2\cdot\dfrac{1}{m}+m^2p^2\cdot\dfrac{1}{p^2}}{m^2p^2\cdot\dfrac{2}{m^2}-m^2p^2\cdot\dfrac{1}{p}}$$ Distributive property

$$= \frac{mp^2+m^2}{2p^2-m^2p}$$ Lowest terms

◀ Work Problem **4** at the Side.

ANSWERS

3. (Both methods give the same answers.)
(a) $\dfrac{5(y-2)}{-3}$ (b) $\dfrac{ab}{b+a}$

4. (a) $\dfrac{s^2-r^2s}{4rs^2+r^2}$ (b) $\dfrac{b}{1+2b^5}$

7.3 Exercises

FOR EXTRA HELP

MyMathLab

Math XL PRACTICE

WATCH

DOWNLOAD

READ

REVIEW

1. Explain in your own words Method 1 for simplifying complex fractions.

   Begin by simplifying the numerator. Then simplify the denominator. Write as a division problem, and proceed.

2. Method 2 for simplifying complex fractions says that we can multiply both the numerator and the denominator of the complex fraction by the same nonzero expression. What property of real numbers from **Section 1.7** justifies this method?

   identity property of multiplication

Use either method to simplify each complex fraction. See Examples 1–3.

3. $\dfrac{\frac{12}{x-1}}{\frac{6}{x}}$ **$\dfrac{2x}{x-1}$**

4. $\dfrac{\frac{24}{t+4}}{\frac{6}{t}}$ **$\dfrac{4t}{t+4}$**

5. $\dfrac{\frac{k+1}{2k}}{\frac{3k-1}{4k}}$ **$\dfrac{2(k+1)}{3k-1}$**

6. $\dfrac{\frac{1-r}{4r}}{\frac{-1-r}{8r}}$ **$\dfrac{2(1-r)}{-1-r}$**

7. $\dfrac{\frac{4z^2x^4}{9}}{\frac{12x^2z^5}{15}}$ **$\dfrac{5x^2}{9z^3}$**

8. $\dfrac{\frac{3y^2x^3}{8}}{\frac{9y^3x^4}{16}}$ **$\dfrac{2}{3yx}$**

9. $\dfrac{\frac{1}{x}+1}{-\frac{1}{x}+1}$ **$\dfrac{1+x}{-1+x}$**

10. $\dfrac{\frac{2}{k}-1}{\frac{2}{k}+1}$ **$\dfrac{2-k}{2+k}$**

11. $\dfrac{\frac{3}{x}+\frac{3}{y}}{\frac{3}{x}-\frac{3}{y}}$ **$\dfrac{y+x}{y-x}$**

12. $\dfrac{\frac{4}{t}-\frac{4}{s}}{\frac{4}{t}+\frac{4}{s}}$ **$\dfrac{s-t}{s+t}$**

13. $\dfrac{\frac{8x-24y}{10}}{\frac{x-3y}{5x}}$ **$4x$**

14. $\dfrac{\frac{10x-5y}{12}}{\frac{2x-y}{6y}}$ **$\dfrac{5y}{2}$**

15. $\dfrac{\frac{x^2-16y^2}{xy}}{\frac{1}{y}-\frac{4}{x}}$ **$x+4y$**

16. $\dfrac{\frac{2}{s}-\frac{3}{t}}{\frac{4t^2-9s^2}{st}}$ **$\dfrac{1}{2t+3s}$**

17. $\dfrac{y-\frac{y-3}{3}}{\frac{4}{9}+\frac{2}{3y}}$ **$\dfrac{3y}{2}$**

18. $\dfrac{p - \dfrac{p+2}{4}}{\dfrac{3}{4} - \dfrac{5}{2p}}$

$\dfrac{3p^2 - 2p}{3p - 10}$

19. $\dfrac{\dfrac{x+2}{x} + \dfrac{1}{x+2}}{\dfrac{5}{x} + \dfrac{x}{x+2}}$

$\dfrac{x^2 + 5x + 4}{x^2 + 5x + 10}$

20. $\dfrac{\dfrac{y+3}{y} - \dfrac{4}{y-1}}{\dfrac{y}{y-1} + \dfrac{1}{y}}$

$\dfrac{y^2 - 2y - 3}{y^2 + y - 1}$

Relating Concepts (Exercises 21–26) For Individual or Group Work

Simplifying a complex fraction by Method 1 is a good way to review the methods of adding, subtracting, multiplying, and dividing rational expressions. Method 2 gives a good review of the fundamental property of rational expressions. Refer to the following complex fraction, and ***work Exercises 21–26 in order.***

$$\frac{\dfrac{4}{m} + \dfrac{m+2}{m-1}}{\dfrac{m+2}{m} - \dfrac{2}{m-1}}$$

21. Add the fractions in the numerator.

$\dfrac{m^2 + 6m - 4}{m(m-1)}$

22. Subtract as indicated in the denominator.

$\dfrac{m^2 - m - 2}{m(m-1)}$

23. Divide your answer from Exercise 21 by your answer from Exercise 22.

$\dfrac{m^2 + 6m - 4}{m^2 - m - 2}$

24. Go back to the original complex fraction and find the least common denominator of all denominators.

$m(m-1)$

25. Multiply the numerator and denominator of the complex fraction by your answer from Exercise 24.

$\dfrac{m^2 + 6m - 4}{m^2 - m - 2}$

26. Your answers for Exercises 23 and 25 should be the same. Write an explanation comparing the two methods. Which method do you prefer? Explain why.

Method 1 involves simplifying the numerator and the denominator separately and then performing a division. Method 2 involves multiplying the fraction by a form of 1, the identity element for multiplication. (Preferences will vary.)

Simplify each expression, using only positive exponents in the answer. See Example 4.

27. $\dfrac{1}{x^{-2} + y^{-2}}$

$\dfrac{x^2y^2}{y^2 + x^2}$

28. $\dfrac{1}{p^{-2} - q^{-2}}$

$\dfrac{p^2q^2}{q^2 - p^2}$, or $\dfrac{p^2q^2}{(q-p)(q+p)}$

29. $\dfrac{x^{-2} + y^{-2}}{x^{-1} + y^{-1}}$

$\dfrac{y^2 + x^2}{xy^2 + x^2y}$, or $\dfrac{y^2 + x^2}{xy(y+x)}$

30. $\dfrac{x^{-1} - y^{-1}}{x^{-2} - y^{-2}}$

$\dfrac{xy}{y + x}$

31. $\dfrac{x^{-1} + 2y^{-1}}{2y + 4x}$

$\dfrac{1}{2xy}$

32. $\dfrac{a^{-2} - 4b^{-2}}{3b - 6a}$

$\dfrac{b + 2a}{3a^2b^2}$

7.4 Equations with Rational Expressions and Graphs

In **Section 7.1,** we defined the domain of a rational function as the set of all possible values of the variable. (We can also refer to this as "the domain of the variable.") Any value that makes the denominator 0 is excluded.

OBJECTIVES

1. **Determine the domain of the variable in a rational equation.**
2. **Solve rational equations.**
3. **Recognize the graph of a rational function.**

OBJECTIVE 1 Determine the domain of the variable in a rational equation. The **domain of the variable in a rational equation** is the intersection (overlap) of the domains of the rational expressions in the equation.

EXAMPLE 1 Determining Domains in Rational Equations

Find the domain of the variable in each equation.

(a) $\frac{2}{x} - \frac{3}{2} = \frac{7}{2x}$

The domains of the three rational terms of the equation are, in order, $\{x \mid x \neq 0\}$, $(-\infty, \infty)$, and $\{x \mid x \neq 0\}$. The intersection of these three domains is all real numbers except 0, which may be written $\{x \mid x \neq 0\}$.

(b) $\frac{2}{x-3} - \frac{3}{x+3} = \frac{12}{x^2-9}$

The domains of these three terms are, respectively, $\{x \mid x \neq 3\}$, $\{x \mid x \neq -3\}$, and $\{x \mid x \neq \pm 3\}$. ($\pm$ is read "positive or negative," or "plus or minus.") The domain of the variable is the intersection of the three domains, all real numbers except 3 and -3, written $\{x \mid x \neq \pm 3\}$.

Work Problem 1 at the Side. ▶

1 Find the domain of the variable in each equation.

(a) $\frac{3}{x} + \frac{1}{2} = \frac{5}{6x}$

(b) $\frac{4}{x-5} - \frac{2}{x+5} = \frac{1}{x^2-25}$

OBJECTIVE 2 Solve rational equations. The easiest way to solve most equations involving rational expressions is to multiply all terms in the equation by the least common denominator. This step will clear the equation of all denominators. ***We can do this only with equations, not expressions.***

CAUTION

When each side of an equation is multiplied by a *variable* expression, the resulting "solutions" may not satisfy the original equation. ***You must either determine and observe the domain or check all proposed solutions in the original equation. It is wise to do both.***

EXAMPLE 2 Solving a Rational Equation

Solve $\frac{2}{x} - \frac{3}{2} = \frac{7}{2x}$.

The domain, which excludes 0, was found in Example 1(a).

$$2x\left(\frac{2}{x} - \frac{3}{2}\right) = 2x\left(\frac{7}{2x}\right) \quad \text{Multiply by the LCD, } 2x.$$

$$2x\left(\frac{2}{x}\right) - 2x\left(\frac{3}{2}\right) = 2x\left(\frac{7}{2x}\right) \quad \text{Distributive property}$$

$$4 - 3x = 7 \quad \text{Multiply.}$$

$$-3x = 3 \quad \text{Subtract 4.}$$

$$\text{Proposed solution} \longrightarrow x = -1 \quad \text{Divide by } -3.$$

Continued on Next Page

ANSWERS

1. **(a)** $\{x \mid x \neq 0\}$ **(b)** $\{x \mid x \neq \pm 5\}$

2 Solve $-\frac{3}{20} + \frac{2}{x} = \frac{5}{4x}$.

3 Solve each equation.

(a) $\frac{3}{x+1} = \frac{1}{x-1} - \frac{2}{x^2-1}$

(b) $\frac{1}{x-3} + \frac{1}{x+3} = \frac{6}{x^2-9}$

Check Replace x with -1 in the original equation.

$$\frac{2}{x} - \frac{3}{2} = \frac{7}{2x} \quad \text{Original equation}$$

$$\frac{2}{-1} - \frac{3}{2} \stackrel{?}{=} \frac{7}{2(-1)} \quad \text{Let } x = -1.$$

$$-2 - \frac{3}{2} \stackrel{?}{=} -\frac{7}{2}$$

$$-\frac{7}{2} = -\frac{7}{2} \quad \text{True}$$

The solution set is $\{-1\}$.

◀ Work Problem **2** at the Side.

EXAMPLE 3 Solving a Rational Equation with No Solution

Solve $\frac{2}{x-3} - \frac{3}{x+3} = \frac{12}{x^2-9}$.

Using the result from Example 1(b), we know that the domain excludes 3 and -3, since these values make one or more of the denominators in the equation equal 0. Multiply each side by the LCD, $(x+3)(x-3)$.

$$(x+3)(x-3)\left(\frac{2}{x-3} - \frac{3}{x+3}\right) = (x+3)(x-3)\left(\frac{12}{x^2-9}\right)$$

$$(x+3)(x-3)\left(\frac{2}{x-3}\right) - (x+3)(x-3)\left(\frac{3}{x+3}\right)$$

$$= (x+3)(x-3)\left(\frac{12}{x^2-9}\right) \quad \text{Distributive property}$$

$$2(x+3) - 3(x-3) = 12 \quad \text{Multiply.}$$

$$2x + 6 - 3x + 9 = 12 \quad \text{Distributive property}$$

$$-x + 15 = 12 \quad \text{Combine like terms.}$$

$$-x = -3 \quad \text{Subtract 15.}$$

$$\text{Proposed solution} \longrightarrow x = 3 \quad \text{Divide by } -1.$$

Since 3 is not in the domain, it cannot be a solution of the equation. Substituting 3 in the original equation shows why.

Check

$$\frac{2}{x-3} - \frac{3}{x+3} = \frac{12}{x^2-9} \quad \text{Original equation}$$

$$\frac{2}{3-3} - \frac{3}{3+3} \stackrel{?}{=} \frac{12}{3^2-9} \quad \text{Let } x = 3.$$

$$\frac{2}{0} - \frac{3}{6} \stackrel{?}{=} \frac{12}{0}$$

Since division by 0 is undefined, the given equation has no solution, and the solution set is $\emptyset$.

◀ Work Problem **3** at the Side.

Answers

2. $\{5\}$

3. (a) $\emptyset$ **(b)** $\emptyset$

EXAMPLE 4 Solving a Rational Equation

Solve $\frac{3}{p^2+p-2} - \frac{1}{p^2-1} = \frac{7}{2(p^2+3p+2)}$.

Factor each denominator to find the LCD, $2(p-1)(p+2)(p+1)$. The domain excludes 1, -2, and -1. Multiply each side by the LCD.

$$2(p-1)(p+2)(p+1)\left(\frac{3}{(p+2)(p-1)} - \frac{1}{(p+1)(p-1)}\right)$$

$$= 2(p-1)(p+2)(p+1)\left(\frac{7}{2(p+2)(p+1)}\right)$$

$$2 \cdot 3(p+1) - 2(p+2) = 7(p-1) \quad \text{Distributive property}$$

$$6p + 6 - 2p - 4 = 7p - 7 \quad \text{Distributive property}$$

$$4p + 2 = 7p - 7 \quad \text{Combine like terms.}$$

$$9 = 3p \quad \text{Subtract } 4p\text{; add 7.}$$

$$\text{Proposed solution} \longrightarrow 3 = p \quad \text{Divide by 3.}$$

Note that 3 is in the domain. Substitute 3 for p in the original equation to check that the solution set is $\{3\}$.

Work Problem 4 at the Side. ▶

EXAMPLE 5 Solving a Rational Equation That Leads to a Quadratic Equation

Solve $\frac{2}{3x+1} = \frac{1}{x} - \frac{6x}{3x+1}$.

Since the denominator $3x + 1$ cannot equal 0, $-\frac{1}{3}$ is excluded from the domain, as is 0. Multiply each side by the LCD, $x(3x+1)$.

$$x(3x+1)\left(\frac{2}{3x+1}\right) = x(3x+1)\left(\frac{1}{x} - \frac{6x}{3x+1}\right)$$

$$x(3x+1)\left(\frac{2}{3x+1}\right) = x(3x+1)\left(\frac{1}{x}\right) - x(3x+1)\left(\frac{6x}{3x+1}\right)$$

Distributive property

$$2x = 3x + 1 - 6x^2$$

Write this quadratic equation in standard form with 0 on the right side.

$$6x^2 - 3x + 2x - 1 = 0$$

$$6x^2 - x - 1 = 0 \quad \text{Standard form}$$

$$(3x+1)(2x-1) = 0 \quad \text{Factor.}$$

$$3x + 1 = 0 \quad \text{or} \quad 2x - 1 = 0 \quad \text{Zero-factor property}$$

$$x = -\frac{1}{3} \quad \text{or} \quad x = \frac{1}{2} \quad \text{Proposed solutions}$$

Because $-\frac{1}{3}$ is not in the domain, it is not a solution. Check that the solution set is $\{\frac{1}{2}\}$.

Work Problem 5 at the Side. ▶

4 Solve
$$\frac{4}{x^2+x-6} - \frac{1}{x^2-4} = \frac{2}{x^2+5x+6}.$$

5 Solve
$$\frac{1}{x+4} + \frac{x}{x-4} = \frac{-8}{x^2-16}.$$

ANSWERS

4. $\{-9\}$

5. $\{-1\}$

OBJECTIVE 3 Recognize the graph of a rational function. As mentioned in **Section 7.1,** a function defined by a quotient of polynomials is a **rational function.** Because one or more values of x may be excluded from the domain of most rational functions, their graphs are often **discontinuous.** That is, there will be one or more breaks in the graph. For example, we use point plotting and observing the domain to graph the simple rational function defined by

$$f(x) = \frac{1}{x}.$$

The domain of this function includes all real numbers except 0. Thus, there will be no point on the graph with $x = 0$. The vertical line with equation $x = 0$ is called a **vertical asymptote** of the graph. The horizontal line with equation $y = 0$ is called a **horizontal asymptote.** We show some typical ordered pairs in the table for both negative and positive x-values.

x	-3	-2	-1	$-\frac{1}{2}$	$-\frac{1}{4}$	$-\frac{1}{10}$	$\frac{1}{10}$	$\frac{1}{4}$	$\frac{1}{2}$	1	2	3
y	$-\frac{1}{3}$	$-\frac{1}{2}$	-1	-2	-4	-10	10	4	2	1	$\frac{1}{2}$	$\frac{1}{3}$

Notice that the closer positive values of x are to 0, the larger y is. Similarly, the closer negative values of x are to 0, the smaller (more negative) y is. Using this observation, excluding 0 from the domain, and plotting the points in the table, we obtain the graph in Figure 2.

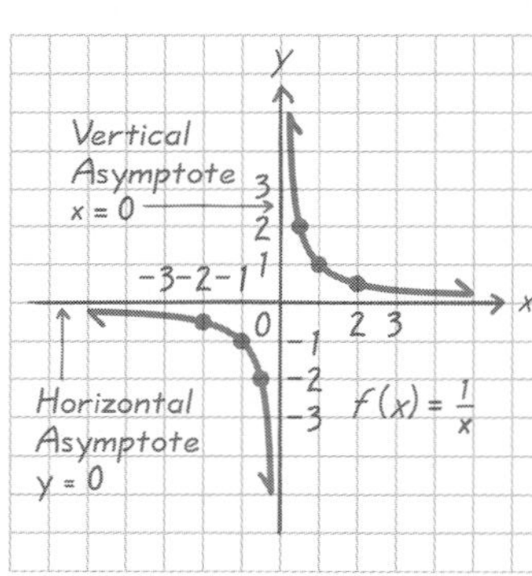

Figure 2

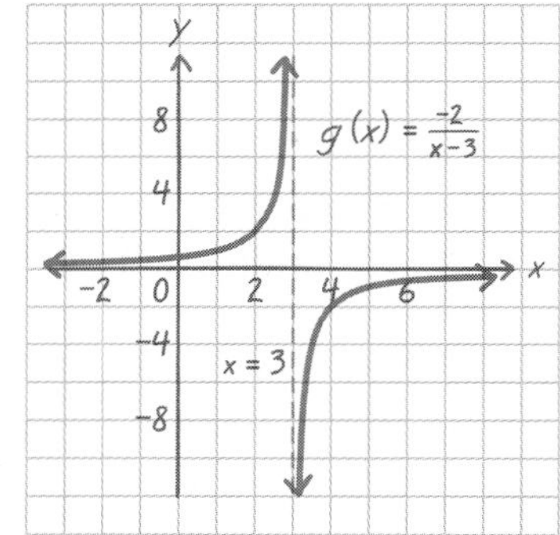

Figure 3

The graph of

$$g(x) = \frac{-2}{x - 3}$$

is shown in Figure 3. Some ordered pairs are shown in the table.

x	-2	-1	0	1	2	2.5	2.75	3.25	3.5	4	5	6
y	$\frac{2}{5}$	$\frac{1}{2}$	$\frac{2}{3}$	1	2	4	8	-8	-4	-2	-1	$-\frac{2}{3}$

There is no point on the graph for $x = 3$ because 3 is excluded from the domain. The dashed line $x = 3$ represents the vertical asymptote and is not part of the graph. As suggested by the points from the table, the graph gets closer to the vertical asymptote as the x-values get closer to 3. Again, $y = 0$ is a horizontal asymptote.

◀ *Work Problem 6 at the Side.*

6 Graph each rational function, and give the equations of the vertical and horizontal asymptotes.

(a) $f(x) = -\frac{1}{x}$

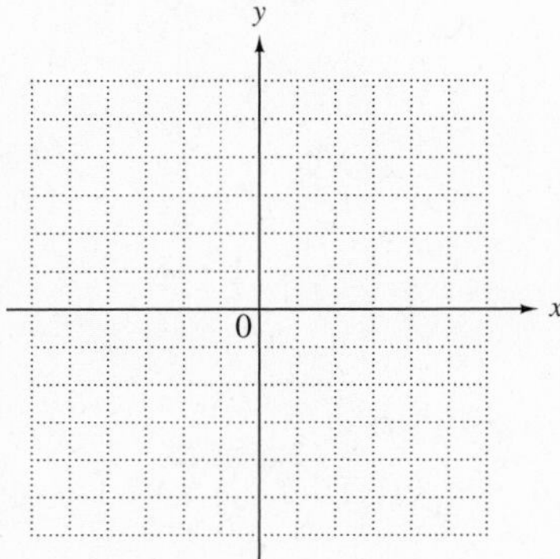

(b) $f(x) = \frac{2}{x + 3}$

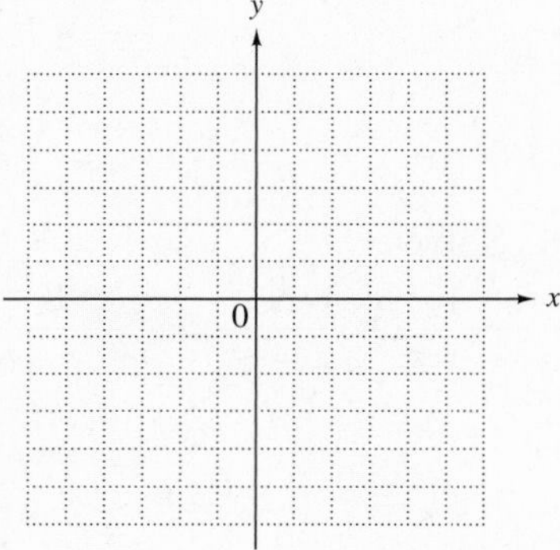

Answers

6. (a) vertical asymptote: $x = 0$; horizontal asymptote: $y = 0$

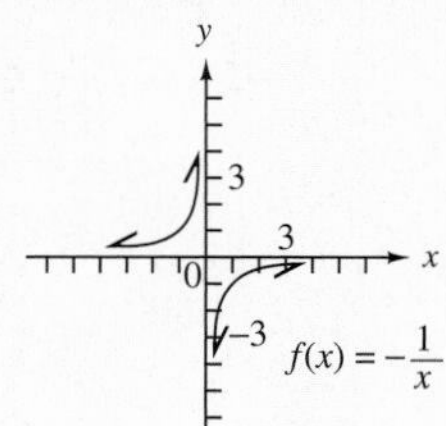

(b) vertical asymptote: $x = -3$; horizontal asymptote: $y = 0$

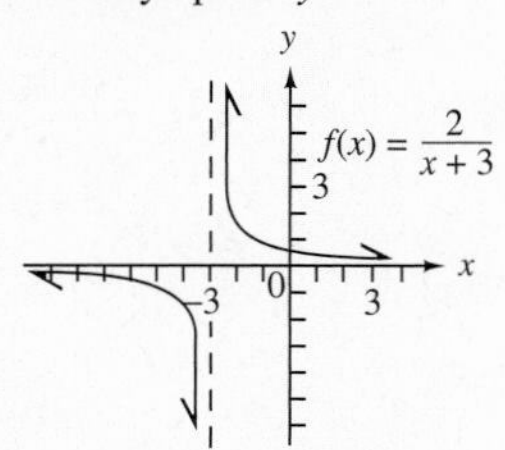

7.4 Exercises

FOR EXTRA HELP

Math XL PRACTICE

WATCH

DOWNLOAD

READ

REVIEW

As explained in this section, any values that would cause a denominator to equal 0 must be excluded from the domain and consequently as solutions of an equation that has variable expressions in the denominators. ***(a)*** *Without actually solving the equation, list all possible numbers that would have to be rejected if they appeared as potential solutions.* ***(b)*** *Then give the domain using set notation. See Example 1.*

1. $\frac{1}{x+1} - \frac{1}{x-2} = 0$

(a) $-1, 2$

(b) $\{x \mid x \neq -1, 2\}$

2. $\frac{3}{x+4} - \frac{2}{x-9} = 0$

(a) $-4, 9$

(b) $\{x \mid x \neq -4, 9\}$

3. $\frac{5}{3x+5} - \frac{1}{x} = \frac{1}{2x+3}$

(a) $-\frac{5}{3}, 0, -\frac{3}{2}$

(b) $\left\{x \mid x \neq -\frac{5}{3}, 0, -\frac{3}{2}\right\}$

4. $\frac{6}{4x+7} - \frac{3}{x} = \frac{5}{6x-13}$

(a) $-\frac{7}{4}, 0, \frac{13}{6}$

(b) $\left\{x \mid x \neq -\frac{7}{4}, 0, \frac{13}{6}\right\}$

5. $\frac{1}{3x} + \frac{1}{2x} = \frac{x}{3}$

(a) 0

(b) $\{x \mid x \neq 0\}$

6. $\frac{5}{6x} - \frac{8}{2x} = \frac{x}{4}$

(a) 0

(b) $\{x \mid x \neq 0\}$

7. $\frac{3x+1}{x-4} = \frac{6x+5}{2x-7}$

(a) $4, \frac{7}{2}$

(b) $\left\{x \mid x \neq 4, \frac{7}{2}\right\}$

8. $\frac{4x-1}{2x+3} = \frac{12x-25}{6x-2}$

(a) $-\frac{3}{2}, \frac{1}{3}$

(b) $\left\{x \mid x \neq -\frac{3}{2}, \frac{1}{3}\right\}$

9. $\frac{2}{x^2-x} + \frac{1}{x+3} = \frac{4}{x-2}$

(a) $0, 1, -3, 2$

(b) $\{x \mid x \neq 0, 1, -3, 2\}$

10. Suppose that in solving the following equation, all of your algebraic steps are correct. Is it possible that your proposed solution would have to be rejected? Explain.

$$\frac{x+7}{4} - \frac{x+3}{3} = \frac{x}{12}$$

Since there are no variables in denominators, there are no proposed solutions that would have to be rejected.

Solve each equation. See Examples 2–5.

11. $\frac{-5}{2x} + \frac{3}{4x} = \frac{-7}{4}$ {1}

12. $\frac{6}{5x} - \frac{2}{3x} = \frac{-8}{45}$ {−3}

13. $x - \frac{24}{x} = -2$ {−6, 4}

14. $p + \frac{15}{p} = -8$ {−5, −3}

15. $\frac{x-4}{x+6} = \frac{2x+3}{2x-1}$ $\left\{-\frac{7}{12}\right\}$

16. $\frac{5x-8}{x+2} = \frac{5x-1}{x+3}$ {−11}

17. $\frac{3x+1}{x-4} = \frac{6x+5}{2x-7}$ ∅

18. $\frac{4x-1}{2x+3} = \frac{12x-25}{6x-2}$ ∅

19. $\frac{1}{y-1} + \frac{5}{12} = \frac{-2}{3y-3}$ {−3}

20. $\frac{4}{m+2} - \frac{11}{9} = \frac{1}{3m+6}$ {1}

21. $\frac{-2}{3t-6} - \frac{1}{36} = \frac{-3}{4t-8}$ {5}

22. $\frac{3}{4m+2} = \frac{17}{2} - \frac{7}{2m+1}$ {0}

23. $\frac{3}{k+2} - \frac{2}{k^2-4} = \frac{1}{k-2}$ {5}

24. $\frac{3}{x-2} + \frac{21}{x^2-4} = \frac{14}{x+2}$ {5}

25. $\frac{1}{y+2} + \frac{3}{y+7} = \frac{5}{y^2+9y+14}$ ∅

26. $\frac{1}{t+3} + \frac{4}{t+5} = \frac{2}{t^2+8t+15}$ ∅

27. $\frac{9}{x} + \frac{4}{6x-3} = \frac{2}{6x-3}$ $\left\{\frac{27}{56}\right\}$

28. $\frac{5}{n} + \frac{4}{6-3n} = \frac{2n}{6-3n}$ $\left\{-\frac{15}{2}\right\}$

29. $\frac{6}{w+3} + \frac{-7}{w-5} = \frac{-48}{w^2-2w-15}$ ∅

30. $\frac{2}{r-5} + \frac{3}{2r+1} = \frac{22}{2r^2-9r-5}$ ∅

31. $\frac{x}{x-3} + \frac{4}{x+3} = \frac{18}{x^2-9}$ {−10}

32. $\frac{2x}{x-3} + \frac{4}{x+3} = \frac{-24}{x^2-9}$ {−2}

33. $\frac{6}{x-4} + \frac{5}{x} = \frac{-20}{x^2-4x}$ ∅

34. $\frac{7}{x-4} + \frac{3}{x} = \frac{-12}{x^2 - 4x}$

∅

35. $\frac{2}{4x+7} + \frac{x}{3} = \frac{6}{12x+21}$

{0}

36. $\frac{5x+14}{x^2-9} = \frac{-2x^2-5x+2}{x^2-9} + \frac{2x+4}{x-3}$

$\{x \mid x \neq -3, 3\}$

37. $\frac{4x-7}{4x^2-9} = \frac{-2x^2+5x-4}{4x^2-9} + \frac{x+1}{2x+3}$

$\left\{x \mid x \neq -\frac{3}{2}, \frac{3}{2}\right\}$

38. What is wrong with the following problem? "Solve $\frac{2x+1}{3x-4} + \frac{1}{2x+3}$."

"Solve" refers to finding the solution set of an equation. What appears here is not an equation, but an expression. "Solve" should be replaced by "Simplify" or "Add."

Graph each rational function. Give the equations of the vertical and horizontal asymptotes. See Objective 3 and Figures 2 and 3.

39. $f(x) = \frac{2}{x}$

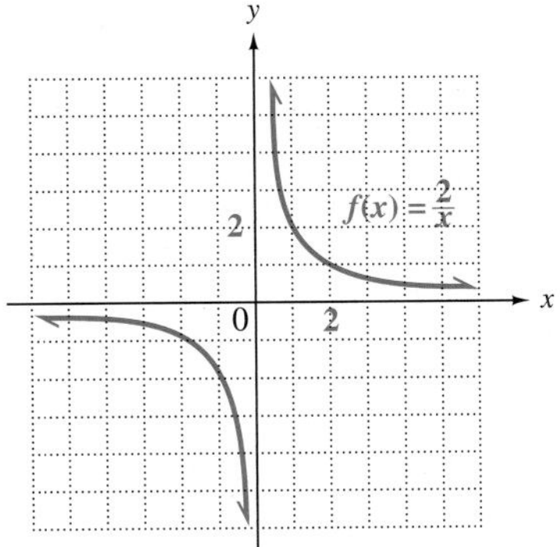

$x = 0; y = 0$

40. $f(x) = \frac{3}{x}$

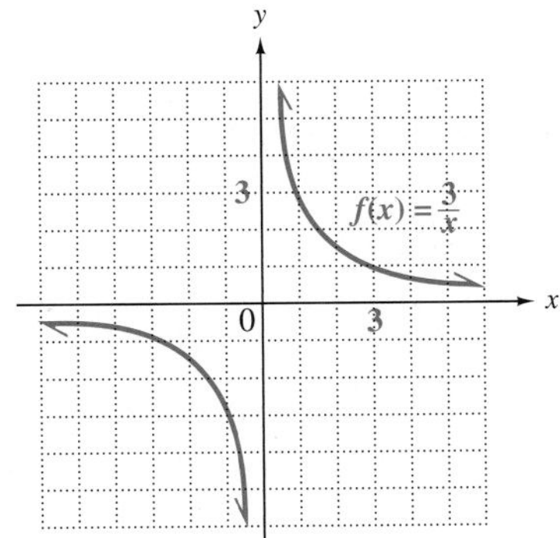

$x = 0; y = 0$

41. $f(x) = \frac{1}{x-2}$

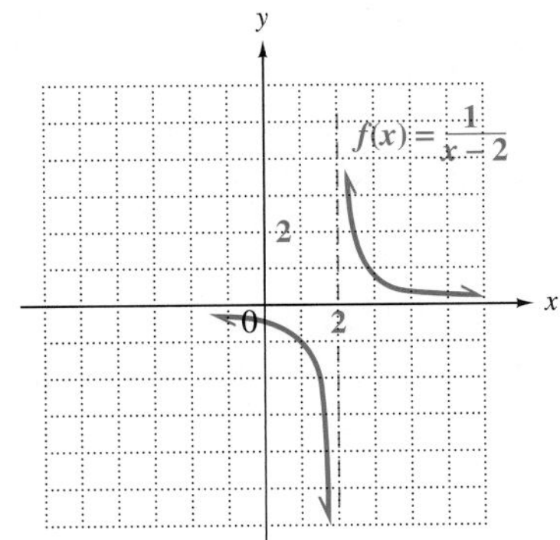

$x = 2; y = 0$

42. $f(x) = \frac{1}{x+2}$

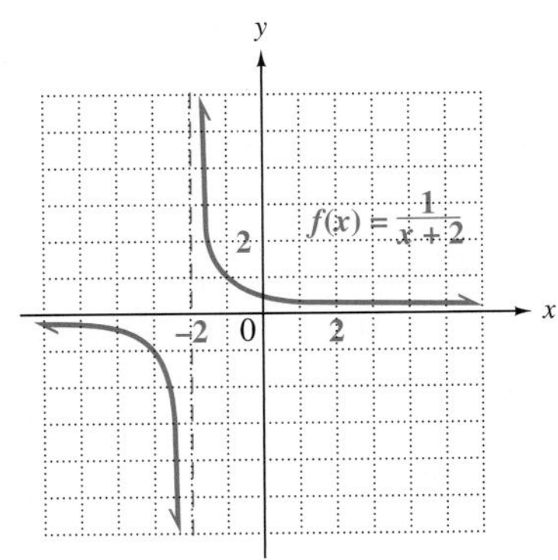

$x = -2; y = 0$

Solve each problem.

43. The average number of vehicles waiting in line to enter a parking area is modeled by the rational function defined by

$$w(x) = \frac{x^2}{2(1 - x)},$$

where x is a quantity between 0 and 1 known as the **traffic intensity.** (*Source*: Mannering, F. and W. Kilareski, *Principles of Highway Engineering and Traffic Control,* John Wiley and Sons, 1990.) To the nearest tenth, find the average number of vehicles waiting for each traffic intensity.

(a) 0.1 **0**

(b) 0.8 **1.6**

(c) 0.9 **4.1**

(d) What happens to waiting time as traffic intensity increases?

The waiting time also increases.

44. The force required to keep a 2000-lb car going 30 mph from skidding on a curve, where r is the radius of the curve in feet, is given by

$$F(r) = \frac{225{,}000}{r}.$$

(a) What radius must a curve have if a force of 450 lb is needed to keep the car from skidding? **500 ft**

(b) As the radius of the curve is lengthened, how is the force affected? **It decreases.**

Relating Concepts (Exercises 45–48) For Individual or Group Work

An equation of the form

$$\frac{A}{x + B} + \frac{x}{x - B} = \frac{C}{x^2 - B^2}$$

will have one rejected solution if the relationship $C = -2AB$ *holds true. (This can be proved using methods not covered in intermediate algebra.) For example, if* $A = 1$ *and* $B = 2$, *then* $C = -2AB = -2(1)(2) = -4$, *and the equation becomes*

$$\frac{1}{x + 2} + \frac{x}{x - 2} = \frac{-4}{x^2 - 4}.$$

This equation has solution set $\{-1\}$; *the potential solution* -2 *must be rejected. To further understand this idea,* ***work Exercises 45–48 in order.***

45. Show that the second equation does indeed have solution set $\{-1\}$ and -2 must be rejected.

Substituting -1 for x gives a true statement, $\frac{4}{3} = \frac{4}{3}$. Substituting -2 for x leads to 0 in the first and third denominators.

46. Let $A = 2$ and let $B = 1$. What is the corresponding value of C? Solve the equation determined by A, B, and C. What is the solution set? What value must be rejected?

$C = -4; \{-2\}; -1$ is rejected.

47. Let $A = 4$ and let $B = -3$. What is the corresponding value of C? Solve the equation determined by A, B, and C. What is the solution set? What value must be rejected?

$C = 24; \{-4\}; 3$ is rejected.

48. Choose two numbers of your own, letting one be A and the other be B. Repeat the process described in Exercises 46 and 47.

Answers will vary.

Summary Exercises on Rational Expressions and Equations

A common student error is to confuse an equation, *such as* $\frac{x}{2} + \frac{x}{3} = -5$, *with an expression, such as* $\frac{x}{2} + \frac{x}{3}$. *Look for the equals sign to distinguish between them. Equations are solved for a numerical answer, while problems involving operations result in simplified expressions.*

Solving an Equation	**Simplifying an Expression Involving an Operation**
Solve: $\frac{x}{2} + \frac{x}{3} = -5.$	Add: $\frac{x}{2} + \frac{x}{3}.$
Multiply each side by the LCD, 6.	Write both fractions with the LCD, 6.
$6\left(\frac{x}{2} + \frac{x}{3}\right) = 6(-5)$	$\frac{x}{2} + \frac{x}{3}$
$6\left(\frac{x}{2}\right) + 6\left(\frac{x}{3}\right) = 6(-5)$	$= \frac{x \cdot 3}{2 \cdot 3} + \frac{x \cdot 2}{3 \cdot 2}$
$3x + 2x = -30$	$= \frac{3x}{6} + \frac{2x}{6}$
$5x = -30$	$= \frac{3x + 2x}{6}$
$x = -6$	$= \frac{5x}{6}$
Check that the solution set is $\{-6\}$.	

Identify each exercise as an expression *or an* equation. *Then simplify the expression by performing the indicated operation, or solve the given equation, as appropriate.*

1. $\frac{x}{2} - \frac{x}{4} = 5$

equation; {20}

2. $\frac{4x - 20}{x^2 - 25} \cdot \frac{(x + 5)^2}{10}$

expression; $\frac{2(x + 5)}{5}$

3. $\frac{6}{7x} - \frac{4}{x}$

expression; $-\frac{22}{7x}$

4. $\dfrac{\frac{1}{x} + \frac{1}{y}}{\frac{1}{x} - \frac{1}{y}}$

expression; $\frac{y + x}{y - x}$

5. $\frac{5}{7t} = \frac{52}{7} - \frac{3}{t}$

equation; $\left\{\frac{1}{2}\right\}$

6. $\frac{x - 5}{3} + \frac{1}{3} = \frac{x - 2}{5}$

equation; {7}

7. $\frac{7}{6x} + \frac{5}{8x}$

expression; $\frac{43}{24x}$

8. $\frac{4}{x} - \frac{8}{x + 1} = 0$

equation; {1}

9. $\dfrac{\frac{6}{x + 1} - \frac{1}{x}}{\frac{2}{x} - \frac{4}{x + 1}}$

expression; $\frac{5x - 1}{-2x + 2}$, **or** $\frac{5x - 1}{-2(x - 1)}$

10. $\frac{8}{r + 2} - \frac{7}{4r + 8}$

expression; $\frac{25}{4(r + 2)}$

11. $\frac{x}{x + y} + \frac{2y}{x - y}$

expression; $\frac{x^2 + xy + 2y^2}{(x + y)(x - y)}$

12. $\frac{3p^2 - 6p}{p + 5} \div \frac{p^2 - 4}{8p + 40}$

expression; $\frac{24p}{p + 2}$

13. $\dfrac{x-2}{9} \cdot \dfrac{5}{8-4x}$

expression; $-\dfrac{5}{36}$

14. $\dfrac{a-4}{3} + \dfrac{11}{6} = \dfrac{a+1}{2}$

equation; $\{0\}$

15. $\dfrac{b^2+b-6}{b^2+2b-8} \cdot \dfrac{b^2+8b+16}{3b+12}$

expression; $\dfrac{b+3}{3}$

16. $\dfrac{10z^2-5z}{3z^3-6z^2} \div \dfrac{2z^2+5z-3}{z^2+z-6}$

expression; $\dfrac{5}{3z}$

17. $\dfrac{5}{x^2-2x} - \dfrac{3}{x^2-4}$

expression; $\dfrac{2x+10}{x(x-2)(x+2)}$

18. $\dfrac{6}{t+1} + \dfrac{4}{5t+5} = \dfrac{34}{15}$

equation; $\{2\}$

19. $\dfrac{\dfrac{5}{x} - \dfrac{3}{y}}{\dfrac{9x^2-25y^2}{x^2y}}$

expression; $\dfrac{-x}{3x+5y}$

20. $\dfrac{-2}{a^2+2a-3} - \dfrac{5}{3-3a} = \dfrac{4}{3a+9}$

equation; $\{-13\}$

21. $\dfrac{4y^2-13y+3}{2y^2-9y+9} \div \dfrac{4y^2+11y-3}{6y^2-5y-6}$

expression; $\dfrac{3y+2}{y+3}$

22. $\dfrac{8}{3k+9} - \dfrac{8}{15} = \dfrac{2}{5k+15}$

equation; $\left\{\dfrac{5}{4}\right\}$

23. $\dfrac{3r}{r-2} = 1 + \dfrac{6}{r-2}$

equation; $\emptyset$

24. $\dfrac{6z^2-5z-6}{6z^2+5z-6} \cdot \dfrac{12z^2-17z+6}{12z^2-z-6}$

expression; $\dfrac{2z-3}{2z+3}$

25. $\dfrac{-1}{3-x} - \dfrac{2}{x-3}$

expression; $\dfrac{-1}{x-3}$, or $\dfrac{1}{3-x}$

26. $\dfrac{\dfrac{t}{4} - \dfrac{1}{t}}{1 + \dfrac{t+4}{t}}$

expression; $\dfrac{t-2}{8}$

27. $\dfrac{2}{y+1} - \dfrac{3}{y^2-y-2} = \dfrac{3}{y-2}$

equation; $\{-10\}$

28. $\dfrac{7}{2x^2-8x} + \dfrac{3}{x^2-16}$

expression; $\dfrac{13x+28}{2x(x+4)(x-4)}$

29. $\dfrac{3}{y-3} - \dfrac{3}{y^2-5y+6} = \dfrac{2}{y-2}$

equation; $\emptyset$

30. $\dfrac{2k + \dfrac{5}{k-1}}{3k - \dfrac{2}{k}}$

expression; $\dfrac{k(2k^2-2k+5)}{(k-1)(3k^2-2)}$

7.5 Applications of Rational Expressions

OBJECTIVES

1. **Find the value of an unknown variable in a formula.**
2. **Solve a formula for a specified variable.**
3. **Solve applications using proportions.**
4. **Solve applications about distance, rate, and time.**
5. **Solve applications about work rates.**

OBJECTIVE 1 Find the value of an unknown variable in a formula. Formulas may contain rational expressions, such as $t = \frac{d}{r}$ and $\frac{1}{f} = \frac{1}{p} + \frac{1}{q}$.

EXAMPLE 1 Finding the Value of a Variable in a Formula

In physics, the focal length, f, of a lens is given by the formula

$$\frac{1}{f} = \frac{1}{p} + \frac{1}{q},$$

where p is the distance from the object to the lens and q is the distance from the lens to the image. See Figure 4. Find q if $p = 20$ cm and $f = 10$ cm.

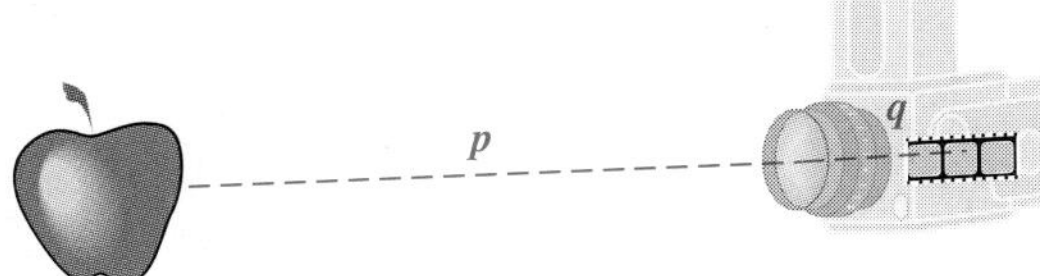

Focal Length of Camera Lens

Figure 4

Replace f with 10 and p with 20.

$$\frac{1}{f} = \frac{1}{p} + \frac{1}{q} \quad \text{Solve this equation for } q.$$

$$\frac{1}{10} = \frac{1}{20} + \frac{1}{q} \qquad \text{Let } f = 10,\ p = 20.$$

$$20q \cdot \frac{1}{10} = 20q\left(\frac{1}{20} + \frac{1}{q}\right) \qquad \text{Multiply by the LCD, } 20q.$$

$$20q \cdot \frac{1}{10} = 20q\left(\frac{1}{20}\right) + 20q\left(\frac{1}{q}\right) \qquad \text{Distributive property.}$$

$$2q = q + 20 \qquad \text{Multiply.}$$

$$q = 20 \qquad \text{Subtract } q.$$

The distance from the lens to the image is 20 cm.

Work Problem 1 *at the Side.* ▶

1 Use the formula given in Example 1 to answer each part.

(a) Find p if $f = 15$ and $q = 25$.

(b) Find f if $p = 6$ and $q = 9$.

(c) Find q if $f = 12$ and $p = 16$.

OBJECTIVE 2 Solve a formula for a specified variable. The goal in solving for a specified variable is to isolate it on one side of the equals sign.

EXAMPLE 2 Solving a Formula for a Specified Variable

Solve $\frac{1}{f} = \frac{1}{p} + \frac{1}{q}$ for p.

$$\frac{1}{f} = \frac{1}{p} + \frac{1}{q}$$

$$fpq \cdot \frac{1}{f} = fpq\left(\frac{1}{p} + \frac{1}{q}\right) \qquad \text{Multiply by the LCD, } fpq.$$

$$pq = fq + fp \qquad \text{Distributive property}$$

Continued on Next Page

ANSWERS

1. (a) $\frac{75}{2}$ (b) $\frac{18}{5}$ (c) 48

Transform the equation so that the terms with p (the specified variable) are on the same side. One way to do this is to subtract fp from each side.

$$pq = fq + fp$$

$$pq - fp = fq \quad \text{Subtract } fp.$$

This is a key step.

$$p(q - f) = fq \quad \text{Factor out } p.$$

$$p = \frac{fq}{q - f} \quad \text{Divide by } q - f.$$

◀ Work Problem **2** at the Side.

2 Solve $\frac{3}{p} + \frac{3}{q} = \frac{5}{r}$ for q.

EXAMPLE 3 Solving a Formula for a Specified Variable

Solve $I = \frac{nE}{R + nr}$ for n.

$$I = \frac{nE}{R + nr}$$

$$(R + nr)I = (R + nr)\frac{nE}{R + nr} \quad \text{Multiply by } R + nr.$$

$$RI + nrI = nE$$

$$RI = nE - nrI \quad \text{Subtract } nrI.$$

$$RI = n(E - rI) \quad \text{Factor out } n.$$

$$\frac{RI}{E - rI} = n \quad \text{Divide by } E - rI.$$

3 Solve $A = \frac{Rr}{R + r}$ for R.

CAUTION

Refer to the steps in Examples 2 and 3 that factor out the desired variable. ***The variable for which you are solving must be a factor on only one side of the equation,*** so that each side can be divided by the remaining factor in the last step.

◀ Work Problem **3** at the Side.

We can now solve problems that translate into equations with rational expressions. To do so, we continue to use the six-step problem-solving method from **Section 2.4.**

OBJECTIVE 3 Solve applications using proportions. Recall from **Section 2.6** that a **ratio** is a comparison of two quantities. The ratio of a to b may be written in any of the following ways:

$$a \text{ to } b, \quad a : b, \quad \text{or} \quad \frac{a}{b}. \quad \text{Ratio of } a \text{ to } b$$

Ratios are usually written as quotients in algebra. A **proportion** is a statement that two ratios are equal, such as

$$\frac{a}{b} = \frac{c}{d}. \quad \text{Proportion}$$

Proportions are a useful and important type of rational equation.

ANSWERS

2. $q = \frac{3rp}{5p - 3r}$, or $q = \frac{-3rp}{3r - 5p}$

3. $R = \frac{-Ar}{A - r}$, or $R = \frac{Ar}{r - A}$

EXAMPLE 4 Solving a Proportion

In 2005, about 15 of every 100 Americans had no health insurance coverage. The population at that time was about 296 million. How many million Americans had no health insurance? (*Source*: U.S. Census Bureau.)

Step 1 **Read** the problem.

Step 2 **Assign a variable.** Let $x =$ the number (in millions) who had no health insurance.

Step 3 **Write an equation.** To get an equation, set up a proportion. The ratio 15 to 100 should equal the ratio x to 296.

$$\frac{15}{100} = \frac{x}{296} \qquad \text{Write a proportion.}$$

Step 4 **Solve.**

$$29{,}600\left(\frac{15}{100}\right) = 29{,}600\left(\frac{x}{296}\right) \qquad \text{Multiply by a common denominator.}$$

$$4440 = 100x \qquad \text{Simplify.}$$

$$x = 44.4 \qquad \text{Divide by 100.}$$

Step 5 **State the answer.** There were about 44.4 million Americans with no health insurance in 2005.

Step 6 **Check** that the ratio of 44.4 million to 296 million equals $\frac{15}{100}$.

Work Problem **4** *at the Side.* ▶

4 Solve the problem.

In 2006, approximately 11.7% (that is, 11.7 of every 100) of the 73,740,000 children under 18 yr of age in the United States had no health insurance. How many such children were uninsured? (*Source*: U.S. Census Bureau.)

EXAMPLE 5 Solving a Proportion Involving Rates

Marissa's car uses 10 gal of gas to travel 210 mi. She has 5 gal of gas in the car, and she still needs to drive 640 mi. If we assume the car continues to use gas at the same rate, how many more gallons will she need?

Step 1 **Read** the problem.

Step 2 **Assign a variable.** Let $x =$ the additional number of gallons of gas.

Step 3 **Write an equation.** To get an equation, set up a proportion.

$$\frac{\text{gallons} \rightarrow 10}{\text{miles} \rightarrow 210} = \frac{5 + x \leftarrow \text{gallons}}{640 \leftarrow \text{miles}}$$

Step 4 **Solve.** We could multiply by the LCD $10 \cdot 21 \cdot 64$. Instead we use an alternative method that involves *cross products:* For $\frac{a}{b} = \frac{c}{d}$ to be true, then the cross products ad and bc must be equal. Thus,

$$10 \cdot 640 = 210(5 + x) \qquad \text{If } \tfrac{a}{b} = \tfrac{c}{d}, \text{ then } ad = bc. \textbf{ (Section 2.6)}$$

$$6400 = 1050 + 210x \qquad \text{Multiply; distributive property}$$

$$5350 = 210x \qquad \text{Subtract 1050.}$$

$$25.5 \approx x. \qquad \text{Divide by 210.}$$

Step 5 **State the answer.** Marissa will need about 25.5 more gallons of gas.

Step 6 **Check.** The 25.5 gal plus the 5 gal equals 30.5 gal.

$$\frac{30.5}{640} \approx \mathbf{0.048} \quad \text{and} \quad \frac{10}{210} \approx \mathbf{0.048}$$

Since the ratios are equal, the answer is correct.

Work Problem **5** *at the Side.* ▶

5 Solve the problem.

Lauren's car uses 15 gal of gasoline to drive 495 mi. She has 6 gal of gasoline in the car, and she wants to know how much more gasoline she will need to drive 600 mi. If we assume that the car continues to use gasoline at the same rate, how many more gallons will she need? (Round your answer to the nearest tenth.)

ANSWERS

4. 8,627,580
5. 12.2 more gallons

OBJECTIVE 4 Solve applications about distance, rate, and time. If an automobile travels at an average rate of 65 mph for 2 hr, then it travels $65 \times 2 = 130$ mi. Recall from **Section 4.4** that this is an example of the basic relationship between distance, rate, and time given by the formula $d = rt$. By solving, in turn, for r and t in the formula, we obtain two other equivalent forms of the formula. The three forms are given below.

Distance, Rate, and Time Relationship

$$d = rt \qquad r = \frac{d}{t} \qquad t = \frac{d}{r}$$

The next example illustrates the uses of these formulas.

EXAMPLE 6 Finding Distance, Rate, or Time

(a) The speed of sound is 1088 ft per sec at sea level at 32°F. In 5 sec, under these conditions, sound travels

$$\underset{\text{Rate}}{\underset{\uparrow}{1088}} \times \underset{\text{Time}}{\underset{\uparrow}{5}} = \underset{\text{Distance}}{\underset{\uparrow}{5440 \text{ ft.}}}$$

Here, we found distance, given rate and time, using $d = rt$.

(b) The winner of the first Indianapolis 500 race (in 1911) was Ray Harroun, driving a Marmon Wasp at an average speed of 74.59 mph. (*Source: Universal Almanac.*)
To complete the 500 mi, it took him

$$\frac{\text{Distance} \rightarrow 500}{\text{Rate} \rightarrow 74.59} \approx 6.70 \text{ hr (rounded).} \leftarrow \text{Time}$$

Here, we found time, given distance and rate, using $t = \frac{d}{r}$. To convert the decimal 0.70 hr to minutes, multiply by 60 to get $0.70(60) = 42$. It took Harroun about 6 hr, 42 min, to complete the race.

(c) At the 2004 Olympic Games in Athens, Greece, Chinese swimmer Luo Xuejuan set an Olympic record of 66.64 sec in the women's 100-m breaststroke swimming event. (*Source: World Almanac and Book of Facts.*)
Her rate was

$$\text{Rate} = \frac{\text{Distance} \rightarrow}{\text{Time} \rightarrow} \frac{100}{66.64} \approx 1.50 \text{ m per sec (rounded).}$$

Here, we found rate, given distance and time, using $r = \frac{d}{t}$.

◀ *Work Problem 6 at the Side.*

Problem-Solving Hint

Many applied problems use the formulas just discussed. The next two examples show how to solve typical applications of the formula $d = rt$.

A helpful strategy for solving such problems is to ***first make a sketch*** showing what is happening in the problem. ***Then make a table*** using the information given, along with the unknown quantities. The table will help organize the information, and the sketch will help set up the equation.

6 Solve each problem.

(a) The world record in the men's 100-m dash was set by Justin Gatlin of the United States in 2006. He ran it in 9.77 sec. What was his speed in meters per second, to the nearest hundredth? (*Source: Guinness World Records.*)

(b) The world record for the women's 3000-m run was set by Julnara Samitova of Russia in 2005. Her speed was 5.539 m per sec. To the nearest second, what was her time? (*Source: Guinness World Records.*)

(c) A small plane flew from Chicago to St. Louis averaging 145 mph. The trip took 2 hr. What is the distance between Chicago and St. Louis?

ANSWERS

6. (a) 10.24 m per sec **(b)** 542 sec **(c)** 290 mi

EXAMPLE 7 **Solving a Problem about Distance, Rate, and Time**

A paddle wheeler goes 10 mi against the current in a river in the same time that it goes 15 mi with the current. If the speed of the current is 3 mph, find the speed of the boat in still water.

Step 1 **Read** the problem. We must find the speed of the boat in still water.

Step 2 **Assign a variable.**

Let $x =$ the speed of the boat in still water.

When the boat is traveling *against* the current, the current slows the boat down, and the speed of the boat is the difference between its speed in still water and the speed of the current. So, the speed against the current is $(x - 3)$ mph.

When the boat is traveling *with* the current, the current speeds the boat up, and the speed of the boat is the sum of its speed in still water and the speed of the current, that is, $(x + 3)$ mph.

Thus, $x - 3 =$ the speed of the boat *against* the current,

and $x + 3 =$ the speed of the boat *with* the current.

Because the time is the same going against the current as with the current, find time in terms of distance and rate (speed) for each situation. Start with the distance formula,

$$d = rt,$$

and divide each side by r to get $t = \frac{d}{r}$. Against the current, the distance is 10 mi and the rate is $(x - 3)$ mph, giving

$$t = \frac{d}{r} = \frac{10}{x - 3}. \quad \text{Time against the current}$$

With the current, the distance is 15 mi and the rate is $(x + 3)$ mph, so

$$t = \frac{d}{r} = \frac{15}{x + 3}. \quad \text{Time with the current}$$

This information is summarized in the following table.

	Distance	Rate	Time
Against Current	10	$x - 3$	$\frac{10}{x - 3}$
With Current	15	$x + 3$	$\frac{15}{x + 3}$

Times are equal.

Step 3 **Write an equation.** Because the times are equal,

$$\frac{10}{x - 3} = \frac{15}{x + 3}.$$

Continued on Next Page

7 Solve the problem.

A plane travels 100 mi against the wind in the same time that it takes to travel 120 mi with the wind. The wind speed is 20 mph.

(a) Complete this table.

	d	r	t
Against Wind	100	$x - 20$	
With Wind	120	$x + 20$	

(b) Find the speed of the plane in still air.

Answers

7. (a) $\frac{100}{x - 20}$; $\frac{120}{x + 20}$ **(b)** 220 mph

Step 4 **Solve.** $\frac{10}{x - 3} = \frac{15}{x + 3}$

$$(x + 3)(x - 3)\left(\frac{10}{x - 3}\right) = (x + 3)(x - 3)\left(\frac{15}{x + 3}\right) \quad \text{Multiply by the LCD.}$$

$$10(x + 3) = 15(x - 3) \quad \text{Multiply.}$$

$$10x + 30 = 15x - 45 \quad \text{Distributive property}$$

$$30 = 5x - 45 \quad \text{Subtract } 10x.$$

$$75 = 5x \quad \text{Add 45.}$$

$$15 = x \quad \text{Divide by 5.}$$

Step 5 **State the answer.** The speed of the boat in still water is 15 mph.

Step 6 **Check** the answer: $\frac{10}{15 - 3} = \frac{15}{15 + 3}$ is true.

◀ Work Problem 7 at the Side.

EXAMPLE 8 Solving a Problem about Distance, Rate, and Time

At O'Hare International Airport in Chicago, Cheryl and Bill are walking to the gate (at the same speed) to catch their flight to Denver. Bill steps onto the moving sidewalk and continues to walk while Cheryl uses the stationary sidewalk. If the sidewalk moves at 1 m per sec and Bill saves 50 sec covering the 300-m distance, what is their walking speed?

Step 1 **Read** the problem. We must find their walking speed.

Step 2 **Assign a variable.** Let x represent their walking speed in meters per second. Thus Cheryl travels at x meters per second and Bill travels at $(x + 1)$ meters per second. Express their times in terms of the known distances and the variable rates. As in Example 6, start with $d = rt$ and divide each side by r to get $t = \frac{d}{r}$. For Cheryl, the distance is 300 m and the rate is x, so Cheryl's time is

$$t = \frac{d}{r} = \frac{300}{x}. \quad \text{Cheryl's time}$$

Bill travels 300 m at a rate of $x + 1$, so his time is

$$t = \frac{d}{r} = \frac{300}{x + 1}. \quad \text{Bill's time}$$

This information is summarized in the following table.

	Distance	Rate	Time
Cheryl	300	x	$\frac{300}{x}$
Bill	300	$x + 1$	$\frac{300}{x + 1}$

Step 3 **Write an equation** using the times from the table.

Bill's time is Cheryl's time less 50 seconds.

$$\frac{300}{x + 1} = \frac{300}{x} - 50$$

Continued on Next Page

Step 4 **Solve.**

$$\frac{300}{x+1} = \frac{300}{x} - 50$$

$$x(x+1)\left(\frac{300}{x+1}\right) = x(x+1)\left(\frac{300}{x} - 50\right) \quad \text{Multiply by the LCD, } x(x+1).$$

$$x(x+1)\left(\frac{300}{x+1}\right) = x(x+1)\left(\frac{300}{x}\right) - x(x+1)(50) \quad \text{Distributive property}$$

$$300x = 300(x+1) - 50x(x+1) \quad \text{Multiply.}$$

$$300x = 300x + 300 - 50x^2 - 50x \quad \text{Distributive property}$$

$$50x^2 + 50x - 300 = 0 \quad \text{Standard form}$$

$$x^2 + x - 6 = 0 \quad \text{Divide by 50.}$$

$$(x+3)(x-2) = 0 \quad \text{Factor.}$$

$$x + 3 = 0 \quad \text{or} \quad x - 2 = 0 \quad \text{Zero-factor property}$$

$$x = -3 \quad \text{or} \quad x = 2 \quad \text{Solve each equation.}$$

Discard the negative answer, since speed cannot be negative.

Step 5 **State the answer.** Their walking speed is 2 m per sec.

Step 6 **Check** the answer in the words of the original problem.

Work Problem 8 at the Side. ▶

8 Solve the problem.

Kathy Manley drove 300 mi north from San Antonio, mostly on the freeway. She usually averaged 55 mph, but an accident slowed her speed through Dallas to 15 mph. If her trip took 6 hr, how many miles did she drive at reduced speed?

	d	r	t
Normal Speed	$300 - x$	55	
Reduced Speed	x	15	

OBJECTIVE 5 Solve applications about work rates. Suppose that you can mow your lawn in 4 hr. Then after 1 hr, you will have mowed $\frac{1}{4}$ of the lawn. After 2 hr, you will have mowed $\frac{2}{4}$, or $\frac{1}{2}$, of the lawn, and so on. This idea is generalized as follows.

Rate of Work

If a job can be completed in t units of time, then the rate of work is

$$\frac{1}{t} \text{ job per unit of time.}$$

Problem-Solving Hint

The relationship between problems involving work and problems involving distance is a very close one. Recall that the formula $d = rt$ says that distance traveled is equal to rate of travel multiplied by time traveled. Similarly, the fractional part of a job accomplished is equal to the rate of work multiplied by the time worked. In the lawn mowing example, after 3 hr, the fractional part of the job done is

$$\underbrace{\frac{1}{4}}_{\text{Rate of work}} \cdot \underbrace{3}_{\text{Time worked}} = \underbrace{\frac{3}{4}}_{\text{Fractional part of job done}}.$$

After 4 hr, $\frac{1}{4}(4) = 1$ whole job has been done.

ANSWER

8. $11\frac{1}{4}$ mi

EXAMPLE 9 Solving a Problem about Work Rates

With spraying equipment, Mateo can paint the woodwork in a small house in 8 hr. His assistant, Chet, needs 14 hr to complete the same job painting by hand. If both Mateo and Chet work together, how long will it take them to paint the woodwork?

Step 1 **Read** the problem again. We are looking for time working together.

Step 2 **Assign a variable.** Let $x =$ the number of hours it will take for Mateo and Chet to paint the woodwork, working together.

Begin by making a table. Based on the previous discussion, Mateo's rate alone is $\frac{1}{8}$ job per hour, and Chet's rate is $\frac{1}{14}$ job per hour.

	Rate	Time Working Together	Fractional Part of the Job Done When Working Together
Mateo	$\frac{1}{8}$	x	$\frac{1}{8}x$
Chet	$\frac{1}{14}$	x	$\frac{1}{14}x$

Sum is 1 whole job.

Step 3 **Write an equation.** Together Mateo and Chet complete 1 whole job. We must add the fractional parts and set the sum equal to 1.

$$\underbrace{\text{Fractional part done by Mateo}}_{\frac{1}{8}x} + \underbrace{\text{Fractional part done by Chet}}_{\frac{1}{14}x} = \underbrace{\text{1 whole job}}_{1}$$

Step 4 **Solve.**

$$56\left(\frac{1}{8}x + \frac{1}{14}x\right) = 56(1) \quad \text{Multiply by the LCD, 56.}$$

$$56\left(\frac{1}{8}x\right) + 56\left(\frac{1}{14}x\right) = 56(1) \quad \text{Distributive property}$$

$$7x + 4x = 56$$

$$11x = 56 \quad \text{Combine like terms.}$$

$$x = \frac{56}{11} \quad \text{Divide by 11.}$$

Step 5 **State the answer.** Working together, Mateo and Chet can paint the woodwork in $\frac{56}{11}$ hr, or $5\frac{1}{11}$ hr.

Step 6 **Check.** Substitute $\frac{56}{11}$ for x in the equation from Step 3.

$$\frac{1}{8}x + \frac{1}{14}x = 1 \quad \text{Equation from Step 3}$$

$$\frac{1}{8}\left(\frac{56}{11}\right) + \frac{1}{14}\left(\frac{56}{11}\right) \stackrel{?}{=} 1 \quad \text{Let } x = \tfrac{56}{11}.$$

$$\frac{7}{11} + \frac{4}{11} = 1 \quad \text{True}$$

Our answer, $\frac{56}{11}$ hr, or $5\frac{1}{11}$ hr, seems correct. See the Problem-Solving Hint on the next page for additional strategies for checking.

Problem-Solving Hint

A common error students make when solving a work problem like that in Example 9 is to add the two times, 8 hr and 14 hr, to get an answer of 22 hr. We reason, however, that x, the time it will take Mateo and Chet working together, must be *less than* 8 hr, since Mateo can complete the job by himself in 8 hr.

Another common error students make is to try to split the job in half between the two workers so that Mateo would work $\frac{1}{2}(8)$, or 4 hr, and Chet would work $\frac{1}{2}(14)$, or 7 hr. In this case, Mateo finishes 3 hr before Chet and they have not worked together to get the entire job done as quickly as possible. If Mateo, when he finishes, helps Chet, the job should actually be completed in a time between 4 hr and 7 hr.

Based on this reasoning, does our answer of $5\frac{1}{11}$ hr in Example 9

Note

An alternative approach in work problems is to consider the part of the job that can be done in 1 hr. For instance, in Example 9 Mateo can do the entire job in 8 hr, and Chet can do it in 14 hr. Thus, their work rates, as we saw in Example 9, are $\frac{1}{8}$ and $\frac{1}{14}$, respectively. Since it takes them x hours to complete the job when working together, in 1 hr they can paint $\frac{1}{x}$ of the woodwork. The amount painted by Mateo in 1 hr plus the amount painted by Chet in 1 hr must equal the amount they can do together. This leads to the equation

$$\underset{\text{Amount by Mateo}}{\frac{1}{8}} + \underset{\text{Amount by Chet}}{\frac{1}{14}} = \underset{\text{Amount together}}{\frac{1}{x}}.$$

Compare this with the equation in Step 3 of Example 9. Multiplying each side by $56x$ leads to

$$7x + 4x = 56,$$

the same equation found in the third line of Step 4 in the example. The same solution results.

Work Problem 9 at the Side. ▶

9 Solve each problem.

(a) Stan needs 45 min to do the dishes, while Deb can do them in 30 min. How long will it take them if they work together?

	Rate	Time Working Together	Fractional Part of the Job Done
Stan	$\frac{1}{45}$	x	
Deb	$\frac{1}{30}$	x	

(b) Suppose it takes Stan 35 min to do the dishes, and together they can do them in 15 min. How long will it take Deb to do them alone?

ANSWERS

9. (a) 18 min **(b)** $26\frac{1}{4}$ min

Math in the Media

TO PLAY BASEBALL, YOU MUST *WORK* AT IT!

In the 1994 movie *Little Big League,* the young Billy Heywood inherits the Minnesota Twins baseball team and becomes its manager. He leads the team to the Division Championship and then to the playoffs. But before the final playoff game, the biggest game of the year, he can't keep his mind on his job because a homework problem is giving him trouble.

If Joe can paint a house in 3 hours, and Sam can paint the same house in 5 hours, how long does it take for them to do it together?

With the help of one of his players, he is able to solve the problem, and the team goes on to victory.

1. Use the method described in Example 9 of **Section 7.5** to solve this problem.

$\frac{15}{8}$ hr, or $1\frac{7}{8}$ hr

2. Before the player was able to solve the problem correctly, Billy got "help" from some of the other players. The incorrect answers they gave him were

(a) 15 hr **(b)** 8 hr **(c)** 4 hr.

Explain the faulty reasoning behind each of these incorrect answers.

(a) The player multiplied: $5 \times 3 = 15$. (b) The player added: $5 + 3 = 8$. (c) The player divided the part (b) incorrect answer by 2, because there were two painters: $\frac{8}{2} = 4$.

3. The player who gave Billy the correct answer solved the problem as follows:

Using the simple formula a times b over a plus b, we get our answer of one and seven-eighths.

Show that if it takes one person a hours to complete one job and another b hours to complete the same job, then the expression stated by the player,

$$\frac{a \cdot b}{a + b}$$

actually does give the number of hours it would take them to do the job together. (*Hint:* Refer to Example 9 and use a and b rather than 8 and 14. Then solve the resulting formula for x.)

$\frac{1}{a}x + \frac{1}{b}x = 1$; $bx + ax = ab$; $x(b + a) = ab$;

$x = \frac{ab}{b + a}$, or $x = \frac{a \cdot b}{a + b}$

7.5 Exercises

FOR EXTRA HELP MyMathLab Math XL PRACTICE  WATCH  DOWNLOAD  READ  REVIEW

In Exercises 1–4, a familiar formula is given. Give the letter of the choice that is an equivalent form of the given formula.

1. $p = br$ (percent)

A. $b = \frac{p}{r}$ **B.** $r = \frac{b}{p}$

C. $b = \frac{r}{p}$ **D.** $p = \frac{r}{b}$

A

2. $V = LWH$ (geometry)

A. $H = \frac{LW}{V}$ **B.** $L = \frac{V}{WH}$

C. $L = \frac{WH}{V}$ **D.** $W = \frac{H}{VL}$

B

3. $m = \frac{F}{a}$ (physics)

A. $a = mF$ **B.** $F = \frac{m}{a}$

C. $F = \frac{a}{m}$ **D.** $F = ma$

D

4. $I = \frac{E}{R}$ (electricity)

A. $R = \frac{I}{E}$ **B.** $R = IE$

C. $E = \frac{I}{R}$ **D.** $E = RI$

D

Solve each problem. See Example 1.

5. A gas law in chemistry says that

$$\frac{PV}{T} = \frac{pv}{t}.$$

Suppose that $T = 300$, $t = 350$, $V = 9$, $P = 50$, and $v = 8$. Find p.

65.625

6. In work with electric circuits, the formula

$$\frac{1}{a} = \frac{1}{b} + \frac{1}{c}$$

occurs. Find b if $a = 8$ and $c = 12$.

24

7. A formula from anthropology says that

$$c = \frac{100b}{L}.$$

Find L if $c = 80$ and $b = 5$.

$\frac{25}{4}$

8. The gravitational force between two masses is given by

$$F = \frac{GMm}{d^2}.$$

Find M to the nearest thousandth if $F = 10$, $G = 6.67 \times 10^{-11}$, $m = 1$, and $d = 3 \times 10^{-6}$.

1.349

Solve each formula for the specified variable. See Examples 2 and 3.

9. $F = \frac{GMm}{d^2}$ for G (physics)

$G = \frac{Fd^2}{Mm}$

10. $F = \frac{GMm}{d^2}$ for M (physics)

$M = \frac{Fd^2}{Gm}$

11. $\frac{1}{a} = \frac{1}{b} + \frac{1}{c}$ for a (electricity)

$a = \frac{bc}{c + b}$

12. $\frac{1}{a} = \frac{1}{b} + \frac{1}{c}$ for b (electricity)

$b = \frac{ac}{c - a}$, or $b = \frac{-ac}{a - c}$

13. $\frac{PV}{T} = \frac{pv}{t}$ for v (chemistry)

$v = \frac{PVt}{pT}$

14. $\frac{PV}{T} = \frac{pv}{t}$ for T (chemistry)

$T = \frac{PVt}{pv}$

15. $I = \frac{nE}{R + nr}$ for r (engineering)

$r = \frac{nE - IR}{In}$

16. $a = \frac{V - v}{t}$ for V (physics)

$V = at + v$

17. $A = \frac{1}{2}h(b + B)$ for b (mathematics)

$b = \frac{2A}{h} - B$, or $b = \frac{2A - Bh}{h}$

18. $S = \frac{n}{2}(a + \ell)d$ for n (mathematics)

$n = \frac{2S}{(a + \ell)d}$

19. $\frac{E}{e} = \frac{R + r}{r}$ for r (engineering)

$r = \frac{eR}{E - e}$

20. $y = \frac{x + z}{a - x}$ for x

$x = \frac{ay - z}{1 + y}$

21. To solve the equation $m = \frac{ab}{a - b}$ for a, what is the first step?

Multiply each side by $a - b$.

22. Suppose you are asked to solve the equation

$$rp - rq = p + q$$

for r. What is the first step?

Factor out r on the left.

Solve each problem. (Give answers to the nearest tenth if an approximation is needed.) See Examples 4 and 5.

23. On May 23, 2008, the Boston Red Sox were in first place in the East Division of the American League, having won 31 of their first 50 regular season games. If the team continued to win the same fraction of its games, how many games would the Red Sox win for the complete 162-game season? Round your answer to the nearest whole number. (*Source:* www.mlb.com)

100 games

24. During 2004–2005, the ratio of teachers to students in public elementary and secondary schools was approximately 1 to 16. If a public school had 846 students, how many teachers would be at the school if this ratio was valid for that school? Round your answer to the nearest whole number. (*Source*: U.S. National Center for Education Statistics.)

53 teachers

25. Biologists tagged 500 fish in a lake on January 1. On February 1 they returned and collected a random sample of 400 fish, 8 of which had been previously tagged. Approximately how many fish does the lake have based on this experiment?

25,000 fish

26. Suppose that in the experiment of Exercise 25, 10 of the previously tagged fish were collected on February 1. What would be the estimate of the fish population?

20,000 fish

27. On a map of the United States, the distance between Seattle and Durango is 4.125 in. The two cities are actually 1238 miles apart. On this same map, what would be the distance between Chicago and El Paso, two cities that are actually 1606 mi apart? (Source: Universal Map Atlas.) **5.4 in.**

28. On a map of the United States, the distance between Reno and Phoenix is 2.5 in. The two cities are actually 768 miles apart. On this same map, what would be the distance between St. Louis and Jacksonville, two cities that are actually 919 mi apart? (*Source:* Universal Map Atlas.) **3.0 in.**

29. Bruce Johnston's Shelby Cobra uses 5 gal of gasoline to drive 156 mi. He has 3 gal of gasoline in the car, and he wants to know how much more gasoline he will need to drive 300 mi. If we assume that the car continues to use gasoline at the same rate, how many more gallons will he need? **6.6 more gallons**

30. Mike Love's T-bird uses 6 gal of gasoline to drive 141 miles. He has 4 gal of gasoline in the car, and he wants to know how much more gasoline he will need to drive 275 mi. If we assume that the car continues to use gasoline at the same rate, how many more gallons will he need? **7.7 more gallons**

Nurses use proportions to determine the amount of a drug to administer when the dose of the drug is measured in milligrams but the drug is packaged in a diluted form in milliliters. (*Source:* Hoyles, Celia, Richard Noss, and Stefano Pozzi, "Proportional Reasoning in Nursing Practice," *Journal for Research in Mathematics Education,* January 2001.) *For example, to find the number of milliliters of fluid needed to administer* 300 mg *of a drug that comes packaged as* 120 mg *in* 2 mL *of fluid, a nurse sets up the proportion*

$$\frac{120 \text{ mg}}{2 \text{ mL}} = \frac{300 \text{ mg}}{x \text{ mL}},$$

where x represents the amount to administer in milliliters. Use this method to find the correct dose for each prescription.

31. 120 mg of Amakacine packaged as 100 mg in 2-mL vials

2.4 mL

32. 1.5 mg of morphine packaged as 20 mg ampules diluted in 10 mL of fluid

0.75 mL

In geometry, it is shown that two triangles with corresponding angle measures equal, called ***similar triangles,*** *have corresponding sides proportional. For example, in the figure, angle A = angle D, angle B = angle E, and angle C = angle F, so the triangles are similar. Then the following ratios of corresponding sides are equal.*

$$\frac{4}{6} = \frac{6}{9} = \frac{2x + 1}{2x + 5}$$

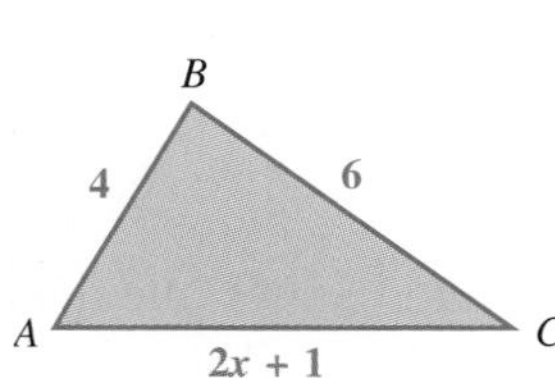

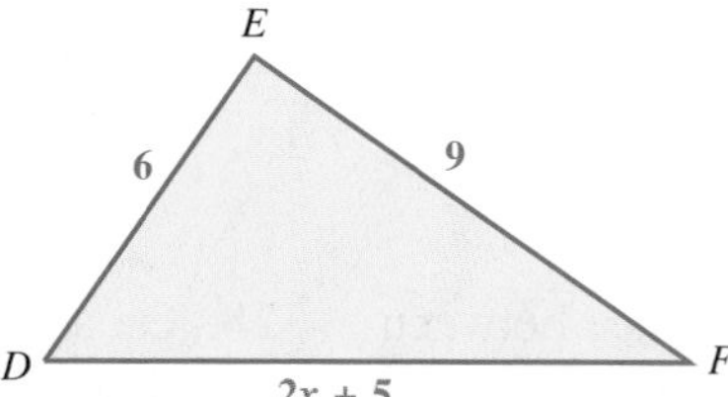

33. Solve for x using the given proportion to find the lengths of the third sides of the triangles.

$x = \frac{7}{2}$; $AC = 8$; $DF = 12$

34. Suppose the following triangles are similar. Find y and the lengths of the two longest sides of each triangle.

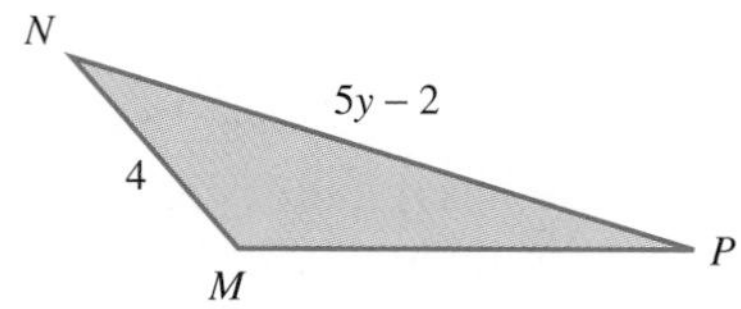

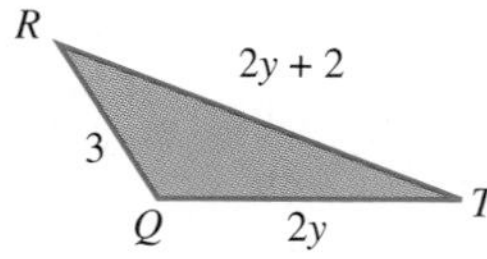

$y = 2$; side $NP = 8$; side $MP = \frac{16}{3}$; side $RT = 6$; side $QT = 4$

Solve each problem. See Example 6.

35. In July 2007, British explorer and endurance swimmer Lewis Gordon Pugh became the first person to swim at the North Pole. To highlight climate change, he swam 0.6 mi in 18.833 min in waters created by melted sea ice. What was his rate (to three decimal places)? (*Source: The Gazette,* July 16, 2007.)

0.032 mi per min

36. In the 2004 Summer Olympics in Athens, Greece, Jody Henry of Australia won the women's 100-m freestyle swimming event. Her rate was 1.854 m per sec. What was her time (to the nearest hundredth of a second)? (*Source: World Almanac and Book of Facts.*)

53.94 sec

37. In 2008, Scott Dixon drove his Target Chip Ganassi Racing car to victory in the Indianapolis 500 (mile) race. His rate was 143.567 mph. What was his time (to the nearest thousandth of an hour)? (*Source:* www.indy500.com) **3.483 hr**

38. In the 2006 Winter Olympics in Torino, Italy, Svetlana Zhurova of Russia won the 500-m speed skating event for women. Her time was 76.57 sec. What was her rate (to three decimal places)? (*Source:* www.espn.com) **6.530 m per sec**

Set up the equation you would use to solve each problem. Do not actually solve the equation. See Examples 7 and 8.

39. Luvenia can row 4 mph in still water. She takes as long to row 8 mi upstream as 24 mi downstream. How fast is the current? (Let x = speed of the current.)

	d	r	t
Upstream	8	$4 - x$	
Downstream	24	$4 + x$	

$$\frac{8}{4 - x} = \frac{24}{4 + x}$$

40. Julio flew his airplane 500 mi against the wind in the same time it took him to fly it 600 mi with the wind. If the speed of the wind was 10 mph, what was the average speed of his plane in still air? (Let x = speed of the plane in still air.)

	d	r	t
Against the Wind	500	$x - 10$	
With the Wind	600	$x + 10$	

$$\frac{500}{x - 10} = \frac{600}{x + 10}$$

Solve each problem. See Examples 7 and 8.

41. Kellen's boat goes 12 mph. Find the rate of the current of the river if she can go 6 mi upstream in the same amount of time she can go 10 mi downstream.

	Distance	Rate	Time
Downstream	10	$12 + x$	
Upstream	6	$12 - x$	

3 mph

42. Kasey can travel 8 mi upstream in the same time it takes her to go 12 mi downstream. Her boat goes 15 mph in still water. What is the rate of the current?

	Distance	Rate	Time
Downstream			
Upstream			

3 mph

43. On his drive from Montpelier, Vermont, to Columbia, South Carolina, Dylan Davis averaged 51 mph. If he had been able to average 60 mph, he would have reached his destination 3 hr earlier. What is the driving distance between Montpelier and Columbia?

1020 mi

44. Leah drove from her apartment to her parents' house for the weekend. Driving to their house on Saturday morning, she was able to average 60 mph because traffic was light. However, returning on Sunday night, she was able to average only 45 mph on the same route, because traffic was heavy. The drive on Sunday took her 1.5 hr longer than the drive on Saturday. What is the distance between Leah's apartment and her parents' house?

270 mi

45. A private plane traveled from San Francisco to a secret rendezvous. It averaged 200 mph. On the return trip, the average speed was 300 mph. If the total traveling time was 4 hr, how far from San Francisco was the secret rendezvous?

480 mi

46. Johnny averages 30 mph when he drives on the old highway to his favorite fishing hole, and he averages 50 mph when most of his route is on the interstate. If both routes are the same length, and he saves 2 hr by traveling on the interstate, how far away is the fishing hole?

150 mi

47. On the first part of a trip to Carmel traveling on the freeway, Marge averaged 60 mph. On the rest of the trip, which was 10 mi longer than the first part, she averaged 50 mph. Find the total distance to Carmel if the second part of the trip took 30 min more than the first part.

190 mi

48. While on vacation, Jim and Annie decided to drive all day. During the first part of their trip on the highway, they averaged 60 mph. When they got to Houston, traffic caused them to average only 30 mph. The distance they drove in Houston was 100 mi less than their distance on the highway. What was their total driving distance if they spent 50 min more on the highway than they did in Houston?

200 mi

Solve each problem. See Example 9.

49. Butch and Peggy want to pick up the mess that their grandson, Grant, has made in his playroom. Butch could do it in 15 min working alone. Peggy, working alone, could clean it in 12 min. How long will it take them if they work together?

	Rate	Time Working Together	Fractional Part of the Job Done
Butch	$\frac{1}{15}$	x	
Peggy	$\frac{1}{12}$	x	

$6\frac{2}{3}$ min

50. Lou can groom Jay Beckenstein's dogs in 8 hr, but it takes his business partner, Janet, only 5 hr to groom the same dogs. How long will it take them to groom Jay's dogs if they work together?

	Rate	Time Working Together	Fractional Part of the Job Done
Lou	$\frac{1}{8}$	x	
Janet	$\frac{1}{5}$	x	

$3\frac{1}{13}$ hr

51. Jerry and Kuba are laying a hardwood floor. Working alone, Jerry can do the job in 20 hr. If the two of them work together, they can complete the job in 12 hr. How long would it take Kuba to lay the floor working alone?

30 hr

52. Mrs. Disher is a high school mathematics teacher. She can grade a set of chapter tests in 5 hr working alone. If her student teacher Mr. Howes helps her, it will take 3 hr to grade the tests. How long would it take Mr. Howes to grade the tests if he worked alone?

$7\frac{1}{2}$ hr

53. If a vat of acid can be filled by an inlet pipe in 10 hr and emptied by an outlet pipe in 20 hr, how long will it take to fill the vat if both pipes are open?

20 hr

54. A winery has a vat to hold Chardonnay. An inlet pipe can fill the vat in 9 hr, while an outlet pipe can empty it in 12 hr. How long will it take to fill the vat if both the outlet and the inlet pipes are open?

36 hr

55. Suppose that Hortense and Mort can clean their entire house in 7 hr, while their toddler, Mimi, just by being around, can completely mess it up in only 2 hr. If Hortense and Mort clean the house while Mimi is at her grandma's, and then start cleaning up after Mimi the minute she gets home, how long does it take from the time Mimi gets home until the whole place is a shambles?

$2\frac{4}{5}$ hr

56. An inlet pipe can fill an artificial lily pond in 60 min, while an outlet pipe can empty it in 80 min. Through an error, both pipes are left open. How long will it take for the pond to fill?

240 min, or 4 hr

7.6 Variation

Certain types of functions are very common, especially in business and the physical sciences. These are functions where y depends on a multiple of x, or y depends on a number divided by x. In such situations, y is said to *vary directly as x* (in the first case) or *vary inversely as x* (in the second case). For example, by the distance formula, the distance traveled varies directly as the rate (or speed) and the time. The simple interest formula and the formulas for area and volume are other familiar examples of *direct variation*.

By contrast, the force required to keep a car from skidding on a curve varies inversely as the radius of the curve. Another example of *inverse variation* is how travel time is inversely proportional to rate or speed.

OBJECTIVES

1. **Write an equation expressing direct variation.**
2. **Find the constant of variation, and solve direct variation problems.**
3. **Solve inverse variation problems.**
4. **Solve joint variation problems.**
5. **Solve combined variation problems.**

OBJECTIVE 1 Write an equation expressing direct variation. The circumference of a circle is given by the formula $C = 2\pi r$, where r is the radius of the circle. See the figure. Circumference is always a constant multiple of the radius. (C is always found by multiplying r by the constant 2π.) Thus,

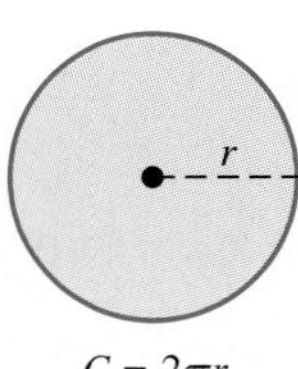

$C = 2\pi r$

As the *radius increases,* the *circumference increases.*

The reverse is also true.

As the *radius decreases,* the *circumference decreases.*

Because of this, the circumference is said to *vary directly* as the radius.

Direct Variation

y varies directly as x if there exists a real number k such that

$$y = kx.$$

Also, y is said to be **proportional to** x. The number k is called the **constant of variation.** In direct variation, for $k > 0$, as the value of x increases, the value of y also increases. Similarly, as x decreases, y decreases.

OBJECTIVE 2 Find the constant of variation, and solve direct variation problems. ***The direct variation equation $y = kx$ defines a linear function, where the constant of variation k is the slope of the line.*** For example, we wrote the equation

$$y = 4.50x$$

to describe the cost y to buy x gallons of gas in Example 8 of **Section 3.4.** The cost varies directly as, or is proportional to, the number of gallons of gas purchased. That is,

As the *number* of gallons of gas *increases, cost increases.*

The reverse is also true.

As the *number* of gallons of gas *decreases, cost decreases.*

The constant of variation k is 4.50, the cost of 1 gallon of gas.

1 Find the constant of variation, and write a direct variation equation.

(a) Ginny Michaud is paid a daily wage. One month she worked 17 days and earned $1334.50.

(b) Distance varies directly as time (at a constant speed). A car travels 100 mi at a constant speed in 2 hr.

2 The charge (in dollars) to customers for electricity (in kilowatt-hours) varies directly as the number of kilowatt-hours used. It costs $52 to use 800 kilowatt-hours. Find the cost to use 1000 kilowatt-hours.

Answers

1. (a) $k = 78.50$; Let E represent her earnings for d days. Then $E = 78.50d$.
   (b) $k = 50$; Let d represent the distance traveled in h hours. Then $d = 50h$.
2. $65

EXAMPLE 1 Finding the Constant of Variation and the Variation Equation

Stella Frolick is paid an hourly wage. One week she worked 43 hr and was paid $795.50. How much does she earn per hour?

Let h represent the number of hours she works and P represent her corresponding pay. Then, P **varies directly as** h, so

$$P = kh.$$

Here k represents Stella's hourly wage. Since $P = 795.50$ when $h = 43$,

$$795.50 = 43k$$

$$k = \mathbf{18.50}. \quad \text{Use a calculator.}$$

This is the constant of variation.

Her hourly wage is $18.50, and P and h are related by

$$P = \mathbf{18.50}h.$$

◀ Work Problem 1 at the Side.

EXAMPLE 2 Solving a Direct Variation Problem

Hooke's law for an elastic spring states that the distance a spring stretches is proportional to the force applied. If a force of 150 newtons* stretches a certain spring 8 cm, how much will a force of 400 newtons stretch the spring?

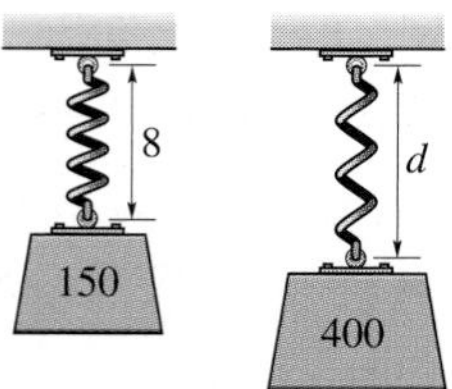

Figure 5

See Figure 5. If d is the distance the spring stretches and f is the force applied, then $d = kf$ for some constant k. Since a force of 150 newtons stretches the spring 8 cm, we can use these values to find k.

$$d = kf \quad \text{Variation equation}$$

$$8 = k \cdot 150 \quad \text{Let } d = 8 \text{ and } f = 150.$$

$$k = \frac{8}{150} \quad \text{Solve for } k.$$

$$k = \frac{4}{75} \quad \text{Lowest terms}$$

Substitute $\frac{4}{75}$ for k in the variation equation $d = kf$ to get

$$d = \frac{4}{75}f.$$

For a force of 400 newtons,

$$d = \frac{4}{75}(400) = \frac{64}{3}. \quad \text{Let } f = 400.$$

The spring will stretch $\frac{64}{3}$ cm, or $21\frac{1}{3}$ cm, if a force of 400 newtons is applied.

◀ Work Problem 2 at the Side.

*A newton is a unit of measure of force used in physics.

In summary, use the following steps to solve a variation problem.

Solving a Variation Problem

Step 1 Write the variation equation.

Step 2 Substitute the initial values and solve for k.

Step 3 Rewrite the variation equation with the value of k from Step 2.

Step 4 Substitute the remaining values, solve for the unknown, and find the required answer.

The direct variation equation $y = kx$ is a linear equation. However, other kinds of variation involve other types of equations. For example, one variable can be proportional to a power of another variable.

Direct Variation as a Power

y* varies directly as the *n*th power of *x if there exists a real number k such that

$$y = kx^n.$$

An example of direct variation as a power is the formula for the area of a circle, $A = \pi r^2$. Here, π is the constant of variation, and the area varies directly as the square of the radius.

EXAMPLE 3 **Solving a Direct Variation Problem**

The distance a body falls from rest varies directly as the square of the time it falls (disregarding air resistance). If a skydiver falls 64 ft in 2 sec, how far will she fall in 8 sec?

Step 1 If d represents the distance the skydiver falls and t the time it takes to fall, then d is a function of t, and

$$d = kt^2$$

for some constant k.

Step 2 To find the value of k, use the fact that the skydiver falls 64 ft in 2 sec.

$$d = kt^2 \quad \text{Variation equation}$$

$$64 = k(2)^2 \quad \text{Let } d = 64 \text{ and } t = 2.$$

$$k = 16 \quad \text{Find } k.$$

Step 3 Using 16 for k, the variation equation becomes

$$d = 16t^2.$$

Step 4 Let $t = 8$ to find the number of feet the skydiver will fall in 8 sec.

$$d = 16(8)^2 = 1024 \quad \text{Let } t = 8.$$

The skydiver will fall 1024 ft in 8 sec.

Work Problem **3** *at the Side.* ▶

3 The area of a circle varies directly as the square of its radius. A circle with radius 3 in. has area 28.278 in.^{2}.

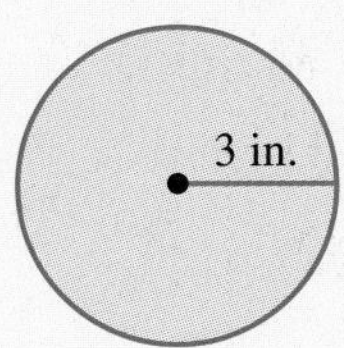

(a) Write a variation equation and give the value of k.

(b) What is the area of a circle with radius 4.1 in.?

Answers

3. (a) $A = kr^2$; 3.142
(b) 52.817 in.^{2} (to the nearest thousandth)

OBJECTIVE 3 Solve inverse variation problems. In direct variation, where $k > 0$, as x increases, y increases. Similarly, as x decreases, y decreases. Another type of variation is *inverse variation*. ***With inverse variation, where k > 0, as one variable increases, the other variable decreases.***

For example, in a closed space, volume decreases as pressure increases, as illustrated by a trash compactor. See Figure 6. As the compactor presses down, the pressure on the trash increases; in turn, the trash occupies a smaller space.

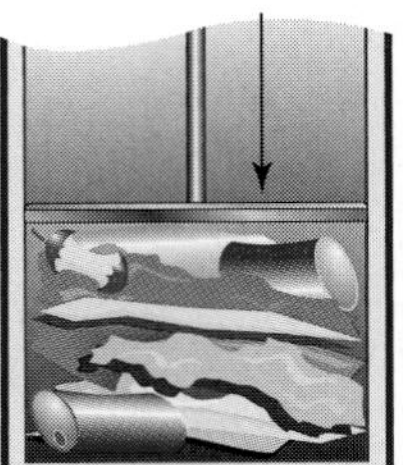

As pressure on trash increases, volume of trash decreases.

Figure 6

Inverse Variation

y* varies inversely as *x if there exists a real number k such that

$$y = \frac{k}{x}.$$

Also, ***y* varies inversely as the *n*th power of *x*** if there exists a real number k such that

$$y = \frac{k}{x^n}.$$

The inverse variation equation also defines a function. Since x is in the denominator, these functions are rational functions, as seen in **Section 7.1.** Another example of inverse variation comes from the distance formula. In its usual form, the formula is

$$d = rt.$$

Dividing each side by r gives

$$t = \frac{d}{r}.$$

Here, t (time) varies inversely as r (rate or speed), with d (distance) serving as the constant of variation. For example, if the distance between Chicago and Des Moines is 300 mi, then

$$t = \frac{300}{r},$$

and the values of r and t might be any of the following.

$$\left.\begin{aligned} r = 50, t = 6 \\ r = 60, t = 5 \\ r = 75, t = 4 \end{aligned}\right\} \text{As } r \text{ increases, } t \text{ decreases.} \qquad \left.\begin{aligned} r = 30, t = 10 \\ r = 25, t = 12 \\ r = 20, t = 15 \end{aligned}\right\} \text{As } r \text{ decreases, } t \text{ increases.}$$

If we *increase* the rate (speed) we drive, time *decreases*. If we *decrease* the rate (speed) we drive, time *increases.*

EXAMPLE 4 **Solving an Inverse Variation Problem**

The weight of an object above Earth varies inversely as the square of its distance from the center of Earth. A space shuttle in an elliptical orbit has a maximum distance from the center of Earth **(apogee)** of 6700 mi. Its minimum distance from the center of Earth **(perigee)** is 4090 mi. See Figure 7. If an astronaut in the shuttle weighs 57 lb at its apogee, what does the astronaut weigh at its perigee?

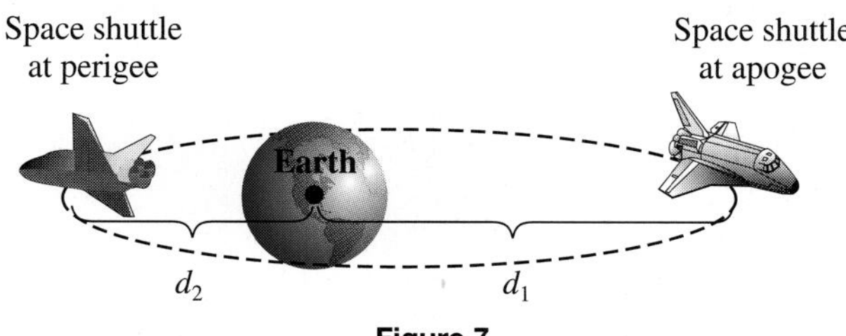

Figure 7

If w is the weight and d is the distance from the center of Earth, then

$$w = \frac{k}{d^2}$$

for some constant k. At the apogee the astronaut weighs 57 lb, and the distance from the center of Earth is 6700 mi. Use these values to find k.

$$57 = \frac{k}{(6700)^2} \quad \text{Let } w = 57 \text{ and } d = 6700.$$

$$k = 57(6700)^2 \quad \text{Solve for } k.$$

Then the weight at the perigee with $d = 4090$ mi is

$$w = \frac{57(6700)^2}{(4090)^2} \approx 153 \text{ lb.} \quad \text{Use a calculator.}$$

Work Problem (4) *at the Side.* ▶

OBJECTIVE 4 **Solve joint variation problems.** It is possible for one variable to depend on several others. If one variable varies directly as the *product* of several other variables (perhaps raised to powers), the first variable is said to *vary jointly* as the others.

Joint Variation

y* varies jointly as *x* and *z if there exists a real number k such that

$$y = kxz.$$

CAUTION

Note that *and* in the expression "y varies jointly as ***x and z***" translates as the product

$$y = kxz.$$

The word *and* does not indicate addition here.

4 If the temperature is constant, the volume of a gas varies inversely as the pressure. For a certain gas, the volume is 10 cm^3 when the pressure is 6 kg per cm^2.

(a) Find the variation equation.

(b) Find the volume when the pressure is 12 kg per cm^2.

ANSWERS

4. (a) $V = \frac{60}{P}$ **(b)** 5 cm^3

5 The volume of a rectangular box of a given height is proportional to its width and length. A box with width 2 ft and length 4 ft has volume 12 ft^3. Find the volume of a box with the same height that is 3 ft wide and 5 ft long.

6 The maximum load that a cylindrical column with a circular cross section can hold varies directly as the fourth power of the diameter of the cross section and inversely as the square of the height. A 9-m column 1 m in diameter will support 8 metric tons. How many metric tons can be supported by a column 12 m high and $\frac{2}{3}$ m in diameter?

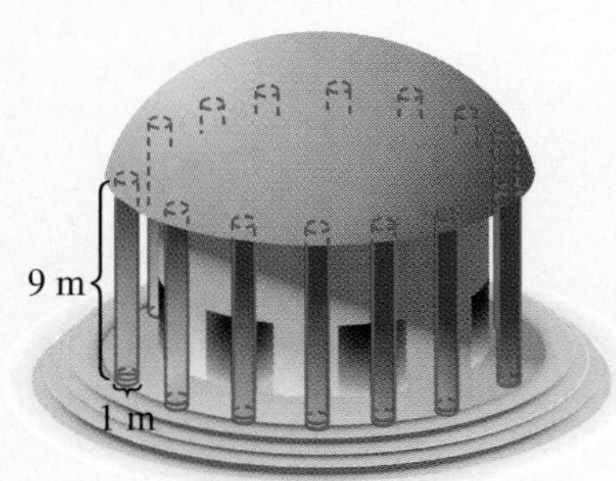

Load = 8 metric tons

EXAMPLE 5 Solving a Joint Variation Problem

The interest on a loan or an investment is given by the formula $I = prt$. Here, for a given principal p, the interest earned I varies jointly as the interest rate r and the time t that the principal is left at interest. If an investment earns \$100 interest at 5% for 2 yr, how much interest will the same principal earn at 4.5% for 3 yr?

We use the formula $I = prt$, where p is the constant of variation because it is the same for both investments. For the first investment,

$$I = prt$$

$$100 = p(0.05)(2) \quad \text{Let } I = 100,\ r = 0.05, \text{ and } t = 2.$$

$$100 = 0.1p$$

$$p = 1000. \quad \text{Divide by } 0.1.$$

Now we find I when $p = 1000$, $r = 0.045$, and $t = 3$.

$$I = 1000(0.045)(3) = 135 \quad \text{Let } p = 1000,\ r = 0.045, \text{ and } t = 3.$$

The interest will be \$135.

◀ Work Problem **5** at the Side.

OBJECTIVE 5 Solve combined variation problems. There are many combinations of direct and inverse variation, called **combined variation.**

EXAMPLE 6 Solving a Combined Variation Problem

Body mass index, or BMI, is used by physicians to assess a person's level of fatness. A BMI from 19 through 25 is considered desirable. BMI varies directly as an individual's weight in pounds and inversely as the square of the individual's height in inches. A man who weighs 118 lb and is 64 in. tall has a BMI of 20. (The BMI is rounded to the nearest whole number.) Find the BMI of a man who weighs 165 lb with a height of 70 in.

Let B represent the BMI, w the weight, and h the height. Then

$$B = \frac{kw}{h^2}. \quad \begin{array}{l}\leftarrow \text{BMI varies directly as the weight.}\\ \leftarrow \text{BMI varies inversely as the square of the height.}\end{array}$$

To find k, let $B = 20$, $w = 118$, and $h = 64$.

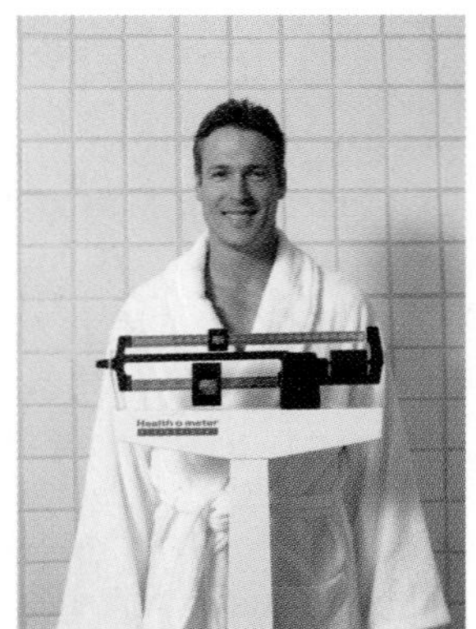

$$20 = \frac{k(118)}{64^2}$$

$$k = \frac{20(64^2)}{118} \quad \text{Multiply by } 64^2; \text{ divide by } 118.$$

$$k \approx 694 \quad \text{Use a calculator.}$$

Now find B when $k = 694$, $w = 165$, and $h = 70$.

$$B = \frac{694(165)}{70^2} \approx 23 \quad \text{Nearest whole number}$$

The man's BMI is 23.

◀ Work Problem **6** at the Side.

ANSWERS

5. 22.5 ft^3
6. $\frac{8}{9}$ metric ton

7.6 Exercises

FOR EXTRA HELP MyMathLab Math XL PRACTICE WATCH DOWNLOAD READ REVIEW

1. **(a)** If the constant of variation is positive and y varies directly as x, then as x increases, y **increases**. (increases/decreases)

   (b) If the constant of variation is positive and y varies inversely as x, then as x increases, y **decreases**. (increases/decreases)

2. Bill Veeck was the owner of several major league baseball teams in the 1950s and 1960s. He was known to often sit in the stands and enjoy games with his paying customers. Here is a quote attributed to him:

   "I have discovered in 20 years of moving around a ballpark, that the knowledge of the game is usually in inverse proportion to the price of the seats."

   Explain in your own words the meaning of this statement. (To prove his point, Veeck once allowed the fans to vote on managerial decisions.)

   The customers in the lower-priced seats know more about the game than those in the higher-priced seats.

Determine whether each equation represents direct, inverse, joint, *or* combined *variation.*

3. $y = \frac{3}{x}$ **inverse**

4. $y = \frac{8}{x}$ **inverse**

5. $y = 10x^2$ **direct**

6. $y = 2x^3$ **direct**

7. $y = 3xz^4$ **joint**

8. $y = 6x^3z^2$ **joint**

9. $y = \frac{4x}{wz}$ **combined**

10. $y = \frac{6x}{st}$ **combined**

Solve each problem. See Examples 2–5.

11. If z varies directly as x, and $z = 30$ when $x = 8$, find z when $x = 4$.

    15

12. If y varies directly as x, and $x = 27$ when $y = 6$, find x when $y = 2$.

    9

13. If d varies directly as r, and $d = 200$ when $r = 40$, find d when $r = 60$.

    300

14. If d varies directly as t, and $d = 150$ when $t = 3$, find d when $t = 5$.

    250

15. If z varies inversely as x, and $z = 50$ when $x = 2$, find z when $x = 25$.

    4

16. If x varies inversely as y, and $x = 3$ when $y = 8$, find y when $x = 4$.

    6

17. If m varies inversely as r, and $m = 12$ when $r = 8$, find m when $r = 16$.

6

18. If p varies inversely as q, and $p = 7$ when $q = 6$, find p when $q = 2$.

21

19. p varies jointly as q and r^2, and $p = 200$ when $q = 2$ and $r = 3$. Find p when $q = 5$ and $r = 2$.

$222\frac{2}{9}$

20. f varies jointly as g^2 and h, and $f = 50$ when $g = 4$ and $h = 2$. Find f when $g = 3$ and $h = 6$.

$84\frac{3}{8}$

*Use personal experience or intuition to determine whether the situation suggests direct or inverse variation.**

21. The rate and the distance traveled by a pickup truck in 3 hr

direct

22. The number of different lottery tickets you buy and your probability of winning that lottery

direct

23. The number of days from now until December 25 and the magnitude of the frenzy of Christmas shopping

inverse

24. Your age and the probability that you believe in Santa Claus

inverse

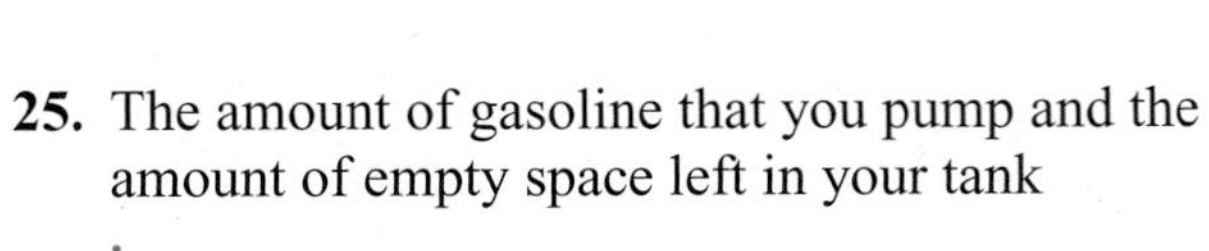

25. The amount of gasoline that you pump and the amount of empty space left in your tank

inverse

26. The surface area of a balloon and its diameter

direct

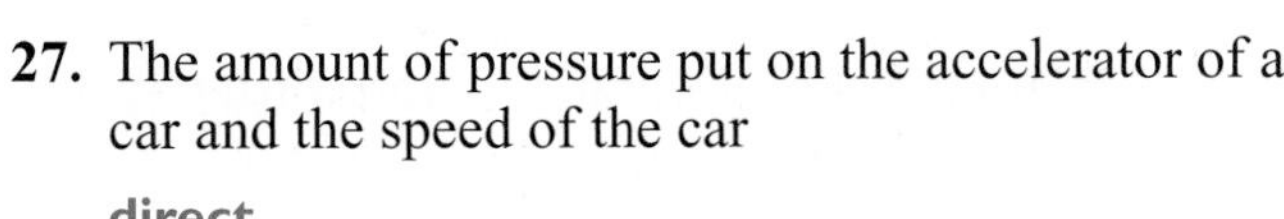

27. The amount of pressure put on the accelerator of a car and the speed of the car

direct

28. The number of days until the end of the baseball season and the number of home runs that Alex Rodriguez has

inverse

---

*The authors thank Linda Kodama of Kapi'olani Community College for suggesting the inclusion of exercises of this type.

Solve each problem involving variation. See Examples 1–6.

29. Matt bought 8 gal of gasoline and paid $36.79. To the nearest tenth of a cent, what is the price of gasoline per gallon?

4.59\frac{9}{10}$

30. Nora gives horseback rides at Shadow Mountain Ranch. A 2.5-hr ride costs $50.00. What is the price per hour?

$20

31. The volume of a can of tomatoes is proportional to the height of the can. If the volume of the can is 300 cm^3 when its height is 10.62 cm, find the volume of a can with height 15.92 cm.

about 450 cm^3

32. The weight of an object on Earth is directly proportional to the weight of that same object on the moon. A 200-lb astronaut would weigh 32 lb on the moon. How much would a 50-lb dog weigh on the moon? **8 lb**

33. For a body falling freely from rest (disregarding air resistance), the distance the body falls varies directly as the square of the time. If an object is dropped from the top of a tower 576 ft high and hits the ground in 6 sec, how far did it fall in the first 4 sec?

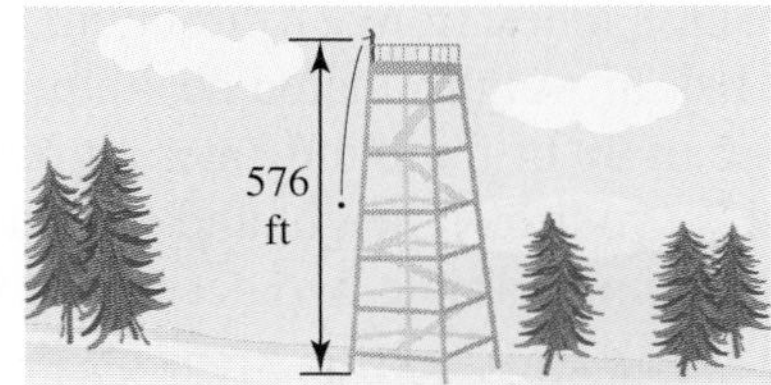

256 ft

34. The amount of water emptied by a pipe varies directly as the square of the diameter of the pipe. For a certain constant water flow, a pipe emptying into a canal will allow 200 gal of water to escape in an hour. The diameter of the pipe is 6 in. How much water would a 12-in. pipe empty into the canal in an hour, assuming the same water flow?

800 gal

35. The current in a simple electrical circuit is inversely proportional to the resistance. If the current is 20 amperes (an **ampere** is a unit for measuring current) when the resistance is 5 ohms, find the current when the resistance is 7.5 ohms.

$13\frac{1}{3}$ amperes

36. The frequency (number of vibrations per second) of a vibrating guitar string varies inversely as its length. That is, a longer string vibrates fewer times in a second than a shorter string. Suppose a guitar string 0.65 m long vibrates 4.3 times per sec. What frequency would a string 0.5 m long have?

5.59 vibrations per sec

37. The amount of light (measured in foot-candles) produced by a light source varies inversely as the square of the distance from the source. If the illumination produced 1 m from a light source is 768 foot-candles, find the illumination produced 6 m from the same source.

$21\frac{1}{3}$ foot-candles

38. The force with which Earth attracts an object above Earth's surface varies inversely with the square of the distance of the object from the center of Earth. If an object 4000 mi from the center of Earth is attracted with a force of 160 lb, find the force of attraction if the object were 6000 mi from the center of Earth.

$71\frac{1}{9}$ lb

39. For a given interest rate, simple interest varies jointly as principal and time. If $2000 left in an account for 4 yr earned interest of $280, how much interest would be earned in 6 yr?

$420

40. The collision impact of an automobile varies jointly as its weight and the square of its speed. Suppose a 2000-lb car traveling at 55 mph has a collision impact of 6.1. What is the collision impact of the same car at 65 mph?

approximately 8.5

41. The force needed to keep a car from skidding on a curve varies inversely as the radius of the curve and jointly as the weight of the car and the square of the speed. If 242 lb of force keep a 2000-lb car from skidding on a curve of radius 500 ft at 30 mph, what force (to the nearest tenth) would keep the same car from skidding on a curve of radius 750 ft at 50 mph?

448.1 lb

42. Almost 70% of the new single-family homes sold in the United States in 2006 used natural gas as the primary heating fuel. (*Source:* U.S. Census Bureau.) The volume of gas varies inversely as the pressure and directly as the temperature. (Temperature must be measured in *Kelvin* (K), a unit of measurement used in physics.) If a certain gas occupies a volume of 1.3 L at 300 K and a pressure of 18 newtons per cm^2, find the volume at 340 K and a pressure of 24 newtons per cm^2.

1.105 L

43. The number of long-distance phone calls between two cities in a certain time period varies jointly as the populations of the cities, p_1 and p_2, and inversely as the distance between them. If 80,000 calls are made between two cities 400 mi apart, with populations of 70,000 and 100,000, how many calls are made between cities with populations of 50,000 and 75,000 that are 250 mi apart?

approximately 68,600 calls

44. A body mass index from 27 through 29 carries a slight risk of weight-related health problems, while one of 30 or more indicates a great increase in risk. Use your own height and weight and the information in Example 6 to determine whether you are at risk.

Answers will vary.

Exercises 45 and 46 describe weight-estimation formulas that fishermen have used over the years. Girth *is the distance around the body of the fish. Give answers to the nearest tenth. (Source: Sacramento Bee,* November 9, 2000.)

45. The weight of a bass varies jointly as its girth and the square of its length. A prize-winning bass weighed in at 22.7 lb and measured 36 in. long with 21 in. girth. How much would a bass 28 in. long with 18 in. girth weigh?

11.8 lb

46. The weight of a trout varies jointly as its length and the square of its girth. One angler caught a trout that weighed 10.5 lb and measured 26 in. long with 18 in. girth. Find the weight of a trout that is 22 in. long with 15 in. girth.

6.2 lb

Chapter 7 Summary

Key Terms

7.1 rational expression A rational expression is the quotient of two polynomials with denominator not 0.

rational function A rational function is a function that is defined by a rational expression in the form $f(x) = \frac{P(x)}{Q(x)}$, where $Q(x) \neq 0$.

7.2 least common denominator (LCD) The least common denominator in a group of denominators is the product of all different factors from each denominator, with each factor raised to the greatest power that occurs in any denominator.

7.3 complex fraction A complex fraction is an expression having a fraction in the numerator, denominator, or both.

7.4 domain of the variable in a rational equation The domain of the variable in a rational equation is the intersection (overlap) of the domains of the rational expressions in the equation.

discontinuous A graph of a function is discontinuous if there are one or more breaks in the graph.

vertical asymptote A rational function in simplest form $f(x) = \frac{P(x)}{x - a}$ has the line $x = a$ as a vertical asymptote; the graph approaches the line on each side but does not intersect it.

horizontal asymptote A horizontal line that a graph approaches as $|x|$ gets larger and larger without bound is called a horizontal asymptote.

7.5 ratio A ratio is a comparison of two quantities using a quotient.

proportion A proportion is a statement that two ratios are equal.

7.6 varies directly y varies directly as x if there exists a real number k such that $y = kx$.

varies inversely y varies inversely as x if there exists a real number k such that $y = \frac{k}{x}$.

constant of variation In the equations for direct and inverse variation, k is the constant of variation.

Test Your Word Power

See how well you have learned the vocabulary in this chapter. Answers, with examples, follow the Quick Review.

1. A **rational expression** is
   A. an algebraic expression made up of a term or the sum of a finite number of terms with real coefficients and integer exponents
   B. a polynomial equation of degree 2
   C. an expression with one or more fractions in the numerator, denominator, or both
   D. the quotient of two polynomials with denominator not zero.

2. In a given set of fractions, the **least common denominator** is
   A. the smallest denominator of all the denominators
   B. the smallest expression that is divisible by all the denominators
   C. the largest integer that evenly divides the numerator and denominator of all the fractions
   D. the largest denominator of all the denominators.

3. A **complex fraction** is
   A. an algebraic expression made up of a term or the sum of a finite number of terms with real coefficients and integer exponents
   B. a polynomial equation of degree 2
   C. an expression with one or more fractions in the numerator, denominator, or both
   D. the quotient of two polynomials with denominator not zero.

4. A **ratio**
   A. compares two quantities using a quotient
   B. says that two quotients are equal
   C. is a product of two quantities
   D. is a difference between two quantities.

5. A **proportion**
   A. compares two quantities using a quotient
   B. says that two ratios are equal
   C. is a product of two quantities
   D. is a difference between two quantities.

▶ Quick Review

Concepts	Examples

7.1 Rational Expressions and Functions; Multiplying and Dividing

Fundamental Property of Rational Numbers

If $\frac{a}{b}$ is a rational number and if c is any nonzero real number, then

$$\frac{a}{b} = \frac{ac}{bc}.$$

$$\frac{3}{4} = \frac{3 \cdot 5}{4 \cdot 5} = \frac{15}{20}$$

Writing a Rational Expression in Lowest Terms

Step 1 Factor the numerator and the denominator completely.

Step 2 Apply the fundamental property.

Write in lowest terms.

$$\frac{2x + 8}{x^2 - 16}$$

$$= \frac{2(x + 4)}{(x - 4)(x + 4)} \quad \text{Factor.}$$

$$= \frac{2}{x - 4} \quad \text{Lowest terms}$$

Multiplying Rational Expressions

Step 1 Factor numerators and denominators.

Step 2 Apply the fundamental property.

Step 3 Multiply the remaining factors in the numerator and in the denominator.

Step 4 Check that the product is in lowest terms.

Multiply.

$$\frac{x^2 + 2x + 1}{x^2 - 1} \cdot \frac{5}{3x + 3}$$

$$= \frac{(x + 1)^2}{(x - 1)(x + 1)} \cdot \frac{5}{3(x + 1)} \quad \text{Factor.}$$

$$= \frac{5}{3(x - 1)} \quad \text{Multiply; lowest terms}$$

Dividing Rational Expressions

Multiply the first rational expression (the dividend) by the reciprocal of the second (the divisor).

Divide.

$$\frac{2x + 5}{x - 3} \div \frac{2x^2 + 3x - 5}{x^2 - 9}$$

$$= \frac{2x + 5}{x - 3} \cdot \frac{x^2 - 9}{2x^2 + 3x - 5} \quad \text{Multiply by the reciprocal.}$$

$$= \frac{2x + 5}{x - 3} \cdot \frac{(x + 3)(x - 3)}{(2x + 5)(x - 1)} \quad \text{Factor.}$$

$$= \frac{x + 3}{x - 1} \quad \text{Multiply; lowest terms}$$

7.2 Adding and Subtracting Rational Expressions

Adding or Subtracting Rational Expressions

Step 1 If the denominators are the same, add or subtract the numerators. Place the result over the common denominator.

If the denominators are different, write all rational expressions with the LCD. Then add or subtract the numerators, and place the result over the common denominator.

Step 2 Make sure that the answer is in lowest terms.

Subtract.

$$\frac{1}{x + 6} - \frac{3}{x + 2}$$

$$= \frac{x + 2}{(x + 6)(x + 2)} - \frac{3(x + 6)}{(x + 6)(x + 2)}$$

$$= \frac{x + 2 - 3(x + 6)}{(x + 6)(x + 2)}$$

$$= \frac{x + 2 - 3x - 18}{(x + 6)(x + 2)}$$

$$= \frac{-2x - 16}{(x + 6)(x + 2)}$$

Concepts	Examples

7.3 Complex Fractions

Simplifying a Complex Fraction

Simplify the complex fraction.

Method 1 Simplify the numerator and denominator separately, as much as possible. Then multiply the numerator by the reciprocal of the denominator. Write the answer in lowest terms.

Method 1

$$\frac{\frac{1}{x^2}-\frac{1}{y^2}}{\frac{1}{x}+\frac{1}{y}} = \frac{\frac{y^2}{x^2y^2}-\frac{x^2}{x^2y^2}}{\frac{y}{xy}+\frac{x}{xy}}$$

$$= \frac{\frac{y^2-x^2}{x^2y^2}}{\frac{y+x}{xy}} = \frac{y^2-x^2}{x^2y^2} \div \frac{y+x}{xy}$$

$$= \frac{(y+x)(y-x)}{x^2y^2} \cdot \frac{xy}{y+x}$$

$$= \frac{y-x}{xy}$$

Method 2 Multiply the numerator and denominator of the complex fraction by the least common denominator of all fractions appearing in the complex fraction. Then simplify the result.

Method 2

$$\frac{\frac{1}{x^2}-\frac{1}{y^2}}{\frac{1}{x}+\frac{1}{y}} = \frac{x^2y^2\left(\frac{1}{x^2}-\frac{1}{y^2}\right)}{x^2y^2\left(\frac{1}{x}+\frac{1}{y}\right)}$$

$$= \frac{y^2-x^2}{xy^2+x^2y}$$

$$= \frac{(y-x)(y+x)}{xy(y+x)}$$

$$= \frac{y-x}{xy}$$

7.4 Equations with Rational Expressions and Graphs

Solving an Equation with Rational Expressions
To solve an equation involving rational expressions, first determine the domain of the variable. Then multiply all the terms in the equation by the least common denominator. Solve the resulting equation. ***Each proposed solution must be checked to see that it is in the domain of the variable in the equation.***

Solve. $\frac{1}{x} + x = \frac{26}{5}$ Note that 0 is excluded from the domain.

$5 + 5x^2 = 26x$ Multiply by $5x$.

$5x^2 - 26x + 5 = 0$ Subtract $26x$.

$(5x - 1)(x - 5) = 0$ Factor.

$5x - 1 = 0$ or $x - 5 = 0$ Zero-factor property

$x = \frac{1}{5}$ or $x = 5$ Solve each equation.

Both check. The solution set is $\{\frac{1}{5}, 5\}$.

The graph of a rational function of the type covered in this section may have one or more breaks. At such points, the graph will approach an asymptote.

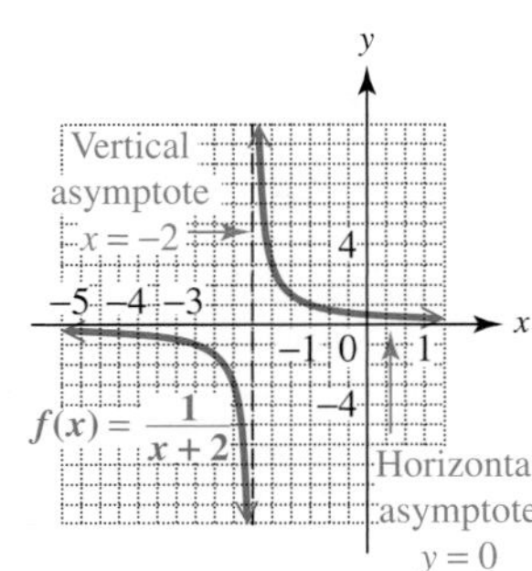

Concepts	Examples

7.5 Applications of Rational Expressions

To solve a motion problem, use the formula

$$d = rt$$

or one of its equivalents,

$$t = \frac{d}{r} \quad \text{or} \quad r = \frac{d}{t}.$$

Solve.

A canal has a current of 2 mph. Find the speed of Amy's boat in still water if it goes 11 mi downstream in the same time that it goes 8 mi upstream.

Let x represent the speed of the boat in still water.

	Distance	Rate	Time
Downstream	11	$x + 2$	$\frac{11}{x+2}$
Upstream	8	$x - 2$	$\frac{8}{x-2}$

Because the times are the same, the equation is

$$\frac{11}{x+2} = \frac{8}{x-2}. \qquad \text{Use } t = \tfrac{d}{r}.$$

$$11(x-2) = 8(x+2) \qquad \text{Multiply by the LCD.}$$

$$11x - 22 = 8x + 16 \qquad \text{Distributive property}$$

$$3x = 38 \qquad \text{Subtract } 8x \text{ and add 22.}$$

$$x = 12\frac{2}{3} \qquad \text{Divide by 3.}$$

The speed in still water is $12\frac{2}{3}$ mph.

To solve a work problem, use the fact that if a complete job is done in t units of time, the rate of work is $\frac{1}{t}$ job per unit of time.

7.6 Variation

If there is some constant k such that

$y = kx^n$, then y varies directly as x^n.

$y = \frac{k}{x^n}$, then y varies inversely as x^n.

$y = kxz$, then y varies jointly as x and z.

The area of a circle **varies directly as** the square of the radius.

$$A = kr^2 \qquad \text{Here, } k = \pi.$$

Pressure **varies inversely as** volume.

$$p = \frac{k}{V}$$

For a given principal, interest **varies jointly as** interest rate and time.

$$I = krt \qquad k \text{ is the given principal.}$$

ANSWERS TO TEST YOUR WORD POWER

1. D; *Examples:* $-\frac{3}{4y^2}, \frac{5x^3}{x+2}, \frac{a+3}{a^2-4a-5}$

2. B; *Example:* The LCD of $\frac{1}{x}, \frac{2}{3},$ and $\frac{5}{x+1}$ is $3x(x+1)$.

3. C; *Examples:* $\frac{\frac{2}{3}}{\frac{4}{7}}, \frac{x-\frac{1}{x}}{x+\frac{1}{y}}, \frac{\frac{2}{a+1}}{a^2-1}$

4. A; *Example:* $\frac{7\text{ in.}}{12\text{ in.}}$ compares two quantities.

5. B; *Example:* The proportion $\frac{2}{3} = \frac{8}{12}$ states that the two ratios are equal.

Chapter 7 Review Exercises

[7.1] ***(a)*** *Find all real numbers that are excluded from the domain.* ***(b)*** *Give the domain using set notation.*

1. $f(x) = \frac{-7}{3x + 18}$

(a) -6

(b) $\{x \mid x \neq -6\}$

2. $f(x) = \frac{5x + 17}{x^2 - 7x + 10}$

(a) $2, 5$

(b) $\{x \mid x \neq 2, 5\}$

3. $f(x) = \frac{9}{x^2 - 18x + 81}$

(a) 9

(b) $\{x \mid x \neq 9\}$

Write in lowest terms.

4. $\frac{12x^2 + 6x}{24x + 12}$

$\frac{x}{2}$

5. $\frac{25m^2 - n^2}{25m^2 - 10mn + n^2}$

$\frac{5m + n}{5m - n}$

6. $\frac{r - 2}{4 - r^2}$

$\frac{-1}{2 + r}$

7. What is meant by the reciprocal of a rational expression?

The reciprocal of a rational expression is another rational expression such that the two rational expressions have a product of 1.

Multiply or divide. Write the answer in lowest terms.

8. $\frac{(2y + 3)^2}{5y} \cdot \frac{15y^3}{4y^2 - 9}$

$\frac{3y^2(2y + 3)}{2y - 3}$

9. $\frac{w^2 - 16}{w} \cdot \frac{3}{4 - w}$

$\frac{-3(w + 4)}{w}$

10. $\frac{z^2 - z - 6}{z - 6} \div \frac{z^2 + 2z - 15}{z^2 - 6z}$

$\frac{z(z + 2)}{z + 5}$

11. $\frac{m^3 - n^3}{m^2 - n^2} \div \frac{m^2 + mn + n^2}{m + n}$

1

[7.2] *Assume that each expression is the denominator of a rational expression. Find the least common denominator for each group.*

12. $32b^3, \quad 24b^5$

$96b^5$

13. $9r^2, \quad 3r + 1$

$9r^2(3r + 1)$

14. $6x^2 + 13x - 5, \quad 9x^2 + 9x - 4$

$(3x - 1)(2x + 5)(3x + 4)$

Add or subtract as indicated.

15. $\frac{8}{z} - \frac{3}{2z^2}$

$\frac{16z - 3}{2z^2}$

16. $\frac{5y + 13}{y + 1} - \frac{1 - 7y}{y + 1}$

12

17. $\frac{6}{5a + 10} + \frac{7}{6a + 12}$

$\frac{71}{30(a + 2)}$

18. $\frac{3r}{10r^2 - 3rs - s^2} + \frac{2r}{2r^2 + rs - s^2}$

$\frac{13r^2 + 5rs}{(5r + s)(2r - s)(r + s)}$

[7.3] *Simplify each complex fraction.*

19. $\dfrac{\frac{3}{t}+2}{\frac{4}{t}-7}$

$\dfrac{3+2t}{4-7t}$

20. $\dfrac{\frac{2}{m-3n}}{\frac{1}{3n-m}}$

-2

21. $\dfrac{\frac{3}{p}-\frac{2}{q}}{\frac{9q^2-4p^2}{qp}}$

$\dfrac{1}{3q+2p}$

22. $\dfrac{x^{-2}-y^{-2}}{x^{-1}-y^{-1}}$

$\dfrac{y+x}{xy}$

[7.4] *Solve each equation.*

23. $\dfrac{1}{t+4}+\dfrac{1}{2}=\dfrac{3}{2t+8}$

$\{-3\}$

24. $\dfrac{-5m}{m+1}+\dfrac{m}{3m+3}=\dfrac{56}{6m+6}$

$\{-2\}$

25. $\dfrac{2}{k-1}-\dfrac{4k+1}{k^2-1}=\dfrac{-1}{k+1}$

$\{0\}$

26. $\dfrac{5}{x+2}+\dfrac{3}{x+3}=\dfrac{x}{x^2+5x+6}$

$\emptyset$

27. After solving the equation

$$\frac{3}{x-3}-\frac{2}{x-2}=\frac{3}{x^2-5x+6},$$

a student got $x=3$ as her final step. She could not understand why the answer in the back of the book was "$\emptyset$," because she checked her algebra several times and was sure that all her algebraic work was correct. Was she wrong or was the answer in the back of the book wrong? Explain.

Although her algebra was correct, 3 is not a solution because it is not in the domain of the variable in the equation. Thus, $\emptyset$ is correct.

28. Explain the difference between simplifying the expression

$$\frac{4}{x}+\frac{1}{2}-\frac{1}{3}$$

and solving the equation

$$\frac{4}{x}+\frac{1}{2}=\frac{1}{3}.$$

In simplifying the expression, we are combining terms to get a single fraction with a denominator of 6*x*, while in solving the equation, we are finding a value for *x* that makes the equation true.

29. Which is the graph of a rational function? Give the equations of its vertical and horizontal asymptotes?

A.

B.

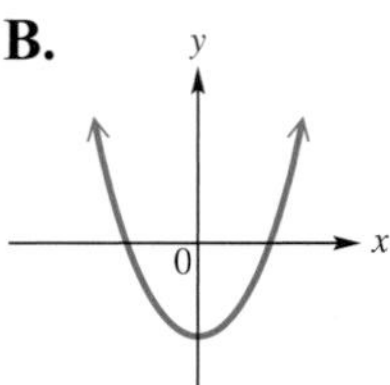

C.

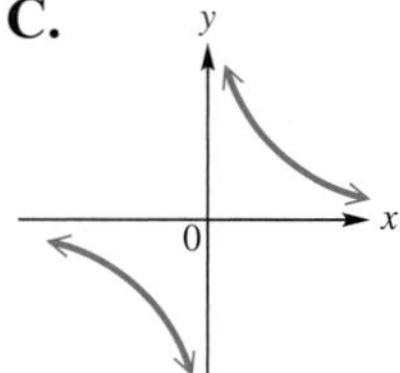

D.

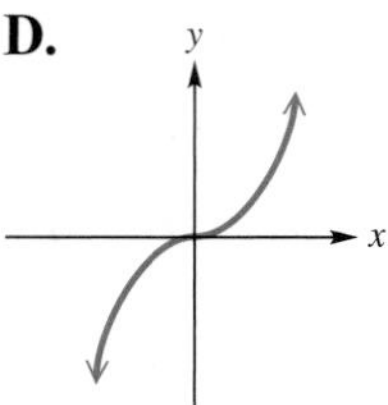

C; $x=0$; $y=0$

[7.5]

30. According to a law from physics, $\dfrac{1}{A}=\dfrac{1}{B}+\dfrac{1}{C}$. Find A if $B=30$ and $C=10$.

$\dfrac{15}{2}$

Solve each formula for the specified variable.

31. $V = \frac{1}{3}\pi r^2 h$ for h (mathematics)

$h = \frac{3V}{\pi r^2}$

32. $\mu = \frac{Mv}{M + m}$ for M (electronics)

$M = \frac{m\mu}{v - \mu}$

Solve each problem.

33. A river has a current of 4 km per hr. Find the speed of Lynn McTernan's boat in still water if it goes 40 km downstream in the same time that it takes to go 24 km upstream.

	d	r	t
Upstream	24	$x - 4$	
Downstream	40		

16 km per hr

34. A sink can be filled by a cold-water tap in 8 min, and filled by the hot-water tap in 12 min. How long would it take to fill the sink with both taps open?

$4\frac{4}{5}$ **min**

[7.6] *Solve each variation problem.*

35. In which one of the following does y vary inversely as x? **C**

A. $y = 2x$ **B.** $y = \frac{x}{3}$ **C.** $y = \frac{3}{x}$ **D.** $y = x^2$

36. If m varies inversely as p^2, and $m = 20$ when $p = 2$, find m when $p = 5$.

$\frac{16}{5}$

37. For the subject in a photograph to appear in the same perspective in the photograph as in real life, the viewing distance must be properly related to the amount of enlargement. For a particular camera, the viewing distance varies directly as the amount of enlargement. A picture taken with this camera that is enlarged 5 times should be viewed from a distance of 250 mm. Suppose a print 8.6 times the size of the negative is made. From what distance should it be viewed? **430 mm**

38. The volume of a rectangular box of a given height is proportional to its width and length. A box with width 4 ft and length 8 ft has volume 64 ft^3. Find the volume of a box with the same height that is 3 ft wide and 6 ft long.

36 ft^3

Mixed Review Exercises

Write in lowest terms.

39. $\frac{x + 2y}{x^2 - 4y^2}$

$\frac{1}{x - 2y}$

40. $\frac{x^2 + 2x - 15}{x^2 - x - 6}$

$\frac{x + 5}{x + 2}$

Perform the indicated operations.

41. $\frac{2}{m} + \frac{5}{3m^2}$

$\frac{6m + 5}{3m^2}$

42. $\frac{k^2 - 6k + 9}{1 - 216k^3} \cdot \frac{6k^2 + 17k - 3}{9 - k^2}$

$\frac{k - 3}{36k^2 + 6k + 1}$

43. $\dfrac{\dfrac{-3}{x} + \dfrac{x}{2}}{1 + \dfrac{x+1}{x}}$

$\dfrac{x^2 - 6}{2(2x + 1)}$

44. $\dfrac{9x^2 + 46x + 5}{3x^2 - 2x - 1} \div \dfrac{x^2 + 11x + 30}{x^3 + 5x^2 - 6x}$

$\dfrac{x(9x + 1)}{3x + 1}$

45. $\dfrac{\dfrac{3}{x} - 5}{6 + \dfrac{1}{x}}$

$\dfrac{3 - 5x}{6x + 1}$

46. $\dfrac{9}{3 - x} - \dfrac{2}{x - 3}$

$\dfrac{11}{3 - x}$, or $\dfrac{-11}{x - 3}$

47. $\dfrac{4y + 16}{30} \div \dfrac{2y + 8}{5}$

$\dfrac{1}{3}$

48. $\dfrac{t^{-2} + s^{-2}}{t^{-1} - s^{-1}}$

$\dfrac{s^2 + t^2}{st(s - t)}$

49. $\dfrac{4a}{a^2 - ab - 2b^2} - \dfrac{6b - a}{a^2 + 4ab + 3b^2}$

$\dfrac{5a^2 + 4ab + 12b^2}{(a + 3b)(a - 2b)(a + b)}$

50. $\dfrac{a}{b} + \dfrac{b}{c} + \dfrac{c}{d}$

$\dfrac{acd + b^2d + bc^2}{bcd}$

Solve each equation.

51. $\dfrac{x + 3}{x^2 - 5x + 4} - \dfrac{1}{x} = \dfrac{2}{x^2 - 4x}$ **$\left\{\dfrac{1}{3}\right\}$**

52. $A = \dfrac{Rr}{R + r}$ for r **$r = \dfrac{AR}{R - A}$, or $r = \dfrac{-AR}{A - R}$**

53. $1 - \dfrac{5}{r} = \dfrac{-4}{r^2}$ **$\{1, 4\}$**

54. $\dfrac{3x}{x - 4} + \dfrac{2}{x} = \dfrac{48}{x^2 - 4x}$ **$\left\{-\dfrac{14}{3}\right\}$**

Solve each problem.

55. Anna and Matthew Sudak need to sort a pile of bottles at the recycling center. Working alone, Anna could do the entire job in 9 hr, while Matthew could do the entire job in 6 hr. How long will it take them if they work together?

$3\frac{3}{5}$ hr

56. Rebecca Song is a college student who lives in an off-campus apartment. Some days she rides her bike to campus, while other days she walks. When she rides her bike, she gets to her first classroom building 36 min faster than when she walks. If her average walking speed is 3 mph and her average biking speed is 12 mph, how far is it from her apartment to the classroom building? **2.4 mi**

57. The frequency (number of vibrations per second) of a vibrating guitar string varies inversely as its length. That is, a longer string vibrates fewer times in a second than a shorter string. Suppose a guitar string 0.65 m long vibrates 4.3 times per sec. What frequency would a string 0.5 m long have?

5.59 vibrations per sec

58. The area of a triangle varies jointly as the lengths of the base and height. A triangle with base 10 ft and height 4 ft has area 20 ft^2. Find the area of a triangle with base 3 ft and height 8 ft.

12 ft^2

Chapter 7 Test

Use the Chapter Test Prep Video CD to see fully worked-out solutions to any of the exercises you want to review.

1. Find all real numbers excluded from the domain of $f(x) = \frac{x+3}{3x^2+2x-8}$. Then give the domain using set notation.

2. Write $\frac{6x^2 - 13x - 5}{9x^3 - x}$ in lowest terms.

Multiply or divide.

3. $\frac{(x+3)^2}{4} \cdot \frac{6}{2x+6}$

4. $\frac{y^2-16}{y^2-25} \cdot \frac{y^2+2y-15}{y^2-7y+12}$

5. $\frac{x^2-9}{x^3+3x^2} \div \frac{x^2+x-12}{x^3+9x^2+20x}$

6. Find the least common denominator for the following group of denominators: $t^2+t-6,\quad t^2+3t,\quad t^2$.

Add or subtract as indicated.

7. $\frac{7}{6t^2} - \frac{1}{3t}$

8. $\frac{9}{x-7} + \frac{4}{x+7}$

9. $\frac{6}{x+4} + \frac{1}{x+2} - \frac{3x}{x^2+6x+8}$

Simplify each complex fraction.

10. $\frac{\frac{12}{r+4}}{\frac{11}{6r+24}}$

11. $\frac{\frac{1}{a}-\frac{1}{b}}{\frac{a}{b}-\frac{b}{a}}$

12. $\frac{2x^{-2}+y^{-2}}{x^{-1}-y^{-1}}$

13. Identify each of the following as an *expression* to be simplified or an *equation* to be solved. Then simplify the one that is an expression, and solve the one that is an equation.

(a) $\frac{2x}{3} + \frac{x}{4} - \frac{11}{2}$

(b) $\frac{2x}{3} + \frac{x}{4} = \frac{11}{2}$

Solve each equation.

14. $\frac{1}{x} - \frac{4}{3x} = \frac{1}{x-2}$

15. $\frac{y}{y+2} - \frac{1}{y-2} = \frac{8}{y^2-4}$

Answers

1. $-2, \frac{4}{3}; \left\{x \mid x \neq -2, \frac{4}{3}\right\}$

2. $\frac{2x-5}{x(3x-1)}$

3. $\frac{3(x+3)}{4}$

4. $\frac{y+4}{y-5}$

5. $\frac{x+5}{x}$

6. $t^2(t+3)(t-2)$

7. $\frac{7-2t}{6t^2}$

8. $\frac{13x+35}{(x-7)(x+7)}$

9. $\frac{4}{x+2}$

10. $\frac{72}{11}$

11. $-\frac{1}{a+b}$

12. $\frac{2y^2+x^2}{xy(y-x)}$

13. (a) expression; $\frac{11(x-6)}{12}$

(b) equation; $\{6\}$

14. $\left\{\frac{1}{2}\right\}$

15. $\{5\}$

16. $x = -1; y = 0$

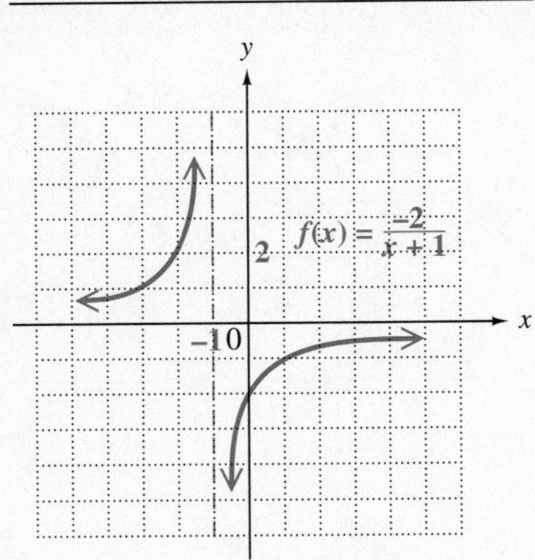

16. Sketch the graph of the function defined by $f(x) = \frac{-2}{x+1}$. Give the equations of its vertical and horizontal asymptotes.

Solve each problem.

17. $3\frac{3}{14}$ hr

17. Wayne can do a job in 9 hr, while Sandra can do the same job in 5 hr. How long would it take them to do the job if they worked together?

18. 15 mph

18. The rate of the current in a stream is 3 mph. Danielle Lalezhar's boat can go 36 mi downstream in the same time that it takes to go 24 mi upstream. Find the rate of her boat in still water.

19. 48,000 fish

19. Biologists collected a sample of 600 fish from Lake Linda on May 1 and tagged each of them. When they returned on June 1, a new sample of 800 fish was collected, and 10 of these had been previously tagged. Use this experiment to determine the approximate fish population of Lake Linda.

20. (a) 3 units

(b) 0

20. In biology, the function defined by

$$g(x) = \frac{5x}{2+x}$$

gives the growth rate g of a population for x units of available food. (*Source:* Smith, J. Maynard, *Models in Ecology,* Cambridge University Press, 1974.)

(a) What amount of food (in appropriate units) would produce a growth rate of 3 units of growth per unit of food?

(b) What is the growth rate if no food is available?

21. 200 amps

21. The current in a simple electrical circuit is inversely proportional to the resistance. If the current is 80 amps when the resistance is 30 ohms, find the current when the resistance is 12 ohms.

22. 0.8 lb

22. The force of the wind blowing on a vertical surface varies jointly as the area of the surface and the square of the velocity. If a wind blowing at 40 mph exerts a force of 50 lb on a surface of 500 ft^2, how much force will a wind of 80 mph place on a surface of 2 ft^2?

Cumulative Review Exercises Chapters R–7

Solve each equation or inequality.

1. $7(2x + 3) - 4(2x + 1) = 2(x + 1)$

$\left\{-\frac{15}{4}\right\}$

2. $0.04x + 0.06(x - 1) = 1.04$

$\{11\}$

3. $\frac{2}{3}x + \frac{5}{12}x \le 20$

$\left(-\infty, \frac{240}{13}\right]$

Solve each problem.

4. Otis Taylor invested some money at 4% interest and twice as much at 3% interest. His interest for the first year was \$400. How much did he invest at each rate?

\$4000 at 4%; \$8000 at 3%

5. A triangle has an area of 42 m^2. The base is 14 m long. Find the height of the triangle.

6 m

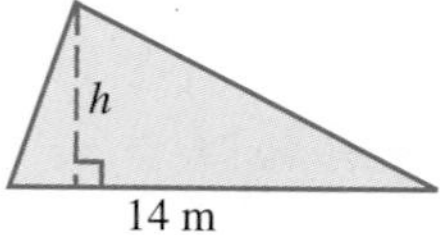

Find the slope of each line.

6. (a) Through $(-5, 8)$ and $(-1, 2)$

$-\frac{3}{2}$

(b) Perpendicular to $4x + 3y = 12$, through $(5, 2)$

$\frac{3}{4}$

7. Write an equation of each line in Exercise 6 in the form $y = mx + b$.

(a) $y = -\frac{3}{2}x + \frac{1}{2}$

(b) $y = \frac{3}{4}x - \frac{7}{4}$

Graph.

8. $-4x + 2y = 8$

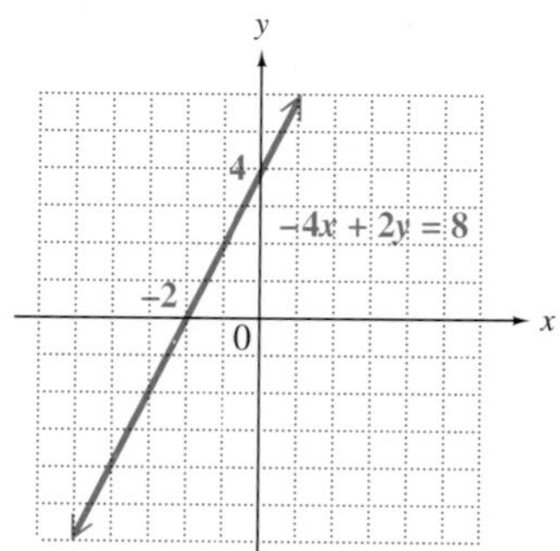

9. $2x + 5y > 10$

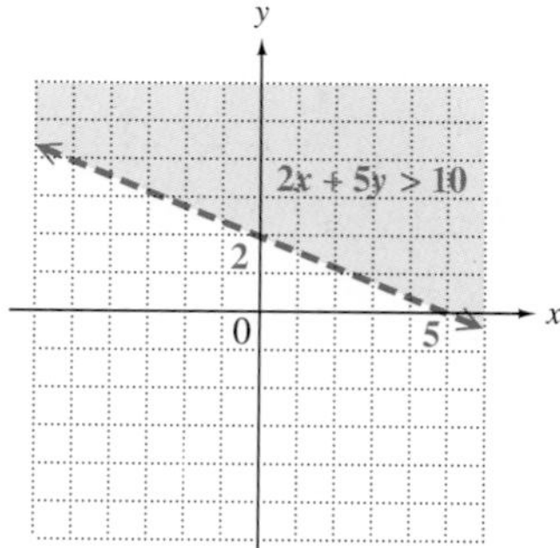

10. Consider the equation $5x - 3y = 8$.

(a) Write y as a function f of x, using function notation $f(x)$.

$f(x) = \frac{5x - 8}{3}$, or $f(x) = \frac{5}{3}x - \frac{8}{3}$

(b) Find $f(1)$. -1

11. If $f(x) = 3x + 6$, what is $f(x + 3)$? $3x + 15$

Solve each system.

12. $4x - y = -7$
$5x + 2y = 1$

$\{(-1, 3)\}$

13. $3x + 4y = 5$
$6x + 7y = 8$

$\{(-1, 2)\}$

14. $2x - y = 6$
$3y = 6x - 18$

$\{(x, y) \mid 2x - y = 6\}$

Simplify. Write each answer with only positive exponents. Assume that all variables represent nonzero real numbers.

15. $\left(\dfrac{a^{-3}b^{4}}{a^{2}b^{-1}}\right)^{-2}$

$\dfrac{a^{10}}{b^{10}}$

16. $\left(\dfrac{m^{-4}n^{2}}{m^{2}n^{-3}}\right) \cdot \left(\dfrac{m^{5}n^{-1}}{m^{-2}n^{5}}\right)$

$\dfrac{m}{n}$

Perform the indicated operations.

17. $(3y^2 - 2y + 6) - (-y^2 + 5y + 12)$

$4y^2 - 7y - 6$

18. $(3x^3 + 13x^2 - 17x - 7) \div (3x + 1)$

$x^2 + 4x - 7$

19. $(4f + 3)(3f - 1)$

$12f^2 + 5f - 3$

20. $(7t^3 + 8)(7t^3 - 8)$

$49t^6 - 64$

21. $\left(\dfrac{1}{4}x + 5\right)^2$

$\dfrac{1}{16}x^2 + \dfrac{5}{2}x + 25$

Factor each polynomial completely.

22. $2x^2 - 13x - 45$

$(2x + 5)(x - 9)$

23. $100t^2 - 25$

$25(2t + 1)(2t - 1)$

24. $8p^3 + 125$

$(2p + 5)(4p^2 - 10p + 25)$

Perform the indicated operations. Express the answer in lowest terms.

25. $\dfrac{2a^2}{a + b} \cdot \dfrac{a - b}{4a}$

$\dfrac{a(a - b)}{2(a + b)}$

26. $\dfrac{x + 4}{x - 2} + \dfrac{2x - 10}{x - 2}$

3

27. $\dfrac{2x}{2x - 1} + \dfrac{4}{2x + 1} + \dfrac{8}{4x^2 - 1}$

$\dfrac{2(x + 2)}{2x - 1}$

Solve.

28. $3x^2 + 4x = 7$

$\left\{-\dfrac{7}{3}, 1\right\}$

29. $\dfrac{-3x}{x + 1} + \dfrac{4x + 1}{x} = \dfrac{-3}{x^2 + x}$

$\{-4\}$

30. $\dfrac{1}{f} = \dfrac{1}{p} + \dfrac{1}{q}$ for q

$q = \dfrac{fp}{p - f}$, or $q = \dfrac{-fp}{f - p}$

8

Equations, Inequalities, and Systems Revisited

The cost of a college education has risen rapidly in the last 15 years. Average higher education tuition and fees for students at public 2-year institutions increased 135% from the 1990–1991 school year to the 2005–2006 school year. (*Source:* National Center for Education Statistics, U.S. Department of Education.)

In Exercises 65–68 in Section 8.2, we use *set operations* and *compound inequalities* to work problems related to college student expenses.

8.1 Review of Solving Linear Equations and Inequalities

OBJECTIVES

1. Solve linear equations.
2. Solve linear inequalities.
3. Solve three-part inequalities.

This section reviews solving linear equations and inequalities in one variable, first introduced in **Chapter 2.**

Recall from **Section 1.2** that an **equation** is a statement that two algebraic expressions are equal. A *linear equation in one variable* involves only real numbers and one variable raised to the first power.

Linear Equation in One Variable

A **linear equation in one variable** can be written in the form

$$Ax + B = C,$$

where A, B, and C are real numbers, with $A \neq 0$.

A linear equation is a **first-degree equation** since the greatest power on the variable is one.

$x + 1 = -2$, $x - 3 = 5$, and $2k + 5 = 10$ Linear equations

$x^2 + 3y = 5$, $\frac{8}{x} = -22$, and $\sqrt{x} = 6$ Nonlinear equations

If the variable in an equation can be replaced by a real number that makes the statement true, then that number is a **solution** of the equation. For example, 8 is a solution of the equation $x - 3 = 5$, since replacing x with 8 gives a true statement, $8 - 3 = 5$. An equation is *solved* by finding its **solution set,** the set of all solutions. The solution set of the equation $x - 3 = 5$ is $\{8\}$.

Equivalent equations are equations that have the same solution set. To solve an equation, we usually start with the given equation and replace it with a series of simpler equivalent equations. For example,

$5x + 2 = 17$, $5x = 15$, and $x = 3$ Equivalent equations

are all equivalent since each has the solution set $\{3\}$.

OBJECTIVE 1 Solve linear equations. We use two important properties to produce equivalent equations.

Addition and Multiplication Properties of Equality

Addition Property of Equality

For all real numbers A, B, and C, the equations

$A = B$ and $A + C = B + C$ are equivalent.

In words, ***the same number may be added to each side of an equation without changing the solution set.***

Multiplication Property of Equality

For all real numbers A and B, and for $C \neq 0$, the equations

$A = B$ and $AC = BC$ are equivalent.

In words, ***each side of an equation may be multiplied by the same nonzero number without changing the solution set.***

Because subtraction and division are defined in terms of addition and multiplication, respectively, these properties can be extended:

The same number may be subtracted from each side of an equation, and each side of an equation may be divided by the same nonzero number, without changing the solution set.

EXAMPLE 1 Solving a Linear Equation

Solve $4x - 2x - 5 = 4 + 6x + 3$.

The goal is to get x alone on one side of the equation.

$$2x - 5 = 7 + 6x \quad \text{Combine like terms.}$$

Next, use the addition property to get the terms with x on the same side of the equation and the remaining terms (the numbers) on the other side. One way to do this is to first subtract $6x$ from each side.

$$2x - 5 - \mathbf{6x} = 7 + 6x - \mathbf{6x} \quad \text{Subtract } 6x.$$

$$-4x - 5 = 7 \quad \text{Combine like terms.}$$

$$-4x - 5 + \mathbf{5} = 7 + \mathbf{5} \quad \text{Add 5.}$$

$$-4x = 12 \quad \text{Combine like terms.}$$

$$\frac{-4x}{\mathbf{-4}} = \frac{12}{\mathbf{-4}} \quad \text{Divide by } -4.$$

$$x = -3 \quad \text{Proposed solution}$$

Check by substituting -3 for x in the *original* equation.

Check

$$4x - 2x - 5 = 4 + 6x + 3 \quad \text{Original equation}$$

$$4(-3) - 2(-3) - 5 \stackrel{?}{=} 4 + 6(-3) + 3 \quad \text{Let } x = -3.$$

$$-12 + 6 - 5 \stackrel{?}{=} 4 - 18 + 3 \quad \text{Multiply.}$$

$$-11 = -11 \quad \text{True}$$

Use parentheses around substituted values to avoid errors.

This is *not* the solution.

The true statement indicates that $\{-3\}$ is the solution set.

Work Problem 1 at the Side. ▶

The steps to solve a linear equation in one variable are as follows.

Solving a Linear Equation in One Variable

Step 1 **Simplify each side separately.** Clear (eliminate) parentheses, fractions, and decimals, using the distributive property as needed, and combine like terms.

Step 2 **Isolate the variable terms on one side.** Use the addition property to get all terms with variables on one side of the equation and all numbers on the other.

Step 3 **Isolate the variable.** Use the multiplication property to get an equation with just the variable (with coefficient 1) on one side.

Step 4 **Check.** Substitute the proposed solution into the original equation to see if a true statement results.

---

1 Solve and check.

(a) $3p + 2p + 1 = -24$

(b) $3p = 2p + 4p + 5$

(c) $4x + 8x = 17x - 9 - 1$

(d) $-7 + 3t - 9t = 12t - 5$

Answers

1. (a) $\{-5\}$ **(b)** $\left\{-\frac{5}{3}\right\}$ **(c)** $\{2\}$ **(d)** $\left\{-\frac{1}{9}\right\}$

2 Solve and check.

(a) $5p + 4(3 - 2p) = 2 + p - 10$

(b) $3(z - 2) + 5z = 2$

(c) $-2 + 3(x + 4) = 8x$

3 Solve and check.

(a) $\frac{2p}{7} - \frac{p}{2} = -3$

(b) $\frac{k+1}{2} + \frac{k+3}{4} = \frac{1}{2}$

Answers

2. (a) $\{5\}$ **(b)** $\{1\}$ **(c)** $\{2\}$
3. (a) $\{14\}$ **(b)** $\{-1\}$

EXAMPLE 2 Solving a Linear Equation

Solve $2(k - 5) + 3k = k + 6$.

Step 1 Use the distributive property to simplify on the left.

$$2(k - 5) + 3k = k + 6$$
$$2k - 10 + 3k = k + 6 \qquad 2(k - 5) = 2(k) - 2(5) = 2k - 10$$
$$5k - 10 = k + 6 \qquad \text{Combine like terms.}$$

Step 2 Next, use the addition property of equality.

$$5k - 10 - k = k + 6 - k \qquad \text{Subtract } k.$$
$$4k - 10 = 6 \qquad \text{Combine like terms.}$$
$$4k - 10 + 10 = 6 + 10 \qquad \text{Add 10.}$$
$$4k = 16 \qquad \text{Combine like terms.}$$

Step 3 Use the multiplication property of equality to get just k on the left.

$$\frac{4k}{4} = \frac{16}{4} \qquad \text{Divide by 4.}$$
$$k = 4$$

Step 4 **Check**

$$2(k - 5) + 3k = k + 6 \qquad \text{Original equation}$$
$$2(4 - 5) + 3(4) \stackrel{?}{=} 4 + 6 \qquad \text{Let } k = 4.$$
$$2(-1) + 12 \stackrel{?}{=} 10$$
$$10 = 10 \qquad \text{True}$$

Always check your work.

The solution checks, so the solution set is $\{4\}$.

◀ Work Problem **2** at the Side.

EXAMPLE 3 Solving a Linear Equation with Fractions

Solve $\frac{x+7}{6} + \frac{2x-8}{2} = -4$.

Step 1

$$6\left(\frac{x+7}{6} + \frac{2x-8}{2}\right) = 6(-4) \qquad \text{Eliminate fractions; multiply by the LCD, 6.}$$
$$6\left(\frac{x+7}{6}\right) + 6\left(\frac{2x-8}{2}\right) = 6(-4) \qquad \text{Distributive property}$$
$$(x + 7) + 3(2x - 8) = -24 \qquad \text{Multiply.}$$
$$x + 7 + 3(2x) - 3(8) = -24 \qquad \text{Distributive property}$$
$$x + 7 + 6x - 24 = -24 \qquad \text{Multiply.}$$
$$7x - 17 = -24 \qquad \text{Combine like terms.}$$

Step 2

$$7x - 17 + 17 = -24 + 17 \qquad \text{Add 17.}$$
$$7x = -7 \qquad \text{Combine like terms.}$$

Step 3

$$\frac{7x}{7} = \frac{-7}{7} \qquad \text{Divide by 7.}$$
$$x = -1$$

Step 4 Check that the solution set is $\{-1\}$.

◀ Work Problem **3** at the Side.

EXAMPLE 4 Solving a Linear Equation with Decimals

Solve $0.06x + 0.09(15 - x) = 0.07(15)$.

Because each decimal number is given in hundredths, multiply each side of the equation by 100. A number can be multiplied by 100 by moving the decimal point two places to the right.

$$0.06x + 0.09(15 - x) = 0.07(15)$$

$$0.06x + 0.09(15 - x) = 0.07(15) \quad \text{Multiply by 100.}$$

Move decimal points 2 places to the right.

$$6x + 9(15 - x) = 7(15)$$

$$6x + 9(15) - 9(x) = 7(15) \quad \text{Distributive property}$$

$$6x + 135 - 9x = 105 \quad \text{Multiply.}$$

$$-3x + 135 = 105 \quad \text{Combine like terms.}$$

$$-3x + 135 - 135 = 105 - 135 \quad \text{Subtract 135.}$$

$$-3x = -30 \quad \text{Combine like terms.}$$

$$\frac{-3x}{-3} = \frac{-30}{-3} \quad \text{Divide by } -3.$$

$$x = 10$$

Check by substituting 10 for x in the original equation.

Check

$$0.06x + 0.09(15 - x) = 0.07(15) \quad \text{Original equation}$$

$$0.06(10) + 0.09(15 - 10) \stackrel{?}{=} 0.07(15) \quad \text{Let } x = 10.$$

$$0.06(10) + 0.09(5) \stackrel{?}{=} 0.07(15)$$

$$0.6 + 0.45 \stackrel{?}{=} 1.05$$

$$1.05 = 1.05 \quad \text{True}$$

The solution set is $\{10\}$.

Work Problem 4 at the Side. ▶

4 Solve and check.

$$0.04x + 0.06(20 - x) = 0.05(50)$$

All of the preceding equations had solution sets containing one element; for example,

$$2(k - 5) + 3k = k + 6$$

has solution set $\{4\}$. This is an example of a *conditional equation,* one which is true only for certain values of the variables. Some linear equations, called *contradictions,* have no solution, while others, called *identities,* have an infinite number of solutions. The table below summarizes these types of equations.

Type of Linear Equation	Number of Solutions	Indication When Solving
Conditional	One	Final line is $x =$ a number. (See Example 5(a).)
Identity	Infinite; solution set {all real numbers}	Final line is true, such as $0 = 0$. (See Example 5(b).)
Contradiction	None; solution set $\emptyset$	Final line is false, such as $-15 = -20$. (See Example 5(c).)

ANSWER

4. $\{-65\}$

5 Solve each equation. Decide whether it is a *conditional equation,* an *identity,* or a *contradiction.*

(a) $5(x+2)-2(x+1)$
$= 3x+1$

(b) $\frac{x+1}{3}+\frac{2x}{3}=x+\frac{1}{3}$

(c) $5(3x+1)=x+5$

EXAMPLE 5 Recognizing Conditional Equations, Identities, and Contradictions

Solve each equation. Decide whether it is a *conditional equation,* an *identity,* or a *contradiction.*

(a)

$$
\begin{aligned}
5(2x+6)-2 &= 7(x+4) \\
10x+30-2 &= 7x+28 && \text{Distributive property} \\
10x+28 &= 7x+28 && \text{Combine like terms.} \\
10x+28-7x-28 &= 7x+28-7x-28 && \text{Subtract } 7x\text{; subtract 28.} \\
3x &= 0 && \text{Combine like terms.} \\
\frac{3x}{3} &= \frac{0}{3} && \text{Divide by 3.} \\
x &= 0
\end{aligned}
$$

The solution set, $\{0\}$, has only one element, so $5(2x+6)-2=7(x+4)$ is a conditional equation.

(b)

$$
\begin{aligned}
5x-15 &= 5(x-3) \\
5x-15 &= 5x-15 && \text{Distributive property} \\
5x-15-5x+15 &= 5x-15-5x+15 && \text{Subtract } 5x\text{; add 15.} \\
0 &= 0 && \text{True}
\end{aligned}
$$

The final line, the *true* statement $0=0$, indicates that the solution set is {all real numbers}, and the equation $5x-15=5(x-3)$ is an identity. (Notice that the first step yielded $5x-15=5x-15$, which is true for all values of x. We could have identified the equation as an identity at that point.)

(c)

$$
\begin{aligned}
5x-15 &= 5(x-4) \\
5x-15 &= 5x-20 && \text{Distributive property} \\
5x-15-5x &= 5x-20-5x && \text{Subtract } 5x. \\
-15 &= -20 && \text{False}
\end{aligned}
$$

Since the result, $-15=-20$, is *false,* the equation has no solution. The solution set is $\emptyset$, so the equation $5x-15=5(x-4)$ is a contradiction.

◀ Work Problem 5 at the Side.

Recall from **Section 2.7** that an **inequality** is a statement with algebraic expressions related by $<$, $\leq$, $>$, or $\geq$.

Linear Inequality

A **linear inequality in one variable** can be written in the form

$$Ax+B<C,$$

where A, B, and C are real numbers, with $A \neq 0$.

$x+5<2$, $t-3\geq 5$, and $2k+5\leq 10$ Linear inequalities

(All definitions and rules are also valid for $>$, $\leq$, and $\geq$.)

ANSWERS

5. (a) contradiction; $\emptyset$
(b) identity; {all real numbers}
(c) conditional; $\{0\}$

OBJECTIVE 2 Solve linear inequalities. We solve an inequality by finding all numbers that make the inequality true. Usually, an inequality has an infinite number of solutions. We find these solutions by using the *addition property of inequality* to produce equivalent inequalities.

Addition Property of Inequality

For all real numbers A, B, and C, the inequalities

$$A < B \quad \text{and} \quad A + C < B + C$$

are equivalent.

In words, adding the same number to each side of an inequality does not change the solution set.

EXAMPLE 6 Using the Addition Property of Inequality

Solve $x - 7 < -12$, and graph the solution set.

$$x - 7 < -12$$

$$x - 7 + 7 < -12 + 7 \quad \text{Add 7.}$$

$$x < -5$$

A graph of the solution set, written $(-\infty, -5)$ using interval notation (introduced in **Section 2.7**), is shown in Figure 1.

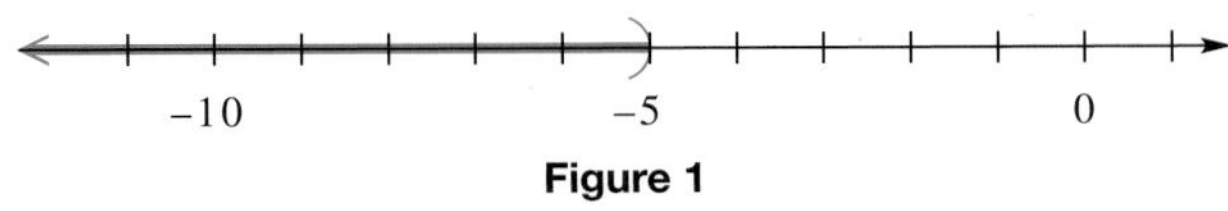

Figure 1

As with equations, the addition property of inequality can be used to *subtract* the same number from each side of an inequality. For example, to solve the inequality $x + 4 > 10$, we subtract 4 from each side to get $x > 6$.

Work Problem 6 at the Side. ▶

The *multiplication property of inequality* has two parts.

Multiplication Property of Inequality

For all real numbers A, B, and C, with $C \neq 0$,

(a) the inequalities

$$A < B \quad \text{and} \quad AC < BC \quad \text{are equivalent } \textbf{if } C > 0;$$

(b) the inequalities

$$A < B \quad \text{and} \quad AC > BC \quad \text{are equivalent } \textbf{if } C < 0.$$

In words, each side of an inequality may be multiplied (or divided) by a *positive* number without changing the direction of the inequality symbol. ***Multiplying (or dividing) by a negative number requires that we reverse the direction of the inequality symbol.***

6 Solve each inequality, check your solutions, and graph the solution set.

(a) $x - 3 < -9$

(b) $p + 6 < 8$

ANSWERS

6. (a) $(-\infty, -6)$

−6 −2 0 2

(b) $(-\infty, 2)$

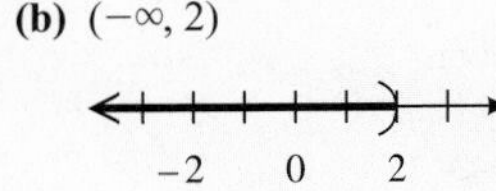

7 Solve, check, and graph the solution set of each inequality.

(a) $2x < -10$

(b) $-7k \geq 8$

(c) $-9m < -81$

EXAMPLE 7 Using the Multiplication Property of Inequality

Solve each inequality, and graph the solution set.

(a) $5m \leq -30$

Use the multiplication property to divide each side by 5. ***Since 5 is positive, do not reverse the direction of inequality symbol.***

$$5m \leq -30$$

$$\frac{5m}{5} \leq \frac{-30}{5} \quad \text{Divide by 5.}$$

$$m \leq -6$$

The solution set is the interval $(-\infty, -6]$, graphed in Figure 2.

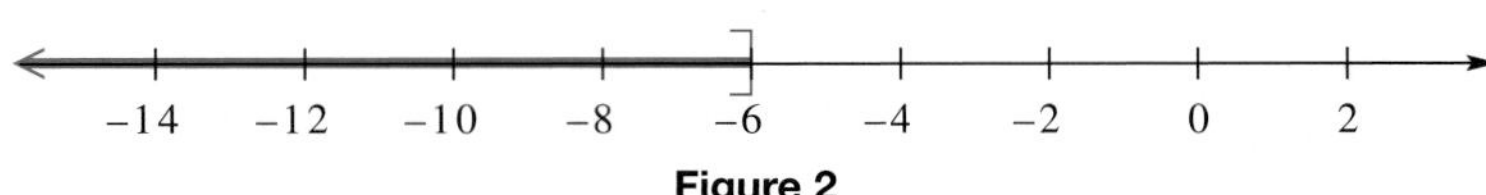

Figure 2

(b) $-4k \leq 32$

Divide each side by -4. ***Since −4 is negative, reverse the direction of inequality symbol.***

$$-4k \leq 32$$

$$\frac{-4k}{-4} \geq \frac{32}{-4} \quad \text{Divide by } -4\text{; reverse the direction of the symbol.}$$

$$k \geq -8$$

Reverse the inequality symbol when dividing by a negative number.

Figure 3 shows the graph of the solution set, $[-8, \infty)$.

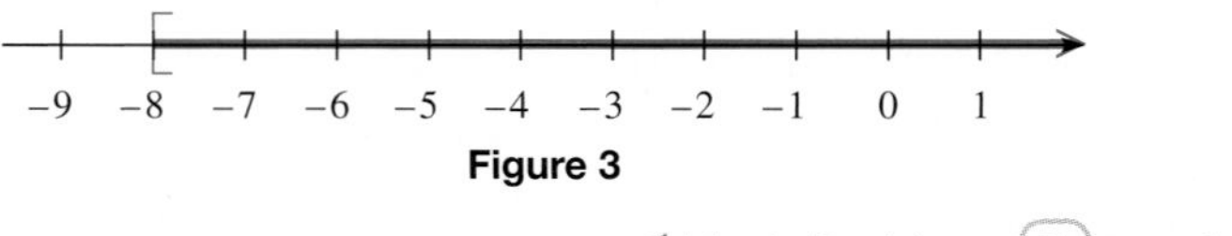

Figure 3

◀ *Work Problem* **7** *at the Side.*

The steps to solve a linear inequality in one variable are as follows.

Solving a Linear Inequality

Step 1 **Simplify each side separately.** Clear parentheses, fractions, and decimals using the distributive property as needed, and combine like terms.

Step 2 **Isolate the variable terms on one side.** Use the addition property of inequality to get all terms with variables on one side of the inequality and all numbers on the other side.

Step 3 **Isolate the variable.** Use the multiplication property of inequality to change the inequality to the form $x < k$ or $x > k$.

CAUTION

Reverse the direction of the inequality symbol only when multiplying or dividing each side of an inequality by a negative number.

ANSWERS

7. (a) $(-\infty, -5)$

−5 −3 −1 0

(b) $\left(-\infty, -\frac{8}{7}\right]$

$-\frac{8}{7}$

−3 −1 0 1

(c) $(9, \infty)$

−3 0 3 6 9

EXAMPLE 8 Solving a Linear Inequality

Solve $-3(x+4)+2 \geq 7-x$, and graph the solution set.

Step 1	$-3(x+4)+2 \geq 7-x$	
	$-3x-12+2 \geq 7-x$	Distributive property
	$-3x-10 \geq 7-x$	Combine like terms.
Step 2	$-3x-10+x \geq 7-x+x$	Add x.
	$-2x-10 \geq 7$	Combine like terms.
	$-2x-10+10 \geq 7+10$	Add 10.
	$-2x \geq 17$	Combine like terms.
Step 3	$\frac{-2x}{-2} \leq \frac{17}{-2}$	Divide by -2; change $\geq$ to $\leq$.
	$x \leq -\frac{17}{2}$	

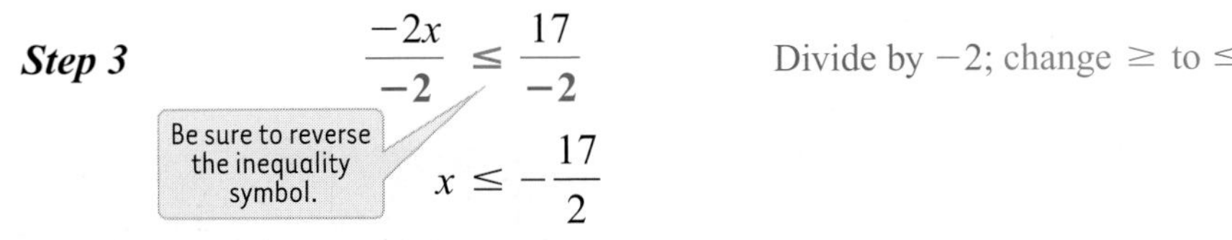

Figure 4 shows the graph of the solution set, $(-\infty, -\frac{17}{2}]$.

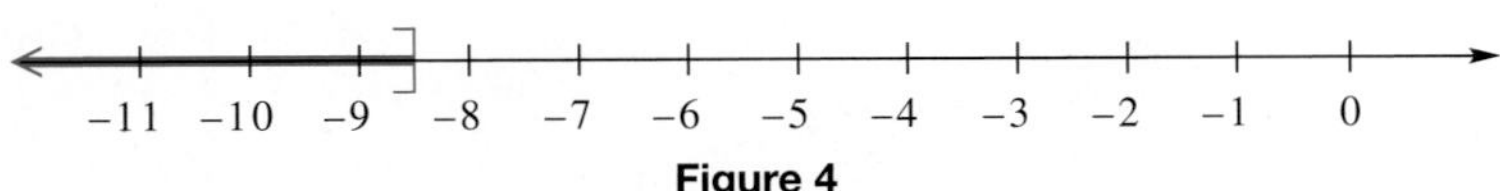

Figure 4

Work Problem 8 at the Side. ▶

OBJECTIVE 3 Solve three-part inequalities. An inequality that says that one number is *between* two other numbers is a *three-part inequality*. (See **Section 2.7.**)

$$-1 < t < 8, \quad 4 \leq 3x - 5 < 6 \qquad \text{Three-part inequalities}$$

To solve a three-part inequality, we work with all three parts at the same time.

EXAMPLE 9 Solving a Three-Part Inequality

Solve $-2 \leq -3k - 1 \leq 5$, and graph the solution set.

Begin by adding 1 to each of the three parts to isolate the variable term in the middle.

$-2+1 \leq -3k-1+1 \leq 5+1$	Add 1 to each part.
$-1 \leq -3k \leq 6$	
$\frac{-1}{-3} \geq \frac{-3k}{-3} \geq \frac{6}{-3}$	Divide each part by -3; reverse the direction of the inequality symbols.
$\frac{1}{3} \geq k \geq -2$	Simplify.
$-2 \leq k \leq \frac{1}{3}$	Rewrite in order based on the number line.

The solution set is the closed interval $[-2, \frac{1}{3}]$. See Figure 5.

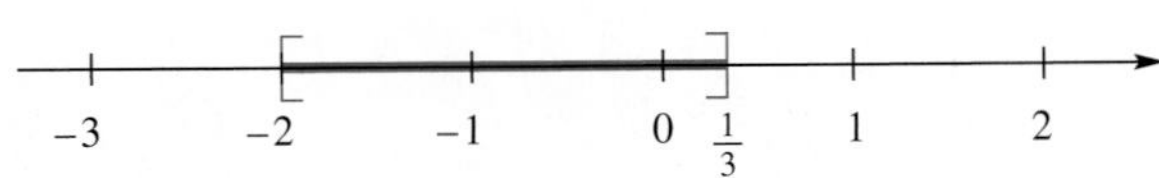

Figure 5

Work Problem 9 at the Side. ▶

---

8 Solve, check, and graph the solution set of each inequality.

(a) $x + 4(2x - 1) \geq x + 2$

(b) $m - 2(m - 4) \leq 3m$

9 Solve, check, and graph the solution set of each inequality.

(a) $-3 \leq x - 1 \leq 7$

(b) $5 < 3x - 4 < 9$

ANSWERS

8. (a) $\left[\frac{3}{4}, \infty\right)$

$\frac{3}{4}$

−1 0 1 2 3

(b) $[2, \infty)$

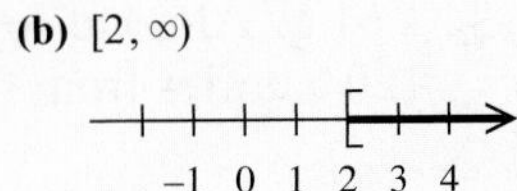

9. (a) $[-2, 8]$

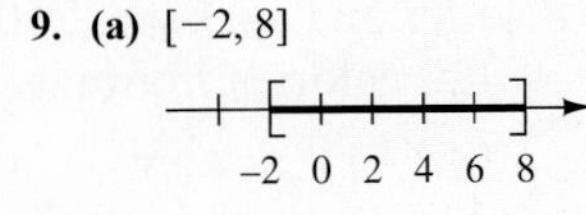

(b) $\left(3, \frac{13}{3}\right)$

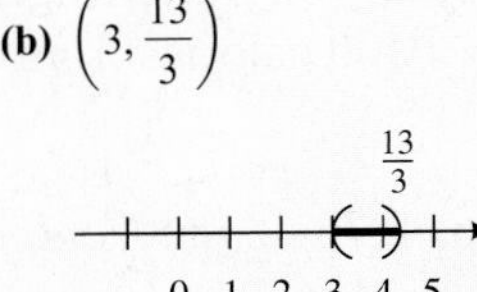

Math in the Media

LEARN MATH, LOSE WEIGHT

The 2004 movie *Mean Girls* stars Lindsay Lohan as Cady Heron, a teenage girl who has been home-schooled until her senior year in high school. A scene in the school cafeteria features her sitting with the Plastics (the "mean girls" of the title). Regina George, played by Rachel McAdams, is reading a candy bar wrapper.

REGINA: *120 calories and 48 calories from fat. What percent is that? I'm only eating food with less than 30% calories from fat.*

CADY: *It's 40%. (Responding to a quizzical look from Regina.) Well, 48 over 120 equals x over 100, and then you cross-multiply and get the value of x.*

REGINA: *Whatever . . . I'm getting cheese fries.*

1. Show that Cady's answer is correct. Let x represent the percent, set up the equation, and solve it. Show all steps.

$$\frac{48}{120} = \frac{x}{100}$$

$$120x = 4800$$

$$x = \frac{4800}{120}$$

$$x = 40$$

The candy bar has $\frac{40}{100}$ = 40% calories from fat.

Use Cady's method to find the percent calories from fat for each of the following candy bars. Round to the nearest percent.

2. 2.05 oz Milky Way: 260 calories, with 90 calories from fat **35%**

3. 1.61 oz Almond Joy: 220 calories, with 120 calories from fat **55%**

4. 1.85 oz PayDay: 250 calories, with 120 calories from fat **48%**

5. 2.1 oz Butterfinger: 270 calories, with 100 calories from fat **37%**

6. 1.76 oz Snickers: 230 calories, with 100 calories from fat **43%**

8.1 Exercises

FOR EXTRA HELP

Math XL PRACTICE

WATCH

DOWNLOAD

READ

REVIEW

1. Which equations are linear equations in x?

A. $3x + x - 2 = 0$ **B.** $12 = x^2$

C. $9x - 4 = 9$ **D.** $\frac{1}{8}x - \frac{1}{x} = 0$

A and C

2. Which of the equations in Exercise 1 are nonlinear equations in x? Explain why.

B is nonlinear because the variable is squared, and D is nonlinear because there is a variable in the second denominator.

3. The following work contains a common student error.

$$8x - 2(2x - 3) = 3x + 7$$

$8x - 4x - 6 = 3x + 7$ Distributive property

$4x - 6 = 3x + 7$ Combine like terms.

$x = 13$ Subtract $3x$; add 6.

WHAT WENT WRONG? Give the correct solution.

The student made a sign error when the distributive property was applied. The left side of the second line should be $8x - 4x + 6$. The correct solution is 1.

4. Identify each as an *expression* or an *equation*.

(a) $5x = 10$ **equation**

(b) $5x + 10$ **expression**

(c) $5x + 6(x - 3) = 12x + 6$ **equation**

(d) $5x + 6(x - 3) - (12x + 6)$ **expression**

Solve and check each equation. See Examples 1 and 2.

5. $9x + 10 = 1$
{−1}

6. $7x - 4 = 31$
{5}

7. $5x + 2 = 3x - 6$
{−4}

8. $9p + 1 = 7p - 9$
{−5}

9. $7x - 5x + 15 = x + 8$
{−7}

10. $2x + 4 - x = 4x - 5$
{3}

11. $12w + 15w - 9 + 5 = -3w + 5 - 9$
{0}

12. $-4t + 5t - 8 + 4 = 6t - 4$
{0}

13. $-5(x + 1) + 3x + 2 = 6x + 4$
$\left\{-\frac{7}{8}\right\}$

14. $5(x + 3) + 4x - 5 = 4 - 2x$
$\left\{-\frac{6}{11}\right\}$

15. $3(2w + 1) - 2(w - 2) = 5$
$\left\{-\frac{1}{2}\right\}$

16. $4(x - 2) + 2(x + 3) = 6$
$\left\{\frac{4}{3}\right\}$

17. $6p - 4(3 - 2p) = 5(p - 4) - 10$
$\{-2\}$

18. $-2k - 3(4 - 2k) = 2(k - 3) + 2$
$\{4\}$

19. $2[w - (2w + 4) + 3] = 2(w + 1)$
$\{-1\}$

20. $4[2t - (3 - t) + 5] = -(2 + 7t)$
$\left\{-\frac{10}{19}\right\}$

Solve and check each equation. See Examples 3 and 4.

21. $\frac{3}{4}x + \frac{5}{2}x = 13$
$\{4\}$

22. $\frac{8}{3}x - \frac{1}{2}x = -13$
$\{-6\}$

23. $\frac{1}{5}x - 2 = \frac{2}{3}x - \frac{2}{5}x$
$\{-30\}$

24. $\frac{3}{4}x - \frac{1}{3}x = \frac{5}{6}x - 5$
$\{12\}$

25. $\frac{3x - 1}{4} + \frac{x + 3}{6} = 3$
$\{3\}$

26. $\frac{3x + 2}{7} - \frac{x + 4}{5} = 2$
$\{11\}$

27. $0.05x + 0.12(x + 5000) = 940$
$\{2000\}$

28. $0.09k + 0.13(k + 300) = 61$
$\{100\}$

29. $0.02(50) + 0.08r = 0.04(50 + r)$
$\{25\}$

30. $0.20(14{,}000) + 0.14t = 0.18(14{,}000 + t)$
$\{7000\}$

31. $0.006(x + 2) = 0.007x + 0.009$
$\{3\}$

32. $0.004x + 0.006(50 - x) = 0.004(68)$
$\{14\}$

Decide whether each equation is a conditional equation, *an* identity, *or a* contradiction. *Give the solution set. See Example 5.*

33. $-x + 4x - 9 = 3(x - 4) - 5$
contradiction; $\emptyset$

34. $-12x + 2x - 11 = -2(5x - 3) + 4$
contradiction; $\emptyset$

35. $-11x + 4(x - 3) + 6x = 4x - 12$
conditional; $\{0\}$

36. $3x - 5(x + 4) + 9 = -11 + 15x$
conditional; $\{0\}$

37. $-2(t + 3) - t - 4 = -3(t + 4) + 2$
identity; {all real numbers}

38. $4(2d + 7) = 2d + 25 + 3(2d + 1)$
identity; {all real numbers}

Match each inequality in Column I with the correct graph or interval in Column II.

I	II
39. $x \leq 3$ D	A. 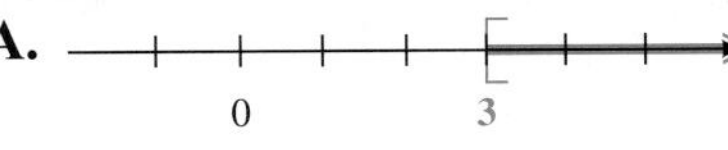
40. $x > 3$ C	B.
41. $x < 3$ B	C. $(3, \infty)$
42. $x \geq 3$ A	D. $(-\infty, 3]$
43. $-3 \leq x \leq 3$ F	E. $(-3, 3)$
44. $-3 < x < 3$ E	F. $[-3, 3]$

Solve each inequality, giving solution sets in both interval and graph forms. Check your answers. See Examples 6–8.

45. $x - 4 \leq 3$

$(-\infty, 7]$

46. $t - 3 \leq 1$

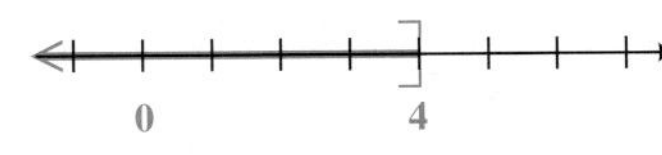

$(-\infty, 4]$

47. $4x + 1 \geq 21$

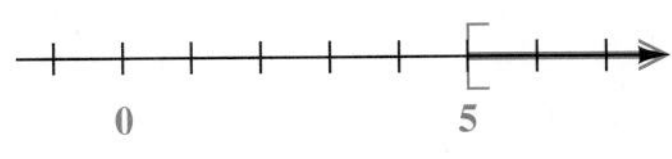

$[5, \infty)$

48. $5t + 2 \geq 52$

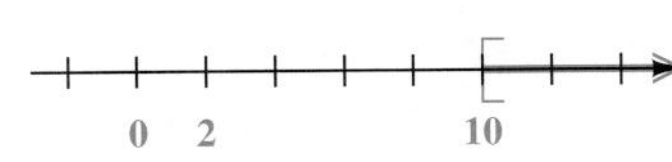

$[10, \infty)$

49. $5x > -25$

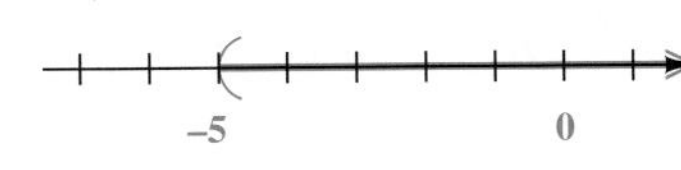

$(-5, \infty)$

50. $7x < -28$

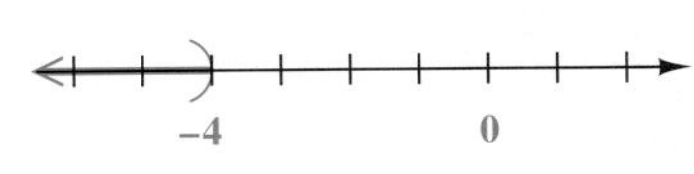

$(-\infty, -4)$

51. $-4x < 16$

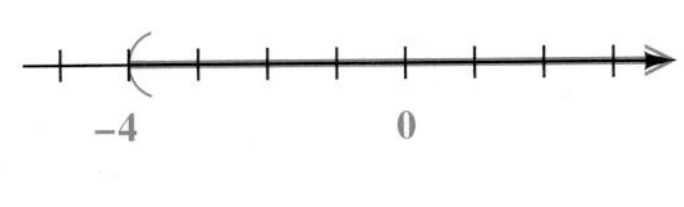

$(-4, \infty)$

52. $-2m > 10$

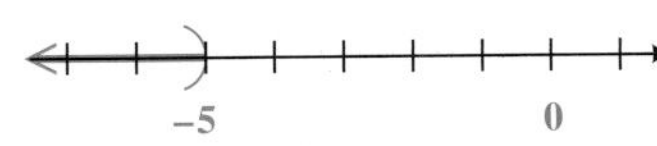

$(-\infty, -5)$

53. $-\frac{3}{4}r \geq 30$

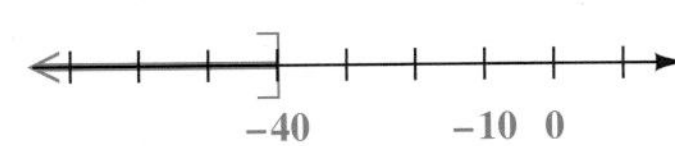

$(-\infty, -40]$

54. $-\frac{2}{3}x \leq 12$

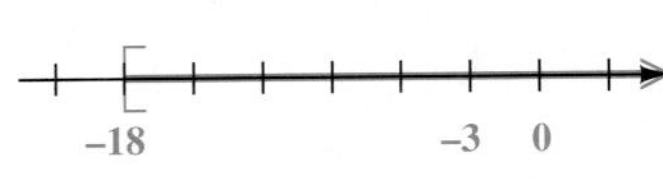

$[-18, \infty)$

55. $\frac{3k - 1}{4} > 5$

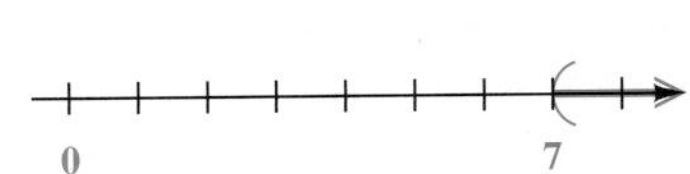

$(7, \infty)$

56. $\frac{5z - 6}{8} < 8$

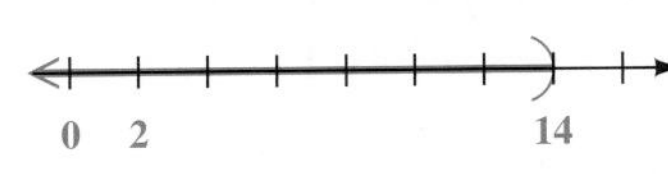

$(-\infty, 14)$

57. $\frac{2k - 5}{-4} > 5$

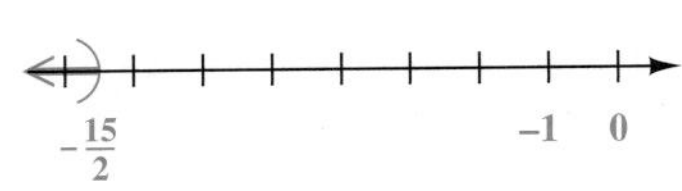

$\left(-\infty, -\frac{15}{2}\right)$

58. $\frac{3z - 2}{-5} < 6$

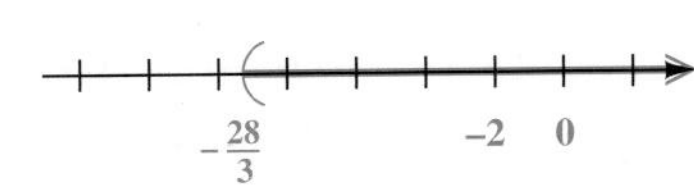

$\left(-\frac{28}{3}, \infty\right)$

59. $3k + 1 < -20$

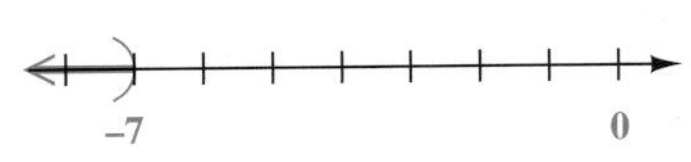

$(-\infty, -7)$

60. $5z + 6 > -29$

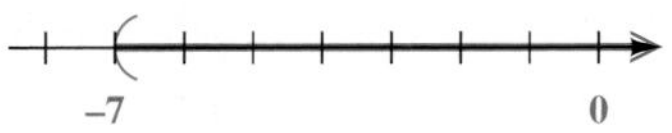

(−7, ∞)

61. $x + 4(2x - 1) \geq x$

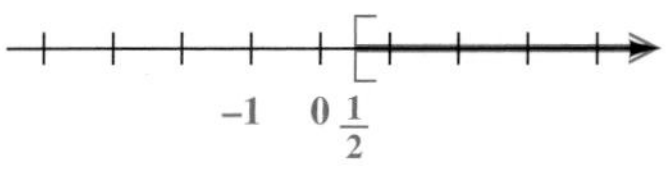

$\left[\frac{1}{2}, \infty\right)$

62. $m - 2(m - 4) \leq 3m$

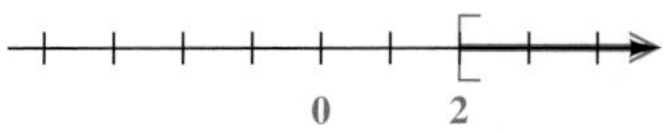

[2, ∞)

63. $-(4 + r) + 2 - 3r < -14$

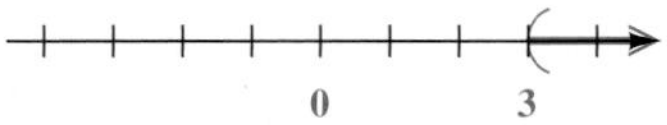

(3, ∞)

64. $-(9 + k) - 5 + 4k \geq 4$

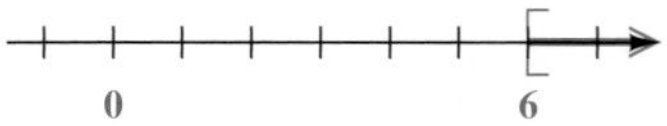

[6, ∞)

65. $-3(z - 6) > 2z - 2$

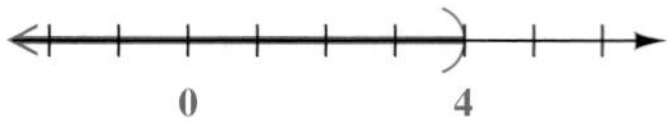

(−∞, 4)

66. $-2(x + 4) \leq 6x + 16$

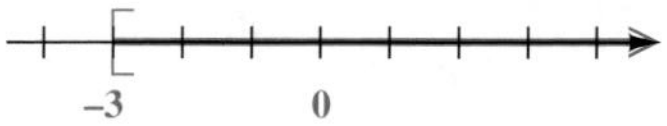

[−3, ∞)

67. $\frac{2}{3}(3k - 1) \geq \frac{3}{2}(2k - 3)$

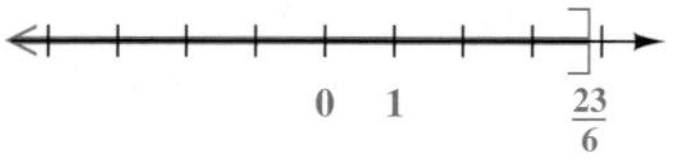

$\left(-\infty, \frac{23}{6}\right]$

68. $\frac{7}{5}(10m - 1) < \frac{2}{3}(6m + 5)$

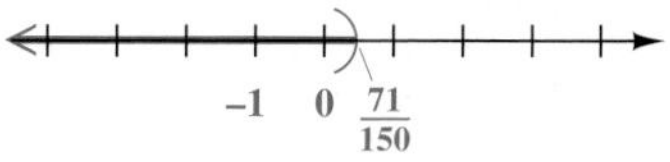

$\left(-\infty, \frac{71}{150}\right)$

69. Which is the graph of $-2 < x$?

A.

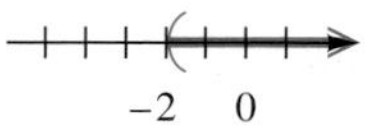

B.

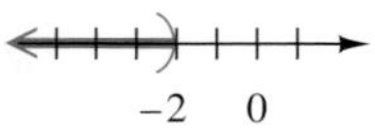

C.

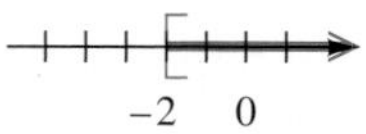

D.

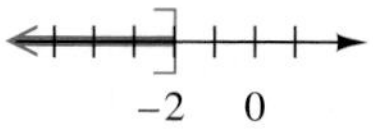

A

Solve each inequality, giving solution sets in both interval and graph forms. Check your answers. See Example 9.

70. $-1 < x + 1 < 8$

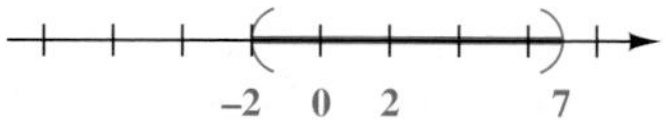

(−2, 7)

71. $-4 < x - 5 < 6$

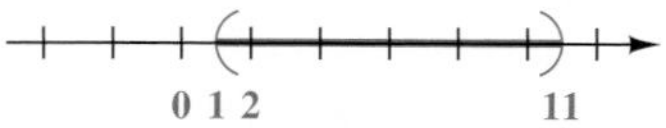

(1, 11)

72. $-15 < 3(p + 2) < 24$

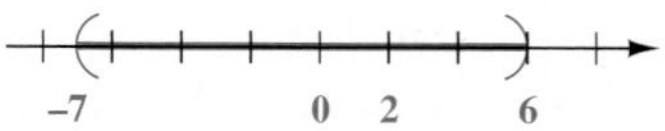

(−7, 6)

73. $-6 \leq 2(z + 2) \leq 16$

[−5, 6]

74. $-19 < 3x - 5 \leq 1$

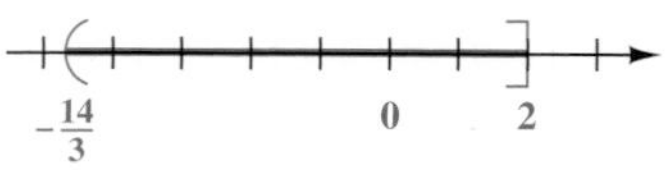

$\left(-\frac{14}{3}, 2\right]$

75. $-16 < 3t + 2 < -10$

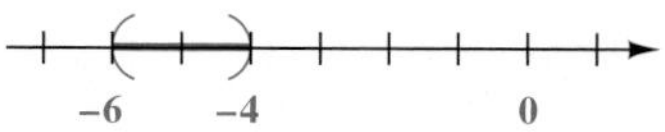

(−6, −4)

8.2 Set Operations and Compound Inequalities

The table shows symptoms of an overactive thyroid and an underactive thyroid.

Underactive Thyroid	Overactive Thyroid
Sleepiness, s	Insomnia, i
Dry hands, d	Moist hands, m
Intolerance of cold, c	Intolerance of heat, h
Goiter, g	Goiter, g

Source: The Merck Manual of Diagnosis and Therapy, 16th Edition, Merck Research Laboratories, 1992.

Let N be the set of symptoms for an underactive thyroid, and let O be the set of symptoms for an overactive thyroid. Suppose we are interested in the set of symptoms that are found in *both* sets N *and* O. In this section, we discuss the use of the words *and* and *or* as they relate to sets and inequalities.

OBJECTIVES

1. **Find the intersection of two sets.**
2. **Solve compound inequalities with the word *and*.**
3. **Find the union of two sets.**
4. **Solve compound inequalities with the word *or*.**

OBJECTIVE 1 Find the intersection of two sets. The intersection of two sets is defined using the word *and*.

Intersection of Sets

For any two sets A and B, the **intersection** of A and B, symbolized $A \cap B$, is defined as follows:

$$A \cap B = \{x \mid x \text{ is an element of } A \textbf{ and } x \text{ is an element of } B\}.$$

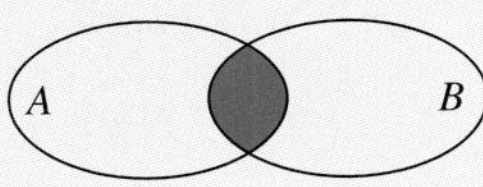

EXAMPLE 1 Finding the Intersection of Two Sets

Let $A = \{1, 2, 3, 4\}$ and $B = \{2, 4, 6\}$. Find $A \cap B$.

The set $A \cap B$ contains those elements that belong to both A *and* B.

$$A \cap B = \{1, 2, 3, 4\} \cap \{2, 4, 6\}$$
$$= \{2, 4\}$$

Work Problem 1 at the Side. ▶

1 List the elements in each set.

(a) $A \cap B$, if $A = \{3, 4, 5, 6\}$ and $B = \{5, 6, 7\}$

(b) $N \cap O$ (Refer to the thyroid table.)

A **compound inequality** consists of two inequalities linked by a connective word such as *and* or *or*. Examples of compound inequalities are

$$x + 1 \le 9 \quad \textbf{and} \quad x - 2 \ge 3$$

and

$$2x > 4 \quad \textbf{or} \quad 3x - 6 < 5.$$

Compound inequalities

OBJECTIVE 2 Solve compound inequalities with the word *and*.

Solving a Compound Inequality with *and*

Step 1 Solve each inequality individually.

Step 2 Since the inequalities are joined with *and,* the solution set of the compound inequality will include all numbers that satisfy both inequalities in Step 1 (the intersection of the solution sets).

ANSWERS

1. (a) $\{5, 6\}$ **(b)** $\{g\}$

2 Solve each compound inequality, and graph the solution set.

(a) $x < 10$ and $x > 2$

(b) $x + 3 \leq 1$ and $x - 4 \geq -12$

3 Solve

$2x \geq x - 1$ and $3x \geq 3 + 2x$,

and graph the solution set.

Answers

2. (a) $(2, 10)$

0 2 4 6 8 10

(b) $[-8, -2]$

−8 −4 −2 0

3. $[3, \infty)$

0 1 2 3 4

EXAMPLE 2 Solving a Compound Inequality with *and*

Solve the compound inequality $x + 1 \leq 9$ and $x - 2 \geq 3$.

Step 1 Solve each inequality individually.

$$x + 1 \leq 9 \quad \text{and} \quad x - 2 \geq 3$$
$$x + 1 - 1 \leq 9 - 1 \quad \text{and} \quad x - 2 + 2 \geq 3 + 2$$
$$x \leq 8 \quad \text{and} \quad x \geq 5$$

Step 2 Because the inequalities are joined with the word *and,* the solution set will include all numbers that satisfy *both* inequalities in Step 1 at the same time. Thus, the compound inequality is true whenever $x \leq 8$ and $x \geq 5$ are both true. The top graph in Figure 6 shows $x \leq 8$, and the bottom graph shows $x \geq 5$.

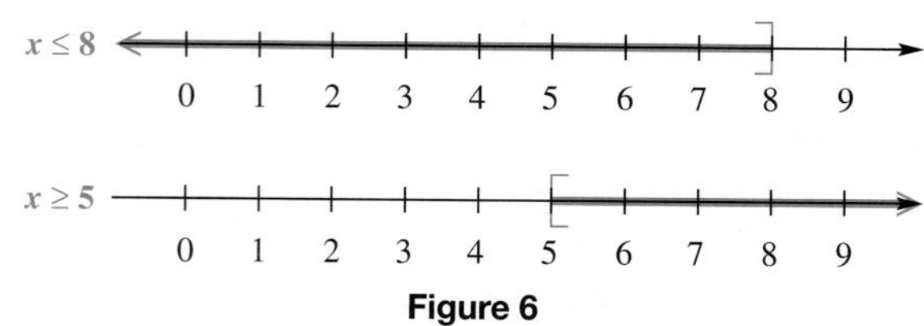

Figure 6

Find the intersection of the two graphs in Figure 6 to get the solution set of the compound inequality. The intersection of the two graphs in Figure 7 shows that the solution set is the closed interval $[5, 8]$.

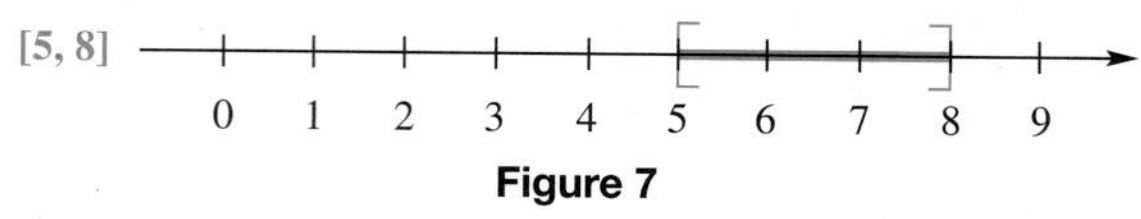

Figure 7

◀ Work Problem **2** at the Side.

EXAMPLE 3 Solving a Compound Inequality with *and*

Solve the compound inequality $-3x - 2 > 5$ and $5x - 1 \leq -21$.

Step 1 Solve each inequality individually.

$$-3x - 2 > 5 \quad \text{and} \quad 5x - 1 \leq -21$$
$$-3x > 7 \quad \text{and} \quad 5x \leq -20$$
$$x < -\frac{7}{3} \quad \text{and} \quad x \leq -4$$

Remember to reverse the inequality symbol.

The graphs of $x < -\frac{7}{3}$ and $x \leq -4$ are shown in Figure 8.

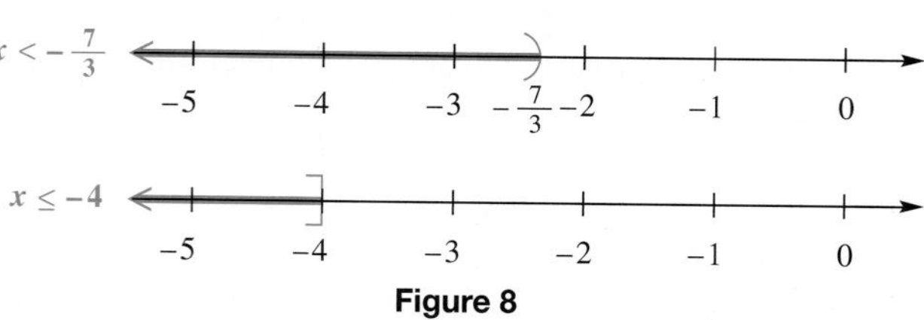

Figure 8

Step 2 Now find all values of x that satisfy both conditions; that is, the real numbers that are less than $-\frac{7}{3}$ and also less than or equal to -4. See Figure 9. The solution set is the infinite interval $(-\infty, -4]$.

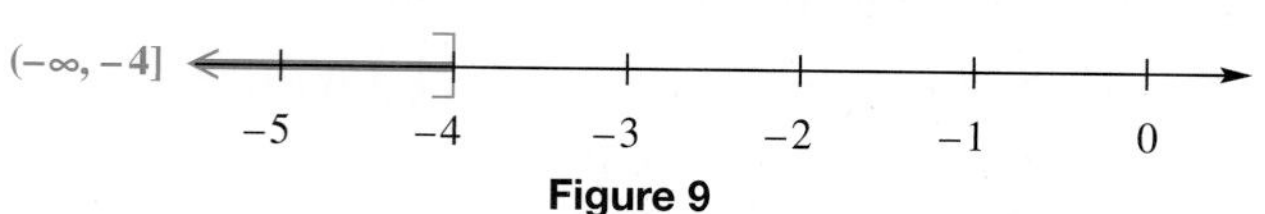

Figure 9

◀ Work Problem **3** at the Side.

EXAMPLE 4 **Solving a Compound Inequality with *and***

Solve $x + 2 < 5$ and $x - 10 > 2$.

First solve each inequality individually.

$$x + 2 < 5 \quad \text{and} \quad x - 10 > 2$$
$$x < 3 \quad \text{and} \quad x > 12$$

The graphs of $x < 3$ and $x > 12$ are shown in Figure 10.

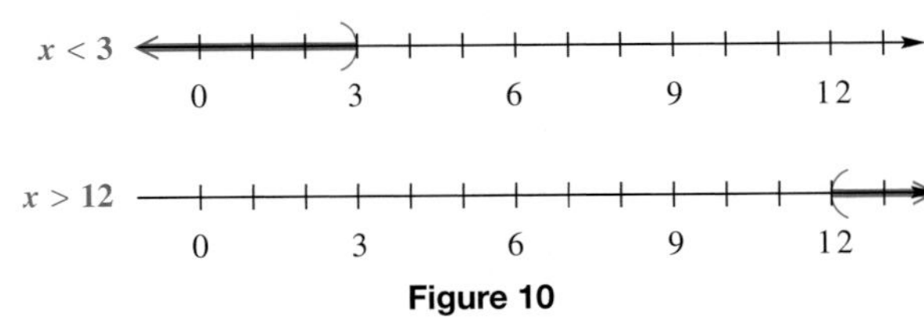

Figure 10

There is no number that is both less than 3 *and* greater than 12, so the given compound inequality has no solution. The solution set is $\emptyset$. See Figure 11.

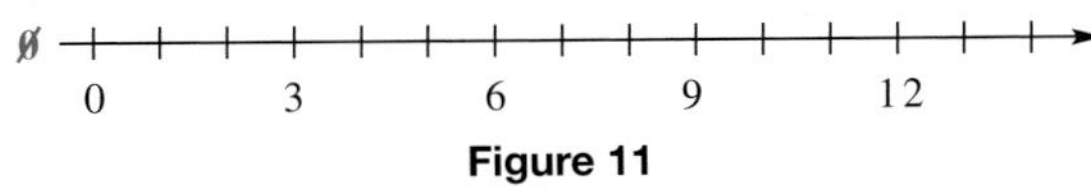

Figure 11

Work Problem 4 at the Side. ▶

4 Solve.

(a) $x < 5$ and $x > 5$

(b) $x + 2 > 3$ and $2x + 1 < -3$

OBJECTIVE 3 **Find the union of two sets.** The union of two sets is defined using the word *or*.

Union of Sets

For any two sets A and B, the **union** of A and B, symbolized $A \cup B$, is defined as follows:

$$A \cup B = \{x \mid x \text{ is an element of } A \textbf{ or } x \text{ is an element of } B\}.$$

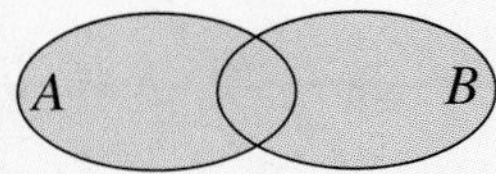

EXAMPLE 5 **Finding the Union of Two Sets**

Let $A = \{1, 2, 3, 4\}$ and $B = \{2, 4, 6\}$. Find $A \cup B$.

Begin by listing all the elements of set A: 1, 2, 3, 4. Then list any additional elements from set B. In this case the elements 2 and 4 are already listed, so the only additional element is 6. Therefore,

$$A \cup B = \{1, 2, 3, 4\} \cup \{2, 4, 6\}$$
$$= \{1, 2, 3, 4, 6\}.$$

The union consists of all elements in either A *or* B (or both).

Note

Although the elements 2 and 4 appeared in both sets A and B, they are written only once in $A \cup B$.

Work Problem 5 at the Side. ▶

5 List the elements in each set.

(a) $A \cup B$, if $A = \{3, 4, 5, 6\}$ and $B = \{5, 6, 7\}$

(b) $N \cup O$ from the thyroid table at the beginning of this section

ANSWERS

4. **(a)** $\emptyset$ **(b)** $\emptyset$

5. **(a)** $\{3, 4, 5, 6, 7\}$ **(b)** $\{s, d, c, g, i, m, h\}$

6 Solve. Give each solution set in both interval and graph forms.

(a) $x + 2 > 3$ or $2x + 1 < -3$

(b) $x - 1 > 2$ or $3x + 5 < 2x + 6$

OBJECTIVE 4 Solve compound inequalities with the word *or*. Use the following steps.

Solving a Compound Inequality with *or*

Step 1 Solve each inequality individually.

Step 2 Since the inequalities are joined with *or,* the solution set includes all numbers that satisfy either one of the two inequalities in Step 1 (the union of the solution sets).

EXAMPLE 6 Solving a Compound Inequality with *or*

Solve $6x - 4 < 2x$ or $-3x \leq -9$.

Step 1 Solve each inequality individually.

$$6x - 4 < 2x \quad \text{or} \quad -3x \leq -9$$
$$4x < 4$$
$$x < 1 \quad \text{or} \quad x \geq 3$$

Remember to reverse the inequality symbol.

The graphs of these two inequalities are shown in Figure 12.

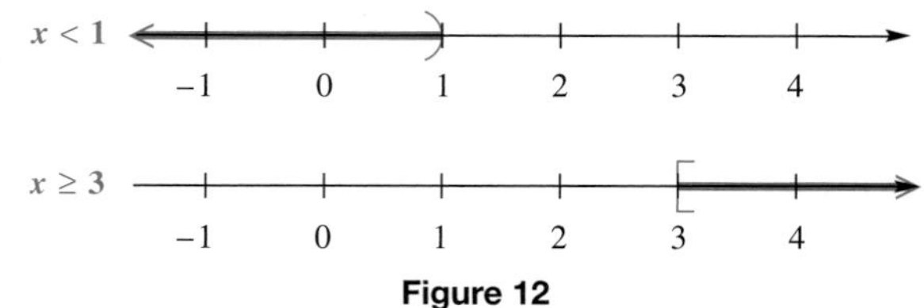

Figure 12

Step 2 Since the inequalities are joined with *or,* we find the union of the two solution sets, as shown in Figure 13. The solution set is the disjoint interval

$$(-\infty, 1) \cup [3, \infty).$$

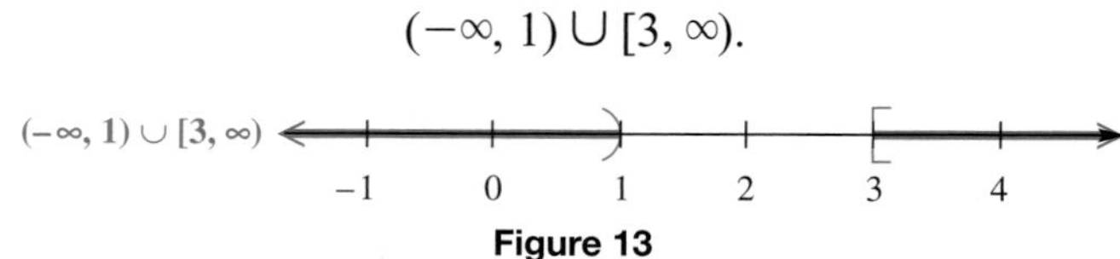

Figure 13

CAUTION
When inequalities are used to write the solution set in Example 6, it *must* be written as

$$x < 1 \quad \text{or} \quad x \geq 3,$$

which keeps the numbers 1 and 3 in their order on the number line. Writing $3 \leq x < 1$ would imply that $3 \leq 1$, which is ***FALSE.*** There is no other way to write the solution set of such a union.

◀ *Work Problem* **6** *at the Side.*

ANSWERS

6. (a) $(-\infty, -2) \cup (1, \infty)$

−2 −1 0 1 2

(b) $(-\infty, 1) \cup (3, \infty)$

0 1 2 3 4

EXAMPLE 7 **Solving a Compound Inequality with *or***

Solve $-4x + 1 \geq 9$ or $5x + 3 \leq -12$.

First, solve each inequality individually.

$$\begin{aligned} -4x + 1 &\geq 9 \quad &\text{or} \quad 5x + 3 &\leq -12 \\ -4x &\geq 8 \quad &\text{or} \quad 5x &\leq -15 \\ x &\leq -2 \quad &\text{or} \quad x &\leq -3 \end{aligned}$$

The graphs of these two inequalities are shown in Figure 14.

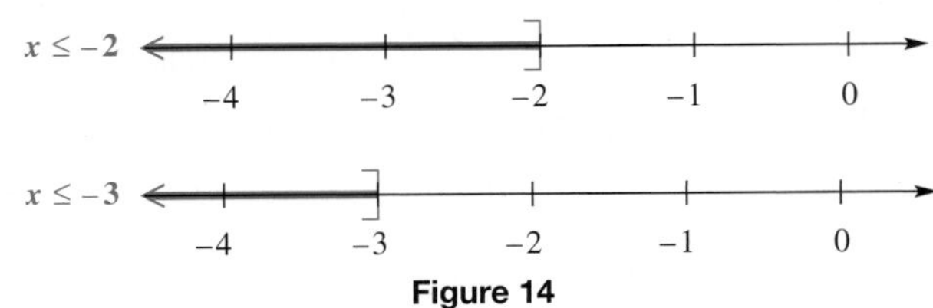

Figure 14

By taking the union, we obtain the interval $(-\infty, -2]$. See Figure 15.

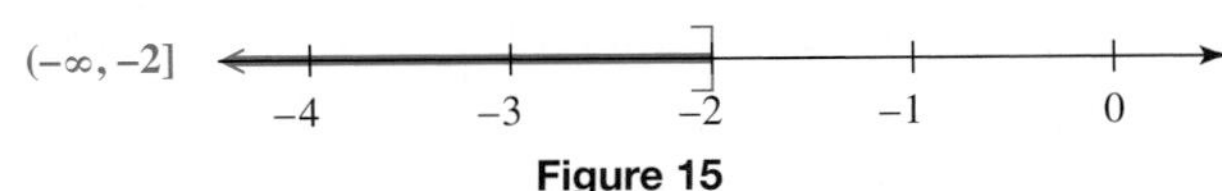

Figure 15

Work Problem **7** *at the Side.* ▶

EXAMPLE 8 **Solving a Compound Inequality with *or***

Solve $-2x + 5 \geq 11$ or $4x - 7 \geq -27$.

Solve each inequality individually.

$$\begin{aligned} -2x + 5 &\geq 11 \quad &\text{or} \quad 4x - 7 &\geq -27 \\ -2x &\geq 6 \quad &\text{or} \quad 4x &\geq -20 \\ x &\leq -3 \quad &\text{or} \quad x &\geq -5 \end{aligned}$$

The graphs of these two inequalities are shown in Figure 16.

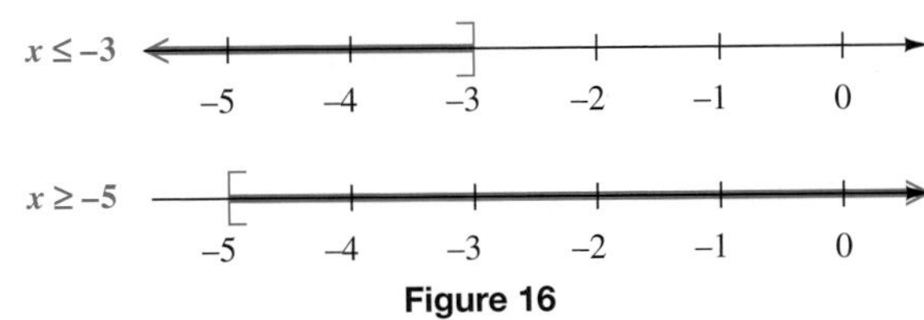

Figure 16

By taking the union, we obtain every real number as a solution, since every real number satisfies at least one of the two inequalities. The set of all real numbers is written in interval notation as $(-\infty, \infty)$ and graphed as in Figure 17.

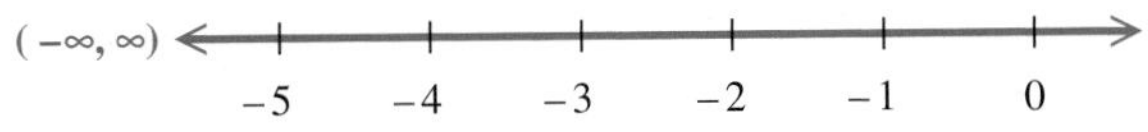

(−∞, ∞)

−5 −4 −3 −2 −1 0

Figure 17

Work Problem **8** *at the Side.* ▶

7 Solve. Give each solution set in both interval and graph forms.

(a) $2x + 1 \leq 9$ or $2x + 3 \leq 5$

(b) $3x - 4 > 2$ or $-2x + 5 < 3$

8 Solve

$3x - 2 \leq 13$ or $x + 5 \geq 7$.

Give the solution set in both interval and graph forms.

ANSWERS

7. (a) $(-\infty, 4]$

(b) $(1, \infty)$

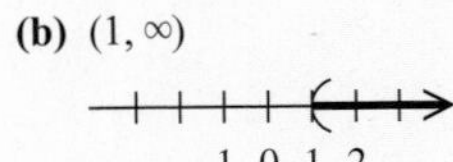

8. $(-\infty, \infty)$

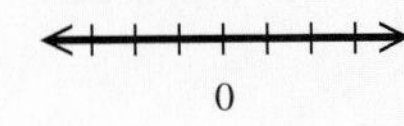

9 From Example 9, list the elements that satisfy each set.

(a) The set of films with admissions greater than 130,000,000 and gross income less than $800,000,000

(b) The set of films with admissions greater than 130,000,000 or gross income less than $800,000,000

EXAMPLE 9 Applying Intersection and Union

The five highest-grossing domestic films (adjusted for inflation) as of July, 2005, are listed in the table.

FIVE ALL-TIME HIGHEST GROSSING DOMESTIC FILMS

Film	Admissions	Gross Income
Gone with the Wind	202,044,569	$1,293,085,000
Star Wars	178,119,595	$1,139,965,000
The Sound of Music	142,415,376	$911,458,000
E.T.	141,925,359	$908,322,298
The Ten Commandments	131,000,000	$838,400,000

Source: Exhibitor Relations Co., Inc.

List the elements of the following sets.

(a) The set of top-five films with admissions greater than 180,000,000 *and* gross income greater than $1,000,000,000

The only film that satisfies both conditions is *Gone with the Wind,* so the set is

{*Gone with the Wind*}.

(b) The set of top-five films with admissions less than 170,000,000 *or* gross income greater than $1,000,000,000

Here, a film that satisfies at least one of the conditions is in the set. This set includes all five films:

{*Gone with the Wind, Star Wars, The Sound of Music, E.T., The Ten Commandments*}.

◀ Work Problem 9 at the Side.

ANSWERS

9. (a) ∅ **(b)** {*Gone with the Wind, Star Wars, The Sound of Music, E.T., The Ten Commandments*}

8.2 Exercises

FOR EXTRA HELP **MyMathLab** Math XL PRACTICE  WATCH  DOWNLOAD  READ  REVIEW

Decide whether each statement is true *or* false. *If it is* false, *explain why.*

1. The union of the solution sets of $2x + 1 = 3$, $2x + 1 > 3$, and $2x + 1 < 3$ is $(-\infty, \infty)$.

true

2. The intersection of the sets $\{x \mid x \geq 5\}$ and $\{x \mid x \leq 5\}$ is $\emptyset$.

false; The intersection is {5}.

3. The union of the sets $(-\infty, 6)$ and $(6, \infty)$ is $\{6\}$.

false; The union is $(-\infty, 6) \cup (6, \infty)$.

4. The intersection of the sets $[6, \infty)$ and $(-\infty, 6]$ is $\{6\}$.

true

Let $A = \{1, 2, 3, 4, 5, 6\}$, $B = \{1, 3, 5\}$, $C = \{1, 6\}$, *and* $D = \{4\}$. *Specify each set. See Examples 1 and 5.*

5. $A \cap D$ — **{4}, or *D***

6. $B \cap C$ — **{1}**

7. $B \cap \emptyset$ — **$\emptyset$**

8. $A \cap \emptyset$ — **$\emptyset$**

9. $A \cup B$ — **{1, 2, 3, 4, 5, 6}, or *A***

10. $B \cup D$ — **{1, 3, 4, 5}**

11. $B \cup C$ — **{1, 3, 5, 6}**

12. $C \cup B$ — **{1, 3, 5, 6}**

Two sets are specified by graphs. Graph the intersection of the two sets.

13.

14.

15.

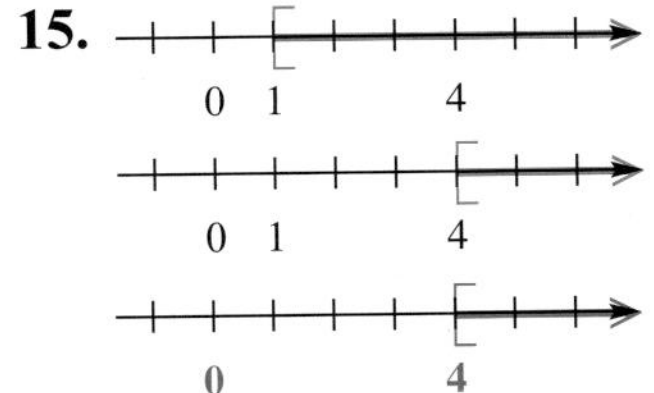

Two sets are specified by graphs. Graph the union of the two sets.

16.

17.

18.

19. Give an example of intersection applied to a real-life situation.

Answers will vary. One example is: The intersection of two streets is the region common to *both* streets.

20. A compound inequality uses one of the words *and* or *or*. Explain how you will determine whether to use *intersection* or *union* when graphing the solution set.

If the word is *and*, use intersection. If the word is *or*, use union.

For each compound inequality, give the solution set in both interval and graph forms. See Examples 2–4.

21. $x < 2$ and $x > -3$

$(-3, 2)$

22. $x < 5$ and $x > 0$

$(0, 5)$

23. $x \leq 2$ and $x \leq 5$

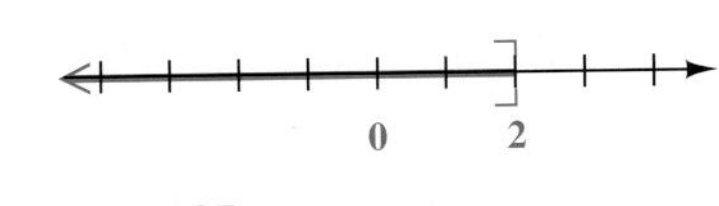

$(-\infty, 2]$

24. $x \geq 3$ and $x \geq 6$

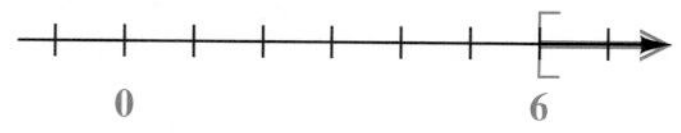

$[6, \infty)$

25. $x \leq 3$ and $x \geq 6$

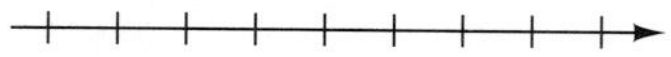

$\emptyset$

26. $x \leq -1$ and $x \geq 3$

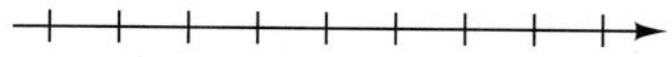

$\emptyset$

27. $x - 3 \leq 6$ and $x + 2 \geq 7$

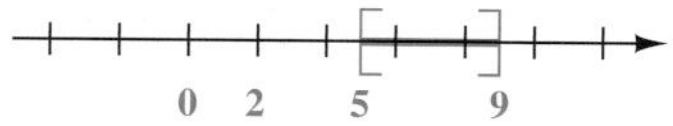

$[5, 9]$

28. $x + 5 \leq 11$ and $x - 3 \geq -1$

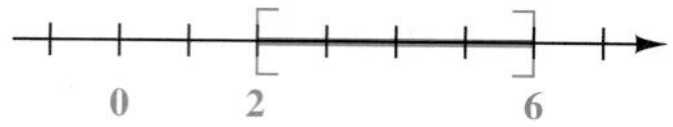

$[2, 6]$

29. $3x - 4 \leq 8$ and $4x - 1 \leq 15$

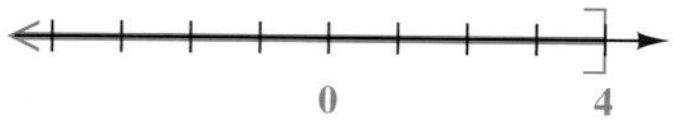

$(-\infty, 4]$

30. $7x + 6 \leq 48$ and $-4x \geq -24$

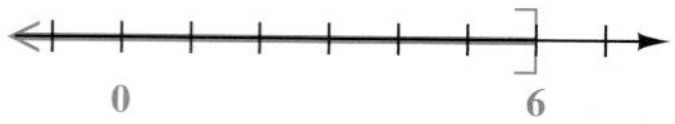

$(-\infty, 6]$

For each compound inequality, give the solution set in both interval and graph forms. See Examples 6–8.

31. $x \leq 1$ or $x \leq 8$

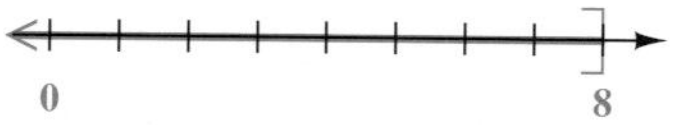

$(-\infty, 8]$

32. $x \geq 1$ or $x \geq 8$

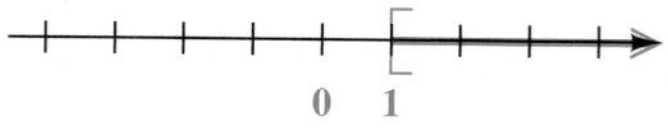

$[1, \infty)$

33. $x \geq -2$ or $x \geq 5$

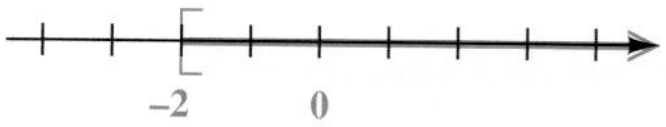

$[-2, \infty)$

34. $x \leq -2$ or $x \leq 6$

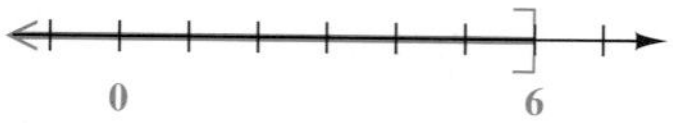

$(-\infty, 6]$

35. $x + 3 \geq 1$ or $x - 8 \leq -4$

$(-\infty, \infty)$

36. $x + 6 \geq 11$ or $x - 4 \leq 3$

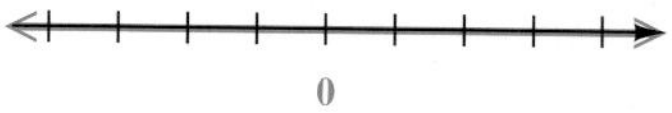

$(-\infty, \infty)$

37. $x + 2 > 7$ or $1 - x > 6$

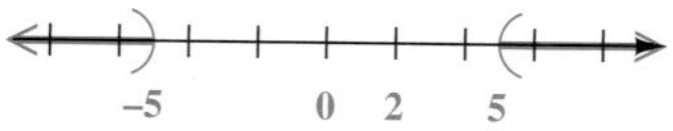

$(-\infty, -5) \cup (5, \infty)$

38. $x + 1 > 3$ or $x + 4 < 2$

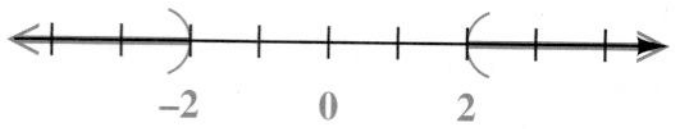

$(-\infty, -2) \cup (2, \infty)$

39. $x + 1 > 3$ or $-4x + 1 \geq 5$

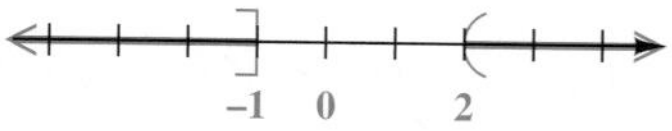

$(-\infty, -1] \cup (2, \infty)$

40. $3x < x + 12$ or $x + 1 > 10$

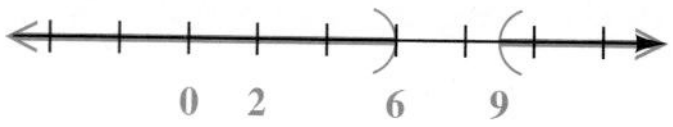

$(-\infty, 6) \cup (9, \infty)$

41. $4x - 8 > 0$ or $4x - 1 < 7$

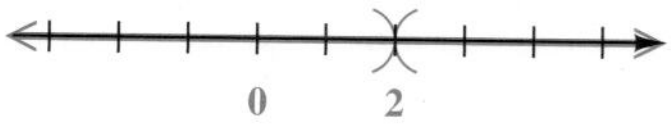

$(-\infty, 2) \cup (2, \infty)$

42. $3x < x + 12$ or $3x - 8 > 10$

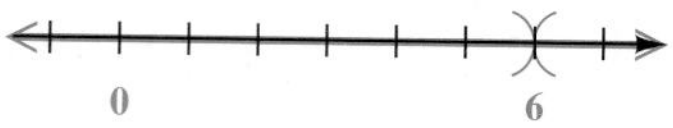

$(-\infty, 6) \cup (6, \infty)$

Express each set in the simplest interval form.

43. $(-\infty, -1] \cap [-4, \infty)$

$[-4, -1]$

44. $[-1, \infty) \cap (-\infty, 9]$

$[-1, 9]$

45. $(-\infty, -6] \cap [-9, \infty)$

$[-9, -6]$

46. $(5, 11] \cap [6, \infty)$

$[6, 11]$

47. $(-\infty, 3) \cup (-\infty, -2)$

$(-\infty, 3)$

48. $[-9, 1] \cup (-\infty, -3)$

$(-\infty, 1]$

49. $[3, 6] \cup (4, 9)$

$[3, 9)$

50. $[-1, 2] \cup (0, 5)$

$[-1, 5)$

For each compound inequality, state whether intersection or union should be used. Then give the solution set in both interval and graph forms. See Examples 2–4 and 6–8.

51. $x < -1$ and $x > -5$

intersection; $(-5, -1)$

52. $x > -1$ and $x < 7$

intersection; $(-1, 7)$

53. $x < 4$ or $x < -2$

union; $(-\infty, 4)$

54. $x < 5$ or $x < -3$

union; $(-\infty, 5)$

55. $x + 1 \geq 5$ and $x - 2 \leq 10$

intersection; $[4, 12]$

56. $2x - 6 \leq -18$ and $2x \geq -18$

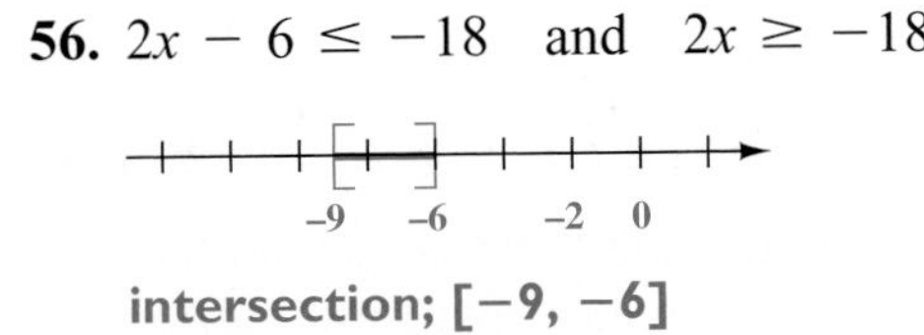

intersection; $[-9, -6]$

57. $-3x \leq -6$ or $-3x \geq 0$

union; $(-\infty, 0] \cup [2, \infty)$

58. $-8x \leq -24$ or $-5x \geq 15$

union; $(-\infty, -3] \cup [3, \infty)$

Relating Concepts (Exercises 59–64) For Individual or Group Work

The figures represent the backyards of neighbors Luigi, Mario, Than, and Joe. Find the area and the perimeter of each yard. Suppose that each resident has 150 ft *of fencing and enough sod to cover* 1400 ft^2 *of lawn.*

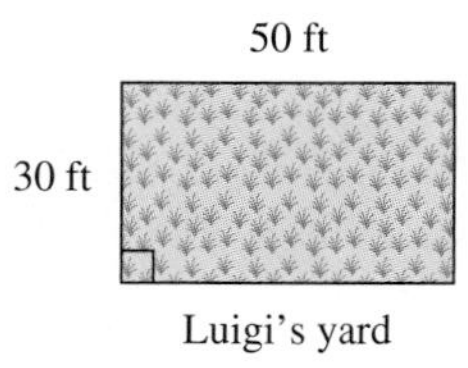

Luigi's yard

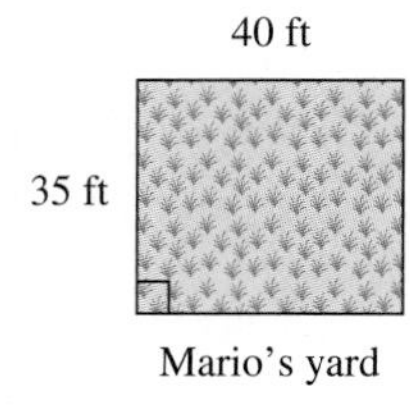

Mario's yard

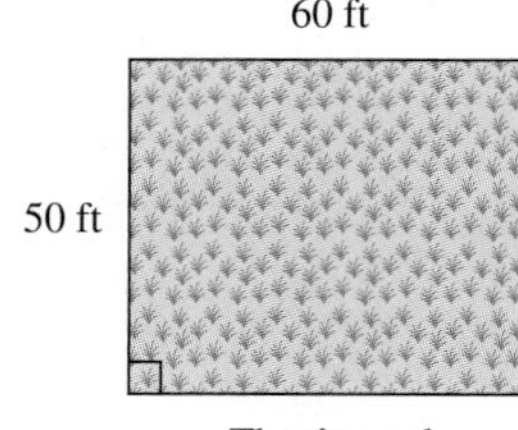

Than's yard

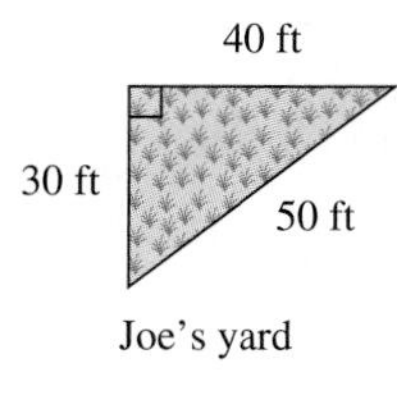

Joe's yard

Give the name or names of the residents whose yards satisfy each description.

59. The yard can be fenced *and* the yard can be sodded.

Mario, Joe

60. The yard can be fenced *and* the yard cannot be sodded.

none of them

61. The yard cannot be fenced *and* the yard can be sodded.

none of them

62. The yard cannot be fenced *and* the yard cannot be sodded.

Luigi, Than

63. The yard can be fenced *or* the yard can be sodded.

Mario, Joe

64. The yard cannot be fenced *or* the yard can be sodded.

all of them

Average expenses for full-time college students at 2-year institutions during the 2005–2006 academic year are shown in the table.

COLLEGE EXPENSES (IN DOLLARS), 2-YEAR INSTITUTIONS

Type of Expense	Public Schools	Private Schools
Tuition and fees	1935	12,450
Board rates	2306	4726
Dormitory charges	2251	3994

Source: National Center for Education Statistics, U.S. Department of Education.

Use the table to list the elements of each set. See Example 9.

65. The set of expenses that are less than \$2500 for public schools *and* are greater than \$5000 for private schools

{Tuition and fees}

66. The set of expenses that are greater than \$2300 for public schools *and* are less than \$12,000 for private schools

{Board rates}

67. The set of expenses that are less than \$2300 for public schools *or* are greater than \$10,000 for private schools

{Tuition and fees, Dormitory charges}

68. The set of expenses that are greater than \$12,000 *or* are less than \$2000

{Tuition and fees}

Relating Concepts (Exercises 69–80) For Individual or Group Work

We can extend the concepts of this section to graph the intersection or union of two linear inequalities in two variables. A pair of inequalities joined with the word *and* is interpreted as the intersection of the solution sets of the inequalities. Thus, ***the graph of the intersection of two linear inequalities in two variables is the region of the plane where all points satisfy both of the inequalities at the same time.***

For example, to graph

$$2x + 4y \geq 5 \quad \textit{and} \quad x \geq 1,$$

we graph each of the two inequalities $2x + 4y \geq 5$ and $x \geq 1$ separately. The graph of $2x + 4y \geq 5$ is shown in Figure A, and the graph of $x \geq 1$ is shown in Figure B.

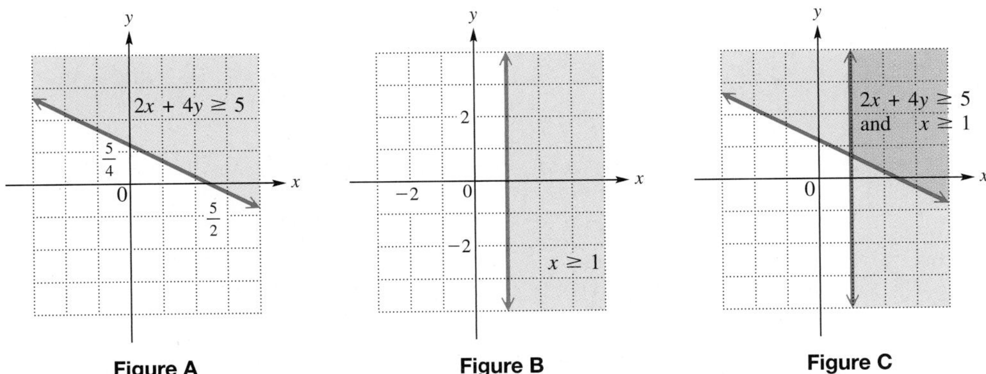

Figure A **Figure B** **Figure C**

In practice, the two graphs in Figures A and B are graphed on the same axes. Then we use heavy shading to indicate the intersection of the graphs, as shown in Figure C. To check, we can use a test point from each of the four regions formed by the intersection of the boundary lines. Verify that only ordered pairs in the heavily shaded region satisfy both inequalities.

When two inequalities are joined by the word *or,* we must find the union of the graphs of the inequalities. ***The graph of the union of two linear inequalities in two variables includes all of the points that satisfy either inequality.***

For example, to graph

$$2x + 4y \geq 5 \quad \textit{or} \quad x \geq 1,$$

we graph both inequalities on the same axes. The graph of the union is the entire shaded region, as shown in Figure D.

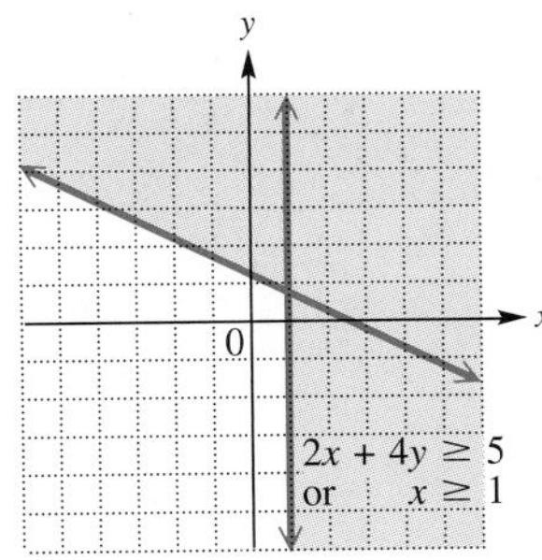

Figure D

(continued)

Graph the intersection of each pair of inequalities.

69. $x + y \leq 1$ and $x \geq 1$

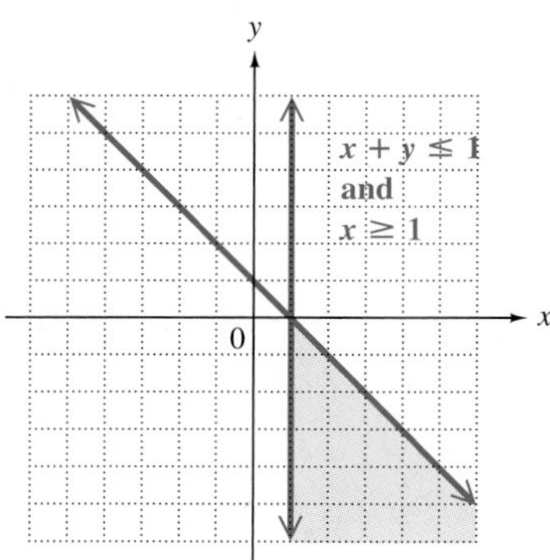

70. $x - y \geq 2$ and $x \geq 3$

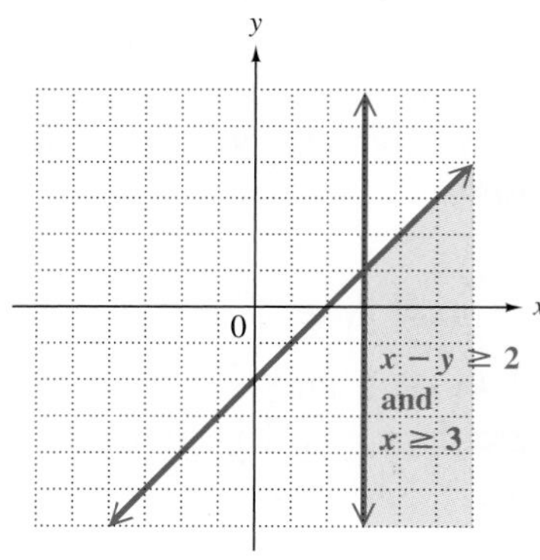

71. $2x - y \geq 2$ and $y < 4$

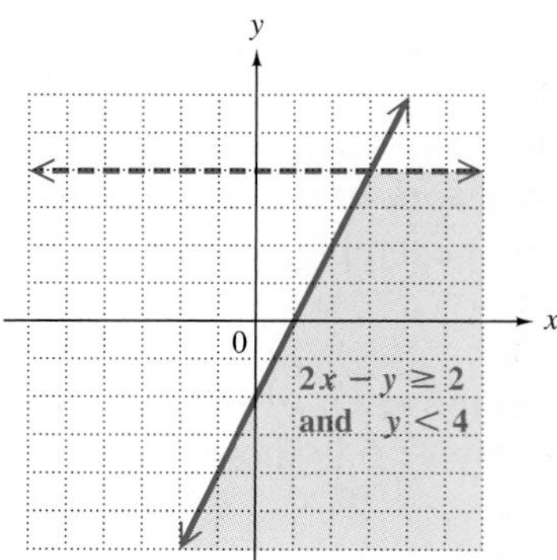

72. $3x - y \geq 3$ and $y < 3$

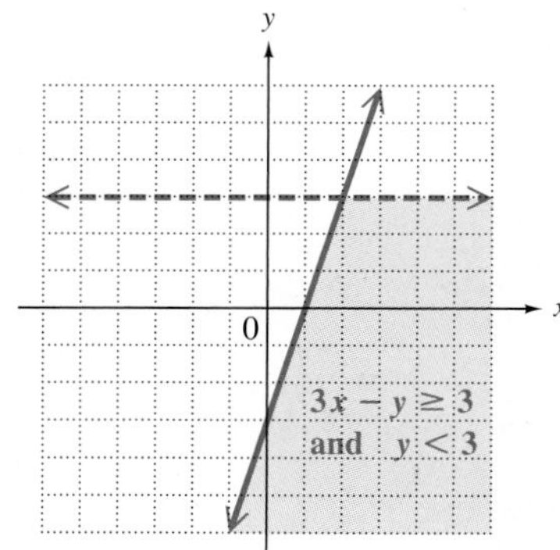

73. $x + y > -5$ and $y < -2$

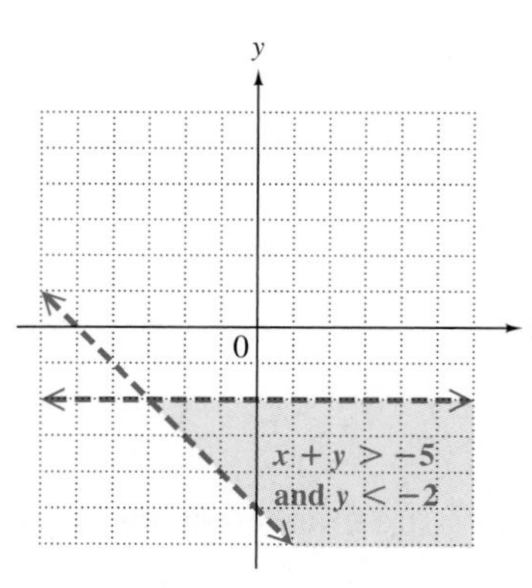

74. $6x - 4y < 10$ and $y > 2$

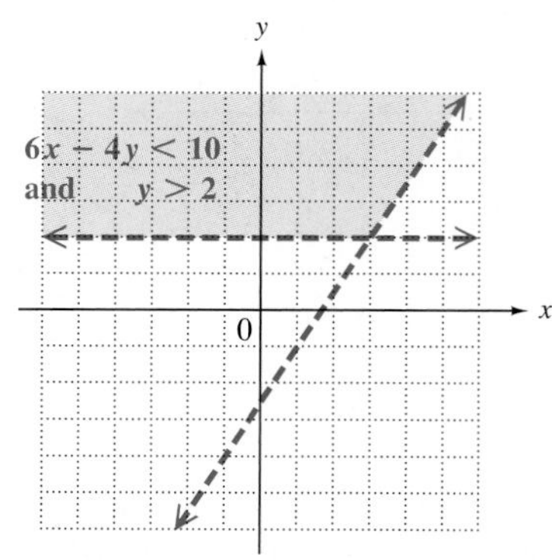

Graph the union of each pair of inequalities.

75. $x - y \geq 1$ or $y \geq 2$

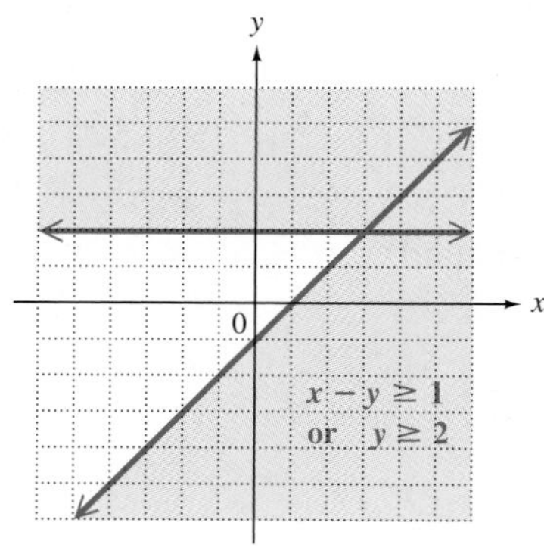

76. $x + y \leq 2$ or $y \geq 3$

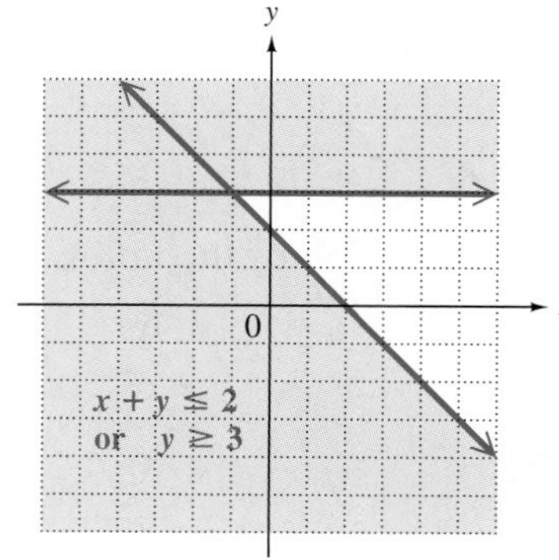

77. $x - 2 > y$ or $x < 1$

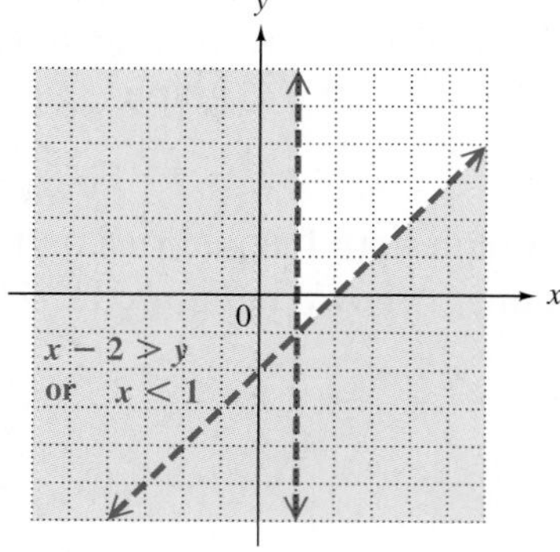

78. $x + 3 < y$ or $x > 3$

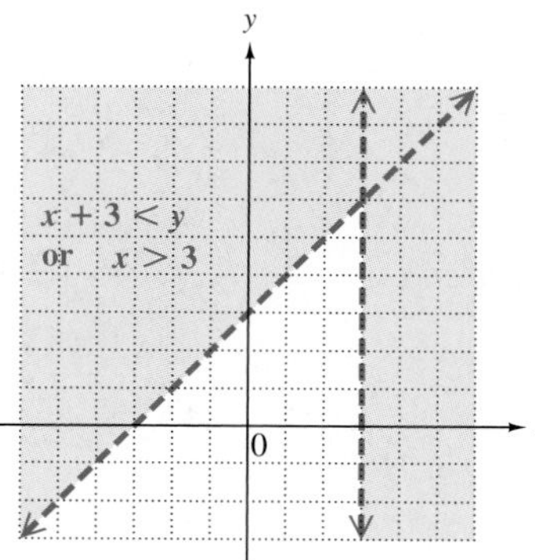

79. $3x + 2y < 6$ or $x - 2y > 2$

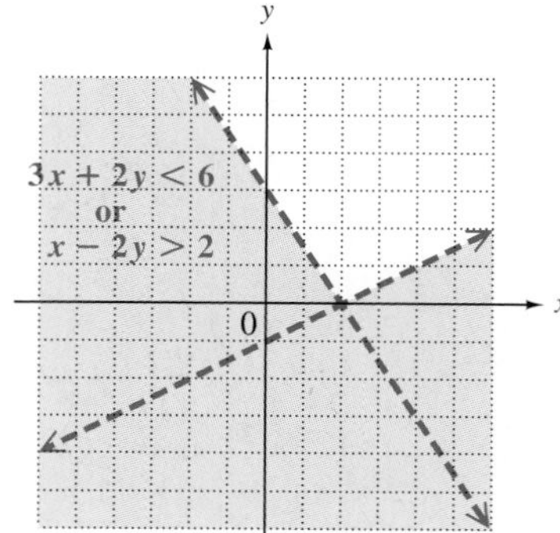

80. $x - y \geq 1$ or $x + y \leq 4$

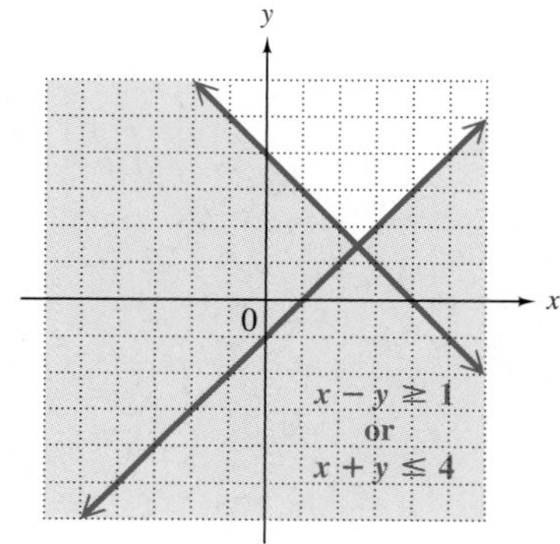

8.3 Absolute Value Equations and Inequalities

OBJECTIVES

1. Use the distance definition of absolute value.
2. Solve equations of the form $|ax + b| = k$, for $k > 0$.
3. Solve inequalities of the form $|ax + b| < k$ and of the form $|ax + b| > k$, for $k > 0$.
4. Solve absolute value equations that involve rewriting.
5. Solve equations of the form $|ax + b| = |cx + d|$.
6. Solve special cases of absolute value equations and inequalities.

In a production line, quality is controlled by randomly choosing items from the line and checking to see how selected measurements vary from the optimum measure. These differences are sometimes positive and sometimes negative, so they are expressed with absolute value. For example, a machine that fills quart milk cartons might be set to release 1 qt (32 oz) plus or minus 2 oz per carton. Then the number of ounces in each carton should satisfy the *absolute value inequality* $|x - 32| \leq 2$, where x is the number of ounces.

OBJECTIVE 1 Use the distance definition of absolute value.

In **Section 1.3,** we saw that the absolute value of a number x, written $|x|$, represents the distance from x to 0 on the number line. For example, the solutions of $|x| = 4$ are 4 and -4, as shown in Figure 18.

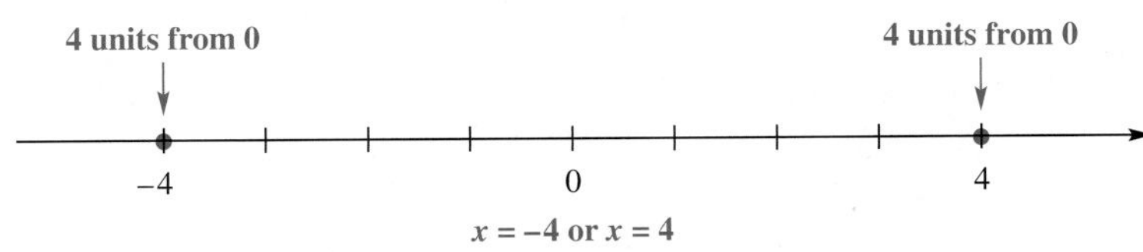

Figure 18

Because absolute value represents distance from 0, it is reasonable to interpret the solutions of $|x| > 4$ to be all numbers that are *more* than 4 units from 0. The set $(-\infty, -4) \cup (4, \infty)$ fits this description. Figure 19 shows the graph of the solution set of $|x| > 4$. Because the graph consists of two separate intervals, the solution set is described using *or* as $x < -4$ or $x > 4$.

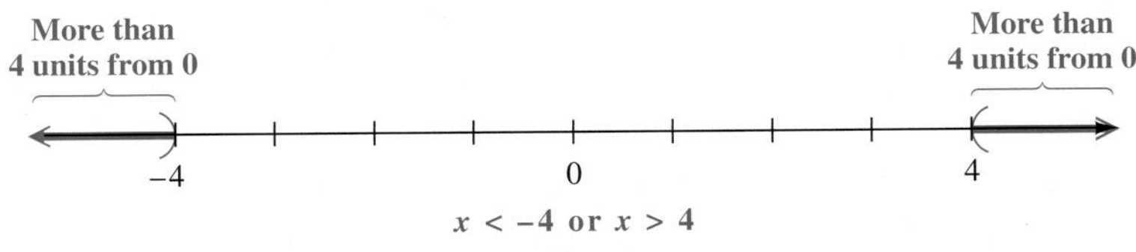

Figure 19

The solution set of $|x| < 4$ consists of all numbers that are *less* than 4 units from 0 on the number line. Another way of thinking of this is to think of all numbers *between* -4 and 4. This set of numbers is given by $(-4, 4)$, as shown in Figure 20. Here, the graph shows that $-4 < x < 4$, which means $x > -4$ *and* $x < 4$.

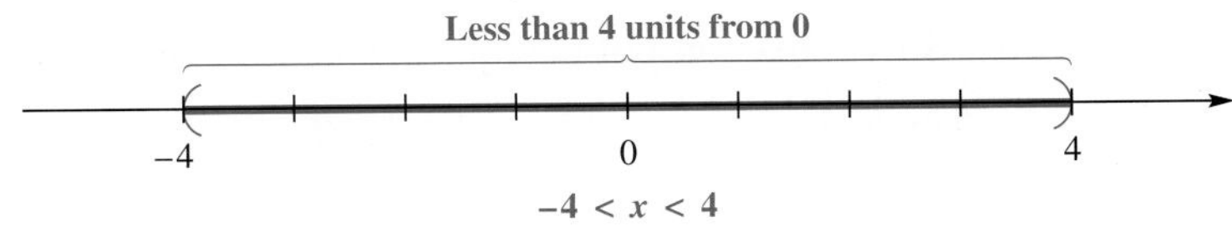

Figure 20

Work Problem 1 at the Side. ▶

1 Graph the solution set of each equation or inequality.

(a) $|x| = 3$

(b) $|x| > 3$

(c) $|x| < 3$

The equation and inequalities just described are examples of **absolute value equations and inequalities.** They involve the absolute value of a variable expression and generally take the form

$$|ax + b| = k, \quad |ax + b| > k, \quad \text{or} \quad |ax + b| < k,$$

where k is a positive number. From Figures 18–20, we see that

$|x| = 4$ has the same solution set as $x = -4$ **or** $x = 4$,

$|x| > 4$ has the same solution set as $x < -4$ **or** $x > 4$,

$|x| < 4$ has the same solution set as $x > -4$ **and** $x < 4$.

ANSWERS

1. (a), (b), (c)

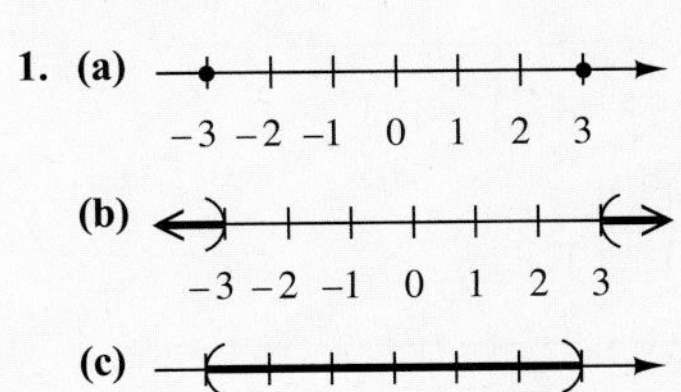

2 Solve, check, and graph the solution set of each equation.

(a) $|x + 2| = 3$

(b) $|3x - 4| = 11$

Thus, we can solve an absolute value equation or inequality by solving the appropriate compound equation or inequality.

Solving Absolute Value Equations and Inequalities

Let k be a positive real number, and p and q be real numbers.

Case 1 To solve $|ax + b| = k$, solve the compound equation

$$ax + b = k \quad \text{or} \quad ax + b = -k.$$

The solution set is usually of the form $\{p, q\}$, which includes two numbers.

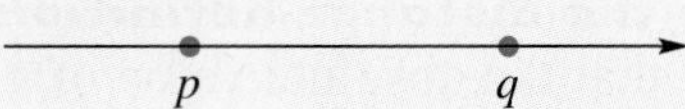

Case 2 To solve $|ax + b| > k$, solve the compound inequality

$$ax + b > k \quad \text{or} \quad ax + b < -k.$$

The solution set is of the form $(-\infty, p) \cup (q, \infty)$, which is a disjoint interval.

Case 3 To solve $|ax + b| < k$, solve the three-part inequality

$$-k < ax + b < k.$$

The solution set is of the form (p, q), an open interval.

OBJECTIVE 2 Solve equations of the form $|ax + b| = k$, for $k > 0$. ***Remember that because absolute value refers to distance from the origin, an absolute value equation will have two parts.***

EXAMPLE 1 Solving an Absolute Value Equation

Solve $|2x + 1| = 7$.

For $|2x + 1|$ to equal 7, $2x + 1$ must be 7 units from 0 on the number line. This can happen only when $2x + 1 = 7$ or $2x + 1 = -7$. This is Case 1 in the preceding box. Solve this compound equation as follows.

$$\begin{aligned} 2x + 1 &= 7 \quad \text{or} \quad & 2x + 1 &= -7 \\ 2x &= 6 \quad \text{or} \quad & 2x &= -8 \\ x &= 3 \quad \text{or} \quad & x &= -4 \end{aligned}$$

Check by substituting 3 and then -4 in the original absolute value equation to verify that the solution set is $\{-4, 3\}$. The graph is shown in Figure 21.

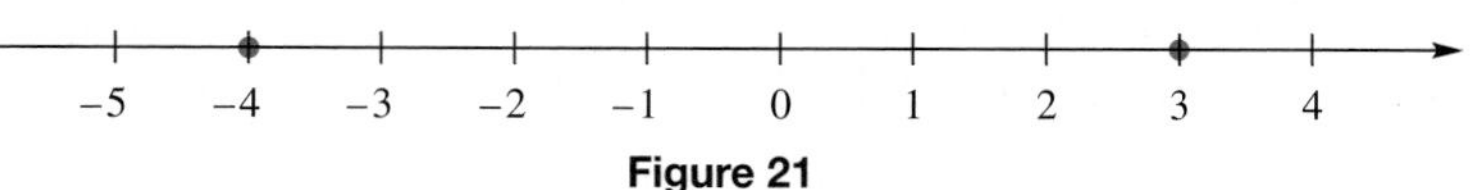

Figure 21

◀ *Work Problem* 2 *at the Side.*

ANSWERS

2. (a) $\{-5, 1\}$

−5 −4 −3 −2 −1 0 1

(b) $\left\{-\frac{7}{3}, 5\right\}$

$-\frac{7}{3}$ 0 2 4 5

Note

Some people prefer to write the compound statements in Cases 1 and 2 of the box on the previous page as the equivalent forms

$$ax + b = k \quad \text{or} \quad -(ax + b) = k$$

and $$ax + b > k \quad \text{or} \quad -(ax + b) > k.$$

These forms produce the same results.

OBJECTIVE 3 Solve inequalities of the form $|ax + b| < k$ and of the form $|ax + b| > k$, for $k > 0$.

EXAMPLE 2 Solving an Absolute Value Inequality with >

Solve $|2x + 1| > 7$.

By Case 2 in the box on the preceding page, this absolute value inequality is rewritten as

$$2x + 1 > 7 \quad \text{or} \quad 2x + 1 < -7,$$

because $2x + 1$ must represent a number that is *more* than 7 units from 0 on either side of the number line. Now, solve the compound inequality.

$$\begin{aligned} 2x + 1 &> 7 \quad \text{or} \quad & 2x + 1 &< -7 \\ 2x &> 6 \quad \text{or} \quad & 2x &< -8 \\ x &> 3 \quad \text{or} \quad & x &< -4 \end{aligned}$$

Check these solutions. The solution set is $(-\infty, -4) \cup (3, \infty)$, a disjoint interval. See the graph in Figure 22.

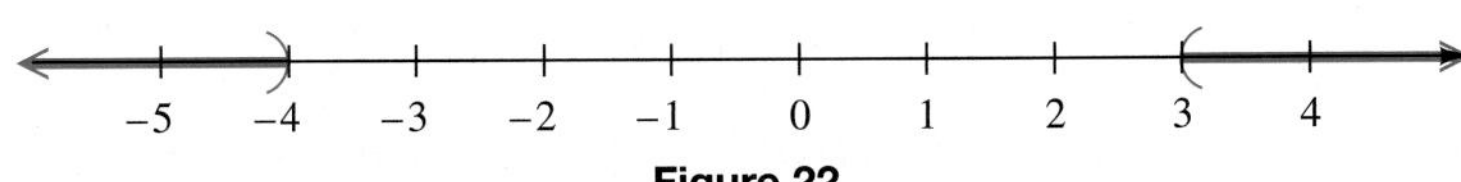

Figure 22

Work Problem 3 at the Side. ▶

EXAMPLE 3 Solving an Absolute Value Inequality with <

Solve $|2x + 1| < 7$.

The expression $2x + 1$ must represent a number that is less than 7 units from 0 on either side of the number line. Another way of thinking of this is to realize that $2x + 1$ must be between -7 and 7. As Case 3 in the box on the preceding page shows, this is written as the three-part inequality

$$-7 < 2x + 1 < 7.$$

$$-8 < 2x < 6 \qquad \text{Subtract 1 from each part.}$$

$$-4 < x < 3 \qquad \text{Divide each part by 2.}$$

Check that the solution set is $(-4, 3)$, so the graph consists of the open interval shown in Figure 23.

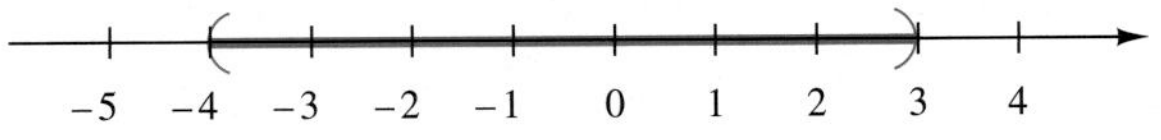

Figure 23

Work Problem 4 at the Side. ▶

3 Solve, check, and graph the solution set of each inequality.

(a) $|x + 2| > 3$

(b) $|3x - 4| \geq 11$

4 Solve, check, and graph the solution set of each inequality.

(a) $|x + 2| < 3$

(b) $|3x - 4| \leq 11$

ANSWERS

3. (a) $(-\infty, -5) \cup (1, \infty)$

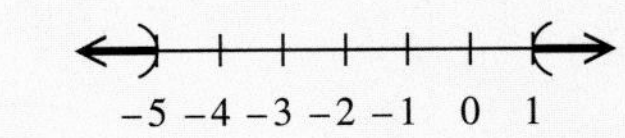

(b) $\left(-\infty, -\frac{7}{3}\right] \cup [5, \infty)$

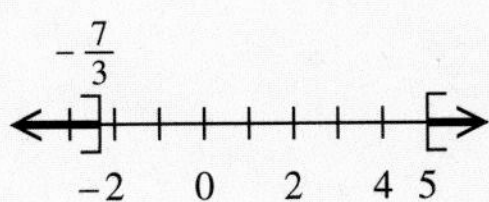

4. (a) $(-5, 1)$

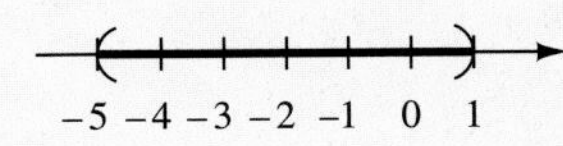

(b) $\left[-\frac{7}{3}, 5\right]$

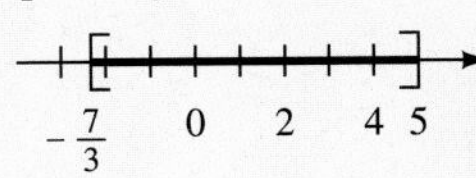

5 Solve $|5x + 2| - 9 = -7$. Check your solutions.

Look back at Figures 21, 22, and 23, with the graphs of $|2x + 1| = 7$, $|2x + 1| > 7$, and $|2x + 1| < 7$. If we find the union of the three sets, we get the set of all real numbers. This is because, for any value of x, $|2x + 1|$ will satisfy one and only one of the following: It is either equal to 7, greater than 7, or less than 7.

CAUTION

When solving absolute value equations and inequalities of the types in Examples 1, 2, and 3, remember the following.

1. The methods described apply when the constant is alone on one side of the equation or inequality and is *positive.*
2. Absolute value equations and absolute value inequalities of the form $|ax + b| > k$ translate into "or" compound statements.
3. Absolute value inequalities of the form $|ax + b| < k$ translate into "and" compound statements, which may be written as three-part inequalities.
4. An "or" statement *cannot* be written in three parts. It would be incorrect to use $-7 > 2x + 1 > 7$ in Example 2, because this would imply that $-7 > 7$, which is *false.*

OBJECTIVE 4 Solve absolute value equations that involve rewriting. Sometimes an absolute value equation or inequality requires some rewriting before it can be set up as a compound statement.

EXAMPLE 4 Solving an Absolute Value Equation That Requires Rewriting

Solve $|x + 3| + 5 = 12$.

First, get the absolute value alone on one side of the equals sign.

$$|x + 3| + 5 = 12$$

$$|x + 3| + 5 - \mathbf{5} = 12 - \mathbf{5} \quad \text{Subtract 5.}$$

$$|x + 3| = 7$$

Now use the method shown in Example 1.

$$x + 3 = 7 \quad \text{or} \quad x + 3 = -7$$

$$x = 4 \quad \text{or} \quad x = -10$$

Check these solutions by substituting each one in the original equation.

Check $|x + 3| + 5 = 12$

$\lvert 4 + 3\rvert + 5 \stackrel{?}{=} 12$ Let $x = 4$.	$\lvert -10 + 3\rvert + 5 \stackrel{?}{=} 12$ Let $x = -10$.
$\lvert 7\rvert + 5 \stackrel{?}{=} 12$	$\lvert -7\rvert + 5 \stackrel{?}{=} 12$
$12 = 12$ True	$12 = 12$ True

The check confirms that the solution set is $\{-10, 4\}$.

◀ *Work Problem* **5** *at the Side.*

CAUTION

When solving an equation like the one in Example 4, do *not* simply drop the absolute value bars.

ANSWER

5. $\left\{-\frac{4}{5}, 0\right\}$

We use a method similar to that used in Example 4 to solve an absolute value *inequality* that requires rewriting:

$\lvert x+3\rvert + 5 \geq 12$	$\lvert x+3\rvert + 5 \leq 12$
$\lvert x+3\rvert \geq 7$	$\lvert x+3\rvert \leq 7$
$x+3 \geq 7$ or $x+3 \leq -7$	$-7 \leq x+3 \leq 7$
$x \geq 4$ or $x \leq -10.$	$-10 \leq x \leq 4.$
Solution set: $(-\infty, -10] \cup [4, \infty)$	Solution set: $[-10, 4]$

Work Problem 6 at the Side. ▶

OBJECTIVE 5 Solve equations of the form $\lvert ax + b\rvert = \lvert cx + d\rvert$. By definition, for two expressions to have the same absolute value, they must either be equal or be negatives of each other.

Solving $\lvert ax + b\rvert = \lvert cx + d\rvert$

To solve an absolute value equation of the form

$$\lvert ax + b\rvert = \lvert cx + d\rvert,$$

solve the compound equation

$$ax + b = cx + d \quad \text{or} \quad ax + b = -(cx + d).$$

EXAMPLE 5 Solving an Equation with Two Absolute Values

Solve $\lvert z+6\rvert = \lvert 2z-3\rvert$.

This equation is satisfied either if $z+6$ and $2z-3$ are equal to each other, or if $z+6$ and $2z-3$ are negatives of each other. Thus,

$$z+6 = 2z-3 \quad \text{or} \quad z+6 = -(2z-3)$$
$$z+9 = 2z \quad \text{or} \quad z+6 = -2z+3$$
$$9 = z \quad \text{or} \quad 3z = -3$$
$$z = 9 \quad \text{or} \quad z = -1.$$

Check that the solution set is $\{9, -1\}$.

Work Problem 7 at the Side. ▶

OBJECTIVE 6 Solve special cases of absolute value equations and inequalities. When a typical absolute value equation or inequality involves a *negative* constant or 0 alone on one side, we use the properties of absolute value to solve. Keep the following in mind.

Special Cases of Absolute Value

Case 1 The absolute value of an expression can never be negative—that is, $\lvert a\rvert \geq 0$ for all real numbers a.

Case 2 The absolute value of an expression equals 0 only when the expression is equal to 0.

6 Solve each inequality, and graph the solution set.

(a) $\lvert x+2\rvert - 3 > 2$

(b) $\lvert 3x+2\rvert + 4 \leq 15$

7 Solve each equation.

(a) $\lvert k-1\rvert = \lvert 5k+7\rvert$

(b) $\lvert 4r-1\rvert = \lvert 3r+5\rvert$

ANSWERS

6. (a) $(-\infty, -7) \cup (3, \infty)$

−7 −4 −2 0 3

(b) $\left[-\frac{13}{3}, 3\right]$

$-\frac{13}{3}$ −2 0 2 3

7. (a) $\{-1, -2\}$ **(b)** $\left\{-\frac{4}{7}, 6\right\}$

8 Solve each equation.

(a) $|6x + 7| = -5$

(b) $\left|\frac{1}{4}x - 3\right| = 0$

9 Solve.

(a) $|x| > -1$

(b) $|x| < -5$

(c) $|x + 2| \leq 0$

(d) $|t - 10| - 2 \leq -3$

EXAMPLE 6 Solving Special Cases of Absolute Value Equations

Solve each equation.

(a) $|5r - 3| = -4$

See Case 1 in the box on the preceding page. ***The absolute value of an expression can never be negative,*** so there are no solutions for this equation. The solution set is $\emptyset$.

(b) $|7x - 3| = 0$

See Case 2 in the box on the preceding page. The expression $7x - 3$ will equal 0 *only* if

$$7x - 3 = 0$$

$$7x = 3 \quad \text{Add 3.}$$

$$x = \frac{3}{7}. \quad \text{Divide by 7.}$$

Thus, the solution set of the original equation is $\left\{\frac{3}{7}\right\}$, with just one element. Check this solution by substituting it in the original equation.

*Work Problem **8** at the Side.*

EXAMPLE 7 Solving Special Cases of Absolute Value Inequalities

Solve each inequality.

(a) $|x| \geq -4$

The absolute value of a number is always greater than or equal to 0. Thus, $|x| \geq -4$ is true for *all* real numbers. The solution set is $(-\infty, \infty)$.

(b) $|x + 6| - 3 < -5$

$$|x + 6| < -2 \quad \text{Add 3 to each side.}$$

There is no number whose absolute value is less than -2, so this inequality has no solution. The solution set is $\emptyset$.

(c) $|x - 7| + 4 \leq 4$

$$|x - 7| \leq 0 \quad \text{Subtract 4 from each side.}$$

The value of $|x - 7|$ will never be less than 0. However, $|x - 7|$ will *equal* 0 when $x = 7$. Therefore, the solution set is $\{7\}$.

*Work Problem **9** at the Side.*

Answers

8. (a) $\emptyset$ **(b)** $\{12\}$
9. (a) $(-\infty, \infty)$ **(b)** $\emptyset$ **(c)** $\{-2\}$ **(d)** $\emptyset$

8.3 Exercises

FOR EXTRA HELP

MyMathLab

Math XL PRACTICE

WATCH

DOWNLOAD

READ

REVIEW

Match each absolute value equation or inequality in Column I with the graph of its solution set in Column II.

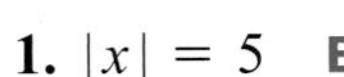

I		II
1. $\lvert x\rvert = 5$	E	A. -5 0 5
$\lvert x\rvert < 5$	C	B. -5 0 5
$\lvert x\rvert > 5$	D	C. -5 0 5
$\lvert x\rvert \leq 5$	B	D. -5 0 5
$\lvert x\rvert \geq 5$	A	E. -5 0 5

I		II
2. $\lvert x\rvert = 9$	E	A. -9 0 9
$\lvert x\rvert > 9$	D	B. -9 0 9
$\lvert x\rvert \geq 9$	A	C. -9 0 9
$\lvert x\rvert < 9$	C	D. -9 0 9
$\lvert x\rvert \leq 9$	B	E. -9 0 9

3. How many solutions will $|ax + b| = k$ have if **(a)** $k = 0$; **(b)** $k > 0$; **(c)** $k < 0$?

one two none

4. Explain when to use *and* and when to use *or* if you are solving an absolute value equation or inequality of the form $|ax + b| = k$, $|ax + b| < k$, or $|ax + b| > k$, where k is a positive number.

Use *or* for the = statement and the > statement. Use *and* for the < statement.

Solve each equation. See Example 1.

5. $|x| = 12$ $\{-12, 12\}$

6. $|x| = 14$ $\{-14, 14\}$

7. $|4x| = 20$ $\{-5, 5\}$

8. $|5x| = 30$ $\{-6, 6\}$

9. $|x - 3| = 9$ $\{-6, 12\}$

10. $|p - 5| = 13$ $\{-8, 18\}$

11. $|2x + 1| = 9$ $\{-5, 4\}$

12. $|2x + 3| = 19$ $\{-11, 8\}$

13. $|4r - 5| = 17$ $\left\{-3, \frac{11}{2}\right\}$

14. $|5t - 1| = 21$ $\left\{-4, \frac{22}{5}\right\}$

15. $|2x + 5| = 14$ $\left\{-\frac{19}{2}, \frac{9}{2}\right\}$

16. $|2x - 9| = 18$ $\left\{-\frac{9}{2}, \frac{27}{2}\right\}$

17. $\left|\frac{1}{2}x + 3\right| = 2$ $\{-10, -2\}$

18. $\left|\frac{2}{3}q - 1\right| = 5$ $\{-6, 9\}$

19. $\left|1 - \frac{3}{4}k\right| = 7$ $\left\{-8, \frac{32}{3}\right\}$

20. $\left|2 - \frac{5}{2}m\right| = 14$ $\left\{-\frac{24}{5}, \frac{32}{5}\right\}$

Solve each inequality, and graph the solution set. See Example 2.

21. $|x| > 3$

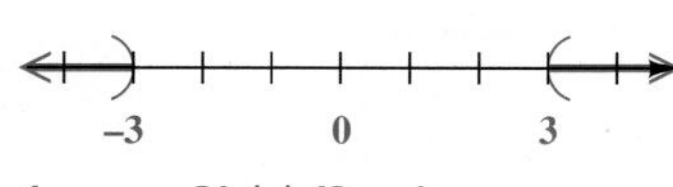

$(-\infty, -3) \cup (3, \infty)$

22. $|x| > 2$

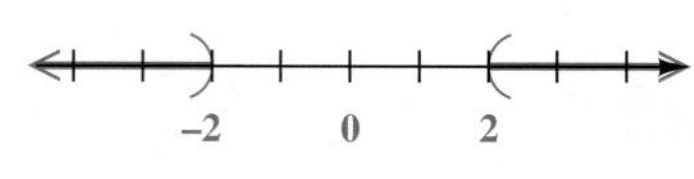

$(-\infty, -2) \cup (2, \infty)$

23. $|k| \geq 4$

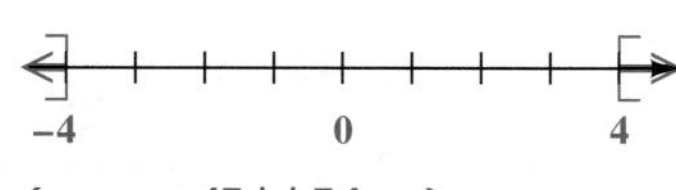

$(-\infty, -4] \cup [4, \infty)$

24. $|r| \geq 1$

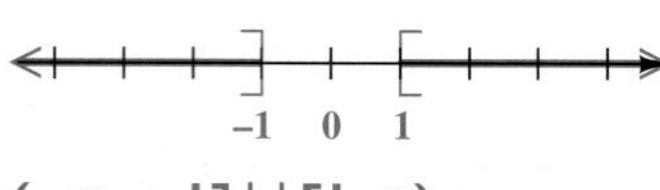

$(-\infty, -1] \cup [1, \infty)$

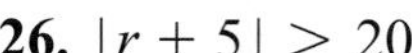

25. $|t + 2| > 8$

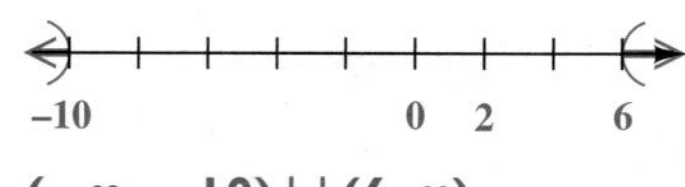

$(-\infty, -10) \cup (6, \infty)$

26. $|r + 5| > 20$

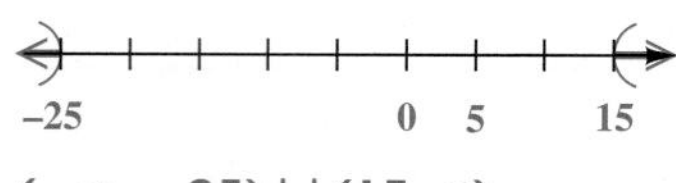

$(-\infty, -25) \cup (15, \infty)$

27. $|3x - 1| \geq 8$

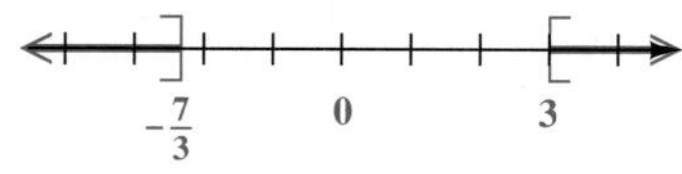

$\left(-\infty, -\frac{7}{3}\right] \cup [3, \infty)$

28. $|4x + 1| \geq 21$

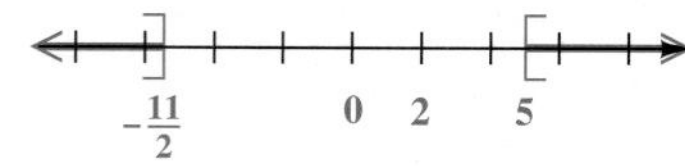

$\left(-\infty, -\frac{11}{2}\right] \cup [5, \infty)$

29. $|3 - x| > 5$

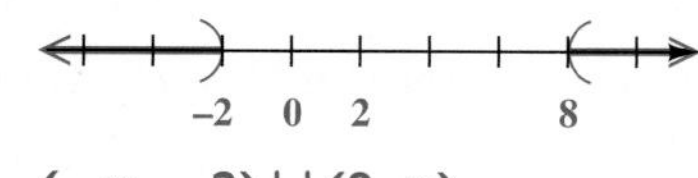

$(-\infty, -2) \cup (8, \infty)$

30. $|5 - x| > 3$

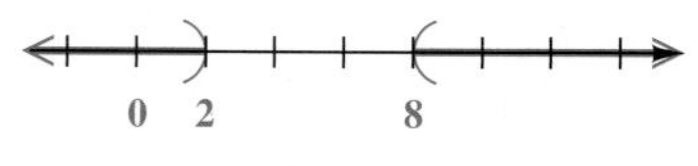

$(-\infty, 2) \cup (8, \infty)$

31. The graph of the solution set of $|2x + 1| = 9$ is given here.

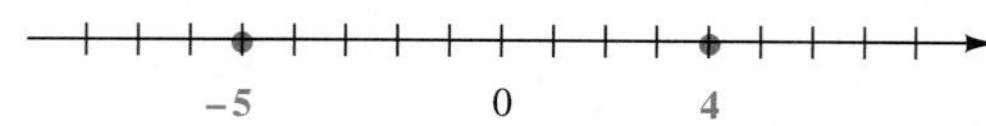

Without actually doing the algebraic work, graph the solution set of each inequality, referring to the graph above.

(a) $|2x + 1| < 9$

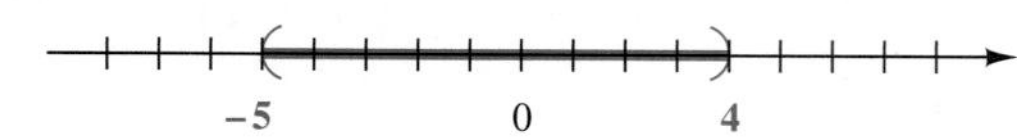

(b) $|2x + 1| > 9$

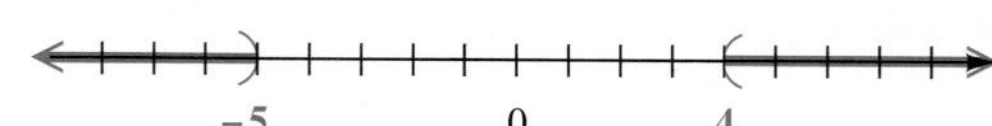

32. The graph of the solution set of $|3x - 4| < 5$ is given here.

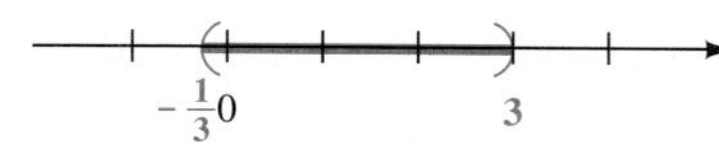

Without actually doing the algebraic work, graph the solution set of the equation and the inequality, referring to the graph above.

(a) $|3x - 4| = 5$

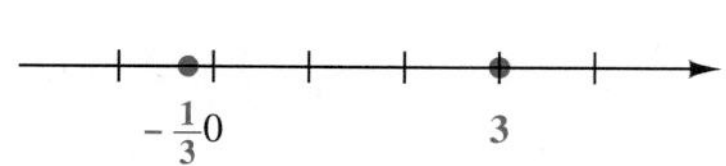

(b) $|3x - 4| > 5$

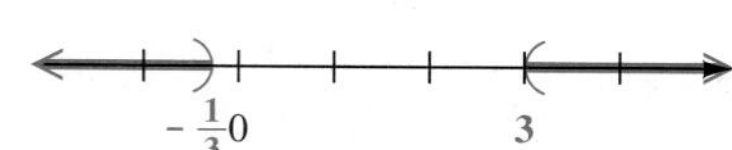

Solve each inequality, and graph the solution set. See Example 3. (Hint: Compare your answers to those in Exercises 21–30.)

33. $|x| \le 3$

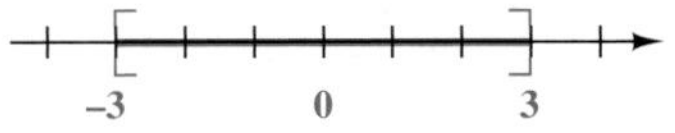

[−3, 3]

34. $|x| \le 2$

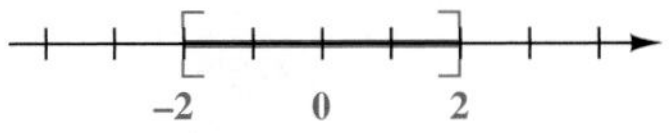

[−2, 2]

35. $|k| < 4$

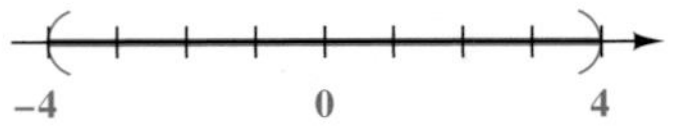

(−4, 4)

36. $|r| < 1$

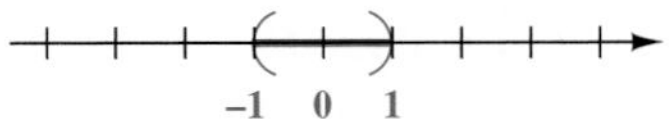

(−1, 1)

37. $|t + 2| \le 8$

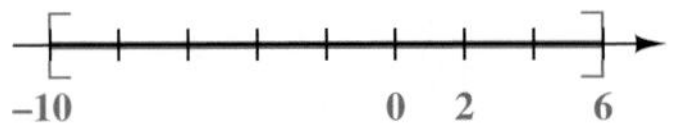

[−10, 6]

38. $|r + 5| \le 20$

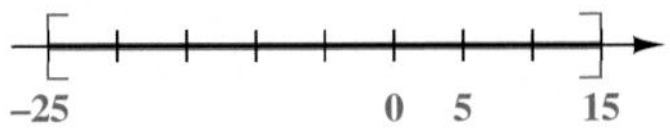

[−25, 15]

39. $|3x - 1| < 8$

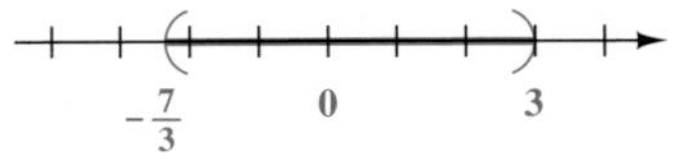

$\left(-\frac{7}{3}, 3\right)$

40. $|4x + 1| < 21$

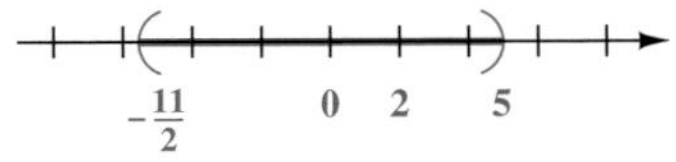

$\left(-\frac{11}{2}, 5\right)$

41. $|3 - x| \le 5$

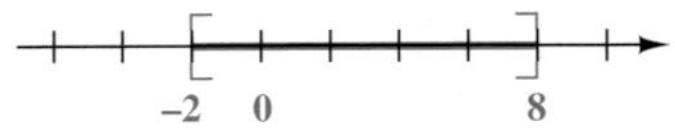

[−2, 8]

42. $|5 - x| \le 3$

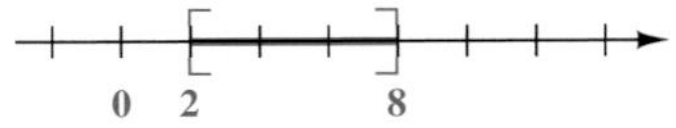

[2, 8]

Exercises 43–50 represent a sampling of the various types of absolute value equations and inequalities. Decide which method of solution applies, find the solution set, and graph. See Examples 1–3.

43. $|-4 + k| > 6$

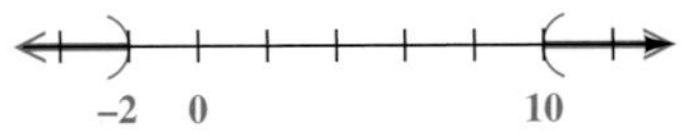

(−∞, −2) ∪ (10, ∞)

44. $|-3 + t| > 5$

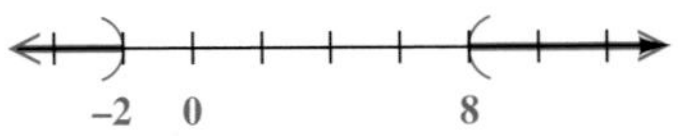

(−∞, −2) ∪ (8, ∞)

45. $|7 + 2z| = 5$

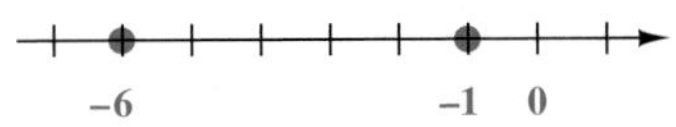

{−6, −1}

46. $|9 - 3p| = 3$

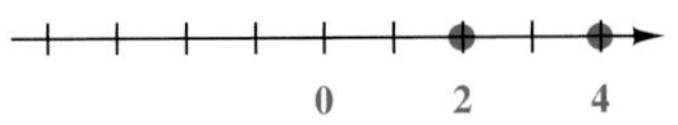

{2, 4}

47. $|3r - 1| \le 11$

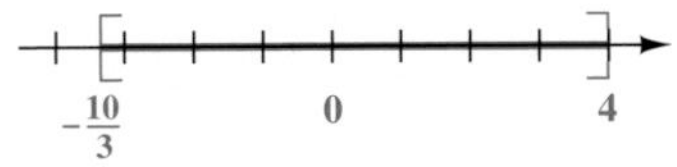

$\left[-\frac{10}{3}, 4\right]$

48. $|2s - 6| \le 6$

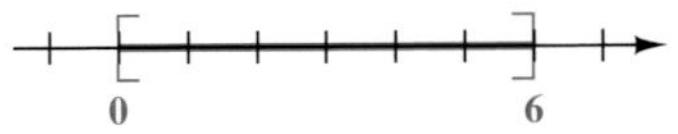

[0, 6]

49. $|-3x - 8| \le 4$

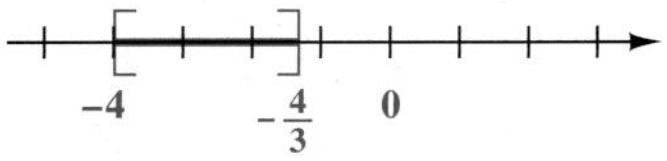

$\left[-4, -\frac{4}{3}\right]$

50. $|-2x - 6| \le 5$

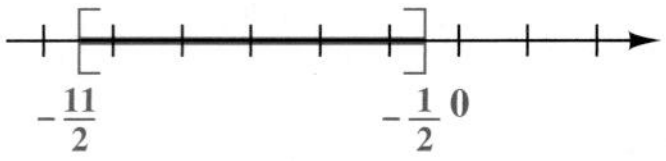

$\left[-\frac{11}{2}, -\frac{1}{2}\right]$

Solve each equation or inequality. Give the solution set using set notation for equations and interval notation for inequalities. See Example 4.

51. $|x| - 1 = 4$

$\{-5, 5\}$

52. $|x| + 3 = 10$

$\{-7, 7\}$

53. $|x + 4| + 1 = 2$

$\{-5, -3\}$

54. $|x + 5| - 2 = 12$

$\{-19, 9\}$

55. $|2x + 1| + 3 > 8$

$(-\infty, -3) \cup (2, \infty)$

56. $|6x - 1| - 2 > 6$

$\left(-\infty, -\frac{7}{6}\right) \cup \left(\frac{3}{2}, \infty\right)$

57. $|x + 5| - 6 \le -1$

$[-10, 0]$

58. $|r - 2| - 3 \le 4$

$[-5, 9]$

Solve each equation. See Example 5.

59. $|3x + 1| = |2x + 4|$

$\{-1, 3\}$

60. $|7x + 12| = |x - 8|$

$\left\{-\frac{10}{3}, -\frac{1}{2}\right\}$

61. $\left|m - \frac{1}{2}\right| = \left|\frac{1}{2}m - 2\right|$

$\left\{-3, \frac{5}{3}\right\}$

62. $\left|\frac{2}{3}r - 2\right| = \left|\frac{1}{3}r + 3\right|$

$\{-1, 15\}$

63. $|6x| = |9x + 1|$

$\left\{-\frac{1}{3}, -\frac{1}{15}\right\}$

64. $|13x| = |2x + 1|$

$\left\{-\frac{1}{15}, \frac{1}{11}\right\}$

65. $|2p - 6| = |2p + 11|$

$\left\{-\frac{5}{4}\right\}$

66. $|3x - 1| = |3x + 9|$

$\left\{-\frac{4}{3}\right\}$

Solve each equation or inequality. See Examples 6 and 7.

67. $|x| \geq -10$
$(-\infty, \infty)$

68. $|x| \geq -15$
$(-\infty, \infty)$

69. $|12t - 3| = -8$
$\emptyset$

70. $|13w + 1| = -3$
$\emptyset$

71. $|4x + 1| = 0$
$\left\{-\frac{1}{4}\right\}$

72. $|6r - 2| = 0$
$\left\{\frac{1}{3}\right\}$

73. $|2q - 1| < -6$
$\emptyset$

74. $|8n + 4| < -4$
$\emptyset$

75. $|x + 5| > -9$
$(-\infty, \infty)$

76. $|x + 9| > -3$
$(-\infty, \infty)$

77. $|7x + 3| \leq 0$
$\left\{-\frac{3}{7}\right\}$

78. $|4x - 1| \leq 0$
$\left\{\frac{1}{4}\right\}$

79. $|5x - 2| \geq 0$
$(-\infty, \infty)$

80. $|4 + 7x| \geq 0$
$(-\infty, \infty)$

81. $|10z + 7| > 0$
$\left(-\infty, -\frac{7}{10}\right) \cup \left(-\frac{7}{10}, \infty\right)$

82. $|4x + 1| > 0$
$\left(-\infty, -\frac{1}{4}\right) \cup \left(-\frac{1}{4}, \infty\right)$

83. $|x - 2| + 3 \geq 2$
$(-\infty, \infty)$

84. $|k - 4| + 5 \geq 4$
$(-\infty, \infty)$

85. The 2007 recommended daily intake (RDI) of calcium for females aged 19–50 is 1000 mg/day. Actual vitamin needs vary from person to person. Write an absolute value inequality to express the RDI plus or minus 100 mg and solve it. (*Source:* Food and Nutrition Board, National Academy of Sciences Institute of Medicine.)
$|x - 1000| \leq 100$; $900 \leq x \leq 1100$

86. The average clotting time of blood is 7.45 sec with a variation of plus or minus 3.6 sec. Write this statement as an absolute value inequality and solve it.
$|x - 7.45| \leq 3.6$; $3.85 \leq x \leq 11.05$

Relating Concepts (Exercises 87–90) For Individual or Group Work

The 10 tallest buildings in Kansas City, Missouri, are listed along with their heights.

Building	Height (in feet)
One Kansas City Place	632
Town Pavilion	591
Hyatt Regency Crown Center	504
Kansas City Power and Light	481
Fidelity Bank and Trust Building	454
City Hall	443
1201 Walnut	427
Federal Office Building	413
Commerce Tower	407
City Center Square	404

Source: World Almanac and Book of Facts.

Use this information to ***work Exercises 87–90 in order.***

87. To find the average of a group of numbers, we add the numbers and then divide by the number of items added. Use a calculator to find the average of the heights.

475.6 ft

88. Let k represent the average height of these buildings. If a height x satisfies the inequality

$$|x - k| < t,$$

then the height is said to be within t feet of the average. Using your result from Exercise 87, list the buildings that are within 50 ft of the average.

1201 Walnut, City Hall, Fidelity Bank and Trust Building, Kansas City Power and Light, Hyatt Regency Crown Center

89. Repeat Exercise 88, but find the buildings that are within 75 ft of the average.

City Center Square, Commerce Tower, Federal Office Building, 1201 Walnut, City Hall, Fidelity Bank and Trust Building, Kansas City Power and Light, Hyatt Regency Crown Center

90. **(a)** Write an absolute value inequality that describes the height of a building that is *not* within 75 ft of the average.

$|x - 475.6| \geq 75$

(b) Solve the inequality you wrote in part (a).

$x \geq 550.6$ **or** $x \leq 400.6$

(c) Use the result of part (b) to find the buildings that are not within 75 ft of the average.

Town Pavilion, One Kansas City Place

(d) Confirm that your answer to part (c) makes sense by comparing it with your answer to Exercise 89.

It makes sense because it includes all buildings *not* listed in the answer to Exercise 89.

Summary Exercises on Solving Linear and Absolute Value Equations and Inequalities

This section of miscellaneous equations and inequalities provides practice in solving all the types introduced in **Chapters 2 and 8.** You might wish to refer to the boxes in these chapters that summarize the various methods of solution.

Solve each equation or inequality. Give the solution set using set notation for equations and interval notation for inequalities.

1. $4z + 1 = 49$

$\{12\}$

2. $|m - 1| = 6$

$\{-5, 7\}$

3. $6q - 9 = 12 + 3q$

$\{7\}$

4. $3p + 7 = 9 + 8p$

$\left\{-\frac{2}{5}\right\}$

5. $|a + 3| = -4$

$\emptyset$

6. $2m + 1 \leq m$

$(-\infty, -1]$

7. $8r + 2 \geq 5r$

$\left[-\frac{2}{3}, \infty\right)$

8. $4(a - 11) + 3a = 20a - 31$

$\{-1\}$

9. $2q - 1 = -7$

$\{-3\}$

10. $|3q - 7| - 4 = 0$

$\left\{1, \frac{11}{3}\right\}$

11. $6z - 5 \leq 3z + 10$

$(-\infty, 5]$

12. $|5z - 8| + 9 \geq 7$

$(-\infty, \infty)$

13. $9x - 3(x + 1) = 8x - 7$

$\{2\}$

14. $|x| \geq 8$

$(-\infty, -8] \cup [8, \infty)$

15. $9x - 5 \geq 9x + 3$

$\emptyset$

16. $13p - 5 > 13p - 8$

$(-\infty, \infty)$

17. $|q| < 5.5$

$(-5.5, 5.5)$

18. $4z - 1 = 12 + z$

$\left\{\frac{13}{3}\right\}$

19. $\frac{2}{3}x + 8 = \frac{1}{4}x$

$\left\{-\frac{96}{5}\right\}$

20. $-\frac{5}{8}x \geq -20$

$(-\infty, 32]$

21. $\frac{1}{4}p < -6$

$(-\infty, -24)$

22. $7z - 3 + 2z = 9z - 8z$

$\left\{\frac{3}{8}\right\}$

23. $\frac{3}{5}q - \frac{1}{10} = 2$

$\left\{\frac{7}{2}\right\}$

24. $|r - 1| < 7$

$(-6, 8)$

25. $r + 9 + 7r = 4(3 + 2r) - 3$
$(-\infty, \infty)$

26. $6 - 3(2 - p) < 2(1 + p) + 3$
$(-\infty, 5)$

27. $|2p - 3| > 11$
$(-\infty, -4) \cup (7, \infty)$

28. $\frac{x}{4} - \frac{2x}{3} = -10$
$\{24\}$

29. $|5a + 1| \leq 0$
$\left\{-\frac{1}{5}\right\}$

30. $5z - (3 + z) \geq 2(3z + 1)$
$\left(-\infty, -\frac{5}{2}\right]$

31. $-2 \leq 3x - 1 \leq 8$
$\left[-\frac{1}{3}, 3\right]$

32. $-1 \leq 6 - x \leq 5$
$[1, 7]$

33. $|7z - 1| = |5z + 3|$
$\left\{-\frac{1}{6}, 2\right\}$

34. $|p + 2| = |p + 4|$
$\{-3\}$

35. $|1 - 3x| \geq 4$
$(-\infty, -1] \cup \left[\frac{5}{3}, \infty\right)$

36. $\frac{1}{2} \leq \frac{2}{3}r \leq \frac{5}{4}$
$\left[\frac{3}{4}, \frac{15}{8}\right]$

37. $-(m + 4) + 2 = 3m + 8$
$\left\{-\frac{5}{2}\right\}$

38. $\frac{p}{6} - \frac{3p}{5} = p - 86$
$\{60\}$

39. $-6 \leq \frac{3}{2} - x \leq 6$
$\left[-\frac{9}{2}, \frac{15}{2}\right]$

40. $|5 - x| < 4$
$(1, 9)$

41. $|x - 1| \geq -6$
$(-\infty, \infty)$

42. $|2r - 5| = |r + 4|$
$\left\{\frac{1}{3}, 9\right\}$

43. $8q - (1 - q) = 3(1 + 3q) - 4$
$(-\infty, \infty)$

44. $8x - (x + 3) = -(2x + 1) - 12$
$\left\{-\frac{10}{9}\right\}$

45. $|r - 5| = |r + 9|$
$\{-2\}$

46. $|r + 2| < -3$
$\emptyset$

47. $2x + 1 > 5$ or $3x + 4 < 1$
$(-\infty, -1) \cup (2, \infty)$

48. $1 - 2x \geq 5$ and $7 + 3x \geq -2$
$[-3, -2]$

8.4 Review of Systems of Linear Equations in Two Variables

This section reviews solving systems of linear equations in two variables.

OBJECTIVES

1 Solve linear systems with two equations and two variables.

2 Solve special systems.

OBJECTIVE 1 Solve linear systems with two equations and two variables. Recall from **Section 4.1** that a **system of linear equations** (often called a **linear system**) consists of two or more linear equations with the same variables. An example of a linear system is

$$\begin{aligned} x + y &= 5 \\ 2x - y &= 4. \end{aligned} \qquad \text{Linear system of equations}$$

The **solution set of a system of linear equations** contains all ordered pairs that satisfy all equations of the system *at the same time*. We introduced three methods for solving linear systems in **Chapter 4:** the *graphing method,* the *substitution method,* and the *elimination method.*

EXAMPLE 1 Solving a System by Graphing

Solve the system of equations by graphing.

$$\begin{aligned} x + y &= 5 \quad (1) \\ 2x - y &= 4 \quad (2) \end{aligned}$$

When we graph these linear equations as shown in Figure 24, the graph suggests that the point of intersection is the ordered pair (3, 2).

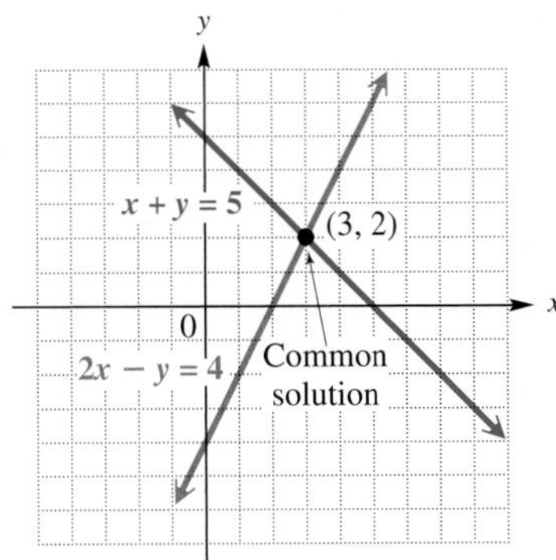

Figure 24

To be sure that (3, 2) is a solution of *both* equations, we check by substituting 3 for x and 2 for y in each equation.

Check

$x + y = 5$ (1)	$2x - y = 4$ (2)
$3 + 2 \stackrel{?}{=} 5$	$2(3) - 2 \stackrel{?}{=} 4$
$5 = 5$ True	$6 - 2 \stackrel{?}{=} 4$
	$4 = 4$ True

Since (3, 2) makes both equations true, $\{(3, 2)\}$ is the solution set of the system.

Work Problem **1** *at the Side.* ▶

1 Solve by graphing.

(a) $\begin{aligned} x - y &= 3 \quad (1) \\ 2x - y &= 4 \quad (2) \end{aligned}$

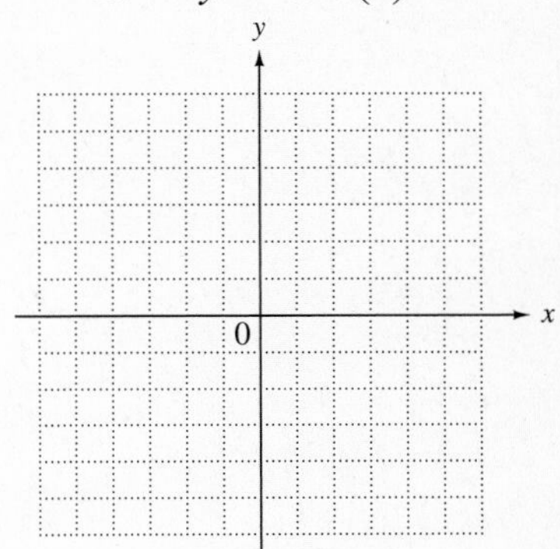

(b) $\begin{aligned} 2x + y &= -5 \quad (1) \\ -x + 3y &= 6 \quad (2) \end{aligned}$

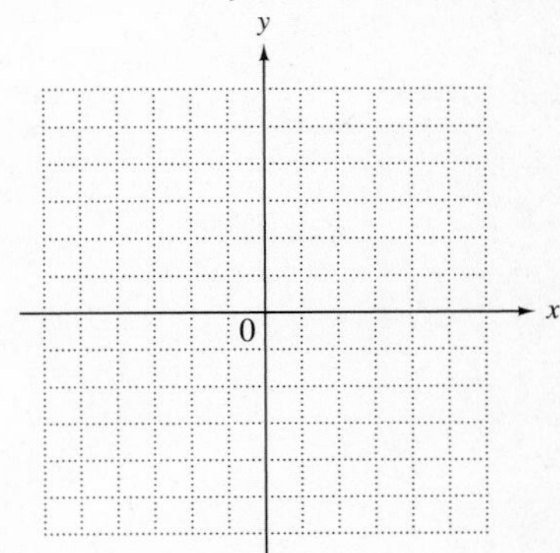

ANSWERS

1. (a) $\{(1, -2)\}$

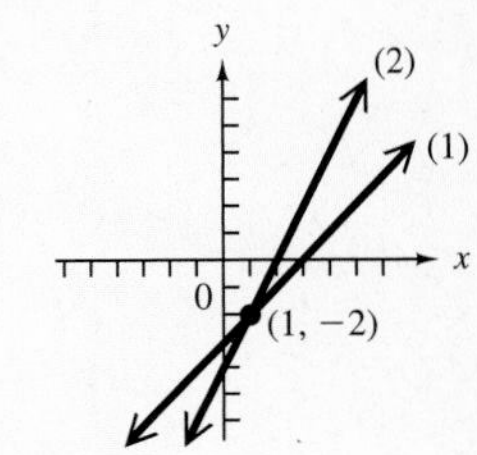

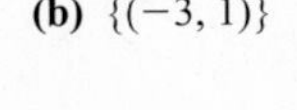
(b) $\{(-3, 1)\}$

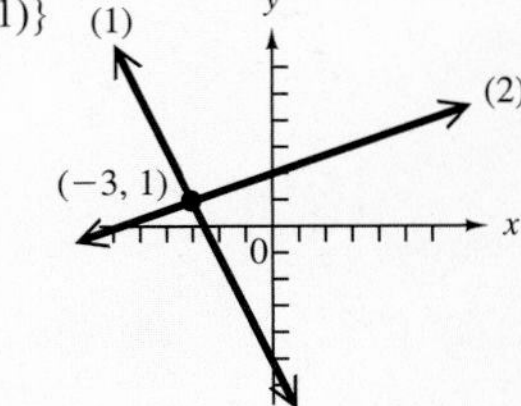

2 Solve by substitution.

(a) $7x - 2y = -2$
$y = 3x$

(b) $5x - 3y = -6$
$x = 2 - y$

Since it can be difficult to read exact coordinates, especially if they are not integers, from a graph, we usually use algebraic methods to solve systems. The **substitution method** is most useful for solving linear systems in which one equation is solved or can be easily solved for one variable in terms of the other.

EXAMPLE 2 Solving a System by Substitution

Solve the system.

$$2x - y = 6 \quad (1)$$
$$x = y + 2 \quad (2)$$

Since equation (2) is solved for x, substitute $y + 2$ for x in equation (1).

$2x - y = 6$	(1)
$2(y + 2) - y = 6$	Let $x = y + 2$.
$2y + 4 - y = 6$	Distributive property
$y + 4 = 6$	Combine like terms.
$y = 2$	Subtract 4.

Be sure to use parentheses here.

We found y. Now find x by substituting 2 for y in equation (2).

$$x = y + 2 = 2 + 2 = 4$$

Write the x-value first in the ordered pair.

Thus, $x = 4$ and $y = 2$, giving the ordered pair (4, 2). Check this solution in both equations of the original system.

Check

$2x - y = 6$	(1)	$x = y + 2$	(2)
$2(4) - 2 \stackrel{?}{=} 6$		$4 \stackrel{?}{=} 2 + 2$	
$8 - 2 \stackrel{?}{=} 6$		$4 = 4$	True
$6 = 6$	True		

Since (4, 2) makes both equations true, the solution set is $\{(4, 2)\}$.

◀ *Work Problem* **2** *at the Side.*

The substitution method is summarized as follows.

Solving a Linear System by Substitution

Step 1 **Solve one of the equations for either variable.** If one of the variable terms has coefficient 1 or -1, choose it, since the substitution method is usually easier this way.

Step 2 **Substitute** for that variable in the other equation. The result should be an equation with just one variable.

Step 3 **Solve** the equation from Step 2.

Step 4 **Find the other value.** Substitute the result from Step 3 into the equation from Step 1 to find the value of the other variable.

Step 5 **Check** the ordered-pair solution in *both* of the *original* equations. Then write the solution set.

ANSWERS

2. (a) $\{(-2, -6)\}$ **(b)** $\{(0, 2)\}$

EXAMPLE 3 Solving a System by Substitution

Solve the system.

$$3x + 2y = 13 \quad (1)$$
$$4x - y = -1 \quad (2)$$

Step 1 First solve one of the equations for x or y. Since the coefficient of y in equation (2) is -1, it is easiest to solve for y in equation (2).

$$4x - y = -1 \quad (2)$$
$$-y = -1 - 4x \quad \text{Subtract } 4x.$$
$$y = 1 + 4x \quad \text{Multiply by } -1.$$

Step 2 Substitute $1 + 4x$ for y in equation (1).

$$3x + 2y = 13 \quad (1)$$
$$3x + 2(1 + 4x) = 13 \quad \text{Let } y = 1 + 4x.$$

Step 3 Solve for x.

$$3x + 2 + 8x = 13 \quad \text{Distributive property}$$
$$11x = 11 \quad \text{Combine like terms; subtract 2.}$$
$$x = 1 \quad \text{Divide by 11.}$$

Step 4 Now solve for y. From Step 1, $y = 1 + 4x$, so if $x = 1$, then

$$y = 1 + 4(1) = 5. \quad \text{Let } x = 1.$$

Step 5 Check the solution (1, 5) in both equations (1) and (2).

Check

$3x + 2y = 13$ (1)	$4x - y = -1$ (2)
$3(1) + 2(5) \stackrel{?}{=} 13$	$4(1) - 5 \stackrel{?}{=} -1$
$3 + 10 \stackrel{?}{=} 13$	$4 - 5 \stackrel{?}{=} -1$
$13 = 13$ True	$-1 = -1$ True

The solution set is $\{(1, 5)\}$.

Work Problem 3 at the Side. ▶

3 Solve by substitution.

(a) $3x - y = 10$
$2x + 5y = 1$

(b) $4x - 5y = -11$
$x + 2y = 7$

The **elimination method** involves combining the two equations in a system so that one variable is eliminated. This is done using the following logic:

If $a = b$ and $c = d$, then $a + c = b + d$.

EXAMPLE 4 Solving a System by Elimination

Solve the system.

$$2x + 3y = -6 \quad (1)$$
$$4x - 3y = 6 \quad (2)$$

Notice that adding the equations together will eliminate the variable y.

$$2x + 3y = -6 \quad (1)$$
$$\underline{4x - 3y = 6} \quad (2)$$
$$6x = 0 \quad \text{Add.}$$
$$x = 0 \quad \text{Solve for } x.$$

Continued on Next Page

ANSWERS

3. **(a)** $\{(3, -1)\}$ **(b)** $\{(1, 3)\}$

To find y, substitute 0 for x in either equation (1) or equation (2).

$$\begin{aligned} 2x + 3y &= -6 && (1) \\ 2(0) + 3y &= -6 && \text{Let } x = 0. \\ 0 + 3y &= -6 && \text{Multiply.} \\ 3y &= -6 && \text{Add.} \\ y &= -2 && \text{Divide by 3.} \end{aligned}$$

The solution of the system is $(0, -2)$. Check by substituting 0 for x and -2 for y in both equations of the original system. The solution set is $\{(0, -2)\}$.

◀ *Work Problem* 4 *at the Side.*

4 Solve by elimination.

(a) $3x - y = -7$
$2x + y = -3$

(b) $-2x + 3y = -10$
$2x + 2y = 5$

By adding the equations in Example 4, we eliminated the variable y because the coefficients of the y-terms were opposites. In many cases the coefficients will *not* be opposites, and we must transform one or both equations so that the coefficients of one pair of variable terms are opposites.

Solving a Linear System by Elimination

Step 1 **Write both equations in standard form $Ax + By = C$.**

Step 2 **Make the coefficients of one pair of variable terms opposites.** Multiply one or both equations by appropriate numbers so that the sum of the coefficients of either the x- or y-terms is 0.

Step 3 **Add** the new equations to eliminate a variable. The sum should be an equation with just one variable.

Step 4 **Solve** the equation from Step 3 for the remaining variable.

Step 5 **Find the other value.** Substitute the result from Step 4 into either of the original equations and solve for the other variable.

Step 6 **Check** the ordered-pair solution in *both* of the *original* equations. Then write the solution set.

EXAMPLE 5 Solving a System by Elimination

Solve the system.

$$\begin{aligned} 5x - 2y &= 4 && (1) \\ 2x + 3y &= 13 && (2) \end{aligned}$$

Step 1 Both equations are in standard form.

Step 2 Suppose that you wish to eliminate the variable x. One way to do this is to multiply equation (1) by 2 and equation (2) by -5.

$$\begin{aligned} 10x - 4y &= 8 && \text{2 times each side of equation (1)} \\ -10x - 15y &= -65 && \text{-5 times each side of equation (2)} \end{aligned}$$

Step 3 Now add.

$$\begin{aligned} 10x - 4y &= 8 \\ -10x - 15y &= -65 \\ \hline -19y &= -57 && \text{Add.} \end{aligned}$$

Step 4 Solve for y. $y = 3$ Divide by -19.

Continued on Next Page

ANSWERS

4. (a) $\{(-2, 1)\}$ (b) $\left\{\left(\frac{7}{2}, -1\right)\right\}$

Step 5 To find x, substitute 3 for y in either equation (1) or (2).

$$\begin{aligned} 2x + 3y &= 13 && (2) \\ 2x + 3(3) &= 13 && \text{Let } y = 3. \\ 2x + 9 &= 13 && \text{Multiply.} \\ 2x &= 4 && \text{Subtract 9.} \\ x &= 2 && \text{Divide by 2.} \end{aligned}$$

Step 6 To check, substitute 2 for x and 3 for y in equations (1) and (2).

Check

$$\begin{aligned} 5x - 2y &= 4 && (1) \\ 5(2) - 2(3) &\stackrel{?}{=} 4 \\ 10 - 6 &\stackrel{?}{=} 4 \\ 4 &= 4 && \text{True} \end{aligned}$$

$$\begin{aligned} 2x + 3y &= 13 && (2) \\ 2(2) + 3(3) &\stackrel{?}{=} 13 \\ 4 + 9 &\stackrel{?}{=} 13 \\ 13 &= 13 && \text{True} \end{aligned}$$

The solution set is $\{(2, 3)\}$.

Work Problem **5** *at the Side.* ▶

5 Solve by elimination.

(a) $x + 3y = 8$
$2x - 5y = -17$

(b) $6x - 2y = -21$
$-3x + 4y = 36$

(c) $2x + 3y = 19$
$3x - 7y = -6$

OBJECTIVE 2 Solve special systems. Since the graph of a linear equation is a straight line, there are three possibilities for the solution set of a linear system in two variables.

Graphs of Linear Systems in Two Variables

Case 1 **The two graphs intersect in a single point.** The coordinates of this point give the only solution of the system. Since the system has a solution, it is **consistent.** The equations are *not* equivalent, so they are **independent.** See Figure 25(a).

Case 2 **The graphs are parallel lines.** There is no solution common to both equations, so the solution set is $\emptyset$ and the system is **inconsistent.** Since the equations are *not* equivalent, they are **independent.** See Figure 25(b).

Case 3 **The graphs are the same line.** Since any solution of one equation of the system is a solution of the other, the solution set is an infinite set of ordered pairs representing the points on the line. This type of system is **consistent** because there is a solution. The equations are equivalent, so they are **dependent.** See Figure 25(c).

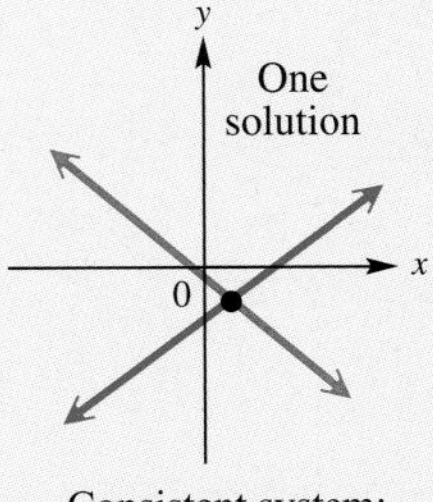

Consistent system; independent equations

(a)

No solution

Inconsistent system; independent equations

(b)

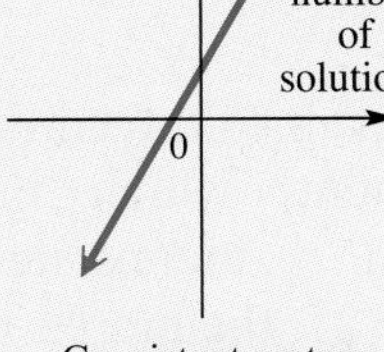

Consistent system; dependent equations

(c)

Figure 25

Case 1 was illustrated in Examples 1–5. We illustrate Cases 2 and 3 in the examples on the next page.

ANSWERS

5. (a) $\{(-1, 3)\}$ **(b)** $\left\{\left(-\frac{2}{3}, \frac{17}{2}\right)\right\}$
(c) $\{(5, 3)\}$

6 Solve the system. Then graph both equations.

$$\begin{aligned} 2x + y &= 6 \quad (1) \\ -8x - 4y &= -24 \quad (2) \end{aligned}$$

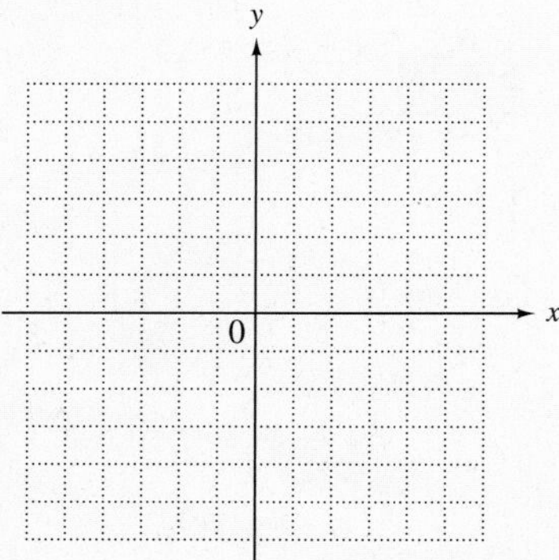

7 Solve the system. Then graph both equations.

$$\begin{aligned} 4x - 3y &= 8 \quad (1) \\ 8x - 6y &= 14 \quad (2) \end{aligned}$$

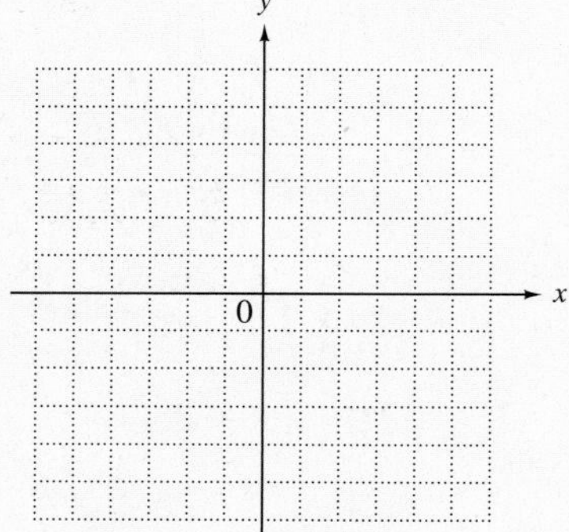

EXAMPLE 6 Solving a System of Dependent Equations

Solve the system.

$$\begin{aligned} 2x - y &= 3 \quad (1) \\ 6x - 3y &= 9 \quad (2) \end{aligned}$$

We multiply equation (1) by -3, and then add the result to equation (2).

$$\begin{aligned} -6x + 3y &= -9 \quad && -3 \text{ times each side of equation (1)} \\ 6x - 3y &= 9 \quad && (2) \\ \hline 0 &= 0 \quad && \text{True} \end{aligned}$$

Adding these equations gives the true statement $0 = 0$. In the original system, we could get equation (2) from equation (1) by multiplying equation (1) by 3. Because of this, equations (1) and (2) are equivalent and have the same graph, as shown in Figure 26. The equations are dependent.

The solution set is the set of all points on the line with equation $2x - y = 3$, written in set-builder notation **(Section 1.3)** as

$$\{(x, y) \mid 2x - y = 3\}$$

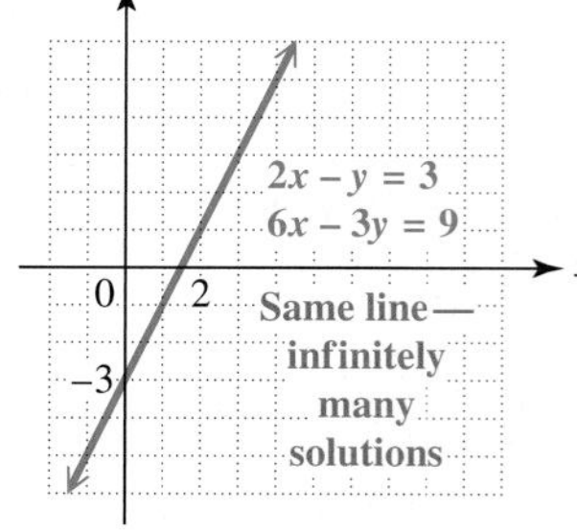

Figure 26

and read "the set of all ordered pairs (x, y), such that $2x - y = 3$."

Work Problem **6** *at the Side.*

Note

To write the solution set in Example 6, ***we use the equation in standard form with coefficients that are integers having greatest common factor 1 and positive coefficient of x.***

EXAMPLE 7 Solving an Inconsistent System

Solve the system.

$$\begin{aligned} x + 3y &= 4 \quad (1) \\ -2x - 6y &= 3 \quad (2) \end{aligned}$$

Multiply equation (1) by 2, and then add the result to equation (2).

$$\begin{aligned} 2x + 6y &= 8 \quad && \text{Equation (1) multiplied by 2} \\ -2x - 6y &= 3 \quad && (2) \\ \hline 0 &= 11 \quad && \text{False} \end{aligned}$$

The result of the addition step is a false statement, which indicates that the system is inconsistent. As shown in Figure 27, the graphs of the equations of the system are parallel lines. There are no ordered pairs that satisfy both equations, so there is no solution for the system. The solution set is $\emptyset$.

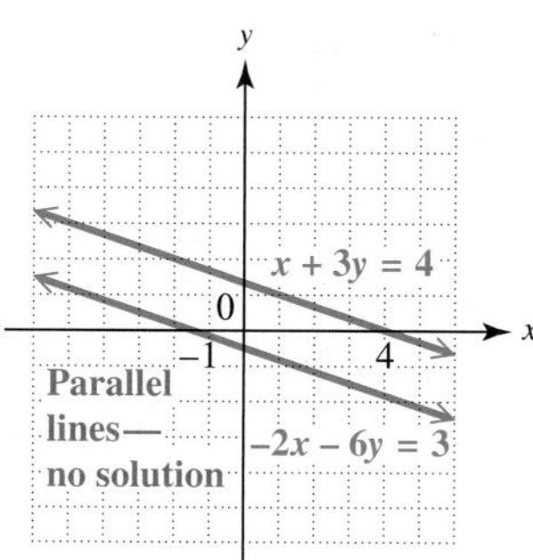

Figure 27

Work Problem **7** *at the Side.*

Answers

6. $\{(x, y) \mid 2x + y = 6\}$

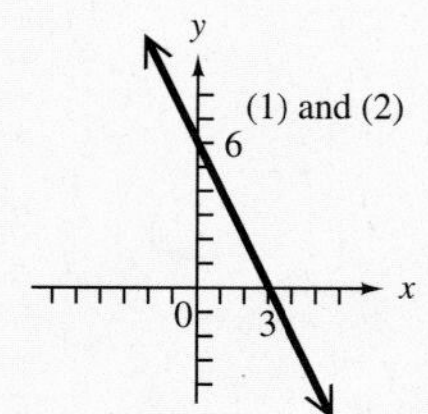

7. $\emptyset$

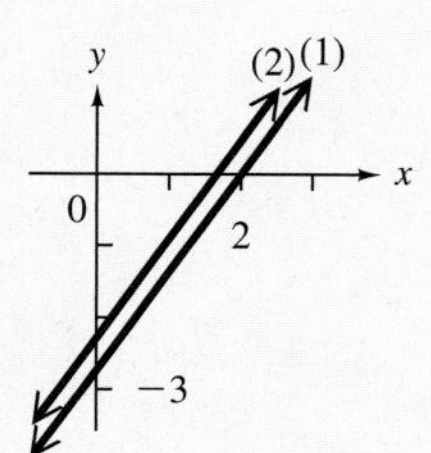

8.4 Exercises

FOR EXTRA HELP

MyMathLab

Math XL PRACTICE  WATCH  DOWNLOAD  READ  REVIEW

1. Which ordered pair could possibly be a solution of the graphed system of equations? Why?

A. (3, 3)
B. (−3, 3)
C. (−3, −3)
D. (3, −3)

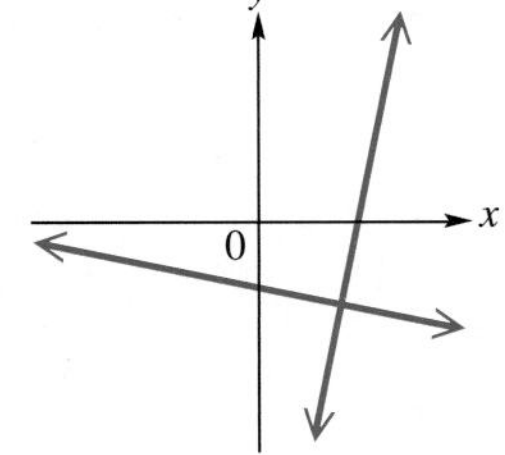

D; The ordered pair solution must be in quadrant IV, since that is where the graphs of the equations intersect.

2. Which ordered pair could possibly be a solution of the graphed system of equations? Why?

A. (3, 0)
B. (−3, 0)
C. (0, 3)
D. (0, −3)

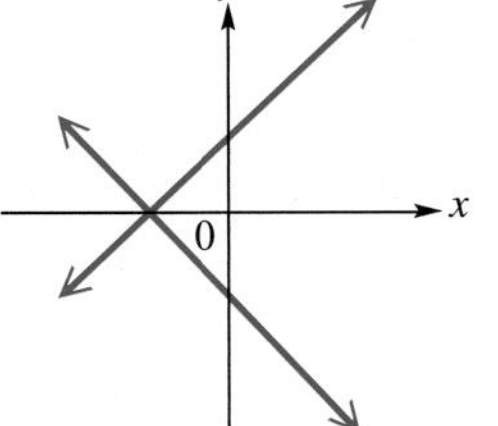

B; The ordered pair solution must be on the x-axis, with $x < 0$, since that is where the graphs of the equations intersect.

3. Match each system in (a)–(d) with the correct graph in A–D.

(a) $x + y = 6$
$x - y = 0$ **B**

(b) $x + y = -6$
$x - y = 0$ **C**

(c) $x + y = 0$
$x - y = -6$ **A**

(d) $x + y = 0$
$x - y = 6$ **D**

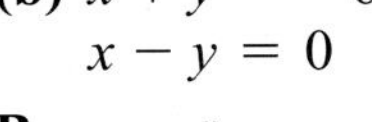

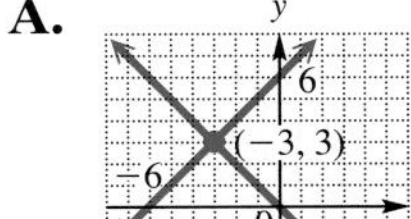

A. 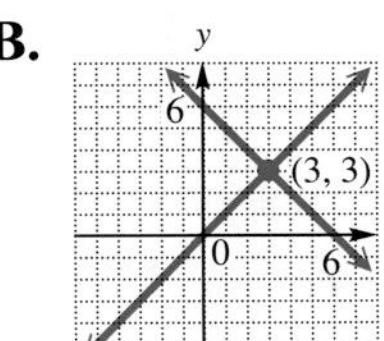

B. 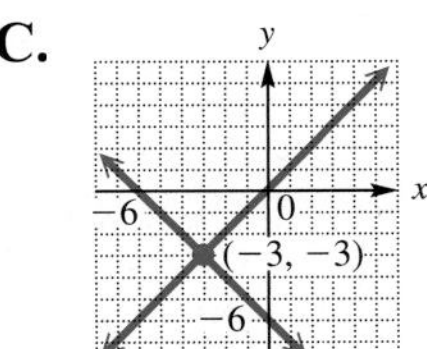

C. 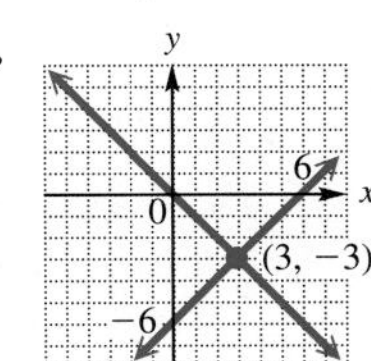

D.

Solve each system by graphing. See Example 1.

4. $x + y = 4$
$2x - y = 2$

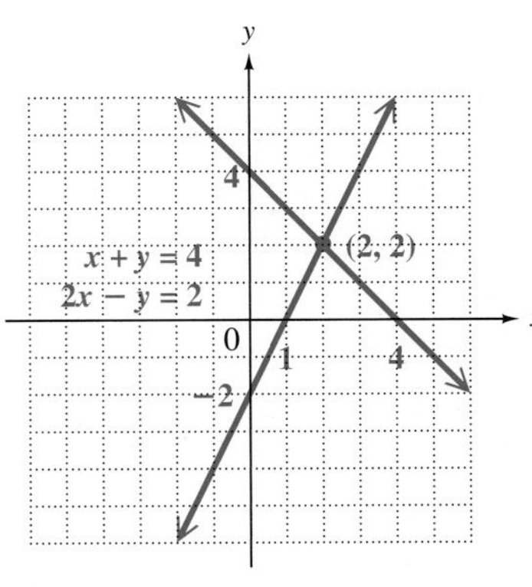

{(2, 2)}

5. $x + y = -5$
$-2x + y = 1$

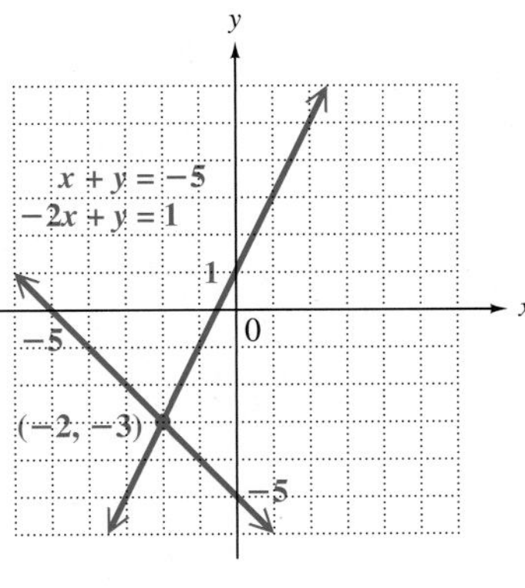

{(−2, −3)}

6. $x - 4y = -4$
$3x + y = 1$

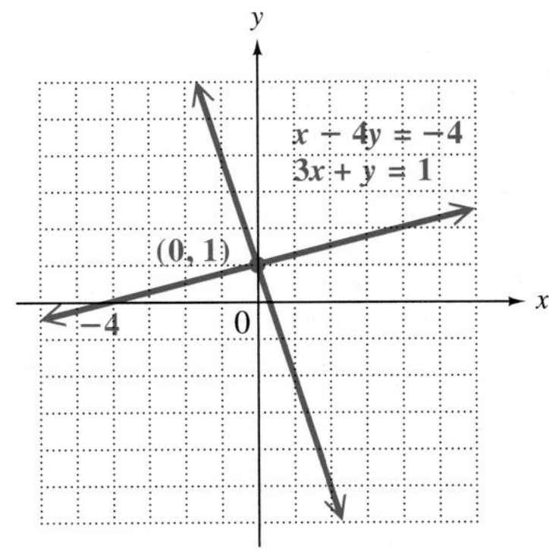

{(0, 1)}

Solve each system by substitution. If the system is inconsistent or has dependent equations, say so. See Examples 2, 3, 6, and 7.

7. $4x + y = 6$
$y = 2x$
{(1, 2)}

8. $2x - y = 6$
$y = 5x$
{(−2, −10)}

9. $-x - 4y = -14$
$y = 2x - 1$
{(2, 3)}

10. $-3x - 5y = -17$
$y = 4x + 8$

$\{(-1, 4)\}$

11. $3x - 4y = -22$
$-3x + y = 0$

$\left\{\left(\frac{22}{9}, \frac{22}{3}\right)\right\}$

12. $-3x + y = -5$
$x + 2y = 0$

$\left\{\left(\frac{10}{7}, -\frac{5}{7}\right)\right\}$

13. $5x - 4y = 9$
$3 - 2y = -x$

$\{(5, 4)\}$

14. $6x - y = -9$
$4 + 7x = -y$

$\{(-1, 3)\}$

15. $x = 3y + 5$
$x = \frac{3}{2}y$

$\left\{\left(-5, -\frac{10}{3}\right)\right\}$

16. $x = 6y - 2$
$x = \frac{3}{4}y$

$\left\{\left(\frac{2}{7}, \frac{8}{21}\right)\right\}$

17. $\frac{1}{2}x + \frac{1}{3}y = 3$
$-3x + y = 0$

$\{(2, 6)\}$

18. $\frac{1}{4}x - \frac{1}{5}y = 9$
$5x - y = 0$

$\{(-12, -60)\}$

19. $y = 2x$
$4x - 2y = 0$

$\{(x, y) \mid 2x - y = 0\}$; dependent equations

20. $x = 3y$
$3x - 9y = 0$

$\{(x, y) \mid x - 3y = 0\}$; dependent equations

21. $5x - 25y = 5$
$x = 5y$

$\emptyset$; inconsistent system

22. $8x + 2y = 4$
$y = -4x$

$\emptyset$; inconsistent system

Solve each system by elimination. If the system is inconsistent or has dependent equations, say so. See Examples 4–7.

23. $-2x + 3y = -16$
$2x - 5y = 24$

$\{(2, -4)\}$

24. $6x + 5y = -7$
$-6x - 11y = 1$

$\{(-2, 1)\}$

25. $2x - 5y = 11$
$3x + y = 8$

$\{(3, -1)\}$

26. $-2x + 3y = 1$
$-4x + y = -3$

$\{(1, 1)\}$

27. $3x + 4y = -6$
$5x + 3y = 1$

$\{(2, -3)\}$

28. $4x + 3y = 1$
$3x + 2y = 2$

$\{(4, -5)\}$

29. $3x + 3y = 0$
$4x + 2y = 3$

$\left\{\left(\frac{3}{2}, -\frac{3}{2}\right)\right\}$

30. $8x + 4y = 0$
$4x - 2y = 2$

$\left\{\left(\frac{1}{4}, -\frac{1}{2}\right)\right\}$

31. $7x + 2y = 6$
$-14x - 4y = -12$

$\{(x, y) \mid 7x + 2y = 6\}$; dependent equations

32. $x - 4y = 2$
$4x - 16y = 8$

$\{(x, y) \mid x - 4y = 2\}$; dependent equations

33. $\frac{x}{2} + \frac{y}{3} = -\frac{1}{3}$
$\frac{x}{2} + 2y = -7$

$\{(2, -4)\}$

34. $\frac{x}{5} + y = \frac{6}{5}$
$\frac{x}{10} + \frac{y}{3} = \frac{5}{6}$

$\left\{\left(13, -\frac{7}{5}\right)\right\}$

35. $5x - 5y = 3$
$x - y = 12$

$\emptyset$; inconsistent system

36. $2x - 3y = 7$
$-4x + 6y = 14$

$\emptyset$; inconsistent system

8.5 Systems of Linear Equations in Three Variables; Applications

OBJECTIVES

1. Understand the geometry of systems of three equations in three variables.
2. Solve linear systems (with three equations and three variables) by elimination.
3. Solve linear systems (with three equations and three variables) in which some of the equations have missing terms.
4. Solve special systems.
5. Solve application problems with three variables using a system of three equations.

A solution of an equation in three variables, such as

$$2x + 3y - z = 4, \quad \text{Linear equation in three variables}$$

is called an **ordered triple** and is written **(x, y, z)**. For example, the ordered triple $(0, 1, -1)$ is a solution of the equation, because

$$2(0) + 3(1) - (-1) = 4$$

is a true statement. Verify that another solution of this equation is $(10, -3, 7)$.

We now extend the term *linear equation* to equations of the form

$$Ax + By + Cz + \ldots + Dw = K,$$

where not all the coefficients $A, B, C, \ldots, D$ equal 0. For example,

$$2x + 3y - 5z = 7 \quad \text{and} \quad x - 2y - z + 3u - 2w = 8$$

are linear equations, the first with three variables and the second with five.

OBJECTIVE 1 Understand the geometry of systems of three equations in three variables. Consider the solution of a system such as

$$\begin{aligned} 4x + 8y + z &= 2 \\ x + 7y - 3z &= -14 \\ 2x - 3y + 2z &= 3. \end{aligned} \quad \text{System of linear equations in three variables}$$

Theoretically, a system of this type can be solved by graphing. However, the graph of a linear equation with three variables is a *plane,* not a line. Since the graph of each equation of the system is a plane, which requires three-dimensional graphing, the graphing method is not practical for solving such systems. However, it does illustrate the number of solutions possible for these systems, as shown in Figure 28.

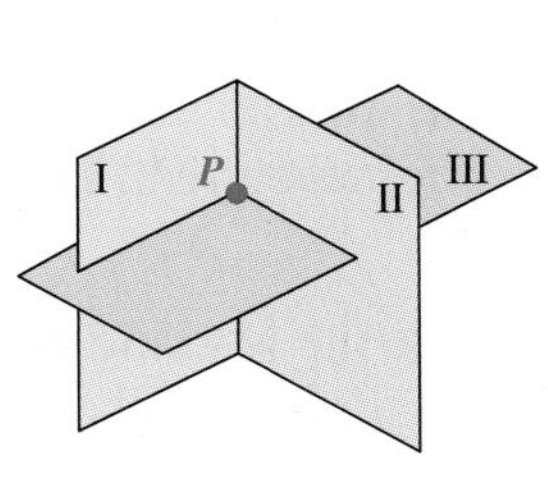

A single solution

(a)

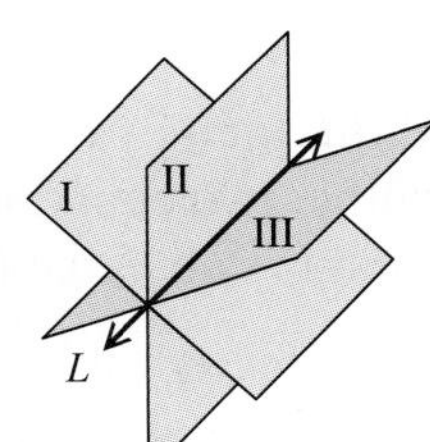

Points of a line in common

(b)

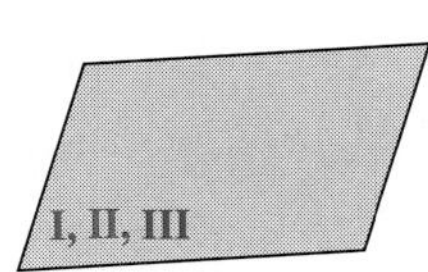

All points in common

(c)

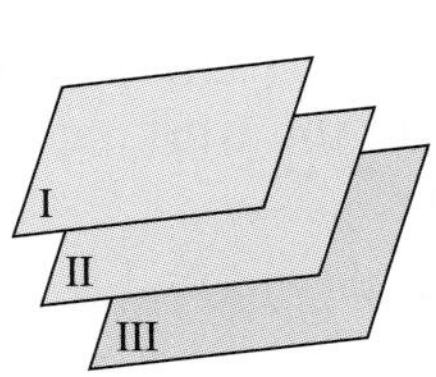

No points in common

(d)

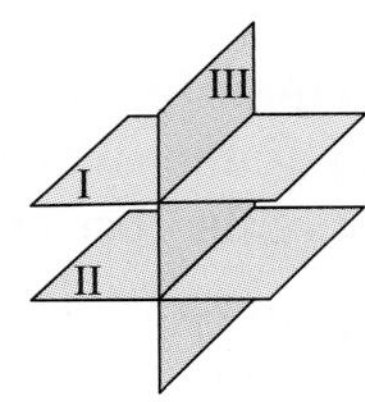

No points in common

(e)

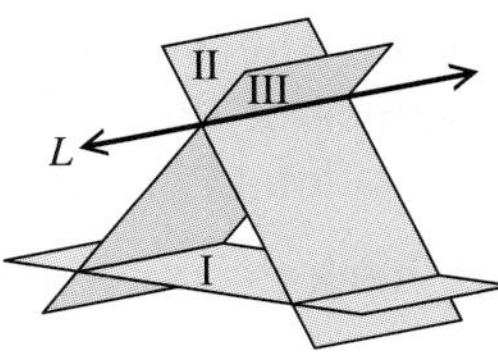

No points in common

(f)

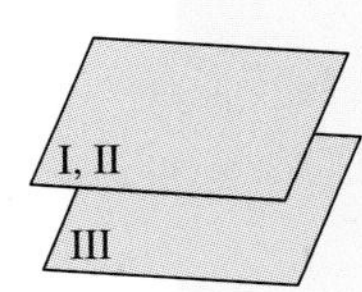

No points in common

(g)

Figure 28

Figure 28 on the preceding page illustrates the following cases.

Graphs of Linear Systems in Three Variables

Case 1 **The three planes may meet at a single, common point** that forms the solution set of the system. See Figure 28(a).

Case 2 **The three planes may have the points of a line in common** so that the infinite set of points that satisfy the equation of the line forms the solution set of the system. See Figure 28(b).

Case 3 **The three planes may coincide** so that the solution set of the system is the set of all points on a plane. See Figure 28(c).

Case 4 **The planes may have no points common to all three** so that there is no solution of the system; the solution set is $\emptyset$. See Figures 28(d)–(g).

OBJECTIVE 2 Solve linear systems (with three equations and three variables) by elimination. Is it possible to solve a system of three equations in three variables such as the one that follows?

$$\begin{aligned} 4x + 8y + z &= 2 \\ x + 7y - 3z &= -14 \\ 2x - 3y + 2z &= 3 \end{aligned}$$

As mentioned, graphing to find the solution set of such a system is impractical. We solve these systems with an extension of the elimination method, reviewed in **Section 8.4.**

Solving a Linear System in Three Variables*

Step 1 **Select a variable and an equation.** A good choice for the variable, which we call the *focus variable,* is one that has coefficient 1 or -1. Then select an equation, usually the one that contains the focus variable, as the *working equation*.

Step 2 **Eliminate the focus variable.** Use the working equation and one of the other two equations of the original system. The result is an equation in two variables.

Step 3 **Eliminate the focus variable again.** Use the working equation and the remaining equation of the original system. The result is another equation in two variables.

Step 4 **Write the equations in two variables that result from Steps 2 and 3 as a system, and solve it.** Doing this gives the values of two of the variables.

Step 5 **Find the value of the remaining variable.** Substitute the values of the two variables found in Step 4 into the working equation to obtain the value of the focus variable.

Step 6 **Check** the ordered-triple solution in *each* of the *original* equations of the system. Then write the solution set.

*The authors wish to thank Christine Heinecke Lehmann of Purdue University North Central for her suggestions here.

EXAMPLE 1 **Solving a System in Three Variables**

Solve the system.

$$\begin{aligned} 4x + 8y + z &= 2 \quad &(1)\\ x + 7y - 3z &= -14 \quad &(2)\\ 2x - 3y + 2z &= 3 \quad &(3) \end{aligned}$$

Step 1 Since z in equation (1) has coefficient 1, we choose z as the focus variable and (1) as the working equation. (Another option would be to choose x as the focus variable, since it also has coefficient 1, and use (2) as the working equation.)

$$4x + 8y + \underset{\text{Focus variable}}{z} = 2 \qquad (1) \leftarrow \text{Working equation}$$

Step 2 Multiply working equation (1) by 3 and add the result to equation (2) to eliminate focus variable z.

$$\begin{aligned} 12x + 24y + 3z &= 6 \qquad &\text{Multiply each side of (1) by 3.}\\ x + 7y - 3z &= -14 \qquad &(2)\\ \hline 13x + 31y \qquad &= -8 \qquad &\text{Add.} \quad (4) \end{aligned}$$

Step 3 Multiply working equation (1) by -2 and add the result to remaining equation (3) to again eliminate focus variable z.

$$\begin{aligned} -8x - 16y - 2z &= -4 \qquad &\text{Multiply each side of (1) by } -2.\\ 2x - 3y + 2z &= 3 \qquad &(3)\\ \hline -6x - 19y \qquad &= -1 \qquad &\text{Add.} \quad (5) \end{aligned}$$

Step 4 Write the equations in two variables that result in Steps 2 and 3 as a system.

Make sure these equations have the same variables.

$$\begin{aligned} 13x + 31y &= -8 \qquad &(4) \quad \text{The result from Step 2}\\ -6x - 19y &= -1 \qquad &(5) \quad \text{The result from Step 3} \end{aligned}$$

Now solve this system. We choose to eliminate x.

$$\begin{aligned} 78x + 186y &= -48 \qquad &\text{Multiply each side of (4) by 6.}\\ -78x - 247y &= -13 \qquad &\text{Multiply each side of (5) by 13.}\\ \hline -61y &= -61 \qquad &\text{Add.}\\ y &= \mathbf{1} \qquad &\text{Divide by } -61. \end{aligned}$$

Substitute 1 for y in either equation (4) or (5) to find x.

$$\begin{aligned} -6x - 19y &= -1 \qquad &(5)\\ -6x - 19(\mathbf{1}) &= -1 \qquad &\text{Let } y = 1.\\ -6x - 19 &= -1 \qquad &\text{Multiply.}\\ -6x &= 18 \qquad &\text{Add 19.}\\ x &= \mathbf{-3} \qquad &\text{Divide by } -6. \end{aligned}$$

Step 5 Now substitute the two values we found in Step 4 in working equation (1) to find the value of the remaining variable, focus variable z.

$$\begin{aligned} 4x + 8y + z &= 2 \qquad &(1)\\ 4(\mathbf{-3}) + 8(\mathbf{1}) + z &= 2 \qquad &\text{Let } x = -3 \text{ and } y = 1.\\ -4 + z &= 2 \qquad &\text{Multiply; add.}\\ z &= \mathbf{6} \qquad &\text{Add 4.} \end{aligned}$$

Continued on Next Page

1 Check that the solution $(-3, 1, 6)$ satisfies equations (2) and (3) of Example 1.

(a) $x + 7y - 3z = -14$ (2)

Does the solution satisfy equation (2)?

(b) $2x - 3y + 2z = 3$ (3)

Does the solution satisfy equation (3)?

2 Solve each system.

(a)
$$\begin{aligned} x + y + z &= 2 \\ x - y + 2z &= 2 \\ -x + 2y - z &= 1 \end{aligned}$$

(b)
$$\begin{aligned} 2x + y + z &= 9 \\ -x - y + z &= 1 \\ 3x - y + z &= 9 \end{aligned}$$

Step 6 It appears that the ordered triple $(-3, 1, 6)$ is the only solution of the system. We must check that the solution satisfies all three original equations of the system. For equation (1),

Check
$$\begin{aligned} 4x + 8y + z &= 2 && (1) \\ 4(-3) + 8(1) + 6 &\stackrel{?}{=} 2 \\ -12 + 8 + 6 &\stackrel{?}{=} 2 \\ 2 &= 2. && \text{True} \end{aligned}$$

◀ Work Problem **1** at the Side.

Because $(-3, 1, 6)$ also satisfies equations (2) and (3), the solution set is $\{(-3, 1, 6)\}$.

◀ Work Problem **2** at the Side.

OBJECTIVE 3 Solve linear systems (with three equations and three variables) in which some of the equations have missing terms. If a linear system has an equation missing a term or terms, one elimination step can be omitted.

EXAMPLE 2 Solving a System of Equations with Missing Terms

Solve the system.

$$\begin{aligned} 6x - 12y &= -5 && (1) \quad \text{Missing } z \\ 8y + z &= 0 && (2) \quad \text{Missing } x \\ 9x - z &= 12 && (3) \quad \text{Missing } y \end{aligned}$$

Since equation (3) is missing the variable y, one way to begin is to eliminate y again using equations (1) and (2).

Leave space for the missing terms.

$$\begin{aligned} 12x - 24y &= -10 && \text{Multiply each side of (1) by 2.} \\ 24y + 3z &= 0 && \text{Multiply each side of (2) by 3.} \\ \hline 12x + 3z &= -10 && \text{Add.} \quad (4) \end{aligned}$$

Use the resulting equation (4) in x and z, together with equation (3), $9x - z = 12$, to eliminate z. Multiply equation (3) by 3.

$$\begin{aligned} 27x - 3z &= 36 && \text{Multiply each side of (3) by 3.} \\ 12x + 3z &= -10 && (4) \\ \hline 39x &= 26 && \text{Add.} \\ x &= \frac{26}{39} = \frac{2}{3} && \text{Divide by 39; lowest terms} \end{aligned}$$

We can find z by substituting this value for x into equation (3).

$$\begin{aligned} 9x - z &= 12 && (3) \\ 9\left(\frac{2}{3}\right) - z &= 12 && \text{Let } x = \tfrac{2}{3}. \\ 6 - z &= 12 && \text{Multiply.} \\ z &= -6 && \text{Subtract 6; multiply by } -1. \end{aligned}$$

Continued on Next Page

ANSWERS

1. **(a)** yes **(b)** yes
2. **(a)** $\{(-1, 1, 2)\}$ **(b)** $\{(2, 1, 4)\}$

We can find y by substituting -6 for z in equation (2).

$$8y + z = 0 \qquad (2)$$
$$8y - 6 = 0 \qquad \text{Let } z = -6.$$
$$8y = 6 \qquad \text{Add 6.}$$
$$y = \frac{6}{8} = \frac{3}{4} \qquad \text{Divide by 8; lowest terms}$$

Thus, $x = \frac{2}{3}$, $y = \frac{3}{4}$, and $z = -6$. Check these values in each of the original equations of the system to verify that the solution set of the system is $\left\{\left(\frac{2}{3}, \frac{3}{4}, -6\right)\right\}$.

Note

Another way to solve the system in Example 2 is to begin by eliminating the variable z from equations (2) and (3). The resulting equation together with equation (1) forms a system of two equations in the variables x and y. Try working Example 2 this way to see that the same solution results.

There are often multiple ways to solve a system of equations. Some ways may involve more work than others.

Work Problem **3** *at the Side.* ▶

OBJECTIVE 4 Solve special systems. Linear systems with three variables may be inconsistent or may include dependent equations.

EXAMPLE 3 Solving an Inconsistent System with Three Variables

Solve the system.

$$2x - 4y + 6z = 5 \qquad (1)$$
$$-x + 3y - 2z = -1 \qquad (2)$$
$$x - 2y + 3z = 1 \qquad (3)$$

Use as the working equation, with focus variable x.

Eliminate x by adding equations (2) and (3) to get the equation

$$y + z = 0.$$

Eliminate x again, using equations (1) and (3).

$$-2x + 4y - 6z = -2 \qquad \text{Multiply each side of (3) by } -2.$$
$$2x - 4y + 6z = 5 \qquad (1)$$
$$0 = 3 \qquad \text{False}$$

This false statement indicates that equations (1) and (3) have no common solution. Thus, the system is inconsistent and the solution set is $\emptyset$. The graph of this system would show these two planes parallel to one another.

Note

If a false statement results when adding as in Example 3, it is not necessary to go any further with the solution. Since two of the three planes are parallel, it is not possible for the three planes to have any common points.

Work Problem **4** *at the Side.* ▶

3 Solve each system.

(a)
$$x - y = 6$$
$$2y + 5z = 1$$
$$3x - 4z = 8$$

(b)
$$5x - y = 26$$
$$4y + 3z = -4$$
$$x + z = 5$$

4 Solve each system.

(a)
$$3x - 5y + 2z = 1$$
$$5x + 8y - z = 4$$
$$-6x + 10y - 4z = 5$$

(b)
$$7x - 9y + 2z = 0$$
$$y + z = 0$$
$$8x - z = 0$$

Answers

3. (a) $\{(4, -2, 1)\}$ **(b)** $\{(5, -1, 0)\}$
4. (a) $\emptyset$ **(b)** $\{(0, 0, 0)\}$

EXAMPLE 4 Solving a System of Dependent Equations with Three Variables

Solve the system.

$$\begin{aligned} 2x - 3y + 4z &= 8 \quad (1) \\ -x + \frac{3}{2}y - 2z &= -4 \quad (2) \\ 6x - 9y + 12z &= 24 \quad (3) \end{aligned}$$

Multiplying each side of equation (1) by 3 gives equation (3). Multiplying each side of equation (2) by -6 also gives equation (3). Because of this, the equations are dependent. All three equations have the same graph, as illustrated in Figure 28(c). The solution set is written

$$\{(x, y, z) \mid 2x - 3y + 4z = 8\}. \qquad \text{Set-builder notation}$$

Although any one of the three equations could be used to write the solution set, we use the equation with coefficients that are integers with greatest common factor 1, as we did in **Section 8.4.**

◀ Work Problem 5 at the Side.

5 Solve the system.

$$\begin{aligned} x - y + z &= 4 \\ -3x + 3y - 3z &= -12 \\ 2x - 2y + 2z &= 8 \end{aligned}$$

EXAMPLE 5 Solving Another Special System

Solve the system.

$$\begin{aligned} 2x - y + 3z &= 6 \quad (1) \\ x - \frac{1}{2}y + \frac{3}{2}z &= 3 \quad (2) \\ 4x - 2y + 6z &= 1 \quad (3) \end{aligned}$$

Multiplying each side of equation (2) by 2 gives equation (1). Thus, these two equations are dependent.

Equations (1) and (3) are not equivalent, however. Multiplying equation (3) by $\frac{1}{2}$ gives

$$2x - y + 3z = \frac{1}{2}, \qquad \text{(3) multiplied by } \tfrac{1}{2}$$

which is *not* equivalent to equation (1). Instead, we obtain two equations with the same coefficients, but with different constant terms. The graphs of equations (1) and (3) have no points in common (that is, the planes are parallel). Thus, the system is inconsistent and the solution set is $\emptyset$, as illustrated in Figure 28(g).

◀ Work Problem 6 at the Side.

6 Solve the system.

$$\begin{aligned} 2x + 3y - z &= 8 \\ \frac{1}{2}x + \frac{3}{4}y - \frac{1}{4}z &= 2 \\ x + \frac{3}{2}y - \frac{1}{2}z &= -6 \end{aligned}$$

OBJECTIVE 5 Solve application problems with three variables using a system of three equations.

Problem-Solving Hint

If an application requires finding *three* unknown quantities, we can use a system of *three* equations to solve it. We extend the method used for two unknowns given in **Section 4.4.**

ANSWERS

5. $\{(x, y, z) \mid x - y + z = 4\}$
6. $\emptyset$

EXAMPLE 6 Solving a Problem Involving Prices

At Panera Bread, a loaf of honey wheat bread costs \$2.95, a loaf of sunflower bread costs \$2.99, and a loaf of French bread costs \$5.79. On a recent day, three times as many loaves of honey wheat bread were sold as sunflower bread. The number of loaves of French bread sold was 5 less than the number of loaves of honey wheat bread sold. Total receipts for these breads were \$87.89. How many loaves of each type of bread were sold? (*Source:* Panera Bread menu.)

Step 1 **Read** the problem again. There are three unknowns in this problem.

Step 2 **Assign variables** to represent the three unknowns.

Let x = the number of loaves of honey wheat bread,
y = the number of loaves of sunflower bread,
and z = the number of loaves of French bread.

Step 3 **Write a system of three equations.** Since three times as many loaves of honey wheat bread were sold as sunflower bread,

$$x = 3y, \quad \text{or} \quad x - 3y = 0. \qquad \text{Subtract } 3y. \quad (1)$$

Also,

Number of loaves of French	equals	5 less than the number of loaves of honey wheat.
↓	↓	↓
z	$=$	$x - 5$,

$$-x + z = -5 \qquad \text{Subtract } x.$$
$$x - z = 5. \qquad \text{Multiply by } -1. \quad (2)$$

Multiplying the cost of a loaf of each kind of bread by the number of loaves of that kind sold and adding gives the total receipts.

$$2.95x + 2.99y + 5.79z = 87.89$$

Multiply each side of this equation by 100 to clear it of decimals.

$$295x + 299y + 579z = 8789 \qquad (3)$$

Step 4 **Solve** the system of three equations,

$$x - 3y = 0 \qquad (1)$$
$$x - z = 5 \qquad (2)$$
$$295x + 299y + 579z = 8789, \qquad (3)$$

using the method shown in this section.

Work Problem 7 at the Side. ▶

Thus, we find that $x = 12$, $y = 4$, and $z = 7$.

Step 5 **State the answer.** The solution set is $\{(12, 4, 7)\}$, meaning that 12 loaves of honey wheat bread, 4 loaves of sunflower bread, and 7 loaves of French bread were sold.

Step 6 **Check.** Since $12 = 3 \cdot 4$, the number of loaves of honey wheat bread is three times the number of loaves of sunflower bread. Also, $12 - 7 = 5$, so the number of loaves of French bread is 5 less than the number of loaves of honey wheat bread. Multiply the appropriate cost per loaf by the number of loaves sold and add the results to check that total receipts were \$87.89.

7 Solve the system of equations from Example 6.

$$x - 3y = 0 \qquad (1)$$
$$x - z = 5 \qquad (2)$$
$$295x + 299y + 579z = 8789 \qquad (3)$$

8 Solve the problem.

A department store display features three kinds of perfume: Felice, Vivid, and Joy. There are 10 more bottles of Felice than Vivid, and 3 fewer bottles of Joy than Vivid. Each bottle of Felice costs \$8, Vivid costs \$15, and Joy costs \$32. The total value of all the perfume is \$589. How many bottles of each are there?

Answers

7. $\{(12, 4, 7)\}$
8. 21 bottles of Felice; 11 of Vivid; 8 of Joy

EXAMPLE 7 Solving a Business Production Problem

A company produces three color television sets, models X, Y, and Z. Each model X set requires 2 hr of electronics work, 2 hr of assembly time, and 1 hr of finishing time. Each model Y requires 1, 3, and 1 hr of electronics, assembly, and finishing time, respectively. Each model Z requires 3, 2, and 2 hr of the same work, respectively. There are 100 hr available for electronics, 100 hr available for assembly, and 65 hr available for finishing per week. How many of each model should be produced each week if all available time must be used?

Step 1 **Read** the problem again. There are three unknowns.

Step 2 **Assign variables.**

Let $x =$ the number of model X produced per week,

$y =$ the number of model Y produced per week,

and $z =$ the number of model Z produced per week.

We organize the information in a table.

	Each Model X	Each Model Y	Each Model Z	Totals	
Hours of Electronics Work	2	1	3	100	→ Gives equation (1)
Hours of Assembly Time	2	3	2	100	→ Gives equation (2)
Hours of Finishing Time	1	1	2	65	→ Gives equation (3)

Step 3 **Write a system of three equations.** The x model X sets require $2x$ hours of electronics, the y model Y sets require $1y$ (or y) hours of electronics, and the z model Z sets require $3z$ hours of electronics. Since 100 hr are available for electronics,

$$2x + y + 3z = 100. \qquad (1)$$

Similarly, from the fact that 100 hr are available for assembly,

$$2x + 3y + 2z = 100, \qquad (2)$$

and the fact that 65 hr are available for finishing leads to the equation

$$x + y + 2z = 65. \qquad (3)$$

Notice that by reading *across* the table, we can quickly determine the coefficients and constants in the equations of the system.

Step 4 **Solve** the system of equations (1), (2), and (3), namely,

$$\begin{aligned} 2x + y + 3z &= 100 \qquad (1)\\ 2x + 3y + 2z &= 100 \qquad (2)\\ x + y + 2z &= 65 \qquad (3) \end{aligned}$$

to find $x = 15$, $y = 10$, and $z = 20$.

Step 5 **State the answer.** The company should produce 15 model X, 10 model Y, and 20 model Z sets per week.

Step 6 **Check** that these values satisfy the conditions of the problem.

◀ Work Problem 9 at the Side.

9 Solve the problem.

A paper mill makes newsprint, bond, and copy machine paper. Each ton of newsprint requires 3 tons of recycled paper and 1 ton of wood pulp. Each ton of bond requires 2 tons of recycled paper, 4 tons of wood pulp, and 3 tons of rags. A ton of copy machine paper requires 2 tons of recycled paper, 3 tons of wood pulp, and 2 tons of rags. The mill has 4200 tons of recycled paper, 5800 tons of wood pulp, and 3900 tons of rags. How much of each kind of paper can be made from these supplies?

ANSWER

9. 400 tons of newsprint; 900 tons of bond; 600 tons of copy machine paper

8.5 Exercises

FOR EXTRA HELP

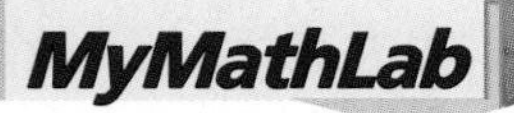

Math XL PRACTICE

WATCH

DOWNLOAD

READ

REVIEW

1. Explain what the following statement means: The solution set of the system

$$\begin{aligned} 2x + y + z &= 3 \\ 3x - y + z &= -2 \\ 4x - y + 2z &= 0 \end{aligned}$$

is $\{(-1, 2, 3)\}$.

The statement means that when −1 is substituted for x, 2 is substituted for y, and 3 is substituted for z in the three equations, the resulting three statements are true.

2. The two equations

$$\begin{aligned} x + y + z &= 6 \\ 2x - y + z &= 3 \end{aligned}$$

have a common solution of $(1, 2, 3)$. Which equation would complete a system of three linear equations in three variables having solution set $\{(1, 2, 3)\}$?

A. $3x + 2y - z = 1$ **B.** $3x + 2y - z = 4$

C. $3x + 2y - z = 5$ **D.** $3x + 2y - z = 6$

B

Solve each system of equations. See Example 1.

3. $\begin{aligned} 2x - 5y + 3z &= -1 \\ x + 4y - 2z &= 9 \\ x - 2y - 4z &= -5 \end{aligned}$

$\{(3, 2, 1)\}$

4. $\begin{aligned} x + 3y - 6z &= 7 \\ 2x - y + z &= 1 \\ x + 2y + 2z &= -1 \end{aligned}$

$\{(1, 0, -1)\}$

5. $\begin{aligned} 3x + 2y + z &= 8 \\ 2x - 3y + 2z &= -16 \\ x + 4y - z &= 20 \end{aligned}$

$\{(1, 4, -3)\}$

6. $\begin{aligned} -3x + y - z &= -10 \\ -4x + 2y + 3z &= -1 \\ 2x + 3y - 2z &= -5 \end{aligned}$

$\{(2, -1, 3)\}$

7. $\begin{aligned} x + 2y + z &= 4 \\ 2x + y - z &= -1 \\ x - y - z &= -2 \end{aligned}$

$\{(1, 0, 3)\}$

8. $\begin{aligned} x - 2y + 5z &= -7 \\ -2x - 3y + 4z &= -14 \\ -3x + 5y - z &= -7 \end{aligned}$

$\{(3, 0, -2)\}$

9. $\begin{aligned} -x + 2y + 6z &= 2 \\ 3x + 2y + 6z &= 6 \\ x + 4y - 3z &= 1 \end{aligned}$

$\left\{\left(1, \frac{3}{10}, \frac{2}{5}\right)\right\}$

10. $\begin{aligned} 2x + y + 2z &= 1 \\ x + 2y + z &= 2 \\ x - y - z &= 0 \end{aligned}$

$\left\{\left(\frac{1}{2}, 1, -\frac{1}{2}\right)\right\}$

11. $\begin{aligned} 2x + 5y + 2z &= 0 \\ 4x - 7y - 3z &= 1 \\ 3x - 8y - 2z &= -6 \end{aligned}$

$\{(0, 2, -5)\}$

12. $\begin{aligned} 5x - 2y + 3z &= -9 \\ 4x + 3y + 5z &= 4 \\ 2x + 4y - 2z &= 14 \end{aligned}$

$\{(0, 3, -1)\}$

13. $\begin{aligned} x + 2y + 3z &= 1 \\ -x - y + 3z &= 2 \\ -6x + y + z &= -2 \end{aligned}$

$\left\{\left(\frac{20}{59}, -\frac{33}{59}, \frac{35}{59}\right)\right\}$

14. $\begin{aligned} x + y - z &= -2 \\ 2x - y + z &= -5 \\ -x + 2y - 3z &= -4 \end{aligned}$

$\left\{\left(-\frac{7}{3}, \frac{22}{3}, 7\right)\right\}$

Solve each system of equations. See Example 2.

15. $\begin{aligned} 2x - 3y + 2z &= -1 \\ x + 2y + z &= 17 \\ 2y - z &= 7 \end{aligned}$

$\{(4, 5, 3)\}$

16. $\begin{aligned} 2x - y + 3z &= 6 \\ x + 2y - z &= 8 \\ 2y + z &= 1 \end{aligned}$

$\{(5, 1, -1)\}$

17. $\begin{aligned} 4x + 2y - 3z &= 6 \\ x - 4y + z &= -4 \\ -x + 2z &= 2 \end{aligned}$

$\{(2, 2, 2)\}$

18. $\begin{aligned} 2x + 3y - 4z &= 4 \\ x - 6y + z &= -16 \\ -x + 3z &= 8 \end{aligned}$

$\{(4, 4, 4)\}$

19. $\begin{aligned} -5x + 2y + z &= 5 \\ -3x - 2y - z &= 3 \\ -x + 6y &= 1 \end{aligned}$

$\{(-1, 0, 0)\}$

20. $\begin{aligned} x + y - z &= 0 \\ 2y - z &= 1 \\ 2x + 3y - 4z &= -4 \end{aligned}$

$\{(1, 2, 3)\}$

21. $\begin{aligned} 2x + y &= 6 \\ 3y - 2z &= -4 \\ 3x - 5z &= -7 \end{aligned}$

$\left\{\left(\frac{8}{3}, \frac{2}{3}, 3\right)\right\}$

22. $\begin{aligned} 4x - 8y &= -7 \\ 4y + z &= 7 \\ -8x + z &= -4 \end{aligned}$

$\left\{\left(\frac{3}{4}, \frac{5}{4}, 2\right)\right\}$

23. Using your immediate surroundings, give an example of three planes that

(a) intersect in a single point;

(b) do not intersect;

(c) intersect in infinitely many points.

Answers will vary. Some possible answers are (a) two perpendicular walls and the ceiling in a normal room, (b) the floors of three different levels of an office building, and (c) three pages of this book (since they intersect in the spine).

24. Suppose that a system has infinitely many ordered triple solutions of the form (x, y, z) such that

$$x + y + 2z = 1.$$

Give three specific ordered triples that are solutions of the system.

Answers will vary. Three possibilities are $\left(1, 1, -\frac{1}{2}\right)$, $\left(0, 0, \frac{1}{2}\right)$, **and** $(2, 5, -3)$.

Solve each system of equations. If the system is inconsistent or has dependent equations, say so. See Examples 1–5.

25. $\begin{aligned} 2x + 2y - 6z &= 5 \\ -3x + y - z &= -2 \\ -x - y + 3z &= 4 \end{aligned}$

∅; inconsistent system

26. $\begin{aligned} -2x + 5y + z &= -3 \\ 5x + 14y - z &= -11 \\ 7x + 9y - 2z &= -5 \end{aligned}$

∅; inconsistent system

27. $\begin{aligned} -5x + 5y - 20z &= -40 \\ x - y + 4z &= 8 \\ 3x - 3y + 12z &= 24 \end{aligned}$

$\{(x, y, z) \mid x - y + 4z = 8\}$; **dependent equations**

28. $x + 4y - z = 3$
$-2x - 8y + 2z = -6$
$3x + 12y - 3z = 9$

$\{(x, y, z) | x + 4y - z = 3\}$; dependent equations

29. $2x + y - z = 6$
$4x + 2y - 2z = 12$
$-x - \frac{1}{2}y + \frac{1}{2}z = -3$

$\{(x, y, z) | 2x + y - z = 6\}$; dependent equations

30. $2x - 8y + 2z = -10$
$-x + 4y - z = 5$
$\frac{1}{8}x - \frac{1}{2}y + \frac{1}{8}z = -\frac{5}{8}$

$\{(x, y, z) | x - 4y + z = -5\}$; dependent equations

31. $x + y - 2z = 0$
$3x - y + z = 0$
$4x + 2y - z = 0$

$\{(0, 0, 0)\}$

32. $2x + 3y - z = 0$
$x - 4y + 2z = 0$
$3x - 5y - z = 0$

$\{(0, 0, 0)\}$

33. $x - 2y + \frac{1}{3}z = 4$
$3x - 6y + z = 12$
$-6x + 12y - 2z = -3$

$\emptyset$; inconsistent system

34. $4x + y - 2z = 3$
$x + \frac{1}{4}y - \frac{1}{2}z = \frac{3}{4}$
$2x + \frac{1}{2}y - z = 1$

$\emptyset$; inconsistent system

35. $x + 5y - 2z = -1$
$-2x + 8y + z = -4$
$3x - y + 5z = 19$

$\{(3, 0, 2)\}$

36. $x + 3y + z = 2$
$4x + y + 2z = -4$
$5x + 2y + 3z = -2$

$\{(-4, 0, 6)\}$

Solve each problem involving three unknowns. See Examples 6 and 7. (In Exercises 37–40, remember that the sum of the measures of the angles of a triangle is 180°.)

37. In the figure, $z = x + 10$ and $x + y = 100$. Determine a third equation involving x, y, and z, and then find the measures of the three angles.

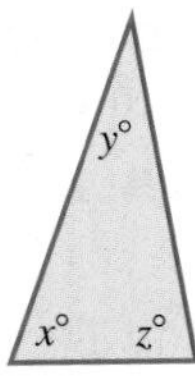

$x + y + z = 180$; angle measures: 70°, 30°, 80°

38. In the figure, x is 10 less than y and x is 20 less than z. Write a system of equations and find the measures of the three angles.

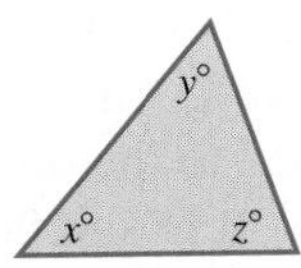

$x = y - 10$, $x = z - 20$, $x + y + z = 180$; 50°, 60°, 70°

39. In a certain triangle, the measure of the second angle is 10° more than three times the first. The third angle measure is equal to the sum of the measures of the other two. Find the measures of the three angles.

first: 20°; second: 70°; third: 90°

40. The measure of the largest angle of a triangle is 12° less than the sum of the measures of the other two. The smallest angle measures 58° less than the largest. Find the measures of the angles.

largest: 84°; middle: 70°; smallest: 26°

41. In a random sample of 100 Americans of voting age, 10 more Americans identify themselves as Independents than Republicans. Six fewer Americans identify themselves as Republicans than Democrats. Assuming that all of those sampled are Republican, Democrat, or Independent, how many of those in the sample identify themselves with each political affiliation? (*Source:* The Gallup Organization.)

Independent: 38; Democrat: 34; Republican: 28

42. In the 2004 Summer Olympics in Athens, Greece, the United States earned 6 more gold medals than bronze. The number of silver medals earned was 19 less than twice the number of bronze medals. The United Stated earned a total of 103 medals. How many of each kind of medal did the United States earn? (*Source: World Almanac and Book of Facts.*)

gold: 35; silver: 39; bronze: 29

43. Tickets for one show on the Harlem Globetrotters' 2006 "Unstoppable" Tour cost \$14, \$20, or, for VIP seats, \$50. Five times as many \$14 tickets were sold as VIP tickets. The number of \$14 tickets was 15 more than the sum of the number of \$20 tickets and the number of VIP tickets. Sales of all three kinds of tickets totaled \$11,700. How many of each kind of ticket were sold? (*Source*: www.ticketmaster.com)

\$14 tickets: 300; \$20 tickets: 225; \$50 tickets: 60

44. Three kinds of tickets are available for a *Third Day* concert: "up close," "in the middle," and "far out." "Up close" tickets cost \$10 more than "in the middle" tickets, while "in the middle" tickets cost \$10 more than "far out" tickets. Twice the cost of an "up close" ticket is \$20 more than 3 times the cost of a "far out" ticket. Find the price of each kind of ticket.

"up close": \$40; "in the middle": \$30; "far out": \$20

45. A wholesaler supplies college T-shirts to three college bookstores: A, B, and C. The wholesaler recently shipped a total of 800 T-shirts to the three bookstores. In order to meet student demand at the three colleges, twice as many T-shirts were shipped to bookstore B as to bookstore A, and the number shipped to bookstore C was 40 less than the sum of the numbers shipped to the other two bookstores. How many T-shirts were shipped to each bookstore?

bookstore A: 140; bookstore B: 280; bookstore C: 380

46. An office supply store sells three models of computer desks: A, B, and C. In January, the store sold a total of 85 computer desks. The number of model B desks was five more than the number of model C desks, and the number of model A desks was four more than twice the number of model C desks. How many of each model did the store sell in January?

model A: 42; model B: 24; model C: 19

47. During the 2005–2006 National Hockey League regular season, the Calgary Flames played 82 games. Their wins and losses totaled 71. They tied 14 fewer games than they lost. How many wins, losses, and ties did they have?

wins: 46; losses: 25; ties: 11

48. (Refer to Exercise 47.) During the same 82-game season, the Minnesota Wild had a total of 44 losses and ties. They had two more wins than losses. How many wins, losses, and ties did they have?

wins: 38; losses: 36; ties: 8

2005-2006 NHL Final Standings

Team	W	L	T	Pts
Calgary	—	—	—	103
Colorado	43	30	9	95
Edmonton	41	28	13	95
Vancouver	42	32	8	92
Minnesota	—	—	—	84

Source: www.sportzdomain.com

8.6 Solving Systems of Linear Equations by Matrix Methods

OBJECTIVES

1. Define a matrix.
2. Write the augmented matrix for a system.
3. Use row operations to solve a system with two equations.
4. Use row operations to solve a system with three equations.
5. Use row operations to solve special systems.

OBJECTIVE 1 Define a matrix. An ordered array of numbers such as

$$\text{Rows} \quad \overset{\text{Columns}}{\begin{bmatrix} 2 & 3 & 5 \\ 7 & 1 & 2 \end{bmatrix}} \quad \text{Matrix}$$

is called a **matrix.** The numbers are called **elements** of the matrix. *Matrices* (the plural of *matrix*) are named according to the number of **rows** and **columns** they contain. The rows are read horizontally, and the columns are read vertically. For example, the first row in the preceding matrix is 2 3 5 and the first column is $\begin{matrix} 2 \\ 7 \end{matrix}$. This matrix is a 2×3 (read "two by three") matrix because it has 2 rows and 3 columns. The number of rows followed by the number of columns gives the **dimensions** of the matrix.

$$\begin{bmatrix} -1 & 0 \\ 1 & -2 \end{bmatrix} \quad \begin{matrix} 2 \times 2 \\ \text{matrix} \end{matrix} \qquad \begin{bmatrix} 8 & -1 & -3 \\ 2 & 1 & 6 \\ 0 & 5 & -3 \\ 5 & 9 & 7 \end{bmatrix} \quad \begin{matrix} 4 \times 3 \\ \text{matrix} \end{matrix}$$

A **square matrix** is one that has the same number of rows as columns. The 2×2 matrix is a square matrix.

Calculator Tip Figure 29 shows how a graphing calculator displays the preceding two matrices. Work with matrices is made much easier by using technology when available. Consult your owner's manual for details.

```
[A]
        [[-1 0 ]
         [1  -2]]
```

```
[B]
      [[8 -1 -3]
       [2 1  6 ]
       [0 5  -3]
       [5 9  7 ]]
```

Figure 29

In this section, we discuss a method of solving linear systems that uses matrices. The advantage of this new method is that it can be done by a graphing calculator or a computer, allowing large systems of equations to be solved easily.

OBJECTIVE 2 Write the augmented matrix for a system. To solve a linear system using matrices, we begin by writing an *augmented matrix* for the system. An **augmented matrix** has a vertical bar that separates the columns of the matrix into two groups. For example, to solve the system

$$x - 3y = 1$$
$$2x + y = -5,$$

we start with the augmented matrix

$$\left[\begin{array}{cc|c} 1 & -3 & 1 \\ 2 & 1 & -5 \end{array}\right]. \quad \text{Augmented matrix}$$

System of equations:

$$\begin{aligned} x - 3y &= 1 \\ 2x + y &= -5 \end{aligned}$$

Augmented matrix:

$$\left[\begin{array}{cc|c} 1 & -3 & 1 \\ 2 & 1 & -5 \end{array}\right]$$

Coefficients of the variables — The bar separates the coefficients from the constants. — Constants

Notice that we place the coefficients of the variables to the left of the bar, and the constants to the right. ***The matrix is just a shorthand way of writing the system of equations, so the rows of the augmented matrix can be treated the same as the equations of a system of equations.***

We know that exchanging the position of two equations in a system does not change the system. Also, multiplying any equation in a system by a nonzero number does not change the system. Comparable changes to the augmented matrix of a system of equations produce new matrices that correspond to systems with the same solutions as the original system.

The following **row operations** produce new matrices that lead to systems having the same solutions as the original system.

Matrix Row Operations

1. Any two rows of the matrix may be interchanged.
2. The numbers in any row may be multiplied by any nonzero real number.
3. Any row may be transformed by adding to the numbers of the row the product of a real number and the corresponding numbers of another row.

Examples of these row operations follow.

Row operation 1:

$$\begin{bmatrix} 2 & 3 & 9 \\ 4 & 8 & -3 \\ 1 & 0 & 7 \end{bmatrix} \text{ becomes } \begin{bmatrix} 1 & 0 & 7 \\ 4 & 8 & -3 \\ 2 & 3 & 9 \end{bmatrix}.$$

Interchange row 1 and row 3.

Row operation 2:

$$\begin{bmatrix} 2 & 3 & 9 \\ 4 & 8 & -3 \\ 1 & 0 & 7 \end{bmatrix} \text{ becomes } \begin{bmatrix} 6 & 9 & 27 \\ 4 & 8 & -3 \\ 1 & 0 & 7 \end{bmatrix}.$$

Multiply the numbers in row 1 by 3.

Row operation 3:

$$\begin{bmatrix} 2 & 3 & 9 \\ 4 & 8 & -3 \\ 1 & 0 & 7 \end{bmatrix} \text{ becomes } \begin{bmatrix} 0 & 3 & -5 \\ 4 & 8 & -3 \\ 1 & 0 & 7 \end{bmatrix}.$$

Multiply the numbers in row 3 by -2; add them to the corresponding numbers in row 1.

The third row operation corresponds to the way we eliminated a variable from a pair of equations in the previous sections.

OBJECTIVE 3 Use row operations to solve a system with two equations. Row operations can be used to rewrite a matrix. The goal is a matrix in the form

$$\left[\begin{array}{cc|c} 1 & a & b \\ 0 & 1 & c \end{array}\right] \quad \text{or} \quad \left[\begin{array}{ccc|c} 1 & a & b & c \\ 0 & 1 & d & e \\ 0 & 0 & 1 & f \end{array}\right]$$

for systems with two or three equations, respectively. Notice that there are 1s down the diagonal from upper left to lower right and 0s below the 1s. A matrix written this way is said to be in **row echelon form.** When these matrices are rewritten as systems of equations, the value of one variable is known, and the rest can be found by substitution. The following examples illustrate this method.

EXAMPLE 1 Using Row Operations to Solve a System with Two Variables

Use row operations to solve the system.

$$\begin{aligned} x - 3y &= 1 \\ 2x + y &= -5 \end{aligned}$$

We start with the augmented matrix of the system.

$$\left[\begin{array}{cc|c} 1 & -3 & 1 \\ 2 & 1 & -5 \end{array}\right] \quad \text{Augmented matrix}$$

Now we use the various row operations to change this matrix into one that leads to a system that is easier to solve.

It is best to work by columns. We start with the first column and make sure that there is a 1 in the first row, first column position. There is already a 1 in this position. Next, we get 0 in every position below the first. To get a 0 in row two, column one, we use the third row operation and add to the numbers in row two the result of multiplying each number in row one by -2. (We abbreviate this as $-2R_1 + R_2$.) Row one remains unchanged.

$$\left[\begin{array}{cc|c} 1 & -3 & 1 \\ 2 + 1(-2) & 1 + -3(-2) & -5 + 1(-2) \end{array}\right]$$

Original number from row two ↑ ↑ -2 times number from row one

$$\left[\begin{array}{cc|c} 1 & -3 & 1 \\ 0 & 7 & -7 \end{array}\right] \quad -2R_1 + R_2$$

The matrix now has a 1 in the first position of column one, with 0 in every position below the first.

Now we go to column two. An entry of 1 is needed in row two, column two. We get this 1 by using the second row operation, multiplying each number of row two by $\frac{1}{7}$.

Stop here—this matrix is in row echelon form.

$$\left[\begin{array}{cc|c} 1 & -3 & 1 \\ 0 & 1 & -1 \end{array}\right] \quad \tfrac{1}{7}R_2$$

This augmented matrix leads to the system of equations

$$\begin{aligned} 1x - 3y &= 1 \\ 0x + 1y &= -1, \end{aligned} \quad \text{or} \quad \begin{aligned} x - 3y &= 1 \\ y &= -1. \end{aligned}$$

From the second equation, $y = -1$. We substitute -1 for y in the first equation to get

$$\begin{aligned} x - 3y &= 1 \\ x - 3(-1) &= 1 && \text{Let } y = -1. \\ x + 3 &= 1 && \text{Multiply.} \\ x &= -2. && \text{Subtract 3.} \end{aligned}$$

Write the values of x and y in the correct order.

The solution set of the system is $\{(-2, -1)\}$. Check this solution by substitution in both equations of the system.

Work Problem **1** *at the Side.* ▶

1 Use row operations to solve the system.

$$\begin{aligned} x - 2y &= 9 \\ 3x + y &= 13 \end{aligned}$$

ANSWER

1. $\{(5, -2)\}$

Calculator Tip If the augmented matrix of the system in Example 1 is entered as matrix A in a graphing calculator (Figure 30(a)) and the row echelon form of the matrix is found (Figure 30(b)), the system becomes

$$x + \frac{1}{2}y = -\frac{5}{2}$$
$$y = -1.$$

While this system looks different from the one we obtained in Example 1, it is equivalent, since its solution set is also $\{(-2, -1)\}$.

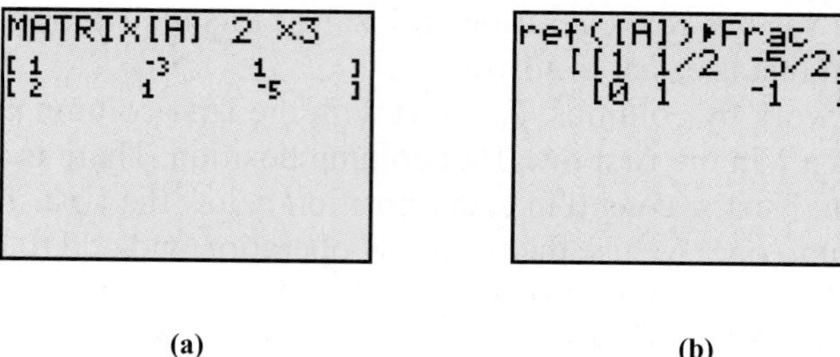

(a) (b)

Figure 30

OBJECTIVE 4 Use row operations to solve a system with three equations. As before, we use row operations to get 1s down the diagonal from left to right and all 0s below each 1.

EXAMPLE 2 Using Row Operations to Solve a System with Three Variables

Use row operations to solve the system.

$$\begin{aligned} x - y + 5z &= -6 \\ 3x + 3y - z &= 10 \\ x + 3y + 2z &= 5 \end{aligned}$$

Start by writing the augmented matrix of the system.

$$\left[\begin{array}{ccc|c} 1 & -1 & 5 & -6 \\ 3 & 3 & -1 & 10 \\ 1 & 3 & 2 & 5 \end{array}\right] \quad \text{Augmented matrix}$$

This matrix already has 1 in row one, column one. Next get 0s in the rest of column one. First, add to row two the results of multiplying each number of row one by -3. This gives the matrix

$$\left[\begin{array}{ccc|c} 1 & -1 & 5 & -6 \\ 0 & 6 & -16 & 28 \\ 1 & 3 & 2 & 5 \end{array}\right]. \quad -3R_1 + R_2$$

Now add to the numbers in row three the results of multiplying each number of row one by -1.

$$\left[\begin{array}{ccc|c} 1 & -1 & 5 & -6 \\ 0 & 6 & -16 & 28 \\ 0 & 4 & -3 & 11 \end{array}\right] \quad -1R_1 + R_3$$

Continued on Next Page

Get 1 in row two, column two by multiplying each number in row two by $\frac{1}{6}$.

$$\left[\begin{array}{ccc|c} 1 & -1 & 5 & -6 \\ 0 & 1 & -\frac{8}{3} & \frac{14}{3} \\ 0 & 4 & -3 & 11 \end{array}\right] \quad \frac{1}{6}R_2$$

Introduce 0 in row three, column two by adding to row three the results of multiplying each number in row two by -4.

$$\left[\begin{array}{ccc|c} 1 & -1 & 5 & -6 \\ 0 & 1 & -\frac{8}{3} & \frac{14}{3} \\ 0 & 0 & \frac{23}{3} & -\frac{23}{3} \end{array}\right] \quad -4R_2 + R_3$$

Finally, obtain 1 in row three, column three by multiplying each number in row three by $\frac{3}{23}$.

$$\left[\begin{array}{ccc|c} 1 & -1 & 5 & -6 \\ 0 & 1 & -\frac{8}{3} & \frac{14}{3} \\ 0 & 0 & 1 & -1 \end{array}\right] \quad \frac{3}{23}R_3$$

This final matrix gives the system of equations

$$\begin{aligned} x - y + 5z &= -6 \\ y - \frac{8}{3}z &= \frac{14}{3} \\ z &= -1. \end{aligned}$$

Substitute -1 for z in the second equation, $y - \frac{8}{3}z = \frac{14}{3}$, to get $y = 2$. Substitute 2 for y and -1 for z in the first equation, $x - y + 5z = -6$, to get $x = 1$. The solution set is $\{(1, 2, -1)\}$. Check by substitution in the original system.

Work Problem **2** *at the Side.* ▶

2 Use row operations to solve the system.

$$\begin{aligned} 2x - y + z &= 7 \\ x - 3y - z &= 7 \\ -x + y - 5z &= -9 \end{aligned}$$

OBJECTIVE 5 Use row operations to solve special systems.

EXAMPLE 3 Recognizing Inconsistent Systems or Dependent Equations

Use row operations to solve each system.

(a) $2x - 3y = 8$
$-6x + 9y = 4$

$$\left[\begin{array}{cc|c} 2 & -3 & 8 \\ -6 & 9 & 4 \end{array}\right] \quad \text{Write the augmented matrix.}$$

$$\left[\begin{array}{cc|c} 1 & -\frac{3}{2} & 4 \\ -6 & 9 & 4 \end{array}\right] \quad \frac{1}{2}R_1$$

$$\left[\begin{array}{cc|c} 1 & -\frac{3}{2} & 4 \\ \mathbf{0} & \mathbf{0} & \mathbf{28} \end{array}\right] \quad 6R_1 + R_2$$

The corresponding system of equations is

$$\begin{aligned} x - \frac{3}{2}y &= 4 \\ 0 &= 28, \quad \text{False} \end{aligned}$$

which has no solution and is inconsistent. The solution set is $\emptyset$.

Continued on Next Page

ANSWER

2. $\{(2, -2, 1)\}$

(b) $-10x + 12y = 30$
$5x - 6y = -15$

$$\left[\begin{array}{rr|r} -10 & 12 & 30 \\ 5 & -6 & -15 \end{array}\right]$$ Write the augmented matrix.

$$\left[\begin{array}{rr|r} 1 & -\frac{6}{5} & -3 \\ 5 & -6 & -15 \end{array}\right]$$ $-\frac{1}{10}R_1$

$$\left[\begin{array}{rr|r} 1 & -\frac{6}{5} & -3 \\ 0 & 0 & 0 \end{array}\right]$$ $-5R_1 + R_2$

The corresponding system is

$$x - \frac{6}{5}y = -3$$

$$0 = 0, \quad \text{True}$$

which has dependent equations. Using the second equation of the original system, we write the solution set as

$$\{(x, y) | 5x - 6y = -15\}.$$

◀ *Work Problem* **3** *at the Side.*

3 Use row operations to solve each system.

(a) $x - y = 2$
$-2x + 2y = 2$

(b) $x - y = 2$
$-2x + 2y = -4$

ANSWERS

3. (a) $\emptyset$ **(b)** $\{(x, y) | x - y = 2\}$

8.6 Exercises

FOR EXTRA HELP

MathXL PRACTICE

WATCH

DOWNLOAD

READ

REVIEW

1. Consider the matrix $\begin{bmatrix} -2 & 3 & 1 \\ 0 & 5 & -3 \\ 1 & 4 & 8 \end{bmatrix}$, and answer the following.

(a) What are the elements of the second row? **0, 5, −3**

(b) What are the elements of the third column? **1, −3, 8**

(c) Is this a square matrix? Explain. **yes; The number of rows is the same as the number of columns (three).**

(d) Give the matrix obtained by interchanging the first and third rows.

(e) Give the matrix obtained by multiplying the first row by $-\frac{1}{2}$.

(f) Give the matrix obtained by multiplying the third row by 3 and adding it to the first row.

(d) $\begin{bmatrix} 1 & 4 & 8 \\ 0 & 5 & -3 \\ -2 & 3 & 1 \end{bmatrix}$ (e) $\begin{bmatrix} 1 & -\frac{3}{2} & -\frac{1}{2} \\ 0 & 5 & -3 \\ 1 & 4 & 8 \end{bmatrix}$ (f) $\begin{bmatrix} 1 & 15 & 25 \\ 0 & 5 & -3 \\ 1 & 4 & 8 \end{bmatrix}$

2. Give the dimensions of each matrix.

(a) $\begin{bmatrix} 3 & -7 \\ 4 & 5 \\ -1 & 0 \end{bmatrix}$ 3×2 **(b)** $\begin{bmatrix} 4 & 9 & 0 \\ -1 & 2 & -4 \end{bmatrix}$ 2×3 **(c)** $\begin{bmatrix} 6 & 3 \\ -2 & 5 \\ 4 & 10 \\ 1 & -11 \end{bmatrix}$ 4×2

Complete the steps in the matrix solution of each system by filling in the blanks. Give the final system and the solution set. See Example 1.

3. $4x + 8y = 44$
$2x - y = -3$

$\left[\begin{array}{cc|c} 4 & 8 & 44 \\ 2 & -1 & -3 \end{array}\right]$

$\left[\begin{array}{cc|c} 1 & \underline{2} & \underline{11} \\ 2 & -1 & -3 \end{array}\right]$ $\frac{1}{4}R_1$

$\left[\begin{array}{cc|c} 1 & 2 & 11 \\ 0 & \underline{-5} & \underline{-25} \end{array}\right]$ $-2R_1 + R_2$

$\left[\begin{array}{cc|c} 1 & 2 & 11 \\ 0 & 1 & \underline{5} \end{array}\right]$ $-\frac{1}{5}R_2$

$x + 2y = 11$
$y = 5$;
$\{(1, 5)\}$

4. $2x - 5y = -1$
$3x + y = 7$

$\left[\begin{array}{cc|c} 2 & -5 & -1 \\ 3 & 1 & 7 \end{array}\right]$

$\left[\begin{array}{cc|c} 1 & -\frac{5}{2} & \underline{-\frac{1}{2}} \\ 3 & 1 & 7 \end{array}\right]$ $\frac{1}{2}R_1$

$\left[\begin{array}{cc|c} 1 & -\frac{5}{2} & -\frac{1}{2} \\ 0 & \underline{\frac{17}{2}} & \underline{\frac{17}{2}} \end{array}\right]$ $-3R_1 + R_2$

$\left[\begin{array}{cc|c} 1 & -\frac{5}{2} & -\frac{1}{2} \\ 0 & 1 & \underline{1} \end{array}\right]$ $\frac{2}{17}R_2$

$x - \frac{5}{2}y = -\frac{1}{2}$
$y = 1$;
$\{(2, 1)\}$

Use row operations to solve each system. See Examples 1 and 3.

5. $x + y = 5$
$x - y = 3$
$\{(4, 1)\}$

6. $x + 2y = 7$
$x - y = -2$
$\{(1, 3)\}$

7. $2x + 4y = 6$
$3x - y = 2$
$\{(1, 1)\}$

8. $4x + 5y = -7$
$x - y = 5$
$\{(2, -3)\}$

9. $3x + 4y = 13$
$2x - 3y = -14$
$\{(-1, 4)\}$

10. $5x + 2y = 8$
$3x - y = 7$
$\{(2, -1)\}$

11. $-4x + 12y = 36$
$x - 3y = 9$
$\emptyset$

12. $2x - 4y = 8$
$-3x + 6y = 5$
$\emptyset$

13. $2x + y = 4$
$4x + 2y = 8$

$\{(x, y) \mid 2x + y = 4\}$

14. $-3x - 4y = 1$
$6x + 8y = -2$

$\{(x, y) \mid 3x + 4y = -1\}$

15. $\frac{1}{2}x + \frac{1}{3}y = 0$
$\frac{2}{3}x + \frac{3}{4}y = 0$

$\{(0, 0)\}$

16. $1.2x + 0.3y = 0$
$2.9x - 0.6y = 0$

$\{(0, 0)\}$

Complete the steps in the matrix solution of each system by filling in the blanks. Give the final system and the solution set. See Example 2.

17. $x + y - z = -3$
$2x + y + z = 4$
$5x - y + 2z = 23$

$$\left[\begin{array}{ccc|c} 1 & 1 & -1 & -3 \\ 2 & 1 & 1 & 4 \\ 5 & -1 & 2 & 23 \end{array}\right]$$

$$\left[\begin{array}{ccc|c} 1 & 1 & -1 & -3 \\ 0 & \underline{-1} & \underline{3} & \underline{10} \\ 0 & \underline{-6} & \underline{7} & \underline{38} \end{array}\right] \quad \begin{array}{l} -2R_1 + R_2 \\ -5R_1 + R_3 \end{array}$$

$$\left[\begin{array}{ccc|c} 1 & 1 & -1 & -3 \\ 0 & 1 & \underline{-3} & \underline{-10} \\ 0 & -6 & 7 & 38 \end{array}\right] \quad -1R_2$$

$$\left[\begin{array}{ccc|c} 1 & 1 & -1 & -3 \\ 0 & 1 & -3 & -10 \\ 0 & 0 & \underline{-11} & \underline{-22} \end{array}\right] \quad 6R_2 + R_3$$

$$\left[\begin{array}{ccc|c} 1 & 1 & -1 & -3 \\ 0 & 1 & -3 & -10 \\ 0 & 0 & 1 & \underline{2} \end{array}\right] \quad -\tfrac{1}{11}R_3$$

$x + y - z = -3$
$y - 3z = -10$
$z = 2;$
$\{(3, -4, 2)\}$

18. $2x + y + 2z = 11$
$2x - y - z = -3$
$3x + 2y + z = 9$

$$\left[\begin{array}{ccc|c} 2 & 1 & 2 & 11 \\ 2 & -1 & -1 & -3 \\ 3 & 2 & 1 & 9 \end{array}\right]$$

$$\left[\begin{array}{ccc|c} 1 & \underline{\tfrac{1}{2}} & \underline{1} & \underline{\tfrac{11}{2}} \\ 2 & -1 & -1 & -3 \\ 3 & 2 & 1 & 9 \end{array}\right] \quad \tfrac{1}{2}R_1$$

$$\left[\begin{array}{ccc|c} 1 & \tfrac{1}{2} & 1 & \tfrac{11}{2} \\ 0 & \underline{-2} & \underline{-3} & \underline{-14} \\ 0 & \underline{\tfrac{1}{2}} & \underline{-2} & \underline{-\tfrac{15}{2}} \end{array}\right] \quad \begin{array}{l} -2R_1 + R_2 \\ -3R_1 + R_3 \end{array}$$

$$\left[\begin{array}{ccc|c} 1 & \tfrac{1}{2} & 1 & \tfrac{11}{2} \\ 0 & 1 & \underline{\tfrac{3}{2}} & \underline{7} \\ 0 & \tfrac{1}{2} & -2 & -\tfrac{15}{2} \end{array}\right] \quad -\tfrac{1}{2}R_2$$

$$\left[\begin{array}{ccc|c} 1 & \tfrac{1}{2} & 1 & \tfrac{11}{2} \\ 0 & 1 & \tfrac{3}{2} & 7 \\ 0 & 0 & \underline{-\tfrac{11}{4}} & \underline{-11} \end{array}\right] \quad -\tfrac{1}{2}R_2 + R_3$$

$$\left[\begin{array}{ccc|c} 1 & \tfrac{1}{2} & 1 & \tfrac{11}{2} \\ 0 & 1 & \tfrac{3}{2} & 7 \\ 0 & 0 & 1 & \underline{4} \end{array}\right] \quad -\tfrac{4}{11}R_3$$

$x + \frac{1}{2}y + z = \frac{11}{2}$
$y + \frac{3}{2}z = 7$
$z = 4;$
$\{(1, 1, 4)\}$

Use row operations to solve each system. See Examples 2 and 3.

19. $x + y - 3z = 1$
$2x - y + z = 9$
$3x + y - 4z = 8$

$\{(4, 0, 1)\}$

20. $2x + 4y - 3z = -18$
$3x + y - z = -5$
$x - 2y + 4z = 14$

$\{(0, -3, 2)\}$

21. $x + y - z = 6$
$2x - y + z = -9$
$x - 2y + 3z = 1$

$\{(-1, 23, 16)\}$

22. $x + 3y - 6z = 7$
$2x - y + 2z = 0$
$x + y + 2z = -1$

$\{(1, 0, -1)\}$

23. $x - y = 1$
$y - z = 6$
$x + z = -1$

$\{(3, 2, -4)\}$

24. $x + y = 1$
$2x - z = 0$
$y + 2z = -2$

$\{(-1, 2, -2)\}$

25. $4x + 8y + 4z = 9$
$x + 3y + 4z = 10$
$5x + 10y + 5z = 12$

$\emptyset$

26. $x + 2y + 3z = -2$
$2x + 4y + 6z = -5$
$x - y + 2z = 6$

$\emptyset$

27. $x - 2y + z = 4$
$3x - 6y + 3z = 12$
$-2x + 4y - 2z = -8$

$\{(x, y, z) \mid x - 2y + z = 4\}$

28. $x + 3y + z = 1$
$2x + 6y + 2z = 2$
$3x + 9y + 3z = 3$

$\{(x, y, z) \mid x + 3y + z = 1\}$

29. $5x + 3y - z = 0$
$2x - 3y + z = 0$
$x + 4y - 2z = 0$

$\{(0, 0, 0)\}$

30. $4x + 5y - z = 0$
$7x - 5y + z = 0$
$x + 3y - 2z = 0$

$\{(0, 0, 0)\}$

Chapter 8 Summary

Key Terms

8.1	**linear (first-degree) equation in one variable**	A linear equation in one variable can be written in the form $Ax + B = C$, where A, B, and C are real numbers, with $A \neq 0$.
	solution	A solution of an equation is a number that makes the equation true when substituted for the variable.
	solution set	The solution set of an equation is the set of all its solutions.
	equivalent equations	Equivalent equations are equations that have the same solution set.
	conditional equation	An equation that is true only for certain value(s) of the variable is called a conditional equation.
	contradiction	An equation that has no solution (that is, its solution set is $\emptyset$) is called a contradiction.
	identity	An equation that is satisfied by every valid replacement of the variable is called an identity.
	linear inequality in one variable	A linear inequality in the variable x can be written in the form $Ax + B < C$, where A, B, and C are real numbers, with $A \neq 0$. (Other inequality symbols may be used.)
8.2	**intersection**	The intersection of two sets A and B is the set of elements that belong to both A and B. 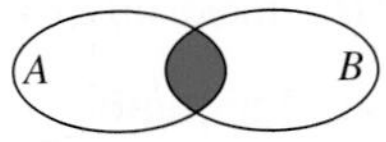
	compound inequality	A compound inequality is formed by joining two inequalities with a connective word such as *and* or *or*.
	union	The union of two sets A and B is the set of elements that belong to either A or B (or both). 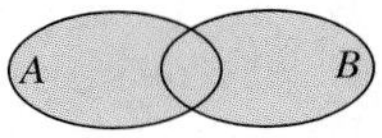
8.3	**absolute value equation; absolute value inequality**	Absolute value equations and inequalities are equations and inequalities that involve the absolute value of a variable expression.
8.4	**system of equations**	Two or more equations that are to be solved at the same time form a system of equations.
	linear system	A linear system is a system of equations that contains only linear equations.
	solution set of a system	All ordered pairs that satisfy all the equations of a system at the same time make up the solution set of the system.
	consistent system	A system is consistent if it has a solution.
	independent equations	Independent equations are equations whose graphs are different lines.
	inconsistent system	A system is inconsistent if it has no solution.
	dependent equations	Dependent equations are equations whose graphs are the same line.
8.6	**matrix**	A matrix is a rectangular array of numbers, consisting of horizontal **rows** and vertical **columns.**
	elements of a matrix	The numbers in a matrix are its elements.
	square matrix	A square matrix is a matrix that has the same number of rows as columns.
	augmented matrix	An augmented matrix is a matrix that has a vertical bar that separates the columns of the matrix into two groups.
	row echelon form	If a matrix is written with 1s down the diagonal from upper left to lower right and 0s below the 1s, it is said to be in row echelon form.

New Symbols

Symbol	Meaning
$\cap$	set intersection
$\cup$	set union
(x, y, z)	ordered triple
$\begin{bmatrix} a & b & c \\ d & e & f \end{bmatrix}$	matrix with 2 rows, 3 columns (2×3)

▶ Test Your Word Power

See how well you have learned the vocabulary in this chapter. Answers, with examples, follow the Quick Review.

1. An **equation** is
   A. an algebraic expression
   B. an expression that contains fractions
   C. an expression that uses any of the four basic operations or the operations of raising to powers or taking roots on any collection of variables and numbers
   D. a statement that two algebraic expressions are equal.

2. A **solution set** is the set of numbers that
   A. make an expression undefined
   B. make an equation false
   C. make an equation true
   D. make an expression equal to 0.

3. An **inequality** is
   A. a statement that two algebraic expressions are equal
   B. a point on a number line
   C. an equation with no solutions
   D. a statement with algebraic expressions related by $<$, $\leq$, $>$, or $\geq$.

4. The **intersection** of two sets *A* and *B* is the set of elements that belong
   A. to both *A* and *B*
   B. to either *A* or *B*, or both
   C. to either *A* or *B*, but not both
   D. to just *A*.

5. The **union** of two sets *A* and *B* is the set of elements that belong
   A. to both *A* and *B*
   B. to either *A* or *B*, or both
   C. to either *A* or *B*, but not both
   D. to just *B*.

6. A **system of equations** consists of
   A. at least two equations with different variables
   B. two or more equations that have an infinite number of solutions
   C. two or more equations that are to be solved at the same time
   D. two or more inequalities that are to be solved.

7. An **inconsistent system** is a system of equations
   A. with one solution
   B. with no solution
   C. with an infinite number of solutions
   D. that have the same graph.

8. **Dependent equations**
   A. have different graphs
   B. have no solution
   C. have one solution
   D. are different forms of the same equation.

9. A **matrix** is
   A. an ordered pair of numbers
   B. an array of numbers with the same number of rows and columns
   C. a pair of numbers written between brackets
   D. a rectangular array of numbers.

▶ Quick Review

Concepts	Examples

8.1 Review of Solving Linear Equations and Inequalities

Solving a Linear Equation in One Variable

Step 1 Simplify each side separately.

Step 2 Isolate the variable terms on one side.

Step 3 Isolate the variable.

Step 4 Check.

Solve the equation.

$$4(8 - 3t) = 32 - 8(t + 2)$$
$$32 - 12t = 32 - 8t - 16 \quad \text{Distributive property}$$
$$32 - 12t = 16 - 8t$$
$$32 - 12t + \mathbf{12t} = 16 - 8t + \mathbf{12t} \quad \text{Add } 12t.$$
$$32 = 16 + 4t$$
$$32 - \mathbf{16} = 16 + 4t - \mathbf{16} \quad \text{Subtract 16.}$$
$$16 = 4t$$
$$\frac{16}{4} = \frac{4t}{4} \quad \text{Divide by 4.}$$
$$4 = t$$

The solution set is $\{4\}$. This can be checked by substituting 4 for *t* in the original equation.

Concepts	Examples
Solving a Linear Inequality in One Variable	Solve $3(x + 2) - 5x \le 12$.
Step 1 Simplify each side of the inequality by clearing parentheses, fractions, and decimals, as needed, and combining like terms.	$3x + 6 - 5x \le 12$ Distributive property $-2x + 6 \le 12$
Step 2 Use the addition property of inequality to get all terms with variables on one side and all terms without variables on the other side.	$-2x + 6 - 6 \le 12 - 6$ Subtract 6. $-2x \le 6$
Step 3 Use the multiplication property of inequality to write the inequality in the form $x < k$ or $x > k$.	$\frac{-2x}{-2} \ge \frac{6}{-2}$ Divide by -2; change $\le$ to $\ge$. $x \ge -3$
If an inequality is multiplied or divided by a negative number, the direction of the inequality symbol must be reversed.	The solution set $[-3, \infty)$ is graphed below. 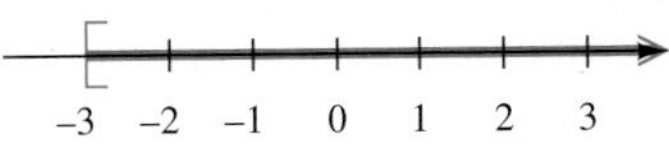

8.2 Set Operations and Compound Inequalities

Concepts	Examples
Solving a Compound Inequality	Solve $x + 1 > 2$ and $2x < 6$.
Step 1 Solve each inequality in the compound inequality individually.	$x + 1 > 2$ and $2x < 6$ $x > 1$ and $x < 3$
Step 2 If the inequalities are joined with *and,* the solution set is the intersection of the two individual solution sets.	The solution set is $(1, 3)$. 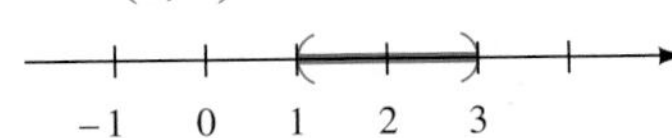
If the inequalities are joined with *or,* the solution set is the union of the two individual solution sets.	Solve $x \ge 4$ or $x \le 0$. The solution set is $(-\infty, 0] \cup [4, \infty)$. 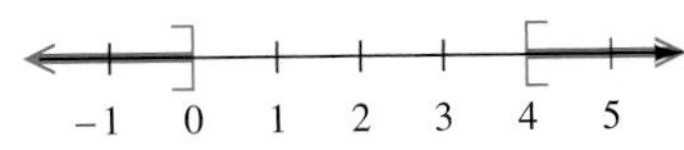

8.3 Absolute Value Equations and Inequalities

Concepts	Examples
Let k be a positive number. To solve $\|ax + b\| = k$, solve the compound equation $ax + b = k$ or $ax + b = -k$.	Solve $\|x - 7\| = 3$. $x - 7 = 3$ or $x - 7 = -3$ $x = 10$ or $x = 4$ The solution set is $\{4, 10\}$. 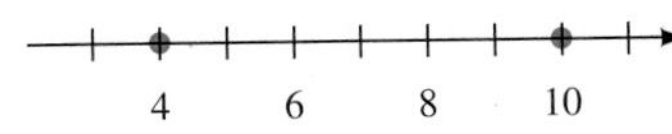
To solve $\|ax + b\| > k$, solve the compound inequality $ax + b > k$ or $ax + b < -k$.	Solve $\|x - 7\| > 3$. $x - 7 > 3$ or $x - 7 < -3$ $x > 10$ or $x < 4$ The solution set is $(-\infty, 4) \cup (10, \infty)$. 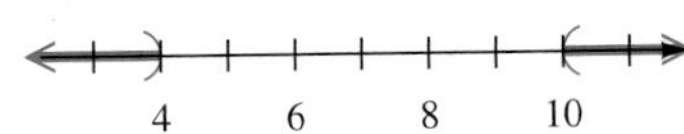

(continued)

8.3 Absolute Value Equations and Inequalities (continued)

Concepts	Examples
To solve $\lvert ax + b \rvert < k$, solve the compound inequality $$-k < ax + b < k.$$	Solve $\lvert x - 7 \rvert < 3$. $$-3 < x - 7 < 3$$ $$4 < x < 10 \quad \text{Add 7 to each part.}$$ The solution set is (4, 10).
To solve an absolute value equation of the form $$\lvert ax + b \rvert = \lvert cx + d \rvert,$$ solve the compound equation $$ax + b = cx + d \quad \text{or} \quad ax + b = -(cx + d).$$	Solve $\lvert x + 2 \rvert = \lvert 2x - 6 \rvert$. $x + 2 = 2x - 6$ or $x + 2 = -(2x - 6)$; $x = 8$; $x + 2 = -2x + 6$; $3x = 4$; $x = \frac{4}{3}$. The solution set is $\{\frac{4}{3}, 8\}$.

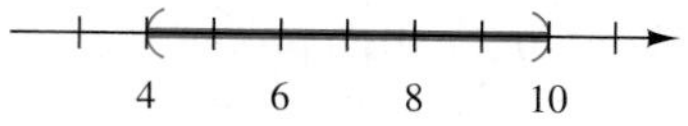

8.4 Review of Systems of Linear Equations in Two Variables

Concepts	Examples
Solving a Linear System by Substitution	Solve by substitution. $$4x - y = 7 \quad (1)$$ $$3x + 2y = 30 \quad (2)$$
Step 1 Solve one of the equations for either variable.	Solve for y in equation (1). $$y = 4x - 7$$
Step 2 Substitute for that variable in the other equation. The result should be an equation with just one variable.	Substitute $4x - 7$ for y in equation (2), and solve for x.
Step 3 Solve the equation from Step 2.	$3x + 2y = 30$ (2); $3x + 2(4x - 7) = 30$ Let $y = 4x - 7$.; $3x + 8x - 14 = 30$ Distributive property; $11x - 14 = 30$ Combine like terms.; $x = 4$ Add 14; divide by 11.
Step 4 Find the value of the other variable by substituting the result from Step 3 into the equation from Step 1.	Substitute 4 for x in the equation $y = 4x - 7$ to find that $y = 9$.
Step 5 Check the ordered-pair solution in *both* of the *original* equations. Then write the solution set.	Check to see that $\{(4, 9)\}$ is the solution set.
Solving a Linear System by Elimination	Solve by elimination.
Step 1 Write both equations in standard form.	$$5x + y = 2 \quad (1)$$ $$2x - 3y = 11 \quad (2)$$
Step 2 Make the coefficients of one pair of variable terms opposites.	To eliminate y, multiply equation (1) by 3, and add the result to equation (2).
Step 3 Add the new equations. The sum should be an equation with just one variable.	$15x + 3y = 6$ 3 times equation (1); $2x - 3y = 11$ (2); $17x = 17$ Add.
Step 4 Solve the equation from Step 3.	$x = 1$ Divide by 17.

Concepts	Examples
Step 5 Find the value of the other variable by substituting the result from Step 4 into either of the original equations.	Let $x = 1$ in equation (1), and solve for y. $$5(1) + y = 2$$ $$y = -3$$
Step 6 Check the ordered-pair solution in *both* of the *original* equations. Then write the solution set.	Check to verify that $\{(1, -3)\}$ is the solution set.
If the result of the addition step (Step 3) is a false statement, such as $0 = 4$, the graphs are parallel lines and *there is no solution. The solution set is* $\emptyset$.	$$\begin{aligned} x - 2y &= 6 \\ -x + 2y &= -2 \\ \hline 0 &= 4 \end{aligned}$$ Solution set: $\emptyset$
If the result is a true statement, such as $0 = 0$, the graphs are the same line, and an *infinite number of ordered pairs are solutions. The solution set is written in set-builder notation as* $\{(x, y) \mid \_\_\_\_\_\}$, *where a form of the equation is written in the blank.*	$$\begin{aligned} x - 2y &= 6 \\ -x + 2y &= -6 \\ \hline 0 &= 0 \end{aligned}$$ Solution set: $\{(x, y) \mid x - 2y = 6\}$

8.5 Systems of Linear Equations in Three Variables; Applications

Concepts	Examples
Solving a Linear System in Three Variables	Solve the system.
Step 1 Select a focus variable, preferably one with coefficient 1 or -1, and a working equation.	$$\begin{aligned} x + 2y - z &= 6 \quad (1) \\ x + y + z &= 6 \quad (2) \\ 2x + y - z &= 7 \quad (3) \end{aligned}$$ We choose z as the focus variable and (2) as the working equation.
Step 2 Eliminate the focus variable, using the working equation and one of the equations of the system.	Add equations (1) and (2) to get $$2x + 3y = 12. \quad (4)$$
Step 3 Eliminate the focus variable again, using the working equation and the remaining equation of the system.	Add equations (2) and (3) to get $$3x + 2y = 13. \quad (5)$$
Step 4 Solve the system of two equations in two variables formed by the equations from Steps 2 and 3.	Use equations (4) and (5) to eliminate x. $$\begin{aligned} -6x - 9y &= -36 && \text{Multiply (4) by } -3. \\ 6x + 4y &= 26 && \text{Multiply (5) by 2.} \\ \hline -5y &= -10 && \text{Add.} \\ y &= 2 && \text{Divide by } -5. \end{aligned}$$ To find x, substitute 2 for y in equation (4). $$\begin{aligned} 2x + 3(2) &= 12 && \text{Let } y = 2 \text{ in (4).} \\ 2x + 6 &= 12 \\ 2x &= 6 \\ x &= 3 \end{aligned}$$
Step 5 Find the value of the remaining variable.	Substitute 3 for x and 2 for y in working equation (2). $$\begin{aligned} x + y + z &= 6 \quad (2) \\ 3 + 2 + z &= 6 \\ z &= 1 \end{aligned}$$
Step 6 Check the ordered-triple solution in each of the original equations of the system. Then write the solution set.	A check of the solution (3, 2, 1) confirms that the solution set is $\{(3, 2, 1)\}$.

Concepts	Examples

8.6 Solving Systems of Linear Equations by Matrix Methods

Matrix Row Operations

1. Any two rows of the matrix may be interchanged.

$$\begin{bmatrix} 1 & 5 & 7 \\ 3 & 9 & -2 \\ 0 & 6 & 4 \end{bmatrix} \text{ becomes } \begin{bmatrix} 3 & 9 & -2 \\ 1 & 5 & 7 \\ 0 & 6 & 4 \end{bmatrix} \quad \text{Interchange } R_1 \text{ and } R_2.$$

2. The numbers in any row may be multiplied by any nonzero real number.

$$\begin{bmatrix} 1 & 5 & 7 \\ 3 & 9 & -2 \\ 0 & 6 & 4 \end{bmatrix} \text{ becomes } \begin{bmatrix} 1 & 5 & 7 \\ 1 & 3 & -\frac{2}{3} \\ 0 & 6 & 4 \end{bmatrix} \quad \tfrac{1}{3}R_2$$

3. Any row may be transformed by adding to the numbers of the row the product of a real number and the numbers of another row.

$$\begin{bmatrix} 1 & 5 & 7 \\ 3 & 9 & -2 \\ 0 & 6 & 4 \end{bmatrix} \text{ becomes } \begin{bmatrix} 1 & 5 & 7 \\ 0 & -6 & -23 \\ 0 & 6 & 4 \end{bmatrix} \quad -3R_1 + R_2$$

A system can be solved by matrix methods. Write the augmented matrix, and use row operations to obtain a matrix in row echelon form.

Solve using row operations.

$$x + 3y = 7$$
$$2x + y = 4$$

$$\left[\begin{array}{cc|c} 1 & 3 & 7 \\ 2 & 1 & 4 \end{array}\right] \quad \text{Augmented matrix}$$

$$\left[\begin{array}{cc|c} 1 & 3 & 7 \\ 0 & -5 & -10 \end{array}\right] \quad -2R_1 + R_2$$

$$\left[\begin{array}{cc|c} 1 & 3 & 7 \\ 0 & 1 & 2 \end{array}\right] \quad -\tfrac{1}{5}R_2$$

$$x + 3y = 7$$
$$y = 2$$

When $y = 2$, $x + 3(2) = 7$, so $x = 1$. The solution set is $\{(1, 2)\}$.

ANSWERS TO TEST YOUR WORD POWER

1. D; *Examples:* $2a + 3 = 7$; $3y = -8$, $x^2 = 4$
2. C; *Example:* $\{8\}$ is the solution set of $2x + 5 = 21$.
3. D; *Examples:* $x < 5$, $7 + 2k \geq 11$, $-5 < 2z - 1 \leq 3$
4. A; *Example:* If $A = \{2, 4, 6, 8\}$ and $B = \{1, 2, 3\}$, $A \cap B = \{2\}$.
5. B; *Example:* Using the preceding sets A and B, $A \cup B = \{1, 2, 3, 4, 6, 8\}$.
6. C; *Example:* $3x - y = 3$, $2x + y = 7$
7. B; *Example:* The equations of two parallel lines form an inconsistent system. Their graphs never intersect, so the system has no solution.
8. D; *Example:* The equations $4x - y = 8$ and $8x - 2y = 16$ are dependent because their graphs are the same line.
9. D; *Examples:* $\begin{bmatrix} 3 & -1 & 0 \\ 4 & 2 & 1 \end{bmatrix}$, $\begin{bmatrix} 1 & 2 \\ 4 & 3 \end{bmatrix}$

Chapter 8  Review Exercises

[8.1] *Solve each equation.*

1. $-(8 + 3x) + 5 = 2x + 6$

$\left\{-\frac{9}{5}\right\}$

2. $-(r + 5) - (2 + 7r) + 8r = 3r - 8$

$\left\{\frac{1}{3}\right\}$

3. $\frac{m - 2}{4} + \frac{m + 2}{2} = 8$

{10}

4. $0.05x + 0.03(1200 - x) = 42$

{300}

Decide whether each equation is a conditional equation, *an* identity, *or a* contradiction. *Give the solution set.*

5. $7r - 3(2r - 5) + 5 + 3r = 4r + 20$

identity; {all real numbers}

6. $8p - 4p - (p - 7) + 9p + 13 = 12p$

contradiction; ∅

7. $-2r + 6(r - 1) + 3r - (4 - r) = -(r + 5) - 5$

conditional; {0}

Solve each inequality. Give the solution set in both interval and graph forms.

8. $-\frac{2}{3}x < 6$

$(-9, \infty)$

9. $-5x - 4 \geq 11$

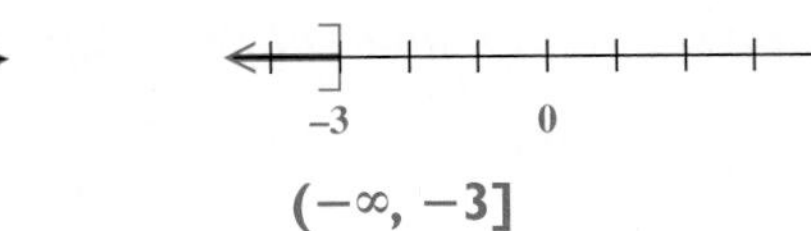

$(-\infty, -3]$

10. $5 - (6 - 4t) \geq 2t - 7$

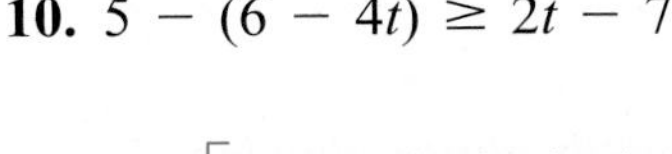

$[-3, \infty)$

11. $-6 \leq 2k \leq 24$

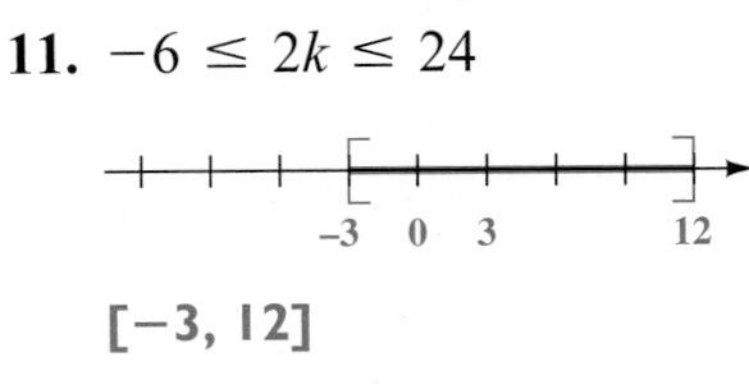

$[-3, 12]$

12. $8 \leq 3x - 1 < 14$

$[3, 5)$

[8.2] *Let* $A = \{a, b, c, d\}$, $B = \{a, c, e, f\}$, *and* $C = \{a, e, f, g\}$. *Find each set.*

13. $A \cap B$

{a, c}

14. $A \cap C$

{a}

15. $B \cup C$

{a, c, e, f, g}

16. $A \cup C$

{a, b, c, d, e, f, g}

Solve each compound inequality. Give the solution set in both interval and graph forms.

17. $x > 4$ and $x < 7$

$(4, 7)$

18. $x + 4 > 12$ and $x - 2 < 12$

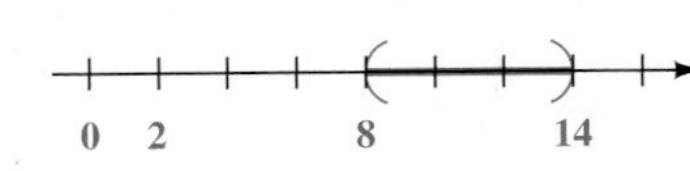

$(8, 14)$

19. $x > 5$ or $x \leq -3$

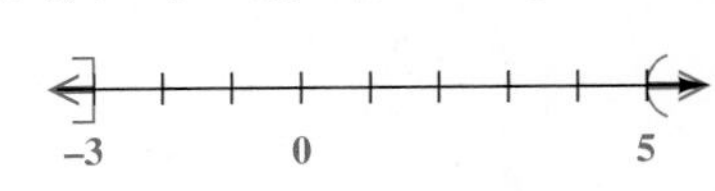

$(-\infty, -3] \cup (5, \infty)$

20. $x \geq -2$ or $x < 2$

$(-\infty, \infty)$

21. $x - 4 > 6$ and $x + 3 \leq 10$

$\emptyset$

22. $-5x + 1 \geq 11$ or $3x + 5 \geq 26$

$(-\infty, -2] \cup [7, \infty)$

Express each union or intersection in simplest interval form.

23. $(-3, \infty) \cap (-\infty, 4)$

$(-3, 4)$

24. $(-\infty, 6) \cap (-\infty, 2)$

$(-\infty, 2)$

25. $(4, \infty) \cup (9, \infty)$

$(4, \infty)$

26. $(1, 2) \cup (1, \infty)$

$(1, \infty)$

[8.3] *Solve each absolute value equation.*

27. $|x| = 7$

$\{-7, 7\}$

28. $|x + 2| = 9$

$\{-11, 7\}$

29. $|3k - 7| = 8$

$\left\{-\frac{1}{3}, 5\right\}$

30. $|z - 4| = -12$

$\emptyset$

31. $|2k - 7| + 4 = 11$

$\{0, 7\}$

32. $|4a + 2| - 7 = -3$

$\left\{-\frac{3}{2}, \frac{1}{2}\right\}$

33. $|3p + 1| = |p + 2|$

$\left\{-\frac{3}{4}, \frac{1}{2}\right\}$

34. $|2m - 1| = |2m + 3|$

$\left\{-\frac{1}{2}\right\}$

Solve each absolute value inequality. Give the solution set in both interval and graph forms.

35. $|x| < 12$

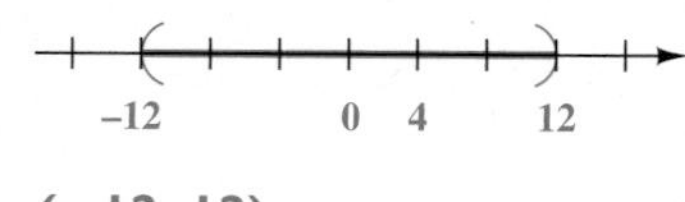

$(-12, 12)$

36. $|-x + 6| \leq 7$

$[-1, 13]$

37. $|2p + 5| \leq 1$

$[-3, -2]$

38. $|x + 1| \geq -3$

$(-\infty, \infty)$

39. $|5r - 1| > 9$

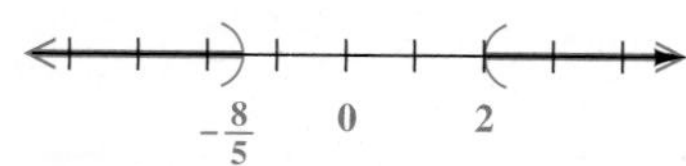

$\left(-\infty, -\frac{8}{5}\right) \cup (2, \infty)$

40. $|3x + 6| \geq 0$

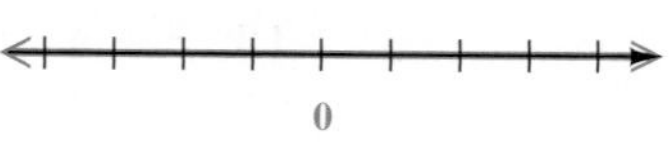

$(-\infty, \infty)$

[8.4] **41.** Solve by graphing: $\begin{aligned} x + 3y &= 8 \\ 2x - y &= 2. \end{aligned}$

{(2, 2)}

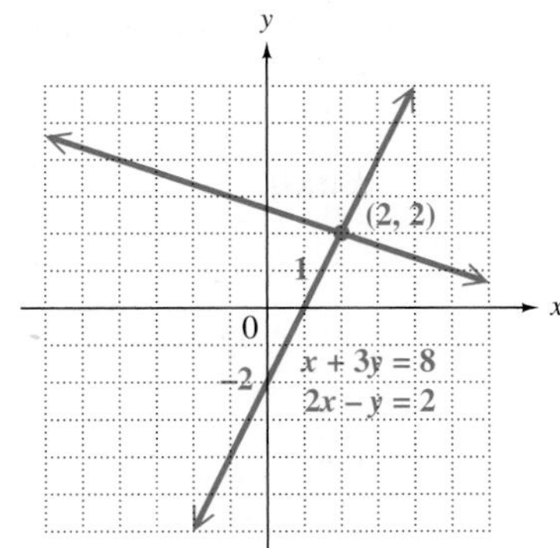

Solve each system using the substitution method.

42. $\begin{aligned} 3x + y &= -4 \\ x &= \frac{2}{3}y \end{aligned}$

$\left\{\left(-\frac{8}{9}, -\frac{4}{3}\right)\right\}$

43. $\begin{aligned} 9x - y &= -4 \\ y &= x + 4 \end{aligned}$

{(0, 4)}

44. $\begin{aligned} -5x + 2y &= -2 \\ x + 6y &= 26 \end{aligned}$

{(2, 4)}

Solve each system using the elimination method. If a system is inconsistent or has dependent equations, say so.

45. $\begin{aligned} 6x + 5y &= 4 \\ -4x + 2y &= 8 \end{aligned}$

{(−1, 2)}

46. $\begin{aligned} \frac{x}{6} + \frac{y}{6} &= -\frac{1}{2} \\ x - y &= -9 \end{aligned}$

{(−6, 3)}

47. $\begin{aligned} 4x + 5y &= 9 \\ 3x + 7y &= -1 \end{aligned}$

$\left\{\left(\frac{68}{13}, -\frac{31}{13}\right)\right\}$

48. $\begin{aligned} -3x + y &= 6 \\ 2y &= 12 + 6x \end{aligned}$

{(x, y) | 3x − y = −6}; dependent equations

49. $\begin{aligned} 5x - 4y &= 2 \\ -10x + 8y &= 7 \end{aligned}$

∅; inconsistent system

50. $\begin{aligned} 3x + 3y &= 0 \\ -2x - y &= 0 \end{aligned}$

{(0, 0)}

[8.5] *Solve each system of equations. If a system is inconsistent or has dependent equations, say so.*

51. $\begin{aligned} 2x + 3y - z &= -16 \\ x + 2y + 2z &= -3 \\ -3x + y + z &= -5 \end{aligned}$

{(1, −5, 3)}

52. $\begin{aligned} 3x - y - z &= -8 \\ 4x + 2y + 3z &= 15 \\ -6x + 2y + 2z &= 10 \end{aligned}$

∅; inconsistent system

53. $\begin{aligned} 4x - y &= 2 \\ 3y + z &= 9 \\ x + 2z &= 7 \end{aligned}$

{(1, 2, 3)}

Solve each problem using a system of equations.

54. The sum of the measures of the angles of a triangle is 180°. The largest angle measures 10° less than the sum of the other two. The measure of the middle-sized angle is the average of the other two. Find the measures of the three angles.

85°, 60°, 35°

55. In the great baseball year of 1961, Yankee teammates Mickey Mantle, Roger Maris, and John Blanchard combined for 136 home runs. Mantle hit 7 fewer than Maris. Maris hit 40 more than Blanchard. What were the home run totals for each player? (*Source:* Neft, David S. and Richard M. Cohen, *The Sports Encyclopedia: Baseball 2003.*)

Mantle: 54; Maris: 61; Blanchard: 21

[8.6] *Solve each system using row operations.*

56. $\begin{aligned} 2x + 5y &= -4 \\ 4x - y &= 14 \end{aligned}$

{(3, −2)}

57. $\begin{aligned} 6x + 3y &= 9 \\ -7x + 2y &= 17 \end{aligned}$

{(−1, 5)}

58. $\begin{aligned} x + 2y - z &= 1 \\ 3x + 4y + 2z &= -2 \\ -2x - y + z &= -1 \end{aligned}$

{(0, 0, −1)}

Mixed Review Exercises

Solve.

59. $(7 - 2x) + 3(5 - 3x) = x + 8$
$\left\{\frac{7}{6}\right\}$

60. $x < 5$ and $x \geq -4$
$[-4, 5)$

61. $\frac{3}{4}(a - 2) - \frac{1}{3}(5 - 2a) < -2$
$\left(-\infty, \frac{14}{17}\right)$

62. $5(2x - 3) = 6(x - 1) + 4x$
$\emptyset$

63. $|7x - 2| > 9$
$(-\infty, -1) \cup \left(\frac{11}{7}, \infty\right)$

64. $|2x - 10| = 20$
$\{-5, 15\}$

65. $|m + 3| \leq 13$
$[-16, 10]$

66. $x \geq -2$ or $x < 4$
$(-\infty, \infty)$

67. $|m - 1| = |2m + 3|$
$\left\{-4, -\frac{2}{3}\right\}$

In Exercises 68 and 69, sketch the graph of each solution set.

68. $x > 6$ and $x < 8$

69. $-5x + 1 \geq 6$ or $3x + 5 \geq 26$

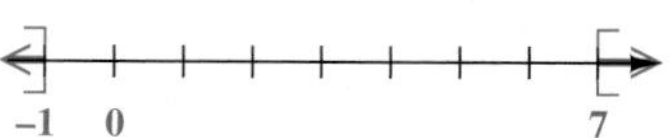

70. If $k < 0$, what is the solution set of

(a) $|5x + 3| < k$, $\emptyset$
(b) $|5x + 3| > k$, $(-\infty, \infty)$
(c) $|5x + 3| = k$? $\emptyset$

71. Which system, A or B, would be easier to solve using the substitution method? Why?

A. $5x - 3y = 7$
$2x + 8y = 3$

B. $7x + 2y = 4$
$y = -3x + 1$

B; The second equation is already solved for y.

Solve by any method.

72. $\frac{2}{3}x + \frac{1}{6}y = \frac{19}{2}$
$\frac{1}{3}x - \frac{2}{9}y = 2$
$\{(12, 9)\}$

73. $2x - 5y = 8$
$3x + 4y = 10$
$\left\{\left(\frac{82}{23}, -\frac{4}{23}\right)\right\}$

74. $x = 7y + 10$
$2x + 3y = 3$
$\{(3, -1)\}$

75. $x + 4y = 17$
$-3x + 2y = -9$
$\{(5, 3)\}$

76. $-7x + 3y = 12$
$5x + 2y = 8$
$\{(0, 4)\}$

77. $2x + 5y - z = 12$
$-x + y - 4z = -10$
$-8x - 20y + 4z = 31$
$\emptyset$

78. In the 2006 Winter Olympics in Turino, Italy, the top medal-winning countries were Germany, the United States, and Canada, with a combined total of 78 medals. Germany won four more medals than the United States, while Canada won one fewer medal than the United States. How many medals did each country win? (*Source:* www.nbcolympics.com)

Germany: 29; U.S.: 25; Canada: 24

Chapter 8 Test

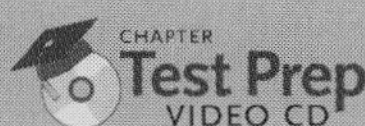

Use the Chapter Test Prep Video CD to see fully worked-out solutions to any of the exercises you want to review

Solve each equation.

1. $3(2x - 2) - 4(x + 6) = 4x + 8$

1. $\{-19\}$

2. $0.08x + 0.06(x + 9) = 1.24$

2. $\{5\}$

3. $\dfrac{x + 6}{10} + \dfrac{x - 4}{15} = 1$

3. $\{4\}$

4. Decide whether each equation is a *conditional equation*, an *identity*, or a *contradiction*. Give its solution set.

(a) $3x - (2 - x) + 4x + 2 = 8x + 3$

(b) $\dfrac{x}{3} + 7 = \dfrac{5x}{6} - 2 - \dfrac{x}{2} + 9$

(c) $-4(2x - 6) = 5x + 24 - 7x$

4. (a) contradiction; $\emptyset$

(b) identity; {all real numbers}

(c) conditional equation; $\{0\}$

Solve each inequality. Give the solution set in both interval and graph forms.

5. $4 - 6(x + 3) \le -2 - 3(x + 6) + 3x$

5. $[1, \infty)$

6. $-6 \le \dfrac{4}{3}x - 2 \le 2$

6. $[-3, 3]$

7. Let $A = \{1, 2, 5, 7\}$ and $B = \{1, 5, 9, 12\}$. Find

(a) $A \cap B$, **(b)** $A \cup B$.

7. (a) $\{1, 5\}$

(b) $\{1, 2, 5, 7, 9, 12\}$

Solve each compound or absolute value inequality. For Exercises 8–11, give the solution set in both interval and graph forms.

8. $3k \ge 6$ and $k - 4 < 5$

8. $[2, 9)$

9. $-4x \le -24$ or $4x - 2 < 10$

9. $(-\infty, 3) \cup [6, \infty)$

10. $|4x + 3| \le 7$

10. $\left[-\dfrac{5}{2}, 1\right]$

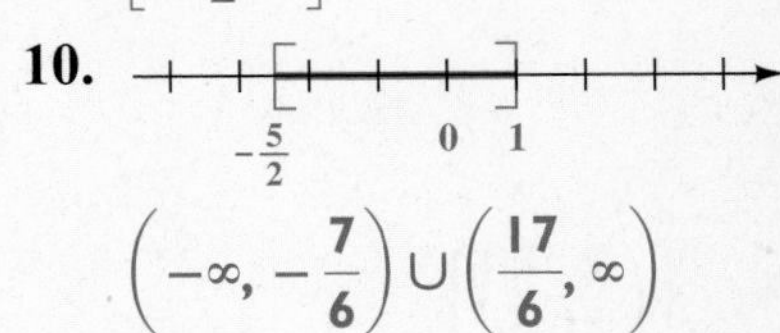

11. $|5 - 6x| > 12$

11. $\left(-\infty, -\dfrac{7}{6}\right) \cup \left(\dfrac{17}{6}, \infty\right)$

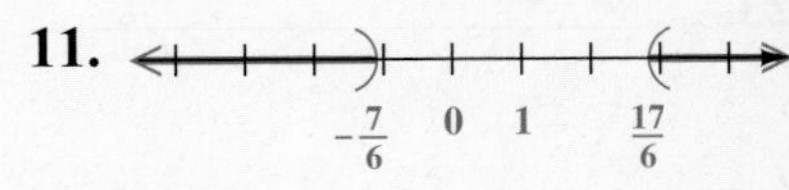

12. $|7 - x| \le -1$

12. $\emptyset$

13. $\left\{-\frac{5}{3}, 3\right\}$

14. $\left\{-\frac{5}{7}, \frac{11}{3}\right\}$

15. $\{(6, 1)\}$

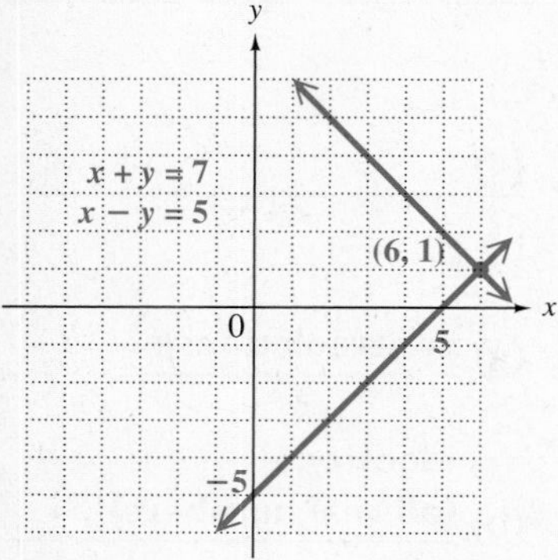

16. $\{(6, -4)\}$

17. $\left\{\left(-\frac{9}{4}, \frac{5}{4}\right)\right\}$

18. $\{(0, -2)\}$

19. $\emptyset$; inconsistent system

20. $\left\{\left(-\frac{2}{3}, \frac{4}{5}, 0\right)\right\}$

21. $\{(3, -2, 1)\}$

22. 60 oz of Orange Pekoe; 30 oz of Irish Breakfast; 10 oz of Earl Grey

23. $\left\{\left(\frac{2}{5}, \frac{7}{5}\right)\right\}$

24. $\{(-1, 2, 3)\}$

Solve each absolute value equation.

13. $|3k - 2| + 1 = 8$

14. $|3 - 5x| = |2x + 8|$

15. Use a graph to solve the system.

$$x + y = 7$$
$$x - y = 5$$

Solve each system by substitution or elimination. If a system is inconsistent or has dependent equations, say so.

16. $2x - 3y = 24$
$y = -\frac{2}{3}x$

17. $3x - y = -8$
$2x + 6y = 3$

18. $-5x + 2y = -4$
$6x + 3y = -6$

19. $3x + 4y = 8$
$8y = 7 - 6x$

20. $3x + 5y + 3z = 2$
$6x + 5y + z = 0$
$3x + 10y - 2z = 6$

21. $4x + y + z = 11$
$x - y - z = 4$
$y + 2z = 0$

22. The owner of a tea shop wants to mix three kinds of tea to make 100 oz of a mixture that will sell for \$0.83 per oz. He uses Orange Pekoe, which sells for \$0.80 per oz, Irish Breakfast, for \$0.85 per oz, and Earl Grey, for \$0.95 per oz. If he wants to use twice as much Orange Pekoe as Irish Breakfast, how much of each kind of tea should he use?

Solve each system using row operations.

23. $3x + 2y = 4$
$5x + 5y = 9$

24. $x + 3y + 2z = 11$
$3x + 7y + 4z = 23$
$5x + 3y - 5z = -14$

Cumulative Review Exercises Chapters R–8

Simplify each expression.

1. $-2(m - 3)$
$-2m + 6$

2. $-(-4m + 3)$
$4m - 3$

3. $3x^2 - 4x + 4 + 9x - x^2$
$2x^2 + 5x + 4$

Evaluate if $p = -4$, $q = -2$, and $r = 5$.

4. $-3(2q - 3p)$
-24

5. $8r^2 + q^2$
204

6. $\dfrac{r}{-p + 2q}$
undefined

7. $\dfrac{rp + 6r^2}{p^2 + q - 1}$
10

Solve.

8. $2z - 5 + 3z = 4 - (z + 2)$
$\left\{\frac{7}{6}\right\}$

9. $\dfrac{3a - 1}{5} + \dfrac{a + 2}{2} = -\dfrac{3}{10}$
$\{-1\}$

10. $-\dfrac{4}{3}d \geq -5$
$\left(-\infty, \frac{15}{4}\right]$

11. $3 - 2(m + 3) < 4m$
$\left(-\frac{1}{2}, \infty\right)$

12. $2k + 4 < 10$ and $3k - 1 > 5$
$(2, 3)$

13. $2k + 4 > 10$ or $3k - 1 < 5$
$(-\infty, 2) \cup (3, \infty)$

14. $|5x + 3| - 10 = 3$
$\left\{-\frac{16}{5}, 2\right\}$

15. $|x + 2| < 9$
$(-11, 7)$

16. $|2y - 5| \geq 9$
$(-\infty, -2] \cup [7, \infty)$

17. $V = lwh$ for h
$h = \frac{V}{lw}$

18. Two planes leave the Dallas-Fort Worth airport at the same time. One travels east at 550 mph, and the other travels west at 500 mph. Assuming no wind, how long will it take for the planes to be 2100 mi apart? **2 hr**

	r	t	d
Eastbound plane	550	x	
Westbound plane	500	x	

19. Graph $4x + 2y = -8$.

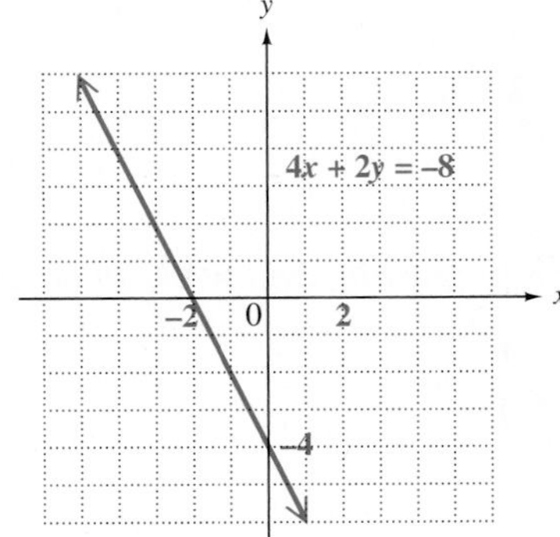

20. Find the slope of the line through the points $(-4, 8)$ and $(-2, 6)$.
-1

21. What is the slope of the line shown here?
0

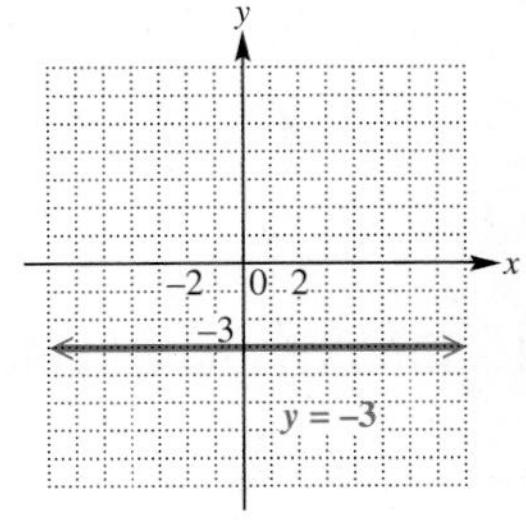

Use the function defined by $f(x) = 2x + 7$ *to find the following.*

22. $f(-4)$

-1

23. The x-intercept of its graph

$\left(-\frac{7}{2}, 0\right)$

24. The y-intercept of its graph

$(0, 7)$

Solve each system.

25. $3x - 2y = -7$
$2x + 3y = 17$

$\{(1, 5)\}$

26. $2x + 3y - 6z = 5$
$8x - y + 3z = 7$
$3x + 4y - 3z = 7$

$\{(1, 1, 0)\}$

Perform the indicated operations. Assume that variables represent nonzero real numbers.

27. $(3x^2y^{-1})^{-2}(2x^{-3}y)^{-1}$

$\frac{y}{18x}$

28. $\frac{5m^{-2}y^3}{3m^{-3}y^{-1}}$

$\frac{5my^4}{3}$

Perform the indicated operations.

29. $(3x^3 + 4x^2 - 7) - (2x^3 - 8x^2 + 3x)$

$x^3 + 12x^2 - 3x - 7$

30. $(7x + 3y)^2$

$49x^2 + 42xy + 9y^2$

31. $(2p + 3)(5p^2 - 4p - 8)$

$10p^3 + 7p^2 - 28p - 24$

Factor.

32. $16w^2 + 50wz - 21z^2$

$(2w + 7z)(8w - 3z)$

33. $4y^2 - 36y + 81$

$(2y - 9)^2$

34. $8p^3 + 27$

$(2p + 3)(4p^2 - 6p + 9)$

Solve.

35. $(p + 4)(2p + 3)(p - 1) = 0$

$\left\{-4, -\frac{3}{2}, 1\right\}$

36. $9q^2 = 6q - 1$

$\left\{\frac{1}{3}\right\}$

37. $\frac{1}{x} = \frac{1}{x + 1} + \frac{1}{2}$

$\{-2, 1\}$

Perform each operation, and write the answer in lowest terms.

38. $\frac{5}{q} - \frac{1}{q}$ $\quad \frac{4}{q}$

39. $\frac{3}{7} + \frac{4}{r}$ $\quad \frac{3r + 28}{7r}$

40. $\frac{4}{5q - 20} - \frac{1}{3q - 12}$ $\quad \frac{7}{15(q - 4)}$

41. $\frac{7z^2 + 49z + 70}{16z^2 + 72z - 40} \div \frac{3z + 6}{4z^2 - 1}$ $\quad \frac{7(2z + 1)}{24}$

42. Simplify the complex fraction $\dfrac{\frac{4}{a} + \frac{5}{2a}}{\frac{7}{6a} - \frac{1}{5a}}$. $\quad \frac{195}{29}$

9

Roots, Radicals, and Root Functions

Tom Skilling is the chief meteorologist for the *Chicago Tribune*. He writes a column titled "Ask Tom Why," in which readers question him on a variety of topics. In the Saturday, August 17, 2002, issue, reader Ted Fleischaker wrote,

> *I cannot remember the formula to calculate the distance to the horizon. I have a stunning view from my 14th-floor condo, 150 ft above the ground. How far can I see?*

Skilling's answer in Section 9.3, Exercise 127, provides a formula for finding the distance to the horizon. The formula includes a *square root*, one of the topics of this chapter.

9.1 Radical Expressions and Graphs

OBJECTIVES

1. Find square roots.
2. Decide whether a given root is rational, irrational, or not a real number.
3. Find cube, fourth, and other roots.
4. Graph functions defined by radical expressions.
5. Find *n*th roots of *n*th powers.
6. Use a calculator to find roots.

In **Section 1.1,** we discussed the idea of the *square* of a number. Recall that squaring a number means multiplying the number by itself.

$$\text{If } a = 8, \quad \text{then} \quad a^2 = 8 \cdot 8 = 64.$$

$$\text{If } a = -4, \quad \text{then} \quad a^2 = (-4)(-4) = 16.$$

$$\text{If } a = -\frac{1}{2}, \quad \text{then} \quad a^2 = \left(-\frac{1}{2}\right)\left(-\frac{1}{2}\right) = \frac{1}{4}.$$

In this chapter, we consider the opposite process.

$$\text{If } a^2 = 64, \quad \text{then} \quad a = ?.$$

$$\text{If } a^2 = 16, \quad \text{then} \quad a = ?.$$

$$\text{If } a^2 = \frac{1}{4}, \quad \text{then} \quad a = ?.$$

OBJECTIVE 1 Find square roots. To find a in the three preceding statements, we must find a number that when multiplied by itself results in the given number. The number a is called a **square root** of the number a^2.

EXAMPLE 1 Finding All Square Roots of a Number

Find all square roots of 49.

To find a square root of 49, think of a number that when multiplied by itself gives 49. One square root is 7 because $7 \cdot 7 = 49$. Another square root of 49 is -7 because $(-7)(-7) = 49$. The number 49 has two square roots, 7 and -7; one is positive, and one is negative.

◀ Work Problem 1 at the Side.

1 Find all square roots.

(a) 100

(b) 25

(c) 36

(d) $\frac{25}{36}$

The **positive** or **principal square root** of a number is written with the symbol $\sqrt{}$. For example, the positive square root of 121 is 11, written

$$\sqrt{121} = 11.$$

The symbol $-\sqrt{}$ is used for the **negative square root** of a number. For example, the negative square root of 121 is -11, written

$$-\sqrt{121} = -11.$$

The symbol $\sqrt{}$, called a **radical sign,** always represents the positive square root (except that $\sqrt{0} = 0$). The number inside the radical sign is called the **radicand,** and the entire expression, radical sign and radicand, is called a **radical.**

Radical sign → $\sqrt{a}$ ← Radicand

Radical

An algebraic expression containing a radical is called a **radical expression.**

Radicals have a long mathematical history. The radical sign $\sqrt{}$ has been used since sixteenth-century Germany and was probably derived from the letter *R*. The radical symbol in the margin comes from the Latin word for root, *radix*. It was first used by Leonardo da Pisa (Fibonnaci) in 1220.

Early radical symbol

ANSWERS

1. (a) $10, -10$ (b) $5, -5$ (c) $6, -6$ (d) $\frac{5}{6}, -\frac{5}{6}$

We summarize our discussion of square roots as follows.

Square Roots of a

If a is a positive real number, then

$\sqrt{a}$ is the positive or principal square root of a,

and $-\sqrt{a}$ is the negative square root of a.

For nonnegative a,

$$\sqrt{a} \cdot \sqrt{a} = \left(\sqrt{a}\right)^2 = a \quad \text{and} \quad -\sqrt{a} \cdot \left(-\sqrt{a}\right) = \left(-\sqrt{a}\right)^2 = a.$$

Also, $\sqrt{0} = 0$.

Calculator Tip Most calculators have a square root key, usually labeled $\sqrt{x}$, that allows us to find the square root of a number. On some models, the square root key must be used in conjunction with the key marked INV or 2nd.

EXAMPLE 2 Finding Square Roots

Find each square root.

(a) $\sqrt{144}$

The radical $\sqrt{144}$ represents the positive or principal square root of 144. Think of a positive number whose square is 144.

$$12^2 = 144, \quad \text{so} \quad \sqrt{144} = 12.$$

(b) $-\sqrt{1024}$

This symbol represents the negative square root of 1024. A calculator with a square root key can be used to find $\sqrt{1024} = 32$. Then, $-\sqrt{1024} = -32$.

(c) $\sqrt{\frac{4}{9}} = \frac{2}{3}$ **(d)** $-\sqrt{\frac{16}{49}} = -\frac{4}{7}$ **(e)** $\sqrt{0.81} = 0.9$

Work Problem 2 at the Side. ▶

As noted above, when the square root of a positive real number is squared, the result is that positive real number. (Also, $\left(\sqrt{0}\right)^2 = 0$.)

EXAMPLE 3 Squaring Radical Expressions

Find the *square* of each radical expression.

(a) $\sqrt{13}$

$\left(\sqrt{13}\right)^2 = 13$ Definition of square root

(b) $-\sqrt{29}$

$\left(-\sqrt{29}\right)^2 = 29$ The square of a *negative* number is positive.

(c) $\sqrt{p^2 + 1}$

$\left(\sqrt{p^2 + 1}\right)^2 = p^2 + 1$

Work Problem 3 at the Side. ▶

2 Find each square root.

(a) $\sqrt{16}$

(b) $-\sqrt{169}$

(c) $-\sqrt{225}$

(d) $\sqrt{729}$

(e) $\sqrt{\frac{36}{25}}$

(f) $\sqrt{0.49}$

3 Find the *square* of each radical expression.

(a) $\sqrt{41}$

(b) $-\sqrt{39}$

(c) $\sqrt{2x^2 + 3}$

ANSWERS

2. (a) 4 **(b)** -13 **(c)** -15 **(d)** 27 **(e)** $\frac{6}{5}$ **(f)** 0.7

3. (a) 41 **(b)** 39 **(c)** $2x^2 + 3$

OBJECTIVE 2 Decide whether a given root is rational, irrational, or not a real number. All numbers with square roots that are rational are called **perfect squares.**

Perfect squares		Rational square roots
25		$\sqrt{25} = 5$
144	are perfect squares since	$\sqrt{144} = 12$
$\frac{4}{9}$		$\sqrt{\frac{4}{9}} = \frac{2}{3}$

A number that is not a perfect square has a square root that is not a rational number. For example, $\sqrt{5}$ is not a rational number because it cannot be written as the ratio of two integers. Its decimal equivalent (or approximation) neither terminates nor repeats. However, $\sqrt{5}$ is a real number and corresponds to a point on the number line. As mentioned in **Section 1.3,** a real number that is not rational is called an **irrational number.** The number $\sqrt{5}$ is irrational. Many square roots of integers are irrational.

> If a is a *positive* real number that is *not* a perfect square, then
> $\sqrt{a}$ is irrational.

Not every number has a real number square root. For example, there is no real number that can be squared to obtain -36. (The square of a real number can never be negative.) Because of this, $\sqrt{-36}$ ***is not a real number.***

> If a is a *negative* real number, then $\sqrt{a}$ is *not* a real number.

CAUTION
Be careful not to confuse $\sqrt{-36}$ and $-\sqrt{36}$. $\sqrt{-36}$ is not a real number since there is no real number that can be squared to obtain -36. However, $-\sqrt{36}$ is the negative square root of 36, which is -6.

EXAMPLE 4 Identifying Types of Square Roots

Tell whether each square root is *rational, irrational,* or *not a real number*.

(a) $\sqrt{17}$
Because 17 is not a perfect square, $\sqrt{17}$ is irrational.

(b) $\sqrt{64}$
The number 64 is a perfect square, 8^2, so $\sqrt{64} = 8$ is a rational number.

(c) $\sqrt{-25}$
There is no real number whose square is -25. Therefore, $\sqrt{-25}$ is not a real number.

◀ Work Problem 4 at the Side.

4 Tell whether each square root is *rational, irrational,* or *not a real number*.

(a) $\sqrt{9}$

(b) $\sqrt{7}$

(c) $\sqrt{\frac{4}{9}}$

(d) $\sqrt{72}$

(e) $\sqrt{-43}$

ANSWERS

4. (a) rational **(b)** irrational **(c)** rational **(d)** irrational **(e)** not a real number

OBJECTIVE 3 Find cube, fourth, and other roots. Finding the square root of a number is the inverse (reverse) of squaring a number. There are inverses to finding the cube of a number, or finding the fourth or higher power of a number. These inverses are the **cube root,** written $\sqrt[3]{a}$, and the **fourth root,** written $\sqrt[4]{a}$. Similar symbols are used for higher roots.

$\sqrt[n]{a}$

The nth root of a is written $\sqrt[n]{a}$.

In general, $\sqrt[n]{a}$ is a number whose n^{th} power equals a. That is,

$$\sqrt[n]{a} = b \quad \textbf{means} \quad b^n = a.$$

In $\sqrt[n]{a}$, the number n is the **index,** or **order,** of the radical.

Index
Radical sign
Radicand
$\sqrt[n]{a}$
Radical

We could write $\sqrt[2]{a}$ instead of $\sqrt{a}$, but the simpler symbol $\sqrt{a}$ is customary since the square root is the most commonly used root.

Calculator Tip A calculator that has a key marked $\sqrt[x]{y}$, x^y, or y^x (again perhaps in conjunction with the INV or 2nd key) can be used to find other roots.

When working with cube roots or fourth roots, it is helpful to memorize the first few *perfect cubes* ($1^3 = 1$, $2^3 = 8$, $3^3 = 27$, and so on) and the first few *perfect fourth powers* ($1^4 = 1$, $2^4 = 16$, $3^4 = 81$, and so on).

Work Problem 5 at the Side. ▶

EXAMPLE 5 Finding Cube Roots

Find each cube root.

(a) $\sqrt[3]{8}$

Look for a number that can be cubed to give 8. Because $2^3 = 8$, $\sqrt[3]{8} = 2$.

(b) $\sqrt[3]{-8} = -2$ because $(-2)^3 = -8$.

(c) $\sqrt[3]{216} = 6$ because $6^3 = 216$.

Notice in Example 5(b) that we can find the cube root of a negative number. (Contrast this with the square root of a negative number, which is not real.) In fact, the cube root of a positive number is positive, and the cube root of a negative number is negative. ***There is only one real number cube root for each real number.***

Work Problem 6 at the Side. ▶

5 Complete the following list of perfect cubes and perfect fourth powers.

Perfect Cubes	Perfect Fourth Powers
$1^3 = 1$	$1^4 = 1$
$2^3 = 8$	$2^4 = 16$
$3^3 = 27$	$3^4 = 81$
$4^3 =$ ____	$4^4 =$ ____
$5^3 =$ ____	$5^4 =$ ____
$6^3 =$ ____	$6^4 =$ ____
$7^3 =$ ____	$7^4 =$ ____
$8^3 =$ ____	$8^4 =$ ____
$9^3 =$ ____	$9^4 =$ ____
$10^3 =$ ____	$10^4 =$ ____

6 Find each cube root.

(a) $\sqrt[3]{27}$

(b) $\sqrt[3]{64}$

(c) $\sqrt[3]{-125}$

ANSWERS

5. Perfect cubes: 64; 125; 216; 343; 512; 729; 1000
Perfect fourth powers: 256; 625; 1296; 2401; 4096; 6561; 10,000

6. (a) 3 **(b)** 4 **(c)** -5

When a radical has an ***even index*** (square root, fourth root, and so on), ***the radicand must be nonnegative*** to yield a real number root. Also,

$$\sqrt{a},\ \sqrt[4]{a},\ \sqrt[6]{a},\ \textbf{and so on are positive (principal) roots;}$$

$$-\sqrt{a},\ -\sqrt[4]{a},\ -\sqrt[6]{a},\ \textbf{and so on are negative roots.}$$

EXAMPLE 6 Finding Other Roots

Find each root.

(a) $\sqrt[4]{16} = 2$ because 2 is positive and $2^4 = 16$.

(b) $-\sqrt[4]{16}$

From part (a), $\sqrt[4]{16} = 2$, so the negative root is $-\sqrt[4]{16} = -2$.

(c) $\sqrt[4]{-16}$

For a real number fourth root, the radicand must be nonnegative. There is no real number that equals $\sqrt[4]{-16}$.

(d) $-\sqrt[5]{32}$

First find $\sqrt[5]{32}$. Because 2 is the number whose fifth power is 32, $\sqrt[5]{32} = 2$. Since $\sqrt[5]{32} = 2$, it follows that

$$-\sqrt[5]{32} = -2.$$

(e) $\sqrt[5]{-32}$

Because $(-2)^5 = -32$, $\sqrt[5]{-32} = -2$.

◀ Work Problem 7 at the Side.

7 Find each root.

(a) $\sqrt[4]{81}$

(b) $\sqrt[4]{-81}$

(c) $-\sqrt[4]{81}$

(d) $\sqrt[5]{243}$

(e) $\sqrt[5]{-243}$

OBJECTIVE 4 Graph functions defined by radical expressions. A **radical expression** is an algebraic expression that contains radicals.

$$3 - \sqrt{x}, \quad \sqrt[3]{x}, \quad \text{and} \quad \sqrt{2x - 1} \qquad \text{Radical expressions}$$

In earlier chapters we graphed functions defined by linear and rational expressions. Now we examine the graphs of functions defined by the radical expressions $f(x) = \sqrt{x}$ and $f(x) = \sqrt[3]{x}$.

Figure 1 shows the graph of the **square root function** defined by $f(x) = \sqrt{x}$, together with a table of selected points. Only nonnegative values can be used for x, so the domain is $[0, \infty)$. Because $\sqrt{x}$ is the principal square root of x, it always has a nonnegative value, so the range is also $[0, \infty)$.

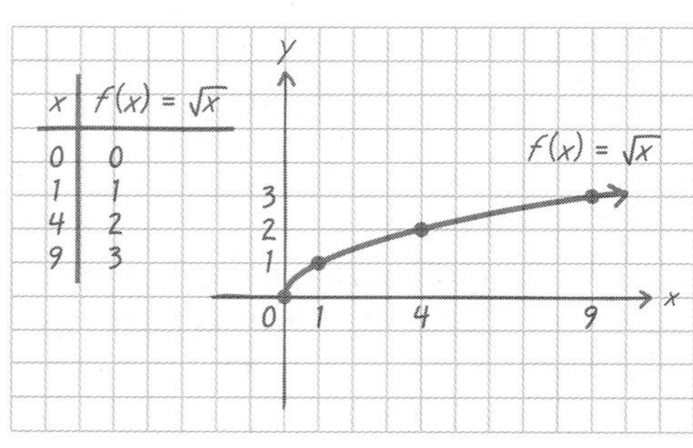

Figure 1

ANSWER

7. (a) 3 (b) not a real number (c) −3 (d) 3 (e) −3

Figure 2 shows the graph of the **cube root function** defined by $f(x) = \sqrt[3]{x}$, together with a table of selected points. Since any real number (positive, negative, or 0) can be used for x in the cube root function, $\sqrt[3]{x}$ can be positive, negative, or 0. Thus, both the domain and the range of the cube root function are $(-\infty, \infty)$.

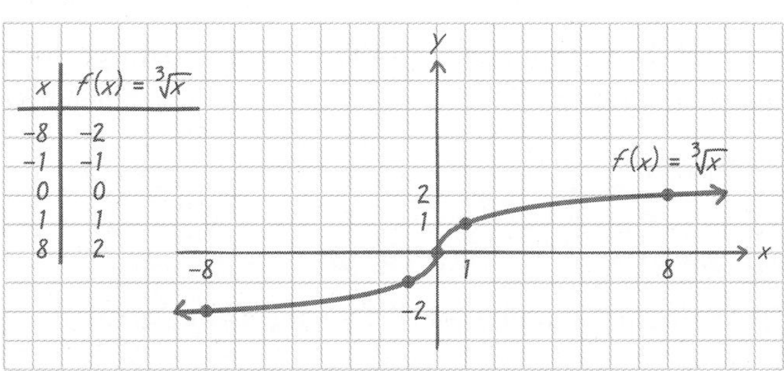

Figure 2

EXAMPLE 7 Graphing Functions Defined with Radicals

Graph each function by creating a table of values. Give the domain and the range.

(a) $f(x) = \sqrt{x - 3}$

A table of values is shown. The x-values were chosen so that the function values are all integers. For the radicand to be nonnegative, we must have $x - 3 \geq 0$, or $x \geq 3$. Therefore, the domain is $[3, \infty)$. Again, function values are positive or 0, so the range is $[0, \infty)$. See the graph in Figure 3.

x	$f(x) = \sqrt{x - 3}$
3	$\sqrt{3 - 3} = 0$
4	$\sqrt{4 - 3} = 1$
7	$\sqrt{7 - 3} = 2$

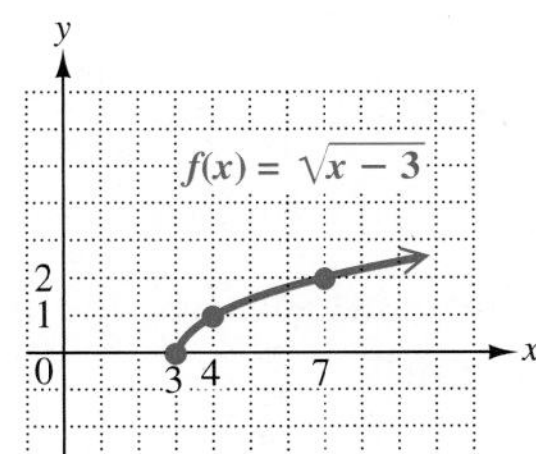

Figure 3

(b) $f(x) = \sqrt[3]{x} + 2$

See the table and Figure 4. Both the domain and the range are $(-\infty, \infty)$.

x	$f(x) = \sqrt[3]{x} + 2$
−8	$\sqrt[3]{-8} + 2 = 0$
−1	$\sqrt[3]{-1} + 2 = 1$
0	$\sqrt[3]{0} + 2 = 2$
1	$\sqrt[3]{1} + 2 = 3$
8	$\sqrt[3]{8} + 2 = 4$

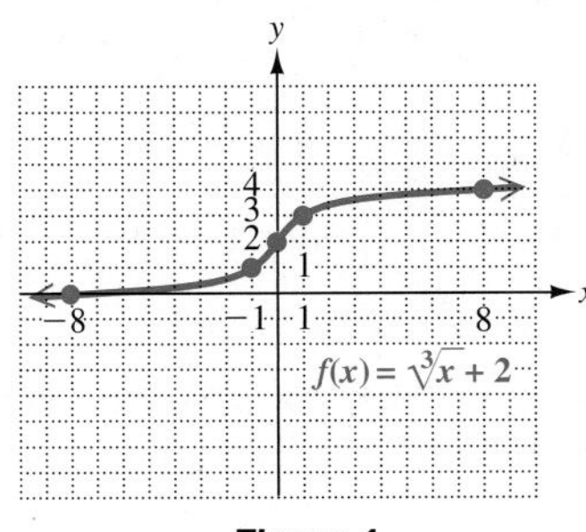

Figure 4

Work Problem **8** at the Side. ▶

8 Graph each function by creating a table of values. Give the domain and range.

(a) $f(x) = \sqrt{x} + 2$

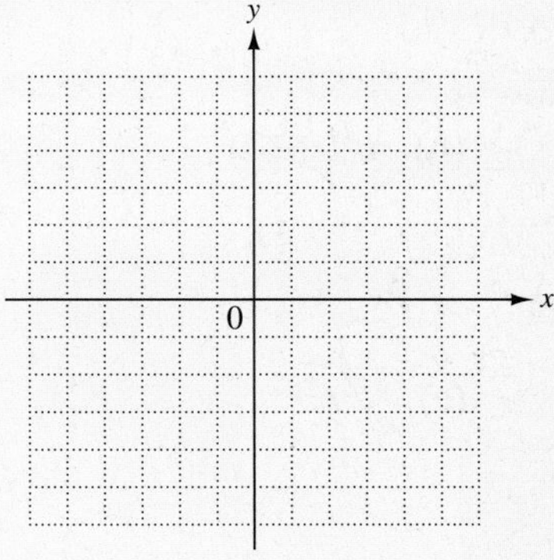

(b) $f(x) = \sqrt[3]{x - 1}$

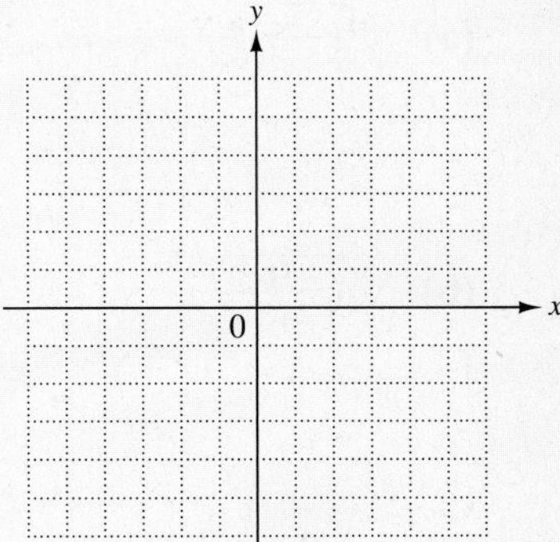

ANSWERS

8. (a) domain: $[0, \infty)$; range: $[2, \infty)$

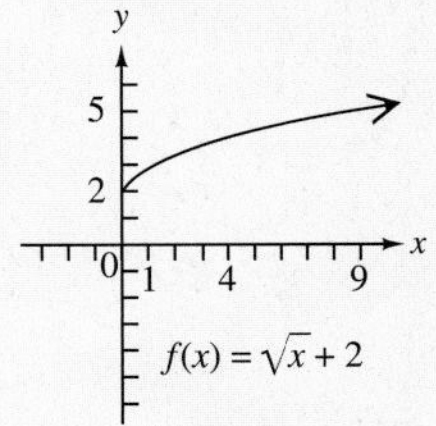

(b) domain: $(-\infty, \infty)$; range: $(-\infty, \infty)$

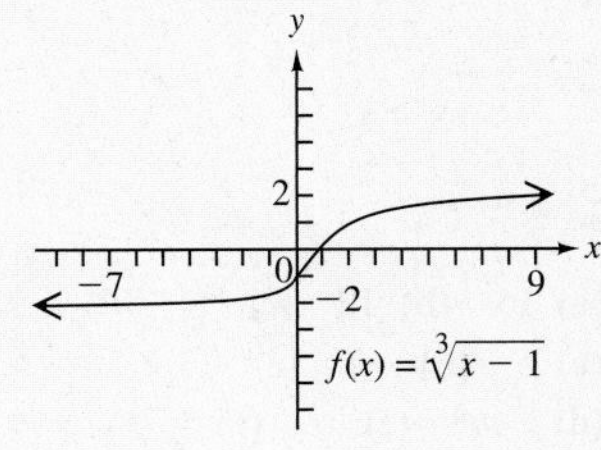

OBJECTIVE 5 Find *n*th roots of *n*th powers. What does $\sqrt{a^2}$ equal? Your first answer might be a, but this is not necessarily true. For example, consider the following:

If $a = 6$, then $\sqrt{a^2} = \sqrt{6^2} = \sqrt{36} = 6$.

If $a = -6$, then $\sqrt{a^2} = \sqrt{(-6)^2} = \sqrt{36} = 6$. ← Instead of -6, we get 6, the *absolute value* of -6.

Since the symbol $\sqrt{a^2}$ represents the *nonnegative* square root, we write $\sqrt{a^2}$ with absolute value bars, as $|a|$, because a may be a negative number.

> **$\sqrt{a^2}$**
>
> For any real number a, $\quad \sqrt{a^2} = |a|$.
>
> In words, the principal square root of a^2 is the absolute value of a.

EXAMPLE 8 Simplifying Square Roots Using Absolute Value

Find each square root that is a real number.

(a) $\sqrt{7^2} = |7| = 7$

(b) $\sqrt{(-7)^2} = |-7| = 7$

(c) $\sqrt{k^2} = |k|$

(d) $\sqrt{(-k)^2} = |-k| = |k|$

◀ Work Problem 9 at the Side.

We can generalize this idea to any nth root.

> **$\sqrt[n]{a^n}$**
>
> If n is an *even* positive integer, then $\quad \sqrt[n]{a^n} = |a|$.
>
> If n is an *odd* positive integer, then $\quad \sqrt[n]{a^n} = a$.
>
> In words, use absolute value when n is even; absolute value is not necessary when n is odd.

EXAMPLE 9 Simplifying Higher Roots Using Absolute Value

Simplify each root.

(a) $\sqrt[6]{(-3)^6} = |-3| = 3$ $\quad$ n is even; use absolute value.

(b) $\sqrt[5]{(-4)^5} = -4$ $\quad$ n is odd.

(c) $-\sqrt[4]{(-9)^4} = -|-9| = -9$ $\quad$ n is even; use absolute value.

(d) $-\sqrt{m^4} = -|m^2| = -m^2$ $\quad$ For all m, $|m^2| = m^2$.

No absolute value bars are needed here because m^2 is nonnegative for any real number value of m.

(e) $\sqrt[3]{a^{12}} = a^4$, because $a^{12} = (a^4)^3$.

(f) $\sqrt[4]{x^{12}} = |x^3|$

We use absolute value bars to guarantee that the result is not negative (because x^3 can be either positive or negative, depending on x). If desired, $|x^3|$ can be written as $x^2 \cdot |x|$.

◀ Work Problem 10 at the Side.

---

9 Find each square root that is a real number.

(a) $\sqrt{15^2}$

(b) $\sqrt{(-12)^2}$

(c) $\sqrt{r^2}$

(d) $\sqrt{(-r)^2}$

10 Simplify.

(a) $\sqrt[4]{(-5)^4}$

(b) $\sqrt[5]{(-7)^5}$

(c) $-\sqrt[6]{(-3)^6}$

(d) $-\sqrt[4]{m^8}$

(e) $\sqrt[3]{x^{24}}$

(f) $\sqrt[6]{y^{18}}$

ANSWERS

9. **(a)** 15 **(b)** 12 **(c)** $|r|$ **(d)** $|r|$

10. **(a)** 5 **(b)** -7 **(c)** -3 **(d)** $-m^2$ **(e)** x^8 **(f)** $|y^3|$

OBJECTIVE 6 Use a calculator to find roots. Radical expressions often represent irrational numbers. To find approximations of such radicals, we usually use a calculator. For example,

$$\sqrt{15} \approx 3.872983346, \quad \sqrt[3]{10} \approx 2.15443469, \quad \text{and} \quad \sqrt[4]{2} \approx 1.189207115,$$

where the symbol $\approx$ means "is approximately equal to." In this book, we often give approximations rounded to three decimal places. Thus,

$$\sqrt{15} \approx 3.873, \quad \sqrt[3]{10} \approx 2.154, \quad \text{and} \quad \sqrt[4]{2} \approx 1.189.$$

> **Calculator Tip** The methods for finding approximations differ among makes and models, and you should always consult your owner's manual for keystroke instructions. Be aware that graphing calculators often differ from scientific calculators in the order in which keystrokes are made.

Figure 5 shows how the preceding approximations are displayed on a TI-83/84 Plus graphing calculator. In Figure 5(a), eight or nine decimal places are shown, while in Figure 5(b), the number of decimal places is fixed at three.

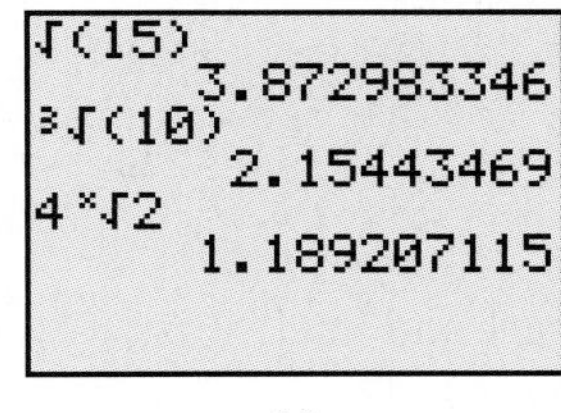

(a)

```
√(15)
             3.873
³√(10)
             2.154
4 ˣ√2
             1.189
```

(b)

Figure 5

There is a simple way to check that a calculator approximation is "in the ballpark." Because 16 is a little larger than 15, $\sqrt{16} = 4$ should be a little larger than $\sqrt{15}$. Thus, 3.873 is a reasonable approximation for $\sqrt{15}$.

EXAMPLE 10 Finding Approximations for Roots

Use a calculator to verify that each approximation is correct.

(a) $\sqrt{39} \approx 6.245$ **(b)** $-\sqrt{72} \approx -8.485$

(c) $\sqrt[3]{93} \approx 4.531$ **(d)** $\sqrt[4]{39} \approx 2.499$

Work Problem 11 at the Side. ▶

11 Use a calculator to approximate each radical to three decimal places.

(a) $\sqrt{17}$

(b) $-\sqrt{362}$

(c) $\sqrt[3]{9482}$

(d) $\sqrt[4]{6825}$

ANSWERS

11. (a) 4.123 **(b)** -19.026 **(c)** 21.166 **(d)** 9.089

Math in the Media

PRIME NUMBERS IN PRIME TIME

The 1997 movie *Contact*, based on the Carl Sagan novel of the same name, portrays Jodie Foster as scientist Ellie Arroway. After years of searching, Ellie makes contact with intelligent life in outer space. Her contact is verified after receiving radio signals that indicate **prime numbers:**

2, 3, 5, 7, 11, 13, and so on.

Her superiors, evidently not familiar with prime numbers, are not convinced and ask her why the aliens just don't speak English? Her response (accompanied by quizzical looks from the bosses) follows.

> *Well, maybe because 70% of the planet speaks other languages. Mathematics is the only universal language, Senator. It's no coincidence they're using primes . . . prime numbers—that would be integers that are divisible only by themselves and 1.*

Integers greater than 1 that are not prime are called **composite numbers,** because they are composed of prime factors in one and only one way. (Factoring is the topic of **Chapter 6,** and we will use it to simplify radical expressions in this chapter.).

1. A prime number is a positive integer greater than 1 whose only factors are 1 and itself. List the first fifteen prime numbers. **2, 3, 5, 7, 11, 13, 17, 19, 23, 29, 31, 37, 41, 43, 47**

2. A recurring feature on the NBC *Today Show* is *Where in the World is Matt Lauer?* Co-host Matt Lauer travels to exotic places during one week, always giving a hint as to where he will be on the following day. In his travels in 2008, Matt gave this clue regarding his next destination: **It's an anagram of a synonym of a homophone of an even prime number.** Where was Matt the next day? **Laos**

3. The 1996 film *The Mirror Has Two Faces* stars Jeff Bridges as mathematician Gregory Larkin, who has written a book on the Twin Prime Conjecture. Search the Internet to discover the statement of this famous unproved conjecture. **There are infinitely many pairs of twin primes. (Twin primes are prime numbers that differ by 2, such as 3 and 5, 5 and 7, 11 and 13, etc.)**

4. Watch the episode "Prime Suspect" from the first season of the CBS television series *NUMB3RS*. The story is based on the premise that a mathematician was very close to proving a famous unsolved problem involving prime numbers, the Riemann Hypothesis, and his daughter was kidnapped as a result. Why did the criminals kidnap the child? **The Riemann Hypothesis concerns how the prime numbers are distributed on the number line. A proof would possibly provide sufficient understanding of primes to break codes that provide security of all Internet commerce (among other things).**

Decide whether each statement is true *or* false. *If* false, *tell why.*

1. Every positive number has two real square roots.

true

2. A negative number has negative square roots.

false; A negative number has no real square roots.

3. Every nonnegative number has two real square roots.

false; Zero has only one square root.

4. The positive square root of a positive number is its principal square root.

true

5. The cube root of every real number has the same sign as the number itself.

true

6. Every positive number has three real cube roots.

false; A positive number has just one real cube root.

Match each expression from Column I with the equivalent choice from Column II. Answers may be used once, more than once, or not at all. See Examples 2 and 4–6.

I		II	
7. $-\sqrt{16}$ E	**8.** $\sqrt{-16}$ F	**A.** 3	**B.** -2
9. $\sqrt[3]{-27}$ D	**10.** $\sqrt[5]{-32}$ B	**C.** 2	**D.** -3
11. $\sqrt[4]{16}$ C	**12.** $-\sqrt[3]{64}$ E	**E.** -4	**F.** Not a real number

Choose the closest approximation of each square root.

13. $\sqrt{123.5}$

A. 9 **B.** 10 **C.** 11 **D.** 12

C

14. $\sqrt{67.8}$

A. 7 **B.** 8 **C.** 9 **D.** 10

B

Refer to the figure to answer the questions in Exercises 15–16.

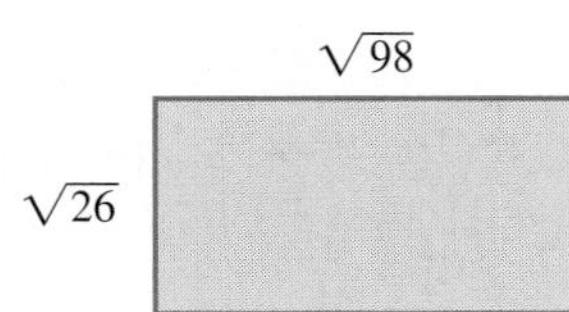

15. Which one of the following is the best estimate of its area?

A. 2500 **B.** 250 **C.** 50 **D.** 100

C

16. Which one of the following is the best estimate of its perimeter?

A. 15 **B.** 250 **C.** 100 **D.** 30

D

17. Consider the expression $-\sqrt{-a}$. Decide whether it is positive, negative, 0, or not a real number if

(a) $a > 0$, **(b)** $a < 0$, **(c)** $a = 0$.

(a) not a real number (b) negative (c) 0

18. If n is odd, under what conditions is $\sqrt[n]{a}$

(a) positive, **(b)** negative, **(c)** 0?

(a) a must be positive ($a > 0$).
(b) a must be negative ($a < 0$).
(c) a must be 0 ($a = 0$).

Find all square roots of each number. See Example 1.

19. 9
$-3, 3$

20. 16
$-4, 4$

21. 64
$-8, 8$

22. 100
$-10, 10$

23. 169
$-13, 13$

24. 225
$-15, 15$

25. $\frac{25}{196}$
$-\frac{5}{14}, \frac{5}{14}$

26. $\frac{81}{400}$
$-\frac{9}{20}, \frac{9}{20}$

27. 900
$-30, 30$

28. 1600
$-40, 40$

Find each square root. See Examples 2 and 4(c).

29. $\sqrt{1}$
1

30. $\sqrt{4}$
2

31. $\sqrt{49}$
7

32. $\sqrt{81}$
9

33. $-\sqrt{256}$
-16

34. $-\sqrt{196}$
-14

35. $-\sqrt{\frac{144}{121}}$
$-\frac{12}{11}$

36. $-\sqrt{\frac{49}{36}}$
$-\frac{7}{6}$

37. $\sqrt{0.64}$
0.8

38. $\sqrt{0.16}$
0.4

39. $\sqrt{-121}$ not a real number

40. $\sqrt{-64}$ not a real number

41. $-\sqrt{-49}$ not a real number

42. $-\sqrt{-100}$ not a real number

Find the square of each radical expression. See Example 3.

43. $\sqrt{100}$ 100

44. $\sqrt{36}$ 36

45. $-\sqrt{19}$ 19

46. $-\sqrt{99}$ 99

47. $\sqrt{\frac{2}{3}}$ $\frac{2}{3}$

48. $\sqrt{\frac{5}{7}}$ $\frac{5}{7}$

49. $\sqrt{3x^2 + 4}$
$3x^2 + 4$

50. $\sqrt{9y^2 + 3}$
$9y^2 + 3$

Write rational, irrational, *or* not a real number *for each number. If a number is rational, give its exact value. If a number is irrational, give a decimal approximation to the nearest thousandth. Use a calculator as necessary. See Examples 4 and 10.*

51. $\sqrt{25}$
rational; 5

52. $\sqrt{169}$
rational; 13

53. $\sqrt{29}$
irrational; 5.385

54. $\sqrt{33}$
irrational; 5.745

55. $-\sqrt{64}$
rational; -8

56. $-\sqrt{81}$
rational; -9

57. $-\sqrt{300}$
irrational; -17.321

58. $-\sqrt{500}$
irrational; -22.361

59. $\sqrt{-29}$
not a real number

60. $\sqrt{-47}$
not a real number

61. $\sqrt{1200}$
irrational; 34.641

62. $\sqrt{1500}$
irrational; 38.730

Find each root that is a real number. Use a calculator as necessary. See Examples 5 and 6.

63. $\sqrt[3]{216}$
6

64. $\sqrt[3]{343}$
7

65. $\sqrt[3]{-64}$
−4

66. $\sqrt[3]{-125}$
−5

67. $-\sqrt[3]{512}$
−8

68. $-\sqrt[3]{1000}$
−10

69. $\sqrt[4]{1296}$
6

70. $\sqrt[4]{625}$
5

71. $-\sqrt[4]{16}$
−2

72. $-\sqrt[4]{256}$
−4

73. $\sqrt[4]{-625}$
not a real number

74. $\sqrt[4]{-256}$
not a real number

75. $\sqrt[6]{729}$
3

76. $\sqrt[6]{64}$
2

77. $\sqrt[6]{-64}$
not a real number

78. $\sqrt[6]{-1}$
not a real number

79. $\sqrt[3]{\frac{64}{27}}$
$\frac{4}{3}$

80. $\sqrt[4]{\frac{81}{16}}$
$\frac{3}{2}$

81. $-\sqrt[6]{\frac{1}{64}}$
$-\frac{1}{2}$

82. $-\sqrt[5]{\frac{1}{32}}$
$-\frac{1}{2}$

83. $\sqrt[3]{0.001}$
0.1

84. $\sqrt[3]{0.125}$
0.5

Graph each function and give its domain and range. See Example 7.

85. $f(x) = \sqrt{x + 3}$

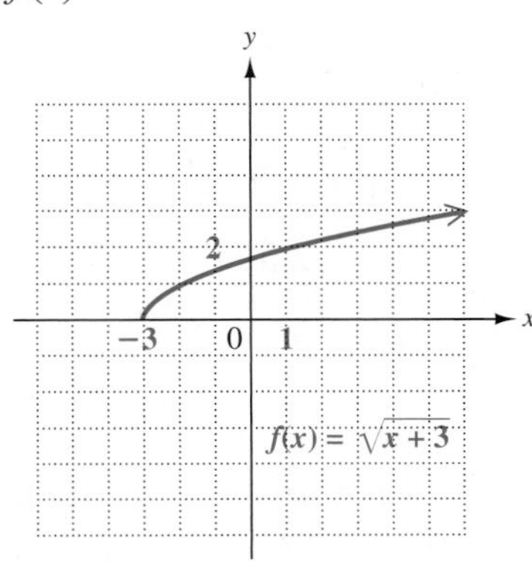

domain: [−3, ∞); range: [0, ∞)

86. $f(x) = \sqrt{x - 5}$

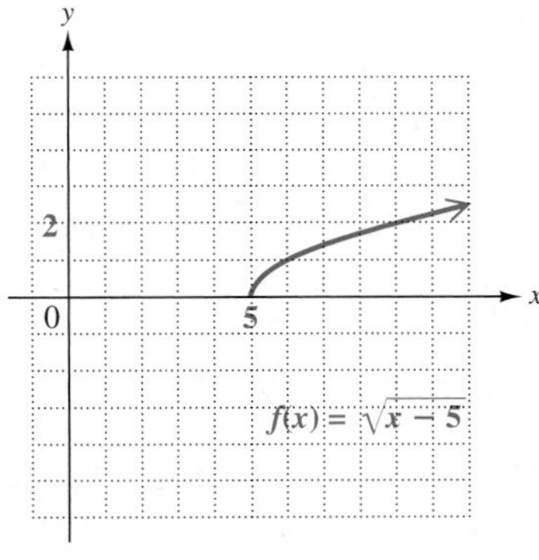

domain: [5, ∞); range: [0, ∞)

87. $f(x) = \sqrt{x} - 2$

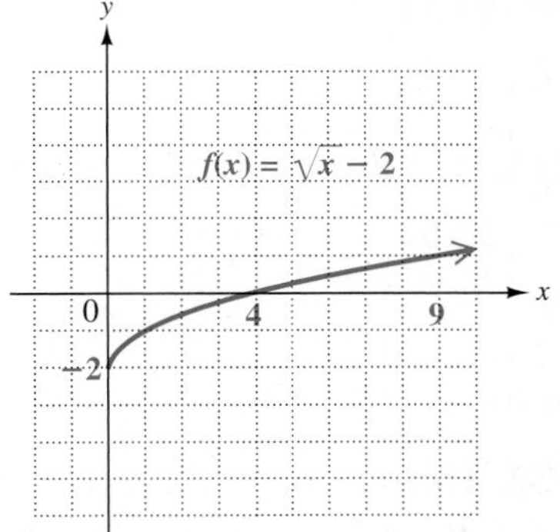

domain: [0, ∞); range: [−2, ∞)

88. $f(x) = \sqrt{x} + 4$

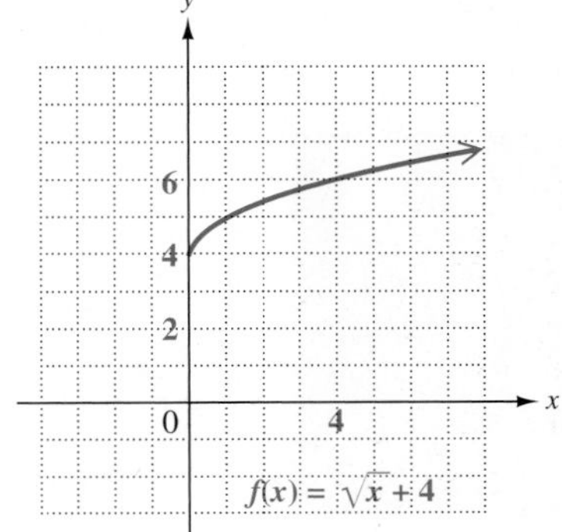

domain: [0, ∞); range: [4, ∞)

89. $f(x) = \sqrt[3]{x} - 3$

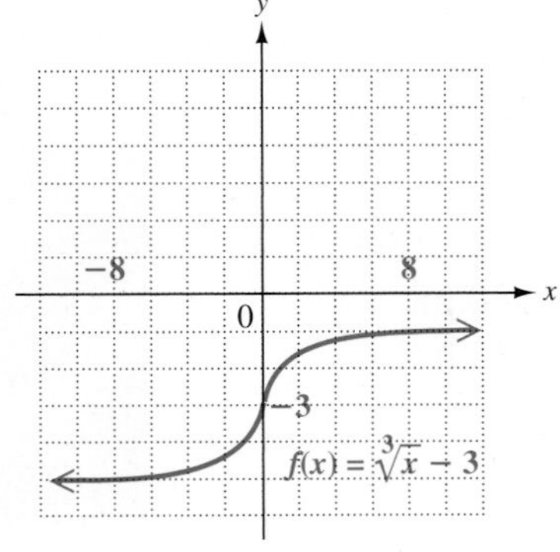

domain: (−∞, ∞); range: (−∞, ∞)

90. $f(x) = \sqrt[3]{x} + 1$

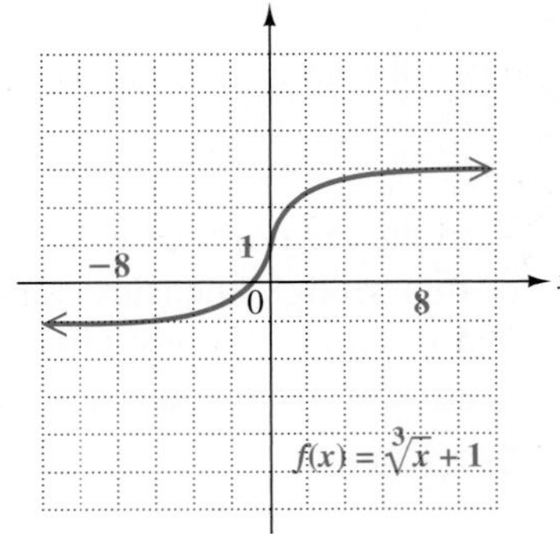

domain: (−∞, ∞); range: (−∞, ∞)

Simplify each root. See Examples 8 and 9.

91. $\sqrt{12^2}$
12

92. $\sqrt{19^2}$
19

93. $\sqrt{(-10)^2}$
10

94. $\sqrt{(-13)^2}$
13

95. $\sqrt[6]{(-2)^6}$
2

96. $\sqrt[6]{(-4)^6}$
4

97. $\sqrt[5]{(-9)^5}$
-9

98. $\sqrt[5]{(-8)^5}$
-8

99. $-\sqrt[6]{(-5)^6}$
-5

100. $-\sqrt[6]{(-7)^6}$
-7

101. $\sqrt{x^2}$
$|x|$

102. $-\sqrt{x^2}$
$-|x|$

103. $\sqrt{(-z)^2}$
$|z|$

104. $\sqrt{(-q)^2}$
$|q|$

105. $\sqrt[3]{x^3}$
x

106. $-\sqrt[3]{x^3}$
$-x$

107. $\sqrt[3]{x^{15}}$
x^5

108. $\sqrt[3]{m^9}$
m^3

109. $\sqrt[6]{x^{30}}$
$|x|^5$ (or $|x^5|$)

110. $\sqrt[4]{k^{20}}$
$|k|^5$ (or $|k^5|$)

Use a calculator to find a decimal approximation for each radical. Round answers to three decimal places. See Example 10.

111. $\sqrt{9483}$
97.381

112. $\sqrt{6825}$
82.614

113. $\sqrt[3]{423}$
7.507

114. $\sqrt[3]{555}$
8.218

115. $\sqrt[4]{100}$
3.162

116. $\sqrt[4]{250}$
3.976

117. $\sqrt[5]{23.8}$
1.885

118. $\sqrt[5]{98.4}$
2.504

Solve each problem.

119. **Heron's formula** gives a method of finding the area of a triangle if the lengths of its sides are known. Suppose that a, b, and c are the lengths of the sides. Let s denote one-half of the perimeter of the triangle (called the **semiperimeter**); that is,

$$s = \frac{1}{2}(a + b + c).$$

Then the area of the triangle is

$$A = \sqrt{s(s - a)(s - b)(s - c)}.$$

Find the area of the Bermuda Triangle, if the "sides" of this triangle measure approximately 850 mi, 925 mi, and 1300 mi. Give your answer to the nearest thousand square miles.

392,000 mi²

120. The Vietnam Veterans' Memorial in Washington, D.C., is in the shape of an unenclosed isosceles triangle with equal sides of length 246.75 ft. If the triangle were enclosed, the third side would have length 438.14 ft. Use Heron's formula from the previous exercise to find the area of this enclosure to the nearest hundred square feet. (*Source:* Information pamphlet obtained at the Vietnam Veterans' Memorial.)

246.75 ft
246.75 ft
438.14 ft
Not to scale

24,900 ft²

9.2 Rational Exponents

OBJECTIVE 1 Use exponential notation for *n*th roots. We now look at exponents that are rational numbers of the form $\frac{1}{n}$, or $1/n$, where n is a natural number.

Consider the product $(3^{1/2})^2 = 3^{1/2} \cdot 3^{1/2}$. Using the rules of exponents from **Section 5.2,** extended to rational exponents, we can simplify this product as follows:

$$
\begin{aligned}
(3^{1/2})^2 &= 3^{1/2} \cdot 3^{1/2} \\
&= 3^{1/2+1/2} && \text{Product rule: } a^m \cdot a^n = a^{m+n} \\
&= 3^1 && \text{Add exponents.} \\
&= 3.
\end{aligned}
$$

Also, by definition,

$$\left(\sqrt{3}\right)^2 = \sqrt{3} \cdot \sqrt{3} = 3.$$

Since both $(3^{1/2})^2$ and $\left(\sqrt{3}\right)^2$ are equal to 3, it seems reasonable to define

$$3^{1/2} = \sqrt{3}.$$

This suggests the following generalization.

> $a^{1/n}$
>
> If $\sqrt[n]{a}$ is a real number, then $\quad \boldsymbol{a^{1/n} = \sqrt[n]{a}}.$

OBJECTIVES

1. Use exponential notation for *n*th roots.
2. Define and use expressions of the form $a^{m/n}$.
3. Convert between radicals and rational exponents.
4. Use the rules for exponents with rational exponents.

EXAMPLE 1 Evaluating Exponentials of the Form $a^{1/n}$

Evaluate each exponential.

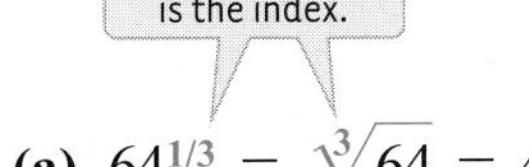

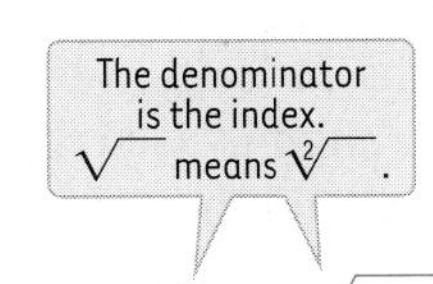

(a) $64^{1/3} = \sqrt[3]{64} = 4$ **(b)** $100^{1/2} = \sqrt{100} = 10$

(c) $-256^{1/4} = -\sqrt[4]{256} = -4$

(d) $(-256)^{1/4} = \sqrt[4]{-256}$ is not a real number, because the radicand, -256, is negative and the index is even.

(e) $(-32)^{1/5} = \sqrt[5]{-32} = -2$ **(f)** $\left(\frac{1}{8}\right)^{1/3} = \sqrt[3]{\frac{1}{8}} = \frac{1}{2}$

> **CAUTION**
>
> Notice the difference between parts (c) and (d) in Example 1. The radical in part (c) is the ***negative fourth root of a positive number,*** while the radical in part (d) is the ***principal fourth root of a negative number, which is not a real number.***

Work Problem 1 at the Side. ▶

1 Evaluate each exponential.

(a) $8^{1/3}$

(b) $9^{1/2}$

(c) $-81^{1/4}$

(d) $(-81)^{1/4}$

(e) $(-64)^{1/3}$

(f) $\left(\frac{1}{32}\right)^{1/5}$

ANSWERS

1. **(a)** 2 **(b)** 3 **(c)** -3 **(d)** not a real number **(e)** -4 **(f)** $\frac{1}{2}$

OBJECTIVE 2 Define and use expressions of the form $a^{m/n}$.

We know that $8^{1/3} = \sqrt[3]{8}$. How should we define a number like $8^{2/3}$? For past rules of exponents to be valid,

$$8^{2/3} = 8^{(1/3)2} = (8^{1/3})^2.$$

Since $8^{1/3} = \sqrt[3]{8}$,

$$8^{2/3} = \left(\sqrt[3]{8}\right)^2 = 2^2 = 4.$$

Generalizing from this example, we define $a^{m/n}$ as follows.

> $\boldsymbol{a^{m/n}}$
>
> If m and n are positive integers with m/n in lowest terms, then
>
> $$a^{m/n} = (a^{1/n})^m,$$
>
> provided that $a^{1/n}$ is a real number. If $a^{1/n}$ is not a real number, then $a^{m/n}$ is not a real number.

EXAMPLE 2 Evaluating Exponentials of the Form $a^{m/n}$

Evaluate each exponential.

Think: $36^{1/2} = \sqrt{36} = 6$

(a) $36^{3/2} = (36^{1/2})^3 = 6^3 = 216$

Think: $125^{1/3} = \sqrt[3]{125} = 5$

(b) $125^{2/3} = (125^{1/3})^2 = 5^2 = 25$

Be careful. The base is 4.

(c) $-4^{5/2} = -(4^{5/2}) = -(4^{1/2})^5 = -(2)^5 = -32$

Because the base here is 4, the negative sign is *not* affected by the exponent.

(d) $(-27)^{2/3} = [(-27)^{1/3}]^2 = (-3)^2 = 9$

Notice how the $-$ sign is used in parts (c) and (d). In part (c), we first evaluate the exponential and then find its negative. In part (d), the $-$ sign is part of the base, -27.

(e) $(-100)^{3/2} = [(-100)^{1/2}]^3$, which is not a real number, since $(-100)^{1/2}$, or $\sqrt{-100}$, is not a real number.

◀ *Work Problem 2 at the Side.*

2 Evaluate each exponential.

(a) $25^{3/2}$

(b) $27^{2/3}$

(c) $-16^{3/2}$

(d) $(-64)^{2/3}$

(e) $(-36)^{3/2}$

Recall from **Section 5.5** that for any natural number n,

$$a^{-n} = \frac{1}{a^n} \quad (a \neq 0).$$

When a rational exponent is negative, we apply this interpretation of negative exponents.

> $\boldsymbol{a^{-m/n}}$
>
> If $a^{m/n}$ is a real number, then
>
> $$a^{-m/n} = \frac{1}{a^{m/n}} \quad (a \neq 0).$$

ANSWERS

2. (a) 125 **(b)** 9 **(c)** -64 **(d)** 16 **(e)** not a real number

EXAMPLE 3 Evaluating Exponentials with Negative Rational Exponents

Evaluate each exponential.

(a) $16^{-3/4} = \frac{1}{16^{3/4}} = \frac{1}{(16^{1/4})^3} = \frac{1}{(\sqrt[4]{16})^3} = \frac{1}{2^3} = \frac{1}{8}$

The denominator of 3/4 is the index and the numerator is the exponent.

(b) $25^{-3/2} = \frac{1}{25^{3/2}} = \frac{1}{(25^{1/2})^3} = \frac{1}{(\sqrt{25})^3} = \frac{1}{5^3} = \frac{1}{125}$

(c) $\left(\frac{8}{27}\right)^{-2/3} = \frac{1}{\left(\frac{8}{27}\right)^{2/3}} = \frac{1}{\left(\sqrt[3]{\frac{8}{27}}\right)^2} = \frac{1}{\left(\frac{2}{3}\right)^2} = \frac{1}{\frac{4}{9}} = \frac{9}{4}$

$\frac{1}{\frac{4}{9}} = 1 \div \frac{4}{9} = 1 \cdot \frac{9}{4}$

We could also use the rule $\left(\frac{b}{a}\right)^{-m} = \left(\frac{a}{b}\right)^m$ here, as follows:

$$\left(\frac{8}{27}\right)^{-2/3} = \left(\frac{27}{8}\right)^{2/3} = \left(\sqrt[3]{\frac{27}{8}}\right)^2 = \left(\frac{3}{2}\right)^2 = \frac{9}{4}.$$

Take the reciprocal only of the base, *not* the exponent.

Work Problem **3** *at the Side.* ▶

CAUTION

Be careful to distinguish between exponential expressions such as

$$16^{-1/4}, \text{ which equals } \frac{1}{2}; \quad -16^{1/4}, \text{ which equals } -2; \quad \text{and}$$
$$-16^{-1/4}, \text{ which equals } -\frac{1}{2}.$$

A negative exponent does not necessarily lead to a negative result. Negative exponents lead to reciprocals, which may be positive.

We obtain an alternative definition of $a^{m/n}$ by applying the power rule a little differently than earlier. If all indicated roots are real numbers, then

$$a^{m/n} = a^{m(1/n)} = (a^m)^{1/n}, \qquad \text{so} \qquad a^{m/n} = (a^m)^{1/n}.$$

$a^{m/n}$

If all indicated roots are real numbers, then

$$a^{m/n} = (a^{1/n})^m = (a^m)^{1/n}.$$

We can now evaluate an expression such as $27^{2/3}$ in two ways:

$27^{2/3} = (27^{1/3})^2 = 3^2 = 9$

or $27^{2/3} = (27^2)^{1/3} = 729^{1/3} = 9.$

The result is the same.

In most cases, it is easier to use $(a^{1/n})^m$.

3 Evaluate each exponential.

(a) $36^{-3/2}$

(b) $32^{-4/5}$

(c) $\left(\frac{4}{9}\right)^{-5/2}$

ANSWERS

3. (a) $\frac{1}{216}$ **(b)** $\frac{1}{16}$ **(c)** $\frac{243}{32}$

4 Write each exponential as a radical. Assume that all variables represent positive real numbers. Use the definition that takes the root first.

(a) $19^{1/2}$

(b) $5^{2/3}$

(c) $4k^{3/5}$

(d) $5x^{3/5} - (2x)^{3/5}$

(e) $x^{-5/7}$

(f) $(m^3 + n^3)^{1/3}$

5 Write each radical as an exponential and simplify. Assume that all variables represent positive real numbers.

(a) $\sqrt{37}$

(b) $\sqrt[4]{9^8}$

(c) $\sqrt[4]{t^4}$

Answers

4. (a) $\sqrt{19}$ **(b)** $\left(\sqrt[3]{5}\right)^2$ **(c)** $4\left(\sqrt[5]{k}\right)^3$
(d) $5\left(\sqrt[5]{x}\right)^3 - \left(\sqrt[5]{2x}\right)^3$
(e) $\dfrac{1}{\left(\sqrt[7]{x}\right)^5}$ **(f)** $\sqrt[3]{m^3 + n^3}$

5. (a) $37^{1/2}$ **(b)** 9^2, or 81 **(c)** t

This rule can also be expressed with radicals as follows.

Radical Form of $a^{m/n}$

If all indicated roots are real numbers, then

$$a^{m/n} = \sqrt[n]{a^m} = \left(\sqrt[n]{a}\right)^m.$$

In words, raise a to the mth power and then take the nth root, or take the nth root of a and then raise to the mth power.

For example,

$$8^{2/3} = \sqrt[3]{8^2} = \sqrt[3]{64} = 4, \quad \text{and} \quad 8^{2/3} = \left(\sqrt[3]{8}\right)^2 = 2^2 = 4,$$

so

$$8^{2/3} = \sqrt[3]{8^2} = \left(\sqrt[3]{8}\right)^2.$$

OBJECTIVE 3 Convert between radicals and rational exponents. Using the definition of rational exponents, we can simplify many problems involving radicals by converting the radicals to numbers with rational exponents. After simplifying, we convert the answer back to radical form.

EXAMPLE 4 Converting between Rational Exponents and Radicals

In (a)–(f), write each exponential as a radical. Assume that all variables represent positive real numbers. Use the definition that takes the root first.

(a) $13^{1/2} = \sqrt{13}$ **(b)** $6^{3/4} = \left(\sqrt[4]{6}\right)^3$ **(c)** $9m^{5/8} = 9\left(\sqrt[8]{m}\right)^5$

(d) $6x^{2/3} - (4x)^{3/5} = 6\left(\sqrt[3]{x}\right)^2 - \left(\sqrt[5]{4x}\right)^3$

(e) $r^{-2/3} = \dfrac{1}{r^{2/3}} = \dfrac{1}{\left(\sqrt[3]{r}\right)^2}$

(f) $(a^2 + b^2)^{1/2} = \sqrt{a^2 + b^2}$ $\quad \sqrt{a^2 + b^2} \neq a + b$

In (g)–(i), write each radical as an exponential. Simplify. Assume that all variables represent positive real numbers.

(g) $\sqrt{10} = 10^{1/2}$

(h) $\sqrt[4]{3^8} = 3^{8/4} = 3^2 = 9$

(i) $\sqrt[6]{z^6} = z^{6/6} = z^1 = z$, since z is positive.

◀ Work Problem **4** at the Side.

Note

In Example 4(i), it was not necessary to use absolute value bars, since the directions specifically stated that the variable represents a positive real number. Because the absolute value of the positive real number z is z itself, the answer is simply z.

◀ Work Problem **5** at the Side.

OBJECTIVE 4 Use the rules for exponents with rational exponents. The definition of rational exponents allows us to apply the rules for exponents first introduced in **Chapter 5.**

Rules for Rational Exponents

Let r and s be rational numbers. For all real numbers a and b for which the indicated expressions exist:

$$a^r \cdot a^s = a^{r+s} \qquad a^{-r} = \frac{1}{a^r} \qquad \frac{a^r}{a^s} = a^{r-s} \qquad \left(\frac{a}{b}\right)^{-r} = \frac{b^r}{a^r}$$

$$(a^r)^s = a^{rs} \qquad (ab)^r = a^r b^r \qquad \left(\frac{a}{b}\right)^r = \frac{a^r}{b^r} \qquad a^{-r} = \left(\frac{1}{a}\right)^r$$

EXAMPLE 5 Applying Rules for Rational Exponents

Write with only positive exponents. Assume that all variables represent positive real numbers.

(a) $2^{1/2} \cdot 2^{1/4}$

$= 2^{1/2+1/4}$ Product rule

$= 2^{3/4}$ Add exponents.

(b) $\dfrac{5^{2/3}}{5^{7/3}}$

$= 5^{2/3-7/3}$ Quotient rule

$= 5^{-5/3}$ Subtract exponents.

$= \dfrac{1}{5^{5/3}}$ $\quad a^{-r} = \frac{1}{a^r}$

(c) $\dfrac{(x^{1/2}y^{2/3})^4}{y}$

$= \dfrac{(x^{1/2})^4(y^{2/3})^4}{y}$ Power rule

$= \dfrac{x^2y^{8/3}}{y^1}$ Power rule

$= x^2y^{8/3-1}$ Quotient rule

$= x^2y^{5/3}$ $\quad \frac{8}{3} - 1 = \frac{8}{3} - \frac{3}{3} = \frac{5}{3}$

(d) $\left(\dfrac{x^4y^{-6}}{x^{-2}y^{1/3}}\right)^{-2/3}$

$= \dfrac{(x^4)^{-2/3}(y^{-6})^{-2/3}}{(x^{-2})^{-2/3}(y^{1/3})^{-2/3}}$ Power rule

$= \dfrac{x^{-8/3}y^4}{x^{4/3}y^{-2/9}}$ Power rule

$= x^{-8/3-4/3}y^{4-(-2/9)}$ Quotient rule

$= x^{-4}y^{38/9}$ Use parentheses to avoid errors. $\quad 4 - \left(-\frac{2}{9}\right) = \frac{36}{9} + \frac{2}{9} = \frac{38}{9}$

$= \dfrac{y^{38/9}}{x^4}$ Definition of negative exponent

Continued on Next Page

6 Write with only positive exponents. Assume that all variables represent positive real numbers.

(a) $11^{3/4} \cdot 11^{5/4}$

(b) $\dfrac{7^{3/4}}{7^{7/4}}$

(c) $\dfrac{9^{2/3}(x^{1/3})^4}{9^{-1/3}}$

(d) $\left(\dfrac{a^3b^{-4}}{a^{-2}b^{1/5}}\right)^{-1/2}$

(e) $a^{2/3}(a^{7/3} + a^{1/3})$

7 Write all radicals as exponentials, and then apply the rules for rational exponents. Give answers in exponential form. Assume that all variables represent positive real numbers.

(a) $\sqrt[5]{m^3} \cdot \sqrt{m}$

(b) $\dfrac{\sqrt[3]{p^5}}{\sqrt{p^3}}$

(c) $\sqrt[4]{\sqrt[3]{x}}$

Answers

6. (a) 11^2, or 121 (b) $\frac{1}{7}$ (c) $9x^{4/3}$ (d) $\frac{b^{21/10}}{a^{5/2}}$ (e) $a^3 + a$

7. (a) $m^{11/10}$ (b) $p^{1/6}$ (c) $x^{1/12}$

The same result is obtained if we simplify within the parentheses first.

$$\left(\frac{x^4y^{-6}}{x^{-2}y^{1/3}}\right)^{-2/3}$$

$$= (x^{4-(-2)}y^{-6-1/3})^{-2/3} \quad \text{Quotient rule}$$

$$= (x^6y^{-19/3})^{-2/3} \quad -6 - \tfrac{1}{3} = -\tfrac{18}{3} - \tfrac{1}{3} = -\tfrac{19}{3}$$

$$= (x^6)^{-2/3}(y^{-19/3})^{-2/3} \quad \text{Power rule}$$

$$= x^{-4}y^{38/9} \quad \text{Power rule}$$

$$= \frac{y^{38/9}}{x^4} \quad \text{Definition of negative exponent}$$

(e) $m^{3/4}(m^{5/4} - m^{1/4})$

$$= m^{3/4}(m^{5/4}) - m^{3/4}(m^{1/4}) \quad \text{Distributive property}$$

$$= m^{3/4+5/4} - m^{3/4+1/4} \quad \text{Product rule}$$

$$= m^{8/4} - m^{4/4}$$

$$= m^2 - m$$

Do not make the common mistake of multiplying exponents in the first step.

◀ Work Problem 6 at the Side.

CAUTION

Use the rules of exponents in problems like those in Example 5. Do not convert the expressions to radical form.

EXAMPLE 6 Applying Rules for Rational Exponents

Write all radicals as exponentials, and then apply the rules for rational exponents. Give answers in exponential form. Assume that all variables represent positive real numbers.

(a) $\sqrt[3]{x^2} \cdot \sqrt[4]{x}$

$$= x^{2/3} \cdot x^{1/4} \quad \text{Convert to rational exponents.}$$

$$= x^{2/3+1/4} \quad \text{Product rule}$$

$$= x^{8/12+3/12} \quad \text{Write exponents with a common denominator.}$$

$$= x^{11/12}$$

(b) $\dfrac{\sqrt{x^3}}{\sqrt[3]{x^2}}$

$$= \frac{x^{3/2}}{x^{2/3}} \quad \text{Convert to rational exponents.}$$

$$= x^{3/2-2/3} \quad \text{Quotient rule}$$

$$= x^{5/6} \quad \tfrac{3}{2} - \tfrac{2}{3} = \tfrac{9}{6} - \tfrac{4}{6} = \tfrac{5}{6}$$

(c) $\sqrt{\sqrt[4]{z}}$

$$= \sqrt{z^{1/4}} \quad \text{Convert the inside radical to rational exponents.}$$

$$= (z^{1/4})^{1/2} \quad \text{Convert to rational exponents.}$$

$$= z^{1/8} \quad \text{Power rule}$$

◀ Work Problem 7 at the Side.

9.2 Exercises

FOR EXTRA HELP

MyMathLab

Math XL PRACTICE  WATCH  DOWNLOAD  READ  REVIEW

Match each expression from Column I with the equivalent choice from Column II.

I		II	
1. $2^{1/2}$ C	**2.** $(-27)^{1/3}$ E	**A.** -4	**B.** 8
3. $-16^{1/2}$ A	**4.** $(-16)^{1/2}$ J	**C.** $\sqrt{2}$	**D.** $-\sqrt{6}$
5. $(-32)^{1/5}$ H	**6.** $(-32)^{2/5}$ G	**E.** -3	**F.** $\sqrt{6}$
7. $4^{3/2}$ B	**8.** $6^{2/4}$ F	**G.** 4	**H.** -2
9. $-6^{2/4}$ D	**10.** $36^{0.5}$ I	**I.** 6	**J.** Not a real number

Evaluate each exponential. See Examples 1–3.

11. $169^{1/2}$ 13

12. $121^{1/2}$ 11

13. $729^{1/3}$ 9

14. $512^{1/3}$ 8

15. $16^{1/4}$ 2

16. $625^{1/4}$ 5

17. $\left(\frac{64}{81}\right)^{1/2}$ $\frac{8}{9}$

18. $\left(\frac{8}{27}\right)^{1/3}$ $\frac{2}{3}$

19. $(-27)^{1/3}$ -3

20. $(-32)^{1/5}$ -2

21. $(-144)^{1/2}$ not a real number

22. $(-36)^{1/2}$ not a real number

23. $100^{3/2}$ 1000

24. $64^{3/2}$ 512

25. $81^{3/4}$ 27

26. $216^{2/3}$ 36

27. $-16^{5/2}$ -1024

28. $-32^{3/5}$ -8

29. $(-8)^{4/3}$ 16

30. $(-243)^{2/5}$ 9

31. $32^{-3/5}$ $\frac{1}{8}$

32. $27^{-4/3}$ $\frac{1}{81}$

33. $64^{-3/2}$ $\frac{1}{512}$

34. $81^{-3/2}$ $\frac{1}{729}$

35. $\left(\frac{125}{27}\right)^{-2/3}$ $\frac{9}{25}$

36. $\left(\frac{64}{125}\right)^{-2/3}$ $\frac{25}{16}$

Write with radicals. Assume that all variables represent positive real numbers. Use the definition that takes the root first. See Example 4.

37. $12^{1/2}$

$\sqrt{12}$

38. $3^{1/2}$

$\sqrt{3}$

39. $8^{3/4}$

$(\sqrt[4]{8})^3$

40. $7^{2/3}$

$(\sqrt[3]{7})^2$

41. $(9q)^{5/8} - (2x)^{2/3}$

$(\sqrt[8]{9q})^5 - (\sqrt[3]{2x})^2$

42. $(3p)^{3/4} + (4x)^{1/3}$

$(\sqrt[4]{3p})^3 + \sqrt[3]{4x}$

43. $(2m)^{-3/2}$

$\frac{1}{(\sqrt{2m})^3}$

44. $(5y)^{-3/5}$

$\frac{1}{(\sqrt[5]{5y})^3}$

45. $(2y + x)^{2/3}$

$(\sqrt[3]{2y + x})^2$

46. $(r + 2z)^{3/2}$

$(\sqrt{r + 2z})^3$

47. $(3m^4 + 2k^2)^{-2/3}$

$\frac{1}{(\sqrt[3]{3m^4 + 2k^2})^2}$

48. $(5x^2 + 3z^3)^{-5/6}$

$\frac{1}{(\sqrt[6]{5x^2 + 3z^3})^5}$

49. Show that, in general, $\sqrt{a^2 + b^2} \neq a + b$ by replacing a with 3 and b with 4.

$\sqrt{a^2 + b^2} = \sqrt{3^2 + 4^2} = 5; a + b = 3 + 4 = 7; 5 \neq 7$

50. Suppose someone claims that $\sqrt[n]{a^n + b^n}$ must equal $a + b$, since when $a = 1$ and $b = 0$, a true statement results:

$$\sqrt[n]{a^n + b^n} = \sqrt[n]{1^n + 0^n} = \sqrt[n]{1^n} = 1 = 1 + 0 = a + b.$$

Explain why this is faulty reasoning.

The statement is true for this particular choice of values for *a* and *b*. However, it is not true *in general*. For example, let *a* = 3, *b* = 4, and *n* = 2.

Simplify by first converting to rational exponents. Assume that all variables represent positive real numbers. See Example 4.

51. $\sqrt{2^{12}}$

64

52. $\sqrt{5^{10}}$

3125

53. $\sqrt[3]{4^9}$

64

54. $\sqrt[4]{6^8}$

36

55. $\sqrt{x^{20}}$

x^{10}

56. $\sqrt{r^{50}}$

r^{25}

57. $\sqrt[3]{x} \cdot \sqrt{x}$

$\sqrt[6]{x^5}$

58. $\sqrt[4]{y} \cdot \sqrt[5]{y^2}$

$\sqrt[20]{y^{13}}$

59. $\frac{\sqrt[3]{t^4}}{\sqrt[5]{t^4}}$

$\sqrt[15]{t^8}$

60. $\frac{\sqrt[4]{w^3}}{\sqrt[6]{w}}$

$\sqrt[12]{w^7}$

Simplify each expression. Write all answers with positive exponents. Assume that all variables represent positive real numbers. See Example 5.

61. $3^{1/2} \cdot 3^{3/2}$

9

62. $6^{4/3} \cdot 6^{2/3}$

36

63. $\frac{64^{5/3}}{64^{4/3}}$

4

64. $\frac{125^{7/3}}{125^{5/3}}$

25

65. $y^{7/3} \cdot y^{-4/3}$

y

66. $r^{-8/9} \cdot r^{17/9}$

r

67. $x^{2/3} \cdot x^{-1/4}$

$x^{5/12}$

68. $x^{2/5} \cdot x^{-1/3}$

$x^{1/15}$

69. $\frac{k^{1/3}}{k^{2/3} \cdot k^{-1}}$

$k^{2/3}$

70. $\frac{z^{3/4}}{z^{5/4} \cdot z^{-2}}$

$z^{3/2}$

71. $\frac{(x^{1/4}y^{2/5})^{20}}{x^2}$

x^3y^8

72. $\dfrac{(r^{1/5}s^{2/3})^{15}}{r^2}$

rs^{10}

73. $\dfrac{(x^{2/3})^2}{(x^2)^{7/3}}$

$\dfrac{1}{x^{10/3}}$

74. $\dfrac{(p^3)^{1/4}}{(p^{5/4})^2}$

$\dfrac{1}{p^{7/4}}$

75. $\dfrac{m^{3/4}n^{-1/4}}{(m^2n)^{1/2}}$

$\dfrac{1}{m^{1/4}n^{3/4}}$

76. $\dfrac{(a^2b^5)^{-1/4}}{(a^{-3}b^2)^{1/6}}$

$\dfrac{1}{b^{19/12}}$

77. $\dfrac{p^{1/5}p^{7/10}p^{1/2}}{(p^3)^{-1/5}}$

p^2

78. $\dfrac{z^{1/3}z^{-2/3}z^{1/6}}{(z^{-1/6})^3}$

$z^{1/3}$

79. $\left(\dfrac{b^{-3/2}}{c^{-5/3}}\right)^2(b^{-1/4}c^{-1/3})^{-1}$

$\dfrac{c^{11/3}}{b^{11/4}}$

80. $\left(\dfrac{m^{-2/3}}{a^{-3/4}}\right)^4(m^{-3/8}a^{1/4})^{-2}$

$\dfrac{a^{5/2}}{m^{23/12}}$

81. $\left(\dfrac{p^{-1/4}q^{-3/2}}{3^{-1}p^{-2}q^{-2/3}}\right)^{-2}$

$\dfrac{q^{5/3}}{9p^{7/2}}$

82. $\left(\dfrac{2^{-2}w^{-3/4}x^{-5/8}}{w^{3/4}x^{-1/2}}\right)^{-3}$

$64w^{9/2}x^{3/8}$

83. $p^{2/3}(p^{1/3} + 2p^{4/3})$

$p + 2p^2$

84. $z^{5/8}(3z^{5/8} + 5z^{11/8})$

$3z^{5/4} + 5z^2$

85. $k^{1/4}(k^{3/2} - k^{1/2})$

$k^{7/4} - k^{3/4}$

86. $r^{3/5}(r^{1/2} + r^{3/4})$

$r^{11/10} + r^{27/20}$

87. $6a^{7/4}(a^{-7/4} + 3a^{-3/4})$

$6 + 18a$

88. $4m^{5/3}(m^{-2/3} - 4m^{-5/3})$

$4m - 16$

89. $5m^{-2/3}(m^{2/3} + m^{-7/3})$

$5 + \dfrac{5}{m^3}$

Write radicals as exponentials, and then apply the rules for rational exponents. Give answers in exponential form. Assume that all radicands represent positive real numbers. See Example 6.

90. $\sqrt[5]{x^3} \cdot \sqrt[4]{x}$

$x^{17/20}$

91. $\sqrt[6]{y^5} \cdot \sqrt[3]{y^2}$

$y^{3/2}$

92. $\dfrac{\sqrt{x^5}}{\sqrt{x^8}}$

$\dfrac{1}{x^{3/2}}$

93. $\dfrac{\sqrt[3]{k^5}}{\sqrt[3]{k^7}}$

$\dfrac{1}{k^{2/3}}$

94. $\sqrt{y} \cdot \sqrt[3]{yz}$

$y^{5/6}z^{1/3}$

95. $\sqrt[3]{xz} \cdot \sqrt{z}$

$x^{1/3}z^{5/6}$

96. $\sqrt[4]{\sqrt[3]{m}}$

$m^{1/12}$

97. $\sqrt[3]{\sqrt{k}}$

$k^{1/6}$

98. $\sqrt{\sqrt[3]{\sqrt[4]{x}}}$

$x^{1/24}$

99. $\sqrt[3]{\sqrt[5]{\sqrt{y}}}$

$y^{1/30}$

100. $\sqrt{y^{5/4}}$

$y^{5/8}$

101. $\sqrt[3]{x^{5/9}}$

$x^{5/27}$

Solve each problem.

102. Meteorologists can determine the duration of a storm by using the function defined by

$$T(D) = 0.07D^{3/2},$$

where D is the diameter of the storm in miles and T is the time in hours. Find the duration of a storm with a diameter of 16 mi. Round your answer to the nearest tenth of an hour.

4.5 hr

103. The threshold weight T, in pounds, for a person is the weight above which the risk of death increases greatly. The threshold weight in pounds for men aged 40–49 is related to height in inches by the function defined by

$$h(T) = (1860.867T)^{1/3}.$$

What height corresponds to a threshold weight of 200 lb for a 46-yr-old man? Round your answer to the nearest inch, and then to the nearest tenth of a foot.

72 in.; 6.0 ft

*The **windchill factor** is a measure of the cooling effect that the wind has on a person's skin. It calculates the equivalent cooling temperature if there were no wind. The National Weather Service uses the formula*

$$\text{Windchill temperature} = 35.74 + 0.6215T - 35.75V^{4/25} + 0.4275TV^{4/25},$$

where T is the temperature in °F and V is the wind speed in miles per hour, to calculate windchill. The chart gives the windchill factor for various wind speeds and temperatures at which frostbite is a risk, and how quickly it may occur.

Wind speed (mph)	Temperature (°F)								
Calm	40	30	20	10	0	−10	−20	−30	−40
5	36	25	13	1	−11	−22	−34	−46	−57
10	34	21	9	−4	−16	−28	−41	−53	−66
15	32	19	6	−7	−19	−32	−45	−58	−71
20	30	17	4	−9	−22	−35	−48	−61	−74
25	29	16	3	−11	−24	−37	−51	−64	−78
30	28	15	1	−12	−26	−39	−53	−67	−80
35	28	14	0	−14	−27	−41	−55	−69	−82
40	27	13	−1	−15	−29	−43	−57	−71	−84

Frostbites times: 30 minutes 10 minutes 5 minutes

Source: National Oceanic and Atmospheric Administration, National Weather Service.

Use the formula to determine the windchill to the nearest tenth of a degree, given the following conditions. Compare your answers with the appropriate entries in the table.

104. 30°F, 15-mph wind

19.0°; The table gives 19°.

105. 10°F, 30-mph wind

−12.3°; The table gives −12°.

9.3 Simplifying Radical Expressions

OBJECTIVES

1. Use the product rule for radicals.
2. Use the quotient rule for radicals.
3. Simplify radicals.
4. Simplify products and quotients of radicals with different indexes.
5. Use the Pythagorean formula.
6. Use the distance formula.

OBJECTIVE 1 Use the product rule for radicals. Is the product of two nth-root radicals equal to the nth root of the product of the radicands? For example, are $\sqrt{36 \cdot 4}$ and $\sqrt{36} \cdot \sqrt{4}$ equal?

$$\sqrt{36 \cdot 4} = \sqrt{144} = 12$$
$$\sqrt{36} \cdot \sqrt{4} = 6 \cdot 2 = 12$$

The result is the same.

This is an example of the **product rule for radicals.**

Product Rule for Radicals

If $\sqrt[n]{a}$ and $\sqrt[n]{b}$ are real numbers and n is a natural number, then

$$\sqrt[n]{a} \cdot \sqrt[n]{b} = \sqrt[n]{ab}.$$

In words, the product of two nth roots is the nth root of the product.

We justify the product rule using the rules for rational exponents. Since $\sqrt[n]{a} = a^{1/n}$ and $\sqrt[n]{b} = b^{1/n}$,

$$\sqrt[n]{a} \cdot \sqrt[n]{b} = a^{1/n} \cdot b^{1/n} = (ab)^{1/n} = \sqrt[n]{ab}.$$

CAUTION

Use the product rule only when the radicals have the same index.

EXAMPLE 1 Using the Product Rule

Multiply. Assume that all variables represent positive real numbers.

(a) $\sqrt{5} \cdot \sqrt{7} = \sqrt{5 \cdot 7} = \sqrt{35}$

(b) $\sqrt{2} \cdot \sqrt{19} = \sqrt{2 \cdot 19} = \sqrt{38}$

(c) $\sqrt{11} \cdot \sqrt{p} = \sqrt{11p}$

(d) $\sqrt{7} \cdot \sqrt{11xyz} = \sqrt{77xyz}$

Work Problem 1 at the Side. ▶

EXAMPLE 2 Using the Product Rule

Multiply. Assume that all variables represent positive real numbers.

(a) $\sqrt[3]{3} \cdot \sqrt[3]{12} = \sqrt[3]{3 \cdot 12} = \sqrt[3]{36}$

(b) $\sqrt[4]{8y} \cdot \sqrt[4]{3r^2} = \sqrt[4]{24yr^2}$

(c) $\sqrt[6]{10m^4} \cdot \sqrt[6]{5m} = \sqrt[6]{50m^5}$

(d) $\sqrt[4]{2} \cdot \sqrt[5]{2}$ cannot be simplified using the product rule for radicals, because the indexes (4 and 5) are different.

Work Problem 2 at the Side. ▶

1 Multiply. Assume that all variables represent positive real numbers.

(a) $\sqrt{5} \cdot \sqrt{13}$

(b) $\sqrt{10y} \cdot \sqrt{3k}$

2 Multiply. Assume that all variables represent positive real numbers.

(a) $\sqrt[3]{2} \cdot \sqrt[3]{7}$

(b) $\sqrt[6]{8r^2} \cdot \sqrt[6]{2r^3}$

(c) $\sqrt[5]{9y^2x} \cdot \sqrt[5]{8xy^2}$

(d) $\sqrt{7} \cdot \sqrt[3]{5}$

ANSWERS

1. (a) $\sqrt{65}$ **(b)** $\sqrt{30yk}$

2. (a) $\sqrt[3]{14}$ **(b)** $\sqrt[6]{16r^5}$ **(c)** $\sqrt[5]{72y^4x^2}$ **(d)** cannot be simplified using the product rule

OBJECTIVE 2 Use the quotient rule for radicals. The **quotient rule for radicals** is similar to the product rule.

Quotient Rule for Radicals

If $\sqrt[n]{a}$ and $\sqrt[n]{b}$ are real numbers, $b \neq 0$, and n is a natural number, then

$$\sqrt[n]{\frac{a}{b}} = \frac{\sqrt[n]{a}}{\sqrt[n]{b}}.$$

In words, the nth root of a quotient is the quotient of the nth roots.

EXAMPLE 3 Using the Quotient Rule

Simplify. Assume that all variables represent positive real numbers.

(a) $\sqrt{\frac{16}{25}} = \frac{\sqrt{16}}{\sqrt{25}} = \frac{4}{5}$

(b) $\sqrt{\frac{7}{36}} = \frac{\sqrt{7}}{\sqrt{36}} = \frac{\sqrt{7}}{6}$

(c) $\sqrt[3]{-\frac{8}{125}} = \sqrt[3]{\frac{-8}{125}} = \frac{\sqrt[3]{-8}}{\sqrt[3]{125}} = \frac{-2}{5} = -\frac{2}{5}$ $\quad \frac{-a}{b} = -\frac{a}{b}$

(d) $\sqrt[3]{\frac{7}{216}} = \frac{\sqrt[3]{7}}{\sqrt[3]{216}} = \frac{\sqrt[3]{7}}{6}$

(e) $\sqrt[5]{\frac{x}{32}} = \frac{\sqrt[5]{x}}{\sqrt[5]{32}} = \frac{\sqrt[5]{x}}{2}$

(f) $-\sqrt[3]{\frac{m^6}{125}} = -\frac{\sqrt[3]{m^6}}{\sqrt[3]{125}} = -\frac{m^2}{5}$ $\quad$ Think: $\sqrt[3]{m^6} = m^{6/3} = m^2$

◀ Work Problem 3 at the Side.

OBJECTIVE 3 Simplify radicals. We use the product and quotient rules to simplify radicals. A radical is **simplified** if the following four conditions are met.

Conditions for a Simplified Radical

1. The radicand has no factor raised to a power greater than or equal to the index.
2. The radicand has no fractions.
3. No denominator has a radical.
4. Exponents in the radicand and the index of the radical have greatest common factor 1.

3 Simplify. Assume that all variables represent positive real numbers.

(a) $\sqrt{\frac{100}{81}}$

(b) $\sqrt{\frac{11}{25}}$

(c) $\sqrt[3]{-\frac{125}{216}}$

(d) $\sqrt{\frac{y^8}{16}}$

(e) $-\sqrt[3]{\frac{x^2}{r^{12}}}$

Answers

3. (a) $\frac{10}{9}$ (b) $\frac{\sqrt{11}}{5}$ (c) $-\frac{5}{6}$ (d) $\frac{y^4}{4}$ (e) $-\frac{\sqrt[3]{x^2}}{r^4}$

EXAMPLE 4 Simplifying Roots of Numbers

Simplify.

(a) $\sqrt{24}$

Check to see whether 24 is divisible by a perfect square (the square of a natural number) such as 4, 9, 16, ... The largest perfect square that divides into 24 is 4.

$$\begin{aligned} &\sqrt{24} \\ &= \sqrt{4 \cdot 6} && \text{Factor; 4 is a perfect square.} \\ &= \sqrt{4} \cdot \sqrt{6} && \text{Product rule} \\ &= 2\sqrt{6} && \sqrt{4} = 2 \end{aligned}$$

(b) $\sqrt{108}$

As shown on the left, the number 108 is divisible by the perfect square 36. If this perfect square is not immediately clear, try factoring 108 into its prime factors, as shown on the right.

$$\begin{aligned} &\sqrt{108} \\ &= \sqrt{36 \cdot 3} \\ &= \sqrt{36} \cdot \sqrt{3} \\ &= 6\sqrt{3} \end{aligned}$$

$$\begin{aligned} &\sqrt{108} \\ &= \sqrt{2^2 \cdot 3^3} \\ &= \sqrt{2^2 \cdot 3^2 \cdot 3} \\ &= \sqrt{2^2} \cdot \sqrt{3^2} \cdot \sqrt{3} && \text{Product rule} \\ &= 2 \cdot 3 \cdot \sqrt{3} && \sqrt{2^2} = 2,\ \sqrt{3^2} = 3 \\ &= 6\sqrt{3} && \text{Multiply.} \end{aligned}$$

(c) $\sqrt{10}$ No perfect square (other than 1) divides into 10, so $\sqrt{10}$ cannot be simplified further.

(d) $\sqrt[3]{16}$

The largest perfect *cube* that divides into 16 is 8, so factor 16 as $8 \cdot 2$.

Remember to write the index.

$$\begin{aligned} &\sqrt[3]{16} \\ &= \sqrt[3]{8 \cdot 2} && \text{8 is a perfect cube.} \\ &= \sqrt[3]{8} \cdot \sqrt[3]{2} && \text{Product rule} \\ &= 2\sqrt[3]{2} && \sqrt[3]{8} = 2 \end{aligned}$$

(e)

Remember the negative sign in each line.

$$\begin{aligned} &-\sqrt[4]{162} \\ &= -\sqrt[4]{81 \cdot 2} && \text{81 is a perfect 4th power.} \\ &= -\sqrt[4]{81} \cdot \sqrt[4]{2} && \text{Product rule} \\ &= -3\sqrt[4]{2} && \sqrt[4]{81} = 3 \end{aligned}$$

Work Problem **4** *at the Side.* ▶

CAUTION

Be careful with which factors belong outside the radical sign and which belong inside. Note in Example 4(b) how $2 \cdot 3$ is written outside because $\sqrt{2^2} = 2$ and $\sqrt{3^2} = 3$. The remaining 3 is left inside the radical.

4 Simplify.

(a) $\sqrt{32}$

(b) $\sqrt{45}$

(c) $\sqrt{300}$

(d) $\sqrt{35}$

(e) $-\sqrt[3]{54}$

(f) $\sqrt[4]{243}$

ANSWERS

4. (a) $4\sqrt{2}$ **(b)** $3\sqrt{5}$ **(c)** $10\sqrt{3}$ **(d)** cannot be simplified further **(e)** $-3\sqrt[3]{2}$ **(f)** $3\sqrt[4]{3}$

5 Simplify. Assume that all variables represent positive real numbers.

(a) $\sqrt{25p^7}$

(b) $\sqrt{72y^3x}$

(c) $\sqrt[3]{-27y^7x^5z^6}$

(d) $-\sqrt[4]{32a^5b^7}$

Answers

5. **(a)** $5p^3\sqrt{p}$ **(b)** $6y\sqrt{2yx}$ **(c)** $-3y^2xz^2\sqrt[3]{yx^2}$ **(d)** $-2ab\sqrt[4]{2ab^3}$

EXAMPLE 5 Simplifying Radicals Involving Variables

Simplify. Assume that all variables represent positive real numbers.

(a) $\sqrt{16m^3}$

$= \sqrt{16m^2 \cdot m}$ Factor.

$= \sqrt{16m^2} \cdot \sqrt{m}$ Product rule

$= 4m\sqrt{m}$

Absolute value bars are not needed around the m in color because of the assumption that all the variables represent *positive* real numbers.

(b) $\sqrt{200k^7q^8}$

$= \sqrt{10^2 \cdot 2 \cdot (k^3)^2 \cdot k \cdot (q^4)^2}$ Factor.

$= 10k^3q^4\sqrt{2k}$ Remove perfect square factors.

(c) $\sqrt[3]{-8x^4y^5}$

$= \sqrt[3]{(-8x^3y^3)(xy^2)}$ Choose $-8x^3y^3$ as the perfect cube that divides into $-8x^4y^5$.

$= \sqrt[3]{-8x^3y^3} \cdot \sqrt[3]{xy^2}$ Product rule

$= -2xy\sqrt[3]{xy^2}$

(d) $-\sqrt[4]{32y^9}$

$= -\sqrt[4]{(16y^8)(2y)}$ $16y^8$ is the largest 4th power that divides into $32y^9$.

$= -\sqrt[4]{16y^8} \cdot \sqrt[4]{2y}$ Product rule

$= -2y^2\sqrt[4]{2y}$

◀ Work Problem **5** at the Side.

Note

From Example 5 we see that if a variable is raised to a power with an exponent divisible by 2, it is a perfect square. If it is raised to a power with an exponent divisible by 3, it is a perfect cube. ***In general, if it is raised to a power with an exponent divisible by n, it is a perfect nth power.***

The conditions for a simplified radical given earlier state that an exponent in the radicand and the index of the radical should have greatest common factor 1. The next example applies this condition.

EXAMPLE 6 Simplifying Radicals by Using Lesser Indexes

Simplify. Assume that all variables represent positive real numbers.

(a) $\sqrt[9]{5^6}$

We can write this radical using rational exponents and then write the exponent in lowest terms. We then express the answer as a radical.

$$\sqrt[9]{5^6} = 5^{6/9} = 5^{2/3} = \sqrt[3]{5^2}, \quad \text{or} \quad \sqrt[3]{25}$$

(b) $\sqrt[4]{p^2} = p^{2/4} = p^{1/2} = \sqrt{p}$ (Recall the assumption that $p > 0$.)

These examples suggest the following rule.

If m is an integer, n and k are natural numbers, and all indicated roots exist, then

$$\sqrt[kn]{a^{km}} = \sqrt[n]{a^m}.$$

Work Problem 6 at the Side. ▶

OBJECTIVE 4 Simplify products and quotients of radicals with different indexes. Since the product and quotient rules for radicals apply only when they have the same index, we multiply and divide radicals with different indexes by using rational exponents.

EXAMPLE 7 Multiplying Radicals with Different Indexes

Simplify $\sqrt{7} \cdot \sqrt[3]{2}$.

Because the different indexes, 2 and 3, have a least common index of 6, we use rational exponents to write each radical as a sixth root.

$$\sqrt{7} = 7^{1/2} = 7^{3/6} = \sqrt[6]{7^3} = \sqrt[6]{343}$$

$$\sqrt[3]{2} = 2^{1/3} = 2^{2/6} = \sqrt[6]{2^2} = \sqrt[6]{4}$$

Therefore,

$$\sqrt{7} \cdot \sqrt[3]{2}$$

$$= \sqrt[6]{343} \cdot \sqrt[6]{4} \quad \text{Substitute; } \sqrt{7} = \sqrt[6]{343},\ \sqrt[3]{2} = \sqrt[6]{4}$$

$$= \sqrt[6]{1372}. \quad \text{Product rule}$$

Work Problem 7 at the Side. ▶

OBJECTIVE 5 Use the Pythagorean formula. The **Pythagorean formula** relates the lengths of the three sides of a right triangle.

Pythagorean Formula

If a and b are the lengths of the shorter sides of a right triangle and c is the length of the longest side, then

$$a^2 + b^2 = c^2.$$

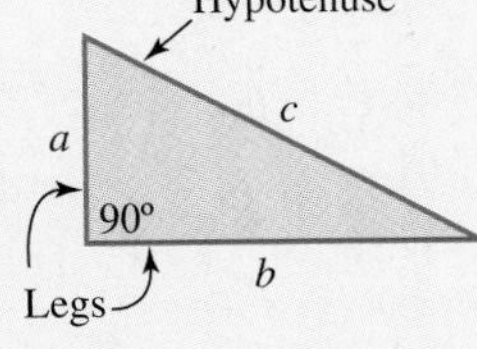

The two shorter sides are the **legs** of the triangle, and the longest side is the **hypotenuse.** The hypotenuse is the side opposite the right angle. Thus,

$$\textbf{leg}^2 + \textbf{leg}^2 = \textbf{hypotenuse}^2.$$

In **Section 10.1,** we will see that an equation such as $x^2 = 7$ has two solutions: $\sqrt{7}$ (the principal, or positive, square root of 7) and $-\sqrt{7}$. Similarly, $c^2 = 52$ has two solutions, $\pm\sqrt{52}$, or $\pm 2\sqrt{13}$. In applications we often choose only the positive square root.

6 Simplify. Assume that all variables represent positive real numbers.

(a) $\sqrt[12]{2^3}$

(b) $\sqrt[6]{t^2}$

7 Simplify $\sqrt{5} \cdot \sqrt[3]{4}$.

ANSWERS

6. **(a)** $\sqrt[4]{2}$ **(b)** $\sqrt[3]{t}$

7. $\sqrt[6]{2000}$

8 Find the length of the unknown side in each triangle.

(a)

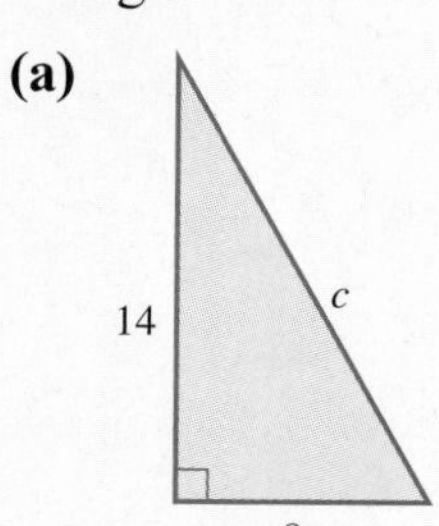

(b)

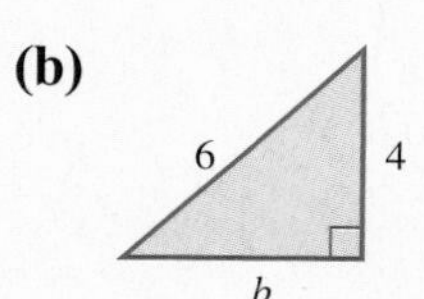

(*Hint:* Write the Pythagorean formula as $b^2 = c^2 - a^2$ here.)

EXAMPLE 8 Using the Pythagorean Formula

Use the Pythagorean formula to find the length of the hypotenuse in the triangle in Figure 6.

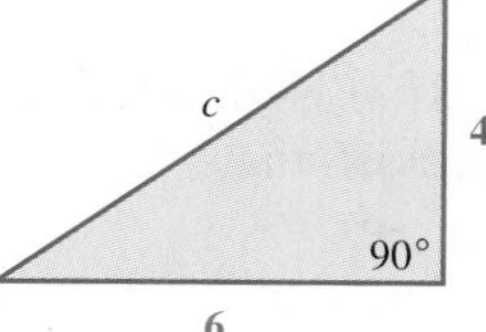

Figure 6

To find the length of the hypotenuse c, let $a = 4$ and $b = 6$. Then, use the formula.

$$a^2 + b^2 = c^2$$

Substitute carefully.

$$4^2 + 6^2 = c^2 \quad \text{Let } a = 4 \text{ and } b = 6.$$

$$16 + 36 = c^2 \quad \text{Apply the exponents.}$$

$$c^2 = 52 \quad \text{Add; interchange sides.}$$

$$c = \sqrt{52} \quad \text{Choose the principal root.}$$

$$c = \sqrt{4 \cdot 13} \quad \text{Factor.}$$

$$c = \sqrt{4} \cdot \sqrt{13} \quad \text{Product rule}$$

$$c = 2\sqrt{13} \quad \text{Simplify.}$$

The length of the hypotenuse is $2\sqrt{13}$.

◀ *Work Problem* **8** *at the Side.*

CAUTION

When substituting in the Pythagorean formula $a^2 + b^2 = c^2$, be sure that the lengths of the legs are substituted for a and b and the length of the hypotenuse is substituted for c.

OBJECTIVE 6 Use the distance formula. The *distance formula* allows us to find the distance between two points in the coordinate plane, or the length of the line segment joining those two points.

Figure 7 shows the points $(3, -4)$ and $(-5, 3)$. The vertical line through $(-5, 3)$ and the horizontal line through $(3, -4)$ intersect at the point $(-5, -4)$. Thus, the point $(-5, -4)$ becomes the vertex of the right angle in a right triangle. By the Pythagorean formula, the sum of the squares of the lengths of the two legs a and b of the right triangle in Figure 7 is equal to the square of the length of the hypotenuse, d:

$$a^2 + b^2 = d^2, \quad \text{or} \quad d^2 = a^2 + b^2.$$

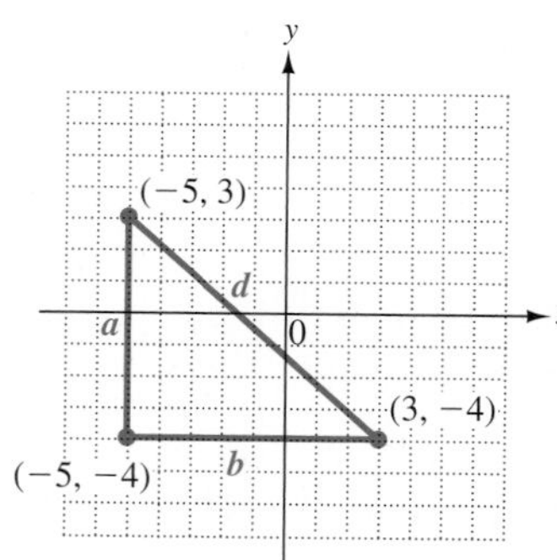

Figure 7

The length a is the difference between the y-coordinates of the endpoints. Since the x-coordinate of both of these points in Figure 7 is -5, the side is vertical, and we can find a by finding the difference between the y-coordinates. We subtract -4 from 3 to get a positive value for a.

$$a = 3 - (-4) = 7$$

ANSWERS

8. (a) $2\sqrt{65}$ **(b)** $2\sqrt{5}$

Similarly, we find b by subtracting -5 from 3.

$$b = 3 - (-5) = 8$$

Substituting these values into the formula, we have

$$d^2 = a^2 + b^2$$

$$d^2 = 7^2 + 8^2 \quad \text{Let } a = 7 \text{ and } b = 8.$$

$$d^2 = 49 + 64 \quad \text{Apply the exponents.}$$

$$d^2 = 113 \quad \text{Add.}$$

$$d = \sqrt{113}. \quad \text{Choose the principal root.}$$

We choose the principal root since distance cannot be negative. Therefore, the distance between $(-5, 3)$ and $(3, -4)$ is $\sqrt{113}$.

Note

It is customary to leave the distance in radical form. Do not use a calculator to get an approximation, unless you are specifically directed to do so.

This result can be generalized. Figure 8 shows the two points (x_1, y_1) and (x_2, y_2). The distance a between (x_1, y_1) and (x_2, y_1) is

$$a = x_2 - x_1,$$

and the distance b between (x_2, y_2) and (x_2, y_1) is

$$b = y_2 - y_1.$$

From the Pythagorean formula,

$$d^2 = a^2 + b^2$$

$$d^2 = (x_2 - x_1)^2 + (y_2 - y_1)^2.$$

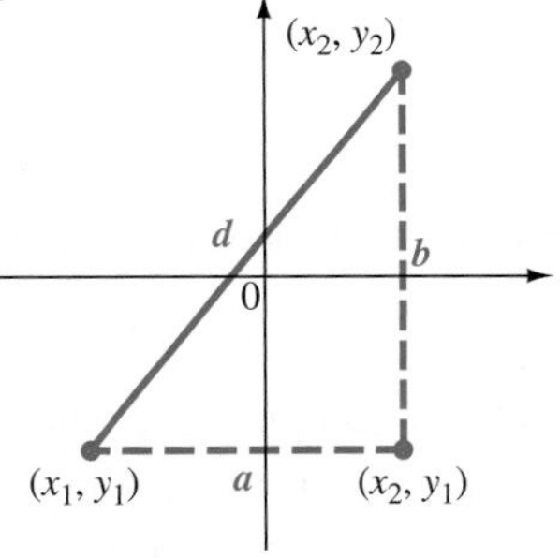

Figure 8

Choosing the principal square root gives the **distance formula.**

Distance Formula

The distance between the points (x_1, y_1) and (x_2, y_2) is

$$d = \sqrt{(x_2 - x_1)^2 + (y_2 - y_1)^2}.$$

EXAMPLE 9 Using the Distance Formula

Find the distance between $(-3, 5)$ and $(6, 4)$.

When using the distance formula to find the distance between two points, designating the points as (x_1, y_1) and (x_2, y_2) is arbitrary. We choose $(x_1, y_1) = (-3, 5)$ and $(x_2, y_2) = (6, 4)$.

$$d = \sqrt{(x_2 - x_1)^2 + (y_2 - y_1)^2} \quad \text{Distance formula}$$

$$d = \sqrt{(6 - (-3))^2 + (4 - 5)^2} \quad x_2 = 6, y_2 = 4, x_1 = -3, y_1 = 5$$

Substitute carefully.

$$d = \sqrt{9^2 + (-1)^2}$$

$$d = \sqrt{82} \quad \text{Leave in radical form.}$$

Work Problem 9 *at the Side.* ▶

9 Find the distance between each pair of points.

(a) $(2, -1)$ and $(5, 3)$

(b) $(-3, 2)$ and $(0, -4)$

Answers

9. **(a)** 5 **(b)** $\sqrt{45}$, or $3\sqrt{5}$

Math in the Media

SO, DID THE SCARECROW REALLY GET A BRAIN?

Probably the most famous mathematical statement in the history of motion pictures is heard in the 1939 classic *The Wizard of Oz.* Ray Bolger's character, the Scarecrow, wants a brain. When the Wizard grants him his "Th.D." (Doctor of Thinkology), the Scarecrow replies with a statement that has made mathematics teachers shudder for 70 years.

Scarecrow: *The sum of the square roots of any two sides of an isosceles triangle is equal to the square root of the remaining side.*

His statement is quite impressive and sounds like the formula for the *Pythagorean Theorem* (page 715). Let's see why it is incorrect.

1. To what kind of triangle does the Scarecrow refer in his statement? To what kind of triangle does the Pythagorean Theorem actually refer?

isosceles triangle; right triangle

2. In the Scarecrow's statement, he refers to square roots. In applying the formula for the Pythagorean Theorem, do you find square roots of the sides? If not, what do you find?

no; The formula for the Pythagorean Theorem involves the squares of the sides, not the square roots.

3. An isosceles triangle has two sides of equal length. Draw an isosceles triangle with two sides of length 9 units and remaining side of length 4 units. Now show that this triangle does not satisfy the Scarecrow's statement.

(This is called a *counterexample* and is sufficient to show that his statement is false in general.)

$\sqrt{9} + \sqrt{9} = 3 + 3 = 6$; $\sqrt{4} = 2$; $6 \neq 2$

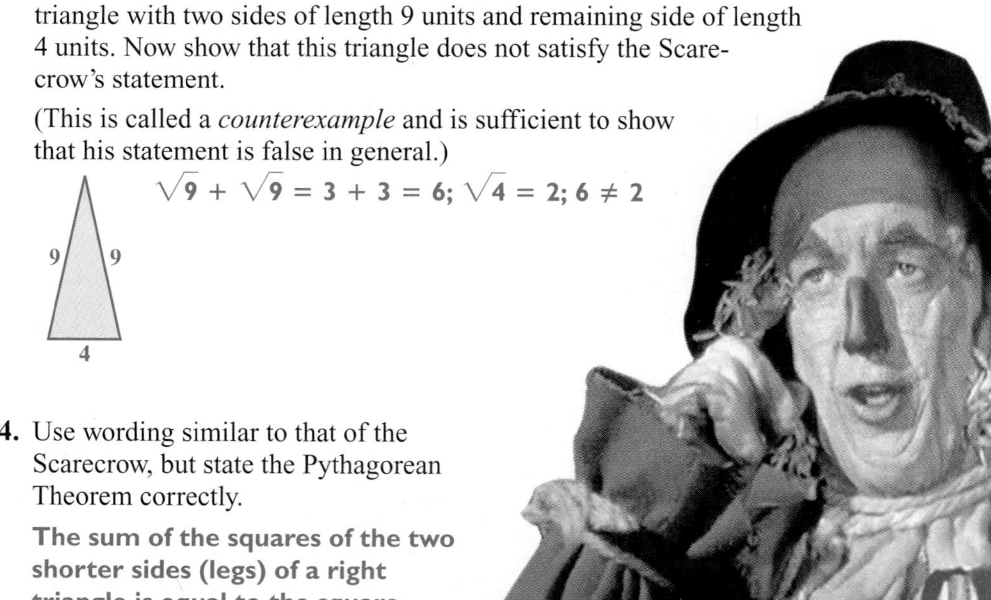

4. Use wording similar to that of the Scarecrow, but state the Pythagorean Theorem correctly.

The sum of the squares of the two shorter sides (legs) of a right triangle is equal to the square of the longest side (hypotenuse).

9.3 Exercises

FOR EXTRA HELP

MyMathLab

Math XL PRACTICE

WATCH

DOWNLOAD

READ

REVIEW

Decide whether each statement is true *or* false *by using the product rule explained in this section. Then support your answer by finding a calculator approximation for each expression.*

1. $2\sqrt{12} = \sqrt{48}$

true; Both are equal to $4\sqrt{3}$ and approximately 6.92820323.

2. $\sqrt{72} = 2\sqrt{18}$

true; Both are equal to $6\sqrt{2}$ and approximately 8.485281374.

3. $3\sqrt{8} = 2\sqrt{18}$

true; Both are equal to $6\sqrt{2}$ and approximately 8.485281374.

4. $5\sqrt{72} = 6\sqrt{50}$

true; Both are equal to $30\sqrt{2}$ and approximately 42.42640687.

5. Explain why $\sqrt[3]{x} \cdot \sqrt[3]{x}$ is not equal to x. What is it equal to?

Because there are only two factors of $\sqrt[3]{x}$, $\sqrt[3]{x} \cdot \sqrt[3]{x} = \left(\sqrt[3]{x}\right)^2$, or $\sqrt[3]{x^2}$.

6. Explain why $\sqrt[4]{x} \cdot \sqrt[4]{x}$ is not equal to x, but *is* equal to $\sqrt{x}$, for $x \geq 0$.

Because there are just two factors, $\sqrt[4]{x} \cdot \sqrt[4]{x} = \left(\sqrt[4]{x}\right)^2 = x^{2/4} = x^{1/2} = \sqrt{x}$, for $x \geq 0$.

7. Which one of the following is *not* equal to $\sqrt{\frac{1}{2}}$? (Do not use calculator approximations.)

A. $\sqrt{0.5}$ **B.** $\sqrt{\frac{2}{4}}$ **C.** $\sqrt{\frac{3}{6}}$ **D.** $\frac{\sqrt{4}}{\sqrt{16}}$

D

8. Use the π key on your calculator to get a value for π. Now find an approximation for $\sqrt[4]{\frac{2143}{22}}$. Does the result mean that π is actually equal to $\sqrt[4]{\frac{2143}{22}}$? Why or why not?

$\sqrt[4]{\frac{2143}{22}} \approx 3.14159265$; It is not actually equal to π. They are only approximations that agree in the first few decimal places.

Multiply using the product rule. Assume all variables represent positive real numbers. See Examples 1 and 2.

9. $\sqrt{5} \cdot \sqrt{6}$

$\sqrt{30}$

10. $\sqrt{10} \cdot \sqrt{3}$

$\sqrt{30}$

11. $\sqrt{14} \cdot \sqrt{x}$

$\sqrt{14x}$

12. $\sqrt{23} \cdot \sqrt{t}$

$\sqrt{23t}$

13. $\sqrt{14} \cdot \sqrt{3pqr}$

$\sqrt{42pqr}$

14. $\sqrt{7} \cdot \sqrt{5xt}$

$\sqrt{35xt}$

15. $\sqrt[3]{7x} \cdot \sqrt[3]{2y}$

$\sqrt[3]{14xy}$

16. $\sqrt[3]{9x} \cdot \sqrt[3]{4y}$

$\sqrt[3]{36xy}$

17. $\sqrt[4]{11} \cdot \sqrt[4]{3}$
$\sqrt[4]{33}$

18. $\sqrt[4]{6} \cdot \sqrt[4]{9}$
$\sqrt[4]{54}$

19. $\sqrt[4]{2x} \cdot \sqrt[4]{3y^2}$
$\sqrt[4]{6xy^2}$

20. $\sqrt[4]{3y^2} \cdot \sqrt[4]{6yz}$
$\sqrt[4]{18y^3z}$

21. $\sqrt[3]{7} \cdot \sqrt[4]{3}$
This product cannot be simplified using the product rule.

22. $\sqrt[5]{8} \cdot \sqrt[6]{12}$
This product cannot be simplified using the product rule.

Simplify. Assume that all variables represent positive real numbers. See Example 3.

23. $\sqrt{\frac{64}{121}}$
$\frac{8}{11}$

24. $\sqrt{\frac{16}{49}}$
$\frac{4}{7}$

25. $\sqrt{\frac{3}{25}}$
$\frac{\sqrt{3}}{5}$

26. $\sqrt{\frac{13}{49}}$
$\frac{\sqrt{13}}{7}$

27. $\sqrt{\frac{x}{25}}$
$\frac{\sqrt{x}}{5}$

28. $\sqrt{\frac{k}{100}}$
$\frac{\sqrt{k}}{10}$

29. $\sqrt{\frac{p^6}{81}}$
$\frac{p^3}{9}$

30. $\sqrt{\frac{w^{10}}{36}}$
$\frac{w^5}{6}$

31. $\sqrt[3]{-\frac{27}{64}}$
$-\frac{3}{4}$

32. $\sqrt[3]{-\frac{216}{125}}$
$-\frac{6}{5}$

33. $\sqrt[3]{\frac{r^2}{8}}$
$\frac{\sqrt[3]{r^2}}{2}$

34. $\sqrt[3]{\frac{t}{125}}$
$\frac{\sqrt[3]{t}}{5}$

35. $-\sqrt[4]{\frac{81}{x^4}}$
$-\frac{3}{x}$

36. $-\sqrt[4]{\frac{625}{y^4}}$
$-\frac{5}{y}$

37. $\sqrt[5]{\frac{1}{x^{15}}}$
$\frac{1}{x^3}$

38. $\sqrt[5]{\frac{32}{y^{20}}}$
$\frac{2}{y^4}$

Express each radical in simplified form. See Example 4.

39. $\sqrt{12}$
$2\sqrt{3}$

40. $\sqrt{18}$
$3\sqrt{2}$

41. $\sqrt{288}$
$12\sqrt{2}$

42. $\sqrt{72}$
$6\sqrt{2}$

43. $-\sqrt{32}$
$-4\sqrt{2}$

44. $-\sqrt{48}$
$-4\sqrt{3}$

45. $-\sqrt{28}$
$-2\sqrt{7}$

46. $-\sqrt{24}$
$-2\sqrt{6}$

47. $\sqrt{30}$
cannot be simplified further

48. $\sqrt{46}$
cannot be simplified further

49. $\sqrt[3]{128}$
$4\sqrt[3]{2}$

50. $\sqrt[3]{24}$
$2\sqrt[3]{3}$

51. $\sqrt[3]{-16}$
$-2\sqrt[3]{2}$

52. $\sqrt[3]{-250}$
$-5\sqrt[3]{2}$

53. $\sqrt[3]{40}$
$2\sqrt[3]{5}$

54. $\sqrt[3]{375}$
$5\sqrt[3]{3}$

55. $-\sqrt[4]{512}$
$-4\sqrt[4]{2}$

56. $-\sqrt[4]{1250}$
$-5\sqrt[4]{2}$

57. $\sqrt[5]{64}$
$2\sqrt[5]{2}$

58. $\sqrt[5]{128}$
$2\sqrt[5]{4}$

59. A student claimed that $\sqrt[3]{14}$ is not in simplified form, since $14 = 8 + 6$, and 8 is a perfect cube. Was his reasoning correct? Why or why not?
His reasoning was incorrect. Here 8 is a term, not a factor.

60. Explain in your own words why $\sqrt[3]{k^4}$ is not a simplified radical.
It is not simplified because the power of k is greater than the index of the radical. The simplified form is $k\sqrt[3]{k}$.

Express each radical in simplified form. Assume that all variables represent positive real numbers. See Example 5.

61. $\sqrt{72k^2}$
$6k\sqrt{2}$

62. $\sqrt{18m^2}$
$3m\sqrt{2}$

63. $\sqrt{144x^3y^9}$
$12xy^4\sqrt{xy}$

64. $\sqrt{169s^5t^{10}}$
$13s^2t^5\sqrt{s}$

65. $\sqrt{121x^6}$
$11x^3$

66. $\sqrt{256z^{12}}$
$16z^6$

67. $-\sqrt[3]{27t^{12}}$
$-3t^4$

68. $-\sqrt[3]{64y^{18}}$
$-4y^6$

69. $-\sqrt{100m^8z^4}$
$-10m^4z^2$

70. $-\sqrt{25t^6s^{20}}$
$-5t^3s^{10}$

71. $-\sqrt[3]{-125a^6b^9c^{12}}$
$5a^2b^3c^4$

72. $-\sqrt[3]{-216y^{15}x^6z^3}$
$6y^5x^2z$

73. $\sqrt[4]{\frac{1}{16}r^8t^{20}}$
$\frac{1}{2}r^2t^5$

74. $\sqrt[4]{\frac{81}{256}t^{12}u^8}$
$\frac{3}{4}t^3u^2$

75. $\sqrt{50x^3}$
$5x\sqrt{2x}$

76. $\sqrt{300z^3}$
$10z\sqrt{3z}$

77. $-\sqrt{500r^{11}}$
$-10r^5\sqrt{5r}$

78. $-\sqrt{200p^{13}}$
$-10p^6\sqrt{2p}$

79. $\sqrt{13x^7y^8}$
$x^3y^4\sqrt{13x}$

80. $\sqrt{23k^9p^{14}}$
$k^4p^7\sqrt{23k}$

81. $\sqrt[3]{8z^6w^9}$
$2z^2w^3$

82. $\sqrt[3]{64a^{15}b^{12}}$
$4a^5b^4$

83. $\sqrt[3]{-16z^5t^7}$
$-2zt^2\sqrt[3]{2z^2t}$

84. $\sqrt[3]{-81m^4n^{10}}$
$-3mn^3\sqrt[3]{3mn}$

85. $\sqrt[4]{81x^{12}y^{16}}$
$3x^3y^4$

86. $\sqrt[4]{81t^8u^{28}}$
$3t^2u^7$

87. $-\sqrt[4]{162r^{15}s^{10}}$
$-3r^3s^2\sqrt[4]{2r^3s^2}$

88. $-\sqrt[4]{32k^5m^{10}}$
$-2km^2\sqrt[4]{2km^2}$

89. $\sqrt{\frac{y^{11}}{36}}$
$\frac{y^5\sqrt{y}}{6}$

90. $\sqrt{\frac{v^{13}}{49}}$
$\frac{v^6\sqrt{v}}{7}$

91. $\sqrt[3]{\frac{x^{16}}{27}}$
$\frac{x^5\sqrt[3]{x}}{3}$

92. $\sqrt[3]{\frac{y^{17}}{125}}$
$\frac{y^5\sqrt[3]{y^2}}{5}$

Simplify. Assume that $x \geq 0$. See Example 6.

93. $\sqrt[4]{48^2}$
$4\sqrt{3}$

94. $\sqrt[4]{50^2}$
$5\sqrt{2}$

95. $\sqrt[4]{25}$
$\sqrt{5}$

96. $\sqrt[6]{8}$
$\sqrt{2}$

97. $\sqrt[10]{x^{25}}$
$x^2\sqrt{x}$

98. $\sqrt[12]{x^{44}}$
$x^3\sqrt[3]{x^2}$

Simplify by first writing the radicals as radicals with the same index. Then multiply. Assume that $x \geq 0$. See Example 7.

99. $\sqrt[3]{4} \cdot \sqrt{3}$
$\sqrt[6]{432}$

100. $\sqrt[3]{5} \cdot \sqrt{6}$
$\sqrt[6]{5400}$

101. $\sqrt[4]{3} \cdot \sqrt[3]{4}$
$\sqrt[12]{6912}$

102. $\sqrt[3]{2} \cdot \sqrt[5]{3}$
$\sqrt[15]{864}$

103. $\sqrt{x} \cdot \sqrt[3]{x}$
$\sqrt[6]{x^5}$

104. $\sqrt[3]{x} \cdot \sqrt[4]{x}$
$\sqrt[12]{x^7}$

Find the unknown length in each right triangle. Simplify the answer if necessary. See Example 8.

105.

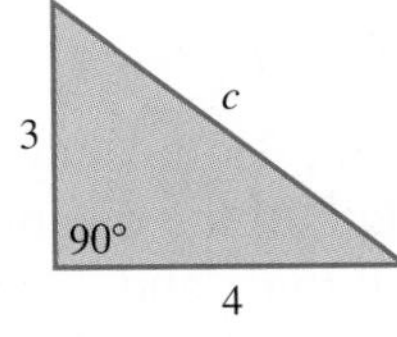

5

106.

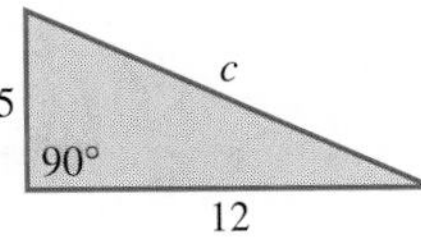

13

107.

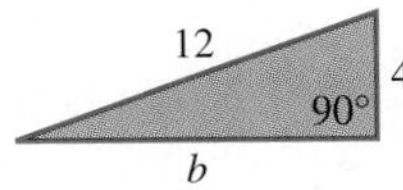

$8\sqrt{2}$

108.

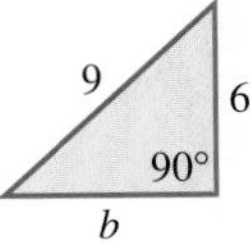

$3\sqrt{5}$

109.

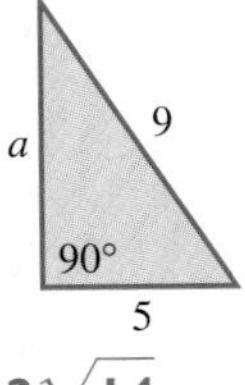

$2\sqrt{14}$

110.

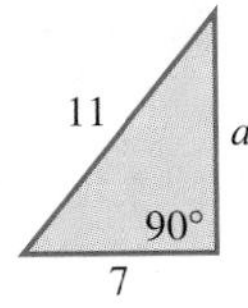

$6\sqrt{2}$

Find the distance between each pair of points. See Example 9.

111. (6, 13) and (1, 1)
13

112. (8, 13) and (2, 5)
10

113. $(-6, 5)$ and $(3, -4)$
$9\sqrt{2}$

114. $(-1, 5)$ and $(-7, 7)$
$2\sqrt{10}$

115. $(-8, 2)$ and $(-4, 1)$
$\sqrt{17}$

116. $(-1, 2)$ and $(5, 3)$
$\sqrt{37}$

117. $(4.7, 2.3)$ and $(1.7, -1.7)$
5

118. $(-2.9, 18.2)$ and $(2.1, 6.2)$
13

119. $(\sqrt{2}, \sqrt{6})$ and $(-2\sqrt{2}, 4\sqrt{6})$
$6\sqrt{2}$

120. $(\sqrt{7}, 9\sqrt{3})$ and $(-\sqrt{7}, 4\sqrt{3})$
$\sqrt{103}$

121. $(x + y, y)$ and $(x - y, x)$
$\sqrt{5y^2 - 2xy + x^2}$

122. $(c, c - d)$ and $(d, c + d)$
$\sqrt{c^2 - 2cd + 5d^2}$

Solve each problem.

123. A Sanyo color television, mode! AVM-2755, has a rectangular screen with a 21.7-in. width. Its height is 16 in. What is the diagonal of the screen to the nearest tenth of an inch? (*Source:* Actual measurements of the author's television.)

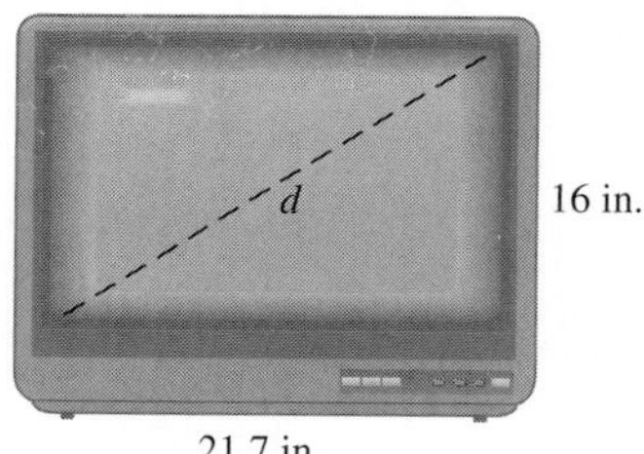

27.0 in.

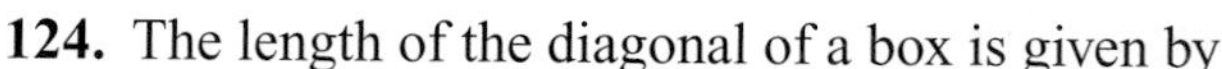

124. The length of the diagonal of a box is given by

$$D = \sqrt{L^2 + W^2 + H^2},$$

where L, W, and H are the length, width, and height of the box. Find the length of the diagonal, D, of a box that is 4 ft long, 3 ft high, and 2 ft wide. Give the exact value, and then round to the nearest tenth of a foot.

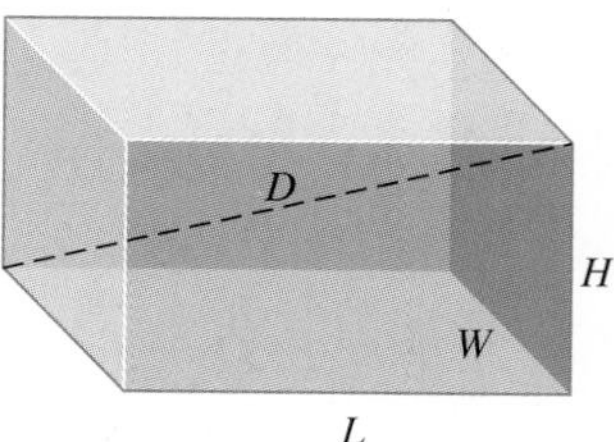

$\sqrt{29}$ ft; 5.4 ft

125. A formula from electronics dealing with impedance of parallel resonant circuits is

$$I = \frac{E}{\sqrt{R^2 + \omega^2 L^2}},$$

where the variables are in appropriate units. Find I if $E = 282$, $R = 100$, $L = 264$, and $\omega = 120\pi$. Give your answer to the nearest thousandth.

0.003

126. In the study of sound, one version of the law of tensions is

$$f_1 = f_2 \sqrt{\frac{F_1}{F_2}}.$$

If $F_1 = 300$, $F_2 = 60$, and $f_2 = 260$, find f_1 to the nearest unit.

581

127. The following letter appeared in the column "Ask Tom Why," written by Tom Skilling of the *Chicago Tribune*.

Dear Tom,

I cannot remember the formula to calculate the distance to the horizon. I have a stunning view from my 14th floor condo, 150 feet above the ground. How far can I see?

Ted Fleischaker; Indianapolis, Ind.

Skilling's answer was as follows.

To find the distance to the horizon in miles, take the square root of the height of your view in feet and multiply that result by 1.224. Your answer will be the number of miles to the horizon. (*Source: Chicago Tribune,* August 17, 2002.)

Assuming Ted's eyes are 6 ft above the ground, the total height from the ground is $150 + 6 = 156$ ft. To the nearest tenth of a mile, how far can he see to the horizon?

15.3 mi

9.4 Adding and Subtracting Radical Expressions

OBJECTIVE 1 Simplify radical expressions involving addition and subtraction. Expressions such as $4\sqrt{2} + 3\sqrt{2}$ and $2\sqrt{3} - 5\sqrt{3}$ can be simplified by using the distributive property.

$$\left.\begin{aligned} &4\sqrt{2} + 3\sqrt{2} \\ &= (4 + 3)\sqrt{2} = 7\sqrt{2} \end{aligned}\right\} \text{This is similar to simplifying } 4x + 3x \text{ as } 7x.$$

$$\left.\begin{aligned} &2\sqrt{3} - 5\sqrt{3} \\ &= (2 - 5)\sqrt{3} = -3\sqrt{3} \end{aligned}\right\} \text{This is similar to simplifying } 2x - 5x \text{ as } -3x.$$

CAUTION
Only radical expressions with the same index and the same radicand may be combined. Expressions such as $5\sqrt{3} + 2\sqrt{2}$ or $3\sqrt{3} + 2\sqrt[3]{3}$ cannot be simplified by combining terms.

EXAMPLE 1 Adding and Subtracting Radicals

Add or subtract to simplify each radical expression.

(a) $3\sqrt{24} + \sqrt{54}$

Simplify each radical; then use the distributive property to combine terms.

$$\begin{aligned} &3\sqrt{24} + \sqrt{54} \\ &= 3\sqrt{4} \cdot \sqrt{6} + \sqrt{9} \cdot \sqrt{6} && \text{Product rule} \\ &= 3 \cdot 2\sqrt{6} + 3\sqrt{6} \\ &= 6\sqrt{6} + 3\sqrt{6} \\ &= 9\sqrt{6} && \text{Combine like terms.} \end{aligned}$$

(b) $2\sqrt{20x} - \sqrt{45x}, \quad x \geq 0$

$$\begin{aligned} &= 2\sqrt{4} \cdot \sqrt{5x} - \sqrt{9} \cdot \sqrt{5x} && \text{Product rule} \\ &= 2 \cdot 2\sqrt{5x} - 3\sqrt{5x} \\ &= 4\sqrt{5x} - 3\sqrt{5x} \\ &= \sqrt{5x} && \text{Combine like terms.} \end{aligned}$$

(c) $2\sqrt{3} - 4\sqrt{5}$

Here the radicals differ and are already simplified, so $2\sqrt{3} - 4\sqrt{5}$ cannot be simplified further.

Work Problem **1** *at the Side.* ▶

CAUTION
Do not confuse the product rule with combining like terms. ***The root of a sum does not equal the sum of the roots.*** For example,

$$\sqrt{9 + 16} \neq \sqrt{9} + \sqrt{16},$$

since $\sqrt{9 + 16} = \sqrt{25} = 5$, but $\sqrt{9} + \sqrt{16} = 3 + 4 = 7$.

OBJECTIVE

1 **Simplify radical expressions involving addition and subtraction.**

1 Add or subtract to simplify each radical expression.

(a) $3\sqrt{5} + 7\sqrt{5}$

(b) $2\sqrt{11} - \sqrt{11} + 3\sqrt{44}$

(c) $5\sqrt{12y} + 6\sqrt{75y}, \; y \geq 0$

(d) $3\sqrt{8} - 6\sqrt{50} + 2\sqrt{200}$

(e) $9\sqrt{5} - 4\sqrt{10}$

ANSWERS

1. (a) $10\sqrt{5}$ **(b)** $7\sqrt{11}$ **(c)** $40\sqrt{3y}$ **(d)** $-4\sqrt{2}$ **(e)** cannot be simplified further

EXAMPLE 2 **Adding and Subtracting Radicals**

Simplify. Assume that all variables represent positive real numbers.

(a) $2\sqrt[3]{16} - 5\sqrt[3]{54}$ — Remember to write the index with each radical.

$= 2\sqrt[3]{8 \cdot 2} - 5\sqrt[3]{27 \cdot 2}$ Factor.

$= 2\sqrt[3]{8} \cdot \sqrt[3]{2} - 5\sqrt[3]{27} \cdot \sqrt[3]{2}$ Product rule

$= 2 \cdot 2 \cdot \sqrt[3]{2} - 5 \cdot 3 \cdot \sqrt[3]{2}$

$= 4\sqrt[3]{2} - 15\sqrt[3]{2}$

$= (4 - 15)\sqrt[3]{2}$ Distributive property

$= -11\sqrt[3]{2}$ Combine like terms.

(b) $2\sqrt[3]{x^2y} + \sqrt[3]{8x^5y^4}$

$= 2\sqrt[3]{x^2y} + \sqrt[3]{(8x^3y^3)x^2y}$ Factor.

$= 2\sqrt[3]{x^2y} + \sqrt[3]{8x^3y^3} \cdot \sqrt[3]{x^2y}$ Product rule

$= 2\sqrt[3]{x^2y} + 2xy\sqrt[3]{x^2y}$

$= (2 + 2xy)\sqrt[3]{x^2y}$ Distributive property — This result cannot be simplified further.

◀ Work Problem 2 at the Side.

2 Simplify. Assume that all variables represent positive real numbers.

(a) $7\sqrt[3]{81} + 3\sqrt[3]{24}$

(b) $-2\sqrt[4]{32} - 7\sqrt[4]{162}$

(c) $\sqrt[3]{p^4q^7} - \sqrt[3]{64pq}$

EXAMPLE 3 **Adding and Subtracting Radicals with Fractions**

Simplify. Assume that all variables represent positive real numbers.

(a) $2\sqrt{\frac{75}{16}} + 4\frac{\sqrt{8}}{\sqrt{32}}$

$= 2\frac{\sqrt{25 \cdot 3}}{\sqrt{16}} + 4\frac{\sqrt{4 \cdot 2}}{\sqrt{16 \cdot 2}}$ Quotient rule; factor.

$= 2\left(\frac{5\sqrt{3}}{4}\right) + 4\left(\frac{2\sqrt{2}}{4\sqrt{2}}\right)$ Product rule; take square roots.

$= \frac{5\sqrt{3}}{2} + 2$ Multiply; $\frac{\sqrt{2}}{\sqrt{2}} = 1$.

$= \frac{5\sqrt{3}}{2} + \frac{4}{2}$ Write with a common denominator.

$= \frac{5\sqrt{3} + 4}{2}$ Add fractions.

(b) $10\sqrt[3]{\frac{5}{x^6}} - 3\sqrt[3]{\frac{4}{x^9}}$

$= 10\frac{\sqrt[3]{5}}{\sqrt[3]{x^6}} - 3\frac{\sqrt[3]{4}}{\sqrt[3]{x^9}}$ Quotient rule

$= \frac{10\sqrt[3]{5}}{x^2} - \frac{3\sqrt[3]{4}}{x^3}$ Simplify denominators.

$= \frac{10\sqrt[3]{5} \cdot x}{x^2 \cdot x} - \frac{3\sqrt[3]{4}}{x^3}$ Write with a common denominator.

$= \frac{10x\sqrt[3]{5} - 3\sqrt[3]{4}}{x^3}$ Subtract fractions.

◀ Work Problem 3 at the Side.

3 Simplify. Assume that all variables represent positive real numbers.

(a) $2\sqrt{\frac{8}{9}} - 2\frac{\sqrt{27}}{\sqrt{108}}$

(b) $\sqrt{\frac{80}{y^4}} + \sqrt{\frac{81}{y^{10}}}$

Answers

2. (a) $27\sqrt[3]{3}$ (b) $-25\sqrt[4]{2}$ (c) $(pq^2 - 4)\sqrt[3]{pq}$

3. (a) $\frac{4\sqrt{2} - 3}{3}$ (b) $\frac{4y^3\sqrt{5} + 9}{y^5}$

9.4 Exercises

FOR EXTRA HELP **MyMathLab**

Math XL PRACTICE  WATCH  DOWNLOAD  READ  REVIEW

1. Which one of the following sums could be simplified without first simplifying the individual radical expressions?

A. $\sqrt{50} + \sqrt{32}$ **B.** $3\sqrt{6} + 9\sqrt{6}$ **C.** $\sqrt[3]{32} - \sqrt[3]{108}$ **D.** $\sqrt[5]{6} - \sqrt[5]{192}$ **B**

2. Let $a = 1$ and $b = 64$.

(a) Evaluate $\sqrt{a} + \sqrt{b}$. Then find $\sqrt{a+b}$. Are they equal? **9; $\sqrt{65}$; no**

(b) Evaluate $\sqrt[3]{a} + \sqrt[3]{b}$. Then find $\sqrt[3]{a+b}$. Are they equal? **5; $\sqrt[3]{65}$; no**

(c) Complete the following: In general, $\sqrt[n]{a} + \sqrt[n]{b} \neq$ **$\sqrt[n]{a+b}$**, based on the observations in parts (a) and (b) of this exercise.

3. Even though the indexes of the terms are not equal, the sum $\sqrt{64} + \sqrt[3]{125} + \sqrt[4]{16}$ can be simplified quite easily. What is this sum? Why can these terms be combined so easily?

15; Each radical expression simplifies to a whole number.

4. Explain why $28 - 4\sqrt{2}$ ***is not equal to*** $24\sqrt{2}$. (This is a common error among algebra students.)

We cannot group 28 − 4 here. Multiplication must take place before subtraction, according to the order of operations. (Only like terms can be combined.)

Simplify. Assume that all variables represent positive real numbers. See Examples 1 and 2.

5. $\sqrt{36} - \sqrt{100}$ **-4**

6. $\sqrt{25} - \sqrt{81}$ **-4**

7. $-2\sqrt{48} + 3\sqrt{75}$ **$7\sqrt{3}$**

8. $4\sqrt{32} - 2\sqrt{8}$ **$12\sqrt{2}$**

9. $\sqrt[3]{16} + 4\sqrt[3]{54}$ **$14\sqrt[3]{2}$**

10. $3\sqrt[3]{24} - 2\sqrt[3]{192}$ **$-2\sqrt[3]{3}$**

11. $\sqrt[4]{32} + 3\sqrt[4]{2}$ **$5\sqrt[4]{2}$**

12. $\sqrt[4]{405} - 2\sqrt[4]{5}$ **$\sqrt[4]{5}$**

13. $6\sqrt{18} - \sqrt{32} + 2\sqrt{50}$ **$24\sqrt{2}$**

14. $5\sqrt{8} + 3\sqrt{72} - 3\sqrt{50}$ **$13\sqrt{2}$**

15. $5\sqrt{6} + 2\sqrt{10}$ **cannot be simplified further**

16. $3\sqrt{11} - 5\sqrt{13}$ **cannot be simplified further**

17. $2\sqrt{5} + 3\sqrt{20} + 4\sqrt{45}$ **$20\sqrt{5}$**

18. $5\sqrt{54} - 2\sqrt{24} - 2\sqrt{96}$ **$3\sqrt{6}$**

19. $8\sqrt{2x} - \sqrt{8x} + \sqrt{72x}$ **$12\sqrt{2x}$**

20. $4\sqrt{18k} - \sqrt{72k} + \sqrt{50k}$ **$11\sqrt{2k}$**

21. $3\sqrt{72m^2} - 5\sqrt{32m^2}$ **$-2m\sqrt{2}$**

22. $9\sqrt{27p^2} - 14\sqrt{108p^2}$ **$-57p\sqrt{3}$**

23. $-\sqrt[3]{54} + 2\sqrt[3]{16}$ **$\sqrt[3]{2}$**

24. $15\sqrt[3]{81} - 4\sqrt[3]{24}$ **$37\sqrt[3]{3}$**

25. $2\sqrt[3]{27x} - 2\sqrt[3]{8x}$ **$2\sqrt[3]{x}$**

26. $6\sqrt[3]{128m} + 3\sqrt[3]{16m}$ **$30\sqrt[3]{2m}$**

27. $\sqrt[3]{x^2y} - \sqrt[3]{8x^2y}$ **$-\sqrt[3]{x^2y}$**

28. $3\sqrt[3]{x^2y^2} - 2\sqrt[3]{64x^2y^2}$ **$-5\sqrt[3]{x^2y^2}$**

29. $3x\sqrt[3]{xy^2} - 2\sqrt[3]{8x^4y^2}$ **$-x\sqrt[3]{xy^2}$**

30. $6q^2\sqrt[3]{5q} - 2q\sqrt[3]{40q^4}$ **$2q^2\sqrt[3]{5q}$**

31. $5\sqrt[4]{32} + 3\sqrt[4]{162}$
$19\sqrt[4]{2}$

32. $2\sqrt[4]{512} + 4\sqrt[4]{32}$
$16\sqrt[4]{2}$

33. $3\sqrt[4]{x^5y} - 2x\sqrt[4]{xy}$
$x\sqrt[4]{xy}$

34. $2\sqrt[4]{m^9p^6} - 3m^2p\sqrt[4]{mp^2}$
$-m^2p\sqrt[4]{mp^2}$

35. $2\sqrt[4]{32a^3} + 5\sqrt[4]{2a^3}$
$9\sqrt[4]{2a^3}$

36. $-\sqrt[4]{16r} + 5\sqrt[4]{r}$
$3\sqrt[4]{r}$

Simplify. Assume that all variables represent positive real numbers. See Example 3.

37. $\frac{2\sqrt{5}}{3} + \frac{\sqrt{5}}{6}$
$\frac{5\sqrt{5}}{6}$

38. $\frac{4\sqrt{3}}{3} + \frac{2\sqrt{3}}{9}$
$\frac{14\sqrt{3}}{9}$

39. $\sqrt{\frac{8}{9}} + \sqrt{\frac{18}{36}}$
$\frac{7\sqrt{2}}{6}$

40. $\sqrt{\frac{12}{16}} + \sqrt{\frac{48}{64}}$
$\sqrt{3}$

41. $\frac{\sqrt{32}}{3} + \frac{2\sqrt{2}}{3} - \frac{\sqrt{2}}{\sqrt{9}}$
$\frac{5\sqrt{2}}{3}$

42. $\frac{\sqrt{27}}{2} - \frac{3\sqrt{3}}{2} + \frac{\sqrt{3}}{\sqrt{4}}$
$\frac{\sqrt{3}}{2}$

43. $3\sqrt{\frac{50}{9}} + 8\frac{\sqrt{2}}{\sqrt{8}}$
$5\sqrt{2} + 4$

44. $9\sqrt{\frac{48}{25}} - 2\frac{\sqrt{2}}{\sqrt{98}}$
$\frac{252\sqrt{3} - 10}{35}$

45. $\sqrt{\frac{25}{x^8}} - \sqrt{\frac{9}{x^6}}$
$\frac{5 - 3x}{x^4}$

46. $\sqrt{\frac{100}{y^4}} + \sqrt{\frac{81}{y^{10}}}$
$\frac{10y^3 + 9}{y^5}$

47. $3\sqrt[3]{\frac{m^5}{27}} - 2m\sqrt[3]{\frac{m^2}{64}}$
$\frac{m\sqrt[3]{m^2}}{2}$

48. $2a\sqrt[4]{\frac{a}{16}} - 5a\sqrt[4]{\frac{a}{81}}$
$-\frac{2a\sqrt[4]{a}}{3}$

49. $3\sqrt[3]{\frac{2}{x^6}} - 4\sqrt[3]{\frac{5}{x^9}}$
$\frac{3x\sqrt[3]{2} - 4\sqrt[3]{5}}{x^3}$

50. $-4\sqrt[3]{\frac{4}{t^9}} + 3\sqrt[3]{\frac{9}{t^{12}}}$
$\frac{-4t\sqrt[3]{4} + 3\sqrt[3]{9}}{t^4}$

Solve each problem. Give answers as simplified radical expressions.

51. Find the perimeter of the triangle.

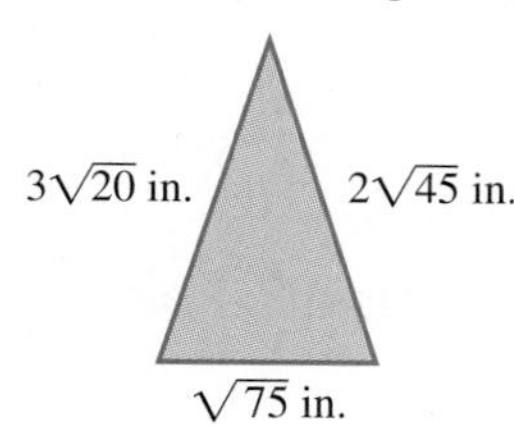

$(12\sqrt{5} + 5\sqrt{3})$ in.

52. Find the perimeter of the rectangle.

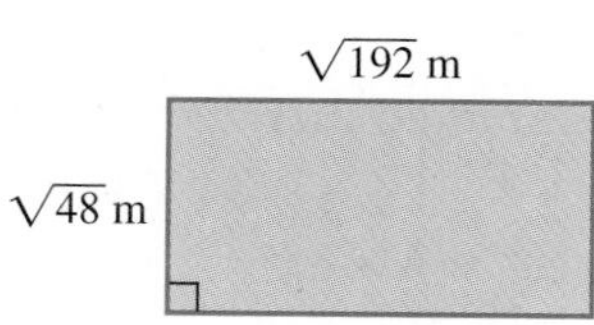

$24\sqrt{3}$ m

53. What is the perimeter of the computer graphic?

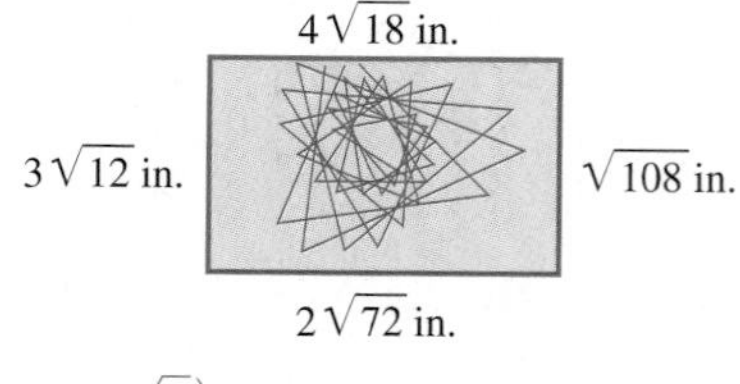

$(24\sqrt{2} + 12\sqrt{3})$ in.

54. Find the area of the trapezoid.

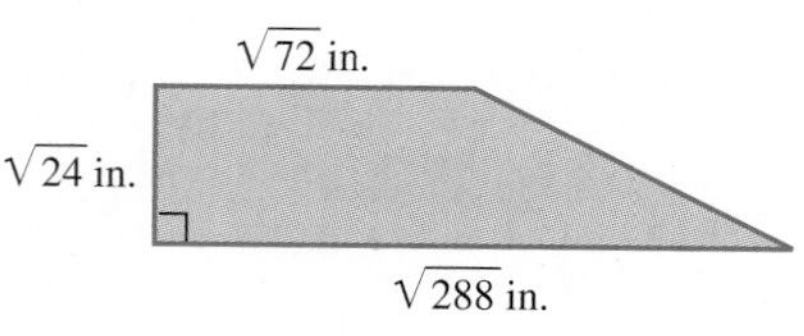

$36\sqrt{3}$ in.^{2}

9.5 Multiplying and Dividing Radical Expressions

OBJECTIVES

1. Multiply radical expressions.
2. Rationalize denominators with one radical term.
3. Rationalize denominators with binomials involving radicals.
4. Write radical quotients in lowest terms.

OBJECTIVE 1 Multiply radical expressions. We multiply binomial expressions involving radicals by using the FOIL method from **Section 5.3.** For example, we find the product of $\sqrt{5}+3$ and $\sqrt{6}+1$ as follows:

$$\left(\sqrt{5}+3\right)\left(\sqrt{6}+1\right)$$

$$= \underbrace{\sqrt{5}\cdot\sqrt{6}}_{\text{First}} + \underbrace{\sqrt{5}\cdot 1}_{\text{Outer}} + \underbrace{3\cdot\sqrt{6}}_{\text{Inner}} + \underbrace{3\cdot 1}_{\text{Last}}$$

$$= \sqrt{30} + \sqrt{5} + 3\sqrt{6} + 3.$$

This result cannot be simplified further.

EXAMPLE 1 Multiplying Binomials Involving Radical Expressions

Multiply, using the FOIL method.

(a) $\left(7-\sqrt{3}\right)\left(\sqrt{5}+\sqrt{2}\right)$

F O I L

$$= 7\sqrt{5} + 7\sqrt{2} - \sqrt{3}\cdot\sqrt{5} - \sqrt{3}\cdot\sqrt{2}$$

$$= 7\sqrt{5} + 7\sqrt{2} - \sqrt{15} - \sqrt{6}$$

(b) $\left(\sqrt{10}+\sqrt{3}\right)\left(\sqrt{10}-\sqrt{3}\right)$

$$= \sqrt{10}\cdot\sqrt{10} - \sqrt{10}\cdot\sqrt{3} + \sqrt{3}\cdot\sqrt{10} - \sqrt{3}\cdot\sqrt{3}$$

$$= 10 - 3$$

$$= 7$$

The product $\left(\sqrt{10}+\sqrt{3}\right)\left(\sqrt{10}-\sqrt{3}\right) = \left(\sqrt{10}\right)^2 - \left(\sqrt{3}\right)^2$ is the difference of squares:

$$(x+y)(x-y) = x^2 - y^2. \quad \text{Here, } x = \sqrt{10} \text{ and } y = \sqrt{3}.$$

(c) $\left(\sqrt{7}-3\right)^2$

$$= \left(\sqrt{7}-3\right)\left(\sqrt{7}-3\right)$$

$$= \sqrt{7}\cdot\sqrt{7} - 3\sqrt{7} - 3\sqrt{7} + 3\cdot 3$$

$$= 7 - 6\sqrt{7} + 9$$

$$= 16 - 6\sqrt{7}$$

Be careful! These terms cannot be combined.

(d) $\left(5-\sqrt[3]{3}\right)\left(5+\sqrt[3]{3}\right)$

$$= 5\cdot 5 + 5\sqrt[3]{3} - 5\sqrt[3]{3} - \sqrt[3]{3}\cdot\sqrt[3]{3}$$

$$= 25 - \sqrt[3]{3^2}$$

$$= 25 - \sqrt[3]{9}$$

Remember to write the index 3 in *each* radical.

(e) $\left(\sqrt{k}+\sqrt{y}\right)\left(\sqrt{k}-\sqrt{y}\right)$

$$= \left(\sqrt{k}\right)^2 - \left(\sqrt{y}\right)^2 \quad \text{Difference of squares}$$

$$= k - y, \quad k \geq 0 \text{ and } y \geq 0$$

1 Multiply, using the FOIL method.

(a) $(2 + \sqrt{3})(1 + \sqrt{5})$

(b) $(4 + \sqrt{3})(4 - \sqrt{3})$

(c) $(\sqrt{13} - 2)^2$

(d) $(4 + \sqrt[3]{7})(4 - \sqrt[3]{7})$

(e) $(\sqrt{p} + \sqrt{s})(\sqrt{p} - \sqrt{s})$, $p \geq 0$ and $s \geq 0$

Note

In Example 1(c) we could have used the formula for the square of a binomial,

$$(x - y)^2 = x^2 - 2xy + y^2,$$

to obtain the same result:

$$\begin{aligned} &(\sqrt{7} - 3)^2 \\ &= (\sqrt{7})^2 - 2(\sqrt{7})(3) + 3^2 \\ &= 7 - 6\sqrt{7} + 9 \\ &= 16 - 6\sqrt{7}. \end{aligned}$$

◀ *Work Problem* **1** *at the Side.*

OBJECTIVE 2 Rationalize denominators with one radical term. As defined earlier, a simplified radical expression will have no radical in the denominator. The origin of this agreement no doubt occurred before the days of high-speed calculation, when computation was a tedious process performed by hand.

For example, consider the radical expression $\frac{1}{\sqrt{2}}$. To find a decimal approximation by hand, it would be necessary to divide 1 by a decimal approximation for $\sqrt{2}$, such as 1.414. It would be much easier if the divisor were a whole number. This can be accomplished by multiplying $\frac{1}{\sqrt{2}}$ by 1 in the form $\frac{\sqrt{2}}{\sqrt{2}}$. ***Multiplying by 1 in any form does not change the value of the original expression.***

$$\frac{1}{\sqrt{2}} \cdot \frac{\sqrt{2}}{\sqrt{2}} = \frac{\sqrt{2}}{2} \qquad \text{Multiply by 1; } \frac{\sqrt{2}}{\sqrt{2}} = 1.$$

Now the computation would require dividing 1.414 by 2 to obtain 0.707, a much easier task.

With current technology, either form of this fraction can be approximated with the same number of keystrokes. See Figure 9, which shows how a calculator gives the same approximation for both forms of the expression.

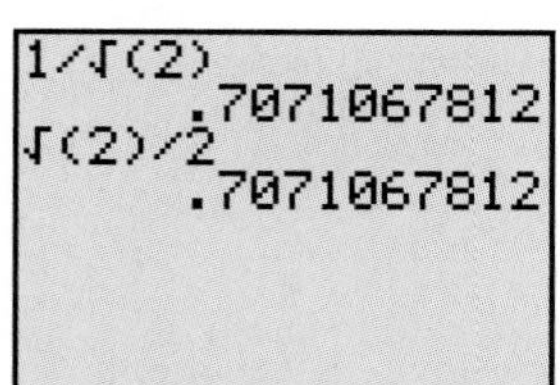

Figure 9

Rationalizing a Denominator

A common way of "standardizing" the form of a radical expression is to have the denominator contain no radicals. The process of removing radicals from a denominator so that the denominator contains only rational numbers is called **rationalizing the denominator.**

ANSWERS

1. **(a)** $2 + 2\sqrt{5} + \sqrt{3} + \sqrt{15}$ **(b)** 13 **(c)** $17 - 4\sqrt{13}$ **(d)** $16 - \sqrt[3]{49}$ **(e)** $p - s$

EXAMPLE 2 Rationalizing Denominators with Square Roots

Rationalize each denominator.

(a) $\frac{3}{\sqrt{7}}$

Multiply by $\frac{\sqrt{7}}{\sqrt{7}}$. This is an application of the multiplicative identity property.

$$\frac{3}{\sqrt{7}} = \frac{3 \cdot \sqrt{7}}{\sqrt{7} \cdot \sqrt{7}} = \frac{3\sqrt{7}}{7}$$

In the denominator, $\sqrt{7} \cdot \sqrt{7} = \sqrt{7 \cdot 7} = \sqrt{49} = 7$. The final denominator is now a rational number.

(b) $\frac{5\sqrt{2}}{\sqrt{5}} = \frac{5\sqrt{2} \cdot \sqrt{5}}{\sqrt{5} \cdot \sqrt{5}} = \frac{5\sqrt{10}}{5} = \sqrt{10}$

(c) $\frac{-6}{\sqrt{12}}$

Less work is involved if the radical in the denominator is simplified first.

$$\frac{-6}{\sqrt{12}} = \frac{-6}{\sqrt{4 \cdot 3}} = \frac{-6}{2\sqrt{3}} = \frac{-3}{\sqrt{3}}$$

Now rationalize the denominator.

$$\frac{-3}{\sqrt{3}} = \frac{-3 \cdot \sqrt{3}}{\sqrt{3} \cdot \sqrt{3}} = \frac{-3\sqrt{3}}{3} = -\sqrt{3}$$

Work Problem 2 at the Side. ▶

2 Rationalize each denominator.

(a) $\frac{8}{\sqrt{3}}$

(b) $\frac{5\sqrt{6}}{\sqrt{5}}$

(c) $\frac{3}{\sqrt{48}}$

(d) $\frac{-16}{\sqrt{32}}$

EXAMPLE 3 Rationalizing Denominators in Roots of Fractions

Simplify each radical. Assume that all variables represent positive real numbers.

(a) $\sqrt{\frac{18}{125}}$

$= \frac{\sqrt{18}}{\sqrt{125}}$ Quotient rule

$= \frac{\sqrt{9 \cdot 2}}{\sqrt{25 \cdot 5}}$ Factor.

$= \frac{3\sqrt{2}}{5\sqrt{5}}$ Product rule

$= \frac{3\sqrt{2} \cdot \sqrt{5}}{5\sqrt{5} \cdot \sqrt{5}}$ Multiply by $\frac{\sqrt{5}}{\sqrt{5}}$.

$= \frac{3\sqrt{10}}{5 \cdot 5}$ Product rule

$= \frac{3\sqrt{10}}{25}$ Multiply.

Continued on Next Page

ANSWERS

2. (a) $\frac{8\sqrt{3}}{3}$ (b) $\sqrt{30}$ (c) $\frac{\sqrt{3}}{4}$ (d) $-2\sqrt{2}$

3 Simplify each radical. Assume that all variables represent positive real numbers.

(a) $\sqrt{\frac{8}{45}}$

(b) $\sqrt{\frac{72}{y}}$

(c) $\sqrt{\frac{200k^6}{y^7}}$

4 Simplify.

(a) $\sqrt[3]{\frac{15}{32}}$

(b) $\sqrt[3]{\frac{m^{12}}{n}}, \quad n \neq 0$

(c) $\sqrt[4]{\frac{6y}{w^2}}, \quad y \geq 0, w \neq 0$

Answers

3. (a) $\frac{2\sqrt{10}}{15}$ **(b)** $\frac{6\sqrt{2y}}{y}$ **(c)** $\frac{10k^3\sqrt{2y}}{y^4}$

4. (a) $\frac{\sqrt[3]{30}}{4}$ **(b)** $\frac{m^4\sqrt[3]{n^2}}{n}$ **(c)** $\frac{\sqrt[4]{6yw^2}}{w}$

(b) $\sqrt{\frac{50m^4}{p^5}}$

$$= \frac{\sqrt{50m^4}}{\sqrt{p^5}} \quad \text{Quotient rule}$$

$$= \frac{\sqrt{25m^4 \cdot 2}}{\sqrt{p^4 \cdot p}} \quad \text{Factor.}$$

$$= \frac{5m^2\sqrt{2}}{p^2\sqrt{p}} \quad \text{Product rule}$$

$$= \frac{5m^2\sqrt{2} \cdot \sqrt{p}}{p^2\sqrt{p} \cdot \sqrt{p}} \quad \text{Multiply by } \frac{\sqrt{p}}{\sqrt{p}}.$$

$$= \frac{5m^2\sqrt{2p}}{p^2 \cdot p} \quad \text{Product rule}$$

$$= \frac{5m^2\sqrt{2p}}{p^3} \quad \text{Multiply.}$$

◀ *Work Problem* **3** *at the Side.*

EXAMPLE 4 Rationalizing Denominators with Higher Roots

Simplify.

(a) $\sqrt[3]{\frac{27}{16}}$

Use the quotient rule and simplify the numerator and denominator.

$$\sqrt[3]{\frac{27}{16}} = \frac{\sqrt[3]{27}}{\sqrt[3]{16}} = \frac{3}{\sqrt[3]{8} \cdot \sqrt[3]{2}} = \frac{3}{2\sqrt[3]{2}}$$

To get a rational denominator, multiply the numerator and denominator by a number that will result in a perfect cube in the radicand in the denominator. Since $2 \cdot 4 = 8$, a perfect cube, apply the multiplicative identity property and multiply the numerator and denominator by $\sqrt[3]{4}$.

$$\sqrt[3]{\frac{27}{16}} = \frac{3}{2\sqrt[3]{2}} = \frac{3 \cdot \sqrt[3]{4}}{2\sqrt[3]{2} \cdot \sqrt[3]{4}} = \frac{3\sqrt[3]{4}}{2\sqrt[3]{8}} = \frac{3\sqrt[3]{4}}{2 \cdot 2} = \frac{3\sqrt[3]{4}}{4}$$

(b) $$\sqrt[4]{\frac{5x}{z}} = \frac{\sqrt[4]{5x} \cdot \sqrt[4]{z^3}}{\sqrt[4]{z} \cdot \sqrt[4]{z^3}} = \frac{\sqrt[4]{5xz^3}}{\sqrt[4]{z^4}} = \frac{\sqrt[4]{5xz^3}}{z}, \quad x \geq 0, z > 0$$

CAUTION

In Example 4(a), a typical error is to multiply the numerator and denominator by $\sqrt[3]{2}$, forgetting that $\sqrt[3]{2} \cdot \sqrt[3]{2} = \sqrt[3]{2^2}$, which does ***not*** equal 2. We need ***three*** factors of 2 to get 2^3 under the radical.

$$\sqrt[3]{2} \cdot \sqrt[3]{2} \cdot \sqrt[3]{2} = \sqrt[3]{2^3}, \quad \text{which does equal} \quad 2.$$

◀ *Work Problem* **4** *at the Side.*

OBJECTIVE 3 Rationalize denominators with binomials involving radicals. Recall the special product

$$(x + y)(x - y) = x^2 - y^2.$$

To rationalize a denominator that contains a binomial expression (one that contains exactly two terms) involving radicals, such as

$$\frac{3}{1 + \sqrt{2}},$$

we must use *conjugates*. The conjugate of $1 + \sqrt{2}$ is $1 - \sqrt{2}$. In general, $x + y$ and $x - y$ are **conjugates.**

Rationalizing a Binomial Denominator

If a radical expression has a sum or difference with square root radicals in the denominator, rationalize the denominator by multiplying both the numerator and denominator by the conjugate of the denominator.

For $\frac{3}{1 + \sqrt{2}}$, we rationalize the denominator by multiplying both the numerator and denominator by $1 - \sqrt{2}$, the conjugate of the denominator.

$$\begin{aligned}
&\frac{3}{1 + \sqrt{2}} \\
&= \frac{3(1 - \sqrt{2})}{(1 + \sqrt{2})(1 - \sqrt{2})} \\
&= \frac{3(1 - \sqrt{2})}{-1} \\
&= \frac{3}{-1}(1 - \sqrt{2}) \\
&= -3(1 - \sqrt{2}), \quad \text{or} \quad -3 + 3\sqrt{2}
\end{aligned}$$

The denominator is now a rational number.

$$\begin{aligned}
&(1 + \sqrt{2})(1 - \sqrt{2}) \\
&= 1^2 - (\sqrt{2})^2 \\
&= 1 - 2 = -1
\end{aligned}$$

EXAMPLE 5 Rationalizing Binomial Denominators

Rationalize each denominator.

(a) $\frac{5}{4 - \sqrt{3}}$

$= \frac{5(4 + \sqrt{3})}{(4 - \sqrt{3})(4 + \sqrt{3})}$ Multiply the numerator and denominator by $4 + \sqrt{3}$.

$= \frac{5(4 + \sqrt{3})}{16 - 3}$ Multiply in the denominator.

$= \frac{5(4 + \sqrt{3})}{13}$ Subtract.

Notice that the numerator is left in factored form. This makes it easier to determine whether the expression is written in lowest terms.

Continued on Next Page

(b) $\dfrac{\sqrt{2}-\sqrt{3}}{\sqrt{5}+\sqrt{3}}$

$= \dfrac{(\sqrt{2}-\sqrt{3})(\sqrt{5}-\sqrt{3})}{(\sqrt{5}+\sqrt{3})(\sqrt{5}-\sqrt{3})}$ Multiply the numerator and denominator by $\sqrt{5}-\sqrt{3}$.

$= \dfrac{\sqrt{10}-\sqrt{6}-\sqrt{15}+3}{5-3}$ Multiply.

$= \dfrac{\sqrt{10}-\sqrt{6}-\sqrt{15}+3}{2}$ Subtract in the denominator.

(c) $\dfrac{3}{\sqrt{5m}-\sqrt{p}}$, $5m \neq p, m > 0, p > 0$

$= \dfrac{3(\sqrt{5m}+\sqrt{p})}{(\sqrt{5m}-\sqrt{p})(\sqrt{5m}+\sqrt{p})}$

$= \dfrac{3(\sqrt{5m}+\sqrt{p})}{5m-p}$

◀ Work Problem **5** at the Side.

OBJECTIVE 4 **Write radical quotients in lowest terms.**

EXAMPLE 6 **Writing Radical Quotients in Lowest Terms**

Write each quotient in lowest terms.

(a) $\dfrac{6+2\sqrt{5}}{4}$

$= \dfrac{2(3+\sqrt{5})}{2 \cdot 2}$ Factor the numerator and denominator. (This is a key step.)

$= \dfrac{3+\sqrt{5}}{2}$ Divide out the common factor.

Here is an alternative method for writing this expression in lowest terms.

$$\frac{6+2\sqrt{5}}{4} = \frac{6}{4} + \frac{2\sqrt{5}}{4} = \frac{3}{2} + \frac{\sqrt{5}}{2} = \frac{3+\sqrt{5}}{2}$$

(b) $\dfrac{5y-\sqrt{8y^2}}{6y}$, $y > 0$

$= \dfrac{5y-2y\sqrt{2}}{6y}$ $\sqrt{8y^2} = \sqrt{4y^2 \cdot 2} = 2y\sqrt{2}$

$= \dfrac{y(5-2\sqrt{2})}{6y}$ Factor the numerator.

$= \dfrac{5-2\sqrt{2}}{6}$ Divide out the common factor.

◀ Work Problem **6** at the Side.

5 Rationalize each denominator.

(a) $\dfrac{-4}{\sqrt{5}+2}$

(b) $\dfrac{15}{\sqrt{7}+\sqrt{2}}$

(c) $\dfrac{\sqrt{3}+\sqrt{5}}{\sqrt{2}-\sqrt{7}}$

(d) $\dfrac{2}{\sqrt{k}+\sqrt{z}}$, $k \neq z, k > 0, z > 0$

6 Write each quotient in lowest terms.

(a) $\dfrac{24-36\sqrt{7}}{16}$

(b) $\dfrac{2x+\sqrt{32x^2}}{6x}$, $x > 0$

Answers

5. (a) $-4(\sqrt{5}-2)$ **(b)** $3(\sqrt{7}-\sqrt{2})$
(c) $\dfrac{-(\sqrt{6}+\sqrt{21}+\sqrt{10}+\sqrt{35})}{5}$
(d) $\dfrac{2(\sqrt{k}-\sqrt{z})}{k-z}$

6. (a) $\dfrac{6-9\sqrt{7}}{4}$ **(b)** $\dfrac{1+2\sqrt{2}}{3}$

9.5 Exercises

FOR EXTRA HELP

MyMathLab

Math XL PRACTICE

WATCH

DOWNLOAD

READ

REVIEW

Match each part of a rule for a special product in Column I with the part it equals in Column II.

I	II
1. $(x+\sqrt{y})(x-\sqrt{y})$ E	**A.** $x-y$
2. $(\sqrt{x}+y)(\sqrt{x}-y)$ C	**B.** $x+2y\sqrt{x}+y^2$
3. $(\sqrt{x}+\sqrt{y})(\sqrt{x}-\sqrt{y})$ A	**C.** $x-y^2$
4. $(\sqrt{x}+\sqrt{y})^2$ F	**D.** $x-2\sqrt{xy}+y$
5. $(\sqrt{x}-\sqrt{y})^2$ D	**E.** x^2-y
6. $(\sqrt{x}+y)^2$ B	**F.** $x+2\sqrt{xy}+y$

Multiply, and then simplify each product. Assume that all variables represent positive real numbers. See Example 1.

7. $\sqrt{3}(\sqrt{12}-4)$
$6-4\sqrt{3}$

8. $\sqrt{5}(\sqrt{125}-6)$
$25-6\sqrt{5}$

9. $\sqrt{2}(\sqrt{18}-\sqrt{3})$
$6-\sqrt{6}$

10. $\sqrt{5}(\sqrt{15}+\sqrt{5})$
$5\sqrt{3}+5$

11. $(\sqrt{6}+2)(\sqrt{6}-2)$
2

12. $(\sqrt{7}+8)(\sqrt{7}-8)$
-57

13. $(\sqrt{12}-\sqrt{3})(\sqrt{12}+\sqrt{3})$
9

14. $(\sqrt{18}+\sqrt{8})(\sqrt{18}-\sqrt{8})$
10

15. $(\sqrt{3}+2)(\sqrt{6}-5)$
$3\sqrt{2}-5\sqrt{3}+2\sqrt{6}-10$

16. $(\sqrt{7}+1)(\sqrt{2}-4)$
$\sqrt{14}-4\sqrt{7}+\sqrt{2}-4$

17. $(\sqrt{3x}+2)(\sqrt{3x}-2)$
$3x-4$

18. $(\sqrt{6y}-4)(\sqrt{6y}+4)$
$6y-16$

19. $(2\sqrt{x}+\sqrt{y})(2\sqrt{x}-\sqrt{y})$
$4x-y$

20. $(\sqrt{p}+5\sqrt{s})(\sqrt{p}-5\sqrt{s})$
$p-25s$

21. $(4\sqrt{x}+3)^2$
$16x+24\sqrt{x}+9$

22. $(5\sqrt{p}-6)^2$
$25p-60\sqrt{p}+36$

23. $(9-\sqrt[3]{2})(9+\sqrt[3]{2})$
$81-\sqrt[3]{4}$

24. $(7+\sqrt[3]{6})(7-\sqrt[3]{6})$
$49-\sqrt[3]{36}$

25. The correct answer to Exercise 7 is $6 - 4\sqrt{3}$. Explain why this is not equal to $2\sqrt{3}$.

Because 6 and $4\sqrt{3}$ are not like terms, they cannot be combined.

26. When we rationalize the denominator in the radical expression $\frac{1}{\sqrt{2}}$, we multiply both the numerator and denominator by $\sqrt{2}$. What property of real numbers covered in **Section 1.7** justifies this procedure?

identity property for multiplication

Rationalize the denominator in each expression. Assume that all variables represent positive real numbers. See Example 2.

27. $\frac{7}{\sqrt{7}}$ **$\sqrt{7}$**

28. $\frac{11}{\sqrt{11}}$ **$\sqrt{11}$**

29. $\frac{15}{\sqrt{3}}$ **$5\sqrt{3}$**

30. $\frac{12}{\sqrt{6}}$ **$2\sqrt{6}$**

31. $\frac{\sqrt{3}}{\sqrt{2}}$ **$\frac{\sqrt{6}}{2}$**

32. $\frac{\sqrt{7}}{\sqrt{6}}$ **$\frac{\sqrt{42}}{6}$**

33. $\frac{9\sqrt{3}}{\sqrt{5}}$ **$\frac{9\sqrt{15}}{5}$**

34. $\frac{3\sqrt{2}}{\sqrt{11}}$ **$\frac{3\sqrt{22}}{11}$**

35. $\frac{-6}{\sqrt{18}}$ **$-\sqrt{2}$**

36. $\frac{-5}{\sqrt{24}}$ **$\frac{-5\sqrt{6}}{12}$**

37. $\frac{-8\sqrt{3}}{\sqrt{k}}$ **$\frac{-8\sqrt{3k}}{k}$**

38. $\frac{-4\sqrt{13}}{\sqrt{m}}$ **$\frac{-4\sqrt{13m}}{m}$**

39. $\frac{6\sqrt{3y}}{\sqrt{y^3}}$ **$\frac{6\sqrt{3}}{y}$**

40. $\frac{-8\sqrt{5y}}{\sqrt{y^5}}$ **$\frac{-8\sqrt{5}}{y^2}$**

41. Explain why $\frac{1}{\sqrt[3]{2}}$ would not be written with the denominator rationalized if you begin by multiplying both the numerator and denominator by $\sqrt[3]{2}$. By what should you multiply them both to achieve the desired result?

To rationalize a cube root, three factors of the quantity under the radical sign are needed. We must multiply by $\sqrt[3]{2^2}$, or $\sqrt[3]{4}$, to rationalize $\sqrt[3]{2}$.

42. Look again at the expression in Exercise 39. Start by multiplying both the numerator and the denominator by $\sqrt{y}$, to obtain the final answer. Then start over, multiplying both the numerator and denominator by $\sqrt{y^3}$, to obtain the same answer. Which method do you prefer? Why?

Both methods lead to the same result, $\frac{6\sqrt{3}}{y}$, but multiplying the numerator and denominator by $\sqrt{y}$ produces this result more directly, with less simplification required.

Simplify. Assume that all variables represent positive real numbers. See Examples 3 and 4.

43. $\sqrt{\frac{7}{2}}$ **$\frac{\sqrt{14}}{2}$**

44. $\sqrt{\frac{10}{3}}$ **$\frac{\sqrt{30}}{3}$**

45. $-\sqrt{\frac{7}{50}}$ **$-\frac{\sqrt{14}}{10}$**

46. $-\sqrt{\frac{13}{75}}$ **$-\frac{\sqrt{39}}{15}$**

47. $\sqrt{\frac{24}{x}}$ **$\frac{2\sqrt{6x}}{x}$**

48. $\sqrt{\frac{52}{y}}$ **$\frac{2\sqrt{13y}}{y}$**

49. $-\sqrt{\frac{98r^3}{s}}$ **$-\frac{7r\sqrt{2rs}}{s}$**

50. $-\sqrt{\frac{150m^5}{n}}$ **$-\frac{5m^2\sqrt{6mn}}{n}$**

51. $\sqrt{\frac{288x^7}{y^9}}$ **$\frac{12x^3\sqrt{2xy}}{y^5}$**

52. $\sqrt{\frac{242t^9}{u^{11}}}$ **$\frac{11t^4\sqrt{2tu}}{u^6}$**

53. $\sqrt[3]{\frac{2}{3}}$ $\frac{\sqrt[3]{18}}{3}$

54. $\sqrt[3]{\frac{4}{5}}$ $\frac{\sqrt[3]{100}}{5}$

55. $\sqrt[3]{\frac{4}{9}}$ $\frac{\sqrt[3]{12}}{3}$

56. $\sqrt[3]{\frac{5}{16}}$ $\frac{\sqrt[3]{20}}{4}$

57. $-\sqrt[3]{\frac{2p}{r^2}}$ $-\frac{\sqrt[3]{2pr}}{r}$

58. $-\sqrt[3]{\frac{6x}{y^2}}$ $-\frac{\sqrt[3]{6xy}}{y}$

59. $\sqrt[4]{\frac{16}{x}}$ $\frac{2\sqrt[4]{x^3}}{x}$

60. $\sqrt[4]{\frac{81}{y}}$ $\frac{3\sqrt[4]{y^3}}{y}$

61. $\sqrt[4]{\frac{2y}{z}}$ $\frac{\sqrt[4]{2yz^3}}{z}$

62. $\sqrt[4]{\frac{7t}{s^2}}$ $\frac{\sqrt[4]{7ts^2}}{s}$

Rationalize the denominator in each expression. Assume that all variables represent positive real numbers and that no denominators are 0. See Example 5.

63. $\frac{2}{4+\sqrt{3}}$ $\frac{2(4-\sqrt{3})}{13}$

64. $\frac{6}{5+\sqrt{2}}$ $\frac{6(5-\sqrt{2})}{23}$

65. $\frac{6}{\sqrt{5}+\sqrt{3}}$ $3(\sqrt{5}-\sqrt{3})$

66. $\frac{12}{\sqrt{6}+\sqrt{3}}$ $4(\sqrt{6}-\sqrt{3})$

67. $\frac{-4}{\sqrt{3}-\sqrt{7}}$ $\sqrt{3}+\sqrt{7}$

68. $\frac{-3}{\sqrt{2}+\sqrt{5}}$ $\sqrt{2}-\sqrt{5}$

69. $\frac{1-\sqrt{2}}{\sqrt{7}+\sqrt{6}}$ $\sqrt{7}-\sqrt{6}-\sqrt{14}+2\sqrt{3}$

70. $\frac{-1-\sqrt{3}}{\sqrt{6}+\sqrt{5}}$ $-\sqrt{6}+\sqrt{5}-3\sqrt{2}+\sqrt{15}$

71. $\frac{\sqrt{2}-\sqrt{3}}{\sqrt{6}-\sqrt{5}}$ $2\sqrt{3}+\sqrt{10}-3\sqrt{2}-\sqrt{15}$

72. $\frac{\sqrt{5}+\sqrt{6}}{\sqrt{3}-\sqrt{2}}$ $\sqrt{15}+\sqrt{10}+3\sqrt{2}+2\sqrt{3}$

73. $\frac{4}{\sqrt{x}-2\sqrt{y}}$ $\frac{4(\sqrt{x}+2\sqrt{y})}{x-4y}$

74. $\frac{5}{3\sqrt{r}+\sqrt{s}}$ $\frac{5(3\sqrt{r}-\sqrt{s})}{9r-s}$

75. $\frac{\sqrt{x}-\sqrt{y}}{\sqrt{2x}+\sqrt{3y}}$ $\frac{x\sqrt{2}-\sqrt{3xy}-\sqrt{2xy}+y\sqrt{3}}{2x-3y}$

76. $\frac{\sqrt{a}+\sqrt{b}}{\sqrt{5a}-\sqrt{2b}}$ $\frac{a\sqrt{5}+\sqrt{2ab}+\sqrt{5ab}+b\sqrt{2}}{5a-2b}$

Write each quotient in lowest terms. Assume that all variables represent positive real numbers. See Example 6.

77. $\frac{25 + 10\sqrt{6}}{20}$

$\frac{5 + 2\sqrt{6}}{4}$

78. $\frac{12 - 6\sqrt{2}}{24}$

$\frac{2 - \sqrt{2}}{4}$

79. $\frac{16 + 4\sqrt{8}}{12}$

$\frac{4 + 2\sqrt{2}}{3}$

80. $\frac{12 + 9\sqrt{72}}{18}$

$\frac{2 + 9\sqrt{2}}{3}$

81. $\frac{6x + \sqrt{24x^3}}{3x}$

$\frac{6 + 2\sqrt{6x}}{3}$

82. $\frac{11y + \sqrt{242y^5}}{22y}$

$\frac{1 + y\sqrt{2y}}{2}$

Relating Concepts (Exercises 83–86) For Individual or Group Work

Sometimes it is desirable to ***rationalize the numerator*** *in an expression. The procedure is similar to rationalizing the denominator. For example, to rationalize the numerator in the following expression, we multiply both the numerator and denominator by the conjugate of the numerator,* $6 + \sqrt{2}$.

$$\frac{6 - \sqrt{2}}{3} = \frac{(6 - \sqrt{2})(6 + \sqrt{2})}{3(6 + \sqrt{2})} = \frac{36 - 2}{3(6 + \sqrt{2})} = \frac{34}{3(6 + \sqrt{2})}$$

In the final expression, the numerator is rationalized. ***Work Exercises 83–86 in order.***

83. Rationalize the numerator of $\frac{8\sqrt{5} - 1}{6}$. $\frac{319}{6(8\sqrt{5} + 1)}$

84. Rationalize the numerator of $\frac{3\sqrt{a} + \sqrt{b}}{\sqrt{b} - \sqrt{a}}$. Assume a and b are positive and $a \neq b$. $\frac{9a - b}{(\sqrt{b} - \sqrt{a})(3\sqrt{a} - \sqrt{b})}$

85. Rationalize the denominator of the expression in Exercise 84. $\frac{(3\sqrt{a} + \sqrt{b})(\sqrt{b} + \sqrt{a})}{b - a}$

86. Describe the difference in the procedures used in Exercises 84 and 85.

In Exercise 84, we multiplied the numerator and denominator by the conjugate of the numerator, while in Exercise 85 we multiplied by the conjugate of the denominator.

Summary Exercises on Operations with Radicals and Rational Exponents

Recall that a simplified radical satisfies the following conditions.

Conditions for a Simplified Radical

1. The radicand has no factor raised to a power greater than or equal to the index.
2. The radicand has no fractions.
3. No denominator has a radical.
4. Exponents in the radicand and the index of the radical have greatest common factor 1.

Perform all indicated operations, and express each answer in simplest form with positive exponents. Assume that all variables represent positive real numbers.

1. $6\sqrt{10} - 12\sqrt{10}$

$-6\sqrt{10}$

2. $\sqrt{7}\left(\sqrt{7} - \sqrt{2}\right)$

$7 - \sqrt{14}$

3. $\left(1 - \sqrt{3}\right)\left(2 + \sqrt{6}\right)$

$2 + \sqrt{6} - 2\sqrt{3} - 3\sqrt{2}$

4. $\sqrt{50} - \sqrt{98} + \sqrt{72}$

$4\sqrt{2}$

5. $\left(3\sqrt{5} + 2\sqrt{7}\right)^2$

$73 + 12\sqrt{35}$

6. $\dfrac{-3}{\sqrt{6}}$

$\dfrac{-\sqrt{6}}{2}$

7. $\dfrac{8}{\sqrt{7} + \sqrt{5}}$

$4\left(\sqrt{7} - \sqrt{5}\right)$

8. $\sqrt[3]{16x^2} - \sqrt[3]{54x^2} + \sqrt[3]{128x^2}$

$3\sqrt[3]{2x^2}$

9. $\dfrac{1 - \sqrt{2}}{1 + \sqrt{2}}$

$-3 + 2\sqrt{2}$

10. $\left(1 - \sqrt[3]{3}\right)\left(1 + \sqrt[3]{3} + \sqrt[3]{9}\right)$

-2

11. $\left(\sqrt{5} + 7\right)\left(\sqrt{5} - 7\right)$

-44

12. $\dfrac{1}{\sqrt{x} - \sqrt{5}}, \quad x \neq 5$

$\dfrac{\sqrt{x} + \sqrt{5}}{x - 5}$

13. $\sqrt[3]{8a^3b^5c^9}$

$2abc^3\sqrt[3]{b^2}$

14. $\dfrac{15}{\sqrt[3]{9}}$

$5\sqrt[3]{3}$

15. $\dfrac{3}{\sqrt{5} + 2}$

$3\left(\sqrt{5} - 2\right)$

16. $\sqrt{\dfrac{3}{5x}}$

$\dfrac{\sqrt{15x}}{5x}$

17. $\dfrac{16\sqrt{3}}{5\sqrt{12}}$

$\dfrac{8}{5}$

18. $\dfrac{2\sqrt{25}}{8\sqrt{50}}$

$\dfrac{\sqrt{2}}{8}$

19. $\dfrac{-10}{\sqrt[3]{10}}$

$-\sqrt[3]{100}$

20. $\dfrac{\sqrt{6}+\sqrt{5}}{\sqrt{6}-\sqrt{5}}$

$11 + 2\sqrt{30}$

21. $\sqrt{12x} - \sqrt{75x}$

$-3\sqrt{3x}$

22. $(5 - 3\sqrt{3})^2$

$52 - 30\sqrt{3}$

23. $(\sqrt{74} - \sqrt{73})(\sqrt{74} + \sqrt{73})$

1

24. $\sqrt[3]{\dfrac{13}{81}}$

$\dfrac{\sqrt[3]{117}}{9}$

25. $-t^2\sqrt[4]{t} + 3\sqrt[4]{t^9} - t\sqrt[4]{t^5}$

$t^2\sqrt[4]{t}$

26. $\dfrac{\sqrt{3}+\sqrt{7}}{\sqrt{6}-\sqrt{5}}$

$3\sqrt{2} + \sqrt{15} + \sqrt{42} + \sqrt{35}$

27. $\dfrac{6}{\sqrt[4]{3}}$

$2\sqrt[4]{27}$

28. $\dfrac{1}{1 - \sqrt[3]{3}}$

$\dfrac{1 + \sqrt[3]{3} + \sqrt[3]{9}}{-2}$

29. $\sqrt[3]{\dfrac{x^2y}{x^{-3}y^4}}$

$\dfrac{x\sqrt[3]{x^2}}{y}$

30. $\sqrt{12} - \sqrt{108} - \sqrt[3]{27}$

$-4\sqrt{3} - 3$

31. $\dfrac{x^{-2/3}y^{4/5}}{x^{-5/3}y^{-2/5}}$

$xy^{6/5}$

32. $\left(\dfrac{x^{3/4}y^{2/3}}{x^{1/3}y^{5/8}}\right)^{24}$

$x^{10}y$

33. $(125x^3)^{-2/3}$

$\dfrac{1}{25x^2}$

34. $(3x^{-2/3}y^{1/2})(-2x^{5/8}y^{-1/3})$

$\dfrac{-6y^{1/6}}{x^{1/24}}$

35. $\dfrac{4^{1/2} + 3^{1/2}}{4^{1/2} - 3^{1/2}}$

$7 + 4 \cdot 3^{1/2}$, or $7 + 4\sqrt{3}$

36. $(\sqrt{6} - \sqrt{5})^2(\sqrt{6} + \sqrt{5})^2$

1

9.6 Solving Equations with Radicals

OBJECTIVES

1. Solve radical equations using the power rule.
2. Solve radical equations that require additional steps.
3. Solve radical equations with indexes greater than 2.

An equation that includes one or more radical expressions with a variable is called a **radical equation.** Some examples of radical equations are

$$\sqrt{x-4}=8, \quad \sqrt{5x+12}=3\sqrt{2x-1}, \quad \text{and} \quad \sqrt[3]{6+x}=27.$$

Radical equations

OBJECTIVE 1 Solve radical equations using the power rule. The equation $x = 1$ has only one solution. Its solution set is $\{1\}$. If we square both sides of this equation, we get $x^2 = 1$. This new equation has two solutions: -1 and 1. Notice that the solution of the original equation is also a solution of the squared equation. However, the squared equation has another solution, -1, that is *not* a solution of the original equation. When solving equations with radicals, we use this idea of raising both sides to a power. This is an application of the **power rule.**

Power Rule for Solving Equations with Radicals

If both sides of an equation are raised to the same power, all solutions of the original equation are also solutions of the new equation.

Read the power rule carefully; it does not say that all solutions of the new equation are solutions of the original equation. They may or may not be. Solutions that do not satisfy the original equation are called **extraneous solutions;** they must be discarded.

CAUTION

When the power rule is used to solve an equation, ***every solution of the new equation* must *be checked in the original equation.***

EXAMPLE 1 Using the Power Rule

Solve $\sqrt{3x+4}=8$.

Use the power rule and square both sides to get

$$\left(\sqrt{3x+4}\right)^2 = 8^2$$

$$3x+4=64$$

$$3x=60 \quad \text{Subtract 4.}$$

$$x=20. \quad \text{Divide by 3.}$$

To check, substitute the proposed solution in the *original* equation.

Check

$$\sqrt{3x+4}=8 \quad \text{Original equation}$$

$$\sqrt{3\cdot 20+4} \stackrel{?}{=} 8 \quad \text{Let } x=20.$$

$$\sqrt{64} \stackrel{?}{=} 8$$

$$8=8 \quad \text{True}$$

Since 20 satisfies the *original* equation, the solution set is $\{20\}$.

Work Problem 1 at the Side. ▶

1 Solve each equation.

(a) $\sqrt{r}=3$

(b) $\sqrt{5x+1}=4$

ANSWERS

1. (a) $\{9\}$ (b) $\{3\}$

The solution of the equation in Example 1 can be generalized.

2 Solve each equation.

(a) $\sqrt{5x+3}+2=0$

(b) $\sqrt{x-9}-3=0$

Solving an Equation with Radicals

Step 1 **Isolate the radical.** Make sure that one radical term is alone on one side of the equation.

Step 2 **Apply the power rule.** Raise both sides of the equation to a power that is the same as the index of the radical.

Step 3 **Solve.** Solve the resulting equation; if it still contains a radical, repeat Steps 1 and 2.

Step 4 **Check** all proposed solutions in the original equation.

CAUTION

Remember to check (Step 4) or you may get an incorrect solution set.

EXAMPLE 2 Using the Power Rule

Solve $\sqrt{5x-1}+3=0$.

Step 1 To isolate the radical on one side, subtract 3 from each side.

$$\sqrt{5x-1}=-3$$

Step 2 Now square both sides.

$$\left(\sqrt{5x-1}\right)^2=(-3)^2$$

Step 3

$$5x-1=9$$

$$5x=10 \quad \text{Add 1.}$$

$$x=2 \quad \text{Divide by 5.}$$

Step 4 Check the proposed solution, 2, by substituting it in the original equation.

Check

$$\sqrt{5x-1}+3=0 \quad \text{Original equation}$$

$$\sqrt{5\cdot 2-1}+3\stackrel{?}{=}0 \quad \text{Let } x=2.$$

$$3+3=0 \quad \text{False}$$

This false result shows that 2 is *not* a solution of the original equation; it is extraneous. The solution set is $\emptyset$.

Note

We could have determined after Step 1 that the equation in Example 2 has no solution because the expression on the left cannot equal a negative number.

◀ Work Problem **2** at the Side.

OBJECTIVE 2 Solve radical equations that require additional steps. The next examples involve finding the square of a binomial. Recall that

$$(x+y)^2=x^2+2xy+y^2.$$

ANSWERS

2. **(a)** $\emptyset$ **(b)** $\{18\}$

EXAMPLE 3 Using the Power Rule; Squaring a Binomial

Solve $\sqrt{4 - x} = x + 2$.

Step 1 The radical is alone on the left side of the equation.

Step 2 Square both sides. On the right, $(x + 2)^2 = x^2 + 2(x)(2) + 2^2$.

$$\left(\sqrt{4 - x}\right)^2 = (x + 2)^2$$

Remember the middle term.

$$4 - x = x^2 + 4x + 4$$

Pay careful attention here. Twice the product of 2 and x

Step 3 The new equation is quadratic, so get 0 on one side.

$0 = x^2 + 5x$ Subtract 4; add x.

$0 = x(x + 5)$ Factor.

Set *each* factor equal to 0. $x = 0$ or $x + 5 = 0$ Zero-factor property

$x = -5$ Solve.

Step 4 Check each proposed solution in the original equation.

Check If $x = 0$, then

$\sqrt{4 - x} = x + 2$

$\sqrt{4 - 0} \stackrel{?}{=} 0 + 2$

$\sqrt{4} \stackrel{?}{=} 2$

$2 = 2$. True

If $x = -5$, then

$\sqrt{4 - x} = x + 2$

$\sqrt{4 - (-5)} \stackrel{?}{=} -5 + 2$

$\sqrt{9} \stackrel{?}{=} -3$

$3 = -3$. False

The solution set is $\{0\}$. The other proposed solution, -5, is extraneous.

Work Problem 3 at the Side. ▶

3 Solve.

(a) $\sqrt{3x - 5} = x - 1$

(b) $x + 1 = \sqrt{-2x - 2}$

EXAMPLE 4 Using the Power Rule; Squaring a Binomial

Solve $\sqrt{x^2 - 4x + 9} = x - 1$.

Square both sides. On the right, $(x - 1)^2 = x^2 - 2(x)(1) + 1^2$.

$$\left(\sqrt{x^2 - 4x + 9}\right)^2 = (x - 1)^2$$

Remember the middle term.

$$x^2 - 4x + 9 = x^2 - 2x + 1$$

Twice the product of x and -1

$-2x = -8$ Subtract x^2 and 9; add $2x$.

$x = 4$ Divide by -2.

Check $\sqrt{x^2 - 4x + 9} = x - 1$ Original equation

$\sqrt{4^2 - 4 \cdot 4 + 9} \stackrel{?}{=} 4 - 1$ Let $x = 4$.

$3 = 3$ True

The solution set of the original equation is $\{4\}$.

Work Problem 4 at the Side. ▶

4 Solve.

$\sqrt{4x^2 + 2x - 3} = 2x + 7$

CAUTION
When a radical equation requires squaring a binomial, as in Examples 3 and 4, ***remember to include the middle term.***

Answers

3. (a) $\{2, 3\}$ **(b)** $\{-1\}$

4. $\{-2\}$

5 **(a)** Verify that 15 is an extraneous solution of the equation in Example 5 and must be discarded.

(b) Solve.

$$\sqrt{2x+3} + \sqrt{x+1} = 1$$

6 Solve each equation.

(a) $\sqrt[3]{2x+7} = \sqrt[3]{3x-2}$

(b) $\sqrt[4]{2x+5} + 1 = 0$

EXAMPLE 5 Using the Power Rule; Squaring Twice

Solve $\sqrt{5x+6} + \sqrt{3x+4} = 2$.

Start by isolating one radical on one side of the equation. Do this by subtracting $\sqrt{3x+4}$ from each side.

$$\sqrt{5x+6} = 2 - \sqrt{3x+4} \qquad \text{Subtract } \sqrt{3x+4}.$$

$$\left(\sqrt{5x+6}\right)^2 = \left(2 - \sqrt{3x+4}\right)^2 \qquad \text{Square both sides.}$$

$$5x + 6 = 4 - 4\sqrt{3x+4} + (3x+4)$$

Be careful here.

Remember the middle term.

Twice the product of 2 and $-\sqrt{3x+4}$

This equation still contains a radical, so isolate the radical term on the right and square both sides again.

$$5x + 6 = 8 + 3x - 4\sqrt{3x+4} \qquad \text{Combine like terms.}$$

$$2x - 2 = -4\sqrt{3x+4} \qquad \text{Subtract 8 and } 3x.$$

Divide *each* term by 2.

$$x - 1 = -2\sqrt{3x+4} \qquad \text{Divide by 2 to make the numbers smaller.}$$

$$(x-1)^2 = \left(-2\sqrt{3x+4}\right)^2 \qquad \text{Square both sides again.}$$

$$x^2 - 2x + 1 = (-2)^2\left(\sqrt{3x+4}\right)^2 \qquad \text{On the right, } (ab)^2 = a^2b^2.$$

$$x^2 - 2x + 1 = 4(3x+4) \qquad \text{Apply the exponents.}$$

$$x^2 - 2x + 1 = 12x + 16 \qquad \text{Distributive property}$$

$$x^2 - 14x - 15 = 0 \qquad \text{Standard form}$$

$$(x+1)(x-15) = 0 \qquad \text{Factor.}$$

$$x + 1 = 0 \quad \text{or} \quad x - 15 = 0 \qquad \text{Zero-factor property}$$

$$x = -1 \quad \text{or} \quad x = 15 \qquad \text{Solve each equation.}$$

Check each of these proposed solutions in the original equation. Only -1 satisfies the equation, so the solution set, $\{-1\}$, has only one element.

◀ Work Problem **5** at the Side.

OBJECTIVE 3 Solve radical equations with indexes greater than 2. The power rule also works for powers greater than 2.

EXAMPLE 6 Using the Power Rule for a Power Greater than 2

Solve $\sqrt[3]{x+5} = \sqrt[3]{2x-6}$.

Raise both sides to the third power.

$$\left(\sqrt[3]{x+5}\right)^3 = \left(\sqrt[3]{2x-6}\right)^3$$

$$x + 5 = 2x - 6$$

$$11 = x \qquad \text{Subtract } x\text{; add 6.}$$

Check $\sqrt[3]{x+5} = \sqrt[3]{2x-6}$ Original equation

$$\sqrt[3]{11+5} \stackrel{?}{=} \sqrt[3]{2 \cdot 11 - 6} \qquad \text{Let } x = 11.$$

$$\sqrt[3]{16} = \sqrt[3]{16} \qquad \text{True}$$

The solution set is $\{11\}$.

◀ Work Problem **6** at the Side.

ANSWERS

5. (a) The final step in the check leads to $16 = 2$, which is false.
(b) $\{-1\}$

6. (a) $\{9\}$ (b) $\emptyset$

9.6 Exercises

FOR EXTRA HELP

MyMathLab

Math XL PRACTICE

WATCH

DOWNLOAD

READ

REVIEW

Check each equation to see if the given value for x is a solution.

1. $\sqrt{3x + 18} = x$

(a) 6 **(b)** -3

yes no

2. $\sqrt{3x - 3} = x - 1$

(a) 1 **(b)** 4

yes yes

3. $\sqrt{x + 2} = \sqrt{9x - 2} - 2\sqrt{x - 1}$

(a) 2 **(b)** 7

yes no

4. $\sqrt{8x - 3} = 2x$

(a) $\frac{3}{2}$ **(b)** $\frac{1}{2}$

yes yes

5. Is 9 a solution of the equation $\sqrt{x} = -3$? If not, what is the solution of this equation? Explain.

no; There is no solution. The radical expression, which is positive, cannot equal a negative number.

6. Before even attempting to solve $\sqrt{3x + 18} = x$, how can you be sure that the equation cannot have a negative solution?

Since the radical on the left side cannot be negative, and it must equal *x*, *x* cannot be negative.

Solve each equation. See Examples 1–4.

7. $\sqrt{x - 2} = 3$

$\{11\}$

8. $\sqrt{x + 1} = 7$

$\{48\}$

9. $\sqrt{6x - 1} = 1$

$\left\{\frac{1}{3}\right\}$

10. $\sqrt{7x - 3} = 5$

$\{4\}$

11. $\sqrt{4x + 3} + 1 = 0$

$\emptyset$

12. $\sqrt{5x - 3} + 2 = 0$

$\emptyset$

13. $\sqrt{3k + 1} - 4 = 0$

$\{5\}$

14. $\sqrt{5z + 1} - 11 = 0$

$\{24\}$

15. $4 - \sqrt{x - 2} = 0$

$\{18\}$

16. $9 - \sqrt{4k + 1} = 0$
{20}

17. $\sqrt{9a - 4} = \sqrt{8a + 1}$
{5}

18. $\sqrt{4p - 2} = \sqrt{3p + 5}$
{7}

19. $2\sqrt{x} = \sqrt{3x + 4}$
{4}

20. $2\sqrt{m} = \sqrt{5m - 16}$
{16}

21. $3\sqrt{z - 1} = 2\sqrt{2z + 2}$
{17}

22. $5\sqrt{4x + 1} = 3\sqrt{10x + 25}$
{20}

23. $k = \sqrt{k^2 + 4k - 20}$
{5}

24. $p = \sqrt{p^2 - 3p + 18}$
{6}

25. $x = \sqrt{x^2 + 3x + 9}$
∅

26. $z = \sqrt{z^2 - 4z - 8}$
∅

27. $\sqrt{9 - x} = x + 3$
{0}

28. $\sqrt{5 - x} = x + 1$
{1}

29. $\sqrt{k^2 + 2k + 9} = k + 3$
{0}

30. $\sqrt{x^2 - 3x + 3} = x - 1$
{2}

31. $\sqrt{r^2 + 9r + 3} = -r$
$\left\{-\frac{1}{3}\right\}$

32. $\sqrt{p^2 - 15p + 15} = p - 5$
∅

33. $\sqrt{z^2 + 12z - 4} + 4 - z = 0$
∅

34. $\sqrt{m^2 + 3m + 12} - m - 2 = 0$
{8}

35. A student wrote the following as his first step in solving $\sqrt{3x + 4} = 8 - x$.

$$3x + 4 = 64 + x^2$$

WHAT WENT WRONG? Solve the given equation correctly.

We cannot just square each term. The right side should be $(8 - x)^2 = 64 - 16x + x^2$. The correct first step is $3x + 4 = 64 - 16x + x^2$, and the solution set is {4}.

36. A student wrote the following as her first step in solving $\sqrt{5x + 6} = \sqrt{x + 3} + 3$.

$$5x + 6 = x + 3 + 9$$

WHAT WENT WRONG? Solve the given equation correctly.

We cannot just square each term. The right side should be $x + 3 + 2\sqrt{x + 3} \cdot 3 + 9$. The correct first step is

$5x + 6 = x + 3 + 2\sqrt{x + 3} \cdot 3 + 9$, and the solution set is {6}.

Solve each equation. See Examples 5 and 6.

37. $\sqrt[3]{2x+5} = \sqrt[3]{6x+1}$
{1}

38. $\sqrt[3]{p-1} = 2$
{9}

39. $\sqrt[3]{a^2+5a+1} = \sqrt[3]{a^2+4a}$
{−1}

40. $\sqrt[3]{r^2+2r+8} = \sqrt[3]{r^2}$
{−4}

41. $\sqrt[3]{2m-1} = \sqrt[3]{m+13}$
{14}

42. $\sqrt[3]{2k-11} - \sqrt[3]{5k+1} = 0$
{−4}

43. $\sqrt[4]{a+8} = \sqrt[4]{2a}$
{8}

44. $\sqrt[4]{z+11} = \sqrt[4]{2z+6}$
{5}

45. $\sqrt[3]{x-8} + 2 = 0$
{0}

46. $\sqrt[3]{r+1} + 1 = 0$
{−2}

47. $\sqrt[4]{2k-5} + 4 = 0$
∅

48. $\sqrt[4]{8z-3} + 2 = 0$
∅

49. $\sqrt{k+2} - \sqrt{k-3} = 1$
{7}

50. $\sqrt{r+6} - \sqrt{r-2} = 2$
{3}

51. $\sqrt{2r+11} - \sqrt{5r+1} = -1$
{7}

52. $\sqrt{3x-2} - \sqrt{x+3} = 1$
{6}

53. $\sqrt{3p+4} - \sqrt{2p-4} = 2$
{4, 20}

54. $\sqrt{4x+5} - \sqrt{2x+2} = 1$
{−1, 1}

55. $\sqrt{3-3p} - 3 = \sqrt{3p+2}$
∅

56. $\sqrt{4x+7} - 4 = \sqrt{4x-1}$
∅

57. $\sqrt{2\sqrt{x+11}} = \sqrt{4x+2}$
$\left\{\frac{5}{4}\right\}$

58. $\sqrt{1+\sqrt{24-10x}} = \sqrt{3x+5}$
$\left\{\frac{2}{9}\right\}$

For each equation, rewrite the expressions with rational exponents as radical expressions, and then solve using the procedures explained in this section.

59. $(2x - 9)^{1/2} = 2 + (x - 8)^{1/2}$

$\{9, 17\}$

60. $(3w + 7)^{1/2} = 1 + (w + 2)^{1/2}$

$\{-2, -1\}$

61. $(2w - 1)^{2/3} - w^{1/3} = 0$

$\left\{\frac{1}{4}, 1\right\}$

62. $(x^2 - 2x)^{1/3} - x^{1/3} = 0$

$\{0, 3\}$

Solve each formula from electricity and radio for the indicated variable. (Source: Cooke, Nelson M., and Joseph B. Orleans, Mathematics Essential to Electricity and Radio, *McGraw-Hill, 1943.)*

63. $V = \sqrt{\frac{2K}{m}}$ for K

$K = \frac{V^2m}{2}$

64. $V = \sqrt{\frac{2K}{m}}$ for m

$m = \frac{2K}{V^2}$

65. $Z = \sqrt{\frac{L}{C}}$ for C

$C = \frac{L}{Z^2}$

66. $Z = \sqrt{\frac{L}{C}}$ for L

$L = CZ^2$

67. $f = \frac{1}{2\pi\sqrt{LC}}$ for L

$L = \frac{1}{4\pi^2 f^2 C}$

68. $r = \sqrt{\frac{Mm}{F}}$ for F

$F = \frac{Mm}{r^2}$

A number of useful formulas involve radicals or radical expressions. Many occur in the mathematics needed for working with objects in space. The formula

$$N = \frac{1}{2\pi}\sqrt{\frac{a}{r}}$$

is used to find the rotational rate N of a space station. Here a is the acceleration and r represents the radius of the space station in meters. To find the value of r that will make N simulate the effect of gravity on Earth, the equation must be solved for r, using the required value of N. (Source: Kastner, Bernice, Space Mathematics, *NASA, 1972.)*

69. Solve the equation for r.

$r = \frac{a}{4\pi^2N^2}$

70. If $a = 9.8$ m per sec^2, find the value of r (to the nearest tenth) using each value of N.

(a) $N = 0.063$ rotation per sec

62.5 m

(b) $N = 0.04$ rotation per sec

155.1 m

9.7 Complex Numbers

OBJECTIVES

1. Simplify numbers of the form $\sqrt{-b}$, where $b > 0$.
2. Recognize subsets of the complex numbers.
3. Add and subtract complex numbers.
4. Multiply complex numbers.
5. Divide complex numbers.
6. Find powers of i.

As we saw in **Section 1.3,** the set of real numbers includes many other number sets (the rational numbers, integers, and natural numbers, for example). In this section, a new set of numbers is introduced that includes the set of real numbers, as well as numbers that are even roots of negative numbers, like $\sqrt{-2}$.

OBJECTIVE 1 Simplify numbers of the form $\sqrt{-b}$, where $b > 0$.

The equation $x^2 + 1 = 0$ has no real number solution since any solution must be a number whose square is -1. In the set of real numbers, all squares are nonnegative numbers because the product of two positive numbers or two negative numbers is positive and $0^2 = 0$. To provide a solution for the equation $x^2 + 1 = 0$, we introduce a new number i.

Imaginary Unit *i*

The **imaginary unit *i*** is defined as

$$i = \sqrt{-1}, \quad \text{where} \quad i^2 = -1.$$

In words, i is the principal square root of -1.

This definition of i makes it possible to define any square root of a negative number as follows.

$\sqrt{-b}$

For any positive number b, $\quad \sqrt{-b} = i\sqrt{b}.$

EXAMPLE 1 Simplifying Square Roots of Negative Numbers

Write each number as a product of a real number and i.

(a) $\sqrt{-100} = i\sqrt{100} = 10i$

(b) $-\sqrt{-36} = -i\sqrt{36} = -6i$

(c) $\sqrt{-2} = i\sqrt{2}$

(d) $\sqrt{-8} = i\sqrt{8} = i\sqrt{4 \cdot 2} = 2i\sqrt{2}$

CAUTION
It is easy to mistake $\sqrt{2}i$ for $\sqrt{2i}$, with the i under the radical. For this reason, we usually write $\sqrt{2}i$ as $i\sqrt{2}$, as in the definition of $\sqrt{-b}$.

Work Problem 1 at the Side. ▶

1 Write each number as a product of a real number and i.

(a) $\sqrt{-16}$

(b) $-\sqrt{-81}$

(c) $\sqrt{-7}$

(d) $\sqrt{-32}$

When finding a product such as $\sqrt{-4} \cdot \sqrt{-9}$, we cannot use the product rule for radicals because it applies only to nonnegative radicands. ***For this reason, we change $\sqrt{-b}$ to the form $i\sqrt{b}$ before performing any multiplications or divisions.***

ANSWERS

1. (a) $4i$ (b) $-9i$ (c) $i\sqrt{7}$ (d) $4i\sqrt{2}$

For example,
$$\sqrt{-4} \cdot \sqrt{-9}$$
$$= i\sqrt{4} \cdot i\sqrt{9} \qquad \sqrt{-b} = i\sqrt{b}$$
$$= i \cdot 2 \cdot i \cdot 3$$
$$= 6i^2$$
$$= 6(-1) \qquad \text{Substitute: } i^2 = -1.$$
$$= -6.$$

First write all square roots in terms of i.

CAUTION
Using the product rule for radicals *before* using the definition of $\sqrt{-b}$ gives a *wrong* answer. The preceding example shows that

$$\sqrt{-4} \cdot \sqrt{-9} = -6, \qquad \text{Correct}$$
but
$$\sqrt{-4(-9)} = \sqrt{36} = 6, \qquad \text{Incorrect}$$
so
$$\sqrt{-4} \cdot \sqrt{-9} \neq \sqrt{-4(-9)}.$$

EXAMPLE 2 Multiplying Square Roots of Negative Numbers

Multiply.

(a)
$$\sqrt{-3} \cdot \sqrt{-7}$$
$$= i\sqrt{3} \cdot i\sqrt{7} \qquad \sqrt{-b} = i\sqrt{b}$$
$$= i^2\sqrt{3 \cdot 7} \qquad \text{Product rule}$$
$$= (-1)\sqrt{21} \qquad \text{Substitute: } i^2 = -1.$$
$$= -\sqrt{21}$$

First write all square roots in terms of i.

(b) $\sqrt{-2} \cdot \sqrt{-8}$
$$= i\sqrt{2} \cdot i\sqrt{8}$$
$$= i^2\sqrt{2 \cdot 8}$$
$$= (-1)\sqrt{16}$$
$$= (-1)4, \quad \text{or} \quad -4$$

(c) $\sqrt{-5} \cdot \sqrt{6}$
$$= i\sqrt{5} \cdot \sqrt{6}$$
$$= i\sqrt{30}$$

◀ Work Problem **2** at the Side.

EXAMPLE 3 Dividing Square Roots of Negative Numbers

Divide.

(a) $\dfrac{\sqrt{-75}}{\sqrt{-3}}$
$$= \frac{i\sqrt{75}}{i\sqrt{3}}$$
$$= \sqrt{\frac{75}{3}} \qquad \text{Quotient rule}$$
$$= \sqrt{25} \qquad \text{Divide.}$$
$$= 5$$

First write all square roots in terms of i.

(b) $\dfrac{\sqrt{-32}}{\sqrt{8}}$
$$= \frac{i\sqrt{32}}{\sqrt{8}}$$
$$= i\sqrt{\frac{32}{8}}$$
$$= i\sqrt{4}$$
$$= 2i$$

◀ Work Problem **3** at the Side.

2 Multiply.

(a) $\sqrt{-7} \cdot \sqrt{-5}$

(b) $\sqrt{-5} \cdot \sqrt{-10}$

(c) $\sqrt{-15} \cdot \sqrt{2}$

3 Divide.

(a) $\dfrac{\sqrt{-32}}{\sqrt{-2}}$

(b) $\dfrac{\sqrt{-27}}{\sqrt{-3}}$

(c) $\dfrac{\sqrt{-40}}{\sqrt{10}}$

ANSWERS

2. **(a)** $-\sqrt{35}$ **(b)** $-5\sqrt{2}$ **(c)** $i\sqrt{30}$
3. **(a)** 4 **(b)** 3 **(c)** $2i$

OBJECTIVE 2 Recognize subsets of the complex numbers. With the imaginary unit i and the real numbers, a new set of numbers can be formed that includes the real numbers as a subset. The *complex numbers* are defined as follows.

Complex Number

If a and b are real numbers, then any number of the form $\boldsymbol{a + bi}$ is called a **complex number.** In the complex number $a + bi$, the number a is called the **real part** and b is called the **imaginary part.***

For a complex number $a + bi$, if $b = 0$, then $a + bi = a$, which is a real number. ***Thus, the set of real numbers is a subset of the set of complex numbers.*** If $a = 0$ and $b \neq 0$, the complex number is said to be a **pure imaginary number.** For example, $3i$ is a pure imaginary number. A number such as $7 + 2i$ is a **nonreal complex number.** These numbers are very useful in applications, particularly in work with electricity.

The relationships among the sets of numbers are shown in Figure 10.

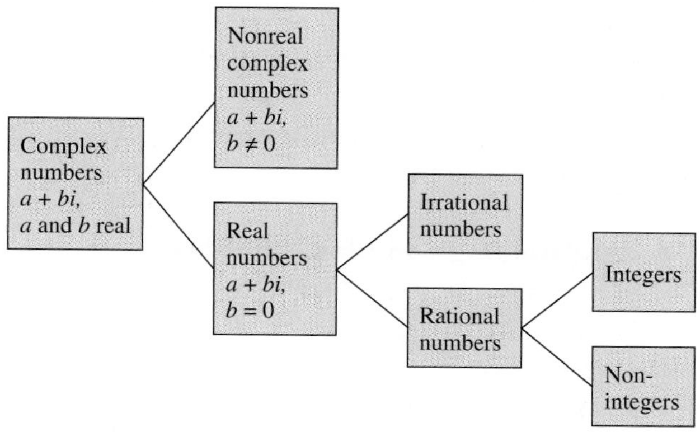

Figure 10

OBJECTIVE 3 Add and subtract complex numbers. The commutative, associative, and distributive properties for real numbers are also valid for complex numbers. ***Thus, to add complex numbers, we add their real parts and add their imaginary parts.***

EXAMPLE 4 Adding Complex Numbers

Add.

(a) $(2 + 3i) + (6 + 4i)$

$= (2 + 6) + (3 + 4)i$ Commutative, associative, and distributive properties

$= 8 + 7i$ Add real parts; add imaginary parts.

(b) $5 + (9 - 3i)$

$= (5 + 9) - 3i$ Associative property

$= 14 - 3i$ Add real parts.

Work Problem 4 at the Side. ▶

4 Add.

(a) $(4 + 6i) + (-3 + 5i)$

(b) $(-1 + 8i) + (9 - 3i)$

*Some texts define bi as the imaginary part of the complex number $a + bi$.

Answers

4. (a) $1 + 11i$ **(b)** $8 + 5i$

5 Subtract.

(a) $(7 + 3i) - (4 + 2i)$

(b) $(-6 - i) - (-5 - 4i)$

(c) $8 - (3 - 2i)$

6 Multiply.

(a) $6i(4 + 3i)$

(b) $(6 - 4i)(2 + 4i)$

(c) $(3 - 2i)(3 + 2i)$

Answers

5. (a) $3 + i$ (b) $-1 + 3i$ (c) $5 + 2i$
6. (a) $-18 + 24i$ (b) $28 + 16i$ (c) 13

To subtract complex numbers, we subtract their real parts and subtract their imaginary parts.

EXAMPLE 5 Subtracting Complex Numbers

Subtract.

(a) $(6 + 5i) - (3 + 2i)$

$= (6 - 3) + (5 - 2)i$ Properties of real numbers

$= 3 + 3i$ Subtract real parts; subtract imaginary parts.

(b) $(7 - 3i) - (8 - 6i)$

$= (7 - 8) + [-3 - (-6)]i$

$= -1 + 3i$

(c) $(-9 + 4i) - (-9 + 8i)$

$= (-9 + 9) + (4 - 8)i$

$= 0 - 4i$

$= -4i$

◀ Work Problem 5 at the Side.

In Example 5(c), the answer was written as $0 - 4i$ and then as just $-4i$. A complex number written in the form $a + bi$, like $0 - 4i$, is in **standard form.** In this section, most answers will be given in standard form, but if a or b is 0, we consider answers such as a or bi to be in standard form.

OBJECTIVE 4 Multiply complex numbers. We multiply complex numbers as we multiply polynomials.

EXAMPLE 6 Multiplying Complex Numbers

Multiply.

(a) $4i(2 + 3i)$

$= 4i(2) + 4i(3i)$ Distributive property

$= 8i + 12i^2$ Multiply.

$= 8i + 12(-1)$ Substitute: $i^2 = -1$.

$= -12 + 8i$ Standard form

(b) $(3 + 5i)(4 - 2i)$

$= \underbrace{3(4)}_{\text{First}} + \underbrace{3(-2i)}_{\text{Outer}} + \underbrace{5i(4)}_{\text{Inner}} + \underbrace{5i(-2i)}_{\text{Last}}$ Use the FOIL method for multiplying binomials. **(Section 5.3)**

$= 12 - 6i + 20i - 10i^2$ Multiply.

$= 12 + 14i - 10(-1)$ Combine imaginary terms; $i^2 = -1$.

$= 12 + 14i + 10$ Multiply.

$= 22 + 14i$ Combine real terms.

(c) $(2 + 3i)(1 - 5i)$

$= 2(1) + 2(-5i) + 3i(1) + 3i(-5i)$ FOIL

$= 2 - 10i + 3i - 15i^2$

$= 2 - 7i - 15(-1)$ Use parentheses around -1 to avoid errors.

$= 2 - 7i + 15$

$= 17 - 7i$

◀ Work Problem 6 at the Side.

The two complex numbers $a + bi$ and $a - bi$ are called *complex conjugates,* or simply *conjugates,* of each other. ***The product of a complex number and its conjugate is always a real number,*** as shown here.

$$(a + bi)(a - bi) = a^2 - abi + abi - b^2i^2$$
$$= a^2 - b^2(-1)$$
$$(a + bi)(a - bi) = a^2 + b^2$$

The product eliminates i.

For example, $(3 + 7i)(3 - 7i) = 3^2 + 7^2 = 9 + 49 = 58.$

OBJECTIVE 5 Divide complex numbers. The quotient of two complex numbers should be a complex number. To write the quotient as a complex number, we need to eliminate i in the denominator. We use conjugates and a process like that for rationalizing a denominator to do this.

EXAMPLE 7 Dividing Complex Numbers

Find each quotient.

(a) $\dfrac{8 + 9i}{5 + 2i}$

Multiply both the numerator and denominator by the conjugate of the denominator. The conjugate of $5 + 2i$ is $5 - 2i$.

$$\frac{8 + 9i}{5 + 2i}$$

$$= \frac{(8 + 9i)(5 - 2i)}{(5 + 2i)(5 - 2i)} \qquad \frac{5 - 2i}{5 - 2i} = 1$$

$$= \frac{40 - 16i + 45i - 18i^2}{5^2 + 2^2} \qquad \text{In the denominator, } (a + bi)(a - bi) = a^2 + b^2.$$

$$= \frac{58 + 29i}{29} \qquad -18i^2 = -18(-1) = 18; \text{ Combine like terms.}$$

$$= \frac{29(2 + i)}{29} \qquad \text{Factor the numerator.}$$

Factor first; then divide out the common factor.

$$= 2 + i \qquad \text{Lowest terms}$$

(b) $\dfrac{1 + i}{i}$

$$= \frac{(1 + i)(-i)}{i(-i)} \qquad \text{Multiply numerator and denominator by } -i, \text{ the conjugate of } i.$$

$$= \frac{-i - i^2}{-i^2} \qquad \text{Distributive property; multiply.}$$

$$= \frac{-i - (-1)}{-(-1)} \qquad \text{Substitute: } i^2 = -1.$$

Use parentheses to avoid errors.

$$= \frac{-i + 1}{1}$$

$$= 1 - i$$

Work Problem 7 at the Side. ▶

7 Find each quotient.

(a) $\dfrac{2 + i}{3 - i}$

(b) $\dfrac{8 - 4i}{1 - i}$

(c) $\dfrac{5}{3 - 2i}$

(d) $\dfrac{5 - i}{i}$

ANSWERS

7. (a) $\frac{1}{2} + \frac{1}{2}i$ (b) $6 + 2i$ (c) $\frac{15}{13} + \frac{10}{13}i$ (d) $-1 - 5i$

8 Find each power of i.

(a) i^{21}

(b) i^{36}

(c) i^{50}

(d) i^{-9}

Calculator Tip In Examples 4–7, we showed how complex numbers can be added, subtracted, multiplied, and divided algebraically. Many current models of graphing calculators can perform these operations. Figure 11 shows how the computations in parts of Examples 4–7 are displayed on a TI-83/84 Plus calculator. Be sure to use parentheses as shown.

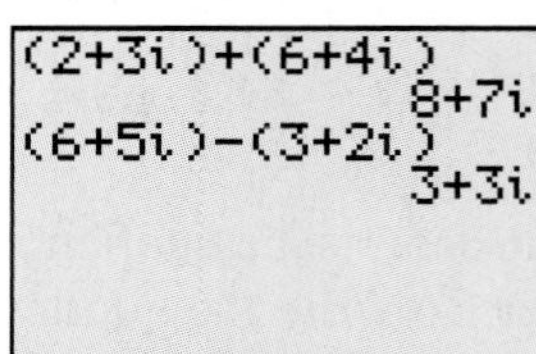

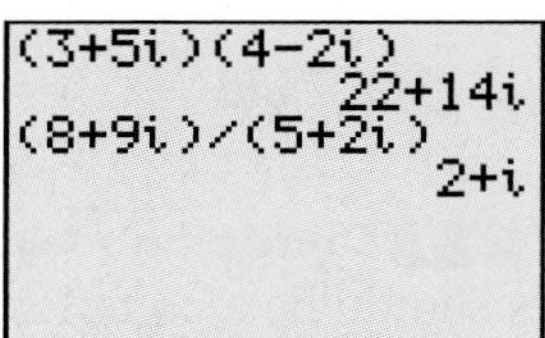

Figure 11

OBJECTIVE 6 Find powers of i. Because i^2 is defined to be -1, we can find higher powers of i as shown in the following examples.

$$i^3 = i \cdot i^2 = i(-1) = -i \qquad i^6 = i^2 \cdot i^4 = (-1) \cdot 1 = -1$$

$$i^4 = i^2 \cdot i^2 = (-1)(-1) = 1 \qquad i^7 = i^3 \cdot i^4 = (-i) \cdot 1 = -i$$

$$i^5 = i \cdot i^4 = i \cdot 1 = i \qquad i^8 = i^4 \cdot i^4 = 1 \cdot 1 = 1$$

As these examples suggest, the powers of i rotate through the four numbers i, -1, $-i$, and 1. Larger powers of i can be simplified by using the fact that $i^4 = 1$. For example,

$$i^{75} = (i^4)^{18} \cdot i^3 = 1^{18} \cdot i^3 = 1 \cdot i^3 = i^3 = -i.$$

EXAMPLE 8 Simplifying Powers of i

Find each power of i.

(a) $i^{12} = (i^4)^3 = 1^3 = 1$

(b) $i^{39} = i^{36} \cdot i^3 = (i^4)^9 \cdot i^3 = 1^9 \cdot (-i) = -i$

(c) $i^{-2} = \frac{1}{i^2} = \frac{1}{-1} = -1$

(d) $i^{-1} = \frac{1}{i} = \frac{1(-i)}{i(-i)} = \frac{-i}{-i^2} = \frac{-i}{-(-1)} = \frac{-i}{1} = -i$

◀ Work Problem 8 at the Side.

Answers

8. (a) i (b) 1 (c) -1 (d) $-i$

9.7 Exercises

FOR EXTRA HELP

PRACTICE

WATCH

DOWNLOAD

READ

REVIEW

Decide whether each expression is equal to $1, -1, i,$ *or* $-i$.

1. $\sqrt{-1}$ i

2. $-i^2$ 1

3. $\frac{1}{i}$ $-i$

4. $(-i)^2$ -1

5. Every real number is a complex number. Explain why this is so.

Any real number a can be written as $a + 0i$, a complex number with imaginary part 0.

6. Not every complex number is a real number. Give an example of this, and explain why this statement is true.

A complex number such as $2 + 3i$, with nonzero imaginary part, is not real.

Write each number as a product of a real number and i. Simplify all radical expressions. See Example 1.

7. $\sqrt{-169}$ $13i$

8. $\sqrt{-225}$ $15i$

9. $-\sqrt{-144}$ $-12i$

10. $-\sqrt{-196}$ $-14i$

11. $\sqrt{-5}$ $i\sqrt{5}$

12. $\sqrt{-21}$ $i\sqrt{21}$

13. $\sqrt{-48}$ $4i\sqrt{3}$

14. $\sqrt{-96}$ $4i\sqrt{6}$

Multiply or divide as indicated. See Examples 2 and 3.

15. $\sqrt{-15} \cdot \sqrt{-15}$ -15

16. $\sqrt{-19} \cdot \sqrt{-19}$ -19

17. $\sqrt{-3} \cdot \sqrt{-19}$ $-\sqrt{57}$

18. $\sqrt{-7} \cdot \sqrt{-15}$ $-\sqrt{105}$

19. $\sqrt{-4} \cdot \sqrt{-25}$ -10

20. $\sqrt{-9} \cdot \sqrt{-81}$ -27

21. $\sqrt{-3} \cdot \sqrt{11}$ $i\sqrt{33}$

22. $\sqrt{-5} \cdot \sqrt{13}$ $i\sqrt{65}$

23. $\frac{\sqrt{-300}}{\sqrt{-100}}$ $\sqrt{3}$

24. $\frac{\sqrt{-40}}{\sqrt{-10}}$ 2

25. $\frac{\sqrt{-75}}{\sqrt{3}}$ $5i$

26. $\frac{\sqrt{-160}}{\sqrt{10}}$ $4i$

Add or subtract as indicated. Write your answers in standard form. See Examples 4 and 5.

27. $(3 + 2i) + (-4 + 5i)$ $-1 + 7i$

28. $(7 + 15i) + (-11 + 14i)$ $-4 + 29i$

29. $(5 - i) + (-5 + i)$ 0

30. $(-2 + 6i) + (2 - 6i)$ 0

31. $(4 + i) - (-3 - 2i)$ $7 + 3i$

32. $(9 + i) - (3 + 2i)$ $6 - i$

33. $(-3 - 4i) - (-1 - 4i)$
-2

34. $(-2 - 3i) - (-5 - 3i)$
3

35. $(-4 + 11i) + (-2 - 4i) + (7 + 6i)$
$1 + 13i$

36. $(-1 + i) + (2 + 5i) + (3 + 2i)$
$4 + 8i$

37. $[(7 + 3i) - (4 - 2i)] + (3 + i)$
$6 + 6i$

38. $[(7 + 2i) + (-4 - i)] - (2 + 5i)$
$1 - 4i$

39. Fill in the blank with the correct response: Because $(4 + 2i) - (3 + i) = 1 + i$, using the definition of subtraction we can check this to find that

$$(1 + i) + (3 + i) = \underline{4 + 2i}.$$

40. Fill in the blank with the correct response: Because $\frac{-5}{2 - i} = -2 - i$, using the definition of division we can check this to find that

$$(-2 - i)(2 - i) = \underline{-5}.$$

Multiply. See Example 6.

41. $(3i)(27i)$
-81

42. $(5i)(125i)$
-625

43. $(-8i)(-2i)$
-16

44. $(-32i)(-2i)$
-64

45. $5i(-6 + 2i)$
$-10 - 30i$

46. $3i(4 + 9i)$
$-27 + 12i$

47. $(4 + 3i)(1 - 2i)$
$10 - 5i$

48. $(7 - 2i)(3 + i)$
$23 + i$

49. $(4 + 5i)^2$
$-9 + 40i$

50. $(3 + 2i)^2$
$5 + 12i$

51. $(12 + 3i)(12 - 3i)$
153

52. $(6 + 7i)(6 - 7i)$
85

53. **(a)** What is the conjugate of $a + bi$? $a - bi$

(b) If we multiply $a + bi$ by its conjugate, we get, $\underline{a^2} + \underline{b^2}$, which is always a real number.

54. Explain the procedure you would use to find the quotient

$$\frac{-1 + 5i}{3 + 2i}.$$

Multiply both the numerator and denominator by the complex conjugate of the denominator, $3 - 2i$. Then simplify the result by multiplying in the numerator and denominator and combining like terms. Write the quotient in the form $a + bi$.

Write each quotient in the form $a + bi$. See Example 7.

55. $\dfrac{2}{1 - i}$
$1 + i$

56. $\dfrac{29}{5 + 2i}$
$5 - 2i$

57. $\dfrac{-7 + 4i}{3 + 2i}$
$-1 + 2i$

58. $\dfrac{-38 - 8i}{7 + 3i}$
$-5 + i$

59. $\dfrac{8i}{2 + 2i}$
$2 + 2i$

60. $\dfrac{-8i}{1 + i}$
$-4 - 4i$

61. $\dfrac{2 - 3i}{2 + 3i}$
$-\dfrac{5}{13} - \dfrac{12}{13}i$

62. $\dfrac{-1 + 5i}{3 + 2i}$
$\dfrac{7}{13} + \dfrac{17}{13}i$

Relating Concepts (Exercises 63–68) For Individual or Group Work

Consider these expressions:

Binomials	***Complex Numbers***
$x + 2, \quad 3x - 1$	$1 + 2i, \quad 3 - i.$

When we add, subtract, or multiply complex numbers in standard form, the rules are the same as those for the corresponding operations on binomials. That is, we add or subtract like terms, and we use FOIL to multiply. Division, however, is comparable to division by the sum or difference of radicals, where we multiply by the conjugate of the denominator to get a rational denominator. To express the quotient of two complex numbers in standard form, we also multiply by the conjugate of the denominator. ***Work Exercises 63–68 in order,*** *to better understand these ideas.*

63. (a) Add the two binomials.
$4x + 1$

(b) Add the two complex numbers.
$4 + i$

64. (a) Subtract the second binomial from the first.
$-2x + 3$

(b) Subtract the second complex number from the first.
$-2 + 3i$

65. (a) Multiply the two binomials.
$3x^2 + 5x - 2$

(b) Multiply the two complex numbers.
$5 + 5i$

66. (a) Rationalize the denominator: $\dfrac{\sqrt{3} - 1}{1 + \sqrt{2}}$.
$-\sqrt{3} + \sqrt{6} + 1 - \sqrt{2}$

(b) Write in standard form: $\dfrac{3 - i}{1 + 2i}$.
$\dfrac{1}{5} - \dfrac{7}{5}i$

67. Explain why the answers for parts (a) and (b) in Exercise 65 do not correspond as the answers in Exercises 63 and 64 do.
Because $i^2 = -1$, two pairs of like terms can be combined in Exercise 65(b).

68. Explain why the answers for parts (a) and (b) in Exercise 66 do not correspond as the answers in Exercises 63 and 64 do.
Because $i^2 = -1$, additional terms can be combined in the numerator and denominator.

69. Recall that if $a \neq 0$, then $\frac{1}{a}$ is called the reciprocal of a. Use this definition to express the reciprocal of $5 - 4i$ in the form $a + bi$.

$\frac{5}{41} + \frac{4}{41}i$

70. Recall that if $a \neq 0$, then a^{-1} is defined to be $\frac{1}{a}$. Use this definition to express $(4 - 3i)^{-1}$ in the form $a + bi$.

$\frac{4}{25} + \frac{3}{25}i$

Find each power of i. See Example 8.

71. i^{18}

-1

72. i^{26}

-1

73. i^{89}

i

74. i^{45}

i

75. i^{96}

1

76. i^{48}

1

77. i^{-5}

$-i$

78. i^{-17}

$-i$

79. A student simplified i^{-18} as follows:

$$i^{-18} = i^{-18} \cdot i^{20} = i^{-18+20} = i^2 = -1.$$

Explain the mathematical justification for this correct work.

Since $i^{20} = (i^4)^5 = 1^5 = 1$, the student multiplied by 1, which is justified by the identity property for multiplication.

80. Explain why

$$(46 + 25i)(3 - 6i) \quad \text{and} \quad (46 + 25i)(3 - 6i)i^{12}$$

must be equal. (Do not actually perform the computation.)

$i^{12} = (i^4)^3 = 1^3 = 1$, so by the identity property for multiplication, the two products must be equal.

***Ohm's law** for the current I in a circuit with voltage E, resistance R, capacitance reactance X_c, and inductive reactance X_L is*

$$I = \frac{E}{R + (X_L - X_c)i}.$$

Use this law to work Exercises 81 and 82.

81. Find I if $E = 2 + 3i$, $R = 5$, $X_L = 4$, and $X_c = 3$.

$\frac{1}{2} + \frac{1}{2}i$

82. Find E if $I = 1 - i$, $R = 2$, $X_L = 3$, and $X_c = 1$.

4

83. Show that $1 + 5i$ is a solution of

$$x^2 - 2x + 26 = 0.$$

$(1 + 5i)^2 - 2(1 + 5i) + 26$ will simplify to 0 when the operations are applied.

84. Show that $3 + 2i$ is a solution of

$$x^2 - 6x + 13 = 0.$$

$(3 + 2i)^2 - 6(3 + 2i) + 13$ will simplify to 0 when the operations are applied.

Chapter 9 Summary

▶ Key Terms

Radical sign → $\sqrt[n]{a}$ ← Radicand; Index → n; Radical

9.1	**square root**	A number a is a square root of b if $a^2 = b$.
	principal square root	The positive square root of a number is its principal square root.
	radicand	The number or expression inside a radical sign is called the radicand.
	radical	A radical sign with a radicand is called a radical.
	radical expression	An algebraic expression containing a radical is called a radical expression.
	perfect square	A number with a rational square root is called a perfect square.
	irrational number	A real number that is not rational is called an irrational number.
	cube root	A number a is a cube root of b if $a^3 = b$.
	index (order)	In a radical of the form $\sqrt[n]{a}$, the number n is the index or order.
9.5	**rationalizing the denominator**	The process of removing radicals from the denominator so that the denominator contains only rational quantities is called rationalizing the denominator.
	conjugate	The conjugate of $a + b$ is $a - b$.
9.6	**radical equation**	A radical equation is an equation that includes one or more radical expressions with variables.
	extraneous solution	An extraneous solution of a radical equation is a solution found after applying the power rule that is not a solution of the original equation.
9.7	**complex number**	A complex number is a number that can be written in the form $a + bi$, where a and b are real numbers.
	real part	The real part of $a + bi$ is a.
	imaginary part	The imaginary part of $a + bi$ is b.
	pure imaginary number	A complex number $a + bi$ with $a = 0$ and $b \neq 0$ is called a pure imaginary number.
	nonreal complex number	A complex number $a + bi$ with $b \neq 0$ is called a nonreal complex number.
	standard form (of a complex number)	A complex number is in standard form if it is written in the form $a + bi$.
	complex conjugates	The complex conjugate of $a + bi$ is $a - bi$.

▶ New Symbols

$\sqrt{}$	radical sign	$a^{1/n}$	a to the power $\frac{1}{n}$
$\sqrt[n]{a}$	radical; principal nth root of a	$a^{m/n}$	a to the power $\frac{m}{n}$
$\pm$	"positive or negative," or "plus or minus"	i	imaginary unit
$\approx$	is approximately equal to		

Test Your Word Power

See how well you have learned the vocabulary in this chapter. Answers, with examples, follow the Quick Review.

1. A **radicand** is
   - **A.** the index of a radical
   - **B.** the number or expression under the radical sign
   - **C.** the positive root of a number
   - **D.** the radical sign.

2. A **hypotenuse** is
   - **A.** either of the two shorter sides of a triangle
   - **B.** the shortest side of a triangle
   - **C.** the side opposite the right angle in a right triangle
   - **D.** the longest side in any triangle.

3. **Rationalizing the denominator** is the process of
   - **A.** eliminating fractions from a radical expression
   - **B.** changing the denominator of a fraction from a radical expression to a rational number
   - **C.** clearing a radical expression of radicals
   - **D.** multiplying radical expressions.

4. An **extraneous solution** is a solution
   - **A.** that does not satisfy the original equation
   - **B.** that makes an equation true
   - **C.** that makes an expression equal 0
   - **D.** that checks in the original equation.

5. A **complex number** is
   - **A.** a real number that includes a complex fraction
   - **B.** a zero multiple of i
   - **C.** a number of the form $a + bi$, where a and b are real numbers
   - **D.** the square root of -1.

Quick Review

Concepts	Examples

9.1 Radical Expressions and Graphs

Concepts

$\sqrt[n]{a} = b$ **means** $b^n = a$.

$\sqrt[n]{a}$ is the principal nth root of a.

$\sqrt[n]{a^n} = |a|$ if n is even; $\sqrt[n]{a^n} = a$ if n is odd.

Functions Defined by Radical Expressions

The square root function defined by $f(x) = \sqrt{x}$ and the cube root function defined by $f(x) = \sqrt[3]{x}$ are two important functions defined by radical expressions.

Examples

The two square roots of 64 are $\sqrt{64} = 8$, the principal square root, and $-\sqrt{64} = -8$.

$$\sqrt[4]{(-2)^4} = |-2| = 2 \qquad \sqrt[3]{-27} = -3$$

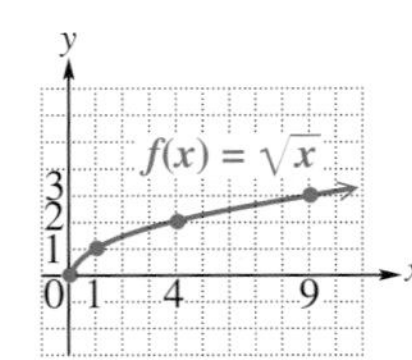

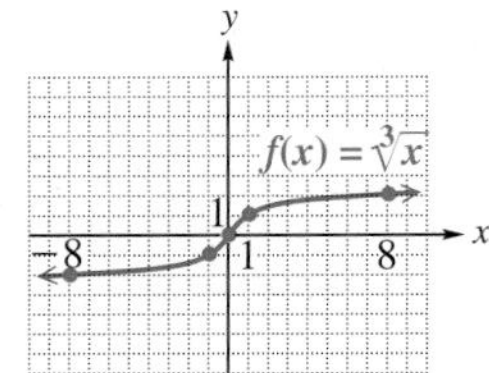

9.2 Rational Exponents

Concepts

$a^{1/n} = \sqrt[n]{a}$ whenever $\sqrt[n]{a}$ exists.

If m and n are positive integers with m/n in lowest terms, then $a^{m/n} = (a^{1/n})^m$, provided that $a^{1/n}$ is a real number.

All of the usual definitions and rules for exponents are valid for rational exponents.

Examples

$$81^{1/2} = \sqrt{81} = 9 \qquad -64^{1/3} = -\sqrt[3]{64} = -4$$

$$8^{5/3} = (8^{1/3})^5 = 2^5 = 32$$

$$5^{-1/2} \cdot 5^{1/4} = 5^{-1/2+1/4} = 5^{-1/4} = \frac{1}{5^{1/4}} \qquad (y^{2/5})^{10} = y^4$$

$$\frac{x^{-1/3}}{x^{-1/2}} = x^{-1/3-(-1/2)} = x^{-1/3+1/2} = x^{1/6}, \quad x > 0$$

9.3 Simplifying Radical Expressions

Concepts

Product and Quotient Rules for Radicals

If $\sqrt[n]{a}$ and $\sqrt[n]{b}$ are real numbers and n is a natural number,

$$\sqrt[n]{a} \cdot \sqrt[n]{b} = \sqrt[n]{ab} \quad \text{and} \quad \sqrt[n]{\frac{a}{b}} = \frac{\sqrt[n]{a}}{\sqrt[n]{b}}, \quad b \neq 0.$$

Examples

$$\sqrt{3} \cdot \sqrt{7} = \sqrt{21} \qquad \sqrt[5]{x^3y} \cdot \sqrt[5]{xy^2} = \sqrt[5]{x^4y^3}$$

$$\frac{\sqrt{x^5}}{\sqrt{x^4}} = \sqrt{\frac{x^5}{x^4}} = \sqrt{x}, \quad x > 0$$

Concepts	Examples

9.3 Simplifying Radical Expressions (*continued*)

Conditions for a Simplified Radical

1. The radicand has no factor raised to a power greater than or equal to the index.
2. The radicand has no fractions.
3. No denominator has a radical.
4. Exponents in the radicand and the index of the radical have greatest common factor 1.

$$\sqrt{18} = \sqrt{9 \cdot 2} = 3\sqrt{2}$$
$$\sqrt[3]{54x^5y^3} = \sqrt[3]{27x^3y^3 \cdot 2x^2} = 3xy\sqrt[3]{2x^2}$$
$$\sqrt{\frac{7}{4}} = \frac{\sqrt{7}}{\sqrt{4}} = \frac{\sqrt{7}}{2}$$
$$\sqrt[9]{x^3} = x^{3/9} = x^{1/3}, \quad \text{or} \quad \sqrt[3]{x}$$

Pythagorean Formula

If a and b are the lengths of the shorter sides of a right triangle and c is the length of the longest side, then

$$a^2 + b^2 = c^2.$$

The two shorter sides are the legs of the triangle, and the longest side is the hypotenuse. The hypotenuse is opposite the right angle.

Find b for the triangle in the figure.

$$10^2 + b^2 = \left(2\sqrt{61}\right)^2$$
$$b^2 = 4(61) - 100$$
$$b^2 = 144$$
$$b = 12$$

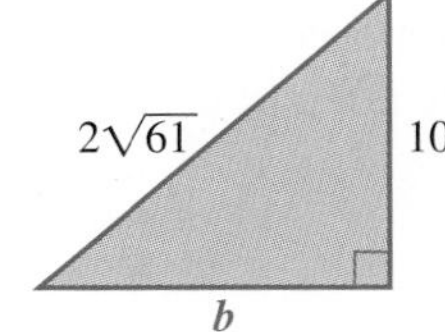

Distance Formula

The distance between (x_1, y_1) and (x_2, y_2) is

$$d = \sqrt{(x_2 - x_1)^2 + (y_2 - y_1)^2}.$$

The distance between $(3, -2)$ and $(-1, 1)$ is

$$\sqrt{(-1 - 3)^2 + [1 - (-2)]^2}$$
$$= \sqrt{(-4)^2 + 3^2}$$
$$= \sqrt{16 + 9}$$
$$= \sqrt{25}, \quad \text{or} \quad 5.$$

9.4 Adding and Subtracting Radical Expressions

Only radical expressions with the same index and the same radicand may be combined.

$$3\sqrt{17} + 2\sqrt{17} - 8\sqrt{17}$$
$$= (3 + 2 - 8)\sqrt{17}$$
$$= -3\sqrt{17}$$

$\sqrt{15} + \sqrt{30}$, $\sqrt{3} + \sqrt[3]{9}$ } cannot be simplified further

9.5 Multiplying and Dividing Radical Expressions

Multiply binomial radical expressions by using the FOIL method. Special products from **Section 5.4** may apply.

$$\left(\sqrt{2} + \sqrt{7}\right)\left(\sqrt{3} - \sqrt{6}\right)$$
$$= \sqrt{6} - 2\sqrt{3} + \sqrt{21} - \sqrt{42} \qquad \sqrt{12} = 2\sqrt{3}$$
$$\left(\sqrt{5} - \sqrt{10}\right)\left(\sqrt{5} + \sqrt{10}\right)$$
$$= 5 - 10, \quad \text{or} \quad -5$$
$$\left(\sqrt{3} - \sqrt{2}\right)^2$$
$$= 3 - 2\sqrt{3} \cdot \sqrt{2} + 2$$
$$= 5 - 2\sqrt{6}$$

Rationalize the denominator by multiplying both the numerator and denominator by the same expression, one that will yield a rational number in the final denominator.

$$\frac{\sqrt{7}}{\sqrt{5}} = \frac{\sqrt{7} \cdot \sqrt{5}}{\sqrt{5} \cdot \sqrt{5}} = \frac{\sqrt{35}}{5}$$

$$\frac{4}{\sqrt{5} - \sqrt{2}} = \frac{4\left(\sqrt{5} + \sqrt{2}\right)}{\left(\sqrt{5} - \sqrt{2}\right)\left(\sqrt{5} + \sqrt{2}\right)}$$
$$= \frac{4\left(\sqrt{5} + \sqrt{2}\right)}{5 - 2} = \frac{4\left(\sqrt{5} + \sqrt{2}\right)}{3}$$

Concepts	Examples

9.6 Solving Equations with Radicals

Solving an Equation with Radicals

Step 1 Isolate one radical on one side of the equation.

Step 2 Raise each side of the equation to a power that is the same as the index of the radical.

Step 3 Solve the resulting equation; if it still contains a radical, repeat Steps 1 and 2.

Step 4 Check all proposed solutions in the *original* equation.

Proposed solutions that do not check are extraneous; they are not part of the solution set.

Solve $\sqrt{2x+3} - x = 0$.

$$\begin{aligned} \sqrt{2x+3} &= x && \text{Add } x. \\ \left(\sqrt{2x+3}\right)^2 &= x^2 && \text{Square both sides.} \\ 2x+3 &= x^2 \\ x^2 - 2x - 3 &= 0 && \text{Standard form} \\ (x+1)(x-3) &= 0 && \text{Factor.} \\ x+1 = 0 \quad \text{or} \quad x-3 &= 0 && \text{Zero-factor property} \\ x = -1 \quad \text{or} \quad x &= 3 && \text{Solve each equation.} \end{aligned}$$

A check shows that 3 is a solution, but -1 is extraneous (as it leads to $2 = 0$, a false statement). The solution set is $\{3\}$.

9.7 Complex Numbers

$i = \sqrt{-1}$, where $i^2 = -1$.

For any positive number b, $\sqrt{-b} = i\sqrt{b}$.

To multiply radicals with negative radicands, first change each factor to the form $i\sqrt{b}$, and then multiply. The same procedure applies to quotients.

$$\sqrt{-25} = i\sqrt{25} = 5i$$

$$\begin{aligned} &\sqrt{-3} \cdot \sqrt{-27} \\ &= i\sqrt{3} \cdot i\sqrt{27} \\ &= i^2\sqrt{81} \\ &= -1 \cdot 9 \\ &= -9 \end{aligned}$$

$$\frac{\sqrt{-18}}{\sqrt{-2}} = \frac{i\sqrt{18}}{i\sqrt{2}} = \sqrt{\frac{18}{2}} = \sqrt{9} = 3$$

Adding and Subtracting Complex Numbers
Add (or subtract) the real parts and add (or subtract) the imaginary parts.

$$(5+3i)+(8-7i) = 13 - 4i \qquad (5+3i)-(8-7i) = -3 + 10i$$

Multiplying Complex Numbers
Multiply complex numbers by using the FOIL method.

$$\begin{aligned} &(2+i)(5-3i) \\ &= 10 - 6i + 5i - 3i^2 && \text{FOIL} \\ &= 10 - i - 3(-1) && i^2 = -1 \\ &= 10 - i + 3 \\ &= 13 - i \end{aligned}$$

Dividing Complex Numbers
Divide complex numbers by multiplying the numerator and the denominator by the conjugate of the denominator.

$$\frac{2}{3+i} = \frac{2(3-i)}{(3+i)(3-i)} = \frac{2(3-i)}{9-i^2} = \frac{2(3-i)}{10} = \frac{3-i}{5} = \frac{3}{5} - \frac{1}{5}i$$

ANSWERS TO TEST YOUR WORD POWER

1. B; *Example:* In $\sqrt{3xy}$, $3xy$ is the radicand.
2. C; *Example:* In a right triangle where the sides measure 9, 12, and 15 units, the hypotenuse is the side with measure 15 units.
3. B; *Example:* To rationalize the denominator of $\frac{5}{\sqrt{3}+1}$, multiply both the numerator and denominator by $\sqrt{3} - 1$ to get $\frac{5(\sqrt{3}-1)}{2}$.
4. A; *Example:* The proposed solution 2 is extraneous in $\sqrt{5x-1} + 3 = 0$, as it leads to $6 = 0$, a false statement.
5. C; *Examples:* -5 (or $-5 + 0i$), $7i$ (or $0 + 7i$), and $\sqrt{2} - 4i$.

Chapter 9 Review Exercises

[9.1] *Find each real number root. Use a calculator as necessary.*

1. $\sqrt{1764}$

   42

2. $-\sqrt{289}$

   −17

3. $-\sqrt{-841}$

   not a real number

4. $\sqrt[3]{216}$

   6

5. $\sqrt[5]{-32}$

   −2

6. $\sqrt{x^2}$

   $|x|$

7. $\sqrt[3]{x^3}$

   x

8. $\sqrt[4]{x^{20}}$

   $|x^5|$

Graph each function. Give the domain and the range.

9. $f(x) = \sqrt{x - 1}$

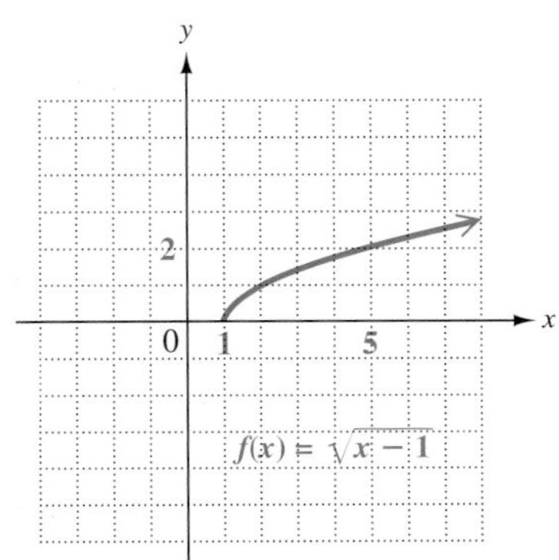

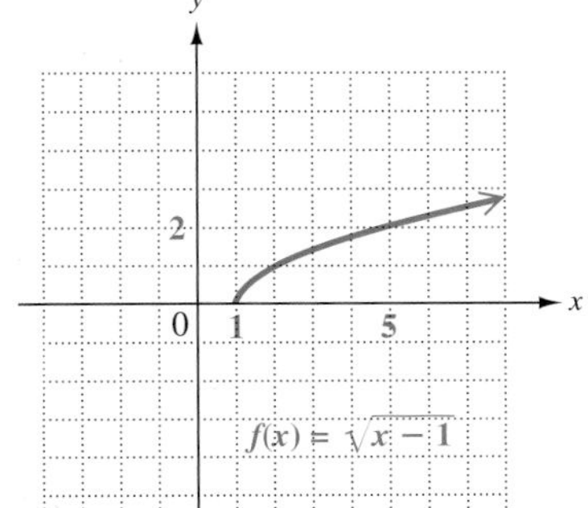
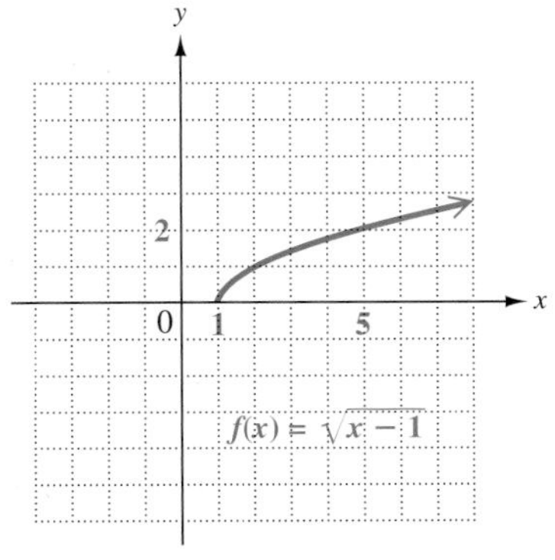

   domain: [1, ∞); range: [0, ∞)

10. $f(x) = \sqrt[3]{x} + 4$

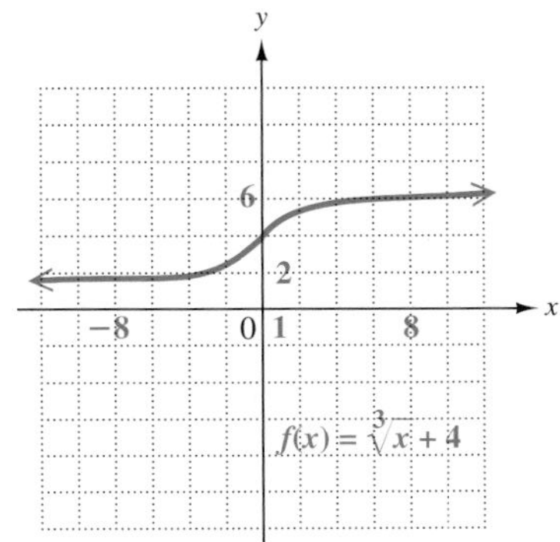

   domain: (−∞, ∞); range: (−∞, ∞)

11. Under what conditions is $\sqrt[n]{a}$ not a real number?

    ***n* must be even, and *a* must be negative.**

12. If a is negative and n is even, what can be said about $a^{1/n}$?

    It is not a real number.

Use a calculator to find a decimal approximation for each radical. Round to the nearest thousandth.

13. $\sqrt{40}$

    6.325

14. $\sqrt{77}$

    8.775

15. $\sqrt{310}$

    17.607

16. Use the formula for the time for one complete swing of a pendulum

$$t = 2\pi\sqrt{\frac{L}{g}}$$

to find the time to the nearest tenth of a second of a complete swing if the pendulum is 3 ft long and g is 32 ft per sec^2.

    1.9 sec

17. Use Heron's formula from Exercise 119 in **Section 9.1,**

$$A = \sqrt{s(s - a)(s - b)(s - c)},$$

where $s = \frac{1}{2}(a + b + c)$, to find the area of a triangle with sides of lengths 11, 13, and 20 in.

    66 in.^{2}

[9.2] *Find each real number root.*

18. $49^{1/2}$ **7**

19. $-8^{1/3}$ **−2**

20. $(-16)^{1/4}$ **not a real number**

21. Explain the relationship between the expressions $a^{m/n}$ and $\sqrt[n]{a^m}$.
By a power rule for exponents and the definition of $x^{1/n}$, $a^{m/n} = (a^m)^{1/n} = \sqrt[n]{a^m}$.

Simplify each expression. Assume that all variables represent positive real numbers.

22. $16^{5/4}$ **32**

23. $-8^{2/3}$ **−4**

24. $-\left(\frac{36}{25}\right)^{3/2}$ $-\frac{216}{125}$

25. $\left(-\frac{1}{8}\right)^{-5/3}$ **−32**

26. $\left(\frac{81}{10{,}000}\right)^{-3/4}$ $\frac{1000}{27}$

27. $7^{1/3} \cdot 7^{5/3}$ **49**

28. $\frac{96^{2/3}}{96^{-1/3}}$ **96**

29. $\frac{k^{2/3}k^{-1/2}k^{3/4}}{2(k^2)^{-1/4}}$ $\frac{k^{17/12}}{2}$

30. Write $2^{4/5}$ as a radical. $\sqrt[5]{2^4}$, or $\sqrt[5]{16}$

Simplify each expression. Write answers in radical form. Assume that all variables represent positive real numbers.

31. $\sqrt{3^{18}}$ 3^9

32. $\sqrt{7^9}$ $7^4\sqrt{7}$

33. $\sqrt[3]{m^5} \cdot \sqrt[3]{m^8}$ $m^4\sqrt[3]{m}$

34. $\sqrt[4]{k^2} \cdot \sqrt[4]{k^7}$ $k^2\sqrt[4]{k}$

35. $\sqrt[3]{\sqrt{m}}$ $\sqrt[6]{m}$

36. $\sqrt[4]{16y^5}$ $2y\sqrt[4]{y}$

37. $\sqrt[5]{y} \cdot \sqrt[3]{y}$ $\sqrt[15]{y^8}$

38. $\frac{\sqrt[3]{y^2}}{\sqrt[4]{y}}$ $\sqrt[12]{y^5}$

[9.3] *Simplify each expression. Assume that all variables represent positive real numbers.*

39. $\sqrt{6} \cdot \sqrt{11}$ $\sqrt{66}$

40. $\sqrt{5} \cdot \sqrt{r}$ $\sqrt{5r}$

41. $\sqrt[3]{6} \cdot \sqrt[3]{5}$ $\sqrt[3]{30}$

42. $\sqrt[4]{7} \cdot \sqrt[4]{3}$ $\sqrt[4]{21}$

43. $\sqrt{20}$ $2\sqrt{5}$

44. $-\sqrt{125}$ $-5\sqrt{5}$

45. $\sqrt[3]{-108x^4y}$ $-3x\sqrt[3]{4xy}$

46. $\sqrt[3]{64p^4q^6}$ $4pq^2\sqrt[3]{p}$

47. $\sqrt{\frac{49}{81}}$ $\frac{7}{9}$

48. $\sqrt{\frac{y^3}{144}}$ $\frac{y\sqrt{y}}{12}$

49. $\sqrt[3]{\frac{m^{15}}{27}}$ $\frac{m^5}{3}$

50. $\sqrt[3]{\frac{r^2}{8}}$ $\frac{\sqrt[3]{r^2}}{2}$

51. $\frac{\sqrt[3]{2^4}}{\sqrt[4]{32}}$ $\sqrt[12]{2}$

52. $\frac{\sqrt{x}}{\sqrt[5]{x}}$ $\sqrt[10]{x^3}$

Find the distance between each pair of points.

53. $(2, 7)$ and $(-1, -4)$ $\sqrt{130}$

54. $(-3, -5)$ and $(4, -3)$ $\sqrt{53}$

[9.4] *Perform the indicated operations. Assume that all variables represent positive real numbers.*

55. $2\sqrt{8} - 3\sqrt{50}$
$-11\sqrt{2}$

56. $8\sqrt{80} - 3\sqrt{45}$
$23\sqrt{5}$

57. $-\sqrt{27y} + 2\sqrt{75y}$
$7\sqrt{3y}$

58. $2\sqrt{54m^3} + 5\sqrt{96m^3}$
$26m\sqrt{6m}$

59. $3\sqrt[3]{54} + 5\sqrt[3]{16}$
$19\sqrt[3]{2}$

60. $-6\sqrt[4]{32} + \sqrt[4]{512}$
$-8\sqrt[4]{2}$

[9.5] *Multiply, and then simplify the products.*

61. $(\sqrt{3} + 1)(\sqrt{3} - 2)$
$1 - \sqrt{3}$

62. $(\sqrt{7} + \sqrt{5})(\sqrt{7} - \sqrt{5})$
2

63. $(3\sqrt{2} + 1)(2\sqrt{2} - 3)$
$9 - 7\sqrt{2}$

64. $(\sqrt{11} + 3\sqrt{5})(\sqrt{11} + 5\sqrt{5})$
$86 + 8\sqrt{55}$

65. $(\sqrt{13} - \sqrt{2})^2$
$15 - 2\sqrt{26}$

66. $(\sqrt{5} - \sqrt{7})^2$
$12 - 2\sqrt{35}$

Rationalize each denominator. Assume that all variables represent positive real numbers.

67. $\dfrac{-6\sqrt{3}}{\sqrt{2}}$
$-3\sqrt{6}$

68. $\dfrac{3\sqrt{7p}}{\sqrt{y}}$
$\dfrac{3\sqrt{7py}}{y}$

69. $-\sqrt[3]{\dfrac{9}{25}}$
$-\dfrac{\sqrt[3]{45}}{5}$

70. $\sqrt[3]{\dfrac{108m^3}{n^5}}$
$\dfrac{3m\sqrt[3]{4n}}{n^2}$

71. $\dfrac{1}{\sqrt{2} + \sqrt{7}}$
$\dfrac{\sqrt{2} - \sqrt{7}}{-5}$

72. $\dfrac{-5}{\sqrt{6} - \sqrt{3}}$
$\dfrac{-5(\sqrt{6} + \sqrt{3})}{3}$

[9.6] *Solve each equation.*

73. $\sqrt{8x + 9} = 5$
$\{2\}$

74. $\sqrt{2z - 3} - 3 = 0$
$\{6\}$

75. $\sqrt{3m + 1} = -1$
$\emptyset$

76. $\sqrt{7z + 1} = z + 1$
$\{0, 5\}$

77. $3\sqrt{m} = \sqrt{10m - 9}$
$\{9\}$

78. $\sqrt{p^2 + 3p + 7} = p + 2$
$\{3\}$

79. $\sqrt{x + 2} - \sqrt{x - 3} = 1$
$\{7\}$

80. $\sqrt[3]{5m - 1} = \sqrt[3]{3m - 2}$
$\left\{-\dfrac{1}{2}\right\}$

81. $\sqrt[4]{x + 6} = \sqrt[4]{2x}$
$\{6\}$

[9.7] *Write as a product of a real number and i.*

82. $\sqrt{-25}$
$5i$

83. $\sqrt{-200}$
$10i\sqrt{2}$

84. $\sqrt{-160}$
$4i\sqrt{10}$

Perform the indicated operations. Write answers in standard form.

85. $(-2 + 5i) + (-8 - 7i)$
$-10 - 2i$

86. $(5 + 4i) - (-9 - 3i)$
$14 + 7i$

87. $\sqrt{-5} \cdot \sqrt{-7}$
$-\sqrt{35}$

88. $\sqrt{-25} \cdot \sqrt{-81}$
-45

89. $\dfrac{\sqrt{-72}}{\sqrt{-8}}$
3

90. $(2 + 3i)(1 - i)$
$5 + i$

91. $(6 - 2i)^2$ $32 - 24i$

92. $\dfrac{3 - i}{2 + i}$ $1 - i$

93. $\dfrac{5 + 14i}{2 + 3i}$ $4 + i$

Find each power of i.

94. i^{11}

$-i$

95. i^{52}

1

96. i^{-13}

$-i$

Mixed Review Exercises

Simplify. Assume that all variables represent positive real numbers.

97. $-\sqrt{169a^2b^4}$ $-13ab^2$

98. $1000^{-2/3}$ $\dfrac{1}{100}$

99. $\dfrac{y^{-1/3} \cdot y^{5/6}}{y}$ $\dfrac{1}{y^{1/2}}$

100. $\dfrac{z^{-1/4}\,x^{1/2}}{z^{1/2}\,x^{-1/4}}$ $\dfrac{x^{3/4}}{z^{3/4}}$

101. $\sqrt[4]{k^{24}}$ k^6

102. $\sqrt[3]{54z^9t^8}$ $3z^3t^2\sqrt[3]{2t^2}$

103. $-5\sqrt{18} + 12\sqrt{72}$

$57\sqrt{2}$

104. $8\sqrt[3]{x^3y^2} - 2x\sqrt[3]{y^2}$

$6x\sqrt[3]{y^2}$

105. $(\sqrt{5} - \sqrt{3})(\sqrt{7} + \sqrt{3})$

$\sqrt{35} + \sqrt{15} - \sqrt{21} - 3$

106. $\dfrac{-1}{\sqrt{12}}$ $-\dfrac{\sqrt{3}}{6}$

107. $\sqrt[3]{\dfrac{12}{25}}$ $\dfrac{\sqrt[3]{60}}{5}$

108. $\dfrac{2\sqrt{z}}{\sqrt{z} - 2}$ $\dfrac{2\sqrt{z}(\sqrt{z} + 2)}{z - 4}$

109. $\sqrt{-49}$

$7i$

110. $(4 - 9i) + (-1 + 2i)$

$3 - 7i$

111. $\dfrac{\sqrt{50}}{\sqrt{-2}}$

$-5i$

Solve each equation.

112. $\sqrt{x + 4} = x - 2$

$\{5\}$

113. $\sqrt{6 + 2x} - 1 = \sqrt{7 - 2x}$

$\left\{\dfrac{3}{2}\right\}$

Solve each problem.

114. Carpenters stabilize wall frames with a diagonal brace as shown in the figure. The length of the brace is given by $L = \sqrt{H^2 + W^2}$. If the bottom of the brace is attached 9 ft from the corner and the brace is 12 ft long, how far up the corner post should it be nailed (to the nearest tenth of a foot)?

7.9 ft

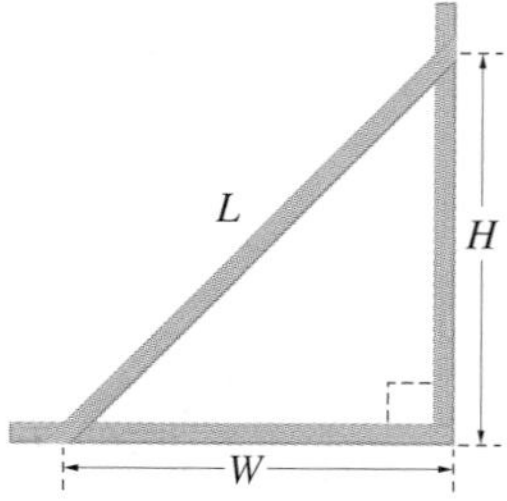

115. Find the perimeter of a triangular electronic highway road sign having the dimensions shown in the figure.

$(12\sqrt{3} + 5\sqrt{2})$ ft

Chapter 9 Test

CHAPTER Test Prep VIDEO CD Use the Chapter Test Prep Video CD to see fully worked-out solutions to any of the exercises you want to review.

Find each root. Use a calculator as necessary.

1. $-\sqrt{841}$ **2.** $\sqrt[3]{-512}$ **3.** $125^{1/3}$

4. For $\sqrt{146.25}$, which choice gives the best estimate?

A. 10 **B.** 11 **C.** 12 **D.** 13

Use a calculator to approximate each root to the nearest thousandth.

5. $\sqrt{478}$ **6.** $\sqrt[3]{-832}$

7. Graph the function defined by $f(x) = \sqrt{x + 6}$, and give the domain and the range.

Simplify each expression. Assume that all variables represent positive real numbers.

8. $(-64)^{-4/3}$

9. $\dfrac{3^{2/5}x^{-1/4}y^{2/5}}{3^{-8/5}x^{7/4}y^{1/10}}$

10. $\sqrt{54x^5y^6}$

11. $\sqrt[4]{32a^7b^{13}}$

12. $\sqrt{2} \cdot \sqrt[3]{5}$

13. $3\sqrt{20} - 5\sqrt{80} + 4\sqrt{500}$

14. $(7\sqrt{5} + 4)(2\sqrt{5} - 1)$

Answers:

1. -29
2. -8
3. 5
4. C
5. 21.863
6. -9.405
7. domain: $[-6, \infty)$; range: $[0, \infty)$

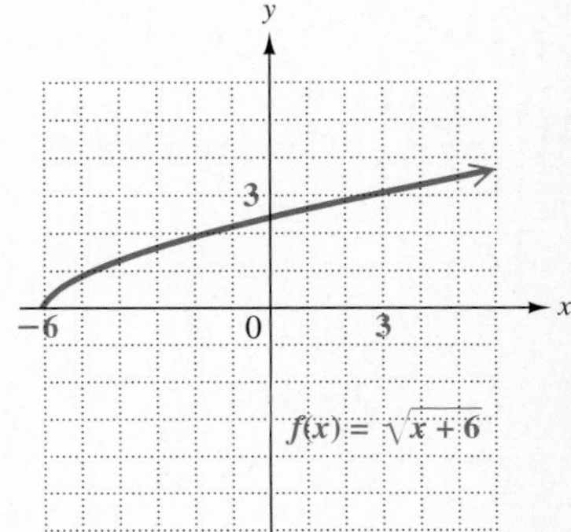

8. $\dfrac{1}{256}$
9. $\dfrac{9y^{3/10}}{x^2}$
10. $3x^2y^3\sqrt{6x}$
11. $2ab^3\sqrt[4]{2a^3b}$
12. $\sqrt[6]{200}$
13. $26\sqrt{5}$
14. $66 + \sqrt{5}$

15. $23 - 4\sqrt{15}$

16. $\frac{-\sqrt{10}}{4}$

17. $\frac{2\sqrt[3]{25}}{5}$

18. $-2(\sqrt{7} - \sqrt{5})$

19. $3 + \sqrt{6}$

20. $\sqrt{26}$

21. $\sqrt{145}$

22. $\{-1\}$

23. $\{6\}$

24. $\{-3\}$

25. $-5 - 8i$

26. $-10 + 10i$

27. $3 + 4i$

28. $-i$

15. $(\sqrt{3} - 2\sqrt{5})^2$

16. $\frac{-5}{\sqrt{40}}$

17. $\frac{2}{\sqrt[3]{5}}$

18. $\frac{-4}{\sqrt{7} + \sqrt{5}}$

19. Write $\frac{6 + \sqrt{24}}{2}$ in lowest terms.

20. Find the distance between the points $(-3, 8)$ and $(2, 7)$.

21. Use the Pythagorean formula to find the exact length of side b in the figure.

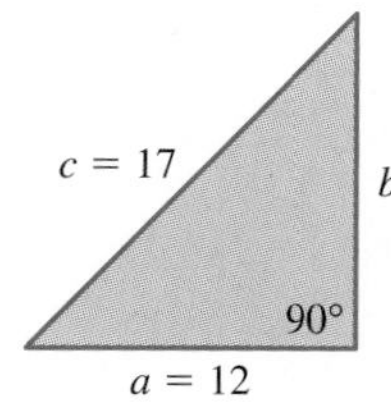

Solve each equation.

22. $\sqrt[3]{5x} = \sqrt[3]{2x - 3}$

23. $\sqrt{7 - x} + 5 = x$

24. $\sqrt{x + 4} - \sqrt{1 - x} = -1$

Perform the indicated operations. Express answers in the form $a + bi$.

25. $(-2 + 5i) - (3 + 6i) - 7i$

26. $(-4 + 2i)(3 - i)$

27. $\frac{7 + i}{1 - i}$

28. Simplify i^{35}.

Cumulative Review Exercises ▶▶▶ Chapters R–9

Solve each equation or inequality.

1. $7 - (4 + 3t) + 2t = -6(t - 2) - 5$ $\left\{\frac{4}{5}\right\}$

2. $\frac{1}{3}x + \frac{1}{4}(x + 8) = x + 7$ $\{-12\}$

3. $|6x - 9| = |-4x + 2|$ $\left\{\frac{11}{10}, \frac{7}{2}\right\}$

4. $-5 - 3(x - 2) < 11 - 2(x + 2)$ $(-6, \infty)$

5. $1 + 4x > 5$ and $-2x > -6$ $(1, 3)$

6. $-2 < 1 - 3x < 7$ $(-2, 1)$

7. Write an equation of the line through the points $(-4, 6)$ and $(7, -6)$. $12x + 11y = 18$

8. Choose the correct response: The lines with equations $2x + 3y = 8$ and $6y = 4x + 16$ are

A. parallel **B.** perpendicular **C.** neither. C

9. For the graph of $f(x) = -3x + 6$,

(a) what is the y-intercept? $(0, 6)$ **(b)** what is the x-intercept? $(2, 0)$

10. Graph the inequality $-2x + y < -6$.

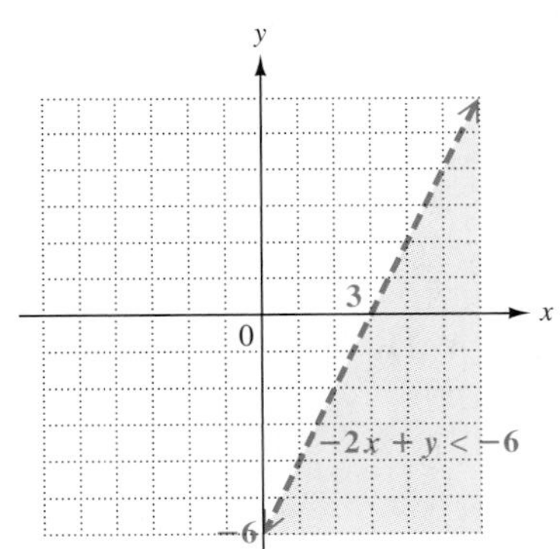

11. Find the measures of the marked angles.

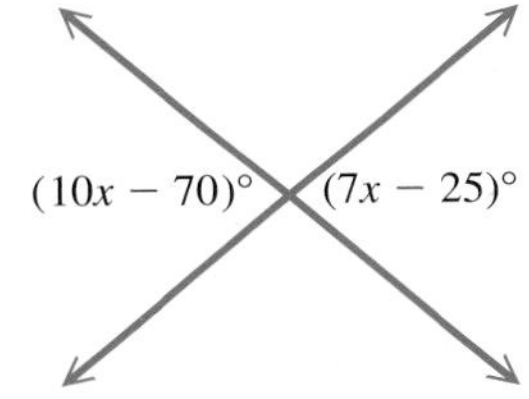

Both angles measure 80°.

Solve each system.

12. $3x - y = 23$
$2x + 3y = 8$

$\{(7, -2)\}$

13. $5x + 2y = 7$
$10x + 4y = 12$

$\emptyset$

14. $2x + y - z = 5$
$6x + 3y - 3z = 15$
$4x + 2y - 2z = 10$

$\{(x, y, z) \mid 2x + y - z = 5\}$

15. In 2006, if you had sent five 2-oz letters and three 3-oz letters by first-class mail, it would have cost \$5.76. Sending three 2-oz letters and five 3-oz letters would have cost \$6.24. What was the 2006 postage rate for one 2-oz letter and for one 3-oz letter? (*Source:* U.S. Postal Service.)

2-oz letter: \$0.63; 3-oz letter: \$0.87

Perform the indicated operations.

16. $(3k^3 - 5k^2 + 8k - 2) - (4k^3 + 11k + 7) + (2k^2 - 5k)$

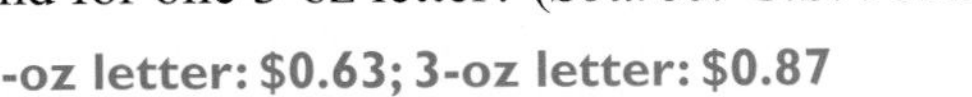

$-k^3 - 3k^2 - 8k - 9$

17. $(8x - 7)(x + 3)$

$8x^2 + 17x - 21$

18. $\frac{8z^3 - 16z^2 + 24z}{8z^2}$

$z - 2 + \frac{3}{z}$

19. $\frac{6y^4 - 3y^3 + 5y^2 + 6y - 9}{2y + 1}$

$3y^3 - 3y^2 + 4y + 1 + \frac{-10}{2y + 1}$

Factor each polynomial completely.

20. $2p^2 - 5pq + 3q^2$
$(2p - 3q)(p - q)$

21. $18k^4 + 9k^2 - 20$
$(3k^2 + 4)(6k^2 - 5)$

22. $x^3 + 512$
$(x + 8)(x^2 - 8x + 64)$

Perform each operation and express answers in lowest terms.

23. $\dfrac{y^2 + y - 12}{y^3 + 9y^2 + 20y} \div \dfrac{y^2 - 9}{y^3 + 3y^2}$
$\dfrac{y}{y + 5}$

24. $\dfrac{1}{x + y} + \dfrac{3}{x - y}$
$\dfrac{4x + 2y}{(x + y)(x - y)}$

Simplify each complex fraction.

25. $\dfrac{\frac{-6}{x - 2}}{\frac{8}{3x - 6}}$
$-\dfrac{9}{4}$

26. $\dfrac{\frac{1}{a} - \frac{1}{b}}{\frac{a}{b} - \frac{b}{a}}$
$-\dfrac{1}{a + b}$

Solve.

27. $2x^2 + 11x + 15 = 0$
$\left\{-3, -\dfrac{5}{2}\right\}$

28. $5t(t - 1) = 2(1 - t)$
$\left\{-\dfrac{2}{5}, 1\right\}$

Simplify.

29. $27^{-5/3}$
$\dfrac{1}{243}$

30. $\dfrac{x^{-2/3}}{x^{-3/4}}$, $x \neq 0$
$x^{1/12}$

31. $8\sqrt{20} + 3\sqrt{80} - 2\sqrt{500}$
$8\sqrt{5}$

32. $\dfrac{-9}{\sqrt{80}}$
$\dfrac{-9\sqrt{5}}{20}$

33. $\dfrac{4}{\sqrt{6} - \sqrt{5}}$
$4(\sqrt{6} + \sqrt{5})$

34. $\dfrac{12}{\sqrt[3]{2}}$
$6\sqrt[3]{4}$

35. Find the distance between the points $(-4, 4)$ and $(-2, 9)$.
$\sqrt{29}$

36. Solve $\sqrt{8x - 4} - \sqrt{7x + 2} = 0$.
$\{6\}$

Solve each problem.

37. The current of a river runs at 3 mph. Brent's boat can go 36 mi downstream in the same time that it takes to go 24 mi upstream. Find the speed of the boat in still water. **15 mph**

38. How many liters of pure alcohol must be mixed with 40 L of 18% alcohol to obtain a 22% alcohol solution?
$\dfrac{80}{39}$, or $2\dfrac{2}{39}$ L

39. A jar containing only dimes and quarters has 29 coins with a face value of \$4.70. How many of each denomination are there?
17 dimes and 12 quarters

40. Brenda rides her bike 4 mph faster than her husband, Chuck. If Brenda can ride 48 mi in the same time that Chuck can ride 24 mi, what are their speeds?
Brenda: 8 mph; Chuck: 4 mph

10

Quadratic Equations, Inequalities, and Functions

In 2008 the prices of food, gasoline, and other products increased throughout the world. In particular, escalating oil prices caused increases in transportation and shipping costs, which trickled down to affect prices of a variety of goods and services.

Although prices tend to go up over time, the rate at which they increase (the inflation rate) varies considerably. The Consumer Price Index (CPI) used by the U.S. government measures changes in prices for goods purchased by typical American families over time. In Example 6 of Section 10.5, we use a *quadratic function* to model the CPI.

10.1 Solving Quadratic Equations by the Square Root Property

OBJECTIVES

1. **Review the zero-factor property.**
2. **Solve equations of the form $x^2 = k$, where $k > 0$.**
3. **Solve equations of the form $(ax + b)^2 = k$, where $k > 0$.**
4. **Solve quadratic equations with nonreal complex solutions.**

We introduced quadratic equations in **Section 6.7.** Recall that a *quadratic equation* is defined as follows.

Quadratic Equation

An equation that can be written in the form

$$ax^2 + bx + c = 0,$$

where a, b, and c are real numbers, with $a \neq 0$, is a **quadratic equation.** The given form is called **standard form.**

A quadratic equation is a *second-degree equation,* that is, an equation with a squared term and no terms of higher degree. For example,

$$4m^2 + 4m - 5 = 0 \quad \text{and} \quad 3x^2 = 4x - 8 \qquad \text{Quadratic equations}$$

are quadratic equations, with the first equation in standard form.

◀ Work Problem 1 at the Side.

OBJECTIVE 1 Review the zero-factor property. In **Section 6.7** we used factoring and the zero-factor property to solve quadratic equations.

Zero-Factor Property

If two numbers have a product of 0, then at least one of the numbers must be 0. That is, if $ab = 0$, then $a = 0$ or $b = 0$.

We solved a quadratic equation such as $3x^2 - 5x - 28 = 0$ using the zero-factor property as follows.

$$3x^2 - 5x - 28 = 0$$

$$(3x + 7)(x - 4) = 0 \qquad \text{Factor.}$$

$$3x + 7 = 0 \quad \text{or} \quad x - 4 = 0 \qquad \text{Zero-factor property}$$

$$3x = -7 \quad \text{or} \quad x = 4 \qquad \text{Solve each equation.}$$

$$x = -\frac{7}{3}$$

The solution set is $\left\{-\frac{7}{3}, 4\right\}$.

◀ Work Problem 2 at the Side.

OBJECTIVE 2 Solve equations of the form $x^2 = k$, where $k > 0$. We can solve an equation such as $x^2 = 9$ by factoring as follows.

$$x^2 = 9$$

$$x^2 - 9 = 0 \qquad \text{Subtract 9.}$$

$$(x + 3)(x - 3) = 0 \qquad \text{Factor.}$$

$$x + 3 = 0 \quad \text{or} \quad x - 3 = 0 \qquad \text{Zero-factor property}$$

$$x = -3 \quad \text{or} \quad x = 3 \qquad \text{Solve each equation.}$$

1 (a) Which of the following are quadratic equations?

A. $x + 2y = 0$

B. $x^2 - 8x + 16 = 0$

C. $2t^2 - 5t = 3$

D. $x^3 + x^2 + 4 = 0$

(b) Which quadratic equation identified in part (a) is in standard form?

2 Solve each equation by factoring.

(a) $x^2 + 3x + 2 = 0$

(b) $3m^2 = 3 - 8m$

(*Hint:* Remember to write the equation in standard form first.)

ANSWERS

1. (a) B, C (b) B

2. (a) $\{-2, -1\}$ (b) $\left\{-3, \frac{1}{3}\right\}$

We might also solve $x^2 = 9$ by noticing that x must be a number whose square is 9. Thus, $x = \sqrt{9} = 3$ or $x = -\sqrt{9} = -3$. This approach is generalized as the **square root property.**

Square Root Property

If k is a complex number and if $x^2 = k$, then

$$x = \sqrt{k} \quad \text{or} \quad x = -\sqrt{k}.$$

CAUTION

If $k \neq 0$, then using the square root property always produces *two* square roots, one positive and one negative.

EXAMPLE 1 Using the Square Root Property

Solve each equation.

(a) $x^2 = 16$

By the square root property, if $x^2 = 16$, then

$$x = \sqrt{16} = 4 \quad \text{or} \quad x = -\sqrt{16} = -4.$$

An abbreviation for $x = 4$ or $x = -4$ is $x = \pm 4$ (read "positive or negative 4," or "plus or minus 4"). Check each solution by substituting it for x in the original equation. The solution set is $\{4, -4\}$, or $\{\pm 4\}$.

(b) $r^2 = 5$

By the square root property, if $r^2 = 5$, then

$$r = \sqrt{5} \quad \text{or} \quad r = -\sqrt{5},$$

Don't forget the negative solution.

and the solution set is $\{\sqrt{5}, -\sqrt{5}\}$, which may be written $\{\pm\sqrt{5}\}$.

(c)

$$m^2 - 20 = 0$$

$$m^2 = 20 \qquad \text{Add 20.}$$

$$m = \sqrt{20} \quad \text{or} \quad m = -\sqrt{20} \qquad \text{Square root property}$$

$$m = 2\sqrt{5} \quad \text{or} \quad m = -2\sqrt{5} \qquad \sqrt{20} = \sqrt{4} \cdot \sqrt{5} = 2\sqrt{5}$$

Check

$$m^2 - 20 = 0 \qquad \text{Original equation}$$

$$(2\sqrt{5})^2 - 20 \stackrel{?}{=} 0 \qquad \text{Let } m = 2\sqrt{5}.$$

$$20 - 20 \stackrel{?}{=} 0 \qquad (2\sqrt{5})^2 = 2^2 \cdot (\sqrt{5})^2 = 4 \cdot 5 = 20$$

$$0 = 0 \qquad \text{True}$$

The check of $-2\sqrt{5}$ is similar. The solution set is $\{2\sqrt{5}, -2\sqrt{5}\}$.

Continued on Next Page

3 Solve each equation.

(a) $m^2 = 64$

(b) $p^2 = 7$

(c) $x^2 - 12 = 0$

(d) $3x^2 - 54 = 0$

(e) $2x^2 + 7 = 55$

4 Solve the problem.

An expert marksman can hold a silver dollar at forehead level, drop it, draw his gun, and shoot the coin as it passes waist level. If the coin falls about 4 ft, use the formula in Example 2 to find the time that elapses between the dropping of the coin and the shot.

Answers

3. **(a)** $\{8, -8\}$ **(b)** $\{\sqrt{7}, -\sqrt{7}\}$
(c) $\{2\sqrt{3}, -2\sqrt{3}\}$ **(d)** $\{3\sqrt{2}, -3\sqrt{2}\}$
(e) $\{2\sqrt{6}, -2\sqrt{6}\}$

4. 0.5 sec

(d)

$$4x^2 - 48 = 0$$
$$4x^2 = 48 \quad \text{Add 48.}$$
$$x^2 = 12 \quad \text{Divide by 4.}$$
$$x = \sqrt{12} \quad \text{or} \quad x = -\sqrt{12} \quad \text{Square root property}$$
$$x = 2\sqrt{3} \quad \text{or} \quad x = -2\sqrt{3} \quad \sqrt{12} = \sqrt{4} \cdot \sqrt{3} = 2\sqrt{3}$$

The solutions are $2\sqrt{3}$ and $-2\sqrt{3}$. Check each in the original equation.

Check $4x^2 - 48 = 0$ Original equation

$4(2\sqrt{3})^2 - 48 \stackrel{?}{=} 0$	$4(-2\sqrt{3})^2 - 48 \stackrel{?}{=} 0$
$4(12) - 48 \stackrel{?}{=} 0$	$4(12) - 48 \stackrel{?}{=} 0$
$48 - 48 \stackrel{?}{=} 0$	$48 - 48 \stackrel{?}{=} 0$
$0 = 0$ True	$0 = 0$ True

The solution set is $\{2\sqrt{3}, -2\sqrt{3}\}$.

(e)

$$5m^2 - 32 = 8$$
$$5m^2 = 40 \quad \text{Add 32.}$$
$$m^2 = 8 \quad \text{Divide by 5.}$$

Don't stop here. Simplify the radicals.

$$m = \sqrt{8} \quad \text{or} \quad m = -\sqrt{8} \quad \text{Square root property}$$
$$m = 2\sqrt{2} \quad \text{or} \quad m = -2\sqrt{2} \quad \sqrt{8} = \sqrt{4} \cdot \sqrt{2} = 2\sqrt{2}$$

The solution set is $\{2\sqrt{2}, -2\sqrt{2}\}$.

◀ Work Problem **3** at the Side.

EXAMPLE 2 **Using the Square Root Property in an Application**

Galileo Galilei (1564–1642) developed a formula for freely falling objects described by

$$d = 16t^2,$$

where d is the distance in feet that an object falls (disregarding air resistance) in t seconds, regardless of weight. Galileo dropped objects from the Leaning Tower of Pisa to develop this formula. If the Leaning Tower is about 180 ft tall, use Galileo's formula to determine how long it would take an object dropped from the tower to fall to the ground. (*Source: Microsoft Encarta Encyclopedia.*)

$$d = 16t^2$$
$$\mathbf{180} = 16t^2 \quad \text{Let } d = 180.$$
$$11.25 = t^2 \quad \text{Divide by 16.}$$
$$t = \sqrt{11.25} \quad \text{or} \quad t = -\sqrt{11.25} \quad \text{Square root property}$$

Time cannot be negative, so we discard $t = -\sqrt{11.25}$. Using a calculator, $\sqrt{11.25} \approx 3.4$ so $t \approx 3.4$. The object would fall to the ground in about 3.4 sec.

◀ Work Problem **4** at the Side.

OBJECTIVE 3 Solve equations of the form $(ax + b)^2 = k$, where $k > 0$. In each equation in Example 1, the exponent 2 had a single variable as its base. We can extend the square root property to solve equations in which the base is a binomial, as shown in the next example.

EXAMPLE 3 Using the Square Root Property

Solve each equation.

(a) Use $x - 3$ as the base. $\quad (x-3)^2 = 16$

$x - 3 = \sqrt{16}$ or $x - 3 = -\sqrt{16}$ — Square root property

$x - 3 = 4$ or $x - 3 = -4$ — $\sqrt{16} = 4$

$x = 7$ or $x = -1$ — Add 3.

Check Substitute each solution in the original equation.

$(x-3)^2 = 16$		$(x-3)^2 = 16$	
$(7-3)^2 \stackrel{?}{=} 16$	Let $x = 7$.	$(-1-3)^2 \stackrel{?}{=} 16$	Let $x = -1$.
$4^2 \stackrel{?}{=} 16$		$(-4)^2 \stackrel{?}{=} 16$	
$16 = 16$	True	$16 = 16$	True

The solutions are 7 and -1, and the solution set is $\{7, -1\}$.

(b) $\quad (x+1)^2 = 6$

$x + 1 = \sqrt{6}$ or $x + 1 = -\sqrt{6}$ — Square root property

$x = -1 + \sqrt{6}$ or $x = -1 - \sqrt{6}$ — Add -1.

Check $\quad \left(-1 + \sqrt{6} + 1\right)^2 = \left(\sqrt{6}\right)^2 = 6;$

$\left(-1 - \sqrt{6} + 1\right)^2 = \left(-\sqrt{6}\right)^2 = 6.$

The solution set is $\left\{-1 + \sqrt{6}, -1 - \sqrt{6}\right\}$.

Work Problem 5 at the Side. ▶

5 Solve each equation.

(a) $(x+2)^2 = 36$

(b) $(x-4)^2 = 3$

EXAMPLE 4 Using the Square Root Property

Solve $(2x-3)^2 = 18$.

$$(2x-3)^2 = 18$$

$2x - 3 = \sqrt{18}$ or $2x - 3 = -\sqrt{18}$ — Square root property

$2x = 3 + \sqrt{18}$ or $2x = 3 - \sqrt{18}$ — Add 3.

$x = \dfrac{3 + \sqrt{18}}{2}$ or $x = \dfrac{3 - \sqrt{18}}{2}$ — Divide by 2.

$x = \dfrac{3 + 3\sqrt{2}}{2}$ or $x = \dfrac{3 - 3\sqrt{2}}{2}$ — $\sqrt{18} = \sqrt{9} \cdot \sqrt{2} = 3\sqrt{2}$

Continued on Next Page

ANSWERS

5. (a) $\{-8, 4\}$ (b) $\left\{4 + \sqrt{3}, 4 - \sqrt{3}\right\}$

We show the check for the first solution. The check for the second solution is similar.

Check

$$(2x - 3)^2 = 18 \quad \text{Original equation}$$

$$\left[2\left(\frac{3 + 3\sqrt{2}}{2}\right) - 3\right]^2 \stackrel{?}{=} 18 \quad \text{Let } x = \frac{3 + 3\sqrt{2}}{2}.$$

$$\left(3 + 3\sqrt{2} - 3\right)^2 \stackrel{?}{=} 18 \quad \text{Multiply.}$$

$$\left(3\sqrt{2}\right)^2 \stackrel{?}{=} 18 \quad \text{Simplify.}$$

$$18 = 18 \quad \text{True}$$

The solution set is $\left\{\frac{3 + 3\sqrt{2}}{2}, \frac{3 - 3\sqrt{2}}{2}\right\}$.

◀ Work Problem 6 at the Side.

6 Solve each equation.

(a) $(x - 3)^2 = 25$

(b) $(3k + 1)^2 = 2$

(c) $(2r + 3)^2 = 8$

OBJECTIVE 4 Solve quadratic equations with nonreal complex solutions. In the equation $x^2 = k$, if $k < 0$, there will be two nonreal complex solutions.

EXAMPLE 5 Solving for Nonreal Complex Solutions

Solve each equation.

(a) $x^2 = -15$

$$x = \sqrt{-15} \quad \text{or} \quad x = -\sqrt{-15} \quad \text{Square root property}$$

$$x = i\sqrt{15} \quad \text{or} \quad x = -i\sqrt{15} \quad \sqrt{-1} = i$$

The solution set is $\{i\sqrt{15}, -i\sqrt{15}\}$.

(b) $(t + 2)^2 = -16$

$$t + 2 = \sqrt{-16} \quad \text{or} \quad t + 2 = -\sqrt{-16} \quad \text{Square root property}$$

$$t + 2 = 4i \quad \text{or} \quad t + 2 = -4i \quad \sqrt{-16} = 4i$$

$$t = -2 + 4i \quad \text{or} \quad t = -2 - 4i \quad \text{Subtract 2.}$$

The solution set is $\{-2 + 4i, -2 - 4i\}$.

◀ Work Problem 7 at the Side.

7 Solve each equation.

(a) $x^2 = -17$

(b) $(k + 5)^2 = -100$

Answers

6. (a) $\{-2, 8\}$
(b) $\left\{\frac{-1 + \sqrt{2}}{3}, \frac{-1 - \sqrt{2}}{3}\right\}$
(c) $\left\{\frac{-3 + 2\sqrt{2}}{2}, \frac{-3 - 2\sqrt{2}}{2}\right\}$

7. (a) $\{i\sqrt{17}, -i\sqrt{17}\}$
(b) $\{-5 + 10i, -5 - 10i\}$

10.1 Exercises

FOR EXTRA HELP

MyMathLab

Math XL PRACTICE

WATCH

DOWNLOAD

READ

REVIEW

1. A student was asked to solve the quadratic equation $x^2 = 16$ and did not get full credit for the solution set $\{4\}$. Why?

   The equation is also true for $x = -4$.

2. Why can't the zero-factor property be used to solve every quadratic equation?

   Some quadratic equations cannot be easily factored.

3. Give a one-sentence description or explanation of each of the following.

   (a) Quadratic equation in standard form

   A quadratic equation in standard form has a second-degree polynomial in decreasing powers equal to 0.

   (b) Square root property

   The square root property states that if the square of a quantity equals a number, then the quantity equals the positive or negative square root of the number.

4. A student tried to solve $x^2 - x - 2 = 5$ as follows.

$$x^2 - x - 2 = 5$$

$$(x - 2)(x + 1) = 5 \quad \text{Factor.}$$

$$x - 2 = 5 \quad \text{or} \quad x + 1 = 5 \quad \text{Zero-factor property}$$

$$x = 7 \quad \text{or} \quad x = 4 \quad \text{Solve each equation.}$$

   This method is incorrect. ***WHAT WENT WRONG?***

   The zero-factor property requires a product equal to 0. The first step should have been to rewrite the equation with 0 on one side.

Solve each equation. See Example 1.

5. $x^2 = 81$
   $\{9, -9\}$

6. $z^2 = 225$
   $\{15, -15\}$

7. $t^2 = 17$
   $\{\sqrt{17}, -\sqrt{17}\}$

8. $k^2 = 19$
   $\{\sqrt{19}, -\sqrt{19}\}$

9. $m^2 = 32$
   $\{4\sqrt{2}, -4\sqrt{2}\}$

10. $x^2 = 54$
   $\{3\sqrt{6}, -3\sqrt{6}\}$

11. $r^2 - 3 = 0$
   $\{\sqrt{3}, -\sqrt{3}\}$

12. $x^2 - 13 = 0$
   $\{\sqrt{13}, -\sqrt{13}\}$

13. $t^2 - 48 = 0$
   $\{4\sqrt{3}, -4\sqrt{3}\}$

14. $p^2 - 50 = 0$
   $\{5\sqrt{2}, -5\sqrt{2}\}$

15. $3n^2 - 72 = 0$
   $\{2\sqrt{6}, -2\sqrt{6}\}$

16. $5z^2 - 200 = 0$
   $\{2\sqrt{10}, -2\sqrt{10}\}$

17. $2x^2 + 7 = 61$
   $\{3\sqrt{3}, -3\sqrt{3}\}$

18. $3x^2 - 8 = 64$
   $\{2\sqrt{6}, -2\sqrt{6}\}$

Solve each equation. See Examples 3 and 4.

19. $(x + 2)^2 = 25$
   $\{-7, 3\}$

20. $(t + 8)^2 = 9$
   $\{-5, -11\}$

21. $(x - 4)^2 = 3$
   $\{4 + \sqrt{3}, 4 - \sqrt{3}\}$

22. $(x + 3)^2 = 11$
$\{-3 + \sqrt{11}, -3 - \sqrt{11}\}$

23. $(t + 5)^2 = 48$
$\{-5 + 4\sqrt{3}, -5 - 4\sqrt{3}\}$

24. $(m - 6)^2 = 27$
$\{6 + 3\sqrt{3}, 6 - 3\sqrt{3}\}$

25. $(3x + 2)^2 = 49$
$\left\{-3, \frac{5}{3}\right\}$

26. $(5x + 3)^2 = 36$
$\left\{-\frac{9}{5}, \frac{3}{5}\right\}$

27. $(3k - 1)^2 = 7$
$\left\{\frac{1 + \sqrt{7}}{3}, \frac{1 - \sqrt{7}}{3}\right\}$

28. $(2x + 4)^2 = 10$
$\left\{\frac{-4 + \sqrt{10}}{2}, \frac{-4 - \sqrt{10}}{2}\right\}$

29. $(4p + 1)^2 = 24$
$\left\{\frac{-1 + 2\sqrt{6}}{4}, \frac{-1 - 2\sqrt{6}}{4}\right\}$

30. $(5k - 2)^2 = 12$
$\left\{\frac{2 + 2\sqrt{3}}{5}, \frac{2 - 2\sqrt{3}}{5}\right\}$

31. $(3x + 1)^2 = 18$
$\left\{\frac{-1 + 3\sqrt{2}}{3}, \frac{-1 - 3\sqrt{2}}{3}\right\}$

32. $(5x + 6)^2 = 75$
$\left\{\frac{-6 + 5\sqrt{3}}{5}, \frac{-6 - 5\sqrt{3}}{5}\right\}$

Find all complex solutions of each equation. See Example 5.

33. $x^2 = -100$
$\{10i, -10i\}$

34. $x^2 = -64$
$\{8i, -8i\}$

35. $x^2 = -12$
$\{2i\sqrt{3}, -2i\sqrt{3}\}$

36. $x^2 = -18$
$\{3i\sqrt{2}, -3i\sqrt{2}\}$

37. $(r - 5)^2 = -3$
$\{5 + i\sqrt{3}, 5 - i\sqrt{3}\}$

38. $(t + 6)^2 = -5$
$\{-6 + i\sqrt{5}, -6 - i\sqrt{5}\}$

39. $(6k - 1)^2 = -8$
$\left\{\frac{1}{6} + \frac{\sqrt{2}}{3}i, \frac{1}{6} - \frac{\sqrt{2}}{3}i\right\}$

40. $(4m - 7)^2 = -27$
$\left\{\frac{7}{4} + \frac{3\sqrt{3}}{4}i, \frac{7}{4} - \frac{3\sqrt{3}}{4}i\right\}$

Solve Exercises 41 and 42 using Galileo's formula, $d = 16t^2$. *Round answers to the nearest tenth. See Example 2.*

41. Mount Rushmore National Memorial in South Dakota features a sculpture of four of America's favorite presidents carved into the rim of the mountain, 500 ft above the valley floor. How long would it take a rock dropped from the top of the sculpture to fall to the ground? (*Source: Microsoft Encarta Encyclopedia.*) **5.6 sec**

42. The Gateway Arch in St. Louis, Missouri, the tallest national monument in the United States, is 630 ft tall. How long would it take an object dropped from the top of it to fall to the ground? (*Source:* www.gatewayarch.com) **6.3 sec**

10.2 Solving Quadratic Equations by Completing the Square

OBJECTIVES

1 Solve quadratic equations by completing the square when the coefficient of the second-degree term is 1.

2 Solve quadratic equations by completing the square when the coefficient of the second-degree term is not 1.

3 Simplify the terms of an equation before solving.

OBJECTIVE 1 Solve quadratic equations by completing the square when the coefficient of the second-degree term is 1. The methods we have studied so far are not enough to solve the equation

$$x^2 + 8x + 10 = 0.$$

If we could write the equation in the form $(x + 4)^2$ equals a constant, we could solve it with the square root property discussed in **Section 10.1.** To do that, we need to have a perfect square trinomial on one side of the equation.

Recall from **Section 6.5** that the perfect square trinomial

$$x^2 + 8x + 16$$

can be factored as $(x + 4)^2$. In the trinomial, the coefficient of x (the first-degree term) is 8 and the constant term is 16. Notice that if we take half of 8 and square it, we get the constant term, 16.

$$\underset{\text{Coefficient of } x}{\left[\frac{1}{2}(8)\right]^2} = 4^2 = \underset{\text{Constant}}{16}$$

Similarly, in

$$x^2 + 12x + 36, \quad \left[\frac{1}{2}(12)\right]^2 = 6^2 = 36,$$

and in

$$m^2 - 6m + 9, \quad \left[\frac{1}{2}(-6)\right]^2 = (-3)^2 = 9.$$

This relationship is true in general and is the idea behind rewriting a quadratic equation so that the square root property can be applied.

Work Problem **1** *at the Side.* ▶

1 Find the constant to be added to get a perfect square trinomial. In each case, take half the coefficient of the first-degree term and square the result.

(a) $x^2 + 4x +$ ___

(b) $t^2 - 2t +$ ___

(c) $m^2 + 5m +$ ___

(d) $x^2 - \frac{2}{3}x +$ ___

EXAMPLE 1 Rewriting to Use the Square Root Property

Solve $x^2 + 8x + 10 = 0$.

This quadratic equation cannot be solved easily by factoring, and it is not in the correct form to solve using the square root property. We need a perfect square trinomial on the left side of the equation. To get this form, we first subtract 10 from each side.

$$x^2 + 8x + 10 = 0 \quad \text{Original equation}$$
$$x^2 + 8x = -10 \quad \text{Subtract 10.}$$

We must add a constant to get a perfect square trinomial on the left.

$$\underbrace{x^2 + 8x + \underline{\ ?\ }}_{\text{Needs to be a perfect square trinomial}}$$

Take half the coefficient of the first-degree term and square the result.

$$\left[\frac{1}{2}(8)\right]^2 = 4^2 = 16 \leftarrow \text{Desired constant}$$

Continued on Next Page

ANSWERS

1. (a) 4 (b) 1 (c) $\frac{25}{4}$ (d) $\frac{1}{9}$

We add this constant, 16, to *each* side of the equation.

$$x^2 + 8x = 10$$

This is a key step.

$$x^2 + 8x + 16 = -10 + 16$$

Next we factor on the left side and add on the right.

$$(x + 4)^2 = 6$$

We can now use the square root property.

$$x + 4 = \sqrt{6} \quad \text{or} \quad x + 4 = -\sqrt{6}$$

$$x = -4 + \sqrt{6} \quad \text{or} \quad x = -4 - \sqrt{6}$$

Check

$$x^2 + 8x + 10 = 0 \qquad \text{Original equation}$$

$$(-4 + \sqrt{6})^2 + 8(-4 + \sqrt{6}) + 10 \stackrel{?}{=} 0 \qquad \text{Let } x = -4 + \sqrt{6}.$$

$$16 - 8\sqrt{6} + 6 - 32 + 8\sqrt{6} + 10 \stackrel{?}{=} 0$$

$$0 = 0 \qquad \text{True}$$

Remember the middle term when squaring $-4 + \sqrt{6}$.

The check of the other solution is similar. Thus, $\{-4 + \sqrt{6}, -4 - \sqrt{6}\}$ is the solution set.

◀ Work Problem **2** at the Side.

2 Solve $n^2 + 6n + 4 = 0$ by completing the square.

The process of changing the form of the equation in Example 1 from

$$x^2 + 8x + 10 = 0 \quad \text{to} \quad (x + 4)^2 = 6$$

is called **completing the square.** Completing the square changes only the form of the equation. To see this, multiply out the left side of $(x + 4)^2 = 6$. Then write the equation in standard form to get $x^2 + 8x + 10 = 0$.

3 Solve each equation by completing the square.

(a) $x^2 + 4x = 1$

(b) $z^2 + 6z - 3 = 0$

EXAMPLE 2 Completing the Square to Solve a Quadratic Equation

Complete the square to solve $x^2 - 6x = 12$.

To complete the square on $x^2 - 6x$, take half of -6, the coefficient of x, and square it.

$$\frac{1}{2}(-6) = -3 \quad \text{and} \quad (-3)^2 = 9$$

Coefficient of x (points to -6)

Add the result, 9, to each side of the equation.

$$x^2 - 6x = 12 \qquad \text{Given equation}$$

$$x^2 - 6x + 9 = 12 + 9 \qquad \text{Add 9.}$$

$$(x - 3)^2 = 21 \qquad \text{Factor; add.}$$

$$x - 3 = \sqrt{21} \quad \text{or} \quad x - 3 = -\sqrt{21} \qquad \text{Square root property}$$

$$x = 3 + \sqrt{21} \quad \text{or} \quad x = 3 - \sqrt{21} \qquad \text{Add 3.}$$

A check indicates that the solution set is $\{3 + \sqrt{21}, 3 - \sqrt{21}\}$.

◀ Work Problem **3** at the Side.

ANSWERS

2. $\{-3 + \sqrt{5}, -3 - \sqrt{5}\}$

3. (a) $\{-2 + \sqrt{5}, -2 - \sqrt{5}\}$

(b) $\{-3 + 2\sqrt{3}, -3 - 2\sqrt{3}\}$

Completing the Square

Solve $ax^2 + bx + c = 0$ $(a \neq 0)$ by completing the square as follows:

Step 1 **Be sure the second-degree (squared) term has coefficient 1.** If the coefficient of the squared term is 1, proceed to Step 2. If the coefficient of the squared term is not 1 but some other nonzero number a, divide each side of the equation by a.

Step 2 **Write the equation in correct form** so that terms with variables are on one side of the equals sign and the constant is on the other side.

Step 3 **Square half the coefficient of the first-degree (linear) term.**

Step 4 **Add the square to each side.**

Step 5 **Factor the perfect square trinomial.** One side should now be a perfect square trinomial. Factor it as the square of a binomial. Simplify the other side.

Step 6 **Solve the equation.** Apply the square root property to complete the solution.

EXAMPLE 3 **Solving a Quadratic Equation by Completing the Square ($a = 1$)**

Solve $k^2 + 5k - 1 = 0$.

Since the coefficient of the squared term is 1, begin with Step 2.

Step 2 $$k^2 + 5k = 1 \quad \text{Add 1 to each side.}$$

Step 3 Take half the coefficient of the first-degree term and square the result.

$$\left[\frac{1}{2}(5)\right]^2 = \left(\frac{5}{2}\right)^2 = \frac{25}{4}$$

Step 4 $$k^2 + 5k + \frac{25}{4} = 1 + \frac{25}{4}$$ Add the square to each side of the equation.

Step 5 $$\left(k + \frac{5}{2}\right)^2 = \frac{29}{4}$$ Factor on the left; add on the right.

Step 6 $$k + \frac{5}{2} = \sqrt{\frac{29}{4}} \quad \text{or} \quad k + \frac{5}{2} = -\sqrt{\frac{29}{4}} \quad \text{Square root property}$$

$$k + \frac{5}{2} = \frac{\sqrt{29}}{2} \quad \text{or} \quad k + \frac{5}{2} = -\frac{\sqrt{29}}{2} \quad \sqrt{\tfrac{a}{b}} = \tfrac{\sqrt{a}}{\sqrt{b}}$$

$$k = -\frac{5}{2} + \frac{\sqrt{29}}{2} \quad \text{or} \quad k = -\frac{5}{2} - \frac{\sqrt{29}}{2} \quad \text{Add } -\tfrac{5}{2}.$$

$$k = \frac{-5 + \sqrt{29}}{2} \quad \text{or} \quad k = \frac{-5 - \sqrt{29}}{2} \quad \tfrac{a}{c} + \tfrac{b}{c} = \tfrac{a+b}{c}$$

Check that the solution set is $\left\{\frac{-5 + \sqrt{29}}{2}, \frac{-5 - \sqrt{29}}{2}\right\}$.

Work Problem **4** at the Side. ▶

4 Solve each equation by completing the square.

(a) $x^2 + 2x - 10 = 0$

(b) $r^2 + 3r - 1 = 0$

Answers

4. (a) $\{-1 + \sqrt{11}, -1 - \sqrt{11}\}$

(b) $\left\{\frac{-3 + \sqrt{13}}{2}, \frac{-3 - \sqrt{13}}{2}\right\}$

OBJECTIVE 2 Solve quadratic equations by completing the square when the coefficient of the second-degree term is not 1. If a quadratic equation has the form

$$ax^2 + bx + c = 0, \quad \text{where } a \neq 1,$$

then to obtain 1 as the coefficient of x^2, we first divide each side of the equation by a.

5 Solve each equation by completing the square.

(a) $9x^2 + 18x = -5$

(b) $4t^2 - 24t + 11 = 0$

EXAMPLE 4 Solving a Quadratic Equation by Completing the Square ($a \neq 1$)

Solve $4x^2 + 16x = 9$.

Before completing the square, the coefficient of x^2 must be 1, not 4. We get 1 as the coefficient of x^2 here by dividing each side by 4.

$$4x^2 + 16x = 9$$

The coefficient of x^2 must be 1.

$$x^2 + 4x = \frac{9}{4} \quad \text{Divide by 4. (Step 1)}$$

The equation is in the correct form, so Step 2 is satisfied. Next, we begin to complete the square by taking half the coefficient of x, and squaring it:

$$\frac{1}{2}(4) = 2 \quad \text{and} \quad 2^2 = 4. \quad \text{(Step 3)}$$

We add the result, 4, to each side of the equation.

$$x^2 + 4x + 4 = \frac{9}{4} + 4 \quad \text{Add 4. (Step 4)}$$

$$(x + 2)^2 = \frac{25}{4} \quad \text{Factor; } \tfrac{9}{4} + 4 = \tfrac{9}{4} + \tfrac{16}{4} = \tfrac{25}{4}. \text{ (Step 5)}$$

$$x + 2 = \sqrt{\frac{25}{4}} \quad \text{or} \quad x + 2 = -\sqrt{\frac{25}{4}} \quad \text{Square root property (Step 6)}$$

$$x + 2 = \frac{5}{2} \quad \text{or} \quad x + 2 = -\frac{5}{2} \quad \text{Take square roots.}$$

$$x = -2 + \frac{5}{2} \quad \text{or} \quad x = -2 - \frac{5}{2} \quad \text{Subtract 2.}$$

$$x = \frac{1}{2} \quad \text{or} \quad x = -\frac{9}{2} \quad -2 = -\tfrac{4}{2}$$

Check

$$4x^2 + 16x = 9$$
$$4\left(\frac{1}{2}\right)^2 + 16\left(\frac{1}{2}\right) \stackrel{?}{=} 9$$
$$4\left(\frac{1}{4}\right) + 8 \stackrel{?}{=} 9$$
$$1 + 8 \stackrel{?}{=} 9$$
$$9 = 9 \quad \text{True}$$

$$4x^2 + 16x = 9$$
$$4\left(-\frac{9}{2}\right)^2 + 16\left(-\frac{9}{2}\right) \stackrel{?}{=} 9$$
$$4\left(\frac{81}{4}\right) - 72 \stackrel{?}{=} 9$$
$$81 - 72 \stackrel{?}{=} 9$$
$$9 = 9 \quad \text{True}$$

The two solutions, $\frac{1}{2}$ and $-\frac{9}{2}$, check, so the solution set is $\left\{-\frac{9}{2}, \frac{1}{2}\right\}$.

◀ Work Problem 5 at the Side.

ANSWERS

5. (a) $\left\{-\frac{1}{3}, -\frac{5}{3}\right\}$ (b) $\left\{\frac{11}{2}, \frac{1}{2}\right\}$

EXAMPLE 5 Solving a Quadratic Equation by Completing the Square ($a \neq 1$)

Solve $2x^2 - 4x - 5 = 0$.

$$x^2 - 2x - \frac{5}{2} = 0 \quad \text{Divide by 2. (Step 1)}$$

$$x^2 - 2x = \frac{5}{2} \quad \text{Add } \tfrac{5}{2}. \text{ (Step 2)}$$

$$\left[\frac{1}{2}(-2)\right]^2 = (-1)^2 = 1 \quad \text{(Step 3)}$$

$$x^2 - 2x + 1 = \frac{5}{2} + 1 \quad \text{Add 1. (Step 4)}$$

$$(x - 1)^2 = \frac{7}{2} \quad \text{Factor; add. (Step 5)}$$

$$x - 1 = \sqrt{\frac{7}{2}} \quad \text{or} \quad x - 1 = -\sqrt{\frac{7}{2}} \quad \text{Square root property (Step 6)}$$

$$x = 1 + \sqrt{\frac{7}{2}} \quad \text{or} \quad x = 1 - \sqrt{\frac{7}{2}} \quad \text{Add 1.}$$

$$x = 1 + \frac{\sqrt{14}}{2} \quad \text{or} \quad x = 1 - \frac{\sqrt{14}}{2} \quad \sqrt{\frac{7}{2}} = \frac{\sqrt{7}}{\sqrt{2}} = \frac{\sqrt{7}}{\sqrt{2}} \cdot \frac{\sqrt{2}}{\sqrt{2}} = \frac{\sqrt{14}}{2}$$

Add the two terms in each solution as follows:

$$1 + \frac{\sqrt{14}}{2} = \frac{2}{2} + \frac{\sqrt{14}}{2} = \frac{2 + \sqrt{14}}{2} \quad 1 = \tfrac{2}{2}$$

$$1 - \frac{\sqrt{14}}{2} = \frac{2}{2} - \frac{\sqrt{14}}{2} = \frac{2 - \sqrt{14}}{2}.$$

Check that the solution set is $\left\{\frac{2 + \sqrt{14}}{2}, \frac{2 - \sqrt{14}}{2}\right\}$.

Work Problem 6 at the Side. ▶

6 Solve each equation by completing the square.

(a) $2r^2 - 4r + 1 = 0$

(b) $3z^2 - 6z - 2 = 0$

(c) $8x^2 - 4x - 2 = 0$

EXAMPLE 6 Solving for Nonreal Complex Solutions

Solve $4p^2 + 8p + 5 = 0$.

$$4p^2 + 8p + 5 = 0$$

The coefficient of the second-degree term must be 1.

$$p^2 + 2p + \frac{5}{4} = 0 \quad \text{Divide by 4.}$$

$$p^2 + 2p = -\frac{5}{4} \quad \text{Subtract } \tfrac{5}{4}.$$

The coefficient of p is 2. Take half of 2; square the result: $\left[\frac{1}{2}(2)\right]^2 = 1^2 = 1$. Add this result to each side. Then write the left side as a perfect square.

$$p^2 + 2p + 1 = -\frac{5}{4} + 1 \quad \text{Add 1.}$$

$$(p + 1)^2 = -\frac{1}{4} \quad \text{Factor; add.}$$

Continued on Next Page

Answers

6. (a) $\left\{\frac{2 + \sqrt{2}}{2}, \frac{2 - \sqrt{2}}{2}\right\}$

(b) $\left\{\frac{3 + \sqrt{15}}{3}, \frac{3 - \sqrt{15}}{3}\right\}$

(c) $\left\{\frac{1 + \sqrt{5}}{4}, \frac{1 - \sqrt{5}}{4}\right\}$

Because the constant on the right side of the equation is negative, this equation will have nonreal complex solutions.

$$p + 1 = \sqrt{-\frac{1}{4}} \quad \text{or} \quad p + 1 = -\sqrt{-\frac{1}{4}} \qquad \text{Square root property}$$

$$p + 1 = \frac{1}{2}i \quad \text{or} \quad p + 1 = -\frac{1}{2}i \qquad \sqrt{-\tfrac{1}{4}} = \tfrac{1}{2}i$$

$$p = -1 + \frac{1}{2}i \quad \text{or} \quad p = -1 - \frac{1}{2}i \qquad \text{Add } -1.$$

Check $4p^2 + 8p + 5 = 0$ Original equation

$$4\left(-1 + \frac{1}{2}i\right)^2 + 8\left(-1 + \frac{1}{2}i\right) + 5 \stackrel{?}{=} 0 \qquad \text{Let } p = -1 + \tfrac{1}{2}i.$$

$$4\left[1 + 2(-1)\left(\frac{1}{2}i\right) - \frac{1}{4}\right] - 8 + 4i + 5 \stackrel{?}{=} 0 \qquad \text{Multiply.}$$

$(x + y)^2 = x^2 + 2xy + y^2$

$$4\left[\frac{3}{4} - i\right] - 3 + 4i \stackrel{?}{=} 0 \qquad \text{Simplify.}$$

$$3 - 4i - 3 + 4i \stackrel{?}{=} 0 \qquad \text{Multiply.}$$

$$0 = 0 \qquad \text{True}$$

The check of the other solution is similar. The solution set is

$$\left\{-1 + \frac{1}{2}i, -1 - \frac{1}{2}i\right\}.$$

◀ Work Problem 7 at the Side.

7 Solve each equation.

(a) $x^2 + 2x + 7 = 0$

(b) $5t^2 - 15t + 12 = 0$

8 Solve each equation.

(a) $r(r - 3) = -1$

(b) $(x + 2)(x + 1) = 5$

OBJECTIVE 3 Simplify the terms of an equation before solving.

EXAMPLE 7 Simplifying before Completing the Square

Solve $(x + 3)(x - 1) = 2$.

$$(x + 3)(x - 1) = 2 \qquad \text{Given equation}$$

$$x^2 + 2x - 3 = 2 \qquad \text{Multiply using the FOIL method.}$$

$$x^2 + 2x = 5 \qquad \text{Add 3.}$$

$$x^2 + 2x + 1 = 5 + 1 \qquad \text{Complete the square—add } [\tfrac{1}{2}(2)]^2 = 1.$$

$$(x + 1)^2 = 6 \qquad \text{Factor on the left; add on the right.}$$

$$x + 1 = \sqrt{6} \quad \text{or} \quad x + 1 = -\sqrt{6} \qquad \text{Square root property}$$

$$x = -1 + \sqrt{6} \quad \text{or} \quad x = -1 - \sqrt{6} \qquad \text{Add } -1.$$

The solution set is $\{-1 + \sqrt{6}, -1 - \sqrt{6}\}$.

◀ Work Problem 8 at the Side.

Note

We will use completing the square in **Section 10.7** when we graph quadratic equations and in **Section 12.2** when we work with circles.

ANSWERS

7. (a) $\{-1 + i\sqrt{6}, -1 - i\sqrt{6}\}$

(b) $\left\{\frac{3}{2} + \frac{\sqrt{15}}{10}i, \frac{3}{2} - \frac{\sqrt{15}}{10}i\right\}$

8. (a) $\left\{\frac{3 + \sqrt{5}}{2}, \frac{3 - \sqrt{5}}{2}\right\}$

(b) $\left\{\frac{-3 + \sqrt{21}}{2}, \frac{-3 - \sqrt{21}}{2}\right\}$

10.2 Exercises

FOR EXTRA HELP

MyMathLab

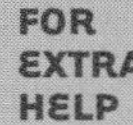

Math XL PRACTICE

WATCH

DOWNLOAD

READ

REVIEW

Complete each trinomial so that it is a perfect square. Then factor the trinomial. See Example 1.

1. $x^2 + 10x +$ ______

$25; (x + 5)^2$

2. $x^2 + 16x +$ ______

$64; (x + 8)^2$

3. $x^2 + 2x +$ ______

$1; (x + 1)^2$

4. $m^2 - 2m +$ ______

$1; (m - 1)^2$

5. $p^2 - 5p +$ ______

$\frac{25}{4}; \left(p - \frac{5}{2}\right)^2$

6. $x^2 + 3x +$ ______

$\frac{9}{4}; \left(x + \frac{3}{2}\right)^2$

Solve each equation by completing the square. See Examples 1–3.

7. $x^2 - 4x = -3$

$\{1, 3\}$

8. $x^2 - 2x = 8$

$\{-2, 4\}$

9. $x^2 + 5x + 6 = 0$

$\{-3, -2\}$

10. $x^2 + 6x + 5 = 0$

$\{-5, -1\}$

11. $x^2 + 2x - 5 = 0$

$\{-1 + \sqrt{6}, -1 - \sqrt{6}\}$

12. $x^2 + 4x + 1 = 0$

$\{-2 + \sqrt{3}, -2 - \sqrt{3}\}$

13. $x^2 + 10x + 18 = 0$

$\{-5 + \sqrt{7}, -5 - \sqrt{7}\}$

14. $x^2 + 8x + 11 = 0$

$\{-4 + \sqrt{5}, -4 - \sqrt{5}\}$

15. $x^2 - 8x = -4$

$\{4 + 2\sqrt{3}, 4 - 2\sqrt{3}\}$

16. $m^2 - 4m = 14$

$\{2 + 3\sqrt{2}, 2 - 3\sqrt{2}\}$

17. $x^2 + x - 1 = 0$

$\left\{\frac{-1 + \sqrt{5}}{2}, \frac{-1 - \sqrt{5}}{2}\right\}$

18. $x^2 + x - 3 = 0$

$\left\{\frac{-1 + \sqrt{13}}{2}, \frac{-1 - \sqrt{13}}{2}\right\}$

19. $r^2 - 3r = 2$

$\left\{\frac{3 + \sqrt{17}}{2}, \frac{3 - \sqrt{17}}{2}\right\}$

20. $x^2 - 3x = 6$

$\left\{\frac{3 + \sqrt{33}}{2}, \frac{3 - \sqrt{33}}{2}\right\}$

Solve each equation by completing the square. See Examples 4, 5, and 7.

21. $4x^2 + 4x - 3 = 0$

$\left\{-\frac{3}{2}, \frac{1}{2}\right\}$

22. $9x^2 + 3x - 2 = 0$

$\left\{-\frac{2}{3}, \frac{1}{3}\right\}$

23. $2x^2 - 4x = 5$

$\left\{\frac{2 + \sqrt{14}}{2}, \frac{2 - \sqrt{14}}{2}\right\}$

24. $2x^2 - 6x = 3$

$\left\{\frac{3 + \sqrt{15}}{2}, \frac{3 - \sqrt{15}}{2}\right\}$

25. $2k^2 + 5k - 2 = 0$

$\left\{\frac{-5 + \sqrt{41}}{4}, \frac{-5 - \sqrt{41}}{4}\right\}$

26. $3r^2 + 2r - 2 = 0$

$\left\{\frac{-1 + \sqrt{7}}{3}, \frac{-1 - \sqrt{7}}{3}\right\}$

27. $3k^2 + 7k = 4$

$\left\{\frac{-7 + \sqrt{97}}{6}, \frac{-7 - \sqrt{97}}{6}\right\}$

28. $2k^2 + 5k = 1$

$\left\{\frac{-5 + \sqrt{33}}{4}, \frac{-5 - \sqrt{33}}{4}\right\}$

29. $(r - 3)(r - 5) = 2$

$\{4 + \sqrt{3}, 4 - \sqrt{3}\}$

30. $(k - 1)(k - 7) = 1$

$\{4 + \sqrt{10}, 4 - \sqrt{10}\}$

31. $-x^2 + 2x = -5$

$\{1 + \sqrt{6}, 1 - \sqrt{6}\}$

32. $-r^2 + 3r = -2$

$\left\{\frac{3 + \sqrt{17}}{2}, \frac{3 - \sqrt{17}}{2}\right\}$

Find all complex solutions of each equation. See Example 6.

33. $m^2 + 4m + 13 = 0$

$\{-2 + 3i, -2 - 3i\}$

34. $t^2 + 6t + 10 = 0$

$\{-3 + i, -3 - i\}$

35. $3r^2 + 4r + 4 = 0$

$\left\{-\frac{2}{3} + \frac{2\sqrt{2}}{3}i, -\frac{2}{3} - \frac{2\sqrt{2}}{3}i\right\}$

36. $4x^2 + 5x + 5 = 0$

$\left\{-\frac{5}{8} + \frac{\sqrt{55}}{8}i, -\frac{5}{8} - \frac{\sqrt{55}}{8}i\right\}$

37. $-m^2 - 6m - 12 = 0$

$\{-3 + i\sqrt{3}, -3 - i\sqrt{3}\}$

38. $-k^2 - 5k - 10 = 0$

$\left\{-\frac{5}{2} + \frac{\sqrt{15}}{2}i, -\frac{5}{2} - \frac{\sqrt{15}}{2}i\right\}$

Relating Concepts (Exercises 39–44) For Individual or Group Work

The Greeks had a method of completing the square geometrically in which they literally changed a figure into a square. For example, to complete the square for $x^2 + 6x$, we begin with a square of side x, as in the figure. We add three rectangles of width 1 to the right side and the bottom to get a region with area $x^2 + 6x$. To fill in the corner (complete the square), we must add 9 1-by-1 squares as shown.

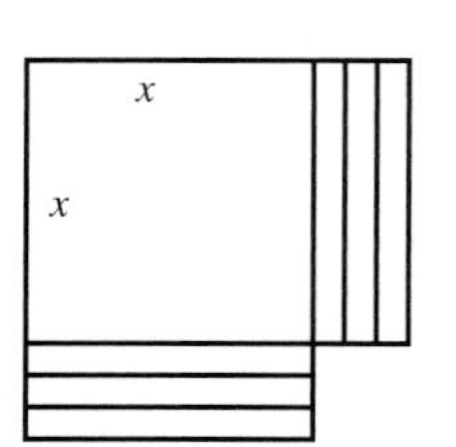

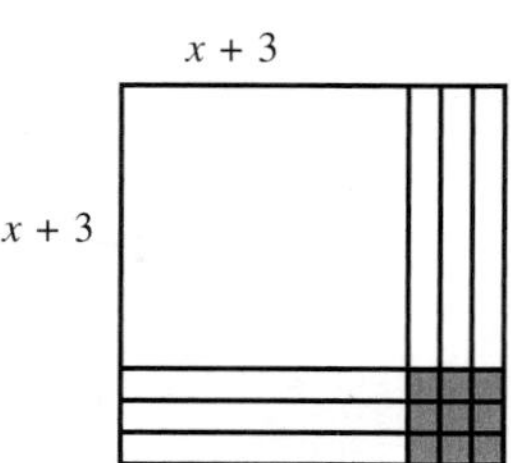

Work Exercises 39–44 in order.

39. What is the area of the original square?

x^2

40. What is the area of each strip?

x

41. What is the total area of the six strips?

$6x$

42. What is the area of each small square in the corner of the second figure?

1

43. What is the total area of the small squares?

9

44. What is the area of the new, larger square?

$(x + 3)^2$, or $x^2 + 6x + 9$

10.3 Solving Quadratic Equations by the Quadratic Formula

OBJECTIVES

1. **Derive the quadratic formula.**
2. **Solve quadratic equations by using the quadratic formula.**
3. **Use the discriminant to determine the number and type of solutions.**

In this section, we complete the square to solve the general quadratic equation

$$ax^2 + bx + c = 0,$$

where a, b, and c are complex numbers and $a \neq 0$. The solution of this general equation gives a formula for finding the solution of *any* specific quadratic equation.

OBJECTIVE 1 Derive the quadratic formula. To solve the general quadratic equation $ax^2 + bx + c = 0$ by completing the square (assuming $a > 0$), we follow the steps given in **Section 10.2.**

$$ax^2 + bx + c = 0$$

$$x^2 + \frac{b}{a}x + \frac{c}{a} = 0 \qquad \text{Divide by } a. \text{ (Step 1)}$$

$$x^2 + \frac{b}{a}x = -\frac{c}{a} \qquad \text{Subtract } \tfrac{c}{a}. \text{ (Step 2)}$$

$$\left[\frac{1}{2}\left(\frac{b}{a}\right)\right]^2 = \left(\frac{b}{2a}\right)^2 = \frac{b^2}{4a^2} \qquad \text{(Step 3)}$$

$$x^2 + \frac{b}{a}x + \frac{b^2}{4a^2} = -\frac{c}{a} + \frac{b^2}{4a^2} \qquad \text{Add } \frac{b^2}{4a^2} \text{ to each side. (Step 4)}$$

Write the left side as a perfect square, and rearrange the right side.

$$\left(x + \frac{b}{2a}\right)^2 = \frac{b^2}{4a^2} + \frac{-c}{a} \qquad \text{(Step 5)}$$

$$\left(x + \frac{b}{2a}\right)^2 = \frac{b^2}{4a^2} + \frac{-4ac}{4a^2} \qquad \text{Write with a common denominator.}$$

$$\left(x + \frac{b}{2a}\right)^2 = \frac{b^2 - 4ac}{4a^2} \qquad \text{Add fractions.}$$

$$x + \frac{b}{2a} = \sqrt{\frac{b^2 - 4ac}{4a^2}} \quad \text{or} \quad x + \frac{b}{2a} = -\sqrt{\frac{b^2 - 4ac}{4a^2}} \qquad \text{Square root property (Step 6)}$$

Since

$$\sqrt{\frac{b^2 - 4ac}{4a^2}} = \frac{\sqrt{b^2 - 4ac}}{\sqrt{4a^2}} = \frac{\sqrt{b^2 - 4ac}}{2a},$$

the right sides of these equations can be expressed as

$$x + \frac{b}{2a} = \frac{\sqrt{b^2 - 4ac}}{2a} \quad \text{or} \quad x + \frac{b}{2a} = \frac{-\sqrt{b^2 - 4ac}}{2a}$$

$$x = \frac{-b}{2a} + \frac{\sqrt{b^2 - 4ac}}{2a} \quad \text{or} \quad x = \frac{-b}{2a} - \frac{\sqrt{b^2 - 4ac}}{2a}$$

$$\boldsymbol{x = \frac{-b + \sqrt{b^2 - 4ac}}{2a}} \quad \text{or} \quad \boldsymbol{x = \frac{-b - \sqrt{b^2 - 4ac}}{2a}}.$$

If $a < 0$, the same two solutions are obtained. The result is the **quadratic formula,** which is abbreviated as shown on the next page.

Quadratic Formula

The solutions of $ax^2 + bx + c = 0$ $(a \neq 0)$ are given by

$$x = \frac{-b \pm \sqrt{b^2 - 4ac}}{2a}.$$

CAUTION

In the quadratic formula, $x = \frac{-b \pm \sqrt{b^2 - 4ac}}{2a}$, ***the square root is added to or subtracted from the value of $-b$ before dividing by $2a$.*** A common student error is to fail to divide ***all*** of the expression $-b \pm \sqrt{b^2 - 4ac}$ by $2a$.

OBJECTIVE 2 Solve quadratic equations by using the quadratic formula. To use the quadratic formula, first write the equation in standard form

$$ax^2 + bx + c = 0.$$

Then identify the values of a, b, and c and substitute them into the quadratic formula.

◀ Work Problem 1 at the Side.

1 Identify the values of a, b, and c. (*Hint:* If necessary, first write the equation in standard form with 0 on the right side.) *Do not actually solve.*

(a) $-3x^2 + 9x - 4 = 0$

(b) $3x^2 = 6x + 2$

EXAMPLE 1 Using the Quadratic Formula (Rational Solutions)

Solve $6x^2 - 5x - 4 = 0$.

Here a, the coefficient of the second-degree term, is 6, while b, the coefficient of the first-degree term, is -5, and the constant c is -4. Substitute these values into the quadratic formula.

$$x = \frac{-b \pm \sqrt{b^2 - 4ac}}{2a} \quad \text{Quadratic formula}$$

$$x = \frac{-(-5) \pm \sqrt{(-5)^2 - 4(6)(-4)}}{2(6)} \quad a = 6, b = -5, c = -4$$

Use parentheses and substitute carefully to avoid errors.

$$x = \frac{5 \pm \sqrt{25 + 96}}{12}$$

$$x = \frac{5 \pm \sqrt{121}}{12}$$

$$x = \frac{5 \pm 11}{12}$$

This last statement leads to two solutions, one from the + sign and one from the − sign.

$$x = \frac{5 + 11}{12} = \frac{16}{12} = \frac{4}{3} \quad \text{or} \quad x = \frac{5 - 11}{12} = \frac{-6}{12} = -\frac{1}{2}$$

Check each solution in the original equation. The solution set is $\left\{-\frac{1}{2}, \frac{4}{3}\right\}$.

◀ Work Problem 2 at the Side.

2 Solve $4x^2 - 11x - 3 = 0$ using the quadratic formula.

ANSWERS

1. (a) -3; 9; -4 (b) 3; -6; -2

2. $\left\{-\frac{1}{4}, 3\right\}$

We could have used factoring to solve the equation in Example 1.

$$6x^2 - 5x - 4 = 0$$

$$(3x - 4)(2x + 1) = 0 \quad \text{Factor.}$$

$$3x - 4 = 0 \quad \text{or} \quad 2x + 1 = 0 \quad \text{Zero-factor property}$$

$$3x = 4 \quad \text{or} \quad 2x = -1 \quad \text{Solve each equation.}$$

$$x = \frac{4}{3} \quad \text{or} \quad x = -\frac{1}{2} \quad \text{Same solutions as in Example 1}$$

When solving quadratic equations, it is a good idea to try factoring first. If the equation cannot be factored or if factoring is difficult, then use the quadratic formula. Later in this section, we will show a way to determine whether factoring can be used to solve a quadratic equation.

EXAMPLE 2 Using the Quadratic Formula (Irrational Solutions)

Solve $4x^2 = 8x - 1$.

Write the equation in standard form as $4x^2 - 8x + 1 = 0$.

$$x = \frac{-b \pm \sqrt{b^2 - 4ac}}{2a} \quad \text{Quadratic formula}$$

$$x = \frac{-(-8) \pm \sqrt{(-8)^2 - 4(4)(1)}}{2(4)} \quad a = 4, b = -8, c = 1$$

$$x = \frac{8 \pm \sqrt{64 - 16}}{8}$$

$$x = \frac{8 \pm \sqrt{48}}{8}$$

$$x = \frac{8 \pm 4\sqrt{3}}{8} \quad \sqrt{48} = \sqrt{16} \cdot \sqrt{3} = 4\sqrt{3}$$

$$x = \frac{4(2 \pm \sqrt{3})}{4(2)} \quad \text{Factor.}$$

Factor first; then divide out the common factor.

$$x = \frac{2 \pm \sqrt{3}}{2} \quad \text{Lowest terms}$$

The solution set is $\left\{\frac{2 + \sqrt{3}}{2}, \frac{2 - \sqrt{3}}{2}\right\}$.

CAUTION

1. ***Every quadratic equation must be written in standard form $ax^2 + bx + c = 0$ before we begin to solve it,*** whether we use factoring or the quadratic formula.
2. ***When writing solutions in lowest terms, be sure to factor first; then divide out the common factor,*** as shown in the last two steps in Example 2.

Work Problem 3 at the Side. ▶

3 Solve each equation using the quadratic formula.

(a) $6x^2 + 4x - 1 = 0$

(b) $2x^2 + 19 = 14x$

Answers

3. (a) $\left\{\frac{-2 + \sqrt{10}}{6}, \frac{-2 - \sqrt{10}}{6}\right\}$

(b) $\left\{\frac{7 + \sqrt{11}}{2}, \frac{7 - \sqrt{11}}{2}\right\}$

EXAMPLE 3 Using the Quadratic Formula (Nonreal Complex Solutions)

Solve $(9x + 3)(x - 1) = -8$.

To write this equation in standard form, we first multiply and collect all nonzero terms on the left.

$$(9x + 3)(x - 1) = -8$$

$$9x^2 - 6x - 3 = -8 \quad \text{Multiply.}$$

$$9x^2 - 6x + 5 = 0 \quad \text{Add 8.}$$

From the equation $9x^2 - 6x + 5 = 0$, we identify $a = 9$, $b = -6$, and $c = 5$.

$$x = \frac{-b \pm \sqrt{b^2 - 4ac}}{2a} \quad \text{Quadratic formula}$$

$$x = \frac{-(-6) \pm \sqrt{(-6)^2 - 4(9)(5)}}{2(9)} \quad \text{Substitute.}$$

$$x = \frac{6 \pm \sqrt{-144}}{18}$$

$$x = \frac{6 \pm 12i}{18} \quad \sqrt{-144} = 12i$$

$$x = \frac{6(1 \pm 2i)}{6(3)} \quad \text{Factor.}$$

$$x = \frac{1 \pm 2i}{3} \quad \text{Lowest terms}$$

$$x = \frac{1}{3} \pm \frac{2}{3}i \quad \text{Standard form } a + bi \text{ for a complex number}$$

The solution set is $\left\{\frac{1}{3} + \frac{2}{3}i, \frac{1}{3} - \frac{2}{3}i\right\}$.

◀ Work Problem **4** at the Side.

4 Solve each equation using the quadratic formula.

(a) $x^2 + x + 1 = 0$

(b) $(x + 2)(x - 6) = -17$

OBJECTIVE 3 Use the discriminant to determine the number and type of solutions. The solutions of the quadratic equation $ax^2 + bx + c = 0$ are given by

$$x = \frac{-b \pm \sqrt{b^2 - 4ac}}{2a}. \quad \leftarrow \text{Discriminant}$$

If a, b, and c are integers, the type of solutions of a quadratic equation—that is, rational, irrational, or nonreal complex—is determined by the expression under the radical sign, $b^2 - 4ac$. Because it distinguishes among the three types of solutions, $b^2 - 4ac$ is called the *discriminant*. By calculating the discriminant before solving a quadratic equation, we can predict whether the solutions will be rational numbers, irrational numbers, or nonreal complex numbers.

ANSWERS

4. (a) $\left\{-\frac{1}{2} + \frac{\sqrt{3}}{2}i, -\frac{1}{2} - \frac{\sqrt{3}}{2}i\right\}$

(b) $\{2 + i, 2 - i\}$

Discriminant

The **discriminant** of $ax^2 + bx + c = 0$ is $\mathbf{b^2 - 4ac}$. If a, b, and c are integers, then the number and type of solutions are determined as follows.

Discriminant	Number and Type of Solutions
Positive, and the square of an integer	Two rational solutions
Positive, but not the square of an integer	Two irrational solutions
Zero	One rational solution
Negative	Two nonreal complex solutions

Calculating the discriminant can also help you decide whether to solve a quadratic equation by factoring or by using the quadratic formula. ***If the discriminant is a perfect square (including 0), then the equation can be solved by factoring. Otherwise, the quadratic formula (or completing the square) should be used.***

EXAMPLE 4 **Using the Discriminant**

Find the discriminant. Use it to predict the number and type of solutions for each equation. Tell whether the equation can be solved by factoring or whether the quadratic formula should be used.

(a) $6x^2 - x - 15 = 0$

First identify the values of a, b, and c. Because $-x = -1x$, the value of b is -1. We find the discriminant by evaluating $b^2 - 4ac$.

$$b^2 - 4ac$$
$$= (-1)^2 - 4(6)(-15) \qquad a = 6, b = -1, c = -15$$
$$= 1 + 360$$
$$= 361$$

Use parentheses and substitute carefully.

A calculator shows that $361 = 19^2$, a perfect square. Since a, b, and c are integers and the discriminant is a perfect square, there will be two rational solutions and the equation can be solved by factoring.

(b) $3x^2 - 4x = 5$

Write the equation in standard form as

$$3x^2 - 4x - 5 = 0 \qquad \text{Subtract 5.}$$

to find $a = 3$, $b = -4$, and $c = -5$. The discriminant is

$$b^2 - 4ac$$
$$= (-4)^2 - 4(3)(-5)$$
$$= 16 + 60$$
$$= 76.$$

Because 76 is positive but *not* the square of an integer and a, b, and c are integers, the equation will have two irrational solutions and is best solved using the quadratic formula.

Continued on Next Page

5 Find the discriminant. Use it to predict the number and type of solutions for each equation.

(a) $2x^2 + 3x = 4$

(b) $2x^2 + 3x + 4 = 0$

(c) $x^2 + 20x + 100 = 0$

(d) $15x^2 + 11x = 14$

(e) Which of the equations in parts (a)–(d) can be solved by factoring?

(c) $4x^2 + x + 1 = 0$

Since $a = 4$, $b = 1$, and $c = 1$, the discriminant is

$$\begin{aligned} &b^2 - 4ac \\ &= 1^2 - 4(4)(1) \\ &= 1 - 16 \\ &= -15. \end{aligned}$$

Since the discriminant is negative and a, b, and c are integers, this quadratic equation will have two nonreal complex solutions. The quadratic formula should be used to solve it.

(d) $4x^2 + 9 = 12x$

Write the equation in standard form as

$$4x^2 - 12x + 9 = 0 \quad \text{Subtract } 12x.$$

to find $a = 4$, $b = -12$, and $c = 9$. The discriminant is

$$\begin{aligned} &b^2 - 4ac \\ &= (-12)^2 - 4(4)(9) \\ &= 144 - 144 \\ &= 0. \end{aligned}$$

Because the discriminant is 0, the quantity under the radical in the quadratic formula is 0, and there is only one rational solution. Again, the equation can be solved by factoring.

◀ *Work Problem* **5** *at the Side.*

Note

In **Section 10.7** we will see how the discriminant can be used to determine the number of x-intercepts of the graph of a quadratic function.

Answers

5. (a) 41; two; irrational
(b) -23; two; nonreal complex
(c) 0; one; rational
(d) 961; two; rational **(e)** (c) and (d)

10.3 Exercises

FOR EXTRA HELP

PRACTICE

WATCH

DOWNLOAD

READ

REVIEW

1. A student wrote the following as the quadratic formula for solving $ax^2 + bx + c = 0, a \neq 0$:

$$x = -b \pm \frac{\sqrt{b^2 - 4ac}}{2a}.$$

This is incorrect. ***WHAT WENT WRONG?***

The fraction bar should extend under the term $-b$.

2. A student attempted to solve the equation $5x^2 - 5x + 1 = 0$ as follows.

$$x = \frac{5 \pm \sqrt{25 - 4(5)(1)}}{2(5)} \qquad a = 5, b = -5, c = 1$$

$$x = \frac{5 \pm \sqrt{5}}{10}$$

$$x = \frac{1}{2} \pm \sqrt{5}$$

This is incorrect. ***WHAT WENT WRONG?***

The last step is wrong. Because 5 is not a common factor in the numerator, the fraction cannot be simplified. The solutions are $\frac{5 \pm \sqrt{5}}{10}$.

Use the quadratic formula to solve each equation. (All solutions for these equations are real numbers.) See Examples 1 and 2.

3. $x^2 - 8x + 15 = 0$

$\{3, 5\}$

4. $x^2 + 3x - 28 = 0$

$\{-7, 4\}$

5. $2x^2 + 4x + 1 = 0$

$\left\{\frac{-2 + \sqrt{2}}{2}, \frac{-2 - \sqrt{2}}{2}\right\}$

6. $2x^2 + 3x - 1 = 0$

$\left\{\frac{-3 + \sqrt{17}}{4}, \frac{-3 - \sqrt{17}}{4}\right\}$

7. $2x^2 - 2x = 1$

$\left\{\frac{1 + \sqrt{3}}{2}, \frac{1 - \sqrt{3}}{2}\right\}$

8. $9x^2 + 6x = 1$

$\left\{\frac{-1 + \sqrt{2}}{3}, \frac{-1 - \sqrt{2}}{3}\right\}$

9. $x^2 + 18 = 10x$

$\{5 + \sqrt{7}, 5 - \sqrt{7}\}$

10. $x^2 - 4 = 2x$

$\{1 + \sqrt{5}, 1 - \sqrt{5}\}$

11. $4k^2 + 4k - 1 = 0$

$\left\{\frac{-1 + \sqrt{2}}{2}, \frac{-1 - \sqrt{2}}{2}\right\}$

12. $4r^2 - 4r - 19 = 0$

$\left\{\frac{1 + 2\sqrt{5}}{2}, \frac{1 - 2\sqrt{5}}{2}\right\}$

13. $2 - 2x = 3x^2$

$\left\{\frac{-1 + \sqrt{7}}{3}, \frac{-1 - \sqrt{7}}{3}\right\}$

14. $26r - 2 = 3r^2$

$\left\{\frac{13 + \sqrt{163}}{3}, \frac{13 - \sqrt{163}}{3}\right\}$

15. $\frac{x^2}{4} - \frac{x}{2} = 1$
(*Hint:* First clear the fractions.)
$\{1 + \sqrt{5}, 1 - \sqrt{5}\}$

16. $p^2 + \frac{p}{3} = \frac{1}{6}$
(*Hint:* First clear the fractions.)
$\left\{\frac{-1 + \sqrt{7}}{6}, \frac{-1 - \sqrt{7}}{6}\right\}$

17. $-2t(t + 2) = -3$
$\left\{\frac{-2 + \sqrt{10}}{2}, \frac{-2 - \sqrt{10}}{2}\right\}$

18. $-3x(x + 2) = -4$
$\left\{\frac{-3 + \sqrt{21}}{3}, \frac{-3 - \sqrt{21}}{3}\right\}$

19. $(r - 3)(r + 5) = 2$
$\{-1 + 3\sqrt{2}, -1 - 3\sqrt{2}\}$

20. $(k + 1)(k - 7) = 1$
$\{3 + \sqrt{17}, 3 - \sqrt{17}\}$

Use the quadratic formula to solve each equation. (All solutions for these equations are nonreal complex numbers.) See Example 3.

21. $x^2 - 3x + 17 = 0$
$\left\{\frac{3}{2} + \frac{\sqrt{59}}{2}i, \frac{3}{2} - \frac{\sqrt{59}}{2}i\right\}$

22. $x^2 - 5x + 20 = 0$
$\left\{\frac{5}{2} + \frac{\sqrt{55}}{2}i, \frac{5}{2} - \frac{\sqrt{55}}{2}i\right\}$

23. $r^2 - 6r + 14 = 0$
$\{3 + i\sqrt{5}, 3 - i\sqrt{5}\}$

24. $t^2 + 4t + 11 = 0$
$\{-2 + i\sqrt{7}, -2 - i\sqrt{7}\}$

25. $4x^2 - 4x = -7$
$\left\{\frac{1}{2} + \frac{\sqrt{6}}{2}i, \frac{1}{2} - \frac{\sqrt{6}}{2}i\right\}$

26. $9x^2 - 6x = -7$
$\left\{\frac{1}{3} + \frac{\sqrt{6}}{3}i, \frac{1}{3} - \frac{\sqrt{6}}{3}i\right\}$

27. $x(3x + 4) = -2$
$\left\{-\frac{2}{3} + \frac{\sqrt{2}}{3}i, -\frac{2}{3} - \frac{\sqrt{2}}{3}i\right\}$

28. $p(2p + 3) = -2$
$\left\{-\frac{3}{4} + \frac{\sqrt{7}}{4}i, -\frac{3}{4} - \frac{\sqrt{7}}{4}i\right\}$

Use the discriminant to determine whether the solutions for each equation are

A. *two rational numbers,* **B.** *one rational number,*
C. *two irrational numbers,* **D.** *two nonreal complex numbers.*

Do not actually solve. See Example 4.

29. $25x^2 + 70x + 49 = 0$
B

30. $4k^2 - 28k + 49 = 0$
B

31. $x^2 + 4x + 2 = 0$
C

32. $9x^2 - 12x - 1 = 0$
C

33. $3x^2 = 5x + 2$
A

34. $4x^2 = 4x + 3$
A

35. $3m^2 - 10m + 15 = 0$
D

36. $18x^2 + 60x + 82 = 0$
D

37. Using the discriminant, which equations in Exercises 29–36 can be solved by factoring?
The equations in Exercises 29, 30, 33, and 34 can be solved by factoring.

38. Based on your answer in Exercise 37, solve the equation given in each exercise.
(a) Exercise 29
$\left\{-\frac{7}{5}\right\}$
(b) Exercise 33
$\left\{-\frac{1}{3}, 2\right\}$

10.4 Equations Quadratic in Form

OBJECTIVES

1. **Solve an equation with fractions by writing it in quadratic form.**
2. **Use quadratic equations to solve applied problems.**
3. **Solve an equation with radicals by writing it in quadratic form.**
4. **Solve an equation that is quadratic in form by substitution.**

OBJECTIVE 1 Solve an equation with fractions by writing it in quadratic form. A variety of nonquadratic equations can be written in the form of a quadratic equation and solved by using one of the methods from **Sections 10.1–10.3.**

EXAMPLE 1 Solving an Equation with Fractions That Leads to a Quadratic Equation

Solve $\frac{1}{x} + \frac{1}{x-1} = \frac{7}{12}$.

Clear fractions by multiplying each side by the least common denominator, $12x(x-1)$. (Note that the domain must be restricted to $x \neq 0$ and $x \neq 1$.)

$$12x(x-1)\left(\frac{1}{x} + \frac{1}{x+1}\right) = 12x(x-1)\left(\frac{7}{12}\right)$$

$$12x(x-1)\frac{1}{x} + 12x(x-1)\frac{1}{x-1} = 12x(x-1)\left(\frac{7}{12}\right) \quad \text{Distributive property}$$

$$12(x-1) + 12x = 7x(x-1)$$

$$12x - 12 + 12x = 7x^2 - 7x \quad \text{Distributive property}$$

$$24x - 12 = 7x^2 - 7x \quad \text{Combine like terms.}$$

$$7x^2 - 31x + 12 = 0 \quad \text{Standard form}$$

$$(7x-3)(x-4) = 0 \quad \text{Factor.}$$

$$7x - 3 = 0 \quad \text{or} \quad x - 4 = 0 \quad \text{Zero-factor property}$$

$$x = \frac{3}{7} \quad \text{or} \quad x = 4 \quad \text{Solve each equation.}$$

Check by substituting these solutions in the original equation. The solution set is $\left\{\frac{3}{7}, 4\right\}$.

Work Problem 1 *at the Side.* ▶

1 Solve each equation. Check your solutions.

(a) $\frac{5}{m} + \frac{12}{m^2} = 2$

(b) $\frac{2}{x} + \frac{1}{x-2} = \frac{5}{3}$

(c) $\frac{4}{m-1} + 9 = -\frac{7}{m}$

OBJECTIVE 2 Use quadratic equations to solve applied problems. In **Sections 4.4 and 7.5** we solved distance-rate-time (or motion) problems that led to linear equations or rational equations. Now we can extend that work to motion problems that lead to quadratic equations. We continue to use the six-step problem-solving method from **Section 2.4.**

EXAMPLE 2 Solving a Motion Problem

A riverboat for tourists averages 12 mph in still water. It takes the boat 1 hr, 4 min to go 6 mi upstream and return. Find the speed of the current.

Step 1 **Read** the problem carefully.

Step 2 **Assign a variable.** Let $x =$ the speed of the current.

The current slows down the boat when it is going upstream, so the rate (or speed) upstream is the speed of the boat in still water less the speed of the current, or $(12 - x)$ mph. See Figure 1 on the next page.

Continued on Next Page

ANSWERS

1. (a) $\left\{-\frac{3}{2}, 4\right\}$ (b) $\left\{\frac{4}{5}, 3\right\}$
(c) $\left\{\frac{7}{9}, -1\right\}$

Riverboat traveling *upstream*—the current slows it down.

Figure 1

Similarly, the current speeds up the boat as it travels downstream, so its speed downstream is $(12 + x)$ mph. Thus,

$12 - x =$ the rate upstream in miles per hour;

$12 + x =$ the rate downstream in miles per hour.

This information can be used to complete a table. We use the distance formula, $d = rt$, solved for time t, $t = \frac{d}{r}$, to write expressions for t.

	d	r	t
Upstream	6	$12 - x$	$\frac{6}{12 - x}$
Downstream	6	$12 + x$	$\frac{6}{12 + x}$

Times in hours

Step 3 **Write an equation.** The total time, 1 hr and 4 min, can be written as

$$1 + \frac{4}{60} = 1 + \frac{1}{15} = \frac{16}{15} \text{ hr.}$$

Because the time upstream plus the time downstream equals $\frac{16}{15}$ hr,

$$\underset{\text{Time upstream}}{\frac{6}{12 - x}} + \underset{\text{time downstream}}{\frac{6}{12 + x}} = \underset{\text{total time.}}{\frac{16}{15}}.$$

Step 4 **Solve** the equation. Multiply each side by $15(12 - x)(12 + x)$, the LCD, and solve the resulting quadratic equation.

$$15(12 + x)6 + 15(12 - x)6 = 16(12 - x)(12 + x)$$

$$90(12 + x) + 90(12 - x) = 16(144 - x^2)$$

$$1080 + 90x + 1080 - 90x = 2304 - 16x^2 \quad \text{Distributive property}$$

$$2160 = 2304 - 16x^2 \quad \text{Combine like terms.}$$

$$16x^2 = 144$$

$$x^2 = 9 \quad \text{Divide by 16.}$$

$$x = 3 \quad \text{or} \quad x = -3 \quad \text{Square root property}$$

Step 5 **State the answer.** The speed of the current cannot be -3, so the answer is 3 mph.

Step 6 **Check** that this value satisfies the original problem.

CAUTION

As shown in Example 2, when a quadratic equation is used to solve an applied problem, sometimes only *one* answer satisfies the application. ***Always check each answer in the words of the original problem.***

Work Problem **2** *at the Side.* ▶

Recall from **Section 7.5** that a person's work rate is $\frac{1}{t}$ part of the job per hour, where t is the time in hours required to do the complete job. Thus, the part of the job the person will do in x hours is $\frac{1}{t}x$.

EXAMPLE 3 Solving a Work Problem

It takes two carpet layers 4 hr to carpet a room. If each worked alone, one of them could do the job in 1 hr less time than the other. How long would it take each carpet layer to complete the job alone?

Step 1 **Read** the problem again. There will be two answers.

Step 2 **Assign a variable.** Let $x =$ the number of hours for the slower carpet layer to complete the job alone. Then the faster carpet layer could do the entire job in $(x - 1)$ hours. The slower person's rate is $\frac{1}{x}$, and the faster person's rate is $\frac{1}{x - 1}$. Together, they can do the job in 4 hr. Complete a table as shown.

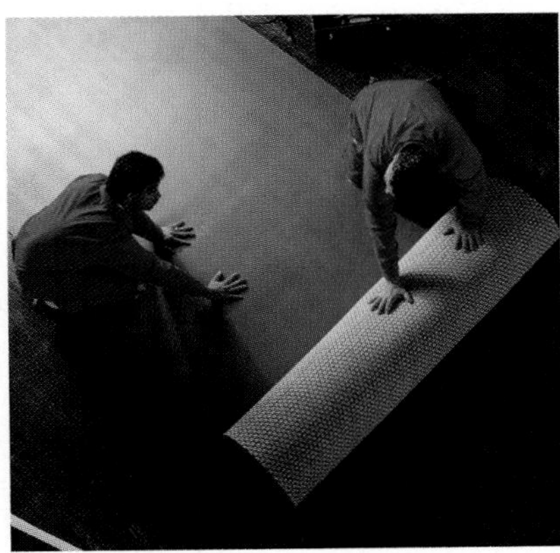

	Rate	Time Working Together	Fractional Part of the Job Done
Slower Worker	$\frac{1}{x}$	4	$\frac{1}{x}(4)$
Faster Worker	$\frac{1}{x-1}$	4	$\frac{1}{x-1}(4)$

Sum is 1 whole job.

Step 3 **Write an equation.** The sum of the fractional parts done by the workers should equal 1 (the whole job).

Part done by slower worker + part done by faster worker = 1 whole job.

$$\frac{4}{x} + \frac{4}{x-1} = 1$$

Step 4 **Solve** the equation. Multiply by $x(x - 1)$, the LCD.

$$x(x-1)\left(\frac{4}{x} + \frac{4}{x-1}\right) = x(x-1)(1) \quad \text{Multiply by the LCD.}$$

$$4(x-1) + 4x = x(x-1) \quad \text{Distributive property}$$

$$4x - 4 + 4x = x^2 - x \quad \text{Distributive property}$$

$$x^2 - 9x + 4 = 0 \quad \text{Standard form}$$

Continued on Next Page

2 Solve each problem.

(a) In 4 hr, Kerrie can go 15 mi upriver and come back. The speed of the current is 5 mph. Complete this table.

	d	*r*	*t*
Up			
Down			

(b) Find the speed of the boat from part (a) in still water.

(c) In $1\frac{3}{4}$ hr, Ken rows his boat 5 mi upriver and comes back. The speed of the current is 3 mph. How fast does Ken row?

Answers

2. (a) row 1: 15; $x - 5$; $\frac{15}{x-5}$

row 2: 15; $x + 5$; $\frac{15}{x+5}$

(b) 10 mph **(c)** 7 mph

3 Solve each problem. Round answers to the nearest tenth.

(a) Carlos can complete a certain lab test in 2 hr less time than Jaime can. If they can finish the job together in 2 hr, how long would it take each of them working alone?

	Rate	Time Working Together	Fractional Part of the Job Done
Carlos			
Jaime			

(b) Two chefs are preparing a banquet. One chef could prepare the banquet in 2 hr less time than the other. Together, they complete the job in 5 hr. How long would it take the faster chef working alone?

The resulting equation $x^2 - 9x + 4 = 0$ cannot be solved by factoring, so use the quadratic formula.

$$x = \frac{-b \pm \sqrt{b^2 - 4ac}}{2a} \quad \text{Quadratic formula}$$

$$x = \frac{-(-9) \pm \sqrt{(-9)^2 - 4(1)(4)}}{2(1)} \quad a = 1, b = -9, c = 4$$

$$x = \frac{9 \pm \sqrt{65}}{2} \quad \text{Simplify.}$$

$$x = \frac{9 + \sqrt{65}}{2} \approx 8.5 \quad \text{or} \quad x = \frac{9 - \sqrt{65}}{2} \approx 0.5 \quad \text{Use a calculator.}$$

Step 5 **State the answer.** Only the solution 8.5 makes sense in the original problem. (Why?) Thus, the slower worker can do the job in about 8.5 hr and the faster in about $8.5 - 1 = 7.5$ hr.

Step 6 **Check** that these results satisfy the original problem.

◀ Work Problem **3** at the Side.

OBJECTIVE 3 Solve an equation with radicals by writing it in quadratic form.

EXAMPLE 4 Solving Radical Equations That Lead to Quadratic Equations

Solve each equation.

(a) $k = \sqrt{6k - 8}$

This equation is not quadratic. However, squaring both sides of the equation gives a quadratic equation that can be solved by factoring.

$$k^2 = 6k - 8 \quad \text{Square each side.}$$

$$k^2 - 6k + 8 = 0 \quad \text{Standard form}$$

$$(k - 4)(k - 2) = 0 \quad \text{Factor.}$$

$$k - 4 = 0 \quad \text{or} \quad k - 2 = 0 \quad \text{Zero-factor property}$$

$$k = 4 \quad \text{or} \quad k = 2 \quad \text{Potential solutions}$$

Recall from **Section 9.6** that squaring both sides of a radical equation can introduce extraneous solutions that do not satisfy the original equation. ***All proposed solutions must be checked in the original (not the squared) equation.***

If $k = 4$, then

$$k = \sqrt{6k - 8}$$
$$4 \stackrel{?}{=} \sqrt{6(4) - 8}$$
$$4 \stackrel{?}{=} \sqrt{16}$$
$$4 = 4. \quad \text{True}$$

If $k = 2$, then

$$k = \sqrt{6k - 8}$$
$$2 \stackrel{?}{=} \sqrt{6(2) - 8}$$
$$2 \stackrel{?}{=} \sqrt{4}$$
$$2 = 2. \quad \text{True}$$

Both solutions check, so the solution set is $\{2, 4\}$.

Continued on Next Page

ANSWERS

3. **(a)** Jaime: 5.2 hr; Carlos: 3.2 hr **(b)** 9.1 hr

(b) $x + \sqrt{x} = 6$

$\sqrt{x} = 6 - x$ Isolate the radical on one side.

$x = 36 - 12x + x^2$ Square each side.

$0 = x^2 - 13x + 36$ Standard form

$0 = (x - 4)(x - 9)$ Factor.

$x - 4 = 0$ or $x - 9 = 0$ Zero-factor property

$x = 4$ or $x = 9$ Proposed solutions

Check both proposed solutions in the *original* equation.

If $x = 4$, then	If $x = 9$, then
$x + \sqrt{x} = 6$	$x + \sqrt{x} = 6$
$4 + \sqrt{4} \stackrel{?}{=} 6$	$9 + \sqrt{9} \stackrel{?}{=} 6$
$6 = 6.$ True	$12 = 6.$ False

Only the solution 4 checks, so the solution set is $\{4\}$.

Work Problem 4 at the Side. ▶

4 Solve each equation. Check your solutions.

(a) $x = \sqrt{7x - 10}$

(b) $2x = \sqrt{x} + 1$

OBJECTIVE 4 Solve an equation that is quadratic in form by substitution. A nonquadratic equation that can be written in the form

$$au^2 + bu + c = 0,$$

for $a \neq 0$ and an algebraic expression u, is called **quadratic in form.**

Many equations that are quadratic in form can be solved more easily by defining and substituting a "temporary" variable u for an expression involving the variable in the original equation. The first step is to define this temporary variable u.

5 Define a variable u, and write each equation in the form $au^2 + bu + c = 0$.

(a) $2x^4 + 5x^2 - 12 = 0$

(b) $2(x + 5)^2 - 7(x + 5) + 6 = 0$

(c) $x^{4/3} - 8x^{2/3} + 16 = 0$

EXAMPLE 5 Defining Substitution Variables

Define a variable u, and write each equation in the form $au^2 + bu + c = 0$.

(a) $x^4 - 13x^2 + 36 = 0$

Look at the two terms involving the variable x, ignoring their coefficients. Try to find one variable expression that is the square of the other. We see that $x^4 = (x^2)^2$, so we define $u = x^2$, and rewrite the original equation as the quadratic equation

$$u^2 - 13u + 36 = 0. \quad \text{Here, } u = x^2.$$

(b) $2(4m - 3)^2 + 7(4m - 3) + 5 = 0$

Because this equation involves both $(4m - 3)^2$ and $(4m - 3)$, choose $u = 4m - 3$. Substituting u for $4m - 3$ gives the quadratic equation

$$2u^2 + 7u + 5 = 0. \quad \text{Here, } u = 4m - 3.$$

(c) $2x^{2/3} - 11x^{1/3} + 12 = 0$

Here we apply one of the power rules for exponents from **Section 5.2:** $(a^m)^n = a^{mn}$. Because $(x^{1/3})^2 = x^{2/3}$, we define $u = x^{1/3}$. With this substitution, the original equation becomes

$$2u^2 - 11u + 12 = 0. \quad \text{Here, } u = x^{1/3}.$$

Work Problem 5 at the Side. ▶

ANSWERS

4. **(a)** $\{2, 5\}$ **(b)** $\{1\}$
5. **(a)** $u = x^2$; $2u^2 + 5u - 12 = 0$
(b) $u = x + 5$; $2u^2 - 7u + 6 = 0$
(c) $u = x^{2/3}$; $u^2 - 8u + 16 = 0$

EXAMPLE 6 Solving Equations That Are Quadratic in Form

Solve each equation.

(a) $x^4 - 13x^2 + 36 = 0$

From Example 5(a), we write this equation in quadratic form by substituting u for x^2.

$$x^4 - 13x^2 + 36 = 0$$

$$(x^2)^2 - 13x^2 + 36 = 0 \qquad x^4 = (x^2)^2$$

$$u^2 - 13u + 36 = 0 \qquad \text{Let } u = x^2.$$

$$(u - 4)(u - 9) = 0 \qquad \text{Factor.}$$

$$u - 4 = 0 \quad \text{or} \quad u - 9 = 0 \qquad \text{Zero-factor property}$$

Don't stop here.

$$u = 4 \quad \text{or} \quad u = 9 \qquad \text{Solve.}$$

$$x^2 = 4 \quad \text{or} \quad x^2 = 9 \qquad \text{Substitute } x^2 \text{ for } u.$$

$$x = \pm 2 \quad \text{or} \quad x = \pm 3 \qquad \text{Square root property}$$

The equation $x^4 - 13x^2 + 36 = 0$, a fourth-degree equation, has four solutions.* The solution set is $\{-3, -2, 2, 3\}$. Check each of the four solutions by substitution.

(b)

$$4x^4 + 1 = 5x^2$$

$$4(x^2)^2 + 1 = 5x^2 \qquad x^4 = (x^2)^2$$

$$4u^2 + 1 = 5u \qquad \text{Let } u = x^2.$$

$$4u^2 - 5u + 1 = 0 \qquad \text{Standard form}$$

$$(4u - 1)(u - 1) = 0 \qquad \text{Factor.}$$

$$4u - 1 = 0 \quad \text{or} \quad u - 1 = 0 \qquad \text{Zero-factor property}$$

$$u = \frac{1}{4} \quad \text{or} \quad u = 1 \qquad \text{Solve.}$$

This is a key step.

$$x^2 = \frac{1}{4} \quad \text{or} \quad x^2 = 1 \qquad \text{Substitute } x^2 \text{ for } u.$$

$$x = \pm\frac{1}{2} \quad \text{or} \quad x = \pm 1 \qquad \text{Square root property}$$

Check that the solution set is $\left\{-1, -\frac{1}{2}, \frac{1}{2}, 1\right\}$.

(c) $x^4 = 6x^2 - 3$

First write the equation as

$$x^4 - 6x^2 + 3 = 0 \quad \text{or} \quad (x^2)^2 - 6x^2 + 3 = 0,$$

which is quadratic in form with $u = x^2$. Substitute u for x^2 and u^2 for x^4 to get

$$u^2 - 6u + 3 = 0.$$

Since this equation cannot be solved by factoring, use the quadratic formula.

Continued on Next Page

*In general, an equation in which an nth-degree polynomial equals 0 has n solutions, although some of them may be repeated.

$$u = \frac{-(-6) \pm \sqrt{(-6)^2 - 4(1)(3)}}{2(1)} \qquad a = 1, b = -6, c = 3$$

$$u = \frac{6 \pm \sqrt{24}}{2} \qquad \text{Simplify.}$$

$$u = \frac{6 \pm 2\sqrt{6}}{2} \qquad \sqrt{24} = \sqrt{4} \cdot \sqrt{6} = 2\sqrt{6}$$

$$u = \frac{2(3 \pm \sqrt{6})}{2} \qquad \text{Factor.}$$

$$u = 3 \pm \sqrt{6} \qquad \text{Lowest terms}$$

$$x^2 = 3 + \sqrt{6} \quad \text{or} \quad x^2 = 3 - \sqrt{6} \qquad \text{Substitute } x^2 \text{ for } u.$$

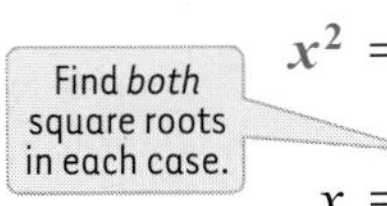

$$x = \pm\sqrt{3 + \sqrt{6}} \quad \text{or} \quad x = \pm\sqrt{3 - \sqrt{6}} \qquad \text{Square root property}$$

The solution set contains four numbers:

$$\left\{\sqrt{3 + \sqrt{6}}, -\sqrt{3 + \sqrt{6}}, \sqrt{3 - \sqrt{6}}, -\sqrt{3 - \sqrt{6}}\right\}.$$

Note

Some students prefer to solve equations like those in Examples 6(a) and (b) by factoring directly. For example,

$$x^4 - 13x^2 + 36 = 0 \qquad \text{Example 6(a) equation}$$

$$(x^2 - 9)(x^2 - 4) = 0 \qquad \text{Factor.}$$

$$(x + 3)(x - 3)(x + 2)(x - 2) = 0. \qquad \text{Factor again.}$$

Using the zero-factor property gives the same solutions obtained in Example 6(a). Equations that cannot be solved by factoring (as in Example 6(c)) must be solved by substitution and the quadratic formula.

Work Problem 6 *at the Side.* ▶

The method used in Example 6 can be generalized.

Solving an Equation That Is Quadratic in Form by Substitution

Step 1 Define a temporary variable u, based on the relationship between the variable expressions in the given equation. Substitute u in the original equation and rewrite the equation in the form $au^2 + bu + c = 0$.

Step 2 **Solve the quadratic equation obtained in Step 1** by factoring or the quadratic formula.

Step 3 **Replace u with the expression it defined in Step 1.**

Step 4 **Solve the resulting equations for the original variable.**

Step 5 **Check** all solutions by substituting them in the original equation.

6 Solve each equation. Check your solutions.

(a) $m^4 - 10m^2 + 9 = 0$

(b) $9k^4 - 37k^2 + 4 = 0$

(c) $x^4 - 4x^2 = -2$

Answers

6. (a) $\{-3, -1, 1, 3\}$ (b) $\left\{-2, -\frac{1}{3}, \frac{1}{3}, 2\right\}$
(c) $\left\{\sqrt{2 + \sqrt{2}}, -\sqrt{2 + \sqrt{2}}, \sqrt{2 - \sqrt{2}}, -\sqrt{2 - \sqrt{2}}\right\}$

EXAMPLE 7 Solving Equations That Are Quadratic in Form

Solve each equation.

(a) $2(4m-3)^2+7(4m-3)+5=0$

Step 1 Because of the repeated quantity $4m-3$, substitute u for $4m-3$ as in Example 5(b).

$$2(4m-3)^2+7(4m-3)+5=0$$

$$2u^2+7u+5=0 \qquad \text{Let } u=4m-3.$$

Step 2

$$(2u+5)(u+1)=0 \qquad \text{Factor.}$$

$$2u+5=0 \quad \text{or} \quad u+1=0 \qquad \text{Zero-factor property}$$

Don't stop here.

$$u=-\frac{5}{2} \quad \text{or} \quad u=-1 \qquad \text{Solve for } u.$$

Step 3

$$4m-3=-\frac{5}{2} \quad \text{or} \quad 4m-3=-1 \qquad \text{Substitute } 4m-3 \text{ for } u.$$

Step 4

$$4m=\frac{1}{2} \quad \text{or} \quad 4m=2 \qquad \text{Solve for } m.$$

$$m=\frac{1}{8} \quad \text{or} \quad m=\frac{1}{2}$$

Step 5 Check that the solution set of the original equation is $\left\{\frac{1}{8}, \frac{1}{2}\right\}$.

(b) $2x^{2/3}-11x^{1/3}+12=0$

From Example 5(c), substitute u for $x^{1/3}$.

$$2u^2-11u+12=0 \qquad \text{Let } x^{1/3}=u;\ x^{2/3}=u^2.$$

$$(2u-3)(u-4)=0 \qquad \text{Factor.}$$

$$2u-3=0 \quad \text{or} \quad u-4=0 \qquad \text{Zero-factor property}$$

$$u=\frac{3}{2} \quad \text{or} \quad u=4 \qquad \text{Solve for } u.$$

$$x^{1/3}=\frac{3}{2} \quad \text{or} \quad x^{1/3}=4 \qquad u=x^{1/3}$$

$$(x^{1/3})^3=\left(\frac{3}{2}\right)^3 \quad \text{or} \quad (x^{1/3})^3=4^3 \qquad \text{Cube each side.}$$

$$x=\frac{27}{8} \quad \text{or} \quad x=64$$

Check that the solution set is $\left\{\frac{27}{8}, 64\right\}$.

CAUTION

A common error when solving problems like those in Examples 6 and 7 is to stop too soon. ***Once you have solved for u, remember to substitute and solve for the values of the original variable.*** Keep in mind that u is just a temporary variable that helps you solve the given equation. As in any equation, you must solve for the variable in the *original* equation.

◀ Work Problem 7 at the Side.

7 Solve each equation. Check your solutions.

(a) $5(r+3)^2+9(r+3)=2$

(b) $4m^{2/3}=3m^{1/3}+1$

ANSWERS

7. (a) $\left\{-5, -\frac{14}{5}\right\}$ (b) $\left\{-\frac{1}{64}, 1\right\}$

10.4 Exercises

FOR EXTRA HELP

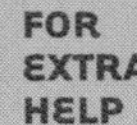

Math XL PRACTICE

WATCH

DOWNLOAD

READ

REVIEW

Based on the discussion and examples of this section, write a sentence describing the first step you would take to solve each equation. Do not actually solve.

1. $\frac{14}{x} = x - 5$

Multiply by the LCD, x.

2. $\sqrt{1 + x} + x = 5$

Isolate the radical term on one side.

3. $(r^2 + r)^2 - 8(r^2 + r) + 12 = 0$

Substitute a variable for $r^2 + r$.

4. $3t = \sqrt{16 - 10t}$

Square both sides.

5. Read this incorrect "solution" carefully. ***WHAT WENT WRONG?***

$$x = \sqrt{3x + 4}$$
$$x^2 = 3x + 4 \quad \text{Square both sides.}$$
$$x^2 - 3x - 4 = 0$$
$$(x - 4)(x + 1) = 0$$
$$x - 4 = 0 \quad \text{or} \quad x + 1 = 0$$
$$x = 4 \quad \text{or} \quad x = -1$$

Solution set: $\{4, -1\}$

The proposed solution -1 does not check. The solution set is $\{4\}$.

6. Read this incorrect "solution" carefully. ***WHAT WENT WRONG?***

$$2(m - 1)^2 - 3(m - 1) + 1 = 0$$
$$2u^2 - 3u + 1 = 0 \quad \text{Let } u = m - 1.$$
$$(2u - 1)(u - 1) = 0$$
$$2u - 1 = 0 \quad \text{or} \quad u - 1 = 0$$
$$u = \frac{1}{2} \quad \text{or} \quad u = 1$$

Solution set: $\left\{\frac{1}{2}, 1\right\}$

The solutions given are for u. Each must be set equal to $m - 1$ and solved for m. The correct solution set is $\left\{\frac{3}{2}, 2\right\}$.

Solve each equation. Check your solutions. See Example 1.

7. $1 - \frac{3}{x} - \frac{28}{x^2} = 0$

$\{-4, 7\}$

8. $4 - \frac{7}{r} - \frac{2}{r^2} = 0$

$\left\{-\frac{1}{4}, 2\right\}$

9. $3 - \frac{1}{t} = \frac{2}{t^2}$

$\left\{-\frac{2}{3}, 1\right\}$

10. $1 + \frac{2}{k} = \frac{3}{k^2}$

$\{-3, 1\}$

11. $\frac{1}{x} + \frac{2}{x + 2} = \frac{17}{35}$

$\left\{-\frac{14}{17}, 5\right\}$

12. $\frac{2}{m} + \frac{3}{m + 9} = \frac{11}{4}$

$\left\{-8, \frac{9}{11}\right\}$

13. $\frac{2}{x + 1} + \frac{3}{x + 2} = \frac{7}{2}$

$\left\{-\frac{11}{7}, 0\right\}$

14. $\frac{4}{3 - p} + \frac{2}{5 - p} = \frac{26}{15}$

$\left\{0, \frac{59}{13}\right\}$

15. $\frac{3}{2x} - \frac{1}{2(x + 2)} = 1$

$\left\{\frac{-1 + \sqrt{13}}{2}, \frac{-1 - \sqrt{13}}{2}\right\}$

16. $\frac{4}{3x} - \frac{1}{2(x + 1)} = 1$

$\left\{\frac{-1 + \sqrt{193}}{12}, \frac{-1 - \sqrt{193}}{12}\right\}$

17. $\frac{6}{p} = 2 + \frac{p}{p + 1}$

$\left\{\frac{2 + \sqrt{22}}{3}, \frac{2 - \sqrt{22}}{3}\right\}$

18. $\frac{k}{2 - k} + \frac{2}{k} = 5$

$\left\{\frac{3 + \sqrt{3}}{3}, \frac{3 - \sqrt{3}}{3}\right\}$

19. A boat goes 20 mph in still water, and the rate of the current is t mph.

(a) What is the rate of the boat when it travels upstream?

$(20 - t)$ mph

(b) What is the rate of the boat when it travels downstream?

$(20 + t)$ mph

20. It takes m hours to grade a set of papers.

(a) What is the grader's rate (in job per hour)?

$\frac{1}{m}$ job per hr

(b) How much of the job will the grader do in 2 hr?

$\frac{2}{m}$ job

Solve each problem. See Examples 2 and 3.

21. On a windy day Yoshiaki found that he could go 16 mi downstream and then 4 mi back upstream at top speed in a total of 48 min. What was the top speed of Yoshiaki's boat if the current was 15 mph?

	d	r	t
Upstream	4	$x - 15$	
Downstream	16		

25 mph

22. Lekesha flew her plane for 6 hr at a constant speed. She traveled 810 mi with the wind, then turned around and traveled 720 mi against the wind. The wind speed was a constant 15 mph. Find the speed of the plane.

	d	r	t
With Wind	810		
Against Wind	720		

255 mph

23. In Canada, Medicine Hat and Cranbrook are 300 km apart. Harry rides his Honda 20 km per hr faster than Yoshi rides his Yamaha. Find Harry's average speed if he travels from Cranbrook to Medicine Hat in $1\frac{1}{4}$ hr less time than Yoshi. (*Source: State Farm Road Atlas.*)

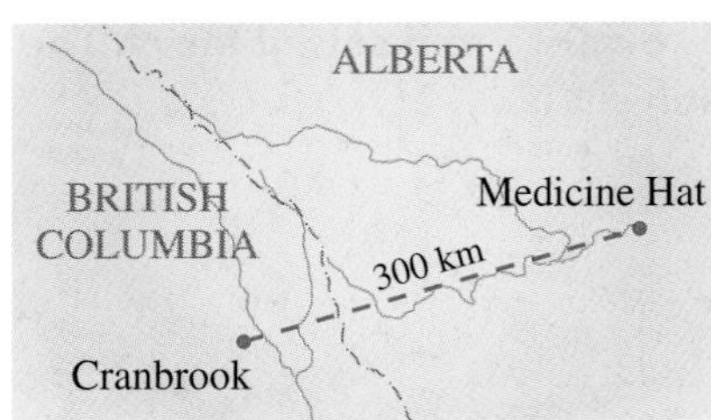

80 km per hr

24. In California, the distance from Jackson to Lodi is about 40 mi, as is the distance from Lodi to Manteca. Rico drove from Jackson to Lodi during the rush hour, stopped in Lodi for a root beer, and then drove on to Manteca at 10 mph faster. Driving time for the entire trip was 88 min. Find his speed from Jackson to Lodi. (*Source: State Farm Road Atlas.*)

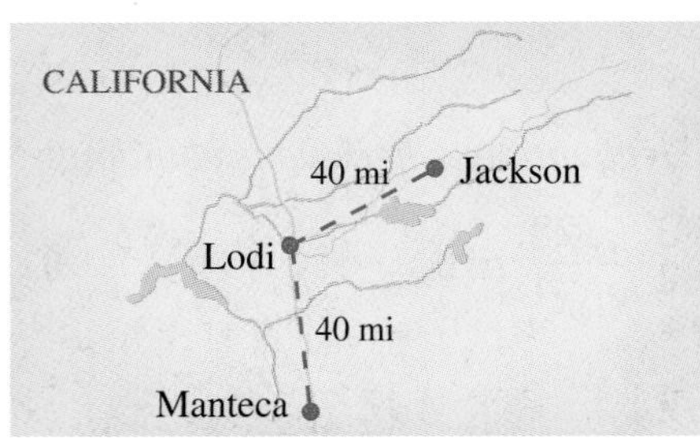

50 mph

25. Working together, two people can cut a large lawn in 2 hr. One person can do the job alone in 1 hr less time than the other. How long (to the nearest tenth) would it take the faster person to do the job? (*Hint:* x is the time of the faster person.)

	Rate	Time Working Together	Fractional Part of the Job Done
Faster Worker	$\frac{1}{x}$	2	
Slower Worker		2	

3.6 hr

26. A janitorial service provides two people to clean an office building. Working together, the two can clean the building in 5 hr. One person is new to the job and would take 2 hr longer than the other person to clean the building alone. How long (to the nearest tenth) would it take the new worker to clean the building alone?

	Rate	Time Working Together	Fractional Part of the Job Done
Faster Worker			
Slower Worker			

11.1 hr

27. A washing machine can be filled in 6 min if both the hot and cold water taps are fully opened. Filling the washer with hot water alone takes 9 min longer than filling it with cold water alone. How long does it take to fill the washer with cold water?

9 min

28. Two pipes together can fill a large tank in 2 hr. One of the pipes, used alone, takes 3 hr longer than the other to fill the tank. How long would each pipe take to fill the tank alone?

3 hr; 6 hr

Solve each equation. Check your solutions. See Example 4.

29. $z = \sqrt{5z - 4}$ $\{1, 4\}$

30. $x = \sqrt{9x - 14}$ $\{2, 7\}$

31. $2x = \sqrt{11x + 3}$ $\{3\}$

32. $4x = \sqrt{6x + 1}$ $\left\{\frac{1}{2}\right\}$

33. $3x = \sqrt{16 - 10x}$ $\left\{\frac{8}{9}\right\}$

34. $4t = \sqrt{8t + 3}$ $\left\{\frac{3}{4}\right\}$

35. $p - 2\sqrt{p} = 8$ $\{16\}$

36. $k + \sqrt{k} = 12$ $\{9\}$

37. $m = \sqrt{\frac{6 - 13m}{5}}$ $\left\{\frac{2}{5}\right\}$

38. $r = \sqrt{\frac{20 - 19r}{6}}$ $\left\{\frac{5}{6}\right\}$

Solve each equation. Check your solutions. See Examples 5–7.

39. $t^4 - 18t^2 + 81 = 0$ $\{-3, 3\}$

40. $x^4 - 8x^2 + 16 = 0$ $\{-2, 2\}$

41. $4k^4 - 13k^2 + 9 = 0$ $\left\{-\frac{3}{2}, -1, 1, \frac{3}{2}\right\}$

42. $9x^4 - 25x^2 + 16 = 0$ $\left\{-\frac{4}{3}, -1, 1, \frac{4}{3}\right\}$

43. $x^4 + 48 = 16x^2$ $\{-2\sqrt{3}, -2, 2, 2\sqrt{3}\}$

44. $z^4 = 17z^2 - 72$ $\{-3, -2\sqrt{2}, 2\sqrt{2}, 3\}$

45. $2x^4 - 9x^2 = -2$ $\left\{\frac{\sqrt{9 + \sqrt{65}}}{2}, -\frac{\sqrt{9 + \sqrt{65}}}{2}, \frac{\sqrt{9 - \sqrt{65}}}{2}, -\frac{\sqrt{9 - \sqrt{65}}}{2}\right\}$

46. $8x^4 + 1 = 11x^2$ $\left\{\frac{\sqrt{11 + \sqrt{89}}}{4}, -\frac{\sqrt{11 + \sqrt{89}}}{4}, \frac{\sqrt{11 - \sqrt{89}}}{4}, -\frac{\sqrt{11 - \sqrt{89}}}{4}\right\}$

47. $(x + 3)^2 + 5(x + 3) + 6 = 0$ $\{-6, -5\}$

48. $(k - 4)^2 + (k - 4) - 20 = 0$ $\{-1, 8\}$

49. $(t + 5)^2 + 6 = 7(t + 5)$ $\{-4, 1\}$

50. $3(m + 4)^2 - 8 = 2(m + 4)$ $\left\{-\frac{16}{3}, -2\right\}$

51. $2 + \frac{5}{3k - 1} = \frac{-2}{(3k - 1)^2}$ $\left\{-\frac{1}{3}, \frac{1}{6}\right\}$

52. $3 - \frac{7}{2p + 2} = \frac{6}{(2p + 2)^2}$ $\left\{-\frac{4}{3}, \frac{1}{2}\right\}$

53. $x^{2/3} + x^{1/3} - 2 = 0$ $\{-8, 1\}$

54. $x^{2/3} - 2x^{1/3} - 3 = 0$

$\{-1, 27\}$

55. $r^{2/3} + r^{1/3} - 12 = 0$

$\{-64, 27\}$

56. $3x^{2/3} - x^{1/3} - 24 = 0$

$\left\{-\frac{512}{27}, 27\right\}$

57. $2\left(1 + \sqrt{r}\right)^2 = 13\left(1 + \sqrt{r}\right) - 6$

$\{25\}$

58. $(k^2 + k)^2 + 12 = 8(k^2 + k)$

$\{-3, -2, 1, 2\}$

Relating Concepts (Exercises 59–64) For Individual or Group Work

Consider the following equation, which contains variable expressions in the denominators. ***Work Exercises 59–64 in order.***

$$\frac{x^2}{(x-3)^2} + \frac{3x}{x-3} - 4 = 0$$

59. Why must 3 be excluded from the domain of this equation?

It would cause both denominators to equal 0, and division by 0 is undefined.

60. Multiply each side of the equation by the LCD, $(x - 3)^2$, and solve. There is only one solution—what is it?

$\frac{12}{5}$

61. Write the equation in a different manner so that it is quadratic in form using the expression $\frac{x}{x-3}$.

$\left(\frac{x}{x-3}\right)^2 + 3\left(\frac{x}{x-3}\right) - 4 = 0$

62. In your own words, explain why the expression $\frac{x}{x-3}$ cannot equal 1.

The numerator can never equal the denominator, since the denominator is 3 less than the numerator.

63. Solve the equation from Exercise 61 by making the substitution $t = \frac{x}{x-3}$. You should get two values for t. Why is one of them impossible for this equation?

$\{\frac{12}{5}\}$**; The values for t are -4 and 1. The value 1 is impossible because it leads to a contradiction (since $\frac{x}{x-3}$ is never equal to 1).**

64. Solve the equation $x^2(x-3)^{-2} + 3x(x-3)^{-1} - 4 = 0$ by letting $s = (x-3)^{-1}$. You should get two values for s. Why is this impossible for this equation?

$\{\frac{12}{5}\}$**; The values for s are $\frac{1}{x}$ and $\frac{-4}{x}$. The value $\frac{1}{x}$ is impossible, since $\frac{1}{x} \neq \frac{1}{x-3}$ for all x.**

Summary Exercises on Solving Quadratic Equations

We have introduced four methods for solving quadratic equations written in standard form $ax^2 + bx + c = 0$. The following table lists some advantages and disadvantages of each method.

METHODS FOR SOLVING QUADRATIC EQUATIONS

Method	Advantages	Disadvantages
Factoring	This is usually the fastest method.	Not all polynomials are factorable; some factorable polynomials are hard to factor.
Square root property	This is the simplest method for solving equations of the form $(ax + b)^2 = c$.	Few equations are given in this form.
Completing the square	This method can always be used, although many people prefer the quadratic formula.	It requires more steps than other methods.
Quadratic formula	This method can always be used.	It is more difficult than factoring because of the square root, although calculators can simplify its use.

Refer to the preceding box. Decide whether factoring, the square root property, *or* the quadratic formula *is most appropriate for solving each quadratic equation. Do not actually solve the equations.*

1. $(2x + 3)^2 = 4$
square root property

2. $4x^2 - 3x = 1$
factoring

3. $z^2 + 5z - 8 = 0$
quadratic formula

4. $2k^2 + 3k = 1$
quadratic formula

5. $3m^2 = 2 - 5m$
factoring

6. $p^2 = 5$
square root property

Solve each quadratic equation by the method of your choice. Check your solutions.

7. $p^2 = 47$
$\{\sqrt{47}, -\sqrt{47}\}$

8. $6x^2 - x - 15 = 0$
$\left\{-\frac{3}{2}, \frac{5}{3}\right\}$

9. $n^2 + 8n + 6 = 0$
$\{-4 + \sqrt{10}, -4 - \sqrt{10}\}$

10. $(x - 4)^2 = 49$
$\{-3, 11\}$

11. $\frac{9}{m} + \frac{5}{m^2} = 2$
$\left\{-\frac{1}{2}, 5\right\}$

12. $3m^2 = 3 - 8m$
$\left\{-3, \frac{1}{3}\right\}$

13. $3x^2 - 9x + 4 = 0$

$\left\{\frac{9 + \sqrt{33}}{6}, \frac{9 - \sqrt{33}}{6}\right\}$

***14.** $x^2 = -12$

$\{2i\sqrt{3}, -2i\sqrt{3}\}$

15. $x\sqrt{2} = \sqrt{5x - 2}$

$\left\{\frac{1}{2}, 2\right\}$

16. $12x^4 - 11x^2 + 2 = 0$

$\left\{-\frac{\sqrt{6}}{3}, -\frac{1}{2}, \frac{1}{2}, \frac{\sqrt{6}}{3}\right\}$

17. $(2k + 5)^2 = 12$

$\left\{\frac{-5 + 2\sqrt{3}}{2}, \frac{-5 - 2\sqrt{3}}{2}\right\}$

18. $\frac{2}{x} + \frac{1}{x - 2} - \frac{5}{3} = 0$

$\left\{\frac{4}{5}, 3\right\}$

19. $t^4 + 14 = 9t^2$

$\{-\sqrt{7}, -\sqrt{2}, \sqrt{2}, \sqrt{7}\}$

20. $2x^2 + 4x = 5$

$\left\{\frac{-2 + \sqrt{14}}{2}, \frac{-2 - \sqrt{14}}{2}\right\}$

***21.** $z^2 + z + 2 = 0$

$\left\{-\frac{1}{2} + \frac{\sqrt{7}}{2}i, -\frac{1}{2} - \frac{\sqrt{7}}{2}i\right\}$

22. $x^4 - 8x^2 = -1$

$\left\{\sqrt{4 + \sqrt{15}}, -\sqrt{4 + \sqrt{15}},\right.$

$\left.\sqrt{4 - \sqrt{15}}, -\sqrt{4 - \sqrt{15}}\right\}$

23. $4t^2 - 12t + 9 = 0$

$\left\{\frac{3}{2}\right\}$

24. $x\sqrt{3} = \sqrt{2 - x}$

$\left\{\frac{2}{3}\right\}$

25. $r^2 - 72 = 0$

$\{6\sqrt{2}, -6\sqrt{2}\}$

26. $-3x^2 + 4x = -4$

$\left\{-\frac{2}{3}, 2\right\}$

27. $x^2 - 5x - 36 = 0$

$\{-4, 9\}$

28. $w^2 = 169$

$\{13, -13\}$

***29.** $3p^2 = 6p - 4$

$\left\{1 + \frac{\sqrt{3}}{3}i, 1 - \frac{\sqrt{3}}{3}i\right\}$

30. $z = \sqrt{\frac{5z + 3}{2}}$

$\{3\}$

31. $2(3k - 1)^2 + 5(3k - 1) = -2$

$\left\{-\frac{1}{3}, \frac{1}{6}\right\}$

***32.** $\frac{4}{r^2} + 3 = \frac{1}{r}$

$\left\{\frac{1}{6} + \frac{\sqrt{47}}{6}i, \frac{1}{6} - \frac{\sqrt{47}}{6}i\right\}$

33. $x - \sqrt{15 - 2x} = 0$

$\{3\}$

34. $3 = \frac{1}{t + 2} + \frac{2}{(t + 2)^2}$

$\left\{-\frac{8}{3}, -1\right\}$

***35.** $4k^4 + 5k^2 + 1 = 0$

$\left\{-i, i, -\frac{1}{2}i, \frac{1}{2}i\right\}$

36. $(x + 1)^{2/3} - (x + 1)^{1/3} = 2$

$\{-2, 7\}$

---

*This exercise requires knowledge of complex numbers.

10.5 Formulas and Further Applications

OBJECTIVE 1 Solve formulas for variables involving squares and square roots. The methods presented earlier in this chapter and the previous one can be used to solve formulas with squares and square roots.

EXAMPLE 1 Solving for Variables Involving Squares or Square Roots

Solve each formula for the given variable.

(a) $w = \frac{kFr}{v^2}$ for v

$w = \frac{kFr}{v^2}$ — The goal is to isolate v on one side.

$v^2 w = kFr$ Multiply by v^2.

$v^2 = \frac{kFr}{w}$ Divide by w.

$v = \pm\sqrt{\frac{kFr}{w}}$ Square root property

$v = \frac{\pm\sqrt{kFr}}{\sqrt{w}} \cdot \frac{\sqrt{w}}{\sqrt{w}} = \frac{\pm\sqrt{kFrw}}{w}$ Rationalize the denominator.

(b) $d = \sqrt{\frac{4A}{\pi}}$ for A

$d = \sqrt{\frac{4A}{\pi}}$ — The goal is to isolate A on one side.

$d^2 = \frac{4A}{\pi}$ Square both sides.

$\pi d^2 = 4A$ Multiply by π.

$\frac{\pi d^2}{4} = A$ Divide by 4.

Work Problem 1 at the Side. ▶

Note

In formulas like $v = \frac{\pm\sqrt{kFrw}}{w}$ in Example 1(a), we will include both positive and negative values.

EXAMPLE 2 Solving for a Second-Degree Variable

Solve $s = 2t^2 + kt$ for t.

Since the equation has terms with t^2 and t, write it in standard form $ax^2 + bx + c = 0$, with t as the variable instead of x.

$s = 2t^2 + kt$

$0 = 2t^2 + kt - s$ Subtract s.

$2t^2 + kt - s = 0$ Standard form

Continued on Next Page

OBJECTIVES

1. Solve formulas for variables involving squares and square roots.
2. Solve applied problems using the Pythagorean formula.
3. Solve applied problems using area formulas.
4. Solve applied problems using quadratic functions as models.

1 Solve each formula for the given variable.

(a) $A = \pi r^2$ for r

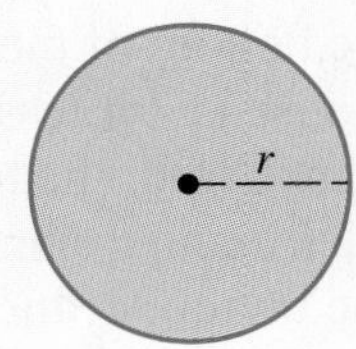

(b) $s = 30\sqrt{\frac{a}{p}}$ for a

ANSWERS

1. (a) $r = \frac{\pm\sqrt{A\pi}}{\pi}$ (b) $a = \frac{ps^2}{900}$

2 Solve $2t^2 - 5t + k = 0$ for t.

Now solve $2t^2 + kt - s = 0$ for t using the quadratic formula.

$$t = \frac{-k \pm \sqrt{k^2 - 4(2)(-s)}}{2(2)} \qquad \text{Let } a = 2,\ b = k, \text{ and } c = -s.$$

$$t = \frac{-k \pm \sqrt{k^2 + 8s}}{4} \qquad \text{Simplify.}$$

The solutions are $t = \dfrac{-k + \sqrt{k^2 + 8s}}{4}$ and $t = \dfrac{-k - \sqrt{k^2 + 8s}}{4}$.

◀ Work Problem **2** at the Side.

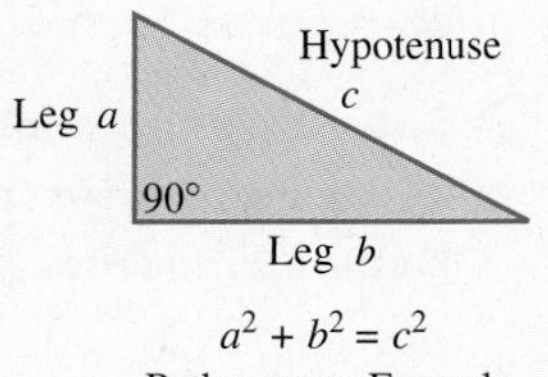

$a^2 + b^2 = c^2$
Pythagorean Formula

OBJECTIVE 2 Solve applied problems using the Pythagorean formula. The Pythagorean formula

$$a^2 + b^2 = c^2,$$

illustrated by the figure in the margin, was introduced in **Section 9.3** and is used to solve applications involving right triangles. Such problems often require solving quadratic equations.

3 Solve the problem.

A 13-ft ladder is leaning against a house. The distance from the bottom of the ladder to the house is 7 ft less than the distance from the top of the ladder to the ground. How far is the bottom of the ladder from the house?

EXAMPLE 3 Using the Pythagorean Formula

Two cars left an intersection at the same time, one heading due north, the other due west. Some time later, they were exactly 100 mi apart. The car headed north had gone 20 mi farther than the car headed west. How far had each car traveled?

Step 1 **Read** the problem carefully.

Step 2 **Assign a variable.**

Let x = the distance traveled by the car headed west.

Then $x + 20$ = the distance traveled by the car headed north.

See Figure 2. The cars are 100 mi apart, so the hypotenuse of the right triangle equals 100.

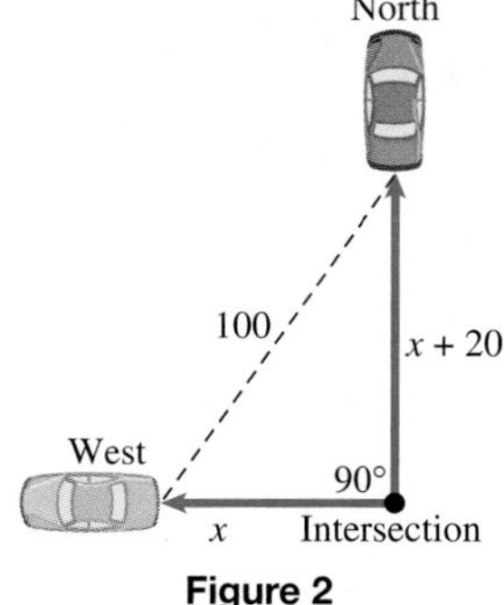

Figure 2

Step 3 **Write an equation.** Use the Pythagorean formula.

$$a^2 + b^2 = c^2$$

$$x^2 + (x + 20)^2 = 100^2 \qquad (x + y)^2 = x^2 + 2xy + y^2$$

Step 4 **Solve.**

$$x^2 + x^2 + 40x + 400 = 10{,}000 \qquad \text{Square the binomial.}$$

$$2x^2 + 40x - 9600 = 0 \qquad \text{Standard form}$$

$$x^2 + 20x - 4800 = 0 \qquad \text{Divide by 2.}$$

$$(x + 80)(x - 60) = 0 \qquad \text{Factor.}$$

$$x + 80 = 0 \quad \text{or} \quad x - 60 = 0 \qquad \text{Zero-factor property}$$

$$x = -80 \quad \text{or} \quad x = 60 \qquad \text{Solve for } x.$$

Step 5 **State the answer.** Distance cannot be negative, so discard the negative solution. The distances are 60 mi and 60 + 20 = 80 mi.

Step 6 **Check.** Since $60^2 + 80^2 = 100^2$, the answers are correct.

◀ Work Problem **3** at the Side.

Answers

2. $t = \dfrac{5 + \sqrt{25 - 8k}}{4}, t = \dfrac{5 - \sqrt{25 - 8k}}{4}$

3. 5 ft

OBJECTIVE 3 Solve applied problems using area formulas.

EXAMPLE 4 Solving an Area Problem

A rectangular reflecting pool in a park is 20 ft wide and 30 ft long. The park gardener wants to plant a strip of grass of uniform width around the edge of the pool. She has enough seed to cover 336 ft^2. How wide will the strip be?

Step 1 **Read** the problem carefully.

Step 2 **Assign a variable.** The pool is shown in Figure 3. If x represents the unknown width of the grass strip, the width of the large rectangle is given by $20 + 2x$ (the width of the pool plus two grass strips), and the length is given by $30 + 2x$.

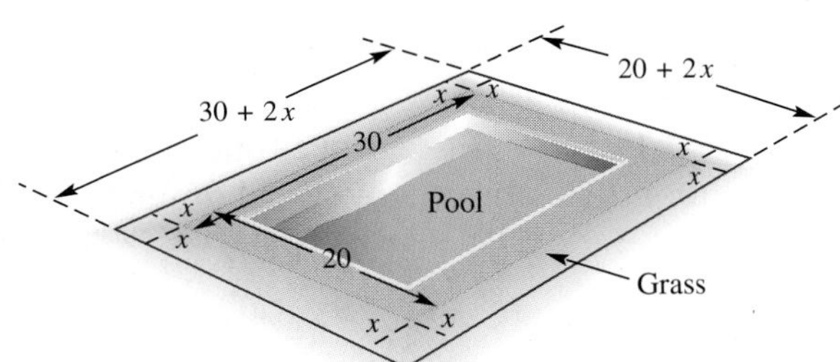

Figure 3

Step 3 **Write an equation.** The area of the large rectangle is given by the product of its length and width, $(30 + 2x)(20 + 2x)$. The area of the pool is $30 \cdot 20 = 600$ ft^2. The area of the large rectangle, minus the area of the pool, should equal the area of the grass strip. Since the area of the grass strip is to be 336 ft^2, the equation is

Area of rectangle − area of pool = area of grass.

$$(30 + 2x)(20 + 2x) - 600 = 336.$$

Step 4 **Solve.**

$$600 + 100x + 4x^2 - 600 = 336 \quad \text{Multiply.}$$

$$4x^2 + 100x - 336 = 0 \quad \text{Standard form}$$

$$x^2 + 25x - 84 = 0 \quad \text{Divide by 4.}$$

$$(x + 28)(x - 3) = 0 \quad \text{Factor.}$$

$$x = -28 \quad \text{or} \quad x = 3 \quad \text{Zero-factor property}$$

Step 5 **State the answer.** The width cannot be −28 ft, so the grass strip should be 3 ft wide.

Step 6 **Check.** If $x = 3$, then the area of the large rectangle is

$$(30 + 2 \cdot 3)(20 + 2 \cdot 3) = 36 \cdot 26 = 936 \text{ ft}^2. \quad \text{Area of pool and strip}$$

The area of the pool is $30 \cdot 20 = 600$ ft^2. So, the area of the grass strip is $936 - 600 = 336$ ft^2, as required. The answer is correct.

Work Problem 4 at the Side. ▶

4 Solve the problem.
Suppose the pool in Example 4 is 20 ft by 40 ft and there is enough seed to cover 700 ft^2. How wide should the grass strip be?

OBJECTIVE 4 Solve applied problems using quadratic functions as models.

Some applied problems can be modeled by *quadratic functions,* which can be written in the form

$$f(x) = ax^2 + bx + c,$$

for real numbers a, b, and c, with $a \neq 0$.

ANSWER

4. 5 ft

EXAMPLE 5 **Solving an Applied Problem Using a Quadratic Function**

If an object is projected upward from the top of a 144-ft building at 112 ft per sec, its position (in feet above the ground) is given by

$$s(t) = -16t^2 + 112t + 144,$$

where t is time in seconds after it was propelled. When does it hit the ground?

When the object hits the ground, its distance above the ground is 0. We must find the value of t that makes $s(t) = 0$.

$0 = -16t^2 + 112t + 144$ Let $s(t) = 0$.

$0 = t^2 - 7t - 9$ Divide by -16.

$$t = \frac{-(-7) \pm \sqrt{(-7)^2 - 4(1)(-9)}}{2(1)} = \frac{7 \pm \sqrt{85}}{2} \approx \frac{7 \pm 9.2}{2}$$ Use the quadratic formula and a calculator.

The solutions are $t \approx 8.1$ or $t \approx -1.1$. Since time cannot be negative, discard the negative solution. The object will hit the ground about 8.1 sec after it is projected.

◀ Work Problem 5 at the Side.

5 Solve the problem.

A ball is projected vertically upward from the ground. Its distance in feet from the ground at t seconds is

$$s(t) = -16t^2 + 64t.$$

At what times will the ball be 32 ft from the ground? Use a calculator and round answers to the nearest tenth. (*Hint:* There are two answers.)

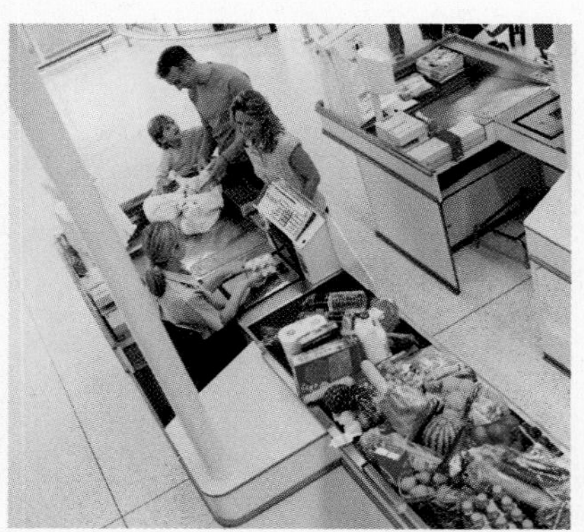

EXAMPLE 6 **Using a Quadratic Function to Model the CPI**

The Consumer Price Index (CPI) is used to measure trends in prices for a "basket" of goods purchased by typical American families. This index uses a base year of 1967, which means that the index number for 1967 is 100. The quadratic function defined by

$$f(x) = -0.065x^2 + 14.8x + 249$$

approximates the CPI for the years 1980–2005, where x is the number of years that have elapsed since 1980. (*Source:* Bureau of Labor Statistics.)

(a) Use the model to approximate the CPI for 1995.

For 1995, $x = 1995 - 1980 = 15$, so find $f(15)$.

$f(x) = -0.065x^2 + 14.8x + 249$ Given model

$f(15) = -0.065(15)^2 + 14.8(15) + 249$ Let $x = 15$.

$f(15) \approx 456$ Nearest whole number

The CPI for 1995 was about 456.

(b) In what year did the CPI reach 550?

Find the value of x that makes $f(x) = 550$.

$f(x) = -0.065x^2 + 14.8x + 249$ Given model

$550 = -0.065x^2 + 14.8x + 249$ Let $f(x) = 550$.

$0 = -0.065x^2 + 14.8x - 301$ Standard form

Now use $a = -0.065$, $b = 14.8$, and $c = -301$ in the quadratic formula.

◀ Work Problem 6 at the Side.

6 Use a calculator to evaluate

$$\frac{-14.8 \pm \sqrt{14.8^2 - 4(-0.065)(-301)}}{2(-0.065)}$$

for both solutions. Round to the nearest whole number. Which solution is valid for this problem?

The first solution is $x \approx 23$. Rounding up to the next whole number, the CPI first reached 550 in $1980 + 23 = 2003$. (Reject the solution $x \approx 205$, as this corresponds to a year far beyond the period covered by the model.)

ANSWERS

5. at 0.6 sec and at 3.4 sec
6. 23; 205; 23

10.5 Exercises

FOR EXTRA HELP

Math XL PRACTICE

WATCH

DOWNLOAD

READ

REVIEW

1. What is the first step in solving a formula like $gw^2 = 2r$ for w?

Solve for w^2 by dividing each side by g.

2. What is the first step in solving a formula like $gw^2 = kw + 24$ for w?

Write it in standard form (with 0 on one side, in decreasing powers of w).

In Exercises 3 and 4, solve for m in terms of the other variables ($m > 0$).

3.

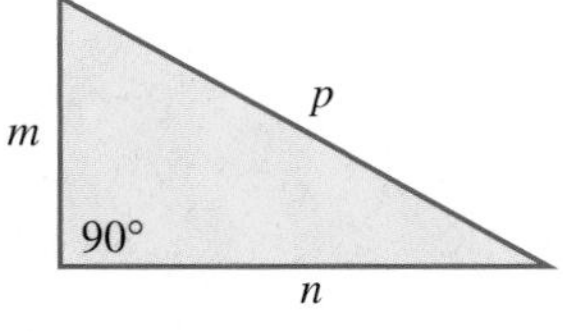

$m = \sqrt{p^2 - n^2}$

4.

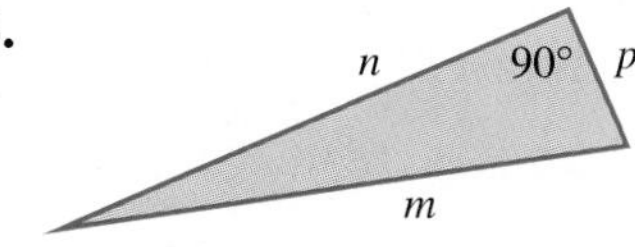

$m = \sqrt{p^2 + n^2}$

Solve each equation for the indicated variable. (Leave $\pm$ *in your answers.) See Examples 1 and 2.*

5. $d = kt^2$ for t

$t = \dfrac{\pm\sqrt{dk}}{k}$

6. $s = kwd^2$ for d

$d = \dfrac{\pm\sqrt{skw}}{kw}$

7. $I = \dfrac{ks}{d^2}$ for d

$d = \dfrac{\pm\sqrt{skI}}{I}$

8. $R = \dfrac{k}{d^2}$ for d

$d = \dfrac{\pm\sqrt{kR}}{R}$

9. $F = \dfrac{kA}{v^2}$ for v

$v = \dfrac{\pm\sqrt{kAF}}{F}$

10. $L = \dfrac{kd^4}{h^2}$ for h

$h = \dfrac{\pm d^2\sqrt{kL}}{L}$

11. $V = \dfrac{1}{3}\pi r^2 h$ for r

$r = \dfrac{\pm\sqrt{3\pi Vh}}{\pi h}$

12. $V = \pi(r^2 + R^2)h$ for r

$r = \dfrac{\pm\sqrt{V\pi h - \pi^2R^2h^2}}{\pi h}$

13. $At^2 + Bt = -C$ for t

$t = \dfrac{-B \pm \sqrt{B^2 - 4AC}}{2A}$

14. $S = 2\pi rh + \pi r^2$ for r

$r = \dfrac{-\pi h \pm \sqrt{\pi^2h^2 + \pi S}}{\pi}$

15. $D = \sqrt{kh}$ for h

$h = \dfrac{D^2}{k}$

16. $F = \dfrac{k}{\sqrt{d}}$ for d

$d = \dfrac{k^2}{F^2}$

17. $p = \sqrt{\dfrac{k\ell}{g}}$ for ℓ

$\ell = \dfrac{p^2g}{k}$

18. $p = \sqrt{\dfrac{k\ell}{g}}$ for g

$g = \dfrac{k\ell}{p^2}$

Solve each problem. When appropriate, round answers to the nearest tenth. See Example 3.

19. Find the lengths of the sides of the triangle.

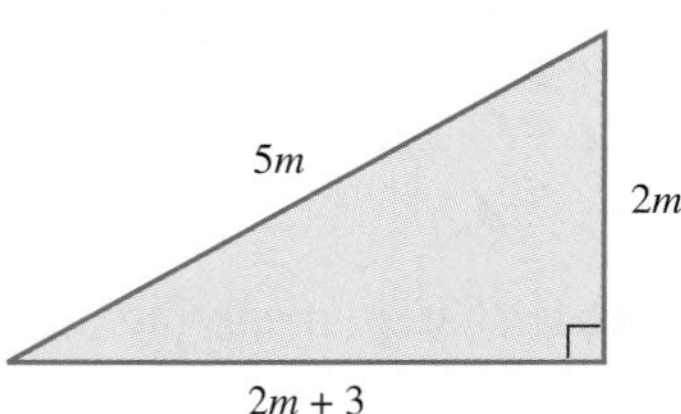

2.3, 5.3, 5.8

20. Find the lengths of the sides of the triangle.

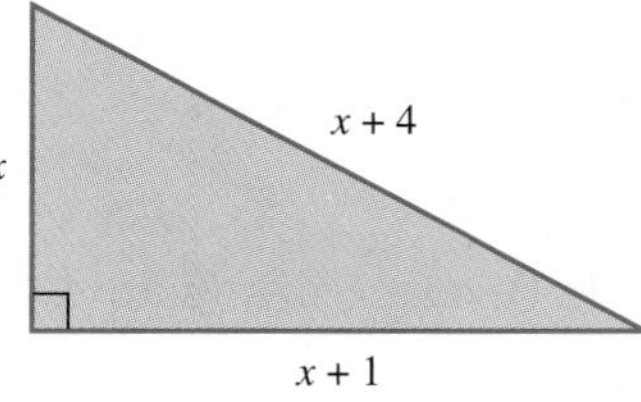

7.9, 8.9, 11.9

21. Two ships leave port at the same time, one heading due south and the other heading due east. Several hours later, they are 170 mi apart. If the ship traveling south traveled 70 mi farther than the other, how many miles did they each travel?

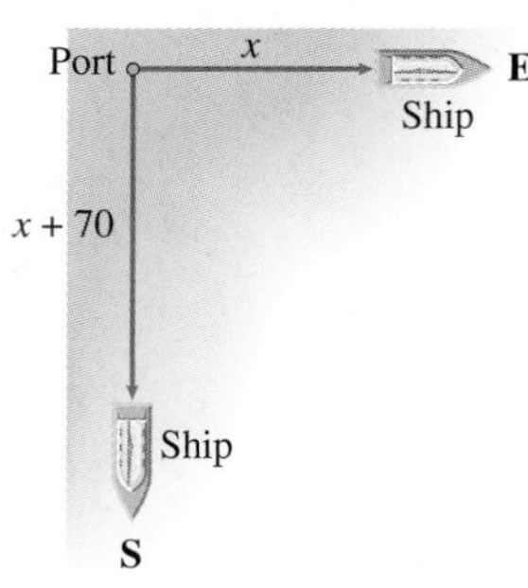

eastbound ship: 80 mi; southbound ship: 150 mi

22. Faith Varnado is flying a kite that is 30 ft farther above her hand than its horizontal distance from her. The string from her hand to the kite is 150 ft long. How high is the kite?

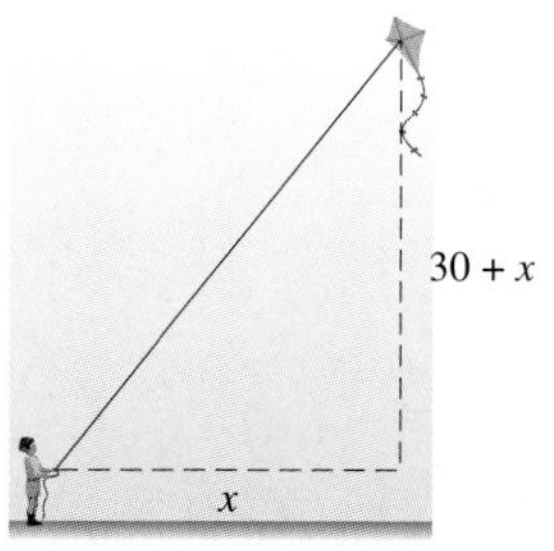

120 ft

Solve each problem. See Example 4.

23. A couple wants to buy a rug for a room that is 20 ft long and 15 ft wide. They want to leave an even strip of flooring uncovered around the edges of the room. How wide a strip will they have if they buy a rug with an area of 234 ft^2?

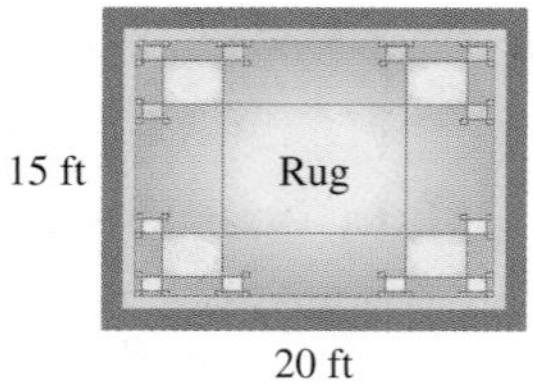

1 ft

24. A club swimming pool is 30 ft wide and 40 ft long. The club members want an exposed aggregate border in a strip of uniform width around the pool. They have enough material for 296 ft^2. How wide can the strip be?

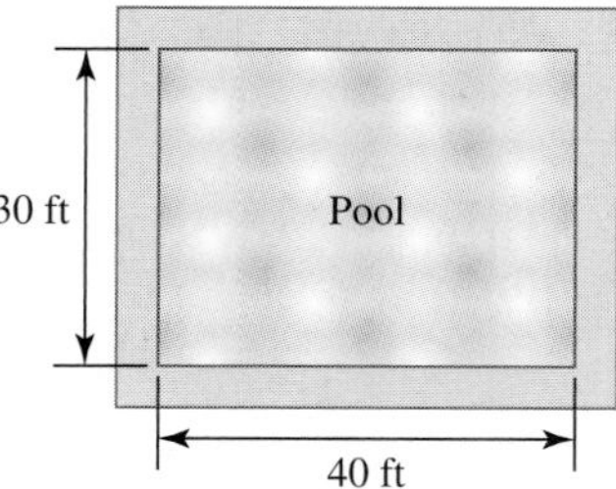

2 ft

25. A rectangular piece of sheet metal has a length that is 4 in. less than twice the width. A square piece 2 in. on a side is cut from each corner. The sides are then turned up to form an uncovered box of volume 256 $in.^3$. Find the length and width of the original piece of metal. **20 in. by 12 in.**

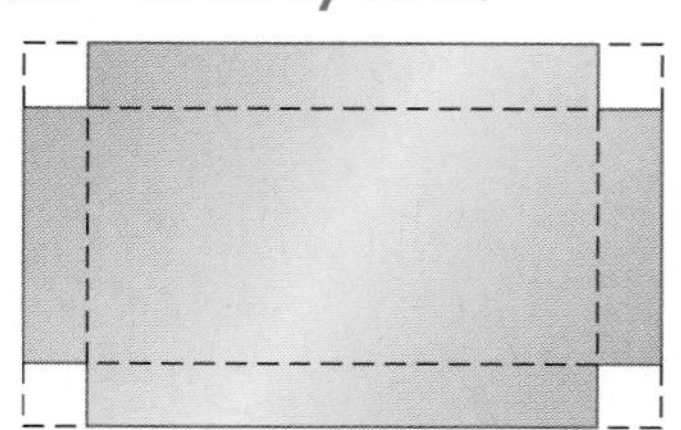

26. Another rectangular piece of sheet metal is 2 in. longer than it is wide. A square piece 3 in. on a side is cut from each corner. The sides are then turned up to form an uncovered box of volume 765 $in.^3$. Find the dimensions of the original piece of metal.

21 in. by 23 in.

Solve each problem. Round answers to the nearest tenth. See Example 5.

27. A ball is projected upward from the ground. Its distance in feet from the ground in t seconds is given by

$$s(t) = -16t^2 + 128t.$$

At what times will the ball be 213 ft from the ground?

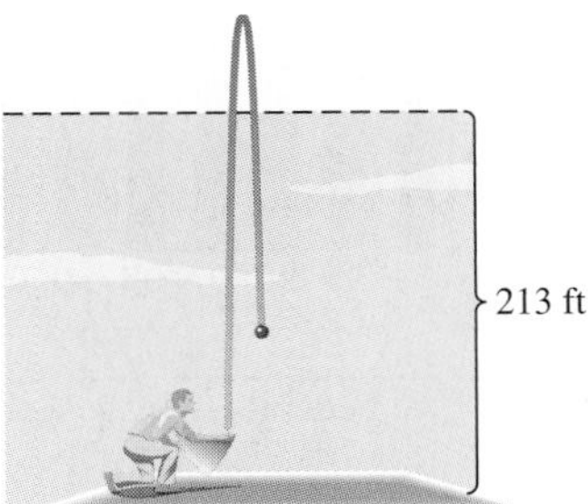

2.4 sec and 5.6 sec

28. A toy rocket is launched from ground level. Its distance in feet from the ground in t seconds is given by

$$s(t) = -16t^2 + 208t.$$

At what times will the rocket be 550 ft from the ground?

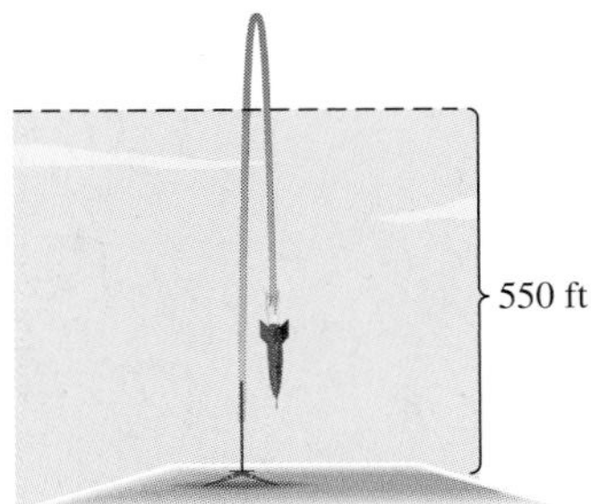

3.7 sec and 9.3 sec

29. The function defined by

$$D(t) = 13t^2 - 100t$$

gives the distance in feet a car going approximately 68 mph will skid in t seconds. Find the time it would take for the car to skid 180 ft.

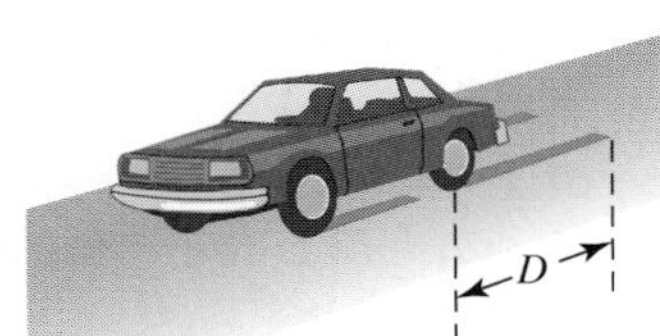

9.2 sec

30. The function given in Exercise 29 becomes

$$D(t) = 13t^2 - 73t$$

for a car going 50 mph. Find the time for this car to skid 218 ft.

7.8 sec

A ball is projected upward from ground level, and its distance in feet from the ground in t seconds is given by $s(t) = -16t^2 + 160t$. Use algebra and a short explanation to answer Exercises 31 and 32.

31. After how many seconds does it reach a height of 400 ft? How would you describe in words its position at this height?

It reaches its *maximum* height at 5 sec because this is the only time it reaches 400 ft.

32. After how many seconds does it reach a height of 425 ft? How would you interpret the mathematical result here?

Because the discriminant is negative, the ball never reaches a height of 425 ft.

Solve each problem using a quadratic equation.

33. A certain bakery has found that the daily demand for blueberry muffins is $\frac{3200}{p}$, where p is the price of a muffin in cents. The daily supply is $3p - 200$. Find the price at which supply and demand are equal. **\$0.80**

34. In one area the demand for compact discs is $\frac{700}{P}$ per day, where P is the price in dollars per disc. The supply is $5P - 1$ per day. At what price, to the nearest cent, does supply equal demand? **\$11.93**

The total number of miles traveled by all motor vehicles in the United States for the years 1994–2003 are shown in the bar graph and can be modeled by the quadratic function defined by

$$f(x) = -1.705x^2 + 75.93x + 2351.$$

Here, $x = 0$ represents 1994, $x = 1$ represents 1995, and so on. Use the graph and the model to work Exercises 35–38. See Example 6.

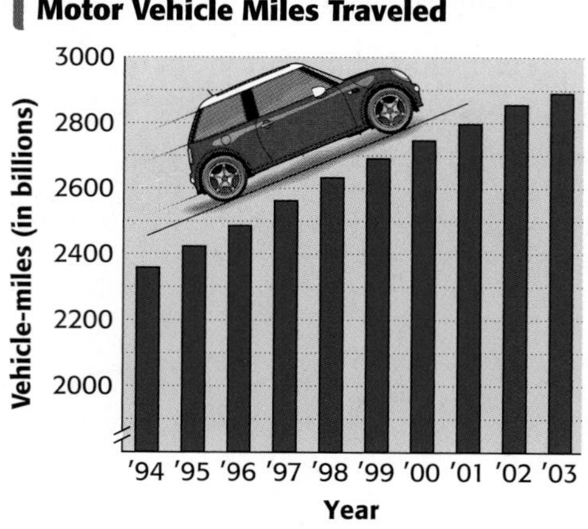

Source: U.S. Federal Highway Administration.

35. (a) Use the graph to estimate miles traveled in 2000 to the nearest ten billion. **2750 billion**

(b) Use the model to approximate miles traveled in 2000 to the nearest ten billion. How does this result compare to your estimate from part (a)?
2750 billion; They are the same.

36. Based on the model, in what year did miles traveled reach 2600 billion? (Round down to the nearest year.) How does this result compare to the vehicle-miles shown in the graph?
1997; The graph indicates that vehicle-miles reached 2600 in 1998.

37. Based on the model, in what year did miles traveled reach 2800 billion? (Round down to the nearest year.) How does this result compare to the vehicle-miles shown in the graph?
2001; The graph indicates that vehicle-miles reached 2800 in 2001.

38. If these data were modeled by a *linear* function defined by $f(x) = ax + b$, would the value of a be positive or negative? Explain.
The value of a would be positive, because it would be the slope of a line that rises from left to right.

William Froude was a 19th-century naval architect who used the expression

$$\frac{v^2}{g\ell}$$

in shipbuilding. This expression, known as the **Froude number,** *was also used by R. McNeill Alexander in his research on dinosaurs.* (Source: *"How Dinosaurs Ran,"* Scientific American, *April 1991.) In Exercises 39 and 40, find to the nearest tenth the value of v (in meters per second), given that* $g = 9.8$ m per sec^2.

39. Rhinoceros: $\ell = 1.2$; Froude number $= 2.57$
5.5 m per sec

40. Triceratops: $\ell = 2.8$; Froude number $= 0.16$
2.1 m per sec

Recall from the **Section 7.5** *exercises that corresponding sides of similar triangles are proportional. Use this fact to find the lengths of the indicated sides of each pair of similar triangles. Check all possible solutions in both triangles. Sides of a triangle cannot be negative (and are not drawn to scale here).*

41. Side AC
5 or 14

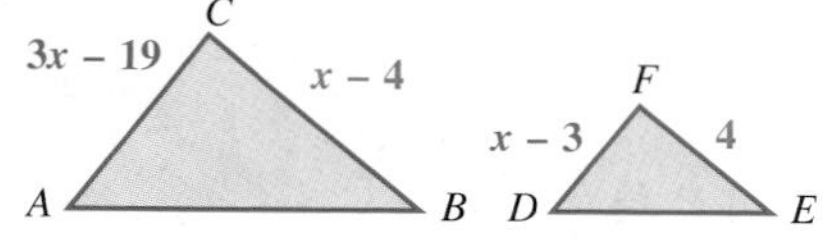

42. Side RQ
4

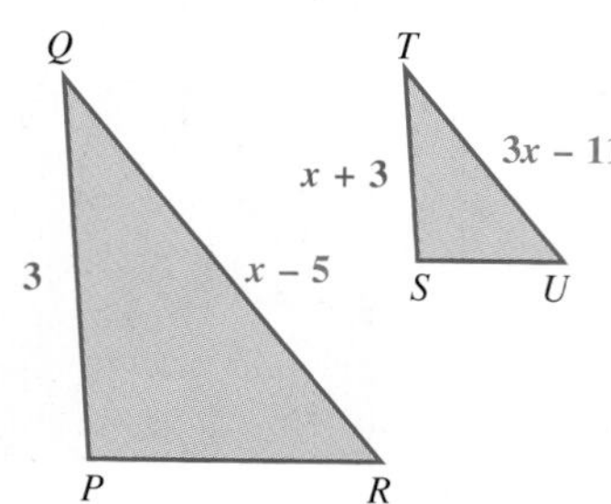

10.6 Graphs of Quadratic Functions

OBJECTIVES

1. **Graph a quadratic function.**
2. **Graph parabolas with horizontal and vertical shifts.**
3. **Use the coefficient of x^2 to predict the shape and direction in which a parabola opens.**
4. **Find a quadratic function to model data.**

OBJECTIVE 1 Graph a quadratic function. Figure 4 gives a graph of the simplest *quadratic function,* defined by $y = x^2$.

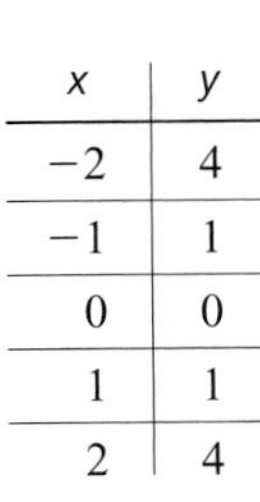

x	y
−2	4
−1	1
0	0
1	1
2	4

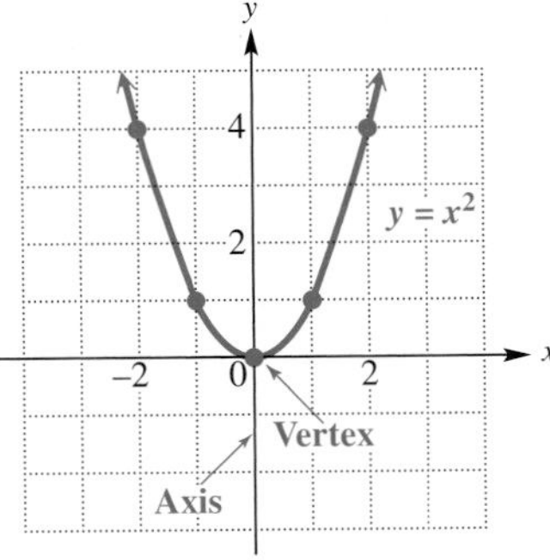

Figure 4

This graph is called a **parabola.** The point (0, 0), the lowest point on the curve, is the **vertex** of this parabola. The vertical line through the vertex is the **axis** of the parabola, here $x = 0$. A parabola is **symmetric about its axis;** that is, if the graph were folded along the axis, the two portions of the curve would coincide. As Figure 4 suggests, x can be any real number, so the domain of the function defined by $y = x^2$ is $(-\infty, \infty)$. Since y is always nonnegative, the range is $[0, \infty)$.

In **Section 10.5,** we solved applications modeled by quadratic functions.

Quadratic Function

A function that can be written in the form

$$f(x) = ax^2 + bx + c$$

for real numbers a, b, and c, with $a \neq 0$, is a **quadratic function.**

The graph of any quadratic function is a parabola with a vertical axis.

Note

We use the variable y and function notation $f(x)$ interchangeably. Although we use the letter f most often to name quadratic functions, other letters can be used. We use the capital letter F to distinguish between different parabolas graphed on the same coordinate axes.

Parabolas, which are a type of *conic section* **(Chapter 12),** have many applications. Cross sections of telescopes, satellite dishes and automobile headlights form parabolas, as do the cables that support suspension bridges.

OBJECTIVE 2 Graph parabolas with horizontal and vertical shifts. Parabolas need not have their vertices at the origin, as does the graph of $f(x) = x^2$. To graph a parabola of the form

$$F(x) = x^2 + k, \quad \text{Vertical shift } k$$

select sample values of x like those that were used to graph $f(x) = x^2$. The corresponding values of $F(x)$ in $F(x) = x^2 + k$ differ by k from those of $f(x) = x^2$. For this reason, the graph of $F(x) = x^2 + k$ is *shifted,* or *translated,* k units vertically compared with that of $f(x) = x^2$.

1 Graph each parabola. Give the vertex, domain, and range.

(a) $f(x) = x^2 + 3$

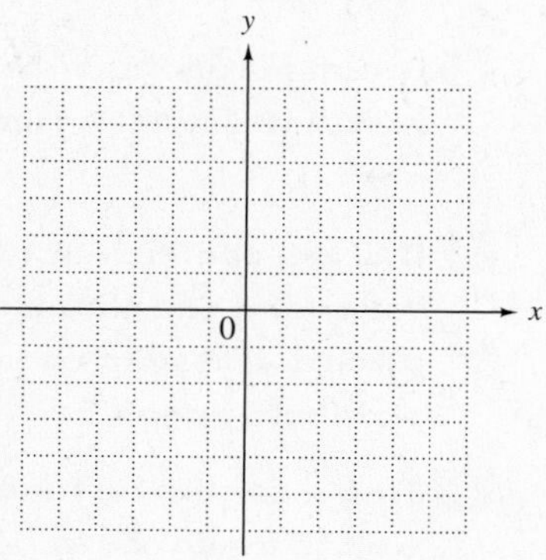

(b) $f(x) = x^2 - 1$

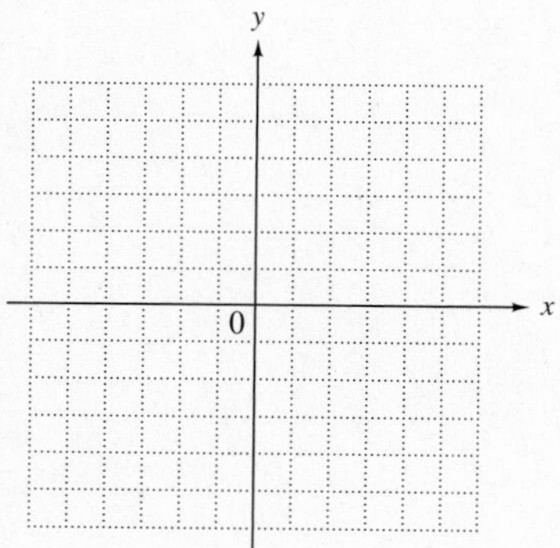

EXAMPLE 1 Graphing a Parabola with a Vertical Shift

Graph $F(x) = x^2 - 2$.

This graph has the same shape as that of $f(x) = x^2$, but since k here is -2, the graph is shifted 2 units down, with vertex $(0, -2)$. Every function value is 2 less than the corresponding function value of $f(x) = x^2$. Plotting points on both sides of the vertex gives the graph in Figure 5.

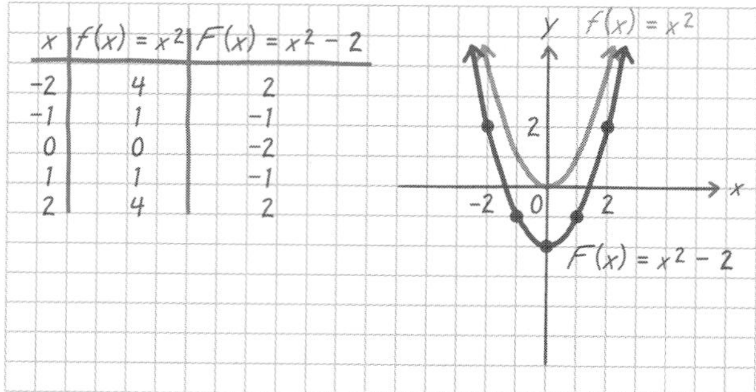

x	$f(x) = x^2$	$F(x) = x^2 - 2$
−2	4	2
−1	1	−1
0	0	−2
1	1	−1
2	4	2

Figure 5

Notice that since the parabola is symmetric about its axis $x = 0$, the plotted points are "mirror images" of each other. Since x can be any real number, the domain is still $(-\infty, \infty)$. The value of y (or $F(x)$) is always greater than or equal to -2, so the range is $[-2, \infty)$. The graph of $f(x) = x^2$ is shown for comparison.

Vertical Shift

The graph of $\boldsymbol{F(x) = x^2 + k}$ is a parabola with the same shape as the graph of $f(x) = x^2$. The parabola is shifted vertically: k units up if $k > 0$, and $|k|$ units down if $k < 0$. The vertex is $(0, k)$.

◀ Work Problem **1** at the Side.

The graph of the function defined by

$$F(x) = (x - h)^2 \qquad \text{Horizontal shift } h$$

is also a parabola with the same shape as that of $f(x) = x^2$. Because $(x - h)^2 \geq 0$ for all x, the vertex of $F(x) = (x - h)^2$ is the lowest point on the parabola. The lowest point occurs here when $F(x)$ is 0. To get $F(x)$ equal to 0, let $x = h$ so the vertex of $F(x) = (x - h)^2$ is $(h, 0)$. Based on this, the graph of $F(x) = (x - h)^2$ is shifted h units horizontally compared with that of $f(x) = x^2$.

EXAMPLE 2 Graphing a Parabola with a Horizontal Shift

Graph $F(x) = (x - 2)^2$.

If $x = 2$, then $F(x) = 0$, which gives the vertex $(2, 0)$. The graph of $F(x) = (x - 2)^2$ has the same shape as that of $f(x) = x^2$ but is shifted 2 units to the right. Plotting several points on one side of the vertex and using symmetry about the axis $x = 2$ to find corresponding points on the other side of the vertex gives the graph in Figure 6 on the next page. Again, the domain is $(-\infty, \infty)$; the range is $[0, \infty)$.

Continued on Next Page

ANSWERS

1. (a)

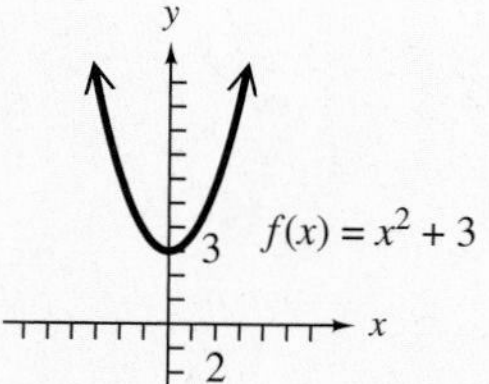

vertex: $(0, 3)$; domain: $(-\infty, \infty)$; range: $[3, \infty)$

(b)

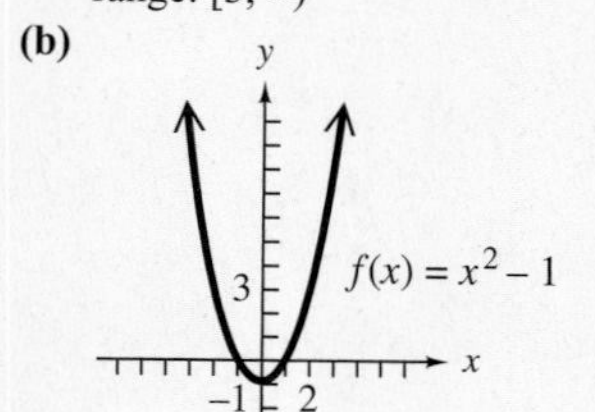

vertex: $(0, -1)$; domain: $(-\infty, \infty)$; range: $[-1, \infty)$

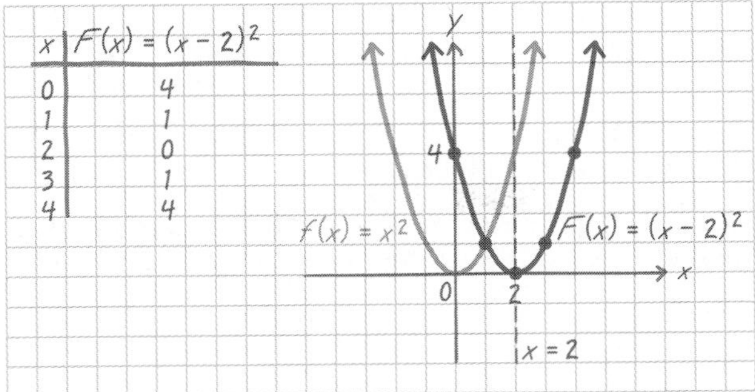

Figure 6

Horizontal Shift

The graph of $\boldsymbol{F(x) = (x - h)^2}$ is a parabola with the same shape as the graph of $f(x) = x^2$. The parabola is shifted h units horizontally: h units to the right if $h > 0$, and $|h|$ units to the left if $h < 0$. The vertex is $(h, 0)$.

CAUTION

Errors frequently occur when horizontal shifts are involved. To determine the direction and magnitude of a horizontal shift, find the value that would cause the expression $x - h$ to equal 0. For example, the graph of $F(x) = (x - 5)^2$ would be shifted 5 units to the *right*, because $+5$ would cause $x - 5$ to equal 0. On the other hand, the graph of $F(x) = (x + 5)^2$ would be shifted 5 units to the *left*, because -5 would cause $x + 5$ to equal 0.

Work Problem 2 at the Side. ▶

A parabola can have both horizontal and vertical shifts.

EXAMPLE 3 Graphing a Parabola with Horizontal and Vertical Shifts

Graph $F(x) = (x + 3)^2 - 2$.

This graph has the same shape as that of $f(x) = x^2$, but is shifted 3 units to the left (since $x + 3 = 0$ if $x = -3$) and 2 units down (because of the -2). See Figure 7. The vertex is $(-3, -2)$, with axis $x = -3$. This function has domain $(-\infty, \infty)$ and range $[-2, \infty)$.

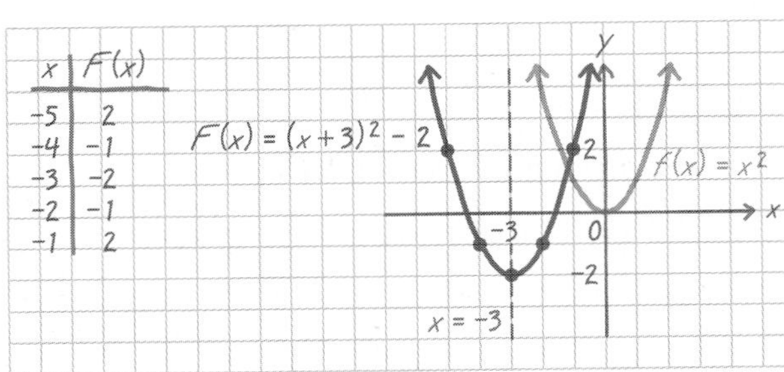

Figure 7

2 Graph each parabola. Give the vertex, axis, domain, and range.

(a) $f(x) = (x - 3)^2$

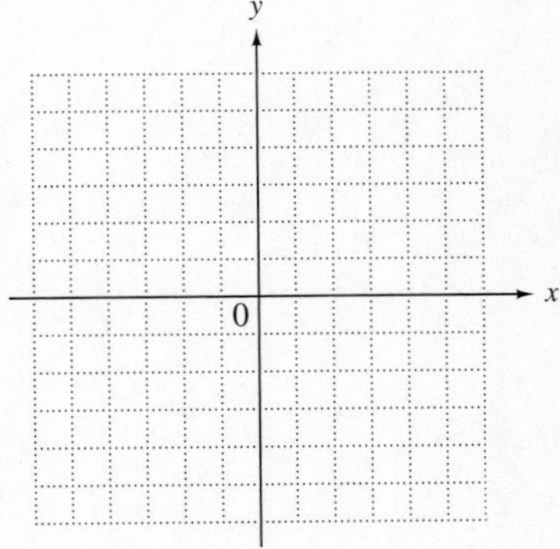

(b) $f(x) = (x + 2)^2$

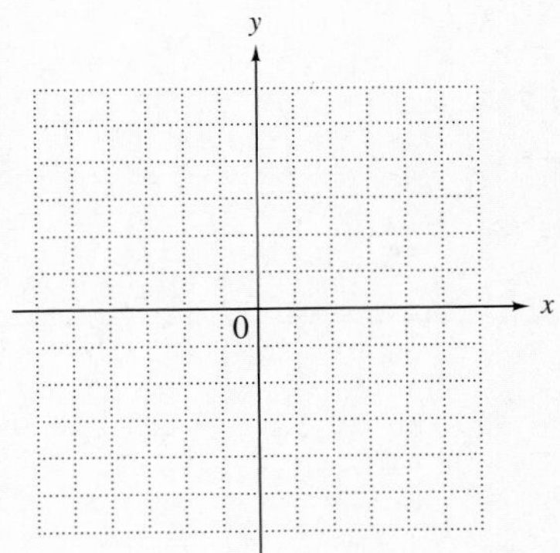

ANSWERS

2. (a)

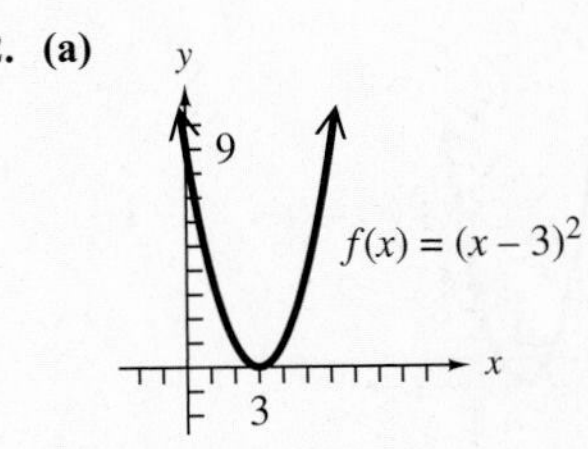

vertex: $(3, 0)$; axis: $x = 3$; domain: $(-\infty, \infty)$; range: $[0, \infty)$

(b)

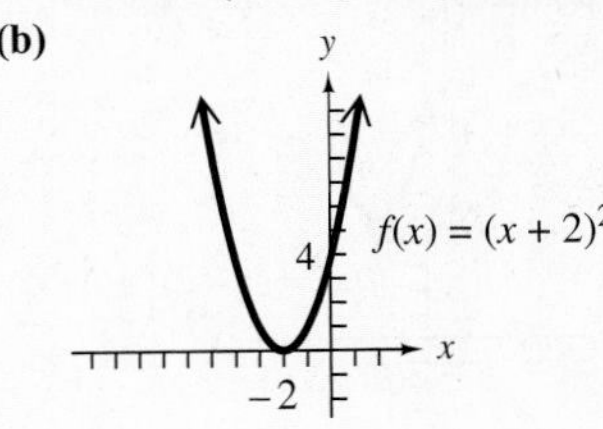

vertex: $(-2, 0)$; axis: $x = -2$; domain: $(-\infty, \infty)$; range: $[0, \infty)$

3 Graph each parabola. Give the vertex, axis, domain, and range.

(a) $f(x) = (x + 2)^2 - 1$

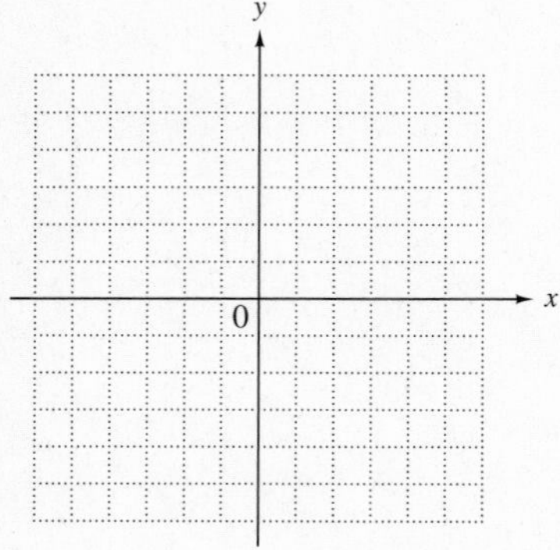

(b) $f(x) = (x - 2)^2 + 5$

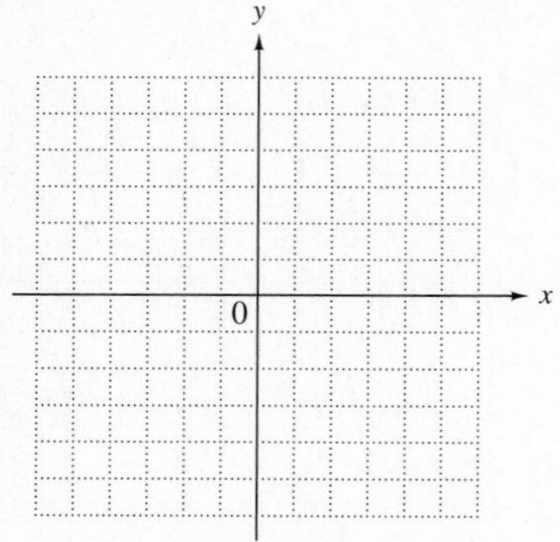

The characteristics of the graph of a parabola with equation of the form $F(x) = (x - h)^2 + k$ are summarized as follows.

Vertex and Axis of a Parabola

The graph of $F(x) = (x - h)^2 + k$ is a parabola with the same shape as the graph of $f(x) = x^2$ with vertex (h, k). The axis is the vertical line $x = h$.

◀ *Work Problem* **3** *at the Side.*

OBJECTIVE 3 Use the coefficient of x^2 to predict the shape and direction in which a parabola opens. Not all parabolas open up, and not all parabolas have the same shape as the graph of $f(x) = x^2$.

EXAMPLE 4 Graphing a Parabola That Opens Down

Graph $f(x) = -\frac{1}{2}x^2$.

This parabola is shown in Figure 8. The coefficient $-\frac{1}{2}$ affects the shape of the graph; the $\frac{1}{2}$ makes the parabola wider (since the values of $\frac{1}{2}x^2$ increase more slowly than those of x^2), and the negative sign makes the parabola open down. The graph is not shifted in any direction; the vertex is still $(0, 0)$ and the axis has equation $x = 0$. Unlike the parabolas graphed in Examples 1–3, the vertex here has the *greatest* function value of any point on the graph. The domain is $(-\infty, \infty)$; the range is $(-\infty, 0]$.

x	$f(x)$
-2	-2
-1	$-\frac{1}{2}$
0	0
1	$-\frac{1}{2}$
2	-2

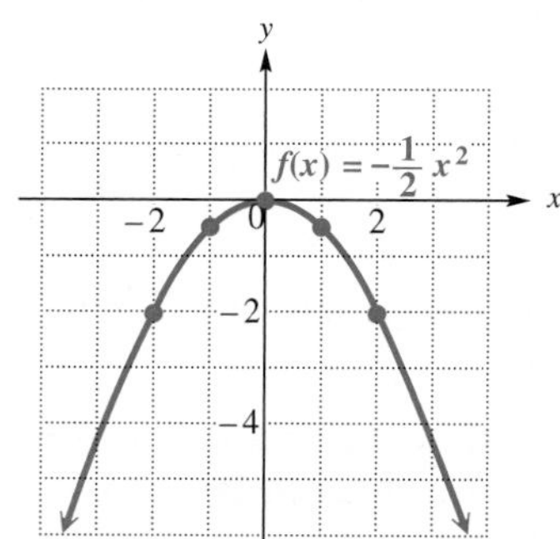

Figure 8

Some general principles concerning the graph of $F(x) = a(x - h)^2 + k$ are summarized as follows.

General Principles

1. The graph of the quadratic function defined by
$$F(x) = a(x - h)^2 + k, \quad a \neq 0,$$
is a parabola with vertex (h, k) and the vertical line $x = h$ as axis.
2. The graph opens up if a is positive and down if a is negative.
3. The graph is wider than that of $f(x) = x^2$ if $0 < |a| < 1$. The graph is narrower than that of $f(x) = x^2$ if $|a| > 1$.

Answers

3. (a)

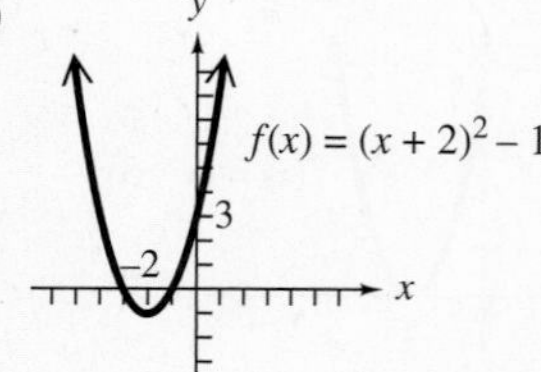

vertex: $(-2, -1)$; axis: $x = -2$; domain: $(-\infty, \infty)$; range: $[-1, \infty)$

(b)

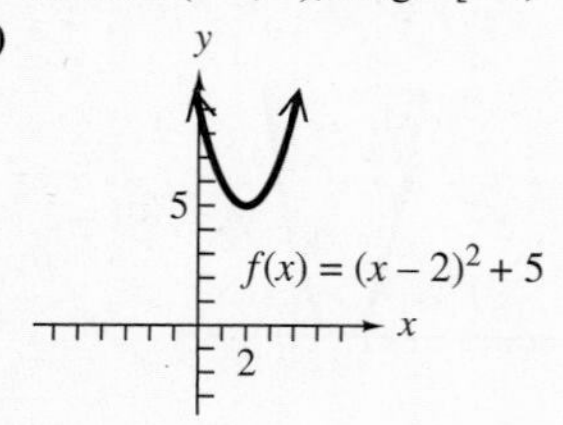

vertex: $(2, 5)$; axis: $x = 2$; domain: $(-\infty, \infty)$; range: $[5, \infty)$

Work Problems **4** *and* **5** *at the Side.* ▶

EXAMPLE 5 Using the General Principles to Graph a Parabola

Graph $F(x) = -2(x + 3)^2 + 4$. Give the domain and the range.

The parabola opens down (because $a < 0$), and is narrower than the graph of $f(x) = x^2$, since $|-2| = 2 > 1$, causing values of $F(x)$ to decrease more quickly than those of $f(x) = -x^2$. This parabola has vertex $(-3, 4)$ as shown in Figure 9. To complete the graph, we plotted the ordered pairs $(-4, 2)$ and, by symmetry, $(-2, 2)$. Symmetry can be used to find additional ordered pairs that satisfy the equation, if desired. The domain is $(-\infty, \infty)$ and the range is $(-\infty, 4]$.

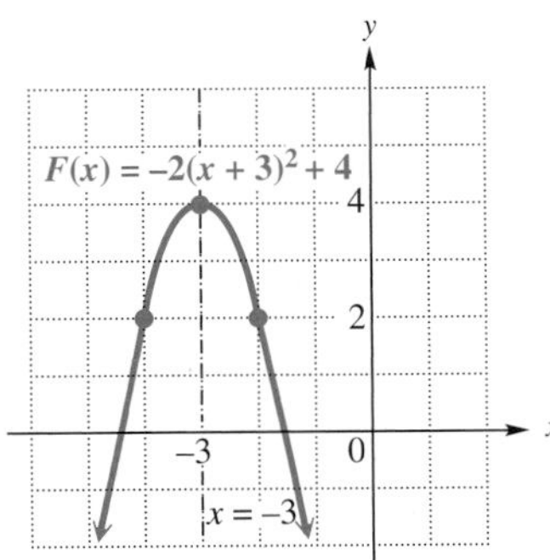

Figure 9

Work Problem **6** *at the Side.* ▶

OBJECTIVE 4 Find a quadratic function to model data.

EXAMPLE 6 Modeling the Number of Multiple Births

After rising steadily over several decades, the number of higher-order multiple births (triplets or more) in the United States started to decline during the first decade of the 21st century. Let x represent the number of years since 1995 and y represent the number of higher-order multiple births. Data for selected years are shown in the table.

Year	x	y
1995	0	4973
1996	1	5939
1997	2	6737
1999	4	7321
2001	6	7471
2003	8	7663
2004	9	7275
2005	10	6694

Source: National Center for Health Statistics.

Find a quadratic function that models the data.

A scatter diagram of the ordered pairs (x, y) is shown in Figure 10 on the next page. The general shape suggested by the scatter diagram indicates that a parabola should approximate these points, as shown by the dashed curve in Figure 11. The equation for such a parabola would have a negative coefficient for x^2 since the graph opens down.

Continued on Next Page

4 Decide whether each parabola opens up or down.

(a) $f(x) = -\frac{2}{3}x^2$

(b) $f(x) = \frac{3}{4}x^2 + 1$

(c) $f(x) = -2x^2 - 3$

(d) $f(x) = 3x^2 + 2$

5 Decide whether each parabola in Problem 4 is wider or narrower than the graph of $f(x) = x^2$.

6 Graph

$$f(x) = \frac{1}{2}(x - 2)^2 + 1.$$

Give the domain and the range.

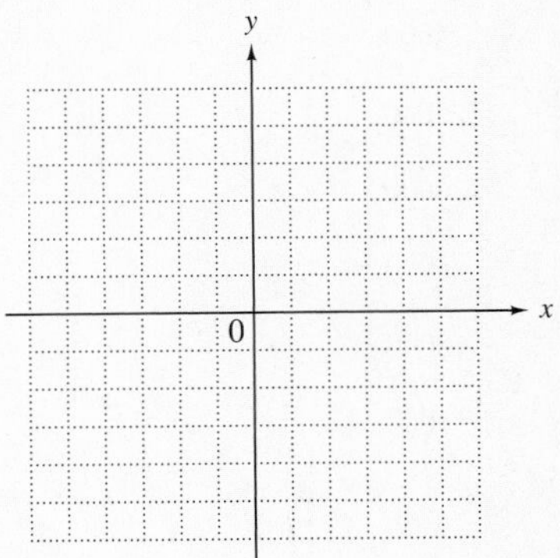

ANSWERS

4. **(a)** down **(b)** up **(c)** down **(d)** up

5. **(a)** wider **(b)** wider **(c)** narrower **(d)** narrower

6.

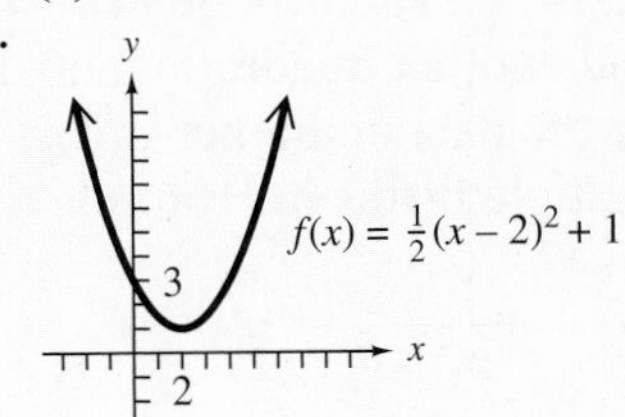

domain: $(-\infty, \infty)$
range: $[1, \infty)$

7 Tell whether a linear or quadratic function would be a more appropriate model for each set of graphed data. If linear, tell whether the slope should be positive or negative. If quadratic, tell whether the coefficient a of x^2 should be positive or negative.

(a)

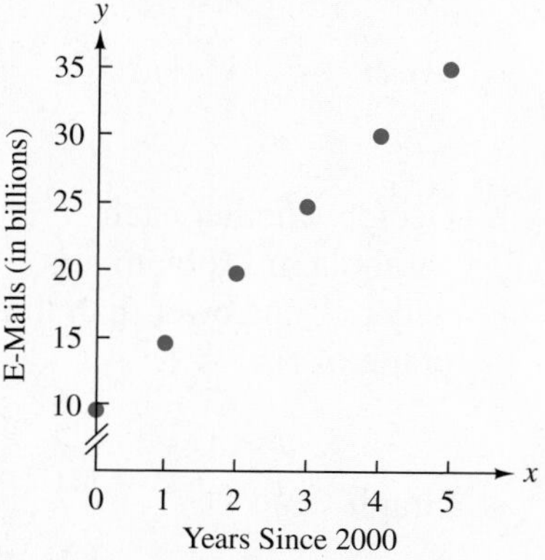

Source: General Accounting Office.

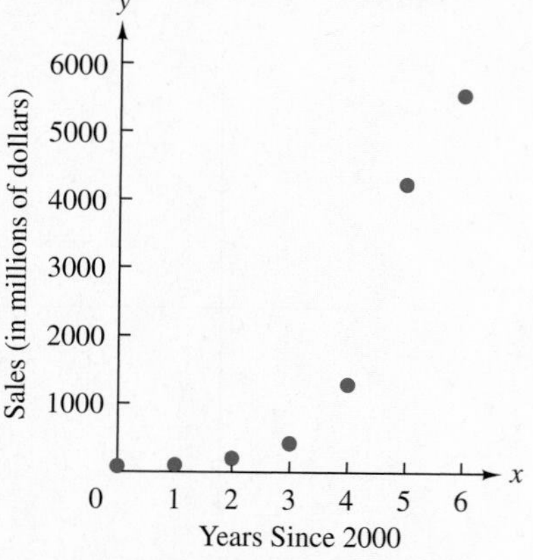

Source: Consumer Electronics Association.

8 Using the points (1, 5939), (6, 7471), and (10, 6694), find another quadratic model for the data on higher-order multiple births in Example 6.

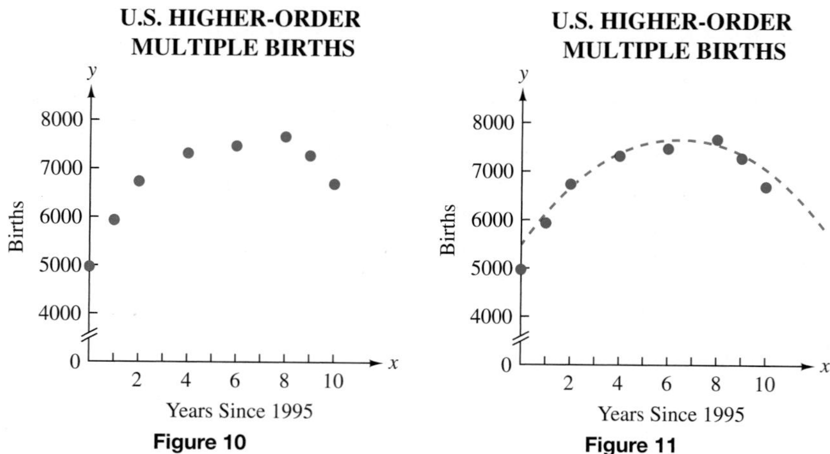

Figure 10 **Figure 11**

To find a quadratic function of the form

$$y = ax^2 + bx + c$$

that models, or *fits,* these data, we choose three representative ordered pairs and use them to write a system of three equations. Using (0, 4973), (4, 7321), and (10, 6694), we substitute the x- and y-values from the ordered pairs into the quadratic form $y = ax^2 + bx + c$ to get the three equations

$$a(0)^2 + b(0) + c = \mathbf{4973} \quad \text{or} \quad c = 4973 \quad (1)$$
$$a(4)^2 + b(4) + c = \mathbf{7321} \quad \text{or} \quad 16a + 4b + c = 7321 \quad (2)$$
$$a(10)^2 + b(10) + c = \mathbf{6694} \quad \text{or} \quad 100a + 10b + c = 6694. \quad (3)$$

We can find the values of a, b, and c by solving this system of three equations in three variables using the methods of **Section 8.5.** From equation (1), $c = \mathbf{4973}$. Substitute 4973 for c in equations (2) and (3) to obtain the equations

$$16a + 4b + \mathbf{4973} = 7321, \quad \text{or} \quad 16a + 4b = 2348 \quad (4)$$
$$100a + 10b + \mathbf{4973} = 6694, \quad \text{or} \quad 100a + 10b = 1721. \quad (5)$$

We can eliminate b from this system of equations in two variables by multiplying equation (4) by -5 and equation (5) by 2, and adding the results to get

$$120a = -8298$$
$$a = \mathbf{-69.15}. \quad \text{Divide by 120; use a calculator.}$$

We substitute -69.15 for a in equation (4) or (5) to find that $b = \mathbf{863.6}$. Using the values we have found for a, b, and c, our model is defined by

$$y = \mathbf{-69.15x^2 + 863.6x + 4973}.$$

◀ Work Problems **7** and **8** at the Side.

Note

If we had chosen three different ordered pairs of data in Example 6, a slightly different model would have resulted, as in Problem 8 at the side.

Calculator Tip The *quadratic regression* feature on a graphing calculator can be used to generate a quadratic model that fits given data. See your owner's manual for details on how to do this.

Answers

7. (a) linear; positive **(b)** quadratic; positive
8. $y = -55.63x^2 + 695.80x + 5299$

10.6 Exercises

FOR EXTRA HELP

MyMathLab

Math XL PRACTICE

WATCH

DOWNLOAD

READ

REVIEW

1. Match each quadratic function with its graph from choices A–D.

(a) $f(x) = (x + 2)^2 - 1$ **B**

(b) $f(x) = (x + 2)^2 + 1$ **C**

(c) $f(x) = (x - 2)^2 - 1$ **A**

(d) $f(x) = (x - 2)^2 + 1$ **D**

A.
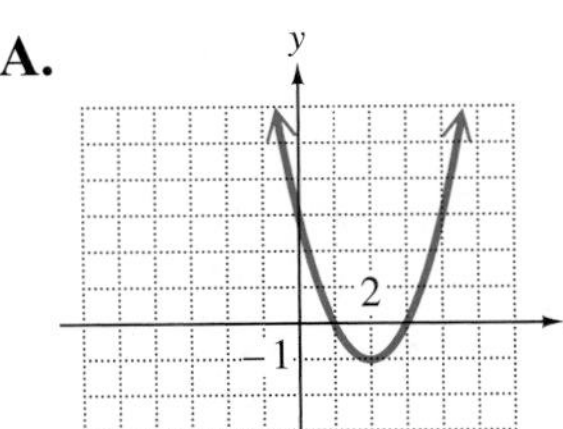

B.
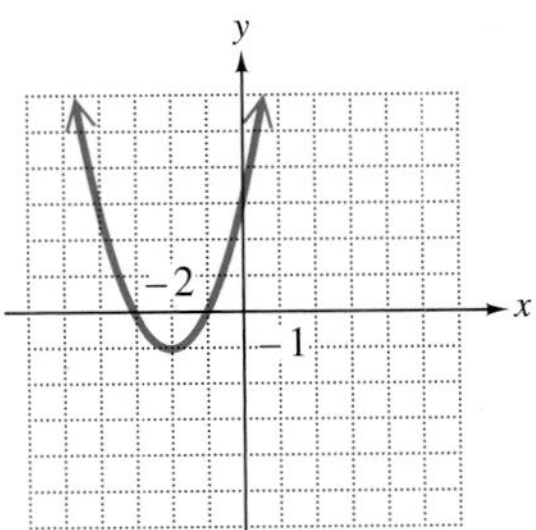

C.
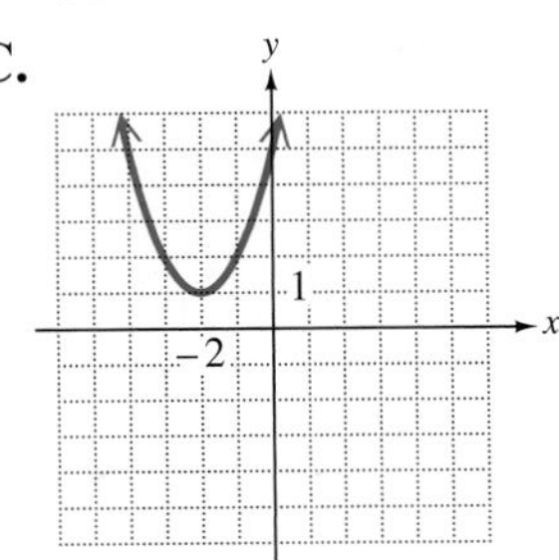

D.
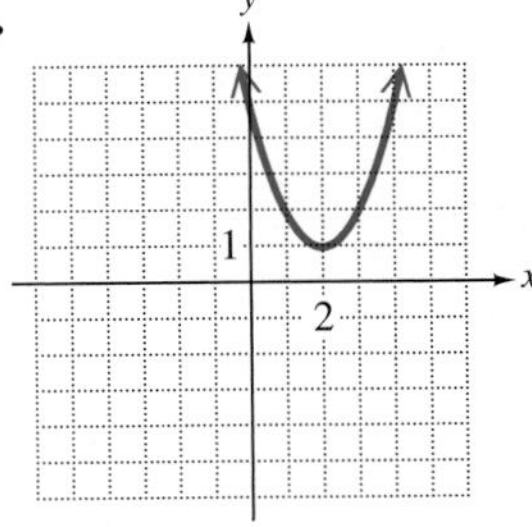

2. Match each quadratic function with its graph from choices A–D.

(a) $f(x) = -x^2 + 2$ **D**

(b) $f(x) = -x^2 - 2$ **C**

(c) $f(x) = -(x + 2)^2$ **B**

(d) $f(x) = -(x - 2)^2$ **A**

A.
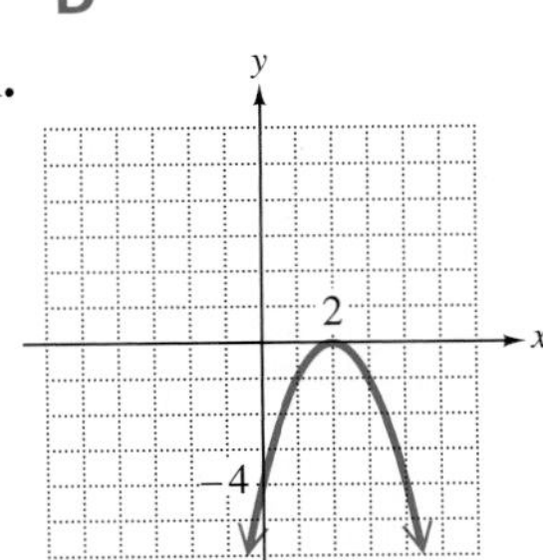

B.
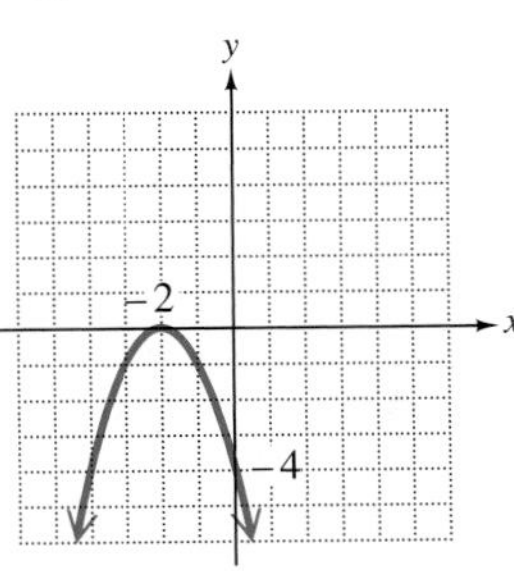

C.
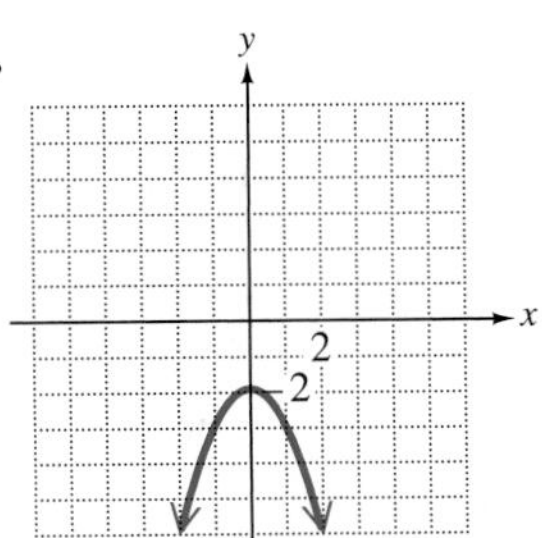

D.
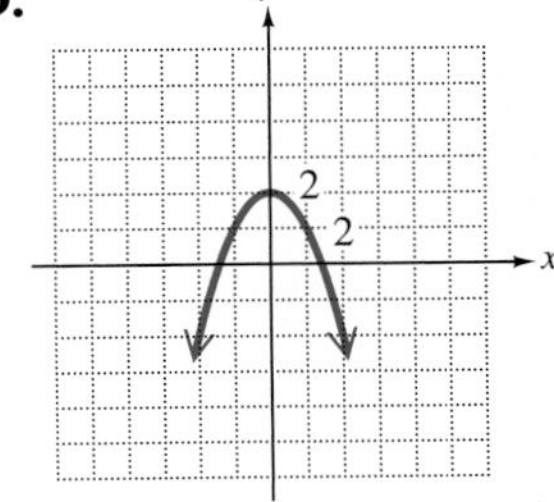

Identify the vertex of each parabola. See Examples 1–4.

3. $f(x) = -3x^2$ **(0, 0)**

4. $f(x) = \frac{1}{2}x^2$ **(0, 0)**

5. $f(x) = x^2 + 4$ **(0, 4)**

6. $f(x) = x^2 - 4$ **(0, −4)**

7. $f(x) = (x - 1)^2$ **(1, 0)**

8. $f(x) = (x + 3)^2$ **(−3, 0)**

9. $f(x) = (x + 3)^2 - 4$ **(−3, −4)**

10. $f(x) = (x - 5)^2 - 8$ **(5, −8)**

11. Describe how each of the parabolas in Exercises 9 and 10 is shifted compared to the graph of $f(x) = x^2$.

In Exercise 9, the parabola is shifted 3 units to the left and 4 units down. The parabola in Exercise 10 is shifted 5 units to the right and 8 units down.

12. What does the value of a in $F(x) = a(x - h)^2 + k$ tell you about the graph of the equation compared to the graph of $f(x) = x^2$?

If $|a| > 1$, the graph of $F(x) = a(x - h)^2 + k$ is narrower than the graph of $f(x) = x^2$. If $0 < |a| < 1$, the graph is wider than the graph of $f(x) = x^2$. If a is negative, the graph opens down; the graph of $f(x) = x^2$ opens up.

For each quadratic function, tell whether the graph opens up or down and whether the graph is wider, narrower, or the same shape as the graph of $f(x) = x^2$. See Examples 4 and 5.

13. $f(x) = -\frac{2}{5}x^2$

down; wider

14. $f(x) = -2x^2$

down; narrower

15. $f(x) = 3x^2 + 1$

up; narrower

16. $f(x) = \frac{2}{3}x^2 - 4$

up; wider

17. For $f(x) = a(x - h)^2 + k$, in what quadrant is the vertex if

(a) $h > 0, k > 0$; **I**
(b) $h > 0, k < 0$; **IV**
(c) $h < 0, k > 0$; **II**
(d) $h < 0, k < 0$? **III**

18. Match each quadratic function with the description of the parabola that is its graph.

(a) $f(x) = (x - 4)^2 - 2$ **D**
(b) $f(x) = (x - 2)^2 - 4$ **B**
(c) $f(x) = -(x - 4)^2 - 2$ **C**
(d) $f(x) = -(x - 2)^2 - 4$ **A**

A. Vertex $(2, -4)$, opens down
B. Vertex $(2, -4)$, opens up
C. Vertex $(4, -2)$, opens down
D. Vertex $(4, -2)$, opens up

Sketch the graph of each parabola. Plot at least two points in addition to the vertex. In Exercises 25–32, give the vertex, axis, domain, and range of the parabola. See Examples 1–5.

19. $f(x) = -2x^2$

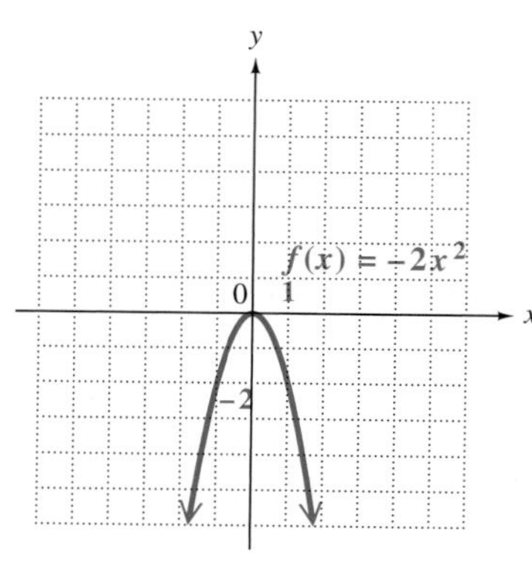

20. $f(x) = \frac{1}{3}x^2$

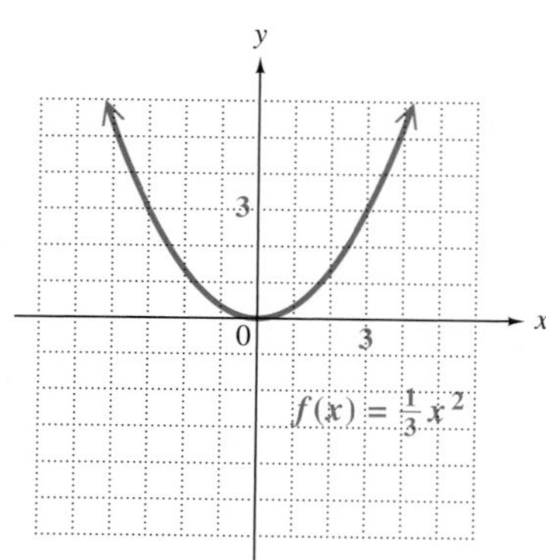

21. $f(x) = x^2 - 1$

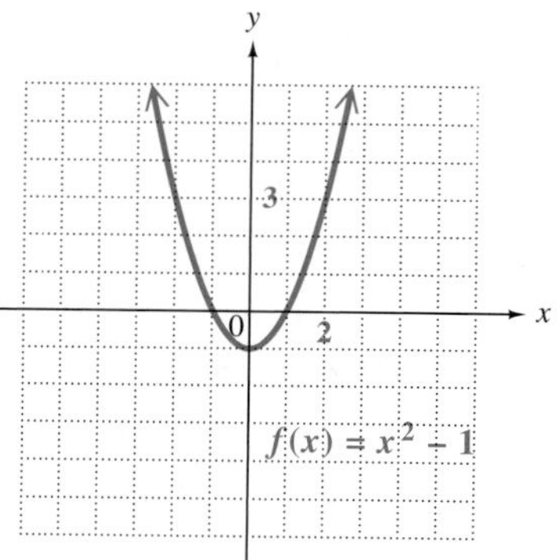

22. $f(x) = x^2 + 3$

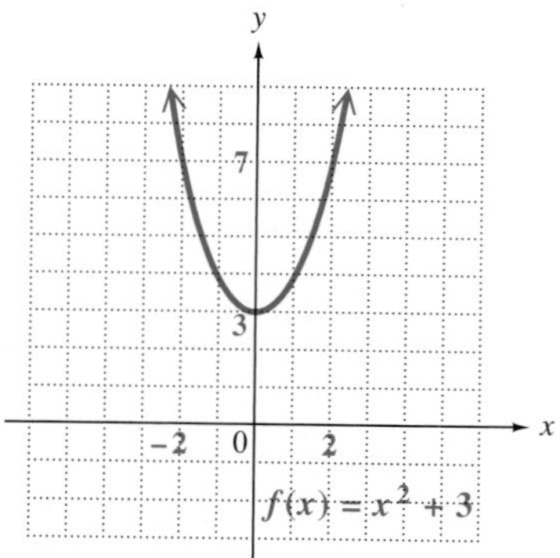

23. $f(x) = -x^2 + 2$

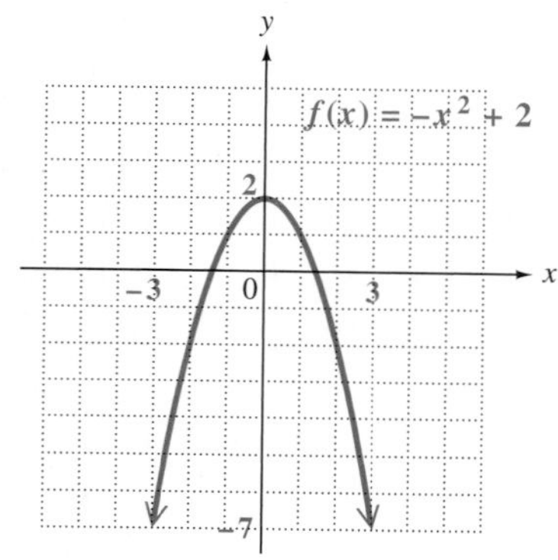

24. $f(x) = 2x^2 - 2$

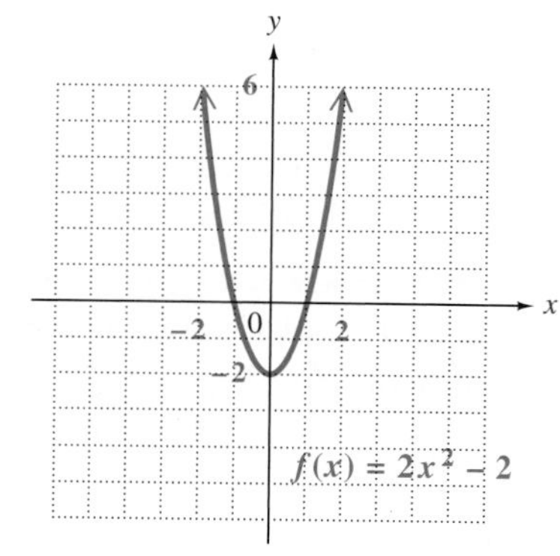

25. $f(x) = \frac{1}{2}(x - 4)^2$

vertex: **(4, 0)**
axis: $\mathbf{x = 4}$
domain: $(-\infty, \infty)$
range: $[0, \infty)$

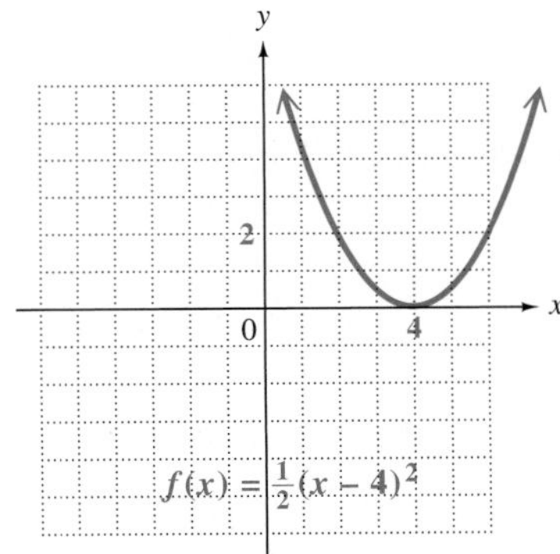

26. $f(x) = -2(x + 1)^2$

vertex: **(−1, 0)**
axis: $\mathbf{x = -1}$
domain: $(-\infty, \infty)$
range: $(-\infty, 0]$

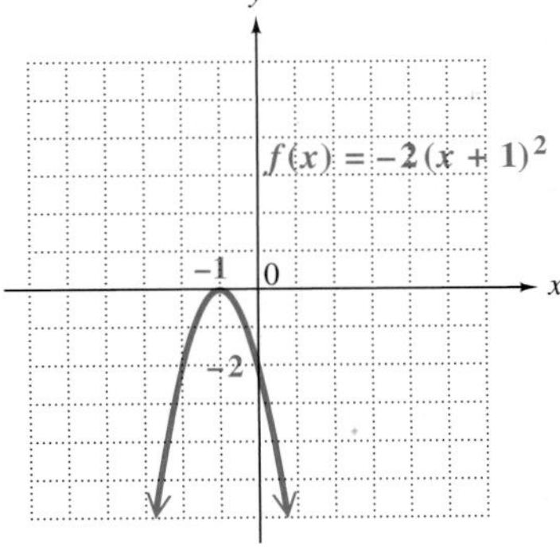

27. $f(x) = (x + 2)^2 - 1$

vertex: **(−2, −1)**
axis: $\mathbf{x = -2}$
domain: $(-\infty, \infty)$
range: $[-1, \infty)$

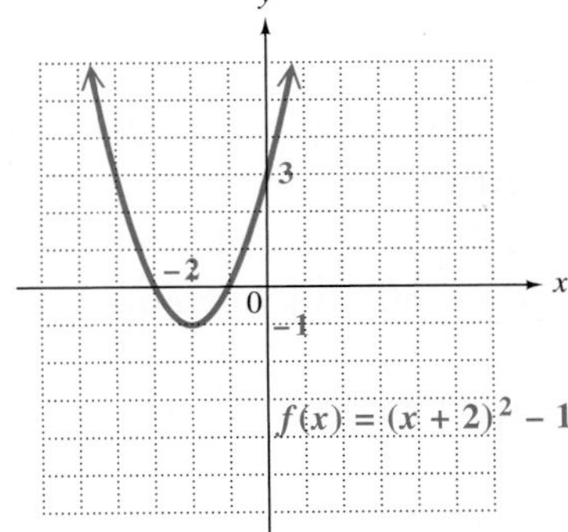

28. $f(x) = (x - 1)^2 + 2$

vertex: **(1, 2)**
axis: $\mathbf{x = 1}$
domain: $(-\infty, \infty)$
range: $[2, \infty)$

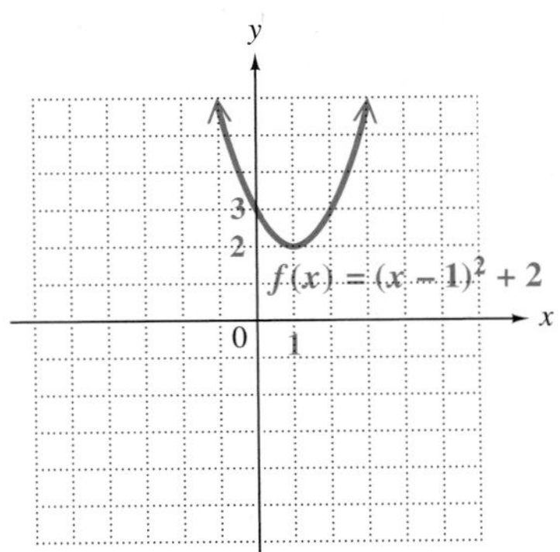

29. $f(x) = -2(x + 3)^2 + 4$

vertex: **(−3, 4)**
axis: $\mathbf{x = -3}$
domain: $(-\infty, \infty)$
range: $(-\infty, 4]$

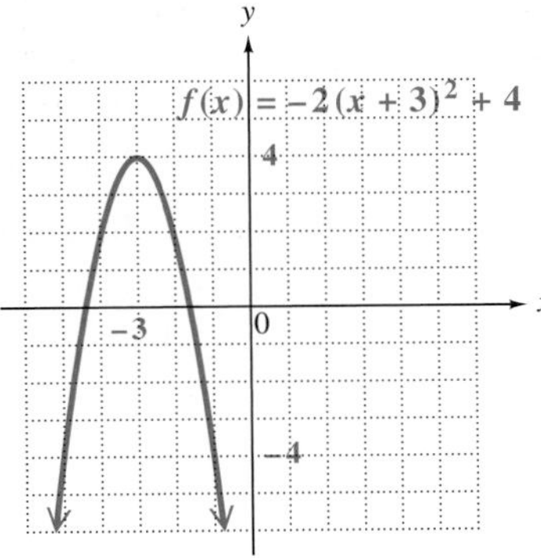

30. $f(x) = 2(x - 2)^2 - 3$

vertex: **(2, −3)**
axis: $\mathbf{x = 2}$
domain: $(-\infty, \infty)$
range: $[-3, \infty)$

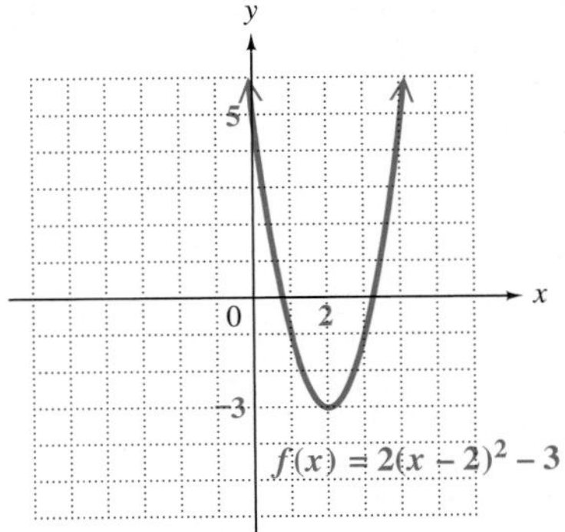

31. $f(x) = -\frac{2}{3}(x + 2)^2 + 1$

vertex: **(−2, 1)**
axis: $\mathbf{x = -2}$
domain: $(-\infty, \infty)$
range: $(-\infty, 1]$

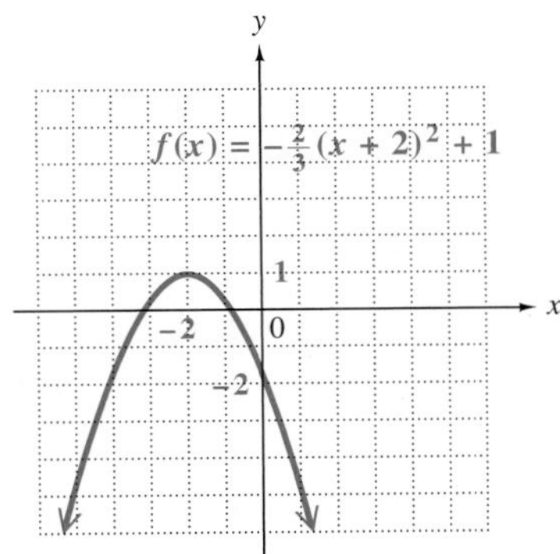

32. $f(x) = -\frac{1}{2}(x + 1)^2 + 2$

vertex: **(−1, 2)**
axis: $\mathbf{x = -1}$
domain: $(-\infty, \infty)$
range: $(-\infty, 2]$

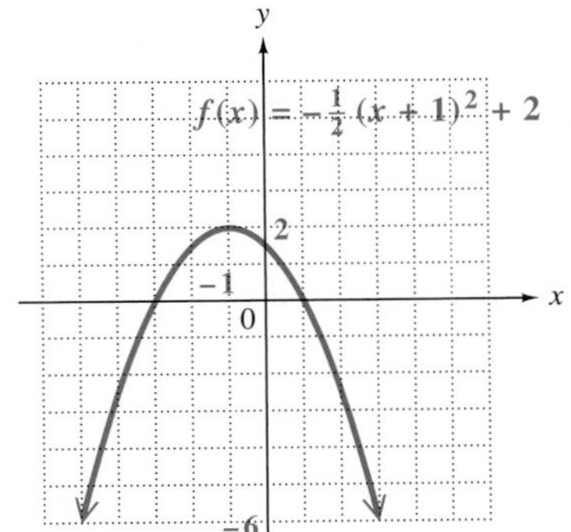

Relating Concepts (Exercises 33–38) For Individual or Group Work

The procedures described in this section that allow the graph of $f(x) = x^2$ to be shifted vertically and horizontally are applicable to other types of functions. In **Section 3.6** *we introduced linear functions of the form $g(x) = ax + b$. Consider the graph of the simplest linear function defined by $g(x) = x$, shown here, and then* **work Exercises 33–38 in order.**

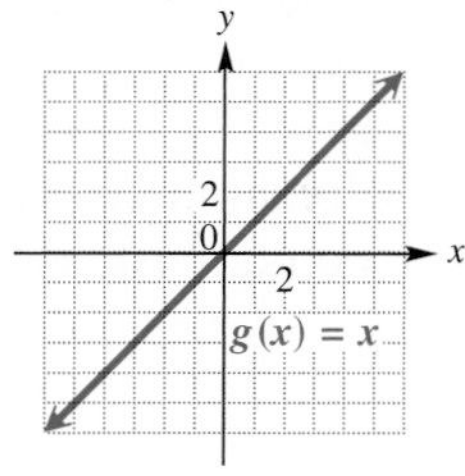

33. Based on the concepts of this section, how does the graph of $F(x) = x^2 + 6$ compare to the graph of $f(x) = x^2$ if a *vertical* shift is considered?

It is shifted 6 units up.

34. Graph the linear function defined by $G(x) = x + 6$.

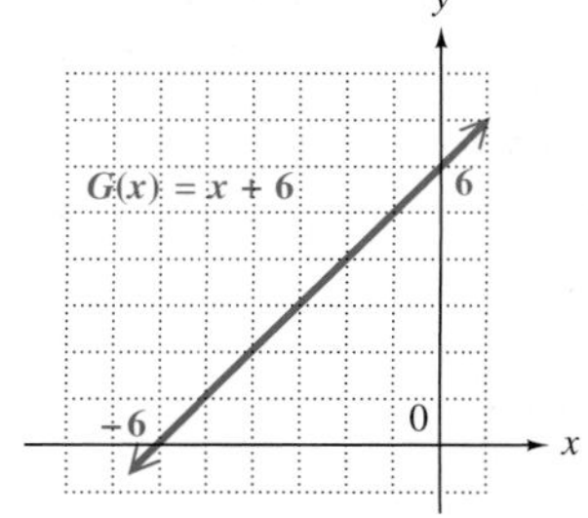

35. Based on the concepts of **Chapter 3,** how does the graph of $G(x) = x + 6$ compare to the graph of $g(x) = x$ if a *vertical* shift is considered? (*Hint:* Look at the y-intercept.) **It is shifted 6 units up.**

36. Based on the concepts of this section, how does the graph of $F(x) = (x - 6)^2$ compare to the graph of $f(x) = x^2$ if a *horizontal* shift is considered?

It is shifted 6 units to the right.

37. Graph the linear function defined by $G(x) = x - 6$.

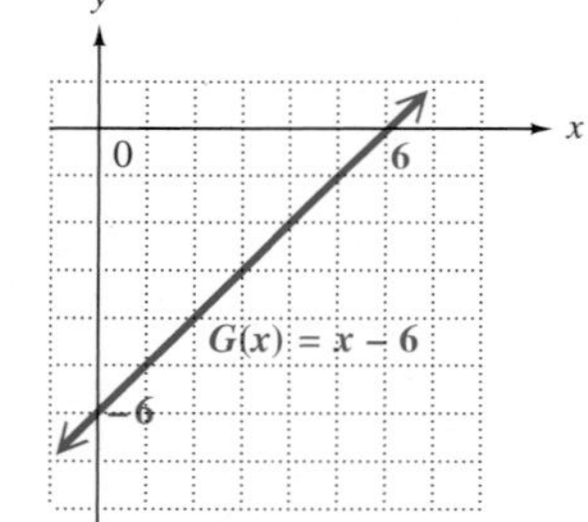

38. Based on the concepts of **Chapter 3,** how does the graph of $G(x) = x - 6$ compare to the graph of $g(x) = x$ if a *horizontal* shift is considered? (*Hint:* Look at the x-intercept.)

It is shifted 6 units to the right.

In Exercises 39–44, tell whether a linear or quadratic function would be a more appropriate model for each set of graphed data. If linear, tell whether the slope should be positive or negative. If quadratic, tell whether the coefficient of x^2 should be positive or negative. See Example 6.

39. **PLASMA TV SALES IN U.S.**

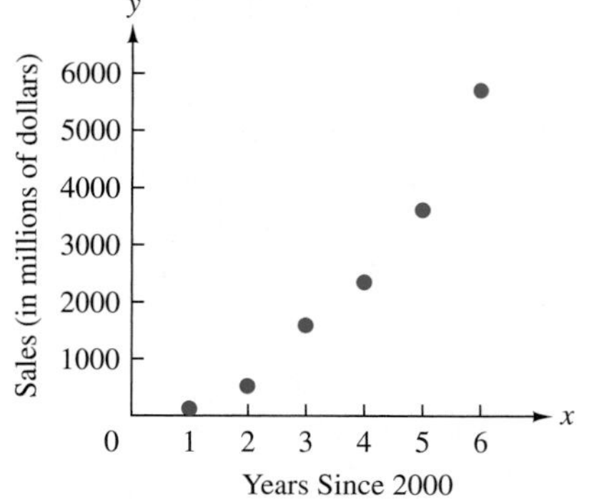

Source: Consumer Electronics Association.

quadratic; positive

40. **AVERAGE DAILY VOLUME OF FIRST-CLASS MAIL**

Source: General Accounting Office.

linear; negative

41. **SOCIAL SECURITY ASSETS***

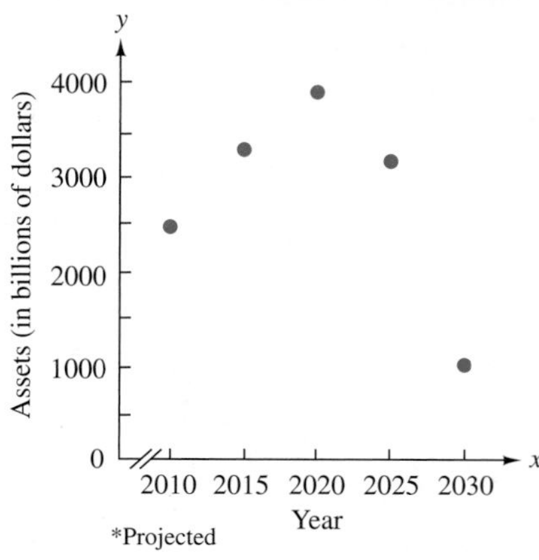

Source: Social Security Administration.

quadratic; negative

42. **FOOD ASSISTANCE SPENDING IN IOWA**

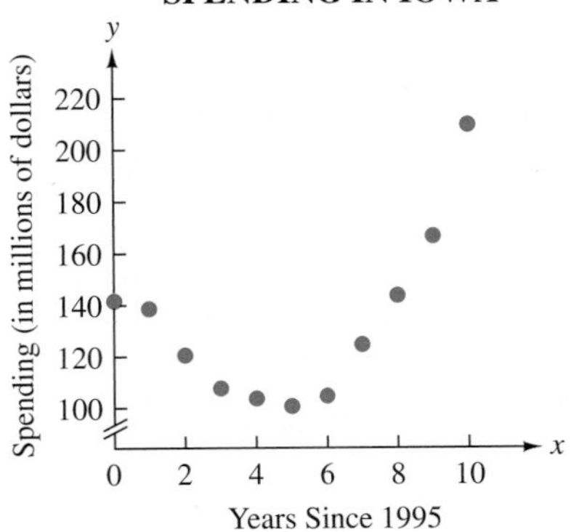

Source: Iowa Department of Human Services.

quadratic; positive

43. **TIME SPENT PLAYING VIDEO GAMES***

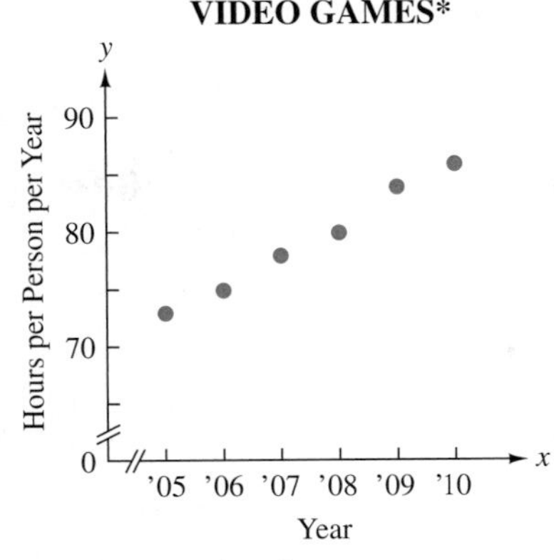

*Later years projected

Source: Veronis Suhler Stevenson.

linear; positive

44. **SALES OF MUSIC CASSETTE TAPES**

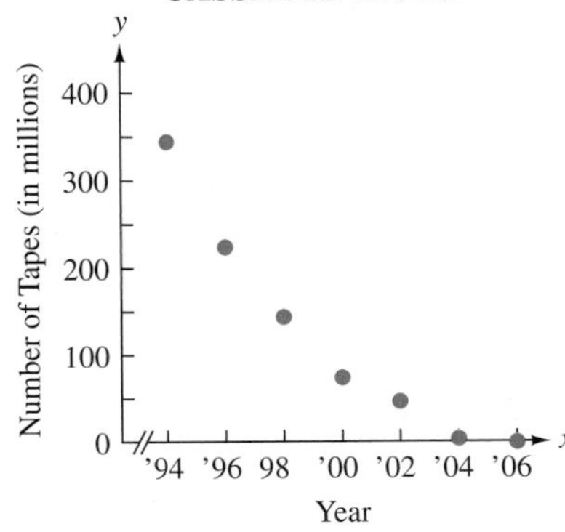

Source: Recording Industry Association of America.

quadratic; positive

Solve each problem. See Example 6.

45. Sales of digital cameras in the United States (in millions of dollars) between 2000 and 2006 are shown in the table. In the year column, 0 represents 2000, 1 represents 2001, and so on.

SALES OF DIGITAL CAMERAS

Year	Sales
0	1825
1	1972
2	2794
3	3921
4	4739
5	5611
6	7805

Source: Consumer Electronics Association.

(a) Use the ordered pairs (year, sales) to make a scatter diagram of the data.

DIGITAL CAMERA SALES IN U.S.

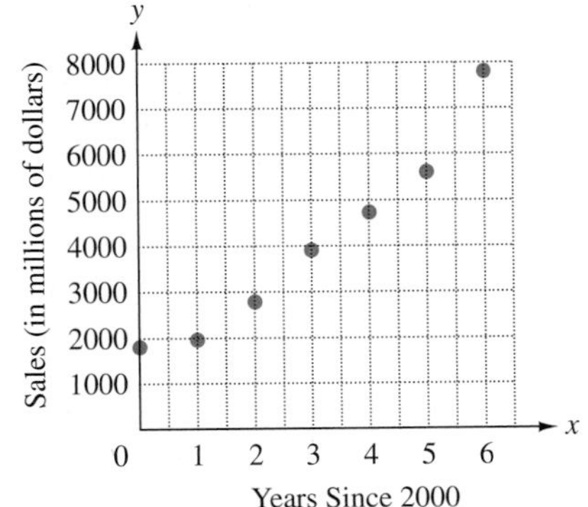

(b) Use the scatter diagram to decide whether a linear or quadratic function would better model the data. If quadratic, should the coefficient a of x^2 be positive or negative? **quadratic; positive**

(c) Use the ordered pairs (0, 1825), (3, 3921), and (6, 7805) to find a quadratic function that models the data. Round the values of a, b, and c in your model to the nearest tenth, as necessary.

$f(x) = 99.3x^2 + 400.7x + 1825$

(d) Use your model from part (c) to approximate the sales of digital cameras in the United States in 2007. Round your answer to the nearest whole number (of millions). **\$9496 million**

(e) Sales of digital cameras were projected to be \$6945 million in 2007. Based on this, is the model valid for 2007? Explain.

No. The number of digital cameras sold in 2007 is far below the number approximated by the model. Rather than continuing to increase, sales of digital cameras fell in 2007.

46. The percent of U.S. high school students in grades 9–12 who smoke is shown in the table for selected years. In the year column, 1 represents 1991, 3 represents 1993, and so on.

HIGH SCHOOL STUDENTS WHO SMOKE

Year	Percent of Students
1	28
3	31
5	35
7	36
9	35
11	29
13	22

Source: Centers for Disease Control and Prevention.

(a) Use the ordered pairs (year, percent of students) to make a scatter diagram of the data.

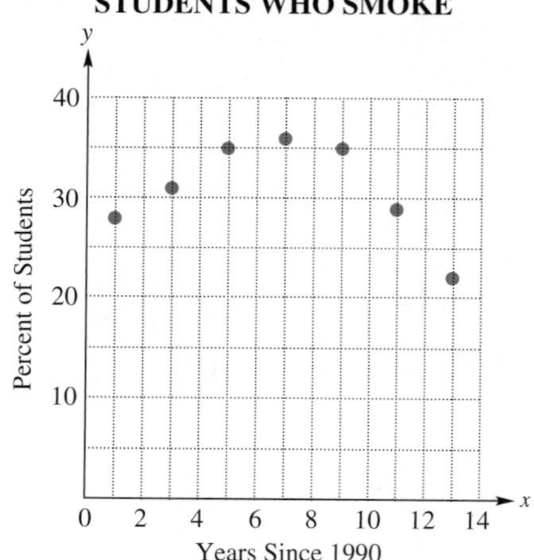

(b) Would a linear or quadratic function better model the data? **quadratic**

(c) Should the coefficient a of x^2 in a quadratic model be positive or negative? **negative**

(d) Use the ordered pairs (1, 28), (7, 36), and (11, 29) to find a quadratic function that models the data. Round the values of a, b, and c in your model to the nearest tenth, as necessary.

$y = -0.3x^2 + 3.8x + 24.5$ (Answers may vary due to roundoff.)

(e) Use your model from part (d) to approximate the percent of high school students who smoked during 1995 and 2003 to the nearest percent. How well does the model approximate the actual data from the table?

1995: 36%; 2003: 23%; The model approximates the data very well.

47. In Exercise 46(d), we determined that the quadratic function defined by

$$y = -0.3x^2 + 3.8x + 24.5$$

modeled the percent of U.S. high school students who smoked in the years 1991–2003.

(a) Use this model to approximate the number of high school students who smoked in 2005 and 2007.

2005: 14%; 2007: 2.4%

(b) The actual smoking rates for high school students were 23% in 2005 and 20% in 2007. (*Source:* Centers for Disease Control and Prevention.) How do the approximations using the model compare to the actual rates for these two years?

The approximations using the model are far too low.

48. Should the model from Exercise 46(d) be used to approximate the smoking rate for high school students in years after 2003? Explain.

No. The model approximates the rates for 2005 and 2007 very poorly, so it should not be used for later years. From the actual data, the teenage smoking rate seemed to level off.

10.7 More about Parabolas and Their Applications

OBJECTIVES

1 Find the vertex of a vertical parabola.

2 Graph a quadratic function.

3 Use the discriminant to find the number of x-intercepts of a parabola with a vertical axis.

4 Use quadratic functions to solve problems involving maximum or minimum value.

5 Graph parabolas with horizontal axes.

OBJECTIVE 1 Find the vertex of a vertical parabola. When the equation of a parabola is given in the form $f(x) = ax^2 + bx + c$, we need to locate the vertex in order to sketch an accurate graph. There are two ways to do this:

1. Complete the square, as shown in Examples 1 and 2, or
2. Use a formula derived by completing the square, as shown in Example 3.

EXAMPLE 1 Completing the Square to Find the Vertex ($a = 1$)

Find the vertex of the graph of $f(x) = x^2 - 4x + 5$.

To find the vertex, we need to write the expression $x^2 - 4x + 5$ in the form $(x - h)^2 + k$. We do this by completing the square on $x^2 - 4x$, as in **Section 10.2.** The process is slightly different here because we want to keep $f(x)$ alone on one side of the equation. Instead of adding the appropriate number to each side, we *add and subtract* it on the right. This is equivalent to adding 0.

$f(x) = x^2 - 4x + 5$

$f(x) = (x^2 - 4x \quad) + 5$ Group the variable terms.

$\left[\frac{1}{2}(-4)\right]^2 = (-2)^2 = 4$

$f(x) = (x^2 - 4x + 4 - 4) + 5$ Add and subtract 4.

$f(x) = (x^2 - 4x + 4) - 4 + 5$ Bring -4 outside the parentheses.

$f(x) = (x - 2)^2 + 1$ Factor; combine like terms.

The vertex of this parabola is (2, 1).

Work Problem 1 at the Side. ▶

EXAMPLE 2 Completing the Square to Find the Vertex ($a \neq 1$)

Find the vertex of the graph of $f(x) = -3x^2 + 6x - 1$.

We must complete the square on $-3x^2 + 6x$. Because the x^2-term has a coefficient other than 1, we factor that coefficient out of the first two terms and then proceed as in Example 1.

$f(x) = -3x^2 + 6x - 1$

$f(x) = -3(x^2 - 2x) - 1$ Factor out -3.

$\left[\frac{1}{2}(-2)\right]^2 = (-1)^2 = 1$

$f(x) = -3(x^2 - 2x + 1 - 1) - 1$ Add and subtract 1.

Now bring -1 ***outside the parentheses; be sure to multiply it by*** -3.

$f(x) = -3(x^2 - 2x + 1) + (-3)(-1) - 1$ Distributive property

This is a key step.

$f(x) = -3(x^2 - 2x + 1) + 3 - 1$

$f(x) = -3(x - 1)^2 + 2$ Factor; combine like terms.

The vertex is (1, 2).

Work Problem 2 at the Side. ▶

1 Find the vertex of the graph of each quadratic function.

(a) $f(x) = x^2 - 6x + 7$

(b) $f(x) = x^2 + 4x - 9$

2 Find the vertex of the graph of each quadratic function.

(a) $f(x) = 2x^2 - 4x + 1$

(b) $f(x) = -\frac{1}{2}x^2 + 2x - 3$

ANSWERS

1. (a) $(3, -2)$ (b) $(-2, -13)$
2. (a) $(1, -1)$ (b) $(2, -1)$

3 Use the formula to find the vertex of the graph of each quadratic function.

(a) $f(x) = -2x^2 + 3x - 1$

(b) $f(x) = 4x^2 - x + 5$

To derive a formula for the vertex of the graph of the quadratic function defined by $f(x) = ax^2 + bx + c$, complete the square.

$$f(x) = ax^2 + bx + c \quad (a \neq 0) \qquad \text{Standard form}$$

$$f(x) = a\left(x^2 + \frac{b}{a}x\right) + c \qquad \text{Factor } a \text{ from the first two terms.}$$

$$\left[\frac{1}{2}\left(\frac{b}{a}\right)\right]^2 = \left(\frac{b}{2a}\right)^2 = \frac{b^2}{4a^2}$$

$$f(x) = a\left(x^2 + \frac{b}{a}x + \frac{b^2}{4a^2} - \frac{b^2}{4a^2}\right) + c \qquad \text{Add and subtract } \frac{b^2}{4a^2}.$$

$$f(x) = a\left(x^2 + \frac{b}{a}x + \frac{b^2}{4a^2}\right) + a\left(-\frac{b^2}{4a^2}\right) + c \qquad \text{Distributive property}$$

$$f(x) = a\left(x^2 + \frac{b}{a}x + \frac{b^2}{4a^2}\right) - \frac{b^2}{4a} + c \qquad -\frac{ab^2}{4a^2} = -\frac{b^2}{4a}$$

$$f(x) = a\left(x + \frac{b}{2a}\right)^2 + \frac{4ac - b^2}{4a} \qquad \text{Factor; rewrite terms with a common denominator.}$$

$$f(x) = a\left[x - \underbrace{\left(\frac{-b}{2a}\right)}_{h}\right]^2 + \underbrace{\frac{4ac - b^2}{4a}}_{k} \qquad f(x) = (x - h)^2 + k$$

Thus, the vertex (h, k) can be expressed in terms of a, b, and c. It is not necessary to remember the expression for k, since it can be found by replacing x with $\frac{-b}{2a}$. Using function notation, if $y = f(x)$, then the y-value of the vertex is $f\left(\frac{-b}{2a}\right)$.

Vertex Formula

The graph of the quadratic function defined by $f(x) = ax^2 + bx + c$ has vertex

$$\left(\frac{-b}{2a}, f\left(\frac{-b}{2a}\right)\right).$$

The axis of the parabola is the line

$$x = \frac{-b}{2a}.$$

EXAMPLE 3 Using the Formula to Find the Vertex

Use the vertex formula to find the vertex of the graph of $f(x) = x^2 - x - 6$.

For this function, $a = 1$, $b = -1$, and $c = -6$. The x-coordinate of the vertex of the parabola is given by

$$\frac{-b}{2a} = \frac{-(-1)}{2(1)} = \frac{1}{2}.$$

The y-coordinate is $f\left(\frac{-b}{2a}\right) = f\left(\frac{1}{2}\right)$.

$$f\left(\frac{1}{2}\right) = \left(\frac{1}{2}\right)^2 - \frac{1}{2} - 6 = \frac{1}{4} - \frac{1}{2} - 6 = -\frac{25}{4}$$

The vertex is $\left(\frac{1}{2}, -\frac{25}{4}\right)$.

◀ Work Problem 3 at the Side.

ANSWERS

3. (a) $\left(\frac{3}{4}, \frac{1}{8}\right)$ (b) $\left(\frac{1}{8}, \frac{79}{16}\right)$

OBJECTIVE 2 Graph a quadratic function.

Graphing a Quadratic Function f

Step 1 **Determine whether the graph opens up or down.** If $a > 0$, the parabola opens up; if $a < 0$, it opens down.

Step 2 **Find the vertex.** Use either the vertex formula or completing the square.

Step 3 **Find any intercepts.** To find the x-intercepts (if any), solve $f(x) = 0$. To find the y-intercept, evaluate $f(0)$.

Step 4 **Complete the graph.** Plot the points found so far. Find and plot additional points as needed, using symmetry about the axis.

EXAMPLE 4 Graphing a Quadratic Function

Graph the quadratic function defined by $f(x) = x^2 - x - 6$.

Step 1 From the equation, $a = 1$, so the graph of the function opens up.

Step 2 The vertex, $\left(\frac{1}{2}, -\frac{25}{4}\right)$, was found in Example 3 by substituting the values $a = 1$, $b = -1$, and $c = -6$ in the vertex formula.

Step 3 Now find any intercepts. Since the vertex, $\left(\frac{1}{2}, -\frac{25}{4}\right)$, is in quadrant IV and the graph opens up, there will be two x-intercepts. To find them, let $f(x) = 0$ and solve the equation.

$$f(x) = x^2 - x - 6$$

$$0 = x^2 - x - 6 \qquad \text{Let } f(x) = 0.$$

$$0 = (x - 3)(x + 2) \qquad \text{Factor.}$$

$$x - 3 = 0 \quad \text{or} \quad x + 2 = 0 \qquad \text{Zero-factor property}$$

$$x = 3 \quad \text{or} \quad x = -2 \qquad \text{Solve each equation.}$$

The x-intercepts are $(3, 0)$ and $(-2, 0)$. Find the y-intercept.

$$f(x) = x^2 - x - 6$$

$$f(0) = 0^2 - 0 - 6 \qquad \text{Let } x = 0.$$

$$f(0) = -6$$

The y-intercept is $(0, -6)$.

Step 4 Plot the points found so far and additional points as needed using symmetry about the axis $x = \frac{1}{2}$. The graph is shown in Figure 12. The domain is $(-\infty, \infty)$, and the range is $\left[-\frac{25}{4}, \infty\right)$.

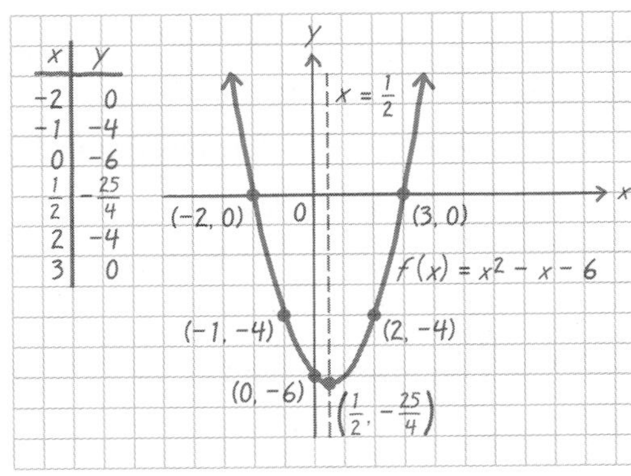

Figure 12

Work Problem 4 at the Side. ▶

4 Graph the quadratic function defined by

$$f(x) = x^2 - 6x + 5.$$

Give the vertex, axis, domain, and range.

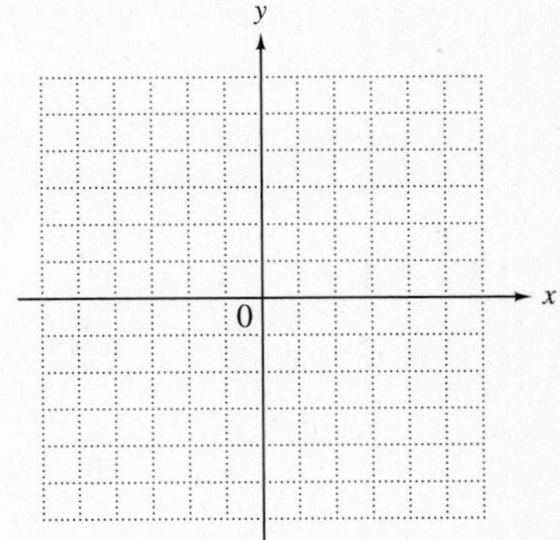

ANSWER

4.

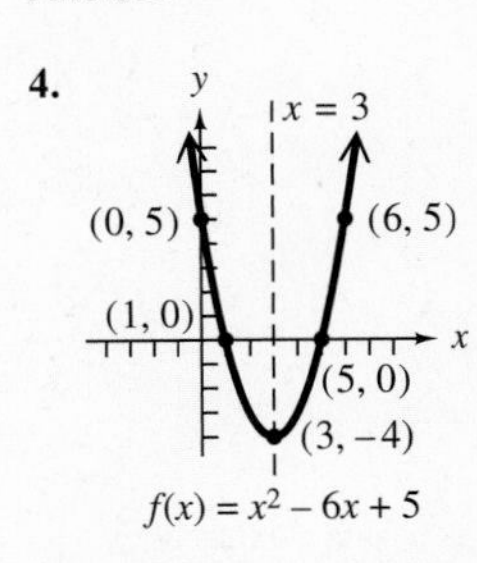

vertex: $(3, -4)$; axis: $x = 3$; domain: $(-\infty, \infty)$; range: $[-4, \infty)$

OBJECTIVE 3 Use the discriminant to find the number of x-intercepts of a parabola with a vertical axis. Recall from **Section 10.3** that the expression $b^2 - 4ac$ is called the **discriminant** of the quadratic *equation* $ax^2 + bx + c = 0$ and that we can use it to determine the number of real solutions of a quadratic equation.

In a similar way, we can use the discriminant of a quadratic *function* to determine the number of x-intercepts of its graph. See Figure 13. If the discriminant is positive, the parabola will have two x-intercepts. If the discriminant is 0, there will be only one x-intercept, and it will be the vertex of the parabola. If the discriminant is negative, the graph will have no x-intercepts.

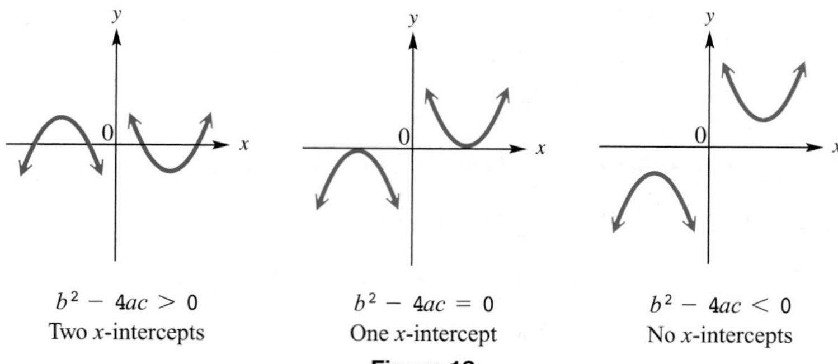

Figure 13

5 Use the discriminant to determine the number of x-intercepts of the graph of each quadratic function.

(a) $f(x) = 4x^2 - 20x + 25$

(b) $f(x) = 2x^2 + 3x + 5$

(c) $f(x) = -3x^2 - x + 2$

EXAMPLE 5 Using the Discriminant to Determine the Number of x-Intercepts

Use the discriminant to determine the number of x-intercepts of the graph of each quadratic function.

(a) $f(x) = 2x^2 + 3x - 5$

The discriminant is $b^2 - 4ac$. Here $a = 2$, $b = 3$, and $c = -5$, so

$$b^2 - 4ac$$
$$= 3^2 - 4(2)(-5) \quad \text{Substitute.}$$
$$= 9 - (-40) \quad \text{Apply the exponent; multiply.}$$
$$= \mathbf{49}. \quad \text{Subtract.}$$

Since the discriminant is positive, the parabola has two x-intercepts.

(b) $f(x) = -3x^2 - 1$

In this equation, $a = -3$, $b = 0$, and $c = -1$. The discriminant is

$$b^2 - 4ac$$
$$= 0^2 - 4(-3)(-1)$$
$$= \mathbf{-12}.$$

The discriminant is negative, so the graph has no x-intercepts.

(c) $f(x) = 9x^2 + 6x + 1$

Here, $a = 9$, $b = 6$, and $c = 1$. The discriminant is

$$b^2 - 4ac$$
$$= 6^2 - 4(9)(1)$$
$$= \mathbf{0}.$$

The parabola has only one x-intercept (its vertex) because the value of the discriminant is 0.

◀ *Work Problem* **5** *at the Side.*

ANSWERS

5. **(a)** discriminant is 0; one x-intercept
(b) discriminant is -31; no x-intercepts
(c) discriminant is 25; two x-intercepts

OBJECTIVE 4 Use quadratic functions to solve problems involving maximum or minimum value. The vertex of a parabola is either the highest or the lowest point on the parabola. The y-value of the vertex gives the maximum or minimum value of y, while the x-value tells where that maximum or minimum occurs.

Problem-Solving Hint

In many applied problems we must find the least or greatest value of some quantity. When we can express that quantity as a quadratic function, the value of k in the vertex (h, k) gives that optimum value.

EXAMPLE 6 Finding the Maximum Area of a Rectangular Region

A farmer has 120 ft of fencing to enclose a rectangular area next to a building. See Figure 14. Find the maximum area he can enclose and the width required to produce this maximum area.

Figure 14

Let x represent the width of the rectangle. Since he has 120 ft of fencing,

$$x + x + \text{length} = 120 \quad \text{Sum of the sides is 120 ft.}$$

$$2x + \text{length} = 120 \quad \text{Combine like terms.}$$

$$\text{length} = 120 - 2x. \quad \text{Subtract } 2x.$$

The area $A(x)$ is given by the product of the width and length, so

$$A(x) = x(120 - 2x)$$

$$A(x) = 120x - 2x^2.$$

To determine the maximum area, find the vertex of the parabola given by $A(x) = 120x - 2x^2$ using the vertex formula. Writing the equation in standard form as

$$A(x) = -2x^2 + 120x$$

gives $a = -2$, $b = 120$, and $c = 0$, so

$$h = \frac{-b}{2a} = \frac{-120}{2(-2)} = \frac{-120}{-4} = 30;$$

$$A(30) = -2(30)^2 + 120(30) = -2(900) + 3600 = 1800.$$

The graph is a parabola that opens down, and its vertex is $(30, 1800)$. Thus, the maximum area will be 1800 ft^2. This area will occur if x, the width of the rectangle, is 30 ft.

Work Problem 6 at the Side. ▶

6 Solve Example 6 if the farmer has only 100 ft of fencing.

Answer

6. The rectangle should be 25 ft by 50 ft with a maximum area of 1250 ft^2.

7 Solve the problem.

A toy rocket is launched from the ground so that its distance in feet above the ground after t seconds is

$$s(t) = -16t^2 + 208t.$$

Find the maximum height it reaches and the number of seconds it takes to reach that height.

CAUTION

Be careful when interpreting the meanings of the coordinates of the vertex. The first coordinate, x, gives the value for which the *function value* is a maximum or a minimum. Be sure to read the problem carefully to determine whether you are asked to find the value of the independent variable, the function value, or both.

EXAMPLE 7 Finding the Maximum Height Attained by a Projectile

If air resistance is neglected, a projectile on Earth shot straight upward with an initial velocity of 40 m per sec will be at a height s in meters given by

$$s(t) = -4.9t^2 + 40t,$$

where t is the number of seconds elapsed after projection. After how many seconds will it reach its maximum height, and what is this maximum height?

For this function, $a = -4.9$, $b = 40$, and $c = 0$. Use the vertex formula.

$$h = \frac{-b}{2a} = \frac{-40}{2(-4.9)} \approx \mathbf{4.1} \qquad \text{Use a calculator.}$$

Thus, the maximum height is attained at 4.1 sec. To find this maximum height, calculate $s(4.1)$.

$$s(\mathbf{4.1}) = -4.9(\mathbf{4.1})^2 + 40(\mathbf{4.1}) \approx \mathbf{81.6} \qquad \text{Use a calculator.}$$

The projectile will attain a maximum height of approximately 81.6 m.

◀ *Work Problem* **7** *at the Side.*

OBJECTIVE 5 Graph parabolas with horizontal axes. If x and y are interchanged in the equation $y = ax^2 + bx + c$, the equation becomes

$$x = ay^2 + by + c.$$

Because of the interchange of the roles of x and y, these parabolas are horizontal (with horizontal lines as axes).

Graph of a Horizontal Parabola

The graph of

$$\boldsymbol{x = ay^2 + by + c} \quad \text{or} \quad \boldsymbol{x = a(y - k)^2 + h}$$

is a parabola with vertex (h, k) and the horizontal line $y = k$ as axis. The graph opens to the right if $a > 0$ and to the left if $a < 0$.

EXAMPLE 8 Graphing a Horizontal Parabola

Graph $x = (y - 2)^2 - 3$. Give the vertex, axis, domain, and range.

This graph has its vertex at $(-3, 2)$, since the roles of x and y are reversed. It opens to the right, the positive x-direction, and has the same shape as $y = x^2$. Plotting a few additional points gives the graph shown in Figure 15 on the next page. Note that the graph is symmetric about its axis, $y = 2$. The domain is $[-3, \infty)$, and the range is $(-\infty, \infty)$.

Continued on Next Page

ANSWER

7. 676 ft; 6.5 sec

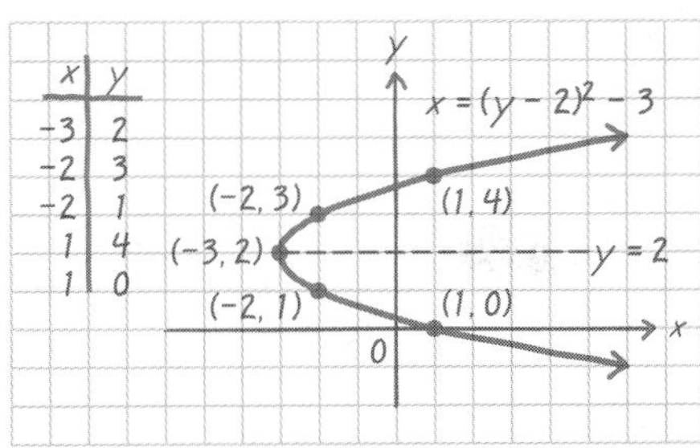

Figure 15

Work Problem 8 at the Side. ▶

When a quadratic equation is given in the form $x = ay^2 + by + c$, completing the square on y will allow us to find the vertex.

EXAMPLE 9 Completing the Square to Graph a Horizontal Parabola

Graph $x = -2y^2 + 4y - 3$. Give the vertex, axis, domain, and range.

$x = -2y^2 + 4y - 3$

$x = -2(y^2 - 2y) - 3$ Factor out -2.

$x = -2(y^2 - 2y + 1 - 1) - 3$ Complete the square within the parentheses; add and subtract 1.

$x = -2(y^2 - 2y + 1) + (-2)(-1) - 3$ Distributive property

$x = -2(y - 1)^2 - 1$ Factor; simplify.

Because of the negative coefficient (-2) in $x = -2(y - 1)^2 - 1$, the graph opens to the left (the negative x-direction). The graph is narrower than the graph of $y = x^2$ because $|-2| > 1$. As shown in Figure 16, the vertex is $(-1, 1)$ and the axis is $y = 1$. The domain is $(-\infty, -1]$, and the range is $(-\infty, \infty)$.

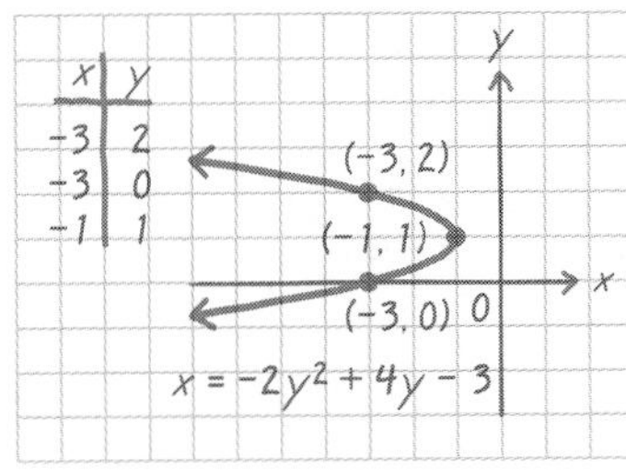

Figure 16

Work Problem 9 at the Side. ▶

CAUTION

Only quadratic equations solved for y (whose graphs are vertical parabolas) are examples of functions. The horizontal parabolas in Examples 8 and 9 are ***not*** graphs of functions, because they do not satisfy the vertical line test.

8 Graph $x = (y + 1)^2 - 4$. Give the vertex, axis, domain, and range.

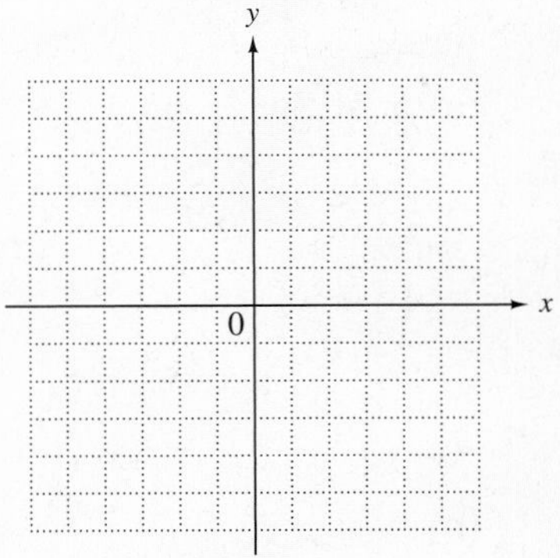

9 Graph $x = -y^2 + 2y + 5$. Give the vertex, axis, domain, and range.

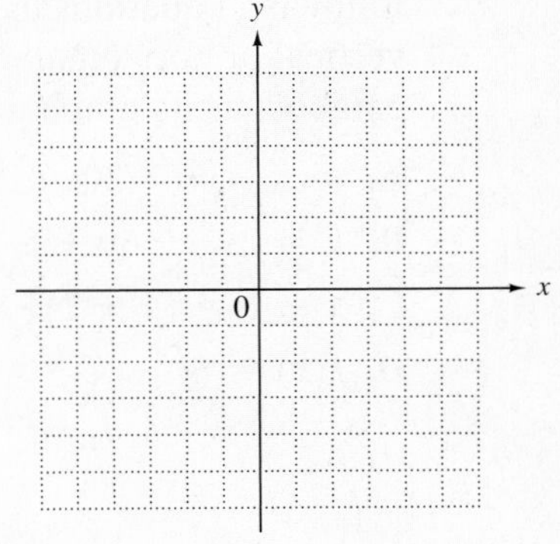

ANSWERS

8.

$x = (y + 1)^2 - 4$

vertex: $(-4, -1)$; axis: $y = -1$; domain: $[-4, \infty)$; range: $(-\infty, \infty)$

9.

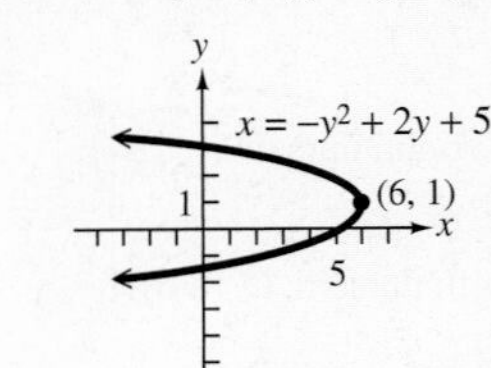

vertex: $(6, 1)$; axis: $y = 1$; domain: $(-\infty, 6]$; range: $(-\infty, \infty)$

In summary, the graphs of parabolas studied in **Sections 10.6 and 10.7** fall into the following categories.

GRAPHS OF PARABOLAS

Equation	Graph
$y = ax^2 + bx + c$ $y = a(x - h)^2 + k$	(h, k), $a > 0$; (h, k), $a < 0$. These graphs represent functions.
$x = ay^2 + by + c$ $x = a(y - k)^2 + h$	(h, k), $a > 0$; (h, k), $a < 0$. These graphs are not graphs of functions.

◀ *Work Problems* **10** *and* **11** *at the Side.*

10 Find the vertex of each parabola. Tell whether the graph opens to the right or to the left. Give the domain and range.

(a) $x = 2y^2 - 6y + 5$

(b) $x = -y^2 + 2y + 5$

11 **(a)** Tell whether each of the following equations has a vertical or horizontal parabola as its graph.

A. $y = -x^2 + 20x + 80$

B. $x = 2y^2 + 6y + 5$

C. $x + 1 = (y + 2)^2$

D. $f(x) = (x - 4)^2$

(b) Which of the equations in part (a) represent functions?

ANSWERS

10. **(a)** $\left(\frac{1}{2}, \frac{3}{2}\right)$; right; domain: $\left[\frac{1}{2}, \infty\right)$; range: $(-\infty, \infty)$

(b) $(6, 1)$; left; domain; $(-\infty, 6]$; range: $(-\infty, \infty)$

11. **(a)** A, D are vertical parabolas; B, C are horizontal parabolas.

(b) A, D

10.7 Exercises

FOR EXTRA HELP MyMathLab

 Math XL PRACTICE WATCH DOWNLOAD  READ  REVIEW

1. How can you determine just by looking at the equation of a parabola whether it has a vertical or a horizontal axis?

   If x is squared, it has a vertical axis; if y is squared, it has a horizontal axis.

2. Why can't the graph of a quadratic function be a horizontal parabola?

   A horizontal parabola fails the vertical line test.

3. How can you determine the number of x-intercepts of the graph of a quadratic function without graphing the function?

   Use the discriminant of the corresponding quadratic equation. If it is positive, there are two x-intercepts. If it is 0, there is just one x-intercept (the vertex), and if it is negative, there are no x-intercepts.

4. If the vertex of the graph of a quadratic function is $(1, -3)$ and the graph opens down, how many x-intercepts does the graph have?

   none

Find the vertex of each parabola. For each equation, decide whether the graph opens up, down, to the left, or to the right, and whether it is wider, narrower, or the same shape as the graph of $y = x^2$. If it is a vertical parabola, use the discriminant to determine the number of x-intercepts. See Examples 1–3, 5, 8, and 9.

5. $y = 2x^2 + 4x + 5$

   $(-1, 3)$; up; narrower; no x-intercepts

6. $y = 3x^2 - 6x + 4$

   $(1, 1)$; up; narrower; no x-intercepts

7. $y = -x^2 + 5x + 3$

   $\left(\frac{5}{2}, \frac{37}{4}\right)$; down; same; two x-intercepts

8. $x = -y^2 + 7y - 2$

   $\left(\frac{41}{4}, \frac{7}{2}\right)$; to the left; same

9. $x = \frac{1}{3}y^2 + 6y + 24$

   $(-3, -9)$; to the right; wider

10. $x = \frac{1}{2}y^2 + 10y - 5$

    $(-55, -10)$; to the right; wider

Graph each parabola. Give the vertex, axis, domain, and range. See Examples 4, 8, and 9.

11. $f(x) = x^2 + 4x + 3$
    vertex: **$(-2, -1)$**
    axis: **$x = -2$**
    domain: **$(-\infty, \infty)$**
    range: **$[-1, \infty)$**

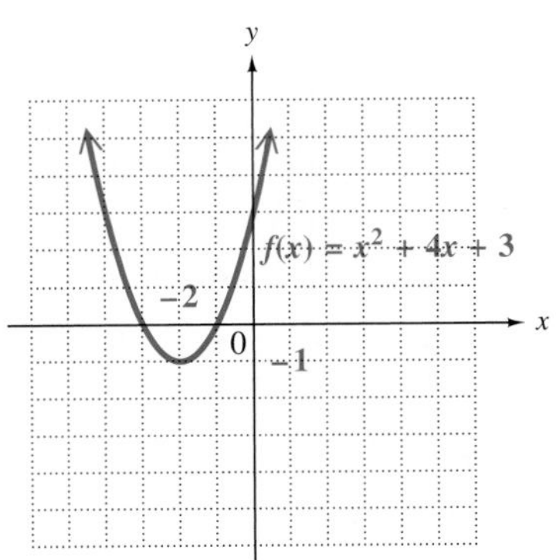

12. $f(x) = x^2 + 2x - 2$
    vertex: **$(-1, -3)$**
    axis: **$x = -1$**
    domain: **$(-\infty, \infty)$**
    range: **$[-3, \infty)$**

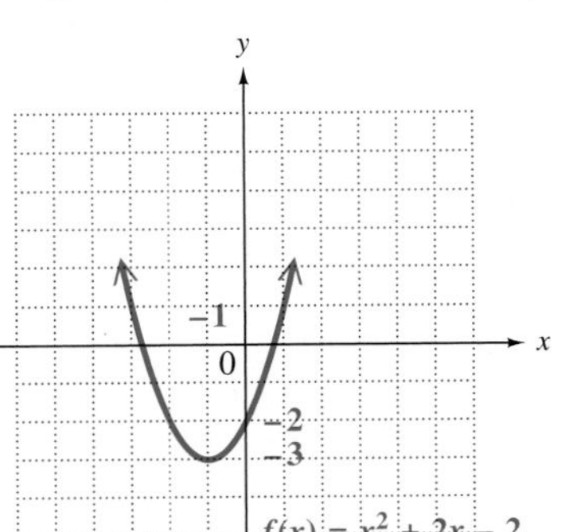

13. $f(x) = -2x^2 + 4x - 5$
    vertex: **$(1, -3)$**
    axis: **$x = 1$**
    domain: **$(-\infty, \infty)$**
    range: **$(-\infty, -3]$**

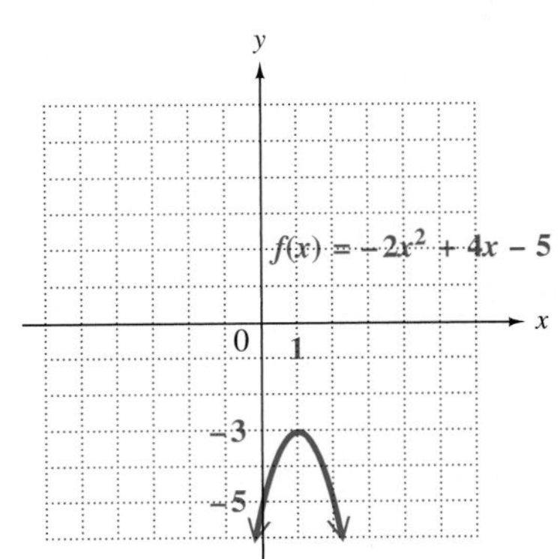

14. $f(x) = -3x^2 + 12x - 8$
vertex: **(2, 4)**
axis: **$x = 2$**
domain: **$(-\infty, \infty)$**
range: **$(-\infty, 4]$**

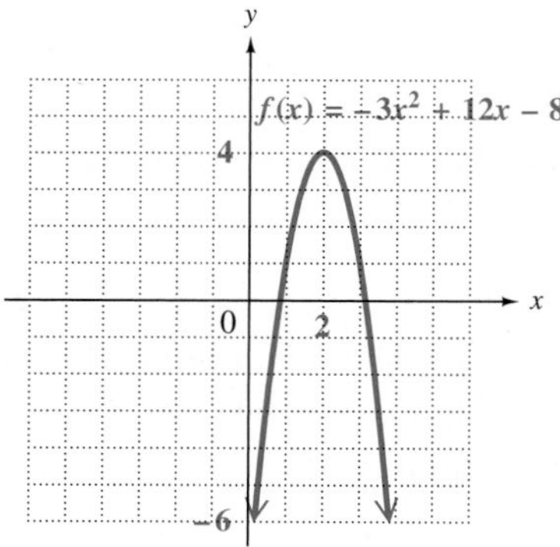

15. $x = -\frac{1}{5}y^2 + 2y - 4$
vertex: **(1, 5)**
axis: **$y = 5$**
domain: **$(-\infty, 1]$**
range: **$(-\infty, \infty)$**

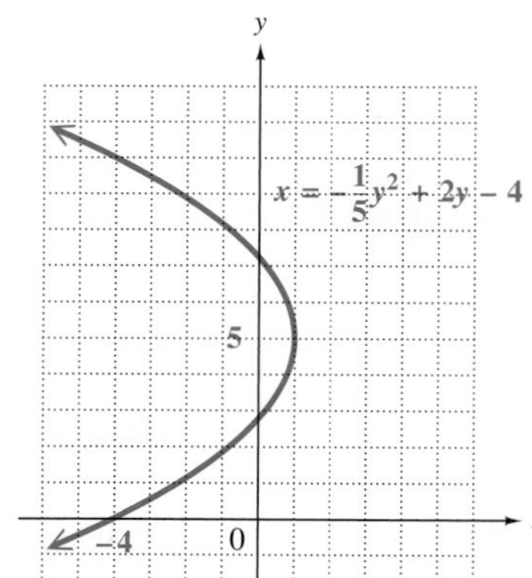

16. $x = -\frac{1}{2}y^2 - 4y - 6$
vertex: **(2, −4)**
axis: **$y = -4$**
domain: **$(-\infty, 2]$**
range: **$(-\infty, \infty)$**

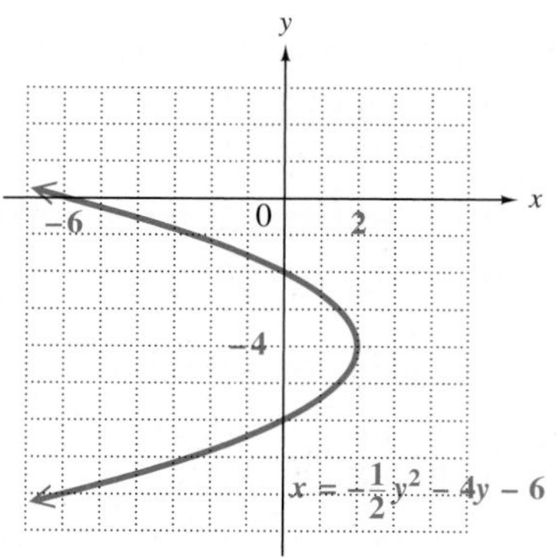

17. $x = 3y^2 + 12y + 5$
vertex: **(−7, −2)**
axis: **$y = -2$**
domain: **$[-7, \infty)$**
range: **$(-\infty, \infty)$**

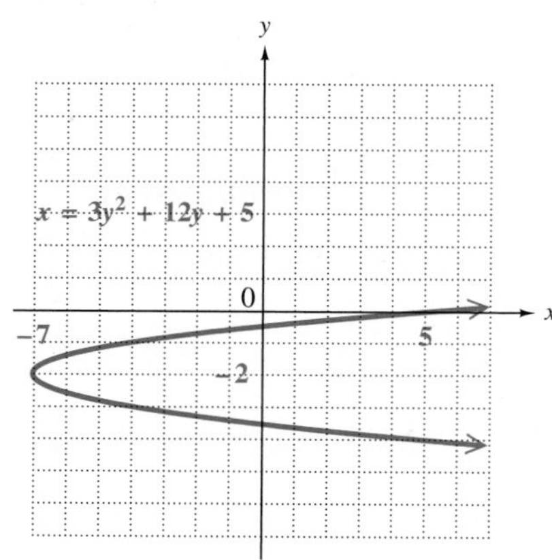

18. $x = 4y^2 + 16y + 11$
vertex: **(−5, −2)**
axis: **$y = -2$**
domain: **$[-5, \infty)$**
range: **$(-\infty, \infty)$**

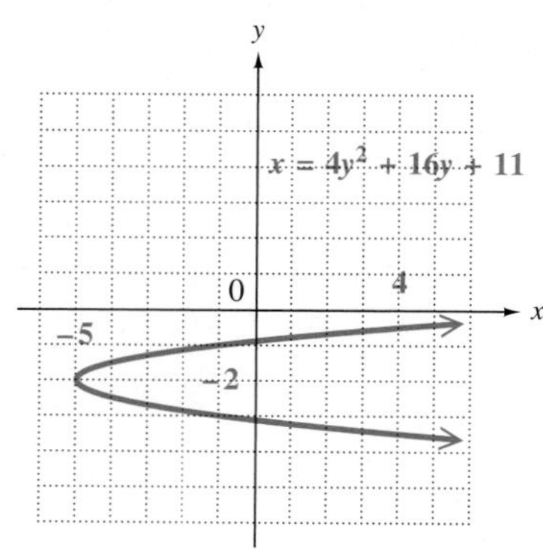

Use the concepts of this section to match each equation with its graph.

19. $y = 2x^2 + 4x - 3$ **F**

20. $y = -x^2 + 3x + 5$ **A**

21. $y = -\frac{1}{2}x^2 - x + 1$ **C**

22. $x = y^2 + 6y + 3$ **B**

23. $x = -y^2 - 2y + 4$ **D**

24. $x = 3y^2 + 6y + 5$ **E**

A.

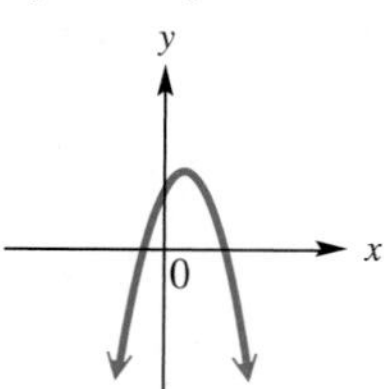

B.

y
0
x

C.

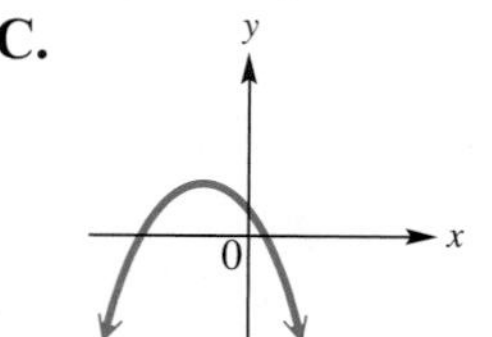

D.

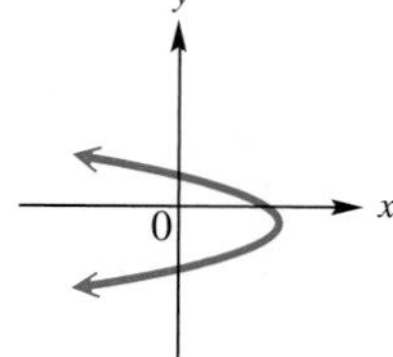

E.

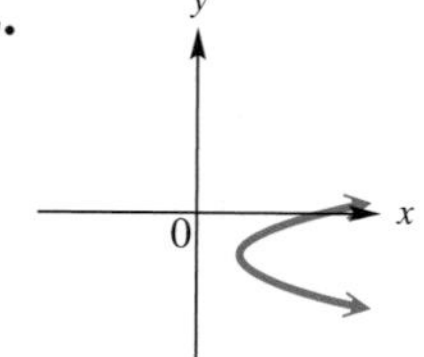

F.

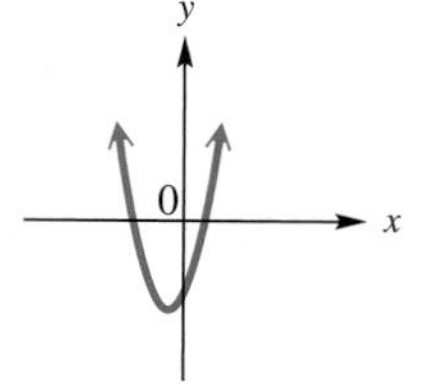

Solve each problem. See Examples 6 and 7.

25. Find the pair of numbers whose sum is 60 and whose product is a maximum. (*Hint:* Let x and $60 - x$ represent the two numbers.)

30 and 30

26. Find the pair of numbers whose sum is 10 and whose product is a maximum.

5 and 5

27. Palo Alto College is planning to construct a rectangular parking lot on land bordered on one side by a highway. The plan is to use 640 ft of fencing to fence off the other three sides. What should the dimensions of the lot be if the enclosed area is to be a maximum?

160 ft by 320 ft

28. Keisha Hughes has 100 m of fencing material to enclose a rectangular exercise run for her dog. What width will give the enclosure the maximum area?

25 m

29. If an object on Earth is projected upward with an initial velocity of 32 ft per sec, then its height (in feet) after t seconds is given by

$$h(t) = 32t - 16t^2.$$

Find the maximum height attained by the object and the number of seconds it takes to hit the ground.

16 ft; 2 sec

30. A projectile on Earth is fired straight upward so that its distance (in feet) above the ground t seconds after firing is given by

$$s(t) = -16t^2 + 400t.$$

Find the maximum height it reaches and the number of seconds it takes to reach that height.

2500 ft; 12.5 sec

31. A charter flight charges a fare of $200 per person, plus $4 per person for each unsold seat on the plane. If the plane holds 100 passengers and if x represents the number of unsold seats, find the following.

(a) A function defined by $R(x)$ that describes the total revenue received for the flight (*Hint:* Multiply the number of people flying, $100 - x$, by the price per ticket, $200 + 4x$.)

$R(x) = 20{,}000 + 200x - 4x^2$

(b) The number of unsold seats that will produce the maximum revenue

25

(c) The maximum revenue

$22,500

32. For a trip, a charter bus company charges a fare of $48 per person, plus $2 per person for each unsold seat on the bus. If the bus has 42 seats and x represents the number of unsold seats, find the following.

(a) A function defined by $R(x)$ that describes the total revenue from the trip (*Hint:* Multiply the total number riding, $42 - x$, by the price per ticket, $48 + 2x$.)

$R(x) = -2x^2 + 36x + 2016$

(b) The number of unsold seats that produces the maximum revenue

9

(c) The maximum revenue

$2178

33. The percent of births in the United States to teenage mothers in the years 1990–2002 can be modeled by the quadratic function defined by

$$f(x) = -0.0334x^2 + 0.2351x + 12.79,$$

where $x = 0$ represents 1990, $x = 1$ represents 1991, and so on. (*Source:* U.S. National Center for Health Statistics.)

(a) Since the coefficient of x^2 in the model is negative, the graph of this quadratic function is a parabola that opens down. Will the y-value of the vertex of this graph be a maximum or a minimum? **maximum**

(b) In what year during this period was the percent of births in the United States to teenage mothers a maximum? (Round down to the nearest year.) Use the actual y-value of the vertex, to the nearest tenth, to find this percent. **1993; 13.2%**

34. The total receipts from individual income taxes by the U.S. Treasury in the years 2002–2006 can be modeled by the quadratic function defined by

$$f(x) = 33.79x^2 - 83.44x + 1036,$$

where $x = 0$ represents 2002, $x = 1$ represents 2003, and so on, and $f(x)$ is in billions of dollars. (*Source:* Internal Revenue Service.)

(a) Since the coefficient of x^2 given in the model is positive, the graph of this quadratic function is a parabola that opens up. Will the y-value of the vertex of this graph be a maximum or minimum? **minimum**

(b) In what year during this period were total receipts from individual taxes a minimum? Use the actual x-value of the vertex, to the nearest tenth, to find this amount. **2003; \$984.5 billion**

35. The graph shows how Social Security assets are expected to change as the number of retirees receiving benefits increases.

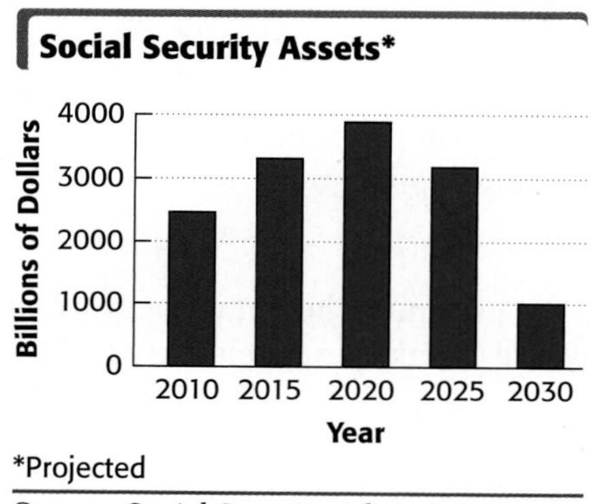

The graph suggests that a quadratic function would be a good fit to the data. The data are approximated by the function defined by

$$f(x) = -20.57x^2 + 758.9x - 3140.$$

In the model, $x = 10$ represents 2010, $x = 15$ represents 2015, and so on, and $f(x)$ is in billions of dollars.

(a) Explain why the coefficient of x^2 in the model is negative, based on the graph.

The coefficient of x^2 is negative because the parabola opens down.

(b) Algebraically determine the vertex of the graph, with coordinates to four significant digits.

(18.45, 3860)

(c) Interpret the answer to part (b) as it applies to the application.

In 2018 Social Security assets will reach their maximum value of \$3860 billion.

36. The graph shows the performance of investment portfolios with different mixtures of U.S. and foreign investments over a 25-yr period.

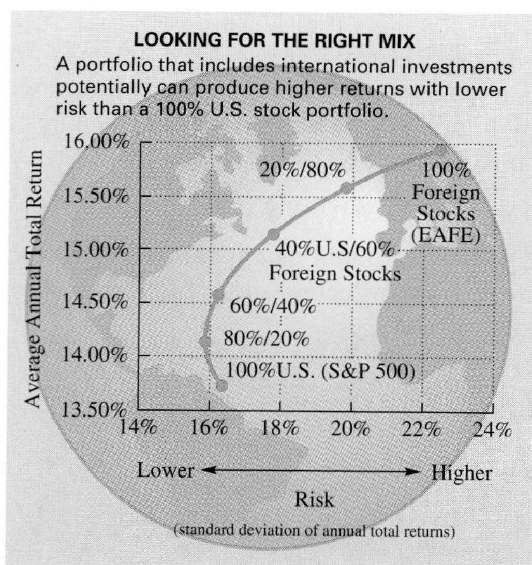

Source: *Financial Ink* Newsletter, Investment Management and Research, Inc., Feb. 1998. Thanks to David Van Geffen for this information.

(a) Is this the graph of a function? Explain.

No. It fails the vertical line test.

(b) What investment mixture shown on the graph appears to represent the vertex? What relative amount of risk does this point represent? What return on investment does it provide?

80% U.S. and 20% foreign; about 15.8%; about 14.2%

(c) Which point on the graph represents the riskiest investment mixture? What return on investment does it provide?

the point corresponding to 100% foreign stocks; 16%

10.8 Polynomial and Rational Inequalities

Now we combine methods of solving linear inequalities and methods of solving quadratic equations to solve *quadratic inequalities*.

OBJECTIVES

1. **Solve quadratic inequalities.**
2. **Solve polynomial inequalities of degree 3 or more.**
3. **Solve rational inequalities.**

Quadratic Inequality

A **quadratic inequality** can be written in the form

$$ax^2 + bx + c < 0 \quad \text{or} \quad ax^2 + bx + c > 0,$$

where a, b, and c are real numbers, with $a \neq 0$.

As before, $<$ and $>$ may be replaced with $\leq$ and $\geq$.

OBJECTIVE 1 Solve quadratic inequalities. One method for solving a quadratic inequality is by graphing the related quadratic function.

EXAMPLE 1 Solving Quadratic Inequalities by Graphing

Solve each inequality.

(a) $x^2 - x - 12 > 0$

To solve the inequality, we graph the related quadratic function defined by $f(x) = x^2 - x - 12$. We are particularly interested in the x-intercepts, which are found as in **Section 10.7** by letting $f(x) = 0$ and solving the quadratic equation

$$x^2 - x - 12 = 0.$$

$$(x - 4)(x + 3) = 0 \quad \text{Factor.}$$

$$x - 4 = 0 \quad \text{or} \quad x + 3 = 0 \quad \text{Zero-factor property}$$

$$x = 4 \quad \text{or} \quad x = -3 \quad \text{Solve each equation.}$$

Thus, the x-intercepts are $(4, 0)$ and $(-3, 0)$. The graph, which opens up since the coefficient of x^2 is positive, is shown in Figure 17(a). Notice from this graph that x-values less than -3 or greater than 4 result in y-values *greater than* 0. Therefore, the solution set of $x^2 - x - 12 > 0$, written in interval notation, is $(-\infty, -3) \cup (4, \infty)$.

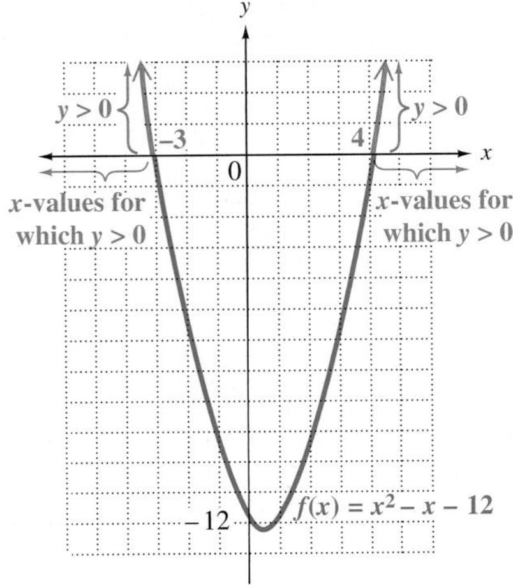

The graph is *above* the x-axis for $(-\infty, -3) \cup (4, \infty)$.

(a)

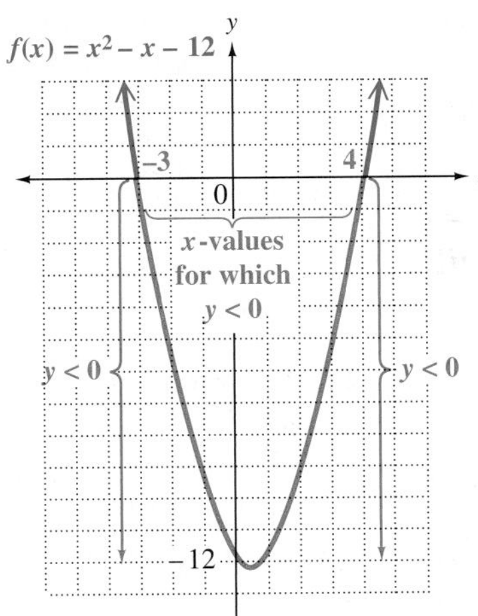

The graph is *below* the x-axis for $(-3, 4)$.

(b)

Figure 17

Continued on Next Page

(b) $x^2 - x - 12 < 0$

Here we want values of y that are *less than* 0. Referring to Figure 17(b) on the previous page, we notice from the graph that x-values between -3 and 4 result in y-values less than 0. Therefore, the solution set of the inequality $x^2 - x - 12 < 0$, written in interval notation, is $(-3, 4)$.

Note

If the inequalities in Example 1 had used $\geq$ and $\leq$, the solution sets would have included the x-values of the intercepts, which make the quadratic expression equal to 0, and been written in interval notation as $(-\infty, -3] \cup [4, \infty)$ for Example 1(a) and $[-3, 4]$ for Example 1(b). Square brackets would indicate that the endpoints -3 and 4 are *included* in the solution sets.

◀ Work Problems 1 and 2 at the Side.

In Example 1, we used graphing to divide the x-axis into intervals. Then using the graphs in Figure 17, we determined which x-values resulted in y-values that were either greater than or less than 0. Another method for solving a quadratic inequality uses these basic ideas without actually graphing the related quadratic function.

EXAMPLE 2 Solving a Quadratic Inequality Using Test Numbers

Solve $x^2 - x - 12 > 0$.

First solve the quadratic equation $x^2 - x - 12 = 0$ by factoring, as in Example 1(a).

$$x^2 - x - 12 = 0$$

$$(x - 4)(x + 3) = 0 \quad \text{Factor.}$$

$$x - 4 = 0 \quad \text{or} \quad x + 3 = 0 \quad \text{Zero-factor property}$$

$$x = 4 \quad \text{or} \quad x = -3 \quad \text{Solve each equation.}$$

The numbers 4 and -3 divide the number line into the three intervals shown in Figure 18. ***Be careful to put the lesser number on the left.*** (Notice the similarity between Figure 18 and the x-axis with intercepts $(-3, 0)$ and $(4, 0)$ in Figure 17(a).)

Interval A		Interval B		Interval C
T	-3	F	4	T

Figure 18

The numbers 4 and -3 are the only numbers that make the expression $x^2 - x - 12$ equal to 0. All other numbers make the expression either positive or negative. The sign of the expression can change from positive to negative or from negative to positive only at a number that makes it 0. Therefore, if one number in an interval satisfies the inequality, then all the numbers in that interval will satisfy the inequality.

To see if the numbers in Interval A satisfy the inequality, choose any number from Interval A in Figure 18 (that is, any number less than -3). Substitute this test number for x in the original inequality $x^2 - x - 12 > 0$. If the result is *true,* then all numbers in Interval A satisfy the inequality.

Continued on Next Page

1 Use the graph to solve each quadratic inequality.

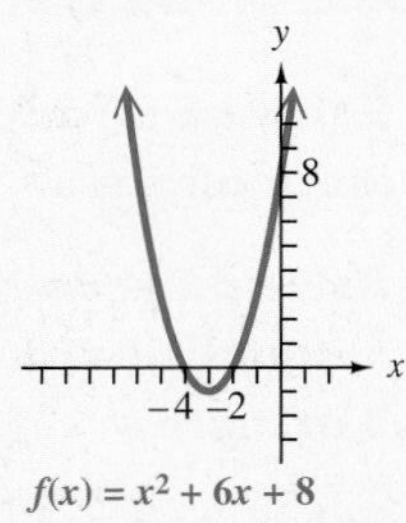

(a) $x^2 + 6x + 8 > 0$

(b) $x^2 + 6x + 8 < 0$

2 Graph $f(x) = x^2 + 3x - 4$ and use the graph to solve each quadratic inequality.

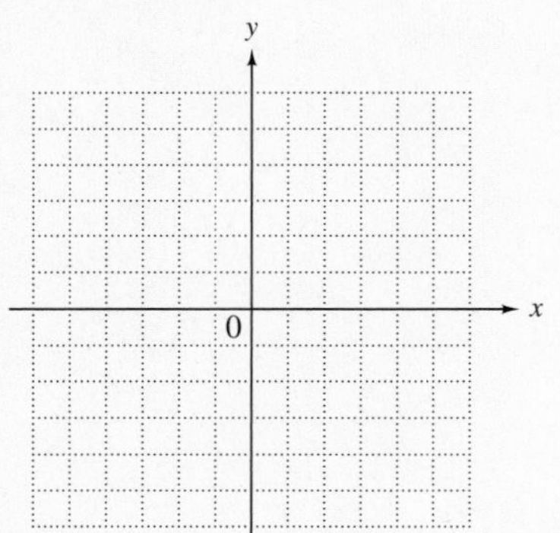

(a) $x^2 + 3x - 4 \geq 0$

(b) $x^2 + 3x - 4 \leq 0$

Answers

1. (a) $(-\infty, -4) \cup (-2, \infty)$ (b) $(-4, -2)$
2. (a) $(-\infty, -4] \cup [1, \infty)$ (b) $[-4, 1]$

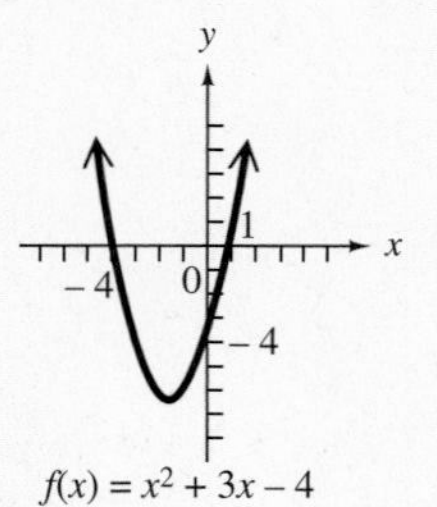

We try -5 from Interval A. Substitute -5 for x.

$$x^2 - x - 12 > 0 \quad \text{Original inequality}$$

Use parentheses to avoid sign errors.

$$(-5)^2 - (-5) - 12 \overset{?}{>} 0 \quad \text{Let } x = -5.$$

$$25 + 5 - 12 \overset{?}{>} 0 \quad \text{Simplify.}$$

$$18 > 0 \quad \text{True}$$

Because -5 from Interval A satisfies the inequality, all numbers from Interval A are solutions.

Try 0 from Interval B. If $x = 0$, then

$$0^2 - 0 - 12 \overset{?}{>} 0 \quad \text{Let } x = 0.$$

$$-12 > 0 \quad \text{False}$$

The numbers in Interval B are *not* solutions.

Work Problem **3** *at the Side.* ▶

In Problem 3 at the side, the test number 5 satisfies the inequality, so the numbers in Interval C are also solutions.

Based on these results (shown by the colored letters in Figure 18), the solution set includes the numbers in Intervals A and C, as shown on the graph in Figure 19. The solution set is written in interval notation as

$$(-\infty, -3) \cup (4, \infty).$$

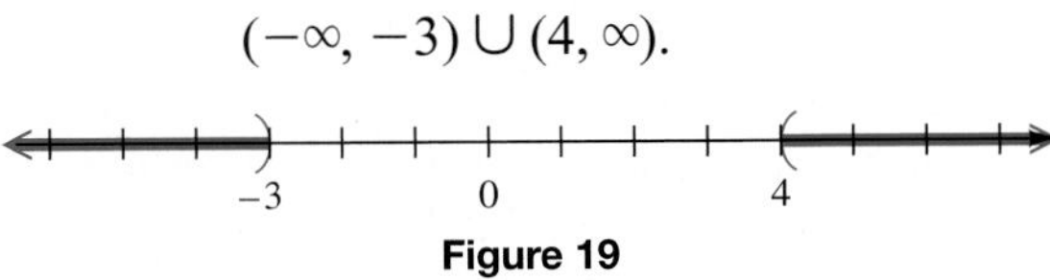

Figure 19

This agrees with the solution set we found by graphing the related quadratic function in Example 1(a).

In summary, a quadratic inequality is solved by following these steps.

Solving a Quadratic Inequality

Step 1 **Write the inequality as an equation and solve it.**

Step 2 **Use the solutions from Step 1 to determine intervals.** Graph the numbers found in Step 1 on a number line. These numbers divide the number line into intervals.

Step 3 **Find the intervals that satisfy the inequality.** Substitute a test number from each interval into the original inequality to determine the intervals that satisfy the inequality. All numbers in those intervals are in the solution set. A graph of the solution set will usually look like one of these. (Square brackets might be used instead of parentheses.)

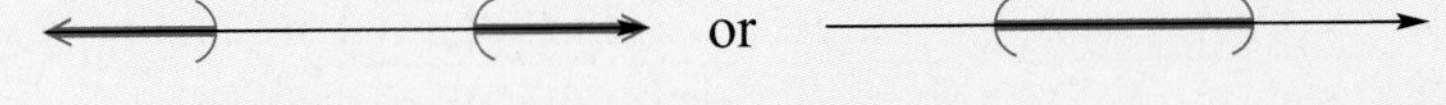

Step 4 **Consider the endpoints separately.** The numbers from Step 1 are included in the solution set if the inequality is $\leq$ or $\geq$; they are not included if it is $<$ or $>$.

Work Problem **4** *at the Side.* ▶

3 Does the number 5 from Interval C satisfy $x^2 - x - 12 > 0$?

4 Solve each inequality, and graph the solution set.

(a) $x^2 + x - 6 > 0$

(b) $3m^2 - 13m - 10 \leq 0$

Answers

3. yes

4. (a) $(-\infty, -3) \cup (2, \infty)$

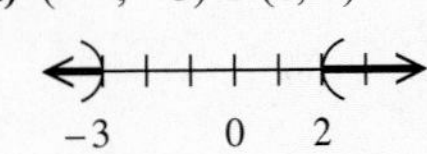

(b) $\left[-\frac{2}{3}, 5\right]$

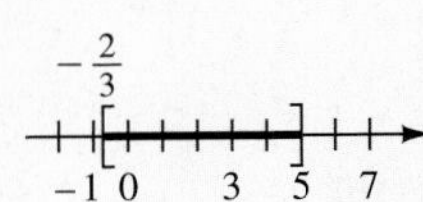

Special cases of quadratic inequalities may occur, as in the next example.

5 Solve each inequality.

(a) $(3x - 2)^2 > -2$

(b) $(3x - 2)^2 < -2$

EXAMPLE 3 Solving Special Cases

Solve each inequality.

(a) $(2x - 3)^2 > -1$

Because $(2x - 3)^2$ is never negative, it is greater than -1 for all replacements for x. Thus, the solution set is all real numbers, $(-\infty, \infty)$.

(b) $(2x - 3)^2 < -1$

Using the same reasoning as in part (a), there is no solution for this inequality. The solution set is $\emptyset$.

◀ *Work Problem* **5** *at the Side.*

OBJECTIVE 2 Solve polynomial inequalities of degree 3 or more. Higher-degree inequalities that have factorable polynomials are solved using a method similar to that of solving quadratic inequalities.

EXAMPLE 4 Solving a Third-Degree Polynomial Inequality

Solve $(x - 1)(x + 2)(x - 4) \leq 0$.

This is a *cubic* (third-degree) inequality rather than a quadratic inequality, but it can be solved using the method shown in the box by extending the zero-factor property to more than two factors. Begin by setting the factored polynomial *equal* to 0 and solving the equation (Step 1).

$$(x - 1)(x + 2)(x - 4) = 0$$

$$x - 1 = 0 \quad \text{or} \quad x + 2 = 0 \quad \text{or} \quad x - 4 = 0$$

$$x = 1 \quad \text{or} \quad x = -2 \quad \text{or} \quad x = 4$$

Locate the numbers -2, 1, and 4 on a number line, as in Figure 20, to determine the Intervals A, B, C, and D (Step 2).

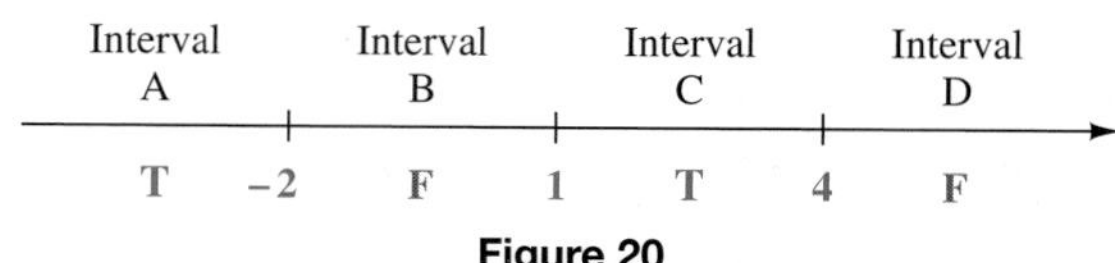

Figure 20

Substitute a test number from each interval in the *original* inequality to determine which intervals satisfy the inequality (Step 3). It is helpful to organize this information in a table.

Interval	Test Number	Test of Inequality	True or False?
A	-3	$-28 \leq 0$	T
B	0	$8 \leq 0$	F
C	2	$-8 \leq 0$	T
D	5	$28 \leq 0$	F

Verify the information given in the table. The numbers in Intervals A and C are in the solution set, which is written as

$$(-\infty, -2] \cup [1, 4].$$

The three endpoints are included since the inequality symbol is $\leq$ (Step 4).

Continued on Next Page

ANSWERS

5. (a) $(-\infty, \infty)$ (b) $\emptyset$

The solution set is graphed in Figure 21.

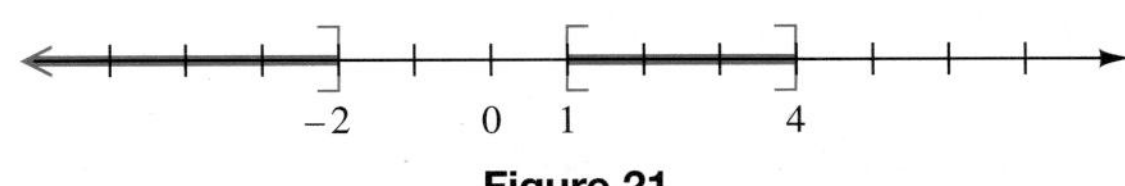

Figure 21

Work Problem 6 at the Side. ▶

OBJECTIVE 3 Solve rational inequalities. Inequalities that involve rational expressions, called **rational inequalities,** are solved similarly using the following steps.

Solving a Rational Inequality

Step 1 **Write the inequality** so that 0 is on one side and there is a single fraction on the other side.

Step 2 **Determine the numbers that make the numerator and denominator equal to 0.**

Step 3 **Divide a number line into intervals.** Use the numbers from Step 2.

Step 4 **Find the intervals that satisfy the inequality.** Test a number from each interval by substituting it into the *original* inequality.

Step 5 **Consider the endpoints separately.** Exclude any values that make the denominator 0.

EXAMPLE 5 Solving a Rational Inequality

Solve $\dfrac{-1}{x-3} > 1$.

Write the inequality so that 0 is on one side (Step 1).

$$\frac{-1}{x-3} - 1 > 0 \quad \text{Subtract 1.}$$

$$\frac{-1}{x-3} - \frac{x-3}{x-3} > 0 \quad \text{Use } x - 3 \text{ as the common denominator.}$$

Be careful with signs.

$$\frac{-1-x+3}{x-3} > 0 \quad \text{Write the left side as a single fraction.}$$

$$\frac{-x+2}{x-3} > 0 \quad \text{Combine like terms in the numerator.}$$

The sign of the rational expression $\frac{-x+2}{x-3}$ will change from positive to negative or negative to positive only at those numbers that make the numerator or denominator 0. The number 2 makes the numerator 0, and 3 makes the denominator 0 (Step 2). These two numbers, 2 and 3, divide a number line into three intervals. See Figure 22 (Step 3).

Interval A		Interval B		Interval C
F	2	T	3	F

Figure 22

Continued on Next Page

6 Solve each inequality, and graph the solution set.

(a) $(x-3)(x+2)(x+1) > 0$

(b) $(x-5)(x+1)(x-3) \le 0$

ANSWERS

6. **(a)** $(-2, -1) \cup (3, \infty)$

−2 −1 0 1 2 3 4

(b) $(-\infty, -1] \cup [3, 5]$

−1 0 1 3 5

Testing a number from each interval in the *original* inequality, $\frac{-1}{x-3} > 1$, gives the results shown in the table (Step 4).

Interval	Test Number	Test of Inequality	True or False?
A	0	$\frac{1}{3} > 1$	F
B	2.5	$2 > 1$	T
C	4	$-1 > 1$	F

The solution set of $\frac{-1}{x-3} > 1$ is the interval $(2, 3)$. This interval does not include 3 since it would make the denominator of the original inequality 0; 2 is not included either since the inequality symbol is $>$, which does not involve equality (Step 5). A graph of the solution set is given in Figure 23.

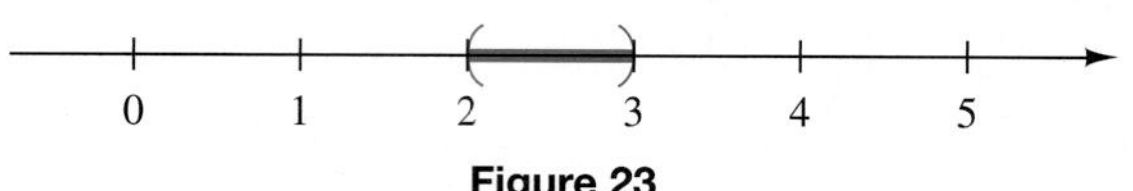

Figure 23

◀ Work Problem **7** at the Side.

CAUTION

When solving a rational inequality, any number that makes the denominator 0 must be excluded from the solution set.

EXAMPLE 6 Solving a Rational Inequality

Solve $\frac{x-2}{x+2} \le 2$.

Write the inequality so that 0 is on one side (Step 1).

$$\frac{x-2}{x+2} - 2 \le 0 \quad \text{Subtract 2.}$$

$$\frac{x-2}{x+2} - \frac{2(x+2)}{x+2} \le 0 \quad \text{Use } x+2 \text{ as the common denominator.}$$

$$\frac{x-2}{x+2} - \frac{2x+4}{x+2} \le 0 \quad \text{Distributive property}$$

$$\frac{x-2-2x-4}{x+2} \le 0 \quad \text{Write as a single fraction.}$$

$$\frac{-x-6}{x+2} \le 0 \quad \text{Combine like terms in the numerator.}$$

The number -6 makes the numerator 0, and -2 makes the denominator 0 (Step 2). These two numbers determine three intervals (Step 3). Test one number from each interval (Step 4) to see that the solution set is the interval

$$(-\infty, -6] \cup (-2, \infty).$$

The number -6 satisfies the original inequality, but -2 cannot be used as a solution since it makes the denominator 0 (Step 5). A graph of the solution set is shown in Figure 24.

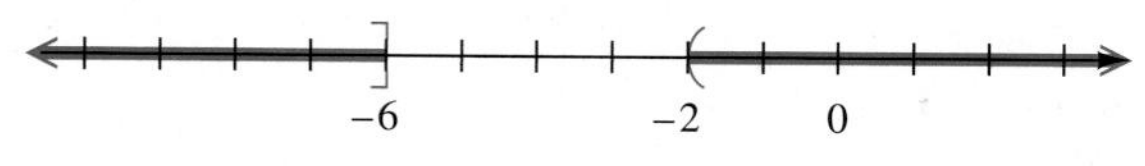

Figure 24

◀ Work Problem **8** at the Side.

7 Solve each inequality, and graph the solution set.

(a) $\frac{2}{x-4} < 3$

(b) $\frac{5}{x+1} > 4$

8 Solve $\frac{x+2}{x-1} \le 5$, and graph the solution set.

ANSWERS

7. (a) $(-\infty, 4) \cup \left(\frac{14}{3}, \infty\right)$

(b) $\left(-1, \frac{1}{4}\right)$

8. $(-\infty, 1) \cup \left[\frac{7}{4}, \infty\right)$

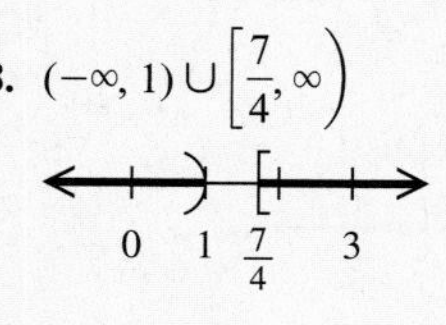

10.8 Exercises

FOR EXTRA HELP

Math XL PRACTICE

WATCH

DOWNLOAD

READ

REVIEW

In Example 1, we determined the solution sets of the quadratic inequalities $x^2 - x - 12 > 0$ *and* $x^2 - x - 12 < 0$ *by graphing* $f(x) = x^2 - x - 12$. *The x-intercepts of this graph indicated the solutions of the equation* $x^2 - x - 12 = 0$. *The x-values of the points on the graph that were* **above** *the x-axis formed the solution set of* $x^2 - x - 12 > 0$, *and the x-values of the points on the graph that were* **below** *the x-axis formed the solution set of* $x^2 - x - 12 < 0$.

In Exercises 1–4, the graph of a quadratic function f is given. Use the graph to find the solution set of each equation or inequality. See Example 1.

1. **(a)** $x^2 - 4x + 3 = 0$
(b) $x^2 - 4x + 3 > 0$
(c) $x^2 - 4x + 3 < 0$

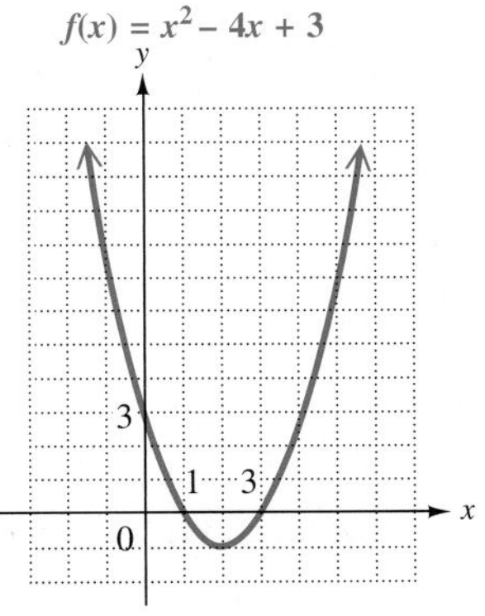

(a) {1, 3} **(b)** $(-\infty, 1) \cup (3, \infty)$ **(c)** (1, 3)

2. **(a)** $3x^2 + 10x - 8 = 0$
(b) $3x^2 + 10x - 8 \geq 0$
(c) $3x^2 + 10x - 8 < 0$

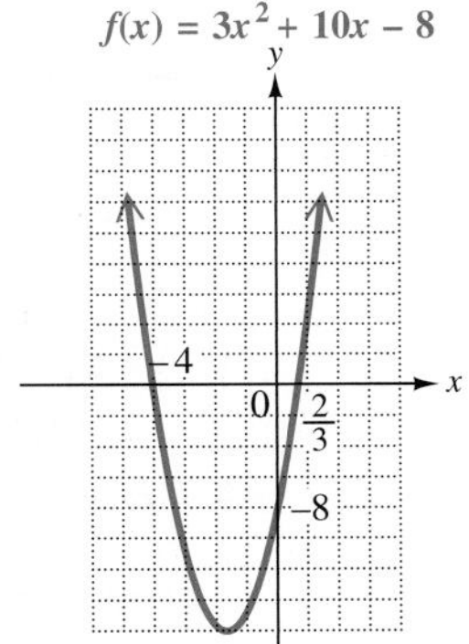

(a) $\left\{-4, \frac{2}{3}\right\}$ **(b)** $(-\infty, -4] \cup \left[\frac{2}{3}, \infty\right)$ **(c)** $\left(-4, \frac{2}{3}\right)$

3. **(a)** $-2x^2 - x + 15 = 0$
(b) $-2x^2 - x + 15 \geq 0$
(c) $-2x^2 - x + 15 \leq 0$

$f(x) = -2x^2 - x + 15$

(a) $\left\{-3, \frac{5}{2}\right\}$ **(b)** $\left[-3, \frac{5}{2}\right]$ **(c)** $(-\infty, -3] \cup \left[\frac{5}{2}, \infty\right)$

4. **(a)** $-x^2 + 3x + 10 = 0$
(b) $-x^2 + 3x + 10 \geq 0$
(c) $-x^2 + 3x + 10 \leq 0$

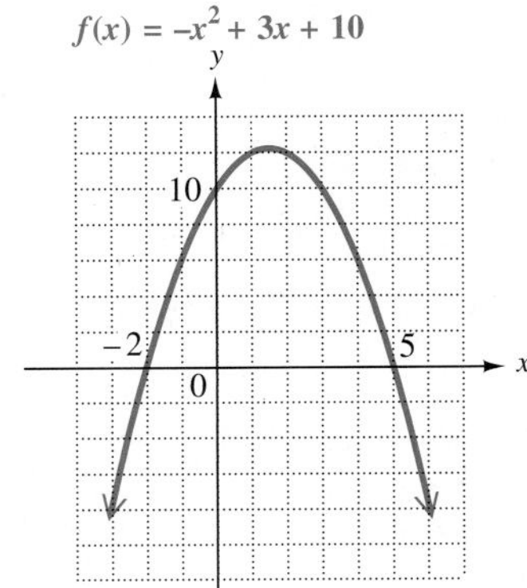

(a) {−2, 5} **(b)** [−2, 5] **(c)** $(-\infty, -2] \cup [5, \infty)$

5. Explain how to determine whether to include or exclude endpoints when solving a quadratic or higher-degree inequality.

Include the endpoints if the symbol is $\geq$ or $\leq$. Exclude the endpoints if the symbol is $>$ or $<$.

6. The solution set of the inequality $x^2 + x - 12 < 0$ is the interval $(-4, 3)$. Without actually performing any work, give the solution set of the inequality $x^2 + x - 12 \geq 0$.

$(-\infty, -4] \cup [3, \infty)$

Solve each inequality, and graph the solution set. See Example 2.

7. $(x + 1)(x - 5) > 0$

$(-\infty, -1) \cup (5, \infty)$

8. $(m + 6)(m - 2) > 0$

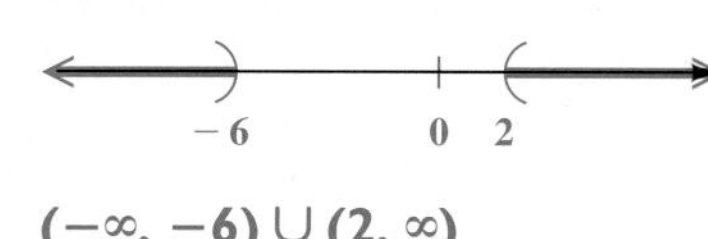

$(-\infty, -6) \cup (2, \infty)$

9. $(r + 4)(r - 6) < 0$

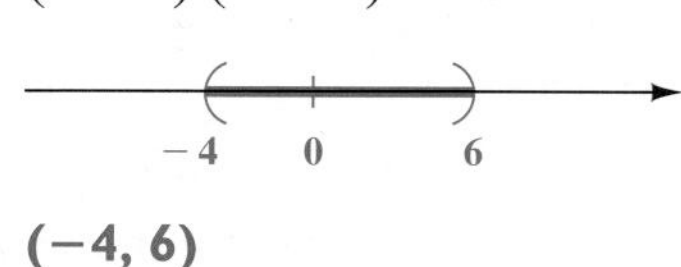

$(-4, 6)$

10. $(x + 4)(x - 8) < 0$

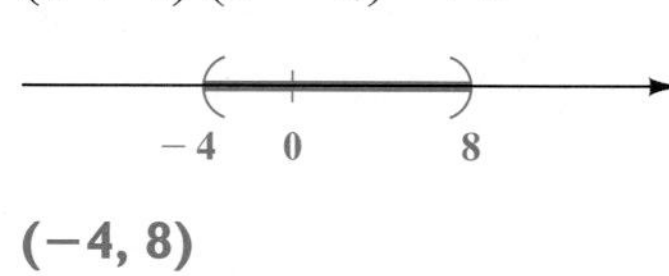

$(-4, 8)$

11. $x^2 - 4x + 3 \geq 0$

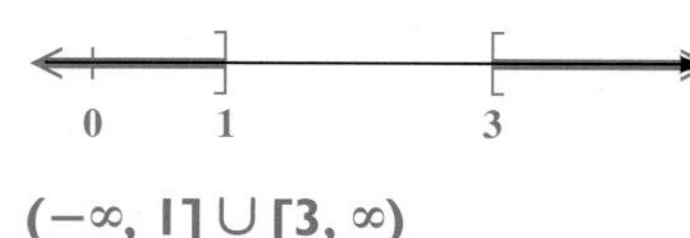

$(-\infty, 1] \cup [3, \infty)$

12. $m^2 - 3m - 10 \geq 0$

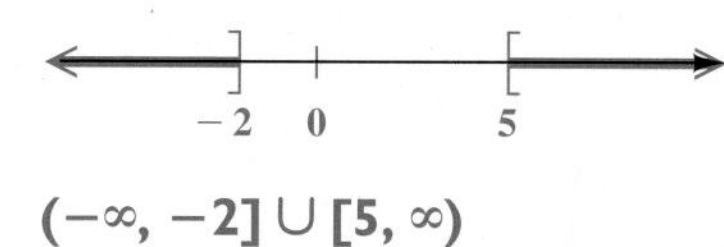

$(-\infty, -2] \cup [5, \infty)$

13. $10t^2 + 9t \geq 9$

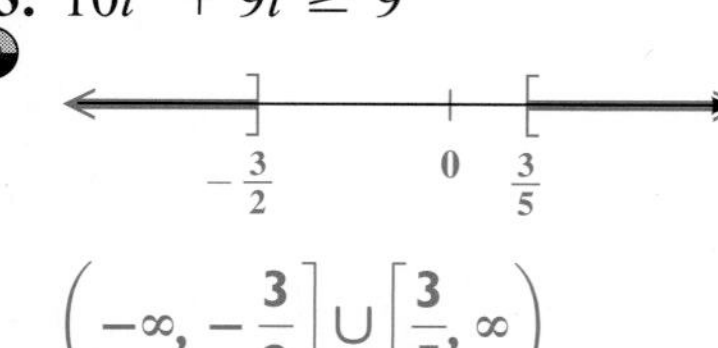

$\left(-\infty, -\frac{3}{2}\right] \cup \left[\frac{3}{5}, \infty\right)$

14. $3r^2 + 10r \geq 8$

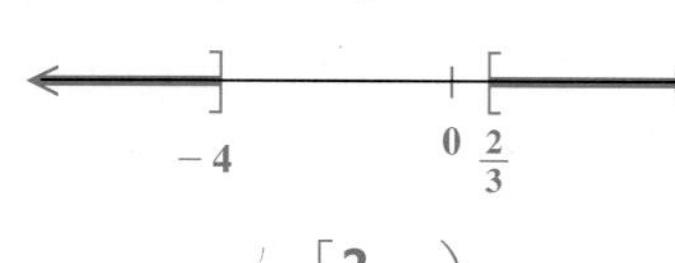

$(-\infty, -4] \cup \left[\frac{2}{3}, \infty\right)$

15. $9p^2 + 3p < 2$

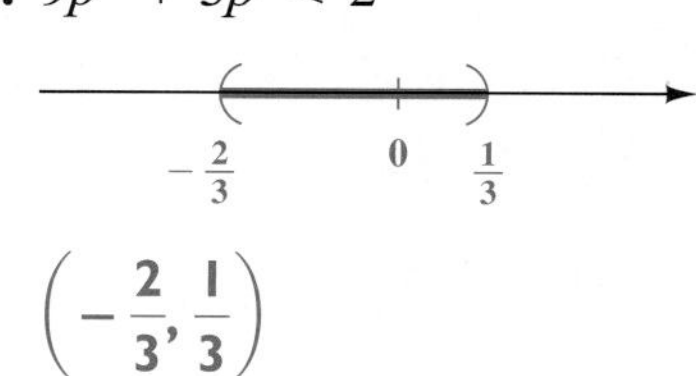

$\left(-\frac{2}{3}, \frac{1}{3}\right)$

16. $2x^2 + x < 15$

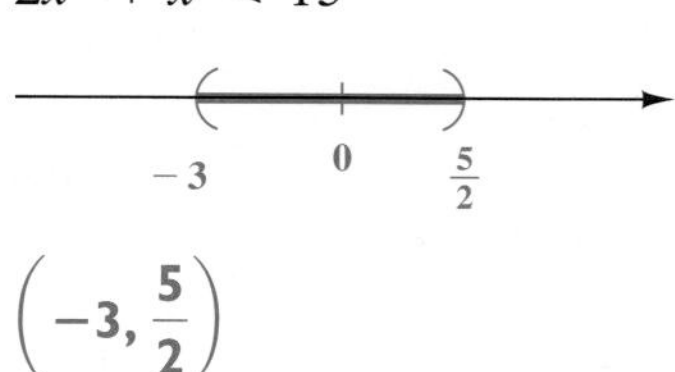

$\left(-3, \frac{5}{2}\right)$

17. $6x^2 + x \geq 1$

$\left(-\infty, -\frac{1}{2}\right] \cup \left[\frac{1}{3}, \infty\right)$

18. $4m^2 + 7m \geq -3$

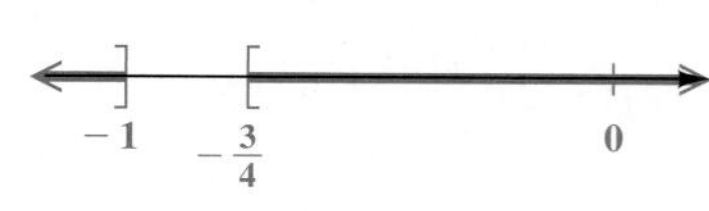

$(-\infty, -1] \cup \left[-\frac{3}{4}, \infty\right)$

19. $x^2 - 6x + 6 \geq 0$
(*Hint:* Use the quadratic formula.)

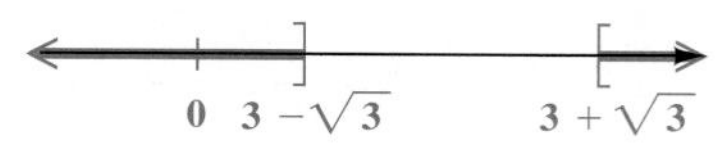

$(-\infty, 3 - \sqrt{3}] \cup [3 + \sqrt{3}, \infty)$

20. $3k^2 - 6k + 2 \leq 0$
(*Hint:* Use the quadratic formula.)

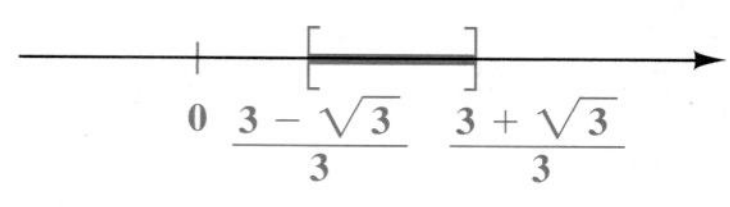

$\left[\frac{3 - \sqrt{3}}{3}, \frac{3 + \sqrt{3}}{3}\right]$

Solve each inequality. See Example 3.

21. $(4 - 3x)^2 \geq -2$
$(-\infty, \infty)$

22. $(6p + 7)^2 \geq -1$
$(-\infty, \infty)$

23. $(3x + 5)^2 \leq -4$
$\emptyset$

24. $(8t + 5)^2 \leq -5$
$\emptyset$

Solve each inequality, and graph the solution set. See Example 4.

25. $(p - 1)(p - 2)(p - 4) < 0$

$(-\infty, 1) \cup (2, 4)$

26. $(2r + 1)(3r - 2)(4r + 7) < 0$

$\left(-\infty, -\frac{7}{4}\right) \cup \left(-\frac{1}{2}, \frac{2}{3}\right)$

27. $(x - 4)(2x + 3)(3x - 1) \geq 0$

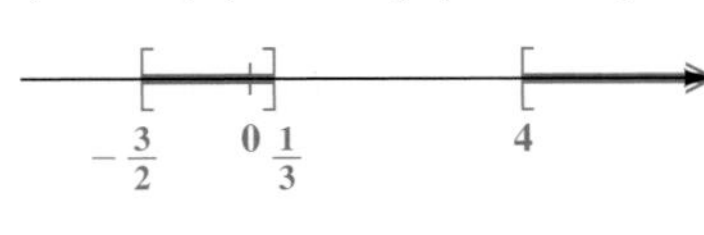

$\left[-\frac{3}{2}, \frac{1}{3}\right] \cup [4, \infty)$

28. $(z + 2)(4z - 3)(2z + 7) \geq 0$

$\left[-\frac{7}{2}, -2\right] \cup \left[\frac{3}{4}, \infty\right)$

Solve each inequality, and graph the solution set. See Examples 5 and 6.

29. $\frac{x - 1}{x - 4} > 0$

$(-\infty, 1) \cup (4, \infty)$

30. $\frac{x + 1}{x - 5} > 0$

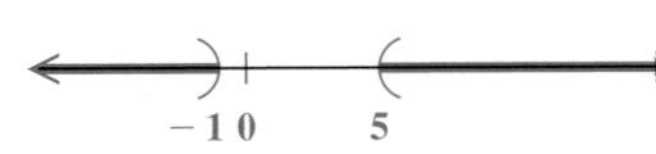

$(-\infty, -1) \cup (5, \infty)$

31. $\frac{2n + 3}{n - 5} \leq 0$

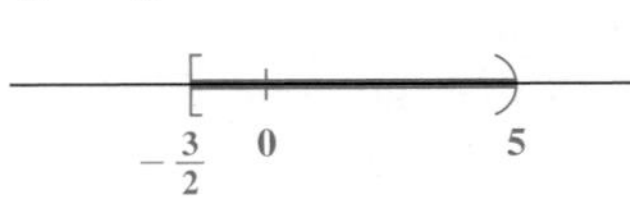

$\left[-\frac{3}{2}, 5\right)$

32. $\frac{3t + 7}{t - 3} \leq 0$

$\left[-\frac{7}{3}, 3\right)$

33. $\frac{8}{x - 2} \geq 2$

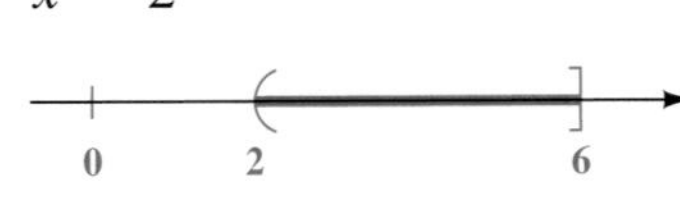

$(2, 6]$

34. $\frac{20}{x - 1} \geq 1$

$(1, 21]$

35. $\frac{3}{2t - 1} < 2$

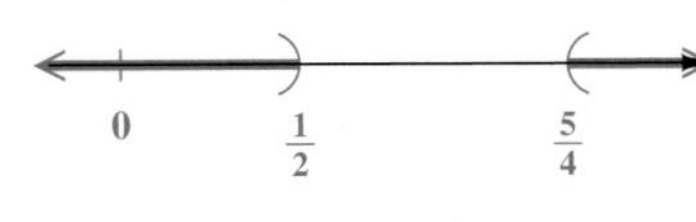

$\left(-\infty, \frac{1}{2}\right) \cup \left(\frac{5}{4}, \infty\right)$

36. $\frac{6}{m - 1} < 1$

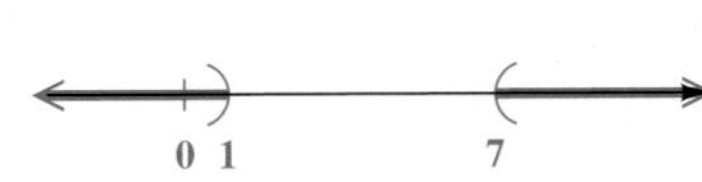

$(-\infty, 1) \cup (7, \infty)$

37. $\frac{w}{w + 2} \geq 2$

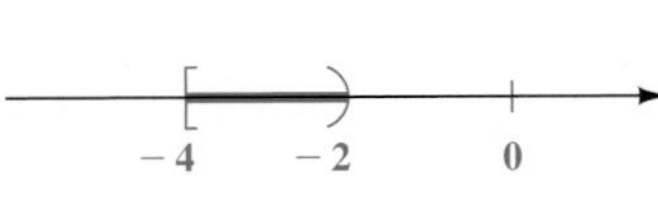

$[-4, -2)$

38. $\dfrac{m}{m+5} \geq 2$

$[-10, -5)$

39. $\dfrac{4k}{2k-1} < k$

$\left(0, \dfrac{1}{2}\right) \cup \left(\dfrac{5}{2}, \infty\right)$

40. $\dfrac{r}{r+2} < 2r$

$\left(-2, -\dfrac{3}{2}\right) \cup (0, \infty)$

41. $\dfrac{x-8}{x-4} \leq 3$

$(-\infty, 2] \cup (4, \infty)$

42. $\dfrac{2t-3}{t+1} \geq 4$

$\left[-\dfrac{7}{2}, -1\right)$

Relating Concepts (Exercises 43–46) For Individual or Group Work

A rock is projected vertically upward from the ground. Its distance s in feet above the ground after t seconds is given by the quadratic function defined by

$$s(t) = -16t^2 + 256t.$$

Work Exercises 43–46 in order, *to see how quadratic equations and inequalities are related.*

43. At what times will the rock be 624 ft above the ground? (*Hint:* Let $s(t) = 624$ and solve the quadratic *equation*.)

3 sec and 13 sec

44. At what times will the rock be more than 624 ft above the ground? (*Hint:* Set $s(t) > 624$ and solve the quadratic *inequality*.)

between 3 sec and 13 sec

45. At what times will the rock be at ground level? (*Hint:* Let $s(t) = 0$ and solve the quadratic *equation*.)

at 0 sec (the time when it is initially projected) and at 16 sec (the time when it hits the ground)

46. At what times will the rock be less than 624 ft above the ground? (*Hint:* Set $s(t) < 624$, solve the quadratic *inequality,* and observe the solutions in Exercises 44 and 45 to determine the least and greatest possible values of t.)

between 0 and 3 sec and between 13 and 16 sec

Chapter 10 Summary

Key Terms

10.1 quadratic equation A quadratic equation is an equation that can be written in the form $ax^2 + bx + c = 0$, where a, b, and c are real numbers, with $a \neq 0$. This form is called standard form.

10.3 quadratic formula The quadratic formula is a formula for solving quadratic equations.

discriminant The discriminant is the expression under the radical in the quadratic formula.

10.4 quadratic in form A nonquadratic equation that can be written as a quadratic equation is called quadratic in form.

10.6 quadratic function A function defined by $f(x) = ax^2 + bx + c$, for real numbers a, b, and c, with $a \neq 0$, is a quadratic function.

parabola The graph of a quadratic function is a parabola.

vertex The point on a parabola that has the least y-value (if the parabola opens up) or the greatest y-value (if the parabola opens down) is called the vertex of the parabola.

axis The vertical (or horizontal) line through the vertex of a vertical (or horizontal) parabola is its axis.

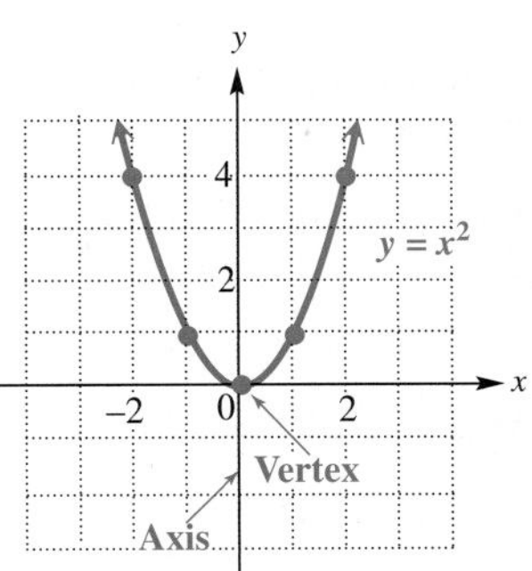

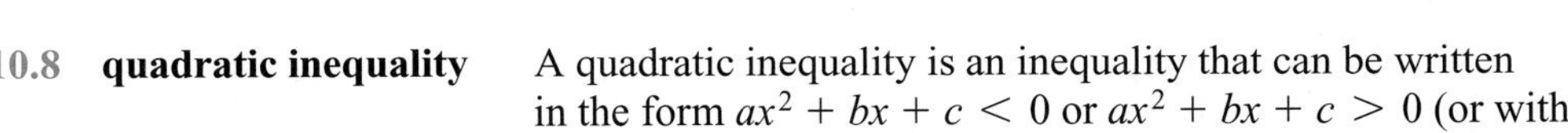

10.8 quadratic inequality A quadratic inequality is an inequality that can be written in the form $ax^2 + bx + c < 0$ or $ax^2 + bx + c > 0$ (or with $\leq$ or $\geq$) where a, b, and c are real numbers, with $a \neq 0$.

rational inequality An inequality that involves a rational expression is a rational inequality.

Test Your Word Power

See how well you have learned the vocabulary in this chapter. Answers, with examples, follow the Quick Review.

1. The **quadratic formula** is
   - **A.** a formula to find the number of solutions of a quadratic equation
   - **B.** a formula to find the type of solutions of a quadratic equation
   - **C.** the standard form of a quadratic equation
   - **D.** a general formula for solving any quadratic equation.

2. A **quadratic function** is a function that can be written in the form
   - **A.** $f(x) = mx + b$ for real numbers m and b
   - **B.** $f(x) = \frac{P(x)}{Q(x)}$, where $Q(x) \neq 0$
   - **C.** $f(x) = ax^2 + bx + c$ for real numbers a, b, and c $(a \neq 0)$
   - **D.** $f(x) = \sqrt{x}$ for $x \geq 0$.

3. A **parabola** is the graph of
   - **A.** any equation in two variables
   - **B.** a linear equation
   - **C.** an equation of degree 3
   - **D.** a quadratic equation in two variables.

4. The **vertex** of a parabola is
   - **A.** the point where the graph intersects the y-axis
   - **B.** the point where the graph intersects the x-axis
   - **C.** the lowest point on a parabola that opens up or the highest point on a parabola that opens down
   - **D.** the origin.

5. The **axis** of a parabola is
   - **A.** either the x-axis or the y-axis
   - **B.** the vertical line (of a vertical parabola) or the horizontal line (of a horizontal parabola) through the vertex
   - **C.** the lowest or highest point on the graph of a parabola
   - **D.** a line through the origin.

6. A parabola is **symmetric about its axis** since
   - **A.** its graph is near the axis
   - **B.** its graph is a mirror image on each side of the axis
   - **C.** its graph looks different on each side of the axis
   - **D.** its graph intersects the axis.

▶ Quick Review

Concepts	Examples

10.1 Solving Quadratic Equations by the Square Root Property

Square Root Property

If x and k are complex numbers and $x^2 = k$, then

$$x = \sqrt{k} \quad \text{or} \quad x = -\sqrt{k}.$$

Solve $(x-1)^2 = 8$.

$$x - 1 = \sqrt{8} \quad \text{or} \quad x - 1 = -\sqrt{8}$$
$$x = 1 + 2\sqrt{2} \quad \text{or} \quad x = 1 - 2\sqrt{2}$$

Solution set: $\{1 + 2\sqrt{2}, 1 - 2\sqrt{2}\}$

10.2 Solving Quadratic Equations by Completing the Square

Completing the Square

To solve $ax^2 + bx + c = 0$ $(a \neq 0)$:

Step 1 If $a \neq 1$, divide each side by a.

Step 2 Write the equation with the variable terms on one side and the constant on the other.

Step 3 Take half the coefficient of x and square it.

Step 4 Add the square to each side.

Step 5 Factor the perfect square trinomial, and write it as the square of a binomial. Simplify the other side.

Step 6 Use the square root property to complete the solution.

Solve $2x^2 - 4x - 18 = 0$.

$x^2 - 2x - 9 = 0$ Divide by 2.

$x^2 - 2x = 9$ Add 9.

$$\left[\frac{1}{2}(-2)\right]^2 = (-1)^2 = 1$$

$x^2 - 2x + 1 = 9 + 1$ Add 1.

$(x-1)^2 = 10$ Factor; add.

$$x - 1 = \sqrt{10} \quad \text{or} \quad x - 1 = -\sqrt{10}$$
$$x = 1 + \sqrt{10} \quad \text{or} \quad x = 1 - \sqrt{10}$$

Solution set: $\{1 + \sqrt{10}, 1 - \sqrt{10}\}$

10.3 Solving Quadratic Equations by the Quadratic Formula

Quadratic Formula

The solutions of $ax^2 + bx + c = 0$ $(a \neq 0)$ are given by

$$x = \frac{-b \pm \sqrt{b^2 - 4ac}}{2a}.$$

Solve $3x^2 + 5x + 2 = 0$.

$$x = \frac{-5 \pm \sqrt{5^2 - 4(3)(2)}}{2(3)}$$
$$x = \frac{-5 \pm 1}{6}$$
$$x = -1 \quad \text{or} \quad x = -\frac{2}{3}$$

Solution set: $\{-1, -\frac{2}{3}\}$

The Discriminant

If a, b, and c are integers, then the discriminant, $b^2 - 4ac$, of $ax^2 + bx + c = 0$ determines the number and type of solutions as follows.

Discriminant	Number and Type of Solutions
Positive, the square of an integer	Two rational solutions
Positive, not the square of an integer	Two irrational solutions
Zero	One rational solution
Negative	Two nonreal complex solutions

For $x^2 + 3x - 10 = 0$, the discriminant is

$3^2 - 4(1)(-10) = 49$. Two rational solutions

For $4x^2 + x + 1 = 0$, the discriminant is

$1^2 - 4(4)(1) = -15$. Two nonreal complex solutions

Concepts	Examples

10.4 Equations Quadratic in Form

A nonquadratic equation that can be written in the form

$$au^2 + bu + c = 0,$$

for $a \neq 0$ and an algebraic expression u, is called quadratic in form. Substitute u for the expression, solve for u, and then solve for the variable in the expression.

Solve $3(x+5)^2 + 7(x+5) + 2 = 0$.

$$3u^2 + 7u + 2 = 0 \quad \text{Let } u = x + 5.$$

$$(3u+1)(u+2) = 0 \quad \text{Factor.}$$

$$u = -\frac{1}{3} \quad \text{or} \quad u = -2$$

$$x + 5 = -\frac{1}{3} \quad \text{or} \quad x + 5 = -2 \quad x + 5 = u$$

$$x = -\frac{16}{3} \quad \text{or} \quad x = -7$$

Solution set: $\left\{-7, -\frac{16}{3}\right\}$

10.5 Formulas and Further Applications

To solve a formula for a second-degree variable, proceed as follows.

(a) If the variable appears only to the second power: Isolate the second-degree variable on one side of the equation, and then use the square root property.

Solve $A = \dfrac{2mp}{r^2}$ for r.

$$r^2A = 2mp \quad \text{Multiply by } r^2.$$

$$r^2 = \frac{2mp}{A} \quad \text{Divide by } A.$$

$$r = \pm\sqrt{\frac{2mp}{A}} \quad \text{Square root property}$$

$$r = \frac{\pm\sqrt{2mpA}}{A} \quad \text{Rationalize the denominator.}$$

(b) If the variable appears to the first and second powers: Write the equation in standard form, and then use the quadratic formula.

Solve $m^2 + rm = t$ for m.

$$m^2 + rm - t = 0 \quad \text{Standard form}$$

$$m = \frac{-r \pm \sqrt{r^2 - 4(1)(-t)}}{2(1)} = \frac{-r \pm \sqrt{r^2 + 4t}}{2}$$

$a = 1, b = r, c = -t$

10.6 Graphs of Quadratic Functions

1. The graph of the quadratic function defined by $F(x) = a(x-h)^2 + k$, $a \neq 0$, is a parabola with vertex at (h, k) and the vertical line $x = h$ as axis.
2. The graph opens up if a is positive and down if a is negative.
3. The graph is wider than the graph of $f(x) = x^2$ if $0 < |a| < 1$ and narrower if $|a| > 1$.

Graph $f(x) = -(x+3)^2 + 1$.

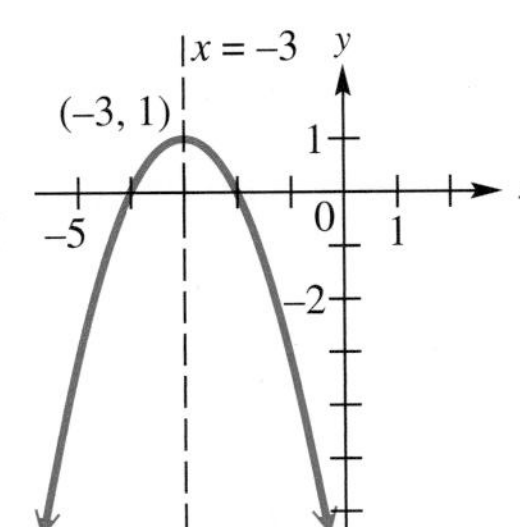

The graph opens down since $a < 0$. It is shifted 3 units left and 1 unit up, so the vertex is $(-3, 1)$, with axis $x = -3$. The domain is $(-\infty, \infty)$; the range is $(-\infty, 1]$.

10.7 More about Parabolas and Their Applications

The vertex of the graph of $f(x) = ax^2 + bx + c$, $a \neq 0$, has coordinates

$$\left(\frac{-b}{2a}, f\left(\frac{-b}{2a}\right)\right).$$

Graphing a Quadratic Function

Step 1 Determine whether the graph opens up or down.
Step 2 Find the vertex.
Step 3 Find the x-intercepts (if any). Find the y-intercept.
Step 4 Find and plot additional points as needed.

Graph $f(x) = x^2 + 4x + 3$.

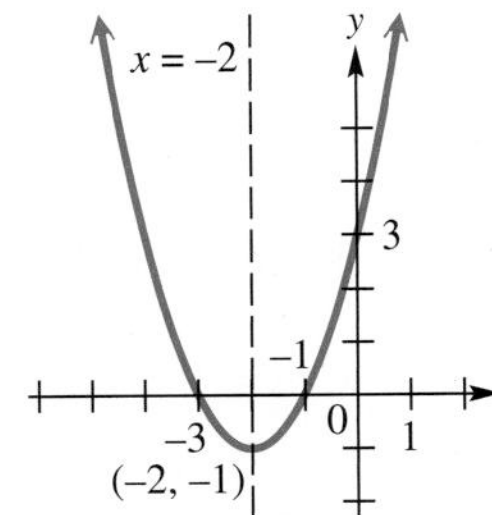

The graph opens up since $a > 0$. The vertex is $(-2, -1)$. The solutions of $x^2 + 4x + 3 = 0$ are -1 and -3, so the x-intercepts are $(-1, 0)$ and $(-3, 0)$. Since $f(0) = 3$, the y-intercept is $(0, 3)$. The domain is $(-\infty, \infty)$; the range is $[-1, \infty)$.

Concepts	Examples

10.7 More about Parabolas and Their Applications (continued)

Horizontal Parabolas

The graph of

$$x = ay^2 + by + c \quad \text{or} \quad x = a(y - k)^2 + h$$

is a horizontal parabola with vertex (h, k) and the horizontal line $y = k$ as axis. The graph opens to the right if $a > 0$ and to the left if $a < 0$.

Horizontal parabolas do not represent functions.

Graph $x = 2y^2 + 6y + 5$.

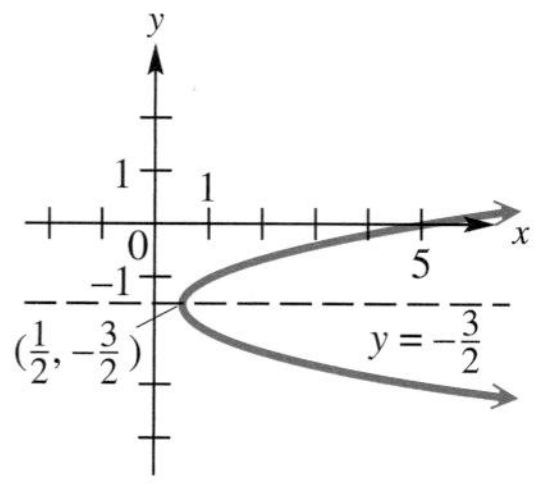

The graph opens to the right since $a > 0$. The vertex is $\left(\frac{1}{2}, -\frac{3}{2}\right)$. The axis is $y = -\frac{3}{2}$. The domain is $\left[\frac{1}{2}, \infty\right)$; the range is $(-\infty, \infty)$.

10.8 Polynomial and Rational Inequalities

Solving a Quadratic Inequality

Step 1 Write the inequality as an equation and solve.

Solve $2x^2 + 5x + 2 < 0$.

$$2x^2 + 5x + 2 = 0$$

$$x = -\frac{1}{2} \quad \text{or} \quad x = -2$$

Step 2 Use the numbers found in Step 1 to divide a number line into intervals.

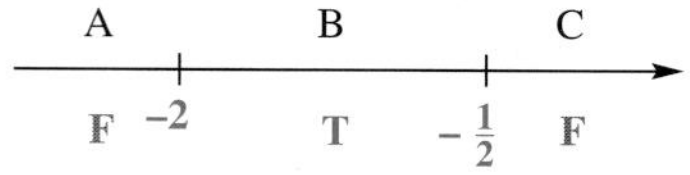

Step 3 Substitute a number from each interval into the original inequality to determine the intervals that belong in the solution set.

Use the method of Example 2 on pages 842–843 to find that $x = -3$ makes the original inequality false, $x = -1$ makes it true, and $x = 0$ makes it false.

Step 4 Consider the endpoints separately.

The solution set is the interval $\left(-2, -\frac{1}{2}\right)$.

Solving a Rational Inequality

Step 1 Write the inequality so that 0 is on one side and there is a single fraction on the other side.

Solve $\frac{x}{x + 2} \geq 4$.

$$\frac{x}{x+2} - 4 \geq 0 \quad \text{Subtract 4.}$$

$$\frac{x}{x+2} - \frac{4(x+2)}{x+2} \geq 0 \quad \text{Write with a common denominator.}$$

$$\frac{-3x - 8}{x+2} \geq 0 \quad \text{Subtract fractions.}$$

Step 2 Determine the numbers that make the numerator and denominator 0.

$-\frac{8}{3}$ makes the numerator 0 and -2 makes the denominator 0.

Step 3 Use the numbers from Step 2 to divide a number line into intervals.

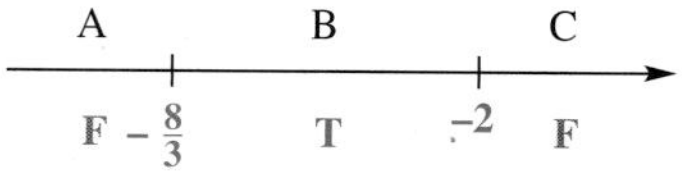

Step 4 Substitute a number from each interval into the original inequality to determine the intervals that belong in the solution set.

Use the method of Example 5 on pages 845–846 to find that $x = -4$ makes the original inequality false, $x = -\frac{7}{3}$ makes it true, and $x = 0$ makes it false.

Step 5 Consider the endpoints separately.

The solution set is the interval $\left[-\frac{8}{3}, -2\right)$. Note that -2 is excluded since it makes the denominator 0.

ANSWERS TO TEST YOUR WORD POWER

1. D; *Example:* The solutions of $ax^2 + bx + c = 0$ $(a \neq 0)$ are given by $x = \frac{-b \pm \sqrt{b^2 - 4ac}}{2a}$. **2.** C; *Examples:* $f(x) = x^2 - 2$, $f(x) = (x + 4)^2 + 1$, $f(x) = x^2 - 4x + 5$ **3.** D; *Examples:* See the figures in the Quick Review for **Sections 10.6 and 10.7.** **4.** C; *Example:* The graph of $y = (x + 3)^2$ has vertex $(-3, 0)$, which is the lowest point on the graph. **5.** B; *Example:* The axis of $y = (x + 3)^2$ is the vertical line $x = -3$. **6.** B; *Example:* Since the graph of $y = (x + 3)^2$ is symmetric about its axis $x = -3$, the points $(-2, 1)$ and $(-4, 1)$ are on this graph.

Chapter 10 Review Exercises

[10.1–10.2] *Solve each equation by using the square root property or completing the square.*

1. $t^2 = 121$

$\{11, -11\}$

2. $p^2 = 3$

$\{\sqrt{3}, -\sqrt{3}\}$

3. $(r - 3)^2 = 10$

$\{3 + \sqrt{10}, 3 - \sqrt{10}\}$

4. $(2p + 1)^2 = 14$

$\left\{\frac{-1 + \sqrt{14}}{2}, \frac{-1 - \sqrt{14}}{2}\right\}$

***5.** $(3k - 2)^2 = -25$

$\left\{\frac{2}{3} + \frac{5}{3}i, \frac{2}{3} - \frac{5}{3}i\right\}$

6. $x^2 + 4x = 15$

$\{-2 + \sqrt{19}, -2 - \sqrt{19}\}$

7. $2m^2 - 3m = -1$

$\left\{\frac{1}{2}, 1\right\}$

8. $-x^2 + 5 = 2x$

$\{-1 + \sqrt{6}, -1 - \sqrt{6}\}$

9. $2x^2 - 3 = -8x$

$\left\{\frac{-4 + \sqrt{22}}{2}, \frac{-4 - \sqrt{22}}{2}\right\}$

10. When it opened in 2008, the Singapore Flyer became the world's largest Ferris wheel, with a height of 165 m. Use the metric version of Galileo's formula, $d = 4.9t^2$ (where d is in meters), to find how long it would take a wallet dropped from the top of the Singapore Flyer to reach the ground. Round your answer to the nearest tenth of a second. (*Source:* www.singaporeflyer.com) **5.8 sec**

[10.3] *Solve each equation using the quadratic formula.*

11. $2x^2 + x - 21 = 0$

$\left\{-\frac{7}{2}, 3\right\}$

12. $k^2 + 5k = 7$

$\left\{\frac{-5 + \sqrt{53}}{2}, \frac{-5 - \sqrt{53}}{2}\right\}$

13. $(t + 3)(t - 4) = -2$

$\left\{\frac{1 + \sqrt{41}}{2}, \frac{1 - \sqrt{41}}{2}\right\}$

***14.** $2x^2 + 3x + 4 = 0$

$\left\{-\frac{3}{4} + \frac{\sqrt{23}}{4}i, -\frac{3}{4} - \frac{\sqrt{23}}{4}i\right\}$

***15.** $3p^2 = 2(2p - 1)$

$\left\{\frac{2}{3} + \frac{\sqrt{2}}{3}i, \frac{2}{3} - \frac{\sqrt{2}}{3}i\right\}$

16. $m(2m - 7) = 3m^2 + 3$

$\left\{\frac{-7 + \sqrt{37}}{2}, \frac{-7 - \sqrt{37}}{2}\right\}$

**Use the discriminant to predict whether the solutions to each equation are*

A. *two rational numbers;* **B.** *one rational number;*
C. *two irrational numbers;* **D.** *two nonreal complex numbers.*

17. $x^2 + 5x + 2 = 0$

C

18. $4t^2 = 3 - 4t$

A

19. $4x^2 = 6x - 8$

D

20. $9z^2 + 30z + 25 = 0$

B

*This exercise requires knowledge of complex numbers.

[10.4] *Solve each equation.*

21. $\frac{15}{x} = 2x - 1$

$\left\{-\frac{5}{2}, 3\right\}$

22. $\frac{1}{n} + \frac{2}{n+1} = 2$

$\left\{-\frac{1}{2}, 1\right\}$

23. $-2r = \sqrt{\frac{48 - 20r}{2}}$

$\{-4\}$

24. $8(3x + 5)^2 + 2(3x + 5) - 1 = 0$

$\left\{-\frac{11}{6}, -\frac{19}{12}\right\}$

25. $2x^{2/3} - x^{1/3} - 28 = 0$

$\left\{-\frac{343}{8}, 64\right\}$

26. $p^4 - 5p^2 + 4 = 0$

$\{-2, -1, 1, 2\}$

Solve each problem. Round answers to the nearest tenth, as necessary.

27. Matthew Sudak drove 8 mi to pick up his cousin Jack, and then drove 11 mi to a mall at a speed 15 mph faster. If Matthew's total travel time was 24 min, what was his speed on the trip to pick up Jack?

40 mph

28. An old machine processes a batch of checks in 1 hr more time than a new one. How long would it take the old machine to process a batch of checks that the two machines together process in 2 hr?

4.6 hr

[10.5] *Solve each formula for the indicated variable. (Give answers with ± when applicable.)*

29. $k = \frac{rF}{wv^2}$ for v

$v = \frac{\pm\sqrt{rFkw}}{kw}$

30. $mt^2 = 3mt + 6$ for t

$t = \frac{3m \pm \sqrt{9m^2 + 24m}}{2m}$

Solve each problem. Round answers to the nearest tenth, as necessary.

31. A large machine requires a part in the shape of a right triangle with a hypotenuse 9 ft less than twice the length of the longer leg. The shorter leg must be $\frac{3}{4}$ the length of the longer leg. Find the lengths of the three sides of the part.

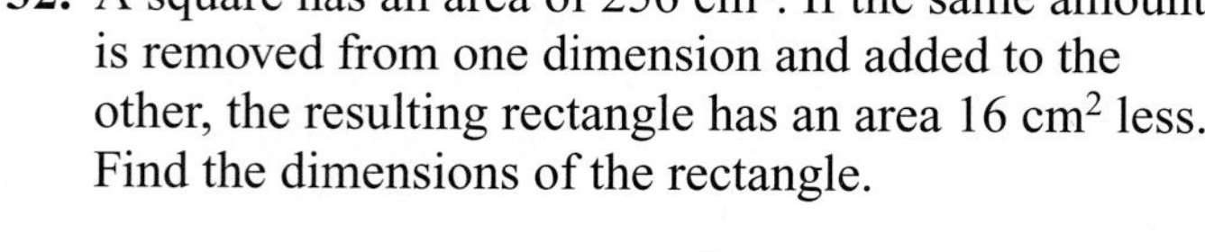

9 ft, 12 ft, 15 ft

32. A square has an area of 256 cm^2. If the same amount is removed from one dimension and added to the other, the resulting rectangle has an area 16 cm^2 less. Find the dimensions of the rectangle.

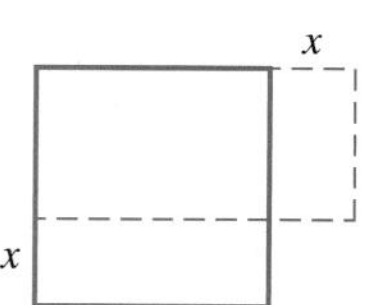

12 cm by 20 cm

33. Nancy wants to buy a mat for a photograph that measures 14 in. by 20 in. She wants to have an even border around the picture when it is mounted on the mat. If the area of the mat she chooses is 352 in.^{2}, how wide will the border be?

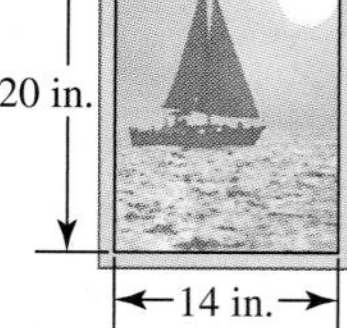

1 in.

34. Lewis Tower, built in Philadelphia, Pennsylvania, in 1929, is 400 ft high. Suppose that a ball is projected upward from the top of the Tower, and its position in feet above the ground is given by the quadratic function defined by

$$f(t) = -16t^2 + 45t + 400,$$

where t is the number of seconds elapsed. How long will it take for the ball to reach a height of 200 ft above the ground? (*Source: World Almanac and Book of Facts.*)

5.2 sec

[10.6–10.7] *Identify the vertex of the graph of each parabola.*

35. $f(x) = -(x - 1)^2$ **36.** $f(x) = (x - 3)^2 + 7$ **37.** $y = -3x^2 + 4x - 2$ **38.** $x = (y - 3)^2 - 4$

(1, 0) **(3, 7)** $\left(\frac{2}{3}, -\frac{2}{3}\right)$ **(−4, 3)**

Graph each parabola. Give the vertex, axis, domain, and range.

39. $y = 2(x - 2)^2 - 3$

vertex: **(2, −3)** domain: **(−∞, ∞)**
axis: **x = 2** range: **[−3, ∞)**

40. $f(x) = -2x^2 + 8x - 5$

vertex: **(2, 3)** domain: **(−∞, ∞)**
axis: **x = 2** range: **(−∞, 3]**

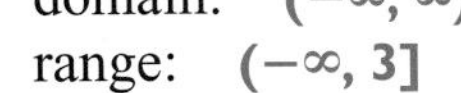

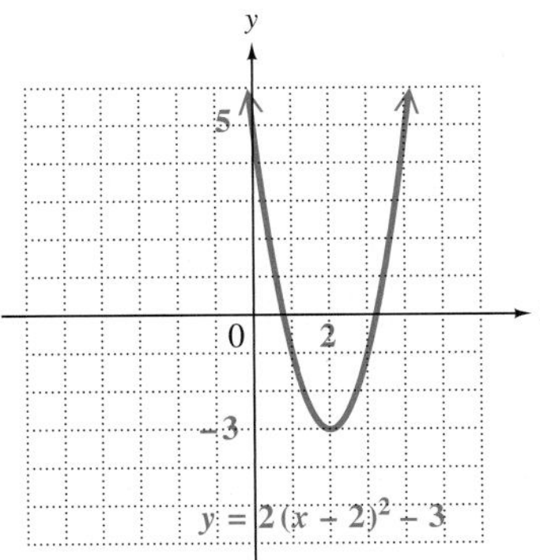

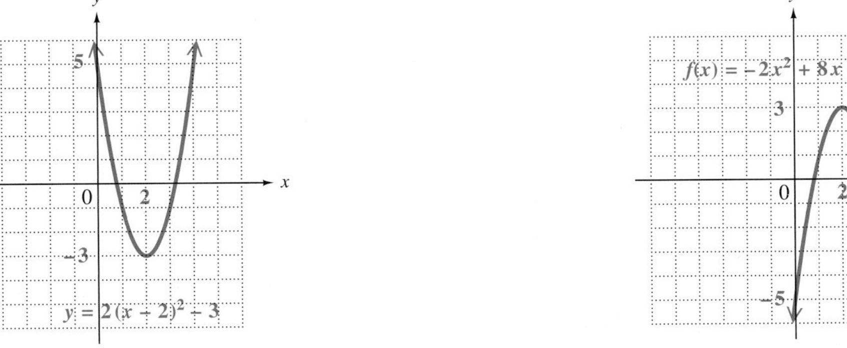

41. $x = 2(y + 3)^2 - 4$

vertex: **(−4, −3)** domain: **[−4, ∞)**
axis: **y = −3** range: **(−∞, ∞)**

42. $x = -\frac{1}{2}y^2 + 6y - 14$

vertex: **(4, 6)** domain: **(−∞, 4]**
axis: **y = 6** range: **(−∞, ∞)**

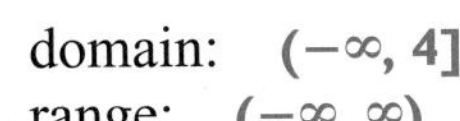

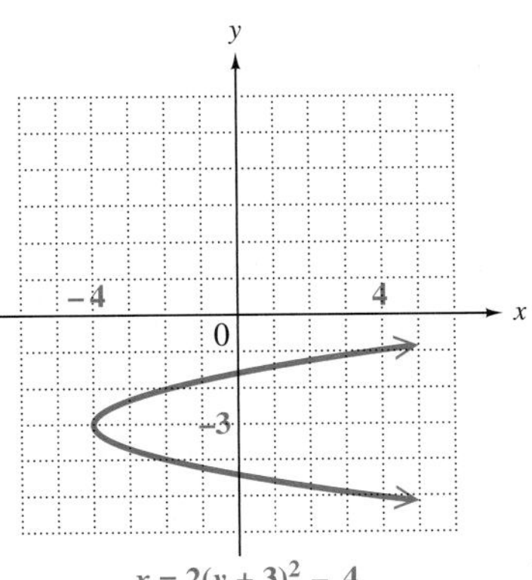

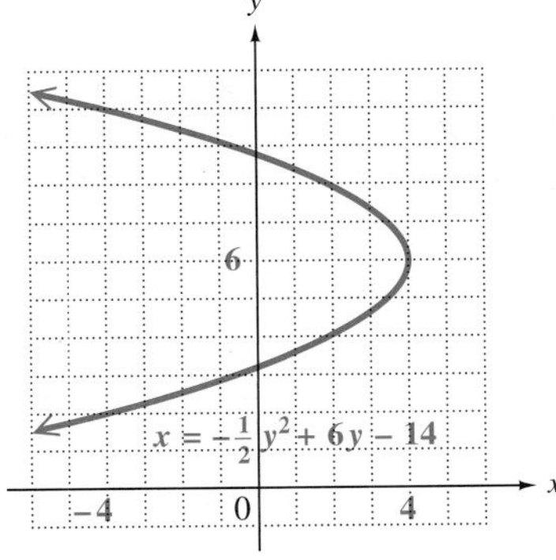

Solve each problem.

43. Total consumer spending on computers, peripherals, and software in the United States for selected years is given in the table. Let $x = 0$ represent 1985, $x = 5$ represent 1990, and so on.

(a) Use the data for 1985, 1995, and 2005 in the quadratic form $ax^2 + bx + c = y$ to write a system of three equations.

$c = 2.9$; $100a + 10b + c = 24.3$; $400a + 20b + c = 56.5$

(b) Solve the system from part (a) to get a quadratic function f that models the data.

$f(x) = 0.054x^2 + 1.6x + 2.9$

(c) Use the model found in part (b) to approximate consumer spending for computers, peripherals, and software games in 2006 to the nearest tenth. How does your answer compare to the actual data from the table?

\$60.3 billion; The result using the model is close, but slightly low.

CONSUMER SPENDING ON COMPUTERS, PERIPHERALS, AND SOFTWARE

Year	Spending (billions of dollars)
1985	2.9
1990	8.9
1995	24.3
2000	43.8
2004	51.6
2005	56.5
2006	61.4

Source: Bureau of Economic Analysis.

44. The height (in feet) of a projectile t seconds after being fired from Earth into the air is given by

$$f(t) = -16t^2 + 160t.$$

Find the number of seconds required for the projectile to reach maximum height. What is the maximum height?

5 sec; 400 ft

45. Find the length and width of a rectangle having a perimeter of 200 m if the area is to be a maximum.

length: 50 m; width: 50 m

[10.8] *Solve each inequality, and graph the solution set.*

46. $(x - 4)(2x + 3) > 0$

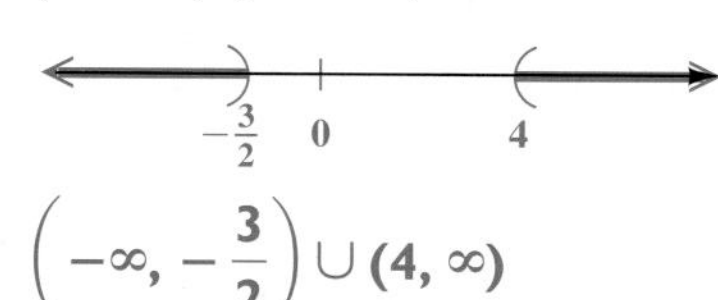

$\left(-\infty, -\frac{3}{2}\right) \cup (4, \infty)$

47. $x^2 + x \leq 12$

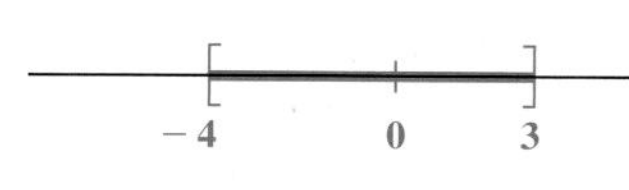

$[-4, 3]$

48. $(x + 2)(x - 3)(x + 5) \leq 0$

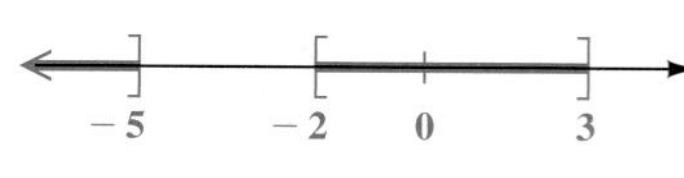

$(-\infty, -5] \cup [-2, 3]$

49. $(4m + 3)^2 \leq -4$

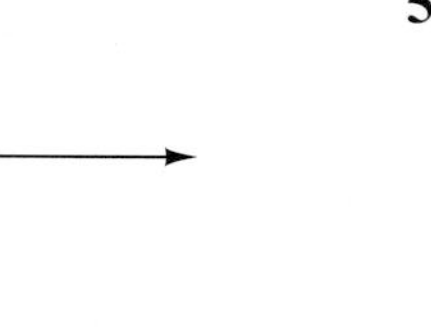

$\emptyset$

50. $\frac{6}{2z - 1} < 2$

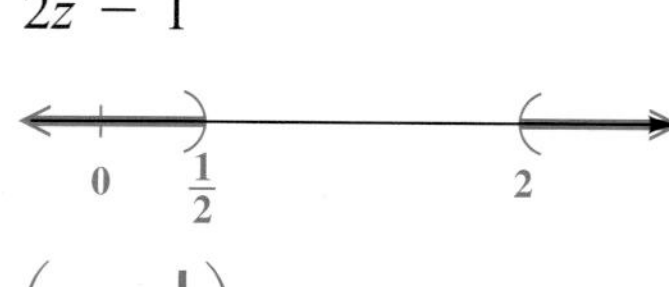

$\left(-\infty, \frac{1}{2}\right) \cup (2, \infty)$

51. $\frac{3t + 4}{t - 2} \leq 1$

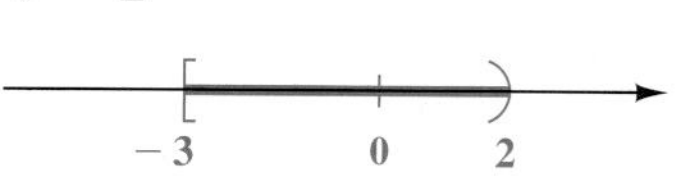

$[-3, 2)$

Mixed Review Exercises

Solve each equation or inequality.

52. $V = r^2 + R^2h$ for R

$R = \frac{\pm\sqrt{Vh - r^2h}}{h}$

***53.** $3t^2 - 6t = -4$

$\left\{1 + \frac{\sqrt{3}}{3}i, 1 - \frac{\sqrt{3}}{3}i\right\}$

54. $(x^2 - 2x)^2 = 11(x^2 - 2x) - 24$

$\{-2, -1, 3, 4\}$

55. $(r - 1)(2r + 3)(r + 6) < 0$

$(-\infty, -6) \cup \left(-\frac{3}{2}, 1\right)$

56. $(3k + 11)^2 = 7$

$\left\{\frac{-11 + \sqrt{7}}{3}, \frac{-11 - \sqrt{7}}{3}\right\}$

57. $S = \frac{Id^2}{k}$ for d

$d = \frac{\pm\sqrt{SkI}}{I}$

58. $2x - \sqrt{x} = 6$

$\{4\}$

59. $6 + \frac{15}{s^2} = -\frac{19}{s}$

$\left\{-\frac{5}{3}, -\frac{3}{2}\right\}$

60. $\frac{-2}{x + 5} \leq -5$

$\left(-5, -\frac{23}{5}\right]$

*This exercise requires knowledge of complex numbers.

Chapter 10 Test

Use the Chapter Test Prep Video CD to see fully worked-out solutions to any of the exercises you want to review.

Solve by using either the square root property or completing the square.

1. $t^2 = 54$

1. $\{3\sqrt{6}, -3\sqrt{6}\}$

2. $(7x + 3)^2 = 25$

2. $\left\{-\frac{8}{7}, \frac{2}{7}\right\}$

3. $x^2 + 2x = 1$

3. $\{-1 + \sqrt{2}, -1 - \sqrt{2}\}$

Solve using the quadratic formula.

4. $2x^2 - 3x - 1 = 0$

4. $\left\{\frac{3 + \sqrt{17}}{4}, \frac{3 - \sqrt{17}}{4}\right\}$

***5.** $3t^2 - 4t = -5$

5. $\left\{\frac{2}{3} + \frac{\sqrt{11}}{3}i, \frac{2}{3} - \frac{\sqrt{11}}{3}i\right\}$

6. $3x = \sqrt{\frac{9x + 2}{2}}$

6. $\left\{\frac{2}{3}\right\}$

***7.** If k is a negative number, then which one of the following equations will have two nonreal complex solutions?

A. $x^2 = 4k$ **B.** $x^2 = -4k$ **C.** $(x + 2)^2 = -k$

7. A

8. What is the discriminant for $2x^2 - 8x - 3 = 0$? How many and what type of solutions does this equation have? (Do not actually solve.)

8. discriminant: 88; two irrational solutions

Solve by any method.

9. $3 - \frac{16}{x} - \frac{12}{x^2} = 0$

9. $\left\{-\frac{2}{3}, 6\right\}$

10. $4x^2 + 7x - 3 = 0$

10. $\left\{\frac{-7 + \sqrt{97}}{8}, \frac{-7 - \sqrt{97}}{8}\right\}$

---

*This exercise requires knowledge of complex numbers.

11. $\left\{-2, -\frac{1}{3}, \frac{1}{3}, 2\right\}$

11. $9x^4 + 4 = 37x^2$

12. $\left\{-\frac{5}{2}, 1\right\}$

12. $12 = (2n + 1)^2 + (2n + 1)$

13. $r = \frac{\pm\sqrt{\pi S}}{2\pi}$

13. Solve for r: $S = 4\pi r^2$. (Leave $\pm$ in your answer.)

Solve each problem.

14. Andrew: 11.1 hr; Kent: 9.1 hr

14. Andrew and Kent do desktop publishing. Kent can prepare a certain prospectus 2 hr faster than Andrew. If they work together, they can do the entire prospectus in 5 hr. How long will it take each of them working alone to prepare the prospectus? Round your answers to the nearest tenth of an hour.

15. 7 mph

15. Bryn Ruhberg paddled her canoe 10 mi upstream, and then paddled back to her starting point. If the rate of the current was 3 mph and the entire trip took $3\frac{1}{2}$ hr, what was Bryn's rate?

16. 2 ft

16. Tyler McGinnis has a pool 24 ft long and 10 ft wide. He wants to construct a concrete walk around the pool. If he plans for the walk to be of uniform width and cover 152 ft^2, what will the width of the walk be?

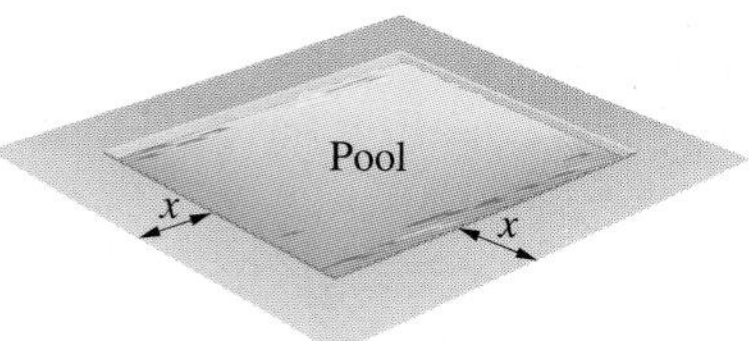

17. 16 m

17. At a point 30 m from the base of a tower, the distance to the top of the tower is 2 m more than twice the height of the tower. Find the height of the tower.

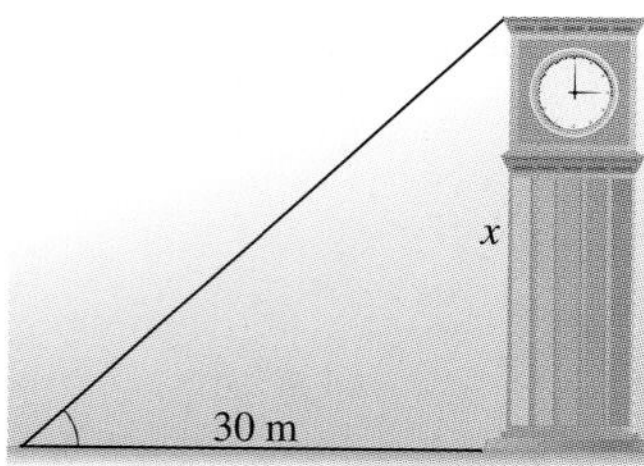

18. Professor Bernstein has found that the number of students attending her intermediate algebra class is approximated by

$$S(x) = -x^2 + 20x + 80,$$

where x is the number of hours that the Campus Center is open daily.

(a) Find the number of hours that the center should be open so that the number of students attending class is a maximum.

(b) What is this maximum number of students?

18. (a) 10 hr

(b) 180 students

19. Which one of the following most closely resembles the graph of $f(x) = a(x - h)^2 + k$ if $a < 0$, $h > 0$, and $k < 0$?

A.

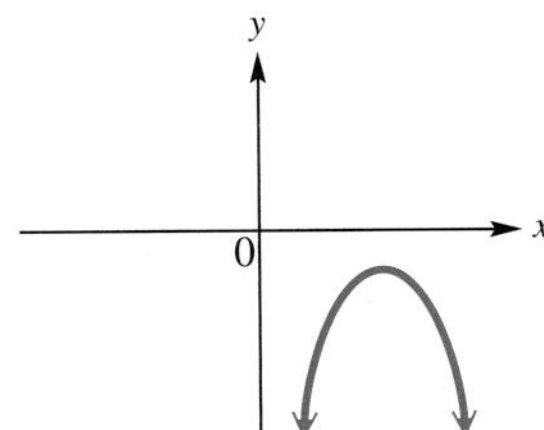

B.

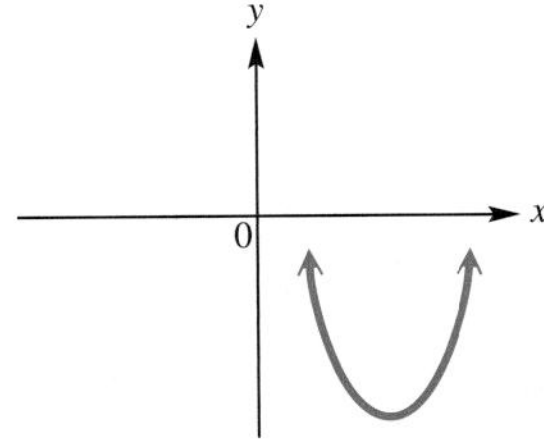

C.

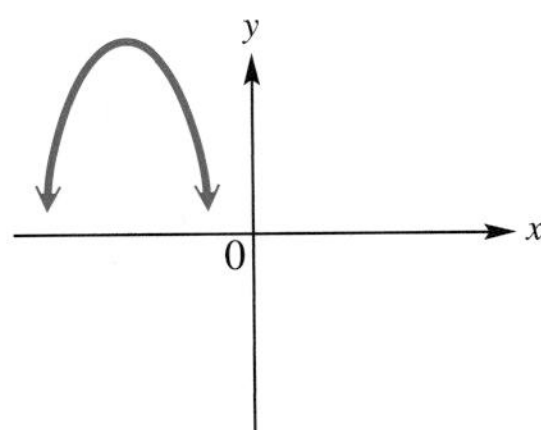

D.

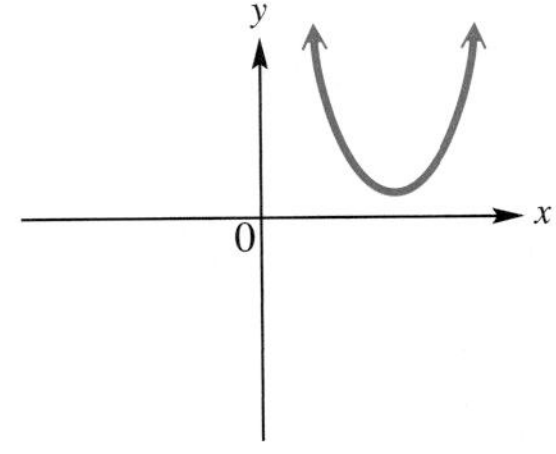

19. A

Graph each parabola. Give the vertex, axis, domain, and range.

20. $f(x) = \frac{1}{2}x^2 - 2$

20. vertex: $(0, -2)$; axis: $x = 0$; domain: $(-\infty, \infty)$; range: $[-2, \infty)$

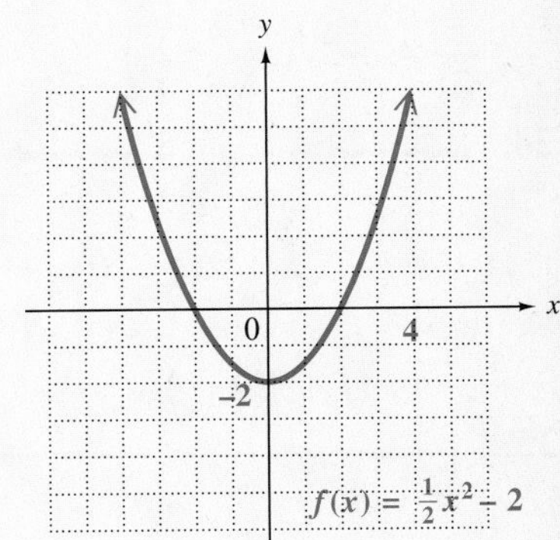

21. **vertex: (2, 3); axis: x = 2; domain: $(-\infty, \infty)$; range: $(-\infty, 3]$**

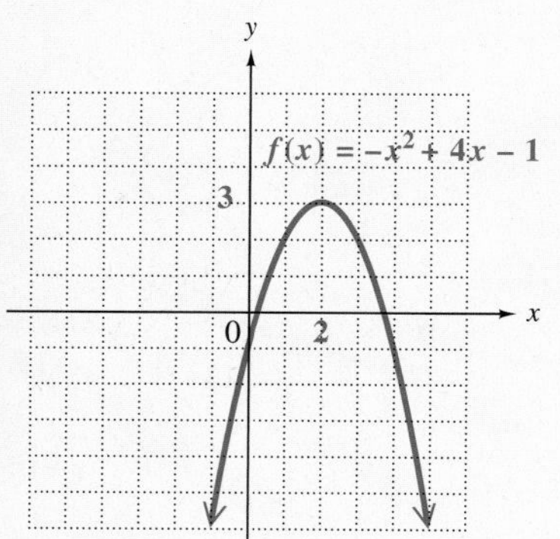

21. 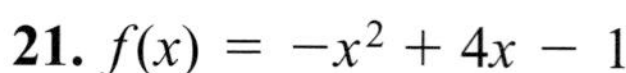
$f(x) = -x^2 + 4x - 1$

22. **vertex: (−5, −2); axis: y = −2; domain: $[-5, \infty)$; range: $(-\infty, \infty)$**

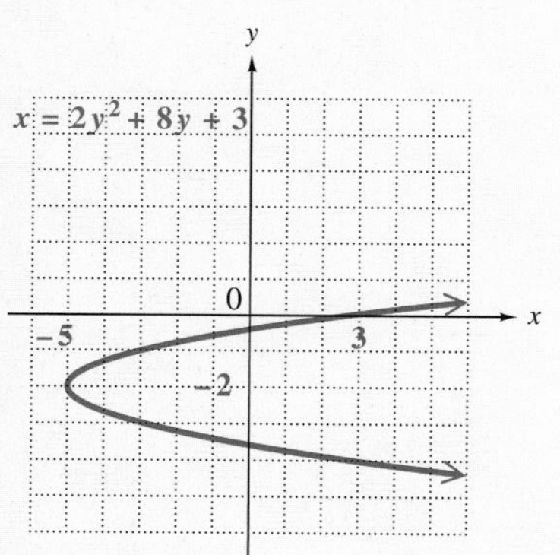

22. 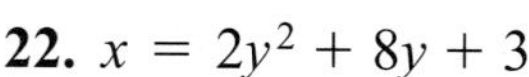
$x = 2y^2 + 8y + 3$

23. **140 ft by 70 ft; 9800 ft²**

23. The manager of Morgan's Department Store wants to construct a rectangular parking lot on land bordered on one side by a highway. The store has 280 ft of fencing that is to be used to fence off the other three sides. What should be the dimensions of the lot if the enclosed area is to be a maximum? What is the maximum area?

Solve. Graph each solution set.

24. **$(-\infty, -5) \cup \left(\frac{3}{2}, \infty\right)$**

−5 0 $\frac{3}{2}$

24. $2x^2 + 7x > 15$

25. **$(-\infty, 4) \cup [9, \infty)$**

0 4 9

25. $\dfrac{5}{t - 4} \leq 1$

Cumulative Review Exercises — Chapters R–10

Solve each equation or inequality.

1. $-2x + 4 = 5(x - 4) + 17$

{1}

2. $-2x + 4 \leq -x + 3$

[1, ∞)

3. $|3x - 7| \leq 1$

$\left[2, \frac{8}{3}\right]$

4. Find the slope and y-intercept of the line with equation $2x - 4y = 7$.

slope: $\frac{1}{2}$**; y-intercept:** $\left(0, -\frac{7}{4}\right)$

5. Write the equation in standard form of the line through $(2, -1)$ and perpendicular to $-3x + y = 5$.

$x + 3y = -1$

Graph each relation. Tell whether or not each is a function, and if it is, give its domain and range.

6. $4x - 5y = 15$

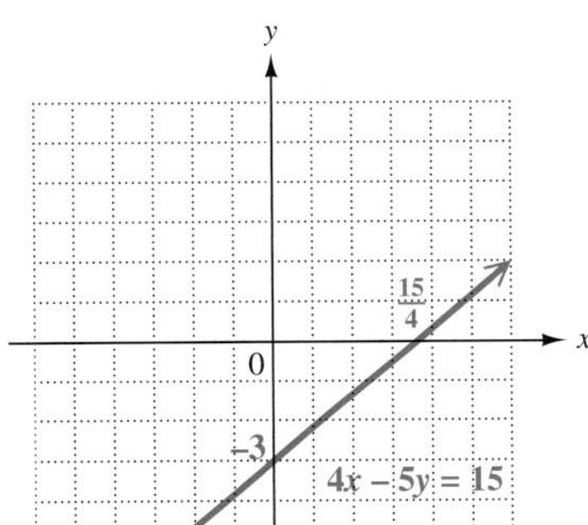

function;
domain: (−∞, ∞);
range: (−∞, ∞)

7. $4x - 5y < 15$

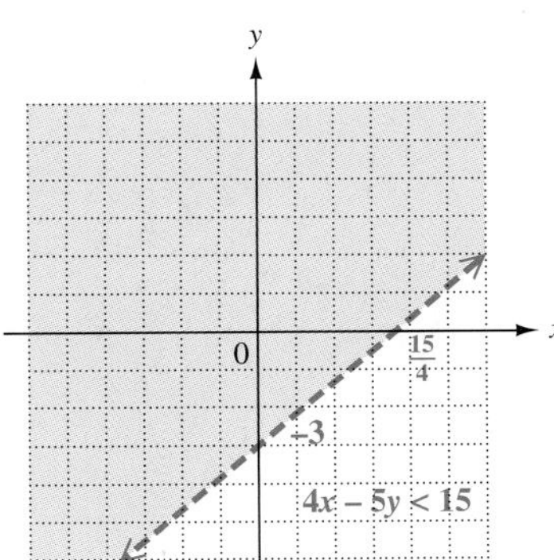

not a function

8. $y = -2(x - 1)^2 + 3$

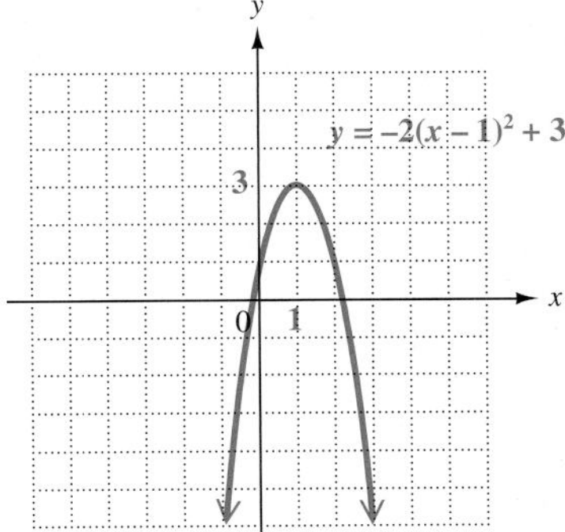

function;
domain: (−∞, ∞);
range: (−∞, 3]

Solve each system of equations.

9. $2x - 4y = 10$
$9x + 3y = 3$

{(1, −2)}

10. $x + y + 2z = 3$
$-x + y + z = -5$
$2x + 3y - z = -8$

{(3, −4, 2)}

Write with positive exponents only. Assume that variables represent positive real numbers.

11. $\left(\frac{x^{-3}y^2}{x^5y^{-2}}\right)^{-1}$

$\frac{x^8}{y^4}$

12. $\frac{(4x^{-2})^2(2y^3)}{8x^{-3}y^5}$

$\frac{4}{xy^2}$

13. Multiply: $\left(\frac{2}{3}t + 9\right)^2$.

$\frac{4}{9}t^2 + 12t + 81$

14. Divide $4x^3 + 2x^2 - x + 26$ by $x + 2$.

$4x^2 - 6x + 11 + \frac{4}{x + 2}$

Factor completely.

15. $16x - x^3$

$x(4 + x)(4 - x)$

16. $24m^2 + 2m - 15$

$(4m - 3)(6m + 5)$

17. $9x^2 - 30xy + 25y^2$

$(3x - 5y)^2$

Perform the operations, and express answers in lowest terms. Assume that denominators represent nonzero real numbers.

18. $\dfrac{5t + 2}{-6} \div \dfrac{15t + 6}{5}$

$-\dfrac{5}{18}$

19. $\dfrac{3}{2 - k} - \dfrac{5}{k} + \dfrac{6}{k^2 - 2k}$

$-\dfrac{8}{k}$

20. $\dfrac{\dfrac{r}{s} - \dfrac{s}{r}}{\dfrac{r}{s} + 1}$

$\dfrac{r - s}{r}$

Simplify each radical expression.

21. $\sqrt[3]{\dfrac{27}{16}}$

$\dfrac{3\sqrt[3]{4}}{4}$

22. $\dfrac{2}{\sqrt{7} - \sqrt{5}}$

$\sqrt{7} + \sqrt{5}$

Solve each equation.

23. $2x = \sqrt{\dfrac{5x + 2}{3}}$

$\left\{\dfrac{2}{3}\right\}$

24. $2x^2 - 4x - 3 = 0$

$\left\{\dfrac{2 + \sqrt{10}}{2}, \dfrac{2 - \sqrt{10}}{2}\right\}$

25. $z^2 - 2z = 15$

$\{-3, 5\}$

26. $\dfrac{3}{x - 3} - \dfrac{2}{x - 2} = \dfrac{3}{x^2 - 5x + 6}$

$\emptyset$

27. $p^4 - 10p^2 + 9 = 0$

$\{-3, -1, 1, 3\}$

28. Two cars left an intersection at the same time, one heading due south and the other due east. Later they were exactly 95 mi apart. The car heading east had gone 38 mi less than twice as far as the car heading south. How far had each car traveled?

southbound car: 57 mi; eastbound car: 76 mi

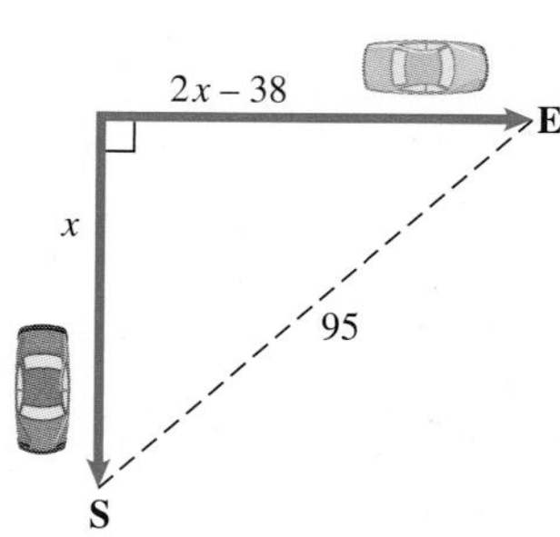

11

Inverse, Exponential, and Logarithmic Functions

In 2001, Apple Computer Inc., introduced the iPod. Since then, the company has sold over 40 million of the popular music players, in spite of warnings by experts that listening to the devices at high volumes may put people at increased risk of hearing loss. In 2006, a federal class-action lawsuit was filed against the company, accusing it of not taking adequate steps to protect the hearing of iPod users. As a result, Apple issued a software update that allows listeners to set maximum volume limits on some of the newer iPod models. (*Source: Sacramento Bee, USA Today.*)

In Example 4 of Section 11.5, we use a *logarithmic function* to calculate the volume level, in *decibels,* of an iPod.

11.1 Inverse Functions

OBJECTIVES

1. **Decide whether a function is one-to-one and, if it is, find its inverse.**
2. **Use the horizontal line test to determine whether a function is one-to-one.**
3. **Find the equation of the inverse of a function.**
4. **Graph f^{-1} from the graph of f.**

In this chapter we will study two important types of functions, *exponential* and *logarithmic*. These functions are related in a special way: They are *inverses* of one another. We begin by discussing inverse functions in general.

Calculator Tip A calculator with the following keys will be essential in this chapter.

We will explain how these keys are used at appropriate places in the chapter.

OBJECTIVE 1 Decide whether a function is one-to-one and, if it is, find its inverse. Suppose we define the function

$$G = \{(-2, 2), (-1, 1), (0, 0), (1, 3), (2, 5)\}.$$

We can form another set of ordered pairs from G by interchanging the x- and y-values of each pair in G. Call this set F, with

$$F = \{(2, -2), (1, -1), (0, 0), (3, 1), (5, 2)\}.$$

To show that these two sets are related, F is called the *inverse* of G. For a function f to have an inverse function, f must be *one-to-one.*

One-to-One Function

In a **one-to-one function,** each x-value corresponds to just one y-value, and each y-value corresponds to just one x-value.

The function shown in Figure 1(a) is not one-to-one because the y-value 7 corresponds to *two* x-values, 2 and 3. That is, the ordered pairs (2, 7) and (3, 7) both appear in the function. The function in Figure 1(b) is one-to-one.

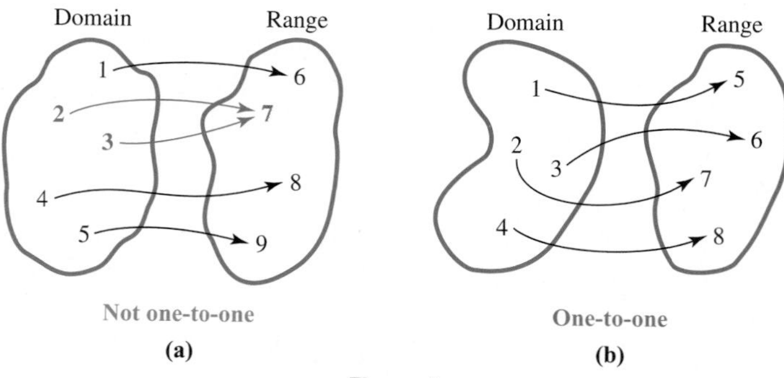

Figure 1

The *inverse* of any one-to-one function f is found by interchanging the components of the ordered pairs of f. The inverse of f is written $\boldsymbol{f^{-1}}$. Read f^{-1} as **"the inverse of f"** or **"f-inverse."**

CAUTION

The symbol $f^{-1}(x)$ does not represent $\dfrac{1}{f(x)}$.

The definition of the inverse of a function follows.

Inverse of a Function

The **inverse** of a one-to-one function f, written f^{-1}, is the set of all ordered pairs of the form (y, x), where (x, y) belongs to f. ***Since the inverse is formed by interchanging x and y, the domain of f becomes the range of f^{-1} and the range of f becomes the domain of f^{-1}.***

For inverses f and f^{-1}, it follows that

$$f(f^{-1}(x)) = x \quad \text{and} \quad f^{-1}(f(x)) = x.$$

EXAMPLE 1 Finding the Inverses of One-to-One Functions

Find the inverse of each function that is one-to-one.

(a) $F = \{(-2, 1), (-1, 0), (0, 1), (1, 2), (2, 2)\}$

Each x-value in F corresponds to just one y-value. However, the y-value 2 corresponds to two x-values, 1 and 2. Also, the y-value 1 corresponds to both -2 and 0. Because some y-values correspond to more than one x-value, F is not one-to-one and does not have an inverse function.

(b) $G = \{(3, 1), (0, 2), (2, 3), (4, 0)\}$

Every x-value in G corresponds to only one y-value, and every y-value corresponds to only one x-value, so G is a one-to-one function. The inverse function is found by interchanging the x- and y-values in each ordered pair.

$$G^{-1} = \{(1, 3), (2, 0), (3, 2), (0, 4)\}$$

Notice how the domain and range of G become the range and domain, respectively, of G^{-1}.

(c) Ozone is the major component in smog. Breathing ozone may cause serious risk of respiratory problems. In 1997, the U.S. Environmental Protection Agency set an 8-hour average ground-level ozone standard of 0.08 parts per million (ppm). The table shows the number of days in which the air in Connecticut exceeded this standard for the years 1995–2006.

Year	Number of Days Exceeding Standard	Year	Number of Days Exceeding Standard
1995	24	2001	26
1996	16	2002	36
1997	27	2003	14
1998	25	2004	6
1999	33	2005	20
2000	13	2006	13

Source: U.S. Environmental Protection Agency.

Let f be the function defined in the table, with the years forming the domain and the numbers of days exceeding the ozone standard forming the range. Then f is not one-to-one, because in two different years (2000 and 2006), the number of days with unacceptable ozone levels was the same, 13.

Work Problem 1 *at the Side.* ▶

1 Find the inverse of each function that is one-to-one.

(a) $\{(1, 2), (2, 4), (3, 3), (4, 5)\}$

(b) $\{(0, 3), (-1, 2), (1, 3)\}$

(c) A Norwegian physiologist has developed a rule for predicting running times based on the time to run 5 km (5K). An example for one runner is shown here. (*Source:* Stephen Seiler, Agder College, Kristiansand, Norway.)

Distance	Time
1.5K	4:22
3K	9:18
5K	16:00
10K	33:40

ANSWERS

1. (a) $\{(2, 1), (4, 2), (3, 3), (5, 4)\}$
   (b) not a one-to-one function
   (c)

Time	Distance
4:22	1.5K
9:18	3K
16:00	5K
33:40	10K

OBJECTIVE 2 Use the horizontal line test to determine whether a function is one-to-one. It may be difficult to decide whether a function is one-to-one just by looking at the equation that defines the function. However, by graphing the function and observing the graph, we can use the *horizontal line test* to tell whether the function is one-to-one.

Horizontal Line Test

A function is one-to-one if every horizontal line intersects the graph of the function at most once.

The horizontal line test follows from the definition of a one-to-one function. Any two points that lie on the same horizontal line have the same y-coordinate. No two ordered pairs that belong to a one-to-one function may have the same y-coordinate, and, therefore, no horizontal line will intersect the graph of a one-to-one function more than once.

EXAMPLE 2 Using the Horizontal Line Test

Use the horizontal line test to determine whether the graphs in Figures 2 and 3 are graphs of one-to-one functions.

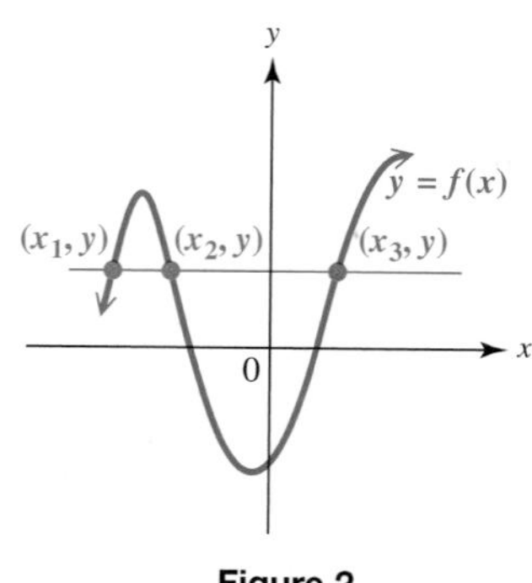

Figure 2

Because the red horizontal line shown in Figure 2 intersects the graph in more than one point (actually three points), the function is not one-to-one.

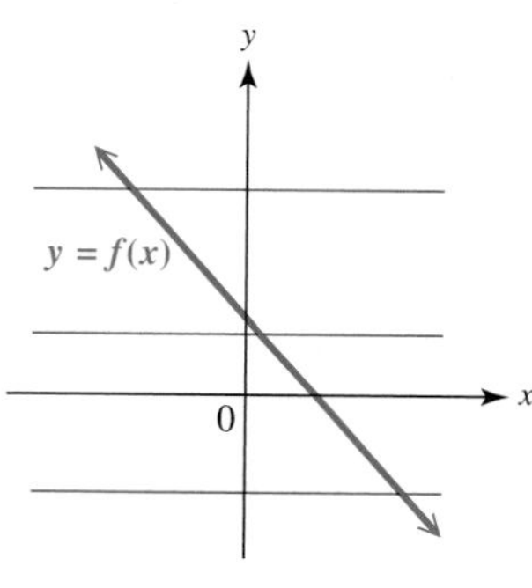

Figure 3

Every horizontal line will intersect the graph in Figure 3 in exactly one point. This function is one-to-one.

◀ Work Problem 2 at the Side.

OBJECTIVE 3 Find the equation of the inverse of a function. By definition, the inverse of a one-to-one function is found by interchanging the x- and y-values of each of its ordered pairs. The equation of the inverse of a function defined by $y = f(x)$ is found in the same way.

Finding the Equation of the Inverse of $y = f(x)$

For a one-to-one function f defined by an equation $y = f(x)$, find the defining equation of the inverse as follows.

Step 1 Interchange x and y.

Step 2 Solve for y.

Step 3 Replace y with $f^{-1}(x)$.

---

2 Use the horizontal line test to determine whether each graph is the graph of a one-to-one function.

(a)

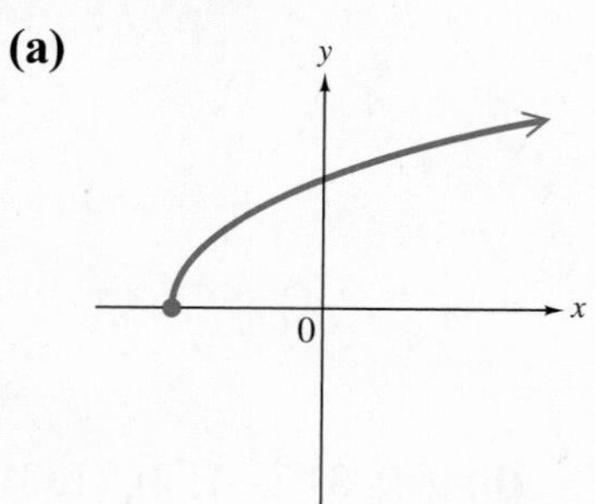

(b)

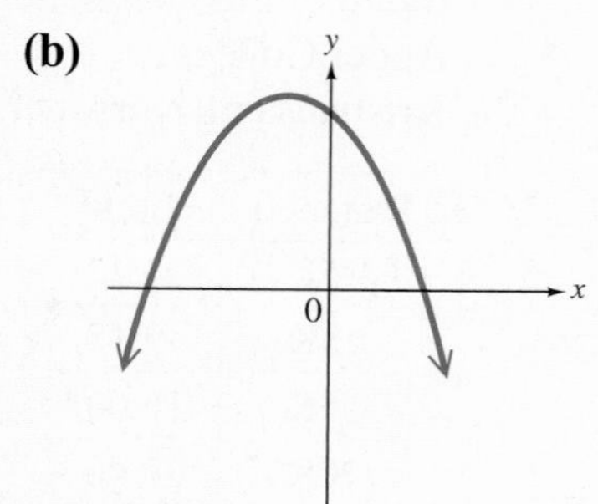

ANSWERS

2. (a) one-to-one **(b)** not one-to-one

EXAMPLE 3 Finding Equations of Inverses

Decide whether each equation defines a one-to-one function. If so, find the equation that defines the inverse.

(a) $f(x) = 2x + 5$

The graph of $y = 2x + 5$ is a nonvertical line, so by the horizontal line test, f is a one-to-one function. To find the inverse, let $y = f(x)$ so that

$$y = 2x + 5$$

$$x = 2y + 5 \quad \text{Interchange } x \text{ and } y. \text{ (Step 1)}$$

$$2y = x - 5 \quad \text{Solve for } y. \text{ (Step 2)}$$

$$y = \frac{x - 5}{2}$$

$$f^{-1}(x) = \frac{x - 5}{2}, \quad \text{Replace } y \text{ with } f^{-1}(x). \text{ (Step 3)}$$

which can also be written

$$f^{-1}(x) = \frac{x}{2} - \frac{5}{2}, \quad \text{or} \quad f^{-1}(x) = \frac{1}{2}x - \frac{5}{2}. \quad \frac{a-b}{c} = \frac{a}{c} - \frac{b}{c}$$

Thus, f^{-1} is a linear function. In the function with $y = 2x + 5$, the value of y is found by starting with a value of x, multiplying by 2, and adding 5. The equation $f^{-1}(x) = \frac{x-5}{2}$ for the inverse has us *subtract* 5, and then *divide* by 2. An inverse is used to "undo" what a function does to the variable x.

(b) $y = x^2 + 2$

This equation has a vertical parabola as its graph, so some horizontal lines will intersect the graph at two points. For example, both $x = 3$ and $x = -3$ correspond to $y = 11$. Because of the x^2-term, there are many pairs of x-values that correspond to the same y-value. This means that the function defined by $y = x^2 + 2$ is not one-to-one and does not have an inverse function.

Following the steps for finding the equation of an inverse leads to

$$y = x^2 + 2$$

$$x = y^2 + 2 \quad \text{Interchange } x \text{ and } y.$$

$$y^2 = x - 2 \quad \text{Solve for } y.$$

$$y = \pm\sqrt{x - 2}. \quad \text{Square root property}$$

The last step shows that there are two y-values for each choice of $x > 2$, so the given function is not one-to-one and cannot have an inverse.

(c) $f(x) = (x - 2)^3$

If we substitute any real number for the range value $f(x)$, take the cube root on both sides, and then add 2, there is one and only one value of x that yields the desired result. So, this is a one-to-one function. (This reasoning holds in general for equations of this type when the exponent is *odd*.)

$$y = (x - 2)^3 \quad \text{Replace } f(x) \text{ with } y.$$

$$x = (y - 2)^3 \quad \text{Interchange } x \text{ and } y.$$

$$\sqrt[3]{x} = \sqrt[3]{(y - 2)^3} \quad \text{Take the cube root on each side.}$$

$$\sqrt[3]{x} = y - 2$$

$$y = \sqrt[3]{x} + 2 \quad \text{Solve for } y.$$

$$f^{-1}(x) = \sqrt[3]{x} + 2 \quad \text{Replace } y \text{ with } f^{-1}(x).$$

Work Problem **3** *at the Side.* ▶

3 Decide whether each equation defines a one-to-one function. If so, find the equation that defines the inverse.

(a) $f(x) = 3x - 4$

(b) $f(x) = x^3 + 1$

(c) $f(x) = (x - 3)^2$

ANSWERS

3. **(a)** one-to-one function; $f^{-1}(x) = \frac{x + 4}{3}$, or $f^{-1}(x) = \frac{1}{3}x + \frac{4}{3}$
(b) one-to-one function; $f^{-1}(x) = \sqrt[3]{x - 1}$
(c) not a one-to-one function

4 Use the given graphs to graph each inverse.

(a)

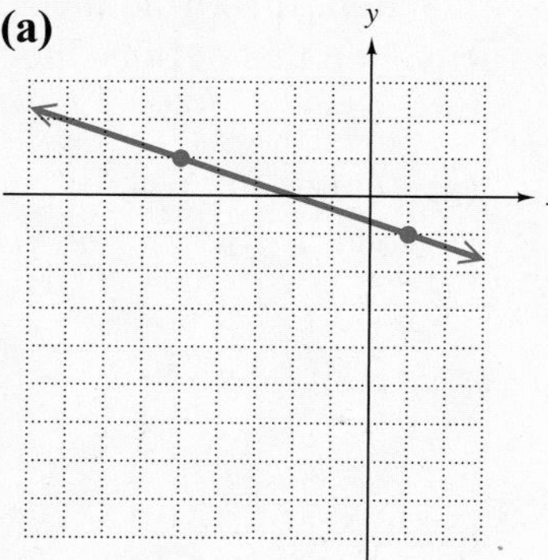

(b)

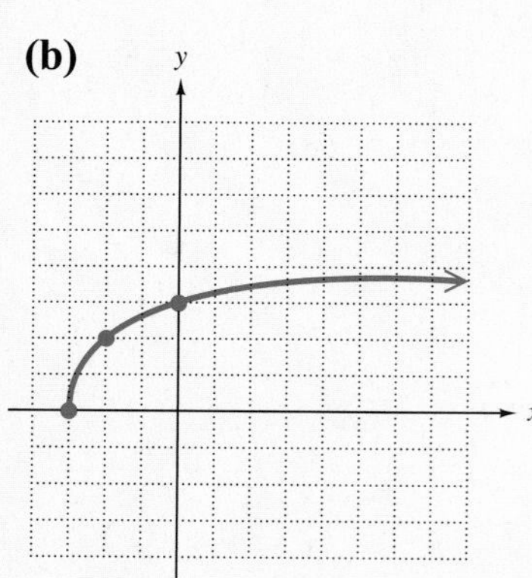

(c)

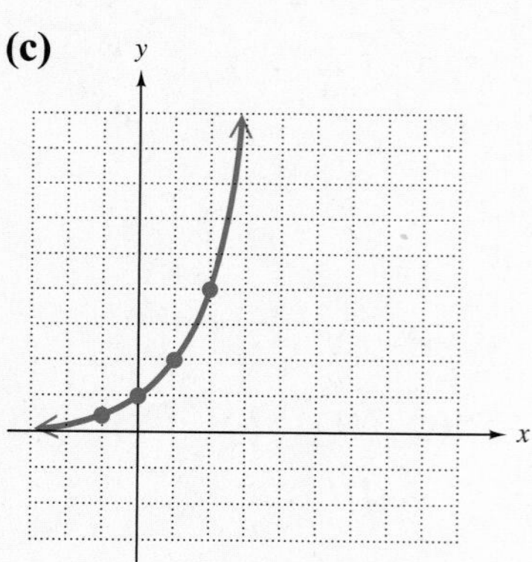

ANSWERS

4. (a) (b) (c)

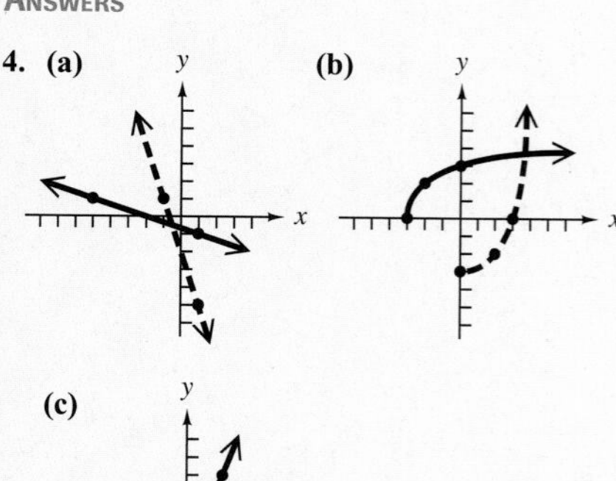

OBJECTIVE 4 Graph f^{-1} from the graph of f. One way to graph the inverse of a function f whose equation is known is to find some ordered pairs that belong to f, interchange x and y to get ordered pairs that belong to f^{-1}, plot those points, and sketch the graph of f^{-1} through the points. A simpler way is to select points on the graph of f and use symmetry to find corresponding points on the graph of f^{-1}.

For example, suppose the point (a, b) shown in Figure 4 belongs to a one-to-one function f. Then the point (b, a) belongs to f^{-1}. The line segment connecting (a, b) and (b, a) is perpendicular to, and cut in half by, the line $y = x$. The points (a, b) and (b, a) are "mirror images" of each other with respect to $y = x$. ***Thus, we can find the graph of f^{-1} from the graph of f by locating the mirror image of each point in f with respect to the line $y = x$.***

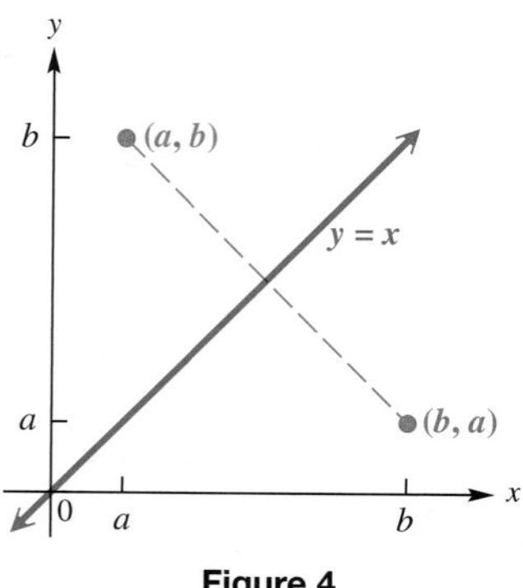

Figure 4

EXAMPLE 4 Graphing the Inverse

Graph the inverses of the functions shown in Figure 5.

In Figure 5 the graphs of two functions are shown in blue. Their inverses are shown in red. In each case, the graph of f^{-1} is symmetric to the graph of f with respect to the line $y = x$.

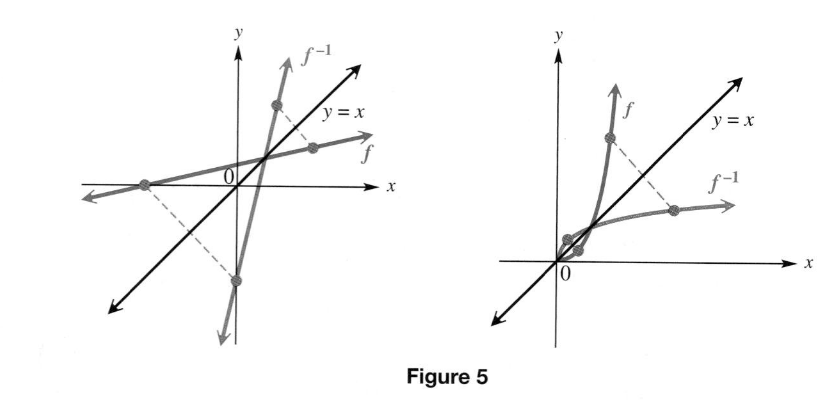

Figure 5

◀ Work Problem 4 at the Side.

11.1 Exercises

FOR EXTRA HELP

MyMathLab | Math XL PRACTICE | WATCH | DOWNLOAD | READ | REVIEW

Answer each question. See Example 1.

1. A new study found that the trans fat content in fast-food products varied widely around the world, based on the type of frying oil used, as shown in the table.

If the set of countries is the domain and the set of trans fat percentages is the range of a function, is it one-to-one? Why or why not?

Country	Percentage of Trans Fat in McDonald's Chicken
Scotland	14
France	11
United States	11
Peru	9
Russia	5
Denmark	1

Source: New England Journal of Medicine.

It is not one-to-one. France and the United States are paired with the same trans fat percentage, 11.

2. The table shows concentrations of a major air pollutant, carbon monoxide, in the United States for the years 2000–2005.

If this correspondence is considered to be a function that pairs each year with its concentration, is it one-to-one? If not, explain why.

Year	Concentration (in parts per million)
2000	3.5
2001	3.2
2002	3.0
2003	2.8
2004	2.6
2005	2.4

Source: E.P.A.

It is one-to-one.

3. Suppose you consider the set of ordered pairs (x, y) such that x represents a person in your mathematics class and y represents that person's mother. How might this function not be a one-to-one function?

Two or more siblings might be in the class. They would be paired with the same mother.

4. The road mileage between Denver, Colorado, and several selected U.S. cities is shown in the table.

If we consider this as a function that pairs each city with a distance, is it one-to-one? How could we change the answer to this question by adding 1 mile to one of the distances shown?

City	Distance to Denver (in miles)
Atlanta	1398
Dallas	781
Indianapolis	1058
Kansas City, MO	600
Los Angeles	1059

Answer to Exercise 4

Yes. By adding 1 to 1058 two distances would be the same, so the function would not be one-to-one.

Choose the correct response from the given list.

5. If a function is made up of ordered pairs in such a way that the same y-value appears in a correspondence with two different x-values, then

A. the function is one-to-one

B. the function is not one-to-one

C. its graph does not pass the vertical line test

D. it has an inverse function associated with it.

B

6. Which equation defines a one-to-one function? Explain why the others are not, using specific examples.

A. $f(x) = x$ **B.** $f(x) = x^2$

C. $f(x) = |x|$ **D.** $f(x) = -x^2 + 2x - 1$

Choice A is one-to-one. In choice B, $f(1) = f(-1) = 1$, and so $f(x) = x^2$ is not one-to-one. In choice C, $f(2) = f(-2) = 2$, and so $f(x) = |x|$ is not one-to-one. Finally, in choice D, $f(-2) = f(4) = -9$, and so $f(x) = -x^2 + 2x - 1$ is also not one-to-one.

7. Only one of the graphs illustrates a one-to-one function. Which one is it? A

A.

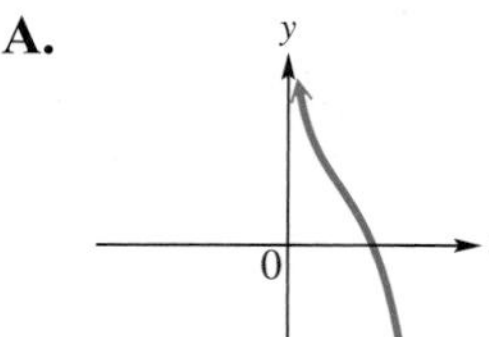

B.

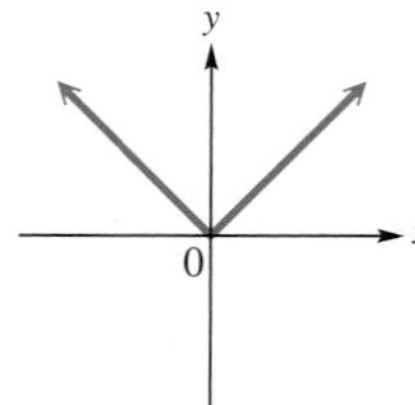

C.

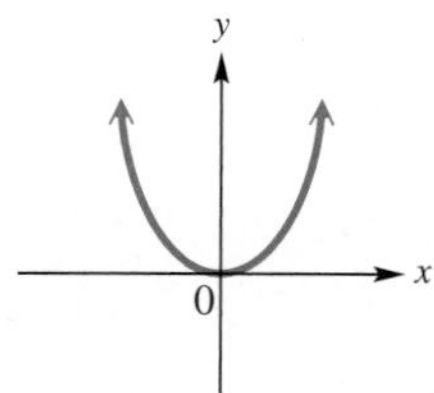

D.

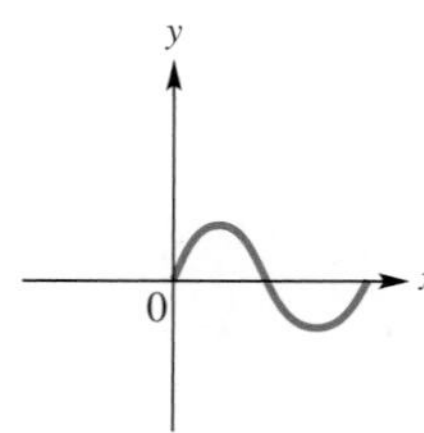

8. If a function f is one-to-one and the point (p, q) lies on the graph of f, then which point *must* lie on the graph of f^{-1}?

A. $(-p, q)$ **B.** $(-q, -p)$

C. $(p, -q)$ **D.** (q, p)

D

If the function is one-to-one, find its inverse. See Examples 1–3.

9. $\{(3, 6), (2, 10), (5, 12)\}$

$\{(6, 3), (10, 2), (12, 5)\}$

10. $\left\{(-1, 3), (0, 5), (5, 0), \left(7, -\frac{1}{2}\right)\right\}$

$\left\{(3, -1), (5, 0), (0, 5), \left(-\frac{1}{2}, 7\right)\right\}$

11. $\{(-1, 3), (2, 7), (4, 3), (5, 8)\}$

not one-to-one

12. $\{(-8, 6), (-4, 3), (0, 6), (5, 10)\}$

not one-to-one

13. $f(x) = 2x + 4$

$f^{-1}(x) = \frac{x - 4}{2}$, or $f^{-1}(x) = \frac{1}{2}x - 2$

14. $f(x) = 3x + 1$

$f^{-1}(x) = \frac{x - 1}{3}$, or $f^{-1}(x) = \frac{1}{3}x - \frac{1}{3}$

15. $g(x) = \sqrt{x - 3}$, $x \geq 3$

$g^{-1}(x) = x^2 + 3$, $x \geq 0$

16. $g(x) = \sqrt{x + 2}$, $x \geq -2$

$g^{-1}(x) = x^2 - 2$, $x \geq 0$

17. $f(x) = 3x^2 + 2$

not one-to-one

18. $f(x) = -4x^2 - 1$

not one-to-one

19. $f(x) = x^3 - 4$

$f^{-1}(x) = \sqrt[3]{x + 4}$

20. $f(x) = x^3 - 3$

$f^{-1}(x) = \sqrt[3]{x + 3}$

Let $f(x) = 2^x$. We will see in the next section that the function f is one-to-one. Find each value, always working part (a) before part (b).

21. (a) $f(3)$ 8

(b) $f^{-1}(8)$ 3

22. (a) $f(4)$ 16

(b) $f^{-1}(16)$ 4

23. (a) $f(0)$ 1

(b) $f^{-1}(1)$ 0

24. (a) $f(-2)$ $\frac{1}{4}$

(b) $f^{-1}\left(\frac{1}{4}\right)$ −2

The graphs of some functions are given in Exercises 25–30. ***(a)*** *Use the horizontal line test to determine whether each function is one-to-one.* ***(b)*** *If the function is one-to-one, graph the inverse of the function with a dashed line (or curve) on the same set of axes. (Remember that if f is one-to-one and $f(a) = b$, then $f^{-1}(b) = a$.) See Example 4.*

25.

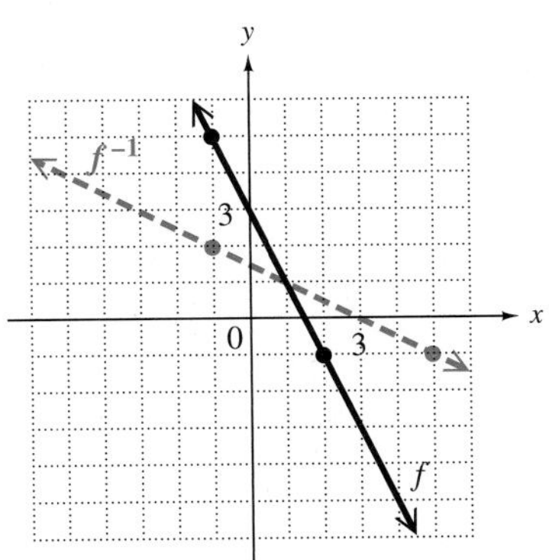

(a) one-to-one

26.

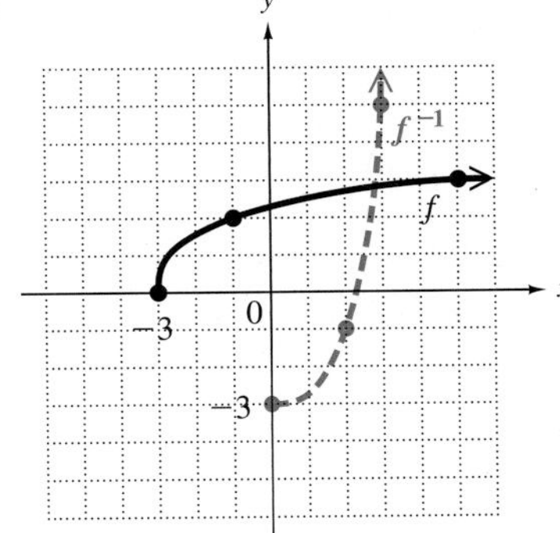

(a) one-to-one

27.

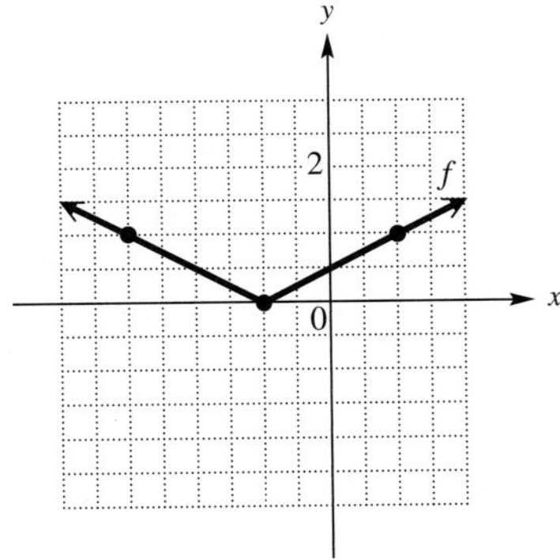

(a) not one-to-one

28.

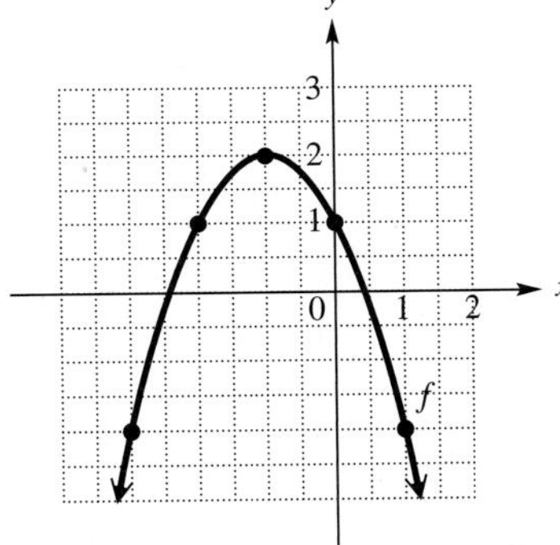

(a) not one-to-one

29.

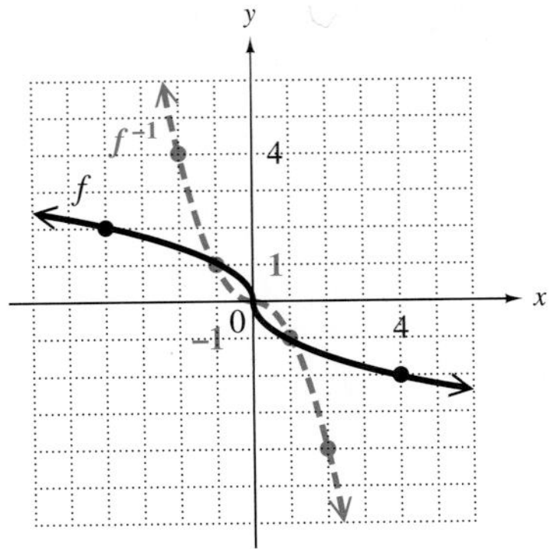

(a) one-to-one

30.

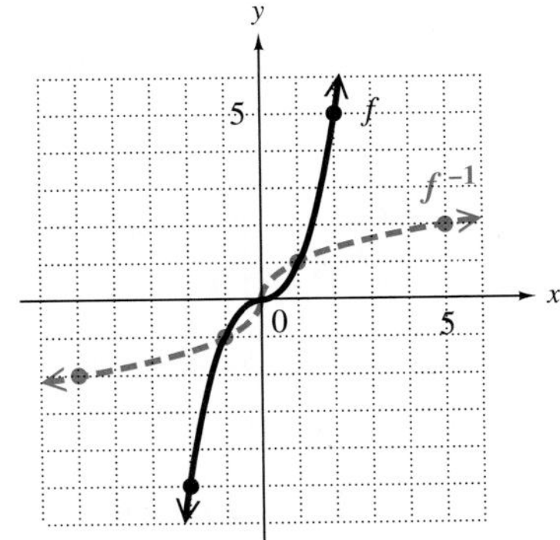

(a) one-to-one

Each function defined in Exercises 31–38 is a one-to-one function. Graph the function as a solid line (or curve), and then graph its inverse on the same set of axes as a dashed line (or curve). In Exercises 35–38 you are given a table to complete so that graphing the function will be easier. See Example 4.

31. $f(x) = 2x - 1$

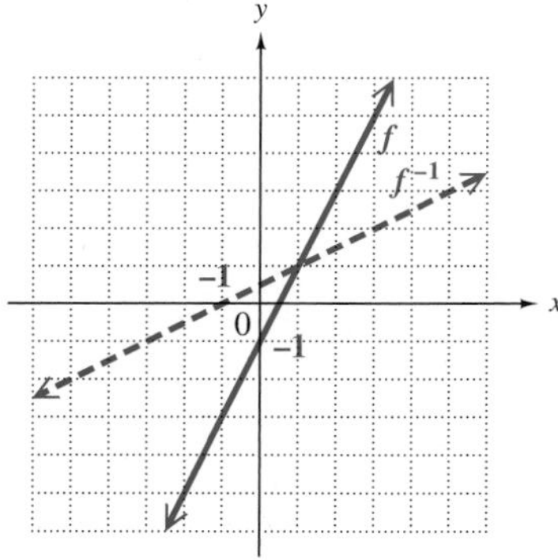

32. $f(x) = 2x + 3$

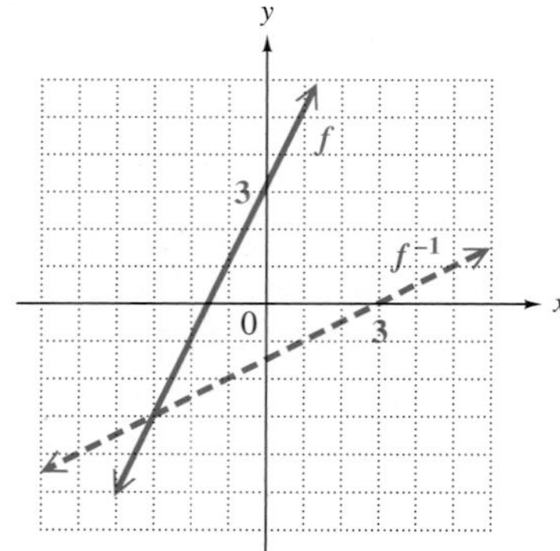

33. $g(x) = -4x$

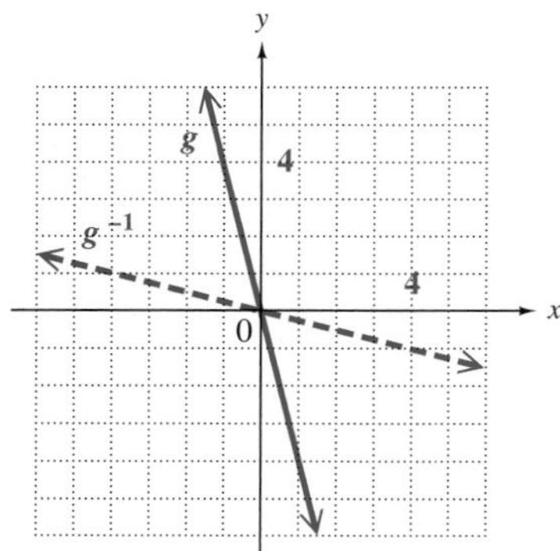

34. $g(x) = -2x$

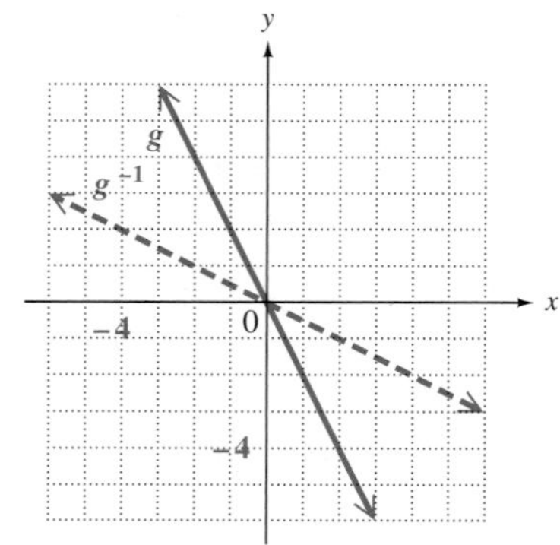

35. $f(x) = y = \sqrt{x}, x \geq 0$

x	f(x) = y
0	**0**
1	**1**
4	**2**

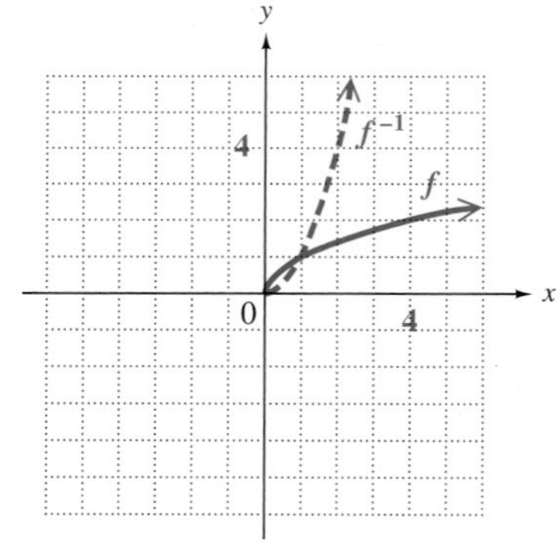

36. $f(x) = y = -\sqrt{x}, x \geq 0$

x	f(x) = y
0	**0**
1	**−1**
4	**−2**

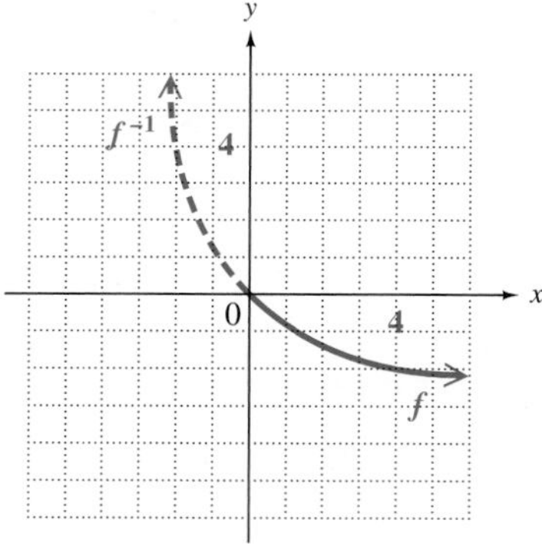

37. $f(x) = y = x^3 - 2$

x	f(x) = y
−1	**−3**
0	**−2**
1	**−1**
2	**6**

38. $f(x) = y = x^3 + 3$

x	f(x) = y
−2	**−5**
−1	**2**
0	**3**
1	**4**

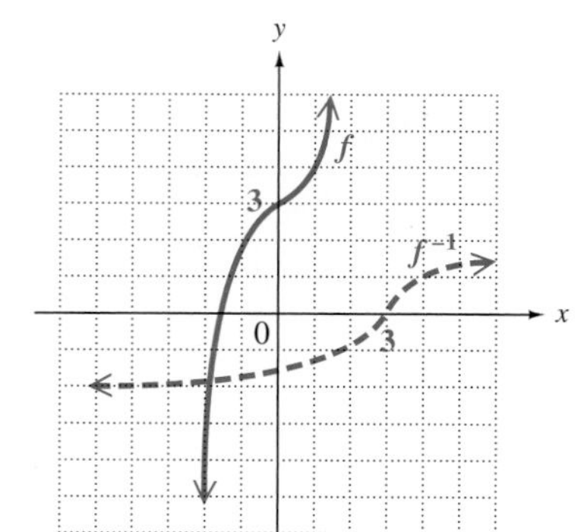

11.2 Exponential Functions

OBJECTIVES

1. **Define exponential functions.**
2. **Graph exponential functions.**
3. **Solve exponential equations of the form $a^x = a^k$ for x.**
4. **Use exponential functions in applications involving growth or decay.**

OBJECTIVE 1 Define exponential functions. In **Section 9.2,** we showed how to evaluate 2^x for rational values of x. For example,

$$2^3 = 8, \qquad 2^{-1} = \frac{1}{2}, \qquad 2^{1/2} = \sqrt{2}, \qquad 2^{3/4} = \sqrt[4]{2^3} = \sqrt[4]{8}.$$

In more advanced courses it is shown that 2^x exists for all real number values of x, both rational and irrational. The following definition of an exponential function assumes that a^x exists for all real numbers x.

Exponential Function

For $a > 0$, $a \neq 1$, and all real numbers x,

$$F(x) = a^x$$

defines the **exponential function with base a.**

Note

The two restrictions on the value of a in the definition of an exponential function $F(x) = a^x$ are important.

1. The restriction $a > 0$ is necessary so that the function can be defined for all real numbers x. Letting a be negative ($a = -2$, for instance) and letting $x = \frac{1}{2}$ would give $(-2)^{1/2}$, which is not real.
2. The restriction $a \neq 1$ is necessary because 1 raised to any power is equal to 1, and the function would then be the linear function defined by $F(x) = 1$.

OBJECTIVE 2 Graph exponential functions. When graphing exponential functions of the form $F(x) = a^x$, pay particular attention to whether $a > 1$ or $a < 1$.

EXAMPLE 1 Graphing an Exponential Function ($a > 1$)

Graph $f(x) = 2^x$.

Choose some values of x, and find the corresponding values of $f(x)$. Plotting these points and drawing a smooth curve through them gives the graph of $f(x) = 2^x$ shown in Figure 6.

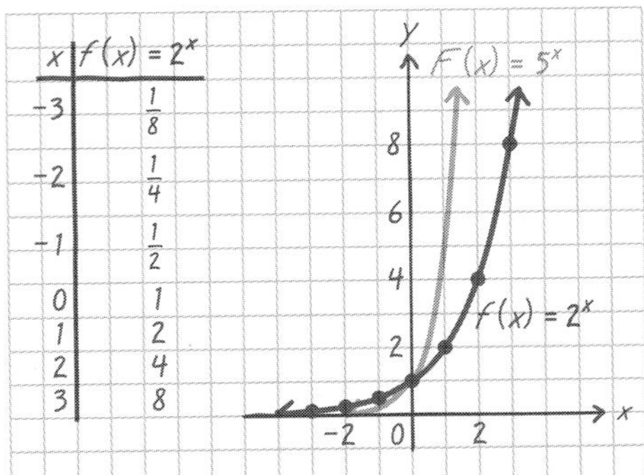

Figure 6

Continued on Next Page

1 Graph.

(a) $f(x) = 10^x$

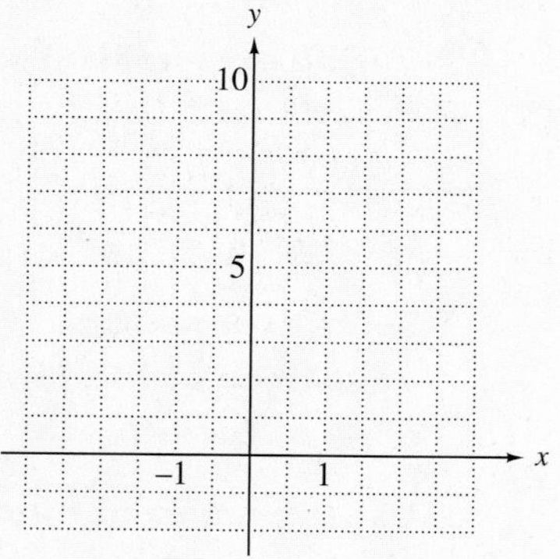

(b) $g(x) = \left(\frac{1}{10}\right)^x$

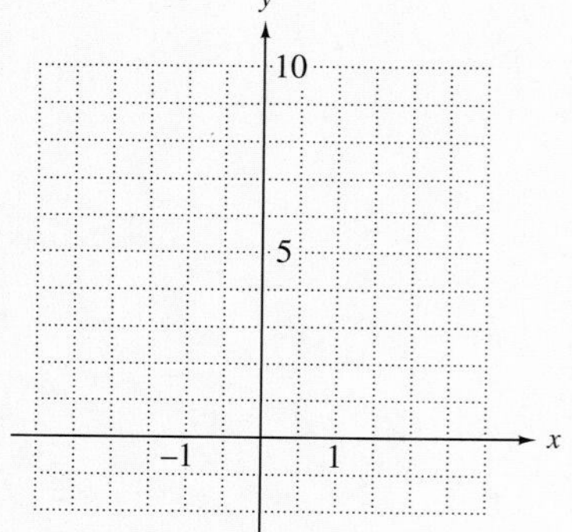

This graph is typical of the graphs of exponential functions of the form $F(x) = a^x$, where $a > 1$. ***The larger the value of a, the faster the graph rises.*** To see this, compare the graph of $F(x) = 5^x$ with the graph of $f(x) = 2^x$ in Figure 6. When graphing such functions, be sure to plot a sufficient number of points to see how rapidly the graph rises.

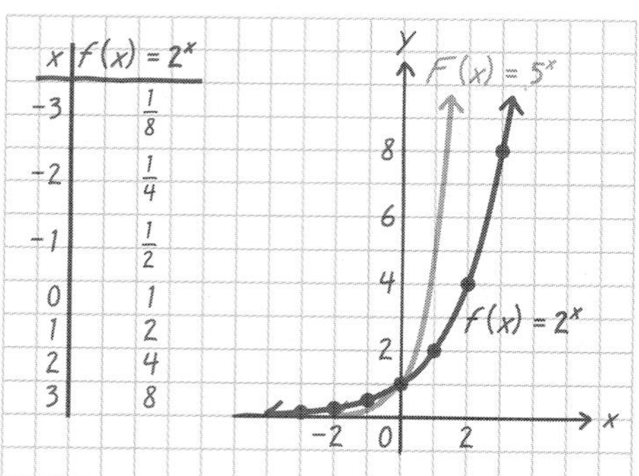

Figure 6 (repeated)

By the vertical line test, the graphs in Figure 6 represent functions. As these graphs suggest, the domain of an exponential function includes all real numbers. Because y is always positive, the range is $(0, \infty)$. Figure 6 also shows an important characteristic of exponential functions where $a > 1$: ***As x gets larger, y increases at a faster and faster rate.***

EXAMPLE 2 Graphing an Exponential Function ($a < 1$)

Graph $g(x) = \left(\frac{1}{2}\right)^x$.

Again, find some points on the graph. The graph, shown in Figure 7, is very similar to that of $f(x) = 2^x$ (Figure 6) with the same domain and range, except that here ***as x gets larger, y decreases.*** This graph is typical of the graphs of exponential functions of the form $F(x) = a^x$, where $0 < a < 1$.

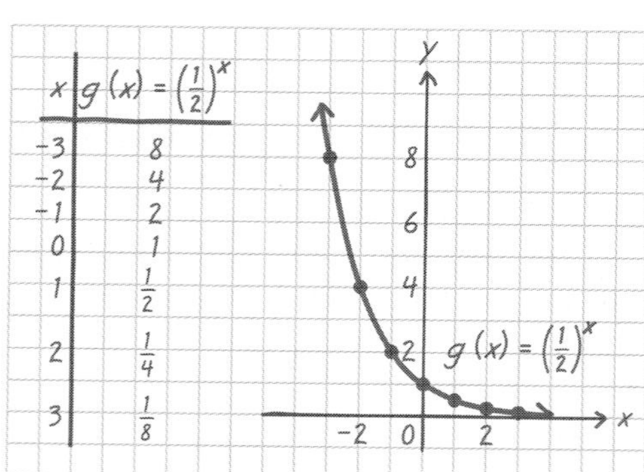

Figure 7

◀ *Work Problem* **1** *at the Side.*

CAUTION

The graph of an exponential function *approaches* the x-axis, but does ***not*** touch it.

Based on Examples 1 and 2, we make the following generalizations about the graphs of exponential functions of the form $F(x) = a^x$.

ANSWERS

1. (a)

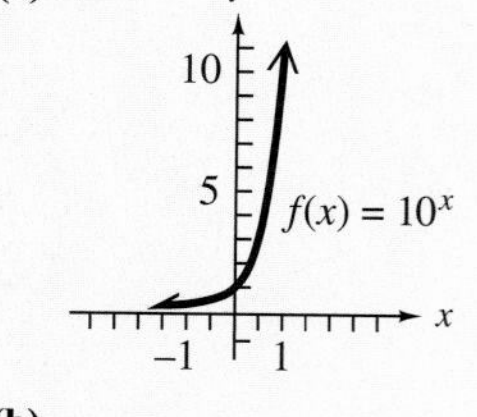

(b)

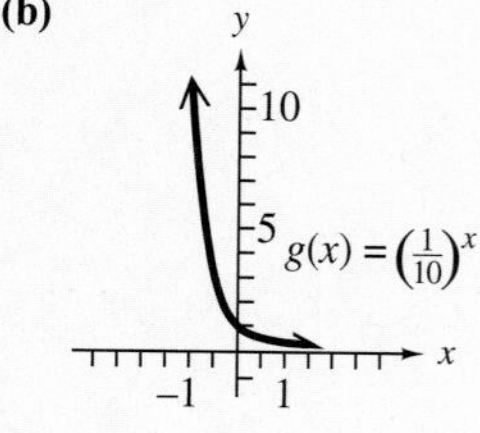

Characteristics of the Graph of $F(x) = a^x$

1. The graph contains the point (0, 1).
2. When $a > 1$, the graph *rises* from left to right. (See Figure 6.) When $0 < a < 1$, the graph *falls* from left to right. (See Figure 7.) In both cases, the graph goes from the second quadrant to the first.
3. The graph approaches the x-axis, but never touches it. (Recall from **Section 7.4** that such a line is called an **asymptote.**)
4. The domain is $(-\infty, \infty)$, and the range is $(0, \infty)$.

EXAMPLE 3 Graphing a More Complicated Exponential Function

Graph $f(x) = 3^{2x-4}$.

Find some ordered pairs.

$$\text{If } x = 0, \text{ then } y = 3^{2(0)-4} = 3^{-4} = \frac{1}{81}.$$

$$\text{If } x = 2, \text{ then } y = 3^{2(2)-4} = 3^{0} = 1.$$

These ordered pairs, $(0, \frac{1}{81})$ and (2, 1), along with the other ordered pairs shown in the table, lead to the graph in Figure 8. The graph is similar to the graph of $f(x) = 3^x$ except that it is shifted to the right and rises more rapidly.

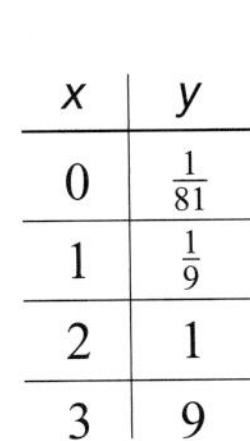

x	y
0	$\frac{1}{81}$
1	$\frac{1}{9}$
2	1
3	9

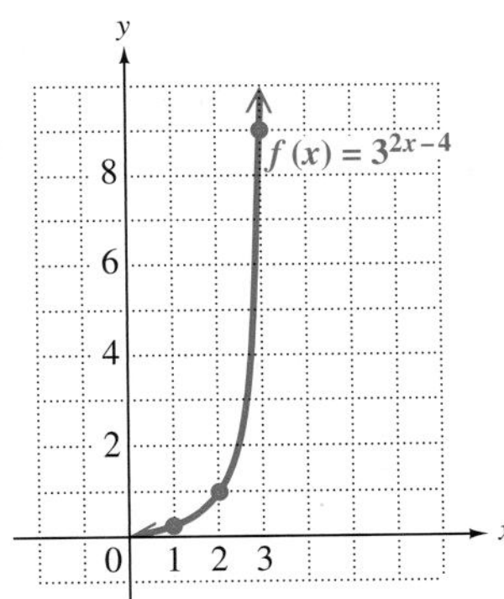

Figure 8

Work Problem 2 at the Side. ▶

OBJECTIVE 3 Solve exponential equations of the form $a^x = a^k$ for x. Until this chapter, we have solved only equations that had the variable as a base, like $x^2 = 8$; all exponents have been constants. An **exponential equation** is an equation that has a variable in an exponent, such as

$$9^x = 27.$$

By the horizontal line test, the exponential function defined by $F(x) = a^x$ is a one-to-one function, so we can use the following property to solve many exponential equations.

Property for Solving an Exponential Equation

For $a > 0$ and $a \neq 1$, if $\quad a^x = a^y \quad$ then $\quad x = y$.

This property would not necessarily be true if $a = 1$.

2 Graph $y = 2^{4x-3}$.

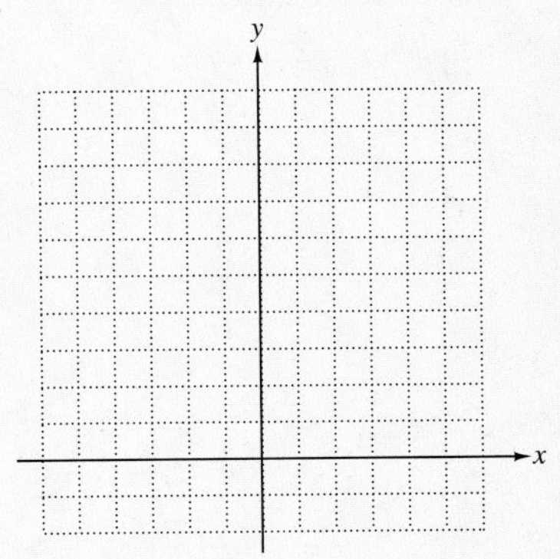

ANSWER

2.

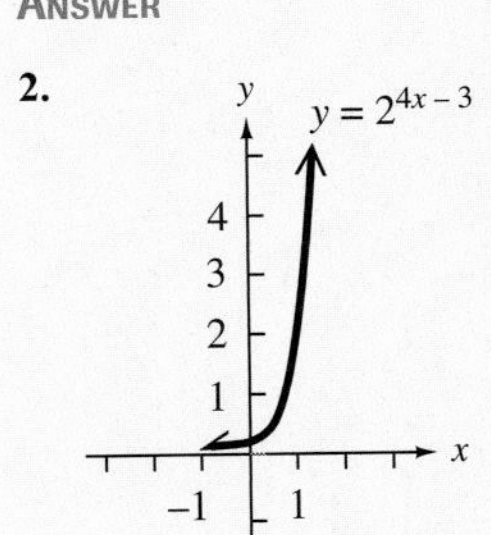

3 Solve each equation and check the solution.

(a) $25^x = 125$

(b) $4^x = 32$

(c) $81^p = 27$

To solve an exponential equation using this property, follow these steps.

Solving an Exponential Equation

Step 1 **Each side must have the same base.** If the two sides of the equation do not have the same base, express each as a power of the same base.

Step 2 **Simplify exponents,** if necessary, using the rules of exponents.

Step 3 **Set exponents equal** using the property given in this section.

Step 4 **Solve** the equation obtained in Step 3.

Note

These steps cannot be applied to an exponential equation like

$$3^x = 12$$

because Step 1 cannot easily be done. A method for solving such equations is given in **Section 11.6.**

EXAMPLE 4 Solving an Exponential Equation

Solve the equation $9^x = 27$.

We can use the property given in the box if both sides are written with the same base. Since $9 = 3^2$ and $27 = 3^3$,

$$9^x = 27$$

$$(3^2)^x = 3^3 \quad \text{Write with the same base. (Step 1)}$$

$$3^{2x} = 3^3 \quad \text{Power rule for exponents (Step 2)}$$

$$2x = 3 \quad \text{If } a^x = a^y \text{, then } x = y. \text{ (Step 3)}$$

$$x = \frac{3}{2}. \quad \text{Solve for } x. \text{ (Step 4)}$$

Check Substitute $\frac{3}{2}$ for x:

$$9^x = 9^{3/2} = (9^{1/2})^3 = 3^3 = 27, \quad \text{as required.}$$

The solution set is $\left\{\frac{3}{2}\right\}$.

◀ *Work Problem* **3** *at the Side.*

EXAMPLE 5 Solving Exponential Equations

Solve each equation.

(a)

$$4^{3x-1} = 16^{x+2}$$

$$4^{3x-1} = (4^2)^{x+2} \quad \text{Write with the same base; } 16 = 4^2.$$

$$4^{3x-1} = 4^{2x+4} \quad \text{Power rule for exponents}$$

$$3x - 1 = 2x + 4 \quad \text{Set exponents equal.}$$

$$x = 5 \quad \text{Subtract } 2x \text{; add 1.}$$

Be careful multiplying the exponents.

Verify that the solution set is $\{5\}$.

Continued on Next Page

Answers

3. (a) $\left\{\frac{3}{2}\right\}$ (b) $\left\{\frac{5}{2}\right\}$ (c) $\left\{\frac{3}{4}\right\}$

(b) $6^x = \frac{1}{216}$

$6^x = \frac{1}{6^3}$ $\quad$ $216 = 6^3$

$6^x = 6^{-3}$ $\quad$ Write with the same base; $\frac{1}{6^3} = 6^{-3}$.

$x = -3$ $\quad$ Set exponents equal.

Check Substitute -3 for x:

$$6^x = 6^{-3} = \frac{1}{6^3} = \frac{1}{216}, \quad \text{as required.}$$

The solution set is $\{-3\}$.

(c) $\left(\frac{2}{3}\right)^x = \frac{9}{4}$

$\left(\frac{2}{3}\right)^x = \left(\frac{4}{9}\right)^{-1}$ $\quad$ $\frac{9}{4} = \left(\frac{4}{9}\right)^{-1}$

$\left(\frac{2}{3}\right)^x = \left[\left(\frac{2}{3}\right)^2\right]^{-1}$ $\quad$ Write with the same base.

$\left(\frac{2}{3}\right)^x = \left(\frac{2}{3}\right)^{-2}$ $\quad$ Power rule for exponents

$x = -2$ $\quad$ Set exponents equal.

Check that the solution set is $\{-2\}$.

Work Problem **4** *at the Side.* ▶

OBJECTIVE 4 Use exponential functions in applications involving growth or decay.

EXAMPLE 6 Solving an Application of Exponential Growth

The graph in Figure 9 shows the concentration of carbon dioxide (in parts per million) in the air. This concentration is increasing exponentially.

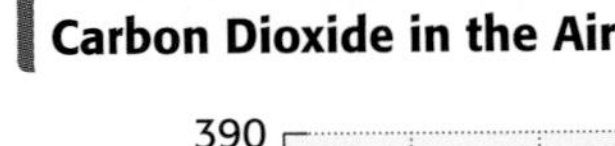

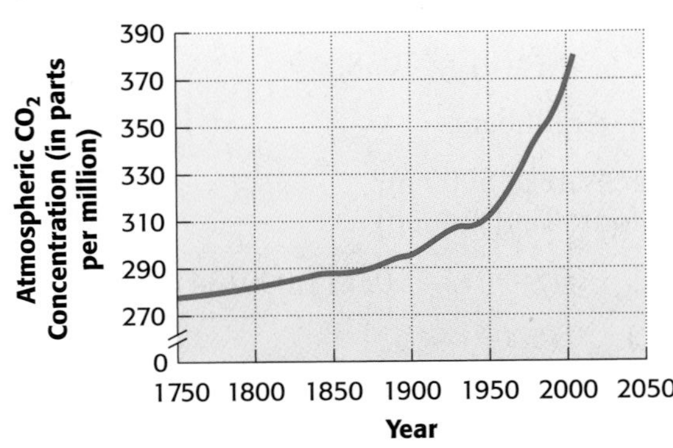

Source: Sacramento Bee; National Oceanic and Atmospheric Administration.

Figure 9

Continued on Next Page

4 Solve each equation and check the solution.

(a) $25^{x-2} = 125^x$

(b) $4^x = \frac{1}{32}$

(c) $\left(\frac{3}{4}\right)^x = \frac{16}{9}$

Answers

4. (a) $\{-4\}$ **(b)** $\left\{-\frac{5}{2}\right\}$ **(c)** $\{-2\}$

5 Use the exponential function in Example 6 to approximate the carbon dioxide concentration in 1925.

The data are approximated by the exponential function defined by

$$f(x) = 266(1.001)^x,$$

where x is the number of years since 1750. Use this function and a calculator to approximate the concentration of carbon dioxide in parts per million for each year.

(a) 1900

Since x represents the number of years since 1750, in this case we have $x = 1900 - 1750 = 150$. Thus, evaluate $f(150)$.

$$f(x) = 266(1.001)^x$$

$$f(150) = 266(1.001)^{150} \quad \text{Let } x = 150.$$

$$f(150) \approx 309 \quad \text{Use a calculator.}$$

The concentration in 1900 was about 309 parts per million.

(b) 1950

Use $x = 1950 - 1750 = 200$.

$$f(200) = 266(1.001)^{200} \quad \text{Let } x = 200.$$

$$f(200) \approx 325 \quad \text{Use a calculator.}$$

The concentration in 1950 was about 325 parts per million.

◀ *Work Problem* **5** *at the Side.*

EXAMPLE 7 Applying an Exponential Decay Function

6 Use the exponential function in Example 7 to find the pressure at 8000 m.

The atmospheric pressure (in millibars) at a given altitude x, in meters, can be approximated by the exponential function defined by

$$f(x) = 1038(1.000134)^{-x},$$

for values of x between 0 and 10,000. Because the base is greater than 1 and the coefficient of x in the exponent is negative, the function values decrease as x increases. This means that as the altitude increases, the atmospheric pressure decreases. (*Source:* Miller, A. and J. Thompson, *Elements of Meteorology,* Fourth Edition, Charles E. Merrill Publishing Company, 1993.)

(a) According to this function, what is the pressure at ground level?

At ground level, $x = 0$, so

$$f(0) = 1038(1.000134)^{-0}$$

$$f(0) = 1038(1)$$

$$f(0) = 1038.$$

The pressure is 1038 millibars.

(b) What is the pressure at 5000 m?

Use a calculator to find $f(5000)$.

$$f(5000) = 1038(1.000134)^{-5000}$$

$$f(5000) \approx 531$$

The pressure is approximately 531 millibars.

◀ *Work Problem* **6** *at the Side.*

ANSWERS

5. 317 parts per million
6. approximately 355 millibars

11.2 Exercises

FOR EXTRA HELP

Math XL PRACTICE

WATCH

DOWNLOAD

READ

REVIEW

Choose the correct response in Exercises 1–4.

1. Which point lies on the graph of $f(x) = 2^x$? C

A. $(1, 0)$ **B.** $(2, 1)$

C. $(0, 1)$ **D.** $\left(\sqrt{2}, \frac{1}{2}\right)$

2. The asymptote of the graph of $F(x) = a^x$ A

A. is the x-axis. **B.** is the y-axis.

C. has equation $x = 1$. **D.** has equation $y = 1$.

3. Which statement is true? C

A. The y-intercept of the graph of $f(x) = 10^x$ is $(0, 10)$.

B. For any $a > 1$, the graph of $f(x) = a^x$ falls from left to right.

C. The point $\left(\frac{1}{2}, \sqrt{5}\right)$ lies on the graph of $f(x) = 5^x$.

D. The graph of $y = 4^x$ rises at a faster rate than the graph of $y = 10^x$.

4. Which equation is graphed here? C

A. $y = 1000\left(\frac{1}{2}\right)^{0.3x}$

B. $y = 1000\left(\frac{1}{2}\right)^{x}$

C. $y = 1000(2)^{0.3x}$

D. $y = 1000^x$

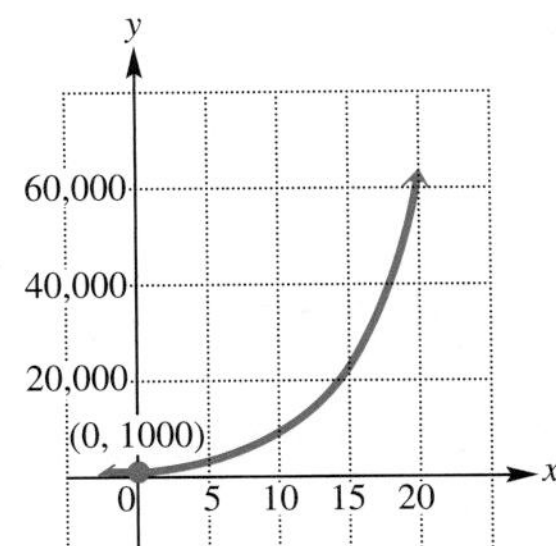

Graph each exponential function. See Examples 1–3.

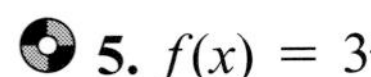

5. $f(x) = 3^x$

6. $f(x) = 5^x$

7. $g(x) = \left(\frac{1}{3}\right)^x$

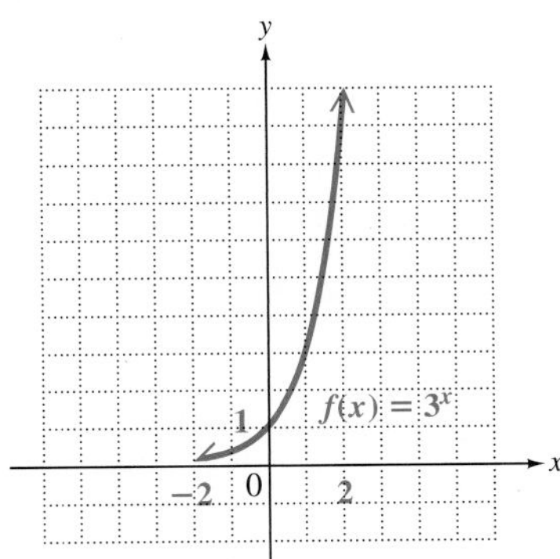

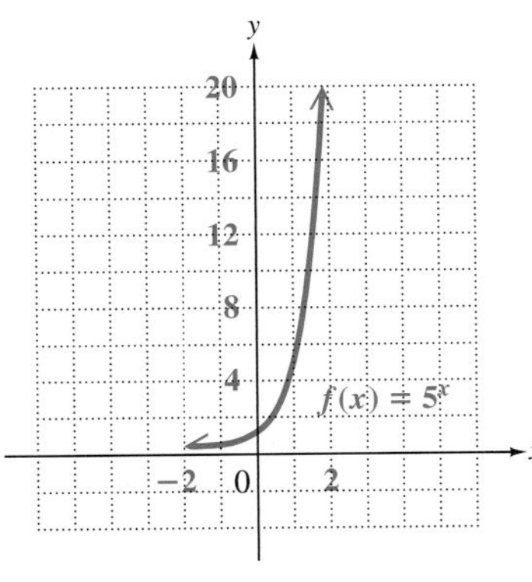

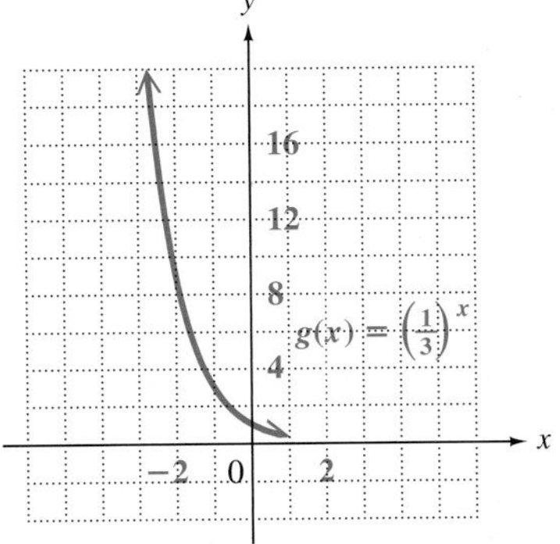

8. $g(x) = \left(\frac{1}{5}\right)^x$

9. $y = 2^{2x-2}$

10. $y = 2^{2x+1}$

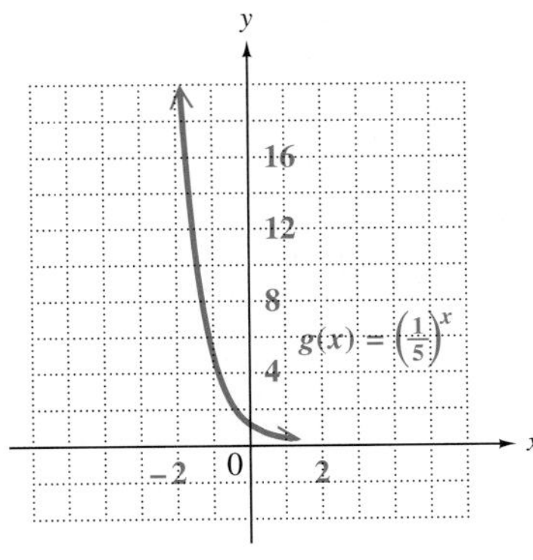

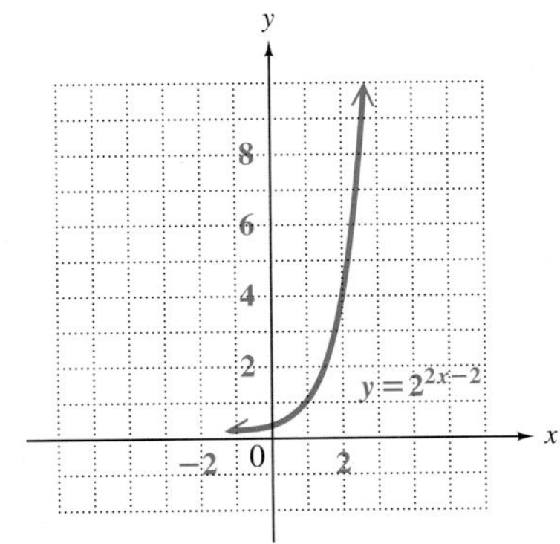

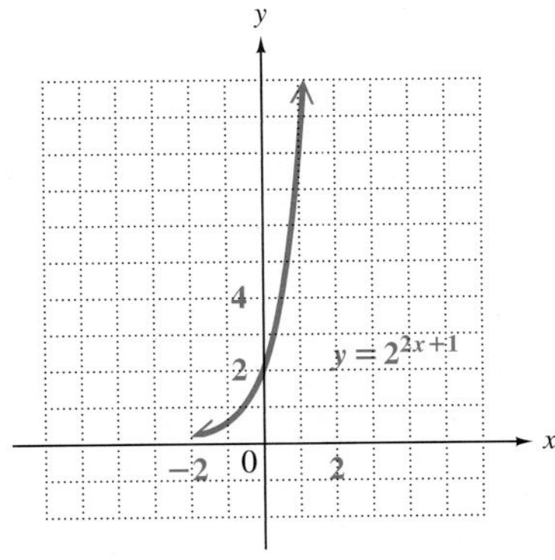

Solve each equation. See Examples 4 and 5.

11. $6^x = 36$ $\{2\}$

12. $8^x = 64$ $\{2\}$

13. $100^x = 1000$ $\left\{\frac{3}{2}\right\}$

14. $8^x = 4$ $\left\{\frac{2}{3}\right\}$

15. $16^{2x+1} = 64^{x+3}$ $\{7\}$

16. $9^{2x-8} = 27^{x-4}$ $\{4\}$

17. $5^x = \frac{1}{125}$ $\{-3\}$

18. $3^x = \frac{1}{81}$ $\{-4\}$

19. $5^x = 0.2$ $\{-1\}$

20. $10^x = 0.1$ $\{-1\}$

21. $\left(\frac{3}{2}\right)^x = \frac{8}{27}$ $\{-3\}$

22. $\left(\frac{4}{3}\right)^x = \frac{27}{64}$ $\{-3\}$

23. **(a)** For an exponential function defined by $f(x) = a^x$, if $a > 1$, then the graph **rises** (rises/falls) from left to right. If $0 < a < 1$, then the graph **falls** (rises/falls) from left to right.

(b) Based on your answers in part (a), make a conjecture (an educated guess) concerning whether an exponential function defined by $f(x) = a^x$ is one-to-one. Then decide whether it has an inverse based on the concepts of **Section 11.1.**

It is one-to-one and thus has an inverse.

Solve each problem. See Examples 6 and 7.

The figure shown here accompanied the article "Is Our World Warming?", which appeared in the October 1990 issue of *National Geographic*. It shows projected temperature increases using two graphs: one an exponential-type curve and the other linear. From the figure, approximate the increase **(a)** for the exponential curve, and **(b)** for the linear graph for each year.

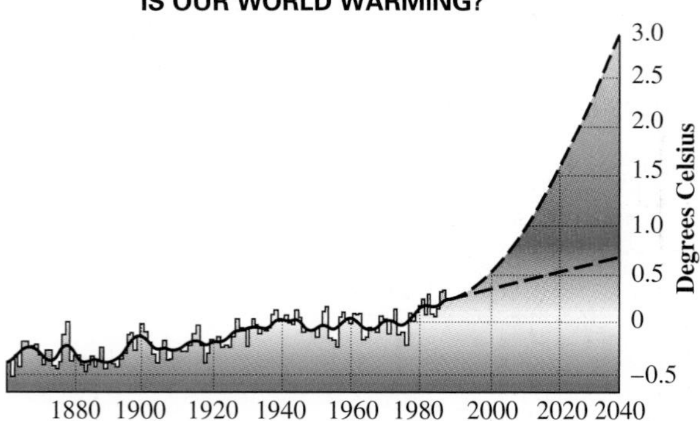

Graph, "Zero Equals Average Global Temperature for the Period 1950–1979." Dale D. Glasgow, © National Geographic Society. Reprinted by permission.

24. 2000 **(a) 0.5°C** **(b) 0.35°C**

25. 2010 **(a) 1.0°C** **(b) 0.4°C**

26. 2020 **(a) 1.6°C** **(b) 0.5°C**

27. 2040 **(a) 3.0°C** **(b) 0.7°C**

28. A small business estimates that the value $V(t)$ of a copy machine is decreasing according to the function defined by

$$V(t) = 5000(2)^{-0.15t},$$

where t is the number of years that have elapsed since the machine was purchased and $V(t)$ is in dollars.

(a) What was the original value of the machine? **$5000**

(b) What is the value of the machine 5 yr after purchase? Give your answer to the nearest dollar. **$2973**

(c) What is the value of the machine 10 yr after purchase? Give your answer to the nearest dollar. **$1768**

(d) Graph the function.

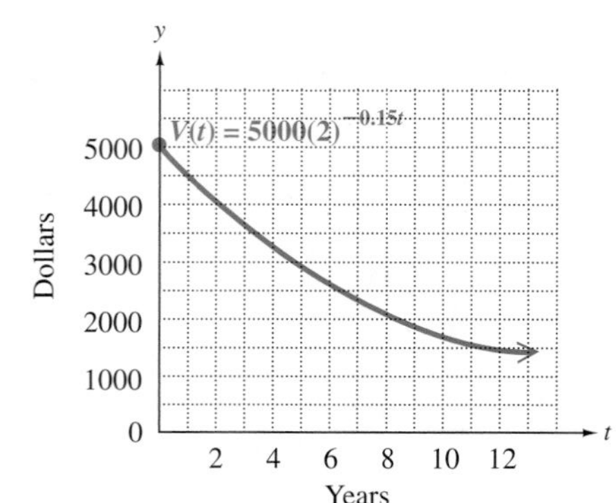

11.3 Logarithmic Functions

The graph of $y = 2^x$ is the curve shown in blue in Figure 10. Because $y = 2^x$ defines a one-to-one function, it has an inverse function. Interchanging x and y gives

$x = 2^y$, the inverse of $y = 2^x$.

As we saw in **Section 11.1,** the graph of the inverse is found by reflecting the graph of $y = 2^x$ about the line $y = x$. The graph of $x = 2^y$ is shown as a red curve in Figure 10.

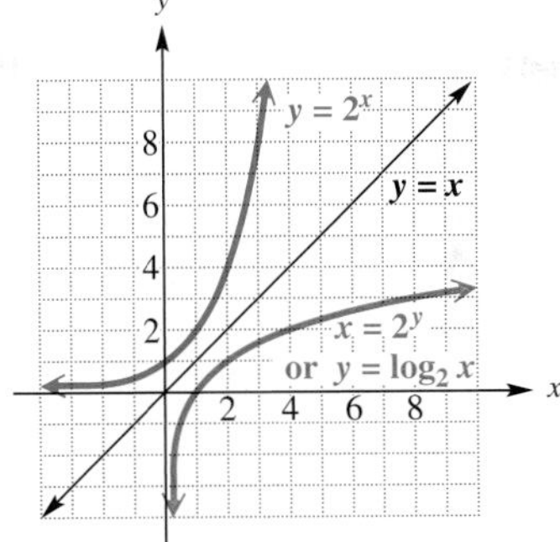

Figure 10

OBJECTIVES

1. **Define a logarithm.**
2. **Convert between exponential and logarithmic forms.**
3. **Solve logarithmic equations of the form $\log_a b = k$ for a, b, or k.**
4. **Define and graph logarithmic functions.**
5. **Use logarithmic functions in applications involving growth or decay.**

OBJECTIVE 1 Define a logarithm. We cannot solve the equation $x = 2^y$ for the dependent variable y with the methods presented up to now. The following definition is used to solve $x = 2^y$ for y.

Logarithm

For all positive numbers a, with $a \neq 1$, and all positive numbers x,

$$y = \log_a x \quad \text{means the same as} \quad x = a^y.$$

This key statement should be memorized. The abbreviation **log** is used for **logarithm.** Read **$\log_a x$** as **"the logarithm of x with base a"** or **"the base a logarithm of x."** To remember the location of the base and the exponent in each form, refer to the following diagrams.

Logarithmic form: $y = \log_a x$ (Exponent: y; Base: a) | Exponential form: $x = a^y$ (Exponent: y; Base: a)

In work with logarithmic and exponential forms, remember the following.

Meaning of $\log_a x$

A logarithm is an exponent. ***The expression $\log_a x$ represents the exponent to which the base a must be raised to obtain x.***

OBJECTIVE 2 Convert between exponential and logarithmic forms. We use the definition of a logarithm to convert between exponential and logarithmic forms. The table shows pairs of equivalent statements.

Exponential Form	Logarithmic Form
$3^2 = 9$	$\log_3 9 = 2$
$\left(\frac{1}{5}\right)^{-2} = 25$	$\log_{1/5} 25 = -2$
$10^5 = 100{,}000$	$\log_{10} 100{,}000 = 5$
$4^{-3} = \frac{1}{64}$	$\log_4 \frac{1}{64} = -3$

Work Problem 1 *at the Side.* ▶

1 Complete the table.

Exponential Form	Logarithmic Form
$2^5 = 32$	
$100^{1/2} = 10$	
	$\log_8 4 = \frac{2}{3}$
	$\log_6 \frac{1}{1296} = -4$

ANSWERS

1. $\log_2 32 = 5$; $\log_{100} 10 = \frac{1}{2}$; $8^{2/3} = 4$; $6^{-4} = \frac{1}{1296}$

OBJECTIVE 3 Solve logarithmic equations of the form $\log_a b = k$ for a, b, or k. A **logarithmic equation** is an equation with a logarithm in at least one term. To solve, first write the equation in exponential form.

EXAMPLE 1 Solving Logarithmic Equations

Solve each equation.

(a) $\log_4 x = -2$

By definition, $\log_4 x = -2$ is equivalent to $x = 4^{-2}$.

$$x = 4^{-2} = \frac{1}{16}$$

The solution set is $\left\{\frac{1}{16}\right\}$.

(b) $\log_{1/2}(3x + 1) = 2$

$3x + 1 = \left(\frac{1}{2}\right)^2$ — This is a key step. Write in exponential form.

$3x + 1 = \frac{1}{4}$ — Apply the exponent.

$12x + 4 = 1$ — Multiply each term by 4.

$12x = -3$ — Subtract 4.

$x = -\frac{1}{4}$ — Divide by 12; lowest terms

Check $\log_{1/2}\left(3\left(-\frac{1}{4}\right) + 1\right) \stackrel{?}{=} 2$ — Let $x = -\frac{1}{4}$.

$\log_{1/2}\frac{1}{4} \stackrel{?}{=} 2$ — Simplify within parentheses

$\left(\frac{1}{2}\right)^2 = \frac{1}{4}$ — Exponential form; true

The solution set is $\left\{-\frac{1}{4}\right\}$.

(c) $\log_x 3 = 2$

$x^2 = 3$ — Write in exponential form.

$x = \pm\sqrt{3}$ — Take square roots.

Notice that only the *principal* square root satisfies the equation, however, since the base x must be a positive number. The solution set is $\{\sqrt{3}\}$.

(d) $\log_{49}\sqrt[3]{7} = x$

$49^x = \sqrt[3]{7}$ — Write in exponential form.

$(7^2)^x = 7^{1/3}$ — Write with the same base.

$7^{2x} = 7^{1/3}$ — Power rule for exponents

$2x = \frac{1}{3}$ — Set exponents equal.

$x = \frac{1}{6}$ — Divide by 2.

The solution set is $\left\{\frac{1}{6}\right\}$.

Work Problem 2 at the Side.

2 Solve each equation.

(a) $\log_3 27 = x$

(b) $\log_5 p = 2$

(c) $\log_m \frac{1}{16} = -4$

(d) $\log_x 12 = 3$

ANSWERS

2. (a) $\{3\}$ (b) $\{25\}$ (c) $\{2\}$ (d) $\{\sqrt[3]{12}\}$

For any real positive number b, we know that $b^1 = b$ and $b^0 = 1$. Writing these two statements in logarithmic form gives the following two properties of logarithms.

Properties of Logarithms

For any positive real number b, with $b \neq 1$,

$$\log_b b = 1 \quad \text{and} \quad \log_b 1 = 0.$$

EXAMPLE 2 Using Properties of Logarithms

Use the preceding two properties of logarithms to evaluate each logarithm.

(a) $\log_7 7 = 1$ **(b)** $\log_{\sqrt{2}} \sqrt{2} = 1$

(c) $\log_9 1 = 0$ **(d)** $\log_{0.2} 1 = 0$

Work Problem 3 at the Side. ▶

3 Evaluate each logarithm.

(a) $\log_{2/5} \frac{2}{5}$

(b) $\log_\pi \pi$

(c) $\log_{0.4} 1$

(d) $\log_6 1$

OBJECTIVE 4 Define and graph logarithmic functions. Now we define the logarithmic function with base a.

Logarithmic Function

If a and x are positive numbers, with $a \neq 1$, then

$$G(x) = \log_a x$$

defines the **logarithmic function with base a.**

EXAMPLE 3 Graphing a Logarithmic Function ($a > 1$)

Graph $f(x) = \log_2 x$.

By writing $y = f(x) = \log_2 x$ in exponential form as $x = 2^y$, we can identify ordered pairs that satisfy the equation. It is easier to choose values for y and find the corresponding values of x. Plotting the points in the table of ordered pairs and connecting them with a smooth curve gives the graph in Figure 11. This graph is typical of logarithmic functions with base $a > 1$.

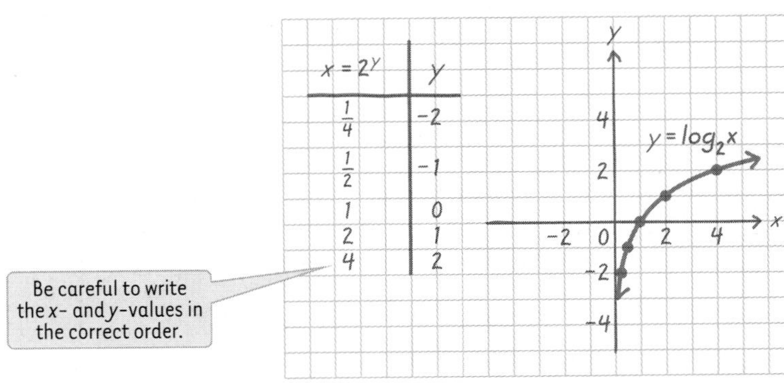

Figure 11

As the table and graph suggest, x is always positive, so the domain of a logarithmic function is $(0, \infty)$. The range includes all real numbers, $(-\infty, \infty)$.

Work Problem 4 at the Side. ▶

4 Graph $y = \log_{10} x$.

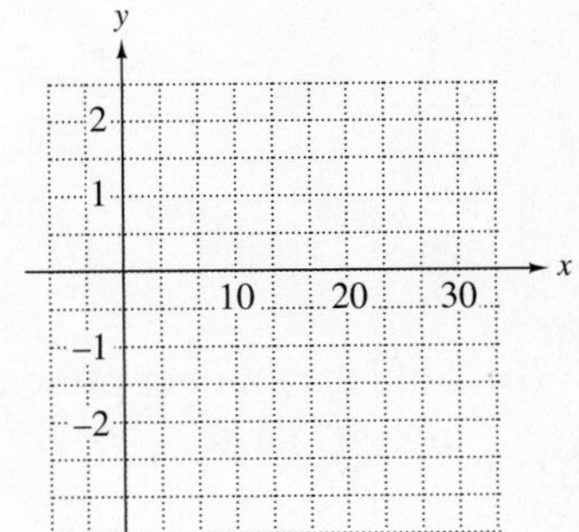

ANSWERS

3. **(a)** 1 **(b)** 1 **(c)** 0 **(d)** 0

4.

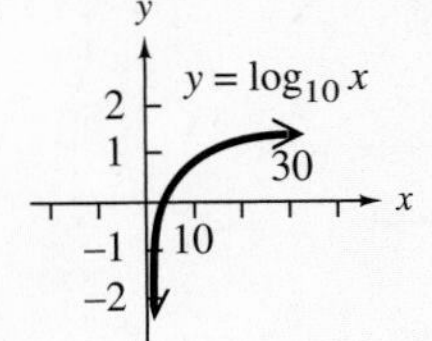

EXAMPLE 4 **Graphing a Logarithmic Function ($0 < a < 1$)**

Graph $g(x) = \log_{1/2} x$.

We write $y = g(x) = \log_{1/2} x$ in exponential form as $x = \left(\frac{1}{2}\right)^y$; then choose values for y and find the corresponding values of x. Plotting these points and connecting them with a smooth curve gives the graph in Figure 12. This graph is typical of logarithmic functions with $0 < a < 1$.

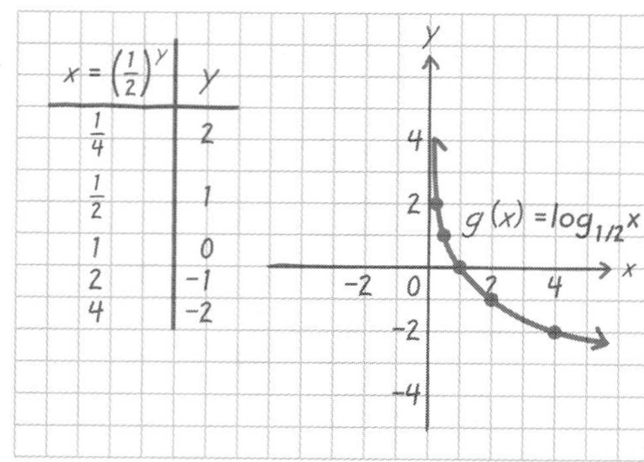

Figure 12

◀ Work Problem 5 at the Side.

Characteristics of the Graph of $G(x) = \log_a x$

1. The graph contains the point $(1, 0)$.
2. When $a > 1$, the graph *rises* from left to right, from the fourth quadrant to the first. (See Figure 11.) When $0 < a < 1$, the graph *falls* from left to right, from the first quadrant to the fourth. (See Figure 12.)
3. The graph approaches the y-axis, but never touches it. (The y-axis is an asymptote.)
4. The domain is $(0, \infty)$, and the range is $(-\infty, \infty)$.

OBJECTIVE 5 Use logarithmic functions in applications involving growth or decay. Logarithmic functions, like exponential functions, can be applied to growth or decay of real-world phenomena.

EXAMPLE 5 **Solving an Application of a Logarithmic Function**

The logarithmic function defined by

$$f(x) = 27 + 1.105 \log_{10} (x + 1)$$

approximates the barometric pressure in inches of mercury at a distance of x miles from the eye of a typical hurricane. (*Source:* Miller, A. and R. Anthes, *Meteorology,* Fifth Edition, Charles E. Merrill Publishing Company, 1985.) Approximate the pressure 9 mi from the eye of the hurricane.

$f(9) = 27 + 1.105 \log_{10} (9 + 1)$ Let $x = 9$.

$f(9) = 27 + 1.105 \log_{10} 10$ Add inside parentheses.

$f(9) = 27 + 1.105(1)$ $\log_{10} 10 = 1$

$f(9) = 28.105$ Add.

The pressure 9 mi from the eye of the hurricane is 28.105 in.

◀ Work Problem 6 at the Side.

5 Graph $y = \log_{1/10} x$.

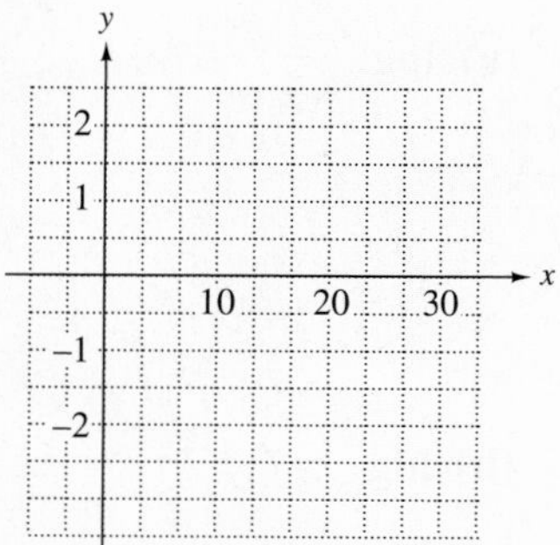

6 Solve the problem.

A population of mites in a laboratory is growing according to the logarithmic function defined by

$$P(t) = 80 \log_{10} (t + 10),$$

where t is the number of days after a study is begun.

(a) Find the number of mites at the beginning of the study.

(b) Find the number present after 90 days.

Answers

5.

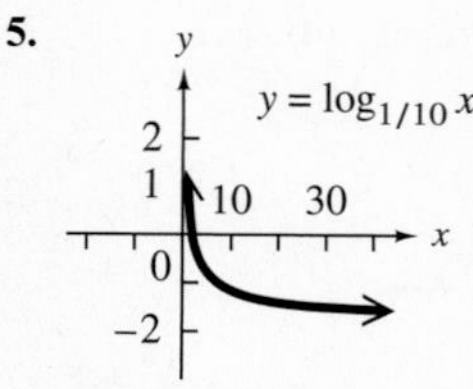

6. (a) 80 **(b)** 160

11.3 Exercises

FOR EXTRA HELP

MyMathLab

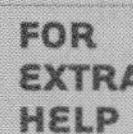
Math XL PRACTICE

WATCH

DOWNLOAD

READ

REVIEW

1. By definition, $\log_a x$ is the exponent to which the base a must be raised in order to obtain x. Use this definition to match the logarithm in Column I with its value in Column II. (*Example:* $\log_3 9$ is equal to 2 because 2 is the exponent to which 3 must be raised in order to obtain 9.)

I		II
(a) $\log_4 16$	C	**A.** -2
(b) $\log_3 81$	F	**B.** -1
(c) $\log_3 \left(\frac{1}{3}\right)$	B	**C.** 2
(d) $\log_{10} 0.01$	A	**D.** 0
(e) $\log_5 \sqrt{5}$	E	**E.** $\frac{1}{2}$
(f) $\log_{13} 1$	D	**F.** 4

2. Match the logarithmic equation in Column I with the corresponding exponential equation from Column II.

I		II
(a) $\log_{1/3} 3 = -1$	B	**A.** $8^{1/3} = \sqrt[3]{8}$
(b) $\log_5 1 = 0$	E	**B.** $\left(\frac{1}{3}\right)^{-1} = 3$
(c) $\log_2 \sqrt{2} = \frac{1}{2}$	D	**C.** $4^1 = 4$
(d) $\log_{10} 1000 = 3$	F	**D.** $2^{1/2} = \sqrt{2}$
(e) $\log_8 \sqrt[3]{8} = \frac{1}{3}$	A	**E.** $5^0 = 1$
(f) $\log_4 4 = 1$	C	**F.** $10^3 = 1000$

Write in logarithmic form. See the table in Objective 2.

3. $4^5 = 1024$

$\log_4 1024 = 5$

4. $3^6 = 729$

$\log_3 729 = 6$

5. $\left(\frac{1}{2}\right)^{-3} = 8$

$\log_{1/2} 8 = -3$

6. $\left(\frac{1}{6}\right)^{-3} = 216$

$\log_{1/6} 216 = -3$

7. $10^{-3} = 0.001$

$\log_{10} 0.001 = -3$

8. $36^{1/2} = 6$

$\log_{36} 6 = \frac{1}{2}$

9. $\sqrt[4]{625} = 5$

$\log_{625} 5 = \frac{1}{4}$

10. $\sqrt[3]{343} = 7$

$\log_{343} 7 = \frac{1}{3}$

Write in exponential form. See the table in Objective 2.

11. $\log_4 64 = 3$

$4^3 = 64$

12. $\log_2 512 = 9$

$2^9 = 512$

13. $\log_{10} \frac{1}{10{,}000} = -4$

$10^{-4} = \frac{1}{10{,}000}$

14. $\log_{100} 100 = 1$

$100^1 = 100$

15. $\log_6 1 = 0$

$6^0 = 1$

16. $\log_\pi 1 = 0$

$\pi^0 = 1$

17. $\log_9 3 = \frac{1}{2}$

$9^{1/2} = 3$

18. $\log_{64} 2 = \frac{1}{6}$

$64^{1/6} = 2$

19. When a student asked his teacher to explain to him how to evaluate $\log_9 3$ without showing any work, his teacher told him, "Think radically." Explain what the teacher meant by this hint.

Since the radical $\sqrt{9} = 9^{1/2} = 3$, the exponent to which 9 must be raised is 1/2.

20. A student told her teacher, "I know that $\log_2 1$ is the exponent to which 2 must be raised in order to obtain 1, but I can't think of any such number." How would you explain to the student that the value of $\log_2 1$ is 0?

Recall that any nonzero number raised to the 0 power equals 1.

Solve each equation for x. See Examples 1 and 2.

21. $x = \log_{27} 3$

$\left\{\frac{1}{3}\right\}$

22. $x = \log_{125} 5$

$\left\{\frac{1}{3}\right\}$

23. $\log_x 9 = \frac{1}{2}$

$\{81\}$

24. $\log_x 5 = \frac{1}{2}$

$\{25\}$

25. $\log_x 125 = -3$

$\left\{\frac{1}{5}\right\}$

26. $\log_x 64 = -6$

$\left\{\frac{1}{2}\right\}$

27. $\log_{12} x = 0$

$\{1\}$

28. $\log_4 x = 0$

$\{1\}$

29. $\log_x x = 1$

$\{x \mid x > 0, x \neq 1\}$

30. $\log_x 1 = 0$

$\{x \mid x > 0, x \neq 1\}$

31. $\log_x \frac{1}{25} = -2$

$\{5\}$

32. $\log_x \frac{1}{10} = -1$

$\{10\}$

33. $\log_8 32 = x$

$\left\{\frac{5}{3}\right\}$

34. $\log_{81} 27 = x$

$\left\{\frac{3}{4}\right\}$

35. $\log_\pi \pi^4 = x$

$\{4\}$

36. $\log_{\sqrt{2}} (\sqrt{2})^9 = x$

$\{9\}$

37. $\log_6 \sqrt{216} = x$

$\left\{\frac{3}{2}\right\}$

38. $\log_4 \sqrt{64} = x$

$\left\{\frac{3}{2}\right\}$

*If the point (p, q) is on the graph of $f(x) = a^x$ (for $a > 0$ and $a \neq 1$), then the point (q, p) is on the graph of $f^{-1}(x) = \log_a x$. Use this fact and refer to the graphs required in Exercises 5–8 in **Section 11.2** to graph each logarithmic function. See Examples 3 and 4.*

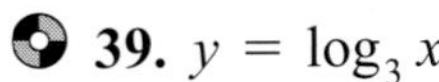

39. $y = \log_3 x$

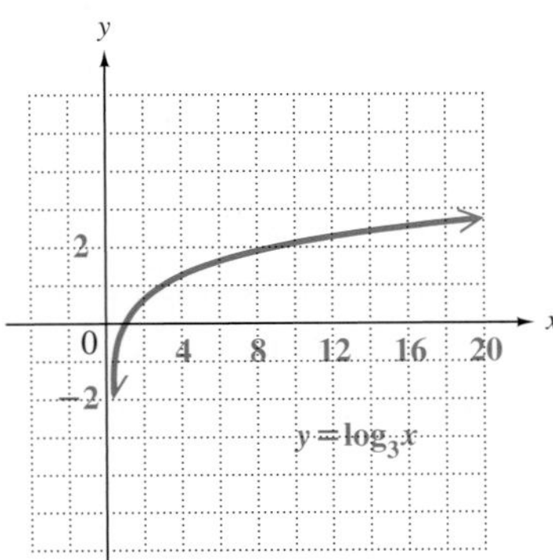

40. $y = \log_5 x$

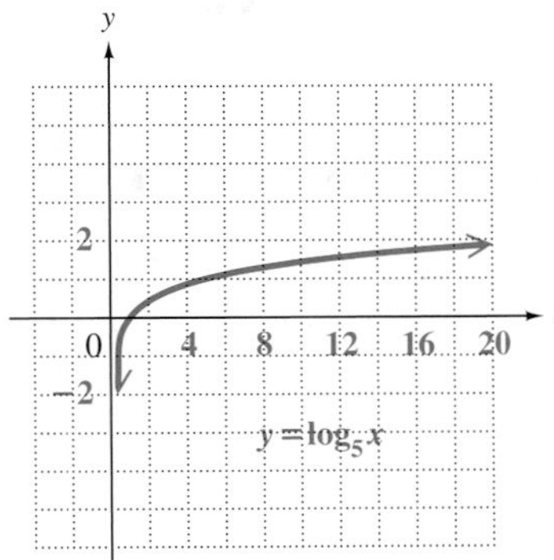

41. $y = \log_{1/3} x$

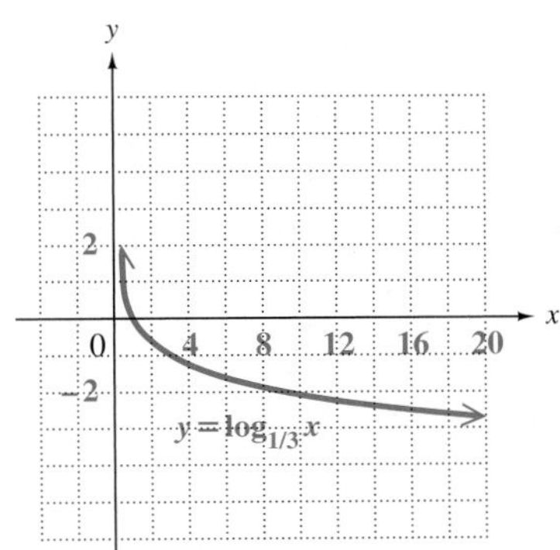

42. $y = \log_{1/5} x$

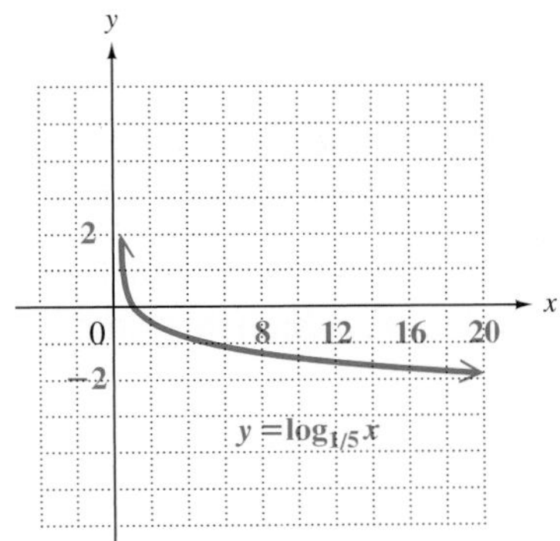

43. Compare the summary of characteristics of the graph of $F(x) = a^x$ in **Section 11.2** with the similar summary of characteristics of the graph of $G(x) = \log_a x$ in this section. Make a list of the characteristics that reinforce the concept that F and G are inverse functions.

Answers will vary.

44. The domain of $F(x) = a^x$ is $(-\infty, \infty)$, while the range is $(0, \infty)$. Therefore, since $G(x) = \log_a x$ defines the inverse of F, the domain of G is **$(0, \infty)$**, while the range of G is **$(-\infty, \infty)$**.

Use the graph to predict the value of $f(t)$ for each value of t.

45. $t = 0$

8

46. $t = 10$

16

47. $t = 60$

24

48. Show that the points determined in Exercises 45–47 lie on the graph of $f(t) = 8 \log_5 (2t + 5)$.

$f(0) = 8$; $f(10) = 16$; $f(60) = 24$

49. Explain why 1 is not allowed as a base for a logarithmic function.

Since every real number power of 1 equals 1, if $y = \log_1 x$, then $x = 1^y$ and so $x = 1$ for every y. This contradicts the definition of a function.

50. Explain why $\log_a 1$ is 0 for any value of a that is allowed as the base of a logarithm. Use a rule of exponents introduced earlier in your explanation.

$x = \log_a 1$ is equivalent to $a^x = 1$. The only value of x that makes $a^x = 1$ is 0. (Recall that $a \neq 1$.)

51. The graphs of both $f(x) = 3^x$ and $g(x) = \log_3 x$ rise from left to right. Which one rises at a faster rate?

$f(x) = 3^x$

52. Use the exponential key of your calculator to find approximations for the expression $(1 + \frac{1}{x})^x$, using x values of 1, 10, 100, 1000, and 10,000. Explain what seems to be happening as x gets larger and larger.

The expression gets closer and closer to a number that is approximately 2.718.

Solve each application of a logarithmic function. See Example 5.

53. According to selected figures from 1981 through 2003, the number of billion cubic feet of natural gas gross withdrawals from crude oil wells in the United States can be approximated by the function defined by

$$f(x) = 3800 + 585 \log_2 x,$$

where $x = 1$ represents 1981, $x = 2$ represents 1982, and so on. (*Source:* Energy Information Administration, Annual Energy Review 2003.) Use this function to approximate the number of cubic feet withdrawn in each of the following years.

(a) 1982 **4385 billion ft³**

(b) 1988 **5555 billion ft³**

(c) 1996 **6140 billion ft³**

54. According to selected figures from the last two decades of the twentieth century, the number of trillion cubic feet of dry natural gas consumed worldwide was approximated by the function defined by

$$f(x) = 51.47 + 6.044 \log_2 (x + 1),$$

where $x = 0$ corresponds to 1980, $x = 1$ to 1981, and so on. (*Source:* Energy Information Administration.) Use the function to approximate consumption in each of the following years.

(a) 1980 **51.47 trillion ft³**

(b) 1987 **69.602 trillion ft³**

(c) 1995 **75.646 trillion ft³**

In the United States, the intensity of an earthquake is rated using the **Richter scale.** *The Richter scale rating of an earthquake of intensity x is given by*

$$R = \log_{10} \frac{x}{x_0},$$

where x_0 is the intensity of an earthquake of a certain (small) size. The figure shows Richter scale ratings for major Southern California earthquakes since 1920. As the figure indicates, earthquakes "come in bunches" and the 1990s were an especially busy time.

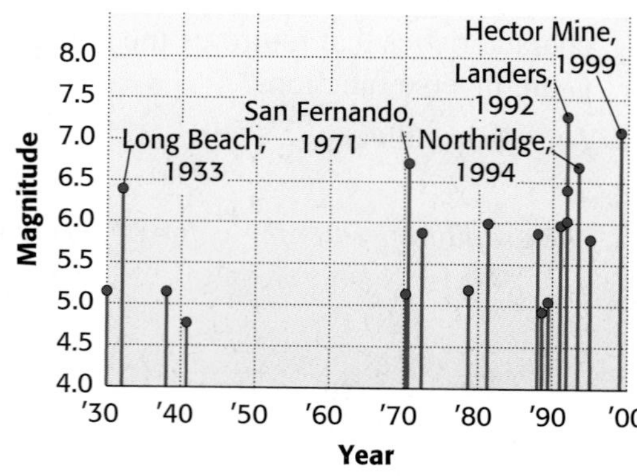

Source: Caltech; U.S. Geological Survey.

55. The 1994 Northridge earthquake had a Richter scale rating of 6.7; the 1992 Landers earthquake had a rating of 7.3. How much more powerful was the Landers earthquake than the Northridge earthquake?

about 4 times as powerful

56. Compare the smallest rated earthquake in the figure (at 4.8) with the Landers quake. How much more powerful was the Landers quake?

about 316 times as powerful

11.4 Properties of Logarithms

Logarithms have been used as an aid to numerical calculation for several hundred years. Today the widespread use of calculators has made the use of logarithms for calculation obsolete. However, logarithms are still very important in applications and in further work in mathematics.

OBJECTIVES

1. **Use the product rule for logarithms.**
2. **Use the quotient rule for logarithms.**
3. **Use the power rule for logarithms.**
4. **Use properties to write alternative forms of logarithmic expressions.**

OBJECTIVE 1 Use the product rule for logarithms. One way in which logarithms simplify problems is by changing a problem of multiplication into one of addition. We know that $\log_2 4 = 2$, $\log_2 8 = 3$, and $\log_2 32 = 5$. Since $2 + 3 = 5$,

$$\log_2 32 = \log_2 4 + \log_2 8$$
$$\log_2 (4 \cdot 8) = \log_2 4 + \log_2 8.$$

This is true in general.

Product Rule for Logarithms

If x, y, and b are positive real numbers, where $b \neq 1$, then

$$\mathbf{\log_b xy = \log_b x + \log_b y.}$$

In words, the logarithm of a product is the sum of the logarithms of the factors.

Note

The word statement of the product rule can be restated by replacing "logarithm" with "exponent." The rule then becomes the familiar rule for multiplying exponential expressions: The *exponent* of a product is equal to the sum of the *exponents* of the factors.

To prove this rule, let $m = \log_b x$ and $n = \log_b y$, and recall that

$$\log_b x = m \quad \text{means} \quad b^m = x.$$
$$\log_b y = n \quad \text{means} \quad b^n = y.$$

Now consider the product xy.

$$xy = b^m \cdot b^n \qquad \text{Substitute.}$$
$$xy = b^{m+n} \qquad \text{Product rule for exponents}$$
$$\log_b xy = m + n \qquad \text{Convert to logarithmic form.}$$
$$\log_b xy = \log_b x + \log_b y \qquad \text{Substitute.}$$

The last statement is the result we wished to prove.

EXAMPLE 1 Using the Product Rule

Use the product rule to rewrite each expression. Assume $x > 0$.

(a) $\log_5 (6 \cdot 9)$
$= \log_5 6 + \log_5 9$

(b) $\log_7 8 + \log_7 12$
$= \log_7 (8 \cdot 12)$
$= \log_7 96$

Continued on Next Page

(c) $\log_3 (3x)$

$= \log_3 3 + \log_3 x$

$= 1 + \log_3 x$ $\quad \log_3 3 = 1$

(d) $\log_4 x^3$

$= \log_4 (x \cdot x \cdot x)$ $\quad x^3 = x \cdot x \cdot x$

$= \log_4 x + \log_4 x + \log_4 x$ $\quad$ Product rule

$= 3 \log_4 x$

◀ Work Problem 1 at the Side.

OBJECTIVE 2 Use the quotient rule for logarithms. The rule for division is similar to the rule for multiplication.

Quotient Rule for Logarithms

If x, y, and b are positive real numbers, where $b \neq 1$, then

$$\log_b \frac{x}{y} = \log_b x - \log_b y.$$

In words, the logarithm of a quotient is the difference between the logarithm of the numerator and the logarithm of the denominator.

The proof of this rule is very similar to the proof of the product rule.

EXAMPLE 2 Using the Quotient Rule

Use the quotient rule to rewrite each expression.

(a) $\log_4 \frac{7}{9}$

$= \log_4 7 - \log_4 9$

(b) $\log_5 6 - \log_5 x$

$= \log_5 \frac{6}{x}, \quad x > 0$

(c) $\log_3 \frac{27}{5}$

$= \log_3 27 - \log_3 5$

$= 3 - \log_3 5$ $\quad \log_3 27 = 3$

CAUTION

There is no property of logarithms to rewrite the logarithm of a sum or difference. For example, we ***cannot*** write $\log_b (x + y)$ in terms of $\log_b x$ and $\log_b y$. Also,

$$\log_b \frac{x}{y} \neq \frac{\log_b x}{\log_b y}.$$

◀ Work Problem 2 at the Side.

1 Use the product rule to rewrite each expression.

(a) $\log_6 (5 \cdot 8)$

(b) $\log_4 3 + \log_4 7$

(c) $\log_8 8k, \quad k > 0$

(d) $\log_5 m^2, \quad m > 0$

2 Use the quotient rule to rewrite each expression.

(a) $\log_7 \frac{9}{4}$

(b) $\log_3 p - \log_3 q, \quad p > 0, \quad q > 0$

(c) $\log_4 \frac{3}{16}$

ANSWERS

1. (a) $\log_6 5 + \log_6 8$ (b) $\log_4 21$ (c) $1 + \log_8 k$ (d) $2 \log_5 m$
2. (a) $\log_7 9 - \log_7 4$ (b) $\log_3 \frac{p}{q}$ (c) $\log_4 3 - 2$

OBJECTIVE 3 Use the power rule for logarithms. An exponential expression such as 2^3 means $2 \cdot 2 \cdot 2$; the base is used as a factor 3 times. Similarly, the product rule can be extended to rewrite the logarithm of a power as the product of the exponent and the logarithm of the base. For example, by the product rule for logarithms,

$$\begin{aligned}&\log_5 2^3\\ &\quad= \log_5 (2 \cdot 2 \cdot 2)\\ &\quad= \log_5 2 + \log_5 2 + \log_5 2\\ &\quad= 3 \log_5 2.\end{aligned}$$

$$\begin{aligned}&\log_2 7^4\\ &\quad= \log_2 (7 \cdot 7 \cdot 7 \cdot 7)\\ &\quad= \log_2 7 + \log_2 7 + \log_2 7 + \log_2 7\\ &\quad= 4 \log_2 7.\end{aligned}$$

Furthermore, we saw in Example 1(d) that $\log_4 x^3 = 3 \log_4 x$. These examples suggest the following rule.

Power Rule for Logarithms

If x and b are positive real numbers, where $b \neq 1$, and if r is any real number, then

$$\log_b x^r = r \log_b x.$$

In words, the logarithm of a number to a power equals the exponent times the logarithm of the number.

As further examples of this rule, we have

$$\log_b m^5 = 5 \log_b m \quad \text{and} \quad \log_3 5^4 = 4 \log_3 5.$$

To prove the power rule, let $\log_b x = m$.

$$\begin{aligned}\log_b x &= m\\ b^m &= x && \text{Convert to exponential form.}\\ (b^m)^r &= x^r && \text{Raise to the power } r.\\ b^{mr} &= x^r && \text{Power rule for exponents}\\ \log_b x^r &= mr && \text{Convert to logarithmic form.}\\ \log_b x^r &= rm && \text{Commutative property}\\ \log_b x^r &= r \log_b x && m = \log_b x\end{aligned}$$

This is the statement to be proved.

As a special case of the power rule, let $r = \frac{1}{p}$, so

$$\log_b \sqrt[p]{x} = \log_b x^{1/p} = \frac{1}{p} \log_b x.$$

For example, using this result, with $x > 0$,

$$\log_b \sqrt[5]{x} = \log_b x^{1/5} = \frac{1}{5} \log_b x \quad \text{and} \quad \log_b \sqrt[3]{x^4} = \log_b x^{4/3} = \frac{4}{3} \log_b x.$$

Another special case is

$$\log_b \frac{1}{x} = \log_b x^{-1} = -\log_b x.$$

Note

For a review of rational exponents, refer to **Section 9.2.**

3 Use the power rule to rewrite each logarithm. Assume that $a > 0, b > 0, x > 0, a \neq 1$, and $b \neq 1$.

(a) $\log_3 5^2$

(b) $\log_a x^4$

(c) $\log_b \sqrt{8}$

(d) $\log_2 \sqrt[3]{2}$

4 Find the value of each logarithmic expression.

(a) $\log_{10} 10^3$

(b) $\log_2 8$

(c) $5^{\log_5 3}$

EXAMPLE 3 Using the Power Rule

Use the power rule to rewrite each logarithm. Assume that $b > 0, x > 0$, and $b \neq 1$.

(a) $\log_5 4^2$
$= 2 \log_5 4$

(b) $\log_b x^5$
$= 5 \log_b x$

(c) $\log_b \sqrt{7}$

When using the power rule with logarithms of expressions involving radicals, begin by rewriting the radical expression with a rational exponent.

$$\log_b \sqrt{7}$$
$$= \log_b 7^{1/2} \qquad \sqrt{x} = x^{1/2}$$
$$= \frac{1}{2} \log_b 7 \qquad \text{Power rule}$$

(d) $\log_2 \sqrt[5]{x^2}$

$$= \log_2 x^{2/5} \qquad \sqrt[5]{x^2} = x^{2/5}$$
$$= \frac{2}{5} \log_2 x \qquad \text{Power rule}$$

◀ Work Problem **3** at the Side.

Two special properties involving both exponential and logarithmic expressions come directly from the fact that logarithmic and exponential functions are inverses of each other.

Special Properties

If $b > 0$ and $b \neq 1$, then

$$b^{\log_b x} = x, \; x > 0 \quad \text{and} \quad \log_b b^x = x.$$

To prove the first statement, let $y = \log_b x$.

$$y = \log_b x$$
$$b^y = x \qquad \text{Convert to exponential form.}$$
$$b^{\log_b x} = x \qquad \text{Replace } y \text{ with } \log_b x.$$

The proof of the second statement is similar.

EXAMPLE 4 Using the Special Properties

Find the value of each logarithmic expression.

(a) $\log_5 5^4 = 4$, since $\log_b b^x = x$.

(b) $\log_3 9$
$= \log_3 3^2 \qquad 9 = 3^2$
$= 2$

(c) $4^{\log_4 10} = 10$, since $b^{\log_b x} = x$.

◀ Work Problem **4** at the Side.

ANSWERS

3. (a) $2 \log_3 5$ (b) $4 \log_a x$ (c) $\frac{1}{2} \log_b 8$ (d) $\frac{1}{3}$

4. (a) 3 (b) 3 (c) 3

Here is a summary of the properties of logarithms.

Properties of Logarithms

If x, y, and b are positive real numbers, where $b \neq 1$, and r is any real number, then

Product Rule $\log_b xy = \log_b x + \log_b y$

Quotient Rule $\log_b \frac{x}{y} = \log_b x - \log_b y$

Power Rule $\log_b x^r = r \log_b x$

Special Properties $b^{\log_b x} = x$ and $\log_b b^x = x$.

OBJECTIVE 4 Use properties to write alternative forms of logarithmic expressions. Applying the properties of logarithms is important for solving equations with logarithms.

EXAMPLE 5 Writing Logarithms in Alternative Forms

Use the properties of logarithms to rewrite each expression. Assume that all variables represent positive real numbers.

(a) $\log_4 4x^3$

$= \log_4 4 + \log_4 x^3$ Product rule

$= 1 + 3 \log_4 x$ $\log_4 4 = 1$; power rule

(b) $\log_7 \sqrt{\frac{m}{n}}$

$= \log_7 \left(\frac{m}{n}\right)^{1/2}$ Write the radical expression with a rational exponent.

$= \frac{1}{2} \log_7 \frac{m}{n}$ Power rule

$= \frac{1}{2}(\log_7 m - \log_7 n)$ Quotient rule

(c) $\log_5 \frac{a^2}{bc}$

$= \log_5 a^2 - \log_5 bc$ Quotient rule

$= 2 \log_5 a - \log_5 bc$ Power rule

$= 2 \log_5 a - (\log_5 b + \log_5 c)$ Product rule

$= 2 \log_5 a - \log_5 b - \log_5 c$

Use parentheses to avoid errors.

(d) $4 \log_b m - \log_b n$

$= \log_b m^4 - \log_b n$ Power rule

$= \log_b \frac{m^4}{n}$ Quotient rule

Continued on Next Page

5 Use the properties of logarithms to rewrite each expression. Assume that all variables represent positive real numbers.

(a) $\log_6 36m^5$

(b) $\log_2 \sqrt{9z}$

(c) $\log_q \dfrac{8r^2}{m-1}$, $m > 1, q \neq 1$

(d) $2 \log_a x + 3 \log_a y$, $a \neq 1$

(e) $\log_4 (3x + y)$

6 Decide whether each statement is *true* or *false*.

(a) $\log_6 36 - \log_6 6 = \log_6 30$

(b) $\log_4 (\log_2 16) = \dfrac{\log_6 6}{\log_6 36}$

(e) $\log_b (x + 1) + \log_b (2x + 1) - \dfrac{2}{3} \log_b x$

$= \log_b (x + 1) + \log_b (2x + 1) - \log_b x^{2/3}$ Power rule

$= \log_b \dfrac{(x + 1)(2x + 1)}{x^{2/3}}$ Product and quotient rules

$= \log_b \dfrac{2x^2 + 3x + 1}{x^{2/3}}$ Multiply in the numerator.

(f) $\log_8 (2p + 3r)$ cannot be rewritten using the properties of logarithms. There is no property of logarithms that allows us to rewrite the logarithm of a sum.

◀ *Work Problem* **5** *at the Side.*

EXAMPLE 6 **Deciding Whether Statements about Logarithms Are True**

Decide whether each statement is *true* or *false*.

(a) $\log_2 8 - \log_2 4 = \log_2 4$

Evaluate both sides.

Left side: $\log_2 8 - \log_2 4 = \log_2 2^3 - \log_2 2^2 = 3 - 2 = 1$

Right side: $\log_2 4 = \log_2 2^2 = 2$

The statement is false because $1 \neq 2$.

(b) $\log_3 (\log_2 8) = \dfrac{\log_7 49}{\log_8 64}$

Evaluate both sides.

Left side: $\log_3 (\log_2 8) = \log_3 3 = 1$

Right side: $\dfrac{\log_7 49}{\log_8 64} = \dfrac{\log_7 7^2}{\log_8 8^2} = \dfrac{2}{2} = 1$

The statement is true because $1 = 1$.

◀ *Work Problem* **6** *at the Side.*

ANSWERS

5. **(a)** $2 + 5 \log_6 m$ **(b)** $\log_2 3 + \frac{1}{2} \log_2 z$
(c) $\log_q 8 + 2 \log_q r - \log_q (m - 1)$
(d) $\log_a x^2 y^3$ **(e)** cannot be rewritten

6. **(a)** false **(b)** false

11.4 Exercises

Decide whether each statement of a logarithmic property is true *or* false. *If it is* false, *correct it by changing the right side of the equation.*

1. $\log_b x + \log_b y = \log_b (x + y)$

false; $\log_b x + \log_b y = \log_b xy$

2. $\log_b \frac{x}{y} = \log_b x - \log_b y$

true

3. $\log_b b^x = x$

true

4. $\log_b x^r = \log_b rx$

false; $\log_b x^r = r \log_b x$

Use the properties of logarithms introduced in this section to express each logarithm as a sum or difference of logarithms, or as a single number if possible. Assume that all variables represent positive real numbers. See Examples 1–5.

5. $\log_7 \frac{4}{5}$

$\log_7 4 - \log_7 5$

6. $\log_8 \frac{9}{11}$

$\log_8 9 - \log_8 11$

7. $\log_2 8^{1/4}$

$\frac{1}{4}\log_2 8$, or $\frac{3}{4}$

8. $\log_3 9^{3/4}$

$\frac{3}{4}\log_3 9$, or $\frac{3}{2}$

9. $\log_4 \frac{3\sqrt{x}}{y}$

$\log_4 3 + \frac{1}{2}\log_4 x - \log_4 y$

10. $\log_5 \frac{6\sqrt{z}}{w}$

$\log_5 6 + \frac{1}{2}\log_5 z - \log_5 w$

11. $\log_3 \frac{\sqrt[3]{4}}{x^2 y}$

$\frac{1}{3}\log_3 4 - 2\log_3 x - \log_3 y$

12. $\log_7 \frac{\sqrt[3]{13}}{pq^2}$

$\frac{1}{3}\log_7 13 - \log_7 p - 2\log_7 q$

13. $\log_3 \sqrt{\frac{xy}{5}}$

$\frac{1}{2}\log_3 x + \frac{1}{2}\log_3 y - \frac{1}{2}\log_3 5$

14. $\log_6 \sqrt{\frac{pq}{7}}$

$\frac{1}{2}\log_6 p + \frac{1}{2}\log_6 q - \frac{1}{2}\log_6 7$

15. $\log_2 \frac{\sqrt[3]{x} \cdot \sqrt[5]{y}}{r^2}$

$\frac{1}{3}\log_2 x + \frac{1}{5}\log_2 y - 2\log_2 r$

16. $\log_4 \frac{\sqrt[4]{z} \cdot \sqrt[5]{w}}{s^2}$

$\frac{1}{4}\log_4 z + \frac{1}{5}\log_4 w - 2\log_4 s$

17. A student erroneously wrote

$$\log_a (x + y) = \log_a x + \log_a y.$$

When his teacher explained that this was wrong, the student claimed he had used the distributive property. Write a few sentences explaining why the distributive property does not apply in this case.

The distributive property tells us that the *product* $a(x + y)$ equals the sum $ax + ay$. In the notation $\log_a (x + y)$, the parentheses do not indicate multiplication. They indicate that $x + y$ is the result of raising a to some power.

18. Write a few sentences explaining how the rules for multiplying and dividing powers of the same base are similar to the rules for finding logarithms of products and quotients.

Logarithms of products and quotients behave exactly as the exponents in products and quotients of exponentials: $a^m \cdot a^n = a^{m+n}$ and $\log_a mn = \log_a m + \log_a n$.

Also, $\frac{a^m}{a^n} = a^{m-n}$ and $\log_a \frac{m}{n} = \log_a m - \log_a n$.

Use the properties of logarithms introduced in this section to rewrite each expression as a single logarithm. Assume that all variables are defined in such a way that the variable expressions are positive, and bases are positive numbers not equal to 1. See Examples 1–5.

19. $\log_b x + \log_b y$

$\log_b xy$

20. $\log_b 2 + \log_b z$

$\log_b 2z$

21. $3 \log_a m - \log_a n$

$\log_a \frac{m^3}{n}$

22. $5 \log_b x - \log_b y$

$\log_b \frac{x^5}{y}$

23. $(\log_a r - \log_a s) + 3 \log_a t$

$\log_a \frac{rt^3}{s}$

24. $(\log_a p - \log_a q) + 2 \log_a r$

$\log_a \frac{pr^2}{q}$

25. $3 \log_a 5 - 4 \log_a 3$

$\log_a \frac{125}{81}$

26. $3 \log_a 5 + \frac{1}{2} \log_a 9$

$\log_a 375$

27. $\log_{10}(x + 3) + \log_{10}(x - 3)$

$\log_{10}(x^2 - 9)$

28. $\log_{10}(y + 4) + \log_{10}(y - 4)$

$\log_{10}(y^2 - 16)$

29. $3 \log_p x + \frac{1}{2} \log_p y - \frac{3}{2} \log_p z - 3 \log_p a$

$\log_p \frac{x^3 y^{1/2}}{z^{3/2} a^3}$

30. $\frac{1}{3} \log_b x + \frac{2}{3} \log_b y - \frac{3}{4} \log_b s - \frac{2}{3} \log_b t$

$\log_b \frac{x^{1/3} y^{2/3}}{s^{3/4} t^{2/3}}$

Decide whether each statement is true *or* false. *See Example 6.*

31. $\log_2(8 + 32) = \log_2 8 + \log_2 32$

false

32. $\log_2(64 - 16) = \log_2 64 - \log_2 16$

false

33. $\log_3 7 + \log_3 7^{-1} = 0$

true

34. $\log_9 14 - \log_{14} 9 = 0$

false

35. $\log_6 60 - \log_6 10 = 1$

true

36. $\log_3 8 + \log_3 \frac{1}{8} = 0$

true

37. $\frac{\log_{10} 7}{\log_{10} 14} = \frac{1}{2}$

false

38. $\frac{\log_{10} 10}{\log_{10} 100} = \frac{1}{10}$

false

Relating Concepts (Exercises 39–44) For Individual or Group Work

Work Exercises 39–44 in order.

39. Evaluate $\log_3 81$.

4

40. Write the *meaning* of the expression $\log_3 81$.

It is the exponent to which 3 must be raised in order to obtain 81.

41. Evaluate $3^{\log_3 81}$.

81

42. Write the *meaning* of the expression $\log_2 19$.

It is the exponent to which 2 must be raised in order to obtain 19.

43. Evaluate $2^{\log_2 19}$.

19

44. Keeping in mind that a logarithm is an exponent, and using the results from Exercises 39–43, what is the simplest form of the expression $k^{\log_k m}$?

m

11.5 Common and Natural Logarithms

Logarithms are important in many applications in biology, engineering, economics, and social science. In this section we find numerical approximations for logarithms. Traditionally, base 10 logarithms were used most often because our number system is base 10. Logarithms to base 10 are called **common logarithms,** and $\log_{10} x$ is abbreviated as simply $\log x$, where the base is understood to be 10.

OBJECTIVES

1. Evaluate common logarithms using a calculator.
2. Use common logarithms in applications.
3. Evaluate natural logarithms using a calculator.
4. Use natural logarithms in applications.

OBJECTIVE 1 Evaluate common logarithms using a calculator.

Calculator Tip In Example 1, we give the results of evaluating some common logarithms using a calculator with a LOG key. (This may be a second function key on some calculators.) For simple scientific calculators, just enter the number, then press the LOG key. For graphing calculators, these steps are reversed. In this section, we give calculator approximations for logarithms to four decimal places.

EXAMPLE 1 Evaluating Common Logarithms

Evaluate each logarithm using a calculator.

(a) $\log 327.1 \approx 2.5147$ **(b)** $\log 437{,}000 \approx 5.6405$

(c) $\log 0.0615 \approx -1.2111$

Notice in part (c) that $\log 0.0615 \approx -1.2111$, a negative result. ***The common logarithm of a number between 0 and 1 is always negative*** because the logarithm is the exponent on 10 that produces the number. Thus,

$$10^{-1.2111} \approx 0.0615.$$

If the exponent (the logarithm) were positive, the result would be greater than 1. See Figure 13.

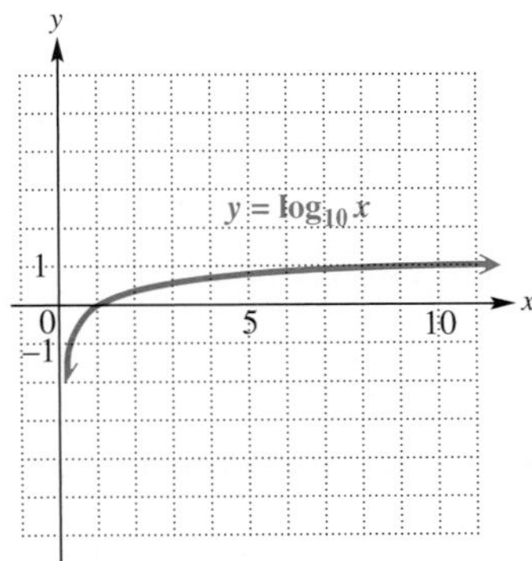

Figure 13

Work Problem 1 at the Side. ▶

1 Evaluate each logarithm to four decimal places using a calculator.

(a) $\log 41{,}600$

(b) $\log 43.5$

(c) $\log 0.442$

OBJECTIVE 2 Use common logarithms in applications. In chemistry, pH is a measure of the acidity or alkalinity of a solution; water, for example, has pH 7. In general, acids have pH numbers less than 7, and alkaline solutions have pH values greater than 7. The **pH** of a solution is defined as

$$\mathbf{pH = -\log[H_3O^+]},$$

where $[H_3O^+]$ is the hydronium ion concentration in moles per liter. It is customary to round pH values to the nearest tenth.

ANSWERS

1. **(a)** 4.6191 **(b)** 1.6385 **(c)** -0.3546

Figure 14 illustrates the pH scale.

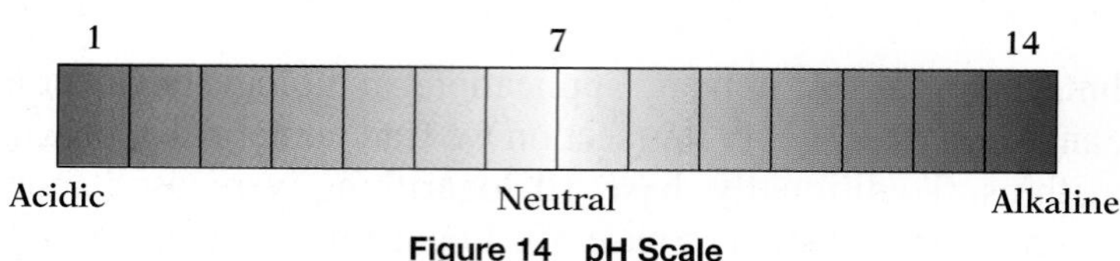

Figure 14 pH Scale

EXAMPLE 2 Using pH in an Application

Wetlands are classified as *bogs, fens, marshes,* and *swamps*. These classifications are based on pH values. A pH value between 6.0 and 7.5, such as that of Summerby Swamp in Michigan's Hiawatha National Forest, indicates that the wetland is a "rich fen." When the pH is between 4.0 and 6.0, the wetland is a "poor fen," and if the pH falls to 3.0 or less, it is a "bog." (*Source:* Mohlenbrock, R., "Summerby Swamp, Michigan," *Natural History,* March 1994.)

Suppose that the hydronium ion concentration of a sample of water from a wetland is 6.3×10^{-3}. How would this wetland be classified?

Use the definition of pH.

$$\begin{aligned} \text{pH} &= -\log(6.3 \times 10^{-3}) && \text{Definition of pH} \\ \text{pH} &= -(\log 6.3 + \log 10^{-3}) && \text{Product rule} \\ \text{pH} &= -[0.7993 - 3(1)] && \text{Use a calculator to find log 6.3.} \\ \text{pH} &= -0.7993 + 3 && \text{Distributive property} \\ \text{pH} &\approx 2.2 \end{aligned}$$

Since the pH is less than 3.0, the wetland is a bog.

◀ Work Problem **2** at the Side.

EXAMPLE 3 Finding Hydronium Ion Concentration

Find the hydronium ion concentration of drinking water with pH 6.5.

$$\begin{aligned} \text{pH} &= -\log[H_3O^+] \\ 6.5 &= -\log[H_3O^+] && \text{Let pH} = 6.5. \\ \log[H_3O^+] &= -6.5 && \text{Multiply by } -1. \end{aligned}$$

Solve for $[H_3O^+]$ by writing the equation in exponential form, remembering that the base is 10.

$$\begin{aligned} [H_3O^+] &= 10^{-6.5} \\ [H_3O^+] &\approx 3.2 \times 10^{-7} && \text{Use a calculator.} \end{aligned}$$

◀ Work Problem **3** at the Side.

2 Solve the problem.

Find the pH of water with a hydronium ion concentration of 1.2×10^{-3}. If this water had been taken from a wetland, is the wetland a rich fen, a poor fen, or a bog?

3 Find the hydronium ion concentrations of solutions with the following pH values.

(a) 4.6

(b) 7.5

ANSWERS

2. 2.9; bog

3. **(a)** 2.5×10^{-5} **(b)** 3.2×10^{-8}

The loudness of sound is measured in a unit called a decibel, abbreviated **dB.** To measure with this unit, we first assign an intensity of I_0 to a very faint sound, called the **threshold sound.** If a particular sound has intensity I, then the decibel level of this louder sound is

$$D = 10 \log \left(\frac{I}{I_0} \right).$$

The table in the margin gives average decibel levels for some common sounds. Any sound over 85 dB exceeds what hearing experts consider safe. Permanent hearing damage can be suffered at levels above 150 dB.

Decibel Level	Example
60	Normal conversation
90	Rush hour traffic, lawn mower
100	Garbage truck, chain saw, pneumatic drill
120	Rock concert, thunderclap
140	Gunshot blast, jet engine
180	Rocket launching pad

Source: Deafness Research Foundation.

EXAMPLE 4 Measuring the Loudness of Sound

If music downloaded from a computer and delivered through the earphones of an iPod has intensity I of $(3.162 \times 10^{11})\, I_0$, find the average decibel level. (*Source: Sacramento Bee.*)

$D = 10 \log \left(\frac{I}{I_0} \right)$ Substitute the given value for I.

$D = 10 \log \left(\frac{(3.162 \times 10^{11}) I_0}{I_0} \right)$

$D = 10 \log (3.162 \times 10^{11})$

$D \approx 115$ dB

Work Problem 4 at the Side. ▶

4 Find the decibel level to the nearest whole number of a whisper with intensity I of $115\, I_0$.

OBJECTIVE 3 Evaluate natural logarithms using a calculator. The most important logarithms used in applications are **natural logarithms,** which have as base the number e. The number $\boldsymbol{e}$ is a fundamental number in our universe. For this reason e, like π, is called a **universal constant.** The letter e is used to honor Leonhard Euler, who published extensive results on the number in 1748. Since it is an irrational number, its decimal expansion never terminates and never repeats. The first few digits of the decimal value of e are **2.718281828.**

Calculator Tip A calculator key e^x or the two keys INV and LN are used to approximate powers of e. For example, a calculator gives

$$e^2 \approx 7.389056099, \quad e^3 \approx 20.08553692, \quad \text{and} \quad e^{0.6} \approx 1.8221188.$$

Logarithms to base e are called natural logarithms because they occur in biology and the social sciences in natural situations that involve growth or decay. The base e logarithm of x is written **ln $\boldsymbol{x}$** (read **"el en $\boldsymbol{x}$"**). A graph of $y = \ln x$, the equation that defines the natural logarithmic function, is given in Figure 15.

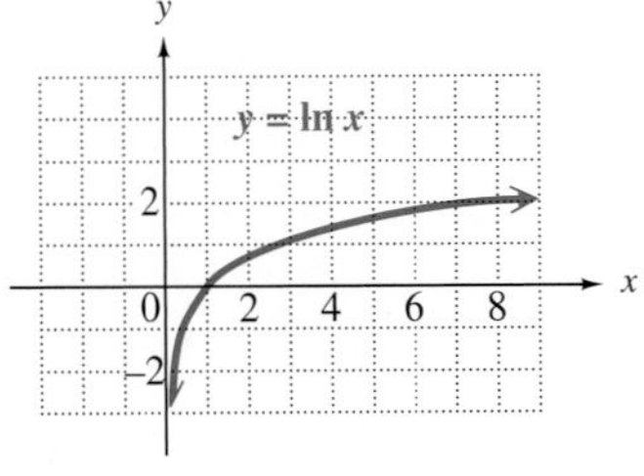

Figure 15

ANSWER

4. 21 dB

5 Find each logarithm to four decimal places.

(a) ln 0.01

(b) ln 27

(c) ln 529

6 Use the natural logarithmic function in Example 6 to approximate the altitude at 700 millibars of pressure.

Calculator Tip A calculator key labeled LN is used to evaluate natural logarithms. If your scientific calculator has an e^x key, but not a key labeled LN, find natural logarithms by entering the number, pressing the INV key, and then pressing the e^x key. This works because $y = e^x$ defines the inverse function of $y = \ln x$ (or $y = \log_e x$).

EXAMPLE 5 Finding Natural Logarithms

Find each logarithm to four decimal places.

(a) $\ln 0.5841 \approx -0.5377$

As with common logarithms, ***a number between 0 and 1 has a negative natural logarithm.***

(b) $\ln 192.7 \approx 5.2611$ **(c)** $\ln 10.84 \approx 2.3832$

Work Problem **5** *at the Side.*

OBJECTIVE 4 Use natural logarithms in applications. A common application of natural logarithmic functions is to express growth or decay of a quantity, as in the next example.

EXAMPLE 6 Applying Natural Logarithms

The altitude in meters that corresponds to an atmospheric pressure of x millibars is given by the natural logarithmic function defined by

$$f(x) = 51{,}600 - 7457 \ln x.$$

(*Source:* Miller, A. and J. Thompson, *Elements of Meteorology,* Fourth Edition, Charles E. Merrill Publishing Company, 1993.) Use this function to find the altitude when atmospheric pressure is 400 millibars.

Let $x = 400$ and substitute in the expression for $f(x)$.

$$f(400) = 51{,}600 - 7457 \ln 400$$

$$f(400) \approx 6900 \text{ (to the nearest hundred)}$$

Atmospheric pressure is 400 millibars at approximately 6900 m.

Calculator Tip In Example 6, the final answer was obtained using a calculator *without* rounding the intermediate values. In general, it is best to wait until the final step to round the answer; otherwise, a buildup of round-off error may cause the final answer to have an incorrect final decimal place digit.

Work Problem **6** *at the Side.*

ANSWERS

5. **(a)** -4.6052 **(b)** 3.2958 **(c)** 6.2710

6. approximately 2700 m

11.5 Exercises

FOR EXTRA HELP MyMathLab Math XL PRACTICE WATCH DOWNLOAD READ REVIEW

Choose the correct response in Exercises 1–4.

1. What is the base in the expression $\log x$?
   A. e **B.** 1 **C.** 10 **D.** x
   C

2. What is the base in the expression $\ln x$?
   A. e **B.** 1 **C.** 10 **D.** x
   A

3. Since $10^0 = 1$ and $10^1 = 10$, between what two consecutive integers is the value of log 5.6?
   A. 5 and 6 **B.** 10 and 11 **C.** 0 and 1 **D.** -1 and 0
   C

4. Since $e^1 \approx 2.718$ and $e^2 \approx 7.389$, between what two consecutive integers is the value of ln 5.6?
   A. 5 and 6 **B.** 2 and 3 **C.** 1 and 2 **D.** 0 and 1
   C

5. Without using a calculator, give the value of $\log 10^{19.2}$.
   19.2

6. Without using a calculator, give the value of $\ln e^{\sqrt{2}}$.
   $\sqrt{2}$

You will need a calculator for the remaining exercises in this set.

Find each logarithm. Give an approximation to four decimal places. See Examples 1 and 5.

7. log 328.4
   2.5164

8. log 457.2
   2.6601

9. log 0.0326
   −1.4868

10. log 0.1741
    −0.7592

11. $\log (4.76 \times 10^9)$
    9.6776

12. $\log (2.13 \times 10^4)$
    4.3284

13. ln 7.84
    2.0592

14. ln 8.32
    2.1187

15. ln 0.0556
    −2.8896

16. ln 0.0217
    −3.8304

17. ln 10
    2.3026

18. $\log e$
    0.4343

Refer to Example 2. In Exercises 19–21, suppose that water from a wetland area is sampled and found to have the given hydronium ion concentration. Determine whether the wetland is a rich fen, *a* poor fen, *or a* bog.

19. 2.5×10^{-5}
    poor fen

20. 2.5×10^{-2}
    bog

21. 2.5×10^{-7}
    rich fen

Find the pH *of the substance with the given hydronium ion concentration. See Example 2.*

22. Ammonia, 2.5×10^{-12}
    11.6

23. Tuna, 1.3×10^{-6}
    5.9

24. Grapes, 5.0×10^{-5}
    4.3

Use the formula for pH *to find the hydronium ion concentration of the substance with the given* pH. *See Example 3.*

25. Human gastric contents, 2.0

1.0 × 10⁻²

26. Human blood plasma, 7.4

4.0×10^{-8}

27. Bananas, 4.6

2.5×10^{-5}

28. Spinach, 5.4

4.0×10^{-6}

Solve each problem. See Examples 4 and 6.

29. The time t in years for an amount increasing at a rate of r (in decimal form) to double is given by

$$t = \frac{\ln 2}{\ln(1 + r)}.$$

This is called **doubling time.** Find the doubling time to the nearest tenth for an investment at each interest rate.

(a) $2\% = 0.02$ **35.0 yr**

(b) $5\% = 0.05$ **14.2 yr**

30. The number of years, $N(r)$, since two independently evolving languages split off from a common ancestral language is approximated by

$$N(r) = -5000 \ln r,$$

where r is the percent of words (as a decimal) from the ancestral language common to both languages now. Find the number of years (to the nearest hundred) since the split for each percent of common words.

(a) 85% (or 0.85) **800 yr**

(b) 35% (or 0.35) **5200 yr**

31. In the central Sierra Nevada of California, the percent of moisture p that falls as snow rather than rain is approximated reasonably well by

$$p(h) = 86.3 \ln h - 680,$$

where h is the altitude in feet.

(a) What percent of the moisture at 5000 ft falls as snow?

55%

(b) What percent at 7500 ft falls as snow?

90%

32. Use the formula from Example 4 to find the decibel level of each sound. (*Source:* The Canadian Society of Otolaryngology.)

(a) noisy restaurant: $I = 10^8 I_0$

80 dB

(b) farm tractor: $I = (6.310 \times 10^9) I_0$

98 dB

(c) snowmobile: $I = 31{,}622{,}776{,}600\, I_0$

105 dB

33. The age in years of a female blue whale is approximated by

$$t = -2.57 \ln\left(\frac{87 - L}{63}\right),$$

where L is its length in feet.

(a) How old is a female blue whale that measures 80 ft?

5.6 yr

(b) The equation that defines t has domain $24 < L < 87$. Explain why.

$t > 0$ and $\frac{87 - L}{63}$ is positive and in the domain of the function only if $24 < L < 87$.

34. The **cost-benefit equation**

$$T = -0.642 - 189 \ln(1 - p)$$

describes the approximate tax T, in dollars per ton, that would result in a $p\%$ (in decimal form) reduction in carbon dioxide emissions.

(a) What tax will reduce emissions 25%?

\$54 per ton

(b) Explain why the equation is not valid for $p = 0$ or $p = 1$.

If $p = 0$, then $\ln(1 - p) = \ln 1 = 0$, so T would be negative. If $p = 1$, then $\ln(1 - p) = \ln 0$, but the domain of $\ln x$ is $(0, \infty)$.

11.6 Exponential and Logarithmic Equations; Further Applications

We solved exponential and logarithmic equations in **Sections 11.2 and 11.3.** General methods for solving these equations depend on the following properties.

Properties for Solving Exponential and Logarithmic Equations

For all real numbers $b > 0$, $b \neq 1$, and any real numbers x and y:

1. If $x = y$, then $b^x = b^y$.

2. If $b^x = b^y$, then $x = y$.

3. If $x = y$, and $x > 0$, $y > 0$, then $\log_b x = \log_b y$.

4. If $x > 0$, $y > 0$, and $\log_b x = \log_b y$, then $x = y$.

We used Property 2 to solve exponential equations in **Section 11.2.**

OBJECTIVES

1. Solve equations involving variables in the exponents.
2. Solve equations involving logarithms.
3. Solve applications of compound interest.
4. Solve applications involving base *e* exponential growth and decay.
5. Use the change-of-base rule.

OBJECTIVE 1 Solve equations involving variables in the exponents. In Examples 1 and 2, we use Property 3.

EXAMPLE 1 Solving an Exponential Equation

Solve $3^x = 12$.

$$3^x = 12$$

$$\log 3^x = \log 12 \quad \text{Property 3 (common logs)}$$

$$x \log 3 = \log 12 \quad \text{Power rule}$$

$$\text{Exact solution} \longrightarrow x = \frac{\log 12}{\log 3} \quad \text{Divide by log 3.}$$

$$\text{Decimal approximation} \longrightarrow x \approx 2.262 \quad \text{Use a calculator.}$$

The solution set is $\{2.262\}$. Check with a calculator that $3^{2.262} \approx 12$.

Work Problem 1 at the Side. ▶

CAUTION

Be careful: $\frac{\log 12}{\log 3}$ is ***not*** equal to log 4. Note that $\log 4 \approx 0.6021$, but $\frac{\log 12}{\log 3} \approx 2.262$.

EXAMPLE 2 Solving an Exponential Equation (Base *e*)

Solve $e^{0.003x} = 40$.

$$\ln e^{0.003x} = \ln 40 \quad \text{Property 3 (natural logs)}$$

$$0.003x \ln e = \ln 40 \quad \text{Power rule}$$

$$0.003x = \ln 40 \quad \ln e = \ln e^1 = 1$$

$$x = \frac{\ln 40}{0.003} \quad \text{Divide by 0.003.}$$

$$x \approx 1230 \quad \text{Use a calculator.}$$

The solution set is $\{1230\}$. Check that $e^{0.003(1230)} \approx 40$.

Work Problem 2 at the Side. ▶

1 Solve each equation and give the decimal approximation to three places.

(a) $2^x = 9$

(b) $10^x = 4$

2 Solve $e^{-0.01t} = 0.38$.

ANSWERS

1. (a) $\{3.170\}$ **(b)** $\{0.602\}$

2. $\{96.8\}$

3 Solve $\log_3 (x + 1)^5 = 3$. Give the exact solution.

General Method for Solving an Exponential Equation

Take logarithms of the same base on both sides of the equation and then use the power rule of logarithms or the special property $\log_b b^x = x$. (See Examples 1 and 2.) As a special case, if both sides can be written as exponentials with the same base, do so, and then set the exponents equal. (See **Section 11.2.**)

OBJECTIVE 2 Solve equations involving logarithms. The properties of logarithms from **Section 11.4** are useful here, as is using the definition of a logarithm to change the equation to exponential form.

EXAMPLE 3 Solving a Logarithmic Equation

Solve $\log_2 (x + 5)^3 = 4$. Give the exact solution.

$$(x + 5)^3 = 2^4 \quad \text{Convert to exponential form.}$$

$$(x + 5)^3 = 16$$

$$x + 5 = \sqrt[3]{16} \quad \text{Take the cube root on each side.}$$

$$x = -5 + \sqrt[3]{16} \quad \text{Subtract 5.}$$

$$x = -5 + 2\sqrt[3]{2} \quad \sqrt[3]{16} = \sqrt[3]{8 \cdot 2} = \sqrt[3]{8} \cdot \sqrt[3]{2} = 2\sqrt[3]{2}$$

Verify that the solution satisfies the equation, so the solution set is $\{-5 + 2\sqrt[3]{2}\}$.

◀ Work Problem **3** at the Side.

4 Solve

$\log_8 (2x + 5) + \log_8 3 = \log_8 33$.

CAUTION

Recall that the domain of $y = \log_b x$ is $(0, \infty)$. For this reason, ***it is always necessary to check that the solution of an equation with logarithms yields only logarithms of positive numbers in the original equation.***

EXAMPLE 4 Solving a Logarithmic Equation

Solve $\log_2 (x + 1) - \log_2 x = \log_2 7$.

$$\log_2 (x + 1) - \log_2 x = \log_2 7$$

Transform the left side to an expression with only *one* logarithm.

$$\log_2 \frac{x + 1}{x} = \log_2 7 \quad \text{Quotient rule}$$

$$\frac{x + 1}{x} = 7 \quad \text{Property 4}$$

$$x + 1 = 7x \quad \text{Multiply by } x.$$

$$1 = 6x \quad \text{Subtract } x.$$

$$\frac{1}{6} = x \quad \text{Divide by 6.}$$

Check this solution by substituting in the original equation. Since we cannot take the logarithm of a *nonpositive* number, both $x + 1$ and x must be positive. If $x = \frac{1}{6}$, this condition is satisfied, so the solution set is $\{\frac{1}{6}\}$.

◀ Work Problem **4** at the Side.

Answers

3. $\{-1 + \sqrt[5]{27}\}$
4. $\{3\}$

EXAMPLE 5 **Solving a Logarithmic Equation**

Solve $\log x + \log(x - 21) = 2$.

$\log x + \log(x - 21) = 2$

$\log x(x - 21) = 2$ Product rule

The base is 10.

$x(x - 21) = 10^2$ Write in exponential form.

$x^2 - 21x = 100$ Distributive property; multiply.

$x^2 - 21x - 100 = 0$ Subtract 100; standard form

$(x - 25)(x + 4) = 0$ Factor.

$x - 25 = 0$ or $x + 4 = 0$ Zero-factor property

$x = 25$ or $x = -4$ Solve each equation.

The value -4 must be rejected as a solution since it leads to the logarithm of at least one negative number in the original equation.

$\log(-4) + \log(-4 - 21) = 2$ The left side is undefined.

Check that the only solution is 25, so the solution set is $\{25\}$.

CAUTION

Do not reject a proposed solution just because it is nonpositive. Reject any value that leads to the logarithm of a nonpositive number.

Work Problem **5** *at the Side.* ▶

5 Solve

$\log_3 2x - \log_3(3x + 15) = -2.$

Solving a Logarithmic Equation

Step 1 **Transform the equation so that a single logarithm appears on one side.** Use the product rule or quotient rule of logarithms to do this.

Step 2 **(a) Use Property 4.** If $\log_b x = \log_b y$, then $x = y$. (See Example 4.)

(b) Write the equation in exponential form. If $\log_b x = k$, then $x = b^k$. (See Examples 3 and 5.)

OBJECTIVE 3 Solve applications of compound interest. So far in this book, applications involving interest have been limited to simple interest using the formula $I = prt$. In most cases, interest paid or charged is **compound interest** (interest paid on both principal and interest). The formula for compound interest is an application of exponential functions.

Compound Interest Formula (for a Finite Number of Periods)

If a principal of P dollars is deposited at an annual rate of interest r compounded (paid) n times per year, the account will contain

$$A = P\left(1 + \frac{r}{n}\right)^{nt}$$

dollars after t years. (In this formula, r is expressed as a decimal.)

ANSWER

5. $\{1\}$

6 Find the value of $2000 deposited at 5% compounded annually for 10 yr.

7 Find the number of years it will take for $500 deposited in an account paying 4% interest compounded semiannually to double.

EXAMPLE 6 Solving a Compound Interest Problem for *A*

How much money will there be in an account at the end of 5 yr if $1000 is deposited at 6% compounded quarterly? (Assume no withdrawals are made.)

Because interest is compounded quarterly, $n = 4$. The other values given in the problem are $P = 1000$, $r = 0.06$ (because $6\% = 0.06$), and $t = 5$.

$$A = P\left(1 + \frac{r}{n}\right)^{nt} \quad \text{Compound interest formula}$$

$$A = 1000\left(1 + \frac{0.06}{4}\right)^{4 \cdot 5} \quad \text{Substitute the given values.}$$

$$A = 1000(1.015)^{20}$$

$$A \approx 1346.86 \quad \text{Use a calculator.}$$

To the nearest cent, the account will contain $1346.86. (The actual amount of interest earned is $1346.86 − $1000 = $346.86. Why?)

◀ Work Problem 6 at the Side.

EXAMPLE 7 Solving a Compound Interest Problem for *t*

Suppose inflation is averaging 3% per year. How many years will it take for prices to double?

We want to find the number of years t for $1 to grow to $2 at a rate of 3% per year. In the compound interest formula, we let $A = 2$, $P = 1$, $r = 0.03$, and $n = 1$.

$$2 = 1\left(1 + \frac{0.03}{1}\right)^{1t} \quad \text{Substitute in the compound interest formula.}$$

$$2 = (1.03)^t \quad \text{Simplify.}$$

$$\log 2 = \log (1.03)^t \quad \text{Property 3}$$

$$\log 2 = t \log 1.03 \quad \text{Power rule}$$

$$t = \frac{\log 2}{\log 1.03} \quad \text{Divide by log 1.03; rewrite.}$$

$$t \approx 23.45 \quad \text{Use a calculator.}$$

Prices will double in about 23 yr. (This is called the **doubling time** of the money.) To check, verify that $1.03^{23.45} \approx 2$.

◀ Work Problem 7 at the Side.

Interest can be compounded annually, semiannually, quarterly, daily, and so on. If the number of compounding periods n is allowed to approach infinity, we have an example of **continuous compounding.** However, the compound interest formula above cannot be used for continuous compounding since there is no finite value for n. The formula for continuous compounding is an example of exponential growth involving the number e.

Continuous Compound Interest Formula

If a principal of P dollars is deposited at an annual rate of interest r compounded continuously for t years, the final amount on deposit is

$$A = Pe^{rt}.$$

Answers

6. about $3257.79
7. about 17.50 yr

EXAMPLE 8 Solving a Continuous Interest Problem

(a) In Example 6, we found that \$1000 invested for 5 yr at 6% interest compounded quarterly would grow to \$1346.86. How much would this same investment grow to if compounded continuously?

$$A = Pe^{rt} \quad \text{Continuous compound interest formula}$$
$$A = 1000e^{(0.06)5} \quad \text{Let } P = 1000,\ r = 0.06, \text{ and } t = 5.$$
$$A \approx 1349.86 \quad \text{Use a calculator; round to two decimal places.}$$

The account will grow to \$1349.86.

(b) How long would it take for the initial investment amount to double?
We must find the value of t that will cause A to be $2(\$1000) = \$2000$.

$$A = Pe^{rt}$$
$$2000 = 1000e^{0.06t} \quad \text{Let } A = 2P = 2000.$$
$$2 = e^{0.06t} \quad \text{Divide by 1000.}$$
$$\ln 2 = 0.06t \quad \text{Take natural logarithms; } \ln e^k = k.$$
$$t = \frac{\ln 2}{0.06} \quad \text{Divide by 0.06; rewrite.}$$
$$t \approx 11.55 \quad \text{Use a calculator.}$$

It would take about 11.55 yr for the original investment to double.

Work Problem 8 *at the Side.* ▶

OBJECTIVE 4 Solve applications involving base *e* exponential growth and decay.

EXAMPLE 9 Solving an Exponential Decay Application

Carbon-14 is a radioactive form of carbon that is found in all living plants and animals. After a plant or animal dies, the amount of radioactive carbon-14 disintegrates according to the natural logarithmic function defined by

$$y = y_0e^{-0.000121t},$$

where t is time in years, y is the amount of the sample at time t, and y_0 is the initial amount present at $t = 0$.

(a) If an initial sample contains $y_0 = 10$ g of carbon-14, how many grams will be present after 3000 yr?
Let $y_0 = 10$ and $t = 3000$ in the formula, and use a calculator.

$$y = 10e^{-0.000121(3000)} \approx 6.96 \text{ g}$$

(b) How long would it take for the initial sample to decay to half of its original amount? (This is called the **half-life.**)
Let $y = \frac{1}{2}(10) = 5$, and solve for t.

$$5 = 10e^{-0.000121t} \quad \text{Substitute.}$$
$$\frac{1}{2} = e^{-0.000121t} \quad \text{Divide by 10.}$$
$$\ln\frac{1}{2} = -0.000121t \quad \text{Take natural logarithms; } \ln e^k = k.$$
$$t = \frac{\ln\frac{1}{2}}{-0.000121} \quad \text{Divide by } -0.000121\text{; rewrite.}$$
$$t \approx 5728 \quad \text{Use a calculator.}$$

The half-life is just over 5700 yr.

Work Problem 9 *at the Side.* ▶

---

8 **(a)** How much will \$2500 grow to at 4% interest compounded continuously for 3 yr?

(b) How long would it take for the initial investment in part (a) to double?

9 Radioactive strontium decays according to the natural logarithmic function defined by

$$y = y_0e^{-0.0239t},$$

where t is time in years.

(a) If an initial sample contains $y_0 = 12$ g of radioactive strontium, how many grams will be present after 35 yr?

(b) What is the half-life of radioactive strontium?

ANSWERS

8. (a) \$2818.74 **(b)** about 17.33 yr
9. (a) 5.20 g **(b)** 29 yr

10 **(a)** Find $\log_3 17$ using common logarithms.

(b) Find $\log_3 17$ using natural logarithms.

OBJECTIVE 5 **Use the change-of-base rule.** In **Section 11.5,** we used a calculator to approximate the values of common logarithms (base 10) or natural logarithms (base e). The following rule is used to convert logarithms from one base to another.

Change-of-Base Rule

If $a > 0, a \neq 1, b > 0, b \neq 1$, and $x > 0$, then

$$\log_a x = \frac{\log_b x}{\log_b a}.$$

Note

Any positive number other than 1 can be used for base b in the change-of-base rule, but usually the only practical bases are e and 10 because calculators give logarithms only for these two bases.

To derive the change-of-base rule, let $\log_a x = m$.

$$\log_a x = m$$

$$a^m = x \quad \text{Change to exponential form.}$$

$$\log_b (a^m) = \log_b x \quad \text{Property 3}$$

$$m \log_b a = \log_b x \quad \text{Power rule}$$

$$(\log_a x)(\log_b a) = \log_b x \quad \text{Substitute for } m.$$

$$\log_a x = \frac{\log_b x}{\log_b a} \quad \text{Divide by } \log_b a.$$

The last step gives the change-of-base rule.

EXAMPLE 10 **Using the Change-of-Base Rule**

Find $\log_5 12$.

Use common logarithms and the change-of-base rule.

$$\log_5 12 = \frac{\log 12}{\log 5}$$

$$\log_5 12 \approx 1.5440 \quad \text{Use a calculator.}$$

Note

Either common or natural logarithms can be used when applying the change-of-base rule. Verify that the same value is found in Example 10 if natural logarithms are used.

*Work Problem **10** at the Side.*

Answers

10. **(a)** 2.5789 **(b)** 2.5789

11.6 Exercises

FOR EXTRA HELP

Math XL PRACTICE

WATCH

DOWNLOAD READ REVIEW

Relating Concepts (Exercises 1–4) For Individual or Group Work

*In **Section 11.2,** we solved an equation such as $5^x = 125$ by writing each side as a power of the same base, setting exponents equal, and then solving the resulting equation. The equation is solved as follows.*

$$5^x = 125 \quad \text{Original equation}$$
$$5^x = 5^3 \quad 125 = 5^3$$
$$x = 3 \quad \text{Set exponents equal.}$$

Solution set: $\{3\}$

The method described in this section can also be used to solve this equation.
Work Exercises 1–4 in order, *to see how this is done.*

1. Take common logarithms on both sides, and write this equation.
$\log 5^x = \log 125$

2. Apply the power rule for logarithms on the left.
$x \log 5 = \log 125$

3. Write the equation so that x is alone on the left.
$x = \dfrac{\log 125}{\log 5}$

4. Use a calculator to find the decimal form of the solution. What is the solution set?
$\dfrac{\log 125}{\log 5} = 3; \{3\}$

Many of the problems in the remaining exercises require a scientific calculator.

Solve each equation. Give solutions to three decimal places. See Example 1.

5. $7^x = 5$
{0.827}

6. $4^x = 3$
{0.792}

7. $9^{-x+2} = 13$
{0.833}

8. $6^{-t+1} = 22$
{−0.725}

9. $3^{2x} = 14$
{1.201}

10. $5^{0.3x} = 11$
{4.966}

11. $2^{y+3} = 5^y$
{2.269}

12. $6^{m+3} = 4^m$
{−13.257}

13. $2^{x+3} = 3^{x-4}$
{15.967}

Solve each equation. Use natural logarithms. Give solutions to three decimal places. See Example 2.

14. $e^{0.006x} = 30$
{566.866}

15. $e^{0.012x} = 23$
{261.291}

16. $e^{-0.103x} = 7$
{−18.892}

17. $e^{-0.205x} = 9$
{−10.718}

18. $\ln e^x = 4$
{4}

19. $\ln e^{3x} = 9$
{3}

20. $\ln e^{0.04x} = \sqrt{3}$
{43.301}

21. $\ln e^{0.45x} = \sqrt{7}$
{5.879}

22. $\ln e^{2x} = \pi$
{1.571}

23. Try solving one of the equations in Exercises 14–17 using common logarithms rather than natural logarithms. (You should get the same solution.) Explain why using natural logarithms is a better choice.

Natural logarithms are a better choice because e is the base.

24. If you were asked to solve

$$10^{0.0025x} = 75,$$

would natural or common logarithms be a better choice? Explain.

Common logarithms are a better choice because 10 is the base.

Solve each equation. Give exact solutions. See Example 3.

25. $\log_3(6x + 5) = 2$
$\left\{\frac{2}{3}\right\}$

26. $\log_5(12x - 8) = 3$
$\left\{\frac{133}{12}\right\}$

27. $\log_2(2x - 1) = 5$
$\left\{\frac{33}{2}\right\}$

28. $\log_6(4x + 2) = 2$
$\left\{\frac{17}{2}\right\}$

29. $\log_7(x + 1)^3 = 2$
$\{-1 + \sqrt[3]{49}\}$

30. $\log_4(x - 3)^3 = 4$
$\{3 + 4\sqrt[3]{4}\}$

31. Suppose that in solving a logarithmic equation having the term $\log(x - 3)$ you obtain a proposed solution of 2. All algebraic work is correct. Explain why you must reject 2 as a solution of the equation.

2 cannot be a solution because $\log(2 - 3) = \log(-1)$, and -1 is not in the domain of $\log x$.

32. Suppose that in solving a logarithmic equation having the term $\log(3 - x)$ you obtain a proposed solution of -4. All algebraic work is correct. Should you reject -4 as a solution of the equation? Explain why or why not.

No, because $3 - (-4) = 7$, and thus -4 is in the domain of $\log(3 - x)$.

Solve each equation. Give exact solutions. See Examples 4 and 5.

33. $\log(6x + 1) = \log 3$
$\left\{\frac{1}{3}\right\}$

34. $\log(7 - x) = \log 12$
{−5}

35. $\log_5(3t + 2) - \log_5 t = \log_5 4$
{2}

36. $\log_2(x + 5) - \log_2(x - 1) = \log_2 3$
{4}

37. $\log 4x - \log(x - 3) = \log 2$

$\emptyset$

38. $\log(-x) + \log 3 = \log(2x - 15)$

$\emptyset$

39. $\log_2 x + \log_2 (x - 7) = 3$

$\{8\}$

40. $\log(2x - 1) + \log 10x = \log 10$

$\{1\}$

41. $\log 5x - \log(2x - 1) = \log 4$

$\left\{\frac{4}{3}\right\}$

42. $\log_3 x + \log_3 (2x + 5) = 1$

$\left\{\frac{1}{2}\right\}$

43. $\log_2 x + \log_2 (x - 6) = 4$

$\{8\}$

44. $\log_2 x + \log_2 (x + 4) = 5$

$\{4\}$

Solve each problem. See Examples 6–8.

45. (a) How much money will there be in an account at the end of 6 yr if $2000 is deposited at 4% compounded quarterly? (Assume no withdrawals are made.)

$2539.47

(b) To one decimal place, how long will it take for the account to grow to $3000?

10.2 yr

46. (a) How much money will there be in an account at the end of 7 yr if $3000 is deposited at 3.5% compounded quarterly? (Assume no withdrawals are made.)

$3828.78

(b) To one decimal place, how long will it take for the account to grow to $5000?

14.7 yr

47. What will be the amount A in an account with initial principal $4000 if interest is compounded continuously at an annual rate of 3.5% for 6 yr?

$4934.71

48. Refer to Exercise 46. Does the money grow to a larger value under those conditions, or when invested for 7 yr at 3% compounded continuously?

The account described in Exercise 46 will amount to $127.75 more, so it is a better deal.

49. How long would it take an initial principal P to double if it is invested at 4.5% compounded continuously?

15.4 yr

50. How long would it take \$4000 to double at 3.25% compounded continuously?

21.3 yr

Solve each problem. See Example 9.

51. A sample of 400 g of lead-210 decays to polonium-210 according to the function defined by

$$A(t) = 400e^{-0.032t},$$

where t is time in years. How much lead will be left in the sample after 25 yr?

about 180 g

52. How long will it take the initial sample of lead in Exercise 51 to decay to half of its original amount?

about 21.7 yr

Use the change-of-base rule (with either common or natural logarithms) to find each logarithm. Give approximations to four decimal places. See Example 10.

53. $\log_6 13$

1.4315

54. $\log_7 19$

1.5131

55. $\log_{\sqrt{2}} \pi$

3.3030

56. $\log_\pi \sqrt{2}$

0.3028

57. $\log_{21} 0.7496$

−0.0947

58. $\log_{19} 0.8325$

−0.0623

59. $\log_{1/2} 5$

−2.3219

60. $\log_{1/3} 7$

−1.7712

61. $\log_{0.3} 12$

−2.0639

One measure of the diversity of the species in an ecological community is the **index of diversity,** *the logarithmic expression*

$$-(p_1 \ln p_1 + p_2 \ln p_2 + \cdots + p_n \ln p_n),$$

where $p_1, p_2, \ldots, p_n$ are the proportions of a sample belonging to each of n species in the sample. (Source: Ludwig, John and James Reynolds, Statistical Ecology: A Primer on Methods and Computing, *New York, Wiley, 1988.) Find the index of diversity to three decimal places if a sample of 100 from a community produces the following numbers.*

62. 90 of one species, 10 of another

0.325

63. 60 of one species, 40 of another

0.673

Chapter 11 Summary

Key Terms

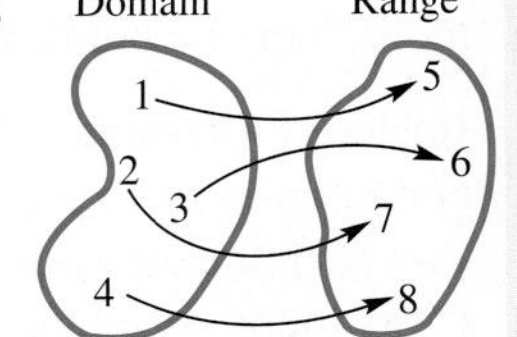

11.1 one-to-one function A one-to-one function is a function in which each x-value corresponds to just one y-value and each y-value corresponds to just one x-value.

inverse of a function f If f is a one-to-one function, the inverse of f is the set of all ordered pairs of the form (y, x), where (x, y) belongs to f.

11.2 exponential equation An equation involving an exponential, where the variable is in the exponent, is an exponential equation.

11.3 logarithm A logarithm is an exponent. The expression $\log_a x$ represents the exponent on the base a that gives the number x.

logarithmic equation A logarithmic equation is an equation with a logarithm in at least one term.

11.5 common logarithm A common logarithm is a logarithm with base 10.

natural logarithm A natural logarithm is a logarithm with base e.

New Symbols

f^{-1}	inverse of f
$\log_a x$	logarithm of x with base a
$\log x$	common (base 10) logarithm of x
$\ln x$	natural (base e) logarithm of x
e	a constant, approximately 2.718281828

Test Your Word Power

See how well you have learned the vocabulary in this chapter. Answers, with examples, follow the Quick Review.

1. In a **one-to-one function**
   A. each x-value corresponds to only one y-value
   B. each x-value corresponds to one or more y-values
   C. each x-value is the same as each y-value
   D. each x-value corresponds to only one y-value and each y-value corresponds to only one x-value.

2. If f is a one-to-one function, then the **inverse** of f is
   A. the set of all solutions of f
   B. the set of all ordered pairs formed by interchanging the coordinates of the ordered pairs of f
   C. an equation involving an exponential expression
   D. the set of all ordered pairs that are the opposite (negative) of the coordinates of the ordered pairs of f.

3. An **exponential function** is a function defined by an expression of the form
   A. $f(x) = ax^2 + bx + c$ for real numbers a, b, c $(a \neq 0)$
   B. $f(x) = \log_a x$, for a and x positive numbers $(a \neq 1)$
   C. $f(x) = a^x$ for all real numbers x $(a > 0, a \neq 1)$
   D. $f(x) = \sqrt{x}$ for $x \geq 0$.

4. A **logarithm** is
   A. an exponent
   B. a base
   C. an equation
   D. a radical expression.

5. A **logarithmic function** is a function defined by an expression of the form
   A. $f(x) = ax^2 + bx + c$ for real numbers a, b, c $(a \neq 0)$
   B. $f(x) = \log_a x$, for a and x positive numbers $(a \neq 1)$
   C. $f(x) = a^x$ for all real numbers x $(a > 0, a \neq 1)$
   D. $f(x) = \sqrt{x}$ for $x \geq 0$.

▶ Quick Review

Concepts | **Examples**

11.1 Inverse Functions

Horizontal Line Test
If a horizontal line intersects the graph of a function in no more than one point, then the function is one-to-one.

Find f^{-1} if $f(x) = 2x - 3$. The graph of f is a straight line, so f is one-to-one by the horizontal line test.

Inverse Functions
For a one-to-one function f defined by an equation $y = f(x)$, the equation that defines the inverse function f^{-1} is found by interchanging x and y, solving for y, and replacing y with $f^{-1}(x)$.

Interchange x and y in the equation $y = 2x - 3$.

$$x = 2y - 3$$

Solve for y to get $\quad y = \dfrac{x+3}{2}.$

Therefore, $\quad f^{-1}(x) = \dfrac{x+3}{2}$, or $f^{-1}(x) = \dfrac{1}{2}x + \dfrac{3}{2}.$

In general, the graph of f^{-1} is the mirror image of the graph of f with respect to the line $y = x$.

The graphs of a nonlinear function f and its inverse f^{-1} are shown here.

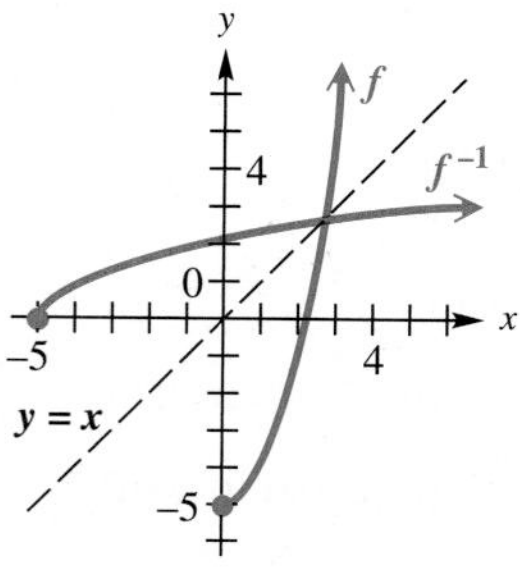

11.2 Exponential Functions

For $a > 0, a \neq 1$, $F(x) = a^x$ defines the exponential function with base a.

$F(x) = 3^x$ defines the exponential function with base 3.

Characteristics of the Graph of $F(x) = a^x$
1. The graph contains the point $(0, 1)$.
2. When $a > 1$, the graph rises from left to right. When $0 < a < 1$, the graph falls from left to right.
3. The x-axis is an asymptote.
4. The domain is $(-\infty, \infty)$; the range is $(0, \infty)$.

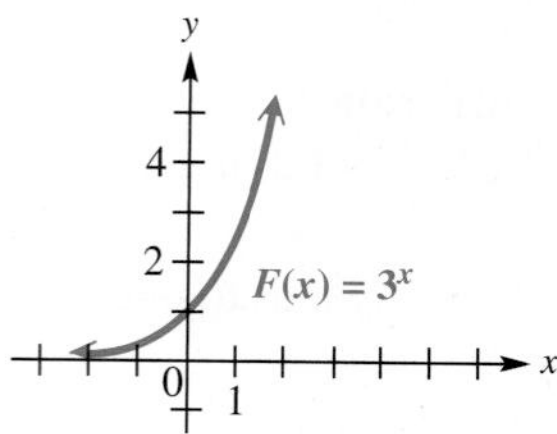

11.3 Logarithmic Functions

$y = \log_a x$ means $x = a^y$.

$y = \log_2 x$ means $x = 2^y$.

For $b > 0, b \neq 1$, $\log_b b = 1$ and $\log_b 1 = 0$.

$\log_3 3 = 1 \qquad \log_5 1 = 0$

For $a > 0, a \neq 1, x > 0$, $G(x) = \log_a x$ defines the logarithmic function with base a.

$G(x) = \log_3 x$ defines the logarithmic function with base 3.

Characteristics of the Graph of $G(x) = \log_a x$
1. The graph contains the point $(1, 0)$.
2. When $a > 1$, the graph rises from left to right. When $0 < a < 1$, the graph falls from left to right.
3. The y-axis is an asymptote.
4. The domain is $(0, \infty)$; the range is $(-\infty, \infty)$.

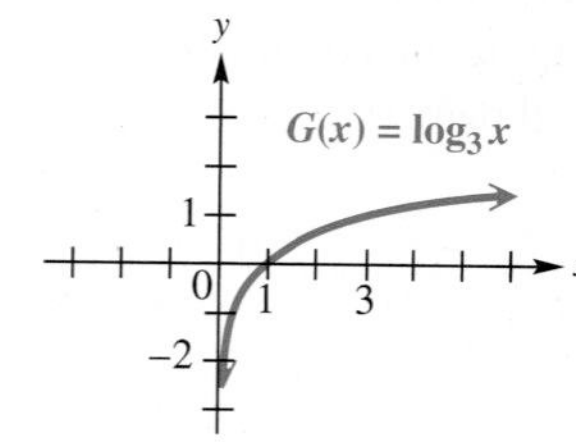

Concepts	Examples

11.4 Properties of Logarithms

Product Rule

$$\log_a xy = \log_a x + \log_a y$$

$$\log_2(3m) = \log_2 3 + \log_2 m$$

Quotient Rule

$$\log_a \frac{x}{y} = \log_a x - \log_a y$$

$$\log_5 \frac{9}{4} = \log_5 9 - \log_5 4$$

Power Rule

$$\log_a x^r = r \log_a x$$

$$\log_{10} 2^3 = 3 \log_{10} 2$$

Special Properties

$$b^{\log_b x} = x \quad \text{and} \quad \log_b b^x = x$$

$$6^{\log_6 10} = 10 \qquad \log_3 3^4 = 4$$

11.5 Common and Natural Logarithms

Common logarithms (base 10) are used in applications such as pH, sound level, and intensity of an earthquake. Use the LOG key of a calculator to evaluate common logarithms.

Use the formula $\text{pH} = -\log[H_3O^+]$ to find the pH (to one decimal place) of grapes with hydronium ion concentration 5.0×10^{-5}.

$\text{pH} = -\log(5.0 \times 10^{-5})$	Substitute.
$\text{pH} = -(\log 5.0 + \log 10^{-5})$	Property of logarithms
$\text{pH} \approx 4.3$	Evaluate.

Natural logarithms (base *e*) are most often used in applications of growth and decay, such as time for money invested to double, decay of chemical compounds, and biological growth. Use the LN key or both the INV and e^x keys to evaluate natural logarithms.

Use the formula for doubling time (in years) $t = \frac{\ln 2}{\ln(1 + r)}$ to find the doubling time to the nearest tenth at an interest rate of 4% compounded annually.

$t = \frac{\ln 2}{\ln(1 + 0.04)}$	Substitute.
$t \approx 17.7$	Evaluate.

The doubling time is about 17.7 yr.

11.6 Exponential and Logarithmic Equations; Further Applications

To solve exponential equations, use these properties ($b > 0, b \neq 1$).

1. If $b^x = b^y$, then $x = y$.

Solve.

$$2^{3x} = 2^5$$
$$3x = 5$$
$$x = \frac{5}{3}$$

The solution set is $\left\{\frac{5}{3}\right\}$.

2. If $x = y$ $(x > 0, y > 0)$, then $\log_b x = \log_b y$.

Solve.

$$5^x = 8$$
$$\log 5^x = \log 8$$
$$x \log 5 = \log 8$$
$$x = \frac{\log 8}{\log 5}$$
$$x \approx 1.2920$$

The solution set is $\{1.2920\}$.

(continued)

Concepts	Examples

11.6 Exponential and Logarithmic Equations; Further Applications *(continued)*

To solve logarithmic equations, use these properties, where $b > 0, b \neq 1, x > 0, y > 0$. First use the properties of **Section 11.4,** if necessary, to write the equation in the proper form.

1. If $\log_b x = \log_b y$, then $x = y$.

Solve.

$$\log_3 2x = \log_3 (x + 1)$$
$$2x = x + 1$$
$$x = 1$$

The solution set is $\{1\}$.

2. If $\log_b x = y$, then $b^y = x$.

Solve.

$$\log_2 (3x - 1) = 4$$
$$3x - 1 = 2^4$$
$$3x - 1 = 16$$
$$3x = 17$$
$$x = \frac{17}{3}$$

The solution set is $\left\{\frac{17}{3}\right\}$.

Change-of-Base Rule

If $a > 0, a \neq 1, b > 0, b \neq 1, x > 0$, then

$$\log_a x = \frac{\log_b x}{\log_b a}.$$

Approximate $\log_3 37$.

$$\log_3 37 = \frac{\ln 37}{\ln 3} = \frac{\log 37}{\log 3} \approx 3.2868$$

ANSWERS TO TEST YOUR WORD POWER

1. D; *Example:* The function $f = \{(0, 2), (1, -1), (3, 5), (-2, 3)\}$ is one-to-one.
2. B; *Example:* The inverse of the one-to-one function f defined in Answer 1 is $f^{-1} = \{(2, 0), (-1, 1), (5, 3), (3, -2)\}$.
3. C; *Examples:* $f(x) = 4^x$, $g(x) = \left(\frac{1}{2}\right)^x$
4. A; *Example:* $\log_a x$ is the exponent to which a must be raised to obtain x; $\log_3 9 = 2$ since $3^2 = 9$.
5. B; *Examples:* $y = \log_3 x$, $y = \log_{1/3} x$

Chapter 11 Review Exercises

[11.1] *Determine whether each graph is the graph of a one-to-one function.*

1.

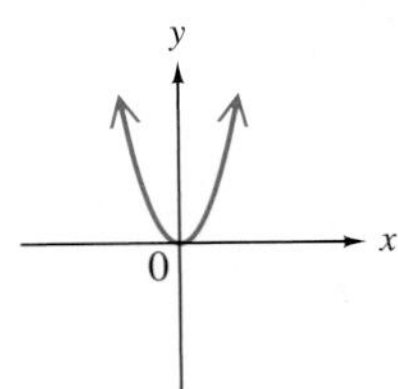

not one-to-one

2.

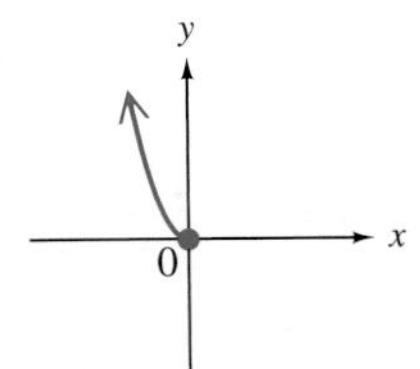

one-to-one

3. The table lists caffeine amounts in several popular 12-oz soft drinks. If the set of sodas is the domain and the set of caffeine amounts is the range of the function consisting of the six pairs listed, is it a one-to-one function? Why or why not?

Soda	Caffeine (mg)
Mountain Dew	55
Diet Coke	45
Dr. Pepper	41
Sunkist Orange Soda	41
Diet Pepsi-Cola	36
Coca-Cola Classic	34

Source: National Soft Drink Association.

This function is not one-to-one because two sodas in the list have 41 mg of caffeine.

Determine whether each function is one-to-one. If it is, find its inverse.

4. $f(x) = -3x + 7$

$f^{-1}(x) = \frac{x - 7}{-3}$, or $f^{-1}(x) = -\frac{1}{3}x + \frac{7}{3}$

5. $f(x) = \sqrt[3]{6x - 4}$

$f^{-1}(x) = \frac{x^3 + 4}{6}$

6. $f(x) = -x^2 + 3$

not one-to-one

Each function graphed is one-to-one. Graph its inverse.

7.

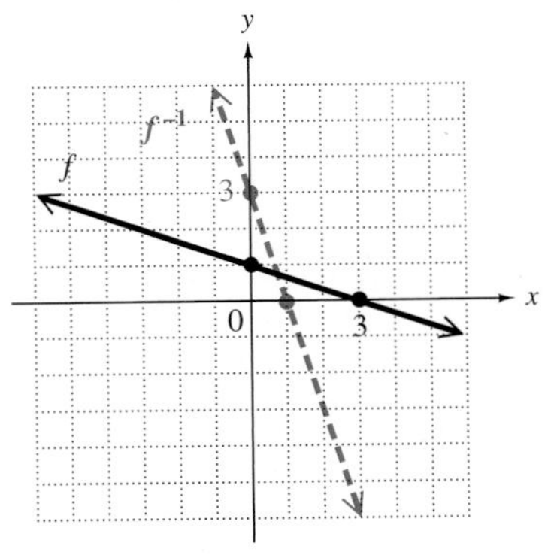

8.

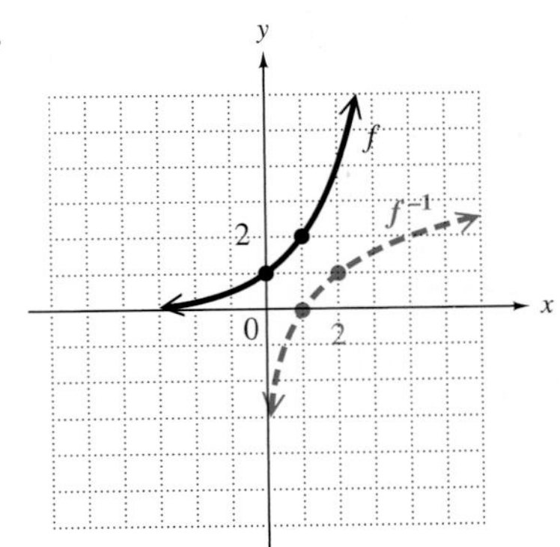

[11.2] *Graph each function.*

9. $f(x) = 4^x$

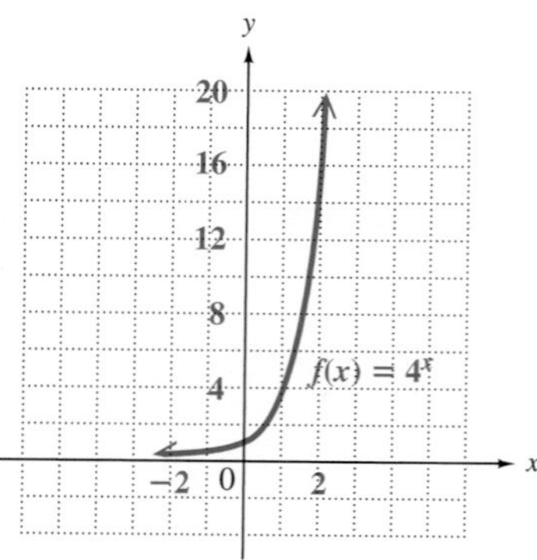

10. $f(x) = \left(\frac{1}{4}\right)^x$

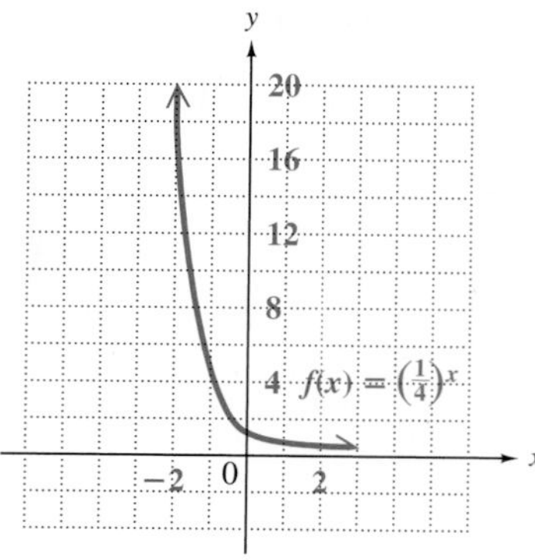

11. $f(x) = 4^{x+1}$

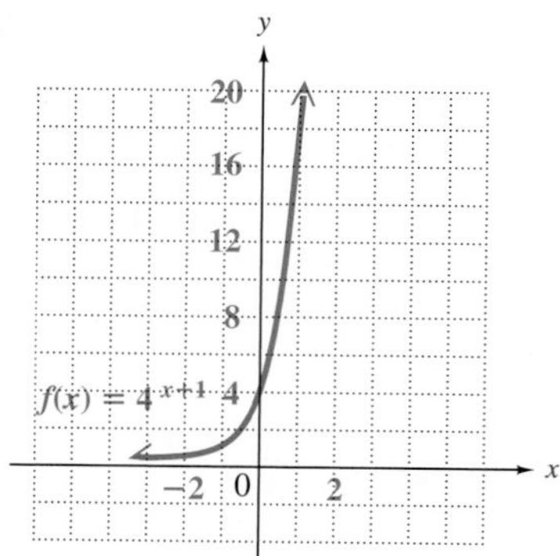

Solve each equation.

12. $4^{3x} = 8^{x+4}$

{4}

13. $\left(\frac{1}{27}\right)^{x-1} = 9^{2x}$

$\left\{\frac{3}{7}\right\}$

14. $5^x = 1$

{0}

 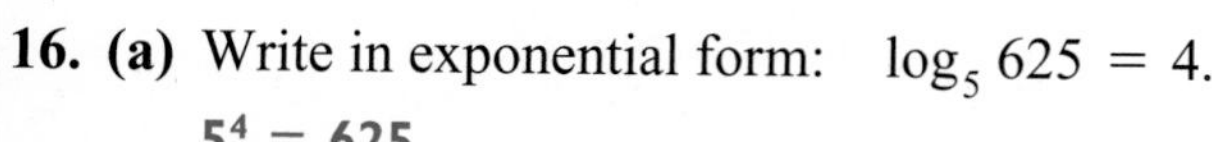

In the remainder of the Chapter Review, many exercises will require a scientific calculator. We do not mark each such exercise.

15. A recent report predicts that the U.S. Hispanic population will increase from 35.6 million in 2000 to 102.8 million in 2050. (*Source:* U.S. Census Bureau.) Assuming an exponential growth pattern, the population is approximated by

$$f(x) = 35.6(2)^{0.0306x},$$

where x represents the number of years since 2000. Use this function to estimate the Hispanic population in each year.

(a) 2015 **48.9 million** **(b)** 2030 **67.3 million**

[11.3]

16. (a) Write in exponential form: $\log_5 625 = 4$.

$5^4 = 625$

(b) Write in logarithmic form: $5^{-2} = 0.04$.

$\log_5 0.04 = -2$

17. (a) In your own words, explain the meaning of $\log_b a$.

$\log_b a$ represents the exponent to which b must be raised to obtain a.

(b) Based on the meaning of $\log_b a$, what is the simplest form of $b^{\log_b a}$? a

Graph each function.

18. $g(x) = \log_4 x$ (*Hint:* See Exercise 9.)

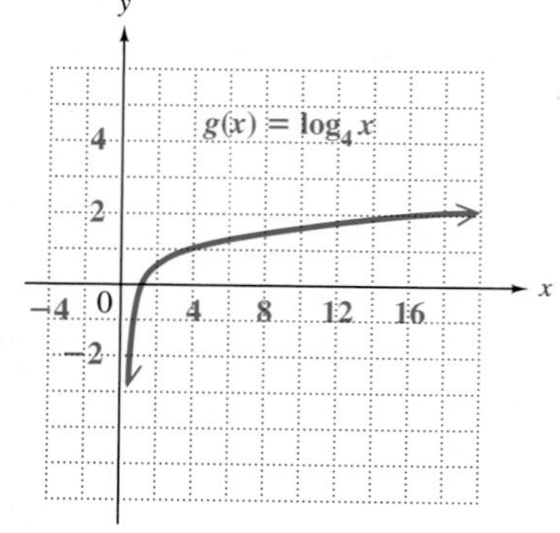

19. $g(x) = \log_{1/4} x$ (*Hint:* See Exercise 10.)

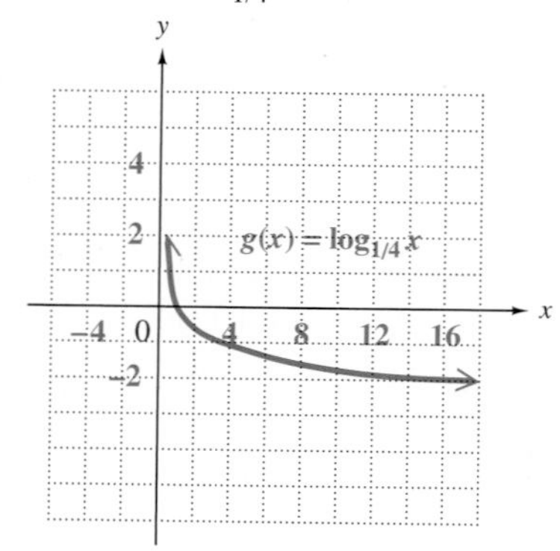

Solve each equation.

20. $\log_8 64 = x$ $\{2\}$

21. $\log_7 \frac{1}{49} = x$ $\{-2\}$

22. $\log_4 x = \frac{3}{2}$ $\{8\}$

23. $\log_b b^2 = 2$ $\{b \mid b > 0, b \neq 1\}$

[11.4] *Apply the properties of logarithms to express each logarithm as a sum or difference of logarithms. Assume that all variables represent positive real numbers.*

24. $\log_4 3x^2$ $\log_4 3 + 2\log_4 x$

25. $\log_2 \frac{p^2 r}{\sqrt{z}}$ $2\log_2 p + \log_2 r - \frac{1}{2}\log_2 z$

Use the properties of logarithms to write each expression as a single logarithm. Assume that all variables represent positive real numbers, $b \neq 1$.

26. $\log_b 3 + \log_b x - 2\log_b y$ $\log_b \frac{3x}{y^2}$

27. $\log_3(x + 7) - \log_3(4x + 6)$ $\log_3 \frac{x + 7}{4x + 6}$

[11.5] *Evaluate each logarithm. Give approximations to four decimal places.*

28. $\log 28.9$ 1.4609

29. $\log 0.257$ −0.5901

30. $\ln 28.9$ 3.3638

31. $\ln 0.257$ −1.3587

Find the pH *of each substance with the given hydronium ion concentration.*

32. Milk, 4.0×10^{-7} 6.4

33. Crackers, 3.8×10^{-9} 8.4

34. If orange juice has pH 4.6, what is its hydronium ion concentration? 2.5×10^{-5}

Solve each problem.

35. **Section 11.5** Exercise 29 introduced the formula for doubling time,

$$t = \frac{\ln 2}{\ln(1 + r)},$$

which gives the number of years required to double your money when it is invested at interest rate r (in decimal form) compounded annually. How long does it take to double your money at each rate? Round answers to the nearest year.

(a) 4% 18 yr **(b)** 6% 12 yr

(c) 10% 7 yr **(d)** 12% 6 yr

(e) Compare each answer in parts (a) – (d) with the following numbers. What do you find?

$$\frac{72}{4}, \frac{72}{6}, \frac{72}{10}, \frac{72}{12}$$

Each comparison shows approximately the same number. For example, in part (a) the doubling time is 18 yr (rounded) and $\frac{72}{4} = 18$. Thus, the formula $t = \frac{72}{100r}$ (called the *Rule of 72*) is an excellent approximation of the doubling time formula.

36. The concentration of a drug injected into the bloodstream decreases with time. The intervals of time T when the drug should be administered are given by

$$T = \frac{1}{k}\ln\frac{C_2}{C_1},$$

where k is a constant determined by the drug in use, C_2 is the concentration at which the drug is harmful, and C_1 is the concentration below which the drug is ineffective. (*Source:* Horelick, Brindell and Sinan Koont, "Applications of Calculus to Medicine: Prescribing Safe and Effective Dosage," *UMAP Module 202,* 1977.) Thus, if $T = 4$, the drug should be administered every 4 hr. For a certain drug, $k = \frac{1}{3}$, $C_2 = 5$, and $C_1 = 2$. How often should the drug be administered? (*Hint:* Round down.)

every 2 hr

[11.6] *Solve each equation. Give solutions to three decimal places.*

37. $3^x = 9.42$

{2.042}

38. $2^{x-1} = 15$

{4.907}

39. $e^{0.06x} = 3$

{18.310}

Solve each equation. Give exact solutions.

40. $\log_3 (9x + 8) = 2$

$\left\{\frac{1}{9}\right\}$

41. $\log_5 (x + 6)^3 = 2$

$\left\{-6 + \sqrt[3]{25}\right\}$

42. $\log_3 (p + 2) - \log_3 p = \log_3 2$

{2}

43. $\log (2x + 3) - \log x = 1$

$\left\{\frac{3}{8}\right\}$

44. $\log_4 x + \log_4 (8 - x) = 2$

{4}

45. $\log_2 x + \log_2 (x + 15) = 4$

{1}

Solve each problem.

46. How much would be in an account after 3 yr if \$6500.00 was invested at 3% annual interest, compounded daily (use $n = 365$)?

\$7112.11

47. Which is a better plan?

Plan A: Invest \$1000.00 at 4% compounded quarterly for 3 yr

Plan B: Invest \$1000.00 at 3.9% compounded monthly for 3 yr

Plan A; it would pay \$2.92 more.

A machine purchased for business use ***depreciates,*** *or loses value, over a period of years. The value of the machine at the end of its useful life is called its* ***scrap value.*** *By one method of depreciation (where it is assumed a constant percentage of the value depreciates annually), the scrap value, S, is given by*

$$S = C(1 - r)^n,$$

where C is the original cost, n is the useful life in years, and r is the constant percent of depreciation.

48. Find the scrap value of a machine costing \$30,000, having a useful life of 12 yr and a constant annual rate of depreciation of 15%.

\$4267

49. A machine has a "half-life" of 6 yr. Find the constant annual rate of depreciation.

about 11%

Use the change-of-base rule (with either common or natural logarithms) to find each logarithm. Give approximations to four decimal places.

50. $\log_{16} 13$

0.9251

51. $\log_4 12$

1.7925

52. $\log_{\sqrt{6}} \sqrt{13}$

1.4315

Mixed Review Exercises

Solve.

53. $\log_3 (x + 9) = 4$

{72}

54. $\log_2 32 = x$

{5}

55. $\log_x \frac{1}{81} = 2$

$\left\{\frac{1}{9}\right\}$

56. $27^x = 81$

$\left\{\frac{4}{3}\right\}$

57. $2^{2x-3} = 8$

{3}

58. $\log_3 (x + 1) - \log_3 x = 2$

$\left\{\frac{1}{8}\right\}$

59. $\log (3x - 1) = \log 10$

$\left\{\frac{11}{3}\right\}$

60. Find the value of n in the equation for Exercise 48 if the scrap value is \$10,000, the cost is \$30,000, and the depreciation rate is 15%.

6.8 yr

Chapter 11 Test

Use the Chapter Test Prep Video CD to see fully worked-out solutions to any of the exercises you want to review.

1. Decide whether each function is one-to-one.

(a) $f(x) = x^2 + 9$

(b)

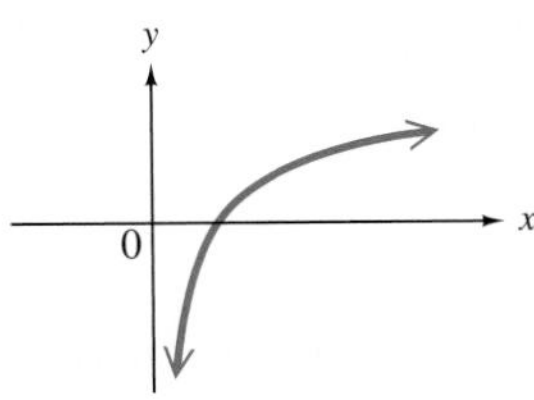

1. (a) not one-to-one

(b) one-to-one

2. Find $f^{-1}(x)$ for the one-to-one function defined by $f(x) = \sqrt[3]{x + 7}$.

2. $f^{-1}(x) = x^3 - 7$

3. Graph the inverse of f, given the graph of f here.

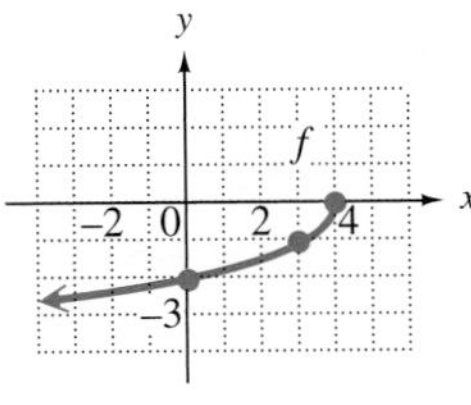

3.

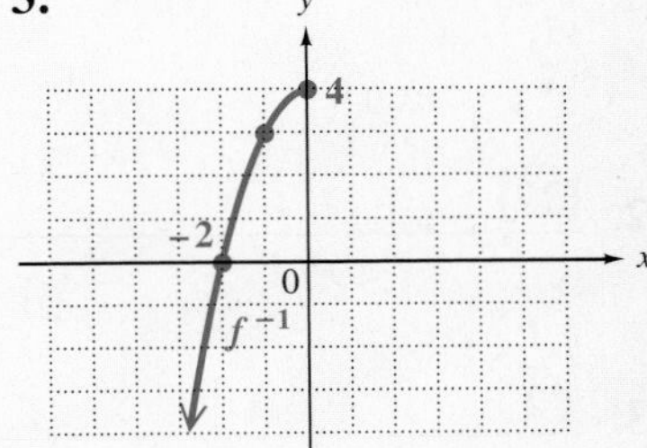

Graph each function.

4. $y = 6^x$

4.

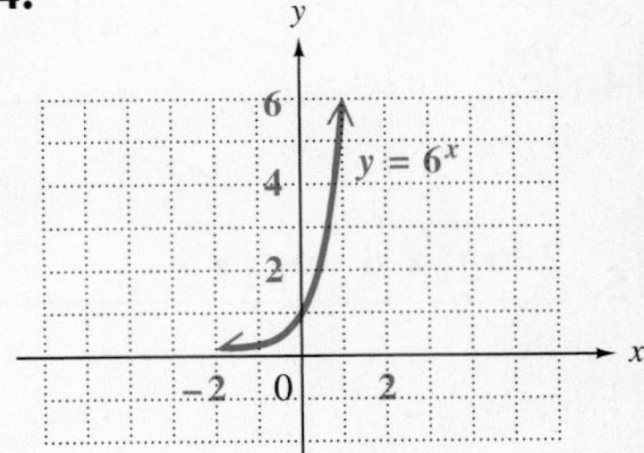

5. $y = \log_6 x$

5.

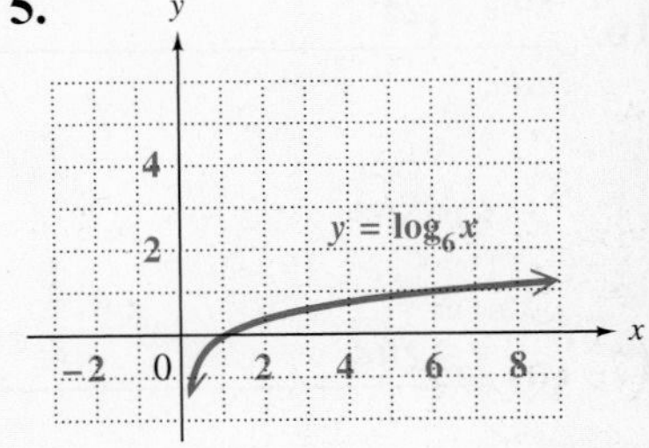

6. Explain how the graph of the function in Exercise 5 can be obtained from the graph of the function in Exercise 4.

6. Interchange the x- and y-values of the ordered pairs, because the functions are inverses.

Solve each equation. Give exact solutions.

7. $5^x = \frac{1}{625}$

8. $2^{3x-7} = 8^{2x+2}$

7. $\{-4\}$

8. $\left\{-\frac{13}{3}\right\}$

9. (a) 775 millibars
(b) 265 millibars

9. The atmospheric pressure (in millibars) at a given altitude x (in meters) is approximated by

$$f(x) = 1013e^{-0.0001341x}.$$

Use this function to approximate the atmospheric pressure at

(a) 2000 m **(b)** 10,000 m.

10. $\log_4 0.0625 = -2$

10. Write in logarithmic form: $4^{-2} = 0.0625$.

11. $7^2 = 49$

11. Write in exponential form: $\log_7 49 = 2$.

Solve each equation.

12. $\{32\}$

12. $\log_{1/2} x = -5$

13. $\left\{\frac{1}{2}\right\}$

13. $x = \log_9 3$

14. $\{2\}$

14. $\log_x 16 = 4$

15. $2 \log_3 x + \log_3 y$

15. Use properties of logarithms to write $\log_3 x^2y$ as a sum or difference of logarithms. Assume that the variables represent positive real numbers.

16. $\log_b \frac{r^{1/4} s^2}{t^{2/3}}$

16. Use properties of logarithms to write $\frac{1}{4} \log_b r + 2 \log_b s - \frac{2}{3} \log_b t$ as a single logarithm. Assume that the variables represent positive real numbers, $b \neq 1$.

17. (a) 1.3284
(b) −0.8440
(c) 2.1245

17. Use a calculator to find an approximation to four decimal places for each logarithm.

(a) $\log 21.3$ **(b)** $\ln 0.43$ **(c)** $\log_6 45$

18. $\{3.9656\}$

18. Solve $3^x = 78$, giving the solution to four decimal places.

19. $\{3\}$

19. Solve $\log_8 (x + 5) + \log_8 (x - 2) = \log_8 8$.

20. (a) $12,507.51
(b) 15.5 yr

20. Suppose that $10,000 is invested at 4.5% annual interest, compounded quarterly.

(a) How much will be in the account in 5 yr if no money is withdrawn?

(b) How long will it take for the initial principal to double?

Cumulative Review Exercises ▶▶▶ Chapters R–11

Let $S = \left\{-\frac{9}{4}, -2, -\sqrt{2}, 0, 0.6, \sqrt{11}, \sqrt{-8}, 6, \frac{30}{3}\right\}$. *List the elements of S that are elements of each set.*

1. Integers

$-2, 0, 6, \frac{30}{3}$ (or 10)

2. Rational numbers

$-\frac{9}{4}, -2, 0, 0.6, 6, \frac{30}{3}$ (or 10)

3. Irrational numbers

$-\sqrt{2}, \sqrt{11}$

Solve each equation or inequality.

4. $7 - (3 + 4x) + 2x = -5(x - 1) - 3$

$\left\{-\frac{2}{3}\right\}$

5. $2x + 2 \leq 5x - 1$

$[1, \infty)$

6. $|2x - 5| = 9$

$\{-2, 7\}$

7. $|4x + 2| > 10$

$(-\infty, -3) \cup (2, \infty)$

8. The graph indicates that the number of international travelers to the United States increased from 41,218 thousand in 2003 to 50,980 thousand in 2006.

(a) Is this the graph of a function?

yes

(b) What is the slope of the line in the graph? Interpret the slope in the context of U.S. travelers to foreign countries.

3254; The number of travelers increased by an average of 3254 thousand per year during the period 2003–2006.

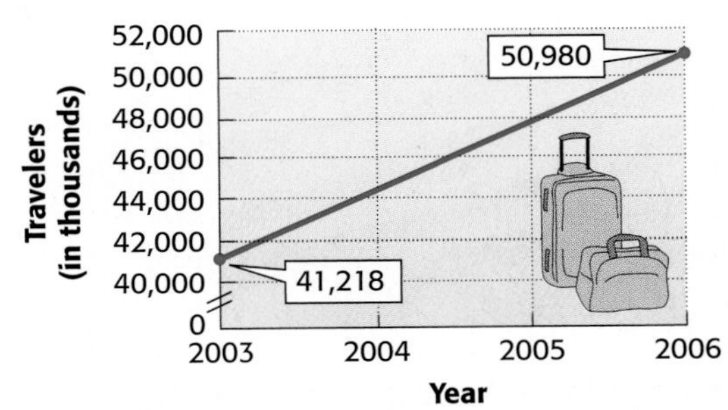

Source: U.S. Department of Commerce.

Solve each system of equations.

9. $5x - 3y = 14$
$2x + 5y = 18$

$\{(4, 2)\}$

10. $x + 2y + 3z = 11$
$3x - y + z = 8$
$2x + 2y - 3z = -12$

$\{(1, -1, 4)\}$

Perform the indicated operations.

11. $(2p + 3)(3p - 1)$

$6p^2 + 7p - 3$

12. $(4k - 3)^2$

$16k^2 - 24k + 9$

13. $(3m^3 + 2m^2 - 5m) - (8m^3 + 2m - 4)$

$-5m^3 + 2m^2 - 7m + 4$

14. Divide $6t^4 + 17t^3 - 4t^2 + 9t + 4$ by $3t + 1$.

$2t^3 + 5t^2 - 3t + 4$

Factor completely.

15. $5z^3 - 19z^2 - 4z$
$z(5z + 1)(z - 4)$

16. $16a^2 - 25b^4$
$(4a + 5b^2)(4a - 5b^2)$

17. $8c^3 + d^3$
$(2c + d)(4c^2 - 2cd + d^2)$

Perform the indicated operations.

18. $\dfrac{(5p^3)^4(-3p^7)}{2p^2(4p^4)}$
$-\dfrac{1875p^{13}}{8}$

19. $\dfrac{x^2 - 9}{x^2 + 7x + 12} \div \dfrac{x - 3}{x + 5}$
$\dfrac{x + 5}{x + 4}$

20. $\dfrac{2}{k + 3} - \dfrac{5}{k - 2}$
$\dfrac{-3k - 19}{(k + 3)(k - 2)}$

Simplify.

21. $\sqrt{288}$
$12\sqrt{2}$

22. $\dfrac{-8^{4/3}}{8^2}$
$-\dfrac{1}{4}$

23. $2\sqrt{32} - 5\sqrt{98}$
$-27\sqrt{2}$

24. Solve $\sqrt{2x + 1} - \sqrt{x} = 1$.
$\{0, 4\}$

25. Multiply $(5 + 4i)(5 - 4i)$.
41

Solve each equation or inequality.

26. $3x^2 = x + 1$
$\left\{\dfrac{1 + \sqrt{13}}{6}, \dfrac{1 - \sqrt{13}}{6}\right\}$

27. $x^2 + 2x - 8 > 0$
$(-\infty, -4) \cup (2, \infty)$

28. $x^4 - 5x^2 + 4 = 0$
$\{-2, -1, 1, 2\}$

Solve.

29. $5^{x+3} = \left(\dfrac{1}{25}\right)^{3x+2}$
$\{-1\}$

30. $\log_5 x + \log_5 (x + 4) = 1$
$\{1\}$

31. Write $\log_5 125 = 3$ in exponential form.
$5^3 = 125$

32. Rewrite the following using the product, quotient, and power rules for logarithms:

$$\log \frac{x^3\sqrt{y}}{z}.$$

$3 \log x + \dfrac{1}{2} \log y - \log z$

Graph.

33. $y = \dfrac{1}{3}(x - 1)^2 + 2$

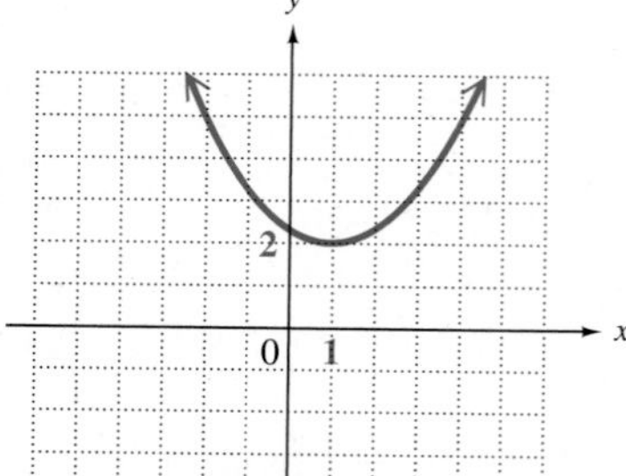

$y = \frac{1}{3}(x - 1)^2 + 2$

34. $f(x) = 2^x$

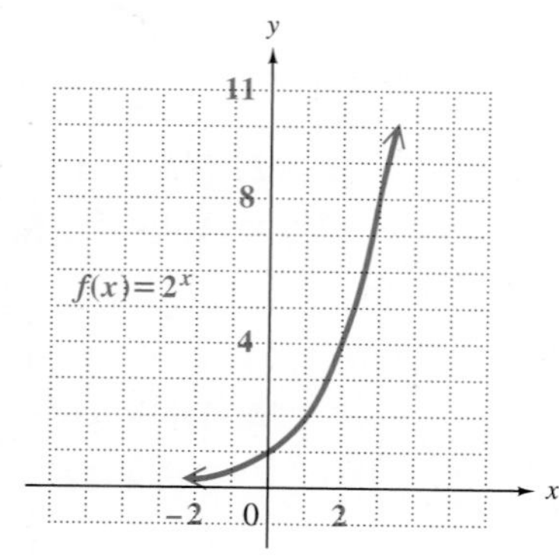

35. $f(x) = \log_3 x$

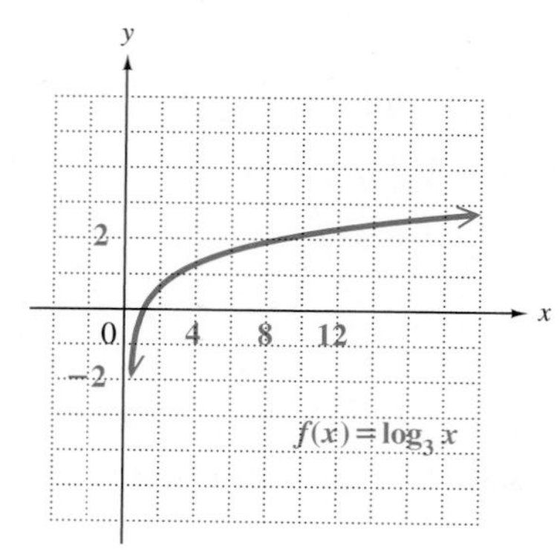

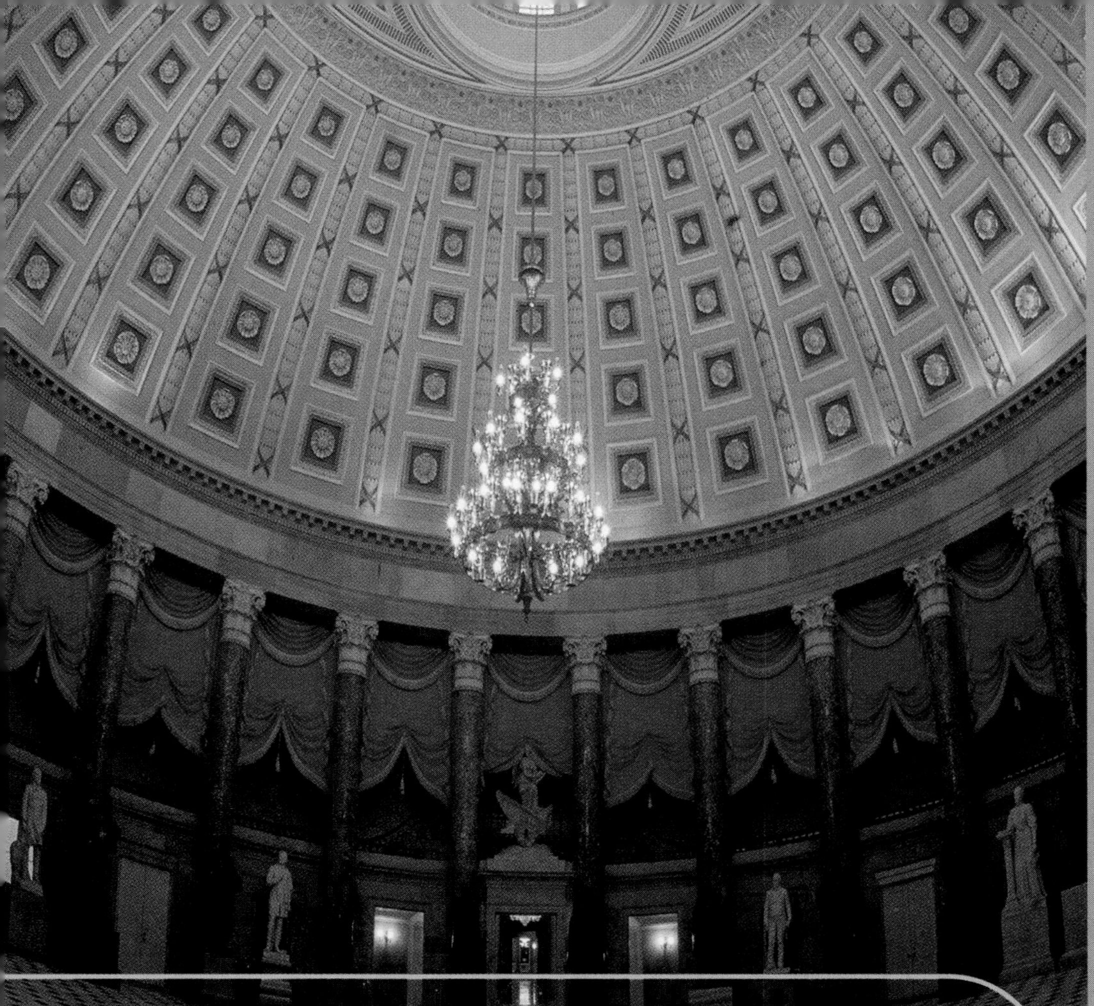

12

Nonlinear Functions, Conic Sections, and Nonlinear Systems

In this chapter, we study a group of curves known as *conic sections.* One conic section, the *ellipse,* has a special reflecting property responsible for "whispering galleries." In a whispering gallery, a person whispering at a certain point in the room can be heard clearly at another point across the room.

The Old House Chamber of the U.S. Capitol, now called Statuary Hall, is a whispering gallery. History has it that John Quincy Adams, whose desk was positioned at exactly the right point beneath the ellipsoidal ceiling, often pretended to sleep there as he listened to political opponents whispering strategies across the room. (*Source: We, the People, The Story of the United States Capitol,* 1991.)

In Section 12.2, we investigate ellipses.

12.1 Additional Graphs of Functions; Operations and Composition

OBJECTIVES

1. **Recognize the graphs of the elementary functions defined by $|x|$, $\frac{1}{x}$, and $\sqrt{x}$, and graph their translations.**
2. **Recognize and graph step functions.**
3. **Perform operations on functions.**
4. **Find the composition of functions.**

In earlier chapters we introduced the function defined by $f(x) = x^2$, sometimes called the **squaring function.** This is one of the most important elementary functions in algebra.

OBJECTIVE 1 Recognize the graphs of the elementary functions defined by $|x|$, $\frac{1}{x}$, and $\sqrt{x}$, and graph their translations. Another one of the elementary functions, defined by $f(x) = |x|$, is called the **absolute value function.** Its graph, along with a table of selected ordered pairs, is shown in Figure 1. Its domain is $(-\infty, \infty)$, and its range is $[0, \infty)$.

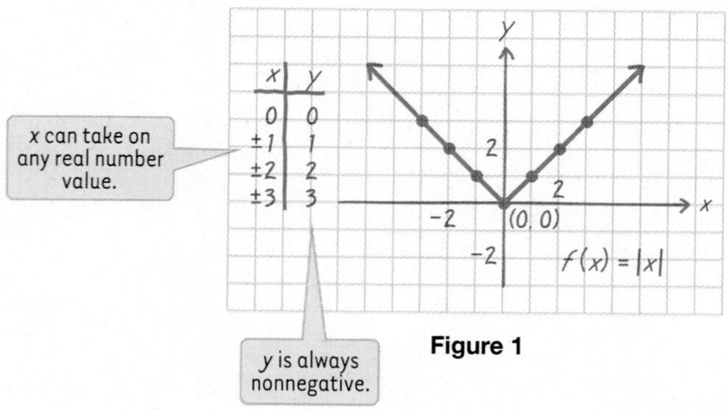

Figure 1

The **reciprocal function,** defined by $f(x) = \frac{1}{x}$, was introduced in **Section 7.4.** Its graph is shown in Figure 2, along with a table of selected ordered pairs. Notice that x can never equal 0 for this function, and as a result, as x gets closer and closer to 0, the graph approaches either ∞ or $-\infty$. Also, $\frac{1}{x}$ can never equal 0, and as x approaches ∞ or $-\infty$, $\frac{1}{x}$ approaches 0. The axes are called **asymptotes** for the function. For the reciprocal function, the domain and the range are both $(-\infty, 0) \cup (0, \infty)$.

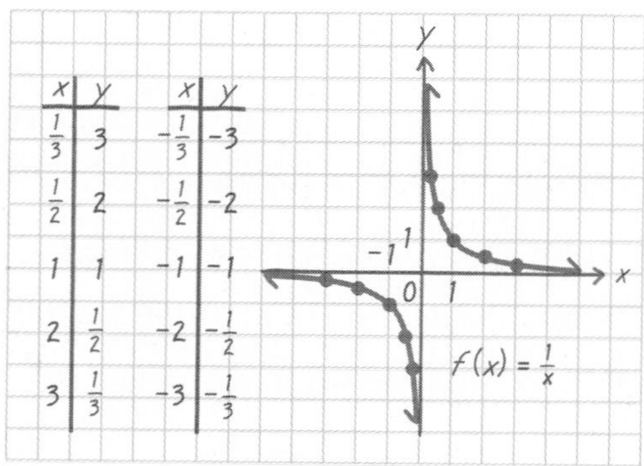

Figure 2

The **square root function**, defined by $f(x) = \sqrt{x}$, was introduced in **Section 9.1.** Its graph is shown in Figure 3 on the next page. Since we restrict function values to be real numbers, x cannot take on negative values. Thus, the domain of the square root function is $[0, \infty)$. Because the principal square root is always nonnegative, the range is also $[0, \infty)$. A table of values is shown along with the graph.

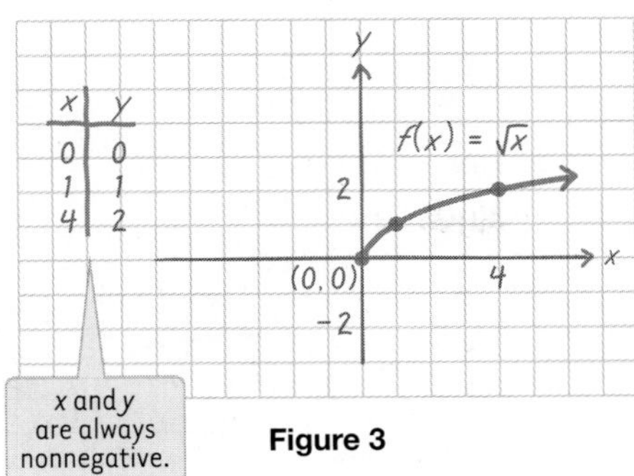

Figure 3

Just as the graph of $f(x) = x^2$ can be shifted, or translated, as we saw in **Section 10.6,** so can the graphs of these other elementary functions.

EXAMPLE 1 Applying a Horizontal Shift

Graph $f(x) = |x - 2|$.

The graph of $y = (x - 2)^2$ is obtained by shifting the graph of $y = x^2$ two units to the right. In a similar manner, the graph of $f(x) = |x - 2|$ is found by shifting the graph of $y = |x|$ two units to the right, as shown in Figure 4. The table of ordered pairs accompanying the graph supports this, as can be seen by comparing it to the table with Figure 1. The domain of this function is $(-\infty, \infty)$, and its range is $[0, \infty)$.

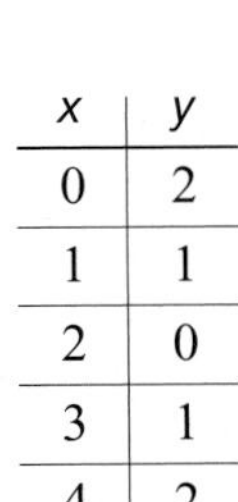

x	y
0	2
1	1
2	0
3	1
4	2

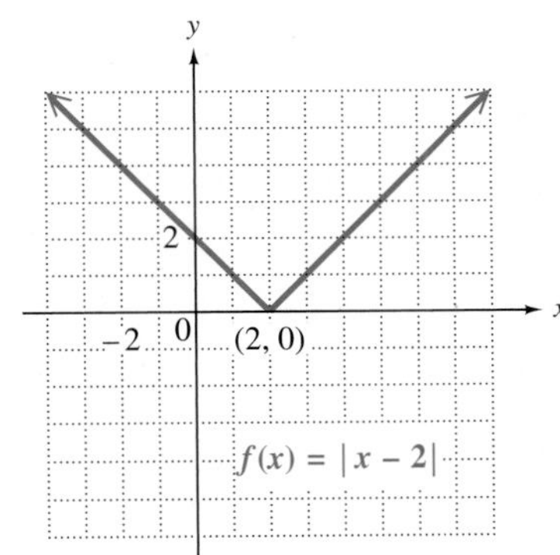

Figure 4

Work Problem 1 at the Side. ▶

EXAMPLE 2 Applying a Vertical Shift

Graph $f(x) = \dfrac{1}{x} + 3$.

The graph of this function is found by shifting the graph of $y = \frac{1}{x}$ three units up. See Figure 5 on the next page. The domain is

$$(-\infty, 0) \cup (0, \infty),$$

and the range is

$$(-\infty, 3) \cup (3, \infty).$$

Continued on Next Page

1 Graph $f(x) = \sqrt{x + 4}$. Give the domain and range.

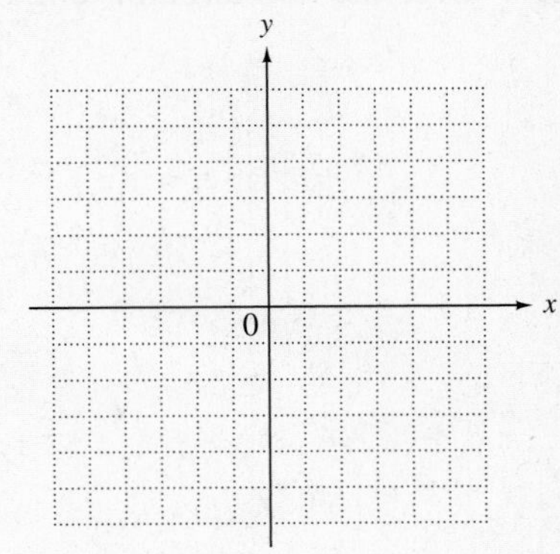

ANSWER

1.

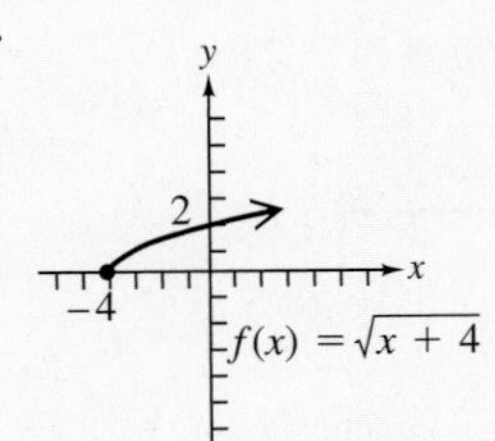

domain: $[-4, \infty)$; range: $[0, \infty)$

2 Graph $f(x) = \frac{1}{x} - 2$.

Give the domain and range.

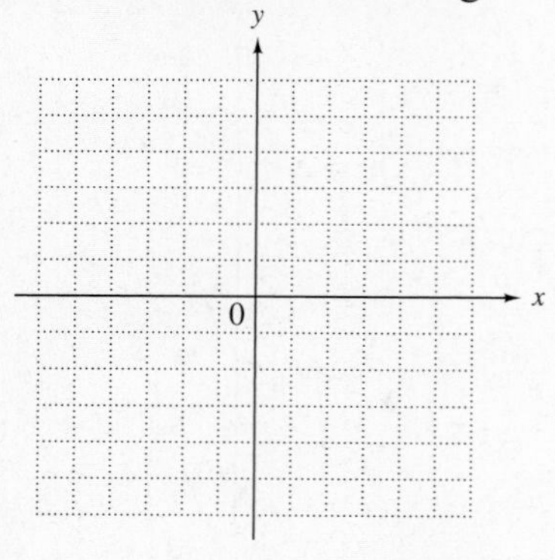

3 Graph $f(x) = |x + 2| + 1$.
Give the domain and range.

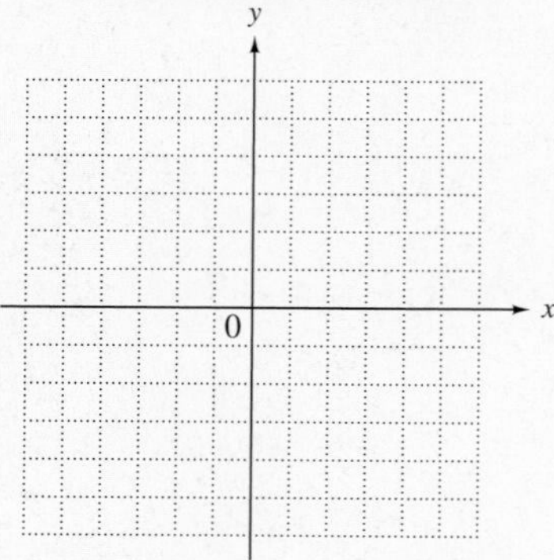

4 Find each of the following.

(a) $[\![18]\!]$ **(b)** $[\![8.7]\!]$

(c) $[\![-5]\!]$ **(d)** $[\![-6.9]\!]$

(e) $\left[\!\!\left[\frac{1}{2}\right]\!\!\right]$

Answers

2.

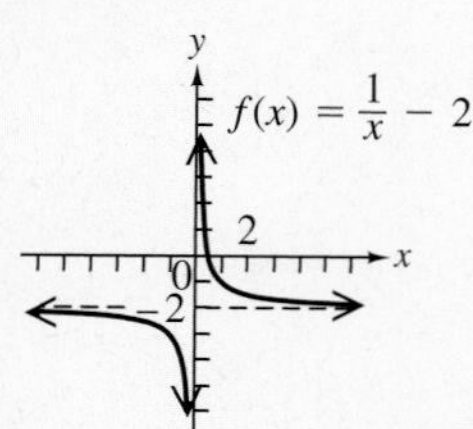

domain: $(-\infty, 0) \cup (0, \infty)$;
range: $(-\infty, -2) \cup (-2, \infty)$

3.

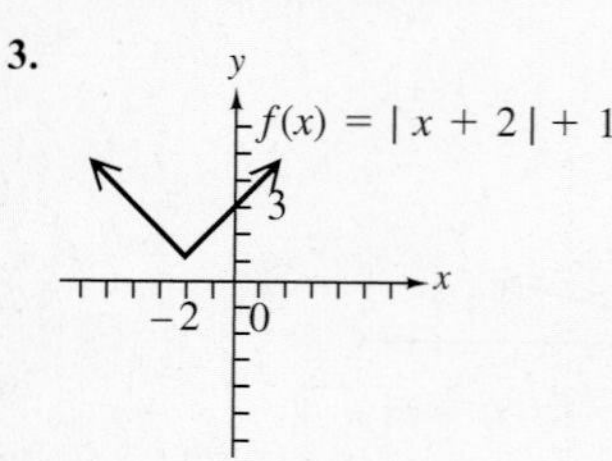

domain: $(-\infty, \infty)$; range: $[1, \infty)$

4. **(a)** 18 **(b)** 8 **(c)** −5 **(d)** −7 **(e)** 0

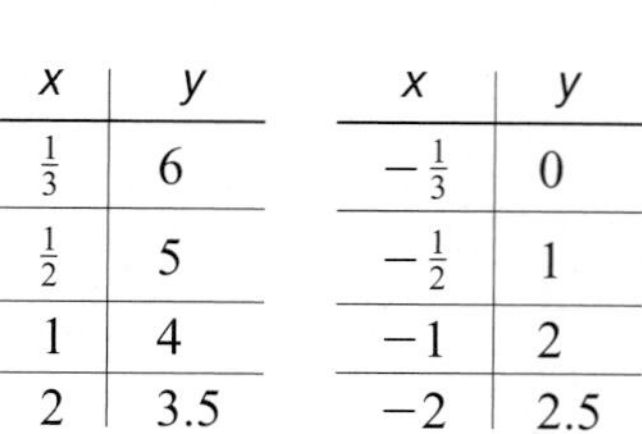

x	y
$\frac{1}{3}$	6
$\frac{1}{2}$	5
1	4
2	3.5

x	y
$-\frac{1}{3}$	0
$-\frac{1}{2}$	1
−1	2
−2	2.5

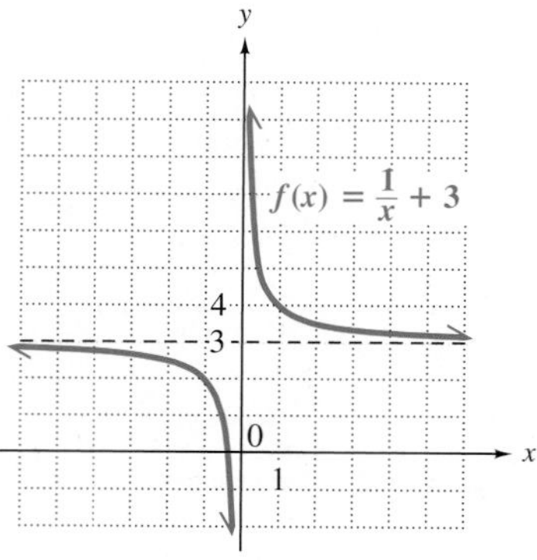

Figure 5

◀ Work Problem **2** at the Side.

EXAMPLE 3 Applying Both Horizontal and Vertical Shifts

Graph $f(x) = \sqrt{x + 1} - 4$.

The graph of $y = (x + 1)^2 - 4$ is obtained by shifting the graph of $y = x^2$ one unit to the left and four units down. Following this pattern here, we shift the graph of $y = \sqrt{x}$ one unit to the left and four units down to get the graph of $f(x) = \sqrt{x + 1} - 4$. See Figure 6. The domain is $[-1, \infty)$, and the range is $[-4, \infty)$.

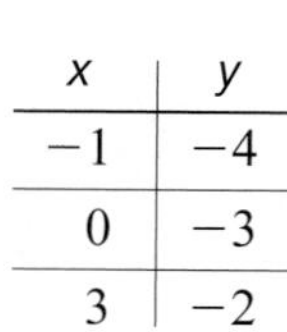

x	y
−1	−4
0	−3
3	−2

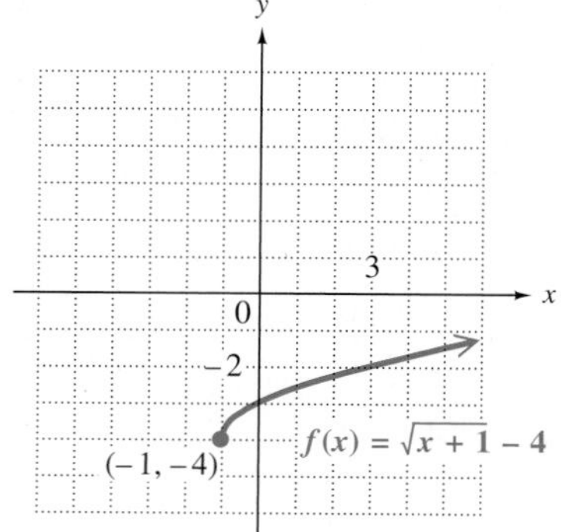

Figure 6

◀ Work Problem **3** at the Side.

OBJECTIVE 2 Recognize and graph step functions. The **greatest integer function,** usually written $f(x) = [\![x]\!]$, is defined as follows:

$[\![x]\!]$ denotes the greatest integer that is less than or equal to x.

For example,

$$[\![8]\!] = 8, \quad [\![7.45]\!] = 7, \quad [\![\pi]\!] = 3, \quad [\![-1]\!] = -1, \quad \text{and} \quad [\![-2.6]\!] = -3.$$

In general, if $f(x) = [\![x]\!]$, then

$$\begin{array}{lll} \text{for} & -2 \le x < -1, & f(x) = -2, \\ \text{for} & -1 \le x < 0, & f(x) = -1, \\ \text{for} & 0 \le x < 1, & f(x) = 0, \\ \text{for} & 1 \le x < 2, & f(x) = 1, \\ \text{for} & 2 \le x < 3, & f(x) = 2, \end{array}$$

and so on.

◀ Work Problem **4** at the Side.

EXAMPLE 4 Graphing the Greatest Integer Function

Graph $f(x) = [\![x]\!]$.

For $[\![x]\!]$,

$$\text{if } -1 \le x < 0, \quad \text{then } [\![x]\!] = -1;$$
$$\text{if } 0 \le x < 1, \quad \text{then } [\![x]\!] = 0;$$
$$\text{if } 1 \le x < 2, \quad \text{then } [\![x]\!] = 1,$$

and so on. Thus, the graph, as shown in Figure 7, consists of a series of horizontal line segments. In each one, the left endpoint is included and the right endpoint is excluded. These segments continue infinitely following this pattern to the left and right. Since x can take any real number value, the domain is $(-\infty, \infty)$. The range is the set of integers $\{\ldots, -4, -3, -2, -1, 0, 1, 2, 3, 4, \ldots\}$. The appearance of the graph is the reason that this function is called a **step function.**

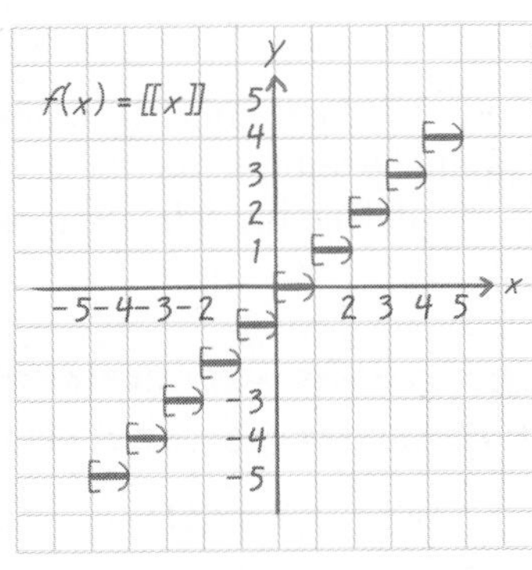

Figure 7

The graph of a step function also may be shifted. For example, the graph of $h(x) = [\![x - 2]\!]$ is the same as the graph of $f(x) = [\![x]\!]$ shifted two units to the right. Similarly, the graph of $g(x) = [\![x]\!] + 2$ is the graph of $f(x)$ shifted two units up.

Work Problem **5** *at the Side.* ▶

5 Graph $f(x) = [\![x + 1]\!]$. Give the domain and range.

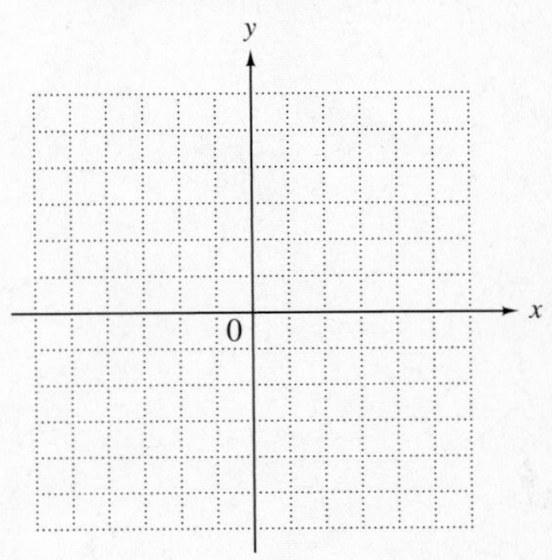

6 Assume that the post office charges 80¢ per oz (or fraction of an ounce) to mail a letter to Europe. Graph the ordered pairs (ounces, cost) for x in the interval $(0, 4]$.

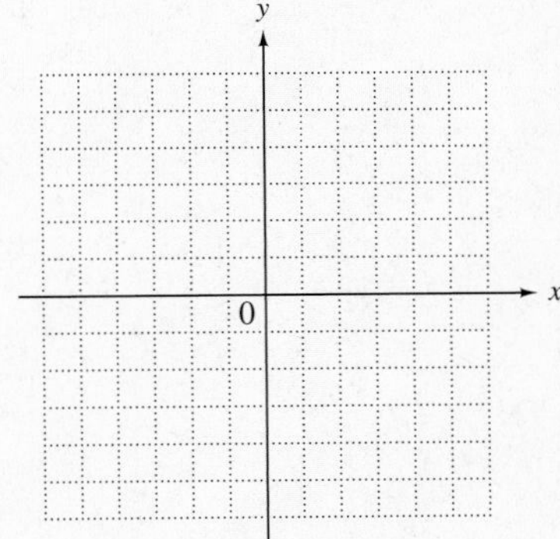

EXAMPLE 5 Applying a Greatest Integer Function

An overnight delivery service charges \$25 for a package weighing up to 2 lb. For each additional pound or fraction of a pound there is an additional charge of \$3. Let $D(x)$ represent the cost to send a package weighing x pounds. Graph $D(x)$ for x in the interval $(0, 6]$.

For x in the interval $(0, 2]$, $\quad y = 25$.

For x in the interval $(2, 3]$, $\quad y = 25 + 3 = 28$.

For x in the interval $(3, 4]$, $\quad y = 28 + 3 = 31$, and so on.

The graph, which is that of a step function, is shown in Figure 8.

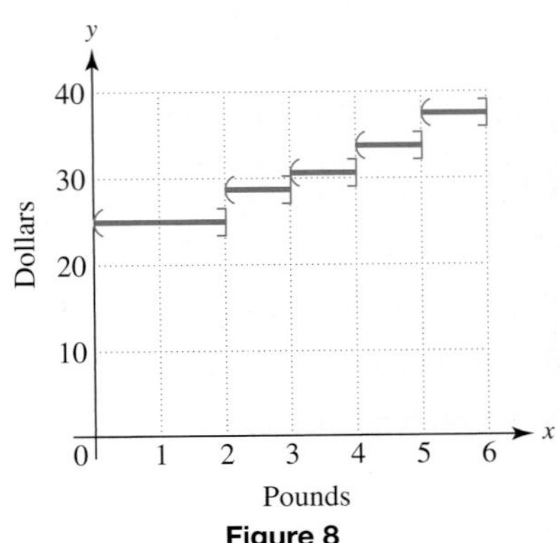

Figure 8

Work Problem **6** *at the Side.* ▶

ANSWERS

5.

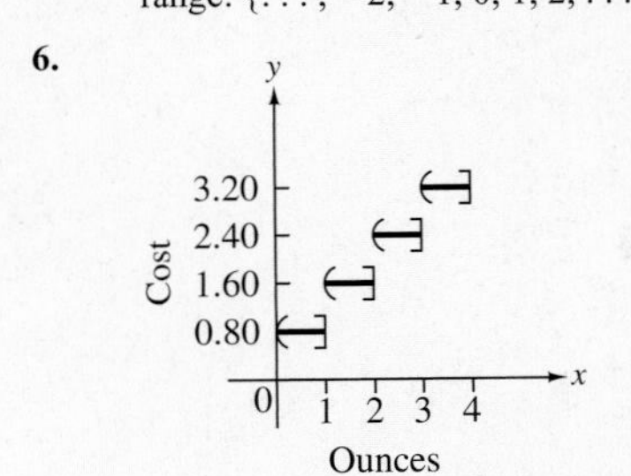

domain: $(-\infty, \infty)$;
range: $\{\ldots, -2, -1, 0, 1, 2, \ldots\}$

6.

OBJECTIVE 3 Perform operations on functions. The operations of addition, subtraction, multiplication, and division are also defined for functions. For example, businesses use the equation "profit equals revenue minus cost," written using function notation as

$$\underset{\substack{\uparrow \\ \text{Profit} \\ \text{function}}}{P(x)} = \underset{\substack{\uparrow \\ \text{Revenue} \\ \text{function}}}{R(x)} - \underset{\substack{\uparrow \\ \text{Cost} \\ \text{function}}}{C(x)},$$

where x is the number of items produced and sold. Thus, the profit function is found by subtracting the cost function from the revenue function.

We define the following **operations on functions.**

Operations on Functions

If $f(x)$ and $g(x)$ define functions, then

$$(f+g)(x) = f(x) + g(x), \quad \text{Sum function}$$

$$(f-g)(x) = f(x) - g(x), \quad \text{Difference function}$$

$$(fg)(x) = f(x) \cdot g(x), \quad \text{Product function}$$

and $$\left(\frac{f}{g}\right)(x) = \frac{f(x)}{g(x)}, \quad g(x) \neq 0. \quad \text{Quotient function}$$

In each case, the domain of the new function is the intersection of the domains of $f(x)$ and $g(x)$. Additionally, the domain of the quotient function must exclude any values of x for which $g(x) = 0$. (Why?)

EXAMPLE 6 Adding and Subtracting Functions

Find each of the following for the polynomial functions defined by

$$f(x) = 10x^2 - 2x \quad \text{and} \quad g(x) = 2x.$$

(a) $(f+g)(2)$

$(f+g)(2) = f(2) + g(2)$ Use the definition.

$= \overbrace{[10(2)^2 - 2(2)]}^{f(x) = 10x^2 - 2x} + \overbrace{2(2)}^{g(x) = 2x}$ Substitute.

This is a key step.

$= [40 - 4] + 4$ Order of operations

$= 40$ Subtract; add.

Alternatively, we could first find $(f+g)(x)$.

$(f+g)(x)$

$= f(x) + g(x)$ Use the definition.

$= (10x^2 - 2x) + 2x$ Substitute.

$= 10x^2$ Combine like terms.

Then, $(f+g)(2)$

$= 10(2)^2$ Substitute.

$= 40.$ The result is the same.

Continued on Next Page

(b) $(f - g)(x)$ and $(f - g)(1)$

$(f - g)(x)$

$= f(x) - g(x)$ Use the definition.

$= (10x^2 - 2x) - 2x$ Substitute.

$= 10x^2 - 4x$ Combine like terms.

Then, $(f - g)(1)$

$= 10(1)^2 - 4(1)$ Substitute.

$= 6.$ $f(1) - g(1)$ gives the same result.

Work Problem 7 at the Side. ▶

EXAMPLE 7 Multiplying Functions

For $f(x) = 3x + 4$ and $g(x) = 2x^2 + x$, find $(fg)(x)$ and $(fg)(-1)$.

$(fg)(x)$

$= f(x) \cdot g(x)$ Use the definition.

$= (3x + 4)(2x^2 + x)$ Substitute.

$= 6x^3 + 3x^2 + 8x^2 + 4x$ FOIL

$= 6x^3 + 11x^2 + 4x$ Combine like terms.

Then $(fg)(-1)$

$= 6(-1)^3 + 11(-1)^2 + 4(-1)$ Let $x = -1$.

$= -6 + 11 - 4$ Be careful with signs.

$= 1.$

(Verify that $f(-1) \cdot g(-1)$ also equals 1, the same result.)

Work Problem 8 at the Side. ▶

EXAMPLE 8 Dividing Functions

For $f(x) = 2x^2 + x - 10$ and $g(x) = x - 2$, find $\left(\frac{f}{g}\right)(x)$ and $\left(\frac{f}{g}\right)(-3)$.

$\left(\frac{f}{g}\right)(x)$

$= \dfrac{f(x)}{g(x)}$ Use the definition.

$= \dfrac{2x^2 + x - 10}{x - 2}$ Substitute.

$= \dfrac{(2x + 5)(x - 2)}{x - 2}$ Factor.

$= 2x + 5, \quad x \neq 2$ $g(x) \neq 0$, so $x \neq 2$.

Then $\left(\frac{f}{g}\right)(-3) = 2(-3) + 5 = -1.$ Let $x = -3$.

(Which is easier to find here: $\left(\frac{f}{g}\right)(-3)$ or $\frac{f(-3)}{g(-3)}$?)

Work Problem 9 at the Side. ▶

7 For

$f(x) = 18x^2 - 24x$

and $g(x) = 3x$,

find each of the following.

(a) $(f + g)(x)$ and $(f + g)(-1)$

(b) $(f - g)(x)$ and $(f - g)(1)$

8 For

$f(x) = 2x + 7$

and $g(x) = x^2 - 4$,

find $(fg)(x)$ and $(fg)(2)$.

9 For

$f(x) = 2x^2 + 17x + 30$

and $g(x) = 2x + 5$,

find $\left(\frac{f}{g}\right)(x)$ and $\left(\frac{f}{g}\right)(-1)$.

Answers

7. (a) $18x^2 - 21x$; 39
(b) $18x^2 - 27x$; -9

8. $2x^3 + 7x^2 - 8x - 28$; 0

9. $x + 6, \quad x \neq -\frac{5}{2}$; 5

OBJECTIVE 4 Find the composition of functions. The diagram in Figure 9 shows a function f that assigns, to each element x of set X, some element y of set Y. Suppose that a function g takes each element of set Y and assigns a value z of set Z. Then f and g together assign an element x in X to an element z in Z. The result of this process is a new function h, which takes an element x in X and assigns it an element z in Z.

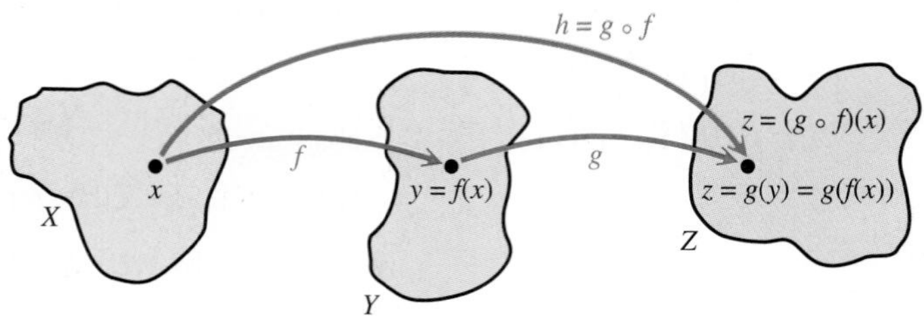

Figure 9

This function h is called the *composition* of functions g and f, written $g \circ f$, and is defined as follows.

Composition of Functions

If f and g are functions, then the **composite function,** or **composition,** of g and f is defined by

$$(g \circ f)(x) = g(f(x))$$

for all x in the domain of f such that $f(x)$ is in the domain of g.

Read $g \circ f$ as "g of f."

As a real-life example of how composite functions occur, consider the following retail situation.

A \$40 pair of blue jeans is on sale for 25% off. If you purchase the jeans before noon, the retailer offers an additional 10% off. What is the final sale price of the blue jeans?

You might be tempted to say that the jeans are 35% off and calculate $40(0.35) = \$14$, giving a final sale price of $\$40 - \$14 = \$26$ for the jeans. ***This is not correct.*** To find the final sale price, we must first find the price after taking 25% off, and then take an additional 10% off that price.

$\$40(0.25) = \$10$, giving a sale price of $\$40 - \$10 = \$30$. Take 25% off original price.

$\$30(0.10) = \$3$, giving a ***final sale price*** of $\$30 - \$3 = \$27$. Take additional 10% off.

This is the idea behind composition of functions.

As another example of composition, suppose an oil well off the Louisiana coast is leaking, with the leak spreading oil in a circular layer over the surface. See Figure 10.

Figure 10

At any time t, in minutes, after the beginning of the leak, the radius of the circular oil slick is given by $r(t) = 5t$ feet. Since $A(r) = \pi r^2$ gives the area of a circle of radius r, the area can be expressed as a function of time by substituting $5t$ for r in $A(r) = \pi r^2$ to get

$$A(r) = \pi r^2$$
$$A(r(t)) = \pi (5t)^2$$
$$A(r(t)) = 25\pi t^2.$$

The function $A(r(t))$ is a composite function of the functions A and r.

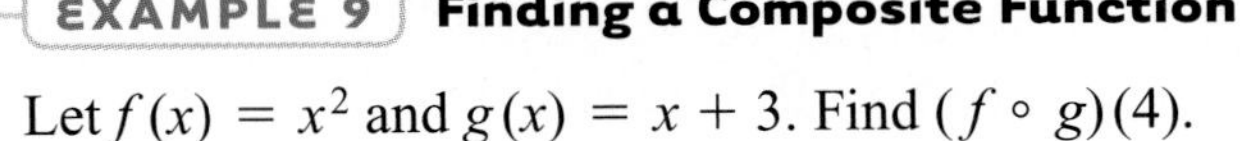

EXAMPLE 9 Finding a Composite Function

Let $f(x) = x^2$ and $g(x) = x + 3$. Find $(f \circ g)(4)$.

$(f \circ g)(4) = f(g(4))$	Definition
$= f(4 + 3)$	Use the rule for $g(x)$; $g(4) = 4 + 3$.
$= f(7)$	Add.
$= 7^2$	Use the rule for $f(x)$; $f(7) = 7^2$.
$= 49$	Square 7.

Evaluate the "inside" function value first.

Now evaluate the "outside" function.

*Work Problem **10** at the Side.* ▶

Notice in Example 9 that if we interchange the order of the functions, the composition of g and f is defined by $g(f(x))$. Once again, letting $x = 4$, we have

$(g \circ f)(4) = g(f(4))$	Definition
$= g(4^2)$	Use the rule for $f(x)$; $f(4) = 4^2$.
$= g(16)$	Square 4.
$= 16 + 3$	Use the rule for $g(x)$; $g(16) = 16 + 3$.
$= 19.$	Add.

Here we see that $(f \circ g)(4) \neq (g \circ f)(4)$ because $49 \neq 19$. In general,

$$(f \circ g)(x) \neq (g \circ f)(x).$$

10 Let $f(x) = x - 4$ and $g(x) = x^2$. Find $(f \circ g)(3)$.

ANSWER

10. 5

11 Let $f(x) = 3x + 6$ and $g(x) = x^3$. Find each of the following.

(a) $(f \circ g)(2)$

(b) $(g \circ f)(2)$

(c) $(f \circ g)(x)$

(d) $(g \circ f)(x)$

EXAMPLE 10 Finding Composite Functions

Let $f(x) = 4x - 1$ and $g(x) = x^2 + 5$. Find each of the following.

(a) $(f \circ g)(2)$

$$\begin{aligned} (f \circ g)(2) &= f(g(2)) \\ &= f(2^2 + 5) && g(x) = x^2 + 5 \\ &= f(9) && \text{Work inside the parentheses.} \\ &= 4(9) - 1 && f(x) = 4x - 1 \\ &= 35 && \text{Multiply; subtract.} \end{aligned}$$

(b) $(f \circ g)(x)$

Here, use $g(x)$ as the input for the function f.

$$\begin{aligned} (f \circ g)(x) &= f(g(x)) \\ &= 4(g(x)) - 1 && \text{Use the rule for } f(x);\ f(x) = 4x - 1. \\ &= 4(x^2 + 5) - 1 && g(x) = x^2 + 5 \\ &= 4x^2 + 20 - 1 && \text{Distributive property} \\ &= 4x^2 + 19 && \text{Combine like terms.} \end{aligned}$$

(c) Find $(f \circ g)(2)$ again, this time using the rule obtained in part (b).

$$\begin{aligned} (f \circ g)(x) &= 4x^2 + 19 && \text{From part (b)} \\ (f \circ g)(2) &= 4(2)^2 + 19 && \text{Let } x = 2. \\ &= 4(4) + 19 && \text{Square 2.} \\ &= 16 + 19 && \text{Multiply.} \\ &= 35 && \text{Add.} \end{aligned}$$

The result, 35, is the same as the result in part (a).

(d) $(g \circ f)(x)$

Here, use $f(x)$ as the input for the function g.

$$\begin{aligned} (g \circ f)(x) &= g(f(x)) \\ &= (f(x))^2 + 5 && \text{Use the rule for } g(x);\ g(x) = x^2 + 5. \\ &= (4x - 1)^2 + 5 && f(x) = 4x - 1 \\ &= 16x^2 - 8x + 1 + 5 && (x - y)^2 = x^2 - 2xy + y^2 \\ &= 16x^2 - 8x + 6 && \text{Combine like terms.} \end{aligned}$$

Compare this result to that in part (b). Again, $(f \circ g)(x) \neq (g \circ f)(x)$.

◀ Work Problem 11 at the Side.

Answers

11. **(a)** 30 **(b)** 1728 **(c)** $3x^3 + 6$ **(d)** $(3x + 6)^3$

12.1 Exercises

FOR EXTRA HELP

Math XL PRACTICE

WATCH

DOWNLOAD

READ

REVIEW

Fill in each blank with the correct response.

1. For the reciprocal function defined by $f(x) = \frac{1}{x}$, **0** is the only real number not in the domain.

2. The range of the square root function, given by $f(x) = \sqrt{x}$, is **$[0, \infty)$**.

3. The lowest point on the graph of $f(x) = |x|$ has coordinates (**0**, **0**).

4. The range of $f(x) = [\![x]\!]$, the greatest integer function, is **$\{\ldots, -2, -1, 0, 1, 2, \ldots\}$**.

Without actually plotting points, match each function defined by the absolute value expression with its graph. See Examples 1–3.

5. $f(x) = |x - 2| + 2$ **B** **6.** $f(x) = |x + 2| + 2$ **C** **7.** $f(x) = |x - 2| - 2$ **A** **8.** $f(x) = |x + 2| - 2$ **D**

A.

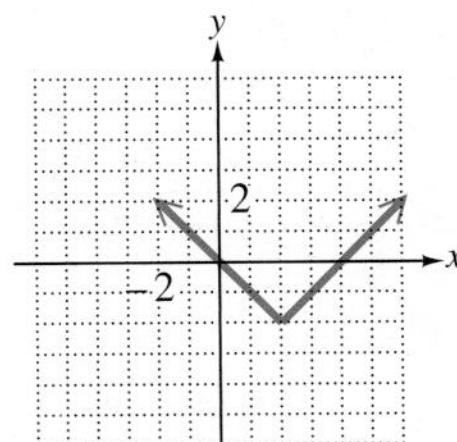

B.

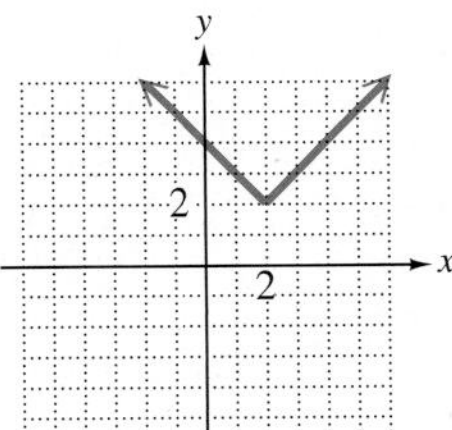

C.

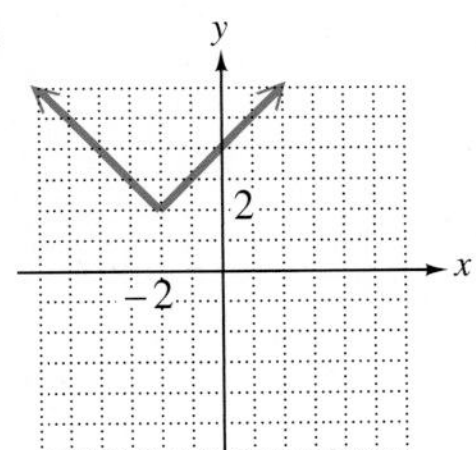

D.

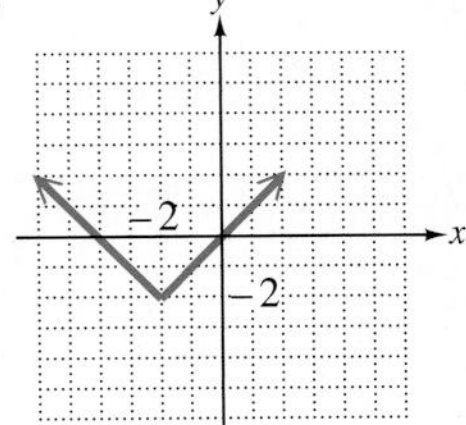

Graph each function. Give the domain and range. See Examples 1–3.

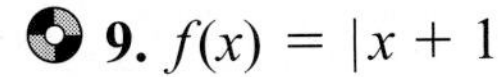
9. $f(x) = |x + 1|$

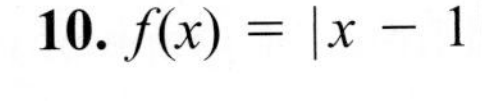
10. $f(x) = |x - 1|$

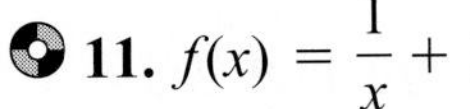
11. $f(x) = \frac{1}{x} + 1$

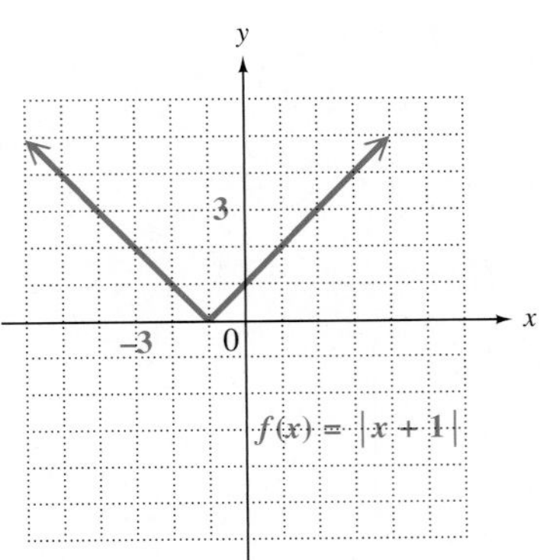

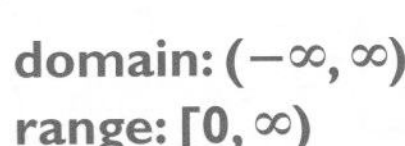
domain: $(-\infty, \infty)$; range: $[0, \infty)$

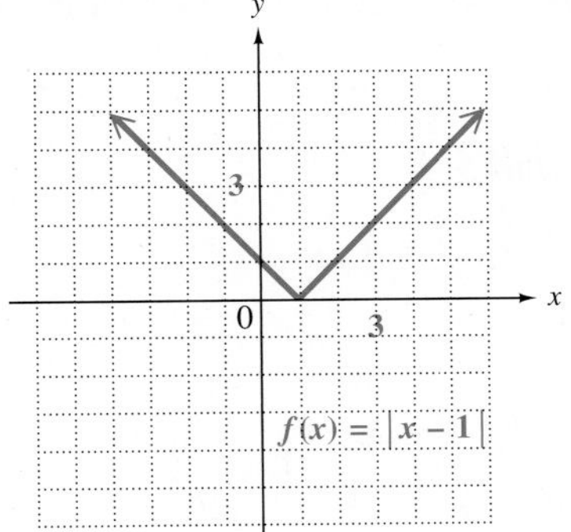

domain: $(-\infty, \infty)$; range: $[0, \infty)$

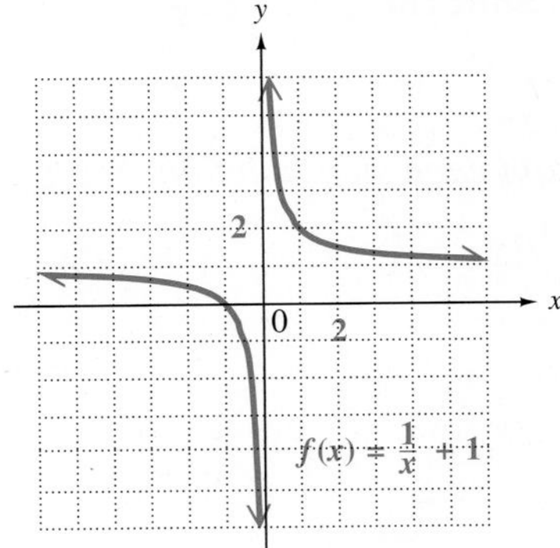

domain: $(-\infty, 0) \cup (0, \infty)$; range: $(-\infty, 1) \cup (1, \infty)$

12. $f(x) = \frac{1}{x} - 1$

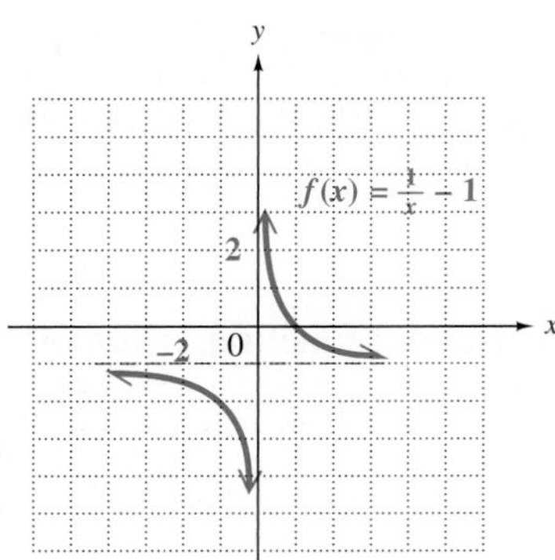

domain: $(-\infty, 0) \cup (0, \infty)$;
range: $(-\infty, -1) \cup (-1, \infty)$

13. $f(x) = \sqrt{x - 2}$

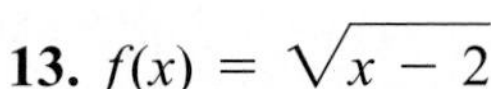

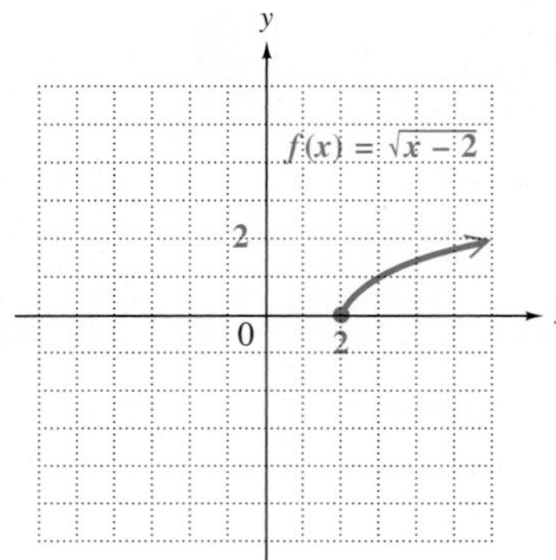

domain: $[2, \infty)$;
range: $[0, \infty)$

14. $f(x) = \sqrt{x + 5}$

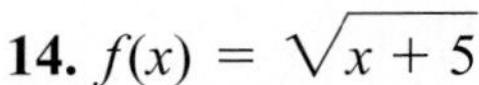

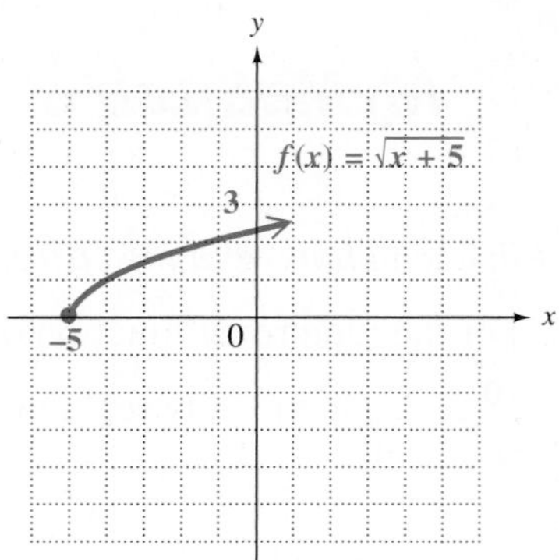

domain: $[-5, \infty)$;
range: $[0, \infty)$

15. $f(x) = \frac{1}{x - 2}$

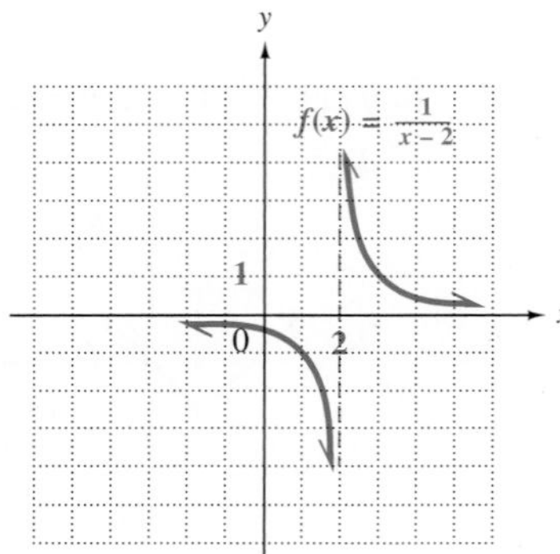

domain: $(-\infty, 2) \cup (2, \infty)$;
range: $(-\infty, 0) \cup (0, \infty)$

16. $f(x) = \frac{1}{x + 2}$

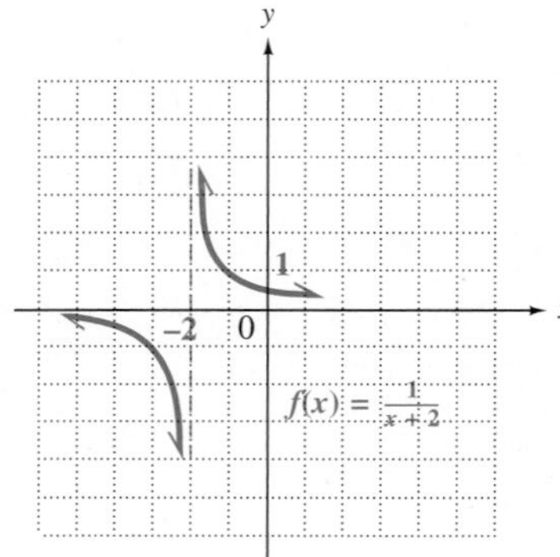

domain: $(-\infty, -2) \cup (-2, \infty)$;
range: $(-\infty, 0) \cup (0, \infty)$

17. $f(x) = \sqrt{x + 3} - 3$

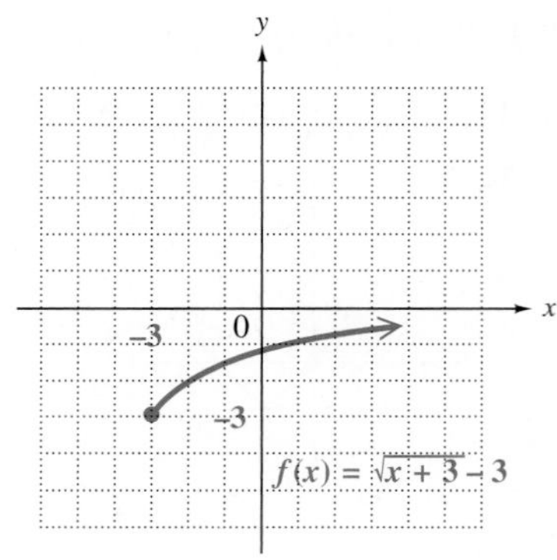

domain: $[-3, \infty)$;
range: $[-3, \infty)$

18. Explain how the graph of $f(x) = \frac{1}{x-3} + 2$ is obtained from the graph of $g(x) = \frac{1}{x}$.

Shift the graph of $g(x) = \frac{1}{x}$ three units to the right and two units up.

Graph each step function. See Examples 4 and 5.

19. $f(x) = [\![-x]\!]$

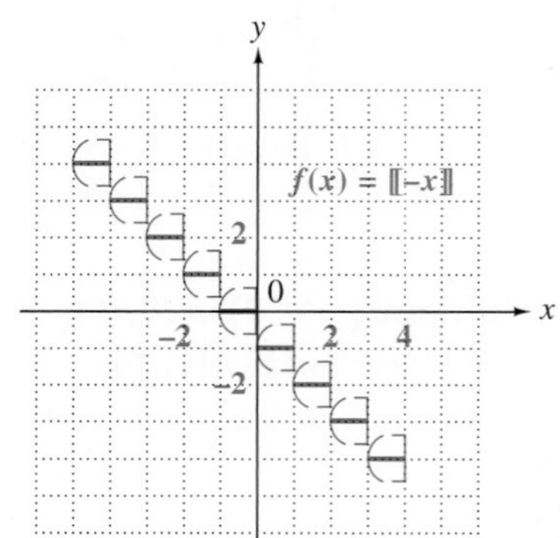

20. $g(x) = [\![x + 2]\!]$

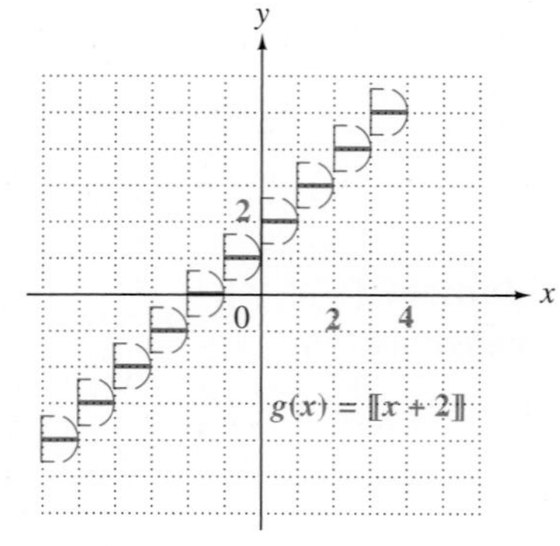

21. $f(x) = [\![x - 3]\!]$

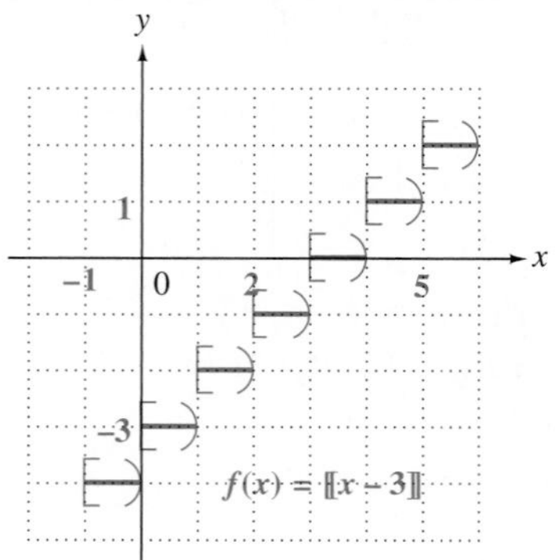

22. The cost of parking a car at an airport hourly parking lot is \$3 for the first half-hour and \$2 for each additional half-hour or fraction thereof. Graph the function defined by $f(x)$ = the cost of parking a car for x hours. Use the interval $(0, 2]$.

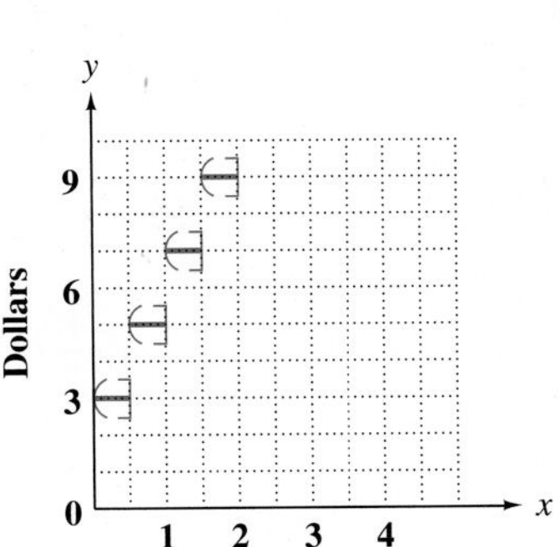

23. Assume that postage rates are 42¢ for the first ounce, plus 17¢ for each additional ounce, and that each letter carries one 42¢ stamp and as many 17¢ stamps as necessary. Graph the function defined by $p(x)$ = the number of stamps on a letter weighing x ounces. Use the interval $(0, 5]$.

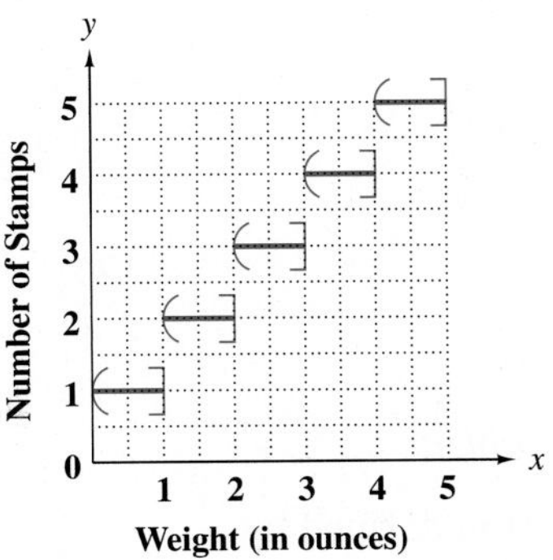

24. A certain long-distance carrier provides service between Podunk and Nowhereville. If x represents the number of minutes for the call, where $x > 0$, then the function f defined by

$$f(x) = 0.40[\![x]\!] + 0.75$$

gives the total cost of the call in dollars. Find the cost of a 5.5-min call. **\$2.75**

Let $f(x) = x^2 - 9$, $g(x) = 2x$, and $h(x) = x - 3$. Find each of the following. See Examples 6–8.

25. $(f + g)(x)$ **$x^2 + 2x - 9$**

26. $(f - g)(x)$ **$x^2 - 2x - 9$**

27. $(fg)(x)$ **$2x^3 - 18x$**

28. $\left(\frac{f}{g}\right)(x)$ **$\frac{x^2 - 9}{2x}$, $x \neq 0$**

29. $(f + g)(3)$ **6**

30. $(f - g)(-3)$ **6**

31. $(fg)(2)$ **-20**

32. $\left(\frac{f}{g}\right)(2)$ **$-\frac{5}{4}$**

33. $(f - h)(x)$ **$x^2 - x - 6$**

34. $(f + h)(x)$ **$x^2 + x - 12$**

35. $\left(\frac{f}{h}\right)(x)$ **$x + 3$, $x \neq 3$**

36. $(fh)(x)$ **$x^3 - 3x^2 - 9x + 27$**

37. $(f - h)(-3)$ **6**

38. $(f + h)(-2)$ **-10**

39. $\left(\frac{f}{h}\right)(1)$ **4**

40. $(fh)(1)$ **16**

Let $f(x) = x^2 + 4$, $g(x) = 2x + 3$, and $h(x) = x + 5$. Find each value or expression. See Examples 9 and 10.

41. $(h \circ g)(4)$ **16**

42. $(f \circ g)(4)$ **125**

43. $(g \circ f)(6)$ **83**

44. $(h \circ f)(6)$ **45**

45. $(f \circ h)(-2)$
13

46. $(h \circ g)(-2)$
4

47. $(f \circ g)(x)$
$4x^2 + 12x + 13$

48. $(g \circ h)(x)$
$2x + 13$

49. $(f \circ h)(x)$
$x^2 + 10x + 29$

50. $(g \circ f)(x)$
$2x^2 + 11$

51. $(h \circ g)(x)$
$2x + 8$

52. $(h \circ f)(x)$
$x^2 + 9$

53. $(f \circ h)\left(\frac{1}{2}\right)$
$\frac{137}{4}$

54. $(h \circ f)\left(\frac{1}{2}\right)$
$\frac{37}{4}$

55. $(f \circ g)\left(-\frac{1}{2}\right)$
8

56. $(g \circ f)\left(-\frac{1}{2}\right)$
$\frac{23}{2}$

Solve each problem.

57. The function defined by

$$f(x) = 12x$$

computes the number of inches in x feet and the function defined by

$$g(x) = 5280x$$

computes the number of feet in x miles. What is $(f \circ g)(x)$ and what does it compute?

$(f \circ g)(x) = 63{,}360x$; It computes the number of inches in x miles.

58. The perimeter x of a square with sides of length s is given by the formula $x = 4s$.

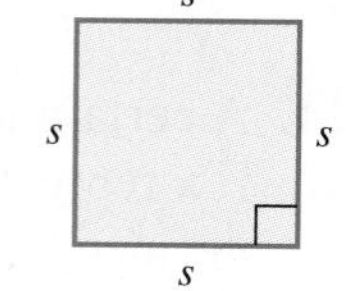

(a) Solve for s in terms of x.

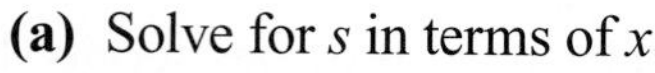

$s = \frac{x}{4}$

(b) If y represents the area of this square, write y as a function of the perimeter x.

$y = \frac{x^2}{16}$

(c) Use the composite function of part (b) to find the area of a square with perimeter 6.

2.25

59. When a thermal inversion layer is over a city (as happens often in Los Angeles), pollutants cannot rise vertically but are trapped below the layer and must disperse horizontally. Assume that a factory smokestack begins emitting a pollutant at 8 A.M. Assume that the pollutant disperses horizontally over a circular area. Suppose that t represents the time, in hours, since the factory began emitting pollutants ($t = 0$ represents 8 A.M.), and assume that the radius of the circle of pollution is $r(t) = 2t$ miles. Let $A(r) = \pi r^2$ represent the area of a circle of radius r. Find and interpret $(A \circ r)(t)$.

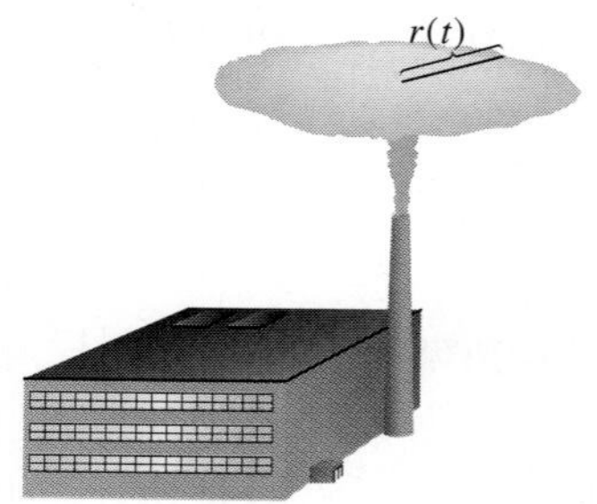

$(A \circ r)(t) = 4\pi t^2$; This is the area of the circular layer as a function of time.

60. An oil well off the Gulf Coast is leaking, with the leak spreading oil over the surface as a circle. At any time t, in minutes, after the beginning of the leak, the radius of the circular oil slick on the surface is $r(t) = 4t$ feet. Let $A(r) = \pi r^2$ represent the area of a circle of radius r. Find and interpret $(A \circ r)(t)$.

$(A \circ r)(t) = 16\pi t^2$; This is the area of the oil slick as a function of time.

12.2 The Circle and the Ellipse

When an infinite cone is intersected by a plane, the resulting figure is called a **conic section.** The parabola is one example of a conic section; *circles, ellipses,* and *hyperbolas* may also result. See Figure 11.

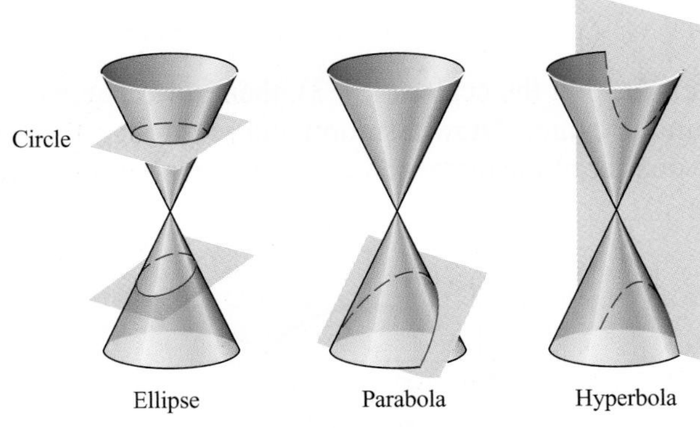

Figure 11

OBJECTIVES

1. Find an equation of a circle given the center and radius.
2. Determine the center and radius of a circle given its equation.
3. Recognize the equation of an ellipse.
4. Graph ellipses.

OBJECTIVE 1 Find an equation of a circle given the center and radius. A **circle** is the set of all points in a plane that lie a fixed distance from a fixed point. The fixed point is called the **center,** and the fixed distance is called the **radius.** We use the distance formula from **Section 9.3** to find an equation of a circle.

EXAMPLE 1 Finding an Equation of a Circle and Graphing It

Find an equation of the circle with radius 3 and center at (0, 0), and graph it.

If the point (x, y) is on the circle, the distance from (x, y) to the center (0, 0) is 3. By the distance formula,

$$\sqrt{(x_2 - x_1)^2 + (y_2 - y_1)^2} = d \qquad \text{Distance formula}$$

$$\sqrt{(x - 0)^2 + (y - 0)^2} = 3 \qquad \text{Let } x_1 = 0, y_1 = 0, \text{ and } d = 3.$$

$$x^2 + y^2 = 9. \qquad \text{Square both sides.}$$

An equation of this circle is $x^2 + y^2 = 9$. The graph is shown in Figure 12.

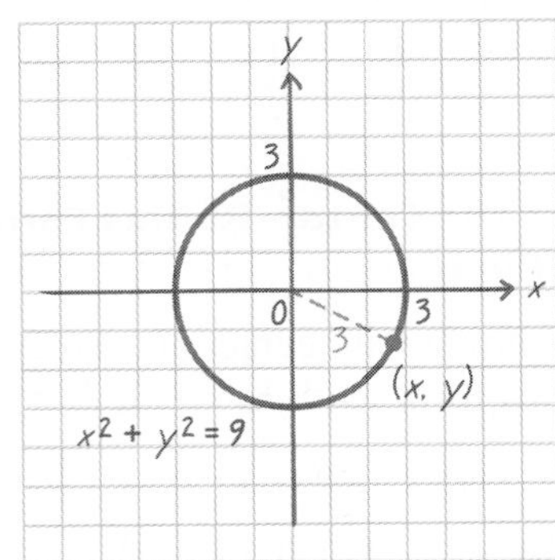

Figure 12

Work Problem 1 at the Side.

1 Find an equation of the circle with radius 4 and center (0, 0). Sketch its graph.

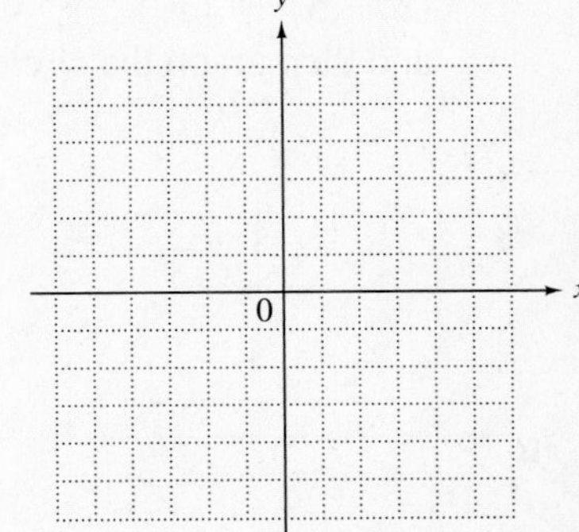

ANSWER

1. $x^2 + y^2 = 16$

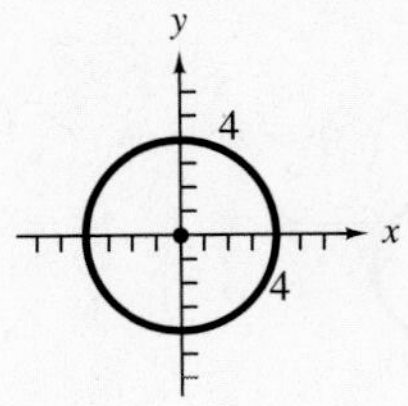

2 **(a)** Find an equation of the circle with center at $(3, -2)$ and radius 4. Graph the circle.

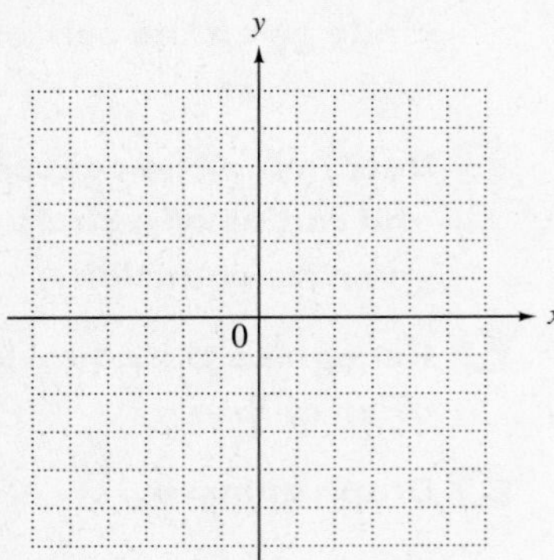

(b) Use the center-radius form to determine the center and radius of $(x - 5)^2 + (y + 2)^2 = 9$, and then graph the circle.

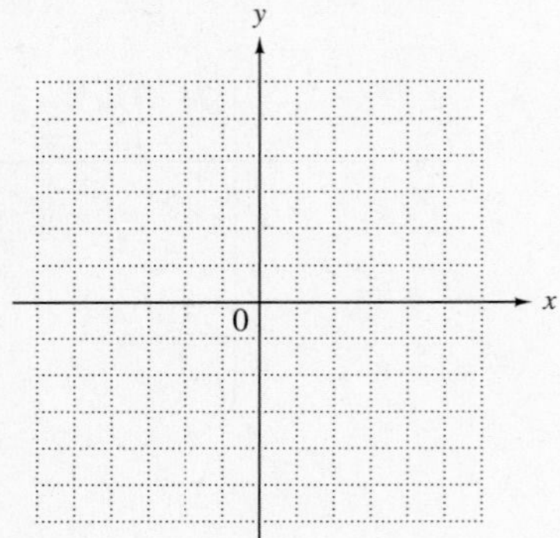

Answers

2. **(a)** $(x - 3)^2 + (y + 2)^2 = 16$

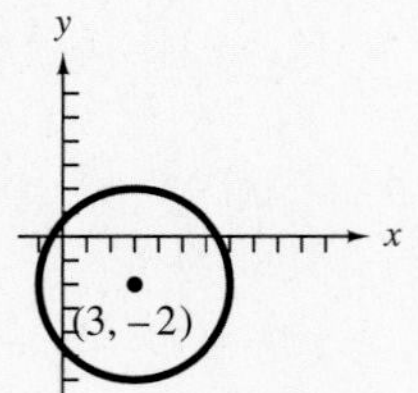

(b) center at $(5, -2)$; radius 3

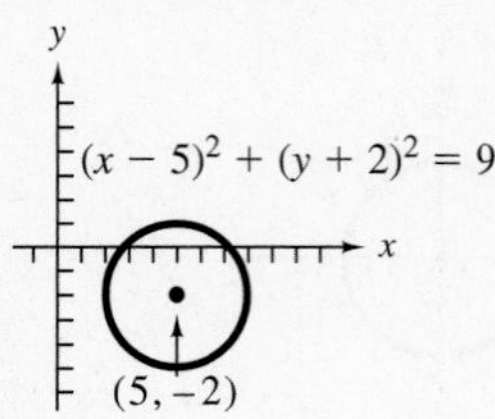

A circle may not be centered at the origin, as seen in the next example.

EXAMPLE 2 **Finding an Equation of a Circle and Graphing It**

Find an equation of the circle with center at $(4, -3)$ and radius 5, and graph it.

$$\sqrt{(x - 4)^2 + [y - (-3)]^2} = 5 \qquad \text{Distance formula}$$

$$(x - 4)^2 + (y + 3)^2 = 25 \qquad \text{Square both sides.}$$

To graph the circle, plot the center $(4, -3)$, then move 5 units right, left, up, and down from the center. Draw a smooth curve through these four points, sketching one quarter of the circle at a time. The graph of this circle is shown in Figure 13.

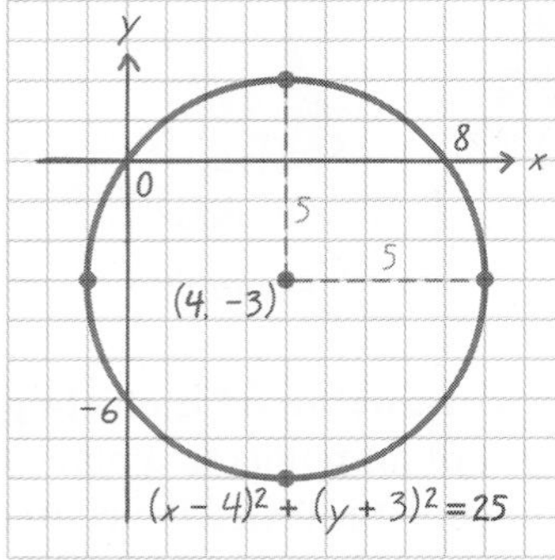

Figure 13

Examples 1 and 2 suggest the form of an equation of a circle with radius r and center at (h, k). If (x, y) is a point on the circle, then the distance from the center (h, k) to the point (x, y) is r. By the distance formula,

$$\sqrt{(x - h)^2 + (y - k)^2} = r.$$

Squaring both sides gives us the **center-radius form** of the equation of a circle.

Equation of a Circle (Center-Radius Form)

An equation of a circle of radius r with center at (h, k) is

$$(x - h)^2 + (y - k)^2 = r^2.$$

EXAMPLE 3 **Using the Center-Radius Form of the Equation of a Circle**

Find an equation of the circle with center at $(-1, 2)$ and radius 4.

Use the center-radius form, with $h = -1$, $k = 2$, and $r = 4$.

$$(x - h)^2 + (y - k)^2 = r^2$$

$$[x - (-1)]^2 + (y - 2)^2 = 4^2$$

Pay attention to signs here.

$$(x + 1)^2 + (y - 2)^2 = 16$$

◀ Work Problem **2** at the Side.

OBJECTIVE 2 Determine the center and radius of a circle given its equation. In the equation found in Example 2, multiplying out $(x - 4)^2$ and $(y + 3)^2$ and then combining like terms gives

$$(x - 4)^2 + (y + 3)^2 = 25$$
$$x^2 - 8x + 16 + y^2 + 6y + 9 = 25$$
$$x^2 + y^2 - 8x + 6y = 0.$$

This general form suggests that an equation with both x^2- and y^2-terms with equal coefficients may represent a circle. The next example shows how to tell, by completing the square. This procedure was introduced in **Section 10.2.**

EXAMPLE 4 Completing the Square to Find the Center and Radius

Graph $x^2 + y^2 + 2x + 6y - 15 = 0$.

Since the equation has x^2- and y^2-terms with equal coefficients, its graph might be that of a circle. To find the center and radius, complete the squares on x and y.

$$x^2 + y^2 + 2x + 6y = 15$$ Transform so that the constant is on the right.

$$(x^2 + 2x \quad) + (y^2 + 6y \quad) = 15$$ Rewrite in anticipation of completing the square.

$$\left[\frac{1}{2}(2)\right]^2 = 1 \qquad \left[\frac{1}{2}(6)\right]^2 = 9$$ Square half the coefficient of each middle term.

$$(x^2 + 2x + 1) + (y^2 + 6y + 9) = 15 + 1 + 9$$ Complete the squares on both x and y.

$$(x + 1)^2 + (y + 3)^2 = 25$$ Factor on the left; add on the right.

$$[x - (-1)]^2 + [y - (-3)]^2 = 5^2$$ Center-radius form

The final equation shows that the graph is a circle with center at $(-1, -3)$ and radius 5. The graph is shown in Figure 14.

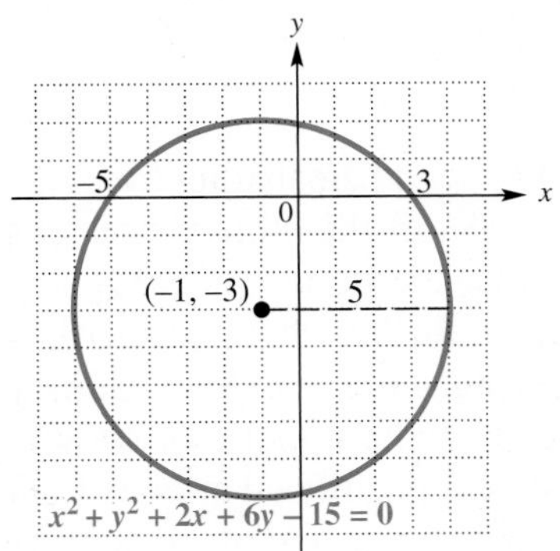

Figure 14

Note

If the procedure of Example 4 leads to an equation of the form $(x - h)^2 + (y - k)^2 = 0$, then the graph is the single point (h, k). If the constant on the right side is negative, then the equation has no graph.

Work Problem 3 at the Side. ▶

3 Find the center and radius of the circle with equation $x^2 + y^2 - 6x + 8y - 11 = 0$.

Answer

3. center at $(3, -4)$; radius 6

OBJECTIVE 3 Recognize the equation of an ellipse. An **ellipse** is the set of all points in a plane the *sum* of whose distances from two fixed points is constant. These fixed points are called **foci** (singular: *focus*). Figure 15 shows an ellipse whose foci are $(c, 0)$ and $(-c, 0)$, with x-intercepts $(a, 0)$ and $(-a, 0)$ and y-intercepts $(0, b)$ and $(0, -b)$. It can be shown in more advanced courses that $c^2 = a^2 - b^2$ for an ellipse of this type. The origin is the **center** of the ellipse.

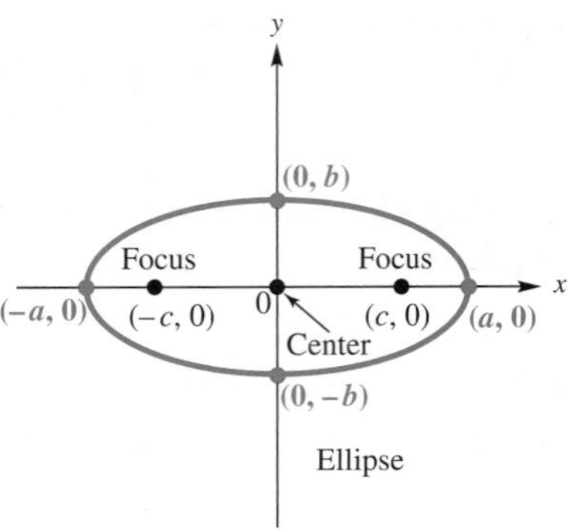

Figure 15

An ellipse has the following equation.

Equation of an Ellipse

The ellipse whose x-intercepts are $(a, 0)$ and $(-a, 0)$ and whose y-intercepts are $(0, b)$ and $(0, -b)$ has an equation of the form

$$\frac{x^2}{a^2} + \frac{y^2}{b^2} = 1.$$

Note

A circle is a special case of an ellipse, where $a^2 = b^2$.

When a ray of light or sound emanating from one focus of an ellipse bounces off the ellipse, it passes through the other focus. See the figure. As mentioned in the chapter introduction, this reflecting property is responsible for whispering galleries. John Quincy Adams was able to listen in on his opponents' conversations because his desk was positioned at one of the foci beneath the ellipsoidal ceiling and his opponents were located across the room at the other focus.

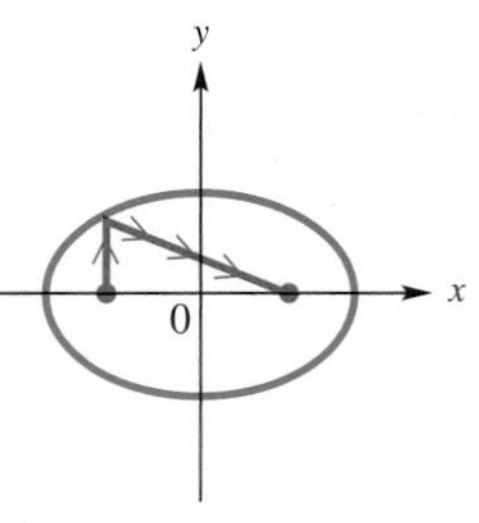

Reflecting property of an ellipse

The paths of Earth and other planets around the sun are approximately ellipses; the sun is at one focus and a point in space is at the other. The orbits of communication satellites and other space vehicles are also elliptical.

Elliptical bicycle gears are designed to respond to the legs' natural strengths and weaknesses. At the top and bottom of the powerstroke, where the legs have the least leverage, the gear offers little resistance, but as the gear rotates, the resistance increases. This allows the legs to apply more power where it is most naturally available. See Figure 16 on the next page.

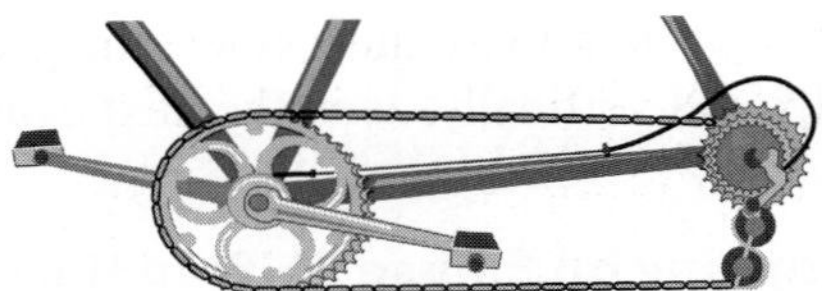

Figure 16

OBJECTIVE 4 Graph ellipses. To graph an ellipse centered at the origin, we plot the four intercepts and then sketch the ellipse through those points.

EXAMPLE 5 Graphing Ellipses

Graph each ellipse.

(a) $\frac{x^2}{49} + \frac{y^2}{36} = 1$

Here, $a^2 = 49$, so $a = 7$, and the x-intercepts for this ellipse are $(7, 0)$ and $(-7, 0)$. Similarly, $b^2 = 36$, so $b = 6$, and the y-intercepts are $(0, 6)$ and $(0, -6)$. Plotting the intercepts and sketching the ellipse through them gives the graph in Figure 17.

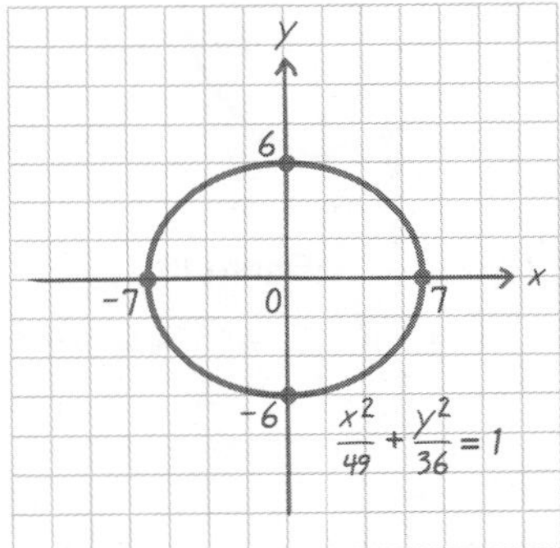

Figure 17

(b) $\frac{x^2}{36} + \frac{y^2}{121} = 1$

The x-intercepts for this ellipse are $(6, 0)$ and $(-6, 0)$, and the y-intercepts are $(0, 11)$ and $(0, -11)$. Join these intercepts with the smooth curve of an ellipse. The graph has been sketched in Figure 18.

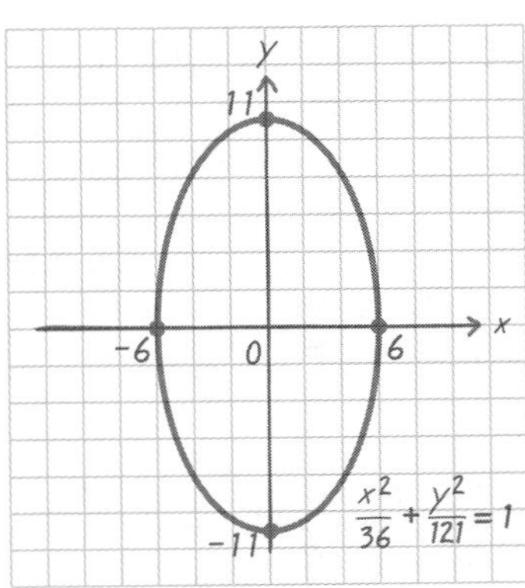

Figure 18

Work Problem 4 at the Side. ▶

4 Graph each ellipse.

(a) $\frac{x^2}{4} + \frac{y^2}{25} = 1$

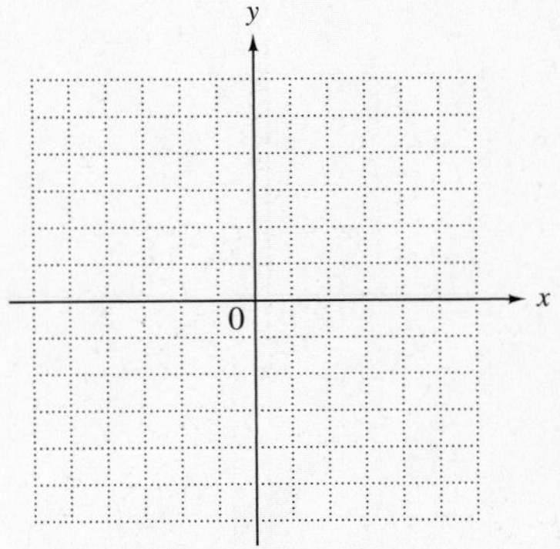

(b) $\frac{x^2}{64} + \frac{y^2}{49} = 1$

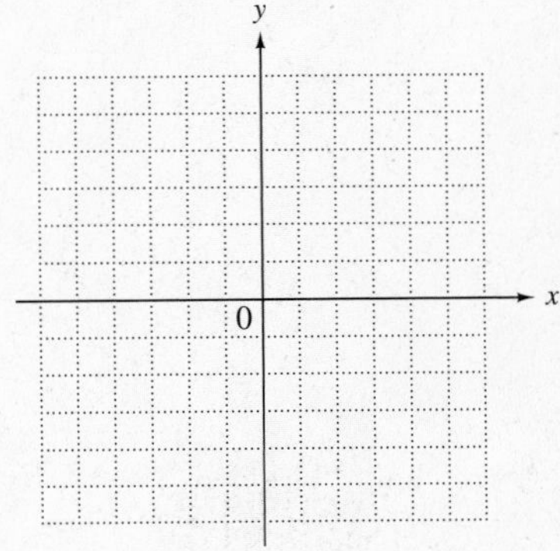

ANSWERS

4. (a) **(b)**

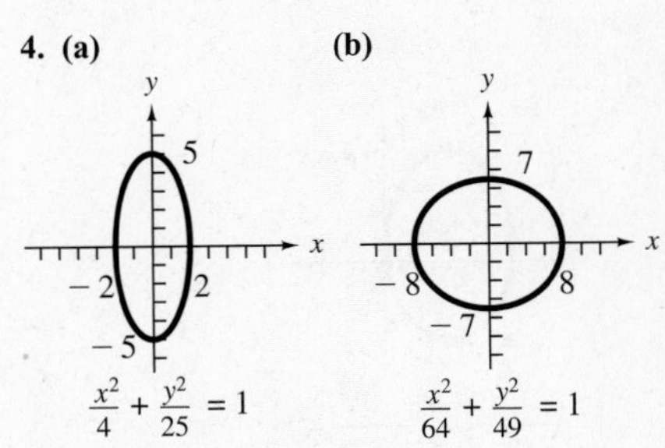

As with the graphs of parabolas and circles, the graph of an ellipse may be shifted horizontally and vertically, as in the next example.

EXAMPLE 6 Graphing an Ellipse Shifted Horizontally and Vertically

Graph $\dfrac{(x-2)^2}{25} + \dfrac{(y+3)^2}{49} = 1.$

Just as $(x-2)^2$ and $(y+3)^2$ would indicate that the center of a circle would be $(2, -3)$, so it is with this ellipse. Figure 19 shows that the graph goes through the four points $(2, 4)$, $(7, -3)$, $(2, -10)$, and $(-3, -3)$. The x-values of these points are found by adding $\pm a = \pm 5$ to 2, and the y-values come from adding $\pm b = \pm 7$ to -3.

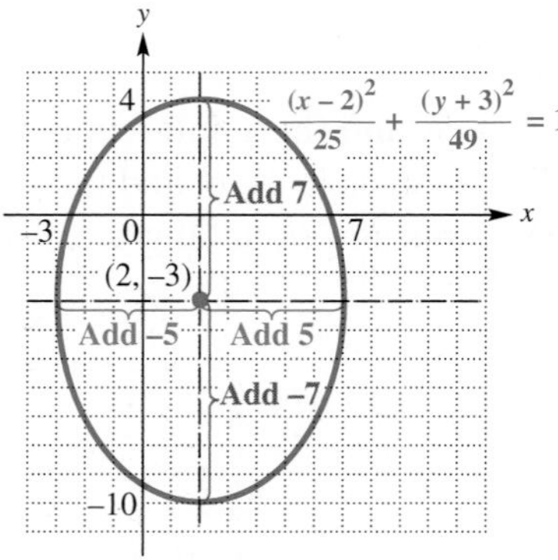

Figure 19

◀ Work Problem 5 at the Side.

Note

The graphs in this section are not graphs of functions. The only conic section whose graph is a function is the vertical parabola with equation $f(x) = ax^2 + bx + c$.

5 Graph

$$\frac{(x+4)^2}{16} + \frac{(y-1)^2}{36} = 1.$$

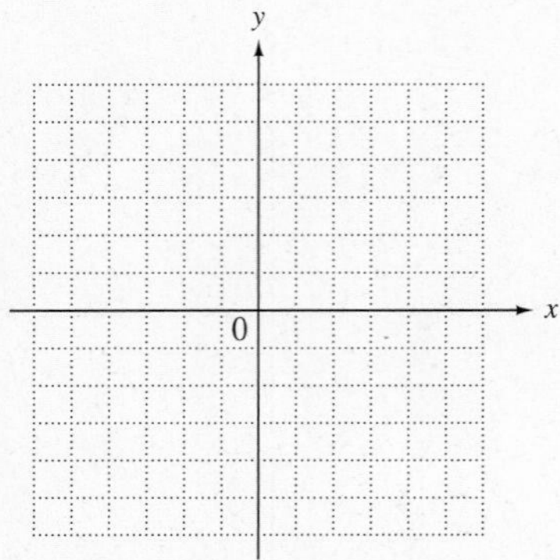

Answer

5. (-4, 7), (-8, 1), (-4, 1), (0, 1), (-4, -5)

$$\frac{(x+4)^2}{16} + \frac{(y-1)^2}{36} = 1$$

12.2 Exercises

FOR EXTRA HELP MyMathLab Math XL PRACTICE  WATCH DOWNLOAD  READ  REVIEW

1. See Example 1. Consider the circle whose equation is $x^2 + y^2 = 25$.
   (a) What are the coordinates of its center? **(0, 0)**
   (b) What is its radius? **5**
   (c) Sketch its graph.

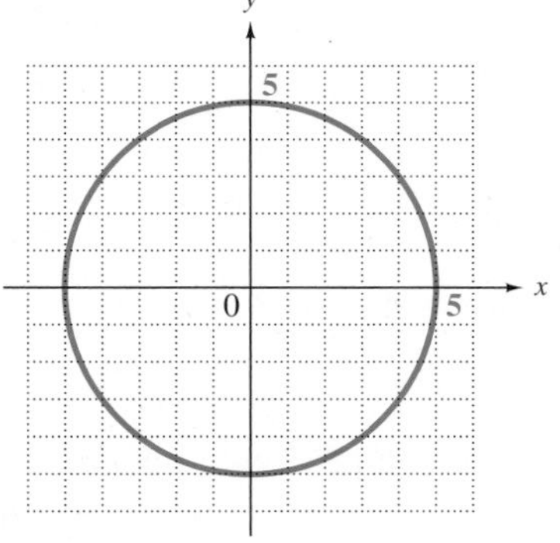

2. Explain why a set of points defined by a circle does not satisfy the definition of a function.
   There will always be domain values that yield more than one range value. A circle fails the vertical line test.

Match each equation with the correct graph. See Examples 1–3.

3. $(x - 3)^2 + (y - 2)^2 = 25$ **B**

4. $(x - 3)^2 + (y + 2)^2 = 25$ **C**

5. $(x + 3)^2 + (y - 2)^2 = 25$ **D**

6. $(x + 3)^2 + (y + 2)^2 = 25$ **A**

A. 

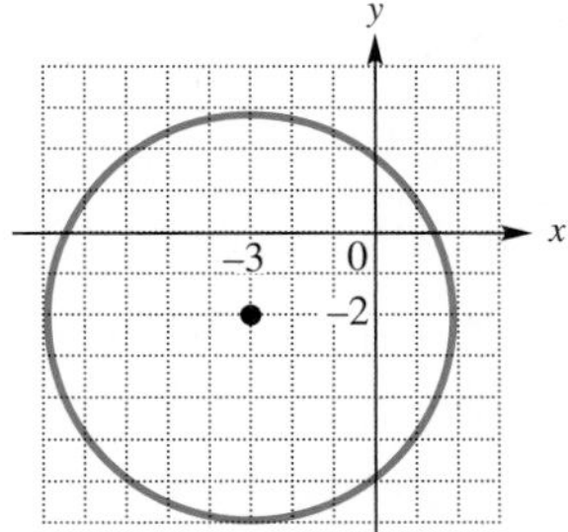

B. 

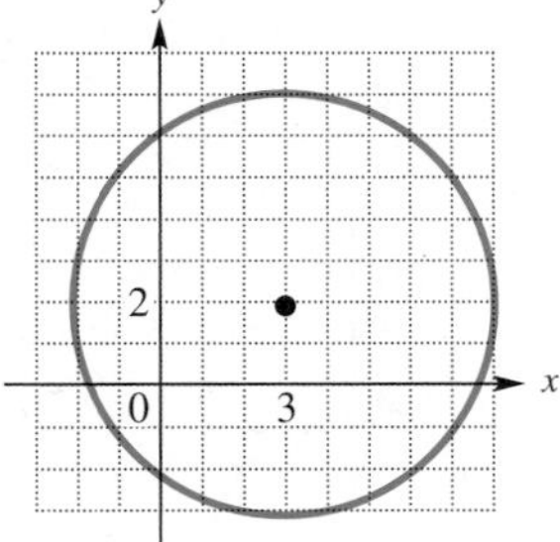

C.

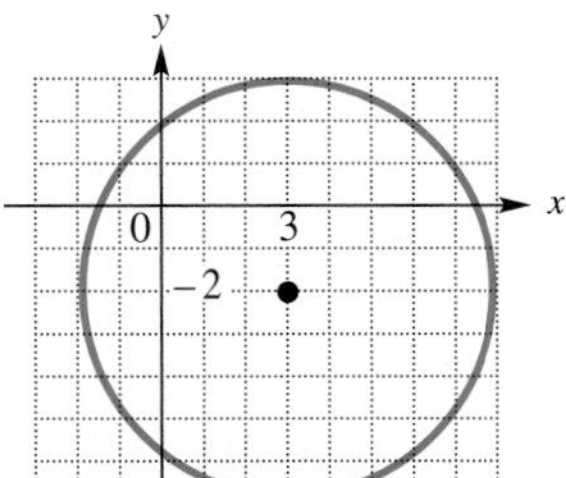

D.

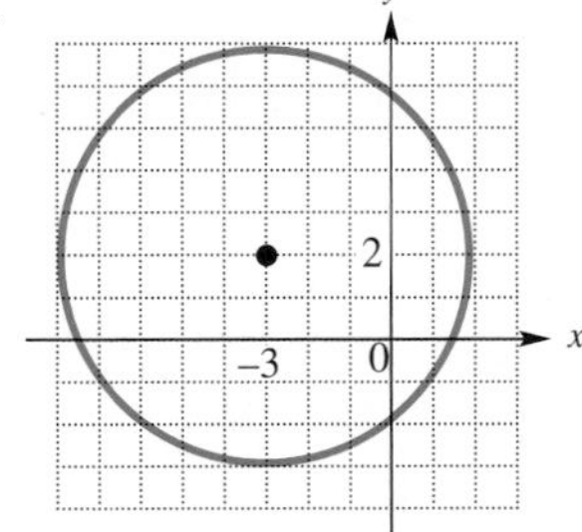

Find the equation of a circle satisfying the given conditions. See Examples 2 and 3.

7. Center: $(-4, 3)$; radius: 2
   $(x + 4)^2 + (y - 3)^2 = 4$

8. Center: $(5, -2)$; radius: 4
   $(x - 5)^2 + (y + 2)^2 = 16$

9. Center: $(-8, -5)$; radius: $\sqrt{5}$
   $(x + 8)^2 + (y + 5)^2 = 5$

10. Center: $(-12, 13)$; radius: $\sqrt{7}$
    $(x + 12)^2 + (y - 13)^2 = 7$

Find the center and radius of each circle. (Hint: In Exercises 15 and 16, divide each side by a common factor.) See Example 4.

11. $x^2 + y^2 + 4x + 6y + 9 = 0$

$(-2, -3)$; $r = 2$

12. $x^2 + y^2 - 8x - 12y + 3 = 0$

$(4, 6)$; $r = 7$

13. $x^2 + y^2 + 10x - 14y - 7 = 0$

$(-5, 7)$; $r = 9$

14. $x^2 + y^2 - 2x + 4y - 4 = 0$

$(1, -2)$; $r = 3$

15. $3x^2 + 3y^2 - 12x - 24y + 12 = 0$

$(2, 4)$; $r = 4$

16. $2x^2 + 2y^2 + 20x + 16y + 10 = 0$

$(-5, -4)$; $r = 6$

17. A circle can be drawn on a piece of posterboard by fastening one end of a string with a thumbtack, pulling the string taut with a pencil, and tracing a curve, as shown in the figure. Explain why this method works.

The thumbtack acts as the center and the length of the string acts as the radius.

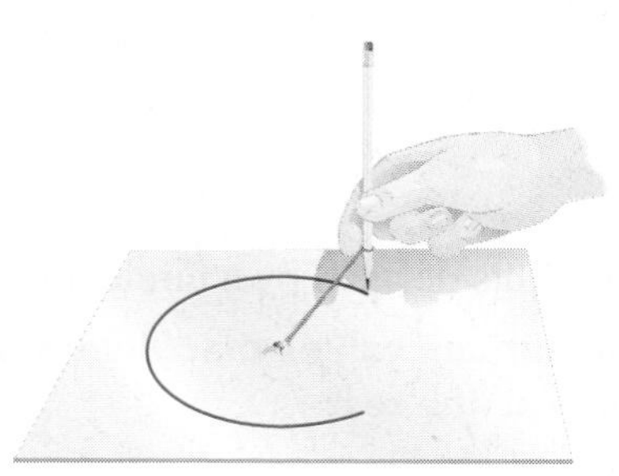

Graph each circle. See Examples 1–4.

18. $x^2 + y^2 = 9$

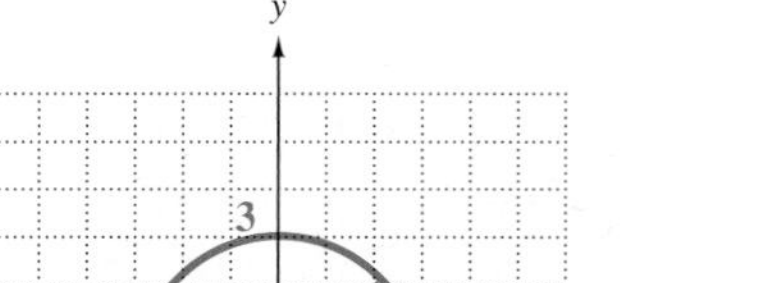

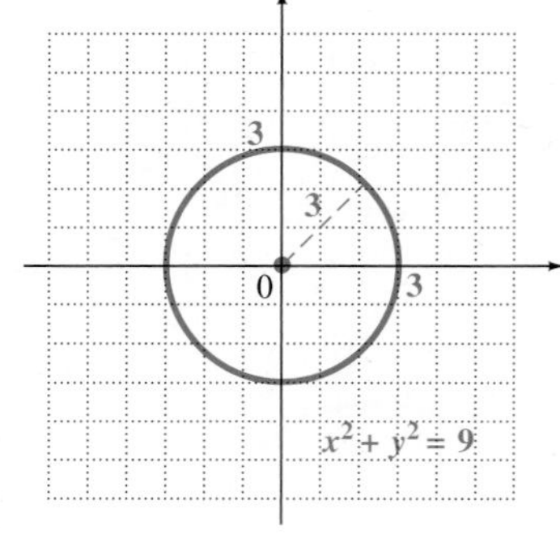

19. $x^2 + y^2 = 4$

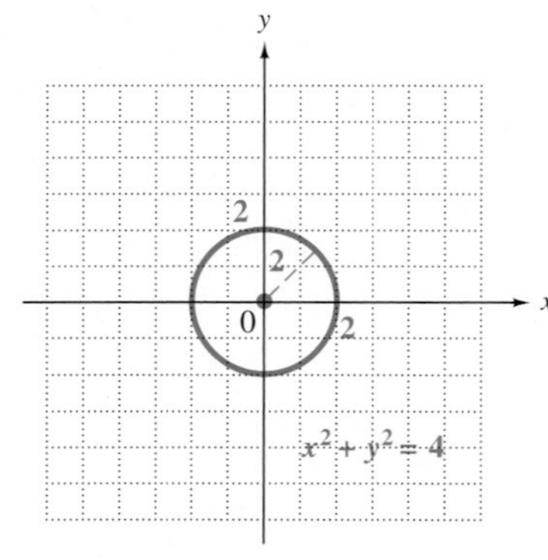

20. $2y^2 = 10 - 2x^2$

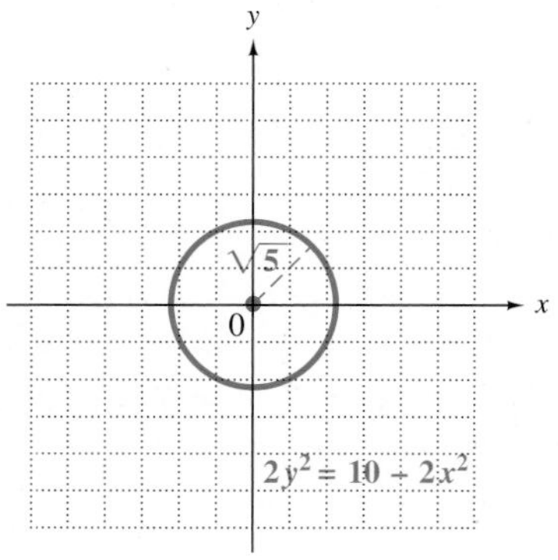

21. $3x^2 = 48 - 3y^2$

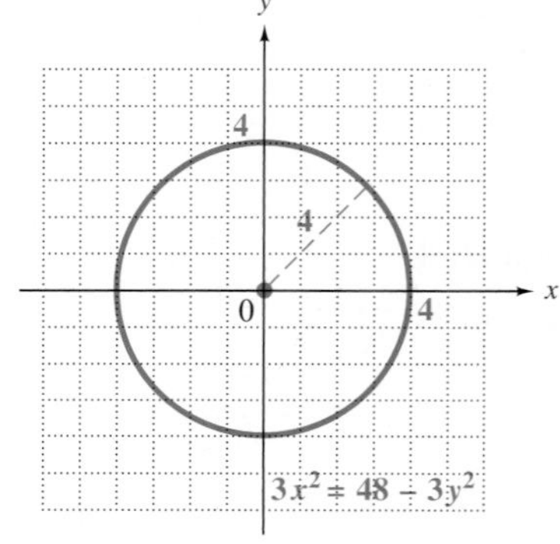

22. $(x + 3)^2 + (y - 2)^2 = 9$

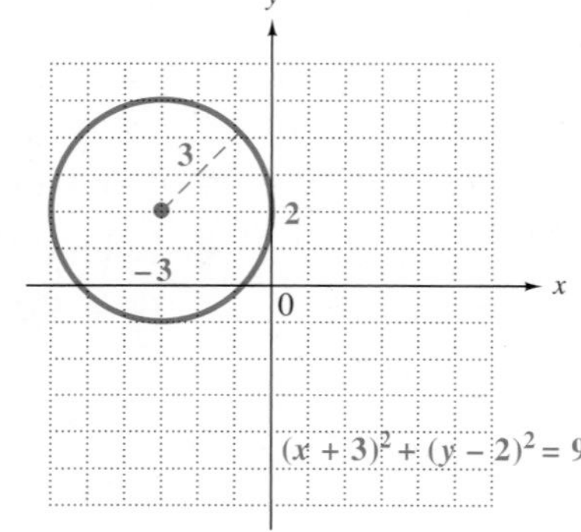

23. $(x - 1)^2 + (y + 3)^2 = 16$

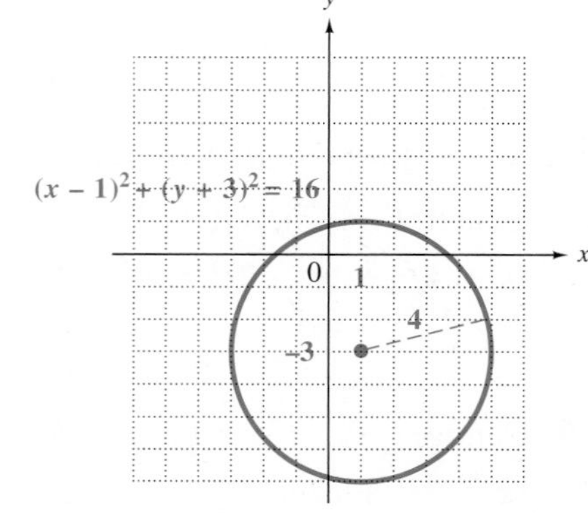

24. $x^2 + y^2 - 4x - 6y + 9 = 0$

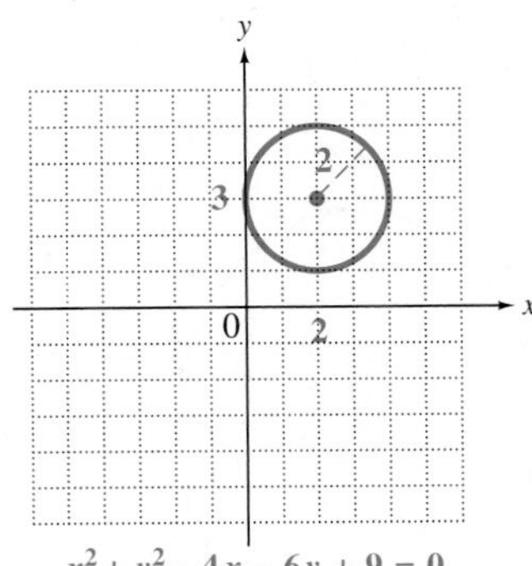

25. $x^2 + y^2 + 8x + 2y - 8 = 0$

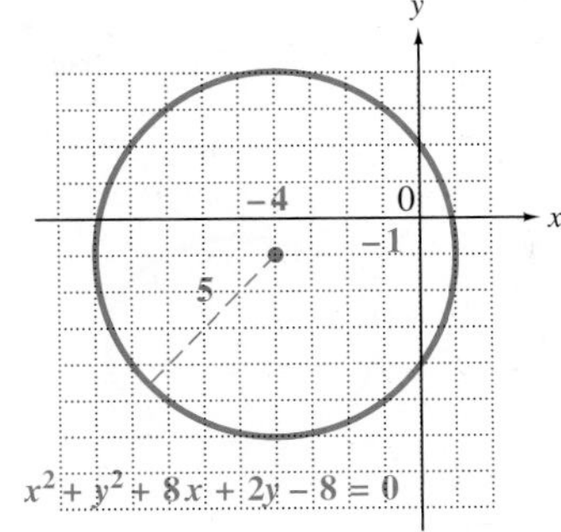

26. $x^2 + y^2 - 4x + 10y + 20 = 0$

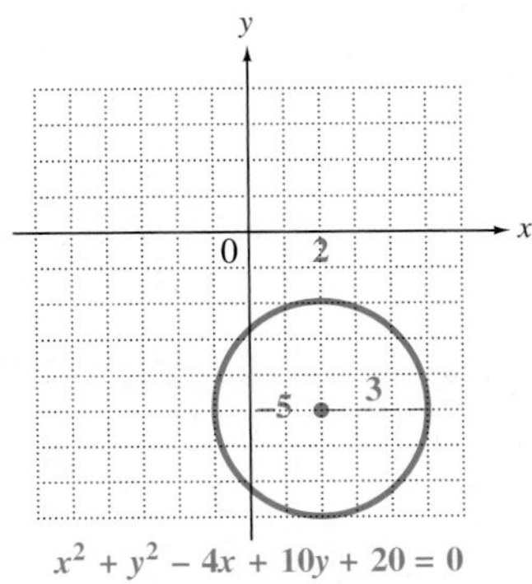

27. An ellipse can be drawn on a piece of posterboard by fastening two ends of a length of string with thumbtacks, pulling the string taut with a pencil, and tracing a curve, as shown in the figure. Explain why this method works.

The two thumbtacks act as foci, and the length of the string is constant, satisfying the requirements of the definition of an ellipse.

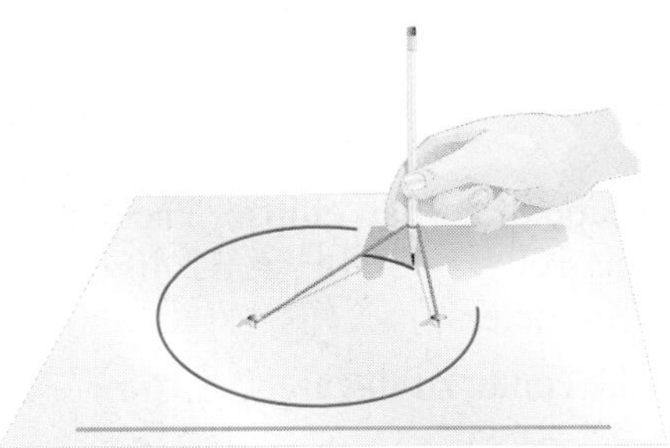

Graph each ellipse. See Examples 5 and 6.

28. $\frac{x^2}{9} + \frac{y^2}{16} = 1$

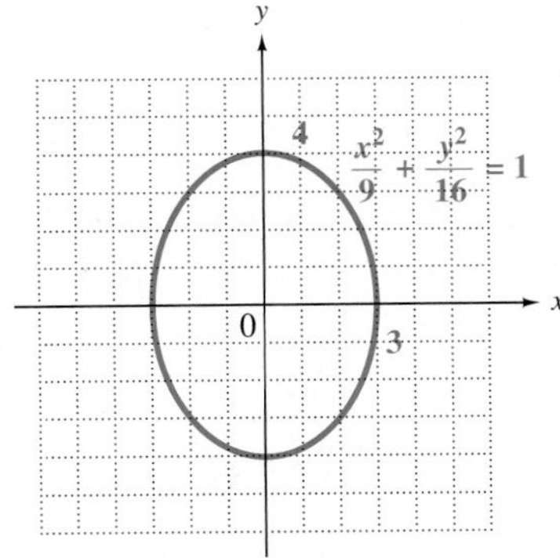

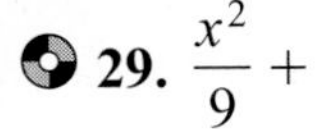

29. $\frac{x^2}{9} + \frac{y^2}{25} = 1$

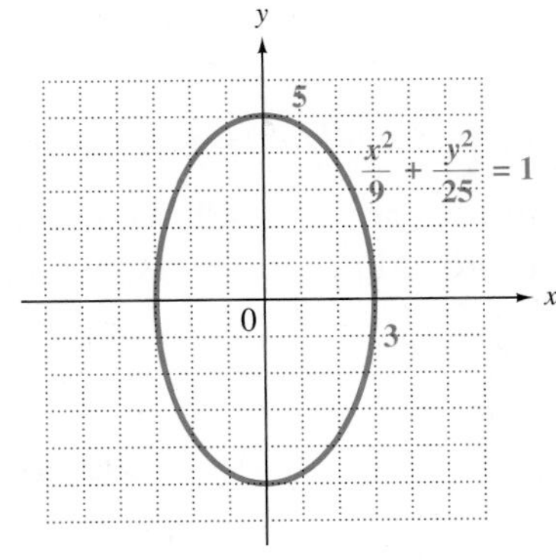

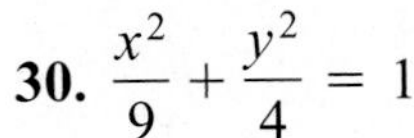

30. $\frac{x^2}{9} + \frac{y^2}{4} = 1$

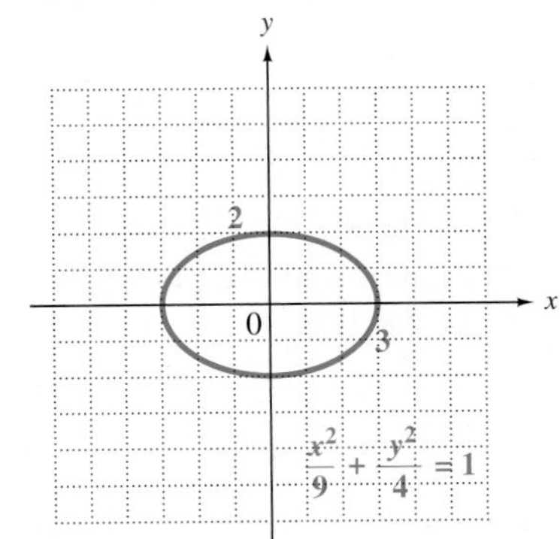

31. $\frac{x^2}{36} + \frac{y^2}{16} = 1$

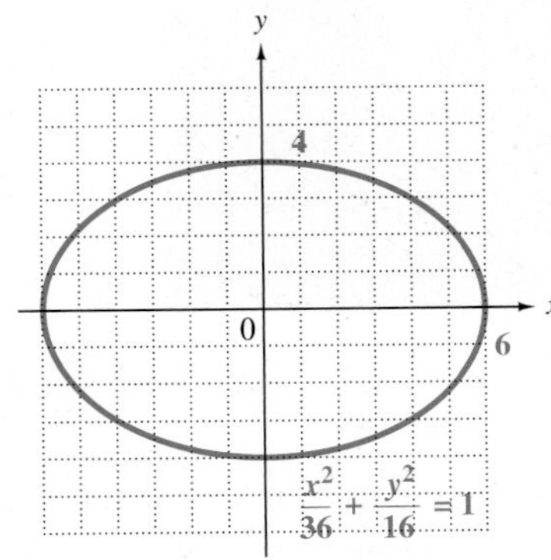

32. $\frac{x^2}{16} + \frac{y^2}{9} = 1$

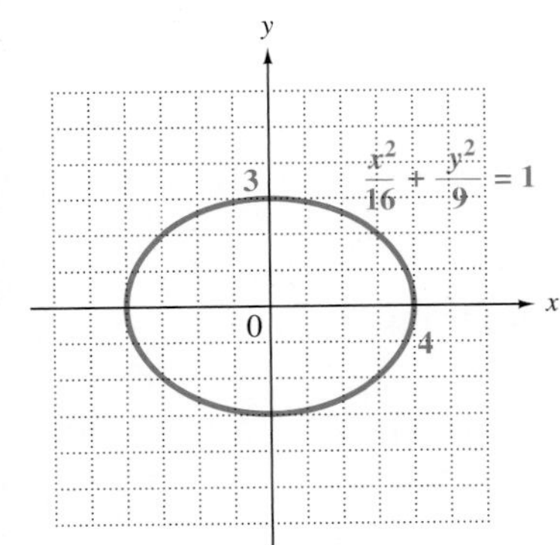

33. $\frac{x^2}{49} + \frac{y^2}{25} = 1$

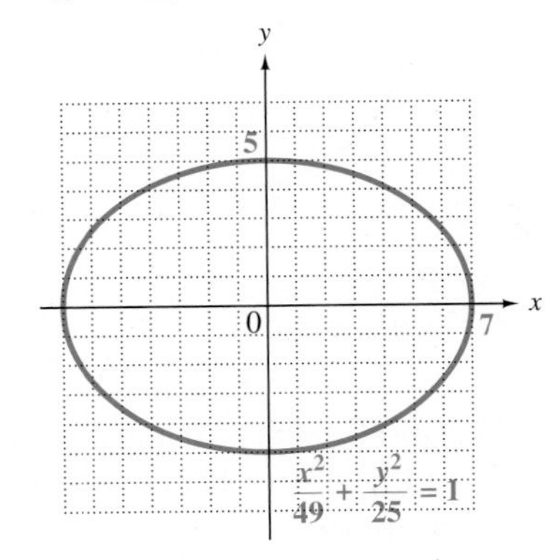

34. $\frac{(x-4)^2}{9}+\frac{(y+2)^2}{4}=1$

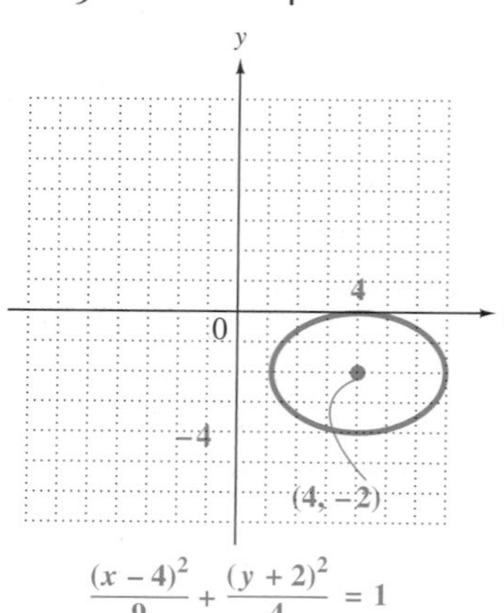

$\frac{(x-4)^2}{9}+\frac{(y+2)^2}{4}=1$

35. $\frac{(x-2)^2}{16}+\frac{(y-1)^2}{9}=1$

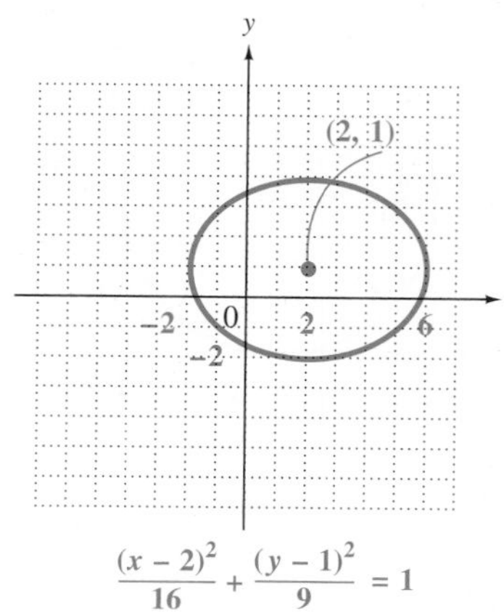

$\frac{(x-2)^2}{16}+\frac{(y-1)^2}{9}=1$

36. $\frac{(x+3)^2}{25}+\frac{(y+2)^2}{36}=1$

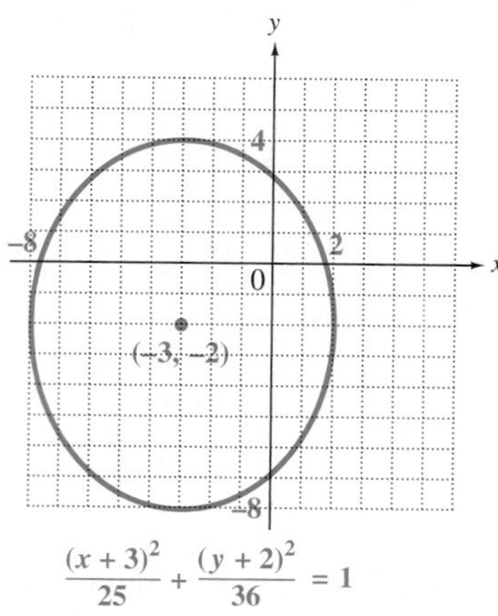

$\frac{(x+3)^2}{25}+\frac{(y+2)^2}{36}=1$

Solve each problem.

37. An arch has the shape of half an ellipse. The equation of the ellipse is $100x^2 + 324y^2 = 32{,}400$, where x and y are in meters.

(a) How high is the center of the arch? **10 m**

(b) How wide is the arch across the bottom? **36 m**

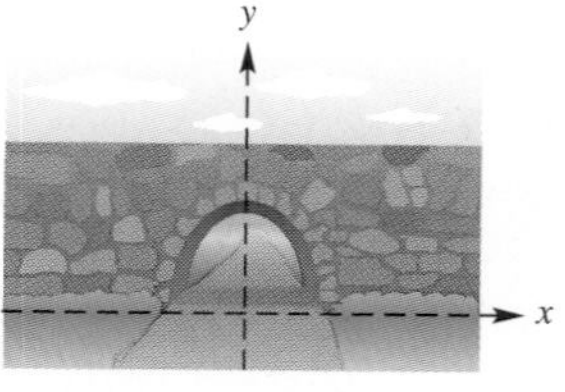

NOT TO SCALE

38. A one-way street passes under an overpass, which is in the form of the top half of an ellipse, as shown in the figure. Suppose that a truck 12 ft wide passes directly under the overpass. What is the maximum possible height of this truck? **12 ft**

NOT TO SCALE

In Exercises 39 and 40, see Figure 15 and use the fact that $c^2 = a^2 - b^2$, where $a^2 > b^2$.

39. The orbit of Mars is an ellipse with the sun at one focus. For x and y in millions of miles, the equation of the orbit is

$$\frac{x^2}{141.7^2}+\frac{y^2}{141.1^2}=1.$$

(*Source:* Kaler, James B., *Astronomy!*, Addison-Wesley, 1997.)

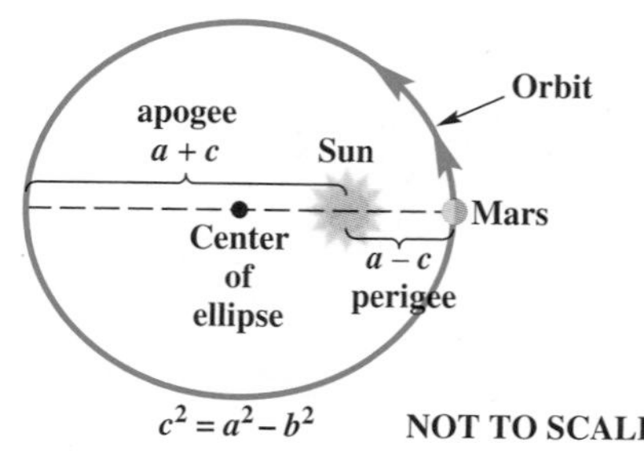

NOT TO SCALE

(a) Find the greatest distance (the **apogee**) from Mars to the sun. **154.7 million mi**

(b) Find the least distance (the **perigee**) from Mars to the sun. **128.7 million mi (Answers are rounded.)**

40. The orbit of Venus around the sun (one of the foci) is an ellipse with equation

$$\frac{x^2}{5013}+\frac{y^2}{4970}=1,$$

where x and y are measured in millions of miles. (*Source:* Kaler, James B., *Astronomy!*, Addison-Wesley, 1997.)

(a) Find the greatest distance between Venus and the sun. **77.4 million mi**

(b) Find the least distance between Venus and the sun. **64.2 million mi (Answers are rounded.)**

12.3 The Hyperbola and Other Functions Defined by Radicals

OBJECTIVES

1. **Recognize the equation of a hyperbola.**
2. **Graph hyperbolas by using asymptotes.**
3. **Identify conic sections by their equations.**
4. **Graph certain square root functions.**

OBJECTIVE 1 Recognize the equation of a hyperbola. A **hyperbola** is the set of all points in a plane such that the absolute value of the *difference* of the distances from two fixed points (called *foci*) is constant. Figure 20 shows a hyperbola. Using the distance formula and the definition above, we can show that this hyperbola has equation $\frac{x^2}{16} - \frac{y^2}{12} = 1$.

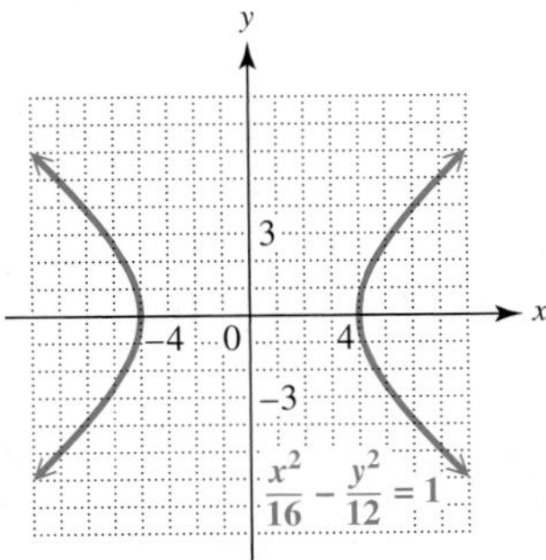

Figure 20

To graph hyperbolas centered at the origin, we need to find their intercepts. For the hyperbola in Figure 20, we proceed as follows.

***x*-Intercepts**

Let $y = 0$.

$$\frac{x^2}{16} - \frac{0^2}{12} = 1 \quad \text{Let } y = 0.$$

$$\frac{x^2}{16} = 1$$

$$x^2 = 16 \quad \text{Multiply by 16.}$$

$$x = \pm 4$$

The x-intercepts of the graph are $(4, 0)$ and $(-4, 0)$.

***y*-Intercepts**

Let $x = 0$.

$$\frac{0^2}{16} - \frac{y^2}{12} = 1 \quad \text{Let } x = 0.$$

$$-\frac{y^2}{12} = 1$$

$$y^2 = -12 \quad \text{Multiply by } -12.$$

Because there are no *real* solutions to $y^2 = -12$, the graph has no y-intercepts.

The graph of $\frac{x^2}{16} - \frac{y^2}{12} = 1$ in Figure 20 has no y-intercepts, while the hyperbola in Figure 21 has no x-intercepts. Its equation is $\frac{y^2}{25} - \frac{x^2}{9} = 1$, with y-intercepts $(0, 5)$ and $(0, -5)$.

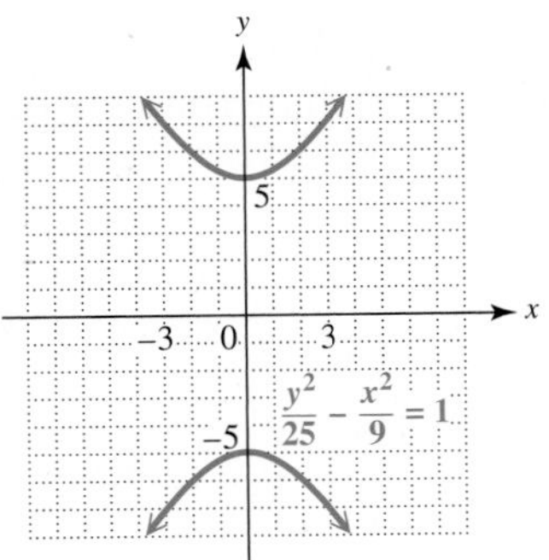

Figure 21

Equations of Hyperbolas

A hyperbola with x-intercepts $(a, 0)$ and $(-a, 0)$ has an equation of the form

$$\frac{x^2}{a^2} - \frac{y^2}{b^2} = 1,$$

and a hyperbola with y-intercepts $(0, b)$ and $(0, -b)$ has an equation of the form

$$\frac{y^2}{b^2} - \frac{x^2}{a^2} = 1.$$

OBJECTIVE 2 Graph hyperbolas by using asymptotes. The two branches of the graph of a hyperbola approach a pair of intersecting straight lines, which are its *asymptotes*. (See Figure 22 on the next page.) The asymptotes are useful for sketching the graph of the hyperbola.

Asymptotes of Hyperbolas

The extended diagonals of the rectangle with vertices (corners) at the points (a, b), $(-a, b)$, $(-a, -b)$, and $(a, -b)$ are the **asymptotes** of the hyperbolas

$$\frac{x^2}{a^2} - \frac{y^2}{b^2} = 1 \quad \text{and} \quad \frac{y^2}{b^2} - \frac{x^2}{a^2} = 1.$$

This rectangle is called the **fundamental rectangle.** Using the methods of **Chapter 3,** we could show that the equations of these asymptotes are

$$y = \frac{b}{a}x \quad \text{and} \quad y = -\frac{b}{a}x.$$

To graph hyperbolas, follow these steps.

Graphing a Hyperbola

Step 1 **Find the intercepts.** Locate the intercepts of the graph at $(a, 0)$ and $(-a, 0)$ if the x^2-term has a positive coefficient, or at $(0, b)$ and $(0, -b)$ if the y^2-term has a positive coefficient.

Step 2 **Find the fundamental rectangle.** Locate the vertices of the fundamental rectangle at (a, b), $(-a, b)$, $(-a, -b)$, and $(a, -b)$.

Step 3 **Sketch the asymptotes.** The extended diagonals of the rectangle are the asymptotes of the hyperbola. They have equations $y = \pm\frac{b}{a}x$.

Step 4 **Draw the graph.** Sketch each branch of the hyperbola through an intercept and approaching (but not touching) the asymptotes.

EXAMPLE 1 Graphing a Horizontal Hyperbola

Graph $\frac{x^2}{16} - \frac{y^2}{25} = 1$.

Step 1 Here $a = 4$ and $b = 5$. The x-intercepts are $(4, 0)$ and $(-4, 0)$.

Step 2 The four points $(4, 5)$, $(-4, 5)$, $(-4, -5)$, and $(4, -5)$ are the vertices of the fundamental rectangle, as shown in Figure 22.

Steps 3 and 4 The equations of the asymptotes are $y = \pm\frac{5}{4}x$. The hyperbola approaches these lines as x and y get larger in absolute value.

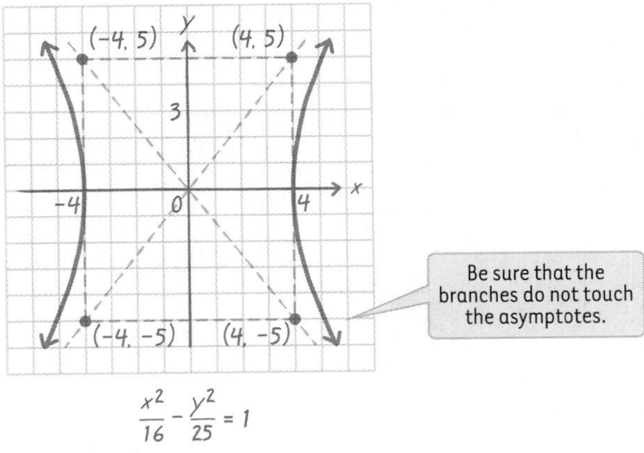

$\frac{x^2}{16} - \frac{y^2}{25} = 1$

Figure 22

Work Problem **1** at the Side. ▶

EXAMPLE 2 Graphing a Vertical Hyperbola

Graph $\frac{y^2}{49} - \frac{x^2}{16} = 1$.

This hyperbola has y-intercepts $(0, 7)$ and $(0, -7)$. The asymptotes are the extended diagonals of the rectangle with vertices at $(4, 7)$, $(-4, 7)$, $(-4, -7)$, and $(4, -7)$. Their equations are $y = \pm\frac{7}{4}x$. See Figure 23.

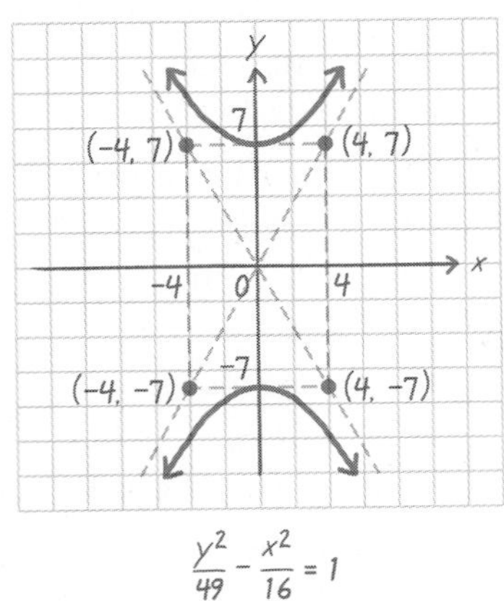

$\frac{y^2}{49} - \frac{x^2}{16} = 1$

Figure 23

Work Problem **2** at the Side. ▶

1 Graph $\frac{x^2}{4} - \frac{y^2}{25} = 1$.

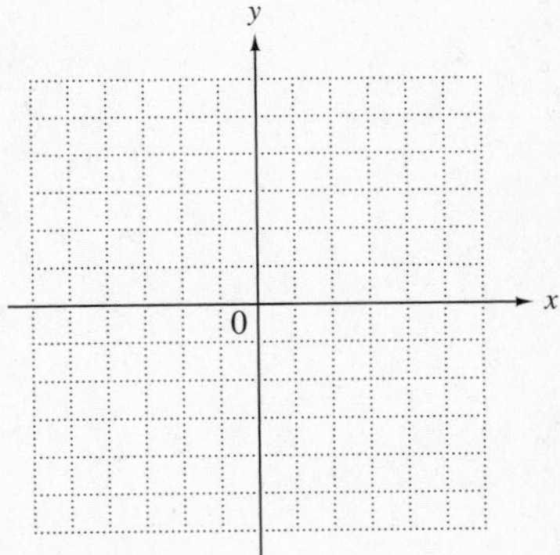

2 Graph $\frac{y^2}{81} - \frac{x^2}{64} = 1$.

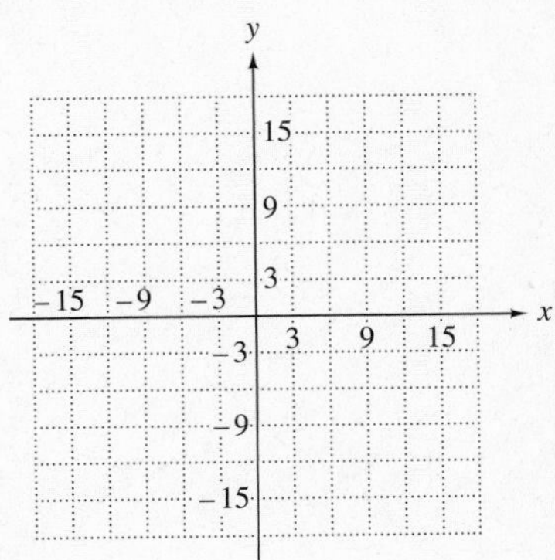

ANSWERS

1.

$\frac{x^2}{4} - \frac{y^2}{25} = 1$

2.

$\frac{y^2}{81} - \frac{x^2}{64} = 1$

OBJECTIVE 3 Identify conic sections by their equations

Rewriting a second-degree equation in one of the forms given for ellipses, hyperbolas, circles, or parabolas makes it possible to identify the graph of the equation.

SUMMARY OF CONIC SECTIONS

Equation	Graph	Description	Identification
$y = ax^2 + bx + c$ or $y = a(x - h)^2 + k$	$a > 0$; (h, k); Parabola	It opens up if $a > 0$, down if $a < 0$. The vertex is (h, k).	It has an x^2-term. y is not squared.
$x = ay^2 + by + c$ or $x = a(y - k)^2 + h$	$a > 0$; (h, k); Parabola	It opens to the right if $a > 0$, to the left if $a < 0$. The vertex is (h, k).	It has a y^2-term. x is not squared.
$(x - h)^2 + (y - k)^2 = r^2$	(h, k); r; Circle	The center is (h, k), and the radius is r.	x^2- and y^2-terms have the same positive coefficient.
$\frac{x^2}{a^2} + \frac{y^2}{b^2} = 1$	$(0, b)$; $(-a, 0)$; $(a, 0)$; $(0, -b)$; Ellipse	The x-intercepts are $(a, 0)$ and $(-a, 0)$. The y-intercepts are $(0, b)$ and $(0, -b)$.	x^2- and y^2-terms have different positive coefficients.
$\frac{x^2}{a^2} - \frac{y^2}{b^2} = 1$	$(-a, 0)$; $(a, 0)$; Hyperbola	The x-intercepts are $(a, 0)$ and $(-a, 0)$. The asymptotes are found from (a, b), $(a, -b)$, $(-a, -b)$, and $(-a, b)$.	x^2 has a positive coefficient. y^2 has a negative coefficient.
$\frac{y^2}{b^2} - \frac{x^2}{a^2} = 1$	$(0, b)$; $(0, -b)$; Hyperbola	The y-intercepts are $(0, b)$ and $(0, -b)$. The asymptotes are found from (a, b), $(a, -b)$, $(-a, -b)$, and $(-a, b)$.	y^2 has a positive coefficient. x^2 has a negative coefficient.

EXAMPLE 3 Identifying the Graphs of Equations

Identify the graph of each equation.

(a) $9x^2 = 108 + 12y^2$

Both variables are squared, so the graph is either an ellipse or a hyperbola. (This situation also occurs for a circle, which is a special case of the ellipse.) To see which one it is, rewrite the equation so that the x^2- and y^2-terms are on one side of the equation and 1 is on the other.

$$9x^2 - 12y^2 = 108 \quad \text{Subtract } 12y^2.$$

$$\frac{x^2}{12} - \frac{y^2}{9} = 1 \quad \text{Divide by 108.}$$

Because of the minus sign, the graph of this equation is a hyperbola.

(b) $x^2 = y - 3$

Only one of the two variables, x, is squared, so this is the vertical parabola $y = x^2 + 3$.

(c) $x^2 = 9 - y^2$

Write the variable terms on the same side of the equation.

$$x^2 + y^2 = 9 \quad \text{Add } y^2.$$

The graph of this equation is a circle with center at the origin and radius 3.

Work Problem 3 at the Side. ▶

3 Identify the graph of each equation.

(a) $3x^2 = 27 - 4y^2$

(b) $6x^2 = 100 + 2y^2$

(c) $3x^2 = 27 - 4y$

(d) $3x^2 = 27 - 3y^2$

OBJECTIVE 4 Graph certain square root functions. Recall from **Section 3.6** that no vertical line will intersect the graph of a function in more than one point. Thus, horizontal parabolas and all circles, ellipses, and hyperbolas are examples of graphs that do not satisfy the conditions of a function. However, by considering only a part of the graph of each of these we have the graph of a function, as seen in Figure 24.

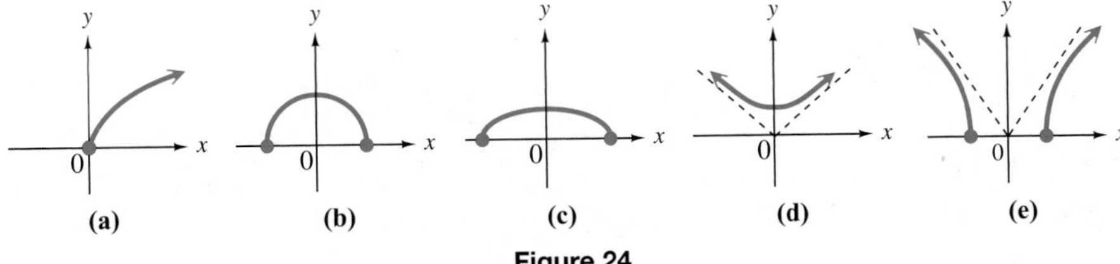

Figure 24

In parts (a)–(d) of Figure 24, the top portion of a conic section is shown (parabola, circle, ellipse, and hyperbola, respectively). In part (e), the top two portions of a hyperbola are shown. In each case, the graph is that of a function since the graph satisfies the conditions of the vertical line test.

In **Sections 9.1** and **12.1** we observed the square root function defined by $f(x) = \sqrt{x}$. To find equations for the types of graphs shown in Figure 24, we extend its definition.

Square Root Function

For an algebraic expression in x defined by u, with $u \geq 0$, a function of the form

$$f(x) = \sqrt{u}$$

is called a **square root function.**

ANSWERS

3. **(a)** ellipse **(b)** hyperbola **(c)** parabola **(d)** circle

EXAMPLE 4 Graphing a Semicircle

Graph $f(x) = \sqrt{25 - x^2}$. Give the domain and range.

Replace $f(x)$ with y and square both sides to get the equation

$$y^2 = 25 - x^2, \quad \text{or} \quad x^2 + y^2 = 25.$$

This is the graph of a circle with center at (0, 0) and radius 5. Since $f(x)$, or y, represents a principal square root in the original equation, $f(x)$ must be nonnegative. This restricts the graph to the upper half of the circle, as shown in Figure 25. Use the graph and the vertical line test to verify that it is indeed a function. The domain is $[-5, 5]$, and the range is $[0, 5]$.

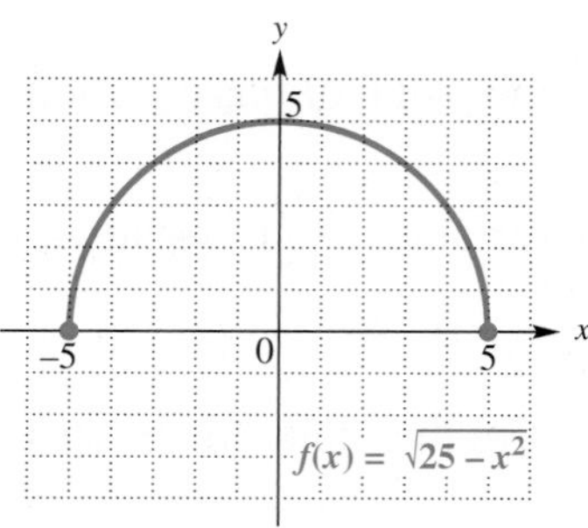

Figure 25

◀ Work Problem 4 at the Side.

EXAMPLE 5 Graphing a Portion of an Ellipse

Graph $\frac{y}{6} = -\sqrt{1 - \frac{x^2}{16}}$. Give the domain and range.

Square both sides to get an equation whose form is known.

$$\frac{y^2}{36} = 1 - \frac{x^2}{16}$$

$$\frac{x^2}{16} + \frac{y^2}{36} = 1 \qquad \text{Add } \frac{x^2}{16}.$$

This is the equation of an ellipse with x-intercepts (4, 0) and (−4, 0) and y-intercepts (0, 6) and (0, −6). Since $\frac{y}{6}$ equals a negative square root in the original equation, y must be nonpositive, restricting the graph to the lower half of the ellipse, as shown in Figure 26. Verify that this is the graph of a function, using the vertical line test. The domain is $[-4, 4]$, and the range is $[-6, 0]$.

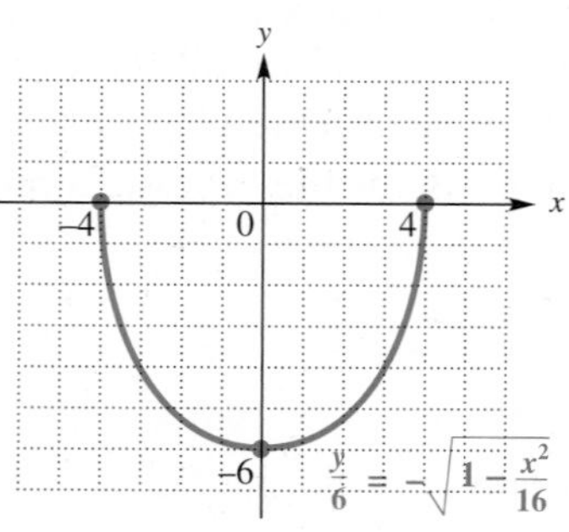

Figure 26

◀ Work Problem 5 at the Side.

---

4 Graph $f(x) = \sqrt{36 - x^2}$. Give the domain and range.

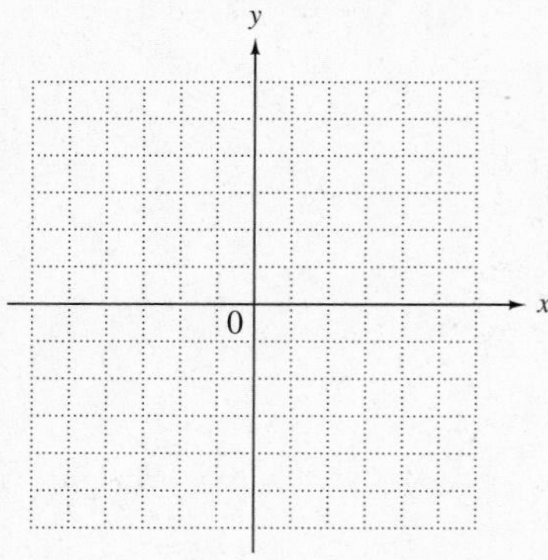

5 Graph

$$\frac{y}{3} = -\sqrt{1 - \frac{x^2}{4}}.$$

Give the domain and range.

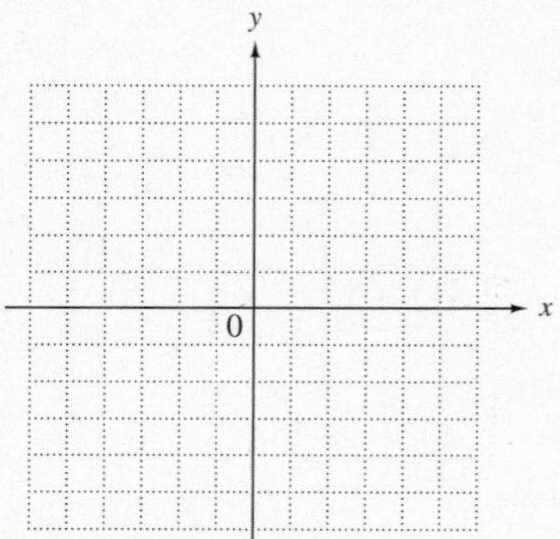

ANSWERS

4.

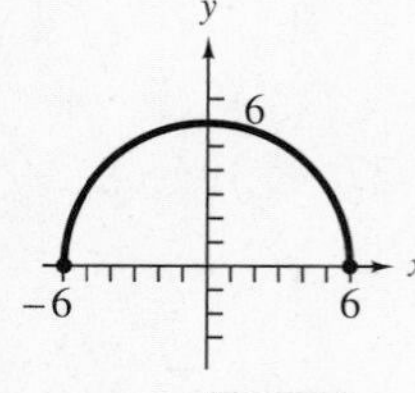

$f(x) = \sqrt{36 - x^2}$

domain: $[-6, 6]$; range: $[0, 6]$

5.

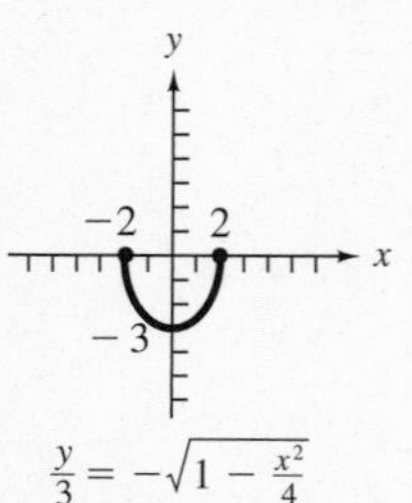

$\frac{y}{3} = -\sqrt{1 - \frac{x^2}{4}}$

domain: $[-2, 2]$; range: $[-3, 0]$

12.3 Exercises

FOR EXTRA HELP

MyMathLab

Math XL PRACTICE | WATCH | DOWNLOAD | READ | REVIEW

Based on the discussions of ellipses in ***Section 12.2*** *and of hyperbolas in this section, match each equation with its graph.*

1. $\frac{x^2}{25} + \frac{y^2}{9} = 1$ **C**

2. $\frac{x^2}{9} + \frac{y^2}{25} = 1$ **B**

3. $\frac{x^2}{9} - \frac{y^2}{25} = 1$ **D**

4. $\frac{x^2}{25} - \frac{y^2}{9} = 1$ **A**

A.

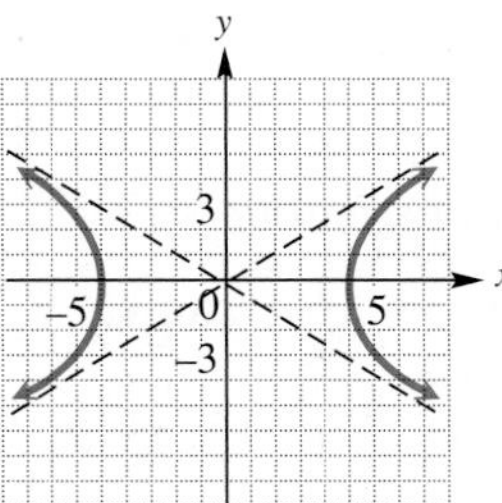

B.

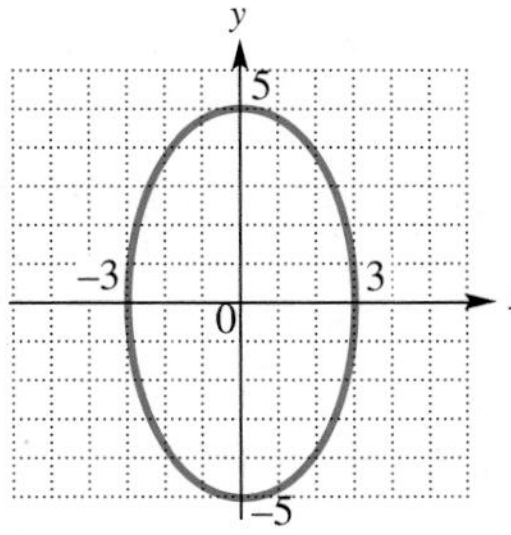

C.

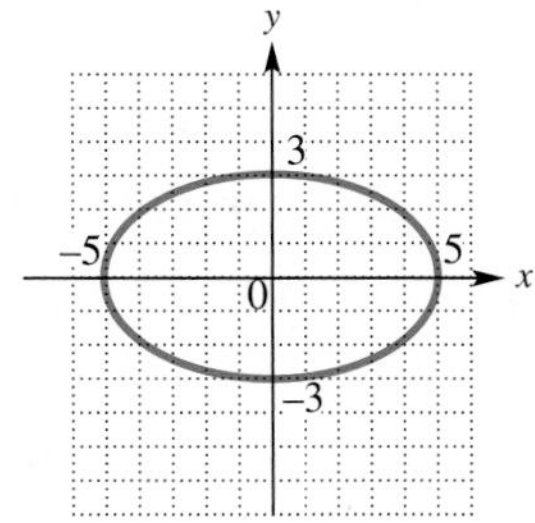

D.

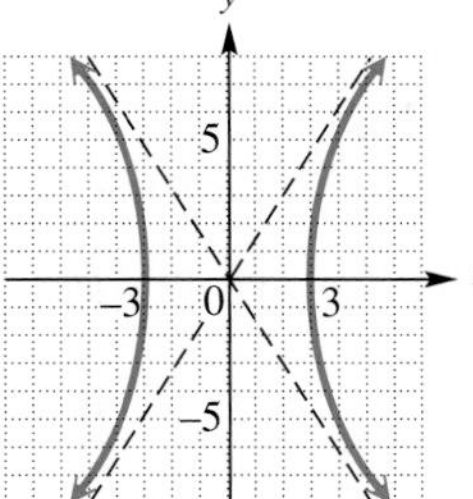

5. Write an explanation of how you can tell from the equation whether the branches of a hyperbola open up and down or left and right.

When written in one of the forms given in the box titled "Equations of Hyperbolas" in this section, it will open up and down if the – sign precedes the x^2-term; it will open left and right if the – sign precedes the y^2-term.

6. Describe how the fundamental rectangle is used to sketch a hyperbola.

The points (a, b), $(a, -b)$, $(-a, -b)$, and $(-a, b)$ are used as vertices of a rectangle. The diagonals of the rectangle are drawn, which are the asymptotes for the hyperbola.

Graph each hyperbola. See Examples 1 and 2.

7. $\frac{x^2}{16} - \frac{y^2}{9} = 1$

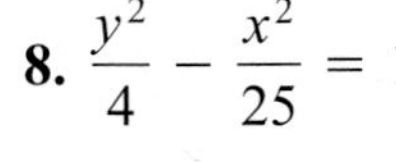

8. $\frac{y^2}{4} - \frac{x^2}{25} = 1$

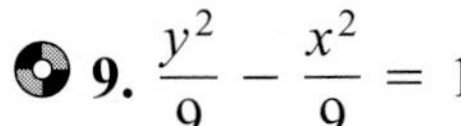

9. $\frac{y^2}{9} - \frac{x^2}{9} = 1$

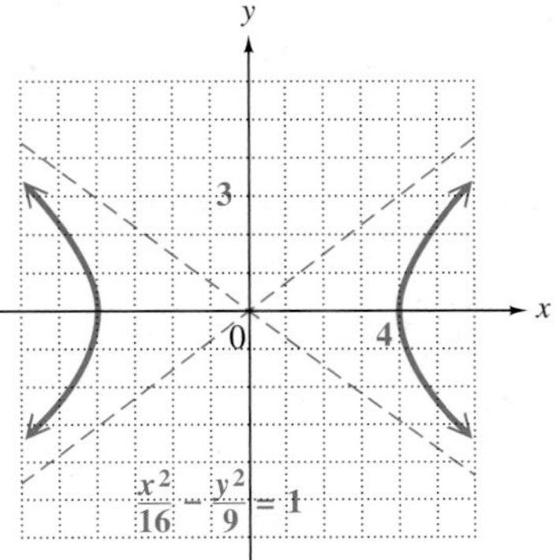

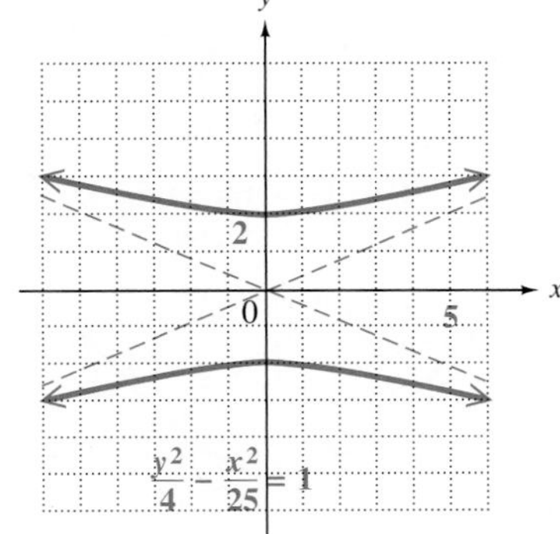

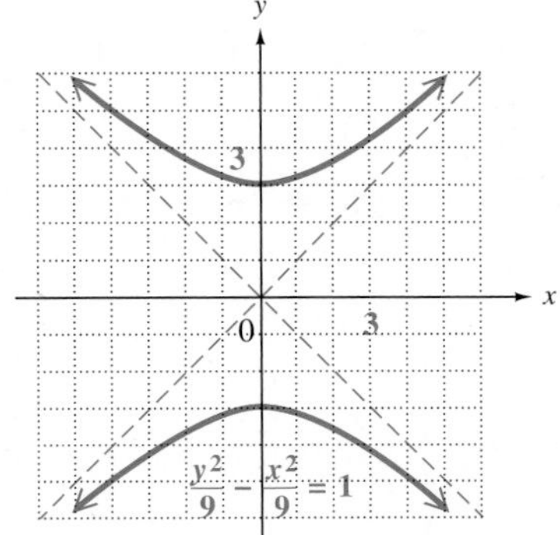

10. $\frac{x^2}{49} - \frac{y^2}{16} = 1$

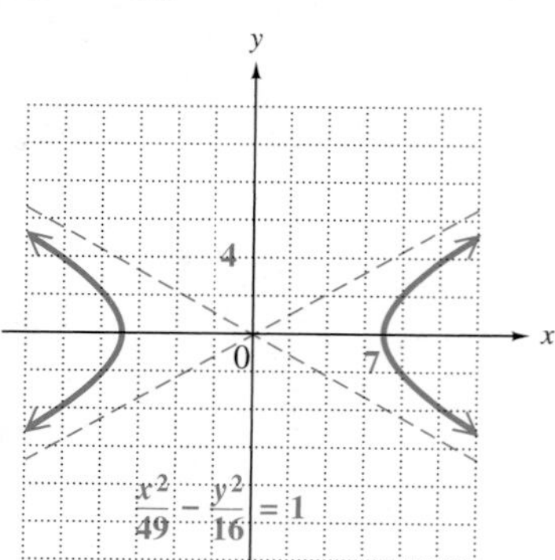

11. $\frac{x^2}{25} - \frac{y^2}{36} = 1$

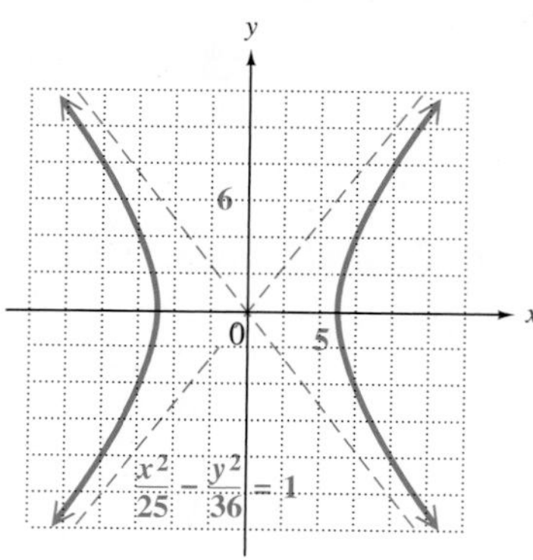

12. $\frac{y^2}{9} - \frac{x^2}{4} = 1$

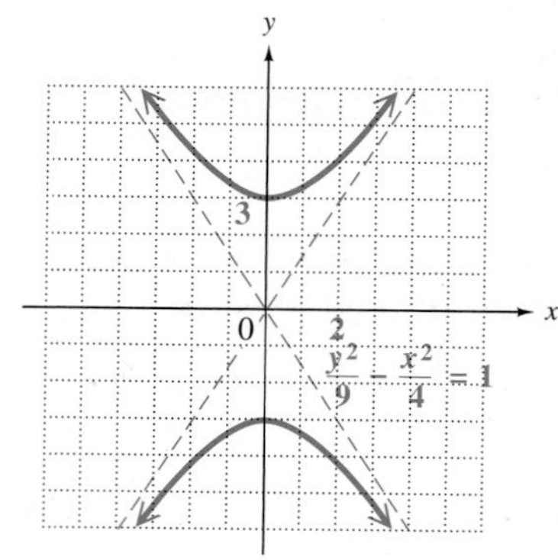

Identify the graph of each equation as a parabola, circle, ellipse, *or* hyperbola, *and sketch it. See Example 3.*

13. $x^2 - y^2 = 16$ **hyperbola**

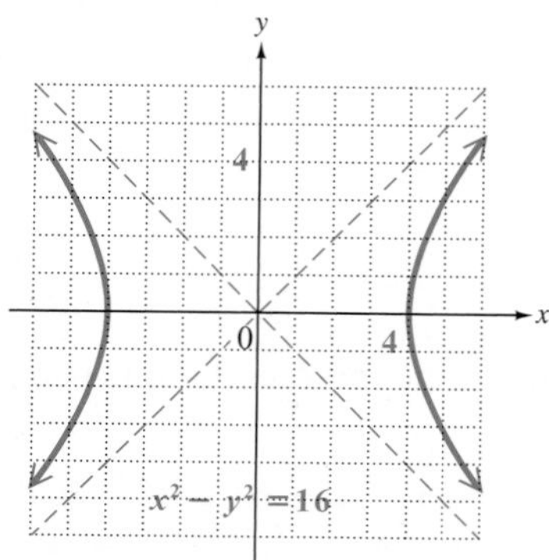

14. $x^2 + y^2 = 16$ **circle**

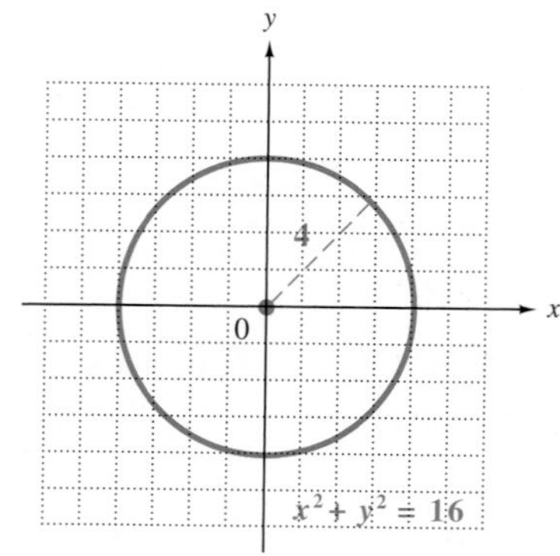

15. $4x^2 + y^2 = 16$ **ellipse**

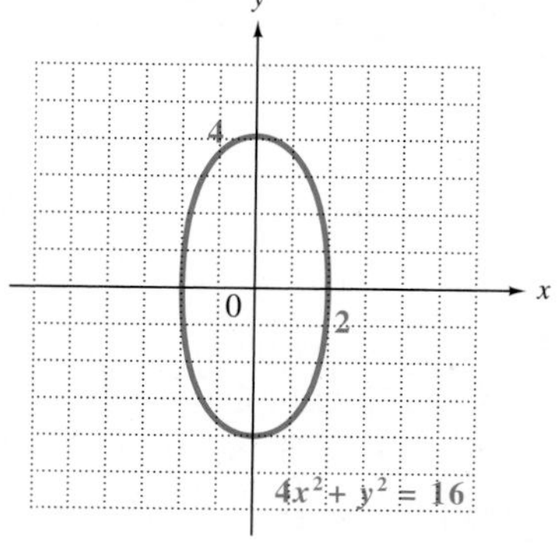

16. $x^2 - 2y = 0$ **parabola**

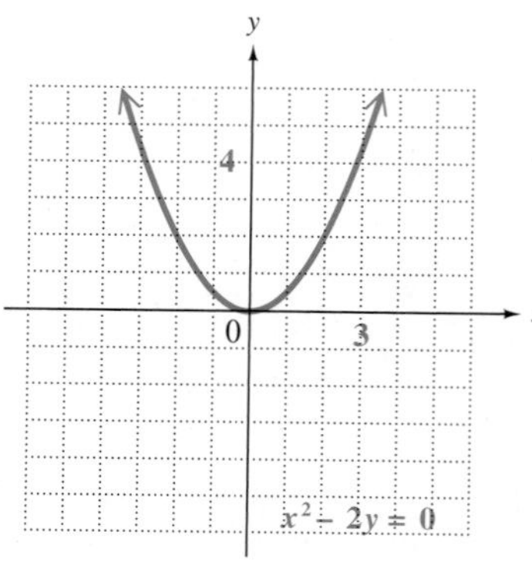

17. $y^2 = 36 - x^2$ **circle**

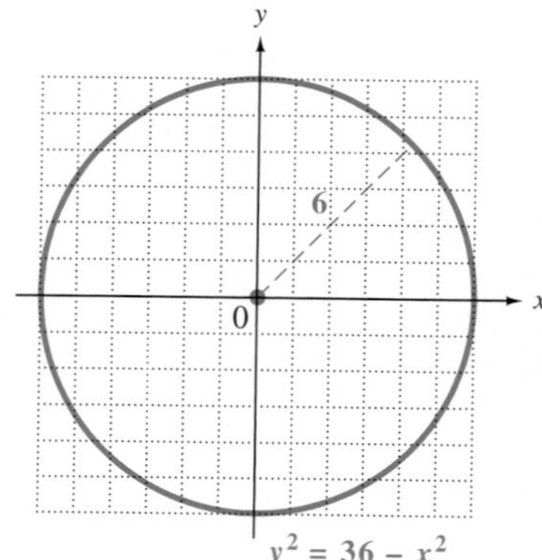

18. $9x^2 + 25y^2 = 225$ **ellipse**

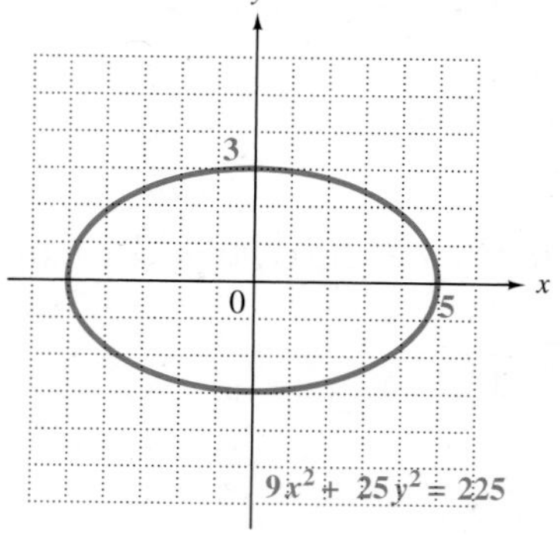

19. $9x^2 = 144 + 16y^2$ **hyperbola**

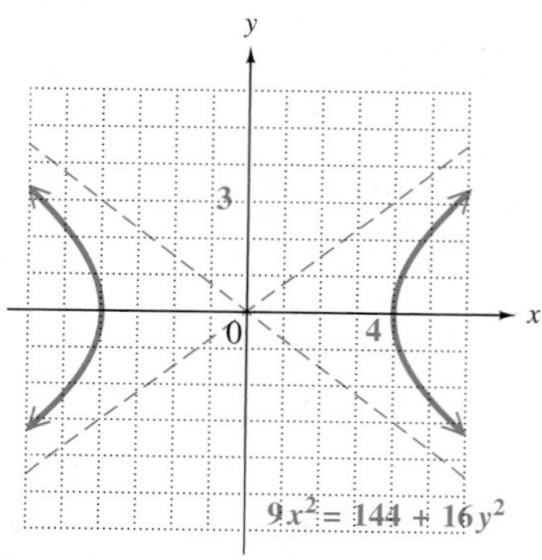

20. $y^2 = 4 + x^2$ **hyperbola**

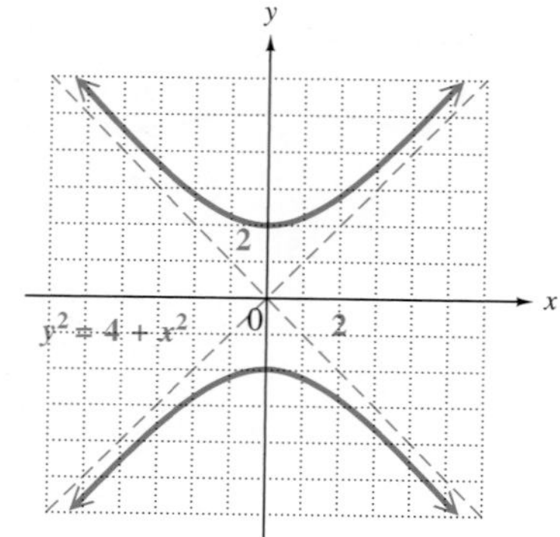

21. $x^2 + 9y^2 = 9$ **ellipse**

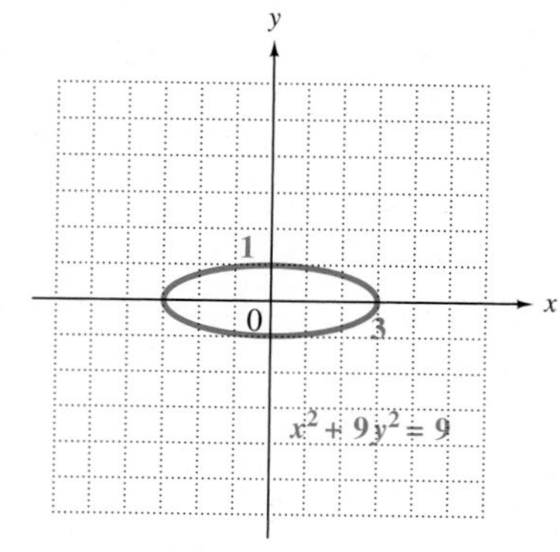

Graph each function defined by a radical expression. Give the domain and range. See Examples 4 and 5.

22. $f(x) = \sqrt{16 - x^2}$

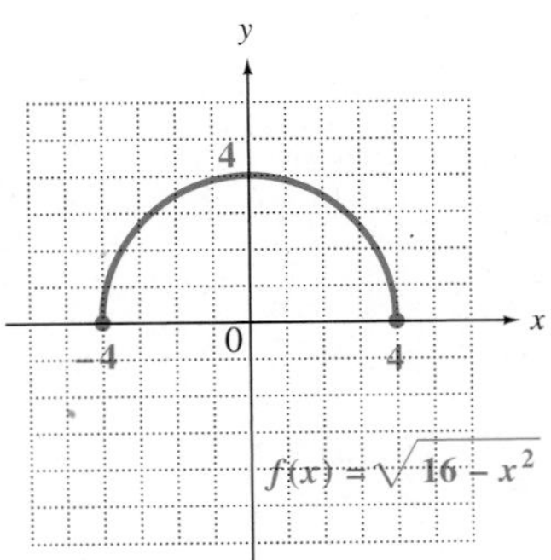

domain: [−4, 4]; range: [0, 4]

23. $f(x) = \sqrt{9 - x^2}$

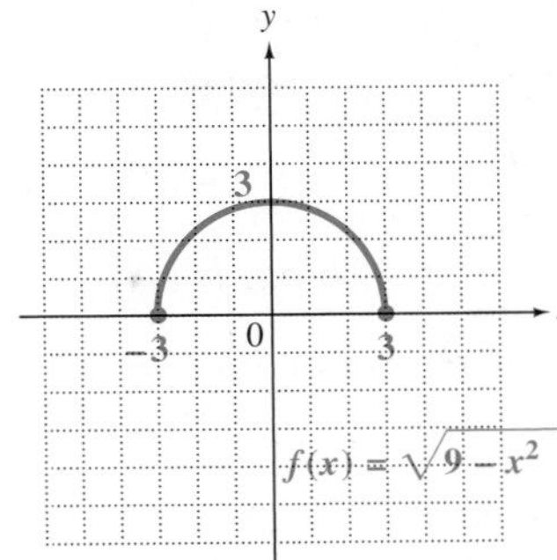

domain: [−3, 3]; range: [0, 3]

24. $f(x) = -\sqrt{36 - x^2}$

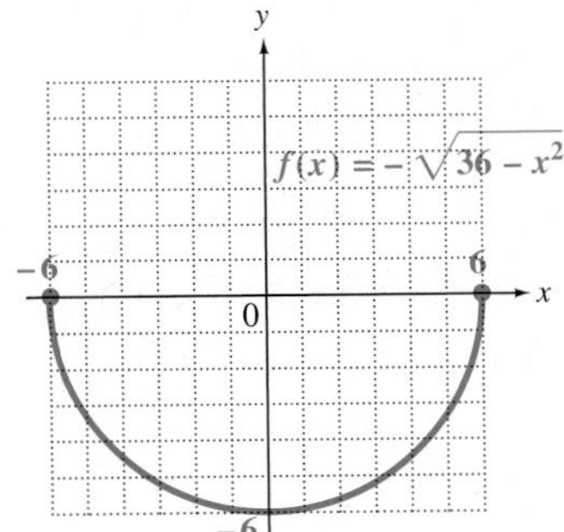

domain: [−6, 6]; range: [−6, 0]

25. $f(x) = -\sqrt{25 - x^2}$

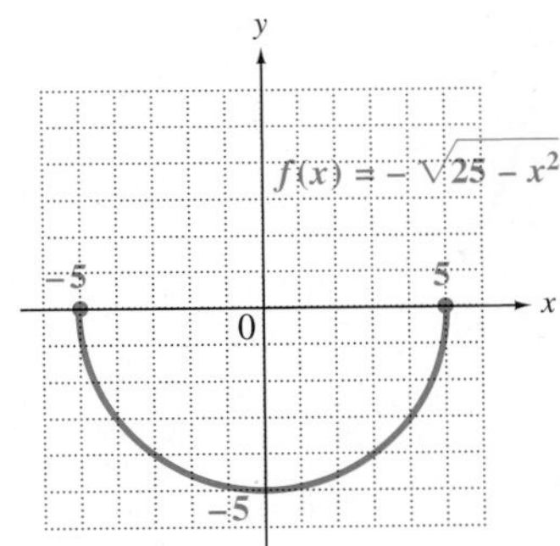

domain: [−5, 5]; range: [−5, 0]

26. $\frac{y}{3} = \sqrt{1 + \frac{x^2}{9}}$

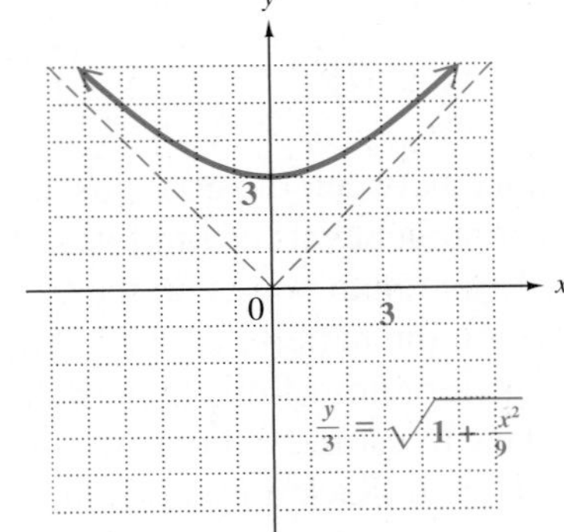

domain: (−∞, ∞); range: [3, ∞)

27. $y = \sqrt{\frac{x + 4}{2}}$

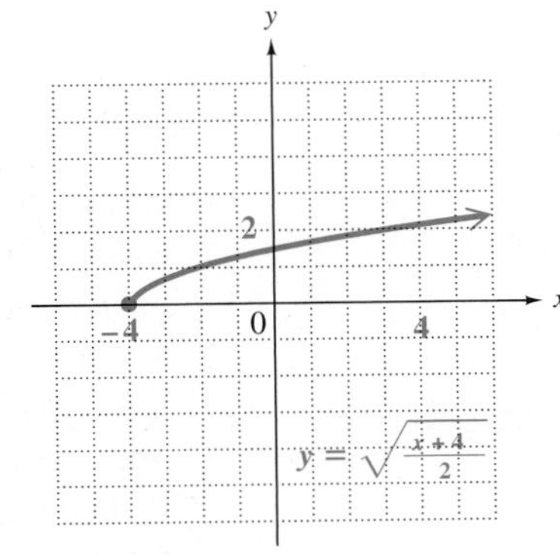

domain: [−4, ∞); range: [0, ∞)

28. $y = -2\sqrt{\frac{9 - x^2}{9}}$

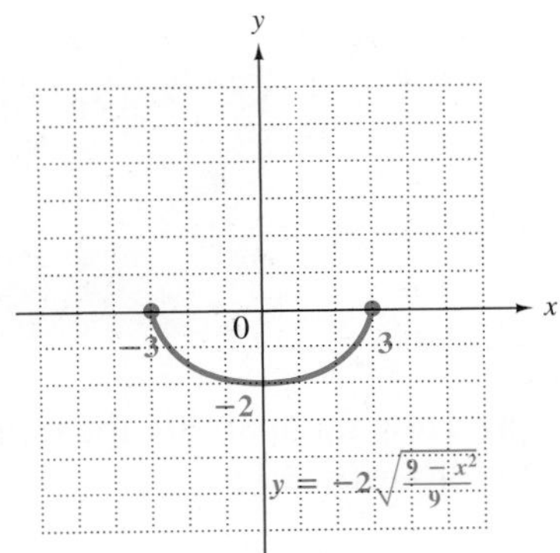

domain: [−3, 3]; range: [−2, 0]

*In **Section 12.2,** Example 6, we saw that the center of an ellipse may be shifted away from the origin. The same process can be applied to hyperbolas. For example, the hyperbola shown at the right,*

$$\frac{(x+5)^2}{4} - \frac{(y-2)^2}{9} = 1,$$

has the same graph as $\frac{x^2}{4} - \frac{y^2}{9} = 1$, *but it is centered at* $(-5, 2)$. *Graph each hyperbola with center shifted away from the origin.*

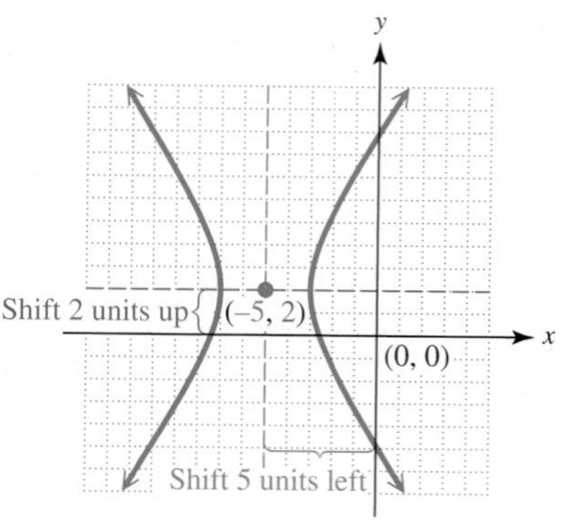

29. 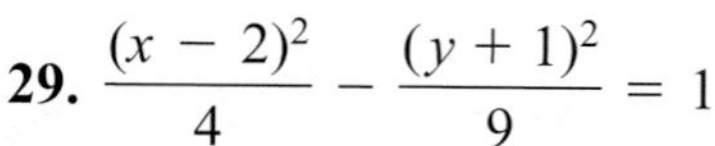

$$\frac{(x-2)^2}{4} - \frac{(y+1)^2}{9} = 1$$

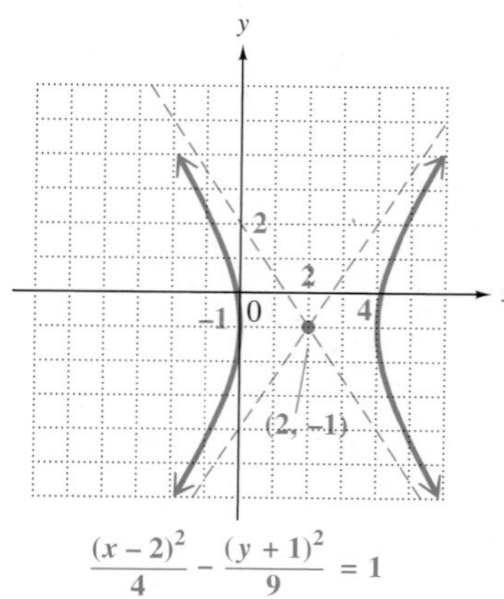

30. $\frac{(x+3)^2}{16} - \frac{(y-2)^2}{4} = 1$

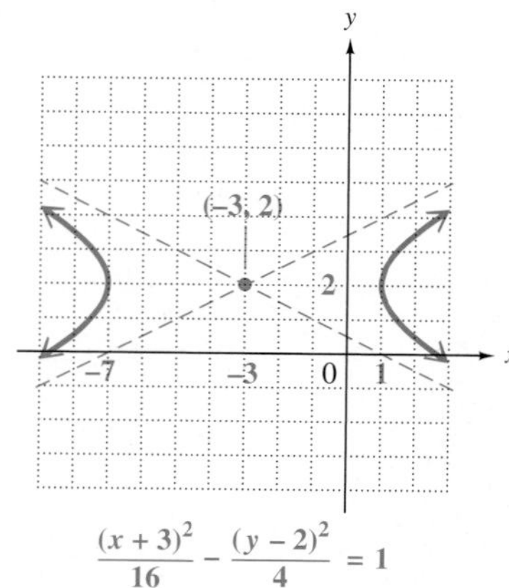

Solve each problem.

31. Two buildings in a sports complex are shaped and positioned like a portion of the branches of the hyperbola with equation

$$400x^2 - 625y^2 = 250{,}000,$$

where x and y are in meters.

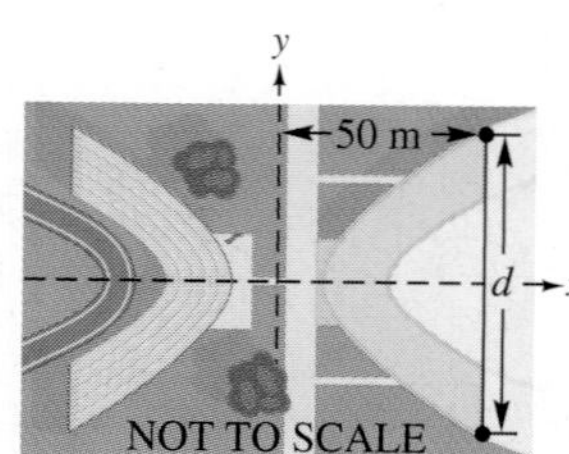

(a) How far apart are the buildings at their closest point? **50 m**

(b) Find the distance d in the figure. **69.3 m**

32. In rugby, after a *try* (similar to a touchdown in American football) the scoring team attempts a kick for extra points. The ball must be kicked from directly behind the point where the try was scored. The kicker can choose the distance but cannot move the ball sideways. It can be shown that the kicker's best choice is on the hyperbola with equation

$$\frac{x^2}{g^2} - \frac{y^2}{g^2} = 1,$$

where $2g$ is the distance between the goal posts. Since the hyperbola approaches its asymptotes, it is easier for the kicker to estimate points on the asymptotes instead of on the hyperbola. What are the equations of the asymptotes of this hyperbola? Why is it relatively easy to estimate them? (*Source:* Isaksen, Daniel C., "How to Kick a Field Goal," *The College Mathematics Journal,* September 1996.)

$y = \pm x$; The lines $y = \pm x$ form a 45° angle with the line through the goal posts. It is fairly easy to estimate a 45° angle.

12.4 Nonlinear Systems of Equations

OBJECTIVES

1. **Solve a nonlinear system by substitution.**
2. **Use the elimination method to solve a system with two second-degree equations.**
3. **Solve a system that requires a combination of methods.**

An equation in which some terms have more than one variable or a variable of degree 2 or greater is called a **nonlinear equation.** A **nonlinear system of equations** includes at least one nonlinear equation.

When solving a nonlinear system, it helps to visualize the types of graphs of the equations of the system to determine the possible number of points of intersection. For example, if a system includes two equations where the graph of one is a circle and the graph of the other is a line, then there may be 0, 1, or 2 points of intersection, as illustrated in Figure 27.

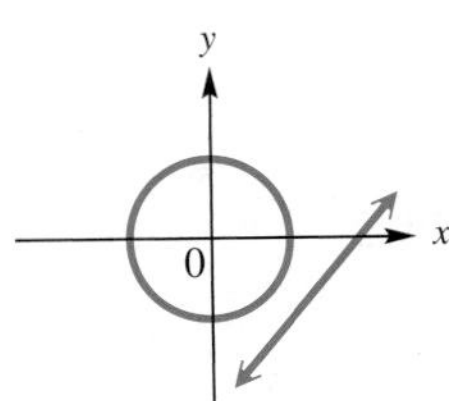

No points of intersection

One point of intersection

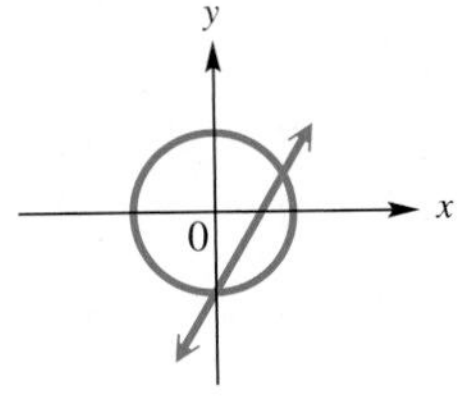

Two points of intersection

Figure 27

OBJECTIVE 1 Solve a nonlinear system by substitution. We solve nonlinear systems by the substitution method, the elimination method, or a combination of the two. The substitution method **(Sections 4.2 and 8.4)** is usually appropriate when one of the equations is linear.

EXAMPLE 1 Solving a Nonlinear System by Substitution

Solve the system.

$$x^2 + y^2 = 9 \quad (1)$$
$$2x - y = 3 \quad (2)$$

The graph of (1) is a circle and the graph of (2) is a line. There may be 0, 1, or 2 points of intersection, as shown in Figure 27. First solve the linear equation for one of the two variables, and then substitute the resulting expression into the nonlinear equation to obtain an equation in one variable.

$$2x - y = 3 \quad (2)$$
$$\mathbf{y = 2x - 3} \quad (3)$$

Substitute $2x - 3$ for y in equation (1).

$$x^2 + y^2 = 9 \quad (1)$$
$$x^2 + (2x - 3)^2 = 9 \quad \text{Let } y = 2x - 3.$$
$$x^2 + 4x^2 - 12x + 9 = 9 \quad \text{Square } 2x - 3.$$
$$5x^2 - 12x = 0 \quad \text{Subtract 9; combine like terms.}$$
$$x(5x - 12) = 0 \quad \text{Factor; GCF is } x.$$
$$x = 0 \quad \text{or} \quad x = \frac{12}{5} \quad \text{Zero-factor property}$$

Set *both* factors equal to 0.

Let $x = 0$ in equation (3) to get $y = -3$. If $x = \frac{12}{5}$ in equation (3), then $y = \frac{9}{5}$. The solution set of the system is $\left\{(0, -3), \left(\frac{12}{5}, \frac{9}{5}\right)\right\}$. The graph in Figure 28 confirms the two points of intersection.

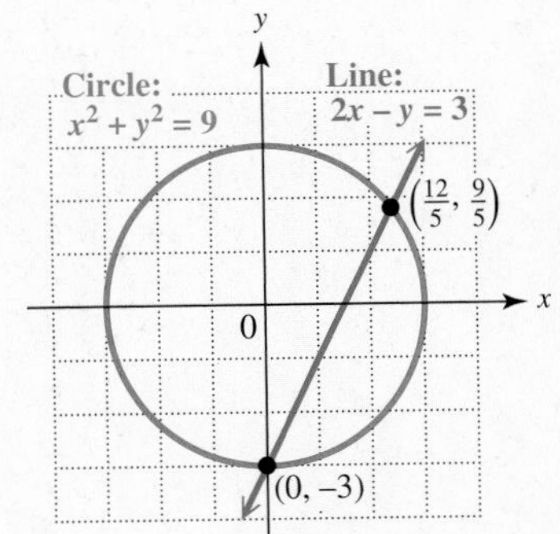

Figure 28

Work Problem 1 at the Side. ▶

1 Solve each system.

(a) $x^2 + y^2 = 10$
$x = y + 2$

(b) $x^2 - 2y^2 = 8$
$y + x = 6$

ANSWERS

1. (a) $\{(3, 1), (-1, -3)\}$
(b) $\{(4, 2), (20, -14)\}$

2 Solve each system.

(a) $xy = 8$
$x + y = 6$

(b) $xy + 10 = 0$
$4x + 9y = -2$

ANSWERS

2. (a) $\{(4, 2), (2, 4)\}$

(b) $\left\{(-5, 2), \left(\frac{9}{2}, -\frac{20}{9}\right)\right\}$

EXAMPLE 2 Solving a Nonlinear System by Substitution

Solve the system.

$$6x - y = 5 \quad (1)$$
$$xy = 4 \quad (2)$$

The graph of (1) is a line. It can be shown by plotting points that the graph of (2) is a hyperbola. Visualizing a line and a hyperbola indicates that there may be 0, 1, or 2 points of intersection. We can solve either equation for one of the variables and then substitute the result into the other equation. Solving $xy = 4$ for x gives $x = \frac{4}{y}$. Substitute $\frac{4}{y}$ for x in equation (1).

$$6\left(\frac{4}{y}\right) - y = 5 \qquad \text{Let } x = \tfrac{4}{y}.$$
$$\frac{24}{y} - y = 5 \qquad \text{Multiply.}$$
$$24 - y^2 = 5y \qquad \text{Multiply by } y,\ y \neq 0.$$
$$0 = y^2 + 5y - 24 \qquad \text{Standard form}$$
$$0 = (y - 3)(y + 8) \qquad \text{Factor.}$$
$$y = 3 \quad \text{or} \quad y = -8 \qquad \text{Zero-factor property}$$

We substitute these results into $x = \frac{4}{y}$ to obtain the corresponding values of x.

$$\text{If } y = 3, \text{ then } x = \frac{4}{3}. \qquad \text{If } y = -8, \text{ then } x = -\frac{1}{2}.$$

The solution set of the system is $\left\{\left(\frac{4}{3}, 3\right), \left(-\frac{1}{2}, -8\right)\right\}$. See Figure 29.

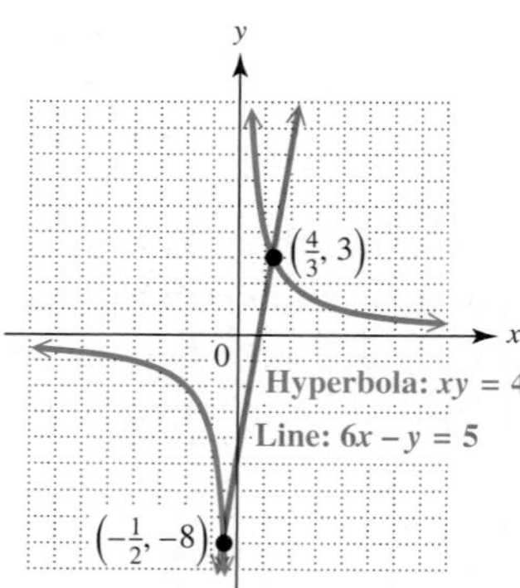

Figure 29

◀ Work Problem **2** at the Side.

OBJECTIVE 2 Use the elimination method to solve a system with two second-degree equations. The elimination method **(Sections 4.3 and 8.4)** is often used when both equations are second degree.

EXAMPLE 3 Solving a Nonlinear System by Elimination

Solve the system.

$$x^2 + y^2 = 9 \quad (1)$$
$$2x^2 - y^2 = -6 \quad (2)$$

The graph of (1) is a circle, while the graph of (2) is a hyperbola. By analyzing the possibilities, we conclude that there may be 0, 1, 2, 3, or 4 points of intersection. Adding the two equations will eliminate y.

Continued on Next Page

$$\begin{aligned} x^2 + y^2 &= 9 && (1)\\ 2x^2 - y^2 &= -6 && (2)\\ \hline 3x^2 \quad &= 3 && \text{Add.}\\ x^2 &= 1 && \text{Divide by 3.}\\ x = 1 \quad \text{or} \quad x &= -1 && \text{Square root property} \end{aligned}$$

Each value of x gives corresponding values for y when substituted into one of the original equations. Using equation (1) is easier since the coefficients of the x^2- and y^2-terms are 1.

If $x = 1$, then	**If $x = -1$, then**
$1^2 + y^2 = 9$	$(-1)^2 + y^2 = 9$
$y^2 = 8$	$y^2 = 8$
$y = \sqrt{8}$ or $y = -\sqrt{8}$	$y = \sqrt{8}$ or $y = -\sqrt{8}$
$y = 2\sqrt{2}$ or $y = -2\sqrt{2}$.	$y = 2\sqrt{2}$ or $y = -2\sqrt{2}$.

The solution set is

$$\{(1, 2\sqrt{2}), (1, -2\sqrt{2}), (-1, 2\sqrt{2}), (-1, -2\sqrt{2})\}.$$

Figure 30 shows the four points of intersection.

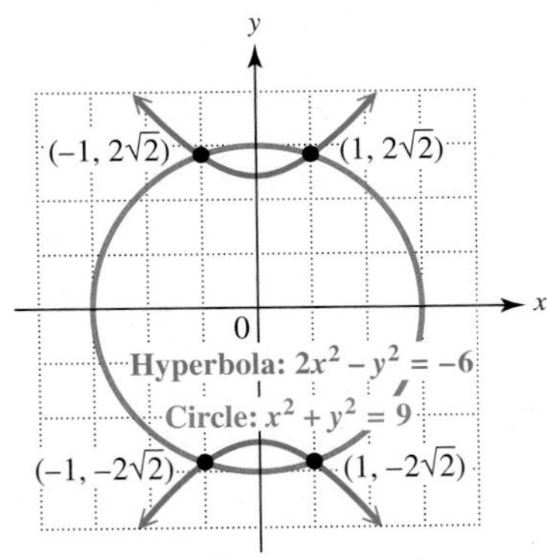

Figure 30

Work Problem 3 at the Side. ▶

3 Solve each system.

(a) $x^2 + y^2 = 41$
$x^2 - y^2 = 9$

(b) $x^2 + 3y^2 = 40$
$4x^2 - y^2 = 4$

OBJECTIVE 3 Solve a system that requires a combination of methods. Solving a system of second-degree equations may require a combination of methods.

EXAMPLE 4 Solving a Nonlinear System by a Combination of Methods

Solve the system.

$$\begin{aligned} x^2 + 2xy - y^2 &= 7 && (1)\\ x^2 - y^2 &= 3 && (2) \end{aligned}$$

While we have not graphed equations like (1), its graph is a hyperbola. The graph of (2) is also a hyperbola. Two hyperbolas may have 0, 1, 2, 3, or 4 points of intersection. We use the elimination method here in combination with the substitution method. We begin by eliminating the squared terms by multiplying each side of equation (2) by -1 and then adding the result to equation (1).

Continued on Next Page

ANSWERS

3. (a) $\{(5, 4), (5, -4), (-5, 4), (-5, -4)\}$
(b) $\{(2, 2\sqrt{3}), (2, -2\sqrt{3}), (-2, 2\sqrt{3}), (-2, -2\sqrt{3})\}$

4 Solve each system.

(a) $x^2 + xy + y^2 = 3$
$x^2 + y^2 = 5$

(b) $x^2 + 7xy - 2y^2 = -8$
$-2x^2 + 4y^2 = 16$

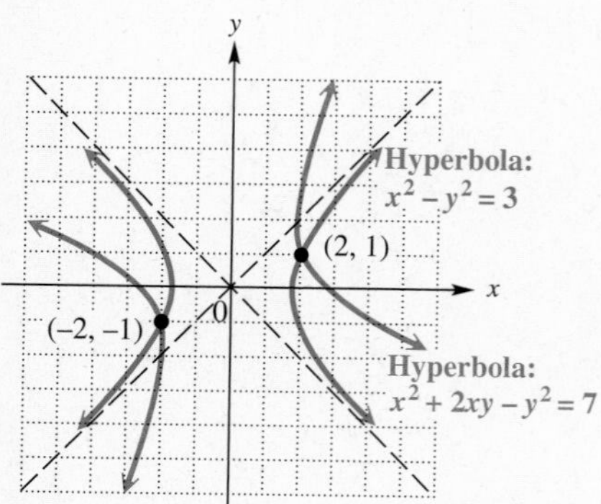

Figure 31

$$\begin{array}{rll} x^2 + 2xy - y^2 = & 7 & (1) \\ -x^2 \qquad + y^2 = & -3 & \text{Multiply (2) by } -1. \\ \hline 2xy \qquad = & 4 & \text{Add.} \end{array}$$

Next, we solve $2xy = 4$ for one of the variables. We choose y.

$$2xy = 4$$

$$y = \frac{2}{x} \qquad (3)$$

Now, we substitute $y = \frac{2}{x}$ into one of the original equations. It is easier to do this with equation (2).

$$x^2 - y^2 = 3 \qquad (2)$$

$$x^2 - \left(\frac{2}{x}\right)^2 = 3 \qquad \text{Let } y = \tfrac{2}{x}.$$

$$x^2 - \frac{4}{x^2} = 3 \qquad \text{Square } \tfrac{2}{x}.$$

$$x^4 - 4 = 3x^2 \qquad \text{Multiply by } x^2, x \neq 0.$$

$$x^4 - 3x^2 - 4 = 0 \qquad \text{Subtract } 3x^2.$$

$$(x^2 - 4)(x^2 + 1) = 0 \qquad \text{Factor.}$$

$$x^2 - 4 = 0 \quad \text{or} \quad x^2 + 1 = 0$$

$$x^2 = 4 \quad \text{or} \quad x^2 = -1$$

$$x = 2 \quad \text{or} \quad x = -2 \qquad x = i \quad \text{or} \quad x = -i$$

Substituting these four values of x into equation (3) gives the corresponding values for y.

If $x = 2$, then $y = 1$. If $x = i$, then $y = -2i$.
If $x = -2$, then $y = -1$. If $x = -i$, then $y = 2i$.

Note that if we substitute the x-values we found into equation (1) or (2) instead of into equation (3), we get extraneous solutions. ***It is always wise to check all solutions in both of the given equations.*** There are four ordered pairs in the solution set, two with real values and two with nonreal complex values. The solution set is

$$\{(2, 1), (-2, -1), (i, -2i), (-i, 2i)\}.$$

The graph of the system, shown in Figure 31, shows only the two real intersection points because the graph is in the real number plane. The two ordered pairs with nonreal complex components are solutions of the system, but do not appear on the graph.

◀ Work Problem 4 at the Side.

Note

In the examples of this section, we analyzed the possible number of points of intersection of the graphs in each system. However, in Examples 2 and 4, we worked with equations whose graphs had not been studied. Keep in mind that it is not absolutely essential to visualize the number of points of intersection in order to solve the system. Furthermore, as in Example 4, there are sometimes nonreal complex solutions to nonlinear systems that do not appear as points of intersection in the real plane. Visualizing the geometry of the graphs is only an aid to solving these systems.

ANSWERS

4. (a) $\{(1, -2), (-1, 2), (2, -1), (-2, 1)\}$
(b) $\{(0, 2), (0, -2), (2i\sqrt{2}, 0), (-2i\sqrt{2}, 0)\}$

12.4 Exercises

FOR EXTRA HELP MyMathLab Math XL PRACTICE WATCH DOWNLOAD READ  REVIEW

Each sketch represents the graphs of a pair of equations in a system. How many ordered pairs of real numbers are in each solution set?

1.

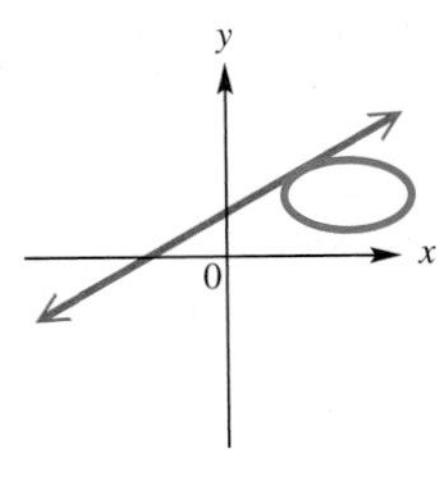

one

2.

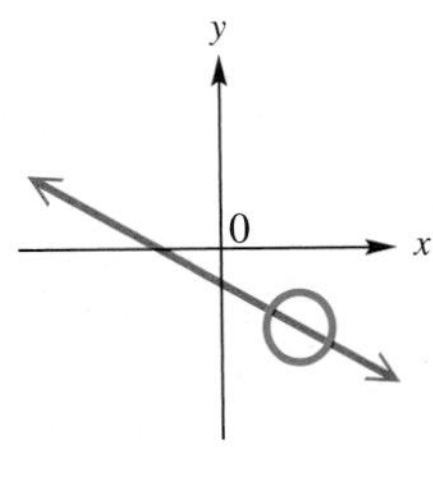

two

3.

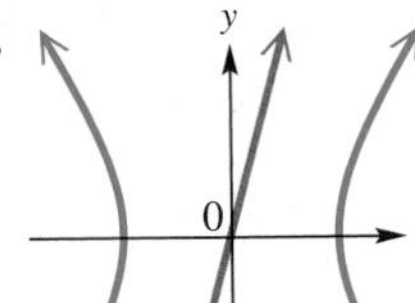

none

4.

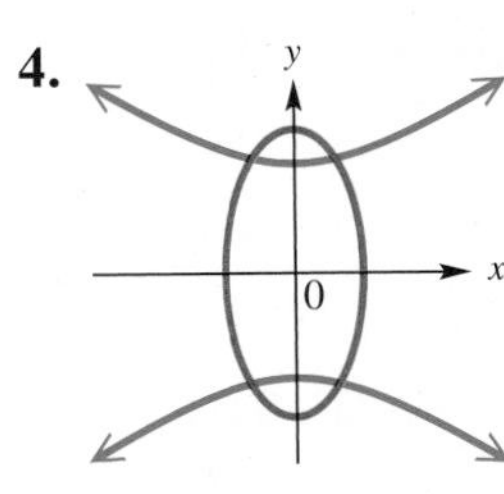

four

Suppose that a nonlinear system is composed of equations whose graphs are those described, and the number of points of intersection of the two graphs is as given. Make a sketch satisfying these conditions. (There may be more than one way to do this.)

5. A line and a circle; no points

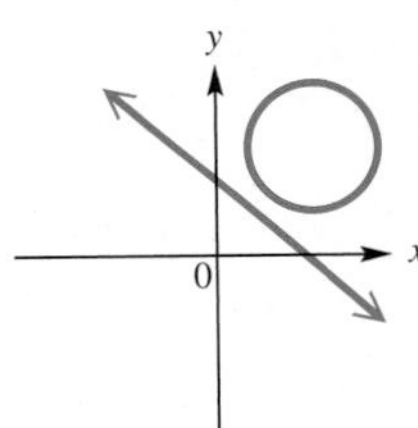

6. A line and a circle; one point

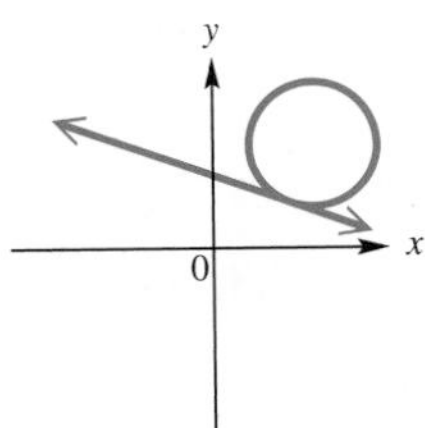

7. A line and an ellipse; two points

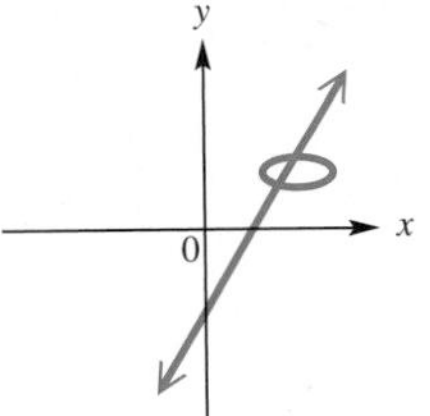

8. A line and a hyperbola; no points

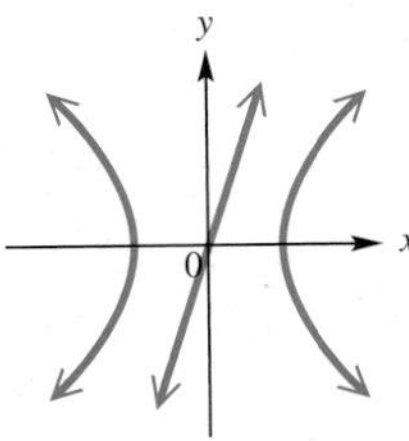

9. A circle and an ellipse; four points

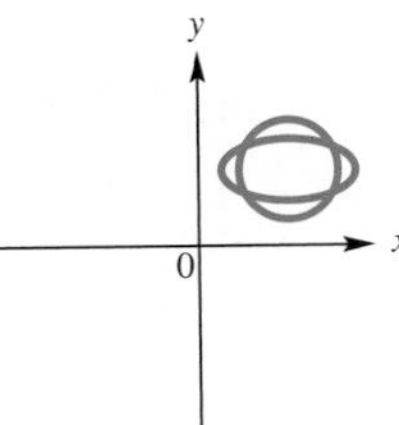

10. A parabola and an ellipse; one point

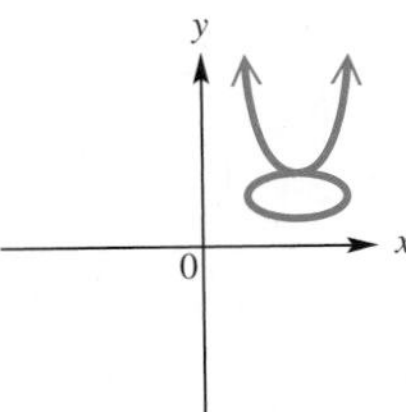

Solve each system by the substitution method. See Examples 1 and 2.

11. $y = 4x^2 - x$
$y = x$

$\left\{(0, 0), \left(\frac{1}{2}, \frac{1}{2}\right)\right\}$

12. $y = x^2 + 6x$
$3y = 12x$

$\{(0, 0), (-2, -8)\}$

13. $y = x^2 + 6x + 9$
$x + y = 3$

$\{(-6, 9), (-1, 4)\}$

14. $y = x^2 + 8x + 16$
$x - y = -4$

$\{(-3, 1), (-4, 0)\}$

15. $x^2 + y^2 = 2$
$2x + y = 1$

$\left\{\left(-\frac{1}{5}, \frac{7}{5}\right), (1, -1)\right\}$

16. $2x^2 + 4y^2 = 4$
$x = 4y$

$\left\{\left(\frac{4}{3}, \frac{1}{3}\right), \left(-\frac{4}{3}, -\frac{1}{3}\right)\right\}$

17. $xy = 4$
$3x + 2y = -10$
$\left\{(-2, -2), \left(-\frac{4}{3}, -3\right)\right\}$

18. $xy = -5$
$2x + y = 3$
$\left\{\left(\frac{5}{2}, -2\right), (-1, 5)\right\}$

19. $xy = -3$
$x + y = -2$
$\{(-3, 1), (1, -3)\}$

20. $xy = 12$
$x + y = 8$
$\{(6, 2), (2, 6)\}$

21. $y = 3x^2 + 6x$
$y = x^2 - x - 6$
$\left\{\left(-\frac{3}{2}, -\frac{9}{4}\right), (-2, 0)\right\}$

22. $y = 2x^2 + 1$
$y = 5x^2 + 2x - 7$
$\left\{\left(\frac{4}{3}, \frac{41}{9}\right), (-2, 9)\right\}$

23. $2x^2 - y^2 = 6$
$y = x^2 - 3$
$\{(-\sqrt{3}, 0), (\sqrt{3}, 0), (-\sqrt{5}, 2), (\sqrt{5}, 2)\}$

24. $x^2 + y^2 = 4$
$y = x^2 - 2$
$\{(0, -2), (\sqrt{3}, 1), (-\sqrt{3}, 1)\}$

Solve each system using the elimination method or a combination of the elimination and substitution methods. See Examples 3 and 4.

25. $3x^2 + 2y^2 = 12$
$x^2 + 2y^2 = 4$
$\{(-2, 0), (2, 0)\}$

26. $2x^2 + y^2 = 28$
$4x^2 - 5y^2 = 28$
$\{(-2\sqrt{3}, -2), (-2\sqrt{3}, 2), (2\sqrt{3}, -2), (2\sqrt{3}, 2)\}$

27. $xy = 6$
$3x^2 - y^2 = 12$
$\{(i\sqrt{2}, -3i\sqrt{2}), (-i\sqrt{2}, 3i\sqrt{2}),$
$(-\sqrt{6}, -\sqrt{6}), (\sqrt{6}, \sqrt{6})\}$

28. $xy = 5$
$2y^2 - x^2 = 5$
$\left\{(\sqrt{5}, \sqrt{5}), (-\sqrt{5}, -\sqrt{5}),\right.$
$\left.\left(i\sqrt{10}, -\frac{i\sqrt{10}}{2}\right), \left(-i\sqrt{10}, \frac{i\sqrt{10}}{2}\right)\right\}$

29. $2x^2 + 2y^2 = 8$
$3x^2 + 4y^2 = 24$
$\{(-2i\sqrt{2}, -2\sqrt{3}), (-2i\sqrt{2}, 2\sqrt{3}),$
$(2i\sqrt{2}, -2\sqrt{3}), (2i\sqrt{2}, 2\sqrt{3})\}$

30. $5x^2 + 5y^2 = 20$
$x^2 + 2y^2 = 2$
$\{(-\sqrt{6}, -i\sqrt{2}), (-\sqrt{6}, i\sqrt{2}),$
$(\sqrt{6}, -i\sqrt{2}), (\sqrt{6}, i\sqrt{2})\}$

31. $x^2 + xy + y^2 = 15$
$x^2 + y^2 = 10$
$\{(-\sqrt{5}, -\sqrt{5}), (\sqrt{5}, \sqrt{5})\}$

32. $2x^2 + 3xy + 2y^2 = 21$
$x^2 + y^2 = 6$
$\{(\sqrt{3}, \sqrt{3}), (-\sqrt{3}, -\sqrt{3})\}$

Solve each problem by using a nonlinear system.

33. The area of a rectangular rug is 84 ft^2 and its perimeter is 38 ft. Find the length and width of the rug.
length: 12 ft; width: 7 ft

34. Find the length and width of a rectangular room whose perimeter is 50 m and whose area is 100 m^2.
length: 20 m; width: 5 m

12.5 Second-Degree Inequalities and Systems of Inequalities

OBJECTIVES

1. Graph second-degree inequalities.
2. Graph the solution set of a system of inequalities.

OBJECTIVE 1 Graph second-degree inequalities. A **second-degree inequality** is an inequality with at least one variable of degree 2 and no variable with degree greater than 2. An example is $x^2 + y^2 \leq 36$. To graph this inequality, we first graph the boundary, as we did with linear inequalities in **Section 3.5.** The boundary of the inequality $x^2 + y^2 \leq 36$ is the graph of the equation $x^2 + y^2 = 36$, a circle with radius 6 and center at the origin, as shown in Figure 32.

The graph of the inequality $x^2 + y^2 \leq 36$ will include either the points outside the circle or the points inside the circle, as well as the boundary. We decide which region to shade by substituting any test point not on the circle, such as (0, 0), into the original inequality. Since $0^2 + 0^2 \leq 36$ is a true statement, the original inequality includes the points inside the circle, the shaded region in Figure 32, and the boundary.

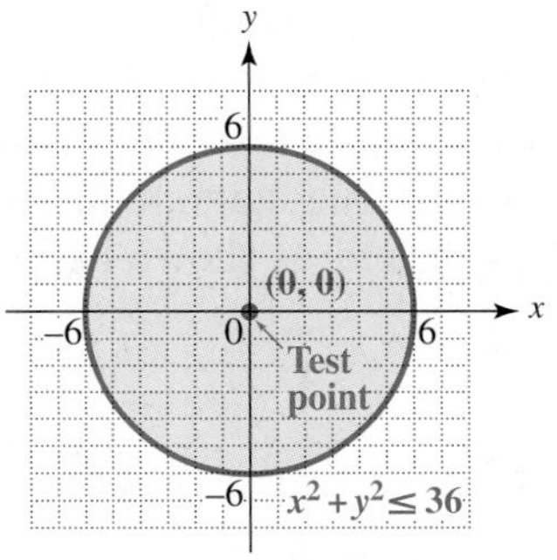

Figure 32

1 Graph $y \geq (x + 1)^2 - 5$.

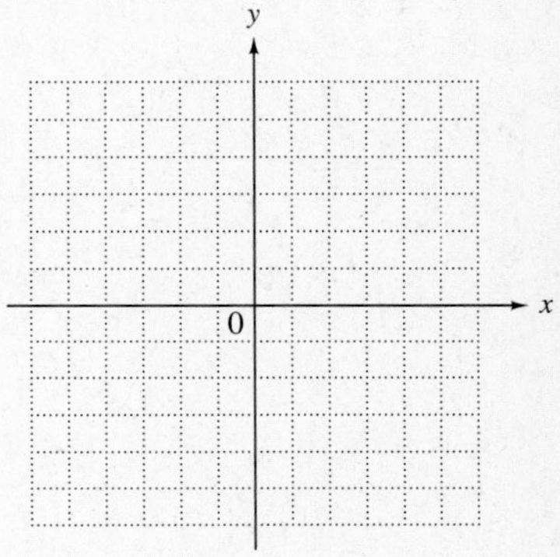

EXAMPLE 1 Graphing a Second-Degree Inequality

Graph $y < -2(x - 4)^2 - 3$.

The boundary, $y = -2(x - 4)^2 - 3$, is a parabola that opens down with vertex at $(4, -3)$. Using (0, 0) as a test point gives

$$0 \stackrel{?}{<} -2(0 - 4)^2 - 3$$

$$0 \stackrel{?}{<} -32 - 3$$

$$0 < -35. \quad \text{False}$$

Because the final inequality is a false statement, the points in the region containing (0, 0) do not satisfy the inequality. Figure 33 shows the final graph. The parabola is drawn as a dashed curve since the points of the parabola itself do not satisfy the inequality, and the region inside (or below) the parabola is shaded.

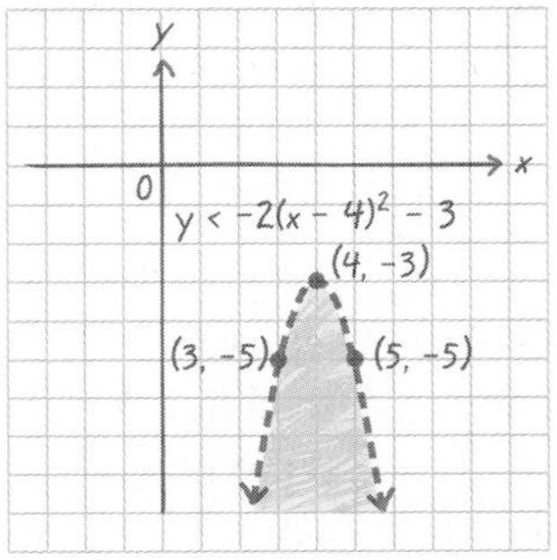

Figure 33

Work Problem **1** *at the Side.* ▶

ANSWER

1.

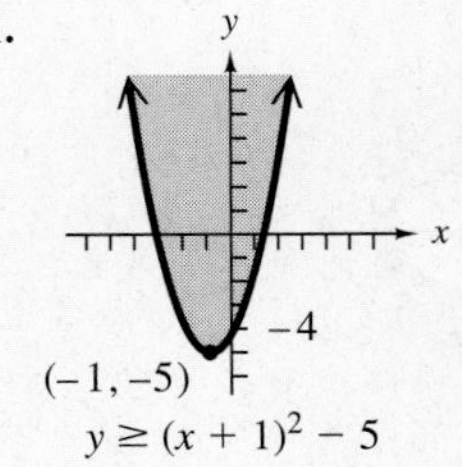

2 Graph $x^2 + 4y^2 > 36$.

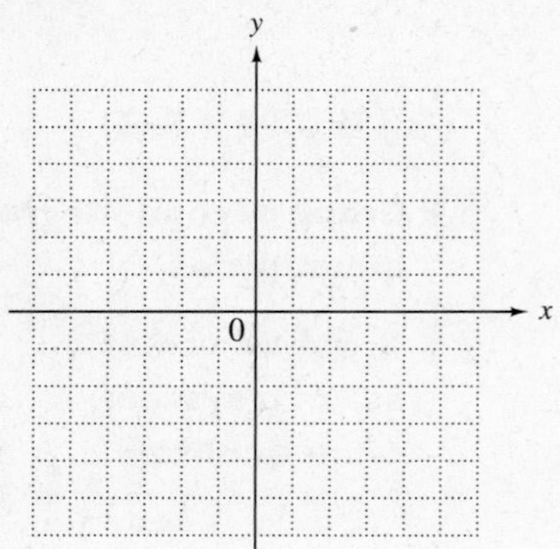

Note

Since the substitution is easy, the origin is the test point of choice when graphing second-degree inequalities unless the graph actually passes through (0, 0).

EXAMPLE 2 Graphing a Second-Degree Inequality

Graph $16y^2 \le 144 + 9x^2$.

First rewrite the inequality as follows.

$$16y^2 - 9x^2 \le 144 \quad \text{Subtract } 9x^2.$$

$$\frac{y^2}{9} - \frac{x^2}{16} \le 1 \quad \text{Divide by 144.}$$

This form shows that the boundary is the hyperbola given by

$$\frac{y^2}{9} - \frac{x^2}{16} = 1.$$

Since the graph is a vertical hyperbola, the desired region will be either the region between the branches or the regions above the top branch and below the bottom branch. Choose (0, 0) as a test point. Substituting into the original inequality leads to $0 \le 144$, a true statement, so the region between the branches containing (0, 0) is shaded, as shown in Figure 34.

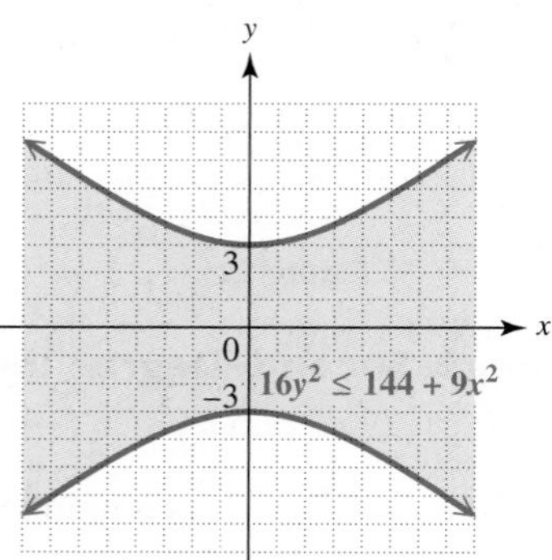

Figure 34

◀ Work Problem **2** at the Side.

OBJECTIVE 2 Graph the solution set of a system of inequalities. If two or more inequalities are considered at the same time, we have a **system of inequalities.** To find the solution set of the system, we find the intersection of the graphs (solution sets) of the inequalities in the system.

EXAMPLE 3 Graphing a System of Two Inequalities

Graph the solution set of the system.

$$2x + 3y > 6$$
$$x^2 + y^2 < 16$$

Begin by graphing the solution set of $2x + 3y > 6$. The boundary line is the graph of $2x + 3y = 6$ and is a dashed line because equality is not included. The test point (0, 0) leads to a false statement in $2x + 3y > 6$, so shade the region above the line, as shown in Figure 35 on the next page.

Continued on Next Page

ANSWER

2.

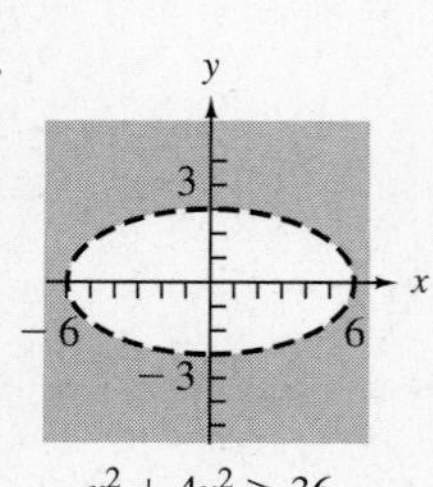

The graph of $x^2 + y^2 < 16$ is the interior of a dashed circle centered at the origin with radius 4. This is shown in Figure 36.

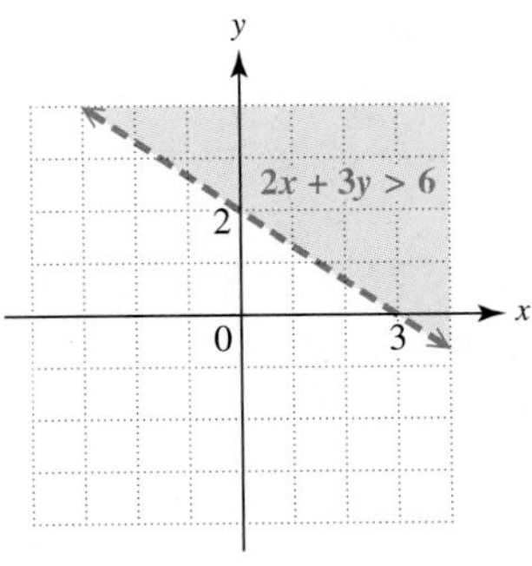

Figure 35

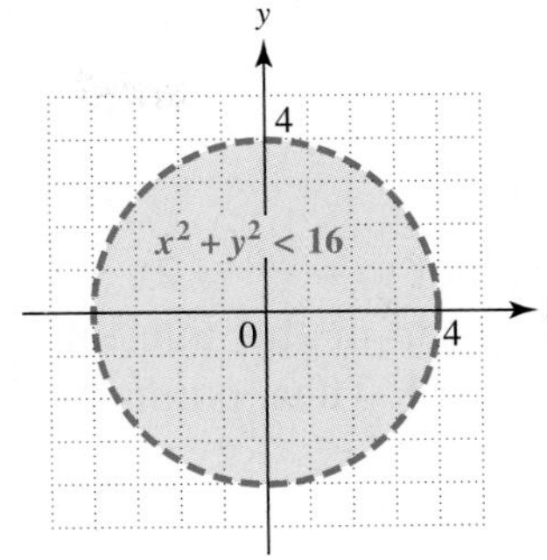

Figure 36

Finally, to show the graph of the solution set of the system, determine the intersection of the graphs of the two inequalities. The overlapping region in Figure 37 is the solution set.

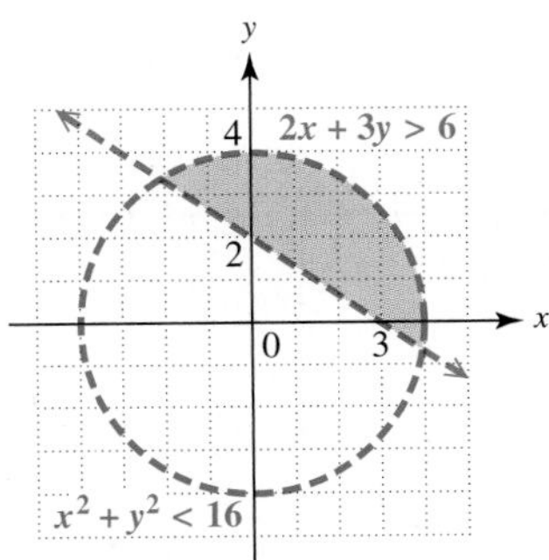

Figure 37

Work Problem **3** at the Side. ▶

EXAMPLE 4 **Graphing a System of Three Inequalities**

Graph the solution set of the system.

$$x + y < 1$$
$$y \le 2x + 3$$
$$y \ge -2$$

Graph each inequality separately, on the same axes. The graph of $x + y < 1$ consists of all points below the dashed line $x + y = 1$. The graph of $y \le 2x + 3$ is the region that lies below the solid line $y = 2x + 3$. Finally, the graph of $y \ge -2$ is the region above the solid horizontal line $y = -2$.

The graph of the system, the intersection of these three graphs, is the triangular region enclosed by the three boundary lines in Figure 38, including two of its boundaries.

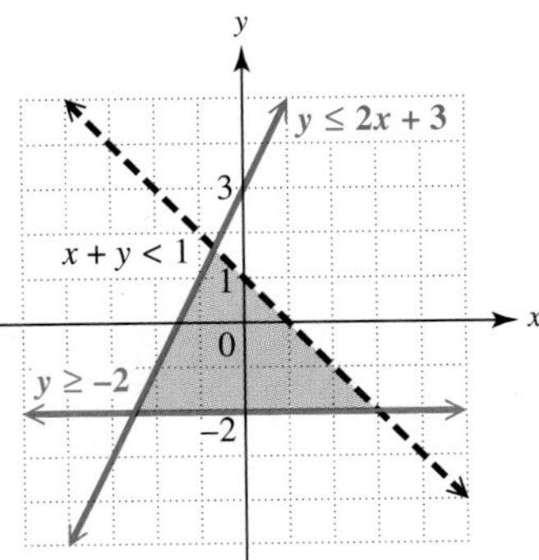

Figure 38

Work Problem **4** at the Side. ▶

3 Graph the solution set of the system.

$$x^2 + y^2 \le 25$$
$$x + y \le 3$$

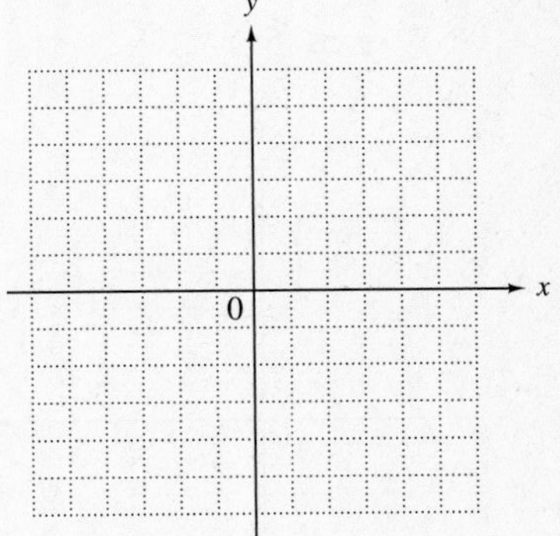

4 Graph the solution set of the system.

$$3x - 4y \ge 12$$
$$x + 3y \ge 6$$
$$y \le 2$$

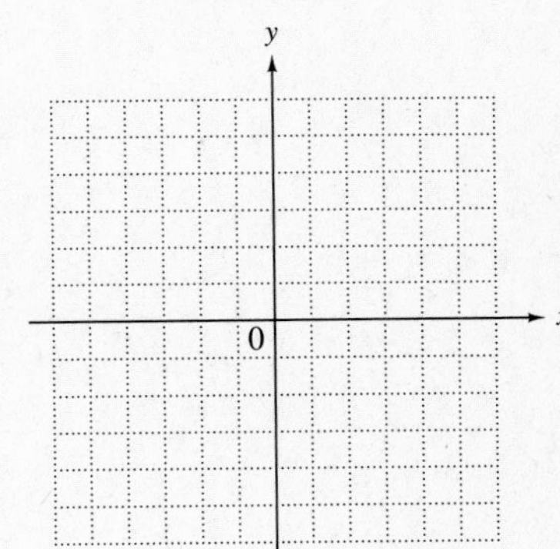

Answers

3.

4.

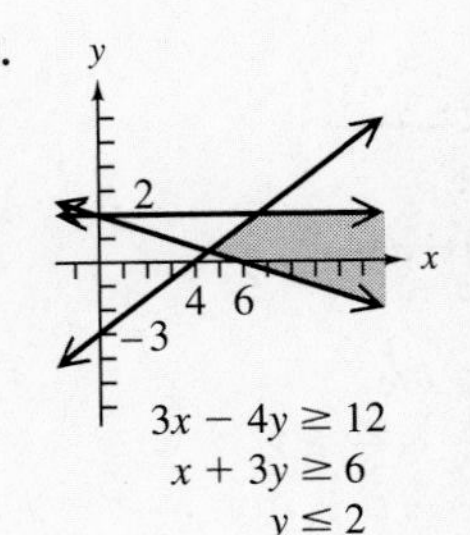

5 Graph the solution set of the system.

$$y \geq x^2 + 1$$
$$\frac{x^2}{9} + \frac{y^2}{4} \geq 1$$
$$y \leq 5$$

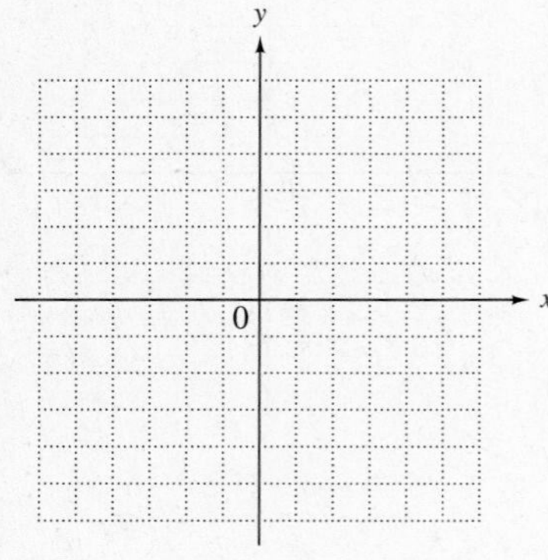

EXAMPLE 5 Graphing a System of Three Inequalities

Graph the solution set of the system.

$$y \geq x^2 - 2x + 1$$
$$2x^2 + y^2 > 4$$
$$y < 4$$

The graph of $y = x^2 - 2x + 1$ is a parabola with vertex at (1, 0). Those points above (or in the interior of) the parabola satisfy the condition $y > x^2 - 2x + 1$. Thus, points on the parabola or in the interior are in the solution set of $y \geq x^2 - 2x + 1$.

The graph of the equation $2x^2 + y^2 = 4$ is an ellipse. We draw it as a dashed curve. To satisfy the inequality $2x^2 + y^2 > 4$, a point must lie outside the ellipse.

The graph of $y < 4$ includes all points below the dashed line $y = 4$. Finally, the graph of the system is the shaded region in Figure 39 that lies outside the ellipse, inside or on the boundary of the parabola, and below the line $y = 4$.

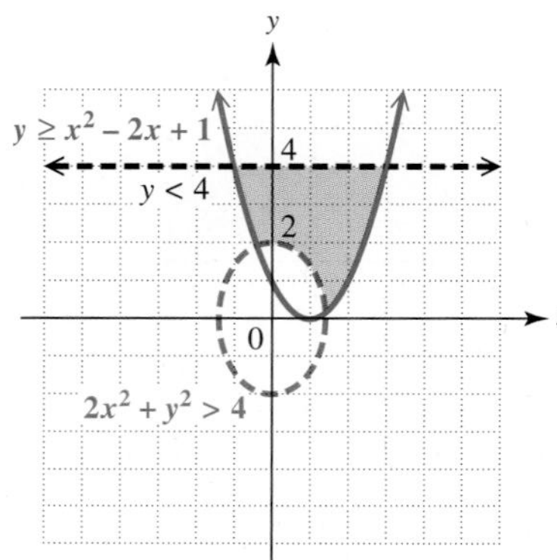

Figure 39

◀ Work Problem **5** at the Side.

ANSWER

5.

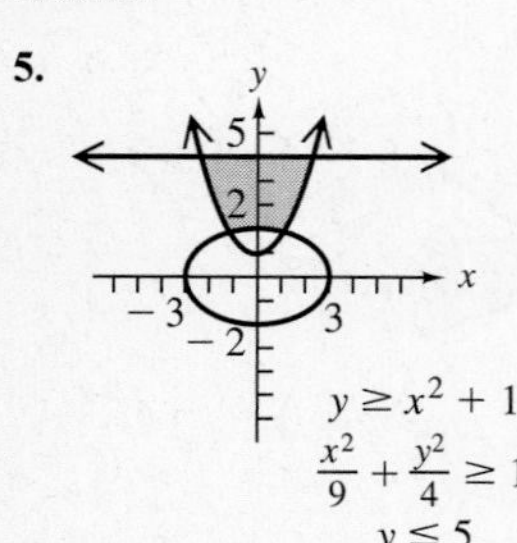

12.5 Exercises

FOR EXTRA HELP

PRACTICE

WATCH

DOWNLOAD

READ

REVIEW

Graph each inequality. See Examples 1 and 2.

1. $y > x^2 - 1$

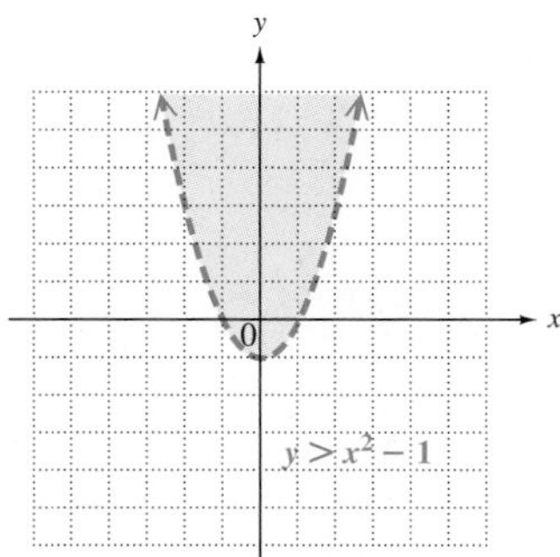

2. $y^2 > 4 + x^2$

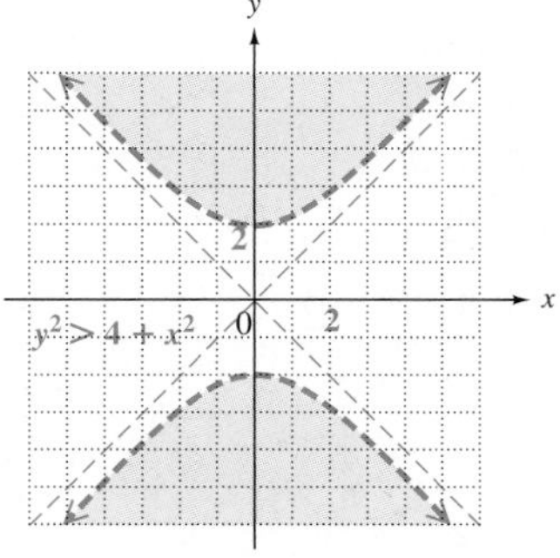

3. $y^2 \leq 4 - 2x^2$

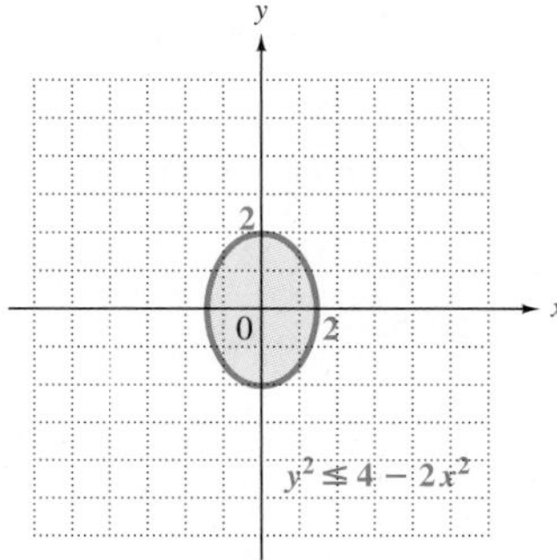

4. $y + 2 \geq x^2$

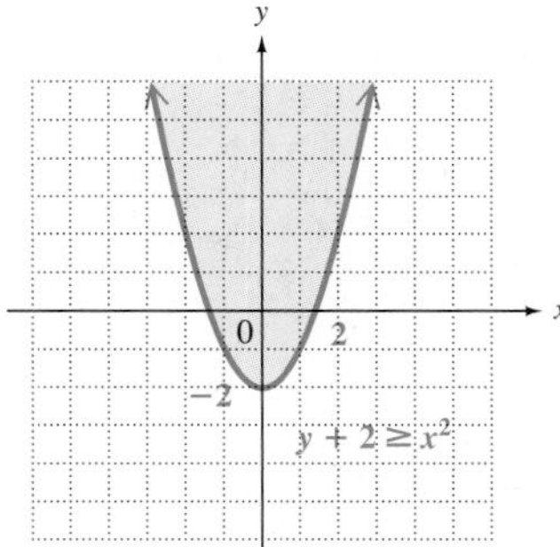

5. $x^2 \leq 16 - y^2$

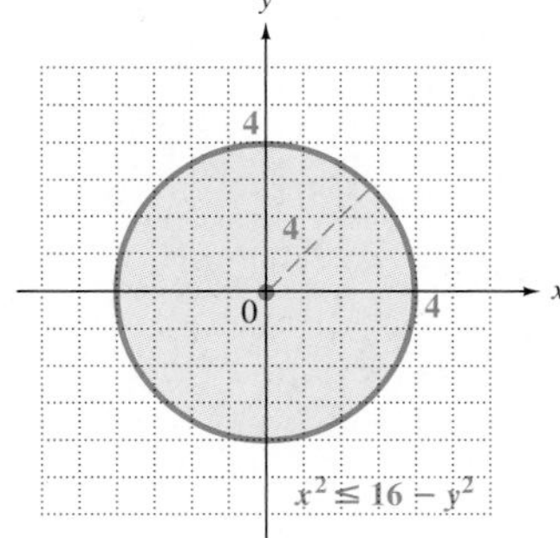

6. $2y^2 \geq 8 - x^2$

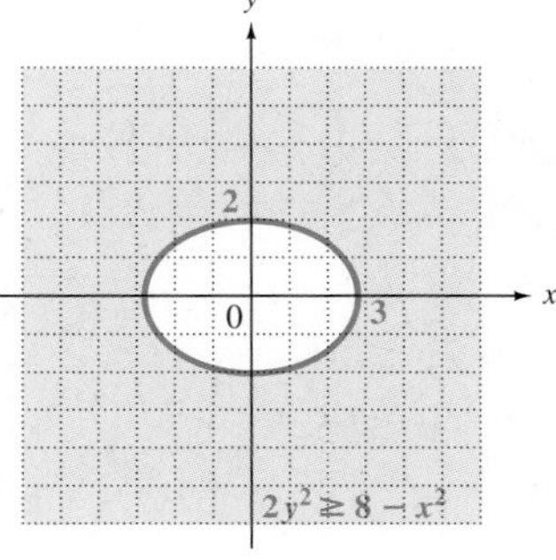

7. $x^2 \leq 16 + 4y^2$

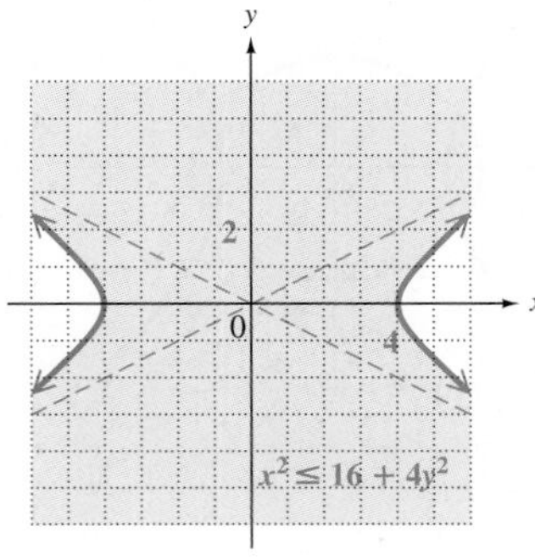

8. $y \leq x^2 + 4x + 2$

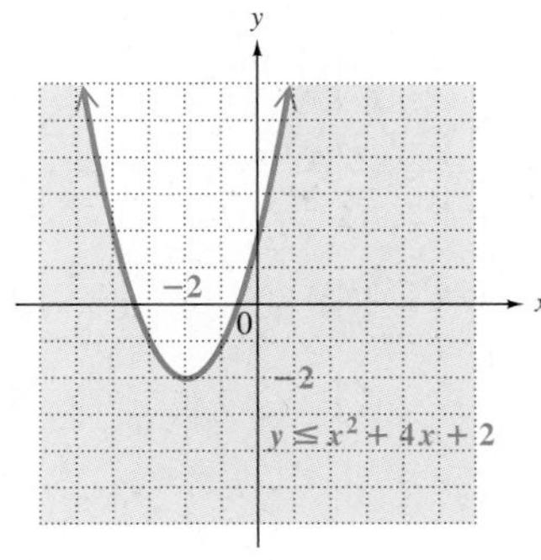

9. $9x^2 < 16y^2 - 144$

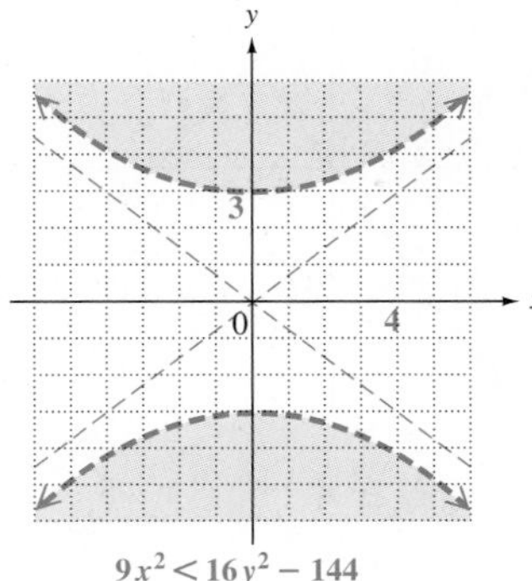

$9x^2 < 16y^2 - 144$

10. $9x^2 > 16y^2 + 144$

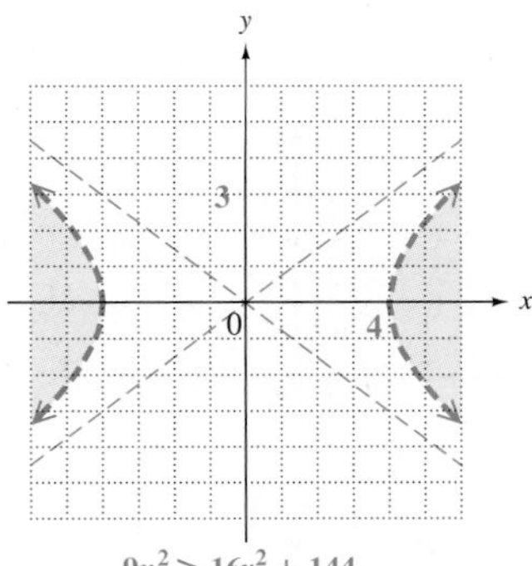

$9x^2 > 16y^2 + 144$

11. $4y^2 \leq 36 - 9x^2$

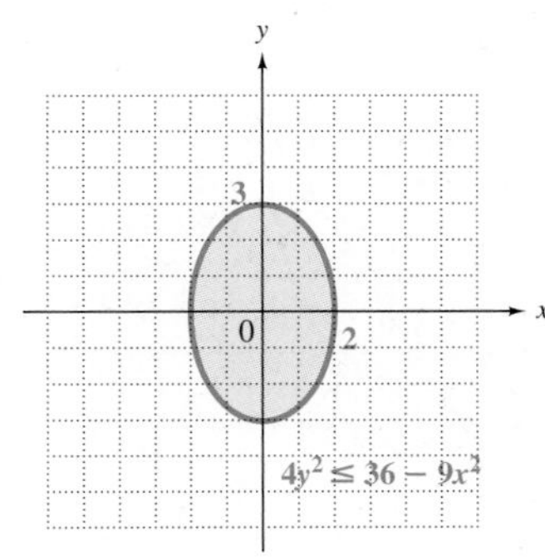

12. $x^2 - 4 \geq -4y^2$

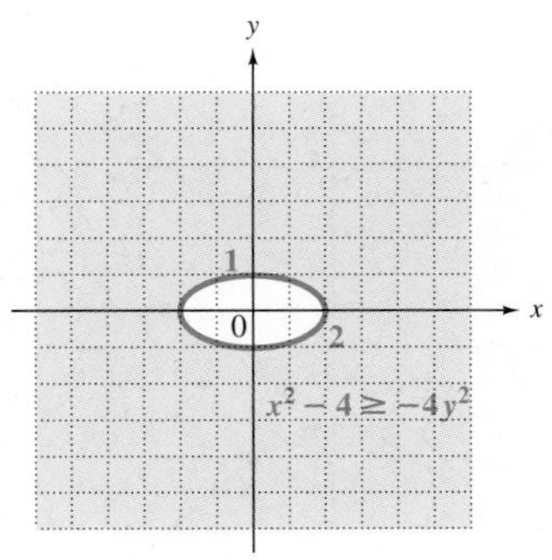

13. $x \geq y^2 - 8y + 14$

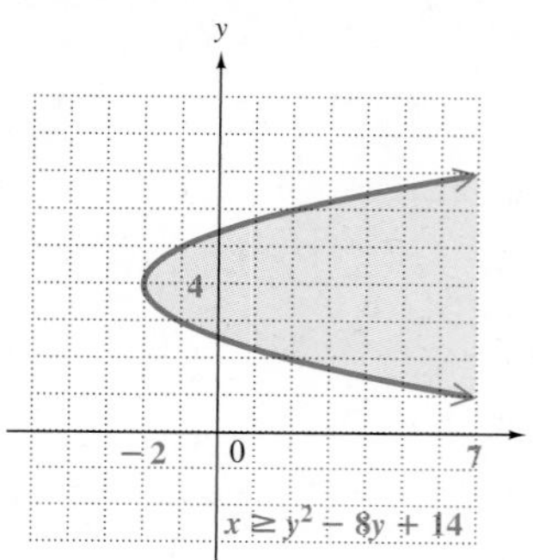

14. $x \leq -y^2 + 6y - 7$

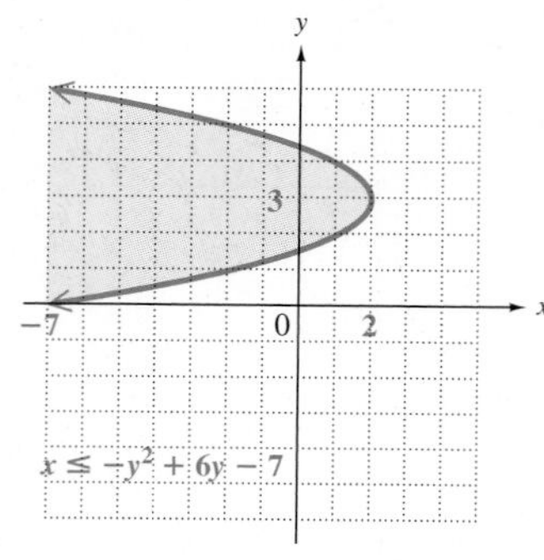

15. $25x^2 \leq 9y^2 + 225$

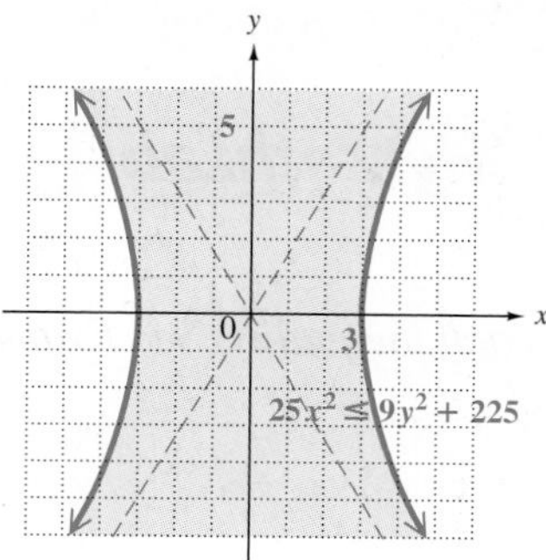

Graph each system of inequalities. See Examples 3–5.

16. $2x + 5y < 10$
$x - 2y < 4$

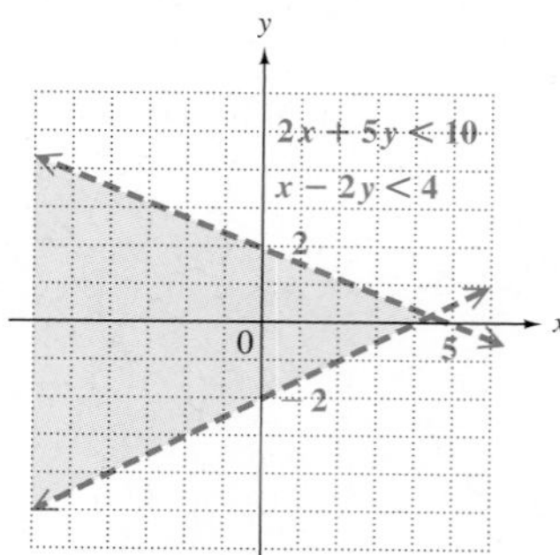

17. $3x - y > -6$
$4x + 3y > 12$

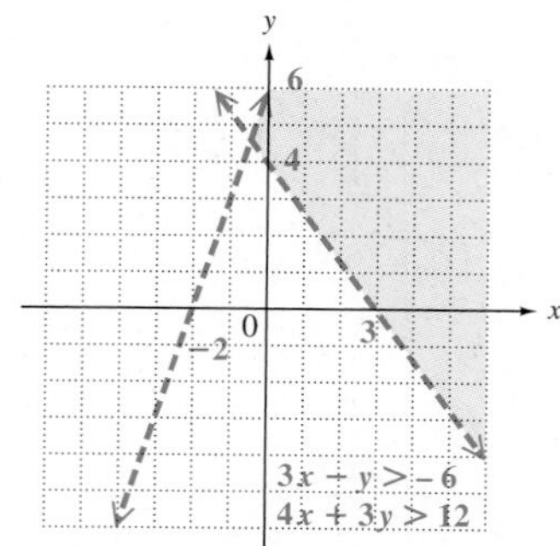

18. $5x - 3y \leq 15$
$4x + y \geq 4$

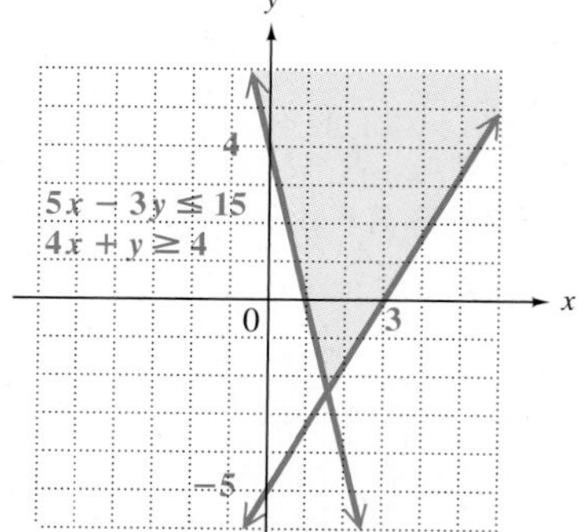

19. $4x - 3y \leq 0$
$x + y \leq 5$

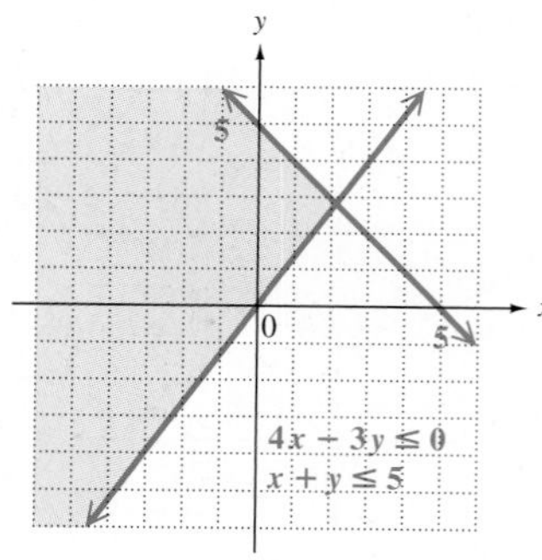

20. $y > x^2 - 4$
$y < -x^2 + 3$

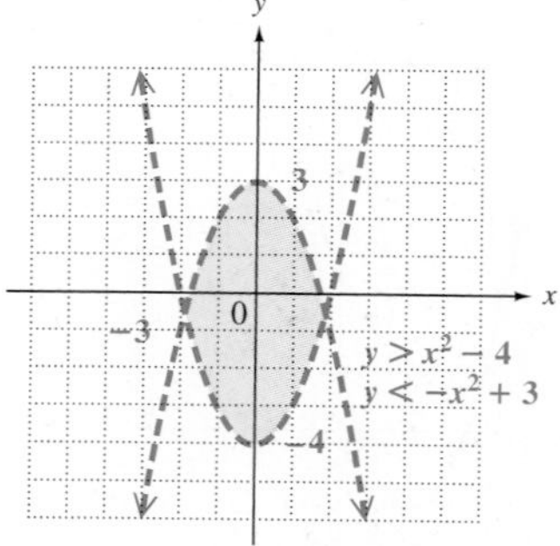

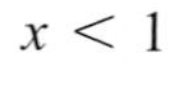 **21.** $x^2 - y^2 \geq 9$
$\frac{x^2}{16} + \frac{y^2}{9} \leq 1$

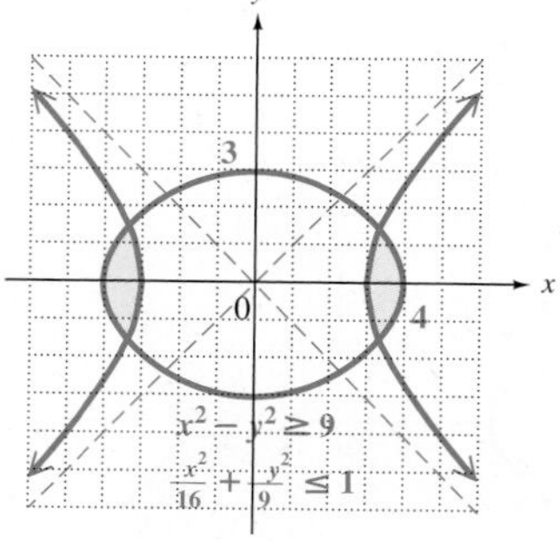

22. $y^2 - x^2 \geq 4$
$-5 \leq y \leq 5$

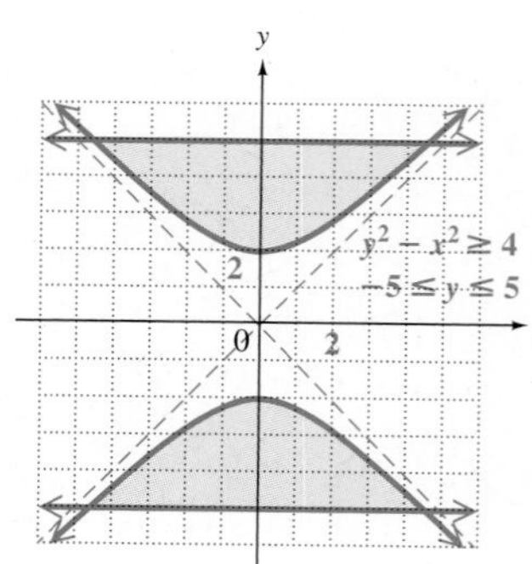

23. $y < x^2$
$y > -2$
$x + y < 3$
$3x - 2y > -6$

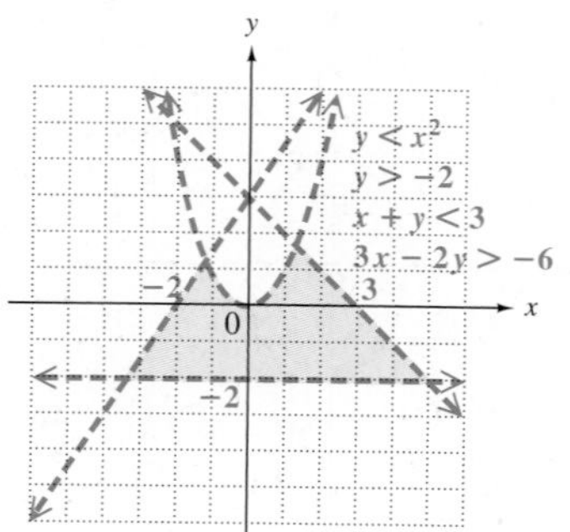

24. $y \leq -x^2$
$y \geq x - 3$
$y \leq -1$
$x < 1$

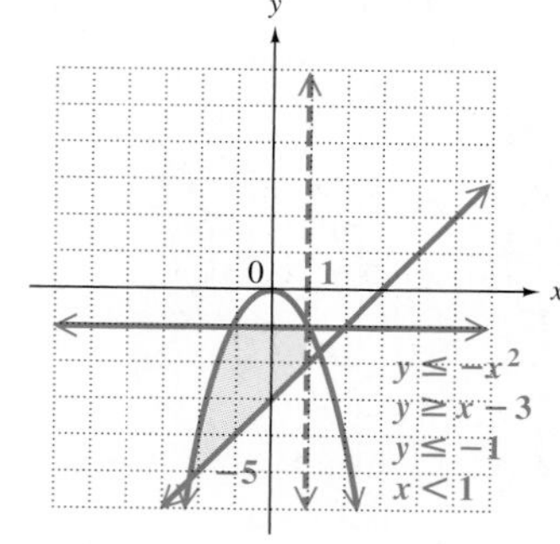

Chapter 12 Summary

Key Terms

12.1	**asymptotes**	Lines that a graph approaches, such as the x- and y-axes for the graph of the reciprocal function, are called asymptotes of the graph.
	greatest integer function	The function defined by $f(x) = [\![x]\!]$, where the symbol $[\![x]\!]$ represents the greatest integer less than or equal to x, is called the greatest integer function.
	step function	A step function is a function with a graph that looks like a series of steps.
	composition (composite function)	If f and g are functions, then the composition of g and f is defined by $(g \circ f)(x) = g(f(x))$ for all x in the domain of f such that $f(x)$ is in the domain of g.
12.2	**conic section**	When a plane intersects an infinite cone at different angles, the figures formed by the intersections are called conic sections.
	circle	A circle is the set of all points in a plane that lie a fixed distance from a fixed point.
	center	The fixed point discussed in the definition of a circle is the center of the circle.
	radius	The radius of a circle is the fixed distance between the center and any point on the circle.
	ellipse	An ellipse is the set of all points in a plane the sum of whose distances from two fixed points **(foci)** is constant.
12.3	**hyperbola**	A hyperbola is the set of all points in a plane such that the absolute value of the difference of the distances from two fixed points (foci) is constant.
	asymptotes of a hyperbola	The two intersecting lines that the branches of a hyperbola approach are called asymptotes of the hyperbola.
	fundamental rectangle	The asymptotes of a hyperbola are the extended diagonals of its fundamental rectangle.
12.4	**nonlinear equation**	An equation in which some terms have more than one variable or a variable of degree 2 or greater is called a nonlinear equation.
	nonlinear system of equations	A nonlinear system of equations is a system with at least one nonlinear equation.
12.5	**second-degree inequality**	A second-degree inequality is an inequality with at least one variable of degree 2 and no variable with degree greater than 2.
	system of inequalities	A system of inequalities consists of two or more inequalities to be solved at the same time.

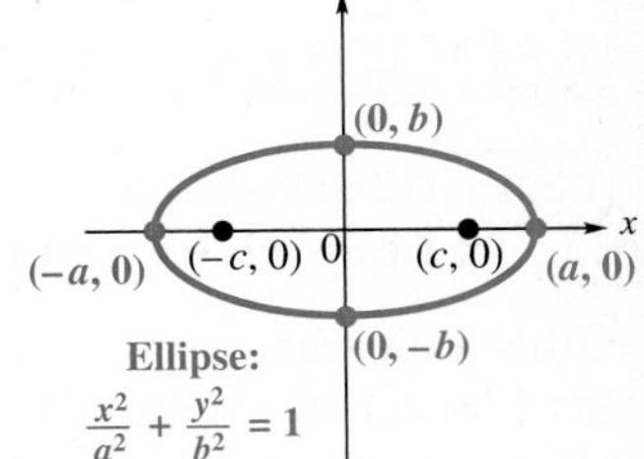

New Symbols

$[\![x]\!]$	greatest integer less than or equal to x	$(f \circ g)(x) = f(g(x))$	composite function

▶ Test Your Word Power

See how well you have learned the vocabulary in this chapter. Answers, with examples, follow the Quick Review.

1. A **circle** is the set of all points in a plane

A. such that the absolute value of the difference of the distances from two fixed points is constant

B. that lie a fixed distance from a fixed point

C. the sum of whose distances from two fixed points is constant

D. that make up the graph of any second-degree equation.

2. An **ellipse** is the set of all points in a plane

A. such that the absolute value of the difference of the distances from two fixed points is constant

B. that lie a fixed distance from a fixed point

C. the sum of whose distances from two fixed points is constant

D. that make up the graph of any second-degree equation.

3. A **hyperbola** is the set of all points in a plane

A. such that the absolute value of the difference of the distances from two fixed points is constant

B. that lie a fixed distance from a fixed point

C. the sum of whose distances from two fixed points is constant

D. that make up the graph of any second-degree equation.

4. A **nonlinear equation** is an equation

A. in which some terms have more than one variable or a variable of degree 2 or greater

B. in which the terms have only one variable

C. of degree 1

D. of a linear function.

▶ Quick Review

Concepts	Examples

12.1 Additional Graphs of Functions; Operations and Composition

Other Functions

In addition to the squaring function, some other important elementary functions in algebra are the absolute value function, defined by $f(x) = |x|$; the reciprocal function, defined by $f(x) = \frac{1}{x}$; the square root function, defined by $f(x) = \sqrt{x}$; and step functions, such as the greatest integer function, defined by $f(x) = [\![x]\!]$. Their graphs can be translated.

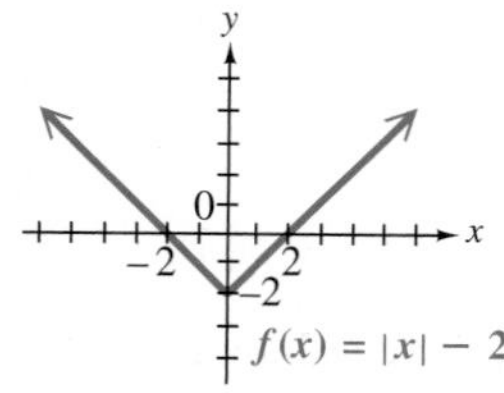

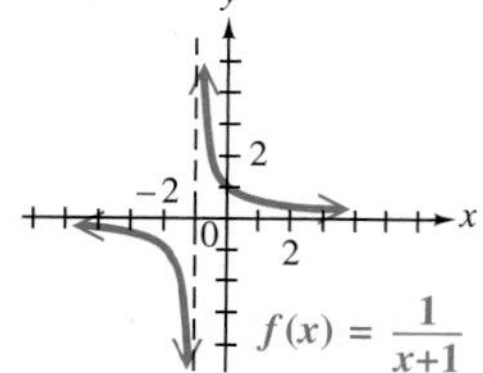

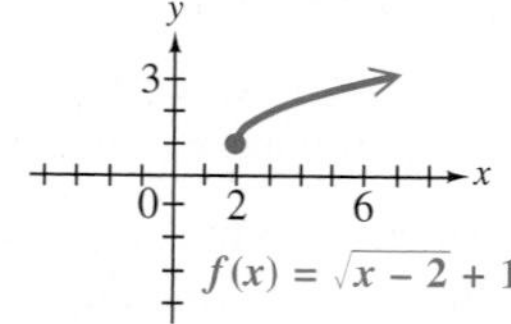

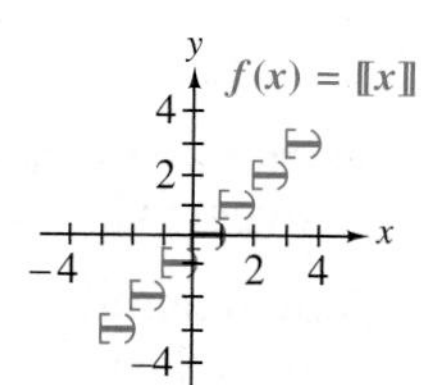

Operations on Functions

If $f(x)$ and $g(x)$ define functions, then

$$(f+g)(x) = f(x) + g(x)$$
$$(f-g)(x) = f(x) - g(x)$$
$$(fg)(x) = f(x) \cdot g(x)$$
$$\left(\frac{f}{g}\right)(x) = \frac{f(x)}{g(x)}, \quad g(x) \neq 0.$$

If $f(x) = x^2$ and $g(x) = 2x + 1$, then

$$(f+g)(x) = f(x) + g(x) = x^2 + 2x + 1,$$
$$(f-g)(x) = f(x) - g(x) = x^2 - 2x - 1,$$
$$(fg)(x) = f(x) \cdot g(x) = 2x^3 + x^2,$$

and $$\left(\frac{f}{g}\right)(x) = \frac{f(x)}{g(x)} = \frac{x^2}{2x+1}, \quad x \neq -\frac{1}{2}.$$

Composition of f and g

$$(f \circ g)(x) = f(g(x))$$

If $f(x) = x^2$ and $g(x) = 2x + 1$, then

$$(f \circ g)(x) = f(g(x)) = (2x+1)^2 = 4x^2 + 4x + 1$$

and $$(g \circ f)(x) = g(f(x)) = 2x^2 + 1.$$

Concepts	Examples

12.2 The Circle and the Ellipse

Circle

The circle with radius r and center at (h, k) has an equation of the form

$$(x - h)^2 + (y - k)^2 = r^2.$$

The circle with equation $(x + 2)^2 + (y - 3)^2 = 25$, which can be written $[x - (-2)^2] + (y - 3)^2 = 5^2$, has center $(-2, 3)$ and radius 5.

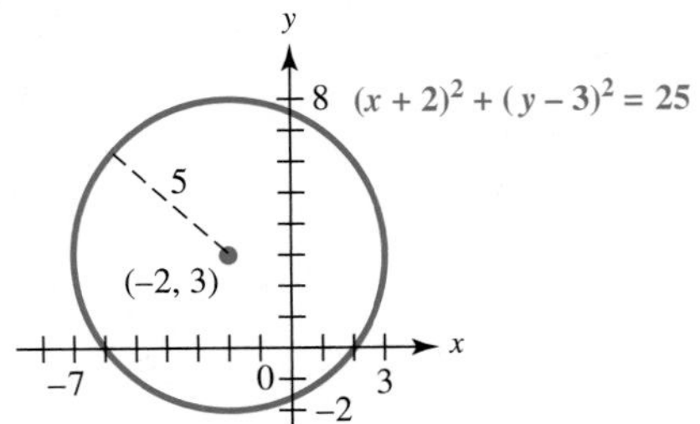

Ellipse

The ellipse whose x-intercepts are $(a, 0)$ and $(-a, 0)$ and whose y-intercepts are $(0, b)$ and $(0, -b)$ has an equation of the form

$$\frac{x^2}{a^2} + \frac{y^2}{b^2} = 1.$$

Graph $\frac{x^2}{9} + \frac{y^2}{4} = 1.$

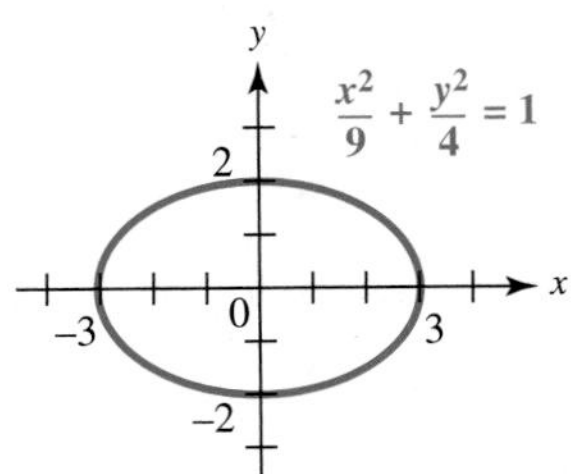

12.3 The Hyperbola and Other Functions Defined by Radicals

Hyperbola

A hyperbola with x-intercepts $(a, 0)$ and $(-a, 0)$ has an equation of the form

$$\frac{x^2}{a^2} - \frac{y^2}{b^2} = 1.$$

and a hyperbola with y-intercepts $(0, b)$ and $(0, -b)$ has an equation of the form

$$\frac{y^2}{b^2} - \frac{x^2}{a^2} = 1.$$

Graph $\frac{x^2}{4} - \frac{y^2}{4} = 1.$

The graph has x-intercepts $(2, 0)$ and $(-2, 0)$.

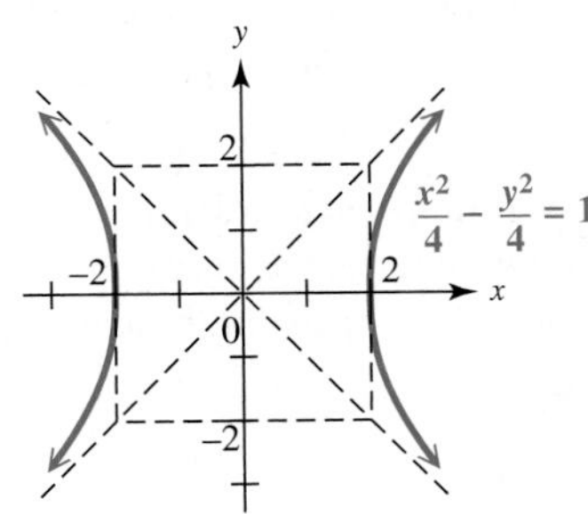

The extended diagonals of the fundamental rectangle with vertices at the points (a, b), $(-a, b)$, $(-a, -b)$, and $(a, -b)$ are the asymptotes of these hyperbolas.

The fundamental rectangle has vertices at $(2, 2)$, $(-2, 2)$, $(-2, -2)$, and $(2, -2)$.

Square Root Function

For an algebraic expression in x defined by u, with $u \geq 0$, a function of the form

$$f(x) = \sqrt{u}$$

is called a square root function.

To graph a square root function, square both sides so that the equation can be easily recognized. Then graph only the part indicated by the original equation.

Graph $y = -\sqrt{4 - x^2}$.

Square both sides and rearrange terms to get

$$x^2 + y^2 = 4.$$

This equation has a circle as its graph. However, graph only the lower half of the circle, since the original equation indicates that y cannot be positive.

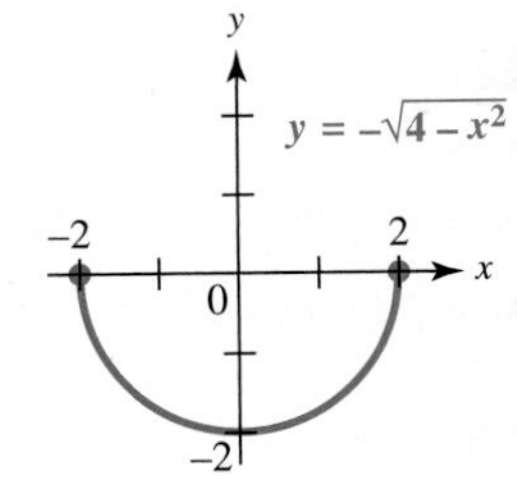

Concepts	Examples

12.4 Nonlinear Systems of Equations

Solving a Nonlinear System

A nonlinear system can be solved by the substitution method, the elimination method, or a combination of the two.

Solve the system.

$$x^2 + 2xy - y^2 = 14 \quad (1)$$
$$x^2 - y^2 = -16 \quad (2)$$

Multiply equation (2) by -1 and use elimination.

$$\begin{aligned} x^2 + 2xy - y^2 &= 14 \\ -x^2 \qquad + y^2 &= 16 \\ \hline 2xy \qquad &= 30 \\ xy &= 15 \end{aligned}$$

Solve for y to obtain $y = \frac{15}{x}$, and substitute into equation (2).

$$x^2 - \left(\frac{15}{x}\right)^2 = -16 \quad (2)$$
$$x^2 - \frac{225}{x^2} = -16$$
$$x^4 + 16x^2 - 225 = 0 \qquad \text{Multiply by } x^2\text{; add } 16x^2.$$
$$(x^2 - 9)(x^2 + 25) = 0 \qquad \text{Factor.}$$
$$x = \pm 3 \quad \text{or} \quad x = \pm 5i \qquad \text{Zero-factor property}$$

Find corresponding y-values to obtain the solution set

$$\{(3, 5), (-3, -5), (5i, -3i), (-5i, 3i)\}.$$

12.5 Second-Degree Inequalities and Systems of Inequalities

Graphing a Second-Degree Inequality

To graph a second-degree inequality, graph the corresponding equation as a boundary and use test points to determine which region(s) form the solution set. Shade the appropriate region(s).

Graphing a System of Inequalities

The solution set of a system of inequalities is the intersection of the solution sets of the individual inequalities.

Graph $y \geq x^2 - 2x + 3$.

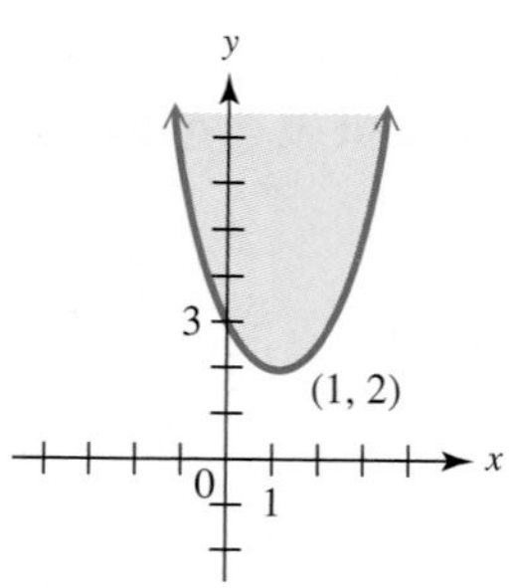

Graph the solution set of the system.

$$3x - 5y > -15$$
$$x^2 + y^2 \leq 25$$

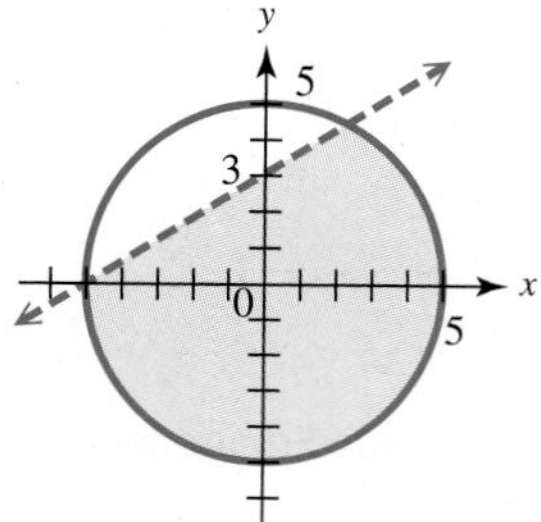

ANSWERS TO TEST YOUR WORD POWER

1. B; *Example*: See the graph of $x^2 + y^2 = 9$ in Figure 12 of **Section 12.2.**
2. C; *Example*: See the graph of $\frac{x^2}{49} + \frac{y^2}{36} = 1$ in Figure 17 of **Section 12.2.**
3. A; *Example*: See the graph of $\frac{x^2}{16} - \frac{y^2}{12} = 1$ in Figure 20 of **Section 12.3.**
4. A; *Examples*: $y = x^2 + 8x + 16$, $xy = 5$, $2x^2 - y^2 = 6$

Chapter 12 Review Exercises

[12.1] *Graph each function. Give the domain and range.*

1. $f(x) = |x + 4|$

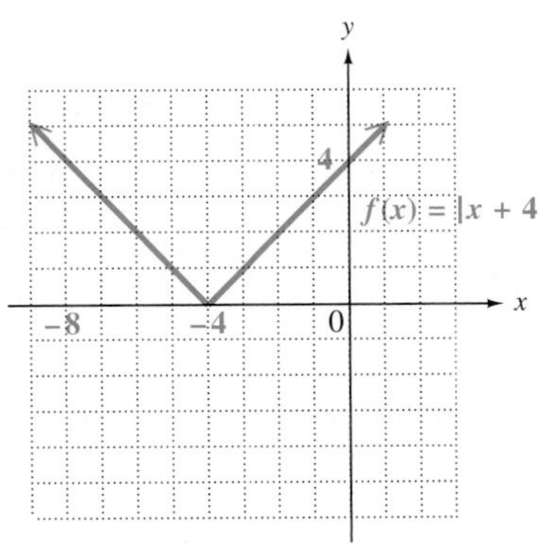

domain: $(-\infty, \infty)$;
range: $[0, \infty)$

2. 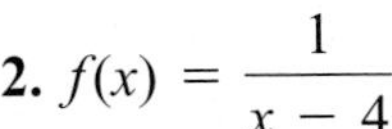 $f(x) = \dfrac{1}{x - 4}$

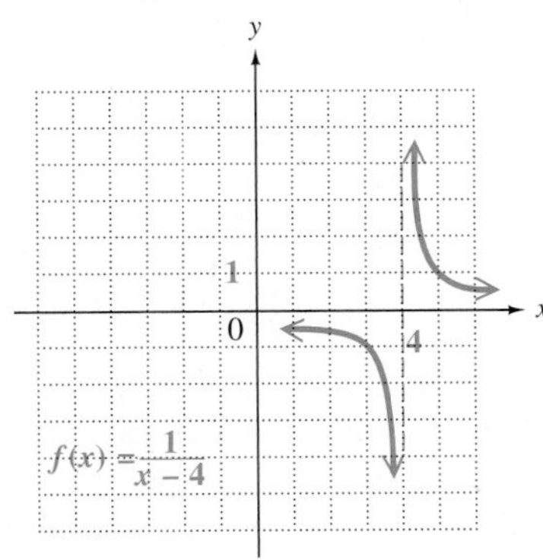

domain: $(-\infty, 4) \cup (4, \infty)$;
range: $(-\infty, 0) \cup (0, \infty)$

3. 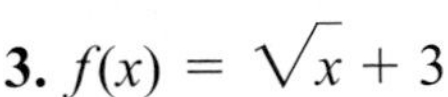 $f(x) = \sqrt{x} + 3$

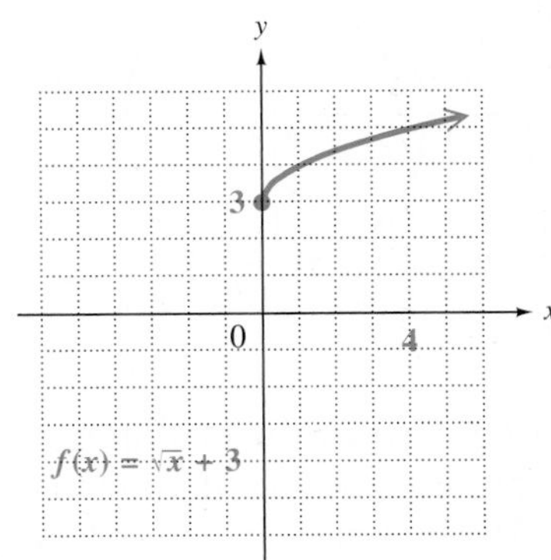

domain: $[0, \infty)$;
range: $[3, \infty)$

4. Find each of the following.

(a) $[\![12]\!]$ **(b)** $\left[\!\left[2\frac{1}{4}\right]\!\right]$ **(c)** $[\![-4.75]\!]$

12 2 −5

For $f(x) = 2x + 3$ *and* $g(x) = 5x^2 - 3x + 2$, *find each of the following.*

5. (a) $(f + g)(x)$ **(b)** $(f - g)(x)$ **6. (a)** $(f + g)(-1)$ **(b)** $(f - g)(-1)$

$5x^2 - x + 5$ $-5x^2 + 5x + 1$ 11 −9

For $f(x) = 12x^2 - 3x$ *and* $g(x) = 3x$, *find each of the following.*

7. (a) $(fg)(x)$ **(b)** $\left(\dfrac{f}{g}\right)(x)$ **8. (a)** $(fg)(-1)$ **(b)** $\left(\dfrac{f}{g}\right)(2)$

$36x^3 - 9x^2$ $4x - 1,\ x \neq 0$ −45 7

Let $f(x) = 3x^2 + 2x - 1$ *and* $g(x) = 5x + 7$. *Find each of the following.*

9. (a) $(f \circ g)(-2)$ **(b)** $(g \circ f)(-2)$ **10. (a)** $(f \circ g)(x)$ **(b)** $(g \circ f)(x)$

20 42 $75x^2 + 220x + 160$ $15x^2 + 10x + 2$

[12.2] *Write an equation for each circle.*

11. Center $(-2, 4)$, $r = 3$
$(x + 2)^2 + (y - 4)^2 = 9$

12. Center $(-1, -3)$, $r = 5$
$(x + 1)^2 + (y + 3)^2 = 25$

13. Center $(4, 2)$, $r = 6$
$(x - 4)^2 + (y - 2)^2 = 36$

Find the center and radius of each circle.

14. $x^2 + y^2 + 6x - 4y - 3 = 0$
$(-3, 2)$; $r = 4$

15. $x^2 + y^2 - 8x - 2y + 13 = 0$
$(4, 1)$; $r = 2$

16. $2x^2 + 2y^2 + 4x + 20y = -34$
$(-1, -5)$; $r = 3$

17. $4x^2 + 4y^2 - 24x + 16y = 48$
$(3, -2)$; $r = 5$

Graph each equation.

18. $x^2 + y^2 = 16$

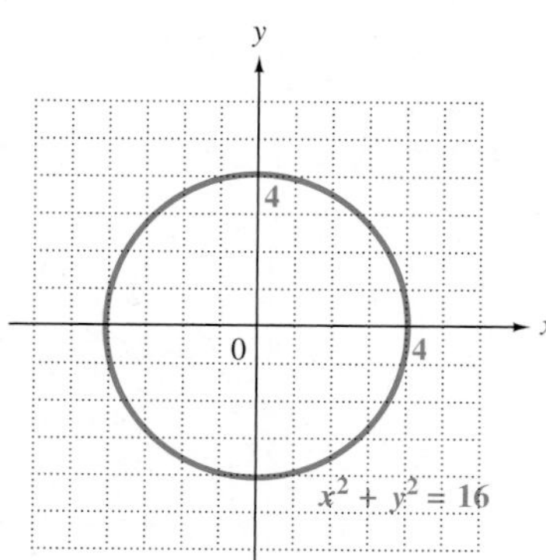

19. $\frac{x^2}{16} + \frac{y^2}{9} = 1$

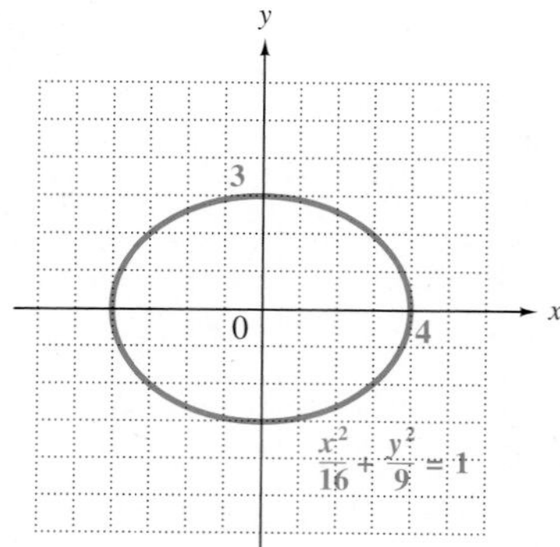

20. $\frac{x^2}{49} + \frac{y^2}{25} = 1$

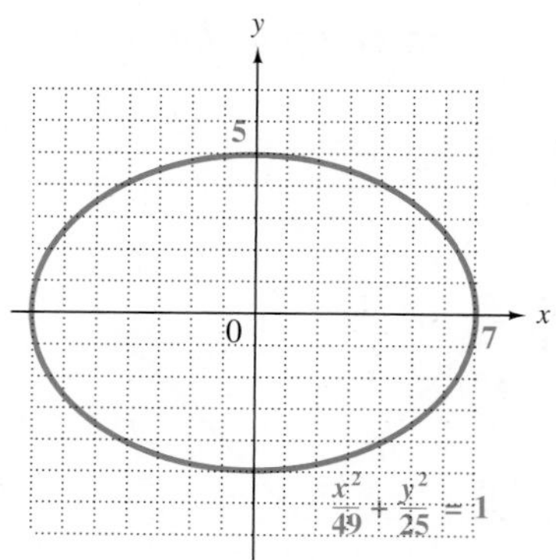

21. A satellite is in an elliptical orbit around Earth with perigee altitude of 160 km and apogee altitude of 16,000 km. See the figure. (*Source*: Kastner, Bernice, *Space Mathematics*, NASA, 1985.) Find the equation of the ellipse.

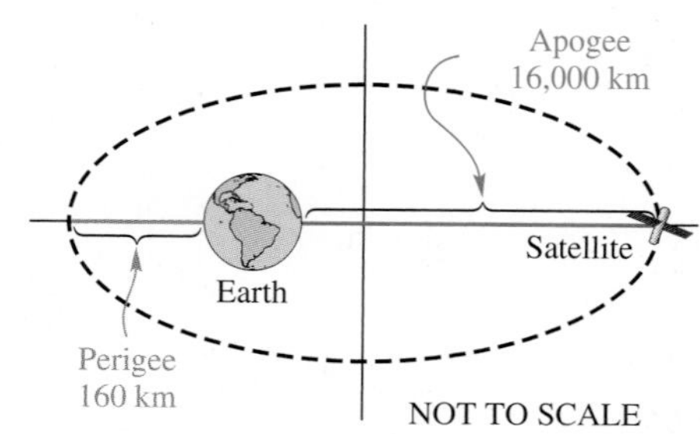

$\frac{x^2}{65{,}286{,}400} + \frac{y^2}{2{,}560{,}000} = 1$

22. (a) The Roman Colosseum is an ellipse with $a = 310$ ft and $b = \frac{513}{2}$ ft. Find the distance between the foci of this ellipse to the nearest tenth of a foot.

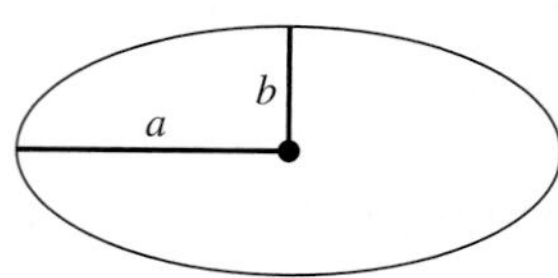

348.2 ft

(b) A formula for the approximate circumference of an ellipse is

$$C \approx 2\pi\sqrt{\frac{a^2 + b^2}{2}},$$

where a and b are the lengths given in part (a). Use this formula to find the approximate circumference of the Roman Colosseum.

1787.6 ft

[12.3] *Graph each equation.*

23. $\frac{x^2}{16} - \frac{y^2}{25} = 1$

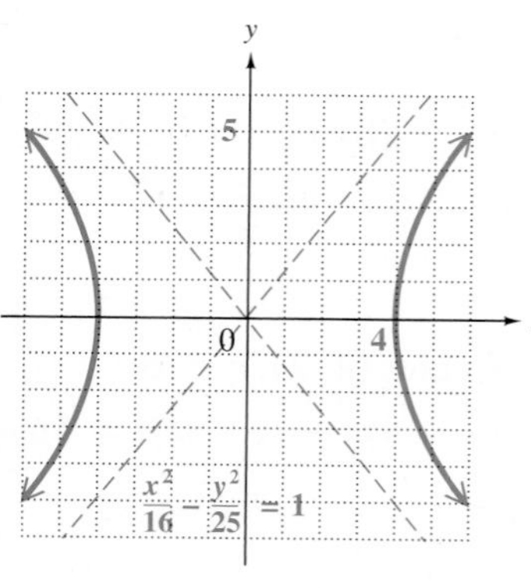

24. $\frac{y^2}{25} - \frac{x^2}{4} = 1$

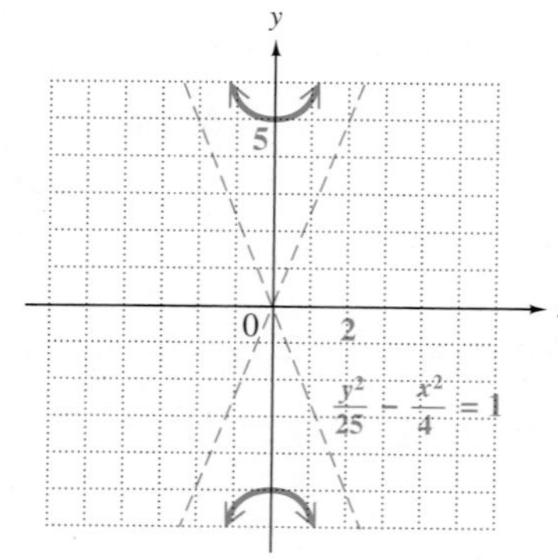

25. $f(x) = -\sqrt{16 - x^2}$

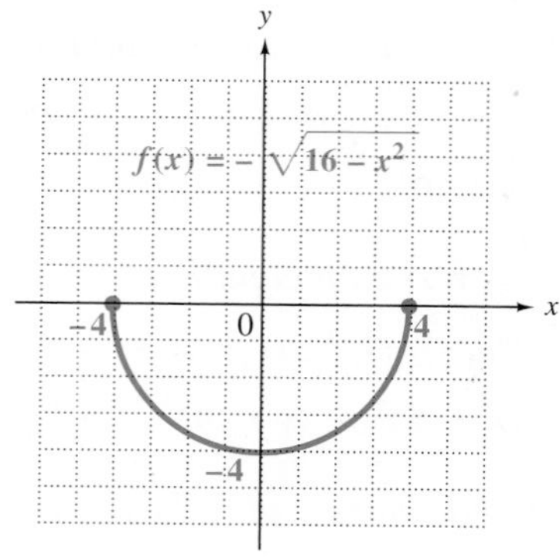

Identify the graph of each equation as a parabola, circle, ellipse, *or* hyperbola.

26. $x^2 + y^2 = 64$ **circle**

27. $y = 2x^2 - 3$ **parabola**

28. $y^2 = 2x^2 - 8$ **hyperbola**

29. $y^2 = 8 - 2x^2$ **ellipse**

30. $x = y^2 + 4$ **parabola**

31. $x^2 - y^2 = 64$ **hyperbola**

32. Ships and planes often use a location-finding system called LORAN. With this system, a radio transmitter at M sends out a series of pulses. (See the figure.) When each pulse is received at transmitter S, it then sends out a pulse. A ship at P receives pulses from both M and S. A receiver on the ship measures the difference in the arrival times of the pulses. A special map gives hyperbolas that correspond to the differences in arrival times (which give the distances d_1 and d_2 in the figure). The ship can then be located as lying on a branch of a particular hyperbola. Suppose $d_1 = 80$ mi and $d_2 = 30$ mi, and the distance between transmitters M and S is 100 mi. Use the definition to find an equation of the hyperbola on which the ship is located.

$\frac{x^2}{625} - \frac{y^2}{1875} = 1$

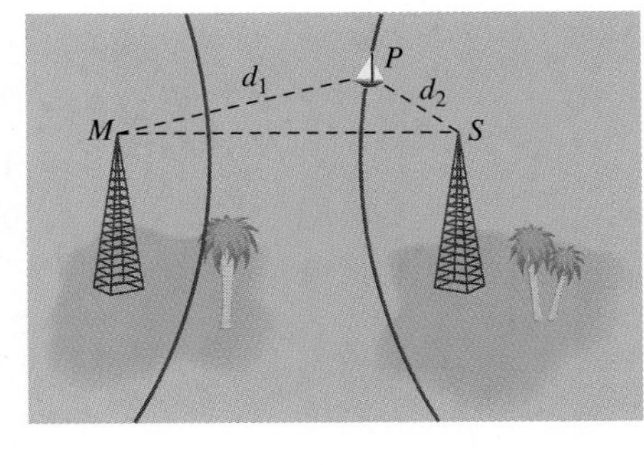

[12.4] *Solve each system.*

33. $2y = 3x - x^2$
$x + 2y = -12$
$\{(6, -9), (-2, -5)\}$

34. $y + 1 = x^2 + 2x$
$y + 2x = 4$
$\{(1, 2), (-5, 14)\}$

35. $x^2 + 3y^2 = 28$
$y - x = -2$
$\{(4, 2), (-1, -3)\}$

36. $xy = 8$
$x - 2y = 6$
$\{(-2, -4), (8, 1)\}$

37. $x^2 + y^2 = 6$
$x^2 - 2y^2 = -6$
$\{(-\sqrt{2}, 2), (-\sqrt{2}, -2), (\sqrt{2}, -2), (\sqrt{2}, 2)\}$

38. $3x^2 - 2y^2 = 12$
$x^2 + 4y^2 = 18$
$\{(-\sqrt{6}, -\sqrt{3}), (-\sqrt{6}, \sqrt{3}), (\sqrt{6}, -\sqrt{3}), (\sqrt{6}, \sqrt{3})\}$

39. How many solutions are possible for a system of two equations whose graphs are a circle and a line?

0, 1, or 2

40. How many solutions are possible for a system of two equations whose graphs are a parabola and a hyperbola?

0, 1, 2, 3, or 4

[12.5] *Graph each inequality.*

41. $9x^2 \geq 16y^2 + 144$

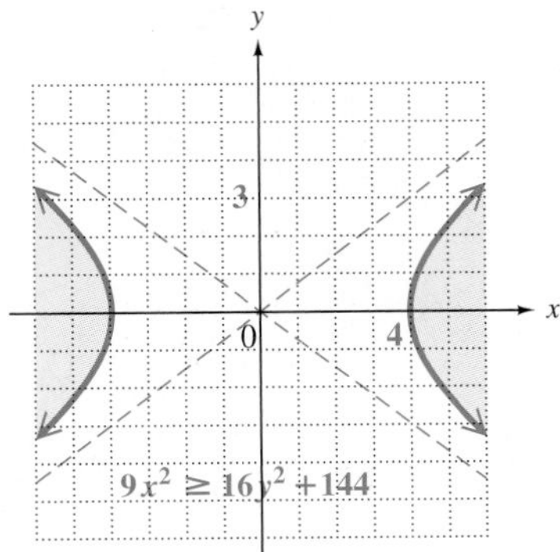

42. $4x^2 + y^2 \geq 16$

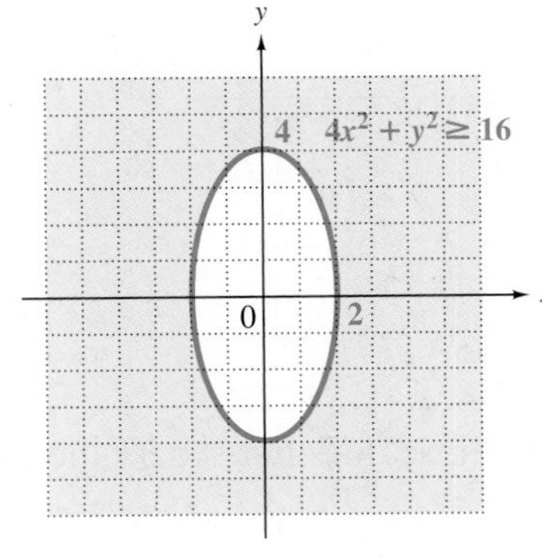

43. $y < -(x + 2)^2 + 1$

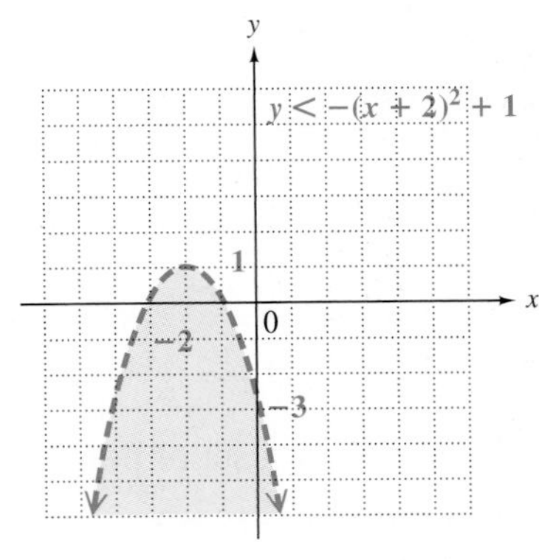

Graph each system of inequalities.

44. $2x + 5y \leq 10$
$3x - y \leq 6$

$2x + 5y \leq 10$
$3x - y \leq 6$

45. $|x| \leq 2$
$|y| > 1$
$4x^2 + 9y^2 \leq 36$

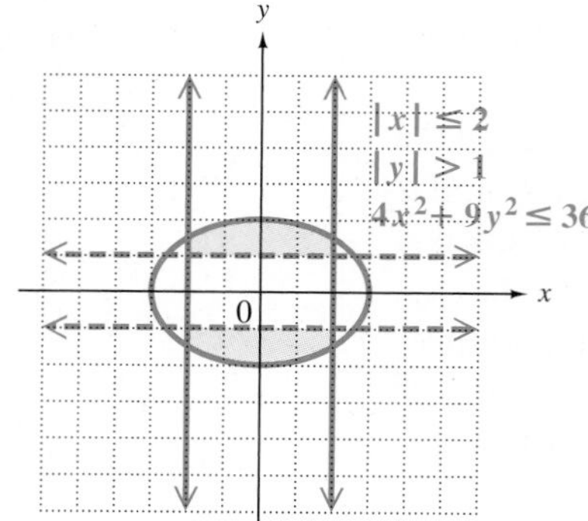

46. $9x^2 \leq 4y^2 + 36$
$x^2 + y^2 \leq 16$

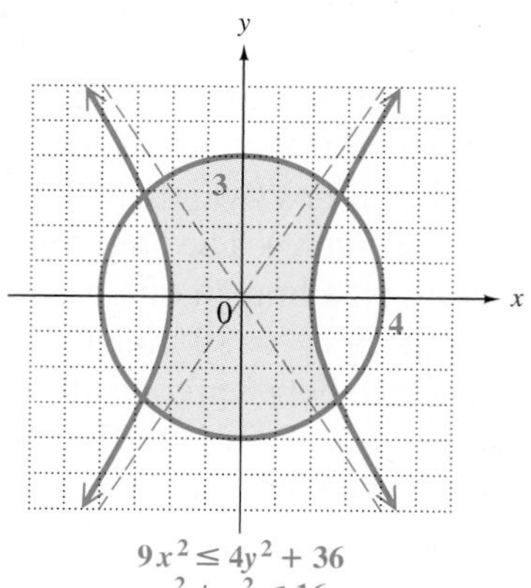

Mixed Review Exercises

Graph.

47. $x^2 + y^2 = 25$

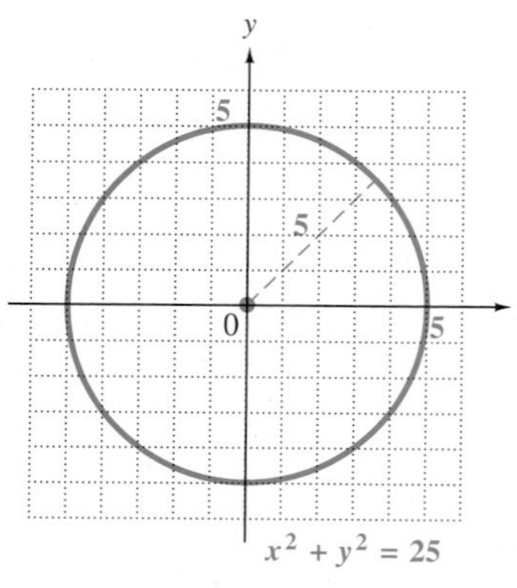

48. $x^2 + 9y^2 = 9$

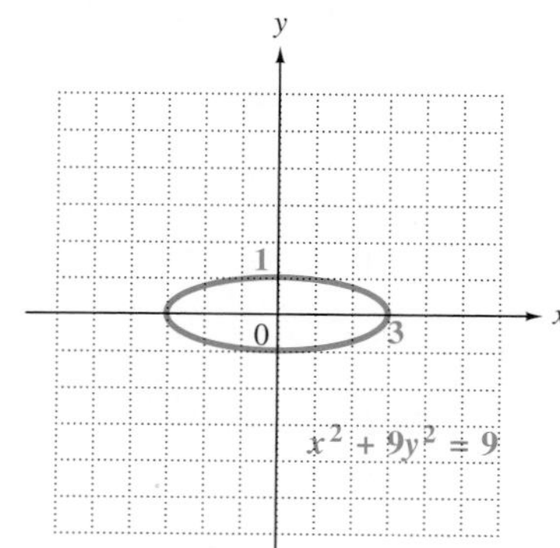

49. $x^2 - 9y^2 = 9$

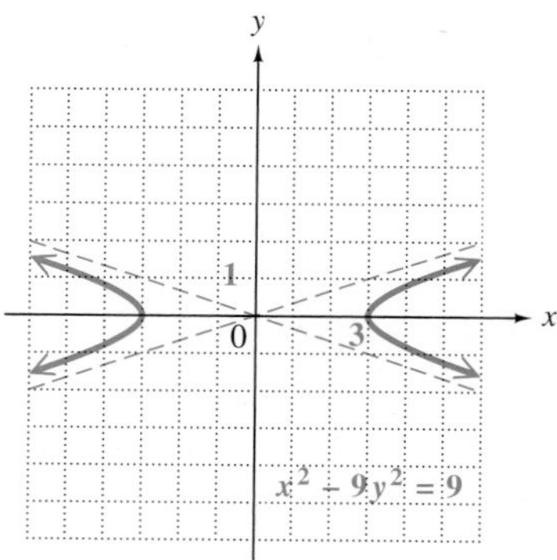

50. $f(x) = \sqrt{4 - x}$

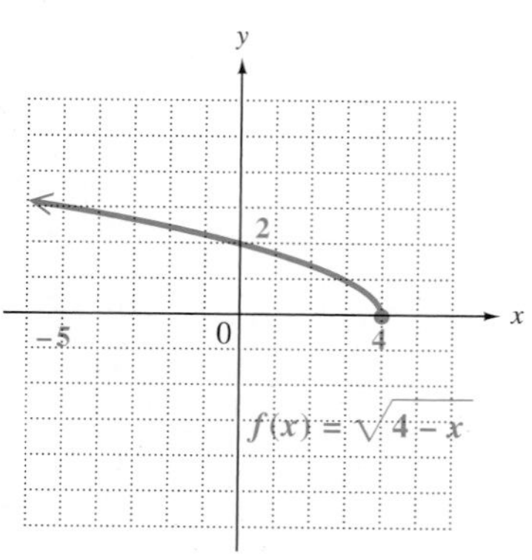

51. $f(x) = [\![x]\!] - 1$

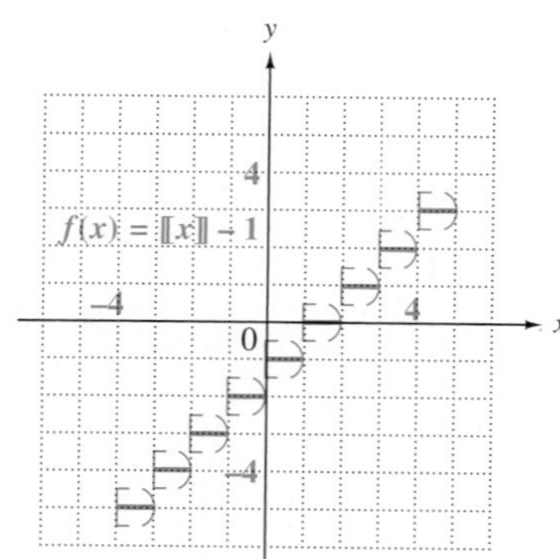

52. $4y > 3x - 12$
$x^2 < 16 - y^2$

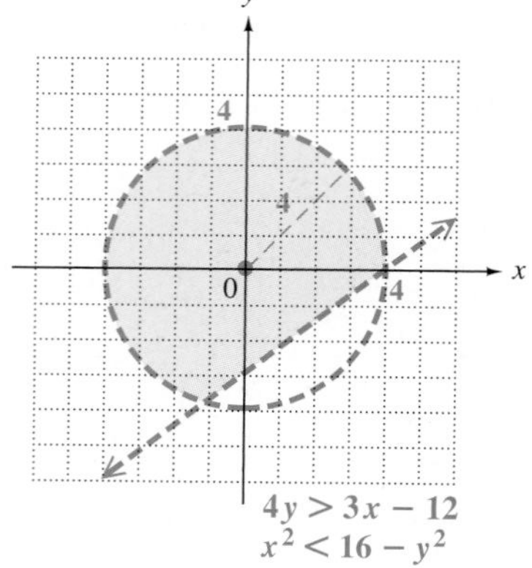

53. Explain why a set of points that form an ellipse does not satisfy the definition of a function.

There are cases where one *x*-value will yield two *y*-values. In a function, every *x* yields one and only one *y*.

Chapter 12 Test

Use the Chapter Test Prep Video CD to see fully worked-out solutions to any of the exercises you want to review.

Match each function with its graph from choices A, B, C, and D.

1. $f(x) = \sqrt{x-2}$
2. $f(x) = \sqrt{x+2}$
3. $f(x) = \sqrt{x} + 2$
4. $f(x) = \sqrt{x} - 2$

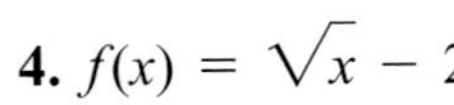

A. 

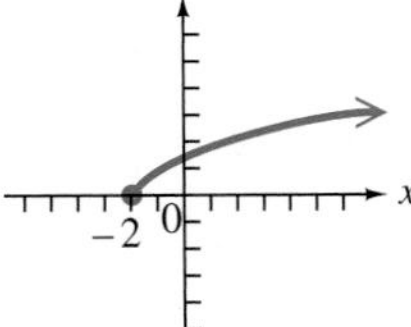

B. 

C.

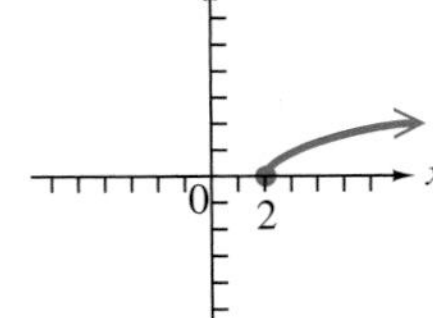

D. 

5. Sketch the graph of $f(x) = |x - 3| + 4$. Give the domain and range.

6. For $f(x) = 3x + 5$ and $g(x) = x^2 + 2$, find each of the following.
   (a) $(f \circ g)(-2)$
   (b) $(f \circ g)(x)$
   (c) $(g \circ f)(x)$

7. Find the center and radius of the circle whose equation is $(x-2)^2 + (y+3)^2 = 16$. Sketch the graph.

Answers

1. C
2. A

3. D
4. B
5. domain: $(-\infty, \infty)$; range: $[4, \infty)$

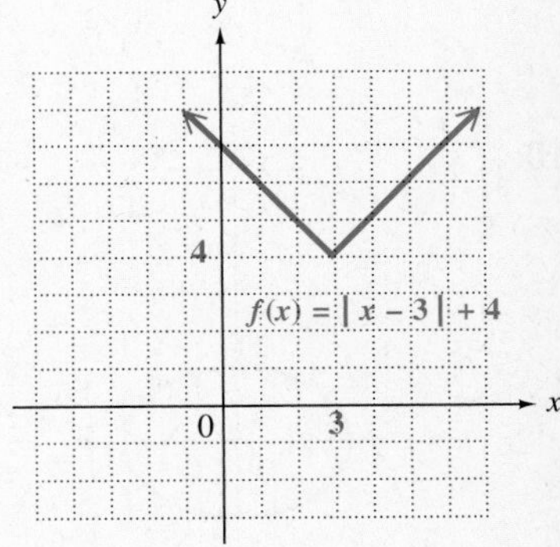

6. (a) 23
   (b) $3x^2 + 11$
   (c) $9x^2 + 30x + 27$
7. center: $(2, -3)$; radius: 4

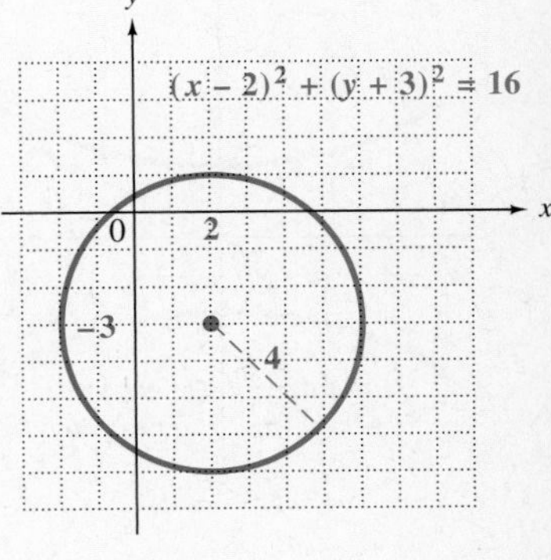

8. center: $(-4, 1)$; radius: 5

8. Find the center and radius of the circle whose equation is $x^2 + y^2 + 8x - 2y = 8$.

Graph.

9.

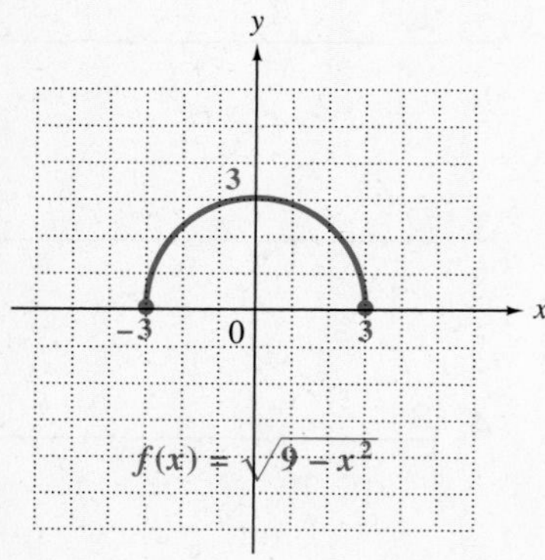

9. $f(x) = \sqrt{9 - x^2}$

10.

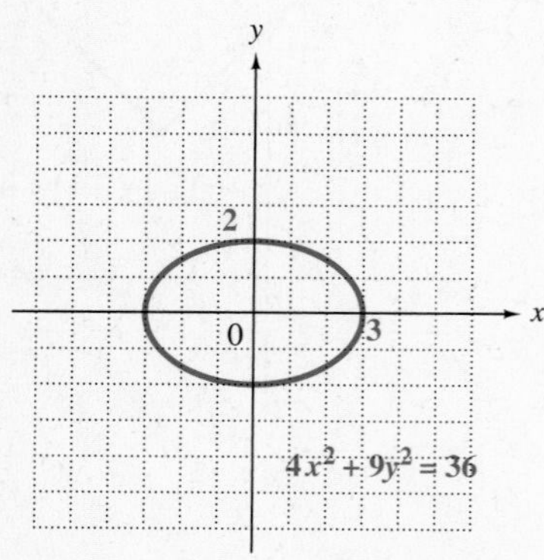

10. $4x^2 + 9y^2 = 36$

11.

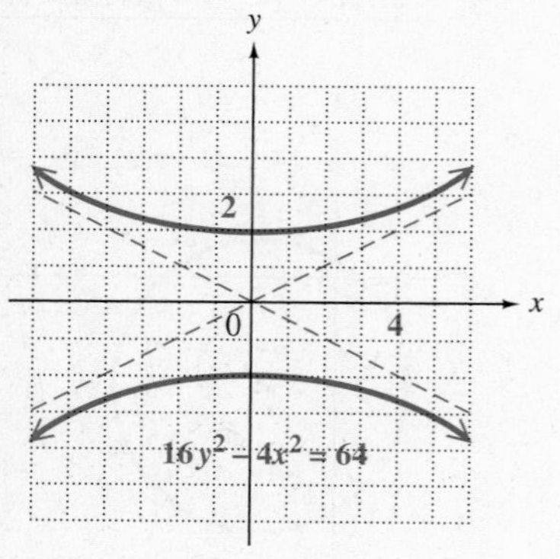

11. $16y^2 - 4x^2 = 64$

12. $\frac{y}{2} = -\sqrt{1 - \frac{x^2}{9}}$

12.

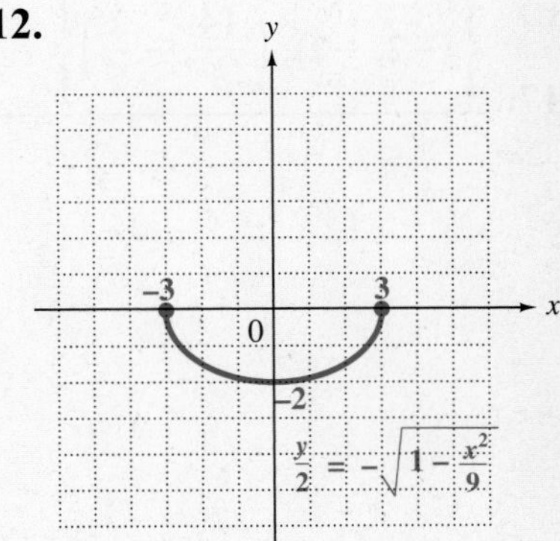

Identify the graph of each equation as a parabola, hyperbola, ellipse, *or* circle.

13. $6x^2 + 4y^2 = 12$

13. **ellipse**

14. $16x^2 = 144 + 9y^2$

14. **hyperbola**

15. $4y^2 + 4x = 9$

15. **parabola**

Solve each nonlinear system.

16. $2x - y = 9$
$xy = 5$

16. $\left\{\left(-\frac{1}{2}, -10\right), (5, 1)\right\}$

17. $\left\{(-2, -2), \left(\frac{14}{5}, -\frac{2}{5}\right)\right\}$

17. $x - 4 = 3y$
$x^2 + y^2 = 8$

18. $\{(-\sqrt{22}, -\sqrt{3}), (-\sqrt{22}, \sqrt{3}), (\sqrt{22}, -\sqrt{3}), (\sqrt{22}, \sqrt{3})\}$

18. $x^2 + y^2 = 25$
$x^2 - 2y^2 = 16$

19.

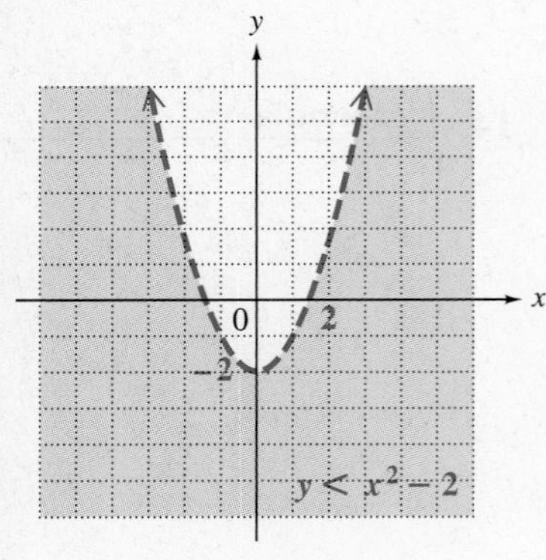

19. Graph the inequality $y < x^2 - 2$.

20.

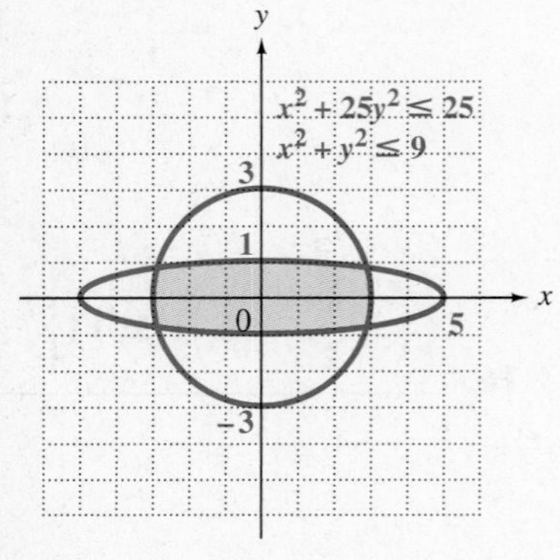

20. Graph the system $\begin{array}{l} x^2 + 25y^2 \leq 25 \\ x^2 + \quad y^2 \leq 9 \end{array}$.

Cumulative Review Exercises Chapters R–12

Solve.

1. $4 - (2x + 3) + x = 5x - 3$

$\left\{\frac{2}{3}\right\}$

2. $-4k + 7 \geq 6k + 1$

$\left(-\infty, \frac{3}{5}\right]$

3. $|5m| - 6 = 14$

$\{-4, 4\}$

4. Find the slope of the line through (2, 5) and (−4, 1).

$\frac{2}{3}$

5. Find the equation of the line through (−3, −2) and perpendicular to the graph of $2x - 3y = 7$.

$3x + 2y = -13$

Solve each system.

6. $3x - y = 12$
$2x + 3y = -3$

$\{(3, -3)\}$

7. $x + y - 2z = 9$
$2x + y + z = 7$
$3x - y - z = 13$

$\{(4, 1, -2)\}$

8. $xy = -5$
$2x + y = 3$

$\left\{(-1, 5), \left(\frac{5}{2}, -2\right)\right\}$

Perform the indicated operations.

9. $(5y - 3)^2$

$25y^2 - 30y + 9$

10. $(2r + 7)(6r - 1)$

$12r^2 + 40r - 7$

11. $\dfrac{8x^4 - 4x^3 + 2x^2 + 13x + 8}{2x + 1}$

$4x^3 - 4x^2 + 3x + 5 + \dfrac{3}{2x + 1}$

Factor.

12. $12x^2 - 7x - 10$

$(3x + 2)(4x - 5)$

13. $z^4 - 1$

$(z^2 + 1)(z + 1)(z - 1)$

14. $a^3 - 27b^3$

$(a - 3b)(a^2 + 3ab + 9b^2)$

Perform each operation.

15. $\dfrac{y^2 - 4}{y^2 - y - 6} \div \dfrac{y^2 - 2y}{y - 1}$

$\dfrac{y - 1}{y(y - 3)}$

16. $\dfrac{5}{c + 5} - \dfrac{2}{c + 3}$

$\dfrac{3c + 5}{(c + 5)(c + 3)}$

17. $\dfrac{p}{p^2 + p} + \dfrac{1}{p^2 + p}$

$\dfrac{1}{p}$

Solve.

18. Kareem and Jamal want to clean their office. Kareem can do the job alone in 3 hr, while Jamal can do it alone in 2 hr. How long will it take them if they work together?

$1\frac{1}{5}$ hr

Simplify. Assume that all variables represent positive real numbers.

19. $\dfrac{(2a)^{-2}a^4}{a^{-3}}$

$\dfrac{a^5}{4}$

20. $4\sqrt[3]{16} - 2\sqrt[3]{54}$

$2\sqrt[3]{2}$

21. $\dfrac{3\sqrt{5x}}{\sqrt{2x}}$

$\dfrac{3\sqrt{10}}{2}$

22. $\dfrac{5 + 3i}{2 - i}$

$\dfrac{7}{5} + \dfrac{11}{5}i$

Solve.

23. $2\sqrt{k} = \sqrt{5k + 3}$

$\emptyset$

24. $10q^2 + 13q = 3$

$\left\{\frac{1}{5}, -\frac{3}{2}\right\}$

25. $3k^2 - 3k - 2 = 0$

$\left\{\frac{3 + \sqrt{33}}{6}, \frac{3 - \sqrt{33}}{6}\right\}$

26. $2(x^2 - 3)^2 - 5(x^2 - 3) = 12$

$\left\{-\frac{\sqrt{6}}{2}, \frac{\sqrt{6}}{2}, -\sqrt{7}, \sqrt{7}\right\}$

27. $\log(x + 2) + \log(x - 1) = 1$

$\{3\}$

28. $F = \frac{kwv^2}{r}$ for v

$v = \frac{\pm\sqrt{rFkw}}{kw}$

29. If $f(x) = x^3 + 4$, find $f^{-1}(x)$.

$f^{-1}(x) = \sqrt[3]{x - 4}$

30. Evaluate.

(a) $3^{\log_3 4}$ 4

(b) $e^{\ln 7}$ 7

31. Use properties of logarithms to write the following as a single logarithm.

$2\log(3x + 7) - \log 4$

$\log\frac{(3x + 7)^2}{4}$

32. If $f(x) = x^2 + 2x - 4$ and $g(x) = 3x + 2$, find **(a)** $(g \circ f)(1)$ **(b)** $(f \circ g)(x)$.

(a) -1 (b) $9x^2 + 18x + 4$

Graph.

33. $f(x) = -3x + 5$

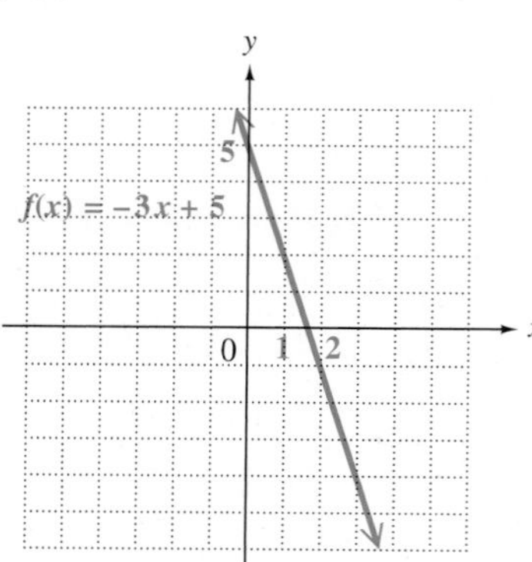

34. $f(x) = -2(x - 1)^2 + 3$

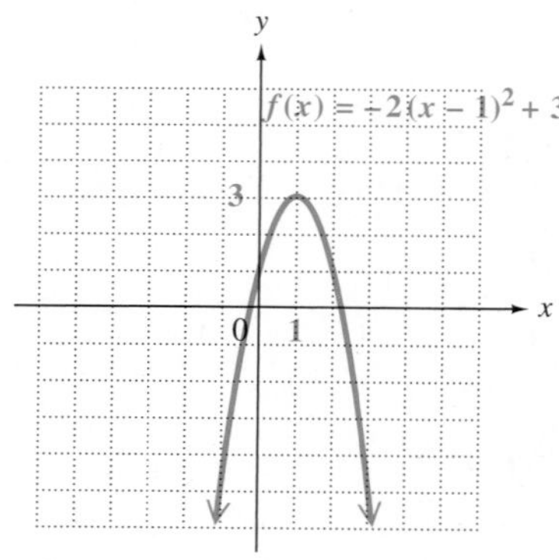

35. $\frac{x^2}{25} + \frac{y^2}{16} \leq 1$

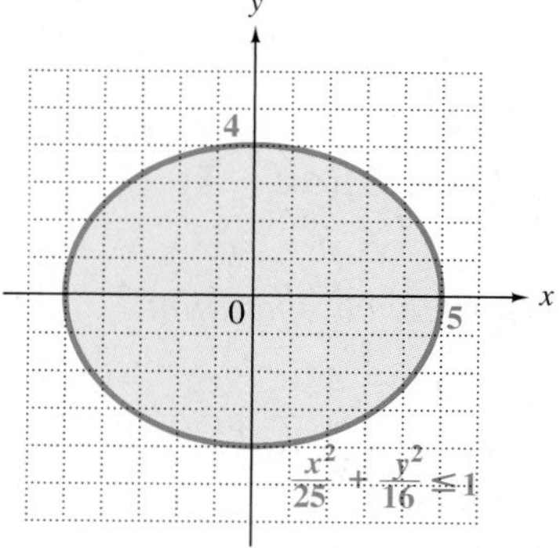

36. $f(x) = \sqrt{x - 2}$

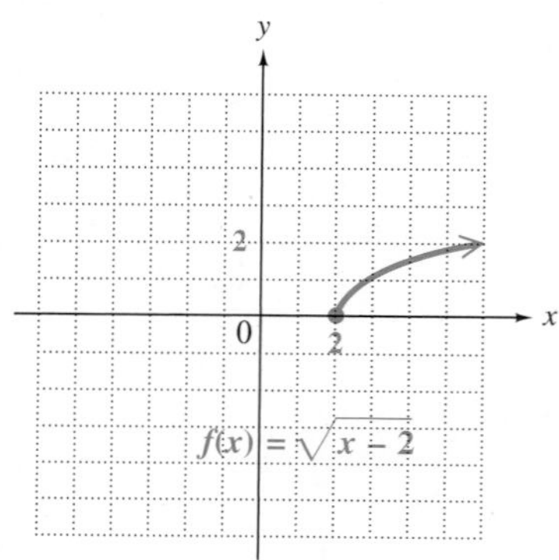

37. $\frac{x^2}{4} - \frac{y^2}{16} = 1$

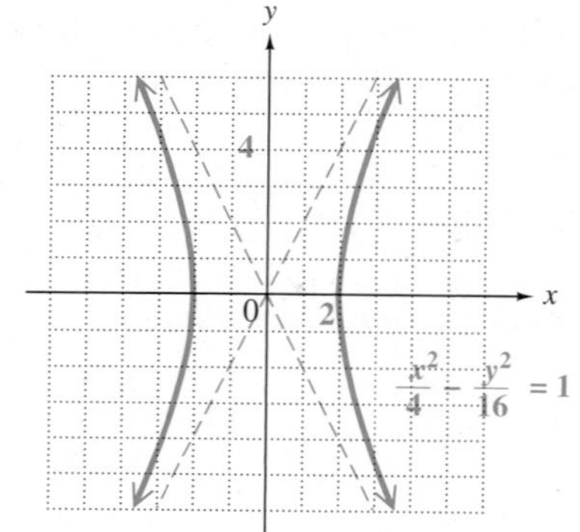

38. $f(x) = 3^x$

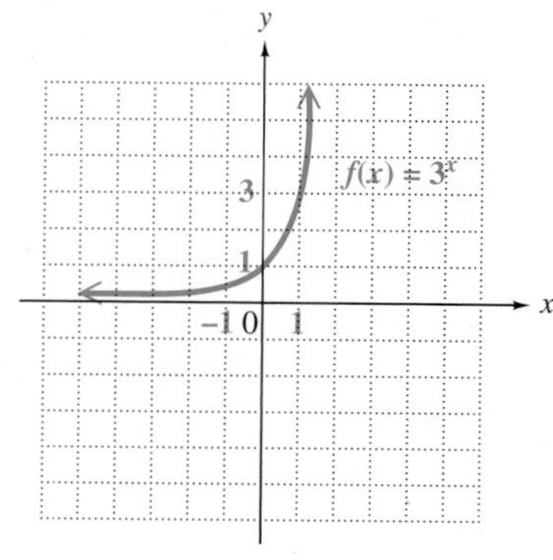

Appendix: Review of Exponents, Polynomials, and Factoring

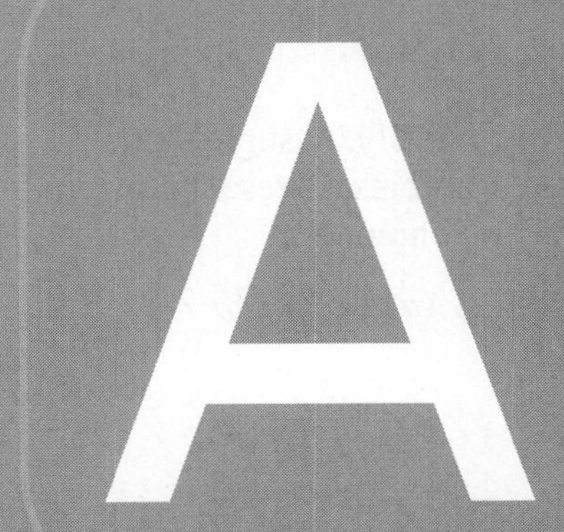

(TRANSITION FROM INTRODUCTORY TO INTERMEDIATE ALGEBRA)

OBJECTIVES

1. Review the basic rules for exponents.
2. Review addition, subtraction, and multiplication of polynomials.
3. Review factoring techniques.

OBJECTIVE 1 Review the basic rules for exponents.

Definitions and Rules for Exponents (Sections 5.2 and 5.5)

If no denominators are 0, for any integers m and n:

			Examples
Product rule		$a^m \cdot a^n = a^{m+n}$	$7^4 \cdot 7^3 = 7^7$
Zero exponent		$a^0 = 1 \quad (a \neq 0)$	$(-3)^0 = 1$
Negative exponent		$a^{-n} = \dfrac{1}{a^n}$	$5^{-3} = \dfrac{1}{5^3}$
Quotient rule		$\dfrac{a^m}{a^n} = a^{m-n}$	$\dfrac{2^2}{2^5} = 2^{-3} = \dfrac{1}{2^3}$
Power rules	(a)	$(a^m)^n = a^{mn}$	$(4^2)^3 = 4^6$
	(b)	$(ab)^m = a^m b^n$	$(3k)^4 = 3^4k^4$
	(c)	$\left(\dfrac{a}{b}\right)^m = \dfrac{a^m}{b^m}$	$\left(\dfrac{2}{3}\right)^{10} = \dfrac{2^{10}}{3^{10}}$
Negative-to-positive rules		$\dfrac{a^{-m}}{b^{-n}} = \dfrac{b^n}{a^m}$	$\dfrac{5^{-3}}{3^{-5}} = \dfrac{3^5}{5^3}$
		$\left(\dfrac{a}{b}\right)^{-m} = \left(\dfrac{b}{a}\right)^m.$	$\left(\dfrac{4}{7}\right)^{-2} = \left(\dfrac{7}{4}\right)^2$

EXAMPLE 1 Applying Definitions and Rules for Exponents

Simplify. Assume that all variables represent nonzero real numbers.

(a) $(x^2y^{-3})(x^{-5}y^7)$

$= (x^{2+(-5)})(y^{-3+7})$ Product rule

$= x^{-3}y^4$

$= \dfrac{1}{x^3}y^4$, or $\dfrac{y^4}{x^3}$

(b) $(-5)^0 + (-5^0)$

$= 1 + (-1)$

$= 0$

$-5^0 = -1 \cdot 5^0$
$= -1 \cdot 1$
$= -1$

Continued on Next Page

1 Simplify. Assume that all variables represent nonzero real numbers.

(a) $(a^{-4}bc^2)(a^2b^{-2}c^4)$

(b) $4^0 + (-4)^0$

(c) $\frac{(x^3y^{-2})^3}{(x^4y^{-3})^2}$

(d) $\left(\frac{2x^2y^{-2}}{x^{-4}y}\right)^{-3}$

2 Add or subtract as indicated.

(a) $(5x^3 - 3x^2 + x + 4) + (2x^3 - x^2 - 3x - 1)$

(b) $(7y^2 - 11y + 8) - (-3y^2 + 4y + 6)$

Answers

1. (a) $\frac{c^6}{a^2b}$ (b) 2 (c) x (d) $\frac{y^9}{8x^{18}}$
2. (a) $7x^3 - 4x^2 - 2x + 3$ (b) $10y^2 - 15y + 2$

(c) $\frac{(t^5s^{-4})^2}{(t^{-3}s^5)^3}$

$= \frac{t^{10}s^{-8}}{t^{-9}s^{15}}$ Power rules (a) and (b)

$= \frac{t^{10}t^9}{s^{15}s^8}$ Definition of negative exponent

$= \frac{t^{19}}{s^{23}}$ Product rule

(d) $\left(\frac{-3x^{-4}y}{x^5y^{-4}}\right)^{-2}$

$= \left(\frac{x^5y^{-4}}{-3x^{-4}y}\right)^2$ Negative-to-positive rule

$= \frac{x^{10}y^{-8}}{9x^{-8}y^2}$ Power rules (a), (b), and (c)

$= \frac{x^{18}}{9y^{10}}$ Quotient rule

(e) $(2x^2y^3z)^2(x^4y^2)^3$

$= (4x^4y^6z^2)(x^{12}y^6)$ Power rules (a) and (b)

$= 4x^{16}y^{12}z^2$ Product rule

◀ Work Problem 1 at the Side.

OBJECTIVE 2 Review addition, subtraction, and multiplication of polynomials.

Adding and Subtracting Polynomials (Section 5.1)

To add polynomials, add like terms. To subtract polynomials, change all signs on the second polynomial and add the result to the first polynomial.

EXAMPLE 2 Adding and Subtracting Polynomials

Add or subtract as indicated.

(a) $(-4x^3 + 3x^2 - 8x + 2) + (5x^3 - 8x^2 + 12x - 3)$

$= (-4 + 5)x^3 + (3 - 8)x^2 + (-8 + 12)x + (2 - 3)$

$= x^3 - 5x^2 + 4x - 1$

(b) $-4(x^2 + 3x - 6) - (2x^2 - 3x + 7)$

$= -4x^2 - 12x + 24 - 2x^2 + 3x - 7$

$= -6x^2 - 9x + 17$

(c) Subtract.

$$\begin{array}{r} 2t^2 - 3t - 4 \\ \underline{-8t^2 + 4t - 1} \end{array} \longrightarrow \begin{array}{rl} 2t^2 - 3t - 4 & \\ \underline{8t^2 - 4t + 1} & \text{Change signs.} \\ 10t^2 - 7t - 3 & \text{Add.} \end{array}$$

◀ Work Problem 2 at the Side.

Multiplying Polynomials (Section 5.3)

To multiply two polynomials, multiply each term of the second polynomial by each term of the first polynomial and add the products. In particular, when multiplying two binomials, use the FOIL method.

There are also several special product rules that are useful when multiplying binomials.

Special Product Rules (Section 5.4)

$$(a + b)^2 = a^2 + 2ab + b^2$$
$$(a - b)^2 = a^2 - 2ab + b^2$$
$$(a + b)(a - b) = a^2 - b^2$$

EXAMPLE 3 Multiplying Polynomials

Find each product.

(a) $(4y - 1)(3y + 2)$

$= 4y(3y) + 4y(2) - 1(3y) - 1(2)$ FOIL

$= 12y^2 + 8y - 3y - 2$ Multiply.

$= 12y^2 + 5y - 2$ Combine like terms.

(b) $(3x + 5y)(3x - 5y)$

$= (3x)^2 - (5y)^2$ $(a + b)(a - b) = a^2 - b^2$

$= 9x^2 - 25y^2$

(c) $(2t + 3)^2$

$= (2t)^2 + 2(2t)(3) + 3^2$
$(a + b)^2 = a^2 + 2ab + b^2$

$= 4t^2 + 12t + 9$

(d) $(5x - 1)^2$

$= (5x)^2 - 2(5x)(1) + 1^2$
$(a - b)^2 = a^2 - 2ab + b^2$

$= 25x^2 - 10x + 1$

(e) $(3x + 2)(9x^2 - 6x + 4)$

$$\begin{array}{rrrrl} & 9x^2 & -\,6x & +\,4 & \\ & & 3x & +\,2 & \text{Multiply vertically.} \\ \hline & 18x^2 & -\,12x & +\,8 & \leftarrow 2(9x^2 - 6x + 4) \\ 27x^3 & -\,18x^2 & +\,12x & & \leftarrow 3x(9x^2 - 6x + 4) \\ \hline 27x^3 & & & +\,8 & \text{Add like terms.} \end{array}$$

The product is a sum of cubes, $27x^3 + 8$.

Work Problem **3** at the Side. ▶

OBJECTIVE 3 Review factoring techniques. Factoring, which involves writing a polynomial as a product, was covered in **Chapter 6.**

Guidelines for Factoring a Polynomial

1. **Is there a common factor?** If so, factor it out.
2. **How many terms are in the polynomial?**

   Two terms: Check to see whether it is a difference of squares or a sum or difference of cubes. If so, factor as in **Section 6.5.**

   Three terms: Is it a perfect square trinomial? If the trinomial is not a perfect square, check to see whether the coefficient of the second-degree term is 1. If so, use the method of **Section 6.2.** If the coefficient of the squared term of the trinomial is not 1, use the general factoring methods of **Sections 6.3 and 6.4.**

   Four terms: Try to factor the polynomial by grouping using the methods of **Sections 6.1 and 6.3.**
3. **Can any factors be factored further?** If so, factor them.

3 Find each product.

(a) $(2x + 5)(3x - 2)$

(b) $(2t + 7y)(2t - 7y)$

(c) $(6r - 5)^2$

(d) $(2x - 3)(4x^2 + 6x + 9)$

ANSWERS

3. (a) $6x^2 + 11x - 10$ **(b)** $4t^2 - 49y^2$ **(c)** $36r^2 - 60r + 25$ **(d)** $8x^3 - 27$

4 Factor each polynomial completely.

(a) $12x^2 - 5x - 2$

(b) $z^2 - 12z + 36$

(c) $8p^3 + 125$

(d) $xy + 6y + xz + 6z$

EXAMPLE 4 Factoring Polynomials

Factor each polynomial completely.

(a) $6x^2y^3 - 12x^3y^2$

$= 6x^2y^2(y - 2x)$ $\quad$ $6x^2y^2$ is the greatest common factor.

(b) $3x^2 - x - 2$

To find the factors, find two terms that multiply to give $3x^2$ ($3x$ and x) and two terms that multiply to give -2 ($+2$ and -1). Make sure that the sum of the outer and inner products in the factored form is $-x$.

$$3x^2 - x - 2 \quad \text{factors as} \quad (3x + 2)(x - 1).$$

Check Multiply the factors using the FOIL method.

(c) $100t^2 - 81$

$= (10t)^2 - 9^2$ $\quad$ Difference of squares

$= (10t + 9)(10t - 9)$ $\quad$ $a^2 - b^2 = (a + b)(a - b)$

(d) $4x^2 + 20xy + 25y^2$

The first and last terms are both perfect squares.

$$4x^2 = (2x)^2 \quad \text{and} \quad 25y^2 = (5y)^2$$

To factor as a perfect square trinomial, take twice the product of the two terms in the binomial $2x + 5y$.

$$2(2x)(5y) = 20xy$$

Twice — First term — Last term

Since $20xy$ is the middle term of the trinomial, the trinomial is a perfect square.

$$4x^2 + 20xy + 25y^2 \quad \text{factors as} \quad (2x + 5y)^2.$$

(e) $1000x^3 - 27$

$= (10x)^3 - 3^3$ $\quad$ Difference of cubes

$= (10x - 3)[(10x)^2 + 10x(3) + 3^2]$ $\quad$ $a^3 - b^3 = (a - b)(a^2 + ab + b^2)$

$= (10x - 3)(100x^2 + 30x + 9)$

(f) $6xy - 3x + 4y - 2$

Since there are four terms, try factoring by grouping.

$6xy - 3x + 4y - 2$

$= (6xy - 3x) + (4y - 2)$ $\quad$ Group the terms.

$= 3x(2y - 1) + 2(2y - 1)$ $\quad$ Factor each group.

$= (2y - 1)(3x + 2)$ $\quad$ Factor out $2y - 1$.

◀ Work Problem **4** at the Side.

ANSWERS

4. **(a)** $(4x + 1)(3x - 2)$ **(b)** $(z - 6)^2$
(c) $(2p + 5)(4p^2 - 10p + 25)$
(d) $(x + 6)(y + z)$

Appendix A Exercises

FOR EXTRA HELP MyMathLab Math XL PRACTICE  WATCH DOWNLOAD READ REVIEW

Apply the definitions and rules for exponents to simplify each expression. Write the final answers using only positive exponents. Assume that all variables represent positive real numbers. See Example 1.

1. $(a^4b^{-3})(a^{-6}b^2)$
$\frac{1}{a^2b}$

2. $(t^{-3}s^{-5})(t^8s^{-2})$
$\frac{t^5}{s^7}$

3. $(5x^{-2}y)^2(2xy^4)^2$
$\frac{100y^{10}}{x^2}$

4. $(7x^{-3}y^4)^3(2x^{-1}y^{-4})^2$
$\frac{1372y^4}{x^{11}}$

5. $-6^0 + (-6)^0$
0

6. $(-12)^0 - 12^0$
0

7. $\frac{(2w^{-1}x^2y^{-1})^3}{(4w^5x^{-2}y)^2}$
$\frac{x^{10}}{2w^{13}y^5}$

8. $\frac{(5p^{-3}q^2r^{-4})^2}{(10p^4q^{-1}r^5)^{-1}}$
$\frac{250q^3}{p^2r^3}$

9. $\left(\frac{-4a^{-2}b^4}{a^3b^{-1}}\right)^{-3}$
$\frac{a^{15}}{-64b^{15}}$

10. $\left(\frac{r^{-3}s^{-8}}{-6r^2s^{-4}}\right)^{-2}$
$36r^{10}s^8$

11. $(7x^{-4}y^2z^{-2})^{-2}(7x^4y^{-1}z^3)^2$
$\frac{x^{16}z^{10}}{y^6}$

12. $(3m^{-5}n^2p^{-4})^3(3m^4n^{-3}p^5)^{-2}$
$\frac{3n^{12}}{m^{23}p^{22}}$

Add or subtract as indicated. See Example 2.

13. $(2a^4 + 3a^3 - 6a^2 + 5a - 12) + (-8a^4 + 8a^3 - 14a^2 + 21a - 3)$
$-6a^4 + 11a^3 - 20a^2 + 26a - 15$

14. $(-6r^4 - 3r^3 + 12r^2 - 9r + 9) + (8r^4 - 13r^3 - 14r^2 - 10r - 3)$
$2r^4 - 16r^3 - 2r^2 - 19r + 6$

15. $(6x^3 - 12x^2 + 3x - 4) - (-2x^3 + 6x^2 - 3x + 12)$

$8x^3 - 18x^2 + 6x - 16$

16. $(10y^3 - 4y^2 + 8y + 7) - (7y^3 + 5y^2 - 2y - 13)$

$3y^3 - 9y^2 + 10y + 20$

17. Add.

$$\begin{array}{r} 5x^2y + 2xy^2 + y^3 \\ \underline{-4x^2y - 3xy^2 + 5y^3} \\ x^2y - xy^2 + 6y^3 \end{array}$$

18. Add.

$$\begin{array}{r} 6ab^3 - 2a^2b^2 + 3b^5 \\ \underline{8ab^3 + 12a^2b^2 - 8b^5} \\ 14ab^3 + 10a^2b^2 - 5b^5 \end{array}$$

19. $3(5x^2 - 12x + 4) - 2(9x^2 + 13x - 10)$

$-3x^2 - 62x + 32$

20. $-4(2t^3 - 3t^2 + 4t - 1) - 3(-8t^3 + 3t^2 - 2t + 9)$

$16t^3 + 3t^2 - 10t - 23$

21. Subtract.

$$\begin{array}{r} 6x^3 - 2x^2 + 3x - 1 \\ \underline{-4x^3 + 2x^2 - 6x + 3} \\ 10x^3 - 4x^2 + 9x - 4 \end{array}$$

22. Subtract.

$$\begin{array}{r} -9y^3 - 2y^2 + 3y - 8 \\ \underline{-8y^3 + 4y^2 + 3y + 1} \\ -y^3 - 6y^2 - 9 \end{array}$$

Find each product. See Example 3.

23. $(3x + 1)(2x - 7)$

$6x^2 - 19x - 7$

24. $(5z + 3)(2z - 3)$

$10z^2 - 9z - 9$

25. $(4x - 1)(x - 2)$

$4x^2 - 9x + 2$

26. $(7t - 3)(t - 4)$
$7t^2 - 31t + 12$

27. $(4t + 3)(4t - 3)$
$16t^2 - 9$

28. $(6x + 1)(6x - 1)$
$36x^2 - 1$

29. $(2y^2 + 4)(2y^2 - 4)$
$4y^4 - 16$

30. $(3b^3 + 2t)(3b^3 - 2t)$
$9b^6 - 4t^2$

31. $(4x - 3)^2$
$16x^2 - 24x + 9$

32. $(9t + 2)^2$
$81t^2 + 36t + 4$

33. $(6r + 5y)^2$
$36r^2 + 60ry + 25y^2$

34. $(8m - 3n)^2$
$64m^2 - 48mn + 9n^2$

35. $(c + 2d)(c^2 - 2cd + 4d^2)$
$c^3 + 8d^3$

36. $(f + 3g)(f^2 - 3fg + 9g^2)$
$f^3 + 27g^3$

37. $(4x - 1)(16x^2 + 4x + 1)$
$64x^3 - 1$

38. $(5r - 2)(25r^2 + 10r + 4)$
$125r^3 - 8$

39. $(7t + 5s)(2t^2 + 5st - s^2)$
$14t^3 + 45st^2 + 18s^2t - 5s^3$

40. $(8p + 3q)(2p^2 - 4pq + q^2)$
$16p^3 - 26p^2q - 4pq^2 + 3q^3$

Factor each polynomial completely. See Example 4.

41. $8x^3y^4 + 12x^2y^3 + 36xy^4$
$4xy^3(2x^2y + 3x + 9y)$

42. $10m^5n + 4m^2n^3 + 18m^3n^2$
$2m^2n(5m^3 + 2n^2 + 9mn)$

43. $x^2 - 2x - 15$
$(x + 3)(x - 5)$

44. $x^2 + x - 12$
$(x + 4)(x - 3)$

45. $2x^2 - 9x - 18$
$(2x + 3)(x - 6)$

46. $3x^2 + 2x - 8$
$(3x - 4)(x + 2)$

47. $36t^2 - 25$

$(6t + 5)(6t - 5)$

48. $49r^2 - 9$

$(7r + 3)(7r - 3)$

49. $16t^2 + 24t + 9$

$(4t + 3)^2$

50. $25t^2 + 90t + 81$

$(5t + 9)^2$

51. $4m^2p - 12mnp + 9n^2p$

$p(2m - 3n)^2$

52. $16p^2r - 40pqr + 25q^2r$

$r(4p - 5q)^2$

53. $x^3 + 1$

$(x + 1)(x^2 - x + 1)$

54. $x^3 + 27$

$(x + 3)(x^2 - 3x + 9)$

55. $8t^3 + 125$

$(2t + 5)(4t^2 - 10t + 25)$

56. $27s^3 + 64$

$(3s + 4)(9s^2 - 12s + 16)$

57. $t^6 - 125$

$(t^2 - 5)(t^4 + 5t^2 + 25)$

58. $w^6 - 27$

$(w^2 - 3)(w^4 + 3w^2 + 9)$

59. $5xt + 15xr + 2yt + 6yr$

$(5x + 2y)(t + 3r)$

60. $3am + 18mb + 2an + 12nb$

$(3m + 2n)(a + 6b)$

61. $6ar + 12br - 5as - 10bs$

$(6r - 5s)(a + 2b)$

62. $7mt + 35ms - 2nt - 10ns$

$(7m - 2n)(t + 5s)$

63. $t^4 - 1$

$(t^2 + 1)(t + 1)(t - 1)$

64. $r^4 - 81$

$(r^2 + 9)(r + 3)(r - 3)$

65. $4x^2 + 12xy + 9y^2 - 1$

$(2x + 3y - 1)(2x + 3y + 1)$

66. $81t^2 + 36ty + 4y^2 - 9$

$(9t + 2y - 3)(9t + 2y + 3)$

Appendix: Strategies for Problem Solving

OBJECTIVE 1 Learn additional problem-solving strategies. In **Section 2.4,** we introduce a six-step method for problem solving that we use with applications throughout this text. This method is based on a problem-solving process developed by Hungarian native George Polya, among whose many publications is the modern classic *How to Solve It*.

Polya's Four-Step Process for Problem Solving

Step 1 **Understand the problem.** You cannot solve a problem if you do not understand what you are asked to find. The problem must be read and analyzed carefully. You may need to read it several times. After you have done so, ask yourself, "What must I find?"

Step 2 **Devise a plan.** There are many ways to attack a problem. Decide what plan is appropriate for the particular problem you are solving.

Step 3 **Carry out the plan.** Once you know how to approach the problem, carry out your plan. You may run into "dead ends" and unforeseen roadblocks, but be persistent. If you are able to solve a problem without a struggle, it isn't much of a problem, is it?

Step 4 **Look back and check.** Check your answer to see that it is reasonable. Does it satisfy the conditions of the problem? Have you answered all the questions the problem asks? Can you solve the problem a different way and come up with the same answer?

Work Problem 1 at the Side.

In Step 2 of Polya's problem-solving process, we are told to devise a plan. The box on the next page lists some strategies that may prove useful.

OBJECTIVE

1 Learn additional problem-solving strategies.

George Polya (1887–1985)

1 Compare the six-step problem-solving method given in **Section 2.4** with Polya's four steps.

ANSWER

1. Step 1 compares to Polya's first step, Steps 2 and 3 compare to his second step, Step 4 compares to his third step, and Step 6 compares to his fourth step.

Problem-Solving Strategies

Make a table or a chart.
Look for a pattern.
Solve a similar simpler problem.
Draw a sketch.
Write an equation and solve it.
If a formula applies, use it.
Work backward.
Guess and check.
Use trial and error.
Use common sense.
Look for a "catch" if an answer seems too obvious or impossible.

The problem in Example 1 first appeared in *Liber Abaci,* a book written by the Italian mathematician Leonardo Pisano (also known as Fibonacci) in the year 1202. We apply Polya's four-step process to solve it.

EXAMPLE 1 Using a Table or a Chart

A man put a pair of rabbits in a cage. During the first month the rabbits produced no offspring, but each month thereafter produced one new pair of rabbits. If each new pair thus produced reproduces in the same manner, how many pairs of rabbits will there be at the end of one year?

Step 1 **Understand the problem.** We can reword it as follows:

How many pairs of rabbits will the man have at the end of one year if he starts with one pair, and they reproduce this way: During the first month of life, each pair produces no new rabbits, but each month thereafter each pair produces one new pair?

Step 2 **Devise a plan.** We can construct a table as shown below.

Month	Number of Pairs at Start	Number of New Pairs Produced	Number of Pairs at End of Month
1st			
2nd			
3rd			
4th			
5th			
6th			
7th			
8th			
9th			
10th			
11th			
12th			

Step 3 **Carry out the plan.** At the start of the first month there is only one pair of rabbits. No new pairs are produced during the first month, so there is $1 + 0 = 1$ pair present at the end of the first month. This pattern continues throughout the table on the next page. We add the number in the first column of numbers to the number in the second column to get the number in the third.

Continued on Next Page

Month	Number of Pairs at Start	+	Number of New Pairs Produced	=	Number of Pairs at End of Month	
1st	1		0		1	$1 + 0 = 1$
2nd	1		1		2	$1 + 1 = 2$
3rd	2		1		3	$2 + 1 = 3$
4th	3		2		5	.
5th	5		3		8	.
6th	8		5		13	.
7th	13		8		21	.
8th	21		13		34	.
9th	34		21		55	.
10th	55		34		89	.
11th	89		55		144	.
12th	144		89		233	$144 + 89 = 233$

There will be 233 pairs of rabbits at the end of one year.

Step 4 **Look back and check.** This problem can be checked by going back and making sure that we have interpreted it correctly, which we have. Double-check the arithmetic. We have answered the question posed by the problem, so the problem is solved.

Work Problem 2 at the Side. ▶

EXAMPLE 2 Working Backward

Rob Zwettler goes to the racetrack with his buddies on a weekly basis. One week he tripled his money, but then lost \$12. He took his money back the next week, doubled it, but then lost \$40. The following week he tried again, taking his money back with him. He quadrupled it, and then played well enough to take that much home with him, a total of \$224. How much did he start with the first week?

This problem asks us to find Rob's starting amount. Since we know his final amount, the method of working backward can be applied. Since his final amount was \$224 and this represents four times the amount he started with on the third week, we *divide* \$224 by 4 to find that he started the third week with \$56. Before he lost \$40 the second week, he had this \$56 plus the \$40 he lost, giving him \$96. This represented double what he started with, so he started with \$96 *divided by* 2, or \$48, the second week. Repeating this process once more for the first week, before his \$12 loss he had

$$\$48 + \$12 = \$60,$$

which represents triple what he started with. Therefore, he started with

$$\$60 \div 3, \quad \text{or} \quad \$20. \longleftarrow \text{Answer}$$

To check our answer, \$20, observe the following equations:

First week: $(3 \times \$20) - \$12 = \$60 - \$12 = \$48$

Second week: $(2 \times \$48) - \$40 = \$96 - \$40 = \$56$

Third week: $(4 \times \$56) = \$224.$ ← His final amount

Work Problem 3 at the Side. ▶

2 Refer to the completed table in Example 1, and observe the sequence of numbers in color, the *Fibonacci sequence*. Choose any four consecutive terms. Multiply the first one chosen by the fourth, and then write the product. Now multiply the two middle terms and write the product. Repeat this process a few more times. What do you notice when the two products are compared?

3 Solve each problem.

(a) Bonnie Boehme bought a book for \$10 and then spent half her remaining money on a train ticket. She then bought lunch for \$4 and spent half her remaining money at a bazaar. She left the bazaar with \$20. How much money did she start with?

(b) If a, b, and c are digits for which

$$\begin{array}{r} 7\ a\ 2 \\ -\ 4\ 8\ b \\ \hline c\ 7\ 3, \end{array}$$

then $a + b + c =$ ____.

A. 14 **B.** 15 **C.** 16
D. 17 **E.** 18

(*Source: Mathematics Teacher* calendar, September 22, 1999.)

Answers

2. The products will always differ by 1.
3. **(a)** \$98 **(b)** D

Recall that $5^2 = 5 \cdot 5 = 25$, that is, 5 squared is 25. Thus, 25 is called a **perfect square.** (See **Section 6.5.**) Other perfect squares include

$$4, \quad 9, \quad 16, \quad 36, \quad \text{and so on.} \qquad \text{Perfect squares}$$

EXAMPLE 3 Using Trial and Error

The mathematician Augustus De Morgan lived in the nineteenth century. He once made the following statement: "I was x years old in the year x^2." In what year was De Morgan born?

We must find the year of De Morgan's birth. The problem tells us that he lived in the nineteenth century, which is another way of saying that he lived during the 1800s. One year of his life was a perfect square, so we must find a number between 1800 and 1900 that is a perfect square. Use trial and error.

$$42^2 = 42 \cdot 42 = 1764$$
$$43^2 = 43 \cdot 43 = \mathbf{1849} \qquad \text{1849 is between 1800 and 1900.}$$
$$44^2 = 44 \cdot 44 = 1936$$

The only natural number whose square is between 1800 and 1900 is 43, since $43^2 = 1849$. Therefore, De Morgan was 43 yr old in 1849. The final step in solving the problem is to subtract 43 from 1849 to find the year of his birth:

$$1849 - 43 = \mathbf{1806}. \qquad \text{He was born in 1806.}$$

While the following check may seem unorthodox, it works: Look up De Morgan's birth date in a book on mathematics history, such as *An Introduction to the History of Mathematics,* Sixth Edition, by Howard W. Eves.

◀ Work Problem 4 at the Side.

As mentioned above, $5^2 = 25$. The inverse (opposite) of squaring a number is called taking the **square root.** (See **Section 9.1.**) We indicate the positive square root using a **radical sign** $\sqrt{\ }$. Thus, $\sqrt{25} = 5$. Also,

$$\sqrt{4} = 2, \quad \sqrt{9} = 3, \quad \sqrt{16} = 4, \quad \text{and so on.} \qquad \text{Square roots}$$

The next problem dates back to Hindu mathematics, circa 850.

EXAMPLE 4 Guessing and Checking

One-fourth of a herd of camels was seen in the forest. Twice the square root of that herd had gone to the mountain slopes, and 3 times 5 camels remained on the riverbank. What is the numerical measure of that herd of camels?

The numerical measure of the herd of camels must be a natural number. Since the problem mentions "one-fourth of a herd" and "the square root of that herd," the number of camels must be both a multiple of 4 and a perfect square, so no fractions will be encountered. The smallest natural number that satisfies both conditions is 4. We write an equation where x represents the numerical measure of the herd.

"one-fourth of the herd" + "twice the square root of that herd" + "3 times 5 camels" = "the numerical measure of the herd"

$$\frac{1}{4}x \quad + \quad 2\sqrt{x} \quad + \quad 3 \cdot 5 \quad = \quad x$$

Continued on Next Page

4 Solve each problem.

(a) Assuming that he lives that long, one of the authors of this book will be 76 yr old in the year x^2, where x is a counting number. In what year was he born?

(b) Place each of the digits 1, 2, 3, 4, 5, 6, 7, and 8 in separate boxes so that boxes that share common corners do not contain successive digits.
(*Source: Mathematics Teacher* calendar, November 29, 1997.)

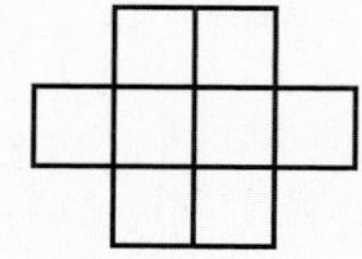

ANSWERS

4. (a) 1949 **(b)** Here is one possible solution.

	3	5	
7	1	8	2
	4	6	

Now substitute 4 for x to see if it is a solution of the equation.

$$\frac{1}{4}(4) + 2\sqrt{4} + 3 \cdot 5 = 4 \qquad \text{Let } x = 4.$$

$$1 + 4 + 15 \stackrel{?}{=} 4 \qquad \sqrt{4} = 2$$

$$20 \neq 4 \qquad \text{Add.}$$

Since 4 is not a solution, try 16, the next perfect square that is a multiple of 4.

$$\frac{1}{4}(16) + 2\sqrt{16} + 3 \cdot 5 = 16 \qquad \text{Let } x = 16.$$

$$4 + 8 + 15 \stackrel{?}{=} 16 \qquad \sqrt{16} = 4$$

$$27 \neq 16 \qquad \text{Add.}$$

Since 16 is not a solution, try 36.

$$\frac{1}{4}(36) + 2\sqrt{36} + 3 \cdot 5 = 36 \qquad \text{Let } x = 36.$$

$$9 + 12 + 15 \stackrel{?}{=} 36 \qquad \sqrt{36} = 6$$

$$36 = 36 \qquad \text{Add.}$$

We see that 36 is the numerical measure of the herd. Check in the words of the problem: "One-fourth of 36, plus twice the square root of 36, plus 3 times 5" gives 9 plus 12 plus 15, which equals 36.

Work Problem 5 at the Side. ▶

5 Solve each problem.

(a) I am thinking of a positive number. If I square it, double the result, take half of that result, and then add 12, I get 21. What is my number?

(b) The same author mentioned in Margin Problem 4(a) graduated from high school in the year that satisfies these conditions:

(1) The sum of the digits is 23;

(2) The hundreds digit is 3 more than the tens digit;

(3) No digit is an 8.

In what year did he graduate?

EXAMPLE 5 Considering a Similar Simpler Problem and Looking for a Pattern

The digit farthest to the right in a natural number is called the *ones* or *units* digit, since it tells how many ones are contained in the number when grouping by tens is considered. What is the ones (or units) digit in 2^{4000}?

Recall that 2^{4000} means that 2 is used as a factor 4000 times:

$$2^{4000} = \underbrace{2 \times 2 \times 2 \times \cdots \times 2.}_{4000 \text{ factors}}$$

Certainly, we are not expected to evaluate this number. To answer the question, we examine some smaller powers of 2 and then look for a pattern. We start with the exponent 1 and look at the first twelve powers of 2.

$$2^1 = 2 \qquad 2^5 = 32 \qquad 2^9 = 512$$

$$2^2 = 4 \qquad 2^6 = 64 \qquad 2^{10} = 1024$$

$$2^3 = 8 \qquad 2^7 = 128 \qquad 2^{11} = 2048$$

$$2^4 = 16 \qquad 2^8 = 256 \qquad 2^{12} = 4096$$

Notice that in each of the four rows above, the ones digit is the same. The final row, which contains the exponents 4, 8, and 12, has the ones digit 6. Each of these exponents is divisible by 4, and since 4000 is divisible by 4, we observe the pattern to predict that the units digit in 2^{4000} is 6.

The units digit for any other power of 2 can be found if we divide the exponent by 4 and compare the remainder to the preceding list of powers. For example, to find the units digit of 2^{543}, we divide 543 by 4 to get a quotient of 135 and a remainder of 3. The units digit is the same as that of 2^3, which is 8.

Work Problem 6 at the Side. ▶

6 Solve each problem.

(a) What is the units digit in 7^{491}?

(b) What is the 103rd digit in the decimal representation for $\frac{1}{11}$?

Answers

5. (a) 3 (b) 1967
6. (a) 3 (b) 0

EXAMPLE 6 Drawing a Sketch

An array of nine dots is arranged in a 3×3 square, as shown in Figure 1. Is it possible to join the dots with exactly four straight lines if you are not allowed to pick up your pencil from the paper and may not trace over a line that has already been drawn? If so, show how.

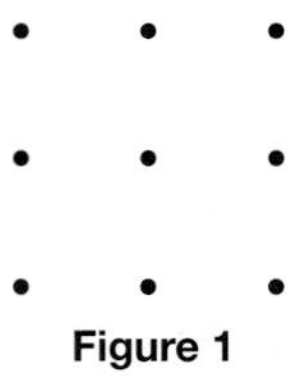

Figure 1

Figure 2 shows three attempts. In each case, something is wrong. In the first sketch, one dot is not joined. In the second, the figure cannot be drawn without picking up your pencil from the paper or tracing over a line that has already been drawn. In the third figure, all dots have been joined, but you have used five lines as well as retraced over the figure.

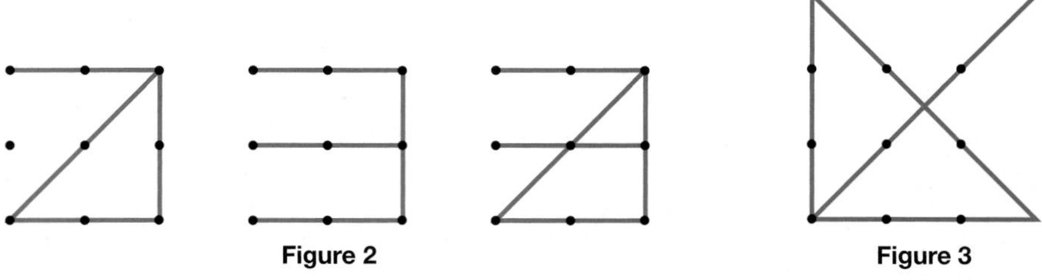

Figure 2 Figure 3

However, the conditions of the problem can be satisfied, as shown in Figure 3. We "went outside of the box," which was not prohibited by the conditions of the problem. This is an example of creative thinking—we used a strategy that is usually not considered at first, since our initial attempts involved "staying within the confines" of the figure.

◀ Work Problem 7 at the Side.

The final example falls into a category of problems that involve a "catch." Some of these problems seem too easy or perhaps impossible at first, because we tend to overlook an obvious situation. We must look carefully at the use of language in such problems. And, of course, we should never forget to use common sense.

EXAMPLE 7 Using Common Sense

Two currently minted U.S. coins together have a total value of $1.05. One is not a dollar. What are the two coins?

Our initial reaction might be, "The only way to have two such coins with a total of $1.05 is to have a nickel and a dollar, but the problem says that one of them is not a dollar." This statement is indeed true. What we must realize here is that the one that is not a dollar is the nickel, and the *other* coin is a dollar! So the two coins are a dollar and a nickel.

◀ Work Problem 8 at the Side.

7 Solve each problem.

(a) What is the maximum number of small squares in which we may place a cross ($\times$) and not have any row, column, or diagonal completely filled with crosses? Illustrate your answer.

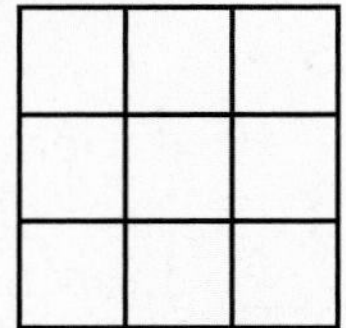

(b) By drawing two straight lines, divide the face of a clock into three regions such that the numbers in the regions have the same total. (*Source: Mathematics Teacher* calendar, October 28, 1998.)

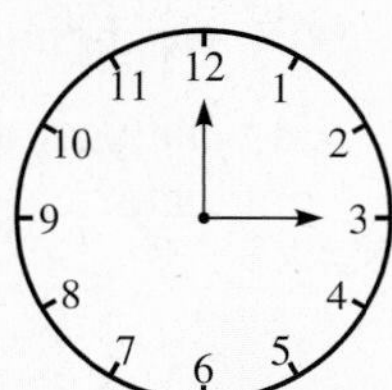

8 Solve each problem.

(a) Which is correct? Three cubed *is* nine or three cubed *are* nine?

(b) If you take 7 bowling pins from 10 bowling pins, what do you have?

ANSWERS

7. (a) 6

	X	X
X		X
X	X	

One of several possibilities

(b) Each region has a sum of 26.

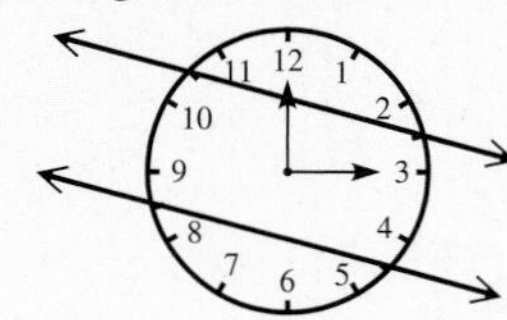

8. (a) Neither is correct, since $3^3 = 27$.
(b) 7 bowling pins

Appendix B Exercises

Exercises 1–9 are from the popular monthly calendar feature in the journal Mathematics Teacher. *The authors wish to thank the many journal contributors for permission to use these problems. Original calendar dates are included.*

Use the various problem-solving strategies to solve each problem. In many cases, there is more than one possible approach, so be creative.

1. You are working in a store that has been very careless with the stock. Three boxes of socks are each incorrectly labeled. The labels say *red socks, green socks,* and *red and green socks*. How can you relabel the boxes correctly by taking only one sock out of one box, without looking inside the boxes? (October 22, 2001)

You should choose a sock from the box labeled *red and green socks.* Since it is mislabeled, it contains only red socks or only green socks, determined by the sock you choose. If the sock is green, relabel this box *green socks.* Since the other two boxes were mislabeled, switch the remaining label to the other box and place the label that says *red and green socks* on the unlabeled box. No other choice guarantees a correct relabeling, since you can remove only one sock.

2. Three dice with faces numbered 1 through 6 are stacked as shown. Seven of the eighteen faces are visible, leaving eleven faces hidden on the back, on the bottom, and between faces. The total number of dots not visible in this view is _____.

A. 21

B. 22

C. 31

D. 41

E. 53

(September 17, 2001)

D

3. You and a friend are playing tick-tack-toe, where three in a row *loses*. You are O. If you want to win, what must your next move be? (October 21, 2001)

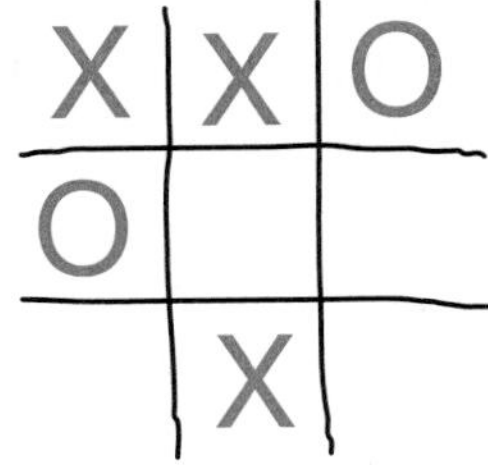

You must place the O in the bottom-left square. No other choice guarantees you a win.

4. How can you connect each square with the triangle that has the same number? Lines cannot cross, enter a square or triangle, or go outside the diagram. (October 15, 1999)

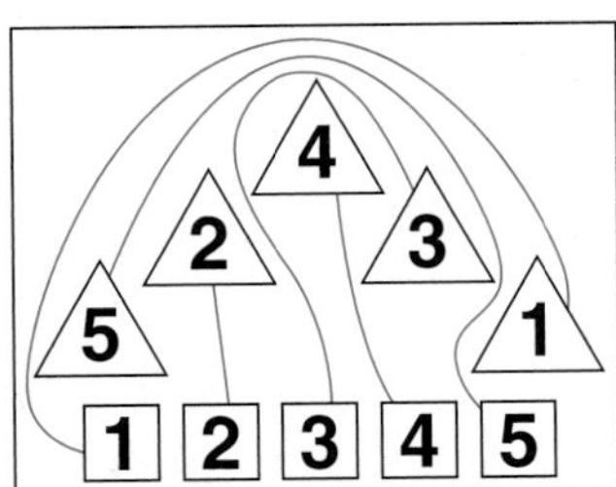

One possible solution is shown.

5. You have brought two unmarked buckets to a stream. The buckets hold 7 gal and 3 gal of water, respectively. How can you obtain exactly 5 gal of water to take home? (October 19, 1997)

One possible sequence is shown here. The numbers represent the number of gallons in each bucket in each successive step.

Big	7	4	4	1	1	0	7	5	5
Small	0	3	0	3	0	1	1	3	0

6. Chip and Dale collected 32 acorns on Monday and stored them with their acorn supply. After Chip fell asleep, Dale ate half the acorns. This pattern continued through Friday night, with 32 acorns being added and half the supply being eaten. On Saturday morning, Chip counted the acorns and found that they had only 35. How many acorns had they started with on Monday morning? (March 12, 1997)

128

7. Pat and Chris have the same birthday. Pat is twice as old as Chris was when Pat was as old as Chris is now. If Pat is now 24 yr old, how old is Chris? (December 3, 2001)

 18

8. Balls numbered 1 through 6 are arranged in a **difference triangle.** Note that in any row, the difference between the larger and the smaller of two successive balls is the number of the ball that appears below them. Arrange balls numbered 1 through 10 in a difference triangle. (May 6, 1998)

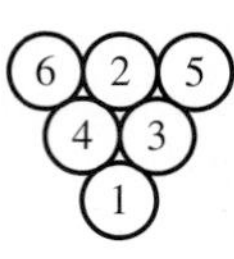

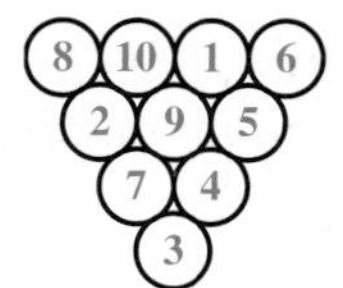

 Here is one solution.

9. What are the final two digits of 7^{1997}? (November 29, 1997)

 07

10. If you raise 3 to the 324th power, what is the units digit of the result?

 1

11. A frog is at the bottom of a 20-ft well. Each day it crawls up 4 ft but each night it slips back 3 ft. After how many days will the frog reach the top of the well?

 17 days

12. A lily pad grows so that each day it doubles its size. On the twentieth day of its life, it completely covers a pond. On what day was the pond half covered?

 the nineteenth day

13. Some children are standing in a circular arrangement. They are evenly spaced and arranged in numerical order. The fourth child is standing directly opposite the twelfth child. How many children are there in the circle? 16

14. A **perfect number** is a natural number that is equal to the sum of all its counting number divisors except itself. For example, 28 is a perfect number, since its divisors other than itself are 1, 2, 4, 7, and 14, and $1 + 2 + 4 + 7 + 14 = 28$. What is the least perfect number? 6

15. A **magic square** is a square array of numbers that has the property that the sum of the numbers in any row, column, or diagonal is the same. Fill in the square below so that it becomes a magic square, and all digits 1, 2, 3, . . . , 9 are used exactly once.

6	1	**8**
7	**5**	3
2	9	**4**

16. Refer to Exercise 15. Complete the magic square below so that all counting numbers 1, 2, 3, . . . , 16 are used exactly once, and the sum in each row, column, or diagonal is 34.

6	12	7	**9**
1	**15**	4	**14**
11	5	**10**	8
16	2	**13**	3

17. What is the minimum number of pitches that a baseball player who pitches a complete game can make in a regulation 9-inning baseball game?

25 pitches (The visiting team's pitcher retires 24 consecutive batters through the first eight innings, using only one pitch per batter. His team does not score either. Going into the bottom of the ninth inning tied 0–0, the first batter for the home team hits his first pitch for a home run. The pitcher threw 25 pitches and loses the game by a score of 1–0.)

18. Draw a square in the following figure so that no two cats share the same region.

19. You have eight coins. Seven are genuine and one is a fake, which weighs a little less than the other seven. You have a balance scale, which you may use only three times. Tell how to locate the bad coin in three weighings. Then show how to detect the bad coin in only *two* weighings.

For three weighings, first balance four against four. Of the lighter four, balance two against the other two. Finally, of the lighter two, balance them one against the other.

To find the bad coin in two weighings, divide the eight coins into groups of 3, 3, 2. Weigh the groups of three against each other on the scale. If the groups weigh the same, the fake is in the two left out and can be found in one additional weighing. If the two groups of three do not weigh the same, pick the lighter group. Choose any two of the coins and weigh them. If one of these is lighter, it is the fake; if they weigh the same, then the third coin is the fake.

20. A person must take a wolf, a goat, and some cabbage across a river. The rowboat to be used has room for the person plus either the wolf, the goat, or the cabbage. If the person takes the cabbage in the boat, the wolf will eat the goat. While the wolf crosses in the boat, the cabbage will be eaten by the goat. The goat and cabbage are safe only when the person is present. Even so, the person gets everything across the river. Explain how. (This problem dates back to around the year 750.)

The person takes the goat across and returns alone. On the second trip, the person takes the wolf across and returns with the goat. On the third trip, the goat is left on the first side while the person takes the cabbage across. Then the person returns alone and brings the goat back across.

21. (This is an ancient Hindu problem.) Beautiful maiden with beaming eyes, tell me . . . which is the number that when multiplied by 3, then increased by $\frac{3}{4}$ the product, then divided by 7, diminished by $\frac{1}{3}$ of the quotient, multiplied by itself, diminished by 52, by the extraction of the square root, addition of 8, and division by 10 gives the number 2?

28

22. A teenager's age increased by 2 gives a perfect square. Her age decreased by 10 gives the square root of that perfect square. She is 5 yr older than her brother. How old is her brother?

9 yr old

23. Draw the following figure without picking up your pencil from the paper and without tracing over a line you have already drawn.

One possible solution is shown.

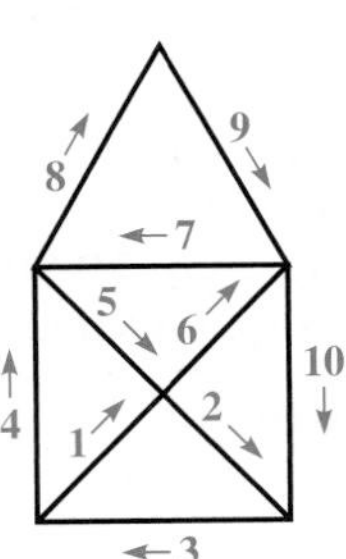

24. Repeat Exercise 23 for the figure shown here.

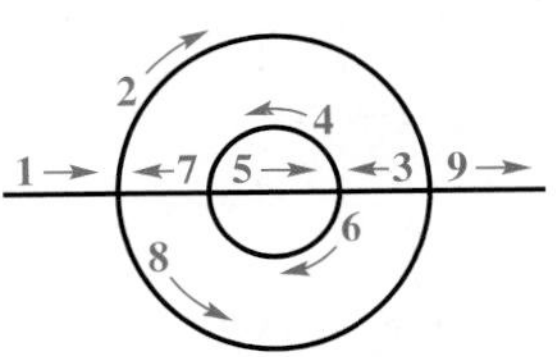

One possible solution is shown.

25. James, Dan, Jessica, and Cathy form a pair of married couples. Their ages are 36, 31, 30, and 29. Jessica is married to the oldest person in the group. James is older than Jessica but younger than Cathy. Who is married to whom, and what are their ages?

Dan (36) is married to Jessica (29); James (30) is married to Cathy (31).

26. If a year has two consecutive months with Friday the thirteenth, what months must they be?

They must be February and March in a non-leap year.

27. How much dirt is there in a cubical hole, 6 ft on each side?

None, since there is no dirt in a hole.

28. Some months have 30 days and some have 31 days. How many months have 28 days?

12; All months have 28 days.

29. Place one of the arithmetic operations $+$, $-$, $\times$, or $\div$ between each pair of successive numbers on the left side of this equation to make it true. Any operation may be used more than once or not at all.

$$1 \quad 2 \quad 3 \quad 4 \quad 5 \quad 6 \quad 7 \quad 8 \quad 9 = 100$$

One solution is
$\mathbf{1 + 2 + 3 + 4 + 5 + 6 + 7 + 8 \times 9 = 100.}$

30. In the addition problem below, some digits are missing as indicated by the blanks. If the problem is done correctly, what is the sum of the missing digits?

$$\begin{array}{cccc} & \underline{4} & 3 & 5 \\ & 8 & \underline{2} & 6 \\ + & 1 & 4 & \underline{7} \\ \hline \underline{1} & 4 & 0 & 8 \end{array}$$

The sum is 14.

31. Fill in the blanks so that the multiplication problem below uses all digits 0, 1, 2, 3, . . . , 9 exactly once, and is correctly worked.

$$\begin{array}{cccccc} & & & \underline{4} & 0 & 2 \\ \times & & & & 3 & \underline{9} \\ \hline \underline{1} & 5, & \underline{6} & \underline{7} & \underline{8} & \end{array}$$

32. Based on your knowledge of elementary arithmetic, describe the pattern that can be observed when the following operations are performed:

$$9 \times 1, \quad 9 \times 2, \quad 9 \times 3, \quad \ldots, \quad 9 \times 9.$$

(*Hint:* Add the digits in the answers. What do you notice?)

The sum of the digits is always 9.

33. How many triangles are in the following figure?

35

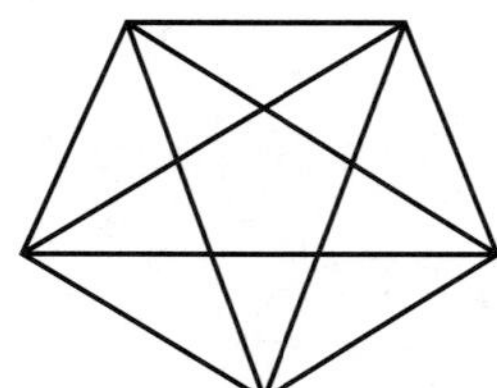

34. How many squares are in the following figure?

14

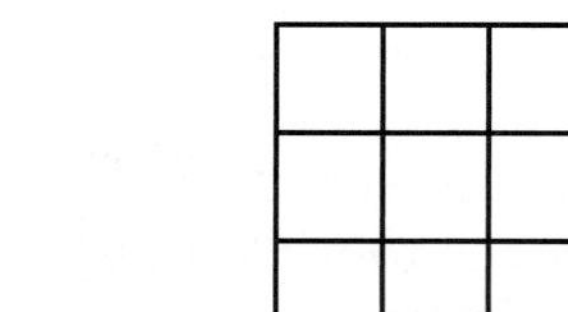

35. If it takes $7\frac{1}{2}$ min to boil an egg, how long does it take to boil 5 eggs?

$7\frac{1}{2}$ min (Boil them all at the same time.)

36. The brother of the chief executive officer (CEO) of a major company died. The man who died had no brother. How is this possible?

The CEO is a woman.

Appendix: Synthetic Division

We begin by reviewing the terminology for the parts of a division problem. The *divisor* is the quantity we are dividing by, the *dividend* is the quantity we are dividing into, and the *quotient* is the result of the division.

$$\begin{array}{rrl} & 1\,562 & \longleftarrow \text{Quotient} \\ \text{Divisor} \longrightarrow 247 & \overline{)385{,}814} & \longleftarrow \text{Dividend} \end{array}$$

OBJECTIVES

1. **Use synthetic division to divide by a polynomial of the form $x - k$.**
2. **Use the remainder theorem to evaluate a polynomial.**
3. **Decide whether a given number is a solution of an equation.**

OBJECTIVE 1 Use synthetic division to divide by a polynomial of the form $x - k$. When one polynomial is divided by a second, if the divisor has the form $x - k$, where the coefficient of the x-term is 1, there is a shortcut method for performing the division. Look at the division of $3x^3 - 2x + 5$ by $x - 3$ on the left below. Notice that we inserted 0 for the missing x^2-term.

$$\begin{array}{r|rrrr} & & 3x^2 & +\ 9x & +\ 25 \\ \hline x-3 & 3x^3 & +\ 0x^2 & -\ 2x & +\ 5 \\ & 3x^3 & -\ 9x^2 & & \\ \hline & & 9x^2 & -\ 2x & \\ & & 9x^2 & -\ 27x & \\ \hline & & & 25x & +\ 5 \\ & & & 25x & -\ 75 \\ \hline & & & & 80 \end{array} \qquad \begin{array}{r|rrrr} & & 3 & 9 & 25 \\ \hline 1-3 & 3 & 0 & -2 & 5 \\ & 3 & -9 & & \\ \hline & & 9 & -2 & \\ & & 9 & -27 & \\ \hline & & & 25 & 5 \\ & & & 25 & -75 \\ \hline & & & & 80 \end{array}$$

On the right, the same division is shown written without the variables. This is why it is *essential* to use 0 as a placeholder in synthetic division. All the numbers in color on the right are repetitions of the numbers directly above them, so they are omitted to condense the work, as shown on the left below.

$$\begin{array}{r|rrrr} & & 3 & 9 & 25 \\ \hline 1-3 & 3 & 0 & -2 & 5 \\ & & -9 & & \\ \hline & & 9 & -2 & \\ & & & -27 & \\ \hline & & & 25 & 5 \\ & & & & -75 \\ \hline & & & & 80 \end{array} \qquad \begin{array}{r|rrrr} & & 3 & 9 & 25 \\ \hline 1-3 & 3 & 0 & -2 & 5 \\ & & -9 & & \\ \hline & & 9 & & \\ & & & -27 & \\ \hline & & & 25 & \\ & & & & -75 \\ \hline & & & & 80 \end{array}$$

The numbers in color on the left are again repetitions of the numbers directly above them; they too are omitted, as shown on the right above. If the 3 in the dividend is brought down to the beginning of the bottom row, the top row can be omitted since it duplicates the bottom row.

$$\begin{array}{r|rrrr} 1-3 & 3 & 0 & -2 & 5 \\ & & -9 & -27 & -75 \\ \hline & 3 & 9 & 25 & 80 \end{array}$$

We omit the 1 at the upper left, since it represents $1x$, which will *always* be the first term in the divisor. Also, to simplify the arithmetic, we replace subtraction in the second row by addition. We compensate for this by changing the -3 at the upper left to its additive inverse, 3.

Additive inverse of -3 ⟶

$$\begin{array}{r|rrrr} 3 & 3 & 0 & -2 & 5 \\ & & 9 & 27 & 75 \\ \hline & 3 & 9 & 25 & 80 \end{array}$$

75 ⟵ Change signs.
80 ⟵ Remainder

The quotient is read from the bottom row.

$$3x^2 + 9x + 25 + \frac{80}{x-3}$$

The first three numbers in the bottom row are the coefficients of the quotient polynomial with degree 1 less than the degree of the dividend. The last number gives the remainder.

Synthetic Division

This shortcut method is called **synthetic division.** It is used *only* when dividing a polynomial by a binomial of the form $x - k$.

1 Divide, using synthetic division.

(a) $\dfrac{3z^2 + 10z - 8}{z + 4}$

(b) $(2x^2 + 3x - 5) \div (x + 1)$

EXAMPLE 1 Using Synthetic Division

Use synthetic division to divide $5x^2 + 16x + 15$ by $x + 2$.

As mentioned above, we use synthetic division only when dividing by a polynomial of the form $x - k$. We change $x + 2$ to this form by writing it as

$$x + 2 = x - (-2),$$

where $k = -2$. Then we write the coefficients of $5x^2 + 16x + 15$.

$x + 2$ leads to -2. ⟶ $-2\overline{)5 \quad 16 \quad 15}$ ⟵ Coefficients

We bring down the 5, and multiply: $-2 \cdot 5 = -10$.

$$\begin{array}{r|rrr} -2 & 5 & 16 & 15 \\ & \downarrow & -10 & \\ \hline & 5 & & \end{array}$$

We add 16 and -10, getting 6. Multiply 6 and -2 to get -12.

$$\begin{array}{r|rrr} -2 & 5 & 16 & 15 \\ & & -10 & -12 \\ \hline & 5 & 6 & \end{array}$$

We add 15 and -12, getting 3.

$$\begin{array}{r|rrr} -2 & 5 & 16 & 15 \\ & & -10 & -12 \\ \hline & 5 & 6 & 3 \end{array}$$

3 ⟵ Remainder

We read the result from the bottom row.

$$\frac{5x^2 + 16x + 15}{x + 2} = 5x + 6 + \frac{3}{x + 2}$$

◀ Work Problem **1** at the Side.

ANSWERS

1. (a) $3z - 2$ (b) $2x + 1 + \dfrac{-6}{x + 1}$

EXAMPLE 2 Using Synthetic Division with a Missing Term

Use synthetic division to find $(-4x^5 + x^4 + 6x^3 + 2x^2 + 50) \div (x - 2)$.

Use the steps given above, inserting a 0 for the missing x-term.

$$\begin{array}{r|rrrrrr} 2 & -4 & 1 & 6 & 2 & 0 & 50 \\ & & -8 & -14 & -16 & -28 & -56 \\ \hline & -4 & -7 & -8 & -14 & -28 & -6 \end{array}$$

Read the result from the bottom row.

$$\frac{-4x^5 + x^4 + 6x^3 + 2x^2 + 50}{x - 2} = -4x^4 - 7x^3 - 8x^2 - 14x - 28 + \frac{-6}{x - 2}$$

Work Problem 2 *at the Side.* ▶

OBJECTIVE 2 Use the remainder theorem to evaluate a polynomial. We can use synthetic division to evaluate polynomials. For example, in the synthetic division of Example 2, where the polynomial was divided by $x - 2$, the remainder was -6.

Replacing x in the polynomial with 2 gives

$$\begin{aligned} &-4x^5 + x^4 + 6x^3 + 2x^2 + 50 \\ &= -4 \cdot 2^5 + 2^4 + 6 \cdot 2^3 + 2 \cdot 2^2 + 50 \\ &= -4 \cdot 32 + 16 + 6 \cdot 8 + 2 \cdot 4 + 50 \\ &= -128 + 16 + 48 + 8 + 50 \\ &= -6, \end{aligned}$$

the same number as the remainder. Thus, dividing by $x - 2$ produced a remainder equal to the result when x is replaced with 2. This always happens, as the following remainder theorem states.

Remainder Theorem

If the polynomial $P(x)$ is divided by $x - k$, then the remainder is equal to $P(k)$.

This result is proved in more advanced courses.

EXAMPLE 3 Using the Remainder Theorem

Let $P(x) = 2x^3 - 5x^2 - 3x + 11$. Find $P(-2)$.

Use the remainder theorem; divide $P(x)$ by $x - (-2)$.

$$\text{Value of } x \rightarrow \begin{array}{r|rrrr} -2 & 2 & -5 & -3 & 11 \\ & & -4 & 18 & -30 \\ \hline & 2 & -9 & 15 & -19 \end{array} \leftarrow \text{Remainder}$$

By this result, $P(-2) = -19$.

Work Problem 3 *at the Side.* ▶

2 Divide, using synthetic division.

(a) $\dfrac{3a^3 - 2a + 21}{a + 2}$

(b) $(-4x^4 + 3x^3 + 18x + 2) \div (x - 2)$

3 Let $P(x) = x^3 - 5x^2 + 7x - 3$. Use synthetic division to find each value.

(a) $P(1)$ (Divide by $x - 1$.)

(b) $P(-2)$

ANSWERS

2. (a) $3a^2 - 6a + 10 + \dfrac{1}{a + 2}$

(b) $-4x^3 - 5x^2 - 10x - 2 + \dfrac{-2}{x - 2}$

3. (a) 0 **(b)** -45

OBJECTIVE 3 Decide whether a given number is a solution of an equation. The remainder theorem can also be used to show that a given number is a solution of an equation.

EXAMPLE 4 Using the Remainder Theorem

Show that -5 is a solution of the equation

$$2x^4 + 12x^3 + 6x^2 - 5x + 75 = 0.$$

One way to show that -5 is a solution is to substitute -5 for x in the equation. However, an easier way is to use synthetic division and the remainder theorem.

$$\begin{array}{r|rrrrr} \text{Proposed solution} \rightarrow -5 & 2 & 12 & 6 & -5 & 75 \\ & & -10 & -10 & 20 & -75 \\ \hline & 2 & 2 & -4 & 15 & 0 \end{array} \leftarrow \text{Remainder}$$

Since the remainder is 0, the polynomial has a value of 0 when $x = -5$, so -5 is a solution of the given equation.

◀ Work Problem 4 at the Side.

The synthetic division in Example 4 shows that $x - (-5)$ divides the polynomial with 0 remainder. Thus $x - (-5) = x + 5$ is a *factor* of the polynomial and

$$2x^4 + 12x^3 + 6x^2 - 5x + 75 = (x + 5)(2x^3 + 2x^2 - 4x + 15).$$

The second factor is the quotient polynomial found in the last row of the synthetic division.

4 Use synthetic division to decide whether 2 is a solution of each equation.

(a) $3x^3 - 11x^2 + 17x - 14 = 0$

(b) $4x^5 - 7x^4 - 11x^2 + 2x + 6 = 0$

ANSWERS

4. **(a)** yes **(b)** no

Appendix C Exercises

FOR EXTRA HELP MyMathLab Math XL PRACTICE

WATCH DOWNLOAD

READ

REVIEW

Choose the letter of the correct setup to perform synthetic division on the indicated quotient.

1. $\dfrac{x^2 + 3x - 6}{x - 2}$

A. $-2\overline{)1 \quad 3 \quad -6}$ B. $-2\overline{)-1 \quad -3 \quad 6}$

C. $2\overline{)1 \quad 3 \quad -6}$ D. $2\overline{)-1 \quad -3 \quad 6}$

C

2. $\dfrac{x^3 - 3x^2 + 2}{x - 1}$

A. $1\overline{)1 \quad -3 \quad 2}$ B. $-1\overline{)1 \quad -3 \quad 2}$

C. $1\overline{)1 \quad -3 \quad 0 \quad 2}$ D. $1\overline{)-1 \quad 3 \quad 0 \quad -2}$

C

Use synthetic division to find each quotient. See Examples 1 and 2.

3. $\dfrac{x^2 - 6x + 5}{x - 1}$

$x - 5$

4. $\dfrac{x^2 - 4x - 21}{x + 3}$

$x - 7$

5. $\dfrac{4m^2 + 19m - 5}{m + 5}$

$4m - 1$

6. $\dfrac{3k^2 - 5k - 12}{k - 3}$

$3k + 4$

7. $\dfrac{2a^2 + 8a + 13}{a + 2}$

$2a + 4 + \dfrac{5}{a + 2}$

8. $\dfrac{4y^2 - 5y - 20}{y - 4}$

$4y + 11 + \dfrac{24}{y - 4}$

9. $(p^2 - 3p + 5) \div (p + 1)$

$p - 4 + \dfrac{9}{p + 1}$

10. $(z^2 + 4z - 6) \div (z - 5)$

$z + 9 + \dfrac{39}{z - 5}$

11. $\dfrac{4a^3 - 3a^2 + 2a - 3}{a - 1}$

$4a^2 + a + 3$

12. $\dfrac{5p^3 - 6p^2 + 3p + 14}{p + 1}$

$5p^2 - 11p + 14$

13. $(x^5 - 2x^3 + 3x^2 - 4x - 2) \div (x - 2)$

$x^4 + 2x^3 + 2x^2 + 7x + 10 + \dfrac{18}{x - 2}$

14. $(2y^5 - 5y^4 - 3y^2 - 6y - 23) \div (y - 3)$

$2y^4 + y^3 + 3y^2 + 6y + 12 + \dfrac{13}{y - 3}$

15. $(-4r^6 - 3r^5 - 3r^4 + 5r^3 - 6r^2 + 3r) \div (r - 1)$

$-4r^5 - 7r^4 - 10r^3 - 5r^2 - 11r - 8 + \dfrac{-8}{r - 1}$

16. $(-3t^5 + 2t^4 - 5t^3 + 6t^2 - 3t - 2) \div (t - 2)$

$-3t^4 - 4t^3 - 13t^2 - 20t - 43 + \dfrac{-88}{t - 2}$

17. $(-3y^5 + 2y^4 - 5y^3 - 6y^2 - 1) \div (y + 2)$

$-3y^4 + 8y^3 - 21y^2 + 36y - 72 + \frac{143}{y + 2}$

18. $(m^6 + 2m^4 - 5m + 11) \div (m - 2)$

$m^5 + 2m^4 + 6m^3 + 12m^2 + 24m + 43 + \frac{97}{m - 2}$

19. $\frac{y^3 + 1}{y - 1}$

$y^2 + y + 1 + \frac{2}{y - 1}$

20. $\frac{z^4 + 81}{z - 3}$

$z^3 + 3z^2 + 9z + 27 + \frac{162}{z - 3}$

Use the remainder theorem to find $P(k)$. See Example 3.

21. $P(x) = 2x^3 - 4x^2 + 5x - 3;\ k = 2$

7

22. $P(x) = x^3 + 3x^2 - x + 5;\ k = -1$

8

23. $P(r) = -r^3 - 5r^2 - 4r - 2;\ k = -4$

-2

24. $P(z) = -z^3 + 5z^2 - 3z + 4;\ k = 3$

13

25. $P(x) = 2x^3 - 4x^2 + 5x - 33;\ k = 3$

0

26. $P(x) = x^3 - 3x^2 + 4x - 4;\ k = 2$

0

Use synthetic division to decide whether the given number is a solution of each equation. See Example 4.

27. $x^3 - 2x^2 - 3x + 10 = 0;\ x = -2$

yes

28. $x^3 - 3x^2 - x + 10 = 0;\ x = -2$

no

29. $m^4 + 2m^3 - 3m^2 + 8m - 8 = 0;\ m = -2$

no

30. $r^4 - r^3 - 6r^2 + 5r + 10 = 0;\ r = -2$

yes

31. $3x^3 + 2x^2 - 2x + 11 = 0;\ x = -2$

no

32. $3z^3 + 10z^2 + 3z - 9 = 0;\ z = -2$

no

33. Explain why it is important to insert 0s as placeholders for missing terms before performing synthetic division.

Since the variables are not present, a missing term will not be noticed in synthetic division, so the quotient will be wrong if placeholders are not inserted.

34. Explain why a 0 remainder in synthetic division of $P(x)$ by k indicates that k is a solution of the equation $P(x) = 0$.

By the remainder theorem, a 0 remainder means that $P(k) = 0$; that is, k is a number that makes $P(x) = 0$.

ANSWERS TO SELECTED EXERCISES

In this section we provide the answers that we think most students will obtain when they work the exercises using the methods explained in the text. If your answer does not look exactly like the one given here, it is not necessarily wrong. In many cases there are equivalent forms of the answer that are correct. For example, if the answer section shows $\frac{3}{4}$ and your answer is 0.75, you have obtained the right answer but written it in a different (yet equivalent) form. Unless the directions specify otherwise, 0.75 is just as valid an answer as $\frac{3}{4}$.

In general, if your answer does not agree with the one given in the text, see whether it can be transformed into the other form. If it can, then it is the correct answer. If you still have doubts, talk with your instructor.

Chapter R Prealgebra Review

Section R.1 (pages 13–16)

1. true **3.** false; The fraction $\frac{17}{51}$ can be simplified to $\frac{1}{3}$. **5.** false; *Product* indicates multiplication, so the product of 8 and 2 is 16. **7.** prime **9.** composite **11.** composite **13.** neither **15.** $2 \cdot 3 \cdot 5$ **17.** $2 \cdot 2 \cdot 3 \cdot 3 \cdot 7$ **19.** $2 \cdot 2 \cdot 31$ **21.** 29 **23.** $\frac{1}{2}$ **25.** $\frac{5}{6}$ **27.** $\frac{1}{5}$ **29.** $\frac{6}{5}$ **31.** $1\frac{5}{7}$ **33.** $6\frac{5}{12}$ **35.** $7\frac{6}{11}$ **37.** $\frac{13}{5}$ **39.** $\frac{83}{8}$ **41.** $\frac{54}{5}$ **43.** A **45.** $\frac{24}{35}$ **47.** $\frac{6}{25}$ **49.** $\frac{6}{5}$, or $1\frac{1}{5}$ **51.** $\frac{232}{15}$, or $15\frac{7}{15}$ **53.** $\frac{10}{3}$, or $3\frac{1}{3}$ **55.** 12 **57.** $\frac{1}{16}$ **59.** $\frac{84}{47}$, or $1\frac{37}{47}$ **61.** $\frac{2}{3}$ **63.** $\frac{8}{9}$ **65.** $\frac{27}{8}$, or $3\frac{3}{8}$ **67.** $\frac{17}{36}$ **69.** $\frac{11}{12}$ **71.** $\frac{4}{3}$, or $1\frac{1}{3}$ **73.** 6 cups **75.** $618\frac{3}{4}$ ft **77.** $\frac{9}{16}$ in. **79.** $\frac{5}{16}$ in. **81.** $16\frac{5}{8}$ yd **83.** $\frac{7}{100}$ **85.** about $4\frac{19}{25}$ million, or 4,760,000

Section R.2 (pages 25–26)

1. (a) 6 **(b)** 9 **(c)** 1 **(d)** 7 **(e)** 4 **3. (a)** 46.25 **(b)** 46.2 **(c)** 46 **(d)** 50 **5.** C **7.** B **9.** $\frac{4}{10}$ **11.** $\frac{64}{100}$ **13.** $\frac{138}{1000}$ **15.** $\frac{3805}{1000}$ **17.** 139; 143.094 **19.** 27; 25.61 **21.** 10; 15.33 **23.** 82; 81.716 **25.** 17; 15.211 **27.** 0.006; 0.006 **29.** 90; 116.48 **31.** 6; 7.15 **33.** 2; 2.05 **35.** 6000; 5711.6 **37.** 0.2; 0.162 **39.** 0.125 **41.** 1.25 **43.** $0.\overline{5}$; 0.556 **45.** $0.1\overline{6}$; 0.167 **47.** To convert a decimal to a percent, move the decimal point two places to the right and attach a percent symbol (%). **49.** 0.54 **51.** 1.17 **53.** 0.024 **55.** 0.0625 **57.** 0.008 **59.** 73% **61.** 0.4% **63.** 128% **65.** 30% **67.** 80% **69.** $18.\overline{18}$% **71.** 175% **73.** $15 **75.** 70%

Chapter 1 The Real Number System

Section 1.1 (pages 35–36)

1. true **3.** false; Using the guidelines for order of operations gives $4 + 3(8 - 2) = 4 + 3(6) = 4 + 18 = 22$. **5.** false; The correct translation is $4 = 16 - 12$. **7.** 49 **9.** 144 **11.** 64 **13.** 1000 **15.** 81 **17.** 1024 **19.** $\frac{16}{81}$ **21.** 0.000064 **23.** The 4 would be applied last because we work first inside the parentheses. **25.** 58 **27.** 13 **29.** 32 **31.** 19 **33.** $\frac{49}{30}$ **35.** 12 **37.** 36.14 **39.** 26 **41.** 4 **43.** 95 **45.** 12 **47.** 14 **49.** $\frac{19}{2}$ **51.** false **53.** true **55.** true **57.** false **59.** false **61.** true **63.** $15 = 5 + 10$ **65.** $9 > 5 - 4$ **67.** $16 \neq 19$ **69.** $2 \leq 3$ **71.** Seven is less than nineteen; true **73.** One-third is not equal to three-tenths; true **75.** Eight is greater than or equal to eleven; false **77.** $30 > 5$ **79.** $3 \leq 12$ **81.** Alaska, Texas, California, Idaho **83.** Alaska, Texas, California, Idaho, Missouri

Section 1.2 (pages 43–46)

1. 11 **3.** $13 + x$; 16 **5.** expression; equation **7.** The equation would be $5x - 9 = 49$. **9.** Answers will vary. Two such pairs are $x = 0, y = 6$ and $x = 1, y = 4$. To find a pair, choose one number, substitute it for a variable, then calculate the value for the other variable. **11. (a)** 64 **(b)** 144 **13. (a)** $\frac{7}{8}$ **(b)** $\frac{13}{12}$ **15. (a)** 9.569 **(b)** 14.353 **17. (a)** 52 **(b)** 114 **19. (a)** 12 **(b)** 33 **21. (a)** 6 **(b)** $\frac{9}{5}$ **23. (a)** $\frac{4}{3}$ **(b)** $\frac{13}{6}$ **25. (a)** $\frac{2}{7}$ **(b)** $\frac{16}{27}$ **27. (a)** 12 **(b)** 55 **29. (a)** 1 **(b)** $\frac{28}{17}$ **31. (a)** 3.684 **(b)** 8.841 **33.** $12x$ **35.** $x - 2$ **37.** $7 - \frac{1}{3}x$ **39.** $2x - 6$ **41.** $\frac{12}{x + 3}$ **43.** $6(x - 4)$ **45.** The word *and* does not signify addition here. In the phrase "the product of a number and 6," *and* connects two quantities to be multiplied. **47.** no **49.** yes **51.** yes **53.** no **55.** yes **57.** yes **59.** $x + 8 = 18$ **61.** $2x + 5 = 5$ **63.** $16 - \frac{3}{4}x = 13$ **65.** $3x = 2x + 8$ **67.** expression **69.** equation **71.** 64.9 yr **72.** 68.5 yr **73.** 72.8 yr **74.** 78.1 yr

Section 1.3 (pages 55–56)

1. 4 **3.** 0 **5.** One example is $\sqrt{13}$. There are others. **7. (a)** 3, 7 **(b)** 0, 3, 7 **(c)** $-9, 0, 3, 7$ **(d)** $-9, -1\frac{1}{4}, -\frac{3}{5}, 0, 3, 5.9, 7$ **(e)** $-\sqrt{7}, \sqrt{5}$ **(f)** All are real numbers. **9.** 2,845,000 **11.** -2809

13. 

15. 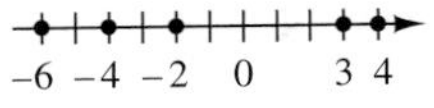

17. 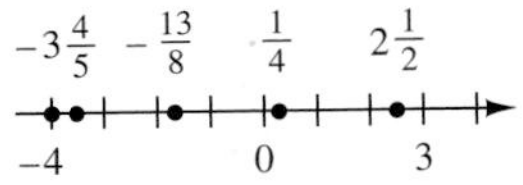

19. -11 **21.** -21 **23.** -100

ANSWERS

25. $-\frac{2}{3}$ 27. false 29. true 31. (a) 2 (b) 2 33. (a) -6 (b) 6 35. (a) $\frac{3}{4}$ (b) $\frac{3}{4}$ 37. 7 39. -12 41. $-\frac{2}{3}$ 43. 9 45. false 47. true 49. No; the statement is false for one number, 0. 51. electronic components and accessories, 2003 to 2004 53. 2002 to 2003

Section 1.4 (pages 63–66)

1. Add -2 and 5. 3. Add -1 and -3. 5. 2 7. -3 9. -10 11. -13 13. -15.9 15. 5 17. 13 19. 0 21. -8 23. $\frac{3}{10}$ 25. $\frac{1}{2}$ 27. $-\frac{3}{4}$ 29. -1.6 31. -8.7 33. -25 35. $-\frac{1}{4}$, or -0.25 37. true 39. false 41. true 43. false 45. true 47. false 49. It must be negative and have the greater absolute value. 50. The sum of a positive number and 5 cannot be -7. 51. It must be positive and have the greater absolute value. 52. The sum of a negative number and -8 cannot be 2. 53. Add the absolute values of the numbers. The sum will be negative. 55. $-5 + 12 + 6$; 13 57. $[-19 + (-4)] + 14$; -9 59. $[-4 + (-10)] + 12$; -2 61. $\left[\frac{5}{7} + \left(-\frac{9}{7}\right)\right] + \frac{2}{7}; -\frac{2}{7}$ 63. $-\$62$ 65. -184 m 67. 37 yd 69. 120°F 71. $-\$107$ 73. -12

Section 1.5 (pages 71–76)

1. -8; -6 3. $7 - 12$; $12 - 7$ 5. -4 7. -10 9. -16 11. 11 13. 19 15. -4 17. 5 19. 0 21. $\frac{3}{4}$ 23. $-\frac{11}{8}$ 25. $\frac{15}{8}$ 27. 13.6 29. -11.9 31. -2.8 33. -6.3 35. -28 37. -18 39. $\frac{37}{12}$ 41. -42.04 43. For example, let $a = 1, b = 1$ or let $a = 2, b = 2$. In general, choose $a = b$. 45. 8 47. For example, $-8 - (-2) = -6$. 49. $4 - (-8)$; 12 51. $-2 - 8$; -10 53. $[9 + (-4)] - 7$; -2 55. $[8 - (-5)] - 12$; 1 57. -58°F 59. 14,776 ft 61. $-\$80$ 63. $105,000 65. 17 67. $1045.55 69. 469 B.C. 71. $323.83 73. 14 ft 75. 40,776 ft 77. 11.3% 79. $649 billion 81. $15,000 83. $2900 85. positive 87. positive

Section 1.6 (pages 85–88)

1. greater than 0 3. less than 0 5. greater than 0 7. -28 9. 30 11. 0 13. $\frac{5}{6}$ 15. -2.38 17. $\frac{3}{2}$ 19. -3 21. -2 23. 16 25. 0 27. undefined 29. $\frac{3}{2}$ 31. C 33. 3 35. 7 37. 4 39. -3 41. -1 43. $\frac{7}{4}$ 45. 68 47. -228 49. 1 51. 0 53. -6 55. undefined 57. $-12 + 4(-7)$; -40 59. $-1 - 2(-8)(2)$; 31 61. $-3[3 - (-7)]$; -30 63. $\frac{3}{10}[-2 + (-28)]$; -9 65. $\frac{-20}{-8 + (-2)}$; 2 67. $\frac{-18 + (-6)}{2(-4)}$; 3 69. $\frac{-\frac{2}{3}\left(-\frac{1}{5}\right)}{\frac{1}{7}}$; $\frac{14}{15}$ 71. $9x = -36$ 73. $\frac{x}{4} = -1$ 75. $x - \frac{9}{11} = 5$ 77. $\frac{6}{x} = -3$ 79. 42 80. 5 81. $8\frac{2}{5}$ 82. $8\frac{2}{5}$ 83. 2 84. $-12\frac{1}{2}$

Summary Exercises on Operations with Real Numbers (pages 89–90)

1. -16 2. 4 3. 0 4. -24 5. -17 6. 76 7. -18 8. 90 9. 38 10. 4 11. -5 12. 5 13. $-\frac{7}{2}$, or $-3\frac{1}{2}$ 14. 4 15. 13 16. $\frac{5}{4}$, or $1\frac{1}{4}$ 17. 9 18. $\frac{37}{10}$, or $3\frac{7}{10}$ 19. 0 20. 25 21. 14 22. 0 23. -4 24. $\frac{6}{5}$, or $1\frac{1}{5}$ 25. -1 26. $\frac{52}{37}$, or $1\frac{15}{37}$ 27. $\frac{17}{16}$, or $1\frac{1}{16}$ 28. $-\frac{2}{3}$ 29. 3.33 30. 1.02 31. -13 32. 0 33. 24 34. -7 35. 37 36. -3 37. -1 38. $\frac{1}{2}$ 39. $-\frac{5}{13}$ 40. 5 41. undefined 42. 0

Section 1.7 (pages 97–100)

1. B 3. C 5. B 7. G 9. commutative property 11. associative property 13. inverse property 15. inverse property 17. identity property 19. commutative property 21. distributive property 23. identity property 25. distributive property 27. (a) 0 (b) 1, -1 29. $25 - (6 - 2) = 25 - 4 = 21$ and $(25 - 6) - 2 = 19 - 2 = 17$. Since these results are different, subtraction is not associative. 31. $7 + r$ 33. s 35. $-6x + (-6)7$; $-6x - 42$ 37. $w + [5 + (-3)]$; $w + 2$ 39. We must multiply $\frac{3}{4}$ by 1 in the form $\frac{3}{3}$: $\frac{3}{4} \cdot \frac{3}{3} = \frac{9}{12}$. 41. 2 43. $4t + 12$ 45. $-8r - 24$ 47. $-5y + 20$ 49. $-16y - 20z$ 51. $8(z + w)$ 53. $5(3 + 17)$; 100 55. $7(2v + 5r)$ 57. $24r + 32s - 40y$ 59. $-24x - 9y - 12z$ 61. $-4t - 5m$ 63. $5c + 4d$ 65. $3q - 5r + 8s$ 67. Answers will vary. For example, "putting on your socks" and "putting on your shoes" 69. false 71. (foreign sales) clerk; foreign (sales clerk) 73. 0 74. $-3(5) + (-3)(-5)$ 75. -15 76. The product $-3(-5)$ must equal 15, since it is the additive inverse of -15.

Section 1.8 (pages 105–106)

1. C 3. A 5. $15x$ 7. $5t$ 9. $4r + 11$ 11. $5 + 2x - 6y$ 13. $-7 + 3p$ 15. -12 17. 5 19. 1 21. -1 23. 74 25. Answers will vary. For example, $-3x$ and $4x$ 27. like 29. unlike 31. like 33. unlike 35. We cannot "add" two unlike terms to obtain a single term, so we must be able to identify like terms in order to combine them. 37. $1 - 2x$ 39. $-\frac{1}{3}t - \frac{28}{3}$ 41. $-4.1r + 4.2$ 43. $-2y^2 + 3y^3$ 45. $-19p + 16$ 47. $-\frac{3}{2}y + 16$ 49. $-16y + 63$ 51. $(x + 3) + 5x$; $6x + 3$ 53. $(13 + 6x) - (-7x)$; $13 + 13x$ 55. $2(3x + 4) - (-4 + 6x)$; 12 57. Wording may vary. One example is "the difference between 9 times a number and the sum of the number and 2." 59. $1000 + 5x$ (dollars) 60. $750 + 3y$ (dollars) 61. $1000 + 5x + 750 + 3y$ (dollars) 62. $1750 + 5x + 3y$ (dollars)

Chapter 1 Review Exercises (pages 113–118)

1. 625 2. 0.00000081 3. 0.009261 4. $\frac{125}{8}$ 5. 27 6. 200 7. 7 8. $\frac{20}{3}$ 9. $13 < 17$ 10. $5 + 2 \neq 10$ 11. Six is less than fifteen. 12. Answers will vary. One example is $2 + 5 \geq \frac{16}{2}$. 13. 30 14. 60

15. 14 **16.** 13 **17.** $x + 6$ **18.** $8 - x$ **19.** $6x - 9$ **20.** $12 + \frac{3}{5}x$ **21.** yes **22.** no **23.** $2x - 6 = 10$ **24.** $4x = 8$ **25.** equation **26.** expression

27. 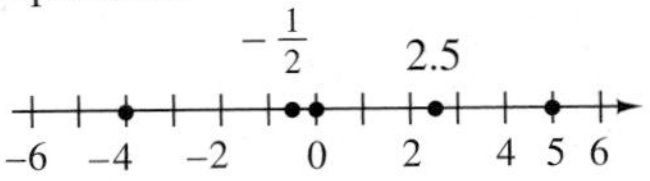

28. 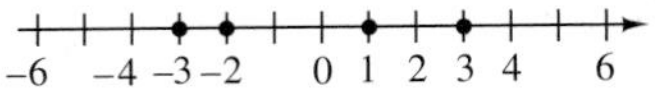

29. 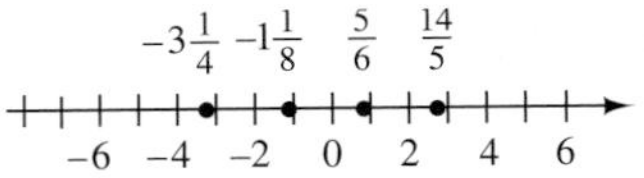

30. 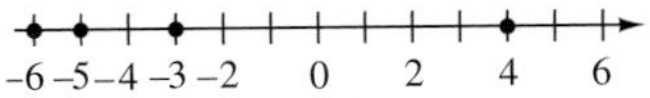

31. -10 **32.** -9 **33.** $-\frac{3}{4}$ **34.** $-|23|$ **35.** true **36.** true **37.** true **38.** false **39.** -3 **40.** -19 **41.** -7 **42.** 9 **43.** -6 **44.** -4 **45.** -17 **46.** $-\frac{29}{36}$ **47.** -10 **48.** -19 **49.** $(-31 + 12) + 19$; 0 **50.** $[-4 + (-8)] + 13$; 1 **51.** \$26.25 **52.** -10°F **53.** -11 **54.** -1 **55.** 7 **56.** $-\frac{43}{35}$ **57.** 10.31 **58.** -12 **59.** 2 **60.** 1 **61.** $-4 - (-6)$; 2 **62.** $[4 + (-8)] - 5$; -9 **63.** $[18 - (-23)] - 15$; 26 **64.** $19 - (-7 - 12)$; 38 **65.** 38 **66.** 12,284.30 **67.** 56 thousand **68.** -293 thousand **69.** -126 thousand **70.** 98 thousand **71.** 36 **72.** -105 **73.** $\frac{1}{2}$ **74.** 10.08 **75.** -20 **76.** -10 **77.** -24 **78.** -35 **79.** 4 **80.** -20 **81.** $-\frac{3}{4}$ **82.** 11.3 **83.** -1 **84.** undefined **85.** 1 **86.** 0 **87.** -18 **88.** -18 **89.** 125 **90.** -423 **91.** $-4(5) - 9$; -29 **92.** $\frac{5}{6}[12 + (-6)]$; 5 **93.** $\frac{12}{8 + (-4)}$; 3 **94.** $\frac{-20(12)}{15 - (-15)}$; -8 **95.** $\frac{x}{x + 5} = -2$ **96.** $8x - 3 = -7$ **97.** identity property **98.** identity property **99.** inverse property **100.** inverse property **101.** associative property **102.** associative property **103.** distributive property **104.** commutative property **105.** $(7 + 1)y$; $8y$ **106.** $-12 \cdot 4 - 12(-t)$; $-48 + 12t$ **107.** $3(2s + 4y)$; $6s + 12y$ **108.** $-1(-4r) + (-1)(5s)$; $4r - 5s$ **109.** $17p^2$ **110.** $16r^2 + 7r$ **111.** $-19k + 54$ **112.** $5s - 6$ **113.** $-45t - 23$ **114.** $-45t^2 - 23.4t$ **115.** $-2(3x) - 7x$; $-13x$ **116.** $\frac{x + 9}{x - 6}$ **117.** No. The use of *and* there indicates the two quantities that are to be multiplied. **118.** Answers may vary. For example, "3 times the difference between 4 times a number and 6" **119.** 16 **120.** $\frac{25}{36}$ **121.** -26 **122.** $\frac{8}{3}$ **123.** $-\frac{1}{24}$ **124.** $\frac{7}{2}$ **125.** 2 **126.** 77.6 **127.** $-1\frac{1}{2}$ **128.** 11 **129.** $-\frac{28}{15}$ **130.** 24 **131.** -11 **132.** -6 **133.** -47°F **134.** 15 points

Chapter 1 Test (pages 119–120)

1. true **2.** false **3.** 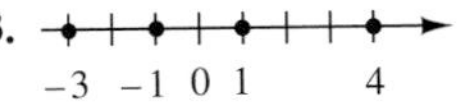

4. $-|-8|$ (or -8) **5.** -1.277 **6.** $\frac{-6}{2 + (-8)}$; 1 **7.** negative **8.** 4 **9.** $-2\frac{5}{6}$ **10.** 2 **11.** 6 **12.** 108 **13.** 11 **14.** $\frac{30}{7}$ **15.** -70 **16.** 3 **17.** 178°F **18.** D **19.** A **20.** E **21.** B **22.** C **23.** $-9x^2 - 6x - 8$ **24.** identity and distributive properties **25.** **(a)** -18 **(b)** -18 **(c)** The distributive property tells us that the two methods produce equal results.

Chapter 2 Equations, Inequalities, and Applications

Section 2.1 (pages 129–130)

1. **(a)** expression; $x + 15$ **(b)** expression; $m + 7$ **(c)** equation; $\{-1\}$ **(d)** equation; $\{-17\}$ **3.** A and B **5.** $\{12\}$ **7.** $\{-3\}$ **9.** $\{4\}$ **11.** $\{-9\}$ **13.** $\left\{-\frac{3}{4}\right\}$ **15.** $\{6.3\}$ **17.** $\{-16.9\}$ **19.** $\{-10\}$ **21.** $\{-13\}$ **23.** $\left\{\frac{4}{15}\right\}$ **25.** $\{7\}$ **27.** $\{-4\}$ **29.** $\{3\}$ **31.** $\{-2\}$ **33.** $\{4\}$ **35.** $\{-16\}$ **37.** $\{2\}$ **39.** $\{2\}$ **41.** $\{-4\}$ **43.** $\{4\}$ **45.** $\{0\}$ **47.** $\left\{\frac{7}{15}\right\}$ **49.** $\{7\}$ **51.** $\{-4\}$ **53.** $\{13\}$ **55.** $\{29\}$ **57.** $\{18\}$ **59.** Answers will vary. One example is $x - 6 = -8$.

Section 2.2 (pages 135–136)

1. $\frac{3}{2}$ **3.** 10 **5.** $-\frac{2}{9}$ **7.** -1 **9.** 6 **11.** -4 **13.** 0.12 **15.** -1 **17.** **(a)** and **(c)**: multiplication property of equality; **(b)** and **(d)**: addition property of equality **19.** B **21.** $\{6\}$ **23.** $\left\{\frac{15}{2}\right\}$ **25.** $\{-5\}$ **27.** $\left\{-\frac{18}{5}\right\}$ **29.** $\{12\}$ **31.** $\{0\}$ **33.** $\{-12\}$ **35.** $\{40\}$ **37.** $\{-12.2\}$ **39.** $\{-48\}$ **41.** $\{72\}$ **43.** $\{-35\}$ **45.** $\{14\}$ **47.** $\left\{-\frac{27}{35}\right\}$ **49.** $\{3\}$ **51.** $\{-5\}$ **53.** $\{20\}$ **55.** $\{-12\}$ **57.** $\{7\}$ **59.** $\{0\}$ **61.** $\{-6\}$ **63.** Answers will vary. One example is $\frac{3}{2}x = -6$. **65.** $-4x = 10$; $-\frac{5}{2}$

Section 2.3 (pages 145–148)

1. Use the addition property of equality to subtract 8 from each side.
3. Clear parentheses by using the distributive property.
5. Clear fractions by multiplying by the LCD, 6.
7. $\{4\}$ **9.** $\{-5\}$ **11.** $\left\{\frac{5}{2}\right\}$ **13.** $\left\{-\frac{1}{2}\right\}$ **15.** $\{5\}$ **17.** $\{1\}$ **19.** $\left\{-\frac{5}{3}\right\}$ **21.** $\{-1\}$ **23.** $\emptyset$ **25.** {all real numbers} **27.** $\{2\}$ **29.** $\left\{-\frac{5}{3}\right\}$ **31.** D **33.** $\{7\}$ **35.** $\{12\}$ **37.** $\{11\}$ **39.** $\{0\}$ **41.** $\left\{\frac{3}{25}\right\}$ **43.** $\{60\}$ **45.** $\{4\}$ **47.** $\{5000\}$ **49.** $\left\{-\frac{72}{11}\right\}$ **51.** $\{0\}$ **53.** $\emptyset$ **55.** {all real numbers} **57.** $\{-6\}$ **59.** $\{15\}$ **61.** $12 - q$ **63.** $\frac{9}{z}$ **65.** $a + 12$; $a - 2$ **67.** $25r$

Summary Exercises on Solving Linear Equations (pages 151–152)

1. $\{-5\}$ **2.** $\{4\}$ **3.** $\{-5.1\}$ **4.** $\{25\}$ **5.** $\{-25\}$ **6.** $\{-6\}$ **7.** $\{-3\}$ **8.** $\{-16\}$ **9.** $\{7\}$ **10.** $\left\{-\frac{96}{5}\right\}$ **11.** $\{5\}$ **12.** $\{23.7\}$ **13.** {all real numbers} **14.** $\{1\}$ **15.** $\{-16\}$ **16.** $\emptyset$ **17.** $\{6\}$

ANSWERS

18. {3} **19.** ∅ **20.** $\left\{\frac{7}{3}\right\}$ **21.** {25} **22.** {−10.8} **23.** {3} **24.** {7} **25.** {2} **26.** {all real numbers} **27.** $\left\{-\frac{2}{7}\right\}$ **28.** {10} **29.** $\left\{\frac{14}{17}\right\}$ **30.** $\left\{-\frac{5}{2}\right\}$ **31.** {all real numbers} **32.** {64}

Section 2.4 (pages 161–166)

1. The procedure should include the following steps: read the problem carefully; assign a variable to represent the unknown to be found; write down variable expressions for any other unknown quantities; translate into an equation; solve the equation; state the answer; check your solution. **3.** D; There cannot be a fractional number of cars. **5.** A; Distance cannot be negative. **7.** 7 **9.** −8 **11.** 6 **13.** −3 **15.** Pennsylvania: 35 screens; Ohio: 33 screens **17.** Democrats: 44; Republicans: 55 **19.** The Police: \$133.2 million; Kenny Chesney: \$71.1 million **21.** wins: 61; losses: 21 **23.** orange: 97 mg; pineapple: 25 mg **25.** 1950 Denver nickel: \$16.00; 1944 Philadelphia nickel: \$12.00 **27.** onions: 81.3 kg; grilled steak: 536.3 kg **29.** 420 lb **31.** American: 18; United: 11; Southwest: 26 **33.** gold: 35; silver: 39; bronze: 29 **35.** 36 million mi **37.** *A* and *B*: 40°; *C*: 100° **39.** yes, 90°; yes, 45° **41.** 18° **43.** 20° **45.** 39° **47.** 50° **49.** 68, 69 **51.** 146, 147 **53.** 10, 12 **55.** 10, 11 **57.** 18 **59.** 2002: \$6.54 billion; 2003: \$6.67 billion; 2004: \$6.77 billion

Section 2.5 (pages 173–178)

1. (a) The perimeter of a plane geometric figure is the distance around the figure. **(b)** The area of a plane geometric figure is the measure of the surface covered or enclosed by the figure. **3.** four **5.** area **7.** perimeter **9.** area **11.** area **13.** $P = 26$ **15.** $A = 64$ **17.** $b = 4$ **19.** $t = 5.6$ **21.** $I = 1575$ **23.** $r = 2.6$ **25.** $A = 50.24$ **27.** $V = 150$ **29.** $V = 52$ **31.** $V = 7234.56$ **33.** length: 18 in.; width: 9 in. **35.** length: 14 m; width: 4 m **37.** shortest: 5 in.; medium: 7 in., longest: 8 in. **39.** two equal sides: 7 m; third side: 10 m **41.** about 154,000 ft^2 **43.** perimeter: 5.4 m; area: 1.8 m^2 **45.** 194.48 ft^2 **47.** 10 ft **49.** 23,800.10 ft^2 **51.** length: 36 in.; maximum volume: 11,664 in.^{3} **53.** 48°, 132° **55.** 55°, 35° **57.** 51°, 51° **59.** 105°, 105° **61.** $t = \frac{d}{r}$ **63.** $H = \frac{V}{LW}$ **65.** $b = P - a - c$ **67.** $r = \frac{C}{2\pi}$ **69.** $r = \frac{I}{pt}$ **71.** $h = \frac{2A}{b}$ **73.** $h = \frac{3V}{\pi r^2}$ **75.** $W = \frac{P - 2L}{2}$ **77.** $m = \frac{y - b}{x}$ **79.** $y = \frac{C - Ax}{B}$ **81.** $r = \frac{M - C}{C}$ **83.** $a = \frac{P - 2b}{2}$

Section 2.6 (pages 185–188)

1. (a) C **(b)** D **(c)** B **(d)** A **3.** $\frac{6}{7}$ **5.** $\frac{18}{55}$ **7.** $\frac{5}{16}$ **9.** $\frac{4}{15}$ **11.** $\frac{6}{5}$ **13.** 10 lb; \$0.429 **15.** 4 oz; \$0.623 **17.** 128 oz; \$0.051 **19.** 36 oz; \$0.049 **21.** {35} **23.** {7} **25.** {2} **27.** {−1} **29.** {5} **31.** $\left\{-\frac{31}{5}\right\}$ **33.** \$30.00 **35.** \$56.85 **37.** 203.3 million vehicles **39.** 50,000 fish **41.** 4 ft **43.** $2\frac{5}{8}$ cups **45.** \$363.84 **47.** 9.234 **49.** 200% **51.** 30,000 **53.** 27% **55.** C **57.** \$304 **59.** 5.1% **61.** 79.4% **63.** 892% **65.** \$237 **67.** \$272 **69.** 30 **70. (a)** $5x = 12$ **(b)** $\left\{\frac{12}{5}\right\}$ **71.** $\left\{\frac{12}{5}\right\}$ **72.** Both methods give the same solution set.

Summary Exercises on Solving Applied Problems (pages 189–190)

1. 48 **2.** 80° **3.** 4 **4.** 104°, 104° **5.** 3 **6.** 18, 20 **7.** 140°, 40° **8.** 36 quart cartons **9.** 24.34 in.; 727.28 in.^{2} **10.** 12.42 cm **11.** $16\frac{2}{3}$% **12.** 510 calories **13.** 4000 calories **14.** Zach Johnson: 289; Phil Mickelson: 299 **15.** \$38.10 **16.** 45.2% **17.** 32 oz; \$0.053 **18.** 8 gold medals

Section 2.7 (pages 201–204)

1. $x > -4$ **3.** $x \le 4$ **5.** $-1 < x \le 2$

7. $(-\infty, 4]$

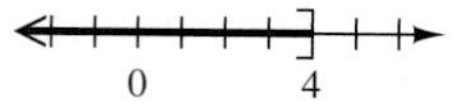

9. $(-3, \infty)$

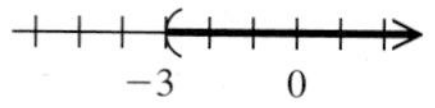

11. $[8, 10]$

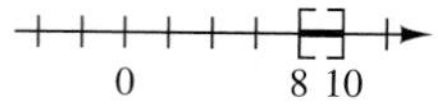

13. $(0, 10]$

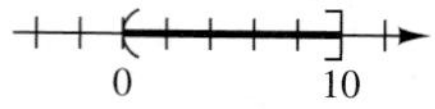

15. $[1, \infty)$

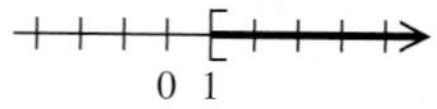

17. $[5, \infty)$

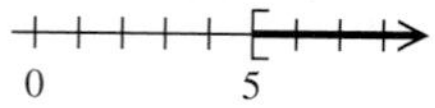

19. $(-\infty, -6)$

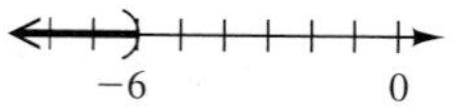

21. It must be reversed when multiplying or dividing by a negative number.

23. $(-\infty, 6)$

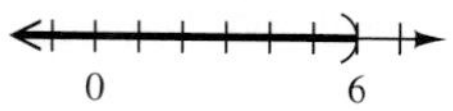

25. $[-10, \infty)$

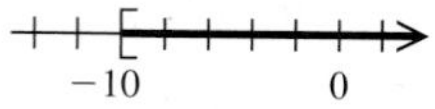

27. $(-\infty, -3)$

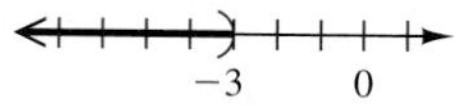

29. $(-\infty, 0]$

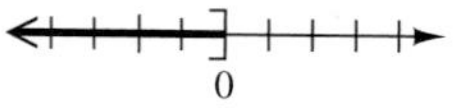

31. $(20, \infty)$

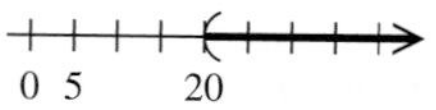

33. $[-3, \infty)$

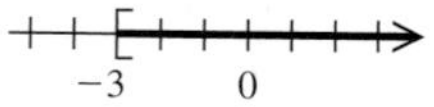

35. $(-\infty, -3]$

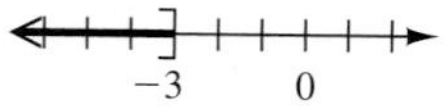

37. $(-1, \infty)$

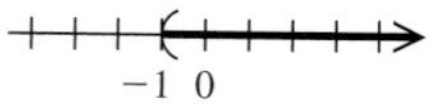

39. $[-5, \infty)$

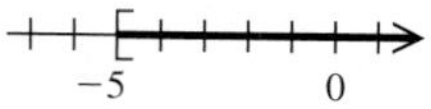

41. $(-\infty, 1)$

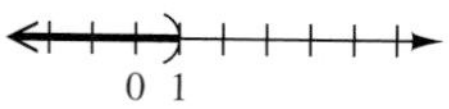

43. $(-\infty, 0]$

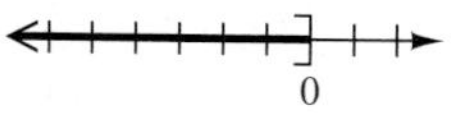

45. $\left(-\frac{1}{2}, \infty\right)$

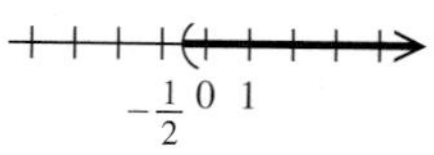

47. $[4, \infty)$

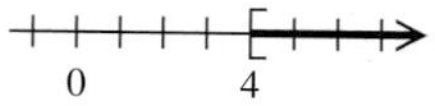

49. $(-\infty, 32)$

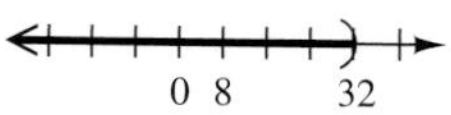

51. $\left(-\infty, \frac{76}{11}\right)$

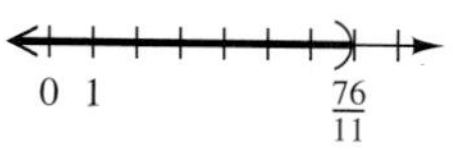

53. $\left[\frac{5}{12}, \infty\right)$ $-1\ 0\ \frac{5}{12}$

55. $[-1, 6]$ $-1\ 0\ \ 6$

57. $(1, 3)$ $0\ 1\ \ 3$

59. $\left(-\frac{11}{6}, -\frac{2}{3}\right]$ $-\frac{11}{6}\ -\frac{2}{3}\ 0$

61. $\left[-\frac{1}{3}, \frac{1}{9}\right)$ $-1\ -\frac{1}{3}\ 0\ \frac{1}{9}\ \ 1$

63. 88 or more **65.** all numbers greater than 16 **67.** It has never exceeded 40°C. **69.** 32 or greater **71.** 15 min

73. 0 4

74. 0 4

75. 0 4

76. It is the set of all real numbers. 0 4

77. The graph would be the set of all real numbers.

Chapter 2 Review Exercises (pages 211–214)

1. $\{9\}$ **2.** $\{4\}$ **3.** $\{-6\}$ **4.** $\left\{\frac{3}{2}\right\}$ **5.** $\{20\}$ **6.** $\left\{-\frac{61}{2}\right\}$ **7.** $\{15\}$ **8.** $\{0\}$ **9.** $\emptyset$ **10.** {all real numbers} **11.** $-\frac{7}{2}$ **12.** 20 **13.** Hawaii: 6425 mi^2; Rhode Island: 1212 mi^2 **14.** Seven Falls: 300 ft; Twin Falls: 120 ft **15.** 80° **16.** 11, 13 **17.** $h = 11$ **18.** $A = 28$ **19.** $r = 4.75$ **20.** $V = 3052.08$ **21.** $h = \frac{A}{b}$ **22.** $h = \frac{2A}{b + B}$ **23.** 135°, 45° **24.** 100°, 100° **25.** perimeter: 326.5 ft; area: 6538.875 ft^2 **26.** diameter: 46.78 ft; radius: 23.39 ft **27.** $\frac{3}{2}$ **28.** $\frac{5}{14}$ **29.** $\frac{3}{4}$ **30.** $\frac{1}{12}$ **31.** $\left\{\frac{7}{2}\right\}$ **32.** $\left\{-\frac{8}{3}\right\}$ **33.** $\left\{\frac{25}{19}\right\}$ **34.** 40% means $\frac{40}{100}$, or $\frac{2}{5}$. It is the same as the ratio of 2 to 5. **35.** $6\frac{2}{3}$ lb **36.** 36 oz **37.** 375 km **38.** 25.5 oz; \$0.137 **39.** 17.48 **40.** 175% **41.** $33\frac{1}{3}$% **42.** 2500 **43.** \$26,595 **44.** \$350.46

45. $[-4, \infty)$ $-4\ \ 0$

46. $(-\infty, 7)$ $0\ \ 7$

47. $[-5, 6)$ $-5\ \ 0\ \ 6$

48. $\left[\frac{1}{2}, \infty\right)$ $-1\ 0\ \frac{1}{2}$

49. $[-3, \infty)$ $-3\ \ 0$

50. $(-\infty, 2)$ $0\ \ 2$

51. $[3, \infty)$ $0\ \ 3$

52. $[46, \infty)$ $0\ 10\ \ 46$

53. $(-\infty, -5)$ $-5\ \ 0$

54. $(-\infty, -37)$ $-37\ \ -10\ 0$

55. $\left[-2, \frac{3}{2}\right)$ $-2\ \ 0\ \frac{3}{2}$

56. $(1, 5]$ $0\ 1\ \ 5$

57. 88 or more **58.** all numbers less than or equal to $-\frac{1}{3}$ **59.** $\{7\}$ **60.** $r = \frac{d}{2}$ **61.** $(-\infty, 2)$ **62.** $\{-9\}$ **63.** $\{70\}$ **64.** $\left\{\frac{13}{4}\right\}$ **65.** $\emptyset$ **66.** {all real numbers} **67.** Pizza Hut: \$9.35 billion; Domino's: \$5.1 billion **68.** 2.0 in. **69.** gold: 11; silver: 12; bronze: 6 **70.** D: 22°; E: 44°; F: 114° **71.** 44 m **72.** 70 ft **73.** 100 oz; \$0.060 **74.** 24°, 66° **75.** 92 or more **76.** \$67.50

Chapter 2 Test (pages 215–216)

1. $\{6\}$ **2.** $\{-6\}$ **3.** $\left\{\frac{13}{4}\right\}$ **4.** $\{-10.8\}$ **5.** $\emptyset$ **6.** $\{21\}$ **7.** $\{30\}$ **8.** {all real numbers} **9.** 67 wins, 15 losses **10.** Hawaii: 4021 mi^2; Maui: 728 mi^2; Kauai: 551 mi^2 **11.** 50° **12. (a)** $W = \frac{P - 2L}{2}$ **(b)** 18 **13.** 100°, 80° **14.** 75°, 75° **15.** $\{6\}$ **16.** $\{-29\}$ **17.** 32 oz; \$0.250 **18.** 2300 mi **19.** 4.2% **20. (a)** $x < 0$ **(b)** $-2 < x \le 3$

21. $(-\infty, 11)$ $0\ \ 11$

22. $[-3, \infty)$ $-3\ \ 0$

23. $(-\infty, 4]$ $0\ \ 4$

24. $(-2, 6]$ $-2\ 0\ \ 6$

25. 83 or more

Cumulative Review Exercises: Chapters R–2 (pages 219–220)

1. $\frac{3}{8}$ **2.** $\frac{3}{4}$ **3.** $\frac{31}{20}$ **4.** $\frac{551}{40}$, or $13\frac{31}{40}$ **5.** 6 **6.** $\frac{6}{5}$ **7.** 34.03 **8.** 27.31 **9.** 30.51 **10.** 56.3 **11.** 35 yd **12.** $7\frac{1}{2}$ cups **13.** $3\frac{3}{8}$ in. **14.** \$2769.88 **15.** true **16.** true **17.** 7 **18.** 1 **19.** 13 **20.** -40 **21.** -12 **22.** undefined **23.** -6 **24.** 28 **25.** 1 **26.** 0 **27.** $\frac{73}{18}$ **28.** -64 **29.** -134 **30.** $-\frac{29}{6}$ **31.** distributive property **32.** commutative property **33.** inverse property **34.** identity property **35.** $7p - 14$ **36.** $2k - 11$ **37.** $\{7\}$ **38.** $\{-4\}$ **39.** $\{-1\}$ **40.** $\left\{-\frac{3}{5}\right\}$ **41.** $\{2\}$ **42.** $\{-13\}$ **43.** $\{26\}$ **44.** $\{-12\}$ **45.** $c = P - a - b - B$ **46.** $s = \frac{P}{4}$

47. $(-\infty, 2]$ $0\ \ 2$

48. $(-\infty, 1)$ $0\ 1$

49. \$2943.00 **50.** \$3750 **51.** \$230.50 **52.** \$98.45 **53.** 30 cm **54.** 16 in.

Chapter 3 Graphs of Linear Equations and Inequalities; Functions

Section 3.1 (pages 231–236)

1. 2005, 2006, 2007 **3.** 2001: about 165 billion lb; 2007: about 185 billion lb **5.** from 2000 to 2005; about \$0.85 **7.** The price of a gallon of gas was decreasing. **9.** does; do not **11.** y **13.** 6 **15.** yes **17.** yes **19.** no **21.** yes **23.** no **25.** No. For two ordered pairs (x, y) to be equal, the x-values must be equal and the y-values must be equal. Here we have $4 \neq -1$ and $-1 \neq 4$. **27.** 11 **29.** $-\frac{7}{2}$ **31.** -4 **33.** -5 **35.** 4; 6; -6; $(0, 4)$; $(6, 0)$; $(-6, 8)$ **37.** 3; -5; -15; $(0, 3)$; $(-5, 0)$; $(-15, -6)$ **39.** -9; -9; -9 **41.** -6; -6; -6 **43.** 8; 8; 8 **45.** $(2, 4)$; I **47.** $(-5, 4)$; II **49.** $(3, 0)$; no quadrant **51.** negative; negative **53.** positive; negative **55.** If $xy < 0$, then either $x < 0$ and $y > 0$ or $x > 0$ and $y < 0$. If $x < 0$ and $y > 0$, then the point lies in quadrant II. If $x > 0$ and $y < 0$, then the point lies in quadrant IV.

57.–68.

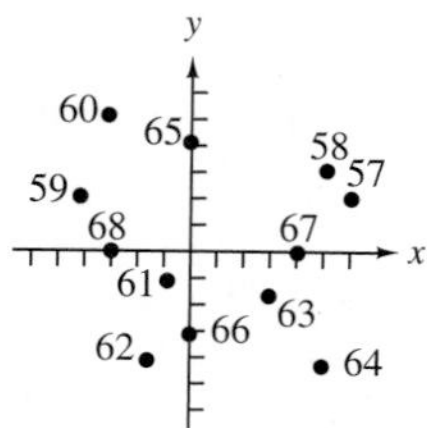

69. -3; 6; -2; 4

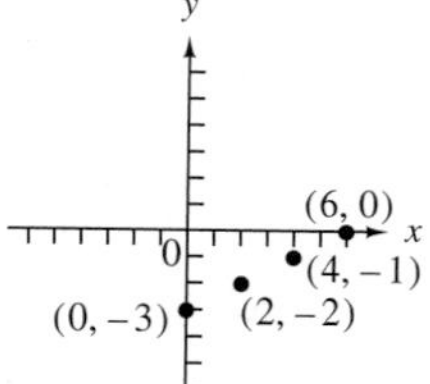

71. -3; 4; -6; $-\frac{4}{3}$

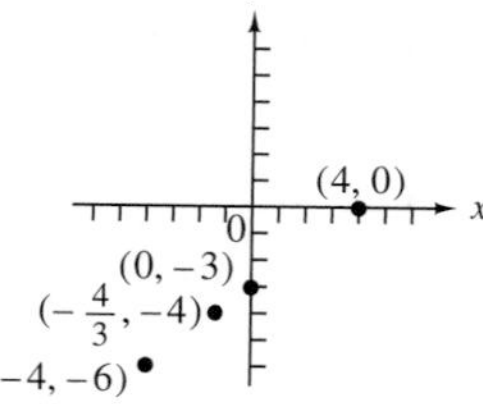

73. -4; -4; -4; -4

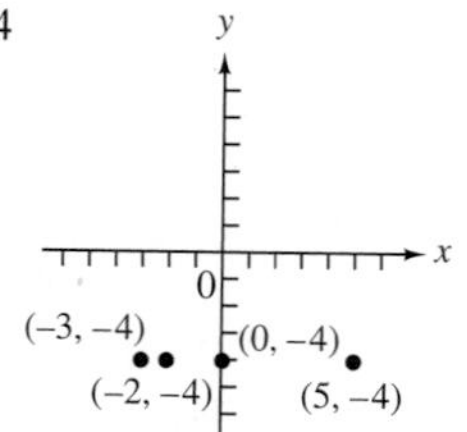

75. The points in each graph appear to lie on a straight line. **77. (a)** $(5, 45)$ **(b)** $(6, 50)$ **79. (a)** $(2000, 32.4)$, $(2001, 31.6)$, $(2002, 31.6)$, $(2003, 30.1)$, $(2004, 29.0)$, $(2005, 27.5)$ **(b)** $(2007, 27.1)$ means that 27.1 percent of 2-year college students in 2007 received a degree within 3 years.

(c)

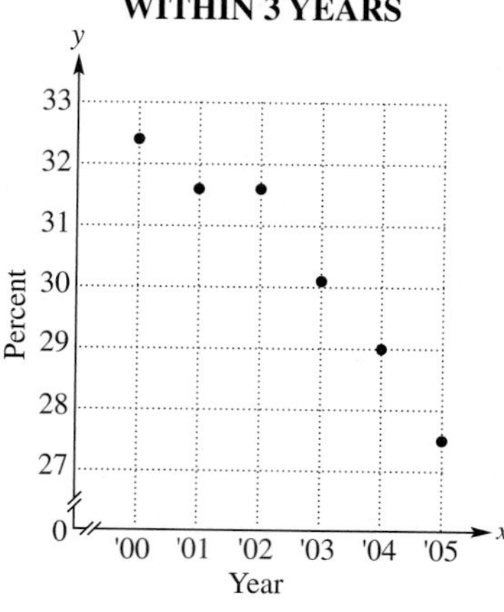

(d) With the exception of the point for 2002, the points lie approximately on a straight line. Rates at which 2-year college students complete a degree within 3 years are generally decreasing.

81. (a) 157, 141, 125, 109 **(b)** $(20, 157)$, $(40, 141)$, $(60, 125)$, $(80, 109)$

(c)

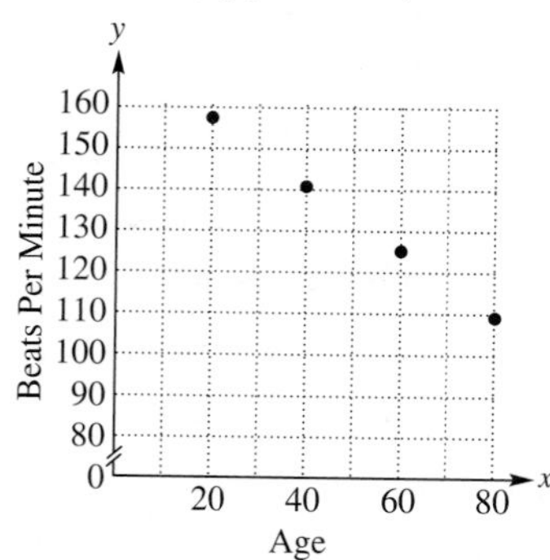

The points lie in a linear pattern.

Section 3.2 (pages 245–250)

1. 5; 5; 3

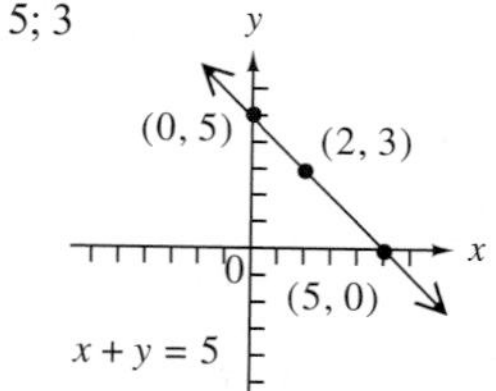

3. 1; 3; -1

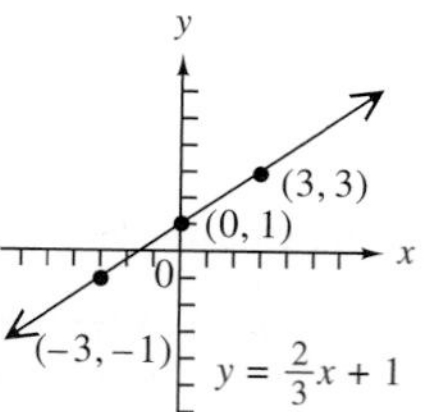

5. -6; -2; -5

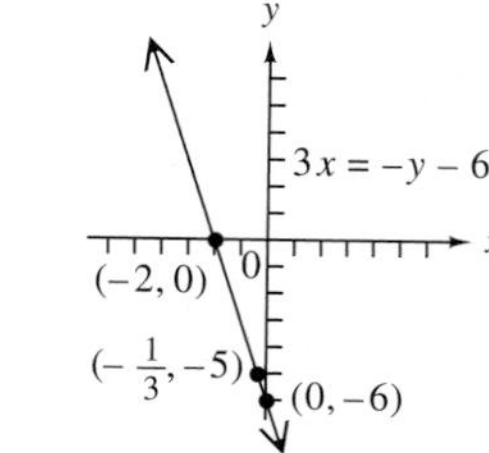

7. (a) A **(b)** C **(c)** D **(d)** B **9.** $(12, 0)$; $(0, -8)$ **11.** $(0, 0)$; $(0, 0)$

13.
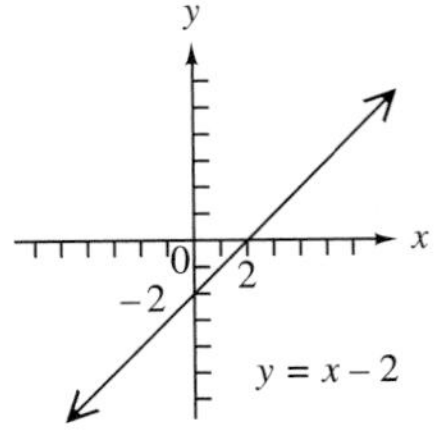

15.
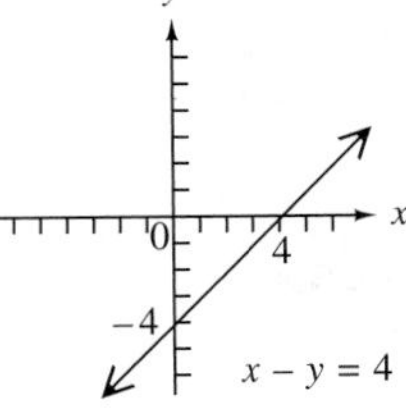

17.
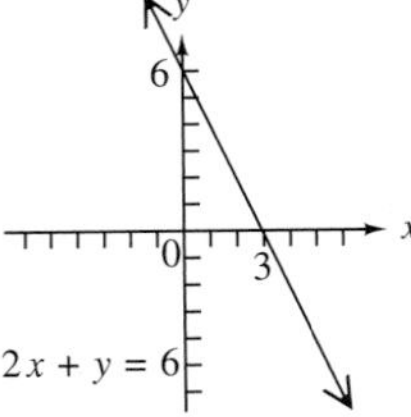

19.
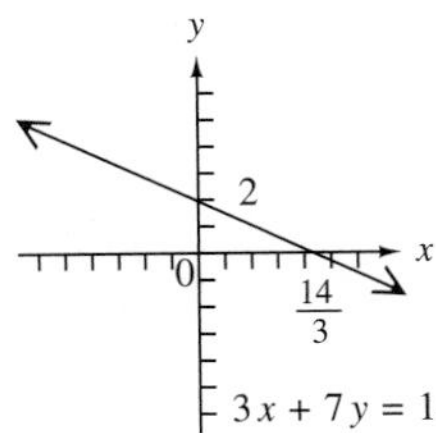

21.
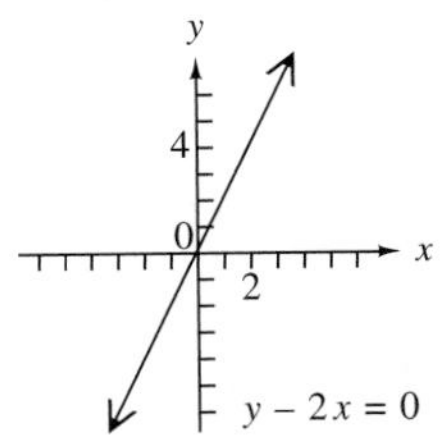

23.
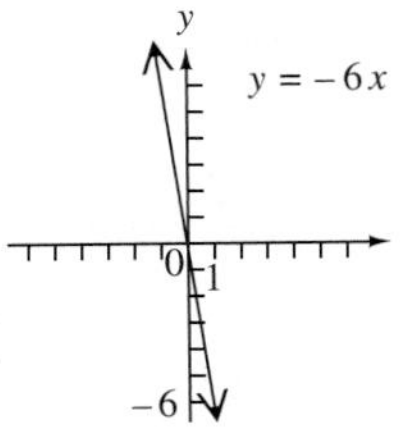

25.
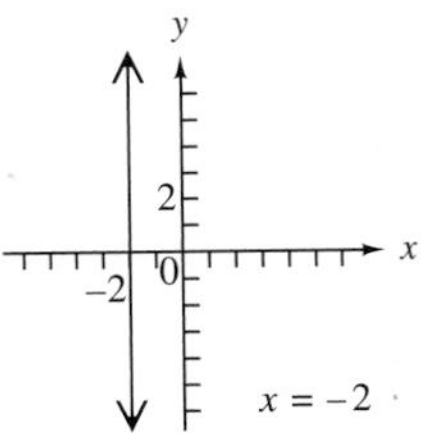

27.
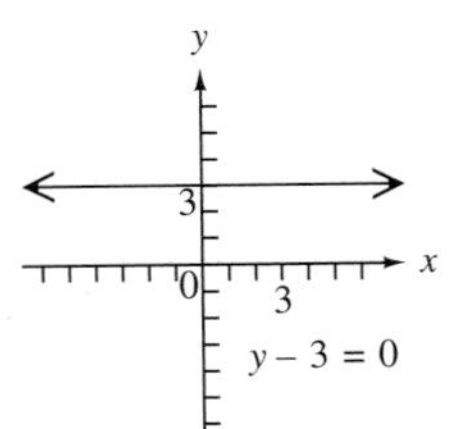

29.
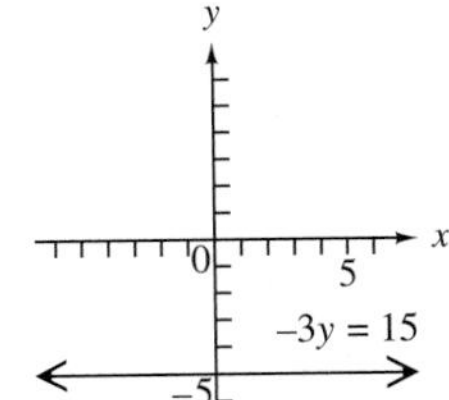

In Exercises 31 and 33, descriptions may vary.

31. The graph is a line with x-intercept $(-3, 0)$ and y-intercept $(0, 9)$. **33.** The graph is a horizontal line with y-intercept $(0, -2)$. **35.** Choose a value *other than* 0 for either x or y. For example, if $x = -5, y = 4$. **37. (a)** 151.5 cm, 159.3 cm, 174.9 cm **(b)** (20, 151.5), (22, 159.3), (26, 174.9)

(c)

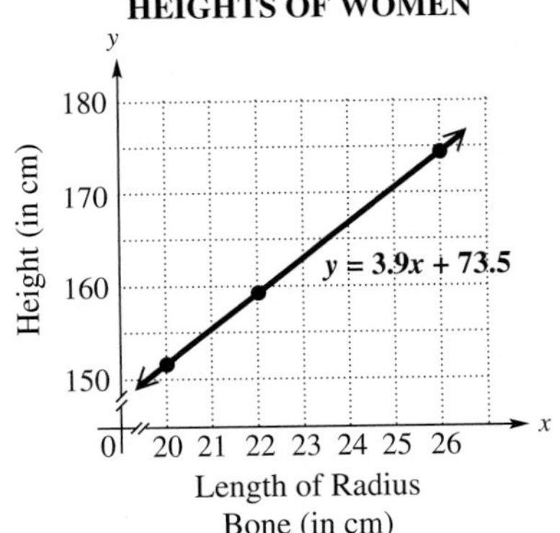

(d) 24 cm; 24 cm

39. (a) \$62.50; \$100 **(b)** 200 **(c)** (50, 62.50), (100, 100), (200, 175)

(d)

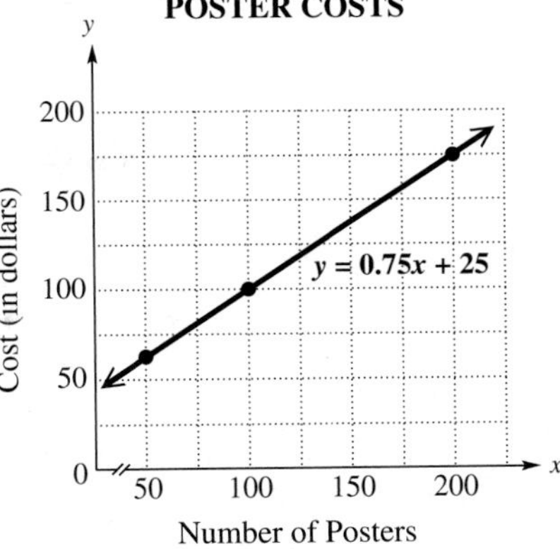

41. (a) \$30,000 **(b)** \$15,000 **(c)** \$5000 **(d)** After 5 yr, the SUV has a value of \$5000. **43. (a)** 1990: 24.1 lb; 2000: 29.5 lb; 2005: 32.2 lb **(b)** 1990: 25 lb; 2000: 30 lb; 2005: 32 lb **(c)** The values are quite close.

Section 3.3 (pages 261–266)

1. $\frac{3}{2}$ **3.** $-\frac{7}{4}$ **5.** 0 **7.** Rise is the vertical change between two different points on a line. Run is the horizontal change between two different points on a line.

9.–12. Answers will vary.

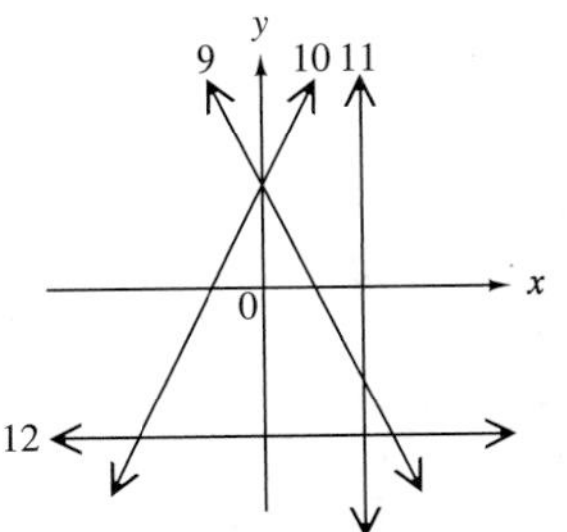

13. (a) falls from left to right **(b)** horizontal **(c)** vertical **(d)** rises from left to right **15.** Because he found the difference $3 - 5 = -2$ in the numerator, he should have subtracted in the same order in the denominator to get $-1 - 2 = -3$. The correct slope is $\frac{-2}{-3} = \frac{2}{3}$.

17. $\frac{5}{4}$ **19.** $\frac{3}{2}$ **21.** -3 **23.** 0 **25.** undefined **27.** $-\frac{1}{2}$ **29.** 5 **31.** $\frac{1}{4}$ **33.** $\frac{3}{2}$ **35.** 0 **37.** undefined **39.** 1 **41. (a)** negative **(b)** 0 **43. (a)** positive **(b)** negative **45. (a)** 0 **(b)** negative **47.** $\frac{4}{3}$; $\frac{4}{3}$; parallel **49.** $\frac{5}{3}$; $\frac{3}{5}$; neither **51.** $\frac{3}{5}$; $-\frac{5}{3}$; perpendicular **53.** $\frac{8}{27}$ **55.** 232 thousand, or 232,000 **56.** positive; increased **57.** 232,000 students **58.** -0.95 **59.** negative; decreased **60.** 0.95 student per computer **61.** 19.5 ft **63.** $-\$4000$ per yr; The value of the machine is decreasing \$4000 each year during these years. **65.** 0% per yr (or no change); The percent of pay raise is not changing—it is 3% each year during these years. **67. (a)** \$18.78 billion per yr **(b)** The positive slope means that personal spending on recreation in the United States *increased* by an average of \$18.78 billion each year from 2000 to 2006. **69.** $-\$1470.67$ million per yr; Sales of analog TVs *decreased* by an average of \$1470.67 million each year from 2003 to 2006. **70.** $\frac{1}{3}$ **71.** $\frac{1}{3}$ **72.** $\frac{1}{3}$ **73.** $\frac{1}{3} = \frac{1}{3} = \frac{1}{3}$ is true. **74.** They are collinear. **75.** They are not collinear.

Section 3.4 (pages 277–282)

1. **(a)** D **(b)** C **(c)** B **(d)** A **3.** $y = 3x - 3$ **5.** $y = -x + 3$ **7.** $y = -\frac{1}{2}x + 2$ **9.** $y = 4x - 3$ **11.** $y = 3$ **13.** **(a)** C **(b)** B **(c)** A **(d)** D

15.

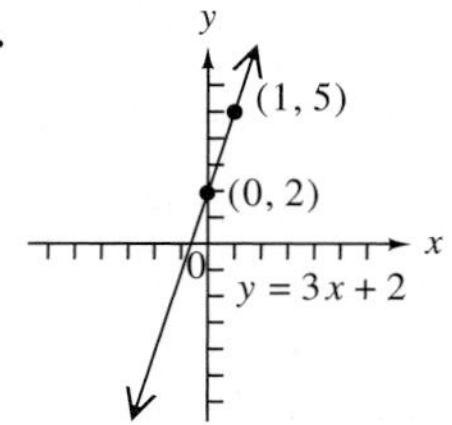

17.

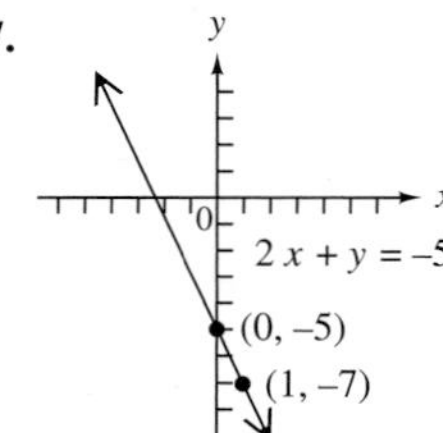

19.

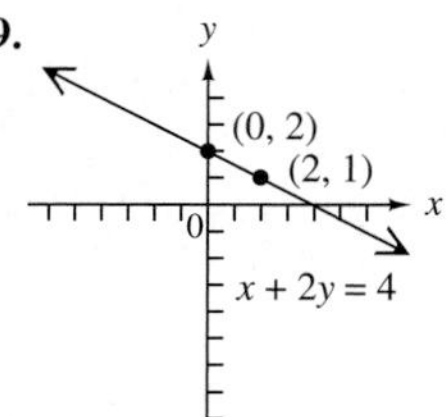

21. $y = \frac{1}{2}x + 4$

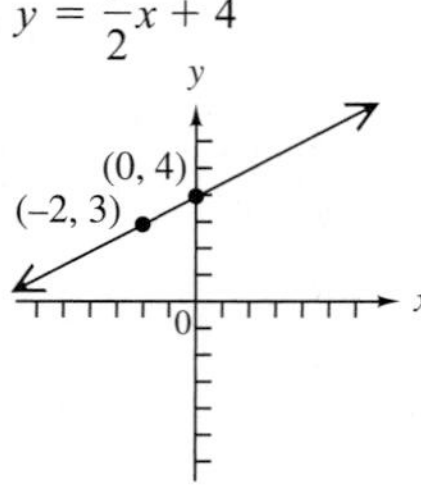

23. $y = -\frac{2}{5}x - \frac{23}{5}$

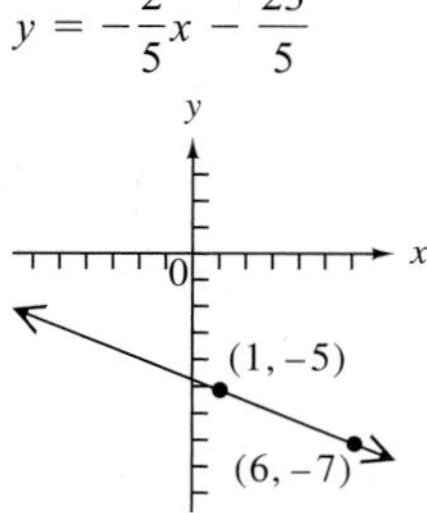

25. $y = 2$

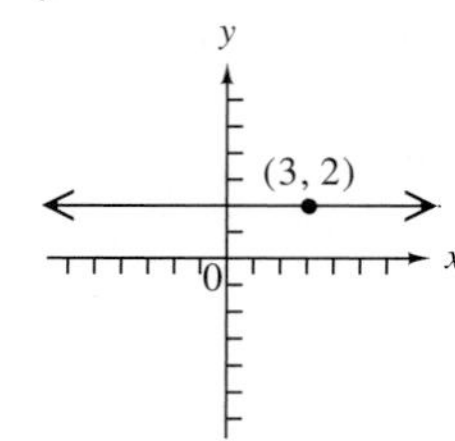

27. $x = 3$ (no slope-intercept form)

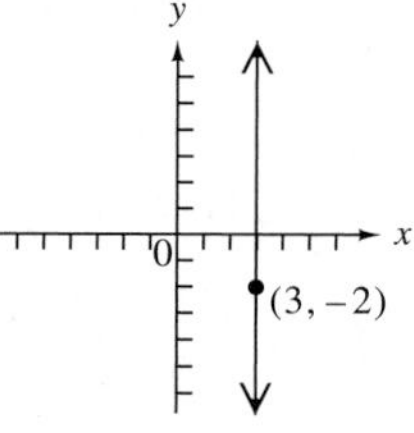

29. $y = \frac{2}{3}x$

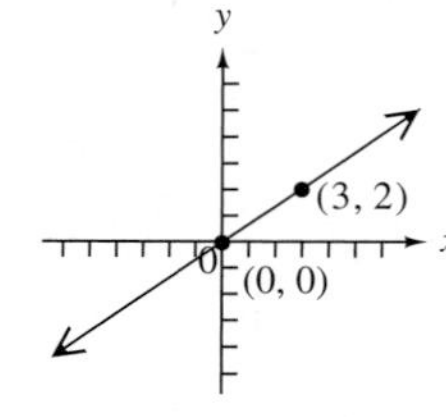

31. $y = 2x - 7$ **33.** $y = -2x - 4$ **35.** $y = \frac{2}{3}x + \frac{19}{3}$ **37.** $y = x - 3$ **39.** $y = -\frac{5}{7}x - \frac{54}{7}$ **41.** $y = -\frac{2}{3}x - 2$ **43.** $x = 3$ (no slope-intercept form) **45.** $y = \frac{1}{3}x + \frac{4}{3}$ **47.** $y = \frac{3}{4}x - \frac{9}{2}$ **49.** $y = -\frac{1}{2}x + 9$ **51.** $y = -2x - 3$ **53.** $y = 45x$; (0, 0), (5, 225), (10, 450) **55.** $y = 5.00x$; (0, 0), (5, 25.00), (10, 50.00) **57.** **(a)** $y = 41x + 99$ **(b)** (5, 304); The cost of a 5-month membership is \$304. **(c)** \$591 **59.** **(a)** $y = 0.20x + 50$ **(b)** (5, 51); The charge for driving 5 mi is \$51. **(c)** 173 mi **61.** **(a)** \$400 **(b)** \$0.25 **(c)** $y = 0.25x + 400$ **(d)** \$425 **(e)** 1500 **63.** **(a)** $y = 1294.7x + 3921$; Sales of digital cameras in the United States increased by \$1294.7 million per yr from 2003 to 2006. **(b)** \$9099.8 million

65. **(a)** (5, 42), (15, 61), (25, 76) **(b)** yes

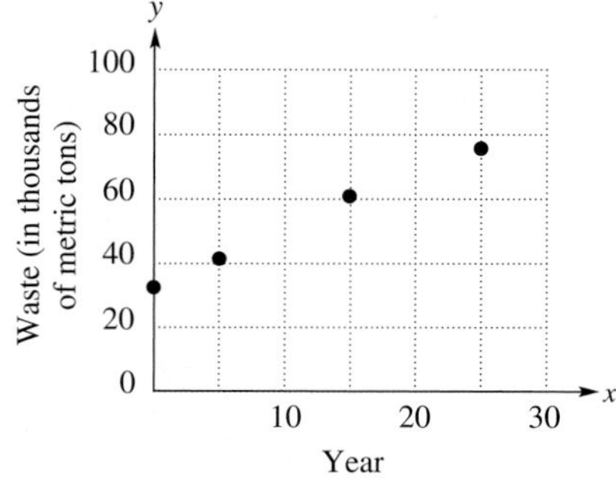

(c) $y = 1.76x + 32$ **(d)** 67.2 thousand (or 67,200) metric tons

Summary Exercises on Linear Equations and Graphs (pages 283–284)

1. -3; $(0, -6)$ **2.** -2; $(0, -4)$ **3.** -4; $(0, -3)$ **4.** -5; $(0, -8)$ **5.** $\frac{3}{2}$; $(0, 6)$ **6.** $\frac{5}{3}$; $(0, 5)$

7.

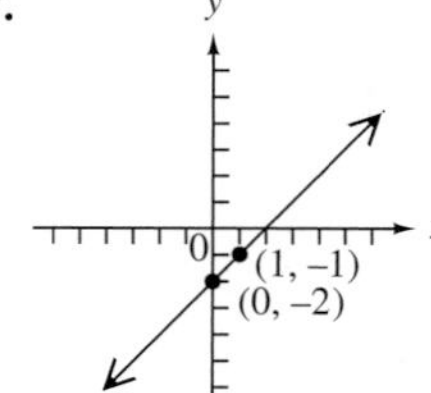

8.

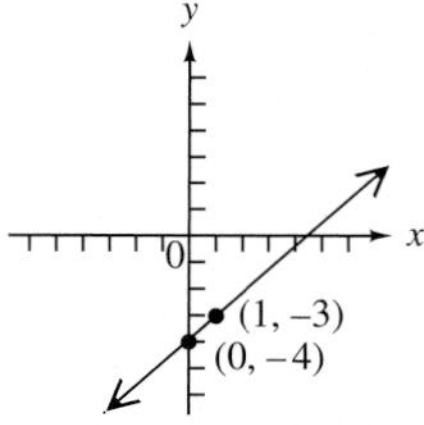

9.

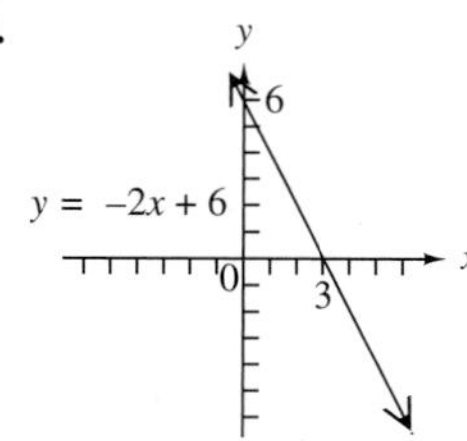

10.

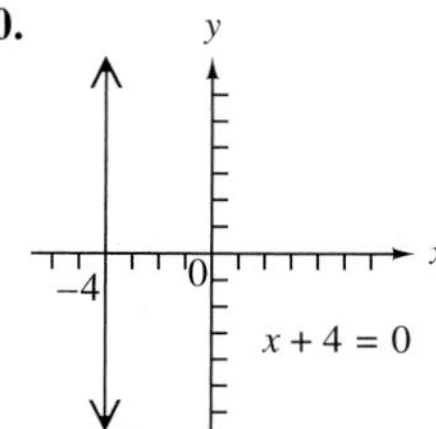

11.

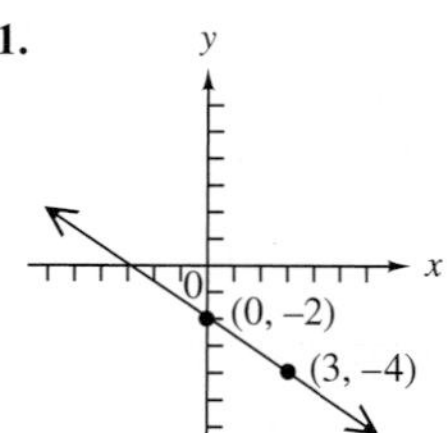

12.

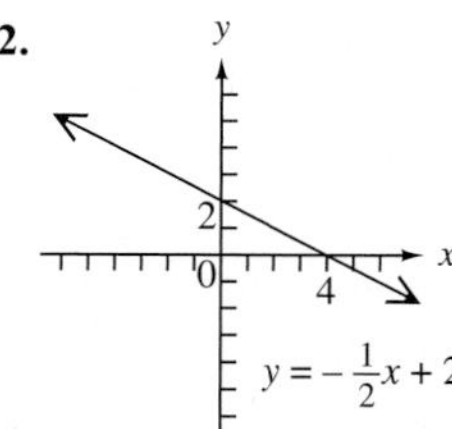

13.

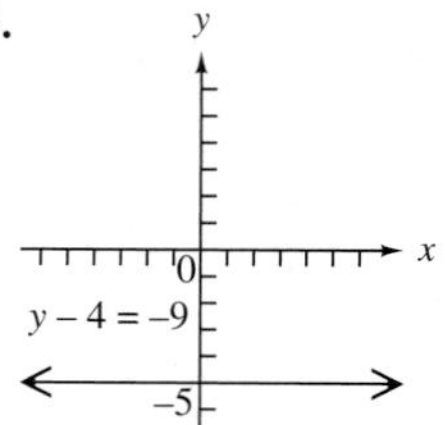

14.

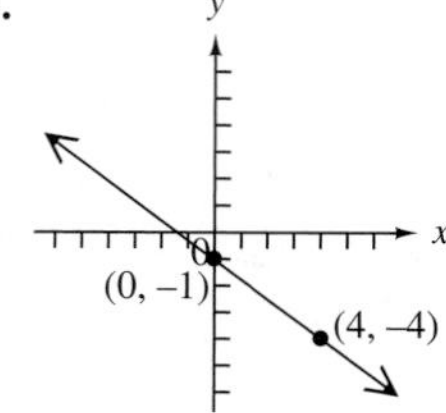

15.

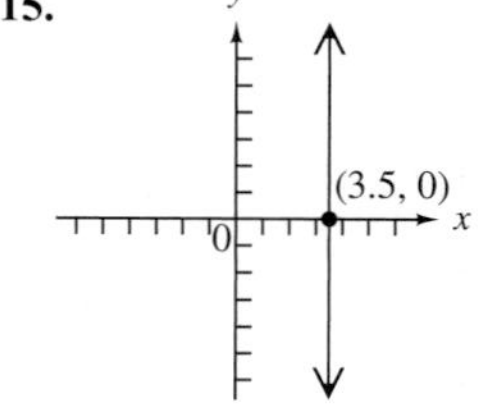

16.

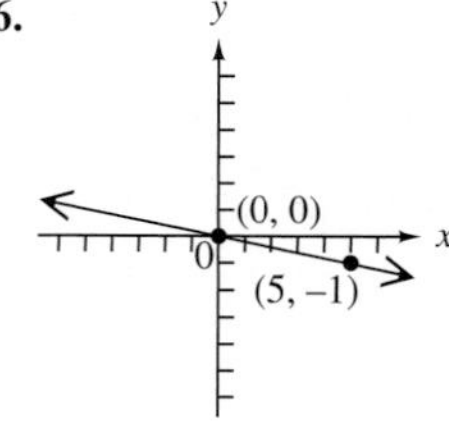

ANSWERS

17.
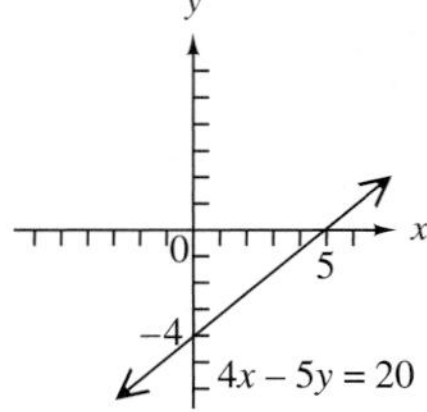

18.
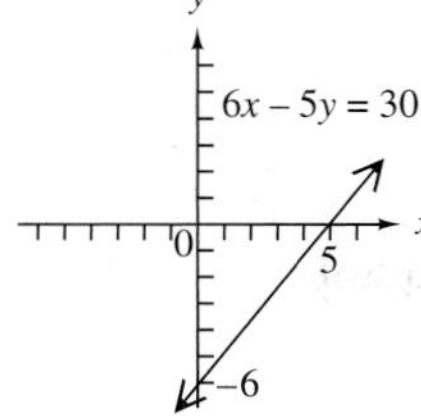

19.
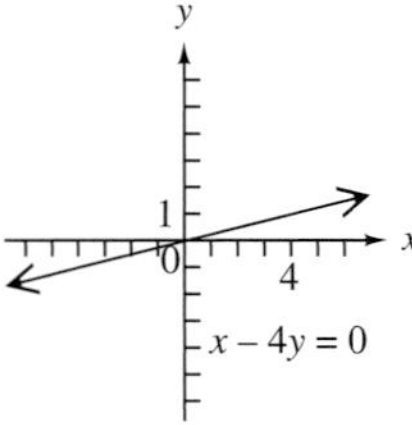

20.
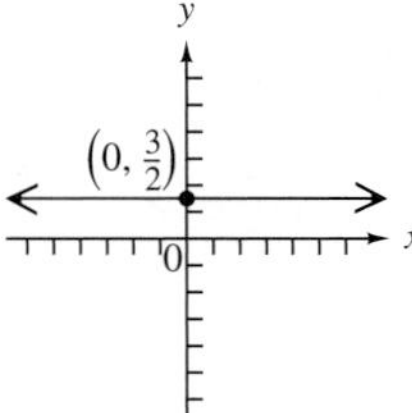

21.
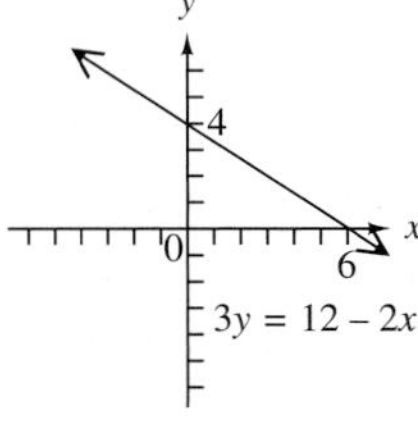

22. $y = -3x - 6$ **23.** $y = -4x - 3$ **24.** $y = \frac{3}{5}x$ **25.** $x = 0$ **26.** $y = 0$ **27.** $y = -2x - 4$ **28.** $y = \frac{5}{3}x + 5$

Section 3.5 (pages 289–290)

1. Use a dashed line if the symbol is $<$ or $>$. Use a solid line if the symbol is $\leq$ or $\geq$.

3.
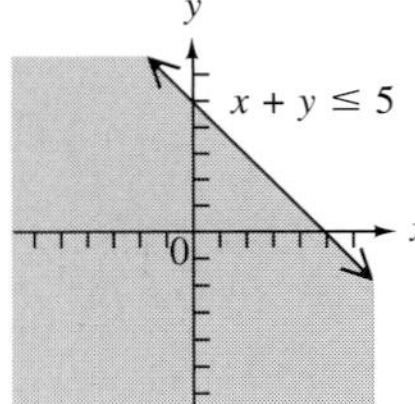

5.
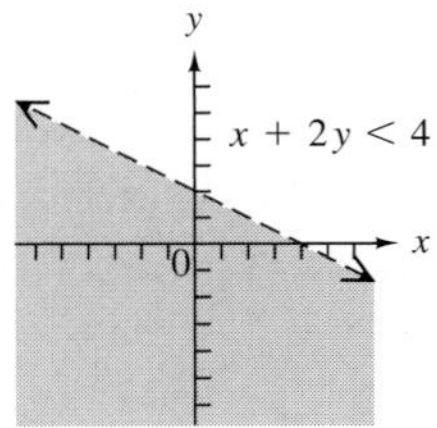

7.
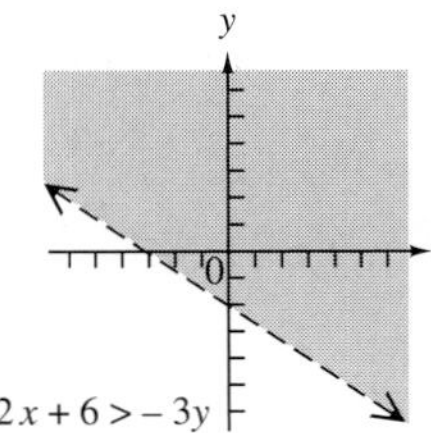

9.
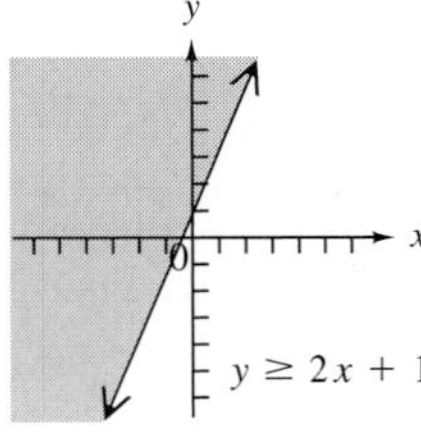

11.
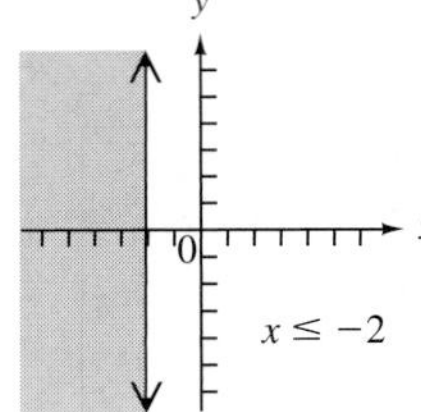

13.
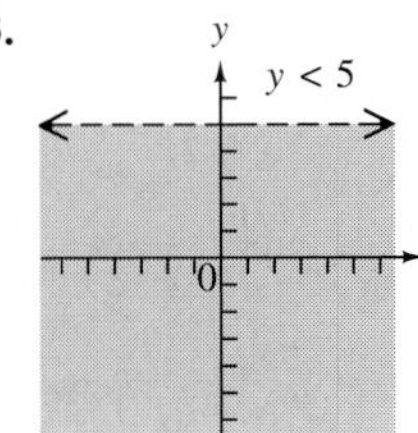

15.
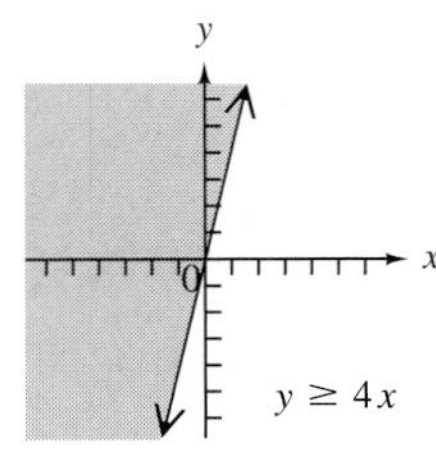

17.
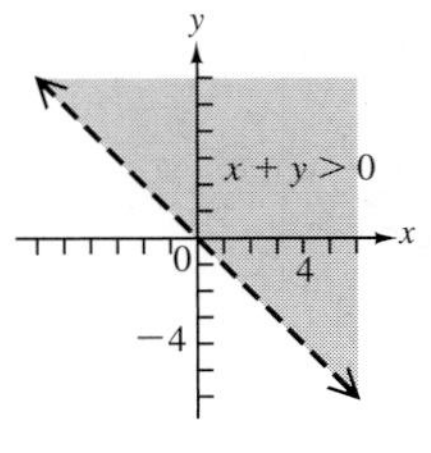

19.
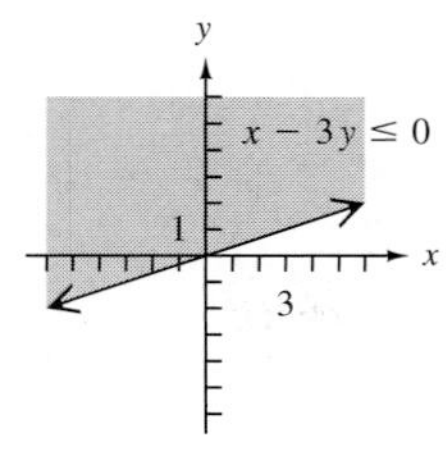

21.
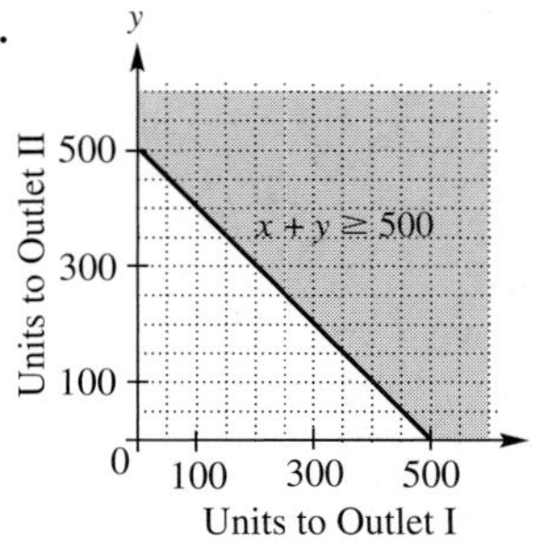

Section 3.6 (pages 301–306)

1. independent variable **3.** **(a)** A relation is a set of ordered pairs. **(b)** The domain is the set of all first components (x-values). **(c)** The range is the set of all second components (y-values). **(d)** A function is a relation in which each domain element is paired with one and only one range element. **5.** function; domain: $\{5, 3, 4, 7\}$; range: $\{1, 2, 9, 3\}$ **7.** not a function; domain: $\{2, 0\}$; range: $\{4, 2, 6\}$ **9.** function; domain: $\{-3, 4, -2\}$; range: $\{1, 7\}$ **11.** not a function; domain: $\{1, 0, 2\}$; range: $\{1, -1, 0, 4, -4\}$ **13.** function; domain: $\{2, 5, 11, 17, 3\}$; range: $\{1, 7, 20\}$ **15.** not a function; domain: $\{1\}$; range: $\{5, 2, -1, -4\}$ **17.** function; domain: $(-\infty, \infty)$; range: $(-\infty, \infty)$ **19.** function; domain: $(-\infty, \infty)$; range: $(-\infty, 4]$ **21.** not a function; domain: $[3, \infty)$; range: $(-\infty, \infty)$ **23.** function; domain: $(-\infty, \infty)$ **25.** not a function; domain: $[0, \infty)$ **27.** function; domain: $(-\infty, \infty)$ **29.** not a function; domain: $(-\infty, \infty)$ **31.** function; domain: $(-\infty, \infty)$ **33.** function; domain: $(-\infty, 0) \cup (0, \infty)$ **35.** function; domain: $(-\infty, 9) \cup (9, \infty)$ **37.** function; domain: $\left(-\infty, -\frac{1}{2}\right) \cup \left(-\frac{1}{2}, \infty\right)$ **39.** **(a)** $[0, 3000]$ **(b)** 25 hr; 25 hr **(c)** 2000 gal **(d)** $g(0) = 0$; The pool is empty at time 0. **41.** Here is one example. The cost of gasoline; number of gallons purchased; cost; number of gallons **43.** 4 **45.** -11 **47.** $-3p + 4$ **49.** $3x + 4$ **51.** $-3x - 2$ **53.** $-\frac{p^2}{9} + \frac{4p}{3} + 1$ **55.** **(a)** 2 **(b)** 3 **57.** **(a)** 15 **(b)** 10 **59.** **(a)** 3 **(b)** -3 **61.** line; -2; linear; $-2x + 4$; -2; 3; -2 **63.** **(a)** $f(x) = -\frac{1}{3}x + 4$ **(b)** 3 **65.** **(a)** $f(x) = 3 - 2x^2$ **(b)** -15 **67.** **(a)** $f(x) = \frac{4}{3}x - \frac{8}{3}$ **(b)** $\frac{4}{3}$

69. domain: $(-\infty, \infty)$; range: $(-\infty, \infty)$

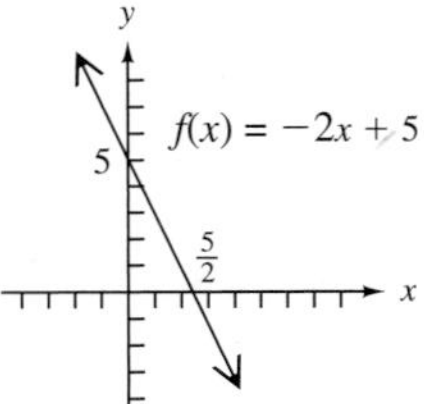

71. domain: $(-\infty, \infty)$; range: $(-\infty, \infty)$

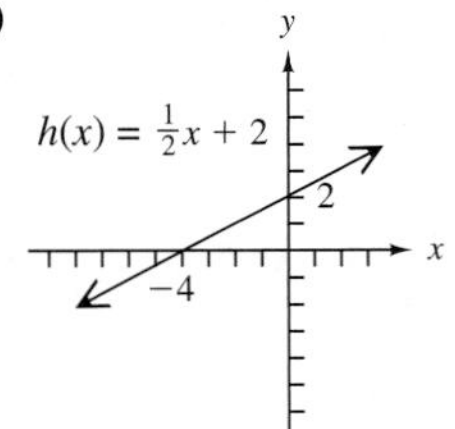

73. domain: $(-\infty, \infty)$; range: $\{-4\}$

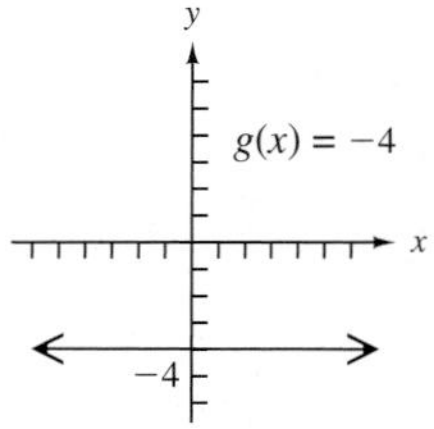

75. (a) \$0; \$2.50; \$5.00; \$7.50 **(b)** $2.50x$
(c)

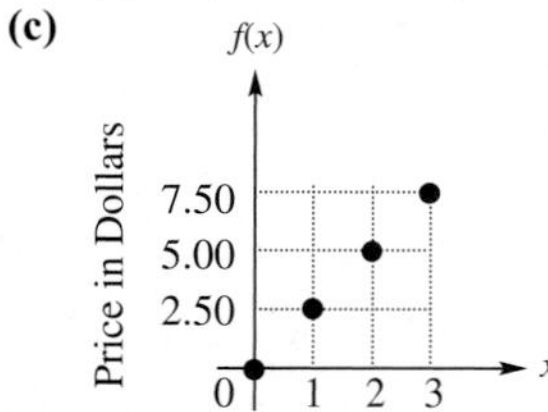

77. 194.53 cm **79.** 177.41 cm

Chapter 3 Review Exercises (pages 313–316)

1. (2001, 51.8), (2002, 51.8), (2003, 52.8), (2004, 51.3), (2005, 51.6), (2006, 52.5), (2007, 51.4) **2.** In the year 2006, 52.5% of first-year college students at two-year public institutions returned for a second year.
3. 2004; 1.5% **4.** 2003; 1.0% **5.** -1; 2; 1 **6.** 2; $\frac{3}{2}$; $\frac{14}{3}$ **7.** no
8. yes **9.** x is positive in quadrants I and IV; y is negative in quadrants III and IV. Thus, if x is positive and y is negative, (x, y) must lie in quadrant IV.
10. In the ordered pair $(k, 0)$, the y-value is 0, so the point lies on the x-axis. In the ordered pair $(0, k)$, the x-value is 0, so the point lies on the y-axis.
11. $\left(-\frac{5}{2}, 0\right)$; $(0, 5)$ **12.** $\left(-\frac{7}{2}, 0\right)$; $(0, -7)$ **13.** $\left(\frac{8}{3}, 0\right)$; $(0, 4)$

14.

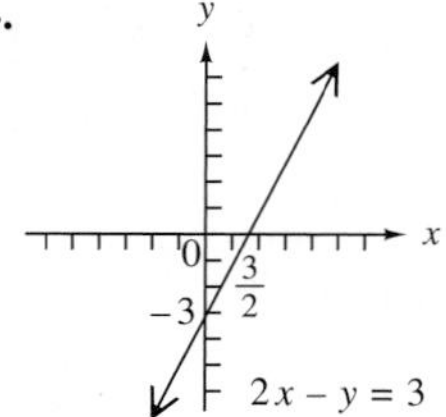

15.

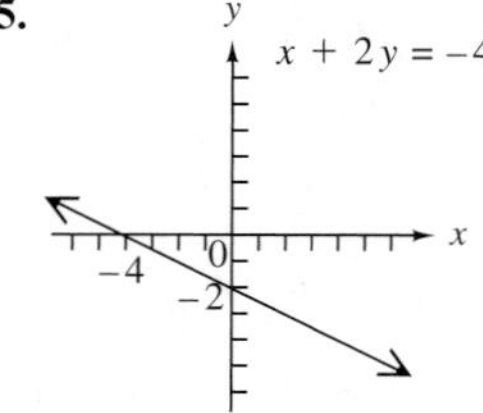

16.

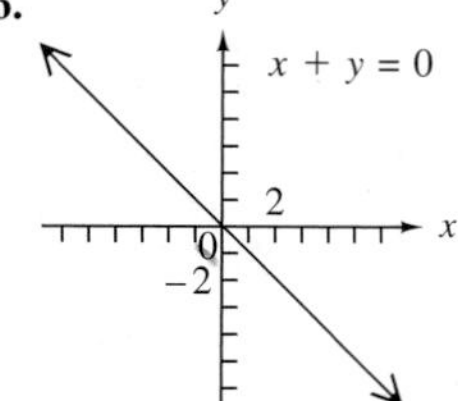

17. $-\frac{1}{2}$ **18.** undefined **19.** 3 **20.** 0 **21.** $-\frac{1}{3}$ **22.** $\frac{3}{2}$ **23. (a)** 2
(b) $\frac{1}{3}$ **24.** parallel **25.** perpendicular **26.** neither **27.** 12 ft
28. \$1408 per yr **29.** $y = -x + \frac{2}{3}$ **30.** $y = -\frac{1}{3}x + 1$
31. $y = \frac{2}{3}x + \frac{14}{3}$ **32.** $y = -\frac{1}{4}x + \frac{3}{2}$ **33.** $y = 1$ **34.** $x = \frac{1}{3}$
35. (a) $y = 4x - 26$ **(b)** $4x - y = 26$
36. (a) $y = -\frac{5}{2}x + 1$ **(b)** $5x + 2y = 2$
37. (a) $y = 57x + 159$; \$843 **(b)** $y = 47x + 159$; \$723

38.

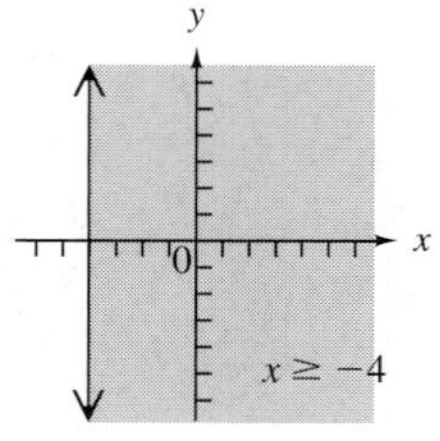

39.

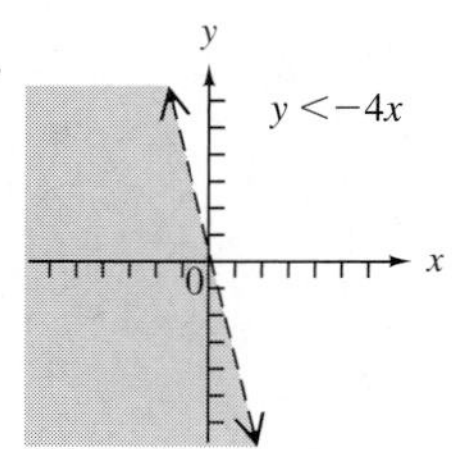

40.

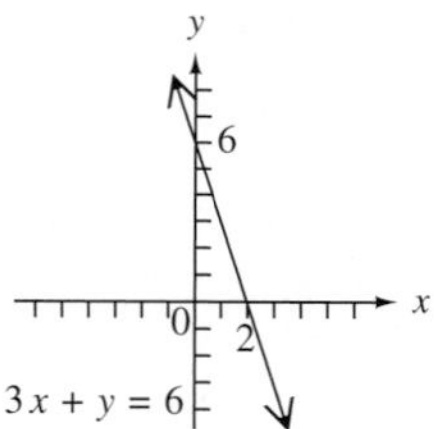

41. domain: $\{-4, 1\}$; range: $\{2, -2, 5, -5\}$; not a function
42. domain: $\{9, 11, 4, 17, 25\}$; range: $\{32, 47, 69, 14\}$; function
43. domain: $[-4, 4]$; range: $[0, 2]$; function **44.** -6 **45.** -15
46. $-2k^2 - 3k - 6$ **47.** function; linear function; domain: $(-\infty, \infty)$
48. not a function; domain: $[0, \infty)$ **49.** function; domain: $(-\infty, 36) \cup (36, \infty)$ **50.** $f(x) = 2x^2$; 18 **51.** It is a horizontal line.
52. (a) A **(b)** C, D **(c)** A, B, D **(d)** D **(e)** C **(f)** B
53. (a) B **(b)** D **(c)** A **(d)** C **54. (a)** yes; domain: {1943, 1953, 1963, 1973, 1983, 1993, 2003}; range: {63.3, 68.8, 69.9, 71.4, 74.6, 75.5, 77.6} **(b)** Answers will vary. Two possible ordered pairs are (1953, 68.8) and (1973, 71.4). **(c)** 77.6; In 2003, life expectancy at birth was 77.6 yr. **(d)** 1993

Chapter 3 Test (pages 317–320)

1. between 1998 and 1999, 1999 and 2000, and 2003 and 2004
2. The unemployment rate was increasing.
3. 2003: 6.0%; 2004: 5.5%; decline: 0.5%
4. x-intercept: (2, 0); y-intercept: (0, 6)

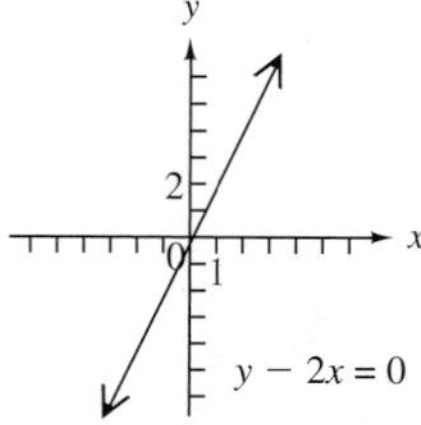

5. x-intercept: (0, 0); y-intercept: (0, 0)

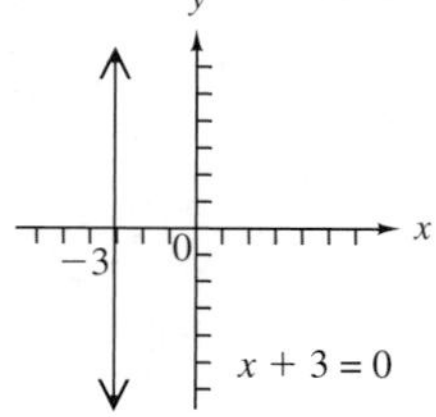

6. x-intercept: $(-3, 0)$; y-intercept: none

7. x-intercept: $(4, 0)$; y-intercept: $(0, -4)$

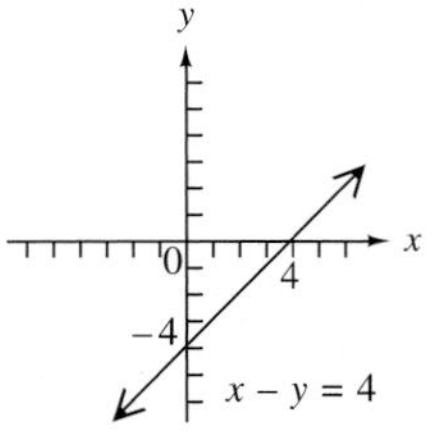

8. $-\frac{8}{3}$ **9.** -2 **10.** undefined **11.** $\frac{5}{2}$ **12.** -1200 farms per yr; The number of farms decreased, on the average, by about 1200 each year from 1980 to 2005. **13.** $y = 2x + 6$ **14.** $y = \frac{5}{2}x - 4$ **15.** $5x - 3y = -57$

16. (a) $y = -\frac{3}{5}x - \frac{11}{5}$ **(b)** $y = -\frac{1}{2}x - \frac{3}{2}$

17.

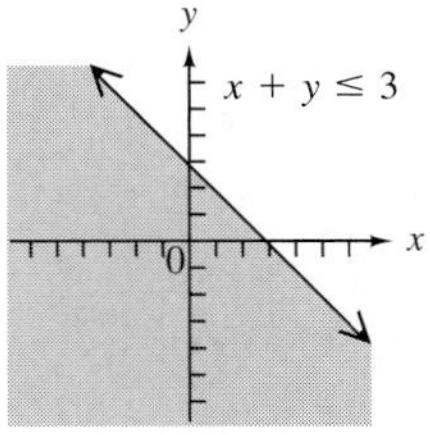

18.

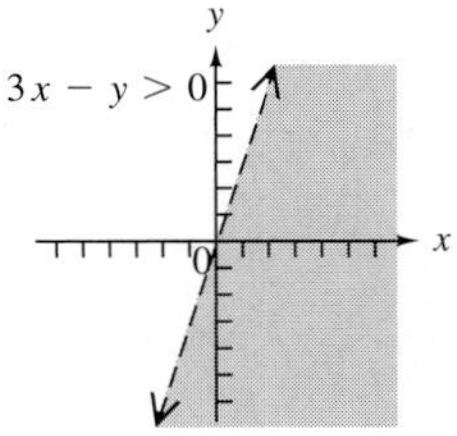

19. C; domain: $(-\infty, \infty)$; range: $[0, \infty)$ **20.** C; domain: $\{0, 3, 6\}$; range: $\{1, 2, 3\}$ **21.** 0; $-a^2 + 2a - 1$ **22.** The slope is negative since sales are decreasing. **23.** $(0, 209)$, $(7, 160)$; -7 **24.** $y = -7x + 209$ **25.** 174 thousand; The equation gives a good approximation of the actual sales.

Cumulative Review Exercises: Chapters R–3 (pages 323–324)

1. $\frac{301}{40}$, or $7\frac{21}{40}$ **2.** 6 **3.** 7 **4.** $\frac{73}{18}$, or $4\frac{1}{18}$ **5.** true **6.** -43

7. distributive property **8.** $-p + 2$ **9.** $h = \frac{3V}{\pi r^2}$ **10.** $\{-1\}$ **11.** $\{2\}$

12. $\{-13\}$

13. $(-2.6, \infty)$

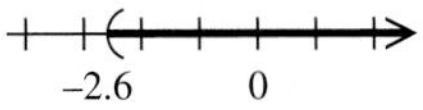

14. $(0, \infty)$

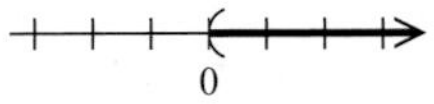

15. $(-\infty, -4]$

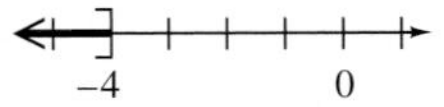

16. high school diploma: \$38,386; bachelor's degree: \$69,506 **17.** 13 mi **18. (a)** 85.53, 76.91, 74.99 **(b)** In 1992, the winning time was approximately 79.78 sec. **19. (a)** \$7000 **(b)** \$10,000 **(c)** about \$30,000

20. $(-4, 0)$; $(0, 3)$ **21.** $\frac{3}{4}$

22.

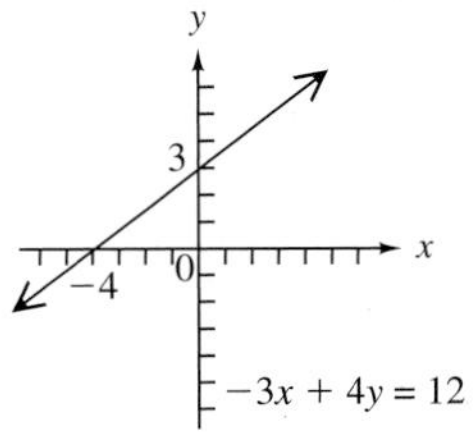

23. 6 **24.** perpendicular **25.** $y = 3x - 11$ **26.** $y = 4$

Chapter 4 Systems of Linear Equations and Inequalities

Section 4.1 (pages 331–334)

1. B, because the ordered pair must be in quadrant II. **3.** There is no way that the sum of two numbers can be both 2 and 4 at the same time. **5.** no **7.** yes **9.** yes **11.** no

We show the graphs here only for Exercises 13–17.

13. $\{(4, 2)\}$

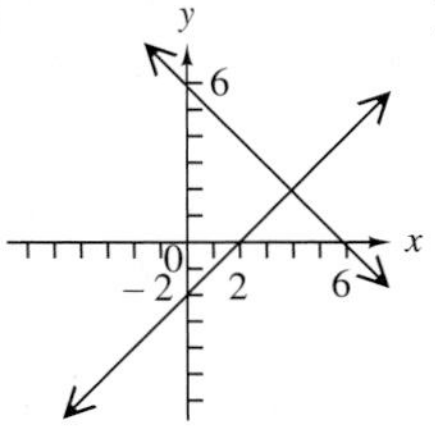

15. $\{(0, 4)\}$

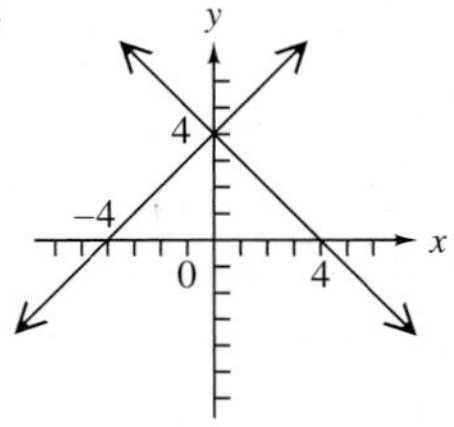

17. $\{(4, -1)\}$

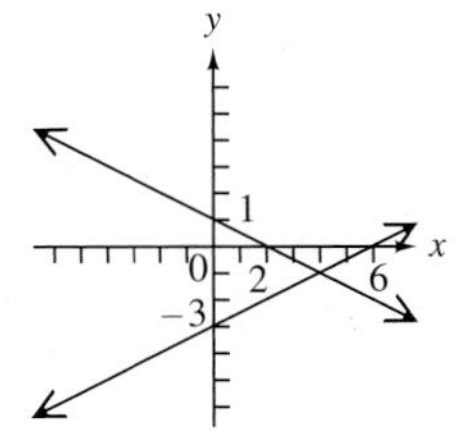

19. $\{(1, 3)\}$ **21.** $\{(0, 2)\}$ **23.** $\emptyset$ (inconsistent system) **25.** $\{(x, y) \mid 2x - y = 4\}$ (dependent equations) **27.** $\{(4, -3)\}$ **29.** $\emptyset$ (inconsistent system) **31. (a)** neither **(b)** intersecting lines **(c)** one solution **33. (a)** dependent **(b)** one line **(c)** infinite number of solutions **35. (a)** inconsistent **(b)** parallel lines **(c)** no solution **37.** 1980–2000 **39. (a)** 1997–2002 **(b)** 2001 **(c)** 2002 **(d)** (2004, 70) (The y-value is approximate.) **(e)** During the period 1997–2004, debit card use went from least popular to most popular of the three methods depicted.

Section 4.2 (pages 341–342)

1. The y-value must also be determined. The solution set is $\{(3, 0)\}$. **3.** $\{(3, 9)\}$ **5.** $\{(7, 3)\}$ **7.** $\{(-2, 4)\}$ **9.** $\{(-4, 8)\}$ **11.** $\{(3, -2)\}$ **13.** $\{(x, y) \mid 3x - y = 5\}$ **15.** $\left\{\left(\frac{1}{3}, -\frac{1}{2}\right)\right\}$ **17.** $\emptyset$ **19.** $\{(x, y) \mid 3x - 4y = 2\}$ **21.** $\{(2, -3)\}$ **23.** $\{(10, -12)\}$ **25.** $\{(-4, 2)\}$ **27.** $\{(7, -3)\}$ **29.** $\{(20, 30)\}$ **30.** To find the total cost, multiply the number of bicycles (x) by the cost per bicycle (400 dollars) and add the fixed cost (5000 dollars). Thus, $y_1 = 400x + 5000$ gives this total cost (in dollars). **31.** $y_2 = 600x$ **32.** $y_1 = 400x + 5000$, $y_2 = 600x$; solution set: $\{(25, 15{,}000)\}$ **33.** 25; 15,000; 15,000

Section 4.3 (pages 347–350)

1. true **3.** true **5.** $\{(-1, 3)\}$ **7.** $\{(-1, -3)\}$ **9.** $\{(-2, 3)\}$ **11.** $\left\{\left(\frac{1}{2}, 4\right)\right\}$ **13.** $\{(3, -6)\}$ **15.** $\{(7, 4)\}$ **17.** $\{(0, 4)\}$ **19.** $\{(-4, 0)\}$ **21.** $\{(0, 0)\}$ **23.** $\emptyset$ **25.** $\{(x, y) \mid x - 3y = -4\}$ **27.** $\{(2, 9)\}$ **29.** $\{(-6, 5)\}$ **31.** $\left\{\left(-\frac{6}{5}, \frac{4}{5}\right)\right\}$ **33.** $\left\{\left(\frac{1}{8}, -\frac{5}{6}\right)\right\}$ **35.** $\{(11, 15)\}$ **37.** $\emptyset$ **39.** $\{(x, y) \mid 2x + y = 0\}$ **41.** $1339 = 1996a + b$ **42.** $1536 = 2004a + b$ **43.** $1996a + b = 1339$, $2004a + b = 1536$; solution set: $\{(24.625, -47{,}812.5)\}$ **44.** $y = 24.625x - 47{,}812.5$

45. 1486.8 (million); This is quite a bit less than the actual figure. **46.** Since the data do not lie in a perfectly straight line, the quantity obtained from an equation determined in this way will probably be "off" a bit. We cannot put too much faith in models such as this one, because not all sets of data points are linear in nature.

Summary Exercises on Solving Systems of Linear Equations (pages 351–352)

1. (a) Use substitution since the second equation is solved for y. **(b)** Use elimination since the coefficients of the y-terms are opposites. **(c)** Use elimination since the equations are in standard form with no coefficients of 1 or -1. Solving by substitution would involve fractions. **2.** The system on the right is easier to solve by substitution because the second equation is already solved for y. **3. (a)** $\{(1, 4)\}$ **(b)** $\{(1, 4)\}$ **(c)** Answers will vary. **4. (a)** $\{(-5, 2)\}$ **(b)** $\{(-5, 2)\}$ **(c)** Answers will vary. **5.** $\{(2, 6)\}$ **6.** $\{(-3, 2)\}$ **7.** $\left\{\left(\frac{1}{3}, \frac{1}{2}\right)\right\}$ **8.** $\emptyset$ **9.** $\{(3, 0)\}$ **10.** $\left\{\left(\frac{3}{2}, -\frac{3}{2}\right)\right\}$ **11.** $\{(x, y) \mid 3x + y = 7\}$ **12.** $\{(9, 4)\}$ **13.** $\left\{\left(-\frac{5}{7}, -\frac{2}{7}\right)\right\}$ **14.** $\{(4, -5)\}$ **15.** $\emptyset$ **16.** $\{(-4, 6)\}$ **17.** $\left\{\left(\frac{19}{3}, -5\right)\right\}$ **18.** $\left\{\left(\frac{22}{13}, -\frac{23}{13}\right)\right\}$ **19.** $\{(-12, -60)\}$ **20.** $\{(2, -4)\}$ **21.** $\{(18, -12)\}$ **22.** $\{(-2, 1)\}$ **23.** $\left\{\left(13, -\frac{7}{5}\right)\right\}$ **24.** $\{(10, -9)\}$ **25.** $\{(0.04, 0.9)\}$

Section 4.4 (pages 359–364)

1. D **3.** B **5.** D **7.** C **9.** the second number; $x - y = 48$; The two numbers are 73 and 25. **11.** *The Phantom of the Opera:* 8197; *Cats:* 7485 **13.** *Spider-Man 3:* \$336.5 million; *Shrek the Third:* \$322.7 million **15.** Terminal Tower: 708 ft; Key Tower: 950 ft **17. (a)** 45 units **(b)** Do not produce; the product will lead to a loss. **19.** 46 ones; 28 tens **21.** 2 DVDs of *Night at the Museum* and 5 Linkin Park CDs **23.** \$2500 at 4%; \$5000 at 5% **25.** The Police: \$107; Van Halen: \$115 **27.** 80 L of 40% solution; 40 L of 70% solution **29.** 30 lb at \$6 per lb; 60 lb at \$3 per lb **31.** nuts: 40 lb; raisins: 20 lb **33.** bicycle: 13.5 mph; car: 49.3 mph **35.** car leaving Cincinnati: 55 mph; car leaving Toledo: 70 mph **37.** Roberto: 17.5 mph; Juana: 12.5 mph **39.** boat: 10 mph; current: 2 mph **41.** plane: 470 mph; wind: 30 mph

Section 4.5 (pages 369–370)

1. C **3.** B

5.

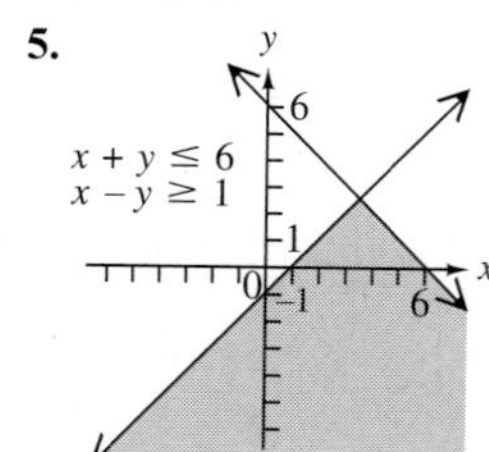

7.

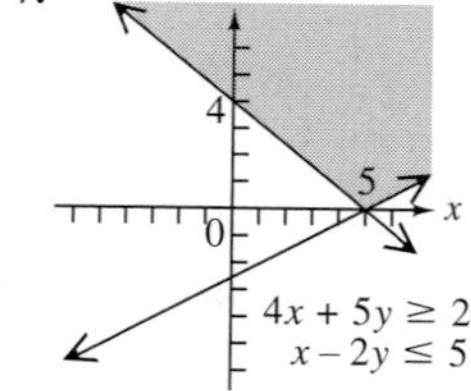

9.

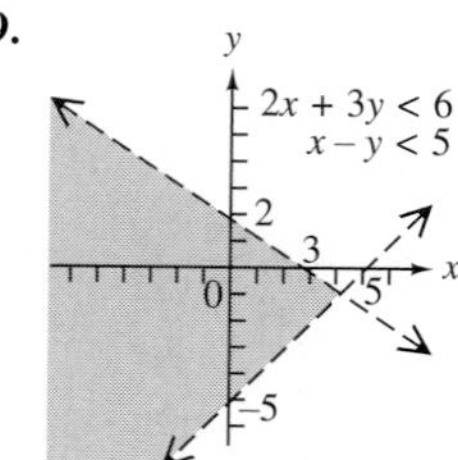

11.

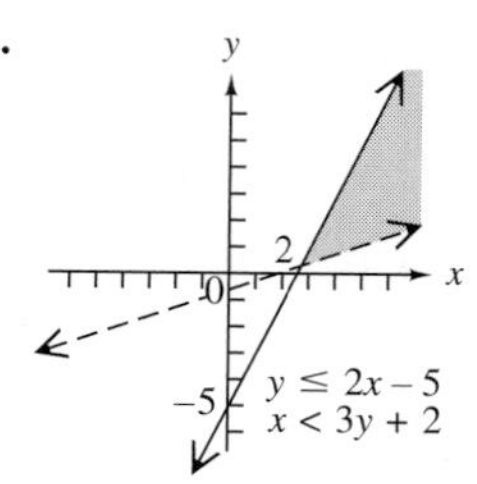

13.

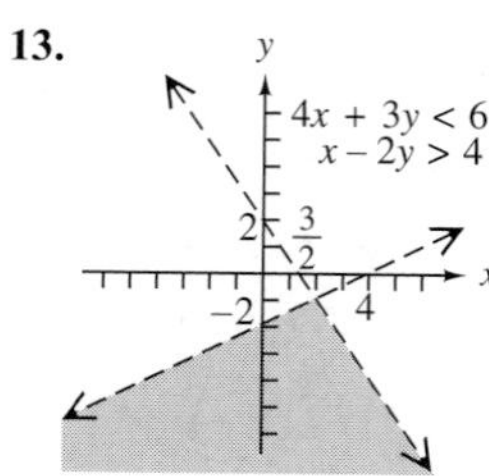

15.

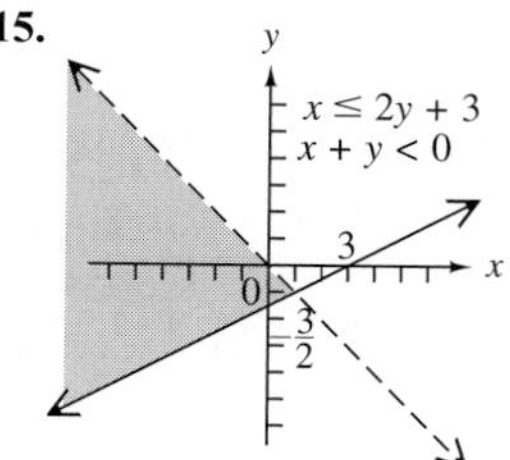

17.

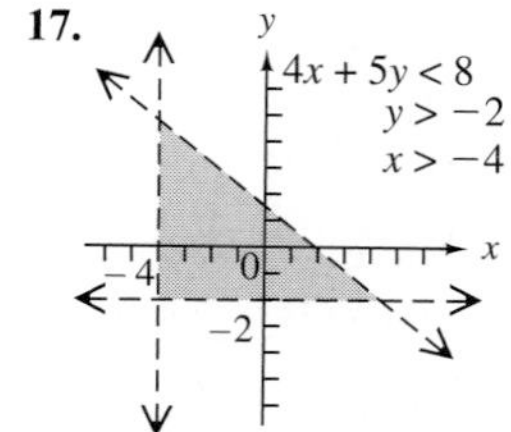

19.

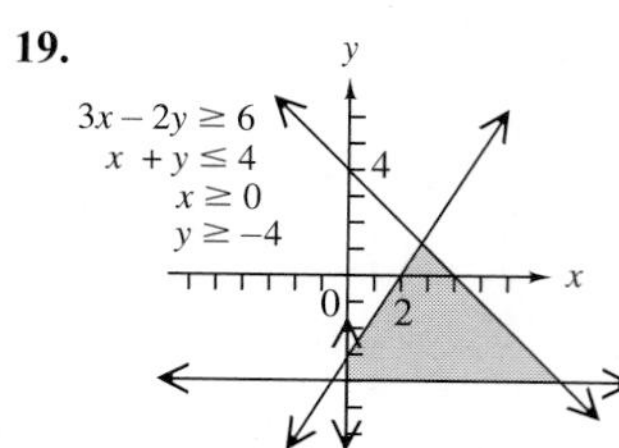

Chapter 4 Review Exercises (pages 375–378)

1. yes **2.** no **3.** $\{(3, 1)\}$ **4.** $\{(0, -2)\}$ **5.** $\{(x, y) \mid x - 2y = 2\}$ **6.** $\emptyset$ **7.** It is not a solution of the system because it is not also a solution of the second equation, $2x + y = 4$. **8.** $\{(2, 1)\}$ **9.** $\{(3, 5)\}$ **10.** $\{(6, 4)\}$ **11.** $\emptyset$ **12.** $\{(7, 1)\}$ **13.** $\{(-5, -2)\}$ **14.** $\{(-4, 3)\}$ **15.** $\{(x, y) \mid 3x - 4y = 9\}$ **16. (a)** 2 **(b)** 9 **17.** $\{(9, 2)\}$ **18.** $\left\{\left(\frac{10}{7}, -\frac{9}{7}\right)\right\}$ **19.** $\{(8, 9)\}$ **20.** $\{(2, 1)\}$ **21.** $\{(6, -4)\}$ **22.** $\{(-8, 5)\}$ **23.** Subway: 20,755; McDonald's: 13,774 **24.** *AARP The Magazine:* 23.4 million; *Reader's Digest:* 10.1 million **25.** length: 27 m; width: 18 m **26.** 13 twenties; 7 tens **27.** 25 lb of \$1.30 candy; 75 lb of \$0.90 candy **28.** plane: 250 mph; wind: 20 mph **29.** \$7000 at 3%; \$11,000 at 4% **30.** 60 L of 40% solution; 30 L of 70% solution

31.

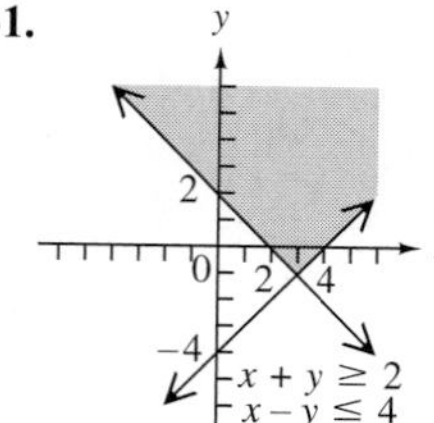

32.

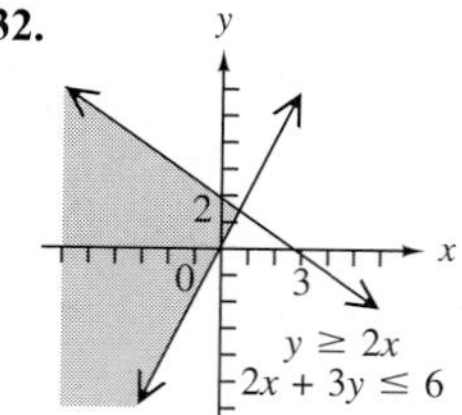

33.

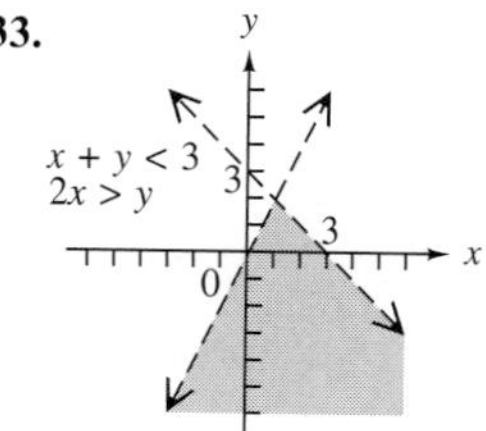

34. B **35.** B **36.** $\{(2, 0)\}$ **37.** $\{(-4, 15)\}$ **38.** $\emptyset$

ANSWERS

39.
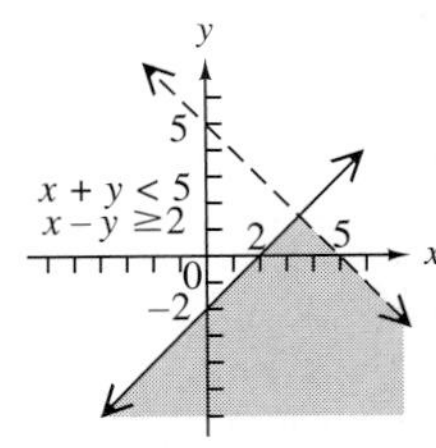

40.
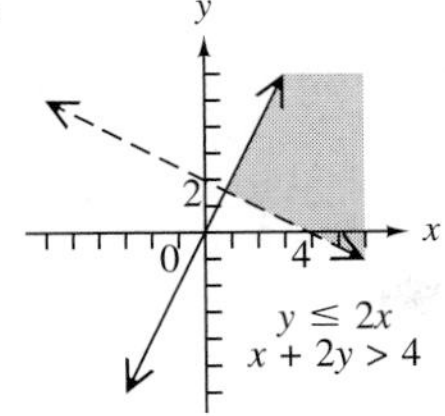

41.
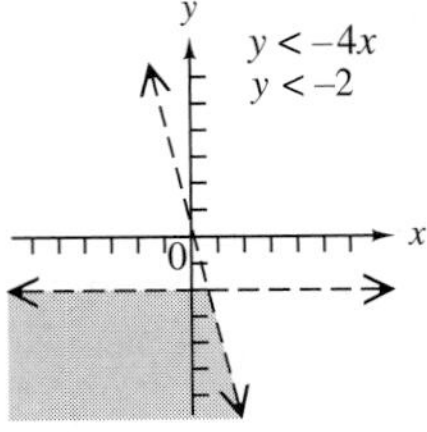

42. 8 in., 8 in., and 13 in. **43.** Giants: 17; Patriots: 14 **44.** **(a)** years 0–6 **(b)** year 6; about $650

Chapter 4 Test (pages 379–380)

1. $\{(2, -3)\}$ **2.** It has no solution. **3.** $\{(1, -6)\}$ **4.** $\{(-35, 35)\}$ **5.** $\{(5, 6)\}$ **6.** $\{(-1, 3)\}$ **7.** $\{(0, 0)\}$ **8.** $\emptyset$ **9.** $\{(x, y) \mid 3x - y = 6\}$ **10.** $\{(12, -4)\}$ **11.** Memphis and Atlanta: 394 mi; Minneapolis and Houston: 1176 mi **12.** Statue of Liberty: 3.6 million; National World War II Memorial: 5.4 million **13.** 20 L of 15% solution; 30 L of 40% solution **14.** slower car: 45 mph; faster car: 60 mph

15.
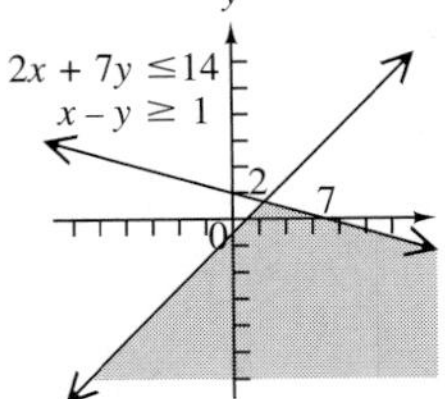

16.
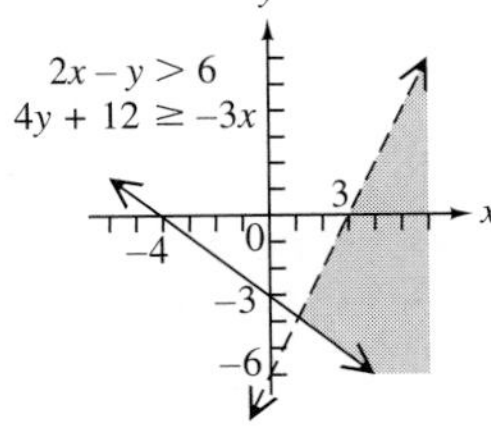

Cumulative Review Exercises: Chapters R–4 (pages 381–382)

1. $-1, 1, -2, 2, -4, 4, -5, 5, -8, 8, -10, 10, -20, 20, -40, 40$ **2.** 1 **3.** commutative property **4.** distributive property **5.** inverse property **6.** 46 **7.** $\left\{-\frac{13}{11}\right\}$ **8.** $\left\{\frac{9}{11}\right\}$ **9.** $(-18, \infty)$ **10.** $\left(-\frac{11}{2}, \infty\right)$ **11.** width: 8.16 in.; length: 10.74 in.

12.
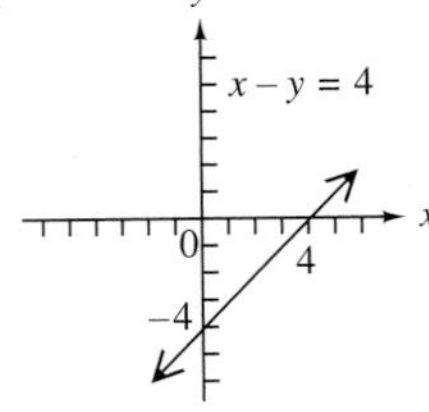

13.
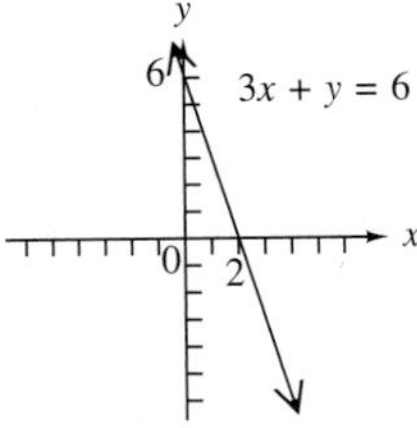

14. $-\frac{4}{3}$ **15.** $-\frac{1}{4}$ **16.** $y = \frac{1}{2}x + 3$ **17.** $y = 2x + 1$ **18.** **(a)** $x = 9$ **(b)** $y = -1$ **19.** $\{(-1, 6)\}$ **20.** $\{(3, -4)\}$ **21.** $\emptyset$ **22.** 405 adults and 49 children **23.** 19 in., 19 in., 15 in.

24.
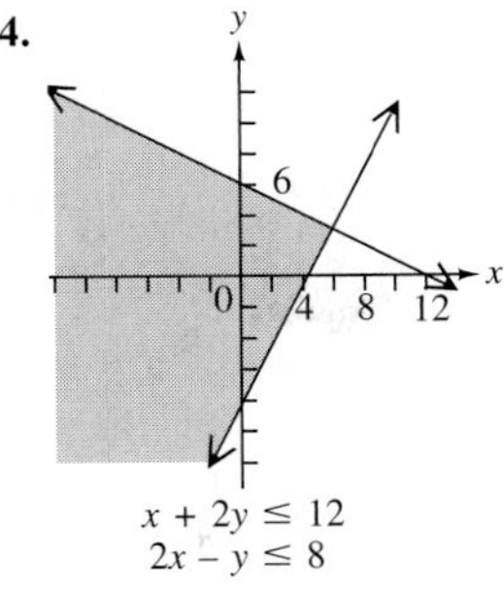

Chapter 5 Exponents and Polynomials

Section 5.1 (pages 389–392)

1. 7; 5 **3.** 8 **5.** 26 **7.** 1; 6 **9.** 1; 1 **11.** $1; \frac{1}{5}$ **13.** $2; -19, -1$ **15.** $3; 1, -8, \frac{2}{3}$ **17.** $2m^5$ **19.** $-r^5$ **21.** $\frac{2}{3}x^4$ **23.** cannot be simplified; $0.2m^5 - 0.5m^2$ **25.** $-5x^5$ **27.** $5p^9 + 4p^7$ **29.** $-2y^2$ **31.** already simplified; 4; binomial **33.** already simplified; $6m^5 + 5m^4 - 7m^3 - 3m^2$; 5; none of these **35.** $x^4 + \frac{1}{3}x^2 - 4$; 4; trinomial **37.** 7; 0; monomial **39.** $1.5x^2 - 0.5x$; 2; binomial **41.** **(a)** -1 **(b)** 5 **43.** **(a)** 19 **(b)** -2 **45.** **(a)** 36 **(b)** -12 **47.** **(a)** -124 **(b)** 5 **48.** 5; 175 **49.** 87 ft; (1, 87) **50.** $16.00 **51.** $27 **52.** 2.5; 130 **53.** $5m^2 + 3m$ **55.** $4x^4 - 4x^2$ **57.** $\frac{7}{6}x^2 - \frac{2}{15}x + \frac{5}{6}$ **59.** $12m^3 - 13m^2 + 6m + 11$ **61.** $2.9x^3 - 3.5x^2 - 1.5x - 9$ **63.** $8r^2 + 5r - 12$ **65.** $5m^2 - 14m + 6$ **67.** $4x^3 + 2x^2 + 5x$ **69.** $-18y^5 + 7y^4 + 5y^3 + 3y^2 + y$ **71.** $-2m^3 + 7m^2 + 8m - 9$ **73.** $-11x^2 - 3x - 3$ **75.** $2x^2 + 8x$ **77.** $8x^2 + 8x + 6$ **79.** $8t^2 + 8t + 13$ **81.** $13a^2b - 7a^2 - b$ **83.** $c^4d - 5c^2d^2 + d^2$ **85.** $12m^3n - 11m^2n^2 - 4mn^2$ **87.** **(a)** $23y + 5t$ **(b)** 25°, 67°, 88°

Section 5.2 (pages 399–400)

1. 1 **3.** false **5.** false **7.** t^7 **9.** $\left(\frac{1}{2}\right)^5$ **11.** $(-8p)^2$ **13.** The expression $(-3)^4$ means $(-3)(-3)(-3)(-3) = 81$, while -3^4 means $-(3 \cdot 3 \cdot 3 \cdot 3) = -81$. **15.** base: 3; exponent: 5; 243 **17.** base: -3; exponent: 5; -243 **19.** base: $-6x$; exponent: 4 **21.** base: x; exponent: 4 **23.** The product rule does not apply to $5^2 + 5^3$ because it is a *sum*, not a product. $5^2 + 5^3 = 25 + 125 = 150$ **25.** 5^8 **27.** 4^{12} **29.** $(-7)^9$ **31.** t^{24} **33.** $-56r^7$ **35.** $42p^{10}$ **37.** The product rule does not apply. **39.** The product rule does not apply. **41.** 4^6 **43.** t^{20} **45.** $343r^3$ **47.** 5^{12} **49.** -8^{15} **51.** $5^5x^5y^5$ **53.** $8q^3r^3$ **55.** $\frac{1}{8}$ **57.** $\frac{a^3}{b^3}$ **59.** $\frac{9^8}{5^8}$ **61.** $-8x^6y^3$ **63.** $9a^6b^4$ **65.** $\frac{5^5}{2^5}$ **67.** $\frac{9^5}{8^3}$ **69.** $2^{12}x^{12}$ **71.** -6^5p^5 **73.** $6^5x^{10}y^{15}$ **75.** x^{21} **77.** $4w^4x^{26}y^7$ **79.** $-r^{18}s^{17}$ **81.** $\frac{125a^6b^{15}}{c^{18}}$ **83.** $25m^6p^{14}q^5$ **85.** $16x^{10}y^{16}z^{10}$ **87.** $30x^7$

Section 5.3 (pages 405–406)

1. $x^2 + 7x + 12$ **3.** $2x^3 + 7x^2 + 7x + 2$ **5.** distributive **7.** $-6m^2 - 4m$ **9.** $6p - \frac{9}{2}p^2 + 9p^4$ **11.** $6y^5 + 4y^6 + 10y^9$ **13.** $12x^3 + 26x^2 + 10x + 1$ **15.** $6r^3 + 5r^2 - 12r + 4$ **17.** $20m^4 - m^3 - 8m^2 - 17m - 15$ **19.** $5x^4 - 13x^3 + 20x^2 + 7x + 5$

21. $m^2 + 12m + 35$ **23.** $n^2 + n - 6$ **25.** $8r^2 - 10r - 3$ **27.** $9x^2 - 4$ **29.** $9q^2 + 6q + 1$ **31.** $15xy - 40x + 21y - 56$ **33.** $6t^2 + 23st + 20s^2$ **35.** $-0.3t^2 + 0.22t + 0.24$ **37.** $x^2 - \frac{5}{12}x - \frac{1}{6}$ **39.** $\frac{15}{16} - \frac{1}{4}r - 2r^2$ **41.** $2x^3 + x^2 - 15x$ **43.** $6y^5 - 21y^4 - 45y^3$ **44.** $(30x + 60)$ yd^2 **45.** $30x + 60 = 600$; $\{18\}$ **46.** **(a)** 10 yd by 60 yd **(b)** 140 yd **47.** \$2100 **48.** \$1260 **49.** The answers are $x^2 - 16$, $y^2 - 4$, and $r^2 - 49$. Each product is the difference of the square of the first term and the square of the last term of the binomials.

Section 5.4 (pages 411–412)

1. **(a)** $4x^2$ **(b)** $12x$ **(c)** 9 **(d)** $4x^2 + 12x + 9$ **3.** $p^2 + 4p + 4$ **5.** $z^2 - 10z + 25$ **7.** $16x^2 - 24x + 9$ **9.** $4p^2 + 20pq + 25q^2$ **11.** $0.64t^2 + 1.12ts + 0.49s^2$ **13.** $25x^2 + 4xy + \frac{4}{25}y^2$ **15.** $9t^3 - 6t^2 + t$ **17.** $-16r^2 + 16r - 4$ **19.** **(a)** $49x^2$ **(b)** 0 **(c)** $-9y^2$ **(d)** $49x^2 - 9y^2$; Because 0 is the identity element for addition, it is not necessary to write "+ 0." **21.** $q^2 - 4$ **23.** $4w^2 - 25$ **25.** $100x^2 - 9y^2$ **27.** $4x^4 - 25$ **29.** $49x^2 - \frac{9}{49}$ **31.** $9p^3 - 49p$ **33.** $(a + b)^2$ **34.** a^2 **35.** $2ab$ **36.** b^2 **37.** $a^2 + 2ab + b^2$ **38.** They both represent the area of the entire large square. **39.** 1225 **40.** $30^2 + 2(30)(5) + 5^2$ **41.** 1225 **42.** They are equal. **43.** $m^3 - 15m^2 + 75m - 125$ **45.** $y^3 + 6y^2 + 12y + 8$ **47.** $8a^3 + 12a^2 + 6a + 1$ **49.** $81r^4 - 216r^3t + 216r^2t^2 - 96rt^3 + 16t^4$ **51.** $3x^5 - 27x^4 + 81x^3 - 81x^2$ **53.** $-8x^6y - 32x^5y^2 - 48x^4y^3 - 32x^3y^4 - 8x^2y^5$ **55.** 512 cu. units

Section 5.5 (pages 421–422)

1. negative **3.** negative **5.** positive **7.** 0 **9.** 1 **11.** 1 **13.** -1 **15.** 0 **17.** 0 **19.** 2 **21.** $\frac{1}{64}$ **23.** 16 **25.** $\frac{49}{36}$ **27.** $\frac{1}{81}$ **29.** $\frac{8}{15}$ **31.** $-\frac{7}{18}$ **33.** 1 **34.** $\frac{5^2}{5^2}$ **35.** $5^{2-2} = 5^0$ **36.** $5^0 = 1$; This supports the definition of a 0 exponent. **37.** $\frac{1}{9}$ **39.** $\frac{1}{6^5}$, or $\frac{1}{7776}$ **41.** 216 **43.** $2r^4$ **45.** $\frac{25}{64}$ **47.** $\frac{p^5}{q^8}$ **49.** r^9 **51.** $\frac{x^5}{6}$ **53.** $3y^2$ **55.** x^3 **57.** $\frac{yz^2}{4x^3}$ **59.** $a + b$ **61.** 343 **63.** $\frac{1}{x^2}$ **65.** $\frac{64x}{9}$ **67.** $\frac{x^2z^4}{y^2}$ **69.** $6x$ **71.** $\frac{1}{m^{10}n^5}$ **73.** $\frac{5}{16x^5}$ **75.** $\frac{36q^2}{m^4p^2}$

Summary Exercises on the Rules for Exponents (pages 423–424)

1. $\frac{6^{12}x^{24}}{5^{12}}$ **2.** $\frac{r^6s^{12}}{729t^6}$ **3.** $100{,}000x^7y^{14}$ **4.** $-128a^{10}b^{15}c^4$ **5.** $\frac{729w^3x^9}{y^{12}}$ **6.** $\frac{x^4y^6}{16}$ **7.** c^{22} **8.** $\frac{1}{k^4t^{12}}$ **9.** $\frac{11}{30}$ **10.** $y^{12}z^3$ **11.** $\frac{x^6}{y^5}$ **12.** 0 **13.** $\frac{1}{z^2}$ **14.** $\frac{9}{r^2s^2t^{10}}$ **15.** $\frac{300x^3}{y^3}$ **16.** $\frac{3}{5x^6}$ **17.** x^8 **18.** $\frac{y^{11}}{x^{11}}$ **19.** $\frac{a^6}{b^4}$ **20.** $6ab$ **21.** $\frac{61}{900}$ **22.** 1 **23.** $\frac{343a^6b^9}{8}$ **24.** 1 **25.** -1 **26.** 0 **27.** $\frac{27y^{18}}{4x^8}$ **28.** $\frac{1}{a^8b^{12}c^{16}}$ **29.** $\frac{x^{15}}{216z^9}$ **30.** $\frac{q}{8p^6r^3}$ **31.** x^6y^6 **32.** 0 **33.** $\frac{343}{x^{15}}$ **34.** $\frac{9}{x^6}$ **35.** $5p^{10}q^9$ **36.** $\frac{7}{24}$ **37.** $\frac{r^{14}t}{2s^2}$ **38.** 1 **39.** $8p^{10}q$ **40.** $\frac{1}{mn^3p^3}$ **41.** -1 **42.** $\frac{3}{40}$ **43.** Using the product rule, simplify as follows: $(10^2)^3 = 10^{2\cdot 3} = 10^6 = 1{,}000{,}000$. **44.** The negative sign is not part of the base: $-5^4 = -1(5)^4 = -1 \cdot 625 = -625$.

Section 5.6 (pages 427–428)

1. $6x^2 + 8$; 2; $3x^2 + 4$ **3.** $3x^2 + 4$; 2 (These may be reversed.); $6x^2 + 8$ **5.** To use the method of this section, the divisor must be just one term. This is true of the first problem, but not the second. **7.** $30x^3 - 10x + 5$ **9.** $-4m^3 + 2m^2 - 1$ **11.** $4t^4 - 2t^2 + 2t$ **13.** $a^4 - a + \frac{2}{a}$ **15.** $-2x^3 + \frac{2x^2}{3} - x$ **17.** $-9x^2 + 5x + 1$ **19.** $\frac{4x^2}{3} + x - \frac{2}{3x}$ **21.** $9r^3 - 12r^2 - 2r + 1 - \frac{2}{3r}$ **23.** $-m^2 + 3m - \frac{4}{m}$ **25.** $\frac{12}{x} - \frac{6}{x^2} + \frac{14}{x^3} - \frac{10}{x^4}$ **27.** $-4b^2 + 3ab - \frac{5}{a}$ **29.** $6x - 2 + \frac{1}{x}$ **31.** $15x^5 - 35x^4 + 35x^3$ **33.** 1423 **34.** $(1 \times 10^3) + (4 \times 10^2) + (2 \times 10^1) + (3 \times 10^0)$ **35.** $x^3 + 4x^2 + 2x + 3$ **36.** They are similar in that the coefficients of the powers of ten are equal to the coefficients of the powers of x. They are different in that one is a number while the other is a polynomial. They are equal if $x = 10$.

Section 5.7 (pages 433–434)

1. The divisor is $2x + 5$; the quotient is $2x^3 - 4x^2 + 3x + 2$. **3.** Divide $12m^2$ by $2m$ to get $6m$. **5.** $x + 2$ **7.** $2y - 5$ **9.** $p - 4 + \frac{44}{p + 6}$ **11.** $r - 5$ **13.** $2a - 14 + \frac{74}{2a + 3}$ **15.** $4x^2 - 7x + 3$ **17.** $3y^2 - 2y + 2$ **19.** $2x^2 - 2x + 3 + \frac{-1}{x + 1}$ **21.** $3k - 4 + \frac{2}{k^2 - 2}$ **23.** $x^2 + 1$ **25.** $x^2 + 1$ **27.** $2p^2 - 5p + 4 + \frac{6}{3p^2 + 1}$ **29.** $x^3 + 6x - 7$ **31.** $2x^2 + \frac{3}{5}x + \frac{1}{5}$ **33.** $(x^2 + x - 3)$ units **35.** 33 **36.** 33 **37.** They are the same. **38.** The answers should agree.

Section 5.8 (pages 439–440)

1. 6.1309×10^9; 5.8689×10^9 **3.** in scientific notation **5.** not in scientific notation; 5.6×10^6 **7.** not in scientific notation; 4×10^{-3} **9.** not in scientific notation; 8×10^1 **11.** A number is written in scientific notation if it is the product of a number whose absolute value is between 1 and 10 (inclusive of 1) and a power of 10. **13.** 5.876×10^9 **15.** 8.235×10^4 **17.** 7×10^{-6} **19.** -2.03×10^{-3} **21.** 750,000 **23.** 5,677,000,000,000 **25.** 1,000,000,000,000 **27.** -6.21 **29.** 0.00078 **31.** 0.000000005134 **33.** 6×10^{11}; 600,000,000,000 **35.** 1.5×10^7; 15,000,000 **37.** 8×10^{-3}; 0.008 **39.** 2.4×10^2; 240 **41.** 6.3×10^{-2}; 0.063 **43.** 6.426×10^4; 64,260 **45.** 3×10^{-4}; 0.0003 **47.** 4×10^1; 40 **49.** 1.3×10^{-5}; 0.000013 **51.** 5×10^2; 500 **53.** 2.6×10^{-3}; 0.0026 **55.** 7.205×10^{-6}; 0.000007205 **57.** 1.5×10^{17} mi **59.** \$3554

Chapter 5 Review Exercises (pages 445–448)

1. $22m^2$; degree 2; monomial **2.** $p^3 - p^2 + 4p + 2$; degree 3; none of these **3.** already in descending powers; degree 5; none of these **4.** $-8y^5 - 7y^4 + 9y$; degree 5; trinomial **5.** $-5a^3 + 4a^2$ **6.** $2r^3 - 3r^2 + 9r$ **7.** $11y^2 - 10y + 9$ **8.** $-13k^4 - 15k^2 - 4k - 6$

9. $10m^3 - 6m^2 - 3$ **10.** $-y^2 - 4y + 26$ **11.** $10p^2 - 3p - 11$ **12.** $7r^4 - 4r^3 - 1$ **13.** 4^{11} **14.** -5^{11} **15.** $-72x^7$ **16.** $10x^{14}$ **17.** 19^5x^5 **18.** -4^7y^7 **19.** $5p^4t^4$ **20.** $\frac{7^6}{5^6}$ **21.** $27x^6y^9$ **22.** t^{42} **23.** $36x^{16}y^4z^{16}$ **24.** $\frac{8m^9n^3}{p^6}$ **25.** $125x^6$ **26.** The product rule for exponents does not apply here because we want the sum of 7^2 and 7^4, not their product. **27.** $10x^2 + 70x$ **28.** $-6p^5 + 15p^4$ **29.** $6r^3 + 8r^2 - 17r + 6$ **30.** $8y^3 + 27$ **31.** $5p^5 - 2p^4 - 3p^3 + 25p^2 + 15p$ **32.** $x^2 + 3x - 18$ **33.** $6k^2 - 9k - 6$ **34.** $12p^2 - 48pq + 21q^2$ **35.** $2m^4 + 5m^3 - 16m^2 - 28m + 9$ **36.** $a^2 + 8a + 16$ **37.** $9p^2 - 12p + 4$ **38.** $4r^2 + 20rs + 25s^2$ **39.** $r^3 + 6r^2 + 12r + 8$ **40.** $8x^3 - 12x^2 + 6x - 1$ **41.** $4z^2 - 49$ **42.** $36m^2 - 25$ **43.** $25a^2 - 36b^2$ **44.** $4x^4 - 25$ **45.** three; two **46.** $(a + b)^2 = (a + b)(a + b) = a^2 + 2ab + b^2$. The term $2ab$ is not in $a^2 + b^2$. **47.** 2 **48.** $\frac{1}{32}$ **49.** $\frac{25}{36}$ **50.** $-\frac{3}{16}$ **51.** 36 **52.** x^2 **53.** $\frac{1}{p^{12}}$ **54.** r^4 **55.** 2^8 **56.** $\frac{1}{9^6}$ **57.** 5^8 **58.** $\frac{1}{8^{12}}$ **59.** $\frac{1}{m^2}$ **60.** y^7 **61.** r^{13} **62.** $25m^6$ **63.** $\frac{y^{12}}{8}$ **64.** $\frac{1}{a^3b^5}$ **65.** $72r^5$ **66.** $\frac{8n^{10}}{3m^{13}}$ **67.** $\frac{5y^2}{3}$ **68.** $-2x^2y$ **69.** $-y^3 + 2y - 3$ **70.** $p - 3 + \frac{5}{2p}$ **71.** $-x^9 + 2x^8 - 4x^3 + 7x$ **72.** $-2m^2n + mn^2 + \frac{6n^3}{5}$ **73.** $2r + 7$ **74.** $4m + 3 + \frac{5}{3m - 5}$ **75.** $2a + 1 + \frac{-8a + 12}{5a^2 - 3}$ **76.** $k^2 + 2k + 4 + \frac{-2k - 12}{2k^2 + 1}$ **77.** 4.8×10^7 **78.** 2.8988×10^{10} **79.** 6.5×10^{-5} **80.** 8.24×10^{-8} **81.** 24,000 **82.** 78,300,000 **83.** 0.000000897 **84.** 0.00000000000995 **85.** 8×10^2; 800 **86.** 4×10^6; 4,000,000 **87.** 2.5×10^{-2}; 0.025 **88.** 1×10^{-2}; 0.01 **89.** 2.796×10^{10} calculations; 1.6776×10^{12} calculations **90.** about 3.3 **91.** 0 **92.** $\frac{243}{p^3}$ **93.** $\frac{1}{49}$ **94.** $49 - 28k + 4k^2$ **95.** $y^2 + 5y + 1$ **96.** $\frac{1296r^8s^4}{625}$ **97.** $-8m^7 - 10m^6 - 6m^5$ **98.** 32 **99.** $5xy^3 - \frac{8y^2}{5} + 3x^2y$ **100.** $\frac{r^2}{6}$ **101.** $8x^3 + 12x^2y + 6xy^2 + y^3$ **102.** $\frac{3}{4}$ **103.** $a^3 - 2a^2 - 7a + 2$ **104.** $8y^3 - 9y^2 + 5$ **105.** $10r^2 + 21r - 10$ **106.** $144a^2 - 1$ **107.** $2x^2 + x - 6$ **108.** $20x^4 + 8x^2$; $25x^8 + 20x^6 + 4x^4$

Chapter 5 Test (pages 449–450)

1. $4t^4 + t^3 - 6t^2 - t$ **2.** $-2y^2 - 9y + 17$ **3.** $-12t^2 + 5t + 8$ **4.** -32 **5.** $\frac{216}{m^6}$ **6.** $-27x^5 + 18x^4 - 6x^3 + 3x^2$ **7.** $2r^3 + r^2 - 16r + 15$ **8.** $t^2 - 5t - 24$ **9.** $8x^2 + 2xy - 3y^2$ **10.** $25x^2 - 20xy + 4y^2$ **11.** $100v^2 - 9w^2$ **12.** $x^3 + 3x^2 + 3x + 1$ **13.** $12x + 36$; $9x^2 + 54x + 81$ **14.** $\frac{1}{625}$ **15.** 2 **16.** $\frac{7}{12}$ **17.** 8^5 **18.** x^2y^6 **19.** $4y^2 - 3y + 2 + \frac{5}{y}$ **20.** $-3xy^2 + 2x^3y^2 + 4y^2$ **21.** $2x + 9$ **22.** $3x^2 + 6x + 11 + \frac{26}{x - 2}$ **23.** **(a)** 3.44×10^{11} **(b)** 5.57×10^{-6} **24.** **(a)** 29,600,000 **(b)** 0.0000000607 **25.** 5.89×10^{15} mi

Cumulative Review Exercises: Chapters R–5 (pages 451–452)

1. $\frac{19}{24}$ **2.** $-\frac{1}{20}$ **3.** 3.72 **4.** 62.006 **5.** \$1836 **6.** -8 **7.** 24 **8.** $\frac{1}{2}$ **9.** -4 **10.** associative property **11.** inverse property **12.** distributive property **13.** $\{10\}$ **14.** $\left\{\frac{13}{4}\right\}$ **15.** $\emptyset$ **16.** $r = \frac{d}{t}$ **17.** $\{-5\}$ **18.** $\{-12\}$ **19.** $\{20\}$ **20.** {all real numbers} **21.** mouse: 160; elephant: 10 **22.** 4 **23.** $[10, \infty)$ **24.** $\left(-\infty, -\frac{14}{5}\right)$ **25.** $[-4, 2)$ **26.** $(0, 2)$ and $(-3, 0)$ **27.**

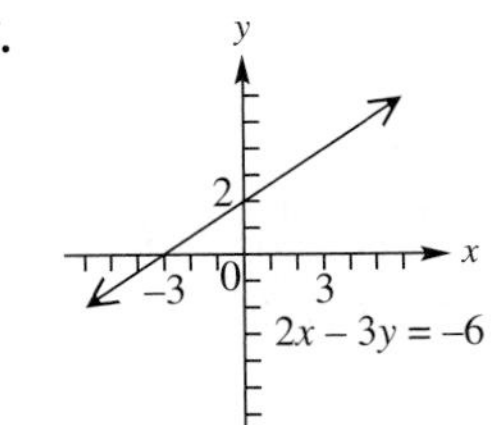

28. $\frac{2}{3}$ **29.** $\{(-3, -1)\}$ **30.** $\{(4, -5)\}$ **31.** $\frac{5}{4}$, or $1\frac{1}{4}$ **32.** 2 **33.** 1 **34.** $\frac{2b}{a^{10}}$ **35.** 3.45×10^4 **36.** $11x^3 - 14x^2 - x + 14$ **37.** $18x^7 - 54x^6 + 60x^5$ **38.** $63x^2 + 57x + 12$ **39.** $25x^2 + 80x + 64$ **40.** $y^2 - 2y + 6$

Chapter 6 Factoring and Applications

Section 6.1 (pages 461–462)

1. 4 **3.** 4 **5.** 6 **7.** 1 **9.** 8 **11.** $10x^3$ **13.** xy^2 **15.** 6 **17.** $3m^2$ **19.** $2z^4$ **21.** $2mn^4$ **23.** $y + 2$ **25.** $a - 2$ **27.** $2 + 3xy$ **29.** $x(x - 4)$ **31.** $3t(2t + 5)$ **33.** $\frac{1}{4}d(d - 3)$ **35.** $-6x^2(2x + 1)$ **37.** $5y^6(13y^4 + 7)$ **39.** no common factor (except 1) **41.** $8m^2n^2(n + 3)$ **43.** $-2x(2x^2 - 5x + 3)$ **45.** $13y^2(y^6 + 2y^2 - 3)$ **47.** $9qp^3(5q^3p^2 + 4p^3 + 9q)$ **49.** $(x + 2)(c + d)$ **51.** $(2a + b)(a^2 - b)$ **53.** $(p + 4)(q - 1)$ **55.** $(5 + n)(m + 4)$ **57.** $(2y - 7)(3x + 4)$ **59.** $(y + 3)(3x + 1)$ **61.** $(z + 2)(7z - a)$ **63.** $(3r + 2y)(6r - x)$ **65.** $(w + 1)(w^2 + 9)$ **67.** $(a + 2)(3a^2 - 2)$ **69.** $(4m - p^2)(4m^2 - p)$ **71.** $(y + 3)(y + x)$ **73.** $(z - 2)(2z - 3w)$ **75.** commutative property **76.** $2x(y - 4) - 3(y - 4)$ **77.** No, because it is not a product. It is the difference between $2x(y - 4)$ and $3(y - 4)$. **78.** $(2x - 3)(y - 4)$; yes

Section 6.2 (pages 467–468)

1. a and b must have different signs. **3.** A prime polynomial is one that cannot be factored using only integers in the factors. **5.** 1 and 12, -1 and -12, 2 and 6, -2 and -6, 3 and 4, -3 and -4; The pair with a sum of 7 is 3 and 4. **7.** 1 and -24, -1 and 24, 2 and -12, -2 and 12, 3 and -8, -3 and 8, 4 and -6, -4 and 6; The pair with a sum of -5 is 3 and -8. **9.** C **11.** $x + 11$ **13.** $x - 8$ **15.** $y - 5$ **17.** $x + 11$ **19.** $y - 9$ **21.** $(y + 8)(y + 1)$ **23.** $(b + 3)(b + 5)$ **25.** $(m + 5)(m - 4)$ **27.** $(x + 8)(x - 5)$ **29.** $(y - 5)(y - 3)$ **31.** $(z - 8)(z - 7)$ **33.** $(r - 6)(r + 5)$ **35.** $(a - 12)(a + 4)$ **37.** prime **39.** $(r + 2a)(r + a)$ **41.** $(x + y)(x + 3y)$ **43.** $(t + 2z)(t - 3z)$ **45.** $(v - 5w)(v - 6w)$ **47.** $4(x + 5)(x - 2)$ **49.** $2t(t + 1)(t + 3)$

51. $-2x^4(x-3)(x+7)$ **53.** $a^3(a+4b)(a-b)$ **55.** $mn(m-6n)(m-4n)$ **57.** The factored form $(2x+4)(x-3)$ is incorrect because $2x+4$ has a common factor of 2, which must be factored out for the trinomial to be *completely* factored.

Section 6.3 (pages 471–472)

1. $(m+6)(m+2)$ **3.** $(a+5)(a-2)$ **5.** $(2t+1)(5t+2)$ **7.** $(3z-2)(5z-3)$ **9.** $(2s-t)(4s+3t)$ **11.** $(3a+2b)(5a+4b)$ **13.** B **15. (a)** 2; 12; 24; 11 **(b)** 3; 8 (Order is irrelevant.) **(c)** $3m$; $8m$ **(d)** $2m^2+3m+8m+12$ **(e)** $(2m+3)(m+4)$ **(f)** $(2m+3)(m+4)=2m^2+11m+12$ **17.** $(2x+1)(x+3)$ **19.** $(4r-3)(r+1)$ **21.** $(4m+1)(2m-3)$ **23.** $(3m+1)(7m+2)$ **25.** $(2b+1)(3b+2)$ **27.** $(4y-3)(3y-1)$ **29.** $3(4x-1)(2x-3)$ **31.** $2m(m-4)(m+5)$ **33.** $-4z^3(z-1)(8z+3)$ **35.** $(3p+4q)(4p-3q)$ **37.** $(3a-5b)(2a+b)$ **39.** $(5-x)(1-x)$ **41.** The student stopped too soon. He needs to factor out the common factor $4x-1$ to get $(4x-1)(4x-5)$ as the correct answer.

Section 6.4 (pages 477–478)

1. B **3.** A **5.** A **7.** $2a+5b$ **9.** x^2+3x-4; $x+4, x-1$, or $x-1, x+4$ **11.** $2z^2-5z-3$; $2z+1, z-3$, or $z-3, 2z+1$ **13.** The binomial $2x-6$ cannot be a factor because it has a common factor of 2, but the polynomial does not. **15.** $(3a+7)(a+1)$ **17.** $(2y+3)(y+2)$ **19.** $(3m-1)(5m+2)$ **21.** $(3s-1)(4s+5)$ **23.** $(5m-4)(2m-3)$ **25.** $(4w-1)(2w-3)$ **27.** $(4y+1)(5y-11)$ **29.** prime **31.** $2(5x+3)(2x+1)$ **33.** $-q(5m+2)(8m-3)$ **35.** $3n^2(5n-3)(n-2)$ **37.** $-y^2(5x-4)(3x+1)$ **39.** $(5a+3b)(a-2b)$ **41.** $(4s+5t)(3s-t)$ **43.** $m^4n(3m+2n)(2m+n)$ **45.** $-1(x+7)(x-3)$ **47.** $-1(3x+4)(x-1)$ **49.** $-1(a+2b)(2a+b)$ **51.** $5\cdot 7$ **52.** $(-5)(-7)$ **53.** The product of $3x-4$ and $2x-1$ is $6x^2-11x+4$. **54.** The product of $4-3x$ and $1-2x$ is $6x^2-11x+4$. **55.** The factors in Exercise 53 are the opposites of the factors in Exercise 54. **56.** $(3-7t)(5-2t)$

Section 6.5 (pages 485–488)

1. 1; 4; 9; 16; 25; 36; 49; 64; 81; 100; 121; 144; 169; 196; 225; 256; 289; 324; 361; 400 **3.** 2 **5.** $(y+5)(y-5)$ **7.** $\left(p+\frac{1}{3}\right)\left(p-\frac{1}{3}\right)$ **9.** prime **11.** $(3r+2)(3r-2)$ **13.** $\left(2m+\frac{3}{5}\right)\left(2m-\frac{3}{5}\right)$ **15.** $4(3x+2)(3x-2)$ **17.** $(14p+15)(14p-15)$ **19.** $(4r+5a)(4r-5a)$ **21.** prime **23.** $(p^2+7)(p^2-7)$ **25.** $(x^2+1)(x+1)(x-1)$ **27.** $(p^2+16)(p+4)(p-4)$ **29.** The teacher was justified, because it was not factored *completely;* x^2-9 can be factored as $(x+3)(x-3)$. The complete factored form is $(x^2+9)(x+3)(x-3)$. **31.** No, it is not a perfect square since the middle term would have to be $30y$. **33.** $(w+1)^2$ **35.** $(x-4)^2$ **37.** $\left(t+\frac{1}{2}\right)^2$ **39.** $(x-0.5)^2$ **41.** $2(x+6)^2$ **43.** $(4x-5)^2$ **45.** $(7x-2y)^2$ **47.** $(8x+3y)^2$ **49.** $-2h(5h-2y)^2$ **51.** 1; 8; 27; 64; 125; 216; 343; 512; 729; 1000 **53.** C, D **55.** $(a+1)(a^2-a+1)$ **57.** $(a-1)(a^2+a+1)$ **59.** $(p+q)(p^2-pq+q^2)$ **61.** $(y-6)(y^2+6y+36)$ **63.** $(k+10)(k^2-10k+100)$ **65.** $(3x-1)(9x^2+3x+1)$ **67.** $(5a+2)(25a^2-10a+4)$ **69.** $(y-2x)(y^2+2xy+4x^2)$ **71.** $(3a-4b)(9a^2+12ab+16b^2)$ **73.** $(2p+9q)(4p^2-18pq+81q^2)$ **75.** $2(2t-1)(4t^2+2t+1)$ **77.** $5(2w+3)(4w^2-6w+9)$ **79.** $(x+y^2)(x^2-xy^2+y^4)$ **81.** $(5k-2m^3)(25k^2+10km^3+4m^6)$ **83.** $(x^3-1)(x^3+1)$ **84.** $(x-1)(x^2+x+1)(x+1)(x^2-x+1)$ **85.** $(x^2-1)(x^4+x^2+1)$ **86.** $(x-1)(x+1)(x^4+x^2+1)$ **87.** The result in Exercise 84 is completely factored. **88.** Show that $x^4+x^2+1=(x^2+x+1)(x^2-x+1)$. **89.** difference of squares **90.** $(x-3)(x^2+3x+9)(x+3)(x^2-3x+9)$

Section 6.6 (pages 491–492)

1. F **3.** A **5.** D **7.** C **9.** H **11.** $(10a+3b)(10a-3b)$ **13.** $6p^3(3p^2-4+2p^3)$ **15.** $(x+7)(x-5)$ **17.** prime **19.** $(6b+1)(b-3)$ **21.** $3mn(3m+2n)(2m-n)$ **23.** $(m+4)(m^2-6)$ **25.** $(2k+7r)^2$ **27.** $(m-2)(n+5)$ **29.** $(x+3)^2(x-3)$ **31.** prime **33.** $(2p-5)(4p^2+10p+25)$ **35.** $(x^2+25)(x+5)(x-5)$ **37.** $(a+6)(b+c)$ **39.** $(7z+2k)(2z-k)$ **41.** $16(4b+5c)(4b-5c)$ **43.** $8(5z+4)(25z^2-20z+16)$ **45.** $(5r-s)(2r+5s)$ **47.** $8x^2(4+2x-3x^3)$ **49.** $(2x-5q)(7x+5q)$ **51.** $(p+2)(4+m)$ **53.** $2(5p+9)(5p-9)$ **55.** $(4a+b)^2$

Section 6.7 (pages 499–500)

1. $\{-5, 2\}$ **3.** $\left\{3, \frac{7}{2}\right\}$ **5.** $\left\{-\frac{5}{6}, 0\right\}$ **7.** $\left\{0, \frac{4}{3}\right\}$ **9.** $\left\{-\frac{1}{2}, \frac{1}{6}\right\}$ **11.** $\{9\}$ **13.** Set each *variable* factor equal to 0, to get $2x=0$ or $3x-4=0$. The solution set is $\left\{0, \frac{4}{3}\right\}$. **15.** $\{-2, -1\}$ **17.** $\{1, 2\}$ **19.** $\{-8, 3\}$ **21.** $\{-1, 3\}$ **23.** $\{-2, -1\}$ **25.** $\{-4\}$ **27.** $\left\{-2, \frac{1}{3}\right\}$ **29.** $\left\{-\frac{4}{3}, \frac{1}{2}\right\}$ **31.** $\left\{-\frac{2}{3}\right\}$ **33.** $\{-3, 3\}$ **35.** $\left\{-\frac{7}{4}, \frac{7}{4}\right\}$ **37.** $\{-11, 11\}$ **39.** $\{0, 7\}$ **41.** $\left\{0, \frac{1}{2}\right\}$ **43.** $\{2, 5\}$ **45.** $\left\{-4, \frac{1}{2}\right\}$ **47.** $\left\{-12, \frac{11}{2}\right\}$ **49.** $\{-2, 0, 2\}$ **51.** $\left\{-\frac{7}{3}, 0, \frac{7}{3}\right\}$ **53.** $\left\{-\frac{5}{2}, \frac{1}{3}, 5\right\}$ **55.** $\left\{-\frac{7}{2}, -3, 1\right\}$ **57. (a)** 64; 144; 4; 6 **(b)** No time has elapsed, so the object hasn't fallen (been released) yet.

Section 6.8 (pages 507–512)

1. Read; variable; equation; Solve; answer; Check; original **3.** *Step 3:* $45=(2x+1)(x+1)$; *Step 4:* $x=4$ or $x=-\frac{11}{2}$; *Step 5:* base: 9 units; height: 5 units; *Step 6:* $9\cdot 5=45$ **5.** *Step 3:* $192=4x(x+2)$; *Step 4:* $x=6$ or $x=-8$; *Step 5:* length: 8 units; width: 6 units; *Step 6:* $8\cdot 6\cdot 4=192$ **7.** length: 14 cm; width: 12 cm **9.** length: 15 in.; width: 12 in. **11.** height: 13 in.; width: 10 in. **13.** mirror: 7 ft; painting: 9 ft **15.** 20, 21 **17.** $-3, -2$ or 4, 5 **19.** $-3, -1$ or 7, 9 **21.** $-2, 0, 2$ or 6, 8, 10 **23.** 12 cm **25.** 12 mi **27.** 8 ft **29. (a)** 1 sec **(b)** $\frac{1}{2}$ sec and $1\frac{1}{2}$ sec **(c)** 3 sec **(d)** The negative solution, -1, does not make sense since t represents time, which cannot be negative. **31. (a)** 46 million; The result using the model is a little more than 44 million, the actual number for 1996.

ANSWERS

(b) 14 **(c)** 184 million; The result is a little more than 182 million, the actual number for 2004. **(d)** 318 million **32.** \$58.6 billion; 16% **33.** 2003: \$522.1 billion; 2004: \$610.8 billion; 2005: \$699.5 billion **34.** The answer using the linear equation is close to the actual data for 2004, but not for the other years. **35.** 2003: \$503.9 billion; 2004: \$601.8 billion; 2005: \$718.2 billion **36.** The answers in Exercise 35 are fairly close to the actual data. The quadratic equation models the data better. **37.** (1, 365.1), (2, 423.7), (3, 496.9), (4, 612.1), (5, 714.4)

38. **U.S. TRADE DEFICIT (Goods and Services)**

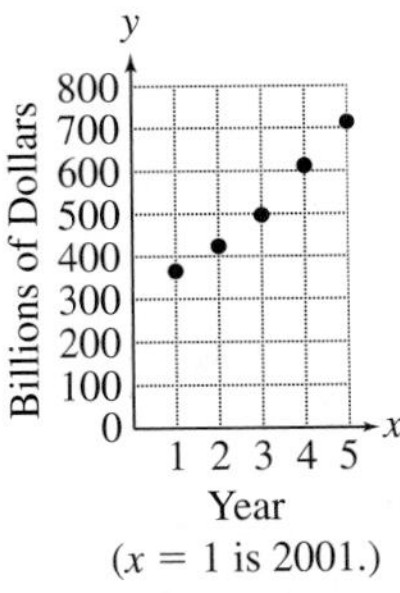

($x = 1$ is 2001.)

39. \$853.1 billion **40. (a)** The actual deficit is quite a bit less than the estimate. **(b)** No, data for later years might not follow the same pattern.

Chapter 6 Review Exercises (pages 517–520)

1. $15(t+3)$ **2.** $30z(2z^2+1)$ **3.** $11x^2(4x+5)$ **4.** $50m^2n^2(2n-mn^2+3)$ **5.** $(x-4)(2y+3)$ **6.** $(2y+3)(3y+2x)$ **7.** $(x+3)(x+7)$ **8.** $(y-5)(y-8)$ **9.** prime **10.** $(r-8)(r+7)$ **11.** $(r+8s)(r-12s)$ **12.** $(p+12q)(p-10q)$ **13.** $-8p(p+2)(p-5)$ **14.** $3x^2(x+2)(x+8)$ **15.** $p^5(p-2q)(p+q)$ **16.** $(3r-1)(r+4)$ **17.** $(3r+2)(2r-3)$ **18.** prime **19.** $4x^3(3x-1)(2x-1)$ **20.** $-mn(3m+5)(m-8)$ **21.** $(2a-5b)(7a+4b)$ **22.** B **23.** D **24.** $(n+8)(n-8)$ **25.** $(5b+11)(5b-11)$ **26.** $36(2p+q)(2p-q)$ **27.** prime **28.** $\left(x+\frac{7}{10}\right)\left(x-\frac{7}{10}\right)$ **29.** $(r-6)^2$ **30.** $(4m+5n)^2$ **31.** $(5x-1)(25x^2+5x+1)$ **32.** $(10p+3)(100p^2-30p+9)$ **33.** $\left\{-\frac{3}{4}, 1\right\}$ **34.** $\left\{0, \frac{5}{2}\right\}$ **35.** $\{-3, -1\}$ **36.** $\{3, 5\}$ **37.** $\left\{-\frac{4}{3}, 5\right\}$ **38.** $\left\{-\frac{8}{9}, \frac{8}{9}\right\}$ **39.** $\{0, 8\}$ **40.** $\{-1, 6\}$ **41.** $\{6\}$ **42.** $\left\{-\frac{2}{5}, -2, -1\right\}$ **43.** $\left\{-\frac{3}{7}, 0, \frac{3}{7}\right\}$ **44.** $\left\{\frac{9}{5}\right\}$ **45.** length: 10 ft; width: 4 ft **46.** 5 ft **47.** 6, 7 or -5, -4 **48.** 26 mi **49.** 112 ft **50.** 192 ft **51.** 256 ft **52.** after 8 sec **53. (a)** \$4.49 billion; The answer using the model is very close. **(b)** \$7.19 billion **54.** D **55.** The factor $2x+8$ has a common factor of 2. The complete factored form is $2(x+4)(3x-4)$. **56.** $(z-x)(z-10x)$ **57.** $(3k+5)(k+2)$ **58.** $(3m+4p)(5m-4)$ **59.** $(y^2+25)(y+5)(y-5)$ **60.** $3m(2m+3)(m-5)$ **61.** $8abc(3b^2c-7ac^2+9ab)$ **62.** prime **63.** $6xyz(2xz^2+2y-5x^2yz^3)$ **64.** $2a^3(a+2)(a-6)$ **65.** $-(2r+3q)(6r-5q)$ **66.** $8(z+2y)(z^2-2zy+4y^2)$ **67.** $(7t+4)^2$ **68.** $\{0, 7\}$ **69.** $\{-5, 2\}$ **70.** $\left\{-\frac{2}{5}\right\}$ **71.** width: 10 m; length: 17 m **72.** 6 m

Chapter 6 Test (pages 521–522)

1. D **2.** $6x(2x-5)$ **3.** $m^2n(2mn+3m-5n)$ **4.** $(2x+y)(a-b)$ **5.** $(x-7)(x-2)$ **6.** $(2x+3)(x-1)$ **7.** $(3x+1)(2x-7)$ **8.** $3(x+1)(x-5)$ **9.** $(5z-1)(2z-3)$ **10.** prime **11.** prime **12.** $(y+7)(y-7)$ **13.** $(9a+11b)(9a-11b)$ **14.** $(x+8)^2$ **15.** $(2x-7y)^2$ **16.** $-2(x+1)^2$ **17.** $(x-8)(x^2+8x+64)$ **18.** $8(k+2)(k^2-2k+4)$ **19.** $4t(t+4)^2$ **20.** $(x^2+9)(x+3)(x-3)$ **21.** $\{-3, 9\}$ **22.** $\left\{\frac{1}{2}, 6\right\}$ **23.** $\left\{-\frac{2}{5}, \frac{2}{5}\right\}$ **24.** $\{10\}$ **25.** $\{0, 3\}$ **26.** $\left\{-8, -\frac{5}{2}, \frac{1}{3}\right\}$ **27.** 6 ft by 9 ft **28.** -2, -1 **29.** 17 ft **30.** 49 million

Cumulative Review Exercises: Chapters R–6 (pages 523–524)

1. $\{0\}$ **2.** $\{0.05\}$ **3.** $\{6\}$ **4.** $t = \frac{A-P}{Pr}$ **5.** 345; 210; 38%; 15% **6.** gold: 11; silver: 12; bronze: 6 **7.** \$600 **8.** 110° and 70° **9. (a)** negative; positive **(b)** negative; negative **10.** $\left(-\frac{1}{4}, 0\right)$, $(0, 3)$ **11.** 12 **12.**

y = 12x + 3

13. (a) 17; A slope of (approximately) 17 means that the retail sales of prescription drugs increased by about \$17 billion per year. **(b)** (2005, 230) **14.** $\{(-1, 2)\}$ **15.** $\emptyset$ **16.** 4 **17.** $\frac{16}{9}$ **18.** 1 **19.** 256 **20.** $\frac{1}{p^2}$ **21.** $\frac{1}{m^6}$ **22.** $-4k^2-4k+8$ **23.** $45x^2+3x-18$ **24.** $9p^2+12p+4$ **25.** $4x^3+6x^2-3x+10$ **26.** $(2a-1)(a+4)$ **27.** $(2m+3)(5m+2)$ **28.** $(4t+3v)(2t+v)$ **29.** $(2p-3)^2$ **30.** $(5r+9t)(5r-9t)$ **31.** $2pq(3p+1)(p+1)$ **32.** $\left\{-\frac{2}{3}, \frac{1}{2}\right\}$ **33.** $\{0, 8\}$ **34.** $\left\{\frac{4}{7}\right\}$ **35.** 5 m, 12 m, 13 m

Chapter 7 Rational Expressions and Functions

Section 7.1 (pages 535–538)

1. C **3.** D **5.** E **7.** Replacing x with 2 makes the denominator 0 and the value of the expression undefined. To find the values excluded from the domain, set the denominator equal to 0 and solve the equation. All solutions of the equation are excluded from the domain. **9.** 7; $\{x \mid x \neq 7\}$ **11.** $-\frac{1}{7}$; $\left\{x \mid x \neq -\frac{1}{7}\right\}$ **13.** 0; $\{x \mid x \neq 0\}$ **15.** $-2, \frac{3}{2}$; $\left\{x \mid x \neq -2, \frac{3}{2}\right\}$ **17.** none; $(-\infty, \infty)$ **19.** none; $(-\infty, \infty)$ **21. (a)** numerator: x^2, $4x$; denominator: x, 4 **(b)** First factor the numerator, getting $x(x+4)$, then divide the numerator and denominator by the common factor of $x+4$ to get $\frac{x}{1}$, or x. **23.** B **25.** x **27.** $\frac{x-3}{x+5}$ **29.** $\frac{x+3}{2x(x-3)}$ **31.** already in lowest terms **33.** $\frac{6}{7}$

35. $\frac{z}{6}$ **37.** $\frac{2}{t-3}$ **39.** $\frac{x-3}{x+1}$ **41.** $\frac{4x+1}{4x+3}$ **43.** $a^2 - ab + b^2$ **45.** $\frac{c+6d}{c-d}$ **47.** $\frac{a+b}{a-b}$ **49.** -1 *In Exercises 51–55, there are other acceptable ways to express each answer.* **51.** $-(x+y)$ **53.** $-\frac{x+y}{x-y}$ **55.** $-\frac{1}{2}$ **57.** already in lowest terms **59.** $\frac{x+4}{x-2}$ **61.** $\frac{2x+3}{x+2}$ **63.** $-\frac{35}{8}$ **65.** $\frac{7x}{6}$ **67.** $-\frac{p+5}{2p}$ (There are other ways.) **69.** $\frac{-m(m+7)}{m+1}$ (There are other ways.) **71.** -2 **73.** $\frac{x+4}{x-4}$ **75.** $\frac{2x+3y}{2x-3y}$ **77.** $\frac{k+5p}{2k+5p}$ **79.** $(k-1)(k-2)$

Section 7.2 (pages 547–550)

1. To add or subtract rational expressions that have a common denominator, first add or subtract the numerators. Then place the result over the common denominator. Write the answer in lowest terms. **3.** $\frac{9}{t}$ **5.** $\frac{2}{x}$ **7.** 1 **9.** $x-5$ **11.** $\frac{1}{p+3}$ **13.** $a-b$ **15.** $72x^4y^5$ **17.** $z(z-2)$ **19.** $2(y+4)$ **21.** $(x+9)^2(x-9)$ **23.** $(m+n)(m-n)$ **25.** $x(x-4)(x+1)$ **27.** $(t+5)(t-2)(2t-3)$ **29.** $2y(y+3)(y-3)$ **31.** Yes, they could both be correct because the expressions are equivalent. Multiplying $\frac{3}{5-y}$ by 1 in the form $\frac{-1}{-1}$ gives $\frac{-3}{y-5}$. **33.** $\frac{31}{3t}$ **35.** $\frac{5-22x}{12x^2y}$ **37.** $\frac{1}{x(x-1)}$ **39.** $\frac{5a^2-7a}{(a+1)(a-3)}$ **41.** 3 **43.** $\frac{3}{x-4}$, or $\frac{-3}{4-x}$ **45.** $\frac{w+z}{w-z}$, or $\frac{-w-z}{z-w}$ **47.** $\frac{-13}{12(3+x)}$ **49.** $\frac{2(2x-1)}{x-1}$ **51.** $\frac{7}{y}$ **53.** $\frac{6}{x-2}$ **55.** $\frac{3x-2}{x-1}$ **57.** $\frac{4x-7}{x^2-x+1}$ **59.** $\frac{2x+1}{x}$ **61.** $\frac{x}{(x-2)^2(x-3)}$ **63.** $\frac{10x+23}{(x+2)^2(x+3)}$ **65.** $\frac{2x(x+12y)}{(x+2y)(x-y)(x+6y)}$ **67.** $c(x) = \frac{10x}{49(101-x)}$ **69.** $\frac{8}{9}$ **70.** $\frac{3}{7}+\frac{5}{9}-\frac{6}{63}$; They are the same. **71.** $\frac{8}{9}$; yes **72.** Answers will vary. Suppose the name is Sosa, so that $x = 4$. The problem is $\frac{3}{2}+\frac{5}{4}-\frac{6}{8}$. The predicted answer is $\frac{8}{4} = 2$, which is correct. **73.** It causes $\frac{3}{x-2}$ and $\frac{6}{x^2-2x}$ to be undefined, since 0 appears in the denominators. **74.** 0

Section 7.3 (pages 555–556)

1. Begin by simplifying the numerator. Then simplify the denominator. Write as a division problem, and proceed. **3.** $\frac{2x}{x-1}$ **5.** $\frac{2(k+1)}{3k-1}$ **7.** $\frac{5x^2}{9z^3}$ **9.** $\frac{1+x}{-1+x}$ **11.** $\frac{y+x}{y-x}$ **13.** $4x$ **15.** $x+4y$ **17.** $\frac{3y}{2}$ **19.** $\frac{x^2+5x+4}{x^2+5x+10}$ **21.** $\frac{m^2+6m-4}{m(m-1)}$ **22.** $\frac{m^2-m-2}{m(m-1)}$ **23.** $\frac{m^2+6m-4}{m^2-m-2}$ **24.** $m(m-1)$ **25.** $\frac{m^2+6m-4}{m^2-m-2}$ **26.** Method 1 involves simplifying the numerator and the denominator separately and then performing a division. Method 2 involves multiplying the fraction by a form of 1, the identity element for multiplication. (Preferences will vary.) **27.** $\frac{x^2y^2}{y^2+x^2}$ **29.** $\frac{y^2+x^2}{xy^2+x^2y}$, or $\frac{y^2+x^2}{xy(y+x)}$ **31.** $\frac{1}{2xy}$

Section 7.4 (pages 561–564)

1. (a) $-1, 2$ **(b)** $\{x \mid x \neq -1, 2\}$ **3. (a)** $-\frac{5}{3}, 0, -\frac{3}{2}$ **(b)** $\left\{x \mid x \neq -\frac{5}{3}, 0, -\frac{3}{2}\right\}$ **5. (a)** 0 **(b)** $\{x \mid x \neq 0\}$ **7. (a)** $4, \frac{7}{2}$ **(b)** $\left\{x \mid x \neq 4, \frac{7}{2}\right\}$ **9. (a)** $0, 1, -3, 2$ **(b)** $\{x \mid x \neq 0, 1, -3, 2\}$ **11.** $\{1\}$ **13.** $\{-6, 4\}$ **15.** $\left\{-\frac{7}{12}\right\}$ **17.** $\emptyset$ **19.** $\{-3\}$ **21.** $\{5\}$ **23.** $\{5\}$ **25.** $\emptyset$ **27.** $\left\{\frac{27}{56}\right\}$ **29.** $\emptyset$ **31.** $\{-10\}$ **33.** $\emptyset$ **35.** $\{0\}$ **37.** $\left\{x \mid x \neq -\frac{3}{2}, \frac{3}{2}\right\}$ **39.** $x = 0; y = 0$ **41.** $x = 2; y = 0$

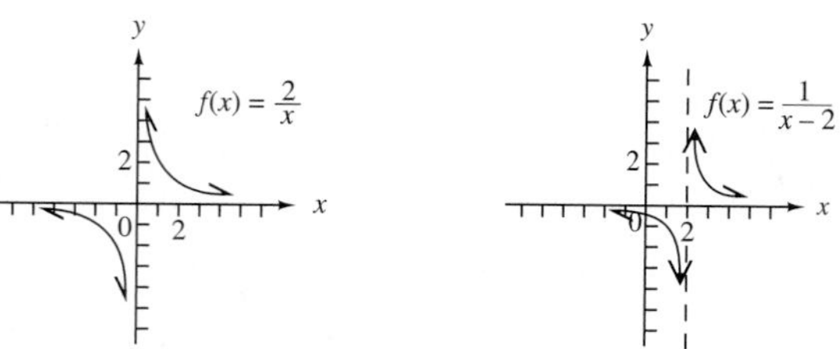

43. (a) 0 **(b)** 1.6 **(c)** 4.1 **(d)** The waiting time also increases. **45.** Substituting -1 for x gives a true statement, $\frac{4}{3} = \frac{4}{3}$. Substituting -2 for x leads to 0 in the first and third denominators. **46.** $C = -4$; $\{-2\}$; -1 is rejected. **47.** $C = 24$; $\{-4\}$; 3 is rejected. **48.** Answers will vary.

Summary Exercises on Rational Expressions and Equations (pages 565–566)

1. equation; $\{20\}$ **2.** expression; $\frac{2(x+5)}{5}$ **3.** expression; $-\frac{22}{7x}$ **4.** expression; $\frac{y+x}{y-x}$ **5.** equation; $\left\{\frac{1}{2}\right\}$ **6.** equation; $\{7\}$ **7.** expression; $\frac{43}{24x}$ **8.** equation; $\{1\}$ **9.** expression; $\frac{5x-1}{-2x+2}$, or $\frac{5x-1}{-2(x-1)}$ **10.** expression; $\frac{25}{4(r+2)}$ **11.** expression; $\frac{x^2+xy+2y^2}{(x+y)(x-y)}$ **12.** expression; $\frac{24p}{p+2}$ **13.** expression; $-\frac{5}{36}$ **14.** equation; $\{0\}$ **15.** expression; $\frac{b+3}{3}$ **16.** expression; $\frac{5}{3z}$ **17.** expression; $\frac{2x+10}{x(x-2)(x+2)}$ **18.** equation; $\{2\}$ **19.** expression; $\frac{-x}{3x+5y}$ **20.** equation; $\{-13\}$ **21.** expression; $\frac{3y+2}{y+3}$ **22.** equation; $\left\{\frac{5}{4}\right\}$ **23.** equation; $\emptyset$ **24.** expression; $\frac{2z-3}{2z+3}$ **25.** expression; $\frac{-1}{x-3}$, or $\frac{1}{3-x}$ **26.** expression; $\frac{t-2}{8}$ **27.** equation; $\{-10\}$ **28.** expression; $\frac{13x+28}{2x(x+4)(x-4)}$ **29.** equation; $\emptyset$ **30.** expression; $\frac{k(2k^2-2k+5)}{(k-1)(3k^2-2)}$

Section 7.5 (pages 577–582)

1. A **3.** D **5.** 65.625 **7.** $\frac{25}{4}$ **9.** $G = \frac{Fd^2}{Mm}$ **11.** $a = \frac{bc}{c+b}$ **13.** $v = \frac{PVt}{pT}$ **15.** $r = \frac{nE - IR}{In}$ **17.** $b = \frac{2A}{h} - B$, or $b = \frac{2A - Bh}{h}$ **19.** $r = \frac{eR}{E - e}$ **21.** Multiply each side by $a - b$. **23.** 100 games **25.** 25,000 fish **27.** 5.4 in. **29.** 6.6 more gallons **31.** 2.4 mL **33.** $x = \frac{7}{2}$; $AC = 8$; $DF = 12$ **35.** 0.032 mi per min **37.** 3.483 hr **39.** $\frac{8}{4-x} = \frac{24}{4+x}$ **41.** 3 mph **43.** 1020 mi **45.** 480 mi **47.** 190 mi **49.** $6\frac{2}{3}$ min **51.** 30 hr **53.** 20 hr **55.** $2\frac{4}{5}$ hr

Section 7.6 (pages 589–592)

1. (a) increases **(b)** decreases **3.** inverse **5.** direct **7.** joint **9.** combined **11.** 15 **13.** 300 **15.** 4 **17.** 6 **19.** $222\frac{2}{9}$ **21.** direct **23.** inverse **25.** inverse **27.** direct **29.** $\$4.59\frac{9}{10}$ **31.** about 450 cm^3 **33.** 256 ft **35.** $13\frac{1}{3}$ amperes **37.** $21\frac{1}{3}$ foot-candles **39.** \$420 **41.** 448.1 lb **43.** approximately 68,600 calls **45.** 11.8 lb

Chapter 7 Review Exercises (pages 597–600)

1. (a) -6 **(b)** $\{x \mid x \neq -6\}$ **2. (a)** 2, 5 **(b)** $\{x \mid x \neq 2, 5\}$ **3. (a)** 9 **(b)** $\{x \mid x \neq 9\}$ **4.** $\frac{x}{2}$ **5.** $\frac{5m+n}{5m-n}$ **6.** $\frac{-1}{2+r}$ **7.** The reciprocal of a rational expression is another rational expression such that the two rational expressions have a product of 1. **8.** $\frac{3y^2(2y+3)}{2y-3}$ **9.** $\frac{-3(w+4)}{w}$ **10.** $\frac{z(z+2)}{z+5}$ **11.** 1 **12.** $96b^5$ **13.** $9r^2(3r+1)$ **14.** $(3x-1)(2x+5)(3x+4)$ **15.** $\frac{16z-3}{2z^2}$ **16.** 12 **17.** $\frac{71}{30(a+2)}$ **18.** $\frac{13r^2+5rs}{(5r+s)(2r-s)(r+s)}$ **19.** $\frac{3+2t}{4-7t}$ **20.** -2 **21.** $\frac{1}{3q+2p}$ **22.** $\frac{y+x}{xy}$ **23.** $\{-3\}$ **24.** $\{-2\}$ **25.** $\{0\}$ **26.** $\emptyset$ **27.** Although her algebra was correct, 3 is not a solution because it is not in the domain of the variable in the equation. Thus, $\emptyset$ is correct. **28.** In simplifying the expression, we are combining terms to get a single fraction with a denominator of $6x$, while in solving the equation, we are finding a value for x that makes the equation true. **29.** C; $x = 0$; $y = 0$ **30.** $\frac{15}{2}$ **31.** $h = \frac{3V}{\pi r^2}$ **32.** $M = \frac{m\mu}{v - \mu}$ **33.** 16 km per hr **34.** $4\frac{4}{5}$ min **35.** C **36.** $\frac{16}{5}$ **37.** 430 mm **38.** 36 ft^3 **39.** $\frac{1}{x-2y}$ **40.** $\frac{x+5}{x+2}$ **41.** $\frac{6m+5}{3m^2}$ **42.** $\frac{k-3}{36k^2+6k+1}$ **43.** $\frac{x^2-6}{2(2x+1)}$ **44.** $\frac{x(9x+1)}{3x+1}$ **45.** $\frac{3-5x}{6x+1}$ **46.** $\frac{11}{3-x}$, or $\frac{-11}{x-3}$ **47.** $\frac{1}{3}$ **48.** $\frac{s^2+t^2}{st(s-t)}$ **49.** $\frac{5a^2+4ab+12b^2}{(a+3b)(a-2b)(a+b)}$ **50.** $\frac{acd+b^2d+bc^2}{bcd}$ **51.** $\left\{\frac{1}{3}\right\}$ **52.** $r = \frac{AR}{R-A}$, or $r = \frac{-AR}{A-R}$ **53.** $\{1, 4\}$ **54.** $\left\{-\frac{14}{3}\right\}$ **55.** $3\frac{3}{5}$ hr **56.** 2.4 mi **57.** 5.59 vibrations per sec **58.** 12 ft^2

Chapter 7 Test (pages 601–602)

1. $-2, \frac{4}{3}$; $\left\{x \mid x \neq -2, \frac{4}{3}\right\}$ **2.** $\frac{2x-5}{x(3x-1)}$ **3.** $\frac{3(x+3)}{4}$ **4.** $\frac{y+4}{y-5}$ **5.** $\frac{x+5}{x}$ **6.** $t^2(t+3)(t-2)$ **7.** $\frac{7-2t}{6t^2}$ **8.** $\frac{13x+35}{(x-7)(x+7)}$ **9.** $\frac{4}{x+2}$ **10.** $\frac{72}{11}$ **11.** $-\frac{1}{a+b}$ **12.** $\frac{2y^2+x^2}{xy(y-x)}$ **13. (a)** expression; $\frac{11(x-6)}{12}$ **(b)** equation; $\{6\}$ **14.** $\left\{\frac{1}{2}\right\}$ **15.** $\{5\}$ **16.** $x = -1$; $y = 0$

$f(x) = \frac{-2}{x+1}$

17. $3\frac{3}{14}$ hr **18.** 15 mph **19.** 48,000 fish **20. (a)** 3 units **(b)** 0 **21.** 200 amps **22.** 0.8 lb

Cumulative Review Exercises: Chapters R–7 (pages 603–604)

1. $\left\{-\frac{15}{4}\right\}$ **2.** $\{11\}$ **3.** $\left(-\infty, \frac{240}{13}\right]$ **4.** \$4000 at 4%; \$8000 at 3% **5.** 6 m **6. (a)** $-\frac{3}{2}$ **(b)** $\frac{3}{4}$ **7. (a)** $y = -\frac{3}{2}x + \frac{1}{2}$ **(b)** $y = \frac{3}{4}x - \frac{7}{4}$

8. **9.**

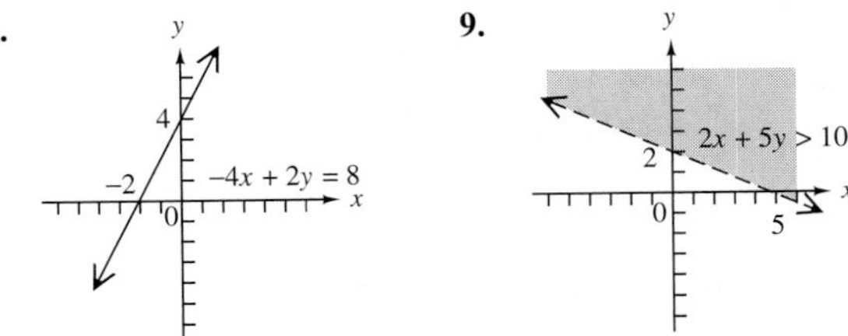

10. (a) $f(x) = \frac{5x-8}{3}$, or $f(x) = \frac{5}{3}x - \frac{8}{3}$ **(b)** -1 **11.** $3x + 15$ **12.** $\{(-1, 3)\}$ **13.** $\{(-1, 2)\}$ **14.** $\{(x, y) \mid 2x - y = 6\}$ **15.** $\frac{a^{10}}{b^{10}}$ **16.** $\frac{m}{n}$ **17.** $4y^2 - 7y - 6$ **18.** $x^2 + 4x - 7$ **19.** $12f^2 + 5f - 3$ **20.** $49t^6 - 64$ **21.** $\frac{1}{16}x^2 + \frac{5}{2}x + 25$ **22.** $(2x+5)(x-9)$ **23.** $25(2t+1)(2t-1)$ **24.** $(2p+5)(4p^2 - 10p + 25)$ **25.** $\frac{a(a-b)}{2(a+b)}$ **26.** 3 **27.** $\frac{2(x+2)}{2x-1}$ **28.** $\left\{-\frac{7}{3}, 1\right\}$ **29.** $\{-4\}$ **30.** $q = \frac{fp}{p-f}$, or $q = \frac{-fp}{f-p}$

Chapter 8 Equations, Inequalities, and Systems Revisited

Section 8.1 (pages 615–618)

1. A and C **3.** The student made a sign error when the distributive property was applied. The left side of the second line should be $8x - 4x + 6$. The correct solution is 1. **5.** $\{-1\}$ **7.** $\{-4\}$

9. $\{-7\}$ **11.** $\{0\}$ **13.** $\left\{-\frac{7}{8}\right\}$ **15.** $\left\{-\frac{1}{2}\right\}$ **17.** $\{-2\}$ **19.** $\{-1\}$

21. $\{4\}$ **23.** $\{-30\}$ **25.** $\{3\}$ **27.** $\{2000\}$ **29.** $\{25\}$ **31.** $\{3\}$

33. contradiction; $\emptyset$ **35.** conditional; $\{0\}$ **37.** identity; {all real numbers}

39. D **41.** B **43.** F

45. $(-\infty, 7]$ 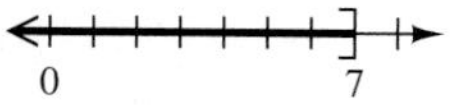

47. $[5, \infty)$ 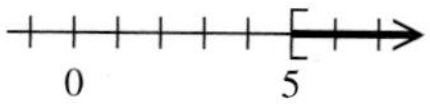

49. $(-5, \infty)$ 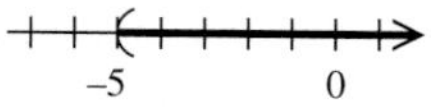

51. $(-4, \infty)$ 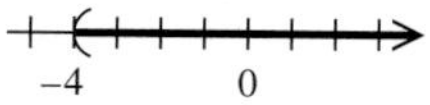

53. $(-\infty, -40]$ 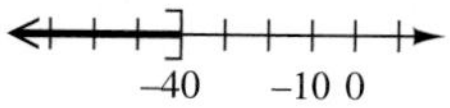

55. $(7, \infty)$ 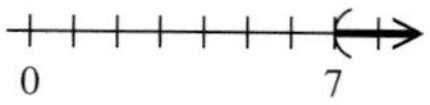

57. $\left(-\infty, -\frac{15}{2}\right)$ 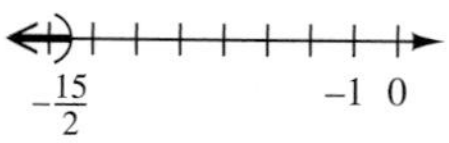

59. $(-\infty, -7)$ 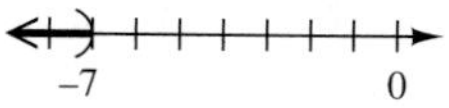

61. $\left[\frac{1}{2}, \infty\right)$ 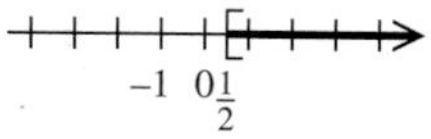

63. $(3, \infty)$ 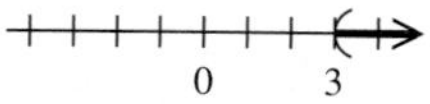

65. $(-\infty, 4)$ 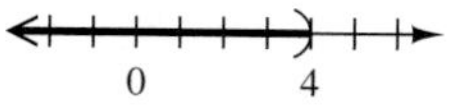

67. $\left(-\infty, \frac{23}{6}\right]$ 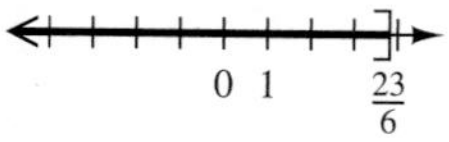

69. A

71. $(1, 11)$ 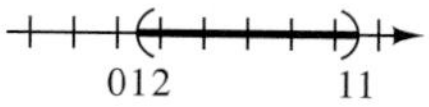

73. $[-5, 6]$ 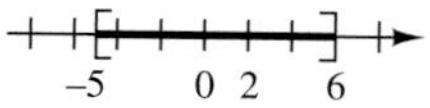

75. $(-6, -4)$ 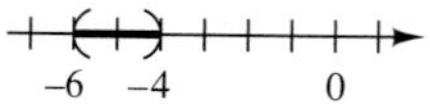

Section 8.2 (pages 625–630)

1. true **3.** false; The union is $(-\infty, 6) \cup (6, \infty)$. **5.** $\{4\}$, or D

7. $\emptyset$ **9.** $\{1, 2, 3, 4, 5, 6\}$, or A **11.** $\{1, 3, 5, 6\}$

13. 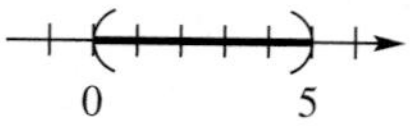

15. 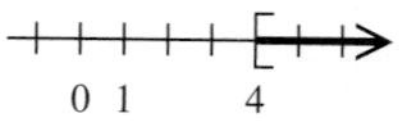

17. 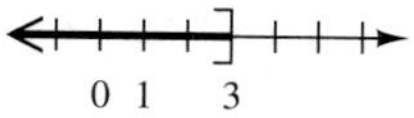

19. Answers will vary. One example is: The intersection of two streets is the region common to *both* streets.

21. $(-3, 2)$ 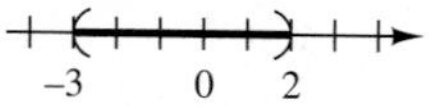

23. $(-\infty, 2]$ 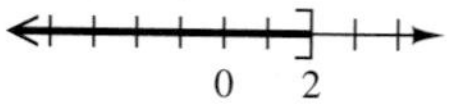

25. $\emptyset$ 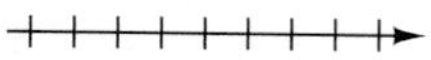

27. $[5, 9]$ 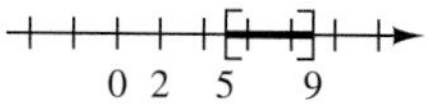

29. $(-\infty, 4]$ 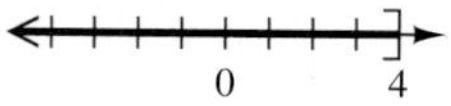

31. $(-\infty, 8]$ 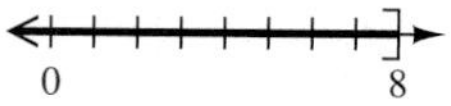

33. $[-2, \infty)$ 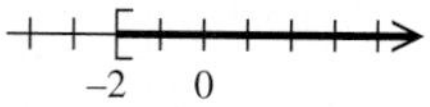

35. $(-\infty, \infty)$ 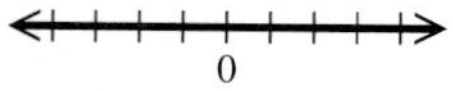

37. $(-\infty, -5) \cup (5, \infty)$ 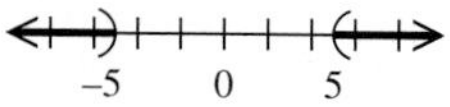

39. $(-\infty, -1] \cup (2, \infty)$ 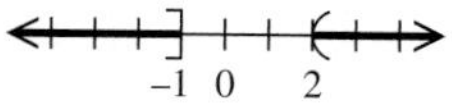

41. $(-\infty, 2) \cup (2, \infty)$ 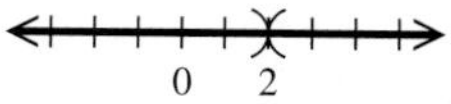

43. $[-4, -1]$ **45.** $[-9, -6]$ **47.** $(-\infty, 3)$ **49.** $[3, 9)$

51. intersection; $(-5, -1)$ 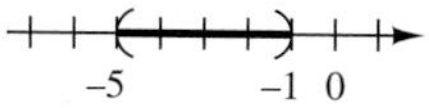

53. union; $(-\infty, 4)$ 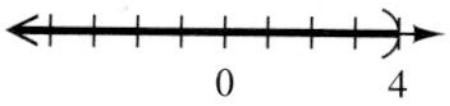

55. intersection; $[4, 12]$ 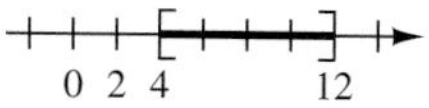

57. union; $(-\infty, 0] \cup [2, \infty)$ 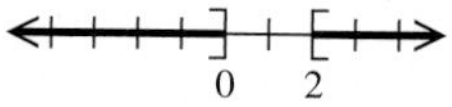

59. Mario, Joe **60.** none of them **61.** none of them **62.** Luigi, Than

63. Mario, Joe **64.** all of them **65.** {Tuition and fees}

67. {Tuition and fees, Dormitory charges}

69.

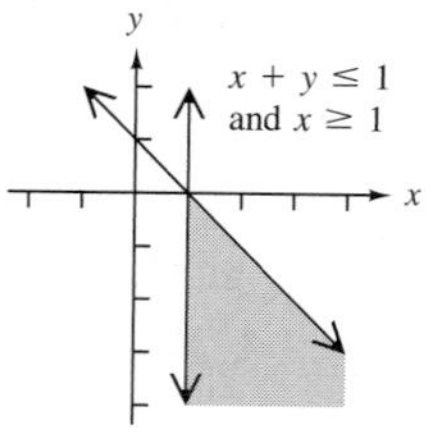

71.

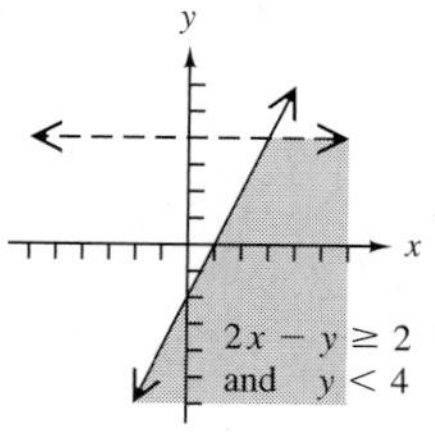

73.

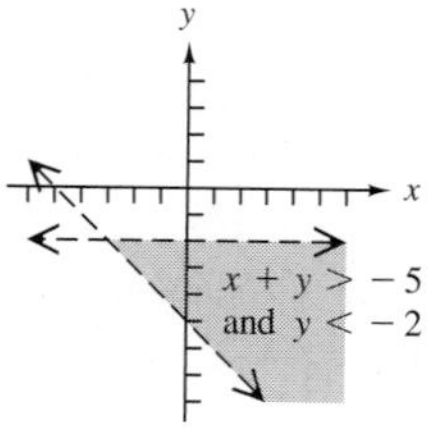

75.

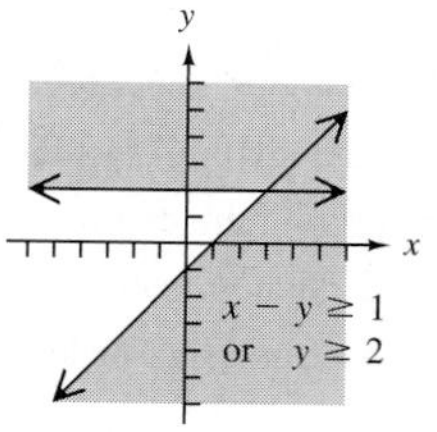

77.

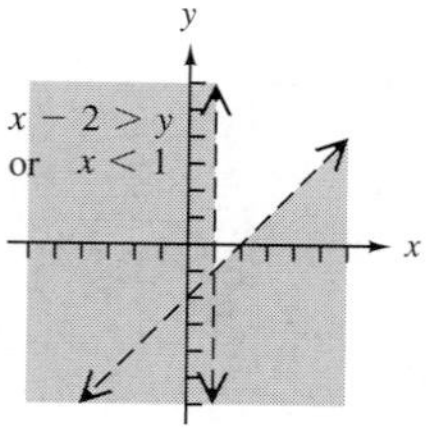

79.

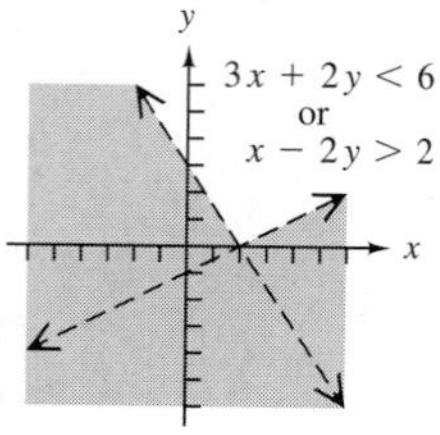

Section 8.3 (pages 637–642)

1. E; C; D; B; A **3.** **(a)** one **(b)** two **(c)** none **5.** $\{-12, 12\}$

7. $\{-5, 5\}$ **9.** $\{-6, 12\}$ **11.** $\{-5, 4\}$ **13.** $\left\{-3, \frac{11}{2}\right\}$

15. $\left\{-\frac{19}{2}, \frac{9}{2}\right\}$ **17.** $\{-10, -2\}$ **19.** $\left\{-8, \frac{32}{3}\right\}$

21. $(-\infty, -3) \cup (3, \infty)$ 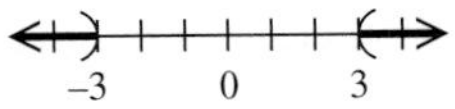

23. $(-\infty, -4] \cup [4, \infty)$ 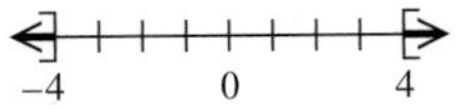

25. $(-\infty, -10) \cup (6, \infty)$ 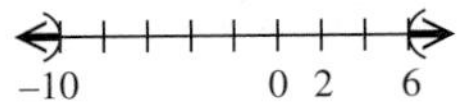

27. $\left(-\infty, -\frac{7}{3}\right] \cup [3, \infty)$ 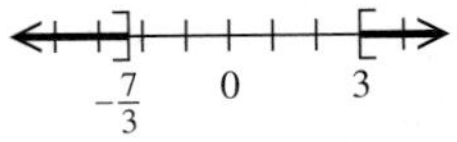

29. $(-\infty, -2) \cup (8, \infty)$ 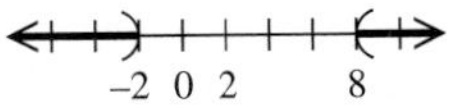

31. (a) 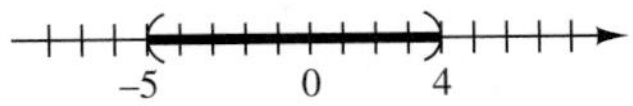

(b) 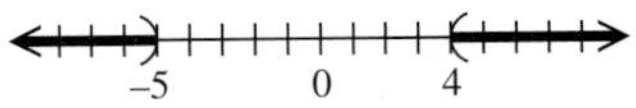

33. $[-3, 3]$ 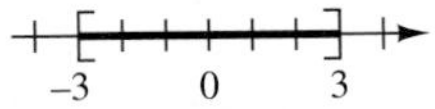

35. $(-4, 4)$ 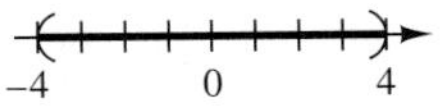

37. $[-10, 6]$ 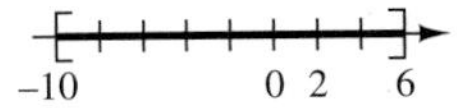

39. $\left(-\frac{7}{3}, 3\right)$ 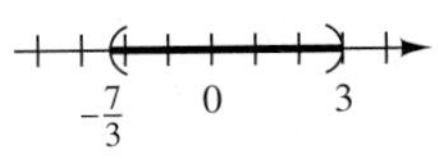

41. $[-2, 8]$ 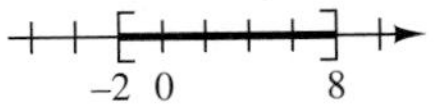

43. $(-\infty, -2) \cup (10, \infty)$ 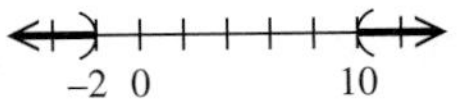

45. $\{-6, -1\}$ 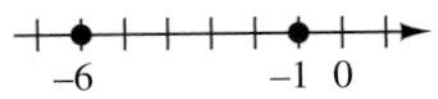

47. $\left[-\frac{10}{3}, 4\right]$ 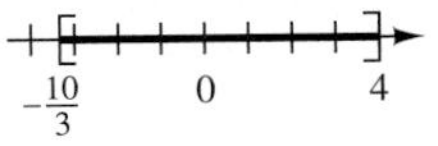

49. $\left[-4, -\frac{4}{3}\right]$ 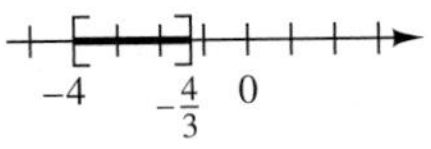

51. $\{-5, 5\}$ **53.** $\{-5, -3\}$ **55.** $(-\infty, -3) \cup (2, \infty)$ **57.** $[-10, 0]$ **59.** $\{-1, 3\}$ **61.** $\left\{-3, \frac{5}{3}\right\}$ **63.** $\left\{-\frac{1}{3}, -\frac{1}{15}\right\}$ **65.** $\left\{-\frac{5}{4}\right\}$ **67.** $(-\infty, \infty)$ **69.** $\emptyset$ **71.** $\left\{-\frac{1}{4}\right\}$ **73.** $\emptyset$ **75.** $(-\infty, \infty)$ **77.** $\left\{-\frac{3}{7}\right\}$ **79.** $(-\infty, \infty)$ **81.** $\left(-\infty, -\frac{7}{10}\right) \cup \left(-\frac{7}{10}, \infty\right)$ **83.** $(-\infty, \infty)$ **85.** $|x - 1000| \le 100$; $900 \le x \le 1100$ **87.** 475.6 ft **88.** 1201 Walnut, City Hall, Fidelity Bank and Trust Building, Kansas City Power and Light, Hyatt Regency Crown Center **89.** City Center Square, Commerce Tower, Federal Office Building, 1201 Walnut, City Hall, Fidelity Bank and Trust Building, Kansas City Power and Light, Hyatt Regency Crown Center **90. (a)** $|x - 475.6| \ge 75$ **(b)** $x \ge 550.6$ or $x \le 400.6$ **(c)** Town Pavilion, One Kansas City Place **(d)** It makes sense because it includes all buildings *not* listed in the answer to Exercise 89.

Summary Exercises on Solving Linear and Absolute Value Equations and Inequalities (pages 643–644)

1. $\{12\}$ **2.** $\{-5, 7\}$ **3.** $\{7\}$ **4.** $\left\{-\frac{2}{5}\right\}$ **5.** $\emptyset$ **6.** $(-\infty, -1]$ **7.** $\left[-\frac{2}{3}, \infty\right)$ **8.** $\{-1\}$ **9.** $\{-3\}$ **10.** $\left\{1, \frac{11}{3}\right\}$ **11.** $(-\infty, 5]$ **12.** $(-\infty, \infty)$ **13.** $\{2\}$ **14.** $(-\infty, -8] \cup [8, \infty)$ **15.** $\emptyset$ **16.** $(-\infty, \infty)$ **17.** $(-5.5, 5.5)$ **18.** $\left\{\frac{13}{3}\right\}$ **19.** $\left\{-\frac{96}{5}\right\}$ **20.** $(-\infty, 32]$ **21.** $(-\infty, -24)$ **22.** $\left\{\frac{3}{8}\right\}$ **23.** $\left\{\frac{7}{2}\right\}$ **24.** $(-6, 8)$ **25.** $(-\infty, \infty)$ **26.** $(-\infty, 5)$ **27.** $(-\infty, -4) \cup (7, \infty)$ **28.** $\{24\}$ **29.** $\left\{-\frac{1}{5}\right\}$ **30.** $\left(-\infty, -\frac{5}{2}\right]$ **31.** $\left[-\frac{1}{3}, 3\right]$ **32.** $[1, 7]$ **33.** $\left\{-\frac{1}{6}, 2\right\}$ **34.** $\{-3\}$ **35.** $(-\infty, -1] \cup \left[\frac{5}{3}, \infty\right)$ **36.** $\left[\frac{3}{4}, \frac{15}{8}\right]$ **37.** $\left\{-\frac{5}{2}\right\}$ **38.** $\{60\}$ **39.** $\left[-\frac{9}{2}, \frac{15}{2}\right]$ **40.** $(1, 9)$ **41.** $(-\infty, \infty)$ **42.** $\left\{\frac{1}{3}, 9\right\}$ **43.** $(-\infty, \infty)$ **44.** $\left\{-\frac{10}{9}\right\}$ **45.** $\{-2\}$ **46.** $\emptyset$ **47.** $(-\infty, -1) \cup (2, \infty)$ **48.** $[-3, -2]$

Section 8.4 (pages 651–652)

1. D; The ordered pair solution must be in quadrant IV, since that is where the graphs of the equations intersect. **3. (a)** B **(b)** C **(c)** A **(d)** D **5.** $\{(-2, -3)\}$

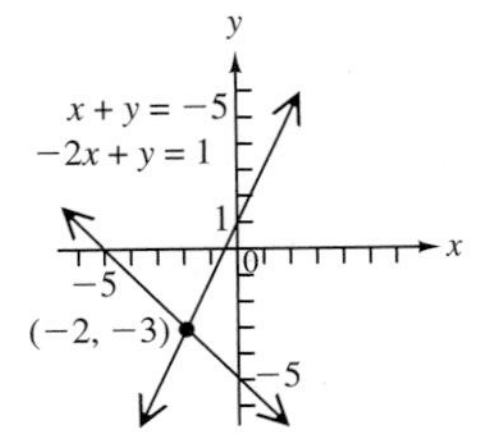

7. $\{(1, 2)\}$ **9.** $\{(2, 3)\}$ **11.** $\left\{\left(\frac{22}{9}, \frac{22}{3}\right)\right\}$ **13.** $\{(5, 4)\}$ **15.** $\left\{\left(-5, -\frac{10}{3}\right)\right\}$ **17.** $\{(2, 6)\}$ **19.** $\{(x, y) \mid 2x - y = 0\}$; dependent equations **21.** $\emptyset$; inconsistent system **23.** $\{(2, -4)\}$ **25.** $\{(3, -1)\}$ **27.** $\{(2, -3)\}$ **29.** $\left\{\left(\frac{3}{2}, -\frac{3}{2}\right)\right\}$ **31.** $\{(x, y) \mid 7x + 2y = 6\}$; dependent equations **33.** $\{(2, -4)\}$ **35.** $\emptyset$; inconsistent system

Section 8.5 (pages 661–664)

1. The statement means that when -1 is substituted for x, 2 is substituted for y, and 3 is substituted for z in the three equations, the resulting three statements are true. **3.** $\{(3, 2, 1)\}$ **5.** $\{(1, 4, -3)\}$ **7.** $\{(1, 0, 3)\}$ **9.** $\left\{\left(1, \frac{3}{10}, \frac{2}{5}\right)\right\}$ **11.** $\{(0, 2, -5)\}$ **13.** $\left\{\left(\frac{20}{59}, -\frac{33}{59}, \frac{35}{59}\right)\right\}$ **15.** $\{(4, 5, 3)\}$ **17.** $\{(2, 2, 2)\}$ **19.** $\{(-1, 0, 0)\}$ **21.** $\left\{\left(\frac{8}{3}, \frac{2}{3}, 3\right)\right\}$ **23.** Answers will vary. Some possible answers are **(a)** two perpendicular walls and the ceiling in a normal room, **(b)** the floors of three different levels of an office building, and **(c)** three pages of this book (since they intersect in the spine). **25.** $\emptyset$; inconsistent system **27.** $\{(x, y, z) \mid x - y + 4z = 8\}$; dependent equations

29. $\{(x, y, z) \mid 2x + y - z = 6\}$; dependent equations **31.** $\{(0, 0, 0)\}$ **33.** $\emptyset$; inconsistent system **35.** $\{(3, 0, 2)\}$ **37.** $x + y + z = 180$; angle measures: 70°, 30°, 80° **39.** first: 20°; second: 70°; third: 90° **41.** Independent: 38; Democrat: 34; Republican: 28 **43.** \$14 tickets: 300; \$20 tickets: 225; \$50 tickets: 60 **45.** bookstore A: 140; bookstore B: 280; bookstore C: 380 **47.** wins: 46; losses: 25; ties: 11

Section 8.6 (pages 671–672)

1. (a) $0, 5, -3$ **(b)** $1, -3, 8$ **(c)** yes; The number of rows is the same as the number of columns (three). **(d)** $\begin{bmatrix} 1 & 4 & 8 \\ 0 & 5 & -3 \\ -2 & 3 & 1 \end{bmatrix}$

(e) $\begin{bmatrix} 1 & -\frac{3}{2} & -\frac{1}{2} \\ 0 & 5 & -3 \\ 1 & 4 & 8 \end{bmatrix}$ **(f)** $\begin{bmatrix} 1 & 15 & 25 \\ 0 & 5 & -3 \\ 1 & 4 & 8 \end{bmatrix}$

3. $\left[\begin{array}{cc|c} 1 & 2 & 11 \\ 2 & -1 & -3 \end{array}\right]$; $\left[\begin{array}{cc|c} 1 & 2 & 11 \\ 0 & -5 & -25 \end{array}\right]$; $\left[\begin{array}{cc|c} 1 & 2 & 11 \\ 0 & 1 & 5 \end{array}\right]$; $x + 2y = 11$; $y = 5$; $\{(1, 5)\}$ **5.** $\{(4, 1)\}$ **7.** $\{(1, 1)\}$ **9.** $\{(-1, 4)\}$ **11.** $\emptyset$ **13.** $\{(x, y) \mid 2x + y = 4\}$ **15.** $\{(0, 0)\}$

17. $\left[\begin{array}{ccc|c} 1 & 1 & -1 & -3 \\ 0 & -1 & 3 & 10 \\ 0 & -6 & 7 & 38 \end{array}\right]$; $\left[\begin{array}{ccc|c} 1 & 1 & -1 & -3 \\ 0 & 1 & -3 & -10 \\ 0 & -6 & 7 & 38 \end{array}\right]$; $\left[\begin{array}{ccc|c} 1 & 1 & -1 & -3 \\ 0 & 1 & -3 & -10 \\ 0 & 0 & -11 & -22 \end{array}\right]$; $\left[\begin{array}{ccc|c} 1 & 1 & -1 & -3 \\ 0 & 1 & -3 & -10 \\ 0 & 0 & 1 & 2 \end{array}\right]$; $x + y - z = -3$; $y - 3z = -10$; $z = 2$; $\{(3, -4, 2)\}$ **19.** $\{(4, 0, 1)\}$ **21.** $\{(-1, 23, 16)\}$ **23.** $\{(3, 2, -4)\}$ **25.** $\emptyset$ **27.** $\{(x, y, z) \mid x - 2y + z = 4\}$ **29.** $\{(0, 0, 0)\}$

Chapter 8 Review Exercises (pages 679–682)

1. $\left\{-\frac{9}{5}\right\}$ **2.** $\left\{\frac{1}{3}\right\}$ **3.** $\{10\}$ **4.** $\{300\}$ **5.** identity; {all real numbers} **6.** contradiction; $\emptyset$ **7.** conditional; $\{0\}$

8. $(-9, \infty)$

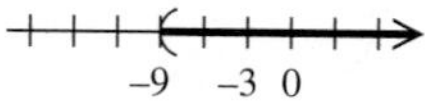

9. $(-\infty, -3]$

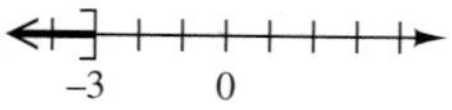

10. $[-3, \infty)$

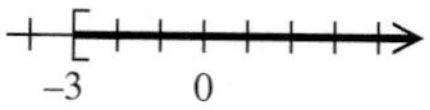

11. $[-3, 12]$

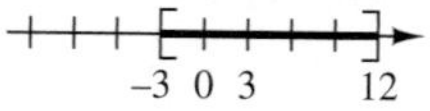

12. $[3, 5)$

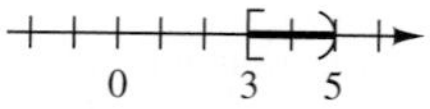

13. {a, c} **14.** {a} **15.** {a, c, e, f, g} **16.** {a, b, c, d, e, f, g}

17. $(4, 7)$

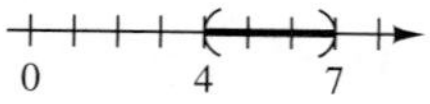

18. $(8, 14)$

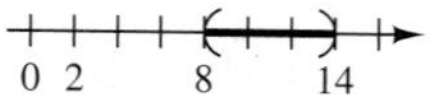

19. $(-\infty, -3] \cup (5, \infty)$

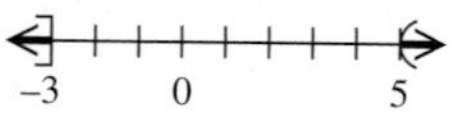

20. $(-\infty, \infty)$

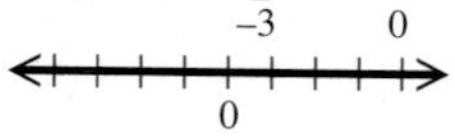

21. $\emptyset$

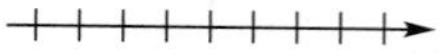

22. $(-\infty, -2] \cup [7, \infty)$

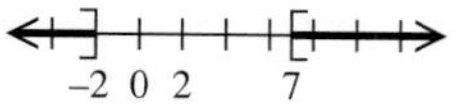

23. $(-3, 4)$ **24.** $(-\infty, 2)$ **25.** $(4, \infty)$ **26.** $(1, \infty)$

27. $\{-7, 7\}$ **28.** $\{-11, 7\}$ **29.** $\left\{-\frac{1}{3}, 5\right\}$ **30.** $\emptyset$ **31.** $\{0, 7\}$

32. $\left\{-\frac{3}{2}, \frac{1}{2}\right\}$ **33.** $\left\{-\frac{3}{4}, \frac{1}{2}\right\}$ **34.** $\left\{-\frac{1}{2}\right\}$

35. $(-12, 12)$

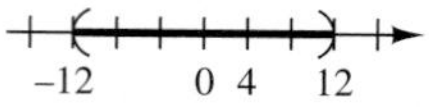

36. $[-1, 13]$

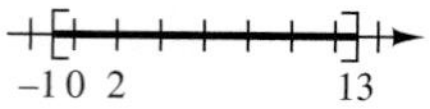

37. $[-3, -2]$

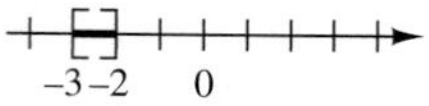

38. $(-\infty, \infty)$

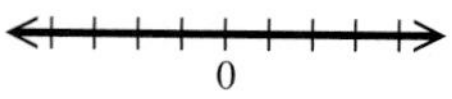

39. $\left(-\infty, -\frac{8}{5}\right) \cup (2, \infty)$

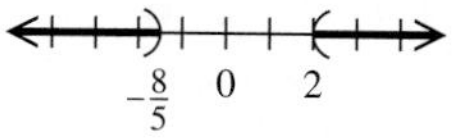

40. $(-\infty, \infty)$

41. $\{(2, 2)\}$ **42.** $\left\{\left(-\frac{8}{9}, -\frac{4}{3}\right)\right\}$ **43.** $\{(0, 4)\}$ **44.** $\{(2, 4)\}$

45. $\{(-1, 2)\}$ **46.** $\{(-6, 3)\}$ **47.** $\left\{\left(\frac{68}{13}, -\frac{31}{13}\right)\right\}$

48. $\{(x, y) \mid 3x - y = -6\}$; dependent equations **49.** $\emptyset$; inconsistent system **50.** $\{(0, 0)\}$ **51.** $\{(1, -5, 3)\}$ **52.** $\emptyset$; inconsistent system **53.** $\{(1, 2, 3)\}$ **54.** 85°, 60°, 35° **55.** Mantle: 54; Maris: 61; Blanchard: 21 **56.** $\{(3, -2)\}$ **57.** $\{(-1, 5)\}$ **58.** $\{(0, 0, -1)\}$ **59.** $\left\{\frac{7}{6}\right\}$ **60.** $[-4, 5)$

61. $\left(-\infty, \frac{14}{17}\right)$ **62.** $\emptyset$ **63.** $(-\infty, -1) \cup \left(\frac{11}{7}, \infty\right)$ **64.** $\{-5, 15\}$

65. $[-16, 10]$ **66.** $(-\infty, \infty)$ **67.** $\left\{-4, -\frac{2}{3}\right\}$

68. 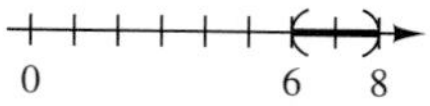 **69.**

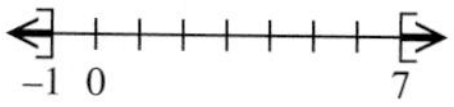

70. (a) $\emptyset$ **(b)** $(-\infty, \infty)$ **(c)** $\emptyset$

71. B; The second equation is already solved for y.

72. $\{(12, 9)\}$ **73.** $\left\{\left(\frac{82}{23}, -\frac{4}{23}\right)\right\}$ **74.** $\{(3, -1)\}$ **75.** $\{(5, 3)\}$ **76.** $\{(0, 4)\}$ **77.** $\emptyset$ **78.** Germany: 29; U.S.: 25; Canada: 24

Chapter 8 Test (pages 683–684)

1. $\{-19\}$ **2.** $\{5\}$ **3.** $\{4\}$ **4. (a)** contradiction; $\emptyset$ **(b)** identity; {all real numbers} **(c)** conditional equation; $\{0\}$

5. $[1, \infty)$

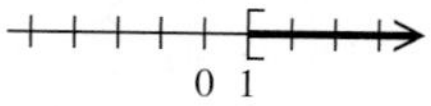

6. $[-3, 3]$

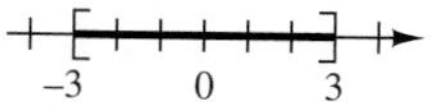

7. (a) $\{1, 5\}$ **(b)** $\{1, 2, 5, 7, 9, 12\}$

8. $[2, 9)$

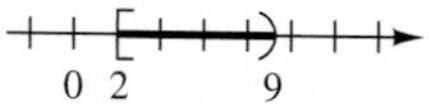

9. $(-\infty, 3) \cup [6, \infty)$

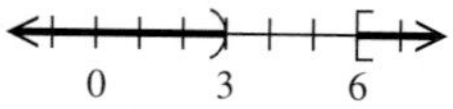

10. $\left[-\frac{5}{2}, 1\right]$

11. $\left(-\infty, -\frac{7}{6}\right) \cup \left(\frac{17}{6}, \infty\right)$

12. $\emptyset$ **13.** $\left\{-\frac{5}{3}, 3\right\}$ **14.** $\left\{-\frac{5}{7}, \frac{11}{3}\right\}$

15. $\{(6, 1)\}$

16. $\{(6, -4)\}$ **17.** $\left\{\left(-\frac{9}{4}, \frac{5}{4}\right)\right\}$ **18.** $\{(0, -2)\}$ **19.** $\emptyset$; inconsistent system **20.** $\left\{\left(-\frac{2}{3}, \frac{4}{5}, 0\right)\right\}$ **21.** $\{(3, -2, 1)\}$

22. 60 oz of Orange Pekoe; 30 oz of Irish Breakfast; 10 oz of Earl Grey

23. $\left\{\left(\frac{2}{5}, \frac{7}{5}\right)\right\}$ **24.** $\{(-1, 2, 3)\}$

Cumulative Review Exercises: Chapters R–8 (pages 685–686)

1. $-2m + 6$ **2.** $4m - 3$ **3.** $2x^2 + 5x + 4$ **4.** -24 **5.** 204

6. undefined **7.** 10 **8.** $\left\{\frac{7}{6}\right\}$ **9.** $\{-1\}$ **10.** $\left(-\infty, \frac{15}{4}\right]$

11. $\left(-\frac{1}{2}, \infty\right)$ **12.** $(2, 3)$ **13.** $(-\infty, 2) \cup (3, \infty)$ **14.** $\left\{-\frac{16}{5}, 2\right\}$

15. $(-11, 7)$ **16.** $(-\infty, -2] \cup [7, \infty)$ **17.** $h = \frac{V}{lw}$ **18.** 2 hr

19.

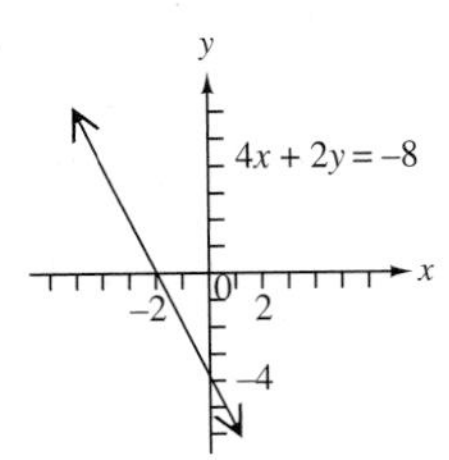

20. -1 **21.** 0 **22.** -1 **23.** $\left(-\frac{7}{2}, 0\right)$ **24.** $(0, 7)$

25. $\{(1, 5)\}$ **26.** $\{(1, 1, 0)\}$ **27.** $\frac{y}{18x}$ **28.** $\frac{5my^4}{3}$

29. $x^3 + 12x^2 - 3x - 7$ **30.** $49x^2 + 42xy + 9y^2$

31. $10p^3 + 7p^2 - 28p - 24$ **32.** $(2w + 7z)(8w - 3z)$

33. $(2y - 9)^2$ **34.** $(2p + 3)(4p^2 - 6p + 9)$

35. $\left\{-4, -\frac{3}{2}, 1\right\}$ **36.** $\left\{\frac{1}{3}\right\}$ **37.** $\{-2, 1\}$ **38.** $\frac{4}{q}$ **39.** $\frac{3r + 28}{7r}$

40. $\frac{7}{15(q - 4)}$ **41.** $\frac{7(2z + 1)}{24}$ **42.** $\frac{195}{29}$

Chapter 9 Roots, Radicals, and Root Functions

Section 9.1 (pages 697–700)

1. true **3.** false; Zero has only one square root. **5.** true **7.** E **9.** D **11.** C **13.** C **15.** C **17.** **(a)** not a real number **(b)** negative **(c)** 0 **19.** $-3, 3$ **21.** $-8, 8$ **23.** $-13, 13$ **25.** $-\frac{5}{14}, \frac{5}{14}$ **27.** $-30, 30$ **29.** 1 **31.** 7 **33.** -16 **35.** $-\frac{12}{11}$ **37.** 0.8 **39.** not a real number **41.** not a real number **43.** 100 **45.** 19 **47.** $\frac{2}{3}$ **49.** $3x^2 + 4$ **51.** rational; 5 **53.** irrational; 5.385 **55.** rational; -8 **57.** irrational; -17.321 **59.** not a real number **61.** irrational; 34.641 **63.** 6 **65.** -4 **67.** -8 **69.** 6 **71.** -2 **73.** not a real number **75.** 3 **77.** not a real number **79.** $\frac{4}{3}$ **81.** $-\frac{1}{2}$ **83.** 0.1

85. domain: $[-3, \infty)$; range: $[0, \infty)$ **87.** domain: $[0, \infty)$; range: $[-2, \infty)$

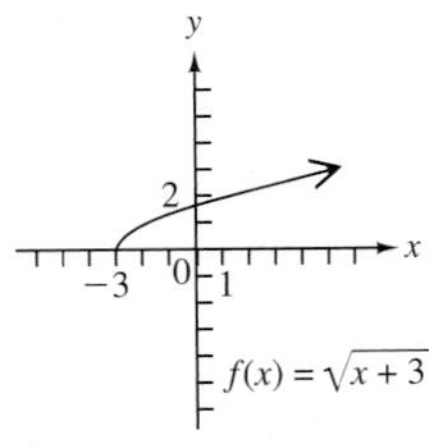

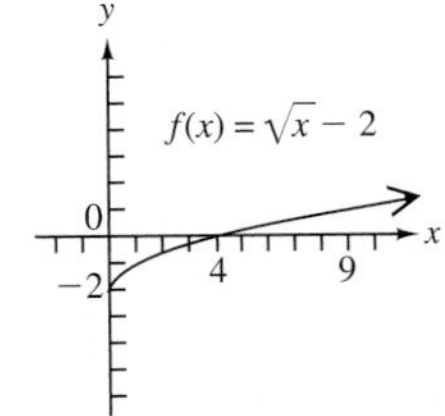

89. domain: $(-\infty, \infty)$; range: $(-\infty, \infty)$

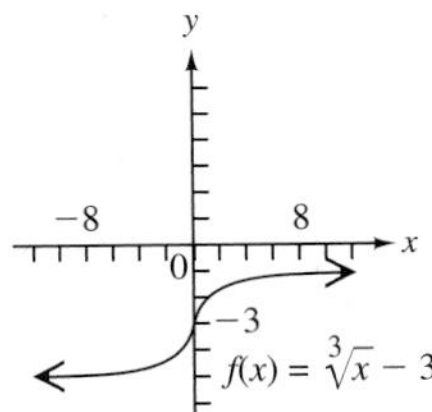

91. 12 **93.** 10 **95.** 2 **97.** -9 **99.** -5 **101.** $|x|$ **103.** $|z|$ **105.** x **107.** x^5 **109.** $|x|^5$ (or $|x^5|$) **111.** 97.381 **113.** 7.507 **115.** 3.162 **117.** 1.885 **119.** 392,000 mi^2

Section 9.2 (pages 707–710)

1. C **3.** A **5.** H **7.** B **9.** D **11.** 13 **13.** 9 **15.** 2 **17.** $\frac{8}{9}$ **19.** -3 **21.** not a real number **23.** 1000 **25.** 27 **27.** -1024 **29.** 16 **31.** $\frac{1}{8}$ **33.** $\frac{1}{512}$ **35.** $\frac{9}{25}$ **37.** $\sqrt{12}$ **39.** $\left(\sqrt[4]{8}\right)^3$ **41.** $\left(\sqrt[8]{9q}\right)^5 - \left(\sqrt[3]{2x}\right)^2$ **43.** $\frac{1}{\left(\sqrt{2m}\right)^3}$ **45.** $\left(\sqrt[3]{2y + x}\right)^2$ **47.** $\frac{1}{\left(\sqrt[3]{3m^4 + 2k^2}\right)^2}$

49. $\sqrt{a^2 + b^2} = \sqrt{3^2 + 4^2} = 5$; $a + b = 3 + 4 = 7$; $5 \neq 7$ **51.** 64 **53.** 64 **55.** x^{10} **57.** $\sqrt[6]{x^5}$ **59.** $\sqrt[15]{t^8}$ **61.** 9 **63.** 4 **65.** y **67.** $x^{5/12}$ **69.** $k^{2/3}$ **71.** x^3y^8 **73.** $\frac{1}{x^{10/3}}$ **75.** $\frac{1}{m^{1/4}n^{3/4}}$ **77.** p^2 **79.** $\frac{c^{11/3}}{b^{11/4}}$ **81.** $\frac{q^{5/3}}{9p^{7/2}}$ **83.** $p + 2p^2$ **85.** $k^{7/4} - k^{3/4}$ **87.** $6 + 18a$

89. $5 + \frac{5}{m^3}$ **91.** $y^{3/2}$ **93.** $\frac{1}{k^{2/3}}$ **95.** $x^{1/3}z^{5/6}$ **97.** $k^{1/6}$ **99.** $y^{1/30}$ **101.** $x^{5/27}$ **103.** 72 in.; 6.0 ft **105.** $-12.3°$; The table gives $-12°$.

Section 9.3 (pages 719–724)

1. true; Both are equal to $4\sqrt{3}$ and approximately 6.92820323. **3.** true; Both are equal to $6\sqrt{2}$ and approximately 8.485281374. **5.** Because there are only two factors of $\sqrt[3]{x}$, $\sqrt[3]{x} \cdot \sqrt[3]{x} = \left(\sqrt[3]{x}\right)^2$, or $\sqrt[3]{x^2}$. **7.** D **9.** $\sqrt{30}$ **11.** $\sqrt{14x}$ **13.** $\sqrt{42pqr}$ **15.** $\sqrt[3]{14xy}$ **17.** $\sqrt[4]{33}$ **19.** $\sqrt[4]{6xy^2}$ **21.** This product cannot be simplified using the product rule. **23.** $\frac{8}{11}$ **25.** $\frac{\sqrt{3}}{5}$ **27.** $\frac{\sqrt{x}}{5}$ **29.** $\frac{p^3}{9}$ **31.** $-\frac{3}{4}$ **33.** $\frac{\sqrt[3]{r^2}}{2}$ **35.** $-\frac{3}{x}$ **37.** $\frac{1}{x^3}$ **39.** $2\sqrt{3}$ **41.** $12\sqrt{2}$ **43.** $-4\sqrt{2}$ **45.** $-2\sqrt{7}$ **47.** cannot be simplified further **49.** $4\sqrt[3]{2}$ **51.** $-2\sqrt[3]{2}$ **53.** $2\sqrt[3]{5}$ **55.** $-4\sqrt[4]{2}$ **57.** $2\sqrt[5]{2}$ **59.** His reasoning was incorrect. Here 8 is a term, not a factor. **61.** $6k\sqrt{2}$ **63.** $12xy^4\sqrt{xy}$ **65.** $11x^3$ **67.** $-3t^4$ **69.** $-10m^4z^2$ **71.** $5a^2b^3c^4$ **73.** $\frac{1}{2}r^2t^5$ **75.** $5x\sqrt{2x}$ **77.** $-10r^5\sqrt{5r}$ **79.** $x^3y^4\sqrt{13x}$ **81.** $2z^2w^3$ **83.** $-2zt^2\sqrt[3]{2z^2t}$ **85.** $3x^3y^4$ **87.** $-3r^3s^2\sqrt[4]{2r^3s^2}$ **89.** $\frac{y^5\sqrt{y}}{6}$ **91.** $\frac{x^5\sqrt[3]{x}}{3}$ **93.** $4\sqrt{3}$ **95.** $\sqrt{5}$ **97.** $x^2\sqrt{x}$ **99.** $\sqrt[6]{432}$ **101.** $\sqrt[12]{6912}$ **103.** $\sqrt[6]{x^5}$ **105.** 5 **107.** $8\sqrt{2}$ **109.** $2\sqrt{14}$ **111.** 13 **113.** $9\sqrt{2}$ **115.** $\sqrt{17}$ **117.** 5 **119.** $6\sqrt{2}$ **121.** $\sqrt{5y^2 - 2xy + x^2}$ **123.** 27.0 in. **125.** 0.003 **127.** 15.3 mi

Section 9.4 (pages 727–728)

1. B **3.** 15; Each radical expression simplifies to a whole number. **5.** -4 **7.** $7\sqrt{3}$ **9.** $14\sqrt[3]{2}$ **11.** $5\sqrt[4]{2}$ **13.** $24\sqrt{2}$ **15.** cannot be simplified further **17.** $20\sqrt{5}$ **19.** $12\sqrt{2x}$ **21.** $-2m\sqrt{2}$ **23.** $\sqrt[3]{2}$ **25.** $2\sqrt[3]{x}$ **27.** $-\sqrt[3]{x^2y}$ **29.** $-x\sqrt[3]{xy^2}$ **31.** $19\sqrt[4]{2}$ **33.** $x\sqrt[4]{xy}$ **35.** $9\sqrt[4]{2a^3}$ **37.** $\frac{5\sqrt{5}}{6}$ **39.** $\frac{7\sqrt{2}}{6}$ **41.** $\frac{5\sqrt{2}}{3}$ **43.** $5\sqrt{2} + 4$ **45.** $\frac{5 - 3x}{x^4}$ **47.** $\frac{m\sqrt[3]{m^2}}{2}$ **49.** $\frac{3x\sqrt[3]{2} - 4\sqrt[3]{5}}{x^3}$ **51.** $\left(12\sqrt{5} + 5\sqrt{3}\right)$ in. **53.** $\left(24\sqrt{2} + 12\sqrt{3}\right)$ in.

Section 9.5 (pages 735–738)

1. E **3.** A **5.** D **7.** $6 - 4\sqrt{3}$ **9.** $6 - \sqrt{6}$ **11.** 2 **13.** 9 **15.** $3\sqrt{2} - 5\sqrt{3} + 2\sqrt{6} - 10$ **17.** $3x - 4$ **19.** $4x - y$ **21.** $16x + 24\sqrt{x} + 9$ **23.** $81 - \sqrt[3]{4}$ **25.** Because 6 and $4\sqrt{3}$ are not like terms, they cannot be combined. **27.** $\sqrt{7}$ **29.** $5\sqrt{3}$ **31.** $\frac{\sqrt{6}}{2}$ **33.** $\frac{9\sqrt{15}}{5}$ **35.** $-\sqrt{2}$ **37.** $\frac{-8\sqrt{3k}}{k}$ **39.** $\frac{6\sqrt{3}}{y}$ **41.** To rationalize a cube root, three factors of the quantity under the radical sign are needed. We must multiply by $\sqrt[3]{2^2}$, or $\sqrt[3]{4}$, to rationalize $\sqrt[3]{2}$. **43.** $\frac{\sqrt{14}}{2}$ **45.** $-\frac{\sqrt{14}}{10}$ **47.** $\frac{2\sqrt{6x}}{x}$ **49.** $-\frac{7r\sqrt{2rs}}{s}$ **51.** $\frac{12x^3\sqrt{2xy}}{y^5}$ **53.** $\frac{\sqrt[3]{18}}{3}$ **55.** $\frac{\sqrt[3]{12}}{3}$ **57.** $-\frac{\sqrt[3]{2pr}}{r}$ **59.** $\frac{2\sqrt[4]{x^3}}{x}$ **61.** $\frac{\sqrt[4]{2yz^3}}{z}$ **63.** $\frac{2\left(4 - \sqrt{3}\right)}{13}$ **65.** $3\left(\sqrt{5} - \sqrt{3}\right)$ **67.** $\sqrt{3} + \sqrt{7}$ **69.** $\sqrt{7} - \sqrt{6} - \sqrt{14} + 2\sqrt{3}$ **71.** $2\sqrt{3} + \sqrt{10} - 3\sqrt{2} - \sqrt{15}$ **73.** $\frac{4\left(\sqrt{x} + 2\sqrt{y}\right)}{x - 4y}$ **75.** $\frac{x\sqrt{2} - \sqrt{3xy} - \sqrt{2xy} + y\sqrt{3}}{2x - 3y}$ **77.** $\frac{5 + 2\sqrt{6}}{4}$ **79.** $\frac{4 + 2\sqrt{2}}{3}$ **81.** $\frac{6 + 2\sqrt{6x}}{3}$ **83.** $\frac{319}{6\left(8\sqrt{5} + 1\right)}$ **84.** $\frac{9a - b}{\left(\sqrt{b} - \sqrt{a}\right)\left(3\sqrt{a} - \sqrt{b}\right)}$ **85.** $\frac{\left(3\sqrt{a} + \sqrt{b}\right)\left(\sqrt{b} + \sqrt{a}\right)}{b - a}$ **86.** In Exercise 84, we multiplied the numerator and denominator by the conjugate of the numerator, while in Exercise 85 we multiplied by the conjugate of the denominator.

Summary Exercises on Operations with Radicals and Rational Exponents (pages 739–740)

1. $-6\sqrt{10}$ **2.** $7 - \sqrt{14}$ **3.** $2 + \sqrt{6} - 2\sqrt{3} - 3\sqrt{2}$ **4.** $4\sqrt{2}$ **5.** $73 + 12\sqrt{35}$ **6.** $\frac{-\sqrt{6}}{2}$ **7.** $4\left(\sqrt{7} - \sqrt{5}\right)$ **8.** $3\sqrt[3]{2x^2}$ **9.** $-3 + 2\sqrt{2}$ **10.** -2 **11.** -44 **12.** $\frac{\sqrt{x} + \sqrt{5}}{x - 5}$ **13.** $2abc^3\sqrt[3]{b^2}$ **14.** $5\sqrt[3]{3}$ **15.** $3\left(\sqrt{5} - 2\right)$ **16.** $\frac{\sqrt{15x}}{5x}$ **17.** $\frac{8}{5}$ **18.** $\frac{\sqrt{2}}{8}$ **19.** $-\sqrt[3]{100}$ **20.** $11 + 2\sqrt{30}$ **21.** $-3\sqrt{3x}$ **22.** $52 - 30\sqrt{3}$ **23.** 1 **24.** $\frac{\sqrt[3]{117}}{9}$ **25.** $t^2\sqrt[4]{t}$ **26.** $3\sqrt{2} + \sqrt{15} + \sqrt{42} + \sqrt{35}$ **27.** $2\sqrt[4]{27}$ **28.** $\frac{1 + \sqrt[3]{3} + \sqrt[3]{9}}{-2}$ **29.** $\frac{x\sqrt[3]{x^2}}{y}$ **30.** $-4\sqrt{3} - 3$ **31.** $xy^{6/5}$ **32.** $x^{10}y$ **33.** $\frac{1}{25x^2}$ **34.** $\frac{-6y^{1/6}}{x^{1/24}}$ **35.** $7 + 4 \cdot 3^{1/2}$, or $7 + 4\sqrt{3}$ **36.** 1

Section 9.6 (pages 745–748)

1. **(a)** yes **(b)** no **3.** **(a)** yes **(b)** no **5.** no; There is no solution. The radical expression, which is positive, cannot equal a negative number. **7.** $\{11\}$ **9.** $\left\{\frac{1}{3}\right\}$ **11.** $\emptyset$ **13.** $\{5\}$ **15.** $\{18\}$ **17.** $\{5\}$ **19.** $\{4\}$ **21.** $\{17\}$ **23.** $\{5\}$ **25.** $\emptyset$ **27.** $\{0\}$ **29.** $\{0\}$ **31.** $\left\{-\frac{1}{3}\right\}$ **33.** $\emptyset$ **35.** We cannot just square each term. The right side should be $(8 - x)^2 = 64 - 16x + x^2$. The correct first step is $3x + 4 = 64 - 16x + x^2$, and the solution set is $\{4\}$. **37.** $\{1\}$ **39.** $\{-1\}$ **41.** $\{14\}$ **43.** $\{8\}$ **45.** $\{0\}$ **47.** $\emptyset$ **49.** $\{7\}$ **51.** $\{7\}$ **53.** $\{4, 20\}$ **55.** $\emptyset$ **57.** $\left\{\frac{5}{4}\right\}$ **59.** $\{9, 17\}$ **61.** $\left\{\frac{1}{4}, 1\right\}$ **63.** $K = \frac{V^2m}{2}$ **65.** $C = \frac{L}{Z^2}$ **67.** $L = \frac{1}{4\pi^2f^2C}$ **69.** $r = \frac{a}{4\pi^2N^2}$

Section 9.7 (pages 755–758)

1. i **3.** $-i$ **5.** Any real number a can be written as $a + 0i$, a complex number with imaginary part 0. **7.** $13i$ **9.** $-12i$ **11.** $i\sqrt{5}$ **13.** $4i\sqrt{3}$ **15.** -15 **17.** $-\sqrt{57}$ **19.** -10 **21.** $i\sqrt{33}$ **23.** $\sqrt{3}$ **25.** $5i$ **27.** $-1 + 7i$ **29.** 0 **31.** $7 + 3i$ **33.** -2 **35.** $1 + 13i$ **37.** $6 + 6i$ **39.** $4 + 2i$ **41.** -81 **43.** -16 **45.** $-10 - 30i$ **47.** $10 - 5i$ **49.** $-9 + 40i$ **51.** 153

53. (a) $a - bi$ (b) $a^2; b^2$ **55.** $1 + i$ **57.** $-1 + 2i$ **59.** $2 + 2i$ **61.** $-\frac{5}{13} - \frac{12}{13}i$ **63.** (a) $4x + 1$ (b) $4 + i$ **64.** (a) $-2x + 3$ (b) $-2 + 3i$ **65.** (a) $3x^2 + 5x - 2$ (b) $5 + 5i$ **66.** (a) $-\sqrt{3} + \sqrt{6} + 1 - \sqrt{2}$ (b) $\frac{1}{5} - \frac{7}{5}i$ **67.** Because $i^2 = -1$, two pairs of like terms can be combined in Exercise 65(b). **68.** Because $i^2 = -1$, additional terms can be combined in the numerator and denominator. **69.** $\frac{5}{41} + \frac{4}{41}i$ **71.** -1 **73.** i **75.** 1 **77.** $-i$ **79.** Since $i^{20} = (i^4)^5 = 1^5 = 1$, the student multiplied by 1, which is justified by the identity property for multiplication. **81.** $\frac{1}{2} + \frac{1}{2}i$ **83.** $(1 + 5i)^2 - 2(1 + 5i) + 26$ will simplify to 0 when the operations are applied.

Chapter 9 Review Exercises (pages 763–766)

1. 42 **2.** -17 **3.** not a real number **4.** 6 **5.** -2 **6.** $|x|$ **7.** x **8.** $|x^5|$
9. domain: $[1, \infty)$; range: $[0, \infty)$

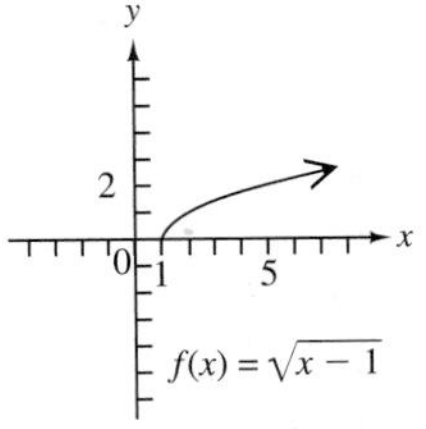

10. domain: $(-\infty, \infty)$; range: $(-\infty, \infty)$

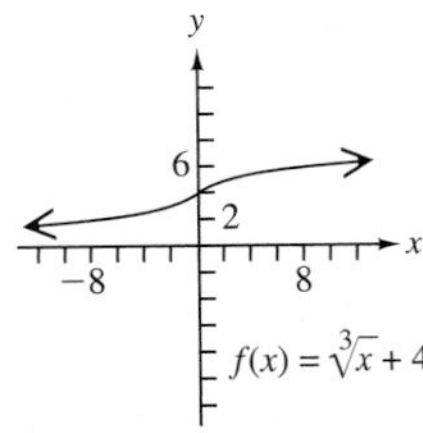

11. n must be even, and a must be negative. **12.** It is not a real number. **13.** 6.325 **14.** 8.775 **15.** 17.607 **16.** 1.9 sec **17.** 66 in.^{2} **18.** 7 **19.** -2 **20.** not a real number **21.** By a power rule for exponents and the definition of $x^{1/n}$, $a^{m/n} = (a^m)^{1/n} = \sqrt[n]{a^m}$. **22.** 32 **23.** -4 **24.** $-\frac{216}{125}$ **25.** -32 **26.** $\frac{1000}{27}$ **27.** 49 **28.** 96 **29.** $\frac{k^{17/12}}{2}$ **30.** $\sqrt[5]{2^4}$, or $\sqrt[5]{16}$ **31.** 3^9 **32.** $7^4\sqrt{7}$ **33.** $m^4\sqrt[3]{m}$ **34.** $k^2\sqrt[4]{k}$ **35.** $\sqrt[6]{m}$ **36.** $2y\sqrt[4]{y}$ **37.** $\sqrt[15]{y^8}$ **38.** $\sqrt[12]{y^5}$ **39.** $\sqrt{66}$ **40.** $\sqrt{5r}$ **41.** $\sqrt[3]{30}$ **42.** $\sqrt[4]{21}$ **43.** $2\sqrt{5}$ **44.** $-5\sqrt{5}$ **45.** $-3x\sqrt[3]{4xy}$ **46.** $4pq^2\sqrt[3]{p}$ **47.** $\frac{7}{9}$ **48.** $\frac{y\sqrt{y}}{12}$ **49.** $\frac{m^5}{3}$ **50.** $\frac{\sqrt[3]{r^2}}{2}$ **51.** $\sqrt[12]{2}$ **52.** $\sqrt[10]{x^3}$ **53.** $\sqrt{130}$ **54.** $\sqrt{53}$ **55.** $-11\sqrt{2}$ **56.** $23\sqrt{5}$ **57.** $7\sqrt{3y}$ **58.** $26m\sqrt{6m}$ **59.** $19\sqrt[3]{2}$ **60.** $-8\sqrt[4]{2}$ **61.** $1 - \sqrt{3}$ **62.** 2 **63.** $9 - 7\sqrt{2}$ **64.** $86 + 8\sqrt{55}$ **65.** $15 - 2\sqrt{26}$ **66.** $12 - 2\sqrt{35}$ **67.** $-3\sqrt{6}$ **68.** $\frac{3\sqrt{7py}}{y}$ **69.** $-\frac{\sqrt[3]{45}}{5}$ **70.** $\frac{3m\sqrt[3]{4n}}{n^2}$ **71.** $\frac{\sqrt{2} - \sqrt{7}}{-5}$ **72.** $\frac{-5(\sqrt{6} + \sqrt{3})}{3}$ **73.** $\{2\}$ **74.** $\{6\}$ **75.** $\emptyset$ **76.** $\{0, 5\}$ **77.** $\{9\}$ **78.** $\{3\}$ **79.** $\{7\}$ **80.** $\left\{-\frac{1}{2}\right\}$ **81.** $\{6\}$ **82.** $5i$ **83.** $10i\sqrt{2}$ **84.** $4i\sqrt{10}$ **85.** $-10 - 2i$ **86.** $14 + 7i$ **87.** $-\sqrt{35}$ **88.** -45 **89.** 3 **90.** $5 + i$ **91.** $32 - 24i$ **92.** $1 - i$ **93.** $4 + i$ **94.** $-i$ **95.** 1 **96.** $-i$ **97.** $-13ab^2$ **98.** $\frac{1}{100}$ **99.** $\frac{1}{y^{1/2}}$ **100.** $\frac{x^{3/4}}{z^{3/4}}$ **101.** k^6 **102.** $3z^3t^2\sqrt[3]{2t^2}$ **103.** $57\sqrt{2}$ **104.** $6x\sqrt[3]{y^2}$ **105.** $\sqrt{35} + \sqrt{15} - \sqrt{21} - 3$ **106.** $-\frac{\sqrt{3}}{6}$ **107.** $\frac{\sqrt[3]{60}}{5}$ **108.** $\frac{2\sqrt{z}(\sqrt{z} + 2)}{z - 4}$ **109.** $7i$ **110.** $3 - 7i$ **111.** $-5i$ **112.** $\{5\}$ **113.** $\left\{\frac{3}{2}\right\}$ **114.** 7.9 ft **115.** $(12\sqrt{3} + 5\sqrt{2})$ ft

Chapter 9 Test (pages 767–768)

1. -29 **2.** -8 **3.** 5 **4.** C **5.** 21.863 **6.** -9.405
7. domain: $[-6, \infty)$; range: $[0, \infty)$

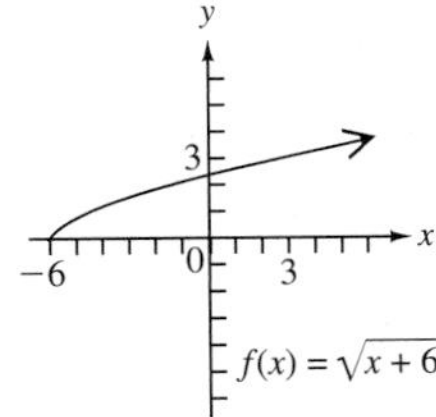

8. $\frac{1}{256}$ **9.** $\frac{9y^{3/10}}{x^2}$ **10.** $3x^2y^3\sqrt{6x}$ **11.** $2ab^3\sqrt[4]{2a^3b}$ **12.** $\sqrt[6]{200}$ **13.** $26\sqrt{5}$ **14.** $66 + \sqrt{5}$ **15.** $23 - 4\sqrt{15}$ **16.** $\frac{-\sqrt{10}}{4}$ **17.** $\frac{2\sqrt[3]{25}}{5}$ **18.** $-2(\sqrt{7} - \sqrt{5})$ **19.** $3 + \sqrt{6}$ **20.** $\sqrt{26}$ **21.** $\sqrt{145}$ **22.** $\{-1\}$ **23.** $\{6\}$ **24.** $\{-3\}$ **25.** $-5 - 8i$ **26.** $-10 + 10i$ **27.** $3 + 4i$ **28.** $-i$

Cumulative Review Exercises: Chapters R–9 (pages 769–770)

1. $\left\{\frac{4}{5}\right\}$ **2.** $\{-12\}$ **3.** $\left\{\frac{11}{10}, \frac{7}{2}\right\}$ **4.** $(-6, \infty)$ **5.** $(1, 3)$ **6.** $(-2, 1)$ **7.** $12x + 11y = 18$ **8.** C **9.** (a) $(0, 6)$ (b) $(2, 0)$
10.

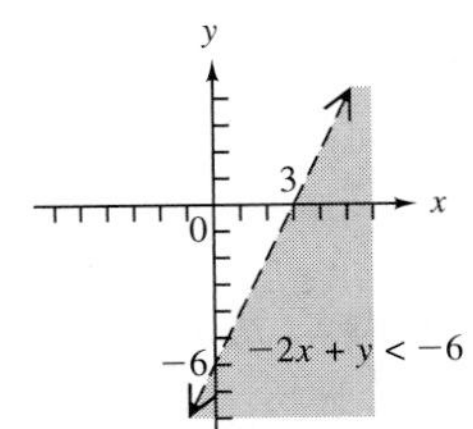

11. Both angles measure 80°. **12.** $\{(7, -2)\}$ **13.** $\emptyset$ **14.** $\{(x, y, z) \mid 2x + y - z = 5\}$ **15.** 2-oz letter: \$0.63; 3-oz letter: \$0.87 **16.** $-k^3 - 3k^2 - 8k - 9$ **17.** $8x^2 + 17x - 21$ **18.** $z - 2 + \frac{3}{z}$ **19.** $3y^3 - 3y^2 + 4y + 1 + \frac{-10}{2y + 1}$ **20.** $(2p - 3q)(p - q)$ **21.** $(3k^2 + 4)(6k^2 - 5)$ **22.** $(x + 8)(x^2 - 8x + 64)$ **23.** $\frac{y}{y + 5}$ **24.** $\frac{4x + 2y}{(x + y)(x - y)}$ **25.** $-\frac{9}{4}$ **26.** $-\frac{1}{a + b}$ **27.** $\left\{-3, -\frac{5}{2}\right\}$ **28.** $\left\{-\frac{2}{5}, 1\right\}$ **29.** $\frac{1}{243}$ **30.** $x^{1/12}$ **31.** $8\sqrt{5}$ **32.** $\frac{-9\sqrt{5}}{20}$ **33.** $4(\sqrt{6} + \sqrt{5})$ **34.** $6\sqrt[3]{4}$ **35.** $\sqrt{29}$ **36.** $\{6\}$ **37.** 15 mph

38. $\frac{80}{39}$, or $2\frac{2}{39}$ L **39.** 17 dimes and 12 quarters
40. Brenda: 8 mph; Chuck: 4 mph

Chapter 10 Quadratic Equations, Inequalities, and Functions

Section 10.1 (pages 777–778)

1. The equation is also true for $x = -4$. **3. (a)** A quadratic equation in standard form has a second-degree polynomial in decreasing powers equal to 0. **(b)** The square root property states that if the square of a quantity equals a number, then the quantity equals the positive or negative square root of the number. **5.** $\{9, -9\}$ **7.** $\{\sqrt{17}, -\sqrt{17}\}$
9. $\{4\sqrt{2}, -4\sqrt{2}\}$ **11.** $\{\sqrt{3}, -\sqrt{3}\}$ **13.** $\{4\sqrt{3}, -4\sqrt{3}\}$
15. $\{2\sqrt{6}, -2\sqrt{6}\}$ **17.** $\{3\sqrt{3}, -3\sqrt{3}\}$ **19.** $\{-7, 3\}$
21. $\{4+\sqrt{3}, 4-\sqrt{3}\}$ **23.** $\{-5+4\sqrt{3}, -5-4\sqrt{3}\}$
25. $\left\{-3, \frac{5}{3}\right\}$ **27.** $\left\{\frac{1+\sqrt{7}}{3}, \frac{1-\sqrt{7}}{3}\right\}$
29. $\left\{\frac{-1+2\sqrt{6}}{4}, \frac{-1-2\sqrt{6}}{4}\right\}$ **31.** $\left\{\frac{-1+3\sqrt{2}}{3}, \frac{-1-3\sqrt{2}}{3}\right\}$
33. $\{10i, -10i\}$ **35.** $\{2i\sqrt{3}, -2i\sqrt{3}\}$ **37.** $\{5+i\sqrt{3}, 5-i\sqrt{3}\}$
39. $\left\{\frac{1}{6}+\frac{\sqrt{2}}{3}i, \frac{1}{6}-\frac{\sqrt{2}}{3}i\right\}$ **41.** 5.6 sec

Section 10.2 (pages 785–786)

1. 25; $(x+5)^2$ **3.** 1; $(x+1)^2$ **5.** $\frac{25}{4}$; $\left(p-\frac{5}{2}\right)^2$
7. $\{1, 3\}$ **9.** $\{-3, -2\}$ **11.** $\{-1+\sqrt{6}, -1-\sqrt{6}\}$
13. $\{-5+\sqrt{7}, -5-\sqrt{7}\}$ **15.** $\{4+2\sqrt{3}, 4-2\sqrt{3}\}$
17. $\left\{\frac{-1+\sqrt{5}}{2}, \frac{-1-\sqrt{5}}{2}\right\}$ **19.** $\left\{\frac{3+\sqrt{17}}{2}, \frac{3-\sqrt{17}}{2}\right\}$
21. $\left\{-\frac{3}{2}, \frac{1}{2}\right\}$ **23.** $\left\{\frac{2+\sqrt{14}}{2}, \frac{2-\sqrt{14}}{2}\right\}$
25. $\left\{\frac{-5+\sqrt{41}}{4}, \frac{-5-\sqrt{41}}{4}\right\}$ **27.** $\left\{\frac{-7+\sqrt{97}}{6}, \frac{-7-\sqrt{97}}{6}\right\}$
29. $\{4+\sqrt{3}, 4-\sqrt{3}\}$ **31.** $\{1+\sqrt{6}, 1-\sqrt{6}\}$
33. $\{-2+3i, -2-3i\}$ **35.** $\left\{-\frac{2}{3}+\frac{2\sqrt{2}}{3}i, -\frac{2}{3}-\frac{2\sqrt{2}}{3}i\right\}$
37. $\{-3+i\sqrt{3}, -3-i\sqrt{3}\}$ **39.** x^2 **40.** x **41.** $6x$ **42.** 1
43. 9 **44.** $(x+3)^2$, or x^2+6x+9

Section 10.3 (pages 793–794)

1. The fraction bar should extend under the term $-b$. **3.** $\{3, 5\}$
5. $\left\{\frac{-2+\sqrt{2}}{2}, \frac{-2-\sqrt{2}}{2}\right\}$ **7.** $\left\{\frac{1+\sqrt{3}}{2}, \frac{1-\sqrt{3}}{2}\right\}$
9. $\{5+\sqrt{7}, 5-\sqrt{7}\}$ **11.** $\left\{\frac{-1+\sqrt{2}}{2}, \frac{-1-\sqrt{2}}{2}\right\}$
13. $\left\{\frac{-1+\sqrt{7}}{3}, \frac{-1-\sqrt{7}}{3}\right\}$ **15.** $\{1+\sqrt{5}, 1-\sqrt{5}\}$
17. $\left\{\frac{-2+\sqrt{10}}{2}, \frac{-2-\sqrt{10}}{2}\right\}$ **19.** $\{-1+3\sqrt{2}, -1-3\sqrt{2}\}$
21. $\left\{\frac{3}{2}+\frac{\sqrt{59}}{2}i, \frac{3}{2}-\frac{\sqrt{59}}{2}i\right\}$ **23.** $\{3+i\sqrt{5}, 3-i\sqrt{5}\}$
25. $\left\{\frac{1}{2}+\frac{\sqrt{6}}{2}i, \frac{1}{2}-\frac{\sqrt{6}}{2}i\right\}$ **27.** $\left\{-\frac{2}{3}+\frac{\sqrt{2}}{3}i, -\frac{2}{3}-\frac{\sqrt{2}}{3}i\right\}$
29. B **31.** C **33.** A **35.** D **37.** The equations in Exercises 29, 30, 33, and 34 can be solved by factoring.

Section 10.4 (pages 803–806)

1. Multiply by the LCD, x. **3.** Substitute a variable for $r^2 + r$.
5. The proposed solution -1 does not check. The solution set is $\{4\}$.
7. $\{-4, 7\}$ **9.** $\left\{-\frac{2}{3}, 1\right\}$ **11.** $\left\{-\frac{14}{17}, 5\right\}$ **13.** $\left\{-\frac{11}{7}, 0\right\}$
15. $\left\{\frac{-1+\sqrt{13}}{2}, \frac{-1-\sqrt{13}}{2}\right\}$ **17.** $\left\{\frac{2+\sqrt{22}}{3}, \frac{2-\sqrt{22}}{3}\right\}$
19. (a) $(20-t)$ mph **(b)** $(20+t)$ mph **21.** 25 mph
23. 80 km per hr **25.** 3.6 hr **27.** 9 min **29.** $\{1, 4\}$
31. $\{3\}$ **33.** $\left\{\frac{8}{9}\right\}$ **35.** $\{16\}$ **37.** $\left\{\frac{2}{5}\right\}$ **39.** $\{-3, 3\}$
41. $\left\{-\frac{3}{2}, -1, 1, \frac{3}{2}\right\}$ **43.** $\{-2\sqrt{3}, -2, 2, 2\sqrt{3}\}$
45. $\left\{\frac{\sqrt{9+\sqrt{65}}}{2}, -\frac{\sqrt{9+\sqrt{65}}}{2}, \frac{\sqrt{9-\sqrt{65}}}{2}, -\frac{\sqrt{9-\sqrt{65}}}{2}\right\}$
47. $\{-6, -5\}$ **49.** $\{-4, 1\}$ **51.** $\left\{-\frac{1}{3}, \frac{1}{6}\right\}$ **53.** $\{-8, 1\}$
55. $\{-64, 27\}$ **57.** $\{25\}$ **59.** It would cause both denominators to equal 0, and division by 0 is undefined. **60.** $\frac{12}{5}$
61. $\left(\frac{x}{x-3}\right)^2 + 3\left(\frac{x}{x-3}\right) - 4 = 0$ **62.** The numerator can never equal the denominator, since the denominator is 3 less than the numerator. **63.** $\left\{\frac{12}{5}\right\}$; The values for t are -4 and 1. The value 1 is impossible because it leads to a contradiction $\left(\text{since } \frac{x}{x-3} \text{ is never equal to } 1\right)$. **64.** $\left\{\frac{12}{5}\right\}$; The values for s are $\frac{1}{x}$ and $\frac{-4}{x}$. The value $\frac{1}{x}$ is impossible, since $\frac{1}{x} \neq \frac{1}{x-3}$ for all x.

Summary Exercises on Solving Quadratic Equations (pages 807–808)

1. square root property **2.** factoring **3.** quadratic formula
4. quadratic formula **5.** factoring **6.** square root property
7. $\{\sqrt{47}, -\sqrt{47}\}$ **8.** $\left\{-\frac{3}{2}, \frac{5}{3}\right\}$ **9.** $\{-4+\sqrt{10}, -4-\sqrt{10}\}$
10. $\{-3, 11\}$ **11.** $\left\{-\frac{1}{2}, 5\right\}$ **12.** $\left\{-3, \frac{1}{3}\right\}$
13. $\left\{\frac{9+\sqrt{33}}{6}, \frac{9-\sqrt{33}}{6}\right\}$ **14.** $\{2i\sqrt{3}, -2i\sqrt{3}\}$ **15.** $\left\{\frac{1}{2}, 2\right\}$

ANSWERS

16. $\left\{-\frac{\sqrt{6}}{3}, -\frac{1}{2}, \frac{1}{2}, \frac{\sqrt{6}}{3}\right\}$ **17.** $\left\{\frac{-5+2\sqrt{3}}{2}, \frac{-5-2\sqrt{3}}{2}\right\}$

18. $\left\{\frac{4}{5}, 3\right\}$ **19.** $\{-\sqrt{7}, -\sqrt{2}, \sqrt{2}, \sqrt{7}\}$

20. $\left\{\frac{-2+\sqrt{14}}{2}, \frac{-2-\sqrt{14}}{2}\right\}$ **21.** $\left\{-\frac{1}{2}+\frac{\sqrt{7}}{2}i, -\frac{1}{2}-\frac{\sqrt{7}}{2}i\right\}$

22. $\left\{\sqrt{4+\sqrt{15}}, -\sqrt{4+\sqrt{15}}, \sqrt{4-\sqrt{15}}, -\sqrt{4-\sqrt{15}}\right\}$

23. $\left\{\frac{3}{2}\right\}$ **24.** $\left\{\frac{2}{3}\right\}$ **25.** $\{6\sqrt{2}, -6\sqrt{2}\}$ **26.** $\left\{-\frac{2}{3}, 2\right\}$

27. $\{-4, 9\}$ **28.** $\{13, -13\}$ **29.** $\left\{1+\frac{\sqrt{3}}{3}i, 1-\frac{\sqrt{3}}{3}i\right\}$

30. $\{3\}$ **31.** $\left\{-\frac{1}{3}, \frac{1}{6}\right\}$ **32.** $\left\{\frac{1}{6}+\frac{\sqrt{47}}{6}i, \frac{1}{6}-\frac{\sqrt{47}}{6}i\right\}$

33. $\{3\}$ **34.** $\left\{-\frac{8}{3}, -1\right\}$ **35.** $\left\{-i, i, -\frac{1}{2}i, \frac{1}{2}i\right\}$ **36.** $\{-2, 7\}$

Section 10.5 (pages 813–816)

1. Solve for w^2 by dividing each side by g. **3.** $m = \sqrt{p^2 - n^2}$
5. $t = \frac{\pm\sqrt{dk}}{k}$ **7.** $d = \frac{\pm\sqrt{skI}}{I}$ **9.** $v = \frac{\pm\sqrt{kAF}}{F}$
11. $r = \frac{\pm\sqrt{3\pi Vh}}{\pi h}$ **13.** $t = \frac{-B \pm \sqrt{B^2 - 4AC}}{2A}$ **15.** $h = \frac{D^2}{k}$
17. $\ell = \frac{p^2 g}{k}$ **19.** 2.3, 5.3, 5.8 **21.** eastbound ship: 80 mi; southbound ship: 150 mi **23.** 1 ft **25.** 20 in. by 12 in.
27. 2.4 sec and 5.6 sec **29.** 9.2 sec **31.** It reaches its *maximum* height at 5 sec because this is the only time it reaches 400 ft. **33.** \$0.80
35. (a) 2750 billion **(b)** 2750 billion; They are the same.
37. 2001; The graph indicates that vehicle-miles reached 2800 in 2001.
39. 5.5 m per sec **41.** 5 or 14

Section 10.6 (pages 823–828)

1. (a) B **(b)** C **(c)** A **(d)** D **3.** (0, 0) **5.** (0, 4) **7.** (1, 0)
9. (−3, −4) **11.** In Exercise 9, the parabola is shifted 3 units to the left and 4 units down. The parabola in Exercise 10 is shifted 5 units to the right and 8 units down. **13.** down; wider **15.** up; narrower
17. (a) I **(b)** IV **(c)** II **(d)** III

19.

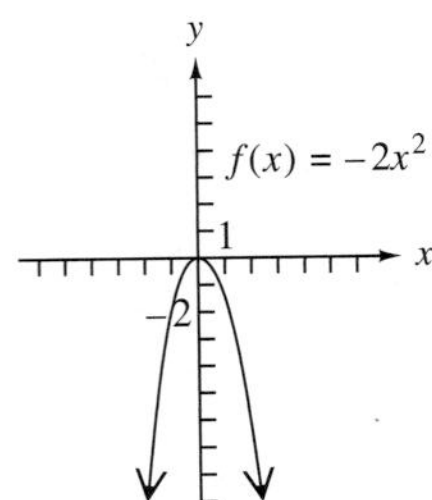

21.

$f(x) = x^2 - 1$

23.

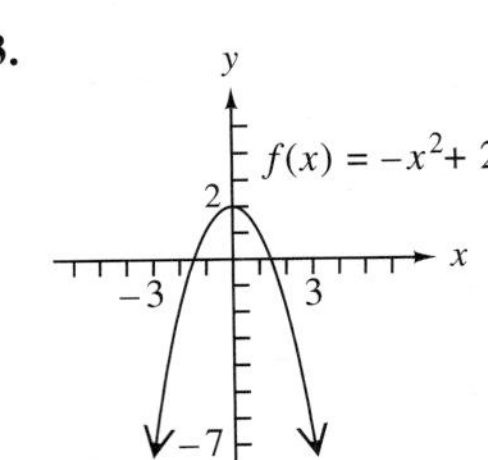

25. vertex: (4, 0); axis: $x = 4$; domain: $(-\infty, \infty)$; range: $[0, \infty)$

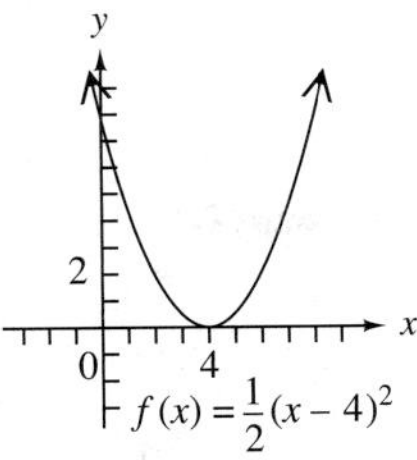

27. vertex: (−2, −1); axis: $x = -2$; domain: $(-\infty, \infty)$; range: $[-1, \infty)$

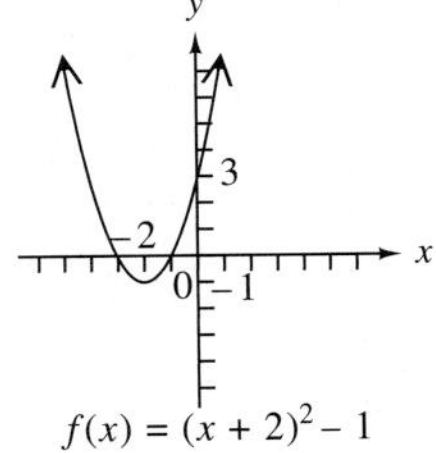

29. vertex: (−3, 4); axis: $x = -3$; domain: $(-\infty, \infty)$; range: $(-\infty, 4]$

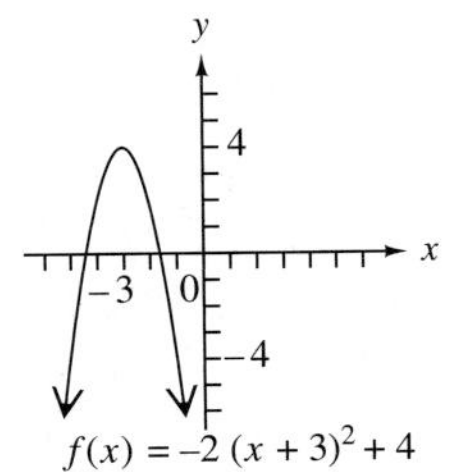

31. vertex: (−2, 1); axis: $x = -2$; domain: $(-\infty, \infty)$; range: $(-\infty, 1]$

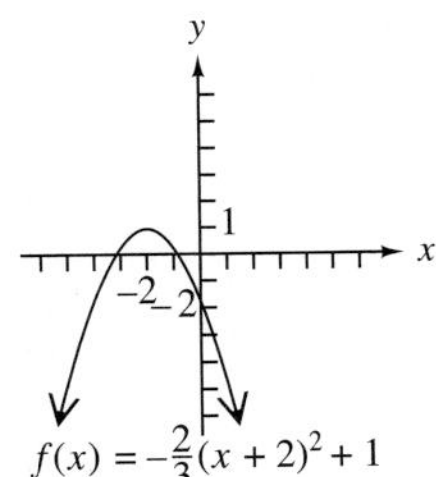

33. It is shifted 6 units up.
34.

$G(x) = x + 6$

35. It is shifted 6 units up. **36.** It is shifted 6 units to the right.
37.

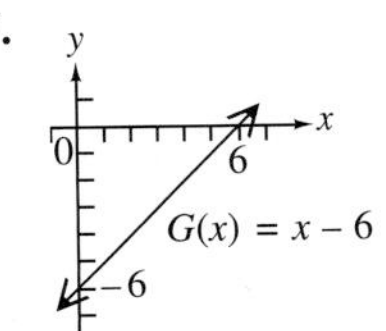

38. It is shifted 6 units to the right. **39.** quadratic; positive
41. quadratic; negative **43.** linear; positive

ANSWERS

45. (a) **DIGITAL CAMERA SALES IN U.S.**

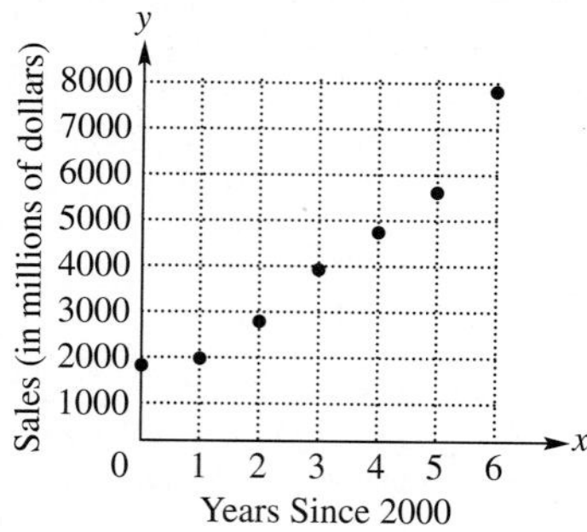

(b) quadratic; positive **(c)** $f(x) = 99.3x^2 + 400.7x + 1825$ **(d)** $9496 million **(e)** No. The number of digital cameras sold in 2007 is far below the number approximated by the model. Rather than continuing to increase, sales of digital cameras fell in 2007.

47. (a) 2005: 14%; 2007: 2.4% **(b)** The approximations using the model are far too low.

Section 10.7 (pages 837–840)

1. If x is squared, it has a vertical axis; if y is squared, it has a horizontal axis. **3.** Use the discriminant of the corresponding quadratic equation. If it is positive, there are two x-intercepts. If it is 0, there is just one x-intercept (the vertex), and if it is negative, there are no x-intercepts. **5.** $(-1, 3)$; up; narrower; no x-intercepts **7.** $\left(\frac{5}{2}, \frac{37}{4}\right)$; down; same; two x-intercepts **9.** $(-3, -9)$; to the right; wider

11. vertex: $(-2, -1)$; axis: $x = -2$; domain: $(-\infty, \infty)$; range: $[-1, \infty)$

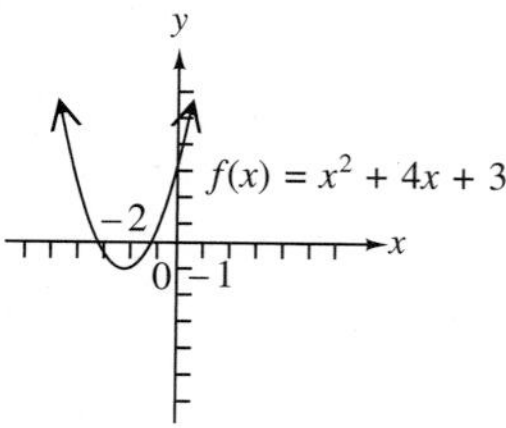

13. vertex: $(1, -3)$; axis: $x = 1$; domain: $(-\infty, \infty)$; range: $(-\infty, -3]$

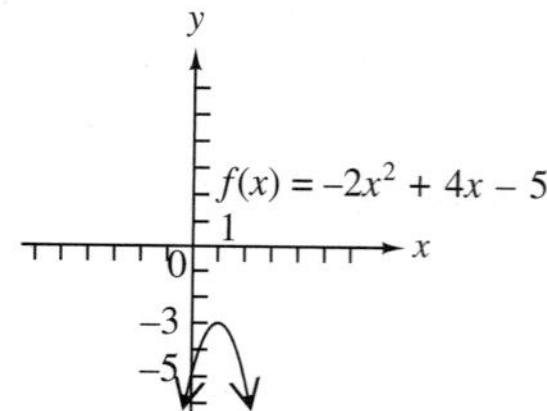

15. vertex: $(1, 5)$; axis: $y = 5$; domain: $(-\infty, 1]$; range: $(-\infty, \infty)$

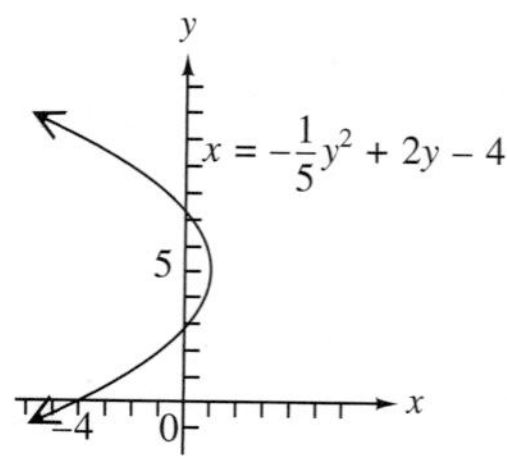

17. vertex: $(-7, -2)$; axis: $y = -2$; domain: $[-7, \infty)$; range: $(-\infty, \infty)$

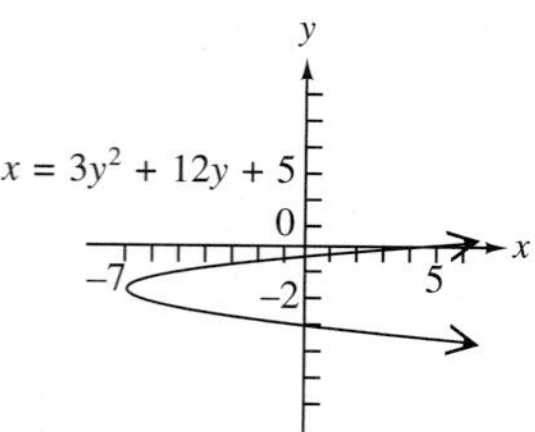

19. F **21.** C **23.** D **25.** 30 and 30 **27.** 160 ft by 320 ft **29.** 16 ft; 2 sec **31. (a)** $R(x) = 20{,}000 + 200x - 4x^2$ **(b)** 25 **(c)** $22,500 **33. (a)** maximum **(b)** 1993; 13.2% **35. (a)** The coefficient of x^2 is negative because the parabola opens down. **(b)** (18.45, 3860) **(c)** In 2018 Social Security assets will reach their maximum value of $3860 billion.

Section 10.8 (pages 847–850)

1. (a) $\{1, 3\}$ **(b)** $(-\infty, 1) \cup (3, \infty)$ **(c)** $(1, 3)$

3. (a) $\left\{-3, \frac{5}{2}\right\}$ **(b)** $\left[-3, \frac{5}{2}\right]$ **(c)** $(-\infty, -3] \cup \left[\frac{5}{2}, \infty\right)$

5. Include the endpoints if the symbol is $\geq$ or $\leq$. Exclude the endpoints if the symbol is $>$ or $<$.

7. $(-\infty, -1) \cup (5, \infty)$

9. $(-4, 6)$

11. $(-\infty, 1] \cup [3, \infty)$

13. $\left(-\infty, -\frac{3}{2}\right] \cup \left[\frac{3}{5}, \infty\right)$

15. $\left(-\frac{2}{3}, \frac{1}{3}\right)$

17. $\left(-\infty, -\frac{1}{2}\right] \cup \left[\frac{1}{3}, \infty\right)$

19. $(-\infty, 3 - \sqrt{3}] \cup [3 + \sqrt{3}, \infty)$

21. $(-\infty, \infty)$ **23.** $\emptyset$

25. $(-\infty, 1) \cup (2, 4)$

27. $\left[-\frac{3}{2}, \frac{1}{3}\right] \cup [4, \infty)$

29. $(-\infty, 1) \cup (4, \infty)$

31. $\left[-\frac{3}{2}, 5\right)$

33. $(2, 6]$

35. $\left(-\infty, \frac{1}{2}\right) \cup \left(\frac{5}{4}, \infty\right)$

37. $[-4, -2)$

39. $\left(0, \frac{1}{2}\right) \cup \left(\frac{5}{2}, \infty\right)$

41. $(-\infty, 2] \cup (4, \infty)$

43. 3 sec and 13 sec **44.** between 3 sec and 13 sec **45.** at 0 sec (the time when it is initially projected) and at 16 sec (the time when it hits the ground) **46.** between 0 and 3 sec and between 13 and 16 sec

Chapter 10 Review Exercises (pages 855–858)

1. $\{11, -11\}$ **2.** $\{\sqrt{3}, -\sqrt{3}\}$ **3.** $\{3 + \sqrt{10}, 3 - \sqrt{10}\}$ **4.** $\left\{\frac{-1 + \sqrt{14}}{2}, \frac{-1 - \sqrt{14}}{2}\right\}$ **5.** $\left\{\frac{2}{3} + \frac{5}{3}i, \frac{2}{3} - \frac{5}{3}i\right\}$ **6.** $\{-2 + \sqrt{19}, -2 - \sqrt{19}\}$ **7.** $\left\{\frac{1}{2}, 1\right\}$ **8.** $\{-1 + \sqrt{6}, -1 - \sqrt{6}\}$. **9.** $\left\{\frac{-4 + \sqrt{22}}{2}, \frac{-4 - \sqrt{22}}{2}\right\}$ **10.** 5.8 sec **11.** $\left\{-\frac{7}{2}, 3\right\}$ **12.** $\left\{\frac{-5 + \sqrt{53}}{2}, \frac{-5 - \sqrt{53}}{2}\right\}$ **13.** $\left\{\frac{1 + \sqrt{41}}{2}, \frac{1 - \sqrt{41}}{2}\right\}$ **14.** $\left\{-\frac{3}{4} + \frac{\sqrt{23}}{4}i, -\frac{3}{4} - \frac{\sqrt{23}}{4}i\right\}$ **15.** $\left\{\frac{2}{3} + \frac{\sqrt{2}}{3}i, \frac{2}{3} - \frac{\sqrt{2}}{3}i\right\}$ **16.** $\left\{\frac{-7 + \sqrt{37}}{2}, \frac{-7 - \sqrt{37}}{2}\right\}$ **17.** C **18.** A **19.** D **20.** B **21.** $\left\{-\frac{5}{2}, 3\right\}$ **22.** $\left\{-\frac{1}{2}, 1\right\}$ **23.** $\{-4\}$ **24.** $\left\{-\frac{11}{6}, -\frac{19}{12}\right\}$ **25.** $\left\{-\frac{343}{8}, 64\right\}$ **26.** $\{-2, -1, 1, 2\}$ **27.** 40 mph **28.** 4.6 hr **29.** $v = \frac{\pm\sqrt{rFkw}}{kw}$ **30.** $t = \frac{3m \pm \sqrt{9m^2 + 24m}}{2m}$ **31.** 9 ft, 12 ft, 15 ft **32.** 12 cm by 20 cm **33.** 1 in. **34.** 5.2 sec **35.** $(1, 0)$ **36.** $(3, 7)$ **37.** $\left(\frac{2}{3}, -\frac{2}{3}\right)$ **38.** $(-4, 3)$

39. vertex: $(2, -3)$; axis: $x = 2$; domain: $(-\infty, \infty)$; range: $[-3, \infty)$

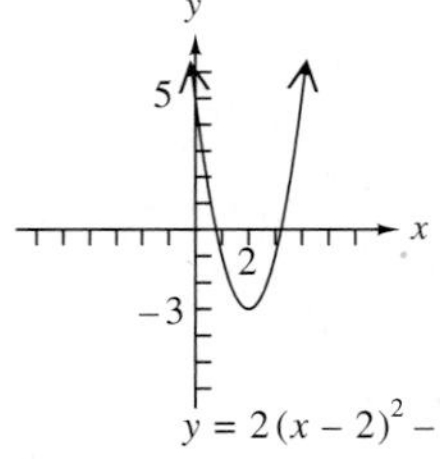

40. vertex: $(2, 3)$; axis: $x = 2$; domain: $(-\infty, \infty)$; range: $(-\infty, 3]$

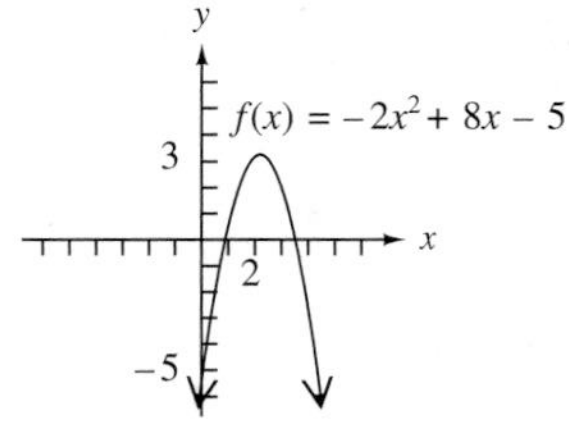

41. vertex: $(-4, -3)$; axis: $y = -3$; domain: $[-4, \infty)$; range: $(-\infty, \infty)$

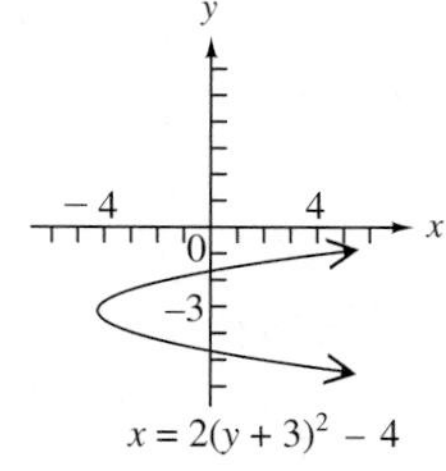

42. vertex: $(4, 6)$; axis: $y = 6$; domain: $(-\infty, 4]$; range: $(-\infty, \infty)$

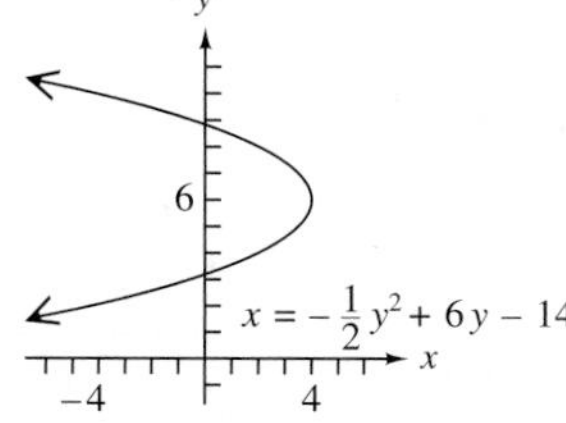

43. **(a)** $c = 2.9$; $100a + 10b + c = 24.3$; $400a + 20b + c = 56.5$ **(b)** $f(x) = 0.054x^2 + 1.6x + 2.9$ **(c)** \$60.3 billion; The result using the model is close but slightly low. **44.** 5 sec; 400 ft **45.** length: 50 m; width: 50 m **46.** $\left(-\infty, -\frac{3}{2}\right) \cup (4, \infty)$ **47.** $[-4, 3]$ **48.** $(-\infty, -5] \cup [-2, 3]$ **49.** $\emptyset$ **50.** $\left(-\infty, \frac{1}{2}\right) \cup (2, \infty)$ **51.** $[-3, 2)$ **52.** $R = \frac{\pm\sqrt{Vh - r^2h}}{h}$ **53.** $\left\{1 + \frac{\sqrt{3}}{3}i, 1 - \frac{\sqrt{3}}{3}i\right\}$ **54.** $\{-2, -1, 3, 4\}$ **55.** $(-\infty, -6) \cup \left(-\frac{3}{2}, 1\right)$ **56.** $\left\{\frac{-11 + \sqrt{7}}{3}, \frac{-11 - \sqrt{7}}{3}\right\}$ **57.** $d = \frac{\pm\sqrt{SkI}}{I}$ **58.** $\{4\}$ **59.** $\left\{-\frac{5}{3}, -\frac{3}{2}\right\}$ **60.** $\left(-5, -\frac{23}{5}\right]$

Chapter 10 Test (pages 859–862)

1. $\{3\sqrt{6}, -3\sqrt{6}\}$ **2.** $\left\{-\frac{8}{7}, \frac{2}{7}\right\}$ **3.** $\{-1 + \sqrt{2}, -1 - \sqrt{2}\}$ **4.** $\left\{\frac{3 + \sqrt{17}}{4}, \frac{3 - \sqrt{17}}{4}\right\}$ **5.** $\left\{\frac{2}{3} + \frac{\sqrt{11}}{3}i, \frac{2}{3} - \frac{\sqrt{11}}{3}i\right\}$ **6.** $\left\{\frac{2}{3}\right\}$ **7.** A **8.** discriminant: 88; two irrational solutions **9.** $\left\{-\frac{2}{3}, 6\right\}$ **10.** $\left\{\frac{-7 + \sqrt{97}}{8}, \frac{-7 - \sqrt{97}}{8}\right\}$ **11.** $\left\{-2, -\frac{1}{3}, \frac{1}{3}, 2\right\}$ **12.** $\left\{-\frac{5}{2}, 1\right\}$ **13.** $r = \frac{\pm\sqrt{\pi S}}{2\pi}$ **14.** Andrew: 11.1 hr; Kent: 9.1 hr **15.** 7 mph **16.** 2 ft **17.** 16 m **18.** **(a)** 10 hr **(b)** 180 students **19.** A

20. vertex: $(0, -2)$; axis: $x = 0$; domain: $(-\infty, \infty)$; range: $[-2, \infty)$

21. vertex: $(2, 3)$; axis: $x = 2$; domain: $(-\infty, \infty)$; range: $(-\infty, 3]$

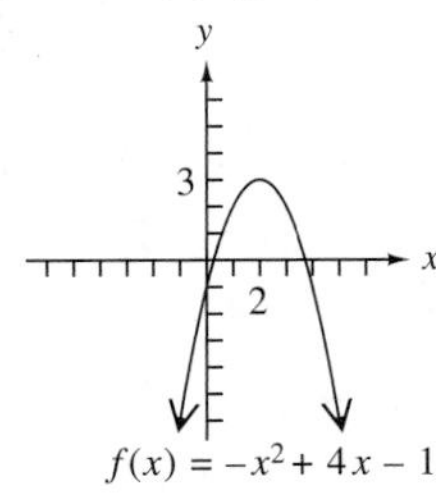

22. vertex: $(-5, -2)$; axis: $y = -2$; domain: $[-5, \infty)$; range: $(-\infty, \infty)$

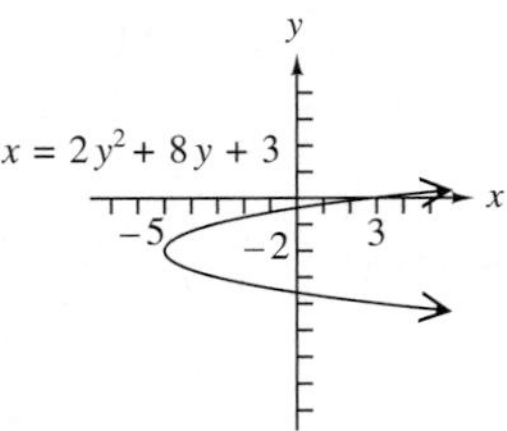

23. 140 ft by 70 ft; 9800 ft^2

24. $(-\infty, -5) \cup \left(\frac{3}{2}, \infty\right)$

25. $(-\infty, 4) \cup [9, \infty)$

Cumulative Review Exercises: Chapters R–10 (pages 863–864)

1. $\{1\}$ **2.** $[1, \infty)$ **3.** $\left[2, \frac{8}{3}\right]$

4. slope: $\frac{1}{2}$; y-intercept: $\left(0, -\frac{7}{4}\right)$ **5.** $x + 3y = -1$

6. function; domain: $(-\infty, \infty)$; range: $(-\infty, \infty)$

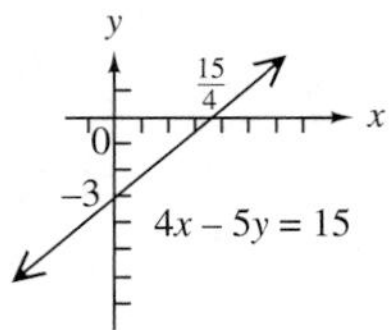

7. not a function

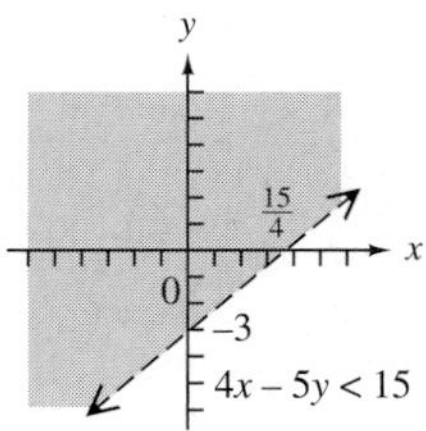

8. function; domain: $(-\infty, \infty)$; range: $(-\infty, 3]$

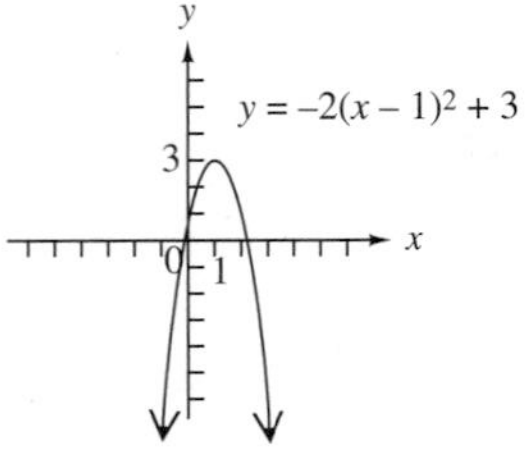

9. $\{(1, -2)\}$ **10.** $\{(3, -4, 2)\}$ **11.** $\frac{x^8}{y^4}$ **12.** $\frac{4}{xy^2}$

13. $\frac{4}{9}t^2 + 12t + 81$ **14.** $4x^2 - 6x + 11 + \frac{4}{x + 2}$

15. $x(4 + x)(4 - x)$ **16.** $(4m - 3)(6m + 5)$ **17.** $(3x - 5y)^2$

18. $-\frac{5}{18}$ **19.** $-\frac{8}{k}$ **20.** $\frac{r - s}{r}$ **21.** $\frac{3\sqrt[3]{4}}{4}$ **22.** $\sqrt{7} + \sqrt{5}$

23. $\left\{\frac{2}{3}\right\}$ **24.** $\left\{\frac{2 + \sqrt{10}}{2}, \frac{2 - \sqrt{10}}{2}\right\}$ **25.** $\{-3, 5\}$ **26.** $\emptyset$

27. $\{-3, -1, 1, 3\}$ **28.** southbound car: 57 mi; eastbound car: 76 mi

Chapter 11 Inverse, Exponential, and Logarithmic Functions

Section 11.1 (pages 871–874)

1. It is not one-to-one. France and the United States are paired with the same trans fat percentage, 11. **3.** Two or more siblings might be in the class. They would be paired with the same mother. **5.** B **7.** A **9.** $\{(6, 3), (10, 2), (12, 5)\}$ **11.** not one-to-one **13.** $f^{-1}(x) = \frac{x - 4}{2}$, or $f^{-1}(x) = \frac{1}{2}x - 2$ **15.** $g^{-1}(x) = x^2 + 3, x \geq 0$ **17.** not one-to-one **19.** $f^{-1}(x) = \sqrt[3]{x + 4}$ **21.** **(a)** 8 **(b)** 3

23. **(a)** 1 **(b)** 0

25. **(a)** one-to-one **27.** **(a)** not one-to-one

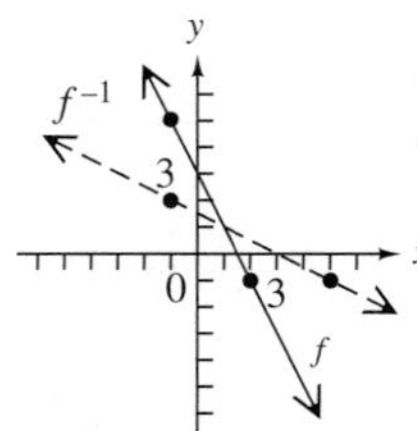

29. **(a)** one-to-one **31.**

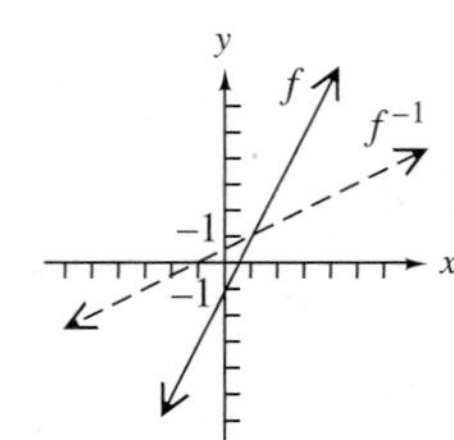

33. **35.** 0, 1, 2

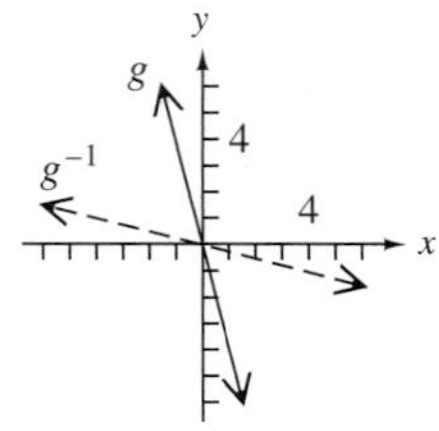

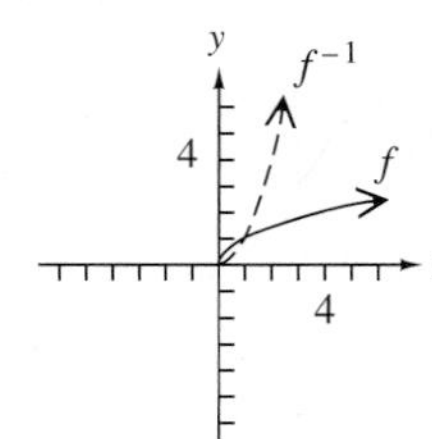

37. $-3, -2, -1, 6$

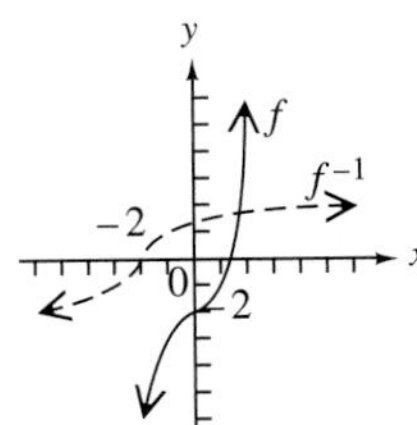

Section 11.2 (pages 881–882)

1. C **3.** C

5.

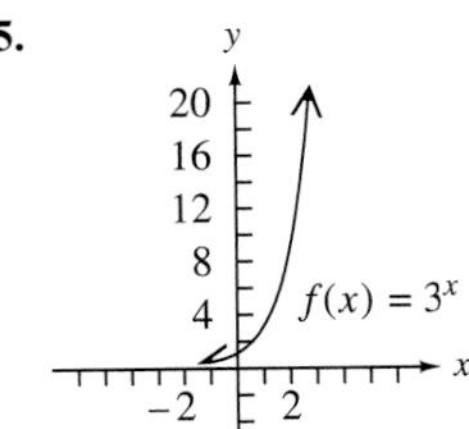

7.

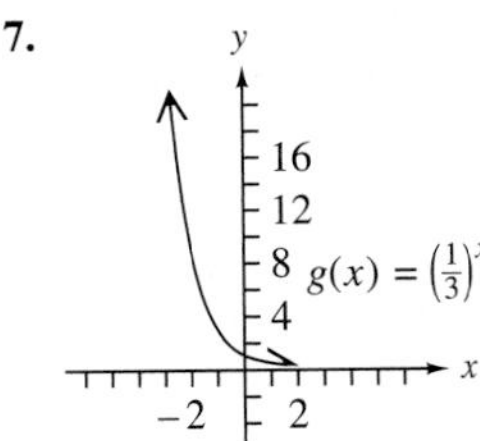

9.

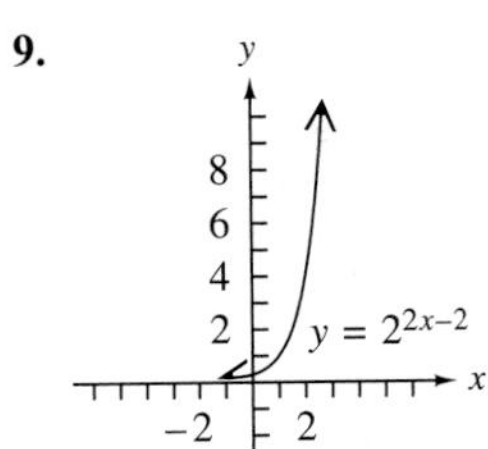

11. $\{2\}$ **13.** $\left\{\frac{3}{2}\right\}$ **15.** $\{7\}$ **17.** $\{-3\}$ **19.** $\{-1\}$ **21.** $\{-3\}$
23. (a) rises; falls **(b)** It is one-to-one and thus has an inverse.
25. (a) 1.0°C **(b)** 0.4°C **27. (a)** 3.0°C **(b)** 0.7°C

Section 11.3 (pages 887–890)

1. (a) C **(b)** F **(c)** B **(d)** A **(e)** E **(f)** D **3.** $\log_4 1024 = 5$
5. $\log_{1/2} 8 = -3$ **7.** $\log_{10} 0.001 = -3$ **9.** $\log_{625} 5 = \frac{1}{4}$
11. $4^3 = 64$ **13.** $10^{-4} = \frac{1}{10{,}000}$ **15.** $6^0 = 1$ **17.** $9^{1/2} = 3$
19. Since the radical $\sqrt{9} = 9^{1/2} = 3$, the exponent to which 9 must be raised is 1/2. **21.** $\left\{\frac{1}{3}\right\}$ **23.** $\{81\}$ **25.** $\left\{\frac{1}{5}\right\}$ **27.** $\{1\}$
29. $\{x \mid x > 0, x \neq 1\}$ **31.** $\{5\}$ **33.** $\left\{\frac{5}{3}\right\}$ **35.** $\{4\}$ **37.** $\left\{\frac{3}{2}\right\}$

39.

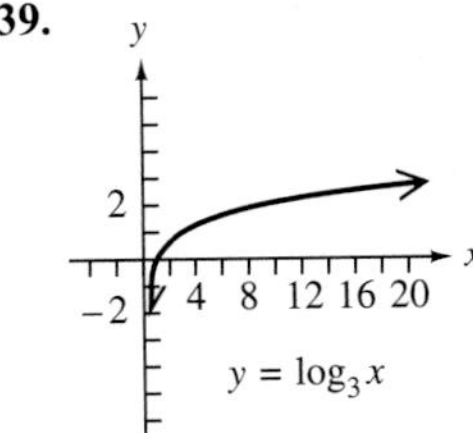

41.

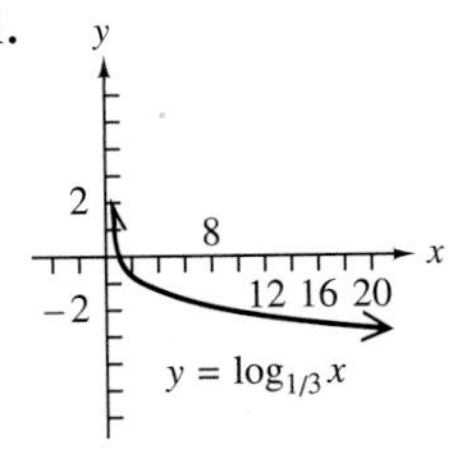

43. Answers will vary. **45.** 8 **47.** 24 **49.** Since every real number power of 1 equals 1, if $y = \log_1 x$, then $x = 1^y$ and so $x = 1$ for every y. This contradicts the definition of a function. **51.** $f(x) = 3^x$
53. (a) 4385 billion ft^3 **(b)** 5555 billion ft^3 **(c)** 6140 billion ft^3
55. about 4 times as powerful

Section 11.4 (pages 897–898)

1. false; $\log_b x + \log_b y = \log_b xy$ **3.** true **5.** $\log_7 4 - \log_7 5$
7. $\frac{1}{4}\log_2 8$, or $\frac{3}{4}$ **9.** $\log_4 3 + \frac{1}{2}\log_4 x - \log_4 y$
11. $\frac{1}{3}\log_3 4 - 2\log_3 x - \log_3 y$ **13.** $\frac{1}{2}\log_3 x + \frac{1}{2}\log_3 y - \frac{1}{2}\log_3 5$
15. $\frac{1}{3}\log_2 x + \frac{1}{5}\log_2 y - 2\log_2 r$ **17.** The distributive property tells us that the *product* $a(x + y)$ equals the sum $ax + ay$. In the notation $\log_a(x + y)$, the parentheses do not indicate multiplication. They indicate that $x + y$ is the result of raising a to some power. **19.** $\log_b xy$
21. $\log_a \frac{m^3}{n}$ **23.** $\log_a \frac{rt^3}{s}$ **25.** $\log_a \frac{125}{81}$ **27.** $\log_{10}(x^2 - 9)$
29. $\log_p \frac{x^3 y^{1/2}}{z^{3/2} a^3}$ **31.** false **33.** true **35.** true **37.** false **39.** 4
40. It is the exponent to which 3 must be raised in order to obtain 81.
41. 81 **42.** It is the exponent to which 2 must be raised in order to obtain 19. **43.** 19 **44.** m

Section 11.5 (pages 903–904)

1. C **3.** C **5.** 19.2 **7.** 2.5164 **9.** -1.4868 **11.** 9.6776
13. 2.0592 **15.** -2.8896 **17.** 2.3026 **19.** poor fen **21.** rich fen
23. 5.9 **25.** 1.0×10^{-2} **27.** 2.5×10^{-5} **29. (a)** 35.0 yr **(b)** 14.2 yr
31. (a) 55% **(b)** 90% **33. (a)** 5.6 yr **(b)** $t > 0$ and $\frac{87 - L}{63}$ is positive and in the domain of the function only if $24 < L < 87$.

Section 11.6 (pages 911–914)

1. $\log 5^x = \log 125$ **2.** $x \log 5 = \log 125$ **3.** $x = \frac{\log 125}{\log 5}$
4. $\frac{\log 125}{\log 5} = 3$; $\{3\}$ **5.** $\{0.827\}$ **7.** $\{0.833\}$ **9.** $\{1.201\}$ **11.** $\{2.269\}$
13. $\{15.967\}$ **15.** $\{261.291\}$ **17.** $\{-10.718\}$ **19.** $\{3\}$ **21.** $\{5.879\}$
23. Natural logarithms are a better choice because e is the base.
25. $\left\{\frac{2}{3}\right\}$ **27.** $\left\{\frac{33}{2}\right\}$ **29.** $\{-1 + \sqrt[3]{49}\}$ **31.** 2 cannot be a solution because $\log(2 - 3) = \log(-1)$, and -1 is not in the domain of $\log x$.
33. $\left\{\frac{1}{3}\right\}$ **35.** $\{2\}$ **37.** $\emptyset$ **39.** $\{8\}$ **41.** $\left\{\frac{4}{3}\right\}$ **43.** $\{8\}$
45. (a) \$2539.47 **(b)** 10.2 yr **47.** \$4934.71 **49.** 15.4 yr
51. about 180 g **53.** 1.4315 **55.** 3.3030 **57.** -0.0947
59. -2.3219 **61.** -2.0639 **63.** 0.673

Chapter 11 Review Exercises (pages 919–922)

1. not one-to-one **2.** one-to-one **3.** This function is not one-to-one because two sodas in the list have 41 mg of caffeine. **4.** $f^{-1}(x) = \frac{x - 7}{-3}$, or $f^{-1}(x) = -\frac{1}{3}x + \frac{7}{3}$ **5.** $f^{-1}(x) = \frac{x^3 + 4}{6}$ **6.** not one-to-one

7.

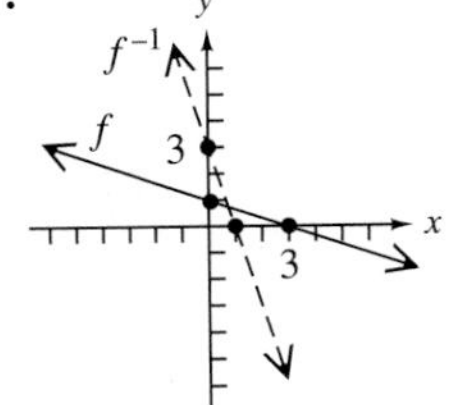

8.

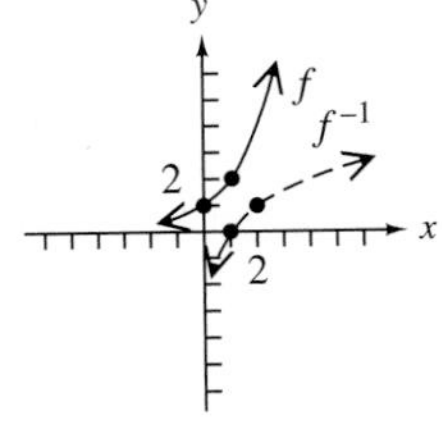

ANSWERS

9.

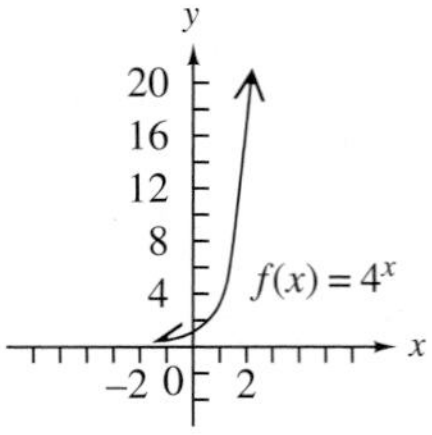

10.

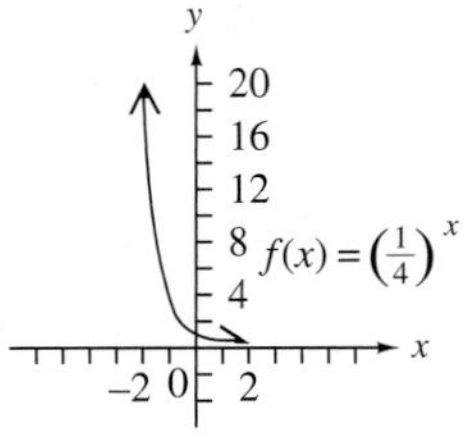

11.

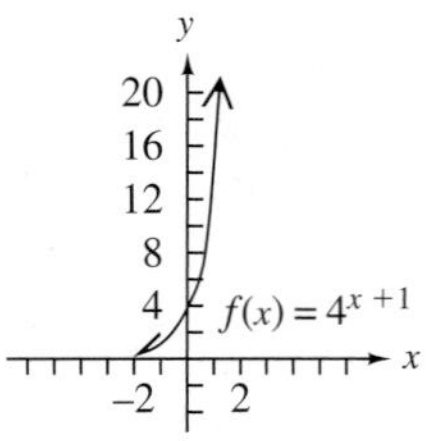

12. $\{4\}$ **13.** $\left\{\frac{3}{7}\right\}$ **14.** $\{0\}$ **15. (a)** 48.9 million **(b)** 67.3 million **16. (a)** $5^4 = 625$ **(b)** $\log_5 0.04 = -2$ **17. (a)** $\log_b a$ represents the exponent to which b must be raised to obtain a. **(b)** a

18.

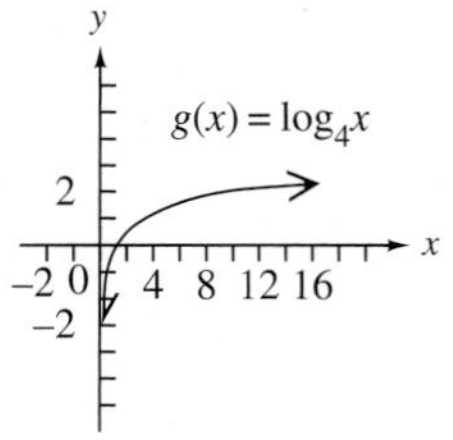

19.

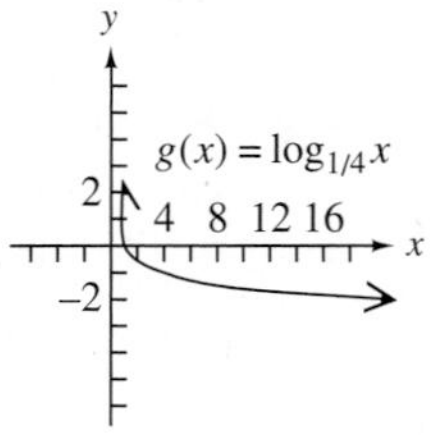

20. $\{2\}$ **21.** $\{-2\}$ **22.** $\{8\}$ **23.** $\{b \mid b > 0, b \neq 1\}$ **24.** $\log_4 3 + 2\log_4 x$ **25.** $2\log_2 p + \log_2 r - \frac{1}{2}\log_2 z$ **26.** $\log_b \frac{3x}{y^2}$ **27.** $\log_3 \frac{x+7}{4x+6}$ **28.** 1.4609 **29.** -0.5901 **30.** 3.3638 **31.** -1.3587 **32.** 6.4 **33.** 8.4 **34.** 2.5×10^{-5} **35. (a)** 18 yr **(b)** 12 yr **(c)** 7 yr **(d)** 6 yr **(e)** Each comparison shows approximately the same number. For example, in part (a) the doubling time is 18 yr (rounded) and $\frac{72}{4} = 18$. Thus, the formula $t = \frac{72}{100r}$ (called the *Rule of 72*) is an excellent approximation of the doubling time formula. **36.** every 2 hr **37.** $\{2.042\}$ **38.** $\{4.907\}$ **39.** $\{18.310\}$ **40.** $\left\{\frac{1}{9}\right\}$ **41.** $\{-6 + \sqrt[3]{25}\}$ **42.** $\{2\}$ **43.** $\left\{\frac{3}{8}\right\}$ **44.** $\{4\}$ **45.** $\{1\}$ **46.** \$7112.11 **47.** Plan A; it would pay \$2.92 more. **48.** \$4267 **49.** about 11% **50.** 0.9251 **51.** 1.7925 **52.** 1.4315 **53.** $\{72\}$ **54.** $\{5\}$ **55.** $\left\{\frac{1}{9}\right\}$ **56.** $\left\{\frac{4}{3}\right\}$ **57.** $\{3\}$ **58.** $\left\{\frac{1}{8}\right\}$ **59.** $\left\{\frac{11}{3}\right\}$ **60.** 6.8 yr

Chapter 11 Test (pages 923–924)

1. (a) not one-to-one **(b)** one-to-one **2.** $f^{-1}(x) = x^3 - 7$

3.

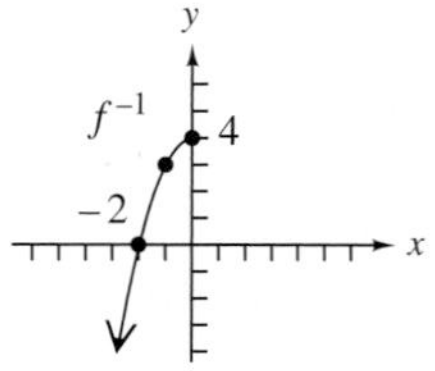

4.

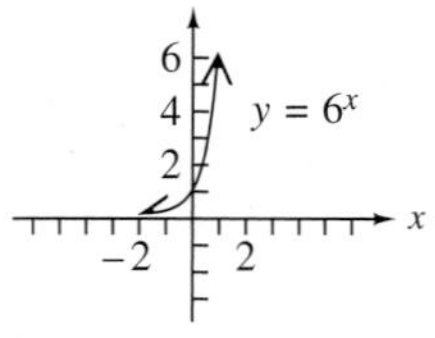

5.

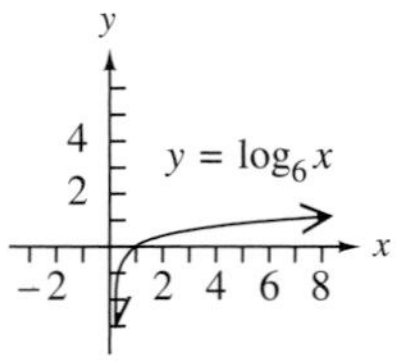

6. Interchange the x- and y-values of the ordered pairs, because the functions are inverses. **7.** $\{-4\}$ **8.** $\left\{-\frac{13}{3}\right\}$ **9. (a)** 775 millibars **(b)** 265 millibars **10.** $\log_4 0.0625 = -2$ **11.** $7^2 = 49$ **12.** $\{32\}$ **13.** $\left\{\frac{1}{2}\right\}$ **14.** $\{2\}$ **15.** $2\log_3 x + \log_3 y$ **16.** $\log_b \frac{r^{1/4}s^2}{t^{2/3}}$ **17. (a)** 1.3284 **(b)** -0.8440 **(c)** 2.1245 **18.** $\{3.9656\}$ **19.** $\{3\}$ **20. (a)** \$12,507.51 **(b)** 15.5 yr

Cumulative Review Exercises: Chapters R–11 (pages 925–926)

1. $-2, 0, 6, \frac{30}{3}$ (or 10) **2.** $-\frac{9}{4}, -2, 0, 0.6, 6, \frac{30}{3}$ (or 10) **3.** $-\sqrt{2}, \sqrt{11}$ **4.** $\left\{-\frac{2}{3}\right\}$ **5.** $[1, \infty)$ **6.** $\{-2, 7\}$ **7.** $(-\infty, -3) \cup (2, \infty)$ **8. (a)** yes **(b)** 3254; The number of travelers increased by an average of 3254 thousand per year during the period 2003–2006. **9.** $\{(4, 2)\}$ **10.** $\{(1, -1, 4)\}$ **11.** $6p^2 + 7p - 3$ **12.** $16k^2 - 24k + 9$ **13.** $-5m^3 + 2m^2 - 7m + 4$ **14.** $2t^3 + 5t^2 - 3t + 4$ **15.** $z(5z + 1)(z - 4)$ **16.** $(4a + 5b^2)(4a - 5b^2)$ **17.** $(2c + d)(4c^2 - 2cd + d^2)$ **18.** $-\frac{1875p^{13}}{8}$ **19.** $\frac{x+5}{x+4}$ **20.** $\frac{-3k-19}{(k+3)(k-2)}$ **21.** $12\sqrt{2}$ **22.** $-\frac{1}{4}$ **23.** $-27\sqrt{2}$ **24.** $\{0, 4\}$ **25.** 41 **26.** $\left\{\frac{1+\sqrt{13}}{6}, \frac{1-\sqrt{13}}{6}\right\}$ **27.** $(-\infty, -4) \cup (2, \infty)$ **28.** $\{-2, -1, 1, 2\}$ **29.** $\{-1\}$ **30.** $\{1\}$ **31.** $5^3 = 125$ **32.** $3\log x + \frac{1}{2}\log y - \log z$

33.

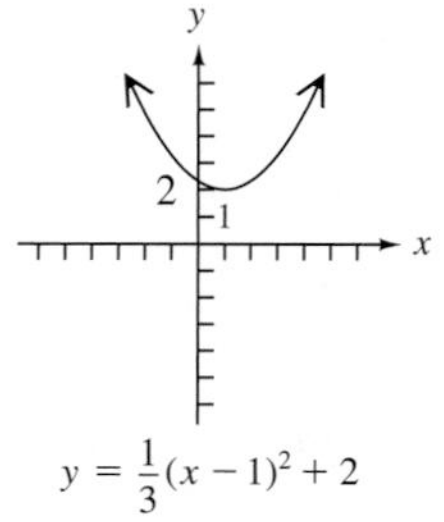

34.

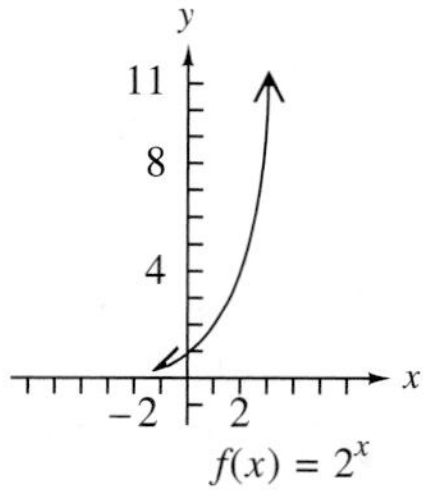

35.

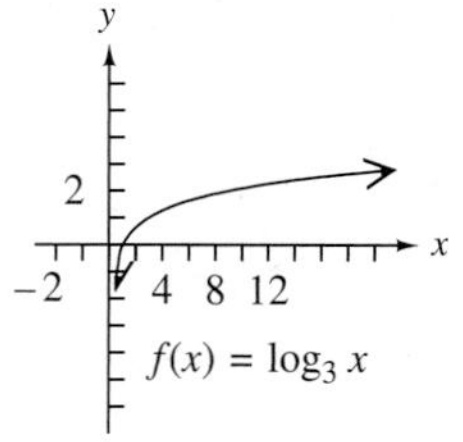

Chapter 12 Nonlinear Functions, Conic Sections, and Nonlinear Systems

Section 12.1 (pages 937–940)

1. 0 **3.** 0; 0 **5.** B **7.** A

9. domain: $(-\infty, \infty)$; range: $[0, \infty)$

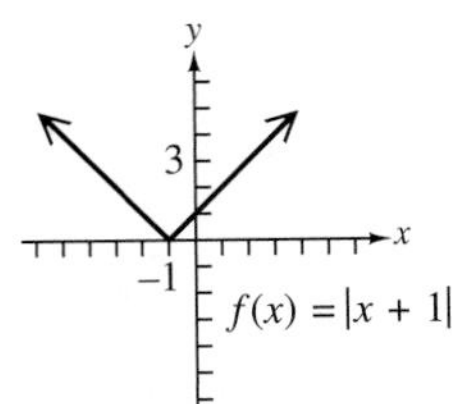

11. domain: $(-\infty, 0) \cup (0, \infty)$; range: $(-\infty, 1) \cup (1, \infty)$

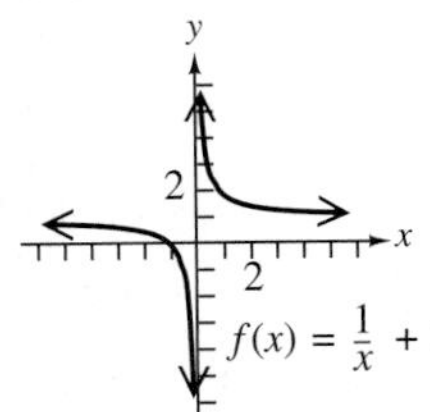

13. domain: $[2, \infty)$; range: $[0, \infty)$

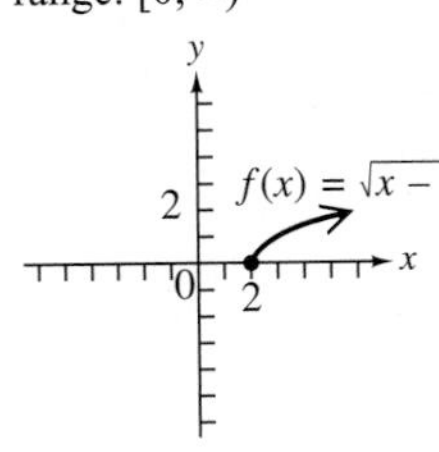

15. domain: $(-\infty, 2) \cup (2, \infty)$; range: $(-\infty, 0) \cup (0, \infty)$

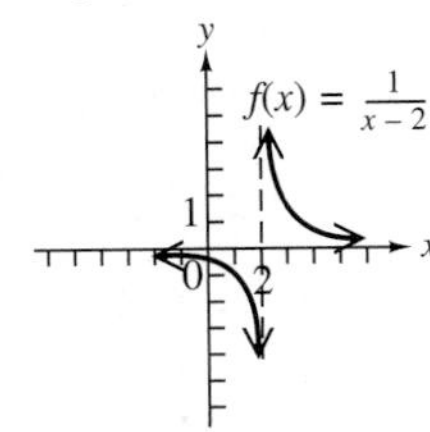

17. domain: $[-3, \infty)$; range: $[-3, \infty)$

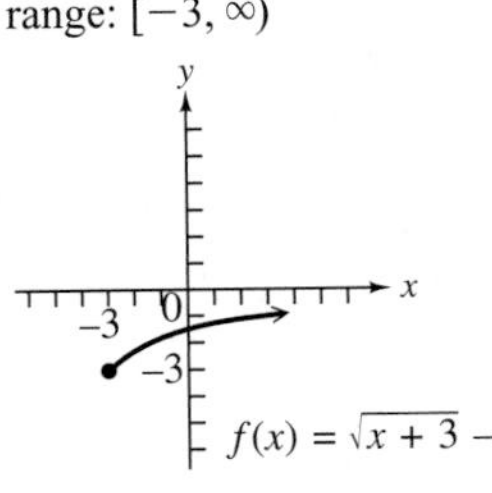

19.

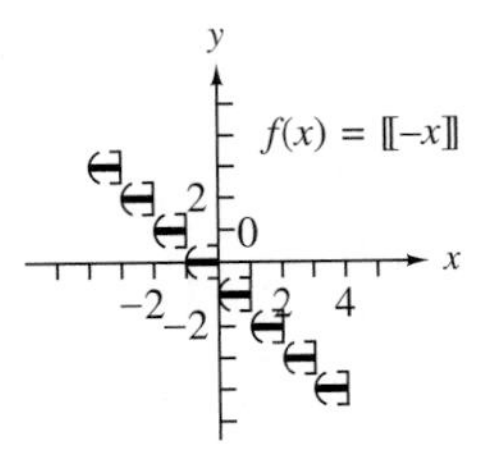

21.

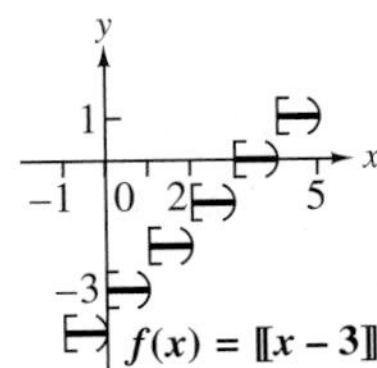

23.

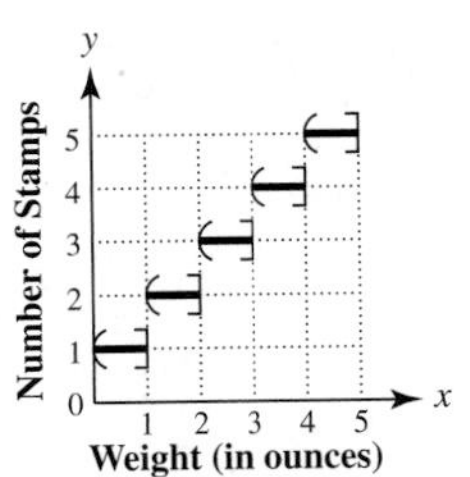

25. $x^2 + 2x - 9$ **27.** $2x^3 - 18x$ **29.** 6 **31.** -20 **33.** $x^2 - x - 6$ **35.** $x + 3,\quad x \neq 3$ **37.** 6 **39.** 4 **41.** 16 **43.** 83 **45.** 13 **47.** $4x^2 + 12x + 13$ **49.** $x^2 + 10x + 29$ **51.** $2x + 8$ **53.** $\frac{137}{4}$ **55.** 8 **57.** $(f \circ g)(x) = 63{,}360x$; It computes the number of inches in x miles. **59.** $(\mathcal{A} \circ r)(t) = 4\pi t^2$; This is the area of the circular layer as a function of time.

Section 12.2 (pages 947–950)

1. **(a)** $(0, 0)$ **(b)** 5 **(c)**

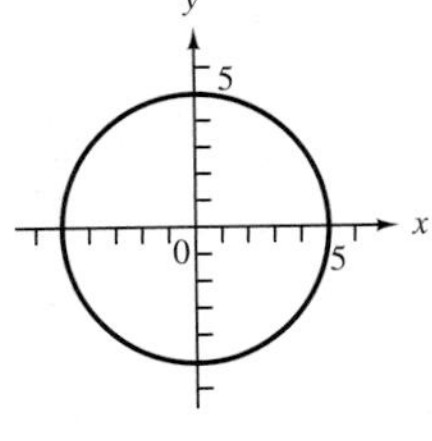

3. B **5.** D **7.** $(x + 4)^2 + (y - 3)^2 = 4$ **9.** $(x + 8)^2 + (y + 5)^2 = 5$ **11.** $(-2, -3)$; $r = 2$ **13.** $(-5, 7)$; $r = 9$ **15.** $(2, 4)$; $r = 4$ **17.** The thumbtack acts as the center and the length of the string acts as the radius.

19.

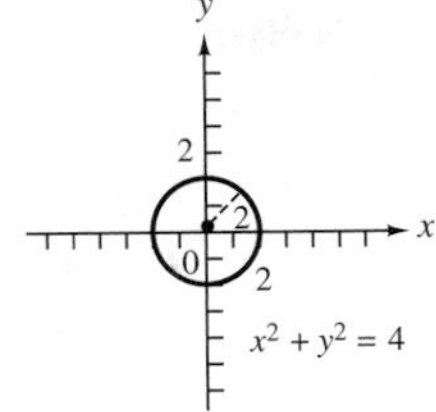

21.

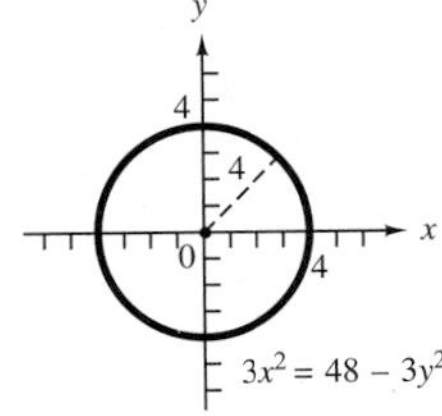

23.

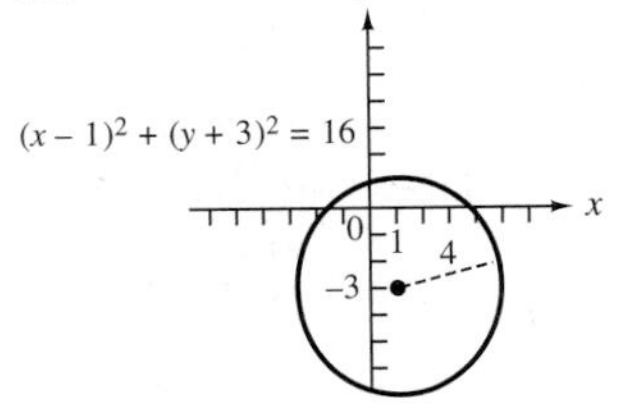

25.

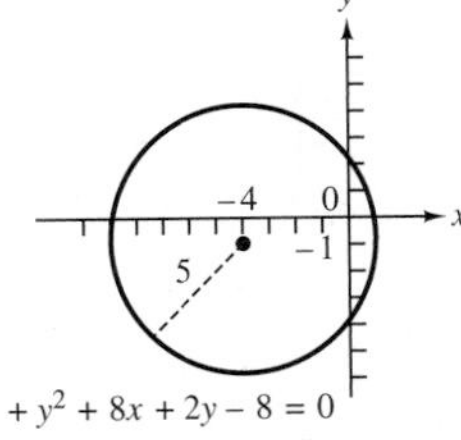

27. The two thumbtacks act as foci, and the length of the string is constant, satisfying the requirements of the definition of an ellipse.

29.

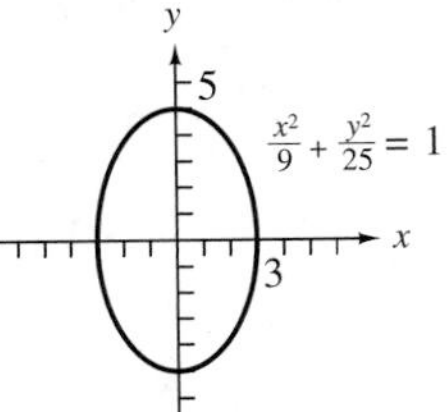

31.

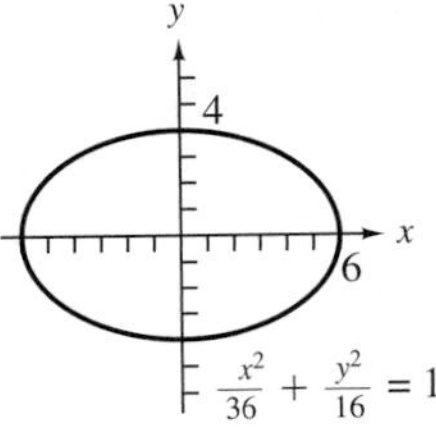

33.

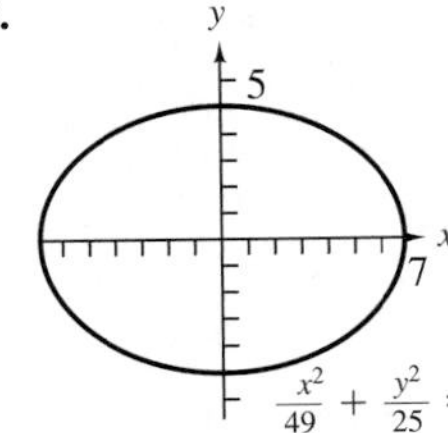

35.

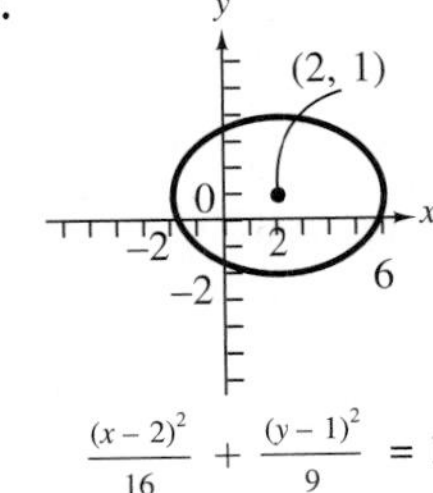

37. **(a)** 10 m **(b)** 36 m **39.** **(a)** 154.7 million mi **(b)** 128.7 million mi (Answers are rounded.)

Section 12.3 (pages 957–960)

1. C **3.** D **5.** When written in one of the forms given in the box titled "Equations of Hyperbolas" in this section, it will open up and down if the $-$ sign precedes the x^2-term; it will open left and right if the $-$ sign precedes the y^2-term.

7.

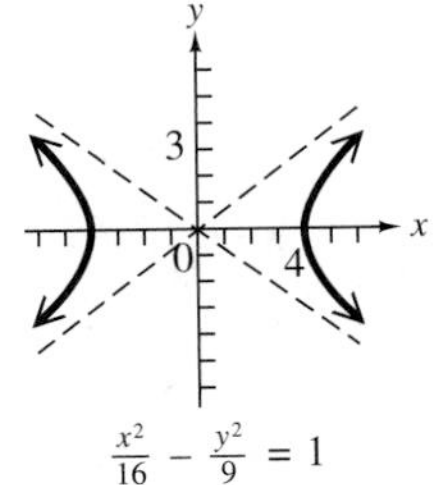

9.

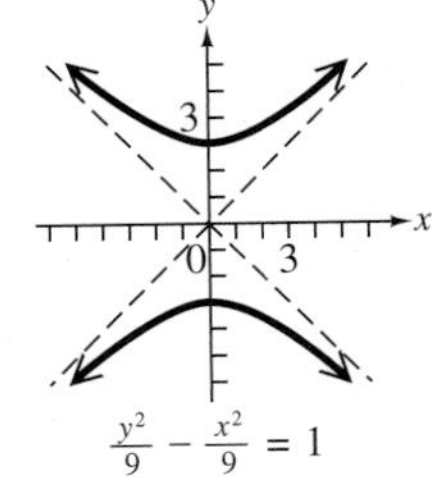

11.

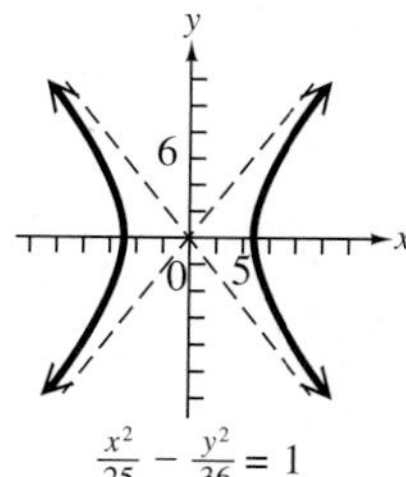
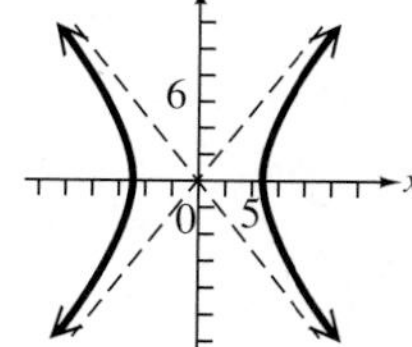

13. hyperbola

$x^2 - y^2 = 16$

15. ellipse

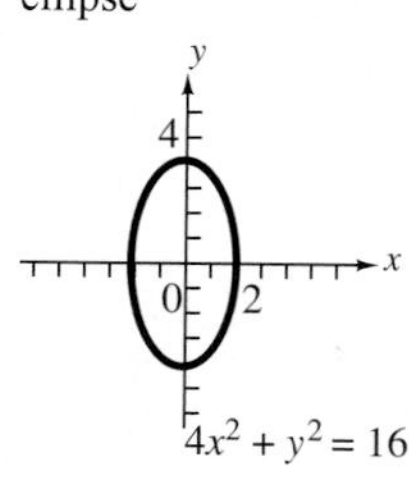

17. circle

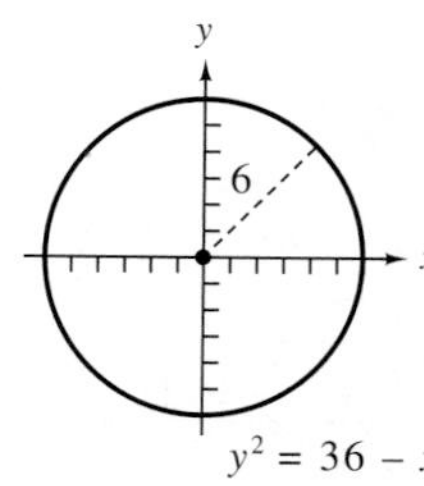

19. hyperbola

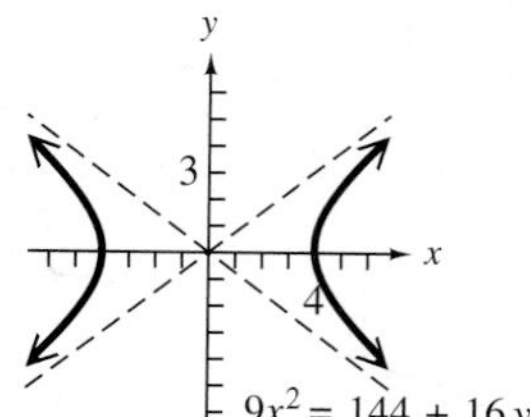

21. ellipse

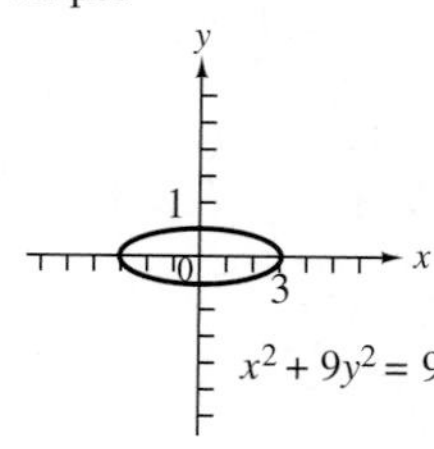

23. domain: $[-3, 3]$; range: $[0, 3]$

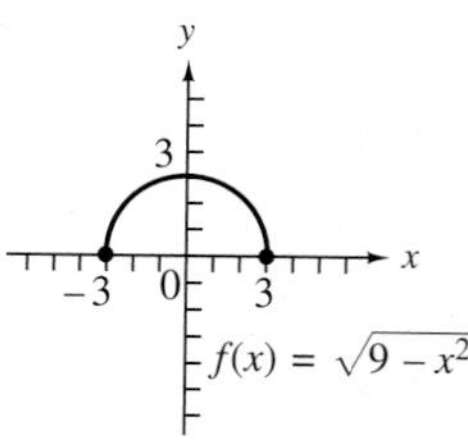

25. domain: $[-5, 5]$; range: $[-5, 0]$

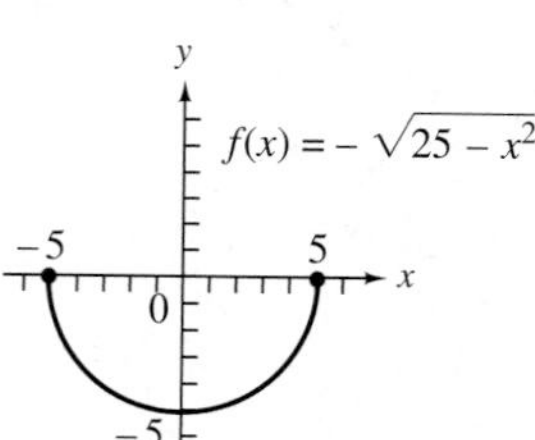

27. domain: $[-4, \infty)$; range: $[0, \infty)$

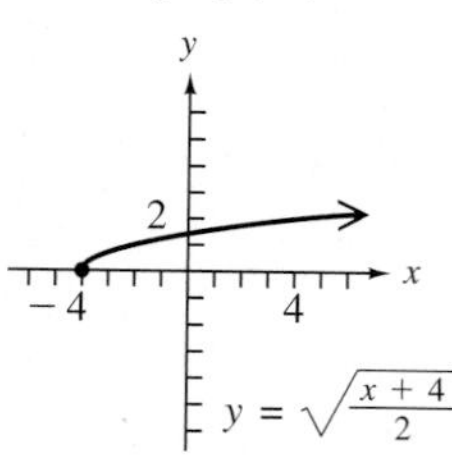

29.

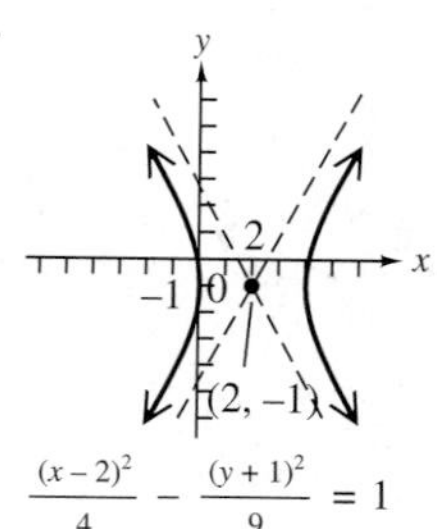

31. **(a)** 50 m **(b)** 69.3 m

Section 12.4 (pages 965–966)

1. one **3.** none

5.

7.

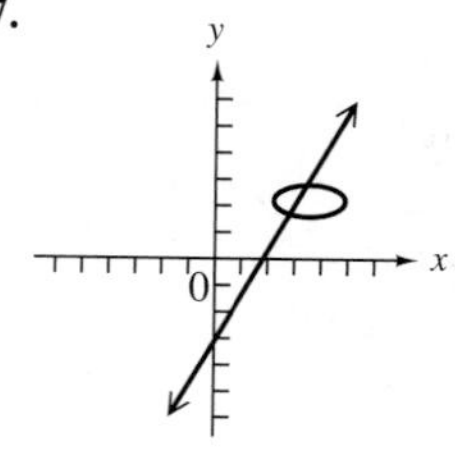

9.

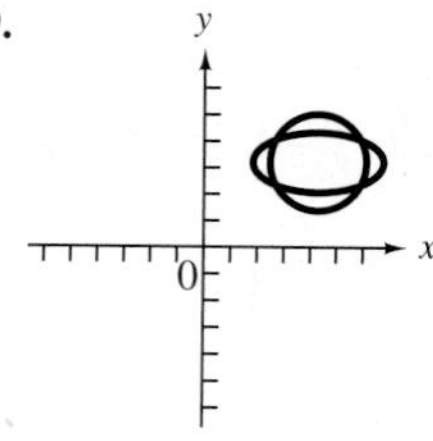

11. $\left\{(0, 0), \left(\frac{1}{2}, \frac{1}{2}\right)\right\}$ **13.** $\{(-6, 9), (-1, 4)\}$

15. $\left\{\left(-\frac{1}{5}, \frac{7}{5}\right), (1, -1)\right\}$ **17.** $\left\{(-2, -2), \left(-\frac{4}{3}, -3\right)\right\}$

19. $\{(-3, 1), (1, -3)\}$ **21.** $\left\{\left(-\frac{3}{2}, -\frac{9}{4}\right), (-2, 0)\right\}$

23. $\{(-\sqrt{3}, 0), (\sqrt{3}, 0), (-\sqrt{5}, 2), (\sqrt{5}, 2)\}$

25. $\{(-2, 0), (2, 0)\}$ **27.** $\{(i\sqrt{2}, -3i\sqrt{2}), (-i\sqrt{2}, 3i\sqrt{2}), (-\sqrt{6}, -\sqrt{6}), (\sqrt{6}, \sqrt{6})\}$ **29.** $\{(-2i\sqrt{2}, -2\sqrt{3}), (-2i\sqrt{2}, 2\sqrt{3}), (2i\sqrt{2}, -2\sqrt{3}), (2i\sqrt{2}, 2\sqrt{3})\}$

31. $\{(-\sqrt{5}, -\sqrt{5}), (\sqrt{5}, \sqrt{5})\}$ **33.** length: 12 ft; width: 7 ft

Section 12.5 (pages 971–972)

1.

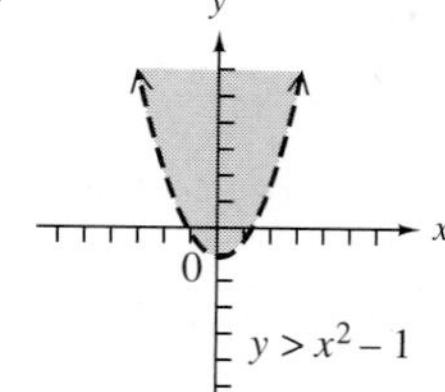

3.

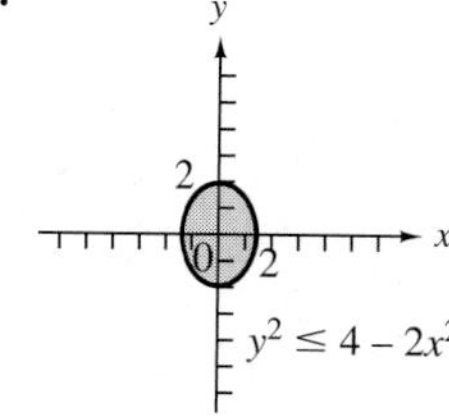

5.

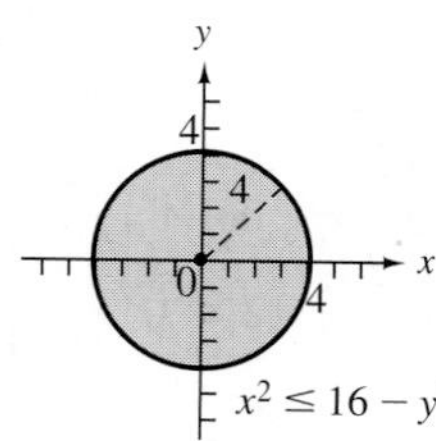

7.

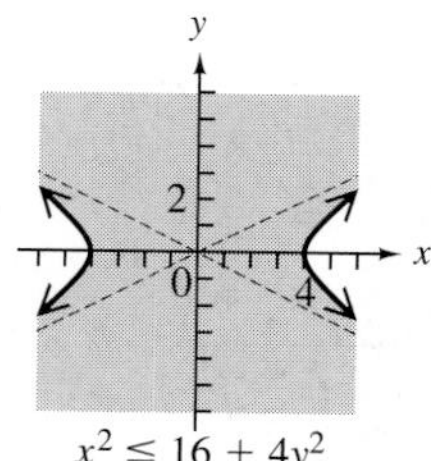

9.

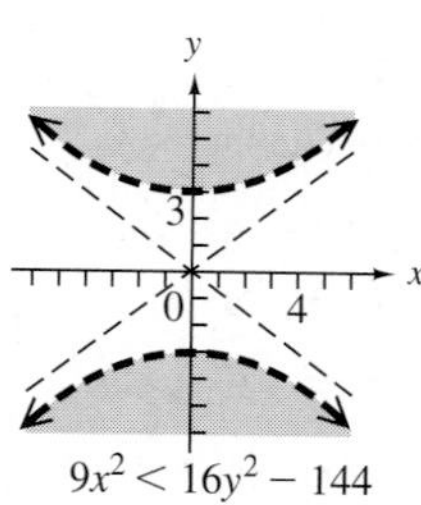

11.

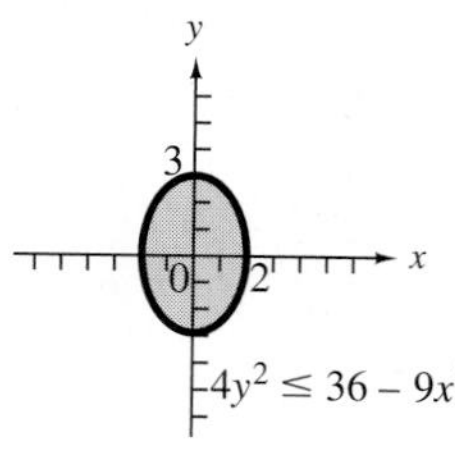

13.

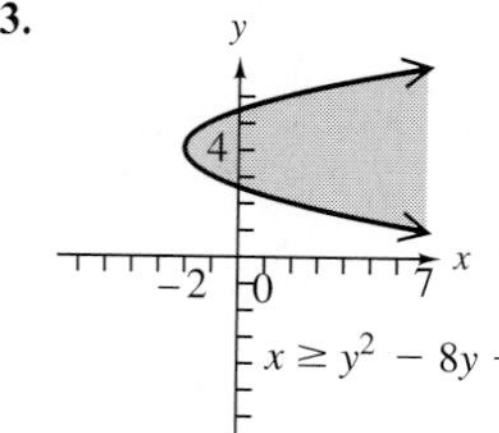

15.

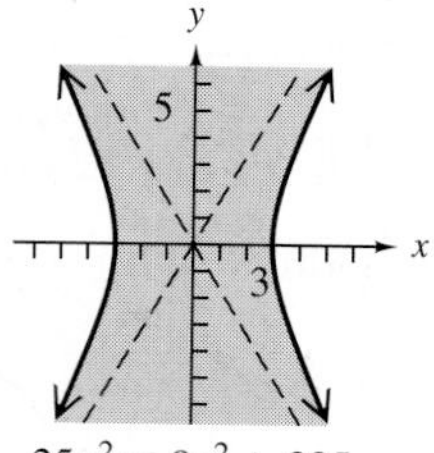

ANSWERS

17.

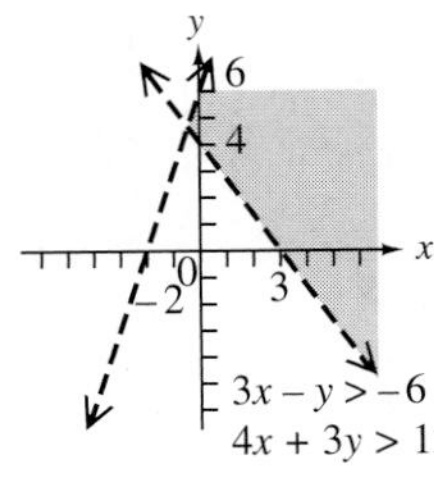

19.

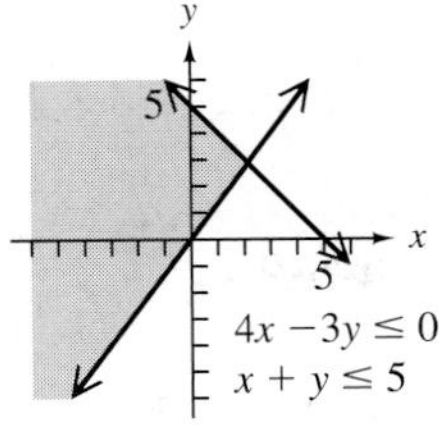

21.

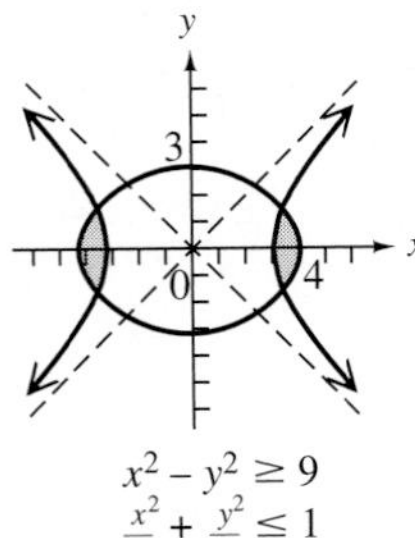

23.

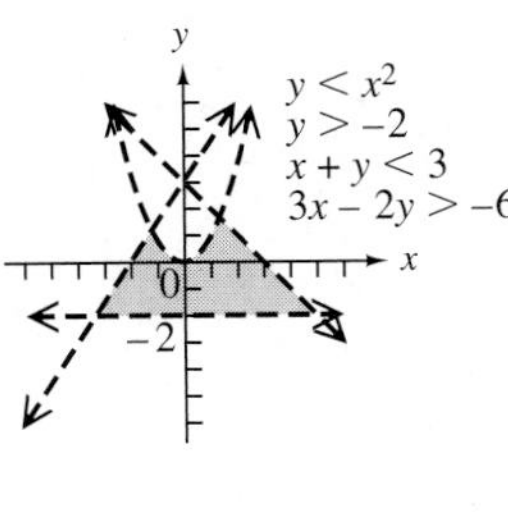

Chapter 12 Review Exercises (pages 977–980)

1. domain: $(-\infty, \infty)$; range: $[0, \infty)$

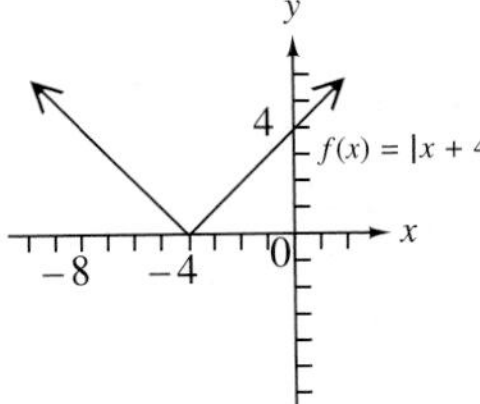

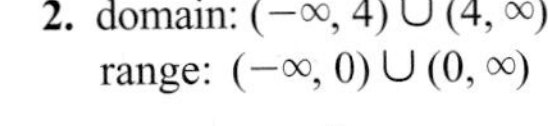

2. domain: $(-\infty, 4) \cup (4, \infty)$; range: $(-\infty, 0) \cup (0, \infty)$

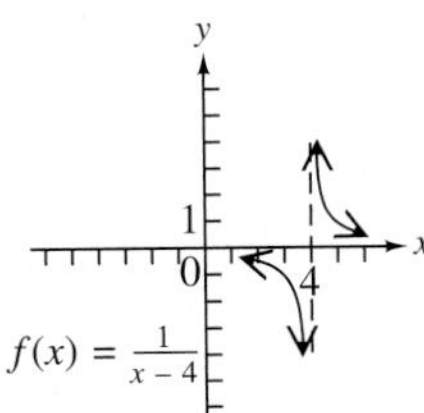

3. domain: $[0, \infty)$; range: $[3, \infty)$

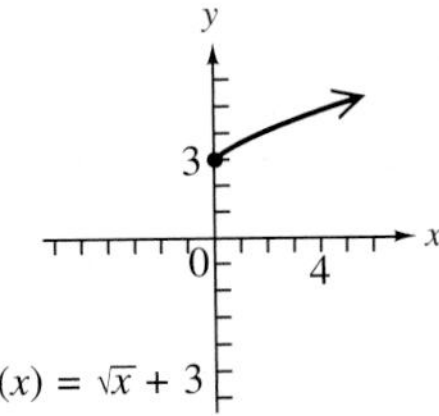

4. (a) 12 **(b)** 2 **(c)** -5 **5. (a)** $5x^2 - x + 5$ **(b)** $-5x^2 + 5x + 1$ **6. (a)** 11 **(b)** -9 **7. (a)** $36x^3 - 9x^2$ **(b)** $4x - 1, \quad x \neq 0$ **8. (a)** -45 **(b)** 7 **9. (a)** 20 **(b)** 42 **10. (a)** $75x^2 + 220x + 160$ **(b)** $15x^2 + 10x + 2$ **11.** $(x + 2)^2 + (y - 4)^2 = 9$ **12.** $(x + 1)^2 + (y + 3)^2 = 25$ **13.** $(x - 4)^2 + (y - 2)^2 = 36$ **14.** $(-3, 2)$; $r = 4$ **15.** $(4, 1)$; $r = 2$ **16.** $(-1, -5)$; $r = 3$ **17.** $(3, -2)$; $r = 5$

18.

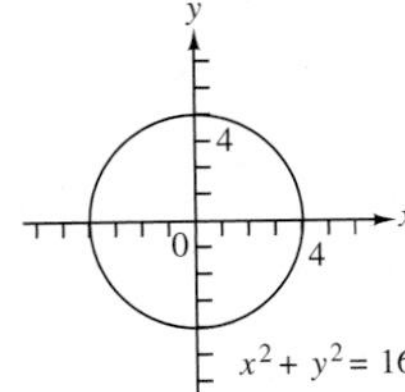

19.

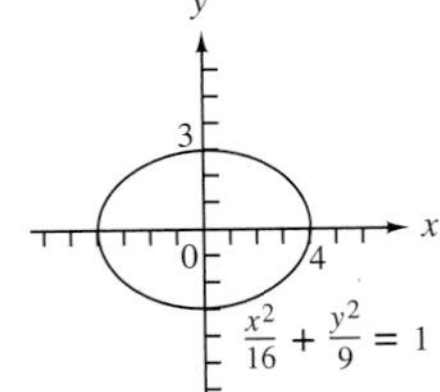

20.

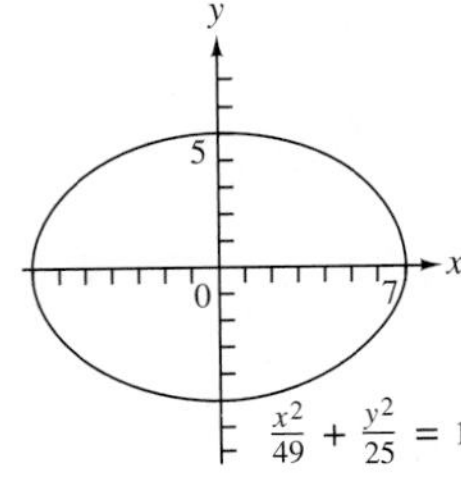

21. $\frac{x^2}{65{,}286{,}400} + \frac{y^2}{2{,}560{,}000} = 1$ **22. (a)** 348.2 ft **(b)** 1787.6 ft

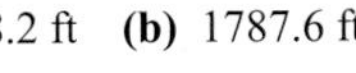

23.

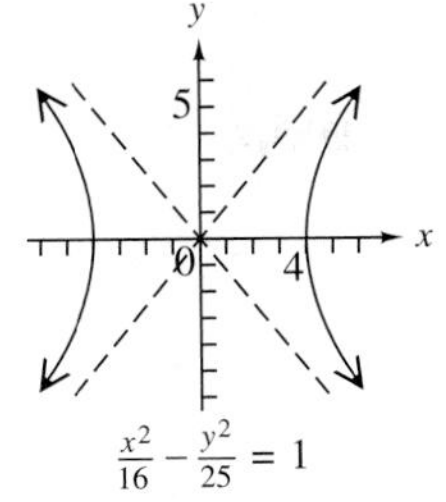

24.

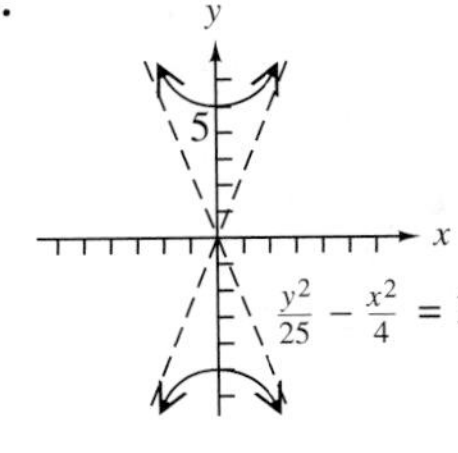

25.

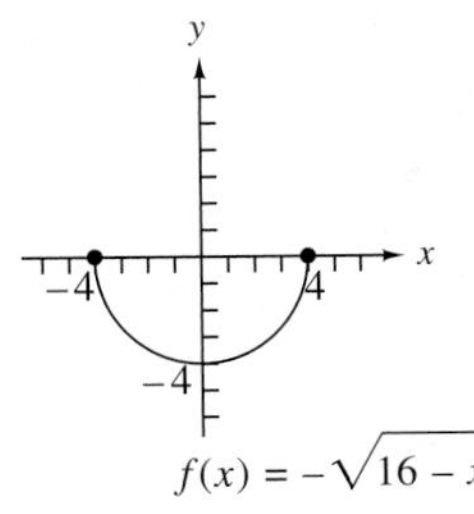

26. circle **27.** parabola **28.** hyperbola **29.** ellipse **30.** parabola **31.** hyperbola **32.** $\frac{x^2}{625} - \frac{y^2}{1875} = 1$ **33.** $\{(6, -9), (-2, -5)\}$ **34.** $\{(1, 2), (-5, 14)\}$ **35.** $\{(4, 2), (-1, -3)\}$ **36.** $\{(-2, -4), (8, 1)\}$

37. $\{(-\sqrt{2}, 2), (-\sqrt{2}, -2), (\sqrt{2}, -2), (\sqrt{2}, 2)\}$

38. $\{(-\sqrt{6}, -\sqrt{3}), (-\sqrt{6}, \sqrt{3}), (\sqrt{6}, -\sqrt{3}), (\sqrt{6}, \sqrt{3})\}$

39. 0, 1, or 2 **40.** 0, 1, 2, 3, or 4

41.

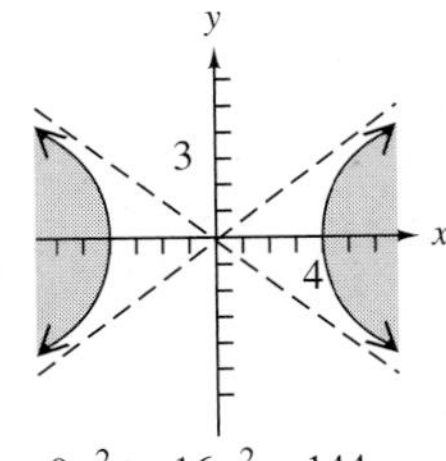

42.

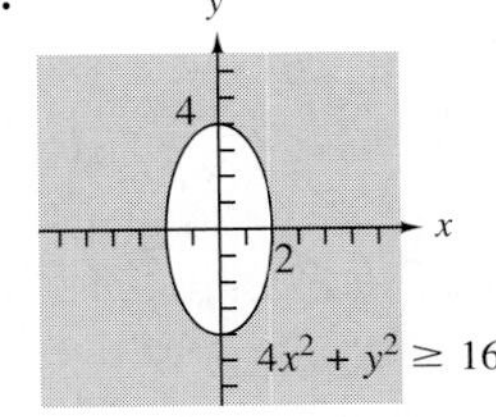

43.

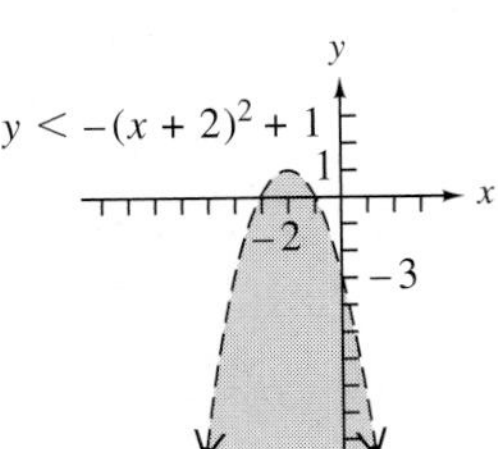

44.

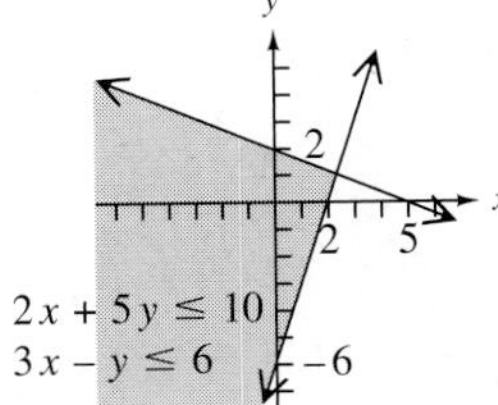

45.

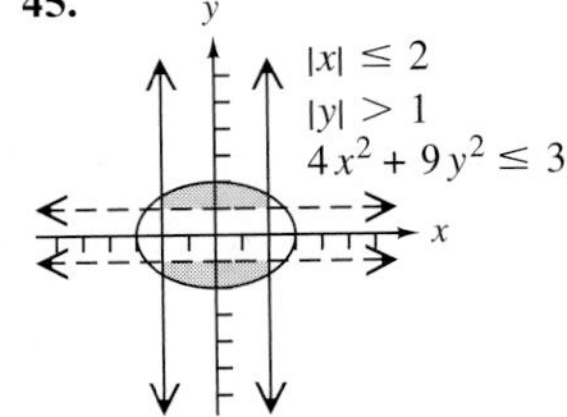

46.

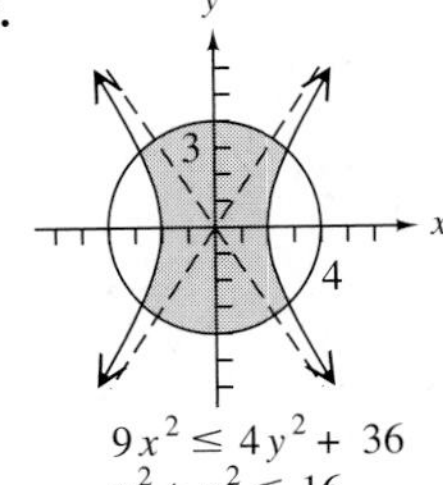

47.

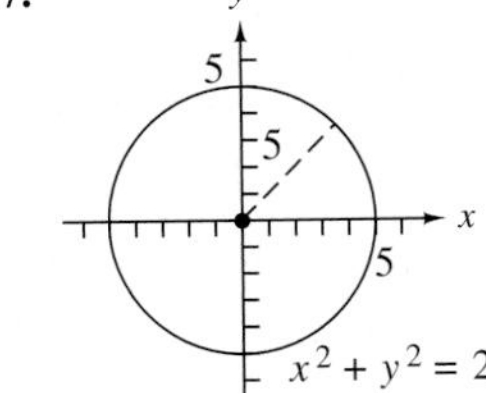

48.

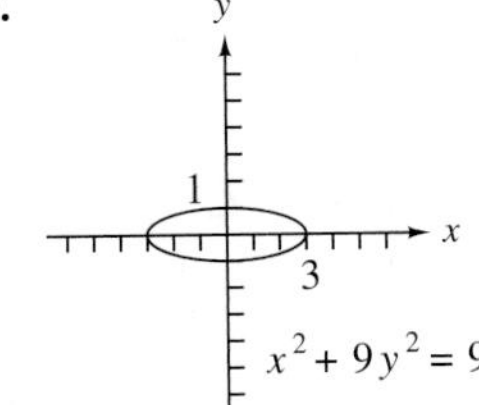

ANSWERS

49.

$x^2 - 9y^2 = 9$

50.

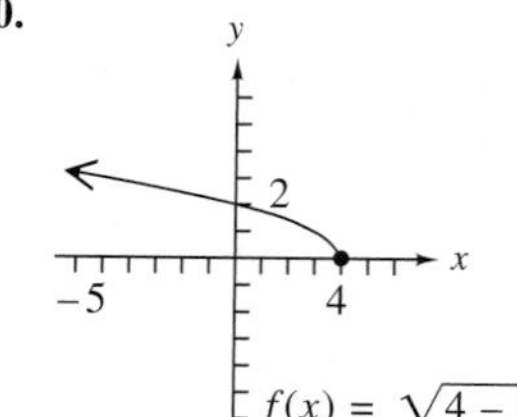

51.

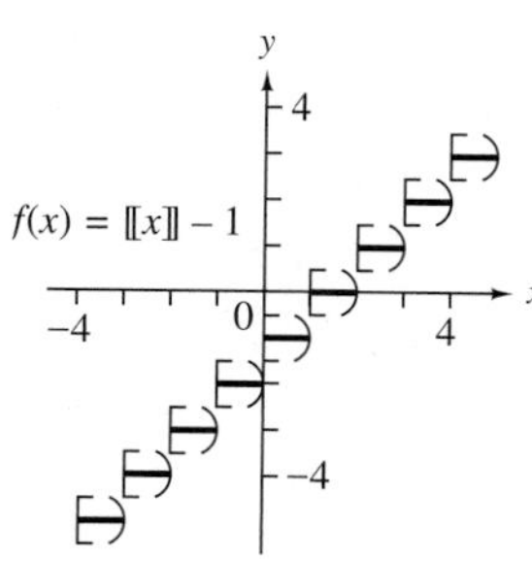

52.

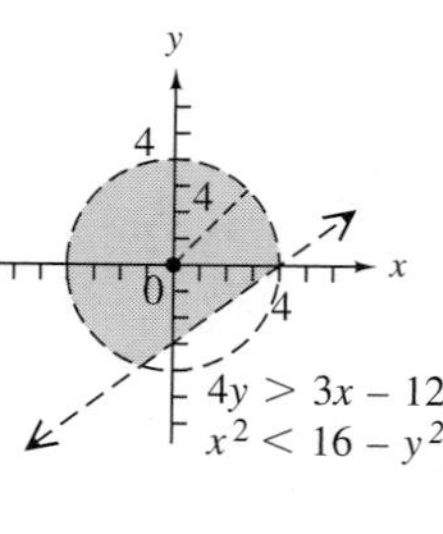

53. There are cases where one x-value will yield two y-values. In a function, every x yields one and only one y.

Chapter 12 Test (pages 981–984)

1. C **2.** A **3.** D **4.** B

5. domain: $(-\infty, \infty)$; range: $[4, \infty)$

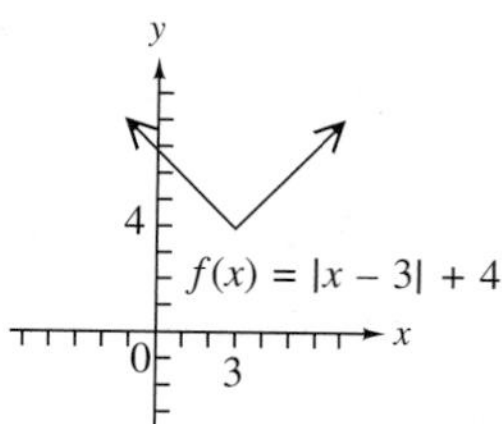

6. (a) 23 **(b)** $3x^2 + 11$ **(c)** $9x^2 + 30x + 27$

7. center: $(2, -3)$; radius: 4

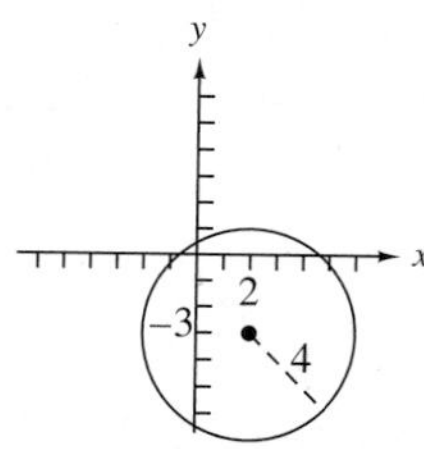

8. center: $(-4, 1)$; radius: 5

9.

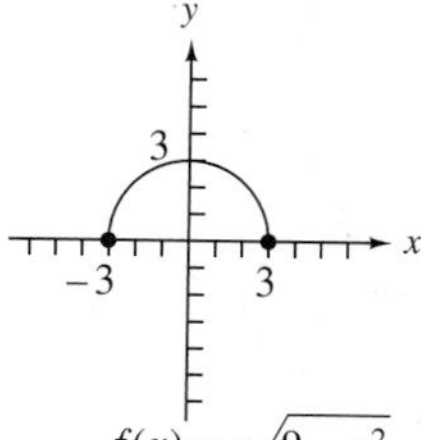

10.

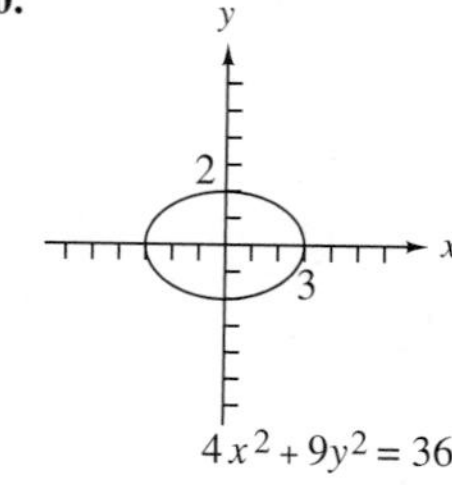

11.

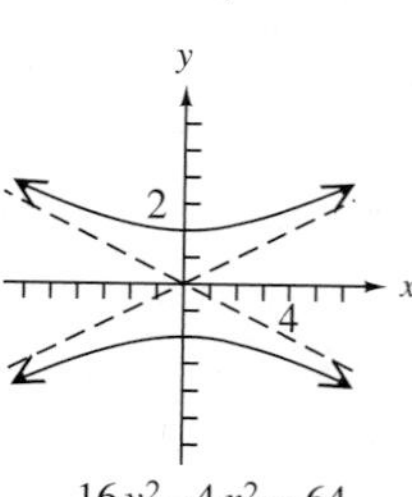

12.

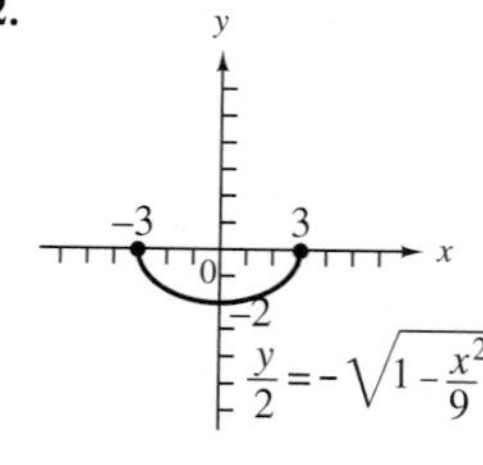

13. ellipse **14.** hyperbola **15.** parabola

16. $\left\{\left(-\frac{1}{2}, -10\right), (5, 1)\right\}$ **17.** $\left\{(-2, -2), \left(\frac{14}{5}, -\frac{2}{5}\right)\right\}$

18. $\left\{(-\sqrt{22}, -\sqrt{3}), (-\sqrt{22}, \sqrt{3}), (\sqrt{22}, -\sqrt{3}), (\sqrt{22}, \sqrt{3})\right\}$

19.

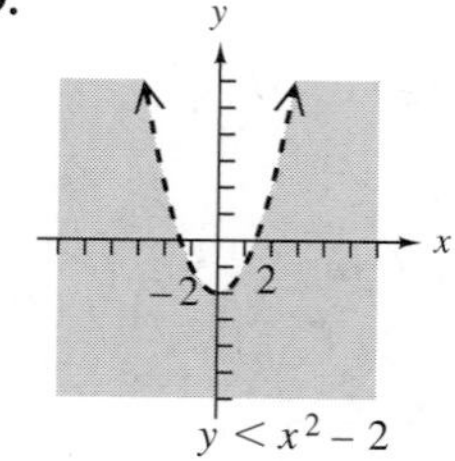

20.

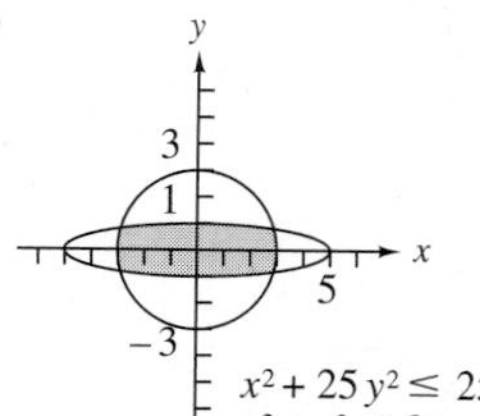

Cumulative Review Exercises: Chapters R–12 (pages 985–986)

1. $\left\{\frac{2}{3}\right\}$ **2.** $\left(-\infty, \frac{3}{5}\right]$ **3.** $\{-4, 4\}$ **4.** $\frac{2}{3}$ **5.** $3x + 2y = -13$

6. $\{(3, -3)\}$ **7.** $\{(4, 1, -2)\}$ **8.** $\left\{(-1, 5), \left(\frac{5}{2}, -2\right)\right\}$

9. $25y^2 - 30y + 9$ **10.** $12r^2 + 40r - 7$

11. $4x^3 - 4x^2 + 3x + 5 + \frac{3}{2x + 1}$ **12.** $(3x + 2)(4x - 5)$

13. $(z^2 + 1)(z + 1)(z - 1)$ **14.** $(a - 3b)(a^2 + 3ab + 9b^2)$

15. $\frac{y - 1}{y(y - 3)}$ **16.** $\frac{3c + 5}{(c + 5)(c + 3)}$ **17.** $\frac{1}{p}$ **18.** $1\frac{1}{5}$ hr

19. $\frac{a^5}{4}$ **20.** $2\sqrt[3]{2}$ **21.** $\frac{3\sqrt{10}}{2}$ **22.** $\frac{7}{5} + \frac{11}{5}i$ **23.** $\emptyset$

24. $\left\{\frac{1}{5}, -\frac{3}{2}\right\}$ **25.** $\left\{\frac{3 + \sqrt{33}}{6}, \frac{3 - \sqrt{33}}{6}\right\}$

26. $\left\{-\frac{\sqrt{6}}{2}, \frac{\sqrt{6}}{2}, -\sqrt{7}, \sqrt{7}\right\}$ **27.** $\{3\}$ **28.** $v = \frac{\pm\sqrt{rFkw}}{kw}$

29. $f^{-1}(x) = \sqrt[3]{x - 4}$ **30. (a)** 4 **(b)** 7 **31.** $\log \frac{(3x + 7)^2}{4}$

32. (a) -1 **(b)** $9x^2 + 18x + 4$

33.

$f(x) = -3x + 5$

34.

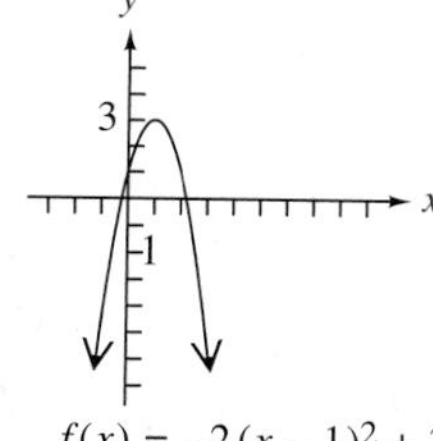

35.

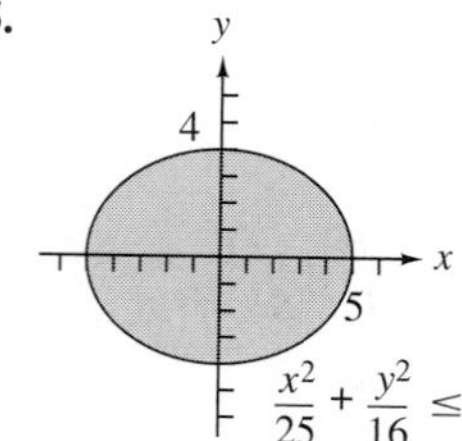

36.

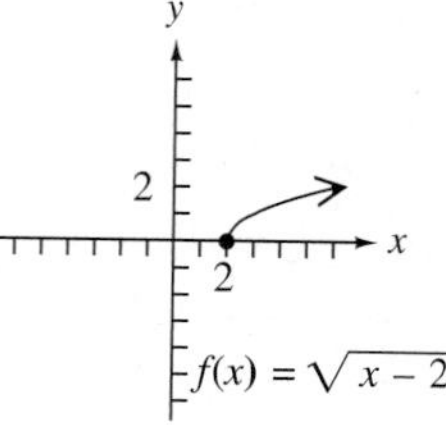

37.

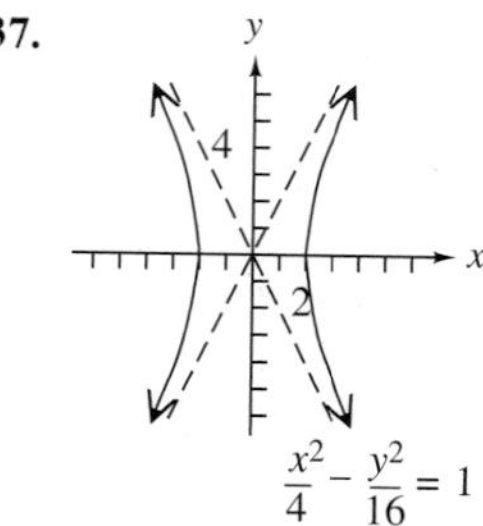

38.

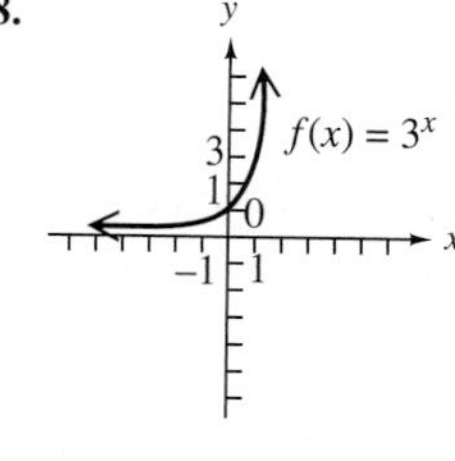

ANSWERS

Appendix A Review of Exponents, Polynomials, and Factoring

(pages 991–994)

1. $\frac{1}{a^2b}$ **3.** $\frac{100y^{10}}{x^2}$ **5.** 0 **7.** $\frac{x^{10}}{2w^{13}y^5}$ **9.** $\frac{a^{15}}{-64b^{15}}$ **11.** $\frac{x^{16}z^{10}}{y^6}$
13. $-6a^4 + 11a^3 - 20a^2 + 26a - 15$ **15.** $8x^3 - 18x^2 + 6x - 16$
17. $x^2y - xy^2 + 6y^3$ **19.** $-3x^2 - 62x + 32$
21. $10x^3 - 4x^2 + 9x - 4$ **23.** $6x^2 - 19x - 7$ **25.** $4x^2 - 9x + 2$
27. $16t^2 - 9$ **29.** $4y^4 - 16$ **31.** $16x^2 - 24x + 9$
33. $36r^2 + 60ry + 25y^2$ **35.** $c^3 + 8d^3$ **37.** $64x^3 - 1$
39. $14t^3 + 45st^2 + 18s^2t - 5s^3$ **41.** $4xy^3(2x^2y + 3x + 9y)$
43. $(x + 3)(x - 5)$ **45.** $(2x + 3)(x - 6)$ **47.** $(6t + 5)(6t - 5)$
49. $(4t + 3)^2$ **51.** $p(2m - 3n)^2$ **53.** $(x + 1)(x^2 - x + 1)$
55. $(2t + 5)(4t^2 - 10t + 25)$ **57.** $(t^2 - 5)(t^4 + 5t^2 + 25)$
59. $(5x + 2y)(t + 3r)$ **61.** $(6r - 5s)(a + 2b)$
63. $(t^2 + 1)(t + 1)(t - 1)$ **65.** $(2x + 3y - 1)(2x + 3y + 1)$

Appendix B Strategies for Problem Solving

(pages 1001–1004)

1. You should choose a sock from the box labeled *red and green socks.* Since it is mislabeled, it contains only red socks or only green socks, determined by the sock you choose. If the sock is green, relabel this box *green socks.* Since the other two boxes were mislabeled, switch the remaining label to the other box and place the label that says *red and green socks* on the unlabeled box. No other choice guarantees a correct relabeling, since you can remove only one sock. **3.** You must place the O in the bottom-left square. No other choice guarantees you a win. **5.** One possible sequence is shown here. The numbers represent the number of gallons in each bucket in each successive step.

Big	7	4	4	1	1	0	7	5	5
Small	0	3	0	3	0	1	1	3	0

7. 18 **9.** 07 **11.** 17 days **13.** 16
15.

6	1	8
7	5	3
2	9	4

17. 25 pitches (The visiting team's pitcher retires 24 consecutive batters through the first eight innings, using only one pitch per batter. His team does not score either. Going into the bottom of the ninth inning tied 0–0, the first batter for the home team hits his first pitch for a home run. The pitcher threw 25 pitches and loses the game by a score of 1–0.)
19. For three weighings, first balance four against four. Of the lighter four, balance two against the other two. Finally, of the lighter two, balance them one against the other. To find the bad coin in two weighings, divide the eight coins into groups of 3, 3, 2. Weigh the groups of three against each other on the scale. If the groups weigh the same, the fake is in the two left out and can be found in one additional weighing. If the two groups of three do not weigh the same, pick the lighter group. Choose any two of the coins and weigh them. If one of these is lighter, it is the fake; if they weigh the same, then the third coin is the fake. **21.** 28

23. Here is one possible solution.

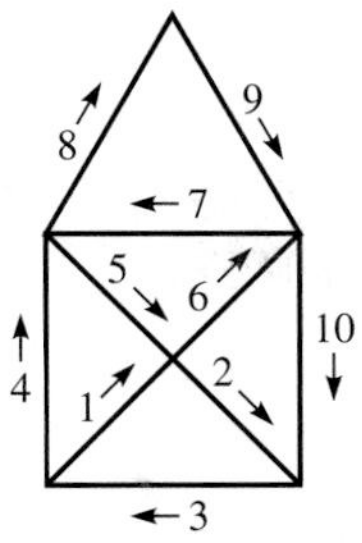

25. Dan (36) is married to Jessica (29); James (30) is married to Cathy (31).
27. None, since there is no dirt in a hole. **29.** One solution is $1 + 2 + 3 + 4 + 5 + 6 + 7 + 8 \times 9 = 100$. **31.** The correct problem follows. $\begin{array}{r} 402 \\ \times \;\; 39 \\ \hline 15{,}678 \end{array}$ **33.** 35 **35.** $7\frac{1}{2}$ min (Boil them all at the same time.)

Appendix C Synthetic Division

(pages 1009–1010)

1. C **3.** $x - 5$ **5.** $4m - 1$ **7.** $2a + 4 + \frac{5}{a + 2}$
9. $p - 4 + \frac{9}{p + 1}$ **11.** $4a^2 + a + 3$
13. $x^4 + 2x^3 + 2x^2 + 7x + 10 + \frac{18}{x - 2}$
15. $-4r^5 - 7r^4 - 10r^3 - 5r^2 - 11r - 8 + \frac{-8}{r - 1}$
17. $-3y^4 + 8y^3 - 21y^2 + 36y - 72 + \frac{143}{y + 2}$
19. $y^2 + y + 1 + \frac{2}{y - 1}$ **21.** 7 **23.** -2 **25.** 0 **27.** yes
29. no **31.** no **33.** Since the variables are not present, a missing term will not be noticed in synthetic division, so the quotient will be wrong if placeholders are not inserted.

SOLUTIONS TO SELECTED EXERCISES

Chapter 1 The Real Number System

Section 1.1 (pages 35–36)

33. $\frac{1}{4} \cdot \frac{2}{3} + \frac{2}{5} \cdot \frac{11}{3}$

$= \frac{1}{6} + \frac{22}{15}$ Multiply; $\frac{2}{12} = \frac{1}{6}$

$= \frac{5}{30} + \frac{44}{30}$ LCD = 30

$= \frac{49}{30}$, or $1\frac{19}{30}$ Add.

45. $\left(\frac{3}{2}\right)^2\left[\left(11 + \frac{1}{3}\right) - 6\right]$

$= \frac{9}{4}\left[\left(11 + \frac{1}{3}\right) - 6\right]$ Apply the exponent.

$= \frac{9}{4}\left[\left(\frac{33}{3} + \frac{1}{3}\right) - 6\right]$ LCD = 3

$= \frac{9}{4}\left(\frac{34}{3} - 6\right)$ Add inside parentheses.

$= \frac{9}{4}\left(\frac{34}{3} - \frac{18}{3}\right)$ LCD = 3

$= \frac{9}{4}\left(\frac{16}{3}\right)$ Subtract inside parentheses.

$= \frac{144}{12}$ Multiply.

$= 12$ Divide.

59. $\frac{9(7 - 1) - 8 \cdot 2}{4(6 - 1)} > 3$

$\frac{9(6) - 8 \cdot 2}{4(5)} > 3$ Work inside parentheses.

$\frac{54 - 16}{20} > 3$ Multiply.

$\frac{38}{20} > 3$ Subtract.

$1\frac{9}{10} > 3$ Write as a mixed number.

Since $1\frac{9}{10} < 3$, the statement is *false*.

Section 1.2 (pages 43–46)

29. **(a)** $\frac{3x + y^2}{2x + 3y}$

$= \frac{3(2) + 1^2}{2(2) + 3(1)}$ Let $x = 2$ and $y = 1$.

$= \frac{3(2) + 1}{4 + 3}$ Work separately above and below the fraction bar.

$= \frac{6 + 1}{7}$

$= \frac{7}{7}$, or 1

(b) $\frac{3x + y^2}{2x + 3y}$

$= \frac{3(1) + 5^2}{2(1) + 3(5)}$ Let $x = 1$ and $y = 5$.

$= \frac{3 + 25}{2 + 15}$

$= \frac{28}{17}$

31. **(a)** $0.841x^2 + 0.32y^2$

$= 0.841 \cdot 2^2 + 0.32 \cdot 1^2$ Let $x = 2$, $y = 1$.

$= 0.841 \cdot 4 + 0.32 \cdot 1$ Apply the exponents.

$= 3.364 + 0.32$ Multiply.

$= 3.684$ Add.

(b) $0.841x^2 + 0.32y^2$

$= 0.841 \cdot 1^2 + 0.32 \cdot 5^2$ Let $x = 1$, $y = 5$.

$= 0.841 \cdot 1 + 0.32 \cdot 25$

$= 0.841 + 8$

$= 8.841$

55. $\frac{z + 4}{2 - z} = \frac{13}{5}$

$\frac{\frac{1}{3} + 4}{2 - \frac{1}{3}} \stackrel{?}{=} \frac{13}{5}$ Let $z = \frac{1}{3}$.

$\frac{\frac{1}{3} + \frac{12}{3}}{\frac{6}{3} - \frac{1}{3}} \stackrel{?}{=} \frac{13}{5}$

$\frac{\frac{13}{3}}{\frac{5}{3}} \stackrel{?}{=} \frac{13}{5}$

$\frac{13}{3} \cdot \frac{3}{5} \stackrel{?}{=} \frac{13}{5}$

$\frac{13}{5} = \frac{13}{5}$ True

The true result shows that $\frac{1}{3}$ is a solution of the equation.

Section 1.3 (pages 55–56)

25. In order to compare these two numbers, write them with a common denominator.

$-\frac{2}{3} = -\frac{8}{12}$ and $-\frac{1}{4} = -\frac{3}{12}$

Since

$\frac{8}{12} > \frac{3}{12}$,

$-\frac{2}{3}$ is farther to the left of 0 on the number line than $-\frac{1}{4}$, so $-\frac{2}{3}$ is the lesser number.

51. A decrease is represented by a negative number, and the amount of decrease is represented by its absolute value. There are three negative numbers in the table, -0.8, -1.7, and -2.2. Of these, -2.2, has the greatest absolute value, 2.2, so electronic components and accessories for 2003 to 2004 represents the greatest decrease.

Section 1.4 (pages 63–66)

35. $\left(-\frac{1}{2} + 0.25\right) - \left(-\frac{3}{4} + 0.75\right)$

$= \left(-\frac{1}{2} + \frac{1}{4}\right) - \left(-\frac{3}{4} + \frac{3}{4}\right)$

$= \left(-\frac{2}{4} + \frac{1}{4}\right) - 0$

$= -\frac{1}{4}$, or -0.25

61. "$\frac{2}{7}$ more than the sum of $\frac{5}{7}$ and $-\frac{9}{7}$" is

$\left[\frac{5}{7} + \left(-\frac{9}{7}\right)\right] + \frac{2}{7}$

$= -\frac{4}{7} + \frac{2}{7}$

$= -\frac{2}{7}$.

Section 1.5 (pages 71–76)

39. $\left(-\frac{3}{8} - \frac{2}{3}\right) - \left(-\frac{9}{8} - 3\right)$

$= \left[-\frac{3}{8} + \left(-\frac{2}{3}\right)\right] - \left[-\frac{9}{8} + (-3)\right]$

$= \left[-\frac{9}{24} + \left(-\frac{16}{24}\right)\right]$

$- \left[-\frac{9}{8} + \left(-\frac{24}{8}\right)\right]$ Get a common denominator for each pair of fractions.

$= -\frac{25}{24} - \left(-\frac{33}{8}\right)$ Add inside the brackets.

$= -\frac{25}{24} - \left(-\frac{99}{24}\right)$ Get a common denominator.

$= -\frac{25}{24} + \frac{99}{24}$

$= \frac{74}{24}$

$= \frac{37}{12}$, or $3\frac{1}{12}$

41. $[-12.25 - (8.34 + 3.57)] - 17.88$
$= [-12.25 - 11.91] - 17.88$
$= [-12.25 + (-11.91)] - 17.88$
$= -24.16 - 17.88$
$= -24.16 + (-17.88)$
$= -42.04$

67. Sum of checks:
\$35.84 + \$26.14 + \$3.12
= \$61.98 + \$3.12
= \$65.10

Sum of deposits:
\$85.00 + \$120.76 = \$205.76

Final balance:
= Beginning balance − checks + deposits
= \$904.89 − \$65.10 + \$205.76
= \$839.79 + \$205.76
= \$1045.55

Her account balance at the end of August was \$1045.55.

Section 1.6 (pages 85–88)

43. $\dfrac{4(2^3 - 5) - 5(-3^3 + 21)}{3[6 - (-2)]}$
$= \dfrac{4(8 - 5) - 5(-27 + 21)}{3[6 - (-2)]}$
$= \dfrac{4(3) - 5(-6)}{3[8]}$
$= \dfrac{12 + 30}{24}$
$= \dfrac{42}{24}$
$= \dfrac{7}{4}$, or $1\frac{3}{4}$

49. $\left(\frac{5}{6}x + \frac{3}{2}y\right)\left(-\frac{1}{3}a\right)$
$= \left[\frac{5}{6}(6) + \frac{3}{2}(-4)\right]\left[-\frac{1}{3}(3)\right]$ Let $x = 6$, $y = -4$, and $a = 3$.
$= [5 + (-6)](-1)$
$= (-1)(-1)$
$= 1$

69. "The product of $-\frac{2}{3}$ and $-\frac{1}{5}$, divided by $\frac{1}{7}$" is
$\dfrac{-\frac{2}{3}\left(-\frac{1}{5}\right)}{\frac{1}{7}}$
$= \dfrac{\frac{2}{15}}{\frac{1}{7}}$
$= \frac{2}{15} \cdot \frac{7}{1}$
$= \frac{14}{15}$.

Section 1.7 (pages 97–100)

59. $-3(8x + 3y + 4z)$
$= -3(8x) + (-3)(3y) + (-3)(4z)$ Distributive property
$= (-3 \cdot 8)x + (-3 \cdot 3)y + (-3 \cdot 4)z$ Associative property
$= -24x - 9y - 12z$ Multiply.

65. $-(-3q + 5r - 8s)$
$= -1(-3q + 5r - 8s)$
$= 3q - 5r + 8s$

Section 1.8 (pages 105–106)

39. $-\frac{4}{3} + 2t + \frac{1}{3}t - 8 - \frac{8}{3}t$
$= \left(2t + \frac{1}{3}t - \frac{8}{3}t\right) + \left(-\frac{4}{3} - 8\right)$
$= \left(2 + \frac{1}{3} - \frac{8}{3}\right)t + \left(-\frac{4}{3} - 8\right)$
$= \left(\frac{6}{3} + \frac{1}{3} - \frac{8}{3}\right)t + \left(-\frac{4}{3} - \frac{24}{3}\right)$
$= -\frac{1}{3}t - \frac{28}{3}$

41. $-5.3r + 4.9 - (2r + 0.7) + 3.2r$
$= -5.3r + 4.9 - 2r - 0.7 + 3.2r$
$= (-5.3r - 2r + 3.2r) + (4.9 - 0.7)$
$= (-5.3 - 2 + 3.2)r + (4.9 - 0.7)$
$= -4.1r + 4.2$

Chapter 2 Equations, Inequalities, and Applications

Section 2.1 (pages 129–130)

27. $10x + 4 = 9x$
$10x + 4 - 9x = 9x - 9x$ Subtract $9x$.
$1x + 4 = 0$
$x + 4 - 4 = 0 - 4$ Subtract 4.
$x = -4$

Check
$10(-4) + 4 \stackrel{?}{=} 9(-4)$ Let $x = -4$.
$-36 = -36$ True
Solution set: $\{-4\}$

47. $\frac{5}{7}x + \frac{1}{3} = \frac{2}{5} - \frac{2}{7}x + \frac{2}{5}$
$\frac{5}{7}x + \frac{1}{3} = \frac{4}{5} - \frac{2}{7}x$
$\frac{5}{7}x + \frac{1}{3} + \frac{2}{7}x = \frac{4}{5} - \frac{2}{7}x + \frac{2}{7}x$ Add $\frac{2}{7}x$.
$\frac{7}{7}x + \frac{1}{3} = \frac{4}{5}$
$1x + \frac{1}{3} - \frac{1}{3} = \frac{4}{5} - \frac{1}{3}$ Subtract $\frac{1}{3}$.
$x = \frac{12}{15} - \frac{5}{15}$ LCD = 15
$x = \frac{7}{15}$

Check
$\frac{5}{7}\left(\frac{7}{15}\right) + \frac{1}{3} \stackrel{?}{=} \frac{2}{5} - \frac{2}{7}\left(\frac{7}{15}\right) + \frac{2}{5}$ Let $x = \frac{7}{15}$.
$\frac{1}{3} + \frac{1}{3} \stackrel{?}{=} \frac{4}{5} - \frac{2}{15}$
$\frac{2}{3} \stackrel{?}{=} \frac{12}{15} - \frac{2}{15}$
$\frac{2}{3} = \frac{10}{15}$, or $\frac{2}{3}$ True

Solution set: $\left\{\frac{7}{15}\right\}$

55. $10(-2x + 1) = -19(x + 1)$
$-20x + 10 = -19x - 19$ Distributive property
$-20x + 10 + 19x = -19x - 19 + 19x$ Add $19x$.
$-x + 10 = -19$ Combine like terms.
$-x + 10 - 10 = -19 - 10$ Subtract 10.
$-x = -29$
$x = 29$

A check confirms this solution.
Solution set: $\{29\}$

Section 2.2 (pages 135–136)

53. $\frac{2}{5}x - \frac{3}{10}x = 2$
$\frac{4}{10}x - \frac{3}{10}x = 2$ LCD = 10
$\frac{1}{10}x = 2$ Combine like terms.
$10 \cdot \frac{1}{10}x = 10 \cdot 2$ Multiply by 10.
$x = 20$

Check
$\frac{2}{5}(20) - \frac{3}{10}(20) \stackrel{?}{=} 2$ Let $x = 20$.
$8 - 6 \stackrel{?}{=} 2$
$2 = 2$ True
Solution set: $\{20\}$

61. $0.9w - 0.5w + 0.1w = -3$
$0.5w = -3$ Combine like terms.
$\frac{0.5w}{0.5} = \frac{-3}{0.5}$ Divide by 0.5.
$w = -6$

A check confirms this solution.
Solution set: $\{-6\}$

SOLUTIONS

Section 2.3 (pages 145–148)

47. $0.02(5000) + 0.03x = 0.025(5000 + x)$

Multiply both sides by 1000, *not* 100.

$$1000[0.02(5000) + 0.03x] = 1000[0.025(5000 + x)]$$

$$20(5000) + 30x = 25(5000 + x)$$

$$100{,}000 + 30x = 125{,}000 + 25x \quad \text{Distributive property}$$

$$5x + 100{,}000 = 125{,}000 \quad \text{Subtract } 25x.$$

$$5x = 25{,}000 \quad \text{Subtract 100,000.}$$

$$x = 5000 \quad \text{Divide by 5.}$$

Check Substituting 5000 for x in the original equation results in a true statement, $250 = 250$.

Solution set: $\{5000\}$

51.

$$-(6k - 5) - (-5k + 8) = -3$$

$$-1(6k - 5) - 1(-5k + 8) = -3$$

$$-6k + 5 + 5k - 8 = -3$$

$$-k - 3 = -3$$

$$-k = 0$$

$$k = 0$$

Check Substituting 0 for k in the original equation results in a true statement, $-3 = -3$.

Solution set: $\{0\}$

Section 2.4 (pages 161–166)

13. ***Step 1***

Read the problem again.

Step 2

Let $x =$ the unknown number. Then $3x$ is three times the number, $x + 7$ is 7 more than the number, and the sum is

$$3x + (x + 7).$$

$2x$ is twice the number, and

$$-11 - 2x$$

is the difference between -11 and twice the number.

Step 3

$$3x + (x + 7) = -11 - 2x$$

Step 4

$$4x + 7 = -11 - 2x$$

$$6x + 7 = -11 \quad \text{Add } 2x.$$

$$6x = -18 \quad \text{Subtract 7.}$$

$$x = -3 \quad \text{Divide by 6.}$$

Step 5

The number is -3.

Step 6

Check that -3 is the correct answer by substituting this result into the words of the original problem. The sum of three times this number and 7 more than the number is

$$3(-3) + (-3 + 7) = -5,$$

and the difference between -11 and twice this number is

$$-11 - 2(-3) = -5.$$

The values are equal, so the number -3 is the correct answer.

37. ***Step 1***

Read the problem again.

Step 2

Let $x =$ the measures of angles A and B.
$x + 60 =$ the measure of angle C.

Step 3

The sum of the measures of the angles of any triangle is 180°, so

$$x + x + (x + 60) = 180.$$

Step 4

$$3x + 60 = 180$$

$$3x = 120 \quad \text{Subtract 60.}$$

$$x = 40 \quad \text{Divide by 3.}$$

Step 5

Angles A and B have measures of 40°, and angle C has a measure of $40 + 60 = 100°$.

Step 6

The answer checks since

$$40 + 40 + 100 = 180.$$

Section 2.5 (pages 173–178)

51. Let L represent the length of the box. The girth is $4 \cdot 18 = 72$ in. Since the length plus the girth is 108, we have

$$L + 72 = 108$$

$$L = 36 \text{ in.} \quad \text{Subtract 72.}$$

The maximum volume of the box is

$$V = LWH$$

$$V = 36(18)(18) \quad \text{Substitute.}$$

$$V = 11{,}664 \text{ in.}^3.$$

55. The two angles are complementary, so the sum of their measures is 90°.

$$(8x - 1) + 5x = 90$$

$$13x - 1 = 90 \quad \text{Combine like terms.}$$

$$13x = 91 \quad \text{Add 1.}$$

$$x = 7 \quad \text{Divide by 13.}$$

Since $x = 7$, we have

$$8x - 1 = 8(7) - 1 = 56 - 1 = 55$$

and

$$5x = 5(7) = 35.$$

The angles measure 55° and 35°.

81.

$$M = C(1 + r) \text{ for } r$$

$$M = C + Cr \quad \text{Distributive property}$$

$$M - C = Cr \quad \text{Subtract } C.$$

$$\frac{M - C}{C} = \frac{Cr}{C} \quad \text{Divide by } C.$$

$$\frac{M - C}{C} = r, \quad \text{or} \quad r = \frac{M - C}{C}$$

Section 2.6 (pages 185–188)

43. Let $x =$ the number of cups of cleaner. Set up a proportion with one ratio involving the number of cups of cleaner and the other involving the number of gallons of water.

$$\frac{x \text{ cups}}{\frac{1}{4} \text{ cup}} = \frac{10\frac{1}{2} \text{ gallons}}{1 \text{ gallon}}$$

$$x \cdot 1 = \frac{1}{4}\left(10\frac{1}{2}\right) \quad \text{Cross products}$$

$$x = \frac{1}{4}\left(\frac{21}{2}\right)$$

$$x = \frac{21}{8}, \quad \text{or} \quad 2\frac{5}{8}$$

The amount of cleaner needed is $2\frac{5}{8}$ cups.

63. The increase in value was

$$\$6200 - \$625 = \$5575.$$

We can state the problem as follows:

"What percent of \$625 is \$5575?"

Substitute $a = 5575$ and $b = 625$ into the percent proportion; then find p.

$$\frac{a}{b} = \frac{p}{100}$$

$$\frac{5575}{625} = \frac{p}{100} \quad \text{Substitute.}$$

$$625p = 100(5575) \quad \text{Cross products}$$

$$p = \frac{100(5575)}{625} \quad \text{Divide by 625.}$$

$$p = 892 \quad \text{Use a calculator.}$$

The percent increase in the value of the coin was 892%.

Section 2.7 (pages 201–204)

49.

$$\frac{2}{3}(p + 3) > \frac{5}{6}(p - 4)$$

$$6\left(\frac{2}{3}\right)(p + 3) > 6\left(\frac{5}{6}\right)(p - 4) \quad \text{Multiply by 6, the LCD.}$$

$$4(p + 3) > 5(p - 4)$$

$$4p + 12 > 5p - 20 \quad \text{Distributive property}$$

$$-p + 12 > -20 \quad \text{Subtract } 5p.$$

$$-p > -32 \quad \text{Subtract 12.}$$

$$\frac{-p}{-1} < \frac{-32}{-1} \quad \text{Divide by } -1\text{; reverse the direction of the symbol.}$$

$$p < 32$$

Solution set: $(-\infty, 32)$

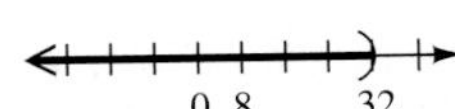

61.
$$4 < -9x + 5 \le 8$$
$$4 - 5 < -9x + 5 - 5 \le 8 - 5$$
$$-1 < -9x \le 3$$
$$\frac{-1}{-9} > x \ge \frac{3}{-9} \quad \text{Divide by } -9\text{; } \textbf{reverse the direction of the symbols.}$$
$$\frac{1}{9} > x \ge -\frac{1}{3}$$
$$-\frac{1}{3} \le x < \frac{1}{9} \quad \text{Rewrite in the order on the number line.}$$

Solution set: $\left[-\frac{1}{3}, \frac{1}{9}\right)$

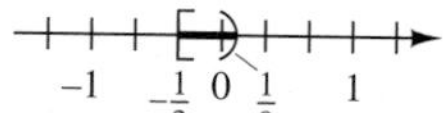

67. The Celsius temperature C must give a Fahrenheit temperature F that has *never exceeded* 104°, which translates as *less than or equal to* 104°.

$$F \le 104$$
$$\frac{9}{5}C + 32 \le 104$$
$$\frac{9}{5}C \le 72 \quad \text{Subtract 32.}$$
$$\frac{5}{9}\left(\frac{9}{5}C\right) \le \frac{5}{9}(72) \quad \text{Multiply by } \frac{5}{9}.$$
$$C \le 40$$

The temperature of Providence, Rhode Island, has never exceeded 40° Celsius.

Chapter 3 Graphs of Linear Equations and Inequalities; Functions

Section 3.1 (pages 231–236)

23. Is $(5, -6)$ a solution of the equation $x = -6$?

Since y does not appear in the equation, we just substitute 5 for x.

$$x = -6$$
$$5 \stackrel{?}{=} -6 \quad \text{Let } x = 5.$$

The result is false, so $(5, -6)$ is not a solution of the equation $x = -6$.

43. The given equation $x - 8 = 0$ may be written as $x = 8$. For any value of y, the value of x will always be 8. The completed table of values follows.

x	y
8	8
8	3
8	0

Section 3.2 (pages 245–250)

29. $-3y = 15$
$$y = \frac{15}{-3}, \text{ or } -5 \quad \text{Divide by } -3.$$

For any value of x, the value of y is -5. Three ordered pairs are $(-2, -5)$, $(0, -5)$, and $(1, -5)$. Plot these points and draw a line through them. The graph, shown at the top of the next column, is a horizontal line.

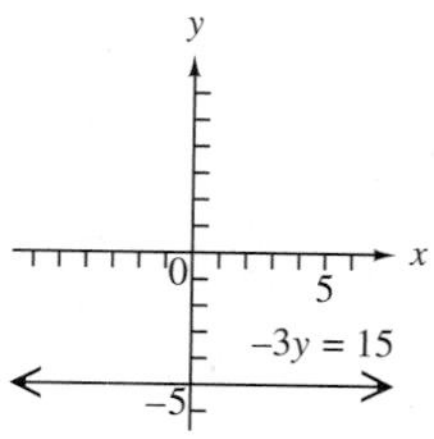

Section 3.3 (pages 261–266)

27. Use the slope formula with

$$\left(-\frac{7}{5}, \frac{3}{10}\right) = (x_1, y_1)$$

and

$$\left(\frac{1}{5}, -\frac{1}{2}\right) = (x_2, y_2).$$

$$\text{slope } m = \frac{\text{change in } y}{\text{change in } x}$$
$$= \frac{y_2 - y_1}{x_2 - x_1}$$
$$= \frac{-\frac{1}{2} - \frac{3}{10}}{\frac{1}{5} - \left(-\frac{7}{5}\right)} \quad \text{Substitute.}$$
$$= \frac{-\frac{5}{10} - \frac{3}{10}}{\frac{1}{5} + \frac{7}{5}} \quad \text{Get a common denominator in the numerator; simplify the denominator.}$$
$$= \frac{-\frac{8}{10}}{\frac{8}{5}} \quad \text{Subtract in the numerator; add in the denominator.}$$
$$= -\frac{8}{10} \div \frac{8}{5} \quad \frac{a}{b} = a \div b$$
$$= -\frac{8}{10} \cdot \frac{5}{8} \quad \text{Multiply by the reciprocal.}$$
$$= -\frac{1}{2} \quad \text{Multiply; lowest terms}$$

53. The slope (or grade) of the hill is the ratio of the rise to the run, or the ratio of the vertical change to the horizontal change. Since the rise is 32 and the run is 108, the slope is

$$\frac{32}{108} = \frac{8 \cdot 4}{27 \cdot 4} = \frac{8}{27}. \quad \text{Lowest terms}$$

63. Use the points $(0, 20)$ and $(4, 4)$ from the graph.

average rate of change
$$= \frac{\text{change in } y}{\text{change in } x} = \frac{4 - 20}{4 - 0} = \frac{-16}{4} = -4$$

The average rate of change is $-\$4000$ per year; that is, the value of the machine is decreasing \$4000 each year during these years.

Section 3.4 (pages 277–282)

3. The rise is 3 and the run is 1, so the slope m is given by

$$m = \frac{\text{rise}}{\text{run}} = \frac{3}{1} = 3.$$

The y-intercept is $(0, -3)$, so $b = -3$. The equation of the line, written in slope-intercept form $y = mx + b$, is

$$y = 3x - 3. \quad m = 3, b = -3$$

45. $\left(\frac{1}{2}, \frac{3}{2}\right), \left(-\frac{1}{4}, \frac{5}{4}\right)$

First, find the slope of the line.

$$m = \frac{\frac{5}{4} - \frac{3}{2}}{-\frac{1}{4} - \frac{1}{2}} = \frac{\frac{5}{4} - \frac{6}{4}}{-\frac{1}{4} - \frac{2}{4}} = \frac{-\frac{1}{4}}{-\frac{3}{4}}$$
$$= -\frac{1}{4} \div \left(-\frac{3}{4}\right) = -\frac{1}{4}\left(-\frac{4}{3}\right) = \frac{1}{3}$$

Now use the point $\left(\frac{1}{2}, \frac{3}{2}\right)$ for (x_1, y_1) and $m = \frac{1}{3}$ in the point-slope form.

$$y - y_1 = m(x - x_1)$$
$$y - \frac{3}{2} = \frac{1}{3}\left(x - \frac{1}{2}\right) \quad \text{Substitute.}$$
$$y - \frac{3}{2} = \frac{1}{3}x - \frac{1}{6} \quad \text{Distributive property}$$
$$y = \frac{1}{3}x - \frac{1}{6} + \frac{9}{6} \quad \text{Add } \tfrac{3}{2} = \tfrac{9}{6}.$$
$$y = \frac{1}{3}x + \frac{4}{3} \quad \text{Combine like terms; } \tfrac{8}{6} = \tfrac{4}{3}$$

Section 3.6 (pages 301–306)

33. $xy = 1$

Divide both sides of the equation by x to rewrite $xy = 1$ as $y = \frac{1}{x}$. Note that x can never equal 0, or the denominator would equal 0. Thus, the domain is

$$(-\infty, 0) \cup (0, \infty).$$

Each nonzero x-value gives exactly one y-value. Therefore, $xy = 1$ defines y as a function of x.

39. Refer to the graph on page 303.

(a) The independent variable is t, the number of hours, and the possible values are in the set $[0, 100]$. The dependent variable is g, the number of gallons, and the possible values are in the set $[0, 3000]$.

(b) The graph rises for the first 25 hr, so the water level increases for 25 hr. The graph falls for $t = 50$ to $t = 75$, so the water level decreases for 25 hr.

(c) There are 2000 gal in the pool when $t = 90$.

(d) $g(0)$ is the number of gallons in the pool at time $t = 0$. Here, $g(0) = 0$, which means the pool is empty at time 0.

53. $g(x) = -x^2 + 4x + 1$
$$g\left(\frac{p}{3}\right) = -\left(\frac{p}{3}\right)^2 + 4\left(\frac{p}{3}\right) + 1$$
$$g\left(\frac{p}{3}\right) = -\frac{p^2}{9} + \frac{4p}{3} + 1$$

67. (a) Solve the given equation for y.

$$4x - 3y = 8$$
$$-3y = -4x + 8$$
$$y = \frac{4}{3}x - \frac{8}{3}$$

Since $y = f(x)$,

$$f(x) = \frac{4}{3}x - \frac{8}{3}.$$

(b) $f(3) = \frac{4}{3}(3) - \frac{8}{3}$

$$f(3) = \frac{12}{3} - \frac{8}{3}$$
$$f(3) = \frac{4}{3}$$

Chapter 4 Systems of Linear Equations and Inequalities

Section 4.1 (pages 331–334)

21. $2x - 3y = -6$

$y = -3x + 2$

To graph $2x - 3y = -6$, find the intercepts.

$$2x - 3(0) = -6 \quad \text{Let } y = 0.$$
$$2x = -6$$
$$x = -3$$

The x-intercept is $(-3, 0)$.

$$2(0) - 3y = -6 \quad \text{Let } x = 0.$$
$$-3y = -6$$
$$y = 2$$

The y-intercept is $(0, 2)$.

Plot the intercepts, $(-3, 0)$ and $(0, 2)$, and draw the line through them.

To graph the second line, start by plotting the y-intercept, $(0, 2)$. From this point, go 3 units down and 1 unit to the right (because the slope is -3) to reach the point $(1, -1)$. Draw the line through $(0, 2)$ and $(1, -1)$.

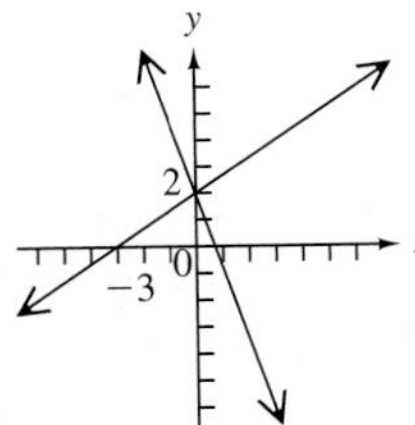

The lines intersect at their common y-intercept, $(0, 2)$.

Solution set: $\{(0, 2)\}$

25. $4x - 2y = 8 \quad (1)$

$2x = y + 4 \quad (2)$

Graph the line $4x - 2y = 8$ using its intercepts, $(2, 0)$ and $(0, -4)$.

Graph the equation $2x = y + 4$ using its intercepts, $(2, 0)$ and $(0, -4)$. Since both equations have the same intercepts, they are equations of the same line. See the graph at the top of the next column.

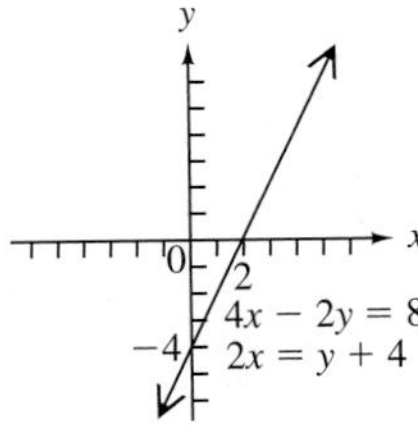

There are an infinite number of solutions, so we use set-builder notation to write the solution set. Rewrite equation (2) in standard form.

$$2x - y = 4$$

The solution set is

$$\{(x, y) \mid 2x - y = 4\}.$$

The given system consists of dependent equations.

Section 4.2 (pages 341–342)

15. $6x - 8y = 6 \quad (1)$

$2y = -2 + 3x \quad (2)$

Solve equation (2) for y.

$$2y = -2 + 3x$$
$$y = \frac{3x - 2}{2} \quad (3)$$

Substitute $\frac{3x - 2}{2}$ for y in equation (1).

$$6x - 8y = 6 \quad (1)$$
$$6x - 8\left(\frac{3x - 2}{2}\right) = 6$$
$$6x - 4(3x - 2) = 6$$
$$6x - 12x + 8 = 6$$
$$-6x + 8 = 6$$
$$-6x = -2$$
$$x = \frac{-2}{-6}, \text{ or } \frac{1}{3}$$

To find y, let $x = \frac{1}{3}$ in equation (3).

$$y = \frac{3x - 2}{2} = \frac{3(\frac{1}{3}) - 2}{2} = \frac{1 - 2}{2} = -\frac{1}{2}$$

Solution set: $\left\{\left(\frac{1}{3}, -\frac{1}{2}\right)\right\}$

19. $12x - 16y = 8 \quad (1)$

$3x = 4y + 2 \quad (2)$

Solve equation (2) for x.

$$3x = 4y + 2$$
$$x = \frac{4y + 2}{3} \quad (3)$$

Substitute $\frac{4y + 2}{3}$ for x in equation (1).

$$12x - 16y = 8 \quad (1)$$
$$12\left(\frac{4y + 2}{3}\right) - 16y = 8$$
$$4(4y + 2) - 16y = 8$$
$$16y + 8 - 16y = 8$$
$$8 = 8 \quad \text{True}$$

This true result means that every solution of one equation is also a solution of the other, so the system has an infinite number of solutions. To write the solution set, rewrite equation (2) in standard form.

$$3x - 4y = 2$$

Solution set: $\{(x, y) \mid 3x - 4y = 2\}$

25. $\frac{x}{5} + 2y = \frac{16}{5} \quad (1)$

$\frac{3x}{5} + \frac{y}{2} = -\frac{7}{5} \quad (2)$

Multiply each side of equation (1) by 5.

$$5\left(\frac{x}{5} + 2y\right) = 5\left(\frac{16}{5}\right)$$
$$x + 10y = 16 \quad (3)$$

Multiply each side of equation (2) by 10.

$$10\left(\frac{3x}{5} + \frac{y}{2}\right) = 10\left(-\frac{7}{5}\right)$$
$$6x + 5y = -14 \quad (4)$$

We now have the simplified system

$$x + 10y = 16 \quad (3)$$
$$6x + 5y = -14. \quad (4)$$

Solve equation (3) for x.

$$x = 16 - 10y \quad (5)$$

Substitute $16 - 10y$ for x in equation (4).

$$6x + 5y = -14 \quad (4)$$
$$6(16 - 10y) + 5y = -14$$
$$96 - 60y + 5y = -14$$
$$-55y = -110$$
$$y = 2$$

To find x, let $y = 2$ in equation (5).

$$x = 16 - 10y \quad (5)$$
$$x = 16 - 10(2)$$
$$x = -4$$

Solution set: $\{(-4, 2)\}$

Section 4.3 (pages 347–350)

25. $-x + 3y = 4 \quad (1)$

$-2x + 6y = 8 \quad (2)$

Multiply equation (1) by -2 and add the result to equation (2).

$$2x - 6y = -8 \quad (3)$$
$$-2x + 6y = 8 \quad (2)$$
$$0 = 0 \quad \text{True}$$

The true statement indicates that the equations of the original system are dependent. To obtain an equation in standard form, with $A > 0$ in $Ax + By = C$, multiply equation (1) by -1.

$$x - 3y = -4$$

Solution set: $\{(x, y) \mid x - 3y = -4\}$

SOLUTIONS

39.
$$6x + 3y = 0 \quad (1)$$
$$-18x - 9y = 0 \quad (2)$$

Multiply equation (1) by 3 and add the result to equation (2).

$$\begin{aligned} 18x + 9y &= 0 \quad (3) \\ -18x - 9y &= 0 \quad (2) \\ \hline 0 &= 0 \quad \text{True} \end{aligned}$$

This true result, $0 = 0$, means that the system has an infinite number of solutions. To obtain an equation for the solution set in which the coefficients have greatest common factor 1, divide both sides of equation (1) by 3 to obtain the equivalent equation

$$2x + y = 0.$$

Solution set: $\{(x, y) \mid 2x + y = 0\}$

Section 4.4 (pages 359–364)

33. ***Step 1***

Read the problem again.

Step 2

Assign variables.

Let $x =$ the average speed of the bicycle;
$y =$ the average speed of the car.

	r	t	d
Bicycle	x	7.5	$7.5x$
Car	y	7.5	$7.5y$

↑ Use $d = rt$.

Step 3

The total distance traveled by the bicycle and the car is 471 mi, so

$$7.5x + 7.5y = 471. \quad (1)$$

The car traveled 35.8 mi faster than the bicycle, so

$$y = x + 35.8. \quad (2)$$

We now have the system

$$7.5x + 7.5y = 471 \quad (1)$$
$$y = x + 35.8. \quad (2)$$

Step 4

Because equation (2) is already solved for y, the substitution method is a good choice. Substitute $x + 35.8$ for y in equation (1).

$$\begin{aligned} 7.5x + 7.5y &= 471 \quad (1) \\ 7.5x + 7.5(x + 35.8) &= 471 \quad \text{Let } y = x + 35.8. \\ 7.5x + 7.5x + 268.5 &= 471 \quad \text{Distributive property} \\ 15x + 268.5 &= 471 \quad \text{Combine like terms.} \\ 15x &= 202.5 \quad \text{Subtract 268.5.} \\ x &= 13.5 \quad \text{Divide by 15.} \end{aligned}$$

To find the value of y substitute 13.5 for x in equation (2).

$$\begin{aligned} y &= x + 35.8 \quad (2) \\ y &= 13.5 + 35.8 \quad \text{Let } x = 13.5. \\ y &= 49.3 \end{aligned}$$

Step 5

The average speed of the bicycle is 13.5 mph, and the average speed of the car is 49.3 mph.

Step 6

In 7.5 hr, the total distance traveled by the bicycle and the car is

$$7.5(13.5) + 7.5(49.3) = 471 \text{ mi.}$$

Since $49.3 - 13.5 = 35.8$, the car traveled 35.8 mph faster than the bicycle, as stated.

Section 4.5 (pages 369–370)

15. $x \le 2y + 3$

$x + y < 0$

Graph $x = 2y + 3$ as a solid line through (3, 0) and (5, 1). Using (0, 0) as a test point will result in the true statement $0 \le 3$, so shade the region containing the origin.

Graph $x + y = 0$ as a dashed line through (0, 0) and (1, −1). Using (1, 0) as a test point will result in the false statement $1 < 0$, so shade the region *not* containing (1, 0).

The solution set of the system is the intersection of the two shaded regions. It includes the portion of the line $x = 2y + 3$ that bounds the region but not the portion of the line $x + y = 0$.

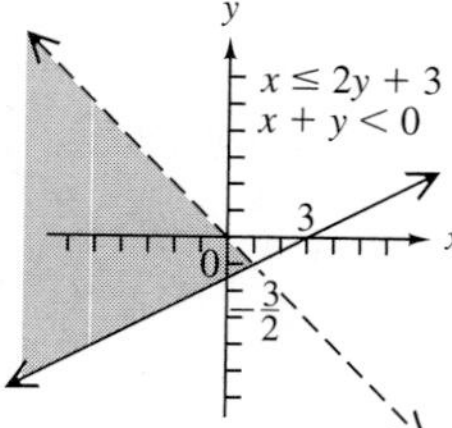

19. $3x - 2y \ge 6$

$x + y \le 4$

$x \ge 0$

$y \ge -4$

Graph $3x - 2y = 6$, $x + y = 4$, $x = 0$, and $y = -4$ as solid lines. All four inequalities are true for (2, −2). Shade the region bounded by the four lines, which contains the test point (2, −2). The solution set is the shaded region.

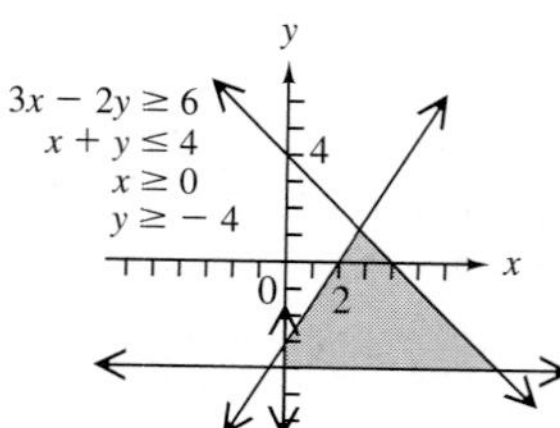

Chapter 5 Exponents and Polynomials

Section 5.1 (pages 389–392)

21. $\frac{1}{2}x^4 + \frac{1}{6}x^4$

$$\begin{aligned} &= \left(\frac{1}{2} + \frac{1}{6}\right)x^4 \quad \text{Distributive property} \\ &= \left(\frac{3}{6} + \frac{1}{6}\right)x^4 \quad \text{Write fractions with a common denominator.} \\ &= \frac{4}{6}x^4, \text{ or } \frac{2}{3}x^4 \quad \text{Add fractions; lowest terms} \end{aligned}$$

37. $0.8x^4 - 0.3x^4 - 0.5x^4 + 7$

$$\begin{aligned} &= (0.8 - 0.3 - 0.5)x^4 + 7 \\ &= 0x^4 + 7 \\ &= 7 \end{aligned}$$

Since 7 can be written as $7x^0$, the degree of the polynomial is 0. The simplified polynomial has one term, so it is a monomial.

57. Add.

$$\begin{array}{r} \frac{2}{3}x^2 + \frac{1}{5}x + \frac{1}{6} \\ \frac{1}{2}x^2 - \frac{1}{3}x + \frac{2}{3} \\ \hline \end{array}$$

Rewrite so that the fractions in each column have a common denominator; then add column by column.

$$\begin{array}{r} \frac{4}{6}x^2 + \frac{3}{15}x + \frac{1}{6} \\ \frac{3}{6}x^2 - \frac{5}{15}x + \frac{4}{6} \\ \hline \frac{7}{6}x^2 - \frac{2}{15}x + \frac{5}{6} \end{array}$$

75. Use the formula for the perimeter of a square, $P = 4s$, with $s = \frac{1}{2}x^2 + 2x$.

$$\begin{aligned} P &= 4s \\ P &= 4\left(\frac{1}{2}x^2 + 2x\right) \quad \text{Substitute.} \\ P &= 4\left(\frac{1}{2}x^2\right) + 4(2x) \quad \text{Distributive property} \\ P &= 2x^2 + 8x \quad \text{Multiply.} \end{aligned}$$

Section 5.2 (pages 399–400)

47. $(-5^2)^6$

$$\begin{aligned} &= (-1 \cdot 5^2)^6 \quad \text{Write the factor } -1. \\ &= (-1)^6 \cdot (5^2)^6 \quad \text{Power rule (b)} \\ &= 1 \cdot 5^{2 \cdot 6} \quad \text{Power rule (a)} \\ &= 5^{12} \end{aligned}$$

81. $\left(\frac{5a^2b^5}{c^6}\right)^3, \quad c \ne 0$

$$\begin{aligned} &= \frac{(5a^2b^5)^3}{(c^6)^3} \quad \text{Power rule (c)} \\ &= \frac{5^3(a^2)^3(b^5)^3}{(c^6)^3} \quad \text{Power rule (b)} \\ &= \frac{125a^6b^{15}}{c^{18}} \quad \text{Power rule (a)} \end{aligned}$$

83. $(-5m^3p^4q)^2(p^2q)^3$

$$\begin{aligned} &= (-1 \cdot 5m^3p^4q)^2(p^2q)^3 \quad \text{Write the factor } -1. \\ &= (-1)^2 \cdot 5^2 \cdot (m^3)^2 \cdot (p^4)^2 \cdot q^2 \cdot (p^2)^3 \cdot q^3 \quad \text{Power rule (b)} \\ &= 1 \cdot 25 \cdot m^6p^8q^2p^6q^3 \quad \text{Power rule (a)} \\ &= 25m^6p^{8+6}q^{2+3} \quad \text{Product rule} \\ &= 25m^6p^{14}q^5 \end{aligned}$$

SOLUTIONS

Section 5.3 (pages 405–406)

37. $\left(x - \frac{2}{3}\right)\left(x + \frac{1}{4}\right)$

F O I L

$= x(x) + x\left(\frac{1}{4}\right) + \left(-\frac{2}{3}\right)x + \left(-\frac{2}{3}\right)\frac{1}{4}$

$= x^2 + \frac{1}{4}x - \frac{2}{3}x - \frac{1}{6}$

$= x^2 + \left(\frac{3}{12}x - \frac{8}{12}x\right) - \frac{1}{6}$

$= x^2 - \frac{5}{12}x - \frac{1}{6}$

Section 5.4 (pages 411–412)

11. $(0.8t + 0.7s)^2$

$= (0.8t)^2 + 2(0.8t)(0.7s) + (0.7s)^2$

$(a + b)^2 = a^2 + 2ab + b^2$

$= 0.64t^2 + 1.12ts + 0.49s^2$

17. $-(4r - 2)^2$

First square the binomial.

$(4r - 2)^2$

$= (4r)^2 - 2(4r)(2) + 2^2$

$= 16r^2 - 16r + 4$

Now multiply by -1.

$-1(16r^2 - 16r + 4)$

$= -16r^2 + 16r - 4$

27. $(2x^2 - 5)(2x^2 + 5)$

$= (2x^2)^2 - 5^2$ $\quad (a+b)(a-b) = a^2 - b^2$

$= 4x^4 - 25$ $\quad (2x^2)^2 = 2^2x^4 = 4x^4$

55. Use the formula for the volume of a cube, $V = e^3$. Here, $V = (x + 2)^3$. If $x = 6$, then

$V = (6 + 2)^3 = 8^3 = 512.$

The volume of the cube is 512 cu. units.

Section 5.5 (pages 421–422)

15. $(-2)^0 - 2^0$

$= 1 - 1$ $\quad a^0 = 1$

$= 0$

31. $-2^{-1} + 3^{-2}$

$= -(2^{-1}) + 3^{-2}$ In -2^{-1}, the base is 2.

$= -\frac{1}{2^1} + \frac{1}{3^2}$ $\quad a^{-n} = \frac{1}{a^n}$

$= -\frac{1}{2} + \frac{1}{9}$ Apply the exponents.

$= -\frac{9}{18} + \frac{2}{18}$ Write fractions with a common denominator.

$= -\frac{7}{18}$ Add.

71. $\frac{(m^7n)^{-2}}{m^{-4}n^3}$

$= \frac{(m^7)^{-2}n^{-2}}{m^{-4}n^3}$ Power rule (b)

$= \frac{m^{7(-2)}n^{-2}}{m^{-4}n^3}$ Power rule (a)

$= \frac{m^{-14}n^{-2}}{m^{-4}n^3}$

$= m^{-14-(-4)}n^{-2-3}$ Quotient rule

$= m^{-10}n^{-5}$

$= \frac{1}{m^{10}n^5}$ $\quad a^{-n} = \frac{1}{a^n}$

Section 5.6 (pages 427–428)

19. $\frac{-3x^3 - 4x^4 + 2x}{-3x^2}$

$= \frac{-4x^4 - 3x^3 + 2x}{-3x^2}$ Write in descending powers.

$= \frac{-4x^4}{-3x^2} - \frac{3x^3}{-3x^2} + \frac{2x}{-3x^2}$ Divide each term by $-3x^2$.

$= \frac{4x^2}{3} + x - \frac{2}{3x}$ Quotient rule; be careful with signs.

Notice how the third term is written with x in the denominator, which is ***not*** the same as $-\frac{2}{3}x$ or $-\frac{2x}{3}$. In $-\frac{2}{3x}$, we are *dividing* by x; in $-\frac{2}{3}x$ and $-\frac{2x}{3}$, we are *multiplying* by x. Applying the quotient rule to the term $\frac{2x}{-3x^2}$ gives

$\frac{2x}{-3x^2} = \frac{2x^1}{-3x^2} = -\frac{2}{3}x^{1-2} = -\frac{2}{3}x^{-1}$

$= -\frac{2}{3}\left(\frac{1}{x}\right) = -\frac{2}{3x}.$

($\frac{4}{3}x^2$ is an acceptable form for the first term, $\frac{4x^2}{3}$. Why?)

29. Use the formula for the area of a rectangle, $A = LW$, with $A = 12x^2 - 4x + 2$ and $W = 2x$.

$A = LW$

$12x^2 - 4x + 2 = L(2x)$ Substitute for A and W.

$\frac{12x^2 - 4x + 2}{2x} = L$ Divide each side by $2x$.

$\frac{12x^2}{2x} - \frac{4x}{2x} + \frac{2}{2x} = L$ Divide each term by $2x$.

$6x - 2 + \frac{1}{x} = L$ $\quad \frac{2}{2x} = \frac{2}{2} \cdot \frac{1}{x} = 1 \cdot \frac{1}{x} = \frac{1}{x}$

$6x - 2 + \frac{1}{x}$ represents the length of the rectangle.

Section 5.7 (pages 433–434)

23. $(x^4 - x^2 - 2) \div (x^2 - 2)$

Use 0 as the coefficient of the missing x^3- and x-terms in the dividend and the missing x-term in the divisor.

$$\begin{array}{r} x^2 \qquad\qquad\quad + 1 \\ x^2 + \mathbf{0x} - 2\overline{)x^4 + \mathbf{0x^3} - x^2 + \mathbf{0x} - 2} \\ \underline{x^4 + 0x^3 - 2x^2} \qquad\qquad \\ x^2 + 0x - 2 \\ \underline{x^2 + 0x - 2} \\ 0 \end{array}$$

The remainder is 0. The answer is the quotient, $x^2 + 1$.

29. $\frac{2x^5 + x^4 + 11x^3 - 8x^2 - 13x + 7}{2x^2 + x - 1}$

$$\begin{array}{r} x^3 \qquad\qquad\qquad + 6x - 7 \\ 2x^2 + x - 1\overline{)2x^5 + x^4 + 11x^3 - 8x^2 - 13x + 7} \\ \underline{2x^5 + x^4 - x^3} \qquad\qquad\qquad\quad \\ 12x^3 - 8x^2 - 13x \quad\; \\ \underline{12x^3 + 6x^2 - 6x} \quad\; \\ -14x^2 - 7x + 7 \\ \underline{-14x^2 - 7x + 7} \\ 0 \end{array}$$

The remainder is 0. The answer is the quotient, $x^3 + 6x - 7$.

Section 5.8 (pages 439–440)

51. $\frac{4 \times 10^5}{8 \times 10^2}$

$= \frac{4}{8} \times \frac{10^5}{10^2}$

$= 0.5 \times 10^3$

Divide; quotient rule

$= (5 \times 10^{-1}) \times 10^3$

Write 0.5 in scientific notation.

$= 5 \times (10^{-1} \times 10^3)$

Associative property

$= 5 \times 10^2$

Product rule

$= 500$ Write without exponents.

55. $\frac{(1.65 \times 10^8)(5.24 \times 10^{-2})}{(6 \times 10^4)(2 \times 10^7)}$

$= \frac{1.65 \times 5.24}{6 \times 2} \times \frac{10^8 \times 10^{-2}}{10^4 \times 10^7}$

Associative and commutative properties

$= 0.7205 \times \frac{10^6}{10^{11}}$ Product rule

$= 0.7205 \times 10^{-5}$ Quotient rule

$= (7.205 \times 10^{-1}) \times 10^{-5}$

Write 0.7205 in scientific notation.

$= 7.205 \times (10^{-1} \times 10^{-5})$

Associative property

$= 7.205 \times 10^{-6}$ Product rule

$= 0.000007205$

Write without exponents.

Chapter 6 Factoring and Applications

Section 6.1 (pages 461–462)

63. $18r^2 + 12ry - 3xr - 2xy$

$= (18r^2 + 12ry) + (-3xr - 2xy)$
Group the terms.

$= 6r(3r + 2y) - x(3r + 2y)$
Factor each group.

$= (3r + 2y)(6r - x)$
Factor out the common factor, $3r + 2y$.

71. $y^2 + 3x + 3y + xy$

$= y^2 + 3y + xy + 3x$
Rearrange terms.

$= (y^2 + 3y) + (xy + 3x)$
Group the terms.

$= y(y + 3) + x(y + 3)$
Factor each group.

$= (y + 3)(y + x)$
Factor out the common factor, $y + 3$.

Section 6.2 (pages 467–468)

35. $a^2 - 8a - 48$

Find two integers whose product is -48 and whose sum is -8. Since c is negative, one integer must be positive and one must be negative.

Factors of -48	Sums of Factors
$-1, 48$	47
$1, -48$	-47
$-2, 24$	22
$2, -24$	-22
$-3, 16$	13
$3, -16$	-13
$-4, 12$	8
$4, -12$	-8 ←
$-6, 8$	2
$6, -8$	-2

Thus,

$a^2 - 8a - 48$ factors as $(a + 4)(a - 12)$.

51. $-2x^6 - 8x^5 + 42x^4$

First, factor out the negative common factor, $-2x^4$.

$$-2x^6 - 8x^5 + 42x^4$$
$$= -2x^4(x^2 + 4x - 21)$$

Now factor $x^2 + 4x - 21$.

Factors of -21	Sums of Factors
$1, -21$	-20
$-1, 21$	20
$3, -7$	-4
$-3, 7$	4 ←

Thus,

$x^2 + 4x - 21$ factors as $(x - 3)(x + 7)$.

The complete factored form is

$$-2x^6 - 8x^5 + 42x^4$$
$$= -2x^4(x - 3)(x + 7).$$

55. $m^3n - 10m^2n^2 + 24mn^3$

First, factor out the GCF, mn.

$$m^3n - 10m^2n^2 + 24mn^3$$
$$= mn(m^2 - 10mn + 24n^2)$$

The expressions $-6n$ and $-4n$ have a product of $24n^2$ and a sum of $-10n$. The complete factored form is

$$m^3n - 10m^2n^2 + 24mn^3$$
$$= mn(m - 6n)(m - 4n).$$

Section 6.3 (pages 471–472)

33. $-32z^5 + 20z^4 + 12z^3$

First factor out the negative common factor, $-4z^3$.

$$-32z^5 + 20z^4 + 12z^3$$
$$= -4z^3(8z^2 - 5z - 3)$$

To factor $8z^2 - 5z - 3$, find two integers whose product is $8(-3) = -24$ and whose sum is -5. These integers are -8 and 3. Now rewrite the given trinomial and factor it.

$-32z^5 + 20z^4 + 12z^3$

$= -4z^3(8z^2 - 5z - 3)$

$= -4z^3(8z^2 - 8z + 3z - 3)$
$-5z = -8z + 3z$

$= -4z^3[(8z^2 - 8z) + (3z - 3)]$
Group the terms.

$= -4z^3[8z(z - 1) + 3(z - 1)]$
Factor each group.

$= -4z^3[(z - 1)(8z + 3)]$
Factor out common factor, $z - 1$.

$= -4z^3(z - 1)(8z + 3)$

39. $5 - 6x + x^2$

Find two integers whose product is $5(1) = 5$ and whose sum is -6. The integers are -1 and -5.

$5 - 6x + x^2$

$= 5 - x - 5x + x^2$

$= (5 - x) + (-5x + x^2)$
Group the terms.

$= 1(5 - x) - x(5 - x)$
Factor each group.

$= (5 - x)(1 - x)$
Factor out the common factor, $5 - x$.

Section 6.4 (pages 477–478)

43. $6m^6n + 7m^5n^2 + 2m^4n^3$

Factor out the GCF, m^4n.

$$6m^6n + 7m^5n^2 + 2m^4n^3$$
$$= m^4n(6m^2 + 7mn + 2n^2)$$

Now factor $6m^2 + 7mn + 2n^2$ by trial and error. Possible factors of $6m^2$ are $6m$ and m or $3m$ and $2m$. Possible factors of $2n^2$ are $2n$ and n.

$(3m + 2n)(2m + n)$
$= 6m^2 + 7mn + 2n^2$ Correct

The complete factored form is

$$6m^6n + 7m^5n^2 + 2m^4n^3$$
$$= m^4n(3m + 2n)(2m + n).$$

Section 6.5 (pages 485–488)

13. $4m^2 - \dfrac{9}{25}$

Because $4m^2 = (2m)^2$ and $\dfrac{9}{25} = \left(\dfrac{3}{5}\right)^2$, $4m^2 - \dfrac{9}{25}$ is a difference of squares.

$$4m^2 - \frac{9}{25}$$
$$= (2m)^2 - \left(\frac{3}{5}\right)^2$$
$$= \left(2m + \frac{3}{5}\right)\left(2m - \frac{3}{5}\right)$$

39. $x^2 - 1.0x + 0.25$

The first and last terms are perfect squares, x^2 and $(-0.5)^2$. The trinomial is a perfect square, since the middle term is

$$2 \cdot x \cdot (-0.5) = -1.0x.$$

Therefore,

$$x^2 - 1.0x + 0.25$$
$$= (x)^2 - 2(x)(0.5) + (0.5)^2$$
$$= (x - 0.5)^2.$$

65. $27x^3 - 1$

Because $27x^3 = (3x)^3$ and $1 = 1^3$, the binomial is a difference of cubes. Use the pattern

$$a^3 - b^3 = (a - b)(a^2 + ab + b^2).$$

$$27x^3 - 1$$
$$= (3x)^3 - 1^3$$
$$= (3x - 1)[(3x)^2 + 3x(1) + 1^2]$$
$$= (3x - 1)(9x^2 + 3x + 1)$$

79. $x^3 + y^6$

Because $y^6 = (y^2)^3$, the binomial is a sum of cubes. Use the pattern

$$a^3 + b^3 = (a + b)(a^2 - ab + b^2).$$

$$x^3 + y^6$$
$$= x^3 + (y^2)^3$$
$$= (x + y^2)[x^2 - x(y^2) + (y^2)^2]$$
$$= (x + y^2)(x^2 - xy^2 + y^4)$$

Section 6.6 (pages 491–492)

21. $18m^3n + 3m^2n^2 - 6mn^3$

$= 3mn(6m^2 + mn - 2n^2)$
Factor out the GCF, $3mn$.

$= 3mn(3m + 2n)(2m - n)$
Factor the trinomial.

29. $x^3 + 3x^2 - 9x - 27$

Factor by grouping.

$$= (x^3 + 3x^2) + (-9x - 27)$$
$$= x^2(x + 3) - 9(x + 3)$$
$$= (x + 3)(x^2 - 9)$$

Since $x^2 - 9$ is a difference of two squares, $x^2 - 3^2$, factor it as $(x + 3)(x - 3)$.

$$= (x + 3)(x + 3)(x - 3)$$
$$= (x + 3)^2(x - 3)$$

SOLUTIONS

35. $x^4 - 625$

Because $x^4 = (x^2)^2$ and $625 = 25^2$, this binomial is a difference of squares.

$$\begin{aligned} &x^4 - 625 \\ &= (x^2)^2 - 25^2 \\ &= (x^2 + 25)(x^2 - 25) \end{aligned}$$

Because $x^2 - 25$ is also a difference of squares, $x^2 - 5^2$, factor it as $(x + 5)(x - 5)$.

$$= (x^2 + 25)(x + 5)(x - 5)$$

43. $1000z^3 + 512$

$$\begin{aligned} &= 8(125z^3 + 64) \quad \text{GCF} = 8 \\ &= 8[(5z)^3 + 4^3] \quad \text{Sum of cubes} \\ &= 8[5z + 4][(5z)^2 - (5z)(4) + 4^2] \\ &= 8(5z + 4)(25z^2 - 20z + 16) \end{aligned}$$

49. $14x^2 - 25xq - 25q^2$

Two integer factors whose product is $14(-25) = -350$ and whose sum is -25 are -35 and 10.

$$\begin{aligned} &14x^2 - 25xq - 25q^2 \\ &= 14x^2 - 35xq + 10xq - 25q^2 \\ &= (14x^2 - 35xq) + (10xq - 25q^2) \\ &= 7x(2x - 5q) + 5q(2x - 5q) \\ &= (2x - 5q)(7x + 5q) \end{aligned}$$

Section 6.7 (pages 499–500)

51. $9y^3 - 49y = 0$

To factor the polynomial, begin by factoring out the greatest common factor.

$$y(9y^2 - 49) = 0$$

Now factor $9y^2 - 49$ as the difference of two squares.

$$y(3y + 7)(3y - 7) = 0$$

Set each of the three factors equal to 0 and solve.

$$\begin{aligned} y = 0 \quad \text{or} \quad 3y + 7 &= 0 \quad \text{or} \quad 3y - 7 = 0 \\ 3y &= -7 \quad \text{or} \quad 3y = 7 \\ y &= -\frac{7}{3} \quad \text{or} \quad y = \frac{7}{3} \end{aligned}$$

Solution set: $\left\{-\frac{7}{3}, 0, \frac{7}{3}\right\}$

53. $(2r + 5)(3r^2 - 16r + 5) = 0$

$(2r + 5)(3r - 1)(r - 5) = 0$

Factor $3r^2 - 16r + 5$.

Set each of the three factors equal to 0, and solve the resulting equations.

$$\begin{aligned} 2r + 5 = 0 \quad &\text{or} \quad 3r - 1 = 0 \quad \text{or} \quad r - 5 = 0 \\ 2r = -5 \quad &\text{or} \quad 3r = 1 \quad \text{or} \quad r = 5 \\ r = -\frac{5}{2} \quad &\text{or} \quad r = \frac{1}{3} \end{aligned}$$

Solution set: $\left\{-\frac{5}{2}, \frac{1}{3}, 5\right\}$

Section 6.8 (pages 507–512)

13. Let $x =$ the length of a side of the square painting.

Then $x - 2 =$ the length of a side of the square mirror.

Since the formula for the area of a square is $A = s^2$, the area of the painting is x^2, and the area of the mirror is $(x - 2)^2$. The difference between their areas is 32, so

$$\begin{aligned} x^2 - (x - 2)^2 &= 32 \\ x^2 - (x^2 - 4x + 4) &= 32 \\ x^2 - x^2 + 4x - 4 &= 32 \\ 4x - 4 &= 32 \\ 4x &= 36 \\ x &= 9. \end{aligned}$$

The length of a side of the painting is 9 ft. The length of a side of the mirror is

$$9 - 2 = 7 \text{ ft}.$$

As a check,

$$9^2 - 7^2 = 81 - 49 = 32, \quad \text{as required.}$$

21. Let $x =$ the least even integer.

Then $x + 2 =$ the next even integer and $x + 4 =$ the third even integer.

$$\begin{aligned} x^2 + (x + 2)^2 &= (x + 4)^2 \\ x^2 + x^2 + 4x + 4 &= x^2 + 8x + 16 \quad \text{Square the binomials.} \\ 2x^2 + 4x + 4 &= x^2 + 8x + 16 \quad \text{Combine like terms.} \\ x^2 - 4x - 12 &= 0 \quad \text{Standard form} \\ (x + 2)(x - 6) &= 0 \quad \text{Factor.} \\ x + 2 = 0 \quad \text{or} \quad x - 6 &= 0 \quad \text{Zero-factor property} \\ x = -2 \quad \text{or} \quad x &= 6 \end{aligned}$$

If $x = -2$, then $x + 2 = 0$ and $x + 4 = 2$.

If $x = 6$, then $x + 2 = 8$ and $x + 4 = 10$.

The integers are -2, 0, and 2, or 6, 8, and 10.

Check

$(-2)^2 + 0^2 \stackrel{?}{=} 2^2$	$6^2 + 8^2 \stackrel{?}{=} 10^2$
$4 + 0 \stackrel{?}{=} 4$	$36 + 64 \stackrel{?}{=} 100$
$4 = 4$	$100 = 100$
True	True

Chapter 7 Rational Expressions and Functions

Section 7.1 (pages 535–538)

15. $f(x) = \dfrac{3x + 1}{2x^2 + x - 6}$

Set the denominator equal to zero and solve.

$$\begin{aligned} 2x^2 + x - 6 &= 0 \\ (x + 2)(2x - 3) &= 0 \quad \text{Factor.} \\ x + 2 = 0 \quad \text{or} \quad 2x - 3 &= 0 \quad \text{Zero-factor property} \\ x = -2 \quad \text{or} \quad x &= \frac{3}{2} \end{aligned}$$

The numbers -2 and $\frac{3}{2}$ are not in the domain of the function. In set notation, the domain is

$$\left\{x \,|\, x \neq -2, \frac{3}{2}\right\}.$$

45. $\dfrac{2c^2 + 2cd - 60d^2}{2c^2 - 12cd + 10d^2}$

$$\begin{aligned} &= \frac{2(c^2 + cd - 30d^2)}{2(c^2 - 6cd + 5d^2)} \quad \text{Factor out the GCF in the numerator and denominator.} \\ &= \frac{2(c + 6d)(c - 5d)}{2(c - d)(c - 5d)} \quad \text{Factor trinomials in the numerator and denominator.} \\ &= \frac{c + 6d}{c - d} \quad \text{Lowest terms} \end{aligned}$$

79. $\left(\dfrac{6k^2 - 13k - 5}{k^2 + 7k} \div \dfrac{2k - 5}{k^3 + 6k^2 - 7k}\right) \cdot \dfrac{k^2 - 5k + 6}{3k^2 - 8k - 3}$

Factor k from the denominator of the divisor; multiply by the reciprocal.

$$\begin{aligned} &= \left(\frac{6k^2 - 13k - 5}{k^2 + 7k} \cdot \frac{k(k^2 + 6k - 7)}{2k - 5}\right) \cdot \frac{k^2 - 5k + 6}{3k^2 - 8k - 3} \\ &= \left(\frac{(3k + 1)(2k - 5)}{k(k + 7)} \cdot \frac{k(k + 7)(k - 1)}{2k - 5}\right) \cdot \frac{(k - 2)(k - 3)}{(3k + 1)(k - 3)} \quad \text{Factor numerators and denominators.} \\ &= \frac{(3k + 1)(k - 1)}{1} \cdot \frac{(k - 2)(k - 3)}{(3k + 1)(k - 3)} \quad \text{Multiply inside the parentheses.} \\ &= (k - 1)(k - 2) \quad \text{Lowest terms} \end{aligned}$$

Section 7.2 (pages 547–550)

13. $\dfrac{a^3}{a^2 + ab + b^2} - \dfrac{b^3}{a^2 + ab + b^2}$

$$\begin{aligned} &= \frac{a^3 - b^3}{a^2 + ab + b^2} \quad \text{Subtract the numerators; keep the common denominator.} \\ &= \frac{(a - b)(a^2 + ab + b^2)}{a^2 + ab + b^2} \quad \text{Factor the difference of cubes in the numerator.} \\ &= a - b \quad \text{Lowest terms} \end{aligned}$$

49. $\dfrac{4x}{x - 1} - \dfrac{2}{x + 1} - \dfrac{4}{x^2 - 1}$

$x^2 - 1 = (x + 1)(x - 1)$, the LCD.

(continued)

SOLUTIONS

$$\frac{4x}{x-1}-\frac{2}{x+1}-\frac{4}{x^2-1}$$

$$=\frac{4x(x+1)}{(x-1)(x+1)}-\frac{2(x-1)}{(x+1)(x-1)}-\frac{4}{(x+1)(x-1)}$$ Fundamental property

$$=\frac{4x(x+1)-2(x-1)-4}{(x+1)(x-1)}$$ Subtract numerators.

$$=\frac{4x^2+4x-2x+2-4}{(x-1)(x+1)}$$ Distributive property

$$=\frac{4x^2+2x-2}{(x-1)(x+1)}$$ Combine like terms in the numerator.

$$=\frac{2(2x^2+x-1)}{(x-1)(x+1)}$$ Factor out the GCF in the numerator.

$$=\frac{2(2x-1)(x+1)}{(x-1)(x+1)}$$ Factor.

$$=\frac{2(2x-1)}{x-1}$$ Lowest terms

57. $\frac{4}{x+1}+\frac{1}{x^2-x+1}-\frac{12}{x^3+1}$

$x^3+1=(x+1)(x^2-x+1)$, the LCD.

$$\frac{4}{x+1}+\frac{1}{x^2-x+1}-\frac{12}{(x+1)(x^2-x+1)}$$

$$=\frac{4(x^2-x+1)}{(x+1)(x^2-x+1)}+\frac{1\cdot(x+1)}{(x^2-x+1)(x+1)}-\frac{12}{(x+1)(x^2-x+1)}$$ Fundamental property

$$=\frac{4(x^2-x+1)+(x+1)-12}{(x+1)(x^2-x+1)}$$ Add and subtract numerators.

$$=\frac{4x^2-4x+4+x+1-12}{(x+1)(x^2-x+1)}$$ Distributive property

$$=\frac{4x^2-3x-7}{(x+1)(x^2-x+1)}$$ Combine like terms in the numerator.

$$=\frac{(4x-7)(x+1)}{(x+1)(x^2-x+1)}$$ Factor the numerator.

$$=\frac{4x-7}{x^2-x+1}$$ Lowest terms

65. $\frac{5x}{x^2+xy-2y^2}-\frac{3x}{x^2+5xy-6y^2}$

Factor each denominator.

$$x^2+xy-2y^2=(x+2y)(x-y)$$
$$x^2+5xy-6y^2=(x+6y)(x-y)$$

The LCD is $(x+2y)(x-y)(x+6y)$.

$$\frac{5x}{(x+2y)(x-y)}-\frac{3x}{(x+6y)(x-y)}$$

$$=\frac{5x(x+6y)}{(x+2y)(x-y)(x+6y)}-\frac{3x(x+2y)}{(x+6y)(x-y)(x+2y)}$$

$$=\frac{5x(x+6y)-3x(x+2y)}{(x+6y)(x-y)(x+2y)}$$

$$=\frac{5x^2+30xy-3x^2-6xy}{(x+2y)(x-y)(x+6y)}$$

$$=\frac{2x^2+24xy}{(x+2y)(x-y)(x+6y)}$$

$$=\frac{2x(x+12y)}{(x+2y)(x-y)(x+6y)}$$

Section 7.3 (pages 555–556)

7. $\dfrac{\frac{4z^2x^4}{9}}{\frac{12x^2z^5}{15}}$

$$=\frac{4z^2x^4}{9}\div\frac{12x^2z^5}{15}$$ Write as a division problem.

$$=\frac{4z^2x^4}{9}\cdot\frac{15}{12x^2z^5}$$ Multiply by the reciprocal of the divisor.

$$=\frac{60z^2x^4}{108x^2z^5}$$ Multiply.

$$=\frac{5\cdot 12\cdot z^2\cdot x^2\cdot x^2}{9\cdot 12\cdot x^2\cdot z^2\cdot z^3}$$ Factor.

$$=\frac{5x^2}{9z^3}$$ Lowest terms

15. $\dfrac{\frac{x^2-16y^2}{xy}}{\frac{1}{y}-\frac{4}{x}}$

Multiply the numerator and denominator by xy, the LCD of all the fractions.

$$=\frac{\left(\frac{x^2-16y^2}{xy}\right)xy}{\left(\frac{1}{y}-\frac{4}{x}\right)xy}$$

$$=\frac{x^2-16y^2}{x-4y}$$ Distributive property

$$=\frac{(x+4y)(x-4y)}{x-4y}$$ Factor the difference of squares.

$$=x+4y$$ Lowest terms

19. $\dfrac{\frac{x+2}{x}+\frac{1}{x+2}}{\frac{5}{x}+\frac{x}{x+2}}$

Multiply the numerator and denominator by $x(x+2)$, the LCD of all the fractions.

$$=\frac{x(x+2)\left(\frac{x+2}{x}+\frac{1}{x+2}\right)}{x(x+2)\left(\frac{5}{x}+\frac{x}{x+2}\right)}$$

$$=\frac{x(x+2)\left(\frac{x+2}{x}\right)+x(x+2)\left(\frac{1}{x+2}\right)}{x(x+2)\left(\frac{5}{x}\right)+x(x+2)\left(\frac{x}{x+2}\right)}$$ Distributive property

$$=\frac{(x+2)(x+2)+x}{5(x+2)+x^2}$$ Multiply.

$$=\frac{x^2+4x+4+x}{5x+10+x^2}$$ Multiply.

$$=\frac{x^2+5x+4}{x^2+5x+10}$$ Combine like terms.

31. $\dfrac{x^{-1}+2y^{-1}}{2y+4x}$

$$=\frac{\frac{1}{x}+\frac{2}{y}}{2y+4x}$$

Multiply the numerator and denominator by xy, the LCD of all the fractions.

$$=\frac{xy\left(\frac{1}{x}+\frac{2}{y}\right)}{xy(2y+4x)}$$

$$=\frac{y+2x}{2xy(y+2x)}$$ Distributive property; factor $2y+4x$ as $2(y+2x)$.

$$=\frac{1}{2xy}$$ Lowest terms

Section 7.4 (pages 561–564)

17. $\frac{3x+1}{x-4}=\frac{6x+5}{2x-7}$

The domain excludes 4 and $\frac{7}{2}$.

Multiply each side by the LCD, $(x-4)(2x-7)$.

$$(x-4)(2x-7)\left(\frac{3x+1}{x-4}\right)=(x-4)(2x-7)\left(\frac{6x+5}{2x-7}\right)$$

$$(2x-7)(3x+1)=(x-4)(6x+5)$$
$$6x^2-19x-7=6x^2-19x-20$$
$$-7=-20 \quad \text{False}$$

The false statement indicates that the original equation has no solution.

Solution set: $\emptyset$

SOLUTIONS

27. $\frac{9}{x}+\frac{4}{6x-3}=\frac{2}{6x-3}$

The domain excludes 0 and $\frac{1}{2}$.

Multiply by the LCD, $x(6x-3)$.

$$x(6x-3)\left(\frac{9}{x}+\frac{4}{6x-3}\right)=x(6x-3)\left(\frac{2}{6x-3}\right)$$

$$9(6x-3)+4x=2x$$

$$54x-27+4x=2x$$

$$56x=27$$

$$x=\frac{27}{56}$$

Note that $\frac{27}{56}$ is in the domain. Substitute $\frac{27}{56}$ for x in the original equation to check the solution.

Solution set: $\left\{\frac{27}{56}\right\}$

37. $\frac{4x-7}{4x^2-9}=\frac{-2x^2+5x-4}{4x^2-9}+\frac{x+1}{2x+3}$

$$\frac{4x-7}{(2x+3)(2x-3)}=\frac{-2x^2+5x-4}{(2x+3)(2x-3)}+\frac{x+1}{2x+3}\quad\text{Factor.}$$

The domain excludes $-\frac{3}{2}$ and $\frac{3}{2}$.

Multiply by the LCD, $(2x+3)(2x-3)$.

$$4x-7=-2x^2+5x-4+(2x-3)(x+1)$$

$$4x-7=-2x^2+5x-4+2x^2-x-3$$

$$4x-7=4x-7\quad\text{True}$$

This equation is true for every real number value of x, but we have already determined that $-\frac{3}{2}$ and $\frac{3}{2}$ are excluded from the domain. Thus, every real number except $-\frac{3}{2}$ and $\frac{3}{2}$ is a solution.

Solution set: $\left\{x \mid x \neq -\frac{3}{2}, \frac{3}{2}\right\}$

Section 7.5 (pages 577–582)

25. Let x = the number of fish in the lake. Write and solve a proportion.

$$\frac{\text{total in lake}}{\text{tagged in lake}}=\frac{\text{total in sample}}{\text{tagged in sample}}$$

$$\frac{x}{500}=\frac{400}{8}$$

$$\frac{x}{500}=50$$

$$x=500(50)$$

$$x=25{,}000$$

There are approximately 25,000 fish.

45. ***Step 1***

Read the problem again.

Step 2

Let x = the distance in miles from San Francisco to the secret rendezvous.

	d	r	t
First Trip	x	200	$\frac{x}{200}$
Return Trip	x	300	$\frac{x}{300}$

Step 3

Time there	plus	time back	equals	4 hr.
$\frac{x}{200}$	$+$	$\frac{x}{300}$	$=$	4

Step 4

Multiply by the LCD, 600.

$$600\left(\frac{x}{200}+\frac{x}{300}\right)=600(4)$$

$$3x+2x=2400$$

$$5x=2400$$

$$x=480$$

Step 5

The distance is 480 mi.

Step 6

Check. 480 mi at 200 mph takes $\frac{480}{200}$, or 2.4 hr; 480 miles at 300 mph takes $\frac{480}{300}$, or 1.6 hr. The total time is 4 hr, as required.

55. Let x = the time from Mimi's arrival home to the time the place is a shambles.

	Rate	Time to Mess up House	Fractional Part of the Job Done
Hortense and Mort	$-\frac{1}{7}$	x	$-\frac{1}{7}x$
Mimi	$\frac{1}{2}$	x	$\frac{1}{2}x$

Notice that Hortense and Mort's rate is negative since they are "undoing" the messing up by cleaning the house.

Part done by Hortense and Mort	$+$	Part done by Mimi	$=$	1 whole job of messing up.
$-\frac{1}{7}x$	$+$	$\frac{1}{2}x$	$=$	1

Multiply by the LCD, 14.

$$14\left(-\frac{1}{7}x\right)+14\left(\frac{1}{2}x\right)=14(1)$$

$$-2x+7x=14$$

$$5x=14$$

$$x=\frac{14}{5},\text{ or } 2\frac{4}{5}$$

It would take $\frac{14}{5}$ hr, or $2\frac{4}{5}$ hr.

Section 7.6 (pages 589–592)

19. Here, p varies jointly as q and r^2, so the variation equation is

$$p=kqr^2,$$

for some constant k. Let $p=200$, $q=2$, and $r=3$ in this equation, and solve for k.

$$200=k\cdot 2\cdot 3^2\quad\text{Substitute.}$$

$$200=18k\quad 3^2=9;\ 2\cdot 9=18$$

$$\frac{200}{18}=k\quad\text{Divide by 18.}$$

$$\frac{100}{9}=k\quad\text{Lowest terms}$$

Now let $k=\frac{100}{9}$, $q=5$, and $r=2$, and find p.

$$p=kqr^2$$

$$p=\frac{100}{9}\cdot 5\cdot 2^2\quad\text{Substitute.}$$

$$p=\frac{100}{9}\cdot 20\quad 2^2=4;\ 5\cdot 4=20$$

$$p=\frac{2000}{9}$$

$$p=222\frac{2}{9}$$

21. For a constant time of 3 hr, if the rate of the pickup truck *increases,* then the distance traveled *increases.* Thus, the variation between the quantities is *direct.*

37. Let I = the illumination produced by a light source and d = the distance from the source. I varies inversely as d^2, so

$$I=\frac{k}{d^2}$$

for some constant k. Since $I=768$ when $d=1$, substitute these values in the equation and solve for k.

$$I=\frac{k}{d^2}$$

$$768=\frac{k}{1^2}$$

$$768=k$$

So $I=\frac{768}{d^2}$. Now let $d=6$.

$$I=\frac{768}{d^2}=\frac{768}{36}=\frac{64}{3},\text{ or } 21\frac{1}{3}$$

The illumination produced by the light source is $21\frac{1}{3}$ foot-candles.

41. Let F = the force, w = the weight of the car, s = the speed, and r = the radius. The force varies inversely as the radius and jointly as the weight and the square of the speed, so

$$F=\frac{kws^2}{r}.$$

(continued)

SOLUTIONS

Let $F = 242$, $w = 2000$, $r = 500$, and $s = 30$, and solve for k.

$$F = \frac{kws^2}{r}$$

$$242 = \frac{k(2000)(30)^2}{500}$$

$$k = \frac{242(500)}{2000(900)}$$

$$k = \frac{121}{1800}$$

So $F = \dfrac{121ws^2}{1800r}$.

Let $r = 750$, $s = 50$, and $w = 2000$.

$$F = \frac{121(2000)(50)^2}{1800(750)} \approx 448.1$$

Approximately 448.1 lb of force would be needed.

Chapter 8 Equations, Inequalities, and Systems Revisited

Section 8.1 (pages 615–618)

19. $2[w - (2w + 4) + 3] = 2(w + 1)$

$2[w - 2w - 4 + 3] = 2(w + 1)$ Distributive property

$2[-w - 1] = 2(w + 1)$ Combine like terms.

$-2w - 2 = 2w + 2$ Distributive property

$-2 = 4w + 2$ Add $2w$.

$-4 = 4w$ Subtract 2.

$-1 = w$ Divide by 4.

Check Substitute -1 for w in the original equation.

Solution set: $\{-1\}$

23. $\dfrac{1}{5}x - 2 = \dfrac{2}{3}x - \dfrac{2}{5}x$

Multiply each side by the LCD, 15, and use the distributive property.

$15\left(\dfrac{1}{5}x\right) - 15(2) = 15\left(\dfrac{2}{3}x\right) - 15\left(\dfrac{2}{5}x\right)$

$3x - 30 = 10x - 6x$ Multiply.

$3x - 30 = 4x$ Combine like terms.

$-30 = x$ Subtract $3x$.

Solution set: $\{-30\}$

57. $\dfrac{2k - 5}{-4} > 5$

Multiply each side by -4 and reverse the direction of the inequality symbol.

$-4\left(\dfrac{2k - 5}{-4}\right) < -4(5)$

$2k - 5 < -20$

$2k < -15$ Add 5.

$k < -\dfrac{15}{2}$ Divide by 2.

Check that the solution set is the interval $\left(-\infty, -\dfrac{15}{2}\right)$.

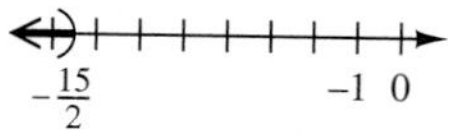

73. $-6 \le 2(z + 2) \le 16$

$-6 \le 2z + 4 \le 16$ Distributive property

$-10 \le 2z \le 12$ Subtract 4.

$-5 \le z \le 6$ Divide by 2.

Check that the solution set is the interval $[-5, 6]$.

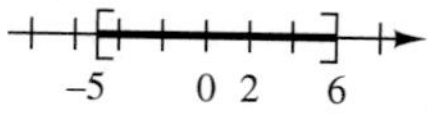

Section 8.2 (pages 625–630)

41. $4x - 8 > 0$ or $4x - 1 < 7$

$4x > 8$ or $4x < 8$

$x > 2$ or $x < 2$

The graph of the solution set is all numbers either greater than 2 or less than 2. This is all real numbers except 2. The solution set is $(-\infty, 2) \cup (2, \infty)$.

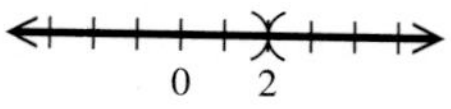

43. $(-\infty, -1] \cap [-4, \infty)$

The intersection is the set of numbers less than or equal to -1 and greater than or equal to -4. The numbers common to *both* original sets are between, and including, -4 and -1. The simplest interval form is $[-4, -1]$.

Section 8.3 (pages 637–642)

19. $\left|1 - \dfrac{3}{4}k\right| = 7$

$1 - \dfrac{3}{4}k = 7$ or $1 - \dfrac{3}{4}k = -7$

$-\dfrac{3}{4}k = 6$ or $-\dfrac{3}{4}k = -8$ Subtract 1.

$k = -8$ or $k = \dfrac{32}{3}$ Multiply by $-\frac{4}{3}$.

Solution set: $\left\{-8, \dfrac{32}{3}\right\}$

49. $|-3x - 8| \le 4$

$-4 \le -3x - 8 \le 4$

$4 \le -3x \le 12$ Add 8.

Divide each part by -3 and reverse the direction of the inequality symbols.

$$\frac{4}{-3} \ge x \ge \frac{12}{-3}$$

$$-\frac{4}{3} \ge x \ge -4$$

Rewrite in order based on a number line.

$$-4 \le x \le -\frac{4}{3}$$

Solution set: $\left[-4, -\dfrac{4}{3}\right]$

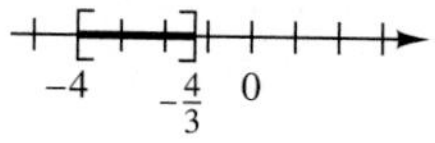

65. $|2p - 6| = |2p + 11|$

$2p - 6 = 2p + 11$

$-6 = 11$ False

No solution

or $2p - 6 = -(2p + 11)$

$2p - 6 = -2p - 11$

$4p = -5$

$p = -\dfrac{5}{4}$

Solution set: $\left\{-\dfrac{5}{4}\right\}$

81. $|10z + 7| > 0$

Since the absolute value of an expression is always nonnegative, there is only one possible value of z that makes this statement false. Solving the equation $10z + 7 = 0$ will give that value of z.

$10z + 7 = 0$

$10z = -7$

$z = -\dfrac{7}{10}$

The solution set of the inequality includes *all values other than* $-\dfrac{7}{10}$, which makes the absolute value expression equal 0.

Solution set: $\left(-\infty, -\dfrac{7}{10}\right) \cup \left(-\dfrac{7}{10}, \infty\right)$

Section 8.4 (pages 651–652)

19. $y = 2x$ (1)

$4x - 2y = 0$ (2)

From equation (1), substitute $2x$ for y in equation (2).

$4x - 2y = 0$ (2)

$4x - 2(2x) = 0$ Let $y = 2x$.

$4x - 4x = 0$

$0 = 0$ True

The equations are dependent, and the solution set is the set of all points on the line. We use one of the equations of the system to write the solution set in set-builder notation. As indicated in the Note on page 650, we give the equation in standard form with coefficients that are integers having greatest common factor 1 and positive coefficient of x.

SOLUTIONS

Thus, we use equation (2) and divide each term by the common factor 2 to get

$$\{(x, y) \mid 2x - y = 0\}$$

as the solution set.

33.
$$\frac{x}{2} + \frac{y}{3} = -\frac{1}{3} \quad (1)$$
$$\frac{x}{2} + 2y = -7 \quad (2)$$

Clear the fractions by multiplying equation (1) by −6 and equation (2) by 6. (We multiply equation (2) by 6 instead of 2 so that when the fractions are cleared, the x-terms are opposites.) Then add the results to eliminate x.

$$\begin{aligned} -3x - 2y &= 2 && \text{Multiply (1) by } -6. \quad (3) \\ 3x + 12y &= -42 && \text{Multiply (2) by 6.} \\ \hline 10y &= -40 && \text{Add.} \\ y &= -4 && \text{Divide by 10.} \end{aligned}$$

To find x, substitute −4 for y in equation (3).

$$\begin{aligned} -3x - 2y &= 2 && (3) \\ -3x - 2(-4) &= 2 && \text{Let } y = -4. \\ -3x + 8 &= 2 \\ -3x &= -6 && \text{Subtract 8.} \\ x &= 2 && \text{Divide by } -3. \end{aligned}$$

The solution $(2, -4)$ checks in both equations (1) and (2).

Solution set: $\{(2, -4)\}$

Section 8.5 (pages 661–664)

13.
$$\begin{aligned} x + 2y + 3z &= 1 \quad (1) \\ -x - y + 3z &= 2 \quad (2) \\ -6x + y + z &= -2 \quad (3) \end{aligned}$$

Step 1

Since x in equation (1) has coefficient 1, we choose it as the focus variable and (1) as the working equation.

Step 2

Add equations (1) and (2) to eliminate x.

$$\begin{aligned} x + 2y + 3z &= 1 && (1) \\ -x - y + 3z &= 2 && (2) \\ \hline y + 6z &= 3 && \text{Add. (4)} \end{aligned}$$

Step 3

Multiply working equation (1) by 6 and add to equation (3) to eliminate x again.

$$\begin{aligned} 6x + 12y + 18z &= 6 && \text{Multiply (1) by 6.} \\ -6x + y + z &= -2 && (3) \\ \hline 13y + 19z &= 4 && \text{Add.} \quad (5) \end{aligned}$$

Step 4

The equations that resulted in Steps 2 and 3 form a system in y and z.

$$\begin{aligned} y + 6z &= 3 \quad (4) \\ 13y + 19z &= 4 \quad (5) \end{aligned}$$

Solve this system of equations (4) and (5).

$$\begin{aligned} -13y - 78z &= -39 && \text{Multiply (4) by } -13. \\ 13y + 19z &= 4 && (5) \\ \hline -59z &= -35 && \text{Add.} \\ z &= \frac{35}{59} \end{aligned}$$

Substitute $\frac{35}{59}$ for z in equation (4) to find y. Be careful—the arithmetic gets messy.

$$\begin{aligned} y + 6z &= 3 && (4) \\ y + 6\left(\frac{35}{59}\right) &= 3 && \text{Let } z = \frac{35}{59}. \\ y + \frac{210}{59} &= 3 \\ y &= \frac{177}{59} - \frac{210}{59} && 3 = \frac{177}{59} \\ y &= -\frac{33}{59} \end{aligned}$$

Step 5

Substitute $-\frac{33}{59}$ for y and $\frac{35}{59}$ for z in working equation (1) to find focus variable x.

$$\begin{aligned} x + 2y + 3z &= 1 \quad (1) \\ x + 2\left(-\frac{33}{59}\right) + 3\left(\frac{35}{59}\right) &= 1 \\ x - \frac{66}{59} + \frac{105}{59} &= 1 \\ x + \frac{39}{59} &= 1 \\ x &= \frac{59}{59} - \frac{39}{59} \\ x &= \frac{20}{59} \end{aligned}$$

Step 6

The solution $\left(\frac{20}{59}, -\frac{33}{59}, \frac{35}{59}\right)$ checks when substituted in equations (1), (2), and (3).

Solution set: $\left\{\left(\frac{20}{59}, -\frac{33}{59}, \frac{35}{59}\right)\right\}$

31.
$$\begin{aligned} x + y - 2z &= 0 \quad (1) \\ 3x - y + z &= 0 \quad (2) \\ 4x + 2y - z &= 0 \quad (3) \end{aligned}$$

We choose z as the focus variable and (2) as the working equation. Eliminate z by adding equations (2) and (3).

$$\begin{aligned} 3x - y + z &= 0 && (2) \\ 4x + 2y - z &= 0 && (3) \\ \hline 7x + y &= 0 && (4) \end{aligned}$$

To get another equation without z, multiply working equation (2) by 2 and add the result to equation (1).

$$\begin{aligned} 6x - 2y + 2z &= 0 && \text{Multiply (2) by 2.} \\ x + y - 2z &= 0 && (1) \\ \hline 7x - y &= 0 && (5) \end{aligned}$$

Add equations (4) and (5) to find x.

$$\begin{aligned} 7x + y &= 0 && (4) \\ 7x - y &= 0 && (5) \\ \hline 14x &= 0 \\ x &= 0 \end{aligned}$$

Substitute 0 for x in equation (4) to find y.

$$\begin{aligned} 7x + y &= 0 \quad (4) \\ 7(0) + y &= 0 \\ 0 + y &= 0 \\ y &= 0 \end{aligned}$$

Substitute 0 for x and 0 for y in working equation (2) to find focus variable z.

$$\begin{aligned} 3x - y + z &= 0 \quad (2) \\ 3(0) - 0 + z &= 0 \\ z &= 0 \end{aligned}$$

The solution $(0, 0, 0)$ checks when substituted in equations (1), (2), and (3).

Solution set: $\{(0, 0, 0)\}$

Chapter 9 Roots, Radicals, and Root Functions

Section 9.1 (pages 697–700)

15. The length $\sqrt{98}$ is closer to $\sqrt{100} = 10$ than to $\sqrt{81} = 9$. The width $\sqrt{26}$ is closer to $\sqrt{25} = 5$ than to $\sqrt{36} = 6$. Use the estimates $L = 10$ and $W = 5$ in the area formula $A = LW$ to find an estimate of the area.

$$A \approx 10 \cdot 5 = 50$$

Choice C is the best estimate.

109.
$$\begin{aligned} &\sqrt[6]{x^{30}} \\ &= \sqrt[6]{(x^5)^6} \\ &= |x^5|, \quad \text{or} \quad |x|^5 \quad \text{(6 is even.)} \end{aligned}$$

119. Let $a = 850$, $b = 925$, and $c = 1300$. First find the semiperimeter s.

$$\begin{aligned} s &= \frac{1}{2}(a + b + c) \\ s &= \frac{1}{2}(850 + 925 + 1300) \\ s &= \frac{3075}{2} \\ s &= 1537.5 \end{aligned}$$

Now find the area A using Heron's formula and a calculator.

$$\begin{aligned} A &= \sqrt{s(s-a)(s-b)(s-c)} \\ A &= \sqrt{1537.5(687.5)(612.5)(237.5)} \\ A &\approx 392{,}128.8 \end{aligned}$$

The area of the Bermuda Triangle is about 392,000 mi^2.

Section 9.2 (pages 707–710)

47. $(3m^4 + 2k^2)^{-2/3}$

$= \dfrac{1}{(3m^4 + 2k^2)^{2/3}}$

$= \dfrac{1}{[(3m^4 + 2k^2)^{1/3}]^2}$

$= \dfrac{1}{\left(\sqrt[3]{3m^4 + 2k^2}\right)^2}$

57. $\sqrt[3]{x} \cdot \sqrt{x}$

$= x^{1/3} \cdot x^{1/2}$ Convert to rational exponents.

$= x^{1/3+1/2}$ Product rule

$= x^{2/6+3/6}$ Get a common denominator.

$= x^{5/6}$

$= \sqrt[6]{x^5}$

79. $\left(\dfrac{b^{-3/2}}{c^{-5/3}}\right)^2 \left(b^{-1/4}c^{-1/3}\right)^{-1}$

$= \left(\dfrac{c^{5/3}}{b^{3/2}}\right)^2 \left(b^{1/4}c^{1/3}\right)$

Definition of negative exponent; power rule

$= \dfrac{c^{10/3}}{b^3}\left(b^{1/4}c^{1/3}\right)$

Power rule

$= \dfrac{c^{10/3}b^{1/4}c^{1/3}}{b^3}$

Multiply.

$= c^{10/3+1/3}b^{1/4-3}$

Product and quotient rules

$= c^{11/3}b^{-11/4}$ $\quad \dfrac{1}{4} - 3 = \dfrac{1}{4} - \dfrac{12}{4} = -\dfrac{11}{4}$

$= \dfrac{c^{11/3}}{b^{11/4}}$

81. $\left(\dfrac{p^{-1/4}q^{-3/2}}{3^{-1}p^{-2}q^{-2/3}}\right)^{-2}$

$= \dfrac{p^{1/2}q^3}{3^2p^4q^{4/3}}$ Power rule

$= \dfrac{p^{1/2-4}q^{3-4/3}}{9}$ Quotient rule

$= \dfrac{p^{1/2-8/2}q^{9/3-4/3}}{9}$ Write exponents with a common denominator.

$= \dfrac{p^{-7/2}q^{5/3}}{9}$

$= \dfrac{q^{5/3}}{9p^{7/2}}$

99. $\sqrt[3]{\sqrt[5]{\sqrt{y}}}$

$= \sqrt[3]{\sqrt[5]{y^{1/2}}}$

$= \sqrt[3]{(y^{1/2})^{1/5}}$

$= (y^{1/10})^{1/3}$

$= y^{1/30}$

Section 9.3 (pages 719–724)

1. Is $2\sqrt{12} = \sqrt{48}$ true?

$2\sqrt{12}$	$\sqrt{48}$
$= 2\sqrt{4 \cdot 3}$	$= \sqrt{16 \cdot 3}$
$= 2\sqrt{4} \cdot \sqrt{3}$	$= \sqrt{16} \cdot \sqrt{3}$
$= 2 \cdot 2 \cdot \sqrt{3}$	$= 4\sqrt{3}$
$= 4\sqrt{3}$	

Each expression equals $4\sqrt{3}$. The calculator approximation for each expression is 6.92820323. The statement is true.

71. $-\sqrt[3]{-125a^6b^9c^{12}}$

$= -\sqrt[3]{(-5a^2b^3c^4)^3}$

$= -(-5a^2b^3c^4)$

$= 5a^2b^3c^4$

87. $-\sqrt[4]{162r^{15}s^{10}}$

$= -\sqrt[4]{81r^{12}s^8(2r^3s^2)}$

$= -\sqrt[4]{81r^{12}s^8} \cdot \sqrt[4]{2r^3s^2}$

$= -3r^3s^2\sqrt[4]{2r^3s^2}$

91. $\sqrt[3]{\dfrac{x^{16}}{27}}$

$= \dfrac{\sqrt[3]{x^{15} \cdot x^1}}{\sqrt[3]{27}}$

$= \dfrac{\sqrt[3]{x^{15}} \cdot \sqrt[3]{x}}{\sqrt[3]{27}}$

$= \dfrac{x^5\sqrt[3]{x}}{3}$

119. Let $(x_1, y_1) = \left(\sqrt{2}, \sqrt{6}\right)$ and $(x_2, y_2) = \left(-2\sqrt{2}, 4\sqrt{6}\right)$.

$d = \sqrt{(x_2 - x_1)^2 + (y_2 - y_1)^2}$

$d = \sqrt{\left(-2\sqrt{2} - \sqrt{2}\right)^2 + \left(4\sqrt{6} - \sqrt{6}\right)^2}$

$d = \sqrt{\left(-3\sqrt{2}\right)^2 + \left(3\sqrt{6}\right)^2}$

$d = \sqrt{9 \cdot 2 + 9 \cdot 6}$

$d = \sqrt{18 + 54}$

$d = \sqrt{72}$

$d = \sqrt{36} \cdot \sqrt{2}$

$d = 6\sqrt{2}$

Section 9.4 (pages 727–728)

13. $6\sqrt{18} - \sqrt{32} + 2\sqrt{50}$

$= 6\sqrt{9 \cdot 2} - \sqrt{16 \cdot 2} + 2\sqrt{25 \cdot 2}$

$= 6\sqrt{9} \cdot \sqrt{2} - \sqrt{16} \cdot \sqrt{2} + 2\sqrt{25} \cdot \sqrt{2}$

$= 6 \cdot 3\sqrt{2} - 4\sqrt{2} + 2 \cdot 5\sqrt{2}$

$= 18\sqrt{2} - 4\sqrt{2} + 10\sqrt{2}$

$= 24\sqrt{2}$

29. $3x\sqrt[3]{xy^2} - 2\sqrt[3]{8x^4y^2}$

$= 3x\sqrt[3]{xy^2} - 2\sqrt[3]{8x^3} \cdot \sqrt[3]{xy^2}$

$= 3x\sqrt[3]{xy^2} - 2 \cdot 2x \cdot \sqrt[3]{xy^2}$

$= 3x\sqrt[3]{xy^2} - 4x\sqrt[3]{xy^2}$

$= (3x - 4x)\sqrt[3]{xy^2}$

$= -x\sqrt[3]{xy^2}$

47. $3\sqrt[3]{\dfrac{m^5}{27}} - 2m\sqrt[3]{\dfrac{m^2}{64}}$

$= \dfrac{3\sqrt[3]{m^5}}{\sqrt[3]{27}} - \dfrac{2m\sqrt[3]{m^2}}{\sqrt[3]{64}}$

$= \dfrac{3\sqrt[3]{m^3} \cdot \sqrt[3]{m^2}}{3} - \dfrac{2m\sqrt[3]{m^2}}{4}$

$= \dfrac{m\sqrt[3]{m^2}}{1} - \dfrac{m\sqrt[3]{m^2}}{2}$

$= \dfrac{2m\sqrt[3]{m^2} - m\sqrt[3]{m^2}}{2}$

$= \dfrac{m\sqrt[3]{m^2}}{2}$

53. To find the perimeter, add the lengths of the sides.

$4\sqrt{18} + \sqrt{108} + 2\sqrt{72} + 3\sqrt{12}$

$= 4\sqrt{9} \cdot \sqrt{2} + \sqrt{36} \cdot \sqrt{3} + 2\sqrt{36} \cdot \sqrt{2} + 3\sqrt{4} \cdot \sqrt{3}$

$= 4 \cdot 3\sqrt{2} + 6\sqrt{3} + 2 \cdot 6\sqrt{2} + 3 \cdot 2\sqrt{3}$

$= 12\sqrt{2} + 6\sqrt{3} + 12\sqrt{2} + 6\sqrt{3}$

$= 24\sqrt{2} + 12\sqrt{3}$

The perimeter is $\left(24\sqrt{2} + 12\sqrt{3}\right)$ in.

Section 9.5 (pages 735–738)

9. $\sqrt{2}\left(\sqrt{18} - \sqrt{3}\right)$

$= \sqrt{2} \cdot \sqrt{18} - \sqrt{2} \cdot \sqrt{3}$

$= \sqrt{36} - \sqrt{6}$

$= 6 - \sqrt{6}$

21. $\left(4\sqrt{x} + 3\right)^2$

$= \left(4\sqrt{x}\right)^2 + 2\left(4\sqrt{x}\right)(3) + 3^2$ $\quad (x + y)^2 = x^2 + 2xy + y^2$

$= 16x + 24\sqrt{x} + 9$

SOLUTIONS

39. $\dfrac{6\sqrt{3y}}{\sqrt{y^3}}$

$= \dfrac{6\sqrt{3y}\cdot\sqrt{y}}{\sqrt{y^3}\cdot\sqrt{y}}$

$= \dfrac{6\sqrt{3y^2}}{\sqrt{y^4}}$

$= \dfrac{6y\sqrt{3}}{y^2}$

$= \dfrac{6\sqrt{3}}{y}$

75. $\dfrac{\sqrt{x}-\sqrt{y}}{\sqrt{2x}+\sqrt{3y}}$

Multiply both the numerator and denominator by the conjugate of the denominator, $\sqrt{2x}-\sqrt{3y}$.

$= \dfrac{(\sqrt{x}-\sqrt{y})(\sqrt{2x}-\sqrt{3y})}{(\sqrt{2x}+\sqrt{3y})(\sqrt{2x}-\sqrt{3y})}$

$= \dfrac{\sqrt{2x^2}-\sqrt{3xy}-\sqrt{2xy}+\sqrt{3y^2}}{(\sqrt{2x})^2-(\sqrt{3y})^2}$

$= \dfrac{x\sqrt{2}-\sqrt{3xy}-\sqrt{2xy}+y\sqrt{3}}{2x-3y}$

Section 9.6 (pages 745–748)

21. $3\sqrt{z-1} = 2\sqrt{2z+2}$

$(3\sqrt{z-1})^2 = (2\sqrt{2z+2})^2$ Square both sides.

$9(z-1) = 4(2z+2)$

$9z-9 = 8z+8$ Distributive property

$z = 17$ Subtract $8z$; add 9.

A check confirms that 17 is a solution of the original equation.

Solution set: $\{17\}$

33. $\sqrt{z^2+12z-4}+4-z = 0$

$\sqrt{z^2+12z-4} = z-4$ Isolate the radical.

$(\sqrt{z^2+12z-4})^2 = (z-4)^2$ Square both sides.

$z^2+12z-4 = z^2-8z+16$ Simplify.

$20z = 20$

$z = 1$

Substituting 1 for z makes the left side of the original equation positive, but the right side is zero, so 1 is not a solution; it is extraneous.

Solution set: $\emptyset$

43. $\sqrt[4]{a+8} = \sqrt[4]{2a}$

Raise each side to the fourth power.

$(\sqrt[4]{a+8})^4 = (\sqrt[4]{2a})^4$

$a+8 = 2a$

$8 = a$

A check confirms that 8 is a solution of the original equation.

Solution set: $\{8\}$

57. $\sqrt{2\sqrt{x+11}} = \sqrt{4x+2}$

$\left(\sqrt{2\sqrt{x+11}}\right)^2 = (\sqrt{4x+2})^2$ Square both sides.

$2\sqrt{x+11} = 4x+2$

$(2\sqrt{x+11})^2 = (4x+2)^2$ Square again.

$4(x+11) = 16x^2+16x+4$

$4x+44 = 16x^2+16x+4$

$16x^2+12x-40 = 0$ Standard form

$4x^2+3x-10 = 0$ Divide by 4.

$(x+2)(4x-5) = 0$ Factor.

$x+2 = 0$ or $4x-5 = 0$ Zero-factor property

$x = -2$ or $x = \dfrac{5}{4}$

Check

Let $x = -2$ in the original equation.

$\sqrt{2\sqrt{-2+11}} \stackrel{?}{=} \sqrt{4(-2)+2}$

$\sqrt{2\sqrt{9}} \stackrel{?}{=} \sqrt{-8+2}$

$\sqrt{6} = \sqrt{-6}$ False

Let $x = \dfrac{5}{4}$ in the original equation.

$\sqrt{2\sqrt{\dfrac{5}{4}+11}} \stackrel{?}{=} \sqrt{4\left(\dfrac{5}{4}\right)+2}$

$\sqrt{2\sqrt{\dfrac{49}{4}}} \stackrel{?}{=} \sqrt{5+2}$

$\sqrt{2\left(\dfrac{7}{2}\right)} = \sqrt{7}$

$\sqrt{7} = \sqrt{7}$ True

Solution set: $\left\{\dfrac{5}{4}\right\}$

61. $(2w-1)^{2/3}-w^{1/3} = 0$

$\sqrt[3]{(2w-1)^2}-\sqrt[3]{w} = 0$ Write with radicals.

$\sqrt[3]{(2w-1)^2} = \sqrt[3]{w}$ Add $\sqrt[3]{w}$.

$\left(\sqrt[3]{(2w-1)^2}\right)^3 = (\sqrt[3]{w})^3$ Cube both sides.

$(2w-1)^2 = w$

$4w^2-4w+1 = w$ Square on the left.

$4w^2-5w+1 = 0$ Standard form

$(4w-1)(w-1) = 0$

$4w-1 = 0$ or $w-1 = 0$

$w = \dfrac{1}{4}$ or $w = 1$

A check confirms that $\dfrac{1}{4}$ and 1 are both solutions of the original equation.

Solution set: $\left\{\dfrac{1}{4}, 1\right\}$

63. Solve $V = \sqrt{\dfrac{2K}{m}}$ for K.

$(V)^2 = \left(\sqrt{\dfrac{2K}{m}}\right)^2$ Square both sides.

$V^2 = \dfrac{2K}{m}$

$\dfrac{V^2m}{2} = K$, or $K = \dfrac{V^2m}{2}$ Multiply by $\dfrac{m}{2}$.

Section 9.7 (pages 755–758)

37. $[(7+3i)-(4-2i)]+(3+i)$ Work inside the brackets first.

$= [(7-4)+(3+2)i]+(3+i)$

$= (3+5i)+(3+i)$

$= (3+3)+(5+1)i$

$= 6+6i$

49. $(4+5i)^2$

$= 4^2+2(4)(5i)+(5i)^2$

$= 16+40i+25i^2$

$= 16+40i+25(-1)$ $i^2 = -1$

$= 16+40i-25$

$= -9+40i$

81. $I = \dfrac{E}{R+(X_L-X_c)i}$

Substitute $2+3i$ for E, 5 for R, 4 for X_L, and 3 for X_c.

$I = \dfrac{2+3i}{5+(4-3)i}$

$I = \dfrac{2+3i}{5+i}$

$I = \dfrac{(2+3i)(5-i)}{(5+i)(5-i)}$

$I = \dfrac{10-2i+15i-3i^2}{5^2-i^2}$

(continued)

SOLUTIONS

$$I = \frac{10 + 13i + 3}{25 + 1}$$

$$I = \frac{13 + 13i}{26}$$

$$I = \frac{13(1 + i)}{13 \cdot 2}$$

$$I = \frac{1 + i}{2}$$

$$I = \frac{1}{2} + \frac{1}{2}i$$

Chapter 10 Quadratic Equations, Inequalities, and Functions

Section 10.1 (pages 777–778)

39. $(6k - 1)^2 = -8$

$6k - 1 = \sqrt{-8}$ or $6k - 1 = -\sqrt{-8}$ Square root property

$6k - 1 = \sqrt{-4} \cdot \sqrt{2}$ or $6k - 1 = -\sqrt{-4} \cdot \sqrt{2}$ Product rule

$6k - 1 = 2i\sqrt{2}$ or $6k - 1 = -2i\sqrt{2}$ $\sqrt{-4} = 2i$

$6k = 1 + 2i\sqrt{2}$ or $6k = 1 - 2i\sqrt{2}$ Add 1.

$k = \frac{1 + 2i\sqrt{2}}{6}$ or $k = \frac{1 - 2i\sqrt{2}}{6}$ Divide by 6.

$k = \frac{1}{6} + \frac{2\sqrt{2}}{6}i$ or $k = \frac{1}{6} - \frac{2\sqrt{2}}{6}i$ Standard form

$k = \frac{1}{6} + \frac{\sqrt{2}}{3}i$ or $k = \frac{1}{6} - \frac{\sqrt{2}}{3}i$ Lowest terms

Solution set: $\left\{\frac{1}{6} + \frac{\sqrt{2}}{3}i, \frac{1}{6} - \frac{\sqrt{2}}{3}i\right\}$

Section 10.2 (pages 785–786)

17. $x^2 + x - 1 = 0$

$x^2 + x = 1$ Add 1.

Take half of 1, the coefficient of x, square it, and add the result to each side.

$x^2 + x + \frac{1}{4} = 1 + \frac{1}{4}$ Add $\left[\frac{1}{2}(1)\right]^2 = \frac{1}{4}$.

$\left(x + \frac{1}{2}\right)^2 = \frac{5}{4}$ Factor; add.

$x + \frac{1}{2} = \sqrt{\frac{5}{4}}$ or $x + \frac{1}{2} = -\sqrt{\frac{5}{4}}$

$x + \frac{1}{2} = \frac{\sqrt{5}}{2}$ or $x + \frac{1}{2} = -\frac{\sqrt{5}}{2}$

$x = -\frac{1}{2} + \frac{\sqrt{5}}{2}$ or $x = -\frac{1}{2} - \frac{\sqrt{5}}{2}$

$x = \frac{-1 + \sqrt{5}}{2}$ or $x = \frac{-1 - \sqrt{5}}{2}$

Solution set: $\left\{\frac{-1 + \sqrt{5}}{2}, \frac{-1 - \sqrt{5}}{2}\right\}$

23. $2x^2 - 4x = 5$

$x^2 - 2x = \frac{5}{2}$ Divide by 2.

$x^2 - 2x + 1 = \frac{5}{2} + 1$ Add $\left[\frac{1}{2}(-2)\right]^2 = 1$.

$(x - 1)^2 = \frac{7}{2}$ Factor; add.

$x - 1 = \sqrt{\frac{7}{2}}$ or $x - 1 = -\sqrt{\frac{7}{2}}$

$x - 1 = \frac{\sqrt{7}}{\sqrt{2}}$ or $x - 1 = -\frac{\sqrt{7}}{\sqrt{2}}$

$x - 1 = \frac{\sqrt{14}}{2}$ or $x - 1 = -\frac{\sqrt{14}}{2}$ Rationalize denominators by multiplying by $\frac{\sqrt{2}}{\sqrt{2}}$.

$x = 1 + \frac{\sqrt{14}}{2}$ or $x = 1 - \frac{\sqrt{14}}{2}$

$x = \frac{2}{2} + \frac{\sqrt{14}}{2}$ or $x = \frac{2}{2} - \frac{\sqrt{14}}{2}$

$x = \frac{2 + \sqrt{14}}{2}$ or $x = \frac{2 - \sqrt{14}}{2}$

Solution set: $\left\{\frac{2 + \sqrt{14}}{2}, \frac{2 - \sqrt{14}}{2}\right\}$

35. $3r^2 + 4r + 4 = 0$

$3r^2 + 4r = -4$ Subtract 4.

$r^2 + \frac{4}{3}r = \frac{-4}{3}$ Divide by 3.

Complete the square.

$$\left[\frac{1}{2}\left(\frac{4}{3}\right)\right]^2 = \left(\frac{2}{3}\right)^2 = \frac{4}{9}$$

Add $\frac{4}{9}$ to each side.

$$r^2 + \frac{4}{3}r + \frac{4}{9} = \frac{-4}{3} + \frac{4}{9}$$

$\left(r + \frac{2}{3}\right)^2 = \frac{-8}{9}$ $\frac{-4}{3} = \frac{-12}{9}$

$r + \frac{2}{3} = \frac{\sqrt{-8}}{\sqrt{9}}$ or $r + \frac{2}{3} = -\frac{\sqrt{-8}}{\sqrt{9}}$

$r + \frac{2}{3} = \frac{2i\sqrt{2}}{3}$ or $r + \frac{2}{3} = \frac{-2i\sqrt{2}}{3}$

$r = -\frac{2}{3} + \frac{2i\sqrt{2}}{3}$ or $r = -\frac{2}{3} - \frac{2i\sqrt{2}}{3}$

Solution set: $\left\{-\frac{2}{3} + \frac{2\sqrt{2}}{3}i, -\frac{2}{3} - \frac{2\sqrt{2}}{3}i\right\}$

37. $-m^2 - 6m - 12 = 0$

Multiply each side by -1.

$m^2 + 6m + 12 = 0$

$m^2 + 6m = -12$ Subtract 12.

Complete the square.

$$\left[\frac{1}{2}(6)\right]^2 = 3^2 = 9$$

$m^2 + 6m + 9 = -12 + 9$

$(m + 3)^2 = -3$

$m + 3 = \sqrt{-3}$ or $m + 3 = -\sqrt{-3}$

$m + 3 = i\sqrt{3}$ or $m + 3 = -i\sqrt{3}$

$m = -3 + i\sqrt{3}$ or $m = -3 - i\sqrt{3}$

Solution set: $\{-3 + i\sqrt{3}, -3 - i\sqrt{3}\}$

Section 10.3 (pages 793–794)

15. $\frac{x^2}{4} - \frac{x}{2} = 1$

First clear fractions by multiplying each side by the LCD, 4.

$4\left(\frac{x^2}{4} - \frac{x}{2}\right) = 4(1)$

$x^2 - 2x = 4$

$x^2 - 2x - 4 = 0$

Here $a = 1$, $b = -2$, and $c = -4$. Substitute in the quadratic formula.

$$x = \frac{-b \pm \sqrt{b^2 - 4ac}}{2a}$$

$$x = \frac{-(-2) \pm \sqrt{(-2)^2 - 4(1)(-4)}}{2(1)}$$

$$x = \frac{2 \pm \sqrt{4 + 16}}{2}$$

$$x = \frac{2 \pm \sqrt{20}}{2}$$

$$x = \frac{2 \pm 2\sqrt{5}}{2}$$

$$x = \frac{2(1 \pm \sqrt{5})}{2}$$

$$x = 1 \pm \sqrt{5}$$

Solution set: $\{1 + \sqrt{5}, 1 - \sqrt{5}\}$

SOLUTIONS

17.
$$-2t(t+2) = -3$$
$$-2t^2 - 4t = -3$$
$$-2t^2 - 4t + 3 = 0$$

Here $a = -2$, $b = -4$, and $c = 3$.

$$t = \frac{-b \pm \sqrt{b^2 - 4ac}}{2a}$$
$$t = \frac{-(-4) \pm \sqrt{(-4)^2 - 4(-2)(3)}}{2(-2)}$$
$$t = \frac{4 \pm \sqrt{16 + 24}}{-4}$$
$$t = \frac{4 \pm \sqrt{40}}{-4}$$
$$t = \frac{4 \pm 2\sqrt{10}}{-4}$$
$$t = \frac{2\left(2 \pm \sqrt{10}\right)}{-2 \cdot 2}$$
$$t = \frac{2 \pm \sqrt{10}}{-2} \cdot \frac{-1}{-1}$$
$$t = \frac{-2 \mp \sqrt{10}}{2}$$
$$t = \frac{-2 \pm \sqrt{10}}{2}$$

Solution set: $\left\{\frac{-2 + \sqrt{10}}{2}, \frac{-2 - \sqrt{10}}{2}\right\}$

27.
$$x(3x+4) = -2$$
$$3x^2 + 4x = -2$$
$$3x^2 + 4x + 2 = 0$$

Here $a = 3$, $b = 4$, and $c = 2$.

$$x = \frac{-b \pm \sqrt{b^2 - 4ac}}{2a}$$
$$x = \frac{-4 \pm \sqrt{4^2 - 4(3)(2)}}{2(3)}$$
$$x = \frac{-4 \pm \sqrt{16 - 24}}{6}$$
$$x = \frac{-4 \pm \sqrt{-8}}{6}$$
$$x = \frac{-4 \pm 2i\sqrt{2}}{6}$$
$$x = \frac{2\left(-2 \pm i\sqrt{2}\right)}{2 \cdot 3}$$
$$x = \frac{-2 \pm i\sqrt{2}}{3}$$

Solution set: $\left\{-\frac{2}{3} + \frac{\sqrt{2}}{3}i, -\frac{2}{3} - \frac{\sqrt{2}}{3}i\right\}$

Section 10.4 (pages 803–806)

15.
$$\frac{3}{2x} - \frac{1}{2(x+2)} = 1$$

Multiply by the LCD, $2x(x+2)$.

$$2x(x+2)\left(\frac{3}{2x} - \frac{1}{2(x+2)}\right) = 2x(x+2) \cdot 1$$
$$3(x+2) - x(1) = 2x(x+2)$$
$$3x + 6 - x = 2x^2 + 4x$$
$$0 = 2x^2 + 2x - 6$$
$$0 = x^2 + x - 3$$

Use $a = 1$, $b = 1$, and $c = -3$ in the quadratic formula.

$$x = \frac{-b \pm \sqrt{b^2 - 4ac}}{2a}$$
$$x = \frac{-1 \pm \sqrt{1^2 - 4(1)(-3)}}{2(1)}$$
$$x = \frac{-1 \pm \sqrt{1 + 12}}{2}$$
$$x = \frac{-1 \pm \sqrt{13}}{2}$$

Use a calculator to check both proposed solutions. Both solutions check.

Solution set: $\left\{\frac{-1 + \sqrt{13}}{2}, \frac{-1 - \sqrt{13}}{2}\right\}$

23. Let $x =$ Harry's average speed.
Then $x - 20 =$ Yoshi's average speed.

	d	r	t
Harry	300	x	$\frac{300}{x}$
Yoshi	300	$x - 20$	$\frac{300}{x - 20}$

It takes Harry $1\frac{1}{4}$ hr, or $\frac{5}{4}$ hr, less time than Yoshi.

$$\frac{300}{x} = \frac{300}{x - 20} - \frac{5}{4}$$

Multiply by the LCD, $4x(x - 20)$.

$$4x(x-20)\left(\frac{300}{x}\right) = 4x(x-20)\left(\frac{300}{x-20} - \frac{5}{4}\right)$$
$$1200(x - 20) = 4x(300) - x(x - 20) \cdot 5$$
$$1200x - 24{,}000 = 1200x - 5x^2 + 100x$$
$$5x^2 - 100x - 24{,}000 = 0$$
$$x^2 - 20x - 4800 = 0$$
$$(x - 80)(x + 60) = 0$$
$$x - 80 = 0 \quad \text{or} \quad x + 60 = 0$$
$$x = 80 \quad \text{or} \quad x = -60$$

Reject $x = -60$. Harry's average speed is 80 km per hr.

37.
$$m = \sqrt{\frac{6 - 13m}{5}}$$
$$m^2 = \frac{6 - 13m}{5} \qquad \text{Square each side.}$$
$$5m^2 = 6 - 13m \qquad \text{Multiply by 5.}$$
$$5m^2 + 13m - 6 = 0 \qquad \text{Standard form}$$
$$(5m - 2)(m + 3) = 0 \qquad \text{Factor.}$$
$$5m - 2 = 0 \quad \text{or} \quad m + 3 = 0 \qquad \text{Zero-factor property}$$
$$m = \frac{2}{5} \quad \text{or} \quad m = -3$$

Check If $m = \frac{2}{5}$, then $\frac{2}{5} = \sqrt{\frac{4}{25}}$. True

If $m = -3$, then $-3 = \sqrt{9}$. False

Solution set: $\left\{\frac{2}{5}\right\}$

51.
$$2 + \frac{5}{3k - 1} = \frac{-2}{(3k - 1)^2}$$

Let $u = 3k - 1$, so $u^2 = (3k - 1)^2$.

$$2 + \frac{5}{u} = -\frac{2}{u^2}$$

Multiply by the LCD, u^2.

$$u^2\left(2 + \frac{5}{u}\right) = u^2\left(-\frac{2}{u^2}\right)$$
$$2u^2 + 5u = -2$$
$$2u^2 + 5u + 2 = 0$$
$$(2u + 1)(u + 2) = 0$$
$$2u + 1 = 0 \quad \text{or} \quad u + 2 = 0$$
$$u = -\frac{1}{2} \quad \text{or} \quad u = -2$$

To find k, substitute $3k - 1$ for u.

$$3k - 1 = -\frac{1}{2} \quad \text{or} \quad 3k - 1 = -2$$
$$3k = \frac{1}{2} \quad \text{or} \quad 3k = -1$$
$$k = \frac{1}{6} \quad \text{or} \quad k = -\frac{1}{3}$$

Check If $k = \frac{1}{6}$, then $2 - 10 = -8$. True

If $k = -\frac{1}{3}$, then $2 - \frac{5}{2} = -\frac{1}{2}$. True

Solution set: $\left\{-\frac{1}{3}, \frac{1}{6}\right\}$

Section 10.5 (pages 813–816)

25. Let x be the width of the sheet metal. Then the length is $2x - 4$.

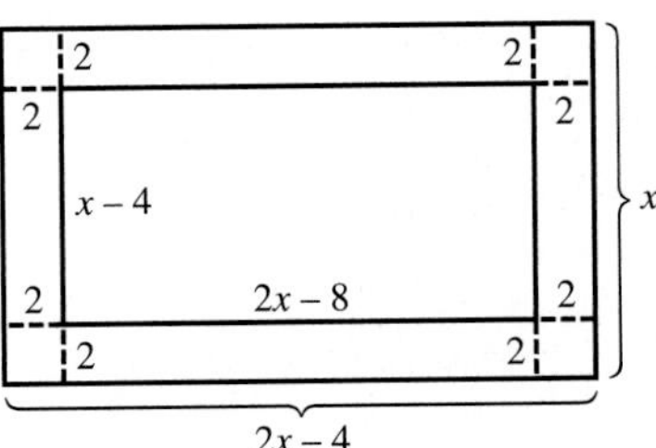

By cutting out 2-in. squares from each corner we get a rectangle with width $(x - 4)$ in. and length

$$(2x - 4) - 4 = (2x - 8) \text{ in.}$$

The uncovered box then has height 2 in., length $(2x - 8)$ in., and width $(x - 4)$ in.

(continued)

Use the formula $V = LWH$ or $V = HLW$.

$$256 = 2(2x - 8)(x - 4)$$
$$256 = 4(x - 4)(x - 4) \quad \text{Factor out 2.}$$
$$64 = (x - 4)^2 \quad \text{Divide by 4.}$$
$$(x - 4)^2 = 64$$

Use the square root property.

$$x - 4 = 8 \quad \text{or} \quad x - 4 = -8$$
$$x = 12 \quad \text{or} \quad x = -4$$

Since x represents width, discard the negative solution. The width is 12 in., and the length is

$$2(12) - 4 = 20 \text{ in.}$$

33. Supply and demand are equal when

$$3p - 200 = \frac{3200}{p}.$$

To solve for p, multiply both sides by p.

$$3p^2 - 200p = 3200$$
$$3p^2 - 200p - 3200 = 0$$

Use the quadratic formula with $a = 3$, $b = -200$, and $c = -3200$.

$$p = \frac{-(-200) \pm \sqrt{(-200)^2 - 4(3)(-3200)}}{2(3)}$$
$$p = \frac{200 \pm \sqrt{40{,}000 + 38{,}400}}{6}$$
$$p = \frac{200 \pm \sqrt{78{,}400}}{6}$$
$$p = \frac{200 \pm 280}{6}$$
$$p = \frac{480}{6} = 80 \quad \text{or} \quad p = \frac{-80}{6} = -\frac{40}{3}$$

Discard the negative solution. The supply and demand are equal when the price is 80 cents, or \$0.80.

41. Write a proportion.

$$\frac{x - 4}{3x - 19} = \frac{4}{x - 3}$$

Multiply by the LCD, $(3x - 19)(x - 3)$.

$$(3x - 19)(x - 3)\left(\frac{x - 4}{3x - 19}\right) = (3x - 19)(x - 3)\left(\frac{4}{x - 3}\right)$$
$$(x - 3)(x - 4) = (3x - 19)4$$
$$x^2 - 7x + 12 = 12x - 76$$
$$x^2 - 19x + 88 = 0$$
$$(x - 8)(x - 11) = 0$$
$$x - 8 = 0 \quad \text{or} \quad x - 11 = 0$$
$$x = 8 \quad \text{or} \quad x = 11$$

If $x = 8$, then

$$3x - 19 = 3(8) - 19 = 5.$$

If $x = 11$, then

$$3x - 19 = 3(11) - 19 = 14.$$

Thus, $AC = 5$ or $AC = 14$.

Section 10.6 (pages 823–828)

31. $f(x) = -\frac{2}{3}(x + 2)^2 + 1$

Because $a = -\frac{2}{3}$, the graph opens down and is wider than the graph of $y = x^2$. Because $h = -2$ and $k = 1$, the graph is shifted 2 units to the left and 1 unit up. The vertex is at $(-2, 1)$ and the axis is $x = -2$. Two other points on the graph are $\left(-4, -\frac{5}{3}\right)$ and $\left(0, -\frac{5}{3}\right)$. We can substitute any value for x, so the domain is $(-\infty, \infty)$. The value of y is always less than or equal to 1, so the range is $(-\infty, 1]$.

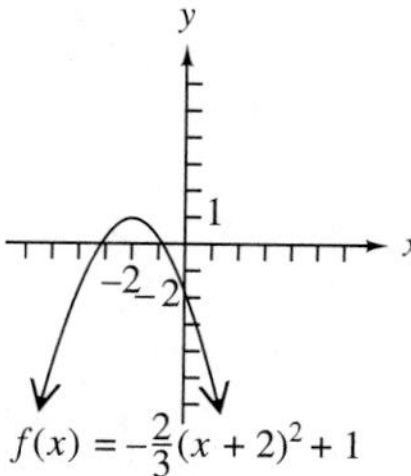

Section 10.7 (pages 837–840)

15. $x = -\frac{1}{5}y^2 + 2y - 4$

The roles of x and y are reversed, so this is a horizontal parabola.

Step 1

Since $a = -\frac{1}{5} < 0$, the graph opens to the left and is wider than the graph of $y = x^2$.

Step 2

The y-coordinate of the vertex is

$$\frac{-b}{2a} = \frac{-2}{2\left(-\frac{1}{5}\right)} = \frac{-2}{-\frac{2}{5}} = -2 \cdot \left(-\frac{5}{2}\right) = 5.$$

The x-coordinate of the vertex is

$$-\frac{1}{5}(5)^2 + 2(5) - 4 = -5 + 10 - 4 = 1.$$

Thus, the vertex is $(1, 5)$. Since the graph opens left, the axis goes through the y-coordinate of the vertex—its equation is $y = 5$.

Step 3

To find the x-intercept, let $y = 0$. If $y = 0$, then $x = -4$, so the x-intercept is $(-4, 0)$.

To find the y-intercepts, let $x = 0$.

$$0 = -\frac{1}{5}y^2 + 2y - 4$$
$$0 = y^2 - 10y + 20 \quad \text{Multiply by } -5.$$
$$y = \frac{-(-10) \pm \sqrt{(-10)^2 - 4(1)(20)}}{2(1)}$$
$$y = \frac{10 \pm \sqrt{20}}{2}$$
$$y = \frac{10 \pm 2\sqrt{5}}{2}$$
$$y = \frac{2\left(5 \pm \sqrt{5}\right)}{2}$$
$$y = 5 \pm \sqrt{5}$$

The y-intercepts are approximately $(0, 7.2)$ and $(0, 2.8)$.

Step 4

For an additional point on the graph, let $y = 7$ (two units above the axis) to get $x = \frac{1}{5}$. So the point $\left(\frac{1}{5}, 7\right)$ is on the graph.

By symmetry, the point $\left(\frac{1}{5}, 3\right)$ (two units below the axis) is on the graph.

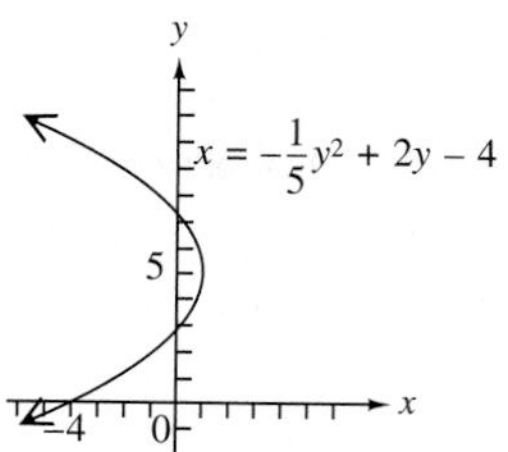

From the graph, we see that the domain is $(-\infty, 1]$ and the range is $(-\infty, \infty)$.

Section 10.8 (pages 847–850)

19. $x^2 - 6x + 6 \geq 0$

Solve the equation

$$x^2 - 6x + 6 = 0.$$

Since $x^2 - 6x + 6$ does not factor, let $a = 1$, $b = -6$, and $c = 6$ in the quadratic formula.

$$x = \frac{-(-6) \pm \sqrt{(-6)^2 - 4(1)(6)}}{2(1)}$$
$$x = \frac{6 \pm \sqrt{12}}{2}$$
$$x = \frac{6 \pm 2\sqrt{3}}{2}$$
$$x = \frac{2\left(3 \pm \sqrt{3}\right)}{2}$$
$$x = 3 \pm \sqrt{3}$$
$$x = 3 + \sqrt{3} \approx 4.7 \text{ or}$$
$$x = 3 - \sqrt{3} \approx 1.3$$

A B C

$3 - \sqrt{3}$ $3 + \sqrt{3}$

Test a number from each interval in the inequality

$$x^2 - 6x + 6 \geq 0.$$

Interval A: Let $x = 0$.

$$0^2 - 6(0) + 6 \stackrel{?}{\geq} 0$$
$$6 \geq 0 \quad \text{True}$$

Interval B: Let $x = 3$.

$$3^2 - 6(3) + 6 \stackrel{?}{\geq} 0$$
$$-3 \geq 0 \quad \text{False}$$

SOLUTIONS

Interval C: Let $x = 5$.

$$5^2 - 6(5) + 6 \overset{?}{\geq} 0$$

$$1 \geq 0 \quad \text{True}$$

The solution set includes the numbers in Intervals A and C, including $3 - \sqrt{3}$ and $3 + \sqrt{3}$ because equality is included in the symbol $\geq$.

Solution set:

$$(-\infty, 3 - \sqrt{3}] \cup [3 + \sqrt{3}, \infty)$$

0 $3-\sqrt{3}$ $3+\sqrt{3}$

37. $$\frac{w}{w+2} \geq 2$$

Write the inequality so that 0 is on one side.

$$\frac{w}{w+2} - 2 \geq 0$$

$$\frac{w}{w+2} - \frac{2(w+2)}{w+2} \geq 0$$

$$\frac{w - 2w - 4}{w+2} \geq 0$$

$$\frac{-w - 4}{w+2} \geq 0$$

The number -4 makes the numerator 0, and -2 makes the denominator 0. These two numbers determine three intervals.

A B C

-4 -2

Test a number from each interval in the inequality

$$\frac{w}{w+2} \geq 2.$$

Interval A: Let $w = -5$.

$$\frac{-5}{-3} \overset{?}{\geq} 2$$

$$\frac{5}{3} \geq 2 \quad \text{False}$$

Interval B: Let $w = -3$.

$$\frac{-3}{-1} \overset{?}{\geq} 2$$

$$3 \geq 2 \quad \text{True}$$

Interval C: Let $w = 0$.

$$\frac{0}{2} \overset{?}{\geq} 2$$

$$0 \geq 2 \quad \text{False}$$

The solution set includes numbers in Interval B, including -4 but excluding -2, which makes the fraction undefined.

Solution set: $[-4, -2)$

-4 -2 0

Chapter 11 Inverse, Exponential, and Logarithmic Functions

Section 11.1 (pages 871–874)

15. Write $g(x) = \sqrt{x - 3}$ as $y = \sqrt{x - 3}$. Since $x \geq 3$, $y \geq 0$. The graph of g is half of a horizontal parabola that opens to the right. The graph passes the horizontal line test, so g is one-to-one. To find the inverse, interchange x and y to get

$$x = \sqrt{y - 3}.$$

Note that now $y \geq 3$, so $x \geq 0$.
Solve for y by squaring both sides.

$$x^2 = y - 3$$

$$x^2 + 3 = y$$

Replace y with $g^{-1}(x)$.

$$g^{-1}(x) = x^2 + 3, \quad x \geq 0$$

23. **(a)** To find $f(0)$, substitute 0 for x.

$$f(x) = 2^x, \text{ so } f(0) = 2^0 = 1.$$

(b) Since f is one-to-one and $f(0) = 1$, it follows that $f^{-1}(1) = 0$.

37. $f(x) = y = x^3 - 2$

Complete the table of values.

x	y
-1	-3
0	-2
1	-1
2	6

Plot these points, and connect them with a solid smooth curve.

Interchange the values of x and y to make a table of values for f^{-1}.

x	y
-3	-1
-2	0
-1	1
6	2

Plot these points, and connect them with a dashed smooth curve. Use the fact that the graph of f^{-1} is symmetric to the graph of f with respect to the line $y = x$.

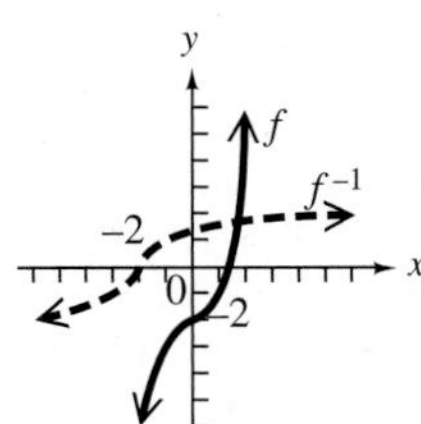

Section 11.2 (pages 881–882)

15. $$16^{2x+1} = 64^{x+3}$$

Write each side as a power of 4.

$$(4^2)^{2x+1} = (4^3)^{x+3}$$

$$4^{4x+2} = 4^{3x+9}$$

Set the exponents equal.

$$4x + 2 = 3x + 9$$

$$x = 7$$

Solution set: $\{7\}$

Check $16^{2x+1} = 64^{x+3}$

$$16^{2(7)+1} \overset{?}{=} 64^{7+3} \quad \text{Let } x = 7.$$

$$16^{15} \overset{?}{=} 64^{10}$$

$$(4^2)^{15} \overset{?}{=} (4^3)^{10}$$

$$4^{30} = 4^{30} \quad \text{True}$$

19. $5^x = 0.2$

$5^x = \frac{2}{10}$ Write the decimal as a fraction.

$5^x = \frac{1}{5}$ Write the fraction in lowest terms.

$5^x = 5^{-1}$ Write with the same base.

$x = -1$ Set exponents equal.

Check by substituting -1 for x in the original equation.

Solution set: $\{-1\}$

Section 11.3 (pages 887–890)

37. $\log_6 \sqrt{216} = x$

$6^x = \sqrt{216}$ Write in exponential form.

$6^x = 216^{1/2}$

$6^x = (6^3)^{1/2}$ Write with the same base.

$6^x = 6^{3/2}$ $(a^m)^n = a^{mn}$

$x = \frac{3}{2}$ Set exponents equal.

Solution set: $\left\{\frac{3}{2}\right\}$

55. $R = \log_{10} \frac{x}{x_0}$

Change to exponential form.

$$10^R = \frac{x}{x_0}, \quad \text{so} \quad x = x_0 10^R.$$

Let $R = 6.7$ for the Northridge earthquake, x_1.

$$x_1 = x_0 10^{6.7}$$

Let $R = 7.3$ for the Landers earthquake, x_2.

$$x_2 = x_0 10^{7.3}$$

The ratio of x_2 to x_1 is

$$\frac{x_2}{x_1} = \frac{x_0 10^{7.3}}{x_0 10^{6.7}} = 10^{0.6} \approx 3.98.$$

The Landers earthquake was about 4 times as powerful as the Northridge earthquake.

Section 11.4 (pages 897–898)

11. $\log_3 \dfrac{\sqrt[3]{4}}{x^2y}$

$= \log_3 \dfrac{4^{1/3}}{x^2y}$ Write the radical expression with a rational exponent.

$= \log_3 4^{1/3} - \log_3 (x^2y)$ Quotient rule

$= \log_3 4^{1/3} - (\log_3 x^2 + \log_3 y)$ Product rule

$= \log_3 4^{1/3} - \log_3 x^2 - \log_3 y$

$= \dfrac{1}{3}\log_3 4 - 2\log_3 x - \log_3 y$ Power rule

29. $3\log_p x + \dfrac{1}{2}\log_p y - \dfrac{3}{2}\log_p z - 3\log_p a$

$= \log_p x^3 + \log_p y^{1/2} - \log_p z^{3/2} - \log_p a^3$ Power rule

$= \left(\log_p x^3 + \log_p y^{1/2}\right) - \left(\log_p z^{3/2} + \log_p a^3\right)$ Group the terms into sums.

$= \log_p x^3y^{1/2} - \log_p z^{3/2}a^3$ Product rule

$= \log_p \dfrac{x^3y^{1/2}}{z^{3/2}a^3}$ Quotient rule

Section 11.5 (pages 903–904)

31. $p(h) = 86.3 \ln h - 680$

(a) $p(5000) = 86.3 \ln 5000 - 680$

$p(5000) \approx 55$ Use a calculator.

The percent of moisture at 5000 ft that falls as snow rather than rain is 55%.

(b) $p(7500) = 86.3 \ln 7500 - 680$

$p(7500) \approx 90$ Use a calculator.

The percent of moisture at 7500 ft that falls as snow rather than rain is 90%.

Section 11.6 (pages 911–914)

13. $2^{x+3} = 3^{x-4}$

$\log 2^{x+3} = \log 3^{x-4}$ Property 3 (common logs)

$(x+3)\log 2 = (x-4)\log 3$ Power rule

$x\log 2 + 3\log 2 = x\log 3 - 4\log 3$ Distributive property

$x\log 2 - x\log 3 = -3\log 2 - 4\log 3$ Get x-terms on one side.

$x(\log 2 - \log 3) = -3\log 2 - 4\log 3$ Factor out x.

$x = \dfrac{-3\log 2 - 4\log 3}{\log 2 - \log 3}$ Divide by $\log 2 - \log 3$.

$x \approx 15.967$ Use a calculator.

Check that $2^{15.967+3} \approx 3^{15.967-4}$.

Solution set: $\{15.967\}$

21. $\ln e^{0.45x} = \sqrt{7}$

$0.45x = \sqrt{7}$ $\ln e^k = k$

$x = \dfrac{\sqrt{7}}{0.45}$ Divide by 0.45.

$x \approx 5.879$ Use a calculator.

Solution set: $\{5.879\}$

37. $\log 4x - \log(x-3) = \log 2$

$\log \dfrac{4x}{x-3} = \log 2$ Quotient rule

$\dfrac{4x}{x-3} = 2$ Property 4

$4x = 2(x-3)$ Multiply by $x - 3$.

$4x = 2x - 6$ Distributive property

$2x = -6$ Subtract $2x$.

$x = -3$ Divide by 2.

Reject $x = -3$, because if $x = -3$, then $4x = -12$ and $x - 3 = -6$, which yield an equation in which the logarithms of negative numbers must be found.

Solution set: $\emptyset$

63. There are 60 of one species and 40 of another, so

$$P_1 = \frac{60}{100} = 0.6 \text{ and } P_2 = \frac{40}{100} = 0.4.$$

Thus, the index of diversity is

$$-(P_1 \ln P_1 + P_2 \ln P_2)$$
$$= -(0.6 \ln 0.6 + 0.4 \ln 0.4)$$
$$\approx 0.673.$$

Chapter 12 Nonlinear Functions, Conic Sections, and Nonlinear Systems

Section 12.1 (pages 937–940)

23. For any portion of the first ounce, the cost will be one 42¢ stamp. If the weight exceeds one ounce (up to two ounces), an additional 17¢ stamp is required. The following table summarizes the weight of a letter, x, and the number of stamps required, $p(x)$, on the interval $(0, 5]$.

x	(0, 1]	(1, 2]	(2, 3]	(3, 4]	(4, 5]
$p(x)$	1	2	3	4	5

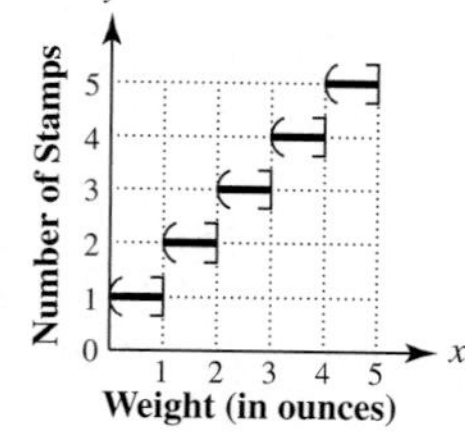

53. $(f \circ h)\left(\dfrac{1}{2}\right) = f\left(h\left(\dfrac{1}{2}\right)\right)$ Definition

$= f\left(\dfrac{1}{2} + 5\right)$ $h(x) = x + 5$

$= f\left(\dfrac{11}{2}\right)$ $5 = \dfrac{10}{2}$

$= \left(\dfrac{11}{2}\right)^2 + 4$ $f(x) = x^2 + 4$

$= \dfrac{121}{4} + \dfrac{16}{4}$

$= \dfrac{137}{4}$

57. $(f \circ g)(x) = f(g(x))$

$= f(5280x)$ $g(x) = 5280x$

$= 12(5280x)$ $f(x) = 12x$

$= 63{,}360x$

$(f \circ g)(x)$ computes the number of inches in x miles.

Section 12.2 (pages 947–950)

21. $3x^2 = 48 - 3y^2$

$3x^2 + 3y^2 = 48$ Add $3y^2$.

$x^2 + y^2 = 16$ Divide by 3.

$x^2 + y^2 = 4^2$

This is an equation of a circle with center (0, 0) and radius 4.

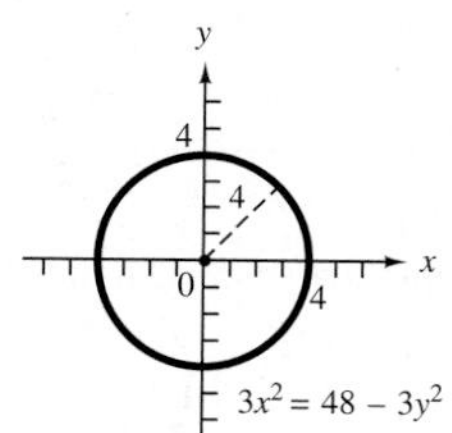

37. **(a)** $100x^2 + 324y^2 = 32{,}400$

$\dfrac{x^2}{324} + \dfrac{y^2}{100} = 1$ Divide by 32,400.

$\dfrac{x^2}{18^2} + \dfrac{y^2}{10^2} = 1$

The height in the center is the y-coordinate of the positive y-intercept. The height is 10 m.

(b) The width of the ellipse is the distance between x-intercepts, $(-18, 0)$ and $(18, 0)$. The width across the bottom of the arch is

$$18 + 18 = 36 \text{ m}.$$

Section 12.3 (pages 957–960)

27. $y = \sqrt{\dfrac{x+4}{2}}$

$y^2 = \dfrac{x+4}{2}$ Square both sides.

$2y^2 = x + 4$ Multiply by 2.

$2y^2 - 4 = x$ Subtract 4.

$2(y-0)^2 - 4 = x$

This is a parabola that opens to the right with vertex $(-4, 0)$. However, y is nonnegative in the original equation, so only the top half of the parabola is included in the graph.

x	y
-4	0
-2	1
0	$\sqrt{2}$
4	2

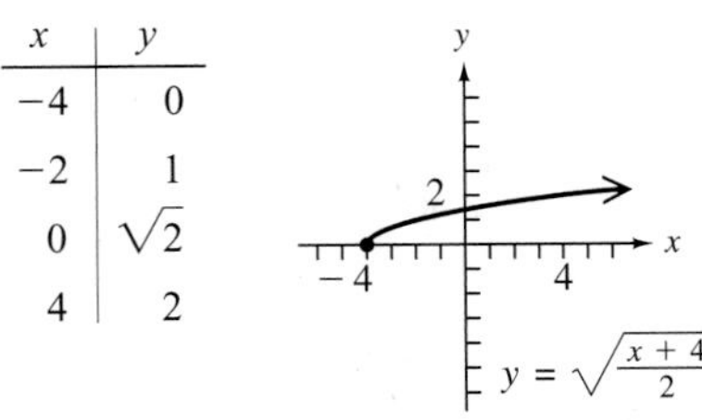

The domain is $[-4, \infty)$, and the range is $[0, \infty)$.

31. (a)

$$400x^2 - 625y^2 = 250{,}000$$

$$\frac{x^2}{625} - \frac{y^2}{400} = 1 \quad \text{Divide by 250,000.}$$

$$\frac{x^2}{25^2} - \frac{y^2}{20^2} = 1$$

The x-intercepts are $(25, 0)$ and $(-25, 0)$. The distance between the buildings is the distance between the x-intercepts. The buildings are

$$25 + 25 = 50 \text{ m}$$

apart at their closest point.

(b) At $x = 50$, $y = \frac{d}{2}$, so $d = 2y$.

$$400(50)^2 - 625y^2 = 250{,}000$$
$$1{,}000{,}000 - 625y^2 = 250{,}000$$
$$-625y^2 = -750{,}000$$
$$y^2 = 1200$$
$$y = \sqrt{1200}$$

The distance d is

$$2\sqrt{1200} \approx 69.3 \text{ m.}$$

Section 12.4 (pages 965–966)

23.

$$2x^2 - y^2 = 6 \quad (1)$$
$$y = x^2 - 3 \quad (2)$$

Solve equation (2) for x^2.

$$x^2 = y + 3 \quad (3)$$

Substitute $y + 3$ for x^2 in equation (1).

$$2x^2 - y^2 = 6 \quad (1)$$
$$2(y + 3) - y^2 = 6$$
$$2y + 6 - y^2 = 6$$
$$0 = y^2 - 2y$$
$$0 = y(y - 2)$$
$$y = 0 \quad \text{or} \quad y = 2$$

From equation (3) with $y = 0$,

$$x^2 = 3, \quad \text{so} \quad x = \pm\sqrt{3}.$$

From equation (3) with $y = 2$,

$$x^2 = 5, \quad \text{so} \quad x = \pm\sqrt{5}.$$

All four ordered pairs check in *both* equations.

Solution set:

$$\{(-\sqrt{3}, 0), (\sqrt{3}, 0), (-\sqrt{5}, 2), (\sqrt{5}, 2)\}$$

33. Let $W =$ the width, and $L =$ the length.

The formula for the area of a rectangle is $A = LW$, or $LW = A$, so

$$LW = 84. \quad (1)$$

The perimeter of a rectangle is given by $P = 2L + 2W$, or $2L + 2W = P$, so

$$2L + 2W = 38. \quad (2)$$

Solve equation (2) for L to get

$$L = 19 - W. \quad (3)$$

Substitute $19 - W$ for L in equation (1).

$$LW = 84 \quad (1)$$
$$(19 - W)W = 84$$
$$19W - W^2 = 84$$
$$-W^2 + 19W - 84 = 0$$
$$W^2 - 19W + 84 = 0 \quad \text{Multiply by } -1.$$
$$(W - 7)(W - 12) = 0$$
$$W - 7 = 0 \quad \text{or} \quad W - 12 = 0$$
$$W = 7 \quad \text{or} \quad W = 12$$

Using equation (3), with $W = 7$,

$$L = 19 - 7 = 12.$$

If $W = 12$, then $L = 7$, which are the same two numbers. Length must be greater than width, so the length is 12 ft and the width is 7 ft.

INDEX

T

PHOTO CREDITS

Antony Dickson/AFP/Getty Images, page 570; **AP,** page 580 left; **AP Photo/Doug Mills,** page 383; **AP Wideworld Photos,** pages 62, 104, 154, 176, 995; **Beth Anderson,** pages 276, 376, 614 bottom, 931; **Bettmann/Corbis,** page 589; **Blend Images/Getty RF,** page 190 bottom, 259, 815 right, 919; **Bureau L.A. Collection/Corbis,** page 576; **Car Cultures/Getty Images,** page 162 left; **Comstock,** page 306 left; **Corbis RF,** pages 440, 590, 664, 778 right, 796, 865, 901; **David Gray/Reuters/Corbis,** page 580 right; **Digital Vision,** pages 179, 199, 214 left, 236, 274, 315, 506, 511 bottom, 771, 812, 821, 861, 871 right, 900, 934; **Courtesy of Ed Walters,** page 188 left; **Everett Collection,** page 624; **Gallo Images ROOTS/Getty RF,** page 163; **Getty Editorial,** pages 115, 118, 123, 362; **Getty Images News,** page 374; **Getty Images Sport,** page 578 left; **Getty RF,** pages 29, 76 top, 93, 190 top left, 323, 342, 350, 522, 588, 605, 628; **Hulton Archives/Getty Editorial,** pages 927, 944; **iStockphoto,** pages 500, 815 left, 828; **Johner Images/Getty RF,** pages 453, 511 top; **Johns Hopkins University Applied Physics Laboratory,** page 439; **MGM/Photofest,** page 718; **NASA,** pages 164, 450, 591, 950; **NASA/Johnson Space Center (PAL AABANEM0),** page 748; **NBC/Photofest,** page 420; **Paramount/The Kobal Collection,** page 614 top; **PhotoDisc,** pages 16 left, 46, 306 right, 579, 778 left, 797, 817, 998; **PhotoDisc Blue,** page 318; **PhotoDisc/Getty RF,** pages 353, 519; **Photofest,** pages 325, 354, 360 top right and bottom; **Photographer's Choice/Getty RF,** page 222; **PictureQuest,** page 710; **Prentice Hall/PAL,** page 816; **Purestock/Getty RF,** pages 221, 244; **Reuters/Corbis,** page 190 top right; **SAU St. Andrew's University MacTutor Archives,** page 493; **Shutterstock,** pages 12, 16 right, 66, 76 bottom, 182, 188 right, 266, 320, 363, 364, 573, 642, 659, 564, 571, 578 right, 687, 700, 724, 774, 827, 855, 940; **Somos/Getty RF,** page 229; **Steve Starr/Corbis,** page 380; **Stockbyte/Getty RF,** pages 214 right, 223; **Stockdisc Premium/Getty Royalty Free,** page 920; **Superstock,** page 189; **20th Century Fox/The Kobal Collection,** page 162 right; **Universal Pictures/Photofest,** page 368; **U.S. Postal Service,** page 769;**Warner Bros./Photofest,** page 360 top left, 696; **Wikipedia/Wolf Meusel,** pages 527 and 550; **Yoshikazu Tsuno/AFP/Getty Images,** page 871 left